FROM ANATOMY TO FUNCTION OF THE CENTRAL NERVOUS SYSTEM

CLINICAL AND NEUROSURGICAL APPLICATIONS

FROM ANATOMY TO FUNCTION OF THE CENTRAL NERVOUS SYSTEM

CLINICAL AND NEUROSURGICAL APPLICATIONS

Edited by

BRANDON MATTEO ASCENZI
Independent Neuroresearcher-Member of Marie Curie Alumni Association (MCAA)

ELSEVIER

ACADEMIC PRESS
An imprint of Elsevier

Academic Press is an imprint of Elsevier
125 London Wall, London EC2Y 5AS, United Kingdom
525 B Street, Suite 1650, San Diego, CA 92101, United States
50 Hampshire Street, 5th Floor, Cambridge, MA 02139, United States

ISBN: 978-0-12-822404-5

For information on all Academic Press publications visit our website at
https://www.elsevier.com/books-and-journals

Publisher: Nikki Levy
Acquisitions Editor: Megan McManus
Editorial Project Manager: Tracy Tufaga
Production Project Manager: Swapna Srinivasan
Cover Designer: Alessandro Basso

Description: In the center, is an artistically reworked Klingler's dissection of the human brain; in the background, is shown an in vitro differentiation of human neural stem cells (hNSCs) into A9 dopaminergic neurons performed by Brandon M. Ascenzi (red: TH positive neurons; green: beta 3 tubulin neuronal marker).

Typeset by TNQ Technologies

Dedication

To my aunt Maria Luisa, in loving memory

Contents

5a. Functional Anatomy of the Brainstem

GIOVANNI PATERNÒ, BARBARA BUCCILLI, GUGLIELMO IESS, SIMON SEBASTEAN VELASCO MONTAÑO, VERONICA VITALI, ORHUN METE ÇEVIK, MERT BABUR, ÖMER ORHUN, AND BRANDON M. ASCENZI

5b. The Reticular Formation

BRANDON M. ASCENZI

6. The Cerebellum

JORGE OJEDA, GIOVANNI PATERNÒ, AND SERGIO JUICA

7. The Diencephalon: Introduction and Epithalamus

LEE ELISEVICH, BRANDON M. ASCENZI, AND KOST ELISEVICH

8. The Diencephalon: Dorsal Thalamus

SRIKANT S. CHAKRAVARTHI, ALEJANDRO MONROY-SOSA, AND KOST ELISEVICH

9. The Diencephalon: Subthalamus or Ventral Thalamus

DANIELE ARMOCIDA, MAURIZIO SALVATI, GIANCARLO D'ANDREA, ALESSANDRO FRATI, AND BRANDON M. ASCENZI

10. The Diencephalon: Hypothalamus

DANIELE ARMOCIDA, MAURO PALMIERI, FULVIO TARTARA, DIEGO GARBOSSA, FABIO COFANO, BRANDON M. ASCENZI, ANTONIO SANTORO, AND ALESSANDRO FRATI

11. The Telencephalon: Neocortex

BRANDON M. ASCENZI

12. The Telencephalon: Cingulate Cortex

HANNA MARKIEWICZ AND BRANDON M. ASCENZI

13. The Telencephalon: Hippocampus and Related Structures

ELHAM HOSSEINI, ALI SEPEHRINEZHAD, JAVAD MOMENI, BRANDON M. ASCENZI, ALI GORJI, AND SAJAD SAHAB-NEGAH

14. The Telencephalon: Amygdala and Claustrum

BARBARA BUCCILLI, SAJAD SAHAB-NEGAH, SADEGH SHIRIAN, ALI GORJI, MARYAM KHALEGHI GHADIRI, AND BRANDON M. ASCENZI

15. The Telencephalon: Basal Ganglia

BRANDON M. ASCENZI, GUILHERME DE CARVALHO CALDAS, ALESSANDRO FRATI, ANTONIO SANTORO, AND MAURO PALMIERI

For ancillary content such as **3D images, podcasts, lectures etc.** consult the following website: *www.ascenzisfunctionalneuroanatomy.com*

Contributors

Daniele Armocida Department of Human Neurosciences, Neurosurgery Division, University of Rome "La Sapienza", Rome, Italy; IRCCS "Neuromed", Experimental Neurosurgery Unit, Pozzilli, Italy

Brandon M. Ascenzi Independent Neuroresearcher-Member of Marie Curie Alumni Association (MCAA)

Mert Babur Department of Neurosurgery, Bakırkoy Training and Research Hospital for Psychiatric and Nervous Diseases, Health Sciences University, Istanbul, Turkey; Center of Advanced Simulation and Education (CASE), Neuroanatomy Lab., Acıbadem Mehmet Ali Aydinlar University, Istanbul, Turkey

Anna Badner Department of Psychiatry and Behavioral Sciences, Stanford University, Palo Alto, CA, USA

Giulia Bechi Gabrielli Neuroimaging Laboratory, IRCCS Santa Lucia Foundation, Rome, Italy; Department of Psychology, University of Rome "La Sapienza", Rome, Italy; Department of Human Neurosciences, University of Rome "La Sapienza", Rome, Italy

Barbara Buccilli University of Rome "La Sapienza", Rome, Italy; Department of Neurosurgery, Cerebrovascular and Neurologic Critical Care, Icahn School of Medicine at Mount Sinai, New York, NY, United States

Guilherme de Carvalho Caldas Neurosurgical Innovations and Training Center, Weill Cornell Medical College, New York City, NY, United States

Srikant S. Chakravarthi Aurora Research Institute, Aurora St. Luke's Medical Center, Milwaukee, WI, United States; Skull Base and Cerebrovascular Lab, Aurora Research Institute, Milwaukee, WI, United States

Fabio Cofano Department of Neuroscience "Rita Levi Montalcini", Neurosurgery Unit, University of Turin, Turin (TO), Italy

Ana Paula Cortes-Contreras Faculty of Medicine, Universidad Nacional Autónoma de México (UNAM), Mexico City, Mexico

Gökşin Nilüfer Demirci Department of Anatomy, School of Medicine, Dokuz Eylul University, Izmir, Turkey

Giancarlo D'Andrea Neurosurgery Department of Fabrizio Spaziani Hospital, Frosinone, Italy

Lee Elisevich Central Michigan University, Mount Pleasant, MI, United States

Kost Elisevich Michigan State University, East Lansing, MI, United States; Department of Neurosurgery, Spectrum Health Medical Group Neurosurgery, Grand Rapids, MI, United States

Leonardo Fogassi Department of Medicine and Surgery, Unit of Neuroscience, University of Parma, Parma, Italy

Alessandro Frati Department of Human Neurosciences, Neurosurgery Division, University of Rome "La Sapienza", Rome, Italy; IRCCS "Neuromed", Experimental Neurosurgery Unit, Pozzilli, Italy

Diego Garbossa Department of Neuroscience "Rita Levi Montalcini", Neurosurgery Unit, University of Turin, Turin (TO), Italy

Isai Garcia Lopez Department of Neurosurgery, National Medical Center 20 de Noviembre, Mexico City, Mexico

Silvia Gigli Department of Radiological, Oncological and Pathological Sciences, University of Rome "La Sapienza", Rome, Italy

Ali Gorji Epilepsy Research Center, Department of Neurosurgery, University of Münster, Münster, Germany; Shefa Neuroscience Research Center, Khatam Alanbia Hospital, Tehran, Iran; Epilepsy Research Center, Department of Neurosurgery and Neurology, Westfälische Wilhelms-Universitat Münster, Münster, Germany

Elham Hosseini Neuroscience Research Center, Mashhad University of Medical Sciences, Mashhad, Iran

Guglielmo Iess Department of Neurosurgery, Istituto Clinico Città Studi, Milan, Italy

Sergio Juica Faculty of Medicine, University of Concepcion and Hospital Regional de Concepción, Concepción, Chile

Hideki Derek Kawai Department of Biosciences, Graduate School of Science and Engineering, Soka University, Tokyo, Japan

Maryam Khaleghi Ghadiri Epilepsy Research Center, Department of Neurosurgery and Neurology, Westfälische Wilhelms-Universität Münster, Münster, Germany

Hanna Markiewicz Warsaw, Poland

Javad Momeni Neuroscience Research Center, Mashhad University of Medical Sciences, Mashhad, Iran

Alejandro Monroy-Sosa Department of Neurosurgery, Star Medica Lomas Verdes Hospital, Mexico City, Mexico; Department of Neurosurgery, Institute for Social Security and Services for State Workers, (ISSSTE) Tlahuac, Center for Research and Innovation in Neuroscience, Instituto Politécnico Nacional, Superior School of Medicine, Mexico City, Mexico; Skull Base and Cerebrovascular Lab, Aurora Research Institute, Milwaukee, WI, United States

Simon Sebastean Velasco Montaño Laboratoire de Recherche, Service de Neurochirurgie, Hôpital Lariboisière, AP-HP, Paris, France

Matthew W. Myers SciLifeLab, Stockholm, Sweden

Jorge Ojeda Biomedical Science Research Laboratory, Basic Sciences Department, Faculty of Medicine, Universidad Católica de la Santísima Concepción, Concepción, Chile

Ömer Orhun Center of Advanced Simulation and Education (CASE), Neuroanatomy Lab., Acıbadem Mehmet Ali Aydinlar University, Istanbul, Turkey; School of Medicine, Acıbadem Mehmet Ali Aydinlar University, Istanbul, Turkey

Mauro Palmieri Department of Human Neurosciences, Neurosurgery Division, University of Rome "La Sapienza", Rome, Italy; Neurosurgical Innovations and Training Center, Weill Cornell Medical College, New York City, NY, United States

Giovanni Paternò Department of Neurosurgery, Istituto Clinico Città Studi, Milan, Italy

Alessandro Pesce Department of Neurosurgery, "Santa Maria Goretti" Hospital, Latina, Italy

Jonathan Ortiz Rafael Department of Neurosurgery, Hospital Juarez de Mexico, Mexico City, Mexico; Center for Research and Innovation in Neuroscience, Superior Medicine School, National Polytechnic Institute, Mexico City, Mexico

Noé Santiago Rea Faculty of Medicine, La Salle University, Mexico City, Mexico

Clelia Rossi-Arnaud Department of Psychology, University of Rome "La Sapienza", Rome, Italy

Stefano Rozzi Department of Medicine and Surgery, Unit of Neuroscience, University of Parma, Parma, Italy

Sajad Sahab-Negah Neuroscience Research Center, Mashhad University of Medical Sciences, Mashhad, Iran

Maurizio Salvati Policlinico Tor Vergata, Neurosurgery Department, University Tor Vergata of Rome, Rome, Italy

Antonio Santoro Department of Human Neurosciences, Neurosurgery Division, University of Rome "La Sapienza", Rome, Italy

Ali Sepehrinezhad Neuroscience Research Center, Mashhad University of Medical Sciences, Mashhad, Iran

Sadegh Shirian Department of Pathology, School of Veterinary Medicine, Shahrekord University, Shahrekord, Iran; Shiraz Molecular Pathology Research Center, Dr. Daneshbod Pathol. Lab., Shiraz, Iran

Pietro Spataro Faculty of Society and Communication Sciences, Universitas Mercatorum, Rome, Italy

Fulvio Tartara IRCCS Fondazione Istituto Neurologico Nazionale C. Mondino, Neurosurgery Unit, Pavia, Italy

Maria Fernanda Tejada Pineda Instituto Tecnológico de Estudios Superiores de Monterrey, School of Medicine and Health Sciences, Mexico City, Mexico

Gülgün Şengül Department of Anatomy, School of Medicine, Ege University, Izmir, Turkey; Department of Neuroscience, Institute of Health Sciences, Ege University, Izmir, Turkey

Nimet Şenoğlu Çiğli Education and Training Hospital Anesthesia and Reanimation, İzmir Bakırçay University Faculty of Medicine, Izmir, Turkey

Orhun Mete Çevik Department of Neurosurgery, Bakırkoy Training and Research Hospital for Psychiatric and Nervous Diseases, Health Sciences University, Istanbul, Turkey; Center of Advanced Simulation and Education (CASE), Neuroanatomy Lab., Acıbadem Mehmet Ali Aydinlar University, Istanbul, Turkey

Yağmur Çınar Anesthesiology and Reanimation Department İzmir Tepecik Training and Research Hospital, İzmir, Turkey

Ariel Ávila Biomedical Sciences Research Laboratory, Faculty of Medicine, Universidad Católica de la Santísima Concepción (UCSC – Chile), Concepción, Chile

Pia Maria Vidal Biomedical Science Research Laboratory, Neuroimmunology and Regeneration of the Central Nervous System Unit, Department of Basic Sciences, Faculty of Medicine, Universidad Católica de la Santísima Concepción, Concepción, Chile

Veronica Vitali Department of Neurosurgery, University of Insubria, Varese, Italy

I. Zoe Chavez Valderrama Department of Neurosurgery, XXI Century National Medical Center, Mexico City, Mexico

About the editor

BRANDON MATTEO ASCENZI, MSc;MD;PhD

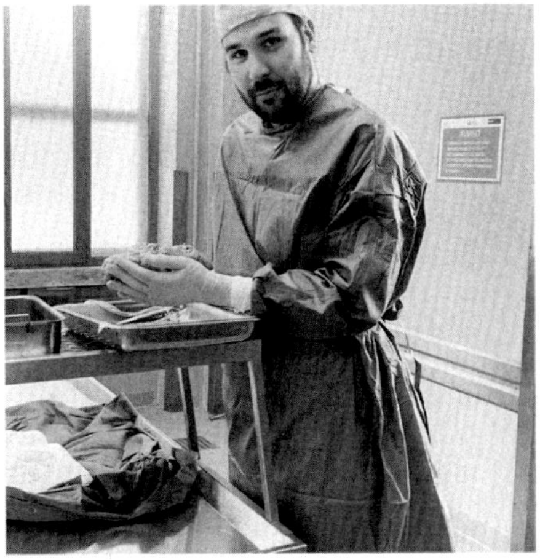

Dr Ascenzi received his bachelor's degree in Techniques of Biomedical Laboratory and master's degree in Neurobiology (Medical Biotechnology) at the School of Medicine and Surgery, University of Rome "Tor Vergata." Over the years, he took part in several research projects about neuromuscular diseases, such as Duchenne's muscular dystrophy, at the Institute of Histology and Embryology of the University of Rome "La Sapienza" and at the School of Sciences (MM.PH.NN) of "Tor Vergata" University.

Since 2015 to middle 2016 he worked at National Research Council "CNR", to a project on the mechanism of maturation and secretion of Nerve Growth Factor (NGF) in a murine model of diabetic and Alzheimer-like encephalopathy. In 2017, he won the European PhD *"Marie Curie"* fellowship and he worked at the *Centro de Biologia Molecular* "Severo Ochoa" of Madrid in collaboration with the University of Lund, Polytechnic University of Milan, Barcelona, Oslo, Munich and others. His fields of research interest mainly focus on the modulation of brain development through neurotransmitters and peptides in the field of neurodevelopmental diseases (ASD, ADHD). Recently, he also started to participate at neurosurgical and radiological works. Currently, he is majoring in medicine; he is 33 years old and this is his first monography.

Preface

This book, whose the writing process required several years of intensive and accurate work, aims to satisfy a wide audience trying to cover virtually all the anatomo-functional aspects of the central nervous system.

The importance to have a textbook as complete as possible is given above all by the increasing number of different professional figures involved in the study of the nervous system from more points of view, having as common need a clear explanation of the anatomy of the nerve structures. Usually, the latter is not of easy understanding due to the complex disposition of brain structures "one inside of other." Indeed, it is extremely important to have in these kinds of texts illustrations that faithfully respect graphically the morphology of the aforementioned structures and above all their real spatial arrangement and relationships with the surrounding areas. In the absence of this, it is very difficult for the reader to get an exhaustive knowledge of neuroanatomy.

In this regard, I wanted to show some brain dissections performed by Klingler's method, an extraordinary tool developed by the scientist of the same name in the early '900s, which allows to isolate the white matter from the gray matter and to realize real "brain sculptures" that clearly show in their entirely the localization and also the morphology of those most hidden brain structures.

All dissections were performed on human brains donated to the laboratories of Prof. Rhoton A. Jr., one of the greatest neurosurgeons in history. However, although the Klingler's method is known for almost a century, it has been and is still used today almost exclusively by neurosurgeons and neuroanatomists and no functional neuroanatomy book had ever been written with the addition of these beautiful images.

Furthermore, I wanted the reader to be able to recognize the same encephalic structures identified through dissection, and so *ex vivo*, also *in vivo* through modern neuroradiological techniques which in turn allow us to study the functionality of the various cortical and subcortical structures.

Some chapters have also used other instrumental methods capable of describing the functional aspects of the nervous system, including electroencephalography, nuclear medicine, and the various computational methods used today in the anatomical study of subcortical structures such as the reticular formation.

The neurobiology of the several neuron types found in the neuroaxis is also discussed together with their neurotransmitter phenotype and synaptological properties.

Finally, with particular attention, the reticular and nonreticular neurochemical systems have been described; they play a fundamental role in vegetative functions such as the sleep—wake rhythm and behavioral aspects of the individual. Some neuropharmacological approaches necessary in some psychiatric and neurological pathologies were also partially covered.

The textbook was realized taking into account the most recent literature produced in the field of neuroscience and presents an appendix that takes into consideration some emerging issues in the neuropsychological field, that is, mnemonic processes (i.e., Attentional Boost Effect) and the neuroanatomical aspects of the anesthetics used in surgery.

All this makes this book a particularly versatile instrument suitable for almost all professionals, especially clinical and neurosurgical ones, who want to get a complete and detailed knowledge of the functional anatomy of the nervous system.

September 9th, 2023

Brandon Matteo Ascenzi

Acknowledgments

First of all, I would like to thank the Rhoton Collection and especially Dr. Spencer Bone of AANS who allowed me to use the beautiful images which represent real brain sculptures which are incomparable in the way they show the structures the brain is made of. I particularly want to thank Dr. Kaan Yagmurlu and Dr. Jeffrey Sorenson, dissectors and collaborators of Professor Rhoton A. Jr. in his lab. I would like to thank all the dissectors who imaged the dissections showed in this book.

My heartfelt thanks go to Dr. Franco Nasella, librarian of the "Paolo Nasella" Biomedical Library of the School of Medicine and Surgery of the University of Rome "Tor Vergata," who, for more than 10 years, has provided me with the bibliographic material necessary to bring forth my studies.

Likewise, I thank Angelo Ippoliti, graphic designer, for assisting me during the graphic processing of the images, and the entire Elsevier team.

In particular, I thank Tracy Tufaga, Editorial Project Manager, who was always available and kind and who continually took care of the entire writing and submission process, finding a solution to any problem that arose.

My warmest thank you goes to my "American mum," Kathie Kessler, who encouraged me emotionally from the beginning, helping me to grow in confidence in undertaking this "long journey."

Of course, I am grateful to my mother and my grandparents who have always supported me in my studies with great efforts and to anyone who loved and believed in me.

Development, topography and neurochemistry of the central nervous system

1

The Development of the Central Nervous System and Human Neural Stem Cells

Ariel Ávila[1], *Brandon M. Ascenzi*[2] *and Pia Maria Vidal*[3]

[1]Biomedical Sciences Research Laboratory, Faculty of Medicine, Universidad Católica de la Santísima Concepción (UCSC – Chile), Concepción, Chile [2]Independent Neuroresearcher-Member of Marie Curie Alumni Association (MCAA) [3]Biomedical Science Research Laboratory, Neuroimmunology and Regeneration of the Central Nervous System Unit, Department of Basic Sciences, Faculty of Medicine, Universidad Católica de la Santísima Concepción, Concepción, Chile

OUTLINE

NEURULATION

The development of the nervous system begins with the formation of the neural tube during the neurulation process. It is at this moment in time when the brain and the spinal cord start to be shaped, as the neural axis of the body and neural crest cells commence detaching to form peripheral structures of the nervous system. The neural tube forms from the bending and folding of the neural plate, ending with the closing of the tube. Thus, the neural plate is the precursor structure and is formed as a thickening of the dorsal surface neural ectoderm bordered by nonneural ectoderm (Wilde et al., 2014).

In humans, the neurulation process occurs between days 20 and 24 during gestation (Carnegie stages 9–11) and proceeds in two stages referred as primary neurulation and secondary neurulation (O'Rahilly and Müller, 1994, 2007). During primary neurulation, the neural plate rolls up and seals fusing in the midline, influenced by the underlying notochord and surrounding mesoderm. With this folding, the neural ectoderm forms the neural tube and the lateral nonneuronal ectoderm connects forming a covering layer of surface ectoderm. As a common feature found in mammals, closing of the neural tube occurs progressing along the spine with more than one initiation point. Distinctively in humans, the initiation points are two, with the first at the hindbrain/cervical boundary and the second at the rostral end of the forebrain (O'Rahilly and Müller, 2002). Thus, while the initiation at the rostral end progresses toward the caudal end, the initiation at the hindbrain/cervical boundary spread bidirectionally toward the rostral and caudal ends. As a consequence, just before total closing of the neural tube, rostral and caudal neuropores can be observed transiently. Regulation of the entire process is cell intrinsic in the neuroepithelium, and also respond to cell extrinsic cues from the nonneural ectoderm, mesoderm, and notochord.

At the cellular level, neurulation includes processes such as convergent extension, apical constriction, and interkinetic nuclear migration. Convergent extension initiates neural tube closure and involves narrowing of

From Anatomy to Function of the Central Nervous System
https://doi.org/10.1016/B978-0-12-822404-5.00001-2

the tissue along its mediolateral axis (convergence) and elongation along its anteroposterior/rostrocaudal axis (extension). This process facilitates neural tube closure by approaching the fusing ends of the neuro ectoderm. Following convergent extension, the margins of the neural plate elevate forming the folds that, after bending, will fuse and form the neural tube. This process occurs in mammals as a progression along the body axis and is based on apical constriction and interkinetic nuclear migration. Apical constriction causes reduction of the apical surface area along the mid line of the neural plate right above notochord. More importantly, since the neural plate is a pseudo stratified epithelium, tissue structure is determined by the positioning of cell nuclei, which varies from apical to basal depending if cells are undergoing mitosis or S-phase, respectively. This is the interkinetic nuclear migration process and, while cells are mostly distributed randomly through the cycle they are in a higher proportion in S-phase along the midline, meaning that the future interior edge of cells is narrowed contributing to bending of the neural plate.

Final closing of the gap formed between neural folds involves interactions between cellular protrusions. Protrusions can display the shape of filopodia or lamellipodia and are actin-rich structures. This is a conserved process found in all studied vertebrates in which neural and nonneural cells make contact sealing the neural tube. Secondary neurulation occurs in the distal part of the spinal cord where the mechanism to extend the neural tube is a canalization where individual cells organize to form the continuing tubular structure.

Key signaling pathways controlling closure of the neural tube include Sonic Hedgehog (SHH) and Wnt, with requirements of retinol and inositol (Copp and Greene, 2010). Convergent extension has been linked to noncanonical Wnt/planar cell polarity (PCP) pathway. Disruption of this pathway in experimental models leads to an open neural plate (Zohn, 2020). This has been observed with the disruption of various PCP genes including Vangl2, Celsr1, Dvl-1 and 2, Fzd-3 and 6, Scrb1, Ptk7, and Sec24b (Juriloff and Harris, 2012; Greene and Copp, 2014). In addition, a human cases study suggests that SHH signaling from the notochord is involved as well in the process (Kirillova et al., 2000). In this line, the relevance of neurulation is manifested in the high severity of the consequences associated with defective closing of the neural tube (Nikolopoulou et al., 2017).

CORTEX ASSEMBLY

Modern humans have a globular brain that has emerged in parallel to the development of craniofacial features and social behaviors. The brain has evolved in volume and shape with an expansion followed by a refinement of its shape (Neubauer et al., 2018). At the onset, each of the structures of the brain increase their size primarily as a result of cell division. This process amplifies a starting pool of progenitors and sequentially originate all cell types for each brain region. Cell death is also present at this time and has a modulatory effect. As the process evolves, further growth is attained by migration of cells born at different places and by the formation of cellular processes in neurons and glial cells. The cerebral cortex is where the higher-order functions of the brain reside. This structure is what largely differentiate humans from individuals of other species an underlies the superior human ability to adapt in response to environmental changes and to display a wide range of unique behaviors. In line with that, dysregulation of the cortical developmental process, due to intrinsic or extrinsic causes, can lead to the development of important cognitive dysfunctions and mental disorders. The cerebral cortex contains around 16 billion neurons, which account for 20% of all brain cells. While this is a small percentage, compared to the contribution of the cerebellum to total brain cells, they represent about 80% of all brain mass owing to their larger size and extensive neuropil (van Essen et al., 2018). Compared to other primates, humans have the larger cortical surface area, with the prefrontal cortex being where we find the biggest difference (Donahue et al., 2018).

Cortical neurogenesis. At the cellular level, there are primarily two types of neuronal cells in the cortex, excitatory neurons and interneurons. The first ones are more abundant and provide excitation to the neuronal circuits by releasing glutamate at synapses. In turn, interneurons are only about 25% of all cortical neurons, but are fundamental for the proper functioning of cortical circuits by providing inhibition and control through the release of gamma aminobutyric acid (GABA) and/or glycine (Brackmann et al., 2004; Aroeira et al., 2011). Both main neuronal types of the brain, oligodendrocytes and astrocytes develop from a small number of cells organized forming the telencephalic vesicles (Silva et al., 2019). These primordial structures are essentially formed by cortical progenitor cells. Importantly, these early progenitors establish a dynamic process characterized by controlled amplification of the progenitor pool, parallel generation of more specialized progenitors, and sequential generation of cortical neurons (Miyata et al., 2001; Noctor et al., 2004; Wilson and Houart, 2004; Sessa et al., 2008; Stancik et al., 2010; Greig et al., 2013; Taverna et al., 2014; Dehay et al., 2015; Lodato and Arlotta, 2015). More in detail, during early stages of embryonic development, the cerebral cortical primordium is almost exclusively formed by radial glial cell progenitors. Radial glia plays a distinctive role, as they give origin to more specialized progenitors and provide a structural scaffold for the migration of cells toward the forming cortical plate. Radial glia cells are also referred as apical progenitors, in attention to their location lining the ventricles, and are characterized for their expression of Pax6. Multiplication of apical progenitors occurs through a series of steps that involve interkinetic nuclear migration during cell cycle progression. From this process, there is an amplification of the apical progenitor pool, but also more specialized progenitors emerge. These new progenitors occupy a more basal position in the developing cortex and characteristically express Tbr2. Consequently, these progenitors are collectively referred as basal progenitors. Distinctively, the developing human SVZ has a massively expanded outer region (OSVZ) where OSVZ progenitor cell types reside. These cells lack contact with the ventricular surface and can undergo proliferative divisions and self-renewing asymmetric divisions (Hansen et al., 2010). During neurogenesis, newly generated neurons sequentially expresses characteristic markers depending on the cortical layer in which neurons will be located. To name some of the most common markers, Tbr1-positive cells are found in layer VI and pre-plate region, Ctip2-positive cells form layer V and Cux1-positive cells take part of layer I–III (Angevine and Sidman, 1961; Rakic, 1972; Anderson, 2008; Marín, 2013; Gallo and Deneen, 2014; Peyre et al., 2015; Lim et al., 2018a).

There are multiple factors that influence the neurogenic process in the developing cerebral cortex. Among them, cell extrinsic factors play important roles, as they generate the ambient in which the neurogenic process takes place. An example that ilustrates how the study of cortical development shed light on the etiology of neurodevelopmental disorders is the research performed on glycine. Evidence has grown supporting the crucial roles for neurotransmitters and their receptors in developmental stages before synapse formation (Herlenius and Lagercrantz, 2004; Ishizuka et al., 2012; Avila et al., 2013; Belinsky et al., 2014; Kemp et al., 2016; Bono et al., 2018). Although all main neurotransmitter systems have been implicated in controlling the neurogenic process in the cortex, most recently we have gained mechanistic insights on how neurotransmitters act at this level by deciphering the role of glycine receptors (GlyRs) during cerebral cortical formation. GlyRs are ligand gated ion channels that share homology with GABA and the nicotinic acetyl choline receptors. As such, they are gated by glycine and contribute to synaptic inhibition in the mature central nervous system. Complementarily, during development, they provide excitation and contribute to the assembly of developing circuits in a similar way as GABA-A receptors. GlyRs, found on ventricular cells and the intermediate zone of the cortex, have been shown to control cortical progenitor cell dynamic affecting their differentiation rate. In particular, we showed that the absence of GlyRs induces progenitor cells to anticipate their differentiation into neurons skyping the amplification process provided by the generation of Tbr2-positive intermediate progenitors. Consequently, more neurons are generated initially at the expense of intermediate progenitors, all of which later impacts on the generation of upper layer neurons in the right numbers (Avila et al., 2014). Thus, GlyR knockout animals at embryonic age (E)11 produce more Tbr1-positive neurons destined to populate the deepest layers of the cortex while, at the end of corticogenesis, they produce less Ctip-2- and Cux-1–positive cells destined to populate the more superior layers of the cortex. This is in line with a moderated microcephaly found in these knockout animals (Avila et al., 2014). Defects induced by the lack of GlyRs in cell numbers are accompanied by morphological abnormalities in dendritic arborizations and later increased susceptibility to epilepsy (Morelli et al., 2016). Remarkably, hereditary mutations at the GlyR gene are associated in humans to the development of autism (Piton et al., 2011; Iossifov et al., 2014; Pilorge et al., 2016; Zhang et al., 2017). Therefore, animal studies have resulted informative in this case providing detailed mechanistic insights in parallel to the description of the human pathology.

Interneuron cell migration. Cortical neurons sequentially form all layers of the cerebral cortex. For this process to occur, it is necessary that the newly generated neurons migrate from their region of origin passing by the older neurons and layers. For excitatory neurons, this is what has been called radial migration and leads to the generation of an inside out patter of organization (Agirman et al., 2017; Silva et al., 2019). On the other hand, cortical interneurons are born in the ventral telencephalon, where they must migrate from the ganglionic eminences to the cortex. Consensus has emerged recognizing that this is a conserved process with interneurons being generated in a higher proportion in the caudal ganglionic eminence in humans comparted to rodents (Hansen et al., 2013; Ma et al., 2013). Interneurons play an important role modulating the activity of cortical networks (Markram et al., 2004). Consequently, disruption of interneuron development has been associated with autism spectrum disorder and Rett syndrome (Chao et al., 2010; Avila et al., 2013; Mariani et al., 2015). Interneurons are a diverse population formed by various cell types that characteristically have defined morphologies, express selected neuronal markers, display unique physiological features, and integrate into selected cortical circuits (Lim et al., 2018b). Interneuron migration takes place from the pallium via several stereotyped streams. This process is referred as tangential migration and occurs mainly through a superficial route in the mantle zone and a deep route in the subventricular zone. As it might be expected, this process is orchestrated by intricated cell intrinsic mechanisms and various cell extrinsic cues. Following the description of the actions of the cellular environment in the control of cortical neurogenesis, the migration process has also been shown to be sensitive to neurotransmitters. In this context, glycinergic neurotransmission has being implicated as well in the control of interneuron cell migration. It was shown that removal of GlyR from interneurons affects interneuron migration decreasing their migration speed. This was found to be the consequence of a diminished cell depolarization that decreased intracellular calcium oscillation and thus affected actomyosin contraction (Avila et al., 2013a,b; Ávila et al., 2020). These findings are somehow descriptive of the complexity in which the migration process takes place, but it also provides insights in the underlying mechanisms of human pathology as this has been discussed to be part of the neurodevelopmental defects linked to autism (Piton et al., 2011; Iossifov et al., 2014; Pilorge et al., 2016; Zhang et al., 2017).

Morphological maturation. In humans and species having a larger brain beyond the threshold of 7 g, the cerebral cortex is gyrencephalic or convoluted (Mota and Herculano-Houzel, 2015). This contrasts with species that have smaller brains, where the cerebral cortex is lissencephalic or with a smooth surface. This difference is a consequence of the evolutionary and developmental process in which a larger cerebral cortex involves a larger surface area. It has been hypothesized that, when the cortex is convoluted, this occurs because proliferation of cortical neurons outpaces that of the underlying subcortical nuclei (van Essen et al., 2018). This hypothesis implies that the cortex in lissencephalic brains have a comparatively smaller number of cells surrounding subcortical structures without needing to fold. Evidence supporting these ideas comes from contrasting the cortex of primates (van Essen, 2007). More in detail, mechanical tension along axons seems to be directing the folding process generating folding patterns that convey the formation of areas and functionality of regions (van Essen, 2007). At the cellular level, the extension of cellular processes is a heterogeneous event that provides enhanced growth to the cerebral cortex by expanding the surface area covered by cells and, by extension, occupied by the whole structure. When cell proliferation slows down, cerebral cortical volume changes dictated by adaptations of the average size of cells. For neuronal cells, this involves the generation of dendrites, axons, and synapses. These specialized structures develop asymmetrically with biased orientation projecting from the cell body toward preferred directions. This developmental process determines changes in the volume of the brain during later stages, although nonneuronal cells have also characteristic shapes and structures that develop over time.

CEREBELLAR DEVELOPMENT

Beyond motor functions, the cerebellum is associated with a number of superior functions such as cognition, affectivity (Stoodley et al., 2012; Salman and Tsai, 2016; Stoodley and Limperopoulos, 2016), articulation of language, working memory, and spacial processing (Sokolov et al., 2017). It is in the cerebellum where most brain neurons are located. In this structure, around 80% of brain neurons correspond to small granule cells packed in a dense cellular layer. The cerebellar cortex is convoluted in all mammals and nonmammalian vertebrate (Yopak et al., 2020). This is hypothesized to be consequence of the cerebellar cortex enveloping deep cerebellar nuclei that have small volumes (van Essen et al., 2018). The cerebellar cortex was at a time suggested to be the best site to understand the developmental process of the brain (Eccles, 1970). However, the early development of the cerebellum differs significantly from the developmental process of the cerebral cortex. The largest and more distinctive neurons in the cerebellum are Purkinje neurons. These cells are born from ventricular progenitors and migrate to form a monolayer below the surface of the cerebellar primordium. They then extend dendrites toward the pial surface. In turn, granule cells, which outnumber by far Purkinje neurons, are born from precursors located at the rhombic lip and migrate passing Purkinje neurons to form the granule cell layer. In this migration, granule cells leave a trailing axon that branch forming parallel fibers. Characteristically in humans, there is a rapid cerebellar expansion during perinatal brain growth. During the first 3 months of life, the cerebellum doubles its volume growing with the highest rate compared with other brain regions (Holland et al., 2014). This growth pattern shows sexual dimorphism with the developmental trajectory of females peaking at 11.8 years, while males reach this peak at 15.6 (Tiemeier et al., 2010).

SPINAL CORD DEVELOPMENT

Following closure of the neural tube, the human spinal cord can be divided into dorsal alar plates, which will give rise to sensory areas, and the ventral basal plates, which will originate motor areas of the spinal cord. These plates surround the neuroepithelium externally. The neuroepithelium of the early spinal cord also has a roof plate and a floor plate, located dorsally and ventrally, respectively, where axons will be crossing during postnatal stages. The marginal layer contains nerve fibers emerging from the immature neurons of the mantle layer. This will later become the white matter of the spinal cord (Ashwell, 2009). There are a number of extracellular signaling molecules responsible for controlling the dorsoventral patterning of the spinal cord, such as bone morphogenetic proteins (BMP) and SHH (discussed in Chapter 4). Dorsal spinal cord neurons arise from six domains (dI1−dI6). The interneuron subtypes dI1-3 migrate ventrally, dI4 and a subset of dI5 populate the deep layers of laminae IV−V, and another subset of dI5 and dI6 interneurons migrate to the ventral horns (ten Donkelaar et al., 2014). Ventral horn neurons arise from four interneuronal (V0, V1, V2, V3) and one motoneuronal (MN) subtypes, organized in five domains from ventral to dorsal axis (p0, 01, p2, pMN, and p3) (Danesin and Soula, 2017) (Table 1.1). There are a number of transcriptional factors that characterize the identity of the neuronal subtype that each domain will generate, either in the dorsal and ventral spinal cord (Dbx1, Dbx2, NKx2.2, NKx6.1, NKx6.2, Irx3, Olig2, Msx1, and Pax2, Pax6, Pax7, TLX3/ISL1, CHX10) (ten Donkelaar et al., 2014). Neuronal precursors in the developing spinal cord have been shown to express cytoskeletal proteins (i.e., vimentin, nestin, and glial fibrillary acidic protein (GFAP), keratin 8 and 18) following a spatiotemporal pattern (Figs. 1.1 and 1.2). For example, keratins and GFAP expressions are detected between 70 and 112 days of gestation in the white matter and have been associated with radial glia development; whereas vimentin expression is detected earlier, at 40 days of gestation in the central canal (Walder et al., 2003; Ferretti et al., 2006). Progenitor cells of the p0, p2, and p3 domains will firstly generate ventral interneurons (V0−V3) and later on will generate ventral astrocytes. Furthermore, progenitor cells of the pMN domain will firstly give rise to motor neurons and later on to oligodendrocytes (ten Donkelaar et al., 2014; Danesin and Soula, 2017). Lumbar motoneurons originate at about 4 weeks postconception in humans (Clowry et al., 2005). The MN group will give rise to medial, lateral, and intermediolateral neurons. Medial motorneurons will innervate axial musculature and are distributed through the entire spinal cord. Lateral motorneurons innervate limb musculature and are located at the cervical and lumbosacral enlargement. Intermediolateral motorneurons innervate viscera and are located in thoracic and upper lumbar segments (Ashwell, 2009). Regarding the glial cells in the human spinal cord, oligodendrocyte precursor cells have been detected in the dorsal spinal cord as early as at 74 days postconception and 83 days postconception in the ventral roots (Hajihosseini et al., 1996). Microglial cells can be detected at 9 weeks postconception and coincide with vascularization and neuronal migration (Verney et al., 2010).

TABLE 1.1 Major interneuron populations in the embryonic spinal cord and their characteristics.

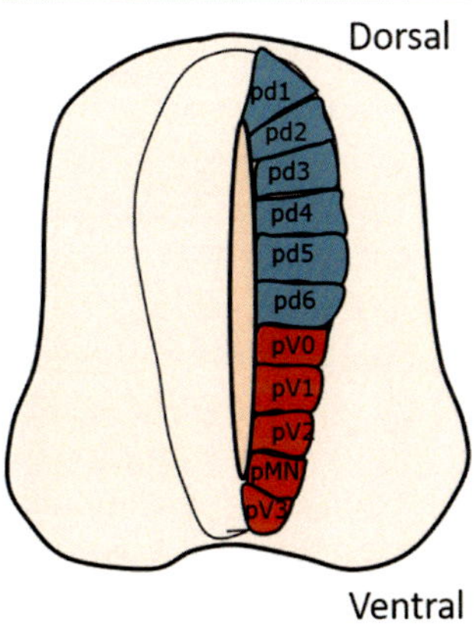

Interneuron domain	Function	Neurotransmitter phenotype	Interneuron population and projection characteristic
pd1	Sensory	Glutamatergic	Contralateral-projecting and ipsilateral-projecting
pd2	Sensory	Glutamatergic	Contralateral-projecting (speculative)
pd3	Sensory	Glutamatergic	Short ipsilateral
pd4	Sensory	GABAergic	Short ipsilateral and short commissural
pd5	Sensory	Glutamatergic	Contralateral-projecting
pd6	Motor	Gabaergic/glycinergic	Short commissural and ipsilateral
pV0	Motor	Gabaergic/glycinergic Acetylcholinergic Glutamatergic	$V0_v$, $V0_D$: long and short commissural $V0_c$: long and short ipsilateral; short ipsilateral and commissural
pV1	Motor	Gabaergic/glycinergic	Ipsilateral rostral interneurons
pV2	Motor	Gabaergic/glycinergic Glutamatergic	$V2_a$, $V2_b$: long and short ipsilateral
pMN	Motor	Acetylcholinergic	Peripheral projections
pV3	Motor	Glutamatergic	$V3_v$, $V3_D$: short commissural and ipsilateral; short ipsilateral

Interneuron domains, function, neurotransmitter type, and characteristics are listed.

Image courtesy of Zavvarian, M.-M., Hong, J., Fehlings, M.G., 2020. The functional role of spinal interneurons following traumatic spinal cord injury. Front. Cell. Neurosci. 14. https://doi.org/10.3389/fncel.2020.00127.

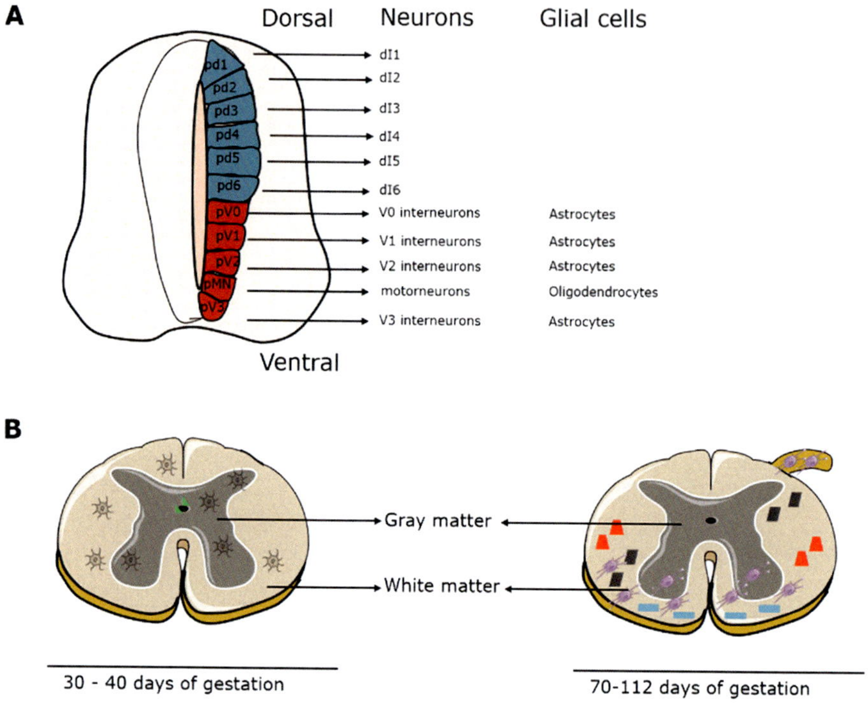

FIGURE 1.1 **Domain organization of dorsal and ventral neural progenitor cells in the spinal cord.** The expression of transcriptional factors preferentially expressed in ventral to dorsal domains (*black color*), intermediate areas (*light blue*), and ventral to dorsal domains (*dark blue*). The combined expression of these factors allows characterization and differentiation of diverse neuronal subpopulations in the spinal cord.

UNDERLYING ALTERATIONS LEADING TO NEURODEVELOPMENTAL DISORDERS

Morphogenic Defects

Neural tube defects (NTDs) are among the most common birth defects in humans and they arise from defective neurulation. Frequency of NTD varies from 0.1 to 12 in 1000 live births (Bhandari and Thada, 2011). Higher prevalence occurs in selected populations affected by environmental factors, but this also depends on genetic factors. The most severe forms of NTDs are fatal and include craniorachischisis and anencephaly, both of them resulting in perinatal death. The first one occurs from a completely open neural plate and the second from failure of closure in the anterior neural plate. A lesser severe NTD results from failure of neural tube closure in the spinal region and is called spina bifida. This is the most common NTD in humans and can manifest as myelomeningocele, meningocele, or spina bifida oculta. Among them, the first one is the most common and severe as it involves protrusion of meninges and spinal cord through the open vertebrae. Meningocele includes only protrusion of meninges and spina occulta occurs when there is malformation of the dorsal part of vertebrae. The consequences of spina bifida are diverse depending on the manifestation and the presence of associated conditions.

The most severe forms of NTDs, mentioned above, occur by defective convergent extension. Abnormal convergent extension in humans leads to fetuses with craniorachischisis (Kirillova et al., 2000). Causes of NTDs are diverse and include genetic predisposition, environmental factors, and medications. Supplementation with folate reduces the occurrence of NTDs by 50% to 70%, in line with folate deficiency being a well stablished risk factor for NTDs (MRC Vitamin Study Research Group, 1991; Czeizel and Dudás, 1992; Berry et al., 1999). Likewise, 82 gene variants in the folate/B12 metabolic pathway are associated with NTDs risk (Pangilinan et al., 2012). Use of valproic acid and carbamazepine is another risk factor that has more recently being established for NTDs (Werler et al., 2011). Regarding cortex formation, human clinical lissencephaly is another neurodevelopmental condition leading to morphological defects that involves an abnormal formation of cortical convolutions leading to smoother brains with thicker cortex. Manifestations of this condition conform a spectrum including agyria, pachygyria, and subcortical band heterotopia. To date, 19 genes have been associated with the condition and an imaging-based classification system with 21 recognizable patterns have been proposed to aid at genetic counseling, as it reliably predicts the most likely causative genes (Di Donato et al., 2017). The two most common causal genes were identified in the 90s and are *LIS1* and *DCX* (Reiner et al., 1993; des Portes et al., 1998). Cerebellar injuries during development can lead to various deleterious consequences such as motor impairments, cognitive outcomes, behavioral dysregulation,

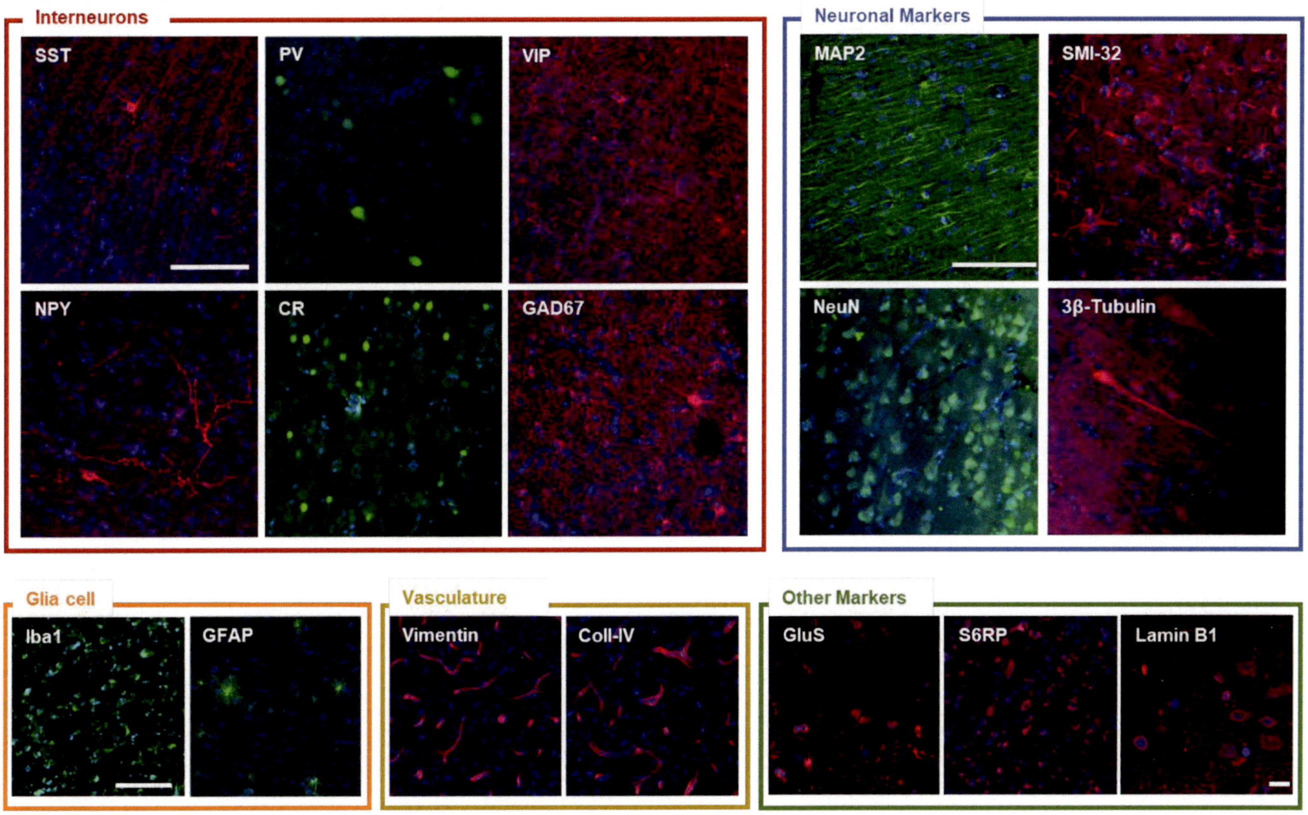

FIGURE 1.2 **Representative confocal images of the principal neural and nonneuronal markers.** *Interneurons*: somatostatin (SST), parvalbumin (PV), vasointestinal peptide (VIP), neuropeptide Y (NPY), calretinin (CR), glutamic acid decarboxylase (GAD67). *Neuronal markers*: microtubule-associated protein 2 (MAP2), nonphosphorylated neurofilament protein (SMI-32 antibody), neuron-specific nuclear protein (NeuN), 3b-Tub (3b-tubulin). *Glial cell*: ionized calcium binding adaptor molecule 1 (Iba1), glial fibrillary acidic protein (GFAP). *Vasculature*: vimentin, collagen IV (Coll-IV). *Other markers*: glutamine synthetase (GluS), phospho-S6 ribosomal protein (S6RP), lamin B1. Pixel size: 0.21 × 0.21 mm. Scale bar 100 um. Scale bar for lamin B1 image: 10 um. *Image courtesy of Pesce, L., Laurino, A., Scardigli, M., Yang, J., Boas, D.A., Hof, P.R., Destrieux, C., Costantini, I., Pavone, F.S., 2022. Exploring the human cerebral cortex using confocal microscopy. Prog. Biophys. Mol. Biol. 168, 3–9. https://doi.org/10.1016/j.pbiomolbio.2021.09.001. Epub 2021 Sep 16. PMID: 34536443; PMCID: PMC8992370.*

and autism-like phenotype (Stoodley and Limperopoulos, 2016). Structural growth trajectories vary across brain structures and thus, magnetic resonance imaging could be used to improve the detection of deviant maturational patterns indicative of neurodevelopmental disorders (Holland et al., 2014).

Environmental Insults

The neurodevelopmental process that leads to the formation of the nervous system is sensitive to many extrinsic factors. Among them, environmental factors associated with life style stand out as their consequences are most preventable.

Alcoholism. Ethanol consumption during pregnancy is one of the most harmful insults for the developing cortex. Alcoholism and alcohol abuse are major health problems affecting the life of millions of people throughout the world. This also has great consequences for the society and the economy. Ethanol is a depressant of the central nervous system (CNS) that can cause addiction and interference with the function of almost every organ. Affected brain functions include awareness, executive functions, cognition, learning, and memory (Niccols, 2007; Spanagel, 2009), among others. Importantly, during pregnancy, alcohol-induced brain damage is a primary source of acquired mental disability. Ethanol exposure during embryonic development can lead to Fetal Alcohol Spectrum Disorder (FASD). The most severe form of this spectrum disorder, the Fetal Alcohol Syndrome (FAS), is characterized by CNS impairment manifested in learning and behavioral disabilities (Driscoll et al., 1990; Spohr et al., 1993; Niccols, 2007; Stade et al., 2011; Denny et al., 2017). In United States, where 18.6% of women aged 35–44 reported drinking during pregnancy (Tan et al., 2015), early estimations quantified the incidence of FAS disorders in 9.1 births per 1000 (Streissguth, 1978). However, more recent studies have suggested that the prevalence can be as high as 1 in 100 live births, with a direct and indirect cost associated with FASD for Canada being approximately 5.3 billion per year (Stade et al., 2009). At the cellular level, ethanol has been shown to stimulate differentiation of embryonic stem cells (Serio et al., 2019), periodontal ligament cells and cementoblasts (Bae et al., 2018), and endothelial progenitors and monocytic cells (Vergori et al., 2016). By contrast, ethanol has been shown to inhibit the differentiation of neural crest cells (Fan et al., 2019), embryonic stem cells (Wang et al., 2017), and osteoclasts (Pan et al., 2016), while it also inhibits the proliferation of some neural stem cells (Kashem et al., 2018). In particular, neurodevelopmental processes affected by ethanol include neurogenesis (Livy et al., 2003), radial cell migration (Delatour et al., 2019), interneuron tangential migration (Skorput et al., 2015), axonal development (Scepanovic and Stewart, 2019), synapse maturation (Montesinos et al., 2018) and function (Lovinger and Roberto, 2010; Abrahao et al., 2017), angiogenesis (Holbrook et al., 2019), cellular metabolism (Stolakis et al., 2019), and cell death, among others (Cuzon et al., 2008). In this regard, various studies have focused on the effect of alcohol exposure on ligand-gated ion channels mediating inhibitory synaptic transmission in the mature central nervous system (Vengeliene et al., 2008; Abrahao et al., 2017). In fact, GABA and glycine receptors (GlyRs) are positively affected by ethanol leading to enhanced synaptic inhibition and accompanying behavioral effects (Burgos et al., 2015; Förstera et al., 2016). Nevertheless, is yet to be known in detail how these neurotransmitter systems mediate all the actions of ethanol during development.

Overnutrition. Besides all the negative consequences for the adult population and young children, overweight and obesity during pregnancy have important consequences for the proper development of the child's brain (Sullivan et al., 2014). In particular, the majority of new born babies in the United States have been exposed to maternal overweight and obesity status (Flegal et al., 2012). This will consequently impact on the programming of behaviors and, in a permissive environment, will likely predispose the infant to become obese later in life (Vucetic and Reyes, 2010). More strikingly, studies have shown that maternal obesity can be linked to the development of mental disorders in the offspring. It is now known for long that maternal obesity alters natural gestational metabolic adjustments and, in consequence, the fetus might be exposed to the consequences of preeclampsia, excessive exposure to fuel sources and altered oxygen and blood supply among other detrimental changes (King, 2006). More recently, neurodevelopmental disorders including autism spectrum disorder (ASD) were found to be associated with metabolic conditions during pregnancy in a case and control study (Krakowiak et al., 2012). In detail, it was found that diabetes, hypertension, and obesity were all maternal conditions associated with higher likelihood of ASD and developmental disorders. In particular, long-term effects for the child of a mother with obesity during pregnancy include increased chances of developing autism, mid intellectual disabilities, attention deficit hyperactivity disorder (ADHD), and schizophrenia in de adult (Krakowiak et al., 2012; Sullivan et al., 2014; Connolly et al., 2016). Likewise, a risk increase of cerebral palsy has been observed in children of obese mothers (Neri and Edlow, 2015). On the other hand, among ASD cases, children of mother with diabetes presented lower expressive language scores compared with children of mothers without metabolic conditions. In line with this, maternal high fat consumption have been found to be able to program offspring behavior (Sullivan et al., 2014). More directly, in a separate study, obesity and diabetes during pregnancy were associated with high risk of ASD and intellectual disabilities unlike others developmental disorders in the offspring (Li et al., 2016). A similar correlation study shows that obesity during pregnancy is associated with lower psychomotor development scores in boys. This is in opposition to what was found in girls at the same age of 3, whose psychomotor development was not affected (Nichols et al., 2019). Although with this result it could be hypothesized that baby boys can be more susceptible to maternal obesity, the mechanisms underlying this sex-specific association remain unclear.

In rodents, high-fat diet during pregnancy and lactation showed to have consequences in the phenotype of the offspring, altering important metabolic parameters and the transcription of key genes related to appetite control (Tozuka et al., 2009; Barrand et al., 2017; Hu et al., 2018). In line with that, high-fat diet has been shown to particularly alter fetal hippocampal development. These alterations, detected in the brain of newborn pups, included higher hippocampal cellular metabolism (Isac et al., 2018), changes in proliferation of neural precursors, decreased apoptosis, and decreased neuronal differentiation within the dentate gyrus (Niculescu and Lupu, 2009; Tozuka et al., 2009; 2010; Nam et al., 2017). Importantly, these cellular defects were found to correlate with later defects on learning processes (Tozuka et al., 2010). Despite all that, it is not clear how higher-order functions of the brain can be affected and lead to more complicated neuro-developmental disorders such as autism. A tentative explanation for this is that maternal high-fat diet modifies the immature hippocampus vulnerability to perinatal insults (Isac et al., 2018). Alternatively, various studies have looked at alterations of the normal synaptic plasticity in the context of maternal obesity (Matikainen-Ankney and Kravitz, 2018). More in detail, long-term potentiation elicited by a single high frequency stimulation train was found to be decreased in 2–3 weeks old pups weaned from mothers under maternal high-fat diet (Ding et al., 2018). Additionally, it has been demonstrated that high-fat diet results in higher concentration of serum leptin, which acts as a neurotrophic factor and modulates certain types of long-term potentiation of synapses (Shanley et al., 2001; Li et al., 2002). On the other hand, important synaptic proteins, such as synaptophysin and NMDA receptors, have been shown to be altered in response to maternal high-fat diet (Nam et al., 2017). Nevertheless, most of the studies on the effects of maternal high-fat diet have focused on the hippocampus disregarding the possible effects on the development of the cerebral cortex. This is particularly important since alterations on the formation and functioning of this structure are most likely to results in neuro-developmental disorders in the offspring.

PERSPECTIVES

Current research has embraced the study of neurodevelopmental disorders with various approaches. This is in expansion with the use of new technologies or renewed strategies. Hereafter some insights linked to the neurodevelopmental conditions described above.

hIPSCs technology. While research methods had advanced enormously in the last years and we have been able to noninvasively obtain important insights on the development of the human brain, animal models continue to be fundamental research tools for validation of experiments (van Essen et al., 2019). However, following up on the concern about the need of advancing toward human-specific mechanistic insights, new experimental strategies are being applied to understand neurodevelopmental processes. Even though there are conserved traits among humans and animal models, the higher-order functions that differentiate humans from other species are achieved by important cortical neural expansion, protracted time of development, particular genetic control, specification of brain areas, and development of extended connectivity (Zhao and Bhattacharyya, 2018). As a consequence, it is not possible to recapitulate all aspects of human brain development and neurological diseases in animal models. Remarkably, the field of neuroscience has been revolutionized by an explosive advance in the experimental strategies linked to the use of human induced pluripotent stem cells (hiPSCs). This has allowed researchers for the first time to model unique events occurring during embryonic human brain development and to explore the early mechanisms that cause synaptopathies (Marchetto et al., 2011; Mariani et al., 2012; Ardhanareeswaran et al., 2017). Since the first time that patients derived cell were obtained (Dimos et al., 2008; Park et al., 2008), this new technology has been widely adopted to study the effects of mutations, to assess the effects of drugs, to attempt cell replacement therapies and to investigate developmental biology processes in a human context. Although obtained from reprogramming of somatic cells, hiPSCs are equivalent to embryonic stem cells keeping their remarkable ability to differentiate into any adult cell. Currently, there are several methods available to differentiate hiPSCs toward specific neuronal subtypes, but essentially, they could be classified in monolayer cell cultures and organoid cultures (Bibel et al., 2007; Farra et al., 2012; Espuny-Camacho et al., 2013; Zhang et al., 2013; Kelava and Lancaster, 2016; Liu et al., 2017; Qian et al., 2018). In both cases, a series of protein markers such as Pax6, Sox2, Sox1, nestin, beta-III tubulin, Map2, and NeuN are used to evidence a neural cell phenotype. Moreover, functional maturation of hiPSCs-derived neurons can be complementarily assessed attending to their electrophysiological properties (Persoon et al., 2018; Rizo and Südhof, 2012; Song et al., 2013; Zhang et al., 2013; Prè et al., 2014; Emperador Melero et al., 2017; Kang et al., 2017). As an example of the power of this approach, the use of human embryoid has helped elucidating the role of FOXG1 gene, which causes a shift toward GABAergic neuronal fate that is associated with autism (Mariani et al., 2015). Nevertheless, modeling a condition involving neurodevelopmental disorders is a complex endeavor because the general nature of such disorders is heterogeneous and its approach requires the efficient characterization of the distinct cellular subtypes and circuits affected in correlation with the hallmark phenotypes of the pathology. Commonly, the clinical manifestations in patients affected by neurodevelopmental actions of ethanol occur at birth or during infancy showing alterations in several brain functions such as learning, memory, cognition, and sensorimotor behavior (Korkmaz, 2011). In this context, ethanol research has also been facilitated by the use of human iPSCs–derived neurons. In these cells, transcriptome studies have shown that several Notch signaling pathway genes that regulate cell fate and synaptic plasticity are affected by alcohol exposure (Jensen et al., 2019). Furthermore, excitatory N-methyl-D-aspartate (NMDA) and inhibitory GABAA receptor subunits genes shown significantly elevated expression on human iPSCs–derived forebrain neurons after acute and chronic treatment, respectively (Lieberman et al., 2018). On the other hand, studies linked to the effect of ethanol as primary cause of depletion of the stem cells pool either in adult neurogenesis or prenatal development have been described observing mixed results (Vangipuram and Lyman, 2010; Larsen et al., 2016; Le Maître et al., 2018). Thus, although using iPSCs exclusively to model the neurodevelopmental effects of ethanol will likely never precisely recapitulate what happen in humans, its use as a complementary approach can contribute to understand human-specific cellular and molecular mechanisms impossible to dissect exclusively using current animal models.

References

Abrahao, K.P., Salinas, A.G., Lovinger, D.M., 2017. Alcohol and the brain: neuronal molecular targets, synapses, and circuits. Neuron 96, 1223–1238.

Agirman, G., Broix, L., Nguyen, L., 2017. Cerebral cortex development: an outside-in perspective. FEBS (Fed. Eur. Biochem. Soc.) Lett. 591, 3978–3992.

Anderson, S.A., 2008. Interneuron Migration from Basal Forebrain to Neocortex: Dependence on Dlx Genes. Science 474, 16–18.

Angevine, J.B., Sidman, R.L., 1961. Autoradiographic study of cell migration during histogenesis of cerebral cortex in the mouse. Nature 192, 766–768.

Ardhanareeswaran, K., Mariani, J., Coppola, G., Abyzov, A., Vaccarino, F.M., 2017. Human induced pluripotent stem cells for modelling neurodevelopmental disorders. Nat. Rev. Neurol. 13, 265–278.

Aroeira, R.I., Ribeiro, J.A., Sebastião, A.M., Valente, C.A., 2011. Age-related changes of glycine receptor at the rat hippocampus: from the embryo to the adult. J. Neurochem. 118, 339–353.

Ashwell, K.W.S., 2009. Development of the spinal cord. Spinal Cord 8–16.

Avila, A., Nguyen, L., Rigo, J.-M., 2013a. Glycine receptors and brain development. Front. Cell. Neurosci. 7, 184.

Avila, A., Vidal, P.M., Dear, T.N., Harvey, R., Rigo, J.M., Nguyen, L., 2013b. Glycine receptor α2 subunit activation promotes cortical interneuron migration. Cell Rep. 4, 738–750.

Avila, A., Vidal, P.M., Tielens, S., Morelli, G., Laguesse, S., Harvey, R.J., Rigo, J., Nguyen, L., 2014. Glycine receptors control the generation of projection neurons in the developing cerebral cortex. Cell Death Differ. 1–13.

Ávila, D., Aedo, E., Sánchez-Hechavarria, M., Ávila, C., Ávila, A., 2020. Glycine receptor inhibition differentially affect selected neuronal populations of the developing embryonic cortex, as evidenced by the analysis of spontaneous calcium oscillations. Int. J. Mol. Sci. 21.

Bae, W.J., Park, J.S., Kang, S.K., Kwon, I.K., Kim, E.C., 2018. Effects of melatonin and its underlying mechanism on ethanol-stimulated senescence and osteoclastic differentiation in human periodontal ligament cells and cementoblasts. Int. J. Mol. Sci. 19.

Barrand, S., Crowley, T.M., Wood-Bradley, R.J., De Jong, K.A., Armitage, J.A., 2017. Impact of maternal high fat diet on hypothalamic transcriptome in neonatal Sprague Dawley rats. PLoS One 12, e0189492.

Belinsky, G.S., Singh, M.B., Oikonomou, K.D., McGovern, M.L., Antic, S.D., 2014. Dopamine receptors in human embryonic stem cell differentiation. In: Dopamine Receptor Technologies, vol 22, pp. 229–249.

Berry, R.J., Li, Z., Erickson, J.D., Li, S., Moore, C.A., Wang, H., Mulinare, J., Zhao, P., Wong, L.-Y.C., Gindler, J., Hong, S.-X., Hao, L., Gunter, E., Correa, A., 1999. Prevention of neural-tube defects with folic acid in China. N. Engl. J. Med. 341, 1485–1490.

Bhandari, J., Thada, P.K., 2011. Neural tube disorders. In: Encyclopedia of Child Behavior and Development. Springer US, p. 1010.

Bibel, M., Richter, J., Lacroix, E., Barde, Y.A., 2007. Generation of a defined and uniform population of CNS progenitors and neurons from mouse embryonic stem cells. Nat. Protoc. 2, 1034–1043.

Bono, F., Savoia, P., Guglielmi, A., Gennarelli, M., Piovani, G., Sigala, S., Leo, D., Espinoza, S., Gainetdinov, R.R., Devoto, P., Spano, P.F., Missale, C., Fiorentini, C., 2018. Role of dopamine D2/D3 receptors in development, plasticity, and neuroprotection in human iPSC-derived midbrain dopaminergic neurons. Mol. Neurobiol. 55, 1054–1067.

Brackmann, M., Zhao, C., Schmieden, V., Braunewell, K.H., 2004. Cellular and subcellular localization of the inhibitory glycine receptor in hippocampal neurons. Biochem. Biophys. Res. Commun. 324, 1137–1142.

Burgos, C.F., Muñoz, B., Guzman, L., Aguayo, L.G., 2015. Ethanol effects on glycinergic transmission: from molecular pharmacology to behavior responses. Pharmacol. Res. 101, 18–29.

Chao, H.T., Chen, H., Samaco, R.C., Xue, M., Chahrour, M., Yoo, J., Neul, J.L., Gong, S., Lu, H.C., Heintz, N., Ekker, M., Rubenstein, J.L., Noebels, J.L., Rosenmund, C., Zoghbi, H.Y., 2010. Dysfunction in GABA signalling mediates autism-like stereotypies and Rett syndrome phenotypes. Nature 468, 263–269.

Clowry, G.J., Moss, J.A., Clough, R.L., 2005. An immunohistochemical study of the development of sensorimotor components of the early fetal human spinal cord. J. Anat. 207, 313–324.

Connolly, N., Anixt, J., Manning, P., Ping-I Lin, D., Marsolo, K.A., Bowers, K., 2016. Maternal metabolic risk factors for autism spectrum disorder-an analysis of electronic medical records and linked birth data. Autism Res. 9, 829–837.

Copp, A.J., Greene, N.D.E., 2010. Genetics and development of neural tube defects. J. Pathol.

Cuzon, V.C., Yeh, P.W.L., Yanagawa, Y., Obata, K., Yeh, H.H., 2008. Ethanol consumption during early pregnancy alters the disposition of tangentially migrating GABAergic interneurons in the fetal cortex. J. Neurosci. 28, 1854–1864.

Czeizel, A.E., Dudás, I., 1992. Prevention of the first occurrence of neural-tube defects by periconceptional vitamin supplementation. N. Engl. J. Med. 327, 1832–1835.

Danesin, C., Soula, C., 2017. Moving the Shh source over time: what impact on neural cell diversification in the developing spinal cord? J. Dev. Biol. 5, 4.

Dehay, C., Kennedy, H., Kosik, K.S., 2015. The outer subventricular zone and primate-specific cortical complexification. Neuron 85, 683–694.

Delatour, L.C., Yeh, P.W., Yeh, H.H., 2019. Ethanol exposure *in utero* disrupts radial migration and pyramidal cell development in the somatosensory cortex. Cerebr. Cortex 29, 2125–2139.

Denny, L., Coles, S., Blitz, R., 2017. Fetal alcohol syndrome and fetal alcohol spectrum disorders. Am. Fam. Physician 96, 515–522.

des Portes, V., Francis, F., Pinard, J.M., Desguerre, I., Moutard, M.L., Snoeck, I., Meiners, L.C., Capron, F., Cusmai, R., Ricci, S., Motte, J., Echenne, B., Ponsot, G., Dulac, O., Chelly, J., Beldjord, C., 1998. Doublecortin is the major gene causing X-linked subcortical laminar heterotopia (SCLH). Hum. Mol. Genet. 7, 1063–1070.

Di Donato, N., Chiari, S., Mirzaa, G.M., Aldinger, K., Parrini, E., Olds, C., Barkovich, A.J., Guerrini, R., Dobyns, W.B., 2017. Lissencephaly: expanded imaging and clinical classification. Am. J. Med. Genet. 173, 1473.

Dimos, J.T., Rodolfa, K.T., Niakan, K.K., Weisenthal, L.M., Mitsumoto, H., Chung, W., Croft, G.F., Saphier, G., Leibel, R., Goland, R., Wichterle, H., Henderson, C.E., Eggan, K., 2008. Induced pluripotent stem cells generated from patients with ALS can be differentiated into motor neurons. Science (New York, NY) 321, 1218–1221.

Ding, Q., Zhao, Y., Yang, Y., Chen, Z., 2018. Cognitive impairment due to leptin withdrawal in rat offspring of dams with maternal diet-induced obesity. Med. Sci. Mon. Int. Med. J. Exp. Clin. Res. 24, 6208–6217.

Donahue, C.J., Glasser, M.F., Preuss, T.M., Rilling, J.K., van Essen, D.C., 2018. Quantitative assessment of prefrontal cortex in humans relative to nonhuman primates. Proc. Natl. Acad. Sci. U. S. A. 115, E5183.

Driscoll, C.D., Streissguth, A.P., Riley, E.P., 1990. Prenatal alcohol exposure: comparability of effects in humans and animal models. Neurotoxicol. Teratol. 12, 231–237.

Eccles, J.C., 1970. Neurogenesis and morphogenesis in the cerebellar cortex. Proc. Natl. Acad. Sci. U. S. A. 66, 294.

Emperador Melero, J., Nadadhur, A.G., Schut, D., Weering, J.V., Heine, V.M., Toonen, R.F., Verhage, M., 2017. Differential maturation of the two regulated secretory pathways in human iPSC-derived neurons. Stem Cell Rep. 8, 659–672.

Espuny-Camacho, I., Michelsen, K.A., Gall, D., Linaro, D., Hasche, A., Bonnefont, J., Bali, C., Orduz, D., Bilheu, A., Herpoel, A., Lambert, N., Gaspard, N., Péron, S., Schiffmann, S.N., Giugliano, M., Gaillard, A., Vanderhaeghen, P., 2013. Pyramidal neurons derived from human pluripotent stem cells integrate efficiently into mouse brain circuits in vivo. Neuron 77, 440–456.

Fan, H., Yuan, F., Yun, Y., Wu, T., Lu, L., Liu, J., Feng, W., Chen, S.Y., 2019. MicroRNA-34a mediates ethanol-induced impairment of neural differentiation of neural crest cells by targeting autophagy-related gene 9a. Exp. Neurol. 320, 112981.

Farra, N., Zhang, W.B., Pasceri, P., Eubanks, J.H., Salter, M.W., Ellis, J., 2012. Rett syndrome induced pluripotent stem cell-derived neurons reveal novel neurophysiological alterations. Mol. Psychiatr. 17, 1261–1271.

Ferretti, P., Mackay, M., Walder, S., 2006. The developing human spinal cord contains distinct populations of neural precursors. Neurodegener. Dis. 3, 38–44.

Flegal, K.M., Carroll, D., Kit, B.K., Ogden, C.L., 2012. Prevalence of obesity and trends in the distribution of body mass index among US adults, 1999–2010. JAMA 307, 491–497.

Förstera, B., Castro, P.A., Moraga-Cid, G., Aguayo, L.G., 2016. Potentiation of gamma aminobutyric acid receptors (GABAAR) by ethanol: how are inhibitory receptors affected? Front. Cell. Neurosci.

Gallo, V., Deneen, B., 2014. Glial development: the crossroads of regeneration and repair in the CNS. Neuron 83, 283–308.

Greene, N.D.E., Copp, A.J., 2014. Neural tube defects. Annu. Rev. Neurosci.

Greig, L.C., Woodworth, M.B., Galazo, M.J., Padmanabhan, H., Macklis, J.D., 2013. Molecular logic of neocortical projection neuron specification, development and diversity. Nat. Rev. Neurosci. 14, 755–769.

Hajihosseini, M., Tham, T.N., Dubois-Dalcq, M., 1996. Origin of oligodendrocytes within the human spinal cord. J. Neurosci. 16, 7981–7994.

Hansen, D.V., Lui, J.H., Parker, P.R., Kriegstein, A.R., 2010. Neurogenic radial glia in the outer subventricular zone of human neocortex. Nature 464, 554–561.

Hansen, D.V., Lui, J.H., Flandin, P., Yoshikawa, K., Rubenstein, J.L., Alvarez-Buylla, A., Kriegstein, A.R., 2013. Non-epithelial stem cells and cortical interneuron production in the human ganglionic eminences. Nat. Neurosci. 16, 1576–1587.

Herlenius, E., Lagercrantz, H., 2004. Development of neurotransmitter systems during critical periods. Exp. Neurol. 190, 8–21.

Holbrook, B.D., Davies, S., Cano, S., Shrestha, S., Jantzie, L.L., Rayburn, W.F., Bakhireva, L.N., Savage, D.D., 2019. The association between prenatal alcohol exposure and protein expression in human placenta. Birth Defects Res. 111, 749–759.

Holland, D., Chang, L., Ernst, T.M., Curran, M., Buchthal, S.D., Alicata, D., Skranes, J., Johansen, H., Hernandez, A., Yamakawa, R., Kuperman, J.M., Dale, A.M., 2014. Structural growth trajectories and rates of change in the first 3 months of infant brain development. JAMA Neurol. 71, 1266–1274.

Hu, S., Wang, L., Yang, D., Li, L., Togo, J., Wu, Y., Liu, Q., Li, B., Li, M., Wang, G., Zhang, X., Niu, C., Li, J., Xu, Y., Couper, E., Whittington-Davies, A., Mazidi, M., Luo, L., Wang, S., Douglas, A., Speakman, J.R., 2018. Dietary fat, but not protein or carbohydrate, regulates energy intake and causes adiposity in mice. Cell Metabol. 28, 415–431.

Iossifov, I., O'Roak, B.J., Sanders, S.J., Ronemus, M., Krumm, N., Levy, D., Stessman, H.A., Witherspoon, K.T., Vives, L., Patterson, K.E., Smith, J.D., Paeper, B., Nickerson, D.A., Dea, J., Dong, S., Gonzalez, L.E., Mandell, J.D., Mane, S.M., Murtha, M.T., Sullivan, C.A., Walker, M.F., Waqar, Z., Wei, L., Willsey, A.J., Yamrom, B., Lee, Y.H., Grabowska, E., Dalkic, E., Wang, Z., Marks, S., Andrews, P., Leotta, A., Kendall, J., Hakker, I., Rosenbaum, J., Ma, B., Rodgers, L., Troge, J., Narzisi, G., Yoon, S., Schatz, M.C., Ye, K., McCombie, W.R., Shendure, J., Eichler, E.E., State, M.W., Wigler, M., 2014. The contribution of de novo coding mutations to autism spectrum disorder. Nature 515, 216–221.

Isac, S., Panaitescu, A.M., Iesanu, M., Grigoras, I.F., Totan, A., Udriste, A., Cucu, N., Peltecu, G., Zagrean, L., Zagrean, A.-M., 2018. Maternal high-fat diet modifies the immature Hippocampus vulnerability to perinatal asphyxia in rats. Neonatology 114, 355–361.

Ishizuka, T., Ozawa, A., Goshima, H., Watanabe, Y., 2012. Involvement of nicotinic acetylcholine receptor in the proliferation of mouse induced pluripotent stem cells. Life Sci. 90, 637–648.

Jensen, K.P., Lieberman, R., Kranzler, H.R., Gelernter, J., Clinton, K., Covault, J., 2019. Alcohol-responsive genes identified in human iPSC-derived neural cultures. Transl. Psychiatry 9, 96.

Juriloff, D.M., Harris, M.J., 2012. A consideration of the evidence that genetic defects in planar cell polarity contribute to the etiology of human neural tube defects. Birth Defects Res. Part A Clin. Mol. Teratol. 94, 824.

Kang, S., Chen, X., Gong, S., Yu, P., Yau, S., Su, Z., Zhou, L., Yu, J., Pan, G., Shi, L., 2017. Characteristic analyses of a neural differentiation model from iPSC-derived neuron according to morphology, physiology, and global gene expression pattern. Sci. Rep. 7, 1–11.

Kashem, M.A., Sultana, N., Balcar, V.J., 2018. Exposure of rat neural stem cells to ethanol affects cell numbers and alters expression of 28 proteins. Neurochem. Res. 43, 1841–1854.

Kelava, I., Lancaster, M.A., 2016. Stem cell models of human brain development. Cell Stem Cell 18, 736–748.

Kemp, P.J., Rushton, D.J., Yarova, P.L., Schnell, C., Geater, C., Hancock, J.M., Wieland, A., Hughes, A., Badder, L., Cope, E., Riccardi, D., Randall, A.D., Brown, J.T., Allen, N.D., Telezhkin, V., 2016. Improving and accelerating the differentiation and functional maturation of human stem cell-derived neurons: role of extracellular calcium and GABA. J. Physiol. 594, 6583–6594.

King, J.C., 2006. Maternal obesity, metabolism, and pregnancy outcomes. Annu. Rev. Nutr. 26, 271–291.

Kirillova, I., Novikova, I., Augé, J., Audollent, S., Esnault, D., Encha-Razavi, F., Lazjuk, G., Attié-Bitach, T., Vekemans, M., 2000. Expression of the Sonic hedgehog gene in human embryos with neural tube defects. Teratology 61, 347–354.

Korkmaz, B., 2011. Theory of mind and neurodevelopmental disorders of childhood. Pediatr. Res. 69, 101R–108R.

Krakowiak, P., Walker, C.K., Bremer, A.A., Baker, A.S., Ozonoff, S., Hansen, R.L., Hertz-Picciotto, I., 2012. Maternal metabolic conditions and risk for autism and other neurodevelopmental disorders. Pediatrics 129, e1121–e1128.

Larsen, Z.H., Chander, P., Joyner, J.A., Floruta, C.M., Demeter, T.L., Weick, J.P., 2016. Effects of ethanol on cellular composition and network excitability of human pluripotent stem cell-derived neurons. Alcohol Clin. Exp. Res. 40, 2339–2350.

Le Maître, T.W., Dhanabalan, G., Bogdanovic, N., Alkass, K., Druid, H., 2018. Effects of alcohol abuse on proliferating cells, stem/

progenitor cells, and immature neurons in the adult human hippocampus. Neuropsychopharmacology 43, 690–699.

Li, X.-L., Aou, S., Oomura, Y., Hori, N., Fukunaga, K., Hori, T., 2002. Impairment of long-term potentiation and spatial memory in leptin receptor-deficient rodents. Neuroscience 113, 607–615.

Li, M., Fallin, M.D., Riley, A., Landa, R., Walker, S.O., Silverstein, M., Caruso, D., Pearson, C., Kiang, S., Dahm, J.L., Hong, X., Wang, G., Wang, M.C., Zuckerman, B., Wang, X., 2016. The association of maternal obesity and diabetes with autism and other developmental disabilities. Pediatrics 137.

Lieberman, R., Kranzler, H.R., Levine, E.S., Covault, J., 2018. Examining the effects of alcohol on GABA A receptor mRNA expression and function in neural cultures generated from control and alcohol dependent donor induced pluripotent stem cells. Alcohol 66, 45–53.

Lim, L., Mi, D., Llorca, A., Marín, O., 2018. Development and functional diversification of cortical interneurons. Neuron 100, 294–313.

Liu, X., Campanac, E., Cheung, H.H., Ziats, M.N., Canterel-Thouennon, L., Raygada, M., Baxendale, V., Pang, A.L.Y., Yang, L., Swedo, S., Thurm, A., Lee, T.L., Fung, K.P., Chan, W.Y., Hoffman, D.A., Rennert, O.M., 2017. Idiopathic autism: cellular and molecular phenotypes in pluripotent stem cell-derived neurons. Mol. Neurobiol. 54, 4507–4523.

Livy, D.J., Miller, E.K., Maier, S.E., West, J.R., 2003. Fetal alcohol exposure and temporal vulnerability: effects of binge-like alcohol exposure on the developing rat hippocampus. Neurotoxicol. Teratol. 25, 447–458.

Lodato, S., Arlotta, P., 2015. Generating neuronal diversity in the mammalian cerebral cortex. Annu. Rev. Cell Dev. Biol. 31, 699–720.

Lovinger, D.M., Roberto, M., 2010. Synaptic effects induced by alcohol. In: Current Topics in Behavioral Neurosciences, pp. 31–86.

Ma, T., Wang, C., Wang, L., Zhou, X., Tian, M., Zhang, Q., Zhang, Y., Li, J., Liu, Z., Cai, Y., Liu, F., You, Y., Chen, C., Campbell, K., Song, H., Ma, L., Rubenstein, J.L., Yang, Z., 2013. Subcortical origins of human and monkey neocortical interneurons. Nat. Neurosci. 16, 1588–1597.

Marchetto, M.C., Brennand, K.J., Boyer, L.F., Gage, F.H., 2011. Induced pluripotent stem cells (iPSCs) and neurological disease modeling: progress and promises. Hum. Mol. Genet. 20, R109–R115.

Mariani, J., Simonini, M.V., Palejev, D., Tomasini, L., Coppola, G., Szekely, A.M., Horvath, T.L., Vaccarino, F.M., 2012. Modeling human cortical development in vitro using induced pluripotent stem cells. Proc. Natl. Acad. Sci. U. S. A. 109, 12770–12775.

Mariani, J., Coppola, G., Zhang, P., Abyzov, A., Provini, L., Tomasini, L., Amenduni, M., Szekely, A., Palejev, D., Wilson, M., Gerstein, M., Grigorenko, E.L., Chawarska, K., Pelphrey, K.A., Howe, J.R., Vaccarino, F.M., 2015. FOXG1-dependent dysregulation of GABA/glutamate neuron differentiation in autism spectrum disorders. Cell 162, 375–390.

Marín, O., 2013. Cellular and molecular mechanisms controlling the migration of neocortical interneurons. Eur. J. Neurosci. 38, 2019–2029.

Markram, H., Toledo-Rodriguez, M., Wang, Y., Gupta, A., Silberberg, G., Wu, C.H., 2004. Interneurons of the neocortical inhibitory system. Nat. Rev. Neurosci. 5, 793–807.

Matikainen-Ankney, B.A., Kravitz, A.V., 2018. Persistent effects of obesity: a neuroplasticity hypothesis. Ann. N. Y. Acad. Sci. 1428, 221–239.

Miyata, T., Kawaguchi, A., Okano, H., Ogawa, M., 2001. Asymmetric inheritance of radial glial fibers by cortical neurons. Neuron 31, 727–741.

Montesinos, J., Pascual, M., Millán-Esteban, D., Guerri, C., 2018. Binge-like ethanol treatment in adolescence impairs autophagy and hinders synaptic maturation: role of TLR4. Neurosci. Lett. 682, 85–91.

Morelli, G., Avila, A., Ravanidis, S., Aourz, N., Neve, R.L., Smolders, I., Harvey, R.J., Rigo, J.-M., Nguyen, L., Brône, B., 2016. Cerebral cortical circuitry formation requires functional glycine receptors. Cerebr. Cortex.

Mota, B., Herculano-Houzel, S., 2015. Cortical folding scales universally with surface area and thickness, not number of neurons. Science (New York, NY) 349, 74–77.

MRC Vitamin Study Research Group, 1991. Prevention of neural tube defects: results of the Medical Research Council Vitamin Study. Lancet 338, 131–137.

Nam, S.M., Kim, J.W., Kwon, H.J., Yoo, D.Y., Jung, H.Y., Kim, D.W., Hwang, I.K., Seong, J.K., Yoon, Y.S., 2017. Differential effects of low- and high-dose zinc supplementation on synaptic plasticity and neurogenesis in the hippocampus of control and high-fat diet-fed mice. Neurochem. Res. 42, 3149–3159.

Neri, C., Edlow, A.G., 2015. Effects of maternal obesity on fetal programming: molecular approaches. Cold Spring Harb. Perspect. Med. 6, a026591.

Neubauer, S., Hublin, J.J., Gunz, P., 2018. The evolution of modern human brain shape. Sci. Adv. 4, eaao5961.

Niccols, A., 2007. Fetal alcohol syndrome and the developing socio-emotional brain. Brain Cognit. 65, 135–142.

Nichols, A.R., Rundle, A.G., Factor-Litvak, P., Insel, B.J., Hoepner, L., Rauh, V., Perera, F., Widen, E.M., 2019. Prepregnancy obesity is associated with lower psychomotor development scores in boys at age 3 in a low-income, minority birth cohort. J. Dev. Orig. Health Dis.

Niculescu, M.D., Lupu, D.S., 2009. High fat diet-induced maternal obesity alters fetal hippocampal development. Int. J. Dev. Neurosci. 27, 627–633.

Nikolopoulou, E., Galea, G.L., Rolo, A., Greene, N.D.E., Copp, A.J., 2017. Neural tube closure: cellular, molecular and biomechanical mechanisms. Development (Cambridge) 144, 552.

Noctor, S.C., Martínez-Cerdeño, V., Ivic, L., Kriegstein, A.R., 2004. Cortical neurons arise in symmetric and asymmetric division zones and migrate through specific phases. Nat. Neurosci. 7, 136–144.

O'Rahilly, R., Müller, F., 1994. Neurulation in the normal human embryo. In: Ciba Foundation Symposium.

O'Rahilly, R., Müller, F., 2002. The two sites of fusion of the neural folds and the two neuropores in the human embryo. Teratology 65, 162–170.

O'Rahilly, R., Müller, F., 2007. Neurulation in the normal human embryo. In: Ciba Foundation Symposium, pp. 70–89.

Pangilinan, F., Molloy, A.M., Mills, J.L., Troendle, J.F., Parle-McDermott, A., Signore, C., O'Leary, V.B., Chines, P., Seay, J.M., Geiler-Samerotte, K., Mitchell, A., VanderMeer, J.E., Krebs, K.M., Sanchez, A., Cornman-Homonoff, J., Stone, N., Conley, M., Kirke, P.N., Shane, B., Scott, J.M., Brody, L.C., 2012. Evaluation of common genetic variants in 82 candidate genes as risk factors for neural tube defects. BMC Med. Genet. 13, 62.

Pan, Z., Zhang, X., Shangguan, Y., Hu, H., Chen, L., Wang, H., 2016. Suppressed osteoclast differentiation at the chondro-osseous junction mediates endochondral ossification retardation in long bones of Wistar fetal rats with prenatal ethanol exposure. Toxicol. Appl. Pharmacol. 305, 234–241.

Park, I.-H., Arora, N., Huo, H., Maherali, N., Ahfeldt, T., Shimamura, A., Lensch, M.W., Cowan, C., Hochedlinger, K., Daley, G.Q., 2008. Disease-specific induced pluripotent stem cells. Cell 134, 877–886.

Persoon, C.M., Moro, A., Nassal, J.P., Farina, M., Broeke, J., Arora, S., Dominguez, N., van Weering, J.R., Toonen, R.F., Verhage, M., 2018. Pool size estimations for dense-core vesicles in mammalian CNS neurons. EMBO J. 37 (20), e99672.

Peyre, E., Silva, C.G., Nguyen, L., 2015. Crosstalk between intracellular and extracellular signals regulating interneuron production, migration and integration into the cortex. Front. Cell. Neurosci. 9, 1–18.

Pilorge, M., Fassier, C., Le Corronc, H., Potey, A., Bai, J., De Gois, S., Delaby, E., Assouline, B., Guinchat, V., Devillard, F., Delorme, R.,

Nygren, G., Rastam, M., Meier, J.C., Otani, S., Cheval, H., James, V.M., Topf, M., Dear, T.N., Gillberg, C., Leboyer, M., Giros, B., Gautron, S., Hazan, J., Harvey, R.J., Legendre, P., Betancur, C., 2016. Genetic and functional analyses demonstrate a role for abnormal glycinergic signaling in autism. Mol. Psychiatr.

Piton, A., Gauthier, J., Hamdan, F.F., Lafrenière, R.G., Yang, Y., Henrion, E., Laurent, S., Noreau, A., Thibodeau, P., Karemera, L., Spiegelman, D., Kuku, F., Duguay, J., Destroismaisons, L., Jolivet, P., Côté, M., Lachapelle, K., Diallo, O., Raymond, A., Marineau, C., Champagne, N., Xiong, L., Gaspar, C., Rivière, J.B., Tarabeux, J., Cossette, P., Krebs, M.O., Rapoport, J.L., Addington, A., Delisi, L.E., Mottron, L., Joober, R., Fombonne, E., Drapeau, P., Rouleau, G.A., 2011. Systematic resequencing of X-chromosome synaptic genes in autism spectrum disorder and schizophrenia. Mol. Psychiatr. 16, 867–880.

Prè, D., Nestor, M.W., Sproul, A.A., Jacob, S., Koppensteiner, P., Chinchalongporn, V., Zimmer, M., Yamamoto, A., Noggle, S.A., Arancio, O., 2014. A time course analysis of the electrophysiological properties of neurons differentiated from human induced Pluripotent Stem Cells (iPSCs). PLoS One 9.

Qian, X., Jacob, F., Song, M.M., Nguyen, H.N., Song, H., Ming, G.-L., 2018. Generation of human brain region-specific organoids using a miniaturized spinning bioreactor. Nat. Protoc. 13, 565–580.

Rakic, P., 1972. Mode of cell migration to the superficial layers of fetal monkey neocortex. J. Comp. Neurol. 145, 61–83.

Reiner, O., Carrozzo, R., Shen, Y., Wehnert, M., Faustinella, F., Dobyns, W.B., Caskey, C.T., Ledbetter, D.H., 1993. Isolation of a Miller-Dieker lissencephaly gene containing G protein beta-subunit-like repeats. Nature 364, 717–721.

Rizo, J., Südhof, T.C., 2012. The membrane fusion enigma: SNAREs, Sec1/Munc18 proteins, and their accomplices—guilty as charged? Annu. Rev. Cell Dev. Biol.

Salman, M.S., Tsai, P., 2016. The Role of the Pediatric Cerebellum in Motor Functions, Cognition, and Behavior: A Clinical Perspective. Neuroimaging Clinics of North America.

Scepanovic, G., Stewart, B.A., 2019. Analysis of *Drosophila* nervous system development following an early, brief exposure to ethanol. Dev. Neurobiol. 79, 780–793.

Serio, R.N., Laursen, K.B., Urvalek, A.M., Gross, S.S., Gudas, L.J., 2019. Ethanol promotes differentiation of embryonic stem cells through retinoic acid receptor-γ. J. Biol. Chem. 294, 5536–5548.

Sessa, A., Mao, C.A., Hadjantonakis, A.K., Klein, W.H., Broccoli, V., 2008. Tbr2 directs conversion of radial glia into basal precursors and guides neuronal amplification by indirect neurogenesis in the developing neocortex. Neuron 60, 56–69.

Shanley, L.J., Irving, A.J., Harvey, J., 2001. Leptin enhances NMDA receptor function and modulates hippocampal synaptic plasticity. J. Neurosci. 21, RC186.

Silva, C.G., Peyre, E., Nguyen, L., 2019. Cell migration promotes dynamic cellular interactions to control cerebral cortex morphogenesis. Nat. Rev. Neurosci. 1.

Skorput, A.G.J., Gupta, V.P., Yeh, P.W.L., Yeh, H.H., 2015. Persistent interneuronopathy in the prefrontal cortex of young adult offspring exposed to ethanol in utero. J. Neurosci. 35, 10977–10988.

Sokolov, A.A., Miall, R.C., Ivry, R.B., 2017. The cerebellum: adaptive prediction for movement and cognition. Trends Cognit. Sci.

Song, M., Mohamad, O., Chen, D., Yu, S.P., 2013. Coordinated development of voltage-gated Na+ and K+ currents regulates functional maturation of forebrain neurons derived from human induced pluripotent stem cells. Stem Cell. Dev. 22, 1551–1563.

Spanagel, R., 2009. Alcoholism: a systems approach from molecular physiology to addictive behavior. Physiol. Rev. 89, 649–705.

Spohr, H.-L., Willms, J., Steinhausen, H.-C., 1993. Prenatal alcohol exposure and long-term developmental consequences. Lancet 341, 907–910.

Stade, B., Ali, A., Bennett, D., Campbell, D., Johnston, M., Lens, C., Tran, S., Koren, G., 2009. The burden of prenatal exposure to alcohol: revised measurement of cost. Can. J. Clin. Pharmacol 16, e91–e102.

Stade, B., Beyene, J., Buller, K., Ross, S., Patterson, K., Stevens, B., Sgro, M., Ungar, W., Watson, W., Koren, G., 2011. Feeling different: the experience of living with fetal alcohol spectrum disorder. J. Popul. Ther. Clin. Pharmacol. 18, e475–e485.

Stancik, E.K., Navarro-Quiroga, I., Sellke, R., Haydar, T.F., 2010. Heterogeneity in ventricular zone neural precursors contributes to neuronal fate diversity in the postnatal neocortex. J. Neurosci. 30, 7028–7036.

Stolakis, V., Liapi, C., Al-Humadi, H., Kalafatakis, K., Gkanti, V., Bimpis, A., Skandali, N., Tsela, S., Theocharis, S., Zarros, A., Tsakiris, S., 2019. Effects of gestational thiamine-deprivation and/or exposure to ethanol on crucial offspring rat brain enzyme activities. J. Matern. Fetal Neonatal Med.

Stoodley, C.J., Limperopoulos, C., 2016. Structure–function relationships in the developing cerebellum: evidence from early-life cerebellar injury and neurodevelopmental disorders. Semin. Fetal Neonatal Med.

Stoodley, C.J., Valera, E.M., Schmahmann, J.D., 2012. Functional topography of the cerebellum for motor and cognitive tasks: an fMRI study. Neuroimage 59, 1560–1570.

Streissguth, A.P., 1978. Fetal alcohol syndrome: an epidemiologic perspective. Am. J. Epidemiol. 107, 467–478.

Sullivan, E.L., Nousen, E.K., Chamlou, K.A., 2014. Maternal high fat diet consumption during the perinatal period programs offspring behavior. Physiol. Behav. 123, 236–242.

Tan, C.H., Denny, C.H., Cheal, N.E., Sniezek, J.E., Kanny, D., 2015. Alcohol use and binge drinking among women of childbearing age – United States, 2011–2013. In: Morbidity and Mortality Weekly Report.

Taverna, E., Götz, M., Huttner, W.B., 2014. The cell biology of neurogenesis: toward an understanding of the development and evolution of the neocortex. Annu. Rev. Cell Dev. Biol.

ten Donkelaar, H.J., Yamada, S., Shiota, K., van der Vliet, T., 2014. Overview of the development of the human brain and spinal cord. Clin. Neuroembryol. 1–52.

Tiemeier, H., Lenroot, R.K., Greenstein, D.K., Tran, L., Pierson, R., Giedd, J.N., 2010. Cerebellum development during childhood and adolescence: a longitudinal morphometric MRI study. Neuroimage 49, 63–70.

Tozuka, Y., Wada, E., Wada, K., 2009. Diet-induced obesity in female mice leads to peroxidized lipid accumulations and impairment of hippocampal neurogenesis during the early life of their offspring. Faseb. J. 23, 1920–1934.

Tozuka, Y., Kumon, M., Wada, E., Onodera, M., Mochizuki, H., Wada, K., 2010. Maternal obesity impairs hippocampal BDNF production and spatial learning performance in young mouse offspring. Neurochem. Int. 57, 235–247.

van Essen, D.C., 2007. Cerebral cortical folding patterns in primates: why they vary and what they signify. Evol. Nerv. Syst. 4, 267–276.

van Essen, D.C., Donahue, C.J., Glasser, M.F., 2018. Development and evolution of cerebral and cerebellar cortex. Brain Behav. Evol. 91, 158.

van Essen, D.C., Donahue, C.J., Coalson, T.S., Kennedy, H., Hayashi, T., Glasser, M.F., 2019. Cerebral cortical folding, parcellation, and connectivity in humans, nonhuman primates, and mice. Proc. Natl. Acad. Sci. U. S. A. 116, 26173.

Vangipuram, S.D., Lyman, W.D., 2010. Ethanol alters cell fate of fetal human brain-derived stem and progenitor cells. Alcohol Clin. Exp. Res. 34, 1574–1583.

Vengeliene, V., Bilbao, A., Molander, A., Spanagel, R., 2008. Neuropharmacology of alcohol addiction. Br. J. Pharmacol. 154, 299–315.

Vergori, L., Lauret, E., Soleti, R., Martinez, M.C., Andriantsitohaina, R., 2016. Low concentration of ethanol favors progenitor cell differentiation and neovascularization in high-fat diet-fed mice model. Int. J. Biochem. Cell Biol. 78, 43–51.

Verney, C., Monier, A., Fallet-Bianco, C., Gressens, P., 2010. Early microglial colonization of the human forebrain and possible involvement in periventricular white-matter injury of preterm infants. J. Anat. 217, 436–448.

Vucetic, Z., Reyes, T.M., 2010. Central dopaminergic circuitry controlling food intake and reward: implications for the regulation of obesity. Wiley Interdiscip. Rev. Syst. Biol. Med. 2, 577–593.

Walder, S., Zhang, F., Ferretti, P., 2003. Up-regulation of neural stem cell markers suggests the occurrence of dedifferentiation in regenerating spinal cord. Dev. Gene. Evol. 213, 625–630.

Wang, Q., Song, J.-W., Liu, Y., Zhao, X.X., 2017. Involvement of Wnt pathway in ethanol-induced inhibition of mouse embryonic stem cell differentiation. Alcohol 58, 13–18.

Werler, M.M., Ahrens, K.A., Bosco, J.L.F., Mitchell, A.A., Anderka, M.T., Gilboa, S.M., Holmes, L.B., 2011. Use of antiepileptic medications in pregnancy in relation to risks of birth defects. Ann. Epidemiol. 21, 842–850.

Wilde, J.J., Petersen, J.R., Niswander, L., 2014. Genetic, epigenetic, and environmental contributions to neural tube closure. Annu. Rev. Genet. 48, 583–611.

Wilson, S.W., Houart, C., 2004. Early steps in the development of the forebrain. Dev. Cell 6, 167–181.

Yopak, K.E., Pakan, J.M.P., Wylie, D., 2020. The cerebellum of nonmammalian vertebrates. Evol. Neurosci. 247–260.

Zhang, Y., Ho, T.N.T., Harvey, R.J., Lynch, J.W., Keramidas, A., 2017. Structure-function analysis of the glyR α2 subunit autism mutation p.R323l reveals a gain-of-function. Front. Mol. Neurosci. 10, 1–13.

Zhang, Y., Pak, C.H., Han, Y., Ahlenius, H., Zhang, Z., Chanda, S., Marro, S., Patzke, C., Acuna, C., Covy, J., Xu, W., Yang, N., Danko, T., Chen, L., Wernig, M., Südhof, T.C., 2013. Rapid single-step induction of functional neurons from human pluripotent stem cells. Neuron 78, 785–798.

Zhao, X., Bhattacharyya, A., 2018. Human models are needed for studying human neurodevelopmental disorders. Am. J. Hum. Genet. 103, 829–857.

Zohn, I.E., 2020. Mouse models of neural tube defects. In: Advances in Experimental Medicine and Biology. Springer, pp. 39–64.

2

Topography of Neuraxis, Vascularization and Ventricular System

Brandon M. Ascenzi[1], *Gökşin Nilüfer Demirci*[2,a], *Silvia Gigli*[3,b] *and Gülgün Şengül*[4,5,a]

[1]Independent Neuroresearcher-Member of Marie Curie Alumni Association (MCAA) [2]Department of Anatomy, School of Medicine, Dokuz Eylul University, Izmir, Turkey [3]Department of Radiological, Oncological and Pathological Sciences, University of Rome "La Sapienza", Rome, Italy [4]Department of Anatomy, School of Medicine, Ege University, Izmir, Turkey [5]Department of Neuroscience, Institute of Health Sciences, Ege University, Izmir, Turkey

OUTLINE

FIRST SECTION: BLOOD SUPPLY, MENINGES AND CSF CIRCULATION

Blood Supply of the Brain

The brain arteries originate from two of the major vessels in the neck: the carotids and the vertebral arteries (VA) (Fig. 2.1). The internal carotid artery (ICA) enters the skull through the petrosal bone in the carotid canal and passes through the cavernous sinus where the ophthalmic artery gives off (Figs. 20.3 and 20.4). Immediately beyond the origin of the posterior communicating artery (PCOM), ICA divides into the middle and anterior cerebral arteries. The vertebral arteries enter the skull through the foramen magnum and, after their passage through the dura mater, they become located within the connective tissue

[a]They contributed to the third section.

[b]She contributed to the second section.

From Anatomy to Function of the Central Nervous System
https://doi.org/10.1016/B978-0-12-822404-5.00022-X

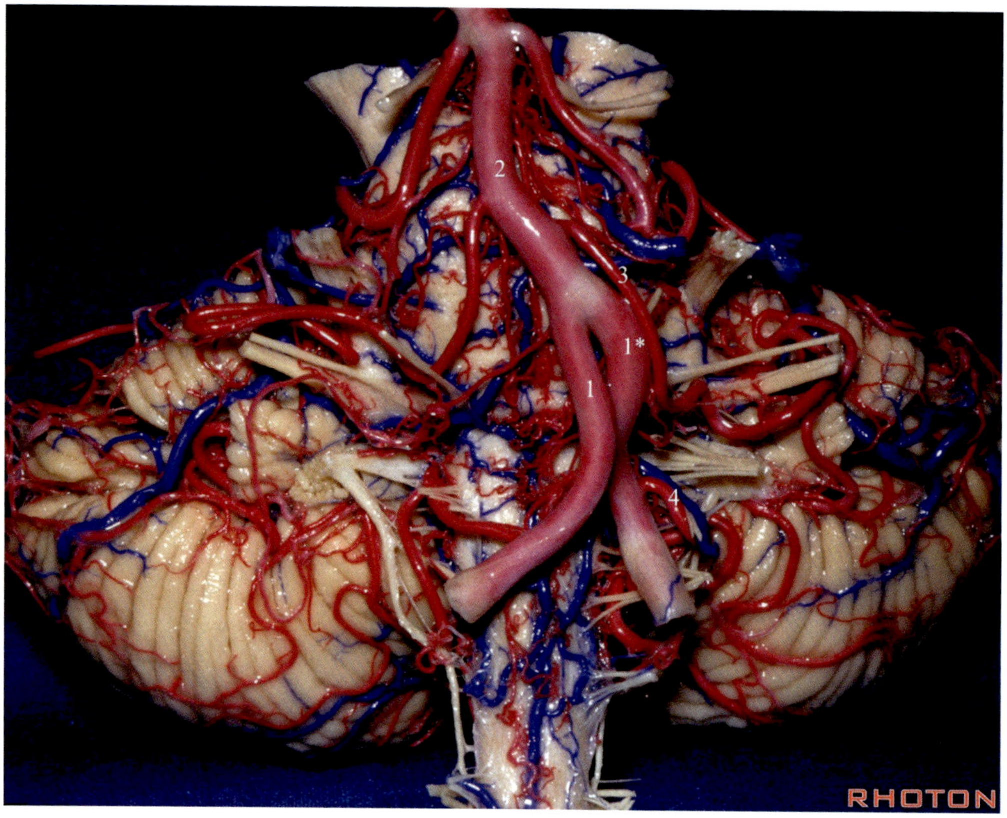

FIGURE 2.1 **Ventral view of the vertebrobasilar system (VBS). It is essentially constituted by:** 1, Vertebral Arteries (VA); and 2, Basilar Artery (BA). It also gives rise: to the Superior Cerebellar Artery (SCA, Figs .2.3A and 2.4); to 3, Anterior Inferior Cerebellar Artery (AICA); and to 4, Posterior Inferior Cerebellar Artery (PICA). *Image courtesy of The Rhoton Collection® http://rhoton.ineurodb.org/.*

derived from the pia mater and the arachnoid (Fig. 2.4).

The middle cerebral artery (MCA) supplies the convexity of the hemisphere (Figs. 2.3A, B, 2.4 and 2.5), and the anterior cerebral artery (ACA) supplies approximately the anterior and superior half of the medial aspect of the hemisphere, up to the precuneus. The vertebral arteries join in the basilar artery (BA) at the ventral aspect of the medulla oblongata. From the BA originate the right and left posterior cerebral arteries (PCA) which supply the posterior, medial, and basal aspects of the cerebral hemisphere. Thus, the vertebrobasilar system (Fig. 2.1), comprised of the aforementioned BA and VA, supply the brainstem and cerebellum and gives rise to the superior cerebrellar artery (SCA, Figs. 2.3A and 2.4), anterior inferior cerebellar artery (AICA, Fig. 2.1) and posterior inferior cerebellar artery (PICA, Fig. 2.1). An anastomotic system of arteries, the Circle of Willis, interconnects the anterior and middle cerebral arteries of both sides with the vertebrobasilar system (Figs. 2.3A, 2.4, 2.12 and 2.14). It is located at the base of the brain and surrounds the infundibulum and the optic chiasm. It includes the anterior communicating artery (ACOM), which interconnects the anterior cerebral arteries, immediately in front of the optic chiasm, and the posterior communicating artery (PCOM), which form an anastomosis between the most distal part of the internal carotid and the posterior cerebral artery, near their origin from the basilar artery. Fig. 2.2 shows the venous drainage of the vertebrobasilar system (lower portion) and cerebello-mesencephalic (upper portion) which both drain into the Galenśystem (see later in "Venus drainage system of the brain").

The initial segments of the middle and anterior cerebral arteries (MCA and ACA, respectively) give rise to central arteries, which enter the brain in the anterior perforated substance (Fig. 2.3B). Together with branches from the posterior communicating artery (PCOM), they supply the basal ganglia, the internal capsule and the thalamus. The middle cerebral artery (MCA) enters the sulcus lateralis (Silvio's scissure; Fig. 2.5) and just before this point, it emits the anterior choroidal artery which also supplies a branch to the globus pallidus. Fig. 2.3C shows the medial and lateral posterior choroidal artery. At the surface of the insula, the middle cerebral artery (MCA) branches into anterior, middle and posterior trunks. The branches of the middle cerebral artery loop over the opercula and ramify over the surface of the cerebral hemisphere to supply the cerebral cortex and the adjacent white matter (Fig. 2.5). The anterior

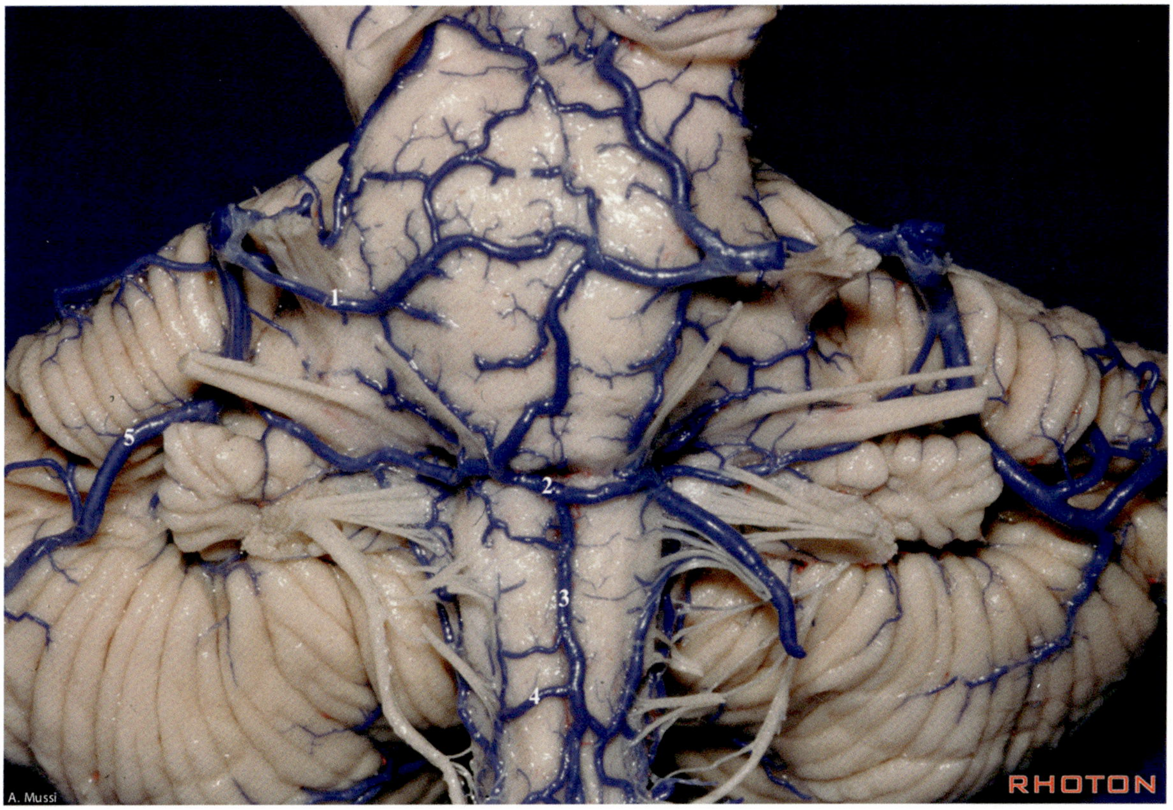

FIGURE 2.2 **Venous drainage of the brainstem and cerebellum.** 1, Transverse pontine vein; 2, vein of pontomedullary sulcus; 3, anteromedian medullary vein; 4, transverse medullary vein; 5, anterior hemispheric vein (middle). *Image courtesy of The Rhoton Collection® http://rhoton.ineurodb.org/. Dissector: A.Mussi.*

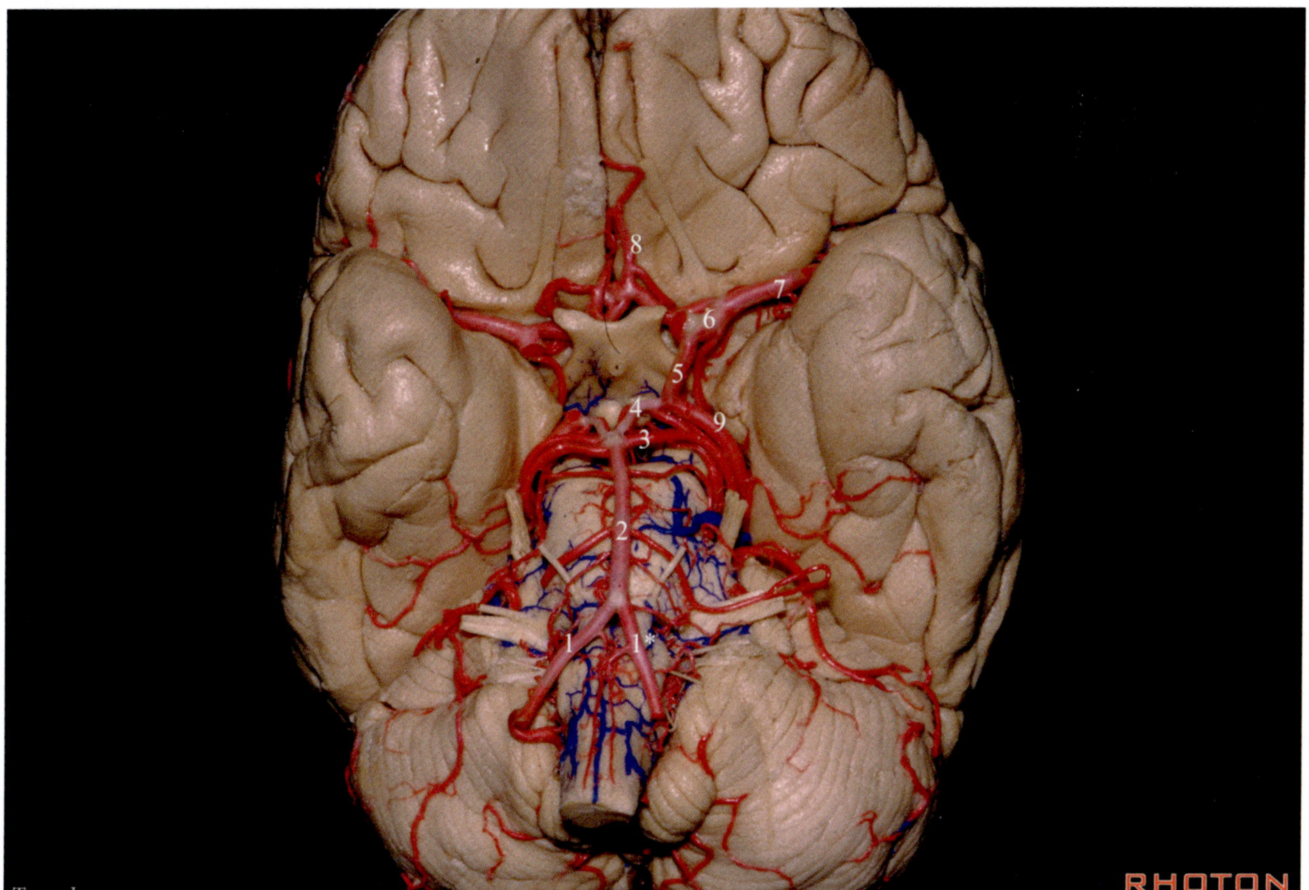

FIGURE 2.3A **Basal arterial circulation of the brain (circle of Willis).** 1, Vertebral arteries (VA); 2, basal artery (BA); 3, superior cerebellar artery (SCA); 4, P1 segment of posterior cerebral artery (PCA); 5, posterior communicating artery (PCOM); 6, internal carotid artery (ICA); 7, middle cerebral artery (MCA); 8, anterior cerebral artery (ACA); 9, P2 segment of posterior cerebral artery (PCA). *Image courtesy of The Rhoton Collection® http://rhoton.ineurodb.org/. Dissector: Tooru Inoue.*

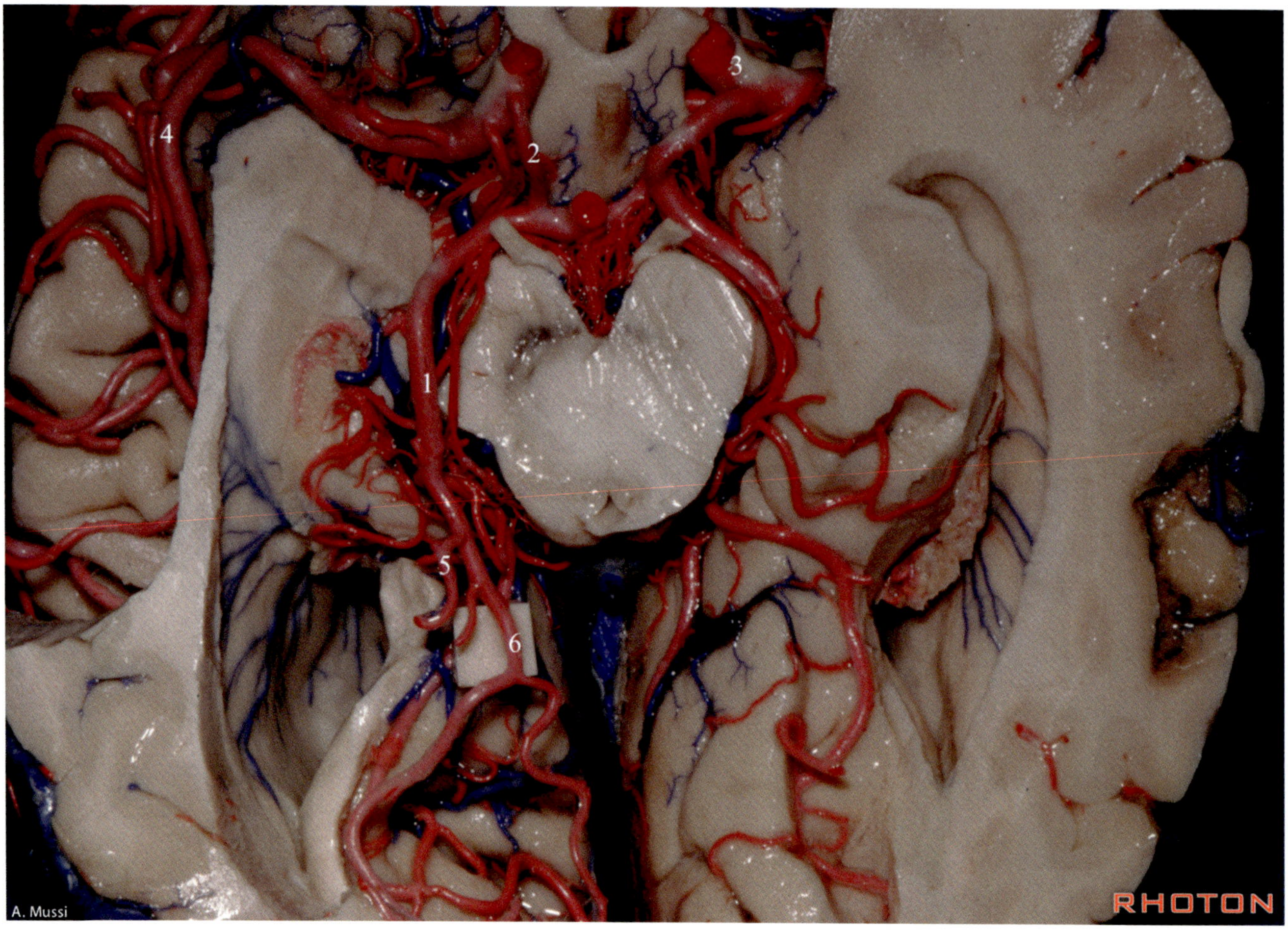

FIGURE 2.3B **Basal arterial circulation of the brain (basal brain dissection).** 1, Posterior cerebral artery (PCA); 2, posterior communicating artery (PCOM); 3, internal carotid artery (ICA); 4, middle cerebral artery (MCA); 5, thalamogeniculate artery (branch of P2 segment of posterior cerebral artery); 6, lateral occipital artery (l OA). *Image courtesy of The Rhoton Collection® http://rhoton.ineurodb.org/. Dissector: A.Mussi.*

cerebral artery (ACA), showed in Figs. 2.3A and 2.4, enters the longitudinal fissure to branch on the medial aspect of the hemisphere. The anterior communicating artery (ACOM), which connects the two anterior cerebral arteries, is located immediately rostral to the optic chiasm (Fig. 2.4). The vertebral arteries (VA) enter the skull through the foramen magnum (Fig. 2.4) and give rise to the anterior spinal artery, which descends in the anterior median fissure of the cord; the vertebral arteries (VA) also emit the posterior inferior cerebellar artery (PICA) showed in Fig. 2.1. The vertebral arteries (VA) unite into the basilar artery (BA) at the ventral aspect of the brainstem (Fig. 2.3A). The basilar artery gives origin to the anterior inferior and superior cerebellar arteries (AICA and SCA, respectively) and splits rostrally into the posterior cerebral arteries (PCA; Figs. 2.3A, B, and 2.4). The oculomotor nerve emerges between the superior cerebellar and the posterior cerebral arteries (SCA and PCA, respectively) and thus marks the bifurcation of the basilar artery (BA). The posterior cerebellar artery (PCA) makes a characteristic, caudally directed curve before it reaches the cerebellum. Both the posterior inferior and superior cerebellar arteries (PICA and SCA, respectively) contribute branches to the dorsolateral brainstem. The posterior cerebral artery (PCA) supplies the medial aspect of the temporal and occipital lobes. The border region of the vascularization territories of the posterior and middle cerebral arteries includes the temporal and occipital poles.

Asymmetries of the brain's arterial system are frequently observed, most often in the caliber of the vertebral, the posterior cerebral, and the posterior communicating arteries. The vertebral, basilar, and posterior cerebral arteries also give rise to smaller branches, which enter the brainstem in the median sulcus and more laterally (Fig. 2.3B). Branches from the basilar and posterior cerebral arteries enter the mesencephalon in the posterior perforated substance, located in the floor of the interpeduncular fossa (see Chapter 5A) The existence of a collateral circulation is of great significance for the vascularization of the CNS. Anastomoses between arteries can be found in relation to three arterial

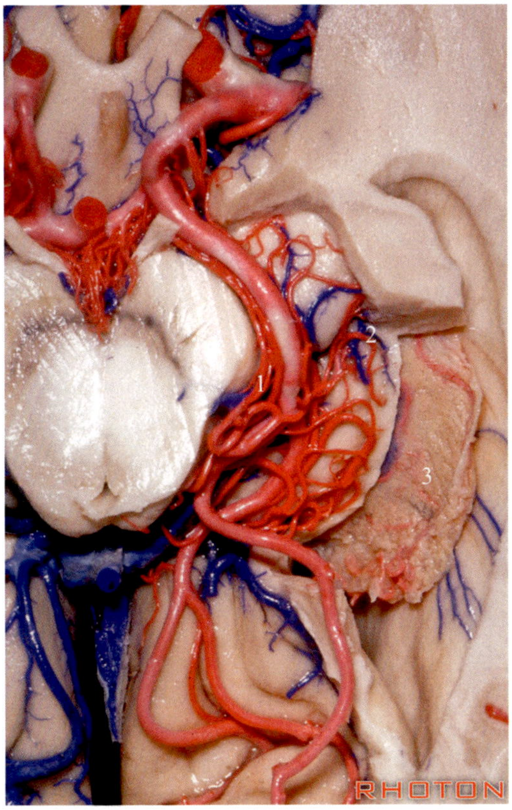

FIGURE 2.3C **Choroidal arteries.** 1, Medial posterior choroidal artery; 2, lateral posterior choroidal artery; 3, choroid plexus. *Image courtesy of The Rhoton Collection® http://rhoton.ineurodb.org/. Dissector: A.Mussi.*

systems: between the two main arterial systems of the carotid and vertebral arteries and between the arterial systems of the brain and the external carotid artery.

VENUS DRAINAGE SYSTEM OF THE BRAIN

The blood of the brain is drained by different systems that mainly converge in the internal jugular vein; drainage ways of smaller caliber are tributaries of the external jugular vein and intravertebral venous plexuses. The venous system of the brain has no valves. Venous sinuses are venous ducts wrapped in a doubling of the dura mater. The major dural venous sinuses include the superior and inferior sagittal sinuses, the straight sinus (or sinus rectus), the superior and inferior petrosal sinuses, the occipital sinus, the transverse sinus, and the sigmoid sinuses (Fig. 2.6). The superior and inferior sagittal sinuses run in the midline along the superior and inferior edges of the *falx cerebri* (Figs. 2.6 and 2.7). The inferior sagittal sinus joins the vein of Galen to form the straight sinus, which then courses along the superior edge of the tentorium cerebelli tentorium cerebelli. The *occipital sinus* originates at the posterior edge of the foramen magnum and meets the straight sinus and superior sagittal sinus at the torcula of Herophili. The superior petrosal sinus joins the cavernous to the sigmoid sinus and runs along the superior ridge of the

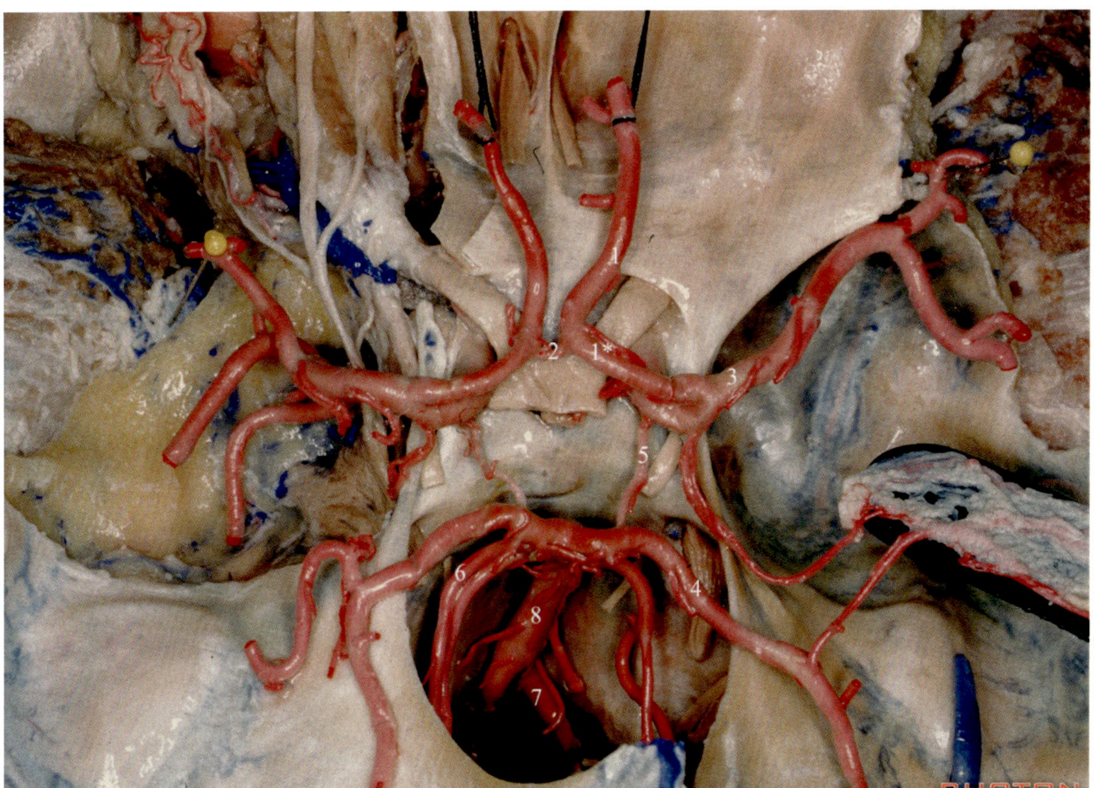

FIGURE 2.4 **Basal skull arteries.** 1, Anterior cerebral artery (ACA); 2, anterior communicating artery (ACOM); 3, middle cerebral artery (MCA); 4, posterior cerebral artery (PCA); 5, posterior communicating artery (PCOM); 6, superior cerebellar artery (SCA); 7, vertebral artery (VA); 8, basilar artery (BA). *Image courtesy of The Rhoton Collection® http://rhoton.ineurodb.org/. Dissector: H.Abe.*

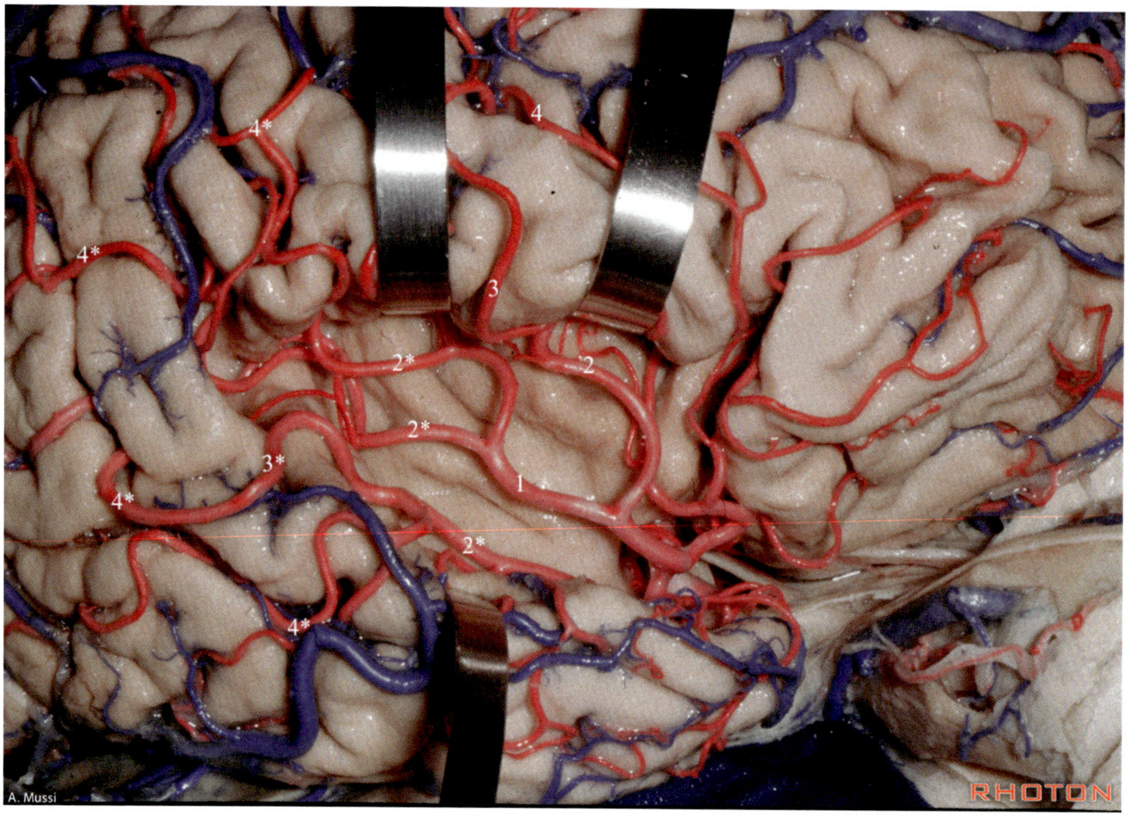

FIGURE 2.5 **Middle cerebral artery.** 1, Middle trunk of the middle cerebral artery (MCA); 2,2*, M2 segment of the MCA; 3,3*, M3 segment of the MCA; 4,4*, M4 segment of the MCA. *Image courtesy of The Rhoton Collection® http://rhoton.ineurodb.org/. Dissector: H.Abe.*

petrous bone. The inferior petrosal sinus has variable tributaries but drains into the internal jugular vein at the jugular bulb. Fig. 2.8 shows the arterovenous system of the brain by CT-angiography 3D reconstruction.

ANATOMOMORPHOLOGICAL DEFECTS OF BRAIN VESSELS

Arteriovenous Malformation

Arteriovenous malformations (AVMs) are an abnormal tangle of small vessels (called nidus), which directly connect brain arteries to brain veins, without any capillaries in-between (as should be normally the case) (Figs. 2.9 and 2.10). This results in a kind of short circuit, whereby blood goes directly from arteries, through the nidus, and into the veins, bypassing brain tissue.

AVS cause problems mostly in two ways:

(1) Bleeding—the vessels that make up the nidus are fragile and can rupture, which results in bleeding into the brain.

(2) AVMs cause problems by "stealing" blood away from the brain. This can cause strokes (not bleeding types, but an ischemic ones) and seizures (epilepsy).

(3) Other problems depend on location of the AVM, but can include paralysis, speech, vision, psychiatric, and other problems, headaches, etc., depending on location of the AVM.

Treatment

Most AVMs are treated by a combination of surgery (Fig. 2.11 shows AVM removal using the Michael Lawton's technique) and preoperative catheter embolization. Embolization decreases blood flow through the AVM and allows the surgery to proceed in a more bloodless, controlled manner.

Aneurysm

A brain aneurysm—also known as a cerebral aneurysm or intracranial aneurysm—is a bulge or ballooning in a blood vessel in the brain (Fig. 2.12). An aneurysm often looks like a berry hanging on a stem. Brain aneurysms form and grow because blood flowing through the blood vessel puts pressure on a weak area of the vessel wall. This can increase the size of the brain aneurysm. If the brain aneurysm leaks or ruptures, it causes bleeding in the brain, known as a hemorrhagic stroke. Most often, a ruptured brain aneurysm occurs in the

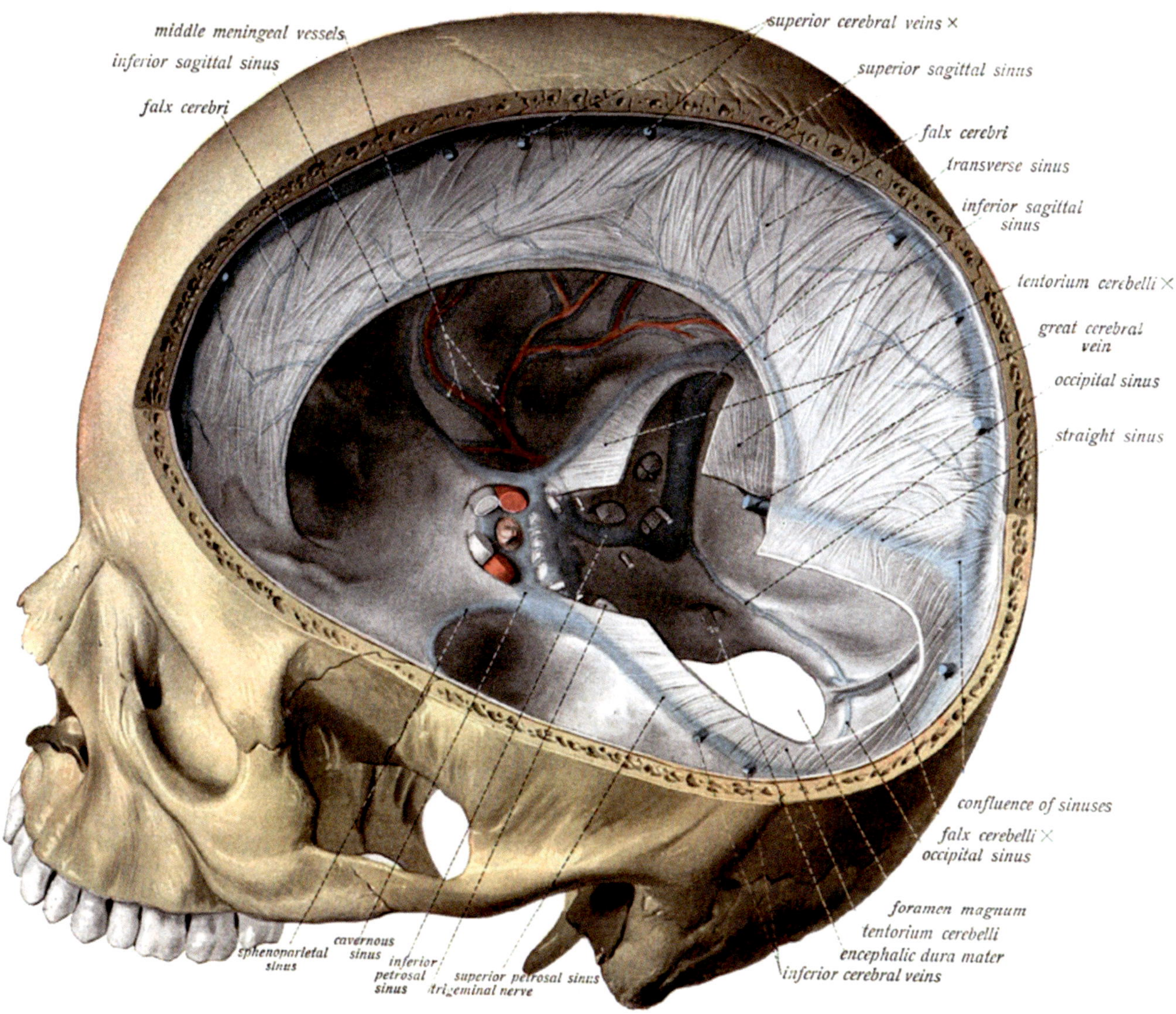

FIGURE 2.6 **Dural venous sinuses.** *From Sobotta's Human Anatomy 1908.*

space between the brain and the thin tissues covering the brain. This type of hemorrhagic stroke is called a subarachnoid hemorrhage (SAH, Fig. 2.13). Aneurysm locations are for grouping purposes. Location is determined by vascular distribution (Fig. 2.14). Aneurysms are named according to the vessel of neck origin, closest branch vessel origin, or next best guess. Use of bony landmarks resulting in terms such as "supraclinoid" or nonspecific vessel locations such as "supraoptic" should be avoided.

Treatment

The two main options are endovascular treatment (which is performed through catheters inserted into arteries under X-ray guidance) and open surgery.

MENINGES

The brain is completely surrounded by three layers of connective tissue called meninges. Starting from the surface of the brain are: the pia mater, arachnoid and dura mater. The dura mater, composed of several layers of collagen fibers, is particularly resistant and thick, this is why it is also called pachymeninx. Fig. 2.15 shows the dura madre and its relations with the middle meningeal artery (MMA), one of the largest branches of the external carotid artery and the most important dural artery because it supplies more than two-thirds of the cranial dura. The thin pia mater, made up of tissue lasso, and the arachnoid make up the leptomeninx.

The cephalic dura mater merges with the periosteum of internal skull plate. Consequently, the dura is firmly attached to the skull, especially along the sutures. The dural

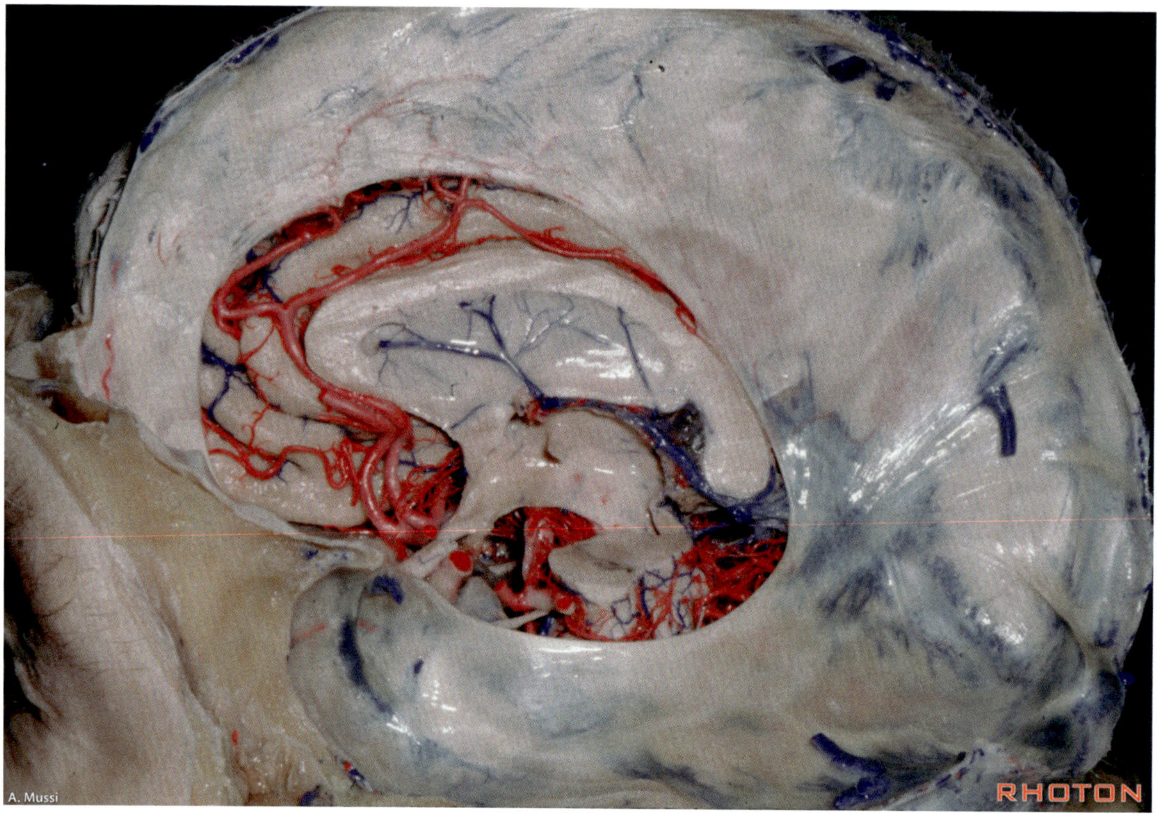

FIGURE 2.7 **Falx cerebri, dural venous sinuses and pericallosal artery.** The broad; glossy; sagittal structure made up of dura mater; crescent-shaped that descends vertically into the longitudinal fissure between the cerebral hemispheres of the cerebrum is the *falx cerebri*. It separates the two hemispheres and supports the *dural venous sinuses* (the superior sagittal sinus (SSS), straight sinuous (SS), and others are clearly visible; for details cf. image above), which provide venous and CSF drainage to the brain. It is attached to the crista galli anteriorly and merges with the tentorium cerebelli posteriorly. Leaning against to the corpus callosum is the *pericallosal artery. Image courtesy of The Rhoton Collection® http://rhoton.ineurodb.org/. Dissector: A. Mussi.*

septa are arranged between the main divisions of the brain. In the midline, the falx cerebri is interposed between the cerebral hemispheres (Fig. 2.16) while the tentorium of the cerebellum (a.k.a. tentorium cerebelli, Figs. 2.6 and 2.7) separates the occipital and temporal lobes of the cerebral hemispheres from the cerebellum and brainstem. This sheet extends in the axial plane over the posterior cranial fossa, divides the cranial cavity into the supratentorial and infratentorial spaces and covers the cerebellum like a tent. The venous sinuses occupy the inner outer margins of the falx (superior and inferior sagittal sinus), the falx junction with the tentorium (senus rectus), and the insertion of the tentorium to the skull (the transverse sinus and the superior and inferior petrosal sinuses and lower) (Figs. 2.6, 2.7 and 2.16).

The pia mater adheres tightly to the surface of the brain and creeps into sulci and depressions. The arachnoid coats the brain at a distance variable, thereby creating a subarachnoid space between the pia and the arachnoid. This space is made up of numerous trabeculae and in the subarachnoid space is contained the cerebrospinal fluid (CSF, see later). The expansions of subarachnoid form the cisternae.

The central nervous system is isolated from the rest of the body by a series of cellular barriers, which limit the flow of hydrophilic molecules between these cells.

These barriers generally consist of extensive tight junctions between the cells, where the outer leaflets of the plasma membranes of two opposing cells are fused. These barriers are found in the epithelium of the choroid plexus (blood–CSF barrier), the outer (barrier) layer of the arachnoid and in the endothelium of capillaries located within arachnoid, and the pia mater and nervous tissue (blood–brain barrier).

CEREBROSPINAL FLUID AND CIRCULATION

In the adult, the total volume of CSF is approximately 150 mL, of which about 125 mL is intracranial and 25 mL is intraspinal. Within the brain, the CSF fills four hollow spaces designated the *left and right lateral ventricles*, the *third ventricle*, and the *fourth ventricle (a schematic and 3D representation are shown in* Figs. 2.18 and 2.19, respectively).

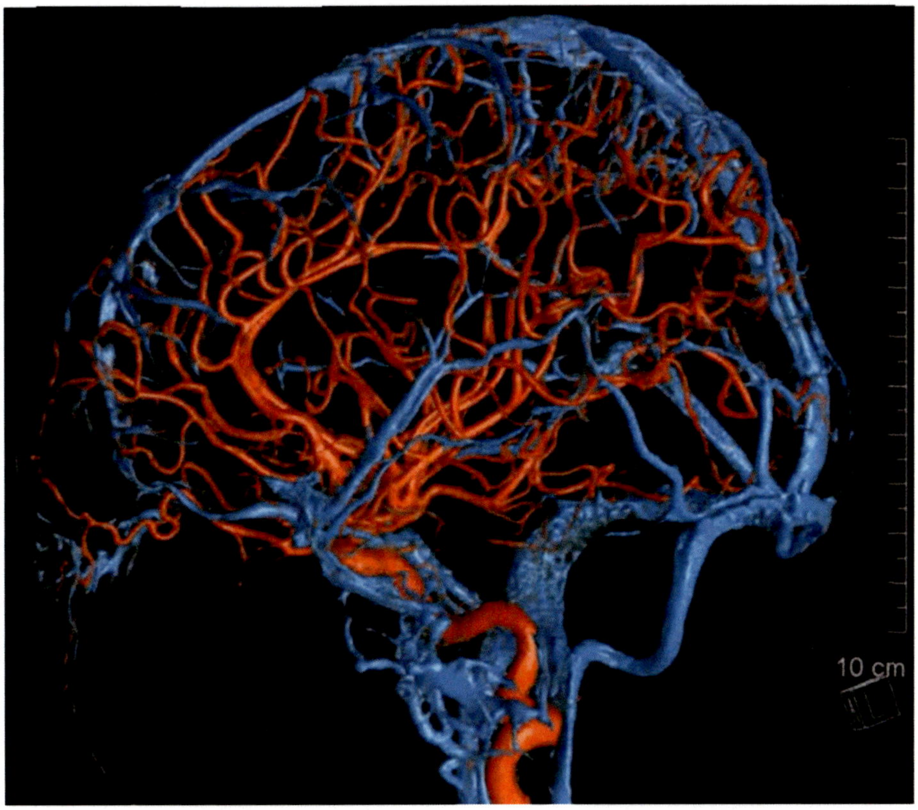

FIGURE 2.8 **CT-angio 3D of the arteriovenous system, in red and blue, respectively.** *Image courtesy of Neuroangio http://neuroangio.org/.*

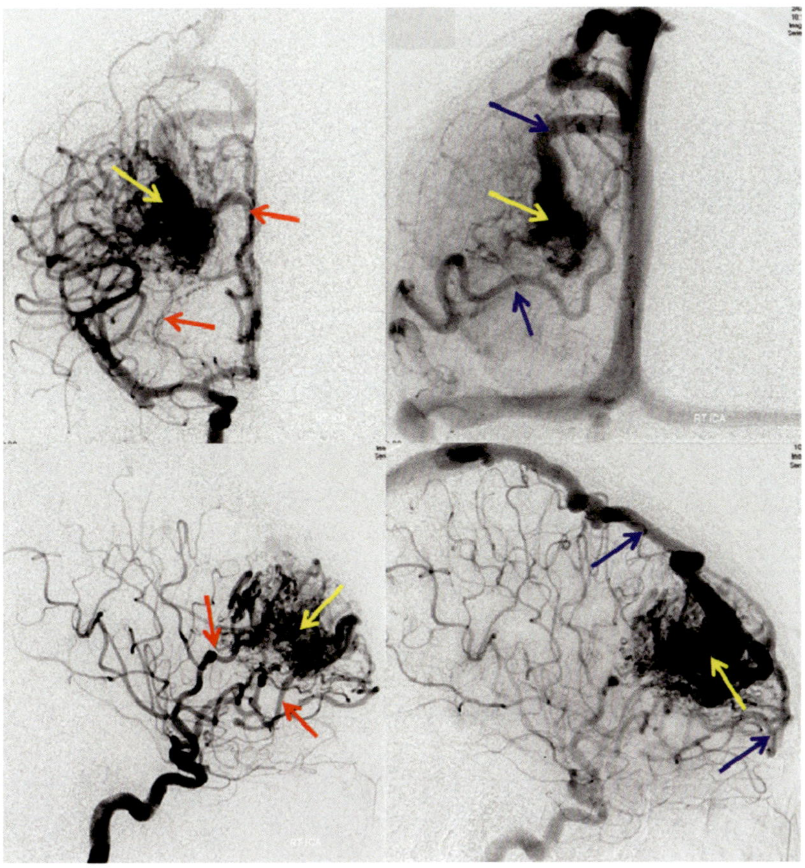

FIGURE 2.9 **Arteriovenous malformation (CT angio).** Frontal (*top row*) and side (*bottom row*) images of the AVM obtained during presurgical angiogram and embolization. The AVM (*yellow arrows*) is supplied by numerous arteries, some of which are shown by *red arrows*, and emptied by one larger and one smaller veins (*blue arrows*). *Image courtesy of Neuroangio http://neuroangio.org/.*

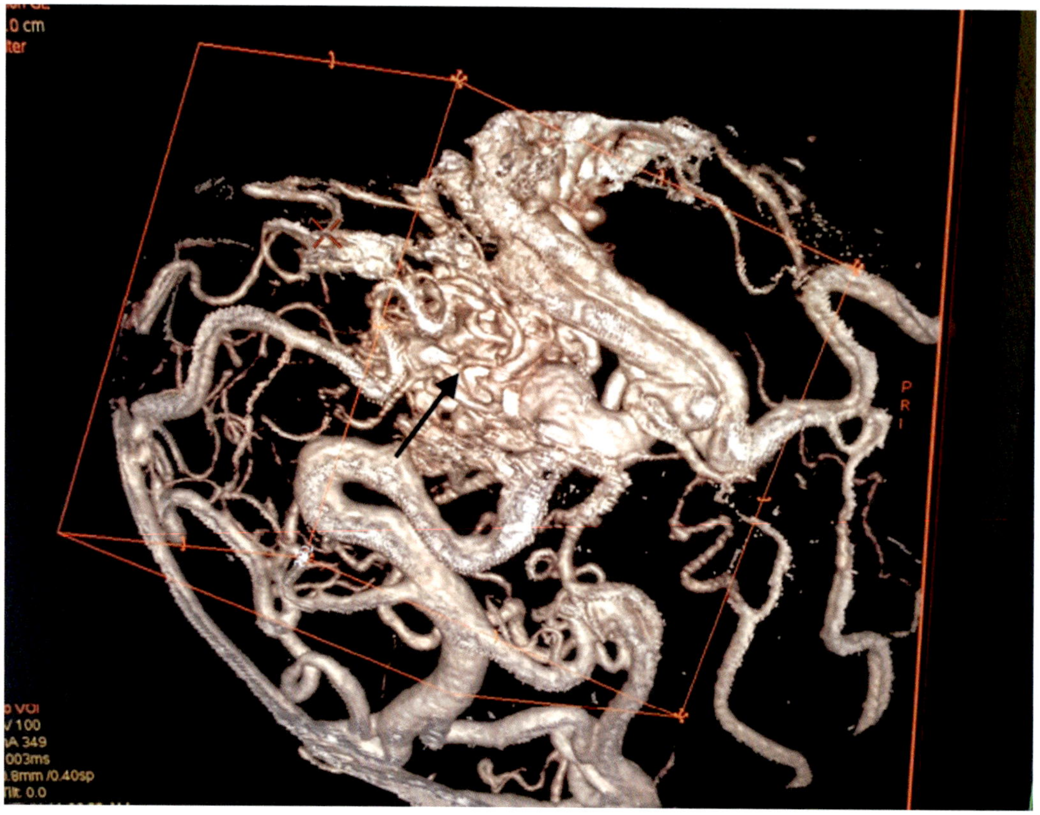

FIGURE 2.10 **CT angio 3D reconstruction of left frontotemporal AVM (*black arrow*) with drainage into MCA and SSS.** *Image courtesy of Dr. Giancarlo D'Andrea, Frosinone Hospital, Lazio, Italy.*

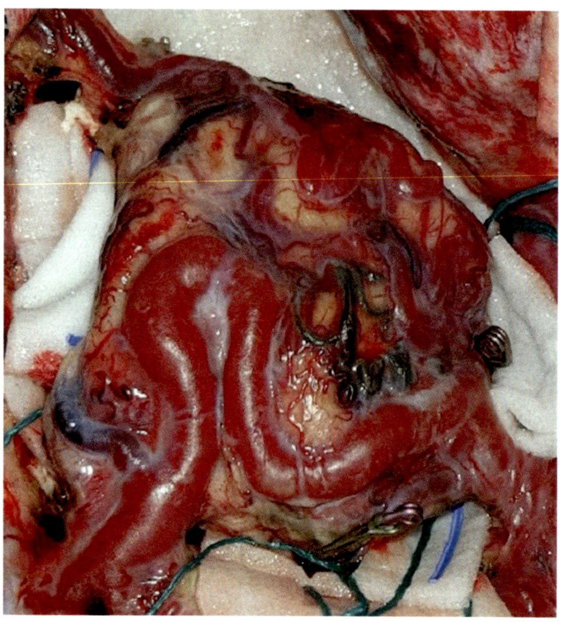

FIGURE 2.11 **AVM removal using the Michael Lawton's technique in which all sides are deafferented before to take it off.** *Image courtesy of Dr. Giancarlo D'Andrea, Frosinone Hospital, Lazio, Italy.*

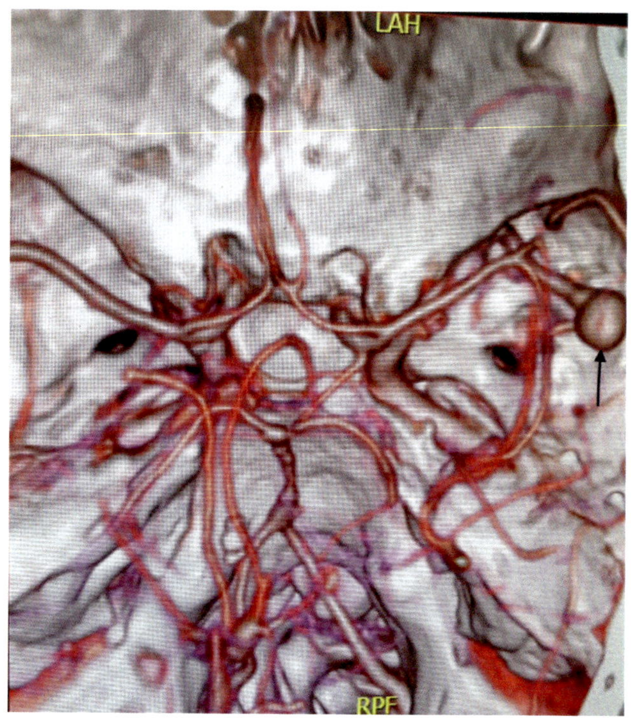

FIGURE 2.12 **CT angio 3D elaborated of circle of Willis and venous aneurysm of AVM (*black arrow*).** *Image courtesy of Dr. Giancarlo D'Andrea, Frosinone Hospital, Lazio, Italy.*

CSF is formed in specialized areas: the aforementioned *choroid* plexuses, which are located in specific regions in the roof of the third, fourth, and lateral

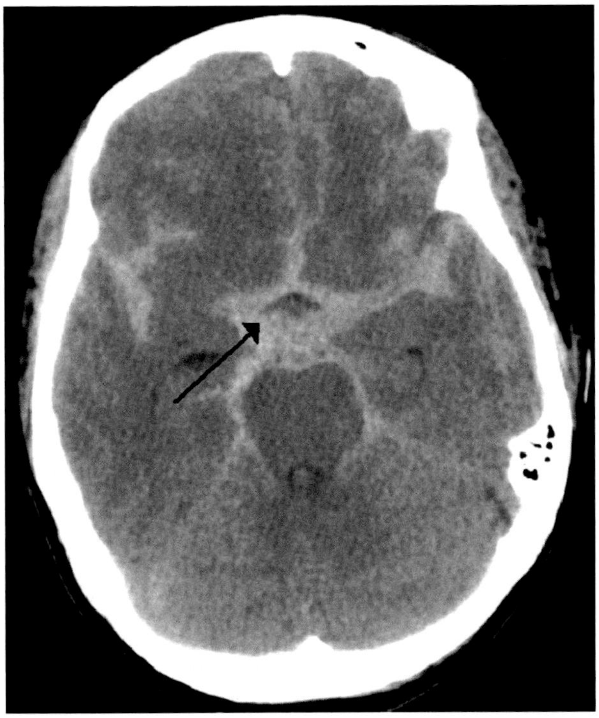

FIGURE 2.13 **Subarachnoid hemorrhage (CT head).** It appears as a white area in the center and extending into the sulci on the sides (marked by *arrow*). *Image courtesy of Dr. James Heilman, MD (Wikipedia).*

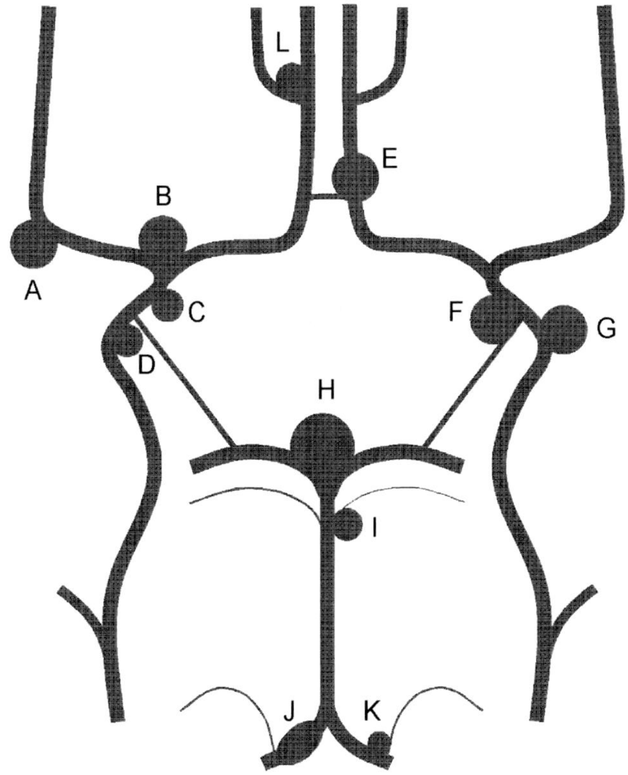

FIGURE 2.14 **Common location of cerebral aneurysms surrounding the circle of Willis.** (A) Middle cerebral (proximal to bifurcation, bifurcation, distal to bifurcation); (B) carotid terminus; (C) anterior choroidal; (D) superior hypophyseal; (E) anterior communicating (proximal to communicating artery, at communicating artery); (F) posterior communicating; (G) ophthalmic; (H) basilar artery (terminus, trunk); (I) superior cerebellar; (J) V4 segment, vertebral; (K) posterior inferior cerebellar; (L) pericallosal artery. *Image courtesy of Meyers, P.M., Schumacher, H.C., Higashida, R.T., Derdeyn, C.P., Nesbit, G.M., Sacks, D., Wechsler, L.R., Bederson, J.B., Lavine, S.D., Rasmussen, P., 2009. Reporting standards for endovascular repair of saccular intracranial cerebral aneurysms. Stroke 40(5), e366-e379. https://doi.org/10.1161/STROKEAHA.108.527572. Epub 2009 Feb 26. PMID: 19246711.*

ventricles. Choroid plexuses are highly vascularized structures that project into the ventricles and secrete CSF into the ventricular system.

The two lateral ventricles lie off the midline (Fig. 2.20), within the left and right cerebral hemispheres, the third ventricle lies in the midline between the two thalami (Fig. 2.21) and the fourth ventricle lies immediately posterior to the brainstem at the center of the cerebellum (Fig. 2.23). The third ventricle is connected to the fourth ventricle through the cerebral aqueduct (a.k.a. aqueduct of Sylvius), a narrow 15 mm conduit (Fig. 2.22). In adults, the mean volume of the two lateral ventricles varies from 9 to 27.6 mL and is slightly larger in men than women. The mean volumes of the left and right lateral ventricles are estimated individually at 14.4 and 13.2 mL, respectively. The combined volume of the third ventricle, aqueduct, and fourth ventricle is estimated at 1.0–1.5 mL. Two additional compartments of CSF situated between the leaves of the septum pellucidum (cavum septi pellucidi) and between the fornices (cavum vergae) are occasionally considered to be fifth and sixth ventricles, respectively.

CSF can be drained into the venous sinuses from the subarachnoidal space by protrusions of the arachnoid known as the arachnoid granulations (or arachnoid villi) (Fig. 2.17) and along the cerebral and spinal nerves into lymph vessels.

Drainage along the arachnoid granulation is pressure-dependent, and they are especially abundant in the superior sagittal sinus and its lateral extensions, the lacunae. Arachnoid cisterns are local widenings of the subarachnoidal space and the most important surround the cerebellum and the brainstem.

The cisterna magna is located between the cerebellum and the medulla oblongata (Fig. 2.18, under the tentorium cerebelli), and it is accessible for puncture through the foramen magnum. The pontine cistern surrounds the pons, and finally, the medullary cistern surrounds the medulla oblongata and continues as the subarachnoidal space surrounding the spinal cord. As before mentioned the vascularization of the meninges is derived from the external carotid artery as the middle meningeal artery, which enters the skull through the foramen spinosum.

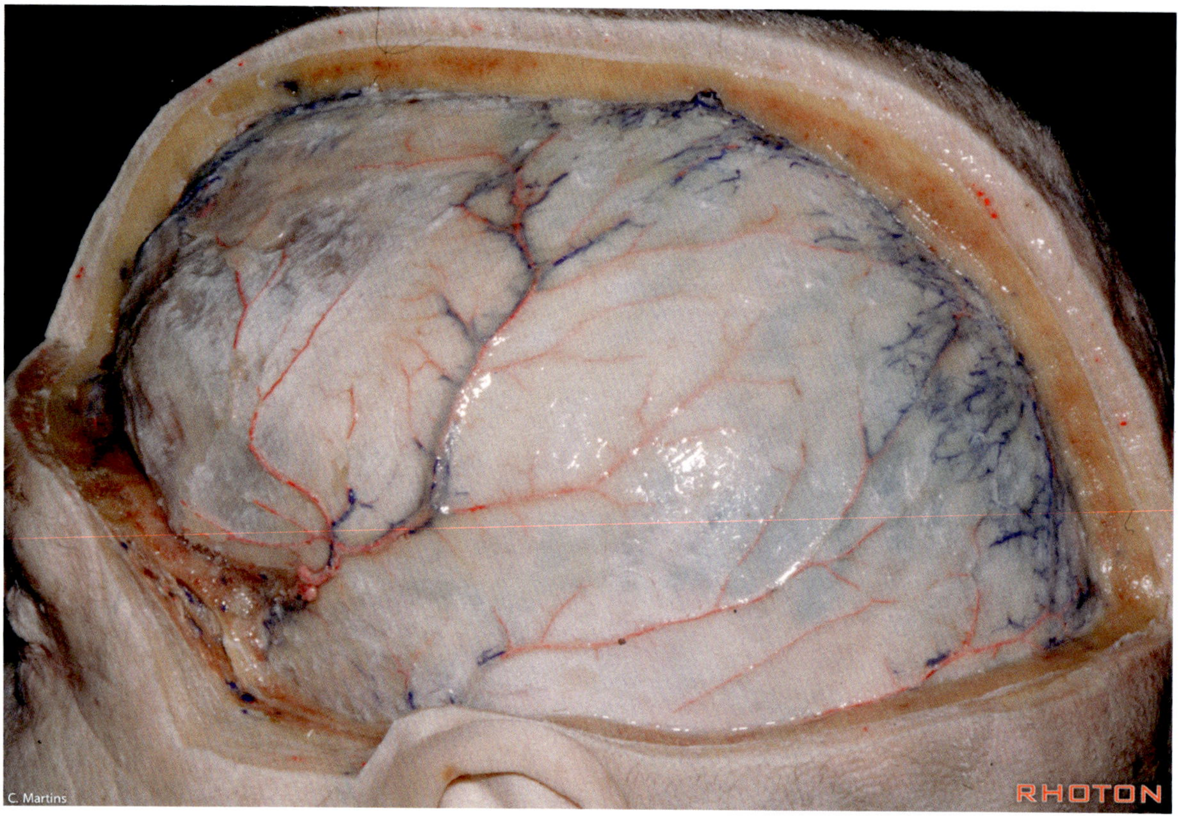

FIGURE 2.15 **Relations of the dura mater and middle meningeal artery (MMA)to the surface of the brain.** *Image courtesy of The Rhoton Collection® http://rhoton.ineurodb.org/. Dissector: C. Martins.*

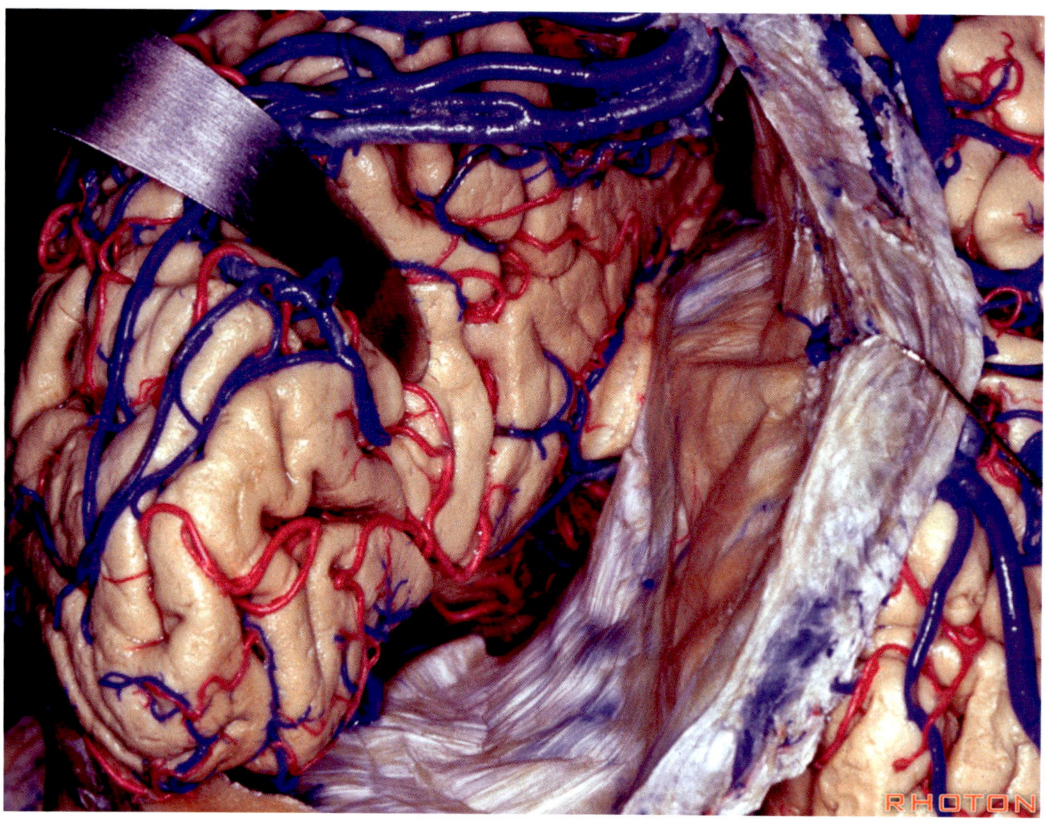

FIGURE 2.16 ***Falx cerebri*, median and attached to the right hemisphere.** *Image courtesy of The Rhoton Collection® http://rhoton.ineurodb.org/.*

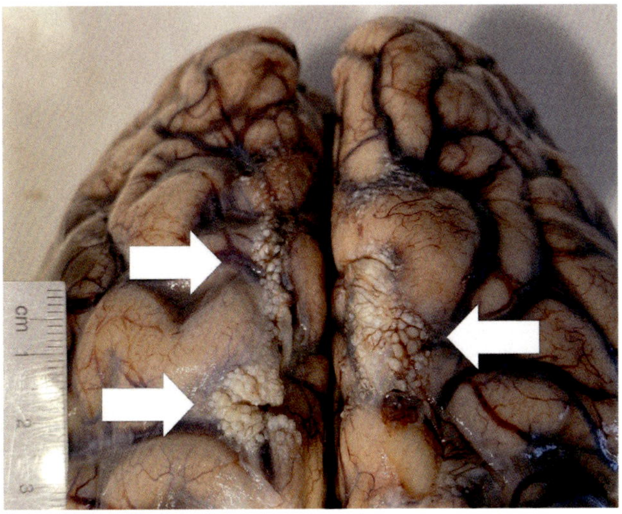

FIGURE 2.17 **Arachnoid granulation (*white arrows*).** *Image by Mikael Häggström, MD.*

Each hemispheric cortex is divided into four lobes based on key fissures and sulci, namely, the Sylvian fissure (lateral sulcus), central sulcus (CS), parietooccipital sulcus and cingulate sulcus. These fissures and sulci are large and readily identifiable between individuals based on their well-defined and characteristic configurations, locations, and depths along the cerebral hemispheres (Fig. 2.24). The four major lobes are divided into smaller sulci and gyri that also exhibit defined patterns that can be recognized and learned through systematic review and practice in identifying these landmarks on routine neuroimaging studies (see later). The normal brain is almost but not quite perfectly symmetric across the interhemispheric fissure, and subtle inherent structural asymmetries may be apparent on neuroimaging, as reviewed recently by Kuo and Massoud.

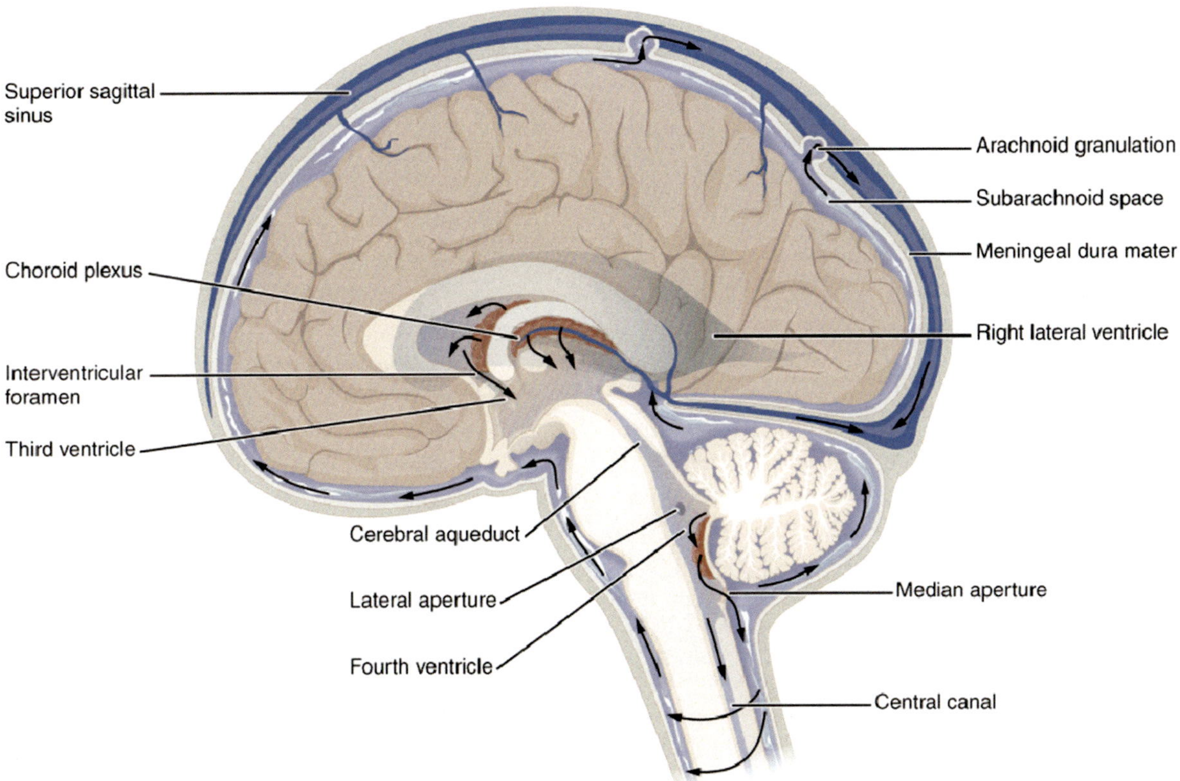

FIGURE 2.18 **Schematic illustration of the ventricular system.** *Image courtesy of "Textbook OpenStax Anatomy and Physiology", 2018.*

SECOND SECTION: TOPOGRAPHY OF THE CEREBRUM: NEW PERSPECTIVES OFFERED BY THE USE OF HIGH-RESOLUTION MAGNETIC RESONANCE IMAGING

The cerebral cortex of each cerebral hemisphere is part of the telencephalon that covers the deeper diencephalic structures of the forebrain (the prosencephalon).

Frontal Lobe

The frontal lobe is the largest lobe within the cerebral hemisphere and is located anterior to the CS and above the Sylvian fissure. The frontal pole represents the most anterior portion of the frontal lobe. The prefrontal cortex corresponds to the anterior part of the frontal lobe. The prefrontal cortex plays a key role in the regulation of higher-order cognitive, emotional, and behavioral

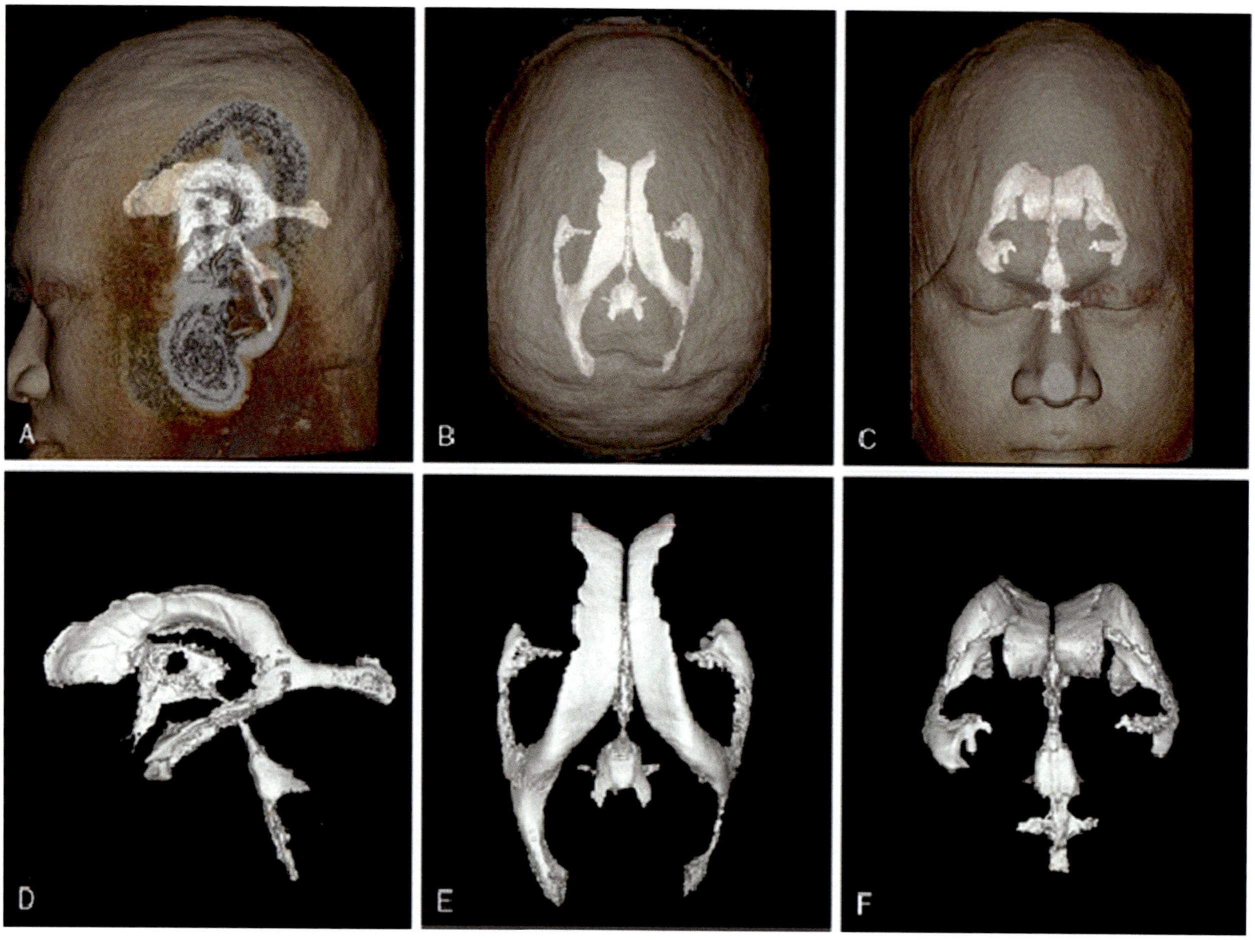

FIGURE 2.19 **Ventricular system.** Three-dimensional (3D) computer reformatted volumetric T2W MR images display the ventricles in lateral (A, D), superior (B, E), and anterior (C, F) views. (A) to (C), Orientation of the ventricular system within the head, seen through the partially transparent skin surface. (D) to (F), Magnified display of the isolated ventricles in the same orientation. *Radiology Key https://radiologykey.com.*

functioning. The frontal lobe is composed of four major gyri: the superior, middle, and inferior frontal gyri and the precentral gyrus. On axial images, the superior frontal gyrus is a vertically oriented gyrus along the superior-most margin of the frontal lobe (Fig. 2.25, SFG). The middle frontal gyrus (see Fig. 2.25, MFG) is a wide bar of tissue that parallels and is separated from the superior frontal gyrus by the superior frontal sulcus (see Fig. 2.25).

The precentral gyrus crosses the superior frontal gyrus along the superior frontal convexity, which is well seen on axial images (see Fig. 2.25, PreCG). The precentral gyrus delineates the most posterior portion of the frontal lobe and extends inferiorly to the Sylvian fissure. The inferior frontal gyrus is part of the lateral and inferior surface of the frontal lobe that overlies the insula anteriorly and is separated from the middle frontal gyrus by the inferior frontal sulcus. The Sylvian fissure lies along its inferior edge; the anterior and ascending rami arising from the Sylvian fissure divide the inferior frontal gyrus into three parts, giving it the shape of the letter "M," known as the "M sign" (Fig. 2.26, yellow line). From anterior to posterior, the three parts of the "M" shape are the pars orbitalis, the pars triangularis, and the pars opercularis (see Fig. 2.26).

Traditionally, the pars opercularis and pars triangularis are considered to form the Broca area (Brodmann areas 44 and 45) in the left hemisphere. The pars opercularis overlies the anterior insula and abuts the orbitofrontal cortex, which runs parallel to the gyrus rectus along the inferior surface of the frontal lobe (Fig. 2.27). The orbitofrontal cortex rests upon the orbital portion of the frontal bone and is composed of four gyri delineated by a distinctive H-shaped sulcus on axial views: the anterior, medial, posterior, and lateral orbital gyri (see Fig. 2.27B). The medial orbital gyrus abuts the vertically oriented olfactory sulcus, which is bordered superiorly by the olfactory tract and divides the orbital gyri from the gyrus rectus (see Fig. 2.27A).

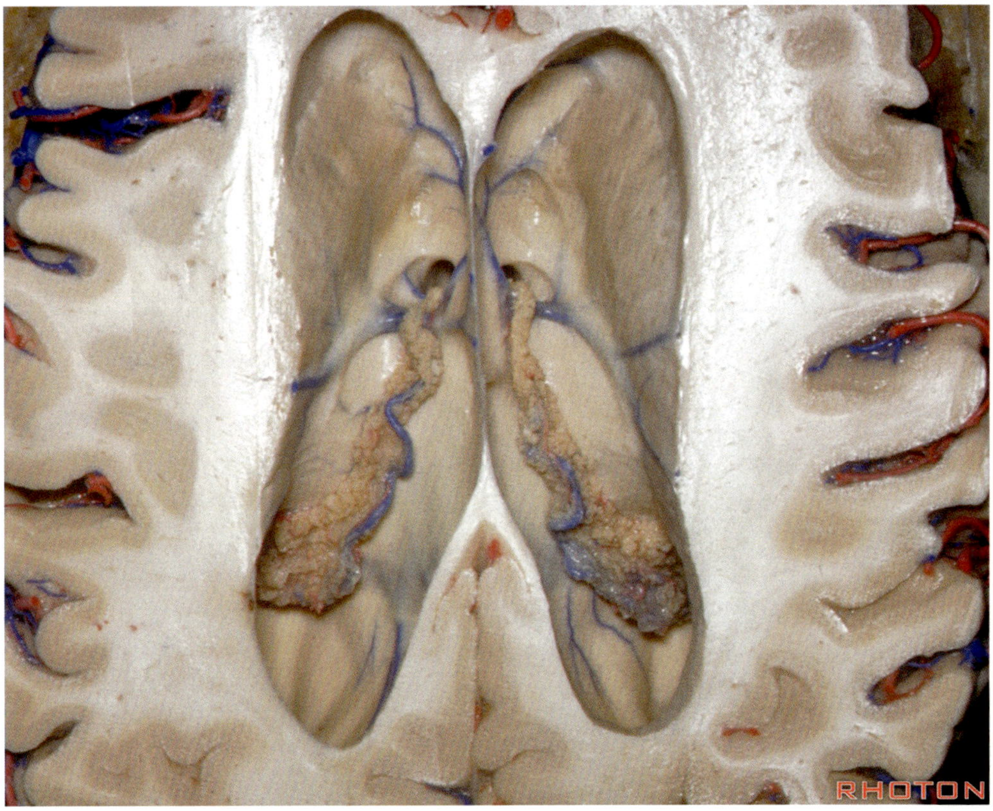

FIGURE 2.20 **Superior view of the lateral ventricles and choroid plexus (granular structures within) in an axial section of the brain.** Anteriorly, it is possible to see the interventricular foramina (of Monro), which communicate, inferiorly, with the third ventricle. *Image courtesy of The Rhoton Collection® http://rhoton.ineurodb.org/.*

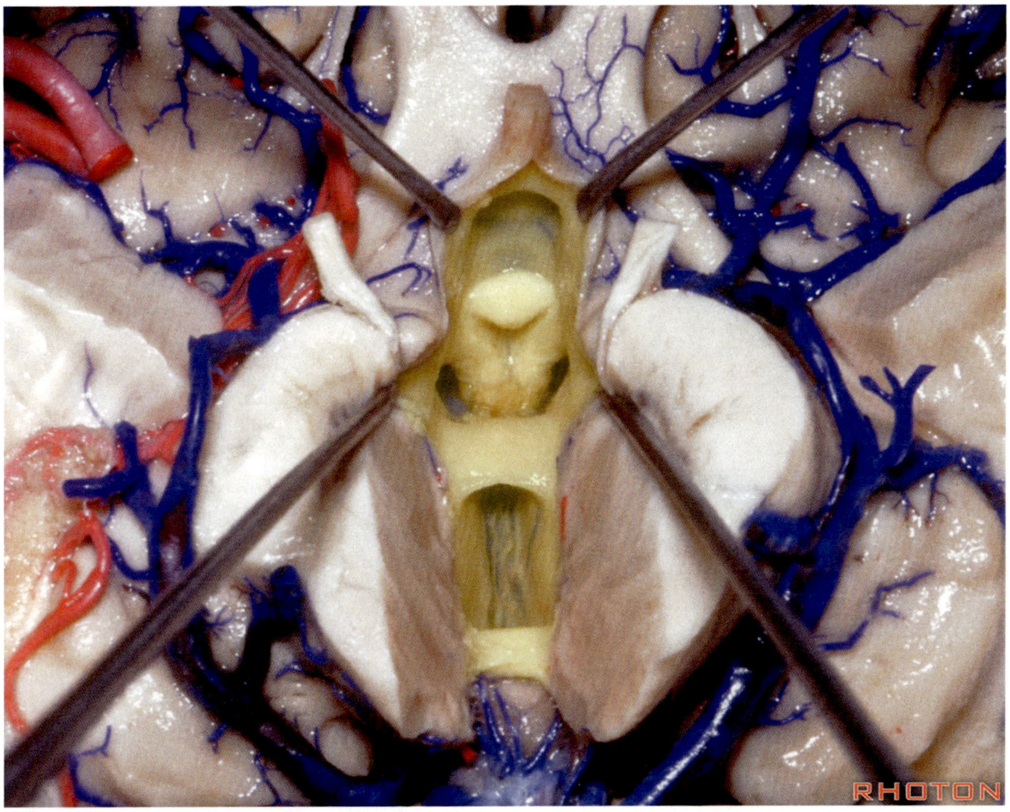

FIGURE 2.21 **Third ventricle.** Basal view of the cerebrum; in yellow is the third ventricle. *Image courtesy of The Rhoton Collection® http://rhoton.ineurodb.org/. Dissector : K. Yagmurlu. Drawing: Jeffrey Sorenson.*

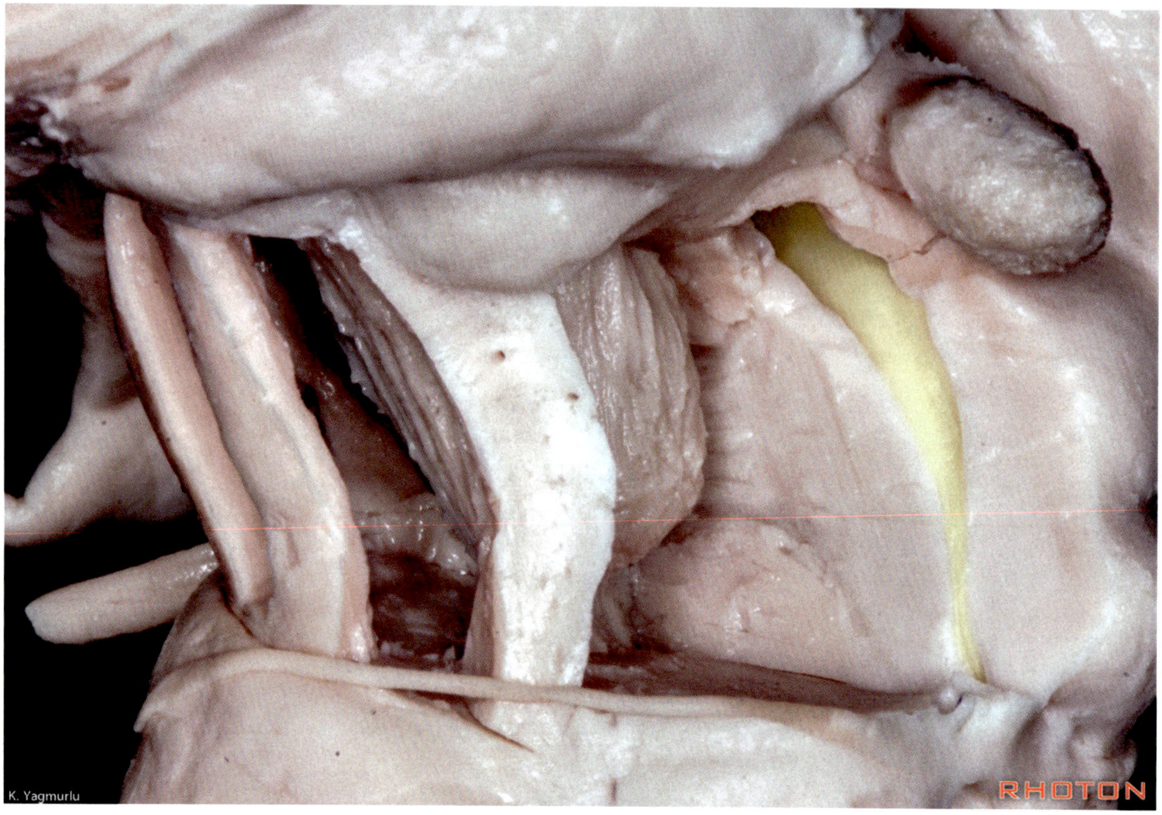

FIGURE 2.22 **Cerebral aqueduct of Silvio (in yellow).** *Image courtesy of The Rhoton Collection® http://rhoton.ineurodb.org/. Dissector: K.Yagmurlu Drawing: Jeffrey Sorenson.*

FIGURE 2.23 **Fourth ventricle (in yellow).** *Image courtesy of The Rhoton Collection® http://rhoton.ineurodb.org/. Dissector: K.Yagmurlu. Drawing: Garrett Venable.*

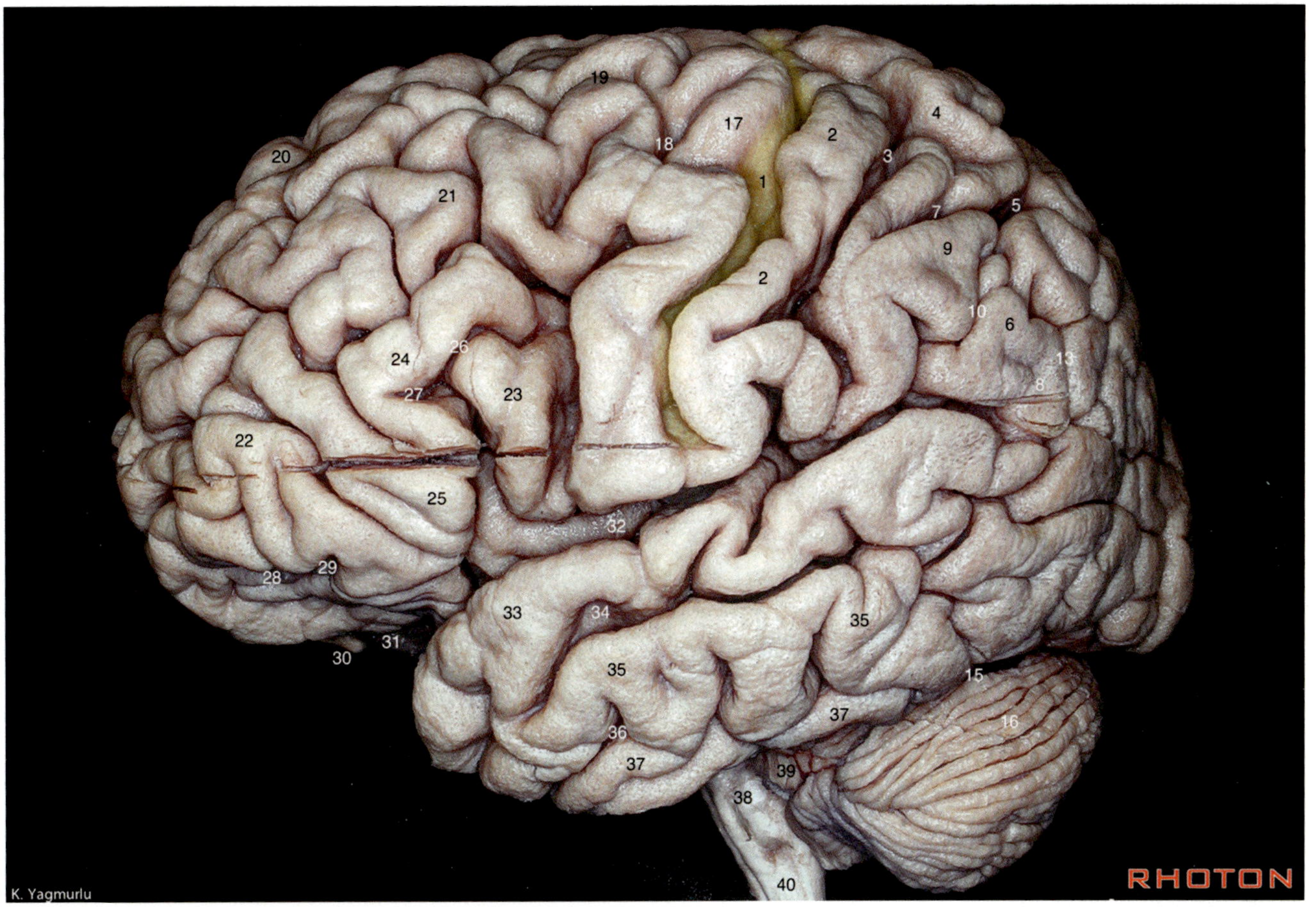

FIGURE 2.24 **Brain topography: lateral view.** 1, Central sulcus (in yellow); 2, Postcentral gyrus; 3, postcentral sulcus; 4, superior parietal lobule; 5, parietooccipital sulcus; 6, inferior parietal lobule; 7, intraparietal sulcus; 8, angular gyrus; 9, supramarginal gyrus; 10, lateral sulcus, posterior branch; 11, occipital gyri; 12, lunate sulcus; 13, anterior occipital sulcus; 14, occipital sulci; 15, preoccipital notch; 16, cerebellar hemisphere; 17, precentral gyrus; 18, precentral sulcus; 19, superior frontal sulcus; 20, superior frontal gyrus; 21, middle frontal gyrus; 22, inferior frontal gyrus; 23, opercular part of the inferior frontal gyrus; 24, triangular part of the inferior frontal gyrus; 25, orbital part of the inferior frontal gyrus; 26, lateral sulcus, ascending branch; 27, lateral sulcus, anterior branch; 28, orbital sulci; 29, orbital gyri; 30, olfactory bulb; 31, olfactory tract; 32, sylvian fissure; 33, superior temporal gyrus; 34, superior temporal sulcus; 35, middle temporal gyrus; 36, inferior temporal sulcus; 37, inferior temporal gyrus; 38, pons; 39, flocculus; 40, medulla oblongata. *Image courtesy of The Rhoton Collection® http://rhoton.ineurodb.org/. Dissector: K.Yagmurlu.*

Central Sulcus

The CS of Rolando separates the frontal and parietal lobes and is an important anatomic landmark when interpreting structural neuroimaging (Naidich et al., 2013; Cho et al. 2002; Su et al., 2019). The CS is flanked by the precentral gyrus, which houses the primary motor cortex anteriorly, and the postcentral gyrus, which houses the primary somatosensory cortex posteriorly (Naidich et al., 2013; Hill et al., 2019). Along the medial surface of the brain, near the interhemispheric fissure at the apex of the CS, the precentral and postcentral gyri fuse to form the paracentral lobule, whereas inferolaterally these gyri fuse to form the subcentral gyrus (Naidich et al., 2013; Ulmer and Jansen, 2020; Naidich and Yousry, 2007; Wagner et al., 2013). This continuous loop of tissue surrounding the CS has been proposed as a separate lobe referred to as the central lobe, although the term has yet to gain wide acceptance. Several signs and landmarks have been described on structural imaging using CT and MRI to identify the CS, most of which rely on relationships and patterns between sulci and gyri (Fig. 2.28). Although most of these signs are 85%–98% reliable in localizing the CS, morphologic variations or changes owing to adjacent cerebral lesions (Fig. 2.29) can be limiting factors, suggesting the need to combine several signs in concert to increase confidence in localization (Naidich et al., 2013). Table 2.1 lists the most commonly described signs for identification of the CS, which are also illustrated in Fig. 2.28.

Cingulate Gyrus

Along the medial surface of the cerebral hemisphere, the cingulate gyrus parallels the corpus callosum, separated from it by the callosal sulcus, as best seen on sagittal imaging (Fig. 2.30). The cingulate gyrus is bordered

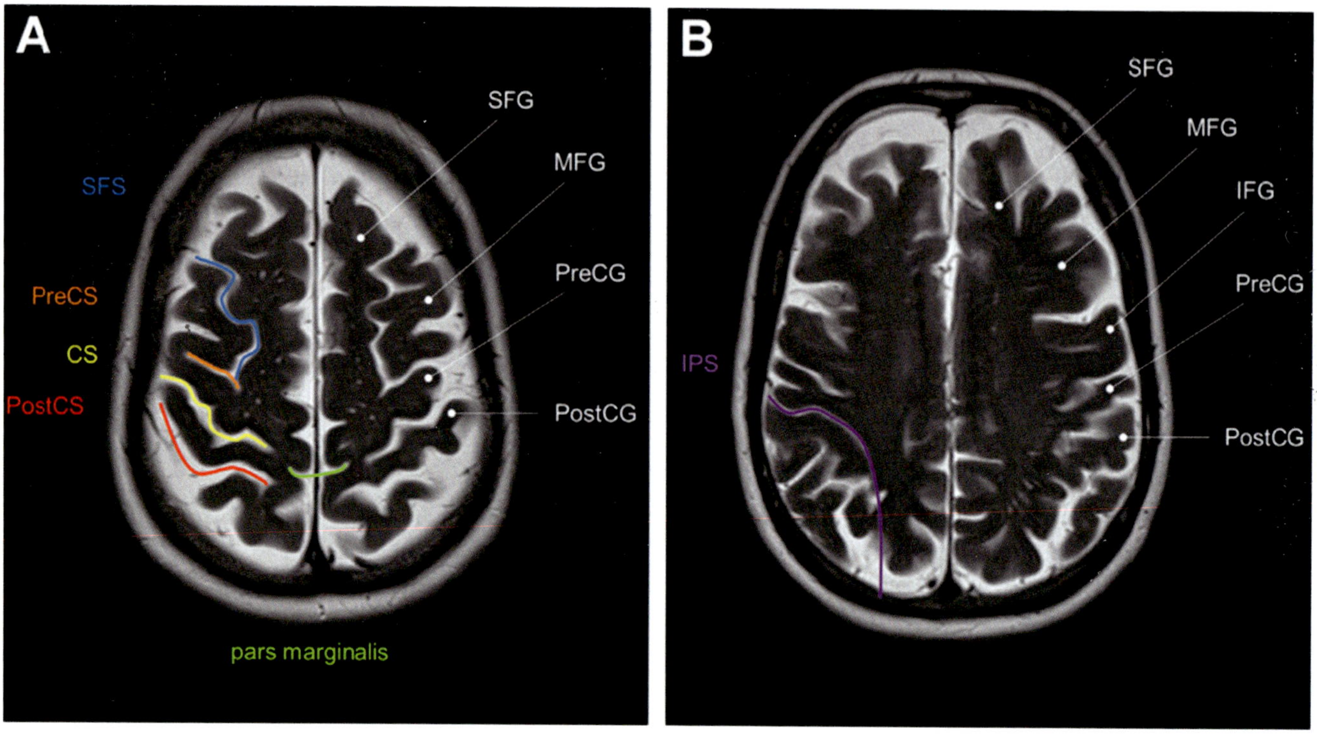

FIGURE 2.25 **Axial T2-weighted spin-echo images at the level of the (A) central sulcus and (B) intraparietal sulcus.** Labeled gyri include the superior frontal gyrus (SFG), middle frontal gyrus (MFG), precentral gyrus (PreCG), and postcentral gyrus (PostCG). Labeled sulci include the superior frontal sulcus (SFS, blue), precentral sulcus (PreCS, orange), central sulcus (CS, yellow), postcentral sulcus (PostCS, red), pars marginalis of the cingulate sulcus (green), and intraparietal sulcus (IPS, purple). *Image courtesy of Vachha, B.A., Massoud, T.F., Huang, S.Y., 2022. Anatomy of the cerebral cortex, lobes, and cerebellum. Neuroimag. Clin. N. Am. 32 (3), 463–473. https://doi.org/10.1016/j.nic.2022.04.008. PMID: 35843656.*

superiorly by the superior frontal gyrus and paracentral lobule. The cingulate gyrus is separated from the superior frontal gyrus and paracentral lobule by the cingulate sulcus (see Fig. 2.30) (Naidich et al., 2013), which runs parallel to the corpus callosum along its anterior portion and bends superiorly to form the pars marginalis. The cingulate gyrus is a major limbic area and part of the Papez circuit (the medial or hippocampal limbic circuit), which begins at the hippocampus, continues as the fimbria and fornix, and ends in the mammillary body. From the mammillary body, the mammillothalamic tract reaches the anterior nuclear group of the thalamus and then the cingulum (the white matter core of the cingulate gyrus), which turns around the splenium of the corpus callosum to end as the radiation of the cingulum into the hippocampus and uncus of the temporal lobe, thus completing the circuit (Fountas et al., 1998); this is in contradistinction to the basolateral or amygdaloid limbic circuit (Shah et al., 2012; Duprez et al., 2005).

Parietal Lobe

The parietal lobe is located posterior to the CS and can be divided into four regions: the postcentral gyrus anteriorly, the superior and inferior parietal lobules posteriorly, and the precuneus medially. The postcentral gyrus is located immediately posterior to and runs parallel to the CS (see Fig. 2.25). The postcentral gyrus is bordered posteriorly by the postcentral sulcus. The superior and inferior parietal lobules are located posterior to the postcentral sulcus and are divided by the intraparietal sulcus (see Fig. 2.25). The precuneus is located posterior to the postcentral sulcus and can be readily identified on the medial surface of the hemisphere on sagittal images (see Fig. 2.30). The precuneus is contiguous with the superior parietal lobule along the superior aspect of the parietal lobe. The parietooccipital sulcus is posterior to the precuneus and divides the parietal lobe from the occipital lobe (see Fig. 2.30). The parietooccipital sulcus is readily identified on sagittal images as a diagonally oriented sulcus that parallels and is located posterior to the ascending ramus of the cingulate sulcus. Between the pars marginalis of the cingulate sulcus and the parietooccipital sulcus lies a sulcus known as the subparietal sulcus (see Fig. 2.30), which separates the precuneus from the posterior portion of the cingulate gyrus. On the lateral surface of the brain, the posterior part of the Sylvian fissure angles superiorly to form the ascending ramus, which terminates in an inverted U-shaped gyrus known as the supramarginal gyrus (Fig. 2.31). The angular gyrus is usually located immediately posterior to the

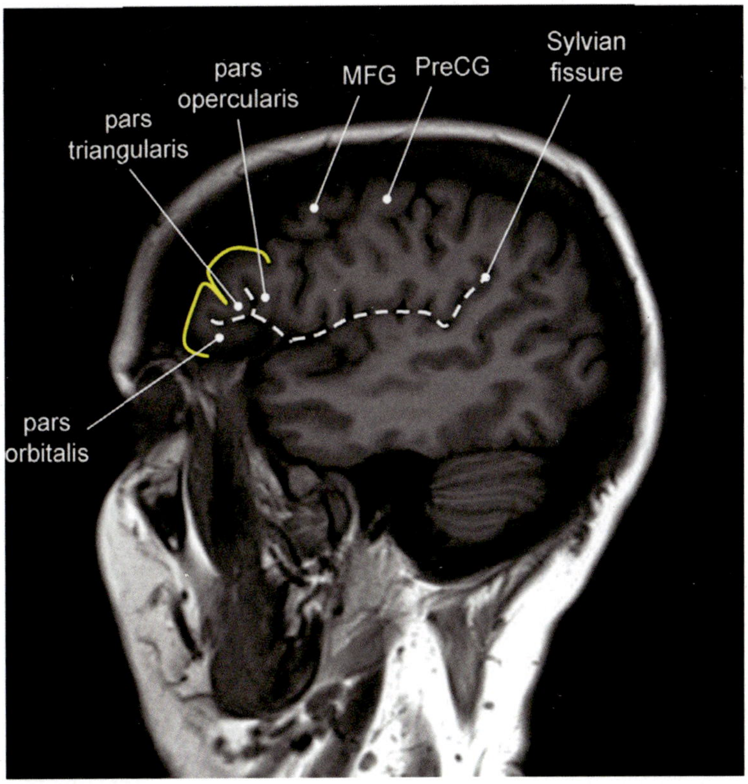

FIGURE 2.26 **Far lateral sagittal T1-weighted image at the level of the Sylvian fissure and frontal operculum.** Labeled gyri include the yellow M-shaped inferior frontal gyrus, consisting of the pars orbitalis, pars triangularis, and pars opercularis; middle frontal gyrus (MFG), and precentral gyrus (PreCG). The Sylvian fissure courses horizontally across the image as denoted by the dashed line, with its anterior and ascending rami separating the pars orbitalis, pars triangularis, and pars opercularis of the inferior frontal gyrus, respectively. *Image courtesy of Vachha, B.A., Massoud, T.F., Huang, S.Y., 2022. Anatomy of the cerebral cortex, lobes, and cerebellum. Neuroimag. Clin. N. Am. 32 (3), 463–473. https://doi.org/10.1016/j.nic.2022.04.008. PMID: 35843656.*

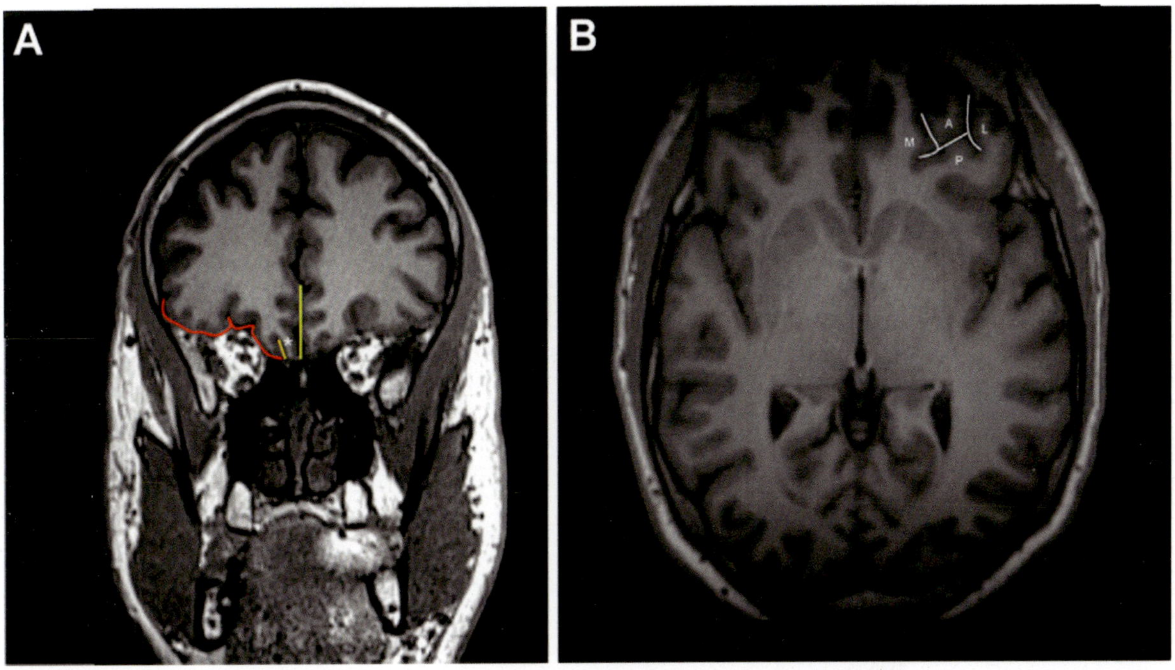

FIGURE 2.27 **Coronal and axial T1-weighted magnetization-prepared gradient echo (MPRAGE) images through the level of the orbitofrontal cortex.** (A) Far anterior coronal image shows the orbitofrontal cortex (red outline), which is separated from the gyrus rectus (asterisk) by the vertically oriented olfactory sulcus (orange line). The interhemispheric fissure is delineated as a yellow line dividing the cerebral hemispheres. (B) Axial image shows the H-shaped orbital sulcus separating the anterior (A), lateral (L), posterior (P), and medial (M) orbital gyri. *Image courtesy of Vachha, B.A., Massoud, T.F., Huang, S.Y., 2022. Anatomy of the cerebral cortex, lobes, and cerebellum. Neuroimag. Clin. N. Am. 32 (3), 463–473. https://doi.org/10.1016/j.nic.2022.04.008. PMID: 35843656.*

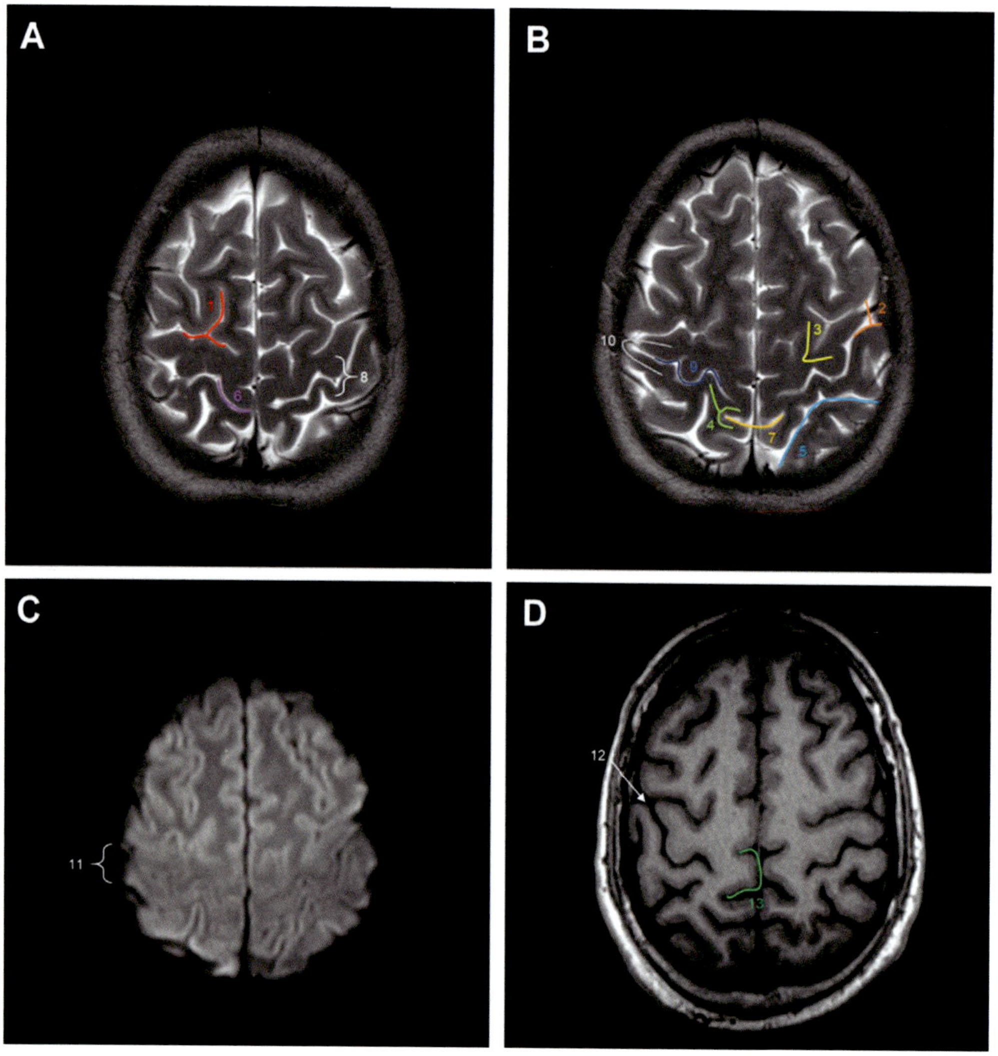

FIGURE 2.28 **Montage of signs indicating the location of the central sulcus on (A, B) axial T2-weighted images, (C) axial diffusion-weighted image, and (D) axial T1-weighted MPRAGE image along the superior convexity.** 1, superior frontal sulcus-precentral sulcus sign (upper T sign); 2, inferior frontal sulcus-precentral sulcus sign (lower T sign); 3, L sign; 4, bifid postcentral gyrus sign; 5, intraparietal sulcus sign; 6, midline sulcus sign; 7, pars bracket sign; 8, thin postcentral gyrus sign; 9, hand knob or reverse-omega sign; 10, U sign; 11, invisible cortex sign; 12, white gray sign; 13, white paracentral lobule sign. *Image courtesy of Vachha, B.A., Massoud, T.F., Huang, S.Y., 2022. Anatomy of the cerebral cortex, lobes, and cerebellum. Neuroimag. Clin. N. Am. 32 (3), 463–473. https://doi.org/10.1016/j.nic.2022.04.008. PMID: 35843656.*

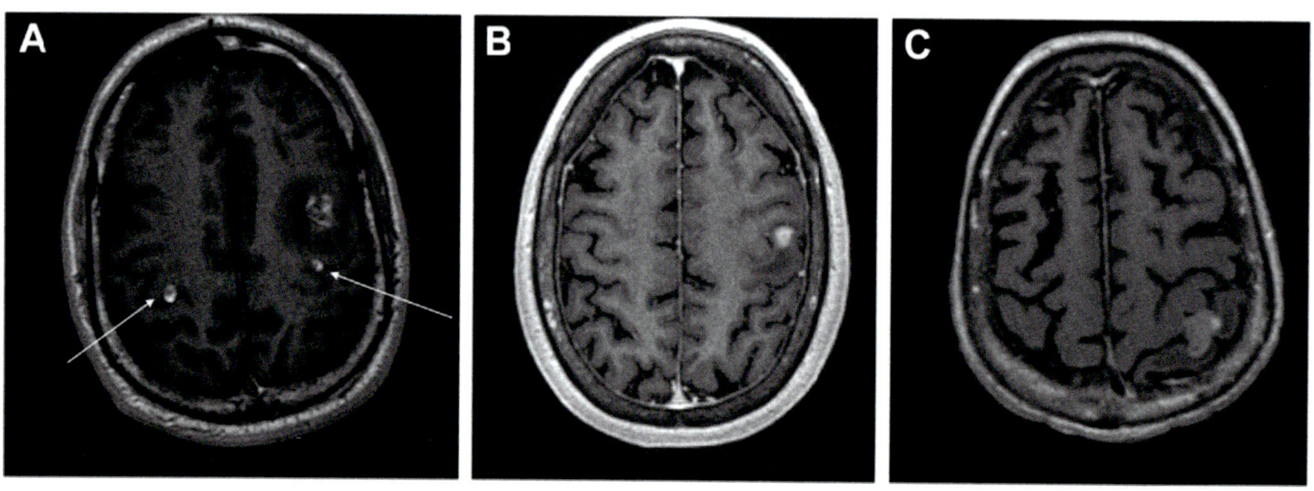

FIGURE 2.29 **Postcontrast axial T1-weighted fast spoiled gradient echo images demonstrating metastatic lesions involving the precentral and postcentral gyri.** (A) Small enhancing metastases (*white arrows*) in the bilateral precentral gyri. A larger enhancing lesion in the posterior aspect of the middle frontal gyrus is associated with marked surrounding edema that extends into the precentral gyrus. (B) Enhancing metastasis involving the left postcentral gyrus with adjacent edema. (C) Lobulated enhancing lesion centered in the left postcentral gyrus extends into the central sulcus.

TABLE 2.1 Central sulcus signs.

Signs	Description
Superior frontal sulcus-precentral sulcus sign (upper T sign)	The easily identifiable superior frontal sulcus joins the precentral sulcus; this intersection has the shape of an upper case T (see Fig. 2.1) and is considered 85% specific (Byrne, 2012; Edwards and Larocca, 1983; Press et al., 1989). The precentral sulcus lies at the posterior end of the superior frontal sulcus, the precentral gyrus is found posterior to the precentral sulcus; and the CS is the next posterior sulcus (Byrne, 2012; Edwards and Larocca, 1983).
L sign	The intersection between the superior frontal gyrus and the precentral gyrus resembles the capital letter "L" with the CS immediately posterior to the precentral gyrus (Edwards and Larocca, 1983; Press et al., 1990).
Bifid postcentral gyrus sign	The pars marginalis of the cingulate sulcus splits the postcentral gyrus into a bifid configuration near the interhemispheric fissure (Edwards and Larocca, 1983; Press et al., 1989).
Intraparietal sulcus sign	The intraparietal sulcus intersects the postcentral sulcus and can be used to identify the central sulcus (Byrne, 2012).
Midline sulcus sign	The CS is identified as the longest sulcus running horizontally and closely approaching the interhemispheric fissure (Edwards and Larocca, 1983).
Pars bracket sign	The bihemispheric pars marginalis formed by the posterior–superior extension of the cingulate sulcus takes the shape of an anteriorly shaped bracket. The CS passes anterior to the pars marginalis on each side and medial to the lateral edge of the pars bracket (Byrne, 2012).
Thin postcentral gyrus sign	The precentral gyrus is thicker than the postcentral gyrus in anteroposterior diameter (Byrne, 2012). In addition, the thickness of the cortical gray matter within the precentral gyrus is thicker than that of the postcentral gyrus (Byrne, 2012).
Hand knob or reverse-omega sign	The motor hand knob area refers to the posterior convex bulge of the precentral gyrus resulting in a symmetric or asymmetrical reverse-omega or epsilon shape in the axial plane (Byrne, 2012; Press et al., 1989, 1990; Schulz et al., 2012). In the sagittal plane, the hand motor area resembles a backward-oriented hook known as the sigmoidal hook sign (Byrne, 2012; Press et al., 1989, 1990).
U sign	The inferolateral aspect of the CS is enclosed by the U-shaped subcentral gyrus (Edwards and Larocca, 1983).
Invisible cortex sign	The cortices abutting the CS appears isointense relative to adjacent white matter on diffusion-weighted MRI sequences (Destrieux et al., 2017). This observation is highly accurate in identifying the inferolateral aspect of the CS (Destrieux et al., 2017).
White gray sign	Increased T1 signal of the anterior and posterior cortices along the CS; the reduced gray–white contrast around the central sulcus was noted to be a reliable sign for identification of the CS on T1-weighted 3D inversion recovery fast-spoiled gradient echo images (Shah et al., 2012).
White paracentral lobule sign	The high signal intensity of the paracentral lobule cortex on T1-weighted imaging compared with that in the superior frontal cortex is useful for CS identification (Courchesne et al., 1989).
Disappearing central sulcus sign	On double inversion recovery sequences, the CS cortex signal intensity is lower than that of the adjacent sulci, and this characteristic can be used to identify the CS (Sherman et al., 2023).

From Vachha, B.A., Massoud, T.F., Huang, S.Y., 2022. Anatomy of the cerebral cortex, lobes, and cerebellum. Neuroimag. Clin. N. Am. 32 (3), 463–473. https://doi.org/10.1016/j.nic.2022.04.008 with permission. For the references mentioned in the table, see the original paper.

supramarginal gyrus. This relationship may vary anatomically even within an individual (see Fig. 2.31), with the interposition of accessory gyri between the supramarginal and angular gyri (see Fig. 2.31B). (Naidich et al., 1995) Taken together, the supramarginal and angular gyri form the inferior parietal lobule. The inferior parietal lobule is separated from the superior parietal lobule by the intraparietal sulcus. The primary intermediate sulcus descends from the intraparietal sulcus to divide the supramarginal gyrus from the angular gyrus.

Temporal Lobe

The Sylvian fissure (also known as the lateral sulcus or lateral fissure) is a key landmark that separates the insula medially and the frontal lobe superiorly from the temporal lobe. The lateral aspect of the Sylvian fissure separates the frontal and parietal opercula from the temporal operculum. The temporal pole lies under the Sylvian fissure at the anterior margin of the temporal lobe, along the most rostral portion of the middle cranial fossa. The circular sulcus outlines the insular cortex and is best visualized on sagittal planes medial to the Sylvian fissure. The circular sulcus consists of anterior, superior, and inferior segments that separate the insula from the frontal, temporal, and parietal opercula (Destrieux et al., 2017). The posterior segment of the Sylvian fissure is contiguous with the superior and inferior segments of the circular sulcus of the insula. The limen insula is located at the junction of the anterior and posterior segments of the Sylvian fissure and delineates the point at which the insular cortex meets the amygdala and superior temporal gyrus.

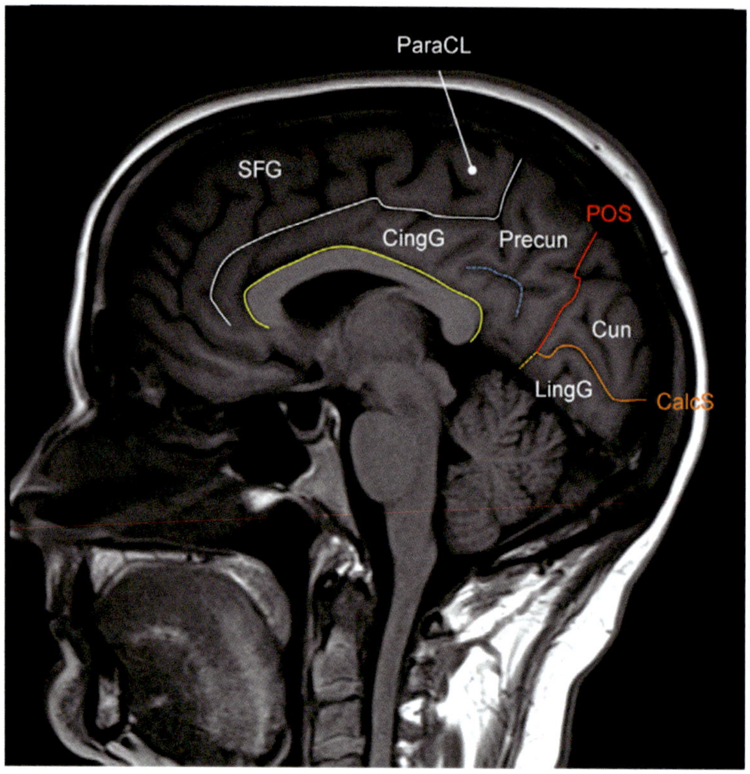

FIGURE 2.30 **Sagittal T1-weighted image demonstrating key gyri and sulci identifiable along the midline.** Labeled gyri include the superior frontal gyrus (SFG), paracentral lobule (ParaCL), cingulate gyrus (CingG), precuneus (Precun), cuneus (Cun), and lingual gyrus (LingG). Outlined sulci include the cingulate sulcus (white line), callosal sulcus (yellow line), subparietal sulcus (blue line), parietooccipital sulcus (red line, POS), and calcarine sulcus (orange line, CalcS). *Image courtesy of Vachha, B.A., Massoud, T.F., Huang, S.Y., 2022. Anatomy of the cerebral cortex, lobes, and cerebellum. Neuroimag. Clin. N. Am. 32 (3), 463–473. https://doi.org/10.1016/j.nic.2022.04.008. PMID: 35843656.*

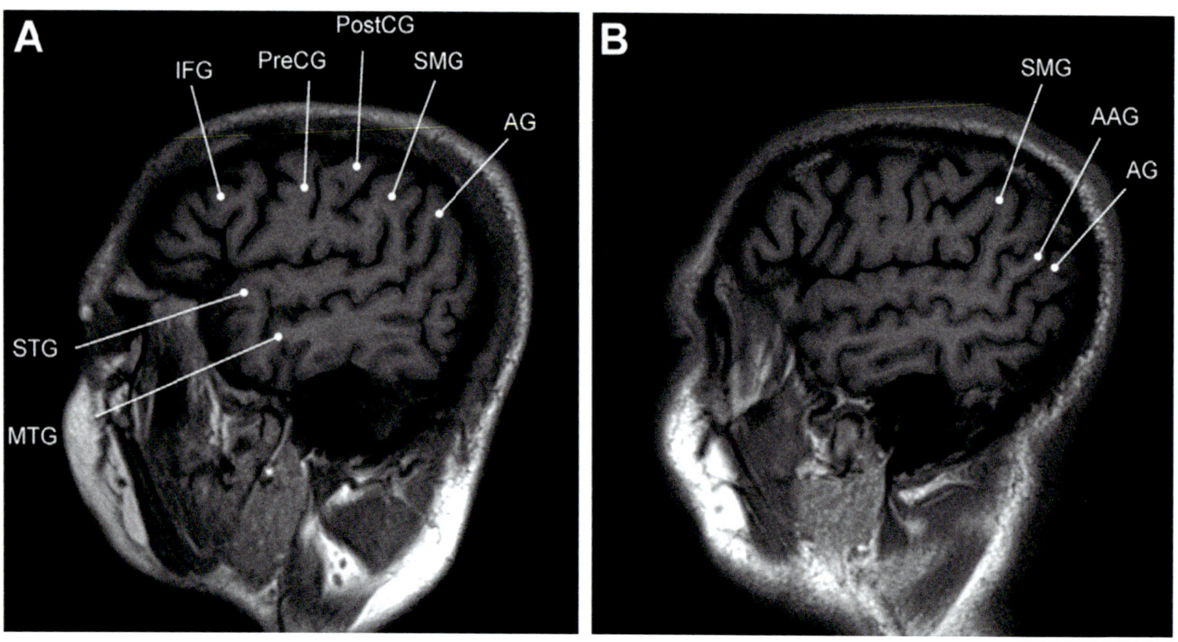

FIGURE 2.31 **Sagittal T1-weighted images along the lateral aspects of the cerebral convexities at the level of the Sylvian fissure illustrating variations in parietal lobe anatomy within an individual.** (A) Right lateral view of the frontal, temporal, and parietal lobes. Labeled gyri include the inferior frontal gyrus (IFG), precentral gyrus (PreCG), postcentral gyrus (PostCG), supramarginal gyrus (SMG), angular gyrus (AG), superior temporal gyrus (STG), and middle temporal gyrus (MTG). (B) Sagittal image of the contralateral hemisphere demonstrates the presence of an accessory preangular gyrus (AAG) interposed between the supramarginal gyrus (SMG) and angular gyrus proper (AG). *Image courtesy of Vachha, B.A., Massoud, T.F., Huang, S.Y., 2022. Anatomy of the cerebral cortex, lobes, and cerebellum. Neuroimag. Clin. N. Am. 32 (3), 463–473. https://doi.org/10.1016/j.nic.2022.04.008. PMID: 35843656.*

The temporal lobe consists of three gyri: the superior, middle, and inferior temporal gyri (Fig. 2.32). The superior temporal sulcus separates the superior and middle temporal gyri, and the inferior temporal sulcus separates the middle and inferior temporal gyri. The traditional Wernicke area (Brodmann Area 22) is considered to lie in the posterior part of the superior temporal gyrus in the dominant hemisphere (Edlow et al., 2019). The inferior temporal gyrus is separated by the lateral occipitotemporal sulcus from the lateral occipitotemporal gyrus (see Fig. 2.32). The lateral occipitotemporal gyrus is separated from the parahippocampal gyrus by the collateral sulcus (see Fig. 2.32). The hippocampal fissure separates the parahippocampal gyrus from the hippocampus (see Fig. 2.32). The temporal stem connects the temporal and frontal lobes and lies in close proximity to the insula. The temporal stem is a key landmark that may be affected in several disorders, including seizures, tumors, and infection. The demarcation between the insula and temporal lobes is readily identified on axial and sagittal views (Fig. 2.33). The insula is generally divided into five gyri. The anterior, middle, and posterior short insular gyri are separated from the anterior and posterior long insular gyri by the CS of the insula, which divides the insula into larger anterior and smaller posterior lobules (Naidich et al., 2004). On axial views, the posterior long insular gyrus is separated from the transverse temporal gyrus or Heschl gyrus (see Fig. 2.33), which is the location of primary auditory cortex. Heschl sulcus is posterior to Heschl gyrus and represents the anterior boundary of the planum temporale, an auditory association area.

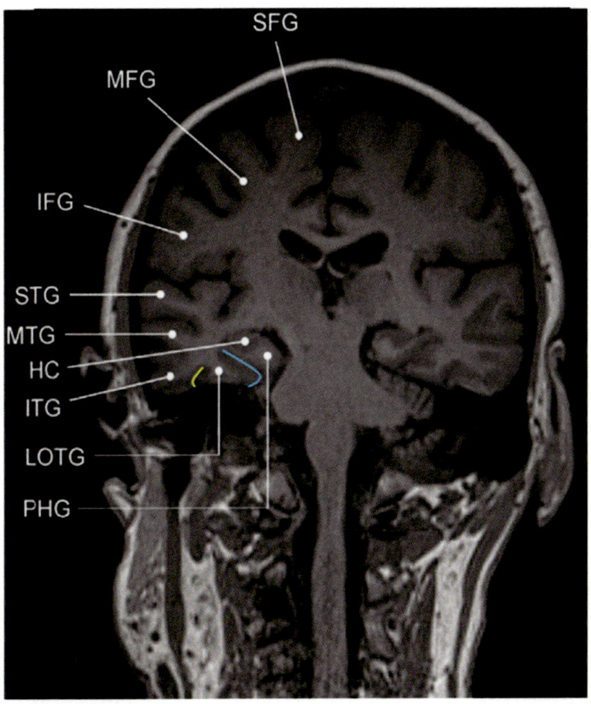

FIGURE 2.32 **Coronal image at the level of the foramina of monro of the lateral ventricles demonstrating the major gyri and sulci of the frontal and temporal lobes.** Labeled gyri include the superior frontal gyrus (SFG), middle frontal gyrus (MFG), inferior frontal gyrus (IFG), superior temporal gyrus (STG), middle temporal gyrus (MTG), hippocampus (HC), inferior temporal gyrus (ITG), lateral occipitotemporal gyrus (LOTG), and parahippocampal gyrus (PHG). Outlined sulci include the lateral occipitotemporal sulcus (yellow line) separating the inferior temporal gyrus and lateral occipitotemporal gyrus and the collateral sulcus (blue line) dividing the lateral occipitotemporal gyrus from the parahippocampal gyrus. *Image courtesy of Vachha, B.A., Massoud, T.F., Huang, S.Y., 2022. Anatomy of the cerebral cortex, lobes, and cerebellum. Neuroimag. Clin. N. Am. 32 (3), 463–473. https://doi.org/10.1016/j.nic.2022.04.008. PMID: 35843656.*

Occipital Lobe

Along the medial surface of the cerebral hemisphere, the parietooccipital sulcus separates the precuneus in the parietal lobe from the cuneus in the occipital lobe (see Fig. 2.30). The calcarine sulcus runs nearly orthogonal to the parietooccipital sulcus and divides the occipital lobe into the triangular-shaped cuneus superiorly and the tongue-shaped lingual gyrus inferiorly as seen on sagittal imaging (see Fig. 2.30). The calcarine sulcus separates the superior and inferior portions of the primary visual cortex. The anterior calcarine sulcus extends anteriorly and inferiorly from the junction of the parietooccipital sulcus and calcarine sulcus to separate the lingual gyrus from the continuation of the cingulate sulcus. Along the lateral surface of the cerebral hemisphere, the occipital lobe is composed of the superior, middle, and inferior occipital gyri, separated by the superior and inferior occipital sulci (Naidich et al., 2013). The superior occipital gyrus drapes over the convexity onto the medial surface to form the cuneus. The superior occipital sulcus or intraoccipital sulcus is contiguous with the intraparietal sulcus and parallels the interhemispheric fissure on axial views. The intraoccipital sulcus runs perpendicular to the parietooccipital sulcus, which is readily identified on axial images as a bifid, barbell-shaped sulcus crossing midline. On axial views, the precuneus sits anterior to the parietooccipital sulcus, and the cuneus is located posterior to the parietooccipital sulcus. The inferior occipital sulcus extends posteriorly from the inferior temporal sulcus. The middle occipital gyrus lies just posterior to the junction of the temporal and parietal lobes and abuts the supramarginal gyrus, the angular gyrus, the superior temporal gyrus, and the middle temporal gyrus. The middle occipital gyrus occupies the most area along the lateral aspect of the occipital lobe and is divided into superior and inferior portions by the horizontally oriented middle occipital sulcus (Naidich et al., 2013).

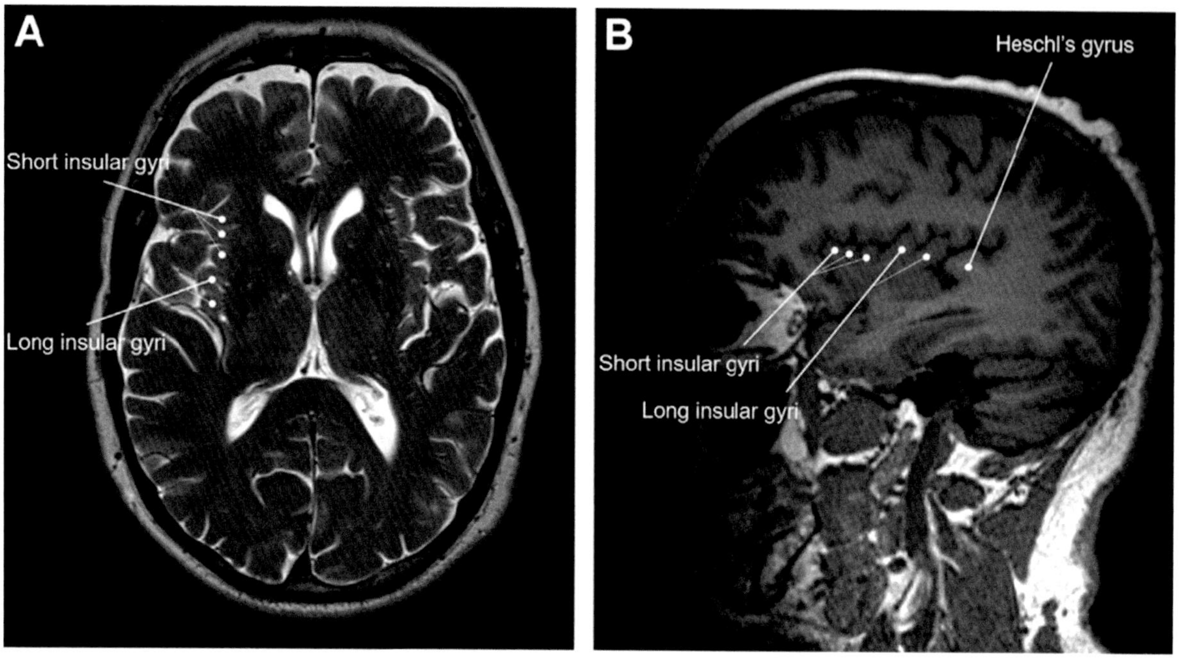

FIGURE 2.33 **Axial and sagittal images of the insula.** (A) Axial T2-weighted spin-echo image demonstrating the division of the insula into the anterior, middle, and posterior short insular gyri anteriorly and the anterior and posterior long insular gyri posteriorly. (B) Sagittal T1-weighted image shows the short and long insular gyri separated by the central sulcus of the insula, and Heschl gyrus posteriorly. *Image courtesy of Vachha, B.A., Massoud, T.F., Huang, S.Y., 2022. Anatomy of the cerebral cortex, lobes, and cerebellum. Neuroimag. Clin. N. Am. 32 (3), 463–473. https://doi.org/10.1016/j.nic.2022.04.008. PMID: 35843656.*

Cerebellum

The cerebellum is located in the posterior cranial fossa and is a major feature of the hindbrain (the rhombencephalon). More specifically, along with the pons, it forms the metencephalon or the upper part of the hindbrain. All cerebellar connections with other parts of the brain pass through the cerebellar peduncles. The cerebellum consists of a highly folded layer of cerebellar gray matter cortex overlying white matter surrounding a fluid-filled fourth ventricle at its base (Fig. 2.34). Four deep cerebellar nuclei are embedded in the white matter. The first detailed descriptions of cerebellar MRI anatomy were provided by Courchesne et al. (1989) and Presset al. (1989, 1990). The cerebellum is divided into two cerebellar hemispheres, connected by a narrow midline zone termed the vermis (see Fig. 2.34).

The vermis has been described as consisting of nine named lobules: lingula, central lobule, culmen, clivus, folium of the vermis, tuber, pyramid, uvula, and nodule or nodulus (Naidich and Yousry, 2007). Three lobes can be distinguished within the cerebellum: the anterior or superior lobe (above the primary or horizontal fissure), the posterior or inferior lobe (below the primary fissure), and the flocculonodular lobe (below the posterior fissure). A set of folds divides the overall structure of the cerebellum into about 10 smaller lobules (I to V in anterior lobe, VI to IX in posterior lobe, and X is considered the flocculonodular lobe). Each surface ridge or gyrus in the cortex is called a folium. Underneath the gray matter folia lies white matter (the arbor vitae, or tree of life, because of its branched appearance on cross section), made up largely of myelinated nerve fibers running to and from the cortex. Embedded within the white matter is a cluster of four deep cerebellar nuclei, composed of gray matter, the dentate, globose, emboliform, and fastigial nuclei. Each of these communicates with different parts of the brain and cerebellar cortex. The dentate nucleus is the largest and is a thin, convoluted layer of gray matter. The flocculus of the flocculonodular lobe is the only part of the cerebellar cortex that does not project to the deep nuclei; its output goes to the vestibular nuclei. Connecting the cerebellum to different parts of the brain and spinal cord are three paired cerebellar peduncles (superior, middle, and inferior). Each superior cerebellar peduncle (brachium conjunctivum) is made of white matter that connects the cerebellum to the midbrain. The middle cerebellar peduncle (brachium pontis) only contains fibers from the pons to the cerebellum arranged in three fasciculi: superior, inferior, and deep. The pontine nuclei (or griseum pontis) in the ventral pons are the nuclei involved in motor activity. Corticopontine fibers carry information from the primary motor cortex to the ipsilateral pontine nucleus, and the pontocerebellar projection then carries that information to the contralateral cerebellum through the middle cerebellar peduncle. The inferior cerebellar

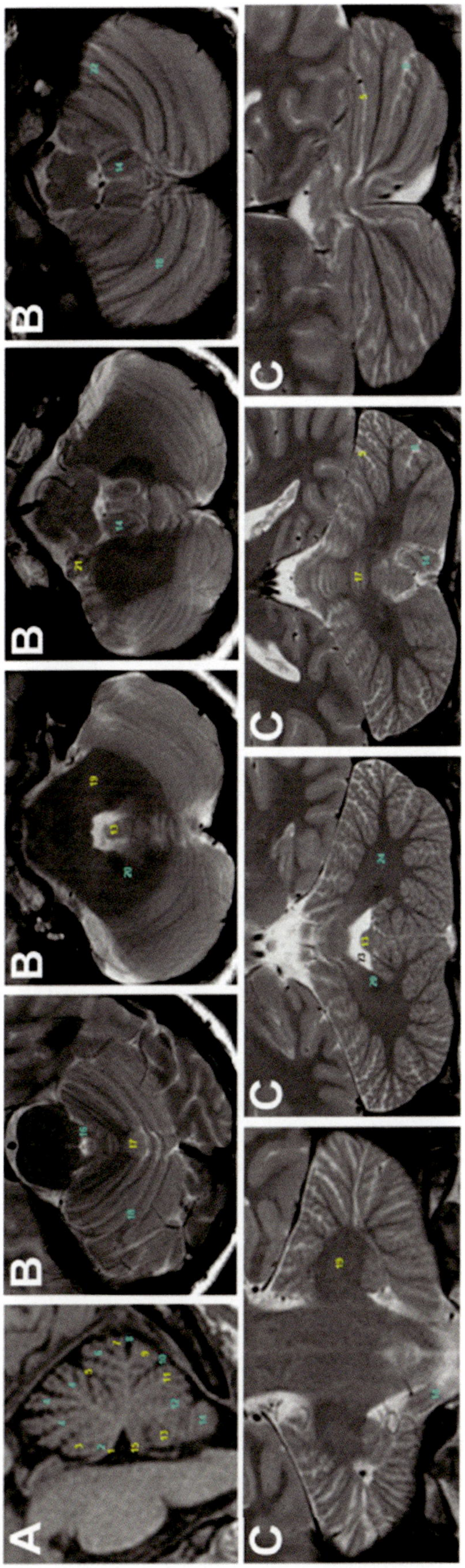

FIGURE 2.34 **(A) Sagittal, (B) axial, and (C) coronal images of the cerebellum.** 1, superior medullary velum; 2, lingula; 3, central lobule; 4, culmen; 5, tentorial fissure; 6, declive; 7, folium; 8, horizontal fissure; 9, tuber; 10, suboccipital fissure; 11, pyramid; 12, uvula; 13, nodulus; 14, tonsil; 15, inferior medullary velum; 16, superior cerebellar peduncle; 17, vermis; 18, cerebellar hemisphere; 19, middle cerebellar peduncle; 20, dentate nucleus; 21, flocculus; 22, cerebellar folia; 23, fourth ventricle; 24, cerebellar white matter (arbor vitae). *Image courtesy of Vachha, B.A., Massoud, T.F., Huang, S.Y., 2022. Anatomy of the cerebral cortex, lobes, and cerebellum. Neuroimag. Clin. N. Am. 32 (3), 463–473. https://doi.org/10.1016/j.nic.2022.04.008. PMID: 35843656.*

peduncle (made of restiform and juxtarestiform bodies) is at the lower part of the fourth ventricle and connects the cerebellum to the spinal cord and medulla oblongata. The inferior cerebellar peduncle carries several input and output fibers for integration of proprioceptive sensory input and motor vestibular functions such as balance and posture maintenance. The cerebellopontine fissure, also called the cerebellopontine angle, is a V-shaped fissure formed by the cerebellum wrapping around the pons and middle cerebellar peduncle. Inferiorly, the cerebellar tonsil is equivalent to a rounded lobule on the undersurface of each cerebellar hemisphere, continuing medially into the uvula of the vermis and superiorly into the flocculonodular lobe.

High-Resolution Neuroimaging: Toward a Better Future

The maximum achievable resolution of an MRI scan is in principle limited only by the time available to scan the subject, as longer scan times are necessary to compensate for the loss of signal-to-noise ratio (SNR) associated with smaller voxels. Recent technological advancements in both hardware and software have allowed MRI to reach increasingly high spatial resolutions, leading to exquisite images of the brain and enabling in vivo visualization of structures such as the stria of Gennari (Fig. 2.42), the brainstem, and substructures of the hippocampus (Zeineh et al., 2014; Wisse et al., 2016). However, a smaller voxel size not only makes the image more susceptible to artifacts associated with small involuntary motion of the subject during the scan, but an extended scan time also means that the total subject motion is likely to be greater. It has recently been shown that a suitable motion-tracking and correction system allows substantial improvements in image quality for very high-resolution imaging, either using a camera-and-marker system (Schulz et al., 2012; Stucht et al., 2015) or additional interleaved lower-resolution MR image "navigators"—even when the subjects being scanned are healthy adults accustomed to the scanner environment. For imaging at the highest resolutions, higher magnetic field strengths are desirable to maximize the available SNR. Recently, several studies demonstrated that ultrahigh-resolution motion-corrected scans can be achieved on a 7T MRI scanner without requiring any additional hardware, delivering images of the whole brain at exceptional quality for various image contrasts (Edlow et al., 2019; Federau and Gallichan, 2016). In the following pages, we will show the brain topography using high-resolution 7T magnetic resonance imaging (Figs. 2.35–2.42).

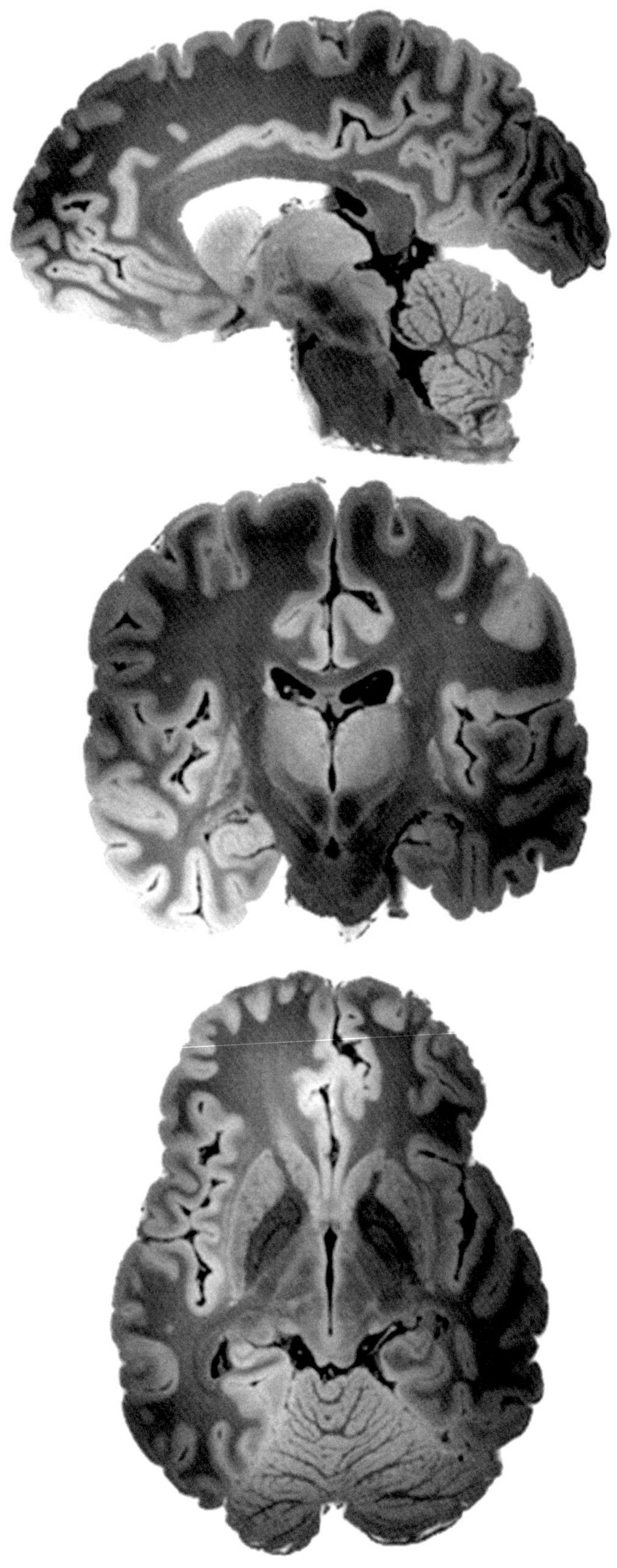

FIGURE 2.35 **High-resolution MRI 7T of brain.** Representative images are displayed in the sagittal (top row), coronal (middle row), and axial (bottom row) planes. For details on the anatomical structures, see later. *Image courtesy of Edlow, B.L., Mareyam, A., Horn, A. et al., 2019. 7 T MRI of the ex vivo human brain at 100 micron resolution. Sci Data 6, 244. https://doi.org/10.1038/s41597-019-0254-8.*

THIRD SECTION: TOPOGRAPHY OF THE SPINAL CORD AND BRAINSTEM: THE IMPORTANCE OF DISSECTION AND IMAGING IN THE NEUROSURGICAL APPROACH

Unlike other parts of the central nervous system, the spinal cord is the only that preserves the embryologic tubular-like appearance. Similar to the developing brain, three embryonic layers, matrix, mantle, and marginal layer, can be clearly recognized in the human cord after the closure of the neural tube. The matrix layer, surrounding the ependymal central canal, originates the neuroblasts that migrate predominantly through a process known as radial migration. In the brain, the massive radial migration of neuroblasts ends in the marginal layer to form the neocortex, whereas in the spinal cord, the migration ends in the mantle layer and remains in direct contact with the matrix layer, to conform a zone of periventricular gray matter. At the same time, the marginal layer in the cord is transformed into a peripheral zone of white matter fibers.

The motor neurons differentiate early in the outermost portions of the anterior (basal plate) mantle layer, and their axons penetrate the marginal layer to conform the ventral root fibers. The spinal ganglia, carrying sensory information and deriving from the neural crest, form bipolar cells with central and peripheral processes. Whereas the central processes constitute the dorsal root fibers that penetrates the marginal layer in the posterior aspect of the cord (alar plate), the peripheral processes join the ventral root fibers to form the mixed (i.e., afferent and efferent) spinal nerves near the intervertebral foramina. Hence, the cord acquires its characteristic H or butterfly-shaped central zone of gray matter (for detail and molecular processes of the development of the central nervous system, see Chapters 1 and 4).

There are usually 31 pairs of spinal nerves; thus, the spinal cord can be subdivided in the same number of segments (8 cervical, 12 thoracic, 5 lumbar, 5 sacral, and 1 coccygeal). The corresponding spinal cord segments are designed accordingly as C1–C8, T1–T12, etc. (Figs. 2.43–2.46) (Bican et al., 2013). With the exception of C1, all segments have ventral and dorsal nerve roots and the conjoined spinal nerve exits through the intervertebral foramen (Figs. 2.47–2.49). The C1 spinal nerve exits between the cranial occipital bone and the first cervical vertebra. Although the presence of ventral nerve roots in the C1 segment is constant, the presence of dorsal nerve roots is variable. Tubbs et al. (2007) found that the C1 segment has no dorsal nerve roots in approximately 54% of cadaveric specimens. When the dorsal nerve roots were present, only 28% had a dorsal root ganglion, and in 50%, the spinal accessory nerve joined the dorsal rootlets. This complex anatomy can be important in cases where rhizotomy is considered for alleviation of pain.

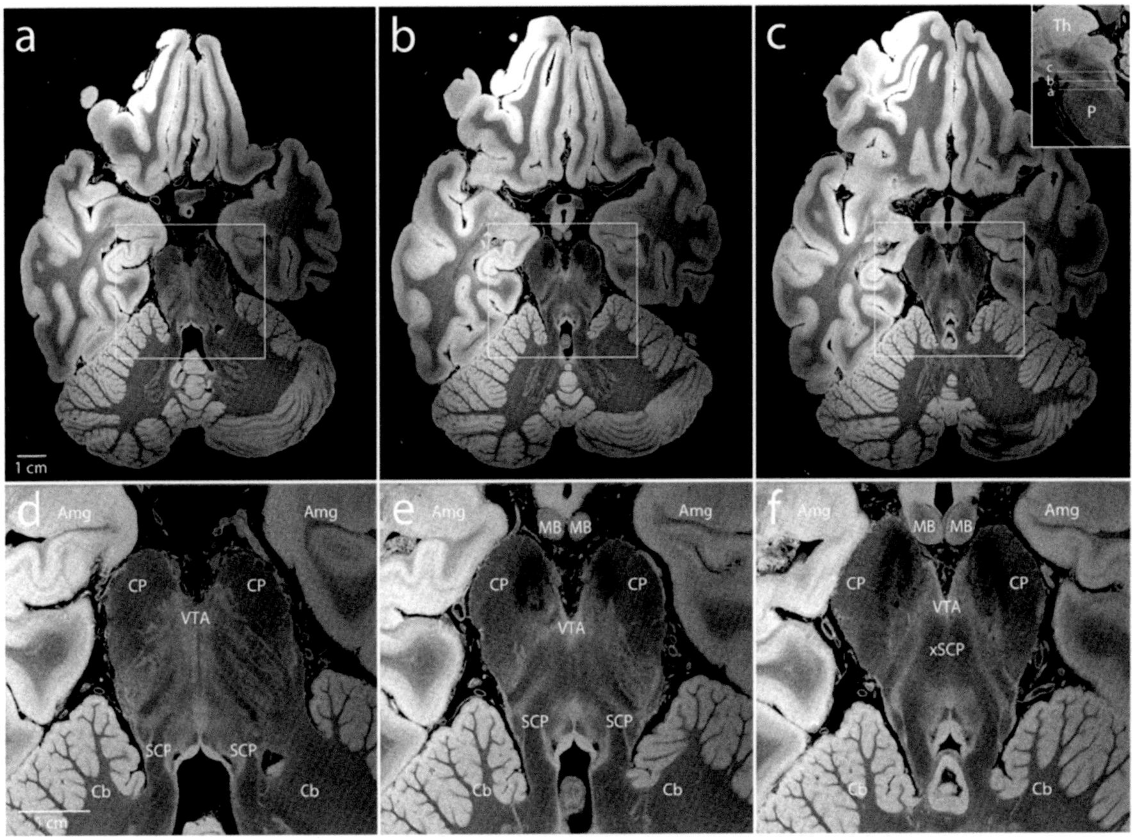

FIGURE 2.36 **Delineation of brainstem neuroanatomy.** Representative axial sections from the synthesized FLASH25 volume are shown at the level of the rostral pons and caudal midbrain (A–C), see inset in panel C). Zoomed views of the brainstem, medial temporal lobe, and anterior cerebellum (within the white rectangles in A–C) are shown in the bottom row (D–F). The anatomic detail that can be visualized in this ex vivo 100 μm resolution MRI dataset is beyond that, which can be seen in typical in vivo MRI datasets. All images are shown in radiologic convention. Neuroanatomic abbreviations: *Amg*, amygdala; *Cb*, cerebellum; *CP*, cerebral peduncle; *MB*, mammillary body; *P*, pons; *SCP*, superior cerebellar peduncle; *Th*, thalamus; *VTA*, ventral tegmental area; *xSCP*, decussation of the superior cerebellar peduncle. *Image courtesy of Edlow, B.L., Mareyam, A., Horn, A. et al., 2019. 7 T MRI of the ex vivo human brain at 100 micron resolution. Sci Data 6, 244. https://doi.org/10.1038/s41597-019-0254-8.*

Until the 12th week of intrauterine life, the cord has the same longitudinal dimension as the spinal canal. Subsequently, there is a discrepancy due to a faster longitudinal growth of the spinal canal as compared with a slower growing of the cord. Eventually, at birth, the *conus medullaris* terminates at approximately L3 vertebral level, and in the adulthood, it terminates at approximately the L1–L2 vertebral level (Fig. 2.43). Congenital anomalies can cause the conus to lay low in the spinal canal, usually below L2–L3 vertebral level.

There are several defining contours along the external portion of the spinal cord: the anterior median fissure, posterior median sulcus, anterolateral sulcus, posterior intermediate sulcus, and the posterolateral sulcus (Fig. 2.44). Of particular importance is a unique bundle of nonmyelinated white matter fibers, named the dorsolateral tract of Lissauer (after a german neurologist who first described it in the late 19th century), which cover the posterior gray horns, run in longitudinal direction, and reach the meningeal surface at the posterolateral sulcus. The other external contours also define the location of other important white matter bundles as described later.

The spinal cord is anchored to the dura mater, like the whole neuraxis, by a saw-toothed fibrous extension of the pia called the denticulate ligament, which is an important surgical landmark. On cross section, the appearance of the adult spinal cord is variable at different levels. Its transverse diameter is bigger than the anteroposterior (AP) diameter at cervical levels, giving the cord an oval shape. In the thoracic and lumbar levels, the cord appears round. The configuration of the gray matter and the gray/white matter ratio vary between spinal cord regions as well (Zhao et al., 2014). Due to the amount of ascending and descending white matter tracts that converge in the cervical region, the size of the cord is the biggest there. This is known as the *cervical enlargement* and is usually located at C4–T1 spinal cord segments (the maximum diameter of the cord is around

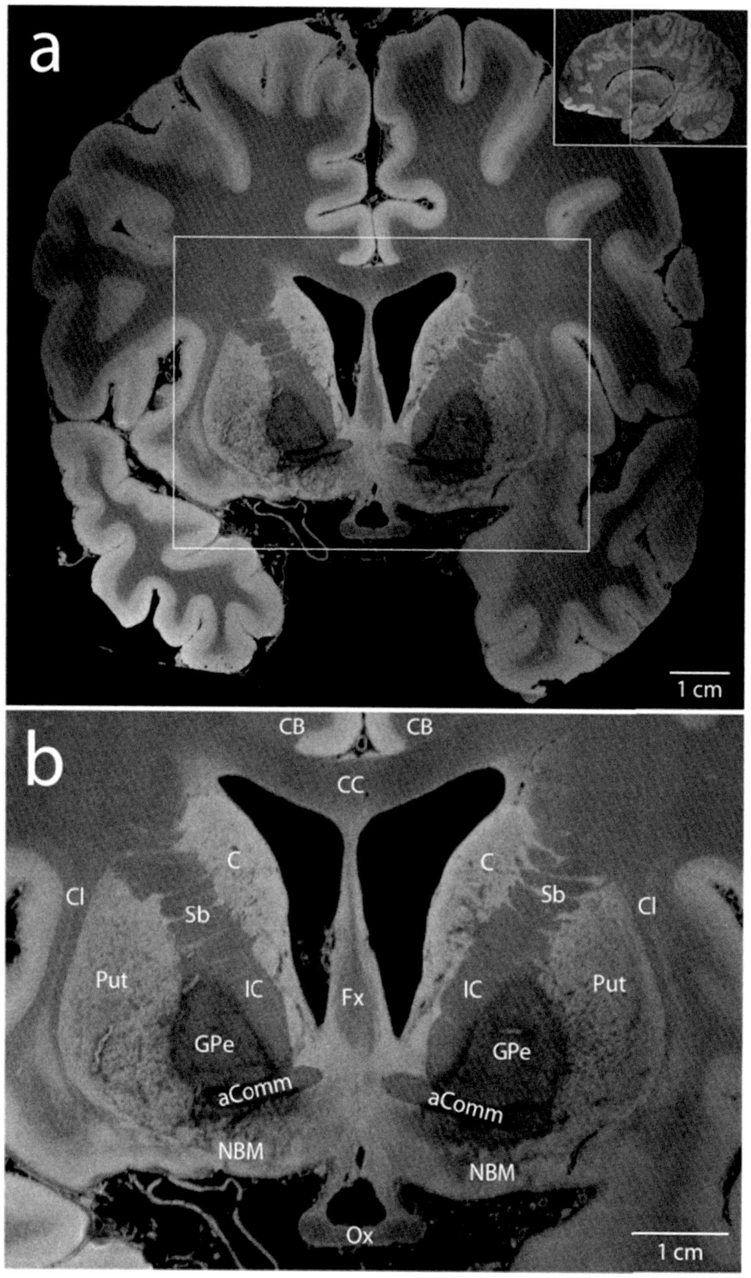

FIGURE 2.37 **Delineation of basal ganglia and basal forebrain neuroanatomy by MRI 7T; coronal section.** A representative coronal section from the synthesized FLASH25 volume is shown in the plane of the anterior commissure (aComm; see inset in A). A zoomed view of the basal ganglia and basal forebrain (within the white rectangle in A) is shown in (B). The anatomic detail that can be visualized in this ex vivo 100 μm resolution MRI dataset is beyond that, which can be seen in standard in vivo MRI datasets. Neuroanatomic abbreviations: *C*, caudate; *CB*, cingulum bundle; *CC*, corpus callosum; *Cl*, claustrum; *Fx*, fornix; *GPe*, globus pallidus externa; *IC*, internal capsule; *NBM*, nucleus basalis of Meynert; *Ox*, optic chiasm; *Put*, putamen; *Sb*, striatal bridges. *Image courtesy of Edlow, B.L., Mareyam, A., Horn, A. et al., 2019. 7 T MRI of the ex vivo human brain at 100 micron resolution.* Sci *Data 6, 244. https://doi.org/10.1038/s41597-019-0254-8.*

C5–C6 measuring about 13 × 7.5 mm in transverse and AP dimensions) (Nieuwenhuys et al., 2008; Mai and Paxinos, 2012).

The thoracic portion of the cord has the smallest diameter (approximately 6.4 mm). Inferiorly, in the lumbosacral segments of the spinal cord, there is once again a slight increase in caliber (9 × 7 mm, transverse and AP dimensions) (Mai and Paxinos, 2012). This is called the *lumbosacral enlargement* and is usually located at L2–S2 spinal cord segments (Fig. 2.46). The average dimensions can have clinical importance, particularly in the setting of central canal stenosis, where a central canal diameter of 10 mm or less is associated with chronic compressive myelopathy in the cervical spine

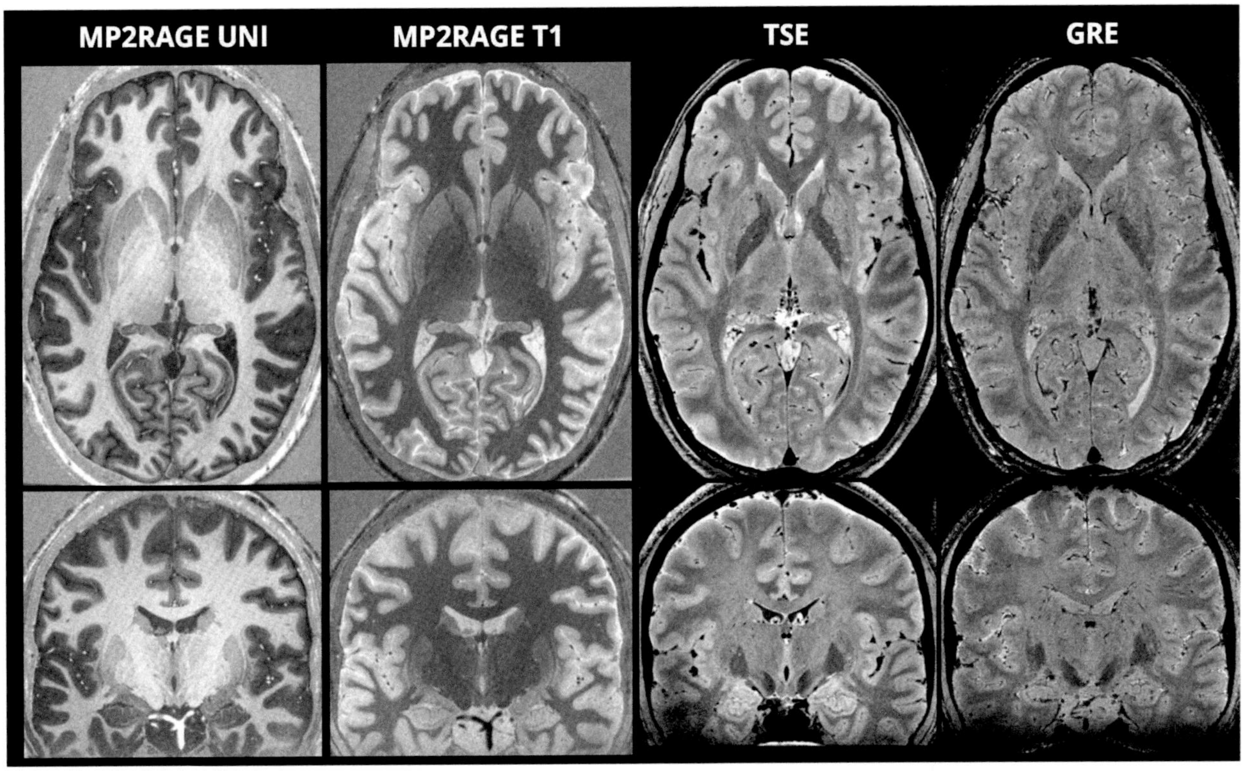

FIGURE 2.38 **Basal ganglia seen by MRI 7T. Overview axial and coronal sections for each of the four contrasts:** uniform T_1-weighted image (MP2RAGE UNI), quantitative T_1 map (MP2RAGE T1), Turbo-spin echo (TSE), and gradient-recalled echo (GRE). ***In this image, the basal ganglia are clearly visible***, see later for details. *Image courtesy of Federau, C., Gallichan, D., 2016. Motion- Correction Enabled Ultra-High Resolution In-Vivo 7T-MRI of the Brain. PLoS One 11 (5): e0154974. https://doi.org/10.1371/journal.pone.0154974.*

(Edwards and Larocca, 1983); or when we are facing the description of an atrophic cord (roughly less the 7 mm in the cervical spine) (Sherman et al., 2023; Fountas et al., 1998).

The longest spinal cord segment is T7. In the cervical cord, the longest segment is C5, and clearly, the shortest segments are in the lumbosacral cord (Fig. 2.43). Due to the length discrepancy between the spinal canal and the spinal cord, the actual longitudinal segmentation of the spinal cord and its correlate with a particular vertebral or clinical segment can be challenging. Recently, an effort to segment the cervical cord and correlate with vertebral bodies/disc levels has shown significant variability between subjects (Cadotte et al., 2015). Each dorsal and ventral nerve root is attached to the cord by a series of rootlets arranged in a longitudinally way (Figs. 2.45, 2.47, 2.48 and 2.49). In humans, each nerve root is composed of six or eight dorsal and ventral rootlets. Posteriorly, these series of rootlets attach to the posterolateral sulcus in an area called dorsal root entry zone (DREZ). Anteriorly, they attach to the anterolateral sulcus through an area called the anterior root entry zone (AREZ). The arrangement of rootlets correlates with the functional segments of the cord and can be evaluated with MRI or DTI when clinical correlation is deemed necessary. However, a coronal acquisition or special protocols with 3D sequences might be necessary. These anatomic landmarks and any variability might prove to be clinically important in the near future, particularly for quantitative and functional imaging of the cord.

The main arterial vascular supply of the cord is provided by the anterior spinal artery (ASA). Superiorly, the ASA arises from two small branches of the intracranial vertebral arteries joining in the midline at the level of the medulla. The ASA additionally is supplied from radiculomedullary branches at multiple segments. The largest radiculomedullary artery, also known as the artery of Adamkiewicz, is located near the lumbosacral enlargement (usually arising from a posterior intercostal artery between T9–T12 vertebral level, more often on the left). The second largest is the artery of the cervical enlargement; it usually enters the central canal at the C5–C6 vertebral level and arises from a cervical vertebral artery or the costocervical trunk. The ASA reaches its largest caliber in the lower thoracic spine caudal to the entry of the artery of Adamkiewicz (Byrne, 2012). Hence, the overall caliber of the thoracic portion of the ASA is the smallest. The ASA, via the sulcal artery, supplies most of the central gray matter and anterior two-

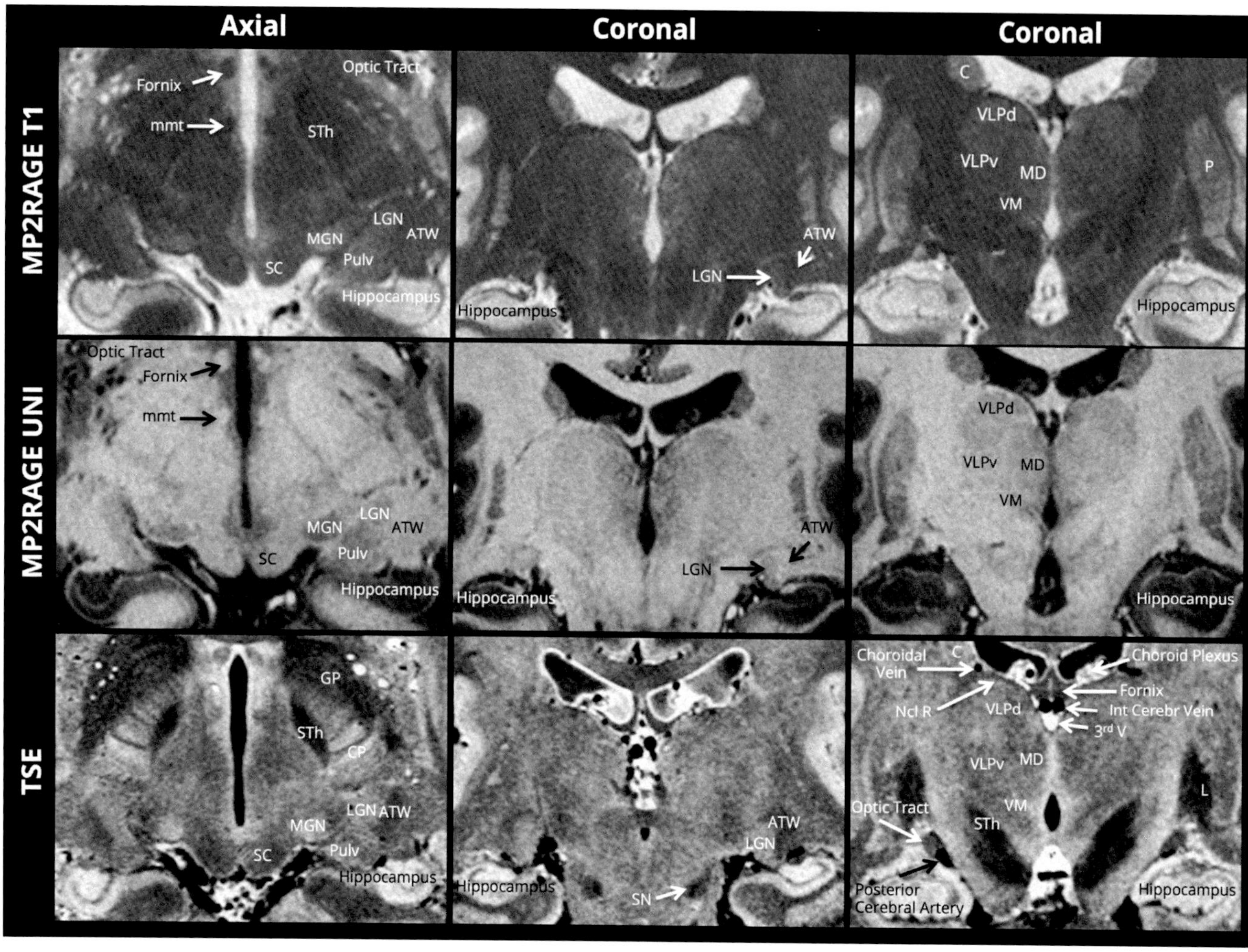

FIGURE 2.39 **Delineation of thalamic nuclei and nearby structures by MRI 7T; axial and coronal section.** *ATW*, area triangularis of Wernicke; *C*, nucleus caudatus; *L*, nucleus lentiformis; *LGN*, lateral geniculate ganglion; *MD*, mediodorsal nucleus; *Ncl R*, nucleus reticularis; *STh*, subthalamic nucleus; *VM*, ventral medial nucleus; *VLPd*, ventral lateral posterior nucleus, dorsal division; *VLPv*, ventral lateral posterior nucleus, ventral division. *Image courtesy of Federau, C., Gallichan, D., 2016. Motion- Correction Enabled Ultra-High Resolution In-Vivo 7T- MRI of the Brain. PLoS One 11 (5), e0154974. https://doi.org/10.1371/journal.pone.0154974.*

thirds of the spinal cord. There are also small paired posterior spinal arteries, which arise from radiculopial branches; they might be longitudinally discontiguous and contribute mainly to the peripheral zone supply of the cord. The white matter can be anatomically subdivided into anterior, posterior, and lateral funiculi (Fig. 2.44). White matter can also be functionally subdivided into ascending, descending and propriospinal pathways; however, this will be the topic discussed in Chapter 4.

TOPOGRAPHY OF THE BRAINSTEM

The brainstem is a part of encephalon, which connects the cerebrum with the spinal cord and the with cerebellum. The brainstem is composed of the parts, caudorostrally: medulla oblongata, pons, and mesencephalon. Located on the basilar part of the occipital bone, it consists of white and gray matter. The gray matter comprises nuclei of cranial nerves (III to XII c.n), nuclei of reticular formation and pontine nuclei. The white matter consists of ascending and descending tracts (see Chapters 4 and 5a).

Three laminae compose the brainstem: tectum, tegmentum, and the basis in a dorsoventral sequence. These laminae extend the entire length of the brainstem, and the composition of each layer varies from zone to zone (Smith and DeMeyer, 2003).

Structures of the brainstem are very diverse with respect to functions they participate in, neurochemicals they contain, and neural pathways they accommodate. As a result, the anatomical organization of the human brainstem is a complex mixture of compact neuronal groups and dispersed cell areas with varying cytoarchitecture (Paxinos et al., 2012).

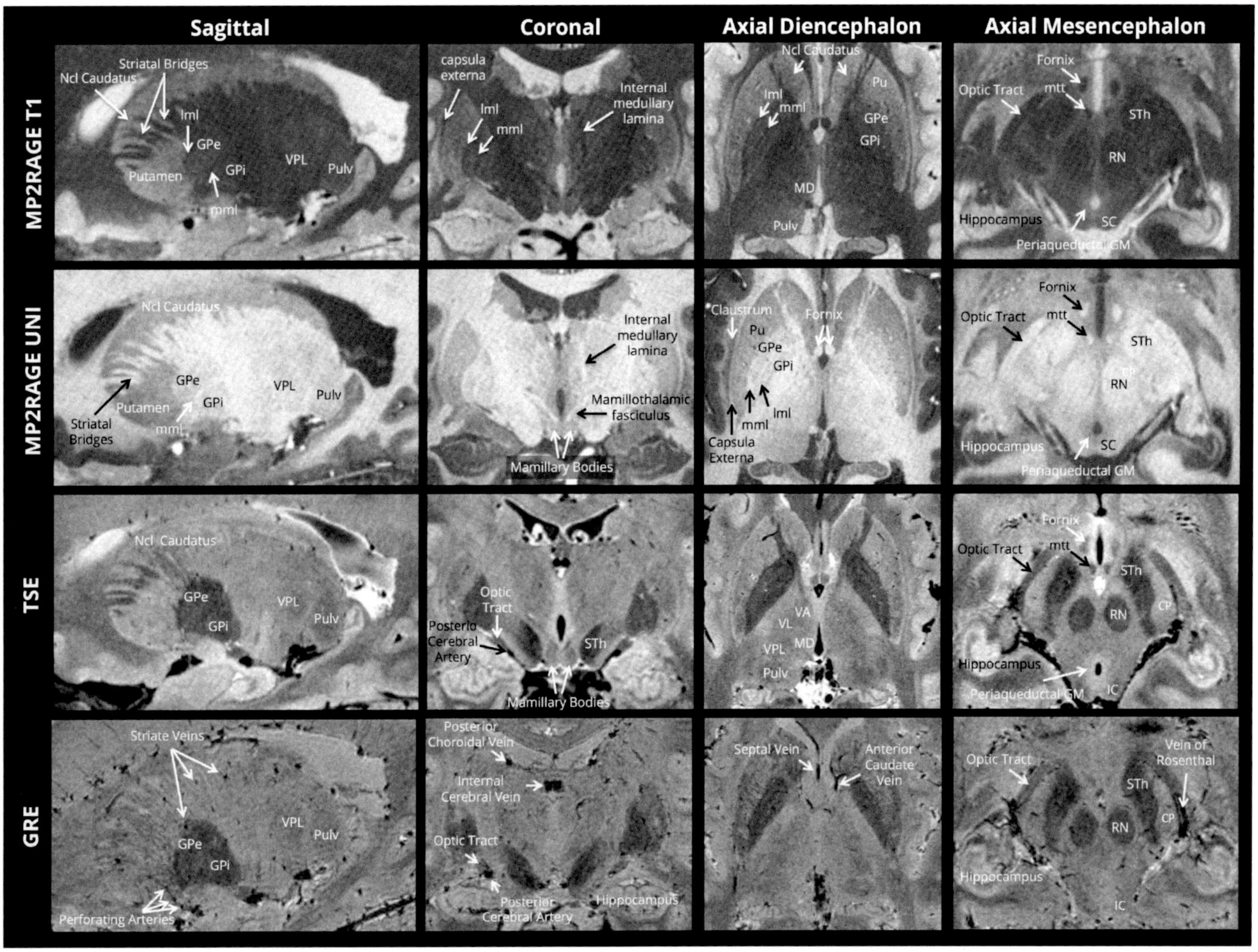

FIGURE 2.40 **Zoomed sections of the basal ganglia and middle brain for each of the four ultrahigh-resolution sequencies (MRI 7T).** *GPe*, globus pallidus externus; *Gpi*, globus pallidus internus; *IC*, inferior colliculus; *lml*, lateral medullary lamina; *MD*, mediodorsal nucleus; *mml*, medial medullary lamina; *mtt*, mamillothalamic tract; *Pulv*, pulvinar; *SC*, superior colliculus; *STh*, subthalamic nucleus; *VA*, ventral anterior nucleus; *VL*, ventral lateral nucleus; *VPL*, ventral posterior lateral nucleus. *Image courtesy of Federau, C., Gallichan, D., 2016. Motion- Correction Enabled Ultra-High Resolution In-Vivo 7T- MRI of the Brain. PLoS One 11 5, e0154974. https://doi.org/10.1371/journal.pone.0154974.*

The dorsal surface of the pons and medulla constitutes the floor of the fourth ventricle and is named the rhomboid fossa, a rhombus-shaped depression resembling a diamond (Fig. 2.23) The rhomboid fossa is delineated by the internal margins of the superior and inferior cerebellar peduncles. The lateral angles of the rhombus are closed by the middle cerebellar peduncles; the inferior angle is the obex, and the superior corner is marked by the frenulum of the superior medullary velum of the fourth ventricle (Szabo et al., 2017).

MEDULLA OBLONGATA

External Features of the Medulla Oblongata

Medulla oblongata is the most caudal part of the brainstem that connects the pons to the spinal cord. It continues with spinal cord approximately at the level of foramen magnum and is about 3 cm long. On the anterior surface of medulla oblongata, the anterior spinal artery sits in the anterior median fissure that is continuous on the spinal cord. On the lateral side of the ventral medial fissure, the pyramids are formed by the descending axons of the corticospinal (pyramidal) tract. The vertical length of the pyramids is 21.36 mm (Kocer et al., 2022). At the distal part of the pyramids, most of the axons (approximately 90%) of the corticospinal tract decussate and form the lateral corticospinal tract. The remaining 10% stay ipsilateral and form the anterior corticospinal tract. Lateral to the pyramids, olivary bodies are situated on both sides. Inferior olivary nuclei underlie the olivary bodies and are important for motor coordination. The transverse length of the olivary bodies is 5.21 mm on the midline and the vertical length is 14.7 mm (Kocer et al., 2022). Preolivary and

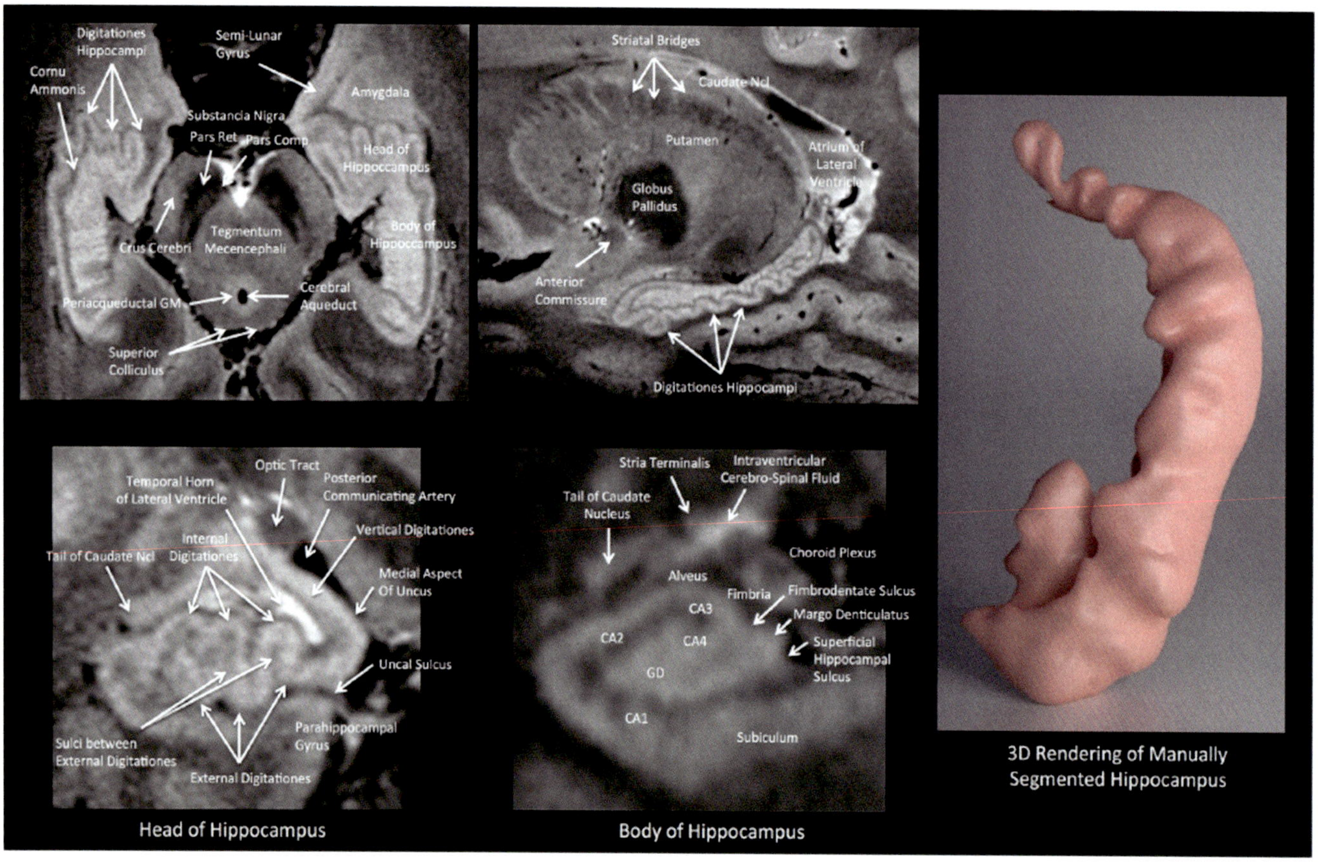

FIGURE 2.41 **(Upper row) oblique axial and sagittal cross sections from the TSE images of the hippocampus, and (lower row) zoomed in coronal section through the hippocampal head and body (MRI 7T).** The digitations hippocampi are particularly clearly depicted. The segments of the cornu ammonis and the dentate gyrus can also be visualized. The contours of the amygdala are also well delimited. The right-hand panel shows a 3D rendering of the hippocampus after manual segmentation from this data. *CA*, cornu ammonis; *GD*, gyrus dentatus. *Image courtesy of Federau, C., Gallichan, D., 2016. Motion- Correction Enabled Ultra-High Resolution In-Vivo 7T- MRI of the Brain. PLoS One 11 (5), e0154974. https://doi.org/10.1371/journal.pone.0154974.*

postolivary grooves appear medially and laterally to each olivary bodies, respectively. The axons of the hypoglossal nerve (XII c.n.) emerge from the preolivary groove. Glossopharyngeal nerve, vagus nerve, and accessory nerve (IX, X and XI c.n.) emerge from the postolivary sulcus from cranially to caudally. Axons of the hypoglossal nerve (XII c.n.) emerge from the preolivary sulcus (Fig. 2.50).

Posteriorly to the postolivary grooves, the medulla oblongata is connected to the cerebellum via the inferior cerebellar peduncles. The posterior surface of the medulla oblongata is examined in two parts: the lower part and the upper part. The lower part meets with spinal cord at the bottom. The posterior median sulcus lies at the midline on the lower part and is continuous on the posterior surface of the spinal cord. Two elongated prominences called the gracile and cuneate tubercle are positioned close to the posterior median sulcus on both sides, from medial to lateral, respectively. These tubercles are formed by the gracile and cuneate nuclei that are the second-order neurons of the dorsal column (see Chapter 4).

The upper part of the dorsal medullary surface is continuous with the dorsal surface of the pons. On the dorsal surface, the upper part of the medulla oblongata and the pons form the floor of the fourth ventricle. The hypoglossal nucleus forms the hypoglossal trigone, and the dorsal vagal nucleus forms the vagal trigone. In the center of the dorsal surface of the medulla oblongata, the posterior median sulcus separating the two gracile tubercles can easily be seen (Fig. 2.51). Lateral to the vagal trigone, there is a vestibular area that contains the underlying vestibular nuclei.

PONS

External Features of the Pons

The pons is continuous with the midbrain cranially and the medulla oblongata caudally. Pons is the Latin name for "bridge" and termed such due to its functional

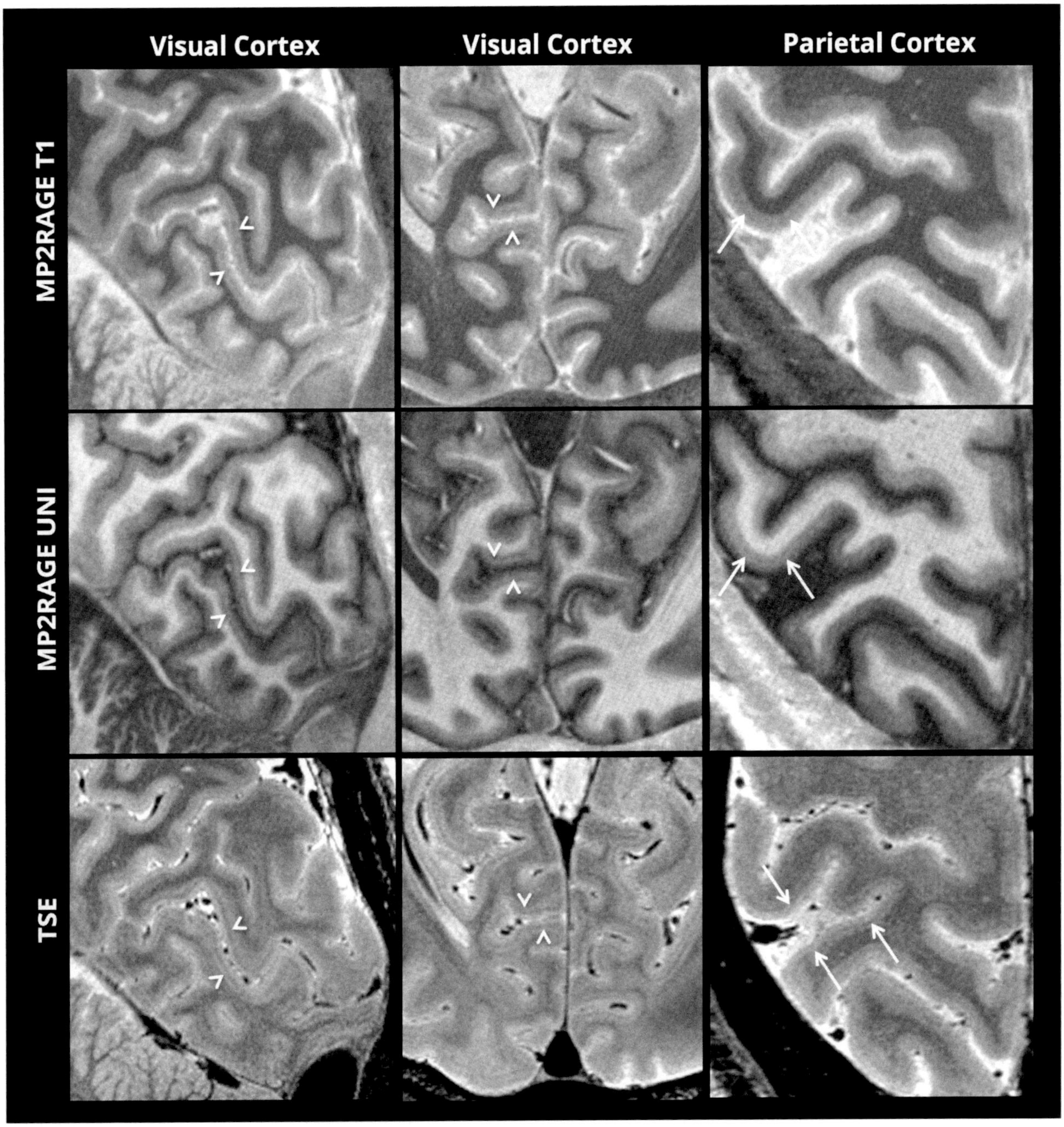

FIGURE 2.42 **MRI 7T:cerebral cortex study.** *Arrow heads*: The stria of gennari in the primary visual cortex. *Arrows*: Intracortical contrast in the parietal cortex. *Image courtesy of Federau, C., Gallichan, D., 2016. Motion- Correction Enabled Ultra-High Resolution In-Vivo 7T- MRI of the Brain. PLoS One 11 (5), e0154974. https://doi.org/10.1371/journal.pone.0154974.*

content of the transverse axons to the right and left cerebellar hemispheres. The pons is about 2.5 cm long.

Anteriorly, the pons is limited by pontomesencephalic sulcus cranially and pontomedullary sulcus (or bulbopontine sulcus) caudally. At its anterior surface, it is close to the clivus; the pontine cistern is located between clivus and pons. A prominent groove in the midline of the anterior surface is called the basilar sulcus, which accommodates the basilar artery. On the right and left ventrolateral sides, the trigeminal nerve fibers emerge as two roots: a small, medial, *motor root* and a large, lateral, *sensory* root. The anterior part of the pons is marked as the anteriorly located part between the emerging trigeminal nerves on both sides and called the basal pons or proper pons. The anterior surface can be easily marked with its convex surface. This convexity has many transverse pontine fibers that originate from the pontine nuclei and cross the midline. These

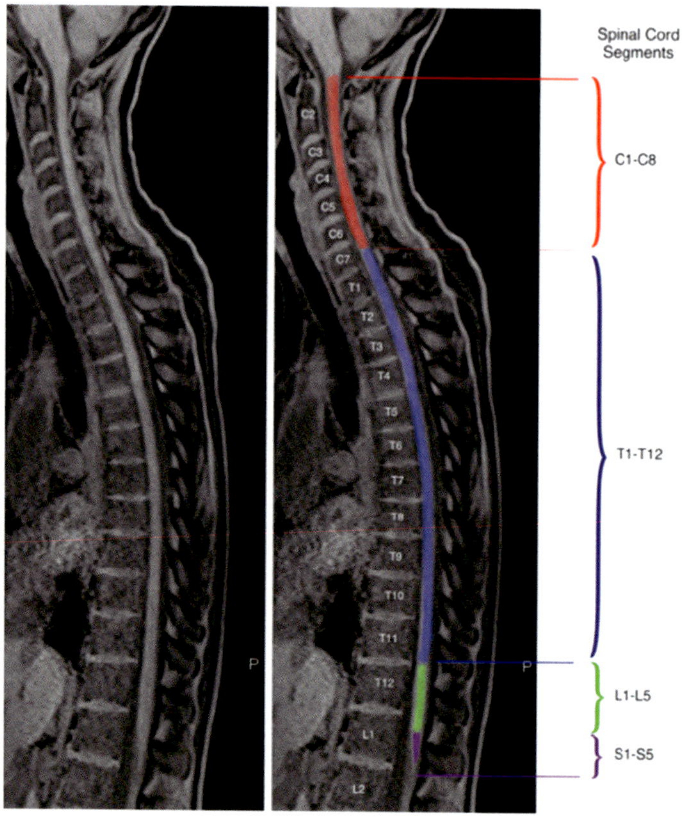

FIGURE 2.43 **MR imaging of the entire spinal cord.** The spinal cord is segmented approximately to the corresponding location of the 31 pair of spinal nerves (8 cervical, 12 thoracic, 5 lumbar, 5 sacral, and 1 coccygeal segments). The last one segment, i.e., the coccygeal spinal cord segment, is not represented here but should be located at the tip of the conus medullaris. In this patient, the conus ends at approximately the L1–L2 vertebral level. The cervical enlargement is located from C4 to T1 spinal cord segments and the lumbar enlargement from L2 to S2 spinal cord segments. *Technique:* Midline sagittal composite section, 1.6 mm thick from overlapping 3D fast Dixon (VIBE) sequence, TR 4.3, TEs = 1.3/~2.5 msec. *MR Image courtesy of Kenneth, L., Weiss, MD; the ImagiCenter and University of Miami, Miller School of Medicine.*

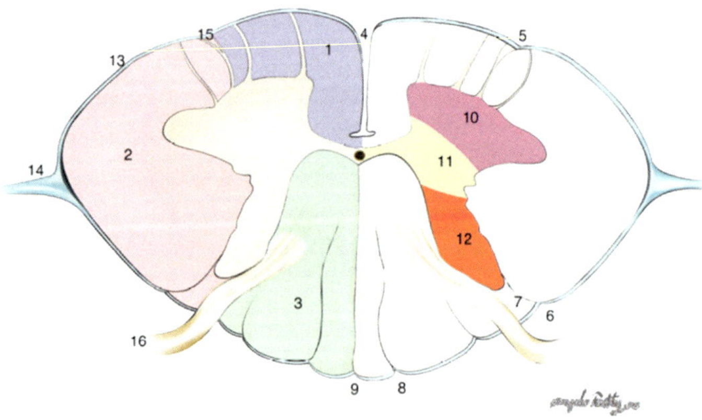

FIGURE 2.44 **Gross anatomy of the spinal cord in adulthood.** Representation of a transverse section of the cervical segment. *Legends:* 1, anterior funicle; 2, lateral funicle; 3, posterior funicle; 4, anterior median fissure; 5, anterolateral sulcus; 6, posterolateral sulcus; 7, Lissauer's tract; 8, posterior intermedian sulcus; 9, posterior median sulcus; 10, anterior horn; 11, intermediate gray zone; 12, posterior horn; 13, pia-arachnoid; 14, denticulate ligament; 15, ventral nerve roots; 16, dorsal nerve root. *Image courtesy of Diaz, E., Morales, H., 2016. Spinal Cord Anatomy and Clinical Syndromes', Seminars in Ultrasound, CT and MRI. 37 (5), 360–371. https://doi.org/10.1053/j.sult.2016.05.002.*

fibers are called the pontocerebellar fibers and form the middle cerebellar peduncle bilaterally. The basal pons contains the pontine nuclei, transverse fibers of pontocerebellar tracts and longitudinal bundles of corticospinal and corticonuclear tracts. The remaining part or the posterior part of the pons is called the pontine

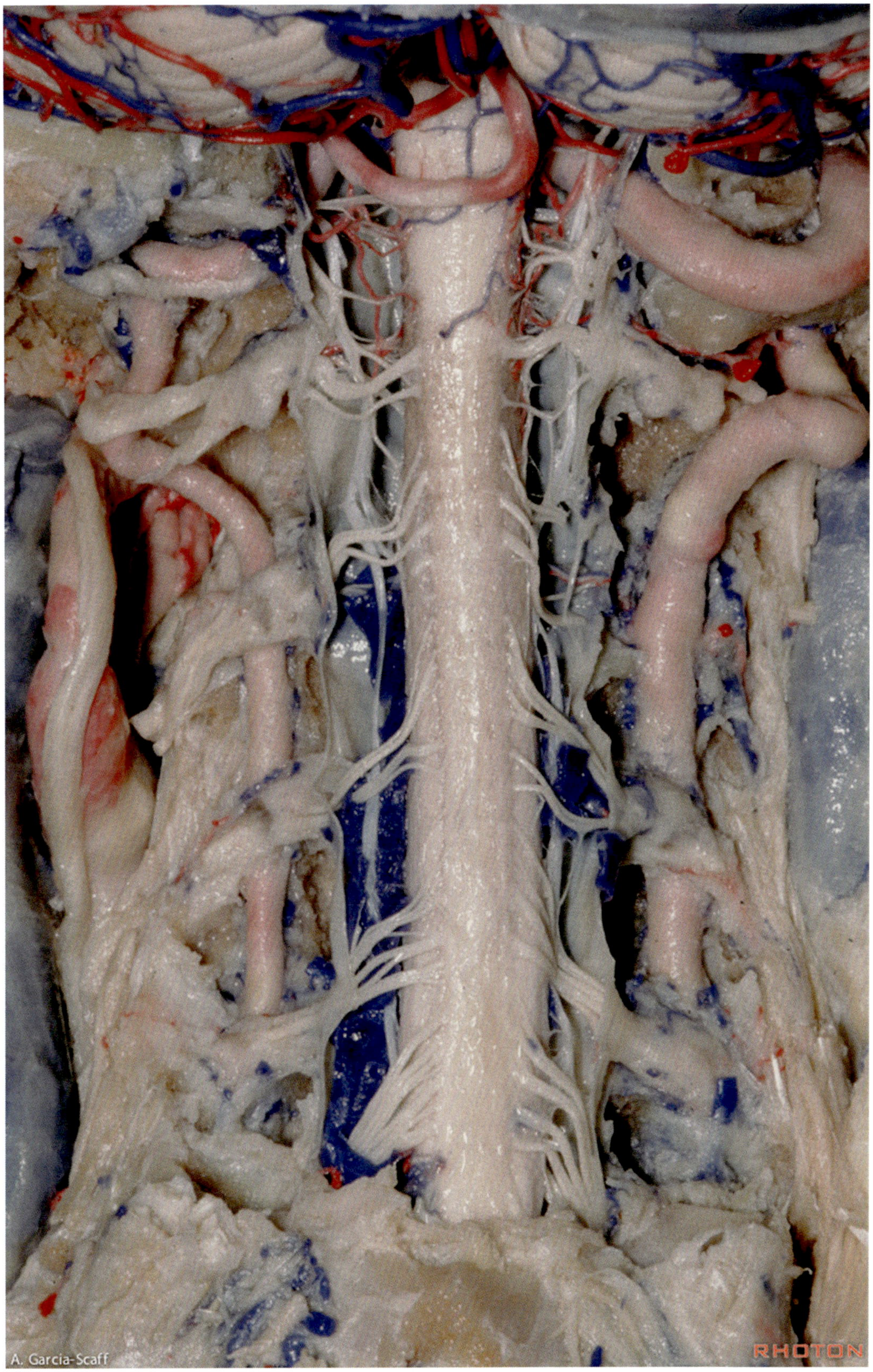

FIGURE 2.45 **Dorsal view of the spinal cord.** The *dorsal roots of the spinal nerves* are easily seen on either side of the dorsal columns. *Image courtesy of The Rhoton Collection® http://rhoton.ineurodb.org/. Dissector: A Garcia-Scaff.*

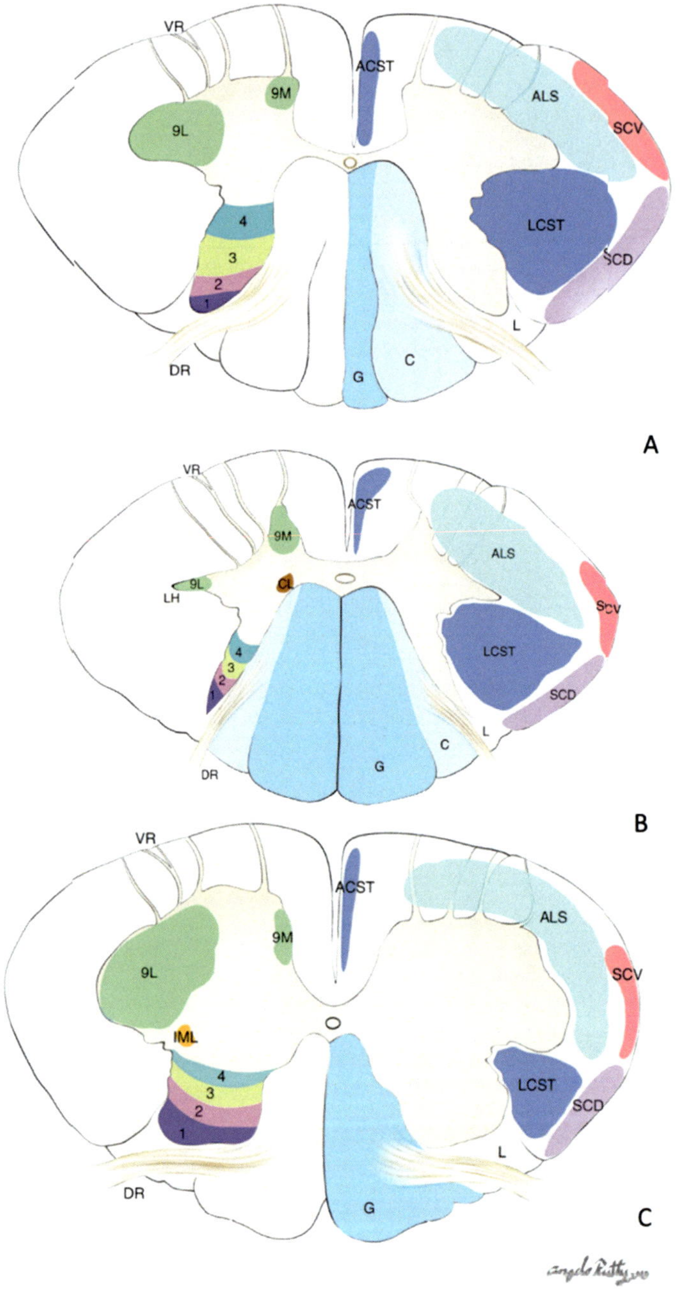

FIGURE 2.46 **Representation of relevant gray and white matter anatomic structures. (A) Cervical segment (C5-cervical enlargement), (B) thoracic segment (T5), and (C) lumbar segment (L5-lumbar enlargement). Note the changes in the gray/white matter relation at different segments.** *Legends:* 1, lamina 1; 2, lamina 2; 3, lamina 3; 4, lamina 4; 9L, lamina 9—lateral motor column (extremities); 9M, lamina 9—medial motor column (axial muscles); *ACST*, anterior corticospinal tract; *ALS*, anterolateral system (spinothalamic tract); *C*, cuneatus; *CL*, Clarke's nucleus (C8–L3); *DR*, dorsal nerve root; *G*, gracilis; *IML*, intermediolateral nucleus (T1–L2); *L*, Lissauer's tract; *LCST*, lateral corticospinal tract; *LH*, lateral horn; *SCD*, dorsal spinocerebellar tract; *SCV*, ventral spinocerebellar tract; *VR*, ventral nerve root. *Image courtesy of Diaz, E., Morales, H., 2016. Spinal Cord Anatomy and Clinical Syndromes', Seminars in Ultrasound, CT and MRI. 37 (5), 360–371. https://doi.org/10.1053/j.sult.2016.05.002.*

tegmentum and accepted as the cranially continuous part of the medulla oblongata. Between the basal pons and the pontine tegmentum, the trapezoid body is located, which is the decussation of the transverse projecting fibers from the cochlear nucleus (especially the anterior cochlear nucleus).

A groove called the bulbopontine sulcus limits caudally the border between the pons and the bulbus on the anterior surface. Roots of abducent (VI c.n.), facial (VII c.n.) and vestibulocochlear (XIII c.n.) nerves emerge from the bulbopontine sulcus from medial to lateral. The facial nerve (VII c.n.) emerges as a motor and sensory

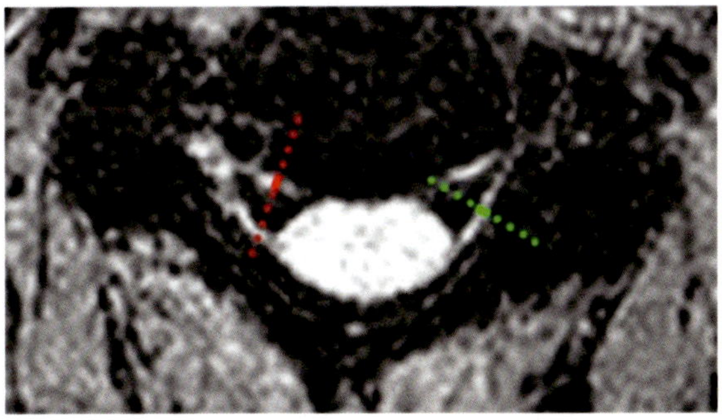

FIGURE 2.47 **Dorsal and ventral nerve root.** MR T2*-weighted gradient echo (T2-FFE). The green line is positioned on the dorsal and the red line on the ventral nerve root. *Image courtesy of Stam, M., et al., 2019. Magnetic resonance imaging of the cervical spinal cord in spinal muscular atrophy. NeuroImage: Clinical 24, 102002. https://doi.org/10.1016/j.nicl.2019.102002.*

root in the same way as the vestibulocochlear nerve (VIII), which emerges as a vestibular and cochlear root. Posteriorly, the pons is cranially limited by an imaginary line just passing below the inferior colliculi and caudally limited by an imaginary line passing from the lateral angles of the fourth ventricle. Its posterior surface forms the floor of the upper half of the fourth ventricle. Additionally, stria medullaris is the transversely crossing fibers that mark the border between the pons and medulla on the posterior surface. These fibers are part of the epithalamus containing afferent axons projecting to the habenula from the septal nuclei, anterior thalamic nuclei, and lateral preoptic–hypothalamic area (Párraga et al., 2016). On the posterior surface, one of the critical landmarks is the medial eminence that is situated bilaterally to the median sulcus. A shallow, longitudinal

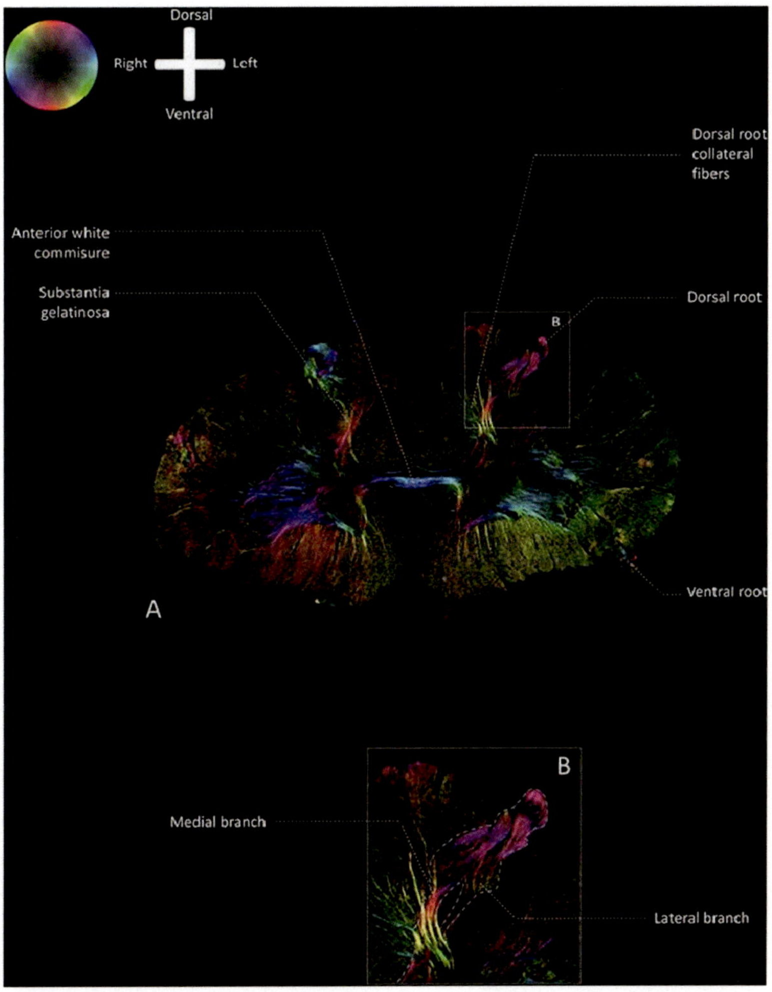

FIGURE 2.48 **Fiber orientation image of the transverse section (100 μm) at level C5–C6 of the cervical human spinal cord.** The fiber orientation is defined by the color sphere and the intensity modulated by the density of in-plane myelinated fibers. (A) Overview of the white matter pathways in the cervical spinal cord. (B) Enlargement of inset in panel A. The medial and lateral bundles of the entering dorsal root can be recognized. The lateral bundle can be seen as a low-intensity area, whereas the medial bundle can be followed up until the deeper layers in the dorsal horn. *Image courtesy of Henssen, D.J.H.A., et al., 2019. Post-mortem 11.7 T Magnetic Resonance Imaging versus Polarized Light Imaging Microscopy to Measure the Angle and Orientation of Dorsal Root Afferents in the Human Cervical Dorsal Root Entry Zone. Frontiers in Neuroanatomy, 13. https://doi.org/10.3389/fnana.2019.00066.*

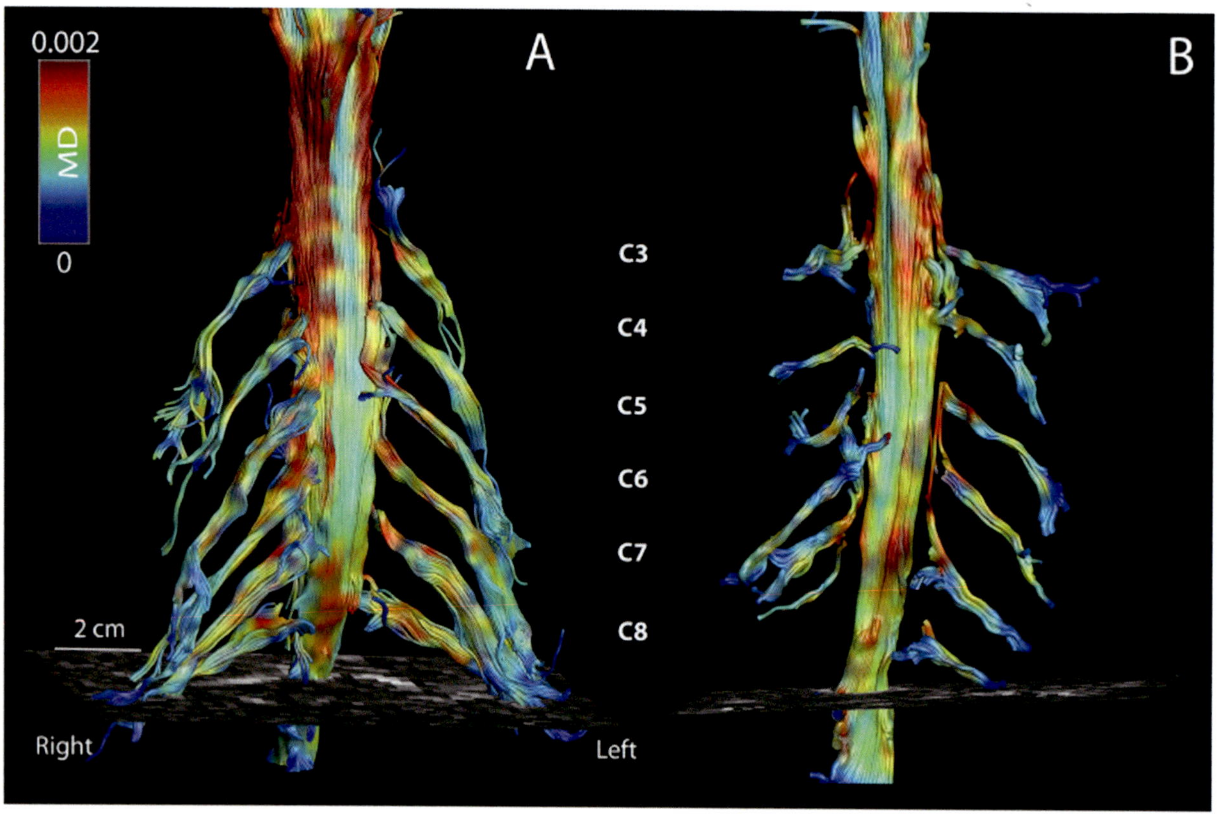

FIGURE 2.49 **Fiber tractography images of cervical tract of the spinal cord and of the cervical nerve roots.** Normal in (A) and atrophic (patient affected by spinal muscular atrophy) in (B). *Image courtesy of Stam, M., et al., 2019. Magnetic resonance imaging of the cervical spinal cord in spinal muscular atrophy. NeuroImage: Clinical 24, 102002. https://doi.org/10.1016/j.nicl.2019.102002.*

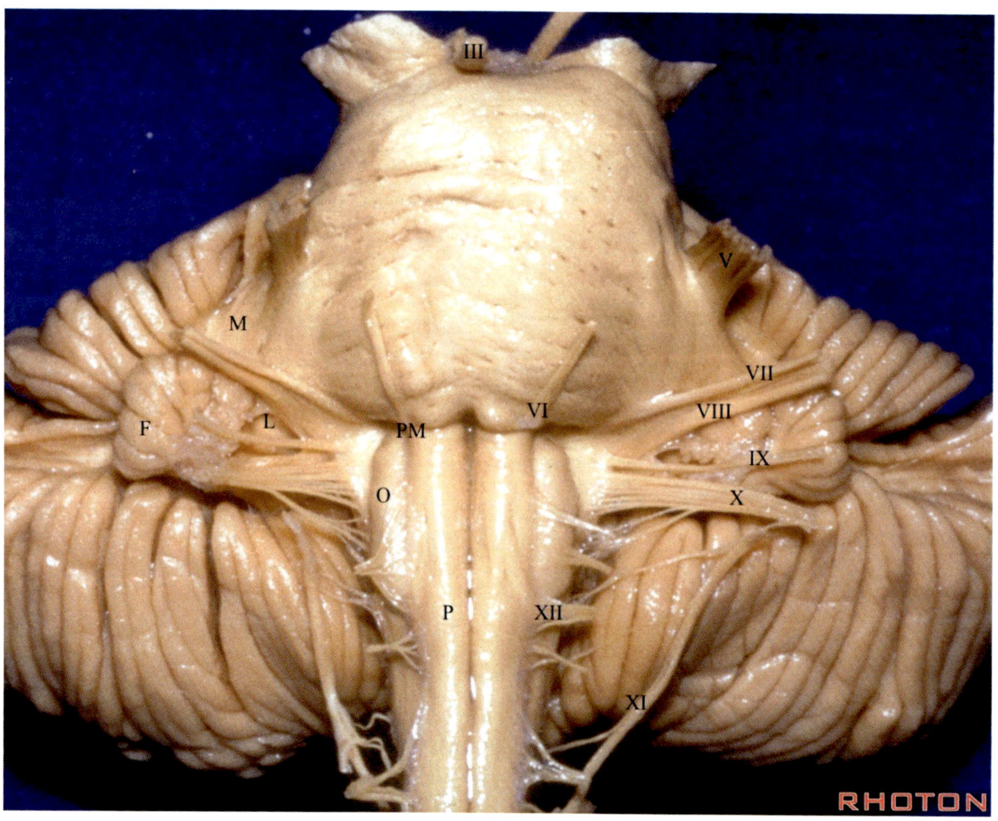

FIGURE 2.50 **Anterior view of the brainstem in all its three components; from the bottom to top: medulla oblongata, pons and mesencephalon. The cranial nerves and some other anatomical components are listed in the following: III, oculomotor nerve; V, trigeminal nerve; VI, abducens nerve; VII, facial nerve; VIII, vestibulocochlear nerve; IX, glossopharyngeal nerve; X, vagus nerve; XI, accessory nerve; XII, hypoglossal nerve; F, flocculus; M, middle cerebellar nerve; L, foramen of luschka; PM, Pontomedullary sulcus; O, olive; P, pyramid.** *Image courtesy of* The Rhoton Collection® *http://rhoton.ineurodb.org/*

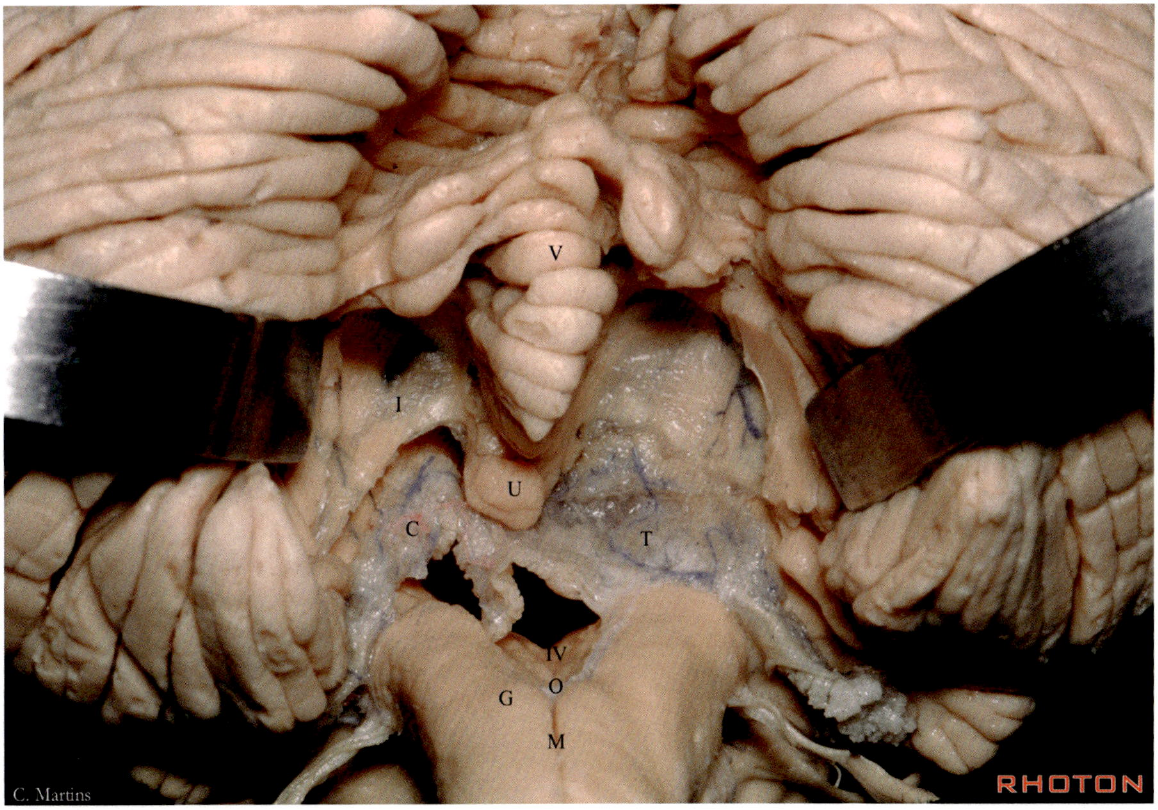

FIGURE 2.51 **Dorsal view of the brainstem.** From the bottom to top: M, posterior median sulcus; O, obex; G, gracile tubercle; IV, fourth ventricle; T, tela choroidea; I, inferior medullary velum; U, uvula; V, vermis; C, choroid plexus. *Image courtesy of The Rhoton Collection® http://rhoton.ineurodb.org/. Dissector: C. Martins.*

groove called the sulcus limitans limits the medial eminence laterally. The lower part of the medial eminence is an elevated area and called the facial colliculus, formed by the abducens nucleus and axons from the motor nucleus of facial nerve looping around the abducens nucleus. Lateral to the sulcus limitans is the vestibular area that consists of the vestibular nuclei that are in the dorsolateral regions of upper medulla oblongata and lower pons.

MESENCEPHALON (MIDBRAIN)

External Features of the Midbrain

The mesencephalon is continuous with the pons caudally and the diencephalon cranially, forming the connection between pons and the diencephalon. It is about 1.5 cm long and is connected to the cerebellum via the superior cerebellar peduncles (Angeles Fernández-Gil, 2010; Camminero and Cascella, 2022).

On the anterior surface of the mesencephalon, there is a deep hollow called the interpeduncular fossa that is limited by the crus cerebri bilaterally. The crus cerebri is composed of corticospinal, corticonuclear, and pontine tracts (Basinger and Hogg, 2023). The floor of the interpeduncular fossa is perforated by small vessels (emerging from the posterior cerebral artery), and it is called the posterior perforated substance. The fibers of the oculomotor nerve (III c.n.) leave the brainstem from the oculomotor sulcus in the interpeduncular fossa. The length of the interpeduncular fossa was measured as 11.26 ± 1.01 mm (Kocer et al., 2022).

The cerebral aqueduct is situated centrally in the midbrain, and the CSF passes from the third ventricle to the fourth ventricle via this passage. The part of the midbrain situated posteriorly to the cerebral aqueduct is the tectum and contains bilateral superior and inferior colliculi. The tectum length was measured as 14.50 ± 2.09 mm (Kocer et al., 2022). On the posterior surface, these colliculi form four rounded eminences also known as corpora quadrigemina. Superior and inferior colliculus thickness were measured 4.36 ± 1.17 mm and 5.06 ± 0.89, respectively (Kocer et al., 2022). The superior colliculus is responsible for visual reflexes. Fibers from each superior colliculus form the superior brachium on both sides and project to the corresponding lateral geniculate body. The inferior colliculus is responsible for auditory information processing. Fibers from each inferior colliculus form the inferior brachium on both sides and project to the corresponding medial geniculate body (Camminero and Cascella, 2022). In the midline of the posterior surface, just below the inferior colliculi, the trochlear nerve (IVc.n.) emerges. The trochlear nerve is the only cranial nerve that emerges from the dorsal surface of the brainstem (Angeles Fernández-Gil, 2010; Camminero and Cascella, 2022).

Acknowledgments

I would like to express my heartfelt gratitude to Dr. Barbara Buccilli for her invaluable assistance and support throughout the process of writing this book chapter. Dr. Buccilli's expertize, guidance and insightful feedback have significantly contributed to the quality and depth of the content. Her dedication and commitment to this project have been truly commendable. I am immensely grateful for her time, expertize and unwavering support.

References

Angeles Fernández-Gil, M., Palacios-Bote, R., Leo-Barahona, M., Mora-Encinas, J.P., 2010. Anatomy of the brainstem: a gaze into the stem of life. Semin. Ultrasound CT MR 31 (3), 196–219. https://doi.org/10.1053/j.sult.2010.03.006.

Basinger, H., Hogg, J.P., 2023. Neuroanatomy, Brainstem. StatPearls. StatPearls Publishing.

Bican, O., Minagar, A., Pruitt, A.A., 2013. The spinal cord. Neurol. Clin. 31 (1), 1–18. https://doi.org/10.1016/j.ncl.2012.09.009.

Byrne, J.V., 2012. Tutorials in Endovascular Neurosurgery and Interventional Neuroradiology. Springer Berlin Heidelberg. https://doi.org/10.1007/978-3-642-19154-1.

Cadotte, D.W., Cadotte, A., Cohen-Adad, J., et al., 2015. Characterizing the location of spinal and vertebral levels in the human cervical spinal cord. Am. J. Neuroradiol. 36 (4), 803–810. https://doi.org/10.3174/ajnr.A4192.

Caminero, F., Cascella, M., 2022. Neuroanatomy, Mesencephalon Midbrain. StatPearls. StatPearls Publishing.

Cho, S., Kurokawa, R., Hagiwara, A., et al., 2022. Localization of the central sulcus using the distinctive high signal intensity of the paracentral lobule on T1-weighted images. Neuroradiology 64 (2), 289–299. https://doi.org/10.1007/s00234-021-02729-4.

Courchesne, E., Press, G., Murakami, J., et al., 1989. The cerebellum in sagittal plane—anatomic-MR correlation: 1. The vermis. Am. J. Roentgenol. 153 (4), 829–835. https://doi.org/10.2214/ajr.153.4.829.

Destrieux, C., Terrier, L.M., Andersson, F., et al., 2017. A practical guide for the identification of major sulcogyral structures of the human cortex. Brain Struct. Funct. 222 (4), 2001–2015. https://doi.org/10.1007/s00429-016-1320-z.

Duprez, T.P., Serieh, B.A., Raftopoulos, C., 2005. Absence of memory dysfunction after bilateral mammillary body and mammillothalamic tract electrode implantation: preliminary experience in three patients. Am. J. Neuroradiol. 26 (1), 195–197 author reply 197–198.

Edlow, B.L., Mareyam, A., Horn, A., et al., 2019. 7 Tesla MRI of the ex vivo human brain at 100 micron resolution. Sci. Data 6 (1), 244. https://doi.org/10.1038/s41597-019-0254-8.

Edwards, W.C., Larocca, H., 1983. The developmental segmental sagittal diameter of the cervical spinal canal in patients with cervical spondylosis. Spine 8, 20.

Federau, C., Gallichan, D., 2016. Motion-correction enabled ultra-high resolution in-vivo 7T-MRI of the brain. PLoS One 11 (5), e0154974. https://doi.org/10.1371/journal.pone.0154974.

Fountas, K.N., Kapsalaki, E.Z., Jackson, J., Vogel, R.L., Robinson, J.S., 1998. Cervical spinal cord-smaller than considered? Spine 23 (14), 1513–1516. https://doi.org/10.1097/00007632-199807150-00001.

Hill, V.B., Cankurtaran, C.Z., Liu, B.P., et al., 2019. A practical review of functional MRI anatomy of the language and motor systems. Am. J. Neuroradiol. 40 (7), 1084–1090. https://doi.org/10.3174/ajnr.A6089.

Kocer, I.B., Demiralin, M.O., Erturk, M., Arslan, D., Sengul, G., 2022. Microsurgical anatomy of safe entry zones on the ventrolateral brainstem: a morphometric study. Neurosurg. Rev. 45 (2), 1363–1370. https://doi.org/10.1007/s10143-021-01644-9.

Mai, J.K., Paxinos, G., 2012. The Human Nervous System. Elsevier Academic Press.

Naidich, T.P., Yousry, T.A., 2007. Functional neuroanatomy. In: Clinical Functional MRI. Springer Berlin Heidelberg, pp. 53–85. https://doi.org/10.1007/978-3-540-49976-3_3.

Naidich, T.P., Valavanis, A.G., Kubik, S., 1995. Anatomic relationships along the low-middle convexity. Neurosurgery 36 (3), 517–532. https://doi.org/10.1227/00006123-199503000-00011.

Naidich, T.P., Tang, C.Y., Ng, J.C., Delman, B.N., 2013. Surface anatomy of the cerebrum. In: Imaging of the Brain. Elsevier, pp. 127–153. https://doi.org/10.1016/B978-1-4160-5009-4.50017-0.

Naidich, T.P., Kang, E., Fatterpekar, G.M., et al., 2004. The insula: anatomic study and MR imaging display at 1.5 T. Am. J. Neuroradiol. 25 (2), 222–232.

Nieuwenhuys, R., Voogd, J., Van Huijzen, C., 2008. The Human Central Nervous System. Springer.

Párraga, R.G., Possatti, L.L., Alves, R.V., Ribas, G.C., Türe, U., de Oliveira, E., 2016. Microsurgical anatomy and internal architecture of the brainstem in 3D images: surgical considerations. J. Neurosurg. 124 (5), 1377–1395. https://doi.org/10.3171/2015.4.JNS132778.

Paxinos, G., Huang, X.F., Sengul, G., Watson, C., 2012. Organization of Brainstem Nuclei.

Press, G., Murakami, J., Courchesne, E., et al., 1989. The cerebellum in sagittal plane—anatomic-MR correlation: 2. The cerebellar hemispheres. Am. J. Roentgenol. 153 (4), 837–846. https://doi.org/10.2214/ajr.153.4.837.

Press, G.A., Murakami, J.W., Courchesne, E., Grafe, M., Hesselink, J.R., 1990. The cerebellum: 3. Anatomic-MR correlation in the coronal plane. Am. J. Roentgenol. 154 (3), 593–602. https://doi.org/10.2214/ajr.154.3.2106226.

Schulz, J., Siegert, T., Reimer, E., et al., 2012. An embedded optical tracking system for motion-corrected magnetic resonance imaging at 7T. Magn. Reson. Mater. Phys. Biol. Med. 25 (6), 443–453. https://doi.org/10.1007/s10334-012-0320-0.

Shah, A., Jhawar, S.S., Goel, A., 2012. Analysis of the anatomy of the Papez circuit and adjoining limbic system by fiber dissection techniques. J. Clin. Neurosci. 19 (2), 289–298. https://doi.org/10.1016/j.jocn.2011.04.039.

Sherman, J.L., Nassaux, Y., Citrin, C.M., 2023. On MR imaging. Measurements of the normal cervical spinal cord. Am. J. Neuroradiol. 11. http://www.ajnr.org/content/11/2/.

Stucht, D., Danishad, K.A., Schulze, P., Godenschweger, F., Zaitsev, M., Speck, O., 2015. Highest resolution in vivo human brain MRI using prospective motion correction. PLoS One 10 (7), e0133921. https://doi.org/10.1371/journal.pone.0133921.

Su, S., Yang, N., Gaillard, F., 2019. Invisible cortex sign: a highly accurate feature to localize the inferolateral central sulcus. J. Med. Imaging Radiat. Oncol. 63 (4), 439–445. https://doi.org/10.1111/1754-9485.12875.

Szabo, B.A., Pascalau, R., Padurean, V.A., 2017. Morphometric study of the human brainstem and its neurovascular relations. Turk. Neurosurg. 20871–20917.

Tubbs, R.S., Loukas, M., Slappey, J.B., Shoja, M.M., Oakes, W.J., Salter, E.G., 2007. Clinical anatomy of the C1 dorsal root, ganglion, and ramus: a review and anatomical study. Clin. Anat. 20 (6), 624–627. https://doi.org/10.1002/ca.20472.

Ulmer, S., Jansen, O., 2020. In: Ulmer, S., Jansen, O. (Eds.), FMRI. Springer International Publishing. https://doi.org/10.1007/978-3-030-41874-8.

Wagner, M., Jurcoane, A., Hattingen, E., 2013. The U sign: tenth landmark to the central region on brain surface reformatted MR imaging. Am. J. Neuroradiol. 34 (2), 323–326. https://doi.org/10.3174/ajnr.A3205.

Wisse, L.E.M., Gerritsen, L., Zwanenburg, J.J.M., et al., 2016. Erratum to "Subfields of the hippocampal formation at 7T MRI: in vivo volumetric assessment". Neuroimage 125, 1176–1177. https://doi.org/10.1016/j.neuroimage.2015.10.054.

Zeineh, M.M., Parekh, M.B., Zaharchuk, G., et al., 2014. Ultrahigh-resolution imaging of the human brain with phase-cycled balanced steady-state free precession at 7 T. Invest. Radiol. 49 (5), 278–289. https://doi.org/10.1097/RLI.0000000000000015.

Zhao, W., Cohen-Adad, J., Polimeni, J.R., et al., 2014. Nineteen-channel receive array and four-channel transmit array coil for cervical spinal cord imaging at 7T. Magn. Reson. Med. 72 (1), 291–300. https://doi.org/10.1002/mrm.24911.

Further Reading

Gallichan, D., Marques, J.P., Gruetter, R., 2016. Retrospective correction of involuntary microscopic head movement using highly accelerated fat image navigators (3D FatNavs) at 7T. Magn. Reson. Med. 75 (3), 1030–1039. https://doi.org/10.1002/mrm.25670.

Maclaren, J., Armstrong, B.S.R., Barrows, R.T., et al., 2012. Measurement and correction of microscopic head motion during magnetic resonance imaging of the brain. PLoS One 7 (11), e48088. https://doi.org/10.1371/journal.pone.0048088.

3

Neurochemistry of the Central Nervous System

Brandon M. Ascenzi[1], *Matthew W. Myers*[2] *and Barbara Buccilli*[3]

[1]Independent Neuroresearcher-Member of Marie Curie Alumni Association (MCAA) [2]SciLifeLab, Stockholm, Sweden
[3]Department of Neurosurgery, Cerebrovascular and Neurologic Critical Care, Icahn School of Medicine at Mount Sinai, New York, NY, United States

OUTLINE

INTRODUCTION

The human brain is a complex and dynamic organ that governs our behavior, thoughts, and emotions. At its core lies a complex and dynamic network of neurons and their interwoven communication systems, primarily governed by a diverse array of neurotransmitters and neuromodulators. These neurochemical systems comprise an extensive spectrum of substances, including amino acids, monoamines, hormones, and other signaling molecules, that orchestrate a broad range of functions, from rudimentary sensory and motor processing to higher cognitive processes such as learning, memory and emotion (Fig. 3.1).

Within this chapter, we undertake an exploration of the most vital neurochemical systems of the brain and their physiological effects. Although we acknowledge that a comprehensive analysis of all systems is untenable, we endeavor to focus on the most significant ones while recommending further neurochemistry resources for inquisitive minds. Furthermore, we undertake a dissection of the pharmacology and neuropsychopharmacology of several psychiatric states and the therapeutic modalities that have been applied to treat them.

In the beginning of this chapter, we delve into the role of acetylcholine in the brain, with particular attention to its cholinergic receptors, including nicotinic and muscarinic receptors, and their diverse subtypes. We then transition to the monoaminergic circuits, encompassing serotonin, dopamine, adrenaline, and noradrenaline, and offer a detailed analysis of their synthesis, receptors, inactivation, and metabolism. Additionally, we explore the functions of amino acids such as glycine, glutamate, and their metabolites, including GABA, within the brain and their influence on neurogenesis and neurodevelopmental disorders. Finally, we critically review the role of opioids and their interplay within the brain, particularly their contributions to pain management and addiction treatment.

Overall, this chapter aims to provide a detailed and nuanced overview of the neurochemical systems that pervade the brain, their corresponding functions, and their potential therapeutic applications. As such, it is intended to serve as a valuable resource for students, researchers, and clinicians with an interest in the intricate workings of the human brain.

ACETYLCHOLINE

Acetylcholine was first synthesized by Adolf von Baeyer in 1867, long before a biological role for this

From Anatomy to Function of the Central Nervous System
https://doi.org/10.1016/B978-0-12-822404-5.00016-4

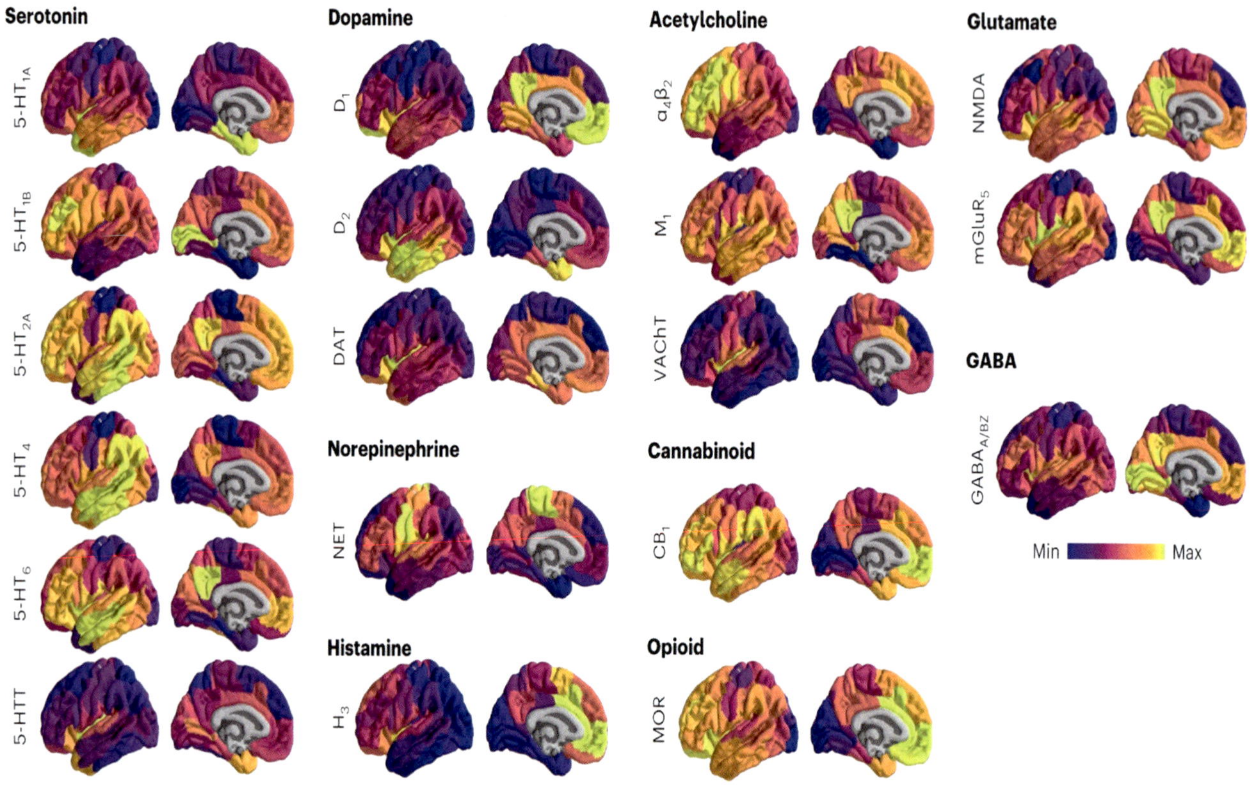

FIGURE 3.1 PET tracer images were collated and averaged to produce mean receptor distribution maps of 19 different neurotransmitter receptors and transporters across nine different neurotransmitter systems and a combined total of more than 1200 healthy participants. *Courtesy of Hansen, J.Y., 2022. Mapping neurotransmitter systems to the structural and functional organization of the human neocortex. Nat. Neurosc. Springer.*

molecule had been established. In 1936, Dale and colleagues chemically identified ACh as the neurotransmitter at the neuromuscular junction (NMJ; Dale et al., 1936). For their work on ACh, both Loewi and Dale received the Nobel Prize in Physiology or Medicine. It is now known that in addition to its widespread role as a transmitter within the CNS (Fig. 3.1) and at the NMJ, ACh also serves as a neurotransmitter at peripheral ganglia and as a mediator of the parasympathetic actions of the autonomic nervous system (Fisher and Wonnacott, 2012). In addition to its pivotal role in the study of chemical transmission, ACh was instrumental in the pharmacological classification of receptors. In 1905, Langley demonstrated that the direct application of nicotine caused skeletal muscle to contract, an observation that led him to postulate the presence of a "receptive substance" in the muscle membrane (for review, see Maehle et al., 2002).

Acetylcholine formation is catalyzed by choline acetyltransferase (ChAT), the enzyme that catalyzes the reaction acetyl-CoA + choline → acetylcholine ChAT is synthesized within the rough endoplasmic reticulum in the cell body and then transported by axoplasmic transport to the nerve terminal, where it becomes highly concentrated within the cytoplasm. Brain regions that possess a high ChAT activity may either be the site of termination of cholinergic projection neurons where ACh is released (e.g., hippocampus) or, alternatively, reflect the presence of a significant number of cholinergic interneurons within the structure (e.g., caudate/putamen). In general, measurement of the activity of ChAT which is an enzyme restricted to presynaptic structures, in a given brain region is a good predictor of the likelihood of cholinergic innervation. In contrast, the activity of acetylcholinesterase (AChE), the degradative enzyme, is not a reliable indicator of cholinergic innervation since the enzyme is found at both pre- and postsynaptic locations. For example, AChE is highly enriched in the substantia nigra even though this structure is only sparsely innervated by cholinergic neurons.

CHOLINERGIC RECEPTORS

Nicotinic Colinergic Receptor (nAChR)

The nicotinic receptor was the first colinergic receptor to be characterized biochemically. nAChRs are pentameric ligand-gated ion channels of four different subunits, designated α, β, γ, and δ in order of increasing MW. The α subunit is present in *Torpedo marmorata*, a marine ray, and muscle nAChRs in two copies. Thus, the nAChR is a pentamer of molecular mass approximately 280 kDa. ACh binding to the extracellular domain leads to opening

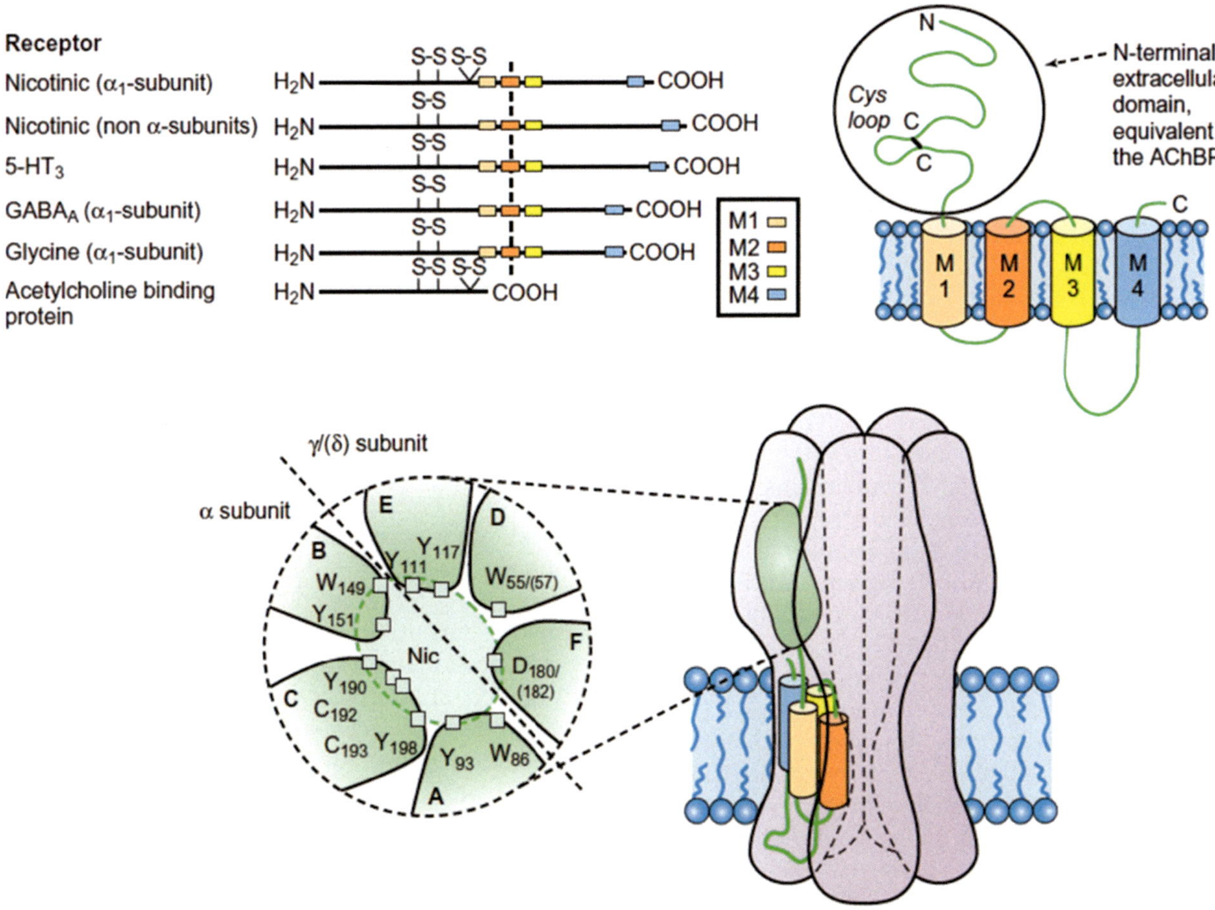

FIGURE 3.2 **Structural features of the nAChR.** *Top left*: Schematic representation of the sequence of various cys-loop receptor subunits including the AChBP, highlighting key conserved features. Reading from the N terminus, the disulfide-bonded cys loop is common to all these subunits and defines the family. The pair of vicinal cysteines close to the first transmembrane domain is a characteristic of nAChR αsubunits and the AChBP only. The colored boxes represent the four transmembrane segments, M1, M2, M3, and M4. The intracellular loop between M3 and M4 is variable in length. *Top right*: Orientation of a nAChR subunit within the membrane. *Bottom*: Schematic of an assembled nAChR, with five subunits arranged to create a central ion channel, lined by M2. The N-terminal ACh-binding site is shown in the insert to be composed of three protein loops from the α subunit (principal face) and three loops from the adjacent (complementary) subunit, in this case the γ subunit of a muscle nAChR. Key amino acid residues involved in ACh binding are indicated. *Courtesy of Fisher, S.K., Wonnacott, S., 2012. Chapter 13: Acetylcholine. In: Brady, S.T., Siegel, G.J., Albers, R.W., Price, D.L. (Eds.), Basic Neurochemistry, eighth ed. Academic Press, pp. 258–282. https://doi.org/10.1016/B978-0-12-374947-5.00013-4.*

of the channel to allow the flux of ions. Under normal conditions Na^+ and Ca^{2+} flow into the muscle cell, producing depolarization, whereas K^+ flows out (Karlin, 2002).

Binding of ACh to the nAChR is communicated to the ion channel extremely rapidly: electrophysiological measurements show that channel opening occurs within tens of microseconds of ACh application, despite a distance of some 50Å between the binding site and the channel (Sine and Engel, 2006). The nAChR channel is made up of the second transmembrane segments (M2) from each subunit (Fig. 3.2).

Muscarinic Cholinergic Receptors

Early studies by Langley and Dale demonstrated that some of the effects of peripherally administered ACh could be mimicked by muscarine, an alkaloid found in the fly agaric mushroom *Amanita muscaria*. Muscarinic cholinergic responses could be readily distinguished from those elicited by nicotinic agonists in that the muscarinic responses could be either excitatory or inhibitory (nicotinic responses are always excitatory) and could be blocked by the inclusion of submicromolar concentrations of two alkaloids, scopolamine or atropine. Moreover, the responses elicited by muscarinic cholinergic ligands had much longer latencies of both onset (~100–250 ms) and offset than nicotinic responses. These observations led to the concept that separate populations of nAChRs and mAChRs mediate the physiological effects of ACh. mAChRs are widely distributed and are present in neurons (and glia) in most regions of the CNS and the PNS as well as in parasympathetically innervated cardiac and smooth muscles (e.g., heart, iris, stomach, bronchioles, intestines, bladder) and salivary, tear, and sweat glands.

Muscarinic Receptor Subtypes Couple to Distinct G-Proteins and Activate Different Effector Mechanisms

M1, M3, and M5 mAChRs preferentially couple to G-proteins of the Gq/G11 family, whereas M2 and M4 receptors typically activate G-proteins of the Gi/Go family. Agonist occupancy of the two groups of mAChRs results in the activation of different downstream effector proteins, as indicated, although some effectors (e.g., mitogen-activated protein kinase) (MAPK) are activated by both groups of receptors. Note that the effects of mAChR activation are mediated by both the α and βγ subunits of the G-proteins. An increase or decrease in the activity of the effector mechanism is indicated by the direction of the arrow. GIRK, G-protein−activated inwardly rectifying K^+ channel; PLCβ, phosphoinositide-specific phospholipase C.

Even if some author finds Mesulam's classification of the major central cholinergic neuron populations and their efferent projections obsolete, in the present text we prefer to describe them on the basis of the studies performed by Kimura et al. (1981), Fibiger (1982), and, indeed, Mesulam et al. (1983a,b, 1984). We know that many descriptions, including the one in question, represent very simple models to explain very important neurochemical circuits, such as the cholinergic one (Ray et al., 2015). However, we believe that to try to better understand complex things, such as neuroanatomy, it is necessary to start from the most basic and simplest descriptions of them and so continue, if necessary, with more detailed ones that fully meet the needs of the general public.

The basal forebrain contains a population of large cholinergic neurons, extending from the septal region rostrally to the subthalamic nucleus caudally. Mesulam et al. (1983a,b, 1984a) subdivided this population into four groups, Ch1−Ch4 (For a charting of these cell masses in the human brain see Perry et al., 1984.)

The Ch1 group corresponds to the medial septal nucleus; about 10% of its neurons are cholinergic.

The Ch2 group corresponds to the vertical limb of the nucleus of the diagonal band of Broca; at least 70% of its neurons are cholinergic.

The Ch3 group most closely corresponds to the horizontal limb of the nucleus of the diagonal band of Broca; only 1% of the neurons of this nucleus are cholinergic.

The Ch4 group, which is very extensive in the human brain, most closely corresponds to the nucleus basalis of (Meynert 1872; Kolliker 1896). This nucleus lies embedded in the substantia innominata, a vaguely defined area situated ventral to the globus pallidus (see Chapter 9 "The subthalamus or ventral thalamus"). At least 90% of the neurons in the nucleus basalis are cholinergic (Price and Stern, 1983). The Ch4 group includes large cholinergic neurons located in the internal and external medullary laminae of the globus pallidus. These cholinergic neurons give rise to the following specific projections.

1. Fibers originating from Ch1, Ch2, and Ch3 travel through the stria medullaris and the tractus habenulo-interpeduncularis to reach the base of the midbrain. Here, they terminate in the interpeduncular nucleus and the area tegmentalis ventralis. Additionally, cholinergic fibers in the stria medullaris project heavily to the medial habenular nucleus. The cholinergic projections from the basal forebrain, passing through the tractus habenulo-interpeduncularis, are further reinforced by fibers originating from the cholinergic elements in the medial habenular nucleus (Bigl et al., 1982; Gottesfeld and Jacobowitz, 1979).
2. Ch1 and Ch2 significantly contribute to the cholinergic projections in the hippocampus. These fibers travel through the fornix and terminate in the cornu ammonis and fascia dentata.
3. Ch2 also sends fibers to the lateral hypothalamus.
4. Cholinergic neurons primarily located in the Ch3 project to the olfactory bulb and terminate in the external layers of the bulb.
5. Cholinergic neurons in the anterolateral portion of Ch4 project, via the ventral amygdalofugal pathway, to the amygdaloid complex, particularly the nucleus basalis.
6. The remaining parts of Ch4 provide major cholinergic projections to the entire neocortex. This projection exhibits a crude topographical organization, with rostral cells projecting to the frontal and parietal cortex, while caudal cells innervate the occipital and temporal cortex. However, the precise organization of this projection is still being investigated. Studies suggest that individual cells in the nucleus basalis may project to specific areas within the cortex, and cortically projecting nucleus basalis neurons display diverse physiological properties (Aston-Jones et al., 1984). These elements yielded spontaneous discharge rates of 0−40 Hz, a variety of impulse amplitudes and waveforms, and a wide range of conduction latencies from the frontal cortex, yielding fiber conduction velocities of about 0.6−8.0 m/s. These values are within the range of speeds for fine myelinated fibers.

The basal forebrain contains a significant population of cholinergic cells that have been implicated in a variety of physiological and behavioral processes (Mesulam and Van Hoesen, 1976; Mesulam et al., 1983a; Divac, 1975). The projection to the olfactory bulb may well play a role comparable to that of the olivocochlear bundle in the auditory system (Mesulam et al., 1983a). The pathway to the hippocampus is possibly involved in memory processes, and that to the amygdala may modulate affective behavior (Mesulam et al., 1983a). The nucleus basalis, suggested to represent a telencephalic extension of the reticular formation of the brain stem, receives afferents from a variety of cortical and subcortical sources (Norgren, 1974; Jones et al., 1976; Leichnetz and Astruc, 1977; Saper et al., 1979; Price and Amaral, 1981; Mesulam et al., 1983a; Mesulam and Mufson, 1984). The cortical input arises from the prepiriform cortex, orbitofrontal cortex, anterior insula, temporal pole, entorhinal cortex, and medial temporal cortex (Mesulam et al., 1983a). Thus, in contrast to its widespread projections to all parts of the neocortex, the nucleus basalis receives reciprocal projections from only a relatively small number of cortical areas (Mesulam et al., 1983a). The subcortical afferents of the nucleus basalis originate from the septal nuclei, nucleus accumbens-ventral pallidum complex, amygdala, hypothalamus, peripeduncular nucleus of the midbrain, and pontine taste area (Mesulam et al., 1983a; Mesulam and Mufson, 1984). In light of its afference, the Ch4 complex, as considered by Mesulam and coworkers, likely acts as a cholinergic relay between limbic plus paralimbic areas and the entire neocortex, potentially influencing complex behavior according to the prevailing emotional and motivational states (Mesulam et al., 1983a; Mesulam and Mufson, 1984).

Regarding neuropathological findings, several studies have reported reduced levels of cholinergic enzymes ChAT and AChE in the cerebral cortex in Alzheimer's disease (Davies, 1979). Neurons in the nucleus basalis of Meynert undergo profound and selective degeneration in Alzheimer's disease and senile dementia (Whitehouse et al., 1981, 1982). Marked decreases in AChE and ChAT have also been observed in the substantia innominata, along with a loss of more than 50% of ChAT-containing neurons in that region (Rossor et al., 1982; Nagai et al., 1983a). These findings have significant implications for understanding the pathophysiology of these disorders (Davies, 1979; Whitehouse et al., 1981, 1982; Rossor et al., 1982; Nagai et al., 1983a). However, the exact relationship between the loss of cholinergic neurons in the nucleus basalis (Ch4) and the loss of cortical cholinergic innervation or potential primary cortical lesions with alterations in the nucleus basalis reflecting retrograde degeneration has yet to be determined (Mesulam et al., 1983a).

All of the groups of cholinergic cells discussed so far consist of long-axoned projection neurons. However, within the neostriatum (nucleus caudatus, putamen, and nucleus accumbens), cholinergic local circuit neurons are found (Kimura et al., 1981; Woolf and Butcher, 1981; Butcher and Woolf, 1982; Fibiger, 1982; **N.B**. see the recent results showed in Fig. 3.3A and B by Deistung et al., 2013). These elements are large and make up approximately 1% of the total population of neostriatal cells (Kimura et al., 1981; Woolf and Butcher, 1981; Butcher and Woolf, 1982; Fibiger, 1982). Notably, they remain unaffected in Alzheimer's disease (Parent et al., 1984). In macaques, the density of cholinergic neurons is higher in the ventral striatum (nucleus accumbens and olfactory tubercle) compared to the dorsal striatum (caudate nucleus and putamen) (Mesulam et al., 1984a).

THE MONOAMINERGIC CIRCUITS

1. *Serotonine*

The structures of serotonin (5-HT) and related compounds are shown in Fig. 3.4. The combination of the hydroxyl group in the 5 position of the indole nucleus and a primary amine nitrogen serving as a proton acceptor at physiological pH, makes 5-HT a hydrophilic substance. The best role of 5-HT is in the brain function. The idea that 5-HT had important behavioral effects stimulated much thought about a central role for 5-HT in mental illness. Subsequently, various theories arose linking abnormalities of 5-HT function in brain to the development of a number of psychiatric disorders, particularly schizophrenia and depression.

Psychotherapeutic drugs are now available that are effective in depression, anxiety disorders, and schizophrenia and have potent and, in some cases selective, effects on serotonergic neurons in brain. Understanding the neuroanatomical organization of serotonergic neurons provides insight into the functions of this neurotransmitter, as well as its possible roles in mental processes and psychiatric disorders. Serotonin-containing neuronal cell bodies are restricted to discrete clusters or groups of cells located along the midline of the brainstem and their axons, innervating nearly every area of the central nervous system. In 1964, Dahlström and Fuxe described nine groups of serotonin-containing cell bodies, which they designated B1 through B9, and which correspond for the most part with the raphe nuclei (Tork, 1990) (see Chapter 5b). Moreover, some serotonergic cell bodies are found outside the raphe nuclei, however not all the cell bodies in the raphe nuclei are serotonergic. In most of the raphe nuclei the majority neurons are nonserotonergic. For example, the dorsal raphe nucleus (B6-DRN) contains the largest number of serotonergic neurons, however, only 40%–50% of the cell bodies in the B6-DRN are serotonergic (Beliveau et al., 2017; Ren et al., 2019). For a detailed description of the neuroanatomy of the serotonergic system see Chapter 5b.

A.

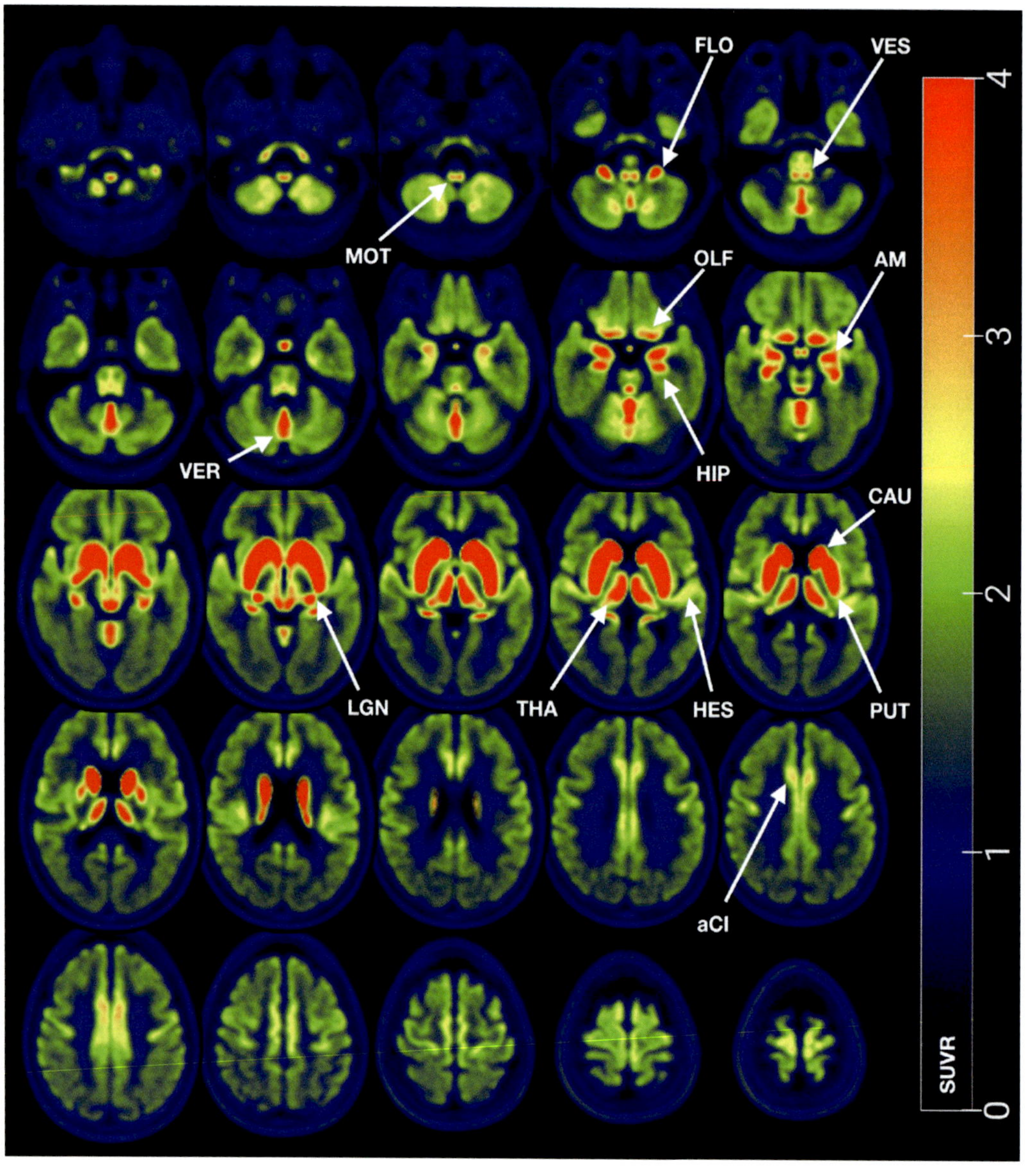

FIGURE 3.3 **Distribution of cholinergic nerve terminals in human brain measured with [^{18}F]FEOBV PET.** The [^{18}F]FEOBV PET image volumes are displayed in a mosaic of axial (A) and coronal/sagittal/axial (B) slices of the brain. Highest tracer binding was present in the putamen (PUT), caudate nucleus (CAU), and nucleus accumbens; the centromedian, lateral geniculate (LGN), and mediodorsal nucleus of the thalamus; and the basal forebrain. Intermediate binding was found in most investigated nuclei of the brainstem, thalamus (THA), and hypothalamus; the vermis (VER) and flocculonodular lobe of the cerebellum; as well as key limbic circuit structures such as the hippocampus (HIP), amygdala (AM), insula, anterior cingulate cortex (aCI), and olfactory cortex (OLF). Lowest binding was present in the cerebellar nuclei and hemispheres, and in most nonlimbic areas of the cerebral cortex and non (HES, Heschl's gyrus; MOT, dorsal motor nucleus of the vagus and hypoglossus nucleus; VES, vestibular nucleus complex). In (B) cross hairs indicate location of (a) olfactory cortex, (b) basal forebrain, (c) Heschl's gyrus, (d) vermis, (e) thalamus, and (f) hypothalamus. Red color indicates SUVR of 4 or higher. *Courtesy of Okkels, N., Horsager, J., Labrador-Espinosa, M.A., Hansen, F.O., Andersen, K.B., Just, M.K., Fedorova, T.D., Skjærbæk, C., Munk, O.L., Hansen, K.V., Gottrup, H., Hansen, A.K., Grothe, M.J., Borghammer, P., 2023. Distribution of cholinergic nerve terminals in the aged human brain measured with [18F] FEOBV PET and its correlation with histological data. Neuroimage 269, 119908. https://doi.org/10.1016/j.neuroimage.2023.119908.*

B.

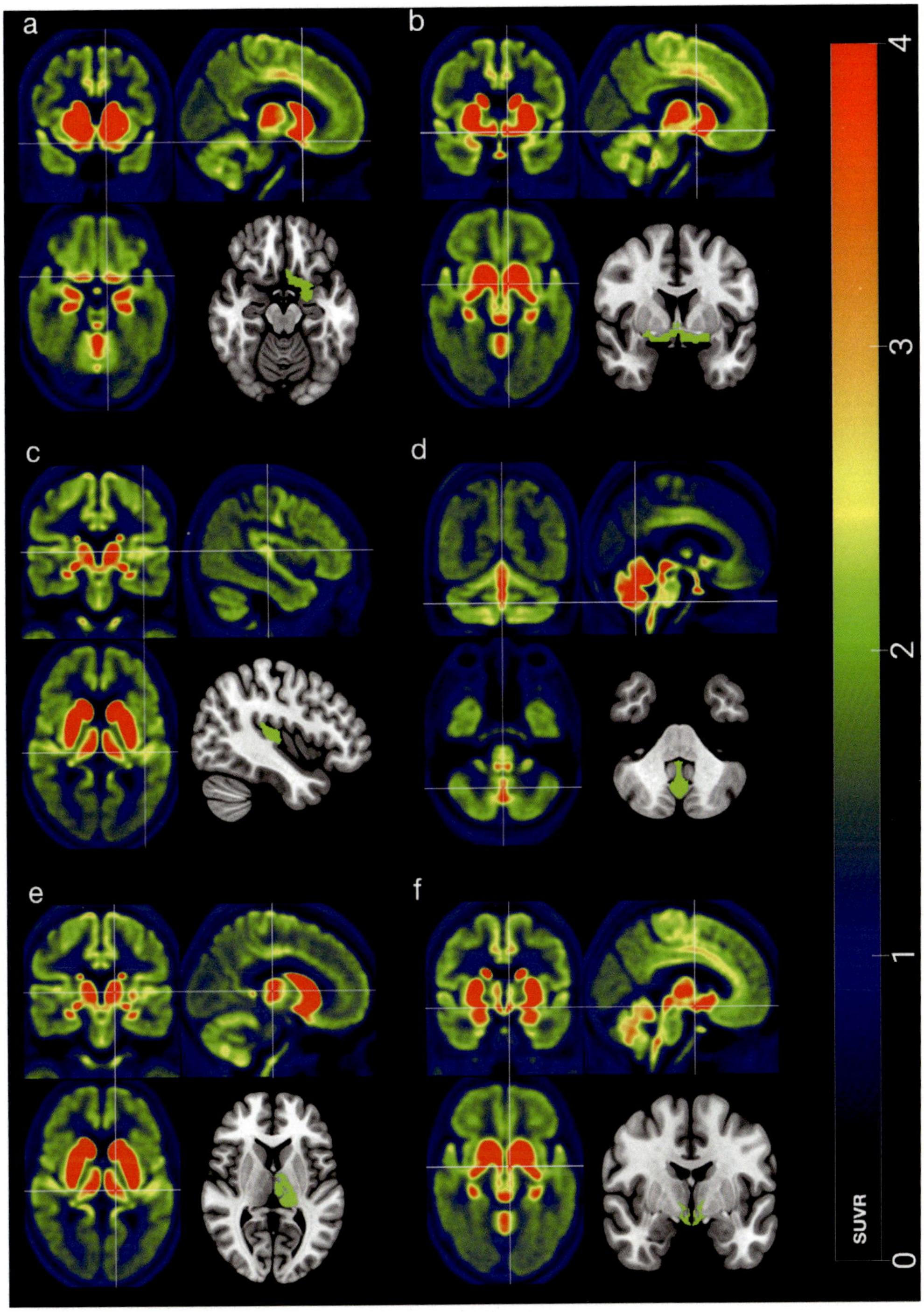

FIGURE 3.3 con'd.

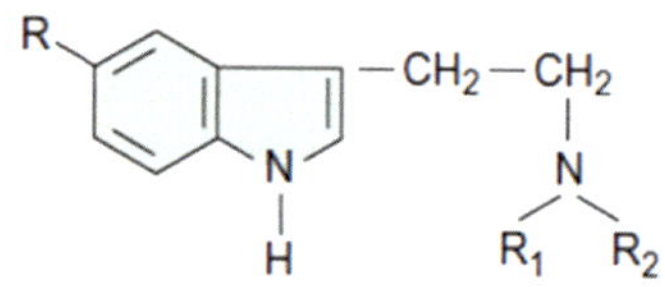

Compound	Position		
	R	R_1	R_2
Tryptamine	H	H	H
Serotonin	OH	H	H
Melatonin	OCH_3	$COCH_3$	H
Diethyltryyptamine (DET)*	H	CH_3CH_2	CH_3CH_2
Dimethyltryptamine (DMT)*	H	CH_3	CH_3
Bufotenine*	OH	CH_3	CH_3

*Psychotropic (modifies mental activity)

FIGURE 3.4 **Chemical structure of 5-hydroxytryptamine (5-HT; serotonin) and related indolealkylamines.** The indole ring structure consists of the benzene ring and the attached five-member ring containing nitrogen. *Courtesy of Hensler, J.G., 2012. Chapter 15: Serotonin. In: Brady, S.T., Siegel, G.J., Albers, R.W., Price, D.L. (Eds.), Basic Neurochemistry, eighth ed. Academic Press, pp. 300–322. https://doi.org/10.1016/B978-0-12-374947-5.00015-8.*

Serotonin Synthesis

L-Tryptophan is the Precursor Serotonin Synthesis

The synthesis and primary metabolic pathways of 5-HT are shown in Fig. 15.5. The initial step in the synthesis of serotonin is the facilitated transport of the amino acid L-tryptophan from blood into brain. The primary source of tryptophan is dietary protein, in turn, the entry of tryptophan into brain is not only related to its concentration in blood, but it is also a function related to its concentration respect to the concentrations of other neutral amino acids. Consequently, lowering the dietary intake of tryptophan while raising the intake of the amino acids with which it competes for transport into brain, lowers the content of 5-HT in brain and changes certain behaviors associated with 5-HT function. This strategy for lowering the brain content of 5-HT has been used clinically to evaluate the importance of brain 5-HT in the mechanism of action of psychotherapeutic drugs.

Serotonergic neurons contain the enzyme L-tryptophan-5-monooxygenase more commonly termed tryptophan hydroxylase (TH), which converts tryptophan to 5-hydroxytryptophan (5-HTP). In brain, this enzyme is synthesized in serotonergic cell bodies of the raphe nuclei and is found only in cells that synthesize 5-HT.

There are two isoforms of tryptophan hydroxylase, referred to as *Tph*1, expressed in the periphery, and *Tph*2, expressed exclusively in brain.

The existence of two tryptophan hydroxylase isoforms may offer new opportunities for affecting central or peripheral 5-HT systems by designing drugs that target the expression and activity of *Tph*1 or *Tph*2 gene. In addition, the duality of 5-HT synthesis changes our approach to peripheral models of psychiatric disorders.

Storage of 5-HT Vesicles and Active Transport from the Cytoplasm

The vesicular "transporter" uses the electrochemical gradient generated by a vesicular H^+-ATPase to drive transport, such that a cytoplasmic amine is exchanged for a luminal proton, i.e., the uptake of 5-HT is coupled to the efflux of H^+.

Vesicles storing 5-HT exhibit some differences from those storing catecholamines in fact in contrast to the latter there is virtually no ATP in serotonin vesicles. Furthermore, serotonergic synaptic vesicles, but not chromaffin granules, contain a specific protein that binds 5-HT with high affinity in the presence of Fe^{2+}. This serotonin-binding protein (SPB) is packaged in secretory vesicles with 5-HT and SPB-5-HT complex is secreted by a calcium-dependent process. Calcium has been reported to stimulate the fusion of vesicular membranes with the plasma membrane and not only in the serotonin release but also in release of others molecules included neurotransmitters.

The serotonin release is also regulated by the firing rate of serotonergic neurons of the raphe nuclei. Numerous studies utilizing a variety of techniques have revealed that an increase in raphe cell firing enhances the release of 5-HT in terminal fields while the opposite effect is observed when raphe cell firing decreases. This means that drugs that change the firing rate of serotonergic soma modify the release of 5-HT as well.

The activity of 5-HT in the synapse is terminated primarily by its reuptake into serotonergic terminals and, in turn, the reuptake of 5-HT is mediated primarily by serotonin transporter (SERT). The SERT is located on serotonergic neurons along the axon rather than within the presynaptic area. The mRNA for the SERT has been localized in brain exclusively to the serotonergic cells in the raphe nuclei. Activity of the SERT regulates the concentration of 5-HT in the extracellular space, thereby influencing synaptic transmission. The activity of 5-HT in the extracellular space is also terminated by its reuptake by various transporter proteins into glia and other neurons. Although mRNA for the SERT has not been detected in glia, it has been showed that primary cultures of astrocytes in vitro can take up 5-HT. Thus, 5-HT uptake into glia and into other nonserotonergic neurons is due to other recently discovered transporters such as organic cation transporter (OCT) and plasma membrane monoamine transporter (PMAT) as well as related transporters such as the norepinephrine transporter (NET) and dopamine transporter (DAT) therefore all of these transporters are capable of removing 5-HT, as well as their native substrates from extracellular fluid in the brain. When SERT function is compromised, for example genetically or pharmacologically, the low-affinity high-capacity transporters for 5-HT play an important role in regulating extracellular 5-HT concentrations (Daws, 2009). It has been known for over 25 years that many of the tricyclic antidepressants (TCAs), e.g., imipramine and amitriptyline, are potent inhibitors of both NE and 5-HT reuptake. Some tricyclic antidepressants, e.g., desipramine, inhibit the reuptake of NE much more potently than the reuptake of 5-HT. Thus, it was unclear for some time whether the inhibition of 5-HT reuptake played any role in the antidepressant action of those TCAs that possessed this pharmacological property. The relatively recent development and use of selective serotonin reuptake inhibitors (SSRIs), e.g., fluoxetine, paroxetine, sertraline, escitalopram (Fig. 3.5) and others, have been shown to be more effective in terms of inhibition of 5-HT reuptake than that of NE (Fig. 3.7A). Thus, selective inhibition of the reuptake of either NE or 5-HT can result in an antidepressant effect. All the aforementioned serotonin reuptake inhibitors (SRIs) in general, which were initially developed as antidepressants, are now the most widely used agents in the treatment of many additional neuropsychiatric disorders (e.g., eating disorders, anxiety, obsessive–compulsive disorder) and at low doses are effective in pain therapy (e.g., one of the most used at this scope is the amitriptyline) (Fig. 3.7A).

Clomipramine, a TCA that is somewhat selective in vivo as an inhibitor of 5-HT reuptake, also produces clinically significant amelioration of the symptoms associated with obsessive–compulsive disorder (OCD). Thus, the mechanism of 5-HT reuptake inhibition not only works in the treatment of the mood disorders, e.g., depression, dysthymic disorder, etc., but it also able to reduces the symptoms of other kinds of psychiatric and nonpsychiatric illness, e.g., erectile disfunction, pain, etc. This clinical effect is not found with drugs such as desipramine that selectively inhibit the reuptake of NE only.

In addition to classical synaptic transmission, 5-HT may relay information by volume transmission or paracrine neurotransmission. Consistent with the idea that a significant mode of transmission for 5-HT is through volume or extra-synaptic transmission is the observation that the SERT is localized perisynaptically along axons and dendrites and is not confined to the terminals (Tao-Cheng and Zhou, 1999). Thus, 5-HT may diffuse away from the site of release before its action is terminated by reuptake into serotonergic neurons. Thus, 5-HT may have a paracrine or neuromodulatory role in brain, adjusting or tuning ongoing activity at synapses some distance from the serotonergic terminal.

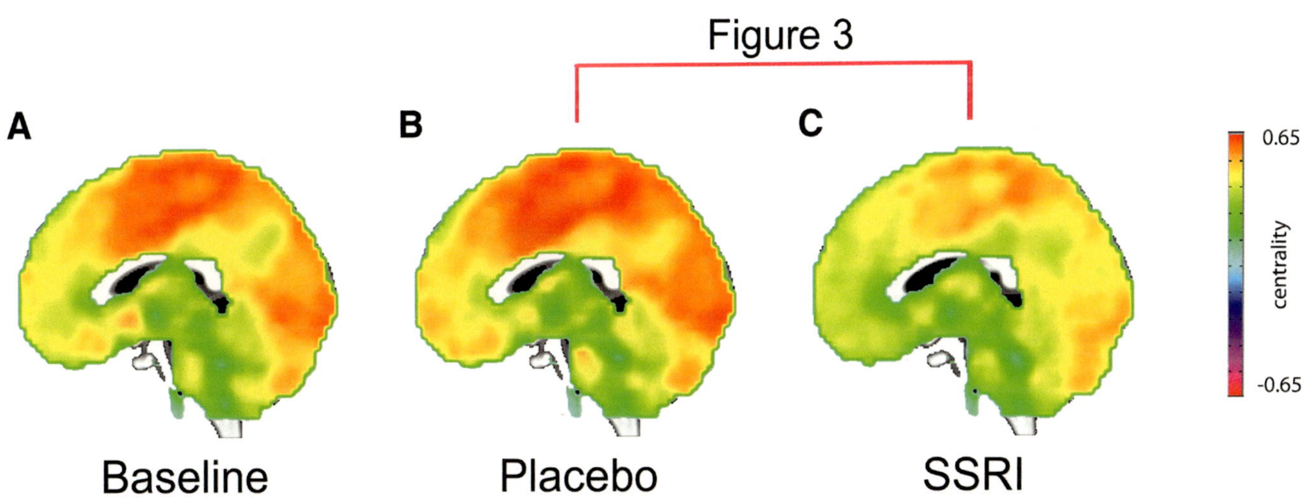

FIGURE 3.5 **Comparison of average degree centrality baseline versus placebo and escitalopram.** Sagittal slices of mean degree centrality of three conditions: (A) baseline, (B) placebo, and (C) escitalopram (20 mg), superimposed on a T1-anatomical standard template. Orange colors indicate higher centrality. Baseline and placebo condition show more similar centrality patterns, whereas degree centrality analysis reveals a global signal change following the administration of the SSRI. Escitalopram (SSRI family drug), influences the availability of the neurotransmitter serotonin and causes major changes in connectivity between functional networks at rest—in other words, the synchronous brain activity in various areas of the resting brain. Escitalopram influences which networks of the brain are activated simultaneously, i.e., fluctuate in synchrony, when the brain is in a resting state. This rapid and brain-encompassing effect of escitalopram is unexpected, as the antidepressant effects of this class of drug usually require 2–3 weeks to unfold. The current study performed by Shaefer A. and coll. suggests that the effect of the serotonin reuptake inhibitors on the brain's resting-state networks begins just a few hours after initial intake of the drug. *Courtesy from Schaefer, A., Burmann, I., Regenthal, R., Arélin, K., Barth, C., Pampel, A., Villringer, A., Margulies, D.S., Sacher, J., 2014. Serotonergic modulation of intrinsic functional connectivity. Curr. Biol. 24 (19), 2314–2318. https://doi.org/10.1016/j.cub.2014.08.024.*

Serotonin may be Involved in a Wide Variety of Behaviors by Setting the Tone of Brain Activity in Relationship to the State of Behavioral Arousal/Activity

Serotonin has been implicated in practically every type of behavior, e.g., appetitive, emotional, motoric, cognitive, and autonomic. However, from a physiologic perspective, it is not clear whether 5-HT affects such behaviors specifically, or does so more generally by coordinating the activity of the nervous system, particularly to set the tone of activity in conjunction with the organism's level of arousal. The primary data that has contributed to the view that 5-HT has a general effect on behavior by modulating the tone of nervous system activity, comes from studies of the firing rate of serotonergic soma in raphe nuclei (Jacobs and Fornal, 1993). Under quiet waking conditions, serotonergic raphe neurons display a slow, clock-like activity of about 1–5 spikes/sec, which shows a gradual decline as the animal becomes drowsy and enters slow wave sleep. A decrease in the regularity of firing accompanies this overall slowing of neuronal activity and, during REM (rapid eye movement) sleep, the activity of these neurons becomes silent. In response to certain types of arousing stimuli, the firing rate of these serotonergic neurons increases and is not surprisingly that some data led to the idea that the activity of serotonergic neurons is related to the level of behavioral arousal/activity. Such data have contributed also to the idea that the activity of serotonergic neurons is associated with motoric output, since atonia of the major skeletal muscle groups occurs during REM sleep (Siegel et al., 1999; Albers and Price, 2012.). Oral–buccal motor activity, such as chewing, biting, licking, or grooming, causes a marked increase in the firing rate of a subgroup of serotonergic soma that are also activated by somatosensory stimuli applied to the head, neck, and face (Frazet and Hensler, 1999). Perturbation of the 5-HT system can elicit changes in a wide variety of behaviors, in fact drugs that target serotonergic neurons and their receptors are used to treat diseases such as depression, anxiety disorders, and schizophrenia.

Serotonin Receptors

The serotonin receptors are a group of G-protein–coupled receptor and ligand-gated ion channels found in the central and peripheral nervous system (Hoyer, 1994; Frazer, 1999; Beliveau, 2017). They are activated by serotonin which acts as their natural ligand mediating both excitatory and inhibitory neurotransmission. They also act as regulators of the release of many other neurotransmitters including glutamate, GABA, dopamine, epinephrine/norepinephrine, and acetylcholine, as well as many hormones such as oxytocin, prolactin, vasopressin, cortisol, corticotropin, and substance P. Serotonin receptors, as mentioned in previously, they influence various biological pathways that underlie numerous behavioral and nonbehavioral processes such as aggression, anxiety, appetite, cognition, learning, memory, mood, nausea, sleep, and thermoregulation. They are the target of a variety of pharmaceutical and recreational drugs including many antidepressants, antipsychotics, anorectics, antiemetics, gastroprokinetic agents, antimigraine agents, hallucinogens, and entactogens (Nichols, 2008). We will talk about their use in neuro- and neuropsychopharmacology later.

Serotonin receptor can be divided into seven families of G-protein–coupled receptors, which activates an intracellular second messenger cascade to produce an excitatory or inhibitory response, except for the 5-HT_3 receptor that is a ligand-gated ion channel and its response depends by the ionic afflux-efflux (see later).

The 5-HT_1 Receptor Family is Composed of the 5-HT_{1A}, 5-HT_{1B}, 5-HT_{1D}, 5-HT_{1E}, and 5-HT_{1F} Receptors

The 5-HT_{1A} Receptor Subtype

The 5-HT_{1A} receptor is present in high density in cortical and limbic structures (Fig. 3.6), i.e., hippocampus, entorhinal cortex, septum, amygdala, frontal cortex (Glikmann-Johnston et al., 2015; Celada et al., 2013). The distribution of this 5-HT_{1A} receptor subtype in the aforementioned brain areas suggest that it can have a role in cognitive or integrative functions, as well as in emotional states (Celada et al., 2004, 2013; Borroto-Escuela et al., 2021). 5-HT1$_A$ receptors being located at the postsynaptic level of the serotonergic neurons are also present in high density on cell body areas and in particular way in the neurons of dorsal- and median raphe nucleus (respectively, B6-DRN and B8-MnR, for details see Chapter 5b). Here they work as somatodendritic autoreceptors involved in the negative feedback modulation of the neuronal activity (Celada et al., 2013; You et al., 2016; McDevitt et al., 2011; Huang et al., 2019; McDevitt and Neumaier, 2011), playing a key role in the regulating serotonin release.

In terminal field areas of serotonergic innervation such as the hippocampus, 5-HT1$_A$ receptors are coupled to the inhibition of adenylyl cyclase activity, as well as the opening of potassium channels, which results in neuronal hyperpolarization (Table 3.1). However, in the dorsal raphe nucleus 5-HT_{1A} receptors are coupled to the opening of potassium channels, but do not appear to be coupled to the inhibition of adenylyl cyclase.

Activation of 5-HT_{1A} receptors in the central nervous system results in a variety of physiological responses such as modulation of food intake, sexual behavior, and body temperature (Altieri et al., 2013; Marcinkiewcz

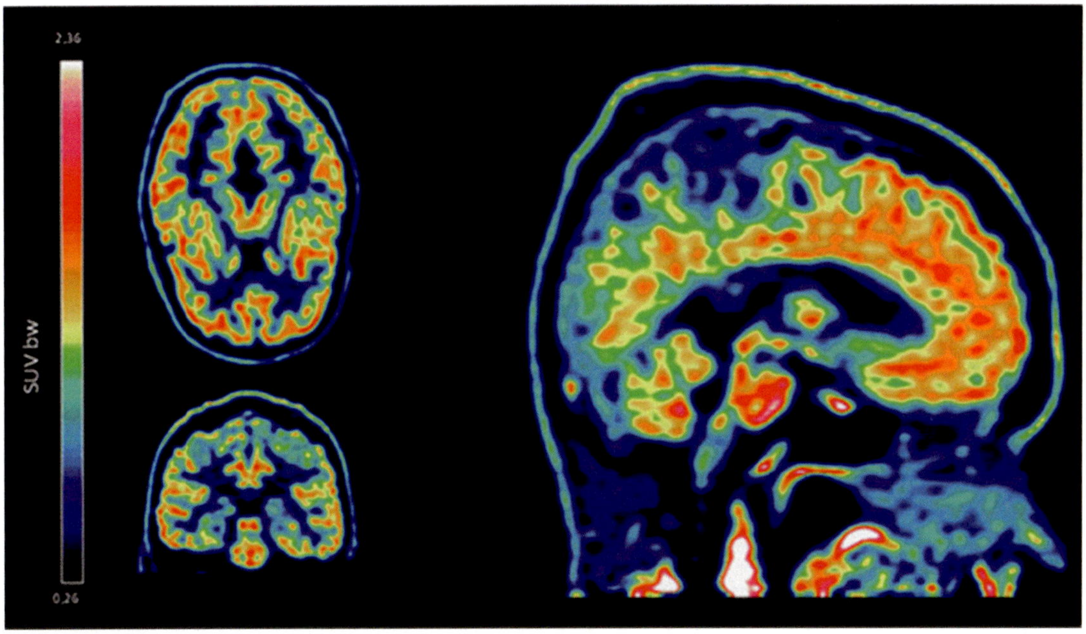

FIGURE 3.6 **Serotonergic system: forebrain innervation.** Here we show the first human brain image obtained in vivo with ^{18}F-F13640, a novel radioligand of 5-HT_{1A} receptor. This is a new molecule able to bind the receptors and, contrarily to the antagonists, being it an agonist preferentially binds to the functionally active state of receptors, a pharmacological property that has not yet been exploited so much for PET imaging of functional 5-HT_{1A} receptors. It is possible to see how the serotoninergic system strongly innervate the forebrain and in particular way the frontal/prefrontal cortex and cingulate cortex, areas particularly implicated in the neurophysiopathology of several psychiatric disorders. *Courtesy from Colom, M., Costes, N., Redouté, J., Dailler, F., Gobert, F., Le Bars, D., Billard, T., Newman-Tancredi, A., Zimmer, L., 2020. 18F-F13640 PET imaging of functional receptors in humans. Eur. J. Nucl. Med. Mol. Imaging 47 (1), 220–221. https://doi.org/10.1007/s00259-019-04473-7.*

TABLE 3.1 Serotonine receptors and distribution.

Receptor	Distribution in brain	Effector mechanism
5-HT_1 receptor family		
5-HT_{1A}	Hippocampus, amygdala, septum entorhinal, cortex, hypothalamus, raphe nuclei	Inhibition of adenylyl cyclase, opening of K^+ channels, activation of MAP kinase pathway
5-HT_{1B}	Substantia nigra, basal ganglia	Inhibition of adenylyl cyclase
(5-$HT_{1D\beta}$)	Frontal cortex, superior colliculus, lateral geniculate, deep nuclei of the cerebellum	
5-HT_{1D}	Globus pallidus, substantia nigra, caudate putamen	Inhibition of adenylyl cyclase
(5-$HT_{1D\alpha}$)	Hippocampus and cortex	
5-ht_{1E}	?	Inhibition of adenylyl cyclase
5-HT_{1F}	Cerebral cortex, striatum, hippocampus, nucleus accumbens, amygdala	Inhibition of adenylyl cyclase
5-HT_2 receptor family		
5-HT_{2A} 13q14–21	Claustrum, cerebral cortex, olfactory tubercle, striatum, nucleus accumbens	Stimulation of phospholipase C closing of K^+ channels
5-HT_{2B} 2q36.3–37.1	Cerebellum, lateral septum, dorsal hypothalamus, medial amygdala	Stimulation of phospholipase C
5-HT_{2C}Xq24	Choroid plexus, globus pallidus, cerebral cortex, hypothalamus, septum, substantia nigra, spinal cord, dorsal and median raphe	Stimulation of phospholipase C

Continued

TABLE 3.1 Serotonine receptors and distribution.—cont'd

Receptor	Distribution in brain	Effector mechanism
5-HT$_3$ receptor family		
5-HT$_3$	Hippocampus, entorhinal cortex	Ligand-gated cation channel
(5-HT$_{3A}$)	Amygdala, nucleus accumbens	
(5-HT$_{3B}$)	Solitary tract nerve, trigeminal nerve, motor nucleus of the dorsal vagal nerve, area postrema, spinal cord	
5-HT$_4$	Hippocampus, striatum, olfactory tubercle, substantia nigra	Stimulation of adenylyl cyclase
5-ht$_{5A}$	?	Inhibition of adenylyl cyclase
5-ht$_{5B}$	?	?
5-HT$_6$	Striatum, nucleus accumbens, olfactory tubercle cortex, amygdala, hypothalamus, hippocampus	Stimulation of adenylyl cyclase
5-HT$_7$	Cerebral cortex, septum, thalamus, hypothalamus, amygdala, hippocampus	Stimulation of adenylyl cyclase

Lower-case appellations are used in some cases because the functions mediated by these receptors in intact tissue are not known.

Courtesy of Hensler, J.G., 2012. Chapter 15: Serotonin. In: Brady, S.T., Siegel, G.J., Albers, R.W., Price, D.L. (Eds.), Basic Neurochemistry, eighth ed. Academic Press, pp. 300–322. https://doi.org/10.1016/B978-0-12-374947-5.00015-8.

et al., 2019; Esquivel-Franco et al., 2020; Tecott et al., 1995; Bombardi et al., 2021; Lee and Clifton, 2020). This receptor has been implicated in affective disorders such as anxiety and depression (Celada et al., 2004, 2013; Bombardi et al., 2021) and in particular seems that its postsynaptic localization mediates the effects of antidepressant drugs in behavioral animal models and on cell proliferation (Huang et al., 2005; Klempin, 2010; Altieri et al., 2013; Albert, 2012; Yohn et al., 2017; Dai et al., 2022) and survival in the hippocampus (Klempin et al., 2010; Dai et al., 2022; Huang et al., 2005). The desensitization of somatodendritic 5-HT$_{1A}$ autoreceptors following chronic administration of drugs used to treat anxiety and depression is also of actual interest, given the critical role of this receptor in regulating the serotonergic neurotransmission (Celada et al., 2004; Assié et al., 2006; Albert et al., 2011; Le Poul et al., 1995).

The 5-HT$_{1B}$ and 5-HT$_{1D}$ Receptor Subtypes

The 5-HT$_{1B}$ and 5-HT$_{1D}$ receptor subtypes are also linked to inhibition of adenylyl cyclase activity (Table 3.1). 5-HT1$_B$ receptors were characterized in certain rodents (rat, mouse, hamster), whereas the 5-HT1$_D$ receptor was characterized using pharmacological criteria in species such as guinea pig, dog, pig, cow, and human (Zgombick et al., 1997; Pullar et al., 2004; Hoyer, 2007, 2019; Mengod et al., 2010; Åhlander-Lüttgen et al., 2003). The 5-HT$_{1B}$ receptor is located in high density in the basal ganglia, particularly in the globus pallidus and the substantia nigra (Tiger et al., 2018; Moret and Briley, 2000.) (Table 3.1). This is turn raises the interesting possibility that this receptor may be involved in the physiopathology of neurodegenerative diseases that involve the basal ganglia as well as Parkinson's disease (Miguelez et al., 2014; Vermeiren et al., 2018). Functional studies indicate that these receptors are located on presynaptic terminals of serotonergic neurons modulating the serotonin release. 5-HT1$_B$ receptors are also found on the postsynaptic terminal where they modulate the release of other neurotransmitters other than serotonin such as acetylcholine in the hippocampus and dopamine in the prefrontal cortex (Tiger et al., 2018; Celada et al., 2013; Stahl, 2015; Ahlander_Lüttgen et al., 2003; Leiser et al., 2015).

Moreover postsynaptic 5-HT$_{1B}$ receptors are also located on cerebral arteries and other vascular tissues (Razzaque et al., 2002; Nilsson et al., 1999) and this plays a role of great importance in the treatment of migraine and cluster headache. Indeed the triptans, a class of 5-HT$_{1B/1D}$ receptors agonists, e.g., sumatriptan, zolmitriptan, interact with 5-HT$_{1B/1D}$ receptors and provoke a vasoconstriction of brain vessels blocking the neurogenic inflammation of the dura mater that gives rise to plasma extravasation presenting with pain and other symptoms that constitute the so-called migraine aura consisting of flashes or bright lights, zigzag lines, blind spots, amaurosis, paresthesia, and many others (Fig. 3.7A).

The release of 5-HT from dorsal raphe nucleus appears to be under the control of 5-HT$_{1D}$ receptors, presumably located on serotonergic terminals. It has been proposed that neurogenic inflammation and trigeminal nociception may be 5-HT$_{1D}$ receptor mediated in fact there is much interest in the 5-HT$_{1D}$ receptor as a useful therapeutic target for migraine and cephalalgia (Spekker et al., 2021; Dussor, 2014; Clemow et al., 2020; Benemei et al., 2017).

The 5-HT_{1E} Receptor Subtype

The 5-HT_{1E} receptor was originally identified in homogenates of human frontal cortex by radioligand binding studies using [^{3}H]-5-HT in the presence of 5-CT to block 5-HT_{1A} and 5-HT_{1D} receptor sites (Klein and Teitler, 2012; Barone et al., 1994.) The 5-HT_{1E} receptor displays a higher degree of homology with the 5-$HT1_D$ receptor (64%) (Hannon and Hoyer, 2008).

The 5-HT_{1F} Receptor Subtype

5-HT_{1F} receptor mRNA is found in cortex, hippocampus, dentate gyrus, nucleus of the solitary tract, spinal cord, uterus, and mesentery. In transfected cells, the 5-HT_{1F} receptor is coupled to the inhibition of adenylyl cyclase.

The 5-HT_{1F} receptor has been the subject of intense research as a target for antimigraine drugs. Activation of 5-HT_{1F} receptors is associated with the inhibition of plasma extravasation in the dura, a component of neurogenic inflammation thought to be a possible cause of migraine. Because activation of 5-HT_{1F} receptors does not mediate constriction of human vasculature (unlike 5-$HT_{1B/1D}$ receptors), 5-HT_{1F} receptor agonists would be devoid of unwanted cardiovascular side effects, an advantage in the treatment of migraine.

The 5-HT2 Receptor Family is Composed of the 5-HT2A, 5-HT2B, and 5HT2C Receptors

The 5-HT_{2A} Receptor Subtype

5-HT_{2A} receptors are localized at postsynaptic level of the serotonergic neurons and are particularly concentrated in the frontal cortex. 5-HT_{2A} receptors are also found in high density in the claustrum, in parts of the limbic system (i.e., amygdala and hippocampus), and in the basal ganglia (Table 3.1). In the cortex, the 5-HT_{2A} receptor is located on local GABAergic interneurons, as well as on pyramidal projection neurons, which are known to be glutamatergic (West et al., 2009).

The high level of 5-HT_{2A} receptor expression throughout the cortex suggests that this 5-HT receptor subtype may play a role in higher cognitive or integrative functions. The 5-HT_{2A} receptor has been implicated in the hallucinogenic effects of 5-HT_2 receptor agonists. There is also considerable interest in the role of the 5-HT_{2A} receptor in antipsychotic drug action. Newer antipsychotic medications (e.g., clozapine, olanzapine) (Fig. 3.7A) are antagonists with high affinity for the 5-$HT_{2A/2C}$ and for D_1−D_4 receptors.

The 5-HT_{2B} Receptor Subtype

The first report of the sensitivity of rat stomach fundus to 5-HT was published by Vane in 1959. This receptor, whose activation results in the contraction of fundus smooth muscle, was originally placed in the 5-HT_1 receptor class because of its sensitivity to serotonin and because responses mediated by this receptor were not blocked by 5-HT_2 or 5-HT_3 receptor antagonists. It has been reclassified as a 5-HT_2 receptor because of its similar pharmacological profile to the 5-HT_{2C} receptor (Table 3.1).

The 5-HT_{2C} Receptor

The 5-HT_{2C} receptor is present in high density in the choroid plexus. High-resolution autoradiography has shown that 5-HT_{2C} receptors are enriched on the epithelial cells of the choroid plexus. It has been proposed that 5-HT-induced activation of 5-HT_{2C} receptors could regulate the composition and volume of the cerebrospinal fluid. 5-HT_{2C} receptors are also found throughout the brain, in particular way in hypothalamus, hippocampus, septum, neocortex, and in areas associated with motor behavior such as substantia nigra and globus pallidus. 5-HT_{2C} receptors are also localized to GABAergic interneurons in the dorsal raphe nucleus (B6-DRN) and the excitation of the latter would result in the suppression of 5-HT cell firing.

5-HT_2 receptors are thought to mediate hallucinogenic properties of some 5-HT receptor agonists, such as LSD (Fig. 3.7A).

In Contrast to the Other 5-HT Receptor Subtypes, the 5-HT3 Receptor is a Ligand-Gated Ion Channel

The 5-HT_3 Receptor

The 5-HT_3 receptor corresponds to the ligand-gated ion channel superfamily like others, e.g., $GABA_A$ receptor. It is made of five subunits and it possesses additional pharmacologically distinct recognition sites where the function of the receptor can be modulated by alcohols and anesthetic agents (Parker et al., 1996).

5-HT_3 receptors have an excitatory effect and are found postsynaptically to serotonergic neurons in the central and peripheral nervous systems. Initially they appeared to be confined to peripheral neurons, where they mediate depolarizing actions of serotonin and modulate neurotransmitters release. Through activation of 5-HT_3 receptors, the serotonin regulates both motility and intestinal secretion throughout the gastrointestinal tract. Moreover, 5-HT_3 receptors are found in high density in peripheral ganglia and nerves (superior cervical ganglion and vagus nerve) as well as in the substantia gelatinosa of Rolando (or lamina II) in the spinal cord. Their localization in spinal cord suggests that 5-HT could modulate nociceptive mechanisms via 5-HT_3 receptors since they facilitate the release of substance P. The localization of 5-HT_3 in cortical and limbic areas of the brain is consistent with behavioral studies in animals, which suggest that 5-HT_3 receptor antagonists (Fig. 3.7A) may have potential anxiolytic,

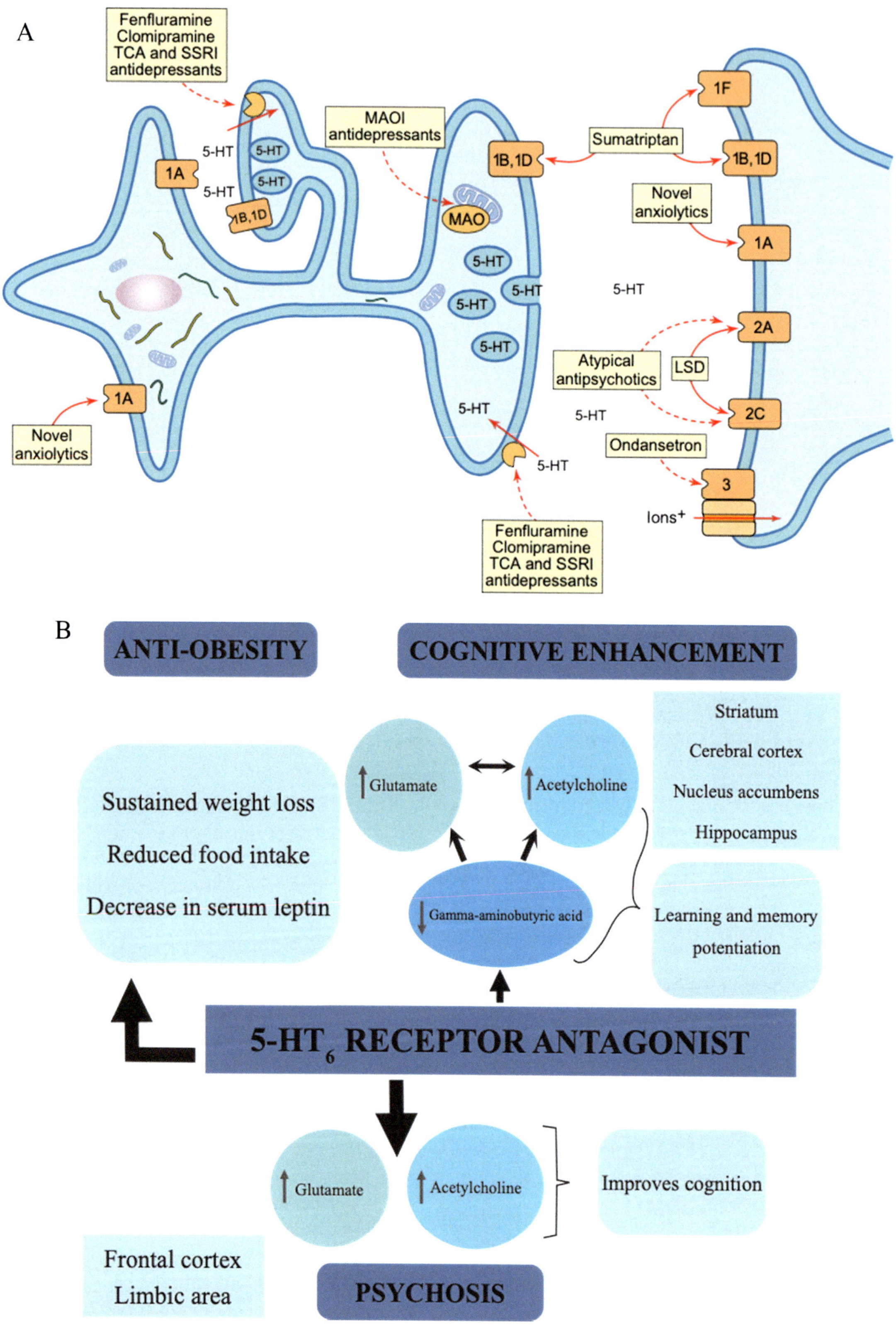

FIGURE 3.7 **(A) Serotonin neurons and receptors are targets for a wide variety of therapeutic drugs.** Drugs that act as agonists to activate receptors are indicated by solid-line arrows, whereas antagonists or inhibitors are shown with broken-line arrows. The 5-HT1A receptor acts as both the somatodendritic autoreceptor and a postsynaptic receptor; anxiolytic drugs such as buspirone are agonists at this receptor. Postsynaptic 5-HT1A receptors are also potential targets of novel antipsychotic drugs. In terminal fields, the autoreceptor is either the 5-HT1B or 5-HT1D subtype; these receptors also function as postsynaptic receptors. The anti-migraine drug sumatriptan is an agonist at these receptors as well as the

antidepressant, and cognitive effects In addition, preclinical findings indicate that antagonism of the 5-HT_3 receptor in the ventral tegmental area, nucleus accumbens, or amygdala reduces alcohol self-administration and/or alcohol-associated effects. Less is known about the effects of 5-HT_3 receptor activity on the self-administration of other drugs of abuse or their associated effects. Clinical findings parallel the preclinical findings such that antagonism of the 5-HT_3 receptor reduces alcohol consumption and some of its subjective effects. The observation that 5-HT_3 receptors modulate the activity of dopaminergic neurons in the VTA has led to the hypothesis that 5-HT_3 receptor antagonists may have potential as antipsychotic drugs and may possess the ability to reduce the rewarding effects of alcohol and certain drugs of abuse.

The highest density of 5-HT_3 receptor sites in the brain is in the area postrema, the site of the chemoreceptor trigger zone (Table 3.1).

5-HT_3 receptor antagonists, such as palonosetron, granisetron, dolasetron, ondansetron, and cilansetron, are commonly used to counteract chemotherapy-induced emesis (Haus et al., 2004; Eleftherios Zarkadas, 2020). Moreover, these agents now appear to be reaching out for newer indications. These drugs can modulate the immune-inflammatory axis through blockade of 5-HT_3 receptors present on immune cells indeed studies have reported neuroprotective and antiinflammatory properties in vitro and in vivo. The effects make 5-HT_3 antagonists' good molecules for the treatment of peripheral inflammatory diseases, including asthma, rheumatoid diseases, inflammatory bowel disease, and sepsis in addition to diabetes and CNS disorders, such as Alzheimer's disease (AD) (Khoury, 2018), seizure, and stroke (Fig. 3.7A).

The 5-HT4, 5-HT6, and 5-HT7 Receptors are Coupled to the Stimulation of Adenylyl-Cyclase

The 5-HT_4 Receptor Family

The 5-HT_4 receptor is associate to adenylyl cyclase stimulation and it is localized with high density in the striatum, substantia nigra, and olfactory tubercle and have been reported in the hippocampus as well (Table 3.1). It is located postsynaptically to serotonergic neurons and modulates the release of several neurotransmitters (i.e., acetylcholine, dopamine, and GABA), as well as indirectly modulating serotonin release. Studies have shown that activating the 5-HT_4 receptor can enhance cognitive performance in rats and monkeys. Furthermore, long-term treatment with a 5-HT_4 receptor agonist has been found to restore learning and memory deficits in a neuroendocrine mouse model of anxiety and depression. Although the role of the 5-HT_4 receptor in memory enhancement in humans remains to be fully established in clinical studies, some evidence suggests that it may be involved in this process (Bockaert et al., 2008).

The 5-HT_6 Receptor Family

The 5-HT_6 receptor provoke adenylyl cyclase stimulation (Heal et al., 2011). Interestingly, this receptor has high affinity for various antipsychotic and antidepressant drugs such as clozapine, amitriptyline, clomipramine, mianserin, and ritanserin (Frazer and Hensler, 1999; Fijal et al., 2014; Boess et al., 1997; Fijał et al., 2014). In striatum, 5-HT_6 receptors mediate an increase in cAMP accumulation, moreover its mRNA has been located in the striatum, nucleus accumbens, amygdala, hippocampus, and cerebral cortex. 5-HT_6 receptors have been found in these brain regions with immunohistochemical and radioligand binding approaches, with a primarily postsynaptic localization (Monsma et al., 1993; Ruat et al., 1993; Ward et al., 1995; Fone et al., 2002; Gerard et al., 1997; Hamon et al., 1999; Roberts et al., 2002).

Studies have demonstrated that cognitive enhancement by 5-HT_6 receptor antagonists involves multiple neurotransmitter systems, including cholinergic, glutamatergic, and GABAergic systems. Specifically, these antagonists enhance cholinergic neurotransmission, leading to positive effects on learning and memory, while their activation modulates the release of other neurotransmitters such as glutamate, GABA, dopamine, and norepinephrine (West, 2009). Consequently, selective ligands of this receptor have emerged as promising candidates for the treatment of neurological and psychiatric disorders such as Alzheimer's disease, schizophrenia, depression, and obesity (Mitchell, 2011; Karila et al., 2015; Morozova et al., 2017; Carr, 2011). It is

5-HT1F receptor. Selective 5-HT1F receptor agonists are of great interest as potential antimigraine drugs. Hallucinogenic drugs such as LSD are agonists at 5-HT2A and 5-HT2C receptors, whereas atypical antipsychotic drugs such as clozapine and olanzapine are antagonists. The 5-HT3 receptor, a ligand-gated ion channel, is blocked by drugs effective in the treatment of chemotherapy-induced nausea and emesis, such as ondansetron. Another important target for psychotherapeutic drugs is the serotonin transporter, which is blocked by drugs effective in the treatment of depression or obsessive–compulsive disorder (clomipramine). The enzyme responsible for the catabolism of serotonin, MAO, is inhibited by another class of antidepressants.**(B) Neurochemical mechanisms involved in blockade of 5-HT_6 receptors depicting its role in memory, cognition, obesity, and psychosis.** *(A) Courtesy of Hensler, J.G., 2012. Chapter 15: Serotonin. In: Brady, S.T., Siegel, G.J., Albers, R.W., Price, D.L. (Eds.), Basic Neurochemistry, eighth ed. Academic Press, pp. 300–322. https://doi.org/10.1016/B978-0-12-374947-5.00015-8. (B) From "Serotonergic 5-HT6 Receptor Antagonists: Heterocyclic Chemistry and Potential Therapeutic Significance" Bali A. 2015, redesigned and adapted to the text.*

noteworthy that several of the 5-HT_6 receptor ligands are indole-based agents bearing structural similarity to the endogenous neurotransmitter serotonin, which may contribute to their pharmacological effects (Fig. 3.7B).

The 5-HT_7 Receptor Family

Recently, analysis of the binding of selective radioligands revealed the presence of high densities of 5-HT_7 receptors in the anterior thalamus and dentate gyrus region of hippocampus, with intermediate levels in septum and hypothalamus, CA1 and CA2 regions of hippocampus and cortex (Hannon and Hoyer, 2008) (Table 3.1). The 5-HT_7 receptor has been implicated in the regulation of sleep, circadian rhythms, and mood. As discussed above, serotonin has been known for some time to induce phase shifts in behavioral circadian rhythms in fact it is able to modulate neuronal activity in the suprachiasmatic nucleus, the likely site of the mammalian circadian clock (Morin, 1999; Pontes et al., 2010; Antle et al., 2003.). There is also increasing evidence that the 5-HT_7 receptor may play a role in psychiatric disorders. This last hypothesis is supported also by its regional distribution in limbic areas and cortex (Yohn et al., 2017; Hedlund, 2009; Thomas and Hagan, 2004; Kołaczkowski et al., 2014; Matthys et al., 2011.). Atypical antipsychotics, such as clozapine and risperidone, and some antidepressants display high affinity for this receptor.

2. Catecholamines (Monoamines): Dopamine, Adrenaline, and Noradrenaline

Catecholamines are part of monoamine class (or biogenic amines) but, differently to the serotonin, they contain a nucleus, the catechol group, that is a benzene group with two adjacent hydroxyl groups, as well as an ethylamine side chain with a single amine group that may have additional substitutions. The predominant catecholamines in the brain are dopamine, norepinephrine, and epinephrine. Catecholamine neurotransmitters were first found in the autonomic nervous system. In 1954 Marthe Vogt demonstrated that norepinephrine was not uniformly distributed in the brain and that its distribution did not necessarily coincide with blood vessel density. This suggested that norepinephrine could function as a central neurotransmitter. Soon after, it was discovered that dopamine had a separate distribution in brain, indicating that it could function independently as a neurotransmitter and not simply be a metabolic intermediate of norepinephrine synthesis. The catecholamines exist in low (micromolar) concentrations in the brain with respect to amino acidic neurotransmitters such as glutamate, glycine, and γ-aminobutyric acid (GABA). However, they have considerable importance in the regulation of multiple aspects of central nervous system function and are vital therapeutic targets.

Tyrosine hydroxylase (TH, tyrosine 3-monooxygenase) is the rate-limiting enzyme in catecholamine biosynthesis. The catecholamines are synthesized in the nerves at the site of their release, both at the terminals and in the cell bodies.

Tyrosine hydroxylase catalyzes the conversion of its substrates tyrosine and molecular oxygen to 3,4-dihydroxy-L-phenylalanine (L-DOPA). It is the most important enzyme in catecholamine synthesis, being the initial and rate-limiting step in the synthesis of dopamine, norepinephrine, and epinephrine (Fig. 3.8). Its importance is also underscored by the fact that disruption of the TH gene is lethal to the embryo.

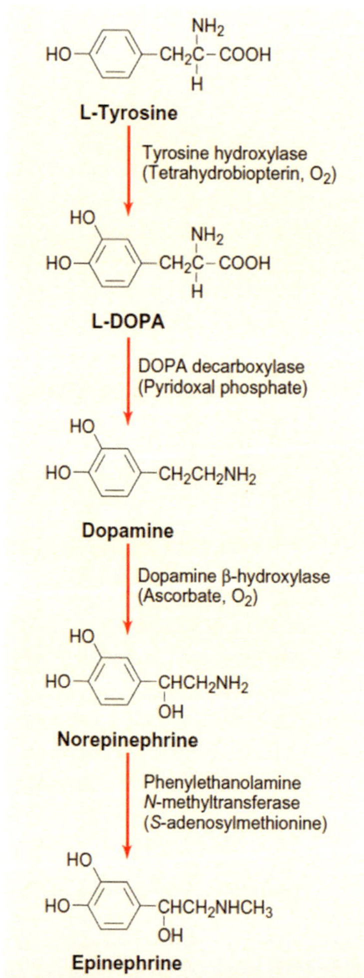

FIGURE 3.8 **Biosynthetic pathway for catecholamines.** *Courtesy of Gnegy, M.E., 2012. Chapter 14: Catecholamines. In: Brady, S.T., et al. (Eds.), Basic Neurochemistry, eighth ed. New York: Academic Press, pp. 283–299. https://doi.org/10.1016/B978-0-12-374947-5.00014-6.*

Dopamine

The levels of tyrosine in the brain are greater than the micromolar affinity of TH for tyrosine, so levels of tyrosine are unlikely to significantly regulate enzyme activity. In an alternate pathway, TH can hydroxylate phenylalanine to form tyrosine, which is converted to L-DOPA. This route is of significance in patients diagnosed with phenylketonuria, a condition in which phenylalanine hydroxylase activity is depressed.

The dopaminergic cell groups have been classified in several groups A8–A16 and they are found in the mesencephalon, the diencephalon, and the telencephalon. The exact localization of these dopaminergic group is discussed in Chapter 5b "The reticular formation," here we only describe the connections of them with the superior centers.

The population of dopaminergic cells in the mesencephalon gives rise to a significant ascending projection, which was categorized into two systems: the nigrostriatal dopaminergic system originating mainly from the substantia nigra or A9 group, and the mesolimbic dopaminergic system arising mainly from the A10 group (Ungerstedt, 1971). Subsequently, the entire forebrain projection was referred to as the mesotelencephalic dopaminergic system (Moore and Bloom, 1978; Lindvall et al., 1983).

For descriptive purposes, the mesotelencephalic dopaminergic system can be further divided into three subsystems: the mesostriatal projection, the mesolimbic projection, and the mesocortical projection. It is important to emphasize that these subdivisions are conceptual and do not have a physical existence. They were established by anatomists based on the functional aspects mediated by the fibers constituting each subsystem.

1. The mesostriatal projection originates from cell groups A8, A9, and A10 (Anden et al., 1964; Lindvall and Björklund, 1974; Lindvall et al., 1983; Moore and Bloom, 1978). The fibers of this projection gather in the medial tegmentum and enter the diencephalon through the dorsal part of the lateral hypothalamic area. They course rostrally through this area, with the majority of fibers diverging dorsally and laterally. Upon traversing the internal capsule, these fibers spread extensively in the caudate nucleus and putamen, forming a dense network of terminals. The nucleus accumbens (NAcc), which is often considered as a ventromedial extension of the caudate nucleus and occupies an intermediate position between the "extrapyramidal" and "limbic" systems, also receives a rich supply of mesostriatal dopaminergic fibers. The cells projecting to the NAcc are primarily located in the A10 group, although some are found in the medial part of the substantia nigra (Simon et al., 1976; Wang, 1981; Swanson, 1982). The dopaminergic innervation of the nucleus caudatus and putamen primarily originates from the compact part of the substantia nigra, but the A8 and A10 groups also contribute to this innervation. Cell group A8, considered as a caudolateral extension of the substantia nigra, has been observed to project to the ventral putamen (Fallon and Moore, 1978b; Beckstead et al., 1979). There is some disagreement in the literature regarding the contribution of the A10 group to the innervation of the neostriatum. Some groups of investigators report that this projection is limited to the most rostromedial portion of the caudate-putamen complex, while another group presents evidence suggesting that efferents from the ventral tegmental area are more widely distributed, extending throughout the ventromedial half of the neostriatum (Carter and Fibiger, 1977; Fallon and Moore, 1978b; Moore and Bloom, 1978; Simon et al., 1976; Beckstead et al., 1979). Collaterals of the mesostriatal projection form a relatively sparse network of terminal axons containing dopamine throughout the globus pallidus (Fallon and Moore, 1978b; Lindvall and Björklund, 1979). There is evidence indicating that the dopaminergic mesostriatal projection is involved in the regulation of complex behavior and plays a crucial role in the organism's ability to adapt to available sensory information in various ways. This specific capacity has been referred to as "the ability to arbitrarily switch motor programs" (see Cools, 1980; Cools et al., 1984b; Jaspers et al., 1984). Parkinson's disease is characterized by the progressive loss of dopaminergic neurons in the substantia nigra, resulting in the degeneration of their ascending projections. This degeneration is believed to underlie the akinesia and rigidity associated with the disease (Hornykiewicz, 1978). The area tegmentalis ventralis, although less developed in humans, also appears to be affected in Parkinson's disease (Bogerts et al., 1983). In line with the functions associated with the dopaminergic mesostriatal projection, it is noteworthy that patients with Parkinson's disease exhibit an impaired ability to switch their behavior, manifesting as a shifting aptitude disorder at both the motor and cognitive levels (Cools et al., 1984a).
2. The mesolimbic projection originates from cell group A10 and the most medial part of the substantia nigra. The ascending fibers of this projection travel alongside the mesostriatal projection within the medial forebrain bundle and distribute to various telencephalic structures (Ungerstedt, 1971; Lindvall, 1975; Moore, 1978; Fallon et al., 1978; Fallon and Moore, 1978a,b; Lindvall et al., 1983; Lindvall and Bjorklund, 1983). These structures include the bulbus olfactorius, anterior olfactory nucleus, olfactory

tubercle or anterior perforated substance, lateral septal nucleus, bed nucleus of stria terminalis, and the amygdaloid complex. The dopaminergic projection from the area tegmentalis ventralis to the nucleus accumbens has been experimentally shown to enhance locomotor activity (Pijnenburg and Van Rossum, 1973; Pijnenburg et al., 1976). It has been proposed that the hyperactivity of the area tegmentalis ventralis and its dopaminergic outflow to limbic structures may play a significant role in the pathophysiology of schizophrenia (Stevens, 1973; Stevens et al., 1979; Wedzony et al., 2007; Murray et al., 2008; Grace, 2010). The dopamine (DA) system's responsiveness requires regulation to facilitate appropriate responses in different contexts. There is a sophisticated mechanism in place to modulate the strength of the DA response, taking into account mood and location-related cues. Disruption of this regulatory system, such as heightened activity in the ventral subiculum, can lead to a mismatch between an individual's responses and environmental cues. This can result in an exaggerated attribution of significance to insignificant stimuli, feeling overwhelmed by attention-demanding cues without the ability to prioritize responses, or displaying behaviors that do not effectively contribute to achieving goals (Grace, 2010; Murray et al., 2008). Consistent with these theories, Bird and colleagues (1979) reported significantly elevated dopamine concentrations in the nucleus accumbens and anterior perforated substance of postmortem brains from schizophrenia patients. Additionally, Reynolds (1983, see also MacKay, 1984) observed a gross asymmetry of dopamine concentrations in the amygdalae of schizophrenic patients, with the left amygdala showing abnormally high levels of this catecholamine.

3. The distinction between the mesolimbic and mesocortical projections is not clear-cut due to several factors. Firstly, both projections originate from cell group A10 in the medial portion of the substantia nigra (Lindvall and Björklund, 1974, 1978; Simon et al., 1976, 1979; Carter and Fibiger, 1977; Fallon et al., 1978; Fallon and Moore, 1978b; Emson and Koob, 1978; Beckstead et al., 1979; Swanson, 1982). Secondly, most of the cortical areas where the mesocortical projection terminates are considered part of the limbic system. These areas include the medial part of the frontal lobe, the prepiriform and piriform cortices, the entorhinal cortex, and the anterior cingulate cortex (Lindvall and Björklund, 1974, 1978; Hökfelt et al., 1974a; Emson and Koob, 1978). These regions exhibit dense and well-defined termination fields of dopaminergic fibers. It's important to note that not all ascending fibers from the compact part of the substantia nigra and the area tegmentalis ventralis to the telencephalon are dopaminergic (Hedreen and Chalmers, 1972; Thierry et al., 1980; Guyenet and Crane, 1981; Van der Kooy et al., 1981; Swanson, 1982). Thus, the mesostriatal projections to the caudatus, putamen, and nucleus accumbens contain a small proportion of nondopaminergic fibers. A higher percentage of nondopaminergic fibers is found in the mesolimbic projection to the lateral septal nucleus, while the mesocortical projection to the medial frontal cortex appears to be about one-third dopaminergic fibers. It's also worth mentioning that in a certain proportion of perikarya in the ventral tegmental area, dopamine coexists with cholecystokinin, and some of these cells project to the nucleus accumbens (Hökfelt et al., 1980b; Studler et al., 1984). In addition to the large mesotelencephalic projection, the mesencephalic dopaminergic neuron population gives rise to other efferent projections. There is strong evidence that group A10 projects via the fasciculus retroflexus to the lateral habenular nucleus (Swanson, 1982; Lindvall et al., 1983; Skagerberg et al., 1984), the subthalamic nucleus contains a plexus of catecholaminergic fibers, possibly from collaterals of the nigrostriatal bundle (Nobin and Björklund, 1973; Lindvall et al., 1983; Guyenet and Cranes, 1981), and the ventral tegmental area projects bilaterally to the locus coeruleus and possibly the lateral parabrachial nucleus (Swanson, 1982). Within the diencephalic dopaminergic cell groups A11–A14, four efferent systems arise: the tubero-infundibular projection, incerto-hypothalamic projection, hypothalamo-septal projection, and hypothalamo-spinal projection.

The dopaminergic tubero-infundibular projection primarily originates from the A12 cell group. The axons of these cells extend to the rostroventral region of the infundibulum and terminate in various layers of the median eminence, with a higher concentration in the external layer. Some axons continue beyond the median eminence and travel through the pituitary stalk to reach the neurointermediate lobe, where they form a dense plexus (Ungerstedt, 1971; Felten, 1976; Hökfelt et al., 1977; Moore and Bloom, 1978). The incerto-hypothalamic projection, as described by Björklund et al. (1975), also contributes to the dopaminergic innervation of the median eminence. Evidence suggests that dopamine released from the tubero-infundibular fibers is transported to the anterior pituitary via the hypothalamo-hypophyseal portal system, inhibiting the release of prolactin-secreting cells (Macleod and Lehmeyer, 1974; Ben-Jonathan et al., 1977). However, the regulation of prolactin release is complex, and dopamine may not be the sole factor involved (McCann et al., 1984). The precise function of the dopaminergic tuberohypophyseal fibers is not fully understood, although recent research has provided some insights.

The incerto-hypothalamic dopaminergic projection originates from the A11 and A13 cell groups and consists of diffusely arranged short fibers within its areas of origin and adjacent dorsal and rostral hypothalamic regions (Björklund et al., 1975). The diencephalo-septal dopaminergic projection comprises fibers from the A11, A13, and A14 groups that project to the lateral septal nucleus (Lindvall and Stenevi, 1978). This indicates that the septal center, primarily supplied with midbrain-originating dopaminergic fibers, also receives an additional dopaminergic projection from the diencephalon.

Experimental evidence has emerged in recent years for the presence of a dopaminergic hypothalamo-spinal projection (Björklund and Skagerberg, 1979; Hökfelt et al., 1979; Skagerberg et al., 1982; Lindvall et al., 1983). In rats, this projection originates from perikarya located in the caudal hypothalamus of the A11 cell group and extends to the ipsilateral half of the spinal cord. The fibers descend within lamina I of the dorsal horn and the adjacent part of the dorsolateral funiculus, as well as along the central canal. The former pathway supplies the lateral portions of the superficial layers of the dorsal horn, while the latter innervates the intermediolateral cell column and related regions in the intermediate zone of the thoracic and upper lumbar segments. The precise course of these descending dopaminergic fibers through the brainstem is not yet known, but it has been suggested that they form part of a dorsal periventricular catecholaminergic bundle, which in turn is a component of the fasciculus longitudinalis dorsalis of Schutz. In rabbits, Blessing and Chalmers (1979) reported that the dopaminergic diencephalospinal fibers originate from the A13 group rather than the A11 group. Additionally, according to Swanson et al. (1981), dopaminergic neurons present in the parvocellular part of the paraventricular nucleus project to the region of the dorsal vagal complex and/or to thoracic levels of the spinal cord (Sawchenko and Swanson, 1982).

Dopamine Receptors

Dopamine receptors primarily exist in the brain but are also found in the kidney and peripheral vascular beds. The first dopamine receptor identified was the D_1 receptor, which was found to couple to G_s (and G_{olf} in the striatum) and stimulate cyclic AMP formation. Subsequently the D_2-like receptor was identified by antagonist binding studies, but was found to have a different pharmacology and location within the synapse from the D_1 receptor (Table 3.2). D_2 receptors are coupled to G_i/G_o GTP-binding proteins and inhibit cAMP formation. Subsequently subtypes within the D_1 and D_2 groups were identified by molecular cloning. In the adult human brain D_1 receptors are expressed at a higher level than any other dopamine receptor. D_1-like receptors have a high affinity for benzazepines, such as SCH23390, a receptor antagonist, and a low affinity for benzamides, such as sulpiride. Through activation of either G_s or G_{olf} in the striatum, the primary action of D1-like receptors is to activate PKA. Prominent substrates of D_1-activated PKA are DARPP-32, a protein phosphatase inhibitor enriched in dopaminoceptive cells, voltage- and ligand-gated ion channels and the cyclic AMP response element–binding protein, CREB. PKA-independent activities of D_1-like receptors have been noted, such as activation of mitogen-activated protein kinase (MAPK) (Neve et al., 2004). D_2-like dopamine receptors are found in a variety of brain regions (Fig. 3.9) and were originally identified by their high affinity for radiolabeled antipsychotic drugs, such as $[^3H]$

TABLE 3.2 Properties of human dopamine receptor subtypes.

	D_1-like		*D_2-like*		
	D_1	D_5	D_2	D_3	d_4
Amino acids	446	477	415/444	400	387
Chromosome	5	4	11	3	11
Effector pathways	↑cAMP	↑cAMP	↓cAMP	↓cAMP	↓cAMP
			↑K^+ channel	↑K^+ channel	↑K^+ channel
			↓Ca^{2+}channel	↓Ca^{2+}channel	↓Ca^{2+}channel
Distribution	Caudate/putamen, nucleus accumbens, olfactory tubercle, cerebral cortex	Hippocampus, hypothalamus, cerebral cortex	Caudate/putamen, nucleus accumbens, midbrain	Olfactory tubercle, hypothalamus	Frontal cortex, medulla, midbrain, nucleus accumbens

Courtesy of Gnegy, M.E., 2012. Chapter 14: Catecholamines. In: Brady, S.T., et al. (Eds.), Basic Neurochemistry, eighth ed. New York: Academic Press, pp. 283–299. https://doi.org/10.1016/B978-0-12-374947-5.00014-6.

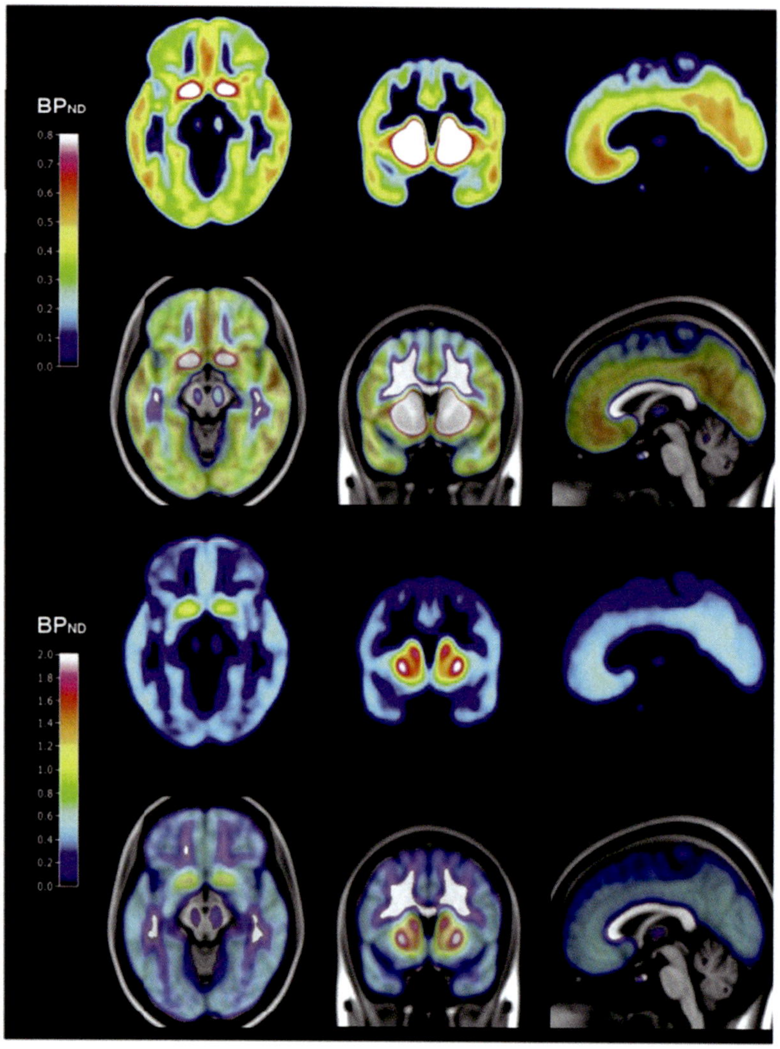

FIGURE 3.9 **The D1-R is widely distributed throughout the brain, in contrast D2/D3-R are found mainly in the striatum.** This figure shows the average BP_{ND} images of D_1 dopamine receptor binding in 15 healthy male subjects, as measured using PET and [^{11}C] SCH 23390. *Courtesy of Cervenka, S., 2019. PET radioligands for the dopamine D1-receptor: Application in psychiatric disorders. Neurosci. Lett. 691, 26–34. https://doi.org/10.1016/j.neulet.2018.03.007.*

haloperidol. The importance of these receptors in schizophrenia was reinforced by the finding that there was a direct relationship between the potency of binding of many therapeutically effective antipsychotic drugs to the D_2-like receptor and their clinical potency (Creese et al., 1976). Although both D_1-like and D_2-like dopamine receptors are important in basal ganglia and frontal cortical dopamine function, the D_2-like receptors are the ones targeted therapeutically by antipsychotic and anti-Parkinson's drugs. D_2, D_3, and D_4 have overlap in their pharmacology, but some differences are emerging with the development of more specific ligands. D_4 receptors have a particularly high affinity for the antipsychotic drug clozapine. All three subtypes have postsynaptic locations in dopamine terminal areas (Fig. 3.9). Both D_2 and D_3 receptors are also located on dopamine neurons and thus would have autoreceptor activity. Evidence in knockout mice demonstrates that the D_2 receptor is the functional dopamine autoreceptor in mice (L'Hirondel et al., 1998).

In the basal ganglia, dopamine autoreceptors reduce both synthesis and release of dopamine, while synthesis-regulating autoreceptors are not evident in the prefrontal cortex. D_2-like receptor signaling is mediated through the activation of heterotrimeric $G_{\alpha i}/_o$, proteins. Through coupling to the $G_{\alpha i}/_o$ proteins, the D_2-like receptors inhibit adenylyl cyclase activity. The behavioral importance of this pathway to the action of D_2-like receptors is demonstrated by the lack of inhibition of locomotor activity by antipsychotic drugs in mice null for adenylyl cyclase 5, the predominant adenylyl cyclase in postsynaptic neurons in the basal ganglia. Inhibition of adenylyl cyclase may also serve as a mechanism by which dopamine synthesis–inhibiting D_2

autoreceptors act, but other mechanisms involving phosphatases may also have a role. D_2-like receptors modulate many signaling pathways independent of adenylyl cyclase, including phospholipases, ion channels, MAP kinases, and the Na^+/H^+ exchanger. Many of these pathways, including those regulating ion channel activity, are mediated by $G_{\beta\gamma}$. Evidence strongly suggests that release inhibiting dopamine autoreceptors act through $G_{\beta\gamma}$-mediated activation of G-protein–regulated inwardly rectifying K^+ channels-2 (GIRK2) channels (Fig. 3.10).

Noradrenaline

Neurons that synthesize noradrenaline are restricted to the pontine and medullary tegmental regions. Seven noradrenergic cell groups, designated as Al–A7, have been described in rodents (Dahlstrom and Fuxe, 1964; Hökfelt et al., 1980a,b). Most of these have also been recognized in primates (Felten et al., 1974; Garver and Sladek, 1975; Hubbard and Di Carlo, 1973, 1974; Jacobowitz and MacLean, 1978; Schofield and Everitt, 1981; Schofield and Dixson, 1982; Felten and Sladek, 1983), including man (Nobin and Björklund, 1973; Olson et al., 1973; Bogerts, 1981). As for the dopamine, the exact localization of the noradrenergic groups is discussed in Chapter 5b "The reticular formation," here we only describe the connections of them with the superior centers.

Somewhat schematically it may be said that the noradrenergic cell group situated within the locus coeruleus, which is the most important noradrenergic center of the brain where all the noradrenergic branches originate, has three extensions: rostral, ventral, and caudolateral. The rostral extension consists of elements

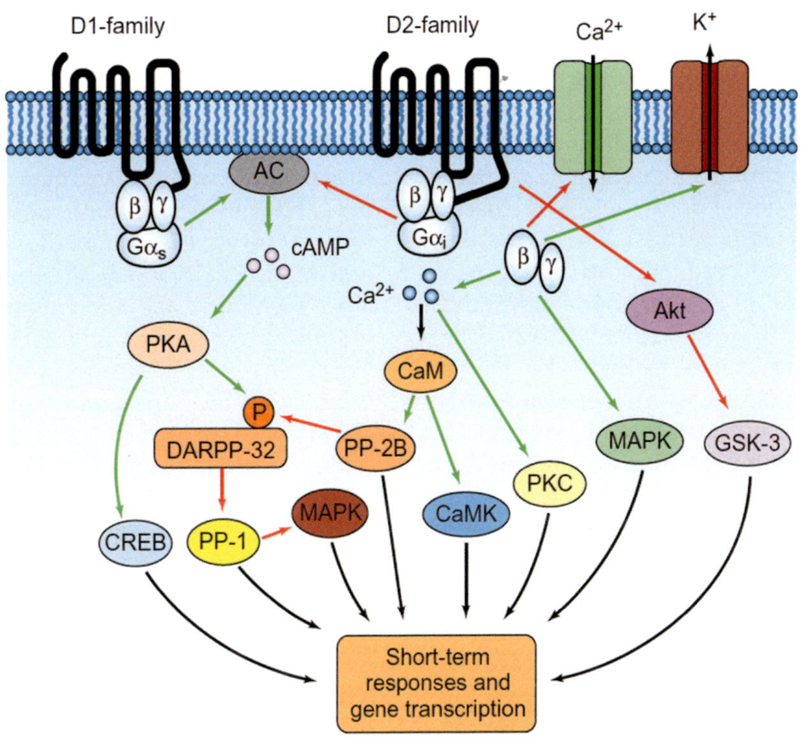

FIGURE 3.10 **Effect of dopamine on intracellular signaling pathways.** Stimulation of receptors by agonists can change enzyme activities as well as gene expression. The D_1 family of receptors (D_1 and D_5) are coupled to adenylyl cyclase (AC) via a stimulatory GTP-binding protein (Gs) which consists of $G_{s\alpha}$, a β and a γ subunit. The D_2 family of receptors (D_2, D_3, and D_4) inhibit adenylyl cyclase activity via coupling to an inhibitory GTP-binding protein (Gi). Activation of adenylyl cyclase leads to formation of cyclic AMP (cAMP) and activation of protein kinase A (PKA). The activated PKA phosphorylates, among other substrates, DARPP-32, which, when phosphorylated, will inhibit protein phosphatase-1. Activation of D_1-family receptors will result in activation of mitogen-activated protein kinase (MAPK). A prominent substrate of PKA that alters gene transcription is CREB (cAMP-response element binding protein). In addition to inhibition of AC by $G_{i\alpha}$, activation of the D_2 family of dopamine receptors results in dissociation of the βγ subunit, which affects numerous activities. When dissociated from the $G_{i\alpha}$ subunit, the βγ subunit inhibits voltage-sensitive Ca^{2+} channels and activates voltage-sensitive K^+ channels. The βγ subunit will also activate a phospholipase C isozyme, leading to an increase in intracellular Ca^{2+}. The Ca^{2+} leads to activation of kinases and phosphatases including MAPK, protein kinase C (PKC) and calmodulin (CaM)-stimulated enzymes such as Ca^{2+}/calmodulin–stimulated protein kinases (CaMK), as well as protein phosphatase-2B (PP-2B, calcineurin). One substrate of PP-2B is DARPP-32. Through a mechanism involving β-arrestin but independent of cAMP, activation of the D_2 family of receptors inhibits Akt activity, leading to an activation of glycogen synthase-kinase 3 (GSK-3) activity, which has direct effects on gene transcription (Beaulieu et al., 2007). *Courtesy of Gnegy, M.E., 2012. Chapter 14: Catecholamines. In: Brady, S.T., et al. (Eds.), Basic Neurochemistry, eighth ed. New York: Academic Press, pp. 283–299. https://doi.org/10.1016/B978-0-12-374947-5.00014-6.*

lying in the caudolateral part of the mesencephalic central gray (A6cg).

- The ventral extension is formed by scattered neurons situated within a cytoarchitectonic entity usually referred to as the subcoeruleus area (A6sc).
- The caudolateral extension of the noradrenergic cell cluster situated within the locus coeruleus is constituted by the A4 group (see later).

The A4, A6cg, and A6sc cell groups are commonly designated together as the (noradrenergic) locus coeruleus complex.

Synthetically, locus coeruleus efferents constitute two ascending fiber systems: the large *dorsal noradrenergic bundle* and the much smaller *rostral limb of the dorsal periventricular pathway*. Other efferents are distributed to the cerebellum and still others descend to the lower medulla oblongata and to the spinal cord.

The *dorsal noradrenergic bundle* or *dorsal tegmental bundle* (Lindvall and Björklund, 1974, 1978), which forms part of the major longitudinal catecholamine bundle, described by Jones and Friedman (1983), traverses the midbrain tegmentum in a position ventrolateral to the periaqueductal gray. At the level of the fasciculus retroflexus the bundle arches rostroventrally and attains the hypothalamus, where it joins the dorsal portion of the medial forebrain bundle complex. Within this complex the dorsal noradrenergic bundle proceeds to the septal region. Along its course, numerous smaller and larger branches emerge to innervate a large number of mesencephalic, diencephalic, and telencephalic grisea. The mesencephalic areas of termination include the central gray substance, the dorsal raphe nucleus, the superior and inferior colliculi, and the interpeduncular nucleus (Levitt and Moore, 1979; Marchand et al., 1980). In the rostral mesencephalon a large group of fibers of the dorsal noradrenergic bundle turns dorsally and passes along the fasciculus retroflexus toward the habenular complex. However, a major component of this group does not reach the epithalamus but rather enters the internal medullary lamina of the thalamus and ascends within it.

A second group of fibers leaving the dorsal bundle at the level of the fasciculus retroflexus ascends within the external medullary lamina. These two fiber groups together provide most of the very large input of the locus coeruleus to the dorsal thalamus. A more rostral branch of the dorsal noradrenergic bundle, the fibers of which join the stria medullaris, also supplies a number of thalamic centers as well as the epithalamus. Within the latter, both the medial and lateral habenular nuclei receive a noradrenergic input (Moore and Bloom, 1979). Practically all areas of the dorsal thalamus receive afferents from the locus coeruleus, most notably, the anterior, ventral, and lateral nuclear complexes and the medial and lateral geniculate bodies. The fibers to the anterior nuclei ascend with the mamillothalamic tract (Lindvall et al., 1974b). Ishikawa and Tanaka (1977) found that in the rhesus monkey, noradrenergic fibers originating from the locus coeruleus distribute to most of the thalamus, except for the midline and medial nuclei. The hypothalamus does receive noradrenergic input from the locus coeruleus (Jacobowitz, 1975; Jones and Moore, 1977; Sawchenko and Swanson, 1982a), but most of these fibers reach this part of the brain via the dorsal periventiicular system to be discussed below. The main telencephalic areas of termination of the dorsal noradrenergic bundle and its branches are: (a) the amygdala (mainly the central and basal nuclei; Tohyama et al., 1974; Jones and Moore, 1977; Fallon et al., 1978); (b) the olfactory tubercle or anterior perforated substance, the anterior olfactory nucleus, and the olfactory bulb (Blackstad et al., 1967; Ungerstedt, 1971; Swanson and Hartman, 1975; Fallon and Moore, 1978a); (c) the nucleus of the diagonal band, the medial septal nucleus, and the bed nucleus of the stria terminalis (Moore, 1978; Lindvall and Stenevi, 1978; Krayliak et al., 1981); (d) the hippocampal formation (gyrus dentatus, cornu Ammonis and subiculum; Blackstad et al., 1967; Ungerstedt, 1971; Lindvall et al., 1974a; Jones and Moore, 1977; Loy et al., 1980); and € the entire neocortex, including the cingulate, retrosplenial, and entorhinal cortical areas (Ungerstedt, 1971; Gatter and Powell, 1977; Lindvall and Björklund, 1978; Levitt and Moore, 1978; Lindvall et al., 1978). In most of the areas mentioned the afferents from the locus coeruleus are mixed with axons originating from other noradrenergic cell groups. However, the noradrenergic fibers innervating the hippocampus and the neocortex seem to originate exclusively from the locus coeruleus. The paths along which the efferents from the locus coeruleus ascend to the various telencephalic centers mentioned, via the dorsal noradrenergic bundle, are complex; several of these centers are approached along two or more different channels. The following simplified survey of the organization of these pathways is based mainly on Moore and Bloom (1979). Along its course through the hypothalamus, a large contingent of noradrenergic fibers leaves the medial forebrain bundle laterally to enter the complex formed by the ansa peduncularis and the ventral amygdalofugal pathway. Some of these fibers enter the amygdala, but others innervate the entorhinal cortex and the hippocampal formation. Upon attaining the septal region, the dorsal noradrenergic bundle breaks up into four major fiber groups. The first of these turns medially into the diagonal band of Broca to innervate the nucleus surrounding that fiber system and the medial septal nucleus. Some of the fibers of this group enter the fornix, along which they pass to the hippocampus. The second group enters the stria terminalis and follows this path to the amygdaloid complex. The third group continues in the medial forebrain bundle as it enters the basal telencephalon. A certain proportion of the fibers of this group innervates the olfactory tubercle, the anterior olfactory nucleus, and the olfactory bulb, but others continue rostrally in the external capsule. These may well correspond partly to the large fiber contingent which, according to Morrison et al. (1981), continues rostrally from the medial forebrain bundle into the frontal pole of the hemisphere, from where they arch caudally to supply the entire frontal, dorsal, and lateral neocortex (see also Shimizu et al., 1974; Tohyama et al., 1974). The fourth group of fibers traverses the rostral septum and encircles the corpus callosum to innervate the neocortex and the hippocampal formation. Ungerstedt (1971) suggested that this group of fibers travels caudally in the cingulum bundle and furnishes laterally directed branches that innervate practically the entire neocortex. Morrison et al. (1981) reported, however, that the fibers arching around the corpus callosum do not follow the cingulum bundle but rather the supracallosal striae, and that these fibers furnish only a marginal innervation of the medial parts of the neocortex. In several earlier studies (e.g., Ungerstedt, 1971; Gatter and Powell, 1977) the noradrenergic innervation of the neocortex is described as diffuse and uniform throughout. However, more recent investigations have revealed regional variations in pattern and density of cortical noradrenergic innervation, adhering to cytoarchitectonic boundaries (e.g., Morrison et al., 1979, 1982a; Levitt et al., 1984). Recent studies such as Nahimi's et al. (2018) demonstrated noradrenergic innervation of the forebrain in healthy people, compared to patients with Parkinson's disease, through the use of PET study (Fig. 3.11). It is noteworthy that most of the centers mentioned above receive, in addition to a projection from the ipsilateral locus coeruleus, a limited number of fibers from the contralateral locus coeruleus. These fibers pass from one dorsal noradrenergic bundle to the other via a number of commissures, among which the posterior commissure, the

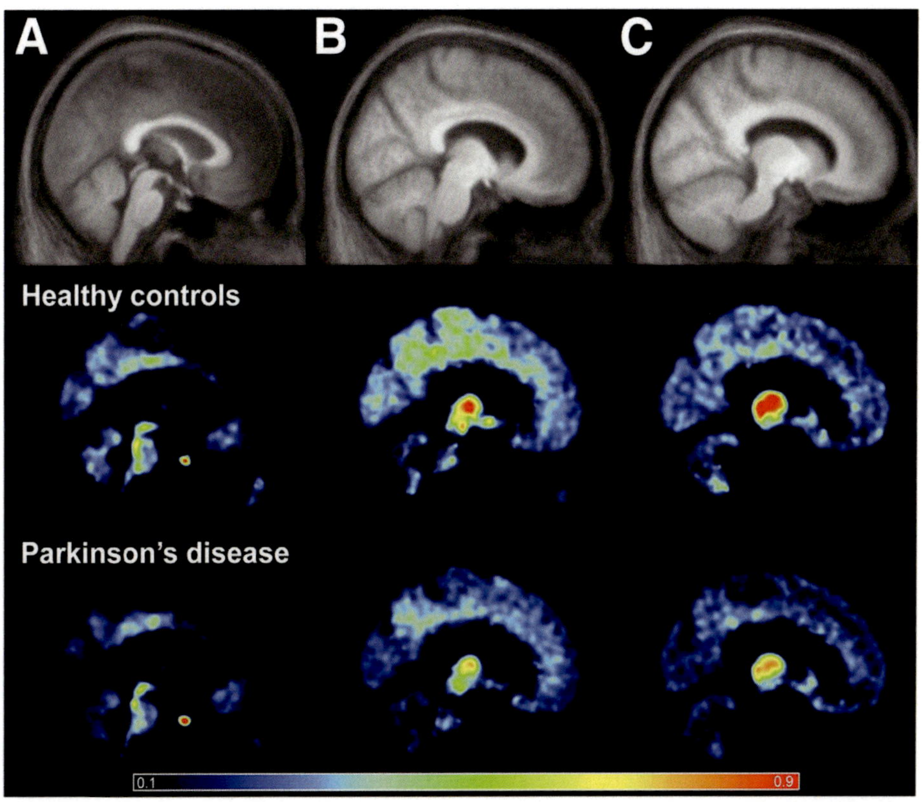

FIGURE 3.11 **Noradrenergic innervation of the forebrain in the healthy people and noradrenergic denervation in Parkinson's disease (PD) patients by use of the radioligand [^{11}C]MeNER BP$_{ND}$.** The radioligand (*S,S*)-^{11}C-2-(α-(2-methoxyphenoxy)benzyl)morpholine (^{11}C-MeNER) is a highly selective inhibitor of noradrenaline transporters, and PET studies suggest that this radioligand is suitable for quantitative neuroimaging of noradrenergic deficits in human brain in vivo. In the present investigation, has been used PET with ^{11}C-MeNER to map the density of noradrenaline transporters in groups of patients with PD and age-matched healthy controls. The maps are scaled from 0.1 (blue) to 0.9 (red). *Upper row*: average of MRI scan of all 25 subjects, *middle row*: BP$_{ND}$ parametric maps of healthy controls, *lower row*: BP$_{ND}$ parametric maps of PD patients. Sections are given in the median plane for the raphe nuclei (A), paramedian for the nucleus ruber, hypothalamus, and part of the thalamus (B), and further lateral in the mid-plane of the thalamus (C). Note also the complete missing tracer binding in the caudate and the low binding in the occipital cortex. *Courtesy of Nahimi, A., Sommerauer, M., Kinnerup, M.B., Østergaard, K., Winterdahl, M., Jacobsen, J. et al., 2018. Noradrenergic deficits in Parkinson disease imaged with 11C-MeNER J. Nucl. Med. 59 (4), 659–664. http://doi.org/10.2967/jnumed.117.190975.*

dorsal supraoptic decussation and the anterior commissure may be mentioned (Jones and Moore, 1977; Moore and Bloom, 1979). The rostral limb of the dorsal periventricular pathway arises mainly from the A6, A6sc, and A6cg cell groups and ascends to the diencephalon within the ventromedial part of the mesencephalic periaqueductal gray, forming part of the fasciculus longitudinalis dorsalis complex (Lindvall and Björklund, 1974; Schofield and Everitt, 1981; Tanaka et al., 1982; Felten and Sladek, 1983). Rostrally, this pathway continues into the diencephalic periventricular fiber plexus. The noradrenergic fibers passing to the cerebellum follow the superior cerebellar peduncle and terminate in the central nuclei as well as in the cortex (Fig. 3.11). These fibers originate mainly from the locus coeruleus (Ungerstedt, 1971; Olson and Fuxe, 1971), but some have been reported to arise from the subcoeruleus area (Pasquier et al., 1980). The descending efferents of the locus coeruleus complex follow two different routes, which may be designated as the *caudal limbs of the dorsal periventricular pathway* and the *dorsal noradrenergic bundle*, respectively. The *caudal limb of the dorsal periventricular pathway*, like the rostral limb of the same system, forms part of the fasciculus longitudinalis dorsalis complex (Felten and Sladek, 1983). It forms, as well, part of the major longitudinal catecholamine bundle, described by Jones and Friedman (1983). It descends through the lateral part of the rhombencephalic reticular core and continues caudally into the lateral funiculus of the spinal cord (Moore and Bloom, 1979; Jones and Friedman, 1983). In primates and man this bundle follows over a certain distance the trajectory of the central tegmental tract (Pearson et al., 1983) system, forms part of the fasciculus longitudinalis dorsalis complex Felten and Sladek (1983). Noradrenergic fibers originating from the locus coeruleus complex supply, partly directly and partly by way of the bundle just discussed, the following rhombencephalic centers (Levitt and Moore, 1979): the nuclei of the lateral lemniscus, the principal sensory trigeminal nucleus, the cochlear nuclei (Kromer and Moore, 1976, 1980), the pontine nuclei, the spinal trigeminal nucleus and the entire rhombencephalic reticular formation. The three centers last mentioned also receive, in addition to fibers from the locus coeruleus, input from other noradrenergic cell groups. The locus coeruleus complex innervates all segments of the spinal cord (Nygren and Olson, 1977; Westlund and Coulter, 1980). This innervation is bilateral, the decussation occurring at spinal levels (Karoum et al., 1980; Commissiong, 1981). The coeruleospinal fibers originate in the ventral part of the locus coeruleus and in the subcoeruleus area (Satoh et al., 1977; Hancock and Fougerousse, 1976; Mason and Fibiger, 1979; Guyenet, 1980; Westlund et al., 1984), descend in the lateral funiculus of the cord (Westlund and Coulter, 1980), and terminate in the ventral parts of the dorsal horn (laminae IV, V, VI), the intermediate gray and the ventral horn (Nygren and Olson, 1977; Westlund and Coulter, 1980). The locus coeruleus complex does not innervate the sympathetic intermediolateral column in the thoracic cord (Nygren and Olson, 1977; Commissiong et al., 1978; Westlund and Coulter, 1980); however, in the rhesus monkey the parasympathetic preganglionic neurons in the sacral cord have been shown to receive major input from the locus coeruleus (Westlund and Coulter, 1980).

Adrenaline

Adrenaline is the final product of the catecholamine chain, dopamine-noradrenaline-adrenaline. The conversion of noradrenaline to adrenaline is specifically catalyzed by the enzyme Phenylethanolamine N-Methyltransferase (PNMT). The presence of adrenaline in the central nervous system was first suggested by Marthe Vogt (1954) and later substantiated by a considerable number of biochemical studies (Gunne, 1962; McGeer and McGeer, 1964; Saavedra et al., 1974; Van der Gugten et al., 1976; Lew et al., 1977; Goldstein et al., 1978, 1980). Immunohistochemical evidence for the existence of adrenaline-containing neurons in the central nervous system has been produced by some groups of investigators (Hökfelt et al., 1974b, 1980b; Goldstein et al., 1978; Howe et al., 1980), all of whom used antibodies against the synthesizing enzyme PNMT. PNMT-containing fields of terminals have been observed in many areas of the central nervous system. The most prominent of these terminal fields include some thalamic midline nuclei; the dorsomedial and paraventricular hypothalamic nuclei, the latter of which shows a high density of terminals; the ventral part of the periaqueductal gray and the lateral periventricular zone of the rostral rhombencephalon; the ventral part of the locus coeruleus; the nucleus solitarius; the nucleus dorsalis of the vagal nerve; and the nucleus intermediolateralis in the spinal cord. The three nuclei last mentioned all show a high density of PNMT-containing terminals (Hökfelt et al., 1974b). It is interesting to note in this context that some groups of investigators who have analyzed the regional distribution of adrenaline (or PNMT) in the brain with biochemical techniques (Saavedra et al., 1974; Van der Gugten et al., 1976; Lew et al., 1977) detected this substance (or the activity of its synthesizing enzyme) not only in all terminal regions described by Hökfelt and colleagues (1974b) but also in several other brain areas, including the basal ganglia, the nucleus accumbens, the amygdala, the septum, the preoptic region, the habenula, the nucleus infundibularis and the median eminence. Only little is known concerning the central actions of the adrenergic neurons. The dense innervation of the paraventricular nucleus suggests that adrenaline might be involved in oxytocin and vasopressin secretion, and the innervation of the dorsomedial hypothalamic nucleus indicates a possible influence on food intake. Effects on the regulation of blood pressure and respiration are suggested by the dense adrenergic innervation of the nucleus solitarius, the dorsal motor vagal nucleus and the sympathetic intermedioventral nucleus in the spinal cord. Physiological experiments (see, e.g., Dampney and Moon, 1980) have shown that the ventrolateral portion of the medulla oblongata contains a circumscribed and highly sensitive vasopressor area, and Goodchild et al. (1984) recently presented evidence suggesting that this center corresponds to the adrenergic C1 group. It has also been reported (Goldstein et al., 1978) that in spontaneously hypertensive rats the PNMT activity in the C1 and C2 regions is notably elevated, and that in these animals the adrenaline level in the C2 region is also much higher than in normotensive rats.

Adrenaline and Noradrenaline Receptors

The word "adrenergic receptors" is referred to adrenaline and noradrenaline receptors (Fig. 3.12). These latter are all metabotropic receptor G-protein coupled. Because these catecholamines function as important peripheral neurotransmitters, adrenergic receptors are widely located throughout the body. It has been identified two principal adrenergic group, α and β, with a total of nine subtypes. A group is divided into α_1-adrenergic and α_2-adrenergic subgroups that are in turn divided respectively in subtypes: α_{1A}, α_{1B}, α_{1D} and α_{2A}, α_{2B}, α_{2C}. β group is divided in β_1, β_2, β_3. A_1, α_2, and β are distinguished by their pharmacology and by their separate signal transduction pathways. α_1-adrenergic receptors are coupled to G_q/G_{11}, which activates phospholipase C, leading to increased Ca^{2+} release and protein kinase C activation in the cell. α_1-adrenergic receptors also activate phospholipase D and mitogen-activated protein (MAP) kinases. α_2-adrenergic receptors are coupled to Gi/Go proteins and primarily inhibit adenylyl cyclase and stimulate phospholipase A_2 activities.

Similar to D_2 dopamine receptors, activation of α_2-adrenergic receptors leads to release of $G_{\beta\gamma}$ resulting in activation of K^+ channels and inhibition of Ca^{2+} channels. All three subtypes of the β-adrenergic receptor are coupled to G_s and activate adenylyl cyclase activity. G-protein–independent activation of MAPK by β-adrenergic receptor activation has been reported. The functional role of this effect is not yet known and it only occurs at high agonist concentrations. All subtypes of adrenergic receptors have postsynaptic locations at the synapse, but the α_2-adrenergic receptor is also located presynaptically, serving as an autoreceptor to inhibit the release of norepinephrine and a presynaptic regulator of the release of other neurotransmitters. Also, α_{2A} and to some extent α_{2C}-adrenergic receptors serve as autoreceptors.

Many β receptors are found on glia, where they have multiple actions, such as reducing the uptake of glutamate, regulating glucose availability, and regulating inflammatory cascades.

Noradrenaline in the CNS plays an important role in attention, arousal, and vigilance. Activation of α_1-adrenergic receptors in the medial preoptic area leads to wakefulness, while blockade of these receptors elicits sedation.

However, α_1-, α_2-, and β-adrenergic receptors in the hypothalamus also are important components of the ascending reticular activating system (Mitchell and Weinshenker, 2010). Norepinephrine enhances long-term memory consolidation in the amygdala and hippocampus through actions at α_1- and β-adrenergic receptors.

Further, α_2-adrenergic receptor agonists enhance prefrontal cortex cognitive function, including working memory. In contrast, release of high concentrations of norepinephrine acting on α_1-adrenergic receptors in the prefrontal cortex impairs prefrontal cortex function. Many antipsychotic drugs block α_1-adrenergic receptors and may ameliorate some prefrontal deficits by this mechanism.

Synaptic Activity of the Catecholamines

After synthesis, catecholamines are stored in synaptic vesicles. This storage is crucial to maintaining a supply of catecholamines that can be released, and also to mediate the process of release through exocytosis. The majority of monoamine-labeled synaptic terminals are characterized by the presence of small, clear vesicles (40–60 nm) and large, granular or dense-core vesicles (70–120 nm) within the terminal (Maley et al., 1990).

In the adrenal medulla, where norephinephine is converted in epinephrine by PNMT, catecholamines are stored in large dense-core chromaffin granules (~300 nm). In both the large, granular vesicles and the chromaffin granules, catecholamines are stored with proteins, notably chromogranin, peptides, and Dopamine-β-hydroxylase (DBH) at sites where norepinephrine is synthesized. Catecholamines are transported into synaptic vesicles and chromaffin granules through vesicular monoamine transporters (VMATs). There are two isoforms of VMAT, 1 and 2. VMAT1 is expressed in adrenal chromaffin granules and intestinal enterochromaffin cells. VMAT2 is expressed in CNS synaptic vesicles, in postganglionic sympathetic cells and in chromaffin granules. VMAT2 has substrate specificity for catecholamines, serotonin and histamine.

As the other neurotransmitters the primary mechanism of inactivation of catecholamine action is reuptake

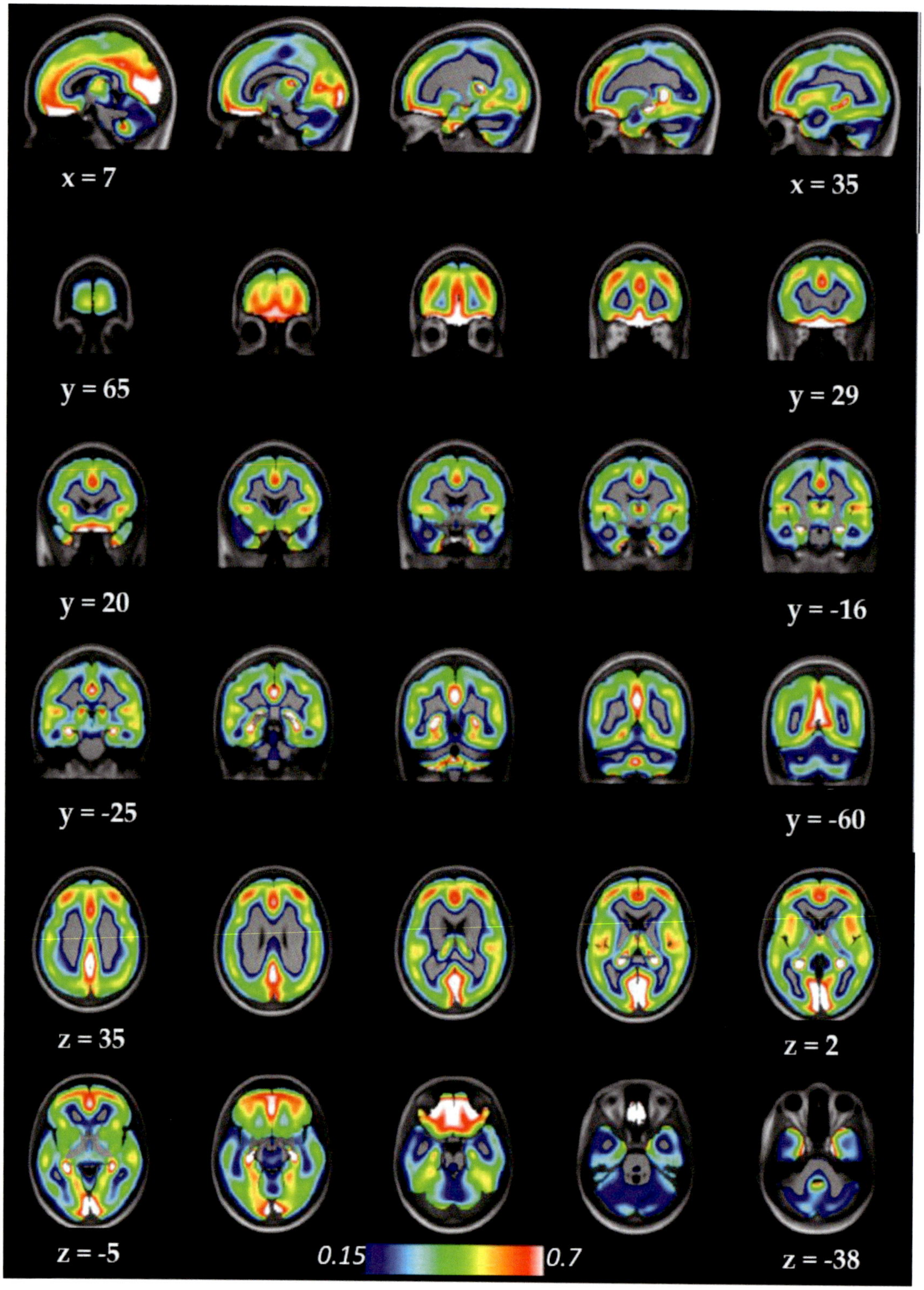

FIGURE 3.12 **Complete human brain mapping of the α_2-ARs *in vivo*.** These data confirm the widespread distribution of these receptors throughout the human brain. The regional distribution of [^{11}C] yohimbine binding is broadly consistent with the major findings of postmortem human brain studies so far performed. Color bar gives estimates of BP_{ND} in units of mL.cm^{-3}. *Courtesy of Laurencin, C., Lancelot, S., Merida, I., Costes, N., Redouté, J., Le Bars, D., Boulinguez, P., Ballanger, B., 2023. Distribution of α_2-adrenergic receptors in the living human brain using [11C]yohimbine PET. Biomolecules 13(5), 843. https://doi.org/10.3390/biom13050843.*

into the terminal via plasmalemmal transporters. The dopamine and norepinephrine transporters (DAT and NET, respectively) belong to the SLC6 family of Na^+, Cl-dependent transporters.

Following exocytotic release, the catecholamine diffuses to the transporter, which exists outside the active zone. Upon binding to the transporter, the catecholamine is taken up into the terminal and subsequently pumped into the synaptic vesicle to protect it from catabolism. Although transporters are generally specific for their substrate, there is overlap between DAT and NET.

However, NET has higher affinity for the endogenous catecholamines than does DAT, but DAT displays greater substrate selectivity for dopamine versus norepinephrine than does NET (Gu et al., 1994). This becomes especially important in an area such as the prefrontal cortex where there are both DAT and NET and there is competition for dopamine uptake. In the rat prefrontal cortex, the DAT content is low and dopamine uptake is primarily through NET. In the basal ganglia, however, an area important for locomotion, cognition and reward, there is very little NET and DAT is a key determinant of the duration and intensity of dopaminergic transmission (Jones et al., 1998).

DAT and NET undergo clathrin-dependent endocytosis and recycling to the surface; substrates, inhibitors and phosphorylation regulate the trafficking. A number of psychoactive and therapeutically useful drugs have their main sites of action at DAT and NET. Amphetamines, such as D-amphetamine and methamphetamine, are substrates at both DAT and NET. Cocaine, on the other hand, is a competitive reuptake blocker with greater affinity for DAT than NET. Both amphetamines and cocaine act to increase synaptic dopamine. Methylphenidate (Ritalin) is an amphetamine by structure, but is a DAT inhibitor like cocaine. Both amphetamine and methylphenidate are therapeutically useful in the treatment of attention deficit hyperactivity disorder (ADHD) and narcolepsy. Blockade of norepinephrine and/or serotonin reuptake has proven to be a successful therapeutic strategy for the treatment of depression. Secondary amine tricyclic antidepressants, such as nortriptyline, have greater selectivity for norepinephrine than serotonin reuptake, and are used to treat depression and neuropathic pain.

Tricyclic antidepressant drugs, especially the tertiary amines, also bind many monoamines, eliciting a number of untoward effects. Drugs have been developed that either are specific for both norepinephrine and serotonin transporters (SNRI) such as venlafaxine (Effexor) and duloxetine (Cymbalta), or are specific only for the norepinephrine transporter such as atomoxetine (Strattera).

Inactivation and Metabolism of Catecholamine

Inactivation of catecholamine is also performed by diffusion. Neither DAT nor NET is located directly in the synaptic cleft, so that dopamine and norepinephrine must diffuse to the reuptake sites. Brain regions differ in their density of reuptake sites (Lewis et al., 2001).

Metabolism of catecholamines takes place locally in the neuron and largely independently of exocytosis (Eisenhofer et al., 2004). The preponderance of catecholamine metabolism occurs in the terminal following leakage of the catecholamine from the vesicle.

Within the terminal, catecholamines are metabolized by monoamine oxidase (MAO) (Fig. 3.13).

MAO is a flavin-containing enzyme that catalyzes the oxidative deamination of a number of monoamines, including the catecholamines, serotonin, and trace amines. There are two isozymes of MAO, termed MAO-A and MAO-B, and they are located on the outer membrane of mitochondria. MAO-A has a higher affinity for serotonin and norepinephrine, while MAO-B prefers β-phenethylamine. In the brain, MAO-A is predominantly found in catecholaminergic neurons, while MAO-B is the form most abundant in serotonergic and histaminergic neurons and especially in glial cells (Shih et al., 1999). MAO-A is concentrated in noradrenergic cell bodies in the *locus coeruleus*, while MAO-B is the predominant form in serotonergic cell bodies in the *raphe nucleus*.

MAO inhibitors (IMAO) have been useful in clinical states, such as depression, where it is advantageous to increase monoamine levels, however nowadays their use is limited because of potentially dangerous side effects such as liver toxicity and interactions with other drugs that elevate catecholamines, causing hypertensive crisis. Selective MAO-B inhibitors, however, notably selegiline and rasagiline, are used therapeutically as adjunct therapy in Parkinson's disease to elevate dopamine levels.

3. Amino Acids

3a. Glutamate

Glutamate is the principal excitatory neurotransmitter. As many as 80%–90% of the neurons of the brain use glutamate as their neurotransmitter and approximately 80%–90% of the synapses in the brain are glutamatergic (Fig. 3.14) (Braitenberg and Schüz, 1998). It is mediator of sensory information, motor coordination, emotions, and cognition, including memory formation and memory retrieval. Glutamate chemically it is one of the most abundant amino acids in the human body formally classified as a nonessential amino acid, because it can be synthesized (in sufficient quantities for health) from α-ketoglutarate, which is produced as part of the

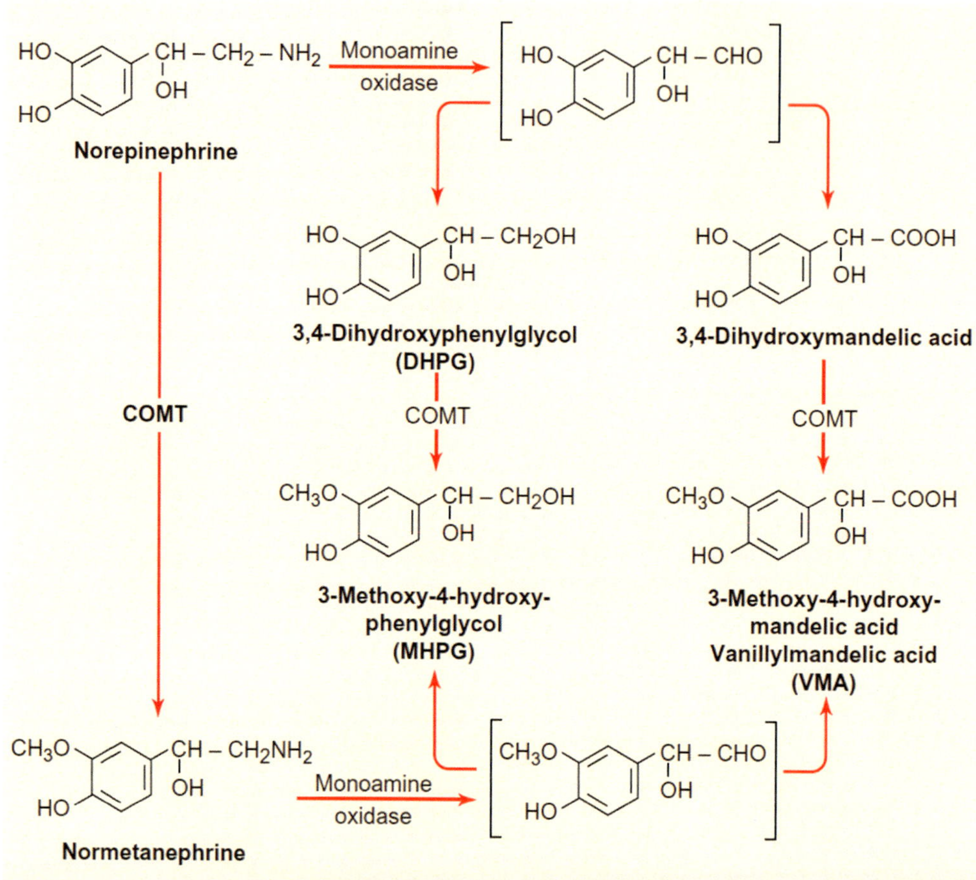

FIGURE 3.13 **Pathways of norepinephrine degradation.** Unstable glycol aldehydes are shown in brackets. COMT, catechol-O-methyltransferase. *Courtesy of Gnegy, M.E., 2012. Chapter 14: Catecholamines. In: Brady, S.T., et al. (Eds.), Basic Neurochemistry, eighth ed. New York: Academic Press, pp. 283–299. https://doi.org/10.1016/B978-0-12-374947-5.00014-6.*

citric acid cycle by a series of reactions whose starting point is citrate (Meldrum, 2000). Glutamate cannot cross the blood–brain brain (BBB) unassisted, but it is actively transported out of the nervous system by a high affinity transport system, which maintains its concentration in brain fluids at a fairly constant level Smith (2000).

Glutamate is synthesized in the central nervous system from glutamine as part of the glutamate-glutamine cycle by the enzyme glutaminase and this can occur in the presynaptic neuron or in neighboring glial cells. Most of the glutamate in the brain is found in glutamatergic neurons. The high concentration of glutamate in these neurons is due to three factors: (1) the high metabolic rate with rapid metabolism of glucose into α-ketoglutarate; (2) the much higher activity of aspartate aminotransferase (which aminates α-ketoglutarate to glutamate) than of tricarboxylic acid cycle (TCA) enzyme α-ketoglutarate dehydrogenase; and (3) the lack of reactions that drain the pool of glutamate much (other than the reversal of the amination and subsequent metabolism of α-ketoglutarate through TCA) (Shashidharan et al., 1997). In GABAergic neurons the glutamate pool is drained by the glutamate decarboxylase reaction, which produces GABA, and in astrocytes it is drained by the glutamine synthetase reaction, which produces glutamine.

Glutamate Receptors

Biochemical receptors for glutamate fall into three major classes, known as AMPA receptor, NMDA receptors and metabotropic receptor (mGlu) (Fig. 3.15). A fourth class, known as kainate receptors, are similar in many respects to AMPA receptors, but much less abundant. Many synapses use multiple types of glutamate receptors. AMPA receptors are ionotropic receptors specialized for fast excitation: in many synapses they produce excitatory electrical responses in their targets a fraction of a millisecond after being stimulated. NMDA receptors are also ionotropic, but they differ from AMPA receptors in being permeable, when activated, to calcium. Their properties make them particularly important for learning and memory (see Chapter 13 "Hippocampus and related structures"). Metabotropic receptors act through second

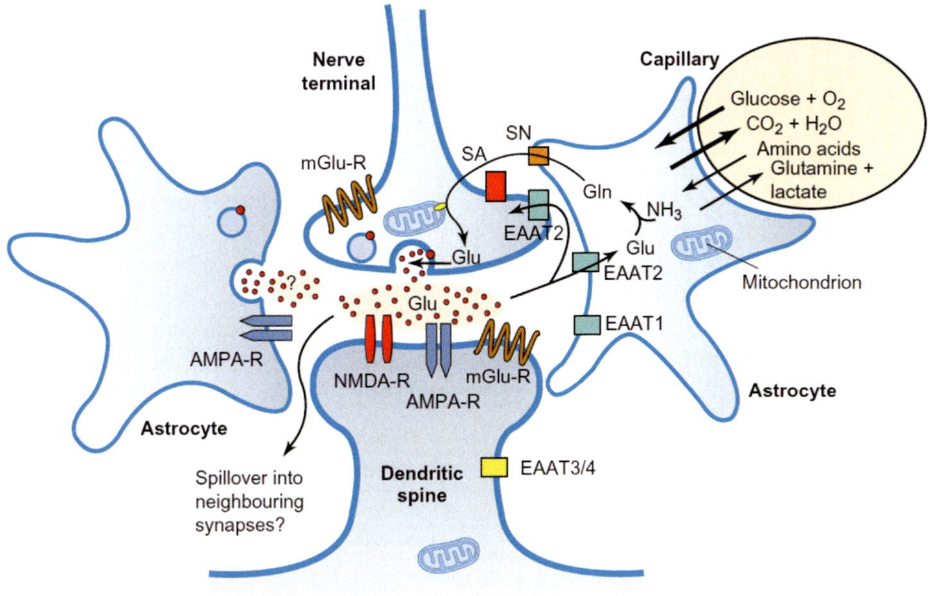

FIGURE 3.14 **A glutamatergic, axodendritic synapse consisting of a presynaptic nerve terminal and a postsynaptic dendritic spine.** The nerve terminal contains synaptic vesicles with glutamate transporters (*red dot*), mitochondria (*blue*) with glutaminase in nerve terminal (*yellow dot*), metabotropic glutamate receptors, and transporters for glutamate (EAAT2) and glutamine (SA transporters). The postsynaptic dendritic spine contains glutamate receptors, both ionotropic (AMPA and NMDA type) and metabotropic, and glutamate transporters (EAAT3 and EAAT4, the latter in cerebellar Purkinje cells). Surrounding the synapse are astrocytic processes with glutamate (EAAT1 and EAAT2) and glutamine transporters (SN transporters), glutamate receptors and even glutamate-filled vesicles. Green rectangles in plasma membrane of the axon terminal represent EAAT2 and of the astrocyte represent EAAT1/EAAT2 glu transporters. Glutamate that escapes out of the synapse without being cleared by transporters may spill over into neighboring synapses. *AMPA-R*, AMPA receptors; *EAAT1–4*, excitatory amino acid transporters 1–4; *Gln*, glutamine; *Glu*, glutamate; *mGlu-R*, metabotropic glutamate receptors; *SA and SN*, system A and system N transporters, respectively (for glutamine). *Courtesy of Hassel, B., Dingledine, R., 2012. Chapter 17: Glutamate and glutamate receptors. In: Brady, S.T., Siegel, G.J., Albers, R.W., Price, D.L. (Eds.), Basic Neurochemistry, eighth ed. Academic Press, pp. 342–366. https://doi.org/10.1016/B978-0-12-374947-5.00017-1.*

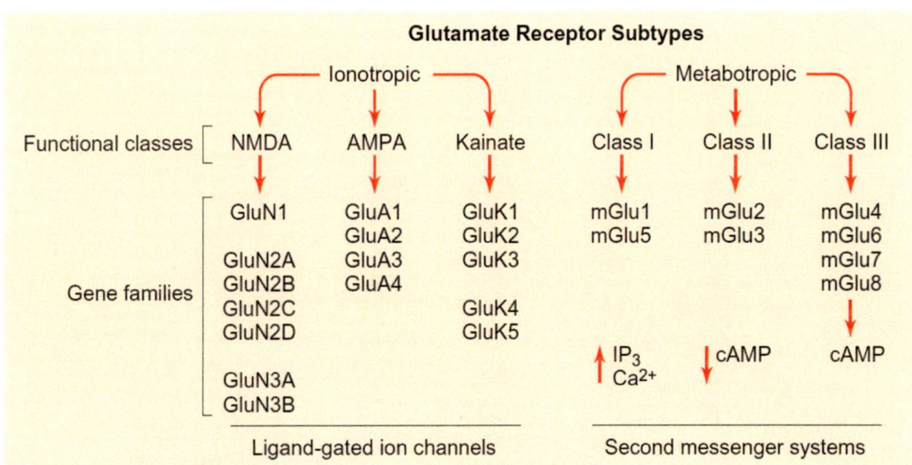

FIGURE 3.15 **Molecular families of glutamate receptors.** Each of the two main glutamate receptor divisions comprises three functionally defined groups (classes) of receptor. These are made up of numerous individual subunits, each encoded by a different gene. *Courtesy of Hassel, B., Dingledine, R., 2012. Chapter 17: Glutamate and glutamate receptors. In: Brady, S.T., Siegel, G.J., Albers, R.W., Price, D.L. (Eds.), Basic Neurochemistry, eighth ed. Academic Press, pp. 342–366. https://doi.org/10.1016/B978-0-12-374947-5.00017-1.*

messenger system to create slow, sustained effects on their targets.

To date, three NMDA receptor subunit families have been identified, one represented by a single gene (GluN1, encoding a protein of ~900 amino acids), the second by four genes (GluN2A-GluN2D, ~1450 amino acids) and the third by two known genes (GluN3A-GluN3B, about 1000 amino acids). Most if not all

NMDA receptors in the brain are heteromeric receptors, as is likely to be the case with AMPA and kainate receptors.

The NMDA receptor is unique among all known neurotransmitter receptors in its requirement for the simultaneous binding of two different agonists for activation. In addition to the conventional agonist binding site typically occupied by glutamate, the binding of glycine to a site on the GluN1 subunit is required for receptor activation (Traynelis et al., 2010; Kleckner and Dingledine, 1988). Because neither glycine nor glutamate acting alone can open this ion channel, they are referred to as "coagonists" of the NMDA receptor (Kleckner and Dingledine, 1988). The glycine site on the NMDA receptor is pharmacologically distinct from the classical inhibitory glycine receptor in that it is not blocked by strychnine and is not activated by β-alanine. Several small analogs of glycine, including serine and alanine, also act as agonists at this site.

Because of its role in synaptic plasticity, glutamate is involved in cognitive functions such as learning and memory in the brain (McEntee and Crook, 1993). The form of plasticity known as long term-potentiation (LTP) takes place at glutamatergic synapses in the hippocampus, neocortex, and other parts of the brain (see Chapter 13 "Hippocampus and related structures"). Glutamate works not only as a point-to-point transmitter, but also through spill-over synaptic crosstalk between synapses in which summation of glutamate released from a neighboring synapse creates extrasynaptic signaling/volume transmission (Okubo et al., 2010).

There is strong electrophysiological and biochemical evidence that the amino acids glutamate is excitatory neurotransmitter used by a large number of neurons in the central nervous system (for review, see Fagg and Foster, 1983; Fonnum, 1984).

Neocortical pyramidal cells contain much glutamate (Storm-Mathisen et al., 1983) and there is evidence suggesting that the following corticofugal projections use this amino acid as a neurotransmitter:

1. The corticostriatal projection, originating from the entire neocortex, extends throughout the ipsilateral nucleus caudatus and putamen (Divac et al., 1977; McGeer et al., 1977; Fonnum et al., 1981). Additionally, a smaller number of fibers from the frontal cortex project to the nucleus accumbens (Walaas, 1981).
2. Corticothalamic fibers originate from the entire ipsilateral neocortex and project to various regions of the thalamus, including the medial nucleus, ventrobasal complex, reticular nucleus, and the dorsal part of the lateral geniculate body (Karlsen and Fonnum, 1978; Fonnum et al., 1981; Young et al., 1981; Rustioni and Cuenod, 1982).
3. Corticotectal fibers originate from the visual cortex and project to the superior colliculus (Karlsen and Fonnum, 1978).
4. Fibers from the rostral and medial regions of the prefrontal area project to the substantia nigra (Carter, 1982).
5. Corticopontine fibers are present (Thangnipon and Storm-Mathisen, 1981; Carter, 1982).
6. Fibers originating from the sensorimotor cortex follow the long corticofugal projection and terminate in the red nucleus, cuneate nucleus, and the cervical and lumbar regions of the spinal cord (Young et al., 1981).

The pathways associated with the hippocampus have been found to contain fibers that utilize glutamate and/or aspartate as neurotransmitters. These pathways include

1. The perforant path, which originates from the lateral part of the entorhinal cortex, passes through the subiculum, and terminates in the cornu ammonis and the fascia dentata (Nadler et al., 1976, 1978; Fonnum et al., 1979).
2. Fibers originating from the cornu ammonis project bilaterally to the lateral septal nucleus via the fornix (Storm-Mathisen and Woxen-OpsahI1978; Fonnum and Walaas, 1978; Storm-Mathisen and Woxen-OpsahI1978; Zaczek et al., 1979; Walaas and Fonnum, 1980b).
3. Fibers, primarily originating from the subiculum, travel through the fornix and terminate in various regions such as the bed nucleus of the stria terminalis, the nucleus of the diagonal band of Broca, the nucleus accumbens, the ventral part of the nucleus caudatus and putamen, the mediobasal hypothalamus, and the mamillary body (Walaas, 1981; Storm-Mathisen and Woxen-Opsahl, 1978; Walaas and Fonnum, 1980b).

In the olfactory system, the terminals of the lateral olfactory tract, which connect the olfactory bulb to the prepiriform cortex, are suggested to utilize glutamate and/or aspartate as neurotransmitters (Harvey et al., 1975; Collins, 1979; Collins et al., 1981).

In the cerebellar cortex, two main excitatory inputs converge on the dendritic trees of Purkinje cells: the parallel fibers from the cerebellar granule cells and the climbing fibers from the inferior olive. Glutamate has been established as the neurotransmitter released by parallel fiber boutons (Young et al., 1974; McBride et al., 1976a,b; Sandoval and Cotman, 1978), while there is evidence suggesting that aspartate is the transmitter of the climbing fibers (Wiklund et al., 1982).

The auditory nerve consists of fibers which peripherally contact the hair cells in the organ of Corti and centrally terminate in the ventral and dorsal cochlear nuclei. Available evidence suggests that at least a certain proportion of the auditory nerve fibers utilize glutamate and/or aspartate as their transmitter(s) (Wenthold and Gulley, 1977; Wenthold, 1978; Altschuler et al., 1981).

It has been suggested that glutamate and/or aspartate may be the neurotransmitter(s) of large, myelinated, non-nociceptive primary afferent fibers entering the central nervous system via the trigeminal nerve and the spinal dorsal roots (for review, see Salt and Hill, 1983). However, the role of these amino acids in primary afferent transmission has been qualified by others Fagg and Foster (1983) as inadequately substantiated.

Glutamate Transporters

Glutamate transporters can be divided into two main classes: excitatory amino acid transporters (EAATs) and vesicular glutamate transporters (VGLUTs). Currently there are five known EAATs (EAAT 1–5). EAAT1-4 are expressed in the brain, however EAAT 1 and 2 are mostly presents in glia and EAAT5 has only been found in the retina (for review see Zhou et al., 2013) EAAT3 can be found in neurons throughout the entire brain and is solely localized to the soma and dendrites (Shashidharan, 1997). EAAT4 is most highly expressed in cerebellar Purkinje cells in mouse, rat and human brains, but is also found throughout the rodent forebrain (Massie, 2008). Neither EAAT3 nor EAAT4 can be detected in the brainstem of adult rodents (Furuta, 1997; Nieoullon et al., 2006).

The second family contains three known types of VGLUTs: VGLUT 1–3. VGLUTs have a much lower affinity for glutamate than EAATs (Shigeri, 2004). VGLUT 1 is predominantly expressed in the cortex, cerebellum and hippocampus whereas VGLUT2 is more restricted to the thalamus, deep cerebellar nuclei, and the brainstem (Fremeau, 2001). VGLUT3 can be detected in the parabrachial complex (Guo et al., 2005).

3b. GABA

GABA is the principal inhibitory neurotransmitter in the adult CNS, Roberts and Awapara discovered it in 1950.Electrophysiological studies between 1950 and 1965 suggested a role for GABA as a neurotransmitter in the mammalian CNS. It is important to underline that before birth, throughout the embryonic period, and in the first 3–4 weeks of postnatal age, GABA has an excitatory action due to the high content of Cl^- in the cytoplasm which in turn is regulated by upregulation of the internal Cl^- transporter NKCC1 and a downregulation of the external Cl^- transporter KCC2 (Ben-Ari, 2012).

GABA synthesis requires the decarboxylation of L-glutamic acid by glutamic acid decarboxylase (GAD); this process is achieved by GAD65 and GAD67. GAD65 is concentrated in the axon terminals where it synthesizes GABA for neurotransmission purposes, instead GAD67 seems able to promote synaptogenesis during early development (for review see Martin and Rimvall, 1993; Waagepetersen et al., 1999) and neuronal migration (as shown for GnRH cells; Lee et al., 2008).

The newly synthesized GABA is then packaged into synaptic vesicles by aid of a vesicular transporter. These proteins, with 10 membrane-spanning regions, are involved in active transport of GABA from the cytosol into secretory vesicles inside the cell, a mechanism energized by an ATP-dependent proton electrochemical gradient, independent of Na^+ (Martin and Olsen, 2000; Gasnier, 2000).

GABA release into the synaptic cleft is stimulated by depolarization of presynaptic neurons. GABA diffuses across the cleft to the target receptors on the postsynaptic surface. The action of GABA at the synapse is terminated by reuptake into both presynaptic nerve terminals and surrounding glial cells (Chen et al., 2004; Madsen et al., 2009). The plasmalemma transport systems mediating reuptake of GABA are both temperature and ion dependent. These transporter molecules are capable of bidirectional neurotransmitter transport. They have an absolute requirement for extracellular Na^+ ions and an additional dependence on Cl^- ions.

GABA taken back up into nerve terminals is available for reutilization but GABA in glia as in neurons is metabolized to succinic semialdehyde by GABA alpha-ketoglutarate transaminase (GABA-T) which in the same reaction converts alpha-ketoglutarate to glutamate. Conversely, GABA cannot be resynthesized from glutamate in this compartment since glia lack GAD.

GABA Rs Physiology and Pharmacology

GABA receptors can be classified into two main groups: ionotropic $GABA_A$Rs and metabotropic $GABA_B$ receptors ($GABA_B$Rs). $GABA_A$Rs are formed by two alpha subunits, two beta subunits and one gamma subunit, while the genome encodes for a total of 19 subunits (Fig. 3.16). There is also a high affinity $GABA_A$-rho receptor—previously known as $GABA_C$ receptor—only composed of rho subunits, especially abundant in the retina (Alakuijala et al., 2005). $GABA_B$R is a metabotropic receptor coupled to a G-protein that inhibit the adenylate cyclase activity (i.e., reducing cAMP production), voltage-gated calcium channels (VGCCs), and activates Kir3 channels (for a complete review see Gassmann and Bettler, 2012). Importantly, GABA-producing enzymes, GABA transporters and GABA receptors have been detected in rodents and primate cortex during embryonic development (Kilb et al., 2013). Furthermore, a large amount of evidence has confirmed that $GABA_A$Rs are functional during development (Ben-Ari et al., 2007). However, it seems that regarding $GABA_B$Rs, only the

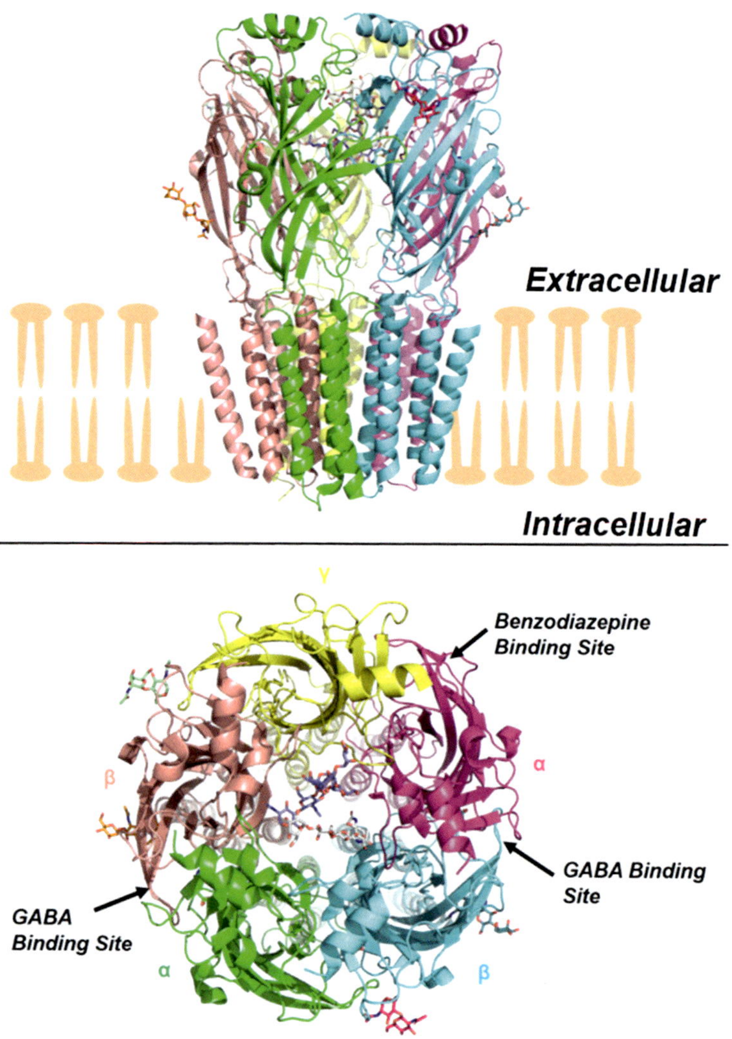

FIGURE 3.16 **Structure of the $GABA_A$receptor ($\alpha1\beta1\gamma2S$: PDB: 6DW1).** *Top*: Side view of the $GABA_A$ receptor embedded in a cell membrana. *Bottom*: View of the receptor from the extracellular face of the membrane. The subunits are labeled according to the $GABA_A$ nomenclature and the approximate locations of the GABA and benzodiazepine (BZ) binding sites are noted (between the α- and β-subunits and between the α- and γ-subunits, respectively). *Courtesy of Wikipedia.*

coupling with adenylate cyclase activity is functional during early corticogenesis in rat (Bony et al., 2013).

The $GABA_AR$ is modulated by several classes of drugs that bind to allosteric sites on the receptor complex: the benzodiazepines, barbiturates, and related intravenous general anesthetics like etomidate and propofol, as well as anesthetic steroids and endogenous neurosteroids. Further, it appears that some types of $GABA_A$ receptors are directly enhanced by ethanol and volatile general anesthetics (Olsen et al., 2008; Jurd et al., 2003; Yamakura et al., 2001; Wallner et al., 2006; Mody and Pearce, 2004; Smith, 2003) (for an in-depth discussion on anesthetics and the nervous system see appendix).

Benzodiazepine agonists represent the currently most useful group of agents in the general class of depressant drugs, which also includes barbiturates, that show anticonvulsant, anxiolytic, and sedative—hypnotic activity. Well-known examples include diazepam, alprazolam, clonazepam, lorazepam, and many others which often are prescribed for their antianxiety effects (Olsen, 2001; Barnard et al., 1998; Whiting, 2006; Olsen and Sieghart, 2008; Rudolph et al., 2001). The mechanism of action of benzodiazepine agonists is to enhance GABAergic transmission. From electrophysiological studies, it is known that these benzodiazepines increase the frequency of channel opening in response to GABA, thus accounting for their pharmacological and therapeutic actions (Martin and Olsen, 2000; Barnard et al., 1998; Olsen and Sieghart, 2008; Olsen and MacDonald, 2002; Polc and Haefely, 1982). In addition, the benzodiazepine site is coupled allosterically to the binding sites for the other modulatory ligands, like barbiturate, steroid, anesthetic, and picrotoxin (Martin and Olsen, 2000).

There is considerable evidence implicating GABA as an important transmitter in the spinal cord (Curtis and Johnson, 1974), used by interneurons to mediate both pre- and postsynaptic inhibition as well as presynaptic facilitation (Barber et al., 1978).

GABA is released by endings that are presynaptic to primary afferent terminals (Barber et al., 1978), and it is presumably also involved in Renshaw cell-mediated recurrent inhibition of motoneurons (Polc and Haefely, 1982; Cullheim and Kellerth, 1981). It has been established that the Purkinje cells in the cerebellar cortex synthesize GABA and release this inhibitory transmitter in their target areas, i.e., the deep cerebellar nuclei and the lateral vestibular nucleus (Obata et al., 1967, 1970; Fonnum et al., 1970; Kawaguchi and Ono, 1973; Saito et al., 1974; McGeer) et al., 1975). GABA has also been suggested to be the transmitter of the inhibitory local circuit neurons (Golgi, stellate and basket cells) in the cerebellar cortex (Kuriyama et al., 1966; McLaughlin et al., 1980; Saito et al., 1974).

The dorsal raphe nucleus (B6-DRN) contains GABA synthesizing elements which exert a tonic inhibitory control on the serotoninergic neurons present in that center (Nanopoulos et al., 1982; Scatton et al., 1984). The superficial zone of the superior colliculus contains two different types of small neurons, which accumulate GABA (Mize et al., 1982). Throughout the reticular nuclei of the thalamus numerous GABA-synthesizing neurons are found. These elements project primarily to other thalamic nuclei on which they probably exert an inhibitory influence (Houser et al., 1980). GABA is a prominent neurotransmitter in the so-called extrapyramidal motor system; within this system it occurs in projection neurons as well as in local circuit elements. The nucleus caudatus and the putamen contain numerous GABAergic neurons, the axons of which traverse the globus pallidus and terminate in the substantia nigra, pars reticulata (Hattori et al., 1973; Ribak et al., 1976; Brownstein et al., 1977; Fonnum et al., 1978; Jessell et al., 1978), where they are believed to influence, either directly or indirectly, the cells of origin of the dopaminergic nigrostriatal pathway (see Chapter 15). Several authors (Brownstein et al., 1977; Fonnum et al., 1978; Jessel et al., 1978) have reported that the GABAergic strionigral fibers arise preferentially from neurons situated in the posterior parts of the striatum. These neurons are of the medium-sized, densely spiny type known as "medium spiny neurons" (MSNs) (Ribak, 1981; but see Bolam et al., 1983a). The globus pallidus contains numerous GABAergic terminals which belong to striopallidal fibers (Fonnum et al., 1978; Jessell et al., 1978). Whether the cells of origin of these fibers are the same as those of the strionigral projection or rather constitute a separate neuron population remains to be determined.

In addition to the strionigral and striopallidal projections, GABA has been shown, to be the neurotransmitter of a pallidonigral pathway (Fonnum et al., 1978; Jessell et al., 1978).

In the nucleus caudatus and the putamen GABA-accumulating local circuit neurons are present (McGeer and McGeer, 1975). These elements, which comprise up to 15% of the total population of striatal neurons, belong to a category characterized as "medium-sized aspiny" (Bolam et al., 1983a). The nucleus accumbens, which according to Heimer and Wilson (1975) constitutes the ventral part of the striatum, contains GABAergic neurons which project to the rostromedial part of the substantia nigra and possibly also to the ventral tegmental area (Walaas and Fonnum, 1980a; Strahlendorf and Barnes, 1983). The pars reticulata of the substantia nigra contains numerous GABA-synthesizing neurons, which project massively to the ventrolateral and intralaminar thalamic nuclei, to the intermediate and deep layers of the superior colliculus and, diffusely, to the ponto-mesencephalic tegmental region (Vincent et al., 1978; Di Chiara et al., 1979; Chevalier et al., 1981; Childs and Gale, 1983). There is evidence suggesting that GABA is a neurotransmitter in the subthalamo-pallidal pathway (Nauta and Cuenod, 1982; Rouzaire-Dubois et al., 1983). The caudal part of the hypothalamus contains some groups of large neurons (Nagai et al., 1983b; Vincent et al., 1983a) which have been reported to project diffusely to the neocortex (Vincent et al., 1983a). GABA plays a role in regulating the release of several hypophysiotropic peptides and anterior pituitary hormones, including luteinizing hormone and prolactin. However, the neuronal organization of these control mechanisms is not known as yet (Racagni et al., 1982; Lamberts et al., 1983; DeFeudis, 1984). The septum and the nucleus of the diagonal band of Broca are rich in GABA-containing neurons (Kohler and Chan-Palay, 1983). A certain proportion of the GABAergic cells present in the medial septal nucleus and in the nucleus of the diagonal band project to the hippocampal region and to the entorhinal area (Kohler et al., 1984a). The medial habenular nucleus has also been reported to receive a GABAergic input from the nucleus of the diagonal band (Contestabile and Fonnum, 1983). A certain proportion of two types of interneurons present in the olfactory bulb, namely the granule cells and the periglomerular cells, use GABA as a neurotransmitter (Ribak et al., 1977; Halasz et al., 1979; Mugnaini et al., 1984a).

There is strong evidence that GABA is concentrated within the basket cells and other inhibitory intrinsic elements of the hippocampal region (Storm-Mathisen and Fonnum, 1971; Storm-Mathisen, 1972; Okada and Shimada, 1975; Storm-Mathisen et al., 1983), and the same holds true for the neocortex (Ribak, 1978; Somogyi et al., 1981; Hendry and Jones, 1981) (Box 3.1).

BOX 3.1

GABA and corticogenesis

In the past, neurotransmitters were not believed to also play a role in the stimulation of downstream signaling after receptor binding but were thought to be responsible only for interneuronal communication. Many results in over the years showed that GABA, as well as other neurotransmitters, plays a crucial role in genes expression thought the regulation of several transcription factor which are stimulated by different agents first of all Ca^{2+}. All these processes are the base of neurogenesis, neuromigration, and neurodifferentiation, which are the most important mechanisms responsible of the brain building therefore it is easy to understand those are at the base of many neurological diseases (e.g., epilepsy, ASD, ADHD, etc.). Below we illustrate how GABA affects the aforementioned processes during brain development (for a complete review see Ascenzi, 2017).

GABA-Induced Neurogenesis: Proliferation and Differentiation

There is an apparent controversy about the exact role of GABA during neurogenesis as some authors have found that it enhances proliferation (Young et al., 2012), while others have found that it decreases it (Nguyen et al., 2003). This discrepancy could be the consequence of a progenitor-dependent effect of GABA, such as ventricular zone (VZ) versus subventricular zone (SVZ) stem cells. In the developing neocortex, it has been proved that exogenous application of GABA on VZ or SVZ progenitors has opposite effects, increasing or decreasing proliferation, respectively (LoTurco et al., 1995; Haydar et al., 2000). These different effects could be both mediated by Ca^{2+}, in fact many authors have shown that the GABA induced depolarization is a means to increase intracellular Ca^{2+}. The latter is a messenger at the core of several cascades leading to the modulation of gene expression (Rosen et al., 1995) and possibly giving different outcomes:

1c. In the VZ, proliferation requires Ca^{2+} waves and connexin hemichannels (Weissman et al., 2004). Recently, it was found that depolarization can propagate through gap junctions, triggering Ca^{2+} oscillations in favor of proliferation (Malmersjo et al., 2013). We could therefore imagine that $GABA_AR$ activation modulate these oscillations of Ca^{2+};

2c. In the SVZ, $GABA_A$-induced depolarization leads to Ca^{2+} increase and inhibition of DNA synthesis (LoTurco et al., 1995).

The excitatory action of GABA in the embryonic period is capable of affecting the expression of several growth factors (e.g., BDNF) and other molecules involved in cellular proliferation (see later).

GABA-Dependent Neuronal Migration

In vitro it was seen that GABA is able to induce neuronal migration from E15 onward via Ca^{2+}-dependent mechanisms and that this process seems to be dose- and age-dependent (Behar et al., 1994). In fact, different concentrations of GABA produce different effects, as it has been shown for example that long-term exposure to $GABA_AR$ agonist leads to a pronounced receptor desensitization (Heck et al., 2007). In addition, different $GABA_A$ subunit compositions hold different effects and there are some evidences indicating a cell-type specific subunit composition (Denter et al., 2010; Cuzon Carlson and Yeh, 2011). Furthermore, GABA seems to promote the migration of both immature pyramidal neurons and immature interneurons (Behar et al., 1996, 1998), through the activation of both $GABA_AR$ and $GABA_BR$ (Behar et al., 1994, 2000, 2001; Lopez-Bendito et al., 2003; Heck et al., 2007; Inada et al., 2011). Surprisingly, it was shown that a high expression of $GABA_AR$ in interneurons of the ferret cortex impairs migration (Abbah and Juliano, 2014); perhaps, this is an indication of the existence of differences between species. Pharmacological manipulations on slices have shown that the radial migration of immature pyramidal neurons and immature interneurons requires $GABA_BR$ (Behar et al., 2000, 2001; Lopez-Bendito et al., 2003). In vivo, the downregulation of $GABA_BRs$ within pyramidal neuron progenitors, delayed their radial migration and many of them remained in deep layers of the cortex (Bony et al., 2013). It is noteworthy that in general, $GABA_BRs$ have been shown to decrease the speed of the neuronal migration in the developing hypothalamic nuclei as well (Davis et al., 2002; McClellan et al., 2008; Stratton et al., 2014).

GABA-Dependent Morphological Maturation of Neurons

The morphological maturation of pyramidal neurons has also been shown to require GABAergic signaling (Sernagor et al., 2010): $GABA_ARs$ through depolarization-dependent mechanisms and $GABA_BRs$ through the modulation of cAMP production. In order to study $GABA_AR$-dependent depolarization on neuronal

BOX 3.1 (*cont'd*)

arborization, the undermentioned studies have reduced the intracellular Cl^- concentration by an increase expression of KCC2 or decrease expression of NKCC1 (Cancedda et al., 2007; Young et al., 2012). These studies showed that premature expression of KCC2 leads to fewer and shorter dendritic processes (Cancedda et al., 2007), likewise, decreasing NKCC1 in immature neurons, strongly reduces the dendritic arborization (Young et al., 2012). This is not specific of pyramidal neurons as $GABA_AR$-induced depolarization also promotes the maturation of immature interneurons in culture (Marty et al., 1996).

The neurotrophic role of $GABA_BR$ has been reviewed a few years ago (Gaiarsa and Porcher, 2013) and we will hereafter refer to few studies relevant to the morphological maturation of neurons. In order to study the effects of $GABA_BR$ activation on morphology, in vitro studies using pharmacological manipulation with $GABA_BR$ agonist (i.e., saclofen and baclofen) and antagonist (i.e., CGP54626) found no effect on dendritic maturation (Gascon et al., 2006) or neuritic outgrowth (Maric et al., 2001).

These studies are somehow surprising considering the number of genes upregulated in the presence of baclofen in culture, such as BDNF (Ghorbel et al., 2005). On the contrary, an in vivo study using genetic manipulation found that a reduction of $GABA_BR$ in pyramidal neuron progenitors induces a multipolar morphology with longer and numerous dendrites, and in vitro, these authors, found a role of $GABA_BR$ in neuronal polarization (Bony et al., 2013). Altogether these findings help to reconsider the model proposed by Behar et al. (2000). It was proposed a role for $GABA_BR$ in migration out of the intermediate zone (IZ) but it seems that $GABA_BR$ contributes to neuronal polarization, a necessary step to pass the IZ. While the model proposed that $GABA_AR$ activation acts as a stop signal in the CP, it seems rather that $GABA_AR$ activation contributes to morphological maturation, an event that would prevent migration.

3c. Glycine

Glycine is widely recognized as a major inhibitory neurotransmitter in the vertebrate CNS, especially the spinal cord and brainstem, where it is crucial for the regulation of motor neuron activity (Lynch, 2009). In addition, glycinergic interneurons are found in the hippocampus, retina, auditory system, and other areas involved in the processing of sensory information. γ-Aminobutyric acid (GABA) and glycine are the two major inhibitory transmitters in the CNS and both are fast acting on ligand-gated chloride ion channels. Like GABA, glycine inhibits neuronal firing by gating Cl^- channels but with a characteristically different pharmacology (Dresbach et al., 2008).

The GlyRs are ionotropic and formed by transmembrane protein complexes formed by the assembling of five subunits arranged around a central pore. Five types of GlyR subunits, four alphas and one beta, have been described so far (Matzenbach et al., 1994; Lynch, 2009). Moreover, alternative splicing can generate additional variants for the alpha 1, 2, and 3 subunits (Kuhse et al., 1991; Malosio et al., 1991a; Nikolic et al., 1998; Miller et al., 2004; Legendre et al., 2009). One study has shown that GlyR subunits alpha 2 and beta are present during prenatal development from E14 onward in rats and that alpha 2 containing GlyRs are characterized by slow kinetics and are abundant in extra-synaptic locations (Malosio et al., 1991b). The presence of such receptors could be an indication of long and tonic activation.

As above mentioned, at embryonal stage both glycine and GABA exert an excitatory action due to the high intracellular chloride (Cl^-) concentration (Avila et al., 2013) due to the high expression of Cl^- importer NKCC1 and the expression of Cl^- exporter KCC2 (Ben-Ari et al., 2007, 2012; Kaila et al., 2014; Luhmann et al., 2014; Watanabe and Fukuda, 2015). Moreover, it is noteworthy that at embryonic day 13 (E13) in mouse, the concentration of glycine in the cortex is about 10 times higher than the GABA one (Benitez-Diaz et al., 2003; Qian et al., 2014). Thus, glycine in adulthood is well established as a major inhibitory neurotransmitter in the spinal cord and lower brain stem (for review, see Pycock and Kerwin, 1981; Fagg and Foster, 1983), where it is released mainly by local circuit neurons (Curtis et al., 1968; Ljungdahl and Hokfelt, 1973). It has been suggested that glycinergic interneurons also occur in two extrapyramidal centers: the substantia nigra and the neostriatum (nucleus caudatus + putamen). In the substantia nigra these neurons would exert a tonic inhibitory influence on the cells of origin of the nigrostriatal dopaminergic pathway and the glycinergic elements would in turn be inhibited by GABA-containing axons descending from the striatum (Pycock and Kerwin, 1981; see above in GABA description and in Chapter 15). In the striatum the glycinergic local circuit neurons would be interposed in an analogous way between GABA and dopamine-containing elements (Pycock and Kerwin, 1981). Finally, it should be mentioned that several authors (Ohta and Oomura, 1979; McGeer and Singh, 1980; Kita and Oomura, 1982) have suggested the existence of a glycinergic cortico-hypothalamic tract. This inhibitory pathway has been reported to originate from the prefrontal cortex, in particular from BA 10 and to terminate in the lateral hypothalamic area (Kita and Oomura, 1982) and in the ventromedial hypothalamic nucleus (Ohta and Oomura, 1979) (Box 3.2).

NEURODEVELOPMENTAL DISORDERS

The development of the CNS requires the markers of the inhibitory system to be functional. Indeed, the lack of glycine, GABA or taurine machinery is known to contribute to several pathologies, as detailed below. We know that glycine is essential for a normal development of the CNS, as for example it was seen that GlyR alpha 2 KO mice display numerous defect in formation of neural circuit that lead to epileptic seizures and excitoxicity (Morelli et al., 2016). In human, the putative role of glycine during development was also pointed out by a genetic study that identified a possible relationship between a single mutation in the GlyR alpha 2 gene and autism (Piton et al., 2011). In addition, a retrospective study of nonketonic hyperglycinemia showed perinatal death in about 2/3 of the patients and seizures in 90% of the patients (Hoover-Fong et al., 2004). Two patients with an atypical variant of nonketonic hyperglycinemia showed mental retardation (Singer et al., 1989; Christodoulou et al., 1993).

There are more evidences regarding the defects in the GABAergic system, especially using rodent models. Indeed, several mutant mouse models affecting the

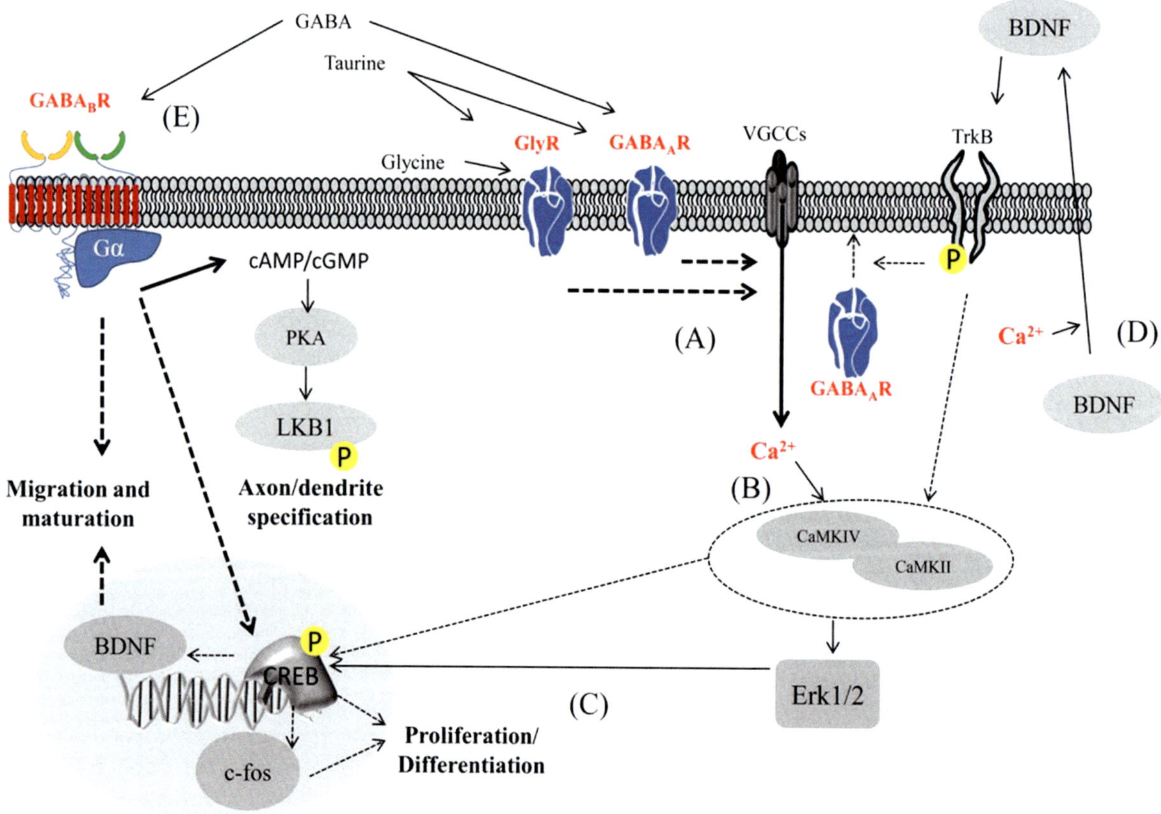

FIGURE 3.17 **Schematic representation of the GABA, glycine, and taurine pathways playing a role in corticogenesis during development.** (A) The activation of ionotropic glycine or $GABA_A$ receptor depolarizes immature neurons leading to the opening of voltage-gated calcium channels (VGCCs). (B) Ca^{2+} through CaMKIV or CaMKII kinases indirectly promotes CREB phosphorylation and subsequently the proliferation or differentiation of neural progenitors. (C) The CREB phosphorylation is partially mediated by Erk1/2 kinase and can lead to the transcription of c-fos and BDNF. (D) Ca^{2+}also promotes BDNF release and, when binding TrkB receptor, it can increase the surface expression of $GABA_A$ receptor and, indirectly, neuronal proliferation and differentiation. (E) The activation of metabotropic GABAB receptor decreases cAMP production and is involved in neuronal polarization, migration, and maturation of neural progenitors, in part via the PKA/LKB1 pathway. *From Ascenzi, M., Bony, G., 2017. The building of the neocortex with non-hyperpolarizing neurotransmitters. Dev. Neurobiol. 77 (9), 1023–1037. https://doi.org/10.1002/dneu.22495.*

BOX 3.2

Glycine-dependent neurogenesis

The role of glycine on the cortical neurogenesis has not been fully explored yet. However, some evidences suggest that GlyRs are expressed by neuronal progenitors and may have an effect on cell proliferation/differentiation. First, functional GlyRs are present on neurosphere primary culture of rat (Nguyen et al., 2002), and in culture of human pluripotent stem cell (James et al., 2014). Second, other authors have shown that a genetic downregulation of GlyRs in zebrafish could reduce the number of interneurons and impair spinal neurons differentiation, without altering the neuronal circuits (McDearmid et al., 2006; Cote and Drapeau, 2012). Likewise, the genetic inactivation of GlyRs in mouse leads to a reduction in the number of neural progenitors in the developing neocortex (Avila et al., 2014). We can hypothesize that the depolarizing effect of glycine would lead to intracellular Ca^{2+} increase that in turn may modulate neural differentiation. Indeed, Ca^{2+} is a well-known modulator of the cell cycle (Hepler, 1994) and Ca^{2+} increase can promote the expression of genes involved in the CNS development and can enhance neurogenesis and interneuron differentiation (Ghosh et al., 1994; Rosen et al., 1995; Bito et al., 1997; Ginty, 1997; Brustein et al., 2013). A recent review about granule cell migration in the cerebellum highlights the role of Ca^{2+} in this process and there are some similarities with the cortex (Komuro et al., 2015). In particular the increase in intracellular Ca^{2+} enhances the speed of migration and it is when Ca^{2+} spikes are lost that the migration is complete (Kumada and Komuro, 2004; Komuro and Kumada, 2005). In the cortex, Ca^{2+} promotes motility (Behar et al., 1996) and the depolarization to hyperpolarization switch could be the stop signal (Marin, 2013).

Glycine-dependent migration and acquisition of the neuronal morphology

The role of glycine in neuronal migration (Fig. 3.17) has been discussed in a recent review (Luhmann et al., 2015), but we would like to highlight two studies that have investigated the putative role of GlyRs in the migration of neocortical neurons. In the first study, the authors used a migration assay on organotypic culture from mice embryo to show that the pharmacological activation of GlyRs slow down the radial migration (Nimmervoll et al., 2011). In the second study, a genetic manipulation of GlyR function in mice showed that these receptors are required in the tangential migration of interneurons (Avila et al., 2013).

At the end of the migration process, the development of the neocortex continues with the maturation of the morphology, mostly at early postnatal age that will set the wiring diagram of the cortex. We know from different models that the glycinergic transmission is able to modulate the maturation of the neuronal morphology. For example, an in vivo study on mice showed that a high glycine concentration enhanced neurite outgrowth in embryonic neurons of the spinal cord via Ca^{2+} influx and synaptic activity (Tapia et al., 2001). The role of glycine in the morphological maturation of neurons from the cerebral neocortex has recently been demonstrated, thanks to the study showing an alteration of the dendritic morphology of neurons in a mutant mouse for GlyR alpha 2 (Morelli et al., 2016).

GABAergic neurotransmission, such as GAD65 KO, GABAA receptor beta 3 subunit KO, or VGAT KO, display higher seizure susceptibility or embryonic lethality (Homanics et al., 1997; Kash et al., 1997; Wojcik et al., 2006). In rats, prenatal exposure to antiepileptic drugs targeting GABA receptors induces hippocampal and neocortical dysplasia, a consequence of neuronal migration defect (Manent et al., 2007). In human, multiple studies have found a high probability of GABA receptor mutations in the brain of autistic patients. For example, a correlation between GABAAR and autism was found for the subunits alpha 4 and beta 1 (Ma et al., 2005; Collins et al., 2006). Others have found brain region specific downregulation of the subunits alpha 4, alpha 5, and beta 1 in autistic patients (Fatemi et al., 2010). More recently, the list of downregulated subunits increased with the identification of alpha 6, beta 2, delta, epsilon, gamma 2, theta, rho 2 (Fatemi et al., 2014). There is also a reduced expression of the GABAB1 subunit of GABABR (Fatemi et al., 2009; Oblak et al., 2010).

These studies link GABA receptors with autism, however, whether the defects happened specifically during the cortical development is not clear. There is almost no evidence regarding the alteration in taurine homeostasis, but we know that the lack of taurine during early life influences the auditory system maturation (Tyson et al., 1989) and its deficiency during pregnancy leads to impaired neurological function in the adult (Sturman et al., 1986). In human, a low plasmatic taurine level is associated with reduced mental development scores (Wharton et al., 2004). If we consider that some effects of taurine are mediated by the activation of GABAARs, we can imagine that taurine would have the same effects as the ones described following the GABAAR-mediated depolarization. For example, an early switch depolarizing to hyperpolarizing GABA through neural-specific overexpression of KCC2 causes an alteration of the neural tube and an abnormal brain development (Horn et al., 2010).

OPIOIDS

Opioids will be covered in Chapter 17 as the latter describes nociception and its control by the peptidergic system.

References

Abbah, J., Juliano, S.L., 2014. Altered migratory behavior of interneurons in a model of cortical dysplasia: the influence of elevated GABAA activity. Cereb. Cortex 24, 2297–2308.

Åhlander-Lüttgen, M., Madjid, N., Schött, P.A., Sandin, J., Ögren, S.O., 2003. Analysis of the role of the 5-HT1B receptor in spatial and aversive learning in the rat. Neuropsychopharmacology 28 (9), 1642–1655. https://doi.org/10.1038/sj.npp.1300235.

Alakuijala, A., Palgi, M., Wegelius, K., Schmidt, M., Enz, R., Paulin, L., Saarma, M., Pasternack, M., 2005. GABA receptor rho subunit expression in the developing rat brain. Brain. Res. Dev. Brain Res. 154, 15–23.

Albers, R.W., Price, D.L., 2012. In: Brady, S.T., Siegel, G.J. (Eds.), Basic Neurochemistry: Principles of Molecular, Cellular, and Medical Neurobiology, eighth ed. Elsevier.

Albert, P.R., 2012. Transcriptional regulation of the 5-HT_{1A} receptor: implications for mental illness. Phil. Trans. Biol. Sci. 367 (1601), 2402–2415. https://doi.org/10.1098/rstb.2011.0376.

Albert, P.R., Le François, B., Millar, A.M., 2011. Transcriptional dysregulation of 5-HT1A autoreceptors in mental illness. Mol. Brain 4 (1), 21. https://doi.org/10.1186/1756-6606-4-21.

Altieri, S.C., Garcia-Garcia, A.L., Leonardo, E.D., Andrews, A.M., 2013. Rethinking 5-HT1A receptors: emerging modes of inhibitory feedback of relevance to emotion-related behavior. ACS Chem. Neurosci. 4 (1), 72–83. https://doi.org/10.1021/cn3002174.

Altschuler, R.A., Neises, G.R., Harmison, G.G., Wenthold, R.J., Fex, J., 1981. Immunocytochemical localization of aspartate aminotransferase immunoreactivity in cochlear nucleus of the guinea pig. Proc. Natl. Acad. Sci. U. S. A. 78, 6553–6557.

Anden, N.E., Carlsson, A., Dahlstroem, A., Fuxe, K., Hillarp, N.A., Larsson, K., 1964. Demonstration and mapping out of nigroneostriatal dopamine neurons. Life Sci. 3, 523–530.

Antle, M.C., Ogilvie, M.D., Pickard, G.E., Mistlberger, R.E., 2003. Response of the mouse circadian system to serotonin 1A/2/7 agonists in vivo: surprisingly little. J. Biol. Rhythms 18 (2), 145–158. https://doi.org/10.1177/0748730403251805.

Assié, M.B., Lomenech, H., Ravailhe, V., Faucillon, V., Newman-Tancredi, A., 2006. Rapid desensitization of somatodendritic 5-HT_{1A} receptors by chronic administration of the high-efficacy 5-HT_{1A} agonist, F13714: a microdialysis study in the rat. Br. J. Pharmacol. 149 (2), 170–178. https://doi.org/10.1038/sj.bjp.0706859.

Aston-Jones, G., Shaver, R., Dinan, T., 1984. Cortically projecting nucleus basalis neurons in rat are physiologically heterogeneous. Neurosci. Lett. 46, 19–24.

Avila, A., Nguyen, L., Rigo, J.M., 2013. Glycine receptors and brain development. Front. Cell. Neurosci. 7, 184.

Avila, A., Vidal, P.M., Dear, T.N., Harvey, R.J., Rigo, J.M., Nguyen, L., 2013. Glycine receptor alpha2 subunit activation promotes cortical interneuron migration. Cell Rep. 4, 738–750.

Avila, A., Vidal, P.M., Tielens, S., Morelli, G., Laguesse, S., Harvey, R.J., Rigo, J.M., Nguyen, L., 2014. Glycine receptors control the generation of projection neurons in the developing cerebral cortex. Cell Death Differ. 21, 1696–1708.

Barber, R.P., Vaughn, J.E., Saito, K., Mclaughlin, B.J., Roberts, E., 1978. GABAergic terminals are presynaptic to primary afferent terminals in the substantia gelatinosa of the rat spinal cord. Brain Res. 141, 35–55.

Barnard, E.A., Skolnick, P., Olsen, R.W., Mohler, H., Sieghart, W., Biggio, G., Braestrup, C., Bateson, A.N., Langer, S.Z., 1998. International Union of Pharmacology. XV. Subtypes of gamma-aminobutyric acidA receptors: classification on the basis of subunit structure and receptor function. Pharmacol. Rev. 50, 291–313.

Barone, P., Jordan, D., Atger, F., Kopp, N., Fillion, G., 1994. Quantitative autoradiography of 5-HT1D and 5-HT1E binding sites labelled by [3H]5-HT, in frontal cortex and the hippocampal region of the human brain. Brain Res. 638 (1–2), 85–94. https://doi.org/10.1016/0006-8993(94)90636-X.

Beaulieu, J.M., Gainetdinov, R.R., Caron, M.G., 2007. The Akt-GSK-3 signaling cascade in the actions of dopamine. Trends Pharmacol. Sci. 28, 166–172.

Beckstead, R.M., Domesick, V.B., Nauta, W.J., 1979. Efferent connections of the substantia nigra and ventral tegmental area in the rat. Brain Res. 175, 191–217.

Behar, T.N., Li, Y.X., Tran, H.T., Ma, W., Dunlap, V., Scott, C., Barker, J.L., 1996. GABA stimulates chemotaxis and chemokinesis of embryonic cortical neurons via calcium-dependent mechanisms. J. Neurosci. 16, 1808–1818.

Behar, T.N., Schaffner, A.E., Colton, C.A., Somogyi, R., Olah, Z., Lehel, C., Barker, J.L., 1994. GABA-induced chemokinesis and NGF-induced chemotaxis of embryonic spinal cord neurons. J. Neurosci. 14, 29–38.

Behar, T.N., Schaffner, A.E., Scott, C.A., Greene, C.L., Barker, J.L., 2000. GABA receptor antagonists modulate postmitotic cell migration in slice cultures of embryonic rat cortex. Cereb. Cortex 10, 899–909.

Behar, T.N., Schaffner, A.E., Scott, C.A., O'Connell, C., Barker, J.L., 1998. Differential response of cortical plate and ventricular zone cells to GABA as a migration stimulus. J. Neurosci. 18, 6378–6387.

Behar, T.N., Smith, S.V., Kennedy, R.T., Mckenzie, J.M., Maric, I., Barker, J.L., 2001. GABA(B) receptors mediate motility signals for migrating embryonic cortical cells. Cereb. Cortex 11, 744–753.

Beliveau, V., Ganz, M., Feng, L., Ozenne, B., Hojgaard, L., Fisher, P.M., Svarer, C., Greve, D.N., Knudsen, G.M., 2017. A high-resolution in vivo atlas of the human brain's serotonin system. J. Neurosci. 37, 120–128.

Ben-Ari, Y., Gaiarsa, J.L., Tyzio, R., Khazipov, R., 2007. GABA: a pioneer transmitter that excites immature neurons and generates primitive oscillations. Physiol. Rev. 87, 1215–1284.

Ben-Ari, Y., Khalilov, I., Kahle, K.T., Cherubini, E., 2012. The GABA excitatory/inhibitory shift in brain maturation and neurological disorders. Neuroscientist 18, 467–486.

Ben-Jonathan, N., Oliver, C., Weiner, H.J., Mical, R.S., Porter, J.C., 1977. Dopamine in hypophysial portal plasma of the rat during the estrous cycle and throughout pregnancy. Endocrinology 100 (2), 452–458. https://doi.org/10.1210/endo-100-2-452.

Benemei, S., Cortese, F., Labastida-Ramírez, A., et al., 2017. Triptans and CGRP blockade—impact on the cranial vasculature. J. Headache Pain 18 (1), 103. https://doi.org/10.1186/s10194-017-0811-5.

Benitez-Diaz, P., Miranda-Contreras, L., Mendoza-Briceno, R.V., Pena-Contreras, Z., Palacios-Pru, E., 2003. Prenatal and postnatal contents of amino acid neurotransmitters in mouse parietal cortex. Dev. Neurosci. 25, 366–374.

Bigl, V., Woolf, N.J., Butcher, L.L., 1982. Cholinergic projections from the basal forebrain to frontal, parietal, temporal, occipital, and cingulate cortices: a combined fluorescent tracer and acetylcholinesterase analysis. Brain Res. Bull. 8, 727–749.

Bird, E.D., Spokes, E.G., Iversen, L.L., 1979. Increased dopamine concentration in limbic areas of brain from patients dying with schizophrenia. Brain 102, 347–360.

Bito, H., Deisseroth, K., Tsien, R.W., 1997. Ca2+-dependent regulation in neuronal gene expression. Curr. Opin. Neurobiol. 7, 419–429.

Björklund, A., Lindvall, O., Nobin, A., 1975. Evidence of an incerto-hypothalamic dopamine neurone system in the rat. Brain Res. 89 (1), 29–42. https://doi.org/10.1016/0006-8993(75)90131-6.

Björklund, A., Skagerberg, G., 1979. Evidence for a major spinal cord projection from the diencephalic A11 dopamine cell group in the rat using transmitter-specific fluorescent retrograde tracing. Brain Res. 177, 170–175.

Blackstad, T.W., Fuxe, K., Hökfelt, T., 1967. Noradrenaline nerve terminals in the hippocampal region of the rat and the guinea pig. Z. Zellforsch. Mikrosk. Anat. 78, 463–473.

Blessing, W.W., Chalmers, J.P., 1979. Direct projection of catecholamine (presumably dopamine)-containing neurons from hypothalamus to spinal cord. Neurosci. Lett. 11, 35–40.

Bockaert, J., Claeysen, S., Compan, V., Dumuis, A., 2008. 5-HT(4) receptors: history, molecular pharmacology and brain functions. Neuropharmacology 55, 922–931.

Boess, F.G., Monsma, F.J., Carolo, C., et al., 1997. Functional and radioligand binding characterization of rat 5-HT 6 receptors stably expressed in HEK293 cells. Neuropharmacology 36 (4–5), 713–720. https://doi.org/10.1016/S0028-3908(97)00019-1.

Bogerts, B., 1981. A brainstem atlas of catecholaminergic neurons in man, using melanin as a natural marker. J. Comp. Neurol. 197, 63–80.

Bogerts, B., Hantsch, J., Herzer, M., 1983. A morphometric study of the dopamine-containing cell groups in the mesencephalon of normals, Parkinson patients, and schizophrenics. Biol. Psychiatry 18, 951–969.

Bolam, J.P., Clarke, D.J., Smith, A.D., Somogyi, P., 1983. A type of aspiny neuron in the rat neostriatum accumulates [3H]gamma-aminobutyric acid: combination of Golgi-staining, autoradiography, and electron microscopy. J. Comp. Neurol. 213, 121–134.

Bombardi, C., Grandis, A., Pivac, N., et al., 2021. Serotonin modulation of hippocampal functions: from anatomy to neurotherapeutics. Prog. Brain Res. 83–158. https://doi.org/10.1016/bs.pbr.2021.01.031.

Bony, G., Szczurkowska, J., Tamagno, I., Shelly, M., Contestabile, A., Cancedda, L., 2013. Non-hyperpolarizing GABAB receptor activation regulates neuronal migration and neurite growth and specification by cAMP/LKB1. Nat. Commun. 4, 1800.

Borroto-Escuela, D.O., Ambrogini, P., Chruścicka, B., et al., 2021. The role of central serotonin neurons and 5-HT heteroreceptor complexes in the pathophysiology of depression: a historical perspective and future prospects. Int. J. Mol. Sci. 22 (4), 1927. https://doi.org/10.3390/ijms22041927.

Braitenberg, V., Schüz, A., 1998. Cortex: statistics and geometry of neuronal connectivity. Springer, Berlin. https://doi.org/10.1007/978-3-662-03733-1N.B.

Brownstein, M.J., Mroz, E.A., Tappaz, M.L., Leeman, S.E., 1977. On the origin of substance P and glutamic acid decarboxylase (GAD) in the substantia nigra. Brain Res. 135, 315–323.

Brustein, E., Cote, S., Ghislain, J., Drapeau, P., 2013. Spontaneous glycine-induced calcium transients in spinal cord progenitors promote neurogenesis. Dev. Neurobiol. 73, 168–175.

Butcher, L.L., Woolf, N.J., 1982. Monoaminergic-cholinergic relationships and the chemical communication matrix of the substantia nigra and neostriatum. Brain Res. Bull. 9, 475–492.

Cancedda, L., Fiumelli, H., Chen, K., Poo, M.M., 2007. Excitatory GABA action is essential for morphological maturation of cortical neurons in vivo. J. Neurosci. 27, 5224–5235.

Carr, G.V., Schechter, L.E., Lucki, I., 2011. Antidepressant and anxiolytic effects of selective 5-HT6 receptor agonists in rats. Psychopharmacology (Berl) 213, 499–507.

Carter, C.J., 1982. Topographical distribution of possible glutamatergic pathways from the frontal cortex to the striatum and substantia nigra in rats. Neuropharmacology 21, 379–383.

Carter, D.A., Fibiger, H.C., 1977. Ascending projections of presumed dopamine-containing neurons in the ventral tegmentum of the rat as demonstrated by horseradish peroxidase. Neuroscience 2, 569–576.

Celada, P., Puig, M., Amargós-Bosch, M., Adell, A., Artigas, F., 2004. The therapeutic role of 5-HT1A and 5-HT2A receptors in depression. J. Psychiatry Neurosci. 29 (4), 252–265.

Celada, P., Puig, M.V., Artigas, F., 2013. Serotonin modulation of cortical neurons and networks. Front. Integr. Neurosci. 7. https://doi.org/10.3389/fnint.2013.00025.

Chen, N.H., Reith, M.E., Quick, M.W., 2004. Synaptic uptake and beyond: the sodium- and chloride-dependent neurotransmitter transporter family SLC6. Pflugers Arch. 447, 519–531.

Chevalier, G., Thierry, A.M., Shibazaki, T., Feger, J., 1981. Evidence for a GABAergic inhibitory nigrotectal pathway in the rat. Neurosci. Lett. 21, 67–70.

Childs, J.A., Gale, K., 1983. Neurochemical evidence for a nigrotegmental GAbAergic projection. Brain Res. 258, 109–114.

Christodoulou, J., Kure, S., Hayasaka, K., Clarke, J.T., 1993. Atypical nonketotic hyperglycinemia confirmed by assay of the glycine cleavage system in lymphoblasts. J. Pediatr. 123 (1), 100–102. https://doi.org/10.1016/s0022-3476(05)81550-1.

Clemow, D.B., Johnson, K.W., Hochstetler, H.M., Ossipov, M.H., Hake, A.M., Blumenfeld, A.M., 2020. Lasmiditan mechanism of action—review of a selective 5-HT1F agonist. J. Headache Pain 21 (1), 71. https://doi.org/10.1186/s10194-020-01132-3.

Collins, A.L., Ma, D., Whitehead, P.L., Martin, E.R., Wright, H.H., Abramson, R.K., Hussman, J.P., Haines, J.L., Cuccaro, M.L., Gilbert, J.R., Pericak-Vance, M.A., 2006. Investigation of autism and GABA receptor subunit genes in multiple ethnic groups. Neurogenetics 7, 167–174.

Collins, G.G., 1979. Evidence of a neurotransmitter role for aspartate and gamma-aminobutyric acid in the rat olfactory cortex. J. Physiol. 291, 51–60.

Collins, G.G., Anson, J., Probett, G.A., 1981. Patterns of endogenous amino acid release from slices of rat and guinea-pig olfactory cortex. Brain Res. 204, 103–120.

Commissiong, J.W., 1981. Evidence that the noradrenergic coerulospinal projection decussates at the spinal level. Brain Res. 212, 145–151.

Commissiong, J.W., Hellström, S.O., Neff, N.H., 1978. A new projection from locus coeruleus to the spinal ventral columns: histochemical and biochemical evidence. Brain Res. 148 (1), 207–213. https://doi.org/10.1016/0006-8993(78)90391-8.

Contestabile, A., Fonnum, F., 1983. Cholinergic and GABAergic forebrain projections to the habenula and nucleus interpeduncularis: surgical and kainic acid lesions. Brain Res. 275, 287–297.

Cools, A.R., 1980. Role of the neostriatal dopaminergic activity in sequencing and selecting behavioural strategies: facilitation of processes involved in selecting the best strategy in a stressful situation. Behav. Brain Res. 1, 361–378.

Cools, A.R., Coolen, J.M., Smit, J.C., Ellenbroek, B.A., 1984. The striato-nigro-collicular pathway and explosive running behaviour: functional interaction between neostriatal dopamine and collicular GABA. Eur. J. Pharmacol. 100, 71–77.

Cools, A.R., Van Den Bercken, J.H., Horstink, M.W., Van Spaendonck, K.P., Berger, H.J., 1984. Cognitive and motor shifting aptitude disorder in Parkinson's disease. J. Neurol. Neurosurg. Psychiatry 47, 443–453.

Cote, S., Drapeau, P., 2012. Regulation of spinal interneuron differentiation by the paracrine action of glycine. Dev. Neurobiol. 72, 208–214.

Creese, I., Burt, D.R., Snyder, S.H., 1976. Dopamine receptor binding predicts clinical and pharmacological potencies of antischizophrenic drugs. Science 192, 481–483.

Cullheim, S., Kellerth, J.O., 1981. Two kinds of recurrent inhibition of cat spinal alpha-motoneurones as differentiated pharmacologically. J. Physiol. 312, 209–224.

Curtis, D.R., Hosli, L., Johnston, G.A., 1968. A pharmacological study of the depression of spinal neurones by glycine and related amino acids. Exp. Brain Res. 6, 1–18.

Curtis, D.R., Johnston, G.A., 1974. Amino acid transmitters in the mammalian central nervous system. Ergeb. Physiol. 69, 97–188.

Cuzon Carlson, V.C., Yeh, H.H., 2011. GABAA receptor subunit profiles of tangentially migrating neurons derived from the medial ganglionic eminence. Cereb. Cortex 21, 1792–1802.

Dahlstroem, A., Fuxe, K., 1964. A method for the demonstration of adrenergic nerve fibres in peripheral nerves. Z. Zellforsch. Mikrosk. Anat. 62, 602–607.

Dai, J., Kim, H., You, Z., et al., 2022. Role of 5-HT1A-mediated upregulation of brain indoleamine 2,3 dioxygenase 1 in the reduced antidepressant and antihyperalgesic effects of fluoxetine during maintenance treatment. Front. Pharmacol. 13. https://doi.org/10.3389/fphar.2022.1084108.

Dale, H.H., Feldberg, W., Vogt, M., 1936. Release of acetylcholine at voluntary motor nerve endings. J. Physiol. 86, 353–380.

Dampney, R.A., Moon, E.A., 1980. Role of ventrolateral medulla in vasomotor response to cerebral ischemia. Am. J. Physiol. 239, H349–H358.

Davies, P., 1979. Neurotransmitter-related enzymes in senile dementia of the Alzheimer type. Brain Res. 171, 319–327.

Davis, A.M., Henion, T.R., Tobet, S.A., 2002. Gamma-aminobutyric acidB receptors and the development of the ventromedial nucleus of the hypothalamus. J. Comp. Neurol. 449, 270–280.

Daws, L.C., 2009. Unfaithful neurotransmitter transporters: focus on serotonin uptake and implications for antidepressant efficacy. Pharmacol. Ther. 121, 89–99.

Defeudis, F.V., 1984. Gamma-Aminobutyric acid-ergic analgesia: implications for gamma-aminobutyric acid-ergic therapy for drug addictions. Drug Alcohol Depend. 14, 101–111.

Deistung, A., Schäfer, A., Schweser, F., Biedermann, U., Turner, R., Reichenbach, J.R., 2013. Toward in vivo histology: a comparison of quantitative susceptibility mapping (QSM) with magnitude-, phase-, and R2*-imaging at ultra-high magnetic field strength. Neuroimage 65, 299–314.

Denter, D.G., Heck, N., Riedemann, T., White, R., Kilb, W., Luhmann, H.J., 2010. GABAC receptors are functionally expressed in the intermediate zone and regulate radial migration in the embryonic mouse neocortex. Neuroscience 167, 124–134.

Di Chiara, G., Porceddu, M.L., Morelli, M., Mulas, M.L., Gessa, G.L., 1979. Evidence for a GABAergic projection from the substantia nigra to the ventromedial thalamus and to the superior colliculus of the rat. Brain Res. 176, 273–284.

Divac, I., 1975. Magnocellular nuclei of the basal forebrain project to neocortex, brain stem, and olfactory bulb. Review of some functional correlates. Brain Res. 93, 385–398.

Divac, I., Fonnum, F., Storm-Mathisen, J., 1977. High affinity uptake of glutamate in terminals of corticostriatal axons. Nature 266, 377–378.

Dresbach, T., Nawrotzki, R., Kremer, T., Schumacher, S., Quinones, D., Kluska, M., Kuhse, J., Kirsch, J., 2008. Molecular architecture of glycinergic synapses. Histochem. Cell Biol. 130, 617–633.

Dussor, G., 2014. Serotonin, 5HT1 agonists, and migraine. Curr. Opin. Support. Palliat. Care 8 (2), 137–142. https://doi.org/10.1097/SPC.0000000000000044.

Eisenhofer, G., Kopin, I.J., Goldstein, D.S., 2004. Catecholamine metabolism: a contemporary view with implications for physiology and medicine. Pharmacol. Rev. 56, 331–349.

Emson, P.C., Koob, G.F., 1978. The origin and distribution of dopamine-containing afferents to the rat frontal cortex. Brain Res. 142, 249–267.

Esquivel-Franco, D.C., de Boer, S.F., Waldinger, M., Olivier, B., Olivier, J.D.A., 2020. Pharmacological studies on the role of 5-HT1A receptors in male sexual behavior of wildtype and serotonin transporter knockout rats. Front. Behav. Neurosci. 14. https://doi.org/10.3389/fnbeh.2020.00040.

Fagg, G.E., Foster, A.C., 1983. Amino acid neurotransmitters and their pathways in the mammalian central nervous system. Neuroscience 9, 701–719.

Fallon, J.H., Koziell, D.A., Moore, R.Y., 1978. Catecholamine innervation of the basal forebrain. II. Amygdala, suprarhinal cortex and entorhinal cortex. J. Comp. Neurol. 180, 509–532.

Fallon, J.H., Moore, R.Y., 1978. Catecholamine innervation of the basal forebrain. III. Olfactory bulb, anterior olfactory nuclei, olfactory tubercle and piriform cortex. J. Comp. Neurol. 180, 533–544.

Fallon, J.H., Moore, R.Y., 1978. Catecholamine innervation of the basal forebrain. IV. Topography of the dopamine projection to the basal forebrain and neostriatum. J. Comp. Neurol. 180, 545–580.

Fatemi, S.H., Reutiman, T.J., Folsom, T.D., Rooney, R.J., Patel, D.H., Thuras, P.D., 2010. mRNA and protein levels for GABAAalpha4, alpha5, beta1 and GABABR1 receptors are altered in brains from subjects with autism. J. Autism Dev. Disord. 40, 743–750.

Fatemi, S.H., Reutiman, T.J., Folsom, T.D., Rustan, O.G., Rooney, R.J., Thuras, P.D., 2014. Downregulation of GABAA receptor protein subunits alpha6, beta2, delta, epsilon, gamma2, theta, and rho2 in superior frontal cortex of subjects with autism. J. Autism Dev. Disord. 44, 1833–1845.

Fatemi, S.H., Reutiman, T.J., Folsom, T.D., Thuras, P.D., 2009. GABA(A) receptor downregulation in brains of subjects with autism. J. Autism Dev. Disord. 39, 223–230.

Felten, D.L., Laties, A.M., Carpenter, M.B., 1974. Monoamine-containing cell bodies in the squirrel monkey brain. Am. J. Anat. 139, 153–165.

Felten, D.L., Sladek Jr., J.R., 1983. Monoamine distribution in primate brain V. Monoaminergic nuclei: anatomy, pathways and local organization. Brain Res. Bull. 10, 171–284.

Fibiger, H.C., 1982. The organization and some projections of cholinergic neurons of the mammalian forebrain. Brain Res. 257, 327–388.

Fijał, K., Popik, P., Nikiforuk, A., 2014. Co-administration of 5-HT6 receptor antagonists with clozapine, risperidone, and a 5-HT2A receptor antagonist: effects on prepulse inhibition in rats. Psychopharmacology (Berl) 231 (1), 269–281. https://doi.org/10.1007/s00213-013-3234-2.

Fisher, S.K., Wonnacott, S., 2012. Acetylcholine. In: Basic Neurochemistry. Elsevier, pp. 258–282. https://doi.org/10.1016/B978-0-12-374947-5.00013-4.

Fone, K.C., Beckett, S.R., Topham, I.A., Swettenham, J., Ball, M., Maddocks, L., 2002. Long-term changes in social interaction and reward following repeated MDMA administration to adolescent rats without accompanying serotonergic neurotoxicity. Psychopharmacology (Berl) 159, 437–444.

Fonnum, F., 1984. Glutamate: a neurotransmitter in mammalian brain. J. Neurochem. 42, 1–11.

Fonnum, F., Gottesfeld, Z., Grofova, I., 1978. Distribution of glutamate decarboxylase, choline acetyl-transferase and aromatic amino acid decarboxylase in the basal ganglia of normal and operated rats. Evidence for striatopallidal, striatoentopeduncular and striatonigral GABAergic fibres. Brain Res. 143, 125–138.

Fonnum, F., Karlsen, R.L., Malthe-Sorenssen, D., Skrede, K.K., Walaas, I., 1979. Localization of neurotransmitters, particularly glutamate, in hippocampus, septum, nucleus accumbens and superior colliculus. Prog. Brain Res. 51, 167–191.

Fonnum, F., Storm-Mathisen, J., Divac, I., 1981. Biochemical evidence for glutamate as neurotransmitter in corticostriatal and corticothalamic fibres in rat brain. Neuroscience 6, 863–873.

Fonnum, F., Storm-Mathisen, J., Walberg, F., 1970. Glutamate decarboxylase in inhibitory neurons. A study of the enzyme in Purkinje cell axons and boutons in the cat. Brain Res. 20, 259–275.

Fonnum, F., Walaas, I., 1978. The effect of intrahippocampal kainic acid injections and surgical lesions on neurotransmitters in hippocampus and septum. J. Neurochem. 31, 1173–1181.

Frazer, A., Gerhardt, G.A., Daws, L.C., 1999. New views of biogenic amine transporter function: implications for neuropsychopharmacology. Int. J. Neuropsychopharmacol. 2, 305–320.

Frazer, A., Hensler, J.G., 1999. Serotonin involvement in physiological function and behavior. In: Basic Neurochemistry: Molecular, Cellular and Medical Aspects, sixth ed. Lippincott-Raven.

Fremeau Jr., R.T., Troyer, M.D., Pahner, I., Nygaard, G.O., Tran, C.H., Reimer, R.J., Bellocchio, E.E., Fortin, D., Storm-Mathisen, J., Edwards, R.H., 2001. The expression of vesicular glutamate transporters defines two classes of excitatory synapse. Neuron 31, 247–260.

Furuta, A., Rothstein, J.D., Martin, L.J., 1997. Glutamate transporter protein subtypes are expressed differentially during rat CNS development. J. Neurosci. 17, 8363–8375.

Gaiarsa, J.L., Porcher, C., 2013. Emerging neurotrophic role of GABAB receptors in neuronal circuit development. Front. Cell. Neurosci. 7, 206.

Garver, D.L., Sladek Jr., J.R., 1975. Monoamine distribution in primate brain. I Catecholamine-containing perikarya in the brain stem of Macaca speciosa. J. Comp. Neurol. 159, 289–304.

Gascon, E., Dayer, A.G., Sauvain, M.O., Potter, G., Jenny, B., De Roo, M., Zgraggen, E., Demaurex, N., Muller, D., Kiss, J.Z., 2006. GABA regulates dendritic growth by stabilizing lamellipodia in newly generated interneurons of the olfactory bulb. J. Neurosci. 26, 12956–12966.

Gasnier, B., 2000. The loading of neurotransmitters into synaptic vesicles. Biochimie 82, 327–337.

Gassmann, M., Bettler, B., 2012. Regulation of neuronal GABA(B) receptor functions by subunit composition. Nat. Rev. Neurosci. 13, 380–394.

Gatter, K., Powell, T., 1977. The projection of the locus coeruleus upon the neocortex in the macaque monkey. Neuroscience 2, 441–445.

Gerard, C., Martres, M.P., Lefevre, K., Miquel, M.C., Verge, D., Lanfumey, L., Doucet, E., Hamon, M., El Mestikawy, S., 1997. Immuno-localization of serotonin 5-HT6 receptor-like material in the rat central nervous system. Brain Res. 746, 207–219.

Ghorbel, M.T., Becker, K.G., Henley, J.M., 2005. Profile of changes in gene expression in cultured hippocampal neurones evoked by the GABAB receptor agonist baclofen. Physiol. Genomics 22, 93–98.

Ghosh, A., Ginty, D.D., Bading, H., Greenberg, M.E., 1994. Calcium regulation of gene expression in neuronal cells. J. Neurobiol. 25, 294–303.

Ginty, D.D., 1997. Calcium regulation of gene expression: isn't that spatial? Neuron 18, 183–186.

Glikmann-Johnston, Y., Saling, M.M., Reutens, D.C., Stout, J.C., 2015. Hippocampal 5-HT1A receptor and spatial learning and memory. Front. Pharmacol. 6. https://doi.org/10.3389/fphar.2015.00289.

Goldstein, M., Lew, J.Y., Hata, F., Lieberman, A., 1978. Binding interactions of ergot alkaloids with monoaminergic receptors in the brain. Gerontology 24 (Suppl. 1), 76–85.

Goldstein, M., Pearson, J., Sauter, A., Ueta, K., Asano, T., Engel, J., Passeltiner, P., Hökfelt, T., Fuxe, K., 1980. Distribution and characterization of PNMT and epinephrine in the CNS. In: Central Adrenaline Neurons. Elsevier.

Goodchild, A., Moon, E., Dampney, R., Howe, P., 1984. Evidence that adrenaline neurons in the rostral ventrolateral medulla have a vasopressor function. Neurosci. Lett. 45, 267–272.

Gottesfeld, Z., Jacobowitz, D.M., 1979. Cholinergic projections from the septal-diagonal band area to the habenular nuclei. Brain Res. 176, 391–394.

Grace, A.A., 2010. Dopamine system dysregulation by the ventral subiculum as the common pathophysiological basis for schizophrenia psychosis, psychostimulant abuse, and stress. Neurotox. Res. 18 (3–4), 367–376. https://doi.org/10.1007/s12640-010-9154-6.

Gu, H., Wall, S.C., Rudnick, G., 1994. Stable expression of biogenic amine transporters reveals differences in inhibitor sensitivity, kinetics, and ion dependence. J. Biol. Chem. 269, 7124–7130.

Gunne, L.M., 1962. Relative adrenaline content in brain tissue. Acta Physiol. Scand. 56, 324–333.

Guo, Z.L., Moazzami, A.R., Longhurst, J.C., 2005. Stimulation of cardiac sympathetic afferents activates glutamatergic neurons in the parabrachial nucleus: relation to neurons containing nNOS. Brain Res. 1053, 97–107.

Guyenet, P.G., 1980. The coeruleospinal noradrenergic neurons: anatomical and electrophysiological studies in the rat. Brain Res. 189, 121–133.

Guyenet, P.G., Cranes, J.K., 1981. Non-dopaminergic nigrostriatal pathway. Brain Res. 213, 291–305.

Halasz, N., Ljungdahl, A., Hökfelt, T., 1979. Transmitter histochemistry of the rat olfactory bulb III. Autoradiographic localization of [3H] GABA. Brain Res. 167, 221–240.

Hamon, M., Doucet, E., Lefevre, K., Miquel, M.C., Lanfumey, L., Insausti, R., Frechilla, D., Del Rio, J., Verge, D., 1999. Antibodies and antisense oligonucleotide for probing the distribution and putative functions of central 5-HT6 receptors. Neuropsychopharmacology 21, 68S–76S.

Hancock, M.B., Fougerousse, C.L., 1976. Spinal projections from the nucleus locus coeruleus and nucleus subcoeruleus in the cat and monkey as demonstrated by the retrograde transport of horseradish peroxidase. Brain Res. Bull. 1, 229–234.

Hannon, J., Hoyer, D., 2008. Molecular biology of 5-HT receptors. Behav. Brain Res. 195, 198–213.

Harvey, J., Scholfield, C., Graham Jr., L., Aprison, M., 1975. Putative transmitters in denervated olfactory cortex. J. Neurochem. 24, 445–449.

Hattori, T., Mcgeer, P.L., Fibiger, H.C., Mcgeer, E.G., 1973. On the source of GABA-containing terminals in the substantia nigra. Electron microscopic autoradiographic and biochemical studies. Brain Res. 54, 103–114.

Haus, U., Spath, M., Farber, L., 2004. Spectrum of use and tolerability of 5-HT3 receptor antagonists. Scand. J. Rheumatol. Suppl. 119, 12–18.

Haydar, T.F., Wang, F., Schwartz, M.L., Rakic, P., 2000. Differential modulation of proliferation in the neocortical ventricular and subventricular zones. J. Neurosci. 20, 5764–5774.

Heal, D., Gosden, J., Smith, S., 2011. The 5-HT6 receptor as a target for developing novel antiobesity drugs. Int. Rev. Neurobiol. 96, 73–109.

Heck, N., Kilb, W., Reiprich, P., Kubota, H., Furukawa, T., Fukuda, A., Luhmann, H.J., 2007. GABA-A receptors regulate neocortical neuronal migration in vitro and in vivo. Cereb. Cortex 17, 138–148.

Hedlund, P.B., 2009. The 5-HT7 receptor and disorders of the nervous system: an overview. Psychopharmacology (Berl) 206 (3), 345–354. https://doi.org/10.1007/s00213-009-1626-0.

Hedreen, J.C., Chalmers, J.P., 1972. Neuronal degeneration in rat brain induced by 6-hydroxydopamine; a histological and biochemical study. Brain Res. 47, 1–36.

Heimer, L., Wilson, R.D., 1975. The subcortical projections of allocortex: similarities in the neuronal associations of the hippocampus, the

piriform cortex and the neocortex. In: Santini, M. (Ed.), Golgi Centennial Symposium Proceedings. Raven, New York, pp. 177–193.
Hendry, S.H., Jones, E.G., 1981. Sizes and distributions of intrinsic neurons incorporating tritiated GABA in monkey sensory-motor cortex. J. Neurosci. 1, 390–408.
Hepler, P.K., 1994. The role of calcium in cell division. Cell Calcium 16, 322–330.
Hökfelt, T., Fuxe, K., Goldstein, M., 1974. Immunohistochemical evidence for the existence of adrenaline neurous in the rat brain. Brain Res. 66, 275–277.
Hökfelt, T., Fuxe, K., Johansson, O., Ljungdahl, A., 1974. Pharmacohistochemical evidence of the existence of dopamine nerve terminals in the limbic cortex. Eur. J. Pharmacol. 25, 108–112.
Hökfelt, T., Goldstein, M., Fuxe, K., Johansson, O., Verhofstad, A., Steinbusch, H., Penke, B., Vargas, J., 1980. Histochemical identification of adrenaline containing cells with special reference to neurons. In: Central Adrenaline Neurons. Elsevier.
Hökfelt, T., Rehfeld, J.F., Skirboll, L., Ivemark, B., Goldstein, M., Markey, K., 1980. Evidence for coexistence of dopamine and CCK in meso-limbic neurones. Nature 285, 476–478.
Hökfelt, T., Johansson, O., Fuxe, K., Elde, R., Goldstein, M., Park, D., Efendic, S., Luft, R., Fraser, H., Jeffcoate, S., 1977. Hypothalamic dopamine neurons and hypothalamic peptides. Adv. Biochem. Psychopharmacol. 16, 99–108.
Hökfelt, T., Phillipson, O., Goldstein, M., 1979. Evidence for a dopaminergic pathway in the rat descending from the A11 cell group to the spinal cord. Acta Physiol. Scand. 107, 393–395.
Homanics, G.E., Delorey, T.M., Firestone, L.L., Quinlan, J.J., Handforth, A., Harrison, N.L., Krasowski, M.D., Rick, C.E., Korpi, E.R., Makela, R., Brilliant, M.H., Hagiwara, N., Ferguson, C., Snyder, K., Olsen, R.W., 1997. Mice devoid of gamma-aminobutyrate type A receptor beta3 subunit have epilepsy, cleft palate, and hypersensitive behavior. Proc. Natl. Acad. Sci. U. S. A. 94, 4143–4148.
Hoover-Fong, J.E., Shah, S., Van Hove, J.L., Applegarth, D., Toone, J., Hamosh, A., 2004. Natural history of nonketotic hyperglycinemia in 65 patients. Neurology 63, 1847–1853.
Horn, Z., Ringstedt, T., Blaesse, P., Kaila, K., Herlenius, E., 2010. Premature expression of KCC2 in embryonic mice perturbs neural development by an ion transport-independent mechanism. Eur. J. Neurosci. 31, 2142–2155.
Hornykiewicz, O., 1978. Psychopharmacological implications of dopamine and dopamine antagonists: a critical evaluation of current evidence. Neuroscience 3, 773–783.
Houser, C.R., Vaughn, J.E., Barber, R.P., Roberts, E., 1980. GABA neurons are the major cell type of the nucleus reticularis thalami. Brain Res. 200, 341–354.
Howe, P.R., Costa, M., Furness, J.B., Chalmers, J.P., 1980. Simultaneous demonstration of phenylethanolamine N-methyltransferase immunofluorescent and catecholamine fluorescent nerve cell bodies in the rat medulla oblongata. Neuroscience 5, 2229–2238.
Hoyer, D., 2007. 5-HT-1D receptor. In: XPharm: The Comprehensive Pharmacology Reference. Elsevier, pp. 1–11. https://doi.org/10.1016/B978-008055232-3.60126-6.
Hoyer, D., 2019. Serotonin receptors nomenclature. In: The Serotonin System. Elsevier, pp. 63–93. https://doi.org/10.1016/B978-0-12-813323-1.00004-9.
Hoyer, D., Clarke, D.E., Fozard, J.R., Hartig, P.R., Martin, G.R., Mylecharane, E.J., Saxena, P.R., Humphrey, P.P., 1994. International Union of Pharmacology classification of receptors for 5-hydroxytryptamine (Serotonin). Pharmacol. Rev. 46, 157–203.
Huang, G.J., Herbert, J., 2005. The role of 5-HT1A receptors in the proliferation and survival of progenitor cells in the dentate gyrus of the adult hippocampus and their regulation by corticoids. Neuroscience 135 (3), 803–813. https://doi.org/10.1016/j.neuroscience.2005.05.056.
Huang, K.W., Ochandarena, N.E., Philson, A.C., et al., 2019. Molecular and anatomical organization of the dorsal raphe nucleus. Elife 8. https://doi.org/10.7554/eLife.46464.
Hubbard, J.E., Di Carlo, V., 1973. Fluorescence histochemistry of monoamine-containing cell bodies in the brain stem of the squirrel monkey (Saimiri sciureus). I. The locus caeruleus. J. Comp. Neurol. 147, 553–566.
Hubbard, J.E., Di Carlo, V., 1974. Fluorescence histochemistry of monoamine-containing cell bodies in the brain stem of the squirrel monkey (Saimiri sciureus). II. Catecholamine-containing groups. J. Comp. Neurol. 153, 369–384.
Inada, H., Watanabe, M., Uchida, T., Ishibashi, H., Wake, H., Nemoto, T., Yanagawa, Y., Fukuda, A., Nabekura, J., 2011. GABA regulates the multidirectional tangential migration of GABAergic interneurons in living neonatal mice. PLoS One 6, e27048.
Ishikawa, M., Tanaka, C., 1977. Morphological organization of catecholamine terminals in the diencephalon of the rhesus monkey. Brain Res. 119, 43–55.
Jacobowitz, D.M., 1975. Fluorescence microscopic mapping of CNS norepinephrine systems in the rat forebrain. In: Anatomical Neuroendocrinology. Karger Publishers.
Jacobowitz, D.M., Maclean, P.D., 1978. A brainstem atlas of catecholaminergic neurons and serotonergic perikarya in a pygmy primate (Cebuella pygmaea). J. Comp. Neurol. 177, 397–416.
Jacobs, B.L., Fornal, C.A., 1993. 5-HT and motor control: a hypothesis. Trends Neurosci. 16, 346–352.
James, O.T., Livesey, M.R., Qiu, J., Dando, O., Bilican, B., Haghi, G., Rajan, R., Burr, K., Hardingham, G.E., Chandran, S., Kind, P.C., Wyllie, D.J., 2014. Ionotropic GABA and glycine receptor subunit composition in human pluripotent stem cell-derived excitatory cortical neurones. J. Physiol. 592, 4353–4363.
Jaspers, R., Schwarz, M., Sontag, K.H., Cools, A.R., 1984. Caudate nucleus and programming behaviour in cats: role of dopamine in switching motor patterns. Behav. Brain Res. 14, 17–28.
Jessell, T.M., Emson, P.C., Paxinos, G., Cuello, A.C., 1978. Topographic projections of substance P and GABA pathways in the striato- and pallido-nigral system: a biochemical and immunohistochemical study. Brain Res. 152, 487–498.
Jones, B.E., Friedman, L., 1983. Atlas of catecholamine perikarya, varicosities and pathways in the brainstem of the cat. J. Comp. Neurol. 215, 382–396.
Jones, B.E., Moore, R.Y., 1977. Ascending projections of the locus coeruleus in the rat. II. Autoradiographic study. Brain Res. 127, 25–53.
Jones, E.G., Burton, H., Saper, C.B., Swanson, L.W., 1976. Midbrain, diencephalic and cortical relationships of the basal nucleus of Meynert and associated structures in primates. J. Comp. Neurol. 167, 385–419.
Jones, S.R., Gainetdinov, R.R., Jaber, M., Giros, B., Wightman, R.M., Caron, M.G., 1998. Profound neuronal plasticity in response to inactivation of the dopamine transporter. Proc. Natl. Acad. Sci. U. S. A. 95, 4029–4034.
Jurd, R., Arras, M., Lambert, S., Drexler, B., Siegwart, R., Crestani, F., Zaugg, M., Vogt, K.E., Ledermann, B., Antkowiak, B., Rudolph, U., 2003. General anesthetic actions in vivo strongly attenuated by a point mutation in the GABA(A) receptor beta3 subunit. FASEB J. 17, 250–252.
Kaila, K., Price, T.J., Payne, J.A., Puskarjov, M., Voipio, J., 2014. Cation-chloride cotransporters in neuronal development, plasticity and disease. Nat. Rev. Neurosci. 15, 637–654.
Karila, D., Freret, T., Bouet, V., Boulouard, M., Dallemagne, P., Rochais, C., 2015. Therapeutic potential of 5-HT6 receptor agonists. J. Med. Chem. 58, 7901–7912.
Karlin, A., 2002. Emerging structure of the nicotinic acetylcholine receptors. Nat. Rev. Neurosci. 3, 102–114.
Karlsen, R.L., Fonnum, F., 1978. Evidence for glutamate as a neurotransmitter in the corticofugal fibres to the dorsal lateral geniculate

body and the superior colliculus in rats. Brain Res. 151 (3), 457–467. https://doi.org/10.1016/0006-8993(78)91079-X.

Karoum, F., Commissiong, J.W., Neff, N.H., Wyatt, R.J., 1980. Biochemical evidence for uncrossed and crossed locus coeruleus projections to the spinal cord. Brain Res. 196, 237–241.

Kash, S.F., Johnson, R.S., Tecott, L.H., Noebels, J.L., Mayfield, R.D., Hanahan, D., Baekkeskov, S., 1997. Epilepsy in mice deficient in the 65-kDa isoform of glutamic acid decarboxylase. Proc. Natl. Acad. Sci. U. S. A. 94, 14060–14065.

Kawaguchi, S., Ono, T., 1973. Bicuculline and picrotoxin sensitive inhibition in interpositus neurones of cat. Brain Res. 58, 260–265.

Khoury, R., Grysman, N., Gold, J., Patel, K., Grossberg, G.T., 2018. The role of 5 HT6-receptor antagonists in Alzheimer's disease: an update. Expet Opin. Invest. Drugs 27 (6), 523–533. https://doi.org/10.1080/13543784.2018.1483334.

Kilb, W., Kirischuk, S., Luhmann, H.J., 2013. Role of tonic GABAergic currents during pre- and early postnatal rodent development. Front. Neural Circuits 7, 139.

Kimura, H., Mcgeer, P.L., Peng, J.H., Mcgeer, E.G., 1981. The central cholinergic system studied by choline acetyltransferase immunohistochemistry in the cat. J. Comp. Neurol. 200, 151–201.

Kita, H., Oomura, Y., 1982. Evidence for a glycinergic cortico-lateral hypothalamic inhibitory pathway in the rat. Brain Res. 235, 131–136.

Kita, H., Oomura, Y., 1982. An HRP study of the afferent connections to rat lateral hypothalamic region. Brain Res. Bull. 8, 63–71.

Kleckner, N.W., Dingledine, R., 1988. Requirement for glycine in activation of NMDA-receptors expressed in Xenopus oocytes. Science 241, 835–837.

Klein, M., Teitler, M., 2012. Distribution of 5-ht1E receptors in the mammalian brain and cerebral vasculature: an immunohistochemical and pharmacological study. Br. J. Pharmacol. 166 (4), 1290–1302. https://doi.org/10.1111/j.1476-5381.2012.01868.x.

Klempin, 2010. Oppositional effects of serotonin receptors 5-HT1a, 2, and 2c in the regulation of adult hippocampal neurogenesis. Front. Mol. Neurosci. https://doi.org/10.3389/fnmol.2010.00014.

Koelliker, A., 1896. Handbuch der Gewebelehre des Menschen, sixth ed., Vol. 2: Nervensystem des Menschen und der Thiere Retrieved from: http://archive.org/details/handbuchdergeweb02kl.

Kohler, C., Chan-Palay, V., 1983. Distribution of gamma aminobutyric acid containing neurons and terminals in the septal area. An immunohistochemical study using antibodies to glutamic acid decarboxylase in the rat brain. Anat. Embryol. 167, 53–65.

Kohler, C., Chan-Palay, V., Wu, J.Y., 1984. Septal neurons containing glutamic acid decarboxylase immunoreactivity project to the hippocampal region in the rat brain. Anat. Embryol. 169, 41–44.

Kołaczkowski, M., Marcinkowska, M., Bucki, A., et al., 2014. Novel arylsulfonamide derivatives with 5-HT$_6$/5-HT$_7$ receptor antagonism targeting behavioral and psychological symptoms of dementia. J. Med. Chem. 57 (11), 4543–4557. https://doi.org/10.1021/jm401895u.

Komuro, H., Kumada, T., 2005. Ca2+ transients control CNS neuronal migration. Cell Calcium 37, 387–393.

Komuro, Y., Galas, L., Lebon, A., Raoult, E., Fahrion, J.K., Tilot, A., Kumada, T., Ohno, N., Vaudry, D., Komuro, H., 2015. The role of calcium and cyclic nucleotide signaling in cerebellar granule cell migration under normal and pathological conditions. Dev. Neurobiol. 75, 369–387.

Kromer, L.F., Moore, R.Y., 1976. Cochlear nucleus innervation by central norepinephrine neurons in the rat. Brain Res. 118, 531–537.

Kromer, L.F., Moore, R.Y., 1980. Norepinephrine innervation of the cochlear nuclei by locus coeruleus neurons in the rat. Anat. Embryol. 158, 227–244.

Kuhse, J., Kuryatov, A., Maulet, Y., Malosio, M.L., Schmieden, V., Betz, H., 1991. Alternative splicing generates two isoforms of the alpha 2 subunit of the inhibitory glycine receptor. FEBS Lett. 283, 73–77.

Kumada, T., Komuro, H., 2004. Completion of neuronal migration regulated by loss of Ca(2+) transients. Proc. Natl. Acad. Sci. U. S. A. 101, 8479–8484.

Kuriyama, K., Haber, B., Sisken, B., Roberts, E., 1966. The gamma-aminobutyric acid system in rabbit cerebellum. Proc. Natl. Acad. Sci. U. S. A. 55, 846–852.

L'hirondel, M., Cheramy, A., Godeheu, G., Artaud, F., Saiardi, A., Borrelli, E., Glowinski, J., 1998. Lack of autoreceptor-mediated inhibitory control of dopamine release in striatal synaptosomes of D2 receptor-deficient mice. Brain Res. 792, 253–262.

Lamberts, R., Vijayan, E., Graf, M., Mansky, T., Wuttke, W., 1983. Involvement of preoptic-anterior hypothalamic GABA neurons in the regulation of pituitary LH and prolactin release. Exp. Brain Res. 52, 356–362.

Le Poul, E., Laaris, N., Doucet, E., Laporte, A.M., Hamon, M., Lanfumey, L., 1995. Early desensitization of somato-dendritic 5-HT1A autoreceptors in rats treated with fluoxetine or paroxetine. Naunyn-Schmiedeberg's Arch. Pharmacol. 352 (2), 141–148. https://doi.org/10.1007/BF00176767.

Lee, M.D., Clifton, P.G., 2020. Role of the serotonergic system in appetite and ingestion control. In: Handbook of Behavioral Neuroscience. Elsevier, pp. 469–487. https://doi.org/10.1016/B978-0-444-64125-0.00027-X.

Lee, J.M., Tiong, J., Maddox, D.M., Condie, B.G., Wray, S., 2008. Temporal migration of gonadotrophin-releasing hormone-1 neurones is modified in GAD67 knockout mice. J. Neuroendocrinol. 20, 93–103.

Legendre, P., Forstera, B., Juttner, R., Meier, J.C., 2009. Glycine receptors caught between genome and proteome—functional implications of RNA editing and splicing. Front. Mol. Neurosci. 2, 23.

Leichnetz, G.R., Astruc, J., 1977. The course of some prefrontal corticofugals to the pallidum, substantia innominata, and amygdaloid complex in monkeys. Exp. Neurol. 54, 104–109.

Leiser, S.C., Li, Y., Pehrson, A.L., Dale, E., Smagin, G., Sanchez, C., 2015. Serotonergic regulation of prefrontal cortical circuitries involved in cognitive processing: a review of individual 5-HT receptor mechanisms and concerted effects of 5-HT receptors exemplified by the multimodal antidepressant vortioxetine. ACS Chem. Neurosci. 6 (7), 970–986. https://doi.org/10.1021/cn500340j.

Levitt, P., Moore, R.Y., 1978. Noradrenaline neuron innervation of the neocortex in the rat. Brain Res. 139, 219–231.

Levitt, P., Moore, R.Y., 1979. Origin and organization of brainstem catecholamine innervation in the rat. J. Comp. Neurol. 186, 505–528.

Levitt, P., Rakic, P., Goldman-Rakic, P., 1984. Region-specific distribution of catecholamine afferents in primate cerebral cortex: a fluorescence histochemical analysis. J. Comp. Neurol. 227, 23–36.

Lew, J.Y., Matsumoto, Y., Pearson, J., Goldstein, M., Hokfelt, T., Fuxe, K., 1977. Localization and characterization of phenylethanolamine N-methyl transferase in the brain of various mammalian species. Brain Res. 119, 199–210.

Lewis, D.A., Melchitzky, D.S., Sesack, S.R., Whitehead, R.E., Auh, S., Sampson, A., 2001. Dopamine transporter immunoreactivity in monkey cerebral cortex: regional, laminar, and ultrastructural localization. J. Comp. Neurol. 432, 119–136.

Lindvall, O., 1975. Mesencephalic dopaminergic afferents to the lateral septal nucleus of the rat. Brain Res. 87, 89–95.

Lindvall, O., Björklund, A., 1974. The organization of the ascending catecholamine neuron systems in the rat brain as revealed by the glyoxylic acid fluorescence method. Acta Physiol. Scand. Suppl. 412, 1–48.

Lindvall, O., Björklund, A., 1978. Anatomy of the dopaminergic neuron systems in the rat brain. Adv. Biochem. Psychopharmacol. 19, 1–23.

Lindvall, O., Björklund, A., 1979. Dopaminergic innervation of the globus pallidus by collaterals from the nigrostriatal pathway. Brain Res. 172, 169–173.

Lindvall, O., Björklund, A., Divac, I., 1978. Organization of catecholamine neurons projecting to the frontal cortex in the rat. Brain Res. 142, 1–24.

Lindvall, O., Bjorklund, A., Moore, R.Y., Stenevi, U., 1974. Mesencephalic dopamine neurons projecting to neocortex. Brain Res. 81, 325–331.

Lindvall, O., Björklund, A., Nobin, A., Stenevi, U., 1974. The adrenergic innervation of the rat thalamus as revealed by the glyoxylic acid fluorescence method. J. Comp. Neurol. 154, 317–347.

Lindvall, O., Björklund, A., Skagerberg, G., 1983. Dopamine-containing neurons in the spinal cord: anatomy and some functional aspects. Ann. Neurol. 14, 255–260.

Lindvall, O., Stenevi, U., 1978. Dopamine and noradrenaline neurons projecting to the septal area in the rat. Cell Tissue Res. 190, 383–407.

Ljungdahl, A., Hokfelt, T., 1973. Autoradiographic uptake patterns of (3H)GABA and (3H)glycine in central nervous tissues with special reference to the cat spinal cord. Brain Res. 62, 587–595.

Lopez-Bendito, G., Lujan, R., Shigemoto, R., Ganter, P., Paulsen, O., Molnar, Z., 2003. Blockade of GABA(B) receptors alters the tangential migration of cortical neurons. Cereb. Cortex 13, 932–942.

Loturco, J.J., Owens, D.F., Heath, M.J., Davis, M.B., Kriegstein, A.R., 1995. GABA and glutamate depolarize cortical progenitor cells and inhibit DNA synthesis. Neuron 15, 1287–1298.

Loy, R., Koziell, D.A., Lindsey, J.D., Moore, R.Y., 1980. Noradrenergic innervation of the adult rat hippocampal formation. J. Comp. Neurol. 189, 699–710.

Luhmann, H.J., Fukuda, A., Kilb, W., 2015. Control of cortical neuronal migration by glutamate and GABA. Front. Cell. Neurosci. 9, 4.

Luhmann, H.J., Kilb, W., Clusmann, H., 2014. Malformations of cortical development and neocortical focus. Int. Rev. Neurobiol. 114, 35–61.

Lynch, J.W., 2009. Native glycine receptor subtypes and their physiological roles. Neuropharmacology 56, 303–309.

Ma, D.Q., Whitehead, P.L., Menold, M.M., Martin, E.R., Ashley-Koch, A.E., Mei, H., Ritchie, M.D., Delong, G.R., Abramson, R.K., Wright, H.H., Cuccaro, M.L., Hussman, J.P., Gilbert, J.R., Pericak-Vance, M.A., 2005. Identification of significant association and gene-gene interaction of GABA receptor subunit genes in autism. Am. J. Hum. Genet. 77, 377–388.

Mackay, A.V., 1984. Sinistral findings in schizophrenia: high dopamine in the left amygdala. Trends Neurosci. 7 (4), 107–108. https://doi.org/10.1016/S0166-2236(84)80234-9.

MacLeod, R.M., Lehmeyer, J.E., 1974. Studies on the mechanism of the dopamine-mediated inhibition of prolactin secretion. Endocrinology 94 (4), 1077–1085. https://doi.org/10.1210/endo-94-4-1077.

Madsen, K.K., Clausen, R.P., Larsson, O.M., Krogsgaard-Larsen, P., Schousboe, A., White, H.S., 2009. Synaptic and extrasynaptic GABA transporters as targets for anti-epileptic drugs. J. Neurochem. 109 (Suppl. 1), 139–144.

Maehle, A.H., Prull, C.R., Halliwell, R.F., 2002. The emergence of the drug receptor theory. Nat. Rev. Drug Discov. 1, 637–641.

Maley, B.E., Engle, M.G., Humphreys, S., Vascik, D.A., Howes, K.A., Newton, B.W., Elde, R.P., 1990. Monoamine synaptic structure and localization in the central nervous system. J. Electron. Microsc. Tech. 15, 20–33.

Malmersjo, S., Rebellato, P., Smedler, E., Planert, H., Kanatani, S., Liste, I., Nanou, E., Sunner, H., Abdelhady, S., Zhang, S., Andang, M., El Manira, A., Silberberg, G., Arenas, E., Uhlen, P., 2013. Neural progenitors organize in small-world networks to promote cell proliferation. Proc. Natl. Acad. Sci. U. S. A. 110, E1524–E1532.

Malosio, M.L., Grenningloh, G., Kuhse, J., Schmieden, V., Schmitt, B., Prior, P., Betz, H., 1991. Alternative splicing generates two variants of the alpha 1 subunit of the inhibitory glycine receptor. J. Biol. Chem. 266, 2048–2053.

Malosio, M.L., Marqueze-Pouey, B., Kuhse, J., Betz, H., 1991. Widespread expression of glycine receptor subunit mRNAs in the adult and developing rat brain. EMBO J. 10, 2401–2409.

Manent, J.B., Represa, A., 2007. Neurotransmitters and brain maturation: early paracrine actions of GABA and glutamate modulate neuronal migration. Neuroscientist 13, 268–279.

Marchand, E.R., Riley, J.N., Moore, R.Y., 1980. Interpeduncular nucleus afferents in the rat. Brain Res. 193, 339–352.

Marcinkiewcz, C.A., Bierlein-De La Rosa, G., Dorrier, C.E., et al., 2019. Sex-dependent modulation of anxiety and fear by 5-HT_{1A} receptors in the bed nucleus of the stria terminalis. ACS Chem. Neurosci. 10 (7), 3154–3166. https://doi.org/10.1021/acschemneuro.8b00594.

Maric, D., Liu, Q.Y., Maric, I., Chaudry, S., Chang, Y.H., Smith, S.V., Sieghart, W., Fritschy, J.M., Barker, J.L., 2001. GABA expression dominates neuronal lineage progression in the embryonic rat neocortex and facilitates neurite outgrowth via GABA(A) autoreceptor/Cl- channels. J. Neurosci. 21, 2343–2360.

Marin, O., 2013. Cellular and molecular mechanisms controlling the migration of neocortical interneurons. Eur. J. Neurosci. 38, 2019–2029.

Martin, D.L., Olsen, R.W., 2000. GABA in the Nervous System: The View at Fifty Years. Lippincott Williams and Wilkins.

Martin, D.L., Rimvall, K., 1993. Regulation of gamma-aminobutyric acid synthesis in the brain. J. Neurochem. 60, 395–407.

Marty, S., Berninger, B., Carroll, P., Thoenen, H., 1996. GABAergic stimulation regulates the phenotype of hippocampal interneurons through the regulation of brain-derived neurotrophic factor. Neuron 16, 565–570.

Mason, S.T., Fibiger, H.C., 1979. Regional topography within noradrenergic locus coeruleus as revealed by retrograde transport of horseradish peroxidase. J. Comp. Neurol. 187, 703–724.

Massie, A., Cnops, L., Smolders, I., McCullumsmith, R., Kooijman, R., Kwak, S., Arckens, L., Michotte, Y., 2008. High-affinity Na+/K+-dependent glutamate transporter EAAT4 is expressed throughout the rat fore- and midbrain. J. Comp. Neurol. 511, 155–172.

Matthys, A., Haegeman, G., Van Craenenbroeck, K., Vanhoenacker, P., 2011. Role of the 5-HT7 receptor in the central nervous system: from current status to future perspectives. Mol. Neurobiol. 43 (3), 228–253. https://doi.org/10.1007/s12035-011-8175-3.

Matzenbach, B., Maulet, Y., Sefton, L., Courtier, B., Avner, P., Guenet, J.L., Betz, H., 1994. Structural analysis of mouse glycine receptor alpha subunit genes. Identification and chromosomal localization of a novel variant. J. Biol. Chem. 269, 2607–2612.

McBride, W.J., Aprison, M.H., Kusano, K., 1976. Contents of several amino acids in the cerebellum, brain stem and cerebrum of the 'staggerer', 'weaver' and 'nervous' neurologically mutant mice. J. Neurochem. 26, 867–870.

Mcbride, W.J., Hingtgen, J.N., Aprison, M.H., 1976. Neurochemical correlates of behavior: levels of amino acids in four areas of the brain of the rat during drug-induced behavioral excitation. Pharmacol. Biochem. Behav. 4, 53–57.

McCann, S.M., Lumpkin, M.D., Mizunuma, H., Khorram, O., Samson, W.K., 1984. Recent studies on the role of brain peptides in control of anterior pituitary hormone secretion. Peptides, 5 (Suppl. 1), 3–7. https://doi.org/10.1016/0196-9781(84)90259-6.

McClellan, K.M., Calver, A.R., Tobet, S.A., 2008. GABAB receptors role in cell migration and positioning within the ventromedial nucleus of the hypothalamus. Neuroscience 151, 1119–1131.

McDearmid, J.R., Liao, M., Drapeau, P., 2006. Glycine receptors regulate interneuron differentiation during spinal network development. Proc. Natl. Acad. Sci. U. S. A. 103, 9679–9684.

McDevitt, R.A., Neumaier, J.F., 2011. Regulation of dorsal raphe nucleus function by serotonin autoreceptors: a behavioral perspective. J. Chem. Neuroanat. 41 (4), 234–246. https://doi.org/10.1016/j.jchemneu.2011.05.001.

McEntee, W.J., Crook, T.H., 1993. Glutamate: its role in learning, memory, and the aging brain. Psychopharmacology (Berl) 111, 391–401.

McGeer, E.G., Singh, E.A., 1980. Possible cortico-hypothalamic glycinergic tract. Neurosci. Lett. 17, 85–87.

McGeer, P., McGeer, E., 1964. Formation of adrenaline by brain tissue. Biochem. Biophys. Res. Commun. 17, 502–507.

Mcgeer, P.L., Hattori, T., McGeer, E.G., 1975. Chemical and autoradiographic analysis of gamma-aminobutyric acid transport in Purkinje cells of the cerebellum. Exp. Neurol. 47, 26–41.

McGeer, P.L., McGeer, E.G., 1975. Evidence for glutamic acid decarboxylase-containing interneurons in the neostriatum. Brain Res. 91, 331–335.

McGeer, P.L., McGeer, E.G., Scherer, U., Singh, K., 1977. A glutamatergic corticostriatal path? Brain Res. 128, 369–373.

Meldrum, B.S., 2000. Glutamate as a neurotransmitter in the brain: review of physiology and pathology. J. Nutr. 130, 15S–1007S.

Mengod, G., Cortés, R., Vilaró, M.T., Hoyer, D., 2010. Distribution of 5-HT receptors in the central nervous system. In: Handbook of Behavioral Neuroscience. Elsevier, pp. 123–138. https://doi.org/10.1016/S1569-7339(10)70074-6.

Mesulam, M.M., Mufson, E.J., 1984. Neural inputs into the nucleus basalis of the substantia innominata (Ch4) in the rhesus monkey. Brain 107 (Pt 1), 253–274.

Mesulam, M.M., Mufson, E.J., Levey, A.I., Wainer, B.H., 1983. Cholinergic innervation of cortex by the basal forebrain: cytochemistry and cortical connections of the septal area, diagonal band nuclei, nucleus basalis (substantia innominata), and hypothalamus in the rhesus monkey. J. Comp. Neurol. 214, 170–197.

Mesulam, M.M., Mufson, E.J., Wainer, B.H., Levey, A.I., 1983. Central cholinergic pathways in the rat: an overview based on an alternative nomenclature (Ch1–Ch6). Neuroscience 10, 1185–1201.

Mesulam, M.M., Rosen, A.D., Mufson, E.J., 1984. Regional variations in cortical cholinergic innervation: chemoarchitectonics of acetylcholinesterase-containing fibers in the macaque brain. Brain Res. 311, 245–258.

Mesulam, M.M., Van Hoesen, G.W., 1976. Acetylcholinesterase-rich projections from the basal forebrain of the rhesus monkey to neocortex. Brain Res. 109, 152–157.

Meynert, T., 1872. Vom Gehirn der Säugetiere. In: Stricker, S. (Ed.), Handbuch der Lehre von den Geweben des Menschen und der Thiere, second ed. W. Engelmann, Leipzig.

Miguelez, C., Morera-Herreras, T., Torrecilla, M., Ruiz-Ortega, J.A., Ugedo, L., 2014. Interaction between the 5-HT system and the basal ganglia: functional implication and therapeutic perspective in Parkinson's disease. Front. Neural Circuits 8. https://doi.org/10.3389/fncir.2014.00021.

Miller, P.S., Harvey, R.J., Smart, T.G., 2004. Differential agonist sensitivity of glycine receptor alpha2 subunit splice variants. Br. J. Pharmacol. 143, 19–26.

Mitchell, E.S., 2011. 5-HT6 receptor ligands as antidementia drugs. Int. Rev. Neurobiol. 96, 163–187.

Mitchell, H.A., Weinshenker, D., 2010. Good night and good luck: norepinephrine in sleep pharmacology. Biochem. Pharmacol. 79, 801–809.

Mize, R.R., Spencer, R.F., Sterling, P., 1982. Two types of GABA-accumulating neurons in the superficial gray layer of the cat superior colliculus. J. Comp. Neurol. 206, 180–192.

Mody, I., Pearce, R.A., 2004. Diversity of inhibitory neurotransmission through GABA(A) receptors. Trends Neurosci. 27, 569–575.

Monsma Jr., F.J., Shen, Y., Ward, R.P., Hamblin, M.W., Sibley, D.R., 1993. Cloning and expression of a novel serotonin receptor with high affinity for tricyclic psychotropic drugs. Mol. Pharmacol. 43, 320–327.

Moore, R.Y., 1978. Catecholamin innervation of the basal forebrain. I. The septal area. J. Comp. Neurol. 177, 665–684.

Moore, R.Y., Bloom, F.E., 1978. Central catecholamine neuron systems: anatomy and physiology of the dopamine systems. Annu. Rev. Neurosci. 1, 129–169.

Moore, R.Y., Bloom, F.E., 1979. Central catecholamine neuron systems: anatomy and physiology of the norepinephrine and epinephrine systems. Annu. Rev. Neurosci. 2, 113–168.

Morelli, G., 2016. Glycine Receptors Alpha 2 Tune Cortical Network Formation.

Moret, C., Briley, M., 2000. The possible role of 5-HT1B/D receptors in psychiatric disorders and their potential as a target for therapy. Eur. J. Pharmacol. 404 (1–2), 1–12. https://doi.org/10.1016/S0014-2999(00)00581-1.

Morin, L.P., 1999. Serotonin and the regulation of mammalian circadian rhythmicity. Ann. Med. 31 (1), 12–33. https://doi.org/10.3109/07853899909019259.

Morozova, M., Burminskiy, D., Rupchev, G., Lepilkina, T., Potanin, S., Beniashvili, A., Lavrovsky, Y., Vostokova, N., Ivaschenko, A., 2017. 5-HT6 receptor antagonist as an adjunct treatment targeting residual symptoms in patients with schizophrenia: unexpected sex-related effects (double-blind placebo-controlled trial). J. Clin. Psychopharmacol. 37, 169–175.

Morrison, J.H., Foote, S.L., Molliver, M.E., Bloom, F.E., Lidov, H.G., 1982. Noradrenergic and serotonergic fibers innervate complementary layers in monkey primary visual cortex: an immunohistochemical study. Proc. Natl. Acad. Sci. U. S. A. 79, 2401–2405.

Morrison, J.H., Molliver, M.E., Grzanna, R., 1979. Noradrenergic innervation of cerebral cortex: widespread effects of local cortical lesions. Science 205, 313–316.

Morrison, J.H., Molliver, M.E., Grzanna, R., Coyle, J.T., 1981. The intracortical trajectory of the coeruleo-cortical projection in the rat: a tangentially organized cortical afferent. Neuroscience 6, 139–158.

Mugnaini, E., Wouterlood, F.G., Dahl, A.L., Oertel, W.H., 1984. Immunocytochemical identification of GABAergic neurons in the main olfactory bulb of the rat. Arch. Ital. Biol. 122, 83–113.

Murray, G.K., Corlett, P.R., Clark, L., et al., 2008. Substantia nigra/ventral tegmental reward prediction error disruption in psychosis. Mol. Psychiatry 13 (3), 267–276. https://doi.org/10.1038/sj.mp.4002058.

Nadler, J.V., Vaca, K.W., White, W.F., Lynch, G.S., Cotman, C.W., 1976. Aspartate and glutamate as possible transmitters of excitatory hippocampal afferents. Nature 260, 538–540.

Nadler, J.V., White, W.F., Vaca, K.W., Perry, B.W., Cotman, C.W., 1978. Biochemical correlates of transmission mediated by glutamate and aspartate. J. Neurochem. 31, 147–155.

Nagai, T., McGeer, P.L., McGeer, E.G., 1983. Distribution of GABA-T-intensive neurons in the rat forebrain and midbrain. J. Comp. Neurol. 218, 220–238.

Nagai, T., McGeer, P.L., Peng, J.H., McGeer, E.G., Dolman, C.E., 1983. Choline acetyltransferase immunohistochemistry in brains of Alzheimer's disease patients and controls. Neurosci. Lett. 36, 195–199.

Nahimi, A., Sommerauer, M., Kinnerup, M.B., Østergaard, K., Winterdahl, M., Jacobsen, J., et al., 2018. Noradrenergic deficits in Parkinson disease imaged with 11C-MeNER. J. Nucl. Med. 59 (4), 659–664. https://doi.org/10.2967/jnumed.117.190975.

Nanopoulos, D., Belin, M.F., Maitre, M., Vincendon, G., Pujol, J.F., 1982. Immunocytochemical evidence for the existence of GABAergic neurons in the nucleus raphe dorsalis. Possible existence of neurons containing serotonin and GABA. Brain Res. 232, 375–389.

Nauta, H.J., Cuenod, M., 1982. Perikaryal cell labeling in the subthalamic nucleus following the injection of 3H-gamma-aminobutyric acid into the pallidal complex: an autoradiographic study in cat. Neuroscience 7, 2725–2734.

Neve, K.A., Seamans, J.K., Trantham-Davidson, H., 2004. Dopamine receptor signaling. J. Recept. Signal Transduct. Res. 24, 165–205.

Nguyen, L., Malgrange, B., Belachew, S., Rogister, B., Rocher, V., Moonen, G., Rigo, J.M., 2002. Functional glycine receptors are expressed by postnatal nestin-positive neural stem/progenitor cells. Eur. J. Neurosci. 15, 1299–1305.

Nguyen, L., Malgrange, B., Breuskin, I., Bettendorff, L., Moonen, G., Belachew, S., Rigo, J.M., 2003. Autocrine/paracrine activation of the GABA(A) receptor inhibits the proliferation of neurogenic polysialylated neural cell adhesion molecule-positive (PSA-NCAM+) precursor cells from postnatal striatum. J. Neurosci. 23, 3278–3294.

Nichols, D.E., Nichols, C.D., 2008. Serotonin receptors. Chem. Rev. 108, 1614–1641.

Nieoullon, A., Canolle, B., Masmejean, F., Guillet, B., Pisano, P., Lortet, S., 2006. The neuronal excitatory amino acid transporter EAAC1/EAAT3: does it represent a major actor at the brain excitatory synapse? J. Neurochem. 98, 1007–1018.

Nikolic, Z., Laube, B., Weber, R.G., Lichter, P., Kioschis, P., Poustka, A., Mulhardt, C., Becker, C.M., 1998. The human glycine receptor subunit alpha3. Glra3 gene structure, chromosomal localization, and functional characterization of alternative transcripts. J. Biol. Chem. 273, 19708–19714.

Nilsson, T., Longmore, J., Shaw, D., Olesen, I.J., Edvinsson, L., 1999. Contractile 5-HT_{1B} receptors in human cerebral arteries: pharmacological characterization and localization with immunocytochemistry. Br. J. Pharmacol. 128 (6), 1133–1140. https://doi.org/10.1038/sj.bjp.0702773.

Nimmervoll, B., Denter, D.G., Sava, I., Kilb, W., Luhmann, H.J., 2011. Glycine receptors influence radial migration in the embryonic mouse neocortex. Neuroreport 22, 509–513.

Nobin, A., Björklund, A., 1973. Topography of the monoamine neuron systems in the human brain as revealed in fetuses. Acta Physiol. Scand. Suppl. 388, 1–40.

Norgren, R., 1974. Gustatory afferents to ventral forebrain. Brain Res. 81, 285–295.

Nygren, L.G., Olson, L., 1977. A new major projection from locus coeruleus: the main source of noradrenergic nerve terminals in the ventral and dorsal columns of the spinal cord. Brain Res. 132, 85–93.

Obata, K., Ito, M., Ochi, R., Sato, N., 1967. Pharmacological properties of the postsynaptic inhibition by Purkinje cell axons and the action of gamma-aminobutyric acid on deiters NEURONES. Exp. Brain Res. 4, 43–57.

Obata, K., Takeda, K., Shinozaki, H., 1970. Further study on pharmacological properties of the cerebellar-induced inhibition of deiters neurones. Exp. Brain Res. 11, 327–342.

Oblak, A.L., Gibbs, T.T., Blatt, G.J., 2010. Decreased GABA(B) receptors in the cingulate cortex and fusiform gyrus in autism. J. Neurochem. 114, 1414–1423.

Ohta, M., Oomura, Y., 1979. Inhibitory pathway from the frontal cortex to the hypothalamic ventromedial nucleus in the rat. Brain Res. Bull. 4, 231–238.

Okada, Y., Shimada, C., 1975. Distribution of gamma-aminobutyric acid (GABA) and glutamate decarboxylase (GAD) activity in the guinea pig hippocampus—microassay method for the determination of GAD activity. Brain Res. 98, 202–206.

Okubo, Y., Sekiya, H., Namiki, S., Sakamoto, H., Iinuma, S., Yamasaki, M., Watanabe, M., Hirose, K., Iino, M., 2010. Imaging extrasynaptic glutamate dynamics in the brain. Proc. Natl. Acad. Sci. U. S. A. 107, 6526–6531.

Olsen, R.W., Sieghart, W., 2008. 'International Union of Pharmacology. LXX. Subtypes of gamma-aminobutyric acid(A) receptors: classification on the basis of subunit composition, pharmacology, and function. Update. Pharmacol. Rev. 60, 243–260.

Olson, L., Boreus, L.O., Seiger, A., 1973. Histochemical demonstration and mapping of 5-hydroxytryptamine- and catecholamine-containing neuron systems in the human fetal brain. Z. Anat. Entwicklungsgesch. 139, 259–282.

Olson, L., Fuxe, K., 1971. On the projections from the locus coeruleus noradrenaline neurons: the cerebellar innervation. Brain Res. 28, 165–171.

Olsen, R.W., Macdonald, R.L., 2002. GABAA receptor complex: structure and function. In: Egebjerg, J., Schousboe, A., Krogsgaard-Larsen, P., (Eds.), Glutamate and GABA receptors and transporters: structure, function, and pharmacology. Taylor and Francis, New York, pp. 203–235 (Chapter 9).

Parent, A., Csonka, C., Etienne, P., 1984. The occurrence of large acetylcholinesterase-containing neurons in human neostriatum as disclosed in normal and Alzheimer-diseased brains. Brain Res. 291, 154–158.

Parker, R.M., Bentley, K.R., Barnes, N.M., 1996. Allosteric modulation of 5-HT3 receptors: focus on alcohols and anaesthetic agents. Trends Pharmacol. Sci. 17, 95–99.

Pasquier, D.A., Gold, M.A., Jacobowitz, D.M., 1980. Noradrenergic perikarya (A5-A7, subcoeruleus) projections to the rat cerebellum. Brain Res. 196 (1), 270–275. https://doi.org/10.1016/0006-8993(80)90737-4.

Pearson, J., Goldstein, N., Markey, K., Brandeis, L., 1983. Human brainstem catecholamine neuronal anatomy as indicated by immunocytochemistry with antibodies to tyrosine hydroxylase. Neuroscience 8, 3–32.

Perry, R.H., Candy, J.M., Perry, E.K., Thompson, J., Oakley, A.E., 1984. The substantia innominata and adjacent regions in the human brain: histochemical and biochemical observations. J. Anat. 138 (Pt 4), 713–732.

Pijnenburg, A.J., Honig, W.M., Van Der Heyden, J.A., Van Rossum, J.M., 1976. Effects of chemical stimulation of the mesolimbic dopamine system upon locomotor activity. Eur. J. Pharmacol. 35, 45–58.

Pijnenburg, A.J., Van Rossum, J.M., 1973. Letter: stimulation of locomotor activity following injection of dopamine into the nucleus accumbens. J. Pharm. Pharmacol. 25, 1003–1005.

Piton, A., Gauthier, J., Hamdan, F.F., Lafreniere, R.G., Yang, Y., Henrion, E., Laurent, S., Noreau, A., Thibodeau, P., Karemera, L., Spiegelman, D., Kuku, F., Duguay, J., Destroismaisons, L., Jolivet, P., Cote, M., Lachapelle, K., Diallo, O., Raymond, A., Marineau, C., Champagne, N., Xiong, L., Gaspar, C., Riviere, J.B., Tarabeux, J., Cossette, P., Krebs, M.O., Rapoport, J.L., Addington, A., Delisi, L.E., Mottron, L., Joober, R., Fombonne, E., Drapeau, P., Rouleau, G.A., 2011. Systematic resequencing of X-chromosome synaptic genes in autism spectrum disorder and schizophrenia. Mol. Psychiatry 16, 867–880.

Polc, P., Haefely, W., 1982. Benzodiazepines enhance the bicuculline-sensitive part of recurrent Renshaw inhibition in the cat spinal cord. Neurosci. Lett. 28, 193–197.

Pontes, ALB de, Engelberth, R.C.G.J., Nascimento, E. da S., et al., 2010. Serotonin and circadian rhythms. Psychol. Neurosci. 3 (2), 217–228. https://doi.org/10.3922/j.psns.2010.2.011.

Price, J.L., Amaral, D.G., 1981. An autoradiographic study of the projections of the central nucleus of the monkey amygdala. J. Neurosci. 1, 1242–1259.

Price, J.L., Stern, R., 1983. Individual cells in the nucleus basalis—diagonal band complex have restricted axonal projections to the cerebral cortex in the rat. Brain Res. 269, 352–356.

Pullar, I.A., Boot, J.R., Broadmore, R.J., et al., 2004. The role of the 5-HT1D receptor as a presynaptic autoreceptor in the guinea pig. Eur. J. Pharmacol. 493 (1–3), 85–93. https://doi.org/10.1016/j.ejphar.2004.04.029.

Pycock, C.J., Kerwin, R.W., 1981. The status of glycine as a supraspinal neurotransmitter. Life Sci. 28, 2679–2686.

Qian, T., Chen, R., Nakamura, M., Furukawa, T., Kumada, T., Akita, T., Kilb, W., Luhmann, H.J., Nakahara, D., Fukuda, A., 2014. Activity-dependent endogenous taurine release facilitates excitatory neurotransmission in the neocortical marginal zone of neonatal rats. Front. Cell. Neurosci. 8, 33.

Racagni, G., Apud, J.A., Cocchi, D., Locatelli, V., Muller, E.E., 1982. GABAergic control of anterior pituitary hormone secretion. Life Sci. 31, 823–838.

Ray, N.J., Metzler-Baddeley, C., Khondoker, M.R., et al., 2015. Cholinergic basal forebrain structure influences the reconfiguration of white matter connections to support residual memory in mild cognitive impairment. J. Neurosci. 35 (2), 739–747. https://doi.org/10.1523/JNEUROSCI.3617-14.2015.

Razzaque, Z., Pickard, J.D., Ma, Q.P., et al., 2002. 5-HT_{1B}-receptors and vascular reactivity in human isolated blood vessels: assessment of the potential craniovascular selectivity of sumatriptan. Br. J. Clin. Pharmacol. 53 (3), 266–274. https://doi.org/10.1046/j.0306-5251.2001.01536.x.

Ren, J., Isakova, A., Friedmann, D., et al., 2019. Single-cell transcriptomes and whole-brain projections of serotonin neurons in the mouse dorsal and median raphe nuclei. Elife 8. https://doi.org/10.7554/eLife.49424.

Reynolds, G.P., 1983. Increased concentrations and lateral asymmetry of amygdala dopamine in schizophrenia. Nature 305, 527–529.

Ribak, C.E., 1978. Aspinous and sparsely-spinous stellate neurons in the visual cortex of rats contain glutamic acid decarboxylase. J. Neurocytol. 7, 461–478.

Ribak, C.E., 1981. The GABAergic neurons of the extrapyramidal system as revealed by immunocytochemistry. Adv. Biochem. Psychopharmacol. 30, 23–36.

Ribak, C.E., Vaughn, J.E., Saito, K., Barber, R., Roberts, E., 1976. Immunocytochemical localization of glutamate decarboxylase in rat substantia nigra. Brain Res. 116, 287–298.

Ribak, C.E., Vaughn, J.E., Saito, K., Barber, R., Roberts, E., 1977. Glutamate decarboxylase localization in neurons of the olfactory bulb. Brain Res. 126, 1–18.

Roberts, J.C., Reavill, C., East, S.Z., Harrison, P.J., Patel, S., Routledge, C., Leslie, R.A., 2002. The distribution of 5-HT(6) receptors in rat brain: an autoradiographic binding study using the radiolabelled 5-HT(6) receptor antagonist [(125)I]SB-258585. Brain Res. 934, 49–57.

Rosen, L.B., Ginty, D.D., Greenberg, M.E., 1995. Calcium regulation of gene expression. Adv. Second Messenger Phosphoprotein Res. 30, 225–253.

Rossor, M.N., Svendsen, C., Hunt, S.P., Mountjoy, C.Q., Roth, M., Iversen, L.L., 1982. The substantia innominata in Alzheimer's disease: an histochemical and biochemical study of cholinergic marker enzymes. Neurosci. Lett. 28, 217–222.

Rouzaire-Dubois, B., Scarnati, E., Hammond, C., Crossman, A.R., Shibazaki, T., 1983. Microiontophoretic studies on the nature of the neurotransmitter in the subthalamo-entopeduncular pathway of the rat. Brain Res. 271, 11–20.

Ruat, M., Traiffort, E., Arrang, J.M., Tardivel-Lacombe, J., Diaz, J., Leurs, R., Schwartz, J.C., 1993. A novel rat serotonin (5-HT6) receptor: molecular cloning, localization and stimulation of cAMP accumulation. Biochem. Biophys. Res. Commun. 193, 268–276.

Rudolph, U., Crestani, F., Mohler, H., 2001. GABA(A) receptor subtypes: dissecting their pharmacological functions. Trends Pharmacol. Sci. 22, 188–194.

Rustioni, A., Cuenod, M., 1982. Selective retrograde transport of D-aspartate in spinal interneurons and cortical neurons of rats. Brain Res. 236, 143–155.

Saavedra, J.M., Palkovits, M., Brownstein, M.J., Axelrod, J., 1974. Localisation of phenylethanolamine N-methyl transferase in the rat brain nuclei. Nature 248, 695–696.

Saito, K., Barber, R., Wu, J., Matsuda, T., Roberts, E., Vaughn, J.E., 1974. Immunohistochemical localization of glutamate decarboxylase in rat cerebellum. Proc. Natl. Acad. Sci. U. S. A. 71, 269–273.

Salt, T.E., Hill, R.G., 1983. Neurotransmitter candidates of somatosensory primary afferent fibres. Neuroscience 10, 1083–1103.

Sandoval, M.E., Cotman, C.W., 1978. Evaluation of glutamate as a neurotransmitter of cerebellar parallel fibers. Neuroscience 3, 199–206.

Saper, C.B., Swanson, L.W., Cowan, W.M., 1979. An autoradiographic study of the efferent connections of the lateral hypothalamic area in the rat. J. Comp. Neurol. 183, 689–706.

Satoh, K., Tohyama, M., Yamamoto, K., Sakumoto, T., Shimizu, N., 1977. Noradrenaline innervation of the spinal cord studied by the horseradish peroxidase method combined with monoamine oxidase staining. Exp. Brain Res. 30, 175–186.

Sawchenko, P.E., Swanson, L.W., 1982. Immunohistochemical identification of neurons in the paraventricular nucleus of the hypothalamus that project to the medulla or to the spinal cord in the rat. J. Comp. Neurol. 205, 260–272.

Scatton, B., Serrano, A., Rivot, J.P., Nishikawa, T., 1984. Inhibitory GABAergic influence on striatal serotonergic transmission exerted in the dorsal raphe as revealed by in vivo voltammetry. Brain Res. 305, 343–352.

Schofield, S.P., Dixson, A.F., 1982. Distribution of catecholamine and indoleamine neurons in the brain of the common marmoset (Callithrix jacchus). J. Anat. 134, 315–338.

Schofield, S.P., Everitt, B.J., 1981. The organisation of catecholamine-containing neurons in the brain of the rhesus monkey (Macaca mulatta). J. Anat. 132, 391–418.

Sernagor, E., Chabrol, F., Bony, G., Cancedda, L., 2010. GABAergic control of neurite outgrowth and remodeling during development and adult neurogenesis: general rules and differences in diverse systems. Front. Cell. Neurosci. 4, 11.

Shashidharan, P., Huntley, G.W., Murray, J.M., Buku, A., Moran, T., Walsh, M.J., Morrison, J.H., Plaitakis, A., 1997. Immunohistochemical localization of the neuron-specific glutamate transporter EAAC1 (EAAT3) in rat brain and spinal cord revealed by a novel monoclonal antibody. Brain Res. 773, 139–148.

Shigeri, Y., Seal, R.P., Shimamoto, K., 2004. Molecular pharmacology of glutamate transporters, EAATs and VGLUTs. Brain Res. Rev. 45, 250–265.

Shih, J.C., Chen, K., Ridd, M.J., 1999. Monoamine oxidase: from genes to behavior. Annu. Rev. Neurosci. 22, 197–217.

Shimizu, N., Onishi, S., Toyama, M., Maeda, T., 1974. Demonstration by degeneration silver method of the ascending projection from the locus ceruleus. Exp. Brain Res. 20, 181–192.

Siegel, G., Agranoff, B., Albers, R., 1999. Basic Neurochemistry: Molecular, Cellular and Medical Aspects, sixth ed. Lippincott-Raven.

Simon, H., Le Moal, M., Calas, A., 1979. Efferents and afferents of the ventral tegmental-A10 region studied after local injection of [3H] leucine and horseradish peroxidase. Brain Res. 178, 17–40.

Simon, H., Le Moal, M., Galey, D., Cardo, B., 1976. Silver impregnation of dopaminergic systems after radiofrequency and 6-OHDA lesions of the rat ventral. Brain Res. 115, 215–231.

Sine, S.M., Engel, A.G., 2006. Recent advances in Cys-loop receptor structure and function. Nature 440, 448–455.

Singer, L.T., Kercsmar, C., Legris, G., Orlowski, J.P., Hill, B.P., Doershuk, C., 1989. Developmental sequelae of long-term infant tracheostomy. Dev. Med. Child Neurol. 31, 224–230.

Skagerberg, G., Björklund, A., Lindvall, O., Schmidt, R.H., 1982. Origin and termination of the diencephalo-spinal dopamine system in the rat. Brain Res. Bull. 9, 237–244.

Skagerberg, G., Lindvall, O., Björklund, A., 1984. Origin, course and termination of the mesohabenular dopamine pathway in the rat. Brain Res. 307, 99–108.

Smith, Q.R., 2000. Transport of glutamate and other amino acids at the blood-brain barrier. J. Nutr. 130, 22S–1016S.

Smith, S.S., 2003. Neurosteroid Effects in the Central Nervous System: The Role of the GABA-A Receptor. CRC Press.

Somogyi, P., Bolam, J.P., Smith, A.D., 1981. Monosynaptic cortical input and local axon collaterals of identified striatonigral neurons. A light and electron microscopic study using the Golgi-peroxidase transport-degeneration procedure. J. Comp. Neurol. 195, 567–584.

Spekker, E., Tanaka, M., Szabó, Á., Vécsei, L., 2021. Neurogenic inflammation: the participant in migraine and recent advancements in translational research. Biomedicines 10 (1), 76. https://doi.org/10.3390/biomedicines10010076.

Stahl, S.M., 2015. Modes and nodes explain the mechanism of action of vortioxetine, a multimodal agent (MMA): actions at serotonin receptors may enhance downstream release of four pro-cognitive neurotransmitters. CNS Spectr. 20 (6), 515–519. https://doi.org/10.1017/S1092852915000358.

Stevens, J.R., 1973. An anatomy of schizophrenia? Arch. Gen. Psychiatry 29, 177–189.

Stevens, J.R., Bigelow, L., Denney, D., et al., 1979. Telemetered EEG-EOG during psychotic behaviors of schizophrenia. Arch. Gen. Psychiatry. 36 (3), 251–262.

Storm-Mathisen, J., 1972. Glutamate decarboxylase in the rat hippocampal region after lesions of the afferent fibre systems. Evidence that the enzyme is localized in intrinsic neurones. Brain Res. 40, 215–235.

Storm-Mathisen, J., Fonnum, F., 1971. Quantitative histochemistry of glutamate decarboxylase in the rat hippocampal region. J. Neurochem. 18, 1105–1111.

Storm-Mathisen, J., Leknes, A.K., Bore, A.T., Vaaland, J.L., Edminson, P., Haug, F.M., Ottersen, O.P., 1983. First visualization of glutamate and GABA in neurones by immunocytochemistry. Nature 301, 517–520.

Storm-Mathisen, J., Woxen Opsahl, M., 1978. Asparate and/or glutamate may be transmitters in hippocampal efferents to septum and hypothalamus. Neurosci. Lett. 9, 65–70.

Strahlendorf, H.K., Barnes, C.D., 1983. Control of substantia nigra pars reticulata neurons by the nucleus accumbens. Brain Res. Bull. 11, 259–263.

Stratton, M.S., Staros, M., Budefeld, T., Searcy, B.T., Nash, C., Eitel, C., Carbone, D., Handa, R.J., Majdic, G., Tobet, S.A., 2014. Embryonic GABA(B) receptor blockade alters cell migration, adult hypothalamic structure, and anxiety- and depression-like behaviors sex specifically in mice. PLoS One 9, e106015.

Studler, J.M., Reibaud, M., Tramu, G., Blanc, G., Glowinski, J., Tassin, J.P., 1984. Pharmacological study on the mixed CCK8/DA meso-nucleus accumbens pathway: evidence for the existence of storage sites containing the two transmitters. Brain Res. 298, 91–97.

Sturman, J.A., Gargano, A.D., Messing, J.M., Imaki, H., 1986. Feline maternal taurine deficiency: effect on mother and offspring. J. Nutr. 116, 655–667.

Swanson, L.W., 1982. The projections of the ventral tegmental area and adjacent regions: a combined fluorescent retrograde tracer and immunofluorescence study in the rat. Brain Res. Bull. 9, 321–353.

Swanson, L.W., Hartman, B.K., 1975. The central adrenergic system. An immunofluorescence study of the location of cell bodies and their efferent connections in the rat utilizing dopamine-beta-hydroxylase as a marker. J. Comp. Neurol. 163, 467–505.

Swanson, L.W., Sawchenko, P.E., Berod, A., Hartman, B.K., Helle, K.B., Vanorden, D.E., 1981. An immunohistochemical study of the organization of catecholaminergic cells and terminal fields in the paraventricular and supraoptic nuclei of the hypothalamus. J. Comp. Neurol. 196, 271–285.

Tanaka, C., Ishikawa, M., Shimada, S., 1982. Histochemical mapping of catecholaminergic neurons and their ascending fiber pathways in the rhesus monkey brain. Brain Res. Bull. 9, 255–270.

Tao-Cheng, J.H., Zhou, F.C., 1999. Differential polarization of serotonin transporters in axons versus soma-dendrites: an immunogold electron microscopy study. Neuroscience 94, 821–830.

Tapia, J.C., Mentis, G.Z., Navarrete, R., Nualart, F., Figueroa, E., Sanchez, A., Aguayo, L.G., 2001. Early expression of glycine and GABA(A) receptors in developing spinal cord neurons. Effects on neurite outgrowth. Neuroscience 108, 493–506.

Tecott, L.H., Sun, L.M., Akana, S.F., et al., 1995. Eating disorder and epilepsy in mice lacking 5-HT2C serotonin receptors. Nature 374 (6522), 542–546. https://doi.org/10.1038/374542a0.

Thangnipon, W., Storm-Mathisen, J., 1981. K+-evoked Ca2+-dependent release of d-[3H] aspartate from terminals of the corticopontine pathway. Neurosci. Lett. 23, 181–186.

Thierry, A.M., Deniau, J.M., Herve, D., Chevalier, G., 1980. Electrophysiological evidence for non-dopaminergic mesocortical and mesolimbic neurons in the rat. Brain Res. 201, 210–214.

Thomas, D., Hagan, J., 2004. 5-HT7 receptors. Current Drug Target -CNS & Neurological Disorders 3 (1), 81–90. https://doi.org/10.2174/1568007043482633.

Tiger, M., Varnäs, K., Okubo, Y., Lundberg, J., 2018. The 5-HT1B receptor—a potential target for antidepressant treatment. Psychopharmacology (Berl) 235 (5), 1317–1334. https://doi.org/10.1007/s00213-018-4872-1.

Tohyama, M., Maeda, T., Hashimoto, J., Shrestha, G.R., Tamura, O., 1974. Comparative anatomy of the locus coeruleus. I. Organization and ascending projections of the catecholamine containing neurons in the pontine region of the bird, Melopsittacus undulatus. J. Hirnforsch. 15, 319–330.

Tork, I., 1990. Anatomy of the serotonergic system. Ann. N. Y. Acad. Sci. 600, 9–34 discussion 34–35.

Traynelis, S.F., Wollmuth, L.P., McBain, C.J., Menniti, F.S., Vance, K.M., Ogden, K.K., Hansen, K.B., Yuan, H., Myers, S.J., Dingledine, R., 2010. Glutamate receptor ion channels: structure, regulation, and function. Pharmacol. Rev. 62, 405–496.

Tyson, J.E., Lasky, R., Flood, D., Mize, C., Picone, T., Paule, C.L., 1989. Randomized trial of taurine supplementation for infants less than or equal to 1300-gram birth weight: effect on auditory brainstem-evoked responses. Pediatrics 83, 406–415.

Ungerstedt, U., 1971. Stereotaxic mapping of the monoamine pathways in the rat brain. Acta Physiol. Scand. Suppl. 367, 1–48.

Van Der Gugten, J., Palkovits, M., Wijnen, H.L., Versteeg, D.H., 1976. Regional distribution of adrenaline in rat brain. Brain Res. 107, 171–175.

Van Der Kooy, D., Coscina, D.V., Hattori, T., 1981. Is there a non-dopaminergic nigrostriatal pathway? Neuroscience 6, 345–357.

Vermeiren, Y., Janssens, J., Van Dam, D., De Deyn, P.P., 2018. Serotonergic dysfunction in amyotrophic lateral sclerosis and Parkinson's disease: similar mechanisms, dissimilar outcomes. Front. Neurosci. 12. https://doi.org/10.3389/fnins.2018.00185.

Vincent, S.R., Hattori, T., McGeer, E.G., 1978. The nigrotectal projection: a biochemical and ultrastructural characterization. Brain Res. 151, 159–164.

Vincent, S.R., Hokfelt, T., Skirboll, L.R., Wu, J.Y., 1983. Hypothalamic gamma-aminobutyric acid neurons project to the neocortex. Science 220, 1309–1311.

Vogt, M., 1954. Norepinephrine and epinephrine in the central nervous system. Pharmacol. Rev. 6, 31–32.

Waagepetersen, H.S., Sonnewald, U., Schousboe, A., 1999. The GABA paradox: multiple roles as metabolite, neurotransmitter, and neurodifferentiative agent. J. Neurochem. 73, 1335–1342.

Walaas, I., 1981. Biochemical evidence for overlapping neocortical and allocortical glutamate projections to the nucleus accumbens and rostral caudatoputamen in the rat brain. Neuroscience 6, 399–405.

Walaas, I., Fonnum, F., 1980. Biochemical evidence for gamma-aminobutyrate containing fibres from the nucleus accumbens to the substantia nigra and ventral tegmental area in the rat. Neuroscience 5, 63–72.

Wallner, M., Hanchar, H.J., Olsen, R.W., 2006. Low dose acute alcohol effects on GABA A receptor subtypes. Pharmacol. Ther. 112, 513–528.

Ward, R.P., Hamblin, M.W., Lachowicz, J.E., Hoffman, B.J., Sibley, D.R., Dorsa, D.M., 1995. Localization of serotonin subtype 6 receptor messenger RNA in the rat brain by in situ hybridization histochemistry. Neuroscience 64, 1105–1111.

Watanabe, M., Fukuda, A., 2015. Development and regulation of chloride homeostasis in the central nervous system. Front. Cell. Neurosci. 9, 371.

Wedzony, K., Chocyk, A., Kolasiewicz, W., Maćkowiak, M., 2007. Glutamatergic neurons of rat medial prefrontal cortex innervating the ventral tegmental area are positive for serotonin 5-HT1A receptor protein. J. Physiol. Pharmacol. 58 (4), 611–624.

Weissman, T.A., Riquelme, P.A., Ivic, L., Flint, A.C., Kriegstein, A.R., 2004. Calcium waves propagate through radial glial cells and modulate proliferation in the developing neocortex. Neuron 43, 647–661.

Wenthold, R.J., 1978. Glutamic acid and aspartic acid in subdivisions of the cochlear nucleus after auditory nerve lesion. Brain Res. 143, 544–548.

Wenthold, R.J., Gulley, R.L., 1977. Aspartic acid and glutamic acid levels in the cochlear nucleus after auditory nerve lesion. Brain Res. 138, 111–123.

West, P.J., Marcy, V.R., Marino, M.J., Schaffhauser, H., 2009. Activation of the 5-HT(6) receptor attenuates long-term potentiation and facilitates GABAergic neurotransmission in rat hippocampus. Neuroscience 164, 692–701.

Westlund, K.N., Bowker, R.M., Ziegler, M.G., Coulter, J.D., 1984. Origins and terminations of descending noradrenergic projections to the spinal cord of monkey. Brain Res. 292, 1–16.

Westlund, K.N., Coulter, J.D., 1980. Descending projections of the locus coeruleus and subcoeruleus/medial parabrachial nuclei in monkey: axonal transport studies and dopamine-beta-hydroxylase immunocytochemistry. Brain Res. 2, 235–264.

Wharton, B.A., Morley, R., Isaacs, E.B., Cole, T.J., Lucas, A., 2004. Low plasma taurine and later neurodevelopment. Arch. Dis. Child. Fetal Neonatal Ed. 89, F497–F498.

Whitehouse, P.J., Price, D.L., Clark, A.W., Coyle, J.T., Delong, M.R., 1981. Alzheimer disease: evidence for selective loss of cholinergic neurons in the nucleus basalis. Ann. Neurol. 10, 122–126.

Whitehouse, P.J., Price, D.L., Struble, R.G., Clark, A.W., Coyle, J.T., Delon, M.R., 1982. Alzheimer's disease and senile dementia: loss of neurons in the basal forebrain. Science 215, 1237–1239.

Whiting, P.J., 2006. GABA-A receptors: a viable target for novel anxiolytics? Curr. Opin. Pharmacol. 6, 24–29.

Wiklund, L., Toggenburger, G., Cuenod, M., 1982. Aspartate: possible neurotransmitter in cerebellar climbing fibers. Science 216, 78–80.

Wojcik, S.M., Katsurabayashi, S., Guillemin, I., Friauf, E., Rosenmund, C., Brose, N., Rhee, J.S., 2006. A shared vesicular carrier allows synaptic corelease of GABA and glycine. Neuron 50, 575–587.

Woolf, N.J., Butcher, L.L., 1981. Cholinergic neurons in the caudate-putamen complex proper are intrinsically organized: a combined Evans blue and acetylcholinesterase analysis. Brain Res. Bull. 7, 487–507.

Yamakura, T., Bertaccini, E., Trudell, J.R., Harris, R.A., 2001. Anesthetics and ion channels: molecular models and sites of action. Annu. Rev. Pharmacol. Toxicol. 41, 23–51.

Yohn, C.N., Gergues, M.M., Samuels, B.A., 2017. The role of 5-HT receptors in depression. Mol. Brain 10 (1), 28. https://doi.org/10.1186/s13041-017-0306-y.

You, I.J., Wright, S.R., Garcia-Garcia, A.L., et al., 2016. 5-HT1A autoreceptors in the dorsal raphe nucleus convey vulnerability to compulsive cocaine seeking. Neuropsychopharmacology 41 (5), 1210–1222. https://doi.org/10.1038/npp.2015.268.

Young, A.B., Bromberg, M.B., Penney JR., J.B., 1981. Decreased glutamate uptake in subcortical areas deafferented by sensorimotor cortical ablation in the cat. J. Neurosci. 1, 241–249.

Young, A.B., Oster-Granite, M.L., Herndon, R.M., Snyder, S.H., 1974. Glutamic acid: selective depletion by viral induced granule cell loss in hamster cerebellum. Brain Res. 73, 1–13.

Young, S.Z., Taylor, M.M., Wu, s., Ikeda-Matsuo, Y., Kubera, C., Bordey, A., 2012. NKCC1 knockdown decreases neuron production through GABA(A)-regulated neural progenitor proliferation and delays dendrite development. J. Neurosci. 32, 13630–13638.

Zaczek, R., Hedreen, J.C., Coyle, J.T., 1979. Evidence for a hippocampal-septal glutamatergic pathway in the rat. Exp. Neurol. 65, 145–156.

Zarkadas, E., Zhang, H., Cai, W., Effantin, G., Perot, J., Neyton, J., Chipot, C., Schoehn, G., Dehez, F., Nury, H., 2020. The binding of palonosetron and other antiemetic drugs to the serotonin 5-HT3 receptor. Structure 28, 1131–1140 e4.

Zgombick, J.M., Bard, J.A., Kucharewicz, S.A., Urquhart, D.A., Weinshank, R.L., Branchek, T.A., 1997. Molecular cloning and pharmacological characterization of guinea pig 5-HT 1B and 5-HT 1D receptors. Neuropharmacology 36 (4–5), 513–524. https://doi.org/10.1016/S0028-3908(97)00023-3.

Zhou, L., Huang, J., Wang, H., Luo, J., Zeng, L., Xiong, K., Chen, D., 2013. Expression of glutamate and GABA during the process of rat retinal synaptic plasticity induced by acute high intraocular pressure. Acta Histochem Cytochem 46, 11–18.

Further Reading

Kievit, J., Kuypers, H.G., 1975. Basal forebrain and hypothalamic connection to frontal and parietal cortex in the Rhesus monkey. Science 187, 660–662.

Lewis, P.R., Shute, C.C., Silver, A., 1967. Confirmation from choline acetylase analyses of a massive cholinergic innervation to the rat hippocampus. J. Physiol. 191, 215–224.

Lüttgen, M., Ove Ogren, S., Meister, B., 2004. Chemical identity of 5-HT2A receptor immunoreactive neurons of the rat septal complex and dorsal hippocampus. Brain Res. 1010 (1–2), 156–165. https://doi.org/10.1016/j.brainres.2004.03.016.

Mayo, W., Dubois, B., Ploska, A., Javoy-Agid, F., Agid, Y., Le Moal, M., Simon, H., 1984. Cortical cholinergic projections from the basal forebrain of the rat, with special reference to the prefrontal cortex innervation. Neurosci. Lett. 47, 149–154.

Vane, J.R., 1959. The relative activities of some tryptamine analogues on the isolated rat stomach strip preparation. Br. J. Pharmacol. Chemother. 14, 87–98.

Villani, L., Contestabile, A., Fonnum, F., 1983. Autoradiographic labeling of the cholinergic habenulo-interpeduncular projection. Neurosci. Lett. 42, 261–266.

Anatomy of the spinal cord and brain

4

The Spinal Cord

Brandon M. Ascenzi[1], *Anna Badner*[2] *and Pia Maria Vidal*[3]

[1]Independent Neuroresearcher-Member of Marie Curie Alumni Association (MCAA) [2]Department of Psychiatry and Behavioral Sciences, Stanford University, Palo Alto, CA, USA [3]Biomedical Science Research Laboratory, Neuroimmunology and Regeneration of the Central Nervous System Unit, Department of Basic Sciences, Faculty of Medicine, Universidad Católica de la Santísima Concepción, Concepción, Chile

OUTLINE

INTRODUCTION

The spinal cord (or medulla spinalis) forms, alongside the brain, the Central Nervous System (CNS) and it lies in the vertebral canal (or spinal canal) with an extension of around 45 cm in man and around 43 cm in long woman. In adults, it has a variable diameter ranging from 13 mm in the cervical and lumbar region and 6.4 mm in the thoracic area. Rostrally, immediately below the foramen magnum it continues with the medulla oblongata, caudally instead ends with the medullary cone which in turn continues with the filum terminalis. The spinal cord is wrapped by three membranes of connective tissue called meninges that are represented, from the outside to inside, by dura mater, arachnoid, and pia mater (Fig. 4.1). They extend into the skull enveloping the brain passing through the foramen magnum. The spinal cord can be considered as a segmental structure from which 31 pairs of spinal nerves originate, which arise from the medulla spinalis by means of ventral- and dorsal roots. The dorsal roots contain the afferent nerve fibers (sensory part of the spinal cord), while the ventral roots contain efferent nerve fibers (motor part of the spinal cord). Along the dorsal root are allocated the Dorsal Root Ganglia (DRG) made by the bodies of the pseudounipolar neurons. DRG are found allocated in the thickness of the intervertebral holes, where the dorsal- and ventral roots associate from each side of the spinal cord, to compose a pair of spinal nerves having a sensitive and motor component. The spinal cord is made up of an external portion of white matter and a central part of gray matter. In cross section, the spinal cord is divided into two symmetrical halves by the posterior median sulcus and by

From Anatomy to Function of the Central Nervous System
https://doi.org/10.1016/B978-0-12-822404-5.00011-5

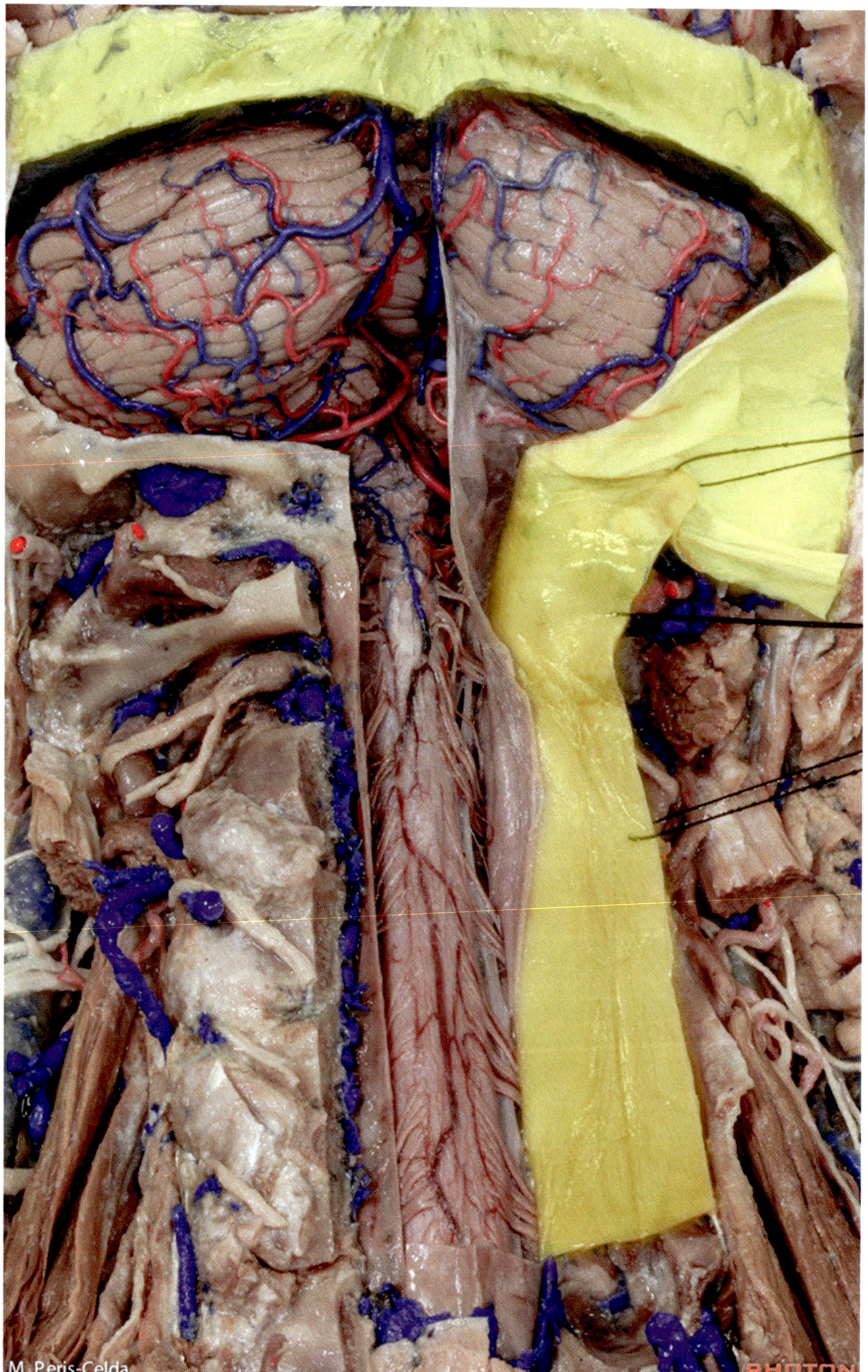

FIGURE 4.1 **Dorsal view of the spinal cord.** In yellow, the dura mater and medially is the arachnoid. *Image courtesy of The Rhoton Collection® http://rhoton.ineurodb.org/. Dissector: M. Peris-Celda. Drawing: Jacob Lepard.*

the anterior median fissure and at cervical and lumbar level presents two enlargements given the greater quantity of gray matter which innervates the upper and lower limbs. At the center of the gray matter and along the entire length of the spinal cord there is a tiny canal, called ependymal canal or central canal, containing the cerebrospinal fluid (CSF). Rostrally, the canal continues into the caudal half of the medulla oblongata to open at the level of the fourth ventricle (see Chapter 5).

SPINAL CORD DEVELOPMENT

During early embryonic development the spinal cord is initially formed from the neural plate (discussed in Chapter 1). The spinal cord can be divided into dorsal and ventral alar plates, which originate by extracellular signaling molecules secreted by the notochord and the adjacent ectoderm (Table 4.1) (Danesin and Soula, 2017; Olmsted and Paluh, 2021; ten Donkelaar et al., 2014). For example, the gradient of Sonic hedgehog (SHH) in the notochord acts as a morphogen factor inducing the formation of the floor plate, controlling the formation of pMN, p3, p2, p1, and p0 progenitor cells, which will originate motorneurons, interneurons, and different glial cells in the basal plate (Danesin and Soula, 2017; Ogura et al., 2018). The roof plate is located dorsally and induces dorsal progenitor domains by production of bone morphogenetic proteins (BMPs) (Ogura et al., 2018). Thanks to the production of a gradient of Shh and BMPs in the floor and roof plates there is activation of transcription factors that will regulate the differentiation and formation spinal interneurons as well as motor neurons. The first synapses appear at approximately 34–36 days post fertilization in the dorsal marginal layer, before this period the synapses are considered transient (Okado, 1981). Work in human-induced pluripotent stem cells (iPSCs) has shed light in the understanding of molecular mechanisms governing motorneurons development (Ho et al., 2016; Sances et al., 2018). For example, motorneurons generation is supported by brain microvascular endothelial cells (BMECs) at

TABLE 4.1 Main signaling pathways involved in spinal cord development.

Signaling pathway	Effect on spinal cord development	References
Sonic hedgehog (SHH)	Responsible for formation the floor plate, p3, p2, p1, and p0 progenitor cells that will generate V3, V2, V1, and V0 interneurons, and pMN progenitor cells that will generate motoneurons.	(Danesin and Soula, 2017; Donkelaar et al., 2014)
Bone morphogenic proteins (BMPs)	BMP 4 and 7 induce formation of the alar and roof plates, the neural crest, and are required for the generation of dorsal interneurons.	(Donkelaar et al., 2014; Ogura et al., 2018)
Retinoic acid (RA)	RA reduces the generation of dorsal interneurons in the spinal cord.	(Ogura et al., 2018)
Notch	Notch signaling prevents premature differentiation of progenitor motoneurons-like, whereas its inhibition promotes differentiation of human embryonic stem cells into motoneurons.	(Ben-Shushan et al., 2015)

approximately 4 weeks post fertilization (Sances et al., 2018). Furthermore, differentiation of interneurons, motorneurons, and primary afferent fibers takes place around 6 weeks post fertilization (ten Donkelaar, 2000).

There is a group of disorders characterized by muscle weakness and/or spastic paralysis as a result of degeneration and/or dysregulation of maturation of a specific neuronal subtype of spinal motor neurons (i.e., amyotrophic lateral sclerosis (ALS), spinal muscular atrophy, primary lateral sclerosis (PLS) (Donkelaar et al., 2014; Ho et al., 2016). The most frequent developmental disorders of the spinal cord are caused by neural tube defects (see Chapter 1).

The evidence suggest that the corticospinal tract reaches all levels of the spinal cord, establishing functional contacts with motor neurons and interneurons in the ventral horn and intermediate nucleus prenatally (Eyre et al., 2000). Dorsal root projections have been reported at 5–8 weeks post fertilization (Donkelaar et al., 2014). The pyramidal tracts reach the caudal medulla oblongata at the end of the embryonic period, and the decussation is formed by 15 weeks post fertilization (ten Donkelaar et al., 2004). However, further development in the number of pyramidal tract fibers that extends down up to the lumbrosacal segments takes place by 27 weeks post fertilization (ten Donkelaar et al., 2004). Furthermore, myelination and maturation of the corticospinal, pyramidal and other descending motor tracts are fully developed after birth (ten Donkelaar et al., 2004; Yeo et al., 2014). Thus, motor development can be divided into gross and fine motor skills, which are related to the descending motor pathways (discussed in motor detail at the end of this chapter).

DEVELOPMENTAL DISORDERS OF THE SPINAL CORD

The most common spinal cord–related pathology during development is the neural tube defect (see Chapter 1). However, there are other less frequent developmental anomalies, such as: (1) duplication of the spinal cord, (2) neuroenteric cysts, (3) syringomyelia, (4) absence or abnormal fiber tracts, among others. There are three types of duplication of the spinal cord: dimyelia (complete duplication of the spinal cord), diplomyelia (accessory spinal cord without roots at the ventral lumbosacral level), and diastematomyelia (complex or typical forms characterized by lateral bifurcation of the spinal cord) (Hori, 1993; Rilliet, 2008). In the case of a neuroenteric cyst (Fig. 4.2), one of its possible origins has been linked with any failure in the notochord to detach itself from the endodermal layer, leading to the formation of a cyst and abnormal gut and vertebral column (ten Donkelaar et al., 2002).

Syringomyelia are tubular cavities in the medulla extending over several segments, with a high incidence in the lower cervical region. It can interrupt spinothalamic fibers, leading to a cape-like distribution of pain. The disorder normally starts after the second decade and slowly progresses (Donkelaar et al., 2014). In the case of absence or abnormal fiber tracts more commonly affects the pyramidal tracts (Miyata et al., 2014; ten Donkelaar et al., 2004). Malformations of the pyramidal tract can take place at various stages of development, and can be caused by abnormal neuronal migration, axon guidance, cell proliferation, defects during the dorsal induction phase of the neural tube, or as a consequence of destructive injuries, such as porencephaly (ten Donkelaar et al., 2004). A potential role of the chromosome 18 into abnormal development of the pyramidal tract has been suggested (Miyata et al., 2014). Eight cases of trisomy 18 aged 16–39 weeks old presented two distinct phenotypic abnormalities: (1) severe hypoplasia of pyramidal tract in female cases and (2) uncrossed pyramidal tract in male cases (Miyata et al., 2014). It has been reported that mutations in the receptor for netrin (DCC), present in the chromosome 18, have been associated with abnormal decussation of pyramidal tracts in a familiar case of congenital mirror movements (Depienne et al., 2011) (Fig. 4.3). A disorder where intentional movements of one side of the body are copied by involuntary movements of the other side. Along the same line, mutations of ROBO3 gene have been identified in patients with horizontal gaze palsy and progressive scoliosis (Jen et al., 2004).

GRAY MATTER OF THE SPINAL CORD

The *gray matter* of the *spinal cord* takes the form of a butterfly, when observed cross-sectionally, where each lateral half can be divided into a *dorsal-* and a *ventral horn* that project respectively toward the dorsolateral and ventrolateral surface of the spinal cord. The gray matter that surrounds the central canal and joins the two halves of the medulla spinalis constitutes the so-called anterior- and posterior gray commissures. The dorsal horn, also known as the posterior horn, is the point where the primary afferent fibers end, while the ventral horn, otherwise referred to as the anterior horn, contains efferent neurons. Between the dorsal and lateral horn, there is also an intermediate column that contains nuclei that innervate visceral and pelvic organs. As suggested by the name, the gray matter is significantly less myelinated than the white matter, containing various prominent nuclei. The gray matter nuclei are organized into nine layers of cellular laminae, commonly defined as Rexed laminae. The dorsal horn is comprised of laminae I–VI, which contains the marginal zone,

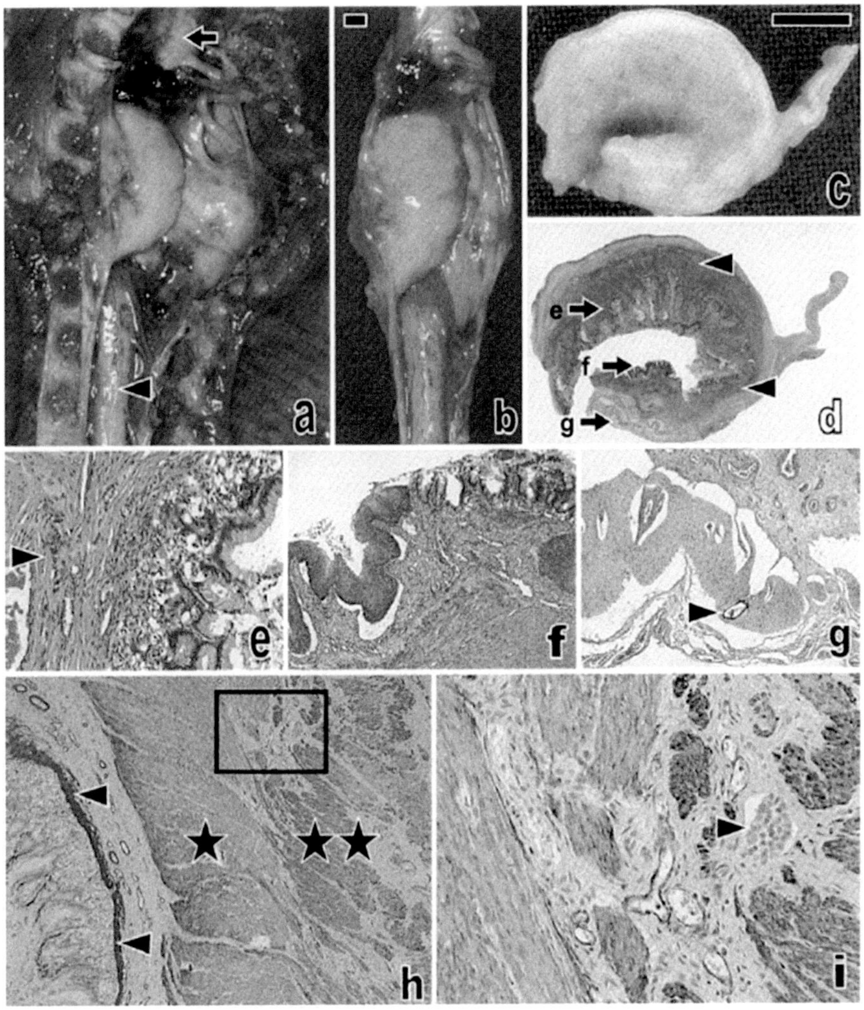

FIGURE 4.2 **Macroscopical (A–C) and microscopical (D–I) neuropathological findings.** In (A) and (B), the in situ position of the cyst and more cranial (*arrow*) and caudal (*arrowhead*) parts of the spinal cord are shown, in (A) after removal of the vertebral bodies and opening of the dural sac, in (B) after removal of the dural sac plus contents from the spinal canal. In (C) and (D), a cross section of the intradural lesion is shown after fixation (C) and in an H&E-stained slice (D). Note the concentric architecture of the lesion with mucosa at the inner surface of the wall, and a prominent surrounding layer of smooth muscle tissue (*arrowheads*) resembling muscularis propria. On the *right*, the transition to the dural sac is seen. In (E–G), higher magnifications of the areas indicated in (D) by *arrows* are shown. Note that the mucosa in (E) resembles fundus-type gastric mucosa with mucus-producing columnar cells in the superficial part and parietal cells (in this image with gray cytoplasm) in the deeper portions of the glands. The *arrowhead* in (E) indicates a cluster of ganglion cells in the submucosa. (F) A transition between mucus-producing mucosa and nonkeratinizing squamous epithelium is seen, the latter resembling the mucosa of the esophagus. (G) Compressed neuroglial tissue in the periphery of the dorsal part of the lesion is seen, the *arrowhead* indicates the central canal with ependymal lining. In (H), an α-smooth muscle actin staining accentuates the muscularis mucosae (*arrowheads*) and the internal and external layer of the muscularis propria (*one* and *two asterisks*, respectively) in the wall of the cystic lesion. The area in the *rectangle* with the transition between those two layers of the muscularis propria in (H) is shown at higher magnification in (I) and contains a cluster of ganglion cells, consistent with plexus myentericus. (D–G) H&E staining. *Bars* (B), (C) 0.5 cm; (E–G), (I) ×200; (H) ×50. *Image courtesy of ten Donkelaar, H.J., Willemsen, M.A., van der Heijden, I., Beems, T., Wesseling, P., 2002. A spinal intradural enterogenous cyst with well-differentiated muscularis propria. Acta Neuropathol. 104 (5), 538–542. https://doi.org/10.1007/s00401-002-0599-4. Epub 2002 Aug 1. PMID: 12410402.*

substantia gelatinosa of Rolando and nucleus proprius (Hochman, 2007). Briefly, lamina I is located at the tip of the dorsal horn, where noxious and thermal information is received in a somatotopically mapped manner. This is also the site of the marginal zone nuclei, which send information to the brain via the lateral spinothalamic tract (see below). Lamina II, corresponding the substantua gelatinosa of Rolando, moderates noxious and nonnoxious sensory information, with direct communication to lamina III and IV. In contrast, lamina III, encompassing the nucleus proprius, controls proprioception as well as light touch. The remainder of the nucleus proprius is found in lamina IV, involved in the reception of nonnoxious sensory input. Lamina V

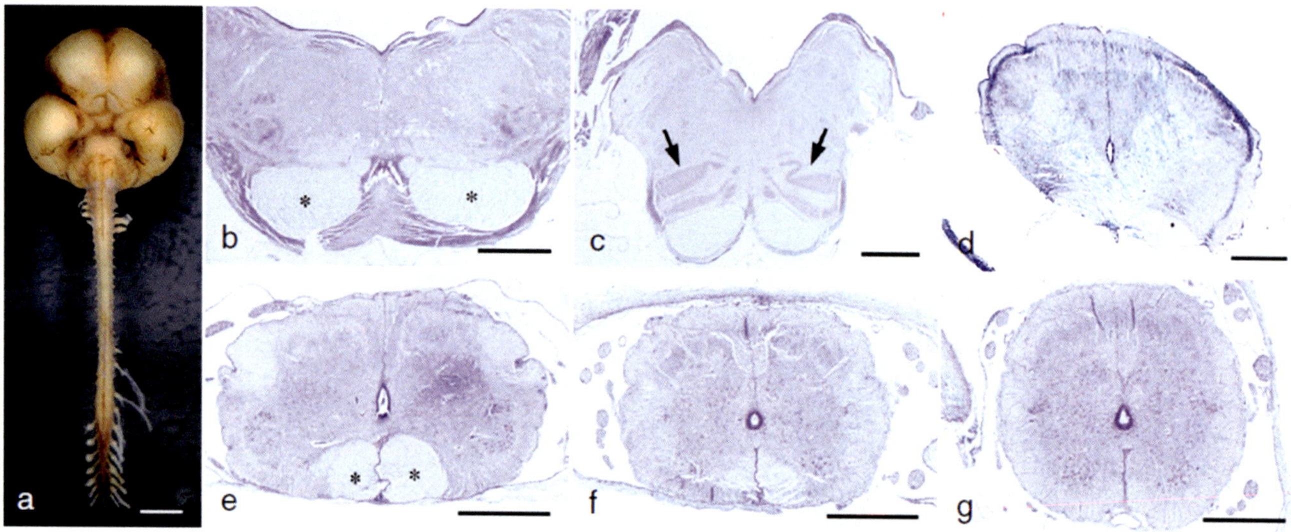

FIGURE 4.3 **Dominant uncrossed pyramidal tract in trisomy 18.** Formalin-fixed brain and spinal cord (panel A) obtained from a 21-week gestation fetus with trisomy 18 having male karyotype (case 4), showing hypoplasia of the pontine base and cerebellum. Cervical and lumbar enlargements appear proportionately developed. The fixed specimen was placed in the water and photographed. Transverse sections of the brainstem and spinal cord show relatively large pyramidal tracts without segmentation in the hypoplastic pontine base (asterisks in panel B), hook-shaped inferior olivary nuclei with dorsal thickening (arrows in panel C), pyramidal decussation (panel D), and dominant anterior corticospinal tracts in the spinal cord at the fourth cervical (asterisks in panel E) and sixth thoracic segments (panel F). The pyramidal tract is not clearly detected in the lower thoracic segment (panel G). KB method, bar in panel A = 10 mm, bars in panels B–G = 1000 μm. *Image courtesy of Miyata, H., Miyata, M., Ohama, E., 2014. Pyramidal tract abnormalities in the human fetus and infant with trisomy 18 syndrome. Neuropathology 34 (3), 219–226. https://doi.org/10.1111/neup.12081. Epub 2013 Dec 9. PMID: 24313853.*

neurons receive modality ambiguous inputs from different pain and innocuous afferents (mediated by Aδ, C, II, and IV sensory axons). This lamina also receives information from descending corticospinal and rubrospinal tracts, allowing for sensory integration of a whole-body response. Only present in the cervical and lumbosacral enlargements, lamina VI contains small interneurons that receive input from muscle spindles and are involved in the modulation of spinal reflexes. This information is further transmitted to the brain via the ipsilateral spinocerebellar pathways. Lamina V and VI are cytologically homogenous, making difficult to clearly distinguish them (Schoenen and Faull, 2004).

The ventral horn is made up of laminae VII–IX, encompassing glutamatergic excitatory and glycinergic inhibitory interneurons. Specifically, lamina VII, which contains the dorsal nucleus of Clarke (see later) as well as some lateral and medial motor neurons, is primarily involved in the regulation of the autonomic nervous system. Lamina VII, both, receives and relays information to the viscera. The central cervical nucleus (CeCv) is located in lamina VII, and received afferent information from upper cervical muscles, cervical joints, lower cervical dorsal roots, and semicircular canals. It is involved in tonic neck reflex, responses to cutaneous, proprioceptive and vestibular stimuli (Heise and Kayalioglu, 2009). Lamina VIII, which houses lateral and medial motor nuclei, is largely confined to the medial part of the anterior gray column and is most prominent at the cervical as well as lumbar vertebral levels. In contrast, the longitudinal columns of lamina IX contain the large α-motor neurons that send axons to innervate skeletal muscle fibers. Although these are largely cholinergic, there are some that apply co-transmitter glutamate. Lastly, the gray matter region surrounding the central canal, lamina X, is responsible for somatosensory processing and visceral nociception, receiving predominately high-threshold Aδ-fiber and C-fiber afferents (Krotov et al., 2019).

WHITE MATTER OF THE SPINAL CORD

The white matter of the spinal cord consists bilaterally of three large components: the dorsal, lateral, and ventral cords made up by myelinated ascending and descending fibers. The dorsal cord consists of primary afferent fibers that enter into the medulla spinalis through the dorsal roots coming from the peripherical nerve (Fig. 4.4), that carries stimuli of vibration, discriminating touch (epicritic), proprioception and weight. Is important classify the afferent fibers that reach the spinal cord based on:

- grade of myelination;
- diameter (or thickness);
- function.

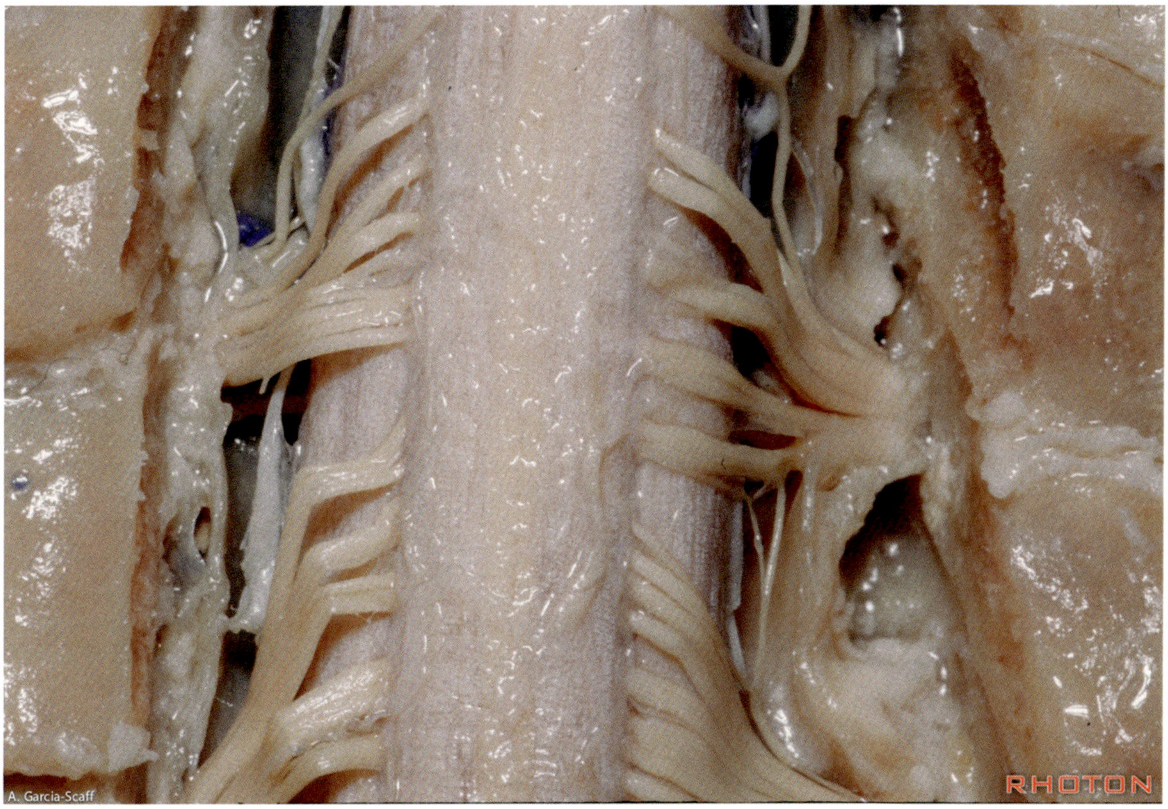

FIGURE 4.4 **Spinal cord and dorsal root of spinal nerves emerging from both posterolateral sulci.** *Image courtesy of The Rhoton Collection® http://rhoton.ineurodb.org/. Dissector A. Garcia-Scaff.*

Fibers are subdivided into A, B, and C groups, on the base of their speed of conduction.

Group A fibers are in turn subclassified into alpha (α), beta (β), gamma (γ), and delta (δ). This additional subdivision is for different degrees of myelination and thickness of the axon, so larger diameter and more myelinated axons transmit the stimulus faster.

Group A fibers are found in both motor and sensory pathways.

Type Aα fibers include type Ia and type Ib sensory fibers of the alternative classification system and they represent the fibers from muscle spindle and the Golgi tendon organs (GTOs), respectively (Oliver et al., 2021).

Type Aβ fiber and type Aγ are included in the type II afferent fibers from stretch receptors. Type Aβ fibers from the skin are mostly dedicated to touch. However, a small fraction of these fast fibers also transmit pain (Nagi et al., 2019). Type Aδ fibers or fibers type III are the afferent fibers of nociceptors. These fibers carry information from peripheral mechanoreceptors and thermoreceptors to the dorsal horn of the spinal cord and terminate in Rexed laminae I–V (Basbaum et al., 2009).

Aδ fibers receive stimuli from the periphery and retransmit them to the somatosensorial cerebral cortex (Brodmann's Areas 3,1,2), primarily relating to acute pain (scharp, immediate, and relatively short lasting). The type of pain results from several nociceptive stimuli: temperature-induced, mechanical and chemical. Aδ fibers being thin (2–5 μm in diameter) and myelinated, send impulses faster than unmyelinated C fibers but more slowly than others more thickly myelinated of the group A of nerve fibers (see above); their conduction speed is moderate. The cell bodies of the A fibers are located in the dorsal root ganglia (DRG) which send a sensitive fiber (a dendrite-like fiber) to the periphery in order to innervate the target organs and a fiber redirected back to the spinal cord (refer to the histology manuals regarding the morphology of the pseudounipolar neuron). Group B nerve fibers are axons moderately myelinated, however less than group A and more than group C. They are usually general visceral afferent fibers and preganglionic nerve fibers part of the autonomic nervous system.

Group C or type IV group fibers are unmyelinated and have a small diameter and low conduction velocity, these latter include postganglionic fibers in the autonomic nervous system (ANS) and some nerve fibers at the dorsal roots.

ASCENDING FIBERS SYSTEM

Dorsal Column (or Dorsal Fasciculi) Medial Lemniscus Pathway (DCML)

All these sensitive fibers once entered into the spinal cord first contact the Lissauer's tract (see later) that covers the dorsal horn to form the dorsal cord contained between dorsal horn and the dorsal median septum.

The dorsal cord is composed of two large ascending tracts of myelinated fibers called gracile fasciculus and cuneate fasciculus (both are also called posterior columns or dorsal columns). The gracile fasciculus carries the sensitive information from the lower extremities of the body while the cuneate fasciculus from the upper extremities. The gracile and cuneate fasciculi carry tactile information, discriminatory touch, vibration, position sense, movement sense, and conscious proprioception.

The gracile fasciculus and the cuneate fasciculus originate, respectively, at the caudal and at the thoracic\upper region of the spinal cord and contain ascending fibers from the primary afferent fibers that enter the spinal cord through the dorsal roots; the primary afferent fibers have their own neuronal body in DRG and constitute the first-order neurons. The axons from the first-order neurons form the ascending tracts of the gracile fasciculus and of the cuneate fasciculus which synapse on the second-order neurons in the gracile nucleus and the cuneate nucleus both located in the lower medulla oblongata; as is possible to see in Figs. 4.5 and 4.6 the gracile fasciculus is located medially to the cuneate fasciculus.

Some axons of second-order neurons of gracile and cuneate fasciculi arise from the laminae IV–VI of the ipsilateral dorsal horn and ascend along the dorsal cord (for some more information see Chapter 17).

From the nucleus gracile and cuneate originate the axons (second order axons) which cross to the other side of the medulla oblongata (sensory decussation) and are so-called because the internal arcuate fibers continue to ascend as the medial lemniscus ending in the posterior ventral lateral nucleus (VPL) of the contralateral thalamus. The neurons of the thalamus, representing the third order neurons, send their axons through the posterior arm of the internal capsule to reach the postcentral gyrus (BA 3,1,2) of the cerebral cortex determining the conscious perception of stimuli. All axons in the DCML pathway are large, myelinated, fast conducting fibers (see above) (O' Sullivan et al., 2019). It Is necessary to specify that the somatosensory perception mediated by DCML carry only somatosensitive information from all the body, excluding the head and face. In fact, sensitive information from the latter is carried by the trigeminal lemniscus that is made up by fibers deriving from the trigeminal ganglion.

SPINOCEREBELLAR TRACT

Another spinal tract that carries exteroceptive and proprioceptive stimuli is represented by the spinocerebellar tract, in turn distinguished in a dorsal and ventral component. Both originate from the peripherical white matter of the spinal cord: the dorsal spinocerebellar tract is located laterally to the lateral corticospinal tract (descending fibers, see later) instead the ventral spinocerebellar tract is located just ventrally to the dorsal one.

The dorsal spinocerebellar tract originates from the neurons of the Clarke's column of the gray matter (a group of interneurons found in the medial part of lamina VII) and ascends in direct way until the medulla oblongata until it enters the cerebellum through the inferior cerebellar peduncle. It has been detected in 20-week-old fetuses. The ventral spinocerebellar tract originates from the neurons of the V–VII laminae of the gray matter and most of these fibers ascend in a decussated way until the mesencephalon where, through the superior cerebellar peduncle, get in the cerebellum where decussate again. The neurons in lamina V receive information from Aβ, Aδ, and C axons which also carry nociceptive information from visceral organs. The neurons cell bodies of the V lamina express neurotransmitters and related protein such as: GABA, dopamine, glycine, somatostatin, and parvalbumin (Heise and Kayalioglu, 2009). The neurons in lamina VI are involved in spinal reflexes and project to the cerebellum, whereas descending brainstem pathways project to the lateral zone of the layer VI. It is the deepest layer of the dorsal horn with the majority of cells either propriospinal neurons or interneurons with short locally projecting axons, targeting the ventral horn motoneurons. Similar to cells in the lamina V, they express GABA, glycine, parvalbumin, and calretinin (Heise and Kayalioglu, 2009). Furthermore, neurons in lamina VII receive information from lamina II to VI along with information from visceral afferent fibers. The neurons in ths region have been shown to have the same neurochemical phenotype (Heise and Kayalioglu, 2009).

The spinocerebellar tracts are organized so that the fibers originating from the lower spinal neurons are more superficial and vice versa. For how concern the dorsal spinocerebellar tract, the neurons of the Clarke's column are activated monosinaptically by the afferent primary fibers Ia and Ib coming, respectively, from neuromuscular spindles and tendon organs (see Chapter 17), but also by the fibers of type II and afferent fibers responsible of the tactile and pressor sensibility. The dorsal spinocerebellar tract therefore transmits specific information for the spatial origin and for the sensory modality used for the fine coordination of the individual muscles of the limbs. The neurons of lamina V–VII, that give rise to the ventral spinocerebellar tract, are activated monosynaptically by

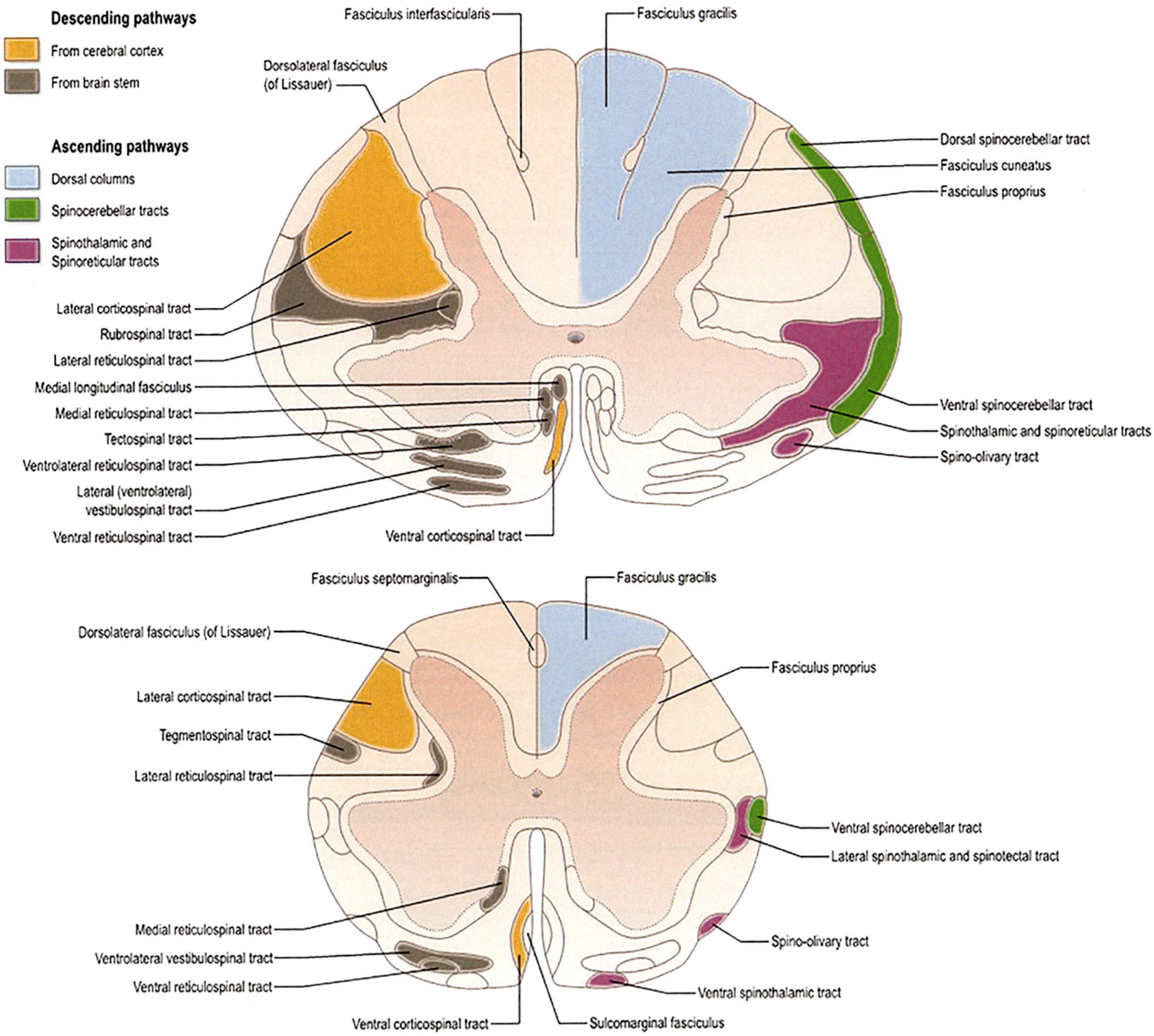

FIGURE 4.5 **Schematic positions of the various bundles of fibers (ascending and descending) in the spinal cord at the midcervical (top) and lumbar (bottom) levels.** *Image courtesy of Adler, J.T., 2010. Gray's Anatomy: The Anatomical Basis of Clinical Practice, 40th edition. J. Surg. Res. 158 (1), 28–29. https://doi.org/10.1016/j.jss.2009.01.035.*

fibers of type Ib and transmit information from large receptor fields that include different segments of a limb. The ventral spinocerebellar tract has no internal subdivisions for the various sensitive modalities and transmits information for the coordination of movement and posture of the entire limb.

SPINOTHALAMIC TRACT

Another very important sensory way making part of the ventrolateral system is the spinothalamic tract which is located in the ventral cord of the medulla spinalis (Fig. 4.7). The first-order neurons of the spinothalamic tract are located in the DRG. They project peripheral processes to tissue in the form of free nerve endings that are sensitive to molecules indicative of cell damage. The central processes enter the spinal cord tract and ascend about two levels there before synapsing on second-order neurons located primarily in laminae I, IV, V, and VI. Therefore, the axons continue to the ventral commissure where they decussate and reach the contralateral site to ascend along the ventral cord of the spinal cord. Some neuroanatomists today tend to no longer consider this tract consisting of two components, one ventral and one lateral, but as a single tract of fibers.

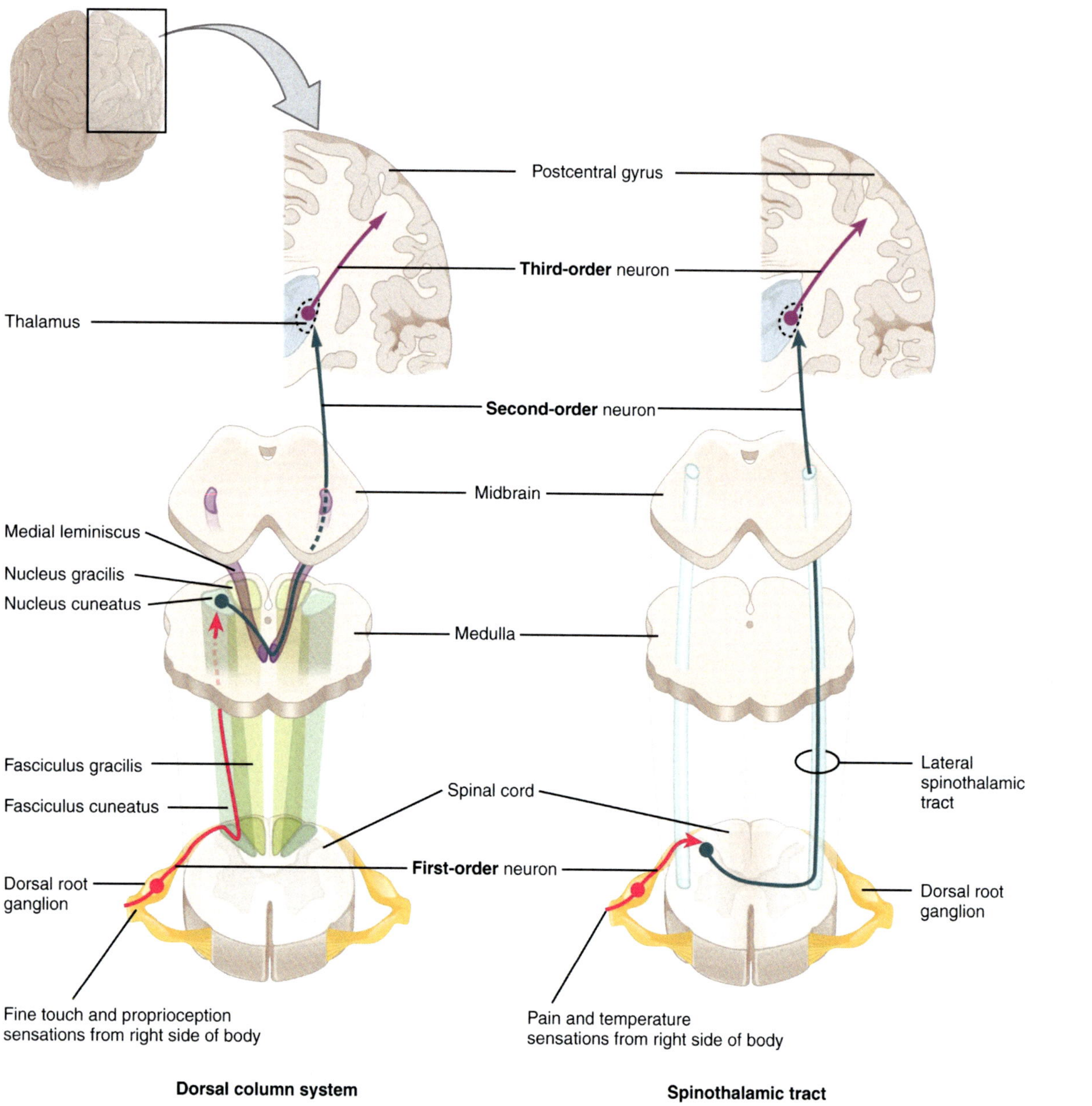

FIGURE 4.6 **Anterior view of ascending sensory pathways of the spinal cord.** The dorsal column system and spinothalamic tract are the major ascending pathways that connect the periphery with the brain. *Image courtesy of OpenStax College - Anatomy and Physiology, Connexions Web site. http://cnx.org/content/col11496/1.6/, June 19, 2013.*

However, we prefer to keep the description of the two components separate to underline that, from a neuro-functional point of view, some specific information is conveyed by a specific component of the spinothalamic tract rather than by the other. The spinothalamic tract is responsible for the tactile perception (not discriminative) and of the thermo-pain perception. The fibers of the lateral component of the spinothalamic tract, ascending along the lateral cord and reaching the brainstem, continue to ascend making synapses with the Ventro-Postero-Lateral Nucleus (VPL) of the thalamus. This tract also terminates on the parietal cortex of the telencephalon sited on the post-central circumvolution (BA 3,1,2) passing through the posterior arm of the internal capsule. Moreover, fibers of this tract, ascending in the spinal cord, make synapses with fibers of the reticular formation of the brainstem taking part in the reticulospinal tract that works mainly in the slow and widespread pain conduction (see later, spinoreticular tract) (Willis et al., 2008).

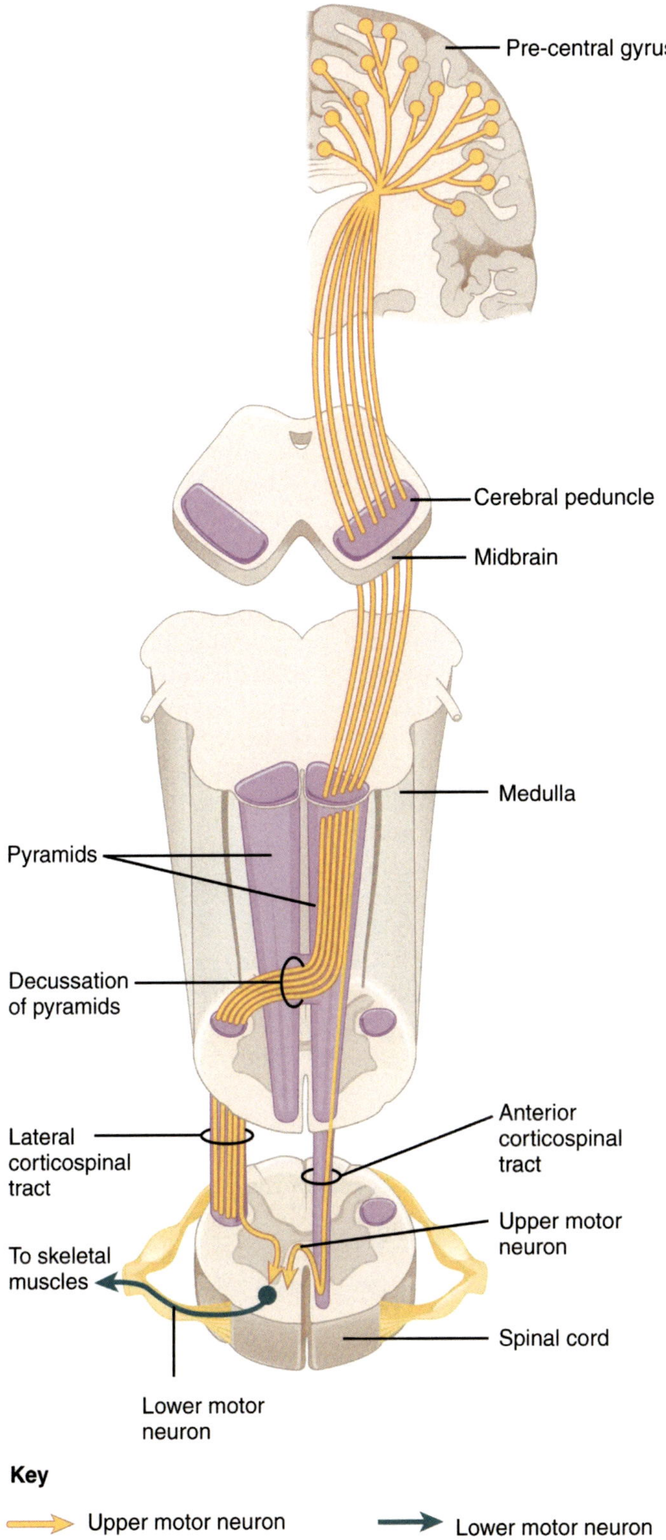

FIGURE 4.7 **The major descending tract that controls skeletal muscle movements is the corticospinal tract.** It is composed of two neurons, the upper motor neuron and the lower motor neuron. The upper motor neuron has its cell body in the primary motor cortex of the frontal lobe and synapses on the lower motor neuron, which is in the ventral horn of the spinal cord and projects to the skeletal muscle in the periphery. *Image courtesy of OpenStax College - Anatomy and Physiology, Connexions Web site. http://cnx.org/content/col11496/1.6/, June 19, 2013.*

The ventral component of the spinothalamic tract, is located in the ventral cord of the spinal cord medially the ventral roots emergence and, after its decussation, proceed its path reaching the VPL and the parietal cortex. These fibers are responsible for the tactile perception (not discriminative) and for the baresthesia, the ability to evaluate the weight of objects or to differentiate objects of different weight, by holding or lifting them. The two components of the spinothalamic tract once reached the brainstem they separate each other, so that the axons of the ventral tract join the medial lemniscus instead the axons of the lateral tract join the spinal lemniscus. Both the lemniscus terminate in the VPL nucleus of the thalamus, this latter in fact is a large mass of gray matter located in the dorsal part of the diencephalon (see Chapter 8) having several functions first of all the one of retransmit the sensory signals and motor signals (see Chapter 15) to the cerebral cortex by the relè nuclei (Scherman et al., 2006). The thalamus also plays an important role in the regulation of consciousness, sleep, and alertness (see in the next chapters). As before said the spinothalamic tract is the most important fibers system involved in the mechanism of pain control therefore it will be described in detail in Chapter 17, above all under neurofunctional point of view. It is important to highlight that other important tract of fibers, as well as the spinoreticular and spinomesencephalic tract, are intercalated in the lateral spinothalamic tract and they will be described in the next pages.

SPINORETICULAR TRACT

The spinoreticular tract begins with first-order neurons which immediately synapse the second-order neurons located in the dorsal horn of the spinal cord and the respective fibers decussate by carrying themselves ventrolaterally, then running along the ventrolateral cord terminating in the reticular formation of the brainstem (Lima et al., 1991).

Sensitive information coming from the reticular formation in turn reaches the intramedian nucleus of the thalamus which belongs to the group of intralaminar nuclei which diffusely project to the entire cerebral cortex where pain reaches the conscious level and promotes behavioral arousal. In fact, it is responsible for the automatic responses to pain, as in the case of an injury, in fact it is believed that the spinoreticular tract projects to neurons with a large receptive field that can cover large areas of the body and play a role in the memory and affective component (emotional) of pain. It is still not quite known whether the spinoreticular tract has ipsilateral fibers in addition to the decussate fibers, however it has been hypothesized that during development the tract was bilateral but that the ipsilateral synapses became ineffective in the postdevelopmental period (Menétrey et al., 1980; Soja et al., 2001). The pathway traveled by the spinoreticular tract explains why when a painful stimulus is perceived in a sleeping subject, he wakes up immediately, in fact, as will be illustrated in Chapter 5b, the reticular formation participates in the ascending reticular activation system (ARES) which projects to the whole cerebral cortex. The spinoreticular tract also projects to the mediodorsal nucleus (MD) of the thalamus which takes part in the structure of the limbic system and sends fibers to the orbitofrontal cortex (OFC, a type of prefrontal cortex) which plays a vital role in mood. At the same time the MD sends fibers to the cingulate cortex, which is why pain has an effect on the emotional system. As far as explained there is a reason why there are neurosurgeries performed on people who have acute and severe pain resulting from cancer or other conditions. This type of intervention essentially destroys the cingulate fascicle in the anterior cingulate cortex (ACC) so that the patient follows by perceiving the somatic sensation of a pain through the spinothalamic tract, but the pain no longer has any effect on the subject's emotionality: the pain somehow continues to be felt by the person but as a feeling of touch or pressure (Sharim et al., 2016). This is explained in more detail in Chapter 12.

SPINOMESENCEPHALIC TRACT

The spinomesencephalic tract is made up of fibers that follow the same path of the last two tracts above described, however other tracts too are mixed to the spinomesencephalic one including the spinotectal tract that projects to the superior colliculus. In addition, the spinomesencephalic fibers synapse also with the periacqueductal gray (PAG) playing a very important role in the modulation of the pain via descending projections (see Chapter 17). The spinomesencephalic tract plays a role especially in the pain perception and likely the spinoreticular tract is involved in emotional elaboration the pain stimuli.

SPINO-OLIVARY TRACT

The fibers that make up the spino-olivary tract come from the DRG and once entered into the spinal cord synapse with the second-order neurons of which axon decussate and ascend along the ventrolateral cord of the medulla spinalis and terminate by synapsing on the inferior olivary nuclei located in the medulla oblongata. Once reached the olivary nuclei the axons of third-order neurons, located in the olivary nuclei decussate again and enter the cerebellum through the inferior cerebellar peduncle. The spino-olivary tract conveys information to the cerebellum from cutaneous and proprioceptive organs providing transmission of unconscious proprioception and balance.

There is little information about the fibers of this tract in primates, but there is evidence in the cadaver that degeneration of axonal terminals in the inferior olivary nuclei can be detected following chordotomy.

DESCENDING FIBERS SYSTEM

Pyramidal System

Descending fibers of the spinal cord are made up of motor fibers mainly originating by cerebral cortex and from brainstem and that are at the base to the control of movement, muscular tone, and posture. From a phylogenetic point of view, there are two tracts of motor fibers that together make up the so-called pyramidal system. These two tracts are:

1. the corticospinal tract;
2. the corticonuclear (or corticobulbar) tract.

The corticospinal tract terminates within the motor neurons of the ventral horn of the spinal cord while the corticonuclear (or corticobulbar) tract, passing through the reticular formation with which they make synapses, terminates within the voluntary motor nuclei of the cranial nerves sited into the brainstem (Fig. 4.7).

For how concern the corticospinal tract it originates from several parts of the cerebral cortex:

- 1/3 of the fibers originates from the precentral gyrus where is located the primary motor cortex (M1, BA4), from a kind of the pyramidal neurons known as *Betz's giant cells* localized in the lamina V. These types of neurons are multipolar and excitatory, they utilize the glutamate as neurotransmitter;
- 1/3 of the fibers originates from the premotor cortex (BA6) lying just anteriorly M1. The medial extension of BA-6, onto the midline surface of the hemisphere, is the site of the supplementary motor area (SMA);
- 1/3 of the fibers have been seen to originate in the postcentral gyrus (BA 3,1,2), that as seen before plays a role mainly in the somatosensorial perception. The fact that corticospinal fibers originate from a sensory area has been seen to have particular importance in integration of the sensibility with the motricity, playing a very important role in the realization of fine movements. Indeed, it is known a pathology that compromises the somatosensory system so that the affected person doesn't know where his body segments are, an inability occurs for how concern the feedback mechanism with the motor system; this condition could configurate in a clinical condition known as sensory ataxia.

Thus, all motor fibers originating from the areas above illustrated head down forming a structure that looks like a fan (in a lateral view) called corona radiata which descending through the posterior arm of the internal capsule is limited medially by the thalamus and laterally by lenticular nucleus. The fibers continuing to descend reaching the brainstem, so they pass through the cerebral peduncles (mesencephalon) continuing to go down until the pons where they tend to be more dispersed because they are between the pontine nuclei having a great anatomy–functional relation with the neocerebellar pathway (see later). At level of the medulla oblongata fibers return to being more compact forming the structures called "pyramids" at the ventral face of the medulla oblongata. When the brainstem and the spinal cord going to unite each other, the most of pyramidal fibers (around the 80%) decussate forming the lateral corticospinal tract that runs long the lateral cord of the medulla spinalis. The remaining 15% or 20% of corticospinal fibers do not decussate and descend in a direct way forming the direct corticospinal tract; the latter, in turn, consists of descending fibers that arise from cells in the motor area of the ipsilateral cerebral hemisphere. The direct corticospinal tract decussates "in situ" with small fibers that pass to the contralateral half of the spinal cord. Both these tracts end on motor neurons of the ventral horn mainly of the laminas VIII and IX, also if some collaterals end on neurons of the dorsal horn in order to have relations with sensitive neurons. In lamina IX there are the alpha and gamma motor neurons. The alpha motor neurons innervate the extrafusal muscle fibers that generating tension by contraction, allow the skeletal movement. The gamma motor neurons innervate intrafusal muscle fibers that serve as specialized sensory organs (proprioceptors) that detect the amount and rate of change in length of a muscle; moreover, these fibers also regulate the muscle tone. A computational approach based on experimental observation of tremor kinematic and electromyography in Parkinson's patients allowed hypothesized that the gamma-dynamic command is integrated with the alpha-dynamic command to generate premotoneuronal outputs. This suggests an alpha and gamma coordination in descending commands, where alpha motor neurons provide the main guide for muscle and gamma-motor neurons develop a more subtle control for maintenance of posture and movement dynamics (Hao et al., 2013; Li et al., 2015). Fig. 4.8 shows the Diffusion Tensor Imaging (tractography) of the cervical cord and brainstem. DTI is one of the most popular and widely used (Assaf et al., 2008) MRI techniques in brain research and diagnostic to describe the orientation of white matter fibers. The process of fiber tracking, called tractography, allows for a virtual dissection and three-dimensional representation of white matter tracts

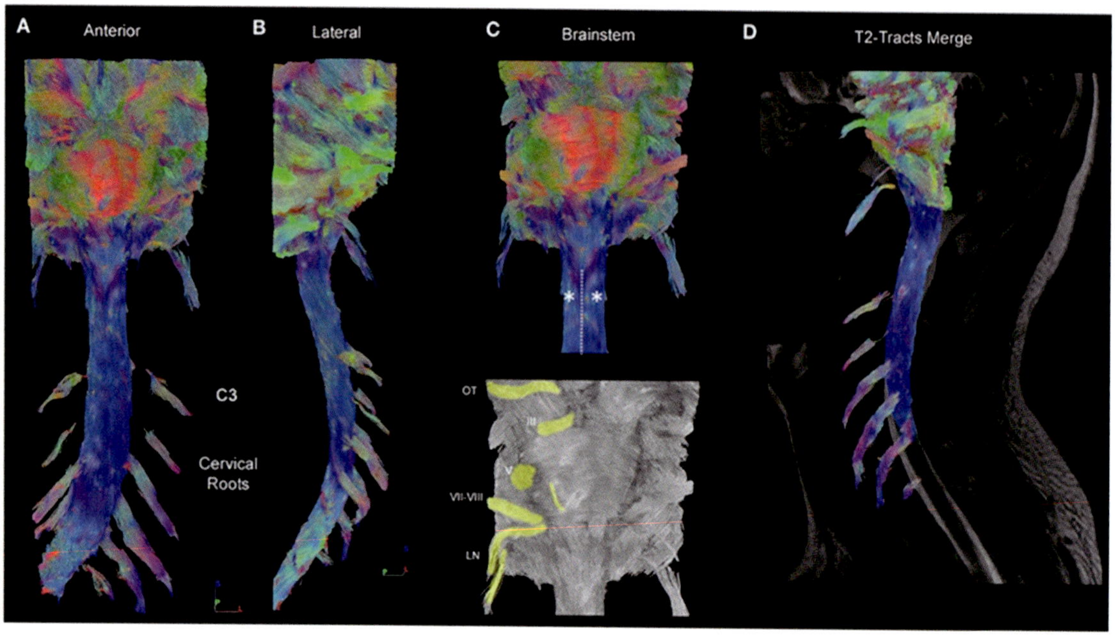

FIGURE 4.8 **Full cervical cord tractography (healthy subject).** (A) Ventral and (B) lateral views of the cervical spinal cord and brainstem tractography showing the ascending/descending pathways with cervical rootlets. The emergence of the rootlets from the spinal cord can be seen laterally (turquoise). Spinal cord fibers (blue) are well reconstructed on the whole length of the spinal cord in a subject who had an important cervical lordosis. (C) From this enlargement of panel A on the brainstem, pontocerebellar fibers are seen (red), cranial nerves (from optic tracts [OT] to lower nerves [LN]), including acoustic-facial bundle [VII–VIII] are highlighted (yellow). The junction between the medulla and spinal cord was identified by its pyramidal bulging (*), and by an invagination that corresponds to the median ventral sulcus (dotted line). (D) The spinal cord tractography is superimposed on the sagittal T2-weighted images. *Image courtesy of Dauleac, C., Frindel, C., Pélissou-Guyotat, I., Nicolas, C., Yeh, F.C., Fernandez Miranda, J., Cotton, F., and Jacquesson, T., 2022. Full cervical cord tractography: A new method for clinical use. Front. Neuroanat. 16. https://doi.org/10.3389/fnana.2022.993464.*

(Catani et al., 2008). As DTI tractography assesses the structural integrity of white matter, it has been widely used for prognostic and/or diagnostic purposes in various brain pathologies such as stroke (Moller et al., 2007), neurodegenerative diseases (Lee et al., 2013), and brain tumors (Berberat et al., 2014).

DESCENDING FIBERS SYSTEM

Extrapyramidal System

Reticulospinal Tract

The reticulospinal tract originates from the reticular formation of the brainstem and in particular way from the pons (medial reticulospinal tract) and from the medulla oblongata (lateral reticulospinal tract). The medial reticulospinal tract goes down running along the ventromedial cord of the medulla spinalis while the lateral reticulospinal tract descends along the lateral cord of the medulla spinalis; both have a direct and a crossed component. The reticulospinal tract ends on neurons of the ventral horn, mostly found in the lamina IX, that innervates the proximal and axial musculature of the body. Motor neurons located in the ventral horn of lamina IX are cholinergic and these cells have been shown to also express dopamine and calmodulin (Heise and Kayalioglu, 2009). The reticulospinal tract projects directly and indirectly (through interneurons) to spinal motor neurons controlling hand muscles (Riddle et al., 2009; Baker et al., 2011). A single reticulospinal neuron exhibits extensive collateralization projecting to multiple motor neuron pools (Alstermark et al., 1986; Matsuyama et al., 1999), suggesting that this tract is capable of activating groups of muscles required for finger manipulations. However, the reticulospinal fibers also end on laminae VII–VIII and the medullary fibers of it (lateral reticulospinal tract), that are the most extended synapse on the VI–VII (Standring, et al., 2008) The majority of cells in lamina VII are premotor interneurons projecting to the motor cells located in lamina IX (Heise and Kayalioglu, 2009). Neuronal populations express GABA, glycine, glutamate, acetylcholine, serotonin, and some other GABA neurons–related proteins such as somatostatin, parvalbumin, calretinin etc. (Heise and Kayalioglu, 2009). Furthermore, in lamina VIII there is more cellular heterogeneity, i.e., size and shape compared to lamina VII. It has been suggested that the majority of cells in lamina VIII are interneurons and express GABA. Regarding connections, short propriospinal fibers are derived from neurons in the lateral halves of laminae V, VI, and VII and normally connect segments nearby. Long propriospinal fibers that connect the cervical and lumbar segments arise from laminae VII and VIII. The reticulospinal tract, in particular way the medial reticulospinal tract, is one of the most important way responsible of the posture maintenance, e.g., enable extension of the legs to maintain postural support (Michielsen et al., 2019). The lateral reticulospinal tract is responsible for flexor motor neurons in fact, inhibiting the medial reticulospinal tract and therefore extensor motor neurons, enable the modulation of stretch reflex (Trompetto et al., 2014). In addition, as in the Chapter 21 will be discussed, the premotor cortex is responsible for the overarching control of locomotion projecting to the brainstem and so to the reticulospinal tract. As a consequence, the reticulospinal tract is able to modulate the control of the locomotion when works with the corticospinal tract (Drew et al., 2004).

In patients who have suffered corticospinal lesion (e.g., stroke survivors), the reticulospinal tract is the most likely candidate descending pathway for functional recovery of hand movements (Zaaimi et al., 2012). Since the reticulospinal tract does activate hand muscles in healthy individuals (albeit weakly), functional recovery would require only the strengthening of this preexisting output, and not the growth of an entirely new category of connections not previously present. Preliminary data show that reticulospinal outputs do indeed strengthen after recovery from corticospinal lesion (Zaaimi et al., 2012).

VESTIBULOSPINAL TRACT

Another extrapyramidal way that interacts with the previous one is the vestibulospinal tract (VST). Two components of the vestibulospinal tract have been described: the medial VST and the lateral VST. The lateral VST originates from the lateral vestibular nuclei in the lower pons, and descends through the ventrolateral region of the medulla oblongata, then terminates at the lateral funiculus of spinal cord. The medial VST originates from the medial vestibular nuclei at the level of lower pons, descends through dorso-medial region of medulla oblongata, then terminates at the ventral funiculus of the spinal cord (Jang et al., 2018). The lateral VST descends ipsilaterally and excites the neurons of laminae VII and VIII. At the same time this tract ends on the interneurons contained in the same laminae inhibiting the contralateral motor neurons. It helps to maintain an upright and balanced posture by stimulating extensor motor neurons in the legs and also innervates muscles of the trunk, thus additionally aiding in body posture.

The medial VST runs down joining to the medial longitudinal tract (mlt), toward the ventral funiculus and terminates on neurons of laminae VII and VIII. Unlike lateral VST, the medial VST innervates muscles that support the head. As a result, medial VST fibers run down only to the cervical segments of the spinal cord (Afifi et al., 1998).

When vestibular nuclei are stimulated, they retransmit the signals to the archicerebellum through the vestibulocerebellar tract. In this way the archicerebellum send back again the information to the vestibular nuclei that trough the vestibulospinal tract goes to stimulate the alpha motor neurons of the ventral horns which send efferent fibers that innervate the neuromuscular junction of the muscles causing a response to the primary stimulus.

RUBROSPINAL TRACT

In humans, one of the major pathways involved in the motor control is the rubrospinal tract. This tract arises from the red nucleus located in the mesencephalon (see Chapter 5) that is formed by two portions: magnocellular and parvocellular. The rubrospinal tract originates from the magnocellular portion and descends in crossed way until the spinal cord. Its axons decussate in the ventral tegmentum crossing the pons and the medulla oblongata ending on laminae V–VII of the gray matter of the spinal cord. The rubrospinal tract runs along the lateral cord of the medulla spinalis and it is very related to corticospinal tract (the decussed component). In the human rubrospinal tract seems to be present almost exclusively in the cervical portion of the spinal cord and playing a role mostly in the contraction of the flexor musculature of the upper limbs. However, it is found to be rudimentary in humans as well as is almost less than half in the bipedal primates than in the quadruped ones (ten Donkelaar et al., 1988). Is interesting from the functional point of view that the rubrospinal tract makes monosynaptic connections to motoneurons innervating distal muscles involved in finger movements (Cheney et al., 1991). A lesion within the region of the red nucleus can result in movement disorders and tremors, but these effects may be more associated with damage to fibers which are associated with the cerebellar and the basal ganglia system.

TECTOSPINAL TRACT

The tectospinal tract originates from the tectum of the mesencephalon mostly by the superior colliculi then, continuing to descend and joining the medial longitudinal fasciculus, projects to the contralateral portion of the first cervical neuromeres on that laminae VI–VIII (Fregosi et al., 2019). The tectospinal tract is well developed in other animals such as the cats, but appears to have a lesser role in humans and is not well understood (Muto et al., 1996). The tectospinal tract has also a tectonuclear component which innervates the oculomotor, trochlear, and abducens nuclei needed for the voluntary movements of the eyes. The function of the tectospinal tract is to mediate reflex postural movements of the head in response to visual and auditory stimuli.

For example, if you were sitting in a quiet room and all of a sudden heard a noise to your right, you would subconsciously turn your head in that direction and orient your eyes toward the direction of the sound, attempting to find the source. In this way, the tectospinal tract will cause head and neck movements with this type of auditory startle response (Okoyama et al., 1991; Yamamoto et al., 2017).

References

Afifi, A., 1998. Functional Neuroanatomy. McGraw Hill, ISBN 978-0-07-001589-0.

Alstermark, B., Sasaki, S., 1986. Integration in descending motor pathways controlling the forelimb in the cat. 14. Differential projection to fast and slow motoneurones from excitatory C3-C4 propriospinal neu rones. Exp. Brain Res. 63 (3), 530–542. https://doi.org/10.1007/BF00237476. PMID: 3019752.

Assaf, Y., et al., 2008. Diffusion tensor imaging (DTI)-based white matter mapping in brain research: a review. J. Mol. Neurosci. https://doi.org/10.1007/s12031-007-0029-0.

Baker, S.N., December 1, 2011. The primate reticulospinal tract, hand function and functional recovery. J. Physiol. 589 (Pt 23), 5603–5612. https://doi.org/10.1113/jphysiol.2011.215160. Epub 2011 Aug 30. PMID: 21878519; PMCID: PMC3249036.

Basbaum, A.I., Bautista, D.M., Scherrer, G., Julius, D., 2009. Cellular and molecular mechanisms of pain. Cell 139 (2), 267–284. https://doi.org/10.1016/j.cell.2009.09.028.

Ben-Shushan, E., Feldman, E., Reubinoff, B.E., 2015. Notch signaling regulates motor neuron differentiation of human embryonic stem cells. Stem Cell. 33 (2), 403–415. https://doi.org/10.1002/stem.1873.

Berberat, G., et al., 2014. Diffusion tensor imaging for target volume definition in glioblastoma multiforme. Strahlenther Onkol. https://doi.org/10.1007/s00066-014-0676-3.

Catani, M., et al., 2008. A diffusion tensor imaging tractography atlas for virtual in vivo dissections. Cortex. https://doi.org/10.1016/j.cortex.2008.05.004.

Cheney, P.D., Fetz, E.E., Mewes, K., 1991. Neural mechanisms underlying corticospinal and rubrospinal control of limb movements. Prog. Brain Res. 87, 213–252. https://doi.org/10.1016/s0079-6123(08)63054-x. PMID: 1866448.

Danesin, C., Soula, C., 2017. Moving the Shh source over time: what impact on neural cell diversification in the developing spinal cord? J. Dev. Biol. 5 (2). https://doi.org/10.3390/jdb5020004.

Depienne, C., Cincotta, M., Billot, S., Bouteiller, D., Groppa, S., Brochard, V., et al., 2011. A novel DCC mutation and genetic heterogeneity in congenital mirror movements. Neurology 76 (3), 260–264. https://doi.org/10.1212/WNL.0b013e318207b1e0.

Donkelaar, H.J.T., Itoh, K., Hori, A., 2014. Development and developmental disorders of the spinal cord. Clinical Neuroembryology: Development and Developmental Disorders of the Human Central Nervous System. Springer Berlin Heidelberg, Berlin, Heidelberg, pp. 271–320.

Drew, T., Prentice, S., Schepens, B., 2004. Cortical and brainstem control of locomotion. Prog. Brain Res. 143, 251–261. https://doi.org/10.1016/S0079-6123(03)43025-2. PMID: 14653170.

Eyre, J.A., Miller, S., Clowry, G.J., Conway, E.A., Watts, C., 2000. Functional corticospinal projections are established prenatally in the human foetus permitting involvement in the development of spinal motor centres. Brain 123 (Pt 1), 51–64. https://doi.org/10.1093/brain/123.1.51.

Fregosi, M., Contestabile, A., Badoud, S., Borgognon, S., Cottet, J., Brunet, J.F., Bloch, J., Schwab, M.E., Rouiller, E.M., May 22, 2019. Corticotectal projections from the premotor or primary motor cortex after cortical lesion or Parkinsonian symptoms in adult macaque monkeys: a pilot tracing study. Front. Neuroanat. 13, 50. https://doi.org/10.3389/fnana.2019.00050. PMID: 31191260; PMCID: PMC6540615.

Hao, M., He, X., Xiao, Q., Alstermark, B., Lan, N., 2013. Corticomuscular transmission of tremor signals by propriospinal neurons in Parkinson's disease. PLoS One 8 (11), e79829. https://doi.org/10.1371/journal.pone.0079829.

Heise, C., Kayalioglu, G., 2009. Chapter 6 - cytoarchitecture of the spinal cord. In: Watson, C., Paxinos, G., Kayalioglu, G. (Eds.), The Spinal Cord, pp. 64–93.

Ho, R., Sances, S., Gowing, G., Amoroso, M.W., O'Rourke, J.G., Sahabian, A., et al., 2016. ALS disrupts spinal motor neuron maturation and aging pathways within gene co-expression networks. Nat. Neurosci. 19 (9), 1256–1267. https://doi.org/10.1038/nn.4345.

Hochman, S., 2007. Spinal cord. Curr. Biol. 17 (22), R950–R955. https://doi.org/10.1016/j.cub.2007.10.014.

Hori, A., 1993. A review of the morphology of spinal cord malformations and their relation to neuro-embryology. Neurosurg. Rev. 16 (4), 259–266. https://doi.org/10.1007/BF00383832.

Jang, S.H., Kwon, Y.H., February 8, 2018. A review of traumatic axonal injury following whiplash injury as demonstrated by diffusion tensor tractography. Front. Neurol. 9, 57. https://doi.org/10.3389/fneur.2018.00057. PMID: 29472891; PMCID: PMC5809420.

Jen, J.C., Chan, W.M., Bosley, T.M., Wan, J., Carr, J.R., Rub, U., et al., 2004. Mutations in a human ROBO gene disrupt hindbrain axon pathway crossing and morphogenesis. Science 304 (5676), 1509–1513. https://doi.org/10.1126/science.1096437.

Krotov, V., Tokhtamysh, A., Safronov, B.V., Belan, P., Voitenko, N., 2019. High-threshold primary afferent supply of spinal lamina X neurons. Pain 160 (9), 1982–1988. https://doi.org/10.1097/j.pain.0000000000001586.

Lee, W., et al., 2013. Classification of diffusion tensor images for the early detection of Alzheimer's disease. Comput. Biol. Med. https://doi.org/10.1016/j.compbiomed.2013.07.004.

Li, S., Zhuang, C., Hao, M., He, X., Marquez, J.C., Niu, C.M., Lan, N., 2015. Coordinated alpha and gamma control of muscles and spindles in movement and posture. Front. Comput. Neurosci. 9, 122. https://doi.org/10.3389/fncom.2015.00122.

Lima, D., Mendes-Ribeiro, J.A., Coimbra, A., 1991. The spino-latero-reticular system of the rat: projections from the superficial dorsal horn and structural characterization of marginal neurons involved. Neuroscience 45 (1), 137–152. https://doi.org/10.1016/0306-4522(91)90110-a. PMID: 1721691.

Matsuyama, K., Mori, F., Kuze, B., Mori, S., August 2, 1999. Morphology of single pontine reticulospinal axons in the lumbar enlargement of the cat: a study using the anterograde tracer PHA-L. J. Comp. Neurol. 410 (3), 413–430. PMID: 10404409.

Menétrey, D., Chaouch, A., Besson, J.M., November 1980. Location and properties of dorsal horn neurons at origin of spinoreticular tract in lumbar enlargement of the rat. J. Neurophysiol. 44 (5), 862–877. https://doi.org/10.1152/jn.1980.44.5.862. PMID: 7441321.

Michielsen, M., Vaughan-Graham, J., Holland, A., Magri, A., Suzuki, M., 2019. The Bobath concept - a model to illustrate clinical practice. Disabil. Rehabil. 41 (17), 2080–2092. https://doi.org/10.1080/09638288.2017.1417496.

Miyata, H., Miyata, M., Ohama, E., 2014. Pyramidal tract abnormalities in the human fetus and infant with trisomy 18 syndrome. Neuropathology 34 (3), 219–226. https://doi.org/10.1111/neup.12081.

Moller, M., et al., 2007. Dynamic changes in corticospinal tracts after stroke detected by fibretracking. J. Neurol. Neurosurg. Psychiatr. https://doi.org/10.1136/jnnp.2006.100248.

Muto, N., Kakei, S., Shinoda, Y., August 12, 1996. Morphology of single axons of tectospinal neurons in the upper cervical spinal cord. J. Comp. Neurol. 372 (1), 9–26. https://doi.org/10.1002/(SICI)1096-9861(19960812)372:1<9::AID-CNE2>3.0.CO;2-7. PMID: 8841918.

Nagi, S.S., Marshall, A.G., Makdani, A., Jarocka, E., Liljencrantz, J., Ridderstrom, M., et al., 2019. An ultrafast system for signaling mechanical pain in human skin. Sci. Adv. 5 (7), eaaw1297. https://doi.org/10.1126/sciadv.aaw1297.

Ogura, T., Sakaguchi, H., Miyamoto, S., Takahashi, J., 2018. Three-dimensional induction of dorsal, intermediate and ventral spinal cord tissues from human pluripotent stem cells. Development 145 (16). https://doi.org/10.1242/dev.162214.

Okado, N., 1981. Onset of synapse formation in the human spinal cord. J. Comp. Neurol. 201 (2), 211–219. https://doi.org/10.1002/cne.902010206.

Oliver, K.M., Florez-Paz, D.M., Badea, T.C., Mentis, G.Z., Menon, V., de Nooij, J.C., March 1, 2021. Molecular correlates of muscle spindle and Golgi tendon organ afferents. Nat. Commun. 12 (1), 1451. https://doi.org/10.1038/s41467-021-21880-3. PMID: 33649316; PMCID: PMC7977083.

Olmsted, Z.T., Paluh, J.L., 2021. Co-development of central and peripheral neurons with trunk mesendoderm in human elongating multi-lineage organized gastruloids. Nat. Commun. 12 (1), 3020. https://doi.org/10.1038/s41467-021-23294-7.

O'Sullivan, S.B., McKibben, R.J., Portney, L.G., 2019. Examination of motor function: motor control and motor learning. In: O'Sullivan, S.B., Schmitz, T.J., Fulk, G. (Eds.), Physical Rehabilitation, 7e. McGraw Hill.

Okoyama, S., 1991. Anatomical plasticity of the tectospinal tract after unilateral lesion of the superior colliculus in the neonatal rat. Exp. Brain Res. 85 (3), 552–558. https://doi.org/10.1007/BF00231739. PMID: 1717304.

Riddle, C.N., Edgley, S.A., Baker, S.N., 2009. Direct and indirect connections with upper limb motoneurons from the primate reticulospinal tract. J. Neurosci. 29 (15), 4993–4999. https://doi.org/10.1523/JNEUROSCI.3720-08.2009.

Rilliet, B., 2008. Diastematomyelia. The Spina Bifida: Management and Outcome. Springer Milan, Milano, pp. 487–513.

Sances, S., Ho, R., Vatine, G., West, D., Laperle, A., Meyer, A., et al., 2018. Human iPSC-derived endothelial cells and microengineered organ-chip enhance neuronal development. Stem Cell Rep. 10 (4), 1222–1236. https://doi.org/10.1016/j.stemcr.2018.02.012.

Schoenen, J., Faull, R., 2004. Spinal cord: cyto- and chemoarchitecture. In: Paxinos, G., Mai, J.K. (Eds.), The Human Nervous System, second ed. Elsevier Academic Press, San Diego.

Sharim, J., Pouratian, N., 2016. Anterior cingulotomy for the treatment of chronic intractable pain: a systematic review. Pain Phys. 19 (8), 537–550. PMID: 27906933.

Sherman, S.M., Guillery, R.W., 2006. Exploring the thalamus and its role in cortical function, second ed. MIT Press.

Soja, P.J., Pang, W., Taepavarapruk, N., McErlane, S.A., February 1, 2001. Spontaneous spike activity of spinoreticular tract neurons during sleep and wakefulness. Sleep 24 (1), 18–25. https://doi.org/10.1093/sleep/24.1.18. PMID: 11204049.

Standring, S., 2008. Gray's Anatomy: The Anatomical Basis of Clinical Practice, 40th Edition. Churchill Livingstone Elsevier, London.

ten Donkelaar, H.J., 1988. Evolution of the red nucleus and rubrospinal tract. Behav. Brain Res. 28 (1–2), 9–20. https://doi.org/10.1016/0166-4328(88)90072-1. PMID: 3289562.

ten Donkelaar, H.J., 2000. Development and regenerative capacity of descending supraspinal pathways in tetrapods: a comparative approach. Adv. Anat. Embryol. Cell Biol. 154 (iii-ix), 1–145. https://doi.org/10.1007/978-3-642-57125-1.

ten Donkelaar, H.J., Lammens, M., Wesseling, P., Hori, A., Keyser, A., Rotteveel, J., 2004. Development and malformations of the human pyramidal tract. J. Neurol. 251 (12), 1429–1442. https://doi.org/10.1007/s00415-004-0653-3.

ten Donkelaar, H.J., Willemsen, M.A., van der Heijden, I., Beems, T., Wesseling, P., 2002. A spinal intradural enterogenous cyst with well-differentiated muscularis propria. Acta Neuropathol. 104 (5), 538–542. https://doi.org/10.1007/s00401-002-0599-4.

ten Donkelaar, H.J., Yamada, S., Shiota, K., van der Vliet, T., 2014. Overview of the Development of the Human Brain and Spinal Cord Clinical Neuroembryology: Development and Developmental Disorders of the Human Central Nervous System. Springer Berlin Heidelberg, Berlin, Heidelberg, pp. 1–52.

Trompetto, C., Marinelli, L., Mori, L., Pelosin, E., Currà, A., Molfetta, L., Abbruzzese, G., 2014. Pathophysiology of spasticity: implications for neurorehabilitation. BioMed Res. Int. 2014, 354906. https://doi.org/10.1155/2014/354906. Epub 2014 Oct 30. PMID: 25530960; PMCID: PMC4229996.

Willis Jr., W.,D., 2008. In: Masland Albright, R.D.T., Gardner, E. (Eds.), Physiological Characteristics of Second-Order Somatosensory Circuits in Spinal Cord and Brainstem, vol. 6. ScienceDirect.

Yamamoto, N., Nakayama, T., Hagio, H., May, 2017. Descending pathways to the spinal cord in teleosts in comparison with mammals, with special attention to rubrospinal pathways. Dev. Growth Differ. 59 (4), 188–193. https://doi.org/10.1111/dgd.12355. Epub 2017 May 16. PMID: 28509386.

Yeo, S.S., Jang, S.H., Son, S.M., 2014. The different maturation of the corticospinal tract and corticoreticular pathway in normal brain development: diffusion tensor imaging study. Front. Hum. Neurosci. 8, 573. https://doi.org/10.3389/fnhum.2014.00573.

Zaaimi, B., Edgley, S.A., Soteropoulos, D.S., Baker, S.N., July, 2012. Changes in descending motor pathway connectivity after corticospinal tract lesion in macaque monkey. Brain 135 (Pt 7), 2277–2289. https://doi.org/10.1093/brain/aws115. Epub 2012 May 11. PMID: 22581799; PMCID: PMC3381720.

Further Reading

Scherman, P., Jenmalm, P., Dahlin, L.B., July, 2016. One-year results of needle fasciotomy and collagenase injection in treatment of Dupuytren's contracture: a two-centre prospective randomized clinical trial. J. Hand Surg. Eur. Vol. 41 (6), 577–582. https://doi.org/10.1177/1753193415617385. Epub 2015 Dec 1. PMID: 26631343.

5a

Functional Anatomy of the Brainstem

Giovanni Paternò[1], *Barbara Buccilli*[2], *Guglielmo Iess*[1], *Simon Sebastean Velasco Montaño*[3], *Veronica Vitali*[4], *Orhun Mete Çevik*[5,6], *Mert Babur*[5,6], *Ömer Orhun*[6,7] *and Brandon M. Ascenzi*[8]

[1]Department of Neurosurgery, Istituto Clinico Città Studi, Milan, Italy [2]Department of Neurosurgery, Cerebrovascular and Neurologic Critical Care, Icahn School of Medicine at Mount Sinai, New York, NY, United States [3]Laboratoire de Recherche, Service de Neurochirurgie, Hôpital Lariboisière, AP-HP, Paris, France [4]Department of Neurosurgery, University of Insubria, Varese, Italy [5]Department of Neurosurgery, Bakırkoy Training and Research Hospital for Psychiatric and Nervous Diseases, Health Sciences University, Istanbul, Turkey [6]Center of Advanced Simulation and Education (CASE), Neuroanatomy Lab., Acıbadem Mehmet Ali Aydinlar University, Istanbul, Turkey [7]School of Medicine, Acıbadem Mehmet Ali Aydinlar University, Istanbul, Turkey [8]Independent Neuroresearcher-Member of Marie Curie Alumni Association (MCAA)

OUTLINE

INTRODUCTION

The brainstem consists of three parts: the medulla oblongata, pons, and midbrain, each of which gives rise to cranial nerves. Posteriorly, the brainstem is connected to the cerebellum. The fourth ventricle is located between the brainstem and the cerebellum.

Some cranial nerves are situated at the level of the cerebral hemisphere, such as the olfactory nerve (including the olfactory bulb and olfactory tract) and the optic nerve, which extends into the optic chiasm, but most of three cranial nerves have their nuclei in the brainstem. In the midbrain we can identify oculomotor, trochlear, and trigeminal (mesencephalic) nuclei. The sensory and motor trigeminal nuclei are in the pons, with the abducens and facial nerve's nuclei. The vestibular nuclei are found at the border between pons and medulla oblongata, in both the aforementioned sections, while the cochlear nuclei can only be found in the medulla, where we can also identify the glossopharyngeal, vagus, and hypoglossal nuclei. The accessory nerve's nucleus has a portion in the medulla and one in the spinal cord. Superiorly and posteriorly to the brainstem, we find the epithalamus, while the mammillary bodies (hypothalamus) are located anteriorly.

In this chapter, the functional anatomy of the brainstem will be described in detail. In the second section,

https://doi.org/10.1016/B978-0-12-822404-5.00023-1

we will take into consideration the new neurosurgical techniques regarding the brainstem safe entry zone (SEZs) and other microsurgical anatomy approaches. Finally, some diseases associated with the brainstem and cerebellum are also mentioned.

EXTERNAL ANATOMY

Medulla Oblongata

The medulla oblongata is the inferior part of the brainstem, extending downward with the spinal cord and upward with the pons. From an external view the medulla oblongata contains several structures topographically important from a macroscopical point of view including the pyramids, the anterior fissure, the corticospinal tract, the olive, anteriorly, and the gracile tubercle, the cuneate tubercle, posteriorly (see Chapter 2).

In details, on the anterior surface of the medulla oblongata, along the midline, there is the anterior fissure; laterally to the anterior fissure are two elongated and paired elevations, the pyramids (Fig. 5a.1A and B). The latter consists of the axons of the corticospinal tract (CST), which descend and cross the midline at the pyramidal decussation. Approximately 75%–90% of the axons in the pyramids cross over to the opposite side of the spinal cord. This crossing is crucial for the control of voluntary movements, so that the left hemisphere of the brain controls the right side of the spinal cord, i.e., the right hemibody, and vice versa.

The CST (Fig. 5a.2) serves as a major pathway for cortical control of spinal cord activity, primarily involved in voluntary movements. The remaining 5%–15% of fibers that do not decussate within the pyramidal decussation form the anterior corticospinal tract, also known as the direct pyramidal tract or anterior cerebrospinal fasciculus. This tract runs close to the anterior median fissure and descends to the lower thoracic cord, gradually decreasing in size.

Laterally to the pyramids, there are ovoidal and paired structures called olives, which correspond to the inferior olivary nuclei. The antero-lateral surface of medulla oblungata is predominantly formed by the olive. The rootlets of the hypoglossal nerve (XII c.n.) originate from the anterolateral sulcus, while the rootlets of the glossopharyngeal nerve (IX c.n.), vagus nerve (X c.n.), and accessory nerve (XI c.n.) are located posteriorly to the olive.

On the posterior surface of the medulla oblongata, along the midline, there is the posterior median sulcus, which differs from the anterior fissure as it is not interrupted by the pyramidal decussation. The sulcus extends upward to the obex, where the medulla oblongata widens into the rhomboid fossa, forming the floor of the fourth ventricle.

The posterior surface of the medulla oblongata can be divided into a lower closed portion and an upper open portion. The caudal (inferior) region of the medulla oblongata, which is continuous with the spinal cord, is referred to as the lower closed portion.

In contrast, the medulla oblongata's upper open region acts as a zone of transition between the spinal cord and the brain's more rostral areas. The distinction between the lower closed and upper open portions of the medulla oblongata helps to define the functional organization of this region. The lower closed portion primarily deals with motor functions and relays information between the brain and the spinal cord, while the upper open portion is involved in sensory processing, cranial nerve control, and integration of vital autonomic functions. The fourth ventricle is located between the posterior surface of the brainstem and the cerebellum (for topography of the brainstem see Chapter 2).

Lateral to the obex, there are the gracile tubercle, cuneate tubercle, and tuberculum cinereum; all these structures are paired. These tubercles correspond to nuclei beneath the surface: the gracile nucleus under the gracile tubercle and the cuneate nucleus under the cuneate tubercle. These nuclei receive the gracile and cuneate fasciculus, respectively, which run in the posterior aspect of the spinal cord. They are involved in transmitting sensations of vibration, fine touch, proprioception, and discrimination between two points and represent the second neuron of the dorsal column medial lemniscus pathway system (DCML) discuss in the previous chapter.

The tuberculum cinereum is formed due to the presence of the spinal trigeminal nucleus, which plays a role in transmitting pain and temperature sensations from the head through the V, VI, IX, and X pair of cranial nerves. This nucleus continues inferiorly with the first and second Rexed laminae of the spinal cord, which contain receptors for substance P and glutamate, neurotransmitters associated with pain transmission.

Pons

The pons is demarcated superiorly by the ponto-mesencephalic sulcus and inferiorly by the ponto-medullary sulcus (Fig. 5a.1A and B). On the anterior surface of the pons, along the midline, we find the basilar sulcus, which may or may not contain the basilar artery, depending on anatomical variability. The lateral surface of the pons is continuous with the middle cerebellar peduncle, and in its upper portion, we can find the posterior (sensory) root of the trigeminal nerve (V c.n.).

On the posterior surface of the pons, within the rhomboid fossa, we can locate the facial colliculus. In this region, we also find the nucleus of the VI c.n. (abducens nerve), specifically within the concavity of the facial nerve's (VII c.n.) first genu. Therefore, when asked about the corresponding structures of the facial colliculus, we can provide two answers: the first genu of the facial nerve and the nucleus of the abducens nerve.

Midbrain

The upper limit of the midbrain consists in the sulcus between the optic tract and the cerebral peduncles and the inferior one by the ponto-mesencephalic sulcus (Fig. 5a.1A and B).

From an external macroscopical examination, we can identify several structures including the superior colliculus and the inferior colliculus (paired structures located on the posterior surface), the cerebral peduncle and the interpeduncular fossa (respectively, paired and impaired structures located on the anterior surface). Performing an axial section at this level we can appreciate postero-anteriorly: the periaqueductal gray (PAG), impaired and median structure inside which the cerebral acqueduct of Silvio runs; laterally the latter and on both sides the two mesencephalic nucleus of trigeminal nerve; the oculomotor nucleus, paired structures located on the ventral edge of the PAG; the medial longitudinal fascicles (mlf, midbrain component); the red nucleus, paired and round structures; laterally to the latters there are, on both of side, the midbrain component of trigeminal, spinal, and medial lemniscus; the substantia nigra (on both side) and the foot of the cerebral peduncle. The anterior surface of the midbrain is related to the diencephalon, to the mammillary bodies.

On the anterior surface of the midbrain, we find the aforementioned cerebral peduncles. The interpeduncular fossa lies between the right and left cerebral peduncles. At the base of the interpeduncular fossa, there are openings through which the thalamoperforating arteries pass, giving rise to the name "posterior perforated substance" for this region. The thalamoperforating arteries supply blood to various parts of the midbrain and thalamus, and they play a crucial role in providing oxygen and nutrients to the Ascending Reticular Activating System (ARAS), which is responsible for maintaining the state of arousal and regulating wakefulness. On the medial border of the cerebral peduncle, we find the emergence of the third cranial nerve, which then continues toward the superior orbital fissure. Beneath the inferior colliculus, we can observe the trochlear nerve (IV c.n.) which is carried anteriorly by enveloping the mesencephalon postero-laterally. This nerve innervates the superior oblique muscle. Located above and behind the superior colliculus, we find the pineal gland, which is part of the epithalamus. Above the pineal gland, there is a large vein known as the great cerebral vein (or Galen vein), which courses toward the straight sinus.

IV Ventricle

The fourth ventricle is a fluid-filled space located between the brainstem anteriorly and the cerebellum posteriorly, playing a crucial role in cerebrospinal fluid circulation. The floor of the fourth ventricle is formed by the rhomboid fossa, while its roof consists of the superior and inferior medullary velum.

The medial aperture, also known as the foramen of Magendie, serves as a connection between the fourth ventricle and the cisterna magna. Through this opening, cerebrospinal fluid produced by the choroidal plexus is drained into the cisterna magna, which is a large subarachnoid space at the base of the skull.

In addition to the medial aperture, we can find the foramina of Luschka, also referred to as the lateral apertures. These apertures provide communication between the fourth ventricle and the prepontine cistern, another subarachnoid space in the brain. The Foramen of Luschka is a small opening located on each side of the fourth ventricle.

The primary function of the Foramen of Luschka is to allow the circulation of cerebrospinal fluid. It provides an additional pathway for the CSF to flow from the fourth ventricle into the subarachnoid space, facilitating the distribution and absorption of CSF throughout the central nervous system. Understanding the anatomy and function of the Foramen of Luschka is important in the context of neurology and neurosurgery, particularly when evaluating conditions related to CSF circulation or when performing procedures that involve the fourth ventricle and its surrounding structures. An obstruction can be a cause of hydrocephalus or increased intracranial pressure.

Cranial Nerves

The first (I) cranial nerve is the olfactory nerve. Its origin is in the olfactory bulb, where the mitral cells are located. The olfactory nerve continues into the olfactory tract, and then the medial and lateral olfactory striae. The medial olfactory stria reaches the subcallosal gyrus; the lateral olfactory stria goes toward the uncus and the insular cortex. Together with the entorhinal cortex and the entorhinal sulcus, they make the primary olfactory area. Between the olfactory striae we find the anterior perforated substance, through which run the lenticulostriate arteries. If we move posteriorly, we find the second cranial nerve (II): the optic nerve. The

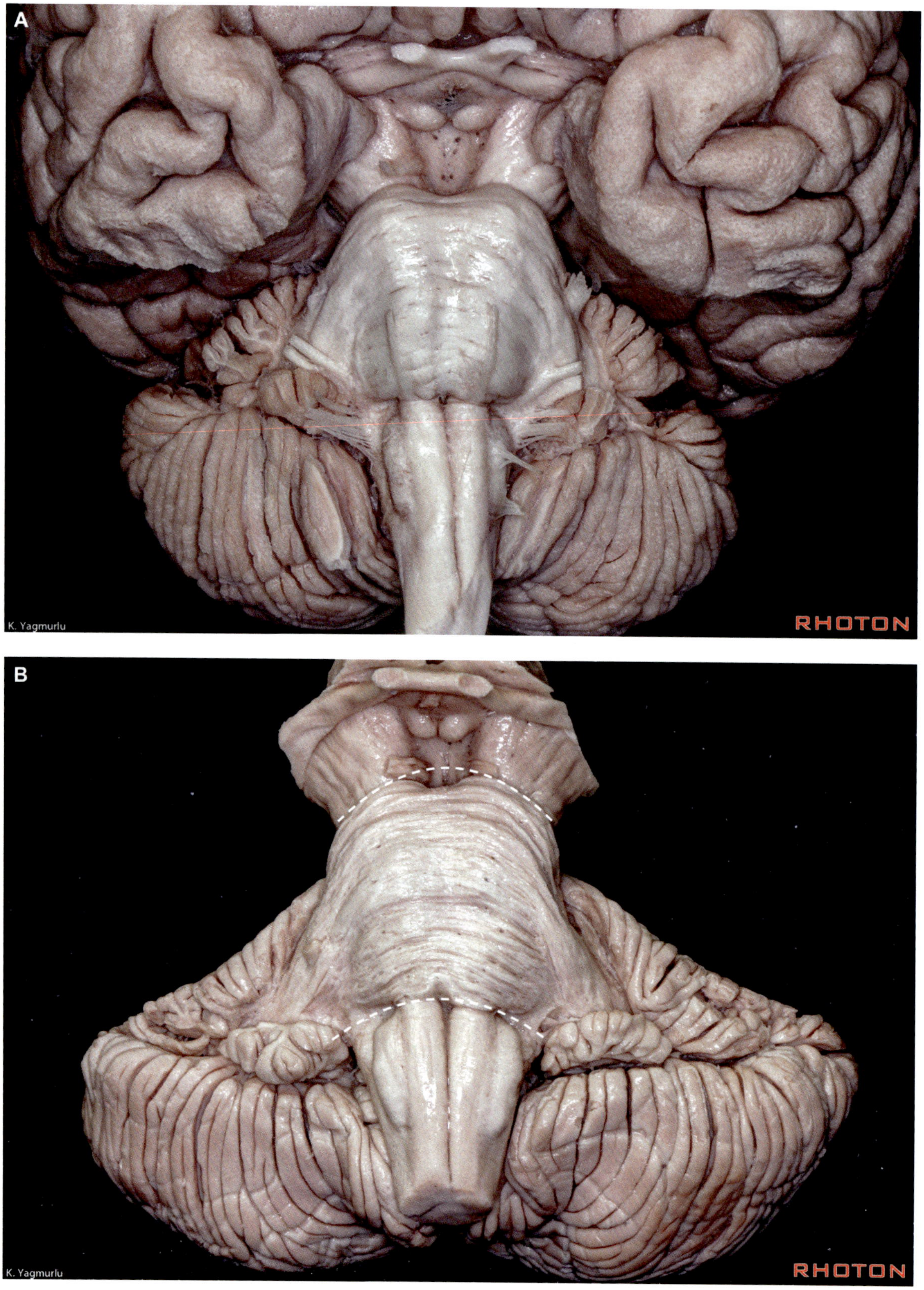

FIGURE 5A.1 **(A) Ventral surface of the brainstem**. The brainstem appears in the foreground consisting of the medulla oblongata, pons, and midbrain, respectively, from *bottom* to *top*. On the anterior surface of the medulla oblongata, the pyramids and the median sulcus that separates them are easily visible. At the level of the medullopontine sulcus, the two emerging nerves are the abducens nerves (VI c.n.). At the level of the pons the transverse pontine fibers are visible on the surface; on both sides the trigeminal nerves can be seen. Above, the two cerebral peduncles

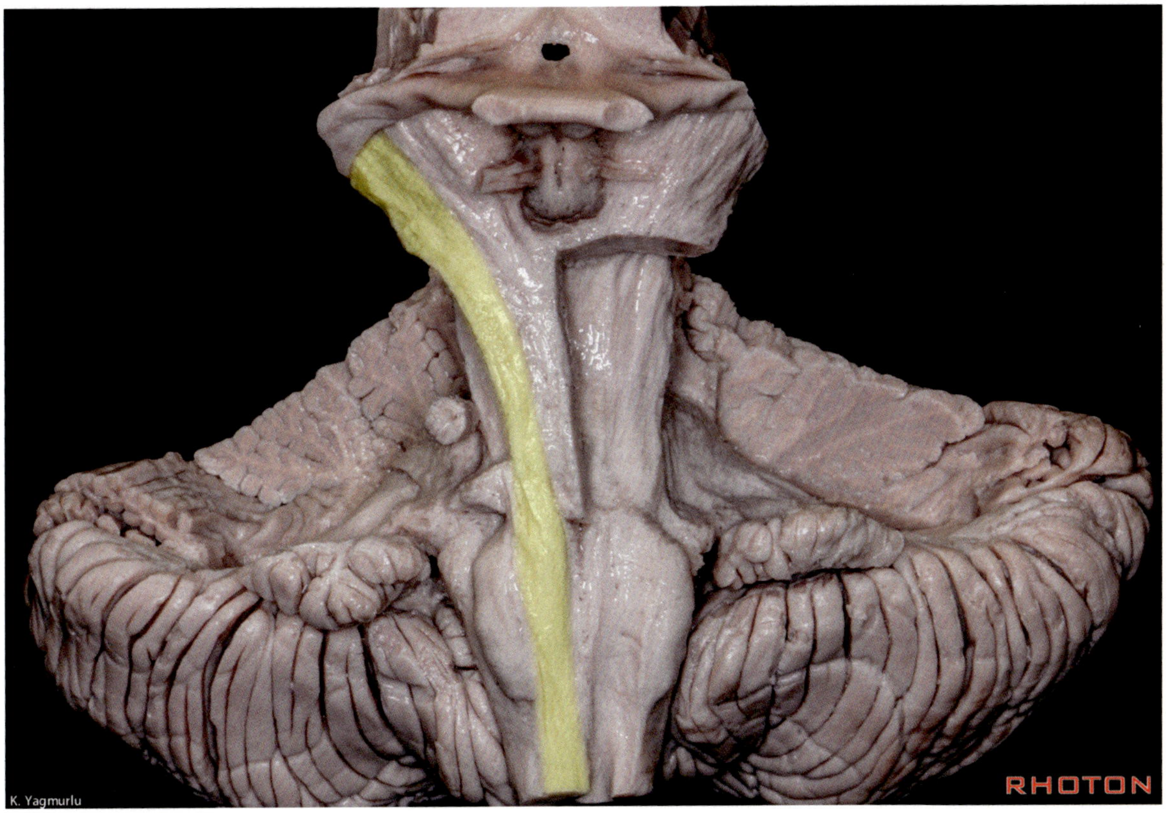

FIGURE 5A.2 **Brainstem: corticospinal tract (superficial dissection).** In *yellow* is the corticospinal tract; medially to this is the corticobulbar tract and even more medially, the frontopontine tract. *Image courtesy of The Rhoton Collection®. http://rhoton.ineurodb.org/. Dissector: K. Yagmurlu. Drawing by Jeffrey Sorenson.*

two optic nerves converge into the optic chiasm; posteriorly to the optic chiasm we find the optic tracts. The optic tract reaches the lateral geniculate nucleus.

At the level of the medial border of the cerebral peduncle is located, as said before, the emergence of the third (III) cranial nerve, the oculomotor.

The small nerve coming from the posterior surface of the brain stem is the fourth (IV) cranial nerve: the trochlear or pathetic nerve.

On the lateral surface of the pons, we see the emergence of the trigeminal nerve. We can distinguish two roots functionally different: a large sensitive root and a smaller motor. The trigeminal nerve innervates the masticatory muscles and carries sensory stimuli from the face, for these reasons it is a mixed nerve (together with the VII, IX, and X c.n., see later).

At the border between the pons and the medulla oblongata (bulbo-pontine sulcus), above the pyramid, we find the sixth (VI) cranial nerve, the abducens. More laterally in the sulcus we find the emergence of the seventh (VII) cranial nerve (the facial nerve) and the intermediate nerve of Wrisberg; also, the latter is part of the VII c.n. and it represents the origin of the parasympathetic and gustative fibers of the facial nerve. Laterally the facial nerve we find the eighth (XIII) cranial nerve (or vestibulocochlear nerve).

At the level of the medulla oblongata, in the posterior collateral sulcus, we have the IX, X, and XI cranial nerves. The XI cranial nerve, the glossopharyngeal nerve, has two components: the cranial portion, that emerges from the posterior collateral sulcus, and the spinal portion comes from the anterior collateral

continue superiorly forming the internal capsule and in the center they describe the fossa interpeduncularis. Here it is possible to see the posterior substantia perforata and the two mammillary bodies further forward. The optic chiasm is the structure located inferior to the cerebral cortex. For a detailed description of the cranial nerves see Chapter 2, Fig. 2.50. **(B) Brainstem: anterior view of the brainstem**. We can observe the two anterior pyramids, divided by the anterior median sulcus. Laterally, we find the inferior olivary nuclei. Higher up, above the first dashed line, we see the anterior surface of the pons, with the pontine transverse fibers. Finally, above the other dashed line, we observe the midbrain, with the two cerebral peduncles. Between them, we see the posterior perforated substance, and anterior to this, the mammillary nuclei. On the side of the olives, the flocculi, together with the cerebellum, are appreciable. The cerebellum envelops the entire postero-dorsal surface of the brainstem. *Image courtesy of The Rhoton Collection®. http://rhoton.ineurodb.org/. Dissector: K. Yagmurlu.*

sulcus. The spinal nucleus of the VIII c.n. can be found in the anterior portion of the segments from C1 to C5/C6. The axons then ascend in the anterior collateral sulcus, and reach the posterior cranial fossa through the foramen magnum. The spinal portion then joins the cranial one. The axons from the spinal portion innervate the sternocleidomastoid muscle and the trapezius muscle.

Anteriorly to the olive, in the preolivary sulcus we find the emergence of the twelfth (XII) c.n., the hypoglossal nerve, which innervates the intrinsic and extrinsic muscles of the tongue.

INTERNAL ANATOMY

Medulla Oblongata

The medulla oblongata serves as a relay center, transmitting information from the spinal cord to the brain. It consists of a ventral portion and a dorsal tegmentum. The dorsal tegmentum contains important structures such as the nuclei of the glossopharyngeal (IX), vagus (X), accessory (XI), and hypoglossal (XII) cranial nerves. Additionally, it houses various white matter tracts including the medial longitudinal fasciculus (Fig. 5a.9), medial lemniscus, spinothalamic tract, central tegmental tract, and spinocerebellar tract (Figs. 5a.3–5a.5 and 5a.9). These tracts play a role in transmitting sensory and motor signals. The inferior olivary nucleus, with its deflated balloon shape, is the most recognizable nucleus of this area and shows connections with the red nucleus and subsequently with the cerebellum. This connectivity forms the rubro-olivo-cerebellar circuit. Notably, this circuit exerts an influence on the ambiguous nucleus, which is located posteriorly to the olive. Interruption or dysfunction of this circuit can result in palatal myoclonus. The ambiguous nucleus, located in the lateral portion of the medullary tegmentum, between the hypoglossal nucleus and the dorsal motor nucleus of the vagus nerve, has a role in the innervation of the pharynx, the larynx, and the soft palate through the IX, X, and XI cranial nerves. Next to the midline we find the axons of the medial lemniscus (Fig. 5a.3), which originates from the gracile and cuneate nuclei. The medial lemniscus is made up of axons of second-order neuron which ascend to the VPL nucleus of the contralateral thalamus.

Behind the medial lemniscus we find the hypoglossal nucleus. This is the real origin of the hypoglossal nerve,

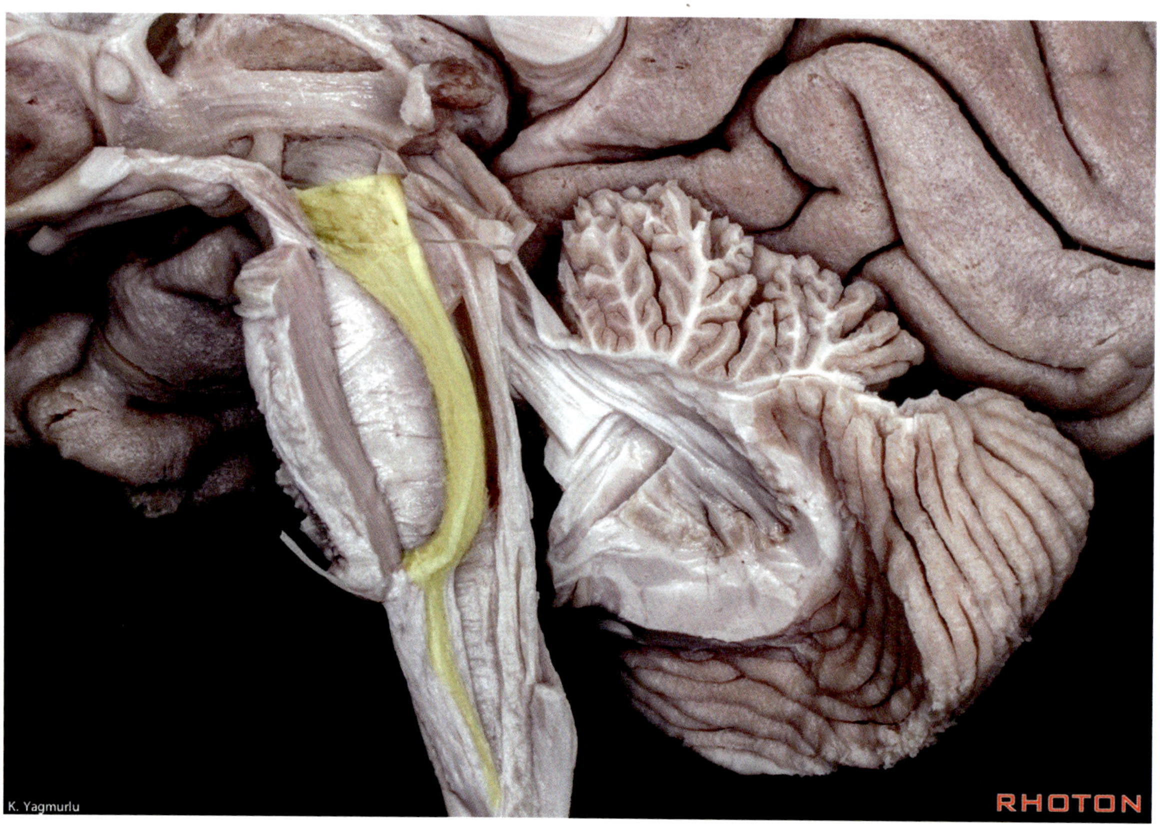

FIGURE 5A.3 **Lateral view of the brainstem (deep dissection).** We can see the **medial lemniscus** (*in yellow*). The corticospinal tract is located anteriorly. Dorsally, we can see the medial longitudinal fasciculus. *Image courtesy of The Rhoton Collection®. http://rhoton.ineurodb.org/. Dissector: K. Yagmurlu. Drawing by Nickalus Khan.*

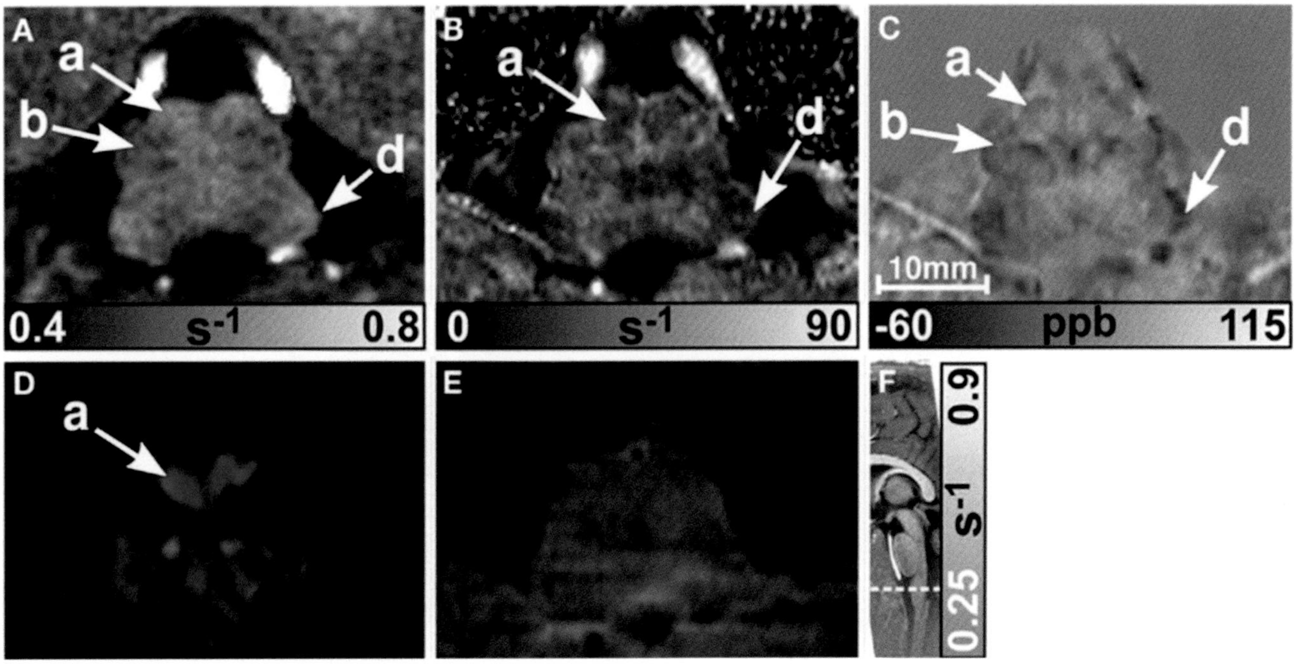

FIGURE 5A.4 **Middle part of the medulla oblongata. hMRI 7T of the same region are presented in axial orientation from (A to E), respectively.** The *arrows* in the images indicate: (a) pyramid, (b) inferior olive, and (d) inferior cerebellar peduncle. The sectional plane of the axial images is indicated by the *dashed line* overlaid on the sagittal R_1 map shown in (F). *Red, blue,* and *green* in the track-density image represent anisotropy along medial-lateral, superior-inferior, and anterior-posterior directions, respectively. *Image courtesy of Deistung, A., et al., 2013. High-resolution MR imaging of the human brainstem in vivo at 7 T. Front. Hum. Neurosci. 7. https://doi.org/10.3389/fnhum.2013.00710.*

which innervates the intrinsic and extrinsic muscles of the tongue.

Anteriorly in the medulla oblongata we find the corticospinal tract, which has a role in fine voluntary movements. Posteriorly and laterally to this tract we find the inferior olivary nucleus, which plays a regulatory role on the ambiguous nucleus. The connections between the inferior olivary nucleus and the ambiguous nucleus are part of a complex network involved in the control of motor functions, particularly in the regulation of movement and coordination. The inferior olivary nucleus provides feedback and modulatory signals to the ambiguous nucleus, which in turn influences the motor output of cranial nerves involved in speech, swallowing, and other voluntary movements of the head and neck. This regulatory role of the inferior olivary nucleus on the ambiguous nucleus helps to ensure the precise coordination and control of motor functions mediated by the cranial nerves that originate from the ambiguous nucleus.

The medulla oblongata has a lower portion that is closed and an upper portion that is open, respectively, below and above the obex, where the fourth ventricle begins.

If we make a horizontal section at the level of the obex, we can study the internal configuration. Anteriorly we have the pyramids, and posterolaterally to them the olives; posteriorly we find the inferior cerebellar peduncle, and between the two peduncles the floor of the fourth ventricle, where we can see the trigones of the hypoglossal and vagus nerves. We should note that the anterior surface of the medulla oblongata seamlessly transitions into the spinal cord. At this junction, we encounter the decussation of the pyramids, which is a distinctive feature. If we make a section above the obex, at the level of the open portion of the medulla oblongata, posteriorly we find the IV ventricle with the inferior medullary velum. The rhomboid fossa forms the floor of the IV ventricle.

We should remember that during embryonic development, the medulla undergoes differentiation into two distinct regions known as the alar plate and basal plate. The alar plate, located in the posterior region, gives rise to sensory nuclei involved in processing touch, proprioception, vibration, and visceral sensory information. On the other hand, the basal plate, found in the anterior region, gives rise to motor nuclei responsible for controlling voluntary movements. When we get to the medulla oblongata and the fourth ventricle, the alar plate is displaced posteriorly, while the basal plate keeps its medial position. Therefore, the nuclei we find in the floor of the fourth ventricle are motor nuclei.

Let's start with the motor nuclei. The first one, resembles a deflated balloon with a concave medial aspect, is

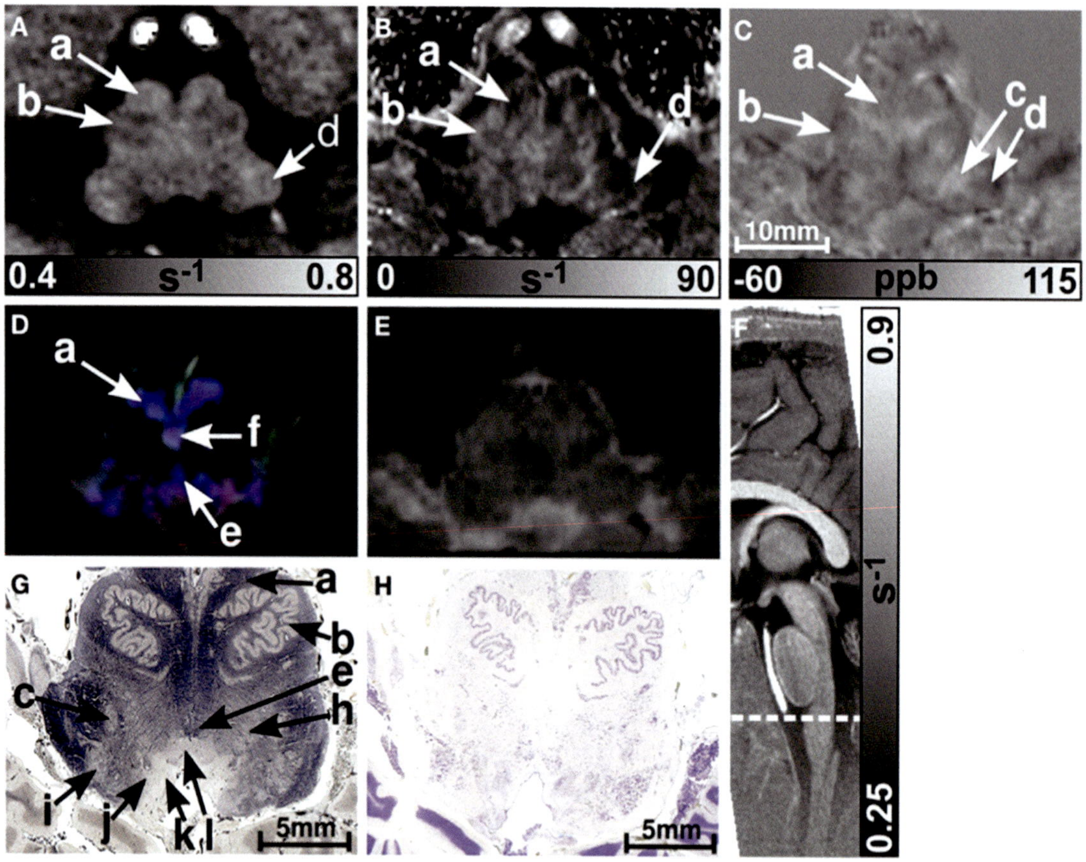

FIGURE 5A.5 **Rostral part of the medulla oblongata. hMRI 7T of the same region are presented in axial orientation from (A to E), respectively.** The *arrows* indicate (a) pyramid, (b) inferior olive, (c) spinal trigeminal nucleus, (d) inferior cerebellar peduncle, (e) medial longitudinal fasciculus, (f) medial lemniscus, (h) reticular formation, (i) accessory cuneate nucleus, (j) solitary nucleus, (k) dorsal motor nucleus of the vagus nerve, and (l) hypoglossal nucleus. The sectional plane of the axial images is indicated by the *dashed line* overlaid on the sagittal R_1 map shown in (F). Axial histological sections from a different subject stained for myelin and cells are illustrated in (G) and (H), respectively. *Image courtesy of Deistung, A., et al., 2013. High-resolution MR imaging of the human brainstem in vivo at 7 T. Front. Hum. Neurosci. 7. https://doi.org/10.3389/fnhum.2013.00710.*

the olivary nucleus, which can be further divided into superior, middle, and inferior portions. It is located in the lower part of the open portion of the medulla oblongata. Moving posteriorly and medially, near the floor of the fourth ventricle, we encounter the hypoglossal nucleus. Lateral to it, we find the dorsal nucleus of the vagus. Dorsal to the inferior olivary nucleus but ventral to the nuclei of the X and XII cranial nerves lies the ambiguous nucleus. These are the primary nuclei derived from the basal plate. Fibers from the hypoglossal nucleus form the hypoglossal nerve, which proceeds ventrally and emerges from the preolivary sulcus.

Laterally we have the dorsal vagal nucleus (or dorsal nucleus of the vagus nerve). The latter gives rise to some of the fibers of the vagus nerve and specifically to the parasympathetic fibers. It is a parasympathetic nucleus (autonomic nervous system), and its fibers are preganglionic fibers. The vagus nerve emerges from the retro-olivary groove. This nucleus is the cardiopneumo-enteric center, because it is the source of the parasympathetic innervation to the thoracic viscera (heart, lungs, bronchus, trachea) and to the abdominal viscera (spleen, transverse colon, splenic flexure), with the exception of the descending colon and the pelvic viscera. Indeed, it is a general, visceral, efferent nucleus.

The ambiguous nucleus innervates the muscles derived from the third and fourth brachial arches. It innervates the palatal muscles, the pharyngeal muscles, the laryngeal muscles, and the voluntary esophageal muscles (superior third of the esophagus). The ambiguous nucleus innervates all these structures through the IX, X, and XI cranial nerves. Its fibers emerge from the retro-olivary groove. This nucleus gives rise only to the cranial portion of the accessory nerve. We should remember that the spinal portion of the accessory nerve originates from the cervical segments of the spinal cord and innervates the sternocleidomastoid muscle and the trapezius muscle. Thus, the cranial portion joins the vagus nerve to innervate the larynx. If there is a lesion of the ambiguous nucleus, the patient presents with dysphagia (difficulty swallowing), dysphonia (hoarseness), and dysarthria (difficulty articulating speech). The muscles derived from the brachial arches with visceral functions have visceral functions in other species as well, like in fish, in which they give rise to gills. The ambiguous nucleus is a special, visceral, efferent nucleus.

We do not know the functions of the olivary nucleus in detail, but we know that, when there is a lesion of this nucleus, the patient presents with movement disorders, due to altered coordination of movements previously learned. The olivary nucleus has connections with the red nucleus and the cerebellum.

Let's now talk about the alar plate, which is laterally displaced because of the presence of the fourth ventricle. As said before sensory nuclei derive from the alar plate. Laterally to the dorsal nucleus of the vagus nerve we find a small nucleus: the nucleus of solitary tract. More laterally, we find nuclei situated at the junction between the medulla oblongata and the pons, which are the vestibular nuclei. We can distinguish three vestibular nuclei at this level: the inferior, lateral, and medial, which are located in the most rostral part of the medulla oblongata. The superior vestibular nucleus is situated in the caudal portion of the pons, and it is related to the floor of the fourth ventricle, in an area which takes the name of vestibular region.

Laterally and posteriorly, we have the inferior cerebellar peduncles, which lead to the cerebellum. Ventrally and dorsally to each peduncle we find two very important nuclei, which are situated laterally to the vestibular region. They are the posterior (or dorsal) and anterior (or ventral) cochlear nuclei. The cochlear nuclei are very close to the junction between medulla oblongata and pons. When their fibers cross the midline, they give rise to the trapezoid body, which is situated in the basal portion of the pons.

In the lateral aspect of the medulla oblongata, posterior and lateral to the olive, lies a significant nucleus known as the spinal trigeminal nucleus. This nucleus is notably large and extends to the first cervical segments of the spinal cord. If we look at it from the side, it looks like a long column with two portions: a superior (or cephalic) portion and an inferior (or caudal) portion. The cephalic portion correspond to the nucleus of solitary tract, a sensory nucleus which receives fibers from neuronal groups coming from outside of the central nervous system, which are ganglionic neurons, whose central (or afferent) fibers enter the CNS and reach the solitary nucleus in its cephalic portion. These ganglionic neurons are situated in the ganglia of three nerves:

- The geniculate ganglion of the facial nerve, in the petrous part of the temporal bone, whose peripheral fibers collect gustative stimuli from the anterior 2/3 of the tongue;
- The glossopharyngeal ganglion, which collects the gustative stimuli from the posterior third of the tongue;
- The vagus ganglion, which collects gustative stimuli from the taste buds in the epiglottis.

The cephalic portion of the nucleus of solitary tract receives gustative stimuli through the VII, IX, and X pairs of cranial nerves.

The caudal portion receives visceral sensory information from the thoracic and abdominal viscera. It collects information from the carotid sinus and the carotid body, situated at the bifurcation of the common carotid artery into the internal and external carotid arteries. In the carotid sinus we find receptors which sense the changes in blood pressure; in the carotid body we find biochemical receptors which sense changes in the CO_2 concentration. These receptors send their information to the glossopharyngeal ganglion, which forwards them to the caudal portion of the solitary nucleus, establishing visceral reflexes, like the vasoconstrictor response after a reduction in the blood pressure. If the blood pressure decreases, the solitary nucleus stimulates the reticular formation in the medulla oblongata. The reticulospinal tract then reaches the sympathetic neurons in the lateral portion of the spinal cord, from C8 to L2, where the preganglionic sympathetic fibers originate. They produce vasoconstriction in response, and increase the blood pressure.

The chemoreceptors of the carotid body send their information to the same ganglion and reach the caudal portion of the solitary nucleus. The sympathetic stimulation of the respiratory center in the reticular formation stimulates the cervical segments C3, C4, and C5 in their anterior horns (see Chapter 5b). From here the phrenic nerve increases the respiratory rate to eliminate the accumulated CO_2.

The caudal portion of the solitary nucleus also receives information from the ganglion of the vagus nerve; the peripheral fibers coming from the thoracic and abdominal viscera. These fibers bring information on organ distension, on changes in pressure and establish visceromotor reflex arcs. For example, the fibers which stimulate the dorsal vagal nucleus (parasympathetic nucleus) have an influence on intestinal peristalsis. If the lungs are distended, the solitary nucleus fibers stimulate the inspiratory center in the reticular formation otherwise, while if the lungs are distended, the solitary nucleus stimulates the expiratory center and inhibits the inspiratory center. All these functions constitute very important reflexes that play a key role in the regulation of the respiratory rate.

Let's now talk about the vestibular nuclei. The vestibular nuclei are situated laterally to the solitary nucleus, in the superior portion of the medulla oblongata, with the exception of the superior vestibular nucleus, which is situated in the caudal portion of the pons. They receive information from the vestibular ganglion (Scarpa's ganglion), situated in the internal ear. This ganglion receives fibers from the hair cells of the cristae ampullares of the semicircular ducts and from the maculae in the utricle and saccule, bringing information about the angular acceleration of the head (when we turn) and the linear acceleration (when we are in a moving car or elevator), respectively. From the vestibular ganglion the vestibular branch of the VIII cranial nerve reaches the medullopontine sulcus and the vestibular nuclei; some fibers reach the inferior cerebellar peduncles directly, heading to the flocculonodular lobe of cerebellum (oldest portion, the archicerebellum) synapsing more precisely the fastigial nucleus. Thus, the majority of the fibers reach the vestibular nuclei and then the archicerebellum though the inferior cerebellar peduncles. These vestibulocerebellar fibers (or vestibulo-olivary fibers) are important for the postural changes of the head. So, the cerebellum forwards the information to the vestibular nuclei and the reticular formation, from which the vestibulospinal and reticulospinal tracts descend to activate the extensor muscles of the neck and torso (for example, if we are falling asleep and our head is falling). This forms a reflex arc to maintain the posture of the head.

The main function of the archicerebellum is to maintain the muscular synergy, assuring that muscular groups work together with good coordination. Through the vestibulospinal tract it signals which groups should contract to maintain thc correct posture and regulates the muscular tone; it controls the direction, strength, and speed of contraction.

More laterally, in the rostral portion of the medulla oblongata, ventrally and dorsally to the inferior cerebellar peduncle, there are the anterior and posterior cochlear nuclei, which receive information from the spiral ganglion (or cochlear ganglion), situated in the modiolus of the cochlea, whose peripheral fibers innervate the hair cells in the organ of Corti, situated in the scala media of the cochlea of the inner ear. Both the cochlear nuclei transmit auditory stimuli (see Chapter 19). The postganglionic fibers reach the medullopontine sulcus, enter the brainstem, and reach the anterior and posterior cochlear nuclei.

Pons

We can divide the pons into two regions, separated by the medial lemniscus: an anterior region, which is the base of the pons, and a posterior region, or pontine tegmentum. For internal topography of the pons (see Figs. 5a.6 and 5a.8).

In the pons we still find the axons of the corticospinal tract, and among its axons the pontocerebellar tract, which reaches the pontine nuclei, then crosses the midline and reaches the contralateral cerebellum. Intermixed with the axons of the corticospinal tract we also find the transverse fibers, which are axons coming from the cerebral cortex and directed to the pontine nuclei. After the pontine nuclei, the axons cross the midline and reach the contralateral cerebellum. This is the ponto-cortico-cerebellar-dento-rubro-spinal circuit (also known as the dentato-rubro-olivary pathway). This is a synergic pathway.

Let's imagine a transversal section of the pons, right above the medullopontine sulcus.

The anterior (or ventral) surface of the pons, on the midline, presents with the basilar sulcus. Laterally the pons blends into the middle cerebellar peduncles, which connect the pons to the cerebellum. On the posterior surface we find the floor of the fourth ventricle; here, laterally to the median eminence, we find the sulcus limitans, which divides the elements derived from the basal plate (medially) from the elements derived from the alar plate (laterally). As we already saw in the medulla oblongata, because of the presence of the fourth ventricle, the sensory nuclei are displaced laterally.

In the basal portion of the pons, near the floor of the fourth ventricle, we find a small nucleus, which is the nucleus of the VI cranial nerve (abducens). Ventrolaterally to it we find the nucleus of the VI c.n. (nucleus of the facial nerve) (Fig. 5a.7).

Medially to the facial nucleus we find some visceral motor nuclei: the salivatory nuclei (Fig. 5a.6). Some authors distinguish a superior and an inferior salivatory nucleus and sometime the former is called "salivary nucleus" to refer to a specific region of it. They both are parasympathetic nuclei. To the facial nucleus we find a nucleus derived from the same column, which is the trigeminal motor nucleus. These are the main nuclei derived from the basal plate, which are located in the medial portion of the pontine tegmentum.

The nucleus of the VI cranial nerve innervates the lateral rectus muscle of the eye. It is a general; somatic; efferent nucleus; because it innervates a voluntary muscle. The fibers originating from the facial nucleus proceed dorsally and around the nucleus of the VI cranial nerve, causing a small protuberance to arise in the floor of the fourth ventricle which corresponds to the facial colliculus. This is the geniculum of the facial nerve. After the geniculum, the fibers proceed anterolaterally toward the medullopontine sulcus, where the nerve emerges. It innervates the muscles derived from the second brachial arch which are the mimic muscles, the posterior belly of

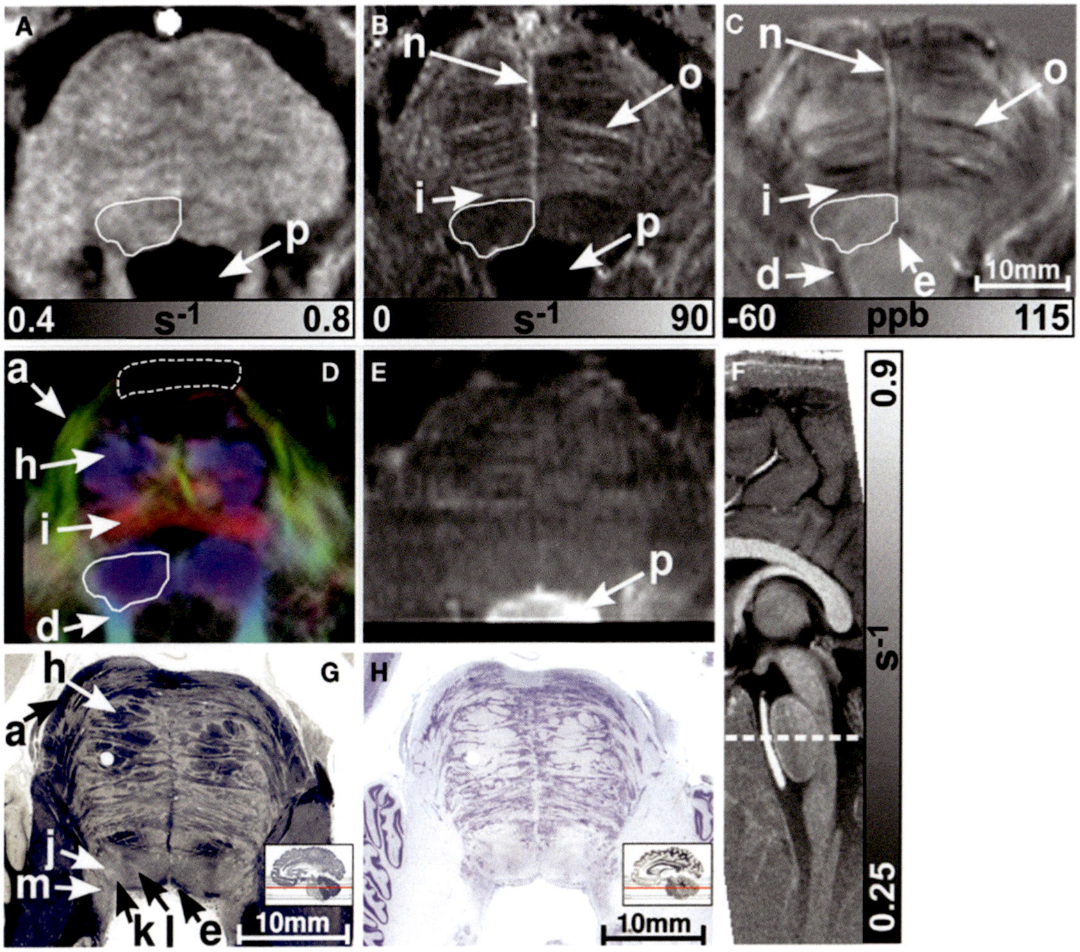

FIGURE 5A.6 **Middle part of the pons. hMRI 7T of the same region are presented in axial orientation from (A to E), respectively.** The *arrows* in the images indicate (a) middle cerebellar peduncle, (d) superior cerebellar peduncle, (e) medial longitudinal fasciculus, (h) corticospinal fibers, (i) pontocerebellar tract, (j) facial nucleus, (k) salivatory nucleus, (l) abducens nucleus, (m) solitary nucleus, (n) vein, (o) fiber bundle that traverses the pons in medial to lateral direction, and (p) fourth ventricle (cerebrospinal fluid). The *solid white outline* in (A–D) summarizes the caudal part of the pontine reticular nucleus, the facial nucleus, salivatory nucleus, and abducens nucleus. The *dashed white outline* in (D) indicates a region between the middle cerebellar peduncle that could not be resolved with TDI. The sectional plane of the axial images is indicated by the *dashed line* overlaid on the sagittal R_1 map shown in (F). Axial histological sections from a different subject stained for myelin and cells are illustrated in (G) and (H), respectively. *Image courtesy of Deistung, A., et al., 2013. High-resolution MR imaging of the human brainstem in vivo at 7 T. Front. Hum. Neurosci. 7. https://doi.org/10.3389/fnhum.2013.00710.*

the digastric muscle, the stylohyoid muscle and the stapedial muscle.

Above the facial nucleus is located the trigeminal motor nucleus, which innervates the muscles derived from the first brachial arch, which are the masticatory muscles such as the temporal muscles, the masseter muscles, the lateral and medial pterygoid muscles, the anterior belly of the digastric muscle, the mylohyoid muscle, and the tensor tympani muscle. This nucleus gives rise to the motor root of the trigeminal nerve. We should remember that, on the lateral surface of the pons, the trigeminal nerve has two roots: a big sensory one, originating from the trigeminal ganglion and, dorsally, a smaller motor one, which will then go through the trigeminal ganglion and join the mandibular branch (third branch) of the trigeminal nerve. Medially to the facial nucleus we find the small salivatory nuclei (Fig. 5a.6).

The superior salivatory nucleus gives rise to preganglionic parasympathetic fibers directed to the sublingual and submandibular glands through the facial nerve, joining the fibers from the facial nucleus. They reach the submandibular ganglion, from which the postganglionic fibers reach the sublingual and submandibular glands.

The inferior salivatory nucleus gives origin to parasympathetic fibers which reach the parotid gland though the glossopharyngeal nerve (VIII n.c.). The preganglionic fibers join this nerve and reach the otic ganglion, situated in the infratemporal fossa, while the postganglionic fibers reach the parotid gland.

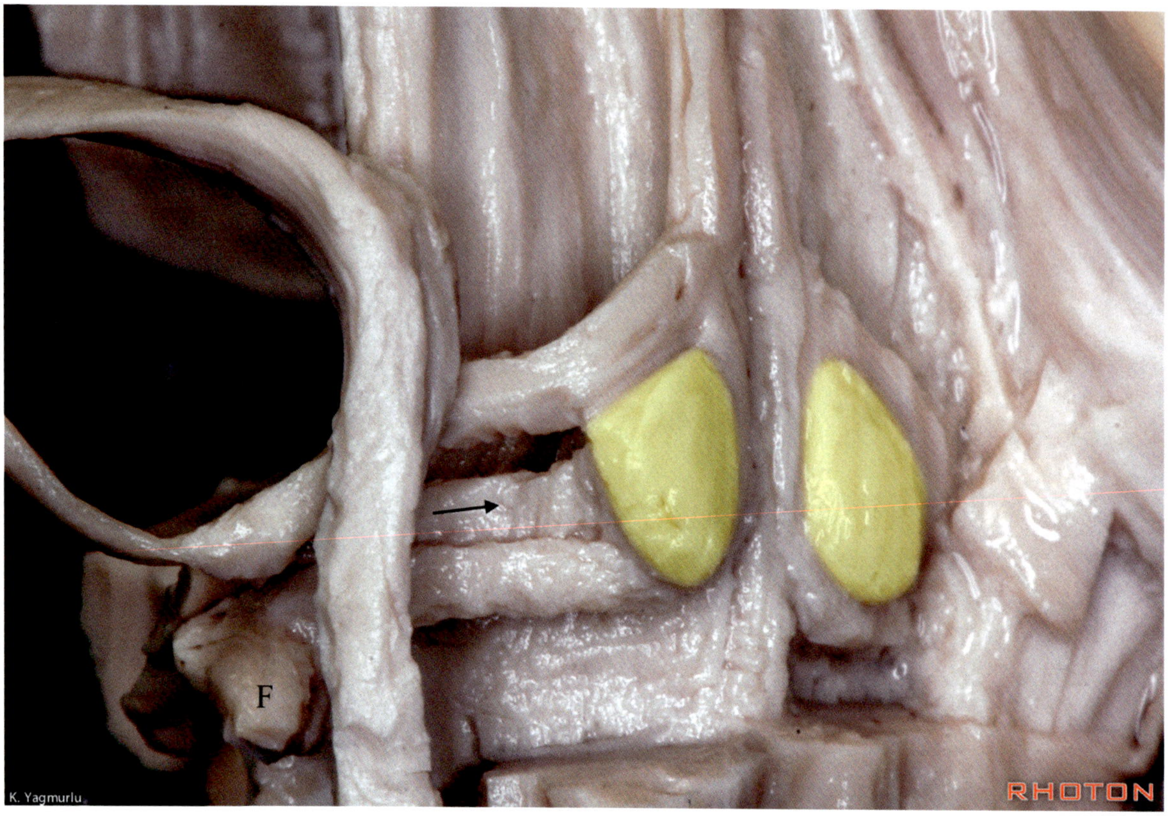

FIGURE 5A.7 **Abducens and facial nerves and nuclei.** It is possible to see the abducens nuclei (*in yellow*), from which the corresponding abducens nerves emerge (*black arrow*). It is also possible to see the facial nerves wrapping around the abducens nuclei and originating from the facial nucleus (F). *Image courtesy of The Rhoton Collection®. http://rhoton.ineurodb.org/. Dissector: K. Yagmurlu. Drawing by Nickalus Khan.*

The last parasympathetic nucleus of this section is the "lacrimal nucleus." Its preganglionic fibers join the facial nerve and then give origin to the greater superficial petrosal nerve, which, after reaching the sphenopalatine ganglion, reaches the lacrimal gland.

The fibers of motor facial nerve provide innervation to the facial mimic muscle while the preganglionic parasympathetic fibers (autonomic) to the submandibular, sublingual, and lacrimal glands. These are the main nuclei derived from the basal plate of the caudal portion of the pontine tegmentum, with the exception of the trigeminal motor nucleus, which is situated in its middle portion.

Now let's talk about the nuclei derived from the alar plate.

Related to the floor of the fourth ventricle we find the superior vestibular nucleus. Before we said the lateral, medial, and inferior vestibular nuclei are situated in the most rostral portion of the medulla oblongata, in the lateral portion of the floor of the fourth ventricle, in the vestibular area. We also said that laterally to them we find the inferior cerebellar peduncle and that the fibers originating from the vestibular nuclei cross the midline forming the trapezoid body. The fibers of the trapezoid body are mostly derived from the ventral cochlear nucleus. The trapezoid body is located in the lower portion of the pons.

Laterally to the facial nucleus is located the main sensory nucleus of the trigeminal nerve, which continues with the spinal nucleus of the trigeminal nerve in the medulla oblongata and in the lateral cervical segments of the spinal cord, at the level of the second Rexed lamina. Rostrally, it continues into the mesencephalic nucleus of the trigeminal nerve. The main sensory nucleus of the trigeminal nerve is important because it is responsible for the epicritic sensation (or fine touch) of the head, through the trigeminal, facial, glossopharyngeal, and vagus nerves. It is involved, for example, in the corneal reflex. The afferent fibers of the corneal reflex are situated in the ophthalmic branch of the trigeminal nerve and reach the ganglionic neurons and then the sensory nucleus. This nucleus is an important integrative center to generate the motor response through the facial nerve, which is the efferent component of this reflex that causes the blinking.

Laterally to the facial nucleus, between it and the sensitive nucleus of the trigeminal nerve, is located a small nucleus: the superior olivary nucleus. We already mentioned this nucleus talking about the cochlear pathway. It receives direct fibers from the homolateral

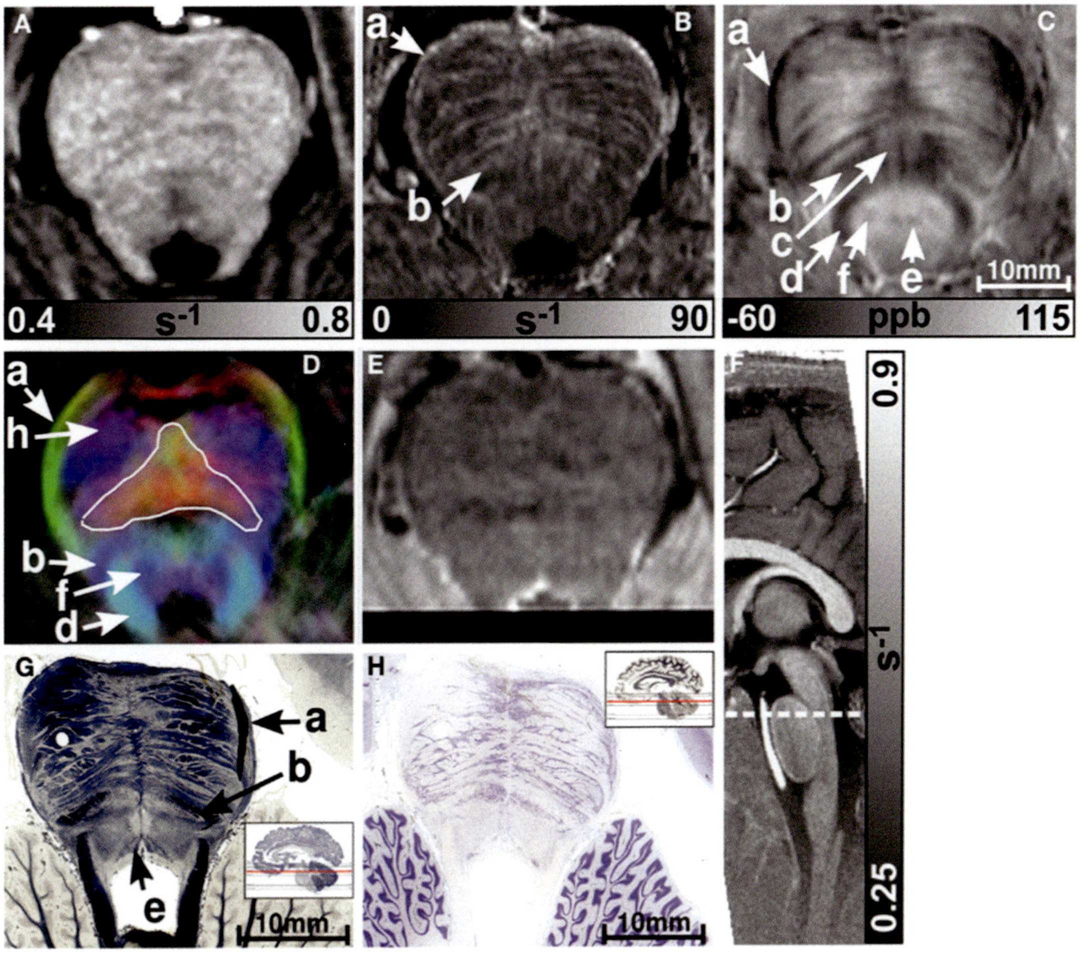

FIGURE 5A.8 **Rostral part of the pons. hMRI 7T of the same region are presented in axial orientation from (A to E), respectively.** The *arrows* in the images indicate (a) middle cerebellar peduncle, (b) medial lemniscus, (c) reticulotegmental nucleus, (d) superior cerebellar peduncle, (e) medial longitudinal fasciculus, (f) central tegmental tract, and (h) corticospinal tract. The solid white outline in (D) indicates the pontocerebellar fibers. The sectional plane of the axial images is indicated by the *dashed line* overlaid on the sagittal R_1 map shown in (F). Axial histological sections from a different subject stained for myelin and cells are illustrated in (G) and (H), respectively. *Image courtesy of Deistung, A., et al., 2013. High-resolution MR imaging of the human brainstem in vivo at 7 T. Front. Hum. Neurosci. 7. https://doi.org/10.3389/fnhum.2013.00710.*

dorsal cochlear nucleus and crossed fibers from the contralateral ventral cochlear nucleus. The superior olivary nucleus then gives rise to the lateral lemniscus, directed to the inferior colliculus, and then the medial geniculate body of the thalamus. It runs in the posterior portion of the posterior branch of the internal capsule, in the sublenticular region, and reaches the superior temporal gyrus (primary auditory area, areas 41 and 42 BA).

These are the main nuclei of the pontine tegmentum.

The small nuclei situated in the base of the pons are the pontine nuclei, which are dispersed in this region. They receive fibers from the corticopontine tract, coming from the frontal, temporal, and parietal lobes and, in a small amount, from the occipital lobe. From the pontine nucleus the fibers cross the midline and reach the contralateral cerebellum through the middle cerebellar peduncle. They mostly reach the neocerebellum synapsing particularly the dentate nucleus. From here the dentatothalamic tract reaches the ventrolateral nucleus (VLN) of the thalamus, in its posterior portion, and then goes back to the cortex. The frontal fibers go back to the precentral cortex and to the premotor area; in this way the cerebellum can exert control over voluntary movements.

In the base of the pons, we can see the fibers of the pyramidal tract, among which we find the corticobulbar fibers, which are crossed fibers directed to the voluntary motor nuclei of the cranial nerves. The corticobulbar fibers directed to the facial nucleus originate in the precentral gyrus, descend toward the thalamus in its ventricular nucleus, then pass though the internal capsule toward the caudal portion of the pons where is located the facial nucleus. It is important to highlight that VII can be divided into two components: the superior component (temporofacial) and the inferior

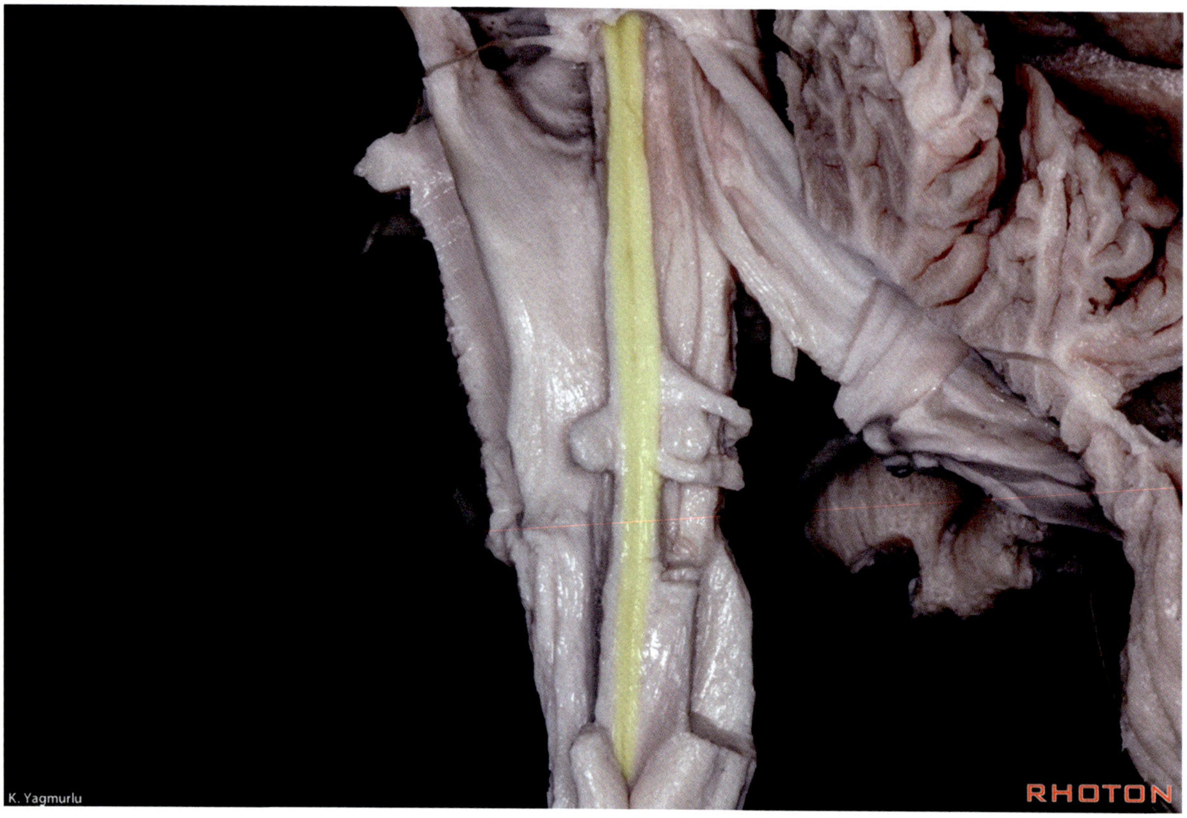

FIGURE 5A.9 **Brainstem: medial longitudinal fasciculus (deep dissection).** In *yellow* is the medial longitudinal fasciculus. Anteriorly, to the latter is the medial lemniscus. Posteriorly we see the central tegmental tract. *Image courtesy of The Rhoton Collection®. http://rhoton.ineurodb.org/. Dissector: K. Yagmurlu. Drawing by Nickalus Khan.*

component (cervicofacial). The fibers from the temporofacial component reach the superior portion of the face; the fibers from the cervicofacial component reach the inferior portion of the face. The temporofacial component receives corticonuclear fibers which are both homo- and contralateral (bihemispheric innervation); the cervicofacial component receives contralateral fibers only. This is important because it helps us understand the clinical presentation of a patient with facial paralysis. We can differentiate between central and peripheral facial paralysis taking into consideration the following. In a peripheral facial paralysis, the lesion is situated at the level of the facial nucleus or the facial nerve. The reader should remember that the facial nerve emerges from the medullopontine sulcus, between the middle cerebellar pedunculus and the cerebellum, in the pontocerebellar angle. It then reaches the internal auditory canal, accompanied by the vestibulocochlear nerve; then it moves toward the Fallopian aqueduct, in its labyrinthic portion, and reaches the vestibulum. It runs behind the tympanic cavity (tympanic portion), and descends toward the inferior surface of the mastoid (mastoid portion); then ascends toward the stylomastoid and enters the parotid gland. Here inside the facial nerve gives rise to two trunks: the temporofacial trunk and the cervicofacial trunk. These trunks further branch out into five ascending branches that innervate the muscles involved in facial expression. After this short review of topographical anatomy of the face, coming back to the peripheral facial paralysis we can say in the latter the lesion is situated at the level of the facial nucleus or the facial nerve therefore the patient presents with paralysis of a whole half of the face. The patient cannot rugate the forehead or close the eye on the healthy side (with supraversion of the open eye), and the nostril and half of the mouth drop.

On the contrary, if we have a central facial paralysis the lesion is located above the facial nucleus. This is possible when there is a lesion in the superior part of the pons or in the midbrain, since the corticonuclear fibers descend in there, or a lesion of the internal capsule or in the cortex of frontal lobe. The lesion affects the fibers contralateral to the affected area; the ipsilateral fibers reaching the cervicofacial portion of the facial nucleus are spared. As a consequence, the superior part of the half-face is spared, and the patient can rugate the forehead and close both eyes. The nostril and half-mouth still drop. Only the inferior part of the contralateral face is paralyzed. However, the platysma muscle in the neck is affected too.

A peripheral paralysis can have benign causes, like a sudden change in temperature, which causes inflammation of the epithelium above the aqueduct, which is small, around 1.5 mm, and can compress the facial nerve, causing Bell's paralysis.

The fibers reaching the hypoglossal and ambiguous nuclei are both homo- and contralateral, like the ones reaching the temporofacial portion of the facial nucleus. The same is true for the trigeminal motor nucleus.

Midbrain

In the midbrain, we find the central gray matter (also known as periacquductal gray—PAG) surrounding the cerebral aqueduct. For internal topography of the midbrain, see Figs. 5a.10 and 5a.11. The neurons within the central gray matter send descending projections through the posterior tract of the spinal cord to inhibit pain signals by releasing endorphins and enkephalins (see Chapter 17). This pathway acts as an opioid system to alleviate pain at the spinal cord level. There are other nuclei involved in pain modulation, such as the raphe nuclei, especially the lower ones. These nuclei are located throughout the brainstem and are part of the reticular formation that will be described in the Chapter 5b.

Ventral to the central gray matter we find the tegmentum of the mesencephalon where are located the red nucleus (paired structure) (Fig. 5a.11). From the red nucleus, axons descend to form the rubrospinal tract, directed to the flexor muscles of the cervical segments. We only find the red nucleus at this level in fact we do not see them at lower sections. This helps us to identify where a transversal section is located in the midbrain.

Ventral to the red nucleus and posteriorly to the cerebral peduncle, we find the substantia nigra (SN). The

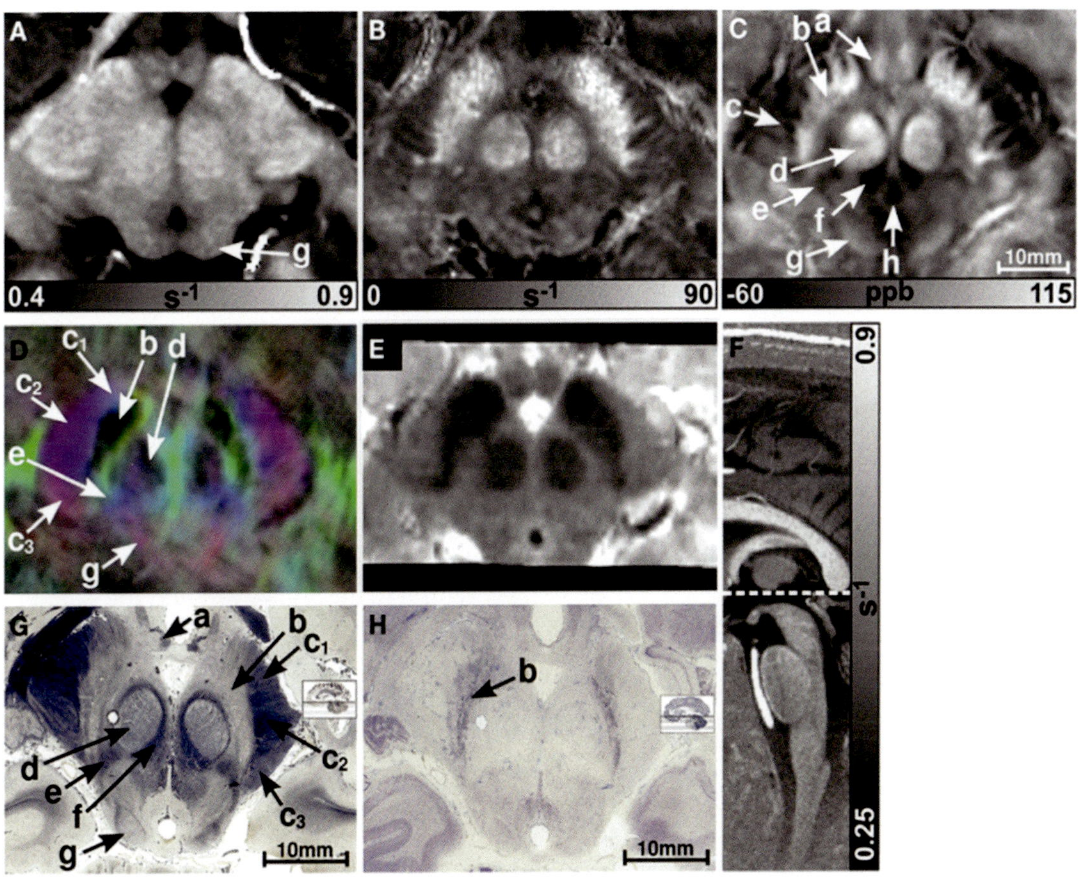

FIGURE 5A.10 **Midbrain. hMRI 7T of the same region are presented in axial orientation from (A to E), respectively.** The *arrows* in the images indicate (a) mammillary body, (b) substantia nigra, (c) crus cerebri, (c_1) corticobulbar fibers, (c_2) corticospinal fibers, (c_3) corticopontine fibers, (d) *red* nucleus, (e) medial lemniscus, (f) central tegmental tract, (g) inferior colliculus, and (h) medial longitudinal fasciculus. The sectional plane of the axial images is indicated by the *dashed line* overlaid on the sagittal R_1 map shown in (F). Axial histological sections from a different subject stained for myelin and cells are illustrated in (G) and (H), respectively. *Red*, *blue*, and *green* in the track-density image represent anisotropy along medial-lateral, superior-inferior, and anterior-posterior directions, respectively. Note that the track-density image is slightly distorted compared to the other image contrasts (A–C, E). *Image courtesy of Deistung, A., et al., 2013. High-resolution MR imaging of the human brainstem in vivo at 7 T. Front. Hum. Neurosci. 7. https://doi.org/10.3389/fnhum.2013.00710.*

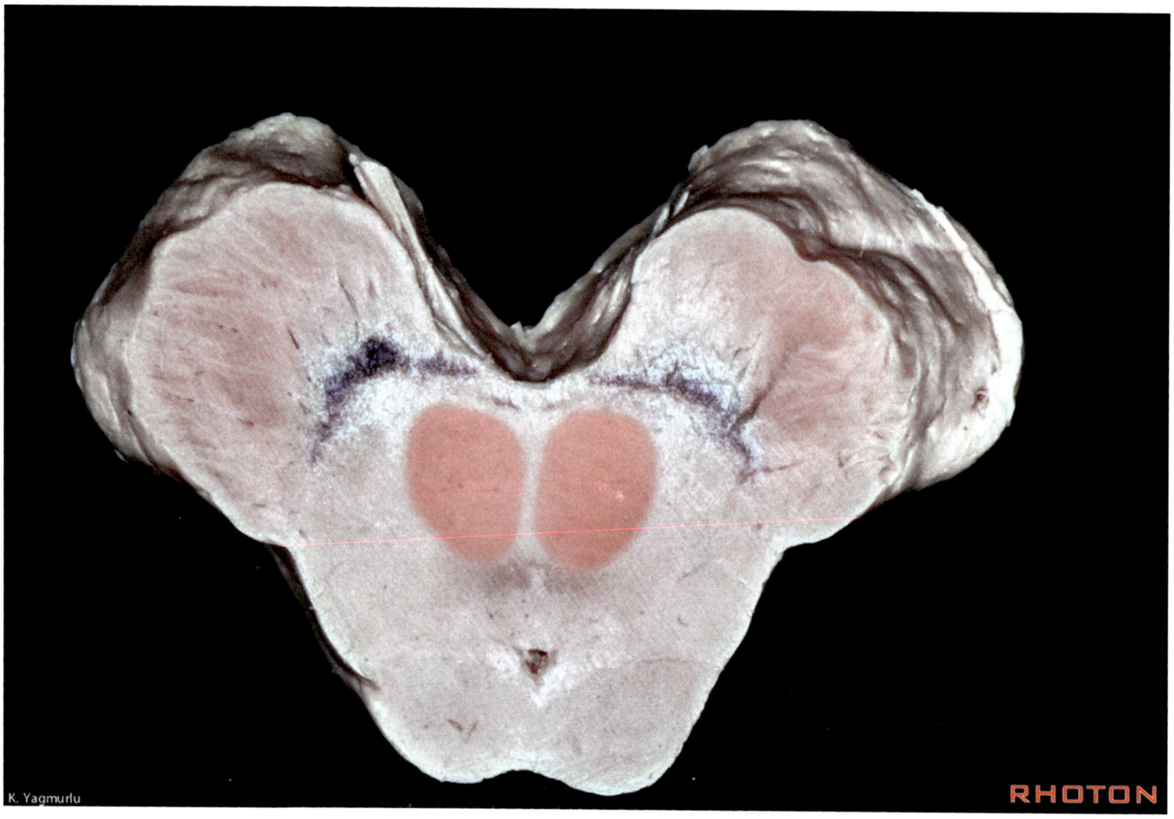

FIGURE 5A.11 **Midbrain.** The tectum is located posteriorly, with the periaqueductal gray (PAG) surrounding the cerebral aqueduct (the small hole in the center). Anteriorly, the two *red* nuclei, indicated in *yellow*, can be seen. In front of the latter is the ventral tegmental area (VTA). The substantia nigra is easily recognized by its dark pigmentation. Finally, still anteriorly, we see the peduncles of the midbrain. *Image courtesy of The Rhoton Collection®. http://rhoton.ineurodb.org/. Dissector: K. Yagmurlu. Drawing by Nickalus Khan.*

substantia nigra is made of neurons with intracytoplasmic neuromelanin inclusions, a product of the synthesis of dopamine. These are dopaminergic neurons, and their number is reduced in some conditions, like Parkinson disease. The substantia nigra is part of the striatum and has an important role in the regulation of motor circuits (see Chapter 15). In turn, we can distinguish two components of the substantia nigra: the pars compacta (SNpc) and the pars reticularis (SNpr).

The pars compacta is the most important, since it is the part which actively produces dopamine. From the *pars compacta*, the nigrostriatal tract reaches the striatum, formed by the putamen and pallidus. The nigrostriatal tract interacts with type 1 and type 2 dopaminergic receptors, which are very important for these circuits.

If there is a lesion in the substantia nigra, dopamine is reduced, and the patient presents with involuntary movements which are very characteristic, like in Parkinson's disease, which presents with tremor at rest, rigidity, bradykinesia, altered postural tone, and a tendency for the patient to fall forward. Parkinson's disease is characterized by degeneration of the *pars compacta* of the substantia nigra, therefore a reduction in dopamine levels in the nigrostriatal tract reaching the caudate and putamen.

The pars reticularis of the substantia nigra is less important in this point of view and has the same connections as the medial portion of the pallidus. They receive striatopallidal and striatonigral fibers and send fibers to the thalamus, to the anteroventral and lateroventral nuclei, which form the circuit of the basal ganglia.

Ventral to the SN we can find the stalk of the cerebral peduncle. The cerebral peduncle has a somatotopic arrangement: the middle 3/5 are occupied by the corticospinal tract, the medial 1/5 is occupied by the frontopontine fibers, and the lateral 1/5 by the temporopontine fibers. The middle 3/5 are the most important, because they constitute the corticospinal tract of which we have already spoke about.

Above the cerebral peduncle we find the optic tract. Anteriorly and laterally to the sylvian aqueduct and behind the substancia nigra, we find another nucleus, which is the nucleus of the third (III) cranial nerve, the oculomotor nerve. The latter ascends along the medial border of the cerebral peduncle to innervate the superior rectus muscle, the inferior rectus muscle, the medial rectus muscle, the inferior oblique muscle, and the levator palpebrae superioris muscle. This nucleus is always accompanied by the accessory oculomotor nucleus, also known as the Edinger–Westphal nucleus. The Edinger–Westphal nucleus gives rise to the parasympathetic component of the oculomotor nerve. Its fibers innervate the ciliary muscle, which controls accommodation of the lens for near vision, and the iris sphincter muscle, which constricts the pupil. Together, these structures play a role in regulating the pupillary light reflex and adjusting the focus of the eye. At the level of the inferior colliculus, we find another nucleus which is the nucleus of the fourth (IV) cranial nerve, which is the only cranial nerve emerging from the posterior surface of the brain stem. It is the trochlear nerve or pathetic nerve. From the colliculi and their nuclei, the tectospinal tract reaches the alpha motoneurons of the cervical segments of the spinal cord and plays a role in the visual and acoustic reflexes. The visual reflexes are important when we see an object in the periphery of our visual field, and we turn toward it; they are a phylogenetic remnant. Before we said that the trigeminal nucleus is a gigantic nucleus, which is divided into multiple portions: spinal, sensory, and mesencephalic. The mesencephalic trigeminal nucleus receives proprioceptive fibers from the masticatory muscles and the temporomandibular joint through the trigeminal ganglion. Its fibers enter the brainstem from the lateral surface of the pons and ascend toward the mesencephalic nucleus, which sends its fibers to the motor trigeminal nucleus, situated behind the facial nucleus. This helps to elicit the masticatory reflex. The stimulus from the masticatory muscles reaches the motor nucleus of the trigeminal nerve, which then initiates the reflex response of muscle relaxation. When we voluntarily contract the masticatory muscles, this reflex relaxation response is elicited, allowing for coordinated and controlled movements during chewing. Behind the substantia nigra, we find the red nucleus, which is anatomically and functionally divided into two portions based on the characteristics of their cells: a magnocellular portion and a parvocellular portion. The fibers originating from the magnocellular portion immediately cross the midline and form the rubrospinal tract, which descends laterally in the spinal cord, accompanied by the corticospinal tract, and reaches the flexor muscles of the superior limbs. The parvocellular portion receives crossed fibers from the contralateral cerebellum, specifically the paleocerebellum, through the superior cerebellar peduncle. It then sends fibers to the inferior olivary nucleus, situated in the medulla oblongata, through the central tegmental tract, which originates at the level of the midbrain and descends laterally in the brainstem to the inferior olivary nucleus. The latter then sends fibers to the cerebellum through the inferior cerebellar peduncle. This is the olivo-cerebello-rubro-olivary circuit.

If there is a lesion affecting the circuit involving the inferior olivary nucleus, red nucleus and cerebellum, the patient may present with a rare condition known as palatal myoclonus. Palatal myoclonus is characterized by rhythmic involuntary jerking movements of the palate. Ventrally to the red nucleus we find the substantia nigra, which can be divided into a *pars compacta* and a *pars reticularis*. Near the red nucleus and the substantia nigra, in an anteromedial position, we find another area rich in dopamine: it is the ventral tegmental area (VTA), situated in the ventral portion of the cerebral pedunculus. This area is part of the mesolimbic system. The mesolimbic fibers are situated in part of the striatum, they are rich in dopamine and reach the amygdala, which is part of the limbic system, and the nucleus accumbens, which is part of the ventral striatum (for details see Chapter 3). The mesolimbic fibers also have relationships with the hippocampus, with which they have a role in memory, and with the anterior portion of the hypothalamus, with which they are involved in visceral functions. The dorsal portion of the anterior perforated substance is involved in the circuits of desire, in which dopamine is also involved.

Many psychiatric drugs, for example most antipsychotics, e.g., haloperidol, block the dopaminergic receptors. The main adverse effect is an iatrogenic parkinsonism presenting by tremor, akinesia, bradykinesia, and alterations of the postural tone.

Posterior to the cerebral pedunculi it runs the medial lemniscus, laterally the spinal lemniscus (which carries temperature and pain from the limbs), and medially the trigeminal lemniscus (which carries sensory stimuli from the head).

The area where the superior, inferior colliculi and their nuclei are located is called mesecephali tectum. The superior colliculus has important connections with the retina, thus a role in the pupillary reflex and also in turning the eyes toward a stimulus. The inferior colliculus plays an important role in the auditory pathway and is reached by the lateral lemniscus, which comes from the cochlear nuclei and the superior olivary nucleus (SON). The inferior colliculus sends connections to the thalamus, specifically the medial geniculate body. From there the fibers reach the temporal lobe and the superior temporal gyrus where is found the primary auditory cortex (see Chapter 19).

PERIAQUEDUCTAL GRAY

The periaqueductal gray (PAG) lines the Sylvius aqueduct in the midbrain and it extends from the posterior commissure all the way down to the locus coeruleus (Fig. 5a.11). The dorsal PAG (dPAG), dorsolateral PAG (dlPAG), lateral PAG (lPAG), and ventrolateral

PAG (vlPAG) have been identified as columns within the PAG. The periaqueductal gray (PAG) is a crucial structure in the transmission and regulation of pain, sympathetic responses, as well as the learning and execution of protective actions. Given that pain is not only dependent on peripheral stimulation, PAG plays a significant role in the descending regulation of pain perception, both in inhibition and facilitation of inhibition of pain. Chronic pain, which can result in depression and anxiety, involves both the suppression and enhancement of pain via these pathways. The PAG also contributes to risk analysis, threat response, and protective behavior. These pathways incorporate sympathetic, motor and emotional reactions, and increasing degrees of awareness to produce learned aversive behaviors. Research has demonstrated that the PAG is active in depressed people due to its intricate link with pain processing. The PAG has only lately been linked to migraines as a potential attack starter (Steiner and Fontebasso, 2002).

The PAG transmits nociceptive and analgesic inputs in a bidirectional way, which can reduce or increase pain perception via ascending and descending projections. The lamina I, II, V receive the first pain impulses. These impulses then climb to the PAG and other structures through the spinothalamic and spinobulbar pathways before being sent to the medial thalamus. The rostral ventromedial medulla (RVMM) and the locus coeruleus (LC) are both significant connections of the PAG. The PAG-LC pathway uses norepinephrine (NE), and has an antinociceptive action on the dorsal horn. One of the main targets for supraspinal opioid analgesia is the serotonergic PAG-RVMM pathway, which is believed to be the main endogenous pain modulator. The PAG stimulates the raphe nuclei with glutamate, and the raphe nuclei send inhibitory signals to pain afferents (Yang and Chang, 2019). It has been demonstrated that during stress reactions, projections from the vlPAG and dPAG to the RVLM, a part of the reticular formation and arousal pathway, cause cardiopulmonary excitement with increased blood pressure, heart rate, and alertness. Certain neuronal subgroups seem to be associated with risk assessment prior to protective responses. This behavioral technique uses anticipation to reduce upcoming pain.

There are two types of defensive behavior: immediate and long-term reactions. Short-term fight or flight is elicited by activation of the dorsolateral and lateral PAG, which causes propagation of nonopioid antinociceptive inputs. Long-term learned behaviors, emerge when the vlPAG is stimulated. Indeed, following the stimulation of the dlPAG and superior colliculus, fear-induced antinociception is elicited. Future avoidance and learning are based on the simultaneous presence of antinociceptive and fear signals.

FUNCTIONAL BRAINSTEM AREAS/ NETWORKS INVOLVED IN SOME NEUROPATHIES

Brainstem Locomotor Networks

Locomotor activity is correlated with a rise in dopaminergic activation. A reduced ascending dopaminergic input to the basal ganglia is often linked to locomotor impairments in pathological diseases where dopamine cells are destroyed, like Parkinson's disease (PD). A recently identified dopaminergic route that connects to the mesencephalic locomotor region (MLR) originates from the substantia nigra pars compacta.

According to Roseberry et al. (2016) and Fahn (2015), a reduction of the ascending dopaminergic projections from the substantia nigra pars compacta (SNc) to the basal ganglia is typically linked to locomotor impairments. The Mesencephalic Locomotor area (MLR), a brainstem area that regulates movement, is reached by projections from the basal ganglia (Ryczko and Dubuc, 2013). The MLR, which includes the PPN (pedunculopontine nucleus) as well as the cuneiform nucleus (CnF), is being investigated as a target for deep brain stimulation to enhance locomotor performance in Parkinson's patients (Plaha and Gill, 2005; Hamani et al., 2016).

The striatum is the primary target of the ascending dopaminergic projections. By raising the activity of D_1 striatal neurons in the direct pathway and decreasing the tonic inhibition delivered to the MLR by the basal ganglia, these projections facilitate the beginning of movement. Dopamine also reduces the activity of striatal neurons in the indirect pathway, that express D_2 receptors. This helps the MLR become less inhibited and promote mobility (Freeze et al., 2013; Roseberry et al., 2016). Following disinhibition, the MLR initiates movement by providing descending excitatory impulses to the reticulospinal neurons.

There is some evidence that dopaminergic cells delivered direct descending projections to brainstem locomotor networks alongside their ascending projections. Radiometric assays and microdialysis have been used to identify dopamine in the MLR's CnF (Saavedra et al., 1979) and PPN (Steiniger and Kretschmer, 2003) in rats. Dopamine is released in the PPN in response to SNc activation.

It is likely that MLR dopamine potentiates glutamatergic signals to the MLR in order to improve locomotor activity. This is supported by the fact that MLR cells rapidly get excitatory inputs when the dopaminergic area is stimulated.

Alterations in the brainstem's locomotor circuitry, especially the MLR, can be blamed for some of the locomotor deficiencies seen in PD. When healthy people are instructed to visualize themselves walking, the peduncolopontine nucleus (PPN) and cuneiform nucleus (CnF), both components of the MLR, are engaged (Peterson et al., 2014; Tattersall et al., 2014). Walking causes a rise in PPN activity, which can be influenced by L-DOPA (Fraix et al., 2013). According to Thevathasan et al. (2012), gait freezing is correlated with lower activity in the PPN. Blood oxygen levels in the MLR are known to be lower during motor arrests (Shine et al., 2013). In the PPN of individuals with PD or progressive supranuclear palsy, neuronal losses have been documented. This involves cholinergic, GABAergic, and glycinergic cells (Pienaar et al., 2013). Neuroimaging demonstrates an association between locomotor deficits in PD patients and different types of MLR anomalies: disrupted connectivity between the MLR, thalamus, and other motor regions (Fling et al., 2014), abnormal microstructure (Vercruysse et al., 2015; Wang et al., 2016), atrophy of the MLR gray matter (Fioravanti et al., 2015), and altered metabolic activity after walking (Tard et al., 2015). Furthermore, Lewy bodies with alpha-synuclein (Seidel et al., 2015) and mitochondrial impairment (Pienaar et al., 2013) were found in the MLR of PD patients, as reported by anatomopathological examinations.

With the extent of PPN injury, the locomotor impairments become more severe.

Both non-Parkinsonian subjects with MLR lesions (Yeo et al., 2012) and older people with midbrain gray matter atrophy (Demain et al., 2014) exhibit locomotor impairments.

Brainstem Involvement in Selective Attention to Speech

People are quite adept at deciphering complicated acoustic configurations. For instance, we can pay attention to one of two conflicting speakers (Middlebrooks et al., 2017). People with impaired hearing have a difficult time comprehending speech in loud settings, and this occurs regardless of whether they are wearing auditory prosthesis (Armstrong et al., 1997). This may be related to the need for selective enhancement to recognize which sound the user wants to focus on. The auditory cortex is primarily responsible for the categorization of various sounds and the creation of related neural representations.

By evaluating the brainstem reaction to running speech, it was recently demonstrated that the auditory brainstem response to speech is also involved in selective attention. Despite having a lesser weight than the auditory cortex, the auditory brainstem activation operates at considerably higher frequencies and has a greater information rate. According to research by Etard et al. (2019), it is possible to determine a listener's attentional focus from their EEG by using the attentional modulation of their brainstem response to speech. EEG data and magnetoencephalography (MEG) data both allow for the detection of a subject's shifting center of attention in a matter of seconds (Miran et al., 2018).

Selective attention reliably modifies subcortical neuronal activity (Forte et al., 2017).

Etard et al. (2019) found that when participants listened to two competing speakers, the brainstem reaction to each speaker's frequency was greater when the speaker was paid attention to as opposed to disregarded. When speech is disregarded, the brainstem responds at a different phase than when speech is attended. The phase shift might also potentially represent the respective relative contributions of various brainstem regions to the reaction measured on the scalp. According to scalp-recorded auditory brainstem activity, the signaling is particularly pronounced in electrodes adjacent to the mastoids and close to the vertex (Bidelman, 2015). Additionally, the 8 ms reaction delay demonstrates a subcortical origin. Recent research using MEG has revealed that the frequency-following response (FFR) to basic auditory signals like long vowels include cerebral components (Coffey et al., 2016). The cortical contributions recorded by EEG come first and are smaller than the subcortical ones.

SOME PATHOLOGIES OF THE BRAINSTEM AND THE CEREBELLUM

Cerebellar Mutism

Cerebellar mutism, also known as posterior fossa syndrome or cerebellar speech disorder, is a rare and temporary condition that can occur after surgery or other interventions involving the posterior fossa of the brain, particularly the cerebellum. It is most commonly seen in children who have undergone surgical removal of tumors in the posterior fossa, such as medulloblastomas.

Cerebellar mutism typically manifests within a few days to weeks after the surgical procedure. It is characterized by a sudden and significant decrease or loss of speech ability. The mutism is often accompanied by other neurological symptoms, such as ataxia (uncoordinated movements), hypotonia (decreased muscle tone), and emotional changes.

The primary feature of cerebellar mutism is the inability to speak or express oneself verbally. Although speech is affected, other cognitive functions, such as understanding language and comprehension, are generally preserved. However, there may be associated difficulties with articulation, voice modulation, and fluency once speech gradually returns. Cerebellar mutism is often associated with other neurological symptoms, like motor deficits (coordination difficulties, balance problems, and muscle weakness). Some patients also suffer from emotional and behavioral changes (irritability, emotional lability, personality alterations). This mutism is usually temporary, and speech gradually returns in weeks to months, but the recovery process and length can vary, and some patients may have persistent deficits. Speech therapy and rehabilitation are essential in the management of this condition and of any lingering deficits. The exact mechanisms underlying cerebellar mutism are not entirely understood. However, it is believed to be related to disruption or damage to the neural connections between the cerebellum and other brain regions involved in speech production, such as the motor cortex, basal ganglia, and brainstem.

Truncal Ataxia and the Parachute Sign

Truncal ataxia and the parachute sign are both related to motor coordination and balance, typically associated with neurological conditions. Truncal ataxia refers to impaired coordination and control of movements involving the trunk. It primarily affects the muscles responsible for maintaining posture and balance. When someone has truncal ataxia, they may exhibit unsteady or wobbly movements, especially when standing or walking. This can result in a swaying or lurching gait and difficulty maintaining an upright posture. Truncal ataxia can be caused by various conditions that affect the cerebellum or its connections, such as cerebellar degeneration, strokes, multiple sclerosis, or certain genetic disorders. The parachute sign is characterized by the involuntary extension of one or both arms in an attempt to protect oneself when falling or losing balance. When an individual with the parachute sign is pushed or loses balance, they instinctively extend their arms forward, as if attempting to deploy a parachute to break a fall. This sign is often associated with corticospinal tract lesions or damage to the motor pathways in the brain, particularly the pyramidal tracts. It can be seen in conditions such as stroke, cerebral palsy, or other upper motor neuron disorders.

Dorsal Vermis Lesions

Dorsal vermis lesions can cause ataxia, a condition characterized by impaired coordination and balance. Individuals may have difficulty with precise movements, balance control, and maintaining steady posture. The coordination difficulties caused by dorsal vermis lesions can result in an abnormal gait pattern. This may include a wide-based or unsteady gait, staggering or stumbling while walking, or a tendency to veer to one side.

Dysmetria refers to difficulties with judging distances and accurately controlling the range of movement. Dorsal vermis lesions can disrupt the fine-tuning of motor movements, leading to overshooting or undershooting when reaching for objects.

Intention tremor is a tremor that occurs during voluntary movements. Dorsal vermis lesions can contribute to the development of this tremor, which typically worsens as an individual approaches a target. The dorsal vermis is involved in controlling eye movements. Lesions in this region can result in oculomotor impairments, such as nystagmus, difficulty with smooth pursuit movements, or impaired saccades.

Speech After Pontocerebellar Ischemia

When a pontocerebellar infarct affects areas involved in speech production, it can lead to speech-related symptoms. The specific manifestations can vary depending on the location and extent of the infarct, as well as individual factors.

Dysarthria refers to a motor speech disorder caused by muscle weakness, incoordination, or difficulty controlling the muscles involved in speech production. In a pontocerebellar infarct, dysarthria may occur due to disruption of motor pathways or damage to the cranial nerves responsible for controlling the muscles of the face, tongue, lips, and vocal cords. Dysarthria can result in slurred or imprecise speech, changes in speech rhythm or speed, and difficulties with articulation or pronunciation.

Apraxia of speech is a disorder characterized by difficulties in planning and coordinating the movements required for speech production. It can result from disruptions to the motor planning areas or connections between the brain regions involved in speech production. In a pontocerebellar infarct, apraxia of speech may be observed, leading to inconsistent errors in speech sound production, difficulty initiating or sequencing speech sounds, and problems with prosody (intonation, stress, and rhythm of speech).

Depending on the location and extent of the infarct, language impairments may also be present. The involvement of adjacent structures or damage to associated white matter tracts can lead to deficits in language comprehension (receptive language) or expression (expressive language).

NEUROSURGICAL APPROACHES TO THE BRAINSTEM

Brainstem Surgery and Safe Entry Zones

Violating the brainstem has been long referred to as a pessimistic chapter of neurosurgery. It was not until

1971 that Lassiter et al. suggested for the first-time the feasibility of removing lesions located deep in the truncus cerebri (Lassiter et al., 1971). In their paper, these authors described their series of 34 patients who underwent brainstem glioma surgical removal, reporting 9 postoperative deaths. A few years later, two other reports were published on the surgical removal of vascular lesions in the brainstem (Chou et al., 1975; Drake et al., 1975). Nevertheless, only in the 1990s and 2000s the number of published studies on the subject witnessed a significant rise.

The high-density of nuclei and fiber tracts carrying motor, autonomic, sensitive, and higher-order information in the brainstem had historically limited the number of attempts of surgeons transgressing its boundaries in order to remove deep-seated lesions (mostly gliomas, cavernous malformations, metastasis, hemangioblastomas). Nevertheless, the wide range of scientific advancements in the last decades in the fields of brain circuits, skull base approaches, neuroimaging techniques, and electrophysiology has paved the way for the concept of "safe entry zones" (SEZ) of the brainstem. Although quite appealing for surgeons, the idea of disposing of a "safe" surgical corridor within the brainstem is (at least partly) misleading. Indeed, no SEZ can be considered as fully safe as demonstrated by two recent published series which have reported rates of morbidity (i.e., new or worsened neurological deficits) following brainstem surgery for cavernous malformations and intrinsic brainstem tumors ranging from 20% to 50% (Cavalheiro et al., 2015; Elhammady et al., 2013). Conversely, it is more appropriate to consider SEZ as trajectories in which eloquent neural elements (i.e., nuclei, white matter tracts) and perforator arteries are sparse, and which can be therefore dissected in order to approach brainstem lesions yielding at the same time limited postoperative morbidity.

Nonetheless, selecting the SEZ associated with the lowest grade of morbidity is not the only challenge for this type of surgery. Other technical difficulties which have to be considered when approaching brainstem lesions are represented by the low threshold for manipulation-induced neurological deficits of the brainstem, the narrow working corridors, the presence of vital neurovascular structures surrounding the truncus cerebri (cranial nerves, cerebral arteries, deep cerebral veins) and the need for wide skull-base approaches with time-intensive bone drilling to expose the basal cisterns. Due to the complexity of brainstem surgery, it is recommended that only centers highly experienced with brainstem and posterior fossa surgery attempt removal of a lesion within the truncus cerebri (Cavalcanti et al., 2016).

Kyoshima et al. were the first in 1993 to describe two SEZ in the floor of the fourth ventricle (the suprafacial and the infrafacial triangles) (Kyoshima et al., 1993). From there, a total of 21 different SEZ for the brainstem have been reported (Yang et al., 2019). However, because most of these SEZ have been used in a limited number of cases and moreover only by single neurosurgical centers, the clinical evidence behind their use is still lacking (Yang et al., 2019). According to a recent systematic review published in 2019, only four SEZ had been utilized in more than 25 clinical cases, namely the supracollicular/infracollicular zones, the suprafacial and intrafacial zones, and the posterior median sulcus (Yang et al., 2019), while vice versa the evidence supporting the use of 10 SEZ was restricted to solely case reports, or even cadaveric dissections (e.g., floccular peduncle, area acustica). Due to the low number of cases and to the inconsistent reporting of postoperative morbidity and mortality rates of the studies, comparisons between SEZ are technically still not possible.

A description of the safe entry zones in the brainstem according to the location and level of evidence follows. A graphical representation of the SEZs is reported in Fig. 5a.12.

Medulla Oblongata

Posterior Median Sulcus

The PMS runs posteriorly below the obex along the midline and is bounded laterally by the gracile fascicles and tubercles (Bricolo et al., 1995). With 47 cases reported as of now, this corridor represents the sole in the medulla oblongata with credible evidence for its use. Because the dorsal vagal and hypoglossal nuclei lie close to the midline, lateral retraction should be strictly limited when using this approach. Of note, the caudal extension of this sulcus in the spinal cord is also considered as the safest route to access intrinsic spinal cord lesions (Iess et al., 2023; Sawada et al., 2017). 47 cases were reported until 2015 using this SEZ.

Lateral Medullary Zone

The LM zone has been proposed by Deshmukh et al., in 2014, as a safe entry to approach rostral dorsolateral medullary lesions. This area is located inferior to the cochlear nuclei and ventral to the foramen of Luschka and the root entry zones of IX and X c.n. After their small case series (which included four patients with excellent outcomes), other 6 cases have been published in literature employing this safe entry zone (Yang et al., 2019). The LM zone exploits a surgical corridor created by opening the foramen of Luschka and by performing a vertical incision in the inferior cerebellar peduncle. Approaches used to gain access to this SEZ include the retrosigmoid, far-lateral, and presigmoid approaches.

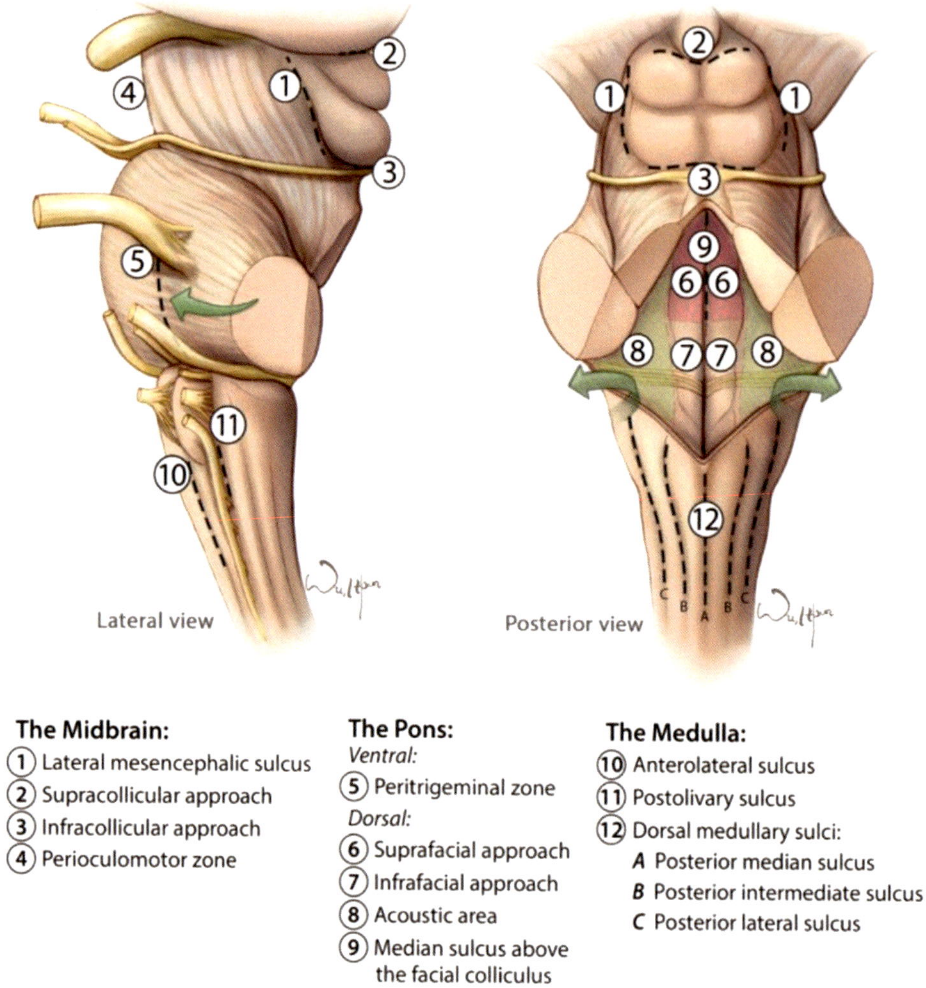

FIGURE 5A.12 **Graphical representation of the safe entry zones (SEZs).** *Image courtesy of* Neurosurgical Atlas *by Aaron Cohen-Gadol, MD.*

Olivary Zone

13 cases have been reported in literature (Yang et al., 2019; Matsushima et al., 2019). The olives are ovoid eminences involved in cerebellar motor-learning and sound processing (D'Angelo et al., 2016). Their functional boundaries are represented medially by the corticospinal tract, the XII c.n. fibers and the medial lemniscus and dorsally by the spinothalamic and tectospinal tracts. A mean distance of 5.5 mm separates the external surface of the olives from the point where white matter tracts coalesce into its hilus (Recalde et al., 2008). Authors describing the use of the OZ adopted the far lateral or retrolabyrinthine approaches (Guberinic et al., 2022).

Posterior Lateral Sulcus

Just posterior to the olive lies the posterior lateral sulcus which represents its dorsal limit. Dissection through this corridor may be justified by exophytic lesions located in the caudal anterolateral medulla (Tayebi et al., 2020). Two case reports have described its use in clinical settings (Yang et al., 2019). The close proximity to the IX and X c.n. fibers may limit its use.

Posterior Intermediate Sulcus

The posterior intermediate sulcus separates the gracile from the cuneate fasciculi. Only one case report can be found in literature which employs this corridor (Gilberto et al., 2010).

Anterolateral Sulcus

The ALS is defined as a limited area between the XII c.n. and C1 nerve. Owing to its superficial position and the proximity to the corticospinal tract, this SEZ may be suited for exophytic lesions. However, evidence for its use is still limited to cadaveric studies (Yagmurlu et al., 2014).

Pons

The Superafacial Zone

This triangle is composed by the superior cerebellar peduncle laterally, the facial colliculus inferiorly, and by the medial longitudinal fasciculus medially (Fig. 5a.9) (Yang et al., 2019; Tayebi et al., 2020).

With 65 cases described starting from 1993, it represents the entry zone with the highest level of evidence regarding its safety (Yang et al., 2019). Most of these cases belong to the series from Cavalheiro et al. (2015) which reported 45 cases of brainstem lesions operated with the SF approach. These authors reported only 1 death (2.2%), 6 (13.3%) medial fascicle longitudinal bundle injuries, and 3 (6.7%) facial paresis in their study.

Although in theory the facial colliculi aids in the topographical localization of this SEZ, in more than one-third of individuals these neural structures are not visually discernible and therefore cannot be utilized as landmarks (Bogucki et al., 1997). In such cases neurophysiological mapping can be very helpful. Care must be taken to avoid extending the cortical incision too lateral as trigeminal mesencephalic tract may become injured by such means. Evidence also suggests it is better to avoid dissecting more than 4.5 mm in depth to protect the medial longitudinal fasciculus (Yagmurlu et al., 2014). The suboccipital craniotomy with telovelar approach should be preferred over the transvermian approach due to the lower rates of cerebellar mutism (Toescu et al., 2021).

The Infrafacial Zone

The infrafacial triangle is a small area abutting the fourth ventricle delimited caudally by the striae medullaris, medially by the medial longitudinal fasciculus, and laterally by the facial nerve fibers the (Cavalcanti et al., 2016). In order to approach the IF the surgeon must first gain access to the interspace (with a mean rostrocaudal distance ranging between 6 and 9 mm) between the intrapontine segment of VII c.n. and the hypoglossal and dorsal vagal nuclei (Strauss and Fahlbusch, 1997). 38 cases have been described in literature, with the largest series belonging to Cavalheiro et al. (2015). To spare the facial colliculus from injury, an ependymal incision no-longer than 6 mm should be performed (Tayebi et al., 2020) just above the striae medullaris. The safety margin of dissection through the pons is represented by the central trigeminal tract which can be found approximately at 5 mm from the floor of the fourth ventricle (Tayebi et al., 2020). Although no deaths were reported, 9 (56.2%) out of 16 patients in the series of Cavalheiro et al. developed postoperative facial paresis with this SEZ (Cavalheiro et al., 2015).

Peritrigeminal Zone Dorsal Approaches

Recalde et al. (2008) were able to precisely estimate a safe pathway anterior to the motor and sensory nuclei of the trigeminal nerve, lateral to the corticospinal tract, employing an approach dissecting the white fibers in front of the trigeminal nerve's emergence. They discovered an average depth of dissection of 11.2 mm to the trigeminal nuclei and an average distance between the V c.n. and the corticospinal tract of 4.64 mm. In front of the trigeminal nuclei, in a downward direction, lie the fibers of VI, VII, and VIII c.n. Directly above the trigeminal root entrance in the middle cerebellar peduncle we find another site that has been employed to treat anteriorly positioned tumors in the pons. Due to the location of the pontine transverse fibers and the middle cerebellar peduncle, it is easy to carefully cut along these fibers behind the corticospinal tract. The intersection between the trigeminal and the facial-vestibulocochlear complex root entrance, and between the middle cerebellar peduncle and the pons, was proposed as a safe corridor by Baghai et al., however it limits vertical manipulation.

Peritrigeminal Zone

The PT zone is delimited cranially by the root of V c.n. and caudally by the complex of VII and VIII c.n. It takes advantage of a surgical corridor lateral to the pyramidal tract and ventral with respect to the motor and sensitive trigeminal nuclei. Recalde et al. (2008) were the first to describe this SEZ. These authors described a mean depth of dissection through the pons of 11.2 mm using this approach, which can be considered fairly deep compared to most SEZ in the brainstem. Vice versa, the interval between the pyramidal tract and V c.n. (4.64 mm on average) limits its extension in an anterior-to-posterior direction. 22 cases can be found in literature using this approach, most of them described in a series of Hauck et al. (Hauck et al., 2010).

Supratrigeminal Zone

First employed by Hebb and Spetzler, the ST zone is situated above the root of V c.n., on the middle cerebellar peduncle and faces the corticospinal tract medially. By dissecting the thick pontine transverse fibers, this approach allows entry to the pons taking care to avoid violating the superior cerebellar peduncle (deeply situated) (Cavalcanti et al., 2016). Limited evidence supports its use, and to date (2023), 8 cases using this approach have been published in the literature (Yang et al., 2019).

Lateral Pontine Zone

Commonly known as the middle cerebellar peduncle zone, the LS is located between the cerebellum and the pons (Yang et al., 2019) and placed laterally to the trigeminal root entry zone. Because of the proximity to the nuclei and the fibers of c.n., only superficial dissection is allowed with this approach. Moreover, vertical manipulation has to be held to the minimum (Cavalcanti et al., 2016). 11 cases have been reported so far using this approach (Yang et al., 2019).

Median Sulcus of the Fourth Ventricle (or Above the Facial Colliculus)

Situated just above facial colliculus, the MS has been used in two clinical cases as an entry point devoid of crossing fibers. Nevertheless, its use is significantly limited by the proximity of the MFL which prohibits even the minimal lateral retraction (Tayebi et al., 2020).

Area Acustica

The area Acustica, which provides access to the brainstem via a breach in the dorsal cochlear nucleus and the peduncular white matter tracts, was first pointed out as a possible safe zone in the early 2000s. Yet so far, no clinical cases have been published in the literature (Yang et al., 2019).

Floccular Peduncle

Similarly to the AA, insights in the safety of the FP (which abuts the lateral recess of the fourth ventricle) as a potential SEZ remain speculative. The potential interval provided by this SEZ is around 8 mm (Akiyama et al., 2018).

Midline Approach

Bricolo et al. (1995) proposed an approach across the midline, which relies on the sparseness of crossing fibers, between the VI and the III c.n. nuclei. Injury to the medial longitudinal fascicle can result in extraocular movement issues. These issues can be triggered by even the smallest amount of lateral retraction. This approach is therefore a risk.

Mesencephalon

Supracollicolar/Infracollicular Zone

Usually grouped together, the sopracollicular and infracollicolar SEZs are located in the quadrigeminal plate and are the most widely used mesencephalic SEZs to approach the dorsal tectal area, just behind to the cerebral aqueduct. In both cases, a small transverse incision usually the length of the inferior or superior collicular complex is made. 49 reported surgeries can be found in literature using this approach, although almost all belong to the case series of Cavalheiro et al. (47/49) (Yang et al., 2019; Cavalheiro et al., 2015). With this SEZ surgeons should be cautious not to injure eloquent structures such as the mesencephalic nucleus and tract of V c.n., III c.n., the trochlear nucleus, the tectospinal tract, the superior cerebral peduncle, the MFL and anterolateral fasciculus (Tayebi et al., 2020). Technically, the SC/IC SEZ can be accessed utilizing the median, paramedian, and extreme-lateral sopracerebellar infratentorial approaches.

The rhomboid fossa conceals a number of important structures. At the level of the facial colliculus the facial nerve wraps around the abducens nerve nucleus at the bottom of the fourth ventricle. The median sulcus and the medial longitudinal fascicle are also located in this area. The striae medullaris lies immediately caudal to the nuclei of X and XII c.n. In this area, Kyoshima et al. (1993) described two safe zones that only slightly disturb the nearby neuronal systems. The soprafacial triangle, which is defined by the facial nerve, the cerebellar peduncles, and the medial longitudinal fascicle, is the first. The striae medullaris, the facial nerve and the medial longitudinal fascicle, form the infrafacial triangle and define the second. The diversity of the striae medullaris and the potential harm to the trigeminal motor nucleus and the lower cranial nerve nuclei were pointed out by Strauss and Fahlbusch (1997) as possible complications in using these approaches.

Anterior Mesencephalic Zone

Also referred to as the perioculomotor region, this area is situated in the lateral two-thirds of the anterior portion of the cerebral peduncle. Its superior and inferior limits are represented by the posterior cerebral artery and by the superior cerebellar artery, respectively (Yang et al., 2019). The proximity of the substantia nigra represents an important dorsal limit when utilizing this approach. Guberinic et al. suggest that the exit zone of III c.n. in the median sulcus could be used to localize the substantia nigra (Guberinic et al., 2022). 13 cases from the year 1995 have been described using this SEZ. Cranial approaches useful to reach the AM zone are represented by the orbitozygomatic, the subtemporal and the anterior petrosectomy.

A small region on the cerebral peduncle, restricted medially by the oculomotor tract and nerve and laterally by the corticospinal tract, can be used to reach lesions affecting the anterior midbrain. The red nucleus and nigrostriatal circuit are located deep medially, while the corticospinal tract is mostly in the intermediate three-fifths of the peduncle. The main trunk of the superior cerebellar artery (SCA) and the posterior cerebral artery (PCA) limit the entrance site in the interpeduncular cistern, respectively, on the superior and inferior sides.

Lateral Mesencephalic Sulcus

The lateral mesencephalic sulcus, which runs from the medial geniculate body to the pontomesencephalic sulcus, divides the peduncular and tegmental regions of the midbrain and is concealed by the lateral mesencephalic vein. The posterior P2 segment (P2P) crosses the sulcus superiorly, the medial posterior choroidal artery crosses it in the middle, and the cerebellomesencephalic sections of the SCA and the trochlear nerve intersect it in the inferior position. The medial lemniscus posteriorly and the substantia nigra anterolaterally define a safe entrance zone. An anteromedial restriction to dissection is imposed by the oculomotor nerve that passes through from the red nucleus to the substantia nigra (Recalde et al., 2008). Bounded by the substantia nigra ventrally and the medial lemniscus dorsally the LMS may be useful for laterally situated lesions in the midbrain. Usually covered by the lateral mesencephalic vein, this SEZ (which has a mean length of 9.6 mm) provides the surgeon with a safe dissection depth of 8 mm on average (recalled). An important anteromedial limit not to be breached is represented by the fibers of III c.n. running from the red nucleus to the substantia nigra. In literature, 13 cases of brainstem surgery using the LMS entry zone are described (Yang et al., 2019).

Intercollicular Region

The rationale behind the use of the intercollicular region SEZ spans from the fact that no white matter tracts crossing this region can be found with the exception of the inferior and superior collicular commissures. Moreover, similarly to the SC/IC, the aqueduct restrains dissection in a ventral direction (Guberinic et al., 2022). One study suggested it may result in less damage to fiber tracts than the other quadrigeminal SEZ, but up to now only one case report using this approach can be found (Yang et al., 2019).

Interpeduncular Zone (or Fossa)

The IPZ is an area located in the caudal mesencephalon where fibers assemble with an orientation that is predominantly diagonal (Guberinic et al., 2022). It is suggested as an alternate surgical corridor for lesions placed anteriorly and midline to the III c.n.'s nucleus. With only three cases described in literature the level of evidence supporting its use is still limited (Yang et al., 2019).

Approaches Division

For a simple understanding of the approaches, we will divide them as follows:

- Lateral Supratentorial
  - Subtemporal "Kawase"
  - Transcavernous
- Lateral Infratentorial
  - Retrosigmoid/Extended Retrosigmoid
  - Far Lateral
- Lateral Supra/Infratentorial
  - Posterior Petrosal (Presigmoid)
- Midline Posterior
  - Supracerebellar Infratentorial
  - Telovelar
- Midline Anterior
  - Extended Endoscopic Endonasal

Lateral Supratentorial

Kawase Approach

A temporal craniotomy is performed using a curvilinear incision and the anterior portion of the temporal muscle is dissected using retrograde dissection technique. Once the bone is exposed, a Burr Hole is placed in the middle of the temporal squama, and a Whirlpool craniotomy is performed around it, centered in the root of the zygoma. Following the completion of the craniotomy, the remaining squamous temporal bone is drilled until it is level with the temporal fossa. Subsequently, careful dural dissection is initiated by tracing the middle meningeal artery until reaching the foramen spinosum, which is then cauterized and cut at its base. Medial to the foramen spinosum lies the GSPN (greater superficial petrosal nerve, Fig. 5a.14). Lateral dural dissection is performed to locate the accurate eminence, which indicates the presence of the foramen of GSPN. The GSPN is gently dissected from its dural attachments in a posterior to anterior direction, serving as the lateral drilling limit. Continuing the dissection along the GSPN anteriorly leads to the junction with the petrous segment of the internal carotid artery (ICA) if it is dehiscent, as well as the gasserian ganglion. Once the GSPN has been safely dissected from its dural attachments, medial dural dissection can be performed to expose the temporal ridge and petrous apex. Tenting sutures can be placed in the detached dura from the temporal fossa to provide better retraction, allowing for dynamic retraction during the procedure. At this point we have fully exposed the anterior petrous apex, and we set the Kawase landmarks:

(A) Lateral GSPN and petrous carotid;
(B) Medial the petrous ridge;
(C) Anterior, Gasserian ganglion; and
(D) Posterior and lateral the arcuate eminence.

The drilling procedure should commence laterally to the gasserian ganglion, as a substantial volume of bone can be efficiently removed in this region. As the dura of the internal auditory canal (IAC) becomes visible, drilling should continue posteriorly and medially. The medial drilling is performed along the edge of the petrous ridge, taking care to avoid the superior petrosal sinus. Upon exposing the dura of the posterior fossa, the entire Kawase rhomboid region must be drilled out to improve the surgical exposure.

To access the intradural space, the dura is opened in an anterior to posterior direction, approximately 2 cm lateral to the superior petrosal sinus. A second incision is made in a lateromedial direction, starting from the initial dural incision and extending toward the superior petrosal sinus. Once the sinus is reached, it should be occluded using clips or cautery and then severed. The dural incision should be extended up to the tentorium, while exercising caution to prevent any damage to the fourth cranial nerve as it enters its respective foramen. At this stage, the anterolateral aspect of the pons becomes exposed. The graphical representation of the exposure limits of this approach is showed in Fig. 5a.13.

Transcavernous Approach

This approach is carried out through a frontotemporozygomatic (FTZ) craniotomy or a pterional approach, depending on the surgeon's preference. It involves performing an anterior clinoidectomy and carefully peeling off the lateral wall of the cavernous sinus. During this step, it is crucial to exercise caution to prevent hemorrhage. The use of hemostatic agents like *Floseal*[1] (a gelatin-based hemostatic matrix) can be valuable in managing bleeding. It is recommended to avoid using bipolar cautery due to the potential risk of cranial nerve damage. However, if cautery is employed, it should be set to the lowest possible power setting.

After completing the FTZ craniotomy, clinoidectomy, and gently retracting the temporal lobe, meticulous dissection of the nerves within the cavernous sinus becomes feasible. The third cranial nerve can serve as a guiding landmark and upper limit of the surgical corridor. Below this nerve lies the cavernous segment of the carotid artery, which serves as the lower boundary of the surgical corridor. By establishing these boundaries, further dissection can be conducted to expose the sella turcica. This may involve drilling the posterior portion of the sella turcica, also known as posterior clinoidectomy, to create a small aperture connecting the middle fossa and posterior fossa. Once the drilling is completed, the interpeduncular fossa becomes visible, and in certain cases, a prominent basilar tip may be observed. The graphical representation of this approach is showed in Fig. 5a.15.

Lateral Infratentorial

Retrosigmoid Approach

As its name suggests, the retrosigmoid approach is centered on the posterior aspect of the sigmoid and lateral sinus, providing a practical window to the posterior fossa, including the lateral portion of the cerebellum and the cerebellopontine angle. The specific type of incision, whether curvilinear, linear, or horseshoe-shaped, can be based on the surgeon's

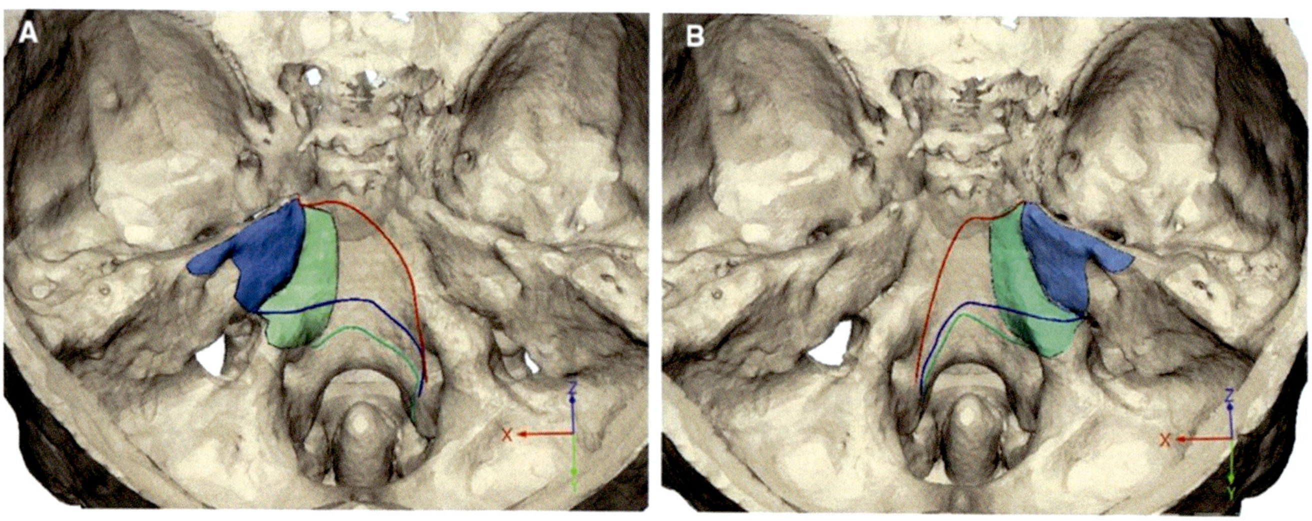

FIGURE 5A.13 **Exposure limits of different approaches: Kawase, subtemporal and extended Kawase.** The inferior limits for subtemporal (*red line*), Kawase (*blue line*), and extended Kawase (*green line*) approaches are delineated in the skull base of group 1 (A, MA < 135 degrees) and group 2 (B, MA ≥ 135 degrees). The debone ranges for the Kawase (*blue area*) and extended Kawase (*green area*) approaches are also marked. *MA*, manipulation angle. *Image courtesy of Yan, Y., Xu, T., Zhao, Y., Mei, Q., Jiang, L., Hou, L., 2023. Exposure region of the Kawase approach and its correlation with skull base anatomy: an evaluation with digital models. Front. Surg. 9, 1047949. https://doi.org/10.3389/fsurg.2022.1047949.*

1 Floseal® (Baxter Healthcare Corporation), a gelatin-based hemostatic matrix.

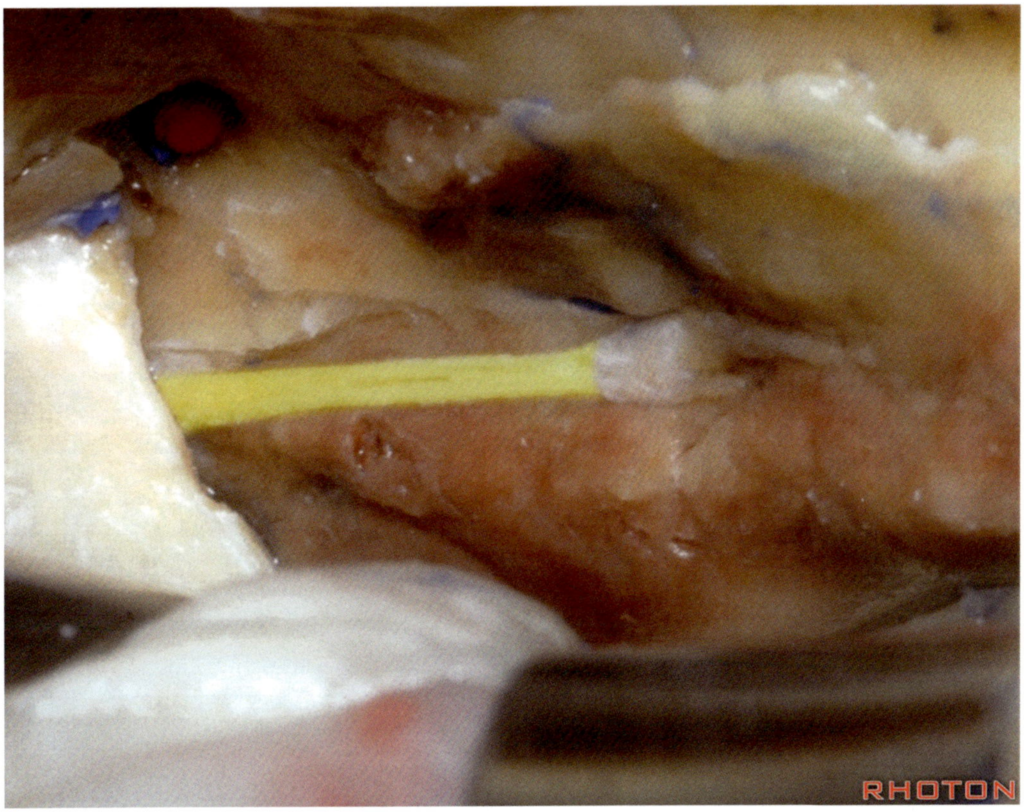

FIGURE 5A.14 **Greater superficial petrosal nerve (*in yellow*).** *Image courtesy of The Rhoton Collection®. http://rhoton.ineurodb.org/. Drawing by Jeffrey Sorenson.*

preference, with consideration given to preserving blood irrigation and innervation. Dissection of soft tissues should be performed meticulously, following distinct planes to ensure the preservation of all structures for proper closure.

To achieve a precise and centered exposure of the junction between the lateral sinus and the sigmoid sinus, neuronavigation or craniometric landmarks can be utilized. Once the targeted area is adequately exposed, a careful drilling procedure with a diamond burr should be carried out through a single Burr hole. Attention must be paid to avoid any injury to the underlying dura, which becomes visible as the bone is being thinned with the diamond burr. Detaching the dura using blunt instruments is important, given the strong attachment of the dura to the bone in this region. After the completion of drilling, the dura is exposed and the sinus can be identified as a purple-colored structure through the dura. If the sinus is not readily exposed, the bone can be further thinned with the diamond burr until the sinus becomes visible, ensuring control in case of injury. Once the drilling is finished, the dura is opened using various techniques, with a practical "T"-shaped incision being one of the options. Sutures are used to reflect the dural flaps out of the surgical field. At this point, the use of a microscope is necessary.

The lateral hemisphere of the cerebellum becomes exposed on its most lateral aspect. By using a Cottonoid® (neurosurgical patties, pads, or micropatties) and performing careful dissection, the foramen magnum cistern can be reached to drain cerebrospinal fluid (CSF) and create space in the cerebellopontine angle. This maneuver allows for easy retraction of the cerebellum, facilitating intradural dissection along the petrous ridge. During the dissection, bridging veins are dissected until reaching the VII–VIII complex before it enters the internal acoustic meatus. The labyrinthine artery may also be encountered in this area. Continuing the exploration in a cephalic direction reveals the V c.n. with the VI c.n. found deeper and underneath it. In a more basal direction, the IX, X, and XI cranial nerves can be observed in some cases. Overall, the retrosigmoid approach provides exposure of multiple cranial nerves, as well as the brainstem, particularly the pons and partially the medulla oblongata in the more caudal region.

Far Lateral

This surgical approach provides access to various skull base structures, with the extent of condylar drilling determining the specific areas exposed. In general, it allows for exposure of the inferior clivus,

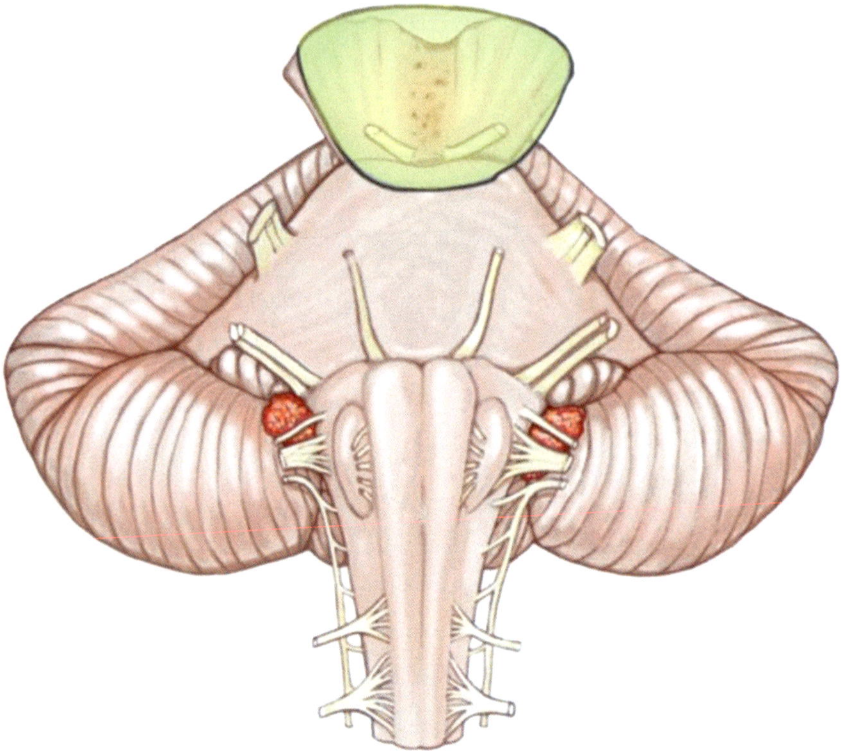

FIGURE 5A.15 **Transcavernous approach (*green area*).** *Image courtesy of The Rhoton Collection®. http://rhoton.ineurodb.org/. Drawing by Jeffrey Sorenson.*

premedullary area, region adjacent to the hypoglossal canal medial to the jugular tubercle and the posterior and lateral aspects of the jugular foramen. Posteriorly, the lateral surface of the pons and medulla oblongata can be fully exposed.

To begin the procedure, a "hockey stick" incision is made, starting in the midline 5 cm below the occipital protuberance, extending 1 cm over the nuchal line, and projecting laterally until reaching the mastoid region. From there, it curves downward to finish around 5 cm below the mastoid tip, typically adjacent to the transverse process of C1. Considering the extent of dissection and the time required for this approach, a musculocutaneous flap is recommended. All muscles are detached from the skull using a Bovie cautery and cut from the sides, creating a single flap that is then reflected downward. Caution must be exercised when reaching the base of the skull near the foramen magnum and C1 to avoid injury to the vertebral artery.

Once the occipital squama and the posterior arch of C1 (and possibly C2) are fully exposed, a suboccipital craniotomy is performed. A Burr hole can be placed lateral and inferior to the torcula, and a second one near the junction of the transverse and sigmoid sinuses. Using a foot plate attachment, a large craniotomy is performed downward until reaching the posterior aspect of the foramen magnum. Soft tissue dissection from C1 and C2, if necessary, is carried out with care to identify and, if needed, reflect the vertebral artery away from the surgical field. Kerrison rongeurs and Bovie cautery are used to gently remove the posterior arches. If required, condylar drilling may be performed to access lesions located closer to the midline of the inferior clivus.

With the bone work completed, the dura is fully exposed, and hemostasis is achieved before opening it. The incision of the dura can be performed in various ways. Here, we propose a paramedian vertical incision extending from C2 to C1 and running up to the most anterior aspect of the suboccipital dura. A second horizontal cut is made starting just a few millimeters above the dura of the foramen magnum, ensuring avoidance of the venous sinus in the area. The dural flaps are reflected and intradural dissection is initiated.

The far lateral approach allows for extensive exposure of the lateral brainstem and the cranial nerves arising from it. The range of exposure extends from the inferior aspect of the VII–VIII complex to the dorsal roots of C1 and C2. Its versatility underscores its importance in neurosurgical practice.

Lateral Supra/Infratentorial

Posterior Petrosal Approach

The PPA (Posterior Petrosectomy Approach) focuses on the angle formed between the dura of the posterior fossa and the superior petrosal sinus over the petrous ridge. The goal is to remove the underlying bone to expose and create a corridor between the temporal lobe and the cerebellum. The preferred incision for this approach is a "question mark" shape. It starts 2 fingerbreadths above the External Auditory Canal (EAC), curves posteriorly to the *asterion* and then proceeds straight down to the mastoid tip. Soft tissue dissection is performed in layers, with the retrograde dissection of the temporal muscle until the root of the zygoma is exposed and retracted anteriorly. The posterior auricular muscle and nuchal muscles are also detached and reflected.

Once the full extent of bone is exposed, a first Burr hole is placed superior to the transverse-sigmoid sinus junction to expose the temporal dura, and a second Burr hole is placed inferior to it to expose the posterior fossa dura. Using a foot plate attachment, a craniotomy is performed, centered on both Burr holes. Cuts are made parallel to the middle fossa, extending to the zygoma, curving up to the full extension of the incision, then curving posteriorly to the Burr holes, and descending to the posterior fossa dura. As it approaches the mastoid tip, the incision curves upward again, ending at the inferior burr hole. The remaining bridge of bone between the burr holes is drilled using diamond burs and the bone is elevated. At this point, the mastoidectomy is initiated by drilling the mastoid bone using cutting and diamond burs. Care must be taken to preserve all inner ear structures. The objective is to completely expose and skeletonize the sigmoid sinus and superior petrosal sinus. Once this is achieved, the superior petrosal sinus is cut, and the sigmoid sinus can be mobilized posteriorly. The dura is incised along the edge of the petrous ridge over the sinus, and the intradural dissection is initiated.

This approach is considered complex and is primarily used for large lesions that extend from the supratentorial to the infratentorial compartment. It provides full exposure of the upper part of the brainstem, from the mesencephalon to the pontomedullary junction, and allows for visualization of all cranial nerves in the region from IV to VII.

Midline Posterior

Supracerebellar Infratentorial Approach

The supracerebellar infratentorial approach offers a surgical route to access the pineal region, tentorial incisura, and structures surrounding the midbrain and upper brainstem. The patient is positioned in the prone position on the operating table, and the head is secured using a three-point Mayfield head holder to ensure stability throughout the procedure. A midline incision is made in the posterior scalp, extending from the inion to a sufficient distance above the hairline, following the natural contour of the skull.

Careful removal of a portion of the skull bone in the suboccipital region, including the occipital bone and part of the posterior fossa, creates a bone window that provides access to the structures of the posterior fossa. The muscles in the suboccipital region, such as the trapezius and suboccipital muscles, are dissected and gently retracted to expose the underlying bone and dura mater. An incision is made in the dura mater allowing visualization of the posterior fossa structures, including the cerebellum and brainstem.

With the dura opened, the surgeon carefully dissects through the cerebellum to create a surgical corridor, which exposes the underlying structures within the posterior fossa. This enables access to the targeted area, such as tumors, vascular malformations, or cysts. The supracerebellar infratentorial approach provides a direct and angled pathway to reach the posterior fossa structures, specifically the pineal region and upper brainstem. It allows surgeons to access deep-seated lesions while minimizing manipulation and retraction of the surrounding brain tissue.

Telovelar Approach

This posterior approach is employed for lesions located in the posterior portion of the brainstem and the fourth ventricle. The goal is to access the area of the fourth ventricle by opening the tela choroidea and inferior medullary velum. The patient's position may vary depending on the center, with options including prone or sitting position.

The incision is initiated at the *inion* and extended downward as necessary until reaching C4. All tissue planes are cauterized using a Bovie cautery, as the sitting position poses a high risk of air embolism. Care must be taken to apply bone wax to control bleeding at all sites when reaching the bone.

Once the full extent of the occipital squama and the posterior part of C1–C2 are exposed, a suboccipital craniotomy is performed, and the posterior arches of C1 and C2 are removed. At this stage, the dura is fully exposed and incised in the midline from C1 to C2. Just before reaching the dura adjacent to the foramen magnum, the incision splits into two diagonal cuts that extend upward to the boundaries of the craniotomy. Dural flaps are secured with sutures, and intradural dissection is initiated.

A brain retractor is utilized to retract the cerebellar vermis, creating an opening in the surgical corridor toward the fourth ventricle. The tela choroidea and inferior medullary velum are cautiously opened and reflected, allowing for a clear view of the full extent of the fourth ventricle.

Midline Anterior

Endoscopic Endonasal Approaches to the Brainstem

The extended endoscopic endonasal approaches (EEEA) are minimally invasive surgical techniques that provide access to the skull base and brainstem through the nasal corridor. These approaches utilize an endoscope, a thin and flexible tube with a camera and light source, to visualize and navigate the surgical field. EEEA can be employed for the treatment of various conditions such as tumors, aneurysms, and vascular malformations.

In the EEEA technique, bony structures of the skull base are drilled to gain access to the brainstem. For most brainstem lesions, a transsphenoidal clival approach is utilized. This involves accessing the sphenoid sinus by opening and removing the rostrum, followed by drilling the posterior aspect of the sinus, which corresponds to the clivus. Once the clivus is drilled, the dura of the posterior fossa is exposed and opened. In this surgical plane, the basilar arteries and their branches can be visualized first, followed by the pons. Anatomically, the advantage of EEEA lies in the fact that the working corridors are medial to all the neuromuscular structures (cranial nerves and arteries), resulting in fewer lesions and less morbidity. However, cerebrospinal fluid leakage is a common complication associated with the procedure.

Compared to traditional open surgery, the EEEA technique offers several advantages, including reduced pain, shorter hospital stays, and faster recovery times. The selection of a specific surgical corridor depends on factors such as the location of the pathology, the patient's anatomy, the surgical objectives, and the surgeon's expertise.

The peduncular approach, also known as the subtemporal transtentorial approach, involves making an incision behind the ear. The patient is positioned on their back with the head slightly turned in the opposite direction of the lesion, and their head is secured using a fixation device. An incision is made following the contour of the temporal bone behind the ear. A bone flap is then created by removing a portion of the skull, with the size and shape of the flap customized based on the specific requirements of the procedure. The dura mater is carefully opened, and gentle retraction of the brain is performed to allow the surgeon to navigate through the brain structures and reach the target area.

During this approach, special attention is given to the structures such as the temporal lobe, tentorium, and midbrain peduncles. Navigating through these structures requires precise dissection and meticulous technique.

On the other hand, the zygomatic approach, also known as the pterional or frontotemporal approach, provides access to the anterior and middle cranial fossae, including the lateral aspects of the brain, sphenoid bone, and cavernous sinus. The zygomatic approach involves making an incision along the temporal region of the skull, extending from above the eyebrow to behind the ear. This incision allows for a broad exposure of the surgical area. An important step in the zygomatic approach is the osteotomy, where a portion of the zygomatic arch (cheekbone) is cut and temporarily removed. This procedure enhances access and visualization of the surgical target. After osteotomizing the zygomatic arch, the temporalis muscle is dissected and gently retracted to expose the underlying skull bone and brain structures. The surgeon then carefully removes a portion of the skull bone, creating a bone window typically extending from the eyebrow region to the temple. Once the bone window is created, an incision is made in the dura mater to expose the brain tissue beneath.

Reaching the interpeduncular cistern, which is a highly complex area requiring meticulous dissection, involves opening the Liliequist membrane. This area houses critical structures such as the basilar artery bifurcation, posterior cerebral arteries (PCAs), superior cerebellar arteries (SCAs), medial posterior choroidal artery, thalamoperforating arteries, and direct perforators of the proximal SCA. Due to the restricted nature of this location, surgical mobility is significantly limited, necessitating utmost precision during dissection.

References

Akiyama, O., Matsushima, K., Nunez, M., et al., 2018. Microsurgical anatomy and approaches around the lateral recess with special reference to entry into the pons. J. Neurosurg. 129 (3), 740–751. https://doi.org/10.3171/2017.5.JNS17251.

Armstrong, M., Pegg, P., James, C., Blamey, P., 1997. Speech perception in noise with implant and hearing aid. Am. J. Otol. 18 (6 Suppl. l), S140–S141.

Bidelman, G.M., 2015. Multichannel recordings of the human brainstem frequency-following response: scalp topography, source generators, and distinctions from the transient ABR. Hear. Res. 323, 68–80. https://doi.org/10.1016/j.heares.2015.01.011.

Bogucki, J., Gielecki, J., Czernicki, Z., 1997. The anatomical aspects of a surgical approach through the floor of the fourth ventricle. Acta Neurochir. 139 (11), 1014–1019. https://doi.org/10.1007/BF01411553.

Bricolo, A., Turazzi, S., 1995. Surgery for gliomas and other mass lesions of the brainstem. In. Adv. Tech. Stand Neurosurg. 261–341. https://doi.org/10.1007/978-3-7091-6898-1_5.

Cavalcanti, D.D., Preul, M.C., Kalani, M.Y.S., Spetzler, R.F., 2016. Microsurgical anatomy of safe entry zones to the brainstem. J. Neurosurg. 124 (5), 1359–1376. https://doi.org/10.3171/2015.4.JNS141945.

Cavalheiro, S., Yagmurlu, K., da Costa, M.D.S., et al., 2015. Surgical approaches for brainstem tumors in pediatric patients. Childs Nerv. Syst. 31 (10), 1815–1840. https://doi.org/10.1007/s00381-015-2799-y.

Chou, S.N., Erickson, D.L., Ortiz-Suarez, H.J., 1975. Surgical treatment of vascular lesions in the brain stem. J. Neurosurg. 42 (1), 23–31. https://doi.org/10.3171/jns.1975.42.1.0023.

Coffey, E.B.J., Herholz, S.C., Chepesiuk, A.M.P., Baillet, S., Zatorre, R.J., 2016. Cortical contributions to the auditory frequency-following response revealed by MEG. Nat. Commun. 7 (1), 11070. https://doi.org/10.1038/ncomms11070.

Demain, A., Westby, G.W.M., Fernandez-Vidal, S., et al., 2014. High-level gait and balance disorders in the elderly: a midbrain disease? J. Neurol. 261 (1), 196–206. https://doi.org/10.1007/s00415-013-7174-x.

Drake, C.G., 1975. Surgical removal of arteriovenous malformations from the brain stem and cerebellopontine angle. J. Neurosurg. 43 (6), 661–670. https://doi.org/10.3171/jns.1975.43.6.0661.

D'Angelo, E., Galliano, E., De Zeeuw, C.I., 2016. Editorial: the olivo-cerebellar system. Front. Neural Circuits 9. https://doi.org/10.3389/fncir.2015.00066.

Elhammady, M.S., Teo, C., 2013. Surgical management of adult intrinsic brainstem tumors. Neurosurgery 60 (Suppl. 1), 131–138. https://doi.org/10.1227/01.neu.0000430770.33467.be.

Etard, O., Kegler, M., Braiman, C., Forte, A.E., Reichenbach, T., 2019. Decoding of selective attention to continuous speech from the human auditory brainstem response. Neuroimage 200, 1–11. https://doi.org/10.1016/j.neuroimage.2019.06.029.

Fahn, S., 2015. The medical treatment of Parkinson disease from James Parkinson to George Cotzias. Mov. Disord. 30 (1), 4–18. https://doi.org/10.1002/mds.26102.

Fioravanti, V., Benuzzi, F., Codeluppi, L., et al., 2015. MRI correlates of Parkinson's disease progression: a voxel based morphometry study. Parkinsons Dis 2015, 1–8. https://doi.org/10.1155/2015/378032.

Fling, B.W., Cohen, R.G., Mancini, M., et al., 2014. Functional reorganization of the locomotor network in Parkinson patients with freezing of gait. PLoS One 9 (6), e100291. https://doi.org/10.1371/journal.pone.0100291.

Forte, A.E., Etard, O., Reichenbach, T., 2017. The human auditory brainstem response to running speech reveals a subcortical mechanism for selective attention. Elife 6. https://doi.org/10.7554/eLife.27203.

Fraix, V., Bastin, J., David, O., et al., 2013. Pedunculopontine nucleus area oscillations during stance, stepping and freezing in Parkinson's disease. PLoS One 8 (12), e83919. https://doi.org/10.1371/journal.pone.0083919.

Freeze, B.S., Kravitz, A.V., Hammack, N., Berke, J.D., Kreitzer, A.C., 2013. Control of basal ganglia output by direct and indirect pathway projection neurons. J. Neurosci. 33 (47), 18531–18539. https://doi.org/10.1523/JNEUROSCI.1278-13.2013.

Giliberto, G., Lanzino, D.J., Diehn, F.E., Factor, D., Flemming, K.D., Lanzino, G., 2010. Brainstem cavernous malformations: anatomical, clinical, and surgical considerations. Neurosurg. Focus 29 (3), E9. https://doi.org/10.3171/2010.6.FOCUS10133.

Guberinic, A., van den Elshout, R., Kozicz, T., Laan, M.T., Henssen, D., 2022. Overview of the microanatomy of the human brainstem in relation to the safe entry zones. J. Neurosurg. 137 (5), 1524–1534. https://doi.org/10.3171/2022.2.JNS211997.

Hamani, C., Lozano, A.M., Mazzone, P.A.M., et al., 2016. Pedunculopontine nucleus region deep brain stimulation in Parkinson disease: surgical techniques, side effects, and postoperative imaging. Stereotact. Funct. Neurosurg. 94 (5), 307–319. https://doi.org/10.1159/000449011.

Hauck, E.F., Barnett, S.L., White, J.A., Samson, D., 2010. The presigmoid approach to anterolateral pontine cavernomas. J. Neurosurg. 113 (4), 701–708. https://doi.org/10.3171/2010.1.JNS08413.

Iess, G., Bonomo, G., Amato, A., et al., 2023. Intraoperative neurophysiologic and angiographic techniques to identify the posterior median sulcus for midline myelotomy. World Neurosurg 171, 103. https://doi.org/10.1016/j.wneu.2022.11.064.

Kyoshima, K., Kobayashi, S., Gibo, H., Kuroyanagi, T., 1993. A study of safe entry zones via the floor of the fourth ventricle for brain-stem lesions. Report of three cases. J. Neurosurg 78 (6), 987–993. https://doi.org/10.3171/jns.1993.78.6.0987.

Lassiter, K.R.L., Alexander, E., Davis, C.H., Kelly, D.L., 1971. Surgical treatment of brain stem gliomas. J. Neurosurg. 34 (6), 719–725. https://doi.org/10.3171/jns.1971.34.6.0719.

Matsushima, K., Kohno, M., Bertalanffy, H., 2019. Microsurgical resection of medullary cavernoma via the olivary zone by the retrosigmoid supracondylar approach. Neurosurg. Focus: Video 1 (2), V7. https://doi.org/10.3171/2019.10.FocusVid.19293.

Middlebrooks, J.C., Simon, J.Z., Popper, A.N., Fay, R.R., 2017. The Springer Handbook of Auditory Research: The Auditory System at the Cocktail Party, vol 60. Springer.

Miran, S., Akram, S., Sheikhattar, A., Simon, J.Z., Zhang, T., Babadi, B., 2018. Real-time tracking of selective auditory attention from M/EEG: a Bayesian filtering approach. Front. Neurosci. 12. https://doi.org/10.3389/fnins.2018.00262.

Peterson, D.S., Pickett, K.A., Duncan, R., Perlmutter, J., Earhart, G.M., 2014. Gait-related brain activity in people with Parkinson disease with freezing of gait. PLoS One 9 (3), e90634. https://doi.org/10.1371/journal.pone.0090634.

Pienaar, I.S., Elson, J.L., Racca, C., Nelson, G., Turnbull, D.M., Morris, C.M., 2013. Mitochondrial abnormality associates with type-specific neuronal loss and cell morphology changes in the pedunculopontine nucleus in Parkinson disease. Am. J. Pathol. 183 (6), 1826–1840. https://doi.org/10.1016/j.ajpath.2013.09.002.

Plaha, P., Gill, S.S., 2005. Bilateral deep brain stimulation of the pedunculopontine nucleus for Parkinson's disease. Neuroreport 16 (17), 1883–1887. https://doi.org/10.1097/01.wnr.0000187637.20771.a0.

Recalde, R.J., Figueiredo, E.G., de Oliveira, E., 2008. Microsurgical anatomy of the safe entry zones on the anterolateral brainstem related to surgical approaches to cavernous malformations. Oper. Neurosurg. 62 (3), 9–17. https://doi.org/10.1227/01.neu.0000317368.69523.40.

Roseberry, T.K., Lee, A.M., Lalive, A.L., Wilbrecht, L., Bonci, A., Kreitzer, A.C., 2016. Cell-type-specific control of brainstem locomotor circuits by basal ganglia. Cell 164 (3), 526–537. https://doi.org/10.1016/j.cell.2015.12.037.

Ryczko, D., Dubuc, R., 2013. The multifunctional mesencephalic locomotor region. Curr. Pharm. Des. 19 (24), 4448–4470. https://doi.org/10.2174/1381612811319240011.

Saavedra, J.M., Kvetnansky, R., Kopin, I.J., 1979. Adrenaline, noradrenaline and dopamine levels in specific brain stem areas of acutely immobilized rats. Brain Res. 160 (2), 271–280. https://doi.org/10.1016/0006-8993(79)90424-4.

Sawada, M., Munemitsu, T., Hojo, M., 2017. Intraoperative FLOW 800 analysis for intramedullary cystic lesion: a technical case report. Oper. Neurosurg. 13 (5), E23–E27. https://doi.org/10.1093/ons/opx044.

Seidel, K., Mahlke, J., Siswanto, S., et al., 2015. The brainstem pathologies of Parkinson's disease and dementia with Lewy bodies. Brain Pathol. 25 (2), 121–135. https://doi.org/10.1111/bpa.12168.

Shine, J.M., Matar, E., Ward, P.B., et al., 2013. Exploring the cortical and subcortical functional magnetic resonance imaging changes associated with freezing in Parkinson's disease. Brain 136 (4), 1204–1215. https://doi.org/10.1093/brain/awt049.

Steiner, T.J., Fontebasso, M., 2002. Headache. BMJ (Clinical Research Ed.) 325 (7369), 881–886.

Steiniger, B., Kretschmer, B.D., 2003. Glutamate and GABA modulate dopamine in the pedunculopontine tegmental nucleus. Exp. Brain Res. 149 (4), 422–430. https://doi.org/10.1007/s00221-003-1382-z.

Strauss, C., Fahlbusch, R., 1997. Anatomical aspects for surgery within the floor of the IVth ventricle. Zentralbl. Neurochir. 58 (1), 7–12.

Tard, C., Delval, A., Devos, D., et al., 2015. Brain metabolic abnormalities during gait with freezing in Parkinson's disease. Neuroscience 307, 281–301. https://doi.org/10.1016/j.neuroscience.2015.08.063.

Tattersall, T.L., Stratton, P.G., Coyne, T.J., et al., 2014. Imagined gait modulates neuronal network dynamics in the human pedunculopontine nucleus. Nat. Neurosci. 17 (3), 449–454. https://doi.org/10.1038/nn.3642.

Tayebi Meybodi, A., Hendricks, B.K., Witten, A.J., Hartman, J., Tomlinson, S.B., Cohen-Gadol, A.A., 2020. Virtual exploration of safe entry zones in the brainstem: comprehensive definition and analysis of the operative approach. World Neurosurg 140, 499–508. https://doi.org/10.1016/j.wneu.2020.05.207.

Thevathasan, W., Pogosyan, A., Hyam, J.A., et al., 2012. Alpha oscillations in the pedunculopontine nucleus correlate with gait performance in parkinsonism. Brain 135 (1), 148–160. https://doi.org/10.1093/brain/awr315.

Toescu, S.M., Samarth, G., Layard Horsfall, H., et al., 2021. Fourth ventricle tumors in children: complications and influence of surgical approach. J. Neurosurg. Pediatr. 27 (1), 52–61. https://doi.org/10.3171/2020.6.PEDS2089.

Vercruysse, S., Leunissen, I., Vervoort, G., Vandenberghe, W., Swinnen, S., Nieuwboer, A., 2015. Microstructural changes in white matter associated with freezing of gait in Parkinson's disease. Mov. Disord. 30 (4), 567–576. https://doi.org/10.1002/mds.26130.

Wang, M., Jiang, S., Yuan, Y., et al., 2016. Alterations of functional and structural connectivity of freezing of gait in Parkinson's disease. J. Neurol. 263 (8), 1583–1592. https://doi.org/10.1007/s00415-016-8174-4.

Yagmurlu, K., Rhoton, A.L., Tanriover, N., Bennett, J.A., 2014. Three-dimensional microsurgical anatomy and the safe entry zones of the brainstem. Oper. Neurosurg. 10 (4), 602–620. https://doi.org/10.1227/NEU.0000000000000466.

Yang, Y., van Niftrik, B., Ma, X., et al., 2019. Analysis of safe entry zones into the brainstem. Neurosurg. Rev. 42 (3), 721–729. https://doi.org/10.1007/s10143-019-01081-9.

Yeo, S.S., Lee, D.G., Choi, B.Y., et al., 2012. Neural connectivity of the pedunculopontine nucleus in relation to walking ability in chronic patients with intracerebral hemorrhage. Eur. Neurol. 67 (4), 226–231. https://doi.org/10.1159/000335248.

Further Reading

Aggarwal, M., Puri, V., Puri, S., 2012. Serotonin and CGRP in migraine. Ann. Neurosci. 19 (2). https://doi.org/10.5214/ans.0972.7531.12190210.

Albright, A.L., Sclabassi, R.J., 1985. Cavitron ultrasonic surgical aspirator and visual evoked potential monitoring for chiasmal gliomas in children. J. Neurosurg. 63 (1), 138–140. https://doi.org/10.3171/jns.1985.63.1.0138.

Back, F.P., Carobrez, A.P., 2018. Periaqueductal gray glutamatergic, cannabinoid and vanilloid receptor interplay in defensive behavior and aversive memory formation. Neuropharmacology 135, 399–411. https://doi.org/10.1016/j.neuropharm.2018.03.032.

Bandler, R., Shipley, M.T., 1994. Columnar organization in the midbrain periaqueductal gray: modules for emotional expression? Trends Neurosci. 17 (9), 379–389. https://doi.org/10.1016/0166-2236(94)90047-7.

Benarroch, E.E., 2008. Descending monoaminergic pain modulation: bidirectional control and clinical relevance. Neurology 71 (3), 217–221. https://doi.org/10.1212/01.wnl.0000318225.51122.63.

Bourbia, N., Pertovaara, A., 2018. Involvement of the periaqueductal gray in the descending antinociceptive effect induced by the central nucleus of amygdala. Physiol. Res. 647–655. https://doi.org/10.33549/physiolres.933699.

Carrive, P., 1993. The periaqueductal gray and defensive behavior: functional representation and neuronal organization. Behav. Brain Res. 58 (1–2), 27–47. https://doi.org/10.1016/0166-4328(93)90088-8.

Deshmukh, V.R., Rangel-Castilla, L., Spetzler, R.F., 2014. Lateral inferior cerebellar peduncle approach to dorsolateral medullary cavernous malformation. J. Neurosurg. 121 (3), 723–729. https://doi.org/10.3171/2014.5.JNS132276.

Freund, W., Wunderlich, A.P., Stuber, G., et al., 2011. The role of periaqueductal gray and cingulate cortex during suppression of pain in complex regional pain syndrome. Clin. J. Pain 27 (9), 796–804. https://doi.org/10.1097/AJP.0b013e31821d9063.

Heinricher, M.M., Tavares, I., Leith, J.L., Lumb, B.M., 2009. Descending control of nociception: specificity, recruitment and plasticity. Brain Res. Rev. 60 (1), 214–225. https://doi.org/10.1016/j.brainresrev.2008.12.009.

Ho, Y.C., Lin, T.B., Hsieh, M.C., et al., 2018. Periaqueductal gray glutamatergic transmission governs chronic stress-induced depression. Neuropsychopharmacology 43 (2), 302–312. https://doi.org/10.1038/npp.2017.199.

Kamson, D.O., Juhász, C., Shin, J., et al., 2014. Patterns of structural reorganization of the corticospinal tract in children with Sturge-Weber syndrome. Pediatr. Neurol. 50 (4), 337–342. https://doi.org/10.1016/j.pediatrneurol.2013.12.012.

Lechanoine, F., Jacquesson, T., Beaujoin, J., et al., 2021. WIKIBrainStem: an online atlas to manually segment the human brainstem at the mesoscopic scale from ultrahigh field MRI. Neuroimage 236, 118080. https://doi.org/10.1016/j.neuroimage.2021.118080.

Lovick, T.A., 1993. The periaqueductal gray-rostral medulla connection in the defence reaction: efferent pathways and descending control mechanisms. Behav. Brain Res. 58 (1–2), 19–25. https://doi.org/10.1016/0166-4328(93)90087-7.

Lu, B., Fan, P., Li, M., et al., 2023. Detection of neuronal defensive discharge information transmission and characteristics in periaqueductal gray double-subregions using PtNP/PEDOT: PSS modified microelectrode arrays. Microsyst. Nanoeng. 9 (1), 70. https://doi.org/10.1038/s41378-023-00546-8.

Menant, O., Andersson, F., Zelena, D., Chaillou, E., 2016. The benefits of magnetic resonance imaging methods to extend the knowledge of the anatomical organisation of the periaqueductal gray in mammals. J. Chem. Neuroanat. 77, 110–120. https://doi.org/10.1016/j.jchemneu.2016.06.003.

Roizen, M.F., Newfield, P., Eger, E.I., Hosobuchi, Y., Adams, J.E., Lamb, S., 1985. Reduced anesthetic requirement after electrical stimulation of periaqueductal gray matter. Anesthesiology 62 (2), 120–123. https://doi.org/10.1097/00000542-198502000-00004.

Ryczko, D., Dubuc, R., 2017. Dopamine and the brainstem locomotor networks: from lamprey to human. Front. Neurosci. 11. https://doi.org/10.3389/fnins.2017.00295.

Sciacca, S., Lynch, J., Davagnanam, I., Barker, R., 2019. Midbrain, pons, and medulla: anatomy and syndromes. Radiographics 39 (4), 1110–1125. https://doi.org/10.1148/rg.2019180126.
Shepherd, T.M., Hoch, M.J., 2022. MRI-visible anatomy of the brainstem. Neuroimaging Clin. N. Am 32 (3), 553–564. https://doi.org/10.1016/j.nic.2022.04.003.
Ullah, F., dos Anjos-Garcia, T., Mendes-Gomes, J., et al., 2017. Connexions between the dorsomedial division of the ventromedial hypothalamus and the dorsal periaqueductal grey matter are critical in the elaboration of hypothalamically mediated panic-like behaviour. Behav. Brain Res. 319, 135–147. https://doi.org/10.1016/j.bbr.2016.11.026.
Welniarz, Q., Dusart, I., Roze, E., 2017. The corticospinal tract: evolution, development, and human disorders. Dev. Neurobiol 77 (7), 810–829. https://doi.org/10.1002/dneu.22455.
Zülch, K.J., 1982. Intracranial tumours of infancy and childhood. In: Tumours of the Central Nervous System in Infancy and Childhood. Springer Berlin Heidelberg, pp. 1–16. https://doi.org/10.1007/978-3-642-95413-9_1.

5b

The Reticular Formation

Brandon M. Ascenzi
Independent Neuroresearcher-Member of Marie Curie Alumni Association (MCAA)

OUTLINE

INTRODUCTION

The *reticular formation* is a phylogenetically ancient portion of the brainstem, which extends longitudinally. Mainly extended from the medulla oblongata to the mesencephalon, it also reaches other structures as well: the first cervical neuromeres of the spinal cord (caudally), the diencephalon and the cerebral cortex (rostrally), and the cerebellum (dorsally). Its name is due to the fact that it is constituted by a diffuse aggregate of neurons and fibers, which often do not allow to identify discrete nuclear aggregates and form kind of a network. It is an important center of sensory—motor integration, whose functional role is still partly uncertain. Currently, on the study of its efferents and afferents, on the basis of electrophysiological data and of those emerged from the study of clinical cases of neurological interest, it is believed that the reticular formation is involved in the control of many functions of the vegetative nervous system, such as cardiocirculatory, respiratory, and gastrointestinal, in the modulation of movement and of nociception, in the regulation of states of consciousness and sleep—wake rhythms.

Neuroanatomically the *reticular formation* is subdivided into three longitudinal columns disposed from both sites:

- Median column
- Medial column
- Lateral column

This subdivision is made by keeping into account the neurochemical, neurophysiologic, and cytologic features of the different types of neurons forming them (Fig. 5b.1A and B).

From Anatomy to Function of the Central Nervous System
https://doi.org/10.1016/B978-0-12-822404-5.00007-3

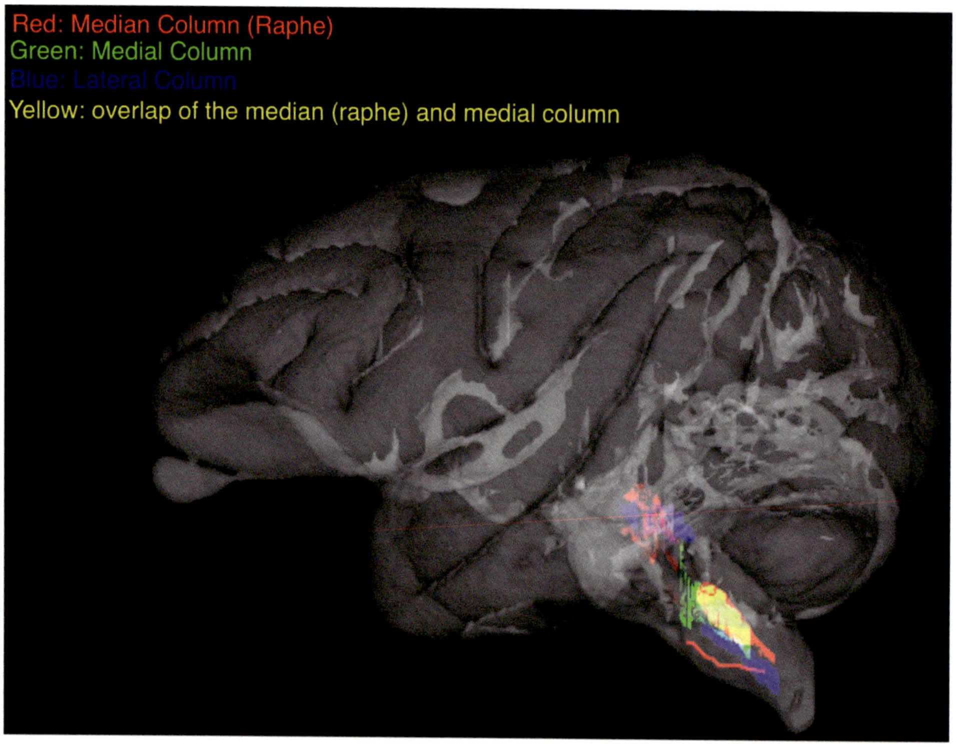

A)

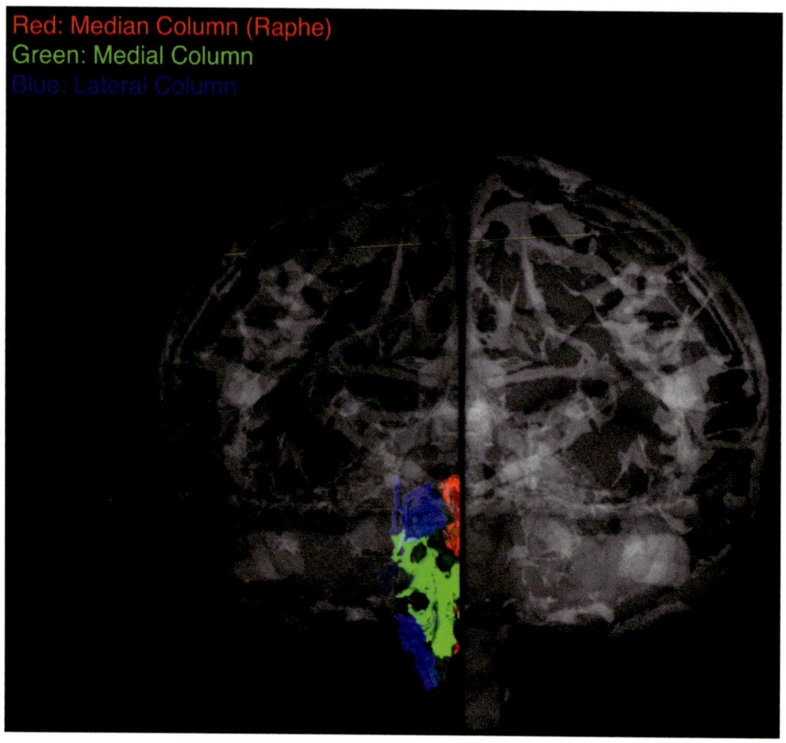

B)

FIGURE 5B.1 **(A) Left-lateral view of the three columns of the reticular formation in *Macaca mulatta* (rhesus macaque) brain model. (B) Back view of the three columns of the reticular formation in *Macaca mulatta* (rhesus macaque) brain model.** *Image created by BrainInfo (1991–present), National Primate Research Center, University of Washington, http://www.braininfo.org.*

THE MEDIAN COLUMN

The median column of the reticular formation consists of numerous nuclei located in the medulla oblongata, the pons, and the midbrain (Figs. 5b.2A,B and 5b.3). These nuclei collectively constitute the raphe nuclei column, which is the primary structure responsible for the origin of the serotonergic system; all the collaterals that reach the overlying structures as well as the forebrain arise from it. The raphe column nuclei primarily consist of the following nuclei: nucleus raphe pallidus (B1-RPa), located in the medulla oblongata, along with nucleus raphe obscurus (B2-ROb) and nucleus raphe magnus (B3-RMg); nucleus raphe pontis (B5-RPn), situated in the pons; and the more rostral nuclei found at the midbrain/midbrain-pons junction, namely the dorsal raphe nucleus (B6-DRN) and superior raphe nucleus (B8-MnR). These raphe nuclei differ in the proportion of serotonergic neurons they contain. Proportions vary from 80% for the dorsal raphe nucleus (DRN) to 10%–20% for the medullary raphe nuclei. Substance P is often colocalized with serotonin, and several other neuropeptides can be found in these neurons.

MEDULLA OBLONGATA NUCLEI

Nucleus Raphe Pallidus (B1-RPa)—Nucleus Raphe Obscurus (B2-ROb)—Nucleus Raphe Magnus (B3-RMg)

The nucleus raphe pallidus (B1-RPa) is positioned ventrally to the nucleus raphe obscurus (B2-ROb), which is located slightly more dorsally. Rostral to these nuclei, there is the nucleus raphe magnus (B3-RMg), which extends its "wings" over the gigantocellular nucleus (Gi) (see the following).

B1-RPa is the smaller nucleus of the raphe nuclei column; its extension corresponds roughly to that of the inferior olivary nucleus and its neurons have an elongated and slim shape, with few dendritic processes oriented predominantly dorsoventrally.

The nucleus raphe obscurus (B2-ROb) has a craniocaudal extension, and at its rostral end, it is at the same level as the abducens nucleus (VI c.n.), positioned just below the medial longitudinal fasciculus (mlf). B2-ROb is located dorsally to the nucleus raphe magnus (RMg-B3), and being located in the median column of the reticular formation, it is between the inferior olivary nucleus and the medial longitudinal fasciculus (mlf). Neurons of the nucleus raphe obscurus (B2-ROb) form two distinct columns at the midline level. These neurons are multipolar, with large to medium size, and their dendrites are oriented dorsoventrally. They are also positive for 5-HT (serotonin). Labeled processes extend in the ventral, dorsal, or lateral directions and range between 7.5 and 12.5 μm in diameter; some of them cross the midline to reach the contralateral B2-Rob; other dendrites are oriented laterally.

The nucleus raphe magnus (B3-RMg) is situated on the midline of the brainstem, specifically in the rostral portion of the medulla oblongata and caudal pons. It is bordered dorsolaterally by the gigantocellular nucleus pars alpha (Giα). Inferomedially, it is positioned relative to the caudal pontine reticular nucleus (PnC) and overlies the medial lemniscus (ml), where its fibers transition from a ventromedial to a more lateral position upon entering the pons. Caudally, the B3-RMg extends until just below the anterior tip of the inferior olivary nuclei (ION). The B3-RMg neurons are serotonergic, bipolar, and multipolar. The bipolar ones are predominant near the midline, whereas the multipolar ones are more abundant in the ventral and lateral divisions of the nucleus. Laterally in B3-RMg, the major axis of dendritic arborization shifts from dorsoventral to mediolateral.

At the level of the pontomedullary junction, serotonergic neurons extend laterally in the neighboring Giα. This population of 5-HT neurons extends both laterally and caudally, arching over the dorsal boundary of the inferior olivary nuclei.

Substance P is found in many B3-RMg neurons of the human brainstem, as well as in the terminals in the neuropil (Halliday et al., 1988a,b; Rikard-Bell et al., 1990). In rodents, substance P was shown to colocalize in B3-RMg serotonergic neurons, as well as thyrotropin-releasing hormones (Poulat et al., 1992, Wu et al., 1993). Several neuropeptides containing neurons (enkephalin, somatostatin, cholecystokinin) (Foster et al., 1985), GABAergic neurons (Reichling and Basbaum, 1990; Inyushkin et al., 2010), and NPY-neurons have been described in the human B3-RMg (Halliday et al., 1988a,b).

PONTINE NUCLEI

Pontine Raphe Nucleus (B5-RPn)[1]

The pontine raphe nucleus (B5-RPn) is situated caudally and dorsolaterally to the oral pontine reticular nucleus (PnO), which is a large structure extending ventrally to the medial longitudinal fasciculus (mlf). This is a small nucleus composed of only a few serotonergic neurons. Rostrally, the column of raphe nuclei

[1]Also known as pontine raphe nucleus, raphe pontis nucleus, nucleus raphes pontis, raphe pontis.

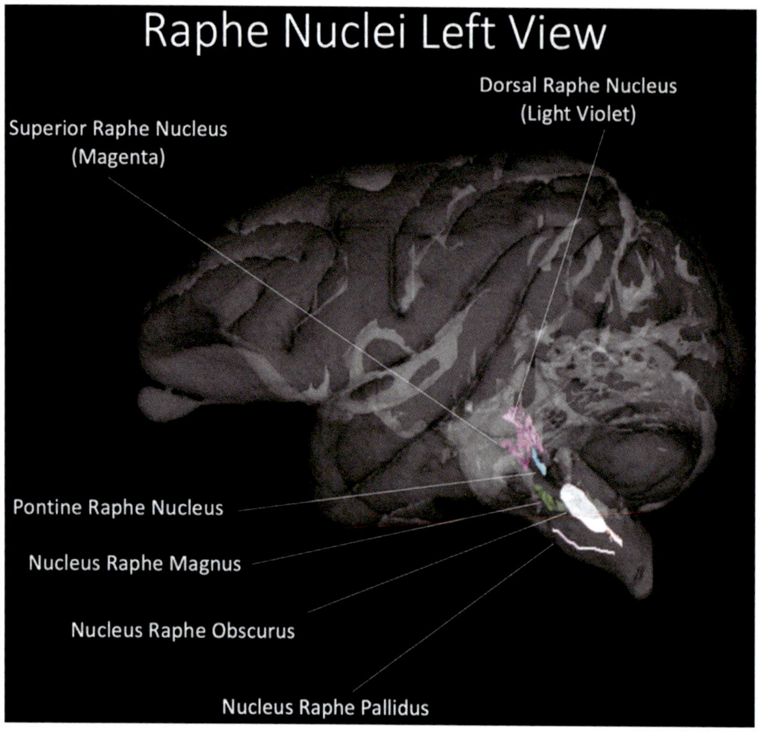

A)

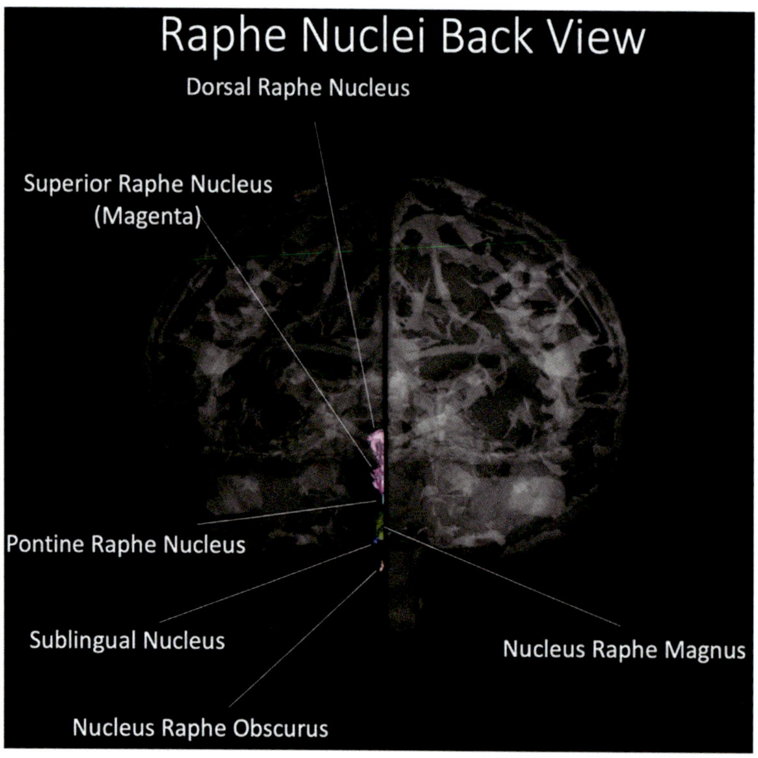

B)

FIGURE 5B.2 **(A) Left-lateral view of the median (raphe) column nuclei of the reticular formation in *Macaca mulatta* (rhesus macaque) brain model. (B) Back view of the median (raphe) column nuclei of the reticular formation in *Macaca mulatta* (rhesus macaque) brain model.** (The sublingual nucleus is not described in the text). *Image created by BrainInfo (1991–present), National Primate Research Center, University of Washington, http://www.braininfo.org.*

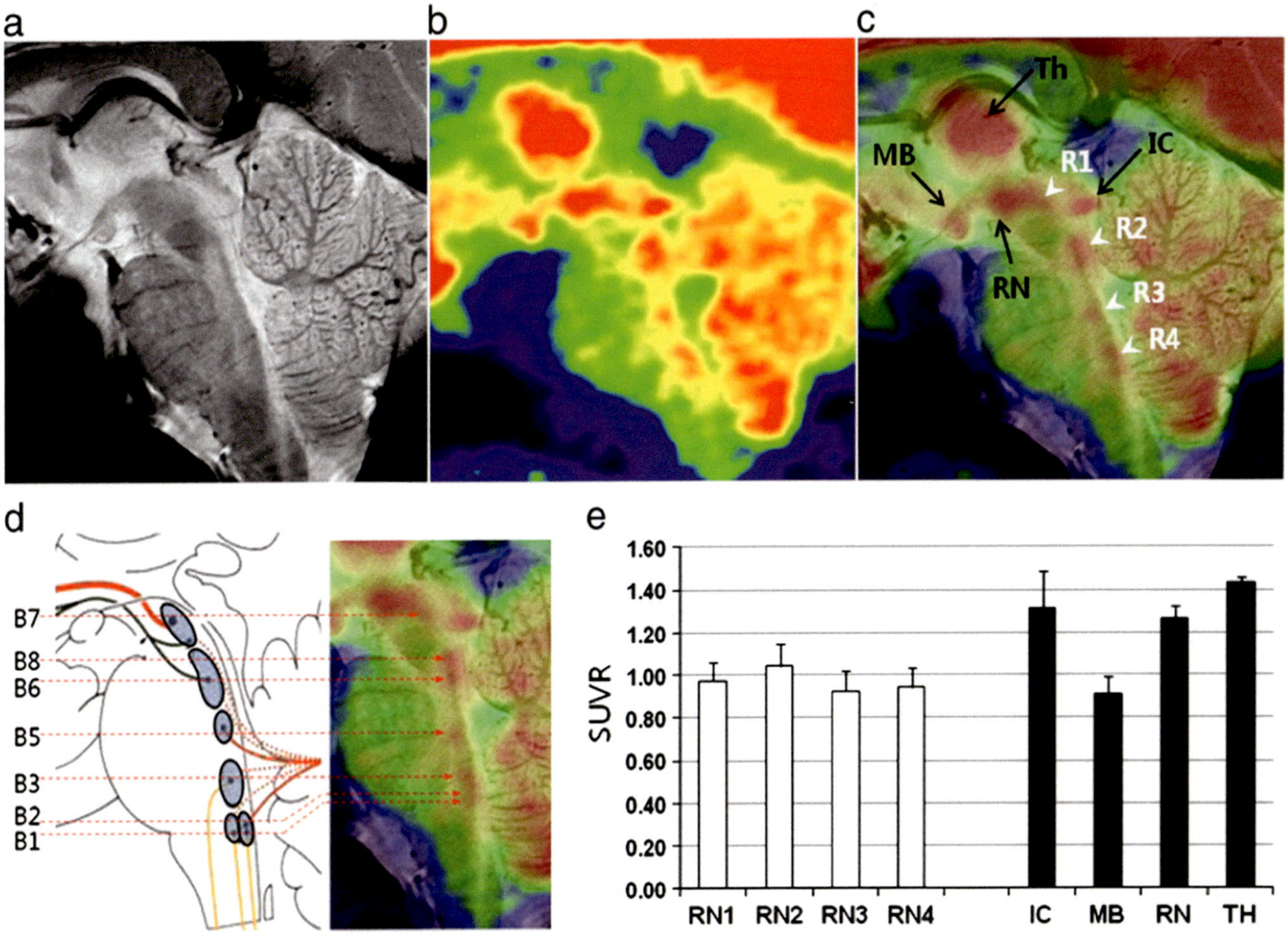

FIGURE 5B.3 **A representative fusion image set of the brainstem area from a single subject obtained by positron emission tomography−magnetic resonance imaging (PET−MRI).** (A) A medial sagittal T2*-weighted image of the brainstem obtained by 7.0 T-MRI. This high-resolution image depicts various landmark structures in the midbrain and brainstem areas such as the mammillary body (MB), thalamus, and red nucleus (RN). (B) [^{18}F]Fluorodeoxyglucose (^{18}F-FDG)-PET image obtained by high-resolution research tomograph (HRRT-PET) of a slice corresponding to (A). Many well-known localized activities are clearly visible, such as the MB, RN, and other structures, together with activity cluster groups that are believed to be the raphe nuclei. (C) Fusion image of (A) and (B). From this fusion image, together with histological information provided by Naidich et al. (2009) and ultrahigh-resolution image of 7.0 T-MRI, it was possible to identify four raphe nucleus groups (arrowheads): dorsal (B6-DNR), superior central (B8-MnR), pontine (B5-RPn), and medullary (B1-RPa-B2-Rob-B3-RMg). (D) Raphe nuclei previously identified in 18 the brain atlas of Naidich et al. (2009), and this new PET−MRI fusion image of the brainstem area obtained by locations and distributions of the raphe nuclei shown in the reference of Naidich et al. and those of the raphe nuclei seen in the PET−MRI fusion image. (E) Relative glucose metabolic activities of each raphe nucleus as well as the other midbrain nuclei measured in seven healthy subjects. R2 shows the highest glucose metabolic activity among the raphe nuclei. The other nonraphe nucleus groups, compared with the raphe nucleus groups, the highest level of activities were found in TH followed by IC, RN, and MB. TH: thalamus, MB: mammillary body, IC: inferior colliculus, RN: red nucleus, R1−R4: raphe nuclei, B1: nucleus raphes pallidus, B2: nucleus raphes obscurus, B3: nucleus raphes magnus, B5: nucleus raphes pontis, B6 and B7: dorsal raphe nucleus (caudal and rostral part, respectively) raphe, B8: nucleus centralis superior (or median raphe nuclei). *Courtesy of Son, Y.D., Cho, Z.H., Kim, H.K., Choi, E.J., Lee, S.Y., Chi, J.G., Park, C.W., Kim, Y.B., 2012. Glucose metabolism of the midline nuclei raphe in the brainstem observed by PET-MRI fusion imaging. Neuroimage 59 (2), 1094−1097. https://doi.org/10.1016/j.neuroimage.2011.09.036.*

continues with the dorsal raphe nucleus (B6-DRN), which is considered one of the most important nuclei from an anatomical and functional perspective.

Dahlström and Füxe in 1964 also described a smaller pontine serotonergic group: the B7. This latter is a group of cells located in the central gray of the pons (Dahlström and Fuxe, 1965; Felten and Sladek, 1983).

MESENCEPHALIC NUCLEI

Dorsal Raphe Nucleus (B6-DRN)−Superior Central Nucleus (B8-MnR)

The *dorsal raphe nucleus* (B6-DRN) is located ventrally to the cerebral aqueduct, surrounded by the periaqueductal gray (PAG), and dorsally to the mlf. This nucleus is divided into a midline region, a ventral region (DRV), located at the bottom of the central gray, and an interfascicular region (DRI), between the trochlear nucleus (IV c.n.) and the mlf. The latter continues anteriorly with the *caudal linear nucleus* (CLi), a subnucleus of the ventral mesencephalic tegmentum, and stops caudally at the level of the rostral opening of the aqueduct into the fourth ventricle (Halliday and Törk, 1986).

The B6-DRN contains the largest population of serotonergic neurons in the human brainstem, estimated around 165,000 neurons (Baker et al., 1991a). Serotonergic neurons are present in all subdivisions of the B6-DRN, although in varying proportions of the total neuronal population as estimated by Nissl-stained sections. There are, in addition, three large subnuclei, central to the periaqueductal gray (PAG): the dorsal (DRD), the ventral (DRV), and the ventrolateral (DRVL) subnuclei. Caudal to the isthmus of the brainstem, the portion of the B6-DRN in the PAG is reduced to two slits of neurons parallel to the ventricular surface of the anterior rhomboid fossa, bordered laterally by the *lateral tegmental nucleus* (see later) and extending caudally up to the level of the trigeminal motor nucleus; this is the caudal division of the B6-DRN (DRC). The largest population of serotonergic neurons is in the DRD (28% of the total number of serotonergic neurons of the B6-DRN, and 61% of all the DRD neurons). The remaining divisions of the B6-DRN amount to 16% −22% of the total serotonergic population each. In the DRI, 100% of the neurons are serotonergic; in the DRV, 82% are; in the other divisions of the B6-DRN, they amount to around 60%. These neurons have small-to-medium bodies, and dendrites oriented parallel to the rostrocaudal axis. The most rostral serotonergic neurons in humans, as in other primates, are located in the rostral part of the interpeduncular nucleus (Törk and Hornung, 1990; Azmitia and Gannon, 1996; Hornung and Fritschy, 1988).

Neurons in the DRVL are multipolar and densely packed, while in the DRD, neurons are loosely arranged, medium-sized, with radiating smooth dendrites. In the DRC, they have smaller somatic diameter, with short dendrites, often oriented parallel with the floor of the fourth ventricle. Several neurotransmitters have been identified in the DRN (Michelsen et al., 2007), and there is a clear difference in the electrophysiological properties of serotonergic and nonserotonergic B6-DRN neurons (Kirby et al., 2003), in particular those using GABA (Allers and Sharp, 2003). Among the GABA neurons in the B6-DRN, very few colocalize with serotonin (Wang and Nakai, 1993; Stamp and Semba, 1995; Charara and Parent, 1998).

Substance P also seems to colocalize in the same neurons (almost 40% of the serotonergic neurons) (Baker et al., 1991a). Furthermore, many serotonergic neurons express the neuropeptide CRF (Commons et al., 2003) and galanin (Melander et al., 1986a,b), and a few neurons also contain the neuropeptides dynorphin, angiotensin, and enkephalin (Björklund and Hökfelt, 1985; Björklund et al., 1990; Tagliaferro et al., 2003; Okere and Waterhouse, 2006). Finally, in DRC, there are dense and restricted groups of neurotensin-containing neurons (Jennes et al., 1982) and cholecystokinin-containing neurons (Otake, 2005). B6-DRN serotonergic neurons express the vesicular glutamate transporter VGLUT3, suggesting a corelease of serotonin and glutamate (Jackson et al., 2009; Hioki et al., 2010).

The *superior central nucleus* (B8-MnR)[2] in its caudal portion is wrapped by the *reticulotegmental nucleus* (RtTg, see later), while rostrally is laterally surrounded by the nuclei of the oral pontine reticular formation.

The B8-MnR extends from the decussation of the superior cerebellar peduncles to the middle pons, at the level of the trigeminal motor nucleus. Dorsally, the B8-MnR terminates below the mlf, while ventrally it extends to the bottom of the midline tegmental region (Baker et al., 1991b). The B8-MnR is divided into dorsal, midline, and paramedian clusters. In the caudal two-thirds of the B8-MnR, the midline division is bordered by a paramedian cell cluster, which lies at the medial border of the lateral cell cluster, and there is a change in structure from rostral to caudal levels of the B8-MnR. Rostrally, the B8-MnR is restricted to a thin dorsoventral slit of neurons near the midline of the mesencephalon. In the rostral pons, the midline group is segregated in a dorsal division, with densely packed neurons, and a ventral one, with more loosely organized neurons.

In the B8-MnR, most neurons have dendrites oriented in the rostrocaudal direction, reminiscent of the ventral division of the DR. Rostrally, the fact that the vicinity of the DRI neurons shares similar morphology makes the separation with the B8-MnR somewhat arbitrary, although at most levels it is clearly delineated by small bundles of the medial

[2]Also known as median raphe nucleus.

longitudinal fasciculus. Neurons in the caudal division are characterized by a round somatic morphology and radially oriented dendrites.

AFFERENT FIBERS TO THE RAPHE NUCLEI

As described thus far, the median column of the reticular formation consists of several serotoninergic cell groups (nuclei) that extend along the brainstem, from the medulla oblongata to the mesencephalon. These nuclei are interconnected and receive numerous afferences from almost the entire brain, sending their efferent fibers to several target structure. We will describe the aforementioned connections in the following paragraphs.

We prefer to categorize the afferent connections of the raphe based on topographical and functional criteria.

AFFERENT CONNECTION FROM THE CEREBRAL CORTEX AND FROM OTHERS BRAIN AREAS TO THE RAPHE NUCLEI (TABLE 5B.1)

Telencephalic afferents to the raphe nuclei originate from the cerebral cortex, primarily from the dorsolateral and dorsomedial prefrontal cortex (dlPFC; dmPFC) (Arnsten et al., 1984; Lee et al., 2003; Gonçalves et al., 2009; Celada et al., 2001), frontopolar and orbitofrontal cortex (Gonçalves et al., 2009; Peyron et al., 1998), cingulate cortex (Behzadi et al., 1990; Peyron et al., 1998), prelimbic and infralimbic cortex, medial and lateral precentral cortex (Hermann, et al., 1997; Peyron et al., 1998), and ventromedial prefrontal cortex (Hajós et al., 1998). It is evident that the majority of afferents originate from the prefrontal cortex (PFC), which explains the association between the serotonergic system and various neurofunctional aspects such as mood, personality, and psychopathological phenomena. Recent anatomical studies performed by Gonçalves et al. (2009) showed that the PFC exerts a powerful top-down control over the B6-DRN and that the largest amount of projection originating from the PFC are from the V layer. Once again, it is worth noting that the regulation of B6-DRN neurons by the medial prefrontal cortex (mPFC) involves the activation of specific receptors, such as the 5-HT_{1A}, $GABA_A$, and glutamate receptors. This indicates the presence of an intricate regulation of B6-DRN neurons by afferent connections from the medial prefrontal cortex (mPFC), and the modulation of various processes (as discussed later). The

TABLE 5B.1 Afferent connection to the raphe nuclei.

Raphe nucleus	Afferences
Superior central nucleus (B8-MnR)	- Septal nucleus - Cingulate cortex - mPFC
Dorsal raphe nucleus (B6-DRN)	- Lateral habenular nuclei - Interpeduncular nuclei - Hypothalamus (preoptic area) - Nucleus of solitary tract - Substantia nigra (SN) - Lateral tegmental field - Locus coeruleus - Tuber cinereum - Amygdala - Entopeduncular nucleus (GPi) - Cingulate cortex - Orbitofrontal cortex - Infralimbic cortex (ventro-mPFC) - Insular cortex - mPFC and 1PFC
Pontine raphe nucleus (B5-RPn)	?
Nucleus raphe magnus (B3-RMg)	- Prelimbic cortex - Infralimbic cortex (ventro-mPFC) - Medial and lateral precentral cortex
Nucleus raphe obscurus (B2-ROb)	?
Nucleus raphe pallidus (Bl-RPa)	- Periaqueductal gray - Hypothalamus (preoptic area) - Prefrontal cortex - Insular and perirhinal cortex

activation of certain serotonergic neurons by descending excitatory fibers triggers the release of serotonin. In turn, serotonin regulates its own secretion through a negative feedback mechanism by interacting with 5-HT_{1A} autoreceptors. Afferences from the mPFC also inhibit the serotonergic neurons through the activation of GABAergic interneurons; in turn, ascending serotoninergic pathways may control the activity of these descending pathways by acting on postsynaptic 5-HT_{1A} receptors (Celada et al., 2001). Lee et al. (2003)'s studies confirmed that the mPFC provides the major glutamatergic afferent projections to each subdivision of the B6-

DRN. The findings indicate that the superior cortical centers are easily involved in the regulation of a broad range of systems, including sensory and motor functions, arousal and sleep–wake cycle, biorhythmic, cognitive, and affective behaviors.

EFFERENT FIBERS OF THE RAPHE NUCLEI (TABLE 5B.2)

The efferences of the raphe nuclei to the cerebral cortex are mainly from the B6-DRN and the B8-MnR. However, a portion also comes from the B9 group (O'Hearn and Molliver, 1984), that in nonhuman primates is located in the ventral part of the B8-MRN and adjacent structures (Felten and Sladek, 1983; Dahlström and Fuxe, 1964b). Neuroanatomical studies performed on primates and other animals, highlighted that B6-DRN and B8-MRN send fibers to frontal areas of the cerebral cortex, such as mPFC (Aston Jones et al., 2001; O'Hearn and Molliver, 1984; Van Bockstaele et al., 1993), dlPFC, and cingulate cortex (Porrino and Goldman-Rakic, 1982), but also to the parietal (O'Hearn and Molliver, 1984) and occipital cortices—(in particular to visive areas 17 and 18 BA—(O'Hearn and Molliver, 1984; Waterhouse et al., 1986, 1993). Efferences to the motor and sensory cortices also have been described (Waterhouse et al., 1986), and this, as well as the connection to the *visive areas*, suggests that these efferences probably play a role in the integration of different kinds of stimuli. Since the efferent pathways described before are connected to the frontal cortical region of the forebrain, this provides an explanation for the association of the serotonergic system with the emotional and physiological aspects of an individual, as mentioned earlier in relation to the afferent connections. These connections rationalize the involvement of the serotoninergic system in various processes such as emotions, arousal, sleep, and more. B6-DRN, locus coeruleus (A6-LC, see later), cholinergic pedunculopontine nucleus (PPTg), and lateral dorsal tegmental nucleus (LDTN) are interconnected, and together, they are involved in sleep/wake control.

The activity of the B6-DRN is high during wake, low during slow wave sleep, and abolished during rapid eye movement (REM) sleep. The raphe nuclei described before also play a crucial role in the etiology of numerous psychiatric disorders, including depression, bipolar disorder, and obsessive-compulsive disorder (OCD) (see later).

Serotonergic axons are distributed extensively throughout the central nervous system. These axons can be categorized into two types based on their appearance: thin axons with small varicosities and coarser axons with large varicosities. The origin of these axons differs depending on the type: axons with small varicosities originate from the B6-DRN and B3-RMg nuclei, while axons with coarse varicosities originate from the B8-MnR and B1-RPa nuclei. Small varicosities lack membrane junctional complexes, the hallmark of the synapse (Descarries et al., 2000). These fibers release their neurotransmitter diffusely (volume transmission). They are also characterized by their sensitivity to amphetamine-derived neurotoxins (Mamounas et al., 1991). Large varicosities make synaptic contact with their targets.

Thin, primarily unmyelinated fibers originating from the B6-DRN nucleus converge to form a dorsal bundle. This system, identified by Forel (1877), consists of fascicles containing fibers of varying sizes. It is situated ventrolaterally to the periaqueductal gray. It contains serotonergic, noradrenergic, and cholinergic elements, and ascending fibers from the *nucleus pontis oralis* and *caudalis* (PnO and PnC, respectively). The area containing Forel's fascicles coincides approximately with the dorsal trigeminothalamic tract. Coarser axons ascend from the B8-MnR in a ventral bundle through the tegmental area. At the mid-forebrain border, the two systems intermingle. They distribute

TABLE 5B.2 Efferent connections of the raphe nuclei.

Raphe nucleus	Efferences
Superior central nucleus (B8-MnR)	- Septal nucleus - Hippocampus - Interpeduncular nucleus - Substantia nigra - Mammillary body - Cingulate cortex - dlPFC - Motor and sensorimotor cortices - Visive cortex (areas 17 and 18a\b)
Dorsal raphe nucleus (B6-DRN)	- Later geniculate body - Striatum - Entorhinal cortex - Olfactory bulb - Amygdala - mPFC - Frontal cortex - Parietal cortex - Visive cortex (areas 17 and 18a\b) - Occipital cortex
Pontine raphe nucleus (B5-RPn)	?
Nucleus raphe magnus (B3-RMg)	?
Nucleus raphe pallidus (B1-RPa)	?

to their cortical targets through the medial forebrain bundle, subthalamus, thalamus, and the internal capsule. The dorsal bundle is the more prominent in primates.

That being said, it becomes clear which are the most significant "target points" in the brain where neuropsychopharmacology exerts its effects. Unlike the afferents originating from the cerebral cortex, which project to almost all raphe nuclei with a few exceptions, the efferents from the raphe nuclei targeting the cerebral cortex are predominantly from the more rostral raphe nuclei groups, particularly the B6-BRN and B8-MnR. Similarly, the inferior centers and the spinal cord also receive afferents from the more caudal raphe nuclei group, primarily B1-RPa, B2-ROb, and B3-RMg. These play a key role in pain modulation by acting on the superficial laminae of the gray matter of the cervical tract of the spinal cord, inhibiting the propagation of the stimulus toward the cortex. B3-RMg in particular influences the marginal layer of the caudal spinal nucleus of the trigeminus (Bowker et al., 1981, 1983), which is responsible for relaying various sensory modalities, including temperature, deep or crude touch, and pain, from the ipsilateral portion of the face, and on laminae I and II of the dorsal horn. The fibers descend in the dorsolateral funiculus. The nuclei of raphe magnus (B3-RMg) and raphe pallidus (B1-RPa) also project to the intermediate gray and innervate the intermediate reticular nucleus (IRt) (Jones et al., 1992; Loewy et al., 1981). The spinal and cranial motoneuronal groups receive a serotonergic input from the nucleus of the raphe pallidus (B1-RPa), from the nucleus of the raphe obscurus (B2-ROb), and from the neighboring reticular formation. This system travels in the lateral part and ventral funiculus (Kojima and Sano, 1983; Kojima et al., 1985). A partial origin of the serotonergic innervation of Onuf's nucleus in the baboon has been described by Rajaofetra et al. (1992a,b).

This creates a boundary zone where mutual control exists between the superior centers and the rostral raphe nuclei in the rostral direction, and between the inferior centers and the caudal raphe nuclei in the caudal direction. However, it is not possible to recognize a clear border between the nuclei connected with the higher centers and those connected with the lower ones, for the following reasons:

(i) Currently, it cannot be ruled out that some thin fibers may be present but have not yet been detected, as most studies have been conducted on animal models other than humans or nonhuman primates;
(ii) Conducting these studies with animal models, including nonhuman primates, is not easily feasible due to numerous ethical reasons.

In the future, it is possible that alternative reciprocal connections between upper and lower regions will be discovered.

The two primary ascending serotonergic pathways from the midbrain raphe nuclei to the forebrain are known as the dorsal periventricular path and the ventral tegmental radiations. Both pathways converge in the caudal hypothalamus, where they join the medial forebrain bundle. Axons of dopaminergic (A8, A9, A10) and noradrenergic (A6) cell body groups also run anteriorly through the medial forebrain bundle (Molliver, 1987).

Ascending projections from the raphe nuclei to forebrain structures are organized in a topographical manner. The B6-DRN and B8-MnR give rise to distinct projections to forebrain regions, which form dissimilar yet partially overlapping patterns of innervation. The B8-MnR projects heavily to the dorsal hippocampus, septum, and hypothalamus, whereas the B6-DRN strongly innervates the ventral hippocampus, amygdala, and striatum. The B6-DRN and B8-MnR send overlapping neuronal projections to the neocortex, which are also organized in a strict topographical manner, and therefore affect different cortical neurons. In both nuclei, cells are organized in particular zones or groups that send axons to specific areas of the brain. For example, the frontal cortex receives a heavy innervation from the rostral and lateral subregions of the B6-DRN. Moreover, raphe neurons send collateral axons to areas of the brain, which are related in function, such as the amygdala and the hippocampus, or the substantia nigra and the caudate-putamen (striatum). The highly organized innervation of forebrain structures by serotonergic neurons of the raphe is fascinating because it suggests that different sets of serotonergic neurons have distinct functions based on their origin and terminal projections. This contrasts with the idea of a nonselective or general role for serotonin (5-HT) in the central nervous system (Molliver, 1987; Hensler, 2006).

Serotonergic axon terminals appear to exhibit morphological differences related to the raphe nucleus of origin. Serotonergic axons from the B8-MnR (type M) look relatively coarse, with large spherical varicosities. By contrast, axons from the B6-DRN (type D) are very fine and typically have small, pleomorphic varicosities. B6-DRN axons appear to be more vulnerable to certain neurotoxic amphetamine derivatives, e.g., 3,4-methylenedioxymethamphetamine (MDMA, or "Ecstasy"), a parachloroamphetamine (PCA). B8-MnR axons appear to be more resistant to the neurotoxic effects of these drugs. Blockade of the serotonin transporter (SERT) prevents the neurotoxic effects of these amphetamine derivatives, indicating that the activity of this transporter is critical for the neurotoxic effects of these drugs.

NEUROPSYCHOPHARMACOLOGY AND SEROTONINERGIC SYSTEM

The release and effects of serotonin are the outcome of intense synaptic activity both within and outside the raphe nuclei. These nuclei, in turn, are regulated by various centers situated in the forebrain; moreover, the influence of other neurochemical systems (e.g., noradrenergic, see later), whose fibers colocalize with the serotonergic ones, plays a crucial role in these processes. One of the most studied factors, which seems to regulate the serotonin activity, is the corticotropin-releasing factor (or hormone) (CRH). The stimulation of CRH receptors (CRFRs), CRF_1 and CRF_2, has an inhibitory and excitatory effect on the 5-HT neurons respectively, causing an inhibition and promotion of serotonin release (Lukkes et al., 2008; Pernar et al., 2004; Valentino and Commons, 2005). As stress is the body's psychological and physiological response to the challenges encountered in various life circumstances, it is also linked to psychiatric disorders characterized by mood and cognitive disturbances (Gold and Chrousos, 2002; Kessler, 1997; McEwen, 2003). The neurobiological processes at the base of mood and cognition alteration are multiples and are usually associated with a dysregulation of the release of noradrenaline (A6-LC) and serotonin (B6-DRN). Recent findings suggest that B6-DRN $5HT_{2A}$ receptors are critical for the modulation of both panic attack-like defensive behavior and unconditioned fear-induced antinociception (da Silva Soares et al., 2019a,b, Fig. 5b.4). Furthermore, it has been observed that the B3-RMg 5HT2A/2C receptors and the gigantocellular nucleus/gigantocellular nucleus pars alpha (Gi/Giα) have the ability to modulate unconditioned fear-induced antinociception (de Oliveira et al., 2016, 2017). Moreover, the alpha- and beta-noradrenergic receptors in the B6-DRN also appear to play a significant role in the pain-like response regulated by medial hypothalamus neurons (Uribe-Mariño et al., 2019). These results also give us a good perspective on the receptor phenotype of some raphe nuclei, and this is of great importance in the study of the possible efficacy of new molecules, especially neuropsychopharmaceuticals.

The regional colocalization of CRF and 5HT IR fibers within the B6-DRN has been reported by Kirby (2000). They showed that CRF-IR fibers were present at all rostrocaudal levels of the B6-DRN and exhibited a topographical distribution. Moreover, at lower doses, CRF predominantly exerts inhibitory effects on B6-DRN discharge, while at higher doses, these effects diminish or become excitatory. These findings provide anatomical and physiological evidence of an impact of CRF on the (B6) DRN-5HT system.

Snyder et al. (2015) found that, in the rat B6-DRN, CRF promotes different forms of behavioral flexibility, depending on social stress history. Thus, CRF appears to play a regulatory role in serotonin release, as demonstrated by blocking the CRF2 receptor in B8-MnR. This blockade resulted in a suppression of serotonin release in the ventral hippocampus and induced fear memory retrieval (Ohmura et al., 2010). Another important mechanism, which actively regulates the 5HT system, is the noradrenergic system, of which A6-LC is the principal component. The two systems are tightly interconnected, and one of the significant roles of norepinephrine (NE) on the raphe nuclei is the tonic activation of serotonin (5HT) neurons by NE neurons (Haddjeri et al., 2004a,b). This also explains why selective serotonin reuptake inhibitors (SSRIs) are frequently coupled with serotonin norepinephrine reuptake inhibitors (SNRIs).

They both act on the regulation of serotonin levels, mainly inhibiting the activity of serotonin transporter (SERT), leading to an increase of the serotonin level in the intersynaptic space. SNRIs, differently from the SSRI, produce their effect also acting on the noradrenergic system.

Regarding the mechanism of action of these drugs on the serotonergic system, SSRIs exert their effects primarily by targeting the rostral raphe nuclei, particularly the B6-DRN neurons, which appear to be a crucial in the regulation of emotions within the serotonergic system. However, recent findings have revealed that the mechanism of action of SSRIs extends beyond solely inhibiting the SERT. It appears that the efficacy of SSRIs is also influenced by the modulation of plasticity in serotonergic functions and their downstream targets (Blier and de Montigny, 1998). This plasticity arises from two key effects of SSRI exposure: increases in extracellular serotonin levels and reduced spontaneous firing of serotonergic neurons (Pineyro and Blier, 1999). The former is the consequence of a functional shift in the release and uptake equilibrium, because of reductions in SERT function; the latter arises because of an increased activation of 5-HT_{1A} serotonin autoreceptors, which inhibit the firing of serotonergic neurons (Blier et al., 1998). Sustained increases in serotonin levels during chronic SSRIs treatment led to functional desensitization of these autoreceptors, an effect associated with reduced 5-HT_{1A} receptor mRNA (Bortolozzi et al., 2012), which leads to reinstatement of normal spontaneous firing, which in turn provokes an increase of serotonin levels in the intersynaptic space (Belmer et al., 2018a,b; Davidson and Stamford, 1995). Studies have shown that this desensitization is a critical component of the antidepressant responses to chronic SSRI treatment in rodent models (Belmer et al., 2018a,b; Bortolozzi et al., 2012; Ferres-Coy et al., 2013). It has recently been discovered that

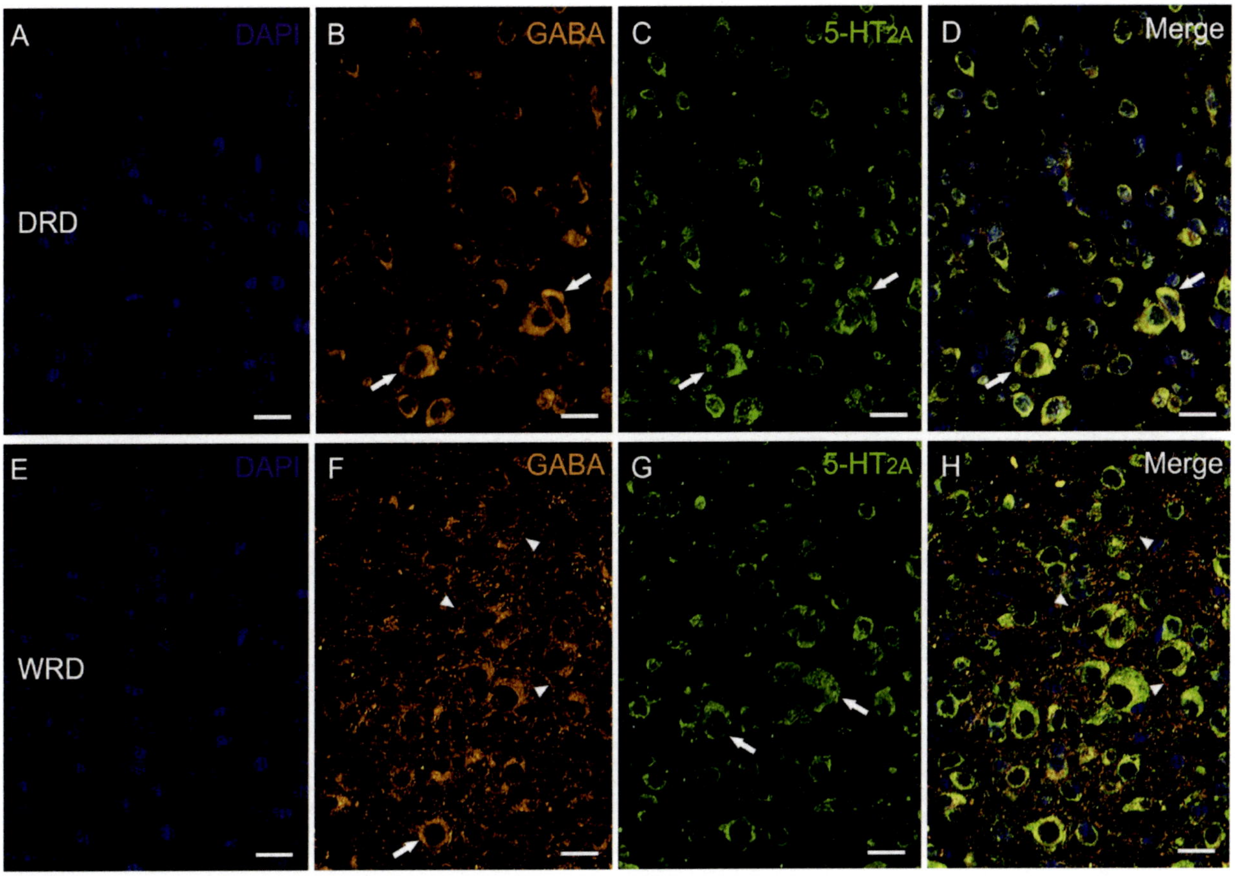

FIGURE 5B.4 **Photomicrographs of transverse sections of the dorsal (DRD) and lateral wings (WRD) of the dorsal raphe nucleus (DRN), showing immunohistochemically labeled gamma-aminobutyric acid (GABA)- and 5-hydroxytryptamine (5-HT)$_{2A}$ receptor-containing fibers and perikarya.** *Image courtesy of da Silva Soares, R., Falconi-Sobrinho, L.L., dos Anjos-Garcia, T., Coimbra, N.C., 2019. 5-Hydroxytryptamine 2A receptors of the dorsal raphe nucleus modulate panic-like behaviors and mediate fear-induced antinociception elicited by neuronal activation in the central nucleus of the inferior colliculus. Behav. Brain Res. 357–358, 71–81. https://doi.org/10.1016/j.bbr.2017.07.016.*

both short- and long-term activation of B6-DRN serotonergic neurons induce antidepressant-like behavioral responses. However, only short-term activation induces anxiogenic-like behaviors (Urban et al., 2016). Recent evidence have also indicated that the antidepressant effects of ketamine are in part mediated by an increase in the AMPA receptor–mediated neurotransmission in depression related areas (B6-DRN, prefrontal and limbic cortex, etc.) Llamosas et al. (2019) demonstrated that ketamine transiently increases spontaneous AMPA receptor–mediated neurotransmission in B6-DRN.

SSRIs and SNRIs treatments are widely used in the treatment of OCD. Recently Minah et al. (2019) showed that the positive response to this kind of therapy also depends on the functional connectivity of the raphe nucleus (the author does not seem to specify precisely which nucleus it is; however, we maintain that it is the rostral nuclei, first of all the B6-DRN). They reported that, compared with the healthy controls (HCs), OCD patients exhibited significantly larger functional connectivity (FC) between the raphe nucleus (i) (RN) and the temporal cortex, including the middle temporal gyrus (MTG), paracingulate gyrus, amygdala, hippocampus, putamen, thalamus and brainstem. Greater RN-left MTG FC was positively correlated with OCD symptoms severity at baseline. In addition, larger FC of the RN-left MTG was also found in SSRI nonresponders, compared with responders, which was a significant predictor of SSRIs response after 16 weeks. The FC of RN may reflect the neurobiological underpinning of OCD and could aid future precision medicine as a differential brain-based biomarker.

THE MEDIAL COLUMN

The medial column or medial tegmental field of the reticular formation is located just medially to the median column we have described so far. One of the most substantial differences compared with the median column is that the medial column presents a more heterogenous composition in terms of neurotransmitter phenotypes; moreover, it contains many large elements.

Let's go to sleep? What the raphe system does about it

The B8-MnR, along with the B6-DRN, is a major source of 5-HT input to the forebrain. By comparison with B6-DRN, however, much less attention has been paid to the role of the B8-MnR in the modulation of states of vigilance/sleep. An early report by Jacobs and colleagues (Rasmussen et al., 1984) described a population of putative 5-HT neurons of B8-MnR that displayed a very similar pattern of activity to "classic" 5-HT B6-DRN neurons. Specifically, these neurons fired at low tonic rates in quite waking and showed progressive decreases in rates in SWS and REM sleep, with a virtual cessation of activity in REM sleep. Subsequent studies, however, have identified a much more heterogenous population of B8-MnR neurons with discharge profiles indicative of serotonergic and GABAergic B8-MnR neurons (see the following).

Possibly, the function most closely associated with B8-MnR and sleep is the modulation of the hippocampal EEG-or the theta rhythm. Specifically, the B8-MnR suppresses (or blocks) theta in NREM sleep, while B8-MnR silencing in REM sleep releases theta during that state (permissive role) (Vertes et al., 2010; Vertes et al., 2004). As is well recognized, theta is a prominent event of REM sleep (Bland and Oddie, 2001; Headley and Paré, 2017; Vertes, 2005; Vertes et al., 2004). As briefly discussed, theta is controlled by ascending reticular activation system (ARES) to septohippocampal system, that is, cells of B5-RPo in the pontine raphe fire at high tonic rates during theta-associated states, transfer this tonic discharge to the supramammillary nucleus (SUM), where it is converted to a (theta) rhythmic pattern of activity that is then relayed to medial septal "pacemaker cells" to generate theta in the hippocampus (Bland and Colom, 1993; Bland et al., 1994; Colom and Bland, 1987; Ford et al., 1989; Bland et al., 1994; Vertes, 1981; Vertes et al., 2010).

By comparison with this ascending theta-generating system, the B8-MnR (and its targets) represents an ascending hippocampal desynchronizing system. B8-MnR disrupts (or blocks) theta by suppressing some or all of the theta-generating sites—B5-RPo, SUM, the medial septum, or the hippocampus. Specifically, B8-MnR stimulation desynchronizes the hippocampal EEG, and B8-MnR lesions produce continuous theta independent of behavior (Assaf and Miller, 1978; Maru et al., 1979; Vertes, 1981; Vertes et al., 2010; Vertes et al., 2004; Yamamoto et al., 1979). Further, the persistent theta produced by B8-MnR lesions can be reversed with infusions of the serotonin precursor, 5-HTP, and the depletion of 5-HT stores with PCPA attenuates the desynchronizing effects of B8-MnR stimulation (McNaughton et al., 1980; Yamamoto et al., 1979). In accord with the foregoing, it has been recently demonstrated that infusions of the $GABA_A$ agonist, muscimol, or the $5\text{-}HT_{1A}$ autoreceptor agonist, 8-OH-DPAT, into B8-MnR produced continuous theta in anesthetized rats (Kinney et al., 1994, 1995, 1996). Finally, Varga et al. (2002) reported that B8-MnR neurons express $GABA_B$ receptors and infusions of the $GABA_B$ agonist, baclofen, into B8-MnR produced long-lasting theta that "resulted from suppression of the serotonergic output from the median raphe."

The discharge properties of MR neurons are also consistent with its role in the modulation of the hippocampal EEG. For instance, it has been demonstrated that about 80% of B8-MnR neurons displayed firing properties associated with changes in theta/hippocampal EEG (Kocsis and Vertes, 1996; Viana Di Prisco et al., 2002). B8-MnR cells consisted of three major types, two of which were putative serotonergic neurons based on their slow, regular rate of discharge. The 5-HT cells were found to be "theta-off" cells, that is, they were active during the nontheta state of NREM sleep and virtually silent during REM sleep—thus permissive to theta of that state. A second major class of B8-MnR cells, putatively GABAergic neurons, fired at high rates during REM sleep (theta-on, REM-on cells) and was thought to inhibit 5-HT B8-MnR (theta-off) cells to generate theta in REM (Vertes, 2010; Viana Di Prisco et al., 2002, Fig. 1). It has further recently been shown that B8-MnR activity is inversely related to hippocampal sharp waves/ripples of NREM sleep (Winterer et al., 2017; ul Haq et al., 2016; Wang et al., 2015). For instance, Wang et al. (2015) reported, in freely behaving mice, that (1) putative serotonergic and nonserotonergic B8-MnR neurons show reduced activity prior to the onset of ripple activity, (2) optogenetic B8-MnR stimulation suppressed ripple activity, and (3) optogenetic inhibition of B8-MnR neurons enhanced sharp waves/ripples during NREM sleep. These findings, together with those on theta, suggest that 5-HT B8-MnR neurons suppress theta and sharp waves/ripples during NREM sleep—with the corollary that the inhibition of B8-MnR releases both types of activity in NREM or REM sleep.

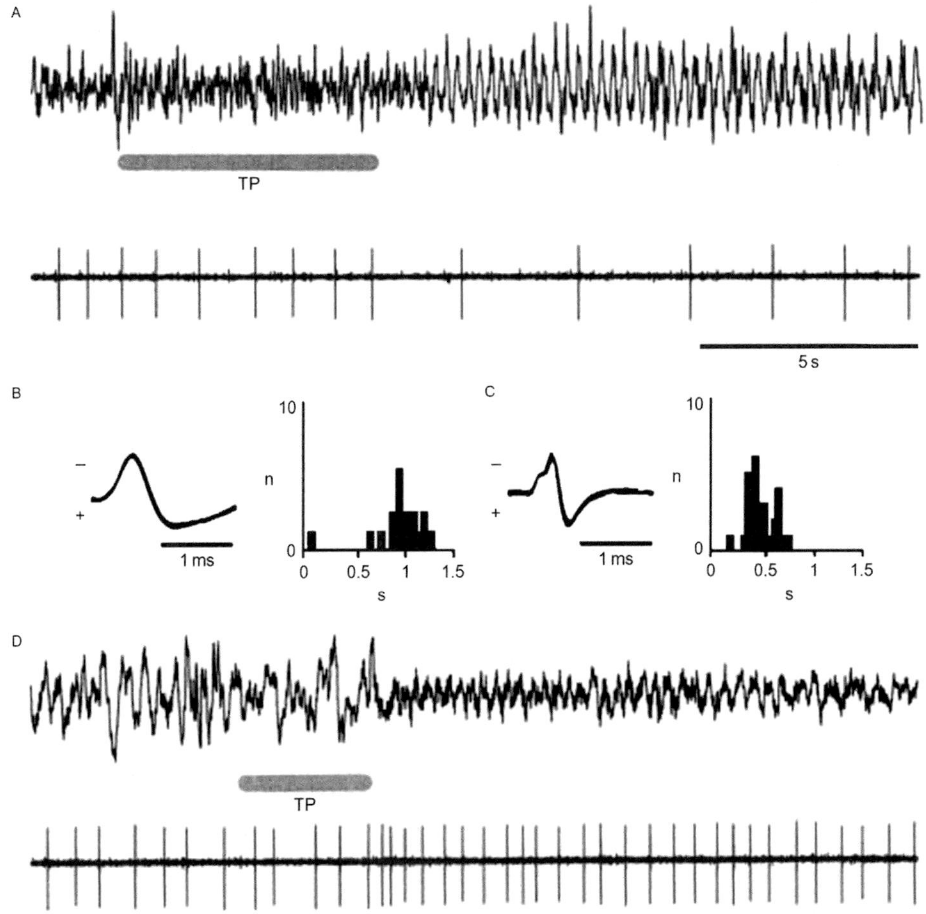

FIGURE 1 **(A)** The discharge characteristics of a slow-firing (SF) cell of the median raphe nucleus that decreased in discharge from nontheta (*left side*) to theta elicited with tail pinch (*TP*, horizontal gray bar). **(B)** Superimposed action potentials of the theta-off cell (A) showing a wide spike (2 ms) and interspike interval histogram (ISIH) demonstrating the slow rate of firing of the cell (1.2 Hz) (0.8 s peak in ISIH) during nontheta conditions. **(C)** Superimposed action potentials of the SF theta-on cell of (D) showing a wide spike (2 ms) and the ISIH demonstrating enhanced firing of the cell to 2.4 Hz (0.4 s peak in ISIH) in the 10 s period following TP-elicited theta. **(D)** The discharge properties of the SF theta-on cell of C showing increased rates of firing during TP-elicited theta. *Courtesy of Linley, S.B., Vertes, R.P., 2019. Chapter 7: Serotonergic systems in sleep and waking. In: Dringenberg, H.C. (Ed.), Handbook of Sleep Research, vol. 30. Elsevier, pp. 101–123. https://doi.org/10.1016/B978-0-12-813743-7.00007-4.*

Like the median column, the medial column also consists of several nuclei located in the medulla oblongata and in the pons (Fig. 5b.5A and B).

The *ventral reticular nuclei* (VRt), A1–A2 and C1–C2, respectively, noradrenergic and adrenergic cellular groups (see later), are located in the caudal part of the medulla oblongata.

In the rostral part of the medulla oblongata, we find the *gigantocellular nucleus* (Gi).

In the caudal pons, there are two nuclei: the *caudal pontine reticular nucleus* (PnC) and the *reticulotegmental pontine nucleus* (RtTg); the latter also belongs to the precerebellar nuclei. Finally, in the rostral pons, we find the *oral pontine reticular nucleus* (PnO).

In the following pages will be described in detail the anatomofunctional characteristics of the nuclei we so far mentioned.

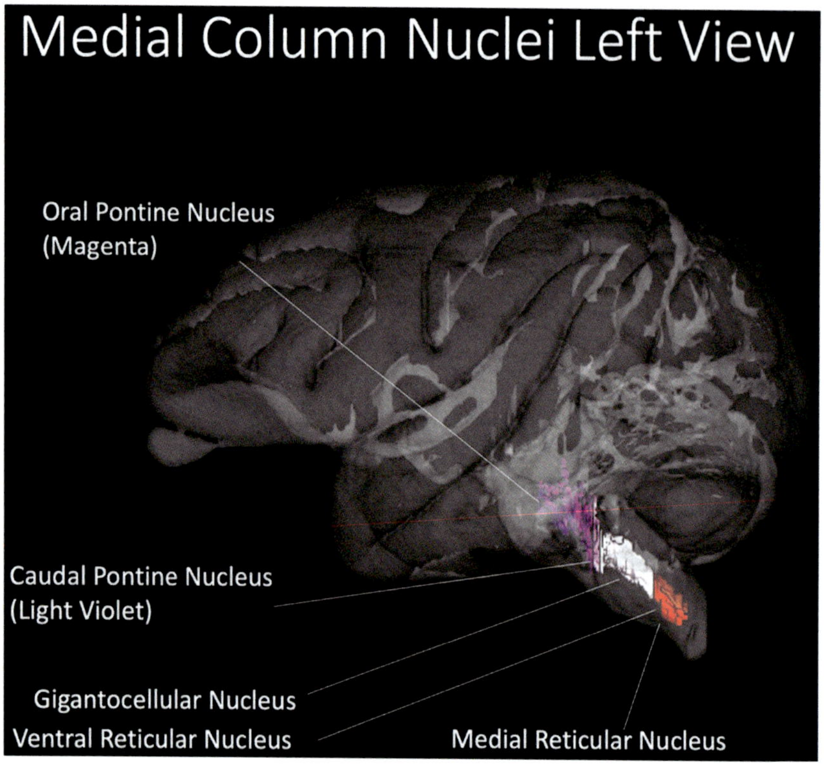

A)

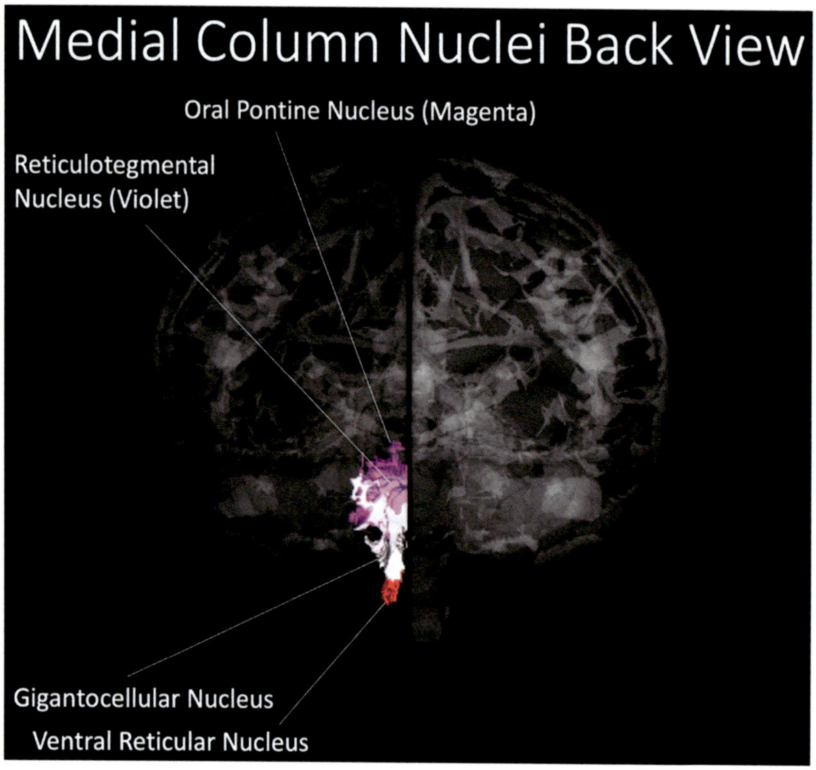

B)

FIGURE 5B.5 **(A) Left-lateral view of the medial column nuclei of the reticular formation in *Macaca mulatta* (rhesus macaque) brain model**. (The medial reticular nucleus is not described in the text). **(B) Back view of the medial column nuclei of the reticular formation in *Macaca mulatta* (rhesus macaque) brain model**. *Image created by BrainInfo (1991–present), National Primate Research Center, University of Washington, http://www.braininfo.org.*

MEDULLA OBLONGATA NUCLEI

Ventral Reticular Nucleus[3]—Gigantocellular Nucleus

The *ventral reticular nucleus* (VRt) is located in the caudal portion of the *medulla oblongata*. Topographically, VRt borders ventrolaterally with the *lateral reticular nucleus* (LRt) and medially with the *nucleus raphe obscurus* (B2-ROb).

It is a continuation of the *parvicellular reticular nucleus* (PCRt), and it has been shown to receive afferent projections from the dentate gyrus of the hippocampus in rabbits (Tang et al., 1987). Immunostaining for tyrosine hydroxylase (TH) revealed large catecholaminergic neurons in VRt (Paxinos et al., 1990).

The gigantocellular nucleus (Gi) is located between the middle olivary levels and the caudal pons, and it extends to the level of the emergence of the facial nerve (VII c.n.) where it is succeeded by the caudal part of the pontine reticular nucleus (PnC) (Fig. 5b.6A and B) As its name implies, the Gi has characteristic large cells, but it also contains many medium- and small-sized cells. It borders medially with the B3-RMg, ventrolaterally with the *nucleus ambiguous* and with the *intermediate reticular zone* (IRt); the latter is interposed between the gigantocellular nucleus (Gi) and the *parvicellular reticular nucleus* (PCRt).

The Gi is essentially subdivided in three parts: Gi ventral pars (GiV), lateral paragigantocellular nucleus (LPGi), and Gi pars alpha (Giα).

The Gi ventral part (GiV) is an AChE-poor area above the dorsal accessory olive, and it borders the LPGi and IRt laterally. Unlike the rat, cat, and monkey, the human GiV does not have giant cells. It continues rostrally into the Gi pars alpha (Giα).

The lateral paragigantocellular nucleus (LPGi) is located in the ventrolateral medulla and is known as a sympathoexcitatory area involved in the control of blood pressure (Guyenet et al., 1993; Dampney et al., 1994). Recent experiments confirmed the role of LPGi in REM sleep, known as paradoxical sleep (PS), and showed that more than two-thirds of LPGi neurons are GABAergic (Sapin et al., 2009), and more than one-fourth send efferent fibers to the wake-active locus coeruleus nucleus (A6-LC).

The LPGi nucleus contains a large number of neurons expressing c-Fos, a marker of neuronal activation, during PS hypersomnia (Verret et al., 2005, 2006). Therefore, since the LC neurons are inhibited by GABA during PS (Gervasoni et al., 1998), it has been postulated that this inhibition could partly arise from the LPGi (Verret et al., 2006).

To get more insight into the role of the LPGi in PS regulation (Sirieix et al., 2012), a combined electrophysiological and anatomical approach in the rat showed that the LPGi contains neurons specifically active during PS (PS-On neurons), neurons inactive during PS (PS-Off neurons), and neurons indifferent to the sleep−wake cycle (Fig. 5b.7)

Furthermore, LPGi activity seems to be regulated by the sublaterodorsal nucleus (SLD), which strongly excites the glycine-GABAergic neurons in the LPGi and the GiV. These nuclei in turn contribute to the inhibition of the local adrenergic(C) and noradrenergic(A) neurons, and the distant A6-LC. In the end, glutamatergic neurons in the PAG may similarly excite the LPGi glycine-GABAergic neurons, while inhibiting the adrenergic/noradrenergic neurons in the LPGi; this would explain the lowering of the blood pressure during sleep (Sirieix et al., 2012).

The LPGi contains the adrenergic cell group C1 and the noradrenergic A1, A2, A4, and A5 cell groups. Studies on experimental animals have shown that the activation of serotonergic cells in the LPGi triggers a solitary nucleus-mediated cardiac baroreflex inhibition, elicited by noxious stimuli (Gau et al., 2009).

Nissl's staining showed the presence of serotonergic cell bodies in LPGi: more than 150 serotonin-positive cells can be seen on each side of a 50-μm section of the caudal hindbrain. The serotonin-positive cells are larger in the LPGi (27 ± 4 μm) than in the caudal part of the intermediate reticular zone (19 ± 4 μm) (Halliday et al., 1988).

Substance P seems to colocalize with serotonine in many of the serotoninergic cells in the LPGi (Halliday et al., 1988b). Zec (2001) examined proximal projections of LPGi using a bidirectional lipophilic fluorescent tracer, 1,1′-dioctadecyl-3,3.3′,3′-tetramethylindocarbocyanine perchlorate (DiI), in postmortem human fetuses, and reported diffusion of DiI to the arcuate nucleus (Ar), the nucleus raphe obscurus (B2-ROb), the hilus of the inferior olive, the bilateral Gi and the intermediate reticular zone (IRt), the vestibular and cochlear nuclei, the cells and fibers at the floor of the fourth ventricle, the medial lemniscus, the lateral lemniscus, the inferior cerebellar peduncle and the cerebellar white matter, the central tegmental tract, and the capsule of the red nucleus.

The Gi pars alpha (Giα) forms a cap over the raphe magnus (B3-RMg). It has small, medium, and large cells, many of which are oriented mediolaterally. It is bordered laterally by the central tegmental tract, as the tract approaches the inferior olive. It is characterized by medium AChE reactivity and has AChE-positive cell bodies. In addition, serotonin-positive cells invade the ventral and lateral part of the Giα.

[3]Also known as nucleus centralis medullae oblongatae, nucleus reticularis ventralis.

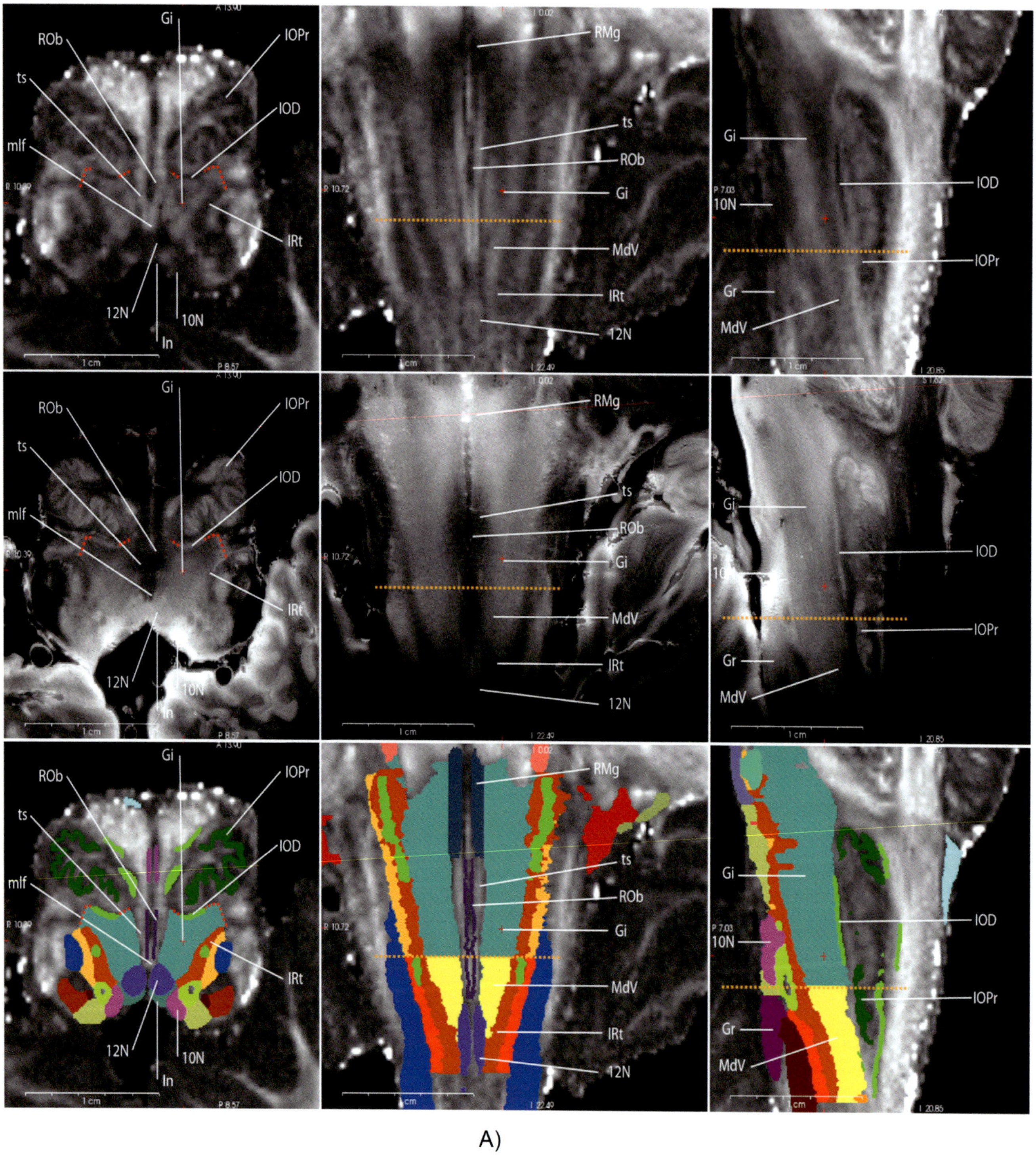

FIGURE 5B.6 **(A) Topography of gigantocellular nucleus (Gi)**. Rostral limit: plane running through the rostral limit of the facial nerve nucleus (7N, *blue dotted line*). Limits at the level of the superior olive: lateral: *straight line* joining the superior olive (SO) to the nucleus of the facial nerve (7N), dorsolateral: intermediate reticular zone (IRt) medial: medial longitudinal fasciculus (mlf), ventrally continued by the tectospinal tract (ts) ventral: *curved line* joining the superior olive to the dorsal aspect of the central tegmental tract (ctg) and medial lemniscus (ml), to finally reach the midline/raphe magnus nucleus (RMg); use the FA map to limit the central tegmental tract (ctg) and medial lemniscus (ml) from the gigantocellular nucleus (Gi). **(B) Topography of gigantocellular nucleus (Gi)**. Limits at the level of the principal inferior olive (IOPr): dorsal-rostrally: praepositus nucleus (Pr) and dorsocaudally: hypoglossal nucleus (12N), intercalatus nucleus (In), and dorsal n. of the vagus nerve (10N); dorsolateral: intermediate reticular zone (IRt); ventral: made of several segments (*red dotted line*): (a) straight sagittal line joining the ventral tip of the intermediate reticular zone (IRt) to the principal inferior olive (IOPr), (b) segment joining the ventral tip of this line to the dorsal accessory olive (IOD), (c) segment joining the IOD to the midline; medial: medial longitudinal fasciculus (mlf) and tectospinal tract (ts), limiting the gigantocellular (Gi) from the raphe obscurus (ROb) nuclei. The theoretical caudal limit of the gigantocellular reticular nucleus from the medullary reticular nucleus is defined by Paxinos as the rostral pole of the linear nucleus, not visible on MRI. It roughly defined by the axial plane running through the rostral pole of the gracile nucleus (Gr, *orange dotted line*). *Courtesy of Fibratlas Group https://fibratlas.univ-tours.fr.*

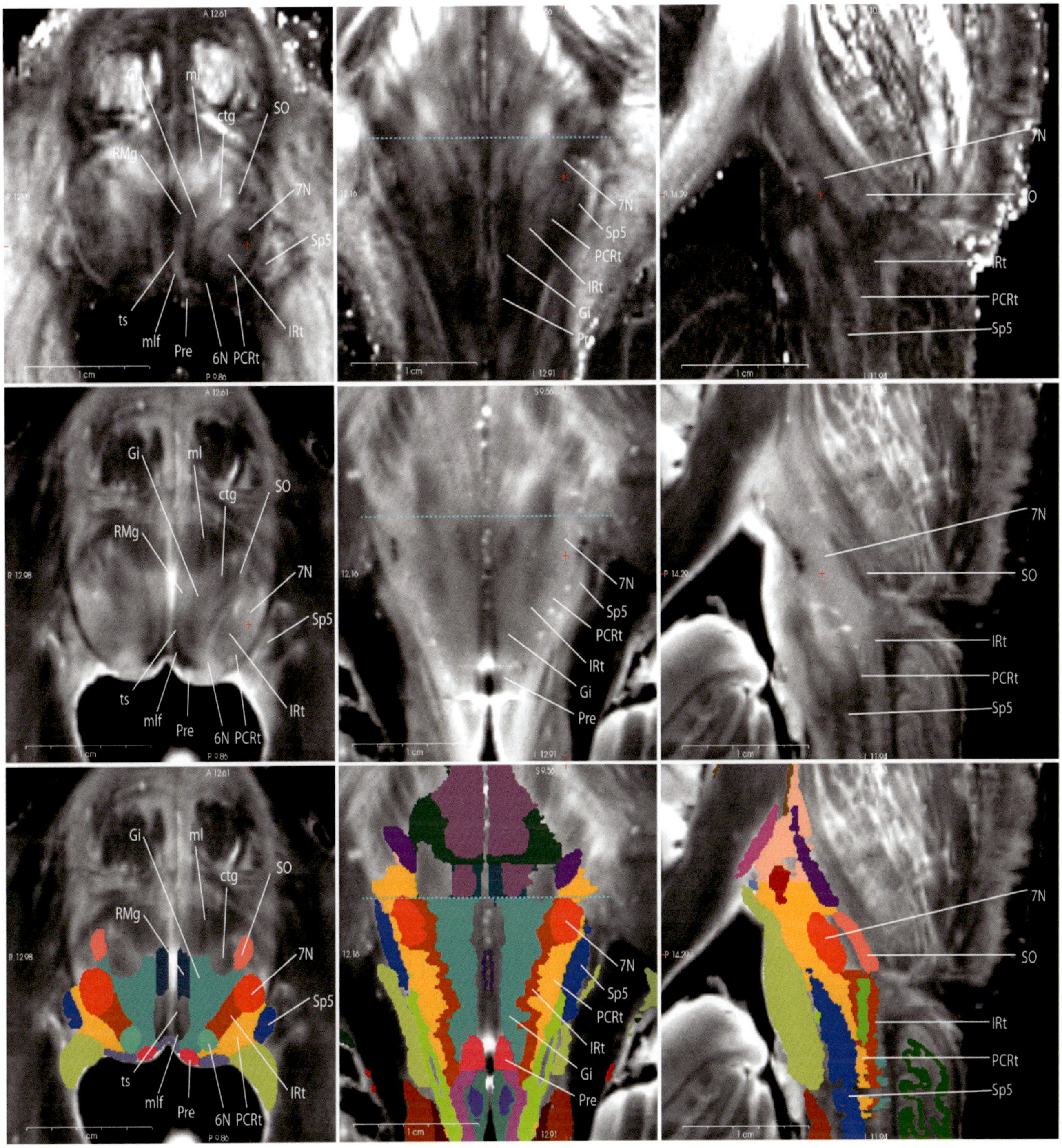

B)

FIGURE 5B.6 cont'd.

The Giα has been identified as a region controlling nociceptive transmission in the caudal spinal nucleus of the trigeminal nerve and the dorsal horn. The serotonergic neurons of the nucleus raphe magnus (B3-RMg) are also involved in this.

Electrophysiological and physiological studies in rats have also shown that the Giα and LPGi provide descending control of pelvic floor organs, specifically by inhibition of sexual reflexes (Hubscher et al., 1996; Johnson and Hubscher, 1998).

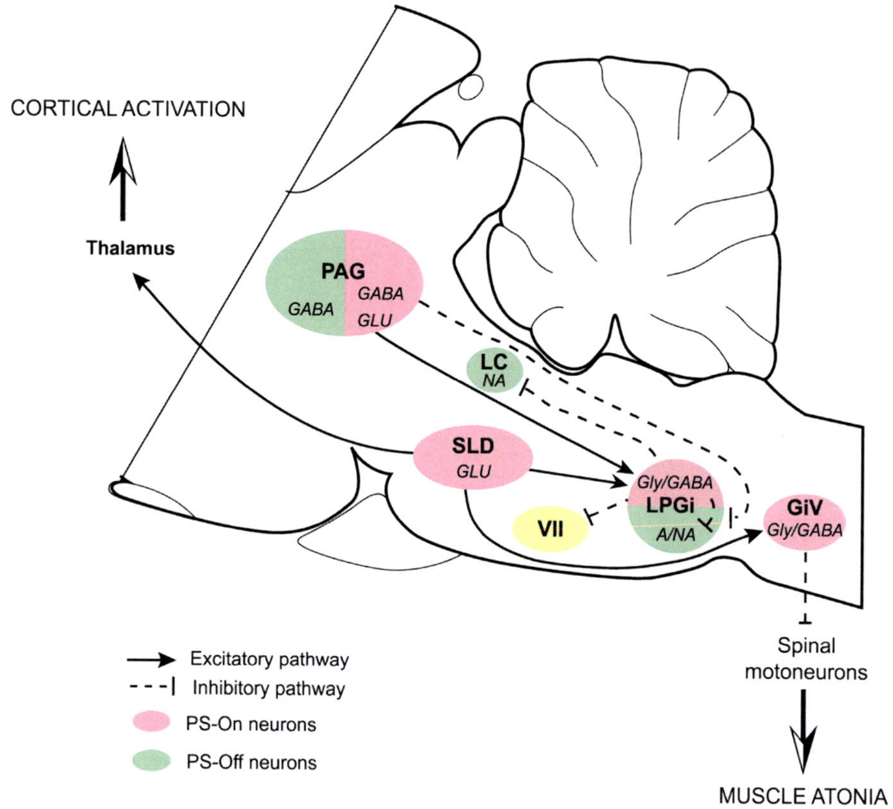

FIGURE 5B.7 **Sagittal representation of the connections of the LPGi activated during PS.** *A/NA*, adrenaline/noradrenalin; *GiV*, gigantocellular reticular nucleus, ventral part; *Glu*, glutamate; *Gly*, glycine; *LC*, locus coeruleus; *LPGi*, lateral paragigantocellular nucleus; *PAG*, periaqueductal gray; *SLD*, sublaterodorsal nucleus; *VII*, facial nucleus.During paradoxical sleep (PS), glutamatergic neurons in the SLD strongly excite Gly/GABAergic neurons in the LPGi and the GiV. In turn, Gly/GABAergic neurons of the LPGi contribute to the hyperpolarization of the facial motoneurons that occurs during PS. They also contribute to the inhibition of the local adrenergic/noradrenergic neurons and the distant LC neurons. Glutamatergic neurons in the PAG may excite the LPGi Gly/GABAergic neurons. Moreover, GABAergic neurons in the PAG would inhibit the adrenergic/noradrenergic neurons in the LPGi. *Courtesy of Sirieix, C., Gervasoni, D., Luppi, P.H., Léger, L., 2012. Role of the lateral paragigantocellular nucleus in the network of paradoxical (REM) sleep: an electrophysiological and anatomical study in the rat. PLoS One 7 (1), e28724. https://doi.org/10.1371/journal.pone.0028724.*

PONS AND MESENCEPHALON–PONS JUNCTION NUCLEI

Caudal Pontine Reticular Nucleus–Oral Pontine Reticular Nucleus–Gigantocellular Nucleus (Gi)–Reticulotegmental Nucleus[4]

The caudal pontine reticular nucleus (PnC) is located in the caudal portion of the pons. Ventrorostrally, it borders with the oral pontine reticular nucleus (PnO), dorsally with the gigantocellular nucleus (Gi) and medially with the pontine raphe nucleus (B4-RPn).

The caudal pontine reticular nucleus (PnC) in rabbits and cats consists exclusively of giant cells. However, in humans, there are both giant cells and normally sized cells. In rodents, it has been shown to play a role in the acoustic startle response (Eggermont et al., 2021). The PnC has been known to mediate head movement, in concert with the gigantocellular reticular nucleus (Gi) and the superior colliculus (Sasaki et al., 2004). The neurons in the dorsal half of this nucleus fire rhythmically during mastication, and in an anesthetized animal, it is possible to induce mastication via electrical stimulation of the PnC or adjacent areas of the gigantocellular nucleus (Scott et al., 2003) The caudal pontine reticular nucleus (PnC) is also thought to play a role in the grinding of teeth during sleep. This region also suppresses muscle tone during REM sleep, activates eye movements, and decreases the sensory input to the cerebral cortex, specifically the primary and somatosensory cortices. The *oral pontine reticular nucleus* (PnO) is located in the rostral portion of the *pons*. It is a large nucleus that

[4]Also known as reticulotegmental nucleus of pons, reticulotegmental pontine nucleus, tegmental reticular nucleus, nucleus reticularis tegmenti pontis.

has a medial border with the superior central nucleus (B8-MnR), a rostral border with the pedunculopontine nucleus (PPTg), a dorsorostral border with the lateral parabrachial nucleus (LPB), and a lateral border with the locus coeruleus (A6-LC).

Regarding the oral pontine reticular nucleus (PnO), it narrows or tapers toward the oral part of the pontine reticular formation, as indicated by its name. It contains occasional or sporadic giant cells. Different neuronal groups of this nucleus have displayed discharge patterns, which coordinate with phasic movements of the eyes to and from paradoxical sleep (REM). From this information, it has been implied that the PnO is involved in the mediation of changing to and from REM sleep (Dergacheva et al., 2002).

The gigantocellular reticular nucleus (Gi), as mentioned before, is one of the largest nuclei located in the medulla oblongata and in the caudal pons. The Gi nucleus has been known to innervate the caudal hypoglossal nucleus and responds to glutamatergic stimuli. The Gi is involved in activating the hypoglossal nucleus and may play a role in the functions and actions of the hypoglossal nerve (Yang et al., 1995).

It additionally receives connections from the periaqueductal gray (PAG), the paraventricular hypothalamic nucleus, the central nucleus of the amygdala, the lateral hypothalamic area, and the parvocellular reticular nucleus (PCRt) (Yang et al., 1995).

Moreover, retrograde studies have shown that the deep mesencephalic reticular formation and the oral pontine reticular nucleus (PnO) project to the Gi (Ohtake et al., 1992).

The dorsal rostral section of the Gi is also involved in mediating expiration along with the parvocellular reticular nucleus (PCRt) (see later in the dedicated section: "Control of the respiratory function by reticular formation").

The *reticulotegmental nucleus* (RtTg) is located between the PnC and the PnO, and it seems to be wrapped by them. Medially, like the PnC, it borders the B4-RPn. The RtTg is medial and slightly ventral to the medial lemniscus. It is made of several cell clusters composed of multipolar neurons, and it has a heterogenous aspect on MRI. The RtTg's principal afferences come from the cerebellum, and it has been proven that they are not only ipsilateral, but they also decussate and reach the contralateral RtTg. This latter, in turn, sends efferences to the vermis (ipsi- and controlateral) and to the cerebellar hemispheres (contralateral only). Therefore, the projections from the pontine tegmentum to the cerebellar lobes are made of crossed fibers only.

Functionally, the RtTg and its related circuits remain largely unknown. Guo et al. (2021) reported that the RtTg is significantly activated during acoustic startle reflexes (ASR), a highly conserved innate defensive behavior, and that chemogenetic inhibition of its glutamatergic neurons decreases the ASR amplitudes. Moreover, viral tracing revealed an ASR neural circuit, which starts from the cochlear nuclei and terminates in the RtTg. This ARS carries information, which are then relayed to the spinal motor neurons. Together, these findings demonstrate how the reticular formation connects the auditory system with motor functions.

PARAMEDIAN PONTINE RETICULAR FORMATION

After having described the pontine and pontomesencephalic nuclei of the reticular formation, we can introduce a region of the latter which plays a fundamental role in the generation of lateral gaze: the paramedian pontine reticular formation (PPRF).

The pontine portion of the horizontal gaze center consists of the medial region of the nucleus reticularis pontis oralis (PnO) and nucleus reticularis pontis caudalis (PnC), which houses medium-lead excitatory burst neurons (EBNs). The EBNs activate the ipsilateral abducens motoneurons causing the ipsilateral eye to abduct (Henn and Cohen, 1976; Hepp and Henn, 1983; Strassman et al., 1986a,b; Hepp et al., 1989). They also contact internuclear neurons in the abducens nucleus (VI c.n.), which activates the medial rectus motoneurons in the contralateral oculomotor nucleus (III c.n), so that a comparable movement is made by the opposite eye, producing conjugate eye movements. The horizontal gaze center also includes the medial region of the nucleus paragigantocellularis dorsalis of the rostral medulla (DPGi), as it houses inhibitory burst neurons (IBNs). IBNs' firing is similar in timing and intensity to that of EBNs during saccades and fixation (Hikosaka and Kawakami, 1977; Hikosaka et al., 1978a,b; Yoshida et al., 1982). IBNs suppress the activity of antagonist muscles (the contralateral lateral rectus and ipsilateral medial rectus) through a glycinergic, crossed inhibitory projection to the motoneurons and internuclear neurons in the abducens nucleus (Yoshida et al., 1982; Strassman et al., 1986a,b).

The PPRF has been hypothesized to generate a command for horizontal saccades by extracting it from saccadic signals, provided by the superior colliculus (SC) (Fuchs et al., 1985; Moschovakis et al., 1996). This role of the PPRF is supported by stimulation studies and evidence that lesions in this region produce horizontal gaze palsies (Goebel et al., 1971; Cohen and Komatsuzaki, 1972; Kato et al., 2007).

The process by which collicular signals are transformed into the required burst neuron firing patterns is still not fully understood (see Moschovakis et al., 1998). For example, there is conflicting evidence in respect to whether the SC directly targets EBNs and IBNs.

Evidence from cats suggests that monosynaptic tectal projections are supplied to PPRF premotor neurons

(Grantyn et al., 1980, 1987a,b; Grantyn and Berthoz, 1987; Izawa et al., 1999), and more specifically to IBNs (Hikosaka and Kawakami, 1977; Grantyn et al., 1979; Takahashi et al., 2005). Chimoto et al. (1996a,b) found evidence of monosynaptic tectal inputs to cats' EBNs when the omnipause inhibition was gated. There is evidence that long lead burst neurons (LLBNs) in the rostral brainstem receive direct inputs from the SC (Luschei and Fuchs, 1972; Hepp and Henn, 1983; Scudder et al., 1996) and send efferents to medium-lead burst neurons (MLBNs). Therefore, it has been suggested that these cells may serve as interneurons between the SC and premotor neurons in primates (Scudder et al., 1996).

The central mesencephalic reticular formation (cMRF) is one of the structures that contains LLBNs (Waitzman et al., 1996; Handel and Glimcher, 1997). While earlier reports of a saccade-related area within the midbrain reticular formation (MRF) exist (Bender and Shanzer, 1964), the cMRF was first described in detail and named by Cohen and Büttner-Ennever (1984) and Cohen et al. (1985). They defined it as an area in the midbrain tegmentum of primates, which produces horizontally directed contraversive saccades following electrical stimulation. They suggested three possible roles for the cMRF: saccade triggering, feedback control of saccadic activity, and feed-forward control of saccade-related activity. Waitzman et al. (1996) recorded from individual cMRF neurons in awake, behaving monkeys, and showed that about three quarters of cMRF neurons discharge before and/or during contraversive visually guided rapid eye movements and during contraversive spontaneous saccades in the dark. The number of spikes appeared to correlate with the size of the horizontal component of the saccade. In later reports, they caused reversible lesions in this structure (Waitzman et al., 2000a,b). Based on their findings, they subdivided the MRF into a caudal region, the cMRF, where inactivation affects the horizontal component of saccades, and a rostral region, the periinterstitial nucleus of Cajal portion of the MRF (piMRF), in which inactivation affects the vertical component of saccades. They suggested that a group of LLBNs located in the piMRF play a role in vertical saccadic eye movements, in contrast to the cMRF, whose LLBN activity is more related to the horizontal eye movements.

The targets of the various cell types described by these studies were not antidromically identified. Only the cells that project back upon the SC have been described physiologically (Moschovakis et al., 1988a,b). The activity of these reticulotectal neurons resembles, in most aspects, the activity of the collicular cells providing them input. More recent studies identified separate classes of cMRF neurons associated with saccade metrics, including amplitude, velocity, and duration (Cromer and Waitzman, 2006, 2007). These authors proposed that cMRF cells, whose firing is most tightly coupled to saccade velocity, may represent an intermediate step in the spatiotemporal transformation needed to convert the firing of output cells within the motor map present in the SC into the temporally coded firing of premotor neurons in the PPRF.

CARDIORESPIRATORY REGULATION BY VLSM

The VLSM is one of the best-studied regions in cardiovascular control (Dampney et al., 1994). Both monoaminergic and nonmonoaminergic neurons with projections to the thoracic visceromotor columns and the ambiguus and dorsal vagus nuclei are involved in these systems. Baroreceptors in the carotid sinus and the ascending aorta connect with the rostral part of the general visceral portion of the nucleus of the solitary tract; fibers from chemoreceptors sensitive to O_2 and CO_2 concentrations terminate more caudally. Short pathways connect the nucleus of the solitary tract with cardiomotor neurons in the dorsal vagal nucleus, subserving slowing down the heart rate. Presympathetic neurons in the VLSM, 50% of which correspond to the adrenergic C1 group, project to the thoracic intermediolateral column (Card et al., 2006; Schreihofer and Guyenet, 1997; Schreihofer et al., 2000) and subserve vasoconstriction. The C1 group is inhibited by caudally located GABAergic neurons in the VLSM (Schreihofer and Guyenet, 2002). These inhibitory neurons receive input from baroreceptor-activated cells in the nucleus of the solitary tract and contribute to the inhibition of vasomotor tone. Ascending projections from the A1 and A2 groups control vasopressin release by the paraventricular hypothalamic nucleus in a complementing manner. Other noradrenergic and raphespinal systems are involved in cardiovascular control, but their pathways have been less well characterized.

CONTROL OF THE RESPIRATORY FUNCTION BY THE RETICULAR FORMATION

The respiratory control system resides in the ventral respiratory column (VRC), located in the ventrolateral medulla (Fig. 5b.8), and is divided into continuous series of four compartments: the Bötzinger (BötC) and pre-Bötzinger complex (preBötC), the rostral ventral respiratory group (rVRG), and the caudal ventral respiratory group (cVRG). Respiratory neurons in this compartment constitute microcircuits that interact within and between these compartments to generate and modulate respiratory patterns.

Each compartment of the VRC exhibits unique characteristic for respiratory-related firing patterns and projections. The BötC and pre-BötC are large groups of

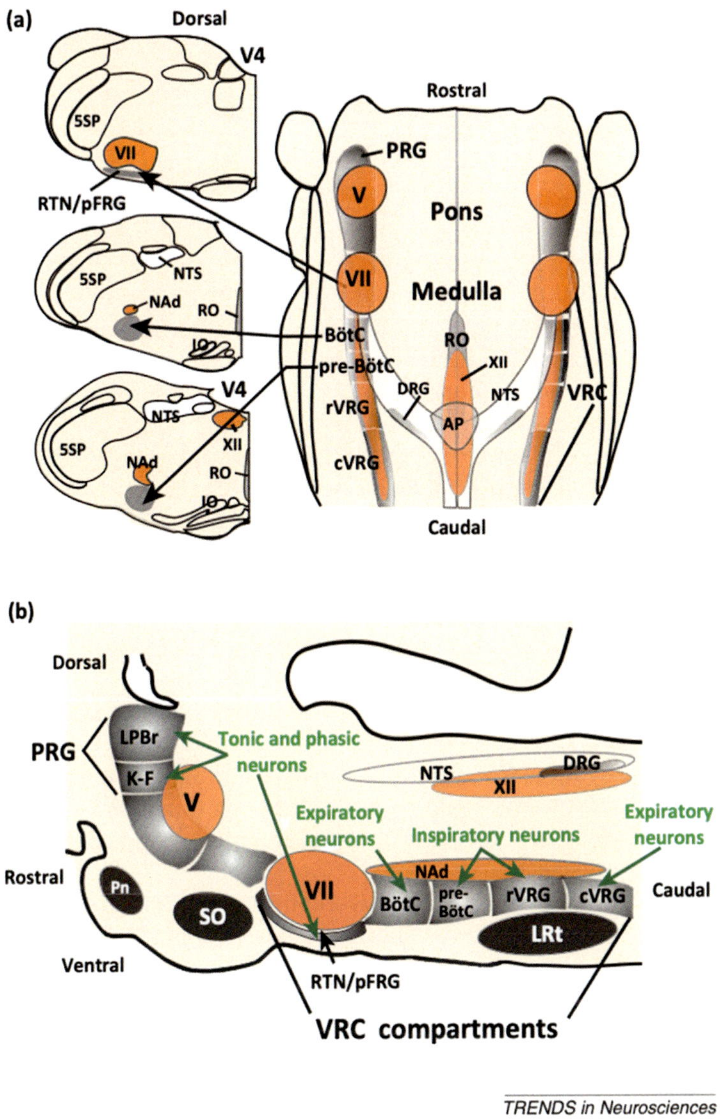

FIGURE 5B.8 **Overview of bilaterally distributed brainstem respiratory compartments of the rat arranged from the rostral pons to the caudal medulla represented neuroanatomically by (A) horizontal (*right*) and several coronal (*left*) brainstem sections and (B) a parasagittal section through one side of the brainstem.** *AP*, area postrema; *LRt*, lateral reticular nucleus; *NAd*, nucleus ambiguus, dorsal division; *Pn*, ventral pontine nucleus; *SO, IO*, superior and inferior olivary complexes; *SP5*, spinal trigeminal nucleus; *V*, motor nucleus of the trigeminal nerve; *V4*, fourth ventricle; *XII*, hypoglossal motor nucleus.Predominant locations of inspiratory, expiratory, tonic, and respiratory-modulated (phasic) interneurons are indicated in (B). *Courtesy of Smith, J.C., Abdala, A.P., Borgmann, A., Rybak, I.A., Paton, J.F., 2013. Brainstem respiratory networks: building blocks and microcircuits. Trends Neurosci. 36 (3),152–162. https://doi.org/10.1016/j.tins.2012.11.004.*

expiratory and inspiratory neurons, respectively, and project to the bulbospinal neurons in the VRG. The BötC is located just rostrally the pre-BötC in the ventrolateral medulla oblongata, at a level immediately caudal to the facial nucleus. The BötC neurons inhibit the inspiratory neurons of the nucleus of the solitary tract (nsl) and the rVRG during expiration, and project bilaterally to the phrenic motor nucleus to inhibit its activity. The BötC inhibitory neurotransmitter seems to be glycine (Schreihofer and Guyenet, 2002).

The pre-BötC is located rostrally to the VRG and the nucleus ambiguous, caudally to the BötC, and it is essential for the generation of the respiratory pattern (pacemaker activity) (Fig. 5b.8).

The rVRG and cVRG are prominent groups of bulbospinal inspiratory and expiratory neurons, respectively. The rVRG (inspiratory neurons) projects to motoneurons of the phrenic nerve in C3/C4, which innervate the diaphragm, the main inspiratory muscle. Conversely to the rVRG, the cVRG (expiratory neurons) projects to the spinal interneurons innervating the motoneurons for the abdominal muscles, which are part of the expiratory muscles. The rVRG is subdivided into the anterior rVRG (arVRG) and the posterior rVRG (prVRG), and recent studies showed that a chemical stimulation of the arVRG—with DL-homocysteic acid (DLH)—induces bradypnea, whereas that of the prVRG does not affect respiration (Monnier et al., 2003).

However, before describing in detail the complete respiratory pattern, it is necessary to take into consideration another important center that regulates the activity of the ones described so far: the pontine respiratory group (PRG), also known as the "pneumotaxic center." This is a group of respiratory neurons located bilaterally in the rostral third of the pons, between the parabrachialis and the Kölliker–Fuse nucleus (KF). Within the PRG, there are inspiratory, expiratory, and inspiratory/expiratory (phase-spanning) neurons, which transmit information to the medullary respiratory centers. It is believed that the primary function of the PRG is to provide a smooth transition from inspiration to expiration. The original name of the PRG, "pneumotaxic center," was given because changes in PRG activity directly influence the rate of breathing. Together with vagal feedback from the pulmonary stretch receptors, which monitor the adequacy of lung inflation, the PRG functions to vary the point of termination of a breath (inspiratory off-switch). Thus, activation of the PRG leads to rapid, shallow breathing, while decreasing PRG output leads to slow, deep breaths. If both of the inspiratory off-switches are ablated (i.e., the vagus nerves are cut and the rostral third of the pons is lesioned), there are very prolonged, deep breaths, separated by short periods of expiration; this pattern of breathing is called apneustic breathing.

The Köelliker–Fuse nucleus is composed of a heterogenous population of neurons, most of which have a respiratory function (Cohen et al., 1959; Dick et al., 1994; Dutschmann and Herbert, 2006; Ezure et al., 2006). Excitatory KF neurons project to the majority of the core respiratory nuclei in the brainstem, including the preBötC, the lateral parafacial nucleus, the BötC, and the rVRG (Geerling et al., 2017; Song et al., 2012). Through these projections, the KF directly participates in the control of respiration by modulating respiratory rate and pattern (Cohen, 1971). As demonstrated by Fung et al. (1995), chemical lesions or pharmacological inhibition of the KF change the rate and pattern of respiration, eliminating the postinspiratory phase and prolonging the inspiration. Conversely, pharmacological excitation of the KF prolongs the duration of the postinspiratory phase (Dutschmann and Herbert, 2006). These studies, in conjunction with others, support the critical role of the KF in the neural control of breathing.

KF neurons also project to a range of cranial nerves, for instance, the vagus and hypoglossal nerves (X and XII c.n., respectively) (Browaldh et al., 2016; Geerling et al., 2017). These projections are important in the coordinated control of the upper airways with breathing during orofacial behaviors involving similar anatomical pathways, such as swallowing, cough, and vocalization (Bautista et al., 2014; Browaldh et al., 2016; Geerling et al., 2017; Jakus et al., 2008). Additionally, through connections with regions mediating the arousal component of hypoxic\hypercapnic chemoreflexes, the KF may also play a role in the state-dependent adjustments of the respiratory rate and pattern (Kaur and Saper, 2019). For instance, the KF receives chemosensory information from the lateral hypothalamus, the retrotrapezoid nucleus (RTN), the medullary raphe, the nucleus of the solitary tract (NTS), and the A6-LC (Guyenet et al., 2015; Hancock et al., 1976; Hebert et al., 1990; Kaur and Saper, 2019; Peyron et al., 1998; Robertson et al., 2013; Yokota et al., 2016).

Now let us describe in detail how these respiratory nuclei work together.

Inspiration. Bulbospinal (aug-I) neurons in the rVRG drive the activity of phrenic motoneurons, which control contraction of the diaphragm to inflate the lungs. The preinspiratory–inspiratory phase (pre-I/I) spiking activity of distinct groups of medullary neurons, which in the pre-BötC are considered critical for rhythm generation and initiation of inspiration activates the tongue protrussor muscles via the hypoglossal nerves. The hypoglossal tone is maintained throughout inspiration to prevent obstruction of the oropharyngeal airway. Branches of the vagus nerve innervate laryngeal muscles, controlling the glottis, and the vocal cords, which are abducted by vagal inspiratory activity to maintain upper airway patency.

Postinspiration. This phase is critical for termination of inspiration. During this phase, the larynx is initially adducted (to stall airflow, thereby prolonging the time for gas exchange) and then slowly abducted to allow exhalation.

Stage 2 expiration. Expiratory bulbospinal neurons in BötC and cVRG are active during this phase and produce low levels of internal intercostal and abdominal muscle activity.

The schematic representation of the functional respiratory patterns is shown in Figs. 5b.9 and 5b.10.

SEROTONERGIC MODULATION OF BREATHING

Nerve fibers with varicosities containing serotonin are densely distributed in the rostral and caudal VRC. In slice preparations at the level of preBötC, burst frequency of hypoglossal neurons, which burst in inspiratory phase, was increased by application of 5-HT, suggesting that 5-HT excites the preBötC and increases respiratory frequency. Thus, 5-HT would be one of the important modulators for respiratory control in the VRC. As stated before, neuronal cell bodies of serotoninergic neurons are distributed in the rostral and caudal raphe nuclei, which send ascending projection to the forebrain and descending projection to the spinal cord, respectively (Steinbusch, 1981; Törk, 1990). Physiological studies reported that electrical and chemical stimulation of both rostral and caudal raphe nuclei, i.e., B8-MnR, B6-DRN, B3-RMg, B2-Rob, and B1-RPa, changed respiratory patterns (Alvarenga et al., 2005; Cao et al., 2006; Doi and Ramirez, 2010; Ptak et al., 2009).

Recently, Morinaga (2019), performing a combination of retrograde tracing techniques, has shed some light on the detailed distribution of serotonergic neurons in the

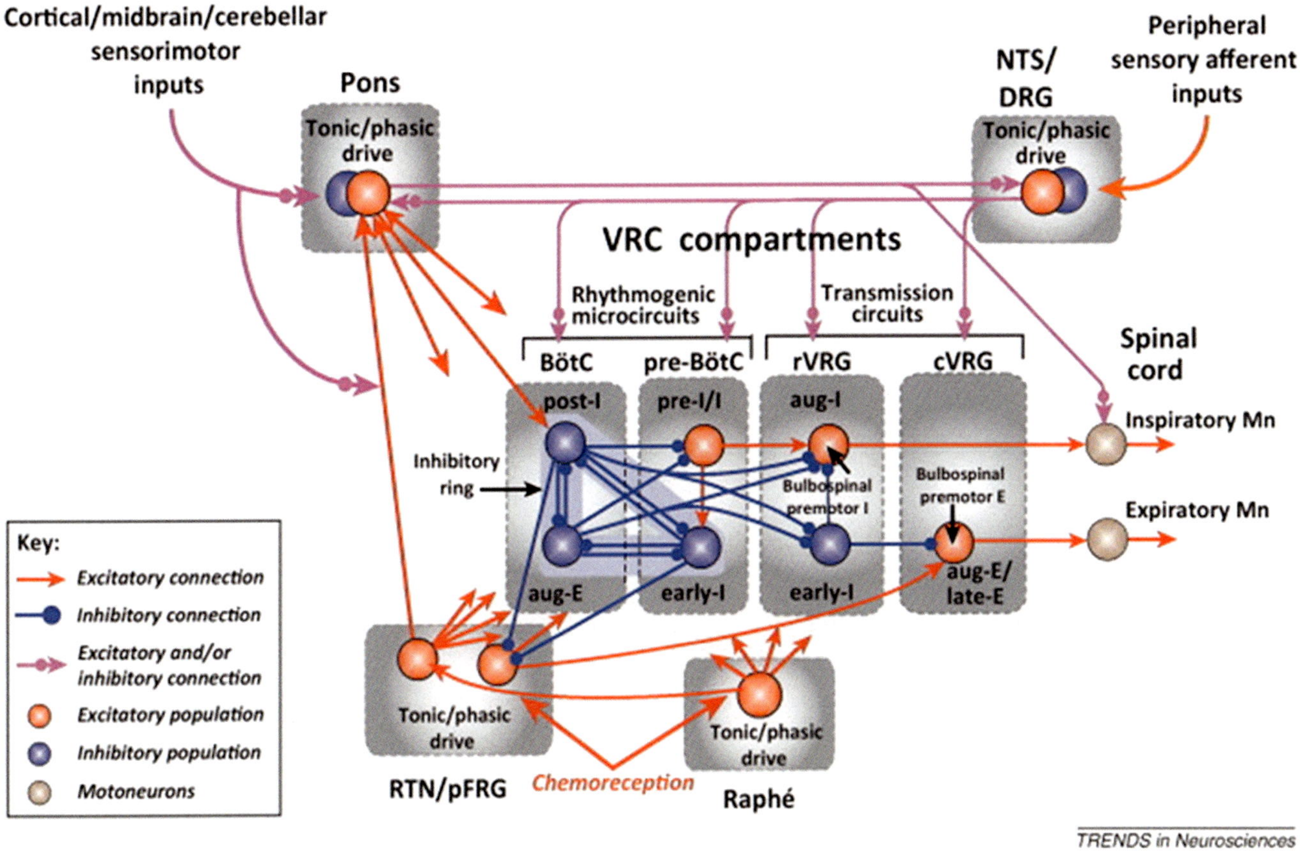

FIGURE 5B.9 **Schematic representation of compartmentalized microcircuits in the brainstem hypothesized to be involved in respiratory rhythm and pattern generation.** Each side of the brainstem contains these compartments with extensive bilateral circuit interconnections (not shown) as well as reciprocal connections, a few of which are indicated. It is hypothesized that the pre-Bötzinger complex (pre-BötC) and Bötzinger complex (BötC) are the major components of the ventral respiratory column (VRC) with rhythmogenic microcircuits generating multiple respiratory patterns as described in the text (see Fig. 5b.10). These circuits, shown in highly simplified form, incorporate the excitatory glutamatergic (*red*) preinspiratory–inspiratory (pre-I/I) population in pre-BötC and three inhibitory GABAergic or glycinergic (*blue*) populations within the pre-BötC–BötC that are postulated (Ben-Tal and Smith, 2008; Smith et al., 2009) to mutually inhibit each other (inhibitory ring of early-I, post-I, and augmenting expiratory [aug-E] neurons; see in the text for respiratory neuron nomenclature). BötC expiratory neurons are a major source of synaptic inhibition of inspiratory neurons during expiration. The pre-I/I excitatory neurons of the pre-BötC, which is the excitatory kernel of the respiratory network, project via premotor circuits to cranial motoneurons (e.g., hypoglossal and vagus) and to rostral ventral respiratory group (rVRG) excitatory (augmenting inspiratory [aug-I]) bulbospinal premotor neurons with projections to spinal cord inspiratory (phrenic) motoneurons (Mn, *brown*). The excitatory bulbospinal expiratory neurons of the caudal VRG (cVRG) project to thoracic and abdominal spinal expiratory motoneurons. Neurons within these compartments are considered to receive tonic, phasic, or rhythmic excitatory drive from the pontine, dorsal respiratory group (DRG), retrotrapezoid nucleus/parafacial respiratory group (RTN/pFRG), and raphe′ compartments. Drives from the latter two compartments are regulated in part by blood or brain CO_2 levels (chemoreception), which may involve modulatory connections from the raphe′ to RTN/pFRG (Guyenet and Mulkey, 2010) as depicted. Some pontine and BötC neurons (not indicated in the schematic) have bulbospinal projections. The spinal cord also contains segmental interneuronal microcircuits (not represented) controlling motoneuronal activity. See the text and Fig. 5b.10 for additional descriptions of the roles of the different populations of excitatory and inhibitory respiratory neurons. *Courtesy of Smith, J.C., Abdala, A.P., Borgmann, A., Rybak, I.A., Paton, J.F., 2013. Brainstem respiratory networks: building blocks and microcircuits. Trends Neurosci. 36 (3):152–162. https://doi.org/10.1016/j.tins.2012.11.004.*

rostral and caudal nuclei projecting to the VRC and analyzed the differences of distribution of labeled neurons. The results of this study indicate that the serotonergic neurons in both the rostral and caudal raphe nuclei projected throughout VRC, and these serotonergic projections may contribute to respiratory responses to various environmental and vital changes. Serotonergic nerve fibers are densely distributed in the rostral to caudal VRC, and serotonin would be one of the important modulators for the respiratory control in the VRC.

Morphologically, these neurons are round, oval, or polygonal, with several dendrites projecting in multiple direction; some of them, situated in the supralemniscal nucleus (SuL), the B3-RMg, the Gi, and the PPy, are elongated horizontally, with dendrites also projecting in a horizontal direction (Figs. 5b.11 and 5b.12).

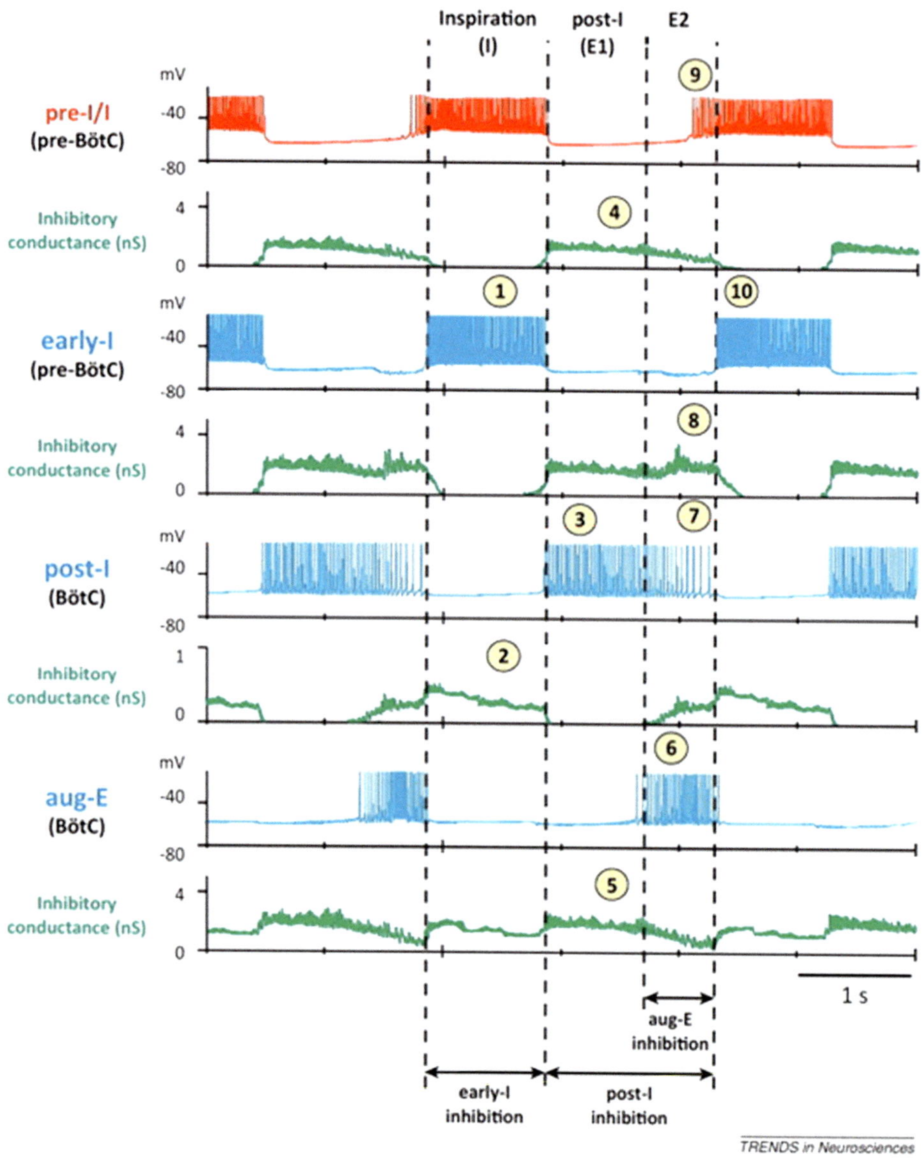

FIGURE 5B.10 **Synaptic inhibition is essential for orchestrating phase transitions to generate the three-phase respiratory pattern.** This figure shows the membrane potential trajectories including spiking patterns and the temporal patterns of postsynaptic inhibitory conductances in single representative neurons from the four main neuron populations in the pre-Bötzinger complex (pre-BötC) and Bötzinger complex (BötC) circuitry from a computational model (Smith et al., 2007; Ribak et al., 2007), which is used here as an aid to explain the possible dynamic operation of these microcircuits. Numbers indicate sequential events, beginning with the inspiratory (I) period. (1) during the I phase, the activity of the pre-BötC early-I inhibitory population declines, in part because of intrinsic spike frequency adaptation. This provides a slow disinhibition of the BötC post-I neurons, which is evident in the trajectory of the post-I neuron inhibitory conductance (2). Once the post-I inhibitory neurons start firing (3), inspiratory activity is inhibited by the abrupt rise in inhibitory conductance in pre-I/I (4) and early-I neurons, which terminates inspiration. Augmenting expiratory (aug-E) neurons are also inhibited (5) and prevented from spiking by the abrupt rise in inhibitory postsynaptic conductance. With the progressive reduction in post-I neuron spiking frequency (adaptation) and inhibition, the activity of the aug-E neuron population (6) develops later in expiration, forming the second expiratory (E2) phase. Aug-E neuron activity produces a rise in inhibitory conductance in post-I neurons, further reducing their spike frequency (7), and in early-I neurons (8), transiently preventing their reactivation. During the E2 phase, the pre-I/I neurons of the pre-BötC, which are probably only weakly inhibited by aug-E neurons and receive high excitatory drive, escape from inhibition and start firing (9). This provides strong excitation of early-I inhibitory neurons. The rapid onset of early-I spiking (10) initiates a wave of inhibition of post-I and aug-E neurons that complete the E-to-I transition. The process is then repeated. All neurons in this model receive tonic excitation, which facilitates the phase transitions by promoting escape from inhibition. These model-based descriptions of the generation of three-phase respiratory activity are generally in accord with the classical descriptions of Büsselberg et al. (2001), updated here with regard to the involvement of specific compartmental microcircuits. *Courtesy of Smith, J.C., Abdala, A.P., Borgmann, A., Rybak, I.A., Paton, J.F., 2013. Brainstem respiratory networks: building blocks and microcircuits. Trends Neurosci. 36 (3):152–162. https://doi.org/10.1016/j.tins.2012.11.004.*

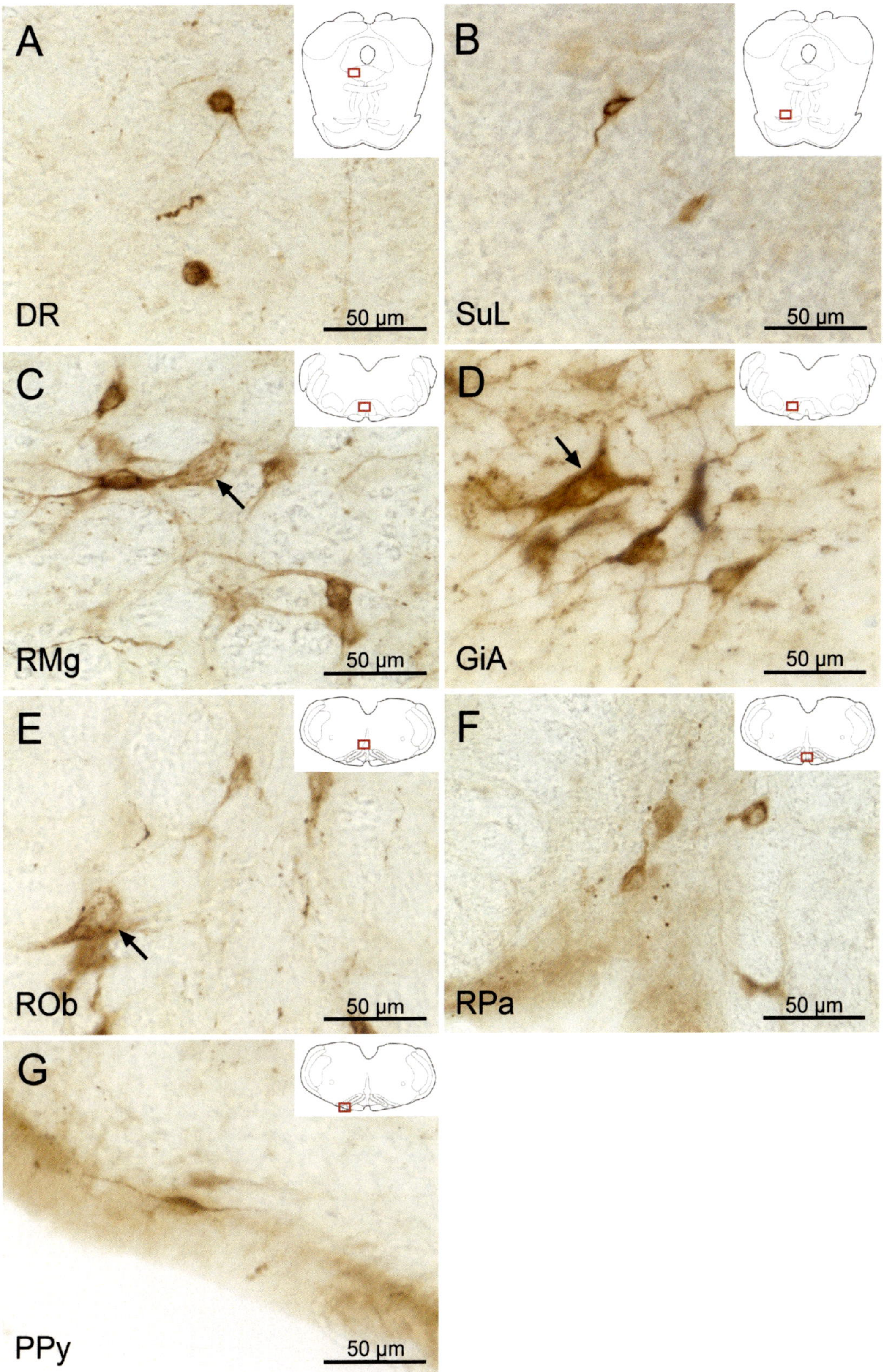

FIGURE 5B.11 **Photomicrographs of CTB-immunoreactive neurons—CTB (cholera toxin b subunit: retrograde tracing)—in each raphe nucleus of #arVRG3**. A–F show high magnification images of CTB-immunoreactive neurons. *Arrows* indicate the large size of CTB-immunoreactive neurons (greater than 30 μm). *Rectangles in insets* show the location of each panel. (A) DR, dorsal raphe nucleus; (B) SuL, supralemniscal nucleus; (C) RMg, nucleus raphe magnus; (D) GiA, nucleus gigantocellularis pars alpha and pars ventralis; (E) ROb, nucleus raphe obscurus; (F) RPa, nucleus raphe pallidus; (G) PPy, parapyramidal area. *Courtesy of Morinaga, R., Nakamuta, N., Yamamoto, Y., 2019. Serotonergic projections to the ventral respiratory column from raphe nuclei in rats. Neurosci. Res. 143, 20–30. https://doi.org/10.1016/j.neures.2018.05.004.*

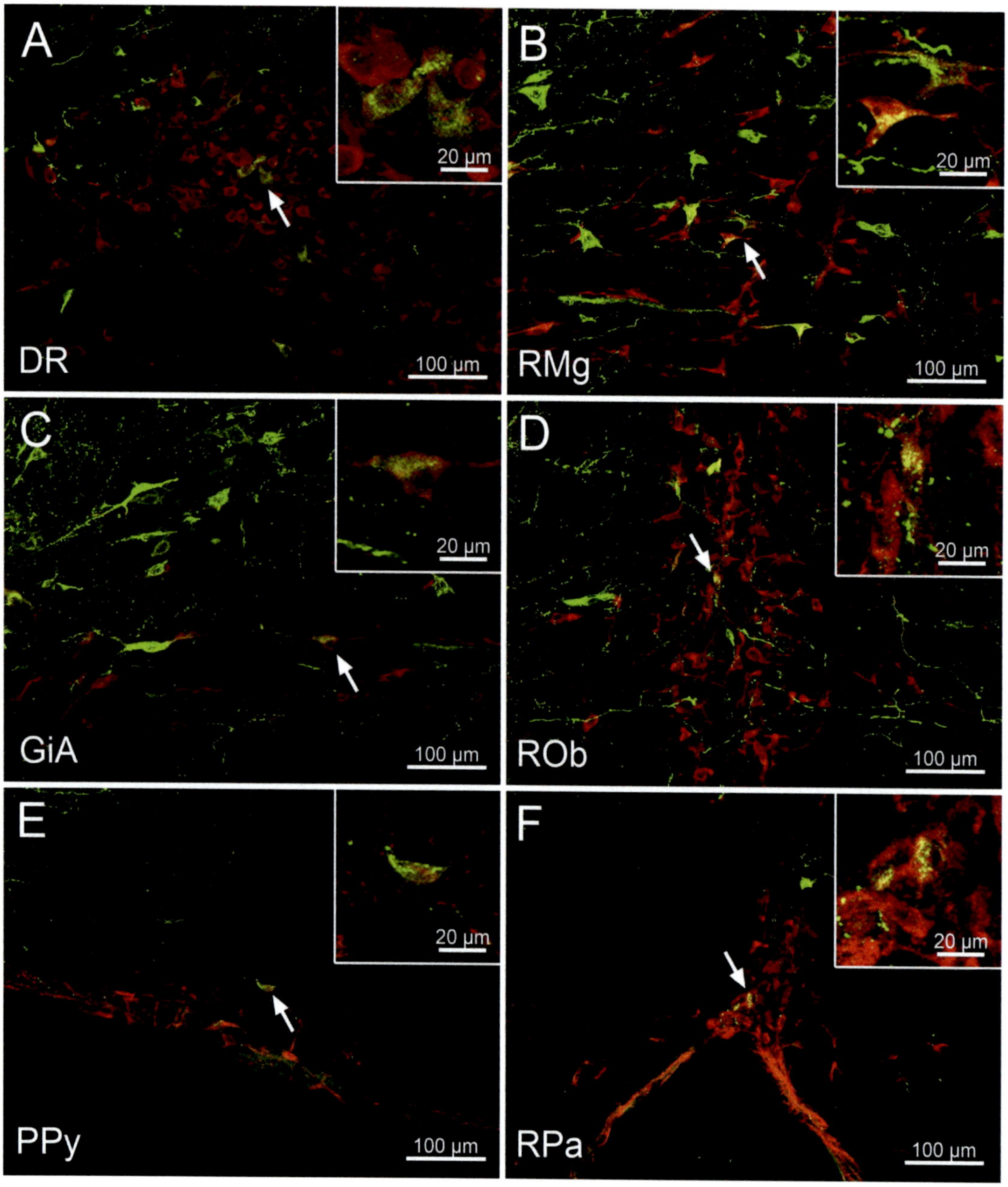

FIGURE 5B.12 **Z-stacks of confocal microscopic images of double immunoreactive neurons with CTB (cholera toxin b subunit: retrograde tracing) and TPH2 (tryptophane hydrohylase 2: serotonin enzyme synthesis) in raphe nuclei of #arVRG3**. (A–F) Low magnification images of double immunoreactive neurons with CTB (*green*) and TPH2 (*red*) in each raphe nucleus. *Insets* show high magnification images of the regions indicated by arrows. (A) DR, dorsal raphe nucleus; (B) RMg, nucleus raphe magnus; (C) GiA, nucleus gigantocellularis pars alpha and pars ventralis; (D) ROb, nucleus raphe obscurus; (E) PPy, parapyramidal area; (F) RPa, nucleus raphe pallidus. *Courtesy of Morinaga, R., Nakamuta, N., Yamamoto, Y., 2019. Serotonergic projections to the ventral respiratory column from raphe nuclei in rats. Neurosci. Res. 143, 20–30. https://doi.org/10.1016/j.neures.2018.05.004.*

Previous immunohistochemical studies revealed that serotonergic neurons in the rostral and caudal raphe nuclei developed independently, and formed ascending projections to the forebrain and descending projections to the spinal cord respectively, right after the appearance of serotonergic neurons (Aitken and Törk, 1988; Wallace and Lauder, 1983) Thus, serotonergic neurons in the caudal raphe nuclei may also form projections to the VRC compartments during the process of the formation of descending projections to the spinal cord.

THE LATERAL COLUMN

The lateral column or lateral tegmental field lies lateral to the medial column described before. Unlike the latter, the lateral column is mostly made of parvicellular elements. The various nuclei that compose it (Fig. 5b.13A and B) have a fairly homogenous neurotransmitter phenotype.

The lateral column nuclei occupy the medulla oblongata, pons, and mesencephalon–pons junction. At the level of medulla oblongata, we find the *lateral reticular nucleus* (LRt) and the *parvicellular reticular nucleus* (PCRt); at the level of the pons, we find the *medial parabrachial nucleus* (MPBN) and the *locus coeruleus* (A6-LC); and at the level of the mesencephalon–pons junction, we find the *lateral parabrachial nucleus* (LPBN) and the *pedunculopontine nucleus* (PPTg). In the following pages, the anatomical–functional characteristics of the nuclei aforementioned will be described in detail.

MEDULLA OBLONGATA NUCLEI

Lateral Reticular Nucleus[5]–Parvocellular Reticular Nucleus[6]

The *lateral reticular nucleus* (LRt) is located caudally to the PCRt and wraps the A1 noradrenergic nucleus. It can be divided into three cytoarchitectonic parts: magnocellular (LRtMC), parvicellular (LRtPC), and subtrigeminal (LRtS5). The magnocellular part is oriented ventromedially within the nucleus, dorsal to the inferior olive; the parvicellular part lies dorsolateral to it. The parvicellular part exhibits reactivity for AchE in islets near the surface of the lateral caudal hindbrain, immediately external to the LRt. The subtrigeminal part consists of large cells that are positive for AChE and tyrosine hydroxylase (TH).The *parvocellular reticular nucleus* (PCRt) is located just above the LTr, laterally to the VRt and to the Gi. It extends caudorostrally and is included between the territory of the medulla oblongata and pons. Throughout its length, it relates to some sensory–motor nuclear groups as well as the nucleus ambiguous, the nucleus of the facial nerve, and the nucleus of the trigeminal nerve. In addition, some adrenergic (C1) and noradrenergic (A5) cell groups are also found in its length. The PCRt contains interneurons that play an important role in various aspects of orofacial motor functions, including suckling, chewing, and swallowing. These interneurons transmit primary oromotor commands to the motoneurons that innervate the masticatory muscles. Several immunohistochemical studies have reported that glutamatergic premotor neurons targeting the trigeminal motor nucleus (MoV) can be found in the rostral PCRt (rPCRt) in rats (Li et al., 1996; Pang et al., 2009; Travers et al., 2005). In addition, electrophysiological reports have shown that premotor neurons, which antidromically respond to MoV stimulation, are also found in the rPCRt (Donga et al., 1991; Westberg et al., 1995).

Nagoya et al. (2017) showed that the majority of *Phox2b*$^+$ neurons—*Phox2b* gene encodes for a transcription factor that is fundamental in the differentiation of the peripheral autonomic nervous system, a part of the cranial nerves, and lower brainstem nuclei (Dauger et al., 2003; Pattyn et al., 1997, 1999)—exhibit a glutamatergic phenotype, little spontaneous activity, low-frequency firing features, and direct projection to the MoV, suggesting that the neurons of the reticular formation dorsal to the MoV (RdV) contains *Phox2b*$^+$ neurons that are candidates for excitatory premotor neurons targeting trigeminal motoneurons. However, the physiological roles and characteristics of *Phox2b*$^+$ rPCRt neurons remain unexplored, even if they seem to contribute to suckling and/or masticatory rhythm generation (Kajiwara et al., 2022). Moreover, it seems that the PCRt also plays a role in facilitating the jaw-opening reflex by stimulation of the red nucleus (RN) (Satoh et al., 2006). Some afferences to the PCRt from the red nucleus (RN) of the mesencephalon have been described (Yasui et al., 2001), and lastly, some descending fibers from the *substantia nigra pars reticulata* (SNpr) to the premotor neurons of the PRCt, which projects to the MoV, have been identified as well (Yasui et al., 1997).

[5]Also known as nucleus reticularis lateralis.

[6]Also known as parvocellular reticular nucleus, nucleus parvocellularis, nucleus reticularis parvicellularis, nucleus reticularis parvocellularis.

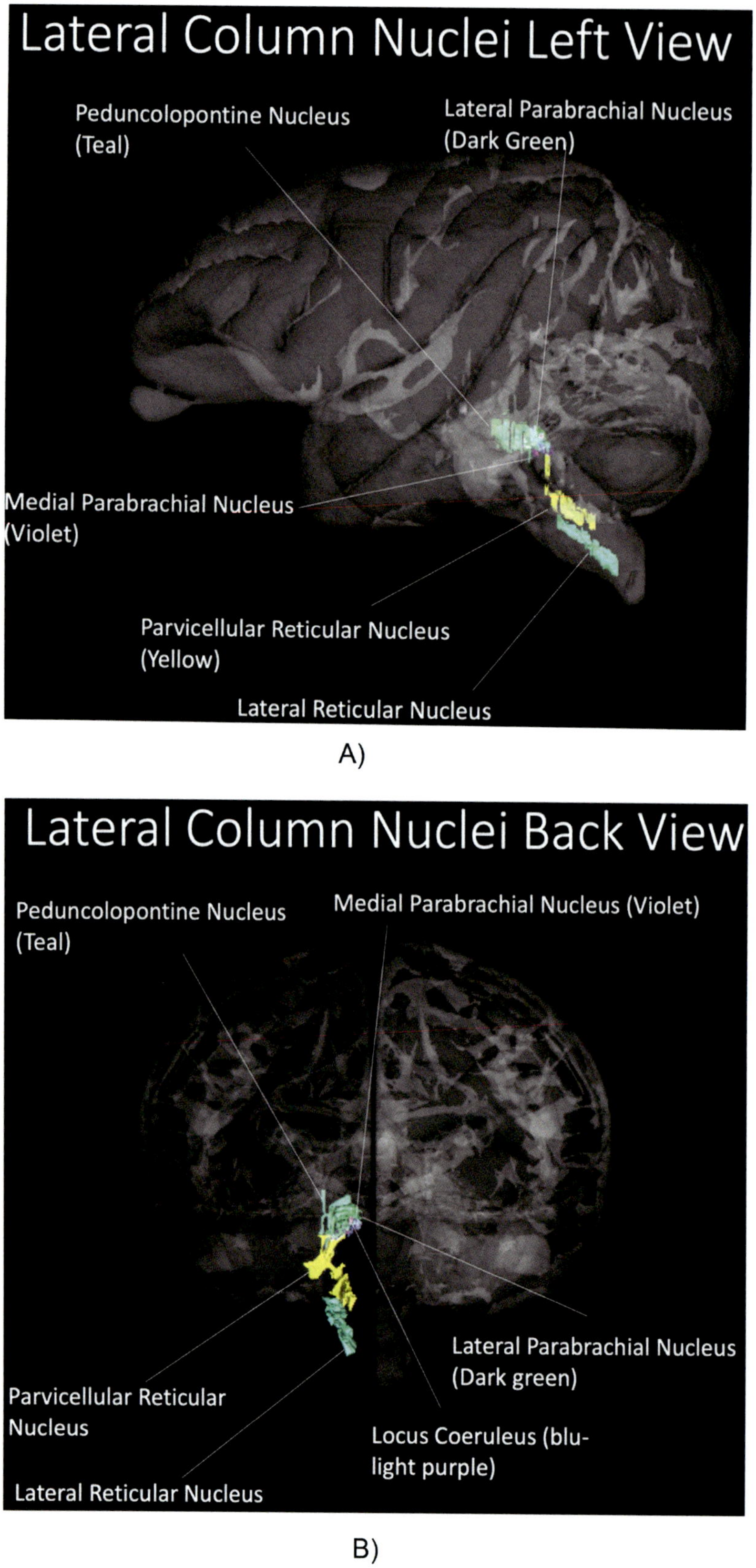

FIGURE 5B.13 **(A) Left-lateral view of the lateral column nuclei of the reticular formation in *Macaca mulatta* (rhesus macaque) brain model. (B) Back view of the lateral column nuclei of the reticular formation in *Macaca mulatta* (rhesus macaque) brain model.** *Image created by BrainInfo (1991–present), National Primate Research Center, University of Washington, http://www.braininfo.org.*

MESENCEPHALON–PONS JUNCTION

Medial Parabrachial Nucleus[7]–Lateral Parabrachial Nucleus[8]–Pedunculopontine Nucleus (PPTg)[9]–Locus Coeruleus (A6-LC)

Brief Note Concerning the Parabrachial Nuclei

The *parabrachial nuclei*, also known as the *parabrachial complex*, are a group of nuclei in the dorsolateral pons and pons–mesencephalon junction that surrounds the superior cerebellar peduncle as it enters the brainstem from the cerebellum. They are named from the Latin term for the superior cerebellar peduncle, the *brachium conjunctivum*. In the human brain, the expansion of the superior cerebellar peduncle expands the parabrachial nuclei, which form a thin strip of gray matter over most of the peduncle. The parabrachial nuclei in humans are typically divided along the lines suggested by Baxter and Olszewski into a medial parabrachial nucleus and a lateral parabrachial nucleus (Olszewski, 1954). These have in turn been subdivided into a dozen subnuclei depending on the experimental model used to perform the anatomical studies. Anyway, in the following, we will describe the three principal parabrachial nuclei: the medial, lateral, and subparabrachial nuclei.

The *medial parabrachial nucleus* (MPBN) is located laterally to the *oral pontine reticular nucleus* (PnO) and caudally to the *lateral parabrachial nucleus* (LPBN). MPBN is one of the major targets of spinal projection neurons in the brain, and it has been shown in rodents that 95% of lamina I projection neurons in the spinal cord project to the MPBN (Todd, 2010). The spinoparabrachial pathway has been implicated in the processing of nociceptive information, and its functional role in pain is further supported by a recent study indicating that excitatory inputs to the MPBN are critical for the affective component of pain (Huang et al., 2019, Bernard et al., 1996; Bester et al., 2000; Jansen and Giesler, 2015; Williams and Ivanusic, 2008). Interestingly, in contrast to many other ascending pathways, the spinal cord sends projections to the bilateral MPBN (Hylden et al., 1989a,b; Li and Li, 2000; Liu et al., 2009; Spike et al., 2003); thus, the ipsilateral and contralateral spinoparabrachial pathways are involved in nociceptive information processing.

Deng et al. (2020) demonstrated that the spinal cord connects with the MPBN in a bilateral manner and that the ipsilateral spinoparabrachial pathway is critical for nocifensive behavior. Moreover, he identified Tacr1-expressing neurons as the major neuronal subtype in the MPBN, which receives direct spinal input, and showed that these neurons are critical for processing nociceptive information.

The functional role of the MPBN has been demonstrated in both electrophysiological and behavioral studies (Barik et al., 2018; Bourgeais et al., 2001a,b; Chiang et al., 2019, 2020; Maeda et al., 2009; Palmiter et al., 2018), and several neuronal subtypes in the MPBN are involved in this process. It has been shown that the activity of calcitonin gene–related peptide (CGRP)–positive neurons in the MPBN increased in response to noxious stimuli (Campos et al., 2018), and tachykinin 1 (Tac1)-positive neurons were found to modulate nocifensive behaviors via a descending pathway (Barik et al., 2018; Roeder et al., 2016). A recent study showed that dynorphin (Pdyn)-positive neurons in the MPBN may convey nociceptive signals from the spinal cord to the lateral parabrachial nucleus (LPBN, see later) via a local circuit (Chiang et al., 2020). Most neurons of the spinal cord that project to the MPBN originate from Rexed's lamina I. These neurons can be morphologically classified as multipolar, fusiform, or pyramidal, based on the shape of their cell bodies and the characteristics of their proximal dendrites in the horizontal plane. There is also evidence that both multipolar and fusiform cells are nociceptive, while pyramidal neurons are nonnociceptive (Almarestani et al., 2007). This same group of investigators wanted to see which types of lamina I neurons belong to the spinoparabrachial tract (SPB) in the rat and characterized them based on the presence or absence of neurokinin 1 receptor (NK-1r) immunoreactivity.

The relative distribution of neurons belonging to the SPB tract was reported as follows: 38.7% multipolar, 36.8% fusiform, 22.7% pyramidal, and 1.9% unclassified (Almarestani et al., 2007).

However, the specific neuronal subtype within the MPBN that is targeted by spinal projections and the precise involvement of these neurons in the processing of nociceptive information is still unknown. The MPBN sends direct projections to multiple targets, including the central nucleus of the amygdala (CeA), the intralaminar thalamic nucleus (ILN), and the ventral tegmental area of the hypothalamus (Chiang et al., 2019; Krout and Loewy, 2000a,b), and it relays information from the gustative area of the solitary nucleus to the ventroposteromedial nucleus (VPM) of the thalamus (Thomas et al., 2009).

[7]Also known as nucleus parabrachialis medialis.

[8]Also known as nucleus parabrachialis lateralis.

[9]Also known as peduncular pontine nucleus, pedunculopontine tegmental nucleus, pedunculotegmental nucleus, nucleus pedunculopontinus, nucleus tegmenti pedunculopontinus.

The CeA is one of the major targets of the MPBN, and it has been proposed that the projection from the MPBN to the CeA is critical for the affective component of pain (Missig et al., 2017; Veinante et al., 2013); this is supported by a behavioral study (Han et al., 2015). The ILN represents another major downstream target of the MPBN in pain processing (Harte et al., 2004; Munn et al., 2009; Thoden et al., 1979), and its spontaneous hyperactivity was detected in patients with deafferentation pain (Rinaldi et al., 1991). Consistently, MPBN neurons, which send projections to the ILN, were shown to be activated by noxious stimuli (Bourgeais et al., 2001a,b).

Furthermore, MPBN neurons receiving spinal input form functional monosynaptic excitatory connections with neurons in the ILN, but not with the amygdala. All of this collectively delineates the neural circuitry underlying nocifensive behavior, offering crucial insights into the circuit-level mechanisms involved in the processing of nociceptive information.

Important: The results of the studies mentioned in the previous explanation do not differentiate between the medial PBN (MPBN) and lateral PBN (LPBN) since such distinction does not seem to exist in mouse models. However, based on the functional characteristics attributed to the medial PBN, it is more appropriate to associate the reported results with the medial PBN rather than the lateral PBN. The lateral PBN appears to be more connected to the hypothalamus and involved in vegetative processes. The *lateral parabrachial nucleus* (LPBN) is one of the three main parabrachial nuclei and is located at the junction of the midbrain and the pons. It receives information from the caudal solitary tract and transmits signals mainly to the medial hypothalamus, but also to the lateral hypothalamus, and many of the nuclei targeted by the MPBN.

The LPBN is positioned dorsally to the pedunculopontine nucleus (PPTg) and is situated rostrally in relation to the A6-LC. The majority of neurons in the LPBN exhibit a distinct fusiform morphology, characterized by a spindle-shaped cell body. These neurons typically extend one to three primary dendrites in opposite directions, resembling bipolar cells. Most dendrites are smooth, with very few, moderate swellings and without any dendritic appendages. These smooth dendrites often extend without branching over long distances, without changes in their diameter. Higher-order dendrites often exhibit varicosities that are linked by very thin segments of the dendrite. The boutons are of different shape and size and can be regularly or irregularly spaced (Herbert and Flügge, 1995).

This region receives different projections from several regions of CNS including the nucleus of solitary tract (NTS) and the area postrema (AP). These inputs convey information arising from peripheral receptors, which are reprojected, after having synapsed with lateral parabrachial nucleus (LPBN), to other hindbrain and forebrain areas, including the paraventricular nucleus of the hypothalamus (PVN), the median preoptic nucleus of the hypothalamus (MnPO), and the central nucleus of the amygdala (CeA) (Ciriello et al., 1984; Fulwiler et al., 1984; Moga et al., 1990; Jhamandas et al., 1996). The lateral parabrachial nucleus (LPBN) is uniquely positioned to integrate and convey signals generated in the peripheral viscera, which then enter the CNS through the area postrema (AP) and the nucleus of the solitary tract (NTS). After their passage through the AP and the NTS, the lateral parabrachial nucleus (LPBN) receives information about the status of the systemic viscera that arise from circulating factors, cardiovascular receptors, peripheral osmoreceptors, and, perhaps, gastrointestinal receptors, all of which influence the amount of sodium and water ingested (De Gobbi et al., 2008; De Luca et al., 2003; Franchini et al., 1999; Kobashi et al., 1993; Menani et al., 1998a,b, 2000, 2004; Menezes et al., 2011).

Recently, Ryan et al. (2017) described how oxitocin receptor–expressing neurons ($Oxtr^{PBN}$ neurons) in the parabrachial nucleus are key regulators of fluid satiation. They are activated by fluid satiation and hypertonic saline injection, and they are directly innervated by oxytocin neurons in the paraventricular hypothalamus (PVH), which mildly attenuate fluid intake. Moreover, activation of the neurons in the nucleus of the solitary tract (nst) substantially suppresses fluid intake and activates $Oxtr^{PBN}$ neurons. These results seem to demonstrate that oxytocin plays a key role in regulating the function of LPN.

Both the parabrachial nuclei (medial and lateral) are involved in the control of sodium and water intake (Johnson and Thunhorst, 1997; Johnson, 2007); in fact, some experiments performed on animal models showed that cells in the lateral parabrachial nucleus (LPBN) are activated after ingestion of sodium solutions by dehydrated rats or rats that received intragastric loads of hypertonic NaCl (Franchini et al., 1999; Kobashi et al., 1993, Yamamoto et al., 1993). The distension of an intravascular balloon positioned at the superior vena cava and right atrial junction increases the expression of c-Fos protein in the LPBN, suggesting that the LPBN is also activated by signals from cardiovascular receptors (De Gobbi et al., 2008). Both warm- and cold-sensitive neurons are found in LPBN, a crucial relay for skin temperature information from the spinal cord to the preoptic area (Xu et al., 2019).

The *pedunculopontine nucleus* (PPTg) is ventral to the LPB and is located at the same level as the DRN. It lies caudal to the substantia nigra and adjacent to the superior cerebellar peduncle. It has two divisions of subnuclei: the *pars compacta,* containing mainly cholinergic neurons, and the *pars dissipata* containing mainly glutamatergic neurons and some noncholinergic neurons (French et al., 2018).

PPTg neurons project axons to a wide range of areas in the brain (Mena-Segovia et al., 2011), particularly parts of the basal ganglia, such as the subthalamic nucleus, the SNpc, and the globus pallidus internus (GPi). It also sends projections to targets in the thalamus, the cerebellum, the basal forebrain, the lower brainstem, and the cerebral cortex, particularly, the supplementary motor area (BA 6 and surrounding areas), the somatosensory area (BA 3,1,2), and the motor cortex (BA 4) (Garcia-Rill et al., 1991; Winn et al., 2006; Tsang et al., 2010; Edley et al., 1983; Woolf et al., 1989; Rye et al., 1988; Mena-Segovia et al., 2004a,b; Moriizumi et al., 1992; Semba et al., 1992; Groenewegen et al., 1993).

PPTg is one of the main components of the ascending reticular activating system (ARES) (Garcia-Rill, 1991; Winn, 2006), and its cholinergic neurons are involved in the regulation of motor control (e.g., locomotion, posture, gaze) and cognitive processes (attention, learning and memory). The PPTg is highly interconnected with several regions of the basal ganglia, and one of its key functions is to regulate their activity. The SNr, through its GABAergic projections to the PPTg, controls the activation of cholinergic and noncholinergic PPTg neurons projecting to different targets. Mixed cholinergic and glutamatergic projections then return from the PPTg to the SNr and GPi, and to the external globus pallidus (GPe) (Clarke et al., 1987; Charara et al., 1994; Lavoie et al., 1994; Saitoh et al., 2003; Kang et al., 1990; Granata et al., 1991). The STN innervates the PPTg through glutamatergic projections, and the PPTg sends both glutamatergic and GABAergic projections back to the STN (Hammond et al., 1983; Bevan et al., 1995). Moreover, the PPTg projection to the SNpc neurons induces the release of dopamine in the striatum, regulating its activity(for details, the reader is referred to Chapter 15).

The PPTg also controls the activity of the striatum through its bilateral projections to the caudal intralaminar nuclei of the thalamus (Erro et al., 1999; Barroso-Chinea et al., 2011).

The PPTg is included in the mesencephalic locomotion region (MLR) that regulates walking and running (Skinner, 1984; Sherman, 2015). Some studies showed that low-threshold electrical stimulation of the PPTg reduces locomotion, and it has been accepted that PPTg acts as a relay from the basal ganglia to the spinal cord (Garcia et al., 1987; Takakusaki, 2004a); new findings show that it can also control the postural muscles tone (Mena-Segovia, 2004; Takakusaki, 2004; Kobayashi, 2004). Regarding this last aspect, projections from the PPTg to the deep cerebellar nucleus have been anatomically identified in rats (Woolf et al., 1989; Newman et al., 1992; Ruggiero et al., 1997). The same connections have been reported in humans too, even if their role is not completely understood (Aravamuthan, 2007), and this raises the possibility that PPTg acts as an interface between the cerebellum and the basal ganglia to influence motor control and cognitive functions. It might relay reward and reinforcement learning information from the basal ganglia to the cerebellum, playing also a crucial role in generating and regulating postural tone and stabilizing posture (Suzuki, 2012).

PPTg neurons are active during walking and REM sleep (Garcia, 2007). Cholinergic activation in the ARAS leads to increased Ach release throughout the reticular formation as well as in the SN, the basal forebrain, the thalamus, and the cerebellum (Garcia- Rill, 1997). These results highlight how the basal ganglia are also involved in regulating the sleep–wake cycle (Mena-Segovia and Cintra, 2002) and how their reciprocal relationship with PPTg is crucial for this function (Mena-Segovia and Giordano, 2003; Takakusaki, 2004a). As a result, it has been seen that PPTg promotes sleep–wake transition as part of the ARAS and works as a crucial element in the generation and maintenance of the rapid rhythms in the cortex, which are associated with wakefulness and REM sleep.

The *locus coeruleus* (A6-LC) is the principal site for brain synthesis of noradrenaline (NA) and its efferences form part of the ascending reticular activating system (ARES). It is responsible for most of the NE activity in the brain and is involved in responses to stress and panic. Studies have shown that sensory stimuli—i.e., smell, hearing, sight, taste and touch, and bilateral masticatory activity (Olszewski, 1954)—accelerate the excitation of the neurons of the A6-LC. This brain area is also closely linked to wakefulness activity and REM sleep. For an in-depth discussion regarding the functional role of this nucleus in the brain, see Chapter 23 *appendix*: "The noradrenergic modulation of attention and memory: The Attentional Boost Effect").

A6-LC is located in the posterior area of the rostral pons, in the lateral floor of the fourth ventricle, and it is lateral to the PnO and caudal to the LPBN. Mature A6-LC neurons are predominantly medium-sized cells (35–45 μm) with a fusiform and multipolar morphology, with relatively sparse ramifications, typically three or four long, thin dendrites with one or two proximal branches (Chan-Palay and Asan, 1989a,b; Patt and Gerhard, 1993; Arendt et al., 1997).

By contrast, A6-LC neuronal axons, particularly those projecting to the forebrain, exhibit extensive bifurcations, suggesting that a single neuron can innervate multiple regions along the neuraxis. In addition to the majority of medium-sized perikarya[10] within the A6-LC, the more sparsely distributed ventrolateral neurons of the caudal A6-LC and subcoeruleus are interspersed with smaller pigmented neurons (15 μm), which are typically spindle-shaped, with two tufts of dendrites (Chan-Palay and Asan, 1989a,b; Patt and Gerhard, 1993).

Collectively, these caudal noradrenergic cells project primarily to the spinal cord and are mostly localized within the brainstem (Westlund and Coulter, 1980).

Consistent with their noradrenergic phenotype, primate's A6-LC neurons have been shown to contain the NE synthetic pathway enzymes tyrosine hydroxylase (TH) and dopamine beta-hydroxylase (DBH), as well as the NE transporter NET (Ordway et al., 1997), the NE catabolic enzyme monoamine oxidase (Westlund et al., 1985), and the α2 adrenoceptor; this latter presumably serves as an autoreceptor (Probst et al., 1984). In addition, A6-LC neurons express a wide range of neuropeptides including neuropeptide (NPY), calcitonin gene–related peptide (CGRP), cholecystokinin (CCK), and somatostatin (SOM) (Palkovits et al., 1982; Fodor et al., 1992; Pau et al., 1998). The neuropeptide galanin is found in the rodent A6-LC neurons (Perez et al., 2001) but not in nonhuman primates (Kordower et al., 1992).

Interestingly, galanin-positive neurons have been reported only in the human subcoeruleus nucleus (Kordower et al., 1992; Miller et al., 1999), indicating a species-specific variation in the distribution of this neuroactive peptide. These neuropeptides potentially serve as noradrenergic cotransmitters, and notably, A6-LC neurons have also been shown to express glial-derived growth factor (GDNF) (Del Fiacco et al., 2002), which plays a critical role in noradrenergic cell survival (Quintero et al., 2004; Pascual et al., 2008).

The main afferent connections of the A6-LC have their origin in the ventrolateral paragigantocellular nucleus, which contains the A5 group, and from a cell group at the medial edge of the nucleus prepositus hypoglossi, in the floor of the fourth ventricle (group EF, the epifascicular nucleus; Koutcherov et al., 2004).

A projection from the lamina I neurons of the dorsal horn was traced to the A6-LC by Westlund and Craig in the monkey (Westlund et al., 1996), and numerous regions project to the peri-LC, included the central nucleus of the amygdala (CeA), the preoptic area (POA), different hypothalamic nuclei, the periaqueductal gray (PAG), the nucleus of the solitary tract (Sol), the prefrontal cortex, the adrenergic cell group C1 (Card et al., 2006), and Barrington's micturition area. The A6-LC is densely innervated by hypocretin/orexin fibers, originating from the perifornical region of the hypothalamus (Espana et al., 2005). An indirect projection from the suprachiasmatic nucleus to the A6-LC was established by transneuronal retrograde tracing with pseudorabies virus in the rat (Aston-Jones et al., 2001; Chen et al., 1999). The projections of the A6-LC cover the entire central nervous system. Its fibers ascend in the dorsal and ventral catecholaminergic bundles (Jones et al., 1983), following a similar course as the serotonergic system. Thus, the main projections in the forebrain include olfactory structures, neocortex, hippocampus, thalamus, basal forebrain, preoptic area, and hypothalamus.

The *subparabrachial nucleus* is also known as the *Kölliker–Fuse nucleus (KF)*. This nucleus is described in this chapter in the paragraph titled "*Control of the respiratory function by reticular formation.*"

ADRENERGIC AND NORADRENERGIC CELLULAR GROUPS

The adrenergic and noradrenergic cell groups are distributed throughout the entire length of the reticular formation, from the medulla oblongata to the mesencephalon. In rodents, both groups are organized into dorsolateral and ventrolateral clusters.

The ventrolateral group consists of the A1 noradrenergic group, as described by Dahlström and Fuxe (1964a), and the C1 adrenergic group, as described by Hökfelt (1974). The A1 group is surrounded by the lateral reticular nucleus (LRt) of the lateral column, while the C1 group extends caudorostrally along the lateral portion of the parvicellular reticular nucleus (PCRt) and encompasses the lower part of the facial motor nucleus (VII c.n.).

The A2 and C2 groups are located between the nucleus of the solitary tract and the parvicellular reticular nucleus (PCRt) and along the borders of the dorsal vagal nucleus, respectively. The C3 group is only present in rodents, but not in primates, and it lays on the median line and the floor of the fourth ventricle, at the rostral level of the medulla oblongata. Rostrally, the noradrenergic neurons are primarily located in the locus coeruleus (A6-LC) and the periaqueductal gray (PAG). However, they are also scattered in the nucleus subcoeruleus and the subependymal extension of the A6-LC, forming the A4 group. This latter is found caudally, at the edge of the lateral recess of the fourth ventricle in the macaque, extending beneath the floor of the ventricle, where its

[10]Perikarya refers to the cell bodies of neurons, which are the central part of the neuron containing the nucleus and other cellular organelles. It is the main region where metabolic processes occur and where signals from other neurons are integrated before being transmitted to other parts of the nervous system.

neurons merge with noradrenergic cells of the A6-LC (Felten and Sladek, 1983).

The A1 group receives afferents from the lamina I neurons of the spinal dorsal horn and, with its projection to the parabrachial nuclei, may serve as a relé in an alternative system mediating emotional and motivational aspects of pain (Craig et al., 1995; Lima et al., 1991; Pan et al., 1999).

Apart from their projections to local structures, the A1, A2, C1, and C2 groups give rise to ascending fibers, which converge in a dorsal noradrenergic bundle (Jones et al., 1983). A periventricular system from the A1 and A2 cell groups ascends toward the medial nuclei of the thalamus.

Both A1 and A2 project to the region of the A5 group, in the ventrolateral paragigantocellular nucleus (PGi). The projection of the A2 group to this region consists of branching axons, which also terminate in the central nucleus of the amygdala (Reyes et al., 2006).

Other neurons of the A1 (and C1) and A2 groups send branching axons to the central nucleus of the amygdala (CeA), to the bed nucleus of the stria terminalis, and to the paraventricular nucleus of the hypothalamus (Petrov et al., 1993; Woulfe et al., 1988, 1990). These connections are part of a more extensive system of projections of the A1 and the A5 groups to the parabrachial nuclei, the Köelliker–Fuse area, the periaqueductal gray (PAG), the hypothalamus, and the vascular organ of the lamina terminalis (OVLT) (Byrum et al., 1987; Chou et al., 2002; Petrov et al., 1993; Woulfe et al., 1988, 1990; Card et al., 2006). Topographically the region of the medulla oblongata containing the aforementioned nuclei is called ventrolateral superficial medulla oblongata (VLSM). This region and the adjoining cell groups in and around the ambiguous nucleus are involved in cardiovascular and respiratory control and may influence sensory transmission in the spinal dorsal horn. In particular, the control of sensory and nociceptive transmission is thought to be mediated by A5 noradrenergic cells group, which give rise to a projection that ends on the dorsal horn neurons and on the intermediolateral cell column, inhibiting the neurons of the laminae I and II through α_2-adrenergic receptors. The analgesic effect produced by the stimulation of the descending noradrenergic system is blocked by intrathecal injection of α_2-, but not α_1-, adrenergic receptor antagonists (Budai et al., 1998; Nuseir and Proudfit, 2000; Yaksh, 1985).

Even if Tavares et al. (1997) sustain that the A5 projection is only inhibitory, several other functions in which A5 can be involved remain largely unknows. Moreover, all this complex system may link nociceptive input to visceromotor functions; however, whether the A5 projection to the thoracic visceromotor columns is inhibitory is not known. The presence of a lamina I projection to the VLSM in monkeys (Craig et al., 1995) suggests that a similar recurrent system may be present in primates.

The projections from A1 and A2 (and from the adrenergic C1 group; Card et al., 2006) to the paraventricular nucleus of the hypothalamus mainly target the parvocellular part of this nucleus (Sawchenko et al., 1982), whose cells secrete vasopressin and corticotropin-releasing hormone (CRH). Moreover, the connection of a population of estrogen-sensitive A1 neurons with gonadotropin-releasing hormone (GnRh) neurons in the preoptic area may be involved in the production of the GnRh/luteinizing hormone surge (Rawson et al., 2001).

Direct projections of the A1 and A2 groups to the spinal cord appear to be absent; however, these spinal projections and/or those from other neurons in the ventrolateral medulla are shunted through the A5 group in the ventrolateral paragigantocellular nucleus. A mainly ipsilateral projection of neurons from the adrenergic C1 group, via the lateral funiculus, to the thoracic intermediomedial nucleus, was described by Card et al. in the rat (Card et al., 2006).

The noradrenergic innervation of the cerebellum is mostly directed to its cortex, and it is present as a randomly oriented plexus in all three cortical layers (Fritschy et al., 1989; Grzanna et al., 1989). Synaptic contacts with Purkinje cell and granule cell dendrites are made by more than 50% of the varicosities (Olschowka et al., 1981).

The A5 group is located in the paragigantocellular nucleus (PGi), while the A7 is located in the pontine reticular formation, ventral to the superior cerebellar peduncle in rodents (Dahlström and Fuxe, 1965) and in primates (Felten and Sladek, 1983).

Strips of noradrenergic cells interconnect A5, A6, and A7.

Neurons of A7 and A5, like those in the locus coeruleus (A6-LC), receive a projection from the PAG, and A7 neurons project to the spinal cord (Bajic et al., 2001; Baker et al., 1991). Both groups receive a projection from the medullary dorsal reticular formation, a supraspinal, pain-modulating area (Leite-Almeida et al., 2006); in addition, A5 receives afferents from the paraventricular nucleus of the hypothalamus (PVN), the area of Köelliker–Fuse, the parabrachial nuclei, and the nucleus of the solitary tract (Byrum et al., 1987; Hosoya et al., 1990).

The locus coeruleus (A6-LC), A5, and A7 give rise to descending systems to the spinal cord: in the monkey, Westlund et al. found a projection from the A6-LC, through the ipsilateral ventrolateral funiculus. and some of its fibers extending in the dorsal funiculus terminated in the dorsal horn. Fibers from A7 descend in the ipsilateral ventrolateral funiculus and the contralateral dorsolateral funiculus; fibers from both nuclei cross the midline at all levels of the cord. Terminations of fibers from the locus coeruleus (A6-LC) and A7 are present in the laminae IV–VIII of the dorsal horn and the intermediate zone, and are heaviest in the motoneuronal cell groups of the lamina IX. Lighter terminations are present around the central canal and in superficial laminae of the dorsal horn. With respect to their terminations in autonomic cell groups, the two nuclei exhibit differences. The locus coeruleus (A6-LC) projects bilaterally to the sacral intermediomedial nucleus and the lateral nuclei of the lateral cord. On the other hand, A7 only projects to the thoracic intermediolateral column. The spinal projection from A5 is a relatively modest one (Hirota et al., 2001; Westlund and Dan Coulter, 1980; Westlund et al., 1982, 1983, 1984), and according to Fritschy and Lyons et al., projections from the A6-LC of the rat to the ventral horn and the thoracic intermediolateral nucleus are scarce; these projections mainly arise from cell groups in A5 and A7 (Fritschy et al., 1987; Lyons et al., 1988, 1989).

In rats, the locus coeruleus (A6-LC) innervates the dorsal horn and the intermediate zone, and in laminae I and II, its fibers descend over the entire length of the spinal cord. A5 and A7 fibers also travel in the ventral and dorsolateral funiculi to terminate in the intermediate zone, the ventral horn, and the intermediolateral cell column (Fritschy et al., 1990). A dense projection of the locus coeruleus to the motoneuronal cell columns with a moderate termination in medial laminae VII and VIII, in lamina X, and the base of the dorsal horn, and a sparse innervation of the superficial dorsal horn, was found in similar experiments in the rat by Proudfit and Clark (Clark et al., 1991; Proudfit et al., 1991). According to these authors, the projection of A7 is restricted to the laminae I–IV of the dorsal horn (Clark et al., 1991), while the A5 projection is directed to the laminae IV–VII of the dorsal horn and to the thoracic intermediolateral nucleus (Clark et al., 1993). In the motoneuronal cell columns (ventral horn) and the intermediolateral nucleus, most synapses are axodendritic, while in the somatosensitive cell column (dorsal horn), the noradrenergic innervation is mainly nonsynaptic (Rajaofetra et al., 1992a,b; Ridet et al., 1993). A projection of the A5 group to the locus coeruleus (A6-LC) appears to be absent, but many neurochemicals systems take their origin from the ventrolateral paragigantocellular nucleus, where neurons were positive for adrenalin, GABA, met-enkephalin, and corticotropin-releasing hormone (CRH). The paragigantocellular region Gi also provides a potent excitatory amino acid input to the A6-LC (Aston-Jones et al., 1991), and met-enkephalin and GABA coexist in certain neurons (Van Bockstaele et al., 1997). Afferents to the ventrolateral paragigantocellular nucleus arise from a wide variety of nuclei with autonomic, visceral, and sensory-related functions (Van Bockstaele et al., 1989). The epifascicular group (EF) inhibits neurons of the A6-LC via a GABAergic pathway (Aston-Jones et al., 1986, 1991; Ennis et al., 1989).

The cerebellar noradrenergic innervation of the cerebellum and the inferior olive stems from the A6-LC and the groups A5 and A7 (Voogd et al., 1998).

The noradrenergic system pairs topographic specificity to the branching of individual noradrenergic axons between widely separated targets. Its varicose axons make synaptic contacts, but are also engaged in volume transmission. In some regions, such as the spinal cord, these two modes of transmission are spatially segregated, with thin fibers without synapses prevailing in the dorsal horn, and synaptic transmission prevailing in the motoneuronal cell columns.

The medullary adrenergic and noradrenergic cell groups, with their spinal output through A5 to the intermediolateral nucleus, their connections with autonomic centers in the brainstem, and their ascending connections to the paraventricular nucleus of the hypothalamus (PVN) and the preoptic area (POA), are involved in a wide range of visceral functions, including cardiovascular control and respiration, and in the control of secretion of different hormones by the hypothalamus.

The locus coeruleus (A6-LC) is implicated in arousal and in other aspects of behavior (see Chapter 3). Connections of the A6-LC and/or A7 with sensory relé nuclei engage these centers in gating mechanisms for sensory input and pain control.

The role of the noradrenergic cell groups in the production of anesthesia was discussed by Hirota and Kushikata (2001). The projections of the rostral noradrenergic cell groups with cranial and spinal motor nuclei subserve level setting in motor control.

DOPAMINERGIC CELLULAR GROUPS

The dopaminergic neurons are located in the mesencephalon and in the diencephalon (Dahlström and Fuxe, 1964; Hökfelt et al., 1984). They are immunoreactive for tyrosine hydroxylase (TH), which is the main enzyme responsible for dopamine synthesis. In mammalian brain, dopaminergic neurons form a semicontinuous population, extending from the mesencephalon to the entire forebrain, and, like the noradrenergic neurons, they are disposed in several nuclei. The first three, A8, A9, and A10, represent the most important groups, as they are the ones from which the nigrostriatal fibers originate (see Chapter 15). Let us describe the aforementioned groups in detail in the following (Fig. 5b.14).

- The A8 group is small and, in rodents (Dahlström and Fuxe, 1965) and in primates (Felten and Sladek, 1983),

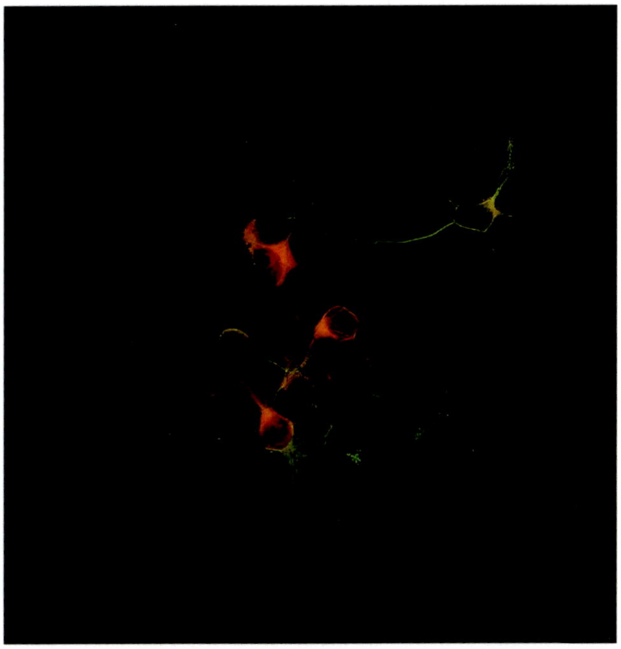

FIGURE 5B.14 **In vitro differentiation of human neural stem cells (hNSCs) into A9 dopaminergic neurons (red: TH positive neurons; green: beta 3 tubulin neuronal marker).** *Image owned by the editor-in-chief of this book.*

is located in the mesencephalic reticular formation, dorsolateral to the *substantia nigra* (SN), at the level of the *red nucleus* (RN) and caudally to it. In mice, it is identified with the retrorubral field and is defined by classical stains (Felten and Sladek, 1983).

- The A9 group is the most densely packed group of dopaminergic cell and, in rodents and primates, is located in the ventrolateral mesencephalon (Dahlström and Fuxe, 1965; Felten and Sladek, 1983) (Fig. 5b.14). It is for the most part identical to the substantia nigra pars compacta(SNpc) (see the Chapter 5a).
- The A10 group is the largest group of dopaminergic neurons located in the *ventral tegmental area* (VTA) in both rodents (Dahlström and Fuxe, 1965) and primates (Felten and Sladek, 1983).
- The A11 group is small and, in the macaque, is located in the *posterior periventricular nucleus* (PPe) and the *intermediate periventricular nucleus of the hypothalamus* (IPe) (Felten and Sladek, 1983). The neurons of this nucleus may be important in the modulation of auditory processing (Nevue et al., 2016).
- The A12 group is small and, in primates, is located in the *arcuate nucleus of the hypothalamus* (ArcH) (Felten and Sladek, 1983).[4] In rats, a few cells belonging to this group are also seen in the anteroventral portion of the *paraventricular nucleus of the hypothalamus* (PVN) (Felten and Sladek, 1983).
- The A13 group is distributed in clusters, which in primates are ventral and medial to the *mammillothalamic tract of the hypothalamus* (mthh), a few extend into the *reuniens nucleus of the thalamus* (Re) (Felten and Sladek, 1983). In mice, the A13 is located ventrally to the mammillothalamic tract of the thalamus (mthh), in the zona incerta (ZI) (Paxinos and Franklin, 2001);
- The A14 group consist of a few cells observed in and near the *preoptic periventricular nucleus* (POP) of primates (Felten and Sladek, 1983). In mice, cells in the *anterodorsal preoptic nucleus* are assigned to this group (Paxinos and Franklin, 2001).
- The A15 group exists in a few species, such as sheep and, maybe, in primates. It is located in the ventral and dorsal aspects of the *preoptic periventricular nucleus* (POP) and in the adjacent part of the *anterior hypothalamic region* (AHR). It continues caudally with the A14 group (Tillet et al., 1994)
- The A16 group is located in the olfactory bulb of vertebrates, including rodents and primates (McGeer, 1995).

CHOLINERGIC CELLULAR GROUPS

Most cholinergic nuclei of the CNS are located in the forebrain, where they establish their connections with many cortical and subcortical structures. The Ch5 and Ch6 cholinergic groups are located in the reticular formation, at the level of the junction of the pons with the mesencephalon, and correspond to the pedunculopontine nucleus (PPTg) and to the lateral dorsal tegmental nucleus (LDTg), which are situated in the periaqueductal gray (PAG) and at the level of the rostral part of the fourth ventricle, respectively (Mesulam et al., 1983a,b; Mufson et al., 1986).

Note to the reader. It should be noted that the discussion of the nuclei of the reticular formation is complex, particularly when it comes to the accurate correspondence between the serotonergic groups B (B1–B9 according to the classification by Dahlström and Fuxe, 1964) and their corresponding nuclei. In the literature, it is sometimes possible to find different correspondences between neuronal groups. Taking into consideration the nucleus raphe dorsalis, we can say that, depending on the topographical meticulousness with which it is described and in which species the nucleus is described, we can find B6-DRN, but also B7-DRN. Faced with this complexity, reader may find himself in a challenging situation. However, to clarify the situation regarding the example of the DRN, it can be said that both neuronal groups B6 and B7 correspond to nuclei within the DRN. The distinction lies in their relative positions, with B6 being located more caudally and B7 being situated more rostrally within the DRN. Furthermore, B7 seems to include both the midrostrocaudal portion and, as previously mentioned, the more rostral one. Another important aspect to consider is that most of the anatomical studies on the reticular formation have been conducted in animal models, particularly in mice, rabbits, and nonhuman primates. These studies often involve the use of retrograde and anterograde tracers, including in situ viral tracers in modern research techniques such as optogenetics and cellular transfection. However, due to ethical reasons, it is challenging to precisely map the corresponding structures in humans. Consequently, the author wants to emphasize that any discrepancies between the descriptive text and the images are a result of the aforementioned limitations.

References

Aitken, A.R., Törk, I., 1988. Early development of serotonin-containing neurons and pathways as seen in wholemount preparations of the fetal rat brain. J. Comp. Neurol. 274 (1), 32–47. https://doi.org/10.1002/cne.902740105.

Allers, K.A., Sharp, T., 2003. Neurochemical and anatomical identification of fast- and slow-firing neurones in the rat dorsal raphe nucleus using juxtacellular labelling methods in vivo. Neuroscience 122 (1), 193–204. https://doi.org/10.1016/S0306-4522(03)00518-9.

Almarestani, L., Waters, S.M., Krause, J.E., Bennett, G.J., Ribeiro-da-Silva, A., 2007. Morphological characterization of spinal cord dorsal horn lamina I neurons projecting to the parabrachial nucleus in the rat. J. Comp. Neurol. 504 (3), 287–297. https://doi.org/10.1002/cne.21410.

Alvarenga, R.M., Pires, J.G.P., Futuro Neto, H.A., 2005. Functional mapping of the cardiorespiratory effects of dorsal and median raphe nuclei in the rat. Braz. J. Med. Biol. Res. 38 (11), 1719–1727. https://doi.org/10.1590/S0100-879X2005001100022.

Aravamuthan, B.R., Muthusamy, K.A., Stein, J.F., Aziz, T.Z., Johansen-Berg, H., 2007. Topography of cortical and subcortical connections of the human pedunculopontine and subthalamic nuclei. Neuroimage 37 (3), 694–705. https://doi.org/10.1016/j.neuroimage.2007.05.050.

Arnsten, A.F.T., Goldman-Rakic, P.S., 1984. Selective prefrontal cortical projections to the region of the locus coeruleus and raphe nuclei in the rhesus monkey. Brain Res. 306 (1–2), 9–18. https://doi.org/10.1016/0006-8993(84)90351-2.

Aston-Jones, G., Ennis, M., Pieribone, V.A., Nickell, W.T., Shipley, M.T., 1986. The brain nucleus locus coeruleus: restricted afferent control of a broad efferent network. Science (1979) 234 (4777), 734–737. https://doi.org/10.1126/science.3775363.

Aston-Jones, G., Shipley, M.T., Chouvet, G., et al., 1991. Afferent regulation of locus coeruleus neurons: anatomy, physiology and pharmacology. Prog. Brain Res. 47–75. https://doi.org/10.1016/S0079-6123(08)63799-1.

Assaf, S.Y., Miller, J.J., 1978. The role of a raphe serotonin system in the control of septal unit activity and hippocampal desynchronization. Neuroscience 3 (6), 539–550. https://doi.org/10.1016/0306-4522(78)90018-0.

Aston-Jones, G., Chen, S., Zhu, Y., Oshinsky, M.L., 2001. A neural circuit for circadian regulation of arousal. Nat. Neurosci. 4 (7), 732–738. https://doi.org/10.1038/89522.

Azmitia, E., Gannon, P., Kheck, N., Whitakerazinitia, P., 1996. Cellular localization of the 5-HT receptor in primate brain neurons and glial cells. Neuropsychopharmacology 14 (1), 35–46. https://doi.org/10.1016/S0893-133X(96)80057-1.

Bajic, D., Van Bockstaele, E.J., Proudfit, H.K., 2001. Ultrastructural analysis of ventrolateral periaqueductal gray projections to the A7 catecholamine cell group. Neuroscience 104 (1), 181–197. https://doi.org/10.1016/S0306-4522(01)00052-5.

Baker, K.G., Halliday, G.M., Halasz, P., et al., 1991a. Cytoarchitecture of serotonin-synthesizing neurons in the pontine tegmentum of the human brain. Synapse 7 (4), 301–320. https://doi.org/10.1002/syn.890070407.

Baker, K.G., Halliday, G.M., Hornung, J.P., Geffen, L.B., Cotton, R.G.H., Törk, I., 1991b. Distribution, morphology and number of monoamine-synthesizing and substance P-containing neurons in the human dorsal raphe nucleus. Neuroscience 42 (3), 757–775. https://doi.org/10.1016/0306-4522(91)90043-N.

Barik, A., Thompson, J.H., Seltzer, M., Ghitani, N., Chesler, A.T., 2018. A brainstem-spinal circuit controlling nocifensive behavior. Neuron 100 (6), 1491–1503.e3. https://doi.org/10.1016/j.neuron.2018.10.037.

Barroso-Chinea, P., Rico, A.J., Conte-Perales, L., et al., 2011. Glutamatergic and cholinergic pedunculopontine neurons innervate the thalamic parafascicular nucleus in rats: changes following experimental parkinsonism. Brain Struct. Funct. 216 (4), 319–330. https://doi.org/10.1007/s00429-011-0317-x.

Bautista, T.G., Dutschmann, M., 2014. Inhibition of the pontine Kölliker-Fuse nucleus abolishes eupneic inspiratory hypoglossal motor discharge in rat. Neuroscience 267, 22–29. https://doi.org/10.1016/j.neuroscience.2014.02.027.

Behzadi, G., Kalén, P., Parvopassu, F., Wiklund, L., 1990. Afferents to the median raphe nucleus of the rat: retrograde cholera toxin and wheat germ conjugated horseradish peroxidase tracing, and selectived-[3H]aspartate labelling of possible excitatory amino acid inputs. Neuroscience 37 (1), 77–100. https://doi.org/10.1016/0306-4522(90)90194-9.

Belmer, A., Patkar, O.L., Lanoue, V., Bartlett, S.E., 2018a. 5-HT1A receptor-dependent modulation of emotional and neurogenic deficits elicited by prolonged consumption of alcohol. Sci. Rep. 8 (1), 2099. https://doi.org/10.1038/s41598-018-20504-z.

Belmer, A., Quentin, E., Diaz, S.L., et al., 2018b. Positive regulation of raphe serotonin neurons by serotonin 2B receptors. Neuropsychopharmacology 43 (7), 1623–1632. https://doi.org/10.1038/s41386-018-0013-0.

Bender, M.B., Shanzer, S., Wagman, I.H., 1964. On the physiologic decussation concerned with head turning. Confin. Neurol. 24, 169–181.

Ben-Tal, A., Smith, J.C., 2008. A model for control of breathing in mammals: coupling neural dynamics to peripheral gas exchange and transport. J. Theor. Biol. 251 (3), 480–497. https://doi.org/10.1016/j.jtbi.2007.12.

Bernard, J.F., Bester, H., Besson, J.M., 1996. Chapter 14: Involvement of the spino-parabrachio -amygdaloid and -hypothalamic pathways in the autonomic and affective emotional aspects of pain. Prog. Brain Res. 243–255. https://doi.org/10.1016/S0079-6123(08)61868-3.

Bester, H., Chapman, V., Besson, J.M., Bernard, J.F., 2000. Physiological properties of the lamina I spinoparabrachial neurons in the rat. J. Neurophysiol. 83 (4), 2239–2259. https://doi.org/10.1152/jn.2000.83.4.2239.

Bevan, M.D., Francis, C.M., Bolam, J.P., 1995. The glutamate-enriched cortical and thalamic input to neurons in the subthalamic nucleus of the rat: convergence with GABA-positive terminals. J. Comp. Neurol. 361 (3), 491–511. https://doi.org/10.1002/cne.903610312.

Björklund, A., Hökfelt, T., 1985. In: Björklund, A., Hökfelt, T. (Eds.), GABA and Neuropeptides in the CNS. Elsevier.

Bland, B.H., Colom, L.V., 1993. Extrinsic and intrinsic properties underlying oscillation and synchrony in limbic cortex. Prog. Neurobiol. 41 (2), 157–208. https://doi.org/10.1016/0301-0082(93)90007-f.

Bland, B.H., Oddie, S.D., 2001. Theta band oscillation and synchrony in the hippocampal formation and associated structures: the case for its role in sensorimotor integration. Behav. Brain Res. 127 (1–2), 119–136. https://doi.org/10.1016/s0166-4328(01)00358-8.

Bland, B.H., Oddie, S.D., Colom, L.V., Vertes, R.P., 1994. Extrinsic modulation of medial septal cell discharges by the ascending brainstem hippocampal synchronizing pathway. Hippocampus 4 (6), 649–660. https://doi.org/10.1002/hipo.450040604.

Blier, P., de Montigny, C., 1998. Possible serotonergic mechanisms underlying the antidepressant and anti-obsessive–compulsive disorder responses. Biol. Psychiatry 44 (5), 313–323. https://doi.org/10.1016/S0006-3223(98)00114-0.

Blier, P., Pineyro, G., El Mansari, M., Bergeron, R., Montigny, C., 1998. Role of somatodendritic 5-HT autoreceptors in modulating 5-HT neurotransmissiona. Ann. N. Y. Acad. Sci. 861 (1), 204–216. https://doi.org/10.1111/j.1749-6632.1998.tb10192.x.

Bortolozzi, A., Castañé, A., Semakova, J., et al., 2012a. New antidepressant strategy based on acute siRNA silencing of 5-HT1A autoreceptors. Mol. Psychiatry 17 (6), 567. https://doi.org/10.1038/mp.2012.52.

Bortolozzi, A., Castañé, A., Semakova, J., et al., 2012b. Selective siRNA-mediated suppression of 5-HT1A autoreceptors evokes strong antidepressant-like effects. Mol. Psychiatry 17 (6), 612–623. https://doi.org/10.1038/mp.2011.92.

Bourgeais, L., Gauriau, C., Bernard, J.F., 2001a. Projections from the nociceptive area of the central nucleus of the amygdala to the forebrain: a PHA-L study in the rat. Eur. J. Neurosci. 14 (2), 229–255. https://doi.org/10.1046/j.0953-816x.2001.01640.x.

Bourgeais, L., Monconduit, L., Villanueva, L., Bernard, J.F., 2001b. Parabrachial internal lateral neurons convey nociceptive messages from the deep laminas of the dorsal horn to the intralaminar thalamus. J. Neurosci. 21 (6), 2159–2165. https://doi.org/10.1523/JNEUROSCI.21-06-02159.2001.

Bowker, R.M., Westlund, K.N., Coulter, J.D., 1981. Origins of serotonergic projections to the spinal cord in rat: an immunocytochemical-retrograde transport study. Brain Res. 226 (1–2), 187–199. https://doi.org/10.1016/0006-8993(81)91092-1.

Bowker, R.M., Westlund, K.N., Sullivan, M.C., Wilber, J.F., Coulter, J.D., 1983. Descending serotonergic, peptidergic and cholinergic pathways from the raphe nuclei: a multiple transmitter complex. Brain Res. 288 (1–2), 33–48. https://doi.org/10.1016/0006-8993(83)90079-3.

Browaldh, N., Bautista, T.G., Dutschmann, M., Berkowitz, R.G., 2016. The Kölliker-Fuse nucleus: a review of animal studies and the implications for cranial nerve function in humans. Eur. Arch. Oto-Rhino-Laryngol. 273 (11), 3505–3510. https://doi.org/10.1007/s00405-015-3861-9.

Budai, D., Harasawa, I., Fields, H.L., 1998. Midbrain periaqueductal gray (PAG) inhibits nociceptive inputs to sacral dorsal horn nociceptive neurons through α_2 -adrenergic receptors. J. Neurophysiol. 80 (5), 2244–2254. https://doi.org/10.1152/jn.1998.80.5.2244.

Büsselberg, D., Bischoff, A.M., Paton, J.F., Richter, D.W., 2001. Reorganisation of respiratory network activity after loss of glycinergic inhibition. Pflugers Archiv. 441 (4), 444–449. https://doi.org/10.1007/s004240000453.

Byrum, C.E., Guyenet, P.G., 1987. Afferent and efferent connections of the A5 noradrenergic cell group in the rat. J. Comp. Neurol. 261 (4), 529–542. https://doi.org/10.1002/cne.902610406.

Campos, C.A., Bowen, A.J., Roman, C.W., Palmiter, R.D., 2018. Encoding of danger by parabrachial CGRP neurons. Nature 555 (7698), 617–622. https://doi.org/10.1038/nature25511.

Cao, Y., Matsuyama, K., Fujito, Y., Aoki, M., 2006. Involvement of medullary GABAergic and serotonergic raphe neurons in respiratory control: electrophysiological and immunohistochemical studies in rats. Neurosci. Res. 56 (3), 322–331. https://doi.org/10.1016/j.neures.2006.08.001.

Card, J.P., Sved, J.C., Craig, B., Raizada, M., Vazquez, J., Sved, A.F., 2006. Efferent projections of rat rostroventrolateral medulla C1 catecholamine neurons: implications for the central control of cardiovascular regulation. J. Comp. Neurol. 499 (5), 840–859. https://doi.org/10.1002/cne.21140.

Celada, P., Puig, M.V., Casanovas, J.M., Guillazo, G., Artigas, F., 2001. Control of dorsal raphe serotonergic neurons by the medial prefrontal cortex: involvement of serotonin-1A, GABA A, and glutamate receptors. J. Neurosci. 21 (24), 9917–9929. https://doi.org/10.1523/JNEUROSCI.21-24-09917.2001.

Chan-Palay, V., Asan, E., 1989a. Alterations in catecholamine neurons of the locus coeruleus in senile dementia of the Alzheimer type and in Parkinson's disease with and without dementia and depression. J. Comp. Neurol. 287 (3), 373–392. https://doi.org/10.1002/cne.902870308.

Chan-Palay, V., Asan, E., 1989b. Quantitation of catecholamine neurons in the locus coeruleus in human brains of normal young and older adults and in depression. J. Comp. Neurol. 287 (3), 357–372. https://doi.org/10.1002/cne.902870307.

Charara, A., Parent, A., 1994. Brainstem dopaminergic, cholinergic and serotoninergic afferents to the pallidum in the squirrel monkey. Brain Res. 640 (1–2), 155–170. https://doi.org/10.1016/0006-8993(94)91870-8.

Charara, A., Parent, A., 1998. Chemoarchitecture of the primate dorsal raphe nucleus. J. Chem. Neuroanat. 15 (2), 111–127. https://doi.org/10.1016/S0891-0618(98)00036-2.

Chen, S., Yang, M., Miselis, R.R., Aston-Jones, G., 1999. Characterization of transsynaptic tracing with central application of pseudorabies virus. Brain Res. 838 (1–2), 171–183. https://doi.org/10.1016/S0006-8993(99)01680-7.

Chiang, M.C., Bowen, A., Schier, L.A., Tupone, D., Uddin, O., Heinricher, M.M., 2019. Parabrachial complex: a hub for pain and aversion. J. Neurosci. 39 (42), 8225–8230. https://doi.org/10.1523/JNEUROSCI.1162-19.2019.

Chiang, M.C., Nguyen, E.K., Canto-Bustos, M., Papale, A.E., Oswald, A.M.M., Ross, S.E., 2020. Divergent neural pathways emanating from the lateral parabrachial nucleus mediate distinct components of the pain response. Neuron 106 (6), 927–939.e5. https://doi.org/10.1016/j.neuron.2020.03.014.

Chimoto, S., Iwamoto, Y., Shimazu, H., Yoshida, K., 1996a. Chapter 11: Functional connectivity of the superior colliculus with saccade-related brain stem neurons in the cat. Prog. Brain Res. 157–165. https://doi.org/10.1016/S0079-6123(08)63327-0.

Chimoto, S., Iwamoto, Y., Shimazu, H., Yoshida, K., 1996b. Monosynaptic activation of medium-lead burst neurons from the superior colliculus in the alert cat. J. Neurophysiol. 75 (6), 2658–2661. https://doi.org/10.1152/jn.1996.75.6.2658.

Chou, T.C., Bjorkum, A.A., Gaus, S.E., Lu, J., Scammell, T.E., Saper, C.B., 2002. Afferents to the ventrolateral preoptic nucleus. J. Neurosci. 22 (3), 977–990. https://doi.org/10.1523/JNEUROSCI.22-03-00977.2002.

Ciriello, J., Kline, R.L., Zhang, T.X., Caverson, M.M., 1984a. Lesions of the paraventricular nucleus alter the development of spontaneous hypertension in the rat. Brain Res. 310 (2), 355–359. https://doi.org/10.1016/0006-8993(84)90159-8.

Ciriello, J., Lawrence, D., Pittman, Q.J., 1984b. Electrophysiological identification of neurons in the parabrachial nucleus projecting directly to the hypothalamus in the rat. Brain Res. 322 (2), 388–392. https://doi.org/10.1016/0006-8993(84)90140-9.

Clark, F.M., Proudfit, H.K., 1991a. The projection of locus coeruleus neurons to the spinal cord in the rat determined by anterograde tracing combined with immunocytochemistry. Brain Res. 538 (2), 231–245. https://doi.org/10.1016/0006-8993(91)90435-X.

Clark, F.M., Proudfit, H.K., 1991b. The projection of noradrenergic neurons in the A7 catecholamine cell group to the spinal cord in the rat demonstrated by anterograde tracing combined with immunocytochemistry. Brain Res. 547 (2), 279–288. https://doi.org/10.1016/0006-8993(91)90972-X.

Clark, F.M., Proudfit, H.K., 1993. The projections of noradrenergic neurons in the A5 catecholamine cell group to the spinal cord in the rat: anatomical evidence that A5 neurons modulate nociception. Brain Res. 616 (1–2), 200–210. https://doi.org/10.1016/0006-8993(93)90210-E.

Clarke, P.B.S., Hommer, D.W., Pert, A., Skirboll, L.R., 1987. Innervation of substantia nigra neurons by cholinergic afferents from pedunculopontine nucleus in the rat: neuroanatomical and electrophysiological evidence. Neuroscience 23 (3), 1011–1019. https://doi.org/10.1016/0306-4522(87)90176-X.

Cohen, M.I., 1971. Switching of the respiratory phases and evoked phrenic responses produced by rostral pontine electrical stimulation. J. Physiol 217 (1), 133–158. https://doi.org/10.1113/jphysiol.1971.sp009563.

Cohen, B., Büttner-Ennever, J.A., 1984. Projections from the superior colliculus to a region of the central mesencephalic reticular formation (cMRF) associated with horizontal saccadic eye movements. Exp. Brain Res. 57 (1), 167–176. https://doi.org/10.1007/BF00231143.

Cohen, B., Komatsuzaki, A., 1972. Eye movements induced by stimulation of the pontine reticular formation: evidence for integration in oculomotor pathways. Exp. Neurol. 36 (1), 101–117. https://doi.org/10.1016/0014-4886(72)90139-2.

Cohen, M.I., Wang, S.C., 1959. Respiratory neuronal activity in pons of cat. J. Neurophysiol. 22 (1), 33–50. https://doi.org/10.1152/jn.1959.22.1.33.

Cohen, B., Matsuo, V., Fradin, J., Raphan, T., 1985. Horizontal saccades induced by stimulation of the central mesencephalic reticular formation. Exp. Brain Res. 57 (3). https://doi.org/10.1007/BF00237847.

Colom, L.V., Bland, B.H., 1987. State-dependent spike train dynamics of hippocampal formation neurons: evidence for theta-on and theta-off cells. Brain Res. 422 (2), 277–286. https://doi.org/10.1016/0006-8993(87)90934-6.

Commons, K.G., Connolley, K.R., Valentino, R.J., 2003. A neurochemically distinct dorsal raphe-limbic circuit with a potential role in affective disorders. Neuropsychopharmacology 28 (2), 206–215. https://doi.org/10.1038/sj.npp.1300045.

Craig, A.D., 1995. Distribution of brainstem projections from spinal lamina I neurons in the cat and the monkey. J. Comp. Neurol. 361 (2), 225–248. https://doi.org/10.1002/cne.903610204.

Cromer, J.A., Waitzman, D.M., 2006. Neurones associated with saccade metrics in the monkey central mesencephalic reticular formation. J. Physiol 570 (3), 507–523. https://doi.org/10.1113/jphysiol.2005.096834.

Cromer, J.A., Waitzman, D.M., 2007. Comparison of saccade-associated neuronal activity in the primate central mesencephalic and paramedian pontine reticular formations. J. Neurophysiol. 98 (2), 835–850. https://doi.org/10.1152/jn.00308.2007.

da Silva Soares, R., Falconi-Sobrinho, L.L., dos Anjos-Garcia, T., Coimbra, N.C., 2019a. 5-Hydroxytryptamine 2A receptors of the dorsal raphe nucleus modulate panic-like behaviours and mediate fear-induced antinociception elicited by neuronal activation in the central nucleus of the inferior colliculus. Behav. Brain Res. 357–358, 71–81. https://doi.org/10.1016/j.bbr.2017.07.016.

da Silva Soares, R., Falconi-Sobrinho, L.L., Almada, R.C., Coimbra, N.C., 2019b. Dorsal raphe nucleus 5-Hydroxytryptamine 2A receptors are critical for the organisation of panic attack-like defensive behaviour and unconditioned fear-induced antinociception elicited by the chemical stimulation of superior colliculus neurons. Eur. Neuropsychopharmacol 29 (7), 858–870. https://doi.org/10.1016/j.euroneuro.2019.05.007.

Dahlström, A., Fuxe, K., 1964a. Evidence for the existence of monoamine-containing neurons in the central nervous system. I. Demonstration of mono- amines in the cell bodies of the brain stem neurons. Acta Physiol. Scand. Suppl. 62 (247), 1–55.

Dahlström, A., Fuxe, K., 1964b. Localization of monoamines in the lower brain stem. Experientia 20 (7), 398–399. https://doi.org/10.1007/BF02147990.

Dahlström, A., Fuxe, K., 1965. Evidence for the existence of monoamine-containing neurons in the central nervous system. Acta Physiol. Scand.

Dampney, R.A., 1994. Functional organization of central pathways regulating the cardiovascular system. Physiol. Rev. 74 (2), 323–364. https://doi.org/10.1152/physrev.1994.74.2.323.

Dauger, S., Pattyn, A., Lofaso, F., et al., 2003. *Phox2b* controls the development of peripheral chemoreceptors and afferent visceral pathways. Development 130 (26), 6635–6642. https://doi.org/10.1242/dev.00866.

Davidson, C., Stamford, J.A., 1995a. Evidence that 5-hydroxytryptamine release in rat dorsal raphé nucleus is controlled by 5-HT1A, 5-HT1B and 5-HT1D autoreceptors. Br. J. Pharmacol. 114 (6), 1107–1109. https://doi.org/10.1111/j.1476-5381.1995.tb13321.x.

Davidson, C., Stamford, J.A., 1995b. The effect of paroxetine on 5-HT efflux in the rat dorsal raphe nucleus is potentiated by both 5-HT1A and receptor antagonists. Neurosci. Lett. 188 (1), 41–44. https://doi.org/10.1016/0304-3940(95)11390-I.

De Gobbi, J.I.F., Menani, J.V., Beltz, T.G., Johnson, R.F., Thunhorst, R.L., Johnson, A.K., 2008. Right atrial stretch alters fore- and hind-brain expression of *c-fos* and inhibits the rapid onset of salt appetite. J. Physiol 586 (15), 3719–3729. https://doi.org/10.1113/jphysiol.2008.151282.

De luca, L.A., Barbosa, S.P., Menani, J.V., 2003. Brain serotonin blockade and paradoxical salt intake in rats. Neuroscience 121 (4), 1055–1061. https://doi.org/10.1016/S0306-4522(03)00316-6.

de Oliveira, R.C., de Oliveira, R., Biagioni, A.F., Falconi-Sobrinho, L.L., dos Anjos-Garcia, T., Coimbra, N.C., 2016. Nicotinic and muscarinic cholinergic receptors are recruited by acetylcholine-mediated neurotransmission within the locus coeruleus during the organisation of post-ictal antinociception. Brain Res. Bull. 127, 74–83. https://doi.org/10.1016/j.brainresbull.2016.08.011.

de Oliveira, R., de Oliveira, R.C., Falconi-Sobrinho, L.L., da Silva Soares, R., Coimbra, N.C., 2017. 5-Hydroxytryptamine 2A/2C receptors of nucleus raphe magnus and gigantocellularis/paragigantocellularis pars α reticular nuclei modulate the unconditioned fear-induced antinociception evoked by electrical stimulation of deep layers of the superior colliculus and dorsal periaqueductal gray matter. Behav. Brain Res. 316, 294–304. https://doi.org/10.1016/j.bbr.2016.09.016.

Del Fiacco, M., Quartu, M., Serra, M.P., Follesa, P., Lai, M.L., Bachis, A., 2002. Topographical localization of glial cell line-derived neurotrophic factor in the human brain stem: an immunohistochemical study of prenatal, neonatal and adult brains. J. Chem. Neuroanat. 23 (1), 29–48. https://doi.org/10.1016/S0891-0618(01)00139-9.

Deng, J., Zhou, H., Lin, J.K., et al., 2020. The parabrachial nucleus directly channels spinal nociceptive signals to the intralaminar thalamic nuclei, but not the amygdala. Neuron 107 (5), 909–923.e6. https://doi.org/10.1016/j.neuron.2020.06.017.

Dergacheva, O.I., Khachikova, I.E., Burikov, A.A., 2002. Impulse activity of neurons in the nucleus pontis oralis in cats during sleep–wakefulness cycle. Ross. Fiziol. Zh. Im. IM Sechenova 88 (12), 1530–1537.

Descarries, L., Mechawar, N., 2000. Ultrastructural evidence for diffuse transmission by monoamine and acetylcholine neurons of the central nervous system. Prog. Brain Res. 27–47. https://doi.org/10.1016/S0079-6123(00)25005-X.

Dick, T.E., Bellingham, M.C., Richter, D.W., 1994. Pontine respiratory neurons in anesthetized cats. Brain Res. 636 (2), 259–269. https://doi.org/10.1016/0006-8993(94)91025-1.

Doi, A., Ramirez, J.M., 2010. State-dependent interactions between excitatory neuromodulators in the neuronal control of breathing.

J. Neurosci. 30 (24), 8251–8262. https://doi.org/10.1523/JNEUROSCI.5361-09.2010.
Donga, R., Lund, J.P., 1991. Discharge patterns of trigeminal commissural last-order interneurons during fictive mastication in the rabbit. J. Neurophysiol. 66 (5), 1564–1578. https://doi.org/10.1152/jn.1991.66.5.1564.
Dutschmann, M., Herbert, H., 2006. The Kölliker-Fuse nucleus gates the postinspiratory phase of the respiratory cycle to control inspiratory off-switch and upper airway resistance in rat. Eur. J. Neurosci. 24 (4), 1071–1084. https://doi.org/10.1111/j.1460-9568.2006.04981.x.
Edley, S.M., Graybiel, A.M., 1983. The afferent and efferent connections of the feline nucleus tegmenti pedunculopontinus, pars compacta. J. Comp. Neurol. 217 (2), 187–215. https://doi.org/10.1002/cne.902170207.
Eggermont, J.J., 2021. Separate auditory pathways for the induction and maintenance of tinnitus and hyperacusis? In. Prog. Brain Res. 101–127. https://doi.org/10.1016/bs.pbr.2020.01.006.
Ennis, M., Aston-Jones, G., 1989. GABA-mediated inhibition of locus coeruleus from the dorsomedial rostral medulla. J. Neurosci. 9 (8), 2973–2981. https://doi.org/10.1523/JNEUROSCI.09-08-02973.1989.
Erro, E., Lanciego, J.L., Giménez-Amaya, J.M., 1999. Relationships between thalamostriatal neurons and pedunculopontine projections to the thalamus: a neuroanatomical tract-tracing study in the rat. Exp. Brain Res. 127 (2), 162–170. https://doi.org/10.1007/s002210050786.
España, R.A., Reis, K.M., Valentino, R.J., Berridge, C.W., 2005. Organization of hypocretin/orexin efferents to locus coeruleus and basal forebrain arousal-related structures. J. Comp. Neurol. 481 (2), 160–178. https://doi.org/10.1002/cne.20369.
Ezure, K., Tanaka, I., 2006. Distribution and medullary projection of respiratory neurons in the dorsolateral pons of the rat. Neuroscience 141 (2), 1011–1023. https://doi.org/10.1016/j.neuroscience.2006.04.020.
Felten, D.L., Sladek, J.R., 1983. Monoamine distribution in primate brain V. Monoaminergic nuclei: anatomy, pathways and local organization. Brain Res. Bull. 10 (2), 171–284. https://doi.org/10.1016/0361-9230(83)90045-X.
Ferrés-Coy, A., Pilar-Cuellar, F., Vidal, R., et al., 2013a. RNAi-mediated serotonin transporter suppression rapidly increases serotonergic neurotransmission and hippocampal neurogenesis. Transl. Psychiatry 3 (1), e211. https://doi.org/10.1038/tp.2012.135.
Ferrés-Coy, A., Santana, N., Castañé, A., et al., 2013b. Acute 5-HT1A autoreceptor knockdown increases antidepressant responses and serotonin release in stressful conditions. Psychopharmacology (Berl) 225 (1), 61–74. https://doi.org/10.1007/s00213-012-2795-9.
Fodor, M., Görcs, T.J., Palkovits, M., 1992. Immunohistochemical study on the distribution of neuropeptides within the pontine tegmentum—particularly the parabrachial nuclei and the locus coeruleus of the human brain. Neuroscience 46 (4), 891–908. https://doi.org/10.1016/0306-4522(92)90192-5.
Ford, R.D., Colom, L.V., Bland, B.H., 1989. The classification of medial septum-diagonal band cells as theta-on or theta-off in relation to hippocampal EEG states. Brain Res. 493 (2), 269–282. https://doi.org/10.1016/0006-8993(89)91162-1.
Forel, A., 1877. Untersuchungen über die Haubenre- gion und ihre oberen Verknüpfungen im Gehirne des Menschen und einiger Säugethiere, mit Beiträgen zu Methoden der Gehirnuntersuchung. Arch f Psychiat u Nervenkrankh 7, 392–495.
Foster, G.A., Schultzberg, M., Gage, F.H., et al., 1985. Transmitter expression and morphological development of embryonic medullary and mesencephalic raphé neurones after transplantation to the adult rat central nervous system—I. Grafts to the spinal cord. Exp. Brain Res. 60 (3). https://doi.org/10.1007/BF00236929.
Franchini, L.F., Vivas, L., 1999. Distribution of Fos immunoreactivity in rat brain after sodium consumption induced by peritoneal dialysis. Am. J. Physiol. Regul. Integr. Comp. Physiol. 276 (4), R1180–R1187. https://doi.org/10.1152/ajpregu.1999.276.4.R1180.
Francis Pau, K.Y., Yu, J.h., Lee, C.J., Spies, H.G., 1998. Topographic localization of neuropeptide Y mRNA in the monkey brainstem. Regul. Pept. 75–76, 145–153. https://doi.org/10.1016/S0167-0115(98)00062-7.
Fritschy, J.M., Grzanna, R., 1989. Immunohistochemical analysis of the neurotoxic effects of DSP-4 identifies two populations of noradrenergic axon terminals. Neuroscience 30 (1), 181–197. https://doi.org/10.1016/0306-4522(89)90364-3.
Fritschy, J.M., Grzanna, R., 1990. Demonstration of two separate descending noradrenergic pathways to the rat spinal cord: evidence for an intragriseal trajectory of locus coeruleus axons in the superficial layers of the dorsal horn. J. Comp. Neurol. 291 (4), 553–582. https://doi.org/10.1002/cne.902910406.
Fritschy, J.M., Lyons, W.E., Mullen, C.A., Kosofsky, B.E., Molliver, M.E., Grzanna, R., 1987. Distribution of locus coeruleus axons in the rat spinal cord: a combined anterograde transport and immunohistochemical study. Brain Res. 437 (1), 176–180. https://doi.org/10.1016/0006-8993(87)91541-1.
Fuchs, A.F., Kaneko, C.R., Scudder, C.A., 1985. Brainstem control of saccadic eye movements. Annu. Rev. Neurosci. 8, 307–337. https://doi.org/10.1146/annurev.ne.08.030185.001515.
Fulwiler, C.E., Saper, C.B., 1984. Subnuclear organization of the efferent connections of the parabrachial nucleus in the rat. Brain Res. Rev. 7 (3), 229–259. https://doi.org/10.1016/0165-0173(84)90012-2.
Fung, M.L., Tomori, Z., St John, W.M., 1995. Medullary neuronal activities in gasping induced by pharyngeal stimulation and hypoxia. Respir. Physiol. 100 (3), 195–202. https://doi.org/10.1016/0034-5687(94)00141-L.
Garcia-Rill, E., 1997. Disorders of the reticular activating system. Med. Hypotheses 49 (5), 379–387. https://doi.org/10.1016/S0306-9877(97)90083-9.
Garciarill, E., 1991. The pedunculopontine nucleus. Prog. Neurobiol. 36 (5), 363–389. https://doi.org/10.1016/0301-0082(91)90016-T.
Gau, R., Sévoz-Couche, C., Laguzzi, R., Hamon, M., Bernard, J.F., 2009. Inhibition of cardiac baroreflex by noxious thermal stimuli: a key role for lateral paragigantocellular serotonergic cells. Pain 146 (3), 315–324. https://doi.org/10.1016/j.pain.2009.09.018.
Geerling, J.C., Yokota, S., Rukhadze, I., Roe, D., Chamberlin, N.L., 2017. Kölliker-Fuse GABAergic and glutamatergic neurons project to distinct targets. J. Comp. Neurol. 525 (8), 1844–1860. https://doi.org/10.1002/cne.24164.
Gervasoni, D., Darracq, L., Fort, P., Soulière, F., Chouvet, G., Luppi, P.H., 1998. Electrophysiological evidence that noradrenergic neurons of the rat locus coeruleus are tonically inhibited by GABA during sleep. Eur. J. Neurosci. 10 (3), 964–970. https://doi.org/10.1046/j.1460-9568.1998.00106.x.
Goebel, H.H., Komatsuzaki, A., Bender, M.B., Cohen, B., 1971. Lesions of the pontine tegmentum and conjugate gaze paralysis. Arch. Neurol. 24 (5), 431–440. https://doi.org/10.1001/archneur.1971.00480350065007.
Gold, P.W., Chrousos, G.P., 2002. Organization of the stress system and its dysregulation in melancholic and atypical depression: high vs low CRH/NE states. Mol. Psychiatry 7 (3), 254–275. https://doi.org/10.1038/sj.mp.4001032.
Gonçalves, L., Nogueira, M.I., Shammah-Lagnado, S.J., Metzger, M., 2009. Prefrontal afferents to the dorsal raphe nucleus in the rat. Brain Res. Bull. 78 (4–5), 240–247. https://doi.org/10.1016/j.brainresbull.2008.11.012.
Granata, A.R., Kitai, S.T., 1991. Inhibitory substantia nigra inputs to the pedunculopontine neurons. Exp. Brain Res. 86 (3). https://doi.org/10.1007/BF00230520.
Grantyn, A., Berthoz, A., 1987. Reticulo-spinal neurons participating in the control of synergic eye and head movements during orienting

in the cat. Exp. Brain Res. 66 (2), 339–354. https://doi.org/10.1007/BF00243309.

Grantyn, A., Grantyn, R., Robiné, K.P., Berthoz, A., 1979. Electroanatomy of tectal efferent connections related to eye movements in the horizontal plane. Exp. Brain Res. 37 (1), 149–172. https://doi.org/10.1007/BF01474261.

Grantyn, R., Baker, R., Grantyn, A., 1980. Morphological and physiological identification of excitatory pontine reticular neurons projecting to the cat abducens nucleus and spinal cord. Brain Res. 198 (1), 221–228. https://doi.org/10.1016/0006-8993(80)90359-5.

Grantyn, A., Ong-Meang Jacques, V., Berthoz, A., 1987a. Reticulo-spinal neurons participating in the control of synergic eye and head movements during orienting in the cat. Exp. Brain Res. 66 (2), 355–377. https://doi.org/10.1007/BF00243310.

Grantyn, R., Perouansky, M., Lux, H.D., Hablitz, J.J., 1987b. Glutamate-induced ionic currents in cultured neurons from the rat superior colliculus. Brain Res. 420 (1), 182–187. https://doi.org/10.1016/0006-8993(87)90258-7.

Groenewegen, H.J., Berendse, H.W., Haber, S.N., 1993. Organization of the output of the ventral striatopallidal system in the rat: ventral pallidal efferents. Neuroscience 57 (1), 113–142. https://doi.org/10.1016/0306-4522(93)90115-V.

Grzanna, R., Berger, U., Fritschy, J.M., Geffard, M., 1989. Acute action of DSP-4 on central norepinephrine axons: biochemical and immunohistochemical evidence for differential effects. J. Histochem. Cytochem. 37 (9), 1435–1442. https://doi.org/10.1177/37.9.2768812.

Guo, W., Fan, S., Xiao, D., et al., 2021. A Brainstem reticulotegmental neural ensemble drives acoustic startle reflexes. Nat. Commun. 12 (1), 6403. https://doi.org/10.1038/s41467-021-26723-9.

Guyenet, P.G., Bayliss, D.A., 2015. Neural control of breathing and CO_2 homeostasis. Neuron 87 (5), 946–961. https://doi.org/10.1016/j.neuron.2015.08.001.

Guyenet, P.G., Koshiya, N., Huangfu, D., Verberne, A.J., Riley, T.A., 1993. Central respiratory control of A5 and A6 pontine noradrenergic neurons. Am. J. Physiol. Regul. Integr. Comp. Physiol. 264 (6), R1035–R1044. https://doi.org/10.1152/ajpregu.1993.264.6.R1035.

Guyenet, P.G., Mulkey, D.K., 2010. Retrotrapezoid nucleus and parafacial respiratory group. Respir. Physiol. Neurobiol. 173 (3), 244–255. https://doi.org/10.1016/j.resp.2010.02.005.

Haddjeri, N., Faure, C., Lucas, G., et al., 2004a. In-vivo modulation of central 5-hydroxytryptamine (5-HT1A) receptor-mediated responses by the cholinergic system. Int. J. Neuropsychopharmacol. 7 (4), 391–399. https://doi.org/10.1017/S1461145704004377.

Haddjeri, N., Lavoie, N., Blier, P., 2004b. Electrophysiological evidence for the tonic activation of 5-HT1A autoreceptors in the rat dorsal raphe nucleus. Neuropsychopharmacology 29 (10), 1800–1806. https://doi.org/10.1038/sj.npp.1300489.

Hajós, M., Richards, C.D., Székely, A.D., Sharp, T., 1998. An electrophysiological and neuroanatomical study of the medial prefrontal cortical projection to the midbrain raphe nuclei in the rat. Neuroscience 87 (1), 95–108. https://doi.org/10.1016/S0306-4522(98)00157-2.

Halliday, G.M., Törk, I., 1986. Comparative anatomy of the ventromedial mesencephalic tegmentum in the rat, cat, monkey and human. J. Comp. Neurol. 252 (4), 423–445. https://doi.org/10.1002/cne.902520402.

Halliday, G.M., Li, Y.W., Joh, T.H., et al., 1988a. Distribution of monoamine-synthesizing neurons in the human medulla oblongata. J. Comp. Neurol. 273 (3), 301–317. https://doi.org/10.1002/cne.902730303.

Halliday, G.M., Li, Y.W., Joh, T.H., et al., 1988b. Distribution of substance P-like immunoreactive neurons in the human medulla oblongata: Co-localization with monoamine-synthesizing neurons. Synapse 2 (4), 353–370. https://doi.org/10.1002/syn.890020403.

Hammond, C., Rouzaire-Dubois, B., Féger, J., Jackson, A., Crossman, A.R., 1983a. Anatomical and electrophysiological studies on the reciprocal projections between the subthalamic nucleus and nucleus tegmenti pedunculopontinus in the rat. Neuroscience 9 (1), 41–52. https://doi.org/10.1016/0306-4522(83)90045-3.

Hammond, C., Shibazaki, T., Rouzaire-Dubois, B., 1983b. Branched output neurons of the rat subthalamic nucleus: electrophysiological study of the synaptic effects on identified cells in the two main target nuclei, the entopeduncular nucleus and the substantia nigra. Neuroscience 9 (3), 511–520. https://doi.org/10.1016/0306-4522(83)90169-0.

Han, S., Soleiman, M.T., Soden, M.E., Zweifel, L.S., Palmiter, R.D., 2015. Elucidating an affective pain circuit that creates a threat memory. Cell 162 (2), 363–374. https://doi.org/10.1016/j.cell.2015.05.057.

Hancock, M.B., Fougerousse, C.L., 1976. Spinal projections from the nucleus locus coeruleus and nucleus subcoeruleus in the cat and monkey as demonstrated by the retrograde transport of horse-radish peroxidase. Brain Res. Bull. 1 (2), 229–234. https://doi.org/10.1016/0361-9230(76)90072-1.

Handel, A., Glimcher, P.W., 1997. Response properties of saccade-related burst neurons in the central mesencephalic Reticular Formation. J. Neurophysiol. 78 (4), 2164–2175. https://doi.org/10.1152/jn.1997.78.4.2164.

Harte, S.E., Hoot, M.R., Borszcz, G.S., 2004. Involvement of the intralaminar parafascicular nucleus in muscarinic-induced antinociception in rats. Brain Res. 1019 (1–2), 152–161. https://doi.org/10.1016/j.brainres.2004.05.096.

Headley, D.B., Paré, D., 2017. Common oscillatory mechanisms across multiple memory systems. NPJ Sci. Learn. 2 (1). https://doi.org/10.1038/s41539-016-0001-2.

Hebert, H., Skriver, E., Kavéus, U., Maunsbach, A.B., 1990. Coexistence of different forms of Na,K-ATPase in two-dimensional membrane crystals. FEBS Lett. 268 (1), 83–87. https://doi.org/10.1016/0014-5793(90)80978-R.

Henn, V., Cohen, B., 1976. Coding of information about rapid eye movements in the pontine reticular formation of alert monkeys. Brain Res. 108 (2), 307–325. https://doi.org/10.1016/0006-8993(76)90188-8.

Hensler, J.G., 2006. Serotonergic modulation of the limbic system. Neurosci. Biobehav. Rev. 30 (2), 203–214. https://doi.org/10.1016/j.neubiorev.2005.06.007.

Hepp, K., Henn, V., 1983. Spatio-temporal recoding of rapid eye movement signals in the monkey paramedian pontine reticular formation (PPRF). Exp. Brain Res. 52 (1). https://doi.org/10.1007/BF00237155.

Hepp, K., Henn, V., Vilis, T., Cohen, B., 1989. Brainstem regions related to saccade generation. Rev. Oculomot. Res. 3, 105–212.

Herbert, H., Flügge, G., 1995. Distribution of alpha2-adrenergic binding sites in the parabrachial complex of the rat. Anat. Embryol. 192 (6), 507–516. https://doi.org/10.1007/BF00187181.

Hermann, D.M., Luppi, P.H., Peyron, C., Hinckel, P., Jouvet, M., 1997. Afferent projections to the rat nuclei raphe magnus, raphe pallidus and reticularis gigantocellularis pars α demonstrated by iontophoretic application of choleratoxin (subunit b). J. Chem. Neuroanat. 13 (1), 1–21. https://doi.org/10.1016/S0891-0618(97)00019-7.

Hikosaka, O., Kawakami, T., 1977. Inhibitory reticular neurons related to the quick phase of vestibular nystagmus? Their location and projection. Exp. Brain Res. (3–4), 27. https://doi.org/10.1007/BF00235511.

Hikosaka, O., Igusa, Y., Imai, H., 1978a. Firing pattern of prepositus hypoglossi and adjacent reticular neurons related to vestibular nystagmus in the cat. Brain Res. 144 (2), 395–403. https://doi.org/10.1016/0006-8993(78)90167-1.

Hikosaka, O., Igusa, Y., Nakao, S., Shimazu, H., 1978b. Direct inhibitory synaptic linkage of pontomedullary reticular burst neurons with abducens motoneurons in the cat. Exp. Brain Res. 33 (3–4). https://doi.org/10.1007/BF00235558.

Hioki, H., Nakamura, H., Ma, Y.F., et al., 2010. Vesicular glutamate transporter 3-expressing nonserotonergic projection neurons constitute a subregion in the rat midbrain raphe nuclei. J. Comp. Neurol. 518 (5), 668–686. https://doi.org/10.1002/cne.22237.

Hirota, K., Kushikata, T., 2001. Central noradrenergic neurones and the mechanism of general anaesthesia. Br. J. Anaesth. 87 (6), 811–813. https://doi.org/10.1093/bja/87.6.811.

Hökfelt, T., Fuxe, K., Goldstein, M., Johansson, O., 1974. Immunohistochemical evidence for the existence of adrenaline neurons in the rat brain. Brain Res. 66 (2), 235–251. https://doi.org/10.1016/0006-8993(74)90143-7.

Hökfelt, T., Mårtensson, R., Björklund, A., Kleinau, S., Goldstein, M., 1984. Distributional maps of tyrosin hydroxylase-immunoreactive neurons in the rat brain. In: Handbook of Chemical Neuroanatomy, vol. 2. Elsevier, pp. 277–379.

Hornung, J.P., Fritschy, J.M., 1988. Serotoninergic system in the brainstem of the marmoset: a combined immunocytochemical and three-dimensional reconstruction study. J. Comp. Neurol. 270 (4), 471–487. https://doi.org/10.1002/cne.902700402.

Hosoya, Y., Sugiura, Y., Ito, R., Kohno, K., 1990. Descending projections from the hypothalamic paraventricular nucleus to the A5 area, including the superior salivatory nucleus, in the rat. Exp. Brain Res. 82 (3). https://doi.org/10.1007/BF00228793.

Huang, T., Lin, S.H., Malewicz, N.M., Zhang, Y., Zhang, Y., Goulding, M., LaMotte, R.H., Ma, Q., 2019. Identifying the pathways required for coping behaviours associated with sustained pain. Nature 565 (7737), 86–90. https://doi.org/10.1038/s41586-018-0793-8.

Hubscher, C.H., Johnson, R.D., 1996. Responses of medullary reticular formation neurons to input from the male genitalia. J. Neurophysiol. 76 (4), 2474–2482. https://doi.org/10.1152/jn.1996.76.4.2474.

Hylden, J.L.K., Anton, F., Nahin, R.L., 1989a. Spinal lamina I projection neurons in the rat: collateral innervation of parabrachial area and thalamus. Neuroscience 28 (1), 27–37. https://doi.org/10.1016/0306-4522(89)90229-7.

Hylden, J.L.K., Nahin, R.L., Traub, R.J., Dubner, R., 1989b. Expansion of receptive fields of spinal lamina I projection neurons in rats with unilateral adjuvant-induced inflammation: the contribution of dorsal horn mechanisms. Pain 37 (2), 229–243. https://doi.org/10.1016/0304-3959(89)90135-8.

Inyushkin, A.N., Merkulova, N.A., Orlova, A.O., Inyushkina, E.M., 2010. Local gabaergic modulation of the activity of serotoninergic neurons in the nucleus raphe magnus. Neurosci. Behav. Physiol. 40 (8), 885–893. https://doi.org/10.1007/s11055-010-9337-x.

Izawa, Y., Sugiuchi, Y., Shinoda, Y., 1999. Neural organization from the superior colliculus to motoneurons in the horizontal oculomotor system of the cat. J. Neurophysiol. 81 (6), 2597–2611. https://doi.org/10.1152/jn.1999.81.6.2597.

Jackson, J., Bland, B.H., Antle, M.C., 2009. Nonserotonergic projection neurons in the midbrain raphe nuclei contain the vesicular glutamate transporter VGLUT3. Synapse 63 (1), 31–41. https://doi.org/10.1002/syn.20581.

Jakus, J., Poliacek, I., Halasova, E., et al., 2008. Brainstem circuitry of tracheal–bronchial cough: c-fos study in anesthetized cats. Respir. Physiol. Neurobiol. 160 (3), 289–300. https://doi.org/10.1016/j.resp.2007.10.014.

Jansen, N.A., Giesler, G.J., 2015. Response characteristics of pruriceptive and nociceptive trigeminoparabrachial tract neurons in the rat. J. Neurophysiol. 113 (1), 58–70. https://doi.org/10.1152/jn.00596.2014.

Jennes, L., Stumpf, W.E., Kalivas, P.W., 1982. Neurotensin: topographical distribution in rat brain by immunohistochemistry. J. Comp. Neurol. 210 (3), 211–224. https://doi.org/10.1002/cne.902100302.

Jhamandas, J.H., Petrov, T., Harris, K.H., Vu, T., Krukoff, T.L., 1996. Parabrachial nucleus projection to the amygdala in the rat: electrophysiological and anatomical observations. Brain Res. Bull. 39 (2), 115–126. https://doi.org/10.1016/0361-9230(95)02084-5.

Johnson, A.K., 2007. The sensory psychobiology of thirst and salt appetite. Med. Sci. Sports Exerc. 39 (8), 1388–1400. https://doi.org/10.1249/mss.0b013e3180686de8.

Johnson, R.D., Hubscher, C.H., 1998. Brainstem microstimulation differentially inhibits pudendal motoneuron reflex inputs. Neuroreport 9 (2), 341–345. https://doi.org/10.1097/00001756-199801260-00030.

Johnson, A.K., Thunhorst, R.L., 1997. The neuroendocrinology of thirst and salt appetite: visceral sensory signals and mechanisms of central integration. Front. Neuroendocrinol. 18 (3), 292–353. https://doi.org/10.1006/frne.1997.0153.

Jones, B.E., Friedman, L., 1983. Atlas of catecholamine perikarya, varicosities and pathways in the brainstem of the cat. J. Comp. Neurol. 215 (4), 382–396. https://doi.org/10.1002/cne.902150404.

Jones, S.L., Light, A.R., 1992. Serotoninergic medullary raphespinal projection to the lumbar spinal cord in the rat: a retrograde immunohistochemical study. J. Comp. Neurol. 322 (4), 599–610. https://doi.org/10.1002/cne.903220413.

Kajiwara, R., Nakamura, S., Ikeda, K., et al., 2022. Intrinsic properties and synaptic connectivity of Phox2b-expressing neurons in rat rostral parvocellular reticular formation. Neurosci. Res. 178, 41–51. https://doi.org/10.1016/j.neures.2021.12.009.

Kang, Y., Kitai, S.T., 1990. Electrophysiological properties of pedunculopontine neurons and their postsynaptic responses following stimulation of substantia nigra reticulata. Brain Res. 535 (1), 79–95. https://doi.org/10.1016/0006-8993(90)91826-3.

Kato, T., Masuda, Y., Morimoto, T., 2007. Patterns of masseter muscle activities during sleep in Guinea pigs. Arch. Oral Biol. 52 (4), 385–386. https://doi.org/10.1016/j.archoralbio.2006.08.004.

Kaur, S., Saper, C.B., 2019. Neural circuitry underlying waking up to hypercapnia. Front. Neurosci. 13. https://doi.org/10.3389/fnins.2019.00401.

Kessler, R.C., 1997. The effects of stressful life events on depression. Annu. Rev. Psychol. 48 (1), 191–214. https://doi.org/10.1146/annurev.psych.48.1.191.

Kinney, G.G., Kocsis, B., Vertes, R.P., 1994. Injections of excitatory amino acid antagonists into the median raphe nucleus produce hippocampal theta rhythm in the urethane-anesthetized rat. Brain Res. 654 (1), 96–104. https://doi.org/10.1016/0006-8993(94)91575-x.

Kinney, G.G., Kocsis, B., Vertes, R.P., 1995. Injections of muscimol into the median raphe nucleus produce hippocampal theta rhythm in the urethane anesthetized rat. Psychopharmacology 120 (3), 244–248. https://doi.org/10.1007/BF02311170.

Kinney, G.G., Kocsis, B., Vertes, R.P., 1996. Medial septal unit firing characteristics following injections of 8-OH-DPAT into the median raphe nucleus. Brain Res. 708 (1–2), 116–122. https://doi.org/10.1016/0006-8993(95)01296-6.

Kirby, L., 2000. Effects of corticotropin-releasing factor on neuronal activity in the serotonergic dorsal raphe nucleus. Neuropsychopharmacology 22 (2), 148–162. https://doi.org/10.1016/S0893-133X(99)00093-7.

Kirby, L.G., Pernar, L., Valentino, R.J., Beck, S.G., 2003. Distinguishing characteristics of serotonin and non-serotonin-containing cells in the dorsal raphe nucleus: electrophysiological and immunohistochemical studies. Neuroscience 116 (3), 669–683. https://doi.org/10.1016/s0306-4522(02)00584-5.

Kobashi, M., Ichikawa, H., Sugimoto, T., Adachi, A., 1993. Response of neurons in the solitary tract nucleus, area postrema and lateral parabrachial nucleus to gastric load of hypertonic saline. Neurosci. Lett. 158 (1), 47–50. https://doi.org/10.1016/0304-3940(93)90609-O.

Kobayashi, T., Good, C., Mamiya, K., Skinner, R.D., Garcia-Rill, E., 2004. Development of REM sleep drive and clinical implications.

J. Appl. Physiol. 96 (2), 735–746. https://doi.org/10.1152/japplphysiol.00908.2003.

Kojima, M., Sano, Y., 1983. The organization of serotonin fibers in the anterior column of the mammalian spinal cord. Anat. Embryol. 167 (1), 1–11. https://doi.org/10.1007/BF00304597.

Kocsis, B., Vertes, R.P., 1996. Midbrain raphe cell firing and hippocampal theta rhythm in urethane-anaesthetized rats. Neuroreport 7 (18), 2867–2872. https://doi.org/10.1097/00001756-199611250-00012.

Kojima, M., Matsuura, T., Tanaka, A., Amagai, T., Imanishi, J., Sano, Y., 1985. Characteristic distribution of noradrenergic terminals on the anterior horn motoneurons innervating the perineal striated muscles in the rat. Anat. Embryol. 171 (3), 267–273. https://doi.org/10.1007/BF00347015.

Kordower, J.H., Le, H.K., Mufson, E.J., 1992. Galanin immunoreactivity in the primate central nervous system. J. Comp. Neurol. 319 (4), 479–500. https://doi.org/10.1002/cne.903190403.

Koutcherov, Y., Huang, X.F., Halliday, G., Paxinos, G., 2004. Organization of human brain stem nuclei. In: The Human Nervous System. Elsevier, pp. 267–320. https://doi.org/10.1016/B978-012547626-3/50011-9.

Krout, K.E., Loewy, A.D., 2000a. Parabrachial nucleus projections to midline and intralaminar thalamic nuclei of the rat. J. Comp. Neurol. 428 (3), 475–494. https://doi.org/10.1002/1096-9861(20001218)428:3<475::AID-CNE6>3.0.CO;2–9.

Krout, K.E., Loewy, A.D., 2000b. Periaqueductal gray matter projections to midline and intralaminar thalamic nuclei of the rat. J. Comp. Neurol. 424 (1), 111–141. https://doi.org/10.1002/1096-9861(20000814)424:1<111::AID-CNE9>3.0.CO;2–3.

Lavoie, B., Parent, A., 1994a. Pedunculopontine nucleus in the squirrel monkey: cholinergic and glutamatergic projections to the substantia nigra. J. Comp. Neurol. 344 (2), 232–241. https://doi.org/10.1002/cne.903440205.

Lavoie, B., Parent, A., 1994b. Pedunculopontine nucleus in the squirrel monkey: distribution of cholinergic and monoaminergic neurons in the mesopontine tegmentum with evidence for the presence of glutamate in cholinergic neurons. J. Comp. Neurol. 344 (2), 190–209. https://doi.org/10.1002/cne.903440203.

Lavoie, B., Parent, A., 1994c. Pedunculopontine nucleus in the squirrel monkey: projections to the basal ganglia as revealed by anterograde tract-tracing methods. J. Comp. Neurol. 344 (2), 210–231. https://doi.org/10.1002/cne.903440204.

Lee, H.S., Kim, M.A., Valentino, R.J., Waterhouse, B.D., 2003. Glutamatergic afferent projections to the dorsal raphe nucleus of the rat. Brain Res. 963 (1–2), 57–71. https://doi.org/10.1016/S0006-8993(02)03841-6.

Leite-Almeida, H., Valle-Fernandes, A., Almeida, A., 2006. Brain projections from the medullary dorsal reticular nucleus: an anterograde and retrograde tracing study in the rat. Neuroscience 140 (2), 577–595. https://doi.org/10.1016/j.neuroscience.2006.02.022.

Li, H., Li, Y.Q., 2000. Collateral projection of substance P receptor expressing neurons in the medullary dorsal horn to bilateral parabrachial nuclei of the rat. Brain Res. Bullet. 53 (2), 163–169. https://doi.org/10.1016/s0361-9230(00)00320-8.

Lima, D., Mendes-Ribeiro, J.A., Coimbra, A., 1991. The spino-latero-reticular system of the rat: projections from the superficial dorsal horn and structural characterization of marginal neurons involved. Neuroscience 45 (1), 137–152. https://doi.org/10.1016/0306-4522(91)90110-A.

Li, Y.Q., Takada, M., Kaneko, T., Mizuno, N., 1996. GABAergic and glycinergic neurons projecting to the trigeminal motor nucleus: a double labeling study in the rat. J. Comparat. Neurol. 373 (4), 498–510. 10.1002/(SICI)1096-9861(19960930)373:4<498::AID-CNE3>3.0.CO;2-x.

Liu, Y., Broman, J., Zhang, M., Edvinsson, L., 2009. Brainstem and thalamic projections from a craniovascular sensory nervous centre in the rostral cervical spinal dorsal horn of rats. Cephalalgia 29 (9), 935–948. https://doi.org/10.1111/j.1468-2982.2008.01829.x.

Llamosas, N., Perez-Caballero, L., Berrocoso, E., Bruzos-Cidon, C., Ugedo, L., Torrecilla, M., 2019. Ketamine promotes rapid and transient activation of AMPA receptor-mediated synaptic transmission in the dorsal raphe nucleus. Prog. Neuro-Psychopharmacol. Biol. Psychiatry 88, 243–252. https://doi.org/10.1016/j.pnpbp.2018.07.022.

Loewy, A.D., McKellar, S., 1981. Serotonergic projections from the ventral medulla to the intermediolateral cell column in the rat. Brain Res. 211 (1), 146–152. https://doi.org/10.1016/0006-8993(81)90074-3.

Lukkes, J.L., Forster, G.L., Renner, K.J., Summers, C.H., 2008. Corticotropin-releasing factor 1 and 2 receptors in the dorsal raphé differentially affect serotonin release in the nucleus accumbens. Eur. J. Pharmacol. 578 (2–3), 185–193. https://doi.org/10.1016/j.ejphar.2007.09.024.

Luschei, E.S., Fuchs, A.F., 1972. Activity of brain stem neurons during eye movements of alert monkeys. J. Neurophysiol. 35 (4), 445–461. https://doi.org/10.1152/jn.1972.35.4.445.

Lyons, W.E., Grzanna, R., 1988. Noradrenergic neurons with divergent projections to the motor trigeminal nucleus and the spinal cord: a double retrograde neuronal labeling study. Neuroscience 26 (2), 681–693. https://doi.org/10.1016/0306-4522(88)90174-1.

Lyons, W., Fritschy, J., Grzanna, R., 1989. The noradrenergic neurotoxin DSP-4 eliminates the coeruleospinal projection but spares projections of the A5 and A7 groups to the ventral horn of the rat spinal cord. J. Neurosci. 9 (5), 1481–1489. https://doi.org/10.1523/JNEUROSCI.09-05-01481.1989.

Maeda, M., Tsuruoka, M., Hayashi, B., Nagasawa, I., Inoue, T., 2009. Descending pathways from activated locus coeruleus/subcoeruleus following unilateral hindpaw inflammation in the rat. Brain Res. Bull. 78 (4–5), 170–174. https://doi.org/10.1016/j.brainresbull.2008.09.005.

Mamounas, L.A., Mullen, C.A., O'hearn, E., Molliver, M.E., 1991. Dual serotoninergic projections to forebrain in the rat: morphologically distinct 5-HT axon terminals exhibit differential vulnerability to neurotoxic amphetamine derivatives. J. Comp. Neurol. 314 (3), 558–586. https://doi.org/10.1002/cne.903140312.

Maru, E., Takahashi, L.K., Iwahara, S., 1979. Effects of median raphe nucleus lesions on hippocampal EEG in the freely moving rat. Brain Res. 163 (2), 223–234. https://doi.org/10.1016/0006-8993(79)90351-2.

McEwen, B.S., 2003. Mood disorders and allostatic load. Biol. Psychiatry 54 (3), 200–207. https://doi.org/10.1016/S0006-3223(03)00177-X.

McGeer, E.G., 1995. Phylogeny and development of catecholamine systems in the CNS of vertebrates. 1994. Edited by W.J.A.J. Smeets and A. Reiner. Published by Cambridge University Press. 488 pages. $C 150.00. Canadian Journal of Neurological Sciences/Journal Canadien des Sciences Neurologiques 22 (4), 339–340. https://doi.org/10.1017/S0317167100039718.

McNaughton, N., Azmitia, E.C., Williams, J.H., Buchan, A., Gray, J.A., 1980. Septal elicitation of hippocampal theta rhythm after localized de-afferentation of serotoninergic fibers. Brain Res. 200 (2), 259–269. https://doi.org/10.1016/0006-8993(80)90918-x.

Melander, T., Hokfelt, T., Rokaeus, A., et al., 1986a. Coexistence of galanin-like immunoreactivity with catecholamines, 5-hydroxytryptamine, GABA and neuropeptides in the rat CNS. J. Neurosci. 6 (12), 3640–3654. https://doi.org/10.1523/JNEUROSCI.06-12-03640.1986.

Melander, T., Hökfelt, T., Rökaeus, A., 1986b. Distribution of galanin-like immunoreactivity in the rat central nervous system. J. Comp. Neurol. 248 (4), 475–517. https://doi.org/10.1002/cne.902480404.

Mena-Segovia, J., Bolam, J.P., 2011. Phasic modulation of cortical high-frequency oscillations by pedunculopontine neurons. Prog.

Brain Res. 85–92. https://doi.org/10.1016/B978-0-444-53839-0.00006-5.

Mena-Segovia, J., Cintra, L., Prospéro-García, O., Giordano, M., 2002. Changes in sleep–waking cycle after striatal excitotoxic lesions. Behav. Brain Res. 136 (2), 475–481. https://doi.org/10.1016/S0166-4328(02)00201-2.

Mena-Segovia, J., Giordano, M., 2003. Striatal dopaminergic stimulation produces c-Fos expression in the PPT and an increase in wakefulness. Brain Res. 986 (1–2), 30–38. https://doi.org/10.1016/S0006-8993(03)03167-6.

Mena-Segovia, J., Bolam, J.P., Magill, P.J., 2004a. Pedunculopontine nucleus and basal ganglia: distant relatives or part of the same family? Trends Neurosci. 27 (10), 585–588. https://doi.org/10.1016/j.tins.2004.07.009.

Mena-Segovia, J., Favila, R., Giordano, M., 2004b. Long-term effects of striatal lesions on c-Fos immunoreactivity in the pedunculopontine nucleus. Eur. J. Neurosci. 20 (9), 2367–2376. https://doi.org/10.1111/j.1460-9568.2004.03696.x.

Menani, J.V., Colombari, D.S.A., Beltz, T.G., Thunhorst, R.L., Johnson, A.K., 1998a. Salt appetite: interaction of forebrain angiotensinergic and hindbrain serotonergic mechanisms. Brain Res. 801 (1–2), 29–35. https://doi.org/10.1016/S0006-8993(98)00530-7.

Menani, J.V., De Luca, L.A., Johnson, A.K., 1998b. Lateral parabrachial nucleus serotonergic mechanisms and salt appetite induced by sodium depletion. Am. J. Physiol. Regul. Integr. Comp. Physiol. 274 (2), R555–R560. https://doi.org/10.1152/ajpregu.1998.274.2.R555.

Menani, J.V., De Luca, L.A., Thunhorst, R.L., Johnson, A.K., 2000. Hindbrain serotonin and the rapid induction of sodium appetite. Am. J. Physiol. Regul. Integr. Comp. Physiol. 279 (1), R126–R131. https://doi.org/10.1152/ajpregu.2000.279.1.R126.

Menani, J.V., Barbosa, S.P., McKinley, M.J., Wade, J.D., De Luca, L.A., 2004. Serotonergic mechanism of the lateral parabrachial nucleus and relaxin-induced sodium intake. Brain Res. 1030 (1), 74–80. https://doi.org/10.1016/j.brainres.2004.09.048.

Menezes, M.F., Barbosa, S.P., De Andrade, C.A.F., Menani, J.V., De Paula, P.M., 2011. Purinergic mechanisms of lateral parabrachial nucleus facilitate sodium depletion-induced NaCl intake. Brain Res. 1372, 49–58. https://doi.org/10.1016/j.brainres.2010.11.075.

Mesulam, M.M., Mufson, E.J., Levey, A.I., Wainer, B.H., 1983a. Cholinergic innervation of cortex by the basal forebrain: cytochemistry and cortical connections of the septal area, diagonal band nuclei, nucleus basalis (Substantia innominata), and hypothalamus in the rhesus monkey. J. Comp. Neurol. 214 (2), 170–197. https://doi.org/10.1002/cne.902140206.

Mesulam, M.M., Mufson, E.J., Wainer, B.H., Levey, A.I., 1983b. Central cholinergic pathways in the rat: an overview based on an alternative nomenclature (Ch1–Ch6). Neuroscience 10 (4), 1185–1201. https://doi.org/10.1016/0306-4522(83)90108-2.

Michelsen, K.A., Schmitz, C., Steinbusch, H.W.M., 2007. The dorsal raphe nucleus—from silver stainings to a role in depression. Brain Res. Rev. 55 (2), 329–342. https://doi.org/10.1016/j.brainresrev.2007.01.002.

Miller, M.A., Kolb, P.E., Leverenz, J.B., Peskind, E.R., Raskind, M.A., 1999. Preservation of noradrenergic neurons in the locus ceruleus that coexpress galanin mRNA in Alzheimer's disease. J. Neurochem. 73 (5), 2028–2036.

Missig, G., Mei, L., Vizzard, M.A., et al., 2017. Parabrachial pituitary adenylate cyclase-activating polypeptide activation of amygdala endosomal extracellular signal–regulated kinase signaling regulates the emotional component of pain. Biol. Psychiatry 81 (8), 671–682. https://doi.org/10.1016/j.biopsych.2016.08.025.

Moga, M.M., Herbert, H., Hurley, K.M., Yasui, Y., Gray, T.S., Saper, C.B., 1990. Organization of cortical, basal forebrain, and hypothalamic afferents to the parabrachial nucleus in the rat. J. Comp. Neurol. 295 (4), 624–661. https://doi.org/10.1002/cne.902950408.

Molliver, M.E., 1987. Serotonergic neuronal systems: what their anatomic organization tells us about function. J. Clin. Psychopharmacol. 7 (6 Suppl. l), 3S–23S.

Monnier, A., Alheid, G.F., McCrimmon, D.R., 2003. Defining ventral medullary respiratory compartments with a glutamate receptor agonist in the rat. J. Physiol. 548 (Pt 3), 859–874. https://doi.org/10.1113/jphysiol.2002.038141.

Moriizumi, T., Hattori, T., 1992. Separate neuronal populations of the rat globus pallidus projecting to the subthalamic nucleus, auditory cortex and pedunculopontine tegmental area. Neuroscience 46 (3), 701–710. https://doi.org/10.1016/0306-4522(92)90156-V.

Morinaga, R., Nakamuta, N., Yamamoto, Y., 2019. Serotonergic projections to the ventral respiratory column from raphe nuclei in rats. Neurosci. Res. 143, 20–30. https://doi.org/10.1016/j.neures.2018.05.004.

Moschovakis, A.K., 1996a. Neural network simulations of the primate oculomotor system. II. Frames of reference. Brain Res. Bull. 40 (5–6), 337–343. https://doi.org/10.1016/0361-9230(96)00124-4.

Moschovakis, A.K., 1996b. The superior colliculus and eye movement control. Curr. Opin. Neurobiol. 6 (6), 811–816. https://doi.org/10.1016/S0959-4388(96)80032-8.

Moschovakis, A.K., Karabelas, A.B., Highstein, S.M., 1988a. Structure-function relationships in the primate superior colliculus. I. Morphological classification of efferent neurons. J. Neurophysiol. 60 (1), 232–262. https://doi.org/10.1152/jn.1988.60.1.232.

Moschovakis, A.K., Karabelas, A.B., Highstein, S.M., 1988b. Structure-function relationships in the primate superior colliculus. II. Morphological identity of presaccadic neurons. J. Neurophysiol. 60 (1), 263–302. https://doi.org/10.1152/jn.1988.60.1.263.

Moschovakis, A., Scudder, C., Highstein, S., 1996. The microscopic anatomy and physiology of the mammalian saccadic system. Prog. Neurobiol. 50 (2–3), 133–254. https://doi.org/10.1016/S0301-0082(96)00034-2.

Moschovakis, A.K., Dalezios, Y., Petit, J., Grantyn, A.A., 1998. New mechanism that accounts for position sensitivity of saccades evoked in response to stimulation of superior colliculus. J. Neurophysiol. 80 (6), 3373–3379. https://doi.org/10.1152/jn.1998.80.6.3373.

Mufson, E.J., Martin, T.L., Mash, D.C., Wainer, B.H., Mesulam, M.M., 1986. Cholinergic projections from the parabigeminal nucleus (Ch8) to the superior colliculus in the mouse: a combined analysis of horseradish peroxidase transport and choline acetyltransferase immunohistochemistry. Brain Res. 370 (1), 144–148. https://doi.org/10.1016/0006-8993(86)91114-5.

Munn, E.M., Harte, S.E., Lagman, A., Borszcz, G.S., 2009. Contribution of the periaqueductal gray to the suppression of pain affect produced by administration of morphine into the intralaminar thalamus of rat. J. Pain 10 (4), 426–435. https://doi.org/10.1016/j.jpain.2008.10.011.

Nagoya, K., Nakamura, S., Ikeda, K., et al., 2017. Distinctive features of Phox2b-expressing neurons in the rat reticular formation dorsal to the trigeminal motor nucleus. Neuroscience 358, 211–226. https://doi.org/10.1016/j.neuroscience.2017.06.035.

Naidich, et al., 2009. Duvernoy's atlas of the human brain stem and cerebellum. Am. J. Neuroradiol. 30 (5), e75. https://doi.org/10.3174/ajnr.A1550. PMCID: PMC7051648.

Nevue, A.A., Felix, R.A., Portfors, C.V., 2016. Dopaminergic projections of the subparafascicular thalamic nucleus to the auditory brainstem. Hear. Res. 341, 202–209. https://doi.org/10.1016/j.heares.2016.09.001.

Nuseir, K., Proudfit, H.K., 2000. Bidirectional modulation of nociception by GABA neurons in the dorsolateral pontine tegmentum that tonically inhibit spinally projecting noradrenergic A7 neurons. Neuroscience 96 (4), 773–783. https://doi.org/10.1016/S0306-4522(99)00603-X.

Ohmura, Y., Izumi, T., Yamaguchi, T., Tsutsui-Kimura, I., Yoshida, T., Yoshioka, M., 2010. The serotonergic projection from the median raphe nucleus to the ventral Hippocampus is involved in the retrieval of fear memory through the corticotropin-releasing factor type 2 receptor. Neuropsychopharmacology 35 (6), 1271–1278. https://doi.org/10.1038/npp.2009.229.

Ohtake, T., 1992. Ascending projections from the gigantocellular reticular and dorsal paragigantocellular nuclei of the medulla oblongata in the rat: an anterograde PHA-L tracing study. Neurosci. Res. 14 (2), 96–116. https://doi.org/10.1016/0168-0102(92)90086-R.

Okere, C.O., Waterhouse, B.D., 2006. Acute restraint increases NADPH-diaphorase staining in distinct subregions of the rat dorsal raphe nucleus: implications for raphe serotonergic and nitrergic transmission. Brain Res. 1119 (1), 174–181. https://doi.org/10.1016/j.brainres.2006.08.058.

Olschowka, J.A., Molliver, M.E., Grzanna, R., Rice, F.L., Coyle, J.T., 1981. Ultrastructural demonstration of noradrenergic synapses in the rat central nervous system by dopamine-beta-hydroxylase immunocytochemistry. J. Histochem. Cytochem. 29 (2), 271–280. https://doi.org/10.1177/29.2.7019303.

Olszewski, J., 1954. Cytoarchitecture of the Human Brainstem. Lippincott, pp. 1–199.

Ordway, G.A., Stockmeier, C.A., Cason, G.W., Klimek, V., 1997. Pharmacology and distribution of norepinephrine transporters in the human locus coeruleus and raphe nuclei. J. Neurosci. 17 (5), 1710–1719. https://doi.org/10.1523/JNEUROSCI.17-05-01710.1997.

Otake, K., 2005. Cholecystokinin and substance P immunoreactive projections to the paraventricular thalamic nucleus in the rat. Neurosci. Res. 51 (4), 383–394. https://doi.org/10.1016/j.neures.2004.12.009.

O'Hearn, E., Molliver, M.E., 1984. Organization of raphe-cortical projections in rat: a quantitative retrograde study. Brain Res. Bull. 13 (6), 709–726. https://doi.org/10.1016/0361-9230(84)90232-6.

Palkovits, M., Epelbaum, J., Tapia-Arancibia, L., Kordon, C., 1982. Somatostatin in catecholamine-rich nuclei of the brainstem. Neuropeptides 3 (2), 139–144. https://doi.org/10.1016/0143-4179(82)90009-9.

Palmiter, R.D., 2018. The parabrachial nucleus: CGRP neurons function as a general alarm. Trends Neurosci. 41 (5), 280–293. https://doi.org/10.1016/j.tins.2018.03.007.

Pan, B., Castro-Lopes, J.M., Coimbra, A., 1999. Central afferent pathways conveying nociceptive input to the hypothalamic paraventricular nucleus as revealed by a combination of retrograde labeling and c-fos activation. J. Comp. Neurol. 413 (1), 129–145.

Pang, Y.W., Ge, S.N., Nakamura, K.C., et al., 2009. Axon terminals expressing vesicular glutamate transporter VGLUT1 or VGLUT2 within the trigeminal motor nucleus of the rat: origins and distribution patterns. J. Comp. Neurol. 512 (5), 595–612. https://doi.org/10.1002/cne.21894.

Pascual, A., Hidalgo-Figueroa, M., Piruat, J.I., Pintado, C.O., Gómez-Díaz, R., López-Barneo, J., 2008. Absolute requirement of GDNF for adult catecholaminergic neuron survival. Nat. Neurosci. 11 (7), 755–761. https://doi.org/10.1038/nn.2136.

Patt, S., Gerhard, L.A., 1993. Golgi study of human locus coeruleus in normal brains and in Parkinson's disease. Neuropathol. Appl. Neurobiol. 19 (6), 519–523. https://doi.org/10.1111/j.1365-2990.1993.tb00480.x.

Pattyn, A., Morin, X., Cremer, H., Goridis, C., Brunet, J.F., 1997. Expression and interactions of the two closely related homeobox genes *Phox2a* and *Phox2b* during neurogenesis. Development 124 (20), 4065–4075. https://doi.org/10.1242/dev.124.20.4065.

Pattyn, A., Morin, X., Cremer, H., Goridis, C., Brunet, J.F., 1999. The homeobox gene Phox2b is essential for the development of autonomic neural crest derivatives. Nature 399 (6734), 366–370. https://doi.org/10.1038/20700.

Paxinos, G., Franklin, K.B.J., 2001. The Mouse Brain in Stereotaxic Coordinates, second ed. Academic Press.

Paxinos, G., Törk, I., 1990. Neuroanatomical nomenclature. Trends Neurosci. 13 (5), 169. https://doi.org/10.1016/0166-2236(90)90041-8.

Pérez, S.E., Wynick, D., Steiner, R.A., Mufson, E.J., 2001. Distribution of galaninergic immunoreactivity in the brain of the mouse. J. Comp. Neurol. 434 (2), 158–185. https://doi.org/10.1002/cne.1171.

Pernar, L., Curtis, A.L., Vale, W.W., Rivier, J.E., Valentino, R.J., 2004. Selective activation of corticotropin-releasing factor-2 receptors on neurochemically identified neurons in the rat dorsal raphe nucleus reveals dual actions. J. Neurosci. 24 (6), 1305–1311. https://doi.org/10.1523/JNEUROSCI.2885-03.2004.

Petrov, T., Krukoff, T.L., Jhamandas, J.H., 1993. Branching projections of catecholaminergic brainstem neurons to the paraventricular hypothalamic nucleus and the central nucleus of the amygdala in the rat. Brain Res. 609 (1–2), 81–92. https://doi.org/10.1016/0006-8993(93)90858-K.

Peyron, C., Tighe, D.K., van den Pol, A.N., de Lecea, L., Heller, H.C., Sutcliffe, J.G., Kilduff, T.S., 1998. Neurons containing hypocretin (orexin) project to multiple neuronal systems. J. Neurosci. 18 (23), 9996–10015. https://doi.org/10.1523/JNEUROSCI.18-23-09996.1998.

Piñeyro, G., Blier, P., 1999. Autoregulation of serotonin neurons: role in antidepressant drug action. Pharmacol. Rev. 51 (3), 533–591.

Porrino, L.J., Goldman-Rakic, P.S., 1982. Brainstem innervation of prefrontal and anterior cingulate cortex in the rhesus monkey revealed by retrograde transport of HRP. J. Comp. Neurol. 205 (1), 63–76. https://doi.org/10.1002/cne.902050107.

Poulat, P., Marlier, L., Rajaofetra, N., Privat, A., 1992. 5-Hydroxytryptamine, substance P and thyrotropin-releasing hormone synapses in the intermediolateral cell column of the rat thoracic spinal cord. Neurosci. Lett. 136 (1), 19–22. https://doi.org/10.1016/0304-3940(92)90637-M.

Probst, A., Cortés, R., Palacios, J.M., 1984. Distribution of alpha 2-adrenergic receptors in the human brainstem: an autoradiographic study using [3H]p-aminoclonidine. Eur. J. Pharmacol. 106 (3), 477–488. https://doi.org/10.1016/0014-2999(84)90051-7.

Proudfit, H.K., Clark, F.M., 1991. The projections of locus coeruleus neurons to the spinal cord. Prog. Brain Res. 123–141. https://doi.org/10.1016/S0079-6123(08)63803-0.

Ptak, K., Yamanishi, T., Aungst, J., et al., 2009. Raphé neurons stimulate respiratory circuit activity by multiple mechanisms via endogenously released serotonin and substance P. J. Neurosci. 29 (12), 3720–3737. https://doi.org/10.1523/JNEUROSCI.5271-08.2009.

Quintero, E.M., Willis, L.M., Zaman, V., et al., 2004. Glial cell line-derived neurotrophic factor is essential for neuronal survival in the locus coeruleus–hippocampal noradrenergic pathway. Neuroscience 124 (1), 137–146. https://doi.org/10.1016/j.neuroscience.2003.11.001.

Rajaofetra, N., Passagia, J.G., Marlier, L., et al., 1992a. Serotoninergic, noradrenergic, and peptidergic innervation of Onuf's nucleus of normal and transected spinal cords of baboons (papio papio). J. Comp. Neurol. 318 (1), 1–17. https://doi.org/10.1002/cne.903180102.

Rajaofetra, N., Ridet, J.L., Poulat, P., et al., 1992b. Immunocytochemical mapping of noradrenergic projections to the rat spinal cord with an antiserum against noradrenaline. J. Neurocytol. 21 (7), 481–494. https://doi.org/10.1007/BF01186952.

Rawson, J.A., Scott, C.J., Pereira, A., Jakubowska, A., Clarke, I.J., 2001. Noradrenergic projections from the A1 field to the preoptic area in the brain of the ewe and Fos responses to oestrogen in the A1 cells. J. Neuroendocrinol. 13 (2), 129–138. https://doi.org/10.1046/j.1365-2826.2001.00588.x.

Reichling, D.B., Basbaum, A.I., 1990. Contribution of brainstem GABAergic circuitry to descending antinociceptive controls: I. GABA-immunoreactive projection neurons in the periaqueductal gray and nucleus raphe magnus. J. Comp. Neurol. 302 (2), 370–377. https://doi.org/10.1002/cne.903020213.

medulla after depletion of C1 cells in rat. J. Physiol. 529 (1), 221–236. https://doi.org/10.1111/j.1469-7793.2000.00221.x.
Reyes, B.A.S., Van Bockstaele, E.J., 2006. Divergent projections of catecholaminergic neurons in the nucleus of the solitary tract to limbic forebrain and medullary autonomic brain regions. Brain Res. 1117 (1), 69–79. https://doi.org/10.1016/j.brainres.2006.08.051.
Ridet, J.L., Rajaofetra, N., Teilhac, J.R., Geffard, M., Privat, A., 1993. Evidence for nonsynaptic serotonergic and noradrenergic innervation of the rat dorsal horn and possible involvement of neuron-glia interactions. Neuroscience 52 (1), 143–157. https://doi.org/10.1016/0306-4522(93)90189-M.
Rikard-Bell, G.C., Törk, L., Sullivan, C., Scheibner, T., 1990. Distribution of substance P-like immunoreactive fibres and terminals in the medulla oblongata of the human infant. Neuroscience 34 (1), 133–148. https://doi.org/10.1016/0306-4522(90)90308-Q.
Rinaldi, P.C., Young, R.F., Albe-Fessard, D., Chodakiewitz, J., 1991. Spontaneous neuronal hyperactivity in the medial and intralaminar thalamic nuclei of patients with deafferentation pain. J. Neurosurg. 74 (3), 415–421. https://doi.org/10.3171/jns.1991.74.3.0415.
Robertson, S.D., Plummer, N.W., de Marchena, J., Jensen, P., 2013. Developmental origins of central norepinephrine neuron diversity. Nat. Neurosci. 16 (8), 1016–1023. https://doi.org/10.1038/nn.3458.
Roeder, Z., Chen, Q., Davis, S., Carlson, J.D., Tupone, D., Heinricher, M.M., 2016. Parabrachial complex links pain transmission to descending pain modulation. Pain 157 (12), 2697–2708. https://doi.org/10.1097/j.pain.0000000000000688.
Ruggiero, D.A., Anwar, M., Golanov, E.V., Reis, D.J., 1997. The pedunculopontine tegmental nucleus issues collaterals to the fastigial nucleus and rostral ventrolateral reticular nucleus in the rat. Brain Res. 760 (1–2), 272–276. https://doi.org/10.1016/S0006-8993(97)00397-1.
Ryan, P.J., Ross, S.I., Campos, C.A., Derkach, V.A., Palmiter, R.D., 2017. Oxytocin-receptor-expressing neurons in the parabrachial nucleus regulate fluid intake. Nat. Neurosci. 20 (12), 1722–1733. https://doi.org/10.1038/s41593-017-0014-z.
Rye, D.B., Lee, H.J., Saper, C.B., Wainer, B.H., 1988. Medullary and spinal efferents of the pedunculopontine tegmental nucleus and adjacent mesopontine tegmentum in the rat. J. Comp. Neurol. 269 (3), 315–341. https://doi.org/10.1002/cne.902690302.
Saitoh, K., Hattori, S., Song, W.J., Isa, T., Takakusaki, K., 2003. Nigral GABAergic inhibition upon cholinergic neurons in the rat pedunculopontine tegmental nucleus. Eur. J. Neurosci. 18 (4), 879–886. https://doi.org/10.1046/j.1460-9568.2003.02825.x.
Sapin, E., Lapray, D., Bérod, A., et al., 2009. Localization of the brainstem GABAergic neurons controlling paradoxical (REM) sleep. PLoS One 4 (1), e4272. https://doi.org/10.1371/journal.pone.0004272.
Sasaki, S., Yoshimura, K., Naito, K., 2004. The neural control of orienting: role of multiple-branching reticulospinal neurons. Prog. Brain Res. 383–389. https://doi.org/10.1016/S0079-6123(03)43036-7.
Satoh, Y., Ishizuka, K., Oskutyte, D., Murakami, T., 2006. Role of the parvicellular reticular formation in facilitating the jaw-opening reflex in rats by stimulation of the red nucleus. Brain Res. 1083 (1), 145–150. https://doi.org/10.1016/j.brainres.2006.02.005.
Sawchenko, P.E., Swanson, L.W., 1982. The organization of noradrenergic pathways from the brainstem to the paraventricular and supraoptic nuclei in the rat. Brain Res. Rev. 4 (3), 275–325. https://doi.org/10.1016/0165-0173(82)90010-8.
Schreihofer, A.M., Guyenet, P.G., 1997. Identification of C1 presympathetic neurons in rat rostral ventrolateral medulla by juxtacellular labeling in vivo. J. Comp. Neurol. 387 (4), 524–536. https://doi.org/10.1002/(SICI)1096-9861(19971103)387:4<524::AID-CNE4>3.0.CO;2-4.
Schreihofer, A.M., Guyenet, P.G., 2002. The baroreflex and beyond: control of sympathetic vasomotor tone by gabaergic neurons in the ventrolateral medulla. Clin. Exp. Pharmacol. Physiol. 29 (5–6), 514–521. https://doi.org/10.1046/j.1440-1681.2002.03665.x.
Schreihofer, A.M., Stornetta, R.L., Guyenet, P.G., 2000. Regulation of sympathetic tone and arterial pressure by rostral ventrolateral medulla after depletion of C1 cells in rat. J. Physiol. 529 (1), 221–236. https://doi.org/10.1111/j.1469-7793.2000.00221.x.
Scott, G., Westberg, K.G., Vrentzos, N., Kolta, A., Lund, J.P., 2003. Effect of lidocaine and NMDA injections into the medial pontobulbar reticular formation on mastication evoked by cortical stimulation in anaesthetized rabbits. Eur. J. Neurosci. 17 (10), 2156–2162. https://doi.org/10.1046/j.1460-9568.2003.02670.x.
Scudder, C.A., Moschovakis, A.K., Karabelas, A.B., Highstein, S.M., 1996. Anatomy and physiology of saccadic long-lead burst neurons recorded in the alert squirrel monkey. II. Pontine neurons. J. Neurophysiol. 76 (1), 353–370. https://doi.org/10.1152/jn.1996.76.1.353.
Semba, K., Fibiger, H.C., 1992. Afferent connections of the laterodorsal and the pedunculopontine tegmental nuclei in the rat: a retro- and antero-grade transport and immunohistochemical study. J. Comp. Neurol. 323 (3), 387–410. https://doi.org/10.1002/cne.903230307.
Sherman, D., Fuller, P.M., Marcus, J., Yu, J., Zhang, P., Chamberlin, N.L., Saper, C.B., Lu, J., 2015. Anatomical location of the mesencephalic locomotor region and its possible role in locomotion, posture, cataplexy, and parkinsonism. Front. Neurol. 6 (140). https://doi.org/10.3389/fneur.2015.00140.
Sirieix, C., Gervasoni, D., Luppi, P.H., Léger, L., 2012. Role of the lateral paragigantocellular nucleus in the network of paradoxical (REM) sleep: an electrophysiological and anatomical study in the rat. PLoS One 7 (1), e28724. https://doi.org/10.1371/journal.pone.0028724.
Skinner, R.D., Garcia-Rill, E., 1984. The mesencephalic locomotor region (MLR) in the rat. Brain Res. 323 (2), 385–389. https://doi.org/10.1016/0006-8993(84)90319-6.
Smith, J.C., Abdala, A.P., Rybak, I.A., Paton, J.F., 2009. Structural and functional architecture of respiratory networks in the mammalian brainstem. Philos. Trans. R. Soc. Lond. B Biol. Sci. 364 (1529), 2577–2587. https://doi.org/10.1098/rstb.2009.0081.
Snyder, K.P., Hill-Smith, T.E., Lucki, I., Valentino, R.J., 2015. Corticotropin-releasing factor in the rat dorsal raphe nucleus promotes different forms of behavioral flexibility depending on social stress history. Neuropsychopharmacology 40 (11), 2517–2525. https://doi.org/10.1038/npp.2015.98.
Song, G., Wang, H., Xu, H., Poon, C.S., 2012. Kölliker–Fuse neurons send collateral projections to multiple hypoxia-activated and nonactivated structures in rat brainstem and spinal cord. Brain Struct. Funct. 217 (4), 835–858. https://doi.org/10.1007/s00429-012-0384-7.
Spike, R.C., Puskár, Z., Andrew, D., Todd, A.J., 2003. A quantitative and morphological study of projection neurons in lamina I of the rat lumbar spinal cord. Eur. J. Neurosci. 18 (9), 2433–2448. https://doi.org/10.1046/j.1460-9568.2003.02981.x.
Stamp, J.A., Semba, K., 1995. Extent of colocalization of serotonin and GABA in the neurons of the rat raphe nuclei. Brain Res. 677 (1), 39–49. https://doi.org/10.1016/0006-8993(95)00119-B.
Steinbusch, H.W.M., 1981. Distribution of serotonin-immunoreactivity in the central nervous system of the rat—cell bodies and terminals. Neuroscience 6 (4), 557–618. https://doi.org/10.1016/0306-4522(81)90146-9.
Strassman, A., Highstein, S.M., McCrea, R.A., 1986a. Anatomy and physiology of saccadic burst neurons in the alert squirrel monkey. I. Excitatory burst neurons. J. Comp. Neurol. 249 (3), 337–357. https://doi.org/10.1002/cne.902490303.
Strassman, A., Highstein, S.M., McCrea, R.A., 1986b. Anatomy and physiology of saccadic burst neurons in the alert squirrel monkey. II. Inhibitory burst neurons. J. Comp. Neurol. 249 (3), 358–380. https://doi.org/10.1002/cne.902490304.
Tagliaferro, R., 2003. In: Apolloni, B., Marinaro, M., Tagliaferri, R. (Eds.), Neural Nets, vol 2859. Springer Berlin Heidelberg. https://doi.org/10.1007/b13826.
Takahashi, M., Sugiuchi, Y., Izawa, Y., Shinoda, Y., 2005. Synaptic inputs and their pathways from fixation and saccade zones of the

superior colliculus to inhibitory burst neurons and pause neurons. Ann. N. Y. Acad. Sci. 1039 (1), 209–219. https://doi.org/10.1196/annals.1325.020.

Takakusaki, K., Habaguchi, T., Saitoh, K., Kohyama, J., 2004a. Changes in the excitability of hindlimb motoneurons during muscular atonia induced by stimulating the pedunculopontine tegmental nucleus in cats. Neuroscience 124 (2), 467–480. https://doi.org/10.1016/j.neuroscience.2003.12.016.

Takakusaki, K., Oohinata-Sugimoto, J., Saitoh, K., Habaguchi, T., 2004b. Role of basal ganglia–brainstem systems in the control of postural muscle tone and locomotion. Prog. Barain Res. 231–237. https://doi.org/10.1016/S0079-6123(03)43023-9.

Takakusaki, K., Saitoh, K., Harada, H., Kashiwayanagi, M., 2004c. Role of basal ganglia–brainstem pathways in the control of motor behaviors. Neurosci. Res. 50 (2), 137–151. https://doi.org/10.1016/j.neures.2004.06.015.

Takakusaki, K., Saitoh, K., Harada, H., Okumura, T., Sakamoto, T., 2004d. Evidence for a role of basal ganglia in the regulation of rapid eye movement sleep by electrical and chemical stimulation for the pedunculopontine tegmental nucleus and the substantia nigra pars reticulata in decerebrate cats. Neuroscience 124 (1), 207–220. https://doi.org/10.1016/j.neuroscience.2003.10.028.

Tang, Z., Zhang, K., Zhang, S., 1987. The fiber projections from the dentate nucleus to the reticular formation of the brain stem in the rabbit. Anat. Embryol. 175 (4), 517–520. https://doi.org/10.1007/BF00309686.

Tavares, L., Lima, D., Coimbra, A., 1997. The pontine A_5 noradrenergic cells which project to the spinal cord dorsal horn are reciprocally connected with the caudal ventrolateral medulla in the rat. Eur. J. Neurosci. 9 (11), 2452–2461. https://doi.org/10.1111/j.1460-9568.1997.tb01662.x.

Thoden, U., Doerr, M., Dieckmann, G., Krainick, J.U., 1979. Medial thalamic permanent electrodes for pain control in man: an electrophysiological and clinical study. Electroencephalogr. Clin. Neurophysiol. 47 (5), 582–591. https://doi.org/10.1016/0013-4694(79)90259-1.

Tillet, Y., Thibault, J., Krieger, M., 1994. Aromatic l-amino acid decarboxylase immunohistochemistry in the suprachiasmatic nucleus of the sheep comparison with tyrosine hydroxylase immunohistochemistry. Brain Res. 648 (2), 319–323. https://doi.org/10.1016/0006-8993(94)91134-7.

Todd, A.J., 2010. Neuronal circuitry for pain processing in the dorsal horn. Nat. Rev. Neurosci. 11 (12), 823–836. https://doi.org/10.1038/nrn2947.

Törk, I., 1990. Anatomy of the serotonergic system. Ann. N. Y. Acad. Sci. 600 (1), 9–34. https://doi.org/10.1111/j.1749-6632.1990.tb16870.x.

Törk, J., Hornung, J., 1990. Raphe nuclei and the serotoninergic system. In: The Human Nervous System. Academic, pp. 1001–1022.

Travers, J.B., Yoo, J.E., Chandran, R., Herman, K., Travers, S.P., 2005. Neurotransmitter phenotypes of intermediate zone reticular formation projections to the motor trigeminal and hypoglossal nuclei in the rat. J. Comp. Neurol. 488 (1), 28–47. https://doi.org/10.1002/cne.20604.

Tsang, E.W., Hamani, C., Moro, E., et al., 2010. Involvement of the human pedunculopontine nucleus region in voluntary movements. Neurology 75 (11), 950–959. https://doi.org/10.1212/WNL.0b013e3181f25b35.

Urban, D.J., Zhu, H., Marcinkiewcz, C.A., et al., 2016. Elucidation of the behavioral program and neuronal network encoded by dorsal raphe serotonergic neurons. Neuropsychopharmacology 41 (5), 1404–1415. https://doi.org/10.1038/npp.2015.293.

Uribe-Mariño, A., Castiblanco-Urbina, M.A., Falconi-Sobrinho, L.L., et al., 2019. The alpha- and beta-noradrenergic receptors blockade in the dorsal raphe nucleus impairs the panic-like response elaborated by medial hypothalamus neurons. Brain Res. 1725, 146468. https://doi.org/10.1016/j.brainres.2019.146468.

ul Haq, R., Anderson, M.L., Hollnagel, J.O., Worschech, F., Sherkheli, M.A., Behrens, C.J., Heinemann, U., 2016. Serotonin dependent masking of hippocampal sharp wave ripples. Neuropharmacology 101, 188–203. https://doi.org/10.1016/j.neuropharm.2015.09.026.

Valentino, R.J., Commons, K.G., 2005. Peptides that fine-tune the serotonin system. Neuropeptides 39 (1), 1–8. https://doi.org/10.1016/j.npep.2004.09.005.

Van Bockstaele, E.J., Chan, J., 1997. Electron microscopic evidence for coexistence of leucine5-enkephalin and γ-aminobutyric acid in a subpopulation of axon terminals in the rat locus coeruleus region. Brain Res. 746 (1–2), 171–182. https://doi.org/10.1016/S0006-8993(96)01194-8.

Van Bockstaele, E.J., Pieribone, V.A., Aston-Jones, G., 1989. Diverse afferents converge on the nucleus paragigantocellularis in the rat ventrolateral medulla: retrograde and anterograde tracing studies. J. Comp. Neurol. 290 (4), 561–584. https://doi.org/10.1002/cne.902900410.

Van Bockstaele, E.J., Biswas, A., Pickel, V.M., 1993. Topography of serotonin neurons in the dorsal raphe nucleus that send axon collaterals to the rat prefrontal cortex and nucleus accumbens. Brain Res. 624 (1–2), 188–198. https://doi.org/10.1016/0006-8993(93)90077-Z.

Varga, V., Sik, A., Freund, T.F., Kocsis, B., 2002. GABA(B) receptors in the median raphe nucleus: distribution and role in the serotonergic control of hippocampal activity. Neuroscience 109 (1), 119–132. https://doi.org/10.1016/s0306-4522(01)00448-1.

Veinante, P., Yalcin, I., Barrot, M., 2013. The amygdala between sensation and affect: a role in pain. J. Mol. Psychiatry 1 (1), 9. https://doi.org/10.1186/2049-9256-1-9.

Verret, L., Léger, L., Fort, P., Luppi, P.H., 2005. Cholinergic and noncholinergic brainstem neurons expressing Fos after paradoxical (REM) sleep deprivation and recovery. Eur. J. Neurosci. 21 (9), 2488–2504. https://doi.org/10.1111/j.1460-9568.2005.04060.x.

Verret, L., Fort, P., Gervasoni, D., Léger, L., Luppi, P.H., 2006. Localization of the neurons active during paradoxical (REM) sleep and projecting to the locus coeruleus noradrenergic neurons in the rat. J. Comp. Neurol. 495 (5), 573–586. https://doi.org/10.1002/cne.20891.

Vertes, R.P., 1981. An analysis of ascending brain stem systems involved in hippocampal synchronization and desynchronization. J. Neurophysiol. 46 (5), 1140–1159. https://doi.org/10.1152/jn.1981.46.5.1140.

Vertes, R.P., 2004. Memory consolidation in sleep; dream or reality. Neuron 44 (1), 135–148. https://doi.org/10.1016/j.neuron.2004.08.034.

Vertes, R.P., 2005. Hippocampal theta rhythm: a tag for short-term memory. Hippocampus 15 (7), 923–935. https://doi.org/10.1002/hipo.20118.

Vertes, R.P., Linley, S.B., Hoover, W.B., 2010. Pattern of distribution of serotonergic fibers to the thalamus of the rat. Brain Struct. Funct. 215 (1), 1–28. https://doi.org/10.1007/s00429-010-0249-x.

Viana Di Prisco, G., Albo, Z., Vertes, R.P., Kocsis, B., 2002. Discharge properties of neurons of the median raphe nucleus during hippocampal theta rhythm in the rat. Exp. Brain Res. 145 (3), 383–394. https://doi.org/10.1007/s00221-002-1123-8.

Voogd, J., Glickstein, M., 1998. The anatomy of the cerebellum. Trends Cogn. Sci. 2 (9), 307–313. https://doi.org/10.1016/S1364-6613(98)01210-8.

Waitzman, D.M., Silakov, V.L., Cohen, B., 1996. Central mesencephalic reticular formation (cMRF) neurons discharging before and during eye movements. J. Neurophysiol. 75 (4), 1546–1572. https://doi.org/10.1152/jn.1996.75.4.1546.

Waitzman, D.M., Silakov, V.L., DePalma-Bowles, S., Ayers, A.S., 2000a. Effects of reversible inactivation of the primate mesencephalic reticular formation. I. Hypermetric goal-directed saccades. J. Neurophysiol. 83 (4), 2260–2284. https://doi.org/10.1152/jn.2000.83.4.2260.

Waitzman, D.M., Silakov, V.L., DePalma-Bowles, S., Ayers, A.S., 2000b. Effects of reversible inactivation of the primate mesencephalic reticular formation. II. Hypometric vertical saccades. J. Neurophysiol. 83 (4), 2285–2299. https://doi.org/10.1152/jn.2000.83.4.2285.

Wallace, J.A., Lauder, J.M., 1983. Development of the serotonergic system in the rat embryo: an immunocytochemical study. Brain Res. Bull. 10 (4), 459–479. https://doi.org/10.1016/0361-9230(83)90144-2.

Wang, Q.P., Ochiai, H., Nakai, Y., 1993. Electron microscopic immunocytochemistry of GABAergic interneurons in the dorsal raphe nucleus of the rat. Showa Univ. J. Medi. Scie. 5 (1), 89–98. https://doi.org/10.15369/sujms1989.5.89.

Wang, Y.Q., Li, R., Zhang, M.Q., Zhang, Z., Qu, W.M., Huang, Z.L., 2015. The neurobiological mechanisms and treatments of REM sleep disturbances in depression. Curr. Neuropharmacol. 13 (4), 543–553. https://doi.org/10.2174/1570159x13666150310002540.

Waterhouse, B.D., Mihailoff, G.A., Baack, J.C., Woodward, D.J., 1986. Topographical distribution of dorsal and median raphe neurons projecting to motor, sensorimotor, and visual cortical areas in the rat. J. Comp. Neurol. 249 (4), 460–476. https://doi.org/10.1002/cne.902490403.

Waterhouse, B.D., Border, B., Wahl, L., Mihailoff, G.A., 1993. Topographic organization of rat locus coeruleus and dorsal raphe nuclei: distribution of cells projecting to visual system structures. J. Comp. Neurol. 336 (3), 345–361. https://doi.org/10.1002/cne.903360304.

Westberg, K.G., Sandström, G., Olsson, K.Å., 1995. Integration in trigeminal premotor interneurones in the cat. Exp. Brain Res. 104 (3). https://doi.org/10.1007/BF00231979.

Westlund, K.N., Craig, A.D., 1996. Association of spinal lamina I projections with brainstem catecholamine neurons in the monkey. Exp. Brain Res. 110 (2). https://doi.org/10.1007/BF00228547.

Westlund, K.N., Dan Coulter, J., 1980. Descending projections of the locus coeruleus and subcoeruleus/medial parabrachial nuclei in monkey: axonal transport studies and dopamine-β-hydroxylase immunocytochemistry. Brain Res. Rev. 2 (1–3), 235–264. https://doi.org/10.1016/0165-0173(80)90009-0.

Westlund, K.N., Bowker, R.M., Ziegler, M.G., Coulter, J.D., 1982. Descending noradrenergic projections and their spinal terminations. Prog. Brain Res. 219–238. https://doi.org/10.1016/S0079-6123(08)64131-X.

Westlund, K.N., Bowker, R.M., Ziegler, M.G., Coulter, J.D., 1983. Noradrenergic projections to the spinal cord of the rat. Brain Res. 263 (1), 15–31. https://doi.org/10.1016/0006-8993(83)91196-4.

Westlund, K.N., Bowker, R.M., Ziegler, M.G., Coulter, J.D., 1984. Origins and terminations of descending noradrenergic projections to the spinal cord of monkey. Brain Res. 292 (1), 1–16. https://doi.org/10.1016/0006-8993(84)90884-9.

Westlund, K.N., Denney, R.M., Kochersperger, L.M., Rose, R.M., Abell, C.W., 1985. Distinct monoamine oxidase A and B populations in primate brain. Science (New York, N.Y.) 230 (4722), 181–183. https://doi.org/10.1126/science.3875898.

Williams, M.C., Ivanusic, J.J., 2008. Evidence for the involvement of the spinoparabrachial pathway, but not the spinothalamic tract or postsynaptic dorsal column, in acute bone nociception. Neurosci. Lett. 443 (3), 246–250. https://doi.org/10.1016/j.neulet.2008.07.076.

Winn, P., 2006. How best to consider the structure and function of the pedunculopontine tegmental nucleus: evidence from animal studies. J. Neurol. Sci. 248 (1–2), 234–250. https://doi.org/10.1016/j.jns.2006.05.036.

Winterer, J., Maier, N., Wozny, C., Beed, P., Breustedt, J., Evangelista, R., Peng, Y., D'Albis, T., Kempter, R., Schmitz, D., 2017. Excitatory microcircuits within superficial layers of the medial entorhinal cortex. Cell Rep. 19 (6), 1110–1116. https://doi.org/10.1016/j.celrep.2017.04.041.

Woolf, N.J., Butcher, L.L., 1989. Cholinergic systems in the rat brain: IV. Descending projections of the pontomesencephalic tegmentum. Brain Res. Bull. 23 (6), 519–540. https://doi.org/10.1016/0361-9230(89)90197-4.

Woulfe, J.M., Hrycyshyn, A.W., Flumerfelt, B.A., 1988. Collateral axonal projection from the A1 noradrenergic cell group to the paraventricular nucleus and bed nucleus of the stria terminalis in the rat. Exp. Neurol. 102 (1), 121–124. https://doi.org/10.1016/0014-4886(88)90084-2.

Woulfe, J.M., Flumerfelt, B.A., Hrycyshyn, A.W., 1990. Efferent connections of the A1 noradrenergic cell group: a DBH immunohistochemical and PHA-L anterograde tracing study. Exp. Neurol. 109 (3), 308–322. https://doi.org/10.1016/S0014-4886(05)80022-6.

Wu, W., Elde, R., Wessendorf, M.W., 1993. Organization of the serotonergic innervation of spinal neurons in rats–III. Differential serotonergic innervation of somatic and parasympathetic preganglionic motoneurons as determined by patterns of co-existing peptides. Neuroscience 55 (1), 223–233. https://doi.org/10.1016/0306-4522(93)90468-u.

Yaksh, T.L., 1985. Pharmacology of spinal adrenergic systems which modulate spinal nociceptive processing. Pharmacol. Biochem. Behav. 22 (5), 845–858. https://doi.org/10.1016/0091-3057(85)90537-4.

Yamamoto, T., Watanabe, S., Oishi, R., Ueki, S., 1979. Effects of midbrain raphe stimulation and lesion on EEG activity in rats. Brain Res. Bull. 4 (4), 491–495. https://doi.org/10.1016/0361-9230(79)90033-9.

Yang, C.C.H., Chan, J.Y.H., Chan, S.H.H., 1995. Excitatory innervation of caudal hypoglossal nucleus from nucleus reticularis gigantocellularis in the rat. Neuroscience 65 (2), 365–374. https://doi.org/10.1016/0306-4522(94)00473-I.

Yasui, Y., Tsumori, T., Ono, K., Kishi, T., 1997. Nigral axon terminals are in contact with parvicellular reticular neurons which project to the motor trigeminal nucleus in the rat. Brain Res. 775 (1–2), 219–224. https://doi.org/10.1016/S0006-8993(97)00957-8.

Yasui, Y., Yokota, S., Ono, K., Tsumori, T., 2001. Projections from the red nucleus to the parvicellular reticular formation and the cervical spinal cord in the rat, with special reference to innervation by branching axons. Brain Res. 923 (1–2), 187–192. https://doi.org/10.1016/S0006-8993(01)03196-1.

Yokota, S., Oka, T., Asano, H., Yasui, Y., 2016. Orexinergic fibers are in contact with Kölliker-Fuse nucleus neurons projecting to the respiration-related nuclei in the medulla oblongata and spinal cord of the rat. Brain Res. 1648, 512–523. https://doi.org/10.1016/j.brainres.2016.08.020.

Yoshida, K., McCrea, R., Berthoz, A., Vidal, P.P., 1982. Morphological and physiological characteristics of inhibitory burst neurons controlling horizontal rapid eye movements in the alert cat. J. Neurophysiol. 48 (3), 761–784. https://doi.org/10.1152/jn.1982.48.3.761.

Further Reading

Bevan, M., Bolam, J., 1995. Cholinergic, GABAergic, and glutamate-enriched inputs from the mesopontine tegmentum to the subthalamic nucleus in the rat. J. Neurosci. 15 (11), 7105–7120. https://doi.org/10.1523/JNEUROSCI.15-11-07105.1995.

Ciriello, J., Caverson, M.M., 1984a. Direct pathway from neurons in the ventrolateral medulla relaying cardiovascular afferent information to the supraoptic nucleus in the cat. Brain Res. 292 (2), 221–228. https://doi.org/10.1016/0006-8993(84)90758-3.

Ciriello, J., Caverson, M.M., 1984b. Ventrolateral medullary neurons relay cardiovascular inputs to the paraventricular nucleus. Am. J. Physiol. Regul. Integr. Comp. Physiol. 246 (6), R968–R978. https://doi.org/10.1152/ajpregu.1984.246.6.R968.

Dahlstroem, A., Fuxe, K., 1964c. Evidence for the existence of monoamine-containing neurons in the central nervous system. I. Demonstration of monoamines in the cell bodies of brain stem neurons. Acta Physiol. Scand. Suppl. 232 (Suppl. L), 1–55.

Fung, M.L., St John, W.M., 1995a. Expiratory neural activities in gasping induced by pharyngeal stimulation and hypoxia. Respir.

Physiol. 100 (2), 119–127. https://doi.org/10.1016/0034-5687(94)00128-M.

Fung, M.L., St John, W.M., 1995b. The functional expression of a pontine pneumotaxic centre in neonatal rats. J Physiol 489 (2), 579–591. https://doi.org/10.1113/jphysiol.1995.sp021074.

Fuxe, K., 1965. Evidence for the existence of monoamine neurons in the central nervous system. IV. Distribution of monoamine nerve terminals in the central nervous system. Acta Physiol. Scand. Suppl. 247 (Suppl. L), 37(+).

García-Gomar, M.G., Videnovic, A., Singh, K., Stauder, M., Lewis, L.D., Wald, L.L., Rosen, B.R., Bianciardi, M., 2022. Disruption of brainstem structural connectivity in REM sleep behavior disorder using 7 Tesla magnetic resonance imaging. Movement disorders. Official Journal of the Movement Disorder Society 37 (4), 847–853. https://doi.org/10.1002/mds.28895.

Guo, W., Fan, S., Xiao, D., He, C., Guan, M., Xiong, W., 2022. A midbrain-reticulotegmental circuit underlies exaggerated startle under fear emotions. Mol. Psychiatry 27 (12), 4881–4892. https://doi.org/10.1038/s41380-022-01782-6.

Herbert, H., Bellintani-Guardia, B., 1995. Morphology and dendritic domains of neurons in the lateral parabrachial nucleus of the rat. J. Comp. Neurol. 354 (3), 377–394. https://doi.org/10.1002/cne.903540307.

Hornung, J.P., 2003. The human raphe nuclei and the serotonergic system. J. Chem. Neuroanat. 26 (4), 331–343. https://doi.org/10.1016/j.jchemneu.2003.10.002.

Hosogai, M., Matsuo, S., Sibahara, T., Kawai, Y., 1998. Projection of respiratory neurons in rat medullary raphé nuclei to the phrenic nucleus. Respir. Physiol. 112 (1), 37–50. https://doi.org/10.1016/S0034-5687(98)00011-5.

Komatsuzaki, A., 1972. Direction of nystagmus and the deviation of the position of the eyeballs when the eyes are closed. Nihon Jibiinkoka Gakkai Kaiho 75 (10), 1054–1055.

Komatsuzaki, A., Alpert, J., Harris, H.E., Cohen, B., 1972. Effects of mesencephalic reticular formation lesions on optokinetic nystagmus. Exp. Neurol. 34 (3), 522–534. https://doi.org/10.1016/0014-4886(72)90047-7.

Pau, F., Lee, C., Cowles, A., Yang, S.P., Hess, D.L., Spies, H., 2002. Possible involvement of norepinephrine transporter activity in he pulsatility of hypothalamic gonadotropin-releasing hormone release: influence of the gonad. J. Neuroendocrinol. 10 (1), 21–29. https://doi.org/10.1046/j.1365-2826.1998.00173.x.

Peyron, C., Petit, J.M., Rampon, C., Jouvet, M., Luppi, P.H., 1997. Forebrain afferents to the rat dorsal raphe nucleus demonstrated by retrograde and anterograde tracing methods. Neuroscience 82 (2), 443–468. https://doi.org/10.1016/S0306-4522(97)00268-6.

Tang, Z., Zhang, S., 1987. The cerebellar projection from the reticular formation of the brain stem in the rabbit. Anat. Embryol. 175 (4), 521–526. https://doi.org/10.1007/BF00309687.

Yaksh, T.L., Noueihed, R., 1985. The physiology and pharmacology of spinal opiates. Annu. Rev. Pharmacol. Toxicol. 25 (1), 433–462. https://doi.org/10.1146/annurev.pa.25.040185.002245.

6

The Cerebellum

Jorge Ojeda[1], *Giovanni Paternò*[2] *and Sergio Juica*[3]

[1]Biomedical Science Research Laboratory, Basic Sciences Department, Faculty of Medicine, Universidad Católica de la Santísima Concepción, Concepción, Chile [2]Department of Neurosurgery, Istituto Clinico Città Studi, Milan, Italy [3]Faculty of Medicine, University of Concepcion and Hospital Regional de Concepción, Concepción, Chile

OUTLINE

THE EMBRYONIC CEREBELLAR DEVELOPMENT

The embryonic cerebellar formation begins during the fourth gestational week (GW) and takes place over an extended period until around 2 years of postnatal age (van Essen et al., 2020). A protuberance on the dorsal alar plate of the rhombencephalon facing the fourth ventricle indicates the onset of the cerebellar development (Cho et al., 2011; Müller & O'Rahilly, 1988). Two germinal zones, the cerebellar primordium and the rhombic lip expand into the basis of the cerebellar hemispheres at the sixth gestational week (GW) (Cho et al., 2011; Marzban et al., 2015; Müller & O'Rahilly, 1988). At the same week, neural progenitors start the migration from the ventricular zone (VZ) (Cho et al., 2011; Müller & O'Rahilly, 1988; Rakic and Sidman, 1970). During the seventh GW, the neural progenitors from the VZ and the rhombic lip continue migrating and expanding to form the dentate nucleus in each cerebellar hemisphere (Cho et al., 2011; Marzban et al., 2015; Müller & O'Rahilly, 1988). Regarding the cerebellar architecture, the superior cerebellar peduncle is formed, and the cerebellar plate thickens at the rostral surface and the vermis becomes evident at the eighth GW. Subsequently, the external granular layer (EGL; also known as external granular zone) extends from the rhombic lip onto the dorsum of the cerebellar protuberance, and the cerebellar plate is arched over the brainstem (Cho et al., 2011; Müller & O'Rahilly, 1988; Rakic and Sidman, 1970). Additionally, the inferior cerebellar peduncle shows an exceptional size connecting the spinal cord and the precerebral nuclei (Müller & O'Rahilly, 1990). At the 10th GW, both hemispheres merge in the midline and the primary fissures appear in the vermis to increase in depth around the 15th GW.

At cellular level, the multipolar and stellate Purkinje cell (PCs) appear at the sixth GW, forming an evident multilayer from the ventricular zone to the molecular layer (ML) since the 10th to the 13th GW (Haldipur et al., 2018; Rakic and Sidman, 1970). At the same time emerges PAX2-expressing cerebellar inhibitory interneuron progenitors (PIPs). Then, in a first stage of maturation, PCs go through a change of dendritic morphology into a spindle shape which is reached at the 16th GW (Haldipur et al., 2018). Then in a second stage of maturation, which reaches the 28th GW, PCs extend causing an increase in the cerebellar volume by

From Anatomy to Function of the Central Nervous System
https://doi.org/10.1016/B978-0-12-822404-5.00008-5

expansion of the EGL. Also, this stage is accompanied by a period of maximum thickness of the EGL between the 20th to 33rd GW and by an increased foliar complexity (Haldipur et al., 2018). Approximately during the same period, the transient lamina dissecans is observed between the maturing PCs and the internal granular layer (IGL) (Rakic and Sidman, 1970). Also within this stage, PCs begin the formation of a monolayer at the 20th to twenty-fourths GW (Rakic and Sidman, 1970). A third stage of PCs maturation starts around the 28th GW when these cells extend long axons and dendritic arborization increasing in complexity which is consistent with the thickening of the molecular layer (ML) that extends since the 38th GW to the first postnatal year (Haldipur et al., 2011; Rakic and Sidman, 1970). On the other hand, the granule cells precursors (GCPs) differentiate into postmitotic granule cells (GCs) in the EGL and migrate inward crossing the Purkinje cell layer (PCL) to establish mature granule cells into the IGL since the 11th GW (Ábrahám et al., 2001; Rakic and Sidman, 1970). A peak of formation is observed around the fourth postnatal month with a subsequent increase in IGL volume (Kiessling et al., 2014). Thus, the proliferation of postmitotic GCs continues in the postnatal period until around eighth months old (Haldipur et al., 2012; Kiessling et al., 2014; Rakic and Sidman, 1970).

The human cerebellar cortex presents four layers at birth but the EGL disappears at the end of the second postnatal year. This cerebellar organization underlies several important events that take place in the last trimester of gestation such as foliation, peak of proliferation, migration of GCPs, interneuron differentiation, late stages of PC differentiation and synapse formation, promoting an increase in size of the cerebellum of fivefold (Haldipur et al., 2018; Rakic and Sidman, 1970). Further, the cerebellar size increases until the adolescence (Tiemeier et al., 2011).

ANATOMY OF THE CEREBELLUM

The human cerebellum is the part of the brain located in the posterior cranial fossa behind the pons and the medulla oblongata, separated by the fourth ventricle (Fig. 6.1). In the lower back, it is related to the skull occipital bone and superiorly with the

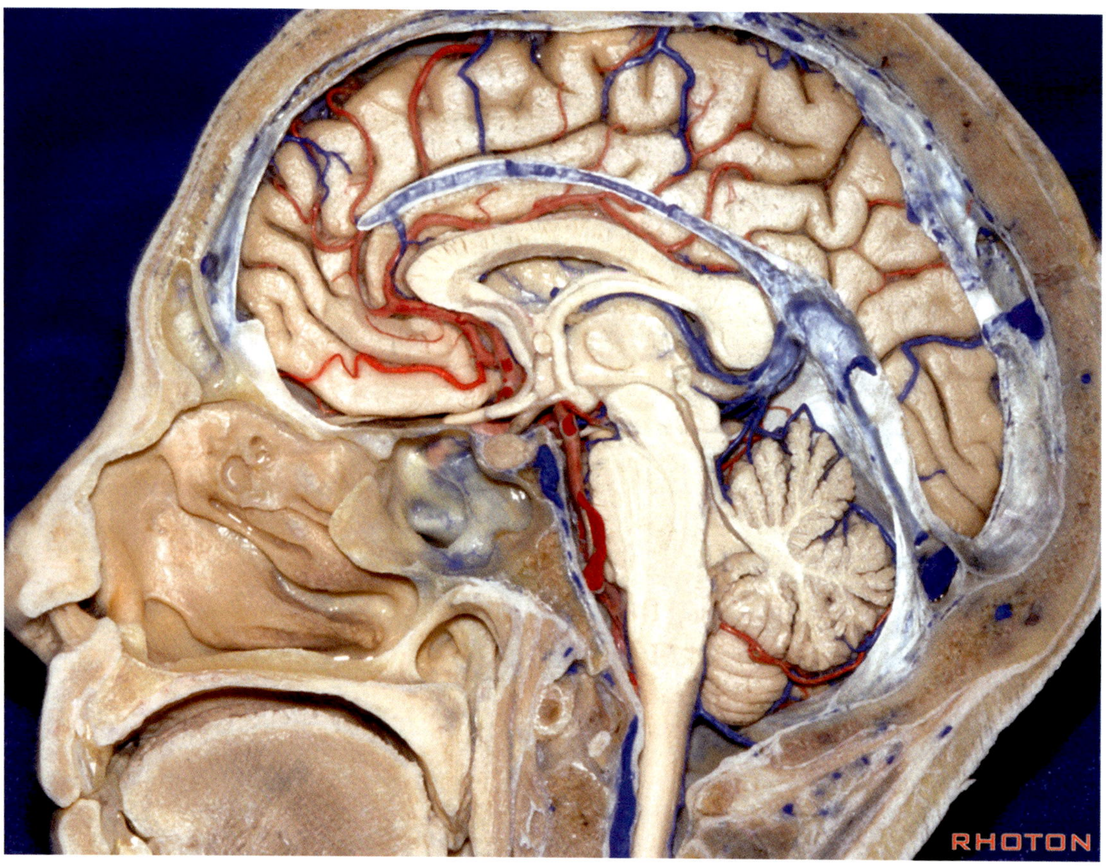

FIGURE 6.1 **Median sagittal section of the skull.** The cerebellum is located in the posterior cerebral fossa and is separated from the occipital lobe by the tentorium cerebelli, a fold of the dura mater. The arbor vitae is easily appreciated in this section; it consists of the white fibers, efferent and afferent from and to the cerebellar cortex. *Image courtesy of The Rhoton Collection® http://rhoton.ineurodb.org/.*

tentorium cerebelli (a dura mater invagination) (Fig. 6.1). The cerebellum has a transverse diameter of 10 cm, an anteroposterior diameter of 6 cm and a vertical diameter of 5 cm (Llinás et al., 2004); weighting approximately from 136 to 169 g and representing around 11% of the adult brain weight (Solov'ev, 2006). Its external organization is like a butterfly, represented by the vermis and two cerebellar hemispheres. The surface is crossed by transverse grooves corresponding to the fissures of the cerebellum that divide it into lobes and lobules which, in turn, further divide into folias (Voogd and Glickstein, 1998).

From a superior view, the superior portion vermis and faces of the cerebellar hemispheres are observed as well as the transverses prima and horizontal fissures that divide three main lobes (Fig. 6.3). From a lower view, it is possible to appreciate a deep cleft called the vallecula and the lower face of the vermis accompanied of both hemispheres, each one representing a dilatation on the internal side called cerebellar tonsil (Fig. 6.4). These structures are in direct contact with the medulla oblongata. The initial portion of the upper vermis as well as the lower vermis from the anterior face is seen, leaving in the middle portion a slit corresponding to the roof of the fourth ventricle. In the internal region of both hemispheres, the hilum, where both of the cerebellar peduncles (upper middle and lower) are distinguished. The upper and lower peduncles are joined by a pons called the superior medullary velum and the inferior medullary velum, respectively (Manni and Petrosini, 2004a).

In front of the vermis there is a rounded structure called nodule that joins each side with the flocculus through a commissure of the flocculus and constituting the flocculonodular lobe (Figs. 6.2 and 6.4). In turn, from the anterior face the beginning of the prima and horizontal fissures are observed (Figs. 6.2 and 6.3).

Likewise, the external organization of the cerebellum can be divided regarding its phylogeny and morphology (Fig. 6.5). In the first one, the phylogenic classification considers the morphology and the function of the cerebellum to describe the archicerebellum, the paleocerebellum, and the neocerebellum (Fig. 6.5) (Manni and Petrosini, 2004a; Roostaei et al., 2014; Voogd and Glickstein, 1998; Voogd et al., 1996). In detail, the archicerebellum is connected with the vestibular nuclei to

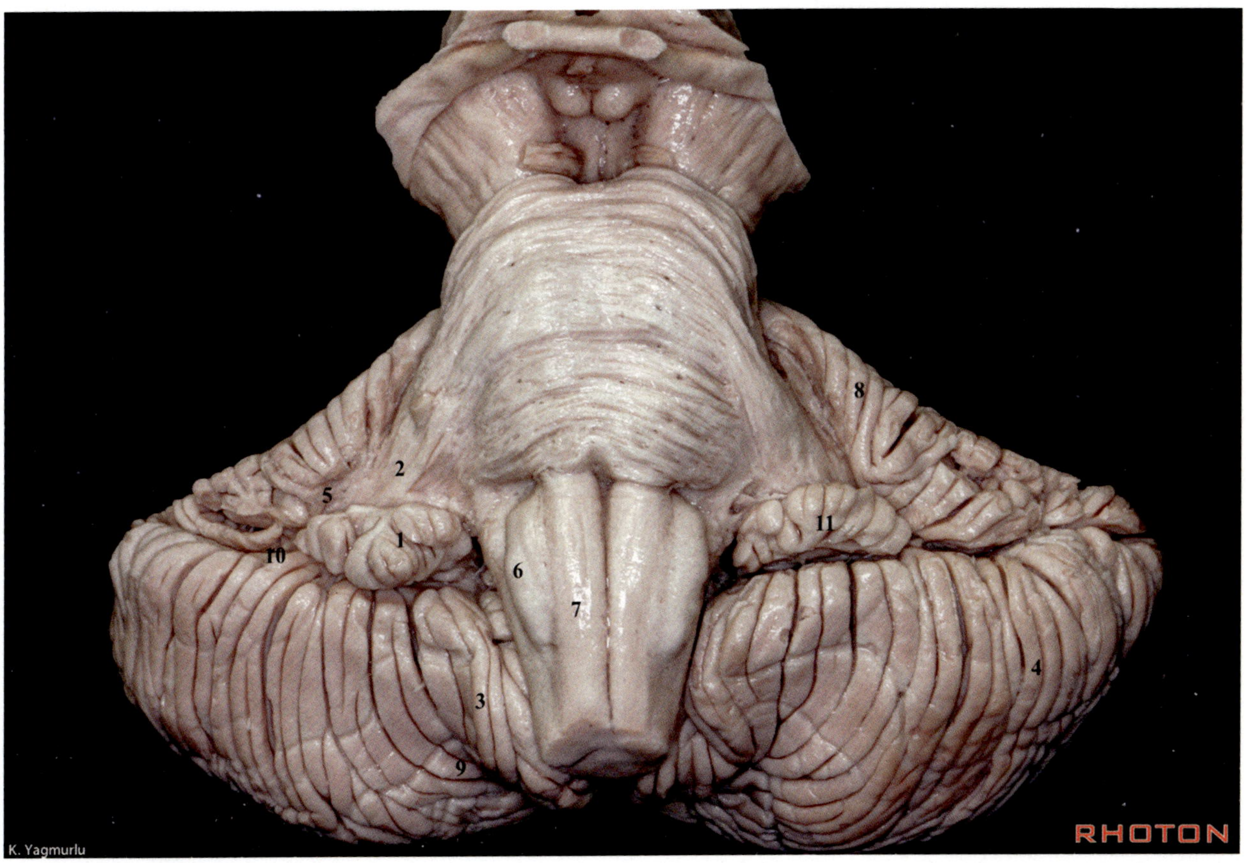

FIGURE 6.2 **Gross anatomy of the cerebellum; anterior view.** 1. Flocculus; 2. Middle cerebellar peduncle; 3. Tonsil; 4. Middle lobe (posterior lobe); 5. Ponto-cerebellar angle; 6. Olive; 7. Pyramid; 8. Anterior lobe; 9. Retrotonsillar fissure; 10. Postero-lateral fissure; 11. Flocculonodular lobe. *Image courtesy of The Rhoton Collection® http://rhoton.ineurodb.org/. Dissector: K. Yagmurlu.*

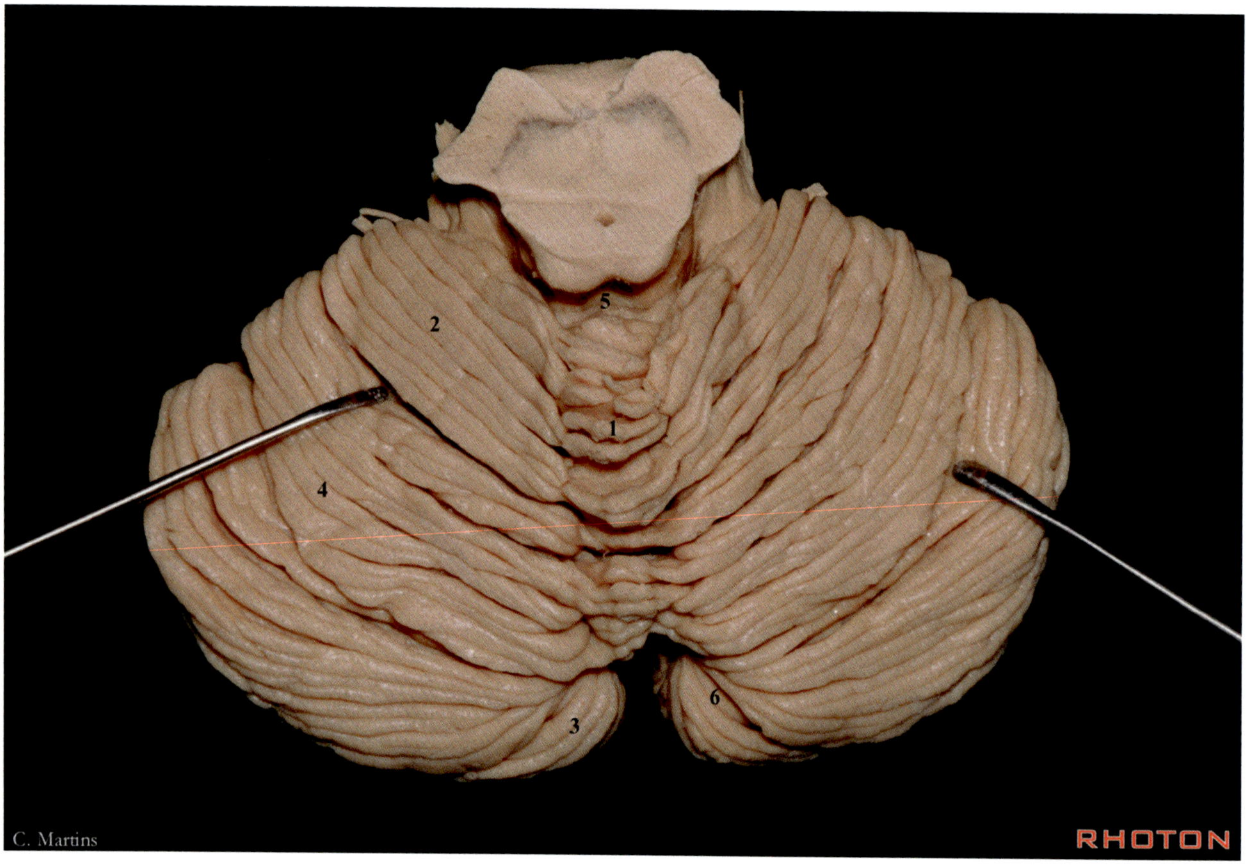

FIGURE 6.3 **Gross anatomy of the cerebellum; superior view.** 1. Vermis; 2. Anterior lobe; 3. Inferior semilunar lobule; 4. Middle lobe (posterior lobe); 5. Anterior cerebellar notch; 6. Horizontal fissure. *Image courtesy of The Rhoton Collection® http://rhoton.ineurodb.org/. Dissector: C. Martins.*

maintain the balance and ocular movements, the paleocerebellum (anterior lobe) is involved in the maintenance of muscular tone and posture, and the neocerebellum (posterior lobe) is related to exploiting the skills learned and the control of voluntary and automatized movements. In the second one, the morphological organization describes two deep fissures, the prima and the posterolateral, which divide the cerebellum into three lobes: anterior, posterior, and flocculonodular. Each lobe contains a portion of the vermis and a portion of the cerebellar hemispheres.

In a classical view of the cerebellum, the portion of the vermis corresponding to each lobe is subdivided into segments with the subsequent association of a pair of lobules located in each hemisphere (Fig. 6.5). The secondary transverse fissures determine the subdivision of each of the lobules. Currently, the cerebellar lobes are divided into 10 lobules identified by Roman numerals (Fig. 6.5). The lobules I–V constitute the anterior lobe, lobules VI–IX the posterior lobe, and the X is the flocculonodular lobe (Fig. 6.5).

The cerebellum is connected to the brainstem through the upper, middle, and lower cerebellar peduncles in each hemisphere. The superior cerebellar peduncles connect with the superior portions of the brainstem, particularly the midbrain. The middle cerebellar peduncles connect to the brainstem at the pons, and the lower cerebellar peduncles connect to the medulla oblongata. Internally, the cerebellum is formed in the same way as the brain where it has an outer tightly folded layer of gray matter corresponding to the cerebellar cortex and an internal mass of white matter interrupted by four intracerebellar nuclei: dentate, fastigial, globose, and emboliform (Figs. 6.6 and 6.7). The white matter formed by axons that go to and from the cerebellum has a medullary body from which branches of axons are projected that reach the different laminae, which is known as the arbor vitae.

Three main arteries originating from the vertebrobasilar system configurate three different arterial territories supply the vascularization of the human cerebellar cortex (see Chapter 2) (Duvernoy et al., 1983; Icardo et al., 1982). The first one is the posterior inferior cerebellar artery supply the inferior cerebellar peduncle, the inferior portion of the vermis and the inferior surface of the cerebellar hemispheres. Subsequently, the posterior inferior cerebellar artery split in the medial o vermis branch and the lateral branch that supply the medial cerebellum as well the dorsolateral area of the bulb,

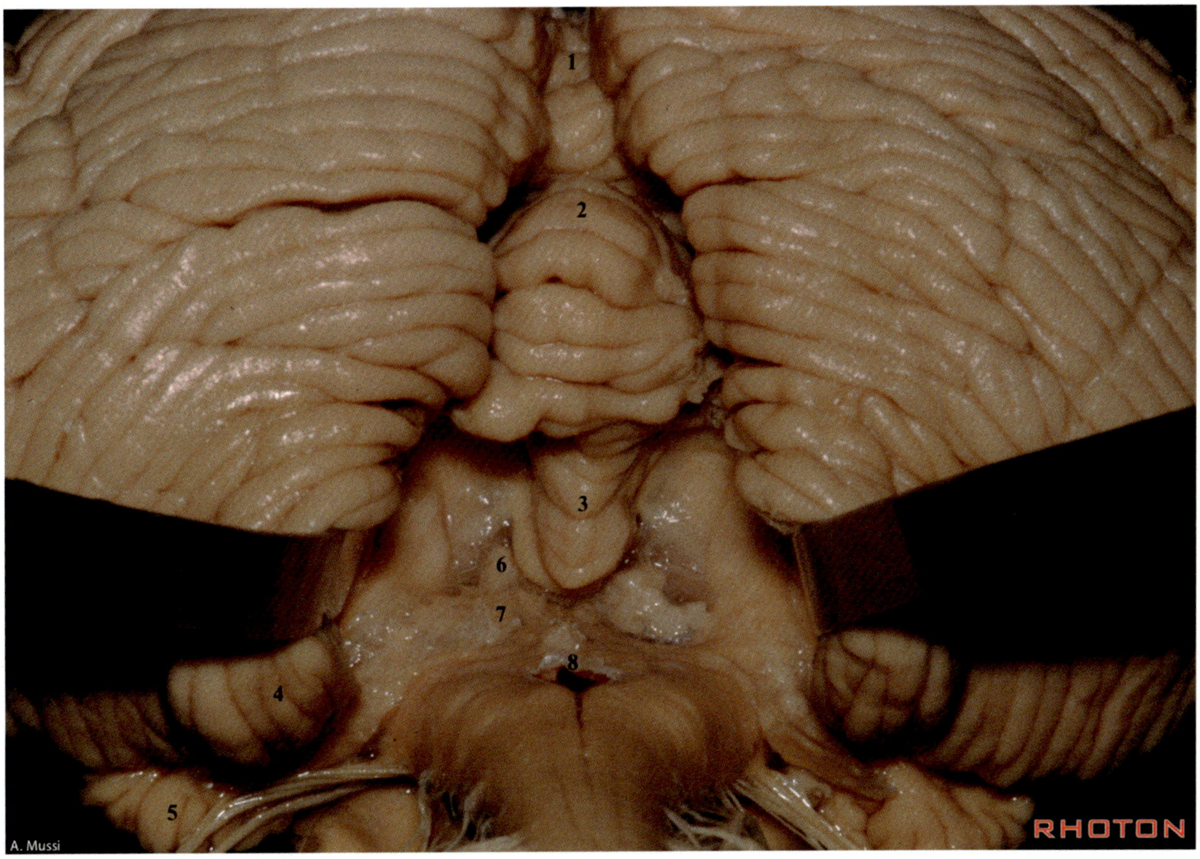

FIGURE 6.4 **Gross anatomy of the cerebellum; lower view.** 1. Vallecula; 2. Vermis; 3. Uvula of the vermis; 4. Tonsil; 5. Flocculonodular lobe; 6. Tela choroidea; 7. Choroid plexus; 8. Fourth ventricle. *Image courtesy of The Rhoton Collection® http://rhoton.ineurodb.org/. Dissector: A. Mussi.*

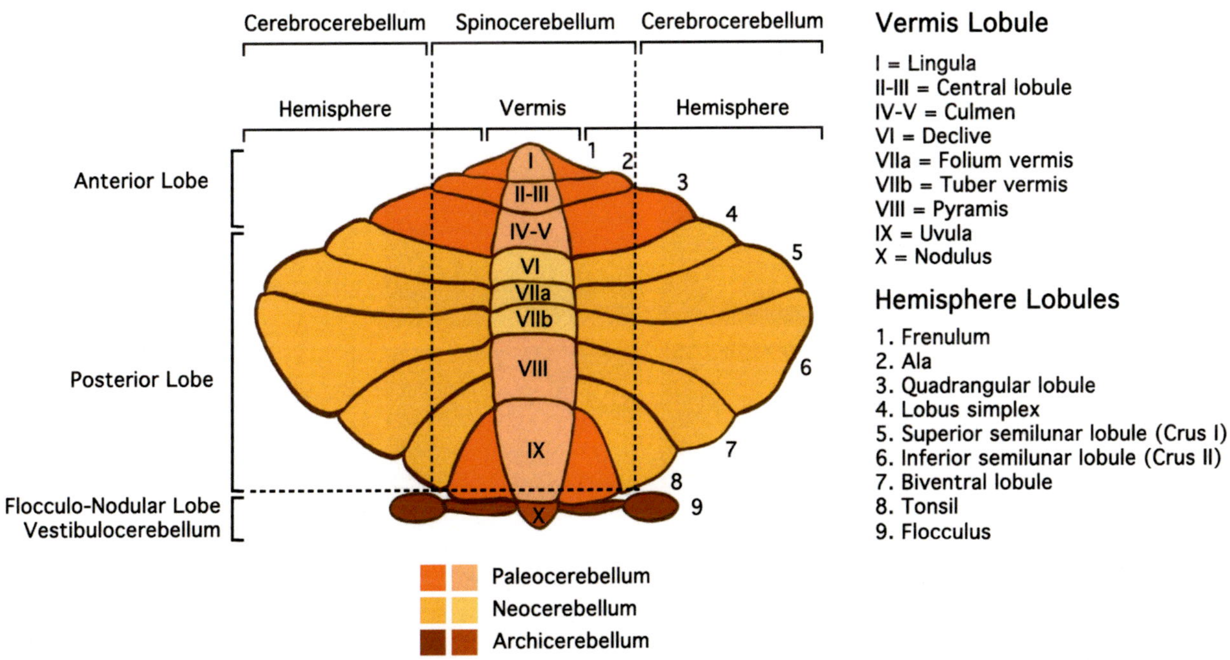

FIGURE 6.5 **The anatomy and functional divisions in schematic view of the cerebellum.** In the anterior–posterior axis, the cerebellum presents three lobes (anterior, posterior, and flocculonodular), each one subdivided in lobules. It is composed of a central part named vermis, and two lateral cerebellar hemispheres. Vermis and paravermial zones form the spinocerebellum that communicates with the spinal cord; the most lateral zones of the hemispheres constitute the cerebrocerebellum that connect with the cerebral cortex. Regarding phylogenesis, the cerebellum can be divided into three parts: the archicerebellum, paleocerebellum, and neocerebellum. *Image of Fredy Díaz, Basic Sciences Department, Faculty of Medicine, Universidad Católica de la Santísima Concepción, Concepción 4090541, Chile.*

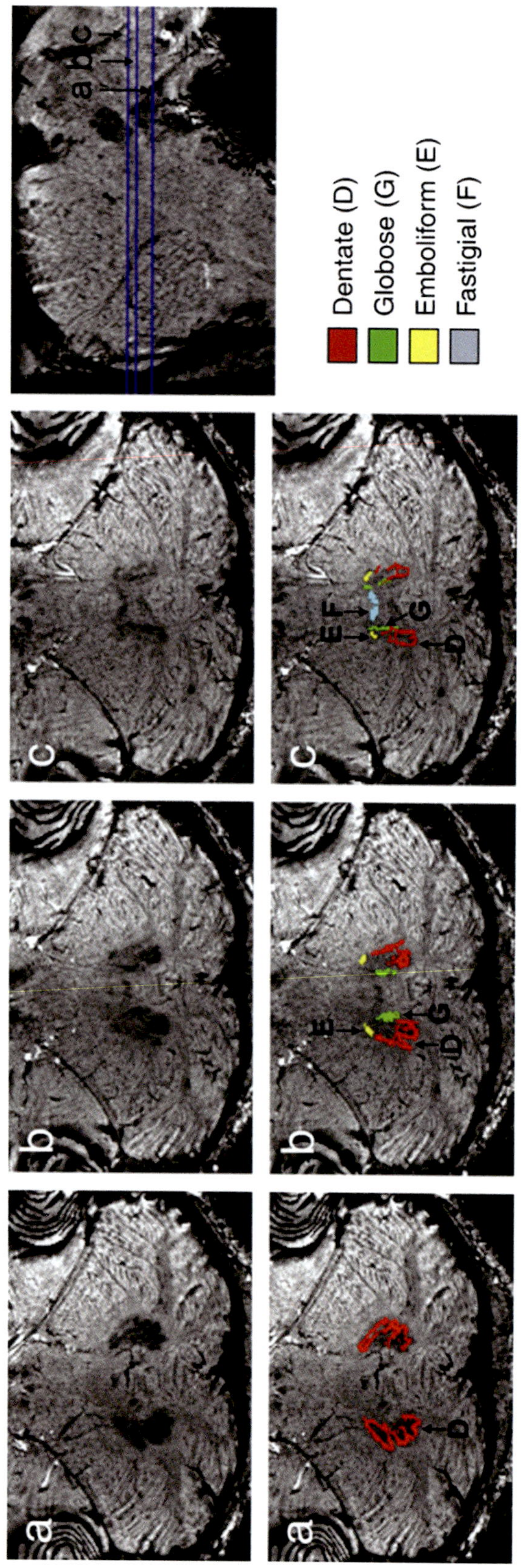

FIGURE 6.6 **Cerebellar nuclei.** ROI drawing of the deep cerebellar nuclei in an individual on the susceptibility-weighted image (7 T, 0.5 mm isotropic resolution). Three horizontal slices are shown without (upper row) and with (lower row) ROIs displayed. *Image courtesy of Diedrichsen, J., Maderwald, S., Küper, M., Thürling, M., Rabe, K., Gizewski, E. R., Ladd, M. E., Timmann, D., 2011. Imaging the deep cerebellar nuclei: a probabilistic atlas and normalization procedure. NeuroImage 54 (3), 1786–1794. https://doi.org/10.1016/j.neuroimage.2010.10.035.*

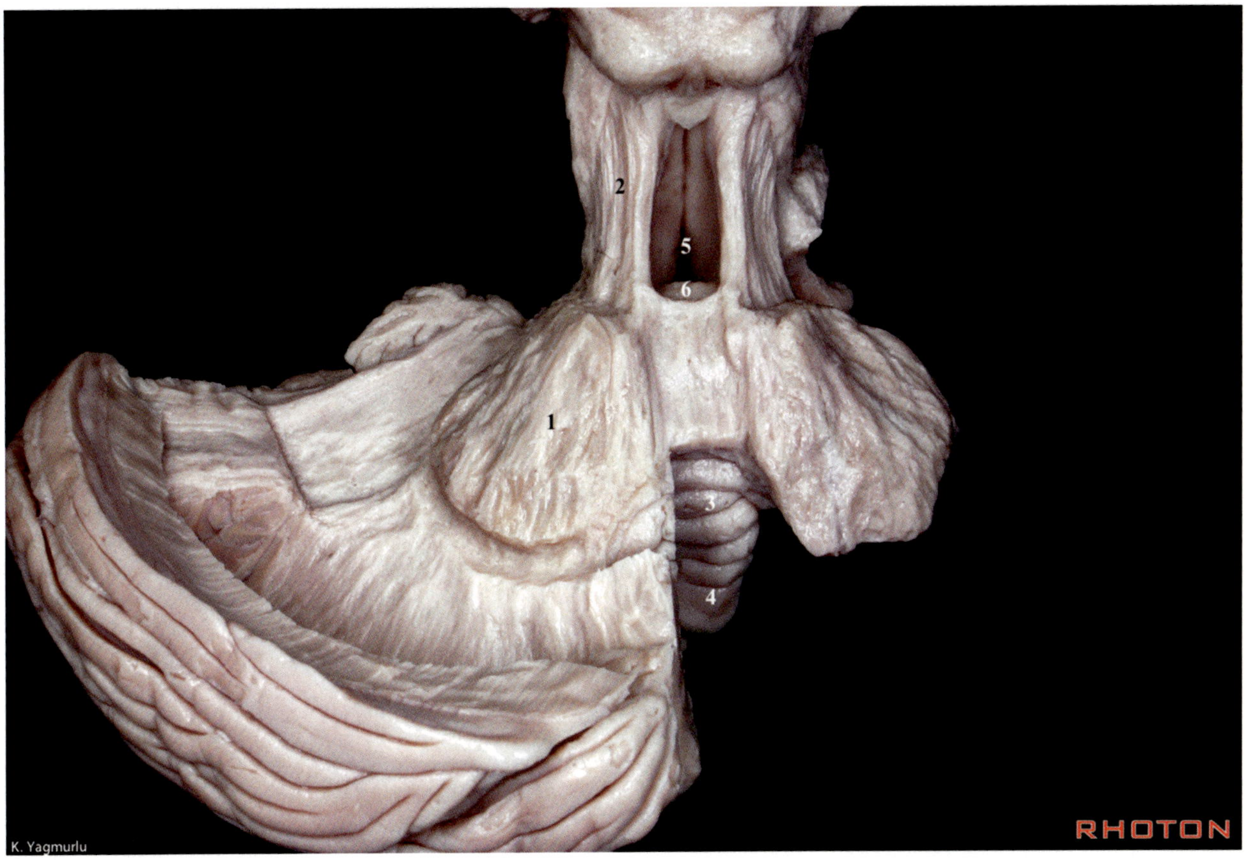

FIGURE 6.7 **Deep dissection of the cerebellum.** 1. Dentate nucleus; 2. Superior cerebellar peduncle; 3. Pyramid vermis; 4. Uvula of vermis; 5. Forth ventricle; 6. Nodule. *Image courtesy of The Rhoton Collection® http://rhoton.ineurodb.org/. Dissector: K. Yagmurlu.*

and the posterior inferolateral portion of the cerebellum, respectively. The second arterial territory corresponds to the anterior inferior cerebellar artery, which is a branch of the basilar artery that supplies anterior petrous surface of the corresponding cerebellar hemisphere, the flocculus, and the inferior portion of the middle peduncle. The third one is other branch of the basilar artery called the superior cerebellar artery that supplies the superior surface of the cerebellar hemispheres, the superior portion of the vermis, the dentate nucleus, as well as the superior cerebellar peduncle (Duvernoy et al., 1983; Duvernoy, 2012; Icardo et al., 1982; Piccinin and Munakomi, 2019). On the other hand, in the venous organization is possible appreciate superficial and deep veins (Delion et al., 2017). In particular, the superficial veins return the blood from the cerebellar cortex and veins transiting on the surface of the cerebellum. The deep veins transiting in the fissures between the cerebellum and the brainstem. All these veins terminate as bridging veins that we can divide in three groups: a superior group emptying into the great vein, a posterior group emptying into the transtentorial sinus, and a lateral group ending into the superior petrosal sinus (Delion et al., 2017).

CELLULAR ORGANIZATION AND CIRCUITRY OF THE CEREBELLUM

The cerebellum is divided into an outer gray matter layer corresponding to the cerebellar cortex, and a deep cerebellar white matter known as arbor vitae (Fig. 6.1). Within the latter are distributed the cerebellar nuclei (Fig. 6.6) (Voogd, 2003). Three layers compose the cerebellar cortex: the superficial molecular layer, the middle Purkinje layer, and the deepest granular layer. In turn, these layers are built by inhibitory cells (stellate, basket, Purkinje, and Golgi cells), excitatory cells (granule and unipolar brush cells), and glial cells (mainly Bergmann glia) (Buffo and Rossi, 2013; Roostaei et al., 2014; Voogd, 2003) (Fig. 6.8). The molecular layer is constituted by the dendritic trees of Purkinje cells, the ascending axons of the granule cells split into two parallel fibers that make synapse with the dendritic trees of PCs, and the inhibitory interneurons stellate cells and basket cells. The Purkinje layer is composed of the cell bodies of PCs along with the Bergmann cells. This layer works as an output of the cerebellar cortex. The granular layer is formed by high density of granule cells and interneurons including Golgi and unipolar brush cells. This layer acts

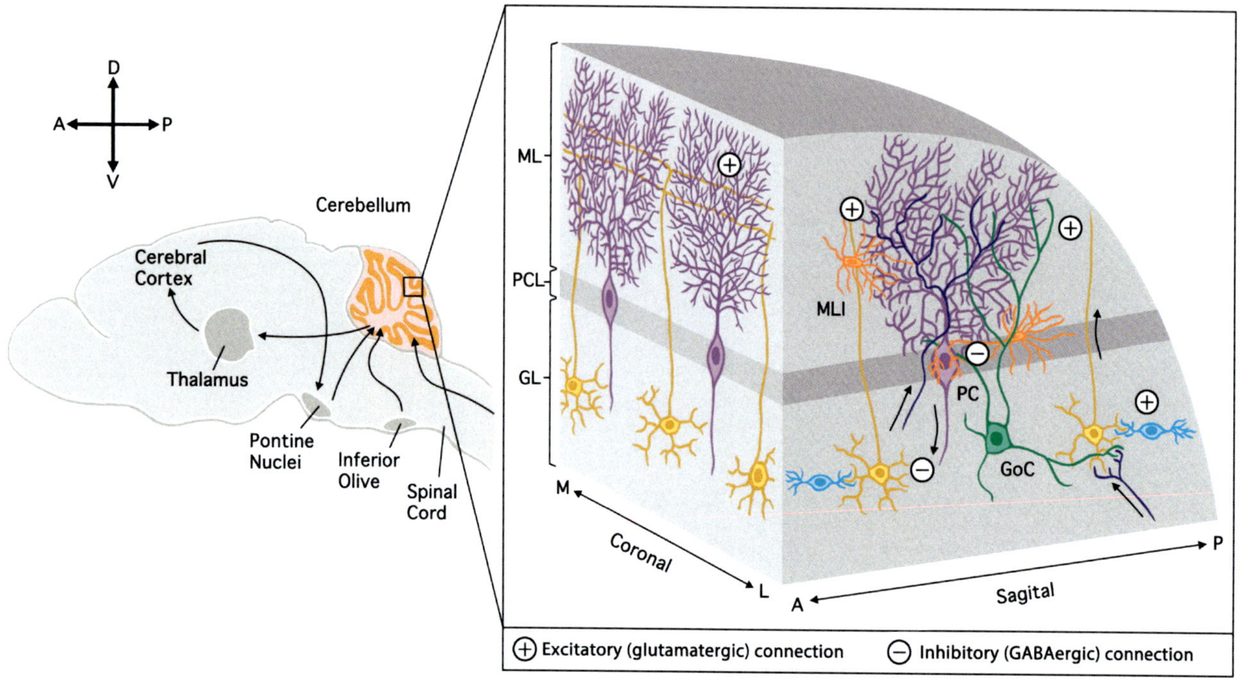

FIGURE 6.8 **Essential characteristics of cellular organization and circuitry of the cerebellum.** Main input and output connections to and from the cerebellum are represented in the brainstem. The main inputs include the spinal cord, inferior olive, and pontine nuclei while main outputs include connections from cerebellar nuclei to the cerebral cortex via the thalamus. The zoomed image shows the cellular distribution in the molecular layer (ML), Purkinje cell layer (PCL), and granular layer (GL), as well as the excitatory and inhibitory circuit connections within the cerebellar cortex (arrows). Purkinje cells (PCs) (purple), granule cells (GCs) (yellow), molecular layer interneurons (MLIs) (orange), Golgi cells (GoCs) (green), and unipolar brush cells interneurons (light blue). *Image of Fredy Díaz, Basic Sciences Department, Faculty of Medicine, Universidad Católica de la Santísima Concepción, Concepción 4090541, Chile.*

as an input layer for the cerebellar cortex because mossy fibers, one of the main cerebellar afferent routes, raise this layer forming the cerebellar glomeruli (Roostaei et al., 2014; Voogd, 2003).

At the cellular level, mossy fibers and climbing fibers are the main input pathways. The first one originates from several structures of CNS such as pontine nuclei, spinal cord, vestibular nuclei, and the reticular formation to synapse indirectly to the Purkinje layer. The second one arises from the contralateral inferior olivary to synapse PCs (Roostaei et al., 2014). Additionally, other neuromodulators influence the cerebellar circuitry such as the aminergic and cholinergic inputs, probably involved in the cerebellar learning and control as well as in the regulation of the speed of supervised learning (Schweighofer et al., 2004). In detail, at the granular layer, mossy fibers send collateral branches to several cerebellar folia forming rosettes to make synapse with dendritic enlargement of the excitatory granule cells in a structure called cerebellar glomeruli. Surrounding these structures are distributed glial cells. Granular cells present a high divergence as they cross and synapse with dendritic trees of PCs at the molecular layer (Roostaei et al., 2014). Its excitatory synaptic transmission is a weak stimulus that propagates small currents to dendrites of PCs. Likewise, granular cells contact Golgi cells in the cerebellar glomeruli to refine the synaptic control contributing with inhibitory inputs, as well as the stellate and basket inhibitory interneurons. Stellate cells make inhibitory GABAergic synapses onto PC dendrites, while basket cells synapse onto PC bodies. Of note is the fact that mossy fibers also synapse with neurons from the deep cerebellar nuclei.

On the other hand, each PCs synapse with a single climbing cell that, in turn, can strongly innervate around 10 dendritic trees of PCs. In turn, the effect of the synaptic transmission of PCs on their target neurons is inhibitory by the release of GABA neurotransmitter. PCs axons project collateral branches close to the cerebellar cortex and, at the same time, they are the only output of the cerebellar cortex extended until the deep cerebellar nuclei. In the Purkinje-deep nuclei pathway, the deep cerebellar nuclei work as the outputs of the cerebellum. Efferents from the deep cerebellar nuclei go through superior and inferior peduncles and make excitatory synapses with their target neurons with the exception of the nucleo-olivary projections (Roostaei et al., 2014).

NEUROANATOMIC CONNECTIONS

The analysis of the pattern of functional activation of the cerebellum and its topographical correlations is essential for the clinical diagnosis to tackle pathological conditions. The construction of detailed maps for a functional topography of the cerebellum has been provided for the progress in neuroimaging studies, neuroanatomy, and clinical data. However, whether the activation of specific areas is relevant in mapped functions is until now an open question whose answer could be predicted, in part, by the type of cerebellar lesion or degeneration. For instance, the more recent association between cerebellar atrophy and cognitive impairment with deficit in executive functions, working memory and language deficits, among others required advanced statistical approaches, as well as to pinpoint the segment of cerebellar lobule involved for each task (Kansal et al., 2017). Nevertheless, cerebellar lesions present a wide spectrum of clinical manifestations that can require the activation of similar patterns. Below is presented an overview of the cerebellar neuroanatomic connections.

The neuroanatomic connections represent fiber tracts carrying input and output between the cerebellum and brainstem via three paired cerebellar peduncles, superior-middle and inferior, that allow the processing of information to and from the periphery (Fig. 6.9).

In this regard, all the inputs that reach the cerebellum synapse in the cerebellar cortex from where axons carry the information to the corresponding deep nuclei, go from the midline vermis projecting to the medial fastigial nuclei, the paravermal regions projecting to the interpositus nucleus, and the lateral hemispheres projecting to the dentate nuclei (Stoodley and Schmahmann, 2018). In turn, axons that are projected from the deep nuclei constitute the output pathways. In detail, the intracerebellar nuclei connected with the archicerebellum portion are the fastigial (committed to sense the balance and the spatial position of the head among others) and the vestibular (that at the level of the brainstem sense the information from the ear, the vestibular organ) (Roostaei et al., 2014; Voogd et al., 1996). Therefore, the inputs correspond to the vestibular cortex portion of the cerebellum. The fiber tracts originated in the vestibular nucleus in the medulla oblongata reach the cerebellar cortex through the inferior cerebellar peduncle and, from there, they synapse with the fastigial nucleus. In this regard, the ipsilateral output pathways are the vestibular fastigial nucleus, which also travel through the inferior cerebellar peduncle to the vestibular nucleus. The reticulo-fastigial fiber tract is crossed fibers

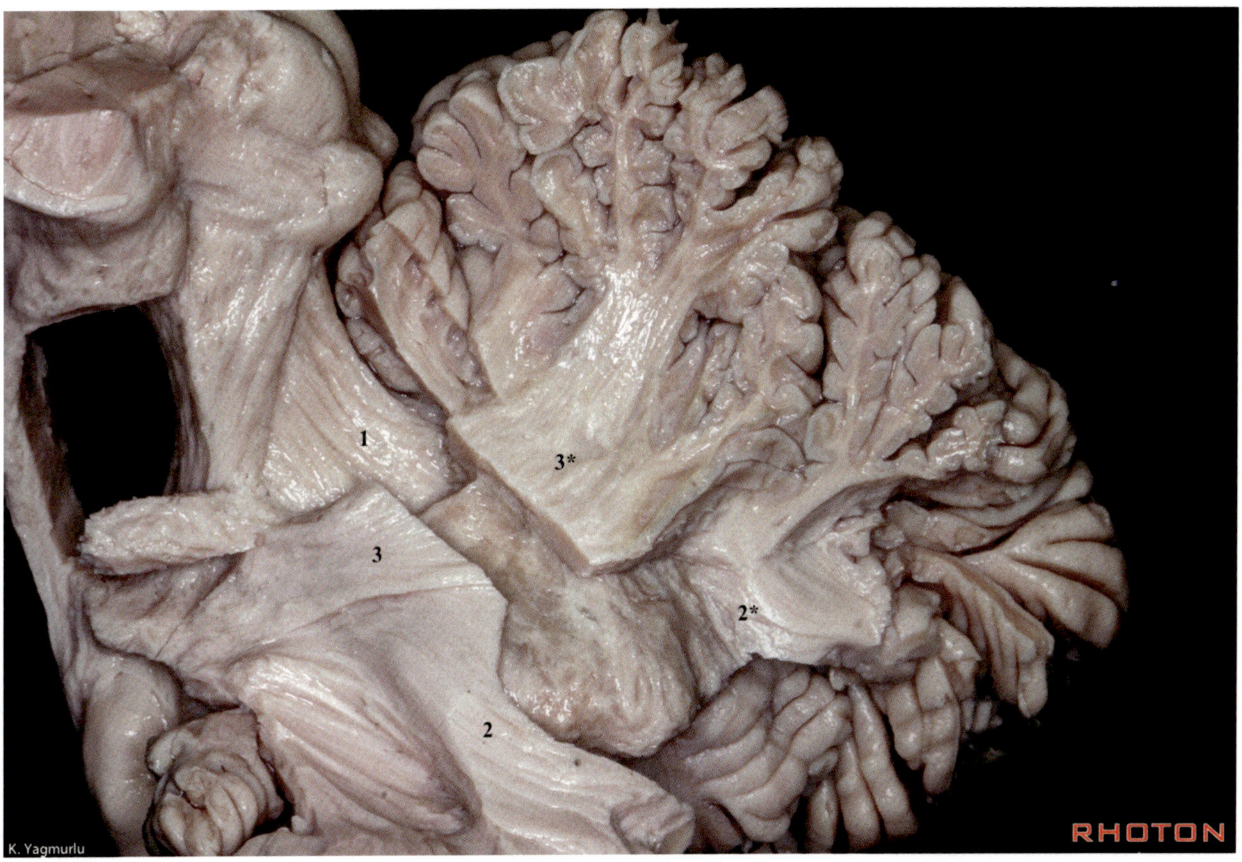

FIGURE 6.9 **Cerebellar peduncles.** The cerebellar white matter is formed by three cerebellar peduncles, which are the superior (1), the middle (2), and the inferior (3). *Image courtesy of The Rhoton Collection® http://rhoton.ineurodb.org/. Dissector: K. Yagmurlu.*

that go around the superior cerebellar peduncles constituting a hook shape known as uncinate fasciculus of Raphe. From these fibers, some collaterals reach into the nucleus of the VI pair in the reticular substance of the medulla oblongata. The flocculus-oculomotor fiber tract is a crossed bundle that runs through the superior cerebellar peduncle and reaches the oculomotor nucleus of the III pair at the level of the peduncles (Manni and Petrosini, 2004b). In addition, the input bundles of the archicerebellum influence the motor neurons of the spinal cord, therefore the vestibulospinal bundles and the spinal reticulum contact neurons in the spinal cord to participate in the control of the ocular and somatic motor response in maintaining balance (Table 6.1).

The paleocerebellum is a proprioceptive regulator, that is, it receives deep unconscious information related to the state of relaxation or contraction of the musculature and the position of the joints to make the necessary adjustments to muscle tone depending on the posture. Its intracerebellar nucleus is the interposal, which is formed by the globose and emboliform nucleus in charge to receive the inputs from the spinal cord and the bulbar oliva. Two nuclei are located in the spinal cord that will give rise to the most important bundles that correspond to the dorsal and ventral spinocerebellar bundles. In detail, the dorsal cerebellar bundle arises in Clarke's column (interneurons found in the medial part of the lamina VII of the spinal cord) arriving at the cerebellum through the superior cerebellar peduncle to send proprioceptive information from the lower extremities and the body trunk. The ventral cerebellar tract works in the same way, but it is originated from the nuclei at the base of the dorsal horns (Fig. 6.10).

Additionally, the olivocerebellar tract brings information from the oliva and paraoliva nucleus housed in the medulla oblongata. Further, the trigemino-cerebellar tract, which is formed by axons of the trigeminal mesencencephalic nucleus that also enters through the superior cerebellar peduncle, transmits the proprioceptive information of the facial skeleton and, therefore, everything regarding the information of the muscle tone of face. Other relevant inputs correspond to the reticulo-cerebellar tract that brings the information of the reticular formation, while from the mesencephalon, the tecto-cerebellar and rubro-cerebellar tracts come from the quadrigeminal tubercles and the red nucleus, respectively (Figs. 6.10 and 6.11).

On the other hand, all the output information propagates from the interpositus nucleus toward the reticular formation, the olive, the red nucleus, and the quadrigeminal tubercles as appropriate. The most important output and main discharge pathway corresponds to the interposito-rubrospinal tract which allows the cerebellum through the superior cerebellar peduncle to reach the contralateral red nucleus from where fibers are distributed to the ventral nucleus of the thalamus which, in turn, reaches the cerebral cortex and the spinal cord through the rubrospinal tract (Table 6.1 and Fig. 6.11).

The neocerebellum is responsible for the synergy between the skills and abilities learned and its intracerebellar nucleus is the dentate that connects with the nuclei of the pons, the bulbar olive, the arcuate nuclei, as well as the red nucleus and the optic thalamus. The inputs correspond to the cerebellopontine, olivocerebellar, and arcuate cerebellar tracts (Manni and Petrosini,

TABLE 6.1 Major circuits and function of the cerebellar cortex.

Phylogenic division	Region	Afferent nerve fibers tract	Efferent nerve fibers tracts	Function
Archicerebellum	Fastigial nuclei Vestibular system	Vestibulo cerebellar	Fastigio-vestibular Fastigio-reticular Flocculus-oculomotor tract Vestibulospinal tract Vestibuloreticular tract	Balance Spatial position of head
Paleocerebellum	Globose nucleus Emboliform nucleus Paraolivar nucleus Spinal cord	Dorsal spinocerebellar Ventral spinocerebellar Olivocerebellar Tectocerebellar Trigeminocerebellar Rubrocerebellar	Reticular formation Interposito-olivar Interposito-rubrospinal Interpositotectal	Muscle tone and posture
Neocerebellum	Dentate nuclei Pontine nuclei Bulbar oliva Arcuate nuclei Red nuclei Thalamic nuclei	Pontocerebellar Olivocerebellar Arcuate cerebellar	Rubrothalamic Dentothalamic	Skills and abilities learned Motor coordination

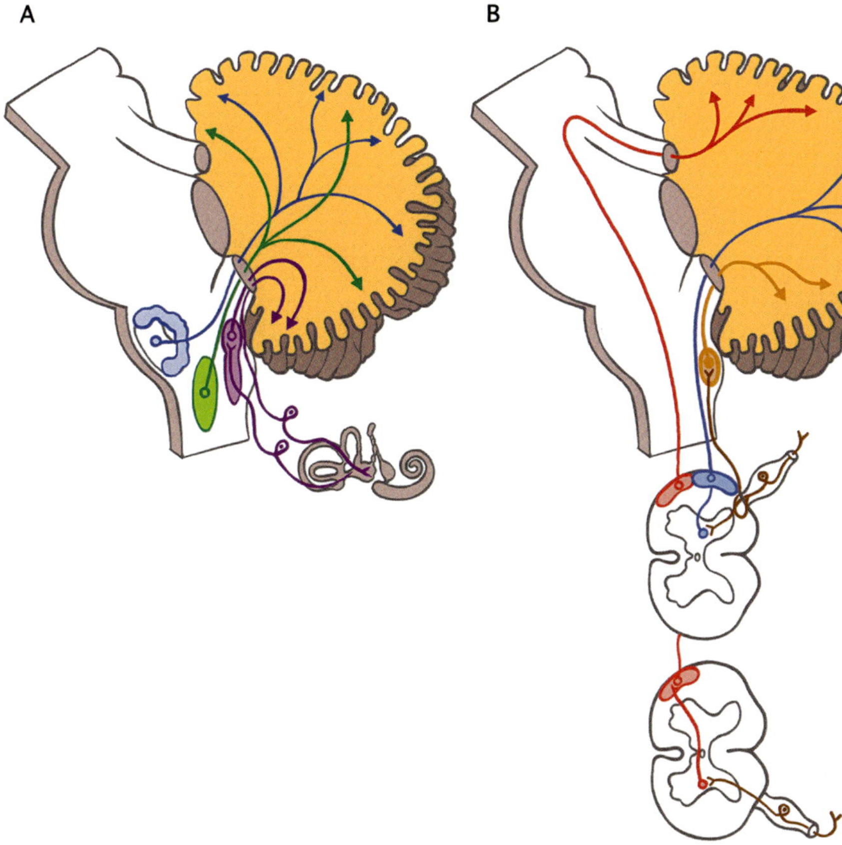

FIGURE 6.10 **Gross maps for a functional topography of phylogenetic division in the cerebellum.** Input signals that go to cerebellum synapse in the cerebellar cortex and then, the message goes to the corresponding deep nuclei. Output signals come from the deep nuclei. Inputs from the inferior olivary nucleus (blue), reticular formation (green), and vestibular nucleus (magenta) are showed (A). The major somatic sensory information to the cerebellum from the spinal cord are the nucleus Cuneate and Clarke's nucleus (B). A summary of the major circuits and functions are presented in Table 6.1. *Credit: Fredy Díaz, Basic Sciences Department, Faculty of Medicine, Universidad Católica de la Santísima Concepción, Concepción 4090541, Chile.*

2004b). The most prominent input connections are those mediated by the cerebellopontine tract that brings the information from the nuclei of the pons, which, in turn, receive the information on movement planning, motor execution, visual and emotional information from the entire cerebral cortex, a function like that of the arcuate cerebellar bundle to the nuclei of the pons. The outputs correspond to the dentatorubral tract; ergo the information propagates from the cerebellar cortex to the dentate nucleus from where dentatorubral fibers emerge to the contralateral red nucleus and, from there, to the thalamus and the cerebral cortex. On the other hand, in the dentatothalamic tract, the fibers go directly to the thalamus without passing through the cerebellar cortex and, from here, they propagate to the regions of the brain cortex from which they originate, that is, the corticopontocerebellar inputs (Table 6.1 and Fig. 6.11).

In the cerebellum, diverse forms of long-term synaptic or nonsynaptic properties of plasticity have been demonstrated to contribute to motor learning and cognitive processing. Thus, there are a rich variety of use-dependent plasticity in the cerebellar cortex and nuclei such as long-term potentiation (LTP), long-term depression (LTD) at excitatory and inhibitory synapses, and continuous modulation of intrinsic excitability (Gruol et al., 2016). For instance, long-term depression (LTD) at excitatory synapses between parallel fibers and PCs under the guidance of climbing fibers is essential in the cellular mechanisms for motor learning (Coesmans et al., 2004). Parallel fiber LTD occurs through the activations of postsynaptic metabotropic glutamate receptors (mGluR) and α-3-hydroxi-5-methyl-4-isoxazolepropionic acid receptor (AMPARs) with the subsequent rise of intracellular calcium and calcium-depending pathways. Parallel fiber LTD results in the clathrin-dependent endocytosis of AMPARs (Wang and Linden, 2000). In turn, the expression of LTP in parallel fiber-PC synapses depends on stimulus frequency. On one hand, 4–8 Hz parallel fiber stimulation induces presynaptic expression of LTP through presynaptic adenylyl cyclase/cyclic adenosine monophosphate (cAMP), and PKA-mediated phosphorylation of

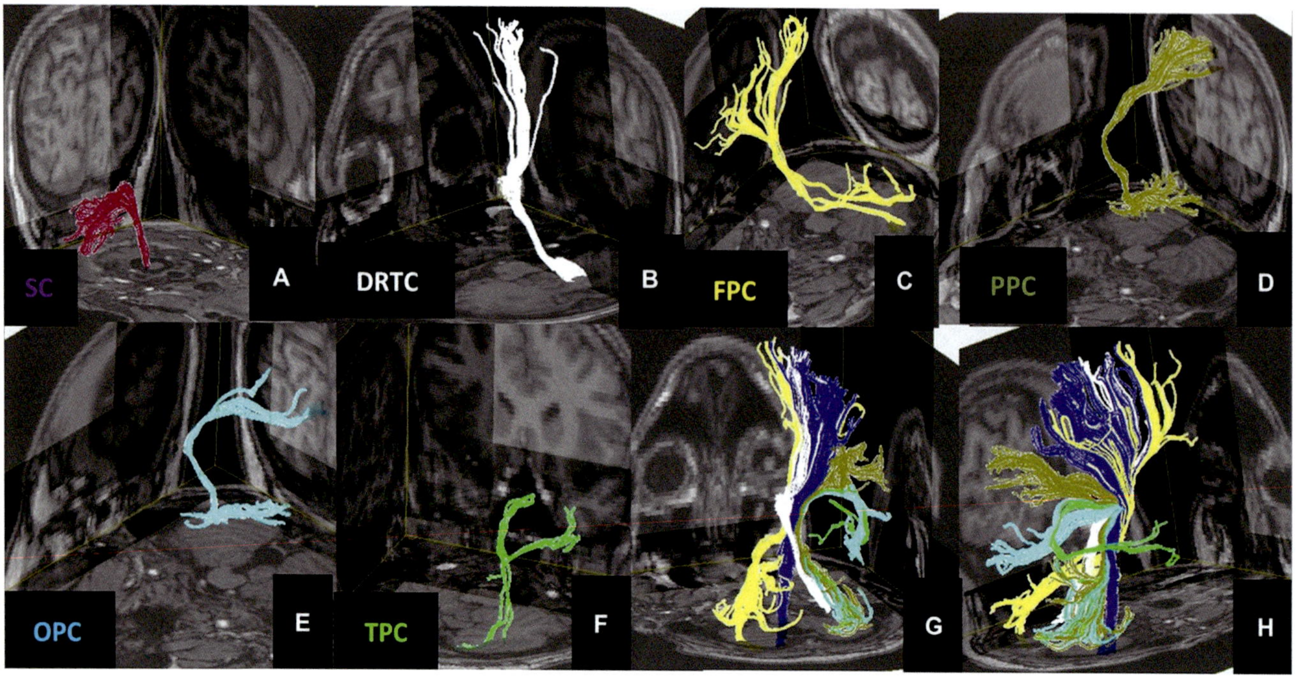

FIGURE 6.11 **3D views of the (A) spinocerebellar, (B) dentate-rubro-Thalamo-cortical, (C) Fronto-Ponto-cerebellar, (D) Parieto-Ponto-cerebellar, (E) Occipito-Ponto-cerebellar, (F) Temporo-Ponto-cerebellar pathways.** All these cerebellar pathways are illustrated together with corticospinal tract (dark blue (dark gray in print version)) in panels (G,H). *Image courtesy of Keser, Z., Hasan, K.M., Mwangi, B.I., Kamali, A., Ucisik-Keser, F.E., Riascos, R.F., Yozbatiran, N., Francisco, G.E., Narayana, P.A., 2015. Diffusion tensor imaging of the human cerebellar pathways and their interplay with cerebral macrostructure. Front Neuroanat. 9, 41. https://doi.org/10.3389/fnana.2015.00041. PMID: 25904851; PMCID: PMC4389543.*

the active zone protein RIM1α. On the other hand, 1 Hz parallel fiber stimulation induces postsynaptically expressed LTP via low levels of calcium in PCs, triggering the release of nitric oxide from parallel fiber terminals to induce the AMPARs traffic (Coesmans et al., 2004; Gruol et al., 2016).

Regarding others neuronal microcircuit regulating information transfer at the cerebellum, climbing fibers (CFs) evoke an LTD at climbing fiber-PC synapse requiring the intracellular calcium, PKC activation and the postsynaptic expression of mGluR1 to control the integrative response of PCs (Shen et al., 2002). Furthermore, the repetitive stimulation of the excitatory climbing fiber-PC synapses is followed by a long-lasting potentiation of γ-aminobutyric acid A ($GABA_A$) receptor like a rebound potentiation. In this phenomenon CFs induce transient increase of postsynaptic calcium from internal stores (Kano et al., 1992). Meanwhile, the mossy fiber synapses provide the cerebellar network with a large reservoir for memory storage (D'Angelo et al., 1999). Mossy fibers trigger the intrinsic excitability in the deep cerebellar nuclei (DCN) (Zhang et al., 2004) and its activation together with postsynaptic granule cells membrane depolarization induce mossy fiber LTP by calcium-depending influx and PKC activation. In this cotext, potentiation was manifested by an increase in NMDA and non-NMDA receptors (D'Angelo et al., 1999; Pugh and Raman, 2006). More recent findings have been shown that mossy fiber theta-burst stimulation (TBS) could induce LTP or LTD at mossy fiber–Golgi cell and granule cell–Golgi cell synapses through calcium channel–depending induction. This forms of long-term plasticity at excitatory Golgi cell synapses could be modulating the response patterns of cerebellar granular layer (Locatelli et al., 2021).

In summary, cerebellar long-term synaptic plasticity occurs in different zones through the cerebellar layers and DCN affecting excitatory and inhibitory synapses and contributing to learning and memory in the cerebellum.

CEREBELLAR DISORDERS

Clinical Manifestations

Given the long period of cerebellum development, a broad variety of malformations could be progress without suspect during the early prenatal period because the prenatal diagnosis of cerebellar or vermian anomalies is complex. Cerebellar malformations often have sporadic onset and progress along with complex syndromes that impair other brain structures besides the cerebellum. The clinical presentation is nonspecific and includes hypotonia, motor delay, nystagmus, and inability for visual fixation. Severely affected patients may also have apnea, feeding difficulties, breathe impairment, spasticity, delayed psychomotor development, and seizures. In some patients, the symptoms are accompanied by cranial nerves dysfunction, including abnormal eye movements, ptosis, facial palsy, hearing impairment, and facial or corneal anesthesia. In other cases, the isolated enlargement of the posterior fossa occurs by mega cisterna magna, posterior fossa arachnoid cyst, and Blake's pouch cyst, all of them with good prognosis (Dror et al., 2009). Cognitive impairment including autism are common but not universal among patients with cerebellar malformations (Aldinger and Doherty, 2016). Some diseases are presented below.

Dandy–Walker Malformation

It is a rare heterogeneous disorder characterized by a hypoplastic vermis with upward rotation, and cystic dilatation of the fourth ventricle, as well as the posterior fossa with torcular-lamboid inversion. The cerebellar hemispheres are generally less affected than the vermis, and the brainstem is normal or moderately hypoplastic. In some cases, Dandy–Walker malformation (DWM) can occur with additional brain abnormalities, including occipital encephalocele, polymicrogyria, agenesis, or dysgenesis of the corpus callosum and heterotopia (Lerman-Sagie et al., 2018). Patients may present a wide range of symptoms from intellectual disability to autism; while in another cases they may be asymptomatic and be diagnosed incidentally (Aldinger and Doherty, 2016). DWM occurs sporadically and in some cases the recurrence may be attributed to the misdiagnosis of autosomal-recessive disorders such as Joubert syndrome (JS).

Cerebellar Hypoplasia

In this condition, the cerebellum does not develop completely or is smaller than the normal size. In comparison with DWM, it does not involve concurrent enlargement of the posterior fossa. The causes are a highly heterogeneous group of diseases that include prenatal infections, exposure to teratogen agents, chromosomal abnormalities, metabolic disorders, genetic syndrome, and malformations.

Initially, the clinical manifestations in children are muscular hypotonia and overall developmental impairment, with the subsequent addition of other signs such as body trunk ataxia (49%–93%), hypotonia (47%–49%), eye movement disorders (40%–46%), dysarthria (38%), intention tremor (9%–35%), microcephaly (20%), and seizures (28%–56%). Intellectual disabilities are present in more than 60% of patients and are severe in 35% of them. In other cases, speech and language disorders are frequent and in a wide range from mild impairment to severe aphasia. Additionally, behavioral disturbances are present and around 5%–20% of patients have characteristics compatible with autism spectrum disorder (Poretti et al., 2014). Diagnosis is obtained using RMI imaging in which several altered patterns affecting majorly the whole cerebellum but also the vermis as well as the cerebellar hemispheres are observed. It is also the most difficult prenatal diagnosis because it is often misdiagnosed as DWM, Blake's pouch cyst or mega cisterna magna.

Cerebellar Vermis Hypoplasia

This cerebellar malformation presents a small complete vermis associated with normal or minimal upward vermis rotation relative to the brainstem (Lerman-Sagie et al., 2018). The retrocerebellar fluid collection is small and communicates directly with the fourth ventricle (Lerman-Sagie et al., 2018; Parisi and Dobyns, 2003).

Unilateral Cerebellar Hypoplasia

Unilateral cerebellar hypoplasia (UCH) could be caused by a vascular damage that affects one cerebellar hemisphere with, in some cases, involvement of vermis which is related with cognitive decline. The clinical presentation of UCH varies from an asymptomatic pathology to a one with severe neurologic damage, and the asymmetry of the cerebellar hemispheres may range in severity from mild to near total aplasia (Massoud et al., 2014; Shaikh et al., 2019).

Pontocerebellar Hypoplasias

Pontocerebellar hypoplasias represent a fetal onset of heterogeneous group of very rare neurodegenerative disorders with reduced volume of pons and cerebellum. It is not necessarily a genetic disorder, but they are listed at least 13 different types related with 19 different genes including tubulinopathies, dystroglycanopathies, some congenital disorders of glycosylation syndromes, among others (Rüsch et al., 2020). Clinical signs include severe motor and developmental delay, and early postnatal lethality.

Rhombencephalosynapsis

It is another rare type of hypoplasia of the cerebellum characterized by the complete or partial absence of the vermis and fusion of the cerebellar hemispheres and dentate nuclei. Additional cerebral abnormalities such as midline defects, aqueductal stenosis leading to hydrocephaleous, corpus callosum dysgenesis may also be manifested (Lerman-Sagie et al., 2018) (Cerebral disorders paper sin autor). The prognosis is poor and in general children die in childhood. Most survivors present neurological delay and motor disorders or ataxia (Poretti et al., 2009).

Chiari Malformations

Chiari malformations constitute a group of entities with very different etiologies, pathophysiology, and clinical features that have in common the presence of anatomical deformities of the brainstem and cerebellum.

The Chiari malformations consist of four types of hindbrain abnormalities, probably unrelated to each other. The majority of Chiari malformations are types 1 or 2; a very limited number of cases comprise the remaining types. Chiari zero is a novel condition.

Type I Chiari Malformation (Fig. 6.12)

- Heterogeneous entity with the common feature of impaired CSF circulation through the foramen magnum
- Caudal descent of the cerebellar tonsils (and brainstem in its subtype, Chiari 1.5) through the foramen magnum >5 mm; more significant than this "number" is the amount of compression of the brainstem at the foramen magnum (best appreciated on T2WI MRI but also in T1W1, see Fig. 6.12)
- Typically diagnosed during second or third decade of life
- Most common symptom is headache, especially in suboccipital region, exacerbated by Valsalva maneuver
- Can be associated with syringomyelia (50%–70%, especially in cervical region) and hydrocephalus (10%)

Type II Chiari Malformation

- Caudal descent of the brainstem, cerebellar tonsils, and vermis through the foramen magnum
- When a child is born with a myelomeningocele, the vast majority (~95%) have an associated Chiari II malformation
- Almost always associated with myelomeningocele, often accompanied by hydrocephalus

Type III Chiari Malformation

- Dislocation of the cerebellum and other neural structures below the foramen magnum into an occipital encephalocele
- Rare condition, usually incompatible with life

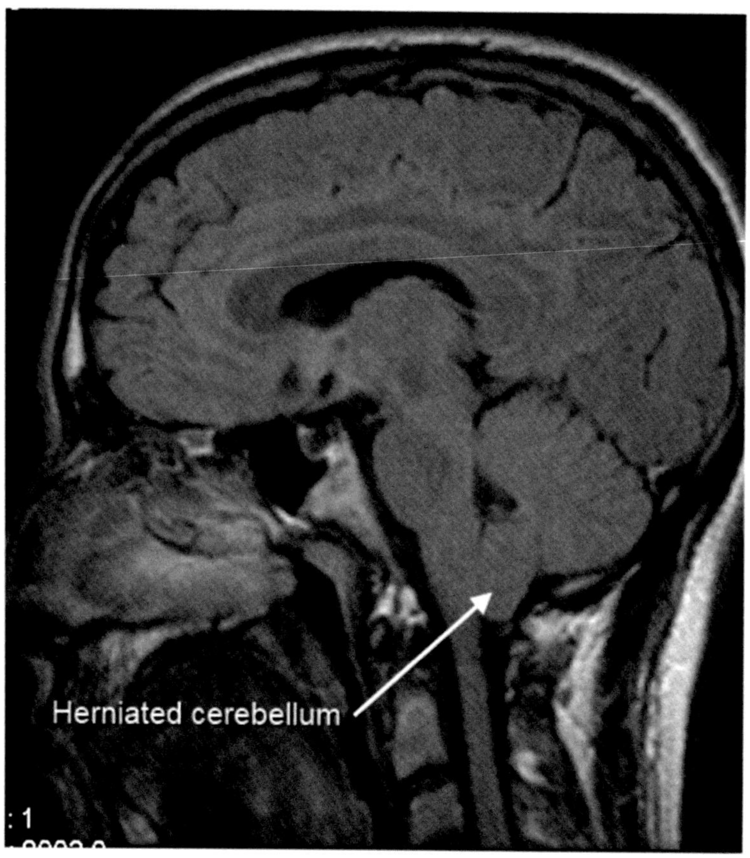

FIGURE 6.12 **Image of a type 1 Arnold-Chiari malformation on T1WI MRI.** The cerebellum has descended 7 mm and there are herniated cerebellar tonsils into the foramen magnum. *Image courtesy of Wikipedia.*

Type IV Chiari Malformation

- Cerebellar hypoplasia (CBLH) without cerebellar herniation and a small posterior fossa.
- Rare condition

Type 0 Chiari Malformation

- Syringohydromyelia without hindbrain herniation that respond to p-fossa decompression

Joubert Syndrome

JS is a ciliopathy with abnormal structure and function of cilia in the brain and retinal photoreceptors, among others, characterized by a distinctive cerebellar and brainstem malformation called molar tooth sign (MTS) on imaging, hypotonia, and developmental delays. Some cases present tachypnea or apnea that improve with age while ataxia develop over time with a correct motor performance delayed. Additionally, abnormal eye movements and facial dysmorphism can be presented. Cognitive disabilities ranging from normal to severe impairment. Currently, more than 30 causative genes have been identified where the most common are the Abelson helper integration site 1 (*AHI1*), ciliogenesis, and planar polarity effector 1 (*CPLANE1*) that codified proteins involved in the normal development (Surisetti et al., 2021).

CBLH-DWM and Vasculogenesis

Several genes could be expressed in altered levels in vascular cell types at the cerebellum. Thus, the disruption of genes working in vasculogenesis might lead cerebellar hemorrhage or altered vascular perfusion (Aldinger et al., 2019). For instance, in prenatal human cerebellar tissue, the gain-of-function of *PDGFRB*, a gene expressed in vascular pericytes and radial glial cells is also highly enriched in pericytes within the cerebellum (Aldinger et al., 2019). In this regard, a pathogenic variant of PDGFRB has been linked with CBLH in addition to posterior fossa cysts as well as DWM (Aldinger et al., 2019; Zarate et al., 2019). However, from previous reports establishes at least 10 cerebellar malformation genes with high expression during vasculogenesis supporting that disruption of vascular-expressed genes impair cerebellar development by a secondary prenatal perfusion defects or hemorrhage and it is a variable feature in CBLH or DWM (Aldinger et al., 2019).

Other Causes of Cerebellar Disorders

Cerebellar infarcts such as hemorrhages can cause lesions in the cerebellum, being one of the main causes of acute-onset ataxic syndrome. Primary brain tumors such as meningioma, gliomas, medulloblastomas, as well as metastatic tumors secondary to melanoma, breast and lung cancer, can also cause damage to the cerebellum. Other less frequently causes such as abscesses or arteriovenous malformations that undergo transformation hemorrhage can manifest with acute ataxia.

THERAPEUTIC MANAGEMENT OF CEREBELLAR DISORDERS

Treatments for cerebellar disorders fall into two categories: specific cause treatments, aimed at halting the progression of the disease, and therapies aimed at enhancing the "cerebellar reserve." Both forms of therapies should be applied as soon as possible, at the time when the cerebellar reserve is still preserved.

Other efficient therapeutical approach is the surgical practice to treat neurological cerebellar disorders that require repair, or readjustment the cerebellar anatomy. In these cases such as cerebral cavernous malformations, Chiari malformation type I or cerebellar arteriovenous malformations, among others, the clinical efficacy of surgery resection and reconstruction of the cerebellar region improve the outcomes in most patients (Lou et al., 2019; Nisson et al., 2018; Wu et al., 2015).

Treatment for Specific Etiology

Most of the acquired cerebellar pathologies have specific treatment related to their etiology that can reverse or at least limit the cerebellum damage, greatly improving the prognosis of the clinical condition. For instance, in the case of infections, it will depend on the causative microorganism being susceptible to treating, especially bacterial and fungal infections. In the case of toxic causes, the mainstay of treatment will be the suspension of drugs or exposure to toxic agents immediately, accompanied by the use of chelators as in the case of heavy metal poisoning.

In the case of vascular pathology, all measures will be applicable, including impact strategies in the case of an ischemic cerebellar infarction (thrombolysis and/or thrombectomy) that are performed as standard procedure. Overall, studies recommend early surgical resection for all surgically accessible cerebellar arteriovenous malformations with the intention to prevent subsequent hemorrhage and resultant poor neurologic outcomes (Tong et al., 2016).

Regarding hereditary causes, the vast majority do not have a specific treatment even though molecular biology has promoted the development of new forms of therapies that could in the future be a real alternative in these pathologies.

Treatment to Enhance the Cerebellar Reserve

Compelling evidence supports that conventional motor rehabilitation is effective in patients with structural injuries such as cerebral infarction, hemorrhage, and trauma. However, in degenerative cerebellar diseases, this type of treatment is uncertain, even though there are different studies that showing positive effects in the intensive motor rehabilitation. In those cases, improving clinical parameters such as stability and coordination of the extremities to delay impact of the disease on the motor performance brings benefits to the daily activities which are maintained in a period of at least 12 months (Ilg et al., 2009; Winfried Ilg et al., 2010).

Studies that are more recent have concluded that noninvasive cerebellar stimulation such as transcranial magnetic stimulation (TMS) and transcranial direct current stimulation (tDCS) improve the motor deficits in cerebellar diseases. The therapeutic effect in a noninvasive cerebellar stimulation is based on the functional reorganization of cerebellar motor control through the lasting restoration of inhibitory modulations of PCs in the dentate–thalamus–cortical pathways that contribute to the stabilization of movement (Di Nuzzo et al., 2018; Ferrucci et al., 2019).

NOVEL TREATMENTS

Gene Therapy

Promising therapies have been evaluated in cell and animal models in cerebellar ataxias of genetic origin. Although clinical studies are required to demonstrate the efficacy and safety of these approaches, it is important to have a general understanding of them since it is very probable that they will become foreseeable therapeutic alternatives in the future. One of the gene therapy approaches is to add a new gene encoding a protein to a target cell or organ, also known as gene replacement therapy, in the context where the dysfunction or lack of a protein is overcome by adding the correct version of the coding gene. Various types of viral or nonviral vectors can be used for this one and a remarkably successful example is the progress made in the spinal muscular atrophy disease. Regarding the cerebellum, animal models of Friederich's Ataxia have been treated with viral vectors that encode Frataxin, preventing cardiomyopathy, sensory neuropathy, and impaired motor coordination. Another promising approach to this disease has been the treatment with viral vectors to increase the levels of brain-derived neurotrophic factor (BDNF), which exerts a protective function as an antiapoptotic, antioxidant, and antiautophagy agent to prevent neurodegeneration and the loss of coordination in preclinical model (Ocana-Santero et al., 2021). On the other hand, in the PolyQ SCAs disease, the focus has been on the incorporation of genes that encode proteins that encode growth factors, neuronal homeostasis, or proteins that activate autophagy as a strategy to counteract the pathological phenotype and delay disease progression (Afonso-Reis et al., 2021). Another approach to be used in PolyQ SCAs would be the silencing of the genes related to pathological RNAs resulting in the lack toxic proteins translated, while eliminating the toxicity exerted by the RNA molecules themselves, which ultimately causes the disease, preventing the appearance or progression of the disease (Afonso-Reis et al., 2021). The endogenous RNA interference pathway is driven by microRNAs (miRNAs), endogenous noncoding RNAs composed of approximately 22 nucleotides that are known as small interfering RNAs. The chemical linkage between an interfering RNA and an mRNA can deactivate the expression of the target mRNA; this mechanism leads to the silencing of the target genes. Taking advantage of this discovery, researchers began to synthesize artificial interference RNA with the aim of inhibiting genes of interest. Another strategy to silence the expression of a gene is the use of antisense oligonucleotides (ASO) that have a structure similar to RNA, the main objective is to decrease the expression of the mutated mRNA through the use of single-stranded ASOs, by blocking translation or through a direct silencing effect. In particular, ASOs bind to complementary mRNA, a process that leads to the recruitment of RNase enzymes. Both strategies mentioned have been studied in animal models, particularly PolyQ SCAs with promising results and improvement in the progression of the condition (Afonso-Reis et al., 2021).

Stem Cell Treatment

Multipotent mesenchymal stem cells are becoming a therapeutic option for various degenerative diseases, since they can generate an exogenous supply of cells capable of promoting neurogenesis and modulating effects, stimulating plasticity and cell differentiation. Positive results have been observed in preclinical models reducing cerebellar degeneration. In this regard, studies have observed that stem cell therapy can slow or stop the progression of spinocerebellar ataxias, with improved motor functions (Appelt et al., 2021).

Currently, there is little evidence regarding the therapeutic usefulness of stem cells in ataxic syndromes. For instance, two clinical studies carried out with cells in patients with ataxic syndromes stand out. In the first of them, mesenchymal stem cells from the umbilical cord were administered intrathecally to 14 patients with ACS showing clinical improvement measured through the scale of daily living activities and the international cooperative scale for the evaluation of ataxia (ICARS), which lasted from 1 to 14 months with an average of 3 months (25). In a second trial, intrathecal and intravenous administration of umbilical cord−derived mesenchymal cells to 16 patients with a genetic diagnosis of SCA1, 2, or 3, resulted in a clinical improvement in scores for the Berg Balance Scale, and ICARS (Jin et al., 2013). The results obtained from these clinical trials carried out in patients affected by some ACS are encouraging; however, a greater number of clinical trials are required to evaluate the effectiveness and safety of stem cell transplantation as a therapeutic option in degenerative cerebellar disorders.

References

Ábrahám, H., Tornóczky, T., Kosztolányi, G., Seress, L., 2001. Cell formation in the cortical layers of the developing human cerebellum. Int. J. Dev. Neurosci. 19 (1), 53−62. https://doi.org/10.1016/S0736-5748(00)00065-4.

Afonso-Reis, R., Afonso, I.T., Nóbrega, C., 2021. Current status of gene therapy research in polyglutamine spinocerebellar ataxias. Int. J. Mol. Sci. 22 (8), 4249.

Aldinger, K.A., Doherty, D., 2016. The genetics of cerebellar malformations. Semin. Fetal Neonatal Med. 21 (5), 321−332. https://doi.org/10.1016/j.siny.2016.04.008.

Aldinger, K.A., Timms, A.E., Thomson, Z., Mirzaa, G.M., Bennett, J.T., Rosenberg, A.B., Haldipur, P., 2019. Redefining the etiologic landscape of cerebellar malformations. Am. J. Hum. Genet. 105 (3), 606−615.

Appelt, P.A., Comella, K., de Souza, L.A.P.S., Luvizutto, G.J., 2021. Effect of stem cell treatment on functional recovery of spinocerebellar ataxia: systematic review and meta-analysis. Cerebellum Ataxias 8 (1), 1−9.

Buffo, A., Rossi, F., 2013. Origin, lineage and function of cerebellar glia. Prog. Neurobiol. 109, 42−63. https://doi.org/10.1016/j.pneurobio.2013.08.001.

Cho, K.H., Rodríguez-Vázquez, J.F., Kim, J.H., Abe, H., Murakami, G., Cho, B.H., 2011. Early fetal development of the human cerebellum. Surg. Radiol. Anat. 33 (6), 523−530. https://doi.org/10.1007/s00276-011-0796-8.

Coesmans, M., Weber, J.T., De Zeeuw, C.I., Hansel, C., 2004. Bidirectional parallel fiber plasticity in the cerebellum under climbing fiber control. Neuron 44 (4), 691−700.

D'Angelo, E., Rossi, P., Armano, S., Taglietti, V., 1999. Evidence for NMDA and mGlu receptor-dependent long-term potentiation of mossy fiber−granule cell transmission in rat cerebellum. J. Neurophysiol. 81 (1), 277−287.

Delion, M., Dinomais, M., Mercier, P., 2017. Arteries and veins of the cerebellum. Cerebellum 16 (5), 880−912. https://doi.org/10.1007/s12311-016-0828-3.

Di Nuzzo, C., Ruggiero, F., Cortese, F., Cova, I., Priori, A., Ferrucci, R., 2018. Non-invasive cerebellar stimulation in cerebellar disorders. CNS Neurol. Disord. Drug Targets 17 (3), 193−198.

Dror, R., Malinger, G., Ben-Sira, L., Lev, D., Pick, C.G., Lerman-Sagie, T., 2009. Developmental outcome of children with enlargement of the cisterna magna identified in utero. J. Child Neurol. 24 (12), 1486−1492.

Duvernoy, H., Delon, S., Vannson, J.L., 1983. The vascularization of the human cerebellar cortex. Brain Res. Bull. 11 (4), 419−480. https://doi.org/10.1016/0361-9230(83)90116-8.

Duvernoy, H.M., 2012. The Human Brain Stem and Cerebellum: Surface, Structure, Vascularization, and Three-Dimensional Sectional Anatomy, with MRI. Springer Science and Business Media.

Ferrucci, R., Bocci, T., Cortese, F., Ruggiero, F., Priori, A., 2019. Noninvasive cerebellar stimulation as a complement tool to pharmacotherapy. Curr. Neuropharmacol. 17 (1), 14−20.

Gruol, D.L., Gruol, D.L., Noriyuki Koibuchi, N.K., Mario Manto, M.M., 2016. Essentials of Cerebellum and Cerebellar Disorders: A Primer for Graduate Students. Springer.

Haldipur, P., Bharti, U., Alberti, C., Sarkar, C., Gulati, G., Iyengar, S., Mani, S., 2011. Preterm delivery disrupts the developmental program of the cerebellum. PLoS One 6 (8). https://doi.org/10.1371/journal.pone.0023449.

Haldipur, P., Bharti, U., Govindan, S., Sarkar, C., Iyengar, S., Gressens, P., Mani, S., 2012. Expression of sonic hedgehog during cell proliferation in the human cerebellum. Stem Cell. Dev. 21 (7), 1059−1068. https://doi.org/10.1089/scd.2011.0206.

Haldipur, P., Dang, D., Millen, K.J., 2018. Embryology. Handb. Clin. Neurol. 154, 29−44. https://doi.org/10.1016/B978-0-444-63956-1.00002-3.

Icardo, J.M., Ojeda, J.L., Garcia-Porrero, J.A., Hurle, J.M., 1982. The cerebellar arteries: cortical patterns and vascularization of the cerebellar nuclei. Cells Tissues Organs 113 (2), 108−116.

Ilg, W., Synofzik, M., Brötz, D., Burkard, S., Giese, M.A., Schöls, L., 2009. Intensive coordinative training improves motor performance in degenerative cerebellar disease. Neurology 73 (22), 1823−1830.

Ilg, W., Brötz, D., Burkard, S., Giese, M.A., Schöls, L., Synofzik, M., 2010. Long-term effects of coordinative training in degenerative cerebellar disease. Mov. Disord. 25 (13), 2239−2246.

Jin, J.-L., Liu, Z., Lu, Z.-J., Guan, D.-N., Wang, C., Chen, Z.-B., Xu, Y., 2013. Safety and efficacy of umbilical cord mesenchymal stem cell therapy in hereditary spinocerebellar ataxia. Curr. Neurovascular Res. 10 (1), 11−20.

Kano, M., Rexhausen, U., Dreessen, J., Konnerth, A., 1992. Synaptic excitation produces a long-lasting rebound potentiation of inhibitory synaptic signals in cerebellar Purkinje cells. Nature 356 (6370), 601−604.

Kansal, K., Yang, Z., Fishman, A.M., Sair, H.I., Ying, S.H., Jedynak, B.M., Onyike, C.U., 2017. Structural cerebellar correlates of cognitive and motor dysfunctions in cerebellar degeneration. Brain 140 (3), 707−720. https://doi.org/10.1093/brain/aww327.

Kiessling, M.C., Büttner, A., Butti, C., Müller-Starck, J., Milz, S., Hof, P.R., Schmitz, C., 2014. Cerebellar granule cells are generated postnatally in humans. Brain Struct. Funct. 219 (4), 1271−1286. https://doi.org/10.1007/s00429-013-0565-z.

Lerman-Sagie, T., Prayer, D., Stöcklein, S., Malinger, G., 2018. Fetal cerebellar disorders. Handb. Clin. Neurol. 155, 3−23.

Llinás, R.R., Walton, K.D., Lang, E.J., 2004. The Synaptic Organization of the Brain. Oxford University Press, New York.

Locatelli, F., Soda, T., Montagna, I., Tritto, S., Botta, L., Prestori, F., D'Angelo, E., 2021. Calcium channel-dependent induction of long-term synaptic plasticity at excitatory Golgi cell synapses of cerebellum. J. Neurosci. 41 (15), 3307−3319.

Lou, Y., Yang, J., Wang, L., Chen, X., Xin, X., Liu, Y., 2019. The clinical efficacy study of treatment to Chiari malformation type I with syringomyelia under the minimally invasive surgery of resection of Submeningeal cerebellar Tonsillar Herniation and reconstruction of Cisterna magna. Saudi J. Biol. Sci. 26 (8), 1927−1931.

Manni, E., Petrosini, L., 2004a. A century of cerebellar somatotopy: a debated representation. Nat. Rev. Neurosci. 5 (3), 241−249.

Manni, E., Petrosini, L., 2004b. ManniE+PetrosiniL-2004, vol. 5, pp. 1–9 (March).

Marzban, H., Del Bigio, M.R., Alizadeh, J., Ghavami, S., Zachariah, R.M., Rastegar, M., 2015. Cellular commitment in the developing cerebellum. Front. Cell. Neurosci. 8 (JAN), 1–26. https://doi.org/10.3389/fncel.2014.00450.

Massoud, M., Cagneaux, M., Garel, C., Varene, N., Moutard, M., Billette, T., Massardier, J., 2014. Prenatal unilateral cerebellar hypoplasia in a series of 26 cases: significance and implications for prenatal diagnosis. Ultrasound Obstet. Gynecol. 44 (4), 447–454.

Müller, F., O'Rahilly, R., 1988. The development of the human brain from a closed neural tube at stage 13. Anat. Embryol. 177 (3), 203–224. https://doi.org/10.1007/BF00321132.

Müller, F., O'Rahilly, R., 1990. The human brain at stages 21-23, with particular reference to the cerebral cortical plate and to the development of the cerebellum. Anat. Embryol. 182 (4), 375–400. https://doi.org/10.1007/BF02433497.

Nisson, P.L., Fard, S.A., Meybodi, A.T., Mooney, M.A., Kim, H., Jahnke, H., Lawton, M.T., 2018. The unique features and outcomes of microsurgically resected cerebellar arteriovenous malformations. World Neurosurg. 120, e940–e949.

Ocana-Santero, G., Díaz-Nido, J., Herranz-Martín, S., 2021. Future prospects of gene therapy for Friedreich's ataxia. Int. J. Mol. Sci. 22 (4), 1815.

Parisi, M.A., Dobyns, W.B., 2003. Human malformations of the midbrain and hindbrain: review and proposed classification scheme. Mol. Genet. Metabol. 80 (1–2), 36–53.

Piccinin, M.A., Munakomi, S., 2019. Neuroanatomy, Vertebrobasilar System.

Poretti, A., Alber, F.D., Bürki, S., Toelle, S.P., Boltshauser, E., 2009. Cognitive outcome in children with rhombencephalosynapsis. Eur. J. Paediatr. Neurol. 13 (1), 28–33.

Poretti, A., Boltshauser, E., Doherty, D., 2014. Cerebellar hypoplasia: differential diagnosis and diagnostic approach. Am. J. Med. Genet. Part C: Seminars in Medical Genetics 166 (2), 211–226 (Wiley Online Library).

Pugh, J.R., Raman, I.M., 2006. Potentiation of mossy fiber EPSCs in the cerebellar nuclei by NMDA receptor activation followed by postinhibitory rebound current. Neuron 51 (1), 113–123.

Rakic, P., Sidman, R.L., 1970. Histogenesis of cortical layers in human cerebellum, particularly the lamina dissecans. J. Comp. Neurol. 139 (4), 473–500. https://doi.org/10.1002/cne.901390407.

Roostaei, T., Nazeri, A., Sahraian, M.A., Minagar, A., 2014. The human cerebellum: a review of physiologic neuroanatomy. Neurol. Clin. 32 (4), 859–869. https://doi.org/10.1016/j.ncl.2014.07.013.

Rüsch, C.T., Bölsterli, B.K., Kottke, R., Steinfeld, R., Boltshauser, E., 2020. Pontocerebellar hypoplasia: a pattern recognition approach. Cerebellum 19 (4), 569–582.

Schweighofer, N., Doya, K., Kuroda, S., 2004. Cerebellar aminergic neuromodulation: towards a functional understanding. Brain Res. Rev. 44 (2–3), 103–116. https://doi.org/10.1016/j.brainresrev.2003.10.004.

Shaikh, M., Khera, P., Panda, S., 2019. Unilateral cerebellar hypoplasia: a rare cause of childhood seizures. J. Pediatr. Neurosci. 14 (4), 236.

Shen, Y., Hansel, C., Linden, D.J., 2002. Glutamate release during LTD at cerebellar climbing fiber–Purkinje cell synapses. Nat. Neurosci. 5 (8), 725–726.

Solov'ev, S.V., 2006. The weight and linear dimensions of the human cerebellum. Neurosci. Behav. Physiol. 36 (5), 479–481. https://doi.org/10.1007/s11055-006-0043-7.

Stoodley, C.J., Schmahmann, J.D., 2018. Functional topography of the human cerebellum. Handb. Clin. Neurol. 154, 59–70. https://doi.org/10.1016/B978-0-444-63956-1.00004-7.

Surisetti, B.K., Holla, V.V., Prasad, S., Neeraja, K., Kamble, N., Yadav, R., Pal, P.K., 2021. Clinical and imaging Profile of patients with Joubert syndrome. J. Mov. Disord. 14 (3), 231.

Tiemeier, H., Lenroot, R.K., Greenstein, D.K., Tran, L., Giedd, J.N., 2011. NIH Public Access 49 (1), 63–70. https://doi.org/10.1016/j.neuroimage.2009.08.016.Cerebellum.

Tong, X., Wu, J., Lin, F., Cao, Y., Zhao, Y., Wang, S., Zhao, J., 2016. Microsurgical outcome of cerebellar arteriovenous malformations: single-center experience. World Neurosurg. 95, 469–479.

van Essen, M.J., Nayler, S., Becker, E.B.E., Jacob, J., 2020. Deconstructing cerebellar development cell by cell. PLoS Genet. 16 (4), 1–14. https://doi.org/10.1371/journal.pgen.1008630.

Voogd, J., 2003. The human cerebellum. J. Chem. Neuroanat. 26 (4), 243–252. https://doi.org/10.1016/j.jchemneu.2003.07.005.

Voogd, J., Glickstein, M., 1998. The anatomy of the cerebellum. Trends Cognit. Sci. 2 (9), 307–313. https://doi.org/10.1016/S1364-6613(98)01210-8.

Voogd, J., Jaarsma, D., Marani, E., 1996. The cerebellum: chemoarchitecture and anatomy. Handbk. Chem. Neuroanat. 12, 1–369.

Wang, Y.T., Linden, D.J., 2000. Expression of cerebellar long-term depression requires postsynaptic clathrin-mediated endocytosis. Neuron 25 (3), 635–647.

Wu, H., Yu, T., Wang, S., Zhao, J., Zhao, Y., 2015. Surgical treatment of cerebellar cavernous malformations: a single-center experience with 58 cases. World Neurosurg. 84 (4), 1103–1111.

Zarate, Y.A., Boccuto, L., Srikanth, S., Pauly, R., Ocal, E., Balmakund, T., Collins, R.T., 2019. Constitutive activation of the PI3K-AKT pathway and cardiovascular abnormalities in an individual with Kosaki overgrowth syndrome. Am. J. Med. Genet. 179 (6), 1047–1052.

Zhang, W., Shin, J.H., Linden, D.J., 2004. Persistent changes in the intrinsic excitability of rat deep cerebellar nuclear neurones induced by EPSP or IPSP bursts. J. Physiol. 561 (3), 703–719.

Further Reading

Mitoma, H., Manto, M., Hadjivassiliou, M., 2021. Immune-mediated cerebellar ataxias: clinical diagnosis and treatment based on immunological and physiological mechanisms. J. Mov. Disord. 14 (1), 10–28. https://doi.org/10.14802/jmd.20040.

7

The Diencephalon: Introduction and Epithalamus

Lee Elisevich[1], *Brandon M. Ascenzi*[3] *and Kost Elisevich*[2]

[1]Central Michigan University, Mount Pleasant, MI, United States [2]Michigan State University, East Lansing, MI, United States [3]Independent Neuroresearcher-Member of Marie Curie Alumni Association (MCAA)

O U T L I N E

INTRODUCTION

The diencephalon constitutes the central core of the cerebrum with the telencephalon forming its bihemispheric crown and includes the thalamus, epithalamus, hypothalamus, and subthalamus. Only its ventral surface is visible on gross inspection of the brain, bounded anteriorly by the optic chiasm, from which the optic tracts extend posterolaterally, and posteriorly by the transition of the internal capsule to the basis pedunculi of the midbrain, slightly caudal to the mammillary bodies (Kiernan, 2008). The surface region itself is identified as the tuber cinereum upon which a midline median eminence is prominent immediately behind the chiasm and from which the neurohypophysial stalk arises. The median eminence itself contains the capillaries of the hypophysial portal system and is considered part of the neurohypophysis.

The diencephalon is, otherwise, a bilateral and symmetrical structure situated on either side of the third ventricle with its anterior margin approximated by a line extending from the anterior commissure to the optic chiasm, defining its connection with the telencephalon, and its posterior extent by a line extending from the posterior commissure to a point caudal to the mammillary bodies defining its transition with the mesencephalon (Fig. 7.1). The remaining dorsal and lateral surfaces of the diencephalon are bounded by the fornix in the immediate parasagittal plane, then outward by the ependymal surface of the floor along the central aspect of the lateral ventricles, and, finally, by the internal capsules in each of the cerebral hemispheres. The latter constitute a massive input–output communication between the cerebral cortex and thalamus on either side.

The tela choroidea is a vascular connective tissue, consisting of both pia matter and ependyma (Fig. 7.2), situated between the right and left fornices and continuous with the choroid plexuses of the lateral and third ventricles. It covers the roof of the diencephalon as a lateral extension from the third ventricle and constitutes the choroid fissure between the fornix and thalamus in the central portion of the lateral ventricles. Its vascular network is supplied predominantly by the anterior choroidal and posteromedial choroidal arteries of the anterior and posterior circulations, respectively, while drainage is managed by the internal cerebral veins that continue to the great cerebral vein (Fig. 7.2).

A number of other features provide some definition to the medial surface of the diencephalon and deserve mention. The massa intermedia (MI) or interthalamic adhesion is apparent in about 70% of individuals and consists of gray matter with fibers that appear largely distributed ipsilaterally. Its normal role is not well defined although its absence has been found more often

From Anatomy to Function of the Central Nervous System
https://doi.org/10.1016/B978-0-12-822404-5.00002-4

in schizophrenia spectrum and bipolar disorders (Damle et al., 2017; Snyder et al., 1998; Takahashi et al., 2008; Trzesniak et al., 2011). A recent study of cortical and subcortical cognitive processing with examination of 402 brains in randomly selected participants identified 42 without an MI (Borghei et al., 2020). Both a greater prevalence and a significantly larger MI were identified in females, but, otherwise, race or ethnicity showed no distinctions. Logistic regression analysis did identify loneliness in addition to a measure of how conflicting information affects choice response speed (Eriksen Flanker Paradigm) as predictors for the presence of a MI. The latter task engages the anterior cingulate cortex in response to processing incongruent stimuli as opposed to congruent stimuli. The presence of subtle neuropsychological effects has therefore been implied by such connections but requires further study. Of some relevance here is the possibility of a more direct limbic functional connection suggested by the finding of fibers of the stria medullaris thalami (SMT) identified passing into and through the MI with putative connections, crossed and uncrossed, between the lateral habenular nuclei (LHb) and, specifically, the right orbitofrontal cortex (Kochanski et al., 2018). Using probabilistic diffusion tensor imaging and seeding of the LHb on either side to define its afferent connections via the SMT, 7/10 subjects were shown with fibers traversing the MI.

The SMT itself forms a visible ridge along the mediodorsal aspect of the diencephalon and is the border from which the ependymal lining of the third ventricle is reflected to form its roof. The ventral surface of this roof is the site from which the forward extension of the tela choroidea is suspended to form the choroid plexus of the third ventricle. More will be mentioned of the SMT in our discussion of the epithalamus and its connections.

Constituent Structures

The thalamus has an egg-shaped appearance in axial section with an apical anterior tubercle and a wider posterior aspect with its anteroposterior axis skewed posterolaterally. It is the largest body within the diencephalon with the left thalamus measuring 7.00 ± 1.12 cc and the

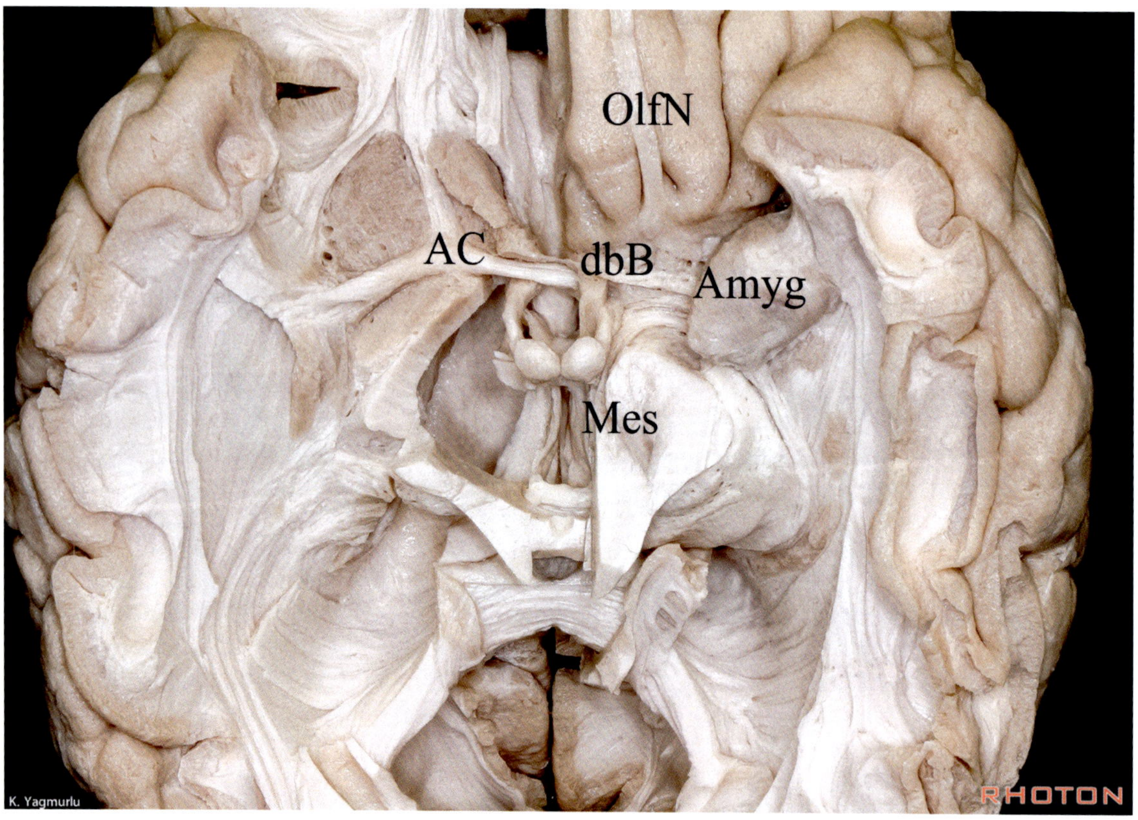

FIGURE 7.1 **Inferior view of the brain.** On the right side (left hemisphere) the mesencephalon (Mes) is apparent, upon which the thalamus is situated. Medioventrally are the two mammillary bodies and the anterior commissure. On the right side, at the level of the anterior commissure (AC) that has been removed, is the thin band of white matter of the diagonal band of Broca (dbB). Laterally, is the almond-shaped amygdala (Amyg). Ventral to the diagonal band is the anterior perforated substance and further ventrally, the olfactory nerve. On the left side, where the thalamus has been removed, at the bottom, is the corpus callosum. *Image courtesy of The Rhoton Collection® http://rhoton.ineurodb.org/. Dissector: K. Yagmurlu.*

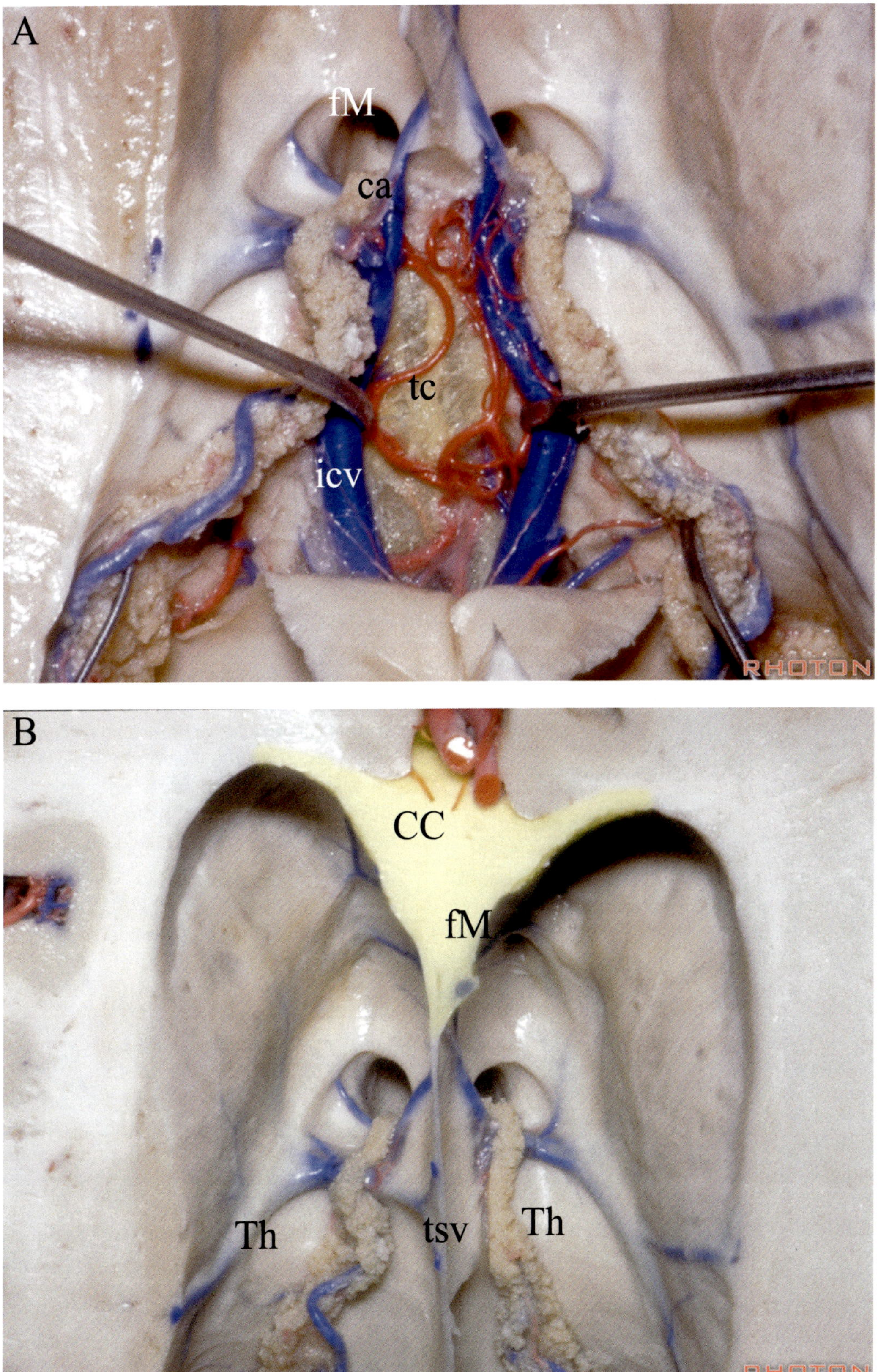

FIGURE 7.2 **Central aspect of the lateral ventricles; anterior views.** (A) The midline tela choroidea (tc) is seen with centrally situated posteromedial choroidal arteries (ca) and bilaterally situated internal cerebral veins (icv). The structure is coveraed by a thinly veiled velum interpositum. (B) The tela choroidea has been removed. The corpus callosum (CC) is highlighted in yellow and the thin midline structure extending behind is the septum pellucidum. The foramen of Monroe (fM) leading to the thrid ventricle presents anteromedially within each of the lateral ventricles rostral to the thalamus. The superior surface of the right and left thalami (Th) form much of the floor of the lateral ventricles. The thalamostriate veins (tsv) are visible in blue. The head and body of the caudate nuclei form part of the floor of the anterior horn of the lateral ventricles. *Image courtesy of The Rhoton Collection® http://rhoton.ineurodb.org/. Drawing of Jeffrey Sorenson.*

right, 6.55 ± 0.94 cc (Spinks et al., 2002). Greater attention has been given more recently to magnetic resonance imaging to provide in vivo volume measures and avoiding potential error attributed to fixation artifact in the case of histological analysis. More importantly, the imaging approach provides a means of clinically distinguishing a variety of disorders impacting the thalamus, including mental health issues, by applying quantitative measures over time to categorize and measure the progress of disease (Hazlett et al., 1999).

The thalamus both processes and transfers sensory information for all modalities, save that for olfaction, to the cerebral cortex, although in this latter modality, it does receive the information from its limbic connection where olfactory input is primary. The integration of locomotor function is facilitated within the thalamus by mediation of cerebrocortical activity through feedback mechanisms that engage both cerebellar and basal ganglia connections. In its role as a coordinating center, each constituent nucleus, save one, projects to the cerebral cortex and receives a reciprocal connection from the respective site that provides the means by which such input may be continuously modulated. The only intrinsic element within the thalamus without a cerebrocorticofugal projection is the thalamic reticular nucleus or zona incerta, which serves to modulate intrinsic thalamic activity in concert with cerebrocortical and brainstem input. Critical to the filtering mechanism that underlies these functions within the thalamus is the rich network of inhibitory interneurons, both GABAergic and peptidergic, along with extrinsic control mediated by serotoninergic and norepinephrinergic afferent input.

The thalamus is divided into three parts—anterior, lateral, and medial—by the internal medullary lamina (IML) that splits rostrally to form a Y-shaped fiber bundle. The anterior thalamic nucleus is situated as the only inhabitant of the anterior part within the split. The dorsal medial nucleus (DM), along with the midline nuclei, constitutes the medial part. The prominent lateral part is divided into dorsal and ventral tiers, each containing a number of nuclei. The dorsal tier has a rostrocaudal arrangement of three nuclei—lateral dorsal (LD), lateral posterior (LP), and pulvinar; likewise, the ventral tier has an arrangement containing the ventral anterior (VA), the ventral lateral (VL), and the ventral posterior that is divided into medial (VPM) and lateral (VPL) segments. Two nuclei—the central median (CM) and parafascicular (PF)—are situated within the IML and are designated the intralaminar nuclei. An external medullary lamina separates the rather thin thalamic reticular nucleus from the rest of the thalamus along its lateral border. Two relay nuclei remain to be mentioned and are situated on the pulvinar surface. The lateral and medial geniculate nuclei subserve the auditory and visual pathways, respectively, and each bear a distinctive topographic arrangement according to their respective modalities.

The subthalamus is somewhat ill-defined because of its preponderance of while matter tracts with a singular well-defined gray matter structure, the subthalamic nucleus. Some of the midbrain gray matter structures, such as the substantia nigra, red nucleus, extend rostrally into the subthalamic area but are considered centered within the midbrain. The midbrain reticular formation also becomes confluent with the zona incerta under the thalamus and between the thalamic and lenticular fasciculi. The fields of Forel constitute three white matter areas or H fields of the subthalamus:

- **field H (prerubal area)** contains several neurons and several fibers of the superior cerebellar peduncle that originate in the dentate nucleus and traverse the red nucleus to end contralaterally in VA and VL of the thalamus. Field H lies medial to the caudal aspect of the zona incerta;
- **field H1 (thalamic fasciculus)** composed of the cerebellothalamic fibers, the ansa lenticularis, and lenticular fasciculus with efferent fibers from the cerebellum and globus pallidus destined to VA and VL of the thalamus;
- **field H2 (lenticular fasciculus)** separated from H1 by the zona incerta and containing fibers of the ansa lenticularis that extend from the globus pallidus to VL of the thalamus.

Three groups of nuclei are recognized within each of the fields and designated, respectively, as the nucleus campi medialis (field H), nucleus campi dorsalis (field H1), and nucleus campi ventralis (field H2). The subthalamic nucleus (corpus Luysi) lies dorsal to the substantia nigra and medial to the internal capsule, measuring 8.5 mm in its maximal dimension (Mavridis et al., 2013). Its width and length change significantly with age, particularly after 60 years. It primarily contains glutamatergic neurons with a small contingent of GABAergic interneurons with glutamatergic afferents arising from a number of sites in the cerebral cortex, particularly the motor area. Its major afferent input is of pallidal origin and is GABAergic with a much smaller dopaminergic input from the substantia nigra pars compacta (SNpc). Efferent projections are primarily to the pallidum and to the substantia nigra pars reticulata (SNpr) (Kita and Kitai, 1987). The subthalamic nucleus is often targeted for deep brain stimulation in Parkinson's disease. Otherwise, it is significantly affected in progressive supranuclear palsy, perhaps predominantly due to oxidative stress often identified in neurodegenerative conditions (Albers et al., 1999; Steele et al., 1964).

The epithalamus presents dorsally and posteriorly relative to the thalamus and consists of the habenular nuclei and its afferent connection, the stria medullaris thalami (SMT), and the pineal gland. For its relatively small dimensions, it plays an outsized role in its influence upon the circadian rhythm and as a regulator of motor activity, memory, and emotions by its connections with the basal ganglia and limbic system. The SMT arises largely from the septal area, projecting limbic influence upon the habenular nuclei with the latter projecting, in turn, to the interpeduncular nucleus centered over the interpeduncular fossa of the midbrain via the habenulointerpeduncular tract (fasciculus retroflexus of Meynert—frM). The pineal gland (epiphysis) attaches to the diencephalon posteriorly by a pineal stalk, which contains a pineal recess lying above the aqueduct of Sylvius along the posterior margin of the third ventricle. A habenular commissure in the dorsal portion of the stalk allows for STM fibers to gain access to the opposite habenular nucleus in addition to acting as a commissural bundle connecting the habenular nuclei.

The hypothalamus, as its name implies, is situated beneath the thalamus and is distinguished from it by the hypothalamic sulcus, which runs along the lateral wall of the third ventricle from the interventricular foramen to the cerebral aqueduct. It is critical to the proper regulation of the basic needs of the body, acting as both a monitor and effector, to afford us optimal circumstances for our well-being. The hypothalamus underlies endocrine, behavioral, metabolic, autonomic, and visceral functions through its intrinsic elements, some of which respond directly to systemic influences via its local vasculature (i.e., hormonal levels, osmotic pressure, temperature) and its connections with the central nervous system including the neurohypophysis and the limbic system; by such influence, it provides the homeostasis needed to adjust to our environment.

The hypothalamus can be divided into distinct domains, each with specializations serving to regulate homeostasis: preoptic, anterior, tuberal, and mammillary. The nuclei of the hypothalamus are typically grouped into lateral and medial zones although it may be difficult to distinguish some nuclei on the basis of distinct function or physiology. Furthermore, recent technological innovation has allowed a much more refined understanding of how neuronal heterogeneity supports the diverse hormonal and synaptic influences that allow adaptation to the environment (Alpar and Harkany, 2018). The lateral zone contains the lateral nucleus and the lateral tuberal nucleus. The medial zone may be divided into dorsal and ventral tiers. In a posteroanterior direction with some overlap, the dorsal tier consists of the posterior, dorsomedial, paraventricular, and anterior nuclei. Similarly, the ventral tier has the mammillary, tuberomammillary, ventromedial, arcuate, supraoptic, and suprachiasmatic nuclei. Other adjacent and associated centers include two circumventricular organs, identified by their permeable vasculature, the neurohypophysis, and the organum vasculosum laminae terminalis imbedded within the lamina terminalis, which extends between the optic chiasm and the anterior commissure to form the anterior wall of the third ventricle. Immediately lateral to and behind the lamina terminalis is the preoptic area. Also designated a telencephalic structure, its connections are not separable from the more anterior aspect of the medial zone of the hypothalamus.

To allow it to perform as an integrative center, somatic, visceral, and special sensory afferent signals combine with those of amygdalohippocampal origin to provide the hypothalamus both extrinsic and intrinsic measures of ourselves and our surroundings as well as simultaneous access to our emotional state and relevant past. A response is automatically executed both rostrally to the thalamus and cerebral cortex and caudally to the brainstem and spinal cord instigating alerting behavior and vasomotor, cardioregulatory, and locomotor activity. Endocrine function is largely mediated by the hypothalamus through the release of neurohypophysial hormones synthesized in the hypothalamus and through the control of hormone release from the anterior hypophysis via releasing factors.

Embryology

Neural development entails considerable coordination and a good deal of time to achieve the complexity required to assume its role. Hence it begins earlier and ends later, long after birth, than other organ systems. Neurulation begins at the trilaminar stage of embryonic development with the central portion of the ectoderm forming the neural plate that invaginates to form a groove, which then folds over to become a neural tube by the third gestational week, closing at either end soon after. Beginning by the fifth gestational week, three primary vesicles form with the prosencephalon (i.e., forebrain), mesencephalon (i.e., midbrain), and rhombencephalon (i.e., hindbrain) forming the rostrocaudal extent of the embryonic brain. Each of these three vesicles has been subdivided into neuromeres representing segments defined by morphological or molecular characteristics with six such segments designated within the prosencephalon, called prosomeres (Ewing and Carstens, 2009). A five vesicle stage is entered when the prosencephalon separates into the diencephalon and telencephalon and the rhombencephalon into the metencephalon and myelencephalon (Ishikawa et al., 2012). The telencephalon then splits into two cerebral hemispheres in the eighth gestational week constituting the cerebral gray and white matter, while the diencephalon gives rise to the pallidum. Neural stem cells within the neuroepithelium (i.e., ventricular zone) of the vesicle walls divide (i.e., neurogenesis) at a considerable rate during this same time to give rise to both neurons and glia, which then differentiate into several cell types, each with distinct specialized functions and that now begin to provide substance to the vesicular assembly.

The diencephalon originates from the median part of the prosencephalon. It consists of two alar plates, which are part of the dorsal aspect of the neural tube and will form the walls and floor of the mature diencephalon. The hypothalamic sulcus, mentioned previously as a longitudinal impression along the third ventricle, divides the alar plate on either side into dorsal and ventral regions to become the thalamus and hypothalamus, respectively. The roof plate consists of a layer of ependymal cells that will combine rostrally with their mesenchymal cover to form the choroid plexus of the third ventricle. The caudal aspect will develop into the pineal gland while some controversy has arisen regarding whether the remainder of the epithalamus has derived from the alar or roof plates. Otherwise, both the thalamus and epithalamus have been identified as arising from a single segment of the diencephalon, prosomere p(2). At times, the roof plate forms an evagination at the site of the interventricular foramina creating a paraphysial cyst within the future third ventricle. With proliferation, the thalami progressively protrude into the vesicle and may fuse to form the interthalamic adhesion. The vesicle itself will become the slender midline third ventricle, limited anteriorly by the lamina terminalis, which is considered to be part of the telencephalon.

The thalamic nuclei will form into dorsal and ventral nuclear groups, each with certain functional characteristics as will be elaborated in the next chapter. The zona limitans intrathalamica (ZLI) is a central morphogenetic organizer of the diencephalon that appears in the alar plate at the neural tube stage. It promotes a rostrocaudal and dorsoventral organization within the plate. Several molecular factors cooperate with one another and act upon the diencephalic epithelium to bring about the regional specialization necessary to establish the complex organization necessary (Lim et al., 2005; Martinez-Ferre and Martinez, 2012). Dorsoventral regionalization has been shown to be codified by signals mediated by retinoic acid, fibroblast growth factor 8 (Fgf8), bone morphogenic proteins (BMPs), and the evolutionarily conserved *cys*-glycoproteins, Wnts, specifying dorsal structures and sonic hedgehog (Shh) acting to do the same with ventral structures. Shh signaling within the ZLI has been one of the most studied and shown to mediate its actions by a concentration gradient that serves to impede directional growth (Sholpp et al., 2006; Rash and Grove, 2011).

These proteins represent a growing number of morphogens responsible for the developmental pathways and nuclear organization involved in the development of the diencephalon. More recently, the Barhl2 homeobox gene, expressed in the p2 progenitor domain and within the ZLI, has been shown to regulate progenitor cell expression through Shh in the developing diencephalon with its loss leading to an absence of thalamocortical projections as well as that of both habenular afferents and efferents (Ding et al., 2017). Shh further mediates the function of another homeobox-containing gene, Pax6, to regulate the partition of the p2 domain into the epithalamus and thalamus (Chatterjee et al., 2014).

In addition to involvement of Fgf activity enabling the positioning of thalamic sensory nuclei in the postnatal brain (Kataoka and Shimogori, 2008), Fgf8 has also been implicated in the emergence of the epithalamus in vertebrates as both an inductive and morphogenic regulator. Its reduction results in a diminished or absent habenular region and pineal in mice (Martinez-Ferre and Martinez, 2009). More recently, a deletion of Foxg1 has been found to result in impairment of development of the epithalamus without any overt changes in Fgf or Shh signaling (Liu et al., 2018). This, in turn, suggests that Foxg1 does not require these two agents in regulating development. Foxg1 deletion resulted in an increase of habenular size along with impairment of innervation via the habenular commissure while mutant forms resulted in reduction of the pineal gland and an increased branching of the choroid plexus. Interestingly, the clinical features of Foxg1-related disorders include sleep disturbance, severe anxiety, cognitive disability, epilepsy, and choroid plexus cysts (Florian et al., 2011), suggesting, to some degree, both habenular and pineal involvement. Finally, an asymmetry of the habenular nuclei occurs because of a preferential leftward migration of parapineal cells and is thought to underlie a lateralization of cognitive function. The left epithalamus expresses Pitx2c, which has been shown to promote the asymmetry by restricting parapineal cell number (Garric et al., 2014).

Although past discussion of the development of the hypothalamus has focused on its origin from the diencephalon, specifically its ventrolateral wall, more recent studies based upon gene expression in the context of the prosomeric model have suggested its origin to be the telencephalon (Rubinstein et al., 1994; Puelles and Rubinstein, 2015). Development here also represents a neural sexual dimorphism with its constitution different in female and male embryos. Hypothalamic cells arise from progenitor cells with those expressing Shh/Fgf10 growing anisotropically to overtake other progenitor populations (Fu et al., 2019). They grow sequentially in a number of directions to ultimately give rise to a rostrally situated anteroventral/tuberal progenitor population, then a caudally situated mammillary progenitor population and, finally, a ventrally situated tuberal infundibular glial population. As this process evolves, stable and separate populations of Shh+ and Fgf10+

progenitors form. The latter, in turn, continue to assert development within each of these populations.

THE EPITHALAMUS

The constituents of the epithalamus include the habenular nuclei and pineal gland with white matter tracts providing primary connection of the limbic system with the habenular nuclei by the stria medullaris thalami (STM). Additional connections are more local and consist of the habenular commissure, within the dorsal pineal stalk, which interconnects the respective nuclei and also includes STM fibers. The posterior commissure lies within the ventral pineal stalk and contains fibers from the superior colliculus, pretectal area, accessory oculomotor nuclei, and habenular nuclei. The posterior commissure therefore appears to have more to do with pupillary reflexes and ocular movement, whereas the habenular commissure subserves aspects of limbic system function.

Habenular Nuclei

The habenulae are a paramidline pair of small nuclei, above the superior colliculi, appearing as minor protuberances within the habenular trigone, a small triangular depression, on each side in the posterosuperior aspect of the third ventricle. They exist in all vertebrates and are involved in circuitry central to adaptive behavior underlying the response to reward as well as those pertaining to addiction and stress (Stephenson-Jones et al., 2011). The habenular nucleus has a medial (MHb) subregion with a relatively high cellular density and a lateral (LHb) more sparsely arranged cellular subregion (Cajal, 1911), each with a particular gene expression (Quina et al., 2009) and particular roles mediated by their respective connections and neurotransmitters. Both have been further divided into subnuclei with 5 identified in the MHb and 10 in the LHb based upon neuronal size and morphology and glia (Andres et al., 1999). Some interaction between the two subregions has been suggested with substance P-positive projections of presumptive MHb origin found in the LHb (Cuello et al., 1978). Moreover, reciprocal connections have been identified by tracing methods between the MHb and the medial aspect of LHb (Lima et al., 2017) although prior electrophysiological study with intracellular biocytin injection showed an asymmetrical arrangement with only the MHb sending projections to the LHb (Kim and Chang, 2005). The latter bore neurons with dendrites much longer than those found among MHb neurons and projected intrinsic axons largely from medial to lateral within the subregion.

Medial Habenula

The less studied of the two subregions, the MHb, has more recently gained recognition as a mediator of mood disorders, drug addiction, and nicotine withdrawal (Lee et al., 2019; McLaughlin et al., 2017). An aversive response to nicotine, in particular, has been an area of study regarding both the MHb and LHb with the former known to both negatively regulate intake and promote aversion (Fowler and Kenny, 2014; Salas et al., 2009; Antolin-Fontes et al., 2015). The MHb is grossly divided into a peptidergic dorsal (dMHb) and a larger cholinergic ventral (vMHb) segment (Aizawa et al., 2012). Substance P (SP)-expressing dMHb neurons are suspected of having a close relationship with vMHb neurons as the latter are partly distinguished by the SP receptor, Tacr1 (Aizawa et al., 2012), along with the gamma aminobutyric acid B ($GABA_B$) receptor and *boutons en passant* noted in the vMHb on efferent fibers of the dMHb (Kim and Chang, 2005). Of interest here is the prominent expression of mu opioid receptors situated on most SP neurons in the MHb and a relatively small subpopulation of cholinergic neurons (Gardon et al., 2014) pointing to a functional interaction in the mediation of pain and addiction (Murtra et al., 2000; Hu et al., 2020). Mu opioid expression is also found within habenular terminals in the interpeduncular nucleus (IPN; Gardon et al., 2014). Finally, some evidence supports a projection from the MHb to the pineal (Ronnekleiv and Moller, 1979; Gugleilmotti and Cristino, 2006).

Afferents of the dMHb arise in the bed nucleus of the anterior commissure (BAC), a dense cellular area situated within the angle formed by the fornix and anterior commissure in the posterior septum; dMHb efferents project to the peripheral part of the interpeduncular nucleus (pIPN) in the ventral midbrain via the frM (Yamaguchi et al., 2013) with sparse efferents suspected also to the VTA (Cuello et al., 1978). A parallel septohabenular pathway from the posterior septum exists involving the vMHb with afferents arising in the triangular septal nucleus (TsN), situated within the ventral hippocampal commissure in the area of the subfornical organ and the interventricular foramen of Monro. The efferent projection of the vMHb is directed to the core part of the interpeduncular nucleus (cIPN). Afferents arriving in both subregions of the MHb otherwise arise in the median raphe nucleus (MRN), nucleus accumbens (NAc), locus coeruleus (LC), superior cervical ganglion (SCG), and the diagonal band nucleus (DBN) (Qin and Luo, 2009; Lecourtier and Kelly, 2007; Gottesfeld, 1983; Sutherland, 1982; Phillipson and Pycock, 1982; Ronnekleiv and Moller, 1979; Herkenham and Nauta, 1977; Conrad et al., 1974). Apart from these more directed inputs to one or both subregions of the MHb, dopaminergic neurons of the midline VTA have been shown to project to the medial aspect of the MHb (Phillipson and Pycock, 1982) although these may reflect, in part, the reciprocal arrangement of the larger communication between the LHb and VTA (Schmidt et al., 2014).

Receptors for SP are found on both glutamatergic and GABAergic neurons in the dorsal raphe nucleus (DRN; Saffroy et al., 1994; Commons and Valentino, 2002), and excitatory SP neuronal projections are presumed to arise from the dMHb (Cuello et al., 1978) although the majority of habenular projections arise in the LHb as part of a dorsal pathway from forebrain to midbrain limbic structures (Aghajanian and Wang, 1977). Stimulation of the Hb exerts an inhibitory effect upon the DRN (Wang and Aghajanian, 1977), an effect ostensibly reflecting the presence of a GABAergic intermediary within the DRN (Zhao et al., 2015).

Neuronal activity in the pIPN is increased by release of SP from the dMHb when rats are exposed to a chronic unpredictable mild stress. This higher SP signal activity appears to promote a degree of anhedonia (Xu et al., 2018) and may even be suitable as a model of depression. A prior study using a knockout mouse model had also affirmed involvement of the dMHb in mediating an anhedonic-like state (Hsu et al., 2016). Some effect upon the fear response may be speculated given such an altered mood state involving the dMHb. Despite selective ablation of the BAC in mice resulting in an enhanced fear response (Yamaguchi et al., 2013), selective ablation of dMHb neurons did not bring about the expected response (Hsu et al., 2016). Further study of the interaction between the dMHb and vMHb may further clarify such an outcome. Elimination of $GABA_B$ receptors on the cholinergic neurons of vMHb neurons enhances conditioned fear (Zhang et al., 2016b). In this circumstance, IPN GABAergic neurons release GABA retrogradely upon presynaptic $GABA_B$ receptors on the terminals of vMHb neurons to promote an enhanced release of glutamate (Koppensteiner et al., 2017). The dynamic of this cholinergic–glutamatergic–GABAergic interaction has not as yet been elucidated. An MHb-IPN synaptic plasticity could be demonstrated through fear conditioning causing activity-dependent potentiation and fear extinction resulting in its reversal. Interestingly, adolescent IPN neurons were found to lack the α-amino-3-hydroxy-5-methyl-4-isoxazolepropionic acid (AMPA) receptor mechanism needed to actuate a GABA release and were incapable of undergoing plasticity.

Cannabinoid type 1 (CB_1) receptors found on presynaptic terminals of vMHb neurons in the IPN are regulators of aversive behavior, and their elimination has been shown to reduce the response to contextual fear conditioning (Soria-Gomez et al., 2015). Inhibition of nicotinic, but not glutamatergic receptors, reverses this effect. Blockade of the CB_1 receptor enhances cholinergic but not glutamatergic transmission, showing that expression of aversive memories can be controlled through selective presynaptic modulation of cholinergic transmission at the vMHb-IPN synapse. Hence, both CB_1 and $GABA_B$ presynaptic activities upon vMHb cholinergic neurons in the IPN play a critical role in the regulation of fear memory.

The cholinergic neurons of the vMHb release both acetylcholine (ACh) and glutamate (Glu) (Frahm et al., 2015; Ren et al., 2011) and harbor the SP receptor, Tacr1, as described before, suggesting a relationship with the peptidergic neurons of the dMHb. Nicotinic receptor (nAChR) subunits are scattered within both the vMHb and IPN with α3 and β4 nAChR subunits together primarily occupying vMHb neurons and the α5 subunit found among GABAergic neurons of the IPN (Wada et al., 1990; Marks et al., 1992; Hsu et al., 2013). There are, however, β4 nAChR subunits located on somatostatin (Som)-positive GABAergic neurons situated in the most dorsal portion of the IPN and are upregulated by glutamate release from vMHb neurons that project monosynaptically onto the site (Zhao-Shea et al., 2013). Blockade of these nAChRs results in somatic withdrawal features in mice chronically exposed to nicotine. These Som + GABAergic IPN neurons project primarily to the serotonin-rich median raphe (MRN)/ paramedian raphe (pMR) and dorsal tegmental area. Somatostatin interneurons are known to inhibit presynaptic glutamate release (Boehm et al., 1997) and will diminish release from vMHb neurons during chronic nicotine exposure. Zhao-Shea et al. (2013) have proposed that nicotine withdrawal will decrease the nAChR signal reducing Som + interneuronal activity and increase glutamate release from the vMHb. Increased excitation of GABAergic neurons will then promote the somatic features of withdrawal.

GABAergic cells are distinguished by glutamic acid decarboxylase (GAD) 1 and 2 markers. Most Som + - GABAergic IPN neurons were found to express GAD1 with comparatively fewer (<25%) expressing GAD2. Activation of predominantly nonSom + GABAergic IPN neurons were also found to trigger somatic withdrawal features in the same study but did not create anxiety, indicating that other pathways and distinct IPN subpopulations of neurons may mediate the affective features of withdrawal (Zhao-Shea et al., 2013). Such a case is supported by the finding that anxiety-related behavior of withdrawal involves ventral tegmental area (VTA)–derived corticotropin releasing factor (CRF) release upon the IPN (Zhao-Shea et al., 2015). Dopaminergic neurons of the VTA projecting onto a number of distinct subregions of the IPN release CRF to activate CRF receptor 1 signaling, which, in turn, modulates glutamate input from the vMHb.

Selective stimulation of α5nAChR-expressing IPN GABAergic neurons does not elicit somatic manifestations of nicotine withdrawal but results in aversion (Morton et al., 2018). Nicotine at high doses promotes aversive behavior through excitatory action of the vMHb upon the IPN and subsequent activation of its GABAergic projections onto the dorsolateral tegmental area (dlTA) (Wolfman et al., 2018). Inhibition of IPN terminals in the dlTA was, however, found to be anxiolytic indicating that the IPN-dlTA connection likely contributes to the expression of anxiety and therefore influences aversive behavior.

Cholinergic activity of the vMHb influences drug addiction after chronic exposure, ostensibly through β4nAChRs, and has been shown to regulate voluntary intake of narcotics and stimulants (Glick et al., 2006) as well as drug-seeking behavior (Lopez et al., 2018). Activation of the dMHb demonstrates reinforcing behavior, while its inhibition elicits an aversive response (Hsu et al., 2014). Moreover, nicotine toxicity diminishes the volume of the MHb and promotes neurodegeneration of the central frM targeting IPN neurons (Ellison, 2002; Ciani et al., 2005).

The MHb exhibits a greater rhythmicity than that within the LHb (Landgraf et al., 2016) with cholinergic neurons, in particular, showing circadian pacemaker behavior, ostensibly from expression of a circadian gene (Zhao and Rusak, 2005; Sakhi et al., 2014a). Both spontaneous firing rate and resting membrane potential show significant circadian variation and indicate that MHb neurons may act separately from the suprachiasmatic nucleus (SCN) possibly shaping diurnal influences upon behaviors such as addiction (Sakhi et al., 2014a). The LHb has also been identified as executing behavioral expression, possibly secondary to influences exerted by the MHb (Paul et al., 2011). Both contain neurons responsive to retinal photic stimulation (Zhao and Rusak, 2005), similar to the SCN. Ultimately, through its influence upon the IPN, the MHb may regulate the sleep—wake cycle (Valjakka et al., 1998), the timing of responsiveness to nicotine (Baldwin et al., 2011; Fowler et al., 2011) as well as fear and anxiety (Yamaguchi et al., 2013).

Lateral Habenula

The LHb receives input from several limbic forebrain sites and those that influence motor activity (Figs. 7.3—7.5). It is involved in assigning negative motivational value toward punishment and, in humans, is seen to guide suppression of behavior to punishment and attraction to reward in situations sometimes so subtle that they seem of little consequence (Lawson et al., 2014; Hu et al., 2020). Moreover, hyperexcitability of the LHb is implicated in major depressive disorder (Browne et al., 2018). This is in keeping with its efferent connections to monoaminergic nuclei of the midbrain, which, in turn, influence motor activity (Lammel et al., 2012). Although LHb neurons display a heterogenous neurotransmitter phenotype, the majority are glutamatergic (Geisler and Trimble, 2008) and exert a short latency, strongly inhibitory influence upon dopaminergic and serotonergic brainstem neurons (Ji and Shepard, 2007; Matsumoto and Hikosaka, 2007), critical for reward-based motor control, particularly behavior underlying prediction error. These sites include the dopaminergic ventral tegmental area (VTA) and substantia nigra pars compacta (SNpc), the serotonergic dorsal (DRN) and median (MRN) raphe, and the rostromedial tegmental area (rmTg) (Herkenham and Nauta, 1979; Araki et al., 1988; Lammel et al., 2012). The latter projects onto both the SNpc and the VTA (Jhou et al., 2009). A hybrid dopaminergic-GABAergic projection from the anterior VTA back to the medial LHb also exists although the connection is not reciprocal (Gruber et al., 2007; Phillipson and Pycock, 1982; Stamatakis et al., 2013) and is likely responsible for modulating activity from the limbic forebrain to monoaminergic midbrain nuclei. These projections, in fact, produce a GABA-mediated inhibitory effect upon LHb neurons suppressing their output under reward conditions (Stamatakis et al., 2013).

Afferent connections of the LHb serve to define a medial and lateral division, which are each composed of a number of subnuclei (Andres et al., 1999). The medial division is both directly and indirectly innervated by the cerebral cortex through the lateral hypothalamic (LHA) and lateral preoptic (LPA) areas, ventral pallidum, substantia innominata, nucleus accumbens, diagonal band of Broca, and amygdala, including the bed nucleus of the stria terminalis (Herkenham and Nauta, 1977; Araki et al., 1984; Garland and Mogenson, 1983). Additional projections have been identified in humans arising from the dorsal and pregenual anterior cingulate cortex, anterior insula cortex, and laterocaudal orbitofrontal cortex using diffusion tensor imaging (Vadovicova, 2014), The lateral division receives input primarily from the basal ganglia and, specifically, the globus pallidus pars interna (GPi) or entopeduncular nucleus (EPN), which receives cerebrocortical input through its striatal intermediary. A topographical arrangement of projection unfolds from the striatum through the EPN of the rat to the habenula. A duality of outflow proceeds from the striosomal network of the striatum to the dorsal EPN and onto the LHb and from the matrix of the striatum to the ventral EPN onto the thalamus (Rajakumar et al., 1993). The former constitutes a subcortical pathway that modulates the activity of the basal ganglia with influence from the ascending serotonergic fibers from the dorsal raphe nucleus (DRN). Dopaminergic inputs to the striatum also

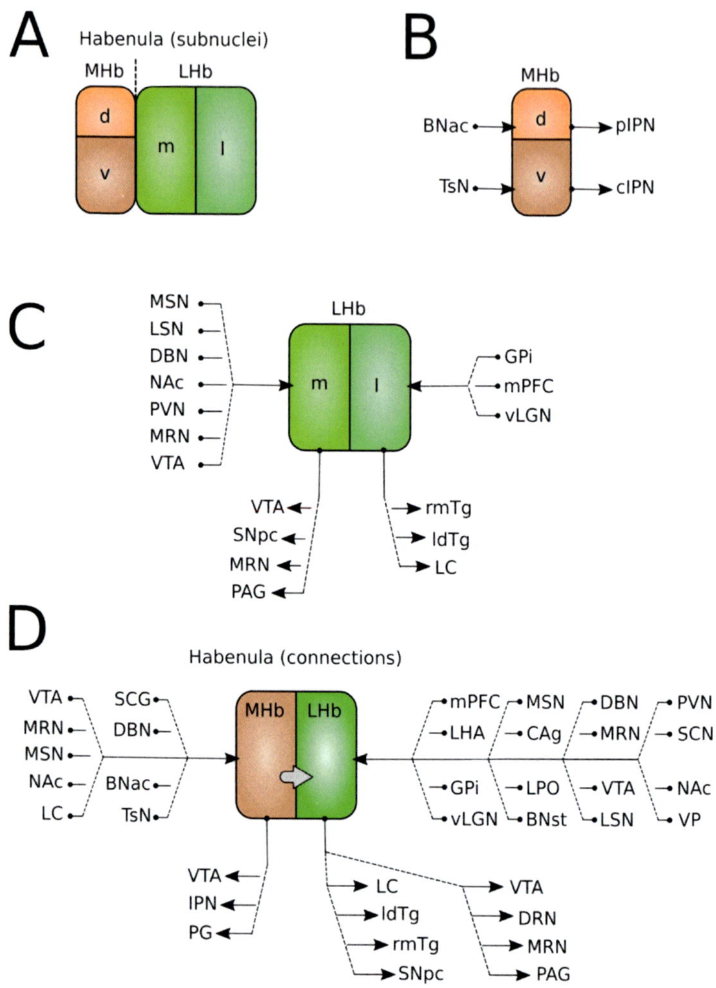

FIGURE 7.3 **Direct connections of the habenular nucleus.** (A) Both the medial (MHb) and lateral (LHb) subnuclei divide into subregions characterized within the MHb by distinct concentrations of peptidergic and cholinergic neurons in dorsal (d) and ventral (v) compartments, respectively, and, within the LHb, by specific afferent input to medial (m) and lateral (l) compartments. (B) The MHb receives preferential input to its dorsal compartment (dMHb) from the bed nucleus of the anterior commissure (BNac) with an efferent projection specifically to the periphery of the interpeduncular nucleus (pIPN). Similarly, the ventral compartment (vMHb) receives preferential input from the triangular septal nucleus (TsN) with an efferent projection descending to the core of the IPN (cIPN). (C) The primary afferent of the lateral compartment of the LHb (lLHb) arises in the globus pallidus pars interna (GPi) with additional minor input from the medial prefrontal cortex (mPF) and a projection from the ventral portion of the lateral geniculate nucleus (vLGN) along with adjacent fibers from the intergeniculate leaflet, which carry visual signals originating in melanopsin-positive retinal ganglion cells. A direct projection from the latter retinal cells arrives within the perihabenular region and is not depicted here. Efferents from the lLHb are distributed to the rostromedial tegmental nucleus (rmTg), laterodorsal tegmental nucleus (ldTg), and locus coeruleus (LC). The medial compartment (mLHb) receives afferents from several limbic areas, specifically, the medial (MSN) and lateral (LSN) septal nuclei, diagonal band of Broca (DBN), nucleus accumbens (NAc), paraventricular nucleus (PVN), median raphe nucleus (MRN), and ventral tegmental nucleus (VTA). Efferents from the mLHb project to the VTA, substantia nigra pars compacta (SNpc), MRN, and the periaqueductal gray (PAG). (D) A variety of afferent inputs at present are considered not to show a predilection toward one or another subregion and can be assembled along with specific compartmental afferents into a general scheme along with efferent projections to various destinations from both MHb and LHb. In addition to those afferent and efferent projections specific to each of the compartments that have already been mentioned, the total breadth of projections to the MHb originate in the VTA, MRN, NAc, LC, MSN, superior cervical ganglion (SCG), DBN, BNac, and TsN while its output remains predominantly to the IPN with minor input to the VTA and the pineal gland (PG). There is a likelihood of input from MHb neurons onto those of the LHb in addition. The LHb receives input from the GPi, mPF, vLGN, lateral hypothalamic area (LHA), lateral preoptic area (LPO), PVN, suprachiasmatic nucleus (SCN), ventral pallidum (VP), NAc, LSN, MSN, central nucleus of the amygdala (CAg), bed nucleus of the stria terminalis (BNst), DBN, MRN, and VTA. Its efferents project to the PAG, MRN, dorsal raphe nucleus (DRN), VTA, SNpc, rmTg, ldTg, and LC.

modulate locomotor activity according to the degree of reward prediction. In this regard, the finding that LHb neurons are more excited by smaller reward prediction than larger (Matsumoto and Hikosaka, 2007) suggests that LHb suppresses dopaminergic neurons (Nakamura and Hikosaka, 2006; Ji and Shepard, 2007) in the SNpc and VTA, both of which project to the striatum.

It is important to recognize that the VTA is differentially activated by inputs from the dlTA and the LHb resulting in locomotor activity reflecting reward and aversion. The dlTA neurons form synapses upon dopaminergic neurons that preferentially project to the nucleus accumbens, whereas the LHb neurons do the same upon those that project onto GABAergic neurons of the rmTA and the medial prefrontal cortex (Lammel et al., 2012). The VTA, as mentioned previously, projects back to the LHb but has been shown by optogenetic activation to suppress LHb activity while increasing the spontaneous activity of VTA dopaminergic neurons (Stamatakis et al., 2013). The connection between the LHb and VTA also pertains specifically to behavior toward nicotine apart from the well-known involvement of the MHb in its negative regulation of nicotine intake (Antolin-Fontes et al., 2015). The LHb appears to be involved with inducing anxiety in acute nicotine exposure (Casarrubea et al., 2015) and with nicotine reinstatement (Khaled et al., 2014) and does so through the expression of α6nAChRs (Zuo et al., 2016). Activation leads to depolarization of LHb neurons with release of glutamate and increased firing. Nicotine also acts upon α4β4nAChRs expressed upon GABAergic input to LHb. This identifies a functionally

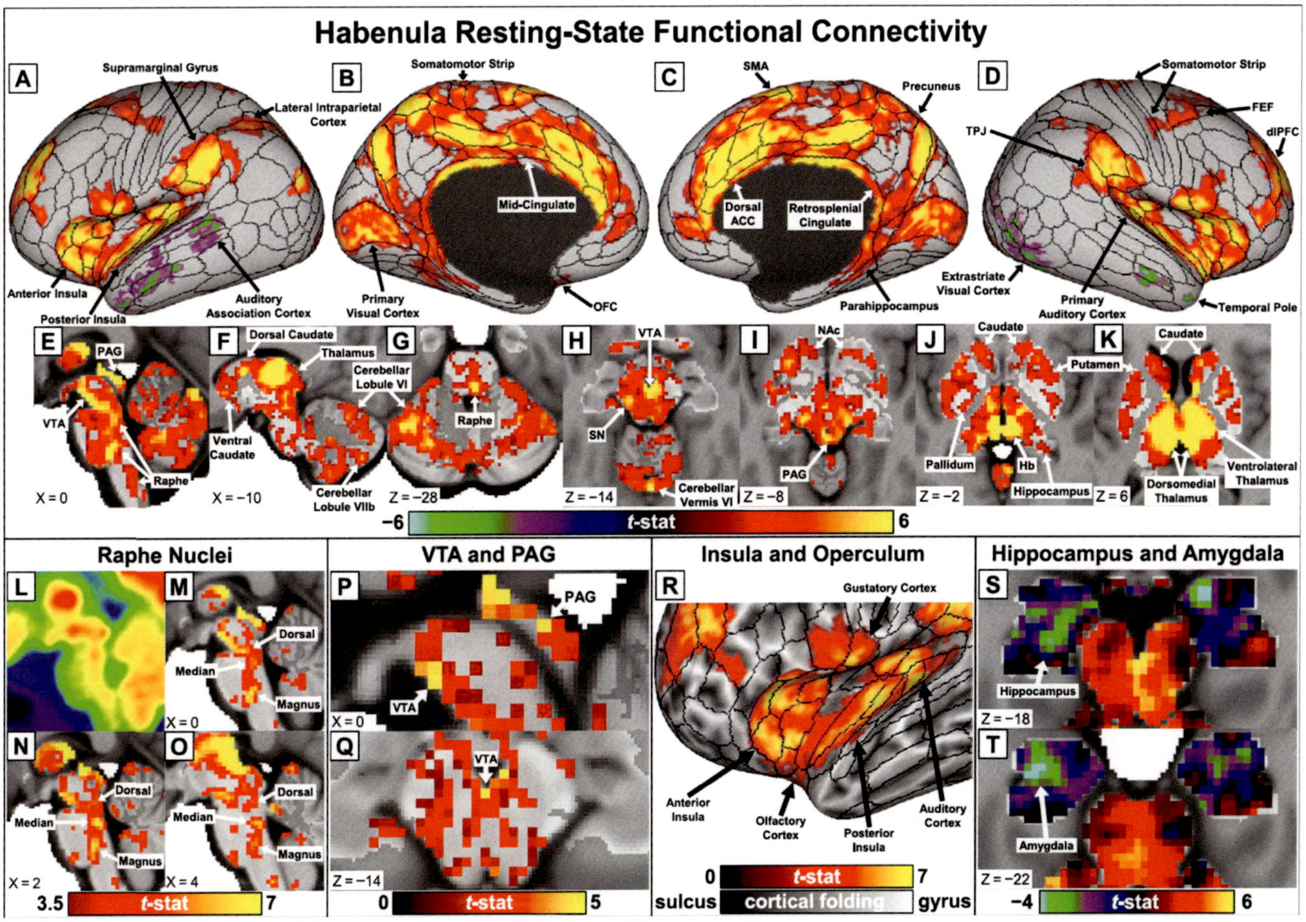

FIGURE 7.4 **The habenula (Hb) inhibits dopaminergic reward signaling in response to negative outcomes and has been linked to numerous functional domains relevant to mental health, including reward prediction, motivation, and aversion processing.** Despite its important neuroscientific and clinical implications, however, the human Hb remains poorly understood due to its small size and the associated technical hurdles to in vivo functional magnetic resonance imaging (fMRI) investigation. Using high-resolution 3T fMRI data from 68 healthy young adults acquired through the Human connectome Project, Ely et al (2019) have recently developed a rigorous approach for mapping the whole-brain resting-state functional connectivity of the human Hb. This study combined an optimized strategy for defining subject-level connectivity seeds to maximize Hb BOLD sensitivity with high-quality surface-based alignment for robust functional localization and cortical sensitivity. *They identified significant positive Hb connectivity with (1) conserved brainstem targets, including the dopaminergic ventral tegmental area, serotonergic raphe nuclei, and periaqueductal gray; (2) subcortical structures related to reward and motor function, including the nucleus accumbens, dorsal striatum, pallidum, thalamus, and cerebellum; and (3) cortical areas associated with the salience network and early sensory processing, including the dorsal anterior cingulate, anterior insula, and primary visual and auditory cortices.* Hb connectivity was strongly biased toward task-positive brain regions, with weak or negative connectivity observed throughout the task-negative default mode network. Essentially, this recent study provides a detailed characterization of Hb resting-state functional connectivity in healthy young adults, demonstrating both the feasibility and clinical potential of studying the human Hb using high-resolution 3T fMRI.Whole-brain Hb resting-state functional connectivity results and expanded views of key findings, displayed on "very inflated" cortical surfaces with areal boundary lines from the multimodal HCP parcellation (Glasser et al., 2016) and in subcortical MNI space. (A–K) Hb connectivity maps for significant ($p_{TFCE\text{-}FWE} < 0.05$) findings in the main analysis. (L) Midsagittal view of the brainstem showing locations of human raphe nuclei as peaks in mean [^{18}F] fluorodeoxyglucose positron emission tomography (PET) images acquired from seven healthy adults, displayed in MNI space. (M–O) Hb connectivity peaks ($t > 3.5$) in similar locations along the medial brainstem were identified as dorsal, median, and magnus raphe nuclei. (P and Q) Hb connectivity with the VTA and PAG remained significant ($p_{TFCE\text{-}FWE} < 0.05$) even in a secondary analysis without spatial smoothing (R) Detail of Hb connectivity with the left anterior insula, posterior insula, and sensory regions of the operculum. (S and T) Unthresholded negative Hb connectivity with the bilateral amygdala and inferior hippocampus. *Images courtesy of Ely, B.A., Stern, E.R., Kim, J., Gabbay, V., Xu, J., 2019. Detailed mapping of human habenula resting-state functional connectivity. NeuroImage 200, 621–634. https://doi.org/10.1016/j.neuroimage.2019.06.015*

complex interaction upon pre- and postsynaptic sites that may facilitate or inhibit activity in such a way as to result in behavior reflecting aversion or reward. In a similar fashion, nicotine has been shown to activate VTA dopaminergic neurons expressing either the α4 or α6 subunit of nAChRs producing sustained increases in firing rate (Liu et al., 2012).

The LHb also projects strongly to the dorsal raphe nucleus (DRN) (Pasquier et al., 1976; Aghajanian and Wang, 1977). Glutamatergic LHb neurons exert an

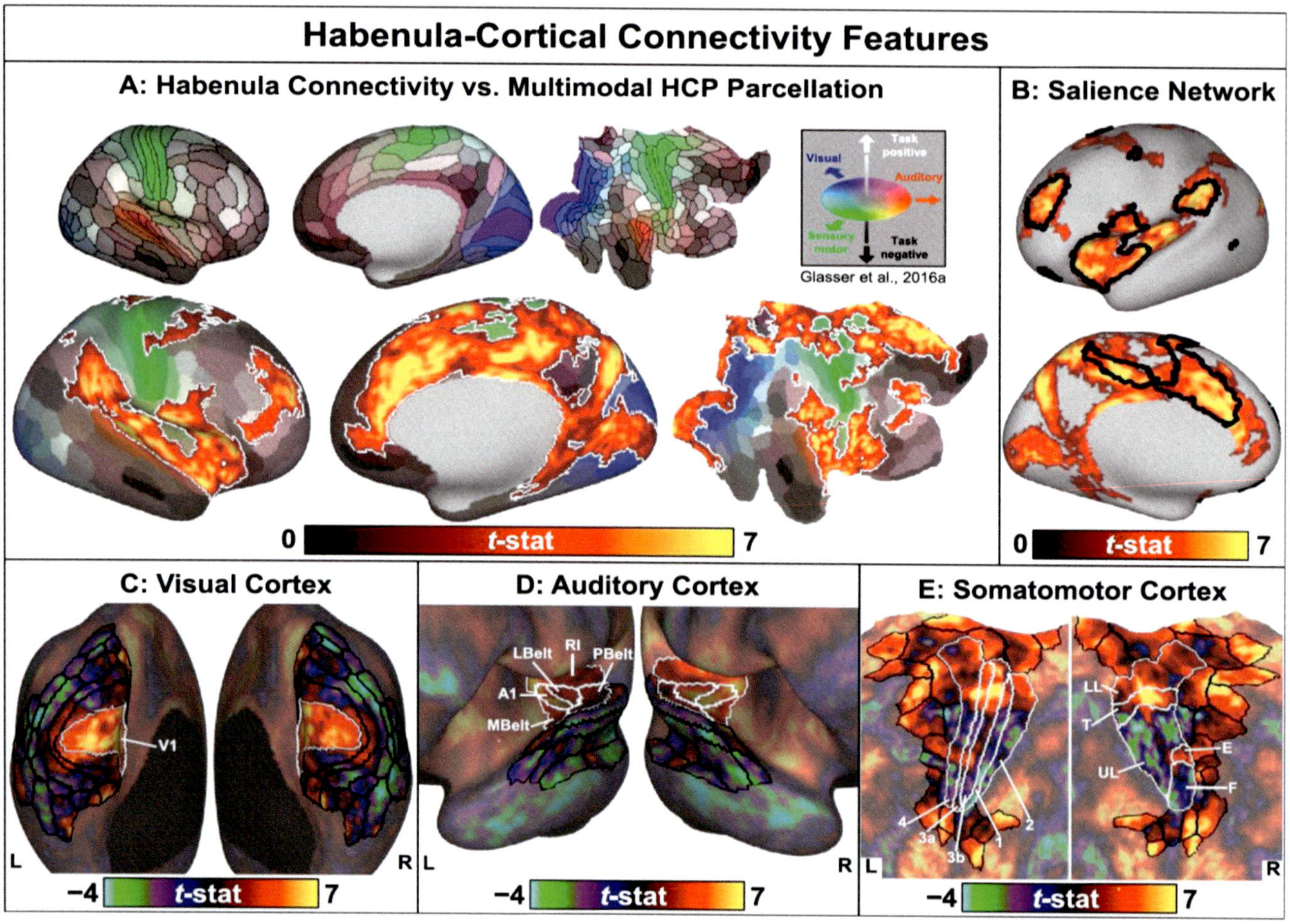

FIGURE 7.5 **Key features of Hb connectivity with the cortex.** (A), *Top*: Composite sensory and task responsiveness map from the multimodal HCP cortical parcellation (Glasser et al., 2016), displayed on "very inflated" and "flat" right cortical surfaces with areal boundary lines. Colors indicate association with major sensory modalities assessed by HCP tasks (blue = visual, red = auditory, green = somatosensory). Shading indicates overall responsiveness to HCP tasks (light = task-positive, dark = task-negative). (A), *Bottom*: Map of significant ($p_{TFCE\text{-}FWE} < 0.05$) positive Hb connectivity (outlined in white) overlaid on the multimodal HCP parcellation. Cortical Hb connectivity included early processing areas of all major sensory modalities and was restricted to task-positive regions, terminating abruptly at boundaries with task-negative areas (e.g., dmPFC, posterior cingulate). (B) Significant ($p_{TFCE\text{-}FWE} < 0.05$) positive Hb connectivity included nearly every part of the cortical ssalience network (outlined in black). Salience network boundaries were created by combining networks 7 and 8 from the 17-network rs-fMRI parcellation published by Yeo et al. (2011) and distributed by the HCP. Displayed on the "very inflated" left cortical surface. (C–E) Unthresholded Hb connectivity maps with left and right visual, auditory, and somatomotor cortices. Areal and subareal boundary lines indicate early sensory cortices (*white borders*) and higher sensory processing areas (*black borders*), as identified in the multimodal HCP parcellation. Areas outside the relevant sensory cortices are displayed at 50% saturation. (C) Hb connectivity transitioned from positive with the primary visual cortex (area V1) to negative for the adjacent extrastriate visual cortex. Displayed on "very inflated" cortical surfaces. (D) Similarly, Hb connectivity transitioned from positive with the early auditory cortex (areas A1, RI, LBelt, MBelt, and PBelt) to negative for adjacent auditory association areas. Displayed on "inflated" cortical surfaces. (E) *Left*: By contrast, Hb connectivity varied between positive and negative within the major areal boundaries of the primary somatomotor cortex (areas 1, 2, 3a, 3b, and 4). (E), *Right*: Instead, these transitions mirrored boundaries between somatotopic subareas corresponding to the lower limbs (LL), trunk (T), upper limbs (UL), eyes (E), and face (F). Unlike the visual and auditory systems, Hb connectivity was also positive with most higher somatosensory processing and motor planning areas. Displayed on "flat" cortical surfaces. *Images courtesy of Ely, B.A., Stern, E.R., Kim, J., Gabbay, V., Xu, J., 2019. Detailed mapping of human habenula resting-state functional connectivity. NeuroImage 200, 621–634. https://doi.org/10.1016/j.neuroimage.2019.06.015*

inhibitory influence upon the serotonergic DRN neurons by acting preferentially upon GABAergic interneurons within the DRN (Stern et al., 1979; Park, 1987; Ferraro et al., 1996; Wang et al., 1992) as is also the case with dopaminergic neurons in the VTA Brinschwitz et al. (2010) via the rmTA. An inhibitory serotonergic projection ascends from the median raphe, largely through the median forebrain bundle, to the LHb (Vertes et al., 1999) and has been shown to act upon presynaptic 5-HT_{1B} receptors situated on afferent terminals in the LHb (Hwang and Chung, 2014). There is also an indirect pathway by which LHb neurons project onto GABAergic neurons within the rmTA that, in turn, project to the DRN (Sego et al., 2014; Lavezzi et al., 2012).

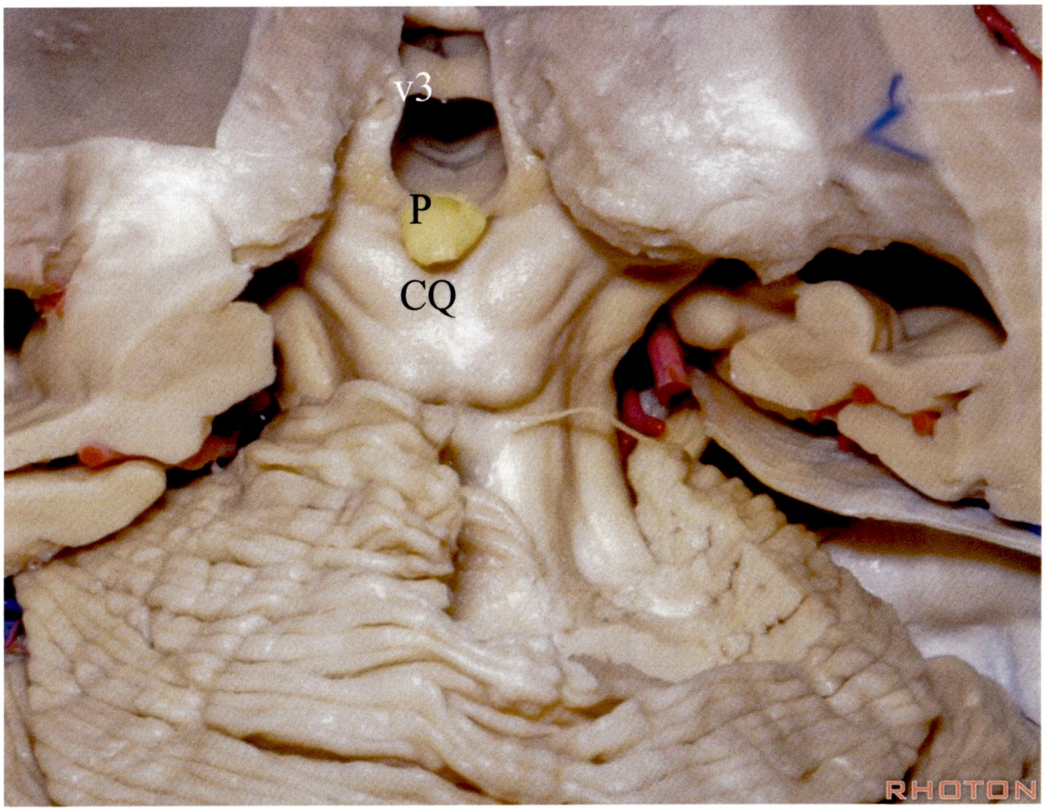

FIGURE 7.6 **Epithalamus; posterior view.** The pineal gland (P) is highlighted in yellow. The two habenular nuclei form minor protuberances within triangular depressions in the posterosuperior part of the third ventricle (v3). Under the pineal gland, is the corpora quadrigemina (CQ; two superior and two inferior colliculi). *Image courtesy of The Rhoton Collection® http://rhoton.ineurodb.org/.*

Serotonergic neurons of the DRN have a wide distribution throughout the central nervous system targeting the hippocampus, amygdala, bed nucleus of the stria terminalis, lateral septal nucleus, LPA, thalamic midline and intralaminar nuclei, medial prefrontal cortex (mPFC), periaqueductal gray, VTA, SNpc, striatum, substantia innominata, and ventral pallidum, among others (Vertes, 1991) indicating that the LHb can modulate several related functions within a number of integrated networks.

Hippocampal pyramidal neurons are activated by the LHb, ostensibly through the suppression of serotonergic neurons in the DRN that project onto them (Ferraro et al., 1997). Median raphe neurons also project onto the hippocampal formation where they play a role in the desynchronization of electroencephalographic activity implicating an effect upon memory function (Vertes et al., 1999). As a consequence of the sort of influence the LHb has upon the hippocampus, cognitive function may be impaired in cases where it has been disturbed in injury or by aging (Villarreal et al., 2002). Moreover, reciprocal connections between the mPFC and DRN (Jacobs and Azmitia, 1992; Peyton et al., 1998) and between the mPFC and hippocampus (Vertes, 2006) give further evidence of the impact that the DRN and, in turn, the LHb may have upon memory and cognition. The mPFC contains a number of 5-HT receptors expressed on both pyramidal neurons and interneurons through which the DRN may suppress activity. In particular, serotonergic projections onto 5-HT_{1A} receptors situated on fast-spiking GABAergic interneurons will dampen gamma rhythm (Puig et al., 2010), which, consequently, will reduce both the ability to encode and retrieve memory as well as to maintain attention (Fries et al., 2001; Montgomery and Buzsaki, 2007).

The mPFC projects directly to both the DRN (Peyton et al., 1998) and the LHb (Thierry et al., 1983) demonstrating an elaborate network of interconnectivity through direct and indirect associations, serving to provide nuance to combined expressions of cognition, emotion, memory, and associated locomotion. The LHb is therefore shown to be a critical element in this design through its input from limbic and basal ganglia to its extensive output largely mediated through projections to the VTA and DRN. As mentioned previously, the LHb has a role in decision-making as it pertains to reward because of its regulatory influence over both dopaminergic and serotonergic neurons while receiving input from both systems as well. Serotonergic input onto LHb neurons suppresses the activity induced by glutamatergic projection from the EPN, suggesting that serotonergic input to the LHb would tend to reduce the aversive response that the basal ganglia would bring about by driving inhibition onto the dopaminergic system via the LHb (Shabel et al., 2012). Serotonergic neurons of raphe origin would be seen to promote a sense of reward therefore. The LHb appears to play a role in depression by suppressing both DRN function (Yang et al., 2008) and dopaminergic activity (Li et al., 2011). Interestingly, an improvement in rat depression-like behavior will also take place when a substance P antagonist is applied to the LHb resulting in increased serotonergic activity in the DRN (Conley et al., 2002; Yang et al., 2014). The LHb is seen to be in a hyperactive state in patients with depression (Morris et al., 1999) with burst-firing glutamatergic neurons, which normally constitute less than 10% of the total number, showing, at times, a threefold increase in rodent models of depression (Yang et al., 2018; Han et al., 2015). This would, in turn, further suppress activity of both downstream dopaminergic and serotonergic neurons (Cui et al., 2019). The maintenance of such bursting activity requires a regulatory environment in which potassium is cleared from the synaptic cleft rapidly and efficiently. This is afforded by the presence of glial Kir4.1 potassium channels situated on astrocytic processes surrounding LHb neurons (Cui et al., 2018; Hu et al., 2020) and their association with gap junctions that provide for a syncytium of astrocytes to form a larger reservoir for the dispersal of potassium from the extracellular space (Du et al., 2018).

Burst behavior in LHb neurons is also dependent upon *N*-methyl-D-aspartate receptors (NMDARs), blockade of which will rapidly stop bursting (Yang et al., 2018). The antidepressant, ketamine, an NMDAR blocker is particularly effective in this regard (Yang et al., 2018; Lodge et al., 1982). A similar effect is mediated by ethosuximide, which blocks low voltage-sensitive T-type calcium channels in LHb neurons to stop burst firing to produce an antidepressant effect (Yang et al., 2018). Consideration has been given to deep brain stimulation of the LHb for the treatment of medically refractory depression (Hauptman et al., 2008; Sartorius et al., 2010; Schneider et al., 2013; Drobisza and Damborská, 2019) with favorable results reported in a rodent model (Meng et al., 2011). Substance abuse and addiction produce symptoms during the drug withdrawal state that are seen in major depressive disorder (Koob and Le Moal, 2001). Hyperactivity of the LHb is evident with drug withdrawal and creates a negative affect (Meye et al., 2016). In both the case of depression and drug withdrawal an asymmetry occurs in the balance of coreleased GABA and glutamate in the rodent EPN-LHb pathway that favors excitation through a disinhibition of EPN-driven LHb activity. A mechanism of metabotropic glutamate receptor–dependent plasticity mediating excitation or inhibition has been proposed to decide upon LHb encoding of opposing motivational states (Valentinova and Mameli, 2016). Although the MHb appears strongly implicated in mediating pain because of the close relationship of mu opioid receptors on SP neurons in the dHb that project to the IPN (Gardon et al., 2014), there is reason to suspect that the LHb, through its efferent projections to both the periaqueductal gray (PAG) and the DRN, may have influence here also (Shelton et al., 2012). Moreover, the LHb ostensibly receives indirect input from lamina I of the spinal cord via the lateral hypothalamus (Craig, 2004; Goto et al., 2005; Herkenham and Nauta, 1977) as well as from the nucleus accumbens in its own projection to the PAG (Yu and Han, 1990). The LHb plays a role in the sleep–wake cycle, possibly that of rapid eye movement (REM) sleep, in particular (Zhang et al., 2016a) with reduced duration of hippocampal theta oscillations (Aizawa et al., 2013). Clock genes within the LHb appear to regulate sleep homeostasis. Lesion of the LHb decreases the amplitude of reduced wake time and increases non-REM sleep following sleep deprivation in rats. Its participation, otherwise, in the circadian circuit is marked by a signature pattern of heightened firing rate during light exposure compared with that during the dark (Sakhi et al., 2014b).

The influence of the LHb on the dopaminergic system has implicated it in the manifestations of schizophrenia, particularly with the lack of appropriate modulation of LHb activity as in the response to unexpected negative outcomes (Shepard et al., 2006). The standard treatment of schizophrenia involves delivery of dopamine D2 receptor antagonists and, more recently, phosphodiesterase inhibitors, which are seen to increase glucose metabolism in the rodent LHb (Pizzolato et al., 1984; Dedeurwaerdere et al., 2011). A glutamatergic signaling dysfunction has been proposed implicating NMDARs (Hu et al., 2015; Snyder and Gao, 2019) as D2 receptor antagonists do not impact the cognitive aspects of the condition and NMDAR antagonists such as ketamine will accentuate both negative and cognitive features of schizophrenia (Krystal et al., 1994). Bilateral deep brain stimulation of the habenula for treatment-resistant schizophrenia was recently reported in two patients with both experiencing improvement during the first 6 months although one subsequently worsened to the point of requiring prolonged hospitalization and withdrawal from the study (Wang et al., 2020). As in depression, there is much yet to discuss although safety and some degree of efficacy for the approach has been shown.

Pineal Gland

Background

The pineal gland (L. "pinea", pine cone) or epiphysis cerebri (Gk. "epi-", upon, "-phusis", growth, "cerebri", cerebrum), captured the attention and imagination of early anatomists and philosophers, when seen as a centrally located solitary entity situated between symmetrically opposed, bilateral structures of the brain, directly behind the third ventricle. The first detailed anatomical description is attributed to Galen of Pergamon (129–200), who speculated incorrectly that it served as a support structure for surrounding vasculature (Lopez-Munoz et al., 2011). Centuries later, French philosopher Rene Descartes (1596–1650) elevated its stature to that of the "seat of the soul," based upon the mistaken belief that it behaved as a mechanical valve controlling the flow of "animal spirits" from the third to fourth ventricle.

In this way, Descartes suggested the pineal gland was responsible for emotional states, madness, and, remarkably, precise control of the human body. Contemporary research does suggest that dysfunction of the pineal gland and its primary end product, melatonin, underlies certain mental conditions, including autism, depression, and schizophrenia, although the nature of these correlations remains unclear (Kulman et al., 2000; Lopez-Munoz et al., 2012). Today the pineal gland is informally referred to as "The Third Eye," partly for its directly photosensitive function and histological resemblance to an eyeball in nonmammalian vertebrates, but most recently for its role in the photoneuroendocrine system, which synchronizes the body's circadian rhythm to the presence of light in the natural environment (Hardeland et al., 2012; Pfeffer et al., 2018).

Anatomy and Histology

The pineal gland is a small, pea-sized (5–8 mm), epithalamic structure, which lies centrally along the midsagittal plane and projects posteriorly from the posterior aspect of the third ventricle, directly above the quadrigeminal plate of the midbrain (Fig. 7.6). Its arterial supply comes through the posterior choroidal arteries (Fig. 7.2A), and it is drained by the dorsally situated internal cerebral veins (Fig. 7.7C). The pineal stalk is comprised of superior and inferior laminae that border the pineal recess within the posterior wall of the third ventricle. The superior lamina includes the habenular commissure and habenular trigone, whereas the inferior lamina includes the posterior commissure.

Morphologically, a well-defined pial capsule encloses a multilobulated gland, septated by connective tissue, with follicles and cords of pinealocytes in a highly variable arrangement (Koshy and Vettivel, 2001). Astrocytes are a significant feature with processes creating an interstitial network surrounding individual pinealocytes and endfeet forming a barrier between the perivascular space and parenchyma (Papasozomenos, 1983). Intercellular communication appears to occur via both synapses and electrotonic junctions (i.e., gap junctions). Sympathetic postganglionic fibers from the superior cervical ganglia (SCG) end upon perivascular spaces and do not form synapses with the pinealocytes themselves. Immunohistochemical staining with neuro-specific enolase (NSE) identifies pinealocytes as neuronal elements, while astrocytes along with less common oligodendrocytes are marked by glial fibrillary acidic protein (GFAP). Corpora arenacea, concretions of calcium and magnesium salts, appear in both the parenchyma and capsule and form either globular arrangements in the vicinity of pinealocytes or lamellar formations where glial fibers predominate (Koshy and Vettivel, 2001). Although evident in some at a very early age, 1% appearing in children under 6 years of age, the prevalence rises quickly to 39% in those aged 8–14 years with dramatic increases beyond 30 years (Tapp, 1979; Doyle and Anderson, 2006). Interestingly, the tendency of the pineal gland to form calcifications likely underlies the misguided medieval belief that madness resulted from "stones" in the brain and the practice of extraction of such "stones of folly" in some of the earliest neurosurgical procedures (Lopez-Munoz et al., 2012).

Neural Connections

Perhaps the best characterized and most pertinent input to the pineal gland conveys photosensory information from retinal ganglion cells through a polysynaptic route. The initial projection involves the monosynaptic retinohypothalamic tract that passes within the optic nerves and chiasm to one of a pair of SCN in the anteromedial ventral hypothalamus. These nuclei lie immediately above the chiasm on either side of the optic recess of the third ventricle and receive bilateral but predominantly contralateral input. Here they partially overlap with input from the geniculohypothalamic tract. Whereas several efferent monosynaptic projections from the SCN are to hypothalamic and thalamic nuclei, the major polysynaptic projection is to the pineal gland. This proceeds through synaptic connections in the hypothalamus, particularly the parvicellular subdivisions of the hypothalamic paraventricular nucleus, through the medial forebrain bundle inferiorly and predominantly to the upper portion (C8–T5) of the intermediolateral cell column (Moore et al., 1968; Larsen et al., 1998; Rando et al., 1981). Preganglionic fibers then ascend to the SCG that will provide sympathetic innervation to the pineal gland in addition to blood vessels of the head and neck, eye, salivary, and thyroid glands and carotid body (Cardinali et al., 1982; Erlich and Apuzzo, 1985). These postganglionic norepinephrinergic fibers proceed from the SCG to the

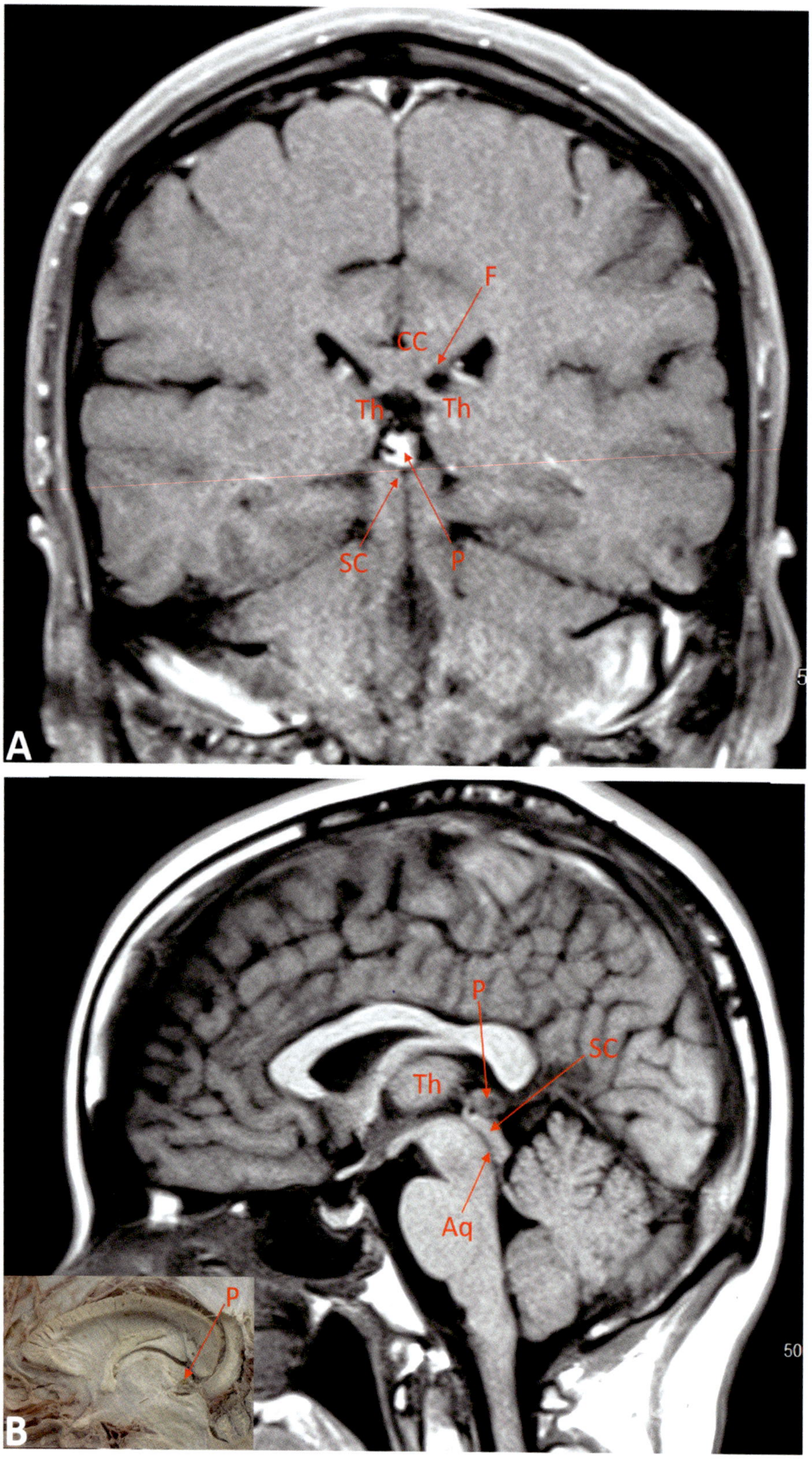

FIGURE 7.7 **Magnetic resonance images of pineal gland (P) presented in coronal (A), sagittal (B), and axial (C) planes.** Both coronal and sagittal profiles identify its location above the quadrigeminal plate. The inset in (B) shows the third ventricular area in a formalin-fixed brain in sagittal view with the pineal gland (P) in both identified in both inset and MRI images with an arrow. Labels identify notable structures in the vicinity as follows: Aq, aqueduct of Sylvius; CC, corpus callosum; ICV, internal cerebral vein; P, pineal gland; SC, superior colliculus; Sp, splenium of corpus callosum; Th, thalamus.

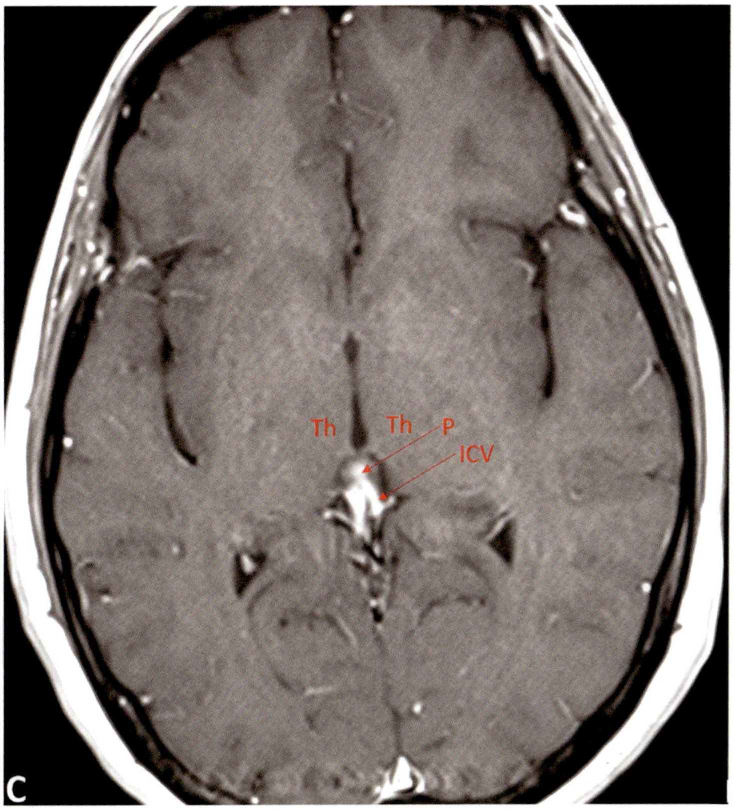

FIGURE 7.7 Cont'd

pineal gland via the carotid plexus and the nervi conarii in the tentorium. Some from either side may fuse with one another before entering the gland to synapse upon the pinealocytes.

Physiology and Biochemistry

The pineal gland has been defined as a photoneuroendocrine transducer that provides a means of coordinating internal function and behavior with the circadian rhythm of the world outside (Axelrod, 1974; Pfeffer et al., 2018; Gheban et al., 2019). It secretes an indoleamine, *N*-acetyl-5-methoxytryptamine or melatonin, with a molecular weight of 232 kDa as its hormone (Tordjman et al., 2017). Its precursor is tryptophan, which is taken up by the pineal and converted to serotonin and finally, melatonin. Release of melatonin is stimulated by darkness, while light inhibits secretion with this process mediated by the postganglionic sympathetic fibers of the SCG (Ostrin, 2019). Apart from this circadian rhythmicity, other cyclical activity is identified by ultradian (Erlich and Apuzzo, 1985; Reuss and Vollrath, 1984), infradian, and seasonal rhythms (Macchi and Bruce, 2004; Reiter, 1980). Melatonin has a short half-life systemically and is conjugated in the liver before excretion in the urine (Kopin et al., 1961). It is lipophilic allowing it to passively diffuse through cell membranes, and its intracellular binding to calmodulin promotes inhibition of neuronal nitric oxide synthase in the brain (Leon et al., 2000). Melatonin receptors, MT1 and MT2, are widely expressed in the CNS, within the cerebral cortex, hippocampus, diencephalon, including the SCN, and mesencephalon (Lacoste et al., 2015). With aging and disease, both melatonin secretion and receptor expression diminish and are ostensibly related to neuropsychiatric dysfunction, impairments in cognition and memory, and neurorestorative capabilities (Bahna and Niles, 2018). The effects of melatonin are surprisingly widespread. They are particularly prominent in sites concerned with reproductive function and equipped with steroid receptors as in the pineal gland, hypothalamus, and the testes (Reiter, 1980). Receptors for testosterone, estradiol, progesterone, and prolactin are all found in the pineal gland. Several cellular effects of melatonin occur within the hypothalamus including protein synthesis, hormone release, serotonin and GABA content, and neurotransmitter uptake (Cardinali, 1981).

Another compound of interest within the pineal is the endogenous powerful hallucinogen, *N,N*-dimethyltryptamine (DMT). It is also synthesized from tryptophan and fulfills the criteria of a naturally occurring neurotransmitter (Christian et al., 1976; Barker, 2018). Because of these attributes, it is thought to play a role in the modulation of perception and consciousness and the manifestations of dreams or spontaneous hallucinatory phenomena such as near-death experiences (Barker, 2018).

Functional Effects

The pineal has a chronobiological function with light as its main *Zeitgeber* or synchronizer, which regulates the phases of melatonin production (Erlich and Apuzzo, 1985), increasing it during the night in the absence of light and inhibiting it otherwise. A circadian rhythm is marked by peak melatonin values during the dark phase but with distinct secretory episodes during the waking day in the presence of light (Weitzman et al., 1978; Touitou et al., 1984) as well as the night. Some indication has been given that spontaneous waking episodes may be associated with melatonin peaks, while REM sleep is associated with melatonin troughs (Birkeland, 1982). Pinealectomy has been shown to abolish the circadian rhythm of REM sleep without causing a change in the total amount of REM sleep in a 24-h cycle (Mouret et al., 1974). Melatonin levels have been noted to become elevated after REM sleep (Sizonenko et al., 1979) due to altered pineal norepinephrine release associated with the increased autonomic activity occurring during this phase of sleep (Erlich and Apuzzo, 1985).

Melatonin reduces hypothalamic gonadotrophin releasing hormone (GnRH) mRNA expression resulting in diminished GnRH and therefore has an antigonadotrophic effect (Erlich and Appuzo, 1985). Its increased secretion has also been shown to inhibit biliary hyperplasia and fibrosis resulting from hepatic cholestasis (Han et al., 2014; Renzi et al., 2011). It may therefore have a role in the treatment of cholestatic liver disease (McMillan et al., 2017). Reduction of melatonin by pharmacologic means (i.e., methysergide) or pinealectomy results in retroperitoneal, cardiac, and pulmonary fibrosis (Cunnane et al., 1979). Both the pineal and the melatonin it secretes inhibit collagen accumulation in healing tissue with the effect not dependent on the type of wound created whether it be a skin wound, myocardial infarction, bone fracture or gastric ulcer (Drobnik and Dabrowski, 2000; Drobnik et al., 2012). Melatonin is involved in all phases of healing from inflammation with the release of mediators, cell proliferation and migration, angiogenesis, fibroblast proliferation, and collagen regulation. It is a free radical scavenger and stimulates antioxidative enzyme activity (Rodriguez et al., 2004) with evidence that it can reduce both ischemia–reperfusion cardiac apoptotic injury (Chen et al., 2009) and ionizing radiation injury (Karbownik and Reiter, 2000). Not surprisingly, it is also produced in several tissues outside the pineal gland and, in particular, within immune cells. Evidence suggests a coordinated action of melatonin synthesis between the pineal gland and the immune system in an immune-pineal axis (Markus et al., 2017). Acute inflammation promotes the transcription factor, NFκB, to switch the synthesis of melatonin from pinealocytes to macrophages returning this back to pinealocytes with resolution of the acute response.

Pineal function has been shown to be regulated by endogenous opioids and melatonin itself has been shown to have analgesic properties (Lowenstein et al., 1984; Kavaliers et al., 1983). Additionally, the pineal modulates diurnal rhythms of analgesia and influences age-related changes in opioid responsiveness. Melatonin has been used effectively in the management of migraines in both the pediatric (Gelfand et al., 2020; Fallah et al., 2015) and adult (Goncalves et al., 2016) populations.

Pathology

Cyst formation within the pineal is found in 25%–40% of cases (Wurtman and Moskowitz, 1977; Pu et al., 2007) and is commonly identified in gliotic areas of the gland, while others are enclosed by parenchymal or ependymal epithelium (Tapp, 1979) and may be uni- or multiloculated (Fig. 7.8). Although rarely symptomatic, intracystic hemorrhages have occurred resulting in acute hydrocephalus and death (Pu et al., 2007). Dermoid and arachnoid cysts may be found as can lipomas within the pineal.

Neoplasms of the pineal can arise from any of its constituent elements and account for less than 1% of primary brain tumors. They are found predominantly in the pediatric age group where they represent 3%–11% of brain tumors (Louis et al., 2007; Villano et al., 2008). One large series of pineal tumors in patients aged 3–73 years identified the following distributions: germinoma (27%), astrocytoma (26%), pineoblastoma (12%), pineocytoma (12%), ependymoma (4.3%), teratoma (4.3%), meningioma (2.7%), metastasis (2.7%), and choriocarcinoma (1.1%) (Regis et al., 1996).

Pineal tumors may bring about mass effect producing aqueductal stenosis with hydrocephalus, headache and obtundation, Parinaud syndrome with vertical gaze palsy and pupillary dysfunction, and/or hypothalamic compression with diabetes insipidus and growth retardation (Carr et al., 2019). There are five categories of pineal tumors (Table 7.1).

Pineal Parenchymal Cell Tumors

These represent 14%–27% of the tumors and include the pineocytoma and pineoblastoma as well as a mixed variety. The slow-growing pineocytoma arises from the pineal epithelium and can be treated by surgery alone. The more aggressive pineoblastoma arises from primitive neuroectoderm. The latter may be recognized by serum and CSF synaptophysin and chromogranin, and

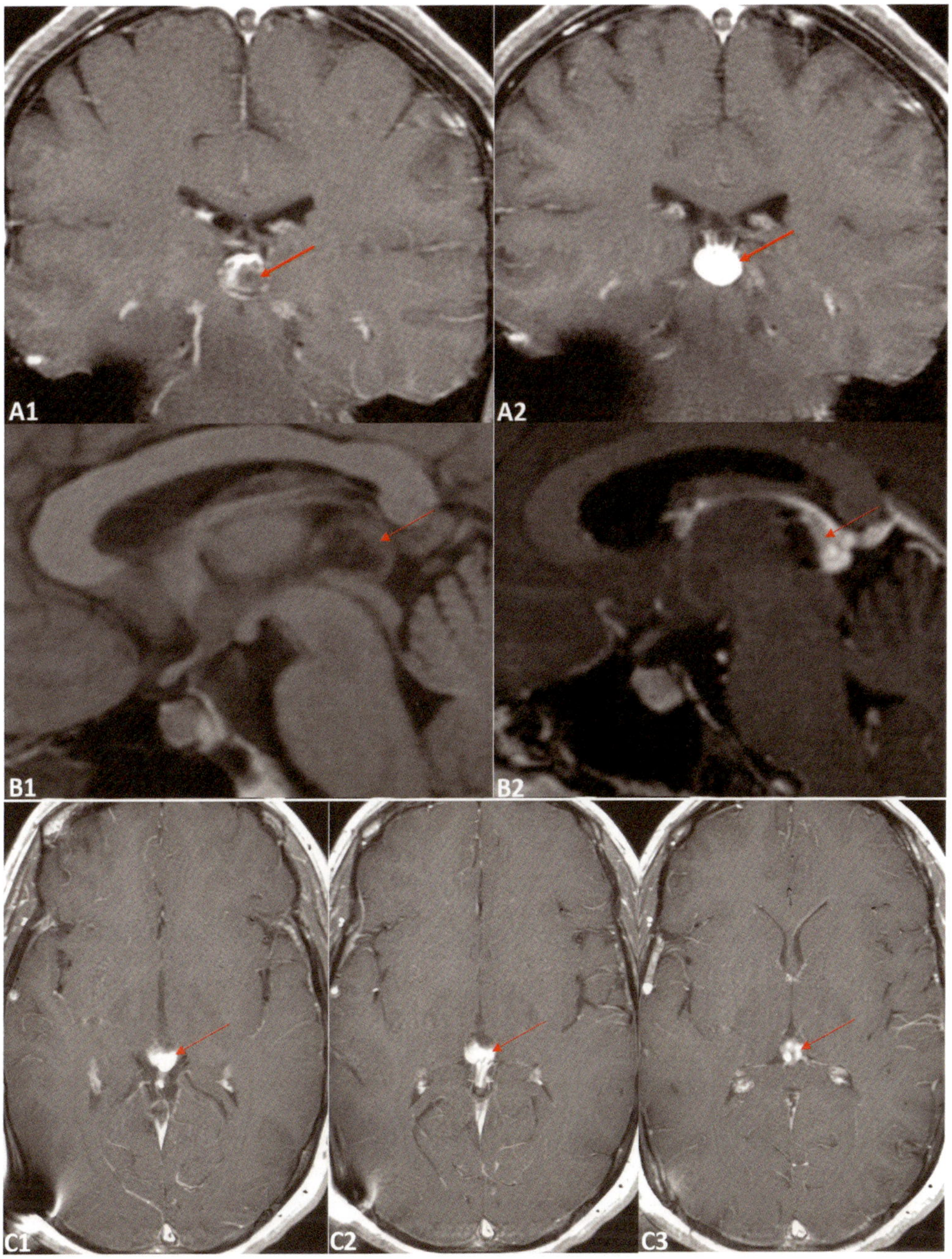

FIGURE 7.8 **Magnetic resonance images of a cystic pineocytoma presented in coronal (A), sagittal (B), and axial (C) display.** (A) The coronal contrast-enhanced images appear in an anterior (*left*) and posterior (*right*) plane identifying the cystic and solid components of the tumor (*arrow*), respectively. (B) The magnified sagittal images appear with noncontrast (*left*) and contrast-enhanced (*right*) features again showing the predominant anterior cystic component and posterior solid tumor. (C) Contrast-enhanced axial images proceed from inferior to superior planes through the pineal region.

treatment has been by surgery, craniospinal irradiation, and chemotherapy with a 5-year survival reaching 50% (Tian et al., 2018).

Germ Cell Tumors

These tumors may be found both in the pineal and in the midline suprasellar regions along the route of primordial germ cell migration (Carr et al., 2019). They comprise 35%–40% of pineal tumors with the great majority of them identified as germinomas (Regis et al., 1996; Villano et al., 2008). Nongerminomatous germ cell tumors include embryonal carcinoma, teratoma, endodermal sinus (i.e., yolk sac) tumor, and choriocarcinoma. Teratomas are the second most common germ cell tumors. Release of beta human chorionic gonadotropin (β-HCG) by germinomas, choriocarcinomas, and

TABLE 7.1 Tumors of and in the vicinity of the pineal gland.

Pineal parenchymal tumors	Pineocytoma Pineoblastoma
Germ cell tumors	Germinoma Teratoma Embryonal carcinoma Endodermal sinus (= yolk sac) Choriocarcinoma
Glial cell tumors	Astrocytoma Anaplastic astrocytoma Glioblastoma Oligodendroglioma Ependymoma
Papillary tumors	
Miscellaneous tumors	Meningioma Metastasis (i.e., melanoma) Craniopharyngioma Chemodectoma

embryonal carcinomas may cause precocious puberty in boys. Embryonal carcinomas, endodermal sinus tumors, and immature teratomas, otherwise, may cause release of alpha-fetoprotein, detectable in serum or CSF (Carr et al., 2019). Germinomas generally respond well to radiation and chemotherapy without surgical intervention, while other germ cell tumors require surgical removal in addition.

Glial Tumors

Fibrillary and pilocytic astrocytomas, ependymomas, oligodendrogliomas, anaplastic astrocytomas, and glioblastomas have all manifested in the pineal, arising from its stroma. Some such as the pilocytic astrocytoma are circumscribed and others infiltrative and poorly resectable. Both chemotherapy and radiation may be implemented as elsewhere within the CNS. Survival is highly dependent upon the grade of the tumor and its response to therapy.

Papillary Tumor

An epithelioid pineal tumor with papillary, epithelial, ependymal, and neuroendocrine differentiation was introduced in the World Health Organization in 2007. Despite cellular pleomorphism and architectural variability, it tends to have a low mitotic rate (Poulgrain et al., 2011). Patients have been treated with surgery and postoperative radiation with recurrences noted.

Miscellaneous Tumors

An assortment of other tumors within and in the vicinity of the pineal has been described and includes meningiomas, metastases, craniopharyngiomas, and chemodectomas. These are relatively uncommon. Most metastases have been identified as melanomas (Tapp, 1979).

Acknowledgment

The authors thank David Tack, PhD, Spectrum Health Office of Research, for graphic design assistance and Karen Poole, PhD, Department of Anatomy, College of Medicine, Central Michigan University for her assistance with anatomical photography.
We extend our heartfelt thanks to Dr. Barbara Buccilli for her valuable contributions and assistance in this chapter. Her expertise and guidance greatly enhanced the content. We are grateful for her support throughout the writing process. We would also like to acknowledge the contributions of our colleagues and mentors, as well as the publishers, for their support and trust in our work.

References

Aghajanian, G.K., Wang, R.Y., 1977. Habenular and other midbrain raphe afferents demonstrated by a modified retrograde tracing technique. Brain Res. 122, 229–242.

Aizawa, H., Kobayashi, M., Tanaka, S., Fukai, T., Okamoto, H., 2012. Molecular characterization of the subnuclei in rat habenula. J. Comp. Neurol. 520, 4051–4066.

Aizawa, H., Yanagihara, S., Kobayashi, M., Niisato, K., Takekawa, T., Harukuni, R., McHugh, T.J., Fukai, T., Isomura, Y., Okamoto, H., 2013. The synchronous activity of lateral habenular neurons is essential for regulating hippocampal theta oscillation. J. Neurosci. 33, 8909–8921.

Albers, D.S., Augood, S.J., Martin, D.M., Standaert, D.G., Vonsattel, J.P.G., Flint Beal, M., 1999. Evidence for oxidative stress in the subthalamic nucleus in progressive supranuclear palsy. J. Neurochem. 73, 881–884.

Alpár, A., Harkany, T., 2018. Novel insights into the spatial and temporal complexity of hypothalamic organization through precision methods allowing nanoscale resolution. J. Intern. Med. 284, 568–580.

Andres, K.H., von During, M., Vey, R.W., 1999. Subnuclear organization of the rat habenular complexes. J. Comp. Neurol. 407, 130–150.

Antolin-Fontes, B., Ables, J.L., Gorlich, A., Ibanez-Tallon, I., 2015. The habenulo-interpeduncular pathway in nicotine aversion and withdrawal. Neuropharmacology 96, 213–222.

Araki, M., McGeer, P.L., McGeer, E.G., 1984. Retrograde HRP tracing combined with a pharmacohistochemical method for GABA transaminase for the identification of presumptive GABAergic projections to the habenula. Brain Res. 304, 271–277.

Araki, M., McGeer, P.L., Kimura, H., 1988. The efferent projections of the rat lateral habenular nucleus revealed by the PHA-L anterograde tracing method. Brain Res. 441, 319–330.

Axelrod, J., 1974. The pineal gland: a neurochemical transducer. Chemical signals from nerves regulate synthesis of melatonin and convey information about internal clocks. Science 184, 1341–1348.

Bahna, S.G., Niles, L.P., 2018. Epigenetic regulation of melatonin receptors in neuropsychiatric disorders. Br. J. Pharmacol. 175, 3209–3219.

Baldwin, P.R., Alanis, R., Salas, R., 2011. The role of the habenula in nicotine addiction. J. Addiction Res. Ther. S1 (1–10), 002. https://doi.org/10.4172/2155-6105.S1-002.

Barker, S.A., 2018. N,N-dimethyltryptamine (DMT), an endogenous hallucinogen: past, present, and future research to determine its role and function. Front. Neurosci. 12 (1–17), 536.

Birkeland, A.J., 1982. Plasma melatonin levels and nocturnal transitions between sleep and wakefulness. Neuroendocrinology 34, 126–131.

Boehm, S., Betz, H, 1997. Somatostatin inhibits excitatory transmission at rat hippocampal synapses via presynaptic receptors. J. Neurosci. 17, 4066–4075.

Borghei, A., Cothran, T., Brahimaj, B., Sani, S., 2020. Role of massa intermedia in human cognitive processing. Brain. Struct. Funct. 225 (3), 985–993. https://doi.org/10.1007/s00429-020-02050-5. PMID: 3212414.

Brinschwitz, K., Dittgen, A., Madai, V.I., Lommel, R., Geisler, S., Veh, R.W., 2010. Glutamatergic axons from the lateral habenula mainly terminate on GABAergic neurons of the ventral midbrain. Neuroscience 168, 463–476.

Browne, C.A., Hammack, R., Lucki, I., 2018. Dysregulation of the lateral habenula in major depressive disorder. Front. Synaptic Neurosci. 10 (1–18), 46. https://doi.org/10.3389/fnsyn.2018.00046.

Cajal S Ramón, y, 1911. Histology of the Nervous System of Man and Vertebrates, vol. 2. Oxford University Press, New York, pp. 342–350.

Cardinali, D.P., Vacas, M.I., Gejman, P.V., Pisarev, M.A., Barontini, M., Boado, R.J., Juvenal, G.J., 1982. The sympathetic superior cervical ganglia as "little neuroendocrine brains". Acta Physiol. Lat. Am. 33, 205–215.

Cardinali, D.P., 1981. Melatonin. A mammalian pineal hormone. Endocr. Rev. 2, 327–346.

Carr, C., O'Neill, B.E., Hochhalter, C.B., Strong, M.J., Ware, M.L., 2019. Biomarkers of pineal region tumors: a review. Ochsner J. 19, 26–31.

Casarrubea, M., Davies, C., Faulisi, F., Pierucci, M., Colangeli, R., Partridge, L., Chambers, S., Cassar, D., Valentino, M., Muscat, R., Benigno, A., Crescimanno, G., Di Giovanni, G., 2015. Acute nicotine induces anxiety and disrupts temporal pattern organization of rat exploratory behavior in hole-board: a potential role for the lateral habenula. Front. Cell. Neurosci. 9 (1–17), 197. https://doi.org/10.3389/fncel.2015.00197.

Chatterjee, M., Guo, Q., Weber, S., Scholpp, S., Li, J.Y., 2014. Pax6 regulates the formation of the habenular nuclei by controlling the temporospatial expression of Shh in the diencephalon in vertebrates. BMC Biol. 12 (1–15), 13.

Chen, Z., Chau, C.C., Gao, J., Chua, K.-W., Ho, Y.-S., Hamdy, R.C., Chua, B.H.L., 2009. Prevention of ischemia/reperfusion-induced cardiac apoptosis and injury by melatonin is independent of glutathione peroxidase 1. J. Pineal Res. 46, 235–241.

Christian, S.T., Harrison, R., Pagel, J., 1976. Evidence for dimethyltryptamine (DMT) as a naturally-occurring transmitter in mammalian brain. Ala. J. Med. Sci. 13, 162–165.

Ciani, E., Severi, S., Bartesaghi, R., Contestabile, A., 2005. Neurochemical correlates of nicotine neurotoxicity on rat habenulo-interpeduncular cholinergic neurons. Neurotoxicology 26, 467–474.

Commons, K.G., Valentino, R.J., 2002. Cellular basis for the effects of substance P in the periaqueductal gray and dorsal raphe nucleus. J. Comp. Neurol. 447, 82–97.

Conley, R.K., Cumberbatch, M.J., Mason, G.S., Williamson, D.J., Harrison, T., Locker, K., Swain, C., Maubach, K., O'Donell, R., Rigby, M., Hewson, L., Smith, D., Rupniak, N.M.J., 2002. Substance P (neurokinin 1) receptor antagonists enhance dorsal raphe neuronal activity. J. Neurosci. 22, 7730–7736.

Conrad, L.C., Leonard, C.M., Pfaff, D.W., 1974. Connections of the median and dorsal raphe nuclei in the rat: an autoradiographic and degeneration study. J. Comp. Neurol. 156, 179–205.

Craig, A.D., 2004. Distribution of trigeminothalamic and spinothalamic lamina I terminations in the macaque monkey. J. Comp. Neurol. 477, 119–148.

Cuello, A.C., Emson, P.C., Paxinos, G., Jessell, T., 1978. Substance P containing and cholinergic projections from the habenula. Brain Res. 149, 413–429.

Cui, Y.H., Yang, Y., Ni, Z., Dong, Y., Cai, G., Foncelle, A., Ma, S., Sang, K., Tang, S., Li, Y., Shen, Y., Berry, H., Wu, S., Hu, H., 2018. Astroglial Kir4.1 in the lateral habenula drives neuronal bursts in depression. Nature 554, 323–327.

Cui, Y.H., Hu, S.H., Hu, H.L., 2019. Lateral habenular burst firing as a target of the rapid antidepressant effects of ketamine. Trends Neurosci. 42, 179–191.

Cunnane, S.C., Manku, M.S., Horrobin, D.F., 1979. The pineal and regulation of fibrosis: pinealectomy as a model of primary biliary cirrhosis: roles of melatonin and prostaglandins in fibrosis and regulation of T lymphocytes. Med. Hypotheses 5, 403–414.

Damle, N.R., Ikuta, T., John, M., Peters, B.D., DeRosse, P., Malhotra, A.K., Szeszko, P.R., 2017. Relationship among interthalamic adhesion size, thalamic anatomy and neuropsychological functions in healthy volunteers. Brain Struct. Funct. 222, 2183–2192.

Dedeurwaerdere, S., Wintmolders, C., Vanhoof, G., Langlois, X., 2011. Patterns of brain glucose metabolism induced by phosphodiesterase 10A inhibitors in the mouse: a potential translational biomarker. J. Pharmacol. Exp. Therapeut. 339, 210–217.

Ding, Q., Balasubramanian, R., Zheng, D., Liang, G., Gan, L., 2017. Barhl2 determines the early patterning of the diencephalon by regulating Shh. Mol. Neurobiol. 54, 4414–4420.

Doyle, A.J., Anderson, G.D., 2006. Physiological calcification of the pineal gland in children on computed tomography: prevalence, observer reliability and association with choroid plexus calcification. Acad. Radiol. 13, 822–826.

Drobisza, D., Damborská, A., 2019. Deep brain stimulation targets for treating depression. Behav. Brain Res. 359, 266–273.

Drobnik, J., Dabrowski, R., 2000. The opposite effect of morning or afternoon application of melatonin on collagen accumulation in the sponge-induced granuloma. Neuroendocrinol. Lett. 21, 209–212.

Drobnik, J., 2012. Wound healing and the effect of pineal gland and melatonin. J. Exp. Integr. Med. 2, 3–14.

Du, M., Li, J., Chen, L., Yu, Y., Wu, Y., 2018. Astrocytic Kir4.1 channels and gap junctions account for spontaneous epileptic seizure. PLoS Comput. Biol. 14, e1005877.

Ellison, G., 2002. Neural degeneration following chronic stimulant abuse reveals a weak link in brain, fasciculus retroflexus, implying the loss of forebrain control circuitry. Eur. Neuropsychopharmacol 12, 287–297.

Erlich, S.S., Apuzzo, L.J., 1985. The pineal gland: anatomy, physiology, and clinical significance. J. Neurosurg. 63, 321–341.

Ewings, E.L., Carstens, M.H., 2009. Neuroembryology and functional anatomy of craniofacial clefts. Indian J. Plast. Surg. 42 (Suppl. l), S19–S34.

Fallah, R., Shoroki, F.F., Ferdosian, F., 2015. Safety and efficacy of melatonin in pediatric migraine prophylaxis. Curr. Drug Saf. 10, 132–135.

Ferraro, G., Montalbano, M.E., Sardo, P., La Grutta, V., 1997. Lateral habenula and hippocampus: a complex interaction raphe cell-mediated. J. Neur. Trans. 104, 615–631. https://doi.org/10.1007/BF01291880.

Ferraro, G., Montalbano, M.E., Sardo, P., La Grutta, V., 1996. Lateral habenular influence on dorsal raphe neurons. Brain Res. Bull. 41, 47–52.

Fires, P., Reynolds, J.H., Rorie, A.E., Desimone, R., 2001. Modulation of oscillatory neuronal synchronization by selective visual attention. Science 291, 1560–1563.

Florian, C., Bahi-Buisson, N., Bienvenu, T., 2011. FOXG1-related disorders: from clinical description to molecular genetics. Mol. Syndromol. 2, 153–163.

Fowler, C.D., Kenny, P.J., 2014. Nicotine aversion: neurobiological mechanisms and relevance to tobacco dependence vulnerability. Neuropharmacology 76 (B), 533–544.

Fowler, C.D., Lu, Q., Johnson, P.M., Marks, M.J., Kenny, P.J., 2011. Habenular alpha5 nicotinic receptor subunit signaling controls nicotine intake. Nature 471, 597–601.

Frahm, S., Antolin-Fontes, B., Gorlich, A., Zander, J.F., Ahnert-Hilger, G., Ibanez-Tallon, 2015. An essential role of acetylcholine-glutamate synergy at habenular synapses in nicotine dependence. Elife 4, e11396. https://doi.org/10.7554/eLife.11396.

Fu, T., Pearson, C., Towers, M., Placzek, M., 2019. Development of the basal hypothalamus through anisotropic growth. J. Neuroen docrinol. 31, e12727.

Gardon, O., Faget, L., Chu Sin Chung, P., Matifas, A., Massotte, D., Kieffer, B.L., 2014. Expression of mu opioid receptor in dorsal diencephalic conduction system: new insights for the medial habenula. Neuroscience 26, 595–609.

Garland, J.C., Mogenson, G.J., 1983. An electrophysiological study of convergence of entopeduncular and lateral preoptic inputs on lateral habenular neurons projecting to the midbrain. Brain Res. 263, 33–41.

Garric, L., Ronsin, B., Roussigné, M., Booton, S., Gamse, J.T., Dufourcq, P., Blader, P., 2014. Pitx2c ensures habenular asymmetry by restricting parapineal cell number. Development 141, 1572–1579.

Geisler, S., Trimble, M., 2008. The lateral habenula: no longer neglected. CNS Spectr. 13, 484–489.

Gelfand, A.A., Ross, A.C., Irwin, S.L., Greene, K.A., Qubty, W.F., Allen, E., 2020. Melatonin for acute treatment of migraine in children and adolescents: a pilot randomized trial. Headache 60, 1712–1721.

Gheban, B.A., Rosca, I.A., Crisan, M., 2019. The morphological and functional characteristics of the pineal gland. Med. Pharm. Rep. 92, 226–234.

Glasser, M.F., Coalson, T.S., Robinson, E.C., Hacker, C.D., Harwell, J., Yacoub, E., Ugurbil, K., Andersson, J., Beckmann, C.F., Jenkinson, M., Smith, S.M., Van Essen, D.C., 2016. A multi-modal parcellation of human cerebral cortex. Nature 536 (7615), 171–178. https://doi.org/10.1038/nature18933. PMID: 27437579.

Glick, S.D., Ramirez, R.L., Livi, J.M., Maisonneuve, I.M., 2006. 18-Methoxycoronaridine acts in the medial habenula and/or interpeduncular nucleus to decrease morphine self-administration in rats. Eur. J. Pharmacol. 537, 94–98.

Gonçalves, A.L., Ferreira, A.M., Ribeiro, R.T., Zukerman, E., Cipolla-Neto, J., Peres, M.F.P., 2016. Randomised clinical trial comparing melatonin 3 mg, amitriptyline 25 mg and placebo for migraine prevention. J. Neurol. Neurosurg. Psychiatry 87, 1127–1132.

Goto, M., Canteras, N.S., Burns, G., Swanson, L.W., 2005. Projections from the subfornical region of the lateral hypothalamic area. J. Comp. Neurol. 493, 412–438.

Gottesfeld, Z., 1983. Origin and distribution of noradrenergic innervation in the habenula: a neurochemical study. Brain Res. 275, 299–304.

Gruber, C., Kahl, A., Lebenheim, L., Kowski, A., Dittgen, A., Veh, R.W., 2007. Dopaminergic projections from the VTA substantially contribute to the mesohabenular pathway in the rat. Neurosci. Lett. 427, 165–170.

Guglielmotti, V., Cristino, L., 2006. The interplay between the pineal complex and the habenular nuclei in lower vertebrates in the context of the evolution of cerebral asymmetry. Brain Res. Bull. 69, 475–488.

Han, Y., Onori, P., Meng, F., DeMorrow, S., Venter, J., Francis, H., Franchitto, A., Ray, D., Kennedy, L., Greene, J., Renzi, A., Mancinelli, R., Gaudio, E., Glaser, S., Alpini, G., 2014. Prolonged exposure of cholestatic rats to complete dark inhibits biliary hyperplasia and liver fibrosis. Am. J. Physiol. Gastrointest. Liver Physiol. 307, G894–G904. https://doi.org/10.1152/ajpgi.00288.2014.

Han, L.-N., Zhang, L., Li, L.-B., Sun, Y.-N., Wang, Y., Chen, L., Guo, Y., Zhang, Y.-M., Zhang, Q.-J., Liu, J., 2015. Activation of serotonin(2C) receptors in the lateral habenular nucleus increases the expression of depression-related behaviors in the hemiparkinsonian rat. Neuropharmacology 93, 68–79.

Hardeland, R., Madrid, J.A., Tan, D.-X., Reiter, R.J., 2012. Melatonin, the circadian multioscillator system and health: the need for detailed analyses of peripheral melatonin signaling. J. Pineal Res. 52, 139–166. https://doi.org/10.1111/j.1600-079X.2011.00934.x.

Hauptman, J.S., DeSalles, A.A., Espinoza, R., Sedrak, M., Ishida, W., 2008. Potential surgical targets for deep brain stimulation in treatment-resistant depression. Neurosurg. Focus 25, E3. https://doi.org/10.3171/FOC/2008/25/7/E3.

Hazlett, E.A., Buchsbaum, M.S., Byne, W., Wei, T.C., Spiegel-Cohen, J., Geneve, C., Kinderlehrer, Haznedar, M.M., Shihabuddin, L., Siever, L.J., 1999. Three-dimensional analysis with MRI and PET of the size, shape and function of the thalamus in the schizophrenia spectrum. Am. J. Psychiatr. 156, 1190–1199.

Herkenham, M., Nauta, W.J., 1977. Afferent connections of the habenular nuclei in the rat. A horseradish peroxidase study, with a note on the fiber-of-passage problem. J. Comp. Neurol. 173, 123–146.

Herkenham, M., Nauta, W.J., 1979. Efferent connections of the habenular nuclei in the rat. J. Comp. Neurol. 187, 19–47.

Hsu, Y.W., Tempest, L., Quina, L.A., Wei, A.D., Zeng, H., Turner, E.E., 2013. Medial habenula output circuit mediated by alpha5 nicotinic receptor-expressing GABAergic neurons in the interpeduncular nucleus. J. Neurosci. 33, 18022–18035.

Hsu, Y.W., Wang, S.D., Wang, S., Morton, G., Zariwala, H.A., de la Iglesia, H.O., Turner, E.E., 2014. Role of the dorsal medial habenula in the regulation of voluntary activity, motor function, hedonic state, and primary reinforcement. J. Neurosci. 34 (34), 11366–11384. https://doi.org/10.1523/JNEUROSCI.1861-14.2014.

Hsu, Y.W., Morton, G., Guy, E.G., Wang, S.D., Turner, E.E., 2016. Dorsal medial habenula regulation of mood-related behaviors and primary reinforcement by tachykinin-expressing habenula neurons. eNeuro 3 (3). https://doi.org/10.1523/ENEURO.0109-16.2016. ENEURO.0109-0116.2016.

Hu, W., MacDonald, M.L., Elswick, D.E., Sweet, R.A., 2015. The glutamate hypothesis of schizophrenia: evidence from human brain tissue studies. Ann. NY Acad. Sci. 1338, 38–57.

Hu, H., Cui, Y., Yang, Y., 2020. Circuits and functions of the lateral habenula in health and in disease. Nature Rev. 21, 277–295.

Hwang, E.-K., Chung, J.-M., 2014. 5HT(1B) receptor-mediated presynaptic depression of excitatory inputs to the rat lateral habenula. Neuropharmacology 81, 153–165.

Ishikawa, Y., Yamamoto, N., Yoshimoto, M., Ito, H., 2012. The primary brain vesicles revisited: are the three primary vesicles (forebrain/midbrain/hindbrain) universal in vertebrates? Brain Behav. Evol. 79, 75–83.

Jacobs, B.L., Azmitia, E.C., 1992. Structure and function of the brain serotonin system. Physiol. Rev. 72, 165–229.

Jhou, T.C., Geisler, S., Marinelli, M., DeGarmo, B.A., Zahm, D.S., 2009. The mesopontine rostromedial tegmental nucleus: a structure targeted by the lateral habenula that projects to the ventral tegmental area of Tsai and substantia nigra compacta. J. Comp. Neurol. 513, 566–596.

Ji, H., Shepard, P.D., 2007. Lateral habenula simulation inhibits rat midbrain dopamine neurons through a $GABA_A$ receptor-mediated mechanism. J. Neurosci. 27, 6923–6930.

Karbownik, M., Reiter, R.J., 2000. Antioxidative effects of melatonin in protection against cellular damage caused by ionizing radiation. Proc. Soc. Exp. Biol. Med. 225, 9–22.

Kataoka, A., Shimogori, T., 2008. Fgf8 controls regional identity in the developing thalamus. Development 135, 2873–2881.

Kavaliers, M., Hirst, M., Teskey, G.C., 1983. Ageing, opioid analgesia and the pineal gland. Life Sci. 32, 2279–2287.

Khaled, M.A., Pushparaj, A., Di Ciano, P., Diaz, J., Le Foll, B., 2014. Dopamine D3 receptors in the basolateral amygdala and the lateral habenula modulate cue-induced reinstatement of nicotine seeking. Neuropsychopharmacology 39, 3049–3058.

Kiernan, J.A., 2008. Barr's The Human Nervous System, ninth ed. Wolters Kluwer, Philadelphia.

Kim, U., Chang, S.-Y., 2005. Dendritic morphology, local circuitry, and intrinsic electrophysiology of neurons in the rat medial and lateral habenular nuclei of the epithalamus. J. Comp. Neurol. 483, 236–250.

Kita, H., Kitai, S.T., 1987. Efferent projections of the subthalamic nucleus in the rat: light and electron microscopic analysis with the PHA-L method. J. Comp. Neurol. 260, 435–452.

Kochanski, R.B., Dawe, R., Kocak, M., Sani, S., 2018. Identification of stria medullaris fibers in the massa intermedia using diffusion tensor imaging. World Neurosurg. 112, e497–e504.

Koob, G.F., Le Moal, M., 2001. Drug addiction, dysregulation of reward, and allostasis. Neuropsychopharmacology 24, 97–129.

Kopin, I.J., Pare, C.M.B., Axelrod, J., Weissbach, H., 1961. The fate of melatonin in animals. J. Biol. Chem. 236, 3072–3075.

Koppensteiner, P., Melani, R., Ninan, I.A., 2017. Cooperative mechanism involving Ca^{2+}-permeable AMPA receptors and retrograde activation of $GABA_B$ receptors in interpeduncular nucleus plasticity. Cell Rep. 20, 1111–1122.

Koshy, S., Vettivel, S.K., 2001. Varying appearances of calcification in human pineal gland: a light microscopic study. J. Anat. Soc. India 50, 17–18.

Krystal, J.H., Karper, L.P., Seibyl, J., Freeman, G.K., Delaney, K., Bremner, J.D., Heninger, G.R., Bowers, M.B., Charney, D., 1994. Subanesthetic effects of the noncompetitive NMDA antagonist, ketamine, in humans. Psychotomimetic, perceptual, cognitive, and neuroendocrine responses. Arch. Gen. Psychiatr. 51, 199–214.

Kulman, G., Lissoni, P., Rovelli, F., Roselli, M.G., Brivio, F., Sequeri, P., 2000. Evidence of pineal endocrine hypofunction in autistic children. Neuroendocrinol. Lett. 21, 31–34.

Lacoste, B., Angeloni, D., Dominguez-Lopez, S., Calderoni, S., Mauro, A., Fraschini, F., Descarries, L., Gobbi, G., 2015. Anatomical and cellular localization of melatonin MT and MT receptors in the adult rat brain. J. Pineal Res. 58, 397–417.

Lammel, S., Lim, B.K., Ran, C., Huang, K.W., Betley, M.J., Tye, K.M., Deisseroth, K., Malenka, R.C., 2012. Input-specific control of reward and aversion in the ventral tegmental area. Nature 491, 212–217.

Landgraf, D., Long, J.E., Welsh, D.K., 2016. Depression-like behavior in mice is associated with disrupted circadian rhythms in nucleus accumbens and periaqueductal grey. Eur. J. Neurosci. 43, 1309–1320.

Larsen, P.J., Enquist, L.W., Card, J.P., 1998. Characterization of the multisynaptic neuronal control of the rat pineal gland using viral transneuronal tracing. Eur. J. Neurosci. 10, 128–145.

Lavezzi, H.N., Parsley, K.P., Zahm, D.S., 2012. Mesopontine rostromedial tegmental nucleus neurons projecting to the dorsal raphe and pedunculopontine tegmental nucleus: psychostimulant-elicited Fos expression and collateralization. Brain Struct. Funct. 217, 719–734.

Lawson, R.P., Seymour, B., Loh, E., Lutti, A., Dolan, R.J., Dayan, P., Weiskopf, N., Poiser, J.P., 2014. The habenula encodes negative motivational value associated with primary punishment in humans. Proc. Natl. Acad. Sci. USA 111, 11858–11863.

Lecourtier, L., Kelly, P.H., 2007. A conductor hidden in the orchestra? Role of the habenular complex in monoamine transmission and cognition. Neurosci. Biobehav. Rev. 31, 658–672.

Lee, H.W., Yang, S.H., Kim, J.Y., Kim, H., 2019. The role of the medial habenula cholinergic system in addiction and emotion-associated behaviors. Front. Psychiatr. 10, 100. https://doi.org/10.3389/fpsyt.2019.00100.

Leon, J., Macias, M., Escames, G., Camacho, E., Khaldy, H., Martin, M., Espinosa, A., Gallo, M.A., Acuna-Castroviejo, D., 2000. Structure related inhibition of calmodulin dependent neuronal nitric-oxide synthase activity by melatonin and synthetic kynurenines. Mol. Pharmacol. 58, 967–975.

Li, B., Piriz, J., Mirrione, M., Chung, C., Proulx, C.D., Schulz, D., Henn, F., Malinow, R., 2011. Synaptic potentiation onto habenula neurons in the learned helplessness model of depression. Nature 470, 535–539.

Lim, Y., Cho, G., Minarcik, J., Golden, J., 2005. Altered BMP signaling disrupts chick diencephalic development. Mech. Dev. 122, 603–620.

Lima, L.B., Bueno, D., Leite, F., Souza, S., Goncalves, L., Furigo, I.C., Donato Jr., J., Metzger, M., 2017. Afferent and efferent connections of the interpeduncular nucleus with special reference to circuits involving the habenula and raphe nuclei. J. Comp. Neurol. 525, 2411–2442.

Liu, L., Zhao-Shea, R., McIntosh, J.M., Gardner, P.D., Tapper, A.R., 2012. Nicotine persistently activates ventral tegmental area dopaminergic neurons via nicotinic acetylcholine receptors containing alpha4 and alpha6 subunits. Mol. Pharmacol. 81, 541–548.

Liu, B., Zhou, K., Wu, X., Zhao, C., 2018. Foxg1 deletion impairs the development of the epithalamus. Mol. Brain 11, 5.

Lodge, D., Anis, N.A., Burton, N.R., 1982. Effects of optical isomers of ketamine on excitation of cat and rat spinal neurons by amino acids and acetylcholine. Neurosci. Lett. 29, 281–286.

Lopez, A.J., Jia, Y., White, A.O., Kwapis, J.L., Espinoza, M., Hwang, P., Campbell, R., Alaghband, Y., Chitnis, O., Matheos, D.P., Lynch, G., Wood, M.A., 2018. Medial habenula cholinergic signaling regulates cocaine-associated relapse-like behavior. Addiction Biol. 24, 403–413.

López-Muñoz, F., Molina, J.D., Rubio, G., Alamo, C., 2011. An historical view of the pineal gland and mental disorders. J. Clin. Neurosci. 18, 1028–3107. https://doi.org/10.1016/j.jocn.2010.11.037.

López-Muñoz, F., Rubio, G., Molina, J.D., Alamo, C., 2012. La glándula pineal como instrumento físico de las facultades del alma: una conexión histórica persistente. Neurologia 27, 161–168. https://doi.org/10.1016/j.nrl.2011.04.018.

Louis, D.N., Ohgaki, H., Wiestler, O.D., 2007. Cavenee WK, Burger PC, Jouvet A, Scheithauer BW, Kleihues P (2007) the 2007 WHO classification of tumours of the central nervous system. Acta Neuropathol. 114, 97–109.

Lowenstein, P.R., Pereyra, E.N., Solveyra, C.G., Cardinali, D.P., 1984. Effect of naloxone on the nocturnal rise of rat pineal melatonin content. Eur. J. Pharmacol. 98, 261–264.

Macchi, M.M., Bruce, J.N., 2004. Human pineal physiology and functional significance of melatonin. Front. Neuroendocrinol. 25, 177–195.

Marks, M.J., Pauly, J.R., Gross, S.D., Deneris, E.S., Hermans-Borgmeyer, I., Heinemann, S.F., Collins, A.C., 1992. Nicotine binding and nicotinic receptor subunit RNA after chronic nicotine treatment. J. Neurosci. 12, 2765–2784.

Markus, R.P., Fernandes, P.A., Kinker, G.S., da Silveira Cruz-Machado, S., Marçola, M., 2017. Immune-pineal axis - acute inflammatory responses coordinate melatonin synthesis by pinealocytes and phagocytes. Br. J. Pharmacol. 175, 3239–3250.

Martinez-Ferre, A., Martinez, S.J., 2009. The development of the thalamic motor learning area is regulated by Fgf8 expression. Neurosciences 29, 13389–13400.

Martinez-Ferre, A., Martinez, S., 2012. Molecular regionalization of the diencephalon. Front. Neurosci. 6, 73. https://doi.org/10.3389/fnins.2012.00073.

Matsumoto, M., Hikosaka, O., 2007. Lateral habenula as a source of negative reward signals in dopamine neurons. Nature 447, 1111–1115.

Mavridis, I., Boviatsis, E., Anagnostopoulou, S., 2013. Anatomy of the human subthalamic nucleus: a combined morphometric study. Anat. Res. Int. 2013, 319710. https://doi.org/10.1155/2013/319710.

McLaughlin, I., Dani, J.A., De Biasi, M., 2017. The medial habenula and interpeduncular nucleus circuitry is critical in addiction, anxiety, and mood regulation. J. Neurochem. 142 (Suppl. 2), 130–143.

McMillin, M., DeMorrow, S., Glaser, S., Venter, J., Kyritsi, K., Zhou, T., Grant, S., Giang, T., Greene, J.F., Wu, N., Jefferson, B., Meng, F., Alpini, G., 2017. Melatonin inhibits hypothalamic gonadotropin-releasing hormone release and reduces biliary hyperplasia and fibrosis in cholestatic rats. Am. J. Physiol. Gastrointest. Liver Physiol. 313, G410–G418. https://doi.org/10.1152/ajpgi.00421.2016.

Meng, H., Wang, Y., Huang, M., Lin, W., Wang, S., Zhang, B., 2011. Chronic deep brain stimulation of the lateral habenula nucleus in a rat model of depression. Brain Res. 1422, 32–38.

Meye, F.J., Soiza-Reilly, M., Smit, T., Diana, M.A., Schwarz, M.K., Mameli, M., 2016. Shifted pallidal co-release of GABA and glutamate in habenula drives cocaine withdrawal and relapse. Nat. Neurosci. 19, 1019–1024.

Montgomery, S.M., Buzsaki, G., 2007. Gamma oscillations dynamically couple hippocampal CA3 and CA1 regions during memory task performance. Proc. Natl. Acad. Sci. U.S.A. 104, 14495–14500.

Moore, R.Y., Heller, A., Bhatnager, R.K., Wurtman, R.J., Axelrod, J., 1968. Central control of the pineal gland: visual pathways. Arch. Neurol. 18, 208–218.

Morris, J.S., Smith, K.A., Cowen, P.J., Friston, K.J., Dolan, R.J., 1999. Covariation of activity in habenula and dorsal raphe nuclei following tryptophan depletion. Neuroimage 10, 163–172.

Morton, G., Nasirova, N., Sparks, D.W., Brodsky, M., Sivakumaran, S., Lambe, E.K., Turner, E.E., 2018. ChRNA5-expressing neurons in the interpeduncular nucleus mediate aversion primed by prior stimulation of nicotine exposure. J. Neurosci. 38, 6900–6920.

Mouret, J., Coindet, J., Chouvet, G., 1974. Effect of pinealectomy on sleep stages and rhythms of the male rat. Brain Res. 81, 97–105.

Murtra, P., Sheasby, A.M., Hunt, S.P., De Felipe, C., 2000. Rewarding effects of opiates are absent in mice lacking the receptor for substance P. Nature 405, 180–183.

Nakamura, K., Hikosaka, O., 2006. Role of dopamine in the primate caudate nucleus in reward modulation of saccades. J. Neurosci. 26, 5360–5369.

Ostrin, L.A., 2019. Ocular and systemic melatonin and the influence of light exposure. Clin. Exp. Optom. 102, 99–108.

Papasozomenos, S.C., 1983. Glial fibrillary acidic (GFA) protein-containing cells in the human pineal gland. J. Neuropathol. Exp. Neurol. 42, 391–408.

Park, M.R., 1987. Monosynaptic inhibitory postsynaptic potentials from lateral habenula recorded in dorsal raphe neurons. Brain Res. Bull. 19, 581–586.

Pasquier, D.A., Anderson, C., Forbes, W.B., Morgane, P.J., 1976. Horseradish peroxidase tracing of the lateral habenular-midbrain raphe nuclei connections in the rat. Brain Res. Bull. 1 (5), 443–451. https://doi.org/10.1016/0361-9230(76)90114-3.

Paul, M.J., Indic, P., Schwartz, W.J., 2011. A role for the habenula in the regulation of locomotor activity cycles. Eur. J. Neurosci. 34, 478–488.

Peyton, C., Petit, J.M., Rampon, C., Jouvet, M., Luppi, P.H., 1998. Forebrain afferents to the rat dorsal raphe nucleus demonstrated by retrograde and anterograde tracing methods. Neurocience 82, 443–468.

Pfeffer, M., Korf, H.W., Wicht, H., 2018. Synchronizing effects of melatonin on diurnal and circadian rhythms. Gen. Comp. Endocrinol. 258, 215–221.

Phillipson, O.T., Pycock, C.J., 1982. Dopamine neurons of the ventral tegmentum project to both medial and lateral habenula. Exp. Brain Res. 45, 89–94.

Pizzolato, G., Soncrant, T.T., Rapoport, S.I., 1984. Haloperidol and cerebral metabolism in the conscious rat: relation to pharmacokinetics. J. Neurochem. 43, 724–732.

Poulgrain, K., Gurgo, R., Winter, C., Ong, B., Lau, Q., 2011. Papillary tumour of the pineal region. J. Clin. Neurosci. 18, 1007–1017.

Pu, Y., Mahankali, S., Hou, J., Li, J., Lancaster, J.L., Gao, J.-H., Appelbaum, D.E., Fox, P.T., 2007. High prevalence of pineal cysts in healthy adults demonstrated by high-resolution, noncontrast brain MR imaging. Am. J. Neuroradiol. 28, 1706–1709.

Puelles, L., Rubenstein, J.L., 2015. A new scenario of hypothalamic organization: rationale of new hypotheses introduced in the updated prosomeric model. Front. Neuroanat. 9, 27.

Puig, M.V., Watakabe, A., Ushimaru, M., Yamamori, T., Kawaguchi, Y., 2010. Serotonin modulates fast-spiking interneuron and synchronous activity in the rat prefrontal cortex through 5-HT_{1A} and 5-HT_{2A} receptors. J. Neurosci. 30, 2211–2222.

Qin, C., Luo, M., 2009. Neurochemical phenotypes of the afferent and efferent projections of the mouse medial habenula. Neuroscience 161, 827–837.

Quina, L.A., Wang, S., Ng, L., Turner, E.E., 2009. Brn3a and Nurr1 mediate a gene regulatory pathway for habenula development. J. Neurosci. 29, 14309–14322.

Rajakumar, N., Elisevich, K., Flumerfelt, B.A., 1993. Compartmental origin of the striato-entopeduncular projection in the rat. J. Comp. Neurol. 331, 286–296.

Rando, T.A., Bowers, C.W., Zigmond, R.E., 1981. Localization of neurons in the rat spinal cord which project to the superior cervical ganglion. J. Comp. Neurol. 196, 73–83.

Rash, B.G., Grove, E.A., 2011. Shh and Gli3 regulate formation of the telencephalic-diencephalic junction and suppress an isthmus-like signaling source in the forebrain. Dev. Biol. (San Diego, CA, U. S.) 359, 242–250.

Regis, J., Bouillot, P., Rouby-Volot, F., Figarella-Branger, D., Dufour, H., Peragut, J.C., 1996. Pineal region tumors and the role of stereotactic biopsy: review of the mortality, morbidity, and diagnostic rates of 370 cases. Neurosurgery 39, 907–914.

Reiter, R.J., 1980. The pineal and its hormones in the control of reproduction in mammals. Endocr. Rev. 1, 109–131.

Ren, J., Qin, C., Hu, F., Tan, J., Qiu, L., Zhao, S., Feng, G., Luo, M., 2011. Habenula "cholinergic" neurons co-release glutamate and acetylcholine and activate postsynaptic neurons via distinct transmission modes. Neuron 69, 445–452.

Renzi, A., Glaser, S., DeMorrow, S., Mancinelli, R., Meng, F., Franchitto, A., Venter, J., White, M., Francis, H., Han, Y., Alvaro, D., Gaudio, E., Carpino, G., Ueno, Y., Onori, P., Alpini, G., 2011. Melatonin inhibits cholangiocyte hyperplasia in cholestatic rats by interaction with MR1 but not MT2 melatonin receptors. Am. J. Physiol. Gastrointest. Liver Physiol. 301, G634–G643. https://doi.org/10.1152/ajpgi.00206.2011.

Reuss, S., Vollrath, L., 1984. Electrophysiological properties of rat pinealocytes: evidence for circadian and ultradian rhythms. Exp. Brain Res. 55, 455–461.

Rodriguez, C., Mayo, J.C., Sainz, R.J., Antolin, I., Herrera, F., Martin, V., Reiter, R.J., 2004. Regulation of antioxidant enzymes: a significant role for melatonin. J. Pineal Res. 36, 1–9.

Ronnekleiv, O.K., Moller, M., 1979. Brain-pineal nervous connections in the rat: an ultrastructure study following habenular lesion. Exp. Brain Res. 37, 551–562.

Rubenstein, J.L., Martinez, S., Shimamura, K., Puelles, L., 1994. The embryonic vertebrate forebrain: the prosomeric model. Science 266, 578–580.

Saffroy, M., Beaujouan, J.C., Petitet, F., Torrens, Y., Glowinski, J., 1994. Differential localization of 3H-[Pro9]SP binding sites in the Guinea pig and rat brain. Brain Res. 633, 317–325.

Sakhi, K., Belle, M.D., Gossan, N., Delagrange, P., Piggins, H.D., 2014a. Daily variation in the electrophysiological activity of mouse medial habenula neurons. J. Physiol. 592, 587–603.

Sakhi, K., Wegner, S., Belle, M.D.C., Howarth, M., Delagrange, P., Brown, T.M., Piggins, H.D.', 2014b. Intrinsic and extrinsic cues regulate the daily profile of mouse lateral habenula neuronal activity. J. Physiol. 592, 5025–5045.

Salas, R., Sturm, R., Boulter, J., De Biasi, M., 2009. Nicotinic receptors in the habenulo-interpeduncular system are necessary for nicotine withdrawal in mice. J. Neurosci. 29, 3014–3018.

Sartorius, A., Kiening, K.L., Kirsch, P., von Gall, C.C., Haberkorn, U., Unterberg, A.W., Henn, F.A., Meyer-Lindenberg, A., 2010. Remission of major depression under deep brain stimulation of the lateral habenula in a therapy-refractory patient. Biol. Psychiatr. 67, e9–e11.

Schmidt, E.R.E., Brignani, S., Adolfs, Y., Lemstra, S., Demmers, J., Vidaki, M., Donahoo, A.-L.S., Lillevali, K., Vasar, E., Richards, L.J., Karagogeos, D., Kolk, S.M., Pasterkamp, R.J., 2014. Subdomain-mediated axon-axon signaling and chemoattraction cooperate to regulate afferent innervation of the lateral habenula. Neuron 83, 372–387.

Schneider, T.M., Beynon, C., Sartorius, A., Unterberg, A.W., Kiening, K.L., 2013. Deep brain stimulation of the lateral habenular complex in treatment-resistant depression: traps and pitfalls of trajectory choice. Neurosurgery 72, 184–193.

Scholpp, S., Wolf, O., Brand, M., Lumsden, A., 2006. Hedgehog signaling from the zona limitans intrathalamica orchestrates patterning of the zebrafish diencephalon. Development 133, 855–864.

Sego, C., Goncalves, L., Lima, L., Furigo, I.C., Donato Jr., J., Metzger, M., 2014. Lateral habenula and the rostromedial tegmental nucleus innervate neurochemically distinct subdivisions of the dorsal raphe nucleus in the rat. J. Comp. Neurol. 522, 1454–1484.

Shabel, S.J., Proulx, C.D., Trias, A., Murphy, R.T., Malinow, R., 2012. Input to the lateral habenula from the basal ganglia is excitatory, aversive, and suppressed by serotonin. Neuron 74, 475–481.

Shelton, L., Pendse, G., Maleki, N., Moulton, E.A., Lebel, A.A., Becerra, L., Borsook, D., 2012. Mapping pain activation and connectivity of the human habenula. J. Neurophysiol. 107, 2633–2648.

Shepard, P.D., Holcomb, H.H., Gold, J.M., 2006. Schizophrenia in transition: the presence of absence: habenular regulation of dopamine neurons and the encoding of negative outcomes. Schizophr. Bull. 32, 417–421.

Sizonenko, P.C., Moore, D.C., Paunier, L., Beaumanoir, A., Nahory, A., 1979. Melatonin secretion in relation to sleep in epileptics. Prog. Brain Res. 52, 549–551.

Snyder, M.A., Gao, W.-J., 2019. NMDA hypofunction for schizophrenia revisited: perspectives from epigenetic mechanisms. Schizophr. Res. 217, 60–70.

Snyder, P.J., Bogerts, B., Wu, H., Bilder, R.M., Deoras, K.S., Lieberman, J.A., 1998. Absence of the adhesio interthalamica as a marker of early developmental neuropathology in schizophrenia: an MRI and postmortem histologic study. J. Neuroimaging 8, 159–163.

Soria-Gomez, E., Busquets-Garcia, A., Hu, F., Mehidi, A., Cannich, A., Roux, L., Louit, I., Alonso, L., Wiesner, T., Georges, F., Verrier, D., Vincent, P., Ferreira, G., Muo, M., Marsicano, 2015. Habenular CB_1 receptors control the expression of aversive memories. Neuron 88, 306–313.

Spinks, R., Magnotta, V.A., Andreasen, N.C., Albright, K.C., Ziebell, S., Nopoulos, P., Cassell, M., 2002. Manual and automated measurement of the whole thalamus and mediodorsal nucleus using magnetic resonance imaging. NeuroImage 17, 631–642.

Stamatakis, A.M., Jennings, J.H., Ung, R.L., Blair, G.A., Weinberg, R.J., Neve, R.L., Boyce, F., Mattis, J., Ramakrishnan, C., Deisseroth, K., Stuber, G.D., 2013. A unique population of ventral tegmental area neurons inhibits the lateral habenula to promote reward. Neuron 80, 1039–1053.

Steele, J., Richardson, J., Olszewski, J., 1964. Progressive supranuclear palsy. A heterogeneous degeneration involving the brainstem, basal ganglia and cerebellum with vertical gaze and pseudobulbar palsy, nuchal dystonia and dementia. Arch. Neurol. 10, 333–359.

Stephenson-Jones, M., Floros, O., Robertson, B., Grillner, 2011. Evolutionary conservation of the habenular nuclei and their circuitry controlling the dopamine and 5-hydroxytryptophan (5-HT) systems. Proc. Natl. Acad. Sci. USA 109, e164–e173.

Stern, W.C., Johnson, A., Bronzino, J.D., Morgane, P.J., 1979. Effects of electrical stimulation of the lateral habenula on single-unit activity of raphe neurons. Exp. Neurol. 65, 326–342.

Sutherland, R.J., 1982. The dorsal diencephalic conduction system: a review of the anatomy and functions of the habenular complex. Neurosci. Biobehav. Rev. 6, 1–13.

Takahashi, T., Suzuki, M., Zhou, S.-Y., Nakamura, K., Tanino, R., Kawasaki, Y., Seal, M.L., Seto, H., Kurachi, M., 2008. Prevalence and length of the adhesio interthalamica in schizophrenia spectrum disorders. Psychiatry Res. Neuroimaging. 164, 90–94.

Tapp, E., 1979. The histology and pathology of the human pineal gland. Prog. Brain Res. 52, 481–500.

Thierry, A.M., Chevalier, G., Ferron, A., Glowinski, J., 1983. Diencephalic and mesencephalic efferents of the medial prefrontal cortex in the rat: electrophysiological evidence for the existence of branched axons. Exp. Brain Res. 50, 275–282.

Tian, Y., Liu, R., Qin, J., Wang, J., Ma, Z., Gong, J., Li, C., 2018. Retrospective analysis of the clinical characteristics, therapeutic aspects, and prognostic factors of 18 cases of childhood pineoblastoma. World Neurosurg. 116, e162–e168.

Tordjman, S., Chokron, S., Delorme, R., Charrier, A., Bellissant, E., Jaafari, N., Fougerou, C., 2017. Melatonin: pharmacology, functions and therapeutic benefits. Curr. Neuropharmacol. 15, 434–443.

Touitou, Y., Fevre, M., Bogdan, R.A., De Prins, J., Beck, H., Touitou, C., 1984. Patterns of plasma melatonin with ageing and mental condition: stability of nyctohemeral rhythms and differences in seasonal variations. Acta Endocrinol. 106, 145–151.

Trzesniak, C., Kempton, M.J., Busatto, G.F., de Oliveira, I.R., Galvao-de Almeida, A., Kambeitz, J., Freitas Ferrari, M.C., Filho, A.S., Chagas, M.H.N., Zuardi, A.W., Hallak, J.E.C., McGuire, P.K., Crippa, J.A.S., 2011. Adhesio interthalamica alterations in schizophrenia spectrum disorders: a systematic review and meta-analysis. Prog. Neuropsychopharmacol. Biol. Psychiatry 35, 877–886.

Vadovicova, K., 2014. Affective and cognitive prefrontal cortex projections to the lateral habenula in humans. Front. Hum. Neurosci. 8, 819. https://doi.org/10.3389/fnhum.2014.00819.

Valentinova, K., Mameli, M., 2016. mGlutR-LTD at excitatory and inhibitory synapses in the lateral habenula tunes neuronal output. Cell Rep. 16, 2298–2307.

Valjakka, A., Vartiainen, J., Tuomisto, L., Tuomisto, J.T., Oikkonen, H., Airaksinen, M.M., 1998. The fasciculus retroflexus controls the integrity of REM sleep by supporting the generation of hippocampal theta rhythm and rapid eye movements in rats. Brain Res. Bull. 47, 171–184.

Vertes, R.P., Fortin, W.J., Crane, A.M., 1999. Projections of the median raphe nucleus in the rat. J. Comp. Neurol. 407, 555–582.

Vertes, R.P., 1991. A PHA-L analysis of ascending projections of the dorsal raphe nucleus in the rat. J. Comp. Neurol. 313, 643–668.

Vertes, R.P., 2006. Interactions among the medial prefrontal cortex, hippocampus and midline thalamus in emotional and cognitive processing in the rat. Neuroscience 142, 1–20.

Villano, J., Propp, J.M., Porter, K.R., Stewart, A.K., Valyi-Nagy, T., Li, X., Engelhard, H.H., McCarthy, B.J., 2008. Malignant pineal germ-cell tumors: an analysis of cases from three tumor registries. Neuro Oncol. 10, 121–130.

Villarreal, J.S., Gonzalez-Lima, F., Berndt, J., Barea-Rodriguez, E.J., 2002. Water maze training in aged rats: effects on brain metabolic capacity and behavior. Brain Res. 939, 43–51.

Wada, E., McKinnon, D., Heinemann, S., Patrick, J., Swanson, L.W., 1990. The distribution of mRNA encoded by a new member of the neuronal nicotinic acetylcholine receptor gene family (alpha 5) in the rat central nervous system. Brain Res. 526, 45–53.

Wang, R.Y., Aghajanian, G.K., 1977. Physiological evidence for habenula as major link between forebrain and midbrain raphe. Science 197, 89–91.

Wang, Q.P., Ochiai, H., Nakai, Y., 1992. GABAergic innervation of serotonergic neurons in the dorsal raphe nucleus of the rat studied by electron microscopy double immunostaining. Brain Res. Bull. 29, 943–948.

Wang, Y., Zhang, C., Zhang, Y., Gong, H., Li, J., Jin, H., Li, D., Liu, D., Sun, B., 2020. Habenula deep brain stimulation for intractable schizophrenia: a pilot study. Neurosurg. Focus 49 (1–7), E9. https://thejns.org/doi/abs/10.3171/2020.4.FOCUS20174.

Weitzman, E.D., Weinberg, U., D'Eletto, R., Lynch, H., Wurtman, R.J., Czeisler, C., Erlich, S., 1978. Studies of the 24 hour rhythm of melatonin in man. J. Neural. Transm. 13, 325–337.

Wolfman, S.L., Gill, D.F., Bogdanic, F., Long, K., Al-Hasani, R., McCall, J.G., Bruchas, M.R., McGehee, D.S., 2018. Nicotine aversion is mediated by GABAergic interpeduncular nucleus inputs to laterodorsal tegmentum. Nat. Commun. 9, 2710. https://doi.org/10.1038/s41467-018-04654-2.

Wurtman, R.J., Moskowitz, M.A., 1977. The pineal organ. N. Engl. J. Med. 296, 1383–1386.

Xu, C., Sun, Y., Cai, X., You, T., Zhao, H., Li, Y., Zhao, H., 2018. Medial habenula-interpeduncular nucleus circuit contributes to anhedonia-like behavior in a rat model of depression. Front. Behav. Neurosci. 12, 238. https://doi.org/10.3389/fnbeh.2018.00238.

Yamaguchi, T., Danjo, T., Pastan, I., Hikida, T., Nakanishi, S., 2013. Distinct roles of segregated transmission of the septo-habenular pathway in anxiety and fear. Neuron 78, 537–544.

Yang, L.-M., Hu, B., Xia, Y.H., Zhang, H., 2008. Lateral habenula lesions improve the behavioral response in depressed rats via increasing the serotonin level in dorsal raphe nucleus. Behav. Brain Res. 188, 84–90.

Yang, L.-M., Yu, L., Jin, H.-J., Zhao, H., 2014. Substance P receptor antagonist in lateral habenula improves rat depression-like behavior. Brain Res. Bull. 100, 22–28.

Yang, Y., Cui, Y., Sang, K., Dong, Y., Ni, Z., Ma, S., Hu, H., 2018. Ketamine blocks bursting in the lateral habenula to rapidly relieve depression. Nature 554, 317–322.

Yeo, B.T.T., Krienen, F.M., Sepulcre, J., Sabuncu, M.R., Lashkari, D., Hollinshead, M., Roffman, J.L., Smoller, J.W., Zollei, L., Polimeni, J.R., Fischl, B., Liu, H., Buckner, R.L., 2011. The organization of the human cerebral cortex estimated by intrinsic functional connectivity. J. Neurophysiol. 106 (3), 1125–1165. https://doi.org/10.1152/jn.00338.2011. PMID: 21653723.

Yu, L.C., Han, J.S., 1990. Habenula as a relay in the descending pathway from nucleus accumbens to periaqueductal grey subserving antinociception. Int. J. Neurosci. 54, 245–251.

Zhang, B., Gao, Y., Li, Y., Yang, J., Zhao, H., 2016a. Sleep deprivation influences circadian gene expression in the lateral habenula. Behav. Neurol. 2016, 7919534. https://doi.org/10.1155/2016/7919534.

Zhang, J., Tan, L., Ren, Y., Liang, J., Lin, R., Feng, Q., Zhou, J., Hu, F., Ren, J., Wei, C., Yu, T., Zhuang, Y., Bettler, B., Wang, F., Luo, M., 2016. Presynaptic excitation via $GABA_B$ receptors in habenula cholinergic neurons regulates fear memory expression. Cell 166, 716–728.

Zhao, H., Rusak, B., 2005. Circadian firing-rate rhythms and light responses of rat habenular nucleus neurons in vivo and in vitro. Neuroscience 132, 519–528.

Zhao, H., Zhang, B.-L., Yang, S.-J., Rusak, B., 2015. The role of lateral habenula-dorsal raphe nucleus circuits in higher brain functions and psychiatric illness. Behav. Brain Res. 277, 89–98.

Zhao-Shea, R., Liu, L., Pang, X., Gardner, P.D., Tapper, A.R., 2013. Activation of GABA neurons in the interpeduncular nucleus triggers physical nicotine withdrawal symptoms. Curr. Biol. 23, 2327–2335.

Zhao-Shea, R., DeGroot, S.R., Liu, L., Vallaster, M., Pang, X., Su, Q., Guangping, G., Rando, O.J., Martin, G.E., Gardner, P.D., Tapper, A.R., 2015. Increased CRF signaling in a ventral tegmental area-interpeduncular nucleus-medial habenula circuit induces anxiety during nicotine withdrawal. Nat. Commun. 6, 6770. https://doi.org/10.1038/ncomms7770.

Zuo, W., Xiao, C., Gao, M., Hopf, F.W., Krnjevic, K., McIntosh, J.M., Fu, R., Wu, J., Bekker, A., Ye, J.-H., 2016. Nicotine regulates activity of lateral habenula neurons via presynaptic and postsynaptic mechanisms. Sci. Rep. 6, 32937. https://doi.org/10.1038/srep32937.

8

The Diencephalon: Dorsal Thalamus

Srikant S. Chakravarthi[1], *Alejandro Monroy-Sosa*[2,3] *and Kost Elisevich*[4]

[1]Aurora Research Institute, Aurora St. Luke's Medical Center, Milwaukee, WI, United States [2]Department of Neurosurgery, Star Medica Lomas Verdes Hospital, Mexico City, Mexico [3]Department of Neurosurgery, Institute for Social Security and Services for State Workers, (ISSSTE) Tlahuac, Center for Research and Innovation in Neuroscience, Instituto Politécnico Nacional, Superior School of Medicine, Mexico City, Mexico [4]Department of Neurosurgery, Spectrum Health Medical Group Neurosurgery, Grand Rapids, MI, United States

OUTLINE

EMBRYOLOGICAL ORIGIN AND DEVELOPMENT OF THE THALAMUS

In order to engage in a thorough discussion on the anatomy and function of the thalamus, we will briefly describe its embryological origin and development along with the structures in its immediate proximity. Between days 28 and 48 of embryological development, both diencephalon and telencephalon arise from the prosencephalon from its basal and alar plates, respectively. The telencephalon develops further into four primary elements—the olfactory system, limbic forebrain, basal ganglia, and cerebral cortex or neopallium. The diencephalon is considered to be the intermediate brain and during its invagination, forms the thalamencephalon, hypothalamus, and subthalamus (posterior forebrain structures) (de Lahunta and Glass, 2009; Puelles et al., 2012). The thalamencephalon, also referred to as the thalamic region or thalamic complex, consists of the thalamus, often referred to as the dorsal thalamus, the epithalamus, and metathalamus (de Lahunta and Glass, 2009).

The diencephalon arises from the medial portion of the prosencephalon and consists of a roof plate, two alar plates, and the third ventricle. The pineal gland and mammillary bodies border the diencephalon posteriorly and ventrally, whereas the optic chiasm is situated anteriorly. The alar plates form the lateral walls and floor of the diencephalon. A distinct groove or sulcus, referred to as the hypothalamic sulcus, extends in an anteroposterior direction and divides the alar plate into dorsal and ventral compartments which eventually become the thalamus and hypothalamus, respectively. Along its sagittal plane, the embryo at week 4 can be first seen invaginating at the level of the isthmus (the segment between the midbrain and hindbrain), with the foramen of Monro also forming in concert. By week 6, an area of the mesencephalon evaginates to form the pineal gland, the cerebral hemispheres start protruding laterally, and the lamina terminalis and hypothalamus become evident. By week 8, the thalamus begins to bulge into the diencephalic lumen and the two fuse along the midline to form the massa intermedia (or the interthalamic adhesion). It is also at this age that the hypothalamic sulcus begins to create a clearer division between the thalamus, hypothalamus, and epithalamus. The mammillary bodies, often considered one of the nuclear groups of the hypothalamus, also begin bulging along the ventral surface of the hypothalamus to eventually connect it to higher centers of the rhinencephalon and lower brainstem nuclear centers. The thalamic nuclear areas divide into two distinct nuclear groups—a dorsal group receiving and

From Anatomy to Function of the Central Nervous System
https://doi.org/10.1016/B978-0-12-822404-5.00013-9

transmitting visual and auditory impulses, and a ventral group relaying information to higher centers. The aforementioned structures and their relative orientation become distinctly visible at week 11 of development (Nakagawa, 2019).

Geographically, the components of the diencephalon are situated along either side of the third ventricle. The hypothalamus and its associated nuclei are found ventral and lateral to the inferior segment of the third ventricle. The floor of the diencephalon eventually gives rise to the neural primordia of the eyes. The infundibulum and neural lobe of the hypophysis are seen to form at week 8. The thalamus eventually becomes the largest structure derived from the embryologic diencephalon, protruding on either side of the third ventricle. Fig. 8.1 highlights the progressive embryological development of the primitive neural tube structures during days 28, 36, and 48, leading to the eventual bulging of the thalamus from the diencephalic lumen during week 8.

THE THALAMUS: LOCATION, STRUCTURE AND FUNCTION

Location

The dorsal thalamus, or *thalamus propria*, is a gray matter nucleated structure and the largest component of the diencephalon, which through development, eventually becomes an ovoid-shaped symmetrical midline structure situated at the center of the lateral ventricle. For the sake of simplicity, the dorsal thalamus—which is the largest component of the overall thalamic system—will be referred to as only the thalamus in the ensuing discussion. At a length of approximately 5.7 cm, it can be considered as paired bulb-shaped halves on either side of the third ventricle (Butler, 2009). The thalamus is strategically positioned above the mesencephalon or midbrain (Figs. 8.2, 8.4 and 8.5), thereby allowing for the relay of nerve fiber connections from the midbrain and below to the cerebral cortex and for the reverse. In relation to the dorsal thalamus, the hypothalamus, with its numerous and distinct nuclei, is situated at the base of the diencephalon, anteroinferior to the thalamus at the 7 o'clock position on sagittal view and superior to the infundibular stalk. The subthalamus is situated ventrally in the diencephalon between the thalamus and midbrain and the epithalamus occupies the dorsal and posterior aspect of the diencephalon.

Components of the Thalamus

The thalamus represents one of the most complex structures in the cerebrum. It serves as a gateway of input and output function, relaying sensory, motor, and limbic stimuli, through functionally specific subnuclei. Although each individual nucleus and their orientation will be described in detail, we can generally conceptualize the thalamus as a concentric system of discrete yet connected nuclei, both from a functional and geographic perspective. Historically, there has been controversy as to the true internal organization and topography of the thalamus; this has been due to several reasons, including innate evolutionary differences in thalamic formation in humans and the progressive effect of age and disease in influencing shape, volume, and neural connectivity. As such, many models of the relative position and orientation of the thalamic nuclei have been proposed based on their presumed function and relative orientation within the thalamus (Hess et al., 2014; Hughes et al., 2012; Mai and Majtanik, 2019). In the ensuing description of the organization and topography of the thalamus, the most current accepted model as suggested by Mai and Majtanik (2019) will be used (Fig. 8.3). With regard to establishing an internal demarcation within the thalamus, Jones (1998) (Jones, 2012; Jones and Friedman, 1982) proposed that the anteroposteriorly directed *internal medullary lamina* (IML), a thin layer of myelinated fibers separating the medial and lateral nuclei, be used as a suitable landmark to organize the nuclei. The proceeding discussion will use this primary landmark, in addition to other internal lamina, to describe the base model of the thalamus.

Structural Organization of the Thalamus

The thalamus, from a structural perspective, represents one of the most complex and convoluted systems in the central nervous system. However, with the incremental exploration of its intrinsic topography, it can be seen that the rationale for this organization bears both functional and connective significance. As previously mentioned, the two halves of the thalamus are connected by the massa intermedia, a white matter bridge between the two gray matter masses. In general, the thalamic nuclei can be structurally organized into three groups: lateral, medial, and anterior (Sheridan and Tadi, 2020). From the dorsal (bird's eye) view, these groups of nuclei can be topographically organized using the IML as an internal demarcation line. The IML divides the ventral nuclei (i.e., ventral anterior, ventral lateral, and ventral posterior) with its anteromedial, dorsal lateral, and dorsal posterior counterparts (i.e., anterior dorsal, lateral dorsal, medial dorsal, and lateral posterior). Lateral to the IML lies a ventral and dorsal tier of nuclei that comprise a considerable bulk of the thalamus. The ventral tier consists of the anterior thalamic nuclei, which are defined by a Y-shaped bifurcation of the

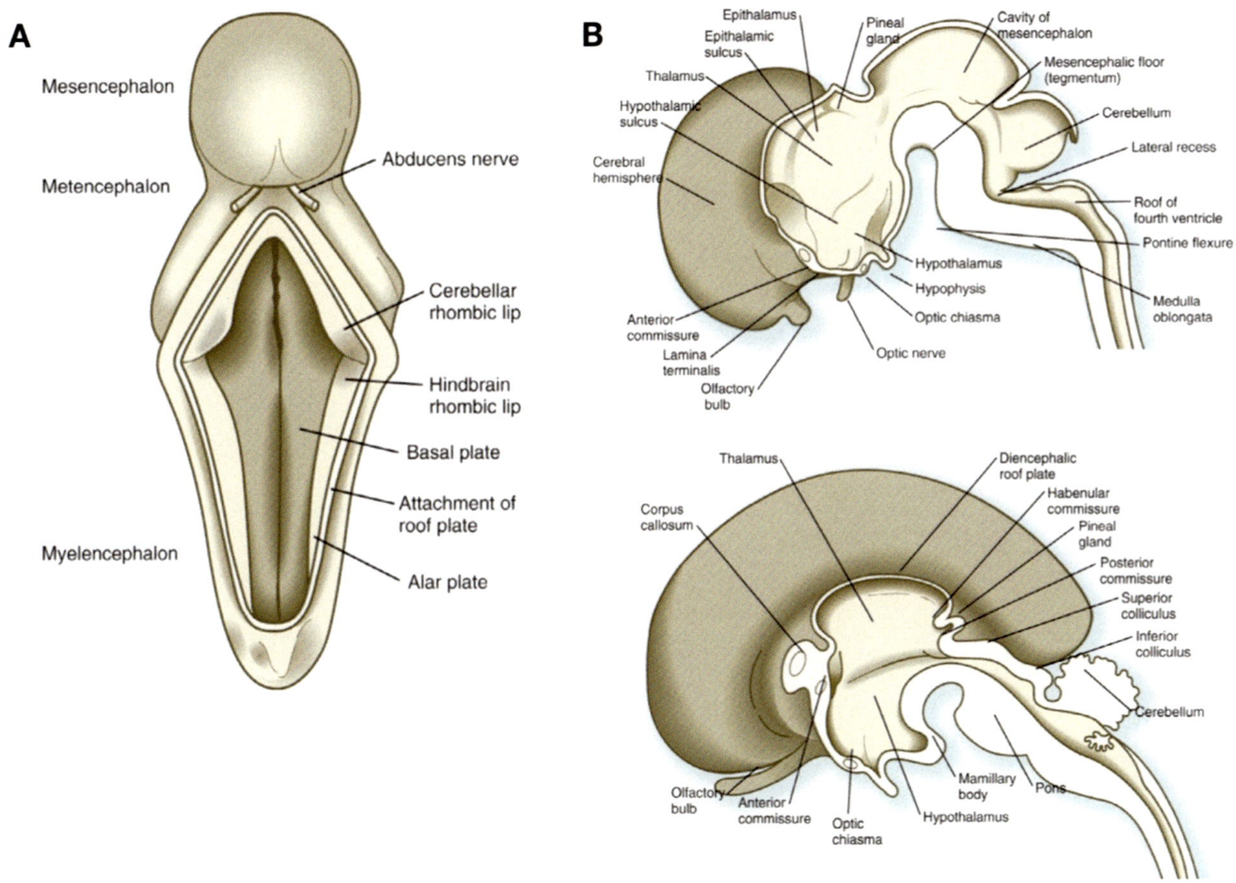

FIGURE 8.1 **Embryological development of the thalamus.** (A) Between days 28 and 48 of embryological development, both diencephalon and telencephalon arise from the prosencephalon from its basal and alar plates, respectively. The diencephalon arises from the medial portion of the prosencephalon and consists of a roof plate, two alar plates, and the third ventricle. Following anterior neuropore closure during day 25 and posterior neuropore closure during day 28, the embryological neural tube, which derives from the neural ectoderm, develops complexity through enlargements of the neural tube called vesicles. (B) The prosencephalon (forebrain), mesencephalon (midbrain), and rhombencephalon (hindbrain) begin forming at day 28. The infundibulum is the first invagination that forms from the anterior–inferior surface of the diencephalon at day 36. These complexities continue to develop at day 48 with the addition of the epiphysis at the superior portion of the diencephalon. By week 8, the thalamus begins to bulge into the diencephalic lumen and the two may fuse along the midline to form the massa intermedia (interthalamic adhesion). The hypothalamic sulcus also begins to create a clearer division between the thalamus, hypothalamus, and epithalamus. The mammillary bodies also begin bulging along the ventral surface of the hypothalamus to eventually connect it to higher centers of the rhinencephalon and lower brainstem nuclear centers. *Courtesy of Sinclair, A.H., 2014. Later development of embryonic central nervous system. In: Reference Module in Biomedical Sciences. Elsevier. https://doi.org/10.1016/B978-0-12-801238-3.05443-X.*

IML as it reaches anteriorly. The pulvinar occupies the posterior pole of the thalamus with the IML veering along its medial side and separating it from the medial and midline nuclei which lie anteromedially in relation. Two additional structures are important to identify from the dorsal view: the external medullary lamina, and the reticular nuclei. Both the external medullary lamina and the reticular nuclei can be seen wrapping around a majority of the lateral nuclei, with the external medullary nuclei in closest proximity to the lateral nuclei. In addition, the reticular nuclei, which is adjacent to the internal capsule laterally, encircles each lateral nucleus, providing its own internal closed-loop pathway; the functional importance of this will be described in the ensuing discussion on each individual nucleus. The following figures show the topographic localization of the thalamus using a coronal and axial section of the brain (Figs. 8.4 and 8.5), respectively. Fig. 8.6 instead shows the thalamus, by a lateral dissection of the third ventricle, and its relationships with the adjacent structures (for details see the caption).

If the thalamic body is cut from its midline (coronal view), one can view the internal medullary lamina curving from a superior to medial direction and also have a Y-shaped bifurcation at its tail. In this view, the internal medullary lamina separates the ventral posterior medial, ventral lateral, and lateral dorsal nuclei from the medial dorsal nuclei. The centromedian

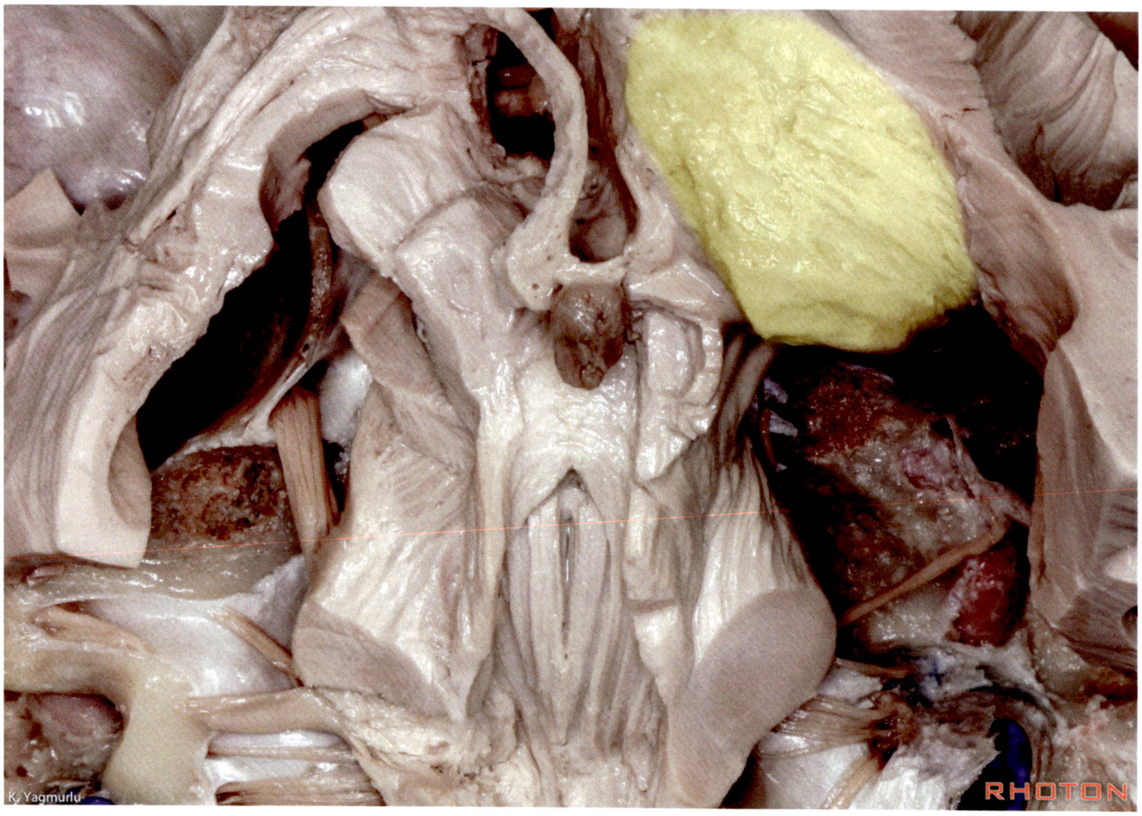

FIGURE 8.2 **Posterior view of the diencephalon.** In yellow, the thalamus and medially to it the habenula bodies. In the center, the dark structure is the pineal gland (epiphysis). *Image courtesy of The Rhoton Collection® http://rhoton.ineurodb.org/. Dissector: K.Yagmurlu. Drawing by Kaan Yagmurlu.*

nucleus is situated in a distinct location as an "island" within the laminar bifurcation, and in between the medial dorsal and ventral posterior medial nuclei. When viewing the thalamus posteriorly, the pulvinar can be seen comprising a majority of this area. In addition, the lateral posterior and ventral posterior nuclei can be seen adjacent to the pulvinar medially and inferiorly, respectively. Two additional small nuclei can be seen from this view, the medial geniculate nucleus (MGN) and the lateral geniculate nucleus (LGN). These two nuclei, which have a primary sensory relay function, can be visualized as buds protruding from the posteroinferior aspect of the thalamus (specifically, from the pulvinar). Taken together, the posterior thalamic region (the pulvinar being the largest component of this) along with the medial and lateral genicular nuclei can be conceptualized as being entirely separate from the remaining anterior, medial, and lateral thalamic nuclei, according to some models (Herrero et al., 2002; Jones, 2012).

Apart from the above-mentioned groups of nuclei, there are two additional nuclei which may not fit this organization schema: midline and intralaminar. The midline nuclei, from a dorsal view, lie in a longitudinal fashion, closest to the interthalamic adhesion, and include the paratenial, paraventricular, reuniens, and rhomboid nuclei. These nuclei encircle the large medial dorsal nucleus and are medial to the internal medullary lamina. From a cross-sectional view and in relation to nonthalamic structures, the midline nuclei along with the medial dorsal nucleus is nestled directly inferior to the lateral ventricle, superior to the third ventricle, and medial to the lateral dorsal nucleus. The intralaminar nuclei are a group of centrally positioned higher order neurons; one of them, the centromedian nucleus, was previously mentioned as sitting between the medial dorsal and ventral posterior nuclei. These groups of neurons are generally divided into two groups: (1) anterior (rostral): central medial, paracentral, and central lateral, and (2) posterior (caudal): centromedian and parafascicular (Fisher and Reynolds, 2014; Saalmann, 2014; Van der Werf et al., 2002). These intralaminar nuclei are essentially located within the internal medullary lamina.

Inter- and Intraconnectivity of the Thalamic Nuclei

Apart from its structural makeup, each thalamic nucleus possesses distinct functions and dynamic

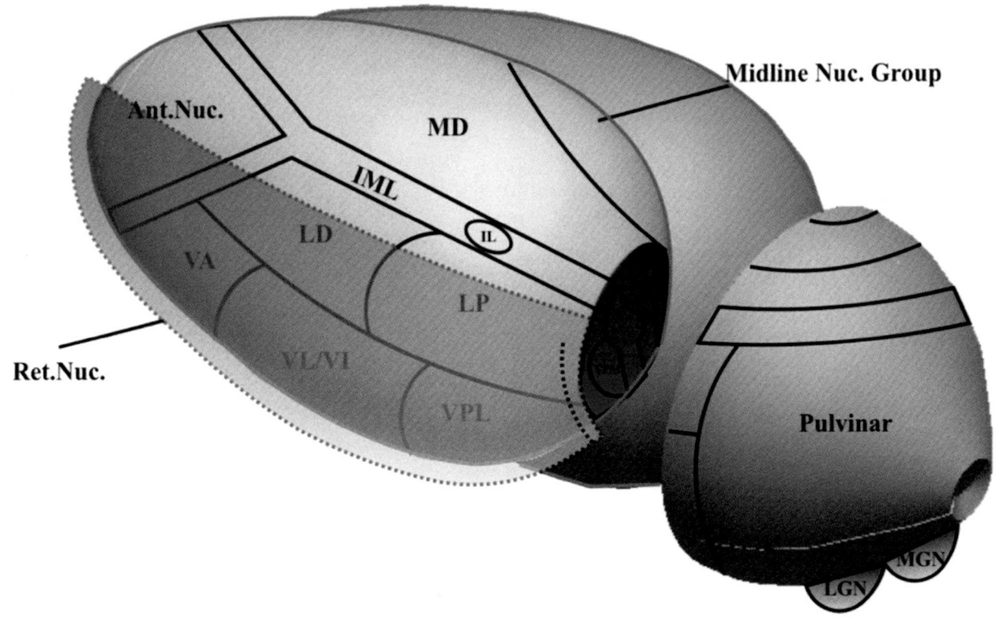

FIGURE 8.3 **Nuclei of the thalamus.** This 3D illustration represents the currently accepted model of the thalamic nuclei. Thalamic nuclei can be structurally organized into three groups, lateral, medial, and anterior. From the dorsal view, these groups of nuclei can be topographically organized using the internal medullary lamina (IML) as an internal demarcation line. The IML divides the ventral nuclei (i.e., ventral anterior (VA), ventral lateral (VL/VI), and ventral posterior (VP)) with its anteromedial, dorsal lateral, and dorsal posterior counterparts (i.e., anterior dorsal (AD), lateral dorsal (LD), medial dorsal (MD), and lateral posterior (LP)). Lateral to the IML lies a ventral and dorsal tier of nuclei that comprise a considerable bulk of the thalamus. The ventral tier consists of the anterior thalamic nuclei, which are defined by a Y-shaped bifurcation of the IML as it reaches anteriorly. The pulvinar occupies the posterior pole of the thalamus with the IML veering along its medial side and separating it from the medial and midline nuclei which lie anteromedially in relation. The pulvinar, along with the lateral (LGB) and medial geniculate bodies (MGB) (seen as bulging from the pulvinar posteroinferiorly), are drawn separately in this illustration in order to show the VPM nucleus. Finally, the reticular nuclei (Ret. Nuc.) can be seen wrapping around each lateral nucleus.

connective relationships. This section will delve into the roles of the major thalamic nuclei and their connections within the thalamus and the cerebrum. The thalamic nuclei can be categorized into three primary nuclear systems: (1) Relay nuclei, consisting of anterior, lateral, and medial groups; (2) Reticular nucleus; and (3) Intralaminar nuclei. It is important to note that with respect to principal neurotransmitter release, the main body of the thalamus (i.e., anterior, lateral, medial, and posterior nuclear groups) is primarily composed of glutamatergic projection neurons and GABAergic interneurons (between output and input signals) (Crabtree, 2018; Jones, 2012; Sherman and Guillery, 2002). Most projection neurons are innervated by glutamatergic-driven inputs. In addition, there are modulator-driven inputs, such as GABAergic neurons of the reticular nucleus and the cholinergic neurons of the brainstem pedunculopontine nucleus, which control the signal produced by these cells.

Relay Nuclei

The relay nuclei represent the majority of the thalamus. The term "relay" refers to the mechanism by which signals or information is received by an entity and then immediately sent to a different location. This rather simple machine-like concept has been attributed to the thalamus as its only major role for centuries; virtually all information sent to the cerebral cortex from peripheral receptors (e.g., retina) first pass through the thalamus (Sherman, 2005). However, more recently, studies have alluded to the processing nature of the thalamus, with the ability to dynamically control and alter information that is being received and sent (e.g., attention and drowsiness). Murray Sherman and Guillery generally defined two sets of relays: (1) "first-order" relays, which receive information from the sensory periphery (i.e., visual, auditory, tactile, etc.) and from other areas, such as the cerebellum and basal ganglia; and (2) "higher-order" relay systems, which receive information from the cerebral cortex itself and in between cortices (Sherman and Guillery, 2002). In general, the projections of each thalamic relay nucleus proceed to more focal locations than those of the interlaminar or reticular nuclei (both of which are considered closed-loop circuits not connecting to the cerebral cortex). There will be a connecting efferent and afferent highway relaying information to specific regions of the cortex. In general, the lateral nuclear group is considered the most clinically important of the relay nuclear groups. For the purposes of simplicity in describing the sending

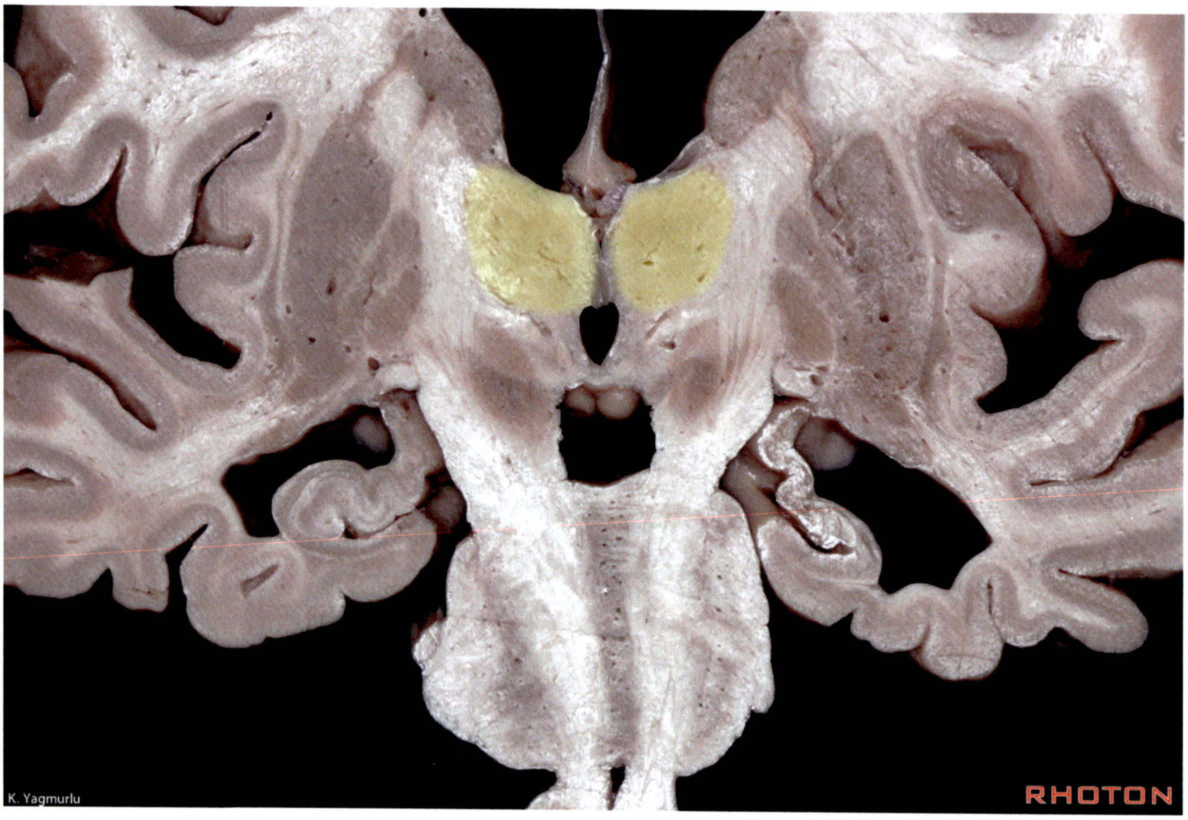

FIGURE 8.4 **Coronal section of the brain.** In yellow the thalamus (*right* and *left*) and above it the lateral ventricles separated by the septum pellucidum. Laterally to the thalamus, the internal capsule (white ribbon-like material) is clearly visible, which descends through both the midbrain and the medulla forming the pyramidal tract. Below the thalamus, the first light gray structure from top to bottom is the subthalamic nucleus and below it is the red nucleus (darker structure). *Image courtesy of The Rhoton Collection® http://rhoton.ineurodb.org/. Dissector: K.Yagmurlu. Drawing by Jeffrey Sorenson.*

and receiving connections of the thalamic nuclei, we will refer to projections sent to specific regions as *outputs* and projections coming to the respective nuclei as *inputs*.

Anterior Nuclear Group

Anterior Nucleus

The anterior nuclear group is nestled in the middle of a complex circuitry of cortical and subcortical connections. In general, it receives a dense concentration of inputs from the mamillary bodies (via the mammillothalamic tract), hypothalamus, subiculum, and retrospenial cortex, and sends outputs to the cingulate gyrus belonging to the Papez circuit (O'Mara, 2013; Wright et al., 2010). The anterior group is generally divided into anteromedial (AM), anterodorsal (AD), and anteroventral (AV) nuclei, each carrying projections which are topographically specific. Together with the lateral dorsal (LD) nucleus, they are considered the "limbic" nuclei, referring to their connections to the limbic structures. Table 8.1 highlights the inputs and outputs of each anterior nucleus, which have been organized and consolidated by Jankowski et al. (2013), O'Mara (2013), as well as their putative function (Schmahmann, 2003). Of note, the hippocampal projections to the mammillary bodies and anterior thalamic nuclei rely on a connection with the fornix; this has important context in understanding the role the fornix plays in cognition and will be described in further detail in the ensuing section (Tsivilis et al., 2008).

As previously mentioned, the Papez circuit plays an important role in the overall integration of the anterior thalamic nucleus. As Papez (1937) described it in 1937, the anterior thalamic nucleus is the central core of a closed-loop circuit system with the following order: hippocampal formation → mammillary bodies → cingulate cortex → parahippocampal gyrus → hippocampal formation. From a functional perspective, it is currently thought that the Papez circuit plays a role in memory and learning as it pertains to emotion (Shah et al., 2012; Vertes et al., 2001). Several other models have functionally divided each individual nucleus and placed them in three separate parallel "subsystems" with the

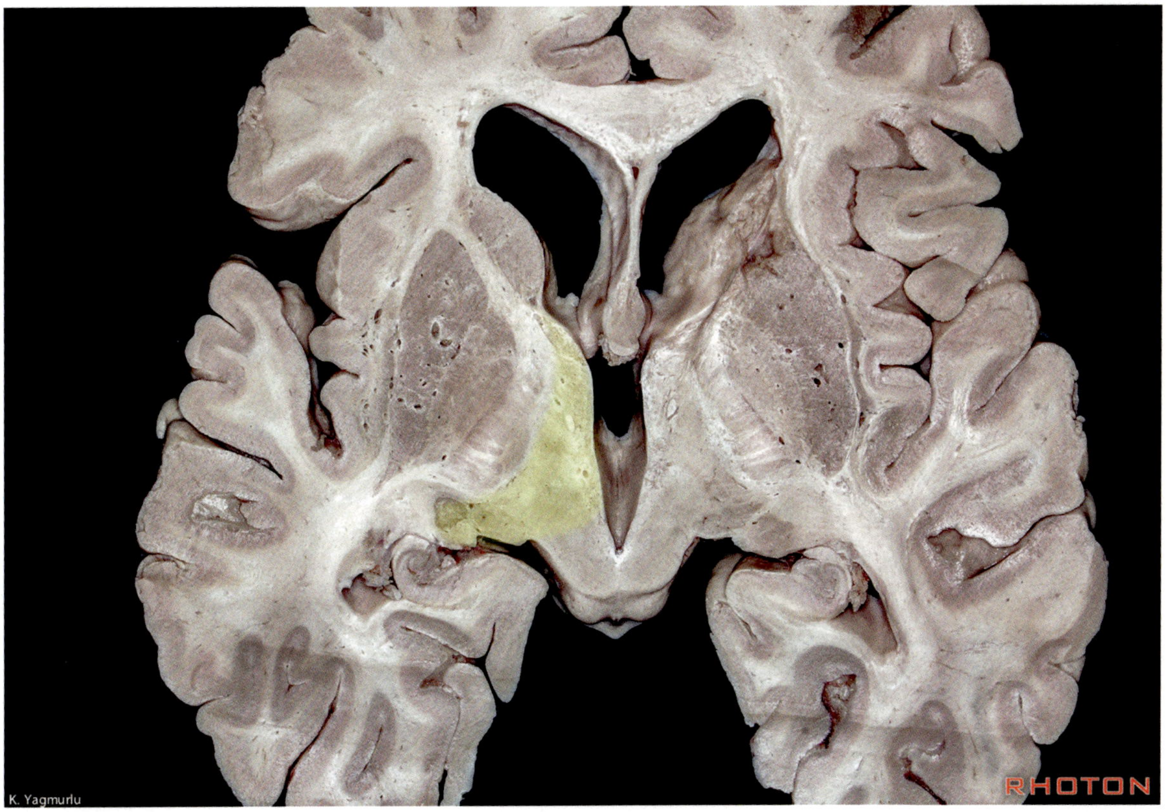

FIGURE 8.5 **Axial section of the brain.** In yellow, the thalamus (*the left*) and in front of it the lateral ventricles separated by the septum pellucidum. Lateral to the thalamus are the posterior limb and genu (more anteriorly) of the internal capsule. Medially the thalamus, the third ventricle is easily seen in the center. The most posterior structure is the roof of mesencephalon with the corpora quadrigemina. *Image courtesy of The Rhoton Collection® http://rhoton.ineurodb.org/. Dissector: K.Yagmurlu. Drawing by Jeffrey Sorenson.*

following schema (Tsivilis et al., 2008; Wright et al., 2010): (1) The anteromedial nucleus forms part of a "forward-feeding" system that conveys information from a hippocampal–diencephalic network to prefrontal areas, thereby playing a part in controlling higher order cognitive and executive functioning; (2) The anteroventral nucleus takes part in a closed-circuit return loop that promotes rhythmic theta wave activity and stimulates the hippocampal formations; and (3) The anterodorsal nucleus provides information for the "head-direction system," which proposes that compass-like spatial and mental navigation (for example, when the head is turned) is due to the activation of these aforementioned circuit loops (Clark and Taube, 2012).

Medial (Medial Dorsal Nucleus)

The medial dorsal thalamus (MD) is situated between the internal medullary lamina and the periventricular gray matter; as such, it is strategically located so that its subdivisions are bidirectionally connected to different prefrontal cortical regions and several critical subcortical regions. It has medial, central, and lateral divisions distinguished by their respective connections noted below. Of note, the MD plays a critical role in a robust system of interconnections, particularly with limbic-driven pathways (Kuroda et al., 1998; Sherman and Guillery, 2002). Explicitly, the MD, along with its surrounding midline interthalamic nuclei, send output signals to the prefrontal cortex, amygdala, basal ganglia, spinothalamic fibers, and olfactory cortex. The primary inputs come from the medial and lateral prefrontal cortices and the orbitofrontal cortex (Georgescu et al., 2020). These interconnections are the essence of the cortico-thalamo-cortical pathways—the medial dorsal-prefrontal and prefrontal-medial dorsal pathways (cortical layers 5 and 6) are glutaminergic excitatory pathways where the cortex plays a strict and higher-order control of MD function. This control plays an important role in facilitating behavioral control as it pertains to guiding inner goals (Miller and Cohen, 2001). In addition, the medial temporal lobe, which is largely responsible for object recognition memory, can send outputs to

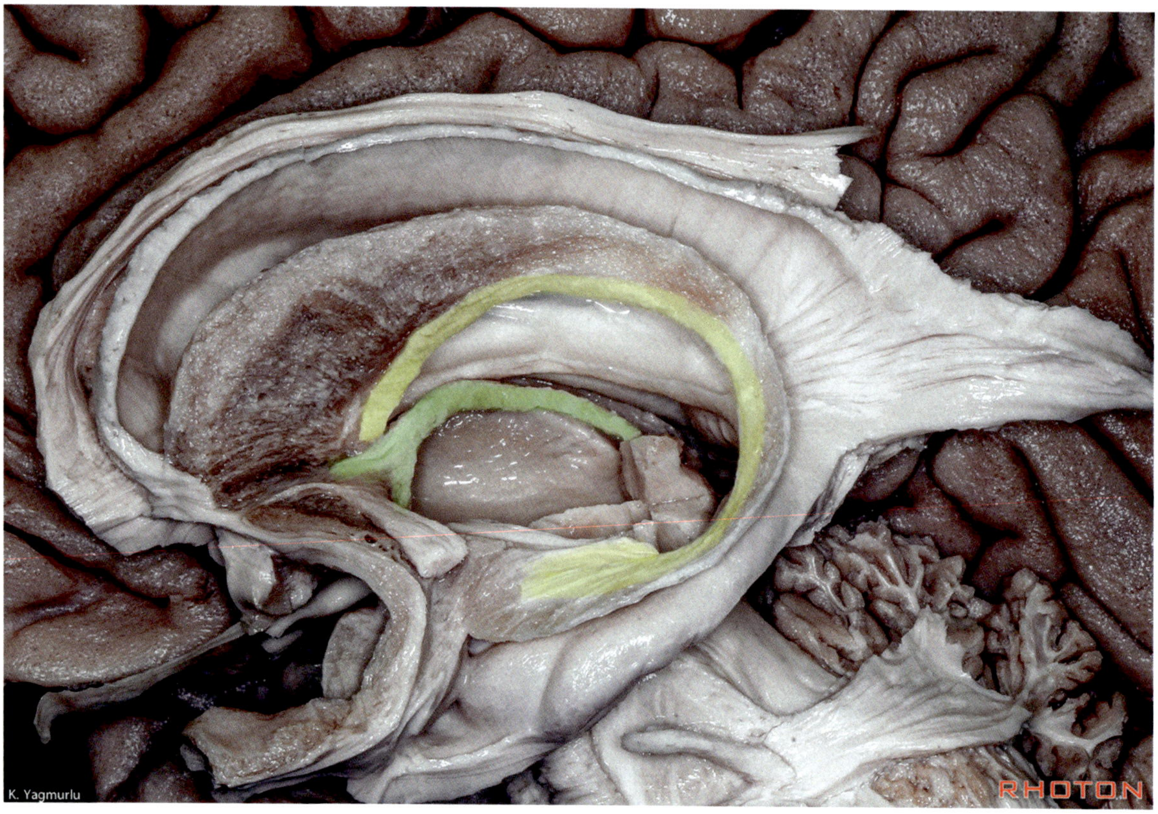

FIGURE 8.6 **Lateral dissection of third ventricle.** Lateral view of the thalamus (egg-shaped structure) in the center, surrounded by the stria medullaris thalami, in green. In front of the latter, the horizontal white structure that moves outward is the anterior commissure. Above the thalamus, the corpus callosum and the fornix (in *yellow*) are visible. The latter is medially adherent to the caudate nucleus (inverted C-shaped structure in *gray*) whose tail is overlying the hippocampus (the "sausage"-shaped structure). *Image courtesy of The Rhoton Collection® http://rhoton.ineurodb.org/. Dissector: K.Yagmurlu. Drawing by Jeffrey Sorenson and Alessandro Basso.*

the prefrontal cortex via the MD (especially in Brodmann areas 14, 24 (anterior cingulate cortex), 25, and 32). The entorhinal cortex and orbitofrontal cortices also play a critical role in this memory function by contributing to the medial temporal lobe-MD-prefrontal circuit by providing direct inputs to the MD.

The MD has important communication with subcortical structures. Specifically, the parvocellular division of the MD receives input from the amygdala (although not as robust a signal), internal segment of the globus pallidus, locus coeruleus, reticular formation, ventral tegmental area, and the reticular component of the substantia nigra. In addition, the lateral part of the MD (which comprises the densocellular and pars multiforms subnuclei) receives input from the internal component of the globus pallidus, locus coeruleus, reticular formation, as well as the reticular component of the substantia nigra and the ventral tegmental area. The magnocellular subnuclei of the MD receives input from the ventral pallidum, locus coeruleus, reticular formation, dorsal raphe, ventral tegmental area, reticular component of the substantia nigra, and basal forebrain (Berendse and Groenewegen, 1990; Giménez-Amaya et al., 1995; Groenewegen, 1988; Groenewegen et al., 1999; Hallanger et al., 1987). It is important to note that, with the exception of the ventral tegmental area and basal forebrain, the neurotransmitter signal is dopaminergic.

When consolidating the aforementioned inputs and outputs to and from the MD, there are generally three distinct close-loop bidirectional cortical-subcortical circuits with several associated inputs (Table 8.2) (Georgescu et al., 2020; Mitchell and Chakraborty, 2013; Parnaudeau et al., 2018).

Although not yet clearly elucidated, some studies have also found a connection between the MD and the cerebellum; specifically, the MD receives two separate inputs, one directly from the deep cerebellar nuclei and the other via the intralaminar nuclei. These inputs are also connected to the prefrontal cortex via output signals, thereby giving rise to a cerebello-thalamo-cortical circuit (Hintzen et al., 2018).

TABLE 8.1 The anterior thalamic nuclear group.

Anterior nuclear group	Inputs	Outputs	Putative function(s)
Anteromedial nucleus	- Medial mamillary bodies - Rostral dorsal reticular nucleus - Prelimbic and medial orbital cortices - Anterior cingulate gyrus - Retrosplenial cortex - Secondary motor cortices - Entorhinal cortex - Subiculum	- Frontal area 2 - Frontopolar cortex - Medial orbital cortex - Anterior cingulate gyrus - Retrosplenial cortex - Entorhinal cortex - Perirhinal cortex - Presubiculum and subiculum - Visual cortex area 18b - Temporal area 2 - Occipital area 1 and 2 - Medial secondary motor cortices	- Learning - Memory - Emotional experience and expression - Drive - Motivation
Anterodorsal nucleus	- Lateral mammillary bodies - Subiculum, including para- and postsubiculum - Retina - Anterior cingulate cortex - Retrosplenial cortex - Dorsal reticular nucleus	- Pre-, para-, and postsubiculum - Hippocampus - Retrosplenial cortex	
Anteroventral nucleus	- Medial mammillary bodies - Dorsal reticular nucleus and laterodorsal tegmental nucleus - Subiculum and postsubiculum - Anterior cingulate cortex - Retrosplenial cortex	- Pre-, para-, and postsubiculum - Entorhinal cortex - Anterior cingulate cortex - Retrosplenial cortex - Secondary motor cortex	

TABLE 8.2 Close-loop bidirectional cortical-subcortical circuits of the MD nucleus.

Bidirectional circuits of MD	Connection	Input(s)
Medial	Connects magnocellular MD to the orbitofrontal and ventromedial prefrontal cortices.	Ventral lateral prefrontal and entorhinal cortices, amygdala, ventral striatum, ventral pallidum.
Central	Connects parvocellular MD to dorsal prefrontal cortex and Brodmann area 10.	Orbitofrontal cortex, dorsal anterior cingulate cortex, globus pallidus, dorsal striatum.
Lateral	Connects intralaminar nuclei (close association to MD) with globus pallidus, dorsal striatum, prefrontal cortex, and frontal eye fields.	No separate inputs.

As one can derive its role from its respective communications and circuitry, the MD is responsible for a host of functions: (1) The frontal lobe, and, in particular, the prefrontal cortex, receives vigorous and reciprocal communication with the MD. It has diffuse connections with and plays a part in the function of the premotor, primary motor, and anterior cingulate cortices (Groenewegen et al., 1997); (2) With the ventromedial prefrontal cortex, orbitofrontal cortex, and medial temporal region, the magnocellular MD, in the central circuit, it plays a role in object recognition and visuospatial memory (Bachevalier et al., 1997; Barbas, 1988; Gower, 1989; Morecraft et al., 1992; Russchen et al., 1987; Tsivilis et al., 2008); (3) The magnocellular MD is potentially involved in a multisensory relay system with the prefrontal cortex via the amygdala, entorhinal cortex, and temporopolar cortex through cortico-amygdaloid and cortico-cortico circuits, mediating cognitive processing of visual and olfactory stimuli; (4) Connection with subcortical structures, such as the basal ganglia creates the lateral circuit (thalamo-gangliar networks) which assists with the higher-order processing of motor tasks (Haber and McFarland, 2001; Haber and Calzavara, 2009); and (5) The cerebello-thalamo-cortical circuits may provide an important connection system for both motor and nonmotor functions, particularly when providing a coordinated response to both internal and external stimuli (Fujita et al., 2020; Zhang et al., 2016).

Lateral Nuclear Group

The lateral nuclear group is the largest of the nuclear groups of the thalamus and clinically is perhaps the most relevant. To reiterate, the lateral relay nuclear group consists of the following nuclei: ventral posterolateral, ventral posteromedial, lateral geniculate, medial geniculate, ventral lateral (also known as ventral intermediate), ventral anterior, pulvinar, lateral dorsal, lateral posterior, ventral medial. Table 8.3 provides a brief overview of the respective inputs, outputs, and putative functions of the lateral nuclear group.

Intralaminar and Midline Nuclei

The midline and intralaminar nuclei are often described together, as they share similarities with respect to connectivity and functionality. Together, the

TABLE 8.3 Lateral thalamic nuclei.

Lateral nuclei	Input(s)	Output(s)	Putative function(s)
Ventral posterolateral	- Medial lemniscus - Spinothalamic tract	- Somatosensory cortex	Relays somatosensory spinal inputs to cortex
Ventral posteromedial	- Trigeminal lemniscus - Trigeminothalamic tract - Taste inputs	- Somatosensory cortex - Taste cortex	Relays somatosensory cranial nerve inputs and taste to cortex
Ventral lateral (ventral intermediate)	- Globus pallidus pars interna - Deep cerebellar nuclei - Substantia nigra pars reticulata	- Motor cortex - Premotor cortex - Supplementary motor cortex	Relays basal ganglia and cerebellar inputs to cortex
Ventral anterior	- Substantia nigra pars reticulata - Internal globus pallidus - Deep cerebellar nuclei	- Diffuse output to frontal lobe, prefrontal, premotor, and supplementary motor cortices	Relays basal ganglia and cerebellar inputs to cortex
Pulvinar	- Tectum (extrageniculate visual pathway) - Other sensory inputs	- Parietotemporooccipital association cortex	Behavioral orientation toward visual and other stimuli
Lateral geniculate	- Retina	- Primary visual cortex	Relays visual inputs to cortex
Medical geniculate	- Inferior colliculus	Primary auditory cortex	Relays auditory inputs to cortex
Lateral dorsal	- Mammillary body - Reticular formation (similar to anterior thalamus)	No known output	Works with anterior thalamus to control limbic-driven pathways
Lateral posterior	- Similar to pulvinar	- Diffuse cortical output	Works with pulvinar to control behavioral orientation toward visual and other stimuli
Ventral medial	- Midbrain reticular formation	- Diffuse cortical output	Theoretical: Maintaining an alert and conscious state

intralaminar and midline nuclei are oddly and conspicuously arranged around the medial dorsal component of the thalamus. The midline nuclei comprise a thin strip of cells tightly encircling the MD and running along the entire length of the thalamus in a dorsal to ventral direction. The intralaminar nuclei, from a cross-sectional view, are located lateral to the MD and situated schematically as "islands" in the internal medullary lamina. In general, both the intralaminar and midline nuclei have, for many decades, been thought to only play a part in nonspecific arousal systems. However, it is now clearer that each nucleus within these systems is involved in more specific and separate cognitive, sensory, and motor functions. From a cognitive perspective, both the intralaminar and midline nuclei seem to have a fundamental role in awareness (Van der Werf et al., 2002). Both of these groups receive strong brainstem inputs. The ascendent reticular activating system (ARES), which is a rostral continuation of the reticular formation, continues upward to communicate with these groups. Following this input, the intralaminar and midline nuclei diffusely relay these signals to cortical centers to mediate vigilance, a necessary component of awareness. The intralaminar and midline nuclei, in fact, do not directly produce "awareness" but produce the necessary arousal of cortical and subcortical regions to induce the components necessary for awareness (Van der Werf et al. 2002). In addition, there has recently been an association with spinothalamic and spinoreticulothalamic projections in receiving tactile and nociceptive information as a component of awareness (Sturm et al., 1999). Tables 8.4 and 8.5, which have been adapted from the work of Van der Werf et al. (2002), provide more detail of each individual nucleus within the midline and intralaminar groups, including their respective inputs, outputs, and putative function.

The subcomponents of the intralaminar nuclei, and their respective inputs and outputs, are described in Table 8.5. Again, the intralaminar nuclei, all of which are located within the white matter confines of the internal medullary lamina, are comprised of a rostral group, consisting of the central medial, paracentral, and central lateral nuclei, and a caudal group, consisting of the parafascicular nuclei. Of note, the outputs of all nuclei of the intralaminar group are the same (Van der Werf et al., 2002).

Reticular Nuclei

The reticular nucleus (RN) is located at the periphery of the lateral thalamus, adjacent to the external medullary lamina, encircling each lateral nucleus; the internal capsule is situated lateral to the reticular nucleus. It takes the form of a sheet of GABAergic neurons, strategically located between the thalamic nuclei and the cortex, and has a modulator activity receiving glutamatergic inputs from both sides (Gentet and Ulrich, 2003; Lam and Sherman, 2015). The RN can also be anatomically organized, along the plane of the nucleus, into four sectors: dorsocaudal visual, ventrocaudal auditory, ventrocentral somatosensory, and rostral motor (Crabtree, 2018; Guillery and Harting, 2003; Sherman and Guillery, 2002).

The RN is unique in that it is the only nuclear group within the thalamus that does not send projections to the cortex; its inputs and outputs are essentially in a closed-loop reciprocal system within the thalamus itself. As such, the RN acts as a regulator of thalamic activity (Crabtree, 2018; Jones, 2012). It is implicated in two key behavioral states: sleep and arousal. These functionally distinct, but connected, subnetworks are, in turn, connected with thalamocortical first-order neurons in sensory (lateral geniculate) and motor (anterior dorsal) nuclei. Each neuron in the RN sector participates in a projection neuron-connected subnetwork, and each subnetwork is synchronously interconnected to lead to an electrically additive effect within the RN system (Lee et al., 2014). At the same time, it must also be noted that each individual RN neuron provides "divergent" inhibitory inputs on neighboring thalamic projection neurons (Copeland et al., 2012; Lam and Sherman, 2015).

There is evidence to suggest that there is an absence of glutamatergic (or excitatory) connections between projection neurons among different thalamic nuclei; as such, it is hypothesized that each of these cells may have their own private lines of communication to the cortex and/or striatum (Crabtree, 1996; Crabtree and Isaac, 2002; Kolmac and Mitrofanis, 1997). The RN, however, is thought to mediate the functional interaction between these projection neurons of different thalamic nuclei through "disynaptic" (connection between two neurons with an intermediate neuron) pathways. Through closed-loop disynaptic thalamocortical-RN-thalamocortical circuits, activation of one thalamic nucleus can inhibit and interrupt firing of a nearby, perhaps overlapping, thalamic nucleus. As such, these activation-inhibition signals have been implicated in controlling responses to sensory stimuli and reward, including attentional orienting, behavioral switching, and sensorimotor learning; many of these functions are in close communication with the various components of the intrathalamic nuclei(Crabtree, 2018; Guillery and Harting, 2003; Minamimoto and Kimura, 2002). Lastly, the RN is implicated in attentional processes that select sensory and motor information through "bottom-up," "top-down," and internuclear mechanisms. "Bottom-up" control refers to the strength of a stimulus while "top-down control" is governed by previous experience

TABLE 8.4 The midline thalamic nuclei.

Midline nuclei	Inputs	Outputs	Putative function(s)
Paraventricular nucleus (Chen and Su, 1990; Kirouac, 2015; Li et al., 2014; Otake and Nakamura, 1995)	- Tuberomammillary nucleus (histamine) - Ventral tegmental area, retrorubral region (dopamine) - Locus ceruleus and nucleus solitarius (noradrenaline) - Dorsal raphe nucleus (serotonin)	- Amygdala - Entorhinal cortex - Stria terminalis - Ventral striatum - Hypothalamus (all connections via ventral forebrain)	Arousal, attention, motivation, affective components of pain
Paratenial nucleus (Bobillier et al., 1979; Chen and Su, 1990; Cornwall and Phillipson, 1988; Newman and Ginsberg, 1992)	- Infralimbic cortex - Dorsal and median raphe nuclei - Central gray nucleus - Locus coeruleus - Parabrachial nucleus - Lateral dorsal tegmental nucleus - Nucleus solitarius	- Medial prefrontal areas - Prelimbic cortex - Medial limbic cortex	
Reuniens nucleus (Berendse and Groenewegen, 1990; Dolleman-Van Der Weel and Witter, 1996; Van der Werf et al., 2002; Witter et al., 1989)	- Infralimbic cortex - Prelimbic cortex - Perirhinal cortex - Cingulate cortex - Dorsal and ventral retrosplenial cortices - Diagonal band nucleus and bed nucleus of the stria terminalis (*basal forebrain*) - Reticular nucleus of the thalamus, lateral geniculate nucleus, medial and lateral preoptic area, medial and lateral hypothalamus, premammillary and supramammillary nuclei - Brainstem: Ventral tegmental area, dorsolateral tegmental nucleus, superior colliculus, central gray, raphe nucleus, parabrachial nucleus	- Inferior thalamic peduncle - Piriform cortex - Claustrum - Olfactory bundle - Rostral and caudal cortical regions	
Rhomboid nucleus (Berendse and Groenewegen, 1991; Groenewegen and Berendse, 1994)	- Raphe nucleus - Supramammillary bodies	- Prelimbic and infralimbic cortices - Ventral and dorsal anterior cingulate cortices - Primary and secondary motor - Primary and secondary somatosensory - Entorhinal cortex - Auditory, gustatory, and visual cortices	

and expectations, with the former being associated with salience and the latter with reward-associated relevance (internuclear control can be associated with a combination of either function) (Buschman and Kastner, 2015; Guillery and Harting, 2003; Lam and Sherman, 2015; Womelsdorf and Everling, 2015). Fig. 8.7 is a circuit representation highlighting the complexity of connections of the thalamic nuclei to and from their respective targets in the cerebrum.

Thalamic Connectivity

As mentioned in the discussion of the individual thalamic nuclei, the thalamus is a multifunctional, dynamic, and interconnected system of both first-order and higher-order relays, modulators, and drivers, with a greater role in subserving deeper and more complex functions of connectivity than previously thought. Crosson (2013), in his treatise on thalamic language

TABLE 8.5 The intralaminar thalamic nuclei.

Intralaminar nuclei	Inputs	Outputs	Putative function(s)
Paracentral nucleus	- Subcortical: Superior colliculus, supramammillary nucleus, reticular formation - Ventral and lateral periaqueductal gray matter - Cortical: Anterior cingulate, retrosplenial, parietal, supplemental motor, somatosensory, and auditory cortices	- Cingulum - Extreme capsule (cortical mantle) - Insular and perirhinal cortices	Arousal, attention, motivation, affective components of pain
Central lateral nucleus	- Brainstem: Mesencephalic, pontine, and medullary reticular formations - Median and dorsal raphe nuclei - Tegmental and pretectal regions - Medial and lateral supramammillary nuclei - Substantia nigra pars reticulata - Superior colliculus - Visual cortices (except primary and secondary visual) - Primary and secondary somatosensory cortices		
Parafascicular (caudal)	- Spinal cord - Medullary, pontine, mesencephalic reticular formations (locus ceruleus, dorsal raphe nucleus) - Brainstem nuclei (vestibular, parabrachial, tegmental, solitarius, ambiguus) - Reticular nucleus of the thalamus, substantia nigra pars reticulata, zona incerta, pretectum, supramammillary nuclei, pretectum		

connectivity, proposed four interthalamic mechanisms: (1) cortico-thalamo-cortical circuits: responsible for passing information from one cortical area to the another via the thalamus; (2) cortico-thalamic circuits: responsible for altering the mode of information transfer (inhibitory or excitatory) for thalamo-cortical relay neurons; (3) cortico-striato-pallido-thalamo-cortico circuits: responsible for enhancing the efficiency of selecting actions for execution; and 4) cortico-thalamic feedback mechanisms: responsible for sharpening the focus of specific information necessary for cognitive processes to properly function. It is now believed that many higher-order functions, apart from language, are regulated by these critical interthalamic mechanisms. Fig. 8.8 describes the thalamus, from a medial surface, in relation to several key tract fibers, including the corpus callosum, corona radiata, and thalamic radiations.

Cortico-Thalamo-Cortical Circuits

As Sherman and Guillery (2002, 2006) stated, the thalamus not only functions to provide first-order relay systems to the brain from the periphery but also acts within higher-order relay systems, mediating communication between one cortical area with another—a cortical hierarchy. Specifically, layer 5 cortico-thalamic projection neurons, normally categorized as 5A (intertelencephalic) and 5B (pyramidal tract) within the cortico-thalamic fiber system, make similar contacts upon thalamic relay neurons as are made by fibers of the optic tract, auditory tecto-thalamic projections, and somatosensory lemniscal pathways (Crosson, 2019). Large terminal processes synapse on these thalamic relay neurons and are in a strategic position to provide a strong influence on relay neuronal firing. As such, layer 5 cortico-thalamic connections are considered drivers of firing patterns. Layer 6 cortical projection neurons,

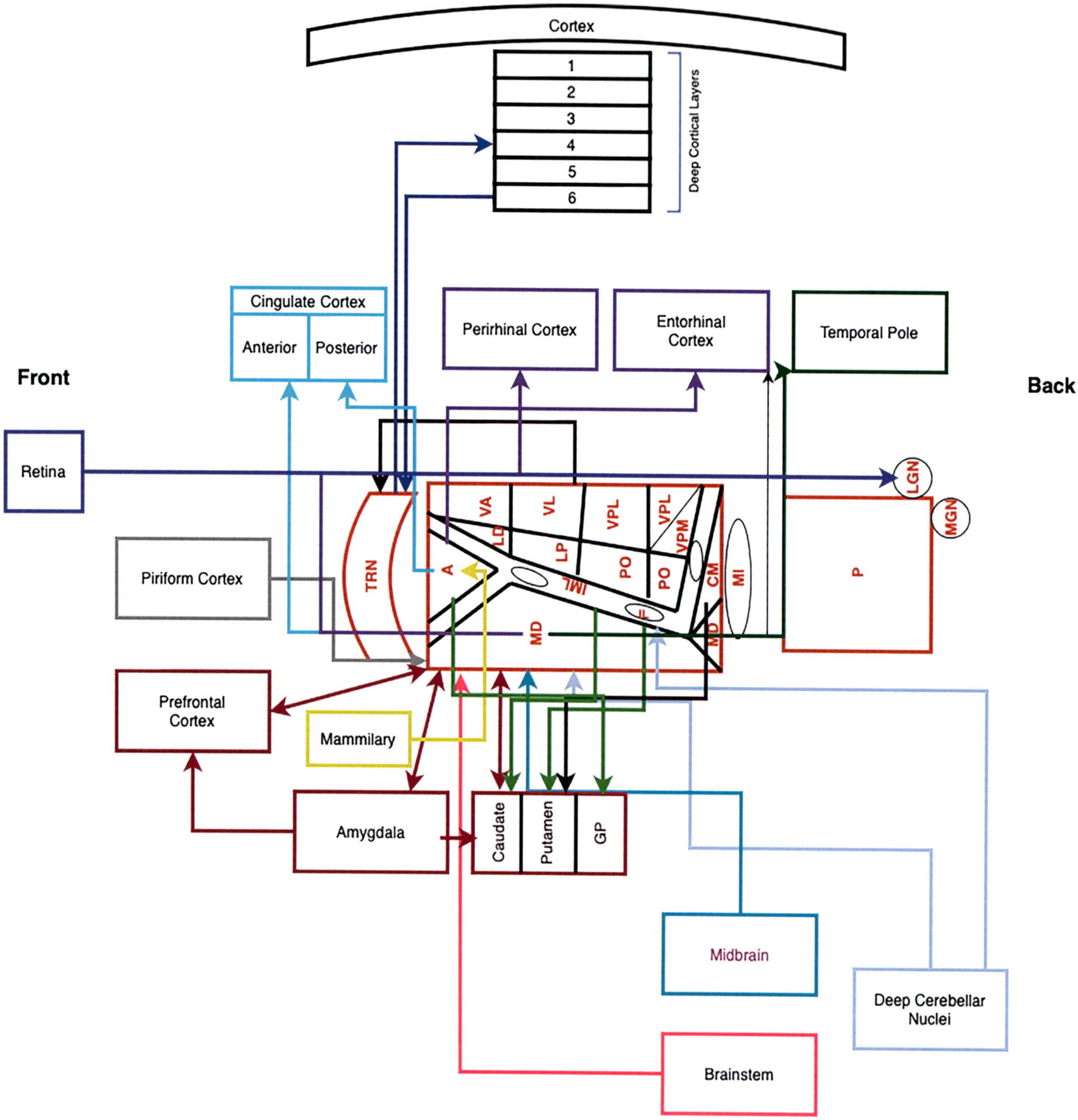

FIGURE 8.7 **The complexity of thalamic circuitry.** The thalamus is set within a dynamic, interconnected system of both first order and higher relays, modulators, and drivers. This schematic provides some of the afferent and efferent connections; the thalamus is topographically located in the center in relation to surrounding structures. Each afferent (input) and efferent (output) signal is color-coded to guide the reader.

contrarily, connect intracortically and function as modulators of relay firing. Within the language paradigm, these layered cortico-thalamo-cortico circuits may play a role in promoting rapidly adapting change in firing patterns. The layer 5 cortical projection neurons are responsible for the altering or modifying of neural code from lower-order impulses in the cortex to higher-order signaling. This is especially important in lexical-semantic processing or translating lexical to semantic code. In addition, information processed by the dorsal and lateral components of the medial geniculate nuclei add to this information by collaborating with primary and secondary auditory cortices, and integrating it with language processing and recognition, by way of the same layer 5 cortical projection neurons (Crosson, 2019; Kawaguchi, 2017). Other complex functions have also

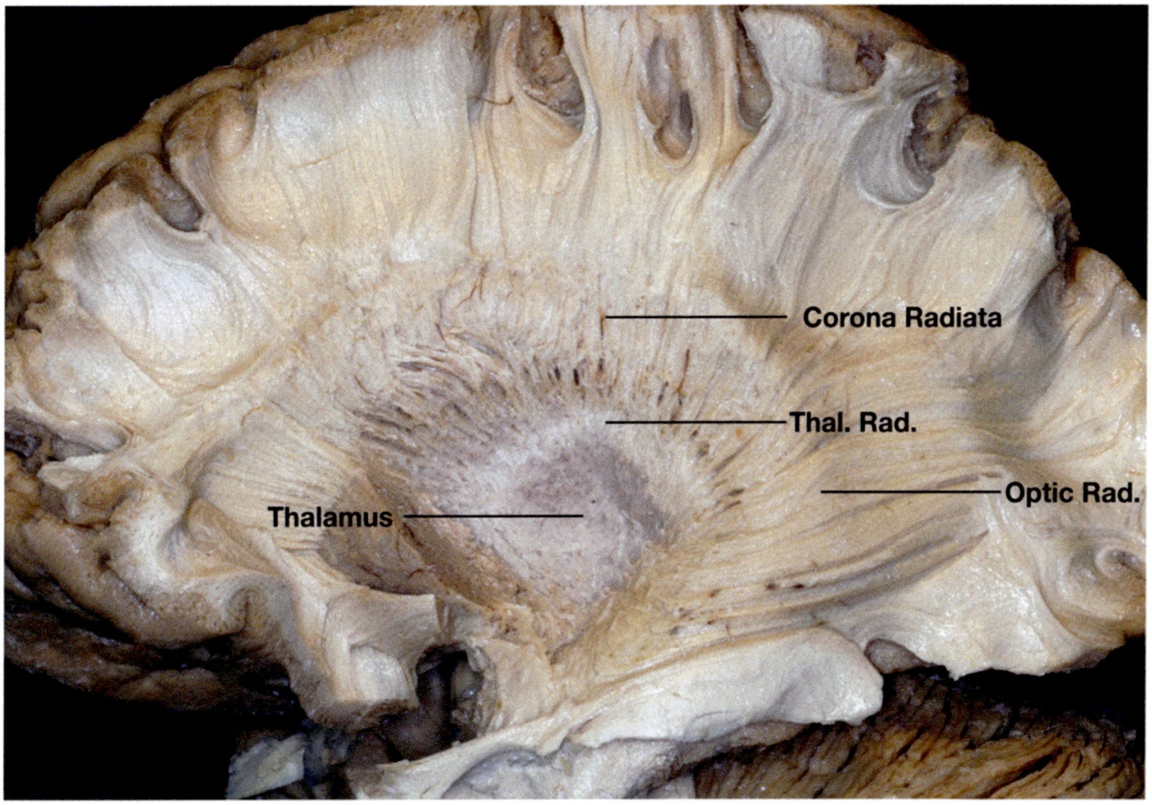

FIGURE 8.8 **Microsurgical anatomy of the thalamus.** *Optic Rad.*, optic radiations; *Thal. Rad.*, thalamic radiations.The sagittal (medial) surface of the thalamus is shown with several related and interconnected white matter tract fibers, including the corpus callosum, optic radiations, corona radiata, and thalamic radiations.

shown cortico-thalamo-cortical circuit modulation and higher-order cortical hierarchy. Through its communication with layer 5B, the somatosensory thalamus directly modulates control of the primary and secondary somatosensory cortices (Theyel et al., 2010).

Anterior Nuclei and Thalamic Radiations

An important aspect of the thalamic connectivity paradigm is the anterior thalamic nucleus and its radiations. Recent evidence has suggested strong hippocampal-anterior thalamic interconnections responsible for episodic (event-related) memory, which is recollective rather than familiar-based. There are numerous parallel temporal-diencephalic pathways which function in a reciprocal, closed-loop manner between the hippocampus and the anterior thalamic nucleus via the anterior thalamic radiations; these connections also extend to the mammillary bodies, retrosplenial cortex, and components of the prefrontal cortex (Aggleton et al., 2010). There exists an alternate indirect counter-opposition pathway from the hippocampal projections via the mammillary bodies, which regulates the aforementioned memory signals. In addition, the anterior medial thalamus participates in interconnections with the prelimbic cortex, anterior cingulate cortex, and retrosplenial cortex (both granular and dysgranular) with direct inputs from the medial mammillary bodies, tegmental nuclei, and subiculum/presubiculum. On the other hand, both the anterior dorsal and anterior ventral nuclei participate in interconnections with only the dysgranular and granular retrosplenial cortices via inputs from the medial and lateral mammillary bodies, tegmental nuclei, and postsubiculum/parasubiculum (Wolff et al., 2015; Wyss et al., 1979). In essence, as proposed by Aggleton et al. (2010), the anterior nuclear groups participate in the following critical functions as it pertains to memory: (1) *Anterior medial "feed-forward" system*: this system is primarily responsible for conveying hippocampal-diencephalic information to prefrontal mechanisms that function in controlling cognitive flexibility, executive function, and immediately occurring judgments; (2) *Anterior ventral "return-loop" system*: through diffuse direct projections *back* to subiculum, presubiculum, and parasubiculum, the anterior ventral cells have an important role in aiding in hippocampal synaptic plasticity; and (3) *Anterior dorsal "head-direction" system*: due to rhythmic theta firing of its cells (Vertes et al., 2001), activity here, along with that of the lateral mammillary bodies, is particularly heightened in the event of head turning, acting as a compass; other roles include mental navigation and imagery manipulation.

Functional Aspects

Relay Systems

As detailed in the previous section on individual thalamic nuclei, a majority of information that is sent to the cortex must first pass through the thalamus; these are located in a strategic position which enables them to process and transfer inputs and outputs within the entire "conscious perceptual" system. It was previously thought that the thalamus only provides a simple mechanical relay of information from the primary "information sources" such as the spinal cord, visual, and auditory systems, to the cerebral cortex. Newer research, such as that by Murray Sherman (2005), reveals that the thalamus actually plays three important functions in the relay process: (1) It provides an interchangeable relay to and from the cortex, adjusting the nature of the information in order to adequately meet certain behavioral demands; (2) It acts as a central link in various looped cortico-thalamo-cortical avenues of information processing; and (3) The thalamocortical relay mechanism carries information of continuously evolving motor instructions, and provides a copy to the cortex of these instructions. Critical relay systems such as sensory, motor, visual, emotion, hearing, and consciousness will be briefly outlined as it pertains to the thalamus.

Thalamic Circuit Properties

In general, the thalamic circuit, regardless of form and function, comprises of three basic cell types: relay cells, interneurons, and reticular cells. Relay cells are glutaminergic (stimulatory) and comprise a majority of the neurons of the thalamus. For example, these cells receive inputs from a primary information source ("driver input"), such as the spinal cord, and relays it to both layers 4 and 6 of the cortex; the lateral geniculate nucleus and ventral posterior nucleus are the primary driver inputs for the retina and the primary somatosensory cortex, respectively. End route, these signals pass through the thalamic reticular nucleus, composed of GABAergic interneurons, providing inhibitory input and thus modulating the relay cells (Sherman, 2005, 2017).

In an effort to control the various stimulatory and inhibitory inputs and outputs contained within every thalamic circuit, there are critical drivers and modulators which control the degree of signal traveling through these circuits. Again, an apt and highly studied example is the communication between retinal receptive field inputs and the visual layer 6 of the cortex that includes the lateral geniculate nucleus. The retinal afferent drivers and cortical input modulators work together to control signals sent to the lateral geniculate nucleus for processing. The drivers differ from modulators in that they predominantly activate ionotropic glutamate receptors (AMPA and NMDA), contain thick axons, and have large terminals on proximal dendrites. Modulators have activating ionotropic and metabotropic glutamate receptors, contain thin axons, and have small terminals on proximal dendrites (Sherman and Guillery, 1996; Sherman, 2017; Theyel et al., 2010). It is also important to note that for the thalamus, drivers are much less numerous than modulators, as in the example provided, where layer 6 inputs have many more synapses than the retinal (driver) inputs (Erisir et al., 1997; Sherman, 2005, 2017; Zhang and Bruno, 2019).

Sensory Relay

The sensory relay system of the thalamus is perhaps the most commonly known and robust of all relay circuits. There are two principal sensory systems that arise from the dorsal roots of the spinal cord. The *spinothalamic*, or ventral lateral, system is responsible for carrying pain, temperature, and two-point discrimination signals to the ventral posterior (VP) nucleus of the thalamus. As it relates to the relative position of innervating axons, this nucleus consists of two divisions: the ventral posterolateral (VPL) nucleus, which receives both spinothalamic and medial lemniscal axons; and the ventral posteromedial (VPM) nucleus, which receives trigeminothalamic axons. Within the VPL, there is somatotopic organization with the contralateral lower and upper limbs represented dorsolaterally and ventromedially, respectively; the opposite side of the head is represented in the VPM. Following its inputs from the spinothalamic axons, the VPL sends its output to the posterior limb of the internal capsule and corona radiata en route to the primary somatosensory area of the cortex. In addition, apart from the contribution of the VPL, the posterior and intralaminar groups of the medial dorsal nucleus also receive inputs from the spinal lemniscus (Al-Chalabi et al., 2020; Willis and Westlund, 1997).

The Motor Thalamus

The motor thalamus (MThal) represents an important and strategic location within the ventral thalamus and shares a strong interconnection with both motor cortices, as well as the cerebellum (dentate and interposed nuclei) and basal ganglia (substantia nigra pars reticulata and internal segment of the globus pallidus), to regulate both cognitive and proprioceptive control of movement. The current system shares many of the features of the "driver-modulator" concept, as proposed by Sherman and Guillery (1996, 1998, 2002) for the sensory thalamus. As described by Bosch-Bouju (2013), the MThal acts as a "super-integrator" of motor information from two functional driver-input subdivisionst: the basal ganglia and cerebral cortex and the cerebellum and cerebral cortex. The key thalamic nuclei involved in this circuit are the ventral anterior (VA), ventral medial (VM), ventral lateral (VL), and ventral lateral posterior (VLp). The VM, VA, and VL receive inputs from the basal ganglia and modulate signals to associative and premotor cortices, whereas VLp receives cerebellar inputs and modulate signals to the primary motor cortices. A vast majority of MThal neurons are glutamatergic (stimulatory) and project to pyramidal neurons of layers 1 and 2 as well as layers 5 and 6. The complex connective relationship of the MThal with its afferents is critical to regulate signals of motor preparation in order to efficiently generate fine motor commands in the motor cortex. In addition, the inputs from the deep cerebellar nuclei allow the thalamus to modulate direction and amplitude as a temporal pattern generator to optimize movement and motor learning assembly (Bosch-Bouju et al., 2013a; Cicirata et al., 1990; Krack et al., 2002; Sommer, 2003). Overall, the open feedback loops involving the basal ganglia, cerebellum, MThal, and cerebral cortex also ensure that the motivational and proprioceptive dimensions of movement are incorporated in the central motor processing unit of the MThal itself (Bosch-Bouju et al., 2013b).

Emotional Circuit and the Thalamus

As more is incrementally being understood about the complex nature of emotion, since its introduction by Papez in 1937 (Papez, 1937), it is now proposed that this critical function belongs to a larger-scale dynamic functionally integrated circuit involving subcortical and cortical structures. From a functional and anatomical perspective, emotion has been proposed to be part of a nonmodular cortical-basal ganglia-thalamo-cortical system which is interconnected, highly accessible, and contains "central-hubs" which act as processing centers. A "closed-loop" circuit exists wherein signals from the striatum pass through the pallidum to the thalamus and feed directly back to the cortex; the thalamus in essence provides a route back to the cortex or cortex-related regions (Papez, 1937). Similarly, the subcortical forebrain regions (i.e., extended amygdala, ventral striatum, and septum), which are caudally connected to structures along the brainstem that mediate autonomic responses, participate in a closed-loop system with the cortex via the thalamus. Ascending systems from the brainstem influence subcortical and cortical processing via a subcortical forebrain-thalamus-cortex system. To reiterate, the subcortical forebrain target regions receive extensive and diverse sensory, cognitive, emotional, motivational, and motor inputs, which project to two different output pathways: (1) a thalamocortical pathway and (2) a descending pathway within the brainstem. The latter descending projection system may project further caudally, loop back into targets within the brainstem, loop back to the subcortical forebrain, or diffusely connect back to subcortical and cortical regions (Parvizi, 2009; Pessoa, 2017). Both inhibitory and stimulatory neurotransmitters—including acetylcholine, dopamine, epinephrine, norepinephrine, and serotonin—are involved in this system.

The mechanism of emotion and its regulation generally follows the aforementioned circuitry, with the thalamus at its core. The significant degree of output and input distribution and integration among the cortico-thalamo-basal ganglia-subcortical-forebrain-brainstem system provides the critical means of intermixing information related to perception, cognition, emotion, motivation, and action. Within the thalamus, the midline nuclei (and, more specifically, the medial dorsal nucleus), plays an important role in this functionally integrated system with dynamic "heterarchical" inputs from the basal, lateral, and central amygdala; ventral striatum; and ventral pallidum. The outputs are to the orbitofrontal, insula, and medial prefrontal cortices (Grossberg, 2017; Pessoa, 2017). From a functional perspective, the cortico-thalamo-amygdala circuit does not strictly mediate a core function such as affect but, as proposed by Pessoa (2017, 2018), its functional state of being at a particular point in time may contribute to multiple mental operations, including arousal, vigilance, novelty, attention, value determination, and decision-making. One interesting example is the functional basis of fear of extinction. There is a top-down role of the medial prefrontal cortex in regulating the amygdala and influencing its signals. In addition, the ventral hippocampus, basolateral amygdala, and medial dorsal thalamus project to multiple areas of the amygdala. Generally speaking, during fear expression and extinction, signals from the basolateral amygdala and medial dorsal thalamus are integrated into the medial prefrontal cortex with signals from multiple sources, including the hippocampus and orbitofrontal cortex, and collectively determine whether to produce a response (Sotres-Bayon and Quirk, 2010; Tovote et al., 2015).

The Language Circuit and the Thalamus

A growing body of knowledge has established the important role of higher-order cortico-thalamo-cortical relays in mediating language function. Specifically, and as proposed by Sherman and colleagues (Sherman, 2005; Sherman and Guillery, 1996, 2002; Sherman, 2017), the terminal processes of layer 5 cortico-thalamic connections (drivers) receive information from the periphery via the thalamus; the information that is processed conveys relevant information to nearby regions in a processing chain (Crosson, 2019).

With respect to language, it had been asserted that the thalamus acts as the cerebrum's "rhythm" and "pulse" generator in information processing; instead, it is now hypothesized that cortical rhythms are generated by both cortico-thalamic and thalamo-cortical responses (Sherman and Guillery, 2002; Steriade et al., 1996; Usrey and Sherman, 2019). Using electrocorticography, Edwards et al. (2010) revealed that language processing was accompanied by large, rapid, temporal increases in amplitude of high gamma rhythms (70–160 Hz) with a normal stimulation frequency being approximately 60 Hz. For example, in their experiments, an auditory stimulus (i.e., noun) for a verb generation task was accompanied by high gamma rhythms in Wernicke's area and left precentral gyrus. As the task was switched to word retrieval, similar high gamma rhythms were now seen in Broca's area (left pars operculis and pars triangularis). In addition, word retrieval, picture naming, and verb generation have all been found to share the same degree of rhythm firing within the cortico-thalamo-cortical circuit. Adding to this information, language information processing seems to serially activate from lower-order systems to higher-order systems, allowing words in a sentence to be successfully and incrementally decoded. This system is especially important in error processing and refining words that are processed and generated; for example, the similarity between the words "knight" and "knife" must be understood and refined, and this is accomplished by feed-forward top-down cortico-thalamo-cortical and thalamic relay circuitries (Bastos et al., 2015; Crosson, 2013, 20195). Finally, with respect to word finding and processing in the lexical-semantic interface, the ventral anterior thalamus plays an integral role by relaying higher-order information from the cortex and sequentially activating lexical, phonemic, and motor planning feed-forward mechanisms. For example, this dynamic mechanism is important in assessing the lexical and semantic differences between "knife," "fork," and "spoon" as it pertains to eating. A thalamic lesion, producing aphasia, would deem the activation of all three words to be equal rather than activating an "error" signal that would differentiate the three words.

Vision and the Thalamus

The lateral geniculate nucleus (LGN), which is a part of the lateral nuclear group, is considered the primary thalamic relay center responsible for visual processing, connecting retinal input from both eyes to the primary visual cortex (V1 or striate cortex, output). More specifically, the LGN has been implicated in mediating binocular processing, in that excitatory retinal input from one eye can be controlled or suppressed by high-contrast visual stimulation of the other eye. In essence, two critical visual functions for binocular arrangement, stereoscopy and hyperacuity, require that inputs from both eyes must be resolved into a singular unified vision, with one eye's excitatory input being suppressed (Dougherty et al., 2021). As with most thalamic nuclei, the LGN contains multiple distinct layers, all receiving multiple direct excitatory inputs from one eye; but evidence suggests that there is congruence among layers to allow for merging of inputs from both eyes to eventually create binocular visual output; this could be due to the constant change in magnitude of visual responses (Dougherty et al., 2019; Erisir et al., 1997; Kaas et al., 1972; Weyand, 2016).

Although the LGN is responsible for receiving and relaying sensory visual inputs, visual perception and visual control of eye movements may rely on higher-order thalamic nuclei. Thalamic pathways originating from different layers of the superior colliculus have been shown to play an important role in controlling and interpreting rapid or saccadic eye movements, termed "active" vision. Such pathways include the following: (1) intermediate layer of the superior colliculus to the parvocellular region of the medial dorsal (MD) nucleus and then to the frontal eye field of the frontal cortex, (2) superficial visual layers of the superior colliculus to the inferior pulvinar nucleus and then to occipital and parietal cortices, (3) intermediate layer of the superior colliculus to the lateral pulvinar and then to the occipital and parietal cortices, and (4) the thalamic reticular nucleus, which passes information through the LGN on its way to the cortex, but does not connect to the cortex directly (Wurtz et al., 2011). Apart from the principal function of the MD nuclei mediating the superior colliculus to frontal cortex pathway in guiding saccadic eye movements, "stable" vision has been found to be mediated via the same MD nuclei by way of the frontal eye fields and reticular formation (Duhamel et al., 1992; Wurtz et al., 2011).

The pulvinar, the largest nucleus of the thalamus, has several subnuclei which also play a role in visual control. Specifically, the ventral visual stream areas of V1, V2, V4, and the inferior temporal (IT) cortex carry outputs from the ventral lateral portions of the pulvinar (Zhou et al., 2016). In addition, cells in layers 5 and 6 of the cortex project topographically to the pulvinar which projects back to the superficial segments of these layers, creating a cortex-thalamus-cortex loop circuit (Sherman and Guillery, 2002). This pathway plays a role in control of attention and relay of sensory information from one area to another. The pulvinar plays an important role in visuospatial attention and control, where damage to this loop circuit can lead to deficits in these functions.

Memory, Cognition, and the Thalamus

Emerging data suggests that the thalamus can play an active role in mediating and processing cognition and memory. Historically, it had been thought that learning and memory were only functions of the hippocampal formations and medial temporal lobe, as evidenced by studies of temporal and diencephalic amnesia; the thalamus was considered only as a passive relay for these areas. The thalamic circuitry models proposed by Sherman and Guillery (1996), however, show that the thalamus plays an active role in cortical function by regulating active participation in driver inputs (i.e., inputs capable of directly eliciting neuronal activity and controlling higher-order relays). In addition, several thalamic nuclei have been proposed to play a limbic role, in connection with the cingulate cortex, to regulate cognition apart from its sensory-motor processing function. Therefore, thalamic nuclei may contribute to cortical functioning and higher-order cognition, playing a role in learning, memory, and flexible adaptation (Wolff and Vann, 2019).

With respect to memory, both the anterior (AT) and medial dorsal (MD) nuclei have been shown to play a role in its regulation. A well-known example of its involvement is Korsakoff syndrome, where impairments of these nuclei result in such symptoms as anterograde (and less commonly, retrograde) amnesia and confabulation. Lesions of the AT nuclei appear to play a role in disrupting the processing of environmental cues as it pertains to memory inputs, as well as monitoring position within the environment (Cassel and de Vasconcelos, 2015; Wolff and Vann, 2019). In addition, AT nuclei-related impairments have been related to the inability to recognize and differentiate between multiple items possessing similar characteristics (Aggleton et al., 2010; Nelson and Vann, 2014; Wolff et al., 2006; Wolff and Vann, 2019). Regarding MD nuclei, it was originally proposed that it plays a role in regulating working memory in conjunction with aspects of executive function (Funahashi, 2013; Halassa and Kastner, 2017; Parnaudeau et al., 2018); rather, it is now being proposed that the MD nuclei may only be involved in task performance, such as delay monitoring or habit formation (Wolff and Vann, 2019). Taken together, both nuclei are thought to play a role in aspects of mental adaptation to external cues and directing attention to task-relevant behavioral features, important for building mental constructs (Wolff and Vann, 2019; Wright et al., 2015). In addition, they appear to contribute an important role in updating "action-outcome" and "stimulus-outcome" associations. Contrary to the previous historical assertion of the cortex being sole contributors to memory adaptation, the thalamus is now considered to play a role in monitoring, maintaining, and updating mental constructs (Wilson et al., 2010; Wolff and Vann, 2019).

As it pertains to memory and cognition, corticothalamic and thalamocortical connections represent reciprocal and parallel circuits involved in adapting cognitive feedback loops in response to environmental cues. The corticothalamic connections, in particular, seem to adapt to dynamic excitatory-inhibitory demands depending on the current behavioral situation, leading to regulation of attention. The reticular nucleus seems also to provide collaterals in this specific network (Guo et al., 2017; Mease et al., 2014). With respect to thalamocortical connections, both MD and AT nuclei appear to provide an active modulating contribution to memory and cognition, rather than serving just as a passive relay. Specifically, the MD nuclei modulate odor-guided behavioral information en route to the piriform cortex (primary olfactory cortex) (Wolff and Vann, 2019). Similarly, the AT nuclei integrate information from the midbrain, diencephalic, hippocampal, and cortical regions (i.e., retrosplenial cortex) (Mathiasen et al., 2017; Vann and Nelson, 2015).

Vascular Supply of the Thalamus

Thalamic blood supply can generally be divided into four territories, each supplying particular groups of nuclei: tuberothalamic, inferolateral, paramedian, posterior choroidal vessels. These territories and their respective branches can vary among individuals. In general, blood supply to the thalamus derives from branches of posterior communicating (PComm) and posterior cerebral arteries (PCA) (Bogousslavsky et al., 1988; Schmahmann, 2003). Briefly, to orient the reader, the inferior, middle, and superior rami (collectively referred to as the paramedian branches) arise from the P1 segment of the PCA, which lies between the PComm and the superior cerebellar artery (SCA). The tuberothalamic artery arises from the middle third of the PComm. The medial posterior choroidal arteries, which generally have either one or two tributaries, arise from the P2 segment of the PCA, which is situated between the lateral posterior choroidal arteries and the PComm. Finally, the lateral posterior choroidal arteries arise from the P3 segment of the PCA. Table 8.6 describes the four main thalamic vascular (arterial) territories and the respective nuclei that are supplied (adapted from Bogousslavsky et al., 1988; Schmahmann, 2003). With respect to the venous system, the thalamostriate, anterior thalamic, superficial thalamic, and superior thalamic veins enter the internal cerebral vein to drain into the basal vein of Rosenthal. The posterior thalamic vein, along with inferior thalamostriate vein, drains directly into the basal vein of Rosenthal (Herrero et al., 2002; Jones, 2012). Fig. 8.9 illustrates the superior, medial, and lateral surfaces of the thalamus and its respective arterial and venous supply.

TABLE 8.6 Territories of thalamic vascular supply.

Thalamic territory	Nuclei supplied
Tuberothalamic artery	Reticular, intralaminar, VA, rostral VL, ventral pole of MD, anterior nuclei (AD, AM, AV), ventral internal medullary lamina, ventral amygdalofugal pathway, mamillothalamic tract
Paramedian artery	MD, intralaminar (CM, Pf, CL), posteromedial VL, ventromedial pulvinar, paraventricular, LD, dorsal internal medullary lamina
Inferolateral artery	Same as paramedian
Principal inferolateral branches	Ventral posterior complex: VPM, VPL, VPI Ventral lateral nucleus, ventral (motor) part
Medial branches	Medial geniculate
Inferolateral pulvinar branches	Rostral and lateral pulvinar, LD nucleus
Posterior choroidal artery Lateral branches Medial branches	LGN, LD, LP, inferolateral parts of pulvinar MGN, posterior parts of CM and CL, pulvinar

Pathologies

As the thalamus is the central passive and active relay and processing hub of the cerebrum, damage to any particular nucleus can significantly impact critical aspects of function (i.e., sensory, motor, and cognitive domains). In an effort to characterize the neurological effects of direct thalamic insults, Fisher (1959) was the first to study the effects of thalamic hemorrhage. He reported sensory deficits such as neglect ("modified anosognosia and hemiasomatognosia"), global dysphasia, confusion, delirium, visual hallucinations, and peduncular hallucinosis (vivid and colorful visions of people and animals). It is worth noting that hemorrhage is rarely confined to the thalamus itself, and symptoms can often be attributed to focal pressure from the hemorrhage mass, surrounding edema affecting adjacent structures, and toxic blood products. Similarly, infarcts have provided useful associations with behavioral manifestations. With respect to thalamic tumors in both adults and children, isolated lesions are rare and often involve adjacent, functionally similar structures (i.e., hypothalamus and mammillary bodies). Therefore, from a clinical standpoint, aligning concrete structure-function correlation with solitary thalamic tumors can be extremely difficult, as the diversity of networks is very complex (Bilginer et al., 2014; Cheek and Taveras, 1966; Nass et al., 2000; Squire et al., 1989; Ziegler et al., 1977). This section will briefly describe several key thalamic pathologies and is not meant to be exhaustive but rather meant to highlight the correlation of structure and function.

Thalamic Infarcts and Hemorrhages

Thalamic infarcts and small-sized hemorrhages (<2 cm in size) can be organized into four categories, based on the nuclear groups affected and the corresponding symptoms (see Table 8.7): (1) Lateral thalamic infarcts and posterolateral hemorrhages often present with severe motor and sensory impairment; (2) Tuberothalamic artery infarcts and anterolateral hemorrhages present with frontal-type neuropsychiatric symptoms (e.g., emotional lability and anger) and mild hemiparesis and hemihyperesthesia; (3) Paramedian artery infarcts and medial hemorrhages often present with amnesia, abulia, vertical- and horizontal-gaze abnormalities, and decreased level of consciousness; and (4) Posterior choroidal artery infarcts and dorsal hemorrhages can present with aphasia, apraxia, amnesia, transient hemiparesis, and hemihyperesthesia (Amici, 2012; Bogousslavsky et al., 1988; Schmahmann, 2003). Table 8.7 more specifically outlines the deficits associated with branches within these four territories (adapted from Schmahmann, 2003). Fig. 8.9 summarizes the vascular supply of the thalamus.

The extent and severity of symptoms of thalamic hemorrhages depends on its size and if small (less than 2 cm in size), mirrors that of the aforementioned vascular territory infarcts. Larger thalamic hemorrhages (greater than 2 cm) can present with overlapping clinical syndromes depending upon the nuclei involved (Amici, 2012; Bogousslavsky et al., 1988). In general, thalamic hemorrhages are rapid in onset and can present with unpredictable and fluctuating impairments in consciousness; fortunately, they present with a relatively good prognosis as compared to hemorrhages of the basal ganglia and pons (Kawahara et al., 1986; Kumral et al., 1995). Large thalamic hemorrhages tend to present with rapidly progressive motor deficits, sensory loss, vertical-gaze, and pupillary abnormalities; contrary to infarcts, hemorrhages cause more pronounced motor deficits than sensory abnormalities (Amici, 2012). In addition, contralateral limb ataxic hemiparesis can also occur. With respect to cognitive deficits, comprehension and verbal repetition are normally preserved, with predominant deficits in semantic and fluency aphasia, such as dysnomia and paraphasic errors (Amici, 2012; Kirshner and Kistler, 1982; Kumral et al., 1995).

Tumors

Solitary thalamic tumors are rare, comprising only 1%–5% off all brain tumors. They can occur in all ages, but do have a greater predilection in children

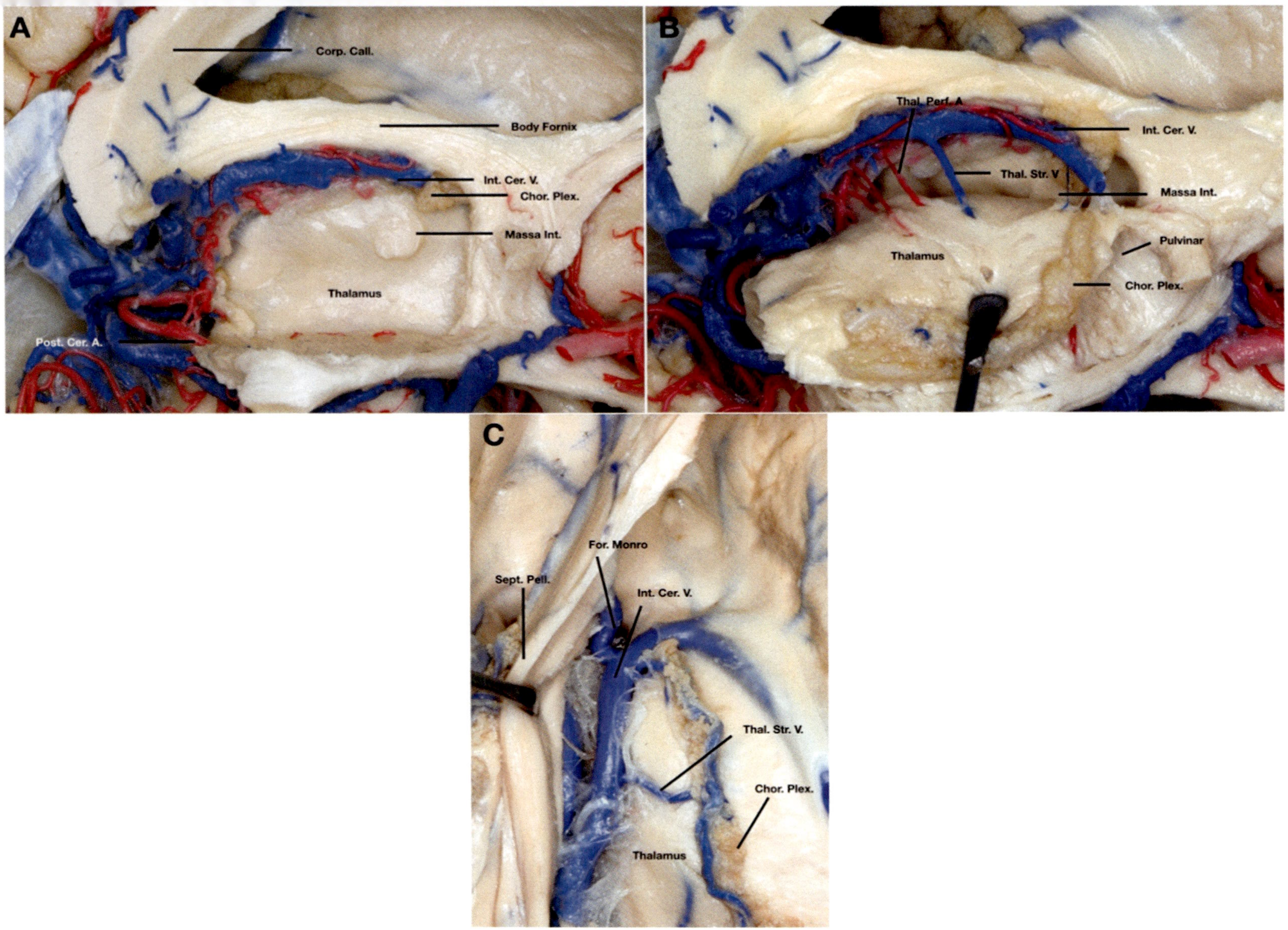

FIGURE 8.9 **Vascular supply of the thalamus.** *Chor. Plex.*, choroid plexus; *Corp. Call.*, corpus callosum; *For. Monro*, foramen of Monro; *Int. Cer. V.*, internal cerebral vein; *Massa Int.*, massa intermedia; *Post. Cer. A.*, posterior cerebral artery; *Sept. Pell.*, septum pellucidum; *Thal. Str. V.*, thalamostriate vein.Medial (A), lateral (B), and superior (C) surfaces of the thalamus are shown. Arterial supply to the thalamus can generally be divided into four territories, each supplying particular groups of nuclei: tuberothalamic, inferolateral, paramedian, posterior choroidal vessels. These territories and their respective branches can vary among individuals. In general, blood supply comes from branches of posterior communicating (PComm) and posterior cerebral arteries (PCA). The inferior, middle, and superior rami (collectively referred to as the paramedian branches) arise from the P1 segment of the PCA, which lies in between PComm and the superior cerebellar artery (SCA). The tuberothalamic artery arises from the middle third of the PComm. The medial posterior choroidal arteries, which generally have either one or two tributaries, arises from the P2 segment of the PCA, which is situated between the lateral posterior choroidal arteries and the PComm. Finally, the lateral posterior choroidal arteries arise from the P3 segment of the PCA. The thalamostriate, anterior thalamic, superficial thalamic, and superior thalamic veins enter the internal cerebral vein to drain into the basal vein of Rosenthal. The posterior thalamic vein, along with inferior thalamostriate vein, drains directly into the basal vein of Rosenthal.

TABLE 8.7 Vascular territories of the thalamus with their branches and corresponding deficits.

Thalamic territory	Associated branches	Deficit
Tuberothalamic artery	Polar artery Premamillary branch of PComm Anterior internal optic artery, inferior thalamic peduncle Anterior thalamic perforating artery Anterior thalamosubthalamic perforating artery	- Fluctuating arousal and attention - Impaired learning, memory, autobiographical memory - Desynchronization of temporal-related information - Personality changes, apathy, abulia - Executive dysfunction, perseveration - Left hemisphere: Language (VL nuclei); right hemisphere: Hemineglect - Acalculia, apraxia, emotional facial expression
Paramedian artery	Retromamillary pedicle Posterior thalamosubthalamic artery Interpeduncular profundus artery Superior ramus of interpeduncular artery	- Impaired arousal - Impaired learning and memory, confabulation, temporal disorientation, autobiographical memory impairment - Same as posterior thalamosubthalamic artery - Left hemisphere: Aphasia - Right hemisphere: Spatial misrepresentation
Inferolateral artery	Thalamogeniculate pedicle Inferolateral branches Medial branches Inferolateral pulvinar branches	- Impaired social skills and personality changes (apathy, aggression, agitation (frontal lobe-related changes)) - Sensory loss, hemiataxia, hemiparesis, postlesion pain (thalamic pain, Dejerine–Roussy) syndrome, right hemisphere - Auditory impairment - Behavioral changes
Posterior choroidal artery	Lateral branches Medial branches	- Hemianopsia, hemiquadrantanopsia - Sensory loss, weakness, aphasia, memory impairment, dystonia, hand tremor

than adults. Thalamic gliomas, which are the most common subtypes of the thalamic tumors, are clinically and anatomically distinct entities. This is mainly due to their unique morphological characteristics, such as segmental growth patterns, circumscribed borders, and expansive compression rather than infiltrative (Ozek and Türe, 2002). In addition, pilocytic astrocytomas, the second most common thalamic tumor, resemble cerebellar astrocytomas in both histological and growth characteristics and are rarely infiltrative (Kelly, 1989). Tumors most often arise along the superior-anterior regions of the thalamus (i.e., anterior nucleus, ventral nucleus, and central portion) (Cheek and Taveras, 1966; Ozek and Türe, 2002). Gazi Yasargil described three modes of extension of thalamic tumors. The first group arise and remain within particular nuclei; they expand locally and compress surrounding structures, such as adjacent nuclei, internal and external capsules, and ventricular wall if medially displaced. The second group expand beyond the confines of the thalamus and compress the surrounding white matter fibers; laterally and superiorly, they may expand into the subpallial white matter of an adjacent lobe or gyrus. Finally, the third group expands into the lateral ventricular cavity without penetrating the ependyma or internal glial membrane (Ozek and Türe, 2002; Yasargil, 1999).

Historically, the outcome of thalamic tumors has been poor; this was mainly due to the deep-seated location of these lesions, its close proximity to other vital structures (i.e., basal ganglia, subthalamus), and relative difficulty of the surgical approach. However, recent advances in surgical technique and other therapeutic approaches have allowed for more accurate and complete resection of these lesions with a lesser degree of morbidity and mortality. MRI and DTI-based imaging have greatly aided in the ability to categorize and assess the extent of thalamic tumors and, consequently, dictate the appropriate surgical approach. It is worth noting that, from an anatomosurgical point of view, the posterior, medial and superior surfaces of the thalamus are considered free surfaces that are more comfortably away from vital structures and more approachable. The ventral lateral border, on the other hand, is intimately associated with the internal capsule and the subthalamus requiring greater caution. This anatomical knowledge as well as the clinical presentation, may be useful in dictating whether to surgically treat or not, which approach to choose, and whether subtotal resection may be more favorable, with or without adjuvant radiotherapy.

A variety of surgical approaches have been developed over the years based on the relative position and extension of the tumor in question. Table 8.8 outlines several commonly used ways to approach the corresponding segments of the thalamus (adapted from Ozek and Türe, 2002). The surgical approach remains challenging, due to the deep location, surrounding white matter tract fibers, proximity to basal ganglia and internal capsule, and complex vascular blood supply. The general consensus for radiologically discrete low-grade tumors and astrocytomas with no peripheral infiltration is for complete resection; biopsy is recommended for infiltrative tumors in order to minimize morbidity and better dictate adjuvant treatment plans. Radical resection of high-grade gliomas has not shown an enhanced benefit in survival. In either case, histopathological analysis of tumor tissue can be useful in better understanding tumor type and behavior (Yasargil, 1998, 1999).

Deep Brain Stimulation of the Thalamus: General Principles and Applications

The utilization of deep brain stimulation (DBS) of the thalamus has shown benefit for a number of chronic neurological and neuropsychiatric conditions; this includes, but is not limited to, essential, Parkinsonian and dystonic tremor, neuropathic and thalamic pain (also known as Dejerine–Roussy syndrome), refractory epilepsy (drug-resistant), and Tourette's syndrome (Nazzaro et al., 2013; Torres et al., 2010; Whiting et al., 2018). The ventral lateral (or ventral intermediate, VIM) nucleus is considered a highly effective target for the treatment of tremor (Fig. 8.10). As described in the previous section on lateral nuclei, the VIM nucleus receives direct input from the globus pallidus pars internus, deep cerebellar nuclei, and substantia nigra pars reticulata, while sending output signals to the motor,

TABLE 8.8 Selected surgical approaches to specific areas of the thalamus.

Surgical approach	Area of thalamus	Advantage of approach	Disadvantages of approach
Anterior transcortical transventricular (middle frontal gyrus)	Anterior and superior thalamus	Easy access, adequate orientation to ventral lateral structures, thalamostriate vein preservation, less injury to pericallosal arteries and draining veins	Risk of postoperative seizures, possible disruption of cortex and white matter, difficulty in maintaining retraction
Transcortical-transtemporal	Ventral posterior thalamus	Lower risk of vessel injury	
Anterior interhemispheric transcallosal	Superior-anterior thalamus	Tumor extension into lateral ventricle, less disruption of cortex, no cortical incision, no need for anticonvulsants	
Posterior interhemispheric parasplenial	Posterior thalamic-pulvinar	Larger exposure of parasplenial, pineal, and pulvinar thalami	
Occipital transventricular	Pulvinar		
Parieto-occipital transventricular	Pulvinar, posterior thalamus		
Infratentorial supracerebellar	Pulvinar	Extra-axial, less invasive	Less visualization when entering between basal veins of Rosenthal; only for small tumors
Pterional transsylvian-transinsular	Ventral posterior	Tumor in close proximity to insula	

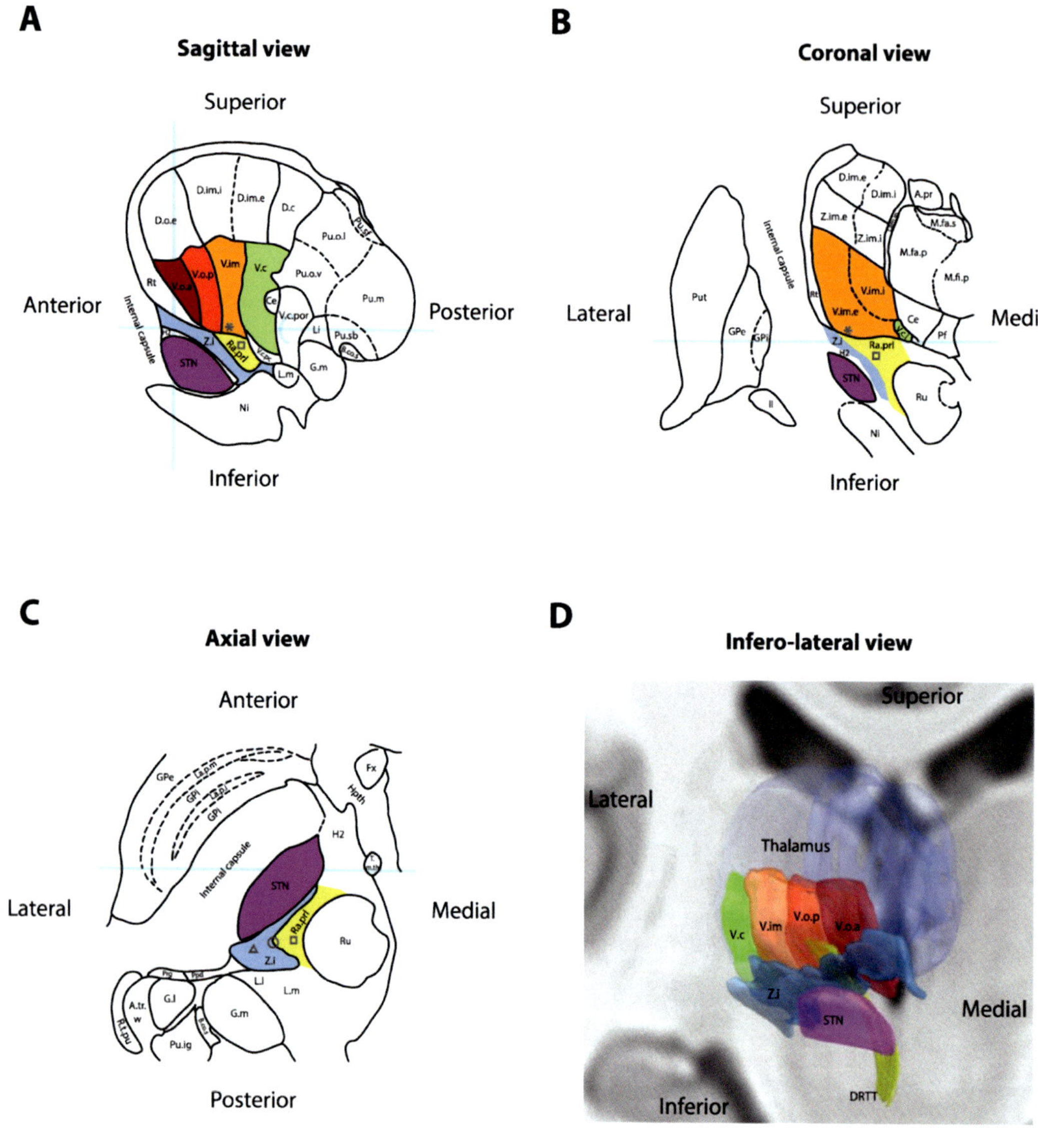

FIGURE 8.10 **Anatomy of the ventral thalamic area: target region for DBS.** *, standard Vim target for deep brain stimulation (DBS); *A.pr*, nucleus anteroprincipalis thalami; *A.tr.w*, area triangularis (Wernicke); *B.co.s*, brachium colliculi superioris; *Ce*, nucleus centralis; *circle*, intermediate PSA target at the interface between Zi and Raprl; *D.c*, nucleus dorso-caudalis; *D.im.e*, nucleus dorso-intermedius externus; *D.im.i*, nucleus dorso-intermedius internus; *D.o.e*, nucleus dorso-oralis externus; *DRTT*, dentato-rubro-thalamic tract; *Fx*, fornix; *G.l.*, corpus geniculatum laterale; *G.m*, corpus geniculatum mediale; *GPe*, globus pallidus externus; *GPi*, globus pallidus internus; *H2*, fasciculus lenticularis; *Hpth*, hypothalamus; *II*, optic tract; *L.l*, lemniscus lateralis; *L.m*, lemniscus medialis; *La.m.ip*, lamina medialis interpolaris; *La.p.i*, lamina pallidi incompleta; *La.p.m*, lamina pallidi medialis; *Li*, nucleus limitans thalami; *M.fa.p*, nucleus medialis fasciculosus posterior; *M.fa.s*, nucleus medialis fasciculosus superior; *M.fi.p*, nucleus medialis fibrosus posterior; *Ni*, substantia nigra; *Pf*, nucleus parafascicularis thalami; *Ppd*, nucleus peripeduncularis; *Prg*, praegeniculatum; *Pu.ig*, nucleus pulvinaris intergeniculatus; *Pu.m*, pulvinar mediale; *Pu.o.l*, nucleus pulvinaris orolateralis; *Pu.o.v*, nucleus pulvinaris oroventralis; *Pu.sb*, nucleus pulvinaris suprabrachialis; *Pu.sf*, nucleus pulvinaris superficialis; *Put*, putamen; *Rt.pu*, right putamen; *Ra.prl*, radiatio praelemniscalis; *Rt*, reticulatum thalami; *Ru*, red nucleus; *square*, standard prelemniscal radiations (Raprl) target for DBS; *STN*, subthalamic nucleus; *T.m.th*, tractus mammillo-thalamicus; *triangle*, standard caudal Zi target for DBS; *V.c*, nucleus ventrocaudalis; *V.c.i*, nucleus ventrocaudalis internus; *V.c.pc*, nucleus ventrocaudalis parvocellularis; *V.c.por*, nucleus ventrocaudalis portae; *V.im*, nucleus ventrointermedius; *V.im.e*, nucleus ventrointermedius externus; *V.im.i*, nucleus ventrointermedius internus; *V.o.a*, nucleus ventrooralis anterior; *V.o.p*, nucleus ventrooralis posterior; *Z.i*, zona incerta; *Z.im.e*, nucleus zentrolateralis intermedius externus; *Z.im.i*, nucleus zentrolateralis intermedius internus.(A) Sagittal cross-section of the thalamus, 14.5 mm lateral to the midline. Light-blue overlays represent the intercommissural plane (horizontally), the midcommissural plane (vertically) and the posterior commissure (half-circle). (B) Coronal cross-section 5 mm posterior to the mid-commissural point. Light-blue overlay represents the intercommissural plane. (C) Axial cross-section 3.5 mm inferior to the intercommissural plane. Light-blue overlay represents the midcommissural plane. Panels A, B, and C redrawn and adapted from Schaltenbrand and Wahren (1977). (D) Three-dimensional reconstruction of the right thalamus seen from an antero-infero-lateral oblique view. Nuclei rendered in Lead-DBS v2 using the DISTAL atlas (ventralis caudalis (Vc), ventral intermediate nucleus of the thalamus (Vim), pallidal receiving area (Vop), Voa and subthalamic nucleus (STN)), zona incerta atlas (Zi), and brainstem connectome atlas (DRTT). *Courtesy from Iorio-Morin, C., Fomenko, A., Kalia, S.K., 2020. Deep-brain stimulation for essential tremor and other tremor syndromes: a narrative review of current targets and clinical outcomes. Brain Sci. 10(12), 925. https://doi.org/10.3390/brainsci10120925.*

premotor, and supplementary motor cortices. Thus, the VIM is considered the "cerebellar nucleus," responsible for relaying basal ganglia and cerebellar inputs to the cortex. It is therefore evident that the VIM can be a particularly important target for controlling severe and chronic movement-related disorders.

In an effort to provide adequate and reversible tremor suppression, Benabid et al. (1987) proposed the implementation of permanent high-frequency electrical stimulation as an adjuvant therapy to contralateral thalamotomy. This method was already being used for the treatment of intractable pain. Prior to this study, bilateral thalamotomies were often performed with the use of acute electrical stimulation as a method of localizing the most appropriate ablative target. The updated method by Benabid essentially avoided the deleterious complications of a bilateral thalamotomy.

Essential tremor (ET) is a symptomatically heterogenous neurodegenerative disorder, which seems to be initiated within the cerebellum and expressed through the corresponding cerebellar relay tracts and VIM. It is characterized by an 8–12 Hz postural and kinetic tremor that is largely bilateral, symmetric, visible, persistent, and involving the hands and forearms. Additional neurological symptoms, such as limb and gait ataxia, eye motion abnormalities, and neuropsychiatric symptoms (e.g., sleep disturbances, pain, dementia, frontal lobe deficits, anxiety, and depression) can also be associated (Lee and Kondziolka, 2005; Whiting et al., 2018). Approximately 50% of patients with ET are refractory to medical therapy. For these patients, DBS is currently a more effective treatment than thalamotomy. The advantages of DBS over thalamotomy are its reversibility, ability to adjust stimulation settings, and the ability to implant the electrodes bilaterally if required. In addition, Yu and Neimat (2008) reported a decreased risk of cognitive deficits compared to thalamotomy. Otherwise, the recent application of focused ultrasound treatment of the VIM has become an alternative for some patients.

The optimal electrode placement for DBS placement for ET is along the anterior margin of the VIM nucleus—the electrode can be placed at the level of the anterior/posterior commissure lines, 6.3 mm anterior to the posterior commissure, and 10 mm lateral to the lateral wall of the third ventricle (Papavassiliou et al., 2008). Depending on the predominant nature of the tremor, that is, head/voice or lower extremity, the electrode can be moved medially or laterally, respectively. The margin of error of electrode placement should be ideally within 2 mm of the anatomically based coordinates. In order to achieve maximal suppression of tremors, it is suggested that a frequency of at least 90 Hz and a voltage range of 1.0–3.5 V be administered (Kuncel et al., 2006). According to Chopra et al. (2013), there are two reasons why DBS may fail in patients with ET; improper electrode placement may result in suboptimal stimulation requiring reimplantation and adequate stimulation may wane over time, despite proper placement. Some of this can be circumvented by the more recent introduction of directional stimulation. Following successful DBS application (unilateral and bilateral), patients have reported significant long-term improvements in tremor, hand motor ability (i.e., writing), and quality of life (Nazzaro et al., 2013; Pahwa et al., 2006).

As previously mentioned, there are a number of other conditions for which thalamic DBS can be executed with some success. In the history of DBS surgery, chronic pain was one of the first conditions treated. Direct ablation of the VPL showed efficacy at 1 year for the treatment of chronic neuropathic pain, particularly following traumatic amputation or brachial plexus avulsion (Pereira et al., 2013). However, several recent multicenter trials failed to meet the prospective criteria for success in treating chronic pain with DBS (Whiting et al., 2018). Apart from chronic pain, DBS has shown reasonable success in medication refractory obsessive compulsive disorder (OCD) when targeting the anterior limb of the internal capsule, in particular, and to a lesser degree, the nucleus accumbens, ventral capsule/ventral striatum, subthalamic nucleus, or inferior thalamic peduncle (Jiménez-Ponce et al., 2009; Kohl et al., 2014). For the management of Tourette's syndrome, the medial thalamus has proven to be an effective target (Rotsides and Mammis, 2013). There has been great interest in the application of DBS and responsive neurostimulation for the treatment of both focal onset epilepsy and generalized epilepsy with the targeting of the anterior and centromedian nuclei as well as the pulvinar (Alcala-Zermeno et al., 2021; Burdette et al., 2020; Warren et al., 2020; Hassler, 1977). This is an evolving field of study that will yield great benefit to these frequently medically refractory conditions.

References

Aggleton, J.P., O'Mara, S.M., Vann, S.D., Wright, N.F., Tsanov, M., Erichsen, J.T., 2010. Hippocampal–anterior thalamic pathways for memory: uncovering a network of direct and indirect actions. Eur. J. Neurosci. 31, 2292–2307. https://doi.org/10.1111/j.1460-9568.2010.07251.x.

Al-Chalabi, M., Reddy, V., Gupta, S., 2020. Neuroanatomy, spinothalamic tract. In: StatPearls. StatPearls Publishing, Treasure Island, FL.

Alcala-Zermeno, J.L., Gregg, N.M., Wirrell, E.C., Stead, M., Worrell, G.A., Van Gompel, J.J., Lundstrom, B.N., 2021. Centromedian thalamic nucleus with or without anterior thalamic nucleus deep brain stimulation for epilepsy in children and adults: a retrospective case series. Seizure 84, 101–107. https://doi.org/10.1016/j.seizure.2020.11.012.

Amici, S., 2012. Thalamic infarcts and hemorrhages. In: Paciaroni, M., Agnelli, G., Caso, V., Bogousslavsky, J. (Eds.), Frontiers of Neurology and Neuroscience. Karger, Basel, pp. 132–136. https://doi.org/10.1159/000333611.

Bachevalier, J., Meunier, M., Lu, M.X., Ungerleider, L.G., 1997. Thalamic and temporal cortex input to medial prefrontal cortex in rhesus monkeys. Exp. Brain Res. 115, 430–444. https://doi.org/10.1007/pl00005713.

Barbas, H., 1988. Anatomic organization of basoventral and mediodorsal visual recipient prefrontal regions in the rhesus monkey. J. Comp. Neurol. 276, 313–342. https://doi.org/10.1002/cne.902760302.

Bastos, A.M., Vezoli, J., Bosman, C.A., Schoffelen, J.-M., Oostenveld, R., Dowdall, J.R., De Weerd, P., Kennedy, H., Fries, P., 2015. Visual areas exert feedforward and feedback influences through distinct frequency channels. Neuron 85, 390–401. https://doi.org/10.1016/j.neuron.2014.12.018.

Benabid, A.L., Pollak, P., Louveau, A., Henry, S., de Rougemont, J., 1987. Combined (thalamotomy and stimulation) stereotactic surgery of the VIM thalamic nucleus for bilateral Parkinson disease. Appl. Neurophysiol. 50, 344–346. https://doi.org/10.1159/000100803.

Berendse, H.W., Groenewegen, H.J., 1990. Organization of the thalamostriatal projections in the rat, with special emphasis on the ventral striatum. J. Comp. Neurol. 299, 187–228. https://doi.org/10.1002/cne.902990206.

Berendse, H.W., Groenewegen, H.J., 1991. Restricted cortical termination fields of the midline and intralaminar thalamic nuclei in the rat. Neuroscience 42, 73–102. https://doi.org/10.1016/0306-4522(91)90151-d.

Bilginer, B., Narin, F., Isıkay, I., Oguz, K.K., Söylemezoglu, F., Akalan, N., 2014. Thalamic tumors in children. Childs Nerv. Syst. 30, 1493–1498. https://doi.org/10.1007/s00381-014-2420-9.

Bobillier, P., Seguin, S., Degueurce, A., Lewis, B.D., Pujol, J.F., 1979. The efferent connections of the nucleus raphe centralis superior in the rat as revealed by radioautography. Brain Res. 166, 1–8. https://doi.org/10.1016/0006-8993(79)90644-9.

Bogousslavsky, J., Regli, F., Uske, A., 1988. Thalamic infarcts: clinical syndromes, etiology, and prognosis. Neurology 38, 837. https://doi.org/10.1212/WNL.38.6.837.

Bosch-Bouju, C., Hyland, B.I., Parr-Brownlie, L.C., 2013. Motor thalamus integration of cortical, cerebellar and basal ganglia information: implications for normal and parkinsonian conditions. Front. Comput. Neurosci. 7. https://doi.org/10.3389/fncom.2013.00163.

Burdette, D.E., Haykal, M.A., Jarosiewicz, B., Fabris, R.R., Heredia, G., Elisevich, K., Patra, S.E., 2020. Brain-responsive corticothalamic stimulation in the centromedian nucleus for the treatment of regional neocortical epilepsy. Epilepsy Behav. 112, 107354. https://doi.org/10.1016/j.yebeh.2020.107354.

Buschman, T.J., Kastner, S., 2015. From behavior to neural dynamics: an integrated theory of attention. Neuron 88, 127–144. https://doi.org/10.1016/j.neuron.2015.09.017.

Butler, A.B., 2009. Thalamus: Evolution in Vertebrates. In: Squire, L.R. (Ed.), Encyclopedia of Neuroscience. Academic Press, Oxford, pp. 955–962. https://doi.org/10.1016/B978-008045046-9.00981-5.

Cassel, J.-C., de Vasconcelos, A.P., 2015. The cognitive thalamus: a bridal chamber not to forget. Neurosci. Biobehav. Rev. 54, 1–2. https://doi.org/10.1016/j.neubiorev.2015.01.017.

Cheek, W.R., Taveras, J.M., 1966. Thalamic tumors. J. Neurosurg. 24, 505–513. https://doi.org/10.3171/jns.1966.24.2.0505.

Chen, S., Su, H.S., 1990. Afferent connections of the thalamic paraventricular and parataenial nuclei in the rat—a retrograde tracing study with iontophoretic application of Fluoro-Gold. Brain Res. 522, 1–6. https://doi.org/10.1016/0006-8993(90)91570-7.

Chopra, A., Klassen, B.T., Stead, M., 2013. Current clinical application of deep-brain stimulation for essential tremor. Neuropsychiatr. Dis. Treat. 9, 1859–1865. https://doi.org/10.2147/NDT.S32342.

Cicirata, F., Angaut, P., Serapide, M.F., Panto, M.R., 1990. Functional organization of the direct and indirect projection via the reticularis thalami nuclear complex from the motor cortex to the thalamic nucleus ventralis lateralis. Exp. Brain Res. 79, 325–337. https://doi.org/10.1007/BF00608242.

Clark, B.J., Taube, J.S., 2012. Vestibular and attractor network basis of the head direction cell signal in subcortical circuits. Front. Neural Circuits 6. https://doi.org/10.3389/fncir.2012.00007.

Copeland, C.S., Neale, S.A., Salt, T.E., 2012. Positive allosteric modulation reveals a specific role for mGlu2 receptors in sensory processing in the thalamus. J. Physiol. 590, 937–951. https://doi.org/10.1113/jphysiol.2011.218065.

Cornwall, J., Phillipson, O.T., 1988. Afferent projections to the dorsal thalamus of the rat as shown by retrograde lectin transport. II. The midline nuclei. Brain Res. Bull. 21, 147–161. https://doi.org/10.1016/0361-9230(88)90227-4.

Crabtree, J.W., 1996. Organization in the somatosensory sector of the cat's thalamic reticular nucleus. J. Comp. Neurol. 366, 207–222. https://doi.org/10.1002/(SICI)1096-9861(19960304)366:2<207::AID-CNE2>3.0.CO;2-9.

Crabtree, J.W., 2018. Functional diversity of thalamic reticular subnetworks. Front. Syst. Neurosci. 12. https://doi.org/10.3389/fnsys.2018.00041.

Crabtree, J.W., Isaac, J.T.R., 2002. New intrathalamic pathways allowing modality-related and cross-modality switching in the dorsal thalamus. J. Neurosci. 22, 8754–8761. https://doi.org/10.1523/JNEUROSCI.22-19-08754.2002.

Crosson, B., 2013. Thalamic mechanisms in language: a reconsideration based on recent findings and concepts. Brain Lang. 126, 73–88. https://doi.org/10.1016/j.bandl.2012.06.011.

Crosson, B., 2019. The role of cortico-thalamo-cortical circuits in language: recurrent circuits revisited. Neuropsychol. Rev. https://doi.org/10.1007/s11065-019-09421-8.

de Lahunta, A., Glass, E., 2009. Chapter 19: Diencephalon. In: de Lahunta, A., Glass, E. (Eds.), Veterinary Neuroanatomy and Clinical Neurology, third ed. W.B. Saunders, Saint Louis, pp. 476–486. https://doi.org/10.1016/B978-0-7216-6706-5.00019-6.

Dolleman-Van Der Weel, M.J., Witter, M.P., 1996. Projections from the nucleus reuniens thalami to the entorhinal cortex, hippocampal field CA1, and the subiculum in the rat arise from different populations of neurons. J. Comp. Neurol. 364, 637–650. https://doi.org/10.1002/(SICI)1096-9861(19960122)364:4<637::AID-CNE3>3.0.CO;2-4.

Dougherty, K., Carlson, B.M., Cox, M.A., Westerberg, J.A., Zinke, W., Schmid, M.C., Martin, P.R., Maier, A., 2021. Binocular suppression in the macaque lateral geniculate nucleus reveals early competitive interactions between the eyes. eNeuro 8. https://doi.org/10.1523/ENEURO.0364-20.2020.

Dougherty, K., Schmid, M.C., Maier, A., 2019. Binocular response modulation in the lateral geniculate nucleus. J. Comp. Neurol. 527, 522–534. https://doi.org/10.1002/cne.24417.

Duhamel, J.R., Colby, C.L., Goldberg, M.E., 1992. The updating of the representation of visual space in parietal cortex by intended eye movements. Science 255, 90–92. https://doi.org/10.1126/science.1553535.

Edwards, E., Nagarajan, S.S., Dallas, S.S., Canolty, R.T., Kirsch, H.E., Barbaro, N.M., Knight, R.T., 2010. Spatiotemporal imaging of cortical activation during verb generation and picture naming. Neuroimage 50 (1), 291–301. https://doi.org/10.1016/j.neuroimage.2009.12.035.

Erisir, A., Van Horn, S.C., Sherman, S.M., 1997. Relative numbers of cortical and brainstem inputs to the lateral geniculate nucleus. Proc. Natl. Acad. Sci. U. S. A. 94, 1517–1520. https://doi.org/10.1073/pnas.94.4.1517.

Fisher, C.M., 1959. The pathologic and clinical aspects of thalamic hemorrhage. Trans. Am. Neurol. Assoc. 84, 56–59.

Fisher, S., Reynolds, J., 2014. The intralaminar thalamus—an expressway linking visual stimuli to circuits determining agency and action selection. Front. Behav. Neurosci. 8. https://doi.org/10.3389/fnbeh.2014.00115.

Fujita, H., Kodama, T., Du Lac, S., 2020. Modular output circuits of the fastigial nucleus for diverse motor and nonmotor functions of the cerebellar vermis. eLife. https://doi.org/10.7554/eLife.58613.

Funahashi, S., 2013. Thalamic mediodorsal nucleus and its participation in spatial working memory processes: comparison with the prefrontal cortex. Front. Syst. Neurosci. 7, 36. https://doi.org/10.3389/fnsys.2013.00036.

Gentet, L.J., Ulrich, D., 2003. Strong, reliable and precise synaptic connections between thalamic relay cells and neurones of the nucleus reticularis in juvenile rats. J. Physiol. 546, 801–811. https://doi.org/10.1113/jphysiol.2002.032730.

Georgescu, I.A., Popa, D., Zagrean, L., 2020. The anatomical and functional heterogeneity of the mediodorsal thalamus. Brain Sci. 10. https://doi.org/10.3390/brainsci10090624.

Giménez-Amaya, J.M., McFarland, N.R., de las Heras, S., Haber, S.N., 1995. Organization of thalamic projections to the ventral striatum in the primate. J. Comp. Neurol. 354, 127–149. https://doi.org/10.1002/cne.903540109.

Gower, E.C., 1989. Efferent projections from limbic cortex of the temporal pole to the magnocellular medial dorsal nucleus in the rhesus monkey. J. Comp. Neurol. 280, 343–358. https://doi.org/10.1002/cne.902800303.

Groenewegen, H.J., 1988. Organization of the afferent connections of the mediodorsal thalamic nucleus in the rat, related to the mediodorsal-prefrontal topography. Neuroscience 24, 379–431. https://doi.org/10.1016/0306-4522(88)90339-9.

Groenewegen, H.J., Berendse, H.W., 1994. The specificity of the "nonspecific" midline and intralaminar thalamic nuclei. Trends Neurosci. 17, 52–57. https://doi.org/10.1016/0166-2236(94)90074-4.

Groenewegen, H.J., Galis-de Graaf, Y., Smeets, W.J., 1999. Integration and segregation of limbic cortico-striatal loops at the thalamic level: an experimental tracing study in rats. J. Chem. Neuroanat. 16, 167–185. https://doi.org/10.1016/s0891-0618(99)00009-5.

Groenewegen, H.J., Wright, C.I., Uylings, H.B., 1997. The anatomical relationships of the prefrontal cortex with limbic structures and the basal ganglia. J. Psychopharmacol. 11, 99–106. https://doi.org/10.1177/026988119701100202.

Grossberg, S., 2017. Towards solving the hard problem of consciousness: the varieties of brain resonances and the conscious experiences that they support. Neural Netw. 87, 38–95. https://doi.org/10.1016/j.neunet.2016.11.003.

Guillery, R.W., Harting, J.K., 2003. Structure and connections of the thalamic reticular nucleus: advancing views over half a century. J. Comp. Neurol. 463, 360–371. https://doi.org/10.1002/cne.10738.

Guo, W., Clause, A.R., Barth-Maron, A., Polley, D.B., 2017. A corticothalamic circuit for dynamic switching between feature detection and discrimination. Neuron 95, 180–194. https://doi.org/10.1016/j.neuron.2017.05.019.

Haber, S., McFarland, N.R., 2001. The place of the thalamus in frontal cortical-basal ganglia circuits. Neuroscientist 7, 315–324. https://doi.org/10.1177/107385840100700408.

Haber, S.N., Calzavara, R., 2009. The cortico-basal ganglia integrative network: the role of the thalamus. Brain Res. Bull. 78, 69–74. https://doi.org/10.1016/j.brainresbull.2008.09.013.

Halassa, M.M., Kastner, S., 2017. Thalamic functions in distributed cognitive control. Nat. Neurosci. 20, 1669–1679. https://doi.org/10.1038/s41593-017-0020-1.

Hallanger, A.E., Levey, A.I., Lee, H.J., Rye, D.B., Wainer, B.H., 1987. The origins of cholinergic and other subcortical afferents to the thalamus in the rat. J. Comp. Neurol. 262, 105–124. https://doi.org/10.1002/cne.902620109.

Hassler, R.G., 1977. Atlas for Stereotaxy of the Human Brain, 2d rev.enl. ed. Thieme, Stuttgart.

Herrero, M.-T., Barcia, C., Navarro, J.M., 2002. Functional anatomy of thalamus and basal ganglia. Childs Nerv. Syst. 18, 386–404. https://doi.org/10.1007/s00381-002-0604-1.

Hess, C.P., Christine, C.W., Apple, A.C., Dillon, W.P., Aminoff, M.J., 2014. Changes in the thalamus in atypical parkinsonism detected using shape analysis and diffusion tensor imaging. Am. J. Neuroradiol. 35, 897–903. https://doi.org/10.3174/ajnr.A3832.

Hintzen, A., Pelzer, E.A., Tittgemeyer, M., 2018. Thalamic interactions of cerebellum and basal ganglia. Brain Struct. Funct. 223, 569–587. https://doi.org/10.1007/s00429-017-1584-y.

Hughes, E.J., Bond, J., Svrckova, P., Makropoulos, A., Ball, G., Sharp, D.J., Edwards, A.D., Hajnal, J.V., Counsell, S.J., 2012. Regional changes in thalamic shape and volume with increasing age. Neuroimage 63, 1134–1142. https://doi.org/10.1016/j.neuroimage.2012.07.043.

Jankowski, M.M., Ronnqvist, K.C., Tsanov, M., Vann, S.D., Wright, N.F., Erichsen, J.T., Aggleton, J.P., O'Mara, S.M., 2013. The anterior thalamus provides a subcortical circuit supporting memory and spatial navigation. Front. Syst. Neurosci. 7, 45. https://doi.org/10.3389/fnsys.2013.00045.

Jiménez-Ponce, F., Velasco-Campos, F., Castro-Farfán, G., Nicolini, H., Velasco, A.L., Salín-Pascual, R., Trejo, D., Criales, J.L., 2009. Preliminary study in patients with obsessive-compulsive disorder treated with electrical stimulation in the inferior thalamic peduncle. Neurosurgery 65, 203–209. https://doi.org/10.1227/01.NEU.0000345938.39199.90 discussion 209.

Jones, E.G., 1998. Viewpoint: the core and matrix of thalamic organization. Neuroscience 85 (2), 331–345. https://doi.org/10.1016/S0306-4522(97)00581-2.

Jones, E.G., 2012. The Thalamus. Springer Science & Business Media.

Jones, E.G., Friedman, D.P., 1982. Projection pattern of functional components of thalamic ventrobasal complex on monkey somatosensory cortex. J. Neurophysiol. 48, 521–544. https://doi.org/10.1152/jn.1982.48.2.521.

Kaas, J.H., Guillery, R.W., Allman, J.M., 1972. Some principles of organization in the dorsal lateral geniculate nucleus. Brain Behav. Evol. 6, 253–299. https://doi.org/10.1159/000123713.

Kawaguchi, Y., 2017. Pyramidal cell subtypes and their synaptic connections in layer 5 of rat frontal cortex. Cereb. Cortex 27, 5755–5771. https://doi.org/10.1093/cercor/bhx252.

Kawahara, N., Sato, K., Muraki, M., Tanaka, K., Kaneko, M., Uemura, K., 1986. CT classification of small thalamic hemorrhages and their clinical implications. Neurology 36, 165–172. https://doi.org/10.1212/wnl.36.2.165.

Kelly, P.J., 1989. Stereotactic biopsy and resection of thalamic astrocytomas. Neurosurgery 25, 185–194 discussion 194–195.

Kirouac, G.J., 2015. Placing the paraventricular nucleus of the thalamus within the brain circuits that control behavior. Neurosci. Biobehav. Rev. 56, 315–329. https://doi.org/10.1016/j.neubiorev.2015.08.005.

Kirshner, H.S., Kistler, K.H., 1982. Aphasia after right thalamic hemorrhage. Arch. Neurol. 39, 667–669. https://doi.org/10.1001/archneur.1982.00510220065018.

Kohl, S., Schönherr, D.M., Luigjes, J., Denys, D., Mueller, U.J., Lenartz, D., Visser-Vandewalle, V., Kuhn, J., 2014. Deep brain stimulation for treatment-refractory obsessive compulsive disorder: a systematic review. BMC Psychiatr. 14, 214. https://doi.org/10.1186/s12888-014-0214-y.

Kolmac, C.I., Mitrofanis, J., 1997. Organisation of the reticular thalamic projection to the intralaminar and midline nuclei in rats. J. Comp. Neurol. 377, 165–178. https://doi.org/10.1002/(SICI)1096-9861(19970113)377:2<165::AID-CNE2>3.0.CO;2-1.

Krack, P., Dostrovsky, J., Ilinsky, I., Kultas-Ilinsky, K., Lenz, F., Lozano, A., Vitek, J., 2002. Surgery of the motor thalamus: problems with the present nomenclatures. Mov. Disord. 17, S2–S8. https://doi.org/10.1002/mds.10136.

Kumral, E., Kocaer, T., Ertübey, N.O., Kumral, K., 1995. Thalamic hemorrhage. A prospective study of 100 patients. Stroke 26, 964–970. https://doi.org/10.1161/01.str.26.6.964.

Kuncel, A.M., Cooper, S.E., Wolgamuth, B.R., Clyde, M.A., Snyder, S.A., Montgomery, E.B., Rezai, A.R., Grill, W.M., 2006. Clinical response to varying the stimulus parameters in deep brain stimulation for essential tremor. Mov. Disord. 21, 1920–1928. https://doi.org/10.1002/mds.21087.

Kuroda, M., Yokofujita, J., Murakami, K., 1998. An ultrastructural study of the neural circuit between the prefrontal cortex and the mediodorsal nucleus of the thalamus. Prog. Neurobiol. 54, 417–458. https://doi.org/10.1016/s0301-0082(97)00070-1.

Lam, Y.-W., Sherman, S.M., 2015. Functional topographic organization of the motor reticulothalamic pathway. J. Neurophysiol. 113, 3090–3097. https://doi.org/10.1152/jn.00847.2014.

Lee, J.Y.K., Kondziolka, D., 2005. Thalamic deep brain stimulation for management of essential tremor. J. Neurosurg. 103, 400–403. https://doi.org/10.3171/jns.2005.103.3.0400.

Lee, S.-C., Patrick, S.L., Richardson, K.A., Connors, B.W., 2014. Two functionally distinct networks of gap junction-coupled inhibitory neurons in the thalamic reticular nucleus. J. Neurosci. 34, 13170–13182. https://doi.org/10.1523/JNEUROSCI.0562-14.2014.

Li, S., Shi, Y., Kirouac, G.J., 2014. The hypothalamus and periaqueductal gray are the sources of dopamine fibers in the paraventricular nucleus of the thalamus in the rat. Front. Neuroanat. 8. https://doi.org/10.3389/fnana.2014.00136.

Mai, J.K., Majtanik, M., 2019. Toward a common terminology for the thalamus. Front. Neuroanat. 12. https://doi.org/10.3389/fnana.2018.00114.

Mathiasen, M.L., Dillingham, C.M., Kinnavane, L., Powell, A.L., Aggleton, J.P., 2017. Asymmetric cross-hemispheric connections link the rat anterior thalamic nuclei with the cortex and hippocampal formation. Neuroscience 349, 128–143. https://doi.org/10.1016/j.neuroscience.2017.02.026.

Mease, R.A., Krieger, P., Groh, A., 2014. Cortical control of adaptation and sensory relay mode in the thalamus. Proc. Natl. Acad. Sci. U. S. A. 111, 6798–6803.

Miller, E.K., Cohen, J.D., 2001. An integrative theory of prefrontal cortex function. Annu. Rev. Neurosci. 24, 167–202. https://doi.org/10.1146/annurev.neuro.24.1.167.

Minamimoto, T., Kimura, M., 2002. Participation of the thalamic CM-Pf complex in attentional orienting. J. Neurophysiol. 87, 3090–3101. https://doi.org/10.1152/jn.2002.87.6.3090.

Mitchell, A.S., Chakraborty, S., 2013. What does the mediodorsal thalamus do? Front. Syst. Neurosci. 7, 37. https://doi.org/10.3389/fnsys.2013.00037.

Morecraft, R.J., Geula, C., Mesulam, M.M., 1992. Cytoarchitecture and neural afferents of orbitofrontal cortex in the brain of the monkey. J. Comp. Neurol. 323, 341–358. https://doi.org/10.1002/cne.903230304.

Nakagawa, Y., 2019. Development of the thalamus: From early patterning to regulation of cortical functions. Wiley Interdiscip. Rev. Dev. Biol. 8, e345. https://doi.org/10.1002/wdev.345.

Nass, R., Boyce, L., Leventhal, F., Levine, B., Allen, J., Maxfield, C., Salsberg, D., Sarno, M., George, A., 2000. Acquired aphasia in children after surgical resection of left-thalamic tumors. Dev. Med. Child Neurol. 42, 580–590. https://doi.org/10.1017/S0012162200001109.

Nazzaro, J.M., Lyons, K.E., Pahwa, R., 2013. Deep brain stimulation for essential tremor. Handb. Clin. Neurol. 116, 155–166. https://doi.org/10.1016/B978-0-444-53497-2.00013-9.

Nelson, A.J.D., Vann, S.D., 2014. Mammilliothalamic tract lesions disrupt tests of visuo-spatial memory. Behav. Neurosci. 128, 494–503. https://doi.org/10.1037/bne0000001.

Newman, D.B., Ginsberg, C.Y., 1992. Brainstem reticular nuclei that project to the cerebellum in rats: a retrograde tracer study. Brain Behav. Evol. 39, 24–68. https://doi.org/10.1159/000114102.

O'Mara, S.M., 2013. The anterior thalamus provides a subcortical circuit supporting memory and spatial navigation. Front. Syst. Neurosci. 7. https://doi.org/10.3389/fnsys.2013.00045.

Otake, K., Nakamura, Y., 1995. Sites of origin of corticotropin-releasing factor-like immunoreactive projection fibers to the paraventricular thalamic nucleus in the rat. Neurosci. Lett. 201, 84–86. https://doi.org/10.1016/0304-3940(95)12148-w.

Ozek, M.M., Türe, U., 2002. Surgical approach to thalamic tumors. Childs Nerv. Syst. 18, 450–456. https://doi.org/10.1007/s00381-002-0608-x.

Pahwa, R., Lyons, K.E., Wilkinson, S.B., Simpson, R.K., Ondo, W.G., Tarsy, D., Norregaard, T., Hubble, J.P., Smith, D.A., Hauser, R.A., Jankovic, J., 2006. Long-term evaluation of deep brain stimulation of the thalamus. J. Neurosurg. 104, 506–512. https://doi.org/10.3171/jns.2006.104.4.506.

Papavassiliou, E., Rau, G., Heath, S., Abosch, A., Barbaro, N.M., Larson, P.S., Lamborn, K., Starr, P.A., 2008. Thalamic deep brain stimulation for essential tremor: relation of lead location to outcome. Neurosurgery 62 (Suppl. 2), 884–894. https://doi.org/10.1227/01.neu.0000316290.83360.7e.

Papez, J.W., 1937. A proposed mechanism of emotion. Arch. Neur. Psych. 38, 725–743. https://doi.org/10.1001/archneurpsyc.1937.02260220069003.

Parnaudeau, S., Bolkan, S.S., Kellendonk, C., 2018. The mediodorsal thalamus: an essential partner of the prefrontal cortex for cognition. Biol. Psychiatry 83, 648–656. https://doi.org/10.1016/j.biopsych.2017.11.008.

Parvizi, J., 2009. Corticocentric myopia: old bias in new cognitive sciences. Trends Cogn. Sci. 13, 354–359. https://doi.org/10.1016/j.tics.2009.04.008.

Pereira, E.A.C., Boccard, S.G., Linhares, P., Chamadoira, C., Rosas, M.J., Abreu, P., Rebelo, V., Vaz, R., Aziz, T.Z., 2013. Thalamic deep brain stimulation for neuropathic pain after amputation or brachial plexus avulsion. Neurosurg. Focus 35, E7. https://doi.org/10.3171/2013.7.FOCUS1346.

Pessoa, L., 2017. A network model of the emotional brain. Trends Cogn. Sci. 21, 357–371. https://doi.org/10.1016/j.tics.2017.03.002.

Pessoa, L., 2018. Emotion and the interactive brain: insights from comparative neuroanatomy and complex systems. Emot. Rev. 10 (3), 204–216. https://doi.org/10.1177/1754073918765675.

Puelles, L., Martinez-de-la-Torre, M., Ferran, J.-L., Watson, C., 2012. Chapter 9: Diencephalon. In: Watson, C., Paxinos, G., Puelles, L. (Eds.), The Mouse Nervous System. Academic Press, San Diego, pp. 313–336. https://doi.org/10.1016/B978-0-12-369497-3.10009-3.

Rotsides, J., Mammis, A., 2013. The use of deep brain stimulation in Tourette's syndrome. Neurosurg. Focus 35, E4. https://doi.org/10.3171/2013.8.FOCUS13292.

Russchen, F.T., Amaral, D.G., Price, J.L., 1987. The afferent input to the magnocellular division of the mediodorsal thalamic nucleus in the monkey, Macaca fascicularis. J. Comp. Neurol. 256, 175–210. https://doi.org/10.1002/cne.902560202.

Saalmann, Y.B., 2014. Intralaminar and medial thalamic influence on cortical synchrony, information transmission and cognition. Front. Syst. Neurosci. 8. https://doi.org/10.3389/fnsys.2014.00083.

Schaltenbrand, G., Wahren, W., 1977. Atlas of Stereotaxy of the Human Brain. Thieme, Stuttgart.

Schmahmann, J.D., 2003. Vascular syndromes of the thalamus. Stroke 34, 2264–2278. https://doi.org/10.1161/01.STR.0000087786.38997.9E.

Shah, A., Jhawar, S.S., Goel, A., 2012. Analysis of the anatomy of the Papez circuit and adjoining limbic system by fiber dissection techniques. J. Clin. Neurosci. 19, 289–298. https://doi.org/10.1016/j.jocn.2011.04.039.

Sheridan, N., Tadi, P., 2020. Neuroanatomy, thalamic nuclei. In: StatPearls. StatPearls Publishing, Treasure Island, FL.

Sherman, S.M., 2005. Thalamic relays and cortical functioning. In: Progress in Brain Research. Elsevier, pp. 107–126. https://doi.org/10.1016/S0079-6123(05)49009-3.

Sherman, S.M., 2017. Functioning of circuits connecting thalamus and cortex. Compr. Physiol. https://doi.org/10.1002/cphy.c160032.

Sherman, S.M., Guillery, R.W., 1996. Functional organization of thalamocortical relays. J. Neurophysiol. 76, 1367–1395. https://doi.org/10.1152/jn.1996.76.3.1367.

Sherman, S.M., Guillery, R.W., 1998. On the actions that one nerve cell can have on another: distinguishing "drivers" from "modulators". Proc Natl Acad Sci USA 95 (12), 7121–7126. https://doi.org/10.1073/pnas.95.12.7121.

Sherman, S.M., Guillery, R.W., 2002. The role of the thalamus in the flow of information to the cortex. Philos. Trans. R. Soc. Lond. B Biol. Sci. 357, 1695–1708. https://doi.org/10.1098/rstb.2002.1161.

Sherman, S.M., Guillery, R.W., 2006. Exploring the Thalamus and Its Role in Cortical Function. MIT Press.

Sommer, M.A., 2003. The role of the thalamus in motor control. Curr. Opin. Neurobiol. 13, 663–670. https://doi.org/10.1016/j.conb.2003.10.014.

Sotres-Bayon, F., Quirk, G.J., 2010. Prefrontal control of fear: more than just extinction. Curr. Opin. Neurobiol. 20, 231–235. https://doi.org/10.1016/j.conb.2010.02.005.

Squire, L.R., Amaral, D.G., Zola-Morgan, S., Kritchevsky, M., Press, G., 1989. Description of brain injury in the amnesic patient N.A. Based on magnetic resonance imaging. Exp. Neurol. 105, 23–35. https://doi.org/10.1016/0014-4886(89)90168-4.

Steriade, M., Contreras, D., Amzica, F., Timofeev, I., 1996. Synchronization of fast (30–40 Hz) spontaneous oscillations in intrathalamic and thalamocortical networks. J. Neurosci. 16, 2788–2808.

Sturm, W., de Simone, A., Krause, B.J., Specht, K., Hesselmann, V., Radermacher, I., Herzog, H., Tellmann, L., Müller-Gärtner, H.W., Willmes, K., 1999. Functional anatomy of intrinsic alertness: evidence for a fronto-parietal-thalamic-brainstem network in the right hemisphere. Neuropsychologia 37, 797–805. https://doi.org/10.1016/s0028-3932(98)00141-9.

Theyel, B.B., Llano, D.A., Sherman, S.M., 2010. The corticothalamocortical circuit drives higher-order cortex in the mouse. Nat. Neurosci. 13, 84–88. https://doi.org/10.1038/nn.2449.

Torres, C.V., Moro, E., Lopez-Rios, A.-L., Hodaie, M., Chen, R., Laxton, A.W., Hutchison, W.D., Dostrovsky, J.O., Lozano, A.M., 2010. Deep brain stimulation of the ventral intermediate nucleus of the thalamus for tremor in patients with multiple sclerosis. Neurosurgery 67, 646–651. https://doi.org/10.1227/01.NEU.0000375506.18902.3E discussion 651.

Tovote, P., Fadok, J.P., Lüthi, A., 2015. Neuronal circuits for fear and anxiety. Nat. Rev. Neurosci. 16, 317–331. https://doi.org/10.1038/nrn3945.

Tsivilis, D., Vann, S.D., Denby, C., Roberts, N., Mayes, A.R., Montaldi, D., Aggleton, J.P., 2008. A disproportionate role for the fornix and mammillary bodies in recall versus recognition memory. Nat. Neurosci. 11, 834–842. https://doi.org/10.1038/nn.2149.

Usrey, W.M., Sherman, S.M., 2019. Corticofugal circuits: communication lines from the cortex to the rest of the brain. J. Comp. Neurol. 527, 640–650. https://doi.org/10.1002/cne.24423.

Van der Werf, Y.D., Witter, M.P., Groenewegen, H.J., 2002. The intralaminar and midline nuclei of the thalamus. Anatomical and functional evidence for participation in processes of arousal and awareness. Brain Res. Rev. 39, 107–140. https://doi.org/10.1016/s0165-0173(02)00181-9.

Vann, S.D., Nelson, A.J.D., 2015. The mammillary bodies and memory: more than a hippocampal relay. Prog. Brain Res. 219, 163–185. https://doi.org/10.1016/bs.pbr.2015.03.006.

Vertes, R.P., Albo, Z., Viana Di Prisco, G., 2001. Theta-rhythmically firing neurons in the anterior thalamus: implications for mnemonic functions of Papez's circuit. Neuroscience 104, 619–625. https://doi.org/10.1016/S0306-4522(01)00131-2.

Warren, A.E.L., Dalic, L.J., Thevathasan, W., Roten, A., Bulluss, K.J., Archer, J., 2020. Targeting the centromedian thalamic nucleus for deep brain stimulation. J. Neurol. Neurosurg. Psychiatry 91, 339–349. https://doi.org/10.1136/jnnp-2019-322030.

Weyand, T.G., 2016. The multifunctional lateral geniculate nucleus. Rev. Neurosci. 27, 135–157. https://doi.org/10.1515/revneuro-2015-0018.

Whiting, B.B., Whiting, A.C., Whiting, D.M., 2018. Thalamic deep brain stimulation. Prog. Neurol. Surg. 33, 198–206. https://doi.org/10.1159/000481104.

Willis, W.D., Westlund, K.N., 1997. Neuroanatomy of the pain system and of the pathways that modulate pain. J. Clin. Neurophysiol. 14, 2–31. https://doi.org/10.1097/00004691-199701000-00002.

Wilson, C.R.E., Gaffan, D., Browning, P.G.F., Baxter, M.G., 2010. Functional localization within the prefrontal cortex: missing the forest for the trees? Trends Neurosci. 33, 533–540. https://doi.org/10.1016/j.tins.2010.08.001.

Witter, M.P., Groenewegen, H.J., Lopes da Silva, F.H., Lohman, A.H., 1989. Functional organization of the extrinsic and intrinsic circuitry of the parahippocampal region. Prog. Neurobiol. 33, 161–253. https://doi.org/10.1016/0301-0082(89)90009-9.

Wolff, M., Vann, S.D., 2019. The cognitive thalamus as a gateway to mental representations. J. Neurosci. 39, 3–14.

Wolff, M., Gibb, S.J., Dalrymple-Alford, J.C., 2006. Beyond spatial memory: the anterior thalamus and memory for the temporal order of a sequence of odor cues. J. Neurosci. 26 (11), 2907–2913. https://doi.org/10.1523/JNEUROSCI.5481-05.2006.

Wolff, M., Alcaraz, F., Marchand, A.R., Coutureau, E., 2015. Functional heterogeneity of the limbic thalamus: from hippocampal to cortical functions. Neurosci. Biobehav. Rev. 54, 120–130. https://doi.org/10.1016/j.neubiorev.2014.11.011.

Womelsdorf, T., Everling, S., 2015. Long-range attention networks: circuit motifs underlying endogenously controlled stimulus selection. Trends Neurosci. 38, 682–700. https://doi.org/10.1016/j.tins.2015.08.009.

Wright, N.F., Erichsen, J.T., Vann, S.D., O'Mara, S.M., Aggleton, J.P., 2010. Parallel but separate inputs from limbic cortices to the mammillary bodies and anterior thalamic nuclei in the rat. J. Comp. Neurol. 518, 2334–2354. https://doi.org/10.1002/cne.22336.

Wright, N.F., Vann, S.D., Aggleton, J.P., Nelson, A.J.D., 2015. A critical role for the anterior thalamus in directing attention to task-relevant stimuli. J. Neurosci. 35, 5480–5488.

Wurtz, R.H., McAlonan, K., Cavanaugh, J., Berman, R.A., 2011. Thalamic pathways for active vision. Trends Cogn. Sci. 15, 177–184. https://doi.org/10.1016/j.tics.2011.02.004.

Wyss, J.M., Swanson, L.W., Cowan, W.M., 1979. A study of subcortical afferents to the hippocampal formation in the rat. Neuroscience 4, 463–476. https://doi.org/10.1016/0306-4522(79)90124-6.

Yasargil, M.G., 1988. Surgical concerns. In: Microneurosurgery, 3B. Thieme, New York, pp. 25–53.

Yasargil, M.G., 1999. A legacy of microneurosurgery: memoirs, lessons, and axioms. Neurosurgery 45.

Yu, H., Neimat, J.S., 2008. The treatment of movement disorders by deep brain stimulation. Neurotherapeutics 5, 26–36. https://doi.org/10.1016/j.nurt.2007.10.072.

Zhang, W., Bruno, R.M., 2019. High-order thalamic inputs to primary somatosensory cortex are stronger and longer lasting than cortical inputs. Elife 8, e44158. https://doi.org/10.7554/eLife.44158.

Zhang, X.Y., Wang, J.J., Zhu, J.N., 2016. Cerebellar fastigial nucleus: from anatomic construction to physiological functions. Cerebellum Ataxias. https://doi.org/10.1186/s40673-016-0047-1.

Zhou, H., Schafer, R.J., Desimone, R., 2016. Pulvinar-cortex interactions in vision and attention. Neuron 89, 209–220. https://doi.org/10.1016/j.neuron.2015.11.034.

Ziegler, D.K., Kaufman, A., Marshall, H.E., 1977. Abrupt memory loss associated with thalamic tumor. Arch. Neurol. 34, 545–548. https://doi.org/10.1001/archneur.1977.00500210047007.

9

The Diencephalon: Subthalamus or Ventral Thalamus

Daniele Armocida[1], *Maurizio Salvati*[2], *Giancarlo D'Andrea*[3], *Alessandro Frati*[1,4] *and Brandon M. Ascenzi*[5]

[1]Department of Human Neurosciences, Neurosurgery Division, University of Rome "La Sapienza", Rome, Italy [2]Policlinico Tor Vergata, Neurosurgery Department, University Tor Vergata of Rome, Rome, Italy [3]Neurosurgery Department of Fabrizio Spaziani Hospital, Frosinone, Italy [4]IRCCS "Neuromed", Experimental Neurosurgery Unit, Pozzilli, Italy [5]Independent Neuroresearcher-Member of Marie Curie Alumni Association (MCAA)

OUTLINE

INTRODUCTION

The portion of the central nervous system (CNS) defined as diencephalon has many areas in which coexist tissues of white and gray substance of difficult macroscopic identification and functional interpretation. Among these, in proximity of the thalamus, the areas identified as epithalamus and subthalamus have been described in different ways since the distinction between them is based on neurofunctional features rather than on purely anatomical boundaries.

DEFINITION AND HISTORICAL CHANGES

The subthalamus is generally identified as the most ventral part of the diencephalon lying between the thalamus and the midbrain (Franklin, 2017); however, the interpretation of this structure as a separate entity was not always accepted. His and later Kuhlenbeck incorporated the entire subthalamic zone in the hypothalamus (Kuhlenbeck, 1951).

Initially the whole ventral diencephalon was included in a region named "region subthalamica" by Forel or "hypothalamus" by Wilhelm His. Herrick made the distinction between the hypothalamus proper, which covers a large collection of nuclei and areas within the ventral margin of the diencephalon and the ventral thalamus (which essentially comprises the reticular thalamic nucleus) the zona incerta and the subthalamic nucleus (STN). According to Kahle in 1956 and Richter in 1965, the subthalamus gives rise to the STN and the internal and external parts of the globus pallidus as well.

This organization model was largely adopted until the end of the 20th century as it seemed to agree with

From Anatomy to Function of the Central Nervous System
https://doi.org/10.1016/B978-0-12-822404-5.00012-7

functional differences: the hypothalamus is involved in the control of neuroendocrine/autonomic responses as well as the expression of instinctive behaviors, while the subthalamus participates in higher cognitive processes or voluntary motor actions by mediating cortico-thalamic interactions or as part of the basal ganglia network (see Chapter 15). However, in the late 20th century, the borders as well as the internal organization of these brain regions were strongly debated once again. The former consensus that both the subthalamus and the hypothalamus belong to the ventral diencephalic vesicle was shaken by evidence that both regions are best regarded as rostral rather than ventral to the thalamus (Puelles, 2019). The borders between the hypothalamus and ventral thalamus were disputed yet again. In 1980, it was believed that the STN undeniably belonged to the ventral thalamus; however, it is now considered to be a part of the subthalamus (Hahn and Swanson, 2012). Furthermore, while the STN ventral thalamic identity was being challenged, organizational analogies between the basal ganglia and the hypothalamic networks were also recognized. Indeed, the systematic study of hypothalamic medial zone nuclei connections led to the conclusion that these nuclei are entangled in loop circuits with the thalamus, cerebral cortex, and cerebral nuclei that parallel similar loops which are representative of the basal ganglia network in which the STN is integrated (Risold et al., 2009).

The boundary of the conventional hypothalamus with the ventral thalamus in amniotes has a tradition burdened with confusion and arbitrary opinions. This was due partly to the various problems posed by the changing concept of subthalamus, which evolved differentially and incoherently relative to its initial synonym, the hypothalamus. The subthalamus was increasingly redefined as a territory centered on the distinct subthalamic nucleus, which was mistakenly thought to be located inside the hypothalamus. This new subthalamus concept tended to pool together the subthalamic nucleus with some ventral thalamic (prethalamic) structures, notably the zona incerta (ZI), the posterior hypothalamus (which is largely a prethalamic and thalamic tegmental field), and the prerubral area (named as Forel's field).

THE MODERN FUNCTIONAL CONSTITUTION OF SUBTHALAMUS

The largest division of the subthalamus is the subthalamic nucleus (STN) that is involved with integration of somatomotor function (Franklin, 2017).

The subthalamus lies beneath the posterior end of the thalamus, behind the hypothalamus and above the cerebral peduncles and links the midbrain to the diencephalon. It is separated from the thalamus by the lateral lemniscus (carrying auditory input) and contains collections of nerve cell bodies from the zona incerta (ZI) and the subthalamic nucleus.

The prerubral area (Forel's field H) is located just rostral to the red nucleus and medial to the subthalamic nucleus. There are scattered neurons in this region and traversing the prerubral area are fibers from the lenticular fasciculus (Forel's field H2) that enter the thalamic fasciculus (Forel's field H1) (Maynard and Downes, 2019).

The zona incerta is located superior to the subthalamic nucleus and is separated from it by the lenticular fasciculus. Superior to the zona incerta are the myelinated axons of the thalamic fasciculus. The zona incerta contains output neurons that project to a variety of locations, including the cerebral cortex, the superior colliculus, the pretectal region, and the basilar pons. Afferent projections arise from the motor cortex and as collaterals from the medial lemniscus. ZI may represent a part of the reticular formation—intralaminar nuclei of the thalamus complex (Mihailoff and Haines, 2018; Puelles and Rubenstein, 2012; Puelles, 2019).

Nowadays, for the area that include subthalamic nucleus, the medially adjacent prerubral area (Forel's field H) and, posteriorly, the zona incerta, both denominations as ventral thalamus and subthalamus are accepted. As the term ventral thalamus (or subthalamus) implies, these cell groups are located ventral (anterior) to the large expanse of the dorsal thalamus (Maynard and Downes, 2019). The subthalamic nucleus receives a major input from the globus pallidus and for this reason is usually included in the basal ganglia but in fact is a dorsally migrated hypothalamic cell mass, which originates from the retromammillary area.

The entopeduncular nucleus (ENa) is usually also attributed to the "subthalamus" rather than to the hypothalamus (where it belongs as well), thus underlining a generally unnoticed, unclear interpretation of the hypothalamic topography of the cerebral peduncle. There is an arbitrary columnar functionalist trend to separate visceromotor ("hypothalamic") from "somatomotor" ("subthalamic") functional entities, imitating the hindbrain columnar prototype (Bittencourt et al., 1992). This explains why the "subthalamus" also tends to be mixed arbitrarily with the concept of "basal ganglia." Furthermore, several authors (Keyser, 1972; Dong et al., 2013; Kahle et al., 1956) expanded the original tegmental subthalamus concept, adding alar domains to the exclusively basal initial ones of Forel and His, in parallel to the conceptual expansion of the hypothalamus.

There are different theories regarding the role of the subthalamus in the development of other basal ganglia structures: some authors (Marchand et al., 1986) described that the pallidal basal ganglia populations, plus the entopeduncular nucleus, derive from the subthalamus (alternatively, the hypothalamus) and thereafter migrate, or are otherwise morphogenetically translocated, into the telencephalon. Other authors (Cobos et al., 2001; Puelles, 2019; Pombero and Martinez, 2009) confuted this hypothesis and stated that these structures do not derive from subthalamus, but directly from telencephalic structures (Kitai and Kita, 1987).

We recommend that this extremely confusing idea of a separate subthalamic region mixed with assorted parts of the hypothalamus, prethalamus and subpallidal telencephalon, should be considered obsolete and wholly abandoned. It should be replaced by the comparatively uncontaminated prosomeric idea of the peduncular hypothalamus, correctly delimited from terminal hypothalamus, diencephalic prethalamus, and telencephalic subpallidum (ten Donkelaar and Vasung, 2015).

SUBTHALAMIC NUCLEUS

The subthalamic nucleus (STN), also known as Corpus of Luys, is an oval-shaped diencephalic structure located ventrally to the thalamus (Figs. 9.1, 9.2, and 9.3, Joel and Weiner, 1997), playing a fundamental role in the circuitry of the basal ganglia. Described for the first time by Jules Bernard Luys in 1865, its functions remained largely unknown until 1927, when J. P. Martin reported the first case of hemi-chorea following lesion of the STN (Lanska, 2010). As a key structure involved in the pathophysiology of several neurological disorders, the STN has also become a target for deep brain stimulation (DBS) in Parkinson's disease (PD) to modulate its firing patterns and improve clinical manifestations (Lozano and Carella, 2002). In PD, the progressive loss of dopaminergic neurons in the substantia nigra pars compacta causes an alteration of the striatopallidal and pallidosubthalamical pathways in the basal ganglia, resulting in an abnormal burst-firing pattern in STN neurons (Magill et al., 2000; Urbain et al., 2002; Temel et al., 2005).

The STN is located in front of the zona incerta and behind the cerebral peduncle ventrally. The lens-shaped cell group of STN is situated rostral and posterior to the substantia nigra and immediately inferior to a distinct myelinated fiber bundle, the lenticular fasciculus. It is the largest nucleus of the subthalamus, which is anatomically a part of the diencephalon (Fig. 9.1).

As a diencephalic structure, the STN has a close topographical relationship also with the nuclei of the basal ganglia and with the structures of the mesencephalon. Dorsally, the STN confines with the zona incerta and the field H2 of Forel (mainly const. by lenticular fasciculus), in particular the lenticular fasciculus defines the dorso-lateral margin of the nucleus while the field H1 of Forel, constituted mainly by the thalamic fasciculus, delineates part of its dorsomedial margin. The field H1 of Forel originates from the conjunction of the ansa lenticularis and part of the adjoining fibers of the lenticular fasciculus (Emmi et al., 2020), which separates the STN from the nucleus of the zona incerta. The ventromedial margin of the STN is delimitated by the ansa lenticularis, a white matter bundle originating from the ventral aspect of the lenticular nucleus which then joins the lenticular fasciculus to form the thalamic fasciculus, circumnavigating the medial aspect of the STN.

Ventrally, the rostral regions of the STN confine with the dorsomedial aspect of the internal capsule as it continues into the base of the cerebral peduncle in the mesencephalon, while the most caudal part of the STN is situated atop the rostral extent of the substantia nigra, in particular the dorsolateral aspect of the SNpr (Figs. 9.1, 9.2, and 9.3). Dopaminergic nigrostriatal fibers pass dorsomedially to the STN through the medial forebrain bundle, where they enter the Fields of Forel and ascend to the striatum (Parent et al., 2001). At this point, the most inferior part of the thalamus is situated lower than the level of the STN. Proceeding more medially, the red nucleus, with its spheroid shape, can be defined adjacent to the midline and medial and posteroinferior to the STN. The substantia nigra, with its darker color compared with neighboring structures, spreads along the subthalamic area inferior to the STN and the red nucleus. The ansa lenticularis, composed of pallidofugal fibers, lies between the GPi and STN within the prerubral region. The ansa lenticularis initially travels inferomedially through the posterior limb of the internal capsule toward the STN, passes inferior to the STN, and stays medial to the nucleus during its final ascent. The zona incerta and substantia nigra overlap in the posterior subthalamic area, just posterior to the STN. The STN is covered by the zona incerta dorsomedially, the substantia nigra inferiorly, the red nucleus medially, and the internal capsule anterolaterally.

The subthalamic nucleus itself is functionally considered a part of the basal ganglia. It receives projections from the globus pallidus, the motor areas of the cerebral cortex, the substantia nigra, and the reticular formation of the pons. The STN sends reciprocally projections to the globus pallidus and the substantia nigra. Input to the basal ganglia is generally considered to be through the caudate nucleus and the putamen that together form the striatum. Received here are afferents from all four lobes of the cortex, thalamic nuclei and part of the substantia nigra pars compacta (SNpc). The output fibers of the basal ganglia usually are from the globus pallidus and another part of the substantia nigra pars reticulata (SNpr). The output of these structures usually is projected to certain thalamic nuclei, the brainstem reticular formation, the superior colliculus, and the cortical motor areas in the frontal lobe (Mihailoff and Haines, 2018).

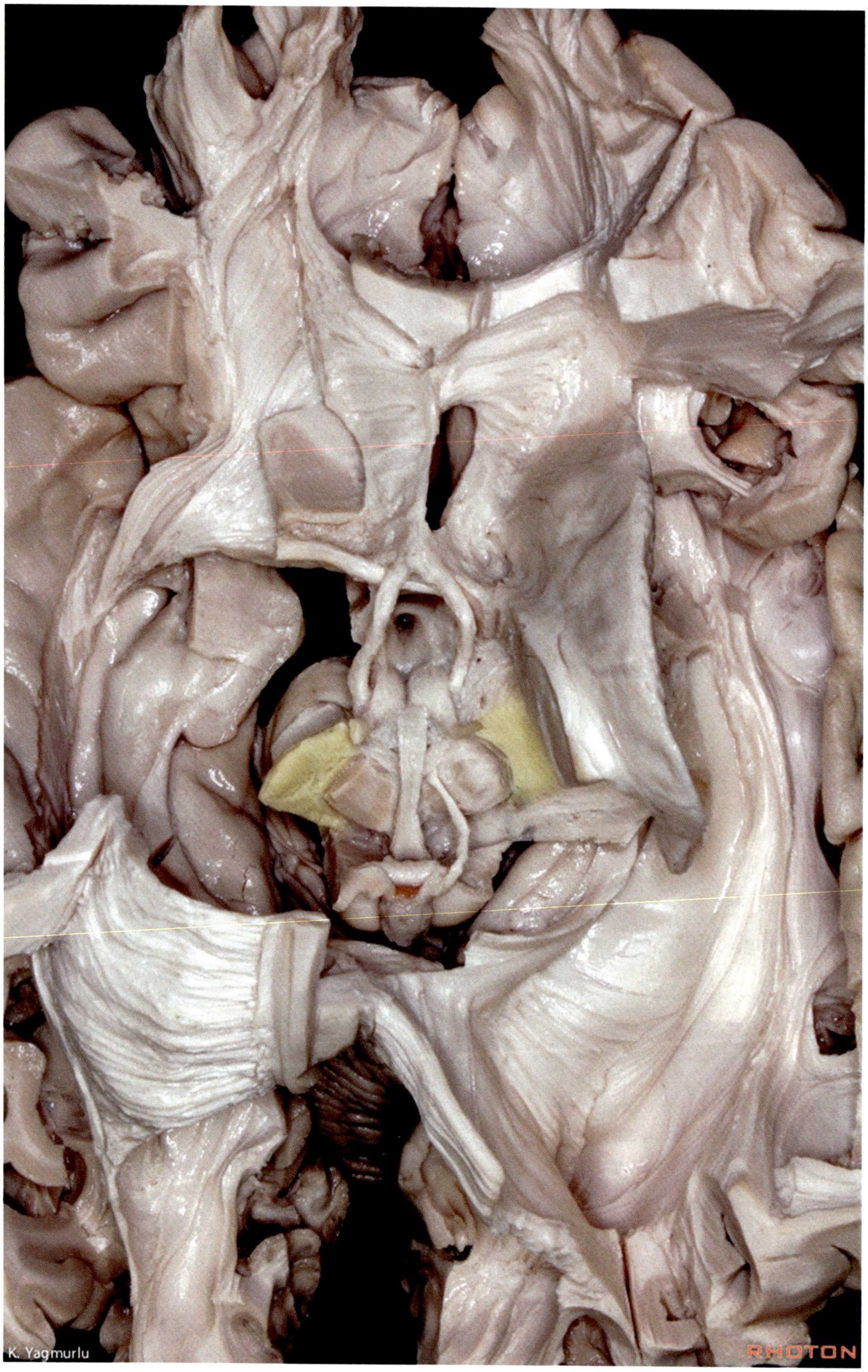

FIGURE 9.1 **Subthalamus; superior view.** Subthalamic nuclei in yellow, lateral to the red nucleus. In a medial anterior direction mammillary bodies and columns of the fornix can be detected. On the midline there is the third ventricle, and posteriorly, the posterior commissure. *Image courtesy of The Rhoton Collection® http://rhoton.ineurodb.org/. Dissector: K.Yagmurlu. Drawing by Jeffrey Sorenson.*

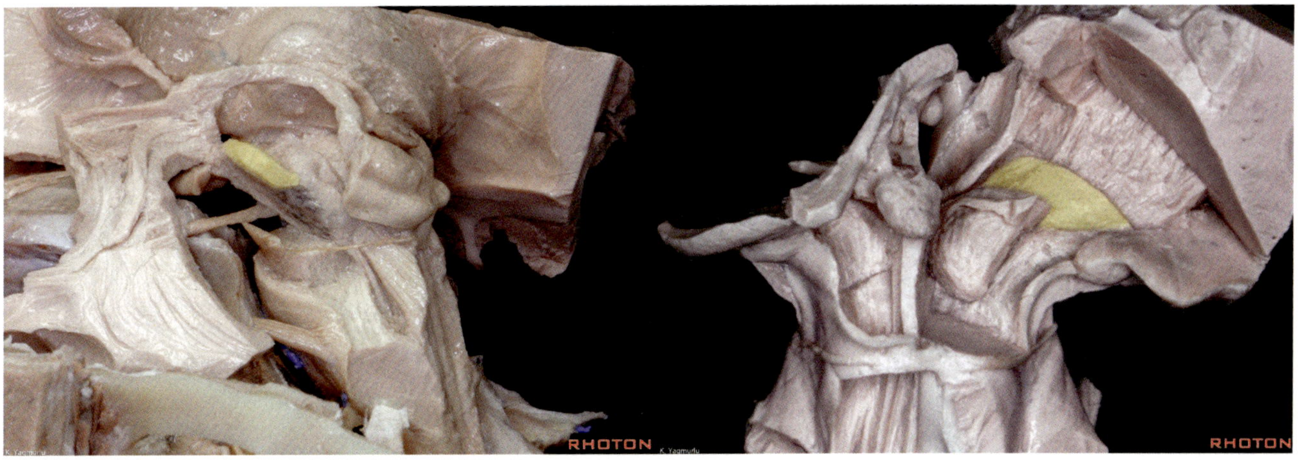

FIGURE 9.2 *Left*. Subthalamus; lateral view of the diencephalic area. Subthalamic nucleus is highlighted in yellow. The substantia nigra is positioned on the ventral surface of the subthalamic nucleus. The red nucleus is located posteriorly. *Right*. Subthalamus; posterior view of the diencephalic area. The subthalamic nucleus is positioned ventrolateral to the red nucleus and medial lemniscus and dorsomedial to the internal capsule. *Images courtesy of The Rhoton Collection® http://rhoton.ineurodb.org/. Dissector: K.Yagmurlu. Drawings by Jeffrey Sorenson.*

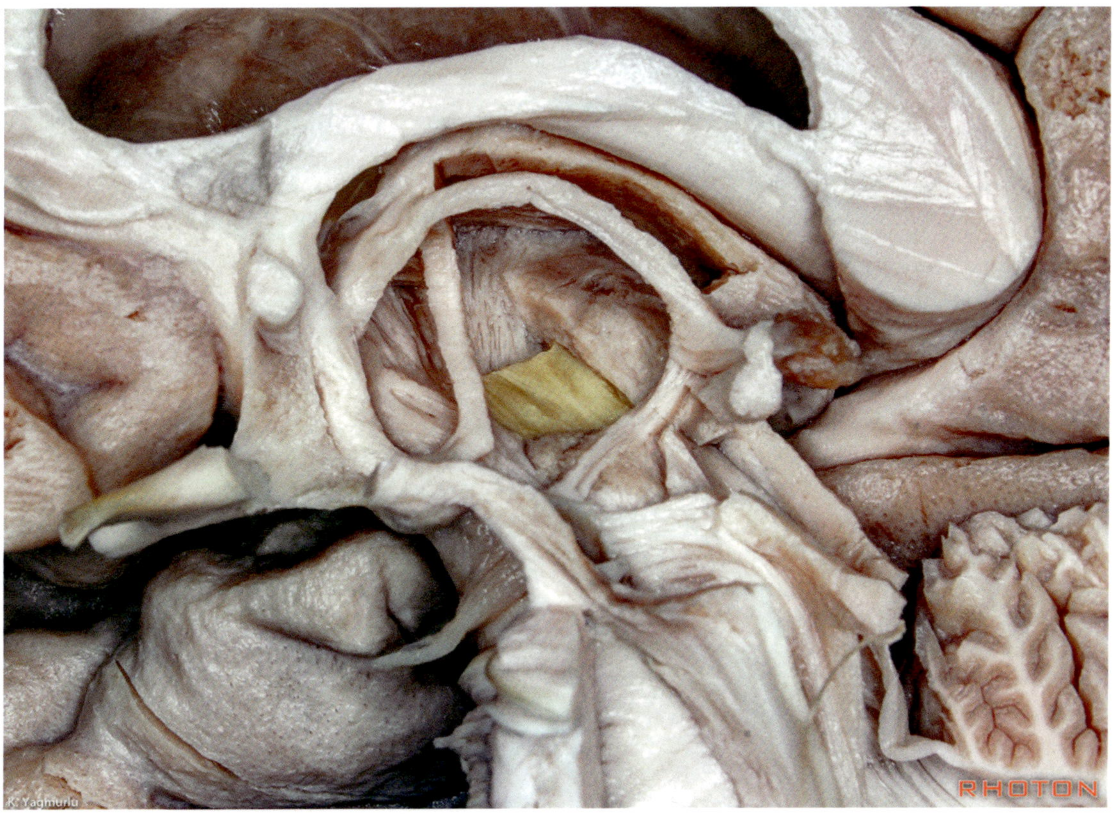

FIGURE 9.3 **Subthalamus; lateral view.** Subthalamic nucleus in yellow, positioned postero-laterally to the internal capsule and supero-medially to the red nucleus. On the midline, both anterior and posterior commissure can be detected. *Image courtesy of The Rhoton Collection® http://rhoton.ineurodb.org/. Dissector: K.Yagmurlu. Drawings by Jeffrey Sorenson.*

The STN is characterized by several functional and regional subdivisions, with each region structurally connected to either the sensorimotor, cognitive, or limbic circuit of the basal ganglia (Parent and Hazrati, 1995; Joel and Weiner, 1997, 2000; Temel et al., 2006). While several subdivisions of the basal ganglia have been proposed (Albin et al., 1989), the most influential model of the organization of these structures views them as components of circuits connecting distinct thalamic, cortical, and subcortical areas in a parallel manner, giving rise to relatively segregated connections, both anatomically and functionally (Alexander and Crutcher, 1990; Parent and Hazrati, 1995; Joel et al., 1997, 2000). More recently, however, the structural segregation of the circuits of the basal ganglia has been questioned by several authors (Percheron and Filion, 1991; Percheron et al., 1994; Reiner and McGeer, 1987), providing evidence on the convergence of different afferent projections on dendritic fields in neurons of these subcortical structures (Shink et al., 1996). According to the classical tripartite organization proposed by Alexander et al., the basal ganglia can be subdivided into

- A motor and oculomotor circuit (forming together the general sensory-motor circuit of the basal ganglia);
- a dorsolateral prefrontal and lateral orbitofrontal circuit (forming the associative circuit); and
- an anterior cingulate circuit (also known as the limbic circuit).

The motor and oculomotor circuit comprises mainly the primary, supplementary, and premotor cortex (BA 4 for primary, BA6 for supplementary and premotor), the dorsolateral part of the caudal putamen (postcommissural), and of the head of the caudate nucleus; the ventrolateral two-thirds of the pallidal complex and part of the lateral substantia nigra; the ventrolateral, ventral anterior, and centromedian nucleus of the thalamus.

The associative circuit comprises mainly the dorsolateral and ventrolateral prefrontal cortex, the rostral regions of the striatum (precommissural), the dorsomedial regions of the pallidal complex and most of the substantia nigra, the parvocellular part of the dorsomedial nucleus of the thalamus.

The limbic circuit comprises the orbitofrontal cortical regions and the anterior cingulate; the nucleus accumbens and the ventral pallidum, the rostral external segment of globus pallidus (GPe) and the rostral ventromedial internal segment of Globus Pallidus (GPi), the medial regions of the substantia nigra and the ventral tegmental area (VTA) of the mesencephalon, the parvocellular part of the ventral anterior nucleus and the magnocellular part of the dorsomedial nucleus of the thalamus (Mai et al., 2008; Aravamuthan et al., 2007; Bahatia and Marsden, 1994; Berney et al.,. 2002; Bostan et al., 2010; Brown et al., 1979).

Returning to STN description, the morphological features of STN neurons have been studied extensively by light and electronic microscopic levels (Rafols and Fox, 1976; Afsharpour, 1985; Chang et al., 1983). Although these studies have proposed different classification schemes for STN neurons, there is an almost complete consensus about the somatic and dendritic morphology of the principal STN neurons in various species (Li et al., 1995).

STN neurons are stellate neurons with radiating and sparsely spined dendrites. Ultrastructure studies have demonstrated that the soma of STN neurons contain abundant organelles with nuclei that have deeply invaginated nuclear envelopes and pale nucleoplasm with little heterochromatin.

The STN is characterized by rich iron deposits (Rutledge et al., 1987) which can be evidenced through Perl's stain in histological sections (Massey and Yousry, 2010); furthermore, these deposits are known to accentuate in pathological conditions, such as PD (Kosta et al., 2006). From a cytoarchitectonic perspective, the STN of rodents, non-human primates (Kuhlenbeck, 1951) and humans is largely characterized by the presence of excitatory glutamatergic neurons (Type I Gray Neurons), even though a small population of γ-aminobutyric acid interneurons (Type II Gray Neurons) has Clarke et al. (1997) been identified (Parent and Hazrati, 1995; Shink et al., 1996; Kearney and Albin, 2000; Tai et al., 2001; Levesque et al., 2005; Marani et al., 2008). In rodents Kita et al. evidenced two morphologically diverse subpopulations of projection neurons in the STh through horseradish peroxidase labeling:

1. Type I STh Neurons, characterized by (a) axon collaterals contacting STh neurons at a local level and by (b) dense dendritic arborizations, especially in proximity of the perykarion.
2. Type II STh Neurons, which possess no axon collaterals at a local level and present fewer dendrites at a proximal level.

Both neuron types found in the rodent STN possess, however, a fundamental common feature which consists in a main axon dividing into two opposite-facing branches; the first branch ascends toward the lenticular nucleus, contacting mainly but not exclusively the globus pallidus; the second branch descends toward the mesencephalon and contacts the neurons of the substantia nigra pars compacta and pars reticulata (Gerfen et al., 1982; Groenewegen et al., 1990).

In primates, glutamatergic neurons in the STN are characterized by a large oval-shaped perykarion with a transverse diameter variable between 25 and 40 μm, while possessing wide dendritic arborizations which extend up to 400 μm from the soma; interneurons are sparser with fewer dendritic arborizations and generally smaller somata. Rafols and Fox (1976), Yelnik et al. (1979) evidenced a branching pattern which varies in relationship to the target of the neurons: neurons contacting both the substantia nigra and the pallidal complex bifurcate in rostral and caudal branches similarly to the neurons identified in rodents by Kita et al., while neurons projecting only to the striatum or the pallidal complex present a single branch which bifurcates in proximity of its target. In particular, the authors identified five subtypes of subthalamic neurons in nonhuman primates based on their axon branching patterns: (1) neurons projecting to the SNr, GPi, and GPe (21.3%); (2) neurons projecting to the SNr and GPe (2.7%); (3) neurons projecting to the GPi and GPe (48%); (4) neurons projecting to the GPe (10.7%); (5) neurons projecting to the striatum (17.3%). It appears evident that, compared to rodents, nonhuman primates present a more diverse population of subthalamic neurons with different axon branching patterns. The typical axonal bifurcation encountered in rodents by Kita et al. is present only in a small portion of primate STh neurons (type I and II STh neurons according to Sato et al. accounting only for the 24% of the total population identified by Sato et al. The remaining neurons present a branching pattern that has no direct correspondence in rodents.

Even though the morphology of subthalamic neurons has been adequately studied in rodents, primates, and even in minipigs (Larsen et al., 2004), only few studies have examined human specimen (Lévesque and Parent, 2005), thus representing a potential field for further investigations.

Lévesque and Parent (2005) studied the morphological characteristics of GABAergic interneurons within the STN compared to the main population projection neurons. The interneurons were identified by a distinct glutamic acid decarboxylase (GAD) immunoreactivity and presented smaller somata than projection neurons (12 vs. 24 μm) with thinner and less numerous dendrites. According to the authors, these neurons appear similar to the interneurons identified in nonhuman primates by Rafols and Fox (1976) through Golgi silver impregnation. Another interesting aspect identified by Lévesque and Parent (2005) is the lack of parvalbumin (PV) and calretinin (CR) immunoreactivity of human GABAergic interneurons within the STN; in fact, only projection neurons appear to be immunoreactive for these calcium-binding proteins. Furthermore PV+ neurons appear more abundant dorsolaterally within the STN, while CR+ neurons were predominant in the ventromedial STh.

While the number of neurons was proportional to the cross-sectional area of the STh across species, the proportional number of neurons expressing parvalbumin (a calcium binding protein) was significantly lower in rodents compared to nonhuman primates and humans.

In the neuropil, at least two types of axon terminals were identified. One type of terminal (Gray's Type I) contained medium-sized round vesicles and formed asymmetrical synapses. The other type of terminal (Gray's Type II) was large and contained both round and flattened vesicles and formed adherens junctions and symmetrical synapses (Lau et al., 2020).

There exists a topographically organized projection from the globus pallidus and ventral pallidum to the subthalamic nucleus and adjacent lateral hypothalamus. The participation of GABA as a neurotransmitter in this projection was evaluated by retrograde labeling of cells in the pallidal area from an iontophoretic deposit of Fluoro-Gold in the subthalamus combined with in situ hybridization for mRNA of the GABA synthetic enzyme, glutamic acid decarboxylase (GAD) (Emmi et al., 2020). A rostrocaudal gradient in the contribution of GABA to the projection was demonstrated with a relatively small percentage of retrogradely labeled cells in the rostral ventral pallidum containing GAD mRNA (7%) compared to the caudal globus pallidus which had over 70% of the Fluoro-Gold containing cells double-labeled for GAD mRNA. Overall, the ventral pallidum contribution to the subthalamic nucleus was less GABAergic than the portion arising from the globus pallidus (35% vs. 61%, respectively) (Emmi et al., 2020).

The neurons of this nucleus use an excitatory neurotransmitter, glutamate. The neurons of the STN are, in normal motor function, usually inhibited from firing by thalamic override. When damage occurs to the basal ganglia pathways and subthalamic or related neurons are not inhibited from firing (i.e., disinhibition occurs), an abnormal increase in involuntary motor movements is seen (Llinas and Jahnsen, 1982; McClelland et al., 2005). This results in disorders associated with hyperkinesia or increased motor movements, such as Huntington's chorea. The vascular supply of the STN derives mainly from the perforating branches of the anterior choroidal artery and the posterior communicating artery, originating from the internal carotid artery and posteromedial choroidal arteries, which derive from the superior cerebellar artery, hence from the vertebrobasilar circulation (Hamani et al., 2004).

So, the STN can also be affected by vascular lesions involving posteromedial branches of the posterior cerebral or posterior communicating arteries, which results in a characteristic clinical condition known as hemiballismus. Patients with this involuntary movement disorder exhibit rapid and forceful flailing movements, which usually involve the contralateral upper extremity. These movements can be very debilitating because the patient has no control over their initiation or duration.

The STN plays a fundamental role within the basal ganglia, thalamocortical motor, associative and limbic circuits. The nucleus has three subdivisions and has been functionally divided into the sensorimotor (dorsolateral), limbic (medial), and cognitive-associative (ventromedial) segments. The dorsolateral sensorimotor region of the STN in Parkinson's disease is the generally accepted target for DBS treatment of motor symptoms, such as tremor and bradykinesia. DBS modulates STN firing, resulting in improvement in motor and nonmotor disability as well as quality of life (Dostrovsky et al., 2002; Grill et al., 2004; McIntyre et al., 2004)

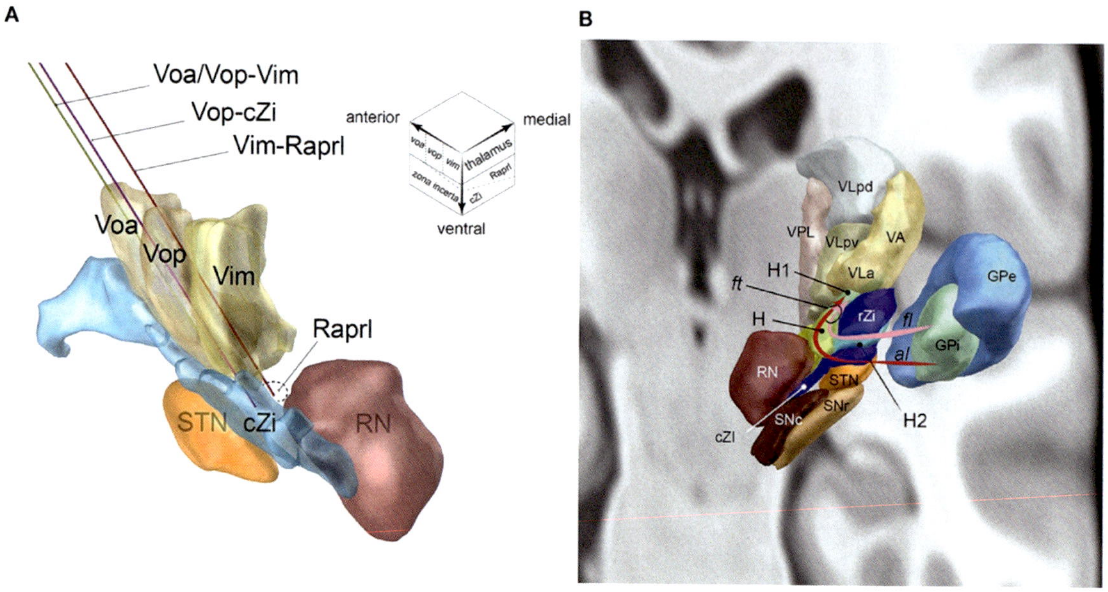

FIGURE 9.4 **Subthalamus Anatomy and Connections (B, see caption) and Deep Brain Stimulation (DBS).** (A) The single DBS-lead dual-target trajectories going through two-story targets made of the ventral thalamic nuclei (upstairs) and posterior subthalamic area (downstairs, inset): Voa/Vop-Vim (yellow), Vop-cZi (blue), and Vop-Raprl (brown). *cZi*, caudal zona incerta; *Vim*, nucleus ventralis intermedius; *Voa*, nucleus ventralis oralis anterior; *Vop*, nucleus ventralis oralis posterior; *Raprl*, prelemniscal radiation; *RN*, nucleus ruber; *STN*, nucleus subthalamicus. The upper right voxel scheme illustrates the two-story structure of ventral motor nuclei of the thalamus (upstairs) and subthalamus (downstairs). (B) Fasciculus lenticularis (*fl*) leaves from dorsal globus pallidus internus (GPi), enters the field H2 of Forel (H2) between rostral zona incerta (rZi) and the subthalamic nucleus (STN), joined with ansa lenticularis(*al*) from ventral GPi in the field H of Forel (H), rostral to the red nucleus (RN), and form the fasciculus thalamicus (*ft*) in the field H1 of Forel (H) then projects to the VA, and VLa nuclei of the thalamus. *GPe*, GP externa; *SNc*, substantia nigra pars compacta; *SNr*, SN pars reticulata; *VLpd and VLpv*, ventrolateral posterior dorsal and ventral nuclei of the thalamus; *VPL*, ventroposterolateral nucleus of the thalamus. *Reproduced from Toda, H., Kambe, D., Shima, A., Nishida, N., Sawamoto, N., 2021. Posterior subthalamic area, pallidothalamic tract, and pedunculopontine nucleus: deep brain stimulation targets for Parkinson's disease and essential tremor. No Shinkei Geka. 49 (4), 820–828 (Japanese). https://doi.org/10.11477/mf.1436204460. PMID: 34376614. Elsevier Inc. The figure components are generated using Lead-DBS (https://www.lead-dbs.org/). Image courtesy from Sugiyama, J., Toda, H., 2022. A single DBS-lead to stimulate the thalamus and subthalamus: two-story targets for tremor disorders. Front. Hum. Neurosci. 16. https://doi.org/10.3389/fnhum.2022.790942.*

STN stimulation has also been explored as an option to treat refractory epilepsy and obsessive-compulsive disorder. Precise and thorough identification of the 3D anatomy of the STN and related structures is a prerequisite for all of these DBS procedures (Fig. 9.4). Safe and effective DBS procedures targeting the STN also depend on a thorough understanding of the 3D anatomy of the region, which is critical for proper assessment of both outcomes and adverse effects. However, even with the best available imaging techniques for displaying brain anatomy in vivo, the peri-STN area is still poorly visualized.

Studies of the effects of stimulation on neural tissue suggest that axons are more readily activated than neuronal cell bodies. Accordingly, capsular effects are commonly observed with high-frequency stimulation. Muscle contractions in the limbs, torso, and head can be detected when the volume of brain tissue activation extends to the corticospinal tract within the internal capsule (IC), which runs from the ventral to the anterolateral aspect of the STN. Facial pulling, dysarthria, and eye deviation may all be related to the activation of the corticobulbar and corticomesencephalic fibers within the IC. Regarding facial muscle contraction during STN DBS, Gorgulho et al. described the most common electrode localization as the anterolateral part of the superior border of the nucleus. Tsuboi et al. reported that dysarthria occurs mainly due to alterations of the laryngeal muscles when the electrode localization is lateral or superior to the STN. Disturbance of frontal eye field projections, running within the posterior limb of the IC and the medial portion of the cerebral peduncle, at the lateral anterosuperior border of the STN is presumably responsible for conjugate eye deviation during stimulation (Turner, 2013).

Noncapsular effects, such as unilateral eye deviation, heat sensation, sweating, and dizziness, are typically observed when stimulation extends medially and posteriorly to the STN. These regions correspond to the hypothalamus, medial lemniscus, and red nucleus (Schmidt et al., 2019; ten Donkelaar and Vasung, 2015; Webb, 2017).

Activation of the medial lemniscus in the posterior subthalamic area, which also includes the ZI and part of the substantia nigra immediately posterior to the STN, can produce contralateral paresthesia (with symptoms such as tingling or electrical sensations (Tamma et al., 2002).

ZONA INCERTA

The zona incerta (ZI) is a large structure made of four neurochemically defined regions. It is globally involved in complex connections with telencephalic and brainstem centers.

The zona incerta (ZI) is a small but diffuse structure in the deep brain first identified by Auguste Forel in 1877, famously described as "an immensely confusing area about which nothing can be said". Forel appreciated that the ZI consisted of gray matter located between the external medullary lamina of the thalamus and the corpus Luysi (or subthalamic nucleus; STN) of otherwise "indefinite" description (Blomstedt et al., 2012). It is telling that Forel found the ZI so difficult to describe given his crucial role in the delineation of surrounding white matter tracts still referred to eponymously as the fields of Forel (Gallay et al., 2008). Since its original description, much has been learned about the ZI and surrounding structures although robust in vivo visualization has remained elusive.

The anatomical boundaries of the ZI have generally been described in the context of its more discrete neighbors rather than based on any consistent feature of the region itself.

During an anatomical dissection removing the thalamus during lateral to medial dissections we can retrieve the ZI, a broad and shallow band of gray matter lying under the thalamus and medial to the IC.

The ZI, a superior continuation of the brainstem reticular formation, is enveloped by efferents from the GPi that pass to the ventrolateral and ventral anterior nuclei of the thalamus. It is difficult to separate the ventral part of the thalamus from the ZI because of their similar external appearance at this stage of dissection, although the ZI emerges as a brighter area of gray matter at higher magnification.

Packed in a small area between the ventral thalamus, STN and red nucleus (RN), the ZI is situated at a complex junction of major white matter pathways including the cerebellothalamic, pallidothalamic, medial lemniscus, and corticospinal tracts. Along its dorsal, ventral, and medial borders, the ZI is surrounded by the thalamic fasciculus (ft; also known as the H1 field of Forel, Fig. 9.4), the fasciculus lenticularis (fl; also known as the H2 field of Forel, see Figs. 9.4 and 9.5), and the H field, which is a convergence of the fasciculus lenticularis and the ansa lenticularis (al) (Figs. 9.4 and 9.5), respectively (Gallay et al., 2008; Nieuwenhuys, 2009). The rostral ZI (rZI) is continuous with the reticular nucleus of the thalamus laterally and with the lateral hypothalamus anteromedially. The caudal ZI (cZI) is laterally bounded by the STN and posterior limb of the internal capsule. To date, most of the details regarding the region are the result of meticulous study of postmortem specimens (Gallay et al., 2008; Morel, 2007; Schaltenbrand and Wahren, 1977).

Cytoarchitectonic and myeloarchitectonic studies in experimental animals (most commonly rodents and primates) have identified the ZI as a gray matter complex consisting of loosely arranged neurons of heterogeneous morphology with a diverse immunohistochemical profile (Nieuwenhuys, 2009) . In Golgi preparations of the ZI, two main neuronal classes have been identified: principal cells and interneurons. Gene expression studies have revealed a common embryological origin along with the reticular nucleus of the thalamus and pregeniculate nucleus of the ventral diencephalon, specifically the prethalamic segment, which predominantly contains GABAergic neurons (Puelles and Rubenstein, 2012; Watson et al., 2014). Through immunohistochemical analysis in experimental animals, a general pattern of at least four component ZI sectors has emerged in the rostral, dorsal, ventral, and caudal directions (Mitrofanis, 2005).

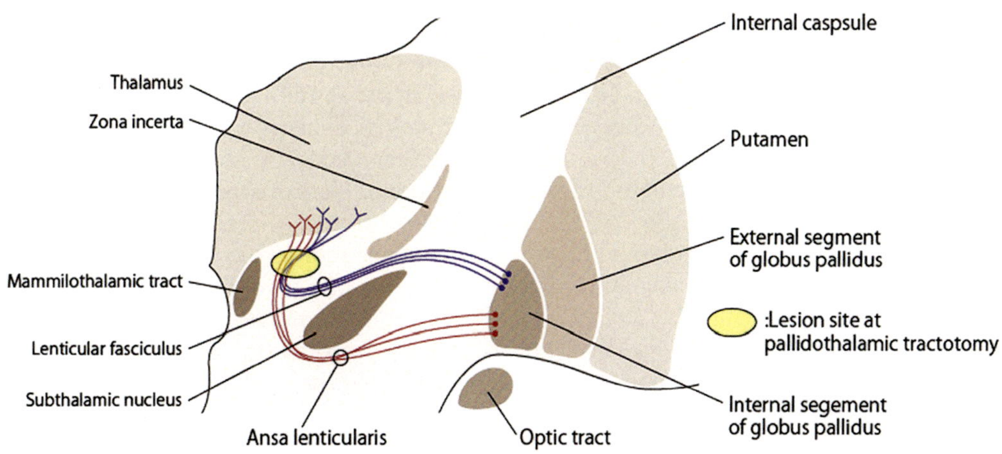

FIGURE 9.5 **Schema of the lesion sites at the pallidothalamic tract and adjacent structures.** The lesion site at the pallidothalamic tract is located far from the internal capsule and optic tract. *Image courtesy of Horisawa, S., Nanke, M., Kawamata, T., Taira, T., 2019. Pallidothalamic tractotomy for Parkinson disease with 1-year follow-up: a case report. World Neurosurg. 121, 193–195. https://doi.org/10.1016/j.wneu.2018.10.055.*

Tract-tracing studies have identified extensive and often bilateral connections between the ZI and the cortex, subcortex, and spinal cord (Mitrofanis, 2005; Watson et al., 2014). At least five functional subsectors within the ZI have been suggested: auditory, limbic, motor, somatosensory, and visual. However, unlike other nearby structures like the STN, no robust immunohistochemical biomarker has been described for the ZI proper.

The zona incerta is a complex region of the brain. First of all, it contains a number of neuronal populations that are distinct in terms of their cytoarchitecture, connections, and pattern of immunomarkers (Kawana and Watanabe, 1981; Kolmac and Mitrofanis, 1999; Mitrofanis, 2005). Secondly, the zona incerta is located at a position where several major tracts converge before turning toward their final localization in the forebrain. These tracts include the superior cerebellar peduncle, the convergence of fasciculus lenticularis and ansa lenticularis (the H field of Forel), the cerebral pedunclens, the nigrostriatal tract, the medial lemniscus, the thalamic fasciculus (the H1 field of Forel), and the lenticular fasciculus (the H2 field of Forel). In fact, the zona incerta was originally defined according to the anatomy of some of these prominent tracts, leading Burton and Jones (1976) to suggest that the neuronal population of the zona incerta would be better named "the nucleus of the zona incerta" to distinguish it from the nearby fiber sheets.

In developmental terms, the zona incerta is a part of prosomere 3 of the true diencephalon. In developmental and gene expression terms, the hypothalamus, diencephalon, and midbrain form a rostrocaudal sequence (Puelles, 2019). Furthermore, gene expression studies show that the diencephalon is divided into three segments from caudal to rostral—prosomere 1 (the pretectum), prosomere 2 (the thalamus), and prosomere 3 (the prethalamus) (Puelles and Rubenstein, 2012). Prosomere 3 consists of the reticular nucleus of the thalamus, the zona incerta, and the pregeniculate nucleus (formerly the ventral lateral geniculate). Prosomere 3 is dominated by GABAergic neurons, whereas prosomere 2 is filled mainly with glutamatergic neurons. Note that GABAergic neurons are scattered through the primate thalamus, but in rodents they are restricted to a small population in the dorsal lateral geniculate nucleus (Jones, 2007; Puelles and Rubenstein, 2012).

Anatomical studies performed on rodents showed that zona incerta receives afferents from several regions of the central nervous system (Mitrofanis, 2005) and it seems likely these same connections are present in the primate brain. Within the neuronal population of the central part of the zona incerta, there is a distinct subdivision into dorsal and ventral sheets and a number of areas are characterized by the expression of particular immunomarkers (Mitrofanis, 2005; Paxinos, 2015; Watson et al., 2014). Up until now the pattern of markers in the caudal part of the zona incerta has not received specific attention.

The zona incerta in rats has been more extensively studied than in other species. The rat zona incerta can be divided into four major regions on the basis of cytoarchitecture (Kawana and Watanabe, 1981; Kolmac and Mitrofanis, 1999; Mitrofanis, 2005). The rat brain atlas of Paxinos (2015) identifies these four regions as (a) rostral part, (b) ventral sheet (ZIV), (c) dorsal sheet (ZID), and (d) caudal part. At the medial border of the ZIV and ZID there is a small group of dopaminergic cells (A13), which belongs to prosomere 3 (Li et al., 2021).

The functions of these regions are still unknown, even if recent works suggest its participation in behavioral expression. Studies about the functional organization of the vibrissal system have provided the first integrated model, illustrating the ZI role in sensory-motor programming. In addition, ZI connections with the superior colliculus and the cerebral cortex as well as recent behavioral studies, point to this region playing a role in cognitive processes related to attention toward salient stimuli (Güngör et al., 2018).

Most of the current knowledge on the ZI was gained from nonspecific manipulations, yet the ZI is a highly heterogenous structure. Histological investigations revealed diverse cell types having the capacity to release chemical signaling molecules such as glutamate, GABA, neuropeptides, and, potentially, dopamine (tyrosine hydroxylase–expressing cells). While the GABAergic neurons are abundant in number and thoroughly expressed across the ZI, the existence of somatostatin (SOM), calretinin (CR), calbindin, parvalbumin (PV), and vasopressin indicate diversity within the GABAergic cells. Such biochemical heterogeneity explains why the ZI has been implicated in quite a wide range of functions (Chometton et al., 2021).

More recently, a stream of research showcased the implication of the ZI in emotionally related experiences and adaptive defensive behaviors. Electrical stimulation in the ZI induced pupil dilation and increased the blood flow to the extremities and the heart rate; all of these are physiological reactions like those related to the acute stress response (Venkataraman et al., 2019; Fig. 9.4). In addition, ZI loss of function impaired mice in their ability to respond to cues predictive of incoming threat. Specifically, mice with ZI electrolytic lesions displayed reduced avoidance in a conditioned avoidance response test (Li et al., 2021). Similarly, ZI tetanus toxin infusion to block synaptic transmission decreased the ability to associate a predictive cue with the negative outcome in a fear conditioning task for mice (Lau et al., 2020). In contrast, optogenetic inactivation of ZI vesicular γ-aminobutyric acid (GABA) transporter (vGAT)-expressing cells during the presentations of loud tones, exacerbated the triggered flight response (Chometton et al., 2021). These previous studies imply that the ZI could be influenced by negative cues and modulate adaptative responses. However, how the ZI senses stressful stimuli when mice experience anxiety-triggering states is not known.

FOREL'S FIELDS

The Forel's fields (FF), already mentioned above, consist of a subthalamic area crossed by the pallidothalamic tracts originating from the globus pallidus internus (GPi). The FF was used as a stereotactic surgical target (ablation and stimulation) to treat cervical dystonia in the 1960s and 1970s. Although recent studies have reappraised the ablation and stimulation of the pallidothalamic tract at FF for Parkinson's disease, the efficacy of deep brain stimulation of FF (FF-DBS) for dystonia has not been well investigated (Bell et al., 1995).

The H Forel's Field constitutes an intricate neuroanatomical structure that occupies a central position within the posterior subthalamus. Anatomically, it features a dense concentration of fiber bundles including corticofugal, pallidothalamic, cerebellothalamic, and other projections that connect functionally relevant areas of the brain. Functionally, the fields of Forel are embedded within the cortico-striato-thalamo-cortical circuit and constitute the main link between the striatopallidal system and the thalamocortical network (Horisawa et al., 2021a, b; Guridi and Gonzalez-Quarante, 2021).

Given the current understanding of basal ganglia involvement in movement disorders and neuropsychiatric disease, Horisawa et al. (2021a, b) sought to investigate the H Forel's field as a potential target in stereotactic functional neurosurgery (Fig. 9.5).

Although historically recognized in the treatment of movement disorders, behavioral disorders and epilepsy, the significance of the H fields is considerably diminished today receiving only little attention (Horisawa et al., 2021a, b). However, H1 Field has been identified as a target for pallidothalamic tractotomy in patients affected by dystonia in a recent report by Horisawa et al. (2021a, b) (Figs. 9.5 and 9.6; for detail see the full article), while older studies report that H2 has been identified as a possible DBS target in Parkinson disease (Story et al., 1965). Furthermore, both H1 and H2 fields have been identified as possible target in the treatment of epilepsy.

NEURORADIOLOGICAL ASSESSMENT

Even though the anatomical features of the subthalamus have been extensively described in literature, in order to detect their precise localization and connections in the living, classical MRI studies must be implemented with fMRI (Barbier and Risold, 2021).

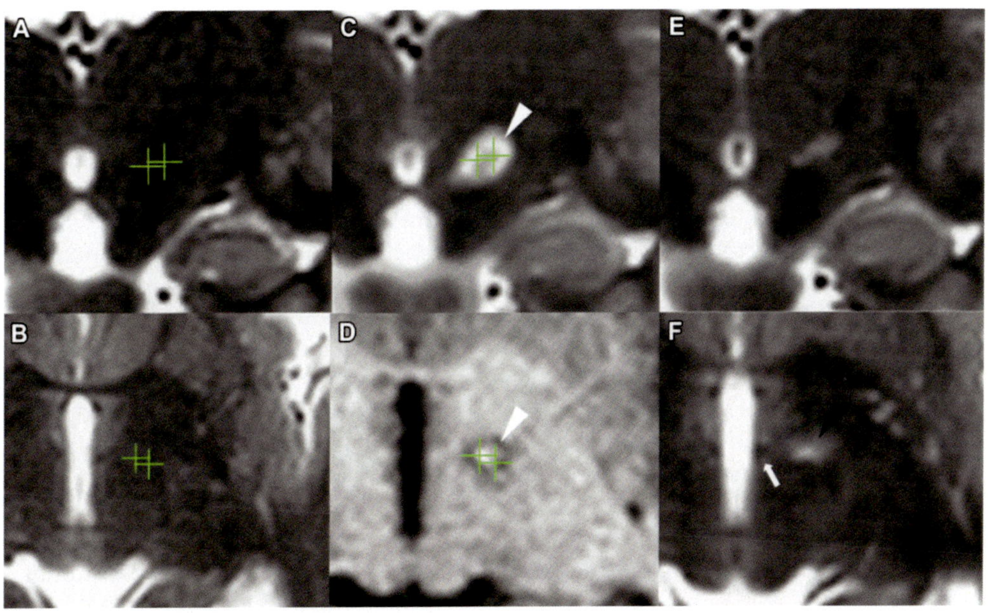

FIGURE 9.6 **Pre- and postoperative T1/2-weighted axial and coronal magnetic resonance imaging (schema of the lesion sites at the pallidothalamic tract and adjacent structures).** (A and B) Preoperative planning coronal (A) and axial (B) images showing two green crosses, which are preplanned targets. (C and D) T2-coronal (C) and T1-axial (D) images on the day of the pallidothalamic tractotomy showing coagulated lesions (arrowhead) at the preplanned targets. (E and F) T2-coronal (E) and axial (F) images 3 months post-surgery showing an old scar (black arrowhead) lateral to the mammillothalamic tract (arrow) and above the subthalamic nucleus (black arrow). *Image courtesy of Horisawa, S., Nanke, M., Kawamata, T., Taira, T., 2019. Pallidothalamic tractotomy for Parkinson disease with 1-year follow-up: a case report. World Neurosurg. 121, 193–195. https://doi.org/10.1016/j.wneu.2018.10.055.*

More specifically fMRI is able to evaluate global brain activation with high temporal and spatial resolution (Obeso et al., 2011), therefore it represents the ideal exam to assess the anatomical and functional boundaries of subthalamic nuclei. Functional imaging has a crucial clinical role since it makes STN DBS more precise. In fact, nonaccurate detection of STN could lead to mispositioning of DBS electrode in adjacent nuclei (Knight et al., 2015) and consequent undesired side effects (Xiao et al., 2014).

Generally, the anatomical landmarks for detecting STN are the anterior (AC) and posterior commissure (PC), retrieved in T1w MRI studies (Xiao et al., 2014). Several authors as well as (Montgomery, 2010; Lanotte et al., 2002; Acar et al., 2007; Guehl et al., 2007) proposed alternative landmarks, such as the mid-point between the AC-PC line, or the middle commissure (MC) point. Another way to detect STN is represented by the indications reported on the two most used atlases, which are the Schaltenbrandt (Schaltenbrand and Wahren, 1977) and the Talairach (Talairach and Tournoux, 1988 Co-planar stereotaxic atlas of the human brain. Thieme, New York.). Many authors infer the location of STN by scaling the atlas, since they both have shorter AC-PC lines (Guehl et al., 2007; Knight et al., 2015). However, even though these anatomical landmarks are easily detected in T1w MRI, STN along with substantia nigra (SN) and red nucleus (RN) are not. Moreover, another important issue must be considered during the radiological assessment of these structures such as the great variability of STN, SN, and RN morphology between individuals, especially in pathological conditions such as PD (Knight et al., 2015; Fig. 9.6). Up to date, the gold standard for detecting STN is still the postmortem histological exam, through tailored neuroanatomical, neurochemical, and neuroarchitectonic analysis. For obvious reasons, several authors attempted to retrieve more accurate MRI studies to detect the exact morphology of STN in vivo (Schaltenbrand and Wahren, 1977; Ashkan et al., 2007; Patel et al., 2008; Daniluk et al., 2010; Richter et al., 2004), by performing T2w studies with 2- or 3-mm axial-slices with 1.5T MRI and then comparing their findings with the data retrieved from the two atlases, but the lack of studies on large groups of patients and the small size of STN did not lead to satisfying results. Moreover, the reconstructions with this methodology were bidimensional, while a tridimensional reconstruction is requested to improve the reliability of the MRI study. The application 3T T1w and T2w MRI with ad hoc automatic segmentation technique (Haegelen et al., 2013) demonstrated the reliability of 3D reconstructions of STN, SN, and RN with this methodology in assessing the variable morphology of these nuclei, even more accurate than those reported in the two atlases (Cuny et al., 2002). The introduction of automatic techniques overcomes the issues related to the "old" manual contour drawing or landmark picking for detecting STN, even though the available results have been collected on small groups of patients and, most-importantly, a modern atlas designed on healthy subjects does not exists.

Therefore, given the actual issues regarding the radiological detection of STN in traditional MRI sequences, functional studies cover a major role in the identification of these nuclei. More specifically, it has been demonstrated that STN is involved both in motor and nonmotor networks (Zhu et al., 2002; Hamani et al., 2004, Fig. 9.7), in fact functional studies demonstrated increase of BOLD signal in motor circuitry (Zhu et al., 2002) (including the primary motor, premotor, and supplementary motor cortices; thalamus; pedunculopontine nucleus; cerebellum) and in the limbic circuitry (including the cingulate and insular cortices) (Knight et al., 2015) (Fig. 9.7).

The results retrieved from the fMRI are then matched with the images retrieved from traditional sequences and with data reported on the two most used atlas (Schaltenbrand and Wahren, 1977; Talairach and Tournoux, 1988). These evidence highlights how the conceptualization of STN anatomy cannot be only descriptive, but it must also consider the functional role that these nuclei have in brain networks. As stated before, the importance of implementing functional with descriptive radiological data is not only theoretical but it is also practical, since detecting STN during DBS procedures is fundamental in order to successfully complete the surgical intervention. One useful methodology to evaluate in vivo effects of STN-DBS is represented by intraoperative high-field 1.5 or 3T fMRI (Hamani et al., 2004; Mallet et al., 2007; Phillips et al., 2006; Jech et al., 2001). In seldom studies the effects of DBS on STN were evaluated by measuring the BOLD effect caused by electrode stimulation during surgery and the results confirmed the activation of both motor and nonmotor circuitry, even though more consistent responses were retrieved from stimulation of motor networks (Obeso et al., 2011).

Another methodology used for detecting STN with more accuracy is based on MRI with 3D reconstructions, but this time the correction made for assessing these nuclei intraoperatively is not radiological, but electrophysiological (Stefurak et al., 2003). After planning the target for STN-DBS, during the surgical procedure an electrophysiological study in the area close to the target is carried out through microrecordings along with clinical assessment (Stefurak et al., 2003). If the recordings coincide with the radiological identification, there is a confirmation that the target for the DBS is correct. On the other hand, discrepancy between 3D MRI and microrecordings can lead to a correction of electrodes' trajectory on the base of the electrophysiological and clinical data.

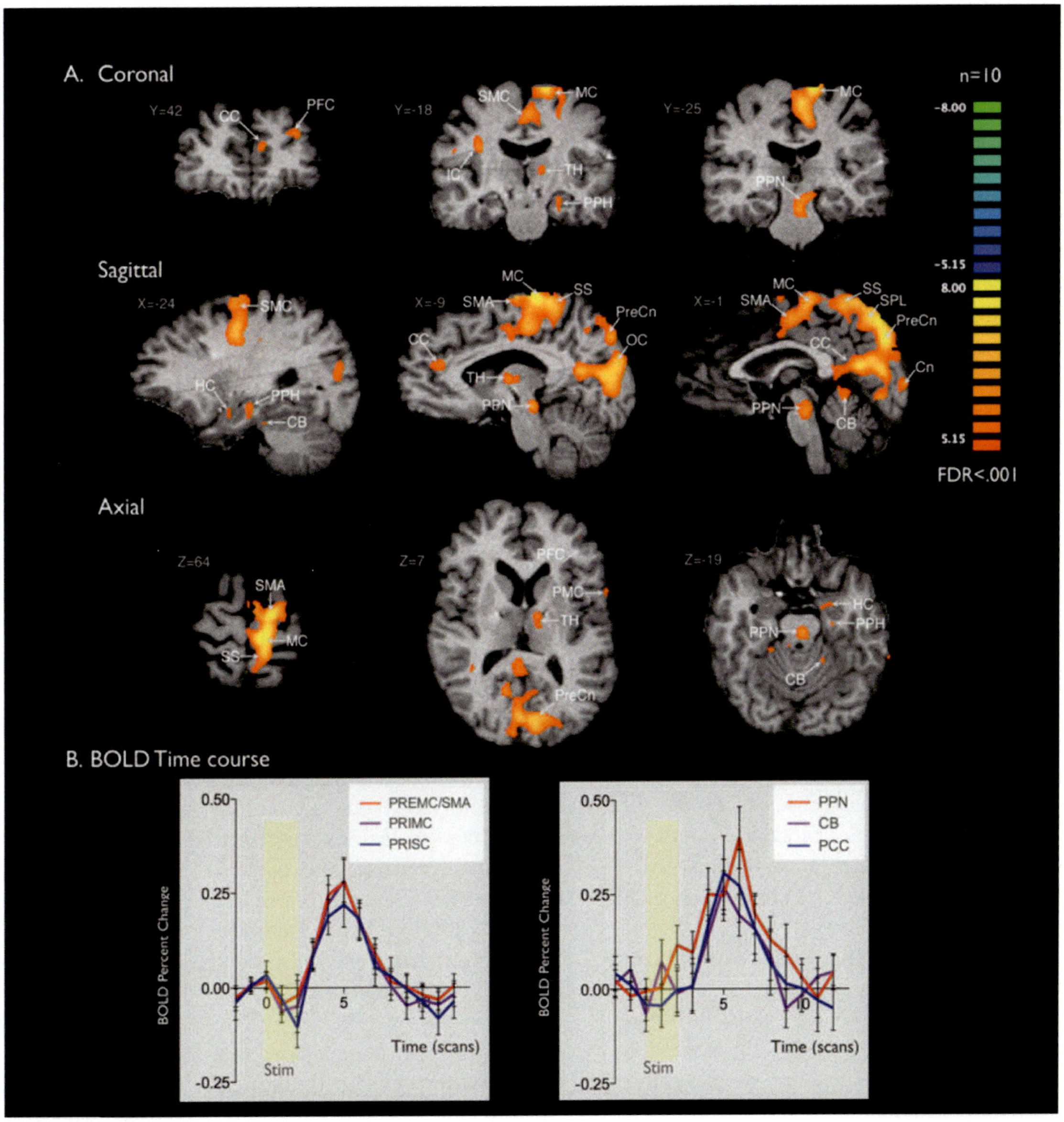

FIGURE 9.7 **Areas of activation with unilateral STN stimulation at 2 V 130–185 Hz 90 μs (n = 10) for PD.** Slice Locations are presented in Talairach coordinates. Significant activation (FDR < .001) was observed in bilateral premotor and primary motor cortices, precuneus, occipital lobes, cerebellum, and anterior and posterior cingulate. Activation of ipsilateral thalamus, pedunculopontine nucleus, parahippocampal gyrus, and hippocampus, and contralateral insula were also observed. (B) The average time courses for five regions of interest were plotted as average percent change in BOLD signal from baseline vs. time (one scan is equal to TR = 3 seconds) using ten frames (30 seconds) prior to stimulation (yellow box) as the baseline. *CB*, cerebellum; *CC*, cingulate cortex; *HC*, hippocampus; *IC*, insula; *MC*, motor cortex; *OC*, occipital cortex; *PFC*, prefrontal cortex; *PPH*, parahippocampal gyrus; *PPN*, pedunculopontine nucleus; *PreCn*, precuneus; *SMA*, supplementary motor area; *SPL*, supramarginal gyrus; *SS*, somatosensory cortex; *TH*, thalamus. *Image courtesy of Knight, E.J., Testini, P., Min, H.K., Gibson, W.S., Gorny, K.R., Favazza, C.P., Felmlee, J.P., Kim, I., Welker, K.M., Clayton, D.A., Klassen, B.T., Chang, S.Y., Lee, K.H., June 2015. Motor and non-motor circuitry activation induced by subthalamic nucleus deep brain stimulation in patients with Parkinson disease: intraoperative functional magnetic resonance imaging for deep brain stimulation. Mayo Clin. Proc. 90(6):773-785. https://doi.org/10.1016/j.mayocp.2015.03.022.*

THERAPEUTICAL NOTICES: DBS IN PD, PSYCHIATRIC DISEASES

Neurosurgeons have been at the forefront of exploration of stereotaxic and other adjacent deep brain regions, due to the clinical impetus for treating a wide range of disorders, e.g., obsessive compulsive disorders and others, for which in past years surgery was the primary option. However, in practice stereotaxic surgery often demonstrated higher risk, less efficacy and more ethical dilemmas than initially expected, along with high targeting variability between patients from ventriculography alone, during the 1960s and 1970s in particular. Concurrently medical treatment was improving, beginning with antipsychotics and better anticonvulsants in the 1950s and L-dopa and carbidopa in the 1960s for movement disturbances, and years later with the use selective serotonin reuptake inhibitor (SSRIs) for the treatment of obsessive-compulsive disorders, so that stereotaxic treatment options were reduced to primarily thalamic treatment of refractory tremor by the 1970s. Thus, very few of the exploratory treatment options refined by neurosurgeons after the advent of stereotaxic approaches in 1949 continued to be valid treatment options, though concepts of human neuroanatomy and interpatient variability were extensively improved by these human exploratory procedures (Deuschl et al., 2006).

Fortunately, much of this rich past has now been resurrected with novel neurobiology concepts and new treatment approaches, as well as adapting the venerable ventriculography guided stereotactic instruments (i.e., Leksell frame in particular) to advanced imaging techniques (i.e., MRI and CT) (Butson and McIntyre, 2005). For example, in the 1980s image-directed stereotaxic biopsy became a significant improvement over open biopsy, then the first, crude version of DBS from the 1970s was extensively adapted to treatment of movement disorders, initially tremor (Deuschl et al., 2011). 1-methyl-4-phenyl-1,2,3,6-tetrahydropyridine (MPTP) was accidently discovered, leading to improved nonhuman primate models of Parkinsonism and hence description of the critical role of subthalamic nucleus (STN) in Parkinson's disease (Parkin et al., 2001). The convergence of these two advances in accurate stereotaxic targeting and neurobiological underpinnings has led to the highly successful approach of STN DBS, now the gold standard of treatment of Parkinson's disease after medical management has started to wane. The knowledge of white matter in the subthalamus also remains relevant, as many neurosurgeons continue to leave an STN DBS contact within the Forel's Fields and lenticular fasciculus superficial to STN because this contact, activating pallidofugal fibers, continues to be useful clinically. Further, the description of CNS growth factors based on the NGF prototype has led in the early 1990s (Hoffert et al., 1994), to development of variously differentiated CNS stem cells and identification of crucial growth factors leading to induction and maintenance of specific CNS cell types, particularly GDNF and Neurturin (NRTN) for dopaminergic cells (Mahato et al., 2020). These specific dopamine neuronal growth factors have now been tried in intraventricular form, direct putaminal infusion, and gene therapy approaches. Further, the possibility of viral-based gene therapy has also been tried in co-opting the primary glutamatergic (excitatory) basis of STN to GABAergic (inhibitory) function using GAD, the synthesizing enzyme for GABA (Buttery and Barker, 2020). Even newer neurobiological approaches have started to define the role of STN in Parkinson's disease, particularly the important role of motor cortex collaterals to STN, using optogenetic approaches. STN may also be useful for implementation of a subcortical brain–machine interface (BMI), due to its optimal location at the convergence of motor systems (Carrillo-Ruiz et al., 2007). Further, continued research on mechanisms of brain function from intraoperative and postoperative studies flourishes, fostering rapid advances in both understanding of the structures being analyzed and treatment routes involving these structures.

In the long quest to both improve subcortical function in a variety of diseases and to analyze function of this difficult to access region, neurosurgeons have continued to take a pivotal role in describing both physiological function of human brain as well as defining therapeutic approaches. In our current search for new therapeutic targets which might transcend the limitations of available therapy sometimes reinvigorating an old site makes considerable sense. This was the case of Laitinen's further description of Leksell's older cases of ventral pallidal lesioning for Parkinson's: these patients quietly did well for many years before a re-examination revealed long-term efficacy in Parkinson's disease (Laitinen et al., 1992). Most available therapeutic targets have either derived from older concepts of brain function (through lesions) or from "designer" surgery based on new scientific approaches. The latter includes the preclinical description of the effects of STN lesions in MPTP nonhuman primates as well as the functional imaging studies pointing to subcallosal frontal lobe involvement (i.e., BA 25) in the treatment of refractory depression (Parkin et al., 2001). Most other targets are based on older lesion data, including thalamic DBS for tremor, anterior thalamic DBS for epilepsy, pallidal DBS for Parkinson's disease, and anterior limb internal capsule DBS for depression/OCD (Follett et al., 2010). Thus, occasionally a study of an older target can be revived with new techniques (i.e., DBS instead of lesions' removal) or new insights into connectivity and refined for future therapeutic approaches (Darmani et al., 2023).

Despite the extensive use of the STN as a DBS target, understanding of the 3D anatomy of the subthalamic region remains challenging due to the variable shape, oblique orientation, relatively small size, and extensive connectivity of the nucleus. As previously mentioned, the STN's intimate relationships with other elusive deep brain structures, such as the zona incerta (ZI), ansa lenticularis, medial forebrain bundle (MFB), medial lemniscus, and substantia nigra (Neudorfer and Maarouf, 2018).

In some publications dealing with human DBS, the term "posterior subthalamic area" (PSA) has been used to describe an area that contains the zona incerta. The term is used to acknowledge the potential current spread into adjacent structures that may partially or totally explain therapeutic effects (Blomstedt et al., 2009a, b). Defined in this way, the PSA is characterized as an area lying inferior to the thalamus, lateral to the red nucleus, and posteromedial to the subthalamic nucleus (Blomstedt et al., 2009a, b). This, in fact, corresponds with Steven Gill's description of the human caudal zona incerta, but PSA is a more inclusive and anatomically accurate term for this purpose. The published PSA DBS target of Gill (Plaha et al., 2006; Heise and Mitrofanis, 2004) is superiorly located in the caudal zona incerta (Fytagoridis et al., 2012). Since 2007, however, that group has shifted their electrode about 1–2 mm more anteriorly, closer to the junction of the posteromedial superior border of the subthalamic nucleus and the caudal zona incerta because of late side effects from stimulating at the original target (S. Gill pers. commun., 2007). Detailed target analysis in the PSA is made possible by using the MRI-directed guide tube technique that enables accurate intraoperative imaging of eventual electrode location (Patel et al., 2008; Thani et al., 2011). Other variations of PSA targeting are difficult to verify anatomically with precision because of technical disadvantages in standard computed tomographic and magnetic resonance imaging of actual brain electrodes as compared with guide tube stylets (Thani et al., 2011).

Bradykinesia in Parkinson's disease is associated with enhanced beta oscillations in the foreband and decrease in the beta frequency oscillations in the subthalamic nucleus and cortex instead has been shown to correlate with a reduction in bradykinesia (Brown, 2007). However, McCarthy et al. (2011) argue that beta oscillations of parkinsonism may primarily arise in the striatum and not in the cortex or subthalamic systems. Their study was based on mathematical modeling of the activity of striatal medium spiny neurons, supported by striatal carbachol injection experiments. Previous to this study, the circuits considered most likely were the subthalamic nucleus/external globus pallidus circuit (Plenz and Kital, 1999; Bevan et al., 2002) and the cortex/subthalamic "patterning" circuit (Magill et al., 2001). In support of the "striatal" hypothesis of McCarthy et al. (2011) is the demonstration by Costa et al. (2006) and Kreitzer (2009) that removal of the dopamine input to the striatum increases the excitability of dopamine receptors. This is consistent with the idea that beta oscillations arise from abnormalities of activity in corticostriatal circuits rather than primary variations in cortical activity. The "striatal" hypothesis fits well with the core finding of loss of striatal dopamine input from the substantia nigra in Parkinson's disease. These studies also support the increasingly popular idea that suppression of beta oscillation may be the mechanism of DBS in Parkinson's disease.

The importance of beta oscillations in DBS in the subthalamic region has recently been challenged. The first and obvious argument against it is that beta oscillations are not evident in essential tremor, despite the fact that the same PSA, thalamic, and subthalamic nucleus targets, that can improve Parkinson's disease, also improve essential tremor. In the context of parkinsonism, Li et al. (2012) showed that beta wave activity can be suppressed by antidromic modification of the corticofugal projection to the subthalamic nucleus. This does not directly dismiss the possibility of an underlying striatal origin for the beta oscillations, but it does argue against a primary role for the neurons of the subthalamic nucleus (Horisawa et al., 2019). An even more fundamental challenge to the beta-oscillation theory of Parkinson's disease was provided by interventions using optogenetics. In these experiments, laser-induced beta oscillations in subthalamic neurons were not sufficient to produce Parkinsonian signs in a transgenic mouse model and the mechanism of DBS-like high-frequency electrical stimulation of the subthalamic nucleus was retrograde modulation of subthalamic neuron afferents (Gradinaru et al., 2009). Optogenetic studies have not yet been published for the caudal zona incerta neurons or their afferents and this technology cannot be used to study primates at this stage. Like the subthalamic nucleus, the zona incerta has extensive connections throughout the brain, leading to the hypothesis that it may have a fundamental regulatory function (Power et al., 1999). Although beta oscillations are present in Parkinson's disease, they appear unlikely to be the underlying cause of symptoms. The extensive connectivity of both the subthalamic nucleus and possibly the zona incerta may yet underpin the comprehensive motor therapeutic effects on rigidity, akinesia, bradykinesia, and tremor when DBS is used in these targets. In terms of the anatomy of the caudal zona incerta in relation to DBS, current concepts and recent studies have not ruled out the possibility that therapeutic effects may be due to one or more of (i) orthodromic efferent nuclear neuron axonal, (ii) antidromic afferent nuclear neuron axonal, or (iii) ortho- or antidromic adjacent tract axonal electrical modulation (Magill et al., 2001; Mai et al., 2008).

In the 1960s and 1970s, stereotactic ablation of Forel's field H (Forel-H-tomy) was performed for the treatment of Parkinson's disease (PD) and cervical dystonia. Almost simultaneously, Jinnai and Mukawa (1970) reported using Forel-H-tomy for treating epilepsy. Although the ablation of Forel's field was forgotten for several decades Gallay's group revived Spiegel's Forel-H-tomy (renamed to "pallidothalamic tractotomy") for the treatment of PD using radiofrequency and focused ultrasound ablation (Gallay et al., 2008).

Lesioning the Forel's field or the subthalamic region is considered a possible treatment for tremoric patients with Parkinson disease, essential tremor and other diseases. This surgical treatment was performed in the 1960s–70s and was an alternative to thalamotomy. Recently, there has been increasing interest in the reappraisal of stimulating and/or lesioning these targets, partly as a result of innovations in imaging and noninvasive ablative technologies, such as magnetic resonance–guided focused ultrasonography.

Nowadays, based on the significant breakthrough that magnetic resonance–guided focused ultrasonography has meant in the neurosurgical world, some classic targets such as the pallidothalamic tract, Forel's fields, and posterior subthalamic area may be reconsidered as surgical alternatives for patients with movement disorders (Horisawa et al., 2021a; Lam et al., 2011).

Forel-H-tomy is also used for the treatment of intractable epilepsy. This indication was introduced by Dennosuke Jinnai in the 1960s. The rationales for Forel-H-tomy in treating epilepsy are interrupting the pathway of epileptic discharge (descending impulse) and elevating the seizure threshold based on experimental animal work (Spiegel et al., 1963). Patients who undergo bilateral Forel-H-tomy may develop serious speech and motor disturbances. To avoid possible complications associated with lesions, deep brain stimulation (DBS) can be a treatment option. DBS disrupts abnormal information flow through the stimulation site (Jinnai and Mukawa, 1970). Therefore, DBS provides similar therapeutic effects to lesioning surgery. The disruption of pathological information flow associated with movement disorders or epilepsy at PTT may achieve symptomatic improvement. The stimulation not only improved the dystonic tremor but also dramatically decreased seizure frequency from two to three seizures each week before surgery to two seizures in 3 years after surgery. Co-stimulation of Forel's field H1 or H2 may be responsible for the dramatic seizure reduction (Guridi and Gonzalez-Quarante, 2021).

Some *Supplementary Material* on the neuroanatomy of the subthalamus are included in the *Appendix B* of this book.

References

Acar, F., Miller, J.P., Berk, M.C., Anderson, G., Burchiel, K.J., 2007. Safety of anterior commissure-posterior commissure-based target calculation of the subthalamic nucleus in functional stereotactic procedures. Stereotact. Funct. Neurosurg. 85 (6), 287–291.

Afsharpour, S., 1985. Topographical projections of the cerebral cortex to the subthalamic nucleus. J. Comp. Neurol. 236 (1), 14-28. doi: 10.1002/cne.902360103. PMID: 2414329.

Albin, R.L., Young, A.B., Penney, J.B., 1989. The functional anatomy of basal ganglia disorders. Trends Neurosci. 12 (10), 366-375. doi: 10.1016/0166-2236(89)90074-x. PMID: 2479133.

Alexander, G.E., Crutcher, M.D., 1990. Functional architecture of basal ganglia circuits: neural substrates of parallel processing. Trends Neurosci. 13 (7), 266-271. doi: 10.1016/0166-2236(90)90107-l. PMID: 1695401.

Aravamuthan, B.R., Muthusamy, K.A., Stein, J.F., Aziz, T.Z., Johansen-Berg, H., 2007. Topography of cortical and subcortical connections of the human pedunculopontine and subthalamic nuclei. Neuroimage 37, 694–705.

Ashkan, K., Blomstedt, P., Zrinzo, L., Tisch, S., Yousry, T., Limousin-Dowsey, P., Hariz, M.I., April 2007. Variability of the subthalamic nucleus: the case for direct MRI guided targeting. Br. J. Neurosurg. 21 (2), 197–200.

Bahatia, K.P., Marsden, C.D., 1994. The behavioral and motor consequences of focal lesions of the basal ganglia in man. Brain 117, 859–876.

Barbier, M., Risold, P.Y., October 4, 2021. Understanding the significance of the hypothalamic Nature of the subthalamic nucleus. eNeuro 8 (5). ENEURO.0116-21.2021.

Bell, K., Churchill, L., Kalivas, P.W., May 1995. GABAergic projection from the ventral pallidum and globus pallidus to the subthalamic nucleus. Synapse 20 (1), 10–18.

Berney, A., Vingerhoets, F., Perrin, A., 2002. Effect on mood of subthalamic DBS for Parkinson's disease: a consecutive series of 24 patients. Neurology 59, 1427–1429.

Bevan, M.D., Magill, P.J., Terman, D., et al., 2002. Move to the rhythm: oscillations in the subthalamic nucleus-external globus pallidus network. Trends Neurosci. 25, 525–531.

Bittencourt, J.C., Presse, F., Arias, C., et al., 1992. The melanin-concentrating hormone system of the rat brain: an immuno-and hybridization histochemical characterization. J. Comp. Neurol. 319, 218–245.

Blomstedt, P., Sandvik, U., Fytagoridis, A., et al., 2009a. The posterior subthalamic area in the treatment of movement disorders: past, present, and future. Neurosurgery 64, 1029–1038.

Blomstedt, P., Fytagoridis, A., Tisch, S., 2009b. Deep brain stimulation of the posterior subthalamic area in the treatment of tremor. Acta Neurochir. 151, 31–36.

Blomstedt, P., Fytagoridis, A., Astrom, M., 2012. Unilateral caudal zona incerta deep brain stimulation for Parkinsonian tremor. Parkinsonism Relat. Disord. 18, 1062–1066.

Bostan, A.C., Dum, R.P., Strick, P.L., 2010. The basal ganglia communicate with the cerebellum. Proc. Natl. Acad. Sci. U.S.A. 107, 8452–8456. https://doi.org/10.1073/pnas.1000496107.

Brown, P., 2007. Abnormal oscillatory synchronisation in the motor system leads to impaired movement. Curr. Opin. Neurobiol. 17, 656–664.

Brown, L.L., Markman, M.H., Wolfson, L.I., Dvorkin, B., Warner, C., Karzman, R., 1979. A direct role of dopamine in the rat subthalamic nucleus and an adjacent intrapeduncular area. Science 206, 1416–1418. https://doi.org/10.1126/science.505015.

Burton, H., Jones, E.G., 1976. The posterior thalamic region and its cortical projection in new world and old world monkeys. J. Comp. Neurol. 168, 249–302.

Butson, C.R., McIntyre, C.C., 2005. Tissue and electrode capacitance reduce neural activation volumes during deep brain stimulation. Clin. Neurophysiol. 116, 2490–2500.

Buttery, P.C., Barker, R.A., 2020. Gene and cell-based therapies for Parkinson's disease: where are we? Neurotherapeutics 17 (4), 1539–1562. https://doi.org/10.1007/s13311-020-00940-4.

Carrillo-Ruiz, J.D., Velasco, F., Jimenez, F., et al., 2007. Neuromodulation of prelemniscal radiations in the treatment of Parkinson's disease. Acta Neurochirurg. 97, 185–190.

Chang, H.T., Kita, H., Kitai, S.T., 1983. The fine structure of the rat subthalamic nucleus: an electron microscopic study. J. Comp. Neurol. 221 (1), 113-123. doi: 10.1002/cne.902210110. PMID: 6643743.

Chometton, S., Barbier, M., Risold, P.Y., 2021. The zona incerta system: involvement in attention and movement. Handb. Clin. Neurol. 180, 173–184.

Clarke, N.P., Bevan, M.D., Cozzari, C., Hartman, B.K., Bolam, J.P., 1997. Glutamate-enriched cholinergic synaptic terminals in the entopeduncular nucleus and subthalamic nucleus of the rat. Neuroscience 81 (2), 371-385. doi: 10.1016/s0306-4522(97)00247-9. PMID: 9300428.

Cobos, I., Shimamura, K., Rubenstein, J.L., Martínez, S., Puelles, L., 2001. Fate map of the avian anterior forebrain at the four-somite stage, based on the analysis of quail-chick chimeras. Dev. Biol. 239 (1), 46-67. doi: 10.1006/dbio.2001.0423. PMID: 11784018.

Costa, R.M., Lin, S.C., Sotnikova, T.D., et al., 2006. Rapid alterations in corticostriatal ensemble coordination during acute dopamine-dependent motor dysfunction. Neuron 52, 359–369.

Cuny, E., Guehl, D., Burbaud, P., Gross, C., Dousset, V., Rougier, A., September 2002. Lack of agreement between direct magnetic resonance imaging and statistical determination of a subthalamic target: the role of electrophysiological guidance. J. Neurosurg. 97 (3), 591–597.

Daniluk, S., G Davies, K., Ellias, S.A., Novak, P., Nazzaro, J.M., February 2010. Assessment of the variability in the anatomical position and size of the subthalamic nucleus among patients with advanced Parkinson's disease using magnetic resonance imaging. Acta Neurochir. 152 (2), 201–210.

Darmani, G., Drummond, N.M., Ramezanpour, H., et al., 2023. Long-term recording of subthalamic aperiodic activities and beta bursts in Parkinson's disease. Mov. Disord. 38 (2), 232–243.

Deuschl, G., Schade-Brittinger, C., Krack, P., et al., 2006. A randomized trial of deep-brain stimulation for Parkinson's disease. N. Engl. J. Med. 355, 896–908 [Erratum appears in N. Engl. J. Med. 2006: 355, 1289].

Deuschl, G., Raethjen, J., Hellriegel, H., et al., 2011. Treatment of patients with essential tremor. Lancet Neurol. 10, 148–161.

Dong, J., Xu, H., Xu, H., Wang, P.F., Cai, G.J., Song, H.F., Wang, C.C., Dong, Z.T., Ju, Y.J., Jiang, Z.Y., 2013. Nesfatin-1 stimulates fatty-acid oxidation by activating AMP-activated protein kinase in STZ-induced type 2 diabetic mice. PLoS One 8 (12), e83397. doi: 10.1371/journal.pone.0083397. PMID: 24391760; PMCID: PMC3877039.

Dostrovsky, J.O., Hutchison, W.D., Lozano, A.M., 2002. The globus pallidus, deep brain stimulation, and Parkinson's disease. Neuroscientist 8 (3), 284-290. doi: 10.1177/1073858402008003014. PMID: 12061508.

Emmi, A., Antonini, A., Macchi, V., Porzionato, A., De Caro, R., April 22, 2020. Anatomy and connectivity of the subthalamic nucleus in humans and non-human primates. Front. Neuroanat. 14, 13.

Follett, K.A., Weaver, F.M., Stern, M., et al., 2010. Pallidal versus subthalamic deep-brain stimulation for Parkinson's disease. N. Engl. J. Med. 362, 2077–2091.

Franklin, S., 2017. The peripheral and central nervous system. In: Conn's Translational Neuroscience. Academic Press.

Fytagoridis, A., Sandvik, U., Astrom, M., et al., 2012. Long term follow-up of deep brain stimulation of the caudal zona incerta for essential tremor. J. Neurol. Neurosurg. Psychiatry 83, 258–262.

Gallay, M.N., Jeanmonod, D., Liu, J., et al., 2008. Human pallidothalamic and cerebellothalamic tracts: anatomical basis for functional stereotactic neurosurgery. Brain Struct. Funct. 212, 443–463.

Gerfen, C.R., Staines, W.A., Arbuthnott, G.W., Fibiger, H.C., 1982. Crossed connections of the substantia nigra in the rat. J. Comp. Neurol. 207 (3), 283-303. doi: 10.1002/cne.902070308. PMID: 7107988.

Gradinaru, V., Mogri, M., Thompson, K.R., et al., 2009. Optical deconstruction of Parkinsonian neural circuitry. Science 324, 354–359.

Grill, W.M., Snyder, A.N., Miocinovic, S., 2004. Deep brain stimulation creates an informational lesion of the stimulated nucleus. Neuroreport 15 (7), 1137-1140. doi: 10.1097/00001756-200405190-00011. PMID: 15129161.

Groenewegen, H.J., Berendse, H.W., Wolters, J.G., Lohman, A.H., 1990. The anatomical relationship of the prefrontal cortex with the striatopallidal system, the thalamus and the amygdala: evidence for a parallel organization. Prog Brain Res. 85, 95-116; discussion 116-8. doi: 10.1016/s0079-6123(08)62677-1. PMID: 2094917.

Guehl, D., Edwards, R., Cuny, E., Burbaud, P., Rougier, A., Modolo, J., Beuter, A., January 2007. Statistical determination of the optimal subthalamic nucleus stimulation site in patients with Parkinson disease. J. Neurosurg. 106 (1), 101–110.

Güngör, A., Baydın, S.S., Holanda, V.M., Middlebrooks, E.H., Isler, C., Tugcu, B., Foote, K., Tanriover, N., May 4, 2018. Microsurgical anatomy of the subthalamic nucleus: correlating fiber dissection results with 3-T magnetic resonance imaging using neuronavigation. J. Neurosurg. 130 (3), 716–732.

Guridi, J., Gonzalez-Quarante, L.H., March 2021. Revisiting Forel field surgery. World Neurosurg. 147, 11–22.

Haegelen, C., Coupé, P., Fonov, V., et al., 2013. Automated segmentation of basal ganglia and deep brain structures in MRI of Parkinson's disease. Int. J. Comput. Assist. Radiol. Surg. 8 (1), 99–110.

Hahn, J.D., Swanson, L.W., 2012. Connections of the lateral hypothalamic area juxtadorsomedial region in the male rat. J. Comp. Neurol. 520 (9), 1831-1890. doi: 10.1002/cne.23064. PMID: 22488503; PMCID: PMC3930178.

Hamani, C., Saint-Cyr, J.A., Fraser, J., Kaplitt, M., Lozano, A.M., 2004. The subthalamic nucleus in the context of movement disorders. Brain 127 (Pt 1), 4-20. doi: 10.1093/brain/awh029. Epub 2003 Nov 7. PMID: 14607789.

Hamani, C., Saint-Cyr, J.A., Fraser, J., Kaplitt, M., Lozano, A.M., January 2004. The subthalamic nucleus in the context of movement disorders. Brain 127 (Pt 1), 4–20.

Heise, C.E., Mitrofanis, J., 2004. Evidence for a glutamatergic projection from the zona incerta to the basal ganglia of rats. J. Comp. Neurol. 468, 482–495.

Hoffer, B.J., Hoffman, A., Bowenkamp, K., et al., 1994. Glial cell line-derived neurotrophic factor reverses toxin-induced injury to midbrain dopaminergic neurons in vivo. Neurosci. Lett. 182, 107–111.

Horisawa, S., Nanke, M., Kawamata, T., Taira, T., 2019. Pallidothalamic tractotomy for Parkinson disease with 1-year follow-up: a case report. World Neurosurg. 121, 193–195.

Horisawa, S., Kohara, K., Murakami, M., Fukui, A., Kawamata, T., Taira, T., 2021a. Deep brain stimulation of the Forel's field for dystonia: preliminary results. Front. Hum. Neurosci. 15, 768057.

Horisawa, S., Miyao, S., Hori, T., Kohara, K., Kawamata, T., Taira, T., 2021b. Comorbid seizure reduction after pallidothalamic tractotomy for movement disorders: revival of Jinnai's Forel-H-tomy. Epilepsia Open 6 (1), 225–229.

Jech, R., Urgosík, D., Tintera, J., Nebuzelský, A., Krásenský, J., Liscák, R., Roth, J., Růzicka, E., November 2001. Functional magnetic resonance imaging during deep brain stimulation: a pilot study in four patients with Parkinson's disease. Mov. Disord. 16 (6), 1126–1132.

Jinnai, D., Mukawa, J., 1970. Forel-H-tomy for the treatment of epilepsy. Confin. Neurol. 32 (2), 307–315.

Joel, D., Weiner, I., 1997. The connections of the primate subthalamic nucleus: indirect pathways and the open-interconnected scheme of basal ganglia-thalamocortical circuitry. Brain Res. Brain Res. Rev. 23 (1-2), 62-78. doi: 10.1016/s0165-0173(96)00018-5. PMID: 9063587.

Joel, D., Weiner, I., 2000. The connections of the dopaminergic system with the striatum in rats and primates: an analysis with respect to the functional and compartmental organization of the striatum. Neuroscience 96 (3), 451-474. doi: 10.1016/s0306-4522(99)00575-8. PMID: 10717427.

Jones, E.G., 2007. The Thalamus, second ed. Cambridge University Press, Cambridge.

Kawana, E., Watanabe, K., 1981. A cytoarchitectonic study of zona incerta in the rat. J. Hirnforsch. 22, 535–541.

Kearney, J.A., Albin, R.L., 2000. Intrasubthalamic nucleus metabotropic glutamate receptor activation: a behavioral, Fos immunohistochemical and [14C]2-deoxyglucose autoradiographic study. Neuroscience 95 (2), 409-416. doi: 10.1016/s0306-4522(99)00439-x. PMID: 10658620.

Kahle, W., 1956. Zur Entwicklung des menschlichen Zwischenhirnes; studien über die Matrixphasen und die örtlichen Reifungsunterschiede im embryonalen menschlichen Gehirn. II [Development of the human diencephalon; studies on the phases of the matrix and the local differences in maturation in the human embryonic brain. II]. Dtsch Z Nervenheilkd 175 (3), 259-318. German. PMID: 13397345.

Keyser, A., 1972. The development of the diencephalon of the Chinese hamster. An investigation of the validity of the criteria of subdivision of the brain. Acta Anat. Suppl. (Basel) 59, 1-178. PMID: 4540788.

Kitai, S.T., Kita, H., 1987. Anatomy and physiology of the subthalamic nucleus: a driving force of the basal ganglia. In: Carpenter, M.B., Jayaraman, A. (Eds.), The Basal Ganglia II, Advances in Behavioral Biology, vol 32. Springer, Boston, MA.

Knight, E.J., Testini, P., Min, H.K., Gibson, W.S., Gorny, K.R., Favazza, C.P., Felmlee, J.P., Kim, I., Welker, K.M., Clayton, D.A., Klassen, B.T., Chang, S.Y., Lee, K.H., June 2015. Motor and nonmotor circuitry activation induced by subthalamic nucleus deep brain stimulation in patients with Parkinson disease: intraoperative functional magnetic resonance imaging for deep brain stimulation. Mayo Clin. Proc. 90 (6), 773−785.

Kolmac, C., Mitrofanis, J., 1999. Distribution of various neurochemicals within the zona incerta: an immunocytochemical and histochemical study. Anat. Embryol. 199, 265−280.

Kosta, P., Argyropoulou, M.I., Markoula, S., Konitsiotis, S., 2006. MRI evaluation of the basal ganglia size and iron content in patients with Parkinson's disease. J. Neurol. 253 (1), 26-32. doi: 10.1007/s00415-005-0914-9. Epub 2005 Jun 29. PMID: 15981079.

Kreitzer, A.C., 2009. Physiology and pharmacology of striatal neurons. Annu. Rev. Neurosci. 32, 127−147.

Kuhlenbeck, H., 1951. The derivatives of thalamus dorsalis and epithalamus in the human brain: their relation to cortical and other centers. Mil. Surg. 108 (3), 205-256. PMID: 14806148.

Laitinen, L.V., Bergenheim, A.T., Hariz, M.I., 1992. Leksell's posteroventral pallidotomy in the treatment of Parkinson's disease. J. Neurosurg. 76, 53−61.

Lam, M.F., Thomas, M.G., Lind, C.R., 2011. Neurosurgical convection-enhanced delivery of treatments for Parkinson's disease. J. Clin. Neurosci. 18, 1163−1167.

Lanotte, M.M., Rizzone, M., Bergamasco, B., Faccani, G., Melcarne, A., Lopiano, L., January 2002. Deep brain stimulation of the subthalamic nucleus: anatomical, neurophysiological, and outcome correlations with the effects of stimulation. J. Neurol. Neurosurg. Psychiatry 72 (1), 53−58.

Lanska, D.J., 2010. Chapter 33: the history of movement disorders. Handb. Clin. Neurol. 95, 501-546. doi: 10.1016/S0072-9752(08)02133-7. PMID: 19892136.

Larsen, M., Bjarkam, C.R., Østergaard, K., West, M.J., Sørensen, J.C., 2004. The anatomy of the porcine subthalamic nucleus evaluated with immunohistochemistry and design-based stereology. Anat. Embryol. (Berl). 208 (3), 239-247. doi: 10.1007/s00429-004-0395-0. Epub 2004 May 27. PMID: 15168115.

Lau, J.C., Xiao, Y., Haast, R.A.M., Gilmore, G., Uludağ, K., MacDougall, K.W., Menon, R.S., Parrent, A.G., Peters, T.M., Khan, A.R., November 2020. Direct visualization and characterization of the human zona incerta and surrounding structures. Hum. Brain Mapp. 41 (16), 4500−4517.

Lévesque, J.C., Parent, A., 2005. GABAergic interneurons in human subthalamic nucleus. Mov. Disord. 20 (5), 574-584. doi: 10.1002/mds.20374. PMID: 15645534.

Li, Z., Decavel, C., Hatton, G.I., 1995. Calbindin-D28k: role in determining intrinsically generated firing patterns in rat supraoptic neurons. J Physiol. 488, 601−608.

Li, Q., Ke, Y., Chan, D.C., et al., 2012. Therapeutic deep brain stimulation in Parkinsonian rats directly influences motor cortex. Neuron 76, 1030−1041.

Li, Z., Rizzi, G., Tan, K.R., September 10, 2021. Zona incerta subpopulations differentially encode and modulate anxiety. Sci. Adv. 7 (37), eabf6709.

Llinas, R., Jahnsen, H., 1982. Electrophysiology of mammalian thalamic neurons in-vitro. Nature 297, 406−408.

Lozano, A.M., Carella, F., 2002. Physiologic studies in the human brain in movement disorders. Parkinsonism Relat. Disord. 8 (6), 455-458. doi: 10.1016/s1353-8020(02)00027-5. PMID: 12217635.

Magill, P.J., Bolam, J.P., Bevan, M.D., 2000. Relationship of activity in the subthalamic nucleus-globus pallidus network to cortical electroencephalogram. J. Neurosci. 20 (2), 820-833. doi: 10.1523/JNEUROSCI.20-02-00820.2000. PMID: 10632612; PMCID: PMC6772398.

Magill, P.J., Bolam, J.P., Bevan, M.D., 2001. Dopamine regulates the impact of the cerebral cortex on the subthalamic nucleus-globus pallidus network. Neuroscience 106, 313−330.

Mahato, A.K., Kopra, J., Renko, J.M., et al., 2020. Glial cell line-derived neurotrophic factor receptor rearranged during transfection agonist supports dopamine neurons in vitro and enhances dopamine release in vivo. Mov. Disord. 35 (2), 245−255.

Mai, J., Paxinos, G., Voss, T., 2008. Atlas of the Human Brain, third ed. Elsevier, Amsterdam.

Mallet, L., Schüpbach, M., N'Diaye, K., Remy, P., Bardinet, E., Czernecki, V., Welter, M.L., Pelissolo, A., Ruberg, M., Agid, Y., Yelnik, J., June 19, 2007. Stimulation of subterritories of the subthalamic nucleus reveals its role in the integration of the emotional and motor aspects of behavior. Proc. Natl. Acad. Sci. U.S.A. 104 (25), 10661−10666.

Marani, E., Heida, T., Lakke, E.A., Usunoff, K.G., 2008. The subthalamic nucleus. Part I: development, cytology, topography and connections. Adv. Anat. Embryol. Cell Biol. 198, 1-113, vii. PMID: 18727483.

Marchand, R., Lajoie, L., Blanchet, C., 1986. Histogenesis at the level of the basal forebrain: the entopeduncular nucleus. Neuroscience 17 (3), 591-607. doi: 10.1016/0306-4522(86)90032-1. PMID: 3703250.

Massey, L.A., Yousry, T.A., 2010. Anatomy of the substantia nigra and subthalamic nucleus on MR imaging. Neuroimaging Clin. N. Am. 20 (1), 7-27. doi: 10.1016/j.nic.2009.10.001. PMID: 19959016.

Maynard, R.L., Downes, N., 2019. Anatomy and Histology of the Laboratory Rat in Toxicology and Biomedical Research. Academic Press.

McClelland 3rd, S., Ford, B., Senatus, P.B., Winfield, L.M., Du, Y.E., Pullman, S.L., Yu, Q., Frucht, S.J., McKhann 2nd, G.M., Goodman, R.R., 2005. Subthalamic stimulation for Parkinson disease: determination of electrode location necessary for clinical efficacy. Neurosurg. Focus 19 (5).

McCarthy, M.M., Moore-Kochlacs, C., Gu, X., Boyden, E.S., Han, X., Kopell, N., 2011. Striatal origin of the pathologic beta oscillations in Parkinson's disease. Proc. Natl. Acad. Sci. USA. 108 (28), 11620-11625. doi: 10.1073/pnas.1107748108. Epub 2011 Jun 22. PMID: 21697509; PMCID: PMC3136295.

McIntyre, C.C., Savasta, M., Kerkerian-Le Goff, L., Vitek, J.L., 2004. Uncovering the mechanism(s) of action of deep brain stimulation:

activation, inhibition, or both. Clin. Neurophysiol. 115 (6), 1239-1248. doi: 10.1016/j.clinph.2003.12.024. PMID: 15134690.

Mihailoff, G.A., Haines, D.E., 2018. Fundamental Neuroscience for Basic and Clinical Applications, fifth ed. Elsevier.

Mitrofanis, J., 2005. Some certainty for the "zone of uncertainty"? Exploring the function of the zona incerta. Neuroscience 130 (1), 1-15. doi: 10.1016/j.neuroscience.2004.08.017. PMID: 15561420.

Montgomery, E.B., 2010. Deep Brain Stimulation Programming: Principles and Practice, vol 17. Oxford University Press, Oxford, p. 179.

Morel, A., 2007. The thalamus and behavior: effects of anatomically distinct strokes. Neurology 68 (19), 1640. doi: 10.1212/01.wnl.0000265608.00944.ff. PMID: 17485660.

Neudorfer, C., Maarouf, M., January 2018. Neuroanatomical background and functional considerations for stereotactic interventions in the H fields of Forel. Brain Struct. Funct. 223 (1), 17–30.

Nieuwenhuys, R., 2009. The forebrain of actinopterygians revisited. Brain Behav. Evol. 73 (4), 229-252. doi: 10.1159/000225622. Epub 2009 Jun 18. PMID: 19546532.

Obeso, I., Ray, N.J., Antonelli, F., Cho, S.S., Strafella, A.P., October 2011. Combining functional imaging with brain stimulation in Parkinson's disease. Int. Rev. Psychiatry 23 (5), 467–475. https://doi.org/10.3109/09540261.2011.621414.

Parent, M., Lévesque, M., Parent, A., 2001. Two types of projection neurons in the internal pallidum of primates: single-axon tracing and three-dimensional reconstruction. J. Comp. Neurol. 439 (2), 162-175. doi: 10.1002/cne.1340. PMID: 11596046.

Parent, A., Hazrati, L.N., 1995. Functional anatomy of the basal ganglia. II. The place of subthalamic nucleus and external pallidum in basal ganglia circuitry. Brain Res. Brain Res. Rev. 20 (1), 128-154. doi: 10.1016/0165-0173(94)00008-d. PMID: 7711765.

Parkin, S., Nandi, D., Giladi, N., et al., 2001. Lesioning the subthalamic nucleus in the treatment of Parkinson's disease. Stereotact. Funct. Neurosurg. 77 (1–4), 68–72.

Patel, N.K., Khan, S., Gill, S.S., 2008. Comparison of atlas- and magnetic-resonance-imaging-based stereotactic targeting of the subthalamic nucleus in the surgical treatment of Parkinson's disease. Stereotact. Funct. Neurosurg. 86 (3), 153–161.

Paxinos, G., 2015. Brain, behaviour and evolution. Adv. Exp. Med. Biol. 821, 1. doi: 10.1007/978-3-319-08939-3_1. PMID: 25416101.

Percheron, G., Filion, M., 1991. Parallel processing in the basal ganglia: up to a point. Trends Neurosci. 14 (2), 55-59. doi: 10.1016/0166-2236(91)90020-u. PMID: 1708537.

Percheron, G., Fénelon, G., Leroux-Hugon, V., Fève, A., 1994. Histoire du système des ganglions de la base. La lente émergence d'un système cérébral majeur [History of the basal ganglia system. Slow development of a major cerebral system]. Rev. Neurol. (Paris) 150 (8-9), 543-54. French. PMID: 7754290.

Phillips, M.D., Baker, K.B., Lowe, M.J., Tkach, J.A., Cooper, S.E., Kopell, B.H., Rezai, A.R., April 2006. Parkinson disease: pattern of functional MR imaging activation during deep brain stimulation of subthalamic nucleus–initial experience. Radiology 239 (1), 209–216.

Plaha, P., Ben-Shlomo, Y., Patel, N.K., Gill, S.S., 2006. Stimulation of the caudal zona incerta is superior to stimulation of the subthalamic nucleus in improving contralateral parkinsonism. Brain 129 (Pt 7), 1732-1747. doi: 10.1093/brain/awl127. Epub 2006 May 23. PMID: 16720681.

Plenz, D., Kital, S.T., 1999. A basal ganglia pacemaker formed by the subthalamic nucleus and external globus pallidus. Nature 400 (6745), 677-682. doi: 10.1038/23281. PMID: 10458164.

Pombero, A., Martinez, S., 2009. Telencephalic morphogenesis during the process of neurulation: an experimental study using quail-chick chimeras. J. Comp. Neurol. 512 (6), 784-797. doi: 10.1002/cne.21933. PMID: 19065633.

Power, B.D., Kolmac, C.I., Mitrofanis, J., 1999. Evidence for a large projection from the zona incerta to the dorsal thalamus. J. Comp. Neurol. 404 (4), 554-565. PMID: 9987997.

Puelles, L., 2019. Survey of midbrain, diencephalon, and hypothalamus neuroanatomic terms whose prosomeric definition conflicts with columnar tradition. Front Neuroanat. 13, 20. https://doi.org/10.3389/fnana.2019.00020. PMID: 30873012; PMCID: PMC6402269.

Puelles, L., Rubenstein, J.L., 2012. Hyphotalamus. In: The Mouse Nervous System. Academic Press.

Rafols, J.A., Fox, C.A., 1976. The neurons in the primate subthalamic nucleus: a Golgi and electron microscopic study. J. Comp. Neurol. 168 (1), 75-111. doi: 10.1002/cne.901680105. PMID: 819471.

Reiner, P.B., McGeer, E.G., 1987. Electrophysiological properties of cortically projecting histamine neurons of the rat hypothalamus. Neurosci. Lett. 73 (1), 43-47. doi: 10.1016/0304-3940(87)90028-0. PMID: 3561856.

Richter, E.O., Hoque, T., Halliday, W., Lozano, A.M., Saint-Cyr, J.A., March 2004. Determining the position and size of the subthalamic nucleus based on magnetic resonance imaging results in patients with advanced Parkinson disease. J. Neurosurg. 100 (3), 541–546.

Risold, P.Y., Croizier, S., Legagneux, K., Brischoux, F., Fellmann, D., Griffond, B., 2009. The development of the MCH system. Peptides 30 (11), 1969-1972. doi: 10.1016/j.peptides.2009.07.016. Epub 2009 Aug 6. PMID: 19647770.

Rutledge, J.N., Hilal, S.K., Silver, A.J., Defendini, R., Fahn, S., 1987. Study of movement disorders and brain iron by MR. Am. J. Roentgenol. 149 (2), 365-379. doi: 10.2214/ajr.149.2.365. PMID: 2440291.

Schaltenbrand, G., Wahren, W., 1977. Atlas for Stereotaxy of the Human Brain, vol 13. Year Book Medical Publishers, Chicago, p. 66.

Schmidt, R., Herrojo Ruiz, M., Kilavik, B.E., Lundqvist, M., Starr, P.A., Aron, A.R., 2019. Beta oscillations in working memory, executive control of movement and thought, and sensorimotor function. J. Neurosci. 39 (42), 8231–8238.

Shink, E., Bevan, M.D., Bolam, J.P., Smith, Y., 1996. The subthalamic nucleus and the external pallidum: two tightly interconnected structures that control the output of the basal ganglia in the monkey. Neuroscience 73 (2), 335-357. doi: 10.1016/0306-4522(96)00022-x. PMID: 8783253.

Spiegel, E.A., Wycis, H.T., Szekely, E.G., Adams, D.J., Flanagan, M., Baird 3rd, H.W., 1963. Campotomy in various extrapyramidal disorders. J. Neurosurg. 20, 871–884.

Stefurak, T., Mikulis, D., Mayberg, H., Lang, A.E., Hevenor, S., Pahapill, P., Saint-Cyr, J., Lozano, A., December 2003. Deep brain stimulation for Parkinson's disease dissociates mood and motor circuits: a functional MRI case study. Mov. Disord. 18 (12), 1508–1516.

Story, J.L., French, L.A., Chou, S.N., Meier, M.J., 1965. Experiences with subthalamic lesions in patients with movement disorders. Confin. Neurol. 26 (3), 218–221.

Tai, L.S., Ng, T.K., Mak, N.K., Yung, K.K., 2001. Co-localization of AMPA-type glutamate receptor immunoreactivity in neurons of the rat subthalamic nucleus. Brain Res. 895 (1-2), 95-103. doi: 10.1016/s0006-8993(01)02036-4. PMID: 11259765.

Talairach, J., Tournoux, P., 1988. Co-planar Stereotaxic Atlas of the Human Brain: 3-Dimensional Proportional System: An Approach to Cerebral Imaging, vol. 8. Thieme Medical Publishers, New York, p. 122.

Tamma, F., Caputo, E., Chiesa, V., Egidi, M., Locatelli, M., Rampini, P., et al., 2002. Anatomo-clinical correlation of intraoperative stimulation-induced side-effects during HF-DBS of the subthalamic nucleus. Neurol. Sci. 23 (Suppl. 2), S109–S110.

Temel, Y., Blokland, A., Steinbusch, H.W., Visser-Vandewalle, V., 2005. The functional role of the subthalamic nucleus in cognitive and limbic circuits. Prog. Neurobiol. 76 (6), 393-413. doi: 10.1016/j.pneurobio.2005.09.005. Epub 2005 Oct 24. PMID: 16249050.

Temel, Y., Kessels, A., Tan, S., Topdag, A., Boon, P., Visser-Vandewalle, V., 2006. Behavioural changes after bilateral subthalamic stimulation in advanced Parkinson disease: a systematic review. Parkinsonism Relat. Disord. 12 (5), 265-272. doi: 10.1016/j.parkreldis.2006.01.004. Epub 2006 Apr 18. PMID: 16621661.

ten Donkelaar, H.J., Vasung, L., 2015. Anatomy and physiology, systems. In: Brain Mapping. Academic Press.

Thani, N.B., Bala, A., Kimber, T.E., Lind, C.R., 2011. High-frequency pallidal stimulation for camptocormia in Parkinson disease: case report. Neurosurgery 68 (5), E1501-E1505. doi: 10.1227/NEU.0b013e318210c859. PMID: 21307785.

Toda, H., Kambe, D., Shima, A., Nishida, N., Sawamoto, N., 2021. Posterior subthalamic area, pallidothalamic tract, and pedunculopontine nucleus: deep brain stimulation targets for Parkinson's disease and essential tremor. No Shinkei Geka. 49 (4), 820-828 (Japanese). doi: 10.11477/mf.1436204460. PMID: 34376614.

Turner, D.A., November 2013. Re-engineering the subthalamus. World Neurosurg. 80 (5), 476–478.

Urbain, N., Rentéro, N., Gervasoni, D., Renaud, B., Chouvet, G., 2002. The switch of subthalamic neurons from an irregular to a bursting pattern does not solely depend on their GABAergic inputs in the anesthetic-free rat. J. Neurosci. 22 (19), 8665-8675. doi: 10.1523/JNEUROSCI.22-19-08665.2002. PMID: 12351741; PMCID: PMC6757798.

Venkataraman, A., Brody, N., Reddi, P., Guo, J., Gordon Rainnie, D., Dias, B.G., 2019. Modulation of fear generalization by the zona incerta. Proc. Natl. Acad. Sci. U.S.A. 116 (18), 9072–9077. https://doi.org/10.1073/pnas.1820541116.

Watson, C., Lind, C.R., Thomas, M.G., February 2014. The anatomy of the caudal zona incerta in rodents and primates. J. Anat. 224 (2), 95–107.

Webb, W.G., 2017. Organization of the nervous system I. In: Neurology for the Speech-Language Pathologist, sixth ed. Mosby.

Xiao, Y., Jannin, P., D'Albis, T., Guizard, N., Haegelen, C., Lalys, F., Vérin, M., Collins, D.L., September 2014. Investigation of morphometric variability of subthalamic nucleus, red nucleus, and substantia nigra in advanced Parkinson's disease patients using automatic segmentation and PCA-based analysis. Hum. Brain Mapp. 35 (9), 4330–4344.

Zhu, X.L., Hamel, W., Schrader, B., Weinert, D., Hedderich, J., Herzog, J., Volkmann, J., Deuschl, G., Müller, D., Mehdorn, H.M., October 2002. Magnetic resonance imaging-based morphometry and landmark correlation of basal ganglia nuclei. Acta Neurochir. 144 (10), 959–969.

10

The Diencephalon: Hypothalamus

Daniele Armocida[1,2], Mauro Palmieri[1], Fulvio Tartara[3], Diego Garbossa[4], Fabio Cofano[4], Brandon M. Ascenzi[5], Antonio Santoro[1] and Alessandro Frati[2]

[1]Department of Human Neurosciences, Neurosurgery Division, University of Rome "La Sapienza", Rome, Italy [2]IRCCS "Neuromed", Experimental Neurosurgery Unit, Pozzilli, Italy [3]IRCCS Fondazione Istituto Neurologico Nazionale C. Mondino, Neurosurgery Unit, Pavia, Italy [4]Department of Neuroscience "Rita Levi Montalcini", Neurosurgery Unit, University of Turin, Turin (TO), Italy [5]Independent Neuroresearcher-Member of Marie Curie Alumni Association (MCAA)

OUTLINE

INTRODUCTION

The hypothalamus represents the ventral-most portion of the diencephalon that surrounds the anterior inferior portion of the third ventricle. Consisting of only 4 cm^3 of nervous tissue (equal to 0.3% of the brain mass), the hypothalamus nonetheless contains the integrative systems which, through the autonomic and endocrine effector systems, control the fluid and electrolyte balance, food intake and energy balance, reproduction, thermoregulation, immune responses, and many emotional responses.

Moreover, although small in size and phylogenetically among the oldest in the human nervous system, hypothalamus play a key role in regulation of mnemonical processes, behavior, emotional and programmatic response to external stimuli (Tony and Joshi, 2021). The number of higher functions in which the hypothalamus appears to be involved is increasing as well as the therapeutic applications and methods by which this structure is being used as a target (Ulrich-Lai et al., 2009).

Significantly, it has been showed that the hypothalamus is implicated in homeostatic and developmental disorders, e.g., sudden infant death syndrome, Prader–Willi's syndrome, disturbance of biological rhythms, infertility, diabetes, and obesity (Kalsbeek et al., 2002), as well as in episodic brain disorders such as migraine, depression, narcolepsy, and cluster headaches (Overeem et al., 2002). Besides these, the hypothalamus is also subject to different types of developmental abnormalities and pathologies such Rathke's cleft cysts, hamartoma, and craniopharyngioma (Yamada and Katagiri, 2007).

ANATOMY

The hypothalamus extends from the lamina terminalis to a vertical plane behind the mammillary bodies, and from the hypothalamic sulcus to the base of the

From Anatomy to Function of the Central Nervous System
https://doi.org/10.1016/B978-0-12-822404-5.00024-3

brain below the third ventricle. It is located inferior to the thalamus and anterior to the tegmental portion of the subthalamus and the midbrain tegmentum (Fig. 10.1). Laterally, it is bounded by the anterior subthalamus, the internal capsule, and the optic tract. The structures of the floor of the third ventricle that reach the pial surface in the interpeduncular fossa, proceeding in an anteroposterior direction, are the optic chiasm, the tuber cinereum, the tuberous eminence and the infundibular peduncle, the mammillary bodies, and the posterior perforated substance (Fig. 10.2). The latter is located at the point of divergence of the cerebral peduncles and is perforated by the small central branches of the PCAs.

It contains the small interpeduncular nucleus on which the fibers of the fasciculus retroflexus on both sides synapse and is connected to the mesencephalic reticular formation and the mammillary bodies. The latter are wrapped in bundles of fibers mainly derived from the fornix.

The tuber cinereum, placed between the mammillary bodies and the optic chiasm, is a convex mass of gray matter. It is a conical and hollow median structure, which continues ventrally with the posterior lobe of hypophysis. At the base of the infundibulum lies the median eminence, delimited by the tuberoinfundibular sulcus (Figs. 10.3 and 10.4).

The hypothalamus contains various neuronal groups classified as nuclei according to phylogenetic, embryogenic, cytoarchitectural, synaptic, and histochemical criteria, many of which are not well defined, especially in the adult. In the hypothalamus, the myelinated connections are few, many of them are diffuse and unmyelinated, and the path of many intrinsic afferent and efferent fibers is still not clear. In the hypothalamus we can recognize four specific regions anteroposteriorly: the preoptic, the supraoptic (or anterior), the tuberous (tuberoinfundibular), and the mamillary. In the medial-lateral sense we found the periventricular, intermediate (or medial), and lateral zone. Between the intermediate and lateral zone there is a paramedian plane, which contains myelinated fibers of the fornix column, the mammillothalamic tract, and the fasciculus retroflexus. It is important to emphasize that these divisions are artificial in any case and functional systems go beyond all this. The PVZ of the hypothalamus delimits the third ventricle and in the anterior wall of this latter lies the *vascular organ of the lamina terminalis* (OVLT) (see Chapter 16), which continues dorsally with the median PON and with the subfornical organ.

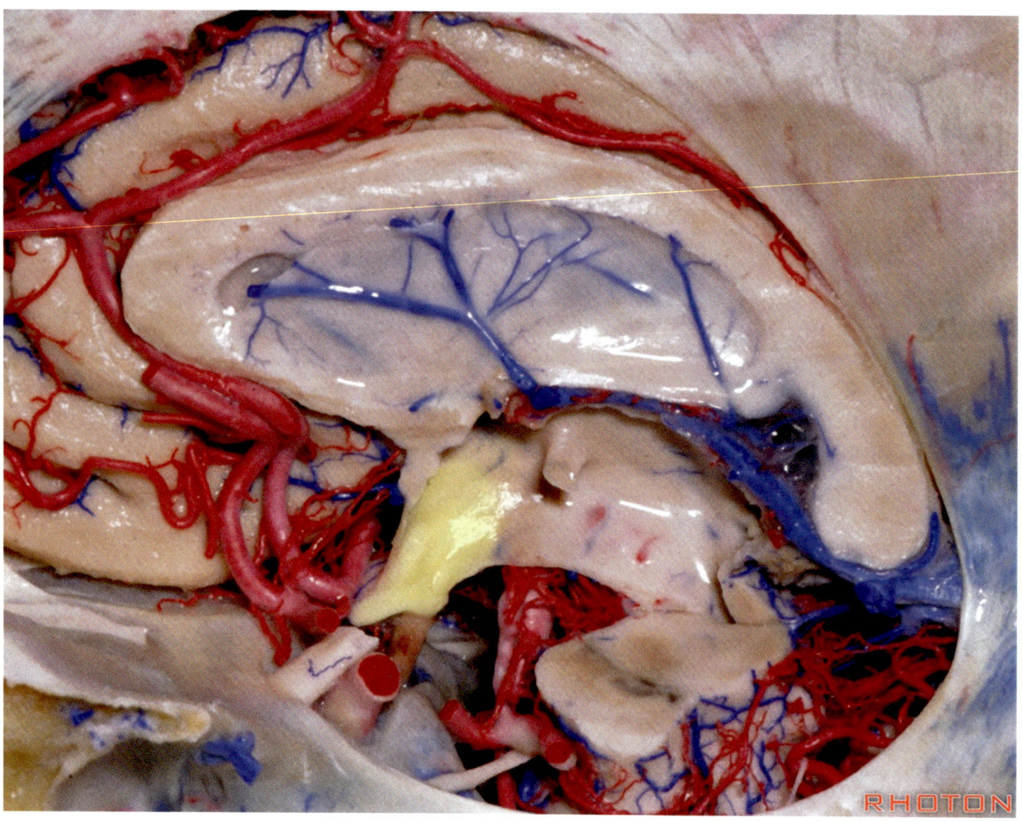

FIGURE 10.1 **Deep brain dissection.** Sagittal view: in yellow, the hypothalamus. *Image courtesy of The Rhoton Collection ®http://rhoton.ineurodb.org/.*

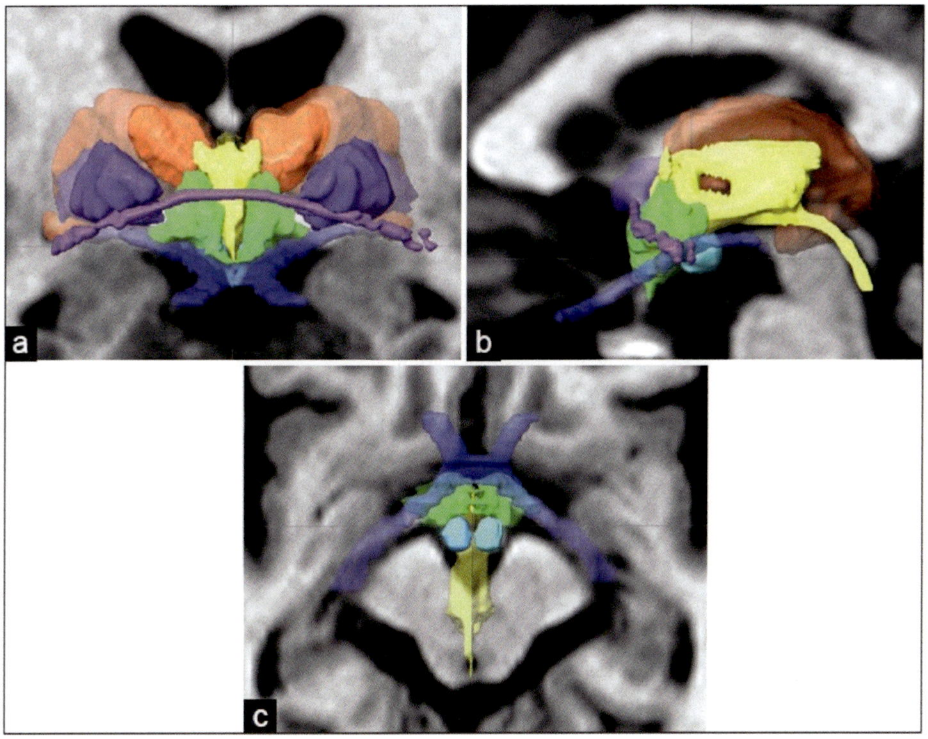

FIGURE 10.2 3D reconstruction of the adult human hypothalamus displayed on three orthogonal ((A) Frontal view; (B) Left lateral view; (C) Inferior view) T1-weighted MRI slices crossing the mammillary bodies (light blue): hypothalamus (green plus light blue), anterior white commissure (purple), thalamus (brown), pallidum (purple), third ventricle (yellow; the brain aqueduct and the ventricular foramen are also reconstructed), and optical system (dark blue). *Image courtesy of Lemaire, J., Nezzar, H., Sakka, L., Boirie, Y., Fontaine, D., Coste, A., et al., 2013. Maps of the adult human hypothalamus. Surg. Neurol. Int. 4:156–163. Available FREE in open access from: http://www.surgicalneurologyint.com/text.asp?2013/4/4/156/110667.*

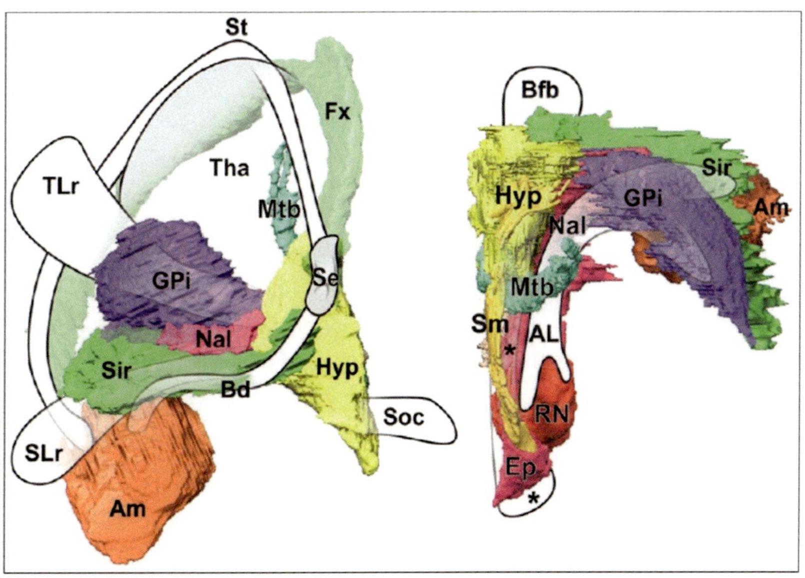

FIGURE 10.3 Overview of structures neighboring the hypothalamus (Hyp), 3D reconstruction from a high-field MRI data set merged with schematic drawing; left, frontal view; right, superior view. *Am*, amygdala; *Bd*, diagonal band of Broca; *Bfb* (and *), basal forebrain bundle; *Ep*, epithalamus; *Fx*, fornix; *GPi*, globus pallidum intern; *Mtb*, mammillothalamic bundle; *Nal*, ansa lenticularis nucleus; *RN*, red nucleus; *Se*, septum; *Sir*, substantia innominata (of Reichert); *SLr*, sublenticular radiations; *Sm*, stria medullaris; *Soc*, supraoptic commissure; *St*, stria terminalis; *Tha*, thalamus; *TLr*, translenticular radiations. *Image courtesy of Lemaire, J., Nezzar, H., Sakka, L., Boirie, Y., Fontaine, D., Coste, A., et al., 2013. Maps of the adult human hypothalamus. Surg. Neurol. Int. 4:156–163. Available FREE in open access from: http://www.surgicalneurologyint.com/text.asp?2013/4/4/156/110667.*

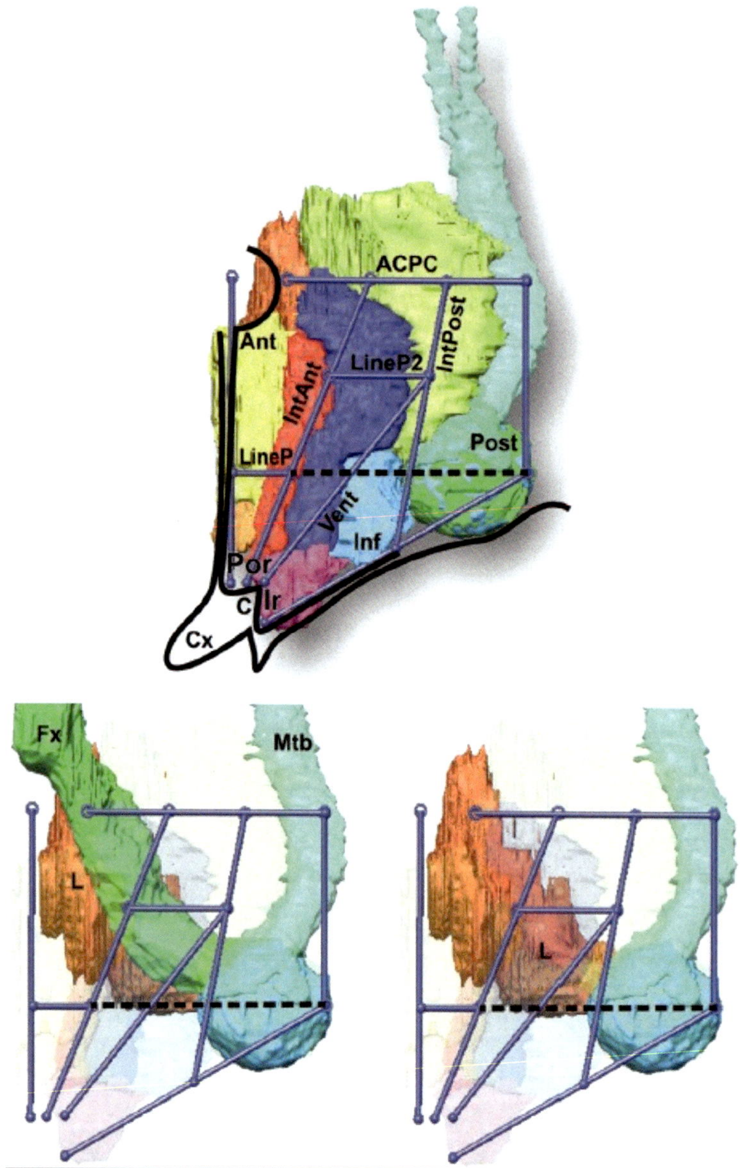

FIGURE 10.4 **MRI landmarks and line used to spot hypothalamic compartments and nuclei (projection on 3D reconstruction from a high-field MRI data set; lateral view).** (A) The hypothalamus is lined posteriorly by the vertical line (Post) placed behind the mammillary body and perpendicular to an anterior commissure posterior commissure (ACPC) segment. The anterior line (Ant) rises from the preoptic recess (Por), perpendicular to ACPC. Two intermediate lines, anterior (IntAnt) and posterior (IntPost), respectively, goes from the dorsal region of the optic chiasma (Cx) to the anterior point of the intermediate third of the ACPC line and from the crest (c) placed between the optical recess and the infundibular recess (Ir), to the posterior point of the intermediate third of the ACPC line. The anterior horizontal line (LineP; doted black line) goes parallel to ACPC (toward the posterior point at the origin of Post). The intermediate horizontal line (LineP2) is parallel to ACPC and origins at the midpoint of the segment of IntAnt between the point where IntAnt crosses LineP and the point where Int Ant crosses ACPC. The ventral line (Vent) goes from C to the intersection of LineP2 and IntPost. (B) The fornix limits the medial border of the lateral nucleus (l), which is visible in the background; the mammillothalamic bundle (Mtb) is displayed. (C) The area limited by the LineP2, LineP, and IntAnt and IntPost projects on the lateral nucleus. *Image courtesy of Lemaire, J., Nezzar, H., Sakka, L., Boirie, Y., Fontaine, D., Coste, A., et al., 2013. Maps of the adult human hypothalamus. Surg. Neurol. Int. 4:156–163. Available FREE in open access from: http://www.surgicalneurologyint.com/text.asp?2013/4/4/156/110667.*

One of the most important hypothalamic nuclei is the SCH. It represents the neuronal substrate of circadian cycles of motor activity, body temperature, plasma hormone levels, renal secretion, sleep-wake rhythm—disturbed by possible suprachiasmatic lesions—and many other variables.

The SCH consists of a ventrolateral and a dorsomedial portion. In the ventrolateral portion, characterized by immunoreactive neurons for *vasoactive intestinal polypeptide* (VIP), the retinal fibers synapse. The ventromedial portion appears to be a generalized input zone, which receives afferents also from the midbrain raphe and parts of the lateral geniculate nucleus. The dorsomedial portion has poor afferents and is characterized by parvocellular neurons immunoreactive for *arginine vasopressin* (AVP). SCH neurons receiving direct retinal afferents do not respond to plots, motion, or color; rather, they function as luminance sensors that respond to the presence or absence of light and their firing rate varies according to light intensity, synchronizing with the light–dark cycle.

The SCH receives glutamatergic afferents from retinal ganglion cells, forming the retinohypothalamic tract, responsible for modulating the light–dark rhythm, which are not essential for the production of the rhythm, which persists in the blind. This nucleus contains many different neurotransmitters (vasopressin, VIP, *neuropeptide Y* [NPY], and neurotensin) and projects to various hypothalamic nuclei (paraventricular, ventromedial, dorsal medial, and arcuate). The SCH also regulates the activity of the preganglionic sympathetic neurons of the more cranial spinal ganglia. These neurons, in turn, fire on the superior cervical ganglion which projects to the pineal gland. In the pineal, which contains modified photoreceptors, circadian variations of sympathetic postganglionic signals parallelly modify the activity of N-acetyltransferase and the consequent production of melatonin.

BLOOD SUPPLY

The hypothalamus is supplied by the circle of Willis, which surrounds it inferiorly, the anteromedial branches of the *anterior cerebral artery* (ACA), the posteromedial branches of the *posterior communicating artery* (pCOA), and the thalamoperforating branches of the *posterior cerebral artery* (PCA). A special note deserves the vascular anatomy of the hypothalamus because not only is its rich vasculature necessary for its supply but also because the vessels in its context play a key role as a clearinghouse for information to and from the rest of the organism (Fitzgerald, 1979).

The key role is undoubtedly determined by its portal system, which determines the main means by which the hypothalamus exerts control over the entire organism. The hypothalamo-neurohypophysial portal system is a capillary plexus that transmits the releasing hormones from the arcuate nucleus of the hypothalamus to the anterior pituitary. Its astroglia podocytes form the blood–brain barrier (BBB) by wrapping podocytes around capillaries. These cells protect the brain from toxins in the blood and facilitate nutrient transport to the neurons. Astroglia also forms a system of microscopic perivascular channels permeating the brain that transmit *cerebrospinal fluid* (CSF)-like lymphatic vessels. The system enables CSF to clear metabolic waste and distribute glucose, amino acids, lipids, and neurotransmitters. This system is most active during sleep, contributing to its restoration function. Arterial pulsation drives lymphatic flow, suggesting that exercise may also enhance it. Aging, brain trauma, and ischemia decrease that CSF flow. Also, larger lymphatic vessels in the meninges help absorb interstitial fluid into the dural venous sinuses (Yao et al., 2021).

HYPOTHALAMUS: BLOOD SUPPLY

The diencephalic vessels consist of the antero-inferior diencephalic branches, which are differently developed and supply especially the telencephalic section of the diencephalon and occasionally the anterior upper thalamic areas. These originate from the distal sections of the precommunicant part of the ACA and the *anterior communicating artery* (AcoA). In addition to the lateral and medial PON, the anterior hypothalamic area contains the *paraventricular nuclei* (PVNs). Vascular alterations with the resulting destruction of these nuclear areas can lead to functional disturbances of the pituitary gland and thus influence the hormone, water, and electrolyte balance and the unequivocally demonstrated centers for temperature sensation and regulation. Since fibers from the PVNs also pass to the spinal nuclei of the sympathetic nervous system and reach it again on returning, this nucleus participates in the regulation of vasodilatation and sweat gland secretion, among others (Haymaker et al., 1969).

A further group of vessels is called the inferior diencephalic branches. These vessels originate especially from the pCOA and internal carotid artery itself. The inferior diencephalic branches also participate in the supply of the anterior and intermediate sections of the hypothalamus and thalamus. In the hypothalamus, the lateral tuberal nuclei and arcuatus are classified to this region. It should be emphasized at this point that ablation of the medial sections of the hypothalamus leads to hyperphagia and the ablation of the lateral areas to hypophagia to aphagia with subsequent leanness. The circadian rhythms for body temperature, plasma components, number of eosinophilic cells, adrenocortical activity, renal secretion mechanisms, sleep and wake centers, emotions, and affections are related to an intact hypothalamus. It is also interesting that these particular control centers are in close connection between the two hypothalamic areas through a large area of decussation, demonstrated experimentally by the fact that in a patient with hyperthermia on the left side of the body, marked destruction of the right hypothalamic area was found (Leclercq et al., 1983).

Posteroinferior Diencephalic Branches

The posterior lower hypothalamic and thalamic areas of the diencephalon are supplied by branches of the precommunicant segment of the PCA, the basilar artery and the superior cerebellar artery. These enter the posterior interpeduncular perforate substance of the midbrain and participate in the perfusion of the most important midbrain nuclei as well. In the posterior hypothalamic area stimulation and destruction leads to changes in blood pressure, heartbeat sequence, pupillary dilatation, as well as movements of the neck and eyes. If the area of the posterior hypothalamus is perfused with noradrenalin, dopamine, apomorphine, among others, then a stimulation-dependent rise in blood pressure can be observed. Therefore, it must be assumed that neural paths from the posterior hypothalamic areas also reach the spinal sympathetic centers.

Inferolateral Diencephalic Branches

These are branches of crucial importance for hypothalamic and lateral geniculate body activity, in fact their functions are well protected by blood flow arising from both hemispheric sides and from different vascular origins. In fact, these important vessels originate mainly from the anterior choroidal artery, and afferent branches have been identified from the middle cerebral artery, deep pallidal branches, and proximal tracts of the acoustic and visual pathways in the internal capsule.

Posterior Diencephalic Branches

On the dorsal surface of the diencephalon (and in the lateral sections of the mesencephalon) branches from various blood vessels are seen. These originate from the posteromedial and posterolateral chorioid branches as well as the PCA itself. For example, we were able to show that five such rami thalamogeniculati on the average originate from the PCA. From the posterolateral chorioid branch run to the lateral corpus geniculatum in 66% and to the pulvinar in 62%, more rarely to the medial corpus geniculatum. The medial posterior chorioid branch (which is double in 24%, triple in 6.2%, and quadrupled in about 1%) supplies especially the corpus geniculatum mediale and its environment, more rarely, however, the thalamic pulvinar (Christ, 1969).

Superior Diencephalic Branches

The upper areas of the thalamus are supplied by the anterior choroid artery as well as by the medial and lateral posterior chorioid branches. Only the upper anterior section of the diencephalon may also receive branches from the ACA and AcoA.

It should be pointed out here that the vessels from the diencephalic branches are of variable diameter, so that the intracerebral length varies and thus the area of perfusion varies as well. Since it is part of the diencephalon, the pineal body is also supplied by the diencephalic branches, particularly by the medial choroidal posterior branch as well as its connection with the artery for the roof of the midbrain (Leclercq et al., 1983; Porter et al., 1978).

TOPOGRAPHIC ANATOMY OF THE HYPOTHALAMUS

The neuroanatomy of the human hypothalamus is often reviewed focusing on its neuroendocrine role. But its functionality is far more complex (Anderson, 2012), and from an anatomical point of view, more new discoveries are being made, thanks to the technological evolution of functional imaging (Baroncini et al., 2012). The hypothalamus encompasses the most ventral part of the diencephalon, of which it forms the floor and contributes to the lateral walls of the third ventricle. Its upper boundary is marked on the ventricular side by a shallow groove defined as the hypothalamic sulcus. Caudally the hypothalamus passes gradually over into the periventricular and tegmental gray matter of the mesencephalon. At least 12 cell clusters or nuclei are recognized within the hypothalamic region. Actually, the usefulness of anatomic subdivisions is limited because dendrites and axons of cell bodies overlap between various nuclei, resulting in overlapping functions; Le Gros Clark (1938), Kuhlenbeck (1973), and Brll (1978) described different classifications of the hypothalamic area.

So, the complex anatomy of the hypothalamus is best represented by a combination of coronal, axial, and parasagittal sections alternatively stained for nerve fibers and cell bodies (Fig. 10.5).

Therefore, it is possible to divide the hypothalamus into three distinct regions in the sagittal plane where we find anteroposteriorly: an anterior region defined as the *preoptic area*, followed immediately posteriorly by the *anterior-chiasmatic area*, the tuberal or *tubero-infundibular part*, and a posterior part defined as the *mammillary area*. This classification is useful because it is easily seen with the naked eye. Moreover, this partition is supplemented by a tripartite and more functional mediolateral (coronal) division represented by: the *periventricular, medial (or nuclear), and lateral zones* (Fig. 10.6), which form parallel zones with a distinctive cellular architecture (Le Gros Clark, 1938).

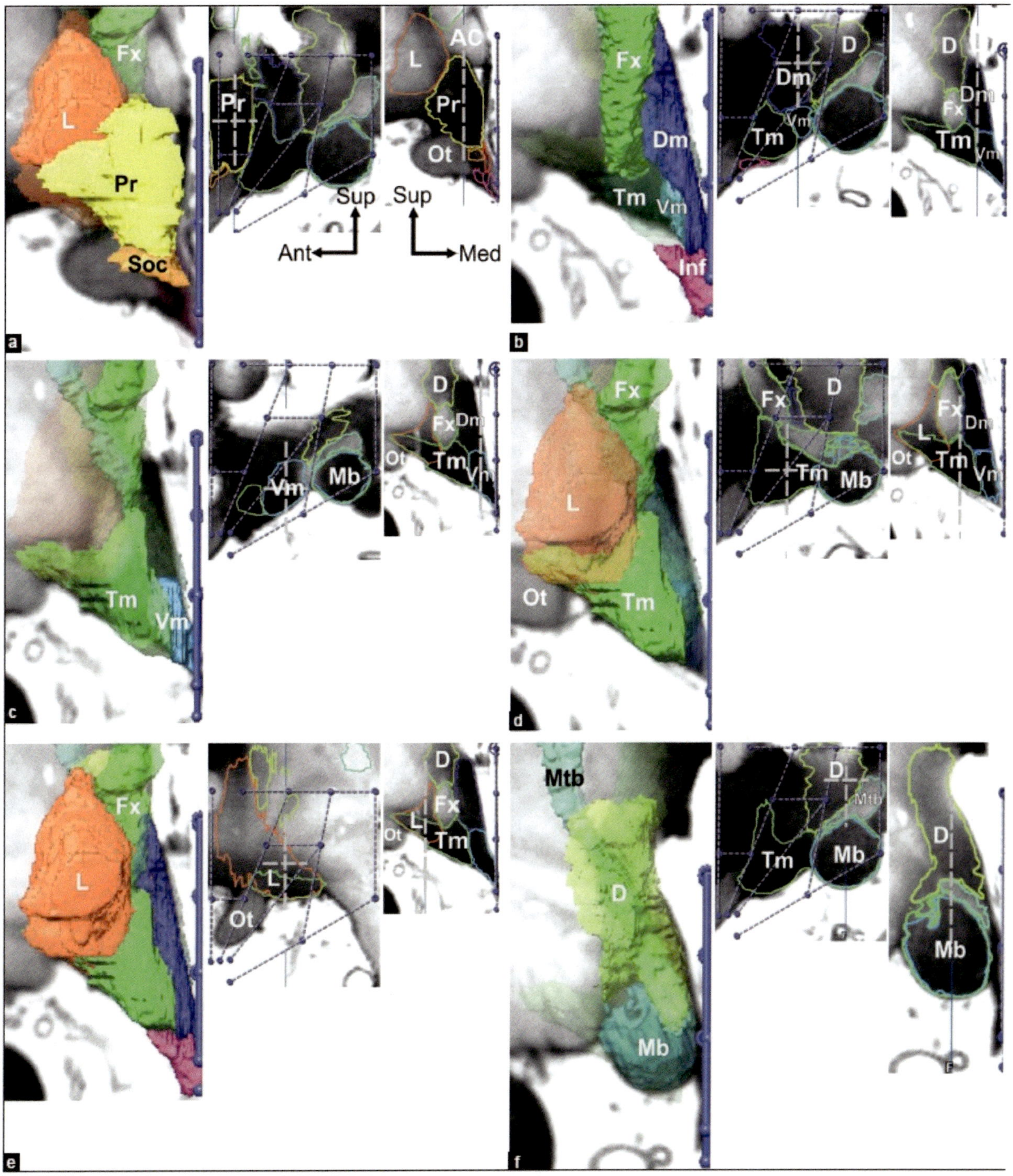

FIGURE 10.5 **MRI landmarks and lines used to spot hypothalamic compartments and nuclei (projection on 3D reconstruction from a high-field MRI data set).** From A to F: left, frontal view; intermediate, sagittal slice, right, coronal slice; gray dotted lines show section plans. (A) Preoptic nuclei (Pr) and supraoptic plus suprachiasmatic nuclei (Soc). (B) Dorsomedial nucleus (Dm) lined laterally by the fornix. (C) Ventromedial nucleus (Vm) lined laterally by the tuberomamillaris nucleus (Tm). (D) Tuberomamillaris nucleus (Tm); placed below the fornix extending laterally to the optical tract (Ot). (E) Lateral nucleus (l), lined medially by the fornix; the inferior region overlaps with the tuberomamillaris nucleus. (F) Dorsal nucleus (d), placed behind the fornix (not visible here) and lined posteriorly by the mammillothalamic bundle (Mtb). *Image courtesy of Lemaire, J., Nezzar, H., Sakka, L., Boirie, Y., Fontaine, D., Coste, A., et al., 2013. Maps of the adult human hypothalamus. Surg. Neurol. Int. 4: 156–163. Available FREE in open access from: http://www.surgicalneurologyint.com/text.asp?2013/4/4/156/110667.*

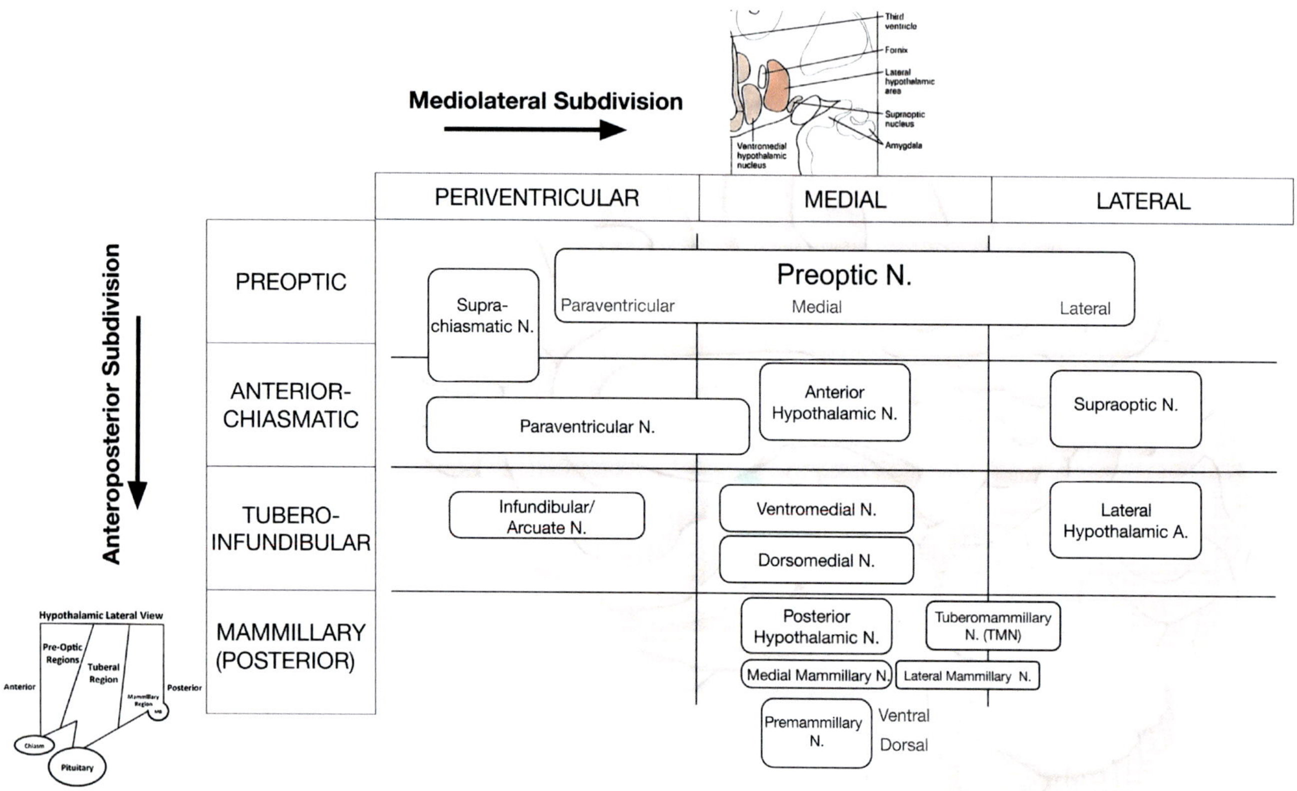

FIGURE 10.6 Schematic representation of the hypothalamic nuclei (for details see in the text).

These two subdivisions on the two planes are necessarily intersected in that is possible to describe the hypothalamus in its subdivision into groups of neurons (cell clusters or nuclei) with specific activity and production that go to make up the different regions in the context of the parenchyma.

For example, the magnocellular neurons in the *supraoptic nucleus* (SON) and PVNs are the site of synthesis of the nonapeptides *antidiuretic hormone* (ADH) and oxytocin are in close relation with the posterior lobe of the pituitary which contains their axonal neurosecretory endings while the parvocellular neurons are scattered around the third ventricle, from the preoptic area toward the infundibulum. This system controls the adenohypophysis by the releasing hormones for *thyrotropin* (TRH), *luteinizing hormone* (LHGR), *growth hormone* (GHRH), and the inhibiting factor for *growth hormone* (somatostatin or SRIF) and *prolactin* (PIH), and anatomically overlap these regions.

The SON and PVNs which, respectively, secrete vasopressin and oxytocin have the most well-defined functions. With the exception of the mammillary nuclei, which are characteristic because of their contour, the individual hypothalamic nuclei cannot be distinctly identified on MRI imaging with standard sequences (Nakane, 1970; Snyder et al., 1977). By convention we therefore describe the various parts of the nucleus using the subdivision, which is more functional in the coronal plane starting from the medial to the most lateral part and from the anterior to posterior part.

Mediolateral Topographic Subdivision (Fig. 10.6)

Periventricular Zone

The periventricular zone (PVZ) consists of a few layers of small cells among which larger, neurosecretory elements are intermingled. The latter form a single concentration, the medial magnocellular part of the PVN. Seems that in its complexity the PVZ is particularly involved in the hypothalamus mood regulation, a fact that may bear important therapeutic and surgical applications. In the PVZ lies the most caudal parts of the *preoptic nucleus* (PON), the *suprachiasmatic nucleus* (SCH), the IFN, the posterior nucleus of the hypothalamus, and PVN.

The importance of the PVN for overall body homeostasis is also shown by its unique relationship with the adenohypophysis at this point. The collection of neurons located near the arcuate and PVNs (the "hypophysiotrophic area") project to the portal system of the median eminence, where they secrete the hypothalamic hormones. Further, at this point, the whole endocrine system is potentially open to all kinds of neural modulation issuing from multiple regions of the brain (Sagar et al., 1986). The PVN is also made up of large cells, which are densely layered particularly in the dorsal sections. PVN, rudimentary in humans, in comparison with that of other mammals, consisting of a 1-mm-thick layer of subependymal neuron clusters without a clear structural organization. The PVN is continuous with the periaqueductal gray substance with which it is connected through the dorsal longitudinal fascicle of Schülz. Despite the comparatively rudimentary structure of the human PVN, its role in behavior is far from simple or unimportant. In addition to the medial part of the PVN, the periventricular preoptic, suprachiasmatic, and infundibular nuclei in the basal hypothalamus represent derivatives of the PVZ.

The dorsocaudal pole of this nucleus usually extends as far as the posterior hypothalamic area with single scattered cells. In the medial area smaller nerve cells are found which have led to a demarcation of the nucleus paraventricularis parvocellularis. A group of smaller nervous cells lies in the ventrocaudal section of the third ventricle. The caudal part of the nucleus is located around the inframamillar recess and corresponds to the posterior periventriculars nucleus (inframammillaris).

In this area, among the fiber tracts of the hypothalamus which include the temporohypothalamic, the thalamohypothalamic, and the pallidosubthalamic fibers, special mention should be made of the ansa lenticularis (Swanon et al., 1981). The ansa lenticularis serves as a connection between the nucleus subthalamicus and the pallidum (Dejerine, 1901). Fibers from the nucleus interstitialis (Cajal) and from the nucleus of the posterior commissure also probably run in this tract (Bucher et al., 1952).

The Medial (or Nuclear) Zone

In contrast to the PVZ, the medial (or nuclear) and lateral zones are larger (relative to overall brain volume) and more complex in humans compared with other mammals. The fornix divides these two zones as it pierces the hypothalamus on its way to the mammillary body. Further this part is divided into an anterior and posterior part by the tuberal region.

The nuclear zone, which is relatively cellular, contains a number of variably discrete cell masses that from anterior to posterior are the *medial PONs*, more diffuse *anterior nucleus*, *paraventricular nucleus*, the *ventromedial nucleus* (DMN) and *dorsomedial nucleus* (VMN), *medial* and *lateral mammillary nuclei*, discrete dorsal and ventral premammillary nuclei (PMV), and diffuse *posterior hypothalamic area*.

Lateral Zone

The lateral zone is partly separated from the medial zone by the postcommissural fornix, a large bundle connecting the hippocampal formation with the mamillary body. The lateral zone contains the magnocellular neurosecretory SON, but most of its territory is occupied by the diffuse *lateral preoptic* and *lateral hypothalamic areas* (LHA), which is less represented in the posterior part of the hypothalamus (Fig. 10.7).

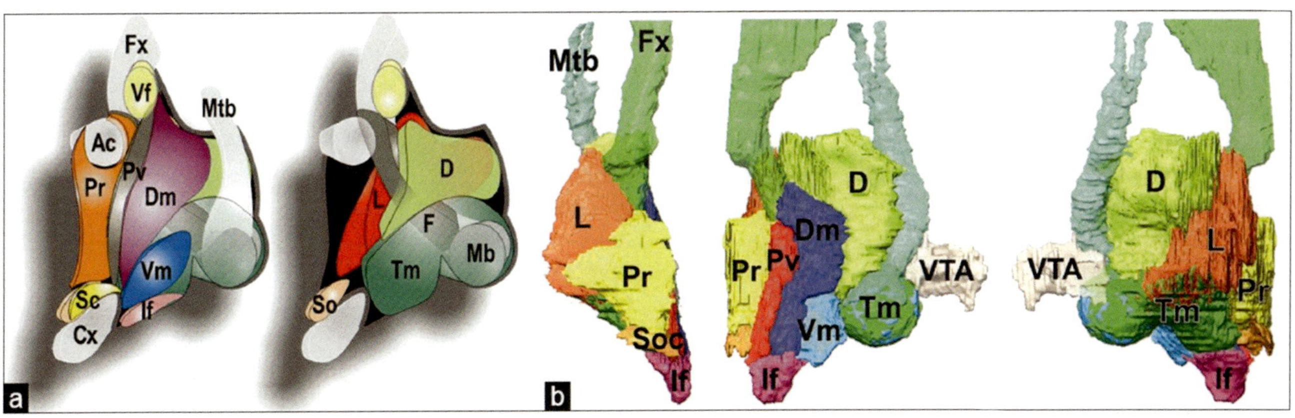

FIGURE 10.7 **Lateral view of the hyphothalamus.** (A) Schematic drawing of hypothalamic nuclei, lateral view from the midline; left, nuclei directly located under the wall of the third ventricle; right, more deeply located nuclei up to the lateral region. (B) 3D overview of hypothalamic nuclei, constructed from a 3D high-field MRI data set; left, frontal view; intermediate, lateral view; right, medial view; the ventral tegmental area (VTA) is located within the retro hypothalamic region.*Ac*, Anterior commissure; *Cx*, optic chiasma ; *D*, dorsal or posterior nucleus; *Dm*, dorsomedial nucleus; *F*, fornix nuclei ; *Fx*, fornix; *If*, infundibular (arcuate) nucleus; *L*, lateral nucleus; *Mb*, mammillary body; *Mtb*, mammillothalamic bundle; *Pr*, preoptic nuclei; *Pv*, paraventricular nucleus; *Sc*, suprachiasmatic (ovoid) nucleus; *So*, supraoptic (tangential) nucleus; *Tm*, tuberomamillaris (mamilloinfundibularis) nucleus; *Vf*, ventricular foramen; *Vm*, ventromedial (tuber principal) nucleus. *Image courtesy of Lemaire, J., Nezzar, H., Sakka, L., Boirie, Y., Fontaine, D., Coste, A., et al., 2013. Maps of the adult human hypothalamus. Surg. Neurol. Int. 4:156–163. Available FREE in open access from: http://www.surgicalneurologyint.com/text.asp?2013/4/4/156/110667.*

The lateral hypothalamus represents the associative zone of the hypothalamus; a number of complex fiber systems make their way through this zone toward and from the hypothalamic nuclei in relation to distant regions of the forebrain and brainstem. The most conspicuous fiber systems that pass through the lateral hypothalamus comprise the medial forebrain bundle, the fornix, and the ansa peduncularis. The LHA also contains the lateral hypothalamic nucleus, which is ventrally and medially limited by the LMN. The lateral and ventromedial nuclei constitute the classical "feeding center" because they exert reciprocal functions in eating behavior and consists of islands of orexinergic neurons (Stuber et al., 2016). Thus, bilateral lesions of the LHA produce weight loss in experimental animals, while electrical stimulation of this region increases appetite and food intake, but numerous researchers had studied the connections between this area and the sympathetic nervous system. By electrophysiological methods the connections between the LHA and the preganglionic sympathetic cells in the spinal cord were demonstrated (Nauta and Haymaker, 1969).

Anteroposterior Topographic Subdivision (Fig. 10.6)

Anterior Part

The most anterior part is located medial to the optic chiasm where lie the PON, the largest nucleus of this area, which is involved in thermal regulation and in the endocrine regulation of sexual behavior. Its delineation is indistinct, particularly dorsolateral from the *nucleus interstitialis of the stria terminalis* but also to the *nucleus pericommissuralis* in the proximity of the anterior commissure. The SON is slightly lateral, with a short extension rostral to the optic tract. In man its major portion is located immediately in front of the chiasm, while the rest lies behind the chiasm at the medial side of the opticus tract.

The nuclei extend as far as the ependymal lining of the optic recess. An unpaired group of nuclei at the front wall of the optic recess is named median PON. In front of it at about the middle of the lamina terminalis, the organum vasculosum laminae terminalis is located (Vann, 2010).

Caudal and posterior to the PONs is the PVN, a vertical column of large cells in close relationship with the wall of the third ventricle (medially) and the column of the fornix (laterally). This nucleus takes part in the secretion of antidiuretic hormone and oxytocin by the posterior pituitary; it is also essential for the integration of the neuroendocrine and autonomic response to stress. The SCH also lies medially, immediately dorsal to the optic chiasm, and plays a role in the regulation of circadian rhythms. The SCH is separated from the SON by the *nucleus hypothalamicus*. The SCH does not reach the level of the supraoptic commissure. According to Knoche (1956) delicate fibers run from the retina over the lamina terminalis into this area and reach the tuber cinereum without interruption and the neurohypophysis.

Tuberal Region

The tuberal (or tubero-infundibular) region is rostrally continuous with the dorsal part of the anterior hypothalamus. The *infundibular (or arcuate) nucleus* is located anteroventrally, behind the optic chiasm, and forms the anterior wall of the infundibulum. This nucleus surrounds the infundibulum in ring-shaped fashion. When the hypophyseal stalk is torn off, this nucleus is also often no longer visible. The nerve cells of this nucleus are small, the nuclei themselves are round to oval. It is involved in food intake, cardiovascular regulation, and the monitoring of adipose tissue fat.

The tuberal region is also the location of the DMN and VMN, the latter being bounded dorsolaterally by the fornix. The VMN is the largest of the tuber cinereum and was called the nucleus principalis tuberis by Cajal (1911). Dorsolateral to this nucleus runs the fornix, occasionally separated from it by the *nucleus perifornicalis* (PFN). The nerve cells are medium to small, but larger than those of the *nucleus infundibularis* (IFN). Lesions to the VMN modulate sexual behavior and lead to overeating and weight gain. As seen below the *deep brain stimulation* (DBS) in this region increases lipolysis and leads to rejection of food and weight loss in animals and humans (Torrealba et al., 2012). The DMN is bounded rostrally by the PVN and has been implicated in daytime feeding and the emotional responses to stress, such as panic-related behavior. This nucleus lies dorsal to the VMN, it possesses a lower cell density with a not clear lateral delineation. In a dorsolateral direction it extends as far as the zona incerta. Its cells are smaller than those of the nucleus ventromedial. Surgical lesions of this nucleus in both male and female patients have caused persistent impotence and loss of libido unrelated to the blood levels of sexual hormones. High-frequency stimulation of this zone leads to sexual arousal.

Mammillary (or Posterior) Part

The mammillary (or posterior) part includes the *posterior hypothalamic nucleus*, the *tuberomammillary nucleus* (TMN), and the mammillary bodies. Historically associated with sympathetic responses and defensive and aggressive behaviors, the posterior nucleus is limited caudally by the mammillothalamic tract and represents one of the largest nuclei of the organ. The posterior nucleus is approximately 7 mm high and 4 mm long at the most from front to back. It lies behind and above the VMN, DMN, PVN nuclei, and in front of the lower part of the mammillothalamic tract. It is located above the TMN, which contains histaminergic neurons that have been implicated in the promotion of wakefulness and motivated behaviors in relation to food, water, sex, and drugs. So, it represents a center through which the efferent impulses of the hypothalamus are sent to the lower levels of the brain and to the spinal cord (Sternson, 2013). The TMN is a rather diffusely organized group of large cells in the caudal tuberal and rostral mamillary regions. It forms part of the medial hypothalamic zone but extends laterally well into the lateral zone. The TMN contains a large population. of histaminergic neurons. This is often an ill-defined area since part of the TMN pertaining to the posterior portion also extends into the lateral area contributing to its constitution.

The undefined posterior margin of the hypothalamus is conventionally represented by the caudal ventral line passing through mamillary bodies plane.

The mammillary bodies constitute one of the major nodes of the classical circuit of Papez, which is essential for episodic memory encoding. The posterior hypothalamus and adjoining brainstem regions are critical for behavioral wakefulness, as originally proposed by von Economo based on clinicopathological cases of lethargic encephalitis. Barely above of this nucleus lies the *premammillaris nucleus*. This nucleus, which is not mentioned in the present nomenclature, consists of a small group of cells in the ventrocaudal hypothalamus which is poorly myelinated. It lies between the caudal pole of the VMN and the mammillary body with whom there are no fibers connections. Behind the last-mentioned nucleus, one finds the *lateral mammillaris nucleus* (LMN) and the TMN in man. These contain large ganglion cells and consist of two or three more or less continuous columns. Because they are usually three spheric ganglia they are described as LMNs.

Hypothalamic Fibers

The magnocellular nuclei anterior hypothalamus, i.e., the supra- and (magnocellular) paraventricular optic nuclei, gives rise to axons that descend through the infundibular stalk to the posterior lobe of the pituitary. These axons, which together form the supraoptico-paraventriculo-hypophyseal tract, transport colloid droplets containing the hormones oxytocin and vasopressin to the posterior pituitary lobe or neurohypophysis, where they are released into the blood. The cells of the infundibular nucleus are involved in controlling secretion of the anterior pituitary hormones by regulating hormones that stimulate or inhibit the liberation of hormones produced in the pituitary. Each of the latter has a corresponding regulating hormone. These regulating hormones pass from the infundibular nucleus along the axons of its constituent cells to the median eminence, where they are released from the axon terminals into capillaries of the hypophyseal portal system.

In addition to the described fibers important for the hormonal control, the hypothalamus is surrounded by a series of fibers and commissures where small areas of gray matter of uncertain functional significance lie in context. Using modern transport methods it was demonstrated that fibers of the hypothalamus reach the spinal cord, more specifically the area of the solitary tract nucleus and the *dorsal motor vagus nucleus*. Most of these neurons originate in the nucleus paraventricularis, others in the zona incerta, the lateral and posterior hypothalamic areas (Greving, 1925). Several cells of the Edinger–Westphal nucleus also send axons in this region. The fibers run within large areas of the midbrain and pons continue in a caudal direction at the ventrolateral edge of the medulla, from which a group of fibers passes in a dorsomedial direction to the nucleus ambiguus, the dorsal motor nucleus, and the solitary tract nucleus (Thompson et al., 2003).

Further, the hypothalamus forms part of the greater limbic system and several hypothalamic fiber systems including the hypothalamo-hypophyseal pathways, medial forebrain bundle, dorsal longitudinal fasciculus of Schütz, mamillothalamic bundle, fornix, fasciculus mammillaris princeps, and mamillary peduncle form part of the circuitry of the greater limbic system.

The *fasciculus mediali telencephali* or *medial forebrain bundle* (MFB) may be considered the central longitudinal pathway of the limbic forebrain–midbrain continuum. It is an assemblage of loosely arranged, mostly thin fibers extending from the septal area to the tegmentum of the midbrain. MFB projections are reciprocated by corresponding ascending projections. Indeed, the MFB is a major descending and ascending link between the forebrain and brain stem.

It traverses the lateral preoptic–hypothalamic area, the scattered neurons of which are collectively designated as the bed nucleus of the medial forebrain bundle. This bundle is highly complex, comprising a variety of short and long ascending and descending links. In the transitional area of the diencephalon and mesencephalon, the MFB fibers are rearranged into a smaller medial and a larger lateral stream. The medial stream roughly maintains the sagittal orientation of the bundle's hypothalamic trajectory. It passes through the medial parts of the mesencephalic and rhombencephalic tegmental areas, just next to the raphe nuclei. The medial stream contains descending fibers, by which several hypothalamic centers project to the raphe nuclei and adjacent part of the medial reticular formation, and also fibers which ascend from the raphe nuclei to the lateral hypothalamus, from where they pass to a variety of diencephalic and telencephalic centers. The lateral stream of fibers from MFB to brainstem sweeps laterally and caudally over the dorsal border of the substantia nigra into a ventrolateral tegmental position immediately caudal to the substantia nigra; at this level it curves dorsomedially and then descends through the central tegmental area to the lateral tegmental field of the pons and medulla oblongata. These descending fibers terminate in a variety of brainstem centers, among which the substantia nigra (pars compacta), parabrachial nuclei, locus coeruleus, nucleus subcoeruleus, the noradrenergic groups, superficial ventrolateral reticular area, and dorsal vagal complex.

The dorsal longitudinal fasciculus of Schütz or periventricular fiber system, much like the medial forebrain bundle, is a composite system consisting of thin ascending and descending fibers. This fascicle extends from the posterior part of the hypothalamus to the caudal medulla oblongata and occupies a periventricular position over its entire length. Rostrally its fibers become incorporated into the hypothalamic periventricular system.

Commissures of the Hypothalamus

In the area of the PVN, the supraoptic dorsal commissura (Ganser) is demonstrated in 2% of cases. Greving (1925) also described the dorsal supraoptic decussation. The fibers of the supraoptic decussation surround the fornix and continue to the other side in the floor of the third ventricle. Further fiber connections are still being discussed today. The ventral supraoptic commissura (Meynert and Gudden) consists of fibers which connect the two optic tracts with each other and extends as far as the entopeduncular midbrain nuclei. They are said to contain a tectopetal and a tectofugal fiber system and extend in part as far as the corpus geniculatum, while other parts reach the colliculus cranialis and the regio pretectalis. The commissura supramammillaris (hypothalamica posterior, subthalamica posterior, retroinfundibularis) connects the nucleus subthalamicus, the zona incerta, and the globus pallidus with each other and with cellular areas of the contralateral tegmentum, as well as possibly with the LMN.

Nonmyelinated commissural fibers are thought to connect nuclear regions of the anterior hypothalamic region and its neighboring structures (septum and diagonal band of Broca) in the dorsal sections of the lamina terminalis. In the chiasmatic section the fibers run between the contralateral nuclei suprachiasmatici and the tuber cinereum (Del Brio et al., 1990).

HYPOTHALAMUS: FUNCTIONAL ASPECTS AND PATHOLOGY

From the anatomy shown above, the hypothalamus presents as a key role that of governing endocrine production for the whole organism in response to stimuli from outside. The hypothalamus is critically involved in the regulation of endocrine functions, control of autonomic reactions, and generation of basic behavioral patterns. The endocrine functions are regulated along the hypothalamo-hypophyseal pathways discussed above. The cells of origin of these pathways may be considered neuroendocrine motoneurons representing the final common pathway for central neural influences on hormone secretion from the anterior and posterior lobes of the pituitary gland.

Further, several hypothalamic centers exert regulatory influences on preganglionic autonomic neurons in the brain stem and spinal cord. Thus, the PVN, referred to as "master controller of the autonomic system," projects directly to the dorsal vagal motor nucleus, nucleus ambiguous and thoracic (sympathetic) and sacral (parasympathetic) preganglionic cell columns. This projection shows a distinct functional differentiation, in that selective clusters of paraventricular neurons address specific sites of preganglionic neurons. Indirect projections from the PVN to the sympathetic and parasympathetic efferent centers are synaptically interrupted in autonomic premotor nuclei such as the A5 noradrenergic cell group and ventrolateral medulla. Apart from the PVN, the DMN and LHA also provide descending connections to the autonomic nuclei of the lower brainstem and spinal cord.

The classic work of Hess and Brügger and others have shown that by electrical stimulation, characteristic behavioral responses can be elicited from different hypothalamic loci. The somatomotor neurons involved in these responses are attained via multisynaptic pathways relayed in, among other structures, the zona incerta, tegmental pedunculopontine nucleus, and reticular formation. In the hypothalamus, these somatomotor responses are integrated with endocrine and autonomic responses into motivated or goal-oriented behaviors that can be divided into two broad classes, one related primarily to survival of the individual and another to survival of the species as a whole. Thermoregulative, ingestive, defensive, and aggressive behaviors belong in the first, sexual and parental behaviors in the second.

Since 1980 a total of 18 peptide agents were found in the brain, all of which were formed in the hypothalamus and exert an effect on the hypophysis or outside the nervous system.

The secretion of hypothalamic releasing and inhibitory hormones from axon terminals of tuberoinfundibular neurons into the portal capillary system is dependent upon several layers of control that can be exerted directly on the cellular bodies or processes of these neurons. For one, neurons of the tuberoinfundibular system can be modulated by substances circulating in the bloodstream that either pass the BBB because they are fat soluble steroids or small molecules, or access tuberoinfundibular neurons via the cerebrospinal fluid due to the periventricular location of many tuberoinfundibular neurons and poor development of tight junctions between ependymal cells in these regions. Feedback effects of thyroid hormone, for example, occur directly on TRH-producing neurons within the PVN as demonstrated by the ability of a microcrystalline implant of *triiodothyronine* (T3) adjacent to the PVN to prevent the hypothyroid-induced increase in TRH biosynthesis on that side but not the opposite side. In addition, tuberoinfundibular neurons receive numerous axosomatic and/or axodendritic contacts from local interneurons and/or other regions in the brain that contain a variety of chemical messengers that contribute to intercommunication between specific neuronal groups or are important in establishing the set point at which the hypophysiotropic substances are secreted in response to hormonal feedback signals.

Historically diencephalon can be divided into two different sectors on the base of their functions: one with parasimpathic maintenance activity (trophotropic) and another one with orthosympatic excitatory activity (ergotropic). Further we can identify two functional self-regulating cycles interconnected with these two sectors in the prosencephalon: a prosencephalic ergotropic cycle and a prosencephalic trophotropic cycle for which the hypothalamus is the most important synaptic center. Nauta and Haymaker (1969) emphasize that the hypothalamus steers not only the ergotropic and the trophotropic cycles but also autonomic and endocrine activities as well as expression and emotions.

Hypothalamus: Neurosecretion

During neurosecretion, the pericaria of the hypothalamic nuclei form certain peptide hormones (releasing hormones or factors), which are directed to the hypophysis by way of a portal circulation where they activate hormone release. On the other hand, the hypothalamus itself is in continuity with the cell bodies of a part of the pituitary gland (the neurohypophysis) for the direct production of hormones in the blood stream, this is the case with the production of vasopressin (antidiuretic hormone, ADH) and oxytocin. In total, the peptides produced are vasopressin, oxytocin, somatostatin, *growth releasing hormone* (GH), and *ACTH-releasing hormone* (ACTRH), *gonadotropin releasing hormone* (GnRH), *thyrotropin releasing hormone* (TRH), *prolactin inhibitor factor* (dopamine). As mentioned before, the parts directly in connection and continuity with hormone production are the PVN, responsible to the oxytocin synthesis, and the SON steers the production of vasopressin. On the basis of different studies is assumed that oxytocin is chiefly synthesized in the PVN and vasopressin in the SON (Fig. 10.8). However, both nuclei contribute to the synthesis of both hormones. Weindl (1973) emphasize that vasopressin, oxytocin, and their peptides (neurophysine) are produced in the magnocellular neurons of the SON and the PVN, and transported through the inner zone of the infundibulum to the posterior lobe of neurohypophysis. There the hormones are stored and passed on to the blood circulation by way of permeable capillaries. Vasopressin and neurophysine fibers end in the outer zone of the infundibulum at permeable nuclei of the hypothalamo-adenohypophyseal circulation.

LRH and GnRH are produced in the neurons of the preoptic area and transported by way of the axons to the portal capillaries in the outer zone of the infundibulum. There they influence the capillaries of the OVLT. Regardless of whether there is direct hormone production or through releasing factors, pathology involving these hypothalamic nuclei, such as a tumor may result in secondary pituitary deficiency, both before and after surgical treatment. Preoperative identification of endocrine deficiency in tumor patients informs selection of the appropriate treatment approach and replacement hormone therapy optimizes patient safety for surgery. In adult and pediatric tumor patients, surgery rarely restores anterior pituitary function in those with preoperative hypopituitarism (Weindl, 1973). We briefly expound here some clinical consequences of a hypothalamic problem responsible for pituitary insufficiency:

Growth Hormone Deficiency

One of the most frequent forms of disturbance in GH production is a primitive failure to release the GH releasing factor (GHRH) occurred in pediatric age. Neither short stature nor normal height are good biomarkers for GH status in children with this disturbance; therefore, we recommend dynamic testing for GHRH. In tumor patients, the phenomenon known as "growth without growth hormone" whereby IGF1 and height can be normal in obese children with sellar tumor mass with GHRH insufficiency, probably due to the action of hyperinsulinemia upon the liver to generate IGF1 or directly via IGF1 receptors. Further, GH replacement therapy is not associated with increased tumor recurrence. The improvement in lipid profile with GH

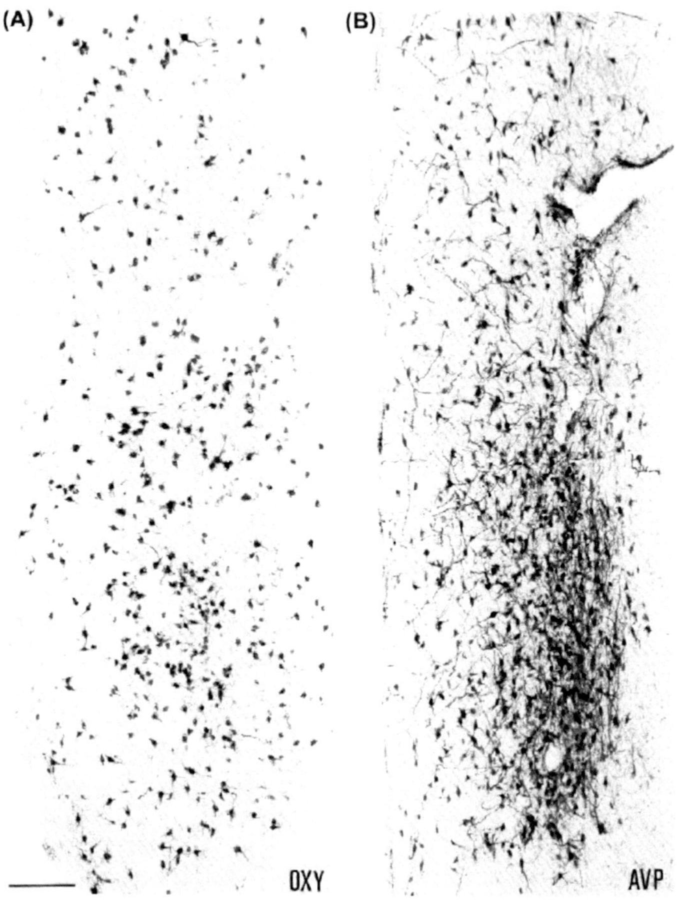

FIGURE 10.8 **A pair of photomicrographs illustrating the distribution of oxytocin (A) and vasopressin (B) neurons in the human paraventricular nucleus.** Note that the vasopressin cells form a dense cluster in the ventrolateral quadrant of the paraventricular nucleus, whereas the oxytocin neurons tend to avoid this region, instead clustering along its borders. Scale 250 mm. *Image courtesy of Saper, C.B., 2012. Chapter 16 – Hypothalamus. In: Mai, J.K., Paxinos, G. (Eds.), The Human Nervous System, third ed. Academic Press, pp. 548–583. https://doi.org/10.1016/B978-0-12-374236-0.10016-1.*

replacement seen in NFPA[1] patients has been shown in some, but not all, studies in these patients. This is a subtle disorder since the pediatrician may be faced with the case of a child with absolutely normal growth and stature with obesity and IGF-1 values in the normal range. If recognized in time the GHRH deficiency can lead to a recovery of the *body mass index* (BMI) with supplementary therapy with an almost complete return to normal values, which, on the other hand, if recognized late this disorder can lead to a situation of irreversibility of the scenario, as well as a severe risk of morbidity and mortality in case it is associated with a tumor with a high proliferation index (Uhlman et al., 2019). Adult CP[2] patients do not demonstrate benefit in reduction in fat mass and have an increase in BMI on GH replacement (0.73 vs. 0.17 kg/m^2 in NFPA). The reason for this difference in the response to GH therapy in CP patients is unclear. The increased mortality rate seen in tumor patients remains high even after GH replacement, although a lower IGF1 measurement at last sampling predeath was associated with greater mortality in a recent KIMS database analysis, suggesting that poor adherence or discontinuation of GH replacement in a terminal illness may have had an impact on the clinical data in this field (Grouselle et al., 2008).

ACTH Deficiency

This is a very rare disorder when isolated; in fact, it is often detected in patients with postoperative hypothalamic-pituitary insufficiency.

Symptomatic hyponatremia secondary to ACTH deficiency may be identified at presentation with tumor patients. ACTH deficiency and higher dose of

1 Non-Functioning Pituitary Adenoma.

2 Cerebral-Palsy.

hydrocortisone replacement therapy (>25 mg/d in acromegaly; >20 mg/d in NFPA patients) in other pituitary disorders are associated with increased mortality (Tuchelt et al., 2000).

Gonadotrophin Deficiency, Fertility, and Pregnancy

Clinical evidence of gonadotrophin deficiency at presentation of hypothalamic disturbance includes menstrual disorders in over half of female with tumor mass patients, impaired sexual function and loss or lack of evolution of secondary sexual characteristics. Pubertal delay can occur, especially in female subjects, if GnRH is not secreted. Pubertal delay is the presenting complaint in almost 20% of tumor patients. A review of pregnancy in hypopituitary women suggested a higher risk of postpartum hemorrhage and babies of small for gestational age. One of the most common cases is when a young girl is underweight. On the other hand, there is interruption of the menstrual cycle during anorexia because, likewise, the hypothalamus notices that the body is deficient in many substances so there is an inhibition of GnRH release. The initiation and maintenance of reproductive capacity in humans is dependent on pulsatile secretion of the hypothalamic hormone GnRH. Congenital hypogonadotropic hypogonadism (CHH) is a rare disorder that results from the failure of the normal episodic GnRH secretion, leading to delayed puberty and infertility. CHH can be associated with an absent sense of smell, also termed Kallmann's syndrome, or with other anomalies. CHH is characterized by rich genetic heterogeneity, with mutations in >30 genes identified to date acting either alone or in combination. CHH can be challenging to diagnose, particularly in early adolescence where the clinical picture mirrors that of constitutional delay of growth and puberty. Timely diagnosis and treatment will induce puberty, leading to improved sexual, bone, metabolic, and psychological health. In most cases, patients require lifelong treatment, yet a notable portion of male patients (~10%–20%) exhibit a spontaneous recovery of their reproductive function (Young et al., 2019).

Given the high prevalence of posterior pituitary disease, it is unsurprising that oxytocin replacement was necessary to manage uterine atony in a patient with failure of uterine involution postpartum; oxytocin deficiency has also been implicated in the high rate of malpresentations (25%) in pregnancies with maternal hypopituitarism (Tobet, 2002).

TSH Deficiency

Very rare, secondary TSH deficiency due to insufficient release of TRH can be present at diagnosis of hypothalamic disfunction and can be unmasked by GH replacement. Replacement of TSH deficiency requires usually oral T4 therapy.

Hyperprolactinemia

Around 50% of adult intrasellar and parasellar tumor patients have evidence of hyperprolactinemia at presentation. Prolactin levels fall after surgery for these patients and prolactin deficiency can be associated with failure of lactation in female tumor patients who pursue fertility (Korkmaz et al., 2021). The most common forms of prolactin disorder mainly involve alterations affecting the release of its inhibitor namely dopamine. In fact, any expansive process that results in kneeling or compression of the pituitary pedicle can result in a reduction of dopamine release and thus a status of hyperprolactinemia (Mercuri et al., 2022). Also, therefore in highly invasive forms of sellar tumor such as craniopharyngioma the patient may present with complete hypothalamic insufficiency and hyperprolactinemia (Rivas et al., 2014).

Diabetes Insipidus

Approximately 20% of patients with a sellar tumor mass may present with DI prior to surgery. Preexisting DI increases the likelihood of perioperative dysnatremia which means that particularly careful monitoring of fluid and electrolyte balance is needed for CP patients. DI is common following surgery for CP and can be temporary or can persist; it is permanent in up to 80% of CP patients. Patients with CP and DI have increased mortality compared to CP patients without DI, an association observed in other hypopituitary cohorts. The reason for this association is unclear but CP patients with DI have been reported to have a higher BMI than those with intact vasopressin (or ADH) secretion (30.4 kg/m^2 compared with 24.6 kg/m^2), which suggests that DI is a marker for hypothalamic damage. Another explanation for the link between mortality and DI in CP patients is the association of dysnatremia with increased mortality. Dysnatremia associated with either DI or SIADH in CP can be exacerbated by abnormalities of thirst (Keil et al., 2008).

Thirst and Adipsia

The somatomotor response to thirst includes general arousal to and the behavior act of drinking. The arousal is believed to be mediated by fibers passing from the SFO to the substantia innominata (SI) and projections from the SFO and AV3V to the LHA. Since the missing of ADH in DI, thirst is almost always normal in patients with this pathology, but the combination of DI and adipsia or hypodipsia is well recognized in tumor patients that involve the hypothalamus. Loss of thirst usually reflects widespread damage to the hypothalamus after attempted total resection or repeated surgeries. Baroregulated vasopressin release in adipsic DI in tumor patients is also absent. In the clinical setting, adipsia should be suspected in a patient who has persistent hypernatremia and yet does not drink in spite of normal cognitive function and free access to water (Pisipati et al., 2011). Diagnosis of adipsic DI is made by use of a visual analogue thirst scale during water deprivation or hypertonic saline infusion test (Contreras et al., 2016). Two hypothalamic cell masses (the SON and paraventricular nuclei), two circumventricular organs (the subfornical and OVLT), and the central projections of the IXth and Xth cranial nerves form essential parts of the neural circuitry associated with regulation of thirst and drinking behavior.

The SON and PVN synthesize the peptide neurohormones vasopressin and oxytocin, give rise to axons descending through the infundibular stalk to the posterior pituitary, thus forming the supraoptico- and paraventriculo-hypophyseal tracts. Vasopressin is responsible for antidiuresis, promoting reabsorption of water by the kidney; oxytocin stimulates smooth muscle in the uterus and mammary glands to contract. Vasopressin and oxytocin are produced by separate neurons which collectively constitute the so-called magnocellular neurosecretory system. Two circumventricular organs (CVOs), the subfornical organ (SFO) and OVLT, are also involved in regulating thirst and drinking behavior (see Chapter 16). The OVLT and immediately adjacent median PON may be designated together as the anteroventral third ventricular area (AV3V). Like the SFO, the OVLT is rich in AII receptors. The SFO and the AV3V are reciprocally connected and, together project to the SO and PVH. Interestingly, the projection from the SFO to the OVLT uses AII as a neurotransmitter. The AV3V is known to be especially sensitive to hyperosmolality of the blood, another stimulus to thirst.

Presumed cognitive inputs from the cerebral cortex are relayed to the PVH by the bed nucleus of the stria terminalis (BST) and ventral lateral septal nucleus (VLS), which in turn receive inputs from the prefrontal cortex (PFC), amygdaloid complex (AM), and hippocampal formation (HF). The control of thirst involves neuroendocrine, autonomic, and somatomotor responses. The neural substrate of the visceromotor response to thirst is formed by axons descending from the PVH to the nucleus of the solitary tract (NST), dorsal motor vagus nucleus (DMX), ventrolateral medulla (VLM), and spinal intermediolateral column (ILC) (Pang et al., 2022).

Hypothalamus: Temperature and Other Centers

Maintaining a stable body temperature is essential for survival, as a hyperthermia or hypothermia condition induced by temperature changes of a few degrees could produce severe debilitating effects. Several studies of body temperature homeostasis have shown that the environment temperature is sensed by skin thermoreceptors, and the information is then routed to the brain through the spinal cord and midbrain, integrated into the preoptic area (POA) of the hypothalamus, and finally transmitted to peripheral tissues for thermogenesis (Tan et al., 2018).

In these studies, were identified specific type of hypothalamic neurons that could elevate thermogenesis in a cold environment. From the temperature center of the hypothalamus, thermoregulating fibers probably cross to the opposite side by way of the pyramids and extend as far as the intermediate nucleus of the spinal cord in a direct link between POA and the body.

The preoptico-hypothalamic continuum plays a prominent role in thermoregulation. The rostral heat control center is located in the preoptic anterior hypothalamic area (POAH). This contains numerous neurons sensitive to increased blood temperature. Many of these warmth-sensitive elements also respond to skin temperature. Stimulation of the POAH induces cutaneous vasodilatation and sweating, and the area also shows a high level of *c-fos* activity following hyperthermia (Zhang et al., 2014a,b). Hyperthermia, hypothermia, and poikilothermia may occur when the POAH is damaged by infarction, trauma, or surgery.

The caudal heat control center is located in the posterior hypothalamic area. This area is temperature-blind, i.e., does not contain neurons that respond to heating or cooling. However, electrical or chemical stimulation there can produce shivering, vasoconstriction, and increased metabolic heat production. Large lesions in the posterior hypothalamus may impair heat production, which results in poikilothermia.

Little is known about the specific circuits involved in thermoregulation. It seems likely that information derived from cutaneous thermoreceptors is relayed by fibers belonging to the spinohypothalamic component of the anterolateral system and its trigeminal equivalent. Cold exposure leads to the secretion of thyrotropin-releasing hormone (TRH) to activate the thyroid axis. In humans the TRH-producing cells are concentrated in the dorsocaudal part of the PVN. As regards the central circuitry involved in relaying autonomic thermoregulatory responses, several direct and indirect projections connect the hypothalamus with the sympathetic and parasympathetic preganglionic cell groups in the lower brain stem and spinal cord. All of these may mediate thermoregulatory impulses.

The discovery of Pacap (pituitary adenylate cyclase activating polypeptide)/BDNF (brain-derived neurotrophic factor)-expressing neurons in POA sheds light to the response to warm temperature in the environment and mediate the decrease of body temperature. Activation of Adcyap1 (adenylate cyclase activating polypeptide 1-expressing or Qrfp (pyroglutamylated RFamide peptide)-expressing hypothalamic neurons is capable of inducing extreme hypothermia, leading to a state of hibernation or torpor (Hrvatin et al., 2020). In contrast, the circuit promoting thermogenesis is less well understood. Although recent studies identified that the *bombesin-like receptor 3* (BRS3)-expressing neurons in POA are capable of inducing thermogenesis (Piñol et al., 2021) and the hypothalamus-projecting glutamatergic prefrontal neurons mediate stress-induced hyperthermia (Kataoka et al., 2020), the cell types involved in thermoregulation outside of POA remain largely unknown.

The efferent fibers of these nuclei proceed through the ventral radices and the communicant branches to the where they synapse with the postganglionic fibers of the cholinergic type. By way of the gray communicant branches, they again join the spinal nerves and reach the sweat glands with these or with the blood vessels. The sweat fibers for face and neck synapse in the superior cervical ganglion and then proceed to the glands with the trigeminal nerve, possibly also with the facial nerve. The limits of the sympathetic dermatomes are often asymmetrical and irregular. In addition to physiological sweating (perspiratio insensibilis) which amount to 800 mL per day and can increase to 3–4 one per day in the tropics (maximum quantity 10), there is cortical emotional sweating with vasoconstriction during psychological stimulation, exhaustion, fear, and terror, in which a strong opposing parasympathetic reaction is involved in which hypotension and sweating are generated. Furthermore, there is gustatory bulbar sweating elicited by taste sensations, spinal sensory sweating in cases of paraplegics, sweating caused by drugs and pathological hyperhidrosis (Thomson et al., 2003).

Sexual Centers

Sexual desire is controlled by brain systems involved in sexual excitation and inhibition. Brain dopamine systems (incertohypothalamic and mesolimbic) that link the hypothalamus and limbic system appear to form the core of the excitatory system. This system also includes melanocortins, oxytocin, and norepinephrine (Pfaus, 2009). According toBauer (1959) who studied 60 autopsy cases, a decrease in the function of the gonads is found chiefly with lesions in the lower and anterior hypothalamic areas. Electrical stimulation of the septum region can lead to erection of the penis and chemical stimulation of this area can evoke an orgasm in women. Further, hypersexuality has been described in cases of deep lying frontotemporal tumors as well as in diseases of the limbic system (Latronico et al., 2016).

In particular, a complex of highly interconnected hypothalamic preoptic and basal telencephalic cell groups are involved in the initiation and coordination of reproductive behavior. This complex, designated here as the telodiencephalic reproductive complex (TDRC), encompasses the hypothalamic ventromedial (VMH), infundibular (INF), ventral premamillary (PMV), and suprachiasmatic (SCH) nuclei, the medial PON and, in the basal telencephalon, the bed nucleus of the stria terminalis (BST), lateral septal nucleus (LS), nucleus of the diagonal band of Broca (NDB), and medial amygdaloid nucleus (MEA).

The considerable differences in sexual behavior between males and females are reflected in differing size, cell number, and structure of several TRDC cell groups, which include the medial preoptic area, BST, VMH, and SCH (Del Abril et al., 1987).

The BST is a telencephalic structure situated at the junction of the septum, nucleus accumbens, and preoptic region. Like all other components of the TDRC, it forms part of the limbic system and is characterized by an extraordinarily large number of neuromediators, particularly neuropeptides.

Phenomena of sexual dimorphism are not confined to structures within the TDRC. The sacral spinal cord contains a distinct, ventrally situated group of motoneurons known as the nucleus of Onuf. The axons emerging from this nucleus supply, among other structures, the muscles of the pelvic floor and the bulbocavernosus and ischiocavernosus muscles. The TDRC contains concentrations of neurons producing gonadotropin-releasing hormone and is an important target of circulating gonadal steroid hormones (Dudas et al., 2000).

Gonadotropin-releasing hormone (GnRH), also known as luteinizing hormone–releasing hormone (LHRH), is a decapeptide essential for mammalian reproduction. It is released from nerve terminals in the median eminence and carried via the hypothalamo-hypophyseal portal system to the anterior pituitary, where it stimulates the synthesis and secretion of the gonadotropic luteinizing hormone (LH) and follicle-stimulating hormone (FSH). The distribution of neurons expressing GnRH in the human brain has been studied with the aid of antibodies against this hormone or utilizing hybridization histochemistry to visualize GnRH mRNA. It appeared that GnRH-producing perikarya are widely scattered over the entire hypothalamus and adjacent basal telencephalic regions, with concentrations in the septum, nuclei of the diagonal band of Broca, dorsal and ventral parts of the medial preoptic area, and infundibular nucleus (Brouwer, 1950).

The MPN and VMH receive highly processed somatosensory information from cortical association areas which attain the hypothalamus via the hippocampal formation and amygdaloid complex. The hippocampo-hypothalamic and amygdalo-hypothalamic projections carrying this information are partly synaptically interrupted in the LS and BST, respectively. The amygdaloid complex also relays to the hypothalamus olfactory information derived from the main and accessory olfactory systems. The accessory olfactory or vomeronasal system comprises the vomeronasal organ or Jacobson's organ, vomeronasal nerve, a special part of the olfactory bulb known as the accessory olfactory bulb and periaqueductal gray (PAG).

PAG or mesencephalic central gray is a conspicuous midline structure which surrounds the cerebral aqueduct. It consists mainly of small, densely packed cells. The PAG receives input from a variety of structures including several limbic cortical fields (prelimbic, infralimbic, anterior cingulate), the amygdaloid complex, MPN, VMH, zona incerta (ZI), ATV, dorsal raphe and central superior nuclei, parabrachial nuclei, pontine and medullary reticular formation, and spinal cord. Many of these regions receive reciprocal projections from the PAG. The PAG plays an important role in a variety of functions and behavioral patterns including defense reactions, reproductive behavior, vocalization, autonomic regulation, and pain modulation (Cantu et al., 2022).

Hypothalamus and Hyperglycemia

Hyperglycemia, caused in part by elevated hepatic glucose production (GP), is a hallmark feature of diabetes and obesity. The hypothalamus responds to hormones and nutrients to regulate hepatic GP and glucose homeostasis. The hypothalamus detects nutrients and hormonal signals, both indicators of the body's energetic status, to regulate peripheral metabolic homeostasis. Signals from gluco-regulatory hormones insulin, leptin, GLP-1 and glucagon are sensed along with nutrient such as fatty acids, glucose and amino acids (Lam, 2010) in the mediobasal hypothalamus (MBH) to initiate feedback mechanisms modulating hepatic GP (Obici et al., 2002).

When conditions such as high-fat feeding disrupt central hormonal and nutrient sensing, the resulting imbalance in peripheral glucose metabolism leads to fasting hyperglycemia and eventually diabetes and obesity. Hyperglycemia and glycosuria occurred following needling of the tuber region as well as destruction of the PVN. Elevated GP contributes to fasting hyperglycemia; thus, dissecting hypothalamic hormonal and nutrient sensing mechanisms that lower GP is an important goal in diabetes and obesity research.

Recent evidence generated via genetic, molecular, and chemical experimental approaches indicates that glucagon and lipid signaling independently trigger complementary hypothalamic mechanisms to lower GP. However, feedback mechanisms can also be altered through disruption of signaling from hypothalamic centers as primum movens in hypothalamic obesity (HO).

HO is a well-recognized complication of both tumor patients and its therapy; more than 80% of tumor patients are overweight or obese. Thus, targeting hypothalamic glucagon or lipid signaling may have therapeutic potential in this type of diabetes and obesity (Lapierre et al., 2014).

If obesity can result from excessive calorie intake (hyperphagia), reduced energy expenditure (low physical activity counts), or excessive energy storage (mediated by hyperinsulinaemia), a loss of the medial zone of the hypothalamus equally leads to hyperphagia, while a loss of the lateral zone results in hypophagia to aphagia. Various hypothalamic nuclei were degenerated: dorsomedial and ventro-medial nuclei of the ventrolateral nucleus and the anterior part of the posterior hypothalamic nucleus. The degeneration of the gray substance around the infundibulum can cause obesity, genital atrophy, or both. Damage to the arcuate nucleus, resulting in leptin resistance and hyperinsulinaemia, is thought to result in hyperphagia; leptin supplementation is not a therapeutic option for these patients, but recent trials have included strategies to reduce insulin secretion (Legradi et al., 1997).

Other Functions of the Hypothalamus

The countless connections that the hypothalamus has with other higher centers and the periphery of the body through hormonal signaling but also through nerve connections has over the decades led to speculation about numerous other functions in which, however, the direct function of this part of the brain remains largely unknown (Haber et al., 1987).

Circadian Rhythms

The small SCH, located in the rostroventral part of the hypothalamus, is considered the major pacemaker of the brain and to coordinate a wide spectrum of endocrine, physiological and behavioral circadian rhythms. Neuron populations are present that are immunoreactive for vasopressin (VP), vasoactive intestinal polypeptide (VIP), neuropeptide-Y (NPY), and neurotensin (NT), all of which show a particular spatial organization. In addition, somatostatin, galanin, and preproenkephalin have been found in the SCH. Individual SCH neurons in isolation have the capacity to generate circadian oscillations in electrical activity and secretion. SCH cells are electrically coupled, promoting synchronization of neural activity and also required for the maintenance of circadian rhythms in behavior. The SCH itself generates biological rhythms with periods of approximately 24 h. These endogenous SCH rhythms are adjusted daily (entrained) to the environmental light/dark cycle through a direct projection from the retina to the SCH known as the retinohypothalamic tract, In a parallel mechanism and only partially mediated by the melatoninergic system (Cammarota et al., 2023). Experimental studies have shown that the SCH in rat, in addition to its retinal input, receives afferents from a number of other sources including the intergeniculate leaflet (the homolog of the primate pregeniculate nucleus), pretectum, midbrain raphe nuclei, ventromedial hypothalamic nucleus, posterior hypothalamic area, lateral septum, ventral subiculum, infralimbic cortex, and paraventricular thalamic nucleus. These SCH afferent connections are topographically organized. The intergeniculate leaflet, midbrain raphe nuclei, and pretectum, which all receive afferents from the retina, like the retinohypothalamic tract project mainly to the ventral part of the SCH. These structures form relays in indirect routes along which photic input from the retina attains the SCH. Except for the paraventricular thalamic nucleus, the remaining cell masses project mainly to the dorsal part of the SCH. The SCH, i.e., the circadian pacemaker, exerts control over the sleep–wake cycle and is also involved in regulating a number of endocrine and behavioral processes (Patton et al., 2023).

Wakefulness and Sleep

The circuitry for regulation of sleep, wakefulness, and arousal includes the following key structures some of which for part of the reticular formation (see Chapter 5b):

a. Two cell masses, the laterodorsal tegmental (LDT) and pedunculopontine tegmental (PPTg) nuclei, both situated in the tegmentum of the caudal midbrain and rostral pons.
b. Noradrenergic neurons clustered in the locus coeruleus (A6-LC), in the rostral pontine tegmentum.
c. Mesencephalic serotoninergic neurons in the dorsal raphe nucleus (B6-DRN) and superior central nucleus or median raphe nucleus (B8-MnR) nuclei. The activity pattern of these neurons closely resembles that of the LC cells.
d. A group of histaminergic cells clustered in the TMN in the posterior hypothalamus.
e. A group of diffusely arranged neurons in the intermediate region of the hypothalamus characterized by presence of the neuropeptide orexin (also known as hypocretin).

Stress Response

The so-called hypothalamo–pituitary–adrenal (HPA) axis plays a crucial role in the response to stress (Bard, 1928). Its structural elements are:

a. Parvocellular neurons in the hypothalamic PVN, which produce corticotropin-releasing hormone (CRH);
b. corticotrope cells in the anterior pituitary gland (APG) that produce adrenocorticotrope hormone (ACTH).

In rats, CRH-producing cells are located in a distinct subnucleus of the PVH, but in humans they are scattered throughout the nucleus (Raadsheer et al., 1993).

These steroid hormones have a wide range of effects, from increasing blood pressure and glucose levels to suppressing inflammatory and immune responses. Most of the CRH-producing neurons in the PVH project to the neurohemal contact zone in the median eminence (ME). Attaining the anterior pituitary via the hypothalamo-hypophyseal portal circulation, CRH stimulates the secretion of ACTH from the corticotropes and ACTH in turn stimulates the release of adrenal glucocorticoids.

Food Intake

Classic studies in the 1940s and 1950s (Brouwer, 1950) revealed that large bilateral electrolytic lesions centered in the region of the ventromedial hypothalamic nucleus (VMH) produce hyperphagia and obesity, while lesions at the same level in the lateral LHA result in aphagia. Moreover, electrical stimulation appeared to elicit feeding behavior from the LHA and feeding suppression from the VMH. Collectively, these findings provided the foundation of a dual-center hypothesis, according to which the VMH contains a "satiety center," whereas the LHA contains a reciprocally interrelated "feeding center." More recent observations have thrown considerable doubt on this hypothesis. Thus, it was found that lesions restricted entirely to the VMH do not lead to obesity, whereas slightly more dorsal lesions, which included the PVN, appeared to be effective. Moreover, evidence suggested that aphagia following LHA lesions is due to damage to fibers passing through the medial forebrain bundle rather than to destruction of specific populations of neurons in the LHA. Finally, it was found that connections between the two feeding control centers are only weakly developed. Thus, the neuroanatomical basis for lesion- and stimulation-induced changes in ingestive behavior remained unclear (Bobrovskikh et al., 2023).

Recent morphological, physiological, and molecular studies have provided new insights on the role of the hypothalamus in regulating food intake and body weight.

It has been shown that interacting populations of peptidergic neurons in the ventromedial hypothalamus and LHA play a crucial role in these regulations and that the PVN is also critically involved.

It has been shown that circulating leptin provides a physiologically important negative feedback signal to control body weight by affecting both ingestive behavior and metabolic activity. Leptin is an adipocyte-derived hormone whose circulating levels vary directly as a function of adipose tissue mass.

The infundibular nucleus has to be considered the principal monitor of circulating leptin-based signals to the brain. It is a major site of leptin receptor expression, and because the BBB is absent locally, its constituent neurons are directly accessible to circulating hormones.

Defensive Behavior

Threatening stimuli may evoke a characteristic pattern of behavioral and autonomic responses commonly designated the affective defense reaction. The autonomic response pattern consists of pupillary dilatation, piloerection, increased respiratory rate and amplitude, and a sympathetic cardiovascular reaction comprising accelerated heart rate, elevated blood pressure, and vasoconstriction in cutaneous, renal, and splanchnic beds but vasodilatation in skeletal muscles (Roeling et al., 1994). This defense response can be elicited by electrical stimulation of a brain area extending from the hypothalamus into the mesencephalic gray matter. Later it appeared that defensive behavior in cats can also be elicited by electrical stimulation of the amygdala, particularly its medial nucleus. Within the hypothalamus, several structures have been reported to be directly involved in defensive behavior. Hess and Brügger found that in the cat this behavior can readily be elicited from a limited zone centered in the perifornical part of the LHA (Wang et al., 2023a,b). A comparable area has been identified in the rat (Lammers et al., 1988). According to Fuchs (Fuchs et al., 1985), the hypothalamic zone from which defensive behavior can be elicited electrically in cats also includes the VMN and anterior hypothalamic nuclei (AH). In rodents an area encompassing the perifornical zone of the LHA plus the ventrolateral pole of the VMN has been designated the hypothalamic "attack area" the cell groups together forming the hypothalamic defensive complex project to the PAG via the dorsal longitudinal fasciculus of Schütz.

IMAGING

The development of new theories on the role and activity of the hypothalamus evident in recent decades has been determined above all thanks to the introduction of the most recent imaging techniques, in fact before of this event most of the studies took place by in vivo analysis using mammalian brain with obviously all the risks of an inaccurate analysis of the higher functions (Bard et al., 1937; Burford et al., 1974).

Nowadays, magnetic resonance imaging (MRI) is the technique of choice for evaluating the anatomy of the hypothalamus in vivo. However, the very small size of this region requires the use of specific high-resolution MRI sequences (Fig. 10.3). The identification of gray or white matter within the hypothalamus, where some nuclei are about 1 mm across, is difficult and must be treated with caution, with frequent reference to sequential anatomical and histological sections. An atlas of the anatomical, histological, and radiological correlates of the human hypothalamus in the coronal plane would thus be very useful for MRI identification of the hypothalamic structures. However, to our knowledge, such an atlas does not exist. In addition, MRI can help establish the relationship of mass lesions to surrounding structures. The diagnosis and management of mass lesions in the sellar and parasellar areas remain challenging. Modern neuroimaging techniques have revealed the presence of lesions and tumors causing structural abnormalities in the hypothalamus of patients with a large variety of disorders (Schindler et al., 2012). The imaging modalities used to characterize hypothalamic and pituitary lesions have significantly evolved over the course of the past several years. Computed tomography (CT) and CT-angiogram play a major role in detecting various sellar lesions, especially in patients who have contraindications to magnetic resonance imaging (MRI) and can also yield important information for surgical planning. However, MRI has become the gold standard for the detection and characterization of hypothalamic and pituitary tumors, infections, cystic, or vascular lesions (Yuh et al., 1994).

A pituitary MRI examination should be obtained if there is concern for mass effect (including visual loss, ophthalmoplegia, headache) or if there is clinical suspicion and laboratory evidence of either hypopituitarism or pituitary hormone excess. The information obtained from MRI images also provides us with assistance in planning surgery (Perosevic et al., 2021). Using intraoperative MRI can be very helpful in assessing the adequacy of tumor resection. In addition, MRI images yield reliable data that allow for noninvasive monitoring of patients postoperatively. When approaching patients with possible sellar or hypothalamic masses, it is important not only to focus on imaging but also detect possible pituitary hormone deficits or excess, in order to establish an appropriate diagnosis and initiate treatment.

SURGICAL APPROACHES

Although rarely, the hypothalamus may be the site of pathology primarily involving its context and for which surgery is required. This is generally complex surgery being the hypothalamus located in one of the deepest regions of the skull, making then surgical approaches varied and with high risks for the patient. Further more recently, the hypothalamus has been used as a stereotactic target for pain syndrome, mood and psychiatric disorders (Buchfelder et al., 2019). The understanding of such anatomic nuances aid in the selection of the most appropriate surgical approach and in the prevention of potential complications.

Different approaches should be considered according to pathology or tumor extension into different compartments. Purely sellar tumors are amenable to endoscopic trans-sellar approaches, whereas those with a suprasellar extension require an extended transtuberculum approach. In some of those patients, a narrow chiasm-pituitary window may block access to the tumor and a transcranial translamina terminalis approach may be favored (Yasargil, 1996).

Craniopharyngioma is the typical tumor that can have suprasellar extension with involvement of the hypothalamus. Craniopharyngiomas (CP) are tumors of the pituitary sella and suprasellar region that are derived from embryonic remnants of Rathke's pouch with an incidence of only 0.13 per 100,000 patient years. CP is classified as a histologically benign tumor; however, a diagnosis of CP is associated with increased morbidity and mortality compared to both the background age- and sex-matched population and to other conditions associated with hypopituitarism; therefore, the diagnosis cannot be considered to follow a benign course. CP patients suffer from high rates of pituitary hormone deficiencies, appetite and thirst disorders, sleep disorders, and impaired temperature regulation as signaled above (Awad et al., 2023).

The suprasellar location predisposes to hypothalamic compression or invasion. These factors contribute to difficulty in achieving gross total surgical resection (GTR) and lead to hypothalamic damage where GTR has been pursued. Hypothalamic disease is postulated to be the main reason why CP patients fare so poorly in comparison with patients with other tumoral causes of hypopituitarism such as NFPA, acromegaly, or even Cushing's disease (Thompson et al., 2019).

Tumors occupying the interpeduncular fossa may pose a significant challenge for an endoscopic endonasal approach and transcranial approaches. Transcavernous approaches and anterior and posterior clinoidectomies may be required for adequate exposure in such patients. Translamina terminalis and/or transcallosal approaches are recommended for resection of purely intraventricular tumors. Tumors extending into the lateral compartment should be considered for transcranial frontotemporal approaches. Because most of occupying space pathologies are midline lesions, the endoscopic endonasal approach represents an excellent approach for most of these. However, transcranial approaches should be considered for lesions with extension into the lateral compartment and for selected tumors involving the ventricular compartment (purely intraventricular tumors and those with extension to the foramen of Monro and/or lateral ventricles).

The distance between the front edge of the brain and the lamina terminalis is of interest for the transfrontal approach to the hypothalamus and the third ventricle with a distance ranged from 52 to 65 mm, and an average of 59.3 mm. Aside from the falx cerebri and the superior sagittal sinus which is usually developed in this area (1 cm above the base of the skull), the veins flowing into this area can also be ligated.

TRANSCALLOSAL APPROACHES TO THE THIRD VENTRICLE

Superior Transcallosal Approach

For the transcallosal approaches from above, attention must be paid to the branches of the ACA, from which occasional fine arteries branch out to the falx cerebri as well as to bridging veins to the superior and inferior sagittal sinuses. After transecting the corpus callosum and the septum pellucidum the two internal cerebral veins are seen to lie close beside each other in the traverse cistern. The postero-medial choroid branches, which supply the upper diencephalic sections can also lie exactly in the midline. After pushing the vessels in the midline aside, the approach is possible through the roof of the third ventricle which may possibly be opened during preparation (Shibuya et al., 1996).

Supra- and Infratentorial Approaches

The third ventricle can be reached through its posterior aspect via supratentorial or infratentorial approaches. At this point it should be emphasized that the angle of the tentorial ridge varies greatly in the material examined.

Bridging veins can come from above and from below and flow into the cerebellar tentorium, possibly forming a tentorial sinus which runs to the sinus rectus or lateral to the transverse sinus. In the region of the splenium the two internal cerebral veins converge as a rule to form the

Vena cerebri magna. In this vein or in the internal cerebral veins empties the basal vein (Rosenthal) and the internal occipital, the posterior pericallosal and from behind the superior vein of the cerebellar vermis. The third ventricle can also be approached by way of the suprapineal recessus which is 4.38 (1.0–9.0) mm on the average.

The transcortical approach to the third ventricle is to be recommended especially in cases of dilated lateral ventricles and is achieved through the frontal gyrus. The thalamostriate vein should be preserved during this approach through the interventricula foramen.

THERAPEUTIC ADVANCES IN FUNCTIONAL NEUROSURGERY

Functional stereotactic neurosurgery of the hypothalamus is currently used for the treatment of cluster headaches, and has recently been shown to have beneficial effects in the treatment of paroxysmal ophthalmic pain in multiple sclerosis patients (Franzini et al., 2007). Excitingly, it has also been shown to activate neural activity in cognitive and memory circuits in patients with mild Alzheimer's disease (Woodard, 2009). Additional applications of hypothalamic DBS may be expected in the field of sleep disorders, thermoregulation, epilepsy (Franzini et al., 2007), disruptive behavior and perhaps certain other diseases involving the autonomic nervous system. Importantly, DBS is also currently being investigated as a weight loss strategy in obesity (Franco et al., 2016). Anatomical knowledge of the hypothalamic structures to be targeted and of the circuitry involved is thus critical to the development of these new neurosurgical treatment strategies.

DBS of the posterior hypothalamus was introduced with impressive clinical results in a refractory group of patients suffering from chronic, medically intractable chronic *cluster headache* (CH).

A comprehensive theory of CH must explain the unilateral headache, as well as the sympathetic impairment and parasympathetic activation. In view of the striking relapsing–remitting pattern, seasonal variation and the clockwise regularity of the attacks, the concept of a central origin for CH has emerged (Holland et al., 2009). Consequently, the posterior hypothalamus was identified early on as a crucial element in the pathogenesis of CH. Functional imaging with positron emission tomography (PET) has shed light on the genesis of several of the trigemino-autonomic syndromes, documenting activation in the hypothalamic gray in CH, short-lasting unilateral neuralgiform headache attacks with *conjunctival injection and tearing* (SUNCT), paroxysmal hemicrania, and hemicrania continua (Domitrz et al., 2022). This area is not involved simply as a response to first division nociceptive pain impulses, but is inherent to each syndrome, probably in some permissive or dysfunctional role. Given that this area is involved in circadian rhythms, sleep–wake cycles and control of the autonomic system, these data suggest involvement of this hypothalamic area as a primum movens in the acute cluster attack (Venner et al., 2020). These findings prompted the use of deep-brain stimulation (DBS) in the posterior hypothalamic gray matter in a patient with intractable CH and led to complete relief from attacks (Bauer, 1959; Toni et al., 2004).

Little is known about the circuits and mechanisms underlying the analgesic effect of DBS; however, it probably involves activation of thalamo-cortical pathways and changes in cortical activity. The uniqueness of the DBS approach is that it allows in vivo investigation of the functional role of the underlying neuronal circuits by switching the hypothalamic stimulator ON/OFF. However, it is unclear whether DBS causes:

1. Local blockade of the hypothalamic trigger activity;
2. A direct antinociceptive effect by activation of the periaqueductal gray (PAG) and/or rostral ventromedial medulla (RVM);
3. Modulation of neuronal pain processing pathways.

Recent findings suggest that hypothalamic stimulation exerts its effect by modulating a complex pattern of activation in some and deactivation in other pain-processing areas distant from the hypothalamus (Kubo et al., 2005). An exclusive focal block, e.g., depolarization effect of the posterior hypothalamic area as the sole therapeutic effector, is therefore unlikely. Following this theory, the constant depolarization would discontinue the biological clock-like pulses from the distant trigeminal and autonomic "executers," involving a certain time lag.

The anatomical basis for the connection between the hypothalamus and the trigeminal system has been described by Malick et al. (2000). They discovered that non-nociceptive input was conveyed directly to the hypothalamus neurons of the trigemino-hypothalamic tract, whereas nociceptive input was relayed via the trigemino-hypothalamic as well as the reticulo-hypothalamic tract. On a functional basis, a bidirectional connection between the two structures exists. Electrical stimulation of the trigeminally innervated superior sagittal sinus in cats leads to activation of the posterior hypothalamus (Benjamin, 2004). The much-discussed missing link between the striking circadian rhythmicity and the excruciating pain attacks could be the orexinergic system. The neuropeptides orexin A and B (or hypocretin-1 and -2) are synthesized in the hypothalamus and have been associated with homeostasis (including the sleep–wake cycle).

Orexinergic projections have been found linking the SCH (regarded as the "master clock" in circadian regulation) with the posterior hypothalamus. The fact that hypothalamic DBS is effective in nearly 60% of otherwise intractable CH, that it has been reported to be effective in *shortlasting, unilateral, neuralgiform headache attacks with conjunctival injection and tearing* (SUNCT) but not in atypical facial pain points toward a highly selective effect in terms of syndromic response. It is not just pain or even trigeminally transmitted pain that will respond to hypothalamic stimulation. For the moment, we have to conclude that CH, and probably trigemino-autonomic headaches as a group, may respond and consequently should be the only target population (Coskun et al., 2014).

Potential Mechanism

Attempts to alleviate medically intractable pain through continuous stimulation of deep-brain structures have been reported for nearly half a century. Over the years, several brain regions, such as the ventroposterolateral nucleus and several other thalamic nuclei, the PAG and periventricular gray (PVG) matter and motor cortex, have been targeted, with varying degrees of success (Zagami et al., 1990). Further, the posterior hypothalamus is able to both decrease and enhance nociceptive responses in the trigeminal nucleus caudalis. In humans, stereotactic thermocoagulation of the postero-medial hypothalamus has been successfully employed to treat otherwise intractable cancer pain (Bartsch et al., 2005).

Based on observations from earlier PET studies, it may be hypothesized that the symptoms of CH are caused by a low threshold "oscillator" that is generated by the hypothalamus and subsequently activates cortical structures of the pain transmitting system, leading to the characteristic short-lived trigemino-autonomic pain (Wang et al., 2023a,b). Nevertheless, it is thus far unclear how the DBS of this precise triggering area prevents CH attacks from occurring (Wang et al., 2023a,b). The fact that artificial hypothalamic stimulation activates not only the hypothalamus but also distinct members of the pain matrix provides evidence that DBS of the hypothalamus not only depolarizes this region (i.e., local depolarization and a local flow response to neuronal firing) but also bidirectionally modulates activity in fundamental structures of the ascending pain pathway. From a clinical point of view, it is noteworthy that it may take weeks or longer before hypothalamic DBS is effective. Similar latencies from implantation to a clinical effect were observed in the recently introduced occipital stimulation of the greater occipital nerve (GON) in intractable CH (Wei et al., 2021). However, similar to the findings in hypothalamic DBS, stimulation of the GON increased the nociceptive blink reflex. In contrast to the central stimulation of the posterior hypothalamus, peripheral GON stimulation probably acts locally in the trigeminocervical complex, which is formed by the caudal trigeminal nucleus and upper cervical afferents. However, the relatively long latency (the headaches did not ameliorate before months of treatment) hints in both stimulation protocols at a more complex and central process, probably involving neuroplastic changes of the CNS, but further studies are needed to evaluate and understand both hypothalamic DBS and GON stimulation more thoroughly (May, 2008).

Neuroresearchers have employed different techniques to reveal the minutiae of the functions of the hypothalamus, ranging from the tracing of fiber pathways to advanced molecular analysis. The hypothalamus mainly exerts an integrative role: it receives a range of afferent inputs and effectuates autonomic, endocrine, and behavioral responses to achieve homeostasis through allostasis. Accordingly, the hypothalamus has been implicated in psychiatric diseases related to feeding, sexual behavior, aggression, and fear (Ricardo, 1983). Stereotactic approaches to specific nuclei or regions of the hypothalamus have provided a unique background for a better understanding of the neural basis of several behavioral disorders (obesity, aggressive behavior, panic disorder, addiction, and paraphilic disorders) (Barbosa et al., 2017). A better understanding of the underlying circuitry of specific behavioral functions—i.e., feeding, drug consumption, aggression, sexuality, and fear—together with the implementation of neuromodulation paradigms (e.g., optogenetic stimulation, radiosurgery, focused ultrasound, intraoperative electrophysiological mapping) and the prediction of the network-level response to different focal stimulations of hypothalamic nuclei or white matter fibers (i.e., with simulation studies) may lead to the development of novel and optimized stereotactic targets and clinical trials of neurosurgery for psychiatric disorders (Zhang et al., 2014a,b).

In conclusion, the hypothalamus still represents one of the most fascinating and mysterious parts of our brain, hiding a real microcosm of activity, control, and management of the entire organism in a sophisticated, precise, and articulated structure only large a few millimeters. From the most primordial instincts such as hunger, thirst, the need to urinate up to the involvement in the centers of memory, coordination and human behavior, the hypothalamus represents both topographically and conceptually one of the centers of our mind.

References

Anderson, D.J., 2012. Optogenetics, sex, and violence in the brain: implications for psychiatry. Biol. Psychiatry 71, 1081–1089.

Awad, M., Butterfield, J.T., Dhawan, S., Tyler, M.A., Venteicher, A.S., 2023. Prognostic factors and treatment impact on overall survival in adult craniopharyngioma. World Neurosurg. 173, e132–e139.

Barbosa, D.A.N., de Oliveira-Souza, R., Monte Santo, F., de Oliveira Faria, A.C., Gorgulho, A.A., De Salles, A.A.F., 2017. The hypothalamus at the crossroads of psychopathology and neurosurgery. Neurosurg. Focus 43 (3), E15.

Bard, P., 1928. A diencephalic mechanism for the expression of rage with special reference to the sympathetic nervous system. Am. J. Physiol. 84, 490–515.

Bard, P., Rioch, D.M., 1937. A Study of four cats deprived of neocortex and additional portions of the forebrain. Bull. Johns Hopkins Hosp. 60, 73–147.

Baroncini, M., Jissendi, P., Balland, E., Besson, P., Pruvo, J.P., Francke, J.P., Dewailly, D., Blond, S., Prevot, V., 2012. MRI atlas of the human hypothalamus. Neuroimage 59 (1), 168–180.

Bartsch, T., Levy, M.J., Knight, Y.E., Goadsby, P.J., 2005. Inhibition of nociceptive dural input in the trigeminal nucleus caudalis by somatostatin receptor blockade in the posterior hypothalamus. Pain 117 (1), 30–39.

Bauer, H.G., 1959. Endocrine and metabolic conditions related to pathology in the hypothalamus. Rev. J. Nerv. Ment. Dis. 128, 323–338.

Benjamin, L., 2004. Hypothalamic activation after stimulation of the superior sagittal sinus in the cat: a Fos study. Neurobiol. Dis. 16, 500–505.

Bobrovskikh, M.A., Gruntenko, N.E., 2023. Mechanisms of neuroendocrine stress response in Drosophila and its effect on carbohydrate and lipid metabolism. Insects 14 (5), 474.

Brll, K., 1978. Circumcriptes Schwitzen. Inaug. Diss. Wiirzburg.

Brouwer, B., 1950. Positive and negative aspects of hypothalamic disorders. J. Neurol. Neurosurg. Psychiatry 13, 16–23.

Bucher, V.M., Birgli, S.M., 1952. Some observations on the fibre connections of the di- and mesencephalon in the cat. Part IV. The ansa lenticularis, pars ascendens mesencephalica, with observations on other systems ascending from and descending to the mesencephalon. J. Comp. Neurol. 99, 415–428.

Buchfelder, M., Schlaffer, S.M., Zhao, Y., 2019. The optimal surgical techniques for pituitary tumors. Best Pract. Res. Clin. Endocrinol. Metabol. 33 (2), 101299.

Burford, G.D., Dyball, R.E.J., Moss, R.L., Pickering, B.T., 1974. Synthesis of both neurohypophysial hormones in both the paraventricular and supraoptic nuclei of the rat. J. Anat. 117, 261–269.

Cajal, S.R.Y., 1911. Histologie du system nerveux de l'homme et des vertebres. Maloine, Paris.

Cammarota, M., Ferlenghi, F., Vacondio, F., Vincenzi, F., Varani, K., Bedini, A., Rivara, S., Mor, M., Boscia, F., 2023. Combined targeting of fatty acid amide hydrolase and melatonin receptors promotes neuroprotection and stimulates inflammation resolution in rats. Br. J. Pharmacol. 180 (10), 1316–1338.

Cantu, D.J., Kaur, S., Murphy, A.Z., Averitt, D.L., 2022. Sex differences in the amygdaloid projections to the ventrolateral periaqueductal gray and their activation during inflammatory pain in the rat. J. Chem. Neuroanat. 124, 102123.

Christ, J.F., 1969. Derivation and boundaries of the hypothalamus, with atlas of hypothalamic grisea. In: Haymaker, W., Anderson, E., Nauta, W.J.H. (Eds.), The Hypothalamus. Charles C. Thomas, pp. 13–60.

Contreras, M., Riveros, M.E., Quispe, M., Sánchez, C., Perdomo, G., Torrealba, F., 2016. The histaminergic tuberomamillary nucleus is involved in appetite for sex, water and amphetamine. PLoS One 11, e0148484.

Coskun, O., Ucar, M., Yildirim, F., Cetinkaya, R., Bolay, H., 2014. EHMTI-0353. MR tractography in short lasting unilateral neuralgiform headache attacks with conjunctival injection and tearing (SUNCT) patients: case reports. J. Headache Pain 15 (S1).

Dejerine, J., 1901. Anatomie des centres nerveux, vol II. J. Rueff et Cie, Paris.

Del Abril, A., Segovia, S., Guillamon, A., 1987. The bed nucleus of the stria terminalis in the rat: regional sex differences controlled by gonadal steroids early after birth. Brain Res. 429, 295–300.

Del Brio, M.A., Riera, P., Gracia, J.M., Alvariz-Uria, M., 1990. Ultrastructural study of the cellular components of the organum vasculosum of the lamina terminalis in the rabbit (Oryctolagus cuniculus). J. Submicr. Cytol. Pathol. 22, 303–309.

Domitrz, I., Piątkowska-Janko, E., Rożniecki, J.J., Hryniewicz, N., Bogorodzki, P., 2022. Paroxysmal hemicrania or short-lasting unilateral neuralgiform headache attacks with trigeminal neuralgia – functional neuroimaging findings. Neurol. Neurochir. Pol. 56 (4), 381–383.

Dudas, B., Mihaly, A., Merchenthaler, I., 2000. Topography and associations of luteinizing hormone- releasing hormone and neuropeptide Y-immunoreactive neuronal systems in the human diencephalon. J. Comp. Neurol. 427, 593–603.

Fitzgerald, K.T., 1979. The structure and function of the pars tuberalis of the vertebrate adenohypophysis. Gen. Comp. Endocrinol. 37, 383–399.

Franco, R., Fonoff, E.T., Alvarenga, P., Lopes, A.C., Miguel, E.C., Teixeira, M.J., Damiani, D., Hamani, C., 2016. DBS for obesity. Brain Sci. 6 (3), 21.

Franzini, A., Marras, C., Tringali, G., Leone, M., Ferroli, P., Bussone, G., 2007. Chronic high frequency stimulation of the posteromedial hypothalamus in facial pain syndromes and behavior disorders. Acta Neurochir. Suppl. 97, 399–406.

Fuchs, S.A.G., Edinger, H.M., Siegel, A., 1985. The organization of the hypothalamic pathways mediating affective defense behavior in the cat. Brain Res. 330, 77–92.

Greving, R., 1925. Beitrige zur Anatomie des Zwischenhirns und seiner Funktion. I. Der anatomische Aufbau der Zwischen- hirnbasis und des anschliegenden Mittelhirngebietes des Men- schen. Z. Anat. Entwickl. Gesch. 75, 579–620.

Grouselle, D., Chaillou, E., Caraty, A., Bluet-Pajot, M.T., Zizzari, P., Tillet, Y., Epelbaum, J., 2008. Pulsatile cerebrospinal fluid and plasma ghrelin in relation to growth hormone secretion and food intake in the sheep. J. Neuroendocrinol. 20, 1138–1146.

Haber, E., Haupert Jr., G.T., 1987. The search for a hypothalamic Na^+, K^+-ATPase inhibitor. Hypertension 9 (4), 315–324.

Haymaker, W., 1969. Blood supply of the human hypothalamus. In: Haymaker, W., Anderson, E., Nauta, W.J.H. (Eds.), The Hypothalamus. Charles C. Thomas, pp. 210–218.

Holland, P.R., Goadsby, P.J., 2009. Cluster headache, hypothalamus, and orexin. Curr. Pain Headache Rep. 13 (2), 147–154.

Hrvatin, S., Sun, S., Wilcox, O.F., Yao, H., Lavin-Peter, A.J., Cicconet, M., Assad, E.G., Palmer, M.E., Aronson, S., Banks, A.S., 2020. Neurons that regulate mouse torpor. Nature 583, 115–121.

Kalsbeek, A., Buijs, R.M., 2002. Output pathways of the mammalian suprachiasmatic nucleus: coding circadian time by transmitter selection and specific targeting. Cell Tissue Res. 309, 109–118.

Kataoka, N., Shima, Y., Nakajima, K., Nakamura, K., 2020. A central master driver of psychosocial stress responses in the rat. Science 367 (6482), 1105–1112. https://doi.org/10.1126/science.aaz4639. PMID: 32139538.

Keil, M.F., Stratakis, C.A., 2008. Pituitary tumors in childhood: update of diagnosis, treatment and molecular genetics. Expert Rev. Neurother. 8 (4), 563–574.

Knoche, H., 1956. Die Verbindung der Retina mit den vegetativen Zentren des Zwischenhirnes und mit der Hypophyse [The connection of the retina with the autonomic centers of the midbrain and pituitary]. Verh Anat. Ges. 53, 140–146; discussion 146-8. German. PMID: 13544013.

Korkmaz, F.N., Gökçay Canpolat, A., Sahin, M., Çorapçioğlu, D., 2021. Determination of the frequency of hyperprolactinemia-related etiologies and the etiology-specific mean prolactin levels. Minerva Endocrinol. https://doi.org/10.23736/S2724-6507.21.03386-8.

Kubo, T., Hagiwara, Y., 2005. Posterior hypothalamus cholinergic stimulation-induced activation of anterior hypothalamic area neurons is enhanced in spontaneously hypertensive rats. Brain Res. 1061 (1), 36–41.

Kuhlenbeck, H., 1973. The central nervous system of Vertebrates. In: Overall Morphologic Pattern, vol 3. Karger, p. 717.

Lam, T.K., 2010. Neuronal regulation of homeostasis by nutrient sensing. Nat. Med. 16, 392–395.

Lammers, J.H., Kruk, M.R., Meelis, W., Van der Poel, A.M., 1988. Hypothalamic substrates for brain stimulation-induced patterns of locomotion and escape jumps in the rat. Brain Res. 449, 294–310.

LaPierre, M.P., Abraham, M.A., Filippi, B.M., Yue, J.T., Lam, T.K., 2014. Glucagon and lipid signaling in the hypothalamus. Mamm. Genome 25 (9–10), 434–441.

Latronico, A.C., Brito, V.N., Carel, J.C., 2016. Causes, diagnosis, and treatment of central precocious puberty. Lancet Diabetes Endocrinol. 4 (3), 265–274.

Le Gros Clark, W.E., 1938. The hypothalamus: morphological, functional, clinical and surgical aspects. In: The Henderson Trust Lectures, Nos. XIII–XVI, Published for the William Ramsy Henderson Trust. Oliver and Boyd, Edinburgh, p. 213.

Leclercq, T., Grisoli, F., 1983. Arterial blood supply of the normal human pituitary gland. J. Neurosurg. 58, 678–681.

Legradi, G., Emerson, C.H., Ahima, R.S., Flier, J.S., Lechan, R.M., 1997. Leptin prevents fasting-induced suppression of prothyrotropin-releasing hormone messenger ribonucleic acid in neurons of the hypothalamic paraventricular nucleus. Endocrinology 138, 2569–2576.

Malick, A., Strassman, R.M., Burstein, R., 2000. Trigeminohypothalamic and reticulohypothalamic tract neurons in the upper cervical spinal cord and caudal medulla of the rat. J. Neurophysiol. 84, 2078–2112.

May, A., 2008. Hypothalamic deep-brain stimulation: target and potential mechanism for the treatment of cluster headache. Cephalalgia 28 (7), 799–803.

Mercuri, V., Armocida, D., Paglia, F., Patrizia, G., Santoro, A., D'Angelo, L., 2022. Giant prolactinoma embedded by pseudoaneurysm of the cavernous carotid artery treated with a tailored therapeutic scheme. J. Neurosci. Rural Pract. 13 (3), 358–369.

Nakane, P.K., 1970. Classifications of anterior pituitary cell types with immunoenzyme histochemistry. J. Histochem. Cytochem. 18, 9–20.

Nauta, S.J.H., Haymaker, W., 1969. Hypothalamic nuclei and fiber connections. In: Haymaker, W., Anderson, E., Nauta, W.J.H. (Eds.), The Hypothalamus. Charles C. Thomas, pp. 136–209.

Obici, S., Feng, Z., Morgan, K., Stein, D., Karkanias, G., Rossetti, L., 2002. Central administration of oleic acid inhibits glucose production and food intake. Diabetes 51, 271–275.

Overeem, S., Vliet, J.A., Lammers, G.J., Zitman, F.G., Swaab, D.F., Ferrari, M.D., 2002. The hypothalamus in episodic brain disorders. Lancet Neurol. 1, 437–444.

Pang, D., Gao, Y., Liao, L., 2022. Functional brain imaging and central control of the bladder in health and disease. Front. Physiol. 12 (13), 914963.

Patton, A.P., Hastings, M.H., 2023. The mammalian circadian timekeeping system. J. Huntingtons Dis. 1–14.

Perosevic, M., Jones, P.S., Tritos, N.A., 2021. Magnetic resonance imaging of the hypothalamo-pituitary region. Handb. Clin. Neurol. 179, 95–112.

Pfaus, J.G., 2009. Pathways of sexual desire. J. Sex. Med. 6 (6), 1506–1533.

Piñol, R.A., Mogul, A.S., Hadley, C.K., Saha, A., Li, C., Škop, V., Province, H.S., Xiao, C., Gavrilova, O., Krashes, M.J., Reitman, M.L., 2021. Preoptic BRS3 neurons increase body temperature and heart rate via multiple pathways. Cell Metab. 33 (7), 1389.e6–1403.e6. https://doi.org/10.1016/j.cmet.2021.05.001. Epub 2021 May 25. PMID: 34038711; PMCID: PMC8266747.

Pisipati, S., Hashim, H., 2011. Vasopressin receptors in voiding dysfunction. Handb. Exp. Pharmacol. 202, 453–483.

Porter, J.C., Kamberi, I.A., Grazia, Y.R., 1978. Pituitary blood flow and portal vesels. In: Martini, L., Ganong, W. (Eds.), Frontiers in Neuroendocrinology. Oxford University Press, pp. 145–175.

Raadsheer, F.C., Sluiter, A.A., Ravid, R., Tilders, F.J., Swaab, D.F., 1993. Localization of corticotropin- releasing hormone (CRH) neurons in the paraventricular nucleus of the human hypothalamus: age-dependent colocalization with vasopressin. Brain Res. 615, 50–62.

Ricardo, J.A., 1983. Hypothalamic pathways involved in metabolic regulatory functions, as identified by track-tracing methods. Adv. Metab. Disord. 10, 1–30.

Rivas, A.M., Sotello, D., Lado-Abeal, J., 2014. Primary and secondary endocrinopathies found in a patient with craniopharyngioma. Am. J. Med. Sci. 348 (6), 534–535.

Roeling, T.A., Veening, J.G., Kruk, M.R., 1994. Efferent connections of the hypothalamic "aggression area" in the rat. Neuroscience 59, 1001–1024.

Sagar, S.M., Martin, J.B., 1986. Hypothalamo-hypophysiotropic peptide systems. In: Bloom, F.E. (Ed.), Handbook of Physiology. Section 1: The Nervous System, Volume IV: Intrinsic Regulatory Systems of the Brain. Oxford University Press, New York, pp. 413–462.

Schindler, S., Geyer, S., Strauß, M., Anwander, A., Hegerl, U., Turner, R., Schönknecht, P., 2012. Structural studies of the hypothalamus and its nuclei in mood disorders. Psychiatr. Res. Neuroimaging 201 (1), 1–9.

Shibuya, M., Takayashu, M., Suzuki, Y., 1996. Bifrontal basal interhemispheric approach to craniopharyngioma resection with or without division of the anterior communicating artery. J. Neurosurg. 84, 951–956.

Snyder, G., Hymer, W.C., Snyder, J., 1977. Functional heterogeneity in somatotrophs isolated from the rat anterior pituitary. Endocrinology 101, 788–799.

Sternson, S.M., 2013. Hypothalamic survival circuits: blueprints for purposive behaviors. Neuron 77, 810–824.

Stuber, G.D., Wise, R.A., 2016. Lateral hypothalamic circuits for feeding and reward. Nat. Neurosci. 19, 198–205.

Swanon, L.W., Mogenson, G.J., 1981. Neural mechanisms for functional coupling of autonomic, endocrine and somatomotor responses in adaptative behavior. Brain Res. Rev. 3, 2–34.

Tan, C.L., Knight, Z.A., 2018. Regulation of body temperature by the nervous system. Neuron 98, 31–48.

Thompson, C.J., Costello, R.W., Crowley, R.K., 2019. Management of hypothalamic disease in patients with craniopharyngioma. Clin. Endocrinol. 90 (4), 506–516.

Thomson, R.H., Swanson, L.W., 2003. Structural characterization of a hypothalamic visceromotor pattern generator network. Brain Res. Rev. 41, 153–202.

Tobet, S.T., 2002. Genes controlling hypothalamic development and sexual differentiation. Eur. J. Neurosci. 16, 373–376.

Toni, R., Malaguti, A., Benfenati, F., Martini, L., 2004. The human hypothalamus: a morphofunctional perspective. J. Endocrinol. Invest. 27 (Supp to n.6), 73–94.

Tony, K.S., Joshi, R., 2021. SIADH as the initial manifestation of acute intermittent porphyria: a case report. Ind. J. Case Rep. 7 (11), 494–496. https://doi.org/10.32677/ijcr.v7i11.3114.

Torrealba, F., Riveros, M.E., Contreras, M., Valdes, J.L., 2012. Histamine and motivation. Front. Syst. Neurosci. 6, 51.

Tuchelt, H., Dekker, K., Bahr, V., Oelkers, W., 2000. Dose-response relationship between plasma ACTH and serum cortisol in the insulin-

hypoglycaemia test in 25 healthy subjects and 109 patients with pituitary disease. Clin. Endocrinol. 53, 301–307.
Uhlman, D., Nguyen, T., Grignol, G., Merchenthaler, I., Dudas, B., 2019. Substance P appears to affect growth via growth hormone-releasing hormone (GHRH) neurons in the human hypothalamus. Brain Struct. Funct. 224 (6), 2079–2085.
Ulrich-Lai, Y.M., Herman, J.P., 2009. Neural regulation of endocrine and autonomic stress responses. Nat. Rev. Neurosci. 10, 397–409.
Vann, S.D., 2010. Re-evaluating the role of the mammillary bodies in memory. Neuropsychologia 48, 2316–2327.
Venner, A., Fuller, P.M., 2020. 0163 Investigating the role of vasoactive intestinal peptide-containing neurons of the ventromedial preoptic area in sleep-wake control. Sleep 43 (Supplement_1), A64.
Wang, D., Lu, Y., Han, Y., Zhang, X., Dong, S., Zhang, H., Wang, G., Wang, G., Wang, J.J., 2023. The influence of etiology and stimulation target on the outcome of deep brain stimulation for chronic neuropathic pain: a systematic review and meta-analysis. Neuromodulation. https://doi.org/10.1016/j.neurom.2022.12.002. Epub ahead of print. PMID: 36697341.
Wang, F., Chen, Y., Lin, Y., Wang, X., Li, K., Han, Y., Wu, J., Shi, X., Zhu, Z., Long, C., Hu, X., Duan, S., Gao, Z., 2023. A parabrachial to hypothalamic pathway mediates defensive behavior. Elife 17 (12), e85450.
Wei, D.Y., Goadsby, P.J., 2021. Cluster headache pathophysiology – insights from current and emerging treatments. Nat. Rev. Neurol. 17 (5), 308–324.
Weindl, A., 1973. Neuroendocrine aspects of circumventricular organs. In: Ganong, W., Martini, L. (Eds.), Frontiers in Neuroendocrinology. Oxford University Press, pp. 3–32.
Woodard, J.L., Seidenberg, M., Nielson, K.A., Antuono, P., Guidotti, L., Durgerian, S., Zhang, Q., Lancaster, M., Hantke, N., Butts, A., Rao, S.M., 2009. Semantic memory activation in amnestic mild cognitive impairment. Brain 132 (Pt 8), 2068–2078.
Yamada, T., Katagiri, H., 2007. Avenues of communication between the brain and tissues/organs involved in energy homeostasis. Endocr. J. 54 (4), 497–505. https://doi.org/10.1507/endocrj.kr-106. Epub 2007 May 18. PMID: 17510499.
Yao, Y., Taub, A.B., LeSauter, J., Silver, R., 2021. Identification of the suprachiasmatic nucleus venous portal system in the mammalian brain. Nat. Commun. 12 (1), 5643.
Yasargil, M.G., 1996. Microneurosurgery. IVB. Microneurosurgery of CNS Tumors. Georg Thieme Verlag, Stuttgart, p. 300.
Young, J., Xu, C., Papadakis, G.E., Acierno, J.S., Maione, L., Hietamäki, J., Raivio, T., Pitteloud, N., 2019. Clinical management of congenital hypogonadotropic hypogonadism. Endocr. Rev. 40 (2), 669–710.
Yuh, W.T.C., Fisher, D.J., Nguyen, H.D., Tali, E.T., Gao, F., Simonson, T.M., Schlechte, J.A., 1994. Sequential MR enhancement pattern in normal pituitary gland and in pituitary adenoma. Am. J. Neuroradiol. 15, 101–108.
Zagami, A.S., Lambert, G.A., 1990. Stimulation of cranial vessels excites nociceptive neurones in several thalamic nuclei of the cat. Exp. Brain Res. 81 (3), 552–566.
Zhang, H., Li, L., Xu, G.Y., Mei, Y.W., Zhang, J.J., Murong, S.X., Sun, S.G., Tong, E.T., 2014. Changes of c-fos, malondialdehyde and lactate in brain tissue after global cerebral ischemia under different brain temperatures. J. Huazhong Univ. Sci. Technol. Med. Sci. 34 (3), 354–358.
Zhang, W., Cline, M.A., Gilbert, E.R., 2014. Hypothalamus-adipose tissue crosstalk: neuropeptide Y and the regulation of energy metabolism. Nutr. Metab. 11, 27.

Further Reading

Bergland, R., Ray, B., Torack, R., 1968. Anatomical variations in the pituitary gland and adjacent structures in 225 human autopsy cases. J. Neurosurg. 28, 93–99.
Brockhaus, H., 1942. Beitrag zur normalen Anatomie des Hypothalamus und der Zona incerta beim Menschen. J. Psychol. Neurol. 51, 96–196.
Brody, I.A., Odom, G.L., Kunkle, E.C., 1960. Pilomotor seizures. Report of a case associated with a cerebral glioma. Neurology 10, 993–997.
Busch, W., 1951. Die Morphologie der Sella turcica und ihre Beziehung zur Hypophyse. Virchows Arch. Path. Anat. 320, 437.
Celesia, G.G., Archer, C.R., Chung, H.D., 1981. Hyperphagie und Adipositas. Beziehungen zu Lisionen des mittleren Hypothalamus. JAMA-D 1 (6), 325–328.
Diepen, R., 1962. Der hypothalamus. In: Handbuch der Mikroskopischen Anatomic des Menschen. 4. Band, Teil 7. Springer, Berlin-Grttingen-Heidelberg.
Dixon, F.W., 1937. A comparative study of the sphenoid sinus. Ann. Otol. Rhinol. Laryngol. 46, 687–698.
Lang, J., 1985. Surgical anatomy of the hypothalamus. Acta Neurochir. 75 (1–4), 5–22.
Stopa, E.G., LeBlanc, V.K., Hill, D.H., Anthony, E.L.P., 1993. A general overview of the anatomy of the neurohypophysis. Ann. N. Y. Acad. Sci. 689, 6–15.
von Dorsche Hahn, H., 1983. Beziehungen zwischen dem Gehirn und dem Endokrinium unter chronobiologischen Aspekten. I. Gehirn, Hypothalamus und Corpus pineale. Anat. Anz. 154, 43–54.
Yamada, S., Sano, T., 1998. Neuropathology of the hypothalamus. In: Barrow, D.L., Selman, W. (Eds.), Neuroendocrinology. Williams & Wilkins, Baltimore, p. 259.

11

The Telencephalon: Neocortex

Brandon M. Ascenzi
Independent Neuroresearcher-Member of Marie Curie Alumni Association (MCAA)

OUTLINE

INTRODUCTION

The neocortex is a very complex structure, composed of six layers that develops from the dorsal sector of the pallium of the telencephalic hemispheres. The term neocortex refers to its late appearance during phylogeny, in comparison with the olfactory cortex representing the "paleocortex" and the hippocampal cortex representing the "archicortex," both present in all amniotes. Regarding the functional differentiation of the neocortex, in several "primitive" mammals (marsupials, insectivores, and rodents), a large part of the neocortex is either occupied by projection areas which receive, through the thalamus, specific impulses relating to the different special senses or are concerned with the steering of motor activity. The sensory fields comprise the somatosensory cortex, receiving impulses from sense organs situated in the skin, the muscles, and the joints, then the primary visual cortex and finally the primary auditory cortex. There is evidence suggesting that already in primitive mammals the various projection areas just mentioned are separated from each other by narrow strips of nonprojection cortex. What we now see is that these strips expand enormously to form a large area known as the parietotemporal association cortex. Another area of association cortex develops in front of the motor cortical areas. This region is known as the prefrontal association cortex. All cortical areas

From Anatomy to Function of the Central Nervous System
https://doi.org/10.1016/B978-0-12-822404-5.00025-5

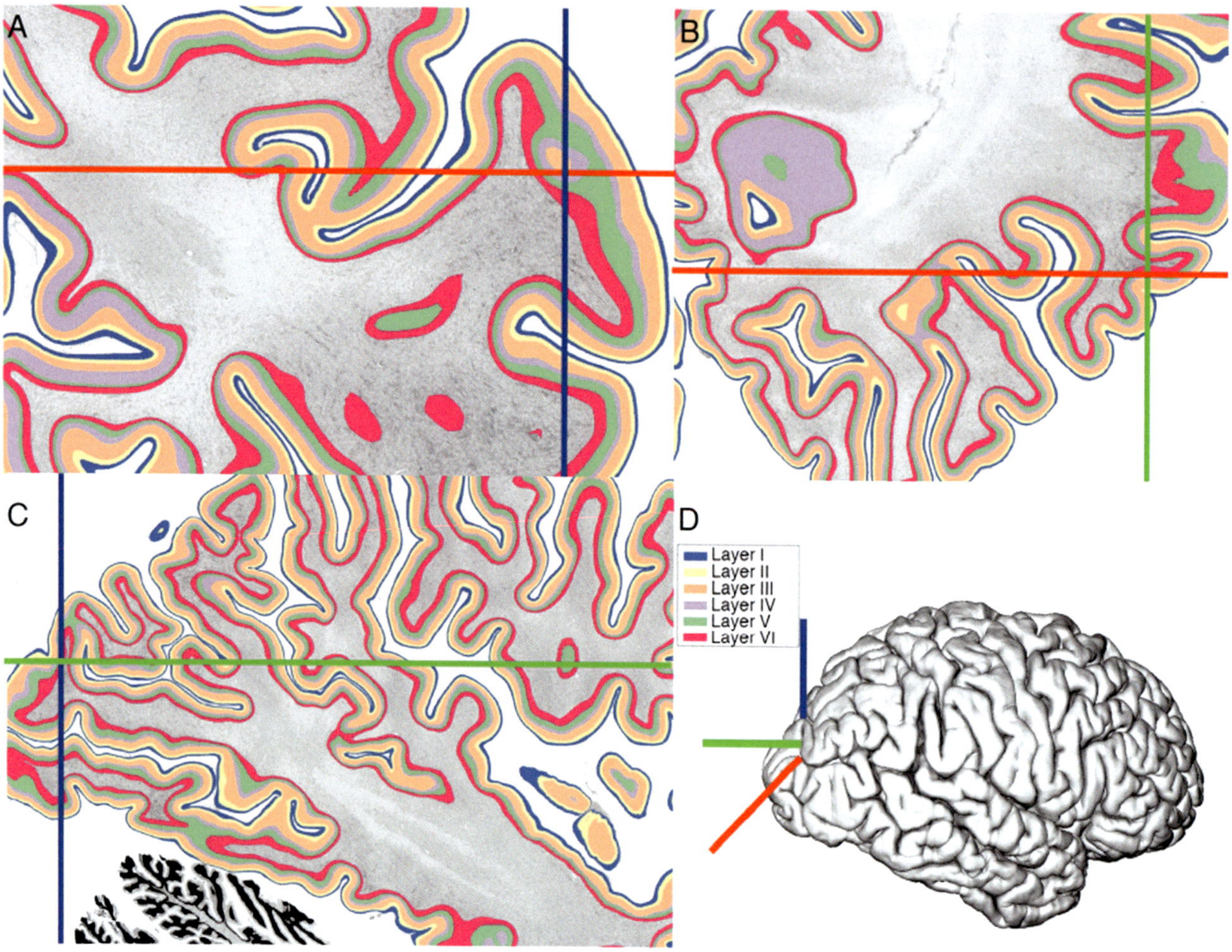

FIGURE 11.1 **Cortical layers in 3D.** Six cortical layers segmented on the 3D volume on 3 orthogonal planes: (A) coronal, (B) axial, (C) sagittal. Panel (D) shows the location of the sections on the reconstructed pial surface of the 3D BigBrain. (A) The coronal plane is the original plane of sectioning. Within this plane, the axes are centered on an area of the cortex where layers would be impossible to segment in 2D because the section only shows part of the gyrus, and most layers are not visible because of the oblique sectioning of the cortex with respect to the gyrus. Underlying data available from https://bigbrain.loris.ca. *Image courtesy from Wagstyl, K., Larocque, S., Cucurull, G., Lepage, C., Cohen, J.P., Bludau, S., et al., 2020. BigBrain 3D atlas of cortical layers: cortical and laminar thickness gradients diverge in sensory and motor cortices. PLoS Biol. 18(4), e3000678. https://doi.org/10.1371/journal.pbio.3000678.*

maintain afferent and efferent connections with subcortical centers. However, these various association cortical areas are primarily connected with other cortical fields.

In cytoarchitectural terms, the neocortex presents a laminar structure throughout its extension. While three layers can be distinguished in the piriform and hippocampal parts of the mammalian cortex, six layers are usually recognizable in the neocortex (Fig. 11.1).

Starting from the surface, the six layers of the neocortex are (I) molecular layer, (II) external granular layer, (III) external pyramidal layer, (IV) internal granular layer, (V) internal pyramidal layer, and (VI) multiform layer.

Analyzing these layers individually, we can say that:

- I layer contains only very few cell bodies;
- II layer is composed of small, densely packed cell bodies. The name of this layer is misleading because most of its constituent somata belong to small pyramidal neurons which, like all typical cortical pyramids, direct their apices toward the surface;
- III layer is a thick layer in which pyramidal somata prevail. These somata increase progressively in size from superficial to deep;
- IV layer consists of small, densely packed, pyramidal and nonpyramidal somata;
- V layer consists mainly of medium-sized and large, loosely arranged pyramidal somata;
- VI layer is composed of relatively tightly packed, spindle-shaped somata. Golgi material reveals that most of these somata belong to modified pyramidal cells.

Lamination of the cerebral cortex is also evident in sections stained for myelin visualization (most axons

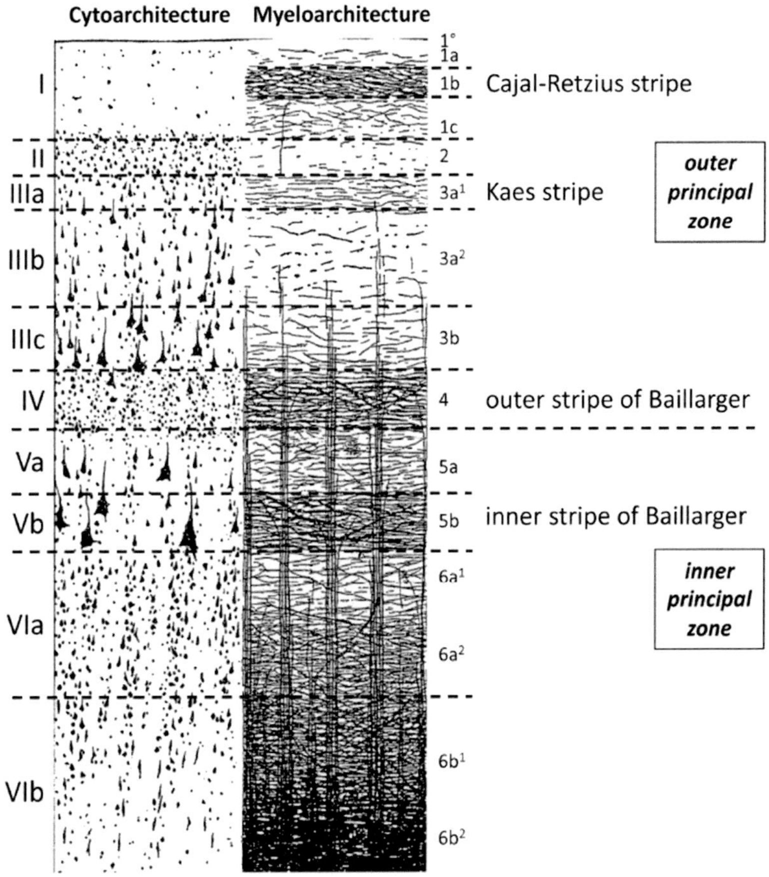

FIGURE 11.2 **Generalized scheme of cortical layering in cyto- and myeloarchitecture.** Roman numbers cytoarchitectonic layers, Arabic numbers myeloarcitectonic layers. *Modified from Zilles, K., Palomero-Gallagher, N., Amunts, K., 2015. Myeloarchitecture and maps of the cerebral cortex. Elsevier eBooks, pp. 137–156. https://doi.org/10.1016/b978-0-12-397025-1.00209-8. Image courtesy from Palomero-Gallagher, N., Zilles, K., 2019. Cortical layers: cyto-, myelo-, receptor- and synaptic architecture in human cortical areas. NeuroImage 197, 716–741. https://doi.org/10.1016/j.neuroimage.2017.08.035.*

of intracortical neurons and major afferent systems are in fact myelinated). Some myelinated fibers have a vertical course and are sometimes grouped in fascicles, others have a horizontal course and in most of the cortical fields three groupings are easily identifiable: one in layer I and two, defined as Baillarger's striae, in layer IV and V (Fig. 11.2).

A slightly more detailed analysis allows us to highlight that, beyond the existence of a common scheme, there are significant differences in the structure of the cerebral cortex: in fact, different regions have different thicknesses and different extensions of the various layers and vary in cell density and also for cell types. On the basis of these differences, at the beginning of the last century, meticulous studies on the architecture of the cerebral cortex were carried out, culminating in the publication of the so-called cytoarchitectural maps. Among them, those by Campbell (1905), Brodmann (1909), and von Economo (1927) must be mentioned, in which the different cortical "areas" were reported. The cytoarchitectural areas thus identified can then be grouped on the basis of common characteristics. Von Economo recognized five basic types: 1. agranular cortex; 2. frontal type cortex; 3. parietal-type cortex; 4. polar cortex; 5. granular cortex (or koniocortex). As is evident from Fig. 11.3, cortical types 2, 3, and 4 have the six layers described above, even if not all have the same degree of development; these types are also termed homotypic cortex. Types 1 and 5, on the other hand, constitute the heterotypic cortices, in which the six canonical layers are indistinguishable. In particular, type 1 cortex is characterized by the absence (or very little relevance) of layers II and IV (therefore, this type of cortex is defined as agranular) and by conspicuous layers III and V; the agranular cortex is characteristic of motor-type cortices (Fig. 11.3). In contrast, type 5 cortex is characterized by rich round cells (granules), well-developed layers II and IV, and poorly developed layers III and V. This type of cortex, defined as granular, is characteristic of cortical areas that receive sensory afferents. The most evident example of granular cortex is represented by 17 BA (or V1), in which layer IV is so developed as to be divided into three sublayers, Iva–c, one of which, IVb, is so rich in fibers to be visible to the naked eye (the so-called Gennari's line). The observation that the distribution of myelinated axons also changes along the horizontal extension of the cortex has represented

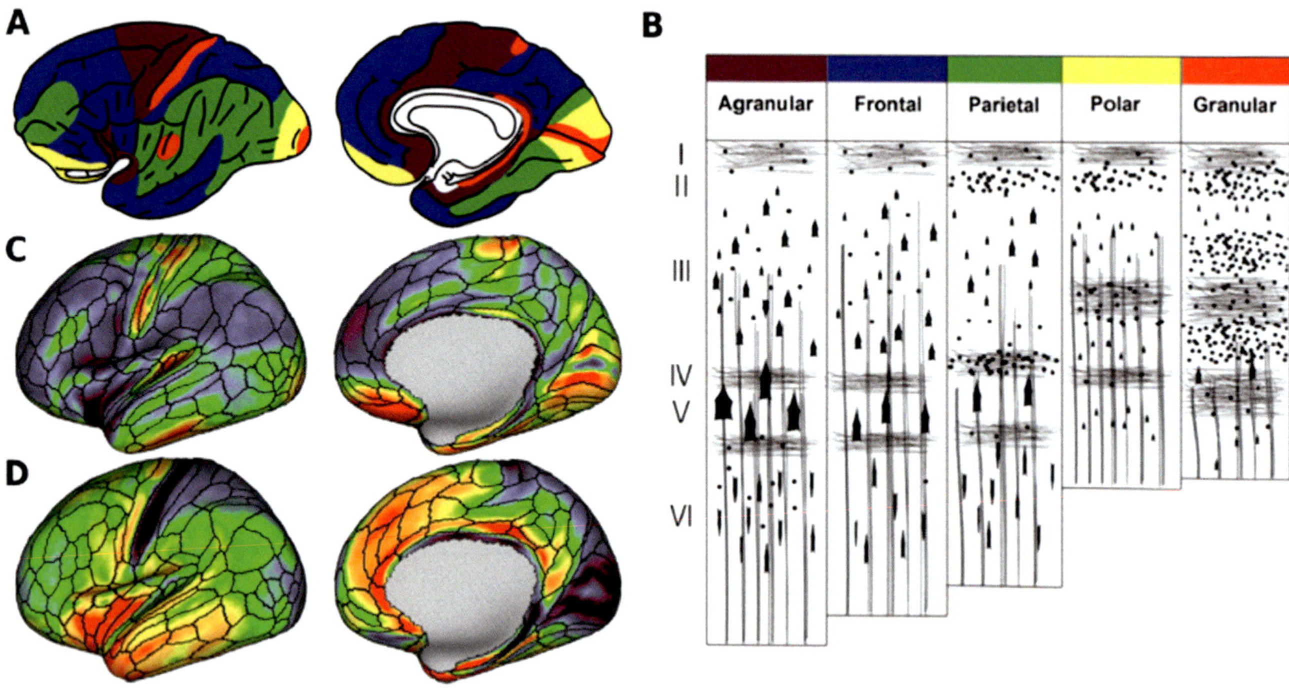

FIGURE 11.3 **Comparison between von Economo–Koskinas atlas (von Economo and Koskinas, 1925) and neurite orientation dispersion and density imaging (NODDI).** (A) Cortical distribution of von Economo's five main cytoarchitectonic types. Color codes are deciphered in the (B) cytoarchitecture type scheme superimposed with tangential and radial myelinated axons (Braak, 2012; Nieuwenhuys, 2013; Triarhou, 2009; von Economo and Koskinas, 1925). The outer and inner stripes of Baillarger are indicated by the red and blue arrowheads, respectively. (C) Orientation dispersion index (ODI) map exhibits relative high values in granular and polar cortices. (D) Cortical thickness map shows relatively low values in granular and polar cortices. Data at https://balsa.wustl.edu/Kz8q. *Image courtesy from Fukutomi, H., et al., 2018. Neurite imaging reveals microstructural variations in human cerebral cortical gray matter. NeuroImage. https://doi.org/10.1016/j.neuroimage.2018.02.017.*

the basis of a series of meticulous studies performed by Cécile and Oskar Vogt (Klatzo, 2003), which have led to the identification of numerous myeloarchitectural areas, largely coinciding with those cytoarchitectural and today generally ignored. Myeloarchitectural studies are of historical importance also as regards the genesis of the concept of associative area (see below).

CELL TYPES

Traditionally the neurons of the cerebral cortex have been classified on the basis of strictly morphological criteria (shape of their body, length of the axon, width and spatial distribution of the dendritic tree), but numerous other parameters have subsequently been used, e.g., the site of axon termination, the type of synaptic terminal, the presence or absence of dendritic spines, the neurotransmitter produced and released, the expression of other molecules active at the synapse (such as neuropeptides) and of receptors for neurotransmitters, and, finally, their biophysical properties. Numerous classifications have derived from this, omitted here because they are too specialized, the comparison of which allows us to highlight interesting correlations. The latter form the basis of the following description, even if it has required some simplifications and in no way represents the whole spectrum of variability and specificity of the morphofunctional characteristics of cortical neurons.

Schematically, all neurons of the cerebral cortex can be divided into two large families: that of *pyramidal neurons* and that of *nonpyramidal neurons*.

PYRAMIDAL NEURONS (OR TYPICAL PYRAMIDAL CELL)

Pyramidal neurons make up about three-quarters of all neurons in the cortex and are located almost exclusively in layers II–III and V–VI (Fig. 11.4). Morphologically, pyramidal neurons are characterized by a triangular body with an inferior base, an apical dendrite of variable length that in most cases reaches the superficial layers and basal dendrites that emerge from the body and head horizontally. Numerous spines emerge from the dendrites of pyramidal neurons which represent sites of synaptic contacts.

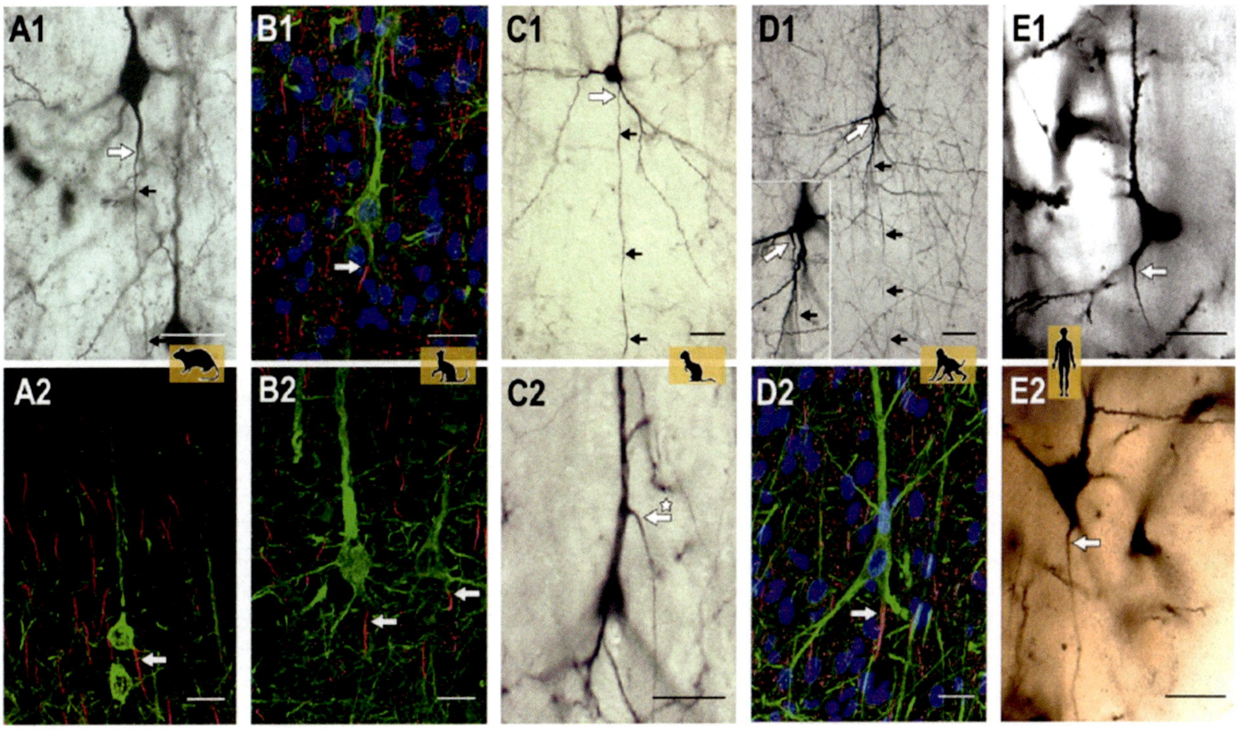

FIGURE 11.4 **The figure shows pyramidal neurons of different cerebral cortex and highlights a recent discovery about the *axon emerging*. Indeed, the canonical view is that axons emerge from the somata and not from dendrites; however the latter is not always necessarily the case: axons may also originate from dendrites.** The terms "axon carrying dendrite" (AcD) and "AcD neurons" have been coined to describe this feature. In rodent hippocampus, AcD cells are shown to be functionally "privileged," since inputs here can circumvent somatic integration and lead to immediate action potential initiation in the axon. Here, we report on the diversity of axon origins in neocortical pyramidal cells of rodent, ungulate, carnivore, and primate. (A1, A2) From rat visual cortex (biocytin, immunofluorescence); (B1, B2) cat visual cortex (immunofluorescence); (C1, C2) ferret visual cortex (biocytin); (D1, D2) macaque premotor cortex (biocytin, immunofluorescence), the inset shows the axon origin at higher magnification; (E1, E2) human auditory cortex (Golgi method; D2 is a montage of two photos). Apical AcDs (asterisk in C2) were rare, less than 10 were detected among the neurons assessed in adult rat, ferret, and macaque, and none in our human material. In all cases, the axon immediately bent down toward the white matter. Axon origins are marked by large arrows, small arrows indicate the course of biocytin-labeled axons. Scale bars 25 μm. *Image courtesy from Wahle, P., Sobierajski, E., Gasterstädt, I., et al., 2022. Neocortical pyramidal neurons with axons emerging from dendrites are frequent in non-primates, but rare in monkey and human. Elife 11, e76101. https://doi.org/10.7554/eLife.76101.*

Pyramidal neurons respond to brief depolarizing stimuli by generating action potentials lasting between 0.5 and 1 ms. Corticocortical neurons produce single action potentials and are termed *regular spiking neurons*, while those projecting to the brainstem and spinal cord generate bursts of 3–5 potentials and are termed *intrinsically bursting neurons*.

The classification of cortical neurons according to their firing properties used in this chapter should not be understood as definitive or as an expression of fixed functional properties. In fact, other classifications exist and, above all, there is a considerable amount of experimental data which demonstrates that the discharge of a given cortical neuron varies following even slight modifications of their membrane potential induced by the application of a depolarizing current or during spontaneous oscillations correlated to behavioral states (for example, during sleep stages). Even the number of neurons exhibiting a given firing pattern can be different in relation to methodological and behavioral variables.

The axons of pyramidal neurons represent the efferent channel of the cerebral cortex. However, this fundamental characteristic of pyramidal neurons should not make us forget that their axons, before leaving the gray matter, emit numerous collateral branches that project both to other cortical layers and to relatively distant cortical regions of the same area (horizontal collaterals or horizontal connections). The influence exerted by these projections is believed to be of no small importance in information processing. The axon terminals of pyramidal neurons form only asymmetric synaptic contacts and use glutamate (Glu) as a neurotransmitter: pyramidal neurons therefore exert an excitatory action on their postsynaptic elements.

The conventional belief is that axons arise from the cell bodies and not from dendrites. However, this assumption is not always accurate. In reality, it is possible for axons to originate from dendrites rather than solely from the cell bodies.

NONPYRAMIDAL NEURONS (OR ATYPICAL PYRAMIDAL CELLS)

Nonpyramidal neurons are further divided into two groups:

1. spiny stellate cells;
2. nonpyramidal aspiny cells (or inhibitory interneurons).

Spiny stellate cells are characterized by the fact that their axon ends in the vicinity of the cell body, otherwise, they have all the characteristics of pyramidal neurons (Fig. 11.5). Indeed: (i) they are rich in spines; (ii) are excitatory (they form asymmetric synapses and use glutamate as a neurotransmitter); and (iii) respond to short depolarizing stimuli by generating action potentials with a duration of between 0.5 and 1 msec. For these reasons, some scholars believe that spiny stellate cells are neurons-modified pyramids.

Nonpyramidal aspiny cells represent about 20%–30% of cortical neurons and constitute a heterogeneous group of cells which share a series of fundamental morpho-functional characteristics: (i) their dendrites are devoid of or poorly equipped with spines (with very rare exceptions); (ii) their axons do not exit the gray matter; (iii) form symmetrical synapses and release GABA as a neurotransmitter. *Nonpyramidal aspiny cells* are interneurons and, in many cases, also produce and release a neuropeptide (e.g., substance P, cholecystokinin, somatostatin, or vasoactive intestinal polypeptide). They respond to brief depolarizing stimuli by generating trains of action potentials of variable duration between 0.2 and 0.3 msec (rapid firing neurons, fast spiking) (Figs. 11.6 and 11.7).

Nonpyramidal neurons are much more frequently electrically coupled than pyramidal neurons. Although electron microscopy studies had long demonstrated the existence of numerous dendrodendritic and dendrosomatic *gap junctions* in the cerebral cortex of primates, the systematic study of electrical synapses in the cerebral cortex has only recently begun. It has been shown that the gap junctions that mediate communication between cortical interneurons require the presence of connexin 36 and that electrical synapses connect only functionally similar interneurons, creating large syncytial networks of interneuron communication. The function of the electrical synapses between cortical interneurons seems to consist in the ability to synchronize the activity of the connected neurons. Despite exhibiting several common features, nonpyramidal neurons exhibit certain elements of specificity, especially as regards the targeting of their axons. It is known, for example, that *chandelier cells*, *bitufted cells*, and *basket cells* form synapses with pyramidal neurons; however, while bitufted cells form synapses with dendrites, basket cells synapse with the cell body, and chandelier cells with the initial segment of the axon. This segregation of inhibitory signals into different portions of pyramidal neurons results in different effects on the generation of action potentials (Figs. 11.6 and 11.7).

Intrinsic Organization of the Cerebral Cortex

Columns

In 1957, V. Mountcastle demonstrated that a microelectrode lowered vertically into the primary somatosensory cortex (S1) of the cat recorded the electrical activity of neurons that always had the same receptive field and always responded to the same submodality. This phenomenon was repeated with a constant periodicity of about 300–500 μm. On the basis of these observations, and of some old anatomical descriptions by

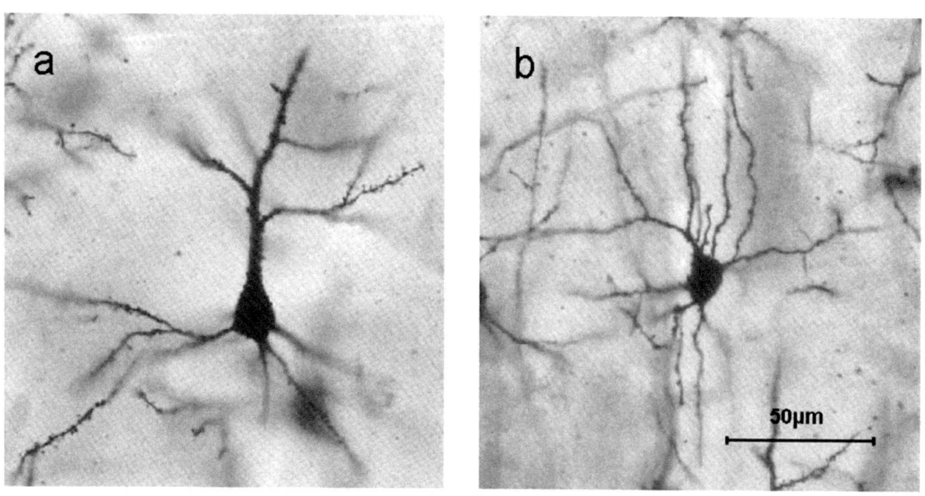

FIGURE 11.5 **Golgi-stained cortical neurons.** (A) Layer II/III pyramidal cell; (B) layer IV spiny stellate cell. *Image courtesy of Wikipedia.*

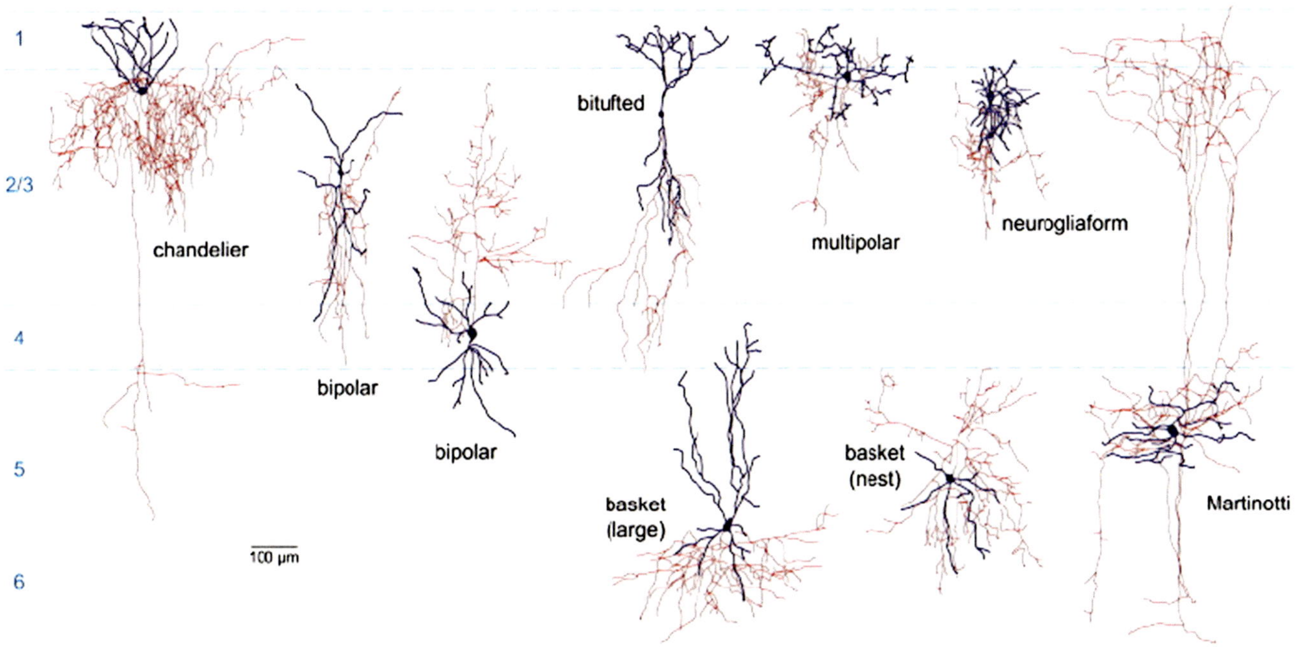

FIGURE 11.6 **Representative morphologies of mouse interneuron types.** Note that the axon (red) and dendrite (blue) arbors are typically less elaborate than in primate cortex. Two variants of bipolar cells and basket cells (large and nest) are shown. Individual cell types adapted from original figures: chandelier cell (Woodruff et al., 2009); bipolar and basket cells (Miyoshi et al., 2007); bitufted, multipolar, and neurogliaform cells (Miyoshi et al., 2010); Martinotti cell (Goldberg et al., 2004). *Image courtesy of Watson, C., Paxinos, G., Puelles, L. (Eds.), 2012. Copyright. In: The Mouse Nervous System. Academic Press, San Diego, p. iv. https://doi.org/10.1016/B978-0-12-369497-3.10035-4.*

Lorente de Nò (1922) which suggested the possibility of a vertical organization of the cerebral cortex, Mountcastle hypothesized that the cerebral cortex was made up of a series of modules, elementary operational units, which he called columns (Fig. 11.8).

Mountcastle's electrophysiological observations were confirmed in all cortex types examined and under different experimental conditions. A notable contribution to the definition of the concept of columnar organization of the cerebral cortex has been provided by numerous other neuroanatomical studies.

The columnar hypothesis postulates that the elementary unit of the mature neocortex is represented by the so-called minicolumn, a narrow chain of neurons that extends vertically through layers II–VI. Each minicolumn contains approximately 80–100 neurons (with the exception of primary visual cortex, 17 BA, where the number is greatest) and is produced by the iterative division of a small group of neuroepithelial progenitor cells. In each column are the main cell types which are densely connected along the vertical axis. According to this hypothesis, a minicolumn represents the smallest processing unit of the neocortex. This latest version of the columnar cortex hypothesis is slightly different from the original one, if only in the dimensions of the columns. This consideration offers the opportunity to underline that, although supported by a considerable amount of experimental data, the columnar organization of the cortex is still a hypothesis, even if very probable and of enormous heuristic value. The large following that this hypothesis has had among scholars of the cerebral column is to be found in the fact that, assuming a modular organization, it is extremely suggestive both from a phylogenetic point of view to explain the increase in cortical development and cerebral capacities (the increase in volume of the cortex is obtained by increasing the number of interconnected modules), both in operational terms, proposing a simplified model (a minicolumn) very useful for trying to understand the flow of information between cortical neurons, in analogy with what has been done, for example, in the case of the cerebellar cortex. In contrast, rigid and schematic readings, such as that according to which the absolute number of neurons and the ratio of pyramidal to nonpyramidal neurons in a strip of cortex 30 μm wide and 25 thick is, with the exception of 17 BA, similar in all cortical areas of all species (including man), neglect undeniable specificities of area and species, specificities that can underlie important functional differences. In this context it is useful to remember that anatomical–functional specificity can be detected at different levels of resolution: perhaps the most striking example in this regard derives from the observation that in the monkey the number of spines of the pyramidal neurons of the prefrontal cortex is about

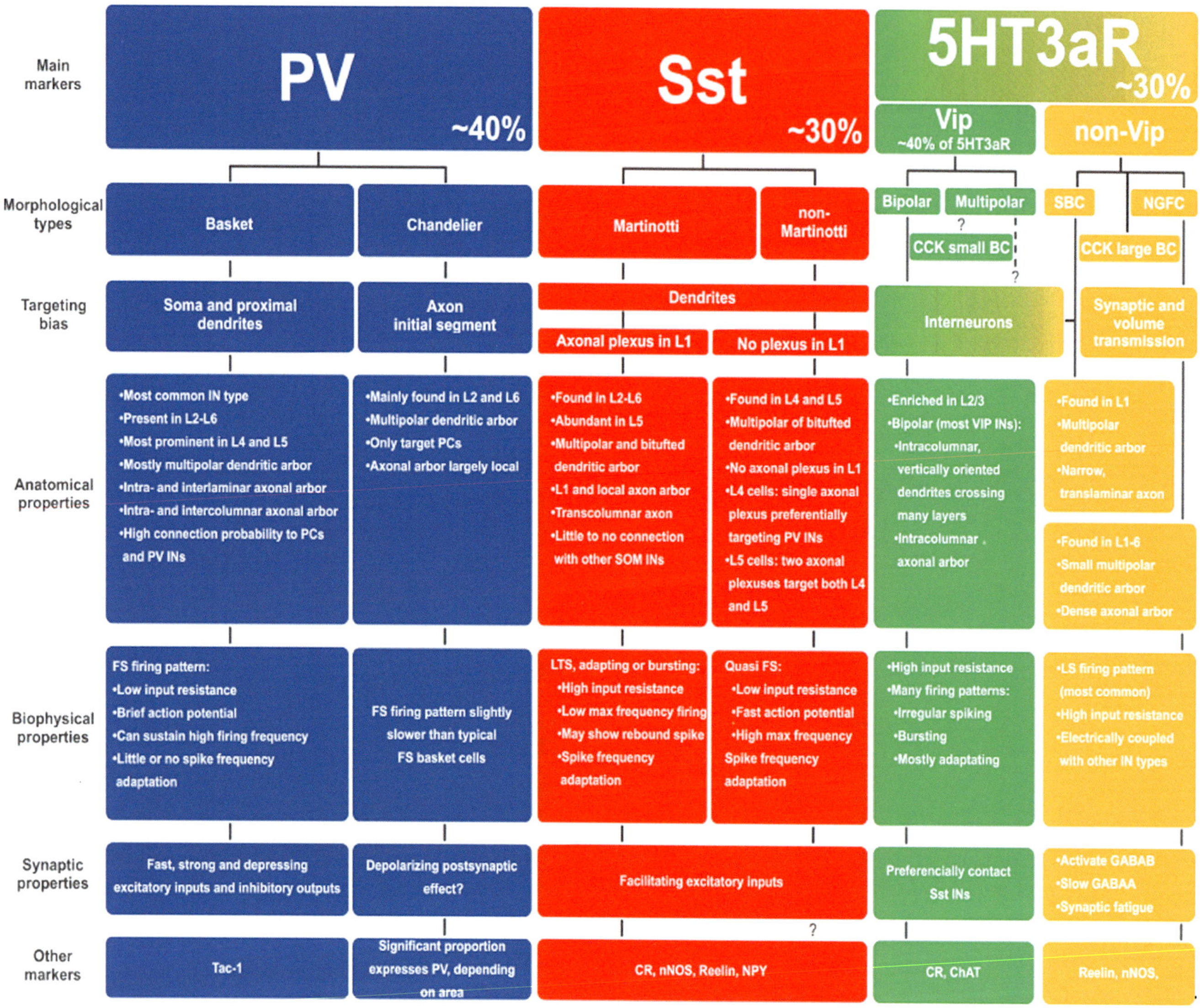

FIGURE 11.7 **Diversity, classification, and properties of neocortical GABAergic interneurons.** Nearly all the INs in neocortex express one of the main three nonoverlapping markers: parvalbumin (PV, blue), somatostatin (Sst, red), and the ionotropic serotonin receptor 5HT3a (5HT3aR, green-yellow). Further subdivisions within each molecular group are revealed by morphological features, cellular and subcellular targeting biases, and expression of other markers, as well as some known anatomical, electrophysiological, and synaptic properties. *Image courtesy form Tremblay, R., Lee, S., Rudy, B., 2016. GABAergic interneurons in the neocortex: from cellular properties to circuits. Neuron 91(2), 260–292. https://doi.org/10.1016/j.neuron.2016.06.033.*

sixteen times greater than that of the pyramidal neurons of 6 BA and that the pyramidal neurons of the human prefrontal cortex have more spines than those of the homologous area of the monkey.

INTRACORTICAL CIRCUITS

At this point, it may be useful to try to simplify the great complexity described in the previous paragraphs and try to identify the synaptic circuit essential for information processing in a minicolumn of any cortical region, a circuit which can therefore be defined as canonical.

The canonical circuit is made up of three populations of neurons that interact with each other: one population is represented by inhibitory interneurons (GABAergic cells), the other two (one located in layers II and III and one in layers V and VI) by excitatory neurons. Spiny stellate cells are homologous to excitatory neurons of layers II and III. Neurons of each subdivision form connections with neurons of the same population. The main interlaminar connections are those existing between the excitatory neurons of the superficial and deep layers; conversely, inhibitory neurons exert their action on both populations of excitatory neurons. The three groups of neurons receive direct inputs from the thalamic fibers, but since the thalamic afferent represents only the 10% of the excitatory

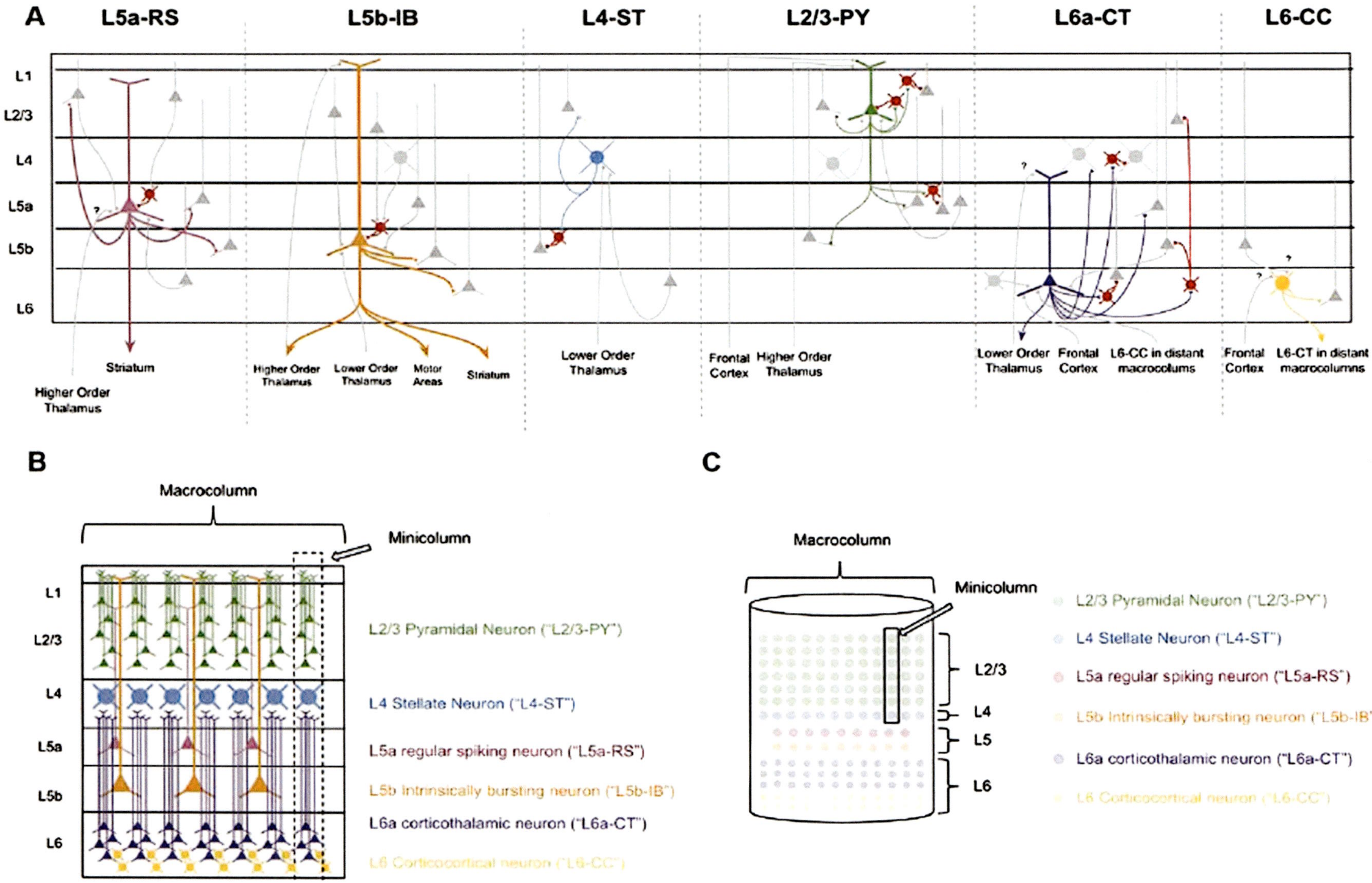

FIGURE 11.8 **The canonical neocortical microcircuit.** (A) The main input/output connectivity of excitatory neuron types within microcircuit. Gray neurons denote excitatory neurons that are either afferent or efferent partners with neuron type in question. Red neurons denote inhibitory interneurons. Question marks ("?") denote hypothesized but unverified connectivity. Triangles denote synapses. (B) Visualization of morphology and laminar distribution of the main types of excitatory neurons within the neocortical microcircuit. Note that there are several known excitatory neuron types not depicted here, including L6b corticothalamic neurons, L6 corticoclaustral neurons, and L2/3 corticocortical pyramidal neurons. Also note that L6-CC neurons are highly varied in morphology, but for simplicity they are only the "star-shaped" morphology is depicted here. (C) A simplified visual model of the neocortical microcircuit. *Image courtesy of Bennett, M., 2020. An attempt at a unified theory of the neocortical microcircuit in sensory cortex. Front. Neural Circuits 14. https://doi.org/10.3389/fncir.2020.00040.*

input, the 90% of the intracortical excitation is generated by the intracortical connections that link the pyramidal neurons together; this very strong recurring excitation appears to be necessary to supply the synaptic current necessary to maintain a reasonable firing rate of action potentials. The inhibition, which prevents recurrent excitation, is so powerful that it is effective even when relatively weak. The simplest and best-known way of information flow in an intracortical circuit is the following: the arrival of potentials from the thalamic afferents activates the neurons of layer IV (and of the lower portion of III) which in turn excite the neurons of layers II and III. The latter activate those of layers V–VI which reactivate in turn the upper layers. Fig. 11.9 shows the distribution of the different types of receptors in the cortical layers.

CORTICAL INTEGRATION: HIERARCHICAL ORGANIZATION AND DISTRIBUTED SYSTEMS

One of the fundamental problems of the physiology of the cerebral cortex has always been that of trying to understand whether a cortical function, for example perception, depends on the convergence in a cortical area (or in a column or in a neuron) of information originating from different sensory channels through specific pathways or if, on the contrary, it is the result of the interaction between different cortical regions. Paraphrasing Sherrington, one could pose the problem in the following terms: is cortical integration the result of a rigidly hierarchical organization or of a democracy?

Answers to this question can be sought both by taking up a fundamental distinction of cortical areas, that between primary areas (sensory or motor) and associative areas, and by studying the organization of the connections that exist between the various cortical areas (corticocortical connections).

ASSOCIATION AREAS OF THE CEREBRAL CORTEX

The distinction between sensorimotor and association areas and the concept that the latter play a particularly important role in the mechanisms that generate higher nervous functions dates back to the second half of the 19th century.

Experimental studies conducted on two groups of connections, thalamocortical and cortico-cortical, have led to very useful functional divisions of the neocortex. It is important to note that, since man, for obvious reasons, cannot be studied using invasive techniques, our knowledge of the connections of the fibers described above is almost completely based on studies conducted on nonhuman primates, in particular the rhesus macaque (Nieuwenhuys et al., 2015).

Association areas can be classified into two main types: modality-specific (unimodal) and higher-order (heteromodal) (Mesulam, 1998). Each primary sensory area is adjoined by a modality-specific sensory association area. These unimodal sensory association areas can be further subdivided into proximal or upstream and distal or downstream components. Proximal areas are only one synapse away from the corresponding primary sensory area, whereas distal areas are at a distance of two or more synapses from the relevant primary area.

In the primate brain at least three unimodal areas have been identified and they are essentially: the *somatosensory*, the *visual*, and the *auditory unimodal association area* (Parra et al., 2022; Bond et al., 2004; Wang et al., 2022).

The high-order heteromodal association areas, which are sometimes also designated as polymodal or supramodal areas, receive their principal cortical afferents from (1) unimodal areas, particularly their distal zones, (2) other heteromodal areas, and (3) paralimbic areas. In the primate brain, three heteromodal association areas have been described and they are the *parietotemporal*, the *medial temporal*, and the *prefrontal* heteromodal association area.

DIFFERENT AREAS AND LOBES IN THE SAME CORTEX

It is known that the cerebral cortex is subdivided principally into five lobes (frontal, parietal, temporal, occipital, and limbic lobe) and in others more restricted areas such as opercula, e.g., parietal, frontal, etc. However, concerning the topographical description of cerebral cortex, we invite the reader to consult the Chapter 2.

Introduction to the Cerebral Connectome

Since the cerebral cortex is divided into different lobes, it would be more accurate to speak of connections in a broad sense and not just those relating to one specifical lobe as can be the frontal lobe. However, it has been known for years now and the emerging results increasingly confirm that the cortex of the frontal lobe, representing more than 60% of the entire neocortex, is the "starting and finishing" point of all commissural, associative, and projective connections coming not only from the frontal cortices but also from the remaining lobar regions. The frontal cortex is the most studied, this is because it is precisely on it that the most important intellectual, planning, organizational, motor skills, etc., depend; all this not to mention all aspects relating to behavior, personality, and psyche. Regarding these last two aspects, we have known for years now how

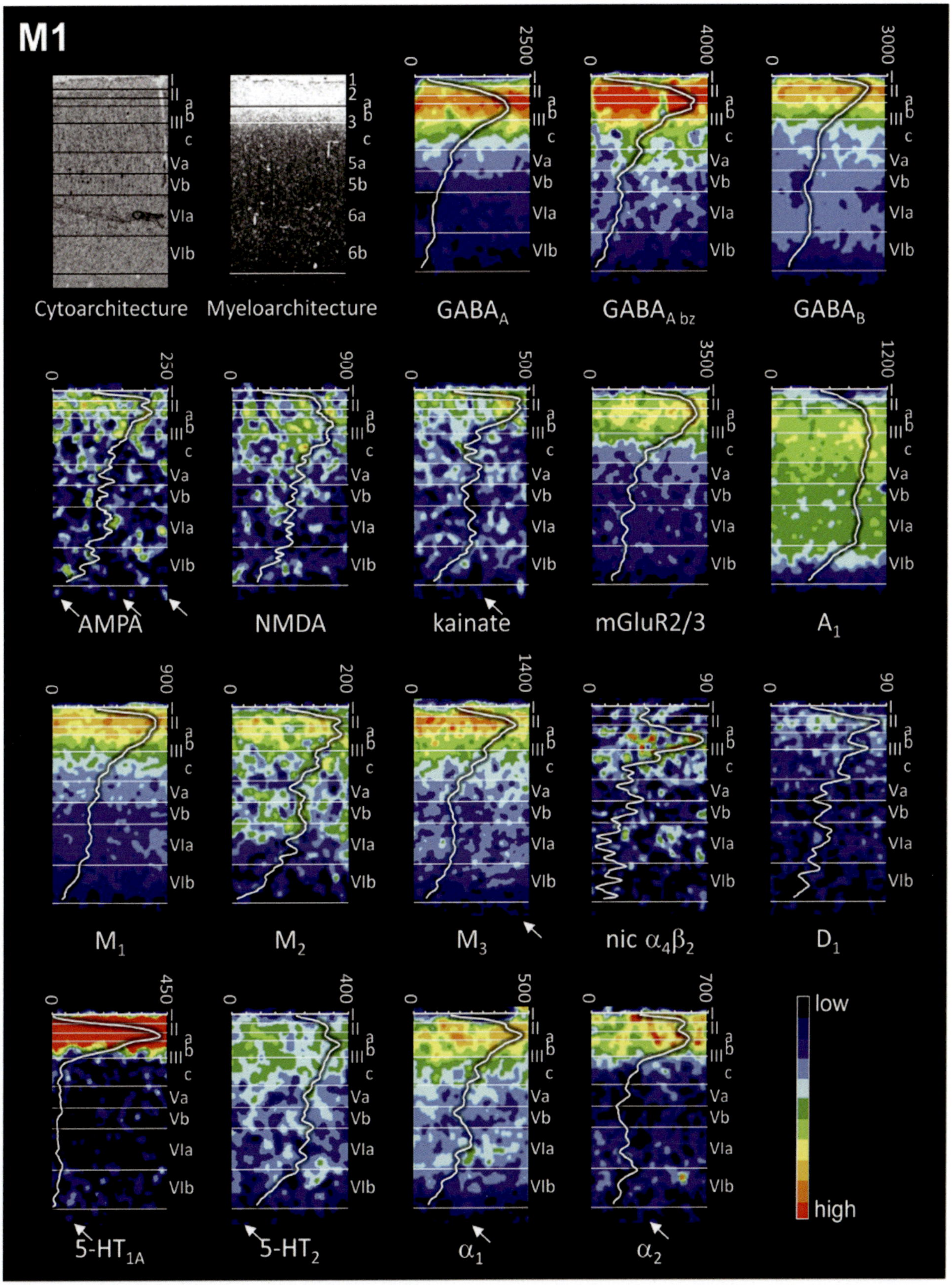

FIGURE 11.9 **Receptor-, cyto-, and myeloarchitecture of the human primary motor cortex (M1, BA4). The absolute receptor concentration (in fmol/mg protein) throughout the cortical depth is provided by the profile curve overlaid onto each receptor autoradiograph.** Note, that the

the frontal cortex has been divided into a more anterior part, known as the prefrontal cortex, and a more posterior part consisting of the motor and premotor cortices. We have seen enough and it can be confirmed that the prefrontal cortex (including 9, 10, 11, 12, 46, 47 BA), is the seat of emotion, thought and personality of the individual and how the latter in particular, it may change following traumatic events affecting these locations (see the case of Pineas Gage (Teles and Ricardo, 2020). The remaining posterior region, which ends with the central sulcus of Rolando, is instead assigned to motricity and is therefore the seat of the motor area (4 BA) and of the supplementary motor areas (6, 8 BA and others). Furthermore, as far as the psychological aspects of the individual are concerned, it has been seen that most of the psychiatric disorders have an "organic" correlation almost always in the aforementioned areas; the last part of this chapter will in fact be dedicated to the physiopathological cortical correlates of some mental disorders.

CONNECTION FIBERS OF THE CEREBRAL CORTEX

The connection fiber tracts can be broadly divided into intrinsic and extrinsic connections according to whether their terminal projections remain confined within the cortex of one same lobe or extend outside of it.

The intrinsic connections are cortico-cortical association fibers between two areas within the same lobe and they vary in length and morphology according to the distance between the areas they connect.

Most of the intrinsic connections are short U-shaped fibers between adjacent gyri. Longer tracts run deeper in the white matter and connect distant cortical regions.

The extrinsic connections, differently from the intrinsic ones, include the association pathways, between one lobe and other lobes within the same hemisphere, and the commissural pathways, that mediate communication mainly between the two hemispheres. Finally, the projection fibers originate from the cortex and terminate in the subcortical regions. The specific arrangement of the intrinsic and extrinsic fibers is unique to each cortical area and determines its functional specialization.

Both intrinsic and extrinsic connections are already at birth, though most of these tracts are poorly myelinated in the first year of life (Deoni et al., 2011).

In short, the principal connections of the cerebral cortex can be subdivided into three classes on the base of their anatomo-functional connotates:

1. commissural fibers;
2. projection fibers;
3. association fibers.

Recent studies highlight that the frontal lobe connections occupy approximately the 40% of the total white matter of the cerebral hemispheres in the monkey brains as compared to the 60% volume occupancy measured in the human brain (Barrett et al., 2020). The relative increase in volume of the human frontal connections was similar for the association, commissural, and projection networks. This suggests a greater direct communication within the frontal cortex and between the frontal cortex and other lobes characterized human brain evolution. The relative greater volume of the frontal lobe connections in the human brain has occurred at the expenses of other tracts that have significantly reduced, such as the anterior commissure. This tract represents only 4% of the telencephalic commissural tracts in the human brain, compared to 28% in the monkey brain (Barrett et al., 2020). When volume measurements were performed for individual frontal association tracts, a greater relative tract volume was observed in humans compared to monkeys for those tracts connecting the frontal lobe to the parietal lobe (superior longitudinal fasciculus—SLF), to temporal lobe (arcuate fasciculus), and to occipital lobe (inferior fronto-occipital fasciculus).

These tracts have an important role in functions that have greatly developed in humans, such as auditory memory (López-Barroso et al., 2013), fine motor skills for tool making (Howells et al., 2018), and language (Catani et al., 2005; Rilling et al., 2008a,b,c). There were no differences between monkey and human brains for fronto-limbic tracts such as the uncinate fasciculus and the cingulum. Overall, these data indicate that humans have benefited from an increased frontal direct top-down control of visual, somatosensory, and auditory information, whereas tracts involved in emotional processing and behavior remained preserved across primate species.

Commissural Pathways

The major commissural fibers lie within the corpus callosum (CC) and they follow a precise topography (Fig. 11.10).

scale has been optimized for each profile to provide the best visualization of changes in receptor densities throughout the cortical ribbon. Roman and Arabic numerals indicate cyto- and myeloarchitectonic layers, respectively. Positions of cytoarchitectonic layers were transferred to the neighboring receptor images. *Image courtesy from Palomero-Gallagher, N., Zilles, K., 2019. Cortical layers: cyto-, myelo-, receptor- and synaptic architecture in human cortical areas. NeuroImage 197, 716–741. https://doi.org/10.1016/j.neuroimage.2017.08.035.*

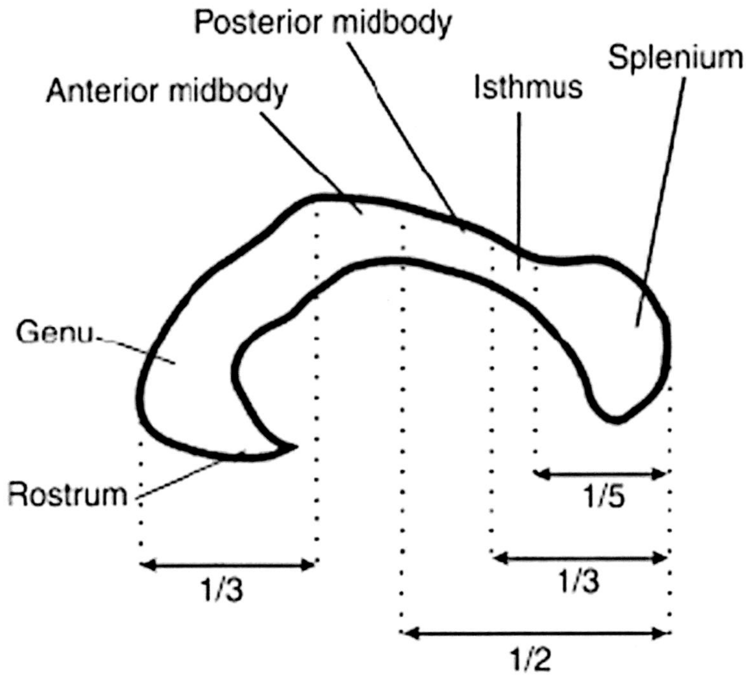

FIGURE 11.10 **Schematic representation of the corpus callosum.** In an adult human, it is approximately 10 cm long, 1 cm wide, and contains approximately 200 million axonal fibers. Standard anatomic divisions are the rostrum, genu, body, isthmus, and splenium. *Image courtesy of Eccher, M., 2014. Corpus callosum. In: Aminoff, M.J., Daroff, R.B. (Eds.), Encyclopedia of the Neurological Sciences, second edition. Academic Press, Oxford, pp. 867–868. https://doi.org/10.1016/B978-0-12-385157-4.01137-4.*

The fibers connecting the prefrontal cortices occupy the most inferior part of the anterior corpus callosum (or anterior third, Figs. 11.10 and 11.11), which is termed rostrum. The fibers that make up the rostrum extend laterally, under the anterior corn of the lateral ventricle. The genu fibers are curved anteriorly and connect the lateral and medial surface of the frontal lobes forming the so-called minor forceps, due to its shape. The fibers of the body of the CC (mid-third) connect the more central cortices of the brain such as precentral and parieto-temporal cortices (motor, somatosensitive, and acoustic cortices, respectively) (Hofer and Frahm, 2006). The isthmus and splenium of the CC (posterior-third) contains visual fibers from the occipital cortex and projections from associative parietal and temporal brain areas. The isthmus contains connecting fibers between superior temporal and inferior parietal areas.

The commissural fibers of the occipital pole form a C-shaped structure similar to but larger than the forceps minor, called the forceps major (Fig. 11.11).

Comparisons between the topography of the midsagittal area of the corpus callosum and its cortical projections in the rhesus monkey (*Macaca mulatta*) (Myers, 1965; Schmahmann and Pandya, 2006) and in vivo spherical deconvolution tractography in the human brain (Catani and Thiebaut de Schotten, 2012) suggest a similar topography of the callosal projections across species, though with some differences, especially for the frontal projections. The commissural connections in general could be omotypichal or eterotypical. The omotypycal connects the correspondent cortical areas of the two hemispheres and vice versa.

Projection Pathways

The cortex of the frontal lobe has the unique feature of establishing extensive direct connections with the major subcortical structures of the diencephalon, basal ganglia, hypothalamus, brainstem, and spinal cord. The three major frontal descending systems include frontostriatal fibers to the caudate and putamen, direct cortico-pontine connections, and corticofugal projections to the nuclei of the motor cranial nerves (corticonuclear tract) and spinal cord (corticospinal tract) (see Chapters 4 and 5a, respectively).

The main ascending projections can be further divided into a larger thalamofrontal group and a smaller nonthalamic component from the amygdala and nucleus accumbens (Nieuwenhuis and Yeung, 2005; Aggleton et al., 2015), mammillary bodies (Allen and Hopkins, 1989), and nucleus basalis of Meynert (Selden et al., 1998). This division into ascending and descending projections should not be interpreted as an absolute segregation of projection fibers, as many tracts contain both ascending and descending fibers.

FRONTOSTRIATAL PROJECTIONS

The frontostriatal projections to the caudate and medial putamen run within the anterior limb of the internal capsule, whereas those directed to the lateral putamen are included in the external capsule. The projections to the basal ganglia and cerebellum are indirectly reciprocal, so that the cortex also receives

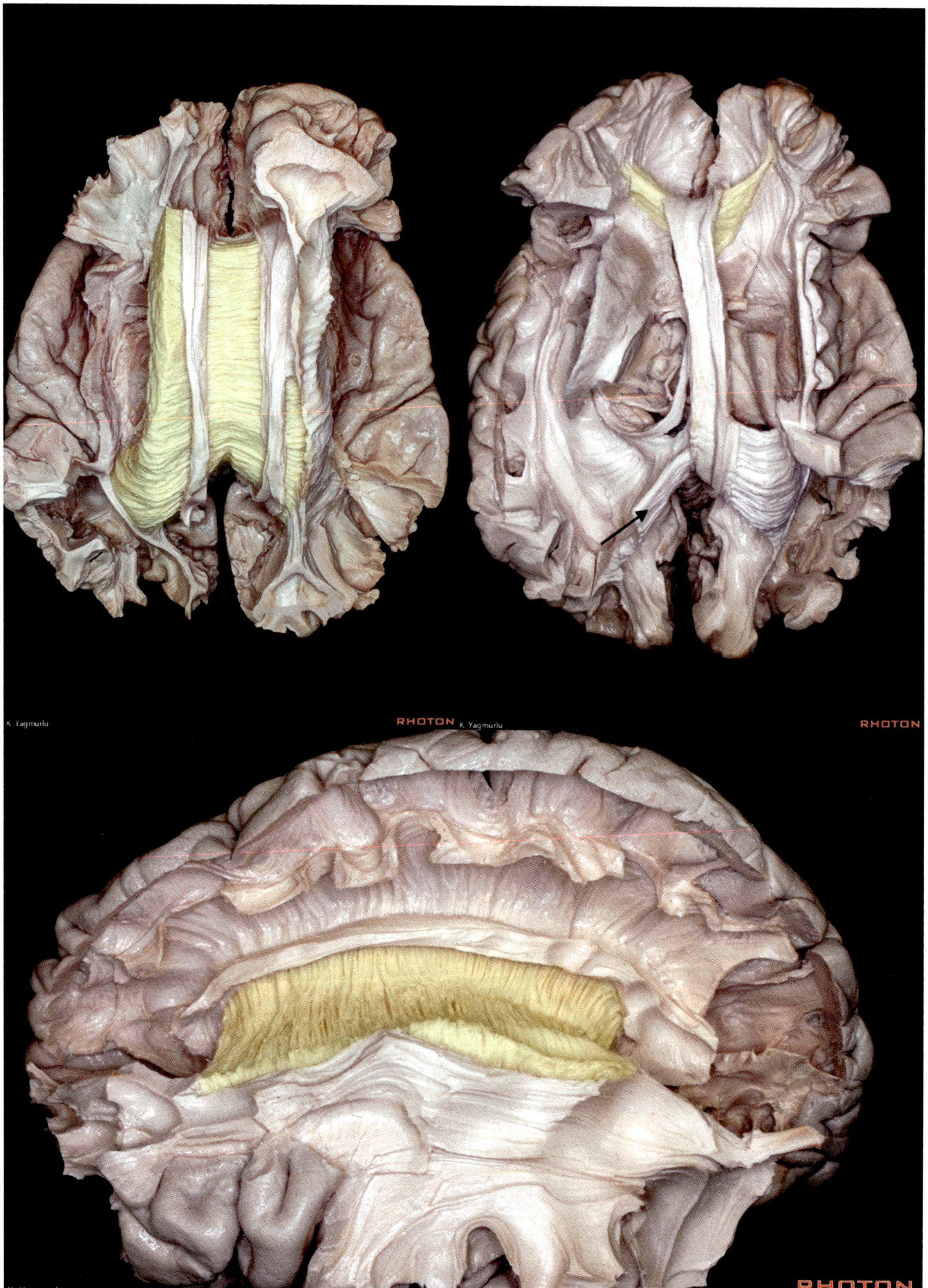

FIGURE 11.11 ***(Top-left)* The corpus callosum (in yellow).** It is the major interhemispheric commissural pathway that connects most neocortical areas. ***(Top Right)* Forceps minor (in yellow) and forceps major (black arrow).** The callosal fibers arising from the genu of the corpus callosum turn in an anterior oblique direction, to interconnect the frontal regions, where they are called the forceps minor. At the level of the

projections from these centers via the thalamus, to create complex corticobasal ganglion and corticocerebellar circuits. According to their cortical origin, the descending frontal fibers contribute to motor, cognitive, affective, and behavioral functions (Schmahmann and Pandya, 2008; Barbas, 2022). Comparative studies of projection fibers are particularly difficult to perform with tractography; nevertheless preliminary findings suggest the corticoponto-cerebellar tract receives a relatively larger contribution from the prefrontal areas in humans compared to monkeys (Ramnani et al., 2006). These data give further support in favor of the fronto-cerebellar participation in human cognition and behavior (Schmahmann and Sherman, 1998).

FRONTOTHALAMIC PROJECTION

Current understanding of the frontothalamic connections is largely based on findings from experimental animal models and occasional human postmortem studies in patients with neurologic or neuropsychiatric disorders. Conversely, decades of structural, metabolic, and functional neuroimaging reports in patients with circumscribed thalamic lesions and in neurosurgical patients with deep brain stimulation represent a valuable source to indirectly validate animal models transposed to humans or propose original interpretations regarding the fronto-thalamic anatomy of the human brain.

In recent years, most contributions from tractography represent replications in humans of previous animal axonal tracing studies. The thalamic radiations originate from the thalamic nuclei and after a short course within the internal capsule enter the corona radiata (Figs. 11.12 and 11.13) to terminate diffusely in the cortex. Diffusion tensor imaging tractography has been used to visualize the thalamic projections and segment the thalamic nuclei according to their cortical projection zones (Behrens et al., 2003). Similar to the fibers of the corticospinal tract, the afferent and efferent thalamic projections are aligned in an orderly manner along the anterior and posterior limbs of the internal capsule, and those connecting to the frontal lobe occupy the most anterior limb (Catani and Thiebaut de Schotten, 2012; Safadi et al., 2018). The efferent thalamofrontal fibers project to the frontal cortex through the anterior and superior thalamic peduncles.

The Association Pathways

The major interlobar association tracts are the superior longitudinal fasciculus (SLF; Fig. 11.17), the *arcuate fasciculus* (AF; it is a subdivision of SLF. For AF see Fig. 11.14), the *cingulum* (Fig. 11.11), the *inferior fronto-occipital fasciculus,* and the *uncinate fasciculus* (Fig. 11.14). These interlobar tracts mediate direct communication between frontal areas and the cortex of other lobes. In addition to these long tracts, short interlobar U-shaped fibers connect the frontal lobe to the parietal lobe (perirolandic fibers), insula (fronto-insular tracts), and cingulate gyrus (Fig. 11.11). Finally, the frontal regions within the same hemisphere communicate among themselves through the fibers that constitute the local intralobar connectivity of the frontal lobes.

The intralobar fibers are strictly confined within the white matter of the frontal lobe and include the frontal aslant tract, the frontal paracingulate tract, the frontal orbitopolar tract, and the two major longitudinal chains of the dorsolateral convexity represented by: the *frontal superior longitudinal* and *frontal inferior longitudinal tract*. Other U-shaped tracts connect the medial and lateral aspects of the frontal pole (i.e., the fronto-marginal tract) and the precentral gyrus to the posterior cortex of the superior, middle, and inferior frontal gyri (Catani et al., 2012b).

LONG AND SHORT INTERLOBAR ASSOCIATION TRACT

The arcuate fasciculus is a dorsal association tract connecting perisylvian regions of the frontal, parietal, and temporal lobe. Within the arcuate fasciculus two parallel pathways have been distinguished:

1. direct pathway (i.e., long segment) connecting the Wernicke's area with Broca's area, localized, respectively, in posterior temporal lobe (BA 41, 42, 22, 37) and in the posterior frontal lobe (BA6, 44, 45), and;

splenium, the fibers turn in a posterior oblique direction, to interconnect the parietooccipital regions, where they are called the forceps major. They cross the midline by passing below the cingulum (arched structure extending anteroposteriorly) to form the roof of the lateral ventricle. ***(Below)* Cingulum (in yellow) over the corpus callosum.** The callosal fibers arising from the splenium, named the tapetal fibers, sweep downward to form the roof and lateral wall of the atrium, temporal, and occipital horns of the lateral ventricle. The anterior commissure crosses the midline at the base of the putamen to connect the orbitofrontal, occipital, and temporal lobes, and especially the amygdala. It has an anterior crus that extends forward, and a posterior crus that extends laterally. The posterior crus is divided into temporal and occipital extensions. From clinical perspective, it was suggested that seizure activity can propagate to the contralateral medial temporal lobe by way of the anterior commissure, so its cutting is part of the hemispherectomy procedure for control of intractable seizures. *Image courtesy of The Rhoton Collection® http://rhoton.ineurodb.org/. Dissector: K.Yagmurlu. Drawings by Jeffrey Sorenson.*

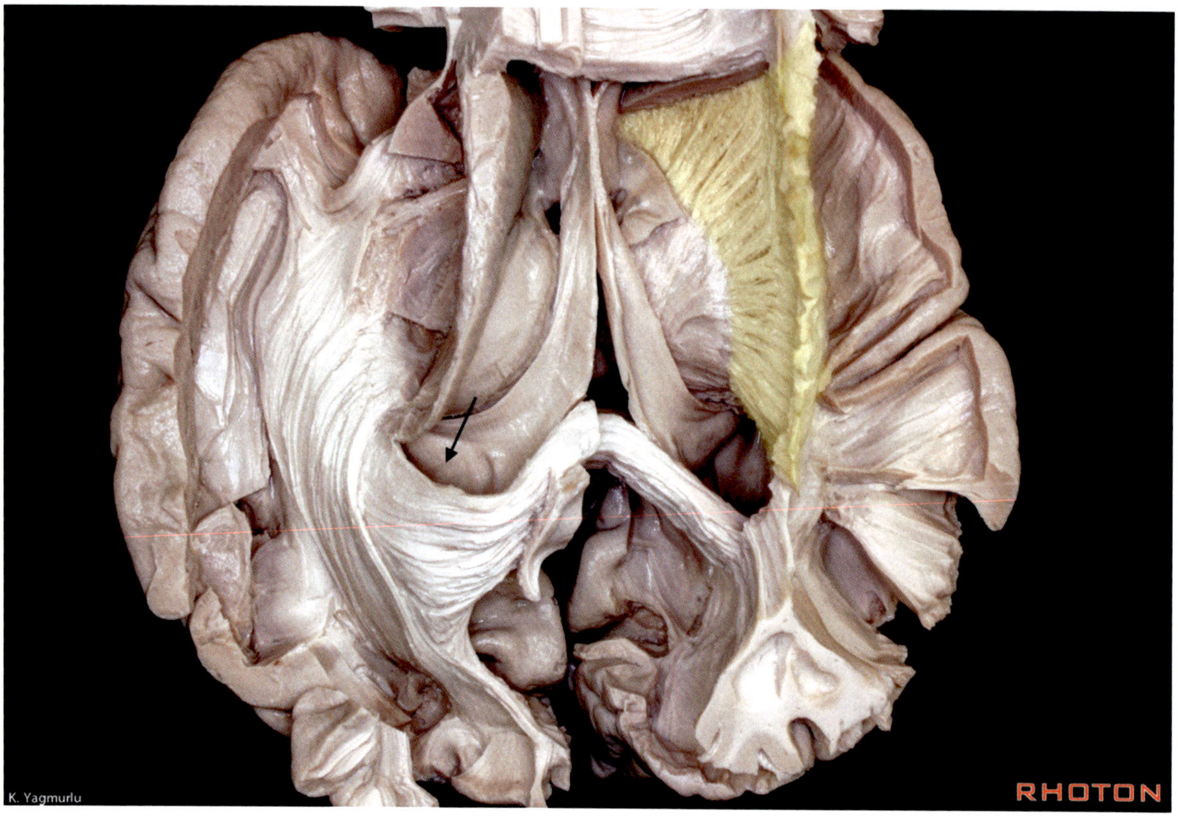

FIGURE 11.12 **The corona radiata (in yellow).** The internal capsule, partially removed, can be seen at the lower end of the corona radiata. The atrium of the lateral ventricle (black arrow) is posterior to the fornix; the latter is easily seen in the center of the brain. *Image courtesy of The Rhoton Collection® http://rhoton.ineurodb.org/. Dissector: K.Yagmurlu. Drawing by Jeffrey Sorenson.*

the indirect pathway, consisting of an anterior segment localized between Broca's region and the inferior parietal lobe (i.e., the Geschwind's region, BA 39, 40) and a posterior segment between Geschwind's and Wernicke's region (Catani et al., 2005) (Fig. 11.14)

The anatomy of the arcuate fasciculus is significantly different between humans and monkeys (Rilling et al., 2008a,b,c; Barrett et al., 2020). In the human brain the fibers of the long segment which originate from the inferior region of the precentral gyrus, i.e., pars opercularis/triangularis, and from the posterior cortex of the middle frontal gyrus reach the posterior areas of the superior and middle temporal gyri.

In monkeys the long segment is significantly reduced in volume (Barrett et al., 2020) and only sparse fibers reach the posterior region of the superior temporal gyrus (Schmahmann and Pandya, 2007a,b), which may explain their poor auditory memory abilities as compared to humans (Schulze et al., 2012). This suggestion has been supported by the study of López-Barroso et al. (2013) who showed that in humans' performance in word learning, correlates with microstructural properties and strength of functional connectivity of the long segment of the arcuate fasciculus. This study demonstrated that the ability to learn new words relies on the direct and fast communication between auditory temporal and motor frontal regions in the left hemisphere; this communication is important for the phoneme categorization of auditory inputs (Chevillet et al., 2013) and the consolidation of the motor codes of the newly learned phonologic sequences (Rauschecker and Scott, 2009; Rodríguez- Fornells et al., 2009).

Another important characteristic of the long segment of the arcuate fasciculus is its variable degree of lateralization in the general population. An extreme degree of leftward lateralization of the long segment has been reported in approximately 60% of the normal population, while the remaining 40% shows either a mild leftward lateralization (20%) or a bilateral, symmetric pattern (20%) (Catani et al., 2007). Overall, females are more likely to have a bilateral pattern compared to males, which may in part explain their superior performance in tasks involving semantic clustering strategies in word learning and overall better recovery of language after left hemisphere stroke. A longitudinal tractography study has indeed confirmed that patients with aphasia due to left hemisphere stroke are more likely to recover normal language at 6 months if they have a larger volume of the long segment in the right

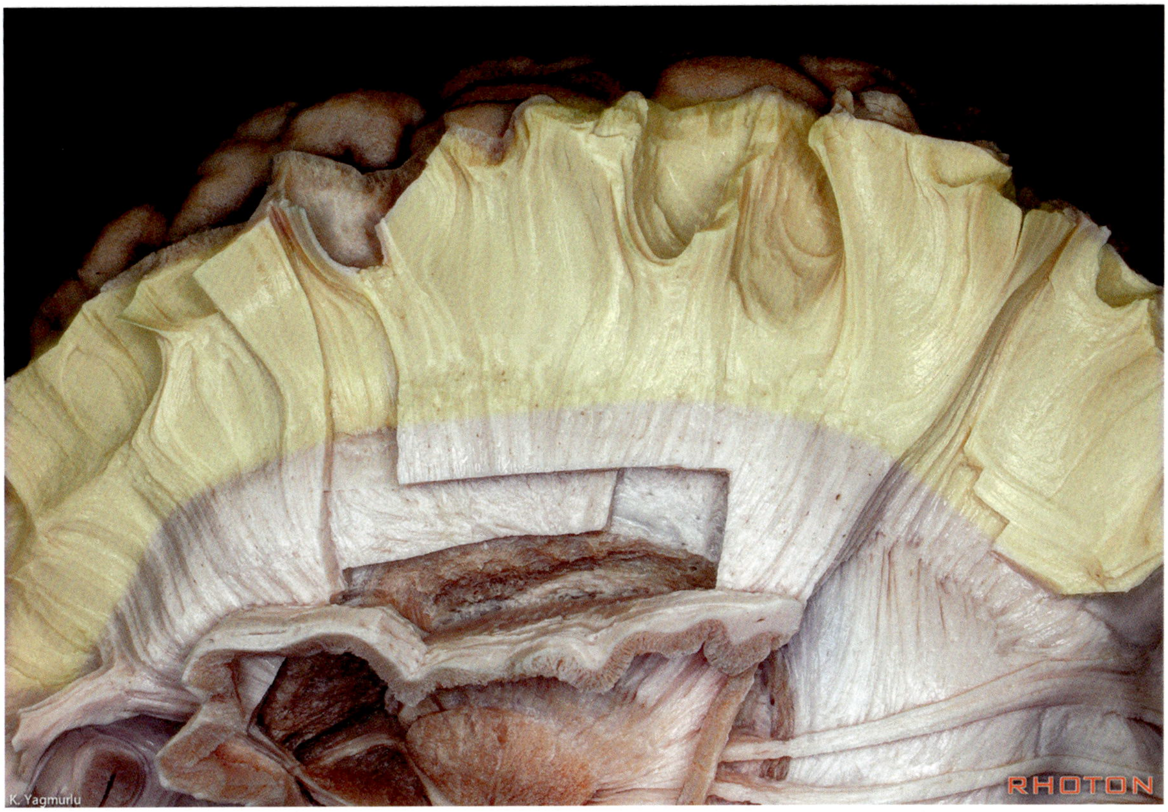

FIGURE 11.13 **The centrum semiovale (in yellow).** It is located above the level of corpus callosum and consists of the superior longitudinal fasciculus association fibers, corona radiata projection fibers and callosal commissural fibers. *Image courtesy of The Rhoton Collection® http://rhoton.ineurodb.org/. Dissector: K.Yagmurlu. Drawing by Jeffrey Sorenson.*

hemisphere (Forkel et al., 2014b). In patients with primary progressive aphasia (PPA) the severity of the deficit in syntactic processing correlates with microstructural damage of the long segment (Wilson et al., 2011). While the direct pathway may support auditory-motor integration and syntactic processing, the role of the indirect pathway could be more complex and related to repetition (Forkel et al., 2020), verbal fluency (Fridriksson et al., 2010a,b), and complex sentence processing (Jacquemot and Scott, 2006; Perani et al., 2011a,b; Wilson et al., 2011). In addition to language, the left arcuate fasciculus has a role in the coordination of goal-directed actions (i.e., praxis) (Heilman and Watson, 2008; Ramayya et al., 2010) and working memory (Jacquemot and Scott, 2006).

In the right hemisphere, normal musical competence depends on the integrity of the right long segment (Loui et al., 2009a,b; Hyde et al., 2011). Impaired ability to sing or to play a previously learned instrument (i.e., expressive amusia) is often seen in patients with inferior frontal lesions (Särkämö et al., 2009a,b). The right arcuate is also important for conveying meaning and emotions through the modulation of speech intonation and rhythm (i.e., prosody), or changes associated with facial, limb, and gestural expression (Ross, 1981).

The arcuate fasciculus and other related fibers that we will consider later are important for language production. Regarding the latter, scientists have described two pathways formed by different fascicles (including the arcuate), which constitute both a conceptual and anatomical division of the language production system, on the basis of their role and their functions; they are the "dorsal" and "ventral" pathway (Fig. 11.15). The former projects toward the inferior parietal and posterior frontal lobe (inferior frontal gyrus and premotor cortex) and it is made of the three segment that constitute the arcuate fasciculus (anterior, direct, and posterior) and it is charge of the translation of sensory\acoustic speech signal into motor–articulatory representations (auditory–motor integration) required for verbal repetition. On the other hand, the "ventral pathway" links the superior and middle temporal gyri, the inferior parietal lobe, and the occipital lobe with the inferior frontal gyrus through the inferior fronto-occipital fasciculus, also called extreme capsule fiber system (Martino et al., 2010a,b; Friederici et al., 2013). Additionally, the most anterior pars of the inferior frontal gyrus and the frontal operculum are connected to the anterior temporal lobe through the uncinate fasciculus (see later) (Friederici et al., 2013). The "ventral pathway" is mainly involved in the mapping of

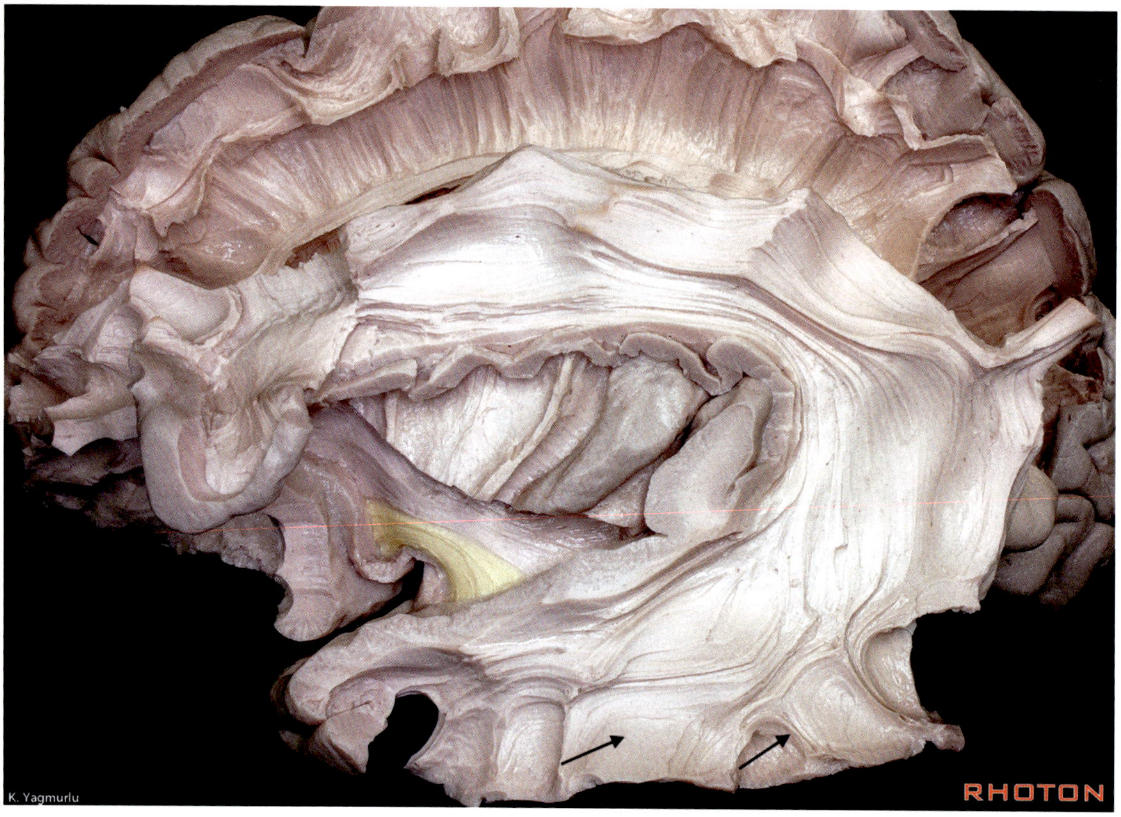

FIGURE 11.14 The **arcuate fasciculus (the reversed C shape structure in the foreground of the image)** is the frontoparieto-temporal long association fibers that connects Wernicke's area (located at the posterior extremity of the fasciculus), which consists of the middle and posterior parts of the superior and middle temporal gyri, to Broca's area (located at the anterior extremity of the fasciculus) situated in the posterior two-thirds of the inferior frontal gyrus. The yellow structure is the **uncinate fasciculus** and the fibers over it is the **inferior frontooccipital fascicle**. Medially the latter, the most anterior white structure, is the **extreme capsule**. The **inferior longitudinal fasciculus** is indicated by the two posterior black arrows. *Image courtesy of The Rhoton Collection® http://rhoton.ineurodb.org/. Dissector: K.Yagmurlu. Drawing by Jeffrey Sorenson.*

sensory\auditory speech signals into conceptual and semantic representations for speech comprehension (for the graphic representation of the "ventral" and "dorsal pathway" see Fig. 11.15).

The postnatal development of the arcuate fasciculus is purely heterogeneous in terms of age-related development of the three segments (i.e., long, anterior and posterior) however it is correlated to the arising and maturation of different abilities related to it such as the oral production of the language, word learning etc. (see later).

Several studies showed a direct correlation anatomofunctional between the development of the arcuate fasciculus and the vocal production in the preterm born children and term born ones. In the former has been seen that they have an impaired linguistic ability when compared with their term-born peers, even in the absence of major disabilities, which can persist as a long-lasting linguistic delay throughout childhood (van Noort-van der Spek, 2012). Thus, the white bundles so far mentioned that play a role in the acquisition of the language ability. Recently (Salvan, 2017) showed that in preterm infants at the time of normal birth, well before the formal emergence of natural language, the microstructural properties of both the left and right arcuate fasciculi are associated with later linguistic abilities. Previous studies in adulthood and adolescence have linked its microstructural properties to: word learning (Lopez-Barroso et al., 2013), development of reading skills (Yeatman et al., 2011, 2012), sentence comprehension performance (Skeide et al., 2015), and to support syntactic processing of language (den Ouden et al., 2012).

Wilkinson et al. (2017) showed that the anterior and long tract have lower apparent diffusion coefficient (ADC) in subject over 3 years old while the posterior tract showed higher ADC. The posterior tract did not show age-related effect in terms of fractional-anisotropy (FA), tract count, length and volume. These results suggest that the posterior AF tract shows a matured state, indexed by most of the used MRI-based measurements, in early postnatal developmental ages and ADC is a measurement that can detect further maturation of the posterior tract (Fig. 11.16).

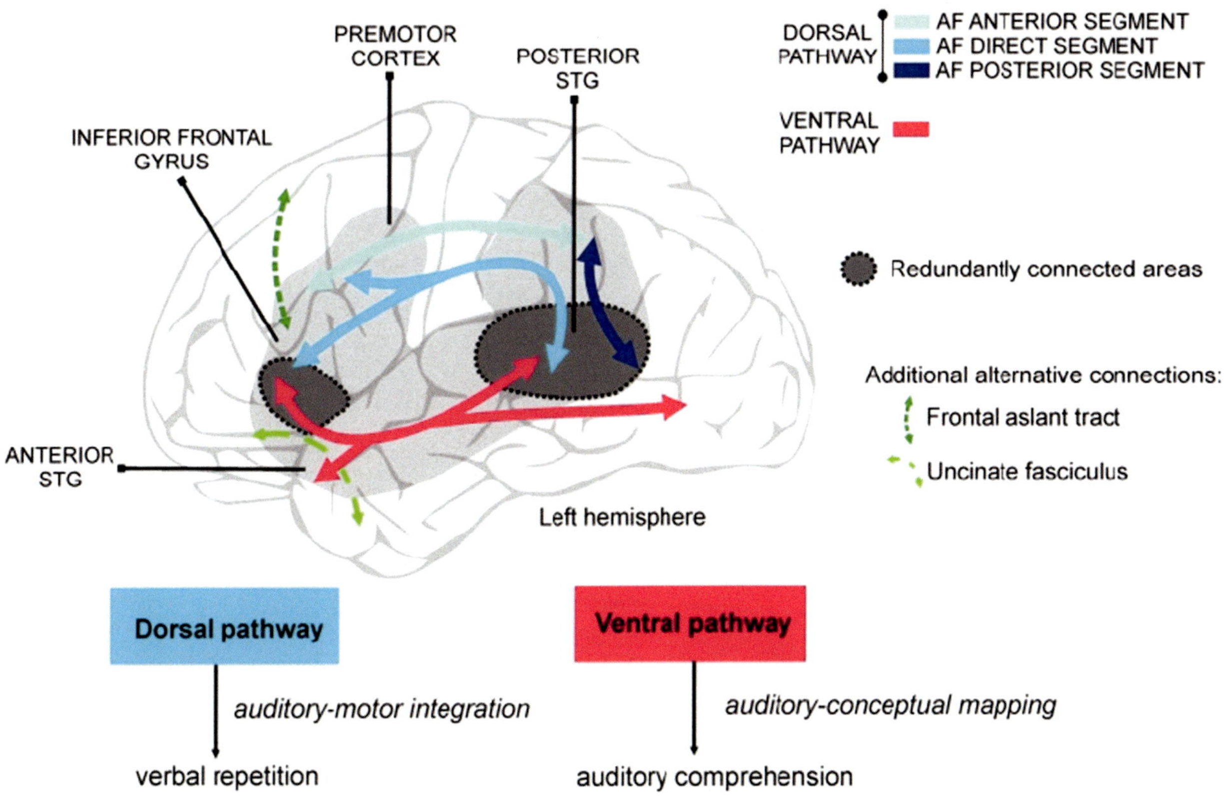

FIGURE 11.15 **Division of labor of the ventral and dorsal pathways for language.** *AF*, arcuate fasciculus; *STG*, superior temporal gyrus.On the top is shown a schematic illustration of the main anatomical connections of the perisylvian areas in the left hemisphere, comprising the ventral and the dorsal pathways. The perisylvian cortex, covering frontal, parietal and temporal cortices, is shaded in gray. Compensation between the dorsal and the ventral streams, normally associated with poorer performance, derives from the partial overlap of the cortical areas they connect and the extra support of additional language-related white-matter pathways (e.g., uncinate fasciculus, frontal aslant tract). Regions showing greater likelihood to be redundantly connected through the ventral and dorsal pathways are colored in dark gray. Compensation mechanisms are crucial in high difficulty language situations and after brain damage. At the bottom, the main functions of the dorsal and ventral streams are illustrated. Although not shown in the figure, the homologous areas in the right hemisphere may also play a crucial role in compensation mechanisms. *Image courtesy from López-Barroso, D., de Diego-Balaguer, R., 2017. Language learning variability within the dorsal and ventral streams as a cue for compensatory mechanisms in aphasia recovery. Front. Hum. Neurosci. 11. https://doi.org/10.3389/fnhum.2017.00476.*

A group of longitudinal fibers connecting the dorsolateral frontal and parietal regions has been described using axonal tracing (Pandya and Kuypers, 1969), cortical electrophysiological recording (Rizzolatti et al., 1998), and more recently diffusion tractography (Thiebaut de Schotten et al., 2011a,b). Petrides and Pandya (1984) used the term superior longitudinal fascicle (SLF) to define this frontoparietal associative system and identified three separate branches. The most dorsal component of the SLF I projects to the superior parietal lobule and superior frontal gyrus. The SLF I is bilateral with lateralization differences between left and right handers (rightward lateralized in left handers) (Howells et al., 2018). The SLF I contributes to processing the spatial coordinates of trunk and limbs and preparatory stages of movement planning (Duffy and Burchfiel, 1971; Leiguarda and Marsden, 2000). It is also involved in oculomotor coordination (Anderson et al., 2012), visual reaching (Johnson et al., 1996), and voluntary orientation of attention (Corbetta and Shulman, 2002).

The second branch of the SLF II projects to the posterior region of the inferior parietal lobule (including the intraparietal sulcus) and the lateral aspect of the superior (i.e., FEF) and middle frontal gyri.

The SLF II is involved in processing the spatial coordinates of the upper limbs, including the hand for reaching and grasping movements (Leiguarda and Marsden, 2000; Goldenberg and Spatt, 2009), and contributes to tasks requiring attention and visuospatial working memory (Levy and Goldman-Rakic, 2000; Corbetta and Shulman, 2002; Thiebaut de Schotten et al., 2005). In humans the SLF II is right lateralized in most but not all subjects (about 70%) and the degree of lateralization is correlated with asymmetry in behavioral performances for visuospatial tasks (Thiebaut de Schotten et al., 2011a,b). This finding indicates that a right hemispheric specialization for visuospatial attention could be related to an unbalanced speed of signal processing along the SLF II of the two hemispheres.

The most ventral branch of the SLF III projects to the inferior parietal lobule (supramarginal and anterior angular gyrus) and the posterior region of the inferior frontal gyrus. The SLF III corresponds to the anterior segment of the arcuate fasciculus and the two terms

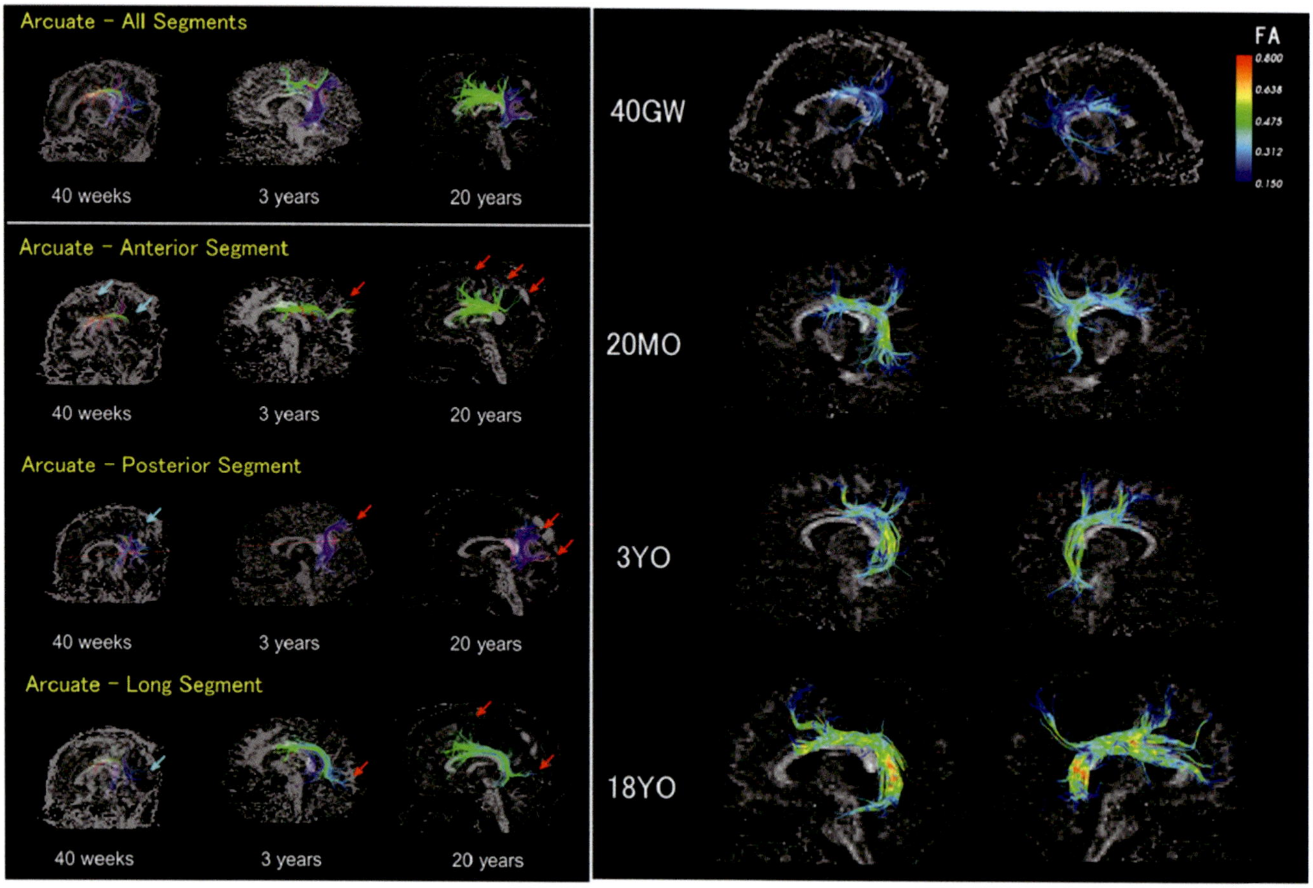

FIGURE 11.16 **(*Left*) Whole arcuate fasciculus pathways (top row), anterior segments (second row), posterior segments (third row), and long segments (bottom row).** In newborns, the anterior, posterior, and long segments of the AF were all thin and sparse, with short pathways running into the cortex that were not detected in regions close to the brain surface (light blue arrows). By 3 years old, such pathways running into the cortex became obvious, with their numbers increasing until adult ages (red arrows). The color coding of fibers is based on a standard RGB code, applied to the vector between the end-points of each fiber (red: Left-right, green: Anterior-posterior, and blue: Dorsal-ventral). HARDI tractography pathways on 40 gestational weeks newborn, 3 years old, and 20 years old subjects. **(*Right*) Color-corded FA values on detected AF pathways in 40 gestational weeks, 20 months old, 3 years old, and 18 years old subjects.** The color coding of fibers is based on FA values. *Image courtesy from Wilkinson, M., Lim, A.R., Cohen, A.H., Galaburda, A.M., Takahashi, E., 2017. Detection and growth pattern of arcuate fasciculus from newborn to adult. Front. Neurosci. 11. https://doi.org/10.3389/fnins.2017.00389.*

are currently used interchangeably. Future tractography studies will establish whether, according to its functional role, the SLF III/anterior segment is to be considered part of the sensory-motor SLF system or the language network (the three SLF are showed in Fig. 11.17).

The uncinate fasciculus connects the anterior temporal lobe to the orbitofrontal region and part of the inferior frontal gyrus and contributes to naming (Papagno et al., 2023; Catani et al., 2013) and face processing (Papagno et al., 2011a,b) (Fig. 11.17). The uncinate fasciculus is severely damaged in those patients with the semantic variant of Primary Progressive Aphasia (PPA)[1] (D'Anna et al., 2016; Montembeault et al., 2018; Catani et al., 2013). In addition to language and memory aspects, the uncinate is involved in several aspects of behavior, including response inhibition, outcome monitoring, object-reward association learning, and multimodal sensory integration, especially for olfactory–gustatory–visceral functions (Catani et al., 2013). Damage to this tract has been correlated with severity of antisocial traits in psychopaths (Craig et al.,

1 Primary progressive aphasia (PPA)is a neurodegenerative clinical syndrome characterized by predominance and insidious onset of language impairments and gradual deterioration of these abilities over time, associated with athrophy of the language network of the brain, including frontal, temporal, and parietal region of the left hemisphere (Mesulam, 2001; Rogalski et al., 2013; Europa et al., 2020; Tippett, 2020). PPA is classified into three variants: logopenic variant PPA(lvPPA), nonfluentagrammatic PPA (nfaPPA), and semantic variant PPA (svPPA), based on clinical (syndromic) characteristic with support from neuroimaging and/or underlying neuropathology.

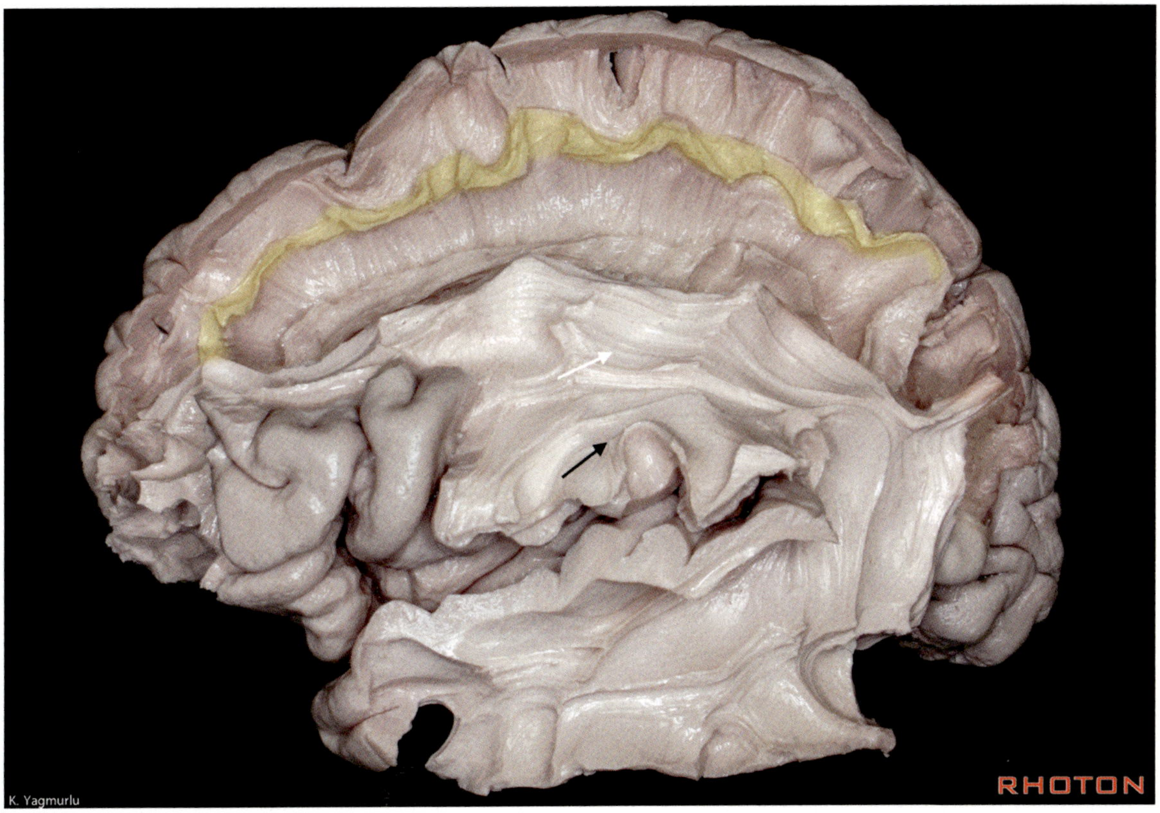

FIGURE 11.17 The **superior longitudinal fasciculus**, the principal long association fiber pathway located in the suprasylvian area, is made up of three parts: SLF I (dorsal pathway), SLF II (middle pathway), and SLF III (ventral pathway). The **SLF I (in yellow)** is positioned within the superior frontal gyrus, the **SLF II (white arrow)** within the middle frontal gyrus, and the **SLF III (black arrow)** within the inferior frontal gyrus. The SLF I lies within the superior bank of the cingulate sulcus and travels just above the cingulum to connect the superior parietal lobe and anterior cingulate cortex. The SLF II extends from the inferior parietal lobe to the middle frontal gyrus. The SLF III courses within the fronto-parietal operculum and extends from the inferior parietal lobe to the inferior frontal gyrus. The SLF is considered a high-order multisensory associative system. *Image courtesy of The Rhoton Collection® http://rhoton.ineurodb.org/. Dissector: K.Yagmurlu. Drawing by Jeffrey Sorenson.*

2009) and behavioral symptoms in frontotemporal degenerative disorders (D'Anna et al., 2016) and traumatic brain injury (Zappalà et al., 2012). The cingulum connects medial regions of the frontal (subgenual cortex, medial frontal gyrus), parietal (precuneus), occipital (cuneus, lingual, and fusiform gyri), and temporal lobes (i.e., parahippocampal gyrus, amygdala) (Vogt, 2016; Catani et al., 2013) (Fig. 11.11). Most of its fibers remain within the cingulate gyrus and connect distant regions of the cingulate cortex (Catani et al., 2002). The anterior cingulate medial prefrontal cortex and the posterior cingulate precuneus form the medial portion of the default-mode network and are interconnected through the dorsal cingulum.

In functional imaging studies, the default-mode network is active during the "resting state" a condition in which the majority of the subjects engage in an introspective, self-directed stream of thought (i.e., similar to daydreaming). A synchronous deactivation of the default network is observed in the transition between the "resting state" and the execution of goal-directed tasks, irrespective of the nature of the task (Raichle et al., 2001). The deactivation of the default-mode network has been linked to a number of functions including working memory, focusing attention to sensorially driven activities, understanding other people's intention (mentalizing or theory of mind), prospective thinking (envisioning the future), and memory for personal events (autobiographic memory) (Amodio and Frith, 2006).

The inferior fronto-occipital fasciculus provides direct connections between occipital and frontal cortex in the human brain. Early human postmortem and recent tractography findings of an inferior fronto-occipital fasciculus in the normal brain (Forkel et al., 2014a) have not been replicated in the macaque brain using axonal tracing methods (Petrides and Pandya, 2007; Schmahmann and Pandya, 2007a,b) (Fig. 11.18). This has led to the hypothesis that the inferior fronto-occipital fasciculus may be a tract unique to the human brain (Catani, 2007). The functions of the inferior fronto-occipital fasciculus remain largely unknown. The occipital areas connected by the inferior fronto-occipital fasciculus include the lingual gyrus and cuneus, which

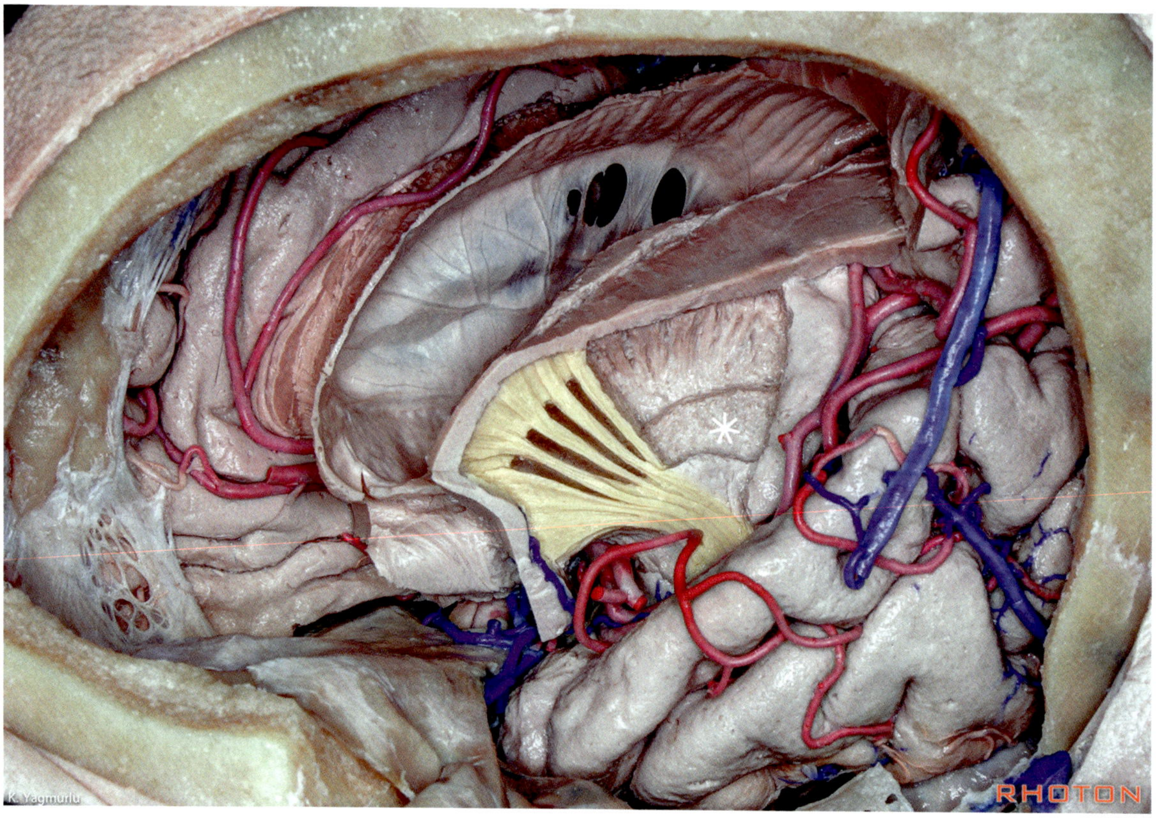

FIGURE 11.18 **Inferior fronto-occipital fasciculus** (in yellow); internal capsule medially; the white star indicates globus pallidus. *Image courtesy of The Rhoton Collection® http://rhoton.ineurodb.org/. Dissector: K.Yagmurlu. Drawing by Kaan Yagmurlu.*

engage in tasks requiring retrieving information from visual memory, spatial attention, and recognition of colors, letters, and faces. The anterior projections of the inferior fronto-occipital fasciculus terminate in the medial and lateral orbitofrontal cortex, which activates in tasks requiring top-down modulation (either activation or inhibition) of visual perception and short-term memory. Also, these frontal regions seem especially important in guiding decision-making and social behavior in relation to the emotional value of the external sensory stimuli (Rudrauf et al., 2008a,b,c). The exact role of the inferior fronto-occipital fasciculus in functions such as mental rotation (Schendan and Stern, 2008), space-directed attention (Urbanski et al., 2008), reading (Shaywitz et al., 2002), and semantic processing (Duffau et al., 2005) remains to be clarified.

A chain of fibers beneath the central sulcus connects the motor cortex of the precentral gyrus to the somatosensory cortex of the postcentral gyrus. In the human brain these U-shaped tracts are not homogeneously distributed along the length of the central sulcus. The existence of these connections in humans is consistent with previous reports supporting the role of somatosensory inputs in motor learning and precision grasping in animals (Hikosaka et al., 1985; Iriki et al., 1991). In monkeys, inactivation of S1 leads to altered finger coordination, such as the inability to oppose the thumb and forefinger and the inaccurate control of grip forces (Hikosaka et al., 1985; Brochier et al., 1999). Furthermore, experimental studies in healthy humans have demonstrated that in conditions of digital anesthesia, where tactile sensation is absent, coordination of thumb and finger movements is impaired due to misalignment of fingers and an imbalance of the pressure applied (Monz et al., 2003). These studies suggest that direct connections between S1 and M1 may play a crucial role in precision finger grasping and articulatory movements of the tongue and mouth. This conclusion is further supported by a tractography study that reported significant correlations between pegboard performance skill and the microstructural properties of white matter connections between S1 and M1 of the left-hand region in a group of healthy adults (Thompson et al., 2017).

The same study also demonstrated that in adults with ASD poor pegboard performance was associated with abnormal diffusion properties of the U-fibers of the hand region.

A system of U-shaped fibers organized around the peri-insular sulcus connects the inferior frontal cortex to the anterior insula of Reil. The anterior insula receives visceral and sensory (mainly gustatory and olfactory) information and integrates it with limbic emotional inputs. It also sends efferent to subcortical nuclei involved in alimentary functions (e.g., salivation, gastric motility) and oropharyngeal control. The frontal operculum controls orofacial movements required for nonverbal facial expression, mastication and deglutition, speech articulation, prosody, and syntactic and semantic in frontotemporal degenerative disorders (D'Anna et al., 2016) and traumatic brain injury (Zappala et al., 2012).

The cingulum connects medial regions of the frontal (subgenual cortex, medial frontal gyrus), parietal (precuneus), occipital (cuneus, lingual, and fusiform gyri), and temporal lobes (i.e., parahippocampal gyrus, amygdala) (Catani et al., 2013). Most of its fibers remain within the cingulate gyrus and connect distant regions of the cingulate cortex (Catani et al., 2002). The anterior cingulate medial prefrontal cortex and the posterior cingulate precuneus form the medial portion of the default-mode network and are interconnected through the dorsal cingulum.

The inferior fronto-occipital fasciculus provides direct connections between occipital and frontal cortex in the human brain (Forkel et al., 2014a). This fascicle could also be a unique trait of the human brain, since it appears not to be present in the macaque (Petrides and Pandya, 2007; Schmahmann and Pandya, 2007a,b) and its functions remain largely undiscovered (Catani, 2007). The occipital areas connected by the inferior fronto-occipital fasciculus include the lingual gyrus and cuneus, which engage in tasks requiring retrieving information from visual memory, spatial attention, and recognition of colors, letters, and faces. The anterior projections of the inferior fronto-occipital fasciculus terminate in the medial and lateral orbitofrontal cortex, which activates in tasks requiring top-down modulation (either activation or inhibition) of visual perception and short-term memory. Also, these frontal regions seem especially important in guiding decision-making and social behavior in relation to the emotional value of the external sensory stimuli (Rudrauf et al., 2008a,b,c). The exact role of the inferior fronto-occipital fasciculus in functions such as mental rotation (Schendan and Stern, 2008), space-directed attention (Urbanski et al., 2008), reading (Shaywitz et al., 2002), and semantic processing (Duffau et al., 2005) remains to be clarified.

A chain of fibers beneath the central sulcus connects the motor cortex of the precentral gyrus to the somatosensory cortex of the postcentral gyrus. In the human brain these U-shaped tracts are not homogeneously distributed along the length of the central sulcus.

A system of U-shaped fibers organized around the peri-insular sulcus connects the inferior frontal cortex to the anterior insula of Reil (Mesulam and Mufson, 1982; Catani et al., 2012a; Cerliani et al., 2012; Yeterian et al., 2012). The anterior insula receives visceral and sensory (especially gustatory and olfactory) information and integrates it with limbic emotional inputs. It also sends efferents to subcortical nuclei involved in salivation, peristalsis, and oropharyngeal control. The frontal operculum controls orofacial movements required for nonverbal facial expression, mastication and deglutition, speech articulation, prosody, and syntactic and semantic aspects of language (Mălîia et al., 2018). Intraoperative stimulation studies and connectivity analysis in humans suggest a topographic organization of the frontal opercular areas. It has been reported that the stimulation of the Rolandic operculum, located beneath the subcentral gyrus, generates oropharyngeal movements and somatosensory perceptual responses of the palate, inner cheeks, pharynx, tongue, inner lips and teeth. It is also interesting that stimulation of more anterior frontal operculum areas is more likely to produce speech problems described as "lack of internal word generation and the other third intrusive feelings or words that could not be ignored and disrupted the verbal sequence" (Mălîia et al., 2018).

Finally, a set of short U-shaped fibers connects the medial aspect of the superior frontal gyrus to the cingulate cortex (Catani and Dell'Acqua, 2013). These very short connections run radially to the cingulate sulcus and are responsible for the humps along the ventral margin of the superior frontal gyrus and the dorsal margin of the cingulate gyrus. Their function is unknown, but they are likely to coordinate the activity of areas distributed across the cingulate sulcus. For example, cortex responsible for pain perception has been identified in both paracentral and dorsal cingulate cortex (Trevisi et al., 2018), which are linked by an oblique cingulate U-tract (Catani and Dell'Acqua, 2013). Other cingulate U-tracts seem to be responsible for connecting areas with different functional specialization, such as the SMA and preSMA that activate, for example, in executive tasks, and dorsal cingulate cortex that activate in survival-relevant goal conflicts (Lieberman and Eisenberg, 2015).

LONG AND SHORT INTRALOBAR ASSOCIATION PATHWAYS

The frontal aslant tract (FAT) is a newly described pathway connecting the cortex of the ventral precentral gyrus and posterior inferior frontal gyrus to the pre-SMA and dorsomedial prefrontal cortex (Lawes et al., 2008; Oishi et al., 2008; Ford et al., 2010; Guevara et al., 2011). This tract is left lateralized in most right-handed subjects, suggesting a role in language (Catani et al., 2012b). Medial regions of the frontal lobe are likely to facilitate speech initiation through direct connections to the pars opercularis and triangularis of the inferior frontal gyrus. Patients with lesions to the FAT or its cortical projections present with various degrees of speech impairment, from a total inability to initiate speech (i.e., mutism) to mild altered fluency (Penfield and Roberts, 1959; Bizzi et al., 2012; Catani et al., 2013).

The first direct evidence of the role of the FAT in language was provided by tractography studies in patients with PPA; this tract is severely damaged in those patients with the nonfluent variant of the condition and its degree of degeneration correlates with the performance in spontaneous speech tasks (Catani et al., 2013; Mandelli et al., 2014). Later studies confirmed an association between the FAT and impaired speech production after stroke (Li et al., 2017) and tumor resection (Chernoff et al., 2018) and in patients with developmental (Kronfeld-Duenias et al., 2016a,b; Neef et al., 2018a,b) and intraoperative (Kemerdere et al., 2016) stuttering. Other studies indicate nonlinguistic functions of the FAT, such as response inhibition control (Bonnelle et al., 2012), executive functions (Dick et al., 2019), and visually guided hand movements (Budisavljevic et al., 2017).

In summary, connections between preSMA and IFG have an important role in voluntary sequential motor planning for speech (Dick et al., 2019) and actions (Budisavljevic et al., 2017). Comparative anatomy studies demonstrated the existence of the FAT in monkeys without significant evolutionary changes across primate species (Barrett et al., 2020). This indicates that the connections between preSMA and IFG are highly preserved in primate brains, which share a common neural substrate for vocalization and gestural communicative intention (Catani and Bambini, 2014). In humans the convergence of FAT and the arcuate fasciculus in the posterior IFG has facilitated the emergence of complex speech characteristic of our species.

The frontal orbitopolar tract is the main associative pathway between posterior orbitofrontal cortex and anterior orbitofrontal and polar regions. Similar connections have been described in the monkey brain between areas 10 and 13 (Price, 2007; Thiebaut de Schotten et al., 2012; Yeterian et al., 2012). The posterior orbital gyrus receives inputs from the limbic regions (i.e., amygdala, hippocampus, nucleus basalis of Meynert, olfactory cortex, and insula) and plays an important role in processing olfactory and gustatory inputs and integration of emotions and memories associated with sensory experiences.

The anterior orbitofrontal cortex receives direct auditory and visual inputs from posterior occipital and temporal cortex through the inferior fronto-occipital and uncinate fasciculus (Rolls, 2002; Price, 2007; Yeterian et al., 2012).

The frontal superior longitudinal and frontal inferior longitudinal tracts are two parallel pathways connecting motor, premotor, and prefrontal regions (Thiebaut de Schotten et al., 2012). These two sets of fibers are likely to play a role in integrating the activity of all local frontal networks in the precentral, premotor, and prefrontal regions, including, for example, the coordination between movement planning, motor execution (carried out by the motor and premotor networks), and the overall goal-directed strategy supervised by the prefrontal networks (Catani et al., 2012b). These networks are responsible for recruiting distributed dorsolateral premotor and prefrontal areas during tasks requiring preforming simultaneous tasks (D'Esposito et al., 1995), switching from one task to another (Dreher and Grafman, 2003), or computing hierarchical structures (Jeon and Friederici, 2016).

ELECTRIC ACTIVITY OF THE CORTEX AND ITS AGE-DEPENDENT ASSESSMENT

The complete development of the cerebral cortex is due not only to obtaining the six layers of the neocortex, the latter in fact needs to be "electrically balanced." What does it mean? We know that brain development begins in the first stage of embryonic life and is complete in most cases by the age of birth, even if the complete development happens around 18 years old (Budday et al., 2015; Guarnieri et al., 2018). However, in the last period of fetal life we witness several neurobiological processes that promote connectivity and synaptogenesis between the different classes of neurons and neuroglia that make up the cortex and all this causes the cerebral cortex to acquire an increasingly mature electrical activity. We recall that neurons communicate through electrical signals which depend on many factors, first of all the activity of neurotransmitters. We saw in Chapter 3 the development and maturation of the various neurotransmitter systems and we also saw first of all that GABA has a completely different activity in the brain before birth in fact, in this period, GABA acts as excitatory neurotransmitter due high of Cl^- ions in the neurons due to overexpression of its internal transporter NKCC1 and low expression of the external Cl^- transporter KCC2 (Ben-Ari et al., 2012). GABA has been shown to remain excitatory during the first 3-4 postnatal weeks (studies performed on rodents) perhaps even longer in humans and this, in turn, plays a particularly important role in the etiopathogenesis of epileptic syndromes and autism (see later in this chapter). Ascenzi et al. (2017) has furthermore emphasized the role of the GABA, glycine, and taurine during the corticogenesis. He particularly highlighted how these inhibitory neurotransmitters are able to promote the migration and differentiation of neurons progenitors toward the pallium by activation of downstream signal pathways at the base of several genes' expression. The electrical activity measured by the EEG recording goes toward a maturation that gets more intense in the last months before the birth and in the first months/years after it. However, the complete settling of the cerebral electrical activity, and therefore of EEG structure, occurs around the age of 22–25 (Matsuura et al., 1985). In the following pages, we will analyze age-dependent EEG patterns, starting from prenatal age and progressing to adulthood. It is important to note that EEG is a highly age-specific and subjective diagnostic tool, and there is no universally defined standard for what is considered normal. This emphasizes the need for electroencephalographic tracings to be interpreted by a skilled neurophysiopathologist, as they possess the expertise required to accurately analyze the results.

Before moving on to discussing the topic introduced, it seems necessary to say a few words about the functioning and the main parameters of electroencephalography. Electroencephalography (EEG) is a method to record the spontaneous electrical activity of the brain. The signals detected by EEG have been shown to represent the postsynaptic potential of pyramidal neurons in the neocortex and allocortex (Amzica et al., 2017) It is typically noninvasive, with the EEG electrodes placed along the scalp using the International 10–20 system or variations of it. Voltage fluctuations measured by the EEG bioamplifer and electrodes allow the evaluation of normal brain activity. As the electrical activity monitored by EEG originates in neurons in the underlying brain tissue, the recordings made by the electrodes on the surface of the scalp vary in accordance with their orientation and distance to the source of the activity. Furthermore, the value recorded is distorted by intermediary tissues and bones, which act in a manner akin to resistors and capacitors in an electrical circuit. This means not all neurons will contribute equally to an EEG signal, with an EEG predominately reflecting the activity of cortical neurons near the electrodes on the scalp. Deep structures within the brain further away from the electrodes will not contribute directly to an EEG; these include the base of the corticali gyrus, mesial walls of the major lobes, hippocampus, thalamus and brainstem (Kandel et al., 2021).

A healthy human EEG will show certain patterns of activity that correlate with how awake a person is. The range of frequencies one observes is between 1 and 30 Hz, and amplitudes will vary between 20 and 100 μV. The observed frequencies are subdivided into various groups: alpha (8–13 Hz), beta (13–30 Hz), theta (4–7 Hz), and delta (0.5–4 Hz). Alpha waves are observed when a person is in a state of relaxed wakefulness and are mostly prominent over the parietal and occipital sites. During intense mental activity, beta waves are more prominent in frontal areas as well as other regions. If a relaxed person is told to open their eyes, one observes alpha activity decreasing and an increase in beta activity; this is happening physiologically. Theta and delta waves are not seen in wakefulness and, if they are, it is a sign of brain dysfunction (Kandel et al., 2021).

EEG General Components of Organization

The most important parameters of an EEG tracing are the organization of the latter concerning to how the waveforms appear across the entirety of the page and include continuity, symmetry, and the anterior posterior gradient.

The *continuity* refers to the waveforms being uninterrupted by periods of flat or very attenuated activity. Healthy children and adults should always have a continuous record, but early neonatal tracings can show periods of discontinuity (this is discussed later).

The *symmetry refers,* in which both the left and right sides appear, largely, the same in terms of both amplitude and frequency. Healthy EEGs should always be symmetric, and intermittent or persistent asymmetries can arise from structural entities such as tumors or bleeds. In the end we find, the *anterior posterior gradient,* in which faster, lower amplitude frequencies are present toward the front of the brain while slower, higher amplitude frequencies are found in the back of the brain. The AP gradient leads into the last component of organization, the posterior dominant rhythm (PDR), which is represented essentially by alpha rhythm.

ELECTROENCEPHALOGRAPHIC FEATURES RELATED TO THE BRAIN DEVELOPMENT

The neonatal EEG contains complex spatiotemporal information that can be difficult to interpret. The precise description of such immature activity and the recognition of normal and pathological patterns have major implications in clinical practice for the diagnosis of cerebral lesions and prognosis of neurological outcome in this highly vulnerable population (André et al., 2010). EEG is certainly more complex than the interpretation of other vital sign signals such as heart rate or respiratory rate. For appropriate assessment of the preterm EEG, multichannel (10–20) conventional EEG recording is essential (Kato et al., 2011; Lloyd et al., 2015; de Vries et al., 2007). Accurate neonatal EEG interpretation requires a thorough appreciation of the appropriate maturational features for neonates of all gestational ages (GA) as baseline EEG patterns evolve in line with the rapid maturational changes taking place in the brain (Wallois, 2010).

The EEG develops in the most immature neonates at 23/24 weeks GA through to full-term age with four majors trends: (i) increasing continuity, with defined periods of EEG quiescence for specific GAs, (ii) the appearance of sleep cycling, (iii) changes in synchrony between hemispheres, (iv) and the appearance of several transient waveforms of prematurity. Assessment of an infant's EEG against these four parameters can indicate whether the maturity of the brain is appropriate for GA.

DISCONTINUOUS VERSUS CONTINUOUS EEG ACTIVITY

In preterm infants the baseline EEG is discontinuous. This discontinuous pattern is characterized by bursts of high amplitude delta–theta activity (sometimes superimposed with faster activity) intermixed with periods of quiescence (interburst intervals—IBIs) (Fig. 11.19). From a physiological perspective, this feature can be explained from animal experiments that have shown that the cortex produces spontaneous and intermittent activity (Khazipov et al., 2006). That is a crucial endogenous driver for the development of brain connectivity before cortical networks are modulated by the exogenous stimuli\sensory input (Colonnese and Khazipov, 2012; Omidvarnia et al., 2014). Furthermore early in development, as mentioned before, GABAergic transmission is not effectively inhibitory and may allow the generation of these endogenous events (Ben-Ari et al., 2012; Vanhatalo and Kaila, 2006).

The activity of early brain networks is characterized by two alternating modes of activity: spontaneous activity transients (SATs or bursts; Fig. 11.20) (Vanhatalo et al., 2005) and the periods of relative inactivity between them (IBIs) (Omidvarnia et al., 2014; Vanhatalo et al., 2005). The duration of IBIs, characterized by low voltage activity, is one of the first parameters generally evaluated, as it has been demonstrated that they reflect brain maturation, being associated with the development of cortical folding (Biagioni et al., 2007). IBIs duration decreases with GA (Niemarkt et al., 2010). With the maturation of normal inhibitory GABAergic transmission, spontaneous events are gradually abolished and "continuous" oscillations emerge at different frequencies due to the increasing influence of exogenous sensory driven input (Vanhatalo and Kaila, 2006). Consequently, the overall amount of discontinuity decreases and continuity increases with GA (Niemarkt et al., 2010).

As sleep cycling evolves (see below), discontinuity becomes primarily a feature of quiet sleep (QS). For example, at 35–36 weeks GA, the EEG should be continuous in active sleep (AS) and waking and discontinuity only occurs in quiet sleep (QS) (Fig. 11.21). Aside from the duration of IBIs, their amplitude also changes over time, becoming less suppressed with increasing GAs.

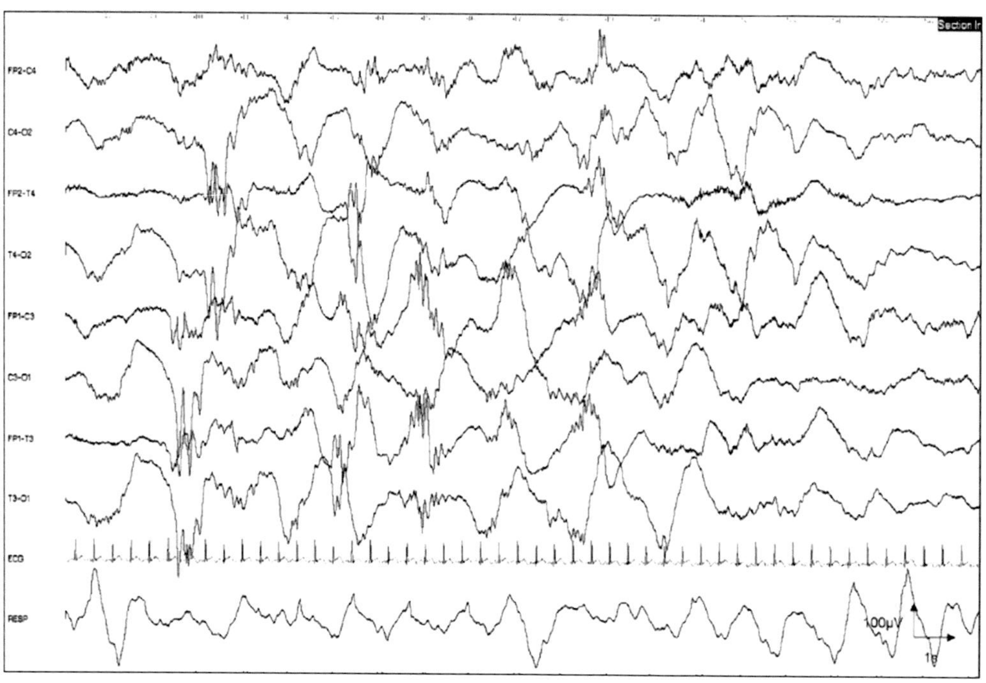

FIGURE 11.19 **EEG in 27 weeks GA.** At this stage sequences of **high-amplitude multifocal delta waves (as in the figure above) are intermixed to periods of quiescence (interburst intervals–IBIs).** The acceptable IBI shortens over time, being up to 60 s at 24 weeks GA, 40 s at 26 weeks, 20 s at 28 weeks, 10 s from 34 to 36 weeks, and 6 s from 37 to 40 weeks; note that these values are generally on the upper limit of normal. After 40 weeks, you can still see some discontinuity in sleep but **by 46 weeks at the most, the tracing should be completely continuous** during both sleep and awake states. *Image courtesy of Levin, K.H., Chauvel, P. (Eds.), 2019. Handbook of Clinical Neurology 3rd Series. Clinical Neurophysiology: Basis and Technical Aspects, vol. 160. Elsevier, pp. v–vi. https://doi.org/10.1016/B978-0-444-64032-1.09997-5.*

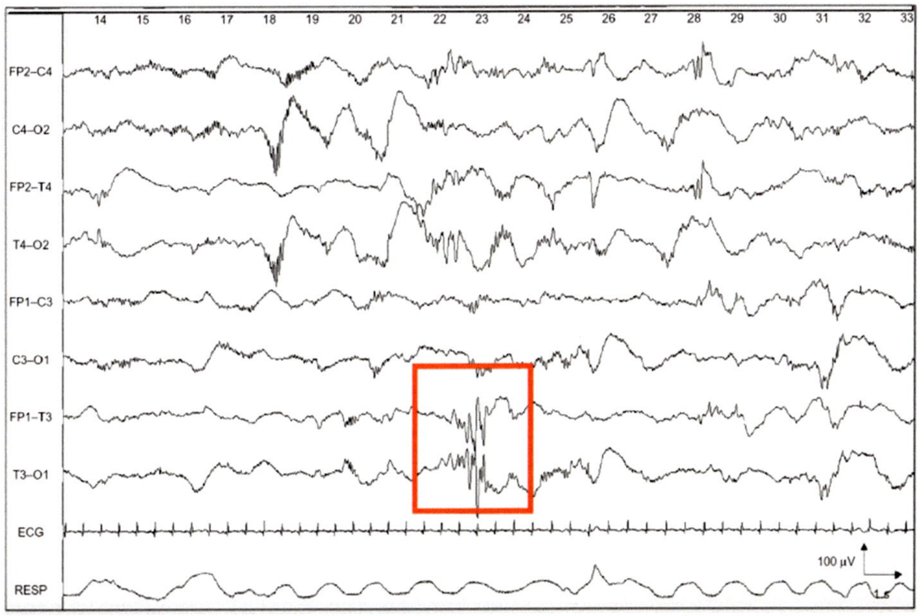

FIGURE 11.20 **Active sleep in 32 weeks GA neonate.** Occipital delta brushes on the right. Burst of theta sharp waves in left temporal area. *Image courtesy of Levin, K.H., Chauvel, P. (Eds.), 2019. Handbook of Clinical Neurology 3rd Series. Clinical Neurophysiology: Basis and Technical Aspects, vol. 160. Elsevier, pp. v–vi. https://doi.org/10.1016/B978-0-444-64032-1.09997-5.*

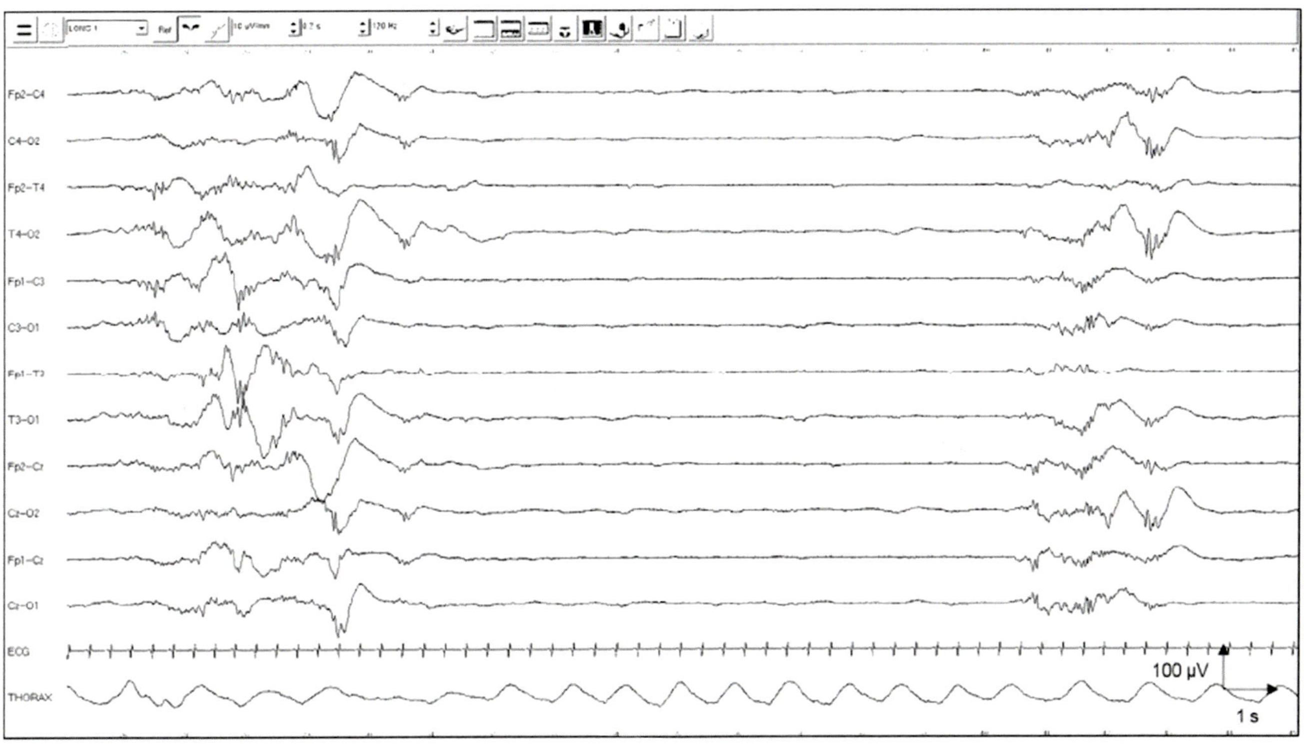

FIGURE 11.21 **Discontinuous EEG in quiet sleep in 34 weeks GA neonate.** *Image courtesy of Levin, K.H., Chauvel, P. (Eds.), 2019. Handbook of Clinical Neurology 3rd Series. Clinical Neurophysiology: Basis and Technical Aspects, vol. 160. Elsevier, pp. v–vi. https://doi.org/10.1016/B978-0-444-64032-1.09997-5.*

Finally, the morphology of burst activity also changes with increasing GAs. Moreover, the amplitude of theta/delta activity bursts decreases with maturity (particularly in the fronto-temporal regions) and superimposed fast activity appears and increases with GA (Wallois, 2010; Jennekens et al., 2012).

Clear agreement between the EEG state and the behavioral state is recognizable after 30 weeks GA;

therefore, in the older literature, it has been reported that proper sleep–wake cycling is clearly visible only after 30 weeks GA (Andre et al., 2010) and that younger infants are in an indeterminate sleep state (Andre et al., 2010). Sleep–wake cycling relies on the maturation of interconnected neural networks located throughout the cortex, diencephalon, and brainstem and is recognizable in younger preterm infants because of the influence of deeper brain structures, before proper thalamo-cortical connectivity has developed (Palmu et al., 2013; Scher and Loparo, 2009).

Synchrony in the EEG is present when all EEG features occur simultaneously in homologous areas over both hemispheres. Although interhemispheric synchrony has been shown to increase with increasing GA (Andre et al., 2010; Scher et al., 1996; Koolen et al., 2014), synchronous bursts/IBI activity between the two hemispheres (paradoxical hypersynchrony) is present in preterm infants <30 weeks GA (Scher et al., 1996). Indeed, high-amplitude bursts of activity are synchronous as early as 24 weeks GA, with 88% of bursts being synchronous between 24 and 27 weeks and almost 100% between 28- and 29-weeks GA (Vecchierini et al., 2003, 2007). High-amplitude delta wave bursts are often synchronous in the occipital areas as early as GA 26–27 weeks (Andre et al., 2010). The high amplitude and synchronization of EEG activity in very early preterm infants might be related to the fact that the cortex is still not fully developed, subcortical (thalamic) control predominates and gradual synaptogenesis of the corpus callosum is ongoing (Vecchierini et al., 2007; Scher et al., 1996; Koolen et al., 2014).

Asynchronous activity after 30 weeks is physiological if seen in <50% of the discontinuous EEG periods (Scher et al., 1996). It persists, with decreasing frequency until 36 weeks, when it is mainly seen at the beginning of QS (Andre et al., 2010, Scher et al., 1996); thereafter asynchrony decreases, disappearing at term age (Scher et al., 1996). Synchrony is an important feature of EEG maturation and reflects the development of the corpus callosum and, therefore, the interconnections between the two hemispheres (Scher et al., 1996; Koolen et al., 2014). However, the visual assessment of synchrony is difficult in the preterm EEG and ambiguity does exist (Koolen et al., 2014; Räsänen et al., 2013).

DELTA ACTIVITY (0–3.5 HZ)

High Amplitude, Diffuse Delta

In earlier GAs (particularly <28 weeks), a common background feature is high amplitude (up to >300 μV), low frequency (0.3–1 Hz) mono- or diphasic, smooth activity (with superimposed faster activity on centro-temporal regions) organized as unilateral or bilateral bursts in centro-occipital or temporal regions (less on frontal), or as short sequences (<80 s) mainly in the occipital regions bilaterally (Andre et al., 2010; Vecchierini et al., 2003, 2007).

Delta brushes are one of the most important features of the preterm EEG. Delta brushes consist of a slow delta wave with fast rhythms superimposed (in the alpha-beta range) mainly on the ascending slope of the slow wave (Andre et al., 2010; Watanabe et al., 1999; Whitehead et al., 2017). They have been reported in all GAs but have a peak between 32 and 35 weeks and tend to disappear between 38 and 42 weeks (Andre et al., 2010; Whitehead et al., 2017) (Fig. 11.22). Their amplitude decreases with maturation and the frequency of both the slow component and the superimposed fast rhythms become faster up to term age (Andre et al., 2010; Vecchierini et al., 2007; Watanabe et al., 1999; Whitehead et al., 2017). Delta brushes are initially seen diffusely over cortical areas, but subsequently become more predominant over central and temporal-occipital regions and finally in the occipital region only (around 36 weeks; Fig. 11.22) (Andre et al., 2010; Vecchierini et al., 2007).

Delta brushes play a central role in early topographic mapping of the primary sensory areas. For example, spontaneous waves of activity in the retina drive cortical delta brushes and mapping of the visual cortex before vision are established (Meister et al., 1991). Similarly, in the somatosensory cortex, spontaneously generated twitches and body movements in immature rats (Khazipov et al., 2004) and human preterm infants (Milh et al., 2007) produce somatosensory feedback, eliciting delta brushes and organize the broad layout of the cortical somatosensory topographic body map, necessary for sensorimotor coordination. Different stimuli evoke delta brushes in the relevant cortical regions: visual inputs evoke delta brushes in occipital regions (Colonnese et al., 2010), auditory inputs in temporal regions (Chipaux et al., 2013), tactile stimuli to hand and foot in lateral and medial regions of the contralateral central cortex, respectively (Milh et al., 2007) and noxious stimuli to the heel can elicit delta brushes in the midtemporal regions (Fabrizi et al., 2011a,b). The somatosensory cortex develops extensively in the fetus both in human\non-human primates and development of delta brushes from 6-month gestation reflects this process (Milh et al., 2007). With the gradual disappearance of delta brushes in general and the progression of background activity from discontinuous to continuous, the sensory elicited delta brushes tend to disappear (around 35 weeks GA) and are replaced by mature evoked potential responses (Colonnese et al., 2010; Fabrizi et al., 2011a,b).

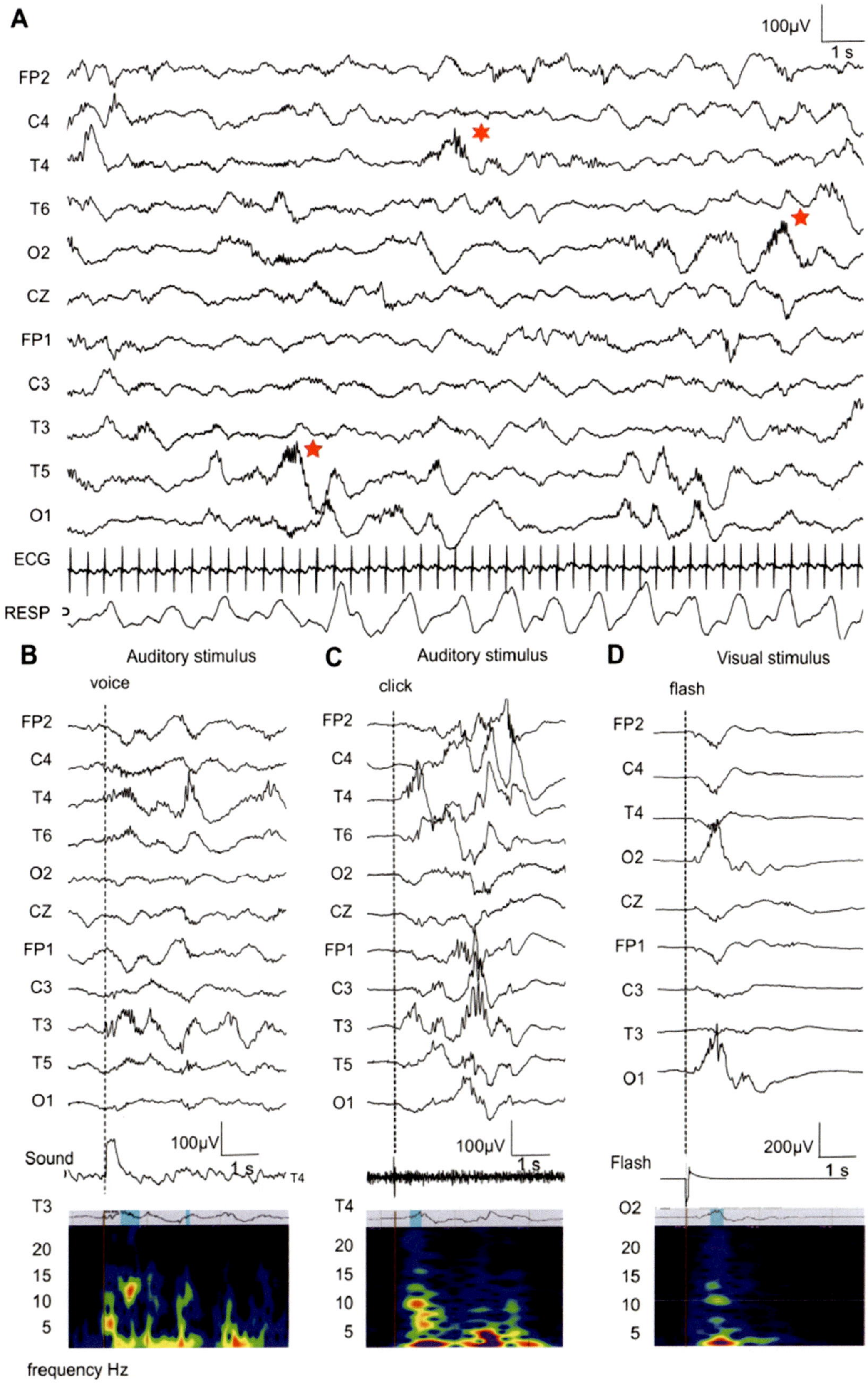

FIGURE 11.22 **Auditory evoked delta-brushes by "click" and voice in temporal cortex.** (A) Electroencephalogram (EEG) in a 31 post-menstrual week (PMW; or week gestational age –GA–) premature infant during active sleep (AS) showing continuous activity with diffuse slow

Overview of Delta Activity Changes

Delta activity is a major characteristic of the preterm EEG that evolves as the infant matures. Several authors have reported that the amplitude of delta waves decreases with GA (Andre et al., 2010; Vecchierini et al., 2007; Watanabe et al., 1999), particularly those ≤1 Hz over occipital areas (Schumacher et al., 2013). Some features about delta activity changes are described below.

Morphology: from smooth or with superimposed theta-alpha rhythms to delta brushes (with superimposed faster rhythms -alpha/beta-).

Spatial organization: from diffuse, to centro-temporal to occipito-temporal to mainly occipital.

Amplitude: it decreases with GA from >300 μV to a range of 50–100 μV at term age. *Frequency*: from the lowest delta range (0–1 Hz) to the higher delta range (2–3.5 Hz).

THETA ACTIVITY (4–7.5 HZ)

Occipital Theta Activity

STOPS is the acronym used for "Sharp theta on the occipitals of premature infants." They are similar to premature temporal theta (PTT), but are faster in frequency, lower in amplitude, more often unilateral, aside from the obvious difference in location. The incidence is higher in the younger ages, with a peak at 25 weeks (Hughes et al., 1990) (Fig. 11.23A).

Occipital Sawtooth

Regular rhythmic high amplitude 4–7 Hz activities that can occur in isolation in the occipital region or spread temporally and last 0.5–3.5 s. The morphology is typically sinusoidal (Biagioni et al., 2000) (indicated with an arrow in Fig. 11.23B).

Temporal Theta Activity

Premature Temporal Theta (PTT) or "Temporal Sawtooth"

It consists of runs of rhythmic theta activity of 4.5–6 Hz, usually bilateral, often asynchronous. Its peak occurrence is at 29–31 weeks GA and this activity disappears in AS at 32 weeks GA and in quiet sleep (QS) at 33–34 weeks GA (Andre et al., 2010; Hughes et al., 1987) (Fig. 11.23C).

Overview of Theta Activity Changes

From the 24 to 28 weeks GA, the *spatial organization* of theta activity is usually diffuse or localized in the occipital regions, then becomes more predominant in the temporal regions between 28 and 30 weeks (Vecchierini et al., 2007). Therefore, premature occipital theta activity is prominent in infants with GA <28 weeks, whereas PTT is prominent between 28 and 30 weeks (Fig. 11.23) (Biagioni et al., 2007). These EEG changes are also associated with changes in cortical folding development, which starts earlier in the occipital lobes and proceeds later to the temporal lobes (Biagioni et al., 2007). The *amplitude* of theta activity decreases with GA, being therefore lower for PTT during the peak at 31 GA (Andre et al., 2010).

ANTERIOR TRANSIENTS

Slow Anterior Dysrhythmia

This feature is characterized by short sequences of delta waves with medium-high amplitude (50–100 μV), in frontal regions. It appears in AS 1, around 36 weeks GA (Andre et al., 2010) (Fig. 11.24A).

Encoches Frontales

These are EEG transients that can be seen from 33 to 34 weeks. Immature frontal transients may be repetitive, often smooth, incomplete, and asymmetrical (unilateral or with difference in amplitude/frequency between the two hemispheres). Mature *encoches frontales* are seen from 35-week GA to term and are diphasic with a small negative deflection followed by a wider and higher amplitude positive deflection. Usually synchronous, they can have a sharp morphology and present a medium-high amplitude (up to 200 μV) (Andre et al., 2010) (Fig. 11.24B). Some authors suggest that these transients are representative of healthy maturation (Andre et al., 2010).

waves and multifocal delta-brushes (red stars). Traces are shown with mean reference montage (high-pass filter 0.53 Hz, notch filter 50 Hz, electrodes placement using 10/20 international system). *ECG*, electrocardiogram; *RESP*, respiration. (B) Example of temporal delta-brushes evoked by single auditory voice stimulus, recorded during quiet sleep (QS) at 31 PMW. Corresponding wavelet analyses of T3-reference derivation is shown below (delta-brushes are outlined in blue on the T3-reference trace above). (C) Example of temporal delta-brushes evoked by single auditory "click" stimulus recorded during QS at 31 PMW and corresponding wavelet analyses on T4-reference derivation. (D) Representative example of occipital delta-brushes evoked by single visual flash stimulus, recorded during QS at 32 PMW with corresponding wavelet analysis on O2-reference derivation. *Image courtesy of Chipaux, M., Colonnese, M.T., Mauguen, A., Fellous, L., Mokhtari, M., et al., 2013. Auditory stimuli mimicking ambient sounds drive temporal "delta-brushes" in premature infants. PLoS One 8 (11), e79028. https://doi.org/10.1371/journal.pone.0079028.*

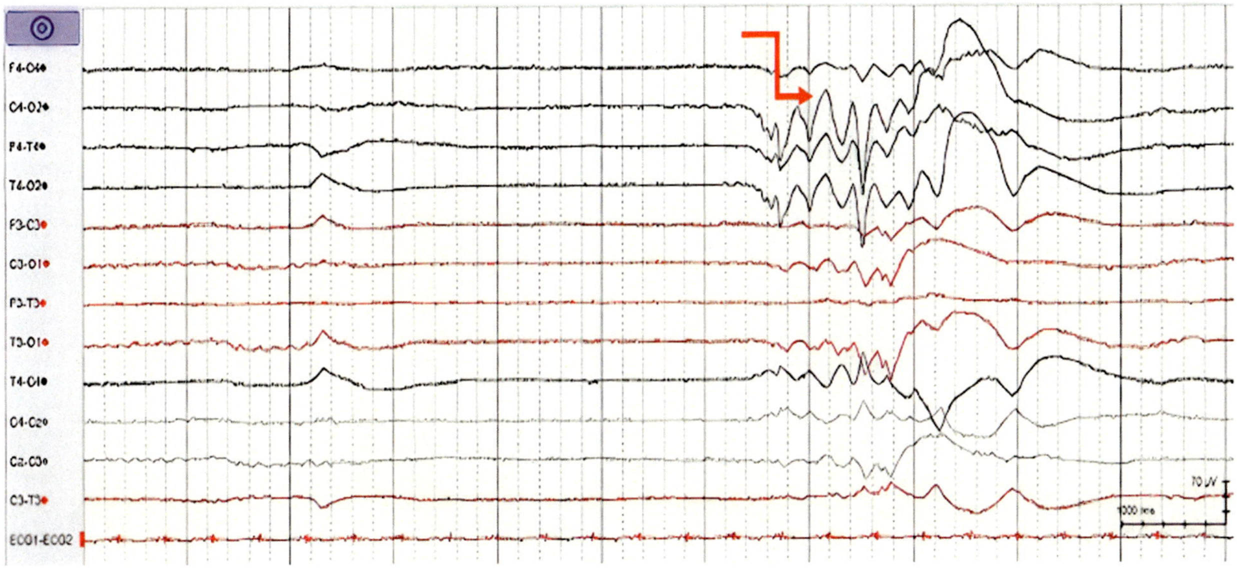

(A)

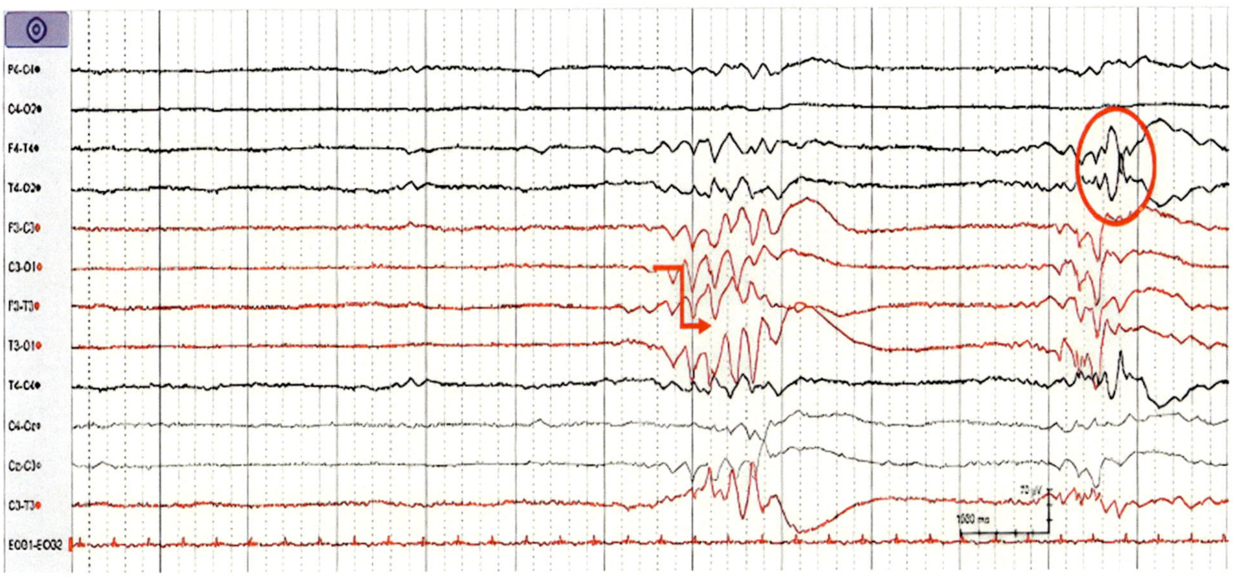

(B)

FIGURE 11.23 (A) Infant 24 + 5 weeks GA: **Theta activity (STOPS)** prominent in occipital regions (arrow), (B) infant 26 weeks GA: Theta activity in centro-temporo-occipital regions (STOPS -arrow- and PTT.-circle-); (C) infant 28 weeks GA: **PTT in temporal regions (circles)** (the time-point of the EEG is indicated with a line on the aEEG; Montage on top right; sensitivity: 70 µV; high cut filter: 70 Hz; low cut filter: 0.500 Hz; time base: 20 mm/s). *Image courtesy of Pavlidis, E., Lloyd, R.O., Mathieson, S., Boylan, G.B., 2017. A review of important electroencephalogram features for the assessment of brain maturation in premature infants. Acta Paediatr. 106 (9):1394–1408. https://doi.org/10.1111/apa.13956. Adapted and modified with permission.*

EEG in the Full-Term Newborn

As in the 36-week GA, the full-term newborn after 38 week GA, falls asleep in AS1, followed by QS and AS2. In contrast to adults who spend only 20% in REM sleep, in the newborn AS represents 50% of the sleep time with a mean sleep cycle duration (including both AS and QS) of approximately 70 min (Curzi-Dascalova et al., 1988).

The background activity is symmetrical and synchronous. Only during the transition from AS to QS can temporary asynchrony be observed for up to several minutes without any pathological significance. Background activity in wake (W) and AS consists of "mixed

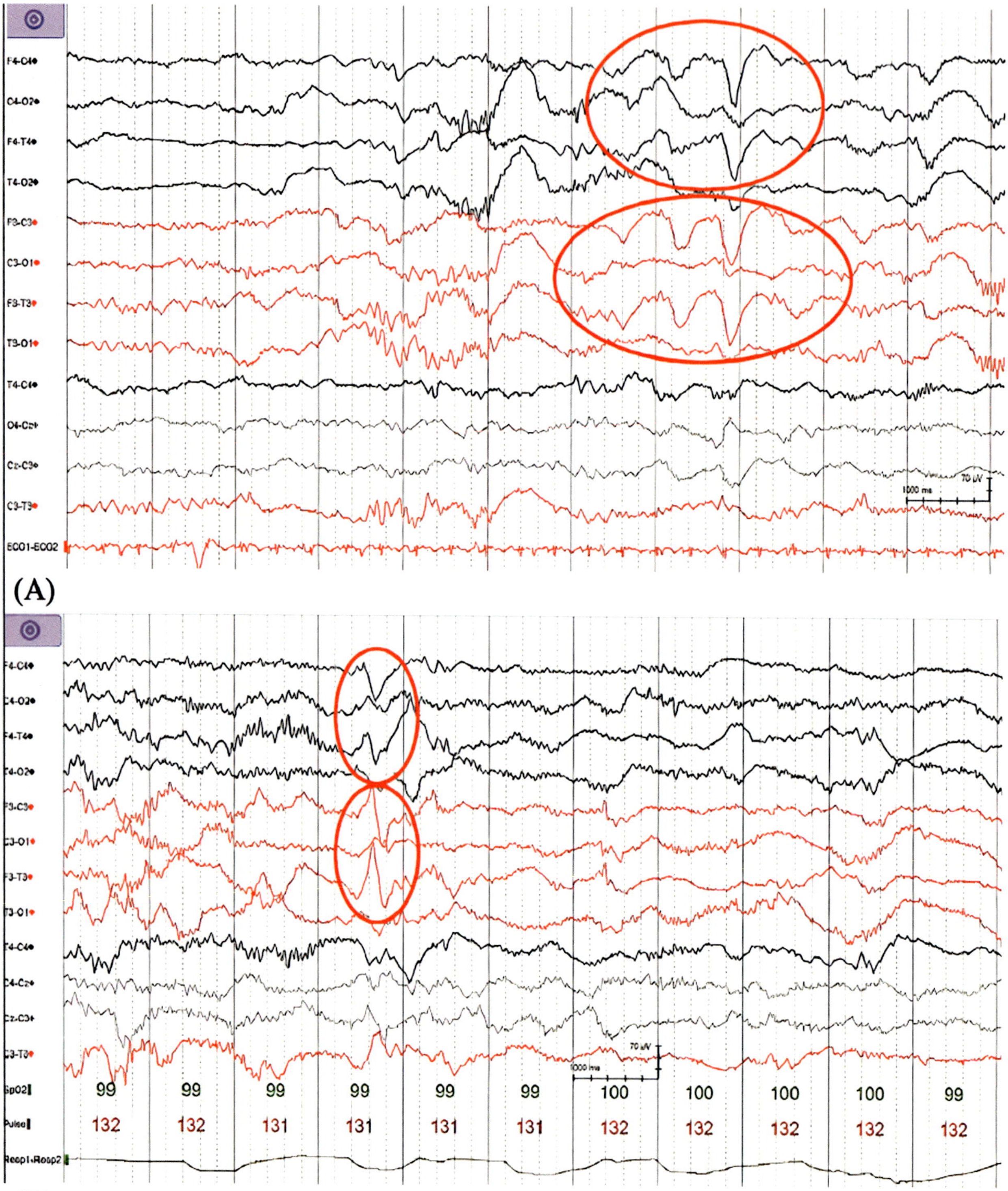

FIGURE 11.24 (A) Infant 35 weeks GA-active sleep: **Slow anterior dysrhythmia,** brief sequence of rhythmic delta activity on both frontal regions (circles); (B) infant 35 + 5 weeks GA—active sleep: **Encoches frontales** on both frontal regions (circles) (the time-point of the EEG is indicated with a line on the aEEG; montage on top right; sensitivity: 70 μV; high cut filter: 70 Hz; low cut filter: 0.500 Hz; time base: 20 mm/s). *Image courtesy of Pavlidis, E., Lloyd, R.O., Mathieson, S., Boylan, G.B., 2017. A review of important electroencephalogram features for the assessment of brain maturation in premature infants. Acta Paediatr. 106 (9):1394–1408. https://doi.org/10.1111/apa.13956. Adapted and modified with permission.*

frequencies," with continuous, irregular, diffuse activity, mainly in the theta (4–7 Hz) band, sometimes mixed with occipital delta (25–50 mV) waves with a Rolandic predominance (Fig. 11.25); frontal transients may occur isolated or in runs of a few seconds; however, they may be absent in strictly normal babies. Anterior slow dysrhythmia, which appears in the newborn from 36 to 44 GA, is mostly present in A1 (André, 2010).

In QS, the background activity is initially continuous: "slow continuous high-voltage activity" (diffuse delta waves delta waves of 1–3 Hz with central or occipital predominance and variable amplitude 50–150 μV).

This slow continuous high-voltage activity is followed by the *tracé alternant*.

It consists of bilateral bursts of delta waves (1–3 Hz) of high amplitude (50–150 μV) mixed with continuous theta (4–7 Hz) activity of lower amplitude (25–50 μV). Bursts last 3–8 s and alternate with low amplitude theta activity of mostly the same duration (Fig. 11.26); this pattern may present high interindividual variability. Bursts also contain frontal transients that are usually more numerous and of higher amplitude than in W and AS.

Unusual but still normal variant patterns can also be observed in the newborn, such as bursts of alpha and theta waves of short duration (<5 s) and variable amplitude, located in rolandic or fronto-rolandic regions mostly in AS1, but also in QS; positive temporal sharp waves that can be isolated or occur in short bursts, lasting less than 400 ms; and rudimentary sleep spindles or prespindles.

The latter are diffuse low amplitude fast activity predominating in frontal or rolandic regions that may be seen in full-term newborns. Their amplitude is much lower, and they occur in longer runs compared to "mature" sleep spindles. Frontal sharp transients (neonatal EEG pattern) and sleep spindles (infantile EEG pattern) can coexist (Kaminska et al., 2019).

EEG in Infancy Between the 1st and 12th Month of Life

The first 3 months of life are characterized by gradual transition from neonatal to infantile EEG patterns, which include gross adult-like outlines. Awake, the mixed frequencies activity present in the neonatal period is gradually replaced by more rhythmic theta waves with increasing frequency, ranging from 3 to 4 Hz at 3 months, to 5 Hz at 5 months, and 6 to 7 Hz by the end of the first year of life (Fig. 11.27).

These rhythms, which precede the occipital alpha rhythm, are initially located in rolandic-occipital regions and can reach an amplitude of 75 μV. A visual blocking response is usually present at the age of 3–4 months. The transition from weakness to sleep is characterized by progressive slowing into the delta frequency range. From the age of 6–8 months on, a strong rhythmicity predominating in centro-parietal or occipital regions is usually observed during drowsiness. At 9 months of age, but sometimes it begins to be recorded

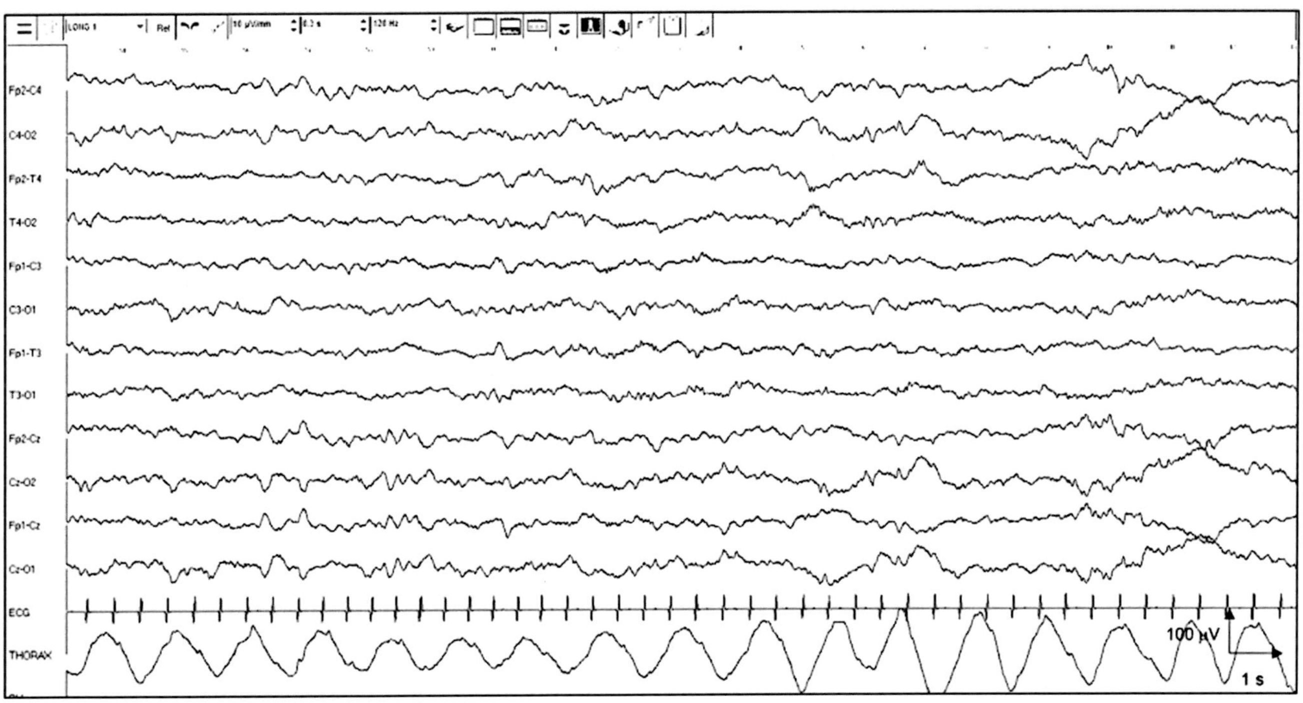

FIGURE 11.25 **Full-term newborn. Quiet wakefulness. Continuous irregular theta activity with some occipital delta waves.** *Image courtesy of Levin, K.H., Chauvel, P. (Eds.), 2019. Handbook of Clinical Neurology 3rd Series. Clinical Neurophysiology: Basis and Technical Aspects, vol. 160, Elsevier, pp. v–vi. https://doi.org/10.1016/B978-0-444-64032-1.09997-5.*

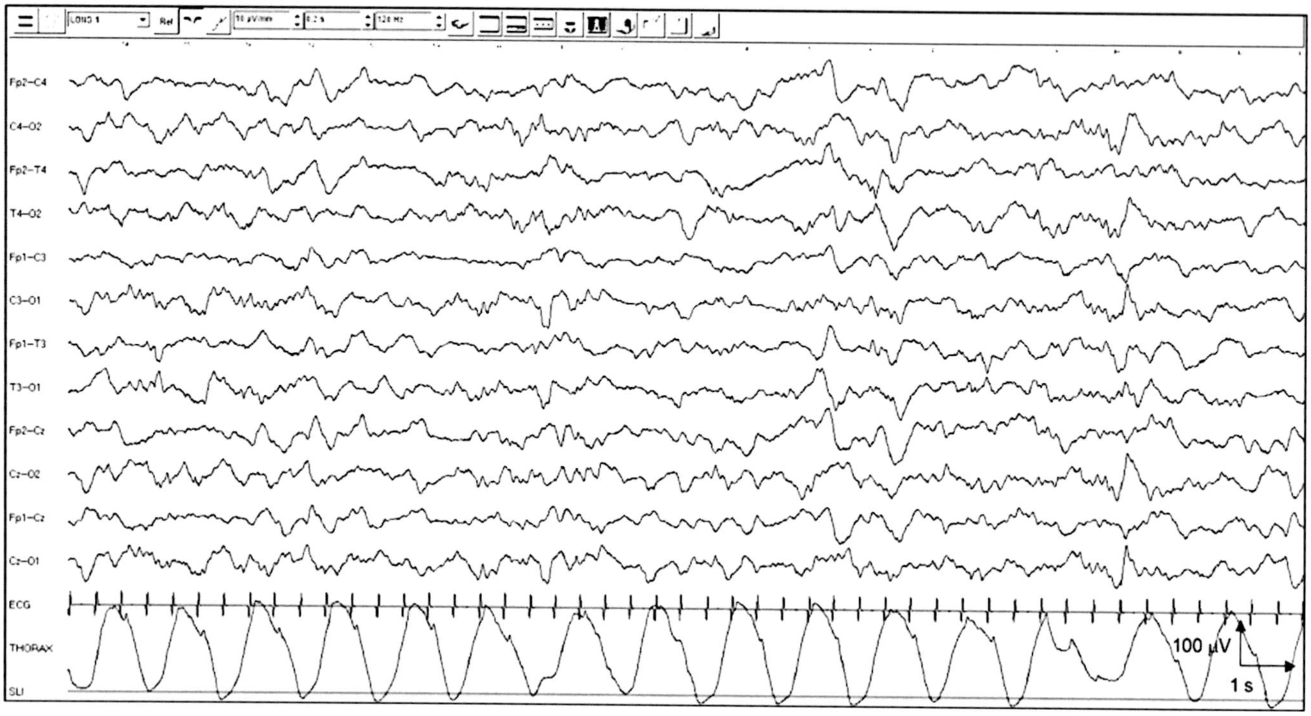

FIGURE 11.26 **Full-term newborn. Quiet sleep. *Tracé alternant* with high-amplitude frontal sharp transients.** *Image courtesy of Levin, K.H., Chauvel, P. (Eds.), 2019. Handbook of Clinical Neurology 3rd Series. Clinical Neurophysiology: Basis and Technical Aspects, vol. 160. Elsevier, pp. v–vi. https://doi.org/10.1016/B978-0-444-64032-1.09997-5.*

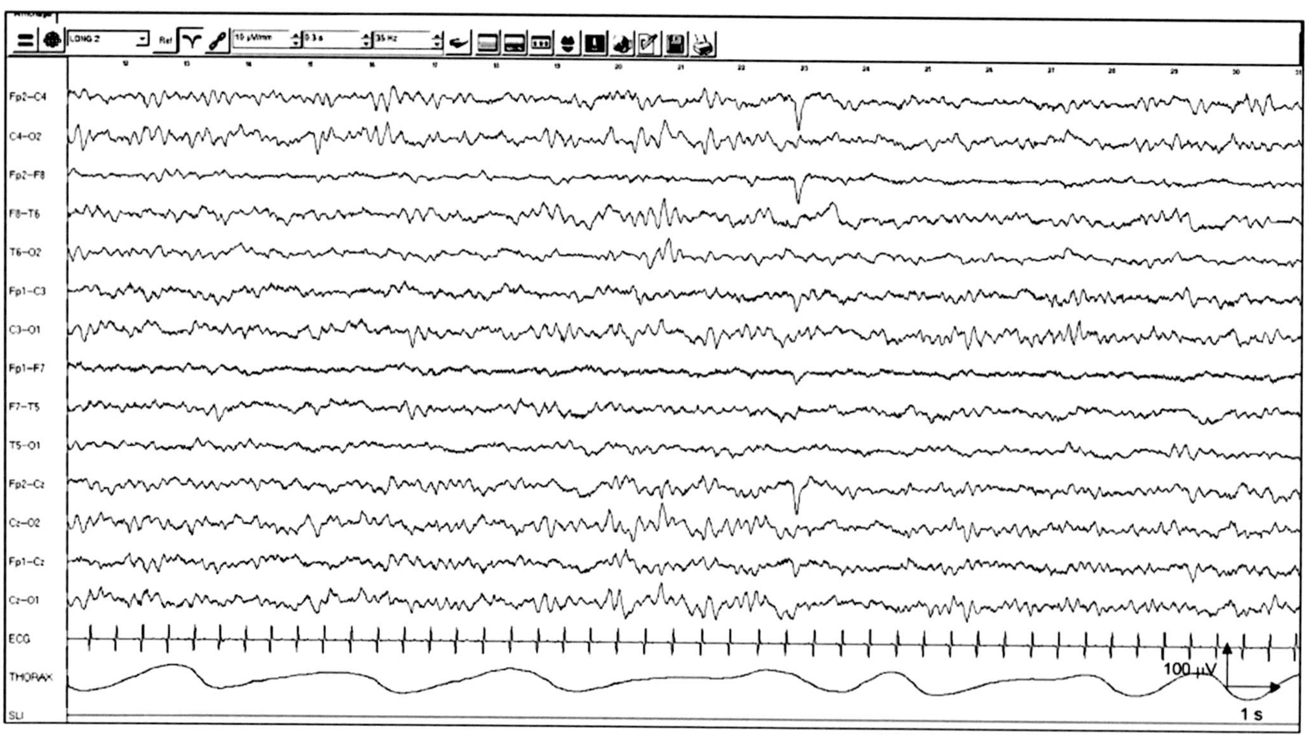

FIGURE 11.27 **Seven-month-old child. Quiet wakefulness. Rolandic-occipital alpha rhythm, symmetrical, low voltage.** *Image courtesy of Levin, K.H., Chauvel, P. (Eds.), 2019. Handbook of Clinical Neurology 3rd Series. Clinical Neurophysiology: Basis and Technical Aspects, vol. 160. Elsevier, pp. v–vi. https://doi.org/10.1016/B978-0-444-64032-1.09997-5.*

even earlier around 3, drowsiness clearly distinguishes between wakefulness and slow sleep. However, it is important to highlight that the drowsiness state, between 3 and 9 years of ages (sometimes recorded up 12–13) is better present and known as *hypnagogic hypersynchrony* (Fig. 11.28). This latter is a normal variant of EEG activity characterized by (sometimes) paroxysmal aspect of 3–5 Hz, high amplitude (75–350 μV), diffuse but prevalent in the fronto-central areas. Hypnagogic hypersynchrony must be distinguished from paroxysmal slow activity consisting of diffuse sequences of slow activity at 3–4 Hz having sharp elements which can be confused with atypical spike-wave complexes. This kind of EEG tracing was already described by (Gibbs, 1964a,b) and defined as "*pseudo petit mal*" having an unclear physiopathogenetic significance, observable during the drowsiness in about 8% of the normal population aged between 1 and 15 years (Eeg-Olofsson et al., 1971). It is also easy to find in children with febrile convulsions but does not constitute an unfavorable prognostic index for the development of future epilepsy.

A visual blocking response is usually present at the age of 3–4 months.

During the first year of life, key stages mark the sleep organization: *tracé alternant* disappears at about 3–4 weeks after birth and is replaced by a continue and diffuse polymorphic delta activity with maximal amplitude (150–200 mV) in the occipital areas. Sleep spindles usually appear during the second month, with a maximal amplitude in central or centro-parietal regions. By 6 months of age, spindles are bilateral but may still be asynchronous and vertex sharp waves plus K-complexes usually appear at the age of 5–6 months. A gradual decrease of REM sleep occurs along the first year of life from around 50% at birth, to 40% at 3–5 months and 30% between 1 and 2 years. REM sleep can present two aspects: diffuse high-voltage delta activity around 2 Hz or diffuse theta activity associated with occipital sharp transients. At 1 year of age, slow sleep can be divided into two different states according to the different patterns: spindles, followed by diffuse delta waves.

EEG From Toddler to Adolescent

Until adolescence the EEG presents a series of modifications which leads to the achievement of the typical pattern of the adult individual. The studies concerning the evolutionary changes of the EEG are very numerous and above all concern the study of the maturation of the three frequency bands typical of the EEG of normal wakefulness (alpha, theta, beta).

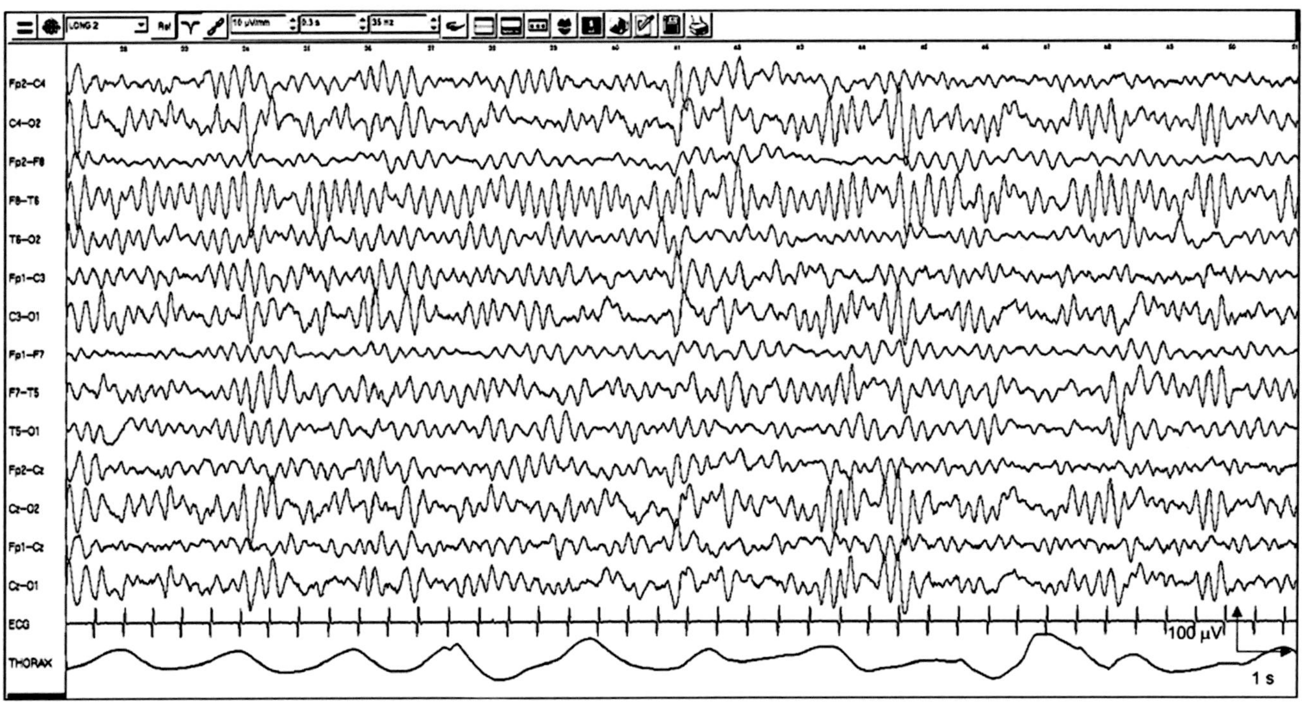

FIGURE 11.28 **Seven-month-old child. Drowsiness. Hypnagogic hypersynchrony.** *Image courtesy of Levin, K.H., Chauvel, P. (Eds.), 2019. Handbook of Clinical Neurology 3rd Series. Clinical Neurophysiology: Basis and Technical Aspects, vol. 160. Elsevier, pp. v–vi. https://doi.org/10.1016/B978-0-444-64032-1.09997-5.*

Maturation of Background Alpha Rhythm

Both classic studies (Kuhlo, 1976) and more modern ones (Eeg-Olofsson et al., 1971, Petersén and Eeg-Olofsson, 1971) demonstrate that the alpha rhythm in the occipital site reaches the average frequency typical of adulthood around 12 years of age and the average amplitude around 15–16 years of age. Some differences between males and females have been reported by various authors (Benninger et al., 1984; Eeg-Olofsson, 1980; La Marche, 1984). In particular, Benninger et al. (1984) state that in girls over 6 years the changes in alpha activity per year occur faster than in boys, with a progressive percentage decrease in theta and a parallel increase before slow alpha (8.0–9.5 Hz) and then fast alpha (9.5–13 Hz). However, interindividual variations are very important in this process. The rhythm is up to 11 Hz at age 10–11 years, whereas its amplitude is generally higher over the nondominant hemisphere (Fig. 11.29). Indeed, in children the amplitude of the alpha rhythm is higher than in adults and some components sometimes have a sharp appearance. This pattern is to be considered perfectly normal and this asymmetry rarely exceeds 20 mV.

Studies with computerized analysis have shown that alpha, both in the occipital, frontal, and central areas, reaches all the typical characteristics of adults only at 22–25 years of age (Matsuura et al., 1985). A particular problem is represented by the presence of slow activities (mainly theta but also delta) in the posterior regions, mixed with the background (or basic) rhythm. At young age the finding of these potentials is considered normal if limited to 10%–20%. A particular type of slow posterior activity is constituted by the appearance, upon closing of the eyes, of rhythmic potentials at 3–4 Hz, of ample voltage, on the occipital regions. Belsh et al. (1983) who have studied this EEG pattern says that it can only be found in childhood and adolescence (6–16 years) and that it may or may not be associated with epileptic-like symptoms but it should not, in any case, be interpreted as epileptiform.

Anyway, from adolescence to adulthood there are no striking changes, with occipital basic alpha rhythm showing a medium frequency of 10 Hz and lower amplitude compared to younger children. Amplitude asymmetry with higher voltage over the nondominant hemisphere does not exceed 20%. Posterior slow activities decrease during adolescence and beta frequencies can be observed in frontal regions.

Maturation of Theta Rhythm

It is common observation that until adolescence the EEGs are very rich in theta activity which progressively

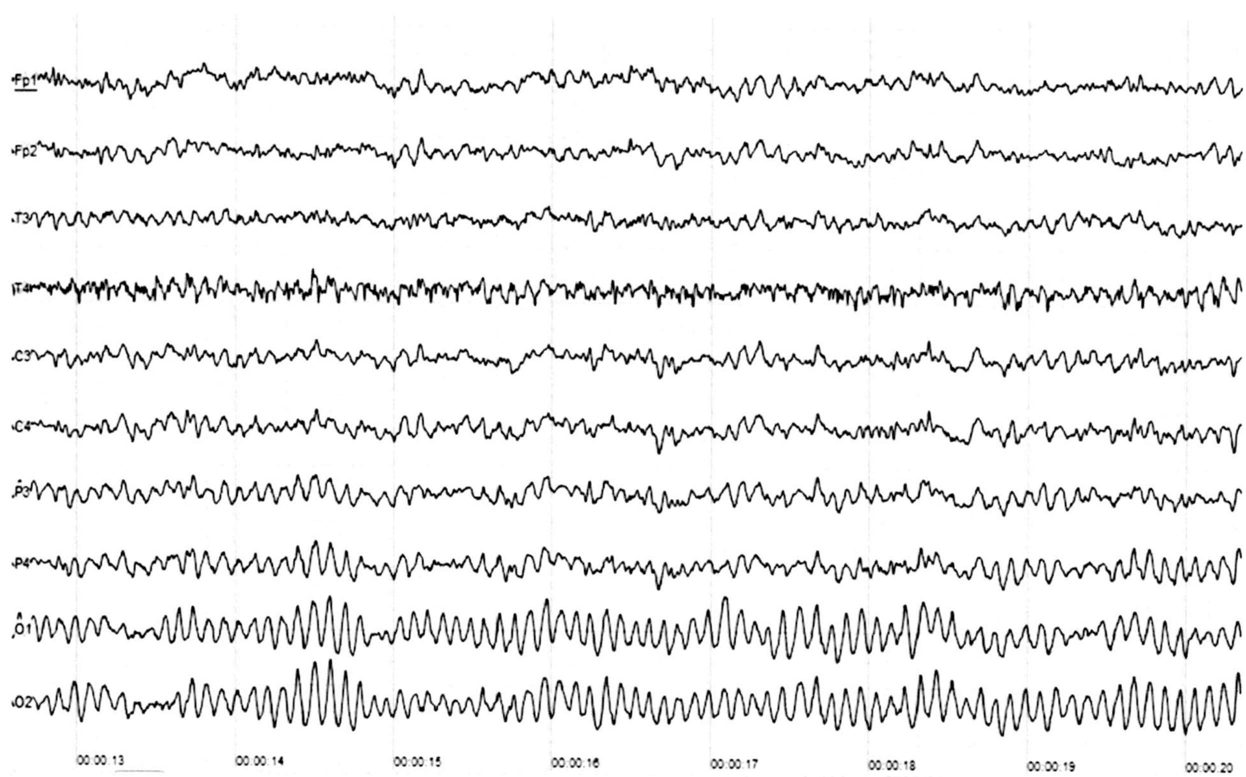

FIGURE 11.29 **Alpha-rhythm 11 Hz in young man more prominent in posterior sites.** Normal alpha range (8–12 Hz). *Image courtesy of Dr. Andrii Cherninskyi (Wikipedia).*

decreases. While studies carried out with visual inspection of the EEG (Eeg-Olofsson et al., 1971) have affirmed that the percentage of theta decreases starting from 8 to 14 years and that this decline ends at 16–21 years, researches performed with computer analysis have shown how theta activity reaches stable "adult" values at 22–25 years in the occipital region and at 26–29 years in the frontal and central region (Matsuura et al., 1985).

Maturation of Beta Rhythm

Scarcely relevant during childhood, beta activity becomes easier to find in adolescence. The maximum percentage of rapid potentials is observed on the frontal areas and on the Rolandic regions especially during the mental activity e.g., attention processes, execution of calculations, etc. Usually, the rapid activity is associated with the presence of rhythm mu, frequently found in adolescence. Computer analysis studies of EEG also demonstrated a higher incidence of rapid activity in females, probably related to endocrinological maturation (Matsuura et al., 1985).

Maturation of Sleep EEG

In younger children, the state of drowsiness is characterized, as already described, by hypnagogic hypersynchrony, while in older children the electroencephalographic changes are increasingly similar to those of the adult subject.

Even the phases of non-REM sleep do not show significant differences compared to the more advanced periods of childhood. In any case, the younger the child, the more the EEG tends to have a higher voltage and up to 2 years it is possible to detect asynchronous and K complexes of greater duration and voltage than observed in adulthood.

Fig. 11.30 shows an EEG of slow wave sleep in adulthood.

Cortex and Neurodevelopmental Disorders

Obsessive Compulsive Disorder

The pathophysiology of obsessive compulsive disorder (OCD) has been widely conceptualized within the cortico-striato-thalamo-cortical (CSTC) model. According to this model, tracts from frontal regions project to the striatum and then travel through direct and indirect pathways to the thalamus and project back to the frontal regions. This model has been corroborated by several reports of structural and functional alterations observed in magnetic resonance imaging (MRI) studies. In particular, volumetric alterations within the orbitofrontal cortex (OFC), anterior cingulate cortex (ACC), and

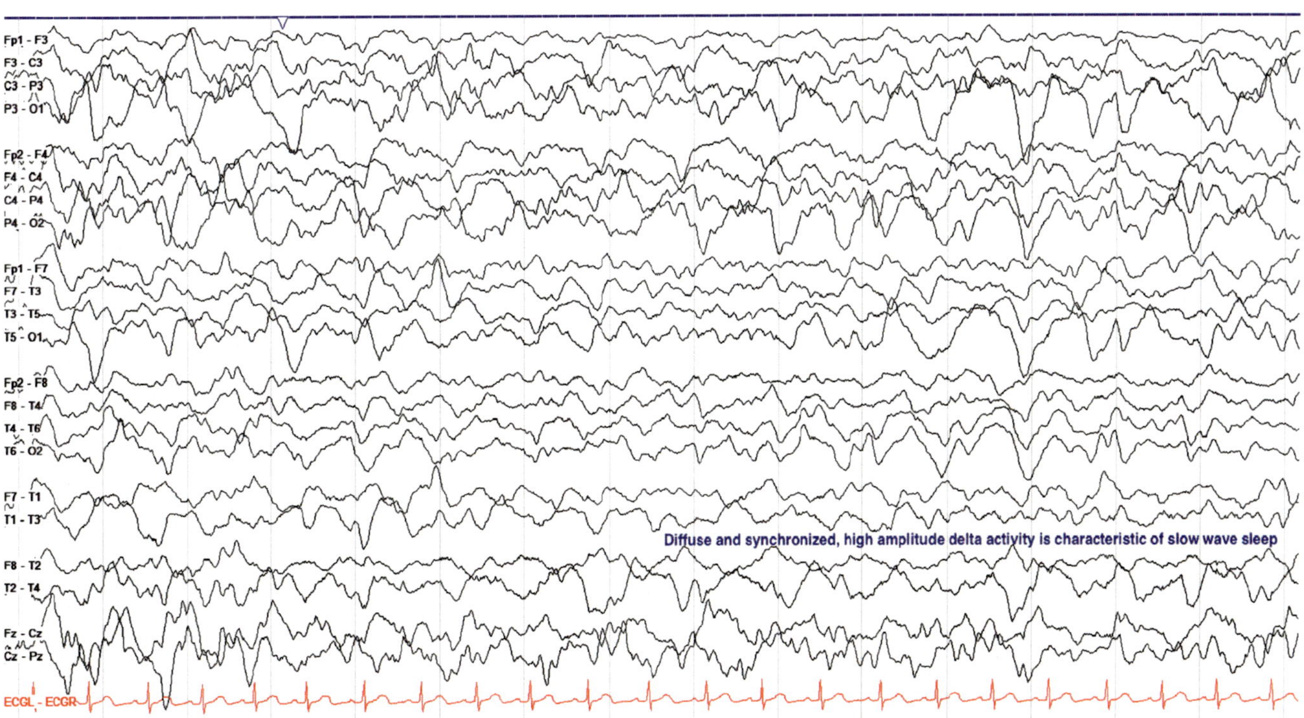

FIGURE 11.30 **Slow wave sleep in adulthood is marked by high amplitude, synchronized delta activity, usually 0.5–2 Hz in frequency.** All of the architecture from earlier sleep phases (vertex waves, POSTS, spindles, and K complexes) can be present in slow wave sleep, but usually become somewhat less organized, and often disappear altogether. Slow wave sleep can appear to be extremely high voltage, particularly in younger patients. Of note, medications such as benzodiazepines and barbiturates reduce the amount of these stages of sleep. *Image courtesy of Dr. David Valentine https://www.learningeeg.com.*

The generation and propagation of the human alpha rhythm

The alpha oscillations (7–13 Hz) (Adrian and Yamagiwa, 1935) are the most salient and the longest-studied brain electroencephalogram (EEG) event during wakefulness and may be fundamental for top-down cognitive processes (Ito et al., 2005; von Stein et al., 2000) such as attention (Saalmann et al., 2012), perception (Samaha and Postle, 2015; Busch et al., 2009), functional inhibition (Jensen and Mazaheri, 2010; Peterson and Voytek, 2017), and working memory (Jensen et al., 2002). However, the underlying neural structure(s) and circuits which generate alpha are intensely controversial. Studies have pointed to the thalamus as the primary alpha pacemaker, with the classic posterior alpha rhythm driven by the pulvinar and/or lateral geniculate nucleus (LGN) (4, 10–12). Within the cortex, instead, is widely assumed that alpha originates from infragranular layers driven by layer V pyramidal cells (van Kerkorle et al., 2014; Silva and Connors, 1991; Buffalo et al., 2011; Mejias et al., 2016; Womelsdorf et al., 2014).

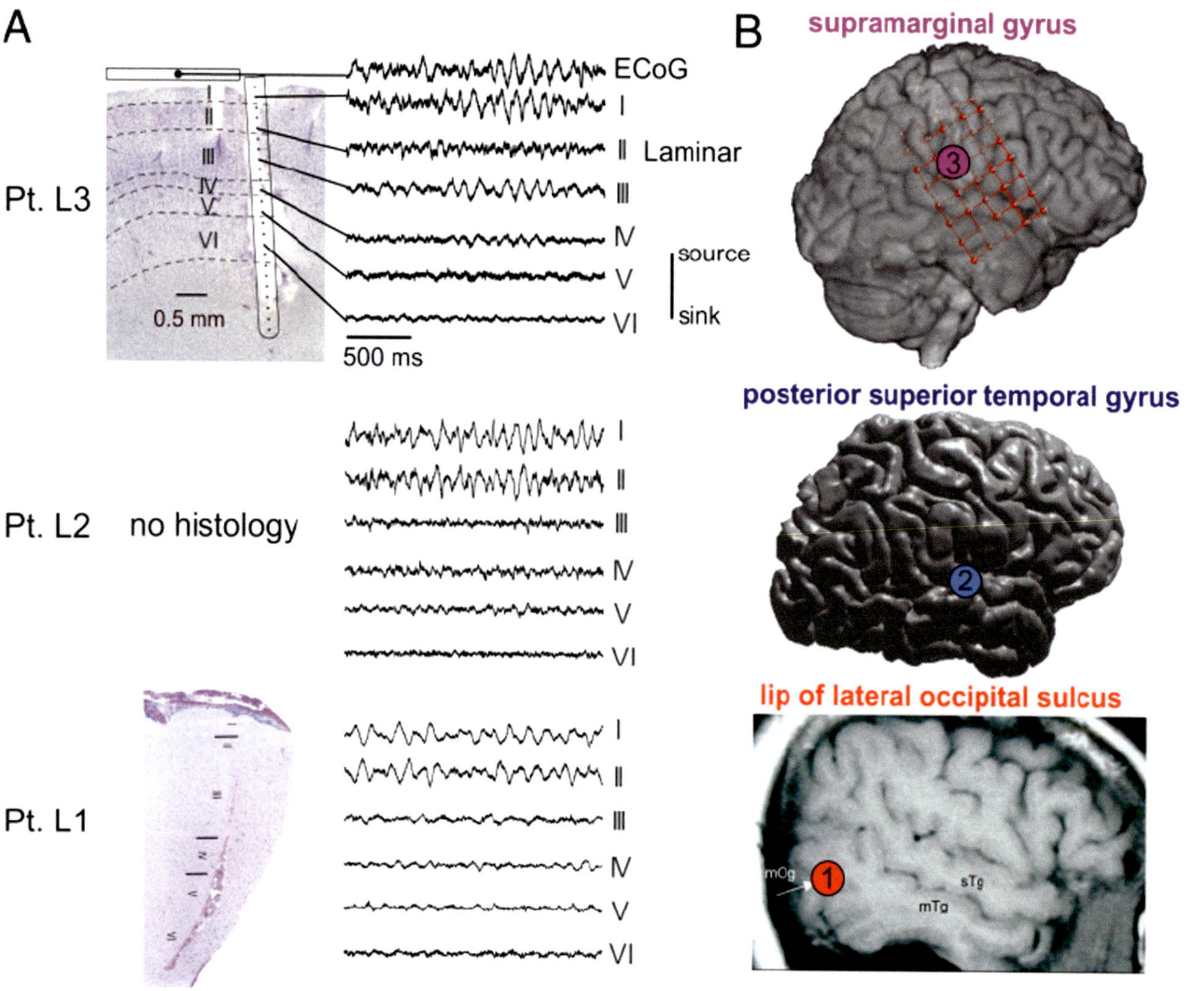

BOX FIGURE 1.1 **Laminar recordings of cortical alpha.** (A) Nissl stains of the explanted tissue surrounding the laminar probe in Pts. L1 and L3, in addition to representative laminar CSD traces from each layer in each subject. Note that despite being made in distinct cortical locations, alpha oscillations were always strongest in layer I/II. Furthermore, in L3, the trace of a simultaneously recorded overlying ECoG contact is near identical to the underlying laminar layer I. (B) Locations of each laminar probe in all patients. *Image courtesy from Halgren, M., Ulbert, I., Bastuji, H., Fabó, D., Erőss, L., Rey, M., Devinsky, O., Doyle, W.K., Mak-McCully, R., Halgren, E., Wittner, L., Chauvel, P., Heit, G., Eskandar, E., Mandell, A., Cash, S.S., 2019. The generation and propagation of the human alpha rhythm. Proc. Natl. Acad. Sci. U.S.A. 116 (47), 23772–23782. https://doi.org/10.1073/pnas.1913092116.*

Despite the prevalence of these hypotheses, the studies used to support them are not definitive and its physiology remains poorly understood; moreover, none of these assumptions have been directly tested via invasive recordings in humans. Recently, Halgren et al. (2019) performed an elegant study using microelectrodes and macroelectrodes in surgical epilepsy patients to measure the intracortical and thalamic generators of the alpha rhythm during quiet wakefulness. They first found that alpha in both visual and somatosensory cortex propagates from higher-order to lower-order areas. In posterior cortex, alpha propagates from higher-order anterosuperior areas toward the occipital pole, whereas in somatosensory cortex it propagates from associative regions toward primary cortex. Several analyses suggest that this cortical alpha leads pulvinar alpha, complicating prevailing theories of a thalamic pacemaker. Interestingly, previous scalp studies of human traveling alpha waves have also found varying propagation directions (Ito et al., 2005; Lozano-Sodevilla and VanRullen, 2019), in contrast to the consistent anterior −posterior directionality Halgren et al. (2019) observed. Alternatively, propagation direction might change with task or behavioral state (Lozano-Sodevilla and VanRullen, 2019; Alamia and VanRullen, 2019). Alpha traveling waves might serve as a mechanism to internally scan the attentional field, tag distinct visual features with different phases (Ermentrout and Kleinfeld, 2001), or facilitate plasticity between upstream and downstream areas (Han et al., 2008; Crunelli, 2018). Simultaneous recordings from human cortex and pulvinar provided several measures suggesting that cortical alpha leads thalamic alpha during quiet wakefulness. Thalamic alpha in the study we are talking about was less common than cortical alpha and discrete alpha bursts in cortex led alpha bursts in the thalamus. Finally, thalamic alpha was Granger caused by cortical alpha. A potential weakness of these bivariate causal analyses is the possibility of a third structure (not recorded from) driving alpha oscillations in both pulvinar and posterior cortex, such as the LGN (Lőrincz et al., 2009). While this is a possibility, the LGN is an unlikely cortical pacemaker, as its major projections are limited to striate and circumstriate cortex (Huijzen et al., 2008) and they found that alpha oscillations propagate toward (and not from) the occipital pole. Because the LGN and pulvinar are the 2 thalamic nuclei with the most robust projections to posterior cortex capable of driving visual alpha, these findings agree with strong cortical influences on thalamic (pulvinar) alpha during quiet wakefulness (though this conclusion only follows for posterior/visual alpha). While this appears to contradict animal studies in which the pulvinar drove cortical alpha (Saalmann et al., 2012), it is consonant with findings that the cortex can still

Continued

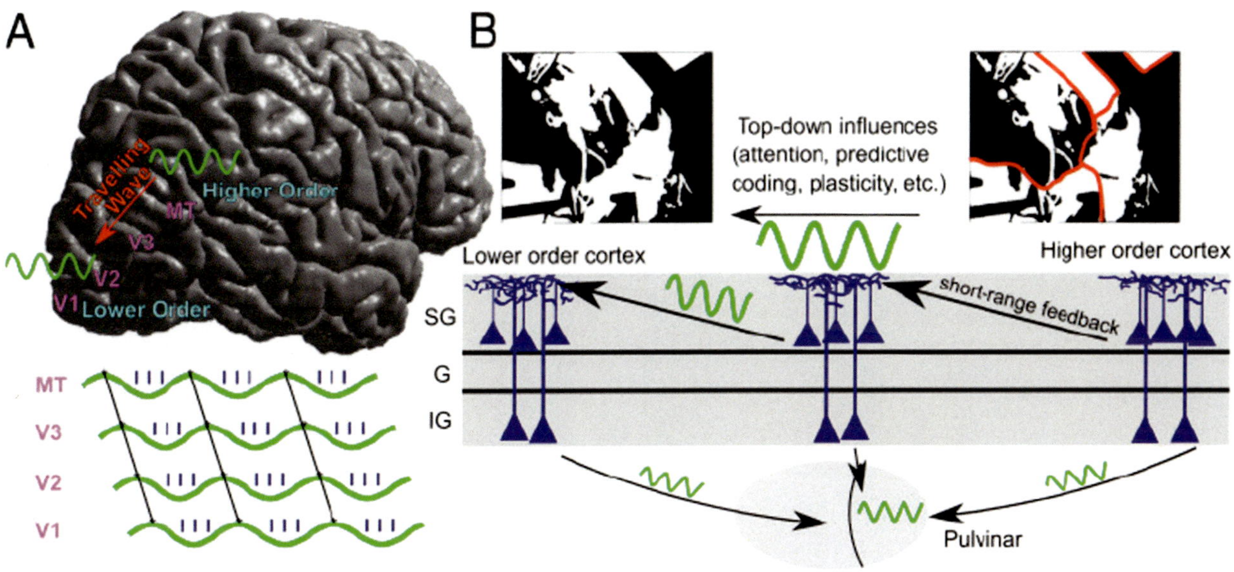

BOX FIGURE 1.2 **Tentative model for how alpha's physiology could mediate feedback.** (A) Alpha propagates as a traveling wave from higher-order (middle temporal, visual area) 3) toward lower-order visual areas 1/2 cortical areas. (B) Alpha is strongest within supragranular cortex and may carry top-down information via short-range feedback connections to constrain lower-level processing; for instance, alpha may play a role in resolving ambiguous visual imagery, such as the picture of a woman and a horse's snout shown above. Cortical alpha in layer VI might influence alpha activity within the pulvinar. *Image courtesy from C. Teufel, S.C. Dakin, P.C., 2018. Fletcher, Prior object-knowledge sharpens properties of early visual feature-detectors. Sci. Rep. 8, 10853.*

generate alpha in vitro (Silva et al., 1991) and actually shows increased alpha-band power when the pulvinar is inactivated (Zhou et al., 2016), as well as alpha coherence within the cortex exceeding thalamocortical alpha coherence (da Silva et al., 1973). There may be separable thalamic and cortical alpha pacemakers which become differentially active and coupled under different behavioral conditions (Saalmann et al., 2012; Fiebelkorn et al., 2019), for instance, this effect may be dependent on eye closure, which was not controlled for in the recordings performed in Halgren et al. (2019)s study. Alternatively, the oscillatory circuit required to generate alpha may require both thalamic and cortical cells, or the pulvinar could enable an intracortical alpha pacemaker with tonic (nonrhythmic) excitation\inhibition without being a direct pacemaker (Quax et al., 2017), perhaps coordinating cortical alpha phase in a task-relevant fashion (Saalmann et al., 2012).

Diffuse thalamocortical matrix projections might mediate this, as matrix projections target superficial

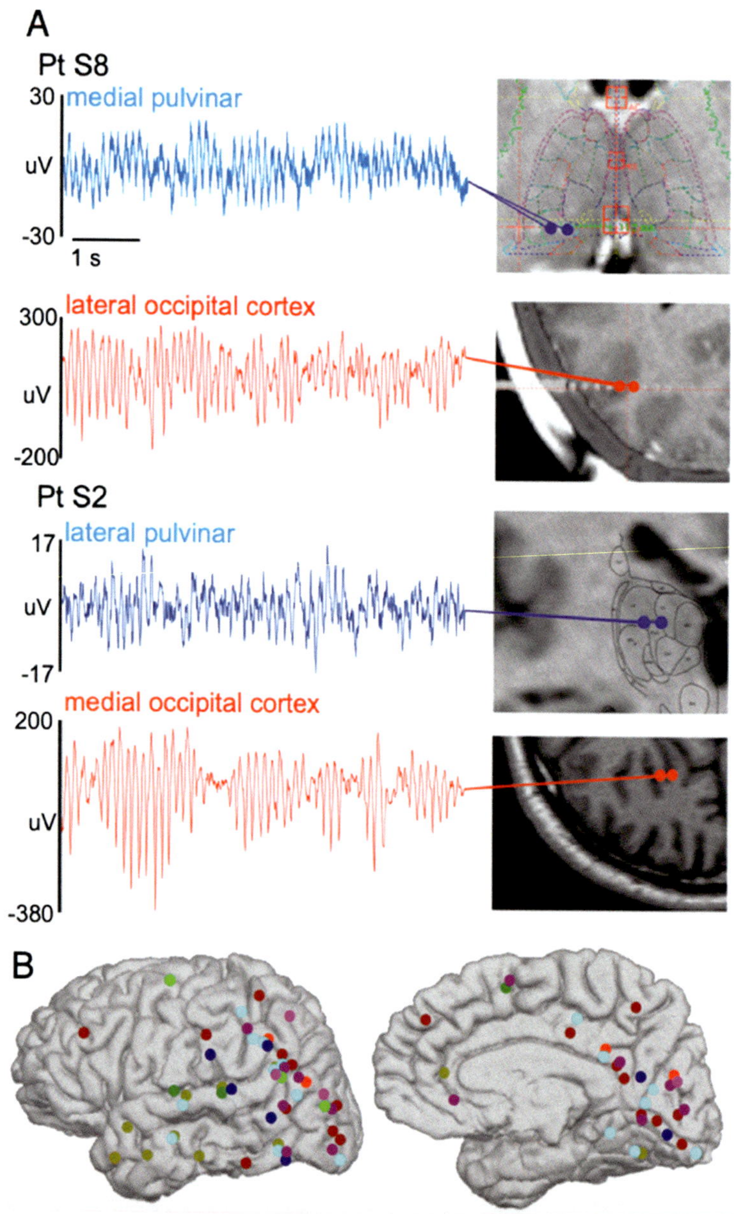

BOX FIGURE 1.3 **Robust alpha rhythms can be recorded in human pulvinar and cortex.** (A) Representative 6-s LFPg traces of simultaneous thalamic and cortical alpha activity. Prominent, largely continuous alpha rhythms can be recorded in various locations within the pulvinar as well as posterior cortex. (B) Cortical implant locations in all SEEG patients displayed on Pt. S3's brain. Each color signifies a different patient. *Image courtesy from Halgren, M., Ulbert, I., Bastuji, H., Fabó, D., Erőss, L., Rey, M., Devinsky, O., Doyle, W.K., Mak-McCully, R., Halgren, E., Wittner, L., Chauvel, P., Heit, G., Eskandar, E., Mandell, A., Cash, S.S., 2019. The generation and propagation of the human alpha rhythm. Proc. Natl. Acad. Sci. U.S.A. 116 (47), 23772–23782. https://doi.org/10.1073/pnas.1913092116.*

cortical layers (Jones, 2002) (in which they found alpha power to be strongest). Nuclei besides pulvinar or LGN, and particularly those with diffuse projections such as the medial dorsal or ventromedial nuclei, could also be involved in generating visual cortical alpha (Zotev et al., 2018). Though direct measurements and causal manipulation of spiking activity in animal models are needed to prove this, the simplest explanation for obtained findings is a leading role for the cortex driving alpha in the pulvinar during quiet wakefulness. Functionally, corticothalamic alpha might inhibit the thalamus to gate feed-forward processing and suppress irrelevant neural assemblies akin to its putative role within the cortex (Jensen and Mazaheri, 2010), as low-frequency corticothalamic activity can inhibit thalamic firing (Crandall et al., 2015). Laminar microelectrode recordings demonstrate that alpha oscillations reflect layer I/II currents [postsynaptic (Buzsáki et al., 2012)] and layer I through III firing (presynaptic), demonstrating that supragranular layers are the source of alpha LFPs and HGP recorded via ECoG and (M)EEG. However, inferring the neural circuit mechanisms which generate the alpha rhythm from Halgren M.'s laminar recordings is more complex. As supragranular pyramidal cells (which is where they observed alpha-phasic firing) are known to make feedback projections to layers I/II (where they recorded driving currents), Halgren M.'s recordings support layer II/III pyramidal cells as the primary alpha generators within the cortex during quiet wakefulness (Box Fig. 1.1). Halgren M.'s ECoG recordings also support this, as the short-range feedback projections subserved by supragranular pyramidal cells (Markov et al., 2014) are a likely intracortical mechanism for mediating the continuously propagating top-down traveling waves measured by using ECoG. These projections would enable oscillations which propagate continuously from high- to low-order cortex (i.e., not in the saltatory manner that might be expected if mediated by long-range feedback) at <1 m/s [the conduction velocity of intracortical fibers (Girard et al., 2001)] (Box Fig. 1.2). Further work employing causal manipulation of infragranular cortex in animal models, as well as controlling for behavioral state and eye closure, will be needed to determine the role of deeper layers in alpha generation. How does this compare to the cortical generation of other rhythms? Slow waves, delta and theta in humans, are also strongest in superficial layers (Csercsa et al., 2010; Halgren et al., 2015, 2018); this may indicate that, in humans, supragranular cortex plays a privileged role in sustaining low-frequency oscillatory activity. A supragranular origin for the alpha rhythm is also in accord with its putative role in neural inhibition (Haegens et al., 2011), as layers I/II contain a dense interneuronal network which strongly inhibits the apical dendrites of excitatory cells throughout the cortical column (Meyer et al., 2011). This short-range inhibition would allow higher-order cortex to modulate the gain of lower-order areas throughout visual cortex, providing a laminar circuit for top-down processes such as attention. Finally, alpha is dominated by currents and firing in supragranular cortical layers. Together, these results suggest that the alpha rhythm likely reflects short-range supragranular feedback, which propagates from higher- to lower-order cortex and cortex to thalamus. These physiological insights suggest how alpha could mediate feedback throughout the thalamocortical system (see also Box Fig. 1.3 for further information).

thalamus have been reported in patients (Goodman et al., 2021; Zhu et al., 2018; Hazari et al., 2019; Jalal et al., 2023; Karas et al., 2019). The left orbitofrontal volumes were found to be significantly smaller in the OCD patients and showed significant negative correlations with obsessive-compulsive symptom severity (Kang et al., 2004; Atmaca et al., 2007) performed a volumetric MRI study in patients who were treatment-naive and healthy controls focusing on the in vivo neuroanatomy of the whole brain. The patients had increased white matter volume than healthy controls. The patient group had significantly smaller left and right OFC volumes and significantly greater left and right thalamus volumes compared with healthy controls. Anterior Cingulate Cortex (ACC) exhibited a near-significant difference between the patients and healthy controls on the left side.

Early evidence from functional imaging studies indicated an increased metabolism and hyperactivity in several brain regions in OCD patients during task performance, including the basal ganglia (Menzies et al., 2008; Friedlander and Desrocher, 2006), OFC (Alptekin et al., 2001), and ACC (Molina et al., 1995a,b). On the opposite, a decreased activation in the dorsolateral prefrontal cortex (dlPFC) (Nakao et al., 2014) and parietal cortex (Maltby et al., 2005; van den Heuvel et al., 2005a,b; Viard et al., 2005; Remijnse et al., 2006) has been described. Emerging evidence suggests a broader cortical dysfunction, involving structural and functional alterations of the anterior insula, lateral and medial temporal lobe regions (Stern et al., 2012). Furthermore, recent multimodal metaanalytic evidence highlights the relevance of the cerebellum and the parietal cortex for the OCD pathophysiology (Eng et al., 2015).

Resting-state fMRI (rs-fMRI) studies have also provided important biomarkers of OCD. For instance, alterations in the normal patterns of functional connectivity (FC) in resting-state networks (RSNs) have been reported in children with OCD, including a significantly increased connectivity between the dorsal striatum and ventromedial frontal cortex, and a decreased FC between dorsal striatum and medial dorsal thalamus to rostral and dorsal ACC, respectively (Fitzgerald et al., 2011). In addition, increased FC within the auditory and cingulate networks was also reported in a pediatric sample (Weber et al., 2014).

Adult OCD individuals exhibited decreased FC of the dorsal striatum and lateral PFC, and of the ventral striatum with ventral tegmental area (Harrison et al., 2009) as well as a decreased dorsal ACC-right anterior operculum FC during rest (Fitzgerald et al., 2010). Altered FC in the default-mode network (DMN) has been reported, particularly its connections with OFC and ACC (Fitzgerald et al., 2010; Jang et al., 2010) and with middle frontal gyrus and putamen (Jang et al., 2010). Moreover, FC alterations between frontoparietal/ventral attention network (VAN) and the structures comprising the DMN, thalamus, lateral frontal cortex, and somatosensory/motor regions were observed (Stern et al., 2012). Using a graph-theory approach, drug-free patients were found to present a diminished FC between the DMN and frontoparietal regions (Shin et al., 2014); interestingly, these alterations were abolished after SSRI treatment (see later).

Altogether, the abovementioned results suggest that a large variety of brain areas and circuits are involved in the pathophysiology of the disease. In particular, FC abnormalities have been mainly observed in orbitofrontal, cingulate, striatal, and default mode regions. Nevertheless, these results were observed using theoretically driven investigations. To the best of our knowledge, the use of whole-brain exploratory approaches to assess FC patterns in OCD patients is scarce, with few notable exceptions (Anticevic et al., 2014; Vaghi et al., 2017a,b). In these studies, OCD patients were characterized by a reduced FC within the lateral prefrontal cortex and an increased FC within the dorsal striatum and thalamus, as well as with hyperconnectivity between basal ganglia and cerebellar regions. Nevertheless, none of the abovementioned strategies integrated structural findings in their analyses. Thus, despite the variety of studies investigating structural and functional MRI patterns in OCD patients, a comprehensive integration of distinct modalities is still unclear. The use of this multimodal/integrative approach may be of upmost relevance, as it may provide useful information on how distinct MRI modalities (that is, brain structure and function) are associated with each other (Uludağ and Roebroeck, 2014).

Furthermore, in accordance with recent developments on the study of OCD, it is anticipated that FC alterations will be observed in networks comprising regions outside of the CSTC model. Others OCD studies showed on one hand a hyperactivation of lOFC and an hypoactivation of mOFC on the other hand the exact contrary hyperactivation of mOFC and hypoactivation of lOFC (Fettes et al., 2017; Milad and Rauch, 2012; Ahmariand and Rauch, 2022; Gowda et al., 2020; Heilbronner et al., 2016).

The reasons for these apparent inconsistencies may be because of several factors including differences in the paradigms and tools used, the heterogeneity of the disorder, comorbidities, and the presence or absence of medications while patients with OCD are participating in the study. Regardless of the apparent discrepancies, all of the above studies suggest that lateral and medial portions of the OFC are indeed dysfunctional in OCD, and that these regions are playing different roles in the disorder. It has been proposed that SSRIs might have their beneficial antiobsessional effects by enhancing serotonergic transmission within OFC, that in turn modulates GABAergic tone within this region (see Blier and de Montigny, 1994; Mansari, 1995; Rauch, 1998b; Rauch et al., 1998a,b,c). Certainly, the OFC has been the most consistently implicated region in neuroimaging studies of OCD; both in terms of between group differences at rest and within group differences over time, where attenuation of OFC hyperactivity has been associated with successful treatment (e.g., Baxter et al., 1992; see Rauch et al., 1998a,b,c for review).

ADHD and OCD both have deficits in inhibitory control (Fineberg et al., 2014; van Velzen et al., 2014; Lipszyc and Schachar, 2010), thought to underlie impulsivity in ADHD and poor control over obsessions and compulsions in OCD (Fineberg et al., 2014). Inhibitory control refers to the ability to inhibit goal-irrelevant behaviors and cognitions and is measured in tasks of response inhibition tasks require the inhibition to no-go stimuli among prepotent go stimuli (go\no-go task) or the withdrawal of already triggered motor responses after stop-signals (stop task) (Aron, 2011; Rubia et al., 2007, 2009). Interference inhibition tasks require the overriding of a prepotent response that interferes with the instructed action (Aron 2011; Rubia et al., 2007, 2009).

Neuroimaging studies show overlapping inhibitory control networks centered on the ventrolateral prefrontal cortex (vlPFC), anterior insula (AI), supplementary motor cortex(SMA), dorsal anterior cingulate cortex (dACC) and the striatal, thalamic, and inferior parietal regions (Aron, 2011; Cai et al., 2014; Hugdahl et al., 2015).

In ADHD, metaanalyses of functional magnetic resonance imaging (fMRI) studies of response and interference inhibition report consistent under activation relative to controls in the right and left vlPFC\AI SMA and caudate (Hart et al., 2013; McCarthy et al., 2014;

McCarthy 2014; Cortese, 2012; Cortese and Castellanos, 2012), while switching also elicits reduced activation in bilateral vlPFC\AI and basal ganglia (Cubillo and Rubia, 2010; Rubia, 2011; Smith et al., 2006). Moreover, poor inhibition performance correlates with decreased gray matter volumes (GMVs) in the vlPFC, AI, anterior cingulate cortex (ACC), striatal and temporoparietal regions (Casey et al., 1997a,b; Depue et al., 2010a,b; McAlonan et al., 2009a,b).

Although structural abnormalities in the prefrontal cortex (Mizuno et al., 2019) have focused the studies exclusively on patients with comorbid ASD and ADHD. While there were no significant differences in prefrontal cortex, cerebellum, and basal ganglia volumes, patients with ASD and ADHD exhibited significantly lower left postcentral gyrus volumes than typically developing controls. They observed significantly lower postcentral gyrus volumes exclusively in children and preadolescent, and not in adolescents. These findings could suggest that abnormal somatosensory, attributed to delayed maturation of the left postcentral gyrus, leads to the core symptoms experienced by patients with comorbid ASD and ADHD (Dougherty et al., 2016; Chantiluke et al., 2014).

LISSENCEPHALY

Lissencephaly (LIS) is a neurodevelopmental disorder caused by a defect in neuronal migration characterized by a smooth cerebral surface, either with agyria (absent gyri) or pachygyria (abnormally broad brain sulci), a thickened cerebral cortex and an increased ratio of gray to white matter. From a macroscopical point of view human lissencephaly, in addition to a lack of sulci and gyri, is characterized by a simplification of neocortical lamination. Typically, the neocortex of lissencephalical patients has three or four layers instead of six. Lissencephaly, like pachygyria, does not necessarily occur across the entire cortex and is often restricted to more frontal or occipital regions, depending on the underlying genetic cause.

Several genes underlie lissencephaly, they are directly or indirectly associated with modulation of microtubules, which are cytoskeletal elements involved in key cellular processes including mitosis, cytokinesis, vesicle transport, and neuronal migration. Heterozygous or hemizygous mutations in the microtubule-associated genes *LIS1* and *DCX* account for 85% of classic LIS (Pilz, 1998). *LIS1*-related lissencephalies tend to exhibit more severe abnormalities over the posterior head regions (P > A gradient) (Dobyns et al., 1999; Saillour et al., 2009). About 12% of LIS patients harbor mutations in the DCX gene, which causes an A > P gradient (Leger et al., 2008).

As mentioned before lissencephaly may be divided into type I (smooth surface) and type II (cobblestone surface), which have distinct mechanisms. Type I lissencephaly is a primary defect of neuronal proliferation and migration features a very thick cortex (10–20 mm vs. normal 4 mm), whereas type II lissencephaly is a primary defect of basal lamina integrity at the pial surface, in which basal lamina breakdown leads to neuronal migration through the basal lamina defects, and disorganization of the cortical structure.

The cytoarchitecture consists of four primitive layers, including an outer marginal layer that contains Cajal-Retzius neurons (layer 1), a superficial cellular layer that contains numerous large and disorganized pyramidal neurons (layer 2) corresponding to the true cortex, a variable cell sparse layer (layer 3), and a deep cellular layer (composed of medium and small neurons) that extends more than half the width of the mantle (layer 4) (Parrini et al., 2009).

Lissencephaly is often associated with additional brain abnormalities including enlarged lateral ventricles, absence of claustrum and external capsules, corpus callosum agenesis, dysplasic basal ganglia, hypoplastic pons, hypoplasia of pyramidal tracts, and cerebellar malformations (Bonneau et al., 2002).

Classical LIS is rare, with a prevalence of about 12 per million births. Some children with LIS have lived more than 20 years, although anecdotal experience suggests that the lifespan is less than 10 years in most patients.

Epilepsy is present in over 90% of LIS children, with onset before 6 months in about 75% of cases and intractability is observed in most children. About 80% of children have infantile spasms in the first year of life although the EEG may not show typical hypsarrhythmia. Subsequently, most children exhibit mixed seizure disorders including persisting spasms, focal motor and generalized tonic seizures, atypical absences, and atonic and myoclonic seizures (Dulac et al., 1983). Mild to severe cognitive/motor impairment is also present (Saillour et al., 2009). A characteristic EEG pattern in LIS consists of diffuse fast rhythms which do not react to eye opening and very high amplitude (more than 400 mV) slow waves, mixed with sharp theta-waves. In late infancy, the EEG reveals a tendency toward bilaterally synchronous discharges of high-amplitude sharp and slow waves that seem to be typical of the most severe forms.

Acknowledgments

I would like to extend my heartfelt appreciation to Dr. Barbara Buccilli MD for her invaluable contribution to the development of this chapter. Her expertise and unwavering support have been instrumental in shaping the chapter.

Dr. Barbara Buccilli MD demonstrated exceptional professionalism and dedication throughout the collaboration process. Her insightful feedback, meticulous attention to detail, and extensive knowledge have significantly enriched the quality of the chapter. I am truly grateful for her time and effort in assisting with its creation.

References

Adrian, E.D., Yamagiwa, K., 1935. The origin of the Berger rhythm. Brain 58, 323–351. https://doi.org/10.1093/brain/58.3.323.

Aggleton, J.P., Wright, N.F., Rosene, D.L., Saunders, R.C., 2015. Complementary patterns of direct amygdala and hippocampal projections to the macaque prefrontal cortex. Cerebr. Cortex 25 (11), 4351–4373. https://doi.org/10.1093/cercor/bhv019.

Ahmari, S.E., Rauch, S.L., 2022. The prefrontal cortex and OCD. Neuropsychopharmacology 47 (1), 211–224. https://doi.org/10.1038/s41386-021-01130-2.

Alamia, A., VanRullen, R., 2019. Alpha oscillations and traveling waves: signatures of predictive coding? PLoS Biol. 17 (10), e3000487. https://doi.org/10.1371/journal.pbio.3000487.

Allen, G.V., Hopkins, D.A., 1989. Mamillary body in the rat: topography and synaptology of projections from the subicular complex, prefrontal cortex, and midbrain tegmentum. J. Comp. Neurol. 286 (3), 311–336. https://doi.org/10.1002/cne.902860303.

Alptekin, K., Degirmenci, B., Kivircik, B., et al., 2001. Tc-99m HMPAO brain perfusion SPECT in drug-free obsessive-compulsive patients without depression. Psychiatry Res. Neuroimaging. 107 (1), 51–56. https://doi.org/10.1016/S0925-4927(01)00086-5.

Amodio, D.M., Frith, C.D., 2006. Meeting of minds: the medial frontal cortex and social cognition. Nat. Rev. Neurosci. 7 (4), 268–277. https://doi.org/10.1038/nrn1884.

Amzica, F., Lopes da Silva, F.H., 2017. In: Schomer, D.L., Lopes da Silva, F.H. (Eds.), Cellular Substrates of Brain Rhythms, vol. 1. Oxford University Press. https://doi.org/10.1093/med/9780190228484.003.0002.

Anderson, B.A., Yantis, S., 2012. Value-driven attentional and oculomotor capture during goal-directed, unconstrained viewing. Atten. Percept. Psychophys. 74 (8), 1644–1653. https://doi.org/10.3758/s13414-012-0348-2.

André, M., Lamblin, M.D., d'Allest, A.M., et al., 2010. Electroencephalography in premature and full-term infants. Developmental features and glossary. Neurophysiol. Clin./Clin. Neurophysiol. 40 (2), 59–124. https://doi.org/10.1016/j.neucli.2010.02.002.

Anticevic, A., Hu, S., Zhang, S., et al., 2014. Global resting-state functional magnetic resonance imaging analysis identifies frontal cortex, striatal, and cerebellar dysconnectivity in obsessive-compulsive disorder. Biol. Psychiatry 75 (8), 595–605. https://doi.org/10.1016/j.biopsych.2013.10.021.

Aron, A.R., 2011. From reactive to proactive and selective control: developing a richer model for stopping inappropriate responses. Biol. Psychiatry 69 (12), e55–e68. https://doi.org/10.1016/j.biopsych.2010.07.024.

Ascenzi, M., Bony, G., 2017. The building of the neocortex with non-hyperpolarizing neurotransmitters. Dev. Neurobiol. 77 (9), 1023–1037. https://doi.org/10.1002/dneu.22495.

Atmaca, M., Yildirim, H., Ozdemir, H., Tezcan, E., Kursad Poyraz, A., 2007. Volumetric MRI study of key brain regions implicated in obsessive–compulsive disorder. Prog. Neuro-Psychopharmacol. Biol. Psychiatry 31 (1), 46–52. https://doi.org/10.1016/j.pnpbp.2006.06.008.

Barbas, H., 2022. Frontal cortex. In: Neuroscience in the 21st Century. Springer International Publishing, pp. 1707–1753. https://doi.org/10.1007/978-3-030-88832-9_45.

Barrett, R.L.C., Dawson, M., Dyrby, T.B., et al., 2020. Differences in frontal network anatomy across primate species. J. Neurosci. 40 (10), 2094–2107. https://doi.org/10.1523/JNEUROSCI.1650-18.2019.

Baxter, L.R., 1992. Neuroimaging studies of obsessive compulsive disorder. Psychiatry Clin. North Am. 15 (4), 871–884.

Behrens, T.E.J., Johansen-Berg, H., Woolrich, M.W., et al., 2003. Non-invasive mapping of connections between human thalamus and cortex using diffusion imaging. Nat. Neurosci. 6 (7), 750–757. https://doi.org/10.1038/nn1075.

Belsh, J.M., Chokroverty, S., Barabas, G., 1983. Posterior rhythmic slow activity in EEG after eye closure. Electroencephalogr. Clin. Neurophysiol. 56 (6), 562–568. https://doi.org/10.1016/0013-4694(83)90024-X.

Ben-Ari, Y., Woodin, M.A., Sernagor, E., et al., 2012. Refuting the challenges of the developmental shift of polarity of GABA actions: GABA more exciting than ever. Front. Cell. Neurosci. 6. https://doi.org/10.3389/fncel.2012.00035.

Bennett, M., 2020. An attempt at a unified theory of the neocortical microcircuit in sensory cortex. Front. Neural Circ. 14. https://doi.org/10.3389/fncir.2020.00040.

Benninger, C., Matthis, P., Scheffner, D., 1984. EEG development of healthy boys and girls. Results of a longitudinal study. Electroencephalogr. Clin. Neurophysiol. 57 (1), 1–12. https://doi.org/10.1016/0013-4694(84)90002-6.

Biagioni, E., Frisone, M.F., Laroche, S., et al., 2000. Occipital sawtooth: a physiological EEG pattern in very premature infants. Clin. Neurophysiol. 111 (12), 2145–2149. https://doi.org/10.1016/S1388-2457(00)00479-X.

Biagioni, E., Frisone, M.F., Laroche, S., et al., 2007. Maturation of cerebral electrical activity and development of cortical folding in young very preterm infants. Clin. Neurophysiol. 118 (1), 53–59. https://doi.org/10.1016/j.clinph.2006.09.018.

Bizzi, A., Nava, S., Ferrè, F., et al., 2012. Aphasia induced by gliomas growing in the ventrolateral frontal region: assessment with diffusion MR tractography, functional MR imaging and neuropsychology. Cortex 48 (2), 255–272. https://doi.org/10.1016/j.cortex.2011.11.015.

Blier, P., de Montigny, C., 1994. Current advances and trends in the treatment of depression. Trends Pharmacol. Sci. 15 (7), 220–226. https://doi.org/10.1016/0165-6147(94)90315-8.

Bond, A.H., 2004. An information-processing analysis of the functional architecture of the primate neocortex. J. Theor. Biol. 227 (1), 51–79. https://doi.org/10.1016/j.jtbi.2003.10.005.

Bonneau, D., Toutain, A., Laquerrière, A., et al., 2002. X-linked lissencephaly with absent corpus callosum and ambiguous genitalia (XLAG): clinical, magnetic resonance imaging, and neuropathological findings. Ann. Neurol. 51 (3), 340–349. https://doi.org/10.1002/ana.10119.

Bonnelle, V., Ham, T.E., Leech, R., et al., 2012. Salience network integrity predicts default mode network function after traumatic brain injury. Proc. Natl. Acad. Sci. 109 (12), 4690–4695. https://doi.org/10.1073/pnas.1113455109.

Braak, H., 2012. Architectonics of the human telencephalic cortex, 4. Springer Science & Business Media.

Brochier, T., Boudreau, M.J., Paré, M., Smith, A.M., 1999. The effects of muscimol inactivation of small regions of motor and somatosensory cortex on independent finger movements and force control in the precision grip. Exp. Brain Res. 128 (1–2), 31–40. https://doi.org/10.1007/s002210050814.

Brodmann, K., 1909. Vergleichende Lokalisationslehre Der Grosshirnrinde in Ihren Prinzipien Dargestellt Auf Grund Des Zellenbaues. Barth.

Budday, S., Steinmann, P., Kuhl, E., 2015. Physical biology of human brain development. Front. Cell. Neurosci. 9. https://doi.org/10.3389/fncel.2015.00257.

Budisavljevic, S., Dell'Acqua, F., Djordjilovic, V., Miotto, D., Motta, R., Castiello, U., 2017. The role of the frontal aslant tract and premotor connections in visually guided hand movements. Neuroimage 146, 419–428. https://doi.org/10.1016/j.neuroimage.2016.10.051.

Buffalo, E.A., Fries, P., Landman, R., Buschman, T.J., Desimone, R., 2011. Laminar differences in gamma and alpha coherence in the ventral stream. Proc. Natl. Acad. Sci. 108 (27), 11262–11267. https://doi.org/10.1073/pnas.1011284108.

Busch, N.A., Dubois, J., VanRullen, R., 2009. The phase of ongoing EEG oscillations predicts visual perception. J. Neurosci. 29 (24), 7869–7876. https://doi.org/10.1523/JNEUROSCI.0113-09.2009.

Buzsáki, G., Anastassiou, C.A., Koch, C., 2012. The origin of extracellular fields and currents — EEG, ECoG, LFP and spikes. Nat. Rev. Neurosci. 13 (6), 407–420. https://doi.org/10.1038/nrn3241.

Cai, W., Ryali, S., Chen, T., Li, C.S.R., Menon, V., 2014. Dissociable roles of right inferior frontal cortex and anterior insula in inhibitory control: evidence from intrinsic and task-related functional parcellation, connectivity, and response profile analyses across multiple datasets. J. Neurosci. 34 (44), 14652–14667. https://doi.org/10.1523/JNEUROSCI.3048-14.2014.

Campbell, A.W., Sherrington, C.S., 1905. Further histological studies on the localisation of cerebral function—the brains of felis, canis , and sus compared with that of homo. Proc. Roy. Soc. Lond. 74 (497–506), 390–392. https://doi.org/10.1098/rspl.1904.0131.

Casey, B.J., Castellanos, F.X., Giedd, J.N., et al., 1997a. Implication of right frontostriatal circuitry in response inhibition and attention-deficit/hyperactivity disorder. J. Am. Acad. Child Adolesc. Psychiatry 36 (3), 374–383. https://doi.org/10.1097/00004583-199703000-00016.

Casey, B.J., Trainor, R.J., Orendi, J.L., et al., 1997b. A developmental functional MRI study of prefrontal activation during performance of a go-no-go task. J. Cogn. Neurosci. 9 (6), 835–847. https://doi.org/10.1162/jocn.1997.9.6.835.

Catani, M., 2007. From hodology to function. Brain 130 (3), 602–605. https://doi.org/10.1093/brain/awm008.

Catani, M., Allin, M.P.G., Husain, M., et al., 2007. Symmetries in human brain language pathways correlate with verbal recall. Proc. Natl. Acad. Sci. 104 (43), 17163–17168. https://doi.org/10.1073/pnas.0702116104.

Catani, M., Bambini, V., 2014. A model for social communication and language evolution and development (SCALED). Curr. Opin. Neurobiol. 28, 165–171. https://doi.org/10.1016/j.conb.2014.07.018.

Catani, M., Bodi, I., Dell'Acqua, F., 2012a. Comment on "The geometric structure of the brain fiber pathways.". Science (1979) 337 (6102), 1605. https://doi.org/10.1126/science.1223425.

Catani, M., Dell'Acqua, F., Bizzi, A., et al., 2012b. Beyond cortical localization in clinico-anatomical correlation. Cortex 48 (10), 1262–1287. https://doi.org/10.1016/j.cortex.2012.07.001.

Catani, M., Dell'Acqua, F., Thiebaut de Schotten, M., 2013. A revised limbic system model for memory, emotion and behaviour. Neurosci. Biobehav. Rev. 37 (8), 1724–1737. https://doi.org/10.1016/j.neubiorev.2013.07.001.

Catani, M., Howard, R.J., Pajevic, S., Jones, D.K., 2002. Virtual in vivo interactive dissection of white matter fasciculi in the human brain. Neuroimage 17 (1), 77–94. https://doi.org/10.1006/nimg.2002.1136.

Catani, M., Jones, D.K., ffytche, D.H., 2005. Perisylvian language networks of the human brain. Ann. Neurol. 57 (1), 8–16. https://doi.org/10.1002/ana.20319.

Catani, M., Thiebaut de Schotten, M., 2012. Atlas of Human Brain Connections. Oxford University Press.

Cerliani, L., Thomas, R.M., Jbabdi, S., et al., 2012. Probabilistic tractography recovers a rostrocaudal trajectory of connectivity variability in the human insular cortex. Hum. Brain Mapp. 33 (9), 2005–2034. https://doi.org/10.1002/hbm.21338.

Chantiluke, K., Christakou, A., Murphy, C.M., et al., 2014. Disorder-specific functional abnormalities during temporal discounting in youth with Attention Deficit Hyperactivity Disorder (ADHD), Autism and comorbid ADHD and Autism. Psychiatry Res. Neuroimaging. 223 (2), 113–120. https://doi.org/10.1016/j.pscychresns.2014.04.006.

Chernoff, B.L., Teghipco, A., Garcea, F.E., et al., 2018. A role for the frontal aslant tract in speech planning: a neurosurgical case study. J. Cogn. Neurosci. 30 (5), 752–769. https://doi.org/10.1162/jocn_a_01244.

Chevillet, M.A., Jiang, X., Rauschecker, J.P., Riesenhuber, M., 2013. Automatic phoneme category selectivity in the dorsal auditory stream. J. Neurosci. 33 (12), 5208–5215. https://doi.org/10.1523/JNEUROSCI.1870-12.2013.

Chipaux, M., Colonnese, M.T., Mauguen, A., et al., 2013. Auditory stimuli mimicking ambient sounds drive temporal "delta-brushes" in premature infants. PLoS One 8 (11), e79028. https://doi.org/10.1371/journal.pone.0079028.

Colonnese, M., Khazipov, R., 2012. Spontaneous activity in developing sensory circuits: implications for resting state fMRI. Neuroimage 62 (4), 2212–2221. https://doi.org/10.1016/j.neuroimage.2012.02.046.

Colonnese, M.T., Kaminska, A., Minlebaev, M., et al., 2010. A conserved switch in sensory processing prepares developing neocortex for vision. Neuron 67 (3), 480–498. https://doi.org/10.1016/j.neuron.2010.07.015.

Corbetta, M., Shulman, G.L., 2002. Control of goal-directed and stimulus-driven attention in the brain. Nat. Rev. Neurosci. 3 (3), 201–215. https://doi.org/10.1038/nrn755.

Cortese, S., 2012. The neurobiology and genetics of Attention-Deficit/Hyperactivity Disorder (ADHD): what every clinician should know. Eur. J. Paediatr. Neurol. 16 (5), 422–433. https://doi.org/10.1016/j.ejpn.2012.01.009.

Cortese, S., Castellanos, F.X., 2012. Neuroimaging of attention-deficit/hyperactivity disorder: current neuroscience-informed perspectives for clinicians. Curr. Psychiatry Rep. 14 (5), 568–578. https://doi.org/10.1007/s11920-012-0310-y.

Craig, M.C., Catani, M., Deeley, Q., et al., 2009. Altered connections on the road to psychopathy. Mol. Psychiatry 14 (10), 946–953. https://doi.org/10.1038/mp.2009.40.

Crandall, S.R., Cruikshank, S.J., Connors, B.W., 2015. A corticothalamic switch: controlling the thalamus with dynamic synapses. Neuron 86 (3), 768–782. https://doi.org/10.1016/j.neuron.2015.03.040.

Crunelli, V., Lőrincz, M.L., Connelly, W.M., et al., 2018. Dual function of thalamic low-vigilance state oscillations: rhythm-regulation and plasticity. Nat. Rev. Neurosci. 19 (2), 107–118. https://doi.org/10.1038/nrn.2017.151.

Csercsa, R., Dombovári, B., Fabó, D., et al., 2010. Laminar analysis of slow wave activity in humans. Brain 133 (9), 2814–2829. https://doi.org/10.1093/brain/awq169.

Cubillo, A., Rubia, K., 2010. Structural and functional brain imaging in adult attention-deficit/hyperactivity disorder. Expert Rev. Neurother. 10 (4), 603–620. https://doi.org/10.1586/ern.10.4.

Curzi-Dascalova, L., Peirano, P., Morel-Kahn, F., 1988. Development of sleep states in normal premature and full-term newborns. Dev. Psychobiol. 21 (5), 431–444. https://doi.org/10.1002/dev.420210503.

D'Anna, L., Mesulam, M.M., Thiebaut de Schotten, M., et al., 2016. Frontotemporal networks and behavioral symptoms in primary progressive aphasia. Neurology 86 (15), 1393–1399. https://doi.org/10.1212/WNL.0000000000002579.

D'Esposito, M., Detre, J.A., Alsop, D.C., Shin, R.K., Atlas, S., Grossman, M., 1995. The neural basis of the central executive system of working memory. Nature 378 (6554), 279–281. https://doi.org/10.1038/378279a0.

de Vries, N.K.S., ter Horst, H.J., Bos, A.F., 2007. The added value of simultaneous EEG and amplitude-integrated EEG recordings in three newborn infants. Neonatology 91 (3), 212–216. https://doi.org/10.1159/000097456.

den Ouden, D.B., Saur, D., Mader, W., et al., 2012. Network modulation during complex syntactic processing. Neuroimage 59 (1), 815–823. https://doi.org/10.1016/j.neuroimage.2011.07.057.

Deoni, S.C.L., Mercure, E., Blasi, A., et al., 2011. Mapping infant brain myelination with magnetic resonance imaging. J. Neurosci. 31 (2), 784–791. https://doi.org/10.1523/JNEUROSCI.2106-10.2011.

Depue, B.E., Burgess, G.C., Bidwell, L.C., Willcutt, E.G., Banich, M.T., 2010a. Behavioral performance predicts grey matter reductions in the right inferior frontal gyrus in young adults with combined type ADHD. Psychiatry Res. Neuroimaging. 182 (3), 231–237. https://doi.org/10.1016/j.pscychresns.2010.01.012.

Depue, B.E., Burgess, G.C., Willcutt, E.G., Ruzic, L., Banich, M.T., 2010b. Inhibitory control of memory retrieval and motor processing associated with the right lateral prefrontal cortex: evidence from deficits in individuals with ADHD. Neuropsychologia 48 (13), 3909–3917. https://doi.org/10.1016/j.neuropsychologia.2010.09.013.

Dick, A.S., Garic, D., Graziano, P., Tremblay, P., 2019. The frontal aslant tract (FAT) and its role in speech, language and executive function. Cortex 111, 148–163. https://doi.org/10.1016/j.cortex.2018.10.015.

Dobyns, W.B., Truwit, C.L., Ross, M.E., et al., 1999. Differences in the gyral pattern distinguish chromosome 17-linked and X-linked lissencephaly. Neurology 53 (2), 270. https://doi.org/10.1212/WNL.53.2.270.

Dougherty, C.C., Evans, D.W., Myers, S.M., Moore, G.J., Michael, A.M., 2016. A comparison of structural brain imaging findings in autism spectrum disorder and attention-deficit hyperactivity disorder. Neuropsychol. Rev. 26 (1), 25–43. https://doi.org/10.1007/s11065-015-9300-2.

Dreher, J.C., 2003. Dissociating the roles of the rostral anterior cingulate and the lateral prefrontal cortices in performing two tasks simultaneously or successively. Cerebr. Cortex 13 (4), 329–339. https://doi.org/10.1093/cercor/13.4.329.

Duffau, H., 2005. New insights into the anatomo-functional connectivity of the semantic system: a study using cortico-subcortical electrostimulations. Brain 128 (4), 797–810. https://doi.org/10.1093/brain/awh423.

Duffy, F.H., Burchfiel, J.L., 1971. Somatosensory system: organizational hierarchy from single units in monkey area 5. Science (1979) 172 (3980), 273–275. https://doi.org/10.1126/science.172.3980.273.

Dulac, O., Plouin, P., Perulli, L., 1983. Electroencephalographic aspects of classic agyria-pachygyria. Rev. Electroencephalogr. Neurophysiol. Clin. 13, 232–239.

Eccher, M., 2014. Corpus callosum. In: Encyclopedia of the Neurological Sciences. Elsevier, pp. 867–868. https://doi.org/10.1016/B978-0-12-385157-4.01137-4.

Eeg-Olofsson, O., 1980. Longitudinal developmental course of electrical activity of brain. Brain Dev. 2 (1), 33–44. https://doi.org/10.1016/S0387-7604(80)80006-4.

Eeg-Olofsson, O., Petersén, I., Selldén, U., 1971. The development of the electroencephalogram in normal children from the age of 1 through 15 Years – paroxysmal activity. Neuropediatrics 2 (04), 375–404. https://doi.org/10.1055/s-0028-1091791.

El Mansari, M., 1995. Alteration of serotonin release in the guinea pig orbito-frontal cortex by selective serotonin reuptake inhibitors relevance to treatment of obsessive-compulsive disorder. Neuropsychopharmacology 13 (2), 117–127. https://doi.org/10.1016/0893-133X(95)00045-F.

Eng, G.K., Sim, K., Chen, S.H.A., 2015. Meta-analytic investigations of structural grey matter, executive domain-related functional activations, and white matter diffusivity in obsessive compulsive disorder: an integrative review. Neurosci. Biobehav. Rev. 52, 233–257. https://doi.org/10.1016/j.neubiorev.2015.03.002.

Ermentrout, G.B., Kleinfeld, D., 2001. Traveling electrical waves in cortex. Neuron 29 (1), 33–44. https://doi.org/10.1016/S0896-6273(01)00178-7.

Europa, E., Iaccarino, L., Perry, D.C., Weis, E., Welch, A.E., Rabinovici, G.D., Miller, B.L., Gorno-Tempini, M.L., Henry, M.L., 2020. Diagnostic assessment in primary progressive aphasia: an illustrative case example. Am. J. Speech Lang. Pathol. 29 (4), 1833–1849. https://doi.org/10.1044/2020_AJSLP-20-00007.

Fabrizi, L., Slater, R., Worley, A., et al., 2011a. A shift in sensory processing that enables the developing human brain to discriminate touch from pain. Curr. Biol. 21 (18), 1552–1558. https://doi.org/10.1016/j.cub.2011.08.010.

Fabrizi, L., Worley, A., Patten, D., et al., 2011b. Electrophysiological measurements and analysis of nociception in human infants. JoVE 58. https://doi.org/10.3791/3118.

Fettes, P., Schulze, L., Downar, J., 2017. Cortico-striatal-thalamic loop circuits of the orbitofrontal cortex: promising therapeutic targets in psychiatric illness. Front. Syst. Neurosci. 11. https://doi.org/10.3389/fnsys.2017.00025.

Fiebelkorn, I.C., Pinsk, M.A., Kastner, S., 2019. The mediodorsal pulvinar coordinates the macaque fronto-parietal network during rhythmic spatial attention. Nat. Commun. 10 (1), 215. https://doi.org/10.1038/s41467-018-08151-4.

Fineberg, N.A., Chamberlain, S.R., Goudriaan, A.E., et al., 2014. New developments in human neurocognition: clinical, genetic, and brain imaging correlates of impulsivity and compulsivity. CNS Spectr. 19 (1), 69–89. https://doi.org/10.1017/S1092852913000801.

Fitzgerald, K.D., Stern, E.R., Angstadt, M., et al., 2010. Altered function and connectivity of the medial frontal cortex in pediatric obsessive-compulsive disorder. Biol. Psychiatry 68 (11), 1039–1047. https://doi.org/10.1016/j.biopsych.2010.08.018.

Fitzgerald, K.D., Welsh, R.C., Stern, E.R., et al., 2011. Developmental alterations of frontal-striatal-thalamic connectivity in obsessive-compulsive disorder. J. Am. Acad. Child Adolesc. Psychiatry 50 (9), 938–948.e3. https://doi.org/10.1016/j.jaac.2011.06.011.

Ford, A., McGregor, K.M., Case, K., Crosson, B., White, K.D., 2010. Structural connectivity of Broca's area and medial frontal cortex. Neuroimage 52 (4), 1230–1237. https://doi.org/10.1016/j.neuroimage.2010.05.018.

Ford, J.H., Verfaellie, M., Giovanello, K.S., 2010. Neural correlates of familiarity-based associative retrieval. Neuropsychologia 48 (10), 3019–3025. https://doi.org/10.1016/j.neuropsychologia.2010.06.010.

Forkel, S.J., Thiebaut de Schotten, M., Dell'Acqua, F., et al., 2014a. Anatomical predictors of aphasia recovery: a tractography study of bilateral perisylvian language networks. Brain 137 (7), 2027–2039. https://doi.org/10.1093/brain/awu113.

Forkel, S.J., Thiebaut de Schotten, M., Kawadler, J.M., Dell'Acqua, F., Danek, A., Catani, M., 2014b. The anatomy of fronto-occipital connections from early blunt dissections to contemporary tractography. Cortex 56, 73–84. https://doi.org/10.1016/j.cortex.2012.09.005.

Forkel, S.J., Rogalski, E., Drossinos Sancho, N., et al., 2020. Anatomical evidence of an indirect pathway for word repetition. Neurology 94 (6), e594–e606. https://doi.org/10.1212/WNL.0000000000008746.

Fridriksson, J., Bonilha, L., Baker, J.M., Moser, D., Rorden, C., 2010a. Activity in preserved left hemisphere regions predicts anomia severity in aphasia. Cerebr. Cortex 20 (5), 1013–1019. https://doi.org/10.1093/cercor/bhp160.

Fridriksson, J., Kjartansson, O., Morgan, P.S., et al., 2010b. Impaired speech repetition and left parietal lobe damage. J. Neurosci. 30 (33), 11057–11061. https://doi.org/10.1523/JNEUROSCI.1120-10.2010.

Friederici, A.D., Mueller, J.L., Sehm, B., Ragert, P., 2013. Language learning without control: the role of the PFC. J. Cogn. Neurosci. 25 (5), 814–821. https://doi.org/10.1162/jocn_a_00350.

Friedlander, L., Desrocher, M., 2006. Neuroimaging studies of obsessive–compulsive disorder in adults and children. Clin. Psychol. Rev. 26 (1), 32–49. https://doi.org/10.1016/j.cpr.2005.06.010.

Fukutomi, H., Glasser, M.F., Zhang, H., et al., 2018. Neurite imaging reveals microstructural variations in human cerebral cortical gray matter. NeuroImage 182, 488–499. https://doi.org/10.1016/j.neuroimage.2018.02.017.

Gibbs, F.A., 1964a. Common types of childhood encephalitis. Arch. Neurol. 10 (1), 1. https://doi.org/10.1001/archneur.1964.00460130005001.

Gibbs, F.A., 1964b. Electroencephalography. Int. J. Neurol. 4, 123–125.

Girard, P., Hupé, J.M., Bullier, J., 2001. Feedforward and feedback connections between areas V1 and V2 of the monkey have similar rapid conduction velocities. J. Neurophysiol. 85 (3), 1328–1331. https://doi.org/10.1152/jn.2001.85.3.1328.

Goldenberg, G., Spatt, J., 2009. The neural basis of tool use. Brain 132 (6), 1645–1655. https://doi.org/10.1093/brain/awp080.

Goodman, W.K., Storch, E.A., Sheth, S.A., 2021. Harmonizing the neurobiology and treatment of obsessive-compulsive disorder. Am. J. Psychiatry 178 (1), 17–29. https://doi.org/10.1176/appi.ajp.2020.20111601.

Gowda, S.M., Shivakumar, V., Narayanaswamy, J.C., Venkatasubramanian, G., 2020. Transcranial direct current stimulation for obsessive–compulsive disorder. In: Non Invasive Brain Stimulation in Psychiatry and Clinical Neurosciences. Springer International Publishing, pp. 249–261. https://doi.org/10.1007/978-3-030-43356-7_18.

Guarnieri, F.C., de Chevigny, A., Falace, A., Cardoso, C., 2018. Disorders of neurogenesis and cortical development. Dialogues Clin. Neurosci. 20 (4), 255–266. https://doi.org/10.31887/DCNS.2018.20.4/ccardoso.

Guevara, P., Poupon, C., Rivière, D., et al., 2011. Robust clustering of massive tractography datasets. Neuroimage 54 (3), 1975–1993. https://doi.org/10.1016/j.neuroimage.2010.10.028.

Haegens, S., Nácher, V., Luna, R., Romo, R., Jensen, O., 2011. α-Oscillations in the monkey sensorimotor network influence discrimination performance by rhythmical inhibition of neuronal spiking. Proc. Natl. Acad. Sci. 108 (48), 19377–19382. https://doi.org/10.1073/pnas.1117190108.

Halgren, E., Kaestner, E., Marinkovic, K., et al., 2015. Laminar profile of spontaneous and evoked theta: rhythmic modulation of cortical processing during word integration. Neuropsychologia 76, 108–124. https://doi.org/10.1016/j.neuropsychologia.2015.03.021.

Halgren, M., Fabó, D., Ulbert, I., et al., 2018. Superficial slow rhythms integrate cortical processing in humans. Sci. Rep. 8 (1), 2055. https://doi.org/10.1038/s41598-018-20662-0.

Halgren, M., Ulbert, I., Bastuji, H., et al., 2019. The generation and propagation of the human alpha rhythm. Proc. Natl. Acad. Sci. U.S.A. 116 (47), 23772–23782. https://doi.org/10.1073/pnas.1913092116.

Han, F., Caporale, N., Dan, Y., 2008. Reverberation of recent visual experience in spontaneous cortical waves. Neuron 60 (2), 321–327. https://doi.org/10.1016/j.neuron.2008.08.026.

Harrison, B.J., Soriano-Mas, C., Pujol, J., et al., 2009. Altered corticostriatal functional connectivity in obsessive-compulsive disorder. Arch. Gen. Psychiatry 66 (11), 1189. https://doi.org/10.1001/archgenpsychiatry.2009.152.

Hart, H., Radua, J., Nakao, T., Mataix-Cols, D., Rubia, K., 2013. Meta-analysis of functional magnetic resonance imaging studies of inhibition and attention in attention-deficit/hyperactivity disorder. JAMA Psychiatry 70 (2), 185. https://doi.org/10.1001/jamapsychiatry.2013.277.

Hazari, N., Narayanaswamy, J., Venkatasubramanian, G., 2019. Neuroimaging findings in obsessive–compulsive disorder: a narrative review to elucidate neurobiological underpinnings. Indian J. Psychiatry 61 (7), 9. https://doi.org/10.4103/psychiatry.IndianJPsychiatry_525_18.

Heilbronner, S.R., Safadi, Z., Haber, S.N., 2016. Neurocircuits commonly involved in psychiatric disorders and their stimulation and lesion therapies. In: Neuromodulation in Psychiatry. John Wiley & Sons, Ltd, pp. 27–48. https://doi.org/10.1002/9781118801086.ch3.

Heilman, K., Watson, R., 2008. The disconnection apraxias. Cortex 44 (8), 975–982. https://doi.org/10.1016/j.cortex.2007.10.010.

Hikosaka, O., Tanaka, M., Sakamoto, M., Iwamura, Y., 1985. Deficits in manipulative behaviors induced by local injections of muscimol in the first somatosensory cortex of the conscious monkey. Brain Res. 325 (1–2), 375–380. https://doi.org/10.1016/0006-8993(85)90344-0.

Hofer, S., Frahm, J., 2006. Topography of the human corpus callosum revisited—comprehensive fiber tractography using diffusion tensor magnetic resonance imaging. Neuroimage 32 (3), 989–994. https://doi.org/10.1016/j.neuroimage.2006.05.044.

Howells, H., Thiebaut de Schotten, M., Dell'Acqua, F., et al., 2018. Frontoparietal tracts linked to lateralized hand preference and manual specialization. Cerebr. Cortex 28 (7), 2482–2494. https://doi.org/10.1093/cercor/bhy040.

Hugdahl, K., Raichle, M.E., Mitra, A., Specht, K., 2015. On the existence of a generalized non-specific task-dependent network. Front. Hum. Neurosci. 9. https://doi.org/10.3389/fnhum.2015.00430.

Hughes, J.R., Fino, J.J., Hart, L.A., 1987. Premature temporal theta (PTθ). Electroencephalogr. Clin. Neurophysiol. 67 (1), 7–15. https://doi.org/10.1016/0013-4694(87)90156-8.

Hughes, J.R., Miller, J.K., Fino, J.J., Hughes, C.A., 1990. The sharp theta rhythm on the occipital areas of prematures (STOP): a newly described waveform. Clin. Electroencephalogr. 21 (2), 77–87. https://doi.org/10.1177/155005949002100207.

Huijzen, C.V., Voogd, J., Nieuwenhuys, R., 2008. The Human Central Nervous System. Springer, New York.

Hyde, K.L., Zatorre, R.J., Peretz, I., 2011. Functional MRI evidence of an abnormal neural network for pitch processing in congenital amusia. Cerebr. Cortex 21 (2), 292–299. https://doi.org/10.1093/cercor/bhq094.

Iriki, A., Pavlides, C., Keller, A., Asanuma, H., 1991. Long-term potentiation of thalamic input to the motor cortex induced by coactivation of thalamocortical and corticocortical afferents. J. Neurophysiol. 65 (6), 1435–1441. https://doi.org/10.1152/jn.1991.65.6.1435.

Ito, J., Nikolaev, A.R., van Leeuwen, C., 2005. Spatial and temporal structure of phase synchronization of spontaneous alpha EEG activity. Biol. Cybern. 92 (1), 54–60. https://doi.org/10.1007/s00422-004-0533-z.

Jacquemot, C., Scott, S.K., 2006. What is the relationship between phonological short-term memory and speech processing? Trends Cogn. Sci. 10 (11), 480–486. https://doi.org/10.1016/j.tics.2006.09.002.

Jalal, B., Chamberlain, S.R., Sahakian, B.J., 2023. Obsessive-compulsive disorder: etiology, neuropathology, and cognitive dysfunction. Brain Behav. 13 (6). https://doi.org/10.1002/brb3.3000.

Jang, J.H., Kim, H.S., Ha, T.H., et al., 2010. Nonverbal memory and organizational dysfunctions are related with distinct symptom dimensions in obsessive-compulsive disorder. Psychiatry Res. 180 (2–3), 93–98. https://doi.org/10.1016/j.psychres.2010.04.016.

Jennekens, W., Niemarkt, H.J., Engels, M., Pasman, J.W., van Pul, C., Andriessen, P., 2012. Topography of maturational changes in EEG burst spectral power of the preterm infant with a normal follow-up at 2 years of age. Clin. Neurophysiol. 123 (11), 2130–2138. https://doi.org/10.1016/j.clinph.2012.03.018.

Jensen, O., 2002. Oscillations in the alpha band (9–12 Hz) increase with memory load during retention in a short-term memory task. Cerebr. Cortex 12 (8), 877–882. https://doi.org/10.1093/cercor/12.8.877.

Jensen, O., Mazaheri, A., 2010. Shaping functional architecture by oscillatory alpha activity: gating by inhibition. Front. Hum. Neurosci. 4. https://doi.org/10.3389/fnhum.2010.00186.

Jeon, H.A., Friederici, A.D., 2016. What does "being an expert" mean to the brain? Functional specificity and connectivity in expertise. Cerebr. Cortex. https://doi.org/10.1093/cercor/bhw329.

Johnson, P.B., Ferraina, S., Bianchi, L., Caminiti, R., 1996. Cortical networks for visual reaching: physiological and anatomical organization of frontal and parietal lobe arm regions. Cerebr. Cortex 6 (2), 102–119. https://doi.org/10.1093/cercor/6.2.102.

Jones, E.G., 2002. Thalamic circuitry and thalamocortical synchrony. Philos. Trans. R. Soc. Lond. B Biol. Sci. 357 (1428), 1659–1673. https://doi.org/10.1098/rstb.2002.1168.

Kaminska, A., 2019. Handbook of Clinical Neurology 3rd Series, pp. v–vi. https://doi.org/10.1016/B978-0-444-64032-1.09997-5.

Kandel, E., Koester, J.D., Mack, S.H., 2021. Principles of Neural Science, sixth ed. McGraw-Hill.

Kang, D.H., Kim, J.J., Choi, J.S., et al., 2004. Volumetric investigation of the frontal-subcortical circuitry in patients with obsessive-compulsive disorder. J. Neuropsychiatry Clin. Neurosci. 16 (3), 342–349. https://doi.org/10.1176/jnp.16.3.342.

Karas, P.J., Lee, S., Jimenez-Shahed, J., Goodman, W.K., Viswanathan, A., Sheth, S.A., 2019. Deep brain stimulation for obsessive compulsive disorder: evolution of surgical stimulation target parallels changing model of dysfunctional brain circuits. Front. Neurosci. 12. https://doi.org/10.3389/fnins.2018.00998.

Kato, T., Okumura, A., Hayakawa, F., Tsuji, T., Natsume, J., Watanabe, K., 2011. Evaluation of brain maturation in pre-term infants using conventional and amplitude-integrated electroencephalograms. Clin. Neurophysiol. 122 (10), 1967–1972. https://doi.org/10.1016/j.clinph.2010.12.063.

Kemerdere, R., de Champfleur, N.M., Deverdun, J., et al., 2016. Role of the left frontal aslant tract in stuttering: a brain stimulation and tractographic study. J. Neurol. 263 (1), 157–167. https://doi.org/10.1007/s00415-015-7949-3.

Khazipov, R., Luhmann, H.J., 2006. Early patterns of electrical activity in the developing cerebral cortex of humans and rodents. Trends Neurosci. 29 (7), 414–418. https://doi.org/10.1016/j.tins.2006.05.007.

Khazipov, R., Sirota, A., Leinekugel, X., Holmes, G.L., Ben-Ari, Y., Buzsáki, G., 2004. Early motor activity drives spindle bursts in the developing somatosensory cortex. Nature 432 (7018), 758–761. https://doi.org/10.1038/nature03132.

Klatzo, I., 2003. Cécile & Oskar Vogt: the significance of their contributions in modern neuroscience. In: Brain Edema XII. Springer Vienna, pp. 29–32. https://doi.org/10.1007/978-3-7091-0651-8_6.

Koolen, N., Dereymaeker, A., Räsänen, O., et al., 2014. Interhemispheric synchrony in the neonatal EEG revisited: activation synchrony index as a promising classifier. Front. Hum. Neurosci. 8. https://doi.org/10.3389/fnhum.2014.01030.

Kronfeld-Duenias, V., Amir, O., Ezrati-Vinacour, R., Civier, O., Ben-Shachar, M., 2016a. Dorsal and ventral language pathways in persistent developmental stuttering. Cortex 81, 79–92. https://doi.org/10.1016/j.cortex.2016.04.001.

Kronfeld-Duenias, V., Amir, O., Ezrati-Vinacour, R., Civier, O., Ben-Shachar, M., 2016b. The frontal aslant tract underlies speech fluency in persistent developmental stuttering. Brain Struct. Funct. 221 (1), 365–381. https://doi.org/10.1007/s00429-014-0912-8.

Kuhlo, W., 1976. The beta rhythms. In: Handbook of Electroencephalography and Clinical Neurophysiology, vol 6a. Elsevier.

La Marche, J.A., 1984. EEG and Evoked Potential Measured of Age and Sex Differences in Central Nervous System Processing. Utah State University.

Lawes, I.N.C., Barrick, T.R., Murugam, V., et al., 2008. Atlas-based segmentation of white matter tracts of the human brain using diffusion tensor tractography and comparison with classical dissection. Neuroimage 39 (1), 62–79. https://doi.org/10.1016/j.neuroimage.2007.06.041.

Leger, P.L., Souville, I., Boddaert, N., et al., 2008. The location of DCX mutations predicts malformation severity in X-linked lissencephaly. Neurogenetics 9 (4), 277–285. https://doi.org/10.1007/s10048-008-0141-5.

Leiguarda, R.C., Marsden, C.D., 2000. Limb apraxias. Brain 123 (5), 860–879. https://doi.org/10.1093/brain/123.5.860.

Levy, R., Goldman-Rakic, P.S., 2000. Segregation of working memory functions within the dorsolateral prefrontal cortex. Exp. Brain Res. 133 (1), 23–32. https://doi.org/10.1007/s002210000397.

Li, M., Zhang, Y., Song, L., et al., 2017. Structural connectivity subserving verbal fluency revealed by lesion-behavior mapping in stroke patients. Neuropsychologia 101, 85–96. https://doi.org/10.1016/j.neuropsychologia.2017.05.008.

Lieberman, M.D., Eisenberger, N.I., 2015. The dorsal anterior cingulate cortex is selective for pain: results from large-scale reverse inference. Proc. Natl. Acad. Sci. 112 (49), 15250–15255. https://doi.org/10.1073/pnas.1515083112.

Lipszyc, J., Schachar, R., 2010. Inhibitory control and psychopathology: a meta-analysis of studies using the stop signal task. J. Int. Neuropsychol. Soc. 16 (6), 1064–1076. https://doi.org/10.1017/S1355617710000895.

Lloyd, R., Goulding, R., Filan, P., Boylan, G., 2015. Overcoming the practical challenges of electroencephalography for very preterm infants in the neonatal intensive care unit. Acta Paediatr. 104 (2), 152–157. https://doi.org/10.1111/apa.12869.

Lopes da Silva, F.H., van Lierop, T.H.M.T., Schrijer, C.F., Storm van Leeuwen, W., 1973. Organization of thalamic and cortical alpha rhythms: spectra and coherences. Electroencephalogr. Clin. Neurophysiol. 35 (6), 627–639. https://doi.org/10.1016/0013-4694(73)90216-2.

López-Barroso, D., Catani, M., Ripollés, P., Dell'Acqua, F., Rodríguez-Fornells, A., de Diego-Balaguer, R., 2013. Word learning is mediated by the left arcuate fasciculus. Proc. Natl. Acad. Sci. 110 (32), 13168–13173. https://doi.org/10.1073/pnas.1301696110.

López-Barroso, D., de Diego-Balaguer, R., 2017. Language learning variability within the dorsal and ventral streams as a cue for compensatory mechanisms in aphasia recovery. Front. Hum. Neurosci. 11. https://doi.org/10.3389/fnhum.2017.00476.

Lorente de Nó, R., 1922. "La corteza cerebral del ratón." Trabajos del Laboratorio de Investigaciones Biológicas de la Universidad de Madrid. Mot. Res 20, 41–78.

Lőrincz, M.L., Kékesi, K.A., Juhász, G., Crunelli, V., Hughes, S.W., 2009. Temporal framing of thalamic relay-mode firing by phasic inhibition during the alpha rhythm. Neuron 63 (5), 683–696. https://doi.org/10.1016/j.neuron.2009.08.012.

Loui, P., Alsop, D., Schlaug, G., 2009a. Tone deafness: a new disconnection syndrome? J. Neurosci. 29 (33), 10215–10220. https://doi.org/10.1523/JNEUROSCI.1701-09.2009.

Loui, P., Wu, E.H., Wessel, D.L., Knight, R.T., 2009b. A generalized mechanism for perception of pitch patterns. J. Neurosci. 29 (2), 454–459. https://doi.org/10.1523/JNEUROSCI.4503-08.2009.

Lozano-Soldevilla, D., VanRullen, R., 2019. The hidden spatial dimension of alpha: 10-Hz perceptual echoes propagate as periodic traveling waves in the human brain. Cell Rep. 26 (2), 374–380.e4. https://doi.org/10.1016/j.celrep.2018.12.058.

Mălîia, M.D., Donos, C., Barborica, A., et al., 2018. Functional mapping and effective connectivity of the human operculum. Cortex 109, 303–321. https://doi.org/10.1016/j.cortex.2018.08.024.

Maltby, N., Tolin, D.F., Worhunsky, P., O'Keefe, T.M., Kiehl, K.A., 2005. Dysfunctional action monitoring hyperactivates frontal–striatal circuits in obsessive–compulsive disorder: an event-related fMRI study. Neuroimage 24 (2), 495–503. https://doi.org/10.1016/j.neuroimage.2004.08.041.

Mandelli, M.L., Caverzasi, E., Binney, R.J., et al., 2014. Frontal white matter tracts sustaining speech production in primary progressive aphasia. J. Neurosci. 34 (29), 9754–9767. https://doi.org/10.1523/JNEUROSCI.3464-13.2014.

Markov, N.T., Vezoli, J., Chameau, P., et al., 2014. Anatomy of hierarchy: feedforward and feedback pathways in macaque visual cortex. J. Comp. Neurol. 522 (1), 225–259. https://doi.org/10.1002/cne.23458.

Martino, J., Brogna, C., Robles, S.G., Vergani, F., Duffau, H., 2010a. Anatomic dissection of the inferior fronto-occipital fasciculus revisited in the lights of brain stimulation data. Cortex 46 (5), 691–699. https://doi.org/10.1016/j.cortex.2009.07.015.

Martino, J., Vergani, F., Robles, S.G., Duffau, H., 2010b. New insights into the anatomic dissection of the temporal stem with special emphasis on the inferior fronto-occipital fasciculus. Oper. Neurosurg. 66 (3), ons4–ons12. https://doi.org/10.1227/01.NEU.0000348564.28415.FA.

Matsuura, M., Yamamoto, K., Fukuzawa, H., et al., 1985. Age development and sex differences of various EEG elements in healthy children and adults — quantification by a computerized wave form recognition method. Electroencephalogr. Clin. Neurophysiol. 60 (5), 394–406. https://doi.org/10.1016/0013-4694(85)91013-2.

McAlonan, G.M., Cheung, C., Cheung, V., Wong, N., Suckling, J., Chua, S.E., 2009a. Differential effects on white-matter systems in high-functioning autism and Asperger's syndrome. Psychol. Med. 39 (11), 1885–1893. https://doi.org/10.1017/S0033291709005728.

McAlonan, G.M., Cheung, V., Chua, S.E., et al., 2009b. Age-related grey matter volume correlates of response inhibition and shifting in attention-deficit hyperactivity disorder. Br. J. Psychiatry 194 (2), 123–129. https://doi.org/10.1192/bjp.bp.108.051359.

McCarthy, H., Skokauskas, N., Frodl, T., 2014. Identifying a consistent pattern of neural function in attention deficit hyperactivity disorder: a meta-analysis. Psychol. Med. 44 (4), 869–880. https://doi.org/10.1017/S0033291713001037.

McCarthy, H.D., 2014. Measuring growth and obesity across childhood and adolescence. Proc. Nutr. Soc. 73 (2), 210–217. https://doi.org/10.1017/S0029665113003868.

Meister, M., Wong, R.O.L., Baylor, D.A., Shatz, C.J., 1991. Synchronous bursts of action potentials in ganglion cells of the developing mammalian retina. Science (1979) 252 (5008), 939–943. https://doi.org/10.1126/science.2035024.

Mejias, J.F., Murray, J.D., Kennedy, H., Wang, X.J., 2016. Feedforward and feedback frequency-dependent interactions in a large-scale laminar network of the primate cortex. Sci. Adv. 2 (11). https://doi.org/10.1126/sciadv.1601335.

Menzies, L., Chamberlain, S.R., Laird, A.R., Thelen, S.M., Sahakian, B.J., Bullmore, E.T., 2008. Integrating evidence from neuroimaging and neuropsychological studies of obsessive-compulsive disorder: the orbitofronto-striatal model revisited. Neurosci. Biobehav. Rev. 32 (3), 525–549. https://doi.org/10.1016/j.neubiorev.2007.09.005.

Mesulam, M., 1998. From sensation to cognition. Brain 121 (6), 1013–1052. https://doi.org/10.1093/brain/121.6.1013.

Mesulam, M.M., 2001. Primary progressive aphasia. Ann. Neurol. 49 (4), 425–432.

Mesulam, M.M., Mufson, E.J., 1982. Insula of the old world monkey. III: efferent cortical output and comments on function. J. Comp. Neurol. 212 (1), 38–52. https://doi.org/10.1002/cne.902120104.

Meyer, H.S., Schwarz, D., Wimmer, V.C., et al., 2011. Inhibitory interneurons in a cortical column form hot zones of inhibition in layers 2 and 5A. Proc. Natl. Acad. Sci. 108 (40), 16807–16812. https://doi.org/10.1073/pnas.1113648108.

Milad, M.R., Rauch, S.L., 2012. Obsessive-compulsive disorder: beyond segregated cortico-striatal pathways. Trends Cogn. Sci. 16 (1), 43–51. https://doi.org/10.1016/j.tics.2011.11.003.

Milh, M., Kaminska, A., Huon, C., Lapillonne, A., Ben-Ari, Y., Khazipov, R., 2007. Rapid cortical oscillations and early motor activity in premature human neonate. Cerebr. Cortex 17 (7), 1582–1594. https://doi.org/10.1093/cercor/bhl069.

Mizuno, Y., Kagitani-Shimono, K., Jung, M., et al., 2019. Structural brain abnormalities in children and adolescents with comorbid autism spectrum disorder and attention-deficit/hyperactivity disorder. Transl. Psychiatry 9 (1), 332. https://doi.org/10.1038/s41398-019-0679-z.

Molina, V., Montz, R., Martín-Loeches, M., Jiménez-Vicioso, A., Carreras, J.L., Rubia, F.J., 1995a. Drug therapy and cerebral perfusion in obsessive-compulsive disorder. J. Nucl. Med. 36 (12), 2234–2238.

Molina, V., Montz, R., Pérez-Castejón, M.J., et al., 1995b. Cerebral perfusion, electrical activity and effects of serotonergic treatment in obsessive-compulsive disorder. Neuropsychobiology 32 (3), 139–148. https://doi.org/10.1159/000119227.

Montembeault, M., Brambati, S.M., Gorno-Tempini, M.L., Migliaccio, R., 2018. Clinical, anatomical, and pathological features in the three variants of primary progressive aphasia: a review. Front. Neurol. 9. https://doi.org/10.3389/fneur.2018.00692.

Monz, B.U.M., König, H.H., Leidl, R., Staib, L., Link, K.H., 2003. Cost effectiveness of adding folinic acid to fluorouracil plus levamisole as adjuvant chemotherapy in patients with colon cancer in Germany. Pharmacoeconomics 21 (10), 709–719. https://doi.org/10.2165/00019053-200321100-00003.

Nakao, T., Okada, K., Kanba, S., 2014. Neurobiological model of obsessive-compulsive disorder: evidence from recent neuropsychological and neuroimaging findings. Psychiatry Clin. Neurosci. 68 (8), 587–605. https://doi.org/10.1111/pcn.12195.

Neef, N.E., Anwander, A., Bütfering, C., et al., 2018a. Structural connectivity of right frontal hyperactive areas scales with stuttering severity. Brain 141 (1), 191–204. https://doi.org/10.1093/brain/awx316.

Neef, N.E., Bütfering, C., Auer, T., et al., 2018b. Altered morphology of the nucleus accumbens in persistent developmental stuttering. J. Fluen. Disord. 55, 84–93. https://doi.org/10.1016/j.jfludis.2017.04.002.

Niemarkt, H.J., Andriessen, P., Peters, C.H.L., Pasman, J.W., Zimmermann, L.J., Bambang Oetomo, S., 2010. Quantitative analysis of maturational changes in EEG background activity in very preterm infants with a normal neurodevelopment at 1year of age. Early Hum. Dev. 86 (4), 219–224. https://doi.org/10.1016/j.earlhumdev.2010.03.003.

Nieuwenhuis, S., Yeung, N., 2005. Neural mechanisms of attention and control: losing our inhibitions? Nat. Neurosci. 8 (12), 1631–1633. https://doi.org/10.1038/nn1205-1631.

Nieuwenhuys, R., 2013. The myeloarchitectonic studies on the human cerebral cortex of the Vogt-Vogt school, and their significance for the interpretation of functional neuroimaging data. In: Microstructural Parcellation of the Human Cerebral Cortex: From Brodmann's Post-Mortem Map to in vivo Mapping with High-Field Magnetic Resonance Imaging, pp. 55–125.

Nieuwenhuys, R., Broere, C.A.J., Cerliani, L., 2015. A new myeloarchitectonic map of the human neocortex based on data from the Vogt–Vogt school. Brain Struct. Funct. 220 (5), 2551–2573. https://doi.org/10.1007/s00429-014-0806-9.

Oishi, K., Zilles, K., Amunts, K., et al., 2008. Human brain white matter atlas: identification and assignment of common anatomical structures in superficial white matter. Neuroimage 43 (3), 447–457. https://doi.org/10.1016/j.neuroimage.2008.07.009.

Omidvarnia, A., Fransson, P., Metsäranta, M., Vanhatalo, S., 2014. Functional bimodality in the brain networks of preterm and term human newborns. Cerebr. Cortex 24 (10), 2657–2668. https://doi.org/10.1093/cercor/bht120.

Palmu, K., Kirjavainen, T., Stjerna, S., Salokivi, T., Vanhatalo, S., 2013. Sleep wake cycling in early preterm infants: comparison of polysomnographic recordings with a novel EEG-based index. Clin. Neurophysiol. 124 (9), 1807–1814. https://doi.org/10.1016/j.clinph.2013.03.010.

Palomero-Gallagher, N., Zilles, K., 2019. Cortical layers: cyto-, myelo-, receptor- and synaptic architecture in human cortical areas. Neuroimage 197, 716–741. https://doi.org/10.1016/j.neuroimage.2017.08.035.

Pavlidis, E., Lloyd, R.O., Mathieson, S., Boylan, G.B., 2017. A review of important electroencephalogram features for the assessment of brain maturation in premature infants. Acta Paediatr. 106 (9), 1394–1408. https://doi.org/10.1111/apa.13956.

Pandya, D.N., Kuypers, H.G.J.M., 1969. Cortico-cortical connections in the rhesus monkey. Brain Res. 13 (1), 13–36. https://doi.org/10.1016/0006-8993(69)90141-3.

Papagno, C., Gallucci, M., Casarotti, A., et al., 2011a. Connectivity constraints on cortical reorganization of neural circuits involved in object naming. Neuroimage 55 (3), 1306–1313. https://doi.org/10.1016/j.neuroimage.2011.01.005.

Papagno, C., Miracapillo, C., Casarotti, A., et al., 2011b. What is the role of the uncinate fasciculus? Surgical removal and proper name retrieval. Brain 134 (2), 405–414. https://doi.org/10.1093/brain/awq283.

Papagno, C., Pascuzzo, R., Ferrante, C., et al., 2023. Deficits in naming pictures of objects are associated with glioma infiltration of the inferior longitudinal fasciculus: a study with diffusion MRI tractography, volumetric MRI, and neuropsychology. Hum. Brain Mapp. 44 (10), 4011–4027. https://doi.org/10.1002/hbm.26325.

Parra, S., Díaz, H., Zainos, A., et al., 2022. Hierarchical unimodal processing within the primary somatosensory cortex during a bimodal detection task. Proc. Natl. Acad. Sci. 119 (52). https://doi.org/10.1073/pnas.2213847119.

Parrini, E., Ferrari, A.R., Dorn, T., Walsh, C.A., Guerrini, R., 2009. Bilateral frontoparietal polymicrogyria, Lennox-Gastaut syndrome, and GPR56 gene mutations. Epilepsia 50 (6), 1344–1353. https://doi.org/10.1111/j.1528-1167.2008.01787.x.

Penfield, W., Roberts, L., 1959. Speech and Brain Mechanisms. Princeton University Press.

Perani, D., Saccuman, M.C., Scifo, P., et al., 2011a. Neural language networks at birth. Proc. Natl. Acad. Sci. 108 (38), 16056–16061. https://doi.org/10.1073/pnas.1102991108.

Perani, D., Tervaniemi, M., Toiviainen, P., 2011b. Tuning the brain for music. Cortex 47 (9), 1023–1025. https://doi.org/10.1016/j.cortex.2011.05.021.

Petersén, I., Eeg-Olofsson, O., 1971. The development of the electroencephalogram in normal children from the age of 1 through 15 years – non-paroxysmal activity. Neuropediatrics 2 (03), 247–304. https://doi.org/10.1055/s-0028-1091786.

Peterson, E.J., Voytek, B., 2017. Alpha oscillations control cortical gain by modulating excitatory-inhibitory background activity. BioRxiv.

Petrides, M., Pandya, D.N., 1984. Projections to the frontal cortex from the posterior parietal region in the rhesus monkey. J. Comp. Neurol. 228 (1), 105–116. https://doi.org/10.1002/cne.902280110.

Petrides, M., Pandya, D.N., 2007. Efferent association pathways from the rostral prefrontal cortex in the macaque monkey. J. Neurosci. 27 (43), 11573–11586. https://doi.org/10.1523/JNEUROSCI.2419-07.2007.

Pilz, D., 1998. LIS1 and XLIS (DCX) mutations cause most classical lissencephaly, but different patterns of malformation. Hum. Mol. Genet. 7 (13), 2029–2037. https://doi.org/10.1093/hmg/7.13.2029.

Price, J.L., 2007. Definition of the orbital cortex in relation to specific connections with limbic and visceral structures and other cortical regions. Ann. N. Y. Acad. Sci. 1121 (1), 54–71. https://doi.org/10.1196/annals.1401.008.

Quax, S., Jensen, O., Tiesinga, P., 2017. Top-down control of cortical gamma-band communication via pulvinar induced phase shifts in the alpha rhythm. PLoS Comput. Biol. 13 (5), e1005519. https://doi.org/10.1371/journal.pcbi.1005519.

Raichle, M.E., MacLeod, A.M., Snyder, A.Z., Powers, W.J., Gusnard, D.A., Shulman, G.L., 2001. A default mode of brain function. Proc. Natl. Acad. Sci. 98 (2), 676–682. https://doi.org/10.1073/pnas.98.2.676.

Ramayya, A.G., Glasser, M.F., Rilling, J.K., 2010. A DTI investigation of neural substrates supporting tool use. Cerebr. Cortex 20 (3), 507–516. https://doi.org/10.1093/cercor/bhp141.

Ramnani, N., Behrens, T.E.J., Johansen-Berg, H., et al., 2006. The evolution of prefrontal inputs to the cortico-pontine system: diffusion imaging evidence from macaque monkeys and humans. Cerebr. Cortex 16 (6), 811–818. https://doi.org/10.1093/cercor/bhj024.

Räsänen, O., Metsäranta, M., Vanhatalo, S., 2013. Development of a novel robust measure for interhemispheric synchrony in the neonatal EEG: activation Synchrony Index (ASI). Neuroimage 69, 256–266. https://doi.org/10.1016/j.neuroimage.2012.12.017.

Rauch, S.L., 1998b. Neuroimaging in OCD: clinical implications. CNS Spectr. 3 (S1), 26–29. https://doi.org/10.1017/S1092852900007185.

Rauch, S.L., Baxter, L.R., Jenike, M.A., Baer, L., Minichiello, W.E., 1998a. Obsessive-Compulsive Disorders: Practical Management. Mosby.

Rauch, S.L., Dougherty, D.D., Shin, L.M., et al., 1998b. Neural correlates of factor-analyzed OCD symptom dimensions: a PET study. CNS Spectr. 3 (7), 37–43. https://doi.org/10.1017/S1092852900006167.

Rauch, S.L., Shin, L.M., Whalen, P.J., Pitman, R.K., 1998c. Neuroimaging and the neuroanatomy of posttraumatic stress disorder. CNS Spectr. 3 (S2), 30–41. https://doi.org/10.1017/S1092852900007306.

Rauschecker, J.P., Scott, S.K., 2009. Maps and streams in the auditory cortex: nonhuman primates illuminate human speech processing. Nat. Neurosci. 12 (6), 718–724. https://doi.org/10.1038/nn.2331.

Remijnse, P.L., Nielen, M.M.A., van Balkom, A.J.L.M., et al., 2006. Reduced orbitofrontal-striatal activity on a reversal learning task in obsessive-compulsive disorder. Arch. Gen. Psychiatry 63 (11), 1225. https://doi.org/10.1001/archpsyc.63.11.1225.

Rilling, J.K., Dagenais, J.E., Goldsmith, D.R., Glenn, A.L., Pagnoni, G., 2008a. Social cognitive neural networks during in-group and out-group interactions. Neuroimage 41 (4), 1447–1461. https://doi.org/10.1016/j.neuroimage.2008.03.044.

Rilling, J.K., Glasser, M.F., Preuss, T.M., et al., 2008b. The evolution of the arcuate fasciculus revealed with comparative DTI. Nat. Neurosci. 11 (4), 426–428. https://doi.org/10.1038/nn2072.

Rilling, J.K., King-Casas, B., Sanfey, A.G., 2008c. The neurobiology of social decision-making. Curr. Opin. Neurobiol. 18 (2), 159–165. https://doi.org/10.1016/j.conb.2008.06.003.

Rizzolatti, G., Luppino, G., Matelli, M., 1998. The organization of the cortical motor system: new concepts. Electroencephalogr. Clin. Neurophysiol. 106 (4), 283–296. https://doi.org/10.1016/S0013-4694(98)00022-4.

Rodríguez-Fornells, A., Cunillera, T., Mestres-Missé, A., de Diego-Balaguer, R., 2009. Neurophysiological mechanisms involved in language learning in adults. Phil. Trans. R. Soc. B Biol. Sci. 364 (1536), 3711–3735. https://doi.org/10.1098/rstb.2009.0130.

Rogalski, E., Weintraub, S., Mesulam, M.M., 2013. Are there susceptibility factors for primary progressive aphasia? Brain Lang 127 (2), 135–138. https://doi.org/10.1016/j.bandl.2013.02.004.

Rolls, E.T., Stringer, S.M., Trappenberg, T.P., 2002. A unified model of spatial and episodic memory. Proc. R. Soc. Lond. B Biol. Sci. 269 (1496), 1087–1093. https://doi.org/10.1098/rspb.2002.2009.

Ross, E.D., 1981. The aprosodias: functional-anatomic organization of the affective components of language in the right hemisphere. Arch. Neurol. 38 (9), 561–569.

Rubia, K., 2011. "Cool" inferior frontostriatal dysfunction in attention-deficit/hyperactivity disorder versus "hot" ventromedial orbitofrontal-limbic dysfunction in conduct disorder: a review. Biol. Psychiatry 69 (12), e69–e87. https://doi.org/10.1016/j.biopsych.2010.09.023.

Rubia, K., Halari, R., Christakou, A., Taylor, E., 2009. Impulsiveness as a timing disturbance: neurocognitive abnormalities in attention-deficit hyperactivity disorder during temporal processes and normalization with methylphenidate. Proc. R. Soc. Lond. B Biol. Sci. 364 (1525), 1919–1931. https://doi.org/10.1098/rstb.2009.0014.

Rubia, K., Smith, A., Taylor, E., 2007. Performance of children with attention deficit hyperactivity disorder (ADHD) on a test battery of impulsiveness. Child Neuropsychol. 13 (3), 276–304. https://doi.org/10.1080/09297040600770761.

Rudrauf, D., David, O., Lachaux, J.P., et al., 2008a. Rapid interactions between the ventral visual stream and emotion-related structures rely on a two-pathway architecture. J. Neurosci. 28 (11), 2793–2803. https://doi.org/10.1523/JNEUROSCI.3476-07.2008.

Rudrauf, D., Mehta, S., Bruss, J., Tranel, D., Damasio, H., Grabowski, T.J., 2008b. Thresholding lesion overlap difference maps: application to category-related naming and recognition deficits. Neuroimage 41 (3), 970–984. https://doi.org/10.1016/j.neuroimage.2007.12.033.

Rudrauf, D., Mehta, S., Grabowski, T., 2008c. Disconnection's renaissance takes shape: formal incorporation in group-level lesion studies. Cortex 44 (8), 1084–1096. https://doi.org/10.1016/j.cortex.2008.05.005.

Saalmann, Y.B., Pinsk, M.A., Wang, L., Li, X., Kastner, S., 2012. The pulvinar regulates information transmission between cortical areas based on attention demands. Science (1979) 337 (6095), 753–756. https://doi.org/10.1126/science.1223082.

Safadi, Z., Grisot, G., Jbabdi, S., et al., 2018. Functional segmentation of the anterior limb of the internal capsule: linking white matter abnormalities to specific connections. J. Neurosci. 38 (8), 2106–2117. https://doi.org/10.1523/JNEUROSCI.2335-17.2017.

Saillour, Y., Carion, N., Quelin, C., et al., 2009. LIS1-Related isolated lissencephaly. Arch. Neurol. 66 (8). https://doi.org/10.1001/archneurol.2009.149.

Salvan, P., Tournier, J.D., Batalle, D., et al., 2017. Language ability in preterm children is associated with arcuate fasciculi microstructure at term. Hum. Brain Mapp. 38 (8), 3836–3847. https://doi.org/10.1002/hbm.23632.

Samaha, J., Postle, B.R., 2015. The speed of alpha-band oscillations predicts the temporal resolution of visual perception. Curr. Biol. 25 (22), 2985–2990. https://doi.org/10.1016/j.cub.2015.10.007.

Särkämö, T., Tervaniemi, M., Soinila, S., et al., 2009a. Amusia and cognitive deficits after stroke. Ann. N. Y. Acad. Sci. 1169 (1), 441–445. https://doi.org/10.1111/j.1749-6632.2009.04765.x.

Särkämö, T., Tervaniemi, M., Soinila, S., et al., 2009b. Cognitive deficits associated with acquired amusia after stroke: a neuropsychological follow-up study. Neuropsychologia 47 (12), 2642–2651. https://doi.org/10.1016/j.neuropsychologia.2009.05.015.

Schendan, H.E., Stern, C.E., 2008. Where vision meets memory: prefrontal–posterior networks for visual object constancy during categorization and recognition. Cerebr. Cortex 18 (7), 1695–1711. https://doi.org/10.1093/cercor/bhm197.

Scher, M.S., Loparo, K.A., 2009. Neonatal EEG/sleep state analyses: a complex phenotype of developmental neural plasticity. Dev. Neurosci. 31 (4), 259–275. https://doi.org/10.1159/000216537.

Scher, M.S., Steppe, D.A., Banks, D.L., 1996. Prediction of lower developmental performances of healthy neonates by neonatal EEG-sleep measures. Pediat. Neurol. 14 (2), 137–144. https://doi.org/10.1016/0887-8994(96)00013-6.

Schmahmann, J., 1998. The cerebellar cognitive affective syndrome. Brain 121 (4), 561–579. https://doi.org/10.1093/brain/121.4.561.

Schmahmann, J., Pandya, D., 2008. Disconnection syndromes of basal ganglia, thalamus, and cerebrocerebellar systems. Cortex 44 (8), 1037–1066. https://doi.org/10.1016/j.cortex.2008.04.004.

Schmahmann, J.D., Pandya, D.N., 2007a. Cerebral white matter — historical evolution of facts and notions concerning the organization of the fiber pathways of the brain. J. Hist. Neurosci. 16 (3), 237–267. https://doi.org/10.1080/09647040500495896.

Schmahmann, J.D., Pandya, D.N., 2007b. The complex history of the fronto-occipital fasciculus. J. Hist. Neurosci. 16 (4), 362–377. https://doi.org/10.1080/09647040600620468.

Schulze, K., Koelsch, S., 2012. Working memory for speech and music. Ann. N. Y. Acad. Sci. 1252 (1), 229–236. https://doi.org/10.1111/j.1749-6632.2012.06447.x.

Schumacher, E.M., Larsson, P.G., Sinding-Larsen, C., et al., 2013. Automated spectral EEG analyses of premature infants during the first three days of life correlated with developmental outcomes at 24 months. Neonatology 103 (3), 205–212. https://doi.org/10.1159/000345923.

Selden, N., 1998. Trajectories of cholinergic pathways within the cerebral hemispheres of the human brain. Brain 121 (12), 2249–2257. https://doi.org/10.1093/brain/121.12.2249.

Shaywitz, B.A., Shaywitz, S.E., Pugh, K.R., et al., 2002. Disruption of posterior brain systems for reading in children with developmental dyslexia. Biol. Psychiatry 52 (2), 101–110. https://doi.org/10.1016/S0006-3223(02)01365-3.

Shin, D.J., Jung, W.H., He, Y., et al., 2014. The effects of pharmacological treatment on functional brain connectome in obsessive-compulsive disorder. Biol. Psychiatry 75 (8), 606–614. https://doi.org/10.1016/j.biopsych.2013.09.002.

Silva, L.R., Amitai, Y., Connors, B.W., 1991. Intrinsic oscillations of neocortex generated by layer 5 pyramidal neurons. Science (1979) 251 (4992), 432–435. https://doi.org/10.1126/science.1824881.

Skeide, M.A., Kirsten, H., Kraft, I., et al., 2015. Genetic dyslexia risk variant is related to neural connectivity patterns underlying phonological awareness in children. Neuroimage 118, 414–421. https://doi.org/10.1016/j.neuroimage.2015.06.024.

Smith, A.B., Taylor, E., Brammer, M., Toone, B., Rubia, K., 2006. Task-specific hypoactivation in prefrontal and temporoparietal brain regions during motor inhibition and task switching in medication-naive children and adolescents with attention deficit hyperactivity disorder. Am. J. Psychiatry 163 (6), 1044–1051. https://doi.org/10.1176/ajp.2006.163.6.1044.

Stern, E.R., Fitzgerald, K.D., Welsh, R.C., Abelson, J.L., Taylor, S.F., 2012. Resting-state functional connectivity between fronto-parietal and default mode networks in obsessive-compulsive disorder. PLoS One 7 (5), e36356. https://doi.org/10.1371/journal.pone.0036356.

Teles, F., Ricardo, V., 2020. Phineas Gage's great legacy. Dement Neuropsychol 14 (4), 419–421. https://doi.org/10.1590/1980-57642020dn14-040013.

Teufel, C., Dakin, S.C., Fletcher, P.C., 2018. Prior object-knowledge sharpens properties of early visual feature-detectors. Sci. Rep. 8 (1), 10853. https://doi.org/10.1038/s41598-018-28845-5.

Thiebaut de Schotten, M., Dell'Acqua, F., Valabregue, R., Catani, M., 2012. Monkey to human comparative anatomy of the frontal lobe association tracts. Cortex 48 (1), 82–96. https://doi.org/10.1016/j.cortex.2011.10.001.

Thiebaut de Schotten, M., Ffytche, D.H., Bizzi, A., et al., 2011a. Atlasing location, asymmetry and inter-subject variability of white matter tracts in the human brain with MR diffusion tractography. Neuroimage 54 (1), 49–59. https://doi.org/10.1016/j.neuroimage.2010.07.055.

Thiebaut de Schotten, M., Urbanski, M., Duffau, H., et al., 2005. Direct evidence for a parietal-frontal pathway subserving spatial awareness in humans. Science (1979) 309 (5744), 2226–2228. https://doi.org/10.1126/science.1116251.

Thiebaut de Schotten, M.T., Dell'Acqua, F., Forkel, S.J., et al., 2011b. A lateralized brain network for visuospatial attention. Nat. Neurosci. 14 (10), 1245–1246. https://doi.org/10.1038/nn.2905.

Thompson, A., Murphy, D., Dell'Acqua, F., et al., 2017. Impaired communication between the motor and somatosensory homunculus is associated with poor manual dexterity in autism spectrum disorder. Biol. Psychiatry 81 (3), 211–219. https://doi.org/10.1016/j.biopsych.2016.06.020.

Tippett, D.C.. Classification of primary progressive aphasia: challenges and complexities. F1000Research 9, F1000 Faculty Rev-64. https://doi.org/10.12688/f1000research.21184.1.
Tremblay, R., Lee, S., Rudy, B., 2016. GABAergic interneurons in the neocortex: from cellular properties to circuits. Neuron 91 (2), 260–292. https://doi.org/10.1016/j.neuron.2016.06.033.
Trevisi, G., Eickhoff, S.B., Chowdhury, F., et al., 2018. Probabilistic electrical stimulation mapping of human medial frontal cortex. Cortex 109, 336–346. https://doi.org/10.1016/j.cortex.2018.06.015.
Triarhou, L.C., (Ed.), 2009. Cellular Structure of the Human Cerebral Cortex (Translated and edited by L.C. Triarhou (Thessaloniki) Plus poster: "The 107 Cortical Cytoarchitectonic Areas of Constantin von Economo and Georg N. Koskinas in the Adult Human Brain", S. Karger AG https://doi.org/10.1159/isbn.978-3-8055-9062-4.
Uludağ, K., Roebroeck, A., 2014. General overview on the merits of multimodal neuroimaging data fusion. Neuroimage 102, 3–10. https://doi.org/10.1016/j.neuroimage.2014.05.018.
Urbanski, M., Thiebaut de Schotten, M., Rodrigo, S., et al., 2008. Brain networks of spatial awareness: evidence from diffusion tensor imaging tractography. J. Neurol. Neurosurg. Psychiatry 79 (5), 598–601. https://doi.org/10.1136/jnnp.2007.126276.
Vaghi, M.M., Luyckx, F., Sule, A., Fineberg, N.A., Robbins, T.W., De Martino, B., 2017a. Compulsivity reveals a novel dissociation between action and confidence. Neuron 96 (2), 348–354.e4. https://doi.org/10.1016/j.neuron.2017.09.006.
Vaghi, M.M., Vértes, P.E., Kitzbichler, M.G., et al., 2017b. Specific frontostriatal circuits for impaired cognitive flexibility and goal-directed planning in obsessive-compulsive disorder: evidence from resting-state functional connectivity. Biol. Psychiatry 81 (8), 708–717. https://doi.org/10.1016/j.biopsych.2016.08.009.
van den Heuvel, O.A., Veltman, D.J., Groenewegen, H.J., et al., 2005a. Disorder-specific neuroanatomical correlates of attentional bias in obsessive-compulsive disorder, panic disorder, and hypochondriasis. Arch. Gen. Psychiatry 62 (8), 922. https://doi.org/10.1001/archpsyc.62.8.922.
van den Heuvel, O.A., Veltman, D.J., Groenewegen, H.J., et al., 2005b. Frontal-striatal dysfunction during planning in obsessive-compulsive disorder. Arch. Gen. Psychiatry 62 (3), 301. https://doi.org/10.1001/archpsyc.62.3.301.
van Kerkoerle, T., Self, M.W., Dagnino, B., et al., 2014. Alpha and gamma oscillations characterize feedback and feedforward processing in monkey visual cortex. Proc. Natl. Acad. Sci. 111 (40), 14332–14341. https://doi.org/10.1073/pnas.1402773111.
van Noort-van der Spek, I.L., Franken, M.C.J.P., Weisglas-Kuperus, N., 2012. Language functions in preterm-born children: a systematic review and meta-analysis. Pediatrics 129 (4), 745–754. https://doi.org/10.1542/peds.2011-1728.
van Velzen, L.S., Vriend, C., de Wit, S.J., van den Heuvel, O.A., 2014. Response inhibition and interference control in obsessive–compulsive spectrum disorders. Front. Hum. Neurosci. 8. https://doi.org/10.3389/fnhum.2014.00419.
Vanhatalo, S., Kaila, K., 2006. Development of neonatal EEG activity: from phenomenology to physiology. Semin. Fetal Neonatal Med. 11 (6), 471–478. https://doi.org/10.1016/j.siny.2006.07.008.
Vanhatalo, S., Palva, J.M., Andersson, S., Rivera, C., Voipio, J., Kaila, K., 2005. Slow endogenous activity transients and developmental expression of K+ -Cl– cotransporter 2 in the immature human cortex. Eur. J. Neurosci. 22 (11), 2799–2804. https://doi.org/10.1111/j.1460-9568.2005.04459.x.
Vecchierini, M.F., André, M., d'Allest, A.M., 2007. Normal EEG of premature infants born between 24 and 30 weeks gestational age: terminology, definitions and maturation aspects. Neurophysiol. Clin./Clin. Neurophysiol. 37 (5), 311–323. https://doi.org/10.1016/j.neucli.2007.10.008.
Vecchierini, M.F., d'Allest, A.M., Verpillat, P., 2003. EEG patterns in 10 extreme premature neonates with normal neurological outcome: qualitative and quantitative data. Brain Dev. 25 (5), 330–337. https://doi.org/10.1016/S0387-7604(03)00007-X.
Viard, A., Flament, M.F., Artiges, E., et al., 2005. Cognitive control in childhood-onset obsessive–compulsive disorder: a functional MRI study. Psychol. Med. 35 (7), 1007–1017. https://doi.org/10.1017/S0033291704004295.
Vogt, B.A., 2016. Midcingulate cortex: structure, connections, homologies, functions and diseases. J. Chem. Neuroanat. 74, 28–46. https://doi.org/10.1016/j.jchemneu.2016.01.010.
von Economo, C., 1927. Zellaufbau Der Grosshirnrinde Des Menschen: Zehn Vorlesungen. Springer.
von Economo, C.F., Koskinas, G.N., 1925. Die cytoarchitektonik der hirnrinde des erwachsenen menschen. Springer.
von Stein, A., Sarnthein, J., 2000. Different frequencies for different scales of cortical integration: from local gamma to long range alpha/theta synchronization. Int. J. Psychophysiol. 38 (3), 301–313. https://doi.org/10.1016/S0167-8760(00)00172-0.
Wagstyl, K., Larocque, S., Cucurull, G., et al., 2020. BigBrain 3D atlas of cortical layers: cortical and laminar thickness gradients diverge in sensory and motor cortices. PLoS Biol. 18 (4), e3000678. https://doi.org/10.1371/journal.pbio.3000678.
Wahle, P., Sobierajski, E., Gasterstädt, I., et al., 2022. Neocortical pyramidal neurons with axons emerging from dendrites are frequent in non-primates, but rare in monkey and human. Elife 11, e76101. https://doi.org/10.7554/eLife.76101.
Wallois, F., 2010. Synopsis of maturation of specific features in EEG of premature neonates. Neurophysiol. Clin./Clin. Neurophysiol. 40 (2), 125–126. https://doi.org/10.1016/j.neucli.2010.02.001.
Wang, L., Li, C., Han, Z., et al., 2022. Spatiotemporal and sensory modality attention processing with domain-specific representations in frontoparietal areas. Cerebr. Cortex 32 (24), 5489–5502. https://doi.org/10.1093/cercor/bhac029.
Watanabe, K., Hayakawa, F., Okumura, A., 1999. Neonatal EEG: a powerful tool in the assessment of brain damage in preterm infants. Brain Dev. 21 (6), 361–372. https://doi.org/10.1016/S0387-7604(99)00034-0.
Weber, A.M., Soreni, N., Noseworthy, M.D., 2014. A preliminary study of functional connectivity of medication naïve children with obsessive–compulsive disorder. Prog. Neuro-Psychopharmacol. Biol. Psychiatry 53, 129–136. https://doi.org/10.1016/j.pnpbp.2014.04.001.
Whitehead, K., Pressler, R., Fabrizi, L., 2017. Characteristics and clinical significance of delta brushes in the EEG of premature infants. Clin. Neurophysiol. Pract. 2, 12–18. https://doi.org/10.1016/j.cnp.2016.11.002.
Wilkinson, M., Lim, A.R., Cohen, A.H., Galaburda, A.M., Takahashi, E., 2017. Detection and growth pattern of arcuate fasciculus from newborn to adult. Front. Neurosci. 11. https://doi.org/10.3389/fnins.2017.00389.
Wilson, S.M., Galantucci, S., Tartaglia, M.C., et al., 2011. Syntactic processing depends on dorsal language tracts. Neuron 72 (2), 397–403. https://doi.org/10.1016/j.neuron.2011.09.014.
Wilson, W.J., Jackson, A., Pender, A., et al., 2011. The CHAPS, SIFTER, and TAPS–R as predictors of (C)AP skills and (C)APD. J. Speech Lang. Hear. Res. 54 (1), 278–291. https://doi.org/10.1044/1092-4388(2010/09-0273.
Womelsdorf, T., Valiante, T.A., Sahin, N.T., Miller, K.J., Tiesinga, P., 2014. Dynamic circuit motifs underlying rhythmic gain control, gating and integration. Nat. Neurosci. 17 (8), 1031–1039. https://doi.org/10.1038/nn.3764.
Yeatman, J.D., Dougherty, R.F., Ben-Shachar, M., Wandell, B.A., 2012. Development of white matter and reading skills. Proc. Natl. Acad. Sci. 109 (44). https://doi.org/10.1073/pnas.1206792109.

Yeatman, J.D., Dougherty, R.F., Rykhlevskaia, E., et al., 2011. Anatomical properties of the arcuate fasciculus predict phonological and reading skills in children. J. Cogn. Neurosci. 23 (11), 3304–3317. https://doi.org/10.1162/jocn_a_00061.

Yeterian, E.H., Pandya, D.N., Tomaiuolo, F., Petrides, M., 2012. The cortical connectivity of the prefrontal cortex in the monkey brain. Cortex 48 (1), 58–81. https://doi.org/10.1016/j.cortex.2011.03.004.

Zappalà, G., Thiebaut de Schotten, M., Eslinger, P.J., 2012. Traumatic brain injury and the frontal lobes: what can we gain with diffusion tensor imaging? Cortex 48 (2), 156–165. https://doi.org/10.1016/j.cortex.2011.06.020.

Zhou, H., Schafer, R.J., Desimone, R., 2016. Pulvinar-cortex interactions in vision and attention. Neuron 89 (1), 209–220. https://doi.org/10.1016/j.neuron.2015.11.034.

Zhu, Y., Jiang, X., Ji, W., 2018. The mechanism of cortico-striato-thalamo-cortical neurocircuitry in response inhibition and emotional responding in attention deficit hyperactivity disorder with comorbid disruptive behavior disorder. Neurosci. Bull. 34 (3), 566–572. https://doi.org/10.1007/s12264-018-0214-x.

Zilles, K., Palomero-Gallagher, N., Amunts, K., 2015. Myeloarchitecture and maps of the cerebral cortex. Elsevier eBooks, pp. 137–156. https://doi.org/10.1016/b978-0-12-397025-1.00209-8.

Zotev, V., Misaki, M., Phillips, R., Wong, C.K., Bodurka, J., 2018. Real-time fMRI neurofeedback of the mediodorsal and anterior thalamus enhances correlation between thalamic BOLD activity and alpha EEG rhythm. Hum. Brain Mapp. 39 (2), 1024–1042. https://doi.org/10.1002/hbm.23902.

Further Reading

Bahi-Buisson, N., Guerrini, R., 2013. Diffuse malformations of cortical development. In: Handbook of Clinical Neurology, pp. 653–665. https://doi.org/10.1016/B978-0-444-52891-9.00068-3.

Budisavljevic, S., Dell'Acqua, F., Zanatto, D., et al., 2016. Asymmetry and structure of the fronto-parietal networks underlie visuomotor processing in humans. Cereb. Cortex bhv348. https://doi.org/10.1093/cercor/bhv348.

Catani, M., Dell'Acqua, F., Vergani, F., et al., 2012c. Short frontal lobe connections of the human brain. Cortex 48 (2), 273–291. https://doi.org/10.1016/j.cortex.2011.12.001.

Corbetta, M., Kincade, J.M., Shulman, G.L., 2002. Neural systems for visual orienting and their relationships to spatial working memory. J. Cogn. Neurosci. 14 (3), 508–523. https://doi.org/10.1162/089892902317362029.

Forkel, S.J., Mahmood, S., Vergani, F., Catani, M., 2015. The white matter of the human cerebrum: part I the occipital lobe by Heinrich Sachs. Cortex 62, 182–202. https://doi.org/10.1016/j.cortex.2014.10.023.

Fridriksson, J., 2010. Preservation and modulation of specific left hemisphere regions is vital for treated recovery from anomia in stroke. J. Neurosci. 30 (35), 11558–11564. https://doi.org/10.1523/JNEUROSCI.2227-10.2010.

Friederici, A.D., Gierhan, S.M., 2013. The language network. Curr. Opin. Neurobiol. 23 (2), 250–254. https://doi.org/10.1016/j.conb.2012.10.002.

Guevara, M.A., Rizo Martínez, L.E., Robles Aguirre, F.A., Hernández González, M., 2012. Prefrontal–parietal correlation during performance of the towers of Hanoi task in male children, adolescents and young adults. Dev. Cogn. Neurosci. 2 (1), 129–138. https://doi.org/10.1016/j.dcn.2011.05.002.

Hans Berger, V., 1875. Nach JBechterew: Die Energie Des Lebenden, vol 102.

Hugdahl, K., 2015. Auditory hallucinations: a review of the ERC "VOICE" project. World J. Psychiatry 5 (2), 193. https://doi.org/10.5498/wjp.v5.i2.193.

Johnson, P.B., Ferraina, S., 1996. Cortical networks for visual reaching: intrinsic frontal lobe connectivity. Eur. J. Neurosci. 8 (7), 1358–1362. https://doi.org/10.1111/j.1460-9568.1996.tb01598.x.

Leiguarda, R., Merello, M., Balej, J., Starkstein, S., Nogues, M., Marsden, C.D., 2000. Disruption of spatial organization and inter-joint coordination in Parkinson's disease, progressive supranuclear palsy, and multiple system atrophy. Mov. Disord. 15 (4), 627–640. https://doi.org/10.1002/1531-8257(200007)15:4<627::aid-mds1006>3.0.co;2-5.

Loui, P., Schlaug, G., 2009. Investigating musical disorders with diffusion tensor imaging. Ann. N. Y. Acad. Sci. 1169 (1), 121–125. https://doi.org/10.1111/j.1749-6632.2009.04781.x.

Mountcastle, V.B., 1957. Modality and topographic properties of single neurons of CAT'S somatic sensory cortex. J. Neurophysiol. 20 (4), 408–434. https://doi.org/10.1152/jn.1957.20.4.408.

Mountcastle, V.B., Davies, P.W., Berman, A.L., 1957. Response properties of neurons of CAT'S somatic sensory cortex to peripheral stimuli. J. Neurophysiol. 20 (4), 374–407. https://doi.org/10.1152/jn.1957.20.4.374.

Nieuwenhuis, S., Heslenfeld, D.J., Alting von Geusau, N.J., Mars, R.B., Holroyd, C.B., Yeung, N., 2005. Activity in human reward-sensitive brain areas is strongly context dependent. Neuroimage 25 (4), 1302–1309. https://doi.org/10.1016/j.neuroimage.2004.12.043.

Norman, L.J., Carlisi, C.O., Christakou, A., et al., 2018. Frontostriatal dysfunction during decision making in attention-deficit/hyperactivity disorder and obsessive-compulsive disorder. Biol. Psychiatry Cogn. Neurosci. Neuroimaging 3 (8), 694–703. https://doi.org/10.1016/j.bpsc.2018.03.009.

Papagno, C., 2011. Naming and the role of the uncinate fasciculus in language function. Curr. Neurol. Neurosci. Rep. 11 (6), 553–559. https://doi.org/10.1007/s11910-011-0219-6.

Raichle, M.E., 2001. Bold insights. Nature 412 (6843), 128–130. https://doi.org/10.1038/35084300.

Ramnani, N., 2006. The primate cortico-cerebellar system: anatomy and function. Nat. Rev. Neurosci. 7 (7), 511–522. https://doi.org/10.1038/nrn1953.

Rauch, S.L., 1998a. Essential papers on obsessive-compulsive disorder. Am. J. Psychiatry 155 (8), 1130–1131. https://doi.org/10.1176/ajp.155.8.1130.

Rizzolatti, G., Arbib, M.A., 1998. Language within our grasp. Trends Neurosci. 21 (5), 188–194. https://doi.org/10.1016/S0166-2236(98)01260-0.

Rizzolatti, G., Fadiga, L., 2007. Grasping Objects and Grasping Action Meanings: The Dual Role of Monkey Rostroventral Premotor Cortex (Area F5), pp. 81–108. https://doi.org/10.1002/9780470515563.ch6.

Rudrauf, D., Lachaux, J.P., Damasio, A., et al., 2009. Enter feelings: somatosensory responses following early stages of visual induction of emotion. Int. J. Psychophysiol. 72 (1), 13–23. https://doi.org/10.1016/j.ijpsycho.2008.03.015.

Scher, M.S., Johnson, M.W., Holditch-Davis, D., 2005. Cyclicity of neonatal sleep behaviors at 25 to 30 weeks' postconceptional age. Pediatr. Res. 57 (6), 879–882. https://doi.org/10.1203/01.PDR.0000157678.84132.A8.

Scher, M.S., 2008. Ontogeny of EEG-sleep from neonatal through infancy periods. Sleep Med. 9 (6), 615–636. https://doi.org/10.1016/j.sleep.2007.08.014.

Skeide, M.A., Friederici, A.D., 2015. Response to Bornkessel-Schlesewsky et al. – towards a nonhuman primate model of language? Trends Cogn. Sci. 19 (9), 483. https://doi.org/10.1016/j.tics.2015.05.011.

Skeide, M.A., Brauer, J., Friederici, A.D., 2016. Brain functional and structural predictors of language performance. Cerebr. Cortex 26 (5), 2127–2139. https://doi.org/10.1093/cercor/bhv042.

Stevens, F.L., Hurley, R.A., Taber, K.H., 2011. Anterior cingulate cortex: unique role in cognition and emotion. J. Neuropsychiatry Clin. Neurosci. 23 (2), 121–125. https://doi.org/10.1176/jnp.23.2.jnp121.

Urbanski, M., Bartolomeo, P., 2008. Line bisection in left neglect: the importance of starting right. Cortex 44 (7), 782–793. https://doi.org/10.1016/j.cortex.2007.04.001.

Vogt, B.A., Nimchinsky, E.A., Vogt, L.J., Hof, P.R., 1995. Human cingulate cortex: surface features, flat maps, and cytoarchitecture. J. Comp. Neurol. 359 (3), 490–506. https://doi.org/10.1002/cne.903590310.

Yeatman, J.D., Feldman, H.M., 2013. Neural plasticity after pre-linguistic injury to the arcuate and superior longitudinal fasciculi. Cortex 49 (1), 301–311. https://doi.org/10.1016/j.cortex.2011.08.006.

Yeatman, J.D., Rauschecker, A.M., Wandell, B.A., 2013. Anatomy of the visual word form area: adjacent cortical circuits and long-range white matter connections. Brain Lang. 125 (2), 146–155. https://doi.org/10.1016/j.bandl.2012.04.010.

12

The Telencephalon: Cingulate Cortex

Hanna Markiewicz[1] *and Brandon M. Ascenzi*[2]

[1]Warsaw, Poland [2]Independent Neuroresearcher-Member of Marie Curie Alumni Association (MCAA)

OUTLINE

CINGULATE GYRUS

The cingulate gyrus (CgG) (or cingulate cortex) is located within the medial surface of the cerebral hemisphere. It has neural connections with many other structures including the contiguous cortices, spinal cord, amygdala, hippocampus, and other regions that are part of the limbic system. The name derives from the Latin word "cingulum," as it is a structure that surrounds the corpus callosum (cc) and appears like a belt as shown in Oszvald et al. (2012). Beneath the surface of the cingulate gyrus lies the cingulum which is a neural tract who connects CgG to the entorhinal cortex and forms one of the largest connections between the components of the limbic system.

Cingulate cortex is located dorsally and adjacent to the corpus callosum at the medial surface of both hemispheres. It consists of the cortical gray matter lining the superior and inferior borders of the cingulate sulcus. It originates beneath the rostrum of the cc, curves around the genu of cc, and extends above the body and courses inferiorly behind the splenium. Then, it connects with the parahippocampal gyrus by the isthmus of the CgG. Isthmus is continuous inferiorly with the parahippocampal gyrus at the temporal lobe. Callosal sulcus separates the CgG from the cc inferiorly and the cingulate sulcus separates it from the superior frontal gyrus, superiorly, described by Palomero-Gallagher et al. (2009), Vogt (2005), and Vogt et al. (2016). Location of cingulate cortex in the brain is showed in Fig. 12.1.

Structural and functional classification of cingulate cortex has been a controversial issue over years. Recently, a four-region neurobiological model was proposed based on structural, circuitry, and functional imaging observations. According to Palomero-Gallagher et al. (2009), this model the cingulate cortex is divided into four parts as anterior cingulate cortex (ACC), midcingulate cortex (MCC), posterior cingulate cortex (PCC), and the retrosplenial cortex (RSC). This division is showed in Fig. 12.2.

The anterior cingulate cortex corresponds to Brodmann's areas 24, 25, 32, and 33. Anteriorly it is separated from the prefrontal cortex by the cingulate sulcus, inferior and posterior boundaries are formed by callosal sulcus, and the imaginary line drawn between the ACC and MCC, respectively. Vogt et al. (2016) suggested that ACC can be further divided into subgenual (sgACC), perigenual (pgACC), and dorsal ACC

From Anatomy to Function of the Central Nervous System
https://doi.org/10.1016/B978-0-12-822404-5.00009-7

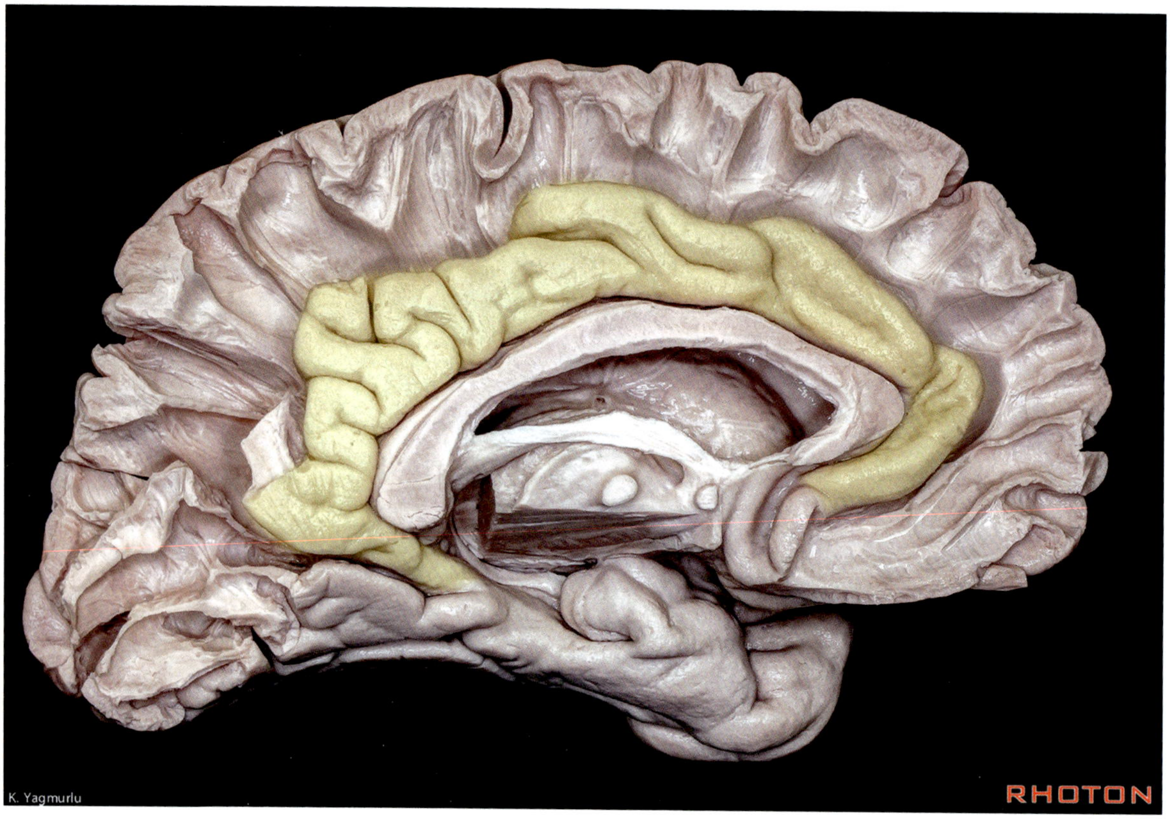

FIGURE 12.1 **Cingulate cortex (yellow).** Inside the lateral ventricle, the caudate nucleus lying on the thalamus is clearly visible (in the center). *Image courtesy of The Rhoton Collection® http://rhoton.ineurodb.org/. Dissector: K.Yagmurlu Drawings: Kaan Yagmurlu.*

(dACC; not included in the previously described neurobiological model). The perigenual anterior cingulate cortex (pgACC) receives information with regards to emotions and processes it and as a result it regulates endocrine and autonomic responses accordingly, as demonstrated by Jackson et al. (2016). It is also involved in affect related to pain. The subgenual anterior cingulate cortex (sgACC) is located below the pgACC, and below genu of the cc. Ge et al. (2020) showed that sgACC is involved in emotion processing and regulation of reward-based social behaviors and has a role in autonomic control, visceral integration, and conditioned learning. SgACC activity correlates with the expected value of options. In previous classifications, the dACC term corresponded to the anterior MCC, however, this is not the case in the recently published neurobiological model by Vogt et al. (2016).

The primary reason for separating the MCC from the ACC is that the connections are quite different with regards to their functions, what is described by Vogt et al. (2016). This separation is also supported by anatomic, cytoarchitectural, and receptor mapping studies conducted by Palomero-Gallagher et al. (2009). ACC projects to the dorsal nucleus of the vagus nerve (DMV) and the solitary nucleus (SN) to regulate the autonomic response. The ACC also have connections with amygdala, periaqueductal gray matter (PAG), dorsomedial and anterior nuclei of the thalamus; with these connections, ACC plays a prominent role in affect-regulation, processing pain and emotions, as demonstrated by Etkin et al. (2011), Tracey et al. (2007), and Ridderinkhof et al. (2004). Simple emotions activate ACC, specifically pgACC is more responsive to happiness while sgACC is responsive to sadness. Furthermore, Margualies et al. (2007) showed that sgACC is activated by emotional conflict regulation and fear-inhibition during extinction stress. Due to its affect-regulation function, the ACC plays a prominent role in psychopathology. Understanding the function of the ACC may help the clinicians provide better treatment options and clinical management.

MCC is located posteriorly to the ACC and anteriorly to the PCC above the cc. It consists of Brodmann's areas 24 and 32. The MCC, according to Vogt (2016), is thought to be responsible for cognitive processing, skeletomotor response to emotions and reward-based decision making. The MCC can be divided into two regions: the anterior MCC and the posterior MCC. Stevens et al. (2011) has showed that, compared to the anterior portion (aMCC), the posterior portion (pMCC) receives stronger projections from inferior parietal cortex, but weaker projections from pain-related thalamic nuclei. In contrast,

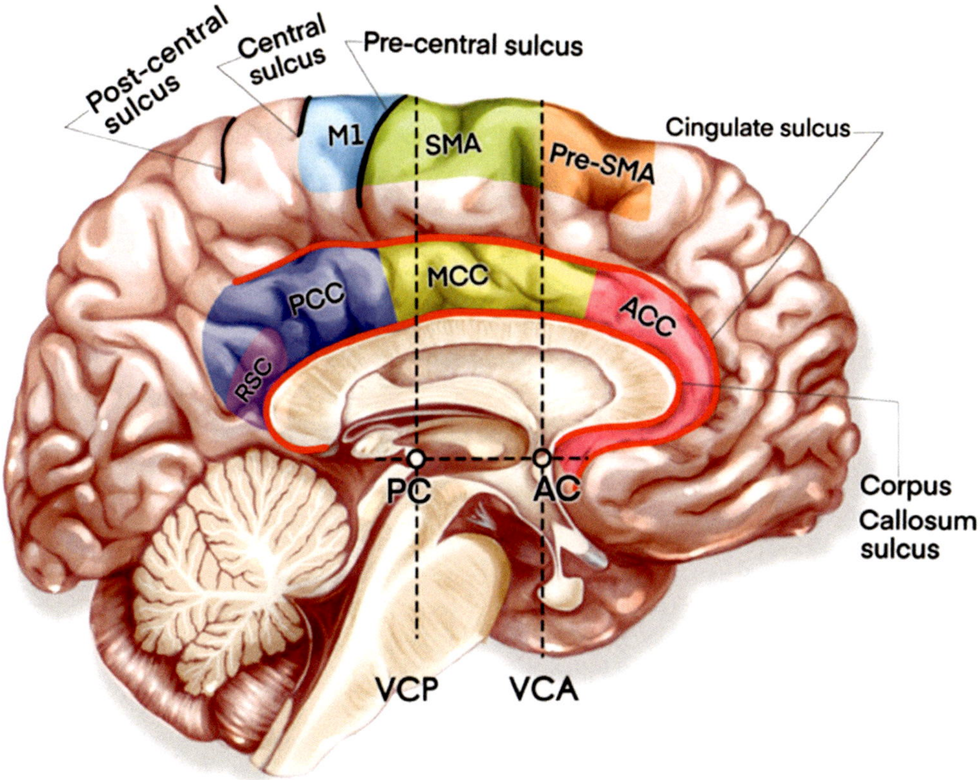

FIGURE 12.2 **The illustration of four cingulate subregions and nearby motor related areas.** Abbreviations: *AC*, anterior commissure; *ACC*, anterior cingulate cortex, *MCC*, midcingulate cortex; *PC*, posterior commissure; *PCC*, posterior cingulate cortex; *RSC*, retrospenial cortex; *VCA*, vertical plane at the AC; *VCP, VCA*, vertical plane at the PC. *Image courtesy of Gong, F., Jin, L., Song, Q., Yang, Z., Chen, H., Wu, J., 2022. Surgical techniques and function outcome for cingulate gyrus glioma, how we do it. Front. Oncol. 12. https://doi.org/10.3389/fonc.2022.986387.*

the aMCC receives strong projections from the pain-related thalamic nuclei and has strong connections with the amygdala, as described by Vogt et al. (2005). According to Stevens et al. (2011), the aMCC is also implicated in emotional appraisal, conflict-monitoring, approach–avoidance decisions, and willed control of actions. PMCC is involved in body-orientation and movement-execution demonstrating its involvement in motor-related tasks. Selection of action and expression of learned fear are other tasks which activate pMCC. Vogt et al. (2016) and Etkin et al. (2007) demonstrated that tasks requiring cognitive control, conflict-monitoring, error-detection or emotion-related (including fear) appraisal activate aMCC. The ACC and the MCC both have skeletomotor functions; however, the ACC is mainly involved in emotion and autonomic regulation, while the MCC has a prominent role in decision making activity. Furthermore, the ACC regulation of motor functions is mainly related to regulation facial expression and vocalization, while the MCC has unique skeletomotor spinal projections, showed by Vogt et al. (2016). Especially the sACC has the highest connectivity with limbic- (amygdala, ventral striatum, hippocampus) and autonomic- (hypothalamus) related areas, whereas MCC has the highest connectivity with cognitive- (dorsal prefrontal cortex, dPFC) and motor-related (premotor and motor cortex, BA 6,4, respectively) areas as demonstrated by Stevens et al. (2011). On the other hand, Vogt et al. (2016) showed that, MCC provides linkage between the pain processing and movement. MCC and pACC are both responsive to pain however, the MCC is more likely to be activated by thermal and pACC by visceral stimulation, according to Vogt et al. (2005).

Jumah et al. (2022) has recently showed that, the cingulate motor areas (CMAs) present within the cingulate sulcus next to the primary and secondary motor areas and are consistent with the location of pMCC, which is activated with execution of motor-related tasks. The CMAs are high-order motor areas that process information from both internal and external conditions. According to Strick et al. (1998), after processing the input such as emotional state signals from the limbic system, the CMAs send motor commands to primary and secondary motor cortices and spinal cord. The posterior cingulate cortex (PCC) contain Brodmann's areas 23, 29, 30, and 31. It is located at the posterior part of the CgG and is adjacent to the medial portion of the inferior parietal lobe. Leech et al. (2014) shows that boundaries of PCC are the parietooccipital sulcus posteriorly, the cingulate sulcus superiorly, the cc inferiorly and the MCC anteriorly.

The PCC is formed by the ventral and dorsal PCC. The ventral part is located superiorly to the splenium and the dorsal part is located posteriorly to the splenium. From studies conducted by Olson et al. (1996), it is known that on a functional level PCC is involved in the topographic and topokinetic memory, it generates the map of head and body in space by using the information of orbital position of the eyes. PCC is most active during the presentation of the familiar places and objects, as described in Sugiura et al. (2005). Furthermore, Ward et al. (2014) showed that, the PCC have reciprocal connections with the medial temporal lobe, particularly with entorhinal cortex and parahippocampal gyrus. It is involved in the associative learning and episodic memory, as demonstrated by Maddock et al. (2001) and Foster et al. (2023). It has connections with lateral intraparietal cortex, precuneus, anterior and lateral thalamic nuclei, caudate nucleus, orbitofrontal cortex, and anterior cingulate cortex. Due to these connections, it is thought to be involved in emotion, social behavior, attention, learning, creativity, spatial navigation, and scene processing' according to Pearson et al. (2011) and Maddock et al. (2003).

The retrosplenial cortex (RSC) is located posterior to the splenium and anteroinferior to the PCC and consists of Brodmann's areas 29 and 30. From studies conducted by Leech et al. (2014) and Mitchell et al. (2018), it is thought to involved in spatial orientation, navigation, landmark processing, visuospatial imagery, and episodic memory. According to paper published by Sugar et al. (2011), it has reciprocal projections with the visual cortex, postsubiculum, anterior thalamic nuclei, and hippocampus.

The RSC especially responds to nonmoving, permanent landmarks and it uses them in spatial judgments as demonstrated by Auger et al. (2013). The RSC damage produces both anterograde and retrograde amnesia indeed patients with lesions involving the RSC produces topographical disorientation where they can recognize the environmental landmarks, but they cannot use them to orient themselves, what was observed during studies conducted by Vann et al. (2009) and Valenstein et al. (1987).

Cilliers et al. (2016) showed that the cingulate cortex is supplied by the anterior cerebral artery due to its localization in the medial surface of the hemisphere. The anterior cerebral artery continues posteriorly as the pericallosal artery within the callosal sulcus and supplies the rostral part of the CgG with small cortical branches. The callosomarginal artery is the largest branch of the anterior cerebral artery and travels within the cingulate sulcus. The portion beneath the paracentral lobule of cingulate cortex is supplied by the callosomarginal artery. The posterior portion of the cingulate cortex is supplied by the precuneal artery, which is a branch of the pericallosal artery (see Chapter 2).

CYTOARCHITECTURE

The two least differentiated areas of anterior cingulate cortex (ACC) are areas 25 and 33 (Fig. 12.3). Area 33 is characterized by almost nonexisting laminar organization when area 25 has a simple laminar cytoarchitecture which contains broad and poorly distributed layers II–III. Neurons from these layers can form aggregations and homogenously distributed pyramids. Layer V is clearly defined, has densely packed neurons and merges with layer VI. Differences in size and density between anterior and posterior divisions of area 25 have been reported by Paxinos et al. (2012), Vogt (1995), and Palomero-Gallagher et al. (2009).

As demonstrated by Vogt (1995) and Paxinos et al. (2012), area 24 is divided based on progressively increasing laminar differentiation. Area 24a contains definable layers II and III, thin layer Va and characterized by neuron sparse, layer Vb. Layer VI is poorly defined. Area 24b consists of well-defined and broad layer III, thickest layer Va in the cingulate gyrus and broad layer Vb containing characteristically shaped pyramids and many spindle pyramidal neurons. Area 24b′ is located on the cingulate gyrus surface caudal to area 24b. Areas 24c, 24c′, and 24c′g are three components of ACC located on ventral bank of cingulate sulcus. Generally, the superficial layers II and III have similar or larger thickness in comparison to layers V and VI.

Area 32 differs from area 24 mainly by the presence of large pyramidal neurons in the deep layer III. It also contains a thick layer II and IV.

Characteristic components of ACC are spindle neurons of von Economo. They are located mainly in layer V and oriented vertically. These neurons have an apical and basal process that express NFPs, an efferent projecting axon and lightly spinous dendrites.

Functions, connections and cytoarchitecture of mid-cingulate cortex (MCC) differ from ACC and PCC and were described by Vogt (1995) and Paxinos et al. (2012). It can be characterized by large neurons in layer Vb of sulcal cortex. There can be observed two main cytoarchitectural differentiations within this region, between gyral and sulcal parts and between anterior and posterior parts. In gyral area the border between ACC and MCC is easily observed due to a number of obvious differences in the structure of area 24 in pACC and 24′ in MCC. In general, the MCC is thicker and has much broader layers II, III, Va, Vb, and VIa than ACC. Sulcal

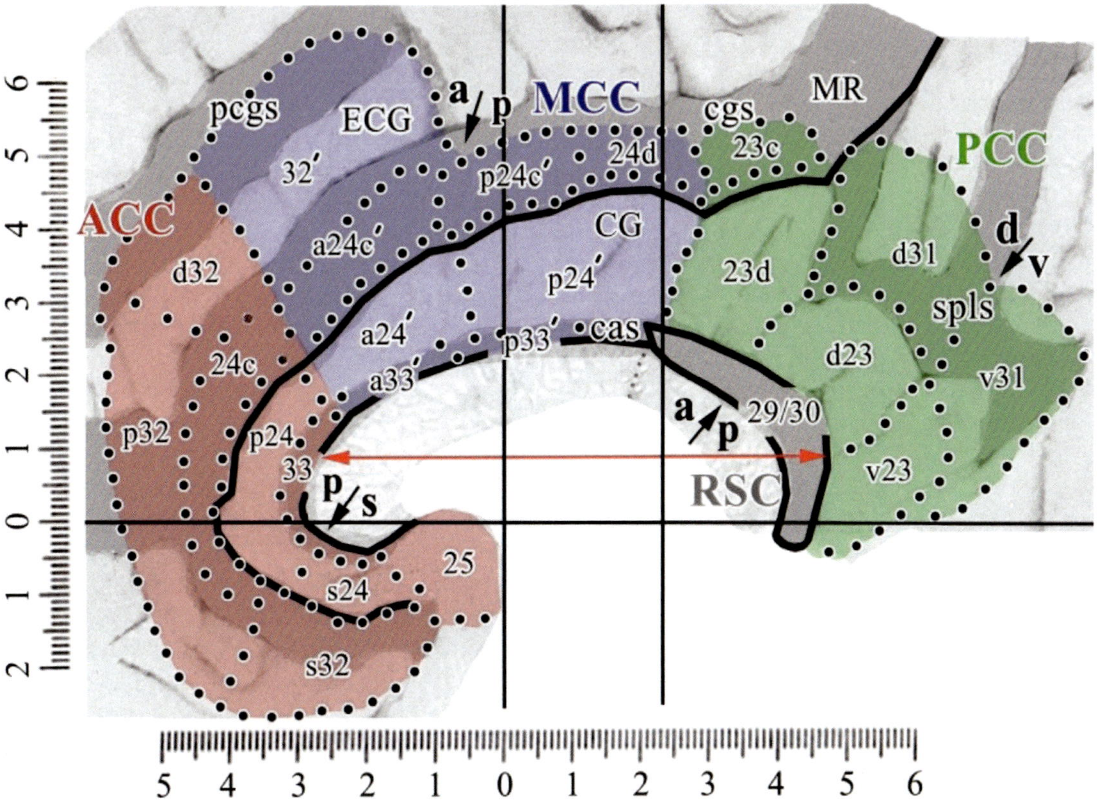

FIGURE 12.3 **Flat maps of the human cingulate cortices numbered second Brodmann's parcellation. The brain has been flipped for easy reading of the areas from left to right.** *Image courtesy of Mai, J.K., Paxinos, G., 2012. Cingulate cortex. In: The Human Nervous System. Elsevier Academic Press, Amsterdam, pp. 943–983. https://doi.org/10.1016/C2009-0-02721-4. Adapted and modified by Vogt B.A. (https://cingulumneurosciences.net).*

area 24c′ of MCC in comparison to its rostral counterpart in ACC is characterized by the decrease in thickness. Area 32 form a belt which is referred as external cingulate gyrus (EGC). The most differentiated layer IV is located in area s32 that progresses to the least differentiated in the area 32′. Well-developed layer III contains large pyramids that shift in position in the ventral-to-dorsal orientation.

The organization of retrosplenial cortex (RSC) is difficult to characterize due to many transitions, so it has to be interpreted in terms of differentiation trends. According to Paxinos et al. (2012), area 29 can be divided into two subdivisions which are unified by the granular layer. The latter is adjusted in lateral area 29l to layer I and it is located deep in the layer of medium-sized pyramidal neurons (layer III) in medial area 29 m. This granular layer is characterized by the presence of a large number of medium-sized pyramids that are NFP-ir. Dysgranular area 30 can be distinguished from adjacent areas by a relatively high density of NFP-ir neurons in layer IIIab. It is layers IIIab and IIIc differentiate from layer III of area 29 m. Layer IV contains aggregations of small neurons and it is often interrupted by bridges of neurons between layers IIIC and Va. Layer IV varies in thickness and has granular neurons either in a separate layer or intermingled with those from layer Va.

Posterior cingulate cortex (PCC) is similar to ACC in terms of homogeneity as demonstrated in Paxinos et al. (2012). Area 23 typically has well-developed layers IIIc, IV, and Va. In area 23a neurons in a layer Va tend to be small and difficult to observe. Even though area 23b is similar to the previously discussed one, there are differences mainly in V layer which contains larger pyramidal neurons. Border between these two areas is clearly indicated by the significant increase in the number of layer Va neurons and the thickness of layer IV. Area 23c is characterized by greater thickness of layers I–IV compared to the features of layers in areas 23a and 23b.

Area 31 forms lobules around the splenial sulci. Layer IV is the thickest on the cingulate gyrus and layer Va is well developed and contain large number of NFP-ir neurons. The main feature of its layer III is the highest density of large, NFP-ir neurons in the entire cingulate gyrus as demonstrated in Vogt et al. (2001) and Paxinos et al. (2012).

Two divisions can be observed in area 29. Vogt et al. (1995) has shown that the lateral division is characterized by high density but poor differentiation of cells. Therefore, there are only two layers: a dense and granular external pyramidal layer and layer V. Medial division contain layers III and IV of the external layers, layer V which can be easily divided into layer Va of large neurons and a sparse layer Va, and a thin layer VI.

Dysgranular belt can be identified based on the differences in organization of cingulate cortex. There is a significant shift in cytoarchitecture from agranular to granular cortices as described in Vogt et al. (1995) and Paxinos et al. (2012). Even though dysgranular belt can be observed in most of cingulate cortex, its cytoarchitecture differs among different dysgranular areas. A progressive differentiation in the caudal to rostral levels of the belt is associated with progressive decreases in the size and densities of neurons in all layers.

CONNECTIONS OF CINGULATE CORTEX

According to Rolls et al. (2019), the anterior cingulate cortex receives inputs from the orbitofrontal cortex (OFC) and amygdala which receive from ventral stream areas. The posterior cingulate cortex receives signals from dorsal stream areas including the parietal cortex and has connections to the hippocampal memory system.

Information related to reward is received by the pregenual anterior cingulate cortex from the medial orbitofrontal cortex (mOFC) while signals related to punishment and not receiving reward come from the lateral orbitofrontal cortex (lOFC). It can be seen on Fig. 12.4. The subgenual cingulate cortex (area 25) probably link rewards and punishers to the autonomic output which was shown by Rolls et al. (2019). Moreover, the orbitofrontal cortex sends inputs to the ACC about the value of stimuli which means the goals including the value of outcomes (the reward received) and the expected value. In combination with the midcingulate motor area, ACC interfaces actions to outcomes with taking into account the cost of actions to obtain the desired goal as demonstrated by Rolls et al. (2019).

Additionally, Rolls et al. (2019) demonstrated that the ACC has connections with the amygdala, some temporal cortical areas, the rostral superior temporal gyrus, the auditory superior temporal gyrus, and the dorsal bank of the superior temporal sulcus (Fig. 12.5).

The PCC has a significant connectivity with parahippocampal areas TF and TH (temporal areas F, H), which in turn project spatial information to the entorhinal cortex and thereby into the hippocampal episodic memory system (Fig. 13.4). The PCC provides a route for spatial including visuospatial information to reach the hippocampus, where it can be combined with object and reward related information to form episodic memory, as described in Rolls et al. (2019). Furthermore, the PCC is also connected with the parietal structures such as the precuneus and lateral parietal areas that, in turn, receive information from the dorsal visual stream and somatosensory areas. Due to these connections, the PCC is involved in spatial processing, action in space and some types of memory as shown in Rolls et al. (2019).

Additionally, Rolls et al. (2019) shows that the PCC is also connected with the orbitofrontal cortex what results in the access to the PCC and by its dorsal route into the hippocampal memory system by the value-related information, as well as by the ventral route via the perirhinal and entorhinal cortex to the hippocampal memory system by object-related information.

Evidence relating the subcallosal cingulate cortex in depression includes evidence that neurons in this region can respond to unpleasant stimuli, meaning that the subgenual cingulate cortex is probably overactive in depression. According to Rolls et al. (2019), the high functional connectivity in depression between the lateral orbitofrontal cortex and the subcallosal ACC may relate to more nonrewarding information transmission to the ACC which can contribute to depression. Due to the complexity of higher cognitive functions and the extensive connections of the cingulate cortex with other brain regions it plays role in the pathogenesis of schizophrenia. In particular, Jumah et al. (2022) have recently shown that the ACC is the area which shows abnormalities (reduced cortical gray matter, decreased gyrus volume and disrupted neuronal arrangement) in schizophrenic patients compared to healthy ones (for detail, see later in the paragraph "Psychopathological aspects associated to the cingulate cortex").

FUNCTIONAL ASPECTS

ACC is closely related to emotion and peripheral autonomic feedback. Different scientists have examined the connection between the activity of ACC and situations when the research subject was exposed to emotional content. According to Paxinos et al. (2012), between sACC and pACC one of the major differences is that the negatively valuable and episodic memories seem to be the most frequently stored in sACC, while pACC stores mostly memories associated with happiness. Hypofunctionality of anterior cingulate cognitive division was reported in ADHD by Bush et al. (1999). Schizophrenia is also characterized by relative underactivity of the rostral ACC during commission of errors as determined by Laurens et al. (2003). Skeletomotor and cognitive functions have been associated with the

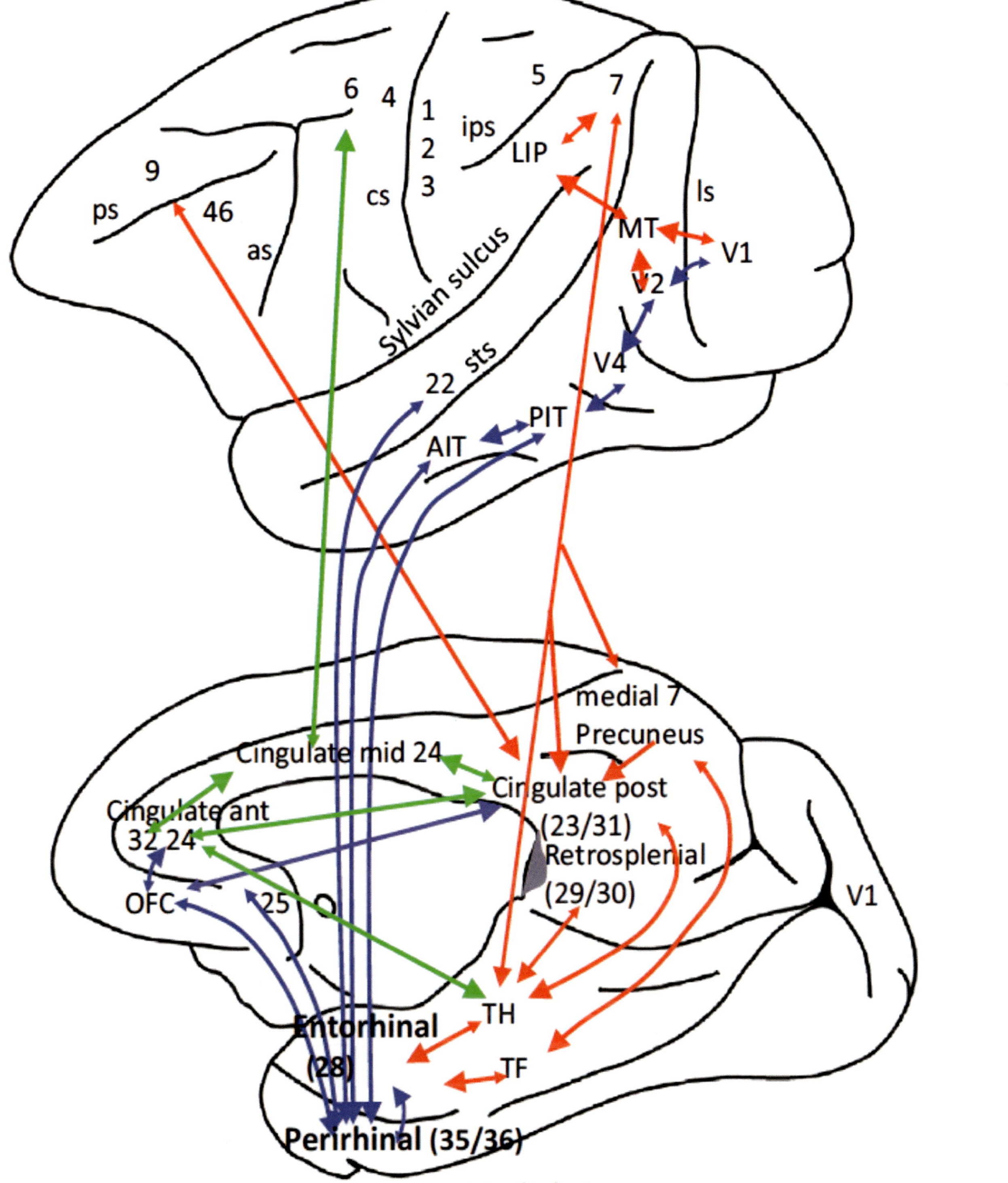

FIGURE 12.4 **The connections of the anterior and posterior cingulate cortex with their input areas, and their outputs to the hippocampal memory system.** A medial view of the macaque brain is shown below, and a lateral view is above. The green arrows show the convergence of reward or outcome information from the ACC and of information about actions from the posterior cingulate cortex to the midcingulate motor area,

activity of MCC. Talaraich et al. (1973) has shown that complex and context-depending gestures such as touching, rubbing, kneading, pressing the fingers or hands together can be evoke by the electrical stimulation of MCC. Environmental influence on this movements have been reported by Posner et al. (1988). Furthermore, sequences of complex finger apposition movements can modify the blood flow in this region which was observed using the human functional imaging by Kwan et al. (2000). Different difficulties arise from paradoxes that are not immediately solvable such as the mechanisms by which cognitive and emotional networks interact in cingulate cortex. MCC provides a cognitive interface with skeletomotor systems via projections to the spinal cord, striatum, supplementary/presupplementary motor cortices, periaqueductal gray (PAG), and pontine/cerebellar systems. Unfortunately, the mechanism by which MCC and sACC interact are not completely understood even though numerous studies on this subject were conducted by Urry et al. (2009) and Critchley et al. (2005). The MCC activity is responsible during decisions selecting between aversive or reward outcomes. Anterior and posterior MCC show different involvement in emotion as determined by Vogt et al. (2003). The aMCC involvement is limited to fear, while pMCC appears to have no activity during simple emotions. The pMCC is associated with the caudal cingulate premotor area (PMD) and neurons located in this area have a shorter duration period between discharge and muscle contractions than the neurons in the rostral cingulate premotor area. The caudal area seems to be more easily driven by passive movements and it has a lesser role in the reward coding of behavior than the rostral area, what was described by Paxinos et al. (2012).

Highest levels of metabolism in the brain were observed in the PCC. Bonhomme et al. (2001) showed that during sleep, metabolism of PCC is significantly regulated while during the altered consciousness under general anesthesia with propofol, its metabolism is reduced. The PCC appears to be involved in topographic and topokinetic memory (orientation of the body in space). Moreover, mental navigation along memorized routes induced the elevation of blood flow in this region, what was shown in studies conducted by Amorim et al. (1997), Ghaem et al. (1997), and Maguire et al. (1998). According to Sugiura et al. (2005), ventral posterior cingulate cortex (vPCC) is particularly active during the exposure to familiar places over objects while dorsal posterior cingulate cortex (dPCC) shows a high level of activity during presentations of familiar places and objects over unfamiliar places and objects. Furthermore, Sugiura et al. (2005) showed that, the dPCC, in contrast to vPCC, has a more profound role in visuospatial functions and it receives input via the dorsal visual stream. The vPCC evaluates emotional and nonemotional sensory inputs by the connections with sACC which is the place where long-term memories of valuable events are stored and RSC where self-references episodic memories are stored. Information is then directed to MCC for motor system output. In opposition, the dPCC shows a closer relation with the MCC and has a closer association with sensorimotor orientation in space and fast adjustments to visuospatial needs as described by Paxinos et al. (2012).

Some problems have arisen around the distinction between the unique contributions to brain functions by the primate retrosplenial cortex (RSC) and PCC since they are adjacent and strongly interconnected. Additionally in human brain RSC has a small size in relation to the spatial resolution of imaging modalities. However, when involvement of RSC was shown in human studies done by Mitchell et al. (2018), it was related with working and long-term memory and visuospatial functions. Furthermore, functional imagining has shown that RSC is a part of network that mediates topokinetic navigation. This part is activated during tasks which require research subject to employ memories in relation to previously learned surrounding environment. It additionally has a role in the retrieval of memories about familiar places and objects. RSC and its thalamic afferents are involved in accessing to stored short- and long-term memories as described by Paxinos et al. (2012).

Hypometabolisms of the cingulate cortex have been recently observed by Hugon et al. (2022) in several neurological and psychiatric diseases including mild cognitive impairment due to Alzheimer's disease, severe depression, and internet gaming disorder.

which then projects to premotor areas including the premotor cortex area 6 and the supplementary motor area. This provides connectivity for action—outcome learning. The anterior cingulate cortex receives reward outcome information from the orbitofrontal cortex (OFC). The posterior cingulate cortex (23 and 31) receives information about actions from the parietal cortex. This cingulate connectivity is compared with that of the hippocampus, which receives information from the ventral "what" processing stream (blue) and the dorsal "where" or "action" processing stream (red), as described in the text. *as*, arcuate sulcus; *cs*, central sulcus; *ios*, inferior occipital sulcus; *ips*, intraparietal sulcus; *ls*, lunate sulcus; retrosplenial cortex (29, 30) is the small region in primates including humans behind the splenium of the corpus callosum shaded gray: it is present in rodents, which do not have a posterior cingulate cortex (Vogt, 2009); *sts*, superior temporal sulcus; 4, 6, motor and premotor cortex. *Image Courtesy of Rolls, E.T., 2019. The cingulate cortex and limbic systems for emotion, action, and memory. Brain Struct. Funct. 224, 3001–3018. https://doi.org/10.1007/s00429-019-01945-2.*

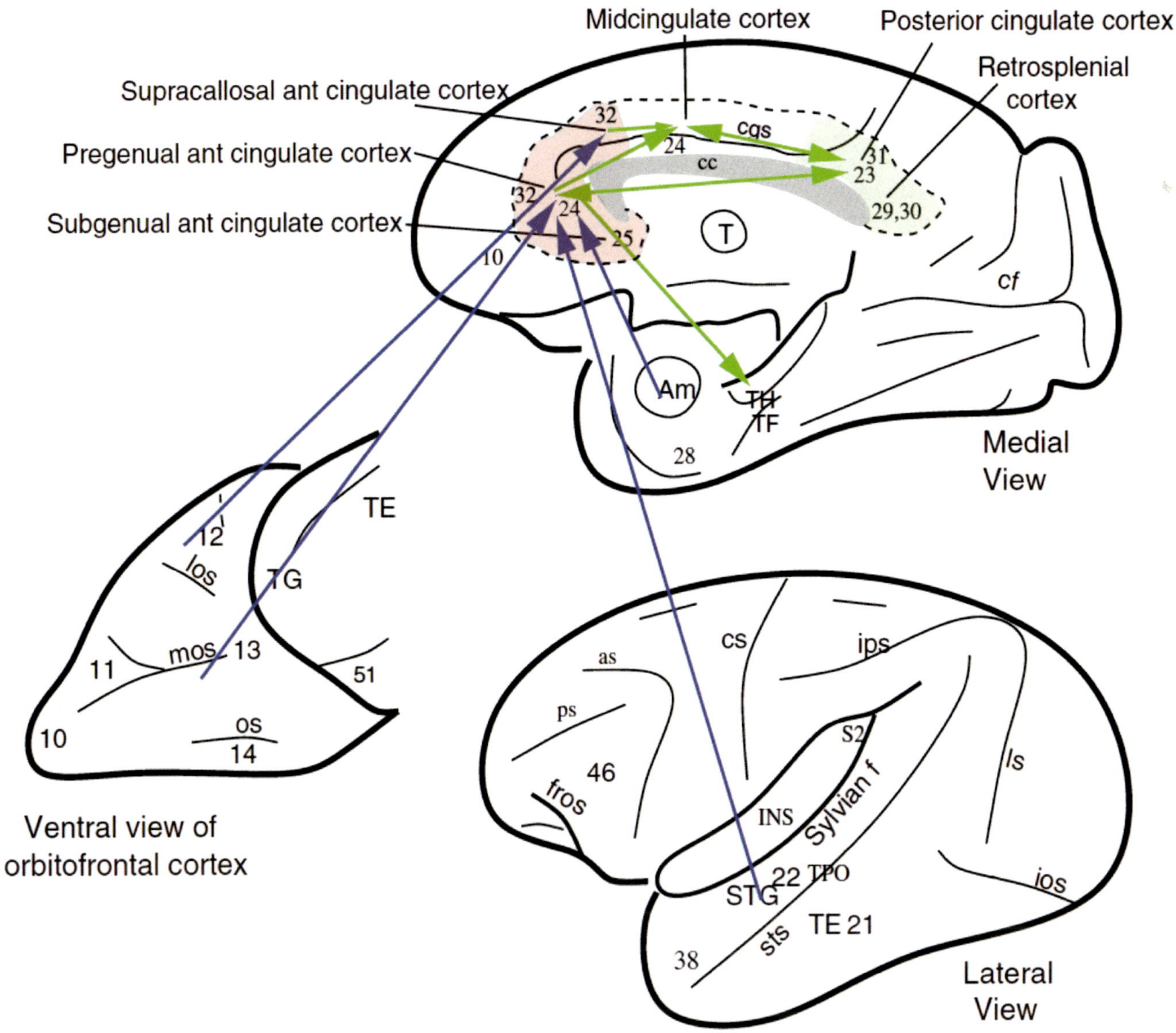

FIGURE 12.5 **Connections of the anterior cingulate cortex shown on views of the primate brain (see text).** The arrows show the main direction of connectivity, but there are connections in both directions. The supracallosal anterior cingulate cortex is also termed the anterior part of the midcingulate cortex, and is distinct from the posterior part of the midcingulate cortex (pMidcingulate). Connections reach the pregenual cingulate cortex especially from the medial/mid-orbitofrontal cortex; and reach the supracallosal anterior cingulate cortex especially from the lateral orbitofrontal cortex. Connections to the anterior cingulate cortex from the temporal lobe are from the (auditory) superior temporal gyrus (STG), from the visual and auditory cortex in the superior temporal sulcus; and from the amygdala. *as*, arcuate sulcus; *cc*, corpus callosum; *cf*, calcarine fissure; *cgs*, cingulate sulcus; *cs*, central sulcus; *ios*, inferior occipital sulcus; *ls*, lunate sulcus; *mos*, medial orbital sulcus; *os*, orbital sulcus; *ps*, principal sulcus; *sts*, superior temporal sulcus; *Sf*, Sylvian (or lateral) fissure (which has been opened to reveal the insula); *Am*, amygdala; *INS*, insula; *TE (21)*, inferior temporal visual cortex; *STG (22)*, superior temporal gyrus auditory association cortex; *TF and TH*, parahippocampal cortex; *TPO*, multimodal cortical area in the superior temporal sulcus; *38, TG*, temporal pole cortex; *13, 11*, medial orbitofrontal cortex; *12*, lateral orbitofrontal cortex; *23, 31*, posterior cingulate cortex; *29, 31*, retrosplenial cortex; *51*, olfactory (prepyriform and periamygdaloid) cortex. A cytoarchitectural map of the human cingulate cortex is provided in Fig. 12.3 of this chapter. *Image courtesy of Rolls, E.T., 2019. The cingulate cortex and limbic systems for emotion, action, and memory. Brain Struct. Funct. 224, 3001–3018. https://doi.org/10.1007/s00429-019-01945-2.*

PSYCHOPATHOLOGICAL ASPECTS ASSOCIATED WITH THE CINGULATE CORTEX

Reduced activation and structural differences in the *cingulate cortex* have been reported in several psychiatric conditions. Emotional numbing and suppression of emotions are common symptoms of these conditions. People with schizophrenia have a smaller *anterior* CgG when compared with controls. The volume of anterior CgG was found to be lower in people with schizophrenia. Furthermore, CgG of people with schizophrenia has reduced cortical gray matter and disrupted neuronal rearrangement as determined by Haznedar et al. (2004), Wang et al. (2007), and Baiano et al. (2007). It has been found by Minzenberg et al. (2009) that in patients with schizophrenia, which often includes a flattening of affect, the aMCC has a reduced activation during executive tasks.

In people with posttraumatic stress disorder (PTSD), it has been found in study conducted by Etkin et al. (2007), that both their sgACC and aMCC activities are reduced during fear conditioning. Due to these findings, Etkin et al. (2007) proposed that reduced activation in sgACC may indicate impairment in emotion-regulation and fear-extinction, whereas the reduced activation in aMCC may relate to reduced experience of negative emotion.

In individuals with psychopathy, absence of the normal activations of ACC, insular and OFC cortices has been reported by Etkin et al. (2007), whereas individuals with social phobia showed an overactivation of these regions. Although results are not consistent across studies, according to Shin et al. (2010), most anxiety disorders besides PTSD show excessive activation in both the pACC and MCC. Patients with Autism Spectrum Disorders (ASD) and Attention Deficit and Hyperactivity Disorder (ADHD) show reduced metabolism, abnormal functional responses, and abnormal functional connectivity in PCC in study conducted by Leech et al. (2014). In Alzheimer's disease especially in prodromal phase, the RSC is one of the regions that undergoes pathological changes. Furthermore, patients with pathologies involving the RSC spatial disorientation, anterograde and retrograde amnesia can be seen in results from studies done by Valenstein et al. (1987) and Tan et al. (2013).

A relationship between **major depressive disorders (MDD)** and subgenual anterior cortex (sgACC) is well known because numerous studies have been conducted on this topic by de Kwaasteniet et al. (2013), Connolly et al. (2013), Schwartz et al. (2019), Hsu et al. (2014), and Chai et al. (2016). During the examinations of sgACC and medial temporal lobe (MTL), two different abnormalities were observed by de Kwaasteniet et al. (2013): the increase in the connection between the sgACC and MTL and the reduced white matter integrity in the uncinate fascicle. Another experiment demonstrates the abnormal state in resting-state functional connectivity (RSFC) of the sgACC in adolescents. Connections between amygdala and the insula also showed high activity in study done by Connolly et al. (2013). Higher hyperactivity of sgACC results in greater the level of sadness of the individuals during MDD as demonstrated by Schwartz et al. (2019). Hsu et al. (2014) has reported that hyperactivity of sgACC is also associated with stress in the individuals with MDD. Study conducted by Chai et al. (2016) showed that children with family members with MDD have abnormalities in the amygdala and the sgACC which was found using the functional magnetic resonance imaging (fMRI).

Previously mentioned studies confirming the relationship between MDD and sgACC have led to studies on MDD treatment. Repetitive transcranial magnetic stimulation was administered to the left dorsolateral prefrontal cortex by Philip et al. (2018), resulting in improved connectivity between the amygdala and ventromedial prefrontal cortex, as well as improved sgACC and default mode network (DMN) connectivity. Further studies to better understand the effects of electroconvulsive therapy (ECT) in MDD were conducted. Liu et al. (2015) established that ECT is an efficient treatment method with its effect on the sgACC. Other studies resulted in the opposite of expectations, the sgACC volumes of pediatric-onset MDD patients were compared with adult-onset MDD patients and it was concluded that the sgACC volumes of pediatric-onset patients were on average the same as healthy individuals as described by Akdoğdu and Erbas (2021). Therefore, the use of sgACC as a biomarker created some doubts, according to Jaworska et al. (2014). Although there was a difference in sgACC volumes between healthy individuals in the younger age group, Gaffrey et al. (2010) found that there are abnormal connections between multiple cortical and subcortical regions of the sgACC in children with a history of preschool-onset MDD. There are many reasons why antidepressants used in patients with MDD do not always provide full benefit. The striking point in the experiments is that the conditions in which the antidepressant can be more effective are neurologically determined. It has been determined by Niida et al. (2014) that regular antidepressant use will be effective in cases of atrophy detected in the sgACC. Self-generated thoughts (SGTs) are the symptoms of MDD. Maltreated children are at greater risk of developing MDD than others and, in these cases, positive SGTs can be associated with depression. Moreover, RSFC with sgACC is reduced in the frontoparietal network in these children as determined by Niida et al. (2014) and Hoffmann et al. (2018).

A study has been conducted by Fiori et al. (2021) to establish the relationship between differential expression in the different parts of brain and MDD. Small RNA sequencing was used to quantify miRNA expression in the anterior cingulate cortex (ACC) and habenula in individuals with MDD and healthy controls. From the obtained results, scientists assessed the expression of the predicted target genes of differentially expressed miRNAs in the brain and found that the expression of Erb-b2 receptor tyrosine kinase 4 (ERBB4), a gene encoding a neuregulin receptor, was downregulated in both regions, and was influenced by miRNA-323a-p in vitro. Furthermore, studies on mice were conducted by Fiori et al. (2021). Mice in which miR-323-3p was overexpressed or knocked-down displayed increased and decreased emotionality, respectively. Additionally, these mice displayed significantly downregulated and upregulated expression of Erb-b4, respectively. These findings indicate that the expression of miRNAs in brain is important in the pathology of MDD and emphasize the involvement of miR-323a-3p and ERBB4 in this phenotype.

In study conducted by Fullana et al. (2019) researchers knocked down the astroglial glutamate transporters GLAST and GLT-1 in infralimbic cortex (IL, rodent equivalent of vACC) using small interfering RNA (siRNA). GLAST and GLT-1 siRNA microinfusion in IL evoked a depressive-like phenotype, associated with a reduced serotonergic function and reduced forebrain BDNF expression. Neither effect occurred after siRNA application in the adjacent prelimbic cortex (PrL), thus emphasizing the critical role of vACC/IL in MDD pathogenesis.

According to Rakel (1999), depression is a chronic condition associated with physical symptoms. The response to the treatments can be monitored by several potential biomarkers, for example, sgACC, amygdala, and anterior insula parts of the brain, as described by Chakrabarty et al. (2016). Studies conducted by He et al. (2019) have shown that there is even a relationship between the expectation of social loss and increase of sgACC activity. Increased blood flow in sgACC has been reported by Yang et al. (2009) in depressed individuals and decreased blood flow can be associated with an improvement in depression. Other studies conducted on this topic, for example, by Lee et al. (2018), report that there are also significant differences in abnormal functional connectivity (FC) in the sgACC and default mode network (DMN). Differences between sexes were reported by Akdoğdu and Erbas (2021) and it has been found that girls in early adolescence are more susceptible to psychological disorders. Reduced gray matter volumes (GMVs) were observed in the sgACC of precocious girls by Okada et al. (2020). Differences in sgACC FC according to sex were also reported. Women are more prone to depression and brain activity higher in negative emotions than men, as determined by Li et al. (2017). Differences can be also observed in the prefrontal-limbic system, which results in different attitudes toward help-seeking. Studies done by Li et al. (2014) have shown that women are more likely to seek help and that they are more receptive to advice.

Liu et al. (2015) have reported that electroconvulsive therapy has been shown to reduce the sgACC activity of depressed individuals.

Bipolar disorder (BD) is characterized by the change of euphoric and unhappy moods. According to Dilsaver et al. (1989), a substantial portion of individuals with this disease commit suicide. Studies conducted on this topic have shown that sgACC is involved in the causes of suicide. Wang et al. (2020) determined that an abnormal relationship between the sgACC and raphe nuclei plays an active role in individuals committing suicide. In patients with BD, the decrease in sgACC metabolism and GMV in the cortex have been reported by Drevets et al. (2008). In patients with this disease also an increased connectivity between the sgACC and amygdala was observed by Rey et al. (2016). The entorhinal cortex and sgACC, which are two significant parts of mood regulation, are important in bipolar patients since the link between these two regions show hyperactivity in patients with BD. The entorhinal cortex is not the only region where the sgACC junction shows hyperactivity; it is also thought to have higher activity in its junctions with sgACC, pregenual ACC, dorsomedial prefrontal cortex, and insula, as described by Apazoglou et al. (2019) and Akdoğdu and Erbas (2021).

Addictions can be divided into two categories as drug-related (cigarette, drug, alcohol addiction) and nonsubstance addiction (gambling, food, internet, mobile phone, and exercise addiction) accordingly to Zou et al. (2017), Weinstein et al. (2014), and Akdoğdu and Erbas (2021). In cigarette addiction, it is believed that the feeling of longing underlines the craving to smoke. Studies have found that sgACC affects the sense of longing which can be used in the treatment. sgACC hyperactivity increases the feeling of longing. Therefore, awareness-based interventions have been found effective in the treatment because these interventions reduce sgACC activity in a study conducted by Westbrook et al. (2013). Orbital and medial prefrontal cortex (OMPFC) network is also linked with the craving triggering smoking. OMPFC region is linked with sgACC which results in the relationship between increased connectivity between sgACC and OMPFC, and the feeling of craving, as demonstrated in study by Janes et al. (2014). ACC activity is seen in early withdrawal. Therefore, according to Allenby et al. (2020) and Akdoğdu and Erbas (2021), if the activity if the ACC can be reduced during the early withdrawal period, it may be easier to quit smoking.

The difference between smoking addiction and internet addiction (IAD) is that IAD shows insufficient ACC activity. It results in decreased connectivity between the ACC and amygdala reported by Cheng et al. (2020). Individuals with IAD have reduced error-monitoring ability, and a study done by Dong et al. (2013) on this revealed that ACC activity is increased during error tracking. Addiction to games has other more significant problems. Individuals addicted to games identify with their characters in the game more than themselves, as described by Choi et al. (2018). Kim et al. (2019) reported that the ACC shows higher activity when thinking about the game and its activity also varies in relation to playing time. In recovery process, a decrease in ACC activity has been reported by Dong et al. (2019). Mobile phone addiction, similar to games addiction, shows lower ACC compared to healthy subjects in study done by Horvath et al. (2020). In drug users, ACC hypoactivity has been observed in tasks requiring emotional attention by Goldstein et al. (2009). The effects of dopamine (DA) on glutamate-mediated excitatory postsynaptic currents (EPSCs) in pyramidal neurons of layer II/III of the ACC in adult mice were reported by Darvish-Ghane et al. (2016). Findings posits a pre- and postsynaptic mode of pyramidal neuron EPSC modulation in mice ACC by DA.

Another study done by Darvish-Ghane et al. (2020) determining DA modulation of synaptic transmission within the ACC and pain related phenotypes was conducted. Data obtained indicate that absence of D_1 receptors (D_1Rs) in the ACC enhanced peripheral sensitivity to mechanical stimuli and D_1R activation decreased glutamatergic synaptic transmission in ACC neurons.

Serotonin 5-HT_{1A} receptors are highly expressed on layer II of cortical neurons and are known to have inhibitory actions. 5-HT_{1A} receptor agonists treatment in autistic cases with SHANK3 mutations and possibly other cases may restore excitatory and inhibitory balance that attenuates core symptoms, as reported by Dunn et al. (2020). A series of experiments investigated the effects of acute tandospirone treatment on spatial learning and self-grooming, subchronic treatment of tandospirone on self-grooming behavior and the effect of tandospirone infusion into the anterior cingulate on self-grooming behavior. These results from study done by Dunn et al. (2020) suggest that stimulation of cortical 5-HT_{1A} receptors may reduce repetitive behaviors and cognitive impairments as observed in autism spectrum disorder, possibly by attenuating an excitation/inhibition imbalance.

Fronto-limbic structures and serotonin 5-HT_{2A} receptors have been implicated in the pathophysiology and treatment of **obsessive-compulsive spectrum disorders** by Mora et al. (2020).

Reduced top-down control by cortical areas is assumed to underlie **pathological forms of aggression**. While the precise underlying molecular mechanisms are still elusive, according to Jager et al. (2020), it seems that balancing the excitatory and inhibitory tones of cortical brain areas has a role in aggression control. Study on mice conducted by Jager et al. (2020) indicate that aggressive behavior can be linked to altered expression of neuroligin-2, a membrane-bound cell adhesion protein that mediates synaptogenesis of mainly inhibitory synapses. A GABA-degrading enzyme ABAT and a potassium-chloride channel involved in $GABA_A$ receptor signaling (KCC2), seem to be involved in modulating aggressive and anxious behaviors observed in BALB/cj mice through affecting GABA signaling in the ACC.

The rostral anterior cingulate cortex (rACC) is a key structure in mediating the negative affective component of chronic pain. Brain-Derived Neurotrophic Factor (BDNF) is known to play a critical role in activity-dependent synaptic plasticity, learning, and memory. It has been shown that BDNF signaling in the rACC might be involved in spontaneous pain—related aversion, but its underlying mechanism is still largely unknown. To address this question, Wang et al. (2017) conducted a study where the mRNA and protein levels of BDNF in the rACC after nerve injury were measured and it has been found that BDNF expression was markedly higher in nerve-injured rats than in controls.

SUBFALCINE HERNIATION

Subfalcine herniation is the most common brain herniation syndrome and this is due to increased intracranial pressure that forces the cingulate gyrus beneath the edge of falx cerebri (see Fig. 12.6). It usually does not produce severe symptomatology but symptoms like nausea, vomiting, confusion may be seen due to intracranial pressure. If the herniation progresses, the anterior cerebral artery may be compressed against the falx cerebri and ipsilateral anterior cerebral artery stroke may present with contralateral lower extremity weakness, as reported by Byard et al. (2013).

CINGULOTOMY

In patients with pain refractory to medications, major depression and obsessive-compulsive disorder (OCD) bilateral cingulotomy can be used. A bilateral cingulotomy involves the cingulotomy via stereotactic ablation or open craniotomy. This procedure's aim is to create a lesion in the *ACC* that damages the supracallosal fibers of the *cingulum*. In patients with refractory pain the cingulotomy is effective, however it does not eliminate the pain but is thought to modulate the perception of it.

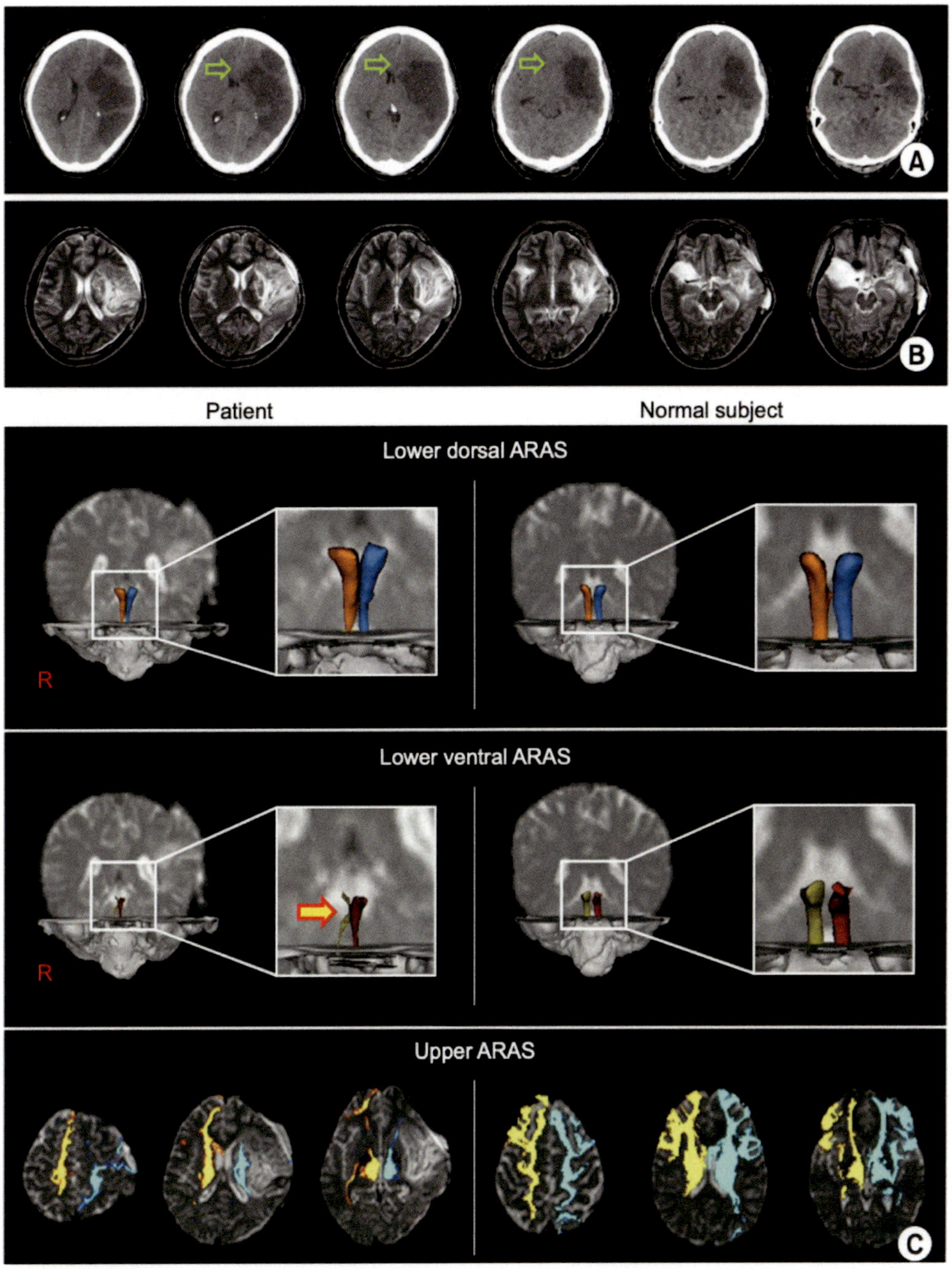

FIGURE 12.6 **Subfalcine herniation.** Brain computed tomography images (A) at 3 days after onset show subfalcine herniation (arrows), and brain magnetic resonance images (B) at 2 weeks after onset show a leukomalactic lesion in the left middle cerebral artery area. (C) Results of diffusion tensor tractography for the ascending reticular activating system (ARAS) of the patient and a normal subject (48-year-old female). The right lower ventral ARAS shows severe narrowing (arrow), whereas the lower portions of the lower dorsal ARAS and the left lower ventral ARAS reveal mild narrowing. The upper ARAS showed decreased neural connectivity in the left hemisphere. *Image courtesy of Jang, S.H., Kwon, Y.H., August 2018. Injury of the lower ascending reticular activating system by subfalcine herniation in a patient with a cerebral infarct. Ann. Rehabil. Med. 42(4): 639–641. https://doi.org/10.5535/arm.2018.42.4.639.*

Then the individual feels the pain but this does not affect his emotional state as the pain feels like a tickle. In patients with refractory OCD bilateral cingulotomy was found effective for the treatment of the refractory OCD by Steele et al. (2008) and Lévêque et al. (2013).

CINGULATE LESIONS

Although gliomas of the CgG are extremely rare, studies on resection of them by Oszvald et al. (2012), Tate et al. (2011), and von Lehe et al. (2009) shed light on the functional dominance of the CgG. Studies regarding the CgG glioma resections have shown the most common preoperative symptom to be seizures, and supplementary motor area (SMA) syndrome as the most common postoperative symptom followed by hemiparesis and aphasia. All studies have shown proof that postoperative SMA syndrome is encountered in lesions located in the anterior CgG. The authors in Oszvald et al. (2012) also argued that the SMA syndrome is caused by the removal of projections rather than the lesion itself.

References

Akdoğdu, B.S., Erbas, O., 2021. Subgenual anterior cingulate cortex and psychiatric disorders, 6, pp. 45–51. https://doi.org/10.5606/dsufnjt.2021.025.

Allenby, C., Falcone, M., Wileyto, E.P., Cao, W., Bernardo, L., Ashare, R.L., et al., 2020. Neural cue reactivity during acute abstinence predicts short-term smoking relapse. Addict. Biol. 25, e12733.

Amorim, M.A., Glasauer, S., Corpinot, K., Berthoz, A., April 1997. Updating an object's orientation and location during nonvisual navigation: a comparison between two processing modes. Percept. Psychophys. 59 (3), 404–418. https://doi.org/10.3758/bf03211907.

Apazoglou, K., Küng, A.L., Cordera, P., Aubry, J.M., Dayer, A., Vuilleumier, P., et al., 2019. Rumination related activity in brain networks mediating attentional switching in euthymic bipolar patients. Int. J. Bipolar Disord. 7, 3.

Auger, S.D., Maguire, E.A., 2013. Assessing the mechanism of response in the retrosplenial cortex of good and poor navigators. Cortex 49 (10), 2904–2913. https://doi.org/10.1016/j.cortex.2013.08.002.

Baiano, M., David, A., Versace, A., Churchill, R., Balestrieri, M., Brambilla, P., 2007. Anterior cingulate volumes in schizophrenia: a systematic review and a meta-analysis of MRI studies. Schizophr. Res. 93 (1–3), 1–12. https://doi.org/10.1016/j.schres.2007.02.012.

Bonhomme, V., Fiset, P., Meuret, P., Backman, S., Plourde, G., Paus, T., Bushnell, M.C., Evans, A.C., March 2001. Propofol anesthesia and cerebral blood flow changes elicited by vibrotactile stimulation: a positron emission tomography study. J. Neurophysiol. 85 (3), 1299–1308. https://doi.org/10.1152/jn.2001.85.3.1299.

Bush, G., Frazier, J.A., Rauch, S.L., Seidman, L.J., Whalen, P.J., Jenike, M.A., Rosen, B.R., Biederman, J., June 15, 1999. Anterior cingulate cortex dysfunction in attention-deficit/hyperactivity disorder revealed by fMRI and the Counting Stroop. Biol. Psychiatry 45 (12), 1542–1552. https://doi.org/10.1016/s0006-3223(99)00083-9.

Byard, R.W., 2013. Patterns of cerebral and cerebellar herniation. Forensic Sci. Med. Pathol. 9 (2), 260–264. https://doi.org/10.1007/s12024-012-9339-9.

Chai, X.J., Hirshfeld-Becker, D., Biederman, J., Uchida, M., Doehrmann, O., Leonard, J.A., et al., 2016. Altered intrinsic functional brain architecture in children at familial risk of major depression. Biol. Psychiatry 80, 849–858.

Chakrabarty, T., Ogrodniczuk, J., Hadjipavlou, G., 2016. Predictive neuroimaging markers of psychotherapy response: a systematic review. Harv. Rev. Psychiatry 24, 396–405.

Cheng, H., Liu, J., 2020. Alterations in amygdala connectivity in internet addiction disorder. Sci. Rep. 10, 2370.

Choi, E.J., Taylor, M.J., Hong, S.B., Kim, C., Kim, J.W., McIntyre, R.S., et al., 2018. Gaming-addicted teens identify more with their cyber-self than their own self: neural evidence. Psychiatry Res. Neuroimaging. 279, 51–59.

Cilliers, K., Page, B.J., 2016. Review of the anatomy of the distal anterior cerebral artery and its anomalies. Turk. Neurosurg. 26 (5), 653–661. https://doi.org/10.5137/1019-5149.JTN.14294-15.

Connolly, C.G., Wu, J., Ho, T.C., Hoeft, F., Wolkowitz, O., Eisendrath, S., et al., 2013. Resting-state functional connectivity of subgenual anterior cingulate cortex in depressed adolescents. Biol. Psychiatry 74, 898–907.

Critchley, H.D., Rotshtein, P., Nagai, Y., O'Doherty, J., Mathias, C.J., Dolan, R.J., 2005. Activity in the human brain predicting differential heart rate responses to emotional facial expressions. Neuroimage 24 (3), 751–762. https://doi.org/10.1016/j.neuroimage.2004.10.013.

Darvish-Ghane, S., Yamanaka, M., Zhuo, M., 2016. Dopaminergic modulation of excitatory transmission in the anterior cingulate cortex of adult mice. Mol. Pain 12. https://doi.org/10.1177/1744806916648153, 1744806916648153.

Darvish-Ghane, S., Quintana, C., Beaulieu, J.M., Martin, L.J., 2020. D1 receptors in the anterior cingulate cortex modulate basal mechanical sensitivity threshold and glutamatergic synaptic transmission. Mol. Brain 13 (1), 121. https://doi.org/10.1186/s13041-020-00661-x.

de Kwaasteniet, B., Ruhe, E., Caan, M., Rive, M., Olabarriaga, S., Groefsema, M., et al., 2013. Relation between structural and functional connectivity in major depressive disorder. Biol. Psychiatry 74, 40–47.

Dilsaver, S.C., 1989. Bipolar disorder. Am. Fam. Physician 40, 156–166.

Dong, G., Shen, Y., Huang, J., Du, X., 2013. Impaired error monitoring function in people with internet addiction disorder: an event-related fMRI study. Eur. Addict. Res. 19, 269–275.

Dong, G.H., Wang, M., Zhang, J., Du, X., Potenza, M.N., 2019. Functional neural changes and altered cortical subcortical connectivity associated with recovery from internet gaming disorder. J. Behav. Addict. 8, 692–702.

Drevets, W.C., Savitz, J., Trimble, M., 2008. The subgenual anterior cingulate cortex in mood disorders. CNS Spectr. 13, 663–681.

Dunn, J.T., Mroczek, J., Patel, H.R., Ragozzino, M.E., 2020. Tandospirone, a partial 5-HT1A receptor agonist, administered systemically or into anterior cingulate attenuates repetitive behaviors in Shank3B mice. Int. J. Neuropsychopharmacol. 23 (8), 533–542. https://doi.org/10.1093/ijnp/pyaa047.

Etkin, A., Wager, T.D., 2007. Functional neuroimaging of anxiety: a meta-analysis of emotional processing in PTSD, social anxiety disorder, and specific phobia. Am. J. Psychiatry 164 (10), 1476–1488. https://doi.org/10.1176/appi.ajp.2007.07030504.

Etkin, A., Egner, T., Kalisch, R., 2011. Emotional processing in anterior cingulate and medial prefrontal cortex. Trends Cognit. Sci. 15 (2), 85–93. https://doi.org/10.1016/j.tics.2010.11.004.

Fiori, L.M., Kos, A., Lin, R., Théroux, J.F., Lopez, J.P., Kühne, C., et al., 2021. miR-323a regulates ERBB4 and is involved in depression. Mol. Psychiatry 26 (8), 4191–4204. https://doi.org/10.1038/s41380-020-00953-7.

Foster, B.L., Koslov, S.R., Aponik-Gremillion, L., Monko, M.E., Hayden, B.Y., Heilbronner, S.R., March 2023. A tripartite view of the posterior cingulate cortex. Nat. Rev. Neurosci. 24 (3), 173–189. https://doi.org/10.1038/s41583-022-00661-x.

Fullana, M.N., Ruiz-Bronchal, E., Ferrés-Coy, A., Juárez-Escoto, E., Artigas, F., Bortolozzi, A., 2019. Regionally selective knockdown of astroglial glutamate transporters in infralimbic cortex induces a depressive phenotype in mice. Glia 67 (6), 1122–1137. https://doi.org/10.1002/glia.23593.

Gaffrey, M.S., Luby, J.L., Repovš, G., Belden, A.C., Botteron, K.N., Luking, K.R., et al., 2010. Subgenual cingulate connectivity in children with a history of preschool-depression. Neuroreport 21, 1182–1188.

Ge, R., Downar, J., Blumberger, D.M., Daskalakis, Z.J., Vila-Rodriguez, F., January–February 2020. Functional connectivity of the anterior cingulate cortex predicts treatment outcome for rTMS in treatment-resistant depression at 3-month follow-up. Brain Stimul. 13 (1), 206–214. https://doi.org/10.1016/j.brs.2019.10.012.

Ghaem, O., Mellet, E., Crivello, F., Tzourio, N., Mazoyer, B., Berthoz, A., Denis, M., 1997. Mental navigation along memorized routes activates the hippocampus, precuneus, and insula. Neuroreport 8 (3), 739–744. https://doi.org/10.1097/00001756-199702100-00032.

Goldstein, R.Z., Alia-Klein, N., Tomasi, D., Carrillo, J.H., Maloney, T., Woicik, P.A., et al., 2009. Anterior cingulate cortex hypoactivations to an emotionally salient task in cocaine addiction. Proc. Natl. Acad. Sci. U. S. A. 106, 9453–9458.

Haznedar, M.M., Buchsbaum, M.S., Hazlett, E.A., Shihabuddin, L., New, A., Siever, L.J., 2004. Cingulate gyrus volume and metabolism in the schizophrenia spectrum. Schizophr. Res. 71 (2–3), 249–262. https://doi.org/10.1016/j.schres.2004.02.025.

He, Z., Zhang, D., Muhlert, N., Elliott, R., 2019. Neural substrates for anticipation and consumption of social and monetary incentives in depression. Soc. Cogn. Affect. Neurosci. 14, 815–826.

Hoffmann, F., Viding, E., Puetz, V.B., Gerin, M.I., Sethi, A., Rankin, G., et al., 2018. Evidence for depressogenic spontaneous thoughts and altered resting-state connectivity in adolescents with a maltreatment history. J. Am. Acad. Child Adolesc. Psychiatry 57, 687–695.e4.

Horvath, J., Mundinger, C., Schmitgen, M.M., Wolf, N.D., Sambataro, F., Hirjak, D., et al., 2020. Structural and functional correlates of smartphone addiction. Addict. Behav. 105, 106334.

Hsu, D.T., Kirouac, G.J., Zubieta, J.K., Bhatnagar, S., 2014. Contributions of the paraventricular thalamic nucleus in the regulation of stress, motivation, and mood. Front. Behav. Neurosci. 8, 73.

Hugon, J., Msika, E.F., Queneau, M., Farid, K., Paquet, C., 2022. Long COVID: cognitive complaints (brain fog) and dysfunction of the cingulate cortex. J. Neurol. 269 (1), 44–46. https://doi.org/10.1007/s00415-021-10655-x.

Jackson, S.A., Horst, N.K., Pears, A., Robbins, T.W., Roberts, A.C., July 2016. Role of the perigenual anterior cingulate and orbitofrontal cortex in contingency learning in the marmoset. Cereb. Cortex 26 (7), 3273–3284. https://doi.org/10.1093/cercor/bhw067.

Jager, A., Amiri, H., Bielczyk, N., van Heukelum, S., Heerschap, A., Aschrafi, A., et al., 2020. Cortical control of aggression: GABA signalling in the anterior cingulate cortex. European neuropsychopharmacology : the journal of the European College of Neuropsychopharmacology 30, 5–16. https://doi.org/10.1016/j.euroneuro.2017.12.007.

Janes, A.C., Farmer, S., Frederick, B., Nickerson, L.D., Lukas, S.E., 2014. An increase in tobacco craving is associated with enhanced medial prefrontal cortex network coupling. PLoS One 9, e88228.

Jaworska, N., MacMaster, F.P., Yang, X.R., Courtright, A., Pradhan, S., Gaxiola, I., et al., 2014. Influence of age of onset on limbic and paralimbic structures in depression. Psychiatry Clin. Neurosci. 68, 812–820.

Jumah, F.R., Dossani, R.H., 2022. Neuroanatomy, cingulate cortex. In: StatPearls. StatPearls Publishing, Treasure Island (FL).

Kim, H., Kim, Y.K., Lee, J.Y., Choi, A.R., Kim, D.J., Choi, J.S., 2019. Hypometabolism and altered metabolic connectivity in patients with internet gaming disorder and alcohol use disorder. Prog. Neuro-Psychopharmacol. Biol. Psychiatry 95, 109680.

Kwan, C.L., Crawley, A.P., Mikulis, D.J., Davis, K.D., April 2000. An fMRI study of the anterior cingulate cortex and surrounding medial wall activations evoked by noxious cutaneous heat and cold stimuli. Pain 85 (3), 359–374. https://doi.org/10.1016/S0304-3959(99)00287-0.

Laurens, K.R., Ngan, E.T., Bates, A.T., Kiehl, K.A., Liddle, P.F., 2003. Rostral anterior cingulate cortex dysfunction during error processing in schizophrenia. Brain 126 (3), 610–622. https://doi.org/10.1093/brain/awg056.

Lee, D., Lee, J., Namkoong, K., Jung, Y.C., 2018. Subregions of the anterior cingulate cortex form distinct functional connectivity patterns in young males with internet gaming disorder with comorbid depression. Front. Psychiatry 9, 380.

Leech, R., Sharp, D.J., 2014. The role of the posterior cingulate cortex in cognition and disease. Brain 137 (Pt 1), 12–32. https://doi.org/10.1093/brain/awt162.

Li, C.R., Zhang, S., Hung, C.C., Chen, C.M., Duann, J.R., Lin, C.P., et al., 2017. Depression in chronic ketamine users: sex differences and neural bases. Psychiatry Res. Neuroimaging. 269, 1–8.

Li, H.J., Sun, J.Z., Zhang, Q.L., Wei, D.T., Li, W.F., Jackson, T., et al., 2014. Neuroanatomical differences between men and women in help-seeking coping strategy. Sci. Rep. 4, 5700.

Liu, Y., Du, L., Li, Y., Liu, H., Zhao, W., Liu, D., et al., 2015. Antidepressant effects of electroconvulsive therapy correlate with subgenual anterior cingulate activity and connectivity in depression. Medicine (Baltim.) 94, e2033.

Lévêque, M., Carron, R., Régis, J., 2013. Radiosurgery for the treatment of psychiatric disorders: a review. World Neurosurg. 80 (3–4), S32.e1-9. https://doi.org/10.1016/j.wneu.2013.07.004.

Maddock, R.J., Garrett, A.S., Buonocore, M.H., 2003. Posterior cingulate cortex activation by emotional words: fMRI evidence from a valence decision task. Hum. Brain Mapp. 18 (1), 30–41. https://doi.org/10.1002/hbm.10075.

Maddock, R.J., Garrett, A.S., Buonocore, M.H., 2001. Remembering familiar people: the posterior cingulate cortex and autobiographical memory retrieval. Neuroscience 104 (3), 667–676. https://doi.org/10.1016/s0306-4522(01)00108-7.

Maguire, E.A., Burgess, N., Donnett, J.G., Frackowiak, R.S., Frith, C.D., O'Keefe, J., 1998. Knowing where and getting there: a human navigation network. Science (New York, N.Y.) 280 (5365), 921–924. https://doi.org/10.1126/science.280.5365.921.

Mai, J.K., Paxinos, G., 2012. Cingulate cortex. In: The Human Nervous System. Elsevier Academic Press, Amsterdam, pp. 943–983. https://doi.org/10.1016/C2009-0-02721-4.

Margulies, D.S., Kelly, A.M.C., Uddin, L.Q., Biswal, B.B., Castellanos, F.X., Milham, M.P., 2007. Mapping the functional connectivity of anterior cingulate cortex. Neuroimage 37 (2), 579–588. https://doi.org/10.1016/j.neuroimage.2007.05.019.

Minzenberg, M.J., Laird, A.R., Thelen, S., Carter, C.S., Glahn, D.C., 2009. Meta-analysis of 41 functional neuroimaging studies of executive function in schizophrenia. Arch. Gen. Psychiatry 66 (8), 811–822. https://doi.org/10.1001/archgenpsychiatry.2009.91.

Mitchell, A.S., Czajkowski, R., Zhang, N., Jeffery, K., Nelson, A.J.D., March 19, 2018. Retrosplenial cortex and its role in spatial cognition. Brain Neurosci. Adv. 2, 2398212818757098. https://doi.org/10.1177/2398212818757098.

Mora, S., Merchán, A., Aznar, S., Flores, P., Moreno, M., 2020. Increased amygdala and decreased hippocampus volume after schedule-induced polydipsia in high drinker compulsive rats. Behav. Brain Res. 390, 112592. https://doi.org/10.1016/j.bbr.2020.112592.

Niida, A., Niida, R., Matsuda, H., Motomura, M., Uechi, A., 2014. Analysis of the presence or absence of atrophy of the subgenual and subcallosal cingulate cortices using voxel-based morphometry on MRI is useful to select prescriptions for patients with depressive symptoms. Int. J. Gen. Med. 7, 513–524.

Okada, N., Yahata, N., Koshiyama, D., Morita, K., Sawada, K., Kanata, S., et al., 2020. Smaller anterior subgenual cingulate volume mediates the effect of girls' early sexual maturation on negative psychobehavioral outcome. Neuroimage 209, 116478.

Olson, C.R., Musil, S.Y., Goldberg, M.E., 1996. Single neurons in posterior cingulate cortex of behaving macaque: eye movement signals. J. Neurophysiol. 76 (5), 3285–3300. https://doi.org/10.1152/jn.1996.76.5.3285.

Oszvald, Á., Quick, J., Franz, K., et al., 2012. Resection of gliomas in the cingulate gyrus: functional outcome and survival. J. Neuro Oncol. 109 (2), 341–348. https://doi.org/10.1007/s11060-012-0898-0.

Palomero-Gallagher, N., Vogt, B.A., Schleicher, A., Mayberg, H.S., Zilles, K., 2009. Receptor architecture of human cingulate cortex: evaluation of the four-region neurobiological model. Hum. Brain Mapp. 30 (8), 2336–2355. https://doi.org/10.1002/hbm.20667.

Pearson, J.M., Heilbronner, S.R., Barack, D.L., Hayden, B.Y., Platt, M.L., 2011. Posterior cingulate cortex: adapting behavior to a changing world. Trends Cogn. Sci. 15 (4), 143–151. https://doi.org/10.1016/j.tics.2011.02.002.

Philip, N.S., Barredo, J., van 't Wout-Frank, M., Tyrka, A.R., Price, L.H., Carpenter, L.L., 2018. Network mechanisms of clinical response to transcranial magnetic stimulation in posttraumatic stress disorder and major depressive disorder. Biol. Psychiatry 83, 263–272.

Posner, M.I., Petersen, S.E., Fox, P.T., Raichle, M.E., 1988. Localization of cognitive operations in the human brain. Science (New York, N.Y.) 240 (4859), 1627–1631. https://doi.org/10.1126/science.3289116.

Rakel, R.E., 1999. Depression. Prim. Care 26, 211–224.

Rey, G., Piguet, C., Benders, A., Favre, S., Eickhoff, S.B., Aubry, J.M., et al., 2016. Resting-state functional connectivity of emotion regulation networks in euthymic and non-euthymic bipolar disorder patients. Eur. Psychiatry 34, 56–63.

Ridderinkhof, K.R., Ullsperger, M., Crone, E.A., Nieuwenhuis, S., 2004. The role of the medial frontal cortex in cognitive control. Science (New York, N.Y.) 306 (5695), 443–447. https://doi.org/10.1126/science.1100301.

Rolls, E.T., 2019. The cingulate cortex and limbic systems for emotion, action, and memory. Brain Struct. Funct. 224 (9), 3001–3018. https://doi.org/10.1007/s00429-019-01945-2.

Schwartz, J., Ordaz, S.J., Kircanski, K., Ho, T.C., Davis, E.G., Camacho, M.C., et al., 2019. Resting-state functional connectivity and inflexibility of daily emotions in major depression. J. Affect. Disord. 249, 26–34.

Shin, L.M., Liberzon, I., 2010. The neurocircuitry of fear, stress, and anxiety disorders. Neuropsychopharmacology 35 (1), 169–191. https://doi.org/10.1038/npp.2009.83.

Steele, J.D., Christmas, D., Eljamel, M.S., Matthews, K., 2008. Anterior cingulotomy for major depression: clinical outcome and relationship to lesion characteristics. Biol Psychiatry 63 (7), 670–677. https://doi.org/10.1016/j.biopsych.2007.07.019.

Stevens, F.L., Hurley, R.A., Taber, K.H., 2011. Anterior cingulate cortex: unique role in cognition and emotion. J Neuropsychiatry Clin. Neurosci. 23 (2), 121–125. https://doi.org/10.1176/jnp.23.2.jnp121.

Strick, P.L., Dum, R.P., Picard, N., 1998. Motor areas on the medial wall of the hemisphere. Novartis Found. Symp. 218, 64–75. https://doi.org/10.1002/9780470515563.ch5.

Sugar, J., Witter, M.P., van Strien, N.M., Cappaert, N.L.M., 2011. The retrosplenial cortex: intrinsic connectivity and connections with the (para)hippocampal region in the rat. An interactive connectome. Front. Neuroinform. 5. https://doi.org/10.3389/fninf.2011.00007.

Sugiura, M., Shah, N.J., Zilles, K., Fink, G.R., 2005. Cortical representations of personally familiar objects and places: functional organization of the human posterior cingulate cortex. J. Cognit. Neurosci. 17 (2), 183–198. https://doi.org/10.1162/0898929053124956.

Talairach, J., Bancaud, J., Geier, S., Bordas-Ferrer, M., Bonis, A., Szikla, G., Rusu, M., 1973. The cingulate gyrus and human behaviour. Electroencephalogr. Clin. Neurophysiol. 34 (1), 45–52. https://doi.org/10.1016/0013-4694(73)90149-1.

Tan, R.H., Wong, S., Hodges, J.R., Halliday, G.M., Hornberger, M., 2013. Retrosplenial cortex (BA 29) volumes in behavioral variant frontotemporal dementia and Alzheimer's disease. Dement. Geriatr. Cogn. Disord 35 (3–4), 177–182. https://doi.org/10.1159/000346392.

Tate, M.C., Kim, C.Y., Chang, E.F., Polley, M.Y., Berger, M.S., 2011. Assessment of morbidity following resection of cingulate gyrus gliomas. Clinical article. J. Neurosurg. 114 (3), 640–647. https://doi.org/10.3171/2010.9.JNS10709.

Tracey, I., Mantyh, P.W., 2007. The cerebral signature for pain perception and its modulation. Neuron 55 (3), 377–391. https://doi.org/10.1016/j.neuron.2007.07.012.

Urry, H.L., van Reekum, C.M., Johnstone, T., Davidson, R.J., September 2009. Individual differences in some (but not all) medial prefrontal regions reflect cognitive demand while regulating unpleasant emotion. Neuroimage 47 (3), 852–863. https://doi.org/10.1016/j.neuroimage.2009.05.069.

Valenstein, E., Bowers, D., Verfaellie, M., Heilman, K.M., Day, A., Watson, R.T., 1987. Retrosplenial amnesia. Brain 110 (Pt 6), 1631–1646. https://doi.org/10.1093/brain/110.6.1631.

Vann, S.D., Aggleton, J.P., Maguire, E.A., 2009. What does the retrosplenial cortex do? Nat. Rev. Neurosci. 10 (11), 792–802. https://doi.org/10.1038/nrn2733.

Vogt, B.A., Berger, G.R., Derbyshire, S.W., December 2003. Structural and functional dichotomy of human midcingulate cortex. Eur. J. Neurosci. 18 (11), 3134–3144. https://doi.org/10.1111/j.1460-9568.2003.03034.x.

Vogt, B.A., 2016. Midcingulate cortex: structure, connections, homologies, functions and diseases. J. Chem. Neuroanat. 74, 28–46. https://doi.org/10.1016/j.jchemneu.2016.01.010.

Vogt, B.A., 2005. Pain and emotion interactions in subregions of the cingulate gyrus. Nat. Rev. Neurosci. 6 (7), 533–544. https://doi.org/10.1038/nrn1704.

Vogt, B. (Ed.), 2009. Cingulate neurobiology and disease. Oxford University Press, USA.

Vogt, B.A., Vogt, L.J., Perl, D.P., Hof, P.R., 2001. Cytology of human caudomedial cingulate, retrosplenial, and caudal parahippocampal cortices. J. Comp. Neurol. 438 (3), 353–376. https://doi.org/10.1002/cne.1320.

Vogt, B.A., Nimchinsky, E.A., Vogt, L.J., Hof, P.R., 1995. Human cingulate cortex: surface features, flat maps, and cytoarchitecture. J. Comp. Neurol. 359 (3), 490–506. https://doi.org/10.1002/cne.903590310.

von Lehe, M., Schramm, J., 2009. Gliomas of the cingulate gyrus: surgical management and functional outcome. Neurosurg. Focus 27 (2), E9. https://doi.org/10.3171/2009.6.FOCUS09104.

Wang, H., Zhu, R., Dai, Z., Tian, S., Shao, J., Wang, X., et al., 2020. Aberrant functional connectivity and graph properties in bipolar II disorder with suicide attempts. J. Affect. Disord. 275, 202–209.

Wang, L., Hosakere, M., Trein, J.C.L., et al., 2007. Abnormalities of cingulate gyrus neuroanatomy in schizophrenia. Schizophr. Res. 93 (1–3), 66–78. https://doi.org/10.1016/j.schres.2007.02.021.

Wang, X., Zhang, L., Zhan, Y., Li, D., Zhang, Y., Wang, G., et al., 2017. Contribution of BDNF/TrkB signalling in the rACC to the development of pain-related aversion via activation of ERK in rats with spared nerve injury. Brain Res 1671, 111–120. https://doi.org/10.1016/j.brainres.2017.07.010.

Ward, A.M., Schultz, A.P., Huijbers, W., Van Dijk, K.R., Hedden, T., Sperling, R.A., March 2014. The parahippocampal gyrus links the default-mode cortical network with the medial temporal lobe memory system. Hum. Brain Mapp. 35 (3), 1061–1073. https://doi.org/10.1002/hbm.22234.

Weinstein, A., Weinstein, Y., 2014. Exercise addiction-diagnosis, biopsychological mechanisms and treatment issues. Curr. Pharm. Des. 20, 4062–4069.

Westbrook, C., Creswell, J.D., Tabibnia, G., Julson, E., Kober, H., Tindle, H.A., 2013. Mindful attention reduces neural and self-reported cue-induced craving in smokers. Soc. Cogn. Affect. Neurosci. 8, 73–84.

Yang, T.T., Simmons, A.N., Matthews, S.C., Tapert, S.F., Frank, G.K., Bischoff-Grethe, A., et al., 2009. Adolescent subgenual anterior cingulate activity is related to harm avoidance. Neuroreport 20, 19–23.

Zou, Z., Wang, H., d'Oleire Uquillas, F., Wang, X., Ding, J., Chen, H., 2017. Definition of substance and non-substance addiction. Adv. Exp. Med. Biol. 1010, 21–41.

13

The Telencephalon: Hippocampus and Related Structures

Elham Hosseini[1], *Ali Sepehrinezhad*[1], *Javad Momeni*[1], *Brandon M. Ascenzi*[2], *Ali Gorji*[3,4] *and Sajad Sahab-Negah*[1]

[1]Neuroscience Research Center, Mashhad University of Medical Sciences, Mashhad, Iran [2]Independent Neuroresearcher-Member of Marie Curie Alumni Association (MCAA) [3]Shefa Neuroscience Research Center, Khatam Alanbia Hospital, Tehran, Iran [4]Epilepsy Research Center, Department of Neurosurgery and Neurology, Westfälische Wilhelms-Universitat Münster, Münster, Germany

OUTLINE

PART I: ANATOMY OF THE HIPPOCAMPUS

Gross anatomy

The hippocampus is a component of the hippocampal formation, which is a larger structure located in the temporal lobe (Fig. 13.1). It extends from the anterior amygdala to the posterior corpus callosum and is a significant part of the limbic system. The hippocampal formation comprises three main parts: the hippocampus proper (or *Cornu Ammonis (CA)*), dentate gyrus, and subiculum. The hippocampal structures extend from the floor and part of the inner wall of the inferior horns of the lateral ventricles (Dekeyzer et al., 2017; Knierim, 2015; Tatu and Vuillier, 2014). The hippocampal formation exhibits its greatest width at its anterior extent, where it curves toward the medial surface of the brain. This medial region is referred to as the uncus and displays intricate images of the hippocampal formation, consisting of various hippocampal fields. The number of rostrocaudal flexures in this bend of the hippocampal formation varies significantly among individuals and can range up to five (Insausti, 2010; Insausti and Amaral, 2004). The uncus comprises a few delicate gyri, known as the hippocampal digitations, which impart a paw-like appearance to this area (Dekeyzer et al., 2017; Oppenheim et al., 1998). As a result, this region is designated as the pes hippocampi ("foot" of hippocampus).

Moving caudally, the hippocampus narrows down and curves upwards around the splenium of the corpus callosum (Schultz and Engelhardt, 2014). During fetal brain development (12–32 weeks postconception), the hippocampus forms through the continuous expansion of the medial edge of the temporal lobe (Ge et al., 2015). This growth pattern results in the hippocampus occupying the floor of the temporal horn of the lateral ventricle (Kiernan and Rajakumar, 2013) (Fig. 13.2). In

From Anatomy to Function of the Central Nervous System
https://doi.org/10.1016/B978-0-12-822404-5.00014-0

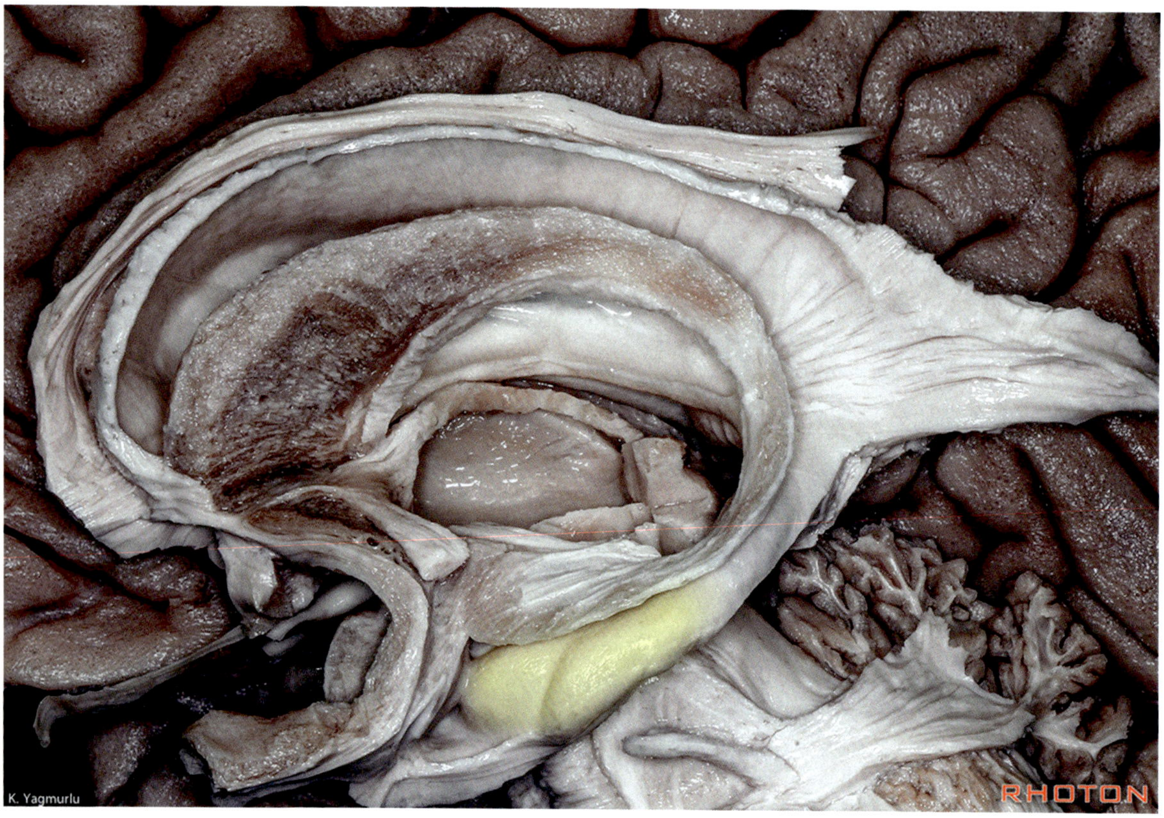

FIGURE 13.1 **Hippocampal formation (in yellow).** *Image courtesy of The Rhoton Collection® http://rhoton.ineurodb.org/. Dissector: K.Yagmurlu Drawing by Alessandro Basso. Stereoscopical Image: use red-blue 3D glasses.*

primates, the hippocampus initially appears above the corpus callosum during embryonic development. As development progresses, the hippocampus migrates toward the temporal lobe, leaving behind a white matter structure called the fornix. The hippocampus internal structure is made up of archicortex, which has fewer cortical layers than both the neocortex (with six layers) and the paleocortex (with either four or five layers) (Chauhan et al., 2021; Knierim, 2015; Siegel and Sapru, 2010). The subiculum is continuous with the six-layered neocortex or para-hippocampal gyrus. The hippocampus contains a region called the alveus, which connects to the ventricles. The alveus is a thin layer of white matter composed of axons of pyramidal cells and is covered by a layer of ependymal cells. The fibers of the alveus converge at the inner margin of the hippocampus, forming the fimbria, which gives rise to the fornix (Chauhan et al., 2021; Dekeyzer et al., 2017). The fimbria covers the dentate gyrus posteriorly and extends to the splenium of the corpus callosum, where it continues as the fornix (Dekeyzer et al., 2017; Insausti and Amaral, 2004; Tatu and Vuillier, 2014). The fornix then curves around the thalamus before separating at the choroidal fissure, which contains the choroid plexus (Dekeyzer et al., 2017; Schultz and Engelhardt, 2014; Kandel et al., 2012). The hippocampal formation can be divided into three parts: the anterior part, known as the head, the intermediate part referred to as the body, and the posterior part, called the tail (Jacobson et al., 2018). Additionally, the hippocampus can be classified into three parts based on their position relative to the corpus callosum: precommissural, supracommissural, and retrocommissural. While the precommissural and supracommissural portions are relatively small, the retrocommissural hippocampus makes up the majority of the hippocampal formation (Ryszard Pluta, 2021; Destrieux et al., 2013).

We can anticipate already now, we will come back to it later in the chapter, that the CA, or hippocampus proper, when observed in a coronal section can be divided into four subregions known as CA1–CA4 (Anand and Dhikav, 2012) (Fig. 13.3). The largest of these subregions is CA1, also referred to as Sommer's sector, which is located at the interface between the subiculum and the hippocampus, and is limited medially by CA2. About 90% of neurons in CA1 are pyramidal cells, while the remaining 10% are interneurons. Input from the supramammillary region of the hypothalamus comes to the mixed CA2 area (Chauhan et al., 2021), which is a small and less important region of the hippocampus, highly resistant to epileptic-induced sclerosis (Anand and Dhikav, 2012). CA2, in

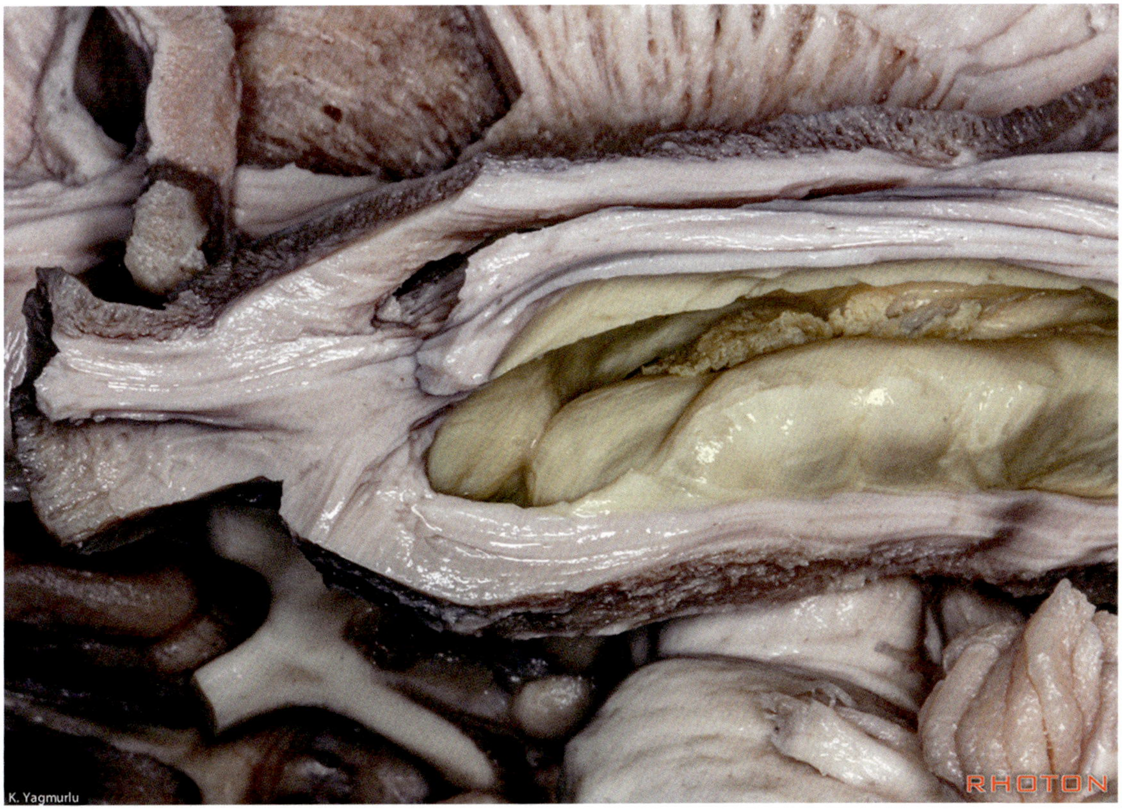

FIGURE 13.2 **Inferior view of the inferior horn of the lateral ventricles (in yellow):** within the latter the hippocampus is easily recognizable. *Image courtesy of The Rhoton Collection® http://rhoton.ineurodb.org/. Dissector: K.Yagmurlu Drawing by Jeffrey Sorenson. Stereoscopical Image: use red-blue 3D glasses.*

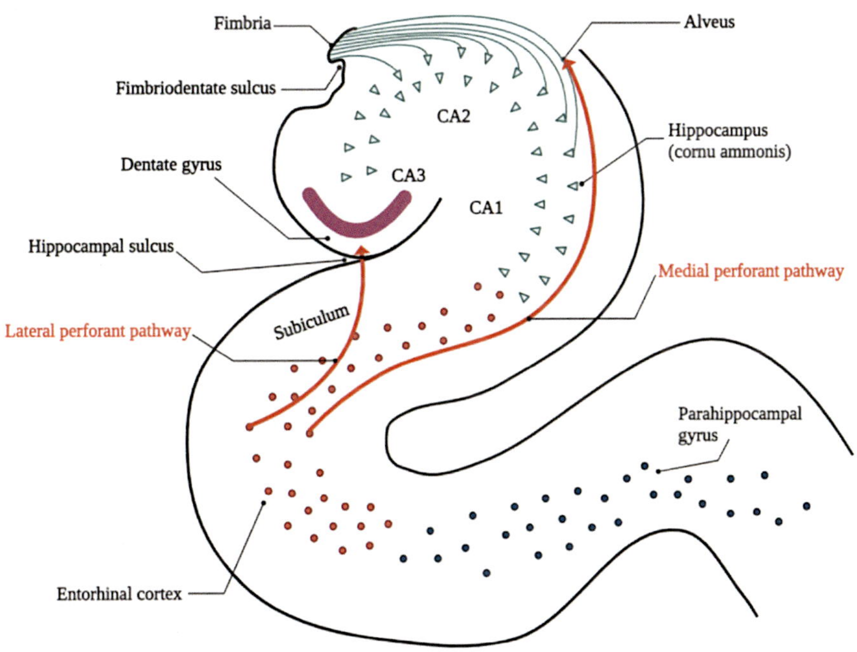

FIGURE 13.3 **Medial and lateral perforant paths (MPP and LPP).** The main way that information from the cortex enters the hippocampus is through two paths called the medial and lateral perforant paths. These paths are formed by the axons of neurons located in the medial and lateral parts of a brain region called the entorhinal cortex (MEC and LEC). *Created with BioRender.com.*

turn, limits the CA3 subregion medially, which is also known as the magnocellular area. A notable feature of the CA3 field is that the collaterals of the axonal processes extending from its pyramidal cells are called recurrent or Schaffer collaterals. Some authors refer to most surface cells of the CA3 area as CA4 , but nowadays CA4 is considered to be part of the hilus of the dentate gyrus (Chauhan et al., 2021). Mossy fibers from the granule cells of the dentate gyrus project to the apical dendrites of the CA3 layer, while axons from the CA3 pyramidal cells contribute to the alveus, fimbria, and fornix. Certain collaterals from these axons (the aforementioned Schaffer's collaterals) synapse with the dendrites of CA1 pyramidal cells. The axons of CA1 pyramidal cells then project to subiculum neurons, while axons from subiculum neurons via the alveolar pathway contribute to the fibers of the fimbria and fornix (Chauhan et al., 2021). We provide a detailed description of the hippocampal fields in the paragraph titled "microscopic structure" (see later).

BLOOD SUPPLY OF THE HIPPOCAMPUS

The anterior choroidal artery runs medially and superiorly to the uncus, between the ambient and semilunar gyrus. It then sends perforating arteries to reach deeper structures. The uncus is closely related to the M1 segment of the middle cerebral arteries and its lenticulostriate arteries. The P2 segment of the posterior cerebral artery and the basal vein supply and drain the caudal part of the head of the hippocampus that faces the crus cerebri and crural cistern. Internal cerebral veins drain into thalamostriatal basal ganglia, thalamus, internal capsule, tela choroidea of third ventricle, and hippocampus. The veins on each side unite to form the internal cerebral vein (Fogwe et al., 2021).

Ventromedial Surface

The ventromedial surface of the temporal lobe is partitioned by two sulci, namely the occipitotemporal and collateral sulci. Toward the front, the collateral sulcus frequently merges with the rhinal sulcus, forming the lateral boundary of the parahippocampal gyrus. This gyrus is divided into an anterior portion consisting of the entorhinal cortex and related perirhinal cortex, and a posterior parahippocampal cortex that encompasses the TF and TH—temporal areas F and H of von Economo and Konskinas and von Bonin and Biley—fields (Blaizot, 2010). The entorhinal cortex terminates along the medial side of the collateral sulcus, while the remaining part of the collateral sulcus is composed of 35 and 36 BA of the perirhinal cortex (Ryszard Pluta, 2021; Sethumadhavan, 2022). The human entorhinal cortex is characterized by wart-like projections on its surface, which correspond to clusters of entorhinal layer II neurons. The dorsomedial aspect of the entorhinal cortex, also known as the anterior parahippocampal gyrus, contains the gyrus ambiens (Maass et al., 2015).

The rostral pole of the uncus is often designated as the gyrus uncinatus. Caudally, it is followed by the most medial bend of the dentate gyrus designated as the tail of the dentate gyrus or limbus of Giacomini. Ventrally, the uncus is limited by the rostral part of the hippocampal fissure (Ryszard Pluta, 2021). Caudally to the uncus, the choroidal fissure marks the medial surface of the dentate gyrus and fimbria. The hippocampal fissure separates the hippocampus from the underlying subiculum (Ryszard Pluta, 2021; Dekeyzer et al., 2017; Zemmoura et al., 2012).

Caudally, the hippocampal formation merges into the retrosplenial region in the form of two obliquely oriented small gyri. The medial one is designated as the fasciola cinerea and contains the caudal part of the dentate gyrus (Destrieux et al., 2013). The lateral one—referred to as the gyrus fasciolaris—represents the terminal portion of CA3 (see later) (Destrieux et al., 2013). Finally, the supracallosal tissue of the hippocampus mainly consists of remnants of the subiculum and CA1 (Destrieux et al., 2013).

Dorsomedial Surface

A number of distinct cytoarchitectonic fields, including the dentate gyrus, hippocampus, and subicular cortices, are included in this region (Schultz and Engelhardt, 2014). Myelinated axons originating from pyramidal neurons of the hippocampus and subiculum travel in the white matter which sheaths the hippocampus (the alveus) and merge into the medially situated fimbria corresponding to a fringe of myelinated fibers along the medial surface of the hippocampus (Matsumoto et al., 2019). The fimbria leaves the hippocampus and coalesces into the fornix which fuses with the ventral surface of the corpus callosum (Fogwe et al., 2021). The part of the fornix fused with the corpus callosum is called the crus fornicis, whereas the major rostral portion is called the body of the fornix. The descending column of the fornix is divided around the anterior commissure forming the two mammillary bodies. The relation with the anterior commissure allows to distinguish the precommissural and the postcommissural fornix (Griffiths et al., 2009). At the posterior corpus callosum (splenium), fibers of the rodent fornix form a prominent hippocampal commissure, which is often referred to as the "psalterium" alluding to an ancient string instrument. Of note, commissural connections in the human hippocampus appear much more limited as compared to rodents (Kandel et al., 2012; Schultz and Engelhardt, 2014).

Subiculum

The subiculum forms the third part of the hippocampal formation (Schultz and Engelhardt, 2014). It is located between CA and the presubiculum (Fig. 13.4). The subiculum can be divided into a superficial molecular layer and a deep pyramidal cell layer. The molecular layer is remarkably wide and harbors the apical tufts of the pyramidal cells. Within the pyramidal layer, separate external and internal sublayers can be distinguished. The pyramidal neurons in the external sublayer are somewhat larger and more widely spaced than those in the internal sublayer. The most superficial zone of the external sublayer contains islands of small, darkly stained cells. Large stellate cells are disseminated throughout the pyramidal layer. The molecular and pyramidal layers both contain numerous GABA-immunoreactive neurons. The subiculum constitutes the main output structure of the hippocampus. It projects to the entorhinal cortex and to several subcortical structures, including the septum, the nucleus accumbens, the anterior thalamus, the hypothalamus and the mamillary nuclei (Insausti and Amaral, 2004; Tatu and Vuillier, 2014).

The three periarchicortical areas situated medially to the subiculum, i.e., the presubiculum, the parasubiculum, and the entorhinal cortex, are so closely and intensively connected with the archicortical structures discussed above that they are commonly considered as integral components of the hippocampal formation (Insausti and Amaral, 2004).

Entorhinal Cortex

The term "entorhinal cortex" corresponds at 28BA, and it extends rostrally to approximately the mid-level of the amygdala and caudally to the level of the anterior limit of the lateral geniculate nucleus. In terms of evolution, the primate entorhinal cortex has undergone significant differentiation. In monkeys, up to seven cytoarchitectonic fields can be distinguished, while up to eight fields can be distinguished in humans (Schultz and Engelhardt, 2014). The deep layers of the entorhinal cortex extend laterally further than the superficial layers, generating a distinct oblique boundary between the entorhinal cortex and adjacent 35BA. This oblique region of transition has also been named the "trans entorhinal zone" (Jacobson et al., 2018). Interestingly, trans entorhinal neurons are highly vulnerable to the earliest preclinical stages of Alzheimer-related neurofibrillary pathology (Amthor et al., 2020). The primate entorhinal cortex

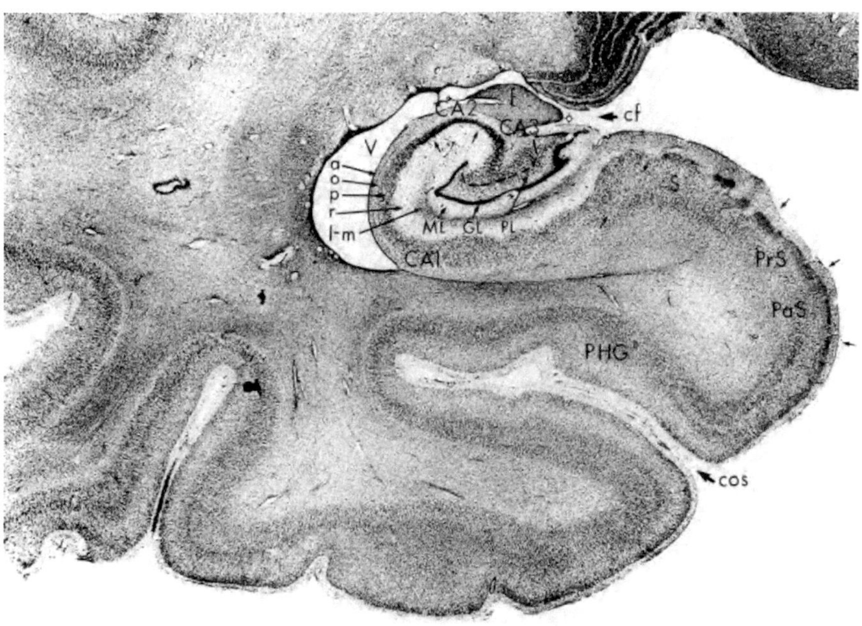

FIGURE 13.4 **Coronal section of the hippocampus.** Dashed line in the dentate gyrus indicates the border between the polymorphic layer of the dentate gyrus (PL) and the CA3 field of the hippocampus. Note that at this level the entorhinal cortex is no longer present and the parasubiculum (PaS) of the hippocampal formation abuts the parahippocampal cortex (PHG), which has replaced the perirhinal cortex in, a caudal continuation of the parahippocampal region. Note also that a small groove (target sign) called the fimbriodentate sulcus separates the molecular layer of the dentate gyrus from the fimbria. *Image courtesy of Insausti, R., Amaral, D.G., 2004. Chapter 23 — Hippocampal formation. In: Paxinos, G., Mai, J.K. (Eds.), The Human Nervous System, second ed. Academic Press, pp. 871–914. https://doi.org/10.1016/B978-012547626-3/50024-7.*

is continuous with the parahippocampal gyrus (fields TH and TF) toward the caudal end (Ryszard Pluta, 2021). The entorhinal cortex has a unique laminar organization that distinguishes it from other neocortical regions. Six distinct layers can be identified in the entorhinal cortex. Layer I is an acellular or plexiform layer, while layer II is highly characteristic of the entorhinal cortex and comprises modified pyramidal and stellate cells organized in islands (Jacobson et al., 2018). These cell islands form small bumps on the entorhinal surface, observable to the naked eye, called "verrucae hippocampi," which are helpful in identifying the entorhinal location. Layer III contains a homogeneous population of pyramidal cells, and their dendrites pass through the space between layer II cell islands (Insausti et al., 2017). Below layer III, a cell-sparse lamina called the lamina dissecans is found. This lamina highlights the lack of a distinct layer IV (internal granule cell layer). Consequently, the cells underlying the lamina dissecans are identified as layer V. Layer VI is not easily distinguishable from layer V in the entorhinal cortex (Insausti et al., 2017).

Fimbria

The fimbria consists of converging fibers of hippocampal nerve cells (alveus), which forms the posterior column of the fornix which extends from the hippocampus to the mammillary body (Dekeyzer et al., 2017; Schultz and Engelhardt, 2014). Posterior column of the fornix curves around the posterior end of the thalamus. The hippocampus ends posteriorly below the splenium of the corpus callosum. The parahippocampal gyrus position is inferolateral respect the hippocampus while dentate gyrus has a denticulated margin which is located between the fimbria and the parahippocampal gyrus (Dekeyzer et al., 2017; Kiernan, 2012).

Dentate Gyrus

The dentate gyrus is the most medial portion of the cerebral cortex, and it is a trilaminar cortical region at the cytoarchitectural level that is composed of the molecular layer, granule cell layer, and polymorphic layer (Amaral et al., 2007). It plays a vital role in pattern separation and associative memory that processes information from the entorhinal cortex to CA3 (Anand and Dhikav, 2012). At the midrostrocaudal levels of the hippocampal formation, the dentate gyrus takes on a typical C- or U-shaped structure and is separated from CA1 and the subiculum ventrally by the hippocampal fissure (Amaral et al., 2007; Hevner, 2016).

The principal cell layer of the dentate gyrus contains a dense packing of granule cell somata (approximately 15×10^6 neurons) (Schultz and Engelhardt, 2014). These granule cells have spiny dendrites that extensively branch in the dentate molecular layer; they are excitatory neurons using glutamate as neurotransmitter. The granule cell layer is the main layer and primarily consists of granule cells, with other neurons located at the border between this layer and the polymorphic layer (such as the soma of the dentate pyramidal basket cell). In humans, dentate granule cells also extend basal dendrites into the subjacent polymorphic layer (Siegel and Sapru, 2010). The molecular layer comprises granule cell dendrites of the dentate gyrus, interneurons, the fibers of the perforant pathway, and some external inputs. Together, the granule cell and molecular layers make up the fascia dentata (Amaral et al., 2007; Hara and Morrison, 2014).

The third and innermost layer of the dentate gyrus is called the polymorphic layer or hilus and is sometimes referred to as the hippocampal field CA4. However, the hilus cells project only to the dentate gyrus. Therefore, the polymorphic layer is usually considered part of the dentate gyrus. Near the hilus, a portion of the pyramidal cell layer is enclosed by the granule cell layer. There is no apparent cytoarchitectonic or connectional difference between these neurons and adjacent CA3 pyramidal neurons. Thus, the term CA4 for these neurons should be avoided, and all pyramidal neurons close to the dentate hilus should be considered as part of CA3 in the hippocampus (Ryszard Pluta, 2021) (Box 13.1).

Microscopic Structure of Hippocampus

The hippocampus proper or CA forms the largest part of the hippocampal formation. In addition to a limited number of small elements, its intermediate cellular layer contains the somata of pyramidal cells, which form by far the most numerous class of neurons in CA. These elements closely resemble their counterparts in other cortices. Each hippocampal pyramidal cell has a stout apical dendrite, which splits up at a shorter or longer distance from the soma into a number of terminal branches (see later). Differences in size and packing density of the pyramidal cells allow a subdivision of CA into separate cytoarchitectonic fields (Insausti and Amaral, 2004) (Fig. 13.4).

Four fields have been designed in the hippocampus: CA1−CA4. CA1 is adjacent to the subiculum, whereas CA4 is situated in the so-called hilar region of the dentate gyrus. However, recently has been proposed to drop the term CA4 and to include the pyramidal cells in the hilar region in CA3. The afferents to the pyramidal neurons in CA show a strict laminar organization. Each

BOX 13.1

Adult neurogenesis and the dentate gyrus

Based on anatomical observations, theoretical work, and lesion experiments, it has been suggested that the dentate gyrus in the mammalian hippocampus serves an essential role in pattern separation (Fares et al., 2019). This neural computation guarantees that newly formed memories are encoded without any interference from previously stored memories that may share similar features. Additionally, the dentate gyrus displays a distinct form of neural plasticity that arises from the continuous integration of newly generated excitatory granule cells, known as adult hippocampal neurogenesis (Tuncdemir et al., 2019). Cumulative evidence suggests that the dentate gyrus and adult-born granule cells play a role in some forms of behavioral discrimination (Christian et al., 2020). However, several factors have been assumed that have effects on neurogenesis, including cognitive-related disorders and aging. For this reason, several studies have investigated this issue. As an example, a study investigated neurogenesis by whole autopsy hippocampi from healthy human individuals aged 14–79 years (Boldrini et al., 2018). It has been reported that there were similar numbers of intermediate neural progenitors and immature neurons in the dentate gyrus (DG), as well as comparable volumes of mature granule neurons and glia across all age groups. However, older individuals showed less angiogenesis and neuroplasticity, as well as a reduced quiescent progenitor pool in the anterior-mid DG, while no changes were observed in the posterior DG. Despite these age-related differences, healthy older individuals without cognitive impairment, neuropsychiatric disease, or treatment still exhibited preserved neurogenesis. This suggests that ongoing hippocampal neurogenesis may play a role in sustaining human-specific cognitive function throughout life, and that declines in neurogenesis could potentially result in compromised cognitive-emotional resilience.

hippocampal field—CA3, CA2, and CA1—has a single cellular layer made up of pyramidal cells. The alveus, formed by the pyramidal cell axons, forms the outermost layer facing the ventricular lumen. The *stratum oriens* lies between the alveus and the pyramidal cell layer and mainly contains the basal dendrites of the pyramidal cells and various types of interneurons. The region above the pyramidal cell layer (toward the hippocampal fissure) contains the apical dendrites of the pyramidal cells and several types of interneurons. Historically, this region has been divided into three parts: *the stratum lucidum*, the *stratum radiatum*, and the *stratum lacunosum-molecular*, corresponding to the most superficial portion (adjacent to the hippocampal fissure) of this region (Dekeyzer et al., 2017; Ryszard Pluta, 2021; Insausti and Amaral, 2004) (Fig. 13.4).

The stratum radiatum primarily consists of the apical dendrites of pyramidal neurons, while the stratum lacunosum-moleculare contains their terminal branches. On the other hand, basal dendrites of pyramidal neurons branch out in the stratum oriens. Axons of pyramidal neurons enter the alveus. In the stratum lucidum of the CA3 region, mossy fibers travel and form synapses with the proximal dendrites that lie just above the pyramidal cell layer of CA3. In the human hippocampus, some mossy fibers also travel within the CA3 pyramidal cell layer. The stratum lucidum is absent in CA2 and CA1, which do not receive mossy fiber input (Ryszard Pluta, 2021; Hara and Morrison, 2014).

The defining characteristic of CA3 neurons is that their proximal dendrites are contacted by mossy fibers, which correspond to the axons of the dentate granule cells. These mossy fibers traverse the stratum lucidum, which is located immediately above the CA3 pyramidal cell layer. Within the stratum radiatum, apical dendrites of CA3 and CA2 neurons receive associational projections from other rostrocaudal levels, as well as cholinergic and histaminergic input from the septal nuclei and supramammillary regions, respectively (Bacciottini, 2001). In CA1, the Schaffer's collaterals, which originate from CA3 terminate in the stratum radiatum and stratum oriens. The perforant pathway projection from the entorhinal cortex to the dentate gyrus travels, in part, in the stratum lacunosum-moleculare of the hippocampus and makes *en passant* synapses on the distal apical dendrites of the hippocampal pyramidal cells. Compared to neighboring fields, particularly CA1, CA3 is relatively insensitive to pathology and is considered the resistant sector of classical neuropathology (Patra et al., 2018; Lee, 2014).

The pyramidal cell layer of the CA2 region is relatively narrow and compact (Dudek, 2016). It can be challenging to establish the borders of CA2 in routine preparations of the human hippocampus. However, the primate CA2 region has a distinct connectivity pattern characterized by prominent histaminergic input from the hypothalamic tuberomammillary nucleus (Kiernan and Rajakumar, 2013). Phylogenetically, CA1 has evolved progressively

(Mtui et al., 2020). The pyramidal cell layer of CA1, in monkeys and humans, becomes thicker and more heterogeneous compared to rodents. For example, the human CA1 pyramidal cell layer can be subdivided into an outer and inner pyramidal cell layer (Deb, 2014). According to stereological studies, the estimated number of CA1 neurons is approximately 14×10^6 (Schultz and Engelhardt, 2014). The CA1 pyramidal cell layer is thinnest close to the CA2 border, where the cells of both sublayers appear most tightly packed. The border of CA1 with CA2 is ill-defined because some CA2 pyramidal cells extend over the emerging CA1 pyramidal cell layer. The pyramidal cell layer of CA1 overlaps that of the subiculum obliquely.

The hippocampus is a circuit of connections that process sensory and spatial information from the superficial layers of the nearby entorhinal cortex, which passes through multiple synapses before arriving at the hippocampal CA1 region, the major output area. The CA1 neurons play a critical role in learning and memory, as evidenced by the severe memory loss observed in patients with lesions in this region, as well as numerous animal studies. Information from the entorhinal cortex reaches the CA1 neurons through two excitatory pathways, one direct and one indirect (Siegel and Sapru, 2010). The direct pathway to CA1 involves layer III neurons in the entorhinal cortex that send their axons through the perforant pathway to make excitatory synapses on the distal regions of the apical dendrites of CA1 neurons. This projection is also known as the monosynaptic pathway (Fig. 13.5).

The interactions between the direct and indirect inputs at each stage of the hippocampal circuit are believed to be crucial for memory storage or recall, although the exact nature of these interactions is yet to be determined (Siegel and Sapru, 2010; Hainmueller and Bartos, 2020)

On the other hand, in the indirect pathway, the axons of neurons in layer II of the entorhinal cortex project through the perforant pathway to excite the granule cells of the dentate gyrus. Then, the axons of the granule cells project through the mossy fiber pathway to excite the pyramidal cells in the CA3 region of the hippocampus. Finally, the axons of the CA3 neurons project, through the Schaffer collateral, to dendrites of the CA1 pyramidal cells. Due to its three successive excitatory synaptic connections, the indirect pathway is often referred to as the trisynaptic pathway (Fig. 13.5). Finally, the CA1 pyramidal cells project back to the deep layers of the entorhinal cortex and forward to the subiculum, that connects the hippocampus with diverse brain regions (Siegel and Sapru, 2010; Witter, 2007).

Aside from the pathways connecting different stages of the hippocampal circuit, there are also strong

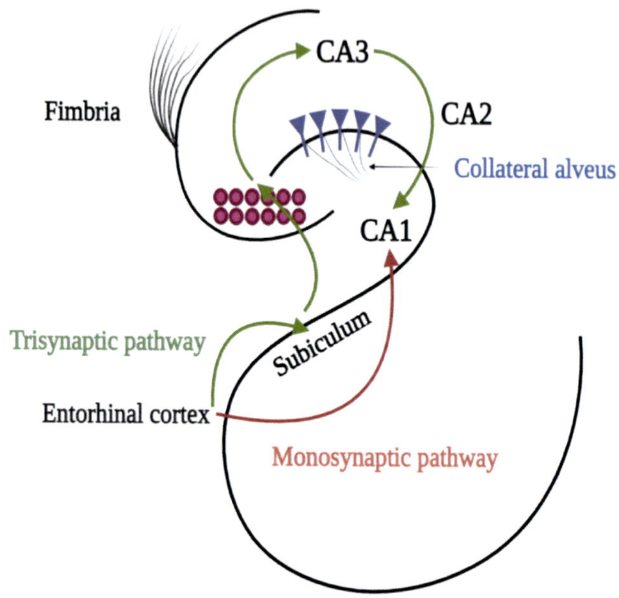

FIGURE 13.5 **Trisynaptic and monosynaptic pathways in hippocampus.** The trisynaptic circuit in the hippocampus involves the transfer of information from the entorhinal cortex to the dentate gyrus via the perforant path, which passes through the subiculum. From there, information flows from the dentate gyrus to CA3 through the mossy fiber pathway, named for the extensive branching of its axons. Finally, information travels from CA3 to CA1 along bundles of axons called Schaffer collaterals. However, there is also a monosynaptic input that skips the dentate gyrus and CA3 and directly transmits information from the entorhinal cortex to CA1. *Created with BioRender.com.*

excitatory connections between CA3 pyramidal neurons. This self-excitation via recurrent collaterals is believed to play a role in the associative aspects of memory storage and recall. However, under pathological conditions, this self-excitation can cause seizures (Siegel and Sapru, 2010; Kempermann et al., 2015). Moreover, neurons in the relatively small CA2 region, situated between CA3 and CA1, receive input from layer II of the entorhinal cortex through both direct and indirect pathways via the dentate gyrus and CA3. Additionally, the CA2 region receives robust input from hypothalamic nuclei that release oxytocin and vasopressin, hormones crucial for social behavior. As a result, CA2 provides CA1 with a third source of excitatory input, in addition to the direct and trisynaptic pathways from the entorhinal cortex (Box 13.2) (Siegel and Sapru, 2010). Finally, CA1, also called Sommer's area, is located closest to the subiculum and contains pyramidal cells. This region is highly vulnerable and is severely impacted by cell damage following hypoglycemia, anoxia, and status epilepticus. However, other parts of the hippocampus may also be affected by this vulnerability (Chauhan et al., 2021; Knierim, 2015). The entorhinal cortex is the primary provider of input to the hippocampus, which is

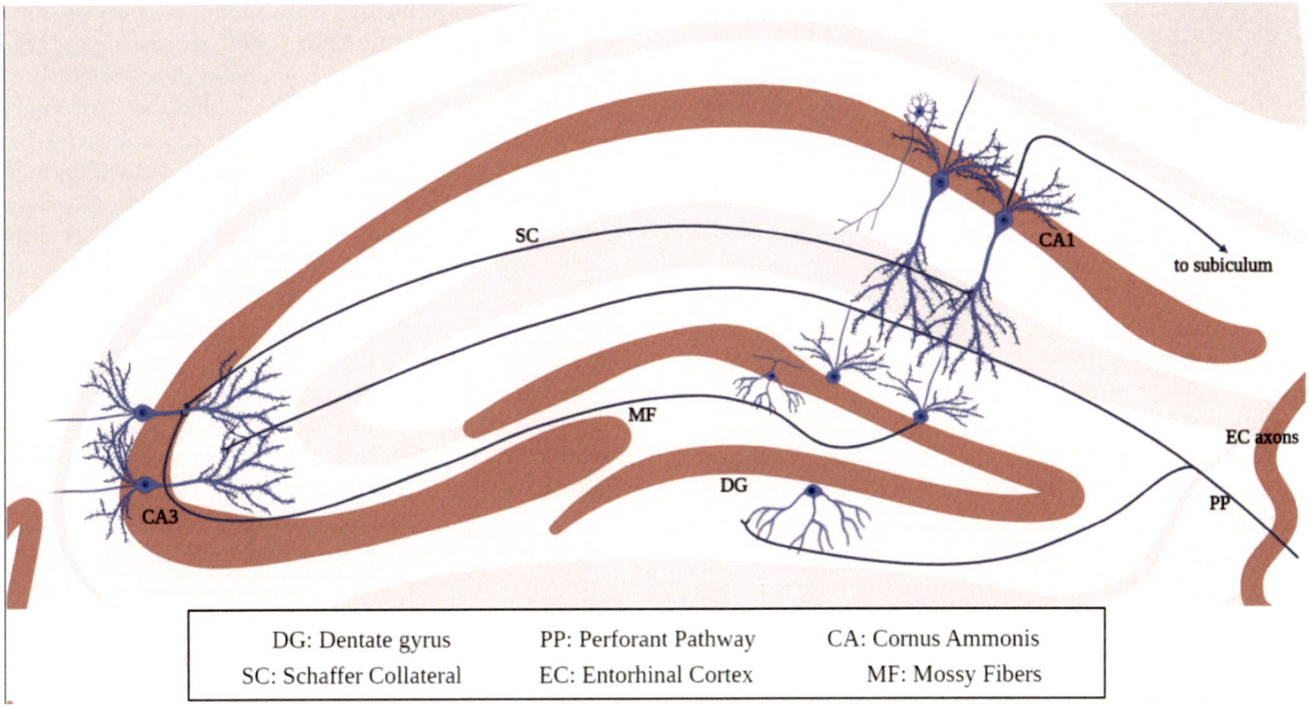

BOX 13.2

The role of Schaffer's Collateral pathway in synaptic plasticity

Schaffer's Collaterals are axons that originate from CA3 and project to CA1 (see figure above, Box 13.2). They are a crucial part of the hippocampal trisynaptic circuit and play a vital role in memory formation and the emotional network of the Papez circuit. These circuits are glutamatergic, meaning they are excitatory and involved in activity-dependent neuronal plasticity (Anand and Dhikav, 2012). For instance, by decreasing the excitability of Schaffer collaterals, it is possible to reduce the transmission of excitatory signals within the hippocampus that are suppressed by cenobamate, an antiepileptic medication (Song et al., 2022). On the other hand, the establishment of synaptic connections is crucial for neural circuit assembly in the brain. To this point, Latrophilin-2 and Latrophilin-3, a group of highly conserved G-protein coupled receptors (GPCRs) from the adhesion G protein–coupled receptor family, are recently shown to be essential for the formation of entorhinal-cortex-derived and Schaffer-collateral-derived synapses in mouse CA1-region neurons, respectively, and their GPCRs signaling pathway is required for synapse formation (Sando and Südhof, 2021). Thus, synaptic plasticity is a cellular mechanism that plays a crucial role in learning and memory. In experimental studies, synaptic plasticity can be induced in CA1 pyramidal neurons of the hippocampus through repeated stimulation of the Schaffer collaterals (Sánchez-Rodríguez et al., 2022). It involves the adjustment of synaptic transmission strength and efficacy in response to specific neural activity patterns (Sánchez-Rodríguez et al., 2022; Dekeyzer et al., 2017; Parmar et al., 2018; Siegel and Sapru, 2010). This process is regulated by a delicate balance between excitatory and inhibitory synaptic transmission. Therefore, Schaffer collaterals axons play important roles in synaptic plasticity for memory in healthy states and disease status.

relayed through the glutamatergic perforant path to the dentate gyrus DG region (synapse 1). From there, the information is transmitted to the CA3 region through the mossy fiber pathway (synapse 2), and then CA3 sends it to the CA1 region via the Schaffer Collateral pathway (synapse 3). Finally, CA1 completes the loop by returning to the entorhinal cortex (Knierim, 2015) (Fig. 13.5).

PART II: FUNCTIONAL CORTICAL CONNECTIONS OF THE HIPPOCAMPUS

To understand how the hippocampus is involved in memory, and its disorders, we need to know its connections to other brain areas, and especially to the cerebral cortex (Aggleton, 2012; Rolls, 2018, 2021). Research in animals has led to the dual stream model, in which ventral cortical processing streams concerned with representations of "what" object or person is present reach the hippocampus by the perirhinal and then lateral entorhinal cortex; and the dorsal cortical stream concerned with "where" representations of spatial locations is connected via the parahippocampal gyrus and medial entorhinal cortex with the hippocampus (Van Hoesen, 1982; Burwell et al., 1995; Burwell, 2000; Knierim et al., 2014; Doan et al., 2019) (shown in schematic and simplified form Fig. 13.6). Then, in the CA3 part of the hippocampus, where there is a single network of neurons, "what" and "where" streams can be combined to form a memory of what happened where (Rolls, 2018, 2021; Kesner and Rolls, 2015). That memory can then be recalled using a partial retrieval cue to the neocortex by corresponding hierarchically organized and segregated backprojection pathways from the hippocampus (Rolls, 2018, 2021; Treves and Rolls, 1994). In macaques, some hippocampal connections have been described that are not included in the conceptual dual stream model shown in (Fig. 13.6). For example, in macaques, the hippocampal CA1 region has some direct connections to the cortex in the anterior part of the superior temporal sulcus (Zhong et al., 2005) and to the orbitofrontal cortex (OFC) and anterior cingulate cortex (ACC) (Cavada et al., 2000; Zhong et al., 2006; Morecraft et al., 2012). It has further been reported in macaques that CA1 neurons have direct connections to a number of temporal cortical areas, including the posterior parahippocampal (areas TF and TH), perirhinal (35 and 36 BA) (Insausti and Munoz, 2001), and ventral inferotemporal areas (areas TEav and TEpv) (Yukie, 2000), with connections from CA1 also to the temporal pole (TG) and subiculum (Blatt and Rosene, 1998), to the pregenual cingulate cortex (Insausti and Munoz, 2001), and to anteroventral TE (Zhong and Rockland, 2004; Ichinohe and Rockland, 2005) and the OFC (Zhong et al., 2006).

Moreover, direct projections to CA1 in macaques have been reported from 7a and 7b BA, area TF, and a region in the occipitotemporal sulcus (Rockland and Van Hoesen, 1999; Ding et al., 2000); and from superior

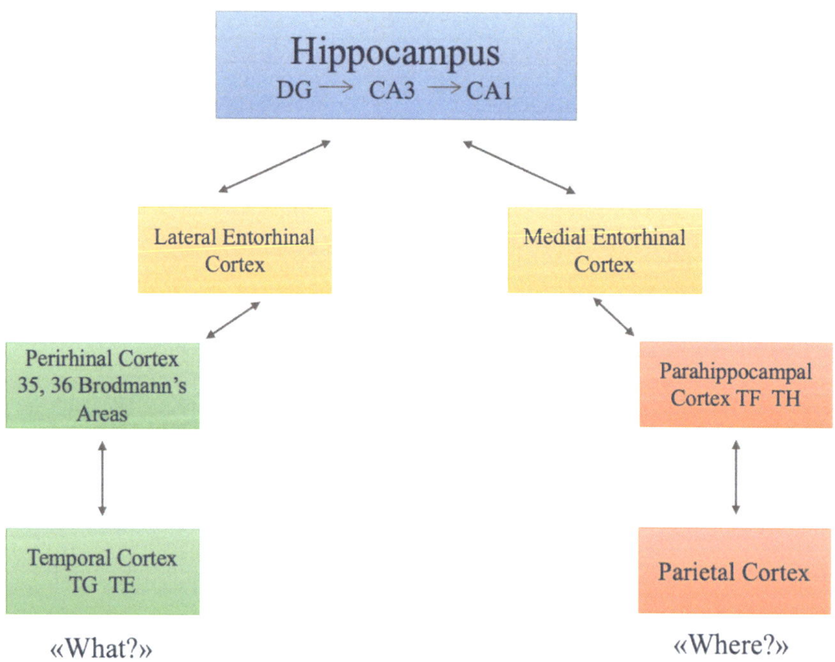

FIGURE 13.6 **A simplified, schematic, conceptual diagram of the dual stream model of hippocampal connectivity.** Neuroanatomical tract-tracing in animals suggests that spatial, "where," representations from the dorsal stream parietal cortex project via the parahippocampal gyrus and then medial entorhinal cortex to the hippocampus, and that "what" object representations from the ventral stream temporal cortex project via the perirhinal cortex and lateral entorhinal cortex to the hippocampus (Aggleton, 2012; Van Hoesen, 1982; Burwell et al., 1995; Burwell, 2000; Knierim et al., 2014; Nilssen et al., 2019; Van Hoesen and Pandya, 1975; Amaral et al., 1983; Saleem and Tanaka, 1996). It is suggested that the CA3 neurons of the hippocampus with their highly developed associative recurrent collateral connections can then associate the inputs to form object-location episodic memories of what happened where on a particular occasion (Rolls, 2018; Kesner and Rolls, 2015). In macaques, the connections are known to be more extensive than shown here. *From Huang, C.C., 2021. Extensive cortical connectivity of the human hippocampal memory system: beyond the "What" and "Where" dual stream model. Cereb. Cortex (in this issue) redesigned by Brandon M. Ascenzi.*

temporal sulcus, the rostral and retrosplenial portions of the cingulate cortex, the agranular insular cortex, and the caudal OFC (Suzuki and Amaral, 1990); and also from anteroventral TE (Zhong and Rockland, 2004). The presubiculum has connections to the cortex in the superior temporal sulcus, the lateral OFC, and the temporal pole (Insausti and Munoz, 2001). In rodents, division of the entorhinal cortex into medial and lateral components as illustrated in Fig. 13.6 is usual, but in at least primates and perhaps rodents too, there may be less segregation than is implied by Fig. 13.6 (Insausti et al., 2017; Nilssen et al., 2019). Thus, reality is more complex at least in primates than the simple dual stream model, as has been recognized (Knierim et al., 2014; Kravitz et al., 2011, 2013; Milner, 2017). A key aim of a recent study performed by Huang (2021) has been precisely to obtain evidence on the direct connections between the human hippocampal system and other cortical areas, as this is fundamental to understanding how the human hippocampus operates in health and disease (Rolls, 2021).

Huang (2021) in this regard showed that effectively the hippocampus has direct connections with many cortical areas beyond the entorhinal cortex, consistent with the hypothesis that in humans these areas are not the exclusive gateway with the hippocampus. Instead, the hippocampus communicates directly with some "what" (temporal cortex) and "where" (parietal and posterior cingulate) cortical areas, and even early sensory cortical areas, which may facilitate the recall of memories with sensory detail. Moreover, they showed that lateral parahippocampal gyrus area TF is, with the perirhinal cortex, strongly connected with the hippocampus and with "what" ventral stream/temporal lobe cortical areas, leading to a reevaluation of the functions of the parahippocampal gyrus. Lastly, they described that medial parahippocampal gyrus area TH is strongly linked with the hippocampus and dorsal visual stream, parietal and posterior cingulate cortical areas in the "where" visual system (Huang, 2021).

In addition, it is known that the amygdala has extensive back-projections to early visual cortical areas (including V1) that end in the superficial layers (Amaral and Price, 1984). In functional terms this leads to the new hypothesis that these back-projections from the perirhinal cortex and parahippocampal gyrus, and it appears from the hippocampus in humans, help to retrieve visual and other sensory details being recalled by the hippocampus. Memory recall is a key function proposed for back-projections to layer 1 of the neocortex (Treves and Rolls, 1994; Rolls, 1989, 2016, 2021). The hippocampal connections with early sensory processing areas for different sensory modalities, including vision, olfaction, somatosensation, and spatial representations in TH, which includes the parahippocampal spatial scene area (PPA) (Sulpizio et al., 2020), may be useful for storing and later recalling sensory details of memories.

Fornix

The fornix (Figs. 13.7 and 13.8) is a compact fiber bundle connecting the hippocampus with the hypothalamus and various other structures (Savransky, 2017). Its fibers first form the alveus, a thin white layer on the ventricular surface of Cornus Ammonis and then converge as the fimbria along the medial aspect of the hippocampus. Anatomically is possible to describe two principal parts of the fornix, the *pre-* and *post-commissural*, and the *commissural connections* (Amaral, 2018). All three structures are below described.

Precommissural fornix. Consists of fibers originate from the hippocampus and subicular cortex travel to the anterior commissure rostrally and terminate in the septal area. The fibers near the anterior and posterior part of the hippocampal formation, respectively, project to the lateral and medial parts of the lateral septal nucleus (Siegel and Sapru, 2010).

Postcommissural fornix. It is made up by fibers that ventrally travel behind the anterior commissure. Postcommissural fibers come from the subicular cortex to terminate in the anterior thalamic nucleus, mammillary bodies, and medial hypothalamus neighboring regions (Christiansen, 2016; Chauhan et al., 2021). In patients with Korsakoff's syndrome, the mammillary bodies are damaged. These patients have a problem in forming new memories (Siegel and Sapru, 2010; Fama et al., 2012).

The axons of the subiculum project to the entorhinal cortex, cingulate cortex, and prefrontal cortex areas. Eventually, these fibers from the entorhinal cortex project to the amygdala and the temporal cortex. Therefore, these connections help the hippocampal formation to send signals throughout the cerebral cortex, including areas which receive and process different types of sensory information (Böhm et al., 2018; Siegel and Sapru, 2010).

Commissural connections. They are fibers that connect the hippocampi on either brain hemisphere to each other. These fibers mainly originate from the CA3–CA4 and terminate in the contralateral hippocampus. These types of fibers are important clinically. It is supposed that these commissural connections may cause spread of seizures from their primary epileptogenic focus to the contralateral hippocampus and form a secondary epileptogenic focus there (Siegel and Sapru, 2010; Postans et al., 2020; Costard et al., 2019) (Box 13.3).

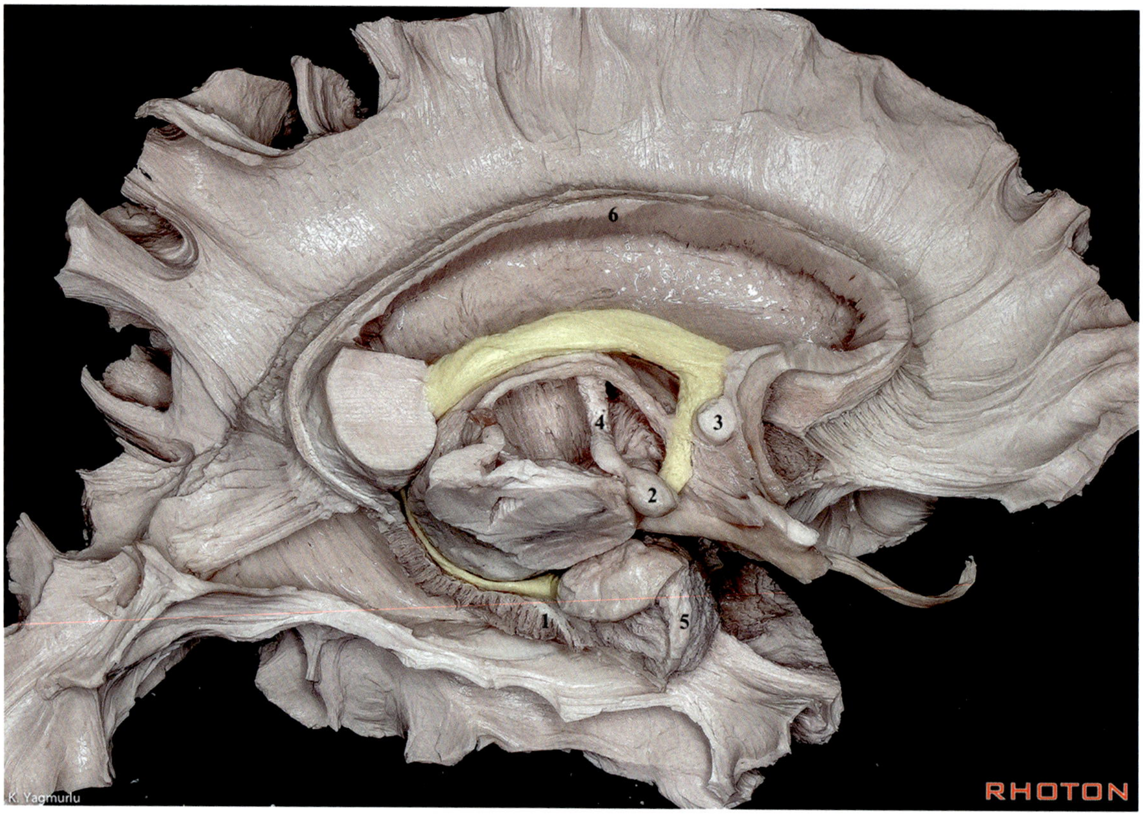

FIGURE 13.7 **Fornix and related structures; lateral deep dissection.** In yellow, the fornix; 1. hippocampus; 2. mammillary body; 3. anterior commissure; 4. mammillothalamic tract; 5. amygdala; 6. corpus callosum. The septum is located anteriorly the anterior commissure. *Image courtesy of The Rhoton Collection® http://rhoton.ineurodb.org/. Dissector: K. Yagmurlu. Drawing by Nickalus Khan. Stereoscopical Image: use red-blue 3D glasses.*

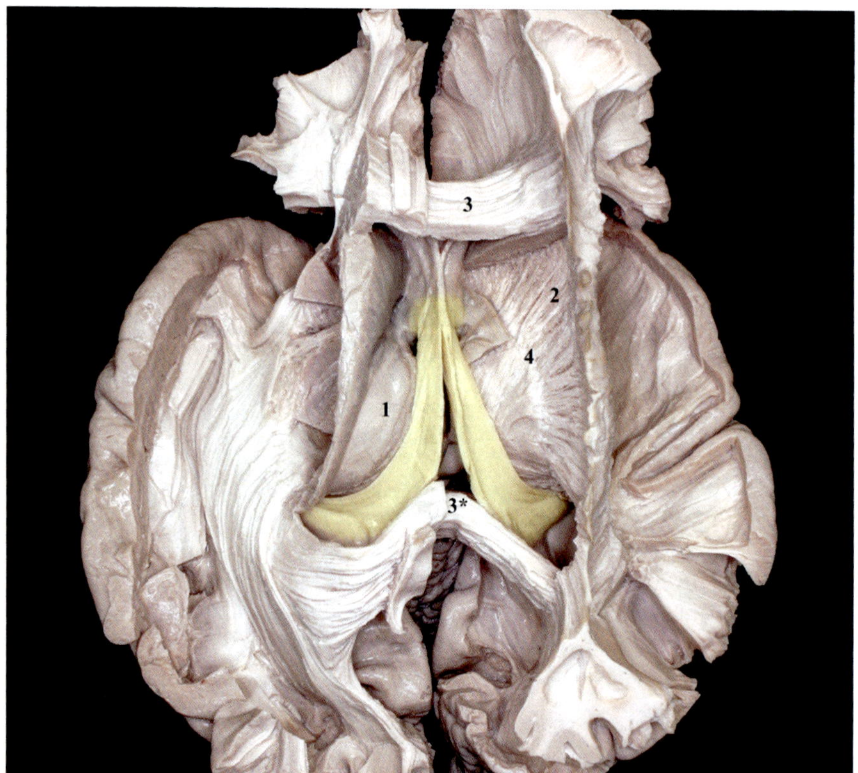

FIGURE 13.8 **Fornix; superior view.** In yellow, the fornix; 1. thalamus; 2. Corona radiata; 3. Corpus callosum (forceps minor, anteriorly (3); forceps major, posteriorly (3*), 4. Internal capsule. *Image courtesy of The Rhoton Collection® http://rhoton.ineurodb.org/. Dissector: K. Yagmurlu. Drawing by Kaan Yagmurlu. Stereoscopical Image: use red-blue 3D glasses.*

BOX 13.3

Recent findings in hippocampal connections

Learning and memory are complex processes that involve multiple brain regions, including the hippocampus, retrosplenial cortex, and lateral septum. A recent investigation has found that the ventral hippocampus sends projections to the basal amygdala, which are responsible for emotional memory. Additionally, the ventral hippocampus sends projections to the nucleus accumbens for social memory and instrumental actions, and to the prefrontal cortex for object-based memory (Bakoyiannis et al., 2023). Furthermore, some of these hippocampal projections also modulate feeding and anxiety-like behaviors (Bakoyiannis et al., 2023). While the dorsal hippocampus is known to be important for acquiring and retrieving episodic memories, the downstream circuitry of this region, specifically the pathways to the retrosplenial cortex and lateral septum, has only recently been explored. A study using optogenetics and contextual fear conditioning in mice investigated the role of these pathways in memory acquisition and expression. The results showed that inhibiting the dorsal hippocampus neuronal terminals in either the retrosplenial cortex or lateral septum impaired memory performance, indicating that both pathways are crucial for memory acquisition. Inhibiting the pathways during memory retrieval had a varying effect on memory performance, highlighting the complexity of memory processing and the differential involvement of hippocampal efferents in physiological and cognitive memory processes (Opalka and Wang, 2020). Overall, new connections in the hippocampus have revealed that the old view of one pathway—one function is outdated and limited. Besides that, new paradigm in medicine has indicated that making connections from structural to biological levels (i.e., cellular, genetic, and molecular levels) is vital for understanding the function of brain regions. In this line, a study investigated to understand how the epigenetic landscape of mammalian brain cells contributes to their diversity (Liu et al., 2021). To do this, the researchers used single-nucleus DNA methylation sequencing to analyze over 100,000 nuclei from different regions of the mouse brain. They identified distinct cell clusters based on epigenetic profiles, signature genes, regulatory elements, and transcription factors. The study found that the DNA methylation landscape varied continuously along spatial gradients in excitatory neurons in the cortex and hippocampus. The researchers developed an artificial neural network model that accurately predicted single neuron cell-type identity and brain area spatial location. By integrating multiple datasets, they were able to annotate the regulatory genome of hundreds of cell types in the mouse brain, which helped establish the epigenetic basis for neuronal diversity and spatial organization (Liu et al., 2021).

THE HIPPOCAMPUS FUNCTIONS THROUGH ITS CIRCUITS

Long-Term Changes in Signaling

The hippocampus is well known for its role in learning and memory. Repeated activity in the hippocampus can cause changes in synaptic efficacy that last longer than those seen at peripheral synapses (Mota et al., 2023). This prolonged synaptic effect may be related to memory. Synaptic plasticity refers to changes in synapses between neurons, which can take two forms: long-term potentiation (LTP) and long-term depression (LTD) (Fig. 13.9) (Pinky and Neha, 2023).

These can exist in early transient or late persistent forms (Bin Ibrahim et al., 2022). LTP is crucial for memory consolidation and is produced by stimulating fibers that make excitatory connections to hippocampal pyramidal cells (Honoré, 2022). This leads to an increase in the intensity of pyramidal cells' response, which lasts for minutes or even weeks (Escobar, 2007). At the end of this process, learning may occur. Glutamate binding to N-methyl-D-aspartate (NMDA) receptors is believed to be the synaptic mechanism responsible for this process. Blocking NMDA receptors results in the loss of LTP and disrupted learning. High frequency stimulation can release magnesium from NMDA receptors, allowing current to flow through both NMDA and non-NMDA channels (Lüscher and Malenka, 2012; Kandel et al., 2012). The increase in calcium concentration in the postsynaptic cell is an important factor in the creation and expression of postsynaptic LTP, and NMDA receptors are highly permeable to calcium. α-Amino-3-hydroxy-5-methyl-4-isoxazole Propionic Acid Receptor (AMPAR) is another receptor that can regulate synaptic transport during LTP (Kandel et al., 2012; Clopath et al., 2008). LTP has been demonstrated at terminals of the perforant pathway in the dentate gyrus and at the synapses of CA3 pyramidal cells on CA1 cells. LTP also occurs in other brain regions, including the cortex, amygdala, and cerebellum (Kandel et al., 2012; Kennedy, 2016).

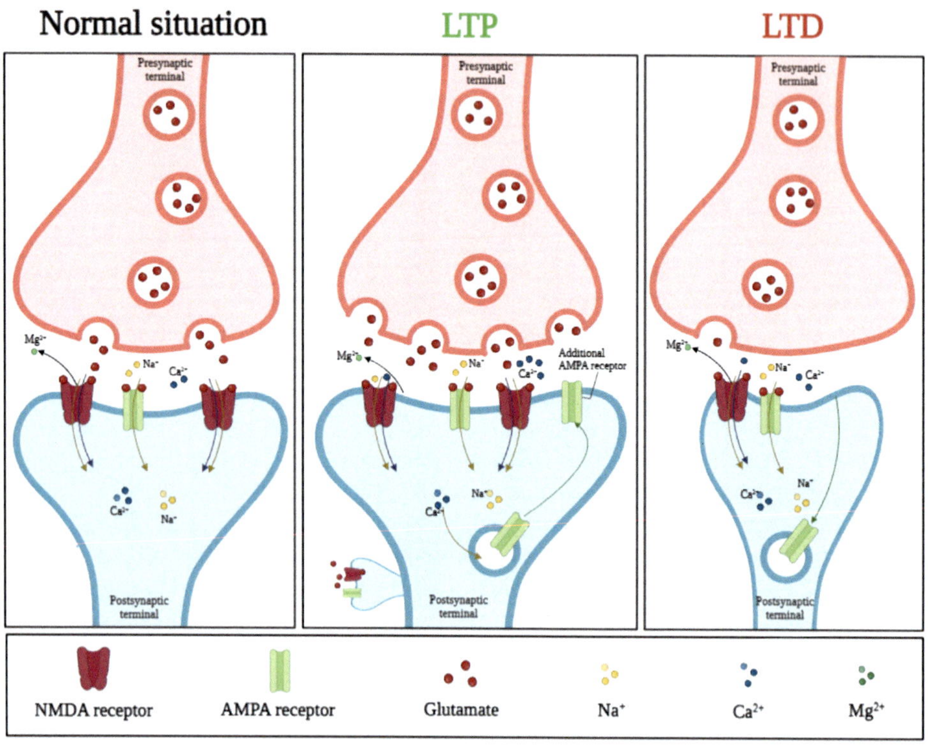

FIGURE 13.9 **The hippocampus uses two mechanisms to trigger learning and memory: long-term potentiation (LTP) and long-term depression (LTD).** LTP is an activity-dependent enhancement in the strength of excitatory synaptic connections that occurs after brief, high-frequency electrical stimulation. In contrast, LTD is a long-lasting decrease in synaptic efficacy that follows low-frequency stimulation. Both LTP and LTD are initiated by the activation of the NMDA receptor and result in changes in synaptic transmission mediated by the AMPA receptor. *Created with BioRender.com.*

LTD is a long-lasting depression of synaptic transmission caused by previous repetitive activity in the same pathway or a different afferent pathway to the same cell (Bin Ibrahim et al., 2022). LTD is necessary for relearning or giving up a concept. There are two types of homosynaptic LTD in the Schaffer collateral–CA1 synapse: one depends on NMDA receptor activation, and the other depends on activation of G-protein–coupled metabotropic glutamate receptors (mGluRs). LTD occurs due to lower levels of NMDA receptor activation and less Ca^{2+} influx compared to LTP. One consequence of LTD induction is a loss of AMPA receptors from the synapse, and one long-term consequence of LTD is synapse elimination. The induction of LTD appears to involve metabotropic glutamate receptors (mGluRs) (Lüscher and Malenka, 2012; Xu et al., 2020). LTP and LTD are important in that learning and memory involve long-term changes in synaptic efficacy (Lüscher and Malenka, 2012).

The hippocampal plasticity changes in patients with depression (Xu et al., 2020; Liu et al., 2017). For example, studies have shown that the hippocampus volume reduces in patients with major depressive disorder (MDD) (Santos et al., 2018). Chronic stress can change the integrity of hippocampal cells by apoptosis in hippocampal subregions, and stress can decrease LTP and increase LTD by increasing the levels of glucocorticoids (Xu et al., 2020; Peters et al., 2018).

Papez Circuit

The hippocampal circuitry postulated by Papez is known as the "Papez circuit" and is believed to function as a memory circuit, responsible for emotional integration and recent memory trace (Aggleton, 2022; Kandel et al., 2012; Siegel and Sapru, 2010). The neocortex and limbic system (responsible for emotional experience) send inputs to the entorhinal cortex and then to the hippocampus. It includes the hippocampus, fimbria, fornix, mammillary body, anterior nucleus of the thalamus, and cingulate gyrus. The entorhinal cortex receives sensory information from the association areas of the frontal, parietal, and temporal lobes, which is then converted into memory through the Papez circuit (Fig. 13.10) (Kamali, 2023; Rajmohan, 2007). Short-term memory, also known

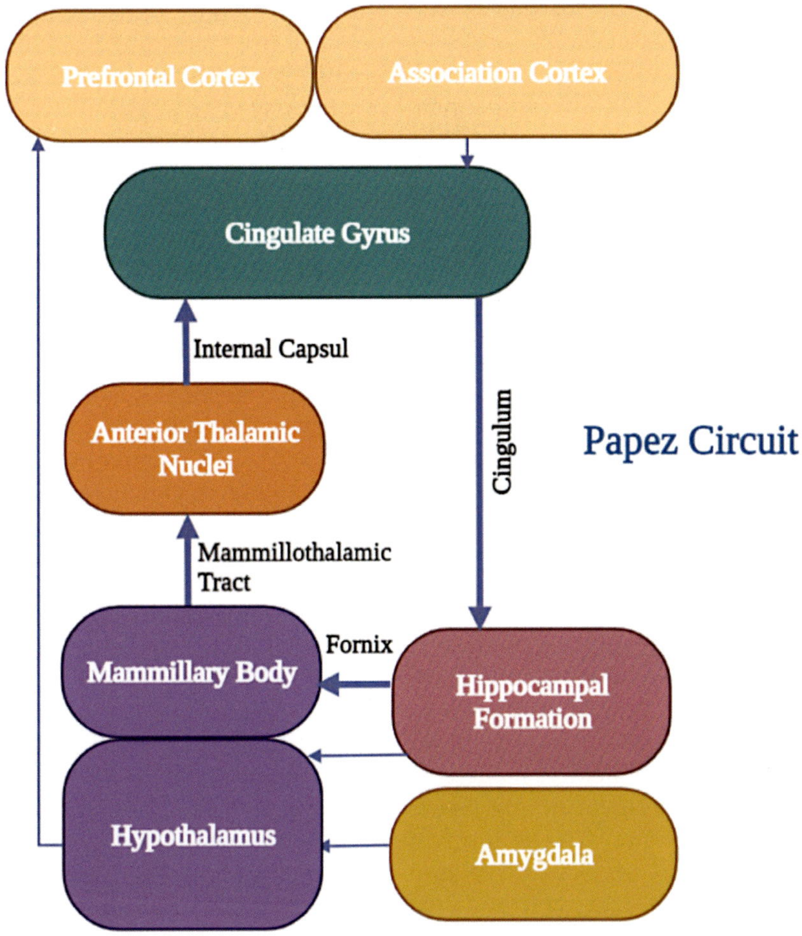

FIGURE 13.10 **The Papez circuit contains several structures of the brain.** It activates and stops with the hippocampus. *Created with BioRender.com.*

as episodic memory, is facilitated by unidirectional activation of synaptic connections, as follows:

- Input from the neocortex and limbic system activates the entorhinal cortex (parahippocampal gyrus);
- Stimuli pass from the entorhinal cortex to the dentate gyrus via the perforant path and then through the CA3 area;
- Schaffer's collaterals from the CA3 transfer stimuli to the CA1, which then efferently synapses at the subiculum;
- Efferent fibers from the subiculum project back to the entorhinal cortex.

Connections between the CA3 and dentate gyrus, and between CA1 and CA3, lack feedback loops. The alveus, fimbria, and fornix, formed by the axon fibers of the CA3 pyramidal cells and neurons of the subiculum, are the major hippocampal output. Fibers from the precommissural fornix (derived from the CA3) are connected to the lateral septal nucleus, and fibers of the postcommissural fornix (derived from the subiculum) are connected to the mammillary bodies and hypothalamic nuclei (Chauhan et al., 2021) (for a more detailed description of limbic system see Chapter 22).

PART III: HIPPOCAMPUS AND CENTRAL NERVOUS SYSTEM DISEASES

Major Anatomical-Related Disorders of Hippocampus

Alzheimer's disease (AD) is a neurodegenerative disease that characterizes by memory loss and cognitive dysfunction and considers as a main cause of dementia worldwide (Scheltens et al., 2016). Increasing evidence proposes that many cognitive impairments, learning deficits, and memory loss that happening in AD are mediated through decrease hippocampal neurogenesis and its atrophy (Demars et al., 2010; Donovan et al., 2006; Lazarov et al., 2010). Abnormal neurogenesis in subgranular zone (decrease in proliferation and abnormal maturation) were also reported in mice model of AD (Donovan et al., 2006; Ren et al., 2023). Similarly, decrease the number of neural progenitor cells and neuronal differentiation were also seen in APPswe/PS1ΔE9 mice model of familial AD (Demars et al., 2010). It seems that dysregulation of hippocampal neurogenesis signaling molecules (i.e., presenilin-1, CREB, Notch 1, and Wnt/β-catenin) may decrease the hippocampal neurogenesis in AD (Hollands et al., 2016). Decreased volume of subregions CA1 and subiculum in moderate AD as well as atrophy of CA3 and CA4 in progressive form has been shown in multishot diffusion-weighted MRI in AD patients (Adachi et al., 2003). Furthermore, a hippocampal analytic technique on data from MRI in AD patients suggested hippocampal atrophy was related to cognitive impairments and considered an indicator for diagnosis of AD (Apostolova et al., 2006a). Hippocampal atrophy especially in subregions CA1 and subiculum in patients with mild cognitive impairment (MCI) increased the risk of conversion of disease to AD (Apostolova et al., 2006b). Importantly, immunohistochemical (IHC) analysis in hippocampal postmortem samples from AD patients indicated neuronal loss and accumulation of Aβ-labeled plaques in CA1 subregions (Furcila et al., 2018). However, Svenningsson et al. (2019) showed that Aβ accumulation and hippocampal atrophy are independently related to memory loss and cognitive impairment in AD (Svenningsson et al., 2019). The subregion CA1 is the first target for injury following AD as previously reported huge aggregation of misfolded proteins and neuronal loss in this region (Furcila et al., 2018; Lace et al., 2009). Synaptic disturbances especially in hippocampal neurons and gliopathy are responsible for cognitive impairments and memory loss in patients with AD (Scheff et al., 2006; Marchetti and Marie, 2011; Rudnitskaya et al., 2022). This is included lower number of synapses in postmortem hippocampal samples (Scheff et al., 2006), downregulated synaptic transmission proteins, such as synaptophysin, synaptogyrin, synaptotagmin, and syntaxin in CA1 subregion (Counts et al., 2014), and disruption of signaling pathways that mediate neurogenesis (Sun et al., 2018). Disruption of functional connectivity in the olfactory bulb-entorhinal cortex-dorsal hippocampus circuit along with neuronal loss in the regions caused recognition memory impairment in AD rats model (Salimi et al., 2022). Furthermore, in APP knock-in mice model of AD, input connectivity strengths to CA1 from other subregions of hippocampus were significantly decreased in compered to control mice (Ye et al., 2022). Dysfunctional inhibitory interneurons and abnormal excitability also seen in dentate gyrus of aged AD mice (Hazra et al., 2013). There were some important mechanisms such as aggregation of misfolding proteins, LTP impairment, decreased hippocampal neurogenesis, impaired cerebral blood flow, and oxidative stress which suggested for hippocampal atrophy following AD (Ondrejcak et al., 2010; Tran et al., 2010; Yassa and Stark, 2011; Molnár, 2011; Hiramatsu et al., 1992; Howland and Wang, 2008).

Parkinson's disease (PD) is another important neurodegenerative disease and is pathologically identified as the progressive destruction of dopaminergic neurons in the midbrain and accumulation of misfolded protein alpha-synuclein (α-syn). There is growing evidence that hippocampal structure and function are pathologically altered in PD and implicated in the induction of cognitive deficits in patients (Tanner et al., 2017; Camicioli et al., 2003; Villar-Conde et al., 2021). In PD patients, hippocampal atrophy especially in CA3 and CA4 and gross changes in dentate gyrus were seen in volumetric analysis using MRI (Tanner et al., 2017; Lenka et al., 2018). Furthermore, in patients with PD, impairment of recognition and verbal memory was well correlated with hippocampal atrophy as compared with control subjects (Camicioli et al., 2003; Brück et al., 2004). Furthermore, decrease in hippocampal volume was associated with depressive symptoms in PD patients (van Mierlo et al., 2015). Postmortem analysis in brain samples from PD patients showed severe expression of α-syn in CA1 subregion of hippocampus and decreased expression of some synaptic proteins (i.e., transmembrane protein 163, receptor accessory protein 2 and caskin-1) in hippocampal sections (Villar-Conde et al., 2021). Importantly, neuronal degeneration was also seen in CA2 and CA3 subregions of hippocampus in postmortem brain samples from patients with PD (Mattila et al., 1999). Atrophy in all subregions of hippocampus were also seen in PD patient using MRI (Pereira et al., 2013). In A53T transgenic mice model of PD, also the expression of α-syn was significantly increased in dentate gyrus and CA2-3 subregions of hippocampus (Flores-Cuadrado et al., 2016).

Temporal lobe epilepsy (TLE) is an intractable seizure disorder that initiates from focal lesions in temporal lobe or induces due to stroke, brain structural changes and glioma (Fang et al., 2020; Chen et al., 2022; Thom et al., 2009). These patients experience some degrees of memory impairments and cognitive deficits, which represent hippocampal involvement in TLE (Mokhothu and Tanaka, 2021). In cases with TLE, fMRI analysis revealed the disruption of hippocampal connectivity with language brain regions played important role in language defects (Nenning et al., 2021). Decreased hippocampal volume in most subregions along with neuronal loss has been seen in MRI data from patients with TLE and hippocampal sclerosis (Jardim et al., 2016). Similarly, the ipsilateral hippocampal atrophy occurred in all subregions (i.e., CA1, CA2, CA3, CA4, dentate gyrus, and subiculum) in MRI data from TLE patients as compared to control cases (Kim et al., 2015). The hippocampal atrophy is associated with cognitive deficits and abnormal electroencephalography in TLE patients (Elsherif and Esmael, 2022). Hippocampal sclerosis was also reported in 57% of children with TLE (Grattan-Smith et al., 1993). Furthermore, autopsy samples from TLE patients indicated severe sclerosis and microglial activation in CA1, CA2 and CA3 subsections of hippocampus (Beach et al., 1995). An imbalance in activation of GABAergic and glutamatergic neurons and neuronal degeneration was seen in CA3 and dentate gyrus of kainic acid (KA)-induced temporal lobe epilepsy model (Cao et al., 2022). Also, reduced neurogenesis, activation of astrocytes and microglia as well as derangement of mossy fibers were shown in hippocampus of KA-induced TLE rats (Upadhya et al., 2019). Importantly, proteomic analysis of hippocampus in pilocarpine-induced TLE rats indicated that 57 proteins associated with synaptic function, cytoskeleton modification, ion channels, and mitochondrial dysfunction were differentially expressed as compared to control rats (Liu et al., 2008). Among these proteins, five proteins including synaptosomal-associated protein 25 (SNAP25; downregulated), voltage-dependent anion channel 2 (VDAC2; upregulated), alpha-internexin (INA; upregulated), synapsin-2 (SYN2; upregulated) and homer protein homolog 2 (HOMER2; upregulated) were validated and quantified by RT-PCR (Liu et al., 2008).

Traumatic brain injury (TBI) is a sudden damage that leading to death and disability in children and adults due to brain injury (Wiles et al., 2023). The hippocampus is a main vulnerable brain structure that affects over weeks after TBI. Hippocampal atrophy and deficit in LTP may mediate impairment of hippocampal learning and memory after TBI (Atkins, 2011). Patients with history of TBI have demonstrated persistent memory and cognitive impairments years after injury that were along with hippocampal atrophy and reduced neuronal activities (Monti et al., 2013; Draper and Ponsford, 2008). The connections between hippocampus and prefrontal cortex are critical for many executive function and memory performances. Diffusion tensor imaging studies in patients with TBI have reported neuronal injuries in some components of limbic system (i.e., corona radiate and cingulum bundle) that responsible for connection of hippocampus to prefrontal cortex (Niogi et al., 2008a,b; Spitz et al., 2013). The memory impairment was also revealed in soldiers with diffuse TBI (Hoge et al., 2008). Also, microscopic postmortem brain analyses on samples taken from patients with persistent memory impairment after mild concussive head injury have shown that axonal injury in the fornices as largest pathway of the hippocampus (Blumbergs et al., 1994). Hippocampal atrophy may consider as a main mechanism that mediates learning and memory deficits following TBI (Atkins, 2011; Ariza et al., 2006). Furthermore, in rats model of TBI, microglial activation, neuroinflammation, and neuronal loss were seen in hippocampus (Komoltsev et al., 2021a). Microglial activation, neuronal loss, decreased synaptogenesis, and synaptic degeneration may cause hippocampal atrophy following TBI (Grady et al., 2003; Scheff et al., 2005; Campbell et al., 2012; Maxwell et al., 2003; Gao et al., 2011; Schumm et al., 2022; Langlois et al., 2022).

Minor Anatomical-Related Disorders of Hippocampus

Multiple sclerosis (MS) is a chronic inflammatory and neurodegenerative condition that involves CNS and occurs when immune system invades the axonal myelin sheath. The existing evidence showed that hippocampal structure and function may be involved in MS that mediates memory impairment and cognitive deficits in these patients (Roosendaal et al., 2010; Geurts et al., 2007; Boscheron et al., 2021). Functional magnetic resonance imaging (fMRI) in MS patients revealed right hippocampal atrophy and impaired normal connectivity of hippocampus with anterior cingulate gyrus, thalamus, and prefrontal cortex in comparison with healthy control subjects (Roosendaal et al., 2010). Furthermore, using high-resolution MRI disclosed moderate CA1 atrophy in patients with relapsing-remitting MS (RRMS) and severe atrophy in all subregions of hippocampus in patients with secondary progressive MS (SPMS) (Sicotte et al., 2008). These anatomical dysfunctions were correlated with impairment of learning and memory tasks in these patients (Sicotte et al., 2008). Importantly, in patients with RRMS, analysis of functional and anatomical connections of hippocampus with other brain regions using diffusion tensor imaging, indicated gross structural changes in hippocampus and these anatomical dysfunctions were correlated with cognitive impairments and disability (Gu et al., 2022). Hulst et al. (2015) showed that hippocampal volume loss decreased its activation during the task and increased the connections to posterior cingulate are implicated in memory impairments and cognitive deficits in MS patients as compared to healthy control (Hulst et al., 2015). Postmortem autopsy analysis on brain samples from MS patients showed several lesions, microglial activation, and neuronal demyelination in CA1 subregion of hippocampus and dentate gyrus as compared to samples from nonneurologic control (Geurts et al., 2007). In experimental autoimmune encephalomyelitis (EAE) mice, decreased GABAergic interneurons, inhibited GABAergic spontaneous inhibitory postsynaptic currents, microglial activation and overexpression of interleukin-1β in hippocampus suggested a new target for cognitive deficits in MS (Nisticò et al., 2013). Furthermore, microglial activation and dendritic damage in hippocampal layers of EAE mice were seen (Crombe et al., 2018). Also, demyelinated hippocampal neurons and loss of neural stem cells in dentate gyrus were identified in cuprizone-induced MS mice (Zhang et al., 2020).

Schizophrenia (SCZ) is a severe mental illness that is accompanied with some symptoms such as hallucinations, delusions, little interest in life, sustain relationships, extreme agitation, disability to do complete tasks or organize their thoughts and slowing of movements (Moya et al., 2023; Nakamura and Takata, 2023; McCutcheon et al., 2020). Many clinical studies have reported gross structural and functional abnormalities in hippocampus in patients with SCZ (Csernansky et al., 1998; Kühn et al., 2012; Schobel et al., 2009; Steen et al., 2006; Bogerts et al., 1990; Heckers and Konradi, 2015; Molina et al., 2003; Medoff et al., 2001; Lieberman et al., 2018; Wegrzyn et al., 2022). A neuroimaging study revealed an abnormal increased in cerebral blood volume in CA1 subregion of hippocampus that was correlated with clinical symptoms in patients with SCZ (Schobel et al., 2009). Decreased hippocampal volume (Bogerts et al., 1990; Prestia et al., 2015; Pujol et al., 2018; Wang et al., 2021; Koolschijn et al., 2010), neuronal loss (Deicken et al., 1998), dysconnectivity of hippocampus with other brain regions (Nelson et al., 2022; Du et al., 2020; Avery et al., 2018), and deformity in subiculum—as the main hippocampal—prefrontal connector path—were also reported in magnetic resonance images of patients with SCZ (Csernansky et al., 1998). Furthermore, reduced hippocampus volume and decreased the ratio of hippocampus to amygdala volume seen in patients with SCZ compared to healthy controls (Prestia et al., 2011). A systematic review and meta-analysis on 1424 patients also demonstrated hippocampal atrophy in SCZ (Steen et al., 2006). Another systematic review and metaanalysis on 413 postmortem brain samples also demonstrated the hippocampal volume and number of neurons in several hippocampal subregions were decreased in patients with SCZ (Roeske et al., 2021). Hyperactivity in subiculum of hippocampus and loss of fast-spiking interneurons were also seen in rats model of SCZ (Lodge and Grace, 2008). Also, injury to ventral hippocampus of neonatal rats have been suggested as an experimental model for induction of SCZ (Andrew Chambers et al., 1996; Tseng et al., 2009). This model impaired cognitive performances and disrupted social interactions (Andrew Chambers et al., 1996; Lipska et al., 2002; Powell et al., 2006).

Huntington's disease (HD) is an autosomal dominant neurodegenerative disorder and described by progressive degeneration of neurons in the striatum. This part of CNS is under study to find exact pathophysiology and therapeutic interventions for HD. Experimental evidence has revealed hippocampal dysfunctions may implicate cognitive impairments in HD patients (Wilkie et al., 2020; de Paula Nascimento-Castro et al., 2022; Ghilan et al., 2014). In a controlled clinical trial study, some hippocampal structure—specific tasks (i.e., virtual Morris water maze and Cambridge Neuropsychological Test Automated Battery) were impaired in HD patients (Begeti et al., 2016). Similar findings were also seen in spatial memory tasks that evaluated hippocampal-related cognition in HD patients (Harris et al., 2019). Furthermore, impaired synaptic transmission processes (decreased paired-pulse depression and increase short-term potentiation) were detected in hippocampus of YAC128 transgenic mice model of HD (Ghilan et al., 2014). These mice also exhibited gross degenerative neurons in dentate gyrus in electron microscopic micrographs (de Paula Nascimento-Castro et al., 2022). Decreased expression of hippocampal CREB-binding protein and lack of histone H3 acetylation was associated with cognitive deficits in $Hdh^{Q7/Q111}$ mice model of HD (Giralt et al., 2012) (Table 13.1).

TABLE 13.1 Evidences of the hippocampal dysfunction in central nervous system diseases.

Disease/ disorder	Case/model	Hippocampal abnormalities	References
AD	26 AD patients/14 healthy control	Atrophy of CA1, subiculum, CA3 and CA4 by multishot diffusion weighted MR images	Adachi et al. (2003)
	34 AD patients	Atrophy of CA1, CA2, and CA3 in structural magnetic resonance hippocampal imaging	Apostolova et al. (2006a)
	6 patients with MCI progressed AD	Hippocampal atrophy in subregions CA1 and subiculum in patients with MCI was considered a risk factor for progression of AD	Apostolova et al. (2006b)
	Postmortem brain tissue obtained from 11 patients with AD	Neuronal loss and accumulation of hyperphosphorylated tau proteins and amyloid-β peptides in CA1	Furcila et al. (2018)

TABLE 13.1 Evidences of the hippocampal dysfunction in central nervous system diseases.—cont'd

Disease/ disorder	Case/model	Hippocampal abnormalities	References
	Postmortem hippocampal samples obtained from AD patients	Decreased number of hippocampal synapses in AD patients	Scheff et al. (2006)
	Postmortem hippocampal samples	Downregulation of genes concerning synaptic transmission (i.e., synaptophysin, synaptogyrin, synaptotagmin, and syntaxin) in CA1 of AD patients	Counts et al. (2014)
	APP/PS1 mouse model of Alzheimer's disease	Severe excitability in CA1 neurons was well correlated with increasing Aβ	Vitale et al. (2021)
	APP/PS1 mouse model of Alzheimer's disease	Impaired LTP and short-term potentiation and increased plaque load in CA1 subregion of hippocampus in old mice	Gengler et al. (2010)
PD	72 patients with PD/48 non-PD cases	CA3 and CA4 hippocampal atrophy and structural changes in dentate gyrus using volumetric MRI	Tanner et al. (2017)
	PD/control	Hippocampal atrophy and cognitive deficits using volumetric MRI	Camicioli et al. (2003)
	20 patients with PD/22 non-PD cases	Left hippocampal atrophy that was correlated with impaired verbal memory in MRI findings	Brück et al. (2004)
	Postmortem brain samples from 20 patients with PD/non-PD cases	A significant positive expression of α-syn in CA2 subregion of hippocampus by IHC and downregulation of several synaptic proteins (i.e., TM163, REEP2, and CSKI1) in hippocampus using proteomic analysis	Villar-Conde et al. (2021)
	Postmortem brain samples from 45 patients with PD	Neuronal degeneration in CA2 and CA3 subsections of hippocampus using IHC	Mattila et al. (1999)
	Postmortem brain samples from 34 PD patients with cognitive deficits	Severe degeneration of cholinergic neurons in hippocampus using IHC	Liu et al. (2019)
	A53T transgenic mice model of PD	Increased expression of α-syn in dentate gyrus and CA2-3 subsections of hippocampus	Flores-Cuadrado et al. (2016)
TLE	62 cases with TLE	Decreased hippocampal volume especially in CA1, CA3 and CA4 subregions as well as neuronal loss using MRI	Jardim et al. (2016)
	32 patients with TLE/14 healthy controls	Impairments of connections between hippocampal subregions with language brain regions	Nenning et al. (2021)
	53 children with TLE	Hippocampal sclerosis is the main lesion in children using MRI	Grattan-Smith et al. (1993)
	47 patients with TLE/41 control	Atrophy in all subregions of hippocampus including CA1, CA2, CA3, CA4, dentate gyrus, and subiculum in MRI analysis	Kim et al. (2015)
	23 patients with TLE/28 healthy controls	Hippocampal atrophy in TLE patients using vertex-based analysis via MRI scanner	Duan et al. (2020)
	45 patients with TLE/45 patients with extra-temporal lobe epilepsy	Hippocampal volume loss that was associated with impaired cognitive functions and abnormal electroencephalography in patients with TLE	Elsherif and Esmael (2022)
	Hippocampal autopsy samples from TLE patients/control autopsy	Hippocampal sclerosis, and microglial activation in CA1, CA2 and CA3 subregions	Beach et al. (1995)
	Kainic acid–induced temporal lobe epilepsy rats/control rats	Decreased neurogenesis, activation of astrocytes and microglia in hippocampus	Upadhya et al. (2019)
	Kainic acid–induced temporal lobe epilepsy mice model/control mice	Impaired activity of GABAergic and glutamatergic neurons, and neuronal loss in CA3 and dentate gyrus via electrophysiological recordings and IHC	Cao et al. (2022)

Continued

TABLE 13.1 Evidences of the hippocampal dysfunction in central nervous system diseases.—cont'd

Disease/disorder	Case/model	Hippocampal abnormalities	References
	Pilocarpine-induced temporal lobe epilepsy rats/control rats	Dysregulation of several proteins that are associated with synaptic function, ion channels and neurotransmission (i.e., SNAP25, SYN2, HOMER2, INA, and VDAC2) in hippocampus	Liu et al. (2008)
TBI	22 patients with mild TBI/22 healthy	Bilateral atrophy and decreased neuronal activity in hippocampus using functional magnetic resonance imaging (fMRI)	Monti et al. (2013)
	109 patients with TBI/healthy control	Hippocampal atrophy that was correlated with verbal memory deficits	Ariza et al. (2006)
	60 patients with mild TBI/43 healthy	Memory and cognitive impairments	Draper and Ponsford (2008)
	34 patients with TBI/26 healthy	Injury in cingulum bundle that was correlated with cognitive impairments using diffusion tensor imaging	Niogi et al. (2008a)
	Postmortem brain sample analyses on 5 patients with mild head injury	Axonal injury in fornices using immunohistochemistry	Blumbergs et al. (1994)
	Lateral fluid percussion cortical injury as model of TBI in rats	Microglia activation, neuroinflammation, and bilaterally neuronal loss in hippocampus	Komoltsev et al. (2021a)
	Lateral hydrodynamic blows −induced TBI in rats	Astrogliosis, and neuronal loss in CA3 subregions and abnormal epileptiform spikes in the hippocampus	Komoltsev et al. (2021b)
MS	25 MS patients/30 healthy control	Right hippocampal atrophy and impaired normal connectivity between hippocampus with anterior cingulate gyrus, thalamus, and frontal cortex in fMRI	Roosendaal et al. (2010)
	23 RRMS patients/18 control	Moderate atrophy in CA1 subregion of hippocampus in RRMS using high-resolution MR	Sicotte et al. (2008)
	11 SPMS patients/18 healthy control	Severe hippocampal volume loss in all subregions in SPMS along with impairment of learning and memory using high-resolution MR	Sicotte et al. (2008)
	20 RRMS patients/16 healthy subjects	An abnormal hippocampal structure that correlated with cognitive impairments and disease disabilities in diffusion tensor imaging	Gu et al. (2022)
	57 MS patients/28 healthy subjects	Hippocampal atrophy reduced hippocampal activation during tasks and increased left hippocampus connections with right posterior cingulate may be responsible for memory impairments and cognitive deficits	Hulst et al. (2015)
	Postmortem brain autopsy from 19 chronic MS patients/7 control	Several lesions, microglial activation, and demyelination in CA1 region of hippocampus and dentate gyrus	Geurts et al. (2007)
	EAE mice	Neuroinflammation, microglial activation, decreased GABAergic interneurons and inhibited GABAergic spontaneous inhibitory postsynaptic currents in hippocampus	Nisticò et al. (2013)
	18 EAE mice/18 control mice	Microglial activation, decreased length of dendrites and structural changes in hippocampal layers using diffusion tensor imaging and the neurite orientation dispersion and density imaging model	Crombe et al. (2018)
	Cuprizone-induced MS mouse model	Demyelinated hippocampal neurons, lack of neurogenesis in the dentate gyrus, change in phenotype of radial glia-like cells from proliferative to mitotically quiescent state	Zhang et al. (2020)
SCZ	18 patients with SCZ/18 healthy subjects	Abnormal hippocampal blood flow that was correlated with symptoms of psychosis	Schobel et al. (2009)
	21 SCZ patients	Bilateral hippocampal atrophy (CA1−CA3 subregions) that was correlated with symptoms	Kühn et al. (2012)

TABLE 13.1 Evidences of the hippocampal dysfunction in central nervous system diseases.—cont'd

Disease/ disorder	Case/model	Hippocampal abnormalities	References
	30 patients with SCZ/18 healthy subjects	Abnormal levels of hippocampal choline in proton magnetic resonance spectroscopic imaging indicated neuronal injury	Deicken et al. (1998)
	15 patients with SCZ/15 healthy subjects	Hippocampal atrophy and axonal injury in hippocampal projections to prefrontal cortex	Csernansky et al. (1998)
	Postmortem meta-analysis on 413 patients with SCZ/415 controls	Hippocampal atrophy and neuronal loss in several subregions of hippocampus	Roeske et al. (2021)
	Rats model of SCZ	Abnormal neuronal activity in ventral subiculum, and loss of hippocampal parvalbumin interneurons	Lodge and Grace (2008)
HD	27 patients with HD/32 healthy subjects	Impaired hippocampal-related cognition in spatial memory task	Harris et al. (2019)
	$Hdh^{Q7/Q111}$ mice model of HD/wild-type mice	Decreased expression of memory-associated genes such as CREB-binding protein, c-fos, Arc, Nr4a2, and lack of histone H3 acetylation in hippocampus	Giralt et al. (2012)
	YAC128 transgenic mice model of HD	Severe neuronal loss in dentate gyrus by light and electron microscopy examinations	de Paula Nascimento-Castro et al. (2022)
	YAC128 transgenic mice model of HD	Disrupted synaptic transmission as seen in reduced paired-pulse depression and increased short-term potentiation in dentate gyrus	Ghilan et al. (2014)

References

Adachi, M., Kawakatsu, S., Hosoya, T., Otani, K., Honma, T., Shibata, A., Sugai, Y., 2003. Morphology of the inner structure of the hippocampal formation in Alzheimer disease. AJNR Am. J. Neuroradiol. 24, 1575–1581.

Aggleton, J.P., 2012. Multiple anatomical systems embedded within the primate medial temporal lobe: implications for hippocampal function. Neurosci. Biobehav. Rev. 36, 1579–1596.

Amaral, D., Insausti, R., Cowan, W., 1983. Evidence for a direct projection from the superior temporal gyrus to the entorhinal cortex in the monkey. Brain Res. 275, 263–277.

Amaral, D., Price, J., 1984. Amygdalo-cortical projections in the monkey (Macaca fascicularis). J. Comp. Neurol. 230, 465–496.

Aggleton, J.P., 2022. Time to retire the serial Papez circuit: implications for space, memory, and attention. Neurosci. Biobehav. Rev. https://doi.org/10.1016/j.neubiorev.2022.104813.

Amaral, R.S.C., 2018. Manual segmentation of the fornix, fimbria, and alveus on high-resolution 3T MRI: application via fully-automated mapping of the human memory circuit white and grey matter in healthy and pathological aging. NeuroImage 170, 132–150. https://doi.org/10.1016/j.neuroimage.2016.10.027.

Amaral, D.G., Dolorfo, C., Alvarez-Royo, P., 1991. Organization of CA1 projections to the subiculum: a PHA-L analysis in the rat. Hippocampus 1, 415–435.

Amaral, D.G., Scharfman, H.E., Lavenex, P., 2007. The dentate gyrus: fundamental neuroanatomical organization (dentate gyrus for dummies). Prog. Brain Res. 163, 3–790.

Amthor, F., Theibert, W.A.B., Standaert, D.G., Roberson, E., 2020. Essentials of Modern Neuroscience. McGraw Hill Professional.

Anand, K.S., Dhikav, V., 2012. Hippocampus in health and disease: an overview. Ann. Indian Acad. Neurol. 15, 239.

Andrew Chambers, R., Moore, J., McEvoy, J.P., Levin, E.D., 1996. Cognitive effects of neonatal hippocampal lesions in a rat model of schizophrenia. Neuropsychopharmacology 15, 587–594.

Apostolova, L.G., Dinov, I.D., Dutton, R.A., Hayashi, K.M., Toga, A.W., Cummings, J.L., Thompson, P.M., 2006. 3D comparison of hippocampal atrophy in amnestic mild cognitive impairment and Alzheimer's disease. Brain 129, 2867–2873.

Apostolova, L.G., Dutton, R.A., Dinov, I.D., Hayashi, K.M., Toga, A.W., Cummings, J.L., Thompson, P.M., 2006. Conversion of mild cognitive impairment to Alzheimer disease predicted by hippocampal atrophy maps. Arch. Neurol. 63, 693–699.

Ariza, M., Serra-Grabulosa, J.M., Junqué, C., Ramírez, B., Mataró, M., Poca, A., Bargalló, N., Sahuquillo, J., 2006. Hippocampal head atrophy after traumatic brain injury. Neuropsychologia 44, 1956–1961.

Atkins, C.M., 2011. Decoding hippocampal signaling deficits after traumatic brain injury. Transl. Stroke Res. 2, 546–555.

Avery, S.N., Rogers, B.P., Heckers, S., 2018. Hippocampal network modularity is associated with relational memory dysfunction in schizophrenia. Biol. Psychiatry Cogn. Neurosci. Neuroimaging 3, 423–432.

Bacciottini, L., et al., 2001. Interactions between histaminergic and cholinergic systems in learning and memory. Behav. Brain Res.

Bakoyiannis, I., Ducourneau, E.-G., Parkes, S.L., Ferreira, G., 2023. Pathway specific interventions reveal the multiple roles of ventral hippocampus projections in cognitive functions. Rev. Neurosci. 34, 825.

Beach, T.G., Woodhurst, W.B., MacDonald, D.B., Jones, M.W., 1995. Reactive microglia in hippocampal sclerosis associated with human temporal lobe epilepsy. Neurosci. Lett. 191, 27–30.

Begeti, F., Schwab, L.C., Mason, S.L., Barker, R.A., 2016. Hippocampal dysfunction defines disease onset in Huntington's disease. J. Neurol. Neurosurg. Psychiatry 87, 975–981.

Bin Ibrahim, M.Z., Benoy, A., Sajikumar, S., 2022. Long-term plasticity in the hippocampus: maintaining within and 'tagging'between synapses. FEBS J. 289, 2176–2201.

Blaizot, X., et al., 2010. The human parahippocampal region: I. Temporal pole cytoarchitectonic and MRI correlation. Cereb. Cortex.

Blatt, G.J., Rosene, D.L., 1998. Organization of direct hippocampal efferent projections to the cerebral cortex of the rhesus monkey: projections from CA1, prosubiculum, and subiculum to the temporal lobe. J. Comp. Neurol. 392, 92–114.

Blumbergs, P.C., Scott, G., Manavis, J., Wainwright, H., Simpson, D.A., McLean, A.J., 1994. Staining of amyloid precursor protein to study axonal damage in mild head injury. Lancet 344, 1055–1056.

Bogerts, B., Ashtari, M., Degreef, G., Alvir, J.M., Bilder, R.M., Lieberman, J.A., 1990. Reduced temporal limbic structure volumes on magnetic resonance images in first episode schizophrenia. Psychiatry Res. 35, 1–13.

Böhm, C., Peng, Y., Geiger, J.R.P., Schmitz, D., 2018. Routes to, from and within the subiculum. Cell Tissue Res 373 (3), 557–563. https://doi.org/10.1007/s00441-018-2848-4. Epub 2018 May 28. PMID: 29808383.

Boldrini, M., Fulmore, C.A., Tartt, A.N., Simeon, L.R., Pavlova, I., Poposka, V., Rosoklija, G.B., Stankov, A., Arango, V., Dwork, A.J., 2018. Human hippocampal neurogenesis persists throughout aging. Cell Stem Cell 22, 589–599.e5.

Boscheron, J., Ruet, A., Deloire, M., Charré-Morin, J., Saubusse, A., Brochet, B., Tourdias, T., Koubiyr, I., 2021. Insights on the relationship between hippocampal connectivity and memory performances at the early stage of multiple sclerosis. Front. Neurol. 12, 667531.

Brück, A., Kurki, T., Kaasinen, V., Vahlberg, T., Rinne, J.O., 2004. Hippocampal and prefrontal atrophy in patients with early non-demented Parkinson's disease is related to cognitive impairment. J. Neurol. Neurosurg. Psychiatry 75, 1467–1469.

Burwell, R.D., 2000. The parahippocampal region: corticocortical connectivity. Ann. N. Y. Acad. Sci. 911, 25–42.

Burwell, R.D., Witter, M.P., Amaral, D.G., 1995. Perirhinal and postrhinal cortices of the rat: a review of the neuroanatomical literature and comparison with findings from the monkey brain. Hippocampus 5, 390–408.

Camicioli, R., Moore, M.M., Kinney, A., Corbridge, E., Glassberg, K., Kaye, J.A., 2003. Parkinson's disease is associated with hippocampal atrophy. Mov. Disord. 18, 784–790.

Campbell, J.N., Register, D., Churn, S.B., 2012. Traumatic brain injury causes an FK506-sensitive loss and an overgrowth of dendritic spines in rat forebrain. J. Neurotrauma 29, 201–217.

Cao, Y., Sun, C., Huang, J., Sun, P., Wang, L., He, S., Liao, J., Lu, Z., Lu, Y., Zhong, C., 2022. Dysfunction of the hippocampal-lateral septal circuit impairs risk assessment in epileptic mice. Front. Mol. Neurosci. 15, 828891.

Cavada, C., Compañy, T., Tejedor, J., Cruz-Rizzolo, R.J., Reinoso-Suárez, F., 2000. The anatomical connections of the macaque monkey orbitofrontal cortex. A review. Cerebr. Cortex 10, 220–242.

Chauhan, P., Jethwa, K., Rathawa, A., Girish, C., Mehra, S., 2021. The anatomy of the hippocampus. In: Pluta, R. (Ed.), Cerebral Ischemia. Exon Publications, Brisbane (AU). Available from: http://www.ncbi.nlm.nih.gov/books/NBK575732/.

Chen, J., Ye, H., Zhang, J., Li, A., Ni, Y., 2022. Pathogenesis of seizures and epilepsy after stroke. Acta Epileptol. 4, 2.

Christian, K.M., Ming, G.-l., Song, H., 2020. Adult neurogenesis and the dentate gyrus: predicting function from form. Behav. Brain Res. 379, 112346.

Christiansen, K., et al., 2016. The status of the precommissural and postcommissural fornix in normal ageing and mild cognitive impairment: An MRI tractography study. NeuroImage. https://doi.org/10.1016/j.neuroimage.2015.12.055.

Clopath, C., Ziegler, L., Vasilaki, E., Büsing, L., Gerstner, W., 2008. Tag-trigger-consolidation: a model of early and late long-term-potentiation and depression. PLoS Comput. Biol. 4, e1000248.

Costard, L.S., et al., 2019. Electrical stimulation of the ventral hippocampal commissure delays experimental epilepsy and is associated with altered microRNA expression. Brain Stimulat. https://doi.org/10.1016/j.brs.2019.06.009.

Counts, S.E., Alldred, M.J., Che, S., Ginsberg, S.D., Mufson, E.J., 2014. Synaptic gene dysregulation within hippocampal CA1 pyramidal neurons in mild cognitive impairment. Neuropharmacology 79, 172–179.

Crombe, A., Planche, V., Raffard, G., Bourel, J., Dubourdieu, N., Panatier, A., Fukutomi, H., Dousset, V., Oliet, S., Hiba, B., Tourdias, T., 2018. Deciphering the microstructure of hippocampal subfields with in vivo DTI and NODDI: applications to experimental multiple sclerosis. Neuroimage 172, 357–368.

Csernansky, J.G., Joshi, S., Wang, L., Haller, J.W., Gado, M., Miller, J.P., Grenander, U., Miller, M.I., 1998. Hippocampal morphometry in schizophrenia by high dimensional brain mapping. Proc. Natl. Acad. Sci. U. S. A. 95, 11406–11411.

de Paula Nascimento-Castro, C., Winkelmann-Duarte, E.C., Mancini, G., Welter, P.G., Plácido, E., Farina, M., Gil-Mohapel, J., Rodrigues, A.L.S., de Bem, A.F., Brocardo, P.S., 2022. Temporal characterization of behavioral and hippocampal dysfunction in the YAC128 mouse model of Huntington's disease. Biomedicines 10, 1433.

Deb, S., 2014. Easy and Interesting Approach to Human Neuroanatomy (Clinically Oriented), First ed. JAYPEE.

Deicken, R.F., Zhou, L., Schuff, N., Fein, G., Weiner, M.W., 1998. Hippocampal neuronal dysfunction in schizophrenia as measured by proton magnetic resonance spectroscopy. Biol. Psychiatry 43, 483–488.

Dekeyzer, S., De Kock, I., Nikoubashman, O., Vanden Bossche, S., Van Eetvelde, R., De Groote, J., Acou, M., Wiesmann, M., Deblaere, K., Achten, E., 2017. "Unforgettable"—a pictorial essay on anatomy and pathology of the hippocampus. Insights Imaging 8, 199–212.

Demars, M., Hu, Y.S., Gadadhar, A., Lazarov, O., 2010. Impaired neurogenesis is an early event in the etiology of familial Alzheimer's disease in transgenic mice. J. Neurosci. Res. 88, 2103–2117.

Destrieux, C., Bourry, D., Velut, S., 2013. Surgical anatomy of the hippocampus. Neurochirurgie 59, 149–158.

Ding, S.L., Van Hoesen, G., Rockland, K.S., 2000. Inferior parietal lobule projections to the presubiculum and neighboring ventromedial temporal cortical areas. J. Comp. Neurol. 425, 510–530.

Doan, T.P., Lagartos-Donate, M.J., Nilssen, E.S., Ohara, S., Witter, M.P., 2019. Convergent projections from perirhinal and postrhinal cortices suggest a multisensory nature of lateral, but not medial, entorhinal cortex. Cell Rep. 29, 617–627.e7.

Donovan, M.H., Yazdani, U., Norris, R.D., Games, D., German, D.C., Eisch, A.J., 2006. Decreased adult hippocampal neurogenesis in the PDAPP mouse model of Alzheimer's disease. J. Comp. Neurol. 495, 70–83.

Draper, K., Ponsford, J., 2008. Cognitive functioning ten years following traumatic brain injury and rehabilitation. Neuropsychology 22, 618–625.

Du, Y., Hao, H., Wang, S., Pearlson, G.D., Calhoun, V.D., 2020. Identifying commonality and specificity across psychosis sub-groups via classification based on features from dynamic connectivity analysis. Neuroimage Clin. 27, 102284.

Duan, Y., Lin, Y., Rosen, D., Du, J., He, L., Wang, Y., 2020. Identifying morphological patterns of hippocampal atrophy in patients with mesial temporal lobe epilepsy and Alzheimer disease. Front. Neurol. 11, 21.

Dudek, S.M., et al., 2016. Rediscovering area CA2: unique properties and functions. Nat. Rev. Neurosci.

Elsherif, M., Esmael, A., 2022. Hippocampal atrophy and quantitative EEG markers in mild cognitive impairment in temporal lobe epilepsy versus extra-temporal lobe epilepsy. Neurol. Sci. 43, 1975–1986.

Escobar, M.L., et al., 2007. Long-term potentiation and depression as putative mechanisms for memory formation. Neural Plasticity and Memory: From Genes to Brain Imaging. Front. Neurosci.

Fama, R., Pitel, A.L., Sullivan, E.V., 2012. Anterograde episodic memory in Korsakoff syndrome. Neuropsychol. Rev. 22 (2), 93–104. https://doi.org/10.1007/s11065-012-9207-0. Epub 2012 May 30. PMID: 22644546; PMCID: PMC4724416.

Fang, S., Zhou, C., Fan, X., Jiang, T., Wang, Y., 2020. Epilepsy-related brain network alterations in patients with temporal lobe glioma in the left hemisphere. Front. Neurol. 11, 684.

Fares, J., Bou Diab, Z., Nabha, S., Fares, Y., 2019. Neurogenesis in the adult hippocampus: history, regulation, and prospective roles. Int. J. Neurosci. 129, 598–611.

Flores-Cuadrado, A., Ubeda-Bañon, I., Saiz-Sanchez, D., de la Rosa-Prieto, C., Martinez-Marcos, A., 2016. Hippocampal α-synuclein and interneurons in Parkinson's disease: data from human and mouse models. Mov. Disord. 31, 979–988.

Fogwe, L., Reddy, V., Mesfin, F., 2021. Neuroanatomy, hippocampus. In: StatPearls. StatPearls Publishing, Treasure Island, FL.

Furcila, D., DeFelipe, J., Alonso-Nanclares, L., 2018. A study of amyloid-β and phosphotau in plaques and neurons in the Hippocampus of Alzheimer's disease patients. J. Alzheimers Dis. 64, 417–435.

Gao, X., Deng, P., Xu, Z.C., Chen, J., 2011. Moderate traumatic brain injury causes acute dendritic and synaptic degeneration in the hippocampal dentate gyrus. PLoS One 6, e24566.

Ge, X., Shi, Y., Li, J., Zhang, Z., Lin, X., Zhan, J., Ge, H., Xu, J., Yu, Q., Leng, Y., 2015. Development of the human fetal hippocampal formation during early second trimester. Neuroimage 119, 33–43.

Gengler, S., Hamilton, A., Hölscher, C., 2010. Synaptic plasticity in the hippocampus of a APP/PS1 mouse model of Alzheimer's disease is impaired in old but not young mice. PLoS One 5, e9764.

Geurts, J.J.G., Bö, L., Roosendaal, S.D., Hazes, T., Daniëls, R., Barkhof, F., Witter, M.P., Huitinga, I., van der Valk, P., 2007. Extensive hippocampal demyelination in multiple sclerosis. J. Neuropathol. Exp. Neurol. 66, 819–827.

Ghilan, M., Bostrom, C.A., Hryciw, B.N., Simpson, J.M., Christie, B.R., Gil-Mohapel, J., 2014. YAC128 Huntington's disease transgenic mice show enhanced short-term hippocampal synaptic plasticity early in the course of the disease. Brain Res. 1581, 117–128.

Giralt, A., Puigdellívol, M., Carretón, O., Paoletti, P., Valero, J., Parra-Damas, A., Saura, C.A., Alberch, J., Ginés, S., 2012. Long-term memory deficits in Huntington's disease are associated with reduced CBP histone acetylase activity. Hum. Mol. Genet. 21, 1203–1216.

Grady, M.S., Charleston, J.S., Maris, D., Witgen, B.M., Lifshitz, J., 2003. Neuronal and glial cell number in the hippocampus after experimental traumatic brain injury: analysis by stereological estimation. J. Neurotrauma 20, 929–941.

Grattan-Smith, J.D., Harvey, A.S., Desmond, P.M., Chow, C.W., 1993. Hippocampal sclerosis in children with intractable temporal lobe epilepsy: detection with MR imaging. AJR Am. J. Roentgenol. 161, 1045–1048.

Griffiths, P.D., Batty, R., Reeves, M.J., Connolly, D.J., 2009. Imaging the corpus callosum, septum pellucidum and fornix in children: normal anatomy and variations of normality. Neuroradiology 51, 337–345.

Gu, X.-Q., Liu, Y., Gu, J.-B., Li, L.-F., Fu, L.-L., Han, X.-M., 2022. Correlations between hippocampal functional connectivity, structural changes, and clinical data in patients with relapsing-remitting multiple sclerosis: a case-control study using multimodal magnetic resonance imaging. Neural Regen. Res. 17, 1115.

Hainmueller, T., Bartos, M., 2020. Dentate gyrus circuits for encoding, retrieval and discrimination of episodic memories. Nat. Rev. Neurosci. 21 (3), 153–168. https://doi.org/10.1038/s41583-019-0260-z. Epub 2020 Feb 10. PMID: 32042144; PMCID: PMC7115869.

Hara, Y., Morrison, J.H., 2014. Synaptic correlates of aging and cognitive decline. In: The Synapse. Elsevier.

Harris, K.L., Armstrong, M., Swain, R., Erzinclioglu, S., Das, T., Burgess, N., Barker, R.A., Mason, S.L., 2019. Huntington's disease patients display progressive deficits in hippocampal-dependent cognition during a task of spatial memory. Cortex 119, 417–427.

Hazra, A., Gu, F., Aulakh, A., Berridge, C., Eriksen, J.L., Žiburkus, J., 2013. Inhibitory neuron and hippocampal circuit dysfunction in an aged mouse model of Alzheimer's disease. PLoS One 8, e64318.

Heckers, S., Konradi, C., 2015. GABAergic mechanisms of hippocampal hyperactivity in schizophrenia. Schizophr. Res. 167, 4–11.

Hevner, R.F., 2016. Evolution of the mammalian dentate gyrus. J. Comp. Neurol. 524, 578–594.

Hiramatsu, M., Edamatsu, R., Mori, A., 1992. Free radicals, lipid peroxidation, SOD activity, neurotransmitters and choline acetyltransferase activity in the aged rat brain. In: Free Radicals and Aging. Springer, pp. 213–218.

Hoge, C.W., McGurk, D., Thomas, J.L., Cox, A.L., Engel, C.C., Castro, C.A., 2008. Mild traumatic brain injury in U.S. Soldiers returning from Iraq. N. Engl. J. Med. 358, 453–463.

Hollands, C., Bartolotti, N., Lazarov, O., 2016. Alzheimer's disease and hippocampal adult neurogenesis; exploring shared mechanisms. Front. Neurosci. 10, 178.

Honoré, E., 2022. Object location learning in mice requires hippocampal somatostatin interneuron activity and is facilitated by mTORC1-mediated long-term potentiation of their excitatory synapses. Mol. Brain.

Howland, J.G., Wang, Y.T., 2008. Synaptic plasticity in learning and memory: stress effects in the hippocampus. Prog. Brain Res. 169, 145–158.

Huang, C.C., 2021. Extensive cortical connectivity of the human hippocampal memory system: beyond the "What" and "Where" dual stream model. Cereb. Cortex. In this issue.

Hulst, H.E., Schoonheim, M.M., Van Geest, Q., Uitdehaag, B.M.J., Barkhof, F., Geurts, J.J.G., 2015. Memory impairment in multiple sclerosis: relevance of hippocampal activation and hippocampal connectivity. Mult. Scler. J. 21, 1705–1712.

Ichinohe, N., Rockland, K.S., 2005. Zinc-enriched amygdalo-and hippocampo-cortical connections to the inferotemporal cortices in macaque monkey. Neurosci. Res. 53, 57–68.

Insausti, R., 2010. Postnatal Development of the Human Hippocampal Formation.

Insausti, R., Amaral, D.G., 2004. Hippocampal formation. Human Nervous Syst. 2, 871–914.

Insausti, R., Muñoz-López, M., Insausti, A.M., Artacho-Pérula, E., 2017. The human periallocortex: layer pattern in presubiculum, parasubiculum and entorhinal cortex. A review. Front. Neuroanat. 11, 84.

Insausti, R., Munoz, M., 2001. Cortical projections of the non-entorhinal hippocampal formation in the cynomolgus monkey (Macaca fascicularis). Eur. J. Neurosci. 14, 435–451.

Jacobson, S., Marcus, E.M., Pugsley, S., 2018. Neuroanatomy for the Neuroscientist. Springer International Publishing.

Jardim, A.P., Corso, J.T., Garcia, M.T.F.C., Gaça, L.B., Comper, S.M., Lancellotti, C.L.P., Centeno, R.S., Carrete, H., Cavalheiro, E.A., Scorza, C.A., Yacubian, E.M.T., 2016. Hippocampal atrophy on MRI is predictive of histopathological patterns and surgical prognosis in mesial temporal lobe epilepsy with hippocampal sclerosis. Epilepsy Res. 128, 169–175.

Kamali, A., 2023. The cortico-limbo-thalamo-cortical circuits: an update to the original Papez circuit of the human limbic system. Brain Topogr. https://doi.org/10.1007/s10548-023-00955-y.

Kandel, E.R., Schwartz, J.H., Jessell, T.M., Siegelbaum, S.A., Hudspeth, A.J., 2012. Principles of Neural Science, fifth ed. McGraw Hill LLC.

Kempermann, G., Song, H., Gage, F.H., 2015. Neurogenesis in the adult hippocampus. Cold Spring Harb. Perspect. Biol. 7 (9), a018812. https://doi.org/10.1101/cshperspect.a018812. PMID: 26330519; PMCID: PMC4563705.

Kennedy, M.B., 2016. Synaptic signaling in learning and memory. Cold Spring Harbor Perspect. Biol. 8, a016824.

Kesner, R.P., Rolls, E.T., 2015. A computational theory of hippocampal function, and tests of the theory: new developments. Neurosci. Biobehav. Rev. 48, 92–147.

Kiernan, J., 2012. Anatomy of the temporal lobe. Epilepsy Res. Treat. 2012.

Kiernan, J., Rajakumar, R., 2013. Barr's the Human Nervous System: An Anatomical Viewpoint. Wolters Kluwer Health.

Kim, J.B., Suh, S.-i., Kim, J.H., 2015. Volumetric and shape analysis of hippocampal subfields in unilateral mesial temporal lobe epilepsy with hippocampal atrophy. Epilepsy Res. 117, 74–81.

Knierim, J.J., 2015. The hippocampus. Curr. Biol. 25, R1116–R1121.

Knierim, J.J., Neunuebel, J.P., Deshmukh, S.S., 2014. Functional correlates of the lateral and medial entorhinal cortex: objects, path integration and local–global reference frames. Phil. Trans. Biol. Sci. 369, 20130369.

Komoltsev, I.G., Frankevich, S.O., Shirobokova, N.I., Volkova, A.A., Onufriev, M.V., Moiseeva, J.V., Novikova, M.R., Gulyaeva, N.V., 2021. Neuroinflammation and neuronal loss in the hippocampus are associated with immediate posttraumatic seizures and corticosterone elevation in rats. Int. J. Mol. Sci. 22, 5883.

Komoltsev, I.G., Volkova, A.A., Levshina, I.P., Novikova, M.R., Gulyaeva, N.V., 2021. Long-term sequelae of traumatic brain injury in rats: a morphological, behavioral, and electrophysiological study. Neurosci. Behav. Physiol. 51, 209–219.

Koolschijn, P.C.M., van Haren, N.E., Cahn, W., Schnack, H.G., Janssen, J., Klumpers, F., Pol, H.E.H., Kahn, R.S., 2010. Hippocampal volume change in schizophrenia. J. Clin. Psychiatry 71, 8248.

Kravitz, D.J., Saleem, K.S., Baker, C.I., Mishkin, M., 2011. A new neural framework for visuospatial processing. Nat. Rev. Neurosci. 12, 217–230.

Kravitz, D.J., Saleem, K.S., Baker, C.I., Ungerleider, L.G., Mishkin, M., 2013. The ventral visual pathway: an expanded neural framework for the processing of object quality. Trends Cognit. Sci. 17, 26–49.

Kühn, S., Musso, F., Mobascher, A., Warbrick, T., Winterer, G., Gallinat, J., 2012. Hippocampal subfields predict positive symptoms in schizophrenia: first evidence from brain morphometry. Transl. Psychiatry 2, e127.

Lace, G., Savva, G.M., Forster, G., de Silva, R., Brayne, C., Matthews, F.E., Barclay, J.J., Dakin, L., Ince, P.G., Wharton, S.B., 2009. Hippocampal tau pathology is related to neuroanatomical connections: an ageing population-based study. Brain 132, 1324–1334.

Langlois, L.D., Selvaraj, P., Simmons, S.C., Gouty, S., Zhang, Y., Nugent, F.S., 2022. Repetitive mild traumatic brain injury induces persistent alterations in spontaneous synaptic activity of hippocampal CA1 pyramidal neurons. IBRO Neurosci. Rep. 12, 157–162.

Lazarov, O., Mattson, M.P., Peterson, D.A., Pimplikar, S.W., van Praag, H., 2010. When neurogenesis encounters aging and disease. Trends Neurosci. 33, 569–579.

Lee, K.J., et al., 2014. Mossy fiber-CA3 synapses mediate homeostatic plasticity in mature hippocampal neurons. Neuron.

Lenka, A., Ingalhalikar, M., Shah, A., Saini, J., Arumugham, S.S., Hegde, S., George, L., Reddy, V., Reddy, Y.C.J., Yadav, R., Pal, P.K., 2018. Hippocampal subfield atrophy in patients with Parkinson's disease and psychosis. J. Neural. Transm. 125, 1361–1372.

Lieberman, J.A., Girgis, R.R., Brucato, G., Moore, H., Provenzano, F., Kegeles, L., Javitt, D., Kantrowitz, J., Wall, M.M., Corcoran, C.M., Schobel, S.A., Small, S.A., 2018. Hippocampal dysfunction in the pathophysiology of schizophrenia: a selective review and hypothesis for early detection and intervention. Mol. Psychiatry 23, 1764–1772.

Lipska, B.K., Aultman, J.M., Verma, A., Weinberger, D.R., Moghaddam, B., 2002. Neonatal damage of the ventral hippocampus impairs working memory in the rat. Neuropsychopharmacology 27, 47–54.

Liu, A.K.L., Chau, T.W., Lim, E.J., Ahmed, I., Chang, R.C.-C., Kalaitzakis, M.E., Graeber, M.B., Gentleman, S.M., Pearce, R.K.B., 2019. Hippocampal CA2 Lewy pathology is associated with cholinergic degeneration in Parkinson's disease with cognitive decline. Acta Neuropathol. Commun. 7, 61.

Liu, H., Zhou, J., Tian, W., Luo, C., Bartlett, A., Aldridge, A., Lucero, J., Osteen, J.K., Nery, J.R., Chen, H., 2021. DNA methylation atlas of the mouse brain at single-cell resolution. Nature 598, 120–128.

Liu, W., Ge, T., Leng, Y., Pan, Z., Fan, J., Yang, W., Cui, R., 2017. The role of neural plasticity in depression: from hippocampus to prefrontal cortex. Neural Plast. 2017.

Liu, X.Y., Yang, J.L., Chen, L.J., Zhang, Y., Yang, M.L., Wu, Y.Y., Li, F.Q., Tang, M.H., Liang, S.F., Wei, Y.Q., 2008. Comparative proteomics and correlated signaling network of rat hippocampus in the pilocarpine model of temporal lobe epilepsy. Proteomics 8, 582–603.

Lodge, D.J., Grace, A.A., 2008. Hippocampal dysfunction and disruption of dopamine system regulation in an animal model of schizophrenia. Neurotox. Res. 14, 97–104.

Lüscher, C., Malenka, R.C., 2012. NMDA receptor-dependent long-term potentiation and long-term depression (LTP/LTD). Cold Spring Harbor Perspect. Biol. 4, a005710.

Maass, A., Berron, D., Libby, L.A., Ranganath, C., Düzel, E., 2015. Functional subregions of the human entorhinal cortex. eLife 4, e06426.

Marchetti, C., Marie, H., 2011. Hippocampal synaptic plasticity in Alzheimer's disease: what have we learned so far from transgenic models? Rev. Neurosci. 22, 373–402.

Matsumoto, N., Kitanishi, T., Mizuseki, K., 2019. The subiculum: unique hippocampal hub and more. Neurosci. Res. 143, 1–12.

Mattila, P.M., Rinne, J.O., Helenius, H., Röyttä, M., 1999. Neuritic degeneration in the hippocampus and amygdala in Parkinson's disease in relation to Alzheimer pathology. Acta Neuropathol. 98, 157–164.

Maxwell, W.L., Dhillon, K., Harper, L., Espin, J., MacIntosh, T.K., Smith, D.H., Graham, D.I., 2003. There is differential loss of pyramidal cells from the human hippocampus with survival after blunt head injury. J. Neuropathol. Exp. Neurol. 62, 272–279.

McCutcheon, R.A., Marques, T.R., Howes, O.D., 2020. Schizophrenia—an overview. JAMA Psychiatry 77, 201–210.

Medoff, D.R., Holcomb, H.H., Lahti, A.C., Tamminga, C.A., 2001. Probing the human hippocampus using rCBF: contrasts in schizophrenia. Hippocampus 11, 543–550.

Milner, A.D., 2017. How do the two visual streams interact with each other? Exp. Brain Res. 235, 1297–1308.

Mokhothu, T.M., Tanaka, K.Z., 2021. Characterizing hippocampal oscillatory signatures underlying seizures in temporal lobe epilepsy. Front. Hum. Neurosci. 15, 785328.

Molina, V., Reig, S., Pascau, J., Sanz, J., Sarramea, F., Gispert, J.D., Luque, R., Benito, C., Palomo, T., Desco, M., 2003. Anatomical and functional cerebral variables associated with basal symptoms but not risperidone response in minimally treated schizophrenia. Psychiatry Res. 124, 163–175.

Molnár, E., 2011. Long-term potentiation in cultured hippocampal neurons. In: Seminars in Cell & Developmental Biology. Elsevier, pp. 506–513.

Monti, J.M., Voss, M.W., Pence, A., McAuley, E., Kramer, A.F., Cohen, N.J., 2013. History of mild traumatic brain injury is associated with deficits in relational memory, reduced hippocampal volume, and less neural activity later in life. Front. Aging Neurosci. 5, 41.

Morecraft, R., Stilwell-Morecraft, K., Cipolloni, P., Ge, J., McNeal, D., Pandya, D., 2012. Cytoarchitecture and cortical connections of the anterior cingulate and adjacent somatomotor fields in the rhesus monkey. Brain Res. Bull. 87, 457–497.

Mota, B., Ramos, M., Marques, S.I., Silva, A., Pereira, P.A., Madeira, M.D., Mateus, N., Cardoso, A., 2023. Effects of high-fat and high-fat high-sugar diets in the anxiety, learning and memory, and in the hippocampus neurogenesis and neuroinflammation of aged rats. Nutrients 15, 1370.

Moya, N.A., Yun, S., Fleps, S.W., Martin, M.M., Nadel, J.A., Beutler, L.R., Zweifel, L.S., Parker, J.G., 2023. The effect of selective nigrostriatal dopamine excess on behaviors linked to the cognitive and negative symptoms of schizophrenia. Neuropsychopharmacology 48, 690–699.

Mtui, E., Gruener, G., Dockery, P., 2020. Fitzgerald's Clinical Neuroanatomy and Neuroscience. Elsevier Health Sciences.

Nakamura, T., Takata, A., 2023. The molecular pathology of schizophrenia: an overview of existing knowledge and new directions for future research. Mol. Psychiatry 1–22.

Nelson, E.A., Kraguljac, N.V., Maximo, J.O., Briend, F., Armstrong, W., Ver Hoef, L.W., Johnson, V., Lahti, A.C., 2022. Hippocampal dysconnectivity and altered glutamatergic modulation of the default mode network: a combined resting-state connectivity and magnetic resonance spectroscopy study in schizophrenia. Biol. Psychiatry Cogn. Neurosci. Neuroimaging 7, 108–118.

Nenning, K.-H., Fösleitner, O., Schwartz, E., Schwarz, M., Schmidbauer, V., Geisl, G., Widmann, C., Pirker, S., Baumgartner, C., Prayer, D., Pataraia, E., Bartha-Doering, L., Langs, G., Kasprian, G., Bonelli, S.B., 2021. The impact of hippocampal impairment on task-positive and task-negative language networks in temporal lobe epilepsy. Clin. Neurophysiol. 132, 404–411.

Nilssen, E.S., Doan, T.P., Nigro, M.J., Ohara, S., Witter, M.P., 2019. Neurons and networks in the entorhinal cortex: a reappraisal of the lateral and medial entorhinal subdivisions mediating parallel cortical pathways. Hippocampus 29, 1238–1254.

Niogi, S.N., Mukherjee, P., Ghajar, J., Johnson, C., Kolster, R.A., Sarkar, R., Lee, H., Meeker, M., Zimmerman, R.D., Manley, G.T., McCandliss, B.D., 2008. Extent of microstructural white matter injury in postconcussive syndrome correlates with impaired cognitive reaction time: a 3T diffusion tensor imaging study of mild traumatic brain injury. AJNR Am. J. Neuroradiol. 29, 967–973.

Niogi, S.N., Mukherjee, P., Ghajar, J., Johnson, C.E., Kolster, R., Lee, H., Suh, M., Zimmerman, R.D., Manley, G.T., McCandliss, B.D., 2008. Structural dissociation of attentional control and memory in adults with and without mild traumatic brain injury. Brain 131, 3209–3221.

Nisticò, R., Mango, D., Mandolesi, G., Piccinin, S., Berretta, N., Pignatelli, M., Feligioni, M., Musella, A., Gentile, A., Mori, F., Bernardi, G., Nicoletti, F., Mercuri, N.B., Centonze, D., 2013. Inflammation subverts hippocampal synaptic plasticity in experimental multiple sclerosis. PLoS One 8, e54666.

Ondrejcak, T., Klyubin, I., Hu, N.-W., Barry, A.E., Cullen, W.K., Rowan, M.J., 2010. Alzheimer's disease amyloid β-protein and synaptic function. NeuroMolecular Med. 12, 13–26.

Opalka, A.N., Wang, D.V., 2020. Hippocampal efferents to retrosplenial cortex and lateral septum are required for memory acquisition. Learn. Mem. 27, 310–318.

Oppenheim, C., Dormont, D., Biondi, A., Lehéricy, S., Hasboun, D., Clémenceau, S., Baulac, M., Marsault, C., 1998. Loss of digitations of the hippocampal head on high-resolution fast spin-echo MR: a sign of mesial temporal sclerosis. Am. J. Neuroradiol. 19, 457–463.

Parmar, S.K., Pruthi, N., Ravindranath, R., Ravindranath, Y., Somanna, S., Philip, M., 2018. Anatomical variations of the temporomesial structures in normal adult brain-A cadaveric study. J. Neurosci. Rural Pract. 9, 317–325.

Patra, D., Tewari, M., Sahni, D., Mathuriya, S., 2018. Microsurgical anatomy of medial temporal lobe in North-West Indian population: cadaveric brain dissection. Asian J. Neurosurg. 13, 674–680.

Pereira, J.B., Junqué, C., Bartrés-Faz, D., Ramírez-Ruiz, B., Marti, M.-J., Tolosa, E., 2013. Regional vulnerability of hippocampal subfields and memory deficits in Parkinson's disease. Hippocampus 23, 720–728.

Peters, A., Reisch, C., Langemann, D., 2018. LTP or LTD? Modeling the influence of stress on synaptic plasticity. Eneuro 5.

Pinky, Neha, 2023. Unravelling of molecular biomarkers in synaptic plasticity of Alzheimer's disease: critical role of the restoration of neuronal circuits. Ageing Res. Rev. 91. https://doi.org/10.1016/j.arr.2023.102069.

Postans, M., Parker, G.D., Lundell, H., Ptito, M., Hamandi, K., Gray, W.P., Aggleton, J.P., Dyrby, T.B., Jones, D.K., Winter, M., 2020. Uncovering a role for the dorsal hippocampal commissure in recognition memory. Cereb. Cortex 30 (3), 1001–1015. https://doi.org/10.1093/cercor/bhz143. PMID: 31364703; PMCID: PMC7132945.

Powell, K.J., Binder, T.L., Hori, S., Nakabeppu, Y., Weinberger, D.R., Lipska, B.K., Robertson, G.S., 2006. Neonatal ventral hippocampal lesions produce an elevation of ΔFosB-like protein(s) in the rodent neocortex. Neuropsychopharmacology 31, 700–711.

Prestia, A., Boccardi, M., Galluzzi, S., Cavedo, E., Adorni, A., Soricelli, A., Bonetti, M., Geroldi, C., Giannakopoulos, P., Thompson, P., Frisoni, G., 2011. Hippocampal and amygdalar volume changes in elderly patients with Alzheimer's disease and schizophrenia. Psychiatry Res. Neuroimaging 192, 77–83.

Prestia, A., Cavedo, E., Boccardi, M., Muscio, C., Adorni, A., Geroldi, C., Bonetti, M., Thompson, P.M., Frisoni, G.B., 2015. Hippocampal and amygdalar local structural differences in elderly patients with schizophrenia. Am. J. Geriatr. Psychiatry 23, 47–58.

Pujol, N., Penadés, R., Junqué, C., Dinov, I., Fu, C.H.Y., Catalán, R., Ibarretxe-Bilbao, N., Bargalló, N., Bernardo, M., Toga, A., Howard, R.J., Costafreda, S.G., 2018. Hippocampal abnormalities and age in chronic schizophrenia: morphometric study across the adult lifespan. Br. J. Psychiatry 205, 369–375.

Rajmohan, V., 2007. The limbic system. Indian J. Psychiatry. https://doi.org/10.4103/0019-5545.33264.

Ren, X.Q., Huang, X., Xing, S.Y., Long, Y., Yuan, D.H., Hong, H., Tang, S.S., 2023. Neuroprotective effects of novel compound FMDB on cognition, neurogenesis and apoptosis in APP/PS1 transgenic mouse model of Alzheimer's disease. Neurochem. Int. 165, 105510.

Rockland, K.S., Van Hoesen, G.W., 1999. Some temporal and parietal cortical connections converge in CA1 of the primate hippocampus. Cerebr. Cortex 9, 232–237.

Roeske, M.J., Konradi, C., Heckers, S., Lewis, A.S., 2021. Hippocampal volume and hippocampal neuron density, number and size in schizophrenia: a systematic review and meta-analysis of postmortem studies. Mol. Psychiatry 26, 3524–3535.

Rolls, E.T., 1989. Functions of neuronal networks in the hippocampus and neocortex in memory. In: Neural Models of Plasticity. Elsevier.

Rolls, E.T., 2016. Cerebral Cortex: Principles of Operation. Oxford University Press.

Rolls, E.T., 2018. The storage and recall of memories in the hippocampo-cortical system. Cell Tissue Res. 373, 577–604.

Rolls, E.T., 2021. Brain Computations: What and How. Oxford University Press, USA.

Roosendaal, S.D., Hulst, H.E., Vrenken, H., Feenstra, H.E., Castelijns, J.A., Pouwels, P.J., Barkhof, F., Geurts, J.J., 2010. Structural and functional hippocampal changes in multiple sclerosis patients with intact memory function. Radiology 255, 595–604.

Rudnitskaya, E.A., Burnyasheva, A.O., Kozlova, T.A., Peunov, D.A., Kolosova, N.G., Stefanova, N.A., 2022. Changes in glial support of the hippocampus during the development of an Alzheimer's disease-like pathology and their correction by mitochondria-targeted antioxidant SkQ1. Int. J. Mol. Sci. 23, 1134.

Ryszard Pluta, M., 2021. In: Cerebral Ischemia. Exon Publications, Australia.

Saleem, K.S., Tanaka, K., 1996. Divergent projections from the anterior inferotemporal area TE to the perirhinal and entorhinal cortices in the macaque monkey. J. Neurosci. 16, 4757–4775.

Salimi, M., Tabasi, F., Abdolsamadi, M., Dehghan, S., Dehdar, K., Nazari, M., Javan, M., Mirnajafi-Zadeh, J., Raoufy, M.R., 2022. Disrupted connectivity in the olfactory bulb-entorhinal cortex-dorsal hippocampus circuit is associated with recognition memory deficit in Alzheimer's disease model. Sci. Rep. 12, 4394.

Sánchez-Rodríguez, I., Temprano-Carazo, S., Jeremic, D., Delgado-Garcia, J.M., Gruart, A., Navarro-López, J.D., Jiménez-Díaz, L., 2022. Recognition memory induces natural LTP-like hippocampal synaptic excitation and inhibition. Int. J. Mol. Sci. 23, 10806.

Sando, R., Südhof, T.C., 2021. Latrophilin GPCR signaling mediates synapse formation. Elife 10, e65717.

Santos, M.A.O., Bezerra, L.S., Carvalho, A.R.M.R., Brainer-Lima, A.M., 2018. Global hippocampal atrophy in major depressive disorder: a meta-analysis of magnetic resonance imaging studies. Trends Psychiatry Psychother. 40, 369–378.

Savransky, A., 2017. Fornix structural connectivity and allostatic load: empirical evidence from Schizophrenia patients and healthy controls. Psychosom. Med. https://doi.org/10.1097/PSY.0000000000000487.

Scheff, S.W., Price, D.A., Hicks, R.R., Baldwin, S.A., Robinson, S., Brackney, C., 2005. Synaptogenesis in the hippocampal CA1 field following traumatic brain injury. J. Neurotrauma 22, 719–732.

Scheff, S.W., Price, D.A., Schmitt, F.A., Mufson, E.J., 2006. Hippocampal synaptic loss in early Alzheimer's disease and mild cognitive impairment. Neurobiol. Aging 27, 1372–1384.

Scheltens, P., Blennow, K., Breteler, M.M.B., de Strooper, B., Frisoni, G.B., Salloway, S., Van der Flier, W.M., 2016. Alzheimer's disease. Lancet 388, 505–517.

Schobel, S.A., Lewandowski, N.M., Corcoran, C.M., Moore, H., Brown, T., Malaspina, D., Small, S.A., 2009. Differential targeting of the CA1 subfield of the hippocampal formation by schizophrenia and related psychotic disorders. Arch. Gen. Psychiatry 66, 938–946.

Schultz, C., Engelhardt, M., 2014. Anatomy of the hippocampal formation. Hippocampus Clin. Neurosci. 34, 6–17.

Schumm, S.N., Gabrieli, D., Meaney, D.F., 2022. Plasticity impairment exposes CA3 vulnerability in a hippocampal network model of mild traumatic brain injury. Hippocampus 32, 231–250.

Sethumadhavan, N., et al., 2022. The perirhinal cortex engages in area and layer-specific encoding of item dimensions. Front. Behav. Neurosci.

Sicotte, N.L., Kern, K.C., Giesser, B.S., Arshanapalli, A., Schultz, A., Montag, M., Wang, H., Bookheimer, S.Y., 2008. Regional hippocampal atrophy in multiple sclerosis. Brain 131, 1134–1141.

Siegel, A., Sapru, H.N., 2010. Essential Neuroscience. Wolters Kluwer Health/Lippincott Williams & Wilkins.

Song, W.S., Cho, Y.S., Oh, S.P., Yoon, S.H., Kim, Y.S., Kim, M.-H., 2022. Cognitive and behavioral effects of the anti-epileptic drug cenobamate (YKP3089) and underlying synaptic and cellular mechanisms. Neuropharmacology 221, 109292.

Spitz, G., Maller, J.J., O'Sullivan, R., Ponsford, J.L., 2013. White matter integrity following traumatic brain injury: the association with severity of injury and cognitive functioning. Brain Topogr. 26, 648–660.

Steen, R.G., Mull, C., McClure, R., Hamer, R.M., Lieberman, J.A., 2006. Brain volume in first-episode schizophrenia: systematic review and meta-analysis of magnetic resonance imaging studies. Br. J. Psychiatry 188, 510–518.

Sulpizio, V., Galati, G., Fattori, P., Galletti, C., Pitzalis, S., 2020. A common neural substrate for processing scenes and egomotion-compatible visual motion. Brain Struct. Funct. 225, 2091–2110.

Sun, L.N., Qi, J.S., Gao, R., 2018. Physical exercise reserved amyloid-beta induced brain dysfunctions by regulating hippocampal neurogenesis and inflammatory response via MAPK signaling. Brain Res. 1697, 1–9.

Suzuki, W.A., Amaral, D.G., 1990. Cortical inputs to the CA1 field of the monkey hippocampus originate from the perirhinal and parahippocampal cortex but not from area TE. Neurosci. Lett. 115, 43–48.

Suzuki, W.A., Amaral, D.G., 1994. Perirhinal and parahippocampal cortices of the macaque monkey: cortical afferents. J. Comp. Neurol. 350, 497–533.

Svenningsson, A.L., Stomrud, E., Insel, P.S., Mattsson, N., Palmqvist, S., Hansson, O., 2019. β-amyloid pathology and hippocampal atrophy are independently associated with memory function in cognitively healthy elderly. Sci. Rep. 9, 11180.

Tanner, J.J., McFarland, N.R., Price, C.C., 2017. Striatal and hippocampal atrophy in idiopathic Parkinson's disease patients without dementia: a morphometric analysis. Front. Neurol. 8, 139.

Tatu, L., Vuillier, F., 2014. Structure and vascularization of the human hippocampus. Hippocampus Clin. Neurosci. 34, 18–25.

Thom, M., Eriksson, S., Martinian, L., Caboclo, L.O., McEvoy, A.W., Duncan, J.S., Sisodiya, S.M., 2009. Temporal lobe sclerosis associated with hippocampal sclerosis in temporal lobe epilepsy: neuropathological features. J. Neuropathol. Exp. Neurol. 68, 928–938.

Tran, T.T., Srivareerat, M., Alkadhi, K.A., 2010. Chronic psychosocial stress triggers cognitive impairment in a novel at-risk model of Alzheimer's disease. Neurobiol. Dis. 37, 756–763.

Treves, A., Rolls, E.T., 1994. Computational analysis of the role of the hippocampus in memory. Hippocampus 4, 374–391.

Tseng, K.Y., Chambers, R.A., Lipska, B.K., 2009. The neonatal ventral hippocampal lesion as a heuristic neurodevelopmental model of schizophrenia. Behav. Brain Res. 204, 295–305.

Tuncdemir, S.N., Lacefield, C.O., Hen, R., 2019. Contributions of adult neurogenesis to dentate gyrus network activity and computations. Behav. Brain Res. 374, 112112.

Upadhya, D., Kodali, M., Gitai, D., Castro, O.W., Zanirati, G., Upadhya, R., Attaluri, S., Mitra, E., Shuai, B., Hattiangady, B., Shetty, A.K., 2019. A model of chronic temporal lobe epilepsy presenting constantly rhythmic and robust spontaneous seizures, co-morbidities and hippocampal neuropathology. Aging Dis. 10, 915–936.

Van Hoesen, G.W., 1982. The parahippocampal gyrus: new observations regarding its cortical connections in the monkey. Trends Neurosci. 5, 345–350.

Van Hoesen, G.W., Pandya, D.N., 1975. Some connections of the entorhinal (area 28) and perirhinal (area 35) cortices of the rhesus monkey. III. Efferent connections. Brain Res. 95, 39–59.

van Mierlo, T.J., Chung, C., Foncke, E.M., Berendse, H.W., van den Heuvel, O.A., 2015. Depressive symptoms in Parkinson's disease are related to decreased hippocampus and amygdala volume. Mov. Disord. 30, 245–252.

Villar-Conde, S., Astillero-Lopez, V., Gonzalez-Rodriguez, M., Villanueva-Anguita, P., Saiz-Sanchez, D., Martinez-Marcos, A., Flores-Cuadrado, A., Ubeda-Bañon, I., 2021. The human hippocampus in Parkinson's disease: an integrative stereological and proteomic study. J. Parkinsons Dis. 11, 1345–1365.

Vitale, P., Salgueiro-Pereira, A.R., Lupascu, C.A., Willem, M., Migliore, R., Migliore, M., Marie, H., 2021. Analysis of age-dependent alterations in excitability properties of CA1 pyramidal neurons in an APPPS1 model of Alzheimer's disease. Front. Aging Neurosci. 13, 668948.

Wang, J., Hart, K.L., Qi, W., Ardekani, B.A., Li, C., Marx, J., Freudenreich, O., Cather, C., Holt, D., Bello, I., Diminich, E.D., Tang, Y., Worthington, M., Zeng, B., Wu, R., Fan, X., Zhao, J., Wang, J., Goff, D.C., 2021. Association of aripiprazole with reduced hippocampal atrophy during maintenance treatment of first-episode schizophrenia. J. Clin. Psychopharmacol. 41, 244–249.

Wegrzyn, D., Juckel, G., Faissner, A., 2022. Structural and functional deviations of the hippocampus in schizophrenia and schizophrenia animal models. Int. J. Mol. Sci. 23, 5482.

Witter, M.P., 2007. The perforant path: projections from the entorhinal cortex to the dentate gyrus. Prog Brain Res 163, 43–61. https://doi.org/10.1016/S0079-6123(07)63003-9. PMID: 17765711.

Wiles, M.D., Braganza, M., Edwards, H., Krause, E., Jackson, J., Tait, F., 2023. Management of traumatic brain injury in the non-neurosurgical intensive care unit: a narrative review of current evidence. Anaesthesia 78, 510–520.

Wilkie, C.M., Barnes, J.R., Benson, C.-L.M., Brymer, K.J., Nafar, F., Parsons, M.P., 2020. Hippocampal synaptic dysfunction in a mouse model of Huntington disease is not alleviated by ceftriaxone treatment. Eneuro 7. ENEURO.0440-19.2020.

Xu, W., Yao, X., Zhao, F., Zhao, H., Cheng, Z., Yang, W., Cui, R., Xu, S., Li, B., 2020. Changes in hippocampal plasticity in depression and therapeutic approaches influencing these changes. Neural Plast. 2020.

Yassa, M.A., Stark, C.E., 2011. Pattern separation in the hippocampus. Trends Neurosci. 34, 515–525.

Ye, Q., Gast, G., Su, X., Saito, T., Saido, T.C., Holmes, T.C., Xu, X., 2022. Hippocampal neural circuit connectivity alterations in an Alzheimer's disease mouse model revealed by monosynaptic rabies virus tracing. Neurobiol. Dis. 172, 105820.

Yukie, M., 2000. Connections between the medial temporal cortex and the CA1 subfield of the hippocampal formation in the Japanese monkey (Macaca fuscata). J. Comp. Neurol. 423, 282–298.

Zemmoura, I., Velut, S., François, P., 2012. The choroidal fissure: anatomy and surgical implications. Adv. Tech. Stand. Neurosurg. 97–113.

Zhang, H., Kim, Y., Ro, E.J., Ho, C., Lee, D., Trapp, B.D., Suh, H., 2020. Hippocampal neurogenesis and neural circuit formation in a cuprizone-induced multiple sclerosis mouse model. J. Neurosci. 40, 447–458.

Zhong, Y.-M., Rockland, K.S., 2004. Connections between the anterior inferotemporal cortex (area TE) and CA1 of the hippocampus in monkey. Exp. Brain Res. 155, 311–319.

Zhong, Y.-M., Yukie, M., Rockland, K.S., 2005. Direct projections from CA1 to the superior temporal sulcus in the monkey, revealed by single axon analysis. Brain Res. 1035, 211–214.

Zhong, Y.-M., Yukie, M., Rockland, K.S., 2006. Distinctive morphology of hippocampal CA1 terminations in orbital and medial frontal cortex in macaque monkeys. Exp. Brain Res. 169, 549–553.

Further Reading

Atoji, Y., Wild, J.M., 2006. Anatomy of the avian hippocampal formation. Rev. Neurosci. 17, 3–16.

14

The Telencephalon: Amygdala and Claustrum

Barbara Buccilli[1,8], *Sajad Sahab-Negah*[2], *Sadegh Shirian*[3,4], *Ali Gorji*[5], *Maryam Khaleghi Ghadiri*[6] *and Brandon M. Ascenzi*[7]

[1]University of Rome "La Sapienza", Rome, Italy [2]Neuroscience Research Center, Mashhad University of Medical Sciences, Mashhad, Iran [3]Department of Pathology, School of Veterinary Medicine, Shahrekord University, Shahrekord, Iran [4]Shiraz Molecular Pathology Research Center, Dr. Daneshbod Pathol. Lab., Shiraz, Iran [5]Epilepsy Research Center, Department of Neurosurgery, University of Münster, Münster, Germany [6]Epilepsy Research Center, Department of Neurosurgery and Neurology, Westfälische Wilhelms-Universität Münster, Münster, Germany [7]Independent Neuroresearcher-Member of Marie Curie Alumni Association (MCAA) [8]Department of Neurosurgery, Cerebrovascular and Neurologic Critical Care, Icahn School of Medicine at Mount Sinai, New York, NY, United States

OUTLINE

INTRODUCTION

The amygdala (Figs. 14.1 and 14.2) is an almond-shaped structure that is situated deep within the temporal lobe. Burdach originally described it as a group of cells that are now known as the basolateral complex. With time, a large number of structures that surround the basolateral complex have been identified in many species and constitute what is now known as the amygdaloid complex. The amygdala is involved in assigning emotional significance to sensory information. In particular, the amygdala has been shown to be an essential component of the circuitry underlying fear-related responses. Disorders of this process are likely to be the underlying cause of anxiety disorders in humans. The amygdaloid complex comprises more than 10 nuclei located in the midtemporal lobe. These nuclei can be distinguished on both cytoarchitectonic and connective grounds, have extensive intranuclear and internuclear connections, and have connections with cortical and subcortical regions.

Analysis of fear conditioning in rats has suggested that long-term synaptic plasticity of inputs to the amygdala underlies the acquisition and storage of fear memory. In agreement with this proposal, synaptic plasticity has been demonstrated at the synapses in the amygdala in both in vitro and in vivo studies.

EMOTION AND COGNITION

William James (1884, 1890) and C. G. Lange (1887) both suggested that emotions are the cognitive

From Anatomy to Function of the Central Nervous System
https://doi.org/10.1016/B978-0-12-822404-5.00006-1

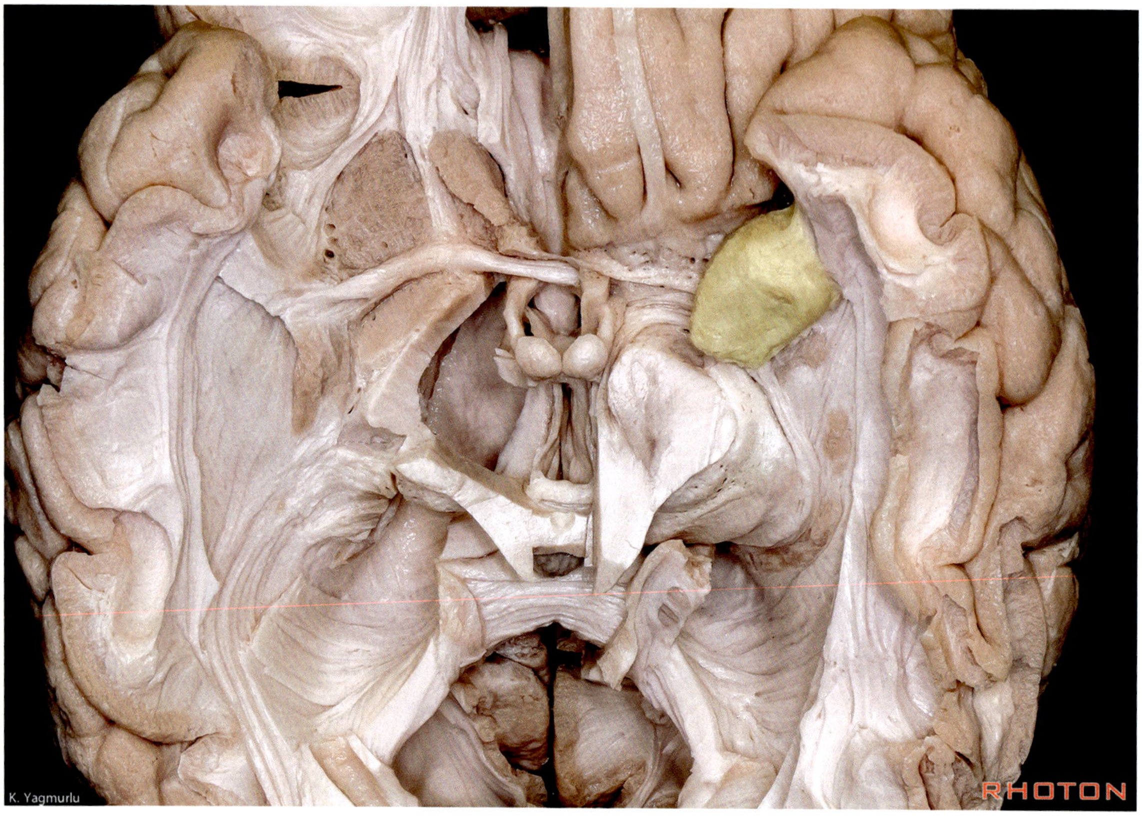

FIGURE 14.1 **Amydgala (*in yellow*).** The *white* ribbon-like structure to the lower of the amygdaloid nucleus that surrounds the thalamus is the stria terminalis. The other two *white* ribbon-like structures (on the *left side* of the amygdala) extending horizontally to the amygdala are, respectively, from *top* to *bottom*, the diagonal band of Broca and the medial forebrain bundle. On the *right* hemisphere (therefore to the *left half* of the image), the ribbon-like structure extending horizontally and posterior to the mammillary bodies is the anterior commissure. *Image courtesy of The Rhoton Collection® http://rhoton.ineurodb.org/. Dissector: K. Yagmurlu. Drawing by Kaan Yagmurlu.*

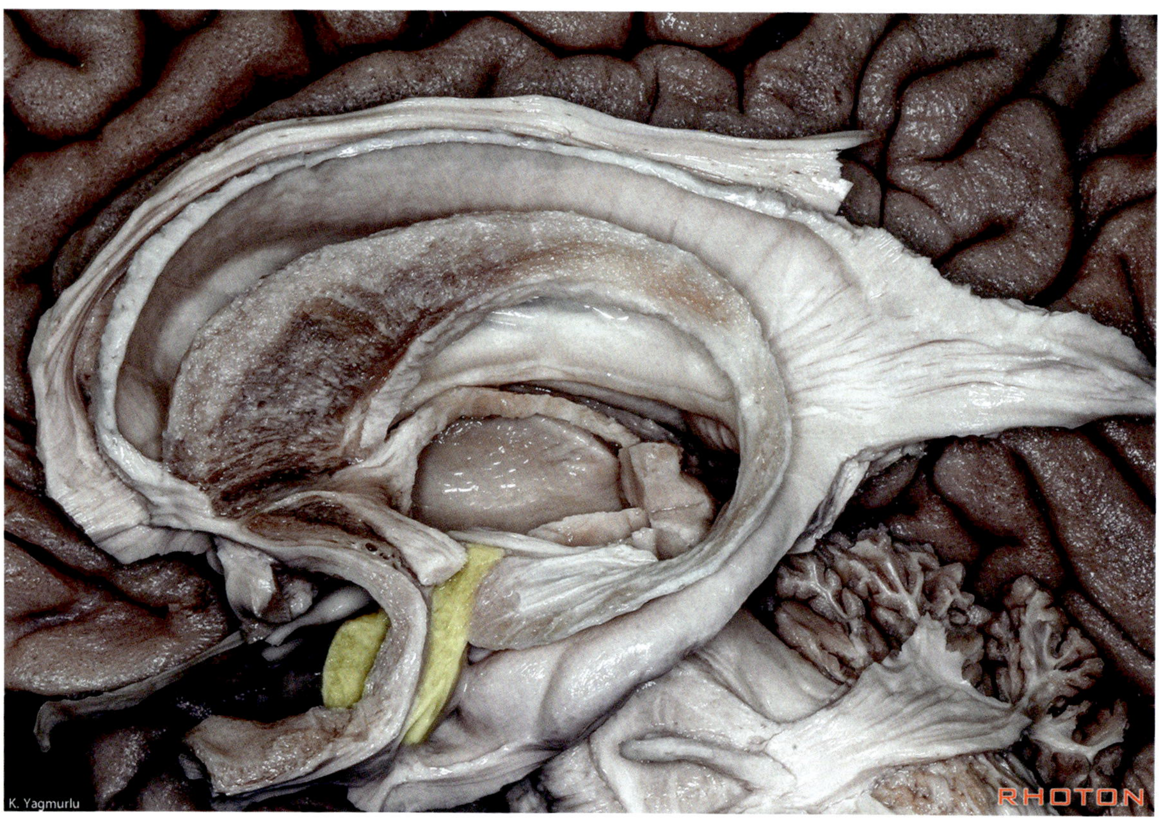

FIGURE 14.2 ***Left view* of the amygdala (*in yellow*),** which is located anterior to the hippocampus (the sausage-like structure). The structure above the amygdala that extends horizontally and latero-medially is the anterior commissure. The *white* C-shaped structure attached medially to the fornix is the stria terminalis. *Image courtesy of The Rhoton Collection® http://rhoton.ineurodb.org/. Dissector: K. Yagmurlu. Drawing by Jeffrey Sorenson.*

responses to physiological responses to external stimuli. This is the James–Lange theory of emotions, which suggests that it is possible to study emotions by examining physiological responses to stimuli.

Cannon (1927) was critical of the James–Lange theory and instead suggested that the hypothalamus and its projections to the cortex and brainstem were the central elements in emotional responses. Papez (1937) then reviewed the anatomical and clinical data and added more medial temporal structures to the circuitry involved in emotional processing. Paul MacLean (1949) called these forebrain circuits the "visceral brain," and introduced the concept of the limbic system. While McLean included the amygdala in the limbic system, the involvement of the amygdala in emotional processing can be attributed to Klüver and Bucy (1997), who examined the behavioral effects of medial temporal lobe lesions in monkeys.

The study of emotions has reached a new level owing to the development of fear conditioning. Fear conditioning is a simple Pavlovian conditioning task, in which a neutral stimulus is paired with an aversive stimulus. Following a relatively small number of such pairings, the neutral stimulus subsequently elicits a behavioral state similar to that evoked by the aversive stimulus.

An increasing number of studies have examined the role of the amygdala in fear, learning, and memory, providing definitive evidence for the role of the amygdala and its projections in fear processing (Adolphs et al., 1998; Davis and Whalen, 2001; Kapp et al., 1992; LeDoux, 1996).

ANATOMY OF THE AMYGDALOID COMPLEX

Thirteen nuclei form the amygdaloid complex and are distinguished by cytoarchitectonics, histochemistry, and connectivity (Krettek and Price, 1978; Pitkänen, 2000).

The amygdala nuclei (Fig. 14.3) can be divided into three groups (Price et al., 1987):

- the deep (or basolateral) group, which includes the lateral nucleus, basal nucleus, and accessory basal nucleus;
- the superficial (or cortical-like) group, which includes the cortical nuclei and nucleus of the lateral olfactory tract;
- the centromedial group, which is composed of the medial and central nuclei.

Outside these groups we also find the intercalated cell masses and the amygdalohippocampal area (Sah et al., 2003).

The basolateral complex is formed by the lateral nucleus (LA), basal nucleus (B), sometimes also called the basolateral nucleus (BLA), and accessory basal nucleus (AB). The LA is situated dorsally, adjacent to the basal nucleus ventrally. It is bounded on the side by the external capsule and on the inside by the central nucleus. It is divided into three sections: the smaller-celled dorsolateral section, the larger-celled ventrolateral section, and the medial section. The basal nucleus is situated beneath the LA and is divided into the rostral magnocellular section and the more caudal intermediate and parvicellular sections. The AB is located ventral to the basal nucleus and is adjacent to the amygdalohippocampal area (AHA). It is composed of the magnocellular subdivision, the intermediate subdivision, and the parvicellular subdivision (Pitkänen, 2000; Pitkänen and Savander, 1997).

Many of the superficial or corticomedial nuclei possess cortical qualities, as they are situated on the surface of the brain and have a stratified structure (Price, 1987). This group includes the nucleus of the lateral olfactory tract (NLOT), bed nucleus of the accessory olfactory tract (BAOT), anterior and posterior cortical nucleus (CoA and CoP, respectively), and the periamygdaloid cortex (PAC). The BAOT is located in the most rostral part of the amygdala, with the CoA situated on its lateral side. The CoA is a layered structure situated lateral to the NLOT, rostrally to the medial nucleus. The CoP has three layers and is situated in the most caudal part of the amygdala, between the AHA dorsally and the PAC laterally. The PAC is located ventrally to the basal nucleus, and is divided into three sections: the periamygdaloid cortex, the medial division, and the sulcal division (Pitkänen, 2000; Price, 1987).

The centromedial nuclear group is located in the dorsomedial area of the amygdaloid complex and is composed of the central (CeA), medial (M), and amygdaloid portions of the bed nucleus of the stria terminalis. The central, medial, and BNST have distinct histochemical and developmental features compared to the cortical nuclei (McDonald, 1998). The CeA is situated dorsomedially and rostrally, with the basolateral complex on its lateral side, the globus pallidus above it, and the stria terminalis on its medial side. The CeA is divided into four sections: the capsular subdivision (CeC), lateral subdivision (CeL), intermediate subdivision (CeI), and medial subdivision (CeM) (Jolkkonen et al., 1998; McDonald et al., 1982). The medial nucleus is located close to the surface, caudally to the NLOT, with the optic tract on its medial side. It is divided into four sections: rostral, central (dorsal and ventral), and caudal.

The remaining amygdala areas are the anterior amygdala area (AAA), amygdalohippocampal area (AHA), and intercalated nuclei (I) (Sah et al., 2003; Pitkänen

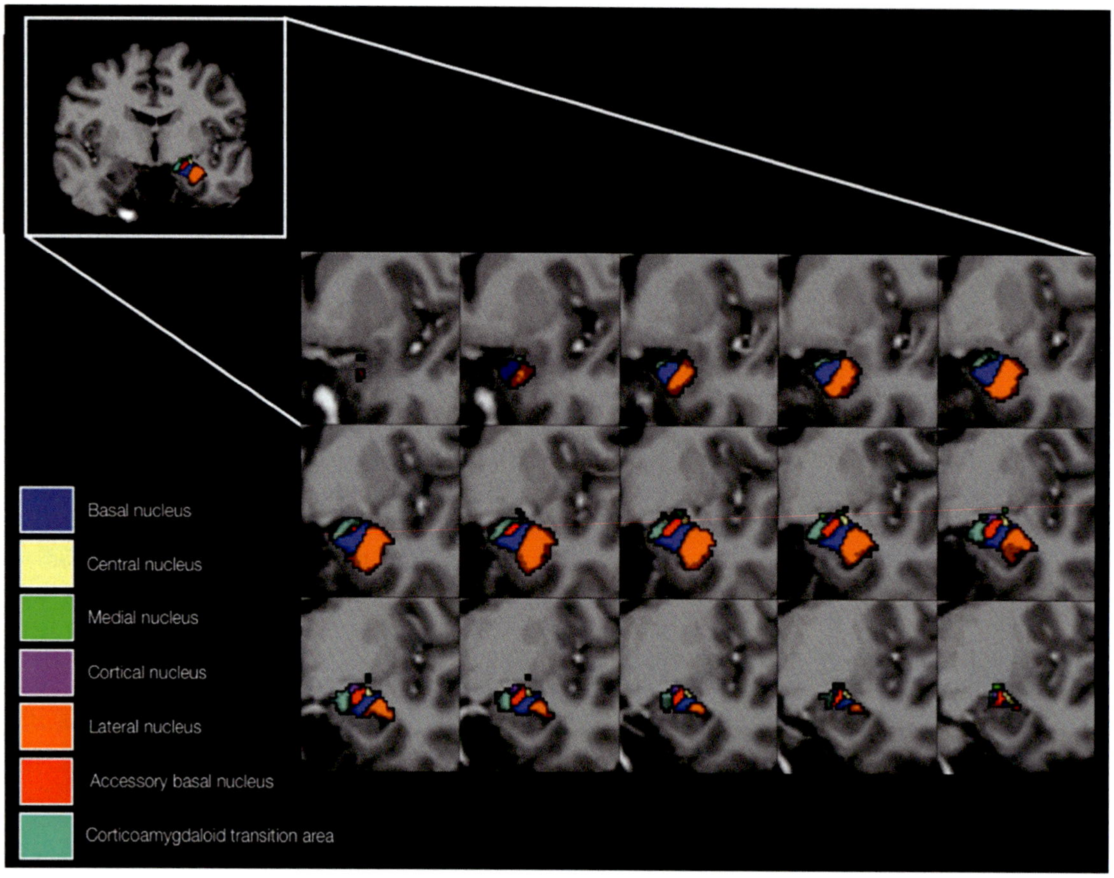

FIGURE 14.3 **An anatomical representation of the segmented amygdala nuclei, including the superficial structures (central, medial, and cortical nuclei), the deep structures (basal, lateral, and accessory basal nuclei) and the corticoamygdaloid transition area.** The study this image was taken from showed that the volumes of the lateral nucleus, cortical nucleus, accessory basal nucleus, and corticoamygdaloid transition area displayed significant negative associations with major depressive disorder severity. *Image courtesy of Brown, S.S.G., Rutland, J.W., Verma, G. et al., 2019. Structural MRI at 7T reveals amygdala nuclei and hippocampal subfield volumetric association with major depressive disorder symptom severity. Sci. Rep. 9, 10166. https://doi.org/10.1038/s41598-019-46687-7.*

et al., 1997; Price et al., 1987). The AHA is the most caudal and has a medial and a lateral subdivision. The intercalated nuclei are small groups of neurons organized in clusters within the fiber bundles (Millhouse et al., 1986).

Some researchers (de Olmos et al., 2004; Alheid and Heimer, 1988) argued that the centromedial amygdala should be extended rostrally and medially. They pointed out that the amygdala innervates the BNST and caudodorsal regions of the substantia inominata (ventral pallidus). Furthermore, these two regions have efferent connections similar to the descending projections of the amygdala. Thus, they argued that these regions were part of the amygdaloid complex. By including these regions they suggested that the centromedial complex should be termed the "extended amygdala."

Swanson and Petrovich (Swanson, 2000; Swanson and Petrovich, 1998) argued that the nuclei of the amygdaloid complex are structurally and functionally heterogeneous and should be divided into four functional systems. These systems are the frontotemporal, autonomic, main olfactory, and accessory olfactory systems.

Sah et al. (2003) proposed a liaison between these classifications, with the basolateral nuclei constituting the frontotemporal group, the centromedial nuclei forming the autonomic group, and the cortical-like nuclei constituting the two olfactory groups.

CONNECTIVITY OF THE AMYGDALA

Multiple brain regions contribute input to each amygdaloid nucleus (Fig. 14.4) (McDonald et al., 1998; Pitkänen, 2000; Price et al., 1987). The amygdala sends efferent projections to several cortical and subcortical areas (Pitkänen, 2000).

There are two main types of inputs to the amygdala: those originating from the brainstem or the hypothalamus, and those originating from cortical and thalamic regions. Sensory information from various regions and structures involved in memory processing is conveyed

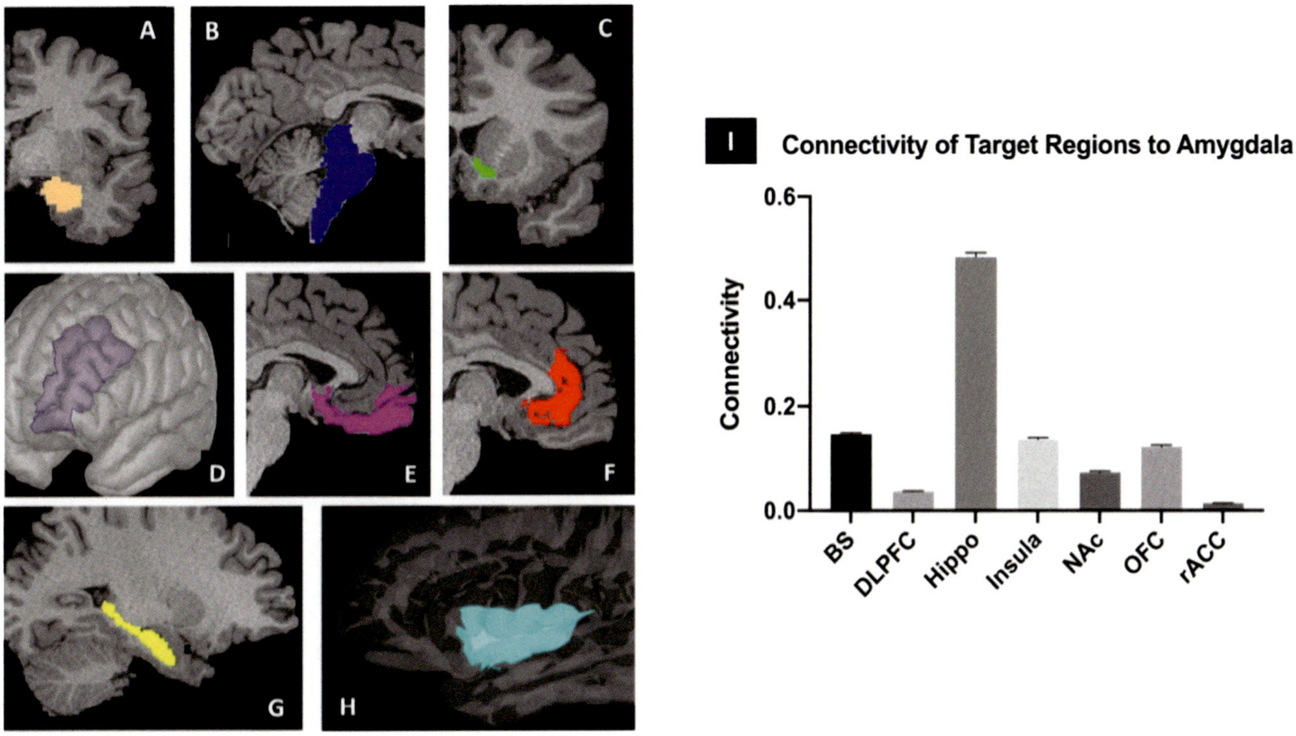

FIGURE 14.4 **Cortical (D, E, F, H) and subcortical (B, C, G) masks of the target brain areas and the amygdala (A, *center*) from a single subject.** (A): Amygdala. (B): Brainstem (BS). (C): Nucleus accumbens (NAc) (D): Dorsolateral prefrontal cortex (dlPFC). (E): Orbitofrontal cortex (OFC). (F): Rostral anterior cingulate cortex (rACC). (G): Hippocampus (Hippo). (H): Insula (I) Probability of connectivity from the amygdala to each target region. The connection probability of each voxel of the amygdala to each of the seven target regions was averaged over all amygdala voxels and this value then averaged across all subjects. Note that the highest probability of connectivity is to the hippocampus (48%). Thus, 48% of all streamlines from the amygdala to the above targets terminated in the hippocampus. *Image courtesy of Avecillas-Chasin, J.M., Levinson, S., Kuhn, T., et al., 2023. Connectivity-based parcellation of the amygdala and identification of its main white matter connections. Sci. Rep. 13, 1305. https://doi.org/10.1038/s41598-023-28100-6.*

through cortical and thalamic inputs. On the other hand, regions associated with behavior and the autonomic nervous system provide input through the hypothalamus and brainstem. The cerebral cortex serves as the primary source of sensory information for the amygdala (McDonald et al., 1998). These projections, predominantly arising from layer V pyramidal neurons, use glutamate as their neurotransmitter (Amaral and Insausti, 1992; Ottersen et al., 1986). The majority of these projections reach the amygdala through the external capsule and are ipsilateral (Mascagni et al., 1993).

The primary sensory cortex is the main origin of cortico-cortical connections that project to the amygdala, transmitting processed information. These inputs can be classified into three groups: those arising from the medial temporal lobe memory system, those that are polymodal, and those that convey sensory information specific to a particular modality (Sah et al., 2003).

Olfactory projections originate from both the main and accessory olfactory bulbs, as well as the primary olfactory cortex. The main olfactory bulb primarily projects to the nucleus of the lateral olfactory tract, anterior cortical nucleus, and periamygdaloid cortex. On the other hand, the accessory olfactory bulb projects to the bed nucleus of the accessory olfactory tract, medial nucleus, and posterior cortical amygdala (Scalia and Winans, 1975). The piriform cortex and anterior olfactory nucleus send their projections to the lateral amygdala, basal nuclei, and accessory basal nuclei (Luskin and Price, 1983). Additionally, the dorsal endopiriform nucleus projects to all cortical nuclei of the amygdala, as well as the nucleus of the lateral olfactory tract, periamydaloid cortex, and medial amygdala (Behan and Haberly, 1999).

Regarding the transmission of somatosensory information to the amygdala, only a small number of direct projections originate from the primary somatosensory areas, while the majority of afferent signals reach the amygdala through the dysgranular parietal insular cortex, located in the parietal lobe (Shi and Cassell, 1998). These projections specifically target the lateral, basal, and central nuclei (McDonald and Jackson, 1987; Shi and Cassell, 1998). While the inputs to the basal nucleus lack a segregated structure, the dorsolateral subdivision of the lateral amygdala shows prominent labeling (Shi and Cassell, 1998). Moreover, somatosensory information reaches the amygdala via projections from the pontine parabrachial nucleus and various thalamic nuclei, including the medial portion of the medial geniculate nucleus and the posterior internuclear nucleus (PIN) (Fig. 14.5). These regions are known to play a role in the transmission of nociceptive information (Bernard et al., 1989; Bordi and LeDoux, 1994; LeDoux et al., 1987). The inputs originating in the PIN target all subdivisions of the lateral amygdala, and also innervate the accessory basal nucleus and the medial subdivision of the central nucleus (Bernard et al., 1993; Linke et al., 2000).

The dorsal subdivision of the lateral amygdala (LA), posterior basal nucleus, and central nucleus receive strong projections from primary gustatory and visceral regions in the anterior and posterior insular cortices (Shi and Cassell, 1998). The LA, basal nucleus (B), and lateral subdivision of the central nucleus (CeL) receive inputs from the posteromedial ventral thalamic nucleus, also known as the thalamic gustatory nucleus (Nakashima et al., 2000; Turner and Herkenham, 1991). Additionally, the parabrachial nucleus, which receives projections from the nucleus of the solitary tract, targets the CeL (Bernard et al., 1993; Ottersen, 1981).

Auditory and visual information is transmitted to the amygdala through association areas rather than from the primary areas, and these inputs play a significant role in fear conditioning. Acoustic inputs from subcortical regions originate from the thalamic medial geniculate nucleus and target corresponding regions within the lateral amygdala (LA) (LeDoux et al. 1990, 1991; Turner et al., 1991). Visual projections to the amygdala originate from both thalamic and higher-order visual areas, following a cascade that involves the temporal cortex (McDonald, 1998; Shi and Davis, 2001). These visual fibers terminate primarily in the dorsal subdivision of the LA, the lateral subdivision of the central nucleus (CeL), and, to a lesser extent, in the magnocellular basal nucleus.

The amygdala receives polymodal sensory data from a variety of sources, including the hippocampus, prefrontal cortex, and perirhinal cortex. The prefrontal cortical regions, many of which are involved in behavior and reward circuits in rats (Rolls, 2000), receive information from all sensory modalities (Ray and Price, 1992). A frontal cortical projection that is both dense and topographically structured has been identified in all species (McDonald, 1998). The LA as well as accessory basal, central, and medial nuclei all receive inputs from the

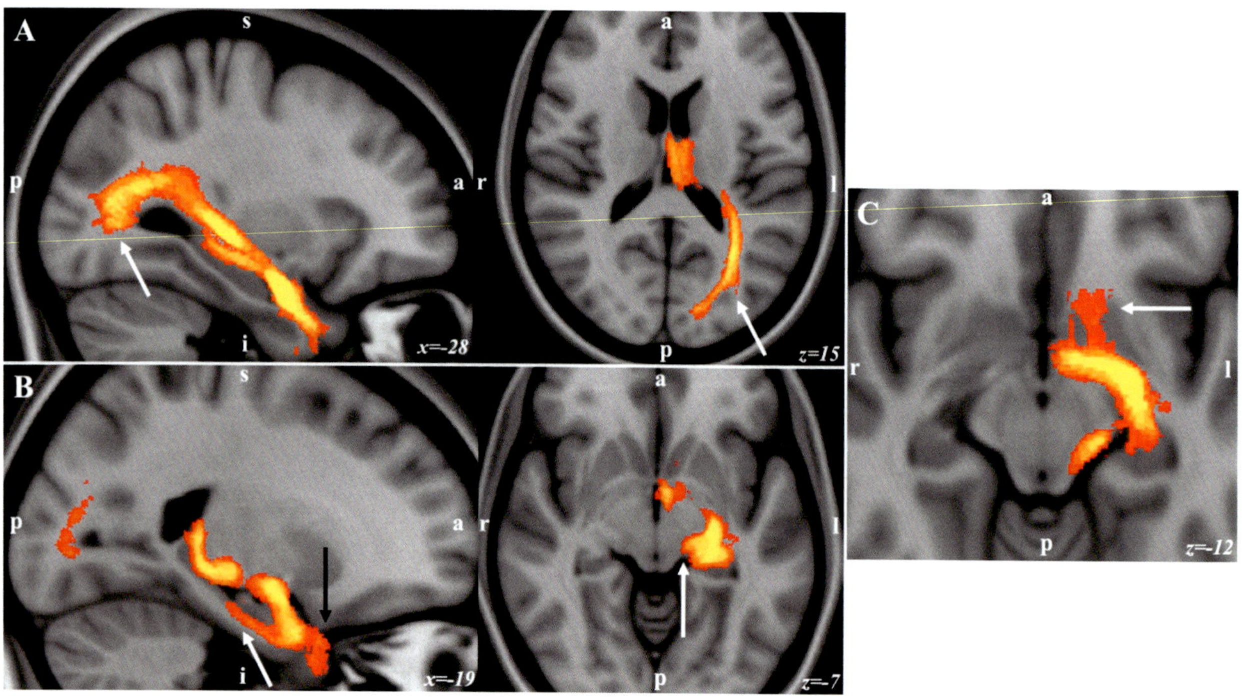

FIGURE 14.5 **Population connectivity maps of the amygdala clusters (I).** (A) The lateral cluster also showed stronger connectivity with sensory areas including parietal, occipital, and posterior temporal areas (*arrows*). (B) The basal and lateral clusters also showed stronger connectivity through the parahippocampal radiation of the cingulate bundle (*white arrow* in sagittal view), with the temporal pole (*black arrow* in sagittal view), and through the amygdalotectal pathways with the colliculi. (C) The lateral cluster had stronger connectivity through the subcaudate white matter and ventromedial part of the uncinate fascicle with the PFC (*arrow*). *Image courtesy of Avecillas-Chasin, J.M., Levinson, S., Kuhn, T., et al., 2023. Connectivity-based parcellation of the amygdala and identification of its main white matter connections. Sci. Rep. 13, 1305. https://doi.org/10.1038/s41598-023-28100-6.*

prefrontal cortex, although the basal nucleus is its primary target (McDonald and Mascagni, 1996).

The perirhinal cortex, entorhinal cortex, parahippocampal cortex, and the hippocampus are regions connected to the long-term declarative memory system (Milner et al., 1998), and have strong and reciprocal connections with the amygdala (McDonald, 1998; Pitkänen, 2000). The perirhinal cortex projects most heavily to the medial division of the LA, but projections to the basal and cortical nuclei have also been observed (Shi and Cassell, 1999). The entorhinal cortex seems to project to the majority of amygdalar nuclei (McDonald and Mascagni, 1997). Although the basal nucleus is the primary target of inputs from the hippocampus to the amygdala, mostly originating in the subicular area (Canteras and Swanson, 1992), many other nuclei are also reached by these fibers.

There are multiple projections originating from the amygdaloid nuclei to various regions, such as the cortex, hypothalamus, and brainstem. Typically, the amygdala's cortical and basolateral regions send projections to cortical sensory centers. The perirhinal area receives reciprocal connections from the lateral amygdala (LA), basal nucleus (B), accessory basal nucleus (AB), medial nucleus (M), and periamygdaloid cortex, as well as other regions of the frontal cortex (Pitkänen, 2000). Significant reciprocal projections are also present between cortical nuclei that receive olfactory projections and the olfactory cortex (Sah et al., 2003).

The basolateral complex of the amygdala has notable projections to the medial temporal lobe memory system, including the hippocampus and perirhinal cortex (Petrovich et al., 2001; Pitkänen, 2000). Additionally, the nucleus accumbens also receives some of these fibers (McDonald, 1991). The basal nucleus, similar to the LA, exhibits substantial projections to the hippocampus, but it also sends significant projections to the prefrontal cortex, nucleus accumbens, and thalamus. The efferent fibers originating from the basolateral complex are believed to be glutamatergic and originate from pyramidal-like neurons (Paré et al., 1995).

When the central nucleus of the amygdala is activated, it can stimulate groups of neurons in the brainstem that regulate the autonomic nervous system, or it can stimulate hypothalamic nuclei that govern these centers (Kapp et al., 1982; LeDoux et al., 1988). The medial subdivision of the central nucleus exhibits multiple projections to nuclei located in the midbrain, pons, and medulla (Veening et al., 1984), as well as to the hypothalamus, bed nucleus of the stria terminalis (Dong et al., 2001), and several other nuclei in these regions.

Projections to the brainstem target three primary areas:

- Periaqueductal gray: responsible for vocalization, startle response, analgesia, and cardiovascular changes (Behbehani, 1995; Rizvi et al., 1991).
- Parabrachial nucleus: involved in pain pathways (Gauriau and Bernard, 2002; Moga et al., 1990).
- Nucleus of the solitary tract (NTS): connected with the vagal system (van der Kooy et al., 1984).

The dorsolateral and caudolateral areas of the hypothalamus, which have projections to autonomic cell groups in the brain stem and spinal cord, receive innervation from the medial and capsular subdivisions of the central nucleus (Swanson and Kuypers, 1980; Petrovich et al., 2001). These regions also receive efferents from the amygdala's olfactory system and the lateral subdivision of the central nucleus. Additionally, the medial nuclei of the hypothalamus, which play a role in behavior control, receive fibers from the amygdala (Petrovich et al., 2001). The amygdala's medial nucleus, posterior basal nucleus, and posterolateral cortical nucleus, which are connected to the olfactory system, innervate the ventromedial nucleus, playing a role in reproductive behavior (Dayas et al., 1999; Petrovich et al., 2001).

Strong projections from the central nucleus of the amygdala (CeA) extend to the bed nucleus of the stria terminalis (BNST), which in turn influences the hypothalamic nuclei. Additionally, both the CeA and BNST heavily target ascending clusters of monoaminergic and cholinergic neurons. These include the serotonergic raphe, cholinergic nucleus basalis, noradrenergic locus coeruleus, dopaminergic substantia nigra, ventral tegmental region, and noradrenergic locus coeruleus (Amaral et al., 1992; Davis and Whalen, 2001; Price et al., 1987). These systems stimulate large portions of the forebrain and temporal lobe memory systems and provide inputs to the amygdaloid complex. Unlike the rapid and direct excitation generated by most glutamatergic afferents, these ascending systems offer modulatory inputs that influence data processing across widespread cell groups, impacting information processing throughout the brain.

The amygdala nuclei display significant intranuclear and internuclear connectivity (Krettek and Price, 1978; Pitkänen, 2000).

Sensory information first enters the amygdala through the basolateral nuclei and then follows a predominantly lateral to medial progression toward the centromedial nuclei, which serve as an output station (Pitkänen et al., 1997).

The lateral nucleus (LA) exhibits extensive rostrocaudal and interdivisional connections (Pitkänen et al., 1995). Specifically, the dorsolateral subdivision projects to both the medial and lateral aspects of the lateral subdivision. Unimodal sensory inputs primarily enter the LA laterally, whereas polymodal afferents and projections from declarative memory systems are predominantly confined to the medial subdivision (Pitkänen, 2000).

The presence of lateral to medial intranuclear connections within the LA suggests that the medial subdivision might serve as a site for integrating sensory information with assessments of past experiences (Sah et al., 2003).

The basal and accessory basal nuclei, along with the capsular region of the central nucleus, receive substantial projections from the lateral nucleus (Pitkänen et al., 1995; Paré et al., 1995; Smith and Paré, 1994). The periamygdaloid cortex also receives projections from the lateral nucleus, and all of these regions exhibit reciprocal connections with the lateral nucleus, except for the central nucleus (Savander et al. 1996, 1997; Jolkkonen et al., 1998). Notably, while the dorsolateral subdivision of the lateral amygdala shows minimal reciprocal projections, there are numerous connections between the medial and ventrolateral subdivisions. Additionally, these reciprocal connections are relatively weaker compared to the intranuclear connections along the lateromedial axis. Most projections to and from the lateral nucleus form asymmetrical synapses, indicating their excitatory nature (Savander et al., 1997; Smith and Paré, 1994). However, some reciprocal connections from the basal nuclei form symmetrical synapses, suggesting a potential inhibitory role (Savander et al., 1997).

All subdivisions of the basal nucleus exhibit robust rostrocaudal connections. The parvicellular subdivision extensively projects to the intermediate and magnocellular subdivisions (Savander et al., 1995). The medial subdivision of the central nucleus is the primary target of the basal nuclei's projections, receiving significant inputs (Paré et al., 1995; Savander et al. 1995, 1996). These projections are believed to be glutamatergic, as they form asymmetric synapses (Paré et al., 1995). The basal nuclei play a crucial role in regulating the output of processed data from the amygdaloid complex through their projections to the central nucleus. The accessory basal nucleus, which possesses strong rostrocaudal connections, sends afferents to the lateral amygdala, central amygdala, and medial divisions (Savander et al., 1996).

The central nucleus, which serves as the primary output region of the amygdala, receives inputs from all other nuclei, although its projections back to them are weak. The medial and capsular subdivisions contribute the majority of amygdaloid inputs to the central nucleus. Within the central nucleus, there are numerous intra- and interdivisional connections, as well as extensive intranuclear connections between each of the four subdivisions (Jolkkonen and Pitkänen, 1998). The lateral subdivision has fewer reciprocal connections compared to the other subdivisions but exhibits prominent projections to them. This suggests that the lateral subdivision may play a role in integrating inputs within the amygdaloid complex (Sah et al., 2003).

Avecillas-Chasin et al. (2023) conducted a tractography study investigating the connectivity patterns of the amygdala with brain regions associated with reward, memory, limbic, and cognitive functions. They identified three spatially contiguous clusters arranged in a medial-to-lateral pattern, which aligned with known amygdala nuclear groups described in histological studies. The medial cluster encompasses the central, medial, and cortical nuclei, whereas the basal cluster corresponds to the basal nuclear complex. The lateral cluster represents the amygdala lateral nuclei of the amygdala.

The study further revealed specific connectivity patterns between the amygdala clusters and other brain regions. The brainstem and the dorsolateral prefrontal cortex (dlPFC) exhibit stronger bilateral connectivity with the lateral cluster. The hippocampus (HC) showed a stronger bilateral connectivity with the basal cluster. The insula demonstrated stronger connectivity with both the medial and lateral clusters. The lateral cluster exhibited robust connectivity with sensory areas, including the parietal, occipital, posterior temporal, and superior colliculi. It also shows strong connectivity with the prefrontal cortex (PFC) through the uncinate fascicle. Both lateral and basal clusters exhibited strong connectivity with the temporal and cingulate cortices.

The basal cluster demonstrated connectivity with the PFC and ventral striatum through amygdalofugal pathways and the internal capsule (Fig. 14.6). Additionally, it displayed stronger connectivity with the HC through the amygdalohippocampal bundle, with the medial thalamus through the inferior thalamic peduncle, and with the cingulum bundle (Fig. 14.8). The medial and basal clusters exhibited stronger connectivity through the ventral amygdalofugal pathway (VAF) and the stria terminalis (ST), which converged in the bed nucleus of the stria terminalis (BNST), hypothalamus, and septal area (Fig. 14.7). The medial cluster showed connectivity with the nucleus accumbens (NAc) and unique connectivity with the ventral tegmental area (VTA) through the medial forebrain bundle, which is involved in reward-related behaviors.

This study's parcellation (Avecillas-Chasin, 2023) of the human amygdala circuitry corresponds to the well-established behavioral model of the amygdala. The findings reveal that the basal and lateral segments serve as the primary input region for sensory and regulatory information, while the basal and medial segments act as the main output region projecting to the hypothalamus, brainstem, septal area, and BNST. These results hold significant implications for understanding behavioral disorders and offer a crucial advancement in mapping the amygdala circuitry in individual patients, which can inform the development of targeted neuromodulatory therapies.

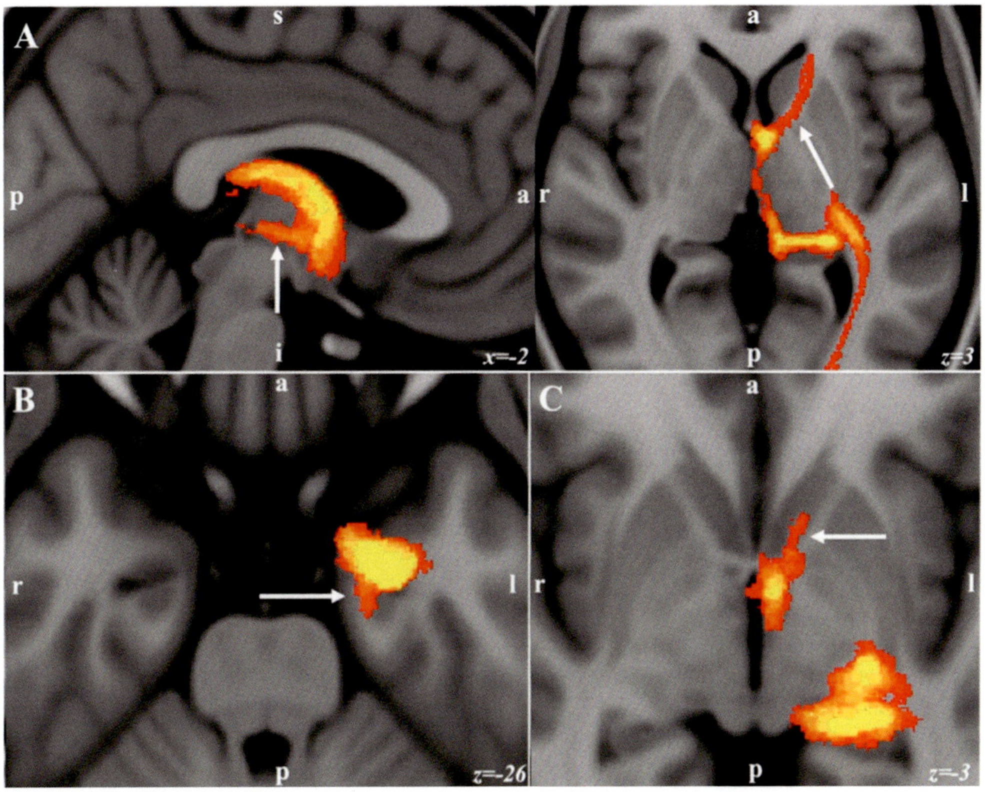

FIGURE 14.6 **Population connectivity maps of the amygdala clusters (II).** (A) The basal cluster had stronger connectivity with the medial thalamus through the inferior thalamic peduncle (*arrow* in sagittal), and through the anterior limb of the internal capsule with the PFC (*arrow* in axial view). (B) With the HC through the amygdalohippocampal bundle and (*arrow*). (C) With the ventral striatum (*arrow* in C). *Image courtesy of Avecillas-Chasin, J.M., Levinson, S., Kuhn, T., et al., 2023. Connectivity-based parcellation of the amygdala and identification of its main white matter connections. Sci. Rep. 13, 1305. https://doi.org/10.1038/s41598-023-28100-6.*

ELECTROPHYSIOLOGIC STUDIES

Neurons within the lateral amygdala (LA) have been categorized into two distinct groups based on their firing properties (Mahanty and Sah, 1998; Paré et al., 1995). The first group, which comprises approximately 95% of the cells, exhibits broad action potentials and varying degrees of spike frequency adaptation in response to sustained depolarizing currents. The extended afterhyperpolarization (AHP) following action potential trains is primarily responsible for this spike frequency adaptation (Faber and Sah, 2002). The second group of neurons responds to sustained depolarization with shorter action potentials and displays minimal frequency adaptation. These two populations have been referred to as pyramidal or projection neurons and interneurons, respectively, due to their similarities to firing patterns observed in cortical and hippocampal neurons (Lang and Paré, 1998; Mahanty and Sah, 1998).

Pyramidal-like neurons within the lateral amygdala (LA) display a range of firing characteristics. Some cells exhibit minimal adaptation and fire repeatedly during a current injection, while others fire only two to three spikes and demonstrate significant spike frequency adaptation (Faber et al., 2001; Funahashi et al., 2000). There are also cells that fall in between these two extremes, firing multiple times with clear spike frequency adaptation. However, the majority of cells in the LA exhibit noticeable adaptation and fire fewer spikes (Faber et al., 2001).

Another cell type, known as a single firer, which stands out from the aforementioned categorization, makes up about 3% of observed cells (Chapman et al., 1990; Faber et al., 2001; Faulkner and Brown, 1999; Yajeya et al., 1997). After current injection, just one action potential is triggered; the excitability can't be increased by applying bigger current injections or by depolarizing the cell (Fig. 14.9).

High levels of glutamate and aspartate immunoreactivity are observed in pyramidal-like neurons within the basolateral complex, indicating their glutamatergic nature (Smith and Paré, 1994). However, these neurons do not exhibit immunoreactivity for glutamic acid decarboxylase, which is involved in GABA synthesis (Carlsen, 1988). Therefore, it is presumed that the output cells of the basolateral complex are predominantly glutamatergic.

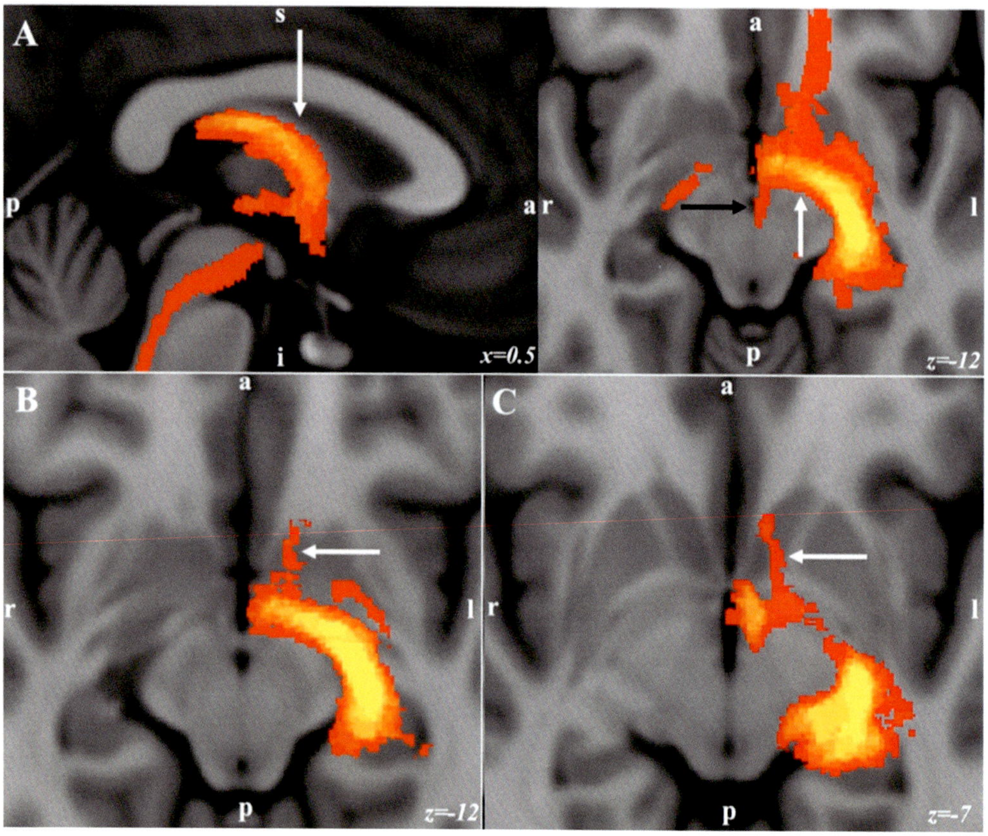

FIGURE 14.7 **Population connectivity maps of the amygdala clusters (III).** (A) The medial and basal clusters showed stronger connectivity through the stria terminalis (*arrow* in sagittal view) and ventral amygdalofugal pathway (*arrow* in axial view). These two pathways converged in the BNST, hypothalamus, and septal area. The medial cluster showed unique connectivity with the ventral tegmental area through the medial forebrain bundle (*black arrow* in axial). (B) and (C) The medial and lateral cluster showed stronger connectivity with ventral and dorsal areas of the NAc respectively. *Image courtesy of Avecillas-Chasin, J.M., Levinson, S., Kuhn, T., et al., 2023. Connectivity-based parcellation of the amygdala and identification of its main white matter connections. Sci. Rep. 13, 1305. https://doi.org/10.1038/s41598-023-28100-6.*

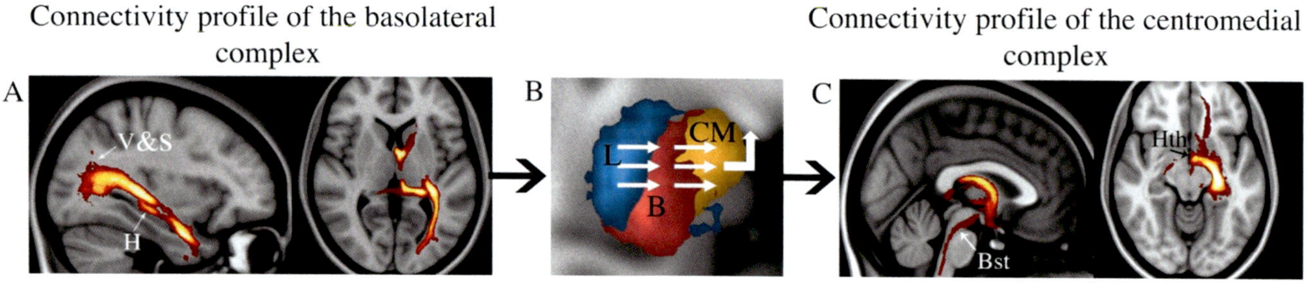

FIGURE 14.8 **Functional model of the amygdala circuit. (A)** The basolateral complex (L, B) receives and integrates information from visual (V), sensory areas (S), hippocampus (H), and prefrontal cortex. **(B)** The centromedial complex (CM) translates this information to behavior through the connections with **(C)** the hypothalamus (Hth) and the brainstem (Bst). *Image courtesy of Avecillas-Chasin, J.M., Levinson, S., Kuhn, T., et al., 2023. Connectivity-based parcellation of the amygdala and identification of its main white matter connections. Sci. Rep. 13, 1305. https://doi.org/10.1038/s41598-023-28100-6.*

The fast-spiking cells found in the basolateral complex are GABAergic interneurons, similar to those found in other regions of the central nervous system (McDonald and Augustine, 1993; Nitecka and Ben-Ari, 1987; Paré and Smith, 1993; Pitkanen and Amaral, 1994). These interneurons likely form local circuits within the complex.

It is hypothesized that excitatory inputs to the interneurons may be the consequence of the activation of axon collaterals of pyramidal cells, suggesting a feedback function for the interneurons.

The most common type of cell found in the central nucleus of the amygdala (CeL) is known as the "medium

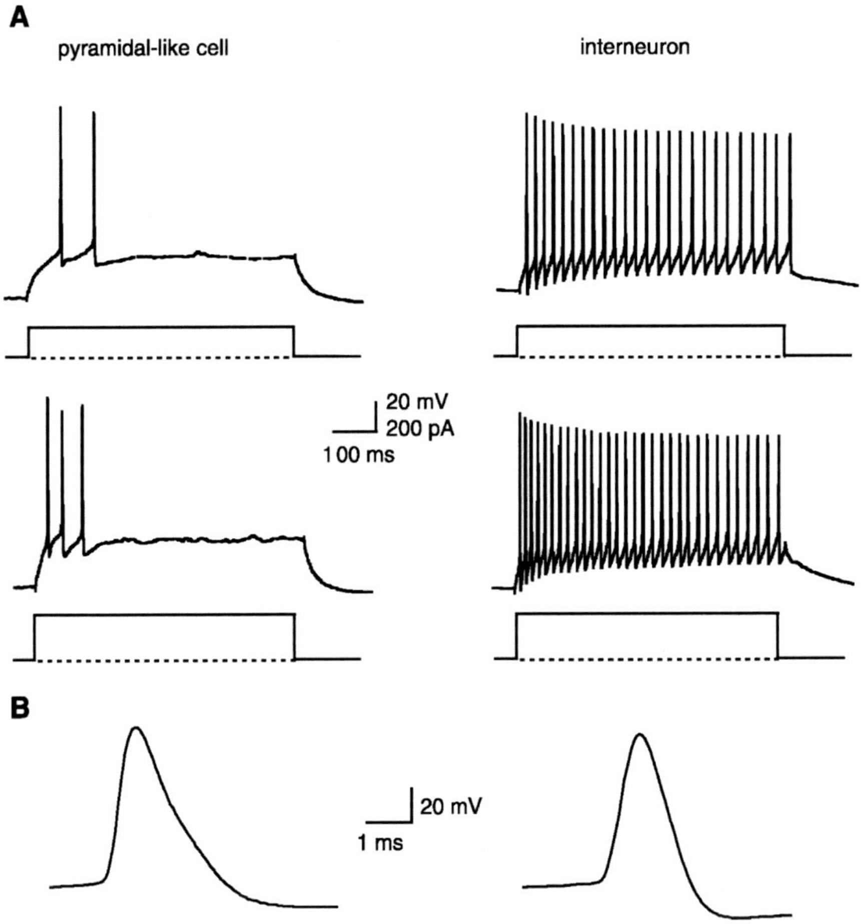

FIGURE 14.9 **Pyramidal-like neurons and interneurons can be distinguished on electrophysiological grounds. Traces show recordings from typical pyramidal-like neuron and interneuron in the basolateral complex.** Traces on the *left* are from a typical pyramidal-like neuron, and those on the *right* are from an interneuron. (A) Injection of a 400-m depolarizing current injection in pyramidal neurons evokes action potentials that show spike frequency adaptation, while similar current injections into interneurons evoke a high-frequency train of action potentials that do not adapt. (B) Action potentials in interneurons have a shorter duration than in pyramidal cells. *Image courtesy of Sah, P., Faber, E.S., Lopez De Armentia, M., Power, J., 2003. The amygdaloid complex: anatomy and physiology. Physiol. Rev. 83 (3), 803–834. https://doi.org/10.1152/physrev.00002.2003.*

spiny neuron" (Hall, 1972; McDonald, 1982). These neurons have an oval or fusiform soma and possess three to five main dendrites that lack spines. Additionally, they have secondary and tertiary dendrites that are moderately spiny and sparsely branched (Martina et al., 1999; McDonald, 1982; Schiess et al., 1999). As these neurons extend their axons out of the nucleus, they release various regional collaterals. Another type of neuron found in the CeL has a slightly larger soma and thick, nonspiny primary dendrites that taper into secondary dendrites with sparse spines. Additionally, a few nonspiny neurons have also been identified (Cassell and Gray, 1989). The CeA contains a fairly equal distribution of these different cell types. Immunohistochemical studies have revealed the presence of numerous peptides within these cells and in the afferents that reach them (Cassell and Gray, 1989; Cassell et al., 1986; Moga and Gray, 1985).

The morphology of neurons in the central nucleus of the amygdala (CeN) differs significantly from those in the basolateral complex (BLA). Neurons in the BLA exhibit a morphology similar to cortical structures, while neurons in the CeN have a more striatal-like morphology. In terms of neurotransmitter projections, the basolateral nuclei predominantly contain glutamatergic projections, whereas the central nuclei primarily have GABAergic projections (Sah et al., 2003).

GABAergic intercalated cells play a crucial role as feed-forward interneurons in transmitting signals between the basolateral complex (BLA) and the central nucleus of the amygdala (CeA) (Delaney and Sah, 2001; Paré and Smith, 1993; Royer et al., 1999). Within the intercalated cell masses, two distinct types of neurons can be identified.

The majority of intercalated cells exhibit ovoid bodies with spiny, predominantly bipolar dendritic trees. Their axons branch into the lateral, basal, and central nuclei of the amygdala (Millhouse, 1986). Another subtype of intercalated cells consists of larger cells with long and thick dendrites that follow the boundaries of the basal, lateral, and central nuclei (Millhouse, 1986; Paré and Smith, 1993). These two cell types bear resemblance to neurons found in the striatum.

Modulating the activity of intercalated neurons in the amygdala has a significant impact on the output of the central nucleus of the amygdala (CeA) due to their inhibitory role in regulating CeA neurons. These intercalated cells form local networks that exhibit a directional flow from the lateral to medial regions, meaning that when they are activated, they preferentially inhibit neurons in the medial direction.

The connectivity pattern within the intercalated cell groups allows for precise regulation of inhibition in the central nucleus, as information transmitted through the basolateral complex activates specific populations of intercalated cells (Royer et al., 2000).

The medial nucleus of the amygdala consists of a single type of cell that bears resemblance to the principal neurons found in the central nucleus's medial subdivision (McDonald, 1983).

In the bed nucleus of the stria terminalis, the neurons exhibit medium-sized somata and multipolar spiny dendrites, which are similar to the major cell types observed in the medial and intermediate subdivisions of the central nucleus (McDonald, 1983).

The anterior amygdaloid region contains neurons with oval somata and three to four main dendrites that have limited branching and few spines. These neurons share similarities with the second class of neurons found in the central nucleus's medial subdivision (Hall, 1972).

The amygdalohippocampal region, cortical nuclei, and lateral olfactory tract nucleus also contain cell types that are comparable to those observed in the basolateral complex of the amygdala (Sah et al., 2003).

FUNCTIONS OF THE AMYGDALAR COMPLEX

The abstract nature of emotions poses challenges in understanding the amygdala and its role in emotion. Bilateral amygdala-specific injuries in humans are rare occurrences. Animal studies, although limited in their ability to determine subjective experiences, have provided valuable insights into the function of the amygdala in emotion (Chapman et al., 1954; Gloor, 1990; Kaada, 1951; LeDoux, 1996).

Both human and animal studies have demonstrated that electrical stimulation of the amygdala elicits fear and anxiety responses. Lesions of the amygdala in both humans and animals selectively impair the expression of certain forms of unconditioned fear, but not all (Chapman et al., 1954; Gloor, 1990; Kaada, 1951; LeDoux, 1996).

As a result, animal research on fear has significantly contributed to our understanding of the amygdala's involvement in emotional processes (LeDoux, 1996).

The amygdala plays a crucial role in various forms of fear-related learning. Lesions in the amygdala disrupt the acquisition of both active avoidance and passive avoidance conditioned responses, while leaving their retention intact (references: Poremba and Gabriel, 1999; Lehmann et al., 2000; Roozendaal et al., 1993; Treit and Menard, 1997).

In addition to fear and aversive stimuli, the amygdala is involved in processing other emotions and is engaged in learning when appetizing stimuli, such as food, are used. Lesions in the amygdala impair gustatory neophobia (Dunn and Everitt, 1988; Sutherland and McDonald, 1990) and affect appetitive Pavlovian conditioning (Gallagher et al., 1990; Sutherland and McDonald, 1990), conditioned location preference, and conditioned taste aversion (Morris et al., 1999; Nachman and Ashe, 1974; Rollins et al., 2001; Welzl et al., 2001).

The amygdala also modulates the formation and retention of memories associated with emotional reactions (McGaugh, 2000; McGaugh et al., 1996; Packard and Cahill, 2001).

The superior temporal sulcus (STS), an important region involved in face processing in primates (Allison et al., 2000; Freiwald et al., 2016; Tsao et al., 2008), exhibits direct anatomical connections with both the amygdala and the OFC. Studies conducted on humans have shown a positive correlation between the size of social networks and these connections (Bickart et al., 2011; Powell et al., 2012).

Further research on humans (Hampton et al., 2016; Sato et al., 2016, 2020) has revealed a significant association between the functional and structural connectivity of the amygdala-OFC pathway and social indicators, including the size of one's social network and perceived social support.

Recent research conducted by Felix-Ortiz and Tye (2014) has demonstrated that inputs from the basolateral amygdala (BLA) to the ventral hippocampus (vHPC) in mice play a role in regulating social interaction. Specifically, their study showed that optogenetic suppression of BLA-vHPC projections led to an increase in social interactions during the home cage intruder test.

The findings of this study highlight the dual impact of the BLA-vHPC circuit on both social and anxiety-related behaviors. Understanding the mechanisms underlying the regulation of social and nonsocial behaviors in this circuit could provide insights into the comorbidity often observed between social deficits and anxiety disorders (Zaboski and Storch, 2018; Gangopadhyay et al., 2021).

Aggression is regulated by complex interactions between various brain regions. The amygdala, in particular, plays a crucial role in modulating aggression. It has been observed that when the prefrontal cortex (PFC) decreases its control over the amygdala, aggression is released, leading to a further potentiation of aggressive behavior. This process involves the involvement of "executive centers" in the hypothalamus and sympathetic centers in the spinal cord (Haller, 2018).

The amygdala, hypothalamus, and brain stem are commonly referred to as the "triggers of aggression" (Bogerts et al., 2018). Experimental studies in animals have shown that stimulation of the medial and basolateral nuclei of the amygdala can elicit various forms of aggressive behavior, and the intensity of aggression is proportional to the degree of activation. Conversely, decreased activity in these amygdalar regions is associated with prosocial and submissive behavior (Haller, 2018).

Interestingly, the amygdala itself exhibits functional diversity in relation to aggression. For example, stimulation of the medial amygdala (ME) increases territorial aggression but decreases predatory aggression. The ME is also involved in mating behavior and protective behavior toward territory and offspring (Haller, 2018; Wang et al., 2013). In contrast, stimulation of the central amygdala (CE) increases predatory aggression, leading to the characterization of the CE as an "independent center of aggression" (Haller, 2018).

It is worth noting that increased activity in the CE has been implicated in pathological aggression associated with reduced emotionality. However, simultaneous activation of both the ME and CE has been linked to violent behavior (Haller, 2018).

In summary, the amygdala, along with other brain regions such as the PFC, hypothalamus, and brain stem, plays a complex role in regulating aggression. The amygdala itself exhibits functional diversity, with different regions contributing to distinct types of aggressive behaviors. These findings highlight the intricate neural mechanisms underlying aggression.

Lower amygdala volume and impaired connectivity between the amygdala and the ventromedial prefrontal cortex (vmPFC) and orbitofrontal cortex (OFC) have been found to be associated with aggressive, violent, and criminal behavior. In individuals displaying such behaviors, there is a decrease in the activity of the vmPFC and OFC (Adebimpe et al., 2019).

PSYCHIATRY AND AMYGDALA

Generalized Anxiety Disorder

The primary characteristic of generalized anxiety disorder (GAD) is the difficulty in differentiating between threatening and nonthreatening stimuli (Madonna et al., 2019). In individuals with GAD, there is an observed increase in the volume, activity, and connectivity of the amygdala with other brain regions, particularly the dorsolateral prefrontal cortex (dlPFC) and anterior cingulate cortex (ACC) (Kolesar et al., 2019). Hyperactivity in the amygdala and hypoactivity in the PFC are the two most commonly observed abnormalities in individuals with GAD (Adebimpe et al., 2019).

Social Phobia

Individuals with social anxiety disorder exhibit heightened amygdala response when exposed to images depicting anger or contempt expressions (Molosh et al., 2020). This heightened response specifically to threatening social cues suggests an abnormal sensitivity of the amygdala in social contexts, while sensitivity to other types of stimuli remains unaffected. Moreover, excessive amygdala activity is associated with reduced activity in the oxytocin (OXT) system (Gottschalk and Domschke, 2017) and impaired functional connectivity between the amygdala and the left hemisphere's ventromedial prefrontal cortex (vmPFC). These findings indicate potential compromises in the structural integrity of the amygdala and its connectivity with the vmPFC in individuals with social anxiety disorder (Šimić et al., 2021).

Post-Traumatic Stress Disorder

The amygdala plays a crucial role in the development of post-traumatic stress disorder (PTSD) by initiating the activation of the hypothalamo-pituitary axis and triggering physiological responses to acute stress (Wahis et al., 2021). Dysregulation of GABAergic interneurons leads to decreased inhibitory control, resulting in heightened activity of glutamatergic neurons and increased excitatory activity in the amygdala, which contributes to the development of anxiety symptoms (Wahis et al., 2021).

The amygdala is involved in both conditioned and unconditioned memory of stressful events, leading to increased activity during recollection. Functional magnetic resonance imaging (fMRI) studies have demonstrated elevated spontaneous amygdala activity and increased amygdala activation when recalling traumatic experiences in individuals with PTSD (Wahis et al., 2021). This explains why individuals with PTSD often exhibit hippocampal atrophy while maintaining normal amygdala volume. Chronic stress exposure is believed to impair hippocampus-dependent memory, while memory stored in the amygdala remains intact (Malikowska-Racia and Salat, 2019).

Furthermore, stress-induced elevation of brain-derived neurotrophic factor (BDNF) levels in the basolateral amygdala (BLA) contributes to the consolidation of memory for stressful events, while simultaneously reducing prefrontal cortex (PFC) control over the amygdala (Fonzo and Etkin, 2017). This mechanism helps explain other characteristics of PTSD, including impaired memory for nonemotionally significant facts but enhanced memory for stressful events.

Panic Attacks and Panic Disorder

Excessive external stimuli can trigger overactivation of the amygdala, leading to spontaneous panic episodes. In individuals with panic disorder, there is a significant reduction in the volume of the right lateral amygdala (LA) and basolateral amygdala (BLA), which also affects the overall right amygdala volume (Asami et al., 2018). The LA recognizes sensory inputs, while the BLA detects potential threats and communicates this information to the central nucleus of the amygdala (CE), resulting in the manifestation of bodily symptoms associated with sympathetic activity. Panic episodes are often preceded by a false sense of danger, particularly when bodily sensations are misinterpreted (Duvarci and Paré, 2007). Furthermore, patients with panic disorder tend to exhibit heightened processing of fearful facial expressions, as evidenced by increased amygdala activity when viewing such pictures compared to healthy individuals (Preter and Klein, 2008). Preclinical models suggest that there is an imbalance between excitation and inhibition within the BLA and CE, further contributing to the pathophysiology of panic disorders (Kunimatsu et al., 2020; Duvarci and Paré, 2007).

Klüver–Bucy Syndrome

A notable study conducted by Klüver and Bucy in 1938 documented a peculiar emotional behavior observed in monkeys following injury to the medial temporal lobe. The main characteristic observed in these cases was a profound lack of fear (Weiskrantz, 1956). This emotional deficit, often referred to as "psychic blindness," hinders individuals from forming emotional connections with people, objects, or events. Strikingly, even after experiencing adverse events, individuals with this condition continue to exhibit their emotional indifference.

The clinical presentation of this condition includes abnormalities in hippocampal development, amygdala dysfunction, and profound global amnesia, which refers to the inability to convert short-term memories into long-term memories. Structural or functional bilateral amygdalectomy, a surgical procedure involving the removal or damage to both amygdalae, can lead to various indications such as loss of fear, heightened compliance, docility, reduced aggression, increased oral exploration, hypersexuality, diminished emotional responsiveness, visual agnosia, blunted affect, and abnormal overeating of inappropriate substances (Weiskrantz, 1956).

Epilepsy and Amygdala

Epilepsy is a common chronic neurological disorder characterized by recurrent and spontaneous seizures (Nelson et al., 2018). The mesial temporal lobe structures, including the amygdala, piriform cortex, and hippocampus, play a significant role in epileptogenesis, particularly in mesial temporal lobe epilepsy (TLE) (Aliashkevich et al., 2003). Both experimental and clinical studies have revealed severe neuropathological changes in the amygdala circuitry in TLE (Pitkänen and Amaral, 1998). TLE patients frequently experience psychiatric problems and emotional disturbances associated with amygdala pathology (Aliashkevich et al., 2003). In patients with epilepsy, the amygdaloid complex exhibits smaller neuronal somata compared to controls. Moreover, these patients show an increased density of dendritic spines per segment in amygdalar neurons. Dendritic changes such as spine bifurcations or focal constrictions are also observed (Farokhian et al., 2017). The alterations in bidirectional connectivity between the amygdala and the hippocampus are thought to play important roles in the epileptogenic network of TLE patients (Guo et al., 2021). Patients with hippocampal sclerosis, a common finding in TLE, exhibit significant reductions in gray and white matter in the mesial temporal lobes (Rhone et al., 2020). Specific regions within the amygdala have been identified as responsible for stimulation-induced apnea in pediatric epilepsy, and imaging studies have revealed abnormalities in connectivity between the temporal lobe and brainstem respiratory sites in these patients (Zikopoulos et al., 2016). Abnormal synaptic function and alterations in receptor densities, particularly GABA and glutamate receptors, have been reported in the basolateral amygdala (BLA) of individuals with epilepsy. These changes may contribute to spontaneous interictal discharges and seizure activity in the amygdala (Aroniadou-Anderjaska et al., 2008). The epileptic amygdala exhibits increased rates of cell injury and apoptosis, as well as downregulation of certain GABAergic receptor subunits (Nelson et al., 2018). Amygdala damage is frequently observed in TLE cases, either in conjunction with damages in other brain regions or as an isolated finding (Lv et al., 2014). Isolated amygdala enlargement has been suggested to be epileptogenic, and it can result from long-lasting inflammatory processes or focal cortical dysplasia (Minami et al., 2015). Histopathologically, clustering hypertrophic neurons associated with

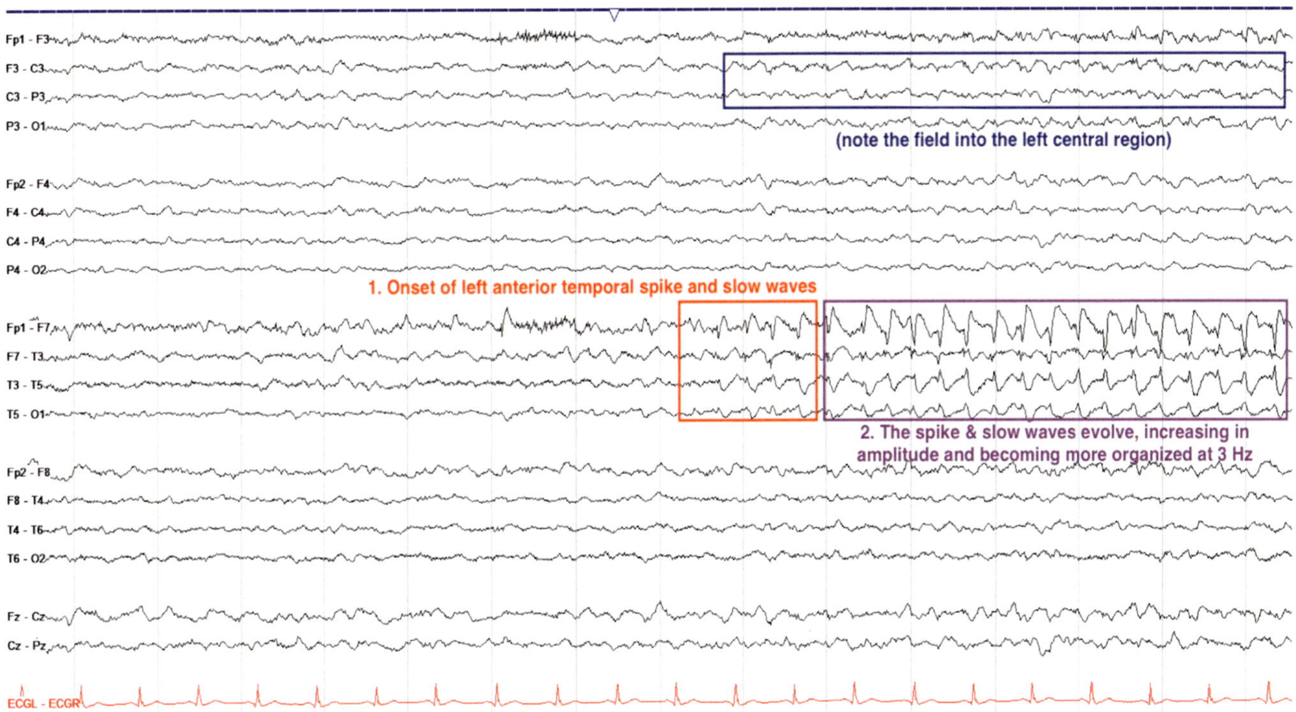

FIGURE 14.10 **EEG of a patient with TLE.** 3Hz spikes and waves localized in the *left* fronto-polar and temporo-occipital regions are easily noticeable. *Image courtesy of Dr. David Valentine https://www.learningeeg.com.*

vacuolation, with or without gliosis, are considered pathological characteristics of amygdala enlargement (Kim et al., 2012). Focal cortical dysplasia is the most frequent pathological finding accompanied by amygdala enlargement in TLE patients, although other conditions such as ganglioglioma, oligodendroglioma, and astrocytoma have also been reported (Fan et al., 2021). Amygdala enlargement in patients with TLE is associated with pathological changes in temporal-insular, temporal-insular-prefrontal, or prefrontal-temporal networks (Makhalova et al., 2022). Detecting amygdala enlargement, even when hidden in neuroimaging evaluations, is crucial in suspected subjects with TLE (Na et al., 2020). The volume change in amygdala enlargement can serve as a prognostic factor in TLE (Kuchukhidze et al., 2022) and may influence treatment outcomes (Chakravarty et al., 2021). Focal atrophy in the basolateral amygdala of the left hemisphere may predict drug responses in patients with medically refractory epilepsy (Li et al., 2021). The activation of emotional neuronal networks involving the amygdala in epilepsy may lead to persistent fear of seizures (Tarrada et al., 2022). Alterations in amygdala-hippocampal networks may also contribute to the pathogenesis of juvenile myoclonic epilepsy, particularly in terms of memory dysfunction (Lee et al., 2017; Jahanbazi Jahan-Abad et al., 2018). Changes in amygdala excitability have been suggested to play a role in the pathogenesis of Dravet syndrome, an epileptic encephalopathy associated with an increased risk of sudden death (Yan et al., 2021). The amygdala may also contribute to epileptic auras in patients with TLE (see Fig. 14.10 below) (Heo et al., 2020).

The periamygdaloid cortex has been extensively implicated in the pathophysiology of focal epilepsy originating from the frontal or temporal lobes. Research conducted on both humans and animals has indicated that the periamygdaloid cortex likely plays a role in amplifying and facilitating human focal epilepsy, and may contribute to the development of intractable epilepsy (Pimontel et al., 2020). Moreover, the anatomical and functional characteristics of the periamygdaloid cortex make it susceptible to epileptic seizures (Pimontel et al., 2020).

THE CLAUSTRUM

The claustrum is a narrow structure composed of neurons and glial cells situated between the external and extreme capsules (Jackson et al., 2020) (Fig. 14.11). It extends rostrocaudally in conjunction with the striatum and dorsoventrally with the insula. Over time, the claustrum has been referred to by various names, including the nucleus taenia formis, insular claustrum, ventral claustrum, endopiriform nucleus, and claustrum

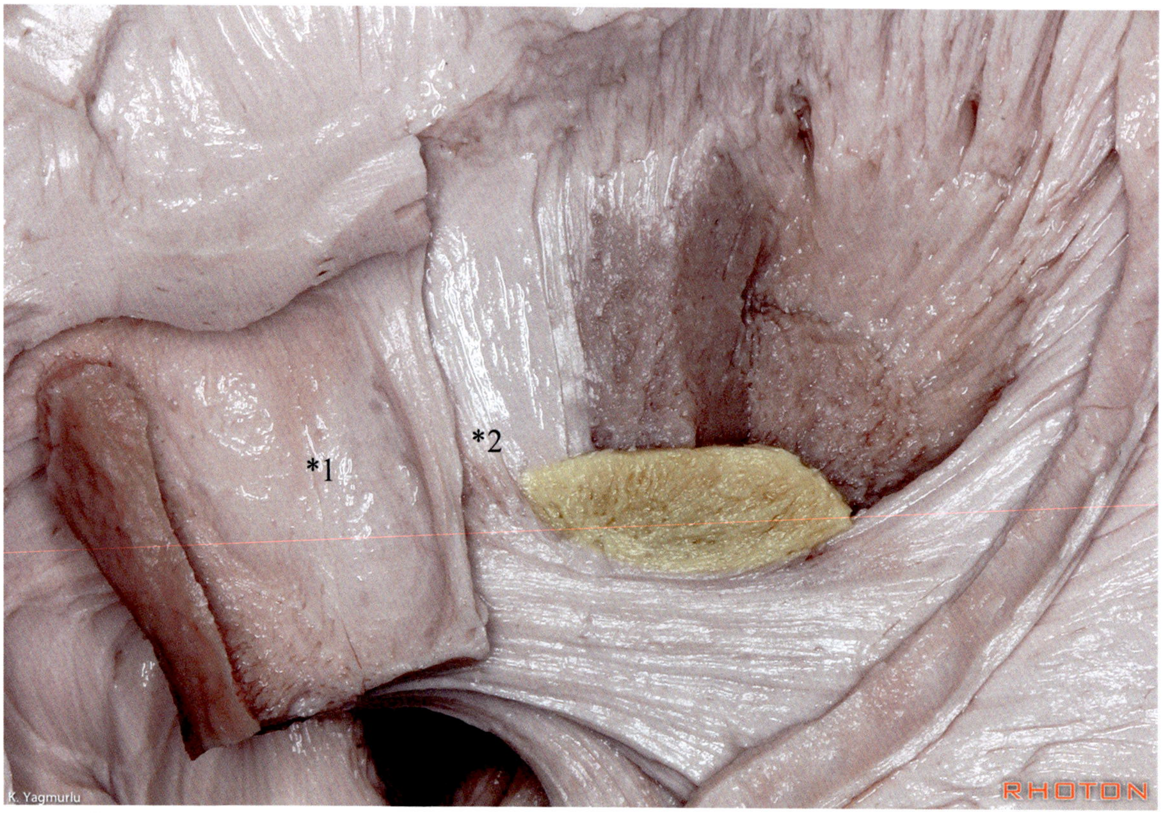

FIGURE 14.11 **The claustrum (*in yellow*) is located between the extreme (*1) and external capsule (*2). Behind the latter is the putamen (*the first gray structure*) and still behind it is the globus pallidus (*the second gray structure*).** *Image courtesy of The Rhoton Collection® http://rhoton.ineurodb.org/. Dissector: K. Yagmurlu. Drawing by Jeffrey Sorenson.*

praefiriforme (Mathur, 2014). However, a comprehensive understanding of its structure and function remains elusive. The claustrum exhibits close connectivity with cortical and subcortical regions, such as the motor, somatosensory, visual, auditory, limbic, associative, and prefrontal cortices, as well as the thalamus, and is implicated in diverse higher-order cognitive processes (Goll et al., 2015). The primary role of the claustrum is thought to involve the integration of inputs from various cortical areas, combining information from touch, color, sound stimuli, and others into unified perceptual experiences rather than isolated outcomes (Benarroch, 2021). Moreover, accumulating evidence suggests that the claustrum is involved in multiple cognitive functions, including learning and memory, sleep, and attention (Barrett et al., 2020). To gain insights into its function, it is crucial to examine the connections of the claustrum, which will be explored by reviewing its inputs and outputs.

AFFERENT AND EFFERENT OF THE CLAUSTRUM

The claustrum exhibits a division into dorsal and ventral regions, with each part further divided into rostral and caudal regions. It establishes connections with various cortical and subcortical structures, and investigations into claustrum connections have predominantly focused on rats, cats, and monkeys. Notably, substantial evidence indicates that a majority of the afferent cortical connections originate from the neocortex (Edelstein and Denaro, 2004). In terms of species differences, robust findings suggest that the dorsal claustrum serves as the primary recipient of inputs from diverse cortices, including the primary motor cortex (BA 4), sensory cortices (BA 3, 1, 2), visual cortex (BA 17), limbic, prelimbic, and paralimbic cortices, as well as the striatum (Nikolenko et al., 2021). Additionally, the ventral claustrum receives projections from the infralimbic cortex, paralimbic cortex, cingulate gyrus, entorhinal cortex, and prepiriform cortex (Smith et al., 2020). In the cat, subcortical afferents arise from the lateral hypothalamus, intrathalamic nuclei, and central nuclei of the thalamus. Reciprocal connections between the lateral intermediate nucleus of the thalamus and the claustrum are also observed in the cat. Conversely, studies in rats have revealed strong projections from the amygdala, specifically from the magnocellular division of the basal nucleus, the accessory basal nucleus, and the lateral division of the amygdalohippocampal area, to both the dorsal and ventral regions of the claustrum (Hoover and Vertes, 2007).

Regarding efferent projections, it is important to note that most connections from the claustrum to cortical areas are bilateral and reciprocal. Furthermore, these efferent connections demonstrate a topographic organization. For example, the most anterior parts of the claustrum project to the prefrontal and motor cortex, while connections to somatosensory, auditory, and visual cortical areas are found in the posterior regions (Jackson et al., 2020). Consequently, two distinct regions can be identified within the claustrum: the anterodorsal portion is associated with the sensorimotor zone, while the posteroventral part corresponds to the visuo-auditory zone. Notably, extensive claustro-cortical connections have been observed in the human brain (Stern, 2017). Consequently, neuronal loss in the central and posterior parts of the claustrum can lead to pathological alterations in the parietal and occipital cortices, suggesting a broad linkage between the human claustrum and the cerebral neocortex, with connections organized topographically similar to those seen in experimental animals.

FUNCTIONS OF THE CLAUSTRUM

Top-down control of attention across modalities (Goll et al., 2015), inhibitory control of the prefrontal cortex (PFC) (Jackson et al., 2018), allocation of attention to reward-related contextual cues (Terem et al., 2020), salience processing (Smith et al., 2019), and regulation of impulsivity (Liu et al., 2019) are among the functions associated with the claustrum. The claustrum is capable of providing robust and persistent feed-forward inhibition to the PFC through interneurons containing neuropeptide Y, effectively suppressing ongoing neural activity (Jackson et al., 2018). This unique circuit organization of the claustrum enables it to exert powerful and rapid control over cortical networks, suggesting its involvement in the regulation of cognitive processes in prefrontal circuits (Jackson et al., 2018). Experimental evidence also indicates that the claustrum plays essential roles in sensorimotor integration (Smith et al., 2012), novelty coding (Kitanishi and Matsuo, 2017), and the modulation of stress-induced anxiety-related behaviors (Seiriki et al., 2017). Furthermore, the claustrum contributes to neocortical activation during wakefulness and hypersomnia in paradoxical sleep (Maciel et al., 2021). Bilateral stimulation of the central core of the claustrum in humans has been shown to induce various sensory symptoms and reflexive movements without altering consciousness or awareness (Bickel and Parvizi, 2019).

CLAUSTRUM AND NEUROLOGICAL DISORDERS

While the claustrum is not specifically associated with any primary disease, it has been implicated in several neurological disorders. Experimental studies have suggested its involvement in focal seizures with altered awareness, as well as motor and cognitive disorders, and encephalitis (Benarroch, 2021). In a cohort study of 34 patients with autoimmune limbic encephalitis and/or epilepsy, imaging investigations revealed bilateral claustrum hyperintensity in four patients. Neuropathological examination of a patient who died during status epilepticus showed marked bilateral astrocytic reaction in the hippocampus and claustrum (Nixon et al., 2001). Abnormal MRI signals have also been observed in the claustrum following acute and chronic phases of repetitive febrile seizures (Yang et al., 2021). In Alzheimer's disease (AD), areas of the claustrum connected with the entorhinal cortex are severely affected, and neurodegeneration and amyloid pathology have been reported in the claustrum of AD patients (Morys et al., 1996). Reduced levels of dopamine and noradrenaline have been observed in the claustrum of patients with Parkinson's disease (PD), indicating its involvement in the motor and nonmotor symptoms of PD (Sitte et al., 2017). PD with dementia has shown evidence of alpha-synuclein and amyloid-beta pathology in the claustrum, and a strong correlation between claustrum pathology and visual hallucinations has been observed in PD patients. Additionally, smaller volume and delayed development of the claustrum have been reported in children with autism, and bilaterally reduced volume of the claustrum has been observed in schizophrenic patients experiencing delusions (Bernstein et al., 2016; Smith et al., 2020). Disruption of binding or synchronization of cognitive, sensory, and motor information in the claustrum has been suggested to contribute to delusions and hallucinations in schizophrenia (Cascella and Sawa, 2014). The presence of a bright claustrum in brain MRI has been proposed as a biomarker for the diagnosis of Wilson's disease (Jacobs et al., 2003). Increased functional connectivity between the amygdala and various brain regions, including the claustrum, has been identified in patients with narcolepsy type 1 (Ballotta et al., 2021). Morphological alterations of the claustrum have also been observed in metabolic diseases such as Niemann–Pick disease, and MRI signal changes within the claustrum and external capsule have been reported in SARS-CoV-2 encephalitis (Mukherjee et al., 2021; Zuhorn et al., 2021).

References

Adebimpe, A., Bassett, D.S., Jamieson, P.E., Romer, D., 2019. Intersubject synchronization of late adolescent brain responses to violent movies: a virtue-ethics approach. Front. Behav. Neurosci. 13. https://doi.org/10.3389/fnbeh.2019.00260.

Adolphs, R., Tranel, D., Damasio, A.R., 1998. The human amygdala in social judgment. Nature 393 (6684), 470–474. https://doi.org/10.1038/30982.

Alheid, G.F., Heimer, L., 1988. New perspectives in basal forebrain organization of special relevance for neuropsychiatric disorders: the striatopallidal, amygdaloid, and corticopetal components of substantia innominata. Neuroscience 27 (1), 1–39. https://doi.org/10.1016/0306-4522(88)90217-5.

Aliashkevich, A.F., Yilmazer-Hanke, D., Van Roost, D., Mundhenk, B., Schramm, J., Blümcke, I., 2003. Cellular pathology of amygdala neurons in human temporal lobe epilepsy. Acta Neuropathol. 106 (2), 99–106. https://doi.org/10.1007/s00401-003-0707-0.

Allison, T., Puce, A., McCarthy, G., 2000. Social perception from visual cues: role of the STS region. Trends Cogn. Sci. 4 (7), 267–278. https://doi.org/10.1016/S1364-6613(00)01501-1.

Amaral, D.G., Insausti, R., 1992. Retrograde transport of D-[3H]-aspartate injected into the monkey amygdaloid complex. Exp. Brain Res. 88 (2), 375–388. https://doi.org/10.1007/BF02259113.

Amaral, D.G., Price, J.L., Pitkäen, A., 1992. Anatomical organisation of the primate amygdaloid complex. In: Aggleton, J.P. (Ed.), The Amyg- Dala: Neurobiological Aspects of Emotion, Memory and Mental Dysfunction. Wiley-Liss, pp. 1–66.

Aroniadou-Anderjaska, V., Fritsch, B., Qashu, F., Braga, M.F.M., 2008. Pathology and pathophysiology of the amygdala in epileptogenesis and epilepsy. Epilepsy Res. 78 (2–3), 102–116. https://doi.org/10.1016/j.eplepsyres.2007.11.011.

Asami, T., Nakamura, R., Takaishi, M., et al., 2018. Smaller volumes in the lateral and basal nuclei of the amygdala in patients with panic disorder. PLoS One 13 (11), e0207163. https://doi.org/10.1371/journal.pone.0207163.

Avecillas-Chasin, J.M., Levinson, S., Kuhn, T., et al., 2023. Connectivity-based parcellation of the amygdala and identification of its main white matter connections. Sci. Rep. 13 (1), 1305. https://doi.org/10.1038/s41598-023-28100-6.

Ballotta, D., Talami, F., Pizza, F., et al., 2021. Hypothalamus and amygdala functional connectivity at rest in narcolepsy type 1. Neuroimage Clin 31, 102748. https://doi.org/10.1016/j.nicl.2021.102748.

Barrett, F.S., Krimmel, S.R., Griffiths, R.R., Seminowicz, D.A., Mathur, B.N., 2020. Psilocybin acutely alters the functional connectivity of the claustrum with brain networks that support perception, memory, and attention. Neuroimage 218, 116980. https://doi.org/10.1016/j.neuroimage.2020.116980.

Behan, M., Haberly, L.B., 1999. Intrinsic and efferent connections of the endopiriform nucleus in rat. J. Comp. Neurol. 408 (4), 532–548.

Behbehani, M.M., 1995. Functional characteristics of the midbrain periaqueductal gray. Prog. Neurobiol. 46 (6), 575–605. https://doi.org/10.1016/0301-0082(95)00009-K.

Benarroch, E.E., 2021. What is the role of the claustrum in cortical function and neurologic disease? Neurology 96 (3), 110–113. https://doi.org/10.1212/WNL.0000000000011280.

Bernard, J.F., Alden, M., Besson, J.M., 1993. The organization of the efferent projections from the pontine parabrachial area to the amygdaloid complex: a Phaseolus vulgaris leucoagglutinin (PHA-L) study in the rat. J. Comp. Neurol. 329 (2), 201–229. https://doi.org/10.1002/cne.903290205.

Bernard, J.F., Peschanski, M., Besson, J.M., 1989. A possible spino (trigemino)-ponto-amygdaloid pathway for pain. Neurosci. Lett. 100 (1–3), 83–88. https://doi.org/10.1016/0304-3940(89)90664-2.

Bernstein, H.G., Ortmann, A., Dobrowolny, H., et al., 2016. Bilaterally reduced claustral volumes in schizophrenia and major depressive disorder: a morphometric postmortem study. Eur. Arch. Psychiatry Clin. Neurosci. 266 (1), 25–33. https://doi.org/10.1007/s00406-015-0597-x.

Bickart, K.C., Wright, C.I., Dautoff, R.J., Dickerson, B.C., Barrett, L.F., 2011. Amygdala volume and social network size in humans. Nat. Neurosci. 14 (2), 163–164. https://doi.org/10.1038/nn.2724.

Bickel, S., Parvizi, J., 2019. Electrical stimulation of the human claustrum. Epilepsy Behav. 97, 296–303. https://doi.org/10.1016/j.yebeh.2019.03.051.

Bogerts, B., Schöne, M., Breitschuh, S., 2018. Brain alterations potentially associated with aggression and terrorism. CNS Spectr. 23 (2), 129–140. https://doi.org/10.1017/S1092852917000463.

Bordi, F., LeDoux, J.E., 1994. Response properties of single units in areas of rat auditory thalamus that project to the amygdala. Exp. Brain Res. 98 (2), 275–286. https://doi.org/10.1007/BF00228415.

Cannon, W.B., 1927. The James-Lange theory of emotions: a critical examination and an alternative theory. Am. J. Psychol. 39 (1/4), 106. https://doi.org/10.2307/1415404.

Canteras, N.S., Swanson, L.W., 1992. Projections of the ventral subiculum to the amygdala, septum, and hypothalamus: a PHAL anterograde tract-tracing study in the rat. J. Comp. Neurol. 324 (2), 180–194. https://doi.org/10.1002/cne.903240204.

Carlsen, J., 1988. Immunocytochemical localization of glutamate decarboxylase in the rat basolateral amygdaloid nucleus, with special reference to GABAergic innervation of amygdalostriatal projection neurons. J. Comp. Neurol. 273 (4), 513–526. https://doi.org/10.1002/cne.902730407.

Cascella, N.G., Sawa, A., 2014. The claustrum in schizophrenia. In: The Claustrum. Elsevier, pp. 237–243. https://doi.org/10.1016/B978-0-12-404566-8.00009-X.

Cassell, M.D., Gray, T.S., 1989. Morphology of peptide-immunoreactive neurons in the rat central nucleus of the amygdala. J. Comp. Neurol. 281 (2), 320–333. https://doi.org/10.1002/cne.902810212.

Cassell, M.D., Gray, T.S., Kiss, J.Z., 1986. Neuronal architecture in the rat central nucleus of the amygdala: a cytological, hodological, and immunocytochemical study. J. Comp. Neurol. 246 (4), 478–499. https://doi.org/10.1002/cne.902460406.

Chakravarty, K., Ray, S., Kharbanda, P.S., Lal, V., Baishya, J., 2021. Temporal lobe epilepsy with amygdala enlargement: a systematic review. Acta Neurol. Scand. 144 (3), 236–250. https://doi.org/10.1111/ane.13455.

Chapman, P.F., Kairiss, E.W., Keenan, C.L., Brown, T.H., 1990. Long-term synaptic potentiation in the amygdala. Synapse 6 (3), 271–278. https://doi.org/10.1002/syn.890060306.

Chapman, W.P., Schroeder, H.R., Geyer, G., et al., 1954. Physiological evidence concerning importance of the amygdaloid nuclear region in the integration of circulatory function and emotion in man. Science (1979) 120 (3127), 949–950. https://doi.org/10.1126/science.120.3127.949.

Davis, M., Whalen, P.J., 2001. The amygdala: vigilance and emotion. Mol. Psychiatry 6 (1), 13–34. https://doi.org/10.1038/sj.mp.4000812.

Dayas, C.V., Buller, K.M., Day, T.A., 1999. Neuroendocrine responses to an emotional stressor: evidence for involvement of the medial but not the central amygdala. Eur. J. Neurosci. 11 (7), 2312–2322. https://doi.org/10.1046/j.1460-9568.1999.00645.x.

Delaney, A.J., Sah, P., 2001. Pathway-specific targeting of GABA a receptor subtypes to somatic and dendritic synapses in the central amygdala. J. Neurophysiol. 86 (2), 717–723. https://doi.org/10.1152/jn.2001.86.2.717.

de Olmos, J.S., Beltramino, C.A., Alheid, G., 2004. Amygdala and extended amygdala of the rat: a cytoarchitectonical, fibroarchitectonical, and chemoarchitectonical survey. In: Paxinos, G. (Ed.), The Rat Nervous System. Elsevier.

Dong, H.W., Petrovich, G.D., Swanson, L.W., 2001. Topography of projections from amygdala to bed nuclei of the stria terminalis. Brain

Res. Rev. 38 (1–2), 192–246. https://doi.org/10.1016/S0165-0173(01)00079-0.
Dunn, L.T., Everitt, B.J., 1988. Double dissociations of the effects of amygdala and insular cortex lesions on conditioned taste aversion, passive avoidance, and neophobia in the rat using the excitotoxin ibotenic acid. Behav. Neurosci. 102 (1), 3–23. https://doi.org/10.1037/0735-7044.102.1.3.
Duvarci, S., Paré, D., 2007. Glucocorticoids enhance the excitability of principal basolateral amygdala neurons. J. Neurosci. 27 (16), 4482–4491. https://doi.org/10.1523/JNEUROSCI.0680-07.2007.
Edelstein, L.R., Denaro, F.J., 2004. The claustrum: a historical review of its anatomy, physiology, cytochemistry and functional significance. Cell. Mol. Biol. (Noisy-le-Grand) 50 (6), 675–702.
Faber, E.S.L., Callister, R.J., Sah, P., 2001. Morphological and electrophysiological properties of principal neurons in the rat lateral amygdala in vitro. J. Neurophysiol. 85 (2), 714–723. https://doi.org/10.1152/jn.2001.85.2.714.
Faber, E.S.L., Sah, P., 2002. Physiological role of calcium-activated potassium currents in the rat lateral amygdala. J. Neurosci. 22 (5), 1618–1628. https://doi.org/10.1523/JNEUROSCI.22-05-01618.2002.
Fan, Z., Sun, B., qin, L.L., et al., 2021. Diagnosis and surgical treatment of non-lesional temporal lobe epilepsy with unilateral amygdala enlargement. Neurol. Sci. 42 (6), 2353–2361. https://doi.org/10.1007/s10072-020-04794-8.
Farokhian, F., Beheshti, I., Sone, D., Matsuda, H., 2017. Comparing CAT12 and VBM8 for detecting brain morphological abnormalities in temporal lobe epilepsy. Front. Neurol. 8. https://doi.org/10.3389/fneur.2017.00428.
Faulkner, B., Brown, T.H., 1999. Morphology and physiology of neurons in the rat perirhinal-lateral amygdala area. J. Comp. Neurol. 411 (4), 613–642.
Felix-Ortiz, A.C., Tye, K.M., 2014. Amygdala inputs to the ventral hippocampus bidirectionally modulate social behavior. J. Neurosci. 34 (2), 586–595. https://doi.org/10.1523/JNEUROSCI.4257-13.2014.
Fonzo, G.A., Etkin, A., 2017. Affective neuroimaging in generalized anxiety disorder: an integrated review. Dialogues Clin. Neurosci. 19 (2), 169–179. https://doi.org/10.31887/DCNS.2017.19.2/gfonzo.
Freiwald, W., Duchaine, B., Yovel, G., 2016. Face processing systems: from neurons to real-world social perception. Annu. Rev. Neurosci. 39 (1), 325–346. https://doi.org/10.1146/annurev-neuro-070815-013934.
Funahashi, M., Matsuo, R., Stewart, M., 2000. Propagation of synchronous burst discharges from entorhinal cortex to morphologically and electrophysiologically identified neurons of rat lateral amygdala. Brain Res. 884 (1–2), 104–115. https://doi.org/10.1016/S0006-8993(00)02854-7.
Gallagher, M., Graham, P., Holland, P., 1990. The amygdala central nucleus and appetitive Pavlovian conditioning: lesions impair one class of conditioned behavior. J. Neurosci. 10 (6), 1906–1911. https://doi.org/10.1523/JNEUROSCI.10-06-01906.1990.
Gangopadhyay, P., Chawla, M., Dal Monte, O., Chang, S.W.C., 2021. Prefrontal–amygdala circuits in social decision-making. Nat. Neurosci. 24 (1), 5–18. https://doi.org/10.1038/s41593-020-00738-9.
Gauriau, C., Bernard, J.F., 2002. Pain pathways and parabrachial circuits in the rat. Exp. Physiol. 87 (2), 251–258. https://doi.org/10.1113/eph8702357.
Gloor, P., 1990. Experiential phenomena of temporal lobe epilepsy. Brain 113 (6), 1673–1694. https://doi.org/10.1093/brain/113.6.1673.
Goll, Y., Atlan, G., Citri, A., 2015. Attention: the claustrum. Trends Neurosci. 38 (8), 486–495. https://doi.org/10.1016/j.tins.2015.05.006.
Gottschalk, M.G., Domschke, K., 2017. Oxytocin and anxiety disorders. Curr. Top. Behav. Neurosci. 467–498. https://doi.org/10.1007/7854_2017_25.
Guo, Z., Zhao, B., Hu, W., et al., 2021. Effective connectivity among the hippocampus, amygdala, and temporal neocortex in epilepsy patients: a cortico-cortical evoked potential study. Epilepsy Behav. 115, 107661. https://doi.org/10.1016/j.yebeh.2020.107661.
Hall, E., 1972. The amygdala of the cat: a Golgi study. Zeitschrift für Zellforschung und Mikroskopische Anatomie 134 (4), 439–458. https://doi.org/10.1007/BF00307668.
Haller, J., 2018. The role of central and medial amygdala in normal and abnormal aggression: a review of classical approaches. Neurosci. Biobehav. Rev. 85, 34–43. https://doi.org/10.1016/j.neubiorev.2017.09.017.
Hampton, W.H., Unger, A., Von Der Heide, R.J., Olson, I.R., 2016. Neural connections foster social connections: a diffusion-weighted imaging study of social networks. Soc. Cogn. Affect Neurosci. 11 (5), 721–727. https://doi.org/10.1093/scan/nsv153.
Heo, K., Kim, K.M., Han, S.M., Cho, K.H., Chu, M.K., 2020. Nasal pain as an aura: amygdala origin? Seizure 83, 13–16. https://doi.org/10.1016/j.seizure.2020.09.028.
Hoover, W.B., Vertes, R.P., 2007. Anatomical analysis of afferent projections to the medial prefrontal cortex in the rat. Brain Struct. Funct. 212 (2),149–179. https://doi.org/10.1007/s00429-007-0150-4 Epub 2007 Jul 27. PMID: 17717690.
Jackson, J., Karnani, M.M., Zemelman, B.V., Burdakov, D., Lee, A.K., 2018. Inhibitory control of prefrontal cortex by the claustrum. Neuron 99 (5), 1029–1039.e4. https://doi.org/10.1016/j.neuron.2018.07.031.
Jackson, J., Smith, J.B., Lee, A.K., 2020. The anatomy and physiology of claustrum-cortex interactions. Annu. Rev. Neurosci. 43 (1), 231–247. https://doi.org/10.1146/annurev-neuro-092519-101637.
Jacobs, D.A., Markowitz, C.E., Liebeskind, D.S., Galetta, S.L., 2003. The "double panda sign" in Wilson's disease. Neurology 61 (7), 969. https://doi.org/10.1212/01.WNL.0000085871.98174.4E.
Jahanbazi Jahan-Abad, A., Sahab Negah, S., Hosseini Ravandi, H., et al., 2018. Human neural stem/progenitor cells derived from epileptic human brain in a self-assembling peptide nanoscaffold improve traumatic brain injury in rats. Mol. Neurobiol. 55 (12), 9122–9138. https://doi.org/10.1007/s12035-018-1050-8.
James, W., 1890. The Principles of Psychology, vol. I. Henry Holt and Co. https://doi.org/10.1037/10538-000.
James, W., 1884. What is an emotion? Mind 9 (34), 188–205.
Jolkkonen, E., Pitkänen, A., 1998. Intrinsic connections of the rat amygdaloid complex: projections originating in the central nucleus. J. Comp. Neurol. 395 (1), 53–72. https://doi.org/10.1002/(SICI)1096-9861(19980525)395:1<53::AID-CNE5>3.0.CO;2-G.
Kaada, B.R., 1951. Somato-motor, autonomic and electrocorticographic responses to electrical stimulation of rhinencephalic and other structures in primates, cat, and dog; a study of responses from the limbic, subcallosal, orbito-insular, piriform and temporal cortex, hippocampus-fornix and amygdala. Acta Physiol. Scand. Suppl. 24 (83), 1–262.
Kapp, B.S., Gallagher, M., Underwood, M.D., McNall, C.L., Whitehorn, D., 1982. Cardiovascular responses elicited by electrical stimulation of the amygdala central nucleus in the rabbit. Brain Res. 234 (2), 251–262. https://doi.org/10.1016/0006-8993(82)90866-6.
Kapp, B.S., Whalen, P.J., Supple, W.F., Pascoe, J.P., 1992. Amygdaloid contributions to conditioned arousal and sensory information processing. In: Aggleton, J.P. (Ed.), The Amygdala: Neurobiological Aspects of Emotion, Memory, and Mental Dysfunction. Wiley-Liss, pp. 229–254.
Kim, D.W., Lee, S.K., Chung, C.K., Koh, Y.C., Choe, G., Lim, S.D., 2012. Clinical features and pathological characteristics of amygdala enlargement in mesial temporal lobe epilepsy. J. Clin. Neurosci. 19 (4), 509–512. https://doi.org/10.1016/j.jocn.2011.05.042.
Kitanishi, T., Matsuo, N., 2017. Organization of the claustrum-to-entorhinal cortical connection in mice. J. Neurosci. 37 (2), 269–280. https://doi.org/10.1523/JNEUROSCI.1360-16.2016.

Klüver, H., Bucy, P.C., 1997. Preliminary analysis of functions of the temporal lobes in monkeys. 1939 [classical article]. J. Neuropsychiatry Clin. Neurosci. 9 (4), 606–620. https://doi.org/10.1176/jnp.9.4.606.

Kolesar, T.A., Bilevicius, E., Wilson, A.D., Kornelsen, J., 2019. Systematic review and meta-analyses of neural structural and functional differences in generalized anxiety disorder and healthy controls using magnetic resonance imaging. Neuroimage Clin 24, 102016. https://doi.org/10.1016/j.nicl.2019.102016.

Krettek, J.E., Price, J.L., 1978. A description of the amygdaloid complex in the rat and cat with observations on intra-amygdaloid axonal connections. J. Comp. Neurol. 178 (2), 255–279. https://doi.org/10.1002/cne.901780205.

Kuchukhidze, G., Unterberger, I., Schmid, E., et al., 2022. Emotional recognition in patients with mesial temporal epilepsy associated with enlarged amygdala. Front. Neurol. 12. https://doi.org/10.3389/fneur.2021.803787.

Kunimatsu, A., Yasaka, K., Akai, H., Kunimatsu, N., Abe, O., 2020. MRI findings in posttraumatic stress disorder. J. Magn. Reson. Imag. 52 (2), 380–396. https://doi.org/10.1002/jmri.26929.

Lang, E.J., Paré, D., 1998. Synaptic responsiveness of interneurons of the cat lateral amygdaloid nucleus. Neuroscience 83 (3), 877–889. https://doi.org/10.1016/S0306-4522(97)00420-X.

Lange, C.G., 1887. The mechanisms of emotion. In: Rand, B. (Ed.), The Classical Psychologists. Houghton Mifflin, pp. 672–684.

LeDoux, J., Farb, C., Ruggiero, D., 1990. Topographic organization of neurons in the acoustic thalamus that project to the amygdala. J. Neurosci. 10 (4), 1043–1054. https://doi.org/10.1523/JNEUROSCI.10-04-01043.1990.

LeDoux, J., Iwata, J., Cicchetti, P., Reis, D., 1988. Different projections of the central amygdaloid nucleus mediate autonomic and behavioral correlates of conditioned fear. J. Neurosci. 8 (7), 2517–2529. https://doi.org/10.1523/JNEUROSCI.08-07-02517.1988.

LeDoux, J.E., 1996. The Emotional Brain: The Mysterious Underpinnings of Emotional Life. Simon & Schuster.

LeDoux, J.E., Farb, C.R., Romanski, L.M., 1991. Overlapping projections to the amygdala and striatum from auditory processing areas of the thalamus and cortex. Neurosci. Lett. 134 (1), 139–144. https://doi.org/10.1016/0304-3940(91)90526-Y.

Ledoux, J.E., Ruggiero, D.A., Forest, R., Stornetta, R., Reis, D.J., 1987. Topographic organization of convergent projections to the thalamus from the inferior colliculus and spinal cord in the rat. J. Comp. Neurol. 264 (1), 123–146. https://doi.org/10.1002/cne.902640110.

Lee, S.C., Amir, A., Haufler, D., Pare, D., 2017. Differential recruitment of competing valence-related amygdala networks during anxiety. Neuron 96 (1), 81–88.e5. https://doi.org/10.1016/j.neuron.2017.09.002.

Lehmann, H., Treit, D., Parent, M.B., 2000. Amygdala lesions do not impair shock-probe avoidance retention performance. Behav. Neurosci. 114 (1), 107–116. https://doi.org/10.1037/0735-7044.114.1.107.

Li, X., Wang, Z., Chen, Q., et al., 2021. Atrophy in the left amygdala predicted drug responses in idiopathic generalized epilepsy patients with tonic–clonic seizures. Front. Neurosci. 15. https://doi.org/10.3389/fnins.2021.640016.

Linke, R., Braune, G., Schwegler, H., 2000. Differential projection of the posterior paralaminar thalamic nuclei to the amygdaloid complex in the rat. Exp. Brain Res. 134 (4), 520–532. https://doi.org/10.1007/s002210000475.

Liu, J., Wu, R., Johnson, B., Vu, J., Bass, C., Li, J.X., 2019. The claustrum-prefrontal cortex pathway regulates impulsive-like behavior. J. Neurosci. 39 (50), 10071–10080. https://doi.org/10.1523/JNEUROSCI.1005-19.2019.

Luskin, M.B., Price, J.L., 1983. The topographic organization of associational fibers of the olfactory system in the rat, including centrifugal fibers to the olfactory bulb. J. Comp. Neurol. 216 (3), 264–291. https://doi.org/10.1002/cne.902160305.

Lv, R.J., Sun, Z.R., Cui, T., Guan, H.Z., Ren, H.T., Shao, X.Q., 2014. Temporal lobe epilepsy with amygdala enlargement: a subtype of temporal lobe epilepsy. BMC Neurol. 14 (1), 194. https://doi.org/10.1186/s12883-014-0194-z.

Maciel, R., Yamazaki, R., Wang, D., et al., 2021. Is REM sleep a paradoxical state?: different neurons are activated in the cingulate cortices and the claustrum during wakefulness and paradoxical sleep hypersomnia. Biochem. Pharmacol. 191, 114514. https://doi.org/10.1016/j.bcp.2021.114514.

MacLean, P.D., 1949. Psychosomatic disease and the "visceral brain.". Psychosom. Med. 11 (6), 338–353. https://doi.org/10.1097/00006842-194911000-00003.

Madonna, D., Delvecchio, G., Soares, J.C., Brambilla, P., 2019. Structural and functional neuroimaging studies in generalized anxiety disorder: a systematic review. Braz. J. Psychiatry 41 (4), 336–362. https://doi.org/10.1590/1516-4446-2018-0108.

Mahanty, N.K., Sah, P., 1998. Calcium-permeable AMPA receptors mediate long-term potentiation in interneurons in the amygdala. Nature 394 (6694), 683–687. https://doi.org/10.1038/29312.

Makhalova, J., Le Troter, A., Aubert-Conil, S., et al., 2022. Epileptogenic networks in drug-resistant epilepsy with amygdala enlargement: assessment with stereo-EEG and 7 T MRI. Clin. Neurophysiol. 133, 94–103. https://doi.org/10.1016/j.clinph.2021.10.012.

Malikowska-Racia, N., Salat, K., 2019. Recent advances in the neurobiology of posttraumatic stress disorder: a review of possible mechanisms underlying an effective pharmacotherapy. Pharmacol. Res. 142, 30–49. https://doi.org/10.1016/j.phrs.2019.02.001.

Martina, M., Royer, S., Paré, D., 1999. Physiological properties of central medial and central lateral amygdala neurons. J. Neurophysiol. 82 (4), 1843–1854. https://doi.org/10.1152/jn.1999.82.4.1843.

Mascagni, F., McDonald, A.J., Coleman, J.R., 1993. Corticoamygdaloid and corticocortical projections of the rat temporal cortex: a Phaseolus vulgaris leucoagglutinin study. Neuroscience 57 (3), 697–715. https://doi.org/10.1016/0306-4522(93)90016-9.

Mathur, B.N., 2014. The claustrum in review. Front. Syst. Neurosci. 8. https://doi.org/10.3389/fnsys.2014.00048.

McDonald, A.J., 1998. Cortical pathways to the mammalian amygdala. Prog. Neurobiol. 55 (3), 257–332. https://doi.org/10.1016/S0301-0082(98)00003-3.

McDonald, A.J., 1982. Cytoarchitecture of the central amygdaloid nucleus of the rat. J. Comp. Neurol. 208 (4), 401–418. https://doi.org/10.1002/cne.902080409.

McDonald, A.J., 1983. Neurons of the bed nucleus of the stria terminalis: a Golgi study in the rat. Brain Res. Bull. 10 (1), 111–120. https://doi.org/10.1016/0361-9230(83)90082-5.

McDonald, A.J., 1991. Topographical organization of amygdaloid projections to the caudatoputamen, nucleus accumbens, and related striatal-like areas of the rat brain. Neuroscience 44 (1), 15–33. https://doi.org/10.1016/0306-4522(91)90248-M.

McDonald, A.J., Augustine, J.R., 1993. Localization of GABA-like immunoreactivity in the monkey amygdala. Neuroscience 52 (2), 281–294. https://doi.org/10.1016/0306-4522(93)90156-A.

McDonald, A.J., Jackson, T.R., 1987. Amygdaloid connections with posterior insular and temporal cortical areas in the rat. J. Comp. Neurol. 262 (1), 59–77. https://doi.org/10.1002/cne.902620106.

McDonald, A.J., Mascagni, F., 1997. Projections of the lateral entorhinal cortex to the amygdala: a Phaseolus vulgaris leucoagglutinin study in the rat. Neuroscience 77 (2), 445–459. https://doi.org/10.1016/S0306-4522(96)00478-2.

McDonald, A.J., Mascagni, F., Guo, L., 1996. Projections of the medial and lateral prefrontal cortices to the amygdala: a Phaseolus vulgaris leucoagglutinin study in the rat. Neuroscience 71 (1), 55–75. https://doi.org/10.1016/0306-4522(95)00417-3.

McGaugh, J.L., 2000. Amygdala: role in modulation of memory storage. In: Aggleton, J.P. (Ed.), The Amygdala: A Functional Analysis. Oxford University Press, pp. 391–424.
McGaugh, J.L., Cahill, L., Roozendaal, B., 1996. Involvement of the amygdala in memory storage: interaction with other brain systems. Proc. Natl. Acad. Sci. USA 93 (24), 13508–13514. https://doi.org/10.1073/pnas.93.24.13508.
Millhouse, O.E., 1986. The intercalated cells of the amygdala. J. Comp. Neurol. 247 (2), 246–271. https://doi.org/10.1002/cne.902470209.
Milner, B., Squire, L.R., Kandel, E.R., 1998. Cognitive neuroscience and the study of memory. Neuron 20 (3), 445–468. https://doi.org/10.1016/S0896-6273(00)80987-3.
Minami, N., Morino, M., Uda, T., et al., 2015. Surgery for amygdala enlargement with mesial temporal lobe epilepsy: pathological findings and seizure outcome. J. Neurol. Neurosurg. Psychiatry 86 (8), 887–894. https://doi.org/10.1136/jnnp-2014-308383.
Moga, M.M., Gray, T.S., 1985. Evidence for corticotropin-releasing factor, neurotensin, and somatostatin in the neural pathway from the central nucleus of the amygdala to the parabrachial nucleus. J. Comp. Neurol. 241 (3), 275–284. https://doi.org/10.1002/cne.902410304.
Moga, M.M., Herbert, H., Hurley, K.M., Yasui, Y., Gray, T.S., Saper, C.B., 1990. Organization of cortical, basal forebrain, and hypothalamic afferents to the parabrachial nucleus in the rat. J. Comp. Neurol. 295 (4), 624–661. https://doi.org/10.1002/cne.902950408.
Molosh, A.I., Dustrude, E.T., Lukkes, J.L., et al., 2020. Panic results in unique molecular and network changes in the amygdala that facilitate fear responses. Mol. Psychiatry 25 (2), 442–460. https://doi.org/10.1038/s41380-018-0119-0.
Morris, R., Frey, S., Kasambira, T., Petrides, M., 1999. Ibotenic acid lesions of the basolateral, but not the central, amygdala interfere with conditioned taste aversion: evidence from a combined behavioral and anatomical tract-tracing investigation. Behav. Neurosci. 113 (2), 291–302. https://doi.org/10.1037/0735-7044.113.2.291.
Morys, J., Bobinski, M., Wegiel, J., Wisniewski, H.M., Narkiewicz, O., 1996. Alzheimer's disease severely affects areas of the claustrum connected with the entorhinal cortex. J. Hirnforsch. 37 (2), 173–180.
Mukherjee, D., Dubey, S., Ganguly, G., Pandit, A., 2021. Claustrum hyperintensity: a rare radiological correlate in Niemann-Pick disease. BMJ Case Rep. 14 (1), e239630. https://doi.org/10.1136/bcr-2020-239630.
Na, H.K., Lee, H., Hong, S., et al., 2020. Volume change in amygdala enlargement as a prognostic factor in patients with temporal lobe epilepsy: a longitudinal study. Epilepsia 61 (1), 70–80. https://doi.org/10.1111/epi.16400.
Nachman, M., Ashe, J.H., 1974. Effects of basolateral amygdala lesions on neophobia, learned taste aversions, and sodium appetite in rats. J. Comp. Physiol. Psychol. 87 (4), 622–643. https://doi.org/10.1037/h0036973.
Nakashima, M., Uemura, M., Yasui, K., Ozaki, H.S., Tabata, S., Taen, A., 2000. An anterograde and retrograde tract-tracing study on the projections from the thalamic gustatory area in the rat: distribution of neurons projecting to the insular cortex and amygdaloid complex. Neurosci. Res. 36 (4), 297–309. https://doi.org/10.1016/S0168-0102(99)00129-7.
Nelson, P.T., Abner, E.L., Patel, E., et al., 2018. The amygdala as a locus of pathologic misfolding in neurodegenerative diseases. J. Neuropathol. Exp. Neurol. 77 (1), 2–20. https://doi.org/10.1093/jnen/nlx099.
Nikolenko, V.N., Rizaeva, N.A., Beeraka, N.M., et al., 2021. The mystery of claustral neural circuits and recent updates on its role in neurodegenerative pathology. Behav. Brain Funct. 17 (1), 8. https://doi.org/10.1186/s12993-021-00181-1.
Nitecka, L., Ben-Ari, Y., 1987. Distribution of GABA-like immunoreactivity in the rat amygdaloid complex. J. Comp. Neurol. 266 (1), 45–55. https://doi.org/10.1002/cne.902660105.
Nixon, J., Bateman, D., Moss, T., 2001. An MRI and neuropathological study of a case of fatal status epilepticus. Seizure 10 (8), 588–591. https://doi.org/10.1053/seiz.2001.0553.
Ottersen, O.P., 1981. Afferent connections to the amygdaloid complex of the rat with some observations in the cat. III. Afferents from the lower brain stem. J. Comp. Neurol. 202 (3), 335–356. https://doi.org/10.1002/cne.902020304.
Ottersen, O.P., Fischer, B.O., Rinvik, E., Storm-Mathisen, J., 1986. Putative amino acid transmitters in the amygdala. Adv. Exp. Med. Biol. 53–66. https://doi.org/10.1007/978-1-4684-7971-3_4.
Packard, M.G., Cahill, L., 2001. Affective modulation of multiple memory systems. Curr. Opin. Neurobiol. 11 (6), 752–756. https://doi.org/10.1016/S0959-4388(01)00280-X.
Papez, J.W., 1937. A proposed mechanism of emotion. Arch. Neurol. Psychiatry 38 (4), 725. https://doi.org/10.1001/archneurpsyc.1937.02260220069003.
Paré, D., Pape, H.C., Dong, J., 1995a. Bursting and oscillating neurons of the cat basolateral amygdaloid complex in vivo: electrophysiological properties and morphological features. J. Neurophysiol. 74 (3), 1179–1191. https://doi.org/10.1152/jn.1995.74.3.1179.
Paré, D., Smith, Y., 1993a. Distribution of GABA immunoreactivity in the amygdaloid complex of the cat. Neuroscience 57 (4), 1061–1076. https://doi.org/10.1016/0306-4522(93)90049-L.
Paré, D., Smith, Y., 1993b. The intercalated cell masses project to the central and medial nuclei of the amygdala in cats. Neuroscience 57 (4), 1077–1090. https://doi.org/10.1016/0306-4522(93)90050-P.
Paré, D., Smith, Y., Paré, J.F., 1995b. Intra-amygdaloid projections of the basolateral and basomedial nuclei in the cat: phaseolus vulgaris-leucoagglutinin anterograde tracing at the light and electron microscopic level. Neuroscience 69 (2), 567–583. https://doi.org/10.1016/0306-4522(95)00272-K.
Petrovich, G.D., Canteras, N.S., Swanson, L.W., 2001. Combinatorial amygdalar inputs to hippocampal domains and hypothalamic behavior systems. Brain Res. Rev. 38 (1–2), 247–289. https://doi.org/10.1016/S0165-0173(01)00080-7.
Pimontel, M.A., Kanellopoulos, D., Gunning, F.M., 2020. Neuroanatomical abnormalities in older depressed adults with apathy: a systematic review. J. Geriatr. Psychiatry Neurol. 33 (5), 289–303. https://doi.org/10.1177/0891988719882100.
Pitkänen, A., 2000. Connectivity of the rat amygdaloid complex. In: Aggleton, J.P. (Ed.), The Amygdala. Oxford University Press, pp. 31–115.
Pitkanen, A., Amaral, D., 1994. The distribution of GABAergic cells, fibers, and terminals in the monkey amygdaloid complex: an immunohistochemical and in situ hybridization study. J. Neurosci. 14 (4), 2200–2224. https://doi.org/10.1523/JNEUROSCI.14-04-02200.1994.
Pitkänen, A., Savander, V., LeDoux, J.E., 1997. Organization of intra-amygdaloid circuitries in the rat: an emerging framework for understanding functions of the amygdala. Trends Neurosci. 20 (11), 517–523. https://doi.org/10.1016/S0166-2236(97)01125-9.
Pitkänen, A., Stefanacci, L., Farb, C.R., Go, G.G., Ledoux, J.E., Amaral, D.G., 1995. Intrinsic connections of the rat amygdaloid complex: projections originating in the lateral nucleus. J. Comp. Neurol. 356 (2), 288–310. https://doi.org/10.1002/cne.903560211.
Pitkänen, A., Amaral, D.G., 1998. Organization of the intrinsic connections of the monkey amygdaloid complex: projections originating in the lateral nucleus. J. Comp. Neurol. 398 (3), 431–458. https://doi.org/10.1002/(SICI)1096-9861(19980831)398:3<431::AID-CNE9>3.0.CO;2-0.
Poremba, A., Gabriel, M., 1999. Amygdala neurons mediate acquisition but not maintenance of instrumental avoidance behavior in rabbits. J. Neurosci. 19 (21), 9635–9641. https://doi.org/10.1523/JNEUROSCI.19-21-09635.1999.

Powell, J., Lewis, P.A., Roberts, N., García-Fiñana, M., Dunbar, R.I.M., 2012. Orbital prefrontal cortex volume predicts social network size: an imaging study of individual differences in humans. Proc. Biol. Sci. 279 (1736), 2157–2162. https://doi.org/10.1098/rspb.2011.2574.

Preter, M., Klein, D.F., 2008. Panic, suffocation false alarms, separation anxiety and endogenous opioids. Prog. Neuro-Psychopharmacol. Biol. Psychiatry 32 (3), 603–612. https://doi.org/10.1016/j.pnpbp.2007.07.029.

Price, J.L., Russchen, E.T., Amaral, D.G., 1987a. The amygdaloid complex. In: Björklund, A., Hökfelt, T., Swanson, L.W. (Eds.), Handbook of Chemical Neuroanatomy, vol. 5. Elsevier.

Price, J.L., Russchen, F.T., Amaral, D.G., 1987b. The limbic region. II: the amygdaloid complex. In: Handbook of Chemical Neuroanatomy, vol. 5. Elsevier.

Ray, J.P., Price, J.L., 1992. The organization of the thalamocortical connections of the mediodorsal thalamic nucleus in the rat, related to the ventral forebrain-prefrontal cortex topography. J. Comp. Neurol. 323 (2), 167–197. https://doi.org/10.1002/cne.903230204.

Rhone, A.E., Kovach, C.K., Harmata, G.I.S., et al., 2020. A human amygdala site that inhibits respiration and elicits apnea in pediatric epilepsy. JCI Insight 5 (6). https://doi.org/10.1172/jci.insight.134852.

Rizvi, T.A., Ennis, M., Behbehani, M.M., Shipley, M.T., 1991. Connections between the central nucleus of the amygdala and the midbrain periaqueductal gray: topography and reciprocity. J. Comp. Neurol. 303 (1), 121–131. https://doi.org/10.1002/cne.903030111.

Rollins, B.L., Stines, S.G., McGuire, H.B., King, B.M., 2001. Effects of amygdala lesions on body weight, conditioned taste aversion, and neophobia. Physiol. Behav. 72 (5), 735–742. https://doi.org/10.1016/S0031-9384(01)00433-4.

Rolls, E.T., 2000. On the brain and emotion. Behav. Brain Sci. 23 (2), 219–228. https://doi.org/10.1017/S0140525X00512424.

Roozendaal, B., Koolhaas, J.M., Bohus, B., 1993. The central amygdala is involved in conditioning but not in retention of active and passive shock avoidance in male rats. Behav. Neural. Biol. 59 (2), 143–149. https://doi.org/10.1016/0163-1047(93)90873-G.

Royer, S., Martina, M., Paré, D., 1999. An inhibitory interface gates impulse traffic between the input and output stations of the amygdala. J. Neurosci. 19 (23), 10575–10583. https://doi.org/10.1523/JNEUROSCI.19-23-10575.1999.

Royer, S., Martina, M., Paré, D., 2000. Polarized synaptic interactions between intercalated neurons of the amygdala. J. Neurophysiol. 83 (6), 3509–3518. https://doi.org/10.1152/jn.2000.83.6.3509.

Sah, P., Faber, E.S.L., Lopez de Armentia, M., Power, J., 2003. The amygdaloid complex: anatomy and physiology. Physiol. Rev. 83 (3), 803–834. https://doi.org/10.1152/physrev.00002.2003.

Sato, W., Kochiyama, T., Kubota, Y., et al., 2016. The association between perceived social support and amygdala structure. Neuropsychologia 85, 237–244. https://doi.org/10.1016/j.neuropsychologia.2016.03.036.

Sato, W., Kochiyama, T., Uono, S., Sawada, R., Yoshikawa, S., 2020. Amygdala activity related to perceived social support. Sci. Rep. 10 (1), 2951. https://doi.org/10.1038/s41598-020-59758-x.

Savander, V., Go, C.G., Ledoux, J.E., Pitkänen, A., 1996a. Intrinsic connections of the rat amygdaloid complex: projections originating in the accessory basal nucleus. J. Comp. Neurol. 374 (2), 291–313. https://doi.org/10.1002/(SICI)1096-9861(19961014)374:2<291::AID-CNE10>3.0.CO;2-Y.

Savander, V., Go, C.G., Ledoux, J.E., Pitkänen, A., 1995. Intrinsic connections of the rat amygdaloid complex: projections originating in the basal nucleus. J. Comp. Neurol. 361 (2), 345–368. https://doi.org/10.1002/cne.903610211.

Savander, V., LeDoux, J.E., Pitkänen, A., 1996b. Topographic projections from the periamygdaloid cortex to select subregions of the lateral nucleus of the amygdala in the rat. Neurosci. Lett. 211 (3), 167–170. https://doi.org/10.1016/0304-3940(96)12750-6.

Savander, V., Miettinen, R., LeDoux, J.E., Pitkänen, A., 1997. Lateral nucleus of the rat amygdala is reciprocally connected with basal and accessory basal nuclei: a light and electron microscopic study. Neuroscience 77 (3), 767–781. https://doi.org/10.1016/S0306-4522(96)00513-1.

Scalia, F., Winans, S.S., 1975. The differential projections of the olfactory bulb and accessory olfactory bulb in mammals. J. Comp. Neurol. 161 (1), 31–55. https://doi.org/10.1002/cne.901610105.

Schiess, M.C., Callahan, P.M., Zheng, H., 1999. Characterization of the electrophysiological and morphological properties of rat central amygdala neurons in vitro. J. Neurosci. Res. 58 (5), 663–673.

Seiriki, K., Kasai, A., Hashimoto, T., et al., 2017. High-speed and scalable whole-brain imaging in rodents and primates. Neuron 94 (6), 1085–1100.e6. https://doi.org/10.1016/j.neuron.2017.05.017.

Shi, C., Davis, M., 2001. Visual pathways involved in fear conditioning measured with fear-potentiated startle: behavioral and anatomic studies. J. Neurosci. 21 (24), 9844–9855. https://doi.org/10.1523/JNEUROSCI.21-24-09844.2001.

Shi, C.J., Cassell, M.D., 1998a. Cascade projections from somatosensory cortex to the rat basolateral amygdala via the parietal insular cortex. J. Comp. Neurol. 399 (4), 469–491. https://doi.org/10.1002/(sici)1096-9861(19981005)399:4<469::aid-cne3>3.0.co;2-#.

Shi, C.J., Cassell, M.D., 1998b. Cortical, thalamic, and amygdaloid connections of the anterior and posterior insular cortices. J. Comp. Neurol. 399 (4), 440–468. https://doi.org/10.1002/(sici)1096-9861(19981005)399:4<440::aid-cne2>3.0.co;2-1.

Shi, C.J., Cassell, M.D., 1999. Perirhinal cortex projections to the amygdaloid complex and hippocampal formation in the rat. J. Comp. Neurol. 406 (3), 299–328. https://doi.org/10.1002/(sici)1096-9861(19990412)406:3<299::aid-cne2>3.0.co;2-9.

Šimić, G., Tkalčić, M., Vukić, V., et al., 2021. Understanding emotions: origins and roles of the amygdala. Biomolecules 11 (6), 823. https://doi.org/10.3390/biom11060823.

Sitte, H.H., Pifl, C., Rajput, A.H., et al., 2017. Dopamine and noradrenaline, but not serotonin, in the human claustrum are greatly reduced in patients with Parkinson's disease: possible functional implications. Eur. J. Neurosci. 45 (1), 192–197. https://doi.org/10.1111/ejn.13435.

Smith, J.B., Lee, A.K., Jackson, J., 2020. The claustrum. Curr. Biol. 30 (23), R1401–R1406. https://doi.org/10.1016/j.cub.2020.09.069.

Smith, J.B., Radhakrishnan, H., Alloway, K.D., 2012. Rat claustrum coordinates but does not integrate somatosensory and motor cortical information. J. Neurosci. 32 (25), 8583–8588. https://doi.org/10.1523/JNEUROSCI.1524-12.2012.

Smith, J.B., Watson, G.D.R., Liang, Z., Liu, Y., Zhang, N., Alloway, K.D., 2019. A role for the claustrum in salience processing? Front. Neuroanat. 13. https://doi.org/10.3389/fnana.2019.00064.

Smith, Y., Paré, D., 1994. Intra-amygdaloid projections of the lateral nucleus in the cat: PHA-L anterograde labeling combined with postembedding GABA and glutamate immunocytochemistry. J. Comp. Neurol. 342 (2), 232–248. https://doi.org/10.1002/cne.903420207.

Stern, P., 2017. Claustro-cortical connectivity. Science (1979) 355 (6323), 364. https://doi.org/10.1126/science.355.6323.364-b.

Sutherland, R.J., McDonald, R.J., 1990. Hippocampus, amygdala, and memory deficits in rats. Behav. Brain Res. 37 (1), 57–79. https://doi.org/10.1016/0166-4328(90)90072-M.

Swanson, L.W., 2000. Cerebral hemisphere regulation of motivated behavior. Brain Res. 886 (1–2), 113–164. https://doi.org/10.1016/S0006-8993(00)02905-X.

Swanson, L.W., Kuypers, H.G.J.M., 1980. The paraventricular nucleus of the hypothalamus: cytoarchitectonic subdivisions and organization of projections to the pituitary, dorsal vagal complex, and spinal cord as demonstrated by retrograde fluorescence double-labeling methods. J. Comp. Neurol. 194 (3), 555–570. https://doi.org/10.1002/cne.901940306.

Swanson, L.W., Petrovich, G.D., 1998. What is the amygdala? Trends Neurosci. 21 (8), 323–331. https://doi.org/10.1016/S0166-2236(98)01265-X.

Tarrada, A., Aron, O., Vignal, J., Ertan, D., Maillard, L., Hingray, C., 2022. Anticipatory anxiety of seizures is associated with ictal emotional distress and amygdala onset seizures. Epilepsia 63 (5), 1130–1140. https://doi.org/10.1111/epi.17215.

Terem, A., Gonzales, B.J., Peretz-Rivlin, N., et al., 2020. Claustral neurons projecting to frontal cortex mediate contextual association of reward. Curr. Biol. 30 (18), 3522–3532.e6. https://doi.org/10.1016/j.cub.2020.06.064.

Treit, D., Menard, J., 1997. Dissociations among the anxiolytic effects of septal, hippocampal, and amygdaloid lesions. Behav. Neurosci. 111 (3), 653–658. https://doi.org/10.1037/0735-7044.111.3.653.

Tsao, D.Y., Moeller, S., Freiwald, W.A., 2008. Comparing face patch systems in macaques and humans. Proc. Natl. Acad. Sci. USA 105 (49), 19514–19519. https://doi.org/10.1073/pnas.0809662105.

Turner, B.H., Herkenham, M., 1991. Thalamoamygdaloid projections in the rat: a test of the amygdala's role in sensory processing. J. Comp. Neurol. 313 (2), 295–325. https://doi.org/10.1002/cne.903130208.

van der Kooy, D., Koda, L.Y., McGinty, J.F., Gerfen, C.R., Bloom, F.E., 1984. The organization of projections from the cortes, amygdala, and hypothalamus to the nucleus of the solitary tract in rat. J. Comp. Neurol. 224 (1), 1–24. https://doi.org/10.1002/cne.902240102.

Veening, J.G., Swanson, L.W., Sawchenko, P.E., 1984. The organization of projections from the central nucleus of the amygdala to brainstem sites involved in central autonomic regulation: a combined retrograde transport-immunohistochemical study. Brain Res. 303 (2), 337–357. https://doi.org/10.1016/0006-8993(84)91220-4.

Wahis, J., Baudon, A., Althammer, F., et al., 2021. Astrocytes mediate the effect of oxytocin in the central amygdala on neuronal activity and affective states in rodents. Nat. Neurosci. 24 (4), 529–541. https://doi.org/10.1038/s41593-021-00800-0.

Wang, Y., He, Z., Zhao, C., Li, L., 2013. Medial amygdala lesions modify aggressive behavior and immediate early gene expression in oxytocin and vasopressin neurons during intermale exposure. Behav. Brain Res. 245, 42–49. https://doi.org/10.1016/j.bbr.2013.02.002.

Weiskrantz, L., 1956. Behavioral changes associated with ablation of the amygdaloid complex in monkeys. J. Comp. Physiol. Psychol. 49 (4), 381–391. https://doi.org/10.1037/h0088009.

Welzl, H., D'Adamo, P., Lipp, H.P., 2001. Conditioned taste aversion as a learning and memory paradigm. Behav. Brain Res. 125 (1–2), 205–213. https://doi.org/10.1016/S0166-4328(01)00302-3.

Yajeya, J., de la Fuente Juan, A., Merchan, M.A., Riolobos, A.S., Heredia, M., Criado, J.M., 1997. Cholinergic responses of morphologically and electrophysiologically characterized neurons of the basolateral complex in rat amygdala slices. Neuroscience 78 (3), 731–743. https://doi.org/10.1016/S0306-4522(96)00614-8.

Yan, W.W., Xia, M., Chiang, J., et al., 2021. Enhanced synaptic transmission in the extended amygdala and altered excitability in an extended amygdala to brainstem circuit in a Dravet syndrome mouse model. eNeuro 8 (3). https://doi.org/10.1523/ENEURO.0306-20.2021.

Yang, F., Sun, L., Li, J., Lin, W., 2021. Repetitive seizures after febrile period exclusively involving bilateral claustrum. Medicine 100 (37), e27129. https://doi.org/10.1097/MD.0000000000027129.

Zaboski, B.A., Storch, E.A., 2018. Comorbid autism spectrum disorder and anxiety disorders: a brief review. Future Neurol. 13 (1), 31–37. https://doi.org/10.2217/fnl-2017-0030.

Zikopoulos, B., John, Y.J., García-Cabezas, M.Á., Bunce, J.G., Barbas, H., 2016. The intercalated nuclear complex of the primate amygdala. Neuroscience 330, 267–290. https://doi.org/10.1016/j.neuroscience.2016.05.052.

Zuhorn, F., Omaimen, H., Ruprecht, B., et al., 2021. Parainfectious encephalitis in COVID-19: the claustrum sign. J. Neurol. 268 (6), 2031–2034. https://doi.org/10.1007/s00415-020-10185-y.

Further Reading

Klüver, H., Bucy, P.C., 1938. An analysis of certain effects of bilateral temporal lobectomy in the rhesus monkey, with special reference to "psychic blindness." J. Psychol. 5 (1), 33–54. https://doi.org/10.1080/00223980.1938.9917551.

Meisner, O.C., Nair, A., Chang, S.W.C., 2022. Amygdala connectivity and implications for social cognition and disorders. In: Handbook of Clinical Neurology. Elsevier, pp. 381–403. https://doi.org/10.1016/B978-0-12-823493-8.00017-1.

15

The Telencephalon: Basal Ganglia

Brandon M. Ascenzi[1], *Guilherme de Carvalho Caldas*[2], *Alessandro Frati*[3], *Antonio Santoro*[3] *and Mauro Palmieri*[2,3]

[1]Independent Neuroresearcher-Member of Marie Curie Alumni Association (MCAA) [2]Neurosurgical Innovations and Training Center, Weill Cornell Medical College, New York City, NY, United States [3]Department of Human Neurosciences, Neurosurgery Division, University of Rome "La Sapienza", Rome, Italy

OUTLINE

INTRODUCTION

Basal ganglia (BG) is the term that refers to a group of masses of gray matter (nuclei) that are closely connected and located deep, at the base (hence "basal") of the forebrain from which the extrapyramidal fibers arise (Figs. 15.1A,B and 15.2). In fact, the extrapyramidal system was thought to be an array of centers which, together with their emerging fibers, constitute a multisynaptic descending system. Striopallidal, pallidoreticular, and reticulospinal pathways were the principal connections in this system. Experimental hodological studies have shown that, although these pathways do indeed exist, the striatum (complex made of caudate nucleus and putamen) and the other "extrapyramidal" centers are not at all interconnected in a unidirectional chain-like fashion. Rather, they and their emerging fiber systems constitute several interrelated loops or circuits, from which output systems emerge at several levels. However, it appears that the idea of independently operating cortical pyramidal and subcortical extrapyramidal motor systems must be abandoned, as the cortex, including its sensory and motor fields, projects strongly to the caudate-putamen complex and the principal output channel of the latter, via the globus pallidus and the thalamus, converges on the motor, premotor, and prefrontal areas of the cortex (Fig. 15.3).

From Anatomy to Function of the Central Nervous System
https://doi.org/10.1016/B978-0-12-822404-5.00021-8

FIGURE 15.1 (A) **Photomicrograph of an acetylcholinesterase-stained coronal section through the rostral, human striatum.** (B) **Photomicrograph of an acetylcholinesterase-stained coronal section through the central, human striatum.** This level includes both the internal and external segments of globus pallidus. *AC*, anterior commissure; *CD*, caudate nucleus; *GPe*, globus pallidus external segment; *GPi*, globus pallidus internal segment; *IC*, internal capsule; *NAc*, accumbens; *P*, putamen; *VS*, ventral striatum. *Image courtesy of Haber, S.N., Adler, A., Bergman, H., 2012. Chapter 20: The basal ganglia. In: Mai, J.K., Paxinos, G. (Eds.), The Human Nervous System, third ed. Academic Press, pp. 678–738. https://doi.org/10.1016/B978-0-12-374236-0.10020-3.*

NEUROANATOMY OF BASAL GANGLIA

Although the basal ganglia (BG) can be mistakenly imagined as separate and distant from each other, they constitute a whole and compact functional structure rather than a morphological entity.

Topographically BG are located between the insular cortex and the lateral ventricles (Fig. 15.4). The insula is hidden deep within the lateral sulcus (or Sylvian fissure) of the cerebral cortex, where the *middle cerebral artery* (M1) with its branches run and it can be seen only when the lips of the fissure are separated (see Chapter 2 "Blood supply, meninges and CSF circulation"). The insular cortex has its central sulcus parallel to the central fissure of the brain (Rolando's fissure) which separates the primary somatosensory cortex (3,1,2 BA) from the primary motor cortex (4 BA). By removing the insular cortex (Fig. 15.5) we immediately see a layer of white matter which is called the *extreme capsule* followed medially by the *claustrum* (Fig. 15.6) which is a thin strip of gray matter that seems to be involved in the integration of the sensory and motor information; it has the projection fibers leading to the cortex (Fernandez-Miranda et al., 2008). Proceeding lateromedial to the claustrum we find the *external capsule* and the *putamen* (Fig. 15.7), the latter is in turn partially separated from the *caudate nucleus* by the *internal capsule* (Yagmurlu et al., 2015). The internal capsule divides the striatal zone into a dorsomedial caudate nucleus and a ventrolateral putamen although the separation of these structures is always incomplete.

Internal and external capsules are both coming from the same projection fibers of the *corona radiata*. This so-called structure is a white matter sheet that continues inferiorly as the internal capsule and superiorly as the *centrum semiovale*. This sheet of both ascending and descending axons carries most of the neural traffic from and to the cerebral cortex. The corona radiata is associated with the corticopontine tract, the corticobulbar tract, and the corticospinal tract (see Chapters 4 and 5a). Therefore, it is worth mentioning when there is little lacunar infarct at the internal capsule, important neurologic problem can occur.

The internal capsule has five parts: anterior and posterior limbs joining at genu part also the sublenticular and retrolenticular parts according to the position of fibers to the lentiform nucleus. Posterior limb contains the corticospinal fibers to the arm nearer to the genu than those coursing to the leg. Anterior limb of the internal capsule separates lentiform nucleus from the caudate head, this region has a striated appearance because gray matter pass through the internal capsule

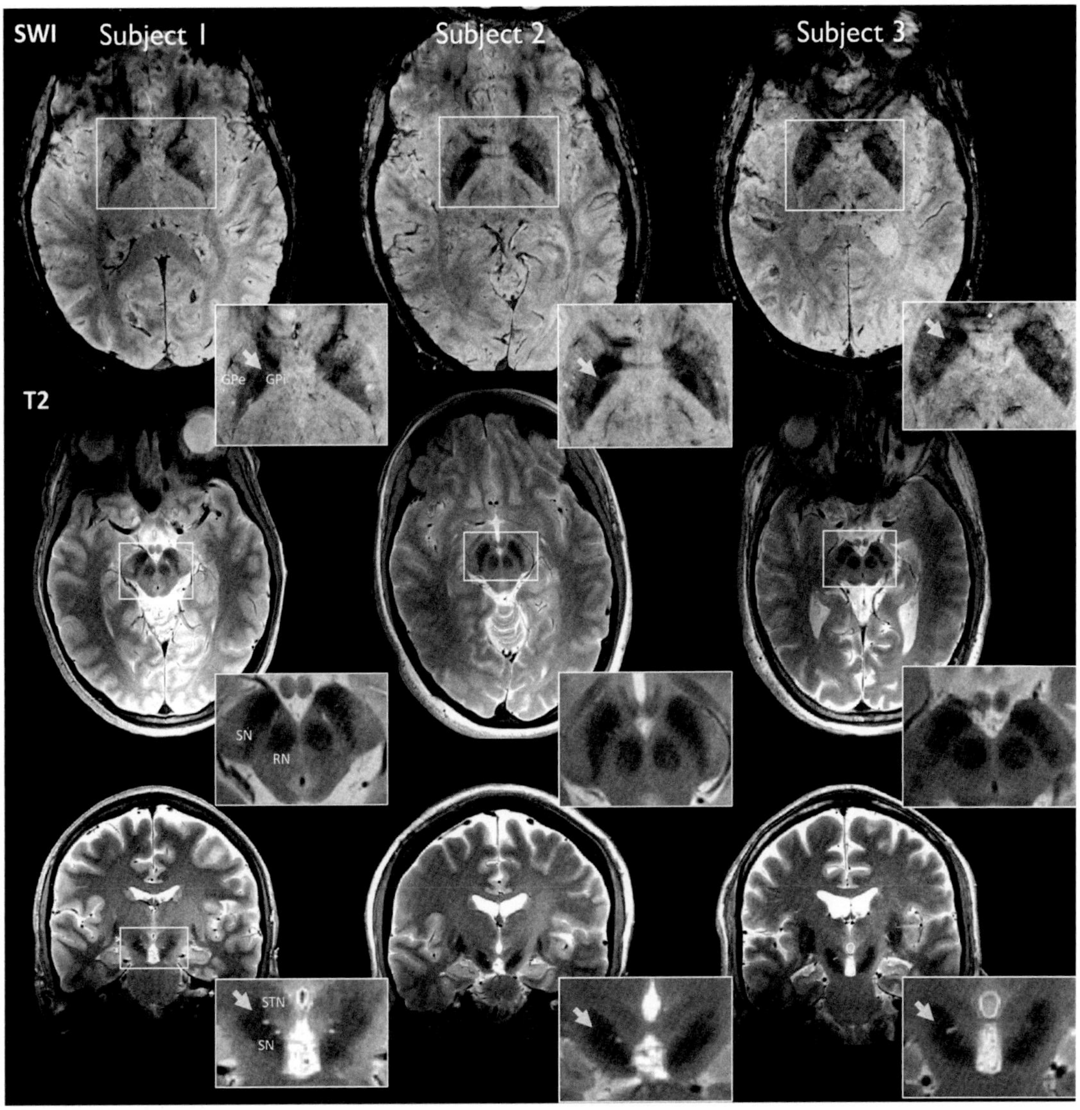

FIGURE 15.2 **Structural imaging of basal ganglia at MRI 7T.** (*Top row*) axial high-resolution susceptibility-weighted images (SWI) in three subjects, at the level of the globus pallidus (GP), putamen and thalamus. 7T SWI provides high contrast between structures such as the external (GPe) and internal (GPi) segments of GP, as can be seen in each *inset* in the *uppermost row*. The *white arrows* indicate the border between GPe and GPi, known as the lamina pallidi medialis (*uppermost insets*). Axial high-resolution T2-weighted images in three subjects at the level of the substantia nigra (SN) and red nucleus (RN) are presented in the *middle row*. Coronal T2-weighted images in three subjects at the level of the subthalamic nucleus (STN) and SN (*bottom row*). Coronal images provide good contrast enabling differentiation between SN and STN along the lateral-medial axis, as indicated by the *white arrow* in each *inset*. SWI and T2-weighted images were co-registered and both used to segment GPe, GPi, SN, and STN. *Image courtesy of Lenglet, C., Abosch, A., Yacoub, E., De Martino, F., Sapiro, G., Harel, N., 2012. Comprehensive in vivo mapping of the human basal ganglia and thalamic connectome in individuals using 7T MRI. PLoS One 7 (1), e29153. https://doi.org/10.1371/journal.pone.0029153.*

(Snell et al., 2006). On the other hand, posterior limb separates it from the thalamus (Rhoton, 2007).

Rostroventrally and caudally, the caudate nucleus and the putamen remain connected and, at intermediate levels, striatal cell bridges are interspersed between the bundles of the internal capsule, this is the reason why this complex is called "striatum."

Morphologically, in the adult human brain, the caudate nucleus is an elongated, curved cell mass, which is in direct contact with the lateral ventricle throughout its extent. It consists of an enlarged rostral part known as the *head of the caudate nucleus*, which protrudes into the anterior horn of the lateral ventricle; a short *intermediate body*, which lies dorsolateral to the thalamus and, a long *tail*, which enters the temporal lobe and terminates in the vicinity of the amygdaloid complex. The *nucleus accumbens* (NAcc) is a medial expansion of the head of the caudate nucleus, which extends around the ventral aspect of the lateral ventricle into the medial or septal wall of the hemisphere (Fig. 15.8).

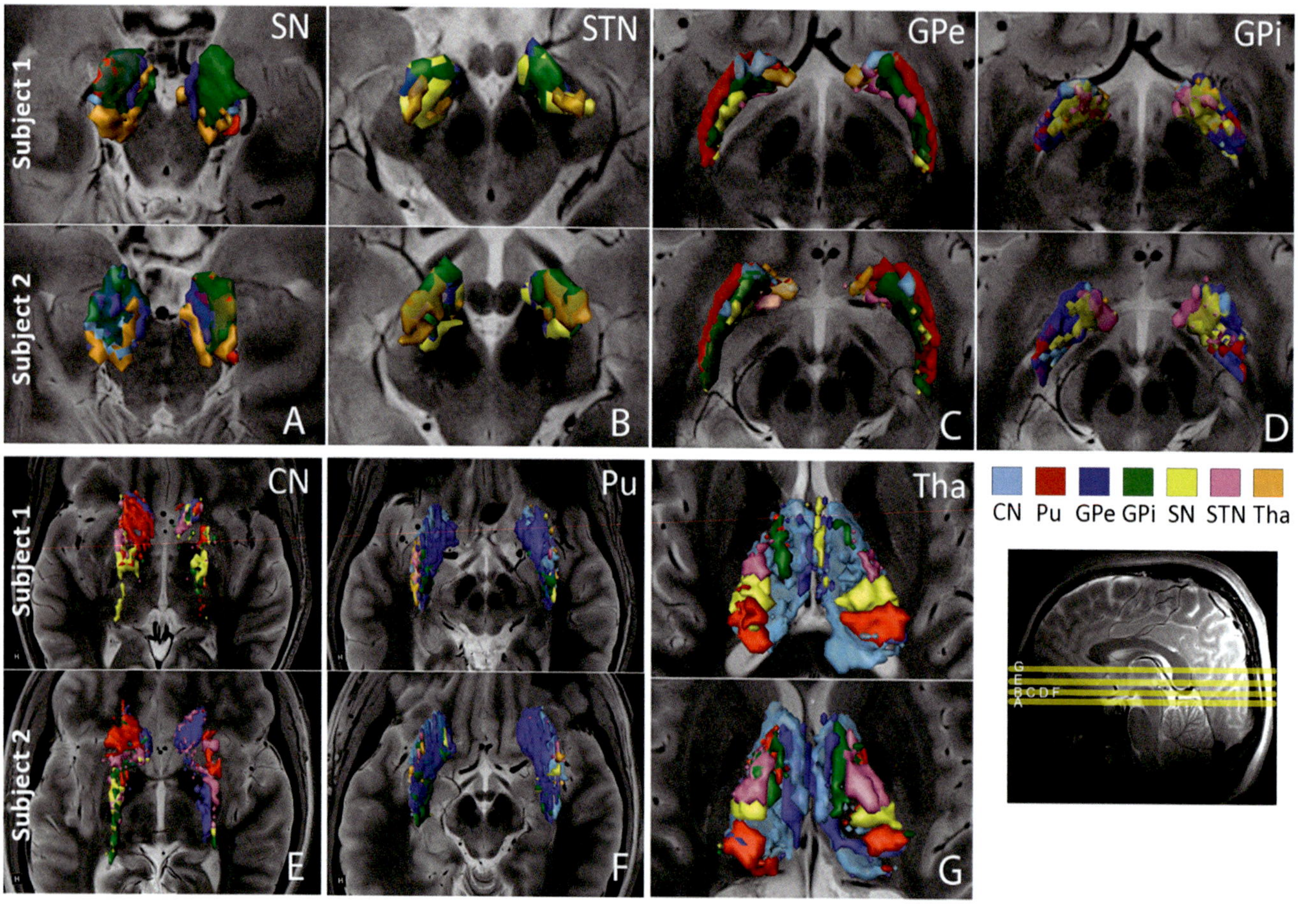

FIGURE 15.3 **Parcellation of the basal ganglia and thalamus based on their white matter projections.** Various subterritories identified in the basal ganglia and thalamus by exploiting the fact that these divisions exhibit distinctively stronger connectivity with other structures. Within each of the seven regions-of-interest, voxels presenting a high probability of connection with another structure are categorized according to the color of that very structure. (*Bottom right inset*) The *yellow lines* depict the location of axial T2-weighted images in each panel. (Panel A) inferior SN; (Panel B, C, D, F) inferior GP; (Panel E) superior GP; (Panel G) superior Tha, inferior-superior view. Color code: Caudate nucleus, CN: light blue; Putamen, Pu: red; external globus pallidus, GPe: dark blue; internal globus pallidus, GPi: green; substantia nigra, SN: yellow; subthalamic nucleus, STN: magenta; thalamus, Tha: orange. *Image courtesy of Lenglet, C., Abosch, A., Yacoub, E., De Martino, F., Sapiro, G., Harel, N. 2012. Comprehensive in vivo mapping of the human basal ganglia and thalamic connectome in individuals using 7T MRI. PLoS One 7 (1), e29153. https://doi.org/10.1371/journal.pone.0029153.*

As introduced ahead the putamen is situated medial to the insula and constitutes, with the *pallidum* or *globus pallidus* (GP), the lentiform nucleus, a cone-shaped complex with the apex directed inwards. The putamen constitutes the outer part and the globus pallidus the inner part of the lentiform nucleus. The globus pallidus can be subdivided into an external (GPe) and an internal segment (GPi) separated from each other via *medial medullary lamina*. The anterior limb of the internal capsule largely separates the caudate nucleus from the putamen, whereas the posterior limb of the internal capsule occupies the space between the lentiform nucleus and the thalamus.

Underneath putamen, anterior commissure passes inferiorly as a thick bundle before axonal branches separating for their destination.

Floor of the BG is the *anterior perforated substance* (APS) from where BG reach the basal surface of the brain. At here *lenticulostriate arteries* enter to supply BG and *inferior striate veins* exit from. Numerous orifices for penetrating blood vessels lead the "perforated" appearance. Lenticulostriate arteries arise predominantly from the first segment of the *middle cerebral artery* (M1) (Yasargil, 1984) and from the first segment of the *anterior cerebral artery* (A1) (Fig. 15.9). Lenticulostriate arteries penetrate through the olfactory tubercle and *substantia innominata* to reach the BG structures. The boundaries of the APS are optic chiasm with optic tract posteriorly then olfactory trigone anteriorly and the uncus laterally.

Fiber dissections (Fig. 15.10) revealed the different pattern of extensions of these arteries which can be divided into three main divisions: anterolateral, anteromedial, and inferolateral.

The anterolateral extension goes toward the limen of the insula to supply the internal capsule and the lateral part of the putamen.

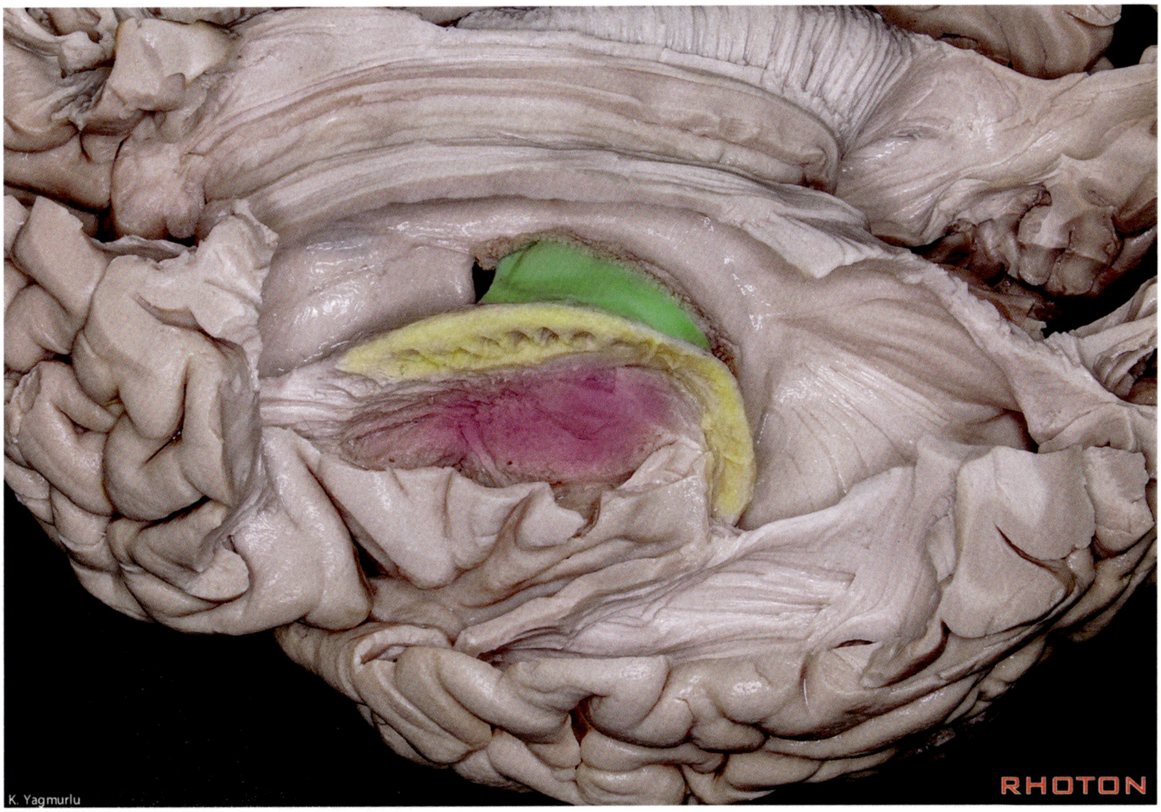

FIGURE 15.4 **Basal ganglia: deep brain dissection.** From inside to outside: *in green*, the thalamus; *in yellow* the caudate nucleus; *in magenta*, the nucleus lenticularis (putamen and globus pallidus). In front of the thalamus, a small hole can be seen: the foramen of Monroe. In front of and above it is the frontal horn of the lateral (*left*) ventricle. The cortex that laterally surrounds this nuclear complex is the insular cortex. *Image courtesy of The Rhoton Collection® http://rhoton.ineurodb.org/. Dissector: K. Yagmurlu. Drawings by Alessandro Basso.*

Most of the anteromedial extension, that is supplied by A1 and *anterior communicating artery*, reaches up to the rostrum of the corpus callosum running as two parallel arches between the lamina terminalis and the posterior margin of the paraterminal gyrus.

Finally, the inferolateral extension follows optic tract posterolateral to the lateral geniculate body. The *recurrent artery of Heubner* is typically the largest of the perforating lenticulostriate arteries arising from anterior communicating artery. This is the biggest supply to the major parts of the BG including the head of the CdN, the medial portion of GB, and anterior crus of the internal capsule (El Falougy et al., 2013).

CELLULAR ANATOMY AND SYNAPTOLOGY OF THE BASAL GANGLIA

Most of the neurons in the striatum belong to the category of *medium spiny neurons* (MSN) also known as *spiny projection neurons* (SPNs) (Fig. 15.11), they are a special type of GABAergic neurons representing ~95% of all neurons in the human striatum. A quantitative analysis has revealed that about 100 million medium-sized and some 600.000 large cells are present in the human striatum. Afferents from the cerebral cortex, the thalamus and the *substantia nigra pars compacta* (SNpc) converge upon these elements, which are all inhibitory and use GABA as their principal neurotransmitter.

The name of "spiny neuron" refers to the fact that numerous spines covering dendrites (Braak et al., 1982; Di Figlia et al., 1976; Graveland et al., 1985; Kemp et al., 1971) which radiate in all directions. Initial dendritic trunk is free of spines then gradually dividing occur from the smooth trunk. This is one of the highest density spine-laden cell in the brain where peak density can be as high as 4–6 per 1 μm in length (Nieuwenhuys et al., 2008).

The principal afferent systems to the striatum are glutamatergic fibers from the cerebral cortex (Fonnum et al., 1981; Kemp et al., 1971; Somogyi et al., 1981) and from the intralaminar thalamic nucleus (Royce, 1987) dopaminergic fibers from the substantia nigra (Kubota et al., 1986) and serotoninergic fibers from the dorsal raphe nucleus (Lavoie et al., 1990; Pasik et al., 1981). All of these extrinsic afferents synapse with the soma

FIGURE 15.5 **Basal ganglia: deep dissection of the insula.** *In yellow,* the caudate nucleus; *in green,* the thalamus. The white fibers surrounding the thalamus make up the corona radiata. *Image courtesy of The Rhoton Collection® http://rhoton.ineurodb.org/. Dissector: K. Yagmurlu. Drawings by Nickalus Khan and Alessandro Basso.*

of MSN and the glutamatergic and dopaminergic fibers mainly contacting the distal portion of the dendritic arbor of these neurons.

Each afferent pathway makes asymmetrical contact with different part of the dendritic spine: the corticostriatal fibers with the head, whereas the thalamostriatal with the shafts (Parent and Hazrati, 1995).

The nigrostriatal dopaminergic fibers make symmetrical synapses and many of them selectively contact the necks of those dendritic spines that receive cortical input (Parent and Hazrati, 1995) (Fig. 14.3), thus they are optimally positioned to specifically modulate cortical influences.

While predominant cell type in the striatum consists of MSNs the remaining consist of interneurons. MSNs are physically active neuron with its phasic bursts of activity and low baseline firing level (Braunlich et al., 2013), whereas interneurons with their high and constant level of baseline activity are tonically active neurons. On the other hand, these latter have smooth dendrites in opposite to the MSNs and their axonal branches are widespread within striatum (Kawaguchi et al., 1995). Their classical axonal branching fashion is that every branch has approximately 120° from each other and the previous trunk. Although functional significance of each type is somewhat unclear, it seems they may prevent decay in learning when there is no reward as a reinforcement (Ashby et al., 2011).

There are so many different interneurons that still cannot be properly categorized, however some of them have been examined in detail.

- *giant cholinergic interneurons*, largest cells of the basal ganglia, had assumed as principal cells until Kölliker recognized that no participation of these axons in the efferent projection in late 1800s. Major input is received by the thalamo-striate system. These cells use Acetylcholine (ACh) as neurotransmitter and are the principal resource of ACh in the striatum (Mallet et al., 2019).

The remaining striatal interneurons are medium to small size and all GABAergic. They can be subdivided into three types:

- Type I, contains calcium binding protein parvalbumin (PV), was characterized in 1980s (Bennet et al., 1993). Its intensely branching axon terminals form basket around the soma of spiny cell. Remarkably, type I interneurons' dendrites are connected with each other

FIGURE 15.6 **Claustrum (middle-deep dissection of the insula).** *In yellow,* the claustrum; the white fibers surrounding the claustrum constitute the external capsule; the dark structure that can be seen between the radial fibers of the external capsule is the putamen. *Image courtesy of The Rhoton Collection® http://rhoton.ineurodb.org/. Dissector: K. Yagmurlu. Drawing by Jeffrey Sorenson.*

via gap junction to make the electrical continuous network (Shepherd, 2004);

- Type II, stains positive for the neuromodulators somatostatin (SOM) and neuropeptide-Y(NPY); also contains nitric oxide (NO) (Nieuwenhuys et al., 2008). Recently, NADPH-diaphorase staining method, which is selective for nitric oxide synthase, has been shown morphologically different (Johannes et al., 2003). These two interneurons type receive their powerful input from the cerebral cortex.
- Type III, distinguishing feature is that it contains calretinin (CR) a calcium binding protein. Yet, little is known about the morphology and connections.

Although the caudate-putamen complex is from a cytoarchitectonic point of view a homogeneous structure, histochemical and particularly immunohistochemical studies have revealed a remarkable heterogeneity within the complex. The first evidence for this chemoarchitectural heterogeneity came from studies in which a staining technique for the enzyme acetylcholinesterase (AChE) was applied. These studies showed that, in the caudate-putamen complex, 300−600 μm wide zones of low AChE activity stand out against an otherwise AChE-rich background (Fig. 15.12).

Graybiel and Ragsdale (Graybiel et al., 1978, 1979, 1983), who first identified these zones, designated them as striatal bodies or striosomes. In recent years, it has gradually become clear that throughout most of the caudate-putamen complex the striosomes and the "matrix" in which they are embedded represent from chemoarchitectonic view distinct tissue compartments. These compartments are related to the intrinsic structure of the complex and to the organization of its afferent and efferent connections.

For how concern the *nucleus accumbens* (NAcc) Heimer and colleagues (Heimer et al., 1975, 1978, 1982) proposed on structural, histochemical, and connectional grounds that together the NAcc and medium-sized cell territories of the olfactory tubercle, it represents the ventral section of the striatum. This concept is corroborated by the fact that the dorsal and ventral sectors of the striatum both receive a robust dopaminergic projection from the ventral midbrain (Lindvall et al., 1983; Moore et al., 1978).

Neurons in the NAcc are mostly MSNs containing mainly D_1 (i.e., DRD1 and DRD5) or D_2 (i.e., DRD2, DRD3, and DRD4) dopamine receptors (see Chapter 3). A subpopulation of MSNs contains both D_1 and D_2 receptors, with approximately 40% of striatal MSNs

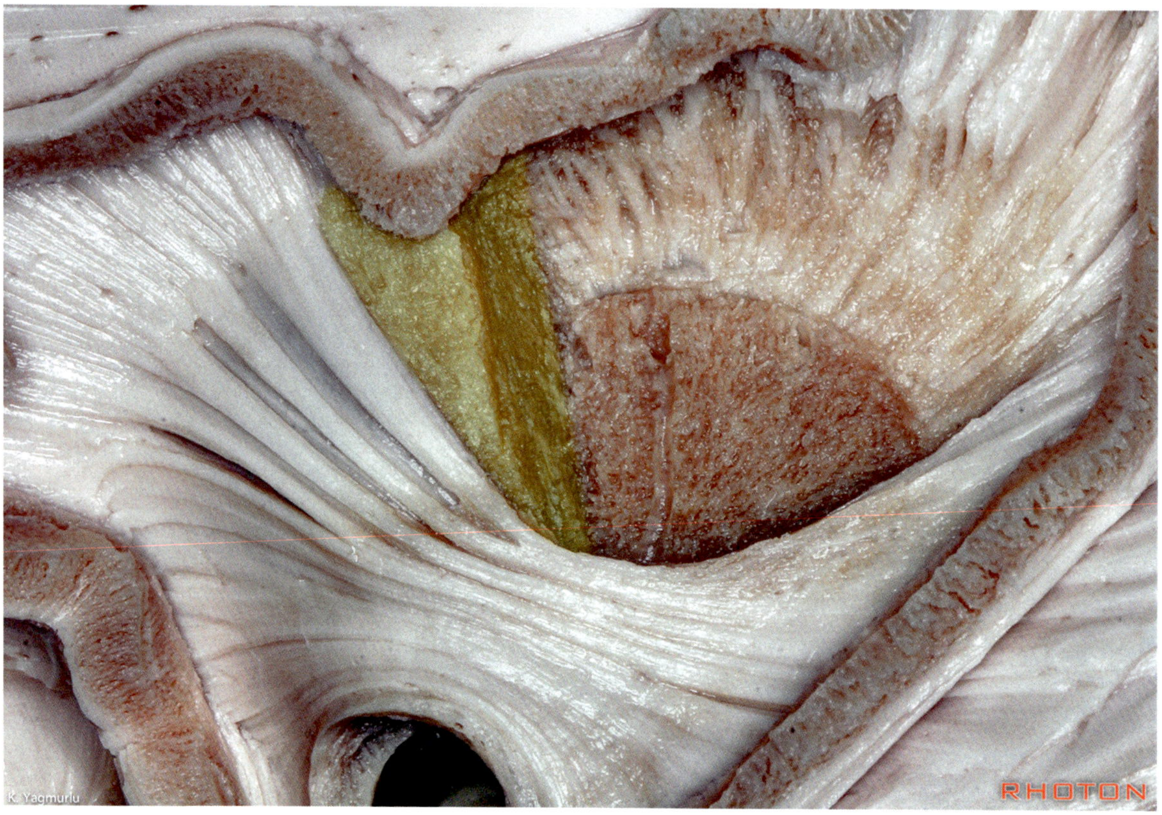

FIGURE 15.7 **Putamen (deep dissection of the insula).** *In yellow*, the putamen; laterally to the latter is the external capsule; medially the putamen, the globus pallidus (the dark structure); the crescent-shaped white fibers surrounding the globus pallidus form the internal capsule. *Image courtesy of The Rhoton Collection® http://rhoton.ineurodb.org/. Dissector: K. Yagmurlu. Drawing by Jeffrey Sorenson.*

expressing both DRD1and DRD2 mRNA (Yager et al., 2015; Ferré et al., 2010). These mixed-type NAcc MSNs with both D_1 and D_2 receptors are mostly confined to the NAcc "shell" (see later) (Yager et al., 2015).

Thus, it appears that each of the major cortical regions projects to a defined, but partially overlapping longitudinal zone of the striatum: the sensorimotor area (BA 3,1,2) projects onto the dorsolateral part of the caudate-putamen; the association areas of the neocortex project to a more medial sector of the caudate-putamen; the mesocortex projects mainly to the medial and ventral regions of the caudate-putamen, but also to the ventral striatum; and, finally, the allocortex projects mainly to the ventral striatum, but also to the medial and ventral parts of the caudate-putamen (McGeorge et al., 1989; Percheron et al., 1984).

Záborszky et al. (1985) showed that the NAcc can be divided into a peripheral and a central subterritory, which they referred to as the "shell" and "core," respectively. Subsequent studies have revealed that such a distinction can also be made with antibodies against substance P and calbindin-28, a calcium-binding protein (for a review see Zahm et al., 1992).

The NAcc "core" region, which merges dorsally with the caudate-putamen, shows a complex striosome/matrix-like compartmentalization. The neurons in the "core," as compared to the neurons in the "shell" (see later), have an increased density of dendritic spines, branch segments, and terminal segments. From the "core," the neurons project to other subcortical areas such as the globus pallidus and the substantia nigra. GABA is one of the main neurotransmitters in the NAcc, and GABA Rs are also abundant (Shirayama et al., 2005; Meredith et al., 1993).

The NAcc "core" is involved in the cognitive processing of motor function related to reward and reinforcement and the regulation of slow–wave sleep (Malenka et al., 2009; Cherasse and Urade, 2017; Valencia Garcia and Fort, 2018; Oishi et al., 2017). Specifically, the "core" encodes new motor programs which facilitate the acquisition of a given reward in the future. The D2-type neurons in the NAcc core which co-express adenosine A_{2A} receptors activation-dependently promote slow–wave sleep (Cherasse and Urade, 2017; Valencia Garcia and Fort, 2018; Oishi et al., 2017).

The "shell" is the outer region of the NAcc, and—unlike the "core"—is considered to be part of the extended amygdala, located at its rostral pole. The neurons in the "shell," as compared to the "core," have a lower density of dendritic spines, less terminal segments

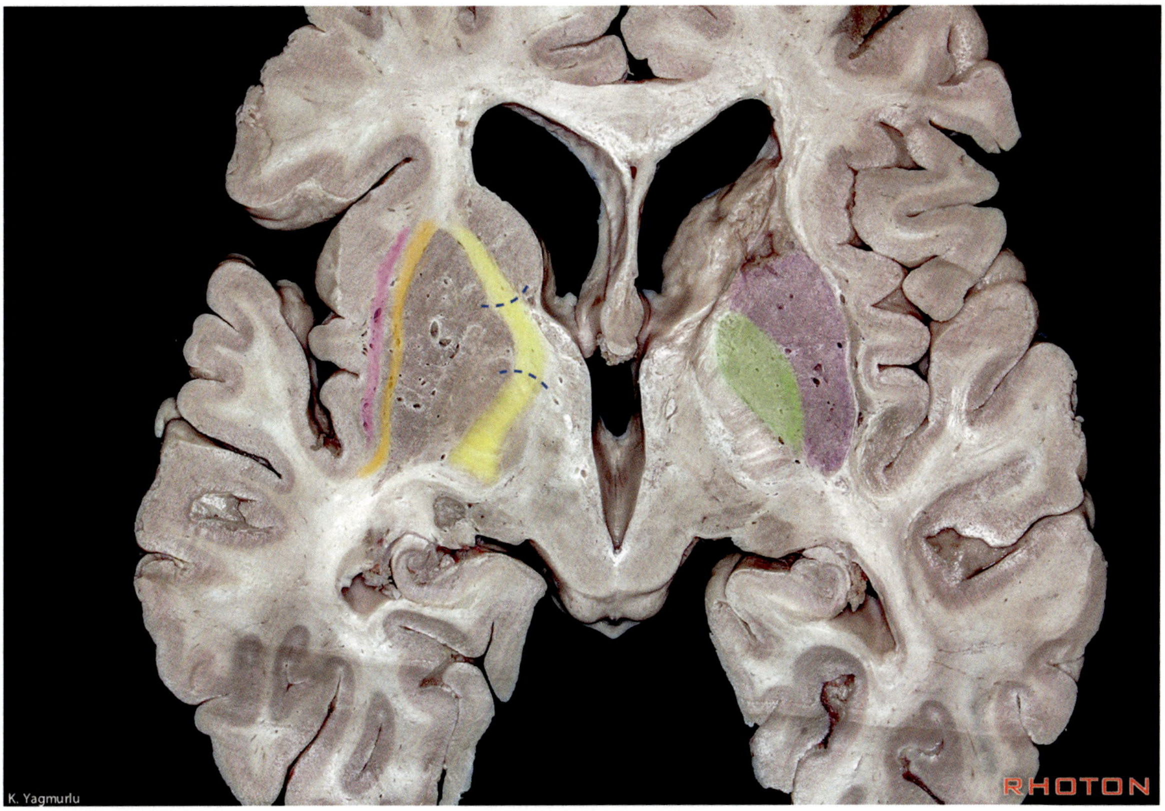

FIGURE 15.8 **Axial section of the brain.** On the *left side* (*left hemisphere*): in yellow, the internal capsule (between the two *dotted lines*, the genu; in front of the *anterior dotted line*, the anterior limb, behind the *posterior dotted line*, the posterior limb); *in orange*, the external capsule; *in magenta*, the extreme capsule. Between the external and extreme capsule is located the claustrum (the thin strip of gray matter). On the *right side* (*right hemisphere*): *in violet*, the putamen; *in green*, the globus pallidus (externus and internus). *Image courtesy from The Rhoton Collection® http://rhoton.ineurodb.org/. Dissector: K. Yagmurlu. Drawings by Jeffrey Sorenson and Alessandro Basso.*

and less branch segments than those in the "core." The "shell" neurons project to the subcommissural part of the ventral pallidum as well as the VTA and to extensive areas in the hypothalamus and extended amygdala (Shirayama et al., 2005; Meredith et al., 1992, 1993).

The *globus pallidus* or *pallidum* (GP) is a triangular mass of cells lying along the medial aspect of the putamen. It is separated from the latter by a fibers sheet, the external medullary lamina. The globus pallidus differs considerably in structure from the striatal nuclei. It is chiefly composed of large, widely spaced, fusiform cells. The total number of pallidal cells in humans is about the same as that of the large striatal neurons, i.e., 600,000 (Thörner et al., 1975).

Many bundles of myelinated fibers traverse the globus pallidus and in fresh preparations give it a paler color than the putamen or the caudate nucleus. An internal medullary lamina divides the globus pallidus into internal and external segments. The pallidal neurons have long, sparsely spiny, poorly branching dendrites. Together, the dendrites of individual cells form flattened, disc-like territories oriented parallel to the external border of the pallidum and perpendicular to the incoming striatal fibers. These striatofugal axons, which are thin and thinly myelinated, issue numerous thin, unmyelinated collaterals parallel to the pallidal dendrites, with which they repeatedly synapse (Di Figlia et al., 1988) (Fig. 15.13). It has already been mentioned that the striatal afferents projecting to the external pallidal segment express enkephalin, whereas those projecting to the internal pallidal segment express substance P.

All large neurons within both pallidal segments are GABAergic, inhibitory projection neurons. The elements in the external segment project to the subthalamic nucleus, whereas those in the internal segment project to the thalamus, the lateral habenular nucleus, and the pedunculopontine nucleus in the caudal mesencephalic tegmentum.

The *substantia innominata* (SI) is a rather ill-defined, flattened cell mass situated directly ventral to the putamen and the globus pallidus; it is partly separated from the latter by the anterior commissure. Its rostral part covers the olfactory tubercle; medially, it borders on the lateral preoptic and lateral hypothalamic areas, and its caudal part is situated between the globus pallidus and the nucleus centralis amygdalae. The

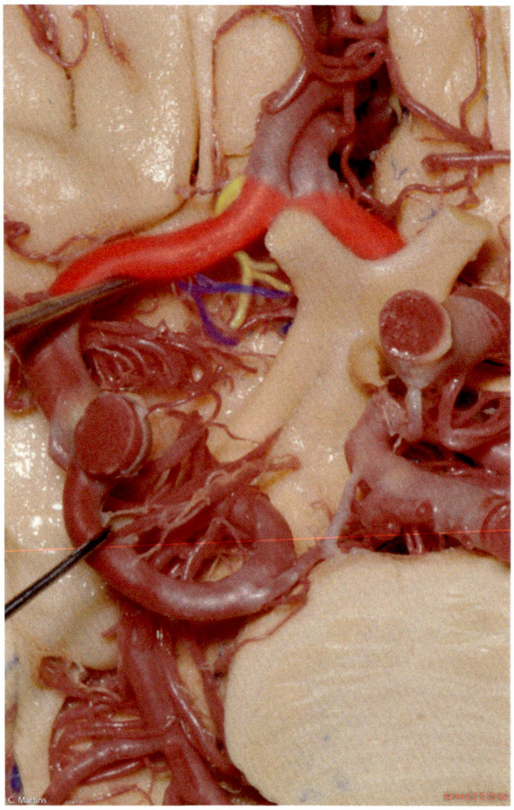

FIGURE 15.9 **Anterior cerebral artery and lenticulostriate arteries.** *In red,* A1 segment of anterior cerebral artery (ACA); *in yellow,* the recurrent artery of Heubner; *in violet* (and the others around it); lenticulostriate arteries. *Image courtesy of The Rhoton Collection® http://rhoton.ineurodb.org/. Dissector: C. Martins. Drawings by Andrew Horowitz, Brian Jimenez and Alessandro Basso.*

substantia innominata is mainly composed of rather loosely arranged, medium-sized cells, but also contains large, deeply stained, fusiform and multipolar neurons. The population of large cells is collectively designated as the *basal nucleus of Meynert* (BNM). It becomes progressively larger and more conspicuous with increasing cerebralization, attaining its greatest development in cetaceans and primates. The BNM is known to project mainly to the neocortex (Divac, 1975; Kievit et al., 1975; Pearson et al., 1982). The mediobasal telencephalon contains groups of large, multipolar neurons, which are collectively denoted as the magnocellular basal forebrain (MBF). This MBF can be topographically subdivided into a rostromedial, an intermediate and a caudolateral part. The rostromedial part is situated in the medial septal nucleus; the intermediate part co-extends roughly with the vertical and horizontal limbs of the *nucleus of the diagonal band of Broca* (DBBv, DBBh), whereas the large caudolateral part is embedded in the substantia innominata and corresponds to the basal nucleus of Meynert.

The MBF is neurochemically heterogeneous and comprises cholinergic, GABAergic, and nitric oxide (NO)-containing cells.

The cholinergic MBF neurons, which project in a topographically ordered fashion to the neocortex and several other telencephalic centers, were subdivided by Mesulam et al. (Mesulam et al., 1984) into four groups. For this topic, please see Chapter 3.

Large GABAergic neurons are distributed over all parts of the MBF, without forming clearly distinct subpopulations (Sarter et al., 2002). A roughly equal number of cholinergic and GABAergic MBF neurons project to the cerebral cortex (Sarter et al., 2002). The NO-containing neurons in the basal forebrain project to microvessels in the neocortex and are supposed to play a role in the regulation of cortical blood flow (Tong et al., 2000). As recently Frederiksen et al. demonstrated (Frederiksen et al., 2019), cerebral blood flow modifications due to vasoconstriction and vasodilatation play a major role in the development of some kind of headaches. The rostral subcommissural region of the substantia innominata seems to be related to the ventral striatum for this reason it is also named as *ventral pallidum*.

Within the ventral pallidum, separate subcommissural and ventral parts have been distinguished by the differential distribution of enkephalin and substance P immunoreactivity and by differences with respect to input and output characteristics (Groenewegen and Uylings, 2000; Haber et al., 1990a,b; Spooren et al., 1991).

In this first section of this chapter, we have described the basal ganglia from a structural and neurochemical point of view and we have also included some neurovascular and neurosurgical elements that are related to some of the more common diseases of the basal ganglia. Among these, ischemia and hemorrhages are the most common vascular pathologies that can affect basal ganglia (Wagner et al., 2009). More specifically, cerebral ischemia affecting perforating branches of the M1 tract of the middle cerebral artery can cause different conditions, from transient ischemic attack (TIA) to contralateral facial-brachio-crural hemiplegia, or in several cases, to coma or death. The same arteries are often affected by small microaneurysms, known as Charcot–Bouchard aneurysms, which rupture can cause hemorrhages of the basal ganglia (Kumar et al., 2005), especially in patients affected by amyloid angiopathy (Fig. 15.14A and B). The treatment of hemorrhages in this area can be only medical, with the administration of mannitol for intracranial hypertension, or also surgical. In this area, hemorrhages can cause hydrocephalus, so an external ventricular drainage can be required, while surgical evacuation of the hematoma cannot be performed always, since the surgical procedure could make more damages to intact basal ganglia (Kumar et al., 2005). However, in case of severe intracranial hypertension, decompressive craniotomy with/without the evacuation of the hematoma is the standard procedure to treat hemorrhages in this area (Fig. 15.15).

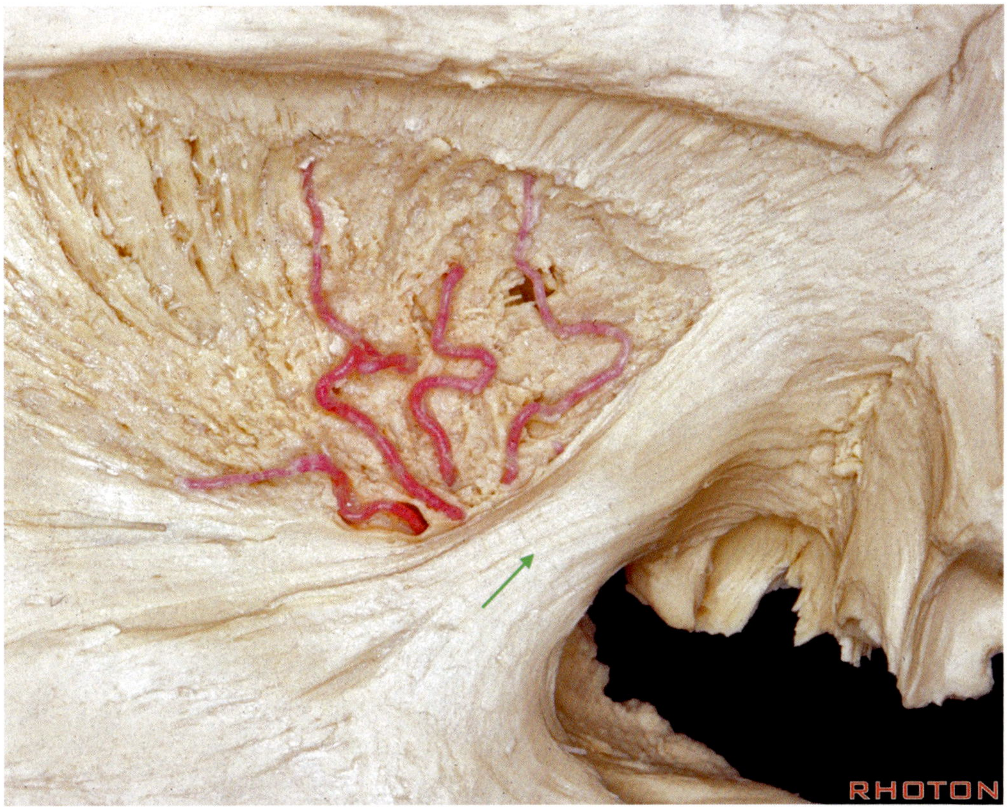

FIGURE 15.10 **Lenticulostriate arteries.** Under the uncinate fascicle (*green arrow*) they go through the anterior perforate substance. The lesion of these arteries, usually given by hypertensive crisis, represents one of the most common causes of hemorrhagic stroke in the putaminal region; see Fig. 15.15. *Image courtesy of The Rhoton Collection® http://rhoton.ineurodb.org/. Drawing by Brandon M. Ascenzi.*

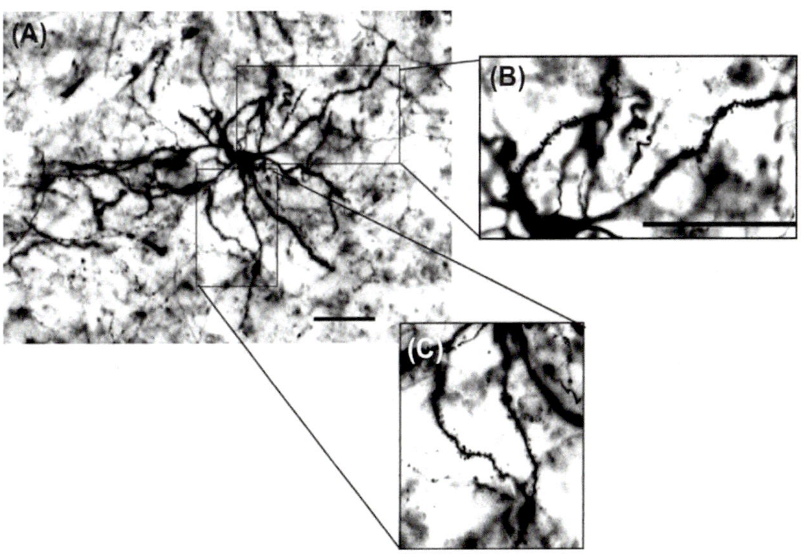

FIGURE 15.11 **Photomicrograph of a medium spiny neuron (MSN) filled with Lucifer Yellow.** Boxes in panel A are enlarged in panels B and C to reveal prominent spines on the dendritic processes. Scale bars = 50 μm. *Image courtesy from Haber, S.N., Adler, A., Bergman, H., 2012. Chapter 20: The basal ganglia. In: Mai, J.K., Paxinos, G. (Eds.), The Human Nervous System, third ed. Academic Press, pp. 678–738. https://doi.org/10.1016/B978-0-12-374236-0.10020-3.*

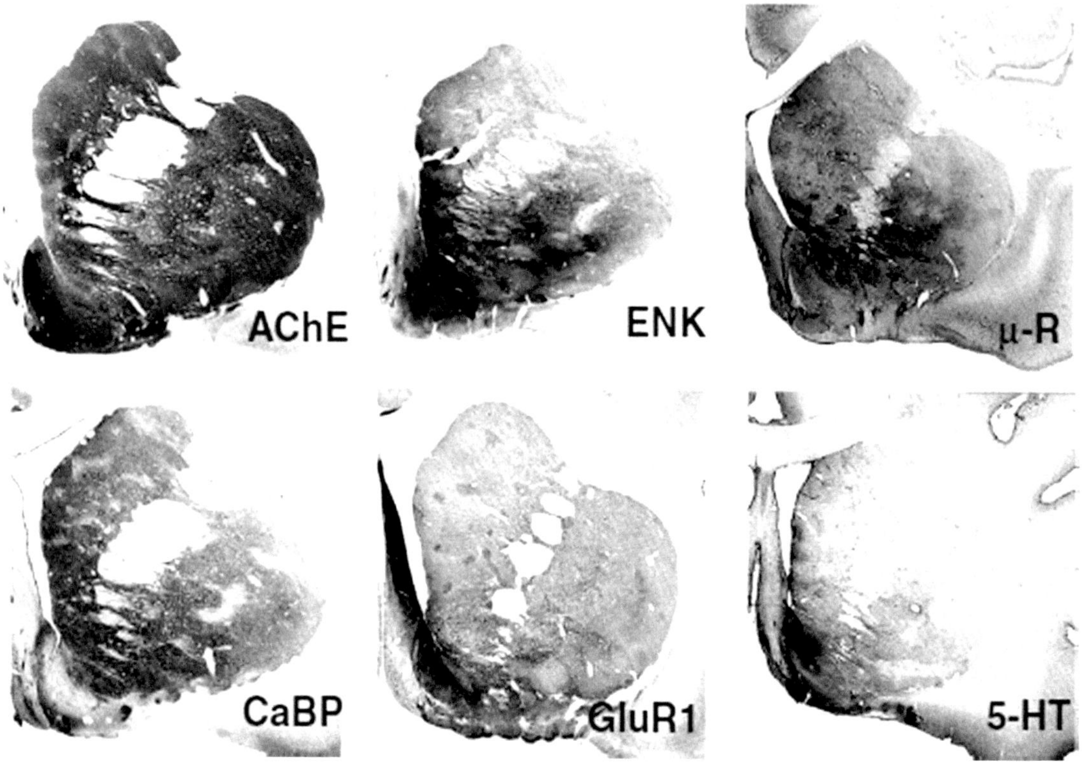

FIGURE 15.12 **Photomicrographs of the striatum at the level of the shell and core, immunostained for various transmitter-related molecules.** *5-HT*, serotonin; *AchE*, acetylcholinesterase; *CaBP*, calbindin-28; *ENK*, enkephalin; *GluR1*, GluR1 AMPA receptor subunit; *m-R*, m-opitate receptor. *Image courtesy from Haber, S.N., Adler, A., Bergman, H., 2012. Chapter 20: The basal ganglia. In: Mai, J.K., Paxinos, G. (Eds.), The Human Nervous System, third ed. Academic Press, pp. 678–738. https://doi.org/10.1016/B978-0-12-374236-0.10020-3.*

In the next section, we will describe how the basal ganglia are interconnected with each other and with the other structures surrounding them therefore we will illustrate their functional role. The basal ganglia represent a large functional unit that works as a "machine" where all the components work together and not separately.

BASAL GANGLIA PATHWAYS—INTRODUCTION

As previously explained, the striatum is constituted by the caudate-putamen complex and the MSNs are the most represented cytotype which, through its extensions, gives rise to the numerous, thin and myelinated fasciculate fibers organized radially between the caudate nucleus and putamen, giving to the entire complex a striated appearance, hence the name.

Two different subpopulations of MSNs that differ from each other for a different neurochemical and receptor phenotype give rise to two pathways that represent the milestone of the functional neuroanatomy of basal ganglia.

These two circuits or pathways are the direct pathway and indirect pathway.

MSNs involved in the direct pathway co-express dopamine receptor D1, substance-P and dynorphin, while striatal projection neurons involved in the indirect pathway co-express dopamine receptor D2 and enkephalin. More recently, clustering of transcriptomic data from mouse striatum revealed specific subsets of genes which expression differentiates between D1-expressing (Drd1a, Tac1, Isl1) and D2-expressing MSNs (Drd2, Adora2a, Penk, Gpr6, and Gp52) (Gerfen, 1989; Graybiel et al., 1991).

The principal neurons of the striatum receive convergent cortical and thalamic input. The density of such afferent synapses has been estimated in rat to be 920.000.000 per mm^3 (Ingham et al., 1998) and of these nearly all are formed on dendritic spines of spiny neurons.

Within the volume of a single MSN dendritic tree, each cortical axon makes at least 40 boutons. The volume of a spiny cell's dendritic tree is about 0.033 cubic millimeter (from the volume of a 0.4 mm-diameter sphere), and so in that volume there should be about 31.000.000 afferent-type synapses formed by cortical and thalamic axons ($0.4^3\pi$ 6*920.000.000).

Single cortical axons can innervate from 1% to 15% of the total volume of the striatum, even though the

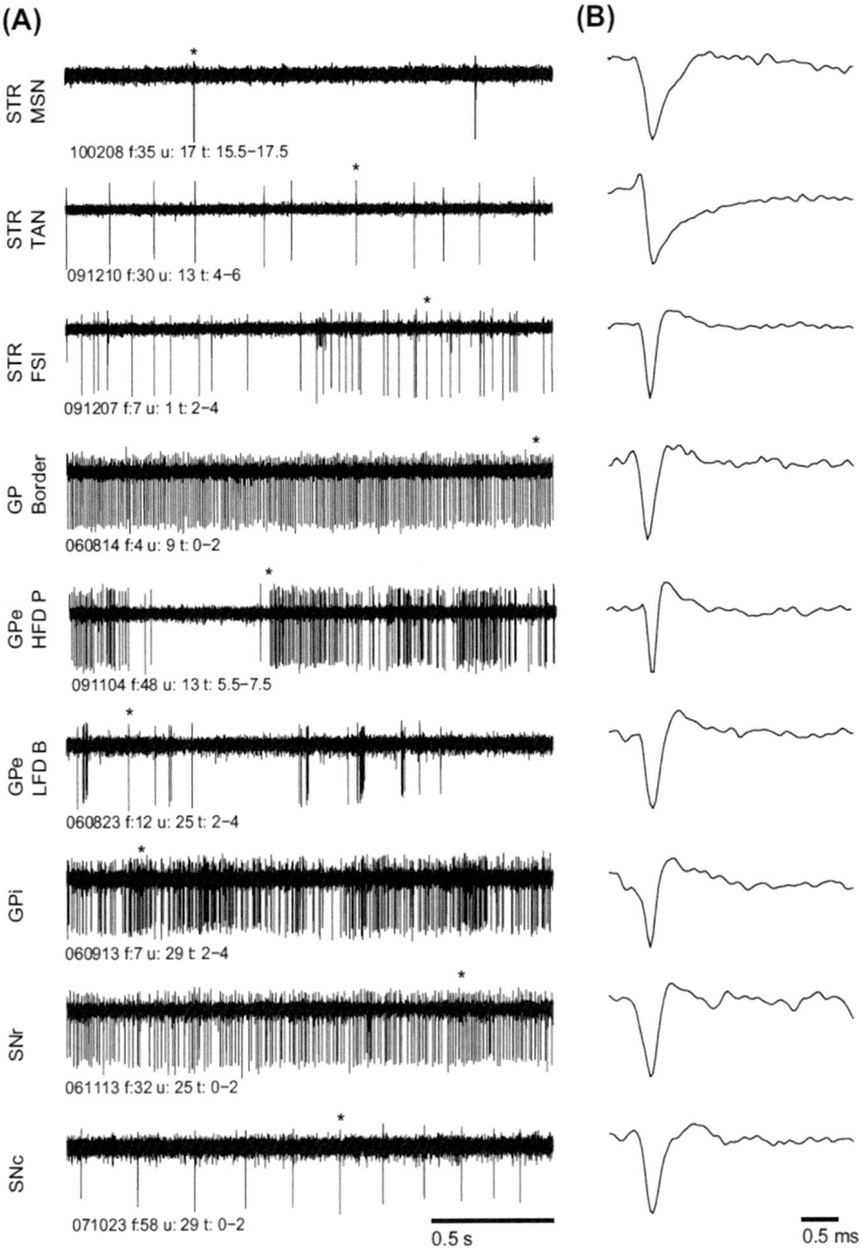

FIGURE 15.13 (A) **Typical examples of 2 s analog traces of extracellular recording filtered between 300 and 6000 Hz**. Asterisk marks the spike displayed in B. (B) **Examples of single spike waveforms taken from the complementary analog trace displayed in A**. *GP border*, globus pallidus border cell; *GPe HFD P*, external segment of globus pallidus high frequency discharge pauser; *GPe LFD B*, external segment of globus pallidus low frequency discharge burster; *GPi*, internal segment of globus pallidus; *SNc*, substantia nigra pars compacta; *SNr*, substantia nigra pars reticulata; *STR FSI*, striatal fast spiking interneuron; *STR MSN*, striatal medium spiny neuron; *STR TAN*, striatal tonically active neuron. *Image courtesy from Haber, S.N., Adler, A., Bergman, H., 2012. Chapter 20: The basal ganglia. In: Mai, J.K., Paxinos, G. (Eds.), The Human Nervous System, third ed. Academic Press, pp. 678–738. https://doi.org/10.1016/B978-0-12-374236-0.10020-3.*

number of cortical axons is about 17 million (Zheng and Wilson, 2002).

The most noticeable feature of the neostriatum as seen in the electron microscope is the large number of small axons forming asymmetrical synapses (Gray's type I) on the heads of dendritic spines. This kind of synapse accounts for about 80% of all synapses in the neostriatum (Ingham et al., 1998). It is the characteristic synaptic type formed by afferents from cerebral cortex and from parts of the intralaminar nuclei of the thalamus (e.g., Kemp et al., 1971). The axons from the cerebral cortex and thalamus are similar in morphology, with both exhibiting small round synaptic vesicles. A smaller number of cortical afferent synapses are made onto dendritic shafts and somata of neostriatal neurons. Some of these are formed on dendritic shafts of MSN projection and others are formed on the somata and dendrites of the aspiny neurons. The intralaminar thalamic nuclei (such as the central lateral nucleus) make synapses on dendritic spines, similarly to the corticostriatal synapses. Another region of the thalamus, the parafascicular nucleus, has been shown to form preferentially axodendritic rather than axospinous synapses, although almost half of the synapses on those cells are still formed onto dendritic spines (Dubé et al., 1988; Xu et al., 1991). It is important to keep in mind that the two types of

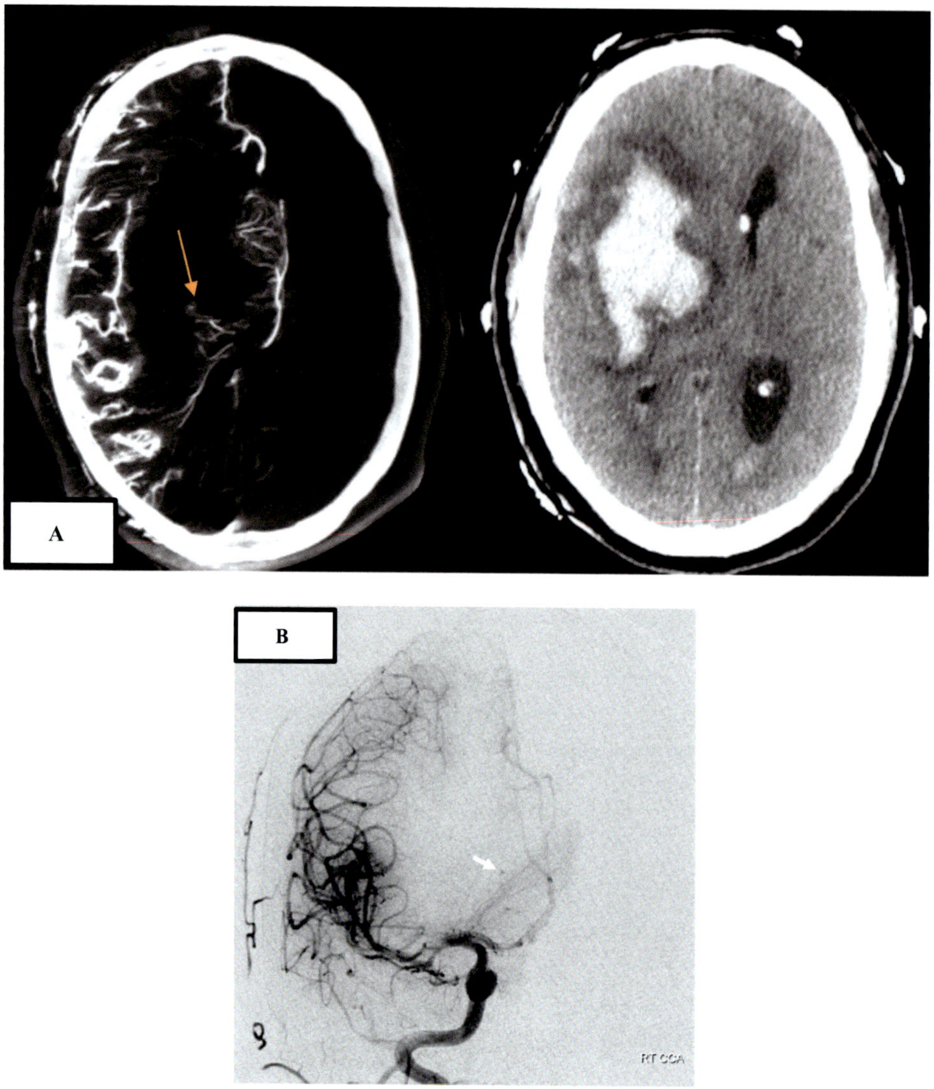

FIGURE 15.14 (A) **Basal ganglia hemorrhage**. Flat CT panel of a large intracerebral basal ganglia hemorrhage. The *orange arrow* points toward a lenticulostriate vessel. As a general consideration, unruptured lenticulostriate artery aneurysms (Charcot–Bouchard microaneurysms) are extremely rare to find in vivo; some have described as an incidental finding on cerebellar hemorrhage imaging studies. (B) **Spot Sign**. Diagnostic angiography of the same case as in A. The *white arrow* shows the focal accumulation/pooling of contrast around lenticulostriate vessels within hematomas. *Image courtesy of Neuroangio website (available on http://neuroangio.org/, with permission).*

pathways or circuits, direct and indirect, are not at all anatomically separated as was believed in the past years but are distinguished from each other based on their neurofunctional point of view (Fig. 15.16).

DIRECT PATHWAY

The direct pathway is made of corticostriatal fibers that come from the cortex and make synapsis with the striatal complex (putamen and caudate nucleus). The striatum sends striatonigral fibers to the *substantia nigra pars reticulata* (SNpr) that in turn sends back their efferences to the *ventral anterior nucleus of the thalamus* (VA) and to the lateral part (paralaminar) of the *mediodorsal nucleus of the thalamus* (MDpl). The loop closure is made by the fiber tracts arising from the VA and from the MDpl, both of which in fact project to the cerebral cortex. The thalamus also receives another quantity of fibers from the striatum that passing through the GPe reaches the GPi and from here gives rise to a pallidothalamic tract which, as well as the nigrothalamic trac, synapses with the VA and MDpl. Both the tracts, the pallidothalamic and nigrothalamic, are inhibitory and use GABA as neurotransmitter.

It is noteworthy that while passing through GP, the fibers constituting the *cortex-striato-SNpr-thalamus-cortex* pathway, or the *cortex-striato-pallido-thalamus-cortex*

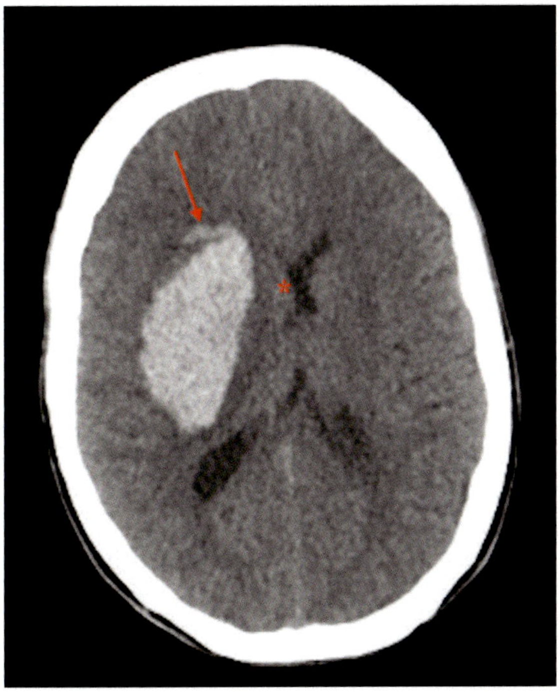

FIGURE 15.15 **CT-scan shows an intracranial hemorrage (*red arrow*)** Affecting *right side* putamen causing compression of the ventricular system (*red asterisk*) and intracranial hypertension. *Image courtesy of Prof. Antonio Santoro, Neurosurgical Department of A.O.U. Policlinico "Umberto I", University of Rome "Sapienza", Rome, Italy.*

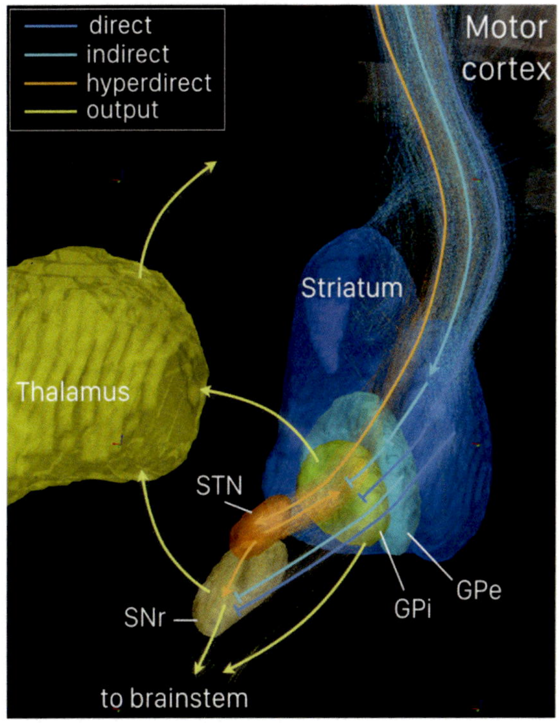

FIGURE 15.16 **Direct, indirect, and hyperdirect circuit.** *GPe*, Globus pallidus externus; *GPi*, Globus pallidus internus; *SNpr*, Substantia nigra pars reticulata; *STN*, Subthalamic nucleus.Depicted are parts of the basal ganglia and their interconnections as revealed by diffusion spectrum imaging based on 30 subjects from the human connectome project (HCP MGH). Direct, indirect, and hyperdirect pathways are visualized in different colors (see legend). Subcortical structures are rendered based on the Harvard–Oxford subcortical (thalamus) as well as the Atlasing of the basal ganglia atlas (other structures). Rendering was generated using TrackVis software. See the text for functional details. *Image courtesy of Andreas Horn, Wikipedia.*

pathway send many collaterals and arborizations to the pallidal neurons. The striatopallidal and striatonigral fibers, as well as pallidothalamic and nigrothalamic fibers, are inhibitory and use GABA as neurotransmitter.

The final station of the direct pathway is, as said before, the cerebral cortex and in particular way: (1) the efferences from the VA are directed to the motor, premotor and supplementary motor areas (4, 6, and 8 BA, respectively); (2) the efferences from MDpl are directed diffusively to the prefrontal cortex and to the frontal eye field (FEF).

Differently to the striatopallidal, striatonigral, pallidothalamic, and nigrothalamic tracts, the talamocortical fibers are excitatory and use the glutamate as neurotransmitter; they make synapsis purely with neurons of I and III layer of neocortex and to lesser extent with V layer (Vaughan et al., 1976).

Regarding the functional mechanism of the direct pathway, it starts with the excitation of striatal MSNs by the corticostriatal fibers that use glutamate as neurotransmitter. The excited MSNs in turn inhibit the pallidal neurons of GPi and\or nigral neurons of SNpr.

This represents a perfect example of "inhibition of an inhibition" ($- \times - = +$) that causes a disinhibition and therefore the excitation of the cortical neurons by the glutamatergic thalamocortical fibers. The direct pathway exerts a facilitating action, contrarily to the indirect pathway (see later), on stimuli originating from the motor cortex that will become physical actions only after they have passed through this very complex system who works as a "positive filter" and have reached the spinal motoneurons.

Recordings of single cell activity in the basal ganglia of awake monkeys have shown that the striatal medium spiny cells fire very rarely and in episodes that only last for about 0.1–3 s, while the pallidal and SNr neurons, on the other hand, fire tonically at very high rates.

In doing so, they keep the thalamic neurons in a permanent state of inhibition. However, when phasic excitatory inputs from the cortex transiently activate the direct pathway, the tonically active pallidal and nigral neurons are briefly suppressed, thus permitting the thalamus and cortex to become activated (DeLong et al., 2000; Wilson, 2004).

INDIRECT PATHWAY

The indirect pathway works in the opposite way than the direct pathway in fact, as the latter has a positive retroaction on the thalamocortical fibers, the indirect pathway exerts a negative retroaction (DeLong et al., 1981).

Essentially the indirect pathway is composed of efferent fibers that from the striatum go to inhibit GPe; the latter normally inhibits the activity of subthalamic nucleus (STN) using GABA. Therefore, if SNT remains uninhibited by GPe action it can excite their glutamatergic fibers the GPi\GPe and SNpr allowing to the GPi fiber that constitute the pallidothalamic fibers to keep the VA\VLa nucleus of thalamus in state of inhibition. It is noteworthy that fibers from GPe project to the STN according to a topographical scheme (Carpenter et al., 1981) (see Chapter 9 for further information).

The STN receive glutamatergic afferences from cerebral cortex and mainly they come from primary motor cortex (BA4), premotor cortex (BA6), and supplemental motor cortex (SMA-BA6) (Nambu et al., 2002). STN is the only excitatory nucleus in the basal ganglia system using the glutamate as neurotransmitter.

At this point is important to introduce a discussion on dopamine that, as we will see, it plays a fundamental role in balancing the entire system of direct and indirect pathway.

As anticipated above the striatum express both D1 and D2 dopamine receptors and they work in opposite way: D1 is excitatory while D2 is inhibitory (see Chapter 3). This means that in the striatum dopamine works differently at the same time: it excites the MSNs expressing D1 who send fibers to the GPi and inhibits the MSNs expressing D2 who send fibers to the GPe. The excitation of MSNs that send fiber to the GPi causes a disinhibition of the GPi-thalamus fibers tract. In contrast, the inhibition of MSNs that send fibers to GPe provokes a disinhibition of GPe-STN fibers and so an increased inhibition of STN.

However, although the synaptic actions are different, the dopaminergic afferent to the direct and indirect pathway provoke the same result: reduce the inhibition of the thalamocortical neurons and facilitate the movements starting in the motor cortex (Alexander and Crutcher, 1990; Wichmann et al., 1996).

CONNECTIONS

Strio-Nigro-Striatal Circuit

The connections between the striatum and substantia nigra represent the most important fibers system that allows a mutual control of the activities of both structures; this fibers system is called strio-nigro-striatal (Fig. 15.17).

Striatonigral fibers from the MSNs of striatum project mainly to the neurons of the SNr (Bunney et al., 1976; Grofova et al., 1975; Kim et al., 1976).

These projections show an inversed dorsoventral topography: the dorsal part of striatum sends fibers to the ventral part of mesencephalon and the ventral part of striatum sends fibers to the dorsal part of mesencephalon. Moreover, the lateral part of the striatum—sensorimotor part—projects to the lateral part of the SNr while the central part of the striatum—associative part—projects to the ventral part of the SNc and SNr; these two projections are largely overlapping (Lindvall et al., 1990). The ventromedial striatum innervates a large part of the dorsomedial region of the mesencephalon that includes a great part of SNc, the medial part of the SNr and the VTA. The striatonigral fibers use GABA and substance P as neurotransmitters and inhibit the neurons of SNc.

The dopaminergic neurons of the SN form the major nigrostriatal projection and their fibers are very thin and topographically organized (Carpenter et al., 1972; Tanaka et al., 1982; Ungerstedt, 1971), it has been seen in fact that each of these connections is directed mainly to a specific region of the striatum but, in the meantime, it also takes part to the innervation of the remaining parts. Specifically, the dorsal tier of the SN neurons mainly projects to the ventral part of the associative striatum with preferential projections of the A8 neurons to the ventrocaudal putamen, A10 to the nucleus accumbens, and of dorsal A9 neurons to both (Fox et al., 1971). The distribution of dopaminergic axons in the striatum is important not only for modulating neuronal activity but also for controlling the plasticity of synaptic circuits as dopamine has been shown to do this in the neostriatum (Fox et al., 1971).

Regarding the effect of dopamine in the modulation of the striatum activity and, in turn, the effect dopamine has on direct and indirect circuits it can be said that: although the synaptic action is different depending on which receptor dopamine binds (D1 is expressed on MSN $_{\text{GABA + substance P}}$ forming the direct pathway and D2 is expressed on MSN $_{\text{GABA + enkephalin}}$ forming the indirect pathway) dopaminergic afferent fibers work in the same way on both circuits in fact they reduce the inhibition of thalamocortical neurons and, ultimately, facilitate movements initiated in the motor cortex.

The mesencephalic dopaminergic neurons establish direct synaptic contacts with MSNs (Freund et al., 1984) and it is known that striatonigral fibers in turn synapse with the somata and dendrites of SNc neurons (Hedreen et al., 1991; Wassef et al., 1981). Axodendritic contacts are found mainly in the SNr, in which many SNc cells extend their long ventrally directed dendrites. As said before the different striatal areas and dopaminergic cells groups are connected each other at different levels and there is also converging evidence indicating that tonic release of

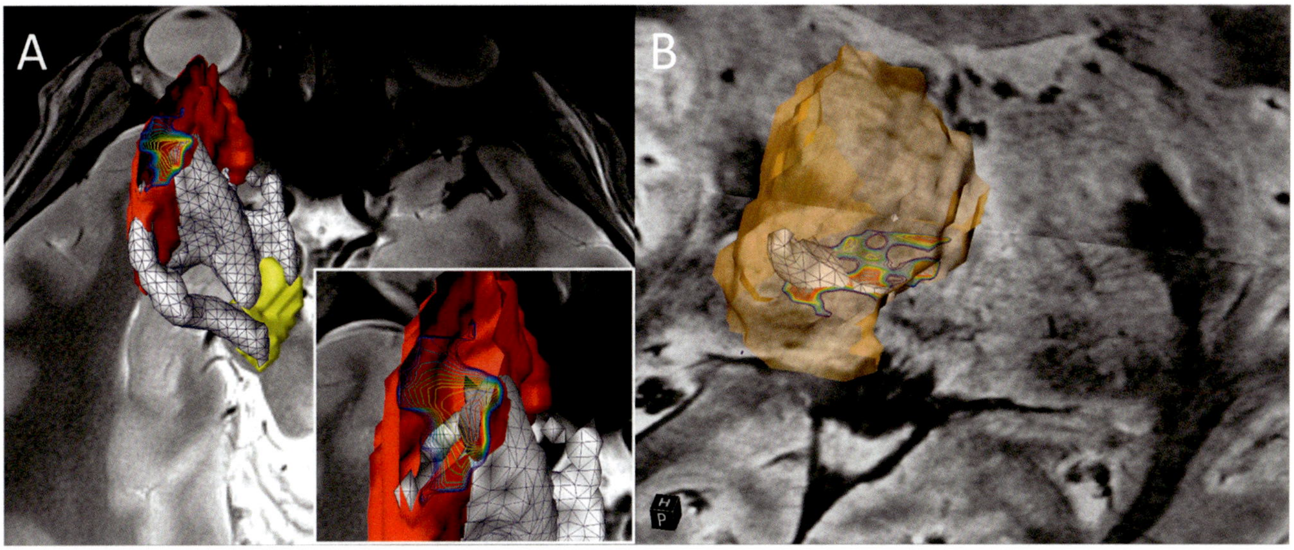

FIGURE 15.17 **Spatial agreement between functional and anatomical connectivity maps.** The figure demonstrates spatial overlap, within the putamen or thalamus, between the areas reached by white matter nigral projections and the resting-state functional activations of the substantia nigra. The white wireframe volumes represent reconstructions of the nigrostriatal (A) and nigrothalamic (B) pathways. R-fMRI activation maps of the substantia nigra were obtained using FDR. They are represented using isolines of equal *P*-values, with *blue lines* corresponding to $P = .01$ and *red lines* to $P \ll .01$. Spatial agreement between anatomical and functional connectivity maps is visible in the dorsal part of the putamen (A) and in thalamus, within the putative Vc nucleus (B). Orientation in (A) is caudo-rostral with T2-weighted image at the level of the substantia nigra. *Inset* in (A) is a magnified view of the overlap area between anatomical and functional connectivity. The *red surface* representing putamen was clipped in its *upper part* to reveal the R-fMRI activation (*isolines*) and endpoint of the nigrostriatal projections. Orientation in (B) is oblique caudo-rostral with SWI images at the level of the inferior and anterior thalamus. *Image courtesy of Lenglet, C., Abosch, A., Yacoub, E., De Martino, F., Sapiro, G., Harel, N., 2012. Comprehensive in vivo mapping of the human basal ganglia and thalamic connectome in individuals using 7T MRI. PLoS One 7 (1), e29153. https://doi.org/10.1371/journal.pone.0029153.*

dopamine attenuates MSN's neuronal response, while phasic release potentiates striatal response (Calabresi et al., 2000). Arborization of nigrostriatal axons have been described by Prensa and Parent (2001). These axons formed spars arborizations, usually composed of several focal subarborization.

While over 90% of the cortical inputs converge on the dendritic spines, as regards the nigrostriatal synapses only half of them are formed on the spines.

Although the dopaminergic input is very important functionally it is not comparable to the cortical and thalamic projections in size. Moreover, most of dopaminergic synapses are symmetrical and only 20% of all striatal synapses are symmetrical.

The boutons that arise from the local collaterals of MSNs contain large synaptic vesicles of variable shape and form symmetrical synapses onto the stalks of dendritic spines, dendritic shafts, somata, and initial segments of axons (Wilson et al., 2004; Somogyi et al., 1981).

Cortical *inputs* to the cholinergic interneurons of striatum are restricted to their distal dendrites, whereas the thalamic *input* makes easily demonstrable synaptic contacts on cholinergic cells (Lapper et al., 1992).

The *output* axons of the MSNs form synapses in the SN and GP; these synapses are morphologically similar to those formed by their local axonal terminals (Chang et al., 1981). In both the GP and SN the dendrites of most neurons are completely encased in a quilt-work of synaptic terminals and the majority of these latter are derived from the neostriatal efferent fibers. A second kind of terminal found on these dendrites, containing small, round vesicles and forming asymmetrical contacts, is derived from the subthalamic nucleus (Kita and Kitai, 1987).

AFFERENCES AND EFFERENCES TO AND FROM THE BASAL GANGLIA

In this section, we will look at the most important afferents and efferences of the basal ganglia. Some of these connections have already been described in the previous paragraphs concerning the two most important circuits, direct and indirect, but we will still take them into consideration from a broader and more functional point of view.

AFFERENCES

Cortico-Striatal Projection

The entire cerebral cortex is known to project topographically to the striatum.

The short-duration monosynaptic component of the corticostriatal EPSP to a single stimulus is mediated by glutamate and primarily by glutamate acting at non-NMDA receptors (e.g., Calabresi et al., 1996; Kita and Kita, 2001).

Cortical inputs are quite heterogeneous from each other based on which cortical region they originate from. Input from the sensorimotor cortex is extensive and bilateral, whereas that from the visual cortex is minimal. Some studies emphasized the high degree of overlap of corticostriatal projections (Braak et al., 2000; Hoover et al., 1999; Schmahmann et al., 1998), while others revealed that the sensorimotor cortex in monkey projects exclusively to the putamen where a somatotopic representation of the leg, arm, and face occurs in the form of obliquely arranged strips (DeLong, 1990, 2000). In contrast, associative areas of the prefrontal, temporal, parietal, and anterior cingulate cortices appear to project almost exclusively to the caudate nucleus in monkeys (Garcia-Rill et al., 2001). These finding suggest that the caudate nucleus in primates is more closely related to complex and associative types of behavior, whereas the putamen appears more directly involved in sensorimotor aspect of forebrain mechanisms. Afferents from limbic and paralimbic cortical areas terminate largely in the ventral portion of the striatum, including the deep layers of the olfactory tubercle and the ventral part of both the caudate nucleus and the putamen.

Thus, based on the organization of its cortical afferents, the striatum can be subdivided into three major functional sectors termed: associative, sensorimotor, and limbic territories (Parent et al., 1990).

The associative territory comprises most of the head of caudate and a large portion of the rostral, precommissural putamen. The sensorimotor includes most of the postcommissural putamen, the dorsolateral rim of the head of the caudate nucleus and the lateral aspect of the body of the caudate nucleus.

Finally, the limbic territory comprises the area of the ventral striatum (Parent et al., 1990).

Terminals for corticostriatal axons are not uniformly distributed in striatal tissue but appear as patches (striosomes) or clusters of various sizes. Furthermore, the cortical areas reciprocally connected via corticocortical connection tend to share common zones of fiber termination in the striatum (Yeterian et al., 1978).

Synaptological Considerations of the Cortico-Striatal Projection

Like most glutamate-mediated synapses, those on neostriatal spiny neurons can undergo long-lasting changes (synaptic plasticity) in effectiveness after tetanic stimulation. Tetanic local stimulation in striatal slices under experimental conditions that minimize NMDA receptor activation leads to long-term depression (LTD) of glutamatergic synaptic transmission.

The same stimulation, applied in the absence of extracellular Mg^{2+}, so that the block of NMDA receptors is removed, will give rise to long-term potentiation (LTP) (Calabresi et al., 1992b). Both of these processes, LTD and LTP, underlie learning and other high-task functions in which the basal ganglia are involved. LTP has been observed in vivo using cortical stimulation at physiological frequencies (Charpier et al., 1999; Reynolds and Wickens, 2000). This relative ease at inducing LTP in vivo may result from the background of depolarizing synaptic activity seen in the spiny neurons in vivo. Both LTP and LTD are induced by trains of high-frequency stimulation, so it is worth asking what might normally determine whether use would increase or decrease the strength of glutamate synapses on spiny neurons. Induction of LTD in the striatum depends upon the presence of dopamine, being prevented by blockade of either D_1 or D_2 receptors (Calabresi et al., 1992a). In addition, LTD requires activation of group I metabotropic glutamate receptors (Sung et al., 2001). AMPA receptor activation is not necessary, but depolarization is required and normally AMPA receptor activation is mainly responsible for the depolarizing synaptic potential at striatal glutamatergic synapses. The effect of depolarization is probably to trigger calcium influx through L-type channels, as LTD induction is also prevented by L-type calcium channel blockers (Strack et al., 1997).

Thus, LTD requires a large number of conditions for its induction in the striatum, including dopaminergic activation, release of NO (presumably by the GABA/SOM/NO interneurons), postsynaptic depolarization and activation of L-channels, and release of glutamate by axons to act at metabotropic glutamate receptors. Once established, LTD reduces the probability of release of glutamate by presynaptic terminals on the spiny neuron (Strack et al., 1997).

Thalamostriatal Projection

The thalamostriatal projection come from centromedian-parafascicular (CM-PF) complex and is directed densely and selectively to the putamen (Fig. 15.18). The CM-PF complex is very enlarged in

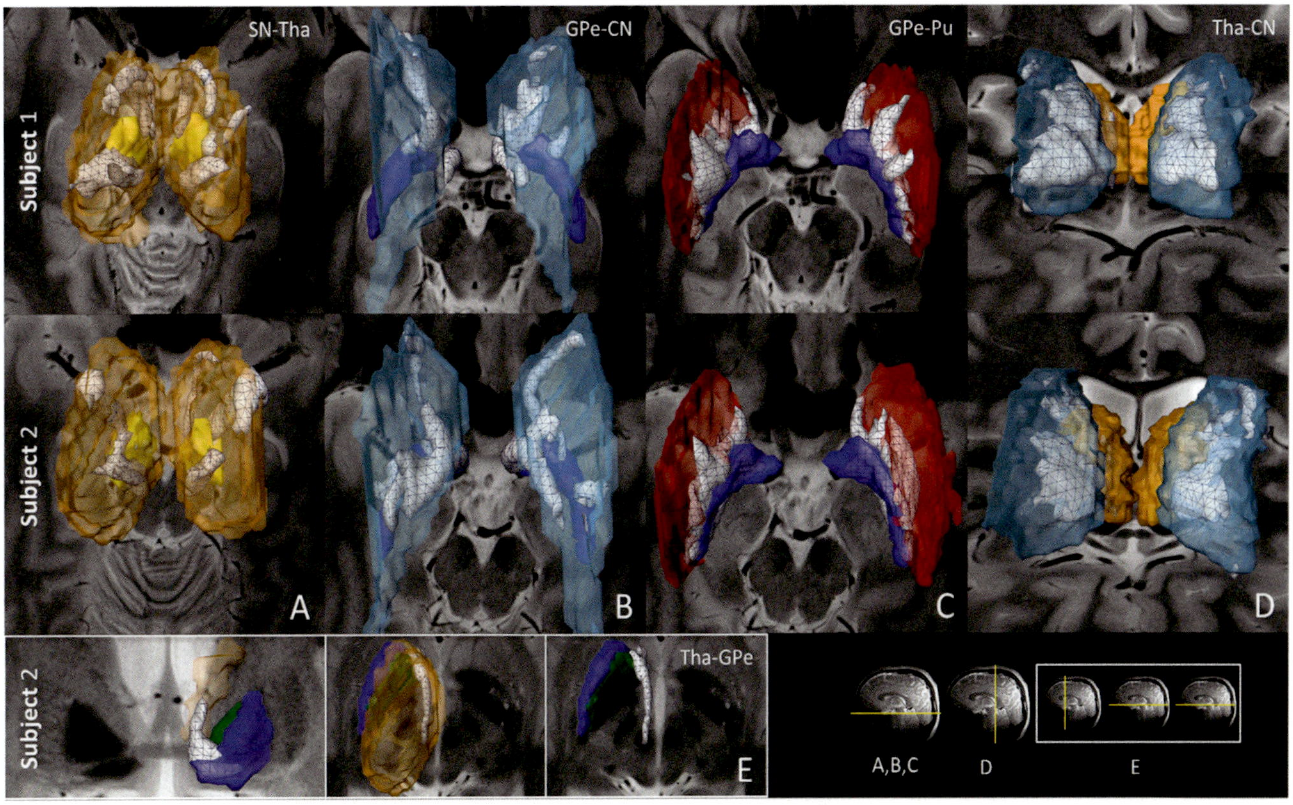

FIGURE 15.18 **Reconstruction of fiber pathways of the basal ganglia and thalamus.** White matter pathways identified using diffusion MRI probabilistic tractography. The white wireframe volumes represent white matter tracts identified as pathways of interest. From *left* to *right*: (Panel A) nigrothalamic pathway, (Panel B) and (Panel C) pallidostriatal pathway, (Panel D) thalamostriatal pathway, (Panel E) thalamopallidal pathway. (*Bottom right inset*) *yellow lines* depict the orientation and location of T2-weighted images in each panel. (Panels A, B, C) axial images at the level of SN; (Panel D) coronal image at the level of the posterior Tha; (Panel E, *left*) coronal image at the level of the anterior thalamus; (Panel E, *middle and right*) axial images at the level of the inferior GP. Color code: Caudate nucleus, CN: light blue; Putamen, Pu: red; external globus pallidus, GPe: dark blue; internal globus pallidus, GPi: green; substantia nigra, SN: yellow; thalamus, Tha: orange. *Image courtesy of Lenglet, C., Abosch, A., Yacoub, E., De Martino, F., Sapiro, G., Harel, N., 2012. Comprehensive in vivo mapping of the human basal ganglia and thalamic connectome in individuals using 7T MRI. PLoS One 7 (1), e29153. https://doi.org/10.1371/journal.pone.0029153.*

monkeys and humans and receives a massive projection from the motor cortex (Mettler, 1947). Anterograde labeling studies revealed that the projections from the centromedian nucleus (CM) terminate principally in the sensorimotor striatal territory, whereas those from the parafascicular nucleus (PF) terminate in the associative territory (Parent et al., 1983; Sadikot et al., 1990).

CM-PF complex is the major source of thalamic afferents to the striatum (Jones, 1985; Parent et al., 1983) in fact the studies further showed that thalamostriatal projection is topographically organized, with a rostrocaudal and dorsoventral correspondence between the intralaminar nuclei and their main determination sites in the striatum. In addition to prominent afferents from the intralaminar nuclei, the striatum receives a significant input for other thalamic nuclei such as the ventral anterior (VA), ventral lateral (VL) and the lateral posterior (LP) nuclei (Parent et al., 1986). The thalamostriatal fibers terminate principally on the shaft and, less abundantly, on the dendritic spines of the MSN (Sadikot et al., 1992a,b). The fibers form asymmetric synapses and are thought to exert an excitatory influence that is mediated probably through glutamate because virtually all excitatory responses evoked in MSN by local stimulation are abolished by application of excitatory amino acid receptor blockers (Jiang et al., 2011; Kita and Kita, 2001).

Mesostriatal Projection

One of the most important systems of afferent fibers directed to the basal ganglia is that coming from reticular formation and in particular way that from mesencephalic reticular formation. These afferent fibers are directed to the intralaminar nuclei of thalamus that in turn projects to the striatum.

The *mesostriatal serotoninergic pathway* that mainly comes from the *dorsal raphe nucleus* (DRN) and maybe represent the most important serotoninergic afference directed to the striatum, and in particular way to its ventral and medial region (Andén et al., 1966; Geyer et al., 1976; Smith et al., 1986).

Striatopallidal

Striatopallidal fibers directed to the GPi and the GABAergic pallidal neurons that project to the thalamus form the largest output system form the basal ganglia. Striatopallidal fibers are topographically organized in both dorsoventral and rostrocaudal sequences and radiate into various parts of the pallidum like spokes of a wheel (Cowan et al., 1994; Szabo, 1967). Them originating from the caudate nucleus and the rostral part of putamen—i.e., the *associative striatum*—pierce the internal capsule and penetrate the globus pallidus thought its dorsal surface. These caudatopallidal fibers arborize in the form of dense plexuses principally on the rostral pole of the GPe and, more caudally, in the dorsomedial third of the both GPe and GPi. Striatopallidal fibers arising from the postcommissural putamen—i.e., *sensorimotor striatum*—from several distinct fascicles that pierce the *external medullary lamina* to reach the GP through its lateral border. In both segments of the pallidum, the putamenopallidal fibers arborize in elongated bands aligned parallel with the medullary lamina. Neuroanatomical studies indicate that axons of striatal neurons located in the associative or sensorimotor striatal territory remain rather well segregated form one another at pallidal levels (Mesulam et al., 1984; Parent et al., 1985).

It has long been assumed that a single striatal efferent system supplies both the globus pallidus and the substantia nigra through axon collaterals (Francois et al., 2002; Kemp et al., 1971). Retrograde double-labeling studies in primates however have revealed that striatal projections to the GPe, to the GPi and to the substantia nigra arise principally from separate neurons (Elliot et al., 2002; Parent et al., 1989). These findings reveal that the striatofugal system is not a monolithic entity but, instead, appears to be composed of distinct striato-GPe, striato-GPi, and striatonigral (see the corresponding paragraph above). These findings agree with data on the distribution of various neuropeptides at both pallidal and nigral levels. The striatopallidal fibers use GABA as neurotransmitter; these projection neurons can, nevertheless, be distinguished from each other by their neuroactive peptide content. It is believed that substance P, enkephalin and dynorphin fibers, distributed differentially in the two segments of the globus pallidus, play different roles in modifying the distinctive efferent system of these segments, which arise from intrinsic GABAergic neurons. Comparison of normal human brains with those from patients neurological and psychiatric disorders as well as Huntington's Disease (HD) indicated substantial reduction of substance P and enkephalin in the globus pallidus and substantia nigra (Emson et al., 1980). More recent studies demonstrated that high levels of substance P might be involved also in depression (Schwarz et al., 2002) and consequentially sleep disturbances (Lieb et al., 2002), in line with the role that globus pallidus play in behavior control.

CONNECTIONS OF THE GLOBUS PALLIDUS

The efferent fibers from the GPe and GPi can be divided into four main bundles:

(a) the ansa lenticularis
(b) the lenticular fasciculus
(c) the pallidotegmental fibers
(d) pallidosubthalamic fibers.

The first three of these arise exclusively from the GPi (Calabresi et al., 2000; Lavoie et al., 1994; Mehler et al., 1960) instead the pallidosubthalamic fibers arise exclusively from the GPe (Carman et al., 1963, 1965).

For how concern the ansa lenticularis and the lenticular fasciculus they are pallidothalamic fibers arising from specific portion of the GPi. Those of the ansa lenticularis arise predominantly from outer part of the GPi (i.e., from that part of the GPi lateral to the accessory medullary lamina) while those of lenticular fasciculus arise from the inner part of GPi (i.e., form the portion medial to the accessory medullary lamina).

Pallidofugal fibers from Forel's field H pass rostrally and laterally along the dorsal surface of the zona incerta where they form part of thalamic fasciculus. The latter projects dorsolaterally over the zona incerta to terminate in specific nuclear subdivision of the rostral ventral tier of the thalamic nuclei (Mehler et al., 1960). For a more detailed description of these structures so far described, see Chapter 9.

In their complex pallidothalamic fibers project to the ventral anterior (VApc, pars principalis) and ventral lateral (VLo, pars oralis) of thalamic nuclei and give off collaterals to the centromedian (CM) nucleus (Fig. 15.19). Pallidal projections to the rostral ventral tier of thalamic nuclei are topographically organized (Huo et al., 1973), and they do not overlap the crossed projection for the deep cerebellar nuclei or the ascending projection for the substantia nigra (Hallanger et al., 1987; Halliday, 2004). Each of the ascending fibers terminate in separate divisions of the thalamic nuclei. The ventral lateral thalamic nucleus, particularly the pars oralis (VLo), project to the supplementary motor area on the medial surface of the hemisphere and to the lateral premotor area (BA 6).

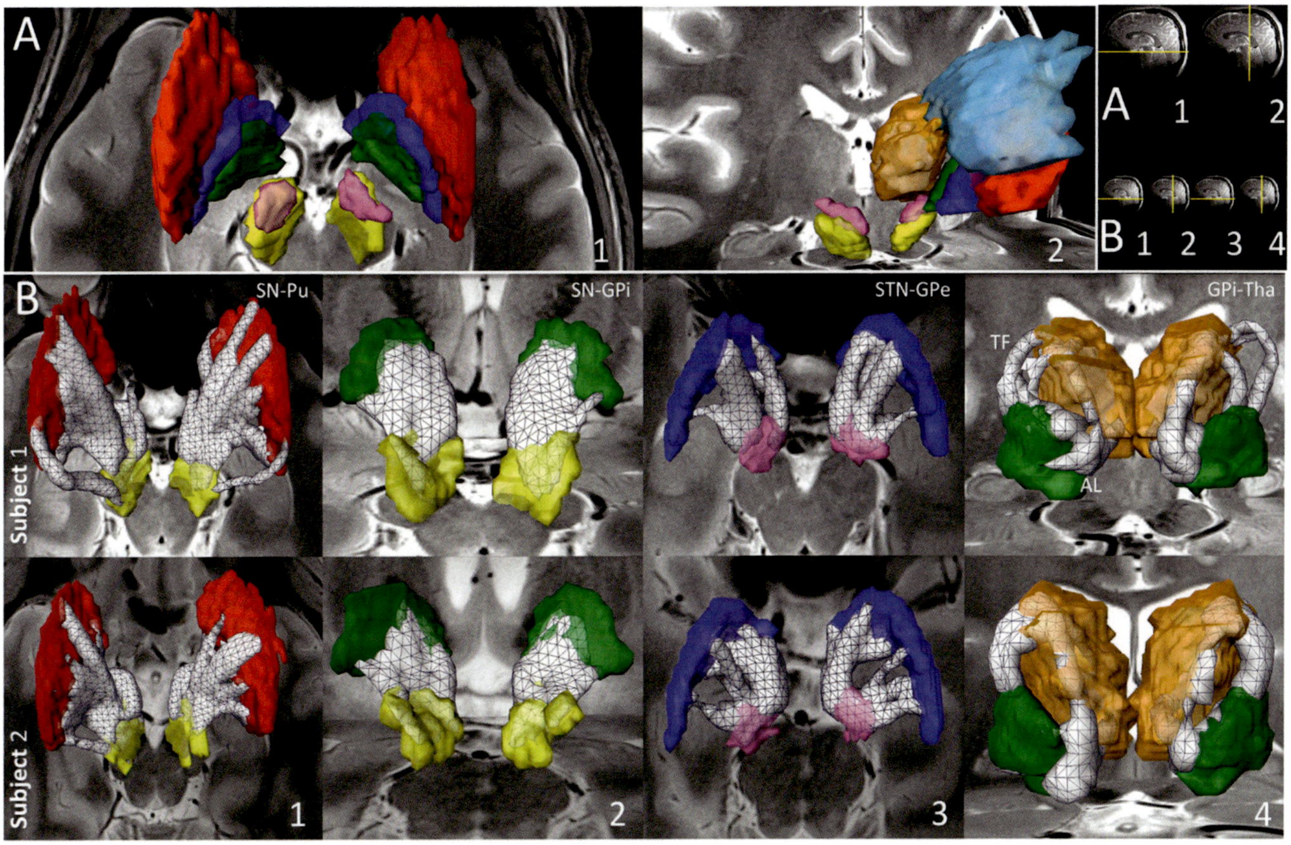

FIGURE 15.19 **Segmentation and reconstruction of fiber pathways of the basal ganglia and thalamus.** (Panel A) three-dimensional visualization of manual segmentations of basal ganglia and thalamus from high-resolution SWI, T2-weighted images and fractional anisotropy maps. Segmentations are superimposed on T2-weighted images. (Panel B) white matter pathways identified using diffusion MRI probabilistic tractography. The white wireframe volumes represent white matter tracts identified as pathways of interest. From *left* to *right*: (1) nigrostriatal, (2) nigropallidal, (3) subthalamopallidal, and (4) pallidothalamic pathways with AL: *ansa lenticularis* and TF: *thalamic fasciculus*. Background images are 7T T2-weighted MR images. (*Top right inset*) *yellow lines* depict the orientation and location of T2-weighted images in each panel. (A1) axial image at the level of SN; (A2) coronal image at the level of the posterior Tha; (B1) axial image at the level of SN, caudo-rostral orientation; (B2) coronal image at the level of the anterior GPi; (B3) axial image slightly inferior to STN to avoid obscuring portions of the tract; (B4) coronal image at the level of the posterior Tha, rostro-caudal orientation. Color code: Caudate nucleus, CN: light blue; Putamen, Pu: red; external globus pallidus, GPe: dark blue; internal globus pallidus, GPi: green; substantia nigra, SN: yellow; subthalamic nucleus, STN: magenta; thalamus, Tha: orange. *Image courtesy of Lenglet, C., Abosch, A., Yacoub, E., De Martino, F., Sapiro, G., Harel, N., 2012. Comprehensive in vivo mapping of the human basal ganglia and thalamic connectome in individuals using 7T MRI. PLoS One 7 (1), e29153. https://doi.org/10.1371/journal.pone.0029153.*

There is a significant proportion, approximately 10%–20%, of the GPi efferent fibers that cross the midline to reach the contralateral ventral tier nuclei of the thalamus (Skinner et al., 1984; Smith et al., 1994). These fibers arborize in patterns that are similar to those disclosed ipsilaterally.

The *pallidotegmental fibers* derived from GPi (Kuo et al., 1973; Mehler et al., 1960) become identifiable as a separate bundle dorsomedial to the subthalamic nucleus. The fibers of this bundle descend along the ventrolateral border of the red nucleus and in the caudal midbrain tegmentum they sweep dorsolaterally to terminate principally upon large cells of the pedunculopontine nucleus (PPN). Although the PPN receives its largest input from the medial pallidal segment, it also receives projections from the cerebral cortex (Grillner et al., 1973) and the substantia nigra (Carman et al., 1963, 1965; Haber et al., 1995; Edley et al., 1983).

Projections of the PPN appear to be mainly ascending, with the largest number of ascending fibers projecting to the SNc, although a significant number of fibers project to the GPi and to the subthalamic nucleus (Carman et al., 1963, 1965; Haber et al., 1995; Edley et al., 1983). Connections between the PPN and the SN are not reciprocal in the strict sense, since nigrotegmental fibers arise from cells of the SNr and ascending projections from the pedunculopontine nucleus end upon cells of the SNc (Berendse et al., 1990; Edley et al., 1983). Large cells in the PPN are strongly cholinergic and in the rat, they seem to project mainly to the ventrolateral thalamic nuclei. Noncholinergic neurons in and around this nucleus are described as projecting to the basal ganglia and related nuclei (Lee et al., 1988). In monkeys, the major projection of the PPN is to the ipsilateral SN and this projection is at least in part cholinergic (Lavoie et al., 1994). Glutamate is also present in neurons of the PPN, including the cholinergic neurons (Lavoie et al., 1994). Thus, the PPN may exert a dual cholinergic and glutamatergic excitatory effect upon nuclei of the basal ganglia.

Also worth noting is the demonstration of a direct projection from the deep cerebellar nuclei to the PPN in monkeys. This projection indicates that the PPN may act as an important functional interface between the two major subcortical systems involved in the control of motor behavior that are: the cerebellum and the basal ganglia.

The *pallidosubthalamic fibers* arising only from the GPe and form, with fibers of subthalamic nucleus that project back to the GP, the *subthalamic fasciculus.*

The pallidosubthalamic fibers arising only from the GPe and are topographically organized (Carman et al., 1963, 1965; Haber et al., 1995; Edley et al., 1983): rostral parts of the GPe project to medial and rostral parts of the subthalamic nucleus. Cells in the central division of the GPe project to the lateral third of the subthalamic nucleus throughout most of its rostrocaudal extent; some of them emit collaterals that arborize in lateral parts of the SNpr. Pallidal neurons projecting to the subthalamic nucleus are GABAergic.

Pallido-Abenular Projection

The dorsal diencephalic conduction (DDC) system is one of two major pathways that interconnect the limbic forebrain and sites in the mid- and hindbrain, the other pathway being the medial forebrain bundle (MFB; Sutherland, 1982). These two pathways appear to represent parallel neural circuits—they share sources of afferent inputs as well as efferent targets and there is an overlap in their physiology and function.

The DDC comprises three core components: the habenular nuclei; the stria medullaris (SM), which is the main fiber tract through which inputs from the forebrain arrive at the habenulae; and the fasciculus retroflexus (FR), a prominent fiber tract that predominantly carries efferent axons from the habenula toward the targets in the midbrain/hindbrain.

In the rat, virtually every entopeduncular neuron appears to project to the LHb, suggesting that the axons are collaterals of the pallido-thalamic pathway (Herkenham et al., 1977). This pallido-habenular pathway also exists in cats and monkeys. However, in monkeys, it appears that while the LHb receives substantial innervation from the internal segment of the globus pallidus, this innervation derives from a different group of pallidal neurons to those that innervate premotor neurons in the thalamus and brainstem (Parent et al., 2001).

While DDC has several functions, several recent reports regarding the pallido-abenular pathway provide evidence for strong functional links between the LHb and globus pallidus, which are involved in modulating motor behaviors and learning new behavioral responses to salient stimuli (Bianco et al., 2009).

CONNECTION (AFFERENCES AND EFFERENCES) OF THE VENTRAL STRIATUM

As previously explained, the ventral striatum is considered the portion of the striatum that is tightly connected with the limbic system for their afferences coming from the allo- and mesocortex and also from the amygdala (Haber et al., 1990a,b). For the anatomical description of the ventral striatum, see the first part of this chapter.

Cortical afferences originating from the limbic cortex, infralimbic, medial orbital cortex, insular agranular cortex, perirhinal and entorhinal cortices, and from the subiculum of the hippocampus (French et al., 2002; Friedman et al., 2002) project bilaterally to the ventral striatum with an ipsilateral predominance. Each of these cortical areas terminate in a striatal field longitudinally oriented which is slightly superimposed on those of adjacent cortical areas (Berendse et al., 1991, 1992).

For how concern the amygdala connections it can be said that amygdalostriatal fibers originate principally from the basal and basal accessory nuclei of the amygdala and that these fibers are topographically organized (Groenewegen et al., 1980; Russchen et al., 1984). Some experiments performed on rats showed that many neurons of the basolateral nucleus of the amygdala send fibers to the nucleus accumbens and to the prefrontal cortex (Shinonaga et al., 1994) (see above). The topographical organization and the projections pattern of these neurons and of the amygdalo-prefrontal and the amygdalo-striatal connection in general, correspond to those of the prefronto-striatal projection.

Ventral striatum receives a strong dopaminergic projection coming from the ventral tier and from the medial zone of the SN. These fibers contrary to the previous ones, are not topographically organized in the ventral mesencephalon (Lynd-Balta et al., 1994).

It is important to highlight that the dopaminergic innervation of the ventral striatum is distributed in inhomogeneous manner that in turn is associated to an inhomogeneous distribution of neuropeptides enkephalin, substance P and dynorphin and of GABA neurotransmitter (Voorn et al., 1986). Moreover, many neurons in the VTA that project to the ventral striatum contain cholecystokinin and that in some cellular elements it colocalize with the dopamine (Heimer et al., 1975, 1978).

The efferences of the ventral striatum are principally directed to the extrapyramidal centers such as ventral pallidum, substantia nigra, and pedunculopontine nucleus (PPT). Also, other structures anatomically and functionally related to the limbic system receive a strong innervation from the ventral striatum, they include the nucleus of the bed of the stria terminalis, lateral hypothalamic area, and the medial part of the mediodorsal nucleus of the thalamus (MD) (Groenewegen et al., 1993; Haber et al., 1990a,b).

The fibers originating from the *nucleus accumbens* (NAcc) and directed to the ventral pallidum are principally GABAergic but fibers containing substance P, enkephalin, and dynorphin also have been described (Haber et al., 1983, 1985). Moreover, it has been seen that the accumbofugal fibers directed to the ventral pallidum project monosynaptically to the efferent cells of this structure (Yang et al., 1985).

These fibers described so far are organized topographically and that the efferences of the ventral striatum, directed to both segments of the globus pallidus are anatomically separated from the efferences of the dorsal striatum equally directed to the globus pallidus (Haber et al., 1990a,b).

Regarding the morphohistological description in toto of the NA as well as the complete description of the macroscopic anatomy of the basal ganglia, the reader is invited to consult the first part of the chapter.

The efferences from the NAcc differ from each other on the base of their origin, "core" and "shell," therefore both the region project to the mesencephalon with only a difference: the "core" projects to the SN and the "shell" to the VTA; moreover the "shell," differently to the "core," projects diffusely to the hypothalamus (Heimer et al., 1991; Zahm et al., 1992).

Considering the roles in which NAcc is involved in we can say that the "core" region plays a great role in the cognitive processing of motor function related to reward and reinforcement and the regulation of slow-wave sleep. Specifically, the "core" encodes new motor programs which facilitate the acquisition of a given reward in the future. The D2-type neurons in the NAcc core which co-express adenosine A_{2A} receptors activation-dependently promote slow—wave sleep (Cherasse and Urade, 2017; Valencia Garcia and Fort, 2018; Oishi et al., 2017).

The "shell," instead, seems to be involved in the cognitive processing of reward, including subjective "liking" reactions to certain pleasurable stimuli, motivational salience, and positive reinforcement (Berridge and Kringelbach, 2015; Malenka et al., 2009). Moreover the "shell" region seems to be a "hedonic center" which is responsible for the pleasurable or "liking" component of some intrinsic rewards (Berridge and Kringelbach, 2015; Malenka et al., 2009). Addictive drugs have a larger effect on dopamine release in the "shell" than in the "core."

Finally, concerning a receptor-associated role, we can say that in the NAcc D1-type MSNs mediate reward-related cognitive processes (Saddoris et al., 2015) whereas the D2-type MSNs mediate aversion-related cognition (Calipari et al., 2016).

Regarding the efferences of the ventral striatum directed to the SN they are distributed rostro-caudally to the whole this structure (Haber et al., 1990a,b); furthermore, these efferences are in direct synaptic contact with the dopaminergic neurons that innervate the dorsal striatum (Somogyi et al., 1981), this is how the ventral striatum is connected with the limbic system (Haber et al., 1990a,b) (see also above the "striato-nigro-striatal" paragraph). The ventral striatum strongly projects to the *nucleus basalis of Meynert* (sigla), contained in the *substantia innominata* (SI), that represents the principal source of cholinergic fibers directed to the cerebral cortex and amygdala.

The complexity and the vast quantity of anatomical connections that the ventral striatum establishes with an equally vast territory of the brain reflects how this system, related to its functional aspects, plays a key role in the behavioral, mood, and other limbic aspects of the individual himself. The above leads us not to extend the anatomical description of the connections of the ventral striatum but rather to describe the neurophysiopathological mechanisms in which the latter plays a key role given its altered activity.

NONMOTOR TASK OF THE BASAL GANGLIA

As widely illustrated so far, the basal ganglia work in concert with the cortex, thalamus, and brainstem to orchestrate and execute planned and motivated behaviors requiring motor, cognitive, and limbic circuits. In addition to their involvement in the expression of goal-directed behaviors through movement, the ventral striatum plays a key role in reward and reinforcement (Haber, 2016) and is important in development of addictive behaviors and habit formation (Koob et al., 1999; Kalivas et al., 1993; Nestler et al., 1993).

What is clear from the recent progress is that this set of subcortical nuclei work in tandem with the cortex (particularly the frontal cortex) via a complex cortico-basal ganglia network to develop and carry out complex behaviors. The concept of parallel cortico-subcortical channels through the basal ganglia has been prominent in how the field has viewed the organizational understanding of basal ganglia circuits (Alexander and Crutcher, 1990).

Since the large number of connections with the cortical areas, specifically with frontal and prefrontal fields, basal ganglia dysfunction is not only linked to neurological disorders but also to psychiatric ones including schizophrenia, depression, obsessive-compulsive disorder (OCD), attention deficit hyperactivity disorder (ADHD), and autism spectrum disorder (ASD). The next paragraph of this chapter will be devoted to the neurophysiopathology of some of these disorders.

BASAL GANGLIA AND NEUROPSYCHIATRIC DISORDERS

It may seem paradoxical but over the years neuroscientists have discovered the function and importance of one or more specific structures or areas of the brain only when these have been damaged and become nonfunctional.

The basal ganglia receive afferent inputs from the entire cerebral cortex but above all from the frontal lobes which are known to be involved in higher-order cognitive tasks, in the personality aspect, in emotional and affective processes.

In parallel, to understand what the specific function of a particular region of the brain was, it was usually necessary to investigate which part the latter was interconnected.

OBSESSIVE COMPULSIVE DISORDER (OCD) AND BASAL GANGLIA

Intrusive thoughts, doubts, or concerns about safety, cleanliness, sex, violence, or symmetry and urges to count or check "to be sure" are common thoughts, feelings, and behaviors that represent universal human experience to 1° or another. People with OCD are unable to control their thoughts and feelings and, consequently, experience obsessions accompanied by anxiety; typically, they also engage in associated compulsions, such as repetitive, stereotyped, or ritualized behaviors, apparently designed to neutralize the anxiety associated with their obsessions. Thus, one perspective conceptualizes OCD as a model disorder of self-regulation and behavioral inhibition (Leisman et al., 2013).

Also in autism, stereotyped, ritualistic, and repetitive behaviors, including compulsive rituals and difficulties in tolerating changes in routine or environment are characteristic in fact the OCD, or at least an obsessive-compulsive tract, it is often a comorbidity of ASD. (For details about ASD see later in the next paragraphs.)

The cortico-basal ganglia circuits linking the orbitofrontal (OFC) and anterior cingulate cortex (ACC) to the caudate nucleus might account for the cardinal features of OCD, they in fact have been suggested to form a neuronal system critical for habit learning and for the routine performance of habits and have specifically been implicated in the acquisition of stereotyped behaviors (Howe et al., 2011; Lingawi et al., 2012).

There is evidence of basal ganglia dysfunction from imaging studies of OCD, with both reduced and increased volumes of caudate nuclei reported by Bush et al. (2005), Harrison et al. (2009), Zarei et al. (2011), Amat et al. (2006) in turn showed that it has been implicated in repetitive actions in monkeys.

Moreover, increased caudate metabolism has been found to be reduced after the pharmacological and psychotherapeutic effective treatment of the OCD (Schwartz et al., 1996; Maia et al., 2008) and, in provoked or activated conditions, patients with OCD have shown increased caudate blood flow (Adams et al., 2005).

The OFC and ACC both project to the ventral part of the caudate nucleus and to the ventral striatum. In the monkey, these regions have been found to send outputs not only to the pallidum but also to a large part of the dopamine containing SNpc, from which the nigrostriatal tract originates. The caudal OFC and the caudal\medial ACC are also major sources of input to striosomal system in the head of the caudate nucleus. Striosomes in this region have been linked to reward effects and may appear to be differentially active under conditions in which animals perform repetitive stereotyped behaviors in response to dopamine receptor agonists (Amat et al., 2006).

These features of the OFC- and ACC-basal ganglia circuits are important not only to understand the OCD symptomatology from a neuroanatomofunctional point of view but also to understand the developmental aspects and the therapeutical approaches of these disorders (for details see also Chapter 11).

The loops running from the neocortex to the basal ganglia and then to the thalamus and back to the neocortex may help to establish cognitive habits, just as they may influence the development of motor habits. If so, the cortico-basal ganglia loop dysfunction in OCD could reflect both sides of basal ganglia function, motor and cognitive, to bring about repetitive thoughts (obsessions) and repetitive actions (compulsions).

Therefore, dysfunction of this cortico-basal ganglia system can contribute to the OCD symptoms and the individuals become stuck in a conceptual framework, unable to shift from one priority set to the next, and thus remaining locked into a specific behavioral output program. Corticothalamic loops are critical for integrative and optimized cortical functioning. The adequacy of basal ganglia function is necessary to facilitate associations among cortical inputs based on context and evaluative signals and thereby promote behavioral automation, normally necessary to reduce the information load on the system. The basal ganglia can relieve the frontal cortex of the substantial computational load in carrying out executive functions. Although it has seen that striatal lesions can induce intense compulsive behaviors and stereotypes, it has not yet found the existence of subtle lesions of the striatum in people with OCD (Litvan et al., 1998).

Magnetic resonance spectroscopy studies have suggested that there exist reduced N-acetylaspartate levels within the striatum of persons with OCD, so that neuronal density there may actually be reduced (Litvan et al., 1996; Canales et al., 2000).

Abnormal brain chemistry in OCD could affect neurotransmission in cortico-basal ganglia circuits leading to the abnormal metabolic activity seen in imaging studies indicated earlier.

Although little is known about the neurochemistry of OCD, the most successful pharmacologic therapy is the treatment with tricyclic antidepressant (TCA) (e.g., clomipramine) which belongs to the class of serotonin reuptake inhibitors (SRIs). However, more recently the use of TCA has been limited to those people who do not respond to a new class of antidepressant: the Selective Serotonin Reuptake Inhibitors (SSRIs) (e.g., paroxetine, fluvoxamine, escitalopram etc.), these latter are preferred to TCA as they have fewer side effects. One suggestion is that SRIs have their beneficial effects via downregulation of $5HT_{1D}$ autoreceptors within the orbitofrontal cortex (Matsunaga et al., 2011). Effective therapy with SRIs or SSRI can reverse the abnormal metabolic activity seen in OCD circuits, suggesting that the modulatory effects of serotonin can act on the cortico-basal ganglia circuit defined in scanning studies (Fan et al., 2010).

However, although clinical results demonstrate strong evidence of primary serotonin abnormality involving other neurochemical systems as well, it is not yet precisely known how neurotransmitter substrates are involved in the pathogenesis of OCD.

The selective serotonin reuptake inhibitors (SSRIs) are the mainstay of the pharmacological treatment of OCD (Pittenger et al., 2014). The tricyclic antidepressant clomipramine was shown to be of benefit in the early 1980s (Insel et al., 1983), but side effects limit its use as a first-line agent. Fluvoxamine was first shown to be beneficial in individuals with OCD by Goodman and colleagues in 1989 (Goodman et al., 1989). Since then, more than 20 blinded, placebo-controlled studies have firmly established the efficacy of SSRI monotherapy in OCD (Soomro et al., 2008). Because of the combination of proven efficacy and a typically benign side effect profile, SSRIs are the first-line pharmacological option for the treatment of OCD (Soomro et al., 2012).

A variety of other agents have been used to augment ineffective SSRI treatment; none are sufficiently well supported by the literature to have entered the standard of care, but small studies provide intriguing evidence of benefit in several cases (Pittenger et al., 2014). Among these, Mirtazapine demonstrated to be helpful in accelerating the initial response to SRI pharmacotherapy but not increasing the ultimate likelihood of response in those who fail initial SRI pharmacotherapy (Pallanti et al., 2002). Moreover, the endogenous opioid system has been postulated to be involved in OCD pathogenesis ever since it was noted that administration of the opioid antagonist naloxone exacerbated OCD symptoms. A double-blind, placebo-controlled crossover trial of 23 treatment refractory OCD patients demonstrated a significantly greater decreased in OCD symptoms in response to weekly oral morphine compared to placebo (Koran et al., 2005). Tramadol hydrochloride, an opioid agonist with lower abuse potential than morphine, has been studied as an augmentation agent for treatment refractory OCD in an open-label trial, with encouraging results (Shapira et al., 1997). Further double-blind studies are needed to establish the efficacy of tramadol as an augmentation agent in OCD.

In case of refractory OCD, deep brain stimulation (DBS) of the ventral anterior limb of the internal capsule (vALIC) (Denys et al., 2020; Widge et al., 2019) showed promising result in reducing symptoms. However, DBS for refractory OCD still lacks consistent data, therefore further studies are required to assess it as a routine treatment.

ATTENTIONAL DEFICIT HYPERACTIVITY DISORDER (ADHD)

ADHD is a highly heritable and prevalent neuropsychiatric disorder estimated to affect 6% of school-aged children (Jahfari et al., 2011). It is manifested by inattention, hyperactivity, and impulsivity, which often respond substantially to treatment with methylphenidate or dextroamphetamine. Etiologic theories suggest a functional deficit in those cortico-striatal circuits modulated by dopamine (see the text later). Teicher et al. (2000) developed a fMRI procedure (T2 relaxometry) to indirectly assess blood volume in the striatum of boys ages 6–12 years in steady-state conditions. Boys with ADHD had higher T2 relaxation time measures in the putamen bilaterally than healthy control subjects. Daily treatment with methylphenidate significantly changed the T2 relaxation times in the putamen of children with ADHD and overall, there was a similar but nonsignificant trend in the right caudate. Teicher et al. (2000) concluded that ADHD symptoms may be closely tied to functional abnormalities in the putamen which is mainly involved in the regulation of motor behavior. Converging evidence implies the involvement of fronto-striatal afferences in ADHD, moreover anatomical studies using MRI have demonstrated subtle reductions in volume of some basal ganglia regions and prefrontal cortex (Teicher et al., 2000; Castellanos et al., 2002). However, cognitive functioning is mildly impaired in this disorder (for review, see Sergeant et al., 2002) and cognitive control of the ability to inhibit inappropriate thoughts and actions is also affected. Several studies have shown that this impairment is related to the reduction in volume in fronto-striatal regions (Semrud-Clikeman et al., 2000) and functional studies have suggested that older children and adults with ADHD may activate these regions less than controls during tasks that require cognitive control (e.g., Vaidya et al., 1998; Bush et al., 1999; Semrud-Clikeman et al., 2000; Durston et al., 2002) showed that the development of this ability is related to the maturation of ventral fronto-striatal circuitry and these results seem therefore to suggest that the ventral fronto-striatal circuits being involved in the control of the activities of the basal ganglia constitute a physical substrate that results in cognitive control of impulses. Volumetric abnormalities have also been associated with the basal ganglia and in turn with ADHD. Qiu et al. (2009) found that boys with ADHD showed significatively smaller basal ganglia volumes compared with typically developing boys. Volume compression was seen bilaterally in the caudate nucleus (head and body) and in anterior putamen as well as in the left anterior globus pallidus and right ventral putamen; volume expansion was most pronounced in the posterior putamen (Qiu et al., 2009). They concluded that the shape compression pattern of basal ganglia in ADHD suggests an atypical brain development involving multiple fronto-subcortical control loops, including circuits with premotor, oculo-motor, and prefrontal cortices.

AUTISM

In Autism Spectrum Disorders (ASD), stereotyped, ritualistic, and repetitive behaviors, including compulsive rituals and difficulties in tolerating changes in routine or environment are characteristic. Moreover, increased risk of Parkinsonism (20%–30%) was observed in individual with ASD compared to the general population (0.1%–0.9%) (Starkstein et al., 2015).

It has been suggested that the behavior features of subjects with ASD may share related pathophysiologic mechanisms (Leisman et al., 2013; Jahfari et al., 2011). Some of the aforementioned authors recorded significant morphological differences in the basal ganglia of ASD people compared with a negative control. They found a significant enlargement of 8% of the total caudate volume in the subjects with autism and this data was proportional to the increased total brain volume and to enlargement of other brain structures earlier reported in the patients with autism (Leisman et al., 2013).

Neuropathological studies of the dorsal and ventral striatum in individuals with autism are extremely sparse. Subramanian et al. (2017), Wegiel et al. (2014) observed in individuals aged from 9 to 60 years a 22% increase in the volume of the caudate nucleus and a 34% increase in the volume of the NAcc. In contrast, these same structures from young individuals with autism aged 4–8 years did not exhibit any differences in volume suggesting that the changes in volumes may be occurring from adolescence into adulthood. Stereological analysis of these dorsal and ventral striatal structures revealed significant decreases in neuronal density in the Nacc and the putamen but normal neuronal density in the caudate nucleus and both segments of the globus pallidus (Wegiel et al., 2014).

Furthermore Sato et al. (2014) reported an increase putamen volume (bilaterally) but it was normal in the caudate nucleus, NAcc, and globus pallidus. If the two studies described are any indication, it suggests that there are indeed differences in the development of striatal circuitry. In addition, limbic areas including the AAC that sends projections to the striatum has been shown in individuals with autism to have a marked reduction in the number of $GABA_ARs$ (Oblak et al., 2009) as well as decreased density of $GABA_B$ Rs (Oblak et al., 2010) and select serotoninergic receptor subtypes (Oblak et al., 2013). This imbalance of neurochemical circuitry within the ACC may contribute to altered excitatory input to striatal areas thus influencing its normal function(s).

An interesting correlation of structural connectivity via diffusion tensor imaging (DTI) of fronto-striatal circuitry and inhibitory control was conducted by Langen et al. (2012). These authors found that adults with autism had reduced total brain white matter volume and decreased functional anisotropy (FA) of white matter tracts connecting the putamen with frontal cortical regions and higher mean diffusivity of white matter tracts connecting NAcc to frontal cortex (Langen et al., 2012). The latter also had worse performance on a go/no go task suggesting that autism may be associated with differences in the anatomy of fronto-striatal white matter tracts. Previously, Di Martino et al., (2011) found hyperconnectivity between the striatum and the pons as well as increased functional connectivity between striatal areas and associative and limbic cortices. A study of basal ganglia morphology and repetitive behavior in children aged 3–4 years with autism compared to a cohort with developmental delay (DD) and with another group of typically developing children (TD) was performed by Estes et al. (2011). In that study, enlargement of striatal structures including left and right putamen and left caudate nucleus was found in the autism group when compared to the other two groups. However, higher levels of repetitive/stereotyped behavior were correlated with decreased volume in specific basal ganglia structures in the autism cohort including the right and left putamen, right globus pallidus and right striatum when not adjusted for total cerebral volume. Decreased volume of the globus pallidus was also demonstrated in a large structural MRI study of individuals with autism compared to controls (Sussman et al., 2015). In a previously reported study contrarily the increased caudate nucleus volume was positively correlated with repetitive/stereotyped behaviors (Hollander et al., 2005).

Basically, it seems that with regard to the neuroanatomical characteristics of the basal ganglia, the ASD has mainly influenced positively or negatively the volume of the striatum in toto.

Furthermore, brain volume also appears to be higher in people with ASD, but this seems to be related to age. It is not yet known what these anatomical differences are involved, or rather, what are the behavioral characteristics caused by these structural differences and much data is still needed. However, this is starting to become a topic beyond the scope of this book, so we invite the interested reader to consult the dedicated papers.

References

Adams, K.H., Hansen, E.S., Pinborg, L.H., Hasselbalch, S.G., Svarer, C., Holm, S., Bolwig, T.G., Knudsen, G.M., September 2005. Patients with obsessive-compulsive disorder have increased 5-HT2A receptor binding in the caudate nuclei. Int. J. Neuropsychopharmacol. 8 (3), 391–401. https://doi.org/10.1017/S1461145705005055.

Alexander, G.E., Crutcher, M.D., 1990. Functional architecture of basal ganglia circuits: neural substrates of parallel processing. Trends Neurosci. 13 (7), 266–271.

Amat, J.A., Bronen, R.A., Saluja, S., Sato, N., Zhu, H., Gorman, D.A., Royal, J., Peterson, B.S., June 2006. Increased number of subcortical hyperintensities on MRI in children and adolescents with Tourette's syndrome, obsessive-compulsive disorder, and attention deficit hyperactivity disorder. Am. J. Psychiatry 163 (6), 1106–1108. https://doi.org/10.1176/ajp.2006.163.6.1106.

Andén, N.E., Dahlström, A., Fuxe, K., et al., 1966. Ascending monoamine neurons to the telencephalon and diencephalon. Acta Physiol. Scand. 67, 313–326.

Ashby, F.G., Crossley, M.J., 2011. A computational model of how cholinergic interneurons protect striatal-dependent learning. J. Cogn. Neurosci. 23 (6), 1549–1566.

Bennett, B.D., Bolam, J.P., 1993. Characterization of calretinin-immunoreactive structures in the striatum of the rat. Brain Res. 609 (1–2), 137–148.

Berendse, H.W., Galis-de Graaf, Y., Groenewegen, H.J., 1992. Topographical organization and relationship with ventral striatal compartments of prefrontal corticostriatal projections. J. Comp. Neurol. 316, 314.

Berendse, H.W., Groenewegen, H.J., 1990. Organization of the thalamostriatal projections in the rat, with special emphasis on the ventral striatum. J. Comp. Neurol. 299, 187–228.

Berendse, H.W., Groenewegen, H.J., 1991. Restricted cortical termination fields of the midline and intralaminar thalamic nuclei in the rat. Neuroscience 42, 73–102.

Berridge, K.C., Kringelbach, M.L., May 2015. Pleasure systems in the brain. Neuron 86 (3), 646–664.

Bianco, I.H., Wilson, S.W., April 12, 2009. The habenular nuclei: a conserved asymmetric relay station in the vertebrate brain. Philos. Trans. R. Soc. Lond. B Biol. Sci. 364 (1519), 1005–1020. https://doi.org/10.1098/rstb.2008.0213.

Braak, H., Braak, E., 1982. Neuronal types in the striatum of man. Cell Tissue Res. 227, 319–342.

Braak, H., Rub, U., Sandmann-Keil, D., et al., 2000. Parkinson's disease: affection of brain stem nuclei controlling premotor and motor neurons of the somatomotor system. Acta Neuropathol. 99, 489–495.

Braunlich, K., Seger, C., 2013. The basal ganglia. Wiley Interdiscip. Rev. Cogn. Sci. 4 (2), 135–148.

Bunney, B.S., Aghajanian, G.K., December 3, 1976. The precise localization of nigral afferents in the rat as determined by a retrograde tracing technique. Brain Res. 117 (3), 423–435. https://doi.org/10.1016/0006-8993(76)90751-4.

Bush, G., Frazier, J.A., Rauch, S.L., Seidman, L.J., Whalen, P.J., Jenike, M.A., Rosen, B.R., Biederman, J., June 15, 1999. Anterior cingulate cortex dysfunction in attention-deficit/hyperactivity disorder revealed by fMRI and the Counting Stroop. Biol. Psychiatry 45 (12), 1542–1552. https://doi.org/10.1016/s0006-3223(99)00083-9.

Bush, G., Valera, E.M., Seidman, L.J., June 1, 2005. Functional neuroimaging of attention-deficit/hyperactivity disorder: a review and suggested future directions. Biol. Psychiatry 57 (11), 1273–1284. https://doi.org/10.1016/j.biopsych.2005.01.034.

Calabresi, P., Centonze, D., Gubellini, P., et al., 2000. Synaptic transmission in the striatum: from plasticity to neurodegeneration. Prog. Neurobiol. 61, 231–265.

Calabresi, P., Maj, R., Mercuri, N.B., Bernardi, G., 1992b. Coactivation of D1 and D2 dopamine receptors is required for long-term synaptic depression in the striatum. Neurosci. Lett. 142 (1), 95–99. https://doi.org/10.1016/0304-3940(92)90628-k.

Calabresi, P., Pisani, A., Mercuri, N.B., Bernardi, G., 1992a. Long-term potentiation in the striatum is unmasked by removing the voltage-dependent magnesium block of NMDA receptor channels. Eur. J. Neurosci. 4 (10), 929–935. https://doi.org/10.1111/j.1460-9568.1992.tb00119.x.

Calabresi, P., Pisani, A., Mercuri, N.B., Bernardi, G., 1996. The corticostriatal projection: from synaptic plasticity to dysfunctions of the basal ganglia. Trends Neurosci 19 (1), 19–24. https://doi.org/10.1016/0166-2236(96)81862-5.

Calipari, E.S., Bagot, R.C., Purushothaman, I., Davidson, T.J., Yorgason, J.T., Peña, C.J., Walker, D.M., Pirpinias, S.T., Guise, K.G., Ramakrishnan, C., Deisseroth, K., Nestler, E.J., March 2016. In vivo imaging identifies temporal signature of D1 and D2 medium spiny neurons in cocaine reward. Proc. Natl. Acad. Sci. U. S. A. 113 (10), 2726–2731.

Canales, J.J., Graybiel, A.M., April 2000. A measure of striatal function predicts motor stereotypy. Nat. Neurosci. 3 (4), 377–383. https://doi.org/10.1038/73949.

Carman, J.B., Cowan, W.M., Powell, T.P., Webster, K.E., 1965. A bilateral cortico-striate projection. J. Neurol. Neurosurg. Psychiatry 28, 71–77.

Carman, J.B., Cowan, W.M., Powell, T.P.S., 1963. The organization of the cortico-striate connexions in the rabbit. Brain 86, 525–562.

Carpenter, M.B., Batton, R.R., Carleton, S.C., Keller, J.T., 1981. Interconnections and organization of pallidal and subthalamic nucleus neurons in the monkey. J. Comp. Neurol. 197, 579–603.

Carpenter, M.B., Peter, P., January 1972. Nigrostriatal and nigrothalamic fibers in the rhesus monkey. J. Comp. Neurol. 144 (1), 93–115. https://doi.org/10.1002/cne.901440105.

Castellanos, F.X., Lee, P.P., Sharp, W., Jeffries, N.O., Greenstein, D.K., Clasen, L.S., Blumenthal, J.D., James, R.S., Ebens, C.L., Walter, J.M., Zijdenbos, A., Evans, A.C., Giedd, J.N., Rapoport, J.L., October 9, 2002. Developmental trajectories of brain volume abnormalities in children and adolescents with attention-deficit/hyperactivity disorder. JAMA 288 (14), 1740–1748. https://doi.org/10.1001/jama.288.14.1740.

Charpier, S., Mahon, S., Deniau, J.M., 1999. In vivo induction of striatal long-term potentiation by low-frequency stimulation of the cerebral cortex. Neuroscience 91 (4), 1209–1222. https://doi.org/10.1016/s0306-4522(98)00719-2.

Cherasse, Y., Urade, Y., November 2017. Dietary zinc acts as a sleep modulator. Int. J. Mol. Sci. 18 (11), 2334.

Cowan, R.L., Wilson, C.J., 1994. Spontaneous firing patterns and axonal projections of single corticostriatal neurons in the rat medial agranular cortex. J. Neurophysiol. 71, 17–32.

DeLong, M.R., 1990. Primate models of movement disorders of basal ganglia origin. Trends Neurosci. 13, 281–285.

DeLong, M.R., 2000. The basal ganglia. In: Kandel, E.R., Schwartz, J.H., Jessel, T.M. (Eds.), Principles of Neural Science, fourth ed. McGraw-Hill, New York, pp. 853–867.

DeLong, M.R., Georgopoulos, A.P., 1981. Motor function of the basal ganglia. In: Brookhart, J.M., Mountcastle, V.B., Brooks, V.B. (Eds.), Handbook of Physiology, Sect 1: The Nervous System, Vol 2: Motor Control, Part 2. American Physiological Society, Bethesda, pp. 1017–1061.

Denys, D., Graat, I., Mocking, R., de Koning, P., Vulink, N., Figee, M., Ooms, P., Mantione, M., van den Munckhof, P., Schuurman, R., March 1, 2020. Efficacy of deep brain stimulation of the ventral anterior limb of the internal capsule for refractory obsessive-compulsive disorder: a clinical cohort of 70 patients. Am. J. Psychiatry 177 (3), 265–271. https://doi.org/10.1176/appi.ajp.2019.19060656.

Di Martino, A., Kelly, C., Grzadzinski, R., Zuo, X.N., Mennes, M., Mairena, M.A., Lord, C., Castellanos, F.X., Milham, M.P., May 1, 2011. Aberrant striatal functional connectivity in children with autism. Biol. Psychiatry 69 (9), 847–856. https://doi.org/10.1016/j.biopsych.2010.10.029.

DiFiglia, M., Pasik, P., Pasik, T., 1976. A Golgi study of neuronal types in the neostriatum of monkeys. Brain Res. 114, 245–256.

DiFiglia, M., Rafols, J.A., 1988. Synaptic organization of the globus pallidus. J. Electron. Microsc. Tech. 10, 247–263.

Divac, I., 1975. Magnocellular nuclei of the basal forebrain project to neocortex, brain stem, and olfactory bulb: review of some functional correlates. Brain Res. 93, 385–398.

Dubé, L., Smith, A.D., Bolam, J.P., 1988. Identification of synaptic terminals of thalamic or cortical origin in contact with distinct medium-size spiny neurons in the rat neostriatum. J. Comp. Neurol. 267 (4), 455–471. https://doi.org/10.1002/cne.902670402.

Durston, S., Thomas, K.M., Worden, M.S., Yang, Y., Casey, B.J., June 2002. The effect of preceding context on inhibition: an event-related fMRI study. Neuroimage 16 (2), 449–453. https://doi.org/10.1006/nimg.2002.1074.

Edley, S.M., Graybiel, A.M., June 20, 1983. The afferent and efferent connections of the feline nucleus tegmenti pedunculopontinus, pars compacta. J. Comp. Neurol. 217 (2), 187–215. https://doi.org/10.1002/cne.902170207.

El Falougy, H., Selmeciova, P., Kubikova, E., Haviarova, Z., 2013. The variable origin of the recurrent artery of Heubner: an anatomical and morphometric study. BioMed Res. Int. 2013, 873434.

Elliott, R., Rubinsztein, J.S., Sahakian, B.J., Dolan, R.J., 2002. The neural basis of mood-congruent processing biases in depression. Arch. Gen. Psychiatry 59, 597–604.

Emson, P.C., Arregui, A., Clement-Jones, V., Sandberg, B.E., Rossor, M., October 13, 1980. Regional distribution of methionine-enkephalin and substance P-like immunoreactivity in normal human brain and in Huntington's disease. Brain Res. 199 (1), 147–160. https://doi.org/10.1016/0006-8993(80)90237-1.

Estes, A., Shaw, D.W., Sparks, B.F., Friedman, S., Giedd, J.N., Dawson, G., Bryan, M., Dager, S.R., June 2011. Basal ganglia morphometry and repetitive behavior in young children with autism spectrum disorder. Autism Res. 4 (3), 212–220. https://doi.org/10.1002/aur.193.

Fan, Q., Tan, L., You, C., Wang, J., Ross, C.A., Wang, X., Zhang, T., Li, J., Chen, K., Xiao, Z., October 2010. Increased N-Acetylaspartate/creatine ratio in the medial prefrontal cortex among unmedicated obsessive-compulsive disorder patients. Psychiatry Clin. Neurosci. 64 (5), 483–490. https://doi.org/10.1111/j.1440-1819.2010.02128.x.

Fernandez-Miranda, J.C., Rhoton Jr., A.L., Kakizawa, Y., Choi, C., Alvarez-Linera, J., 2008. The claustrum and its projection system in the human brain: a microsurgical and tractographic anatomical study. J. Neurosurg. 108 (4), 764–774.

Ferré, S., Lluís, C., Justinova, Z., Quiroz, C., Orru, M., Navarro, G., Canela, E.I., Franco, R., Goldberg, S.R.(J., 2010. Adenosine-cannabinoid receptor interactions. Implications for striatal function. Br. J. Pharmacol. 160 (3), 443–453.

Fonnum, F., Storm-Mathisen, J., Divac, I., 1981. Biochemical evidence for glutamate as neurotransmitter in corticostriatal and corticothalamic fibres in rat brain. Neuroscience 6, 863–873.

Fox, C.A., Andrade, A.N., Schwyn, R.C., Rafols, J.A., 1971-1972. The aspiny neurons and the glia in the primate striatum: a Golgi and electron microscopic study. J. Hirnforsch. 13 (4), 341–362.

Francois, C., Tande, D., Yelnik, J., Hirsch, E.C., 2002. Distribution and morphology of nigral axons projecting to the thalamus in primates. J. Comp. Neurol. 447, 249–260.

Frederiksen, S.D., Haanes, K.A., Warfvinge, K., Edvinsson, L., April 2019. Perivascular neurotransmitters: regulation of cerebral blood flow and role in primary headaches. J. Cereb. Blood Flow Metab. 39 (4), 610–632. https://doi.org/10.1177/0271678X17747188.

French, S.J., Totterdell, S., 2002. Hippocampal and prefrontal cortical inputs monosynaptically converge with individual projection neurons of the nucleus accumbens. J. Comp. Neurol. 446, 151–165.

Freund, T.F., Powell, J.F., Smith, A.D., 1984. Tyrosine hydroxylase-immunoreactive boutons in synaptic contact with identified striatonigral neurons, with particular reference to dentritic spines. Neuroscience 13, 1189–1215.

Friedman, D.P., Aggleton, J.P., Saunders, R.C., 2002. Comparison of hippocampal, amygdala, and perirhinal projections to the nucleus accumbens: combined anterograde and retrograde tracing study in the Macaque brain. J. Comp. Neurol. 450, 345–365.

Garcia-Rill, E., Skinner, R.D., Miyazato, H., Homma, Y., 2001. Pedunculopontine stimulation induces prolonged activation of pontine reticular neurons. Neuroscience 104, 455–465.

Gerfen, C.R., 1989. The neostriatal mosaic: striatal patch-matrix organization is related to cortical lamination. Science 246, 385–388.

Geyer, M.A., Puerto, A., Dawsey, W.I., et al., 1976. Histologic and enzymatic studies of the mesolimbic and mesostriatal serotonergic pathways. Brain Res. 106, 241–256.

Goodman, W.K., et al., 1989. Efficacy of fluvoxamine in obsessive-compulsive disorder. A double-blind comparison with placebo. Arch. Gen. Psychiatry 46 (1), 36–44.

Graveland, G.A., Williams, R.S., DiFiglia, M., 1985. A Golgi study in the human neostriatum: neurons and afferent fibers. J. Comp. Neurol. 234, 317–333.

Graybiel, A.M., Flaherty, A.W., Giménez-Amaya, J.-M., 1991. Striosomes and matrisomes. In: Bernardi, G., Carpenter, M.B., di Chiara, G., Morelli, M., Stanzione, P. (Eds.), The Basal Ganglia, vol. 3. Plenum, New York, pp. 3–12.

Graybiel, A.M., Ragsdale, C.W., 1978. Histochemically distinct compartments in the striatum of human, monkey and cat demonstrated by acetylcholinesterase staining. Proc. Natl. Acad. Sci. U. S. A. 75, 5723–5726.

Graybiel, A.M., Ragsdale Jr., C.W., 1979. Fiber connections of the basal ganglia. In: Bloom, R.E., Kreutzberg, G.W., Cuénod, M. (Eds.), Development and Chemical Specificity of Neurons. Elsevier, Amsterdam, pp. 239–283.

Graybiel, A.M., Ragsdale Jr., C.W., 1983. Biochemical anatomy of the striatum. In: Emson, P.C. (Ed.), Chemical Neuroanatomy. Raven Press, New York, pp. 427–504.

Grillner, S., Shik, M.I., 1973. On the descending control of the lumbosacral spinal cord from the "mesencephalic locomotor region". Acta Physiol. Scand. 87, 320–333.

Groenewegen, H.J., Becker, N.E.K., Lohman, A.H., 1980. Subcortical afferents of the nucleus accumbens septi in the cat, studied with retrograde axonal transport of horseradish peroxidase and bisbenzimide. Neuroscience 5, 1903–1916.

Groenewegen, H.J., Berendse, H.W., Haber, S.N., 1993. Organization of the output of the ventral striatopallidal system in the rat: ventral pallidal efferents. Neuroscience 57, 113–142.

Groenewegen, H.J., Uylings, H.B., 2000. The prefrontal cortex and the integration of sensory, limbic and autonomic information. Progr. Brain Res. 126, 3–28. https://doi.org/10.1016/S0079-6123(00)26003-2.

Grofová, I., June 27, 1975. The identification of striatal and pallidal neurons projecting to substantia nigra. An experimental study by means of retrograde axonal transport of horseradish peroxidase. Brain Res. 91 (2), 286–291. https://doi.org/10.1016/0006-8993(75)90550-8.

Haber, S.N., Kunishio, K., Mizobuchi, M., Lynd-Balta, E., 1995. The orbital and medial prefrontal circuit through the primate basal ganglia. J. Neurosci. 15 (7 Pt 1), 4851–4867. https://doi.org/10.1523/JNEUROSCI.15-07-04851.1995.

Haber, S.N., Lynd, E., Klein, C., Groenewegen, H.J., 1990a. Topographic organization of the ventral striatal efferent projections in the rhesus monkey: a anterograde tracing study. J. Comp. Neurol. 293, 282–298.

Haber, S.N., Nauta, W.J.H., 1983. Ramifications of the globus pallidus in the rat as indicated by patterns of immunohistochemistry. Neuroscience 9, 245–260.

Haber, S.N., Watson, S.J., 1985. The comparative distribution of enkephalin dynorphin and substance P in the human globus pallidus and basal forebrain. Neuroscience 14, 1011–1024.

Haber, S.N., Wolfe, D.P., Groenewegen, H.J., 1990b. The relationship between ventral striatal efferent fibers and the distribution of peptide-positive woolly fibers in the forebrain of the rhesus monkey. Neuroscience 39, 323–338.

Haber, S.N., March 2016. Corticostriatal circuitry. Dialogues Clin. Neurosci. 18 (1), 7–21. https://doi.org/10.31887/DCNS.2016.18.1/shaber.

Hallanger, A.E., Levey, A.I., Lee, H.J., Rye, D.B., Wainer, B.H., 1987. The origins of cholinergic and other subcortical afferents to the thalamus in the rat. J. Comp. Neurol. 262, 105–124.

Halliday, G., 2004. Substantia nigra and locus coeruleus. In: Paxinos, G., Mai, J.K. (Eds.), The Human Nervous System, second ed. Elsevier, Amsterdam, pp. 449–463.

Harrison, B.J., Soriano-Mas, C., Pujol, J., Ortiz, H., López-Solà, M., Hernández-Ribas, R., Deus, J., Alonso, P., Yücel, M., Pantelis, C., Menchon, J.M., Cardoner, N., November 2009. Altered corticostriatal functional connectivity in obsessive-compulsive disorder. Arch. Gen. Psychiatry 66 (11), 1189–1200. https://doi.org/10.1001/archgenpsychiatry.2009.152.

Hedreen, J.C., DeLong, M.R., 1991. Organization of striatopallidal, striatonigral, and nigrostriatal projections in the macaque. J. Comp. Neurol. 304, 569–595.

Heimer, L., 1978. The olfactory cortex and the ventral striatum. In: Livingston, K.E., Hornykiewicz, O. (Eds.), Limbic Mechanisms: The Continuing Evolution of the Limbic System Concept. Plenum, New York, pp. 95–187.

Heimer, L., Switzer, R.D., Van Hoesen, G.W., 1982. Ventral striatum and ventral pallidum. Components of the motor system? Trends Neurosci. 5, 83–87.

Heimer, L., Wilson, R.D., 1975. The subcortical projections of the allocortex: similarities in the neural associations of the hippocampus, the piriform cortex, and the neocortex. In: Santini, M. (Ed.), Golgi Centennial Symposium Proceedings. Raven, New York, pp. 177–193.

Heimer, L., Zahm, D.S., Churchill, L., Kalivas, P.W., Wohltmann, C., 1991. Specificity in the projection patterns of accumbal core and shell in the rat. Neuroscience 41, 89–125.

Herkenham, M., Nauta, W.J., 1977. Afferent connections of the habenular nuclei in the rat. A horseradish peroxidase study, with a note on the fiber-of-passage problem. J. Comp. Neurol. 173, 123–146.

Hollander, E., Anagnostou, E., Chaplin, W., Esposito, K., Haznedar, M.M., Licalzi, E., Wasserman, S., Soorya, L., Buchsbaum, M., August 1, 2005. Striatal volume on magnetic resonance imaging and repetitive behaviors in autism. Biol. Psychiatry 58 (3), 226–232. https://doi.org/10.1016/j.biopsych.2005.03.040.

Hoover, J.E., Strick, P.L., 1999. The organization of cerebellar and basal ganglia outputs to primary motor cortex as revealed by retrograde transneuronal transport of herpes simplex virus type 1. J. Neurosci. 19, 1446–1463.

Howe, M.W., Atallah, H.E., McCool, A., Gibson, D.J., Graybiel, A.M., October 4, 2011. Habit learning is associated with major shifts in frequencies of oscillatory activity and synchronized spike firing in striatum. Proc. Natl. Acad. Sci. U. S. A. 108 (40), 16801–16806. https://doi.org/10.1073/pnas.1113158108.

Ingham, C.A., Hood, S.H., Taggart, P., Arbuthnott, G.W., 1998. Plasticity of synapses in the rat neostriatum after unilateral lesion of the nigrostriatal dopaminergic pathway. J. Neurosci. 18 (12), 4732–4743. https://doi.org/10.1523/JNEUROSCI.18-12-04732.1998.

Insel, T.R., et al., 1983. Obsessive-compulsive disorder. A double-blind trial of clomipramine and clorgyline. Arch. Gen. Psychiatry 40 (6), 605–612.

Jahfari, S., Waldorp, L., van den Wildenberg, W.P., Scholte, H.S., Ridderinkhof, K.R., Forstmann, B.U., May 4, 2011. Effective connectivity reveals important roles for both the hyperdirect (fronto-subthalamic) and the indirect (fronto-striatal-pallidal) fronto-basal ganglia pathways during response inhibition. J. Neurosci. 31 (18), 6891–6899. https://doi.org/10.1523/JNEUROSCI.5253-10.2011.

Jiang, H., Stein, B.E., McHaffie, J.G., December 1, 2011. Physiological evidence for a trans-basal ganglia pathway linking extrastriate visual cortex and the superior colliculus. J. Physiol. 589 (Pt 23), 5785–5799. https://doi.org/10.1113/jphysiol.2011.213553.

Johannes, S., Reif, A., Senitz, D., Riederer, P., Lauer, M., 2003. NADPH-diaphorase staining reveals new types of interneurons in human putamen. Brain Res. 980 (1), 92–99.

Jones, E.G., 1985. The Thalamus. Plenum, New York.

Kalivas, P.W., Churchill, L., Klitenick, M.A., 1993. The circuitry mediating the translation of motivational stimuli into adaptive motor responses. In: Kalivas, P.W., Barnes, C.D. (Eds.), Limbic Motor Circuits and Neuropsychiatry. CRC Press, Inc, Boca Raton, FL, pp. 237–275.

Kawaguchi, Y., Wilson, C.J., Augood, S.J., Emson, P.C., 1995. Striatal interneurones: chemical, physiological and morphological characterization. Trends Neurosci. 18 (12), 527–535.

Kemp, J.M., Powell, T.P., September 30, 1971. The connexions of the striatum and globus pallidus: synthesis and speculation. Philos. Trans. R. Soc. Lond. B Biol. Sci. 262 (845), 441–457. https://doi.org/10.1098/rstb.1971.0106.

Kemp, J.M., Powell, T.P.S., 1971. The site of the termination of afferent fibers in the caudate nucleus. Philos. Trans. R. Soc. Lond. [Biol.] 262, 413–427.

Kemp, J.M., Powell, T.P.S., 1971. The Structure of the Caudate Nucleus of the Cat: Light and Electron.

Kievit, J., Kuypers, H.G.J.M., 1975. Basal forebrain and hypothalamic connections to frontal and parietal cortex in the rhesus monkey. Science 187, 660–662.

Kim, R., Nakano, K., Jayaraman, A., Carpenter, M.B., October 1, 1976. Projections of the globus pallidus and adjacent structures: an autoradiographic study in the monkey. J. Comp. Neurol. 169 (3), 263–290. https://doi.org/10.1002/cne.901690302.

Kita, H., Kita, T., 2001. Number, origins, and chemical types of rat pallidostriatal projection neurons. J. Comp. Neurol. 437, 438–448. https://doi.org/10.1002/cne.1294.

Kita, H., Kitai, S.T., 1987. Efferent projections of the subthalamic nucleus in the rat: light and electron microscopic analysis with the PHA-L method. J. Comp. Neurol. 260 (3), 435–452. https://doi.org/10.1002/cne.902600309.

Koob, G.F., 1999. Drug reward and addiction. In: Fundamental Neuroscience. Academic Press, New York, NY, pp. 1261–1279.

Koran, L.M., et al., 2005. Double-blind treatment with oral morphine in treatment-resistant obsessive-compulsive disorder. J. Clin. Psychiatry 66 (3), 353–359.

Kubota, Y., Inagaki, S., Kito, S., 1986. Innervation of substance P neurons by catecholaminergic terminals in the neostriatum. Brain Res. 375, 163–167.

Kumar, V., Abbas, A.K., Fausto, N. (Eds.), 2005. Robbins and Cotran Pathologic Basis of Disease, seventh ed. Elsevier/Saunders, Philadelphia.

Kuo, J.S., Carpenter, M.B., October 1, 1973. Organization of pallidothalamic projections in the rhesus monkey. J. Comp. Neurol. 151 (3), 201–236. https://doi.org/10.1002/cne.901510302.

Langen, M., Leemans, A., Johnston, P., Ecker, C., Daly, E., Murphy, C.M., Dell'acqua, F., Durston, S., , AIMS Consortium, Murphy, D.G., February 2012. Fronto-striatal circuitry and inhibitory control in autism: findings from diffusion tensor imaging tractography. Cortex 48 (2), 183–193. https://doi.org/10.1016/j.cortex.2011.05.018.

Lapper, S.R., Bolam, J.P., 1992. Input from the frontal cortex and the parafascicular nucleus to cholinergic interneurons in the dorsal striatum of the rat. Neuroscience 51 (3), 533–545.

Lavoie, B., Parent, A., 1990. Immunohistochemical study of the serotoninergic innervation of the basal ganglia in the squirrel monkey. J. Comp. Neurol. 299, 1–16.

Lavoie, B., Parent, A., June 8, 1994. Pedunculopontine nucleus in the squirrel monkey: cholinergic and glutamatergic projections to the substantia nigra. J. Comp. Neurol. 344 (2), 232–241. https://doi.org/10.1002/cne.903440205.

Lee, H.J., Rye, D.B., Hallanger, A.E., Levey, A.I., Wainer, B.H., September 22, 1988. Cholinergic vs. noncholinergic efferents from the mesopontine tegmentum to the extrapyramidal motor system nuclei. J. Comp. Neurol. 275 (4), 469–492. https://doi.org/10.1002/cne.902750402.

Leisman, G., Melillo, R., 2013. The basal ganglia: motor and cognitive relationships in a clinical neurobehavioral context. Rev. Neurosci. 24 (1), 9–25. https://doi.org/10.1515/revneuro-2012-0067.

Lieb, K., Ahlvers, K., Dancker, K., Strohbusch, S., Reincke, M., Feige, B., Berger, M., Riemann, D., Voderholzer, U., December 2002. Effects of the neuropeptide substance P on sleep, mood, and neuroendocrine measures in healthy young men. Neuropsychopharmacology 27 (6), 1041–1049. https://doi.org/10.1016/S0893-133X(02)00369-X.

Lindvall, O., Björklund, A., 1983. Dopamine- and norepinephrine-containing neuron systems: their anatomy in the rat brain. In: Emson, P.C. (Ed.), Clinical Neuroanatomy. Raven, New York, pp. 229–255.

Lindvall, O., Brundin, P., Widner, H., Rehncrona, S., Gustavii, B., Frackowiak, R., Leenders, K.L., Sawle, G., Rothwell, J.C., Marsden, C.D., et al., February 2, 1990. Grafts of fetal dopamine neurons survive and improve motor function in Parkinson's disease. Science 247 (4942), 574–577. https://doi.org/10.1126/science.2105529.

Lingawi, N.W., Balleine, B.W., January 18, 2012. Amygdala central nucleus interacts with dorsolateral striatum to regulate the acquisition of habits. J. Neurosci. 32 (3), 1073–1081. https://doi.org/10.1523/JNEUROSCI.4806-11.2012.

Litvan, I., Mega, M.S., Cummings, J.L., Fairbanks, L., November 1996. Neuropsychiatric aspects of progressive supranuclear palsy. Neurology 47 (5), 1184–1189. https://doi.org/10.1212/wnl.47.5.1184.

Litvan, I., Paulsen, J.S., Mega, M.S., Cummings, J.L., October 1998. Neuropsychiatric assessment of patients with hyperkinetic and hypokinetic movement disorders. Arch. Neurol. 55 (10), 1313–1319. https://doi.org/10.1001/archneur.55.10.1313. Erratum in: Arch. Neurol. 1998 Dec;55(12):1591.

Lynd-Balta, E., Haber, S.N., 1994. The organization of midbrain projections to the ventral striatum in the primate. Neuroscience 59, 609–623.

Maia, T.V., Cooney, R.E., Peterson, B.S., 2008 Fall. The neural bases of obsessive-compulsive disorder in children and adults. Dev. Psychopathol. 20 (4), 1251–1283. https://doi.org/10.1017/S0954579408000606.

Malenka, R.C., Nestler, E.J., Hyman, S.E., 2009. In: Sydor, A., Brown, R.Y. (Eds.), Molecular Neuropharmacology: A Foundation for Clinical Neuroscience, second ed. McGraw-Hill Medical, New York, pp. 147–148. 367, 376.

Mallet, N., Leblois, A., Maurice, N., Beurrier, C., 2019. Striatal cholinergic interneurons: how to elucidate their function in health and disease. Front. Pharmacol. 10, 1488. https://doi.org/10.3389/fphar.2019.01488.

Matsunaga, H., Hayashida, K., Maebayashi, K., Mito, H., Kiriike, N., November 2011. A case series of aripiprazole augmentation of selective serotonin reuptake inhibitors in treatment-refractory obsessive compulsive disorder. Int. J. Psychiatry Clin. Pract. 15 (4), 263–269. https://doi.org/10.3109/13651501.2011.605958.

McGeorge, A.J., Faull, R.L.M., 1989. The organization of the projection from the cerebral cortex to the striatum in the rat. Neuroscience 29, 503–537.

Mehler, W.R., Feferman, M.E., Nauta, W.J., 1960. Ascending axon degeneration following anterolateral cordotomy. An experimental study in the monkey. Brain 83, 718–750. https://doi.org/10.1093/brain/83.4.718.

Meredith, G.E., Agolia, R., Arts, M.P., Groenewegen, H.J., Zahm, D.S., September 1992. Morphological differences between projection neurons of the core and shell in the nucleus accumbens of the rat. Neuroscience 50 (1), 149–162.

Meredith, G.E., Pennartz, C.M., Groenewegen, H.J., 1993. The cellular framework for chemical signalling in the nucleus accumbens. In: Chemical Signalling in the Basal Ganglia. Progress in Brain Research, vol 99, pp. 3–24.

Mesulam, M.M., Mufson, E.J., Levey, A.I., Wainer, B.H., 1984. Atlas of cholinergic neurons in the forebrainvand upper brainstem of the macaque based on monoclonal choline acetyltransferas immunohistochemistry and acetylcholinesterase histochemistry. Neuroscience 12, 669–686.

Mettler, F.A., February 1947. Extracortical connections of the primate frontal cerebral cortex; thalamocortical connections. J. Comp. Neurol. 86 (1), 95–117. https://doi.org/10.1002/cne.900860105.

Moore, R.Y., Bloom, F.E., 1978. Central catecholamine neuron systems: anatomy and physiology of the dopamine systems. Annu. Rev. Neurosci. 1, 129–169.

Nambu, A., Tokuno, H., Takada, M., 2002. Functional significance of the cortico-subthalamo-pallidal 'hyperdirect' pathway. Neurosci. Res. 43, 111–117.

Nestler, E.J., Hope, B.T., Widnell, K.L., 1993. Drug addiction: a model for the molecular basis of neural plasticity. Neuron 11 (6), 995–1006.

Nieuwenhuys, R., Voogd, J., Van Huijzen, C., 2008. The Human Central Nervous System, fourth ed. Springer, New York.
Oblak, A., Gibbs, T.T., Blatt, G.J., August 2009. Decreased GABAA receptors and benzodiazepine binding sites in the anterior cingulate cortex in autism. Autism Res. 2 (4), 205–219. https://doi.org/10.1002/aur.88. Erratum in: Autism Res. 2009 Aug;2(4):237.
Oblak, A., Gibbs, T.T., Blatt, G.J., December 2013. Reduced serotonin receptor subtypes in a limbic and a neocortical region in autism. Autism Res. 6 (6), 571–583. https://doi.org/10.1002/aur.1317.
Oblak, A.L., Gibbs, T.T., Blatt, G.J., September 1, 2010. Decreased GABA(B) receptors in the cingulate cortex and fusiform gyrus in autism. J. Neurochem. 114 (5), 1414–1423. https://doi.org/10.1111/j.1471-4159.2010.06858.x.
Oishi, Y., Xu, Q., Wang, L., Zhang, B.J., Takahashi, K., Takata, Y., Luo, Y.J., Cherasse, Y., Schiffmann, S.N., de Kerchove d'Exaerde, A., Urade, Y., Qu, W.M., Huang, Z.L., Lazarus, M., September 2017. Slow-wave sleep is controlled by a subset of nucleus accumbens core neurons in mice. Nat. Commun. 8 (1), 734.
Pallanti, S., Quercioliand, L., Koran, L.M., 2002. Citalopram intravenous infusion in resistant obsessive-compulsive disorder: an open trial. J. Clin. Psychiatry 63 (9), 796–801.
Parent, A., Hazrati, L.N., 1995. Functional anatomy of the basal ganglia. I. The cortico-basal ganglia-thalamo-cortical loop. Brain Res. Rev. 20 (1), 91–127.
Parent, A., Mackey, A., De Bellefeuille, L., December 1983. The subcortical afferents to caudate nucleus and putamen in primate: a fluorescence retrograde double labeling study. Neuroscience 10 (4), 1137–1150. https://doi.org/10.1016/0306-4522(83)90104-5.
Parent, A., Smith, Y., Filion, M., Dumas, J., January 16, 1989. Distinct afferents to internal and external pallidal segments in the squirrel monkey. Neurosci. Lett. 96 (2), 140–144. https://doi.org/10.1016/0304-3940(89)90047-5.
Parent, A., 1986. Comparative Neurobiology of the Basal Ganglia. John Wiley & Sons, New York.
Parent, A., July 1990. Extrinsic connections of the basal ganglia. Trends Neurosci. 13 (7), 254–258. https://doi.org/10.1016/0166-2236(90)90105-j.
Parent, M., Lévesque, M., Parent, A., October 15, 2001. Two types of projection neurons in the internal pallidum of primates: single-axon tracing and three-dimensional reconstruction. J. Comp. Neurol. 439 (2), 162–175. https://doi.org/10.1002/cne.1340.
Pasik, P., Pasik, T., Pecci Saavedra, J., Holstein, G.R., 1981. Light and electron microscopic immunocytochemical localization of serotonin in the basal ganglia of cats and monkeys. Anat. Rec. 199, 194.
Pearson, R.C.A., Gatter, K.C., Brodal, P., Powell, T.S.P., 1982. The projection of the basal nucleus of Meynert upon the neocortex in the monkey. Brain Res. 259, 132–136.
Percheron, G., Yelnik, J., Francois, C., 1984. A Golgi analysis of the primate globus pallidus. III. Spatial organization of the striato-pallidal complex. J. Comp. Neurol. 227, 214–227.
Pittenger, C., Bloch, M.H., September 2014. Pharmacological treatment of obsessive-compulsive disorder. Psychiatr. Clin. North Am. 37 (3), 375–391. https://doi.org/10.1016/j.psc.2014.05.006.
Prensa, L., Parent, A., 2001. The nigrostriatal pathway in the rat: A single-axon study of the relationship between dorsal and ventral tier nigral neurons and the striosome/matrix striatal compartments. J. Neurosci. 21 (18), 7247–7260. https://doi.org/10.1523/JNEUROSCI.21-18-07247.20019.
Qiu, A., Crocetti, D., Adler, M., Mahone, E.M., Denckla, M.B., Miller, M.I., Mostofsky, S.H., 2009. Basal ganglia volume and shape in children with attention deficit hyperactivity disorder. Am. J. Psychiatr. 166 (1), 74–82. https://doi.org/10.1176/appi.ajp.2008.08030426.
Reynolds, J.N., Wickens, J.R., 2000. Substantia nigra dopamine regulates synaptic plasticity and membrane potential fluctuations in the rat neostriatum, in vivo. Neuroscience 99 (2), 199–203. https://doi.org/10.1016/s0306-4522(00)00273-6.
Rhoton Jr., A.L., 2007. The cerebrum. Anatomy. Neurosurgery 61 (Suppl. l), 37–118 discussion 118–119.
Royce, G.J., 1987. Recent research on the centromedian and parafascicular nuclei. In: The Basal Ganglia II: Structure and Function – Current Concepts. Plenum, New York, pp. 293–319. Advances in behavioral biology, vol. 32.
Russchen, F.T., Price, J.L., 1984. Amygdalostriatal projections in the rat, topographical organization and fiber morphology shown using the lectin PHA-L as an anterograde tracer. Neurosci. Lett. 47, 15–22.
Saddoris, M.P., Cacciapaglia, F., Wightman, R.M., Carelli, R.M., August 2015. Differential dopamine release dynamics in the nucleus accumbens core and shell reveal complementary signals for error prediction and incentive motivation. J. Neurosci. 35 (33), 11572–11582.
Sadikot, A.F., Parent, A., François, C., 1990. The centre médian and parafascicular thalamic nuclei project respectively to the sensorimotor and associative-limbic striatal territories in the squirrel monkey. Brain Res 510 (1), 161–165. https://doi.org/10.1016/0006-8993(90)90746-x.
Sadikot, A.F., Parent, A., François, C., January 8, 1992a. Efferent connections of the centromedian and parafascicular thalamic nuclei in the squirrel monkey: a PHA-L study of subcortical projections. J. Comp. Neurol. 315 (2), 137–159. https://doi.org/10.1002/cne.903150203.
Sadikot, A.F., Parent, A., Smith, Y., Bolam, J.P., June 8, 1992b. Efferent connections of the centromedian and parafascicular thalamic nuclei in the squirrel monkey: a light and electron microscopic study of the thalamostriatal projection in relation to striatal heterogeneity. J. Comp. Neurol. 320 (2), 228–242. https://doi.org/10.1002/cne.903200207.
Sarter, M., Bruno, J.P., 2002. The neglected constituent of the basal forebrain corticopetal projection system: GABAergic projections. Eur. J. Neurosci. 15, 1867–1873.
Sato, W., Kubota, Y., Kochiyama, T., Uono, S., Yoshimura, S., Sawada, R., Sakihama, M., Toichi, M., November 25, 2014. Increased putamen volume in adults with autism spectrum disorder. Front. Hum. Neurosci. 8, 957. https://doi.org/10.3389/fnhum.2014.00957.
Schmahmann, J.D., 1998. Dysmetria of thought: clinical consequences of cerebellar dysfunction on cognition and affect. Trends Cogn. Sci. 2, 362–371.
Schwartz, J.M., Stoessel, P.W., Baxter Jr., L.R., Martin, K.M., Phelps, M.E., February 1996. Systematic changes in cerebral glucose metabolic rate after successful behavior modification treatment of obsessive-compulsive disorder. Arch. Gen. Psychiatry 53 (2), 109–113. https://doi.org/10.1001/archpsyc.1996.01830020023004.
Schwarz, M.J., Ackenheil, M., March 2002. The role of substance P in depression: therapeutic implications. Dialogues Clin. Neurosci. 4 (1), 21–29. https://doi.org/10.31887/DCNS.2002.4.1/mschwarz.
Semrud-Clikeman, M., Steingard, R.J., Filipek, P., Biederman, J., Bekken, K., Renshaw, P.F., April 2000. Using MRI to examine brain-behavior relationships in males with attention deficit disorder with hyperactivity. J. Am. Acad. Child Adolesc. Psychiatry 39 (4), 477–484. https://doi.org/10.1097/00004583-200004000-00017.
Sergeant, J.A., Geurts, H., Oosterlaan, J., March 10, 2002. How specific is a deficit of executive functioning for attention-deficit/hyperactivity disorder? Behav. Brain Res. 130 (1–2), 3–28. https://doi.org/10.1016/s0166-4328(01)00430-2.
Shapira, N.A., et al., 1997. Open-label pilot study of tramadol hydrochloride in treatment-refractory obsessive-compulsive disorder. Depress. Anxiety 6 (4), 170–173.
Shepherd, G.M., 2004. The Synaptic Organization of the Brain. Oxford University Press, New York.

Shinonaga, Y., Takada, M., Mizuno, N., 1994. Topographic organization of collateral projections from the basolateral amygdaloid nucleus to both the prefrontal cortex and nucleus accumbens in the rat. Neuroscience 58, 389–397.

Shirayama, Y., Takahashi, S., Minabe, Y., Ogino, T., 2005. In vitro 1H NMR spectroscopy shows an increase in N-acetylaspartylglutamate and glutamine content in the hippocampus of amygdaloid-kindled rats. J. Neurochem. 92 (6), 1317–1326. https://doi.org/10.1111/j.1471-4159.2004.02958.x.

Skinner, R.D., Garcia-Rill, E., 1984. The mesencephalic locomotor region (MLR) in the rat. Brain Res. 323, 385–389.

Smith, M.L., Hale, B.D., Booze, R.M., 1994. Calbindin- D28k immunoreactivity within the cholinergic and GABAergic projection neurons of the basal forebrain. Exp. Neurol. 130, 230–236.

Smith, Y., Parent, A., 1986. Differential connections of caudate nucleus and putamen in the squirrel monkey (Saimiri sciureus). Neuroscience 18, 347–371.

Snell, R.S., Snell, R.S., 2006. Clinical Uroanatomy, sixth ed. Lippincott Williams & Wilkins, Philadelphia.

Somogyi, P., Bolam, J.P., Smith, A.D., 1981. Monosynaptic cortical input and local axon collaterals of identified striatonigral neurons. A light and electron microscopic study in using the Golgi-peroxidase transport degeneration procedure. J. Comp. Neurol. 195, 567–584.

Soomro, G.M., et al., 2008. Selective serotonin re-uptake inhibitors (SSRIs) versus placebo for obsessive compulsive disorder (OCD) Cochrane. Database Syst. Rev. 1, CD001765.

Soomro, G.M., 2012. Obsessive compulsive disorder. Clin. Evid. 2012.

Spooren, W.P.J.M., Veening, J.G., Groenewegen, H.J., Cools, A.R., 1991. Efferent connections of the striatopallidal and amygdaloid components of the substantia innominata in the cat: projections tothe nucleus accumbens and caudate nucleus. Neuroscience 44, 431–447.

Starkstein, S., Gellar, S., Parlier, M., Payne, L., Piven, J., 2015. High rates of parkinsonism in adults with autism. J. Neurodev. Disorder. 7 (1), 29. https://doi.org/10.1186/s11689-015-9125-6.

Strack, S., Choi, S., Lovinger, D.M., Colbran, R.J., 1997. Translocation of autophosphorylated calcium/calmodulin-dependent protein kinase II to the postsynaptic density. J. Biol. Chem. 272 (21), 13467–13470. https://doi.org/10.1074/jbc.272.21.13467.

Subramanian, K., Brandenburg, C., Orsati, F., Soghomonian, J.J., Hussman, J.P., Blatt, G.J., 2017. Basal ganglia and autism—a translational perspective. Autism Res. 10 (11), 1751–1775. https://doi.org/10.1002/aur.1837.

Sung, K.W., Choi, S., Lovinger, D.M., 2001. Activation of group I mGluRs is necessary for induction of long-term depression at striatal synapses. J. Neurophysiol. 86 (5), 2405–2412. https://doi.org/10.1152/jn.2001.86.5.2405.

Sussman, D., Leung, R.C., Vogan, V.M., Lee, W., Trelle, S., Lin, S., Cassel, D.B., Chakravarty, M.M., Lerch, J.P., Anagnostou, E., Taylor, M.J., April 15, 2015. The autism puzzle: diffuse but not pervasive neuroanatomical abnormalities in children with ASD. Neuroimage Clin. 8, 170–179. https://doi.org/10.1016/j.nicl.2015.04.008.

Sutherland, R.J., 1982. The dorsal diencephalic conduction system: a review of the anatomy and functions of the habenular complex. Neurosci. Biobehav. Rev. 6, 1–13.

Szabo, J., December 1967. The efferent projections of the putamen in the monkey. Exp. Neurol. 19 (4), 463–476. https://doi.org/10.1016/0014-4886(67)90166-5.

Tanaka, C., Ishikawa, M., Shimada, S., 1982. Histochemical mapping of catecholaminergic neurons and their ascending fiber pathways in the rhesus monkey brain. Brain Res. Bull. 9, 255–270.

Teicher, M.H., Anderson, C.M., Polcari, A., Glod, C.A., Maas, L.C., Renshaw, P.F., April 2000. Functional deficits in basal ganglia of children with attention-deficit/hyperactivity disorder shown with functional magnetic resonance imaging relaxometry. Nat. Med. 6 (4), 470–473. https://doi.org/10.1038/74737.

Thörner, G., Lange, H., Hopf, A., 1975. Morphometrisch- statistische Strukturanalysen des Striatum, Pallidum und Nucleus subthalamicus beim Menschen. II. Pallidum. J. Hirnforsch. 16, 404–413.

Tong, C.K., Hamel, E., 2000. Basal forebrain nitric oxide synthase (NOS)-containing neurons project to microvessels and NOS neurons in the rat neocortex: cellular basis for cortical blood flow regulation. Eur. J. Neurosci. 12, 2769.

Ungerstedt, U., 1971. Stereotaxic mapping of the monoamine pathways in the rat brain. Acta Physiol. Scand. Suppl. 367, 1–49.

Vaidya, C.J., Austin, G., Kirkorian, G., Ridlehuber, H.W., Desmond, J.E., Glover, G.H., Gabrieli, J.D., November 24, 1998. Selective effects of methylphenidate in attention deficit hyperactivity disorder: a functional magnetic resonance study. Proc. Natl. Acad. Sci. U. S. A. 95 (24), 14494–14499. https://doi.org/10.1073/pnas.95.24.14494.

Valencia Garcia, S., Fort, P., February 2018. Nucleus Accumbens, a new sleep-regulating area through the integration of motivational stimuli. Acta Pharmacol. Sin. 39 (2), 165–166.

Vaughan, G.M., Pelham, R.W., Pang, S.F., et al., 1976. Nocturnal elevation of plasma melatonin and urinary 5-hydroxyindoleacetic acid in young men: attempts at modification by brief changes in environmental lighting and sleep and by autonomic drugs. J. Clin. Endocr. Metab. 42, 752–764.

Voorn, P., Jorritsma-Byham, B., Van Dijk, C., Buijs, R.M., 1986. The dopaminergic innervation of the ventral striatum in the rat: a light- and electronmicroscopical study with antibodies against dopamine. J. Comp. Neurol. 251, 84–99.

Wagner, S.J., Begaz, T., 2009. Basal ganglion stroke presenting as subtle behavioural change. BMJ Case Rep. https://doi.org/10.1136/bcr.10.2008.1139.

Wassef, M., Berod, A., Sotelo, C., 1981. Dopaminergic dendrites in the pars reticulata of the rat substantia nigra and their striatal input: combined immunocytochemical localization of tyrosine hydroxylase and anterograde degeneration. Neuroscienc 6, 2125–2139.

Wegiel, J., Flory, M., Kuchna, I., Nowicki, K., Ma, S.Y., Imaki, H., Wegiel, J., Cohen, I.L., London, E., Wisniewski, T., Brown, W.T., September 18, 2014. Stereological study of the neuronal number and volume of 38 brain subdivisions of subjects diagnosed with autism reveals significant alterations restricted to the striatum, amygdala and cerebellum. Acta Neuropathol. Commun. 2, 141. https://doi.org/10.1186/s40478-014-0141-7.

Wichmann, T., DeLong, M.R., 1996. Functional and pathophysiological models of the basal ganglia. Curr. Opin. Neurobiol. 6, 751–758.

Widge, A.S., Zorowitz, S., Basu, I., Paulk, A.C., Cash, S.S., Eskandar, E.N., Deckersbach, T., Miller, E.K., Dougherty, D.D., April 4, 2019. Deep brain stimulation of the internal capsule enhances human cognitive control and prefrontal cortex function. Nat. Commun. 10 (1), 1536. https://doi.org/10.1038/s41467-019-09557-4.

Wilson, C.J., 2004. Basal ganglia. In: Shepherd, G.M. (Ed.), The Synaptic Organization of the Brain. Oxford University Press, Oxford, pp. 361–413.

Xu, Z.C., Wilson, C.J., Emson, P.C., 1991. Synaptic potentials evoked in spiny neurons in rat neostriatal grafts by cortical and thalamic stimulation. J. Neurophysiol. 65 (3), 477–493. https://doi.org/10.1152/jn.1991.65.3.477.

Yager, L.M., Garcia, A.F., Wunsch, A.M., Ferguson, S.M., August 2015. The ins and outs of the striatum: role in drug addiction. Neuroscience 301, 529–541.

Yagmurlu, K., Vlasak, A.L., Rhoton Jr., A.L., 2015. Three-dimensional topographic fiber tract anatomy of the cerebrum. Neurosurgery 11 (Suppl. 2), 274–305.

Yang, C.R., Mogenson, G.J., 1985. An electrophysiological study of the neural projections from the hippocampus to the ventral pallidum and the subpallidal areas by way of the nucleus accumbens. Neuroscience 15, 1015–1024.

Yasargil, M.G., 1984. Microneurosurgery in 4 volumes. I, microsurgical anatomy of the basal cisterns and vessels of the brain, diagnostic studies, general operative techniques and pathological considerations of the intracranial aneurysms. In: Thieme Classics. Thieme Stratton, New York.

Yeterian, E.H., Van Hoesen, G.W., January 6, 1978. Cortico-striate projections in the rhesus monkey: the organization of certain cortico-caudate connections. Brain Res. 139 (1), 43–63. https://doi.org/10.1016/0006-8993(78)90059-8.

Zahm, D.S., Brog, J.S., 1992. On the significance of subterritories in the "accumbens" part of the rat ventral striatum. Neuroscience 50, 751–767.

Zarei, M., Mataix-Cols, D., Heyman, I., Hough, M., Doherty, J., Burge, L., Winmill, L., Nijhawan, S., Matthews, P.M., James, A., December 1, 2011. Changes in gray matter volume and white matter microstructure in adolescents with obsessive-compulsive disorder. Biol. Psychiatry 70 (11), 1083–1090. https://doi.org/10.1016/j.biopsych.2011.06.032.

Záborszky, L., Alheid, G.F., Beinfeld, M.C., et al., 1985. Cholecystokinin innervation of the ventral striatum: a morphological and radioimmunological study. Neuroscience 14, 427–453.

Zheng, T., Wilson, C.J., 2002. Corticostriatal combinatorics: the implications of corticostriatal axonal arborizations. J. Neurophysiol. 87 (2), 1007–1017. https://doi.org/10.1152/jn.00519.2001.

Further Reading

Bain, P.G., 2009. Deep Brain Stimulation. Oxford University Press, Oxford.

Baydin, S., Yagmurlu, K., Tanriover, N., Gungor, A., Rhoton Jr., A.L., 2016. Microsurgical and fiber tract anatomy of the nucleus accumbens. Oper. Neurosurg. (Hagerstown). 12 (3), 269–288.

Corbit, V.L., Whalen, T.C., Zitelli, K.T., Crilly, S.Y., Rubin, J.E., Gittis, A.H., May 18, 2016. Pallidostriatal projections promote β oscillations in a dopamine-depleted biophysical network model. J. Neurosci. 36 (20), 5556–5571. https://doi.org/10.1523/JNEUROSCI.0339-16.2016.

Guzman, J.N., Hernandez, A., Galarraga, E., et al., 2003. Dopaminergic modulation of axon collaterals interconnecting spiny neurons of the rat striatum. J. Neurosci. 23 (26), 8931–8940.

Kitai, S.T., Kocsis, J.D., Preston, R.J., Sugimori, M., 1976. Monosynaptic inputs to caudate neurons identified by intracellular injection of horseradish peroxidase. Brain Res. 109 (3), 601–606.

Nauta, W.J., Mehler, W.R., January 1966. Projections of the lentiform nucleus in the monkey. Brain Res. 1 (1), 3–42. https://doi.org/10.1016/0006-8993(66)90103-x.

Shirayama, Y., Chaki, S., October 2006. Neurochemistry of the nucleus accumbens and its relevance to depression and antidepressant action in rodents. Curr. Neuropharmacol. 4 (4), 277–291.

Voorn, P., Vanderschuren, L.J., Groenewegen, H.J., Robbins, T.W., Pennartz, C.M., 2004. Putting a spin on the dorsal-ventral divide of the striatum. Trends Neurosci. 27 (8), 468–474.

16

The Circumventricular Organs

Daniele Armocida[1], *Alessandro Pesce*[2], *Giovanni Paternò*[3], *Mauro Palmieri*[1], *Maurizio Salvati*[5], *Brandon M. Ascenzi*[6] *and Alessandro Frati*[1,4]

[1]Department of Human Neurosciences, Neurosurgery Division, University of Rome "La Sapienza", Rome, Italy [2]Department of Neurosurgery, "Santa Maria Goretti" Hospital, Latina, Italy [3]Department of Neurosurgery, Istituto Clinico Città Studi, Milan, Italy [4]IRCCS "Neuromed", Experimental Neurosurgery Unit, Pozzilli, Italy [5]Policlinico Tor Vergata, Neurosurgery Department, University Tor Vergata of Rome, Rome, Italy [6]Independent Neuroresearcher-Member of Marie Curie Alumni Association (MCAA)

OUTLINE

INTRODUCTION

The central nervous system (CNS) plays an essential role in maintaining constancy of the "milieu interieur" as a consequence of the ability of autonomic control centers in the hypothalamus and medulla to continually monitor and regulate the constituents of this internal environment (Cattaneo and Parmiani, 1982). To achieve this objective these "control centers" must obtain information about the "current state" of all important variables (e.g., glucose, sodium, calcium, temperature, pH, osmolarity), in all regions of the body (Hatton, 1990). Autonomic control centers derive this afferent information from a variety of specialized sensory structures. Traditionally, sensory information was believed to reach the brain either by direct diffusion from the circulation into the CNS (substances that cross the blood–brain barrier, BBB), or as a result of neural input derived from peripherally located sensory structures (e.g., osmoreceptors, baroreceptors, chemoreceptors) transmitted to the CNS through autonomic afferents (e.g., vagus, glossopharyngeal nerves). However, neither of these mechanisms provides adequate explanations for the established roles of many BBB-impermeable circulating signals in controlling the output of autonomic control centers in the brain. Recent works have suggested mechanisms through which such information may reach autonomic control centers behind the BBB, including specific transporters, transendothelial cell signaling, and the focus of this particular chapter, actions at circumventricular organs (CVOs) of the brain, structures that represent the only regions of the CNS that are not protected by the normal BBB.

THE CIRCUMVENTRICULAR ORGANS

The CNS is separated from the peripheral circulation by the BBB, which is formed by specialized endothelial cells: astrocytes and pericytes (Tagami et al., 1990). CVOs are highly vascularized midline structures lacking a complete BBB (Hofer, 1958; Hoffman and Olszewski, 1961) due to the presence of fenestrated vasculature (Fig. 16.1), allowing peripheral circulating factors to penetrate the brain and influence neuronal activity (Bearer and Orci, 1985; Krisch et al., 1987; Shaver et al., 1990; McKinley et al., 2003). Modified ependymal cells known as tanycytes are present in the ependymal lining (Rizzoti and Lovell-Badge, 2017) (see Box 16.1).

Tanycytes are specialized ependymal cells interfacing with the CSF and have a long basal process that terminates

From Anatomy to Function of the Central Nervous System
https://doi.org/10.1016/B978-0-12-822404-5.00010-3

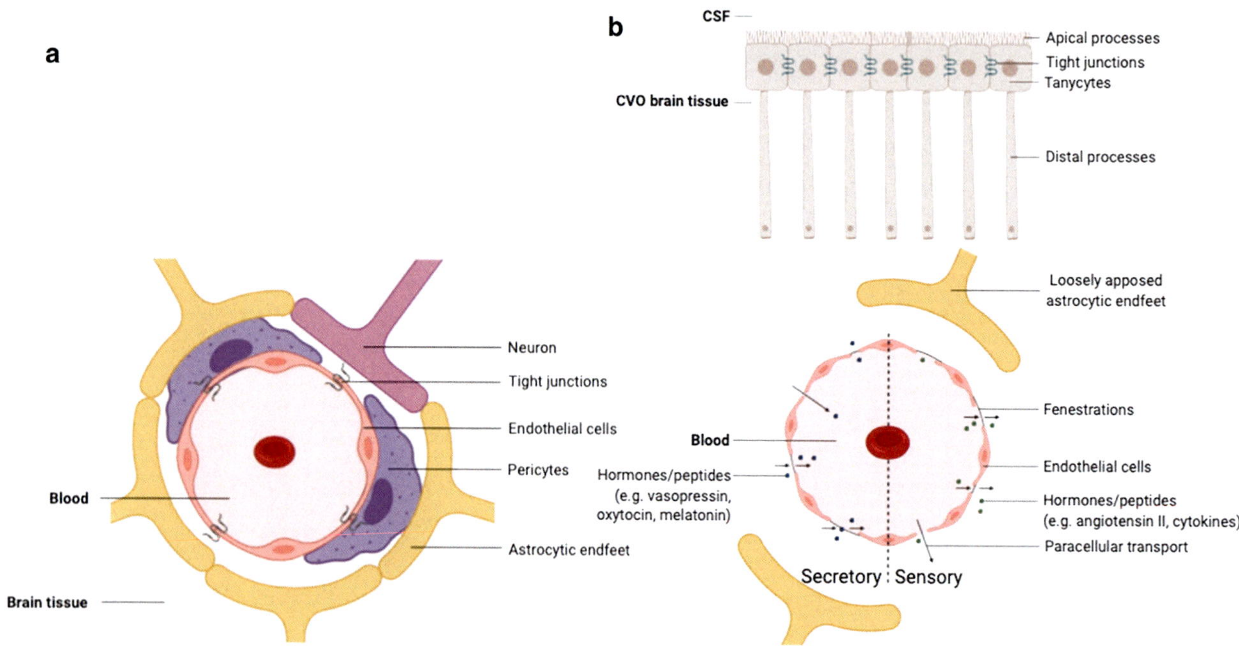

FIGURE 16.1 (A) Cross section of a blood vessel with a normal blood–brain barrier. (B) Circumventricular organ with a cross section of a capillary with a fenestrated endothelium, with on the left half the secretory and on the right half the sensory type. Tanycytes monitor cerebrospinal fluid composition and pass this information to the circumventricular organ. *Figure created with BioRender.com Image courtesy from Verheggen, I.C.M., de Jong, J.J.A., van Boxtel, M.P.J., et al., 2020. Permeability of the windows of the brain: feasibility of dynamic contrast-enhanced MRI of the circumventricular organs. Fluids Barriers CNS 17, 66. https://doi.org/10.1186/s12987-020-00228-x*

at the capillary (Horstmann, 1954; Rodríguez et al., 2005). Ependymal cells in the Arc and ME (median eminence, see later) are the best-known tanycytes and have a single, long cellular process projecting to discrete hypothalamic parenchyma and fenestrated capillaries (Fig. 16.2; Rodríguez et al., 2005). Tanycytes also exist at the ventricular borders of the OVLT and SFO (organum vasculosum of lamina terminalis and subfornical organ, see later) (Mullier et al., 2010; Langlet et al., 2013). A honeycomb distribution pattern of tight junction proteins is observed at tanycytes around the CVO proximal cell bodies, possibly to prevent the diffusion of blood-derived molecules into the CSF (Mullier et al., 2010; Langlet et al., 2013). Hypothalamic tanycytes also reduce the release into circulation of thyrotropin releasing hormone (TRH) by activating the tanycytic receptor for thyrotropin-releasing hormone (Müller-Fielitz et al., 2017).

However, circulating leptin (an important hormone produced by adipose tissue that plays a key role in food intake and in particular in regulating the sense of satiety) regulation is recognized by leptin receptors on tanycytes in the ME, then extracellular signal-regulated kinase signaling of tanycytes promotes leptin transport, activating neurons in the mediobasal hypothalamus (Balland et al., 2014).

Astrocytes and tanycytes often express several kinds of tight junction proteins (Fig. 16.2, in part. see D and E) (Morita et al., 2016), and electron microscopy reveals junctions between, intimately apposed, astrocytic or tanycytic cellular membranes (Fig. 16.2, in part. see F) (Morita et al., 2016). Astrocytes are reported to express ZO-1 and occludin in the periphery of infarct areas during the stroke recovery process to prevent ingress of neurotoxic molecules (Yang et al., 2007, 2013).

Thus, these specialized areas are points of communication between the blood, the brain parenchyma, and the CSF. Neurons and glial cells of the CVOs express a unique repertoire of receptors and ion channels and receive a wide range of chemical signals from the bloodstream (Hendelman, 2005). Via their interconnections with the hypothalamus and brainstem, these sensory CVOs have a critical role in sodium and water balance, cardiovascular regulation, energy metabolism, and immunomodulation (Hatton, 1997; Hatton and Walters, 1973). They are also involved in the pathogenetic mechanisms of fever, vomiting, and other responses to potentially noxious stimuli. Indeed, CVOs are sites of access to the CNS for circulating microorganisms, prion proteins, and autoantibodies (Benarroch, 2011) and studies show that expression of TLR2 and TLR4 (Toll-like receptors 2 and 4, very important in the immune response) is remarkably higher in the CVOs than in other brain regions (Nakano et al., 2015; Muneoka et al., 2019; Murayama et al., 2019).

With the remarkable technological improvements in animal and molecular research in recent years, more

Box 16.1 New "sources" of neural stem cells? The tanycytes of the circumventricular organs are something of this

Tanycytes in the ME and Arc are shown to be a neurogenic niche in adult mammals (Lee et al., 2012; Robins et al., 2013). Tanycytes in the spinal cord are NSCs that, remarkably, proliferate and differentiate into astrocytes and oligodendrocytes upon spinal injury (Meletis et al., 2008). Recently, Miyata et al. have found two types of NSCs in the CVOs: astrocyte- and tanycyte-like NSCs (Hourai and Miyata, 2013; Furube et al., 2020; Nambu, 2021). Radial glial cells in the ventricle zone are NSCs that, during cortical development, produce new neurons and glial cells (Ohtaka-Maruyama and Okado, 2015). Tanycytes in the adult mammalian brain are NSC precursors, ultimately derived from radial glia during development (Goodman and Hajihosseini, 2015). The CSF provides important functions to support brain homeostasis, morphogenesis, and proliferation during development (Johansson, 2014). In adults, the CSF actively synthesizes and secretes factors that promote the proliferation of quiescent and transit-amplifying NSCs in the subventricular zone (Silva-Vargas et al., 2016). Aging speed is controlled primarily by hypothalamic NSCs via levels of exosomal miRNAs in the CSF (Zhang et al., 2018). Moreover, continuous angiogenesis occurs in the CVOs of adult mice, possibly reconstructing neurons, glia, and fenestrated capillaries (Morita et al., 2013; Nakano et al., 2015; Furube et al., 2014; Asami et al., 2020). Thus, tanycytes, throughout the adult mammalian brain, are accepted as NSCs, with their proliferation and differentiation potentially regulated by factors.

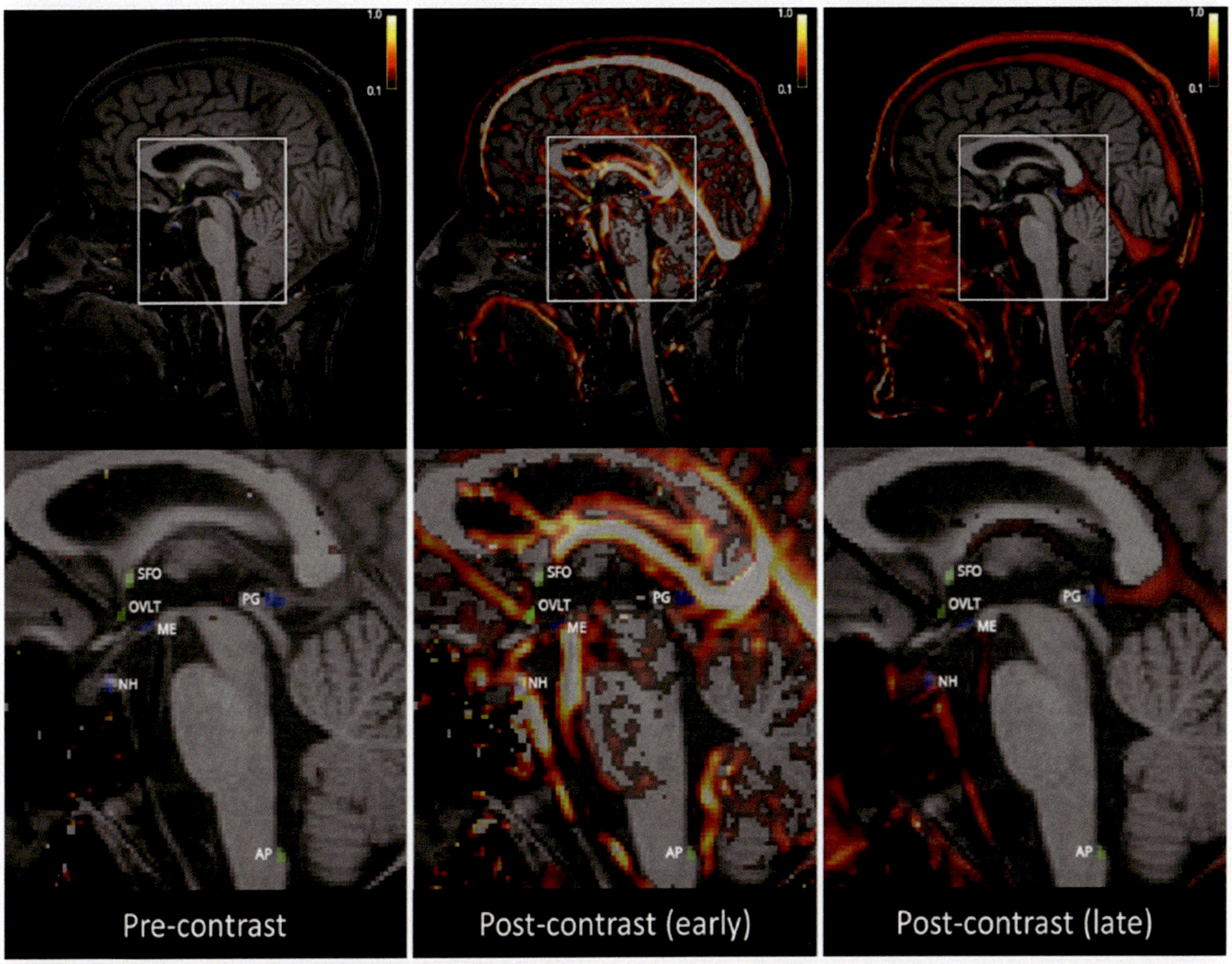

Permeability of the windows of the brain: feasibility of dynamic contrast-enhanced MRI of the circumventricular organs. In detail in this image: the representative contrast agent concentration maps [mM] showing the distribution of contrast agent before (left), shortly after (middle), and approximately 12 min after (right) contrast agent injection. *AP*, area postrema; *ME*, median eminence; *NH*, neurohypophysis; *OVLT*, organum vascolosum of lamina terminalis; *PG*, pineal gland; *SFO*, subfornical organ. *Image courtesy of Verheggen, I.C.M., de Jong, J.J.A., van Boxtel, M.P.J., et al., 2020. Permeability of the windows of the brain: feasibility of dynamic contrast-enhanced MRI of the circumventricular organs. Fluids Barriers CNS 17, 66. https://doi.org/10.1186/s12987-020-00228-x*

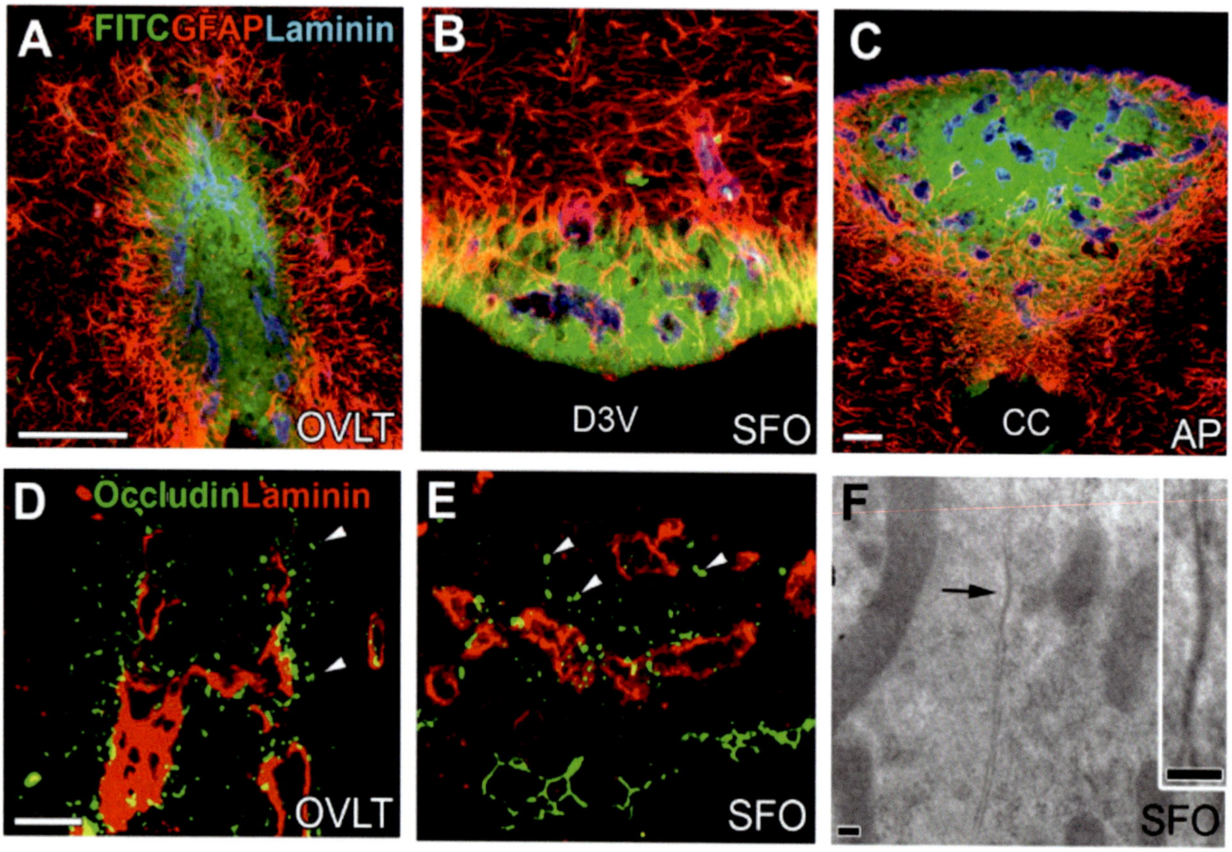

FIGURE 16.2 **Vascular permeability and tight junctions in the CVOs of the adult mouse brain.** Strong fluorescence of blood-derived fluorescein isothiocyanate is seen at the central subdivision of the sensory CVOs, but fluorescence is weak at distal subdivisions and no fluorescence is seen in the neighboring brain region beyond dense astrocytic and tanycytic barriers (A–C). Tight junction protein occludin is seen at parenchyma close to fenestrated capillaries and tanycytic layers (D and E). An electron microscopic image reveals the tight junction-like structure between the plasma membrane of neighboring cells making complete contact (F). Scale bar = 50 μm (A, C, and D) and 100 nm (F). AP, area postrema; CC, central canal; D3V, dorsal third ventricle; FITC, fluorescein isothiocyanate; GFAP, glial fibrillar acidic protein; OVLT, organum vasculosum of the lamina terminalis; SFO, subfornical organ. *Image courtesy of Miyata, S., 2022. Glial functions in the blood-brain communication at the circumventricular organs. Front. Neurosci. 16 . https://doi.org/10.3389/fnins.2022.991779.*

and more circumventricular areas are being discovered that are not directly protected by the BBB and that acquire the status of CVOs (Felten et al., 2015); at present, the main organs described being part of this particular group of substructures are supraoptic nucleus (SON), subfornical organ (SFO), neurohypophysis (NH), median eminence (ME), subcommissural organ (SCO), area postrema (AP), pineal gland (PG), and organum vasculosum of the lamina terminalis (OVLT) (Apuzzo, 1987; Arendt and Aulinas, 2000). Their anatomical arrangement is showed in Fig. 16.3 (murine) and Fig. 1 of Box 16.1 (human).

Glial cells consisting of astrocytes and microglia/macrophages are present in all these organs. The pineal gland, neurohypophysis and the median eminence lack of neurons that are present in the rest of the CVOs. Most of the CVOs are overlined by ependymal cells except the pineal and the neurohypophysis.

These organs are important sites for communication with the cerebrospinal fluid as well as between the brain and peripheral organs via blood-borne products as they lack the BBB (Kaur and Ling, 2017). The CVOs are structures strategically located along the surface of the brain ventricles and, from the structural standpoint, they are subdivided into two groups: ependymal and paraependymal and have been further classified into *sensory* and *secretory*. The CVOs have a unique neural and vascular architecture and consist of neurons, glia, and ependymal cells combined with leptomeningeal and vascular elements. A common salient feature of the CVOs is presence of fenestrated capillaries and loosely apposed astrocytic processes that create relatively large perivascular spaces. These characteristics allow the ready passage of large molecules, such as peptides and polar substances, between the blood to the perivascular spaces, exposing neurons to peripheral signals

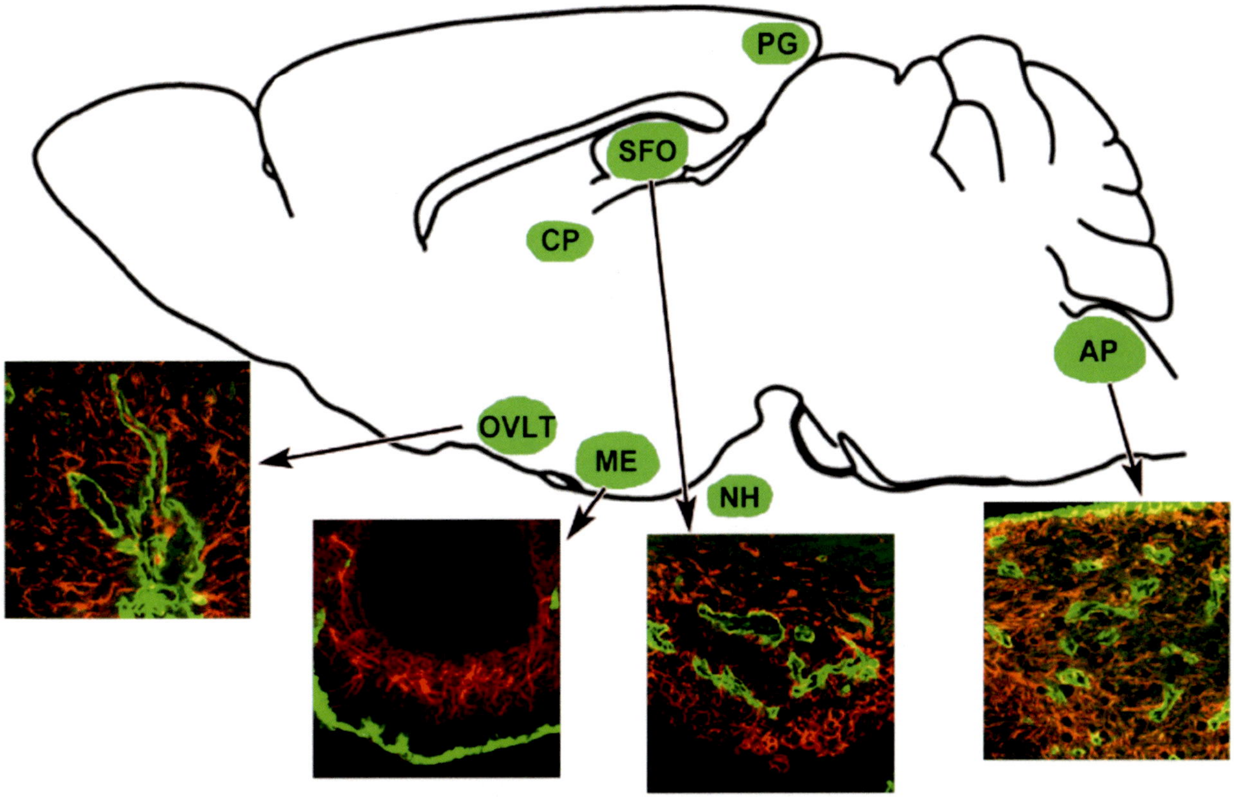

FIGURE 16.3 **Schematic representation revealing CVOs locations in the adult rodent brain.** Insets indicate that the CVOs are characterized by dense and thick laminin^{+}(green) capillaries and dense GFAP^{+} (red) astrocytic and tanycytic networks. Scale bar = 50 μm. AP, area postrema; CP, choroid plexus; GFAP, glial fibrillar acidic protein; ME, median eminence; NH, neurohypophysis; OVLT, organum vasculosum of the lamina terminalis; PG, pineal gland; SFO, subfornical organ. *The illustration is rearranged with permission from Elsevier B.V. Takagi et al. (2019) by Miyata (2022). Image courtesy of Miyata, S., 2022. Glial functions in the blood-brain communication at the circumventricular organs. Front. Neurosci. 16. https://doi.org/10.3389/fnins.2022.991779.*

(McKinley et al., 1998; Benarroch, 2011; Fry and Ferguson, 2007).

The CVOs endothelial intercellular space exhibits a size-limited vascular permeability lower than approximately MW 10,000 (Miyata, 2015, 2017). However, it is often misunderstood that the fenestrated capillaries of the CVOs allow molecules of any size to enter the parenchyma. High-molecular-weight molecules (HMW) cannot pass freely through fenestrate capillaries in the CVOs. HMW molecules such as hormones and cytokines reach the brain parenchyma beyond the BBB via transcytosis (Gonçalves and De Felice, 2021). The unique morphologic and functional characteristics of the CVOs provide the basis for both their important role in homeostasis and their involvement in a wide range of neurologic and medical disorders. The accessibility of peptides and polar substances, such as receptor antagonists and monoclonal antibodies to specific targets in the CVOs, provides an opportunity for pharmacologic therapy of many disorders.

In order to describe the location of CVOs, it is important to remember the location and pivotal role that the hypothalamus plays, with its small size, within the brain and its role in homeostasis. Its upper limit is represented by the hypothalamic sulcus which is a smooth groove separating it from the thalamus; its lower limit is represented by three structures and two different recesses, and they are, respectively: mammillary bodies, tuber cinereum, the optic chiasm, and hypophyseal and chiasmatic recesses. Such structures in the shape of a double tunnel constitute the floor of the third ventricle. The anterior limit is represented by the lamina terminalis of the corpus callosum, while the posterior limit is represented by the subthalamus.

To describe the first analyzed CVO, the supraoptic nucleus, it is necessary to take into consideration that the hypothalamus *sensu stricto* consists of the chiasmatic, tuberal, and mamillary region. These regions form the hypothalamic gray matter, which is composed of easily distinguishable nuclear grays. Prominent components of the chiasmatic region are the magnocellular neurosecretory complex (supraoptic nucleus, paraventricular nucleus, accessory neurosecretory nucleus), the sexually dimorphic intermediate nucleus, retrochiasmatic and supraoptic nuclei (Braak and Braak, 1992).

Supraoptic Nucleus

The supraoptic nucleus (SON) covers the posterior part of the optic chiasm and the proximal part of the optic tract. It consists of three parts:

- a large dorsolateral part which contains 53,000 neurons, 90% of which contain vasopressin and 10% oxytocin (Dierickx and Vandesande, 1977; Fliers et al., 1985);
- a dorsomedial part; and
- a ventromedial part that, together, contain 23,000 neurons; 85% of these contain vasopressin and 15% oxytocin (Dierickx and Vandesande, 1977).

The neurons of the supraoptic form a population of extremely stable cells in normal aging and Alzheimer's disease (Swaab et al., 1993; Swaab, 1997).

SON neurons play a fundamental role in regulating body homeostasis by controlling the secretion of oxytocine (OT) and vasopressine (VP) from the neurohypophysis in response to changes in plasma osmolarity and so in parturition and lactation (Brownstein et al., 1980; Gainer and Wray, 1994; Armstrong, 1995). Furthermore, these neurons provide special structural and functional advantages for the study of the organization and plasticity of nuclear compartments in both physiological conditions and in response to experimental conditions of up- or downregulation of the global transcriptional activity induced by hyperosmolarity versus hypoosmolality (Gainer, 1998). First, SON neurons are large which have a high rate of global transcription, with OT and VP mRNAs being particularly abundant in their respective neuronal phenotypes (Hazell et al., 2012). Second, the ultrastructure of the cellular nucleus shows very well-defined nuclear compartments (Lafarga et al., 1991, 1993; Garcia-Segura et al., 1993). Third, SON neurons can be easily activated or inactivated in vivo by changes in plasma osmolarity which modify the expression level of cell-specific and house-keeping genes and the hormonal biosynthesis specific to these neurons (Glasgow et al., 2000) . Finally, these neurosecretory cells, like all mammalian neurons, are interphase G0 cells, which avoids the influence of cell cycle on the organization of nuclear compartments (Lafarga et al., 1992, 1998).

The magnocellular neuroendocrine system is composed of axons from the large oxytocin- and vasopressin-containing neurons in the supraoptic and paraventricular nuclei that terminate in the posterior lobe of the pituitary.

Magnocellular neurons project a single axon to the neurohypophyseal parenchyma where their hormonal content is secreted directly in the blood stream. In addition, it is now acknowledged that oxytocin and vasopressin can be centrally released in various brain regions including within the paraventricular nucleus (PVN) and supraoptic nucleus (SON) (Ludwig and Pittman, 2003) where they have local modulatory effects (Kombian et al., 2002; Ludwig et al., 2002). Neurohypophysial hormone secretion is dependent upon the electrical activity of magnocellular neurons, which is itself under the influence of synaptic afferent input (Leng et al., 1999). Glutamate and GABA are, respectively, the main excitatory and inhibitory neurotransmitters in the hypothalamus and in the SON, approximately 25% of all synapses are glutamatergic, whereas approximately 40% are GABAergic. Most glial cells in this nucleus are astrocytes with a radial morphology; their cell bodies are located in the ventral portion of the SON from which arise thick and long processes orientated ventrodorsally and processes surrounding neuronal elements (Salm, 2000).

Vascular Organ of Lamina Terminalis

The Organum Vasculosum of Lamina Terminalis (OVLT) is a paraependymal sensory CVO. It is at its greatest extent in the lamina terminalis, approximately midway between the optic chiasm and the anterior commissure (Duvernoy et al., 1969). The OVLT appears to be part of a neural network within the lamina terminalis and hypothalamus, which is involved in the regulation of fluid balance (McKinley et al., 1984, 2004).

It responds to factors such as angiotensin II and regulates the thirst drive via its connections to the median preoptic nucleus (MPN) (McKinley et al., 2015). OVLT along with the median eminence and the subfornical organ is interconnected with midventral hypothalamus and all these structures are in proximity of the anterior wall of the third ventricle: the lamina terminalis. This latter is a thin layer of neuroepithelium that forms, as said above, part of the anterior wall of the third ventricle and where OVLN neurons are located.

Like other CVOs, the OVLT is composed of neurons, glial cells, and tanycytes. Axon terminals containing several neuropeptides and neurotransmitters including somatostatin, angiotensin, dopamine, norepinephrine, serotonin, acetylcholine, oxytocin, AVP, and TRH innervate the OVLT.

The region of the hypothalamus that immediately surrounds the OVLT regulates a diverse array of autonomic processes. However, because the OVLT is potentially involved in the maintenance of so many processes, definitive studies ascribing specific functions to the OVLT are inherently difficult.

The efferent projections of the OVLT are not well defined because of the fundamental difficulty of injecting this small structure with specific neuroanatomic tracers without contaminating surrounding preoptic nuclei.

In early embryos, lamina terminalis progenitors can be found medially in the preoptic area of the anterior neural plate, just in front of the hypothalamus and posterior to the anterior neural border (ANB) (Cobos et al., 2001). Upon neural tube closure, the ANB folds up into the commissural plate, a template for the formation of all telencephalic commissures (Suárez et al., 2014). The ANB/commissural plate also functions as a signaling center that secretes several FGFs and that is involved in the patterning of the cerebral cortex (Shimogori et al., 2004). Thus, the lamina terminalis emerges between the FGF-secreting ANB and the SHH-producing prechordal plate/hypothalamus and there is substantial cross-regulation between these two signaling centers (Ohkubo et al., 2002; Storm et al., 2006). At the same time, BMPs and WNTs from the dorsal midline of the telencephalon antagonize the ventral-anterior domains of FGF and SHH signaling (Ohkubo et al., 2002; Gunhaga et al., 2003; Shimogori et al., 2004; Storm et al., 2006).

Homeobox genes important for the formation of the lamina terminalis are Six3, which appears to integrate FGF and SHH signaling (Lagutin et al., 2003; Storm et al., 2006; Geng et al., 2008; Jeong et al., 2008), and the SHH target gene Nkx2.1, which is essential for hypothalamus and ventral telencephalon development (Dale et al., 1997; Pera and Kessel, 1997; Corbin et al., 2003). Taken together, the lamina terminalis is formed under the control of those signals that pattern the emerging telencephalon more generally, but how exactly the OVLT is established within the preoptic region remains unknown. In a recent reevaluation of hypothalamic regionalization, Puelles and Rubenstein (2015) introduced an "acroterminal hypothalamic domain" (ATD), a medial strip of neuroepithelium that connects the anterior ends of the floor and roof plates. The OVLT, median eminence, and neurohypophysis originate within this ATD. However, as several other structures such as the optic stalk and chiasm and the postoptic commissure also form within the ATD, it is not just a progenitor domain of the anteroventral CVOs.

Subfornical Organ

The subfornical organ (SFO), named due to its proximity to the fornix, is one of the paraependymal sensory CVOs (Dempsey, 1968). Basic anatomical features of the SFO have been outlined in several mammalian species (McKinley et al., 1998). It is located along the anterior wall of the third ventricle and extends into its lumen. It occupies the dorsal extremity of the lamina terminalis and is adjacent to the choroid plexus (McKinley et al., 1998).

In addition, SFO has extensive afferent and efferent neural connections within the hypothalamus, brainstem, thalamus, cortex, and other regions (McKinley et al., 2003).

As in other CVOs, SFO harbors several unique non-neuronal cell populations including specialized endothelial cells containing fenestrations (Sposito and Gross, 1987; Shaver et al., 1990) with extended perivascular space surrounded by basement membranes (Morita et al., 2016) and distinct glia cells called tanycytes (Hicks et al., 2021).

SFO is a heterogeneous nucleus that contains several types of neurons that differ in their pattern of gene expression. Notably, a few subsets of neurons have been associated with specific neural functional pathways, e.g., excitatory nitric-oxide synthase (nNOS)-positive neurons trigger thirst (Zimmerman et al., 2019). In addition, subpopulations of SFO neurons express receptors for hormones associated with regulating energy metabolism, such as amylin (Sexton et al., 1994; Pulman et al., 2006), ghrelin (Pulman et al., 2006), leptin (Smith et al., 2009), cholecystokinin (Ahmed et al., 2014), neuropeptide Y (Kishi et al., 2006), and glucagon-like peptide 1 (Göke et al., 1995) , as well as molecules involved in neuroimmune interactions, such as lipopolysaccharide receptor (Lacroix et al., 1998), prostaglandins (Matsumura and Simon, 1990; Zhang and Rivest, 2000), interleukin 1b and interleukin 6 (Takahashi et al.), and TNFα. Moreover, also express: relaxin receptors (Burazin et al., 2005), vasopressin V1 receptor (Phillips et al., 1988), atrial natriuretic peptide (Quirion et al., 1984; Mendelsohn et al., 1984), estrogen receptor (Somponpun and Sladek, 2003), prolactin receptor (Kamesh et al., 2018), galanin receptor (O'Donnell et al., 1999), and somatostatin (Patel et al., 1986), as well as ion channels such as TRPV1 (Mannari et al., 2013), Nax (Hiyama et al., 2004; Nehmé et al., 2012), and classical neurotransmitter receptors such as glutamate, GABA, and acetylcholine (McKinley et al., 2003). Although it remains unknown whether the expression of these molecules defines distinct neuronal subpopulations within the SFO, these findings are consistent with the involvement of this nucleus in a myriad of centrally regulated homeostatic processes, including energy metabolism and food anticipatory activity (Pulman et al., 2006; Smith et al., 2010), thirst (Hollis et al., 2008), reproduction (Summerlee et al., 1987; Sunn et al., 2002), immune responses (Takahashi et al., 1997), cardiovascular regulation (Cancelliere and Ferguson, 2017; Jeong et al., 2008) and systemic osmoregulation (Oldfield and McKinley, 1994).

One of the best-characterized functions of the SFO is its role in mediating dipsogenic and pressor effects of circulating angiotensin II (Ang II) (Hoffman and Phillips, 1976; Ferguson and Bains). Neurons in this area express Ang II type 1 (AT1) receptors (Song et al., 1992). Ang II causes an increase in action potential firing rate of SFO neurons (Felix and Künzle, 1974; Felix, 1976; Allen et al., 1988), thereby stimulating neural pathways involved in drinking, pressor response, and salt appetite via its projections to the bed nucleus of the stria terminalis, the median preoptic, and supraoptic and paraventricular nuclei (Weisinger et al., 1990 ; Oldfield and McKinley, 1994; Fitts et al., 2005; Hollis et al., 2008).

All these projections regulate the activity of magnocellular neurosecretory neurons releasing antidiuretic hormone (ADH or vasopressin) (Oldfield and McKinley, 1994), as well as contribute to the regulation of the activity of preautonomic neurons associated with cardiovascular control (Cancelliere et al., 2015).

Neurohypophysis

The neurohypophysis forms the posterior lobe of the pituitary gland (hypophysis cerebri), a pea-sized structure protruding ventrally from the hypothalamus via the stalk-like infundibular stem.

The neurohypophysis per se is composed of the *infundibulum*, a conical downgrowth of the tuber cinereum, a cylindrical portion termed the *pituitary stalk* and the *bulbous posterior* or *neural lobe of the pituitary gland*. These structures consist predominantly of nonmyelinated nerve fiber tracts originating from the hypothalamus. The neurohypophysis represents only one portion of a more complex neurosecretory unit which begins with the paired supraoptic and paraventricular nuclei (see Chapter 10).

In the context of its endocrinological function, also the neurohypophysis contains two major neuronal groups: the *magnocellular* and the *parvocellular*. The first are wide neurons connecting the supraoptic and paraventricular nuclei with the posterior hypophyseal lobe. The magnocellular neurons of these hypothalamic nuclei are engaged in the production of vasopressin and oxytocin, substances erroneously referred to as "posterior lobe" hormones. The supraoptico- and paraventriculohypophyseal fiber tracts emanate from these nuclei and, in combination, comprise the majority of the hypothalamo-hypophyseal tract, which in turn makes up the bulk of the pituitary stalk. The parvocellular neurons are smaller, connecting the periventricular nuclei with the median eminence (Scheithauer et al., 1992).

As stated before, the supraoptic and paraventricular nuclei overlie the proximal optic tracts and underlie the subependymal zone of the anterolateral third ventricle, respectively. Magnocellular neurons of hypothalamus produce hormones which are synaptically conveyed to the neurohypophysis through the hypothalamus pituitary bundle, whereas the parvicellular neurons, differently, produce peptide substances, which are poured through the pituitary tuber bundle leading directly to the infundibular capillary network, known as the hypothalamic–hypophysial portal system.

The hypothalamic-hypophysial system is composed of two capillary beds, one located in the median eminence and the other one in the anterior hypophysis. The capillary bed of the median eminence is supplied by the superior hypophysial arteries, and it is divided in two plexuses: external and internal. The first one is directly in contact with the external surface of the median eminence and it is composed by hexagonal units that are filled with axon terminals and glial cells, while the second one is composed by several capillary loops which are supplied by external plexuses and then are connected to anterior hypophysis by portal veins. Once into the anterior hypophysis, blood is then drained into the cavernous and posterior intercavernous sinuses. All capillaries in the portal system are fenestrated, thus molecules that are not blocked by BBB can then be drained in the systemic circulation.

Pathology of the posterior lobe of the pituitary gland (or neurohypophysis) may develop several clinical features such as diabetes insipidus, which is caused by infections, vascular accidents, complications of brain surgery and benign and malignant tumors as well as metastasis (Rhoton, 2019). Inappropriate secretion of vasopressin (Schwartz Bartter's syndrome) is due to excessive vasopressin release from the posterior pituitary or the ectopic production by various tumors, mainly carcinoid of the lung (Felix, 1988).

Median Eminence

The median eminence (ME), one of the paraependymal CVOs, is the structure at the base of the hypothalamus where hypothalamic-releasing and -inhibiting hormones converge onto the portal capillary system that vascularizes the anterior pituitary gland.

The median eminence lies in the center of the tuber cinereum and is composed of an extensive array of blood vessels and nerve endings. It's extremely rich blood supply arises from the superior hypophyseal artery (a branch of the internal carotid artery), which sends off many small branches that form capillary loops. The small capillary loops extend into the internal and external zones of the median eminence, form anastomoses, and drain into sinusoids that become the pituitary portal veins that enter the vascular pool of the pituitary gland; the vessels are fenestrated, allowing diffusion of the peptide-releasing factors to their site of action in the anterior pituitary gland.

This vascular complex in the base of the hypothalamus and its "arteriolized" venous drainage to the pituitary compose a circulatory system analogous to the portal vein system of the liver, it has been termed the "hypophyseal-portal circulation."

This region acts as a key interface between the neural and endocrine systems involved in hypothalamic-adenohypophyseal regulation of reproduction, stress, lactation, growth, and thyroid and metabolic systems (El-Majdoubi and Weiner, 2002). The ME is a very unusual neural structure: although it contains nerve terminals and glial cells, it is virtually devoid of synapses and it has structural properties that distinguish it from other brain regions (Yin and Gore, 2010).

Anatomically, represents a small swelling found at the interface between the ventral hypothalamus and the infundibular recess.

As described before, the magnocellular supraoptic and paraventricular nuclei produce vasopressin and oxytocin. The output from the supraoptic nucleus is primarily restricted to the neurohypophysis, whereas the paraventricular nucleus projects not only to the posterior lobe of the pituitary gland but also to the median eminence and to several extrahypothalamic regions including autonomic centers in the brainstem and the spinal cord.

The neurosecretory cells of the parvocellular secretory system are located throughout the periventricular zone and the preoptic area. There is a distinct localization for each of the hypothalamic-releasing hormones or factors that influence the anterior pituitary gland (or adenohypophysis) (see Swaab, 1997, 2003). These hormones are transported axonally to the median eminence where they are released into the perivascular spaces surrounding the portal capillaries which join into the portal veins through which the hormones are transported to the vascular sinusoids in the adenohypophysis. Here, they influence the secretion of the pituitary hormones: TSH (thyroid-stimulating hormone or thyrotropin), ACTH, FSH, LH, GH (growth hormone or somatotropin), and PRL (prolactin).

The neurohypophysis and the median eminence are usually considered together, since they are connected to form a functional unit and emerge in direct proximity and are thus likely to share many developmental mechanisms (Szarek et al., 2010).

Based on events occurring during the genesis of a preovulatory surge of luteinizing hormone (LH) and an ovulatory failure of central origin, the hypothalamic ME is probably a major neuroendocrine control site for reproductive activity in the men (Advis and Contijoch, 1993).

Subcommissural Organ

The subcommissural organ (SCO) is a small gland located in the dorsocaudal region of the third ventricle, at the entrance of the Sylvian aqueduct. It is a highly differentiated ependyma that covers the ventral aspect of the posterior commissure (Grondona et al., 2012). Its topographical position and its location at the interface between the CSF and a special and circumscribed vascular plexus led to consider the SCO as one of the CVOs (Rodríguez et al., 1998).

The SCO is an ependymal differentiation located in the midline roof plate of the most caudal portion of the diencephalon (prosomere 1), under the Posterior Commissure (PC). Because of the secretory nature of the SCO, it has been regarded as a true brain gland (Stanic et al., 2010).

Concerning its histological structure, two layers of different cell types are present in the SCO: the ependymal and the hypendymal cell layers. The ependymal layer is formed by a tall pseudostratified epithelium in contact with the cerebrospinal fluid of the third ventricle. It resembles the morphology of the Floor Plate epithelium, its cells being quite similar to radial glial cells. On the other hand, the hypendymal cells are located under the ependymal layer of the SCO (Oksche, 1961). While the ependymal layer did not undergo substantial changes throughout phylogeny, the degree of development of the hypendymal layer differs considerably between lower vertebrates, where it is poorly developed, and mammals, in which the hypendyma forms a prominent subependymal layer (Oksche, 1961). The organization pattern of hypendymal cells varies among species. In some species, the hypendymal cells arrange in clusters preferentially situated in a fixed position of the SCO, while in other species like bovine, they form a continuous layer. Occasionally, the hypendymal cells form rosette-like aggregates with a central cavity.

Ependymal SCO secretory cells are tall, cylindrical epithelial cells with a clear regionalization, allowing the recognition of several cytoplasmic regions: perinuclear, intermediate, subapical, and apical. SCO ependymal cells display a basal process that project dorsally and contact with local blood vessels and with the external limiting membrane. In mammals, such processes are produced by both ependymal and hypendymal SCO cells, but in lower vertebrates, they are mainly present in the ependymal cells (Oksche, 1961; Rodríguez et al., 1998). In certain species of elasmobranchs and snakes studied, the endings of these basal processes contain numerous secretory related structures, such as RER, microtubules, and secretory granules, raising the possibility of a local release of the secretion into the leptomeningeal space (Rodriguez et al., 2005).

The subependymal space of the SCO is occupied by a dense capillary network whose vessels run closely to the hypendymal cells (Oksche, 1961). The presence of such vascular network in the basal region of the SCO and its apical contact with the ventricular cavity are both common characteristics of circumventricular organs;

therefore, the SCO is regarded as one of them (Hatton, 1990).

There is a close relationship between the SCO cells and the PC, somehow reminding that between the floor plate (FP) with commissural axons of the spinal cord, or the one between the glial palisade and the axons of the optic chiasm. The pioneer axons forming the PC cross the midline near the basal region and the basal processes of the developing SCO (Caprile et al., 2009; Hoyo-Becerra et al., 2010). In accordance, SCO basal processes (both from ependymal and hypendymal cells) break through the nerve bundle of the PC, and even in some species, the hypendymal cells are located among axons of the PC bundles, proving the intimate association of the two structures.

The SCO secretes glycoproteins of high molecular mass; the bulk of these glycoproteins is released into the ventricle where they condense to form a threadlike structure, the Reissner's fiber (RF). By addition of newly released glycoproteins to its cephalic end, RF grows caudally and extends along the aqueduct, fourth ventricle, and the whole length of the central canal of the spinal cord. At its caudal most end, the central canal becomes dilated, forming a terminal ventricle, also known as ampulla caudal. When reaching the terminal ventricle, RF glycoproteins undergo some chemical changes and form a gel mass, known as massa caudalis.

A detailed immunohistochemical study of the expression of the SCO secretion in the chick SCO and the relationships between the SCO cells and the axons of the PC, allowed to define two regional domains within the SCO with functional implications. A lateral domain, where the SCO cells express its secretion and the PC axons associated to them are highly fasciculated, and a midline domain, where the SCO cells do not express immunoreactive secretory material and the associated axons, are defasciculated (Stanic et al., 2010).

Although the subcommissural organ is not highly permeable and does not have fenestrated capillaries, its role in the functioning and relative pathology is still unclear but new evidence indicates that it may be involved in the hypertension produced by aldosterone acting on the brain (Ganong, 2000).

Pineal Gland

The pineal gland is a small neuroendocrine organ whose main and most conserved function is the nighttime secretion of melatonin. It is considered a CVO due to its lack of BBB allowing it to secrete melatonin into the blood stream and CSF.

In lower vertebrates, the pineal gland is directly photosensitive. In contrast, in higher vertebrates, the direct photosensitivity of the pineal gland had been lost. Rather, the action of this gland as a relay between environmental light conditions and body functions involves reception of light information by the retina. In parallel to this sensory regression, the pineal gland (and its accessory organs) appears to have lost several functions in relation to light and temperature, which are important in lower vertebrate species. In humans, the functions of the pineal gland overlap with the functions of melatonin, including general effects on cell protection, sleep, and immunity (Nordlund and Lerner, 1977; Harman, 1993, 1994; Ames et al., 1993). Recently, the role of melatonin has received a considerable amount of attention due to increased cancer risk in shift workers and the discovery that patients suffering from neurodegenerative diseases, autism, or depression exhibit abnormal melatonin rhythms (Sapède and Cau, 2013).

The pineal gland, regarded as an endocrine gland, is an oval-shaped structure within the mammalian brain. The gland forms during the seventh week after gestation as an "evagination" of the dorsal midline of the diencephalon. Development takes place after the neural tube closes. The thickening of the ependymal forms a small vesicle and, in mammals, it loses a direct connection to the third ventricle. Once it elongates into a thin stall-like structure the shifting of the tissue posteriorly makes it appear connected to the third ventricle.

Due to its ability to secrete hormones, such as melatonin, directly into the blood and cerebrospinal fluid (CSF) without the use of ducts, making it an endocrine gland. The pineal body can be located in the epithalamus which is the geometrical center of the brain. The gland become fully developed around five to 7 years of age and is approximately 0.8 cm long and weighs about 0.1 g.

The pineal gland is bordered by the posterior commissure ventrally, the habenular commissure dorsally, and the corpus callous superiorly. Below it overhangs the depression between the two superior colliculi on the posterior surface of the midbrain. It is connected to the caudal diencephalon by a stalk. Thus, the pineal gland may be grossly divided into an apex, a body, and a stalk. Moreover, the pineal gland, as a diencephalic structure, is connected to the third ventricle. It is observed to be dark brown in color and is covered by the pia mater. Although it is a true diencephalic structure, it is positioned between the diencephalon and mesencephalon, dorsally. Due to its positioning and function, the pineal gland is highly vascularized and the two main arteries are the medial posterior choroidal arteries which anatomically branch from the posterior cerebral artery (Patel et al., 2020).

Our perception of pineal gland function has attained new dimensions during the last decade. The gland is active throughout the life of an individual and secretes melatonin, the major pineal hormone, and many other substances such as indoles and polypeptides (Srinivasan, 1989). The secretion of pineal gland is not only regulated by sympathetic nerve fibers but also by other central pinealopetal projections. Many neurotransmitter-receptor sites have been identified recently in the pineal gland. The gland plays an established role in controlling reproduction, is involved in the control of sexual maturation (Hiller-Sturmhöfel and Bartke, 1998) and has a major influence on the circadian organization of vertebrates including human beings. The hormone melatonin has a potential therapeutic valve in treating disorders that are associated with biological rhythm disturbances like sleep disorders, "jet lag" phenomenon and affective disorders (Srinivasan, 1989). The gland is actively involved in the mechanisms controlling sleep—wakefulness cycle and human mood disorders (Arendt and Aulinas, 2022) and participates in the neuroendocrine mechanism of controlling stress and acts even as an oncostatic gland, so may be considered an "equilibrating-tranquilizing gland" contributing to longevity (Srinivasan, 1989).

Melatonin membrane receptors have been identified in the suprachiasmatic nucleus of the hypothalamus (SCN) of vertebrates (Pevet, 2002) and the signal transduction pathways seemed to be involved in both MT1 and MT2 to induce an increase in the expression of two clock genes, Period 1 (*Per*1) and Period 2 (*Per*2) (Kandalepas et al., 2016). Without the feedback information of melatonin, SCN would not properly interpret the natural photoperiodic changes and would exhibit a free running internal rhythm in which the cycle is longer than 24 h. In this situation the SCN would also instruct the pineal gland to exhibit an unusual melatonin circadian rhythm which is also longer than 24 h (Tan et al., 2018). This phenomenon is apparent in completely blind animals and humans whose eyes, specifically the MRGC, do not appropriately receive environmental photoperiodic information (Nakagawa et al., 1992; Lewy and Newsome, 1983; Palm et al., 1991). Importantly, melatonin administration to blind subjects partially re-entrains their biological rhythms close to normal (Lockley et al., 2000).

Pineal gland is mainly comprised of pinealocytes, microglia, and astrocytes. The lineage of pinealocytes is elusive. Current information suggests that pinealocytes are differentiated from Pax6-expressing neuroepithelial cells and they are specialized to synthesize and release melatonin (and possible some other substances) (Rath et al., 2009). Therefore, given the intense metabolic activity required to synthetize melatonin, mitochondria in pinealocytes show two special characteristics: first, the pinealocytes contain a wider number of mitochondria in respect to those of neuronal cells and, second, the morphologies of these mitochondria exhibit obvious dynamic alterations related to their fission, fusion and mitophagy activities during a 24 h period. Because of the high density of mitochondria, it seems that the latter are the major sites for melatonin synthesis in fact, recently, this was further confirmed by Suofu et al., (2017); they also highlighted that the exact subsite of melatonin synthesis occurred in the matrix of mitochondria. Thus, the numerous mitochondria in pinealocytes relate to their melatonin synthetic function, however this does not naturally exclude the extra-mitochondrial melatonin production. In cytosol, melatonin can also be synthesized in fact red blood cells and platelets which are without mitochondria still produced melatonin (Rosengarten et al., 1972; Launayet al., 1982). Due to the substrate, particularly acetyl coenzyme A availability, melatonin synthesis in the extra-mitochondrial sites would not be as efficient as in the mitochondria since acetyl coenzyme A is concentrated in the latter (Tan et al., 2016).

Current studies have reported that melatonin per se can regulate mitochondrial morphology. Melatonin upregulates the levels of mitochondrial fusion proteins, *mitofusin* 1 (Mfn1) and *Opa*1, to promote mitochondrial fusion and inhibits the nuclear translocation of dynamin-related protein 1 (DrP1). The nuclear translocation of DrP1 increases mitochondrial fission and the inhibition of DrP1 nuclear translocation by melatonin results in suppression of mitochondrial fission. Thus, the net result of melatonin is to promote the mitochondrial fusion and to reduce mitochondrial fission (Zhang et al., 2020).

The effects of melatonin on mitophagy are still elusive. Some reports document that melatonin inhibits mitophagy and others show that melatonin promotes this process depending on the experimental conditions and cell type (Coto-Montes et al., 2012). Currently it is not possible to determine whether the mitochondrial dynamic changes in pinealocytes relate to their functional activity may be controlled by the clock genes, such as *Per* 1 and 2, or it is a result of their melatonin production rhythm. Thus, changes in melatonin levels generated by pinealocytes result in mitochondrial dynamic changes.

In addition to the pinealocytes, the astrocytes and the microglia in the pineal gland also have the capacity to synthesize melatonin with great efficiency. The melatonin synthetic machinery (including genes AANAT/SNAT and HIOMT/ASMT) has been identified and melatonin production has been detected in these cells (Markus et al., 2017) hypothesized that melatonin synthesis was coordinated by both pinealocytes and macrophages/glia and astrocytes for the immunoresponse. For example, an acute inflammatory response drives NFκB to switch melatonin synthesis from pinealocytes to macrophages/microglia and, upon acute inflammatory resolution, back to pinealocytes. A participation of melatonin production by these cells would significantly improve the capacity of pineal gland to generate melatonin as a whole. This is particularly important in the situations in which melatonin is required, for example under the oxidative stress or inflammation (Ikram et al., 2021). It was reported that CSF melatonin and its oxidative metabolites, AFMK, are elevated by several orders of magnitude in patients with meningitis (De Oliveira Silva et al., 2005). This is probably the outcome of a coordinated physiology of all the cells mentioned above. However, the major function of astrocytes and the microglia in the pineal gland is to regulate the pinealocyte melatonin synthesis under normal conditions (Villela et al., 2013). The regulatory mechanisms are well documented, i.e., astrocytes/glia are excited under different conditions which include elevated intracellular calcium concentration, which results in the NFκB activation (Villela et al., 2013) The excited astrocytes/microglia, then, release soluble TNFα which is the signaling molecule to the pinealocytes for inhibition of melatonin synthesis by targeting AANAT. Other regulatory mechanisms may also be involved. For example, purinergic signaling on melatonin synthesis in pineal gland was reported. ATP binding to its receptor in pinealocytes inhibits melatonin synthesis via suppression of the gene expression as well as the activity of the ASMT rather than the AANAT (Costa and Montagense., 2021).

Once melatonin is synthesized in the pineal gland, it is rapidly released and the outlets for melatonin release are several. The classic concept is that pineal melatonin is released into the precapillary spaces, enters the capillaries and then, via surrounding veins and sinuses, reaches the general circulation. However, a more important route for melatonin release from pineal gland was uncovered, i.e., melatonin is directly released to the CSF of the third ventricle of the mammals. Compelling evidence supports this secretory route. Anatomically, a portion of pineal gland is nakedly exposed into the CSF of the third ventricle (bathed by the CSF) and many canaliculi of the pineal gland directly open into the CSF of third ventricle. Pineal melatonin via these canaliculi is directly discharged into the CSF; this results in extremely high melatonin level in the CSF of pineal recess of third ventricle. In the sheep, the melatonin levels in the CSF of pineal recess of third ventricle are several orders of magnitude higher than those in the blood (Tricoire et al., 2002, 2003). A melatonin concentration gradient in the CSF around the pineal recess of third ventricle was observed, indicating that the main source of CSF melatonin originated from the pineal recess of the third ventricle.

Area Postrema

The area postrema (AP) is the most caudal of the sensory CVOs and was the first to be recognized as such in the early part of last century (Wilson, 1906). These early studies showed that the AP, but not surrounding area, was stained by intravenously injected dyes (Wislocki and Leduc, 1952) suggesting the AP had unique access to the circulation. These observations were later confirmed by studies showing that systemic injections of horseradish peroxidase (HRP) resulted in extensive staining of the AP, as well as anatomical studies that clearly demonstrated the existence of the fenestrated capillaries within this CVO (Dempsey, 1973; Krisch, 1978). Functionally, the AP for many years gained notoriety as the "chemoreceptor trigger zone" at which noxious chemical stimulation acts to induce the emetic reflex (Carpenter et al., 1983). Recent studies, however, have suggested additional important roles for the AP in the control of CSF balance, cardiovascular regulation, metabolism, and immune function (for review, see (Borison, 1974). It is now well recognized that the AP expresses several receptors and sensors on neuronal cell bodies the activation of which has been shown to influence the excitability of AP neurons (Fry and Ferguson, 2007). Efferent projections of these AP neurons to important autonomic control centers behind the BBB in turn play important roles in regulation of the autonomic nervous system.

The AP is a hindbrain structure situated in the fourth ventricle on the dorsal surface of the medulla immediately adjacent to the nucleus tractus solitarius (NTS) (Fig. 16.4). In rodents, the AP is a single structure that descends out in to the fourth ventricle (Fig. 16.3). In higher animals such as rabbits and primates, the AP is a bilateral structure positioned on either side of the medullary midline.

The rodent AP is composed of three distinct regions based on neuronal cell body location and projection patterns. Regions of the AP include the mantle zone, the central zone, and the ventral zone (containing a row of helically arranged tanycytes joined together by tight junctions that define the boundary with the NTS) (McKinley et al., 2003). The central and mantle zones are rich with neuronal cell bodies and axons situated next to ependymal cells and the ventral zone contains mostly glia (McKinley et al., 2003).

The AP, along with the other two sensory CVOs (SFO and OVLT), is one of the most highly vascularized regions in the entire mammalian brain (McKinley et al., 2003). Calculations based on plasma flow, neutral amino acid transfer, and capillary permeability suggest that plasma flow in the AP is 50% higher than through surrounding medullary regions, while even more significantly the AP shows a 150-fold increase in surface

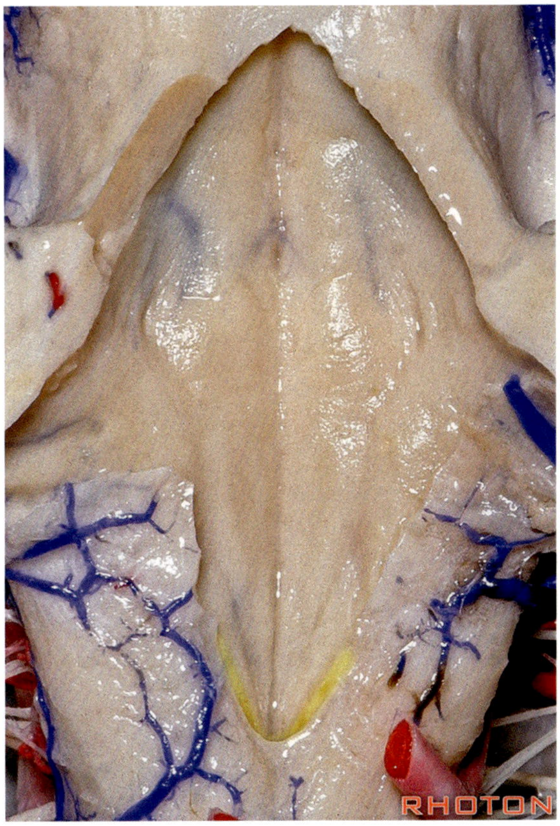

FIGURE 16.4 **Floor of the fourth ventricle.** In yellow, the area postrema (AP). *Image courtesy of The Rhoton Collection® http://rhoton.ineurodb.org/. Drawing by Jeffrey Sorenson.*

area/permeability ratio when compared with the adjacent regions of the dorsomedial medulla (Gross et al., 1991). This is the result of an extensive network of highly fenestrated blood vessels complete with specialized "pockets," called Virchow–Robin spaces, which act to slow blood flow for optimal diffusion into the perivascular space, prolonging exposure of the constituents of the circulation to receptors and sensors on neurons and glia within the AP (Gross et al., 1991). Strategically neurons are present in the perivascular sheaths of capillaries where they are positioned to optimally sense blood-borne agents (Dempsey, 1973).

Ultrastructural analysis of AP neuronal dendritic trees suggests that apical dendrites extend out toward the basal lamina side of endothelial cells and receive limited synaptic inputs and are therefore appropriately positioned to receive blood-borne information from the vasculature (Morita and Finger, 1987). Basal dendrites, however, extend out in the opposite direction and receive neuronal input from vagal afferents of the lung, stomach, and cardiovascular system, therefore making this structure uniquely situated to receive both neuronal and humoral information that can then be integrated by the cell body.

The combined dense vasculature and small size of AP neurons has until recently created significant challenges to detailed analysis of their intrinsic properties using electrophysiological techniques. However, the recent development of in vitro dissociated cell and slice preparations (Yang et al., 2007; Funahashi and Adachi, 1993; Hay and Lindsley, 1995) have allowed a number of groups to begin to examine the intrinsic electrical properties of AP neurons. Both in vivo and in vitro studies illustrate that AP neurons exhibit a relatively low rate of spontaneous pacemaking activity (<1 Hz), which has been suggested to be regulated by a cAMP-dependent hyperpolarization-activated cation current (Ih) (Funahashi et al., 2003). AP neurons have relatively high input resistances (~3 GΩ) indicative of a reduced number of synaptic inputs (Yang et al., 2007; Ferguson and Bains, 1997), although axosomatic and axodendritic connections clearly do exist (Dempsey, 1973). AP neurons express ion channels typical of many other populations of CNS neurons (Funahashi et al., 2002), which can be modulated by a variety of signaling peptides, properties that will be described in more detail.

Anatomical Connections of the Area Postrema

Anatomical studies completed in the mid-1980s described the anatomical connections of the AP (van der Kooy and Phillips, 1977; Shapiro and Miselis, 1985). Retrograde tracing studies using cholera-toxin HRP and wheat germ agglutinin HRP showed that AP sends major and minor efferents to a variety of different nuclei in both the medulla and the hypothalamus. Major projections include those to the adjacent nucleus of the solitary tract (NTS) and the lateral parabrachial nucleus (LPBN), both of which are well recognized as multifunctional integrative brainstem structures (Menani et al., 1996; Miura and Takayama, 1991; Vigier and Gahery, 1970; van der Kooy and Phillips, 1977; Leslie and Gwyn, 1984; Shapiro and Miselis, 1985). Less prominent efferent networks include projections to the nucleus ambiguus, dorsal motor nucleus of the vagus, dorsal regions of the tegmental nucleus, cerebellar vermis, paratrigeminal nucleus, ventrolateral catecholaminergic column in the medulla, and the spinal trigeminal tract (van der Kooy and Phillips, 1977). The AP also receives afferent input from several different and functionally distinct regions of the medulla and hypothalamus including reciprocal connections from the NTS and LPBN in addition to substantial axonal projections arising from the parvocellular regions of the paraventricular nucleus (PVN) and dorsomedial nucleus of the hypothalamus (DMH) (van der Kooy and Phillips, 1977). Intriguingly information from the AP reaches the PVN through both monosynaptic and polysynaptic connections suggesting an integrative capacity with potentially bidirectional information processing. Other major sources of neuronal information to the AP come from vagal and carotid sinus nerve afferents originating from the lungs and gastrointestinal and cardiovascular systems.

Neurochemistry of AP Neurons

In addition to glutamate and GABA (Walberg and Ottersen, 1992), AP neurons have been shown to contain several signaling neuropeptides which have therefore been suggested to play important roles as neurotransmitters at the identified primary termination sites of AP neurons. Immunohistochemical studies have shown the presence of enkephalin (Merchenthaler et al., 1986) , neurotensin (Newton and Maley, 1985), and serotonin (Lança and van der Kooy, 1985; Newton and MAley,1985) containing neurons in the area postrema, although neither the specific function nor projection site of each chemical phenotype of neuron has been completely described. Cholecytokinin (CCK) and catecholamine (part of the A2 noradrenergic group from the dorsolateral medulla) neurons are found in the central zone and project to the NTS (Petrov et al., 1992; Hermann et al., 2005) controlling the emetic reflex and arousal states as the A2 neurons project to oxytocin-containing neurosecretory cells in the supraoptic area and are activated during experimental fear conditioning (Zhu and Onaka, 2002).

In addition to peptidergic neurons described previously, the AP also contains a tremendously rich complement of peptide and hormone receptors including adiponectin, adrenomedullin, amylin, angiotensin II, CCK, endothelin, ghrelin, glucagon-like peptide 1 (GLP-1), interleukins, natriuretic peptides, orexins, prokineticin 2, peptide YY (PYY), and vasopressin (Price et al., 2008).

CLINICAL CORRELATIONS OF THE CVOs

CVOs as a port of entry, trigger, and propagation of pathogenic stimuli. The morphologic and functional features of the CVOs allow the access of circulating cytokines, immunoglobulins, infectious agents, and prion proteins to these areas, leading to exposure of local neurons, glial, and endothelial cells to these signals. Reactive astrocytes and microglia in the CVOs may act as innate immune effectors, producing proinflammatory cytokines and chemokines that recruit and activate perivascular macrophages, lymphocytes (Kaur and Ling, 2017). The CVOs may also recruit circulating inflammatory cells into the brain through upregulation of adhesion molecules in their fenestrated capillaries (Schulz and Engelhardt, 2005). The proinflammatory cytokine signaling initiated at the level of the CVOs may spread to astrocytes, microglia, and neurons in adjacent brain areas via volume transmission and neural connections and, conceivably, to distal brain areas via the CSF. In response to these signals, hypothalamic and brainstem neurons initiate the several responses including fever, somnolence, endocrine and autonomic disturbances, thirst, hunger, and vomiting (Kaur and Ling, 2017).

For example, entry of circulating *Trypanosoma brucei* via the CVOs triggers local production of proinflammatory cytokines that, via action on hypothalamic nuclei, trigger the typical manifestations of human sleeping sickness, including disruption of sleep—wake rhythms (Bentivoglio et al., 2018).

CVOs are a target of AQP4 antibodies in neuromyelitis spectrum disorders (Iorio and Papi, 2021). Neuromyelitis optica (NMO) is a CNS inflammatory demyelinating autoimmune disease characterized by relapsing attacks that target the optic nerves and spinal cord. It is due to the presence of NMO—immunoglobulin G (IgG) autoantibodies that bind specifically to the AQP4 water channel concentrated in astrocytic foot processes. Not surprisingly, NMO also affects the AQP4—enriched CVOs, including the AP (Iorio and Papi, 2021). Consistent with this selective vulnerability, characteristic symptoms of NMO are intractable nausea, vomiting and hiccups, in various combinations, which reflects involvement of the AP, as shown in MRI studies.

In a study performed by Hillebrand et al. (2019) in NMO patients, there was selective loss of AQP4 in floor of the fourth ventricle and AP; this was associated with tissue rarefaction, inflammation, complement deposition, and morphologic alterations in reactive astrocytes with relative preservation of neurons, axons, and myelin (Hillebrand et al., 2019). In these patients, despite loss of AQP4 expression, the abundant reactive astrocytes retained their foot processes and showed morphologic changes that were interpreted as a compensatory strategy to maintain water flux in the absence of AQP4. The lack of necrosis in the AP in the setting of severe destructive demyelinating lesions in the spinal cord or optic nerves in the same cases suggest regional differences in vulnerability to autoantibody-triggered complement deposition (Hillebrand et al., 2019).

Inflammatory responses, particularly those associated with adaptive immunity, have a role in the genesis of hypertension (Lazartigues, 2010; Marvar et al., 2010). Sympathetic activation elicits hypertension and stimulates the accumulation of inflammatory cells, including macrophages and T lymphocytes in peripheral vessels and the kidney. Circulating inflammatory cells and cytokines accessing the brain can also impair central regulation of blood pressure and promote hypertension. The CVOs may provide an interface for the interactions among sympathetic activation, inflammation, and hypertension. For example, angiotensin II, acting via AT1 receptors in the SFO, elicits oxidative stress and sympathetic activation, leading to activation of circulating T cells, vascular inflammation, and hypertension (Marvar et al., 2010).

References

Advis, J.P., Contijoch, A.M., May 1993. The median eminence as a site for neuroendocrine control of reproduction in hens. Poultry Sci. 72 (5), 932–939. https://doi.org/10.3382/ps.0720932.

Ahmed, F.N., Ahmad, R.R., Din, U.K., Noorani, M.S., 2014. Oscillations for neutral functional differential equations. Sci. World J. 2014, 124310. doi: 10.1155/2014/124310. Epub 2014 Aug 18. PMID: 25202713; PMCID: PMC4151387.

Allen, A.M., Chai, S.Y., Clevers, J., McKinley, M.J., Paxinos, G., Mendelsohn, F.A., 1988. Localization and characterization of angiotensin II receptor binding and angiotensin converting enzyme in the human medulla oblongata. J. Comp. Neurol. 269 (2), 249–264. doi: 10.1002/cne.902690209. PMID: 2833536.

Ames, B.N., Shigenaga, M.K., Hagen, T.M., September 1, 1993. Oxidants, antioxidants, and the degenerative diseases of aging. Proc. Natl. Acad. Sci. U.S.A. 90 (17), 7915–7922. https://doi.org/10.1073/pnas.90.17.7915.

Apuzzo, M.J., 1987. Surgery of the Third Ventricle. Radiology of Third Ventriclular Lesions, Baltimore, pp. 268–272.

Arendt, J., Aulinas, A., 2000. Physiology of the pineal gland and melatonin. [Updated 30 October 2022]. In: Feingold, K.R., Anawalt, B., Blackman, M.R., et al. (Eds.), Endotext [Internet]. MDText.com, Inc., South Dartmouth (MA). https://www.ncbi.nlm.nih.gov/books/NBK550972/?report=classic.

Arendt, J., Aulinas, A., 2022. Physiology of the pineal gland and melatonin. In: Feingold, K.R., Anawalt, B., Blackman, M.R., Boyce, A., Chrousos, G., Corpas, E., de Herder, W.W., Dhatariya, K., Dungan, K., Hofland, J., Kalra, S., Kaltsas, G., Kapoor, N., Koch, C., Kopp, P., Korbonits, M., Kovacs, C.S., Kuohung, W., Laferrère, B., Levy, M., McGee, E.A., McLachlan, R., New, M., Purnell, J., Sahay, R., Shah, A.S., Singer, F., Sperling, M.A., Stratakis, C.A., Trence, D.L., Wilson, D.P., (Eds.) Endotext [Internet]. South Dartmouth (MA): MDText.-com, Inc.; 2000–. PMID: 31841296.

Armstrong, W.E., 1995. Morphological and electrophysiological classification of hypothalamic supraoptic neurons. Prog. Neurobiol. 47, 239–291.

Asami, A., Kurganov, E., Miyata, S., 2020. Proliferation of endothelial cells in the choroid plexus of normal and hydrocephalic mice. J. Chem. Neuroanat. 106, 101796. doi: 10.1016/j.jchemneu.2020.101796. Epub 2020 Apr 29. PMID: 32360474.

Balland, E., Dam, J., Langlet, F., Caron, E., Steculorum, S., Messina, A., Rasika, S., Falluel-Morel, A., Anouar, Y., Dehouck, B., Trinquet, E., Jockers, R., Bouret, S.G., Prévot, V., 2014. Hypothalamic tanycytes are an ERK-gated conduit for leptin into the brain. Cell Metab. 19 (2), 293–301. doi: 10.1016/j.cmet.2013.12.015. PMID: 24506870; PMCID: PMC3936883.

Bearer, E.L., Orci, L., 1985. Endothelial fenestral diaphragms: a quick-freeze, deep-etch study. J. Cell Biol. 100, 418–428. https://doi.org/10.1083/jcb.100.2.418.

Benarroch, E.E., September 20, 2011. Circumventricular organs: receptive and homeostatic functions and clinical implications. Neurology 77 (12), 1198–1204. https://doi.org/10.1212/WNL.0b013e31822f04a0.

Bentivoglio, M., Kristensson, K., Rottenberg, M.E., 2018. Circumventricular organs and parasite neurotropism: neglected gates to the brain? Front. Immunol. 9, 2877. https://doi.org/10.3389/fimmu.2018.02877.

Borison, H.L., 1974. Area postrema: chemoreceptor trigger zone for vomiting–is that all? Life Sci. 14 (10), 1807–1817. doi: 10.1016/0024-3205(74)90399-3. PMID: 4603263.

Braak, H., Braak, E., 1992. Anatomy of the human hypothalamus (chiasmatic and tuberal region). Prog. Brain Res. 93, 3–14. https://doi.org/10.1016/s0079-6123(08)64559-8 discussion 14-6.

Brownstein, M.J., Russell, J.T., Gainer, H., 1980. Synthesis, transport, and release of posterior pituitary hormones. Science 207, 373–378.

Cancelliere, N.M., Black, E.A., Ferguson, A.V., 2015. Neurohumoral integration of cardiovascular function by the lamina terminalis. Curr. Hypertens. Rep. 17 (12), 93. doi: 10.1007/s11906-015-0602-9. PMID: 26531751.

Cancelliere, N.M., Ferguson, A.V., 2017. Subfornical organ neurons integrate cardiovascular and metabolic signals. Am. J. Physiol. Regul. Integr. Comp. Physiol. 312 (2), R253–R262. doi: 10.1152/ajpregu.00423.2016. Epub 2016 Dec 21. PMID: 28003212.

Caprile, T., Osorio, G., Henríquez, J.P., Montecinos, H., 2009. Polarized expression of integrin beta1 in diencephalic roof plate during chick development, a possible receptor for SCO-spondin. Dev. Dyn. 238 (10), 2494–2504. doi: 10.1002/dvdy.22070. PMID: 19681158.

Cattaneo, L., Parmiani, G., 1982. Anatomia del sistema nervoso centrale e periferico dell'uomo. Monduzzi.

Carpenter, D.O., Briggs, D.B., Strominger, N., 1983. Responses of neurons of canine area postrema to neurotransmitters and peptides. Cell Mol. Neurobiol. 3 (2), 113–126. doi: 10.1007/BF00735276. PMID: 6140078.

Cobos, I., Shimamura, K., Rubenstein, J.L., et al., 2001. Fate map of the avian anterior forebrain at the four-somite stage, based on the analysis of quail-chick chimeras. Dev. Biol. 239, 46–67.

Corbin, J.G., Rutlin, M., Gaiano, N., et al., 2003. Combinatorial function of the homeodomain proteins Nkx2.1 and Gsh2 in ventral telencephalic patterning. Development 130, 4895–4906.

Costa, R., Montagense, S., 2021. The role of astrocytes in generating circadian rhythmicity in health and disease. J. Neurochem. 157, 42–52.

Coto-Montes, A., Boga, J.A., Rosales- Corral, S., et al., September 2012. Role of melatonin in the regulation autophagy and mitophagy. A review. Mol. Cell. Endocrinol. 361, 12–23.

Dale, J.K., Vesque, C., Lints, T.J., et al., 1997. Cooperation of BMP7 and SHH in the induction of forebrain ventral midline cells by prechordal mesoderm. Cell 90, 257–269.

De Oliveira Silva, S., Ximenes, V.F., Livramento, J.A., Catalani, L.H., Campa, A., 2005. High concentrations of the melatonin metabolite, N1-acetyl-N2-formyl-5-methoxykynuramine, in cerebrospinal fluid of patients with meningitis: a possible immunomodulatory mechanism. J. Pineal Res. 39, 302–306. https://doi.org/10.1111/j.1600-079X.2005.00247.x.

Dempsey, E.W., 1968. Fine-structure of the rat's intercolumnar tubercle and its adjacent ependyma and choroid plexus, with especial reference to the appearance of its sinusoidal vessels in experimental argyria. Exp. Neurol. 22, 568–589. https://doi.org/10.1016/0014-4886(68)90150-7.

Dempsey, E.W., 1973. Neural and vascular ultrastructure of the area postrema in the rat. J. Comp. Neurol. 150 (2), 177–199. doi: 10.1002/cne.901500206. PMID: 4723064.

Dierickx, K., Vandesande, F., 1977. Immunocytochemical localization of the vasopressinergic and oxytocinergic neurons in the human hypothalamus. Cell Tissue Res. 184, 15–27.

Duvernoy, H., Koritké, J.G., Monnier, G., 1969. Sur la vascularisation de la lame terminale humaine. Z. Zellforsch. Mikrosk. Anat. 102, 49–77.

El-Majdoubi, M., Weiner, R.I., 2002. Localization of olfactory cyclic nucleotide-gated channels in rat gonadotropin-releasing hormone neurons. Endocrinology 143 (6), 2441–2444. doi: 10.1210/endo.143.6.8951. PMID: 12021210.

Felix, D., Künzle, H., 1974. Iontophoretic and autoradiographic studies on the role of proline in nervous transmission. Pflugers Arch. 350 (2), 135–144. doi: 10.1007/BF00586233. PMID: 4368845.

Felix, D., 1976. Peptide and acetylcholine action on neurones of the cat subfornical organ. Naunyn Schmiedebergs Arch. Pharmacol. 292 (1), 15–20. doi: 10.1007/BF00506484. PMID: 934351.

Felix, I.A., September 1988. Pathology of the neurohypophysis. Pathol. Res. Pract. 183 (5), 535–537. https://doi.org/10.1016/s0344-0338(88)80003-7.

Felten, D.L., O'Banion, M.K., Maida, M.E., 2015. Netter's Atlas of Neuroscience. Elsevier Health Sciences.
Fliers, E., Swaab, D.F., Pool, C.W., Verwer, R.W.H., 1985. The vasopressin and oxytocin neurons in the human supraoptic and paraventricular nucleus: changes with aging and in senile dementia. Brain Res. 342, 45–53.
Fitts, J.P., Shang, X., Flynn, G.W., Heinz, T.F., Eisenthal, K.B., 2005. Electrostatic surface charge at aqueous/alpha-Al_2O_3 single-crystal interfaces as probed by optical second-harmonic generation. J. Phys. Chem. B. 109 (16), 7981–7986. doi: 10.1021/jp040297d. PMID: 16851933.
Ferguson, A.V., Bains, J.S.,1997. Actions of angiotensin in the subfornical organ and area postrema: implications for long term control of autonomic output. Clin. Exp. Pharmacol. Physiol. 24 (1), 96–101. doi: 10.1111/j.1440-1681.1997.tb01790.x. PMID: 9043813.
Funahashi, M., Adachi, A., 1993. Glucose-responsive neurons exist within the area postrema of the rat: in vitro study on the isolated slice preparation. Brain Res. Bull. 32 (5), 531–535. doi: 10.1016/0361-9230(93)90303-s. PMID: 8221148.
Funahashi, M., Mitoh, Y., Matsuo, R., 2002. Electrophysiological properties of the rat area postrema neurons displaying both the transient outward current and the hyperpolarization-activated inward current. Brain Res. Bull. 58 (3), 337–343. doi: 10.1016/s0361-9230(02)00798-0. PMID: 12128161.
Funahashi, M., Mitoh, Y., Kohjitani, A., Matsuo, R., 2003. Role of the hyperpolarization-activated cation current (Ih) in pacemaker activity in area postrema neurons of rat brain slices. J. Physiol. 552 (Pt 1), 135–148. doi: 10.1113/jphysiol.2003.047191. Epub 2003 Aug 1. PMID: 12897173; PMCID: PMC2343317.
Fry, M., Ferguson, A.V., 2007. The sensory circumventricular organs: brain targets for circulating signals controlling ingestive behavior. Physiol. Behav. 91, 413–423.
Furube, E., Mannari, T., Morita, S., Nishikawa, K., Yoshida, A., Itoh, M., Miyata, S., 2014. VEGF-dependent and PDGF-dependent dynamic neurovascular reconstruction in the neurohypophysis of adult mice. J. Endocrinol. 222 (1), 161–179. doi: 10.1530/JOE-14-0075. Epub 2014 May 23. PMID: 24860149.
Furube, E., Ishii, H., Nambu, Y., Kurganov, E., Nagaoka, S., Morita, M., Miyata, S., 2020. Neural stem cell phenotype of tanycyte-like ependymal cells in the circumventricular organs and central canal of adult mouse brain. Sci. Rep. 10 (1), 2826. doi: 10.1038/s41598-020-59629-5. PMID: 32071335; PMCID: PMC7029029.
Gainer, H., 1998. Cell-specific gene expression in oxytocin and vasopressin magnocellular neurons. Adv. Exp. Med. Biol. 449, 15–27. doi: 10.1007/978-1-4615-4871-3_2. PMID: 10026782.
Gainer, H., Wray, S., 1994. Cellular and molecular biology of oxytocin and vasopressin. In: Knobil, E., Neil, D. (Eds.), The Physiology of Reproduction, second ed. Raven Press, New York, pp. 1099–1129.
Ganong, W.F., 2000 . Circumventricular organs: definition and role in the regulation of endocrine and autonomic function. Clin. Exp. Pharmacol. Physiol. 27 (5–6), 422–427. https://doi.org/10.1046/j.1440-1681.2000.03259.x.
Garcia-Segura, L.M., Berciano, M.T., Lafarga, M., 1993. Nuclear compartmentalization in transcriptionally activated hypothalamic neurons. Biol. Cell. 77, 143–154.
Geng, X., Speirs, C., Lagutin, O., et al., 2008. Haploinsufficiency of Six3 fails to activate Sonic hedgehog expression in the ventral forebrain and causes holoprosencephaly. Dev. Cell 15, 236–247.
Göke, R., Larsen, P.J., Mikkelsen, J.D., Sheikh, S.P., 1995. Identification of specific binding sites for glucagon-like peptide-1 on the posterior lobe of the rat pituitary. Neuroendocrinology 62 (2), 130–134. doi: 10.1159/000126997. PMID: 8584112.
Goodman, T., Hajihosseini, M.K., 2015. Hypothalamic tanycytes-masters and servants of metabolic, neuroendocrine, and neurogenic functions. Front. Neurosci. 9, 387. doi: 10.3389/fnins.2015.00387. PMID: 26578855; PMCID: PMC4624852.
Gonçalves, R.A., De Felice, F.G., 2021. The crosstalk between brain and periphery: implications for brain health and disease. Neuropharmacology 197, 108728. https://doi.org/10.1016/j.neuropharm.2021.108728.
Glasgow, E., Murase, T., Zhang, B., Verbalis, J.G., Gainer, H., 2000. Gene expression in the rat supraoptic nucleus induced by chronic hyperosmolality versus hyposmolality. Am. J. Physiol. Regul. Integr. Comp. Physiol. 279 (4), R1239–R1250. doi: 10.1152/ajpregu.2000.279.4.R1239. PMID: 11003989.
Grondona, J.M., Hoyo-Becerra, C., Visser, R., Fernández-Llebrez, P., López-Ávalos, M.D., 2012. The subcommissural organ and the development of the posterior commissure. Int. Rev. Cell Mol. Biol. 296, 63–137. https://doi.org/10.1016/B978-0-12-394307-1.00002-3.
Gross, P.M., Wall, K.M., Wainman, D.S., Shaver, S.W., 1991. Subregional topography of capillaries in the dorsal vagal complex of rats: II. Physiological properties. J. Comp. Neurol. 306 (1), 83–94. doi: 10.1002/cne.903060107. PMID: 2040731.
Gunhaga, L., Marklund, M., Sjödal, M., et al., 2003. Specification of dorsal telencephalic character by sequential Wnt and FGF signaling. Nat. Neurosci. 6, 701–707.
Harman, D., 1993. Free radical involvement in aging. Pathophysiology and therapeutic implications. Drugs Aging 3 (1), 60–80.
Harman, D., 1994. Free-radical theory of aging. Increasing the functional life span. Ann. N. Y. Acad. Sci. 717, 1–15.
Hatton, G.I., 1990. Emerging concepts of structure-function dynamics in adult brain: the hypothalamus-neurohypophysial system. Prog. Neurobiol. 34, 437–504.
Hatton, G.I., 1997. Function-related plasticity in hypothalamus. Annu. Rev. Neurosci. 20, 375–397.
Hatton, G.I., Walters, J.K., 1973. Induced multiple nucleoli, nucleolar margination, and cell size changes in supraoptic neurons during dehydration and rehydration in the rat. Brain Res. 59, 137–154.
Hay, M., Lindsley, K.A., 1995. Membrane properties of area postrema neurons. Brain Res. 705 (1-2), 199–208. doi: 10.1016/0006-8993(95)01159-5. PMID: 8821750.
Hazell, G.G., Hindmarch, C.C., Pope, G.R., Roper, J.A., Lightman, S.L., Murphy, D., et al., 2012. G protein-coupled receptors in the hypothalamic paraventricular and supraoptic nuclei–serpentine gateways to neuroendocrine homeostasis. Front. Neuroendocrinol. 33 (1), 45–66.
Hendelman, W., 2005. Atlas of Functional Neuroanatomy. CRC press.
Hermann, G.E., Nasse, J.S., Rogers, R.C., 2005. Alpha-1 adrenergic input to solitary nucleus neurones: calcium oscillations, excitation and gastric reflex control. J. Physiol. 562 (Pt 2), 553–568. doi: 10.1113/jphysiol.2004.076919. Epub 2004 Nov 11. PMID: 15539398; PMCID: PMC1665513.
Hicks, A.I., Kobrinsky, S., Zhou, S., Yang, J., Prager-Khoutorsky, M., September 6, 2021. Anatomical organization of the rat subfornical organ. Front. Cell. Neurosci. 15, 691711. https://doi.org/10.3389/fncel.2021.691711.
Hillebrand, S., Schanda, K., Nigritinou, M., et al., 2019. Circulating AQP4-specific auto-antibodies alone can induce neuromyelitis optica spectrum disorder in the rat. Acta Neuropathol. 137 (3), 467–485. https://doi.org/10.1007/s00401-018-1950-8.
Hiller-Sturmhöfel, S., Bartke, A., 1998. The endocrine system: an overview. Alcohol Health Res. World 22 (3), 153–164.
Hiyama, T.Y., Watanabe, E., Okado, H., Noda, M., 2004. The subfornical organ is the primary locus of sodium-level sensing by Na(x) sodium channels for the control of salt-intake behavior. J. Neurosci. 24 (42), 9276–9281. doi: 10.1523/JNEUROSCI.2795-04.2004. PMID: 15496663; PMCID: PMC6730094.
Hofer, H., 1958. Zur Morphologie der circumventrikulären Organe des Zwischenhirns der Säugetiere. Zool. Anz. 22, 202–261.
Hoffman, H.J., Olszewski, J., 1961. Spread of sodium fluorescein in normal brain tissue. A study of the mechanism of the blood-brain

barrier. Neurology 11, 1081–1085. https://doi.org/10.1212/wnl.11.12.1081.

Hoffman, W.E., Phillips, M.I., 1976. The effect of subfornical organ lesions and ventricular blockade on drinking induced by angiotensin II. Brain Res. 108 (1), 59–73. doi: 10.1016/0006-8993(76)90164-5. PMID: 1276893.

Hollis, J.H., McKinley, M.J., D'Souza, M., Kampe, J., Oldfield, B.J., 2008 The trajectory of sensory pathways from the lamina terminalis to the insular and cingulate cortex: a neuroanatomical framework for the generation of thirst. Am. J. Physiol. Regul. Integr. Comp. Physiol. 294 (4), R1390–R401. doi: 10.1152/ajpregu.00869.2007. Epub 2008 Jan 30. PMID: 18234743.

Horstmann, E., 1954. Die Faserglia des Selachiergehirns. Z. Zellforsch. 39, 588–617.

Hourai, A., Miyata, S., 2013. Neurogenesis in the circumventricular organs of adult mouse brains. J. Neurosci. Res. 91 (6), 757–770. doi: 10.1002/jnr.23206. Epub 2013 Mar 22. PMID: 23526379.

Hoyo-Becerra, C., López-Avalos, M.D., Cifuentes, M., Visser, R., Fernández-Llebrez, P., Grondona, J.M., 2010. The subcommissural organ and the development of the posterior commissure in chick embryos. Cell Tissue Res. 339 (2), 383–395. doi: 10.1007/s00441-009-0899-2. Epub 2009 Dec 11. PMID: 20012322.

Ikram, M., Park, H.Y., Ali, T., Kim, M.O., November 27, 2021. Melatonin as a potential regulator of oxidative stress, and neuroinflammation: mechanisms and implications for the management of brain injury-induced neurodegeneration. J. Inflamm. Res. 14, 6251–6264. https://doi.org/10.2147/JIR.S334423.

Iorio, R., Papi, C., 2021. The human hypothalamus. In: Handbook of Clinical Neurology. Elsevier.

Jeong, Y., Leskow, F.C., El-Jaick, K., et al., 2008. Regulation of a remote Shh forebrain enhancer by the Six3 homeoprotein. Nat. Genet. 40, 1348–1353.

Johansson, P.A., 2014. The choroid plexuses and their impact on developmental neurogenesis. Front. Neurosci. 8, 340. doi: 10.3389/fnins.2014.00340. PMID: 25386116; PMCID: PMC4208491.

Kamesh, A., Black, E.A.E., Ferguson, A.V., 2018. The subfornical organ: a novel site for prolactin action. J. Neuroendocrinol. 30 (9), e12613. doi: 10.1111/jne.12613. Epub 2018 Aug 7. PMID: 29862587.

Kandalepas, P.C., Mitchell, J.W., Gillette, M.U., June 30, 2016. Melatonin signal transduction pathways require E-box-mediated transcription of Per1 and Per2 to reset the SCN clock at dusk. PLoS One 11 (6), e0157824. https://doi.org/10.1371/journal.pone.0157824.

Kaur, C., Ling, E.A., September 2017. The circumventricular organs. Histol. Histopathol. 32 (9), 879–892. https://doi.org/10.14670/HH-11-881.

Kishi, R., Bongiovanni, R., de Nadai, T.R., Freitas, R.L., de Oliveira, R., Ferreira, C.M., Coimbra, N.C., 2006. Dorsal raphe nucleus and locus coeruleus neural networks and the elaboration of the sweet-substance-induced antinociception. Neurosci. Lett. 395 (1), 12–27. doi: 10.1016/j.neulet.2005.10.046. Epub 2005 Nov 10. PMID: 16289556.

Kombian, S.B., Hirasawa, M., Mouginot, D., Pittman, Q.J., 2002. Modulation of synaptic transmission by oxytocin and vasopressin in the supraoptic nucleus. Prog. Brain Res. 46, 139–235. https://doi.org/10.1016/s0079-6123(02)39020-4. IQWX 235±246.

Krisch, B., 1978. Altered pattern of vasopressin distribution in the hypothalamus of rats subjected to immobilization stress. An immunohistochemical study. Cell Tissue Res. 189 (2), 267–275. doi: 10.1007/BF00209276. PMID: 350410.

Krisch, B., Leonhardt, H., Oksche, A., 1987. Compartments in the organum vasculosum laminae terminalis of the rat and their delineation against the outer cerebrospinal fluid-containing space. Cell Tissue Res. 250, 331–347.

Lacroix, L., Broersen, L.M., Weiner, I., Feldon, J., 1998. The effects of excitotoxic lesion of the medial prefrontal cortex on latent inhibition, prepulse inhibition, food hoarding, elevated plus maze, active avoidance and locomotor activity in the rat. Neuroscience 84 (2), 431–42. doi: 10.1016/s0306-4522(97)00521-6. PMID: 9539214.

Lafarga, M., Andres, M.A., Berciano, M.T., Maquiera, E., 1991. Organization of nucleoli and nuclear bodies in osmotically stimulated supraoptic neurons of the rat. J. Comp. Neurol. 308, 329–339.

Lafarga, M., Martinez-Guijarro, J.F., Berciano, M.T., Blaco-Ibañez, J.M., Andres, M.A., Mellström, B., Naranjo, J.R., 1993. Nuclear Fos domains in transcriptionally activated supraoptic nucleus neurons. Neuroscience 57, 353–364.

Lafarga, M., Berciano, M.T., Martınez-Guijarro, F.J., Andres, M.A., Mellström, B., Lopez-García, C., Naranjo, J.R., 1992. Fos-like expression and nuclear size in osmotically stimulated supraoptic nucleus neurons. Neuroscience 50, 867–875.

Lafarga, M., Berciano, M.T., Garcia-Segura, L.M., Andres, M.A., Carmo-Fonseca, M., 1998. Acute osmotic/stress stimuli induce a transient decrease of transcriptional activity in the neurosecretory neurons of supraoptic nuclei. J. Neurocytol. 27, 205–217.

Lagutin, O.V., Zhu, C.C., Kobayashi, D., et al., 2003. Six3 repression of Wnt signaling in the anterior neuroectoderm is essential for vertebrate forebrain development. Genes Dev. 17, 368–379.

Lança, A.J., van der Kooy, D., 1985. A serotonin-containing pathway from the area postrema to the parabrachial nucleus in the rat. Neuroscience 14 (4), 1117–1126. doi: 10.1016/0306-4522(85)90281-7. PMID: 2987755.

Langlet, F., Mullier, A., Bouret, S.G., Prevot, V., Dehouck, B., 2013. Tanycyte-like cells form a blood-cerebrospinal fluid barrier in the circumventricular organs of the mouse brain. J. Comp. Neurol. 521, 3389–3405. https://doi.org/10.1002/cne.23355.

Launay, J.M., Lamaître, B.J., Husson, H.P., Dreux, C., Hartmann, L., Da Prada, M., 1982. Melatonin synthesis by rabbit platelets. Life Sci. 31, 1487–1494. https://doi.org/10.1016/0024-3205(82)90010-8.

Lazartigues, E., 2010. Inflammation and neurogenic hypertension: a new role for the circumventricular organs? Circ. Res. 107 (2), 166–167. https://doi.org/10.1161/CIRCRESAHA.110.224873.

Lee, D.A., Bedont, J.L., Pak, T., Wang, H., Song, J., Miranda-Angulo, A., Takiar, V., Charubhumi, V., Balordi, F., Takebayashi, H., Aja, S., Ford, E., Fishell, G., Blackshaw, S., 2012. Tanycytes of the hypothalamic median eminence form a diet-responsive neurogenic niche. Nat. Neurosci.15 (5), 700–702. doi: 10.1038/nn.3079. PMID: 22446882; PMCID: PMC3380241.

Leng, G., Brown, C.H., Russell, J.A., 1999. Physiological pathways regulating the activity of magnocellular neurosecretory cells. Prog. Neurobiol. 57, 625–655. https://doi.org/10.1016/s0301-0082(98)00072-0. SUX 625±655.

Leslie, R.A., Gwyn, D.G., 1984. Neuronal connections of the area postrema. Fed. Proc. 43 (15), 2941–2943. PMID: 6389179.

Lewy, A.J., Newsome, D.A., 1983. Different types of melatonin circadian secretory rhythms in some blind subjects. J. Clin. Endocrinol. Metab. 56, 1103–1107. https://doi.org/10.1210/jcem-56-6-1103.

Lockley, S.W., Skene, D.J., James, K., Thapan, K., Wright, J., Arendt, J., 2000. Melatonin administration can entrain the free-running circadian system of blind subjects. J. Endocrinol. 164, R1–R6. https://doi.org/10.1677/joe.0.164R001.

Ludwig, M., Pittman, Q.J., 2003. Talking back: dendritic neurotransmitter release. Trends Neurosci. 26, 255–261. https://doi.org/10.1016/S0166-2236(03)00072-9. PTX 255±261.

Ludwig, M., Sabatier, N., Dayanithi, G., Russell, J.A., Leng, G., 2002. The active role of dendrites in the regulation of magnocellular neurosecretory cell behavior. Prog. Brain Res. 139, 247–256. https://doi.org/10.1016/s0079-6123(02)39021-6. IQWX 247±256.

Mannari, T., Morita, S., Furube, E., Tominaga, M., Miyata, S., 2013. Astrocytic TRPV1 ion channels detect blood-borne signals in the sensory circumventricular organs of adult mouse brains. Glia 61 (6), 957–971. doi: 10.1002/glia.22488. Epub 2013 Mar 7. PMID: 23468425.

Markus, R.P., Fernandes, P.A., Kinker, G.S., da Silveira Cruz-Machado, S., Marçola, M., 2017. Immune-Pineal Axis—acute inflammatory responses coordinate melatonin synthesis by pinealocytes and phagocytes. Br. J. Pharmacol. https://doi.org/10.1111/bph.14083.

Marvar, P.J., Lob, H., Vinh, A., Zarreen, F., Harrison, D.G., 2010. The central nervous system and inflammation in hypertension. Curr. Opin. Pharmacol. 11, 156–161.

Matsumura, K., Simon, E., 1990. Locations and properties of angiotensin II-responsive neurones in the circumventricular region of the duck brain. J. Physiol. 429, 281–296. doi: 10.1113/jphysiol.1990.sp018256. PMID: 2277348; PMCID: PMC1181699.

McKinley, M.J., Congiu, M., Denton, D.A., Park, R.G., Penschow, J., Simpson, J.B., et al., 1984. The anterior wall of the third ventricle and homeostatic responses to dehydration. J. Physiol. (Paris) 79, 421–427.

McKinley, M.J., Allen, A.M., Burns, P., Colvill, L.M., Oldfield, B.J., 1998. Interaction of circulating hormones with the brain: the roles of the subfornical organ and the organum vasculosum of the lamina terminalis. Clin. Exp. Pharmacol. Physiol. Suppl. 25, S61–S67.

McKinley, M.J., McAllen, R.M., Davern, P., Giles, M.E., Penschow, J., Sunn, N., et al., 2003. The sensory circumventricular organs of the mammalian brain. Adv. Anat. Embryol. Cell Biol. 172, 1–122.

McKinley, M.J., Clarke, I.J., Oldfield, B.J., 2004. Circumventricular organs. In: Paxinos, G., Mai, J.K. (Eds.), The Human Nervous System, second ed. Elsevier, Amsterdam, pp. 562–591.

McKinley, M.J., Yao, S.T., Uschakov, A., McAllen, R.M., Rundgren, M., Martelli, D., 2015. The median preoptic nucleus: front and centre for the regulation of body fluid, sodium, temperature, sleep and cardiovascular homeostasis. Acta Physiol. (Oxf). 214 (1), 8–32. doi: 10.1111/apha.12487. Epub 2015 Apr 1. PMID: 25753944.

Meletis, K., Barnabé-Heider, F., Carlén, M., Evergren, E., Tomilin, N., Shupliakov, O., Frisén, J., 2008. Spinal cord injury reveals multilineage differentiation of ependymal cells. PLoS Biol. 6 (7), e182. doi: 10.1371/journal.pbio.0060182. PMID: 18651793; PMCID: PMC2475541.

Menani, J.V., Colombari, E., Talman, W.T., Johnson, A.K., 1996. Commissural nucleus of the solitary tract lesions reduce food intake and body weight gain in rats. Brain Res. 740 (1-2), 102–108. doi: 10.1016/s0006-8993(96)00850-5. PMID: 8973803.

Mendelsohn, F.A., Quirion, R., Saavedra, J.M., Aguilera, G., Catt, K.J., 1984. Autoradiographic localization of angiotensin II receptors in rat brain. Proc. Natl. Acad. Sci. USA. 1 (5), 1575–1579. doi: 10.1073/pnas.81.5.1575. PMID: 6324205; PMCID: PMC344880.

Merchenthaler, I., Csontos, C., Kallo, I., Arimura, A., 1986. The hypothalamo-infundibular growth hormone-releasing hormone (GH-RH) system of the rat. Brain Res. 378 (2), 297–302. doi: 10.1016/0006-8993(86)90932-7. PMID: 3089538.

Miura, M., Takayama, K., 1991. Circulatory and respiratory responses to glutamate stimulation of the lateral parabrachial nucleus of the cat. J. Auton. Nerv. Syst. 32 (2), 121–133. doi: 10.1016/0165-1838(91)90062-8. PMID: 1709415.

Miyata, S., 2015. New aspects in fenestrated capillary and tissue dynamics in the sensory circumventricular organs of adult brains. Front. Neurosci. 9, 390. https://doi.org/10.3389/fnins.2015.00390.

Miyata, S., 2017. Advances in understanding of structural reorganization in the hypothalamic neurosecretory system. Front. Endocrinol. 8, 275. https://doi.org/10.3389/fendo.2017.00275.

Miyata, S., 2022. Glial functions in the blood-brain communication at the circumventricular organs. Front. Neurosci. 16, 991779. doi: 10.3389/fnins.2022.991779. PMID: 36278020; PMCID: PMC9583022.

Morita, Y., Finger, T.E., 1987. Topographic representation of the sensory and motor roots of the vagus nerve in the medulla of goldfish, *Carassius auratus*. J. Comp. Neurol. 264 (2), 231–249. doi: 10.1002/cne.902640208. PMID: 3680630.

Morita, S., Ukai, S., Miyata, S., 2013. VEGF-dependent continuous angiogenesis in the median eminence of adult mice. Eur. J. Neurosci. 37 (4), 508–518. doi: 10.1111/ejn.12047. Epub 2012 Nov 22. PMID: 23173692.

Morita, S., Furube, E., Mannari, T., Okuda, H., Tatsumi, K., Wanaka, A., et al., 2016. Heterogeneous vascular permeability and alternative diffusion barrier in sensory circumventricular organs of adult mouse brain. Cell Tissue Res. 363, 497–511. https://doi.org/10.1007/s00441-015-2207-7.

Müller-Fielitz, H., Stahr, M., Bernau, M., Richter, M., Abele, S., Krajka, V., Benzin, A., Wenzel, J., Kalies, K., Mittag, J., Heuer, H., Offermanns, S., Schwaninger, M., September 7, 2017. Tanycytes control the hormonal output of the hypothalamic-pituitary-thyroid axis. Nat. Commun. 8 (1), 484. https://doi.org/10.1038/s41467-017-00604-6.

Mullier, A., Bouret, S.G., Prevot, V., Dehouck, B., 2010. Differential distribution of tight junction proteins suggests a role for tanycytes in blood-hypothalamus barrier regulation in the adult mouse brain. J. Comp. Neurol. 518, 943–962. https://doi.org/10.1002/cne.22273.

Muneoka, S., Murayama, S., Nakano, Y., Miyata, S., 2019. TLR4 in circumventricular neural stem cells is a negative regulator for thermogenic pathways in the mouse brain. J. Neuroimmunol. 331, 58–73. https://doi.org/10.1016/j.jneuroim.2018.04.017.

Murayama, S., Kurganov, E., Miyata, S., 2019. Activation of microglia and macrophages in the circumventricular organs of the mouse brain during TLR2-induced fever and sickness responses. J. Neuroimmunol. 334, 576973. https://doi.org/10.1016/j.jneuroim.2019.576973.

Nakagawa, H., Isaki, K., Sack, R.L., Lewy, A.J., 1992. Free-running melatonin, sleep propensity, cortisol and temperature rhythms in a totally blind person. Jpn. J. Psychiatry Neurol. 46, 210–212. https://doi.org/10.1111/j.1440-1819.1992.tb00837.x.

Nakano, Y., et al., 2015. Astrocytic TLR4 expression and LPS-induced nuclear translocation of STAT3 in the sensory circumventricular organs of adult mouse brain. J. Neuroimmunol. 278, 144–158. https://doi.org/10.1016/j.jneuroim.2014.12.013.

Nakano, Y., Furube, E., Morita, S., Wanaka, A., Nakashima, T., Miyata, S., 2015. Astrocytic TLR4 expression and LPS-induced nuclear translocation of STAT3 in the sensory circumventricular organs of adult mouse brain. J. Neuroimmunol. 278, 144–158. doi: 10.1016/j.jneuroim.2014.12.013. Epub 2014 Dec 16. PMID: 25595264.

Nambu, A., 2021. Distinct roles of the dorsal versus ventral subthalamic nucleus. Mov. Disord. 36 (10), 2223. doi: 10.1002/mds.28752. Epub 2021 Aug 18. PMID: 34405440.

Nehmé, B., Henry, M., Mouginot, D., Drolet, G., 2012. The expression pattern of the Na(+) Sensor, Na(X) in the hydromineral homeostatic network: a comparative study between the rat and mouse. Front. Neuroanat. 6, 26. doi: 10.3389/fnana.2012.00026. PMID: 22833716; PMCID: PMC3400090.

Newton, B.W., Maley, B.E., 1985. Distribution of neurotensin-like immunoreactivity in rat and cat area postrema. Peptides 6 (2), 301–306. doi: 10.1016/0196-9781(85)90056-7. PMID: 3898039.

Nordlund, J.J., Lerner, A.B., October 1977. The effects of oral melatonin on skin color and on the release of pituitary hormones. J. Clin. Endocrinol. Metab. 45 (4), 768–774. https://doi.org/10.1210/jcem-45-4-768.

O'Donnell, D., Ahmad, S., Wahlestedt, C., Walker, P., 1999. Expression of the novel galanin receptor subtype GALR2 in the adult rat CNS: distinct distribution from GALR1. J. Comp. Neurol. 409 (3), 469–481. PMID: 10379831.

Ohkubo, Y., Chiang, C., Rubenstein, J.L., 2002. Coordinate regulation and synergistic actions of BMP4, SHH and FGF8 in the rostral prosencephalon regulate morphogenesis of the telencephalic and optic vesicles. Neuroscience 111, 1–17.

Ohtaka-Maruyama, C., Okado, H., 2015. Molecular pathways underlying projection neuron production and migration during cerebral cortical development. Front. Neurosci. 9, 447. doi: 10.3389/fnins.2015.00447. Erratum in: Front Neurosci. 10, 160. PMID: 26733777; PMCID: PMC4682034.

Oksche, A., 1961. Comparative studies on the secretory activity of the subcommissural organ and the glial character of its cells. Z. Zellforsch. Mikrosk. Anat. 54, 549–612.

Oldfield, B.J., McKinley, M.J., 1994. Distribution of Fos in rat brain resulting from endogenously-generated angiotensin II. Kidney Int. 46 (6), 1567–1569. doi: 10.1038/ki.1994.448. PMID: 7700008.

Palm, L., Blennow, G., Wetterberg, L., 1991. Correction of non-24-hour sleep/wake cycle by melatonin in a blind retarded boy. Ann. Neurol. 29, 336–339. https://doi.org/10.1002/ana.410290318.

Patel, S., Dulluc, J., Geffard, M., 1986. Comparison of serotonin and 5-methoxytryptamine immunoreactivity in rat raphe nuclei. Histochemistry 85 (3), 259–263. doi: 10.1007/BF00494813. PMID: 3528081.

Patel, S., Rahmani, B., Gandhi, J., Seyam, O., Joshi, G., Reid, I., Smith, N.L., Waltzer, W.C., Khan, S.A., May 2020. Revisiting the pineal gland: a review of calcification, masses, precocious puberty, and melatonin functions. Int. J. Neurosci. 130 (5), 464–475. https://doi.org/10.1080/00207454.2019.1692838.

Pera, E.M., Kessel, M., 1997. Patterning of the chick forebrain anlage by the prechordal plate. Development 124, 4153–4162.

Petrov, T., Krukoff, T.L., Jhamandas, J.H.,1992. The hypothalamic paraventricular and lateral parabrachial nuclei receive collaterals from raphe nucleus neurons: a combined double retrograde and immunocytochemical study. J. Comp. Neurol. 318 (1), 18–26. doi: 10.1002/cne.903180103. PMID: 1583154.

Pevet, P., March 2002. Melatonin, dialogues. Clin. Neurosci. 4 (1), 57–72.

Phillips, A.G., Blaha, C.D., Fibiger, H.C., Lane, R.F., 1988. Interactions between mesolimbic dopamine neurons, cholecystokinin, and neurotensin: evidence using in vivo voltammetry. Ann. N. Y. Acad. Sci. 537, 347–361. doi: 10.1111/j.1749-6632.1988.tb42119.x. PMID: 3059931.

Price, C.J., Hoyda, T.D., Ferguson, A.V., April 2008. The area postrema: a brain monitor and integrator of systemic autonomic state. Neuroscientist 14 (2), 182–194. https://doi.org/10.1177/1073858407311100.

Puelles, L., Rubenstein, J.L., 2015. A new scenario of hypothalamic organization: rationale of new hypotheses introduced in the updated prosomeric model. Front. Neuroanat. 9, 27.

Pulman, K.J., Fry, W.M., Cottrell, G.T., Ferguson, A.V., 2006. The subfornical organ: a central target for circulating feeding signals. J. Neurosci. 26 (7), 2022–2030. doi: 10.1523/JNEUROSCI.3218-05.2006. PMID: 16481435; PMCID: PMC6674925.

Quirion, R., Dalpé, M., De Lean, A., Gutkowska, J., Cantin, M., Genest, J., 1984. Atrial natriuretic factor (ANF) binding sites in brain and related structures. Peptides. 5 (6), 1167-1172. doi: 10.1016/0196-9781(84)90183-9. PMID: 6241671.

Rath, M.F., Bailey, M.J., Kim, J.S., Coon, S.L., Klein, D.C., Moller, M., January 2009. Developmental and daily expression of the Pax4 and Pax6 homeobox genes in the rat retina, localization of Pax4 in photoreceptor cells. J. Neurochem. 108 (1), 285–294.

Rhoton Jr., A.L., 2019. Rhoton's Cranial Anatomy and Surgical Approaches. Oxford University Press.

Rizzoti, K., Lovell-Badge, R., April 15, 2017. Pivotal role of median eminence tanycytes for hypothalamic function and neurogenesis. Mol. Cell. Endocrinol. 445, 7–13. https://doi.org/10.1016/j.mce.2016.08.020.

Robins, S.C., Trudel, E., Rotondi, O., Liu, X., Djogo, T., Kryzskaya, D., Bourque, C.W., Kokoeva, M.V., 2013. Evidence for NG2-glia derived, adult-born functional neurons in the hypothalamus. PLoS One 8 (10), e78236. doi: 10.1371/journal.pone.0078236. PMID: 24205170; PMCID: PMC3812154.

Rodríguez, E.M., Rodríguez, S., Hein, S., April 15, 1998. The subcommissural organ. Microsc. Res. Tech. 41 (2), 98–123. https://doi.org/10.1002/(SICI)1097-0029(19980415)41:2<98::AID-JEMT2>3.0.CO;2-M.

Rodriguez, E.M., et al., 2005. Hypothalamic tanycytes: a key component of brain-endocrine interaction. Int. Rev. Cytol. 247, 89–164. https://doi.org/10.1016/S0074-7696(05)47003-5.

Rosengarten, H., Meller, E., Friedhoff, A.J., 1972. In vitro enzymatic formation of melatonin by human erythrocytes. Res. Commun. Chem. Pathol. Pharmacol. 4, 457–465.

Salm, A.K., 2000. Mechanisms of glial retraction in the hypothalamo-neurohypophysial system of the rat. Exp. Physiol. 85, 197–202. https://doi.org/10.1111/j.1469-445x.2000.tb00024.x. VSX 197±202.

Sapède, D., Cau, E., 2013. The pineal gland from development to function. Curr. Top. Dev. Biol. 106, 171–215. https://doi.org/10.1016/B978-0-12-416021-7.00005-5.

Scheithauer, B.W., Horvath, E., Kovacs, K., January 15, 1992. Ultrastructure of the neurohypophysis. Microsc. Res. Tech. 20 (2), 177–186. https://doi.org/10.1002/jemt.1070200206.

Schulz, M., Engelhardt, B., September 30, 2005. The circumventricular organs participate in the immunopathogenesis of experimental autoimmune encephalomyelitis. Cerebrospinal Fluid Res. 2, 8. https://doi.org/10.1186/1743-8454-2-8.

Sexton, P.M., Houssami, S., Brady, C.L., Myers, D.E., Findlay, D.M., 1994. Amylin is an agonist of the renal porcine calcitonin receptor. Endocrinology 134 (5), 2103–2107. doi: 10.1210/endo.134.5.8156909. PMID: 8156909.

Shaver, S.W., Sposito, N.M., Gross, P.M., 1990. Quantitative fine structure of capillaries in subregions of the rat subfornical organ. J. Comp. Neurol. 294, 145–152. https://doi.org/10.1002/cne.902940111.

Shimogori, T., Banuchi, V., Ng, H.Y., et al., 2004. Embryonic signaling centers expressing BMP, WNT and FGF proteins interact to pattern the cerebral cortex. Development 131, 5639–5647.

Silva-Vargas, V., Maldonado-Soto, A.R., Mizrak, D., Codega, P., Doetsch, F., 2016. Age-dependent niche signals from the choroid plexus regulate adult neural stem cells. Cell Stem Cell. 19 (5), 643–652. doi: 10.1016/j.stem.2016.06.013. Epub 2016 Jul 21. PMID: 27452173.

Shapiro, R.E., Miselis, R.R., 1985. The central neural connections of the area postrema of the rat. J. Comp. Neurol. 234 (3), 344–364. doi: 10.1002/cne.902340306. PMID: 3988989.

Smith, P.M., Chambers, A.P., Price, C.J., Ho, W., Hopf, C., Sharkey, K.A., Ferguson, A.V., 2009. The subfornical organ: a central nervous system site for actions of circulating leptin. Am. J. Physiol. Regul. Integr. Comp. Physiol. 296 (3), R512–R520. doi: 10.1152/ajpregu.90858.2008. Epub 2008 Nov 19. PMID: 19020290.

Smith, P.M., Rozanski, G., Ferguson, A.V., 2010. Acute electrical stimulation of the subfornical organ induces feeding in satiated rats. Physiol. Behav. 99 (4), 534–547. doi: 10.1016/j.physbeh.2010.01.013. Epub 2010 Jan 22. PMID: 20096716.

Somponpun, S.J., Sladek, C.D., 2003. Osmotic regulation of estrogen receptor-beta in rat vasopressin and oxytocin neurons. J. Neurosci. 23 (10), 4261–4269. doi: 10.1523/JNEUROSCI.23-10-04261.2003. PMID: 12764114; PMCID: PMC6741117.

Song, K., Allen, A.M., Paxinos, G., Mendelsohn, F.A., 1992. Mapping of angiotensin II receptor subtype heterogeneity in rat brain. J. Comp. Neurol. 316 (4), 467–84. doi: 10.1002/cne.903160407. PMID: 1577995.

Sposito, N.M., Gross, P.M., 1987. Topography and morphometry of capillaries in the rat subfornical organ. J. Comp. Neurol. 260 (1), 36–46. doi: 10.1002/cne.902600104. PMID: 3597833.

Srinivasan, V., October–December 1989. The pineal gland: its physiological and pharmacological role. Indian J. Physiol. Pharmacol. 33 (4), 263–272.

Stanic, K., Montecinos, H., Caprile, T., 2010. Subdivisions of chick diencephalic roof plate: implication in the formation of the posterior commissure. Dev. Dynam. 239, 2584–2593.

Storm, E.E., Garel, S., Borello, U., et al., 2006. Dose-dependent functions of Fgf8 in regulating telencephalic patterning centers. Development 133, 1831–1844.

Sunn, N., Egli, M., Burazin, T.C., Burns, P., Colvill, L., Davern, P., Denton, D.A., Oldfield, B.J., Weisinger, R.S., Rauch, M., Schmid, H.A., McKinley, M.J., 2002. Circulating relaxin acts on subfornical organ neurons to stimulate water drinking in the rat. Proc. Natl. Acad. Sci. USA 99 (3), 1701–1706. doi: 10.1073/pnas.022647699. PMID: 11830674; PMCID: PMC122254.

Summerlee, A.J., O'Byrne, K.T., Jones, S.A., Eltringham, L., 1987. The subfornical organ and relaxin-induced inhibition of reflex milk ejection in lactating rats. J. Endocrinol. 115 (2), 347–353. doi: 10.1677/joe.0.1150347. PMID: 3437253.

Suofu, Y., Li, W., Jean-Alphonse, F.G., Jia, J., Khattar, N.K., Li, J., Baranov, S.V., Leronni, D., Mihalik, A.C., He, Y., et al., 2017. Dual role of mitochondria in producing melatonin and driving GPCR signaling to block cytochrome c release. Proc. Natl. Acad. Sci. USA 114, E7997–E8006. https://doi.org/10.1073/pnas.1705768114.

Suárez, R., Gobius, I., Richards, L.J., 2014. Evolution and development of interhemispheric connections in the vertebrate forebrain. Front. Hum. Neurosci. 8, 497. doi: 10.3389/fnhum.2014.00497. PMID: 25071525; PMCID: PMC4094842.

Swaab, D.F., 1997. Neurobiology and neuropathology of the human hypothalamus. In: Handbook of Chemical Neuroanatomy, pp. 39–137.

Swaab, D.F., 2003. Human Hypothalamus: Basic and Clinical Aspects, Part II, ISBN 9780444514905.

Swaab, D.F., Hofman, M.A., Lucassen, P.J., Purba, J.S., Raadsheer, F.C., van de Nes, J.P., 1993. Functional neuroanatomy and neuropathology of the human hypothalamus. Anat. Embryol. 187, 317–330.

Szarek, E., Cheah, P.S., Schwartz, J., Thomas, P., 2010. Molecular genetics of the developing neuroendocrine hypothalamus. Mol. Cell Endocrinol. 323 (1), 115–123. doi: 10.1016/j.mce.2010.04.002. Epub 2010 Apr 10. PMID: 20385202.

Tagami, M., Nara, Y., Kubota, A., Fujino, H., Yamori, Y., 1990. Ultrastructural changes in cerebral pericytes and astrocytes of stroke-prone spontaneously hypertensive rats. Stroke 21, 1064–1071. https://doi.org/10.1161/01.STR.21.7.1064.

Takahashi, S., Fang, J., Kapás, L., Wang, Y., Krueger, J.M.,1997. Inhibition of brain interleukin-1 attenuates sleep rebound after sleep deprivation in rabbits. Am. J. Physiol. 273 (2 Pt 2), R677–R682. doi: 10.1152/ajpregu.1997.273.2.R677. PMID: 9277554.

Takagi, S., Furube, E., Nakano, Y., Morita, M., Miyata, S., 2019. Microglia are continuously activated in the circumventricular organs of mouse brain. J. Neuroimmunol. 331, 74-86. doi: 10.1016/j.jneuroim.2017.10.008. Epub 2017 Oct 19. PMID: 29107327.

Tan, D.-X., Manchester, L.C., Qin, L., Melatonin, R.R.J., 2016. A mitochondrial targeting molecule involving mitochondrial protection and dynamics. Int. J. Mol. Sci. 17, 2124. https://doi.org/10.3390/ijms17122124.

Tan, D.X., Xu, B., Zhou, X., Reiter, R.J., January 31, 2018. Pineal calcification, melatonin production, aging, associated health consequences and rejuvenation of the pineal gland. Molecules 23 (2), 301. https://doi.org/10.3390/molecules23020301.

Tricoire, H., Locatelli, A., Chemineau, P., Malpaux, B., 2002. Melatonin enters the cerebrospinal fluid through the pineal recess. Endocrinology 143, 84–90. https://doi.org/10.1210/endo.143.1.8585.

Tricoire, H., Møller, M., Chemineau, P., Malpaux, B., 2003. Origin of cerebrospinal fluid melatonin and possible function in the integration of photoperiod. Reprod. Suppl. 61, 311–321.

van der Kooy, D., Phillips, A.G., 1977. Trigeminal substrates of intracranial self-stimulation in the brainstem. Science 196 (4288), 447–449. doi: 10.1126/science.850789. PMID: 850789.

Vigier, D., Gahery, Y., 1970. Influence de l'adrénaline sur l'amplitude du potentiel antidromique du nerf radial superficiel provoqué par la stimulation du noyau cunéiforme. Rôle possible de l'area postrema [Influence of adrenaline on the antidromic potential amplitude of the superficial radial nerve provoked by cuneiform nucleus stimulation. Possible role of the area postrema]. J. Physiol. (Paris). 62 (Suppl. 3), 462. French. PMID: 5516033.

Villela, D., Atherino, V.F., Lima Lde, S., et al., 2013. Modulation of pineal melatonin synthesis by glutamate involves paracrine interactions between pinealocytes and astrocytes through NF-κB activation. BioMed Res. Int. 2013, 618432. https://doi.org/10.1155/2013/618432.

Walberg, F., Ottersen, O.P., 1992. Neuroactive amino acids in the area postrema. An immunocytochemical investigation in rat with some observations in cat and monkey (*Macaca fascicularis*). Anat. Embryol. (Berl) 185 (6), 529–545. doi: 10.1007/BF00185613. PMID: 1605365.

Weisinger, R.S., Denton, D.A., Di Nicolantonio, R., Hards, D.K., McKinley, M.J., Oldfield, B., Osborne, P.G., 1990. Subfornical organ lesion decreases sodium appetite in the sodium-depleted rat. Brain Res. 526 (1), 23–30. doi: 10.1016/0006-8993(90)90245-7. PMID: 2078815.

Wilson, J.T., 1906. On the anatomy of the calamus region in the human bulb; with an account of a hitherto undescribed "Nucleus postremus.": Part I. J. Anat. Physiol. 40 (Pt 3), 210–241. PMID: 17232680; PMCID: PMC1287464.

Wislocki, G.B., Leduc, E.H., 1952. The cytology and histochemistry of the subcommissural organ and Reissner's fiber in rodents. J. Comp. Neurol. 97 (3), 515–543. doi: 10.1002/cne.900970305. PMID: 13034932.

Yang, Y., Estrada, E.Y., Thompson, J.F., Liu, W., Rosenberg, G.A., 2007. Matrix metalloproteinase-mediated disruption of tight junction proteins in cerebral vessels is reversed by synthetic matrix metalloproteinase inhibitor in focal ischemia in rat. J. Cerebr. Blood Flow Metabol. 27, 697–709. https://doi.org/10.1038/sj.jcbfm.9600375.

Yang, Y., Thompson, J.F., Taheri, S., Salayandia, V.M., McAvoy, T.A., Hill, J.W., et al., 2013. Early inhibition of MMP activity in ischemic rat brain promotes expression of tight junction proteins and angiogenesis during recovery. J. Cerebr. Blood Flow Metabol. 33, 1104–1114. https://doi.org/10.1038/jcbfm.2013.56.

Yin, W., Gore, A.C., August 2010. The hypothalamic median eminence and its role in reproductive aging. Ann. N. Y. Acad. Sci. 1204, 113–122. https://doi.org/10.1111/j.1749-6632.2010.05518.x.

Zhang, X., Zhao, D., Wu, W., et al., June 15, 2020. Melatonin regulates mitochondrial dynamic and alleviates neuron damage in prion disease. Aging (Albany, NY) 12 (11), 11139–11151.

Zimmerman, C.A., Huey, E.L., Ahn, J.S., Beutler, L.R., Tan, C.L., Kosar, S., et al., 2019. A gut-to-brain signal of fluid osmolarity controls thirst satiation. Nature 568, 98–102. https://doi.org/10.1038/s41586-019-1066-x.

Zhang, J., Rivest, S., 2000. A functional analysis of EP4 receptor-expressing neurons in mediating the action of prostaglandin E2 within specific nuclei of the brain in response to circulating interleukin-1beta. J. Neurochem. 74 (5), 2134-2145. doi: 10.1046/j.1471-4159.2000.0742134.x. PMID: 10800959.

Zhang, H., Zhang, S., Zhang, J., Liu, D., Wei, J., Fang, W., Zhao, W., Chen, Y., Shang, D., 2018. ZO-1 expression is suppressed by GM-CSF via miR-96/ERG in brain microvascular endothelial cells. J. Cereb. Blood Flow Metab. 38 (5), 809–822. doi: 10.1177/0271678X17702668. Epub 2017 Apr 21. PMID: 28430012; PMCID: PMC5987931.

Zhu, L., Onaka, T., 2002. Involvement of medullary A2 noradrenergic neurons in the activation of oxytocin neurons after conditioned fear stimuli. Eur. J. Neurosci. 16 (11), 2186–2198. doi: 10.1046/j.1460-9568.2002.02285.x. PMID: 12473086.

Functional systems

17

General Sensory System and Taste

Mauro Palmieri[1,2], *Guilherme de Carvalho Caldas*[2], *Alejandro Monroy-Sosa*[3], *Isai Garcia Lopez*[4], *Noé Santiago Rea*[5], *Ana Paula Cortes-Contreras*[6] *and Brandon M. Ascenzi*[7]

[1]Department of Human Neurosciences, Neurosurgery Division, University of Rome "La Sapienza", Rome, Italy [2]Neurosurgical Innovations and Training Center, Weill Cornell Medical College, New York City, NY, United States [3]Department of Neurosurgery, Institute for Social Security and Services for State Workers, (ISSSTE) Tlahuac, Center for Research and Innovation in Neuroscience, Instituto Politécnico Nacional, Superior School of Medicine, Mexico City, Mexico [4]Department of Neurosurgery, National Medical Center 20 de Noviembre, Mexico City, Mexico [5]Faculty of Medicine, La Salle University, Mexico City, Mexico [6]Faculty of Medicine, Universidad Nacional Autónoma de México (UNAM), Mexico City, Mexico [7]Independent Neuroresearcher-Member of Marie Curie Alumni Association (MCAA)

OUTLINE

INTRODUCTION

The somatosensory system (SSS) is made up of a complex of neurophysiological abilities, which depend, in turn, on various anatomical structures both external and internal to the body (Freund, 2003). In the peripheral nervous system, the main responsible for processing sensory inputs are highly specialized structures known as receptors, whose function is to transform the inputs into a sensory experience, i.e., perceived by the individual (Mountcastle et al., 1975). All this takes place through the transduction mechanism of the energy intrinsic to the input itself, which in itself does not yet constitute a perceived stimulus, through the aforementioned structures (Srinivasan et al., 1990). The combination of environmental stimuli therefore passes from the entry point, more or less peripheral, up to the highest and most complex cortical levels and delegates to the elaboration, integration, and recognition of that stimulus, which becomes sight, hearing, touch, taste, or smell

From Anatomy to Function of the Central Nervous System
https://doi.org/10.1016/B978-0-12-822404-5.00004-8

depending on the type of receptor primarily involved. In this chapter, we will describe the somatosensory and gustatory systems, those that are, respectively, responsible for tactile and gustatory perception (Salinas et al., 2000). Furthermore, we will carefully analyze the physiology of pain (nociception) as it assumes a fundamental role in preserving the body from harmful stimuli.

MECHANOCEPTION AND PROPRIOCEPTION

The touch is represented by the superficial mechanoception, that is to say, by receptors sensitive to mechanical stimuli, which are found at the skin level. Mechanoception includes three qualities: *pressure* (intensity and time), *touch* in the strict sense (speed), and *vibration* (acceleration) (Birznieks et al., 2009).

In general, mechanoception allows the body to relate to the external environment, to actively explore it and to passively receive information. When it comes to touch in humans, the organ of choice for this modality of the somatosensory system is the hand, the part of the body that allows us to explore the surrounding world and to recognize objects (Gardner, 2008). It should be emphasized that the hand is important in man and primates more generally, while other parts of the body are more important in lower mammals and invertebrates (Mountcastle et al., 1972; Romo et al., 2002). For example, rodents explore their surroundings with whiskers on their snouts, the so-called whiskers, while octopuses can recognize different shapes using their tentacles. In other words, during evolution, there has been a shift from a coarse exploration with the tentacles and whiskers to a finer exploration with the fingers. This change of shape and function of the various parts of the body is accompanied by a modification of different brain areas (Edin et al., 1991; Johansson et al., 1982).

Proprioception includes three qualities: the sense of position, the sense of movement, and the sense of force of the various parts of the body. Proprioception is therefore different from interoception, which is instead understood as the set of sensations that come from the internal organs (intestine, lungs, bladder, etc.) (Gardner et al., 2007; Milner et al., 1995). Proprioception relates the body to the external environment and the different parts of the body with respect to each other. In other words, through proprioception, we are able to know where the limbs are positioned in space, how much force is exerted in lifting a weight, at what speed a limb is moved, and so on. Therefore, at every instant, one has the exact spatial representation of one's body in space (Felleman et al., 1991; Fitzgerald et al., 2006).

Receptors and Primary Afferent Fibers

The need to combine touch and proprioception arises from the common anatomical–functional organization. In fact, the pathways that convey superficial mechanoception and proprioception travel together in the spinal cord to the subcortical areas and higher brain centers (Muniak et al., 2007; Brisben et al., 1999).

Mechanoreceptors

There are four most important superficial mechanoreceptors: Meissner's corpuscles, Pacinian corpuscles, Merkel's disks, and Ruffini's corpuscles (see Table 17.1). According to the characteristics and distribution in the skin, these receptors have different functions such as all these receptors are connected to Aβ-type myelinated fibers with large diameter and high conduction velocity, thus ensuring rapid transmission of information to the central nervous system (Connor et al., 1990). The Aβ-type represent the primary afferent fibers, which are nothing more than the peripheral axons of the T neurons (primary sensory neurons, *pseudounipolar*) of the dorsal root ganglia (or DRG) (Nelsin et al., 1980; Nolano et al., 2003).

Meissner's Corpuscles

They are very superficial, just below the epidermis, and are the most common mechanoreceptors of glabrous (hairless) skin, for example, the fingertips, palms of the hands, and soles of the feet (Sripati et al., 2006; Carlson et al., 1981). Since they are rapidly adapting type 1 receptors (RA1 receptors) and are found on the surface, they are activated by even very light pressure on the skin, for example, just touching an object on the palm of the hand. Meissner's corpuscles account for about 40% of the sensory innervation of the hand. As shown in Table 17.1, RA1 receptors show an increase in firing rate at the onset of stimulation, followed by a sharp reduction in nerve activity (Bolanowski et al., 2003).

Pacinian Corpuscles

They are located deep in the subcutaneous tissues, including the interosseous membranes and intestines. These receptors have a characteristic appearance, with an onion-skin layered concentric capsule and are very rapidly adapting (Pons et al., 1992). This feature is well suited to making Pacinian corpuscles the ideal receptors for the sense of vibration and the recognition of object surfaces. In fact, rubbing a rough surface on the skin induces a high-frequency vibration, which represents the optimal stimulus for this type of receptor. Interestingly,

TABLE 17.1 Mechanoceptors.

Receptor type	Anatomical characteristics	Associated axons (and diameters)	Location	Function	Rate of adaptation	Threshold of activation
Free nerve endings	Minimally specialized *nerve* endings	C, Aδ	All skin	Pain, temperature, crude touch	Slow	High
Meissner's corpuscles	Encapsulated; between dermal papillae	Aβ 6–12 μm	Principally glabrous skin	Touch, pressure (dynamic)	Rapid (RA1)	Low
Pacinian corpuscles	Encapsulated; onion-like covering	Aβ 6–12 μm	Subcutaneous tissue, interosseous membranes, viscera	Deep pressure, vibration (dynamic)	Rapid (RA2)	Low
Merkel's disks	Encapsulated; associated with peptide-releasing cells	Aβ 6–12 μm	All skin, hair follicles	Touch, pressure (static)	Slow (SA1)	Low
Ruffini's corpuscles	Encapsulated; oriented along stretch lines	Aβ 6–12 μm	All skin	Stretching of skin	Slow (SA2)	Low
Muscle spindles	Highly specialized	La and II	Muscles	Muscle length	Both slow and rapid	Low
Krause corpucles	Encapsulated Highly specialized Highly specialized	Aβ 6–12 μm	Glabrous skin and mucous membranes	Light pressure, cold perception	Slow	Low
		lb 12–20 μm	Tendons	Muscle	Slow	Low
Golgi Mazzoni Corpuscles Neuromuscular spindle	Highly specialized (feirly thick capsule)	La and II	Muscle	Tension changes speed and length	Slow	Low
Joint receptors	Minimally specialized	—	Joints	Joint position	Rapid	Low

selective stimulation of afferent fibers of Pacinian corpuscles induces a tickling and vibratory sensation in humans. They represent about 15% of the hand skin receptors (Talbot et al., 1968; Snider et al., 1998).

Merkel Disks

They are superficial, lying just below the epidermis, such as Meissner's corpuscles. They are slow-adapting type 1 receptors (SA1 receptors) and are found particularly on the fingers and lips (DiCarlo et al., 1998). They appear to play an important role in static stimulation of the skin, where the shapes and edges of objects placed on the skin are recognized. Merkel's disks make up about 25% of the sensory innervation of the hand. SA1 show a high firing rate at the onset of stimulation, which slowly decreases over time (see Table 17.1) (Westling et al., 1987; Gardner and Costanzo, 1981).

Ruffini Corpuscles

They are found deep in the skin, ligaments, and tendons. They are classified as slowly adapting type 2 receptors (SA2), and their elongated spindle shape makes them suitable for recording the tension produced by finger and limb movements. They make up about 20% of the hand mechanoreceptors. SA2 receptors have a similar firing pattern to SA1 receptors, i.e., a high initial firing rate that slowly decreases (see Table 17.1) (Yoshioka et al., 2001; Mountcastle et al., 2005).

Proprioceptors (Mechanoreceptors Specialized for Proprioception)

Less clear are the role and function of the proprioceptors, although their relationship with position sense, movement sense, and muscle force sense is well established. However, it is not clear which type of receptor is responsible for each of these functions (Robinson et al., 1980).

The various proprioceptors involved in these functions are the slow- and fast-adapting *neuromuscular spindles*, whose afferent fibers are of type Ia and II, the slow-adapting *Golgi tendon organs*, whose afferent fibers are of type Ib, and the rapidly adapting *joint receptors* (Koch et al., 1989). While the first two seem to be more important for the sense of muscular strength, the third type appears to be more involved in the sense of position and movement of the various parts of the body, especially the limbs (Jones et al., 2000; Ouilliam et al., 1978).

ASCENDING WAYS

The central axons of the T cell neurons of the DRG enter the spinal cord through the posterior roots. Each posterior root has a cutaneous innervation area called the dermatome (Pandya et al., 1982; Saal et al., 2014). The concept of dermatome is very important from both a functional and clinical point of view. It arises from the fact that the fibers that come from different peripheral nerves regroup in the posterior root (Wimmer et al., 2010).

The knowledge of the dermatomal organization is of fundamental importance from the clinical point of view; in fact, on the basis of the cutaneous territory presenting sensitivity disturbances, it is possible to trace the dorsal root responsible for the dermatomal symptoms (Weinstein et al., 1968; Flanagan et al., 2003). Once in the spinal cord, the axons of the T-shaped neurons enter the posterior cords, or dorsal columns, and travel up to the medulla. Therefore, in the spinal cord, tactile and proprioceptive sensitivity is represented ipsilaterally, in contrast to the pain and thermal pathways that cross immediately in the spinal cord (for more details on spinal cord pathways, see Chapter 4) (Kaas et al., 2008). This organization is very important in some clinical scenarios: in fact, the Brown–Sequard Syndrome (BSS), an incomplete spinal cord injury that results from a hemisection to the spinal cord on either side, typically in the cervical and thoracic spinal cord region (Sharms, 2021) causes complete tactile and proprioceptive anesthesia on the same side of the lesion and pain and thermal sensitivity complete impairment on the other side due to the different pathway organization in the spinal cord. This lesion leads to interruption of normal function of nerve tracts in one half of the spinal cord (Ranga, 2014). It typically presents with paralysis (hemiparaplegia) on the side of the lesion due to compromise to the lateral corticospinal tracts; loss of proprioception and vibration sense on the same side from damage to the dorsal column; and a loss of pain and temperature sensation (hemianesthesia) contralateral to the lesion owing to injury to the lateral spinothalamic tract (Fogassi et al., 2005).

As shown in the overview of the ascending tactile and proprioceptive pathways, the axons of the T-shaped neurons, after ascending the spinal cord in the dorsal columns, arrive in the medulla at the gracilis and cuneate nuclei of the dorsal columns where they form the first synapse of the dorsal column–medial lemniscal pathway (DCML) (Weber et al., 2013; Gardner et al., 1988). From the gracilis and cuneate nuclei, the secondary sensory neurons ascend toward the thalamus forming the *medial lemniscus*: the right medial lemniscus crosses with the left one; therefore, tactile and proprioceptive information reaches the contralateral thalamus (Bennet et al., 1997; Oberlaender et al., 2012). The proprioceptive tactile information that comes from the face is instead conveyed by the trigeminal nerve. In this case, the primary sensory neurons are located in the ganglion of Gasser, and their peripheral axons innervate the upper face via the ophthalmic nerve, the middle face via the maxillary nerve, and the lower face via the mandibular nerve (Jones et al., 1986; Jenmalm et al., 2003). The central branch of T-shaped neurons of Gasser's ganglion projects to the midbrain trigeminal nuclei where they synapse with secondary sensory neurons. From here, the axons of secondary sensory neurons join the medial lemniscus from the gracilis and cuneate nuclei to then reach the thalamus (Wing et al., 1996).

THERMOCEPTION

The term thermoception refers to the sensory modality through which we perceive heat and cold (Pruszynski et al., 2014).

There is an indifferent temperature zone for the skin, which does not evoke any sensation of heat or cold and which ranges from 31°C (88°F) to 36°C (97°F). Above 36°C, a sensation of heat appears up to about 45°C (115°F), where a painful sensation appears. Below 31°C, a sensation of cold appears, up to 17°C (63°F), which represents the point at which cold pain appears (Lederman et al., 1987). Alongside these conscious sensations, skin sensitivity to temperature is also involved in thermoregulation, a physiological function outside of consciousness (Romo et al., 2003).

The pathways that carry thermal information will be described in detail with regard to pain, since the thermal and pain pathways ascend in the spinal cord to higher centers with the same organization (Harris et al., 2015; Pons et al., 1991).

By exploring different skin points with a pointed metal probe capable of varying the temperature at the tip, it can be seen that in some points it is possible to evoke only sensations of heat, while in others only sensations of cold (Philips et al., 1990). For example, in 1 cm^2 of the palm, there are up to five points for cold and only half for heat. These correspond to receptors for cold, whose primary afferent fibers are of type Aδ, and for heat, whose primary afferent fibers are of type C (Brochiet et al., 1999).

INTEROCEPTION

The physiological processes that take place within the organism are generally outside of consciousness (Peters et al., 2009). For example, heart rate, blood pressure, and digestive processes are not perceived. However, particularly in some circumstances, these internal organ functions arrive at the conscious state mainly through two mechanisms (Hinkley et al., 2007).

The first is indirect: as an example, just think of the perception of the heartbeat following excessive physical exercise (Hikosaka et al., 1985). In this case, the sensation of the heart beating is due to an excessive increase in volume and a change in shape and position of the heart, which excite muscle and even skin receptors to produce the conscious perception of the heartbeat. There are also direct mechanisms through which internal physiological functions reach consciousness (Johnson et al., 1981; Jones et al., 1969).

For example, excessive stretching of the stomach induces the excitation of mechanoreceptors located in the gastric wall, inducing a feeling of satiety. Similarly, excessive distention of the intestine, for example due to the presence of excess air, produces a sense of swelling, just as distension of the bladder produces a feeling of fullness (LaMotte et al., 1979; Pause et al., 1989). These examples clearly demonstrate how physiological processes of internal organs, normally unconscious, can become conscious in certain conditions. The term interoception brings together all the sensations that come from internal organs (Lieber et al., 2017), with the exception of the locomotor system (muscles, tendons, joints), which is instead involved in proprioception (Chapman et al., 2005). The concept of interoception has developed particularly in recent years in relation to emotions and sensations of well-being and malaise (Warren et al., 1986; Manfredi et al., 2014). In fact, a complex balance between the sensations that come from the whole organism is essential for that overall pleasant sensation, which is defined as well-being. However, it should be noted that the concept of interoception awaits to be defined in more precise terms from a functional point of view (Vega-Bermudez et al., 1999; Murray et al., 1984).

CORTICAL STRUCTURES AND THEIR FUNCTION IN SOMESTHESIS

The neuroanatomical bases of the somatosensory system have been already in part discussed in Chapter 4 and in the first part of this chapter. As we have seen previously, only three relay stations are interposed between the cutaneous receptors and the cerebral cortex. The cutaneous mechanoreceptors send their axons to the caudal part of the bulb, where they terminate at the level of the nucleus gracilis and cuneatus. These second-order neurons project directly to the contralateral thalamus via the medial fasciculus, where they terminate at the level of the VPL nucleus (Dejerine et al., 1906). A parallel path to the latter is that described by the trigeminal lemniscus, which originates from the main sensory nucleus of the trigeminal nerve, where the representation of the face is present, up to ending at the VPM nucleus of the thalamus. Third-order neurons send their axons from the thalamus to the primary somatosensory (SI) cortex (Fig. 17.1), located in the postcentral gyrus of the parietal lobe (Wimmer et al., 2010).

Let us now analyze more deeply the cortical aspects related to the somatosensory system, so we can start by saying that SI is formed by four areas cytoarchitecturally distinguished by Brodmann in the past century: areas 3a, 3b, 1, and 2 (Brochier et al., 1999). Most of the thalamic fibers terminate in areas 3a and 3b whose neurons in turn project to areas 1 and 2 of Brodmann. Thalamic neurons also send modest projections directly to areas 1 and 2. These four cortical regions have different functional properties: areas 3b and 1 receive information from skin receptors, while areas 3a and 2 receive proprioceptive information from muscle and joint receptors. However, these four cortical areas are extensively interconnected with each other, so that both series analysis mechanisms and parallel analysis mechanisms are used in higher-order processing of sensory information. The secondary somatosensory cortex (SII) is located at the level of the upper lip of the lateral fissure and receives projections from all four areas of SI (Costanzo et al., 1980). Projections from SI are indispensable for normal functioning of SII. For example, when projections from the hand area of SI are interrupted,

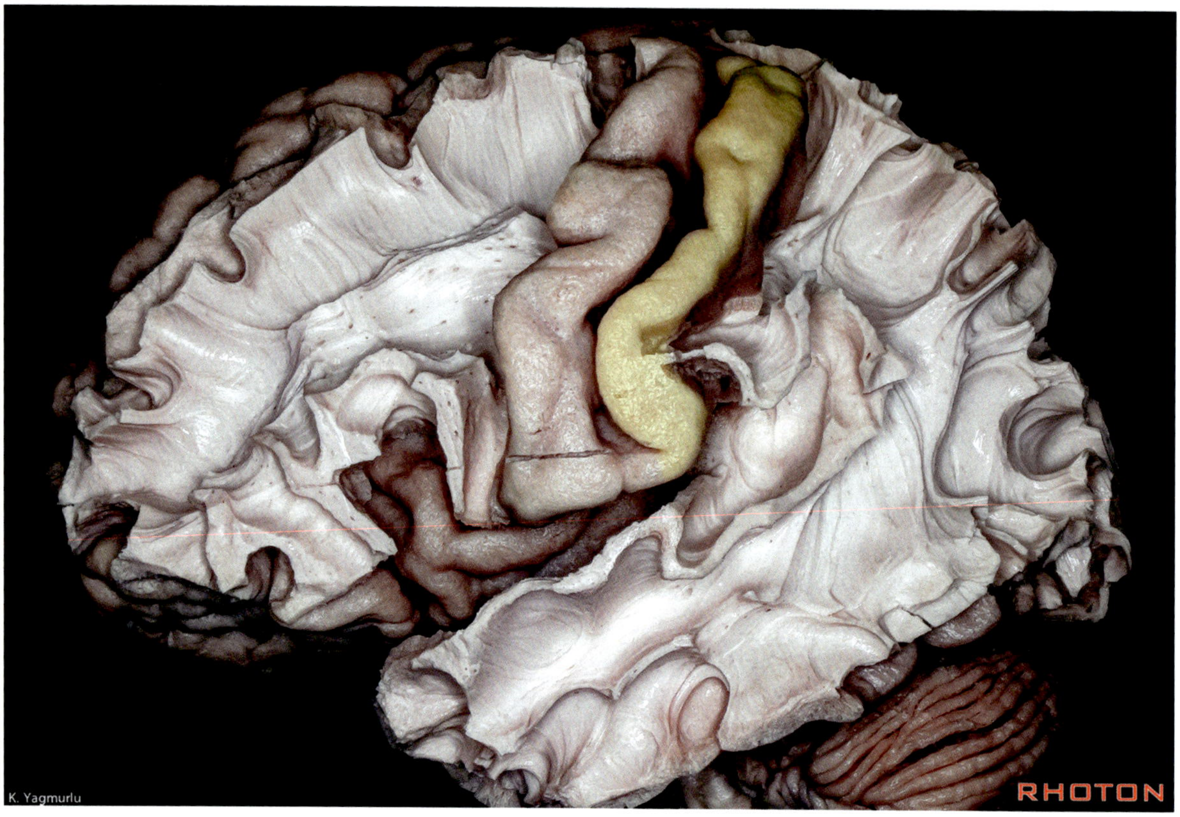

FIGURE 17.1 **The postcentral gyrus (in yellow)** is the site of the primary somesthetic area (S1) made up of Brodmann's areas 3a, 3b, 1, and 2. Anterior to the postcentral gyrus is the precentral gyrus, site of the primary motor cortex (M1) consisting of Brodmann's area 4. Between the precentral and the postcentral gyri is the central sulcus of Rolando. *Image courtesy of The Rhoton Collection® http://rhoton.ineurodb.org/. Dissector: K.Yagmurlu. Drawing by Kaan Yagmurlu.*

the stimuli applied to the skin of the hand no longer activate the neurons of SI, while, on the contrary, the ablation of S2 has no effect on the responses of S1 neurons. Area SII projects to the cortex of the insula, which, in turn, projects to areas of the temporal lobe thought to be important for tactile memory. In the posterior parietal cortex (PPC, areas 5 and 7 of Brodmann), there are other important somatosensitive cortical areas. These last receive afferents from SI and pulvinar and are implicated in common functions with that information. They are also interconnected bilaterally through the corpus callosum (Flanagan et al., 2003). The 5 BA integrates tactile information from the cutaneous mechanoreceptors with proprioceptive information coming from the underlying muscles and joints and also integrates information coming from the two hands. The 7 BA receives visual, tactile, and proprioceptive information and therefore is able to integrate stereognostic information with visual ones (Carlson et al., 1981). The PPC projects to the motor cortex of the frontal lobe (4 BA) and plays an important functional role in the process of initiating and guiding movement. Since each cortical neuron receives afferents from the receptors of a particular skin area, it also possesses a receptive field: therefore, each cortical neuron is defined by its receptive field and its mode of activation. Each point of the skin is represented, at the cortical level, by a population of neurons connected with afferent fibers innervating a particular point on the skin (Nelson et al., 1980). When this point is stimulated, the population of cortical neurons connected with the receptors located at that point is excited; in turn the stimulation of another point of the skin activates another population of cortical neurons and so on. Thus, we perceive contact with a particular skin site as a specific population of brain neurons is activated (Snider et al., 1998). The receptive fields of the sensory neurons that innervate individual fingers cover very small areas of the skin, while those of the cortical neurons that receive inputs from these neurons cover larger areas, such as the entire tip of a finger or several adjacent fingers or even the entire palmar surface of the contralateral hand (Johansson et al., 1982).

The receptive field of an area 3b neuron results from the convergence of about 300–400 mechanoceptive afferents; moreover, the receptive fields of higher-order cortical neurons are even larger. In the PPC, receptive fields are often bilateral, located in symmetrical areas of the ipsilateral and contralateral hands (Merzenich et al., 1993). Cortical neuron receptive fields comprise functional regions of the skin that are activated simultaneously during motor activity, and their size and location are not immutable but can be changed by experience and by injury to sensory nerves. These receptive fields are formed during development and are preserved by the simultaneous activation of the afferent pathways. Although the same receptive fields of cortical neurons cover large areas of the skin, a single neuron is able to discriminate fine details of stimuli as it responds optimally to the stimulation of the center of the receptive field. When stimuli are applied toward the periphery of the receptive field, responses become progressively weaker until they disappear altogether; therefore, a stimulus applied to the tip of the index finger strongly excites some neurons, while others fire weakly or not at all. The ability to locate a stimulus on the skin is based on the information provided by the entire population of excited cells (Raja et al., 1984).

The gray matter thickness of the postcentral gyrus, as determined by cortical morphometry, is approximately 1.8 mm, which varies sharply in contrast to the neighboring precentral gyrus (premotor cortex), which has an average thickness of approximately 2.7 mm. The location and borders of SI and the regions within it can be identified by analyzing the histological changes that occur from one area to the next. An in vitro quantitative receptor autoradiography analysis of various neurotransmitter binding sites across SI, such as muscarinic (M1 and M2) receptors, α_1-adrenoceptors, serotonin receptors (5-HT_{1A}, 5-HT_{1D}, 5-HT_{1E}, and 5-HT_{2A}), L-glutamate receptors, and gamma-amino butyric acid (GABA) receptors (specifically $GABA_A$ receptors), revealed that most receptor densities are located lower in deeper cortical layers and exhibit changes in density patterns at the borders defined by the cytoarchitecture (De Biasi et al., 1990).

SII is in the parietal operculum (OP) and can be divided into four regions: OP1, OP2, OP3, and OP4 (Robinson et al., 1980). Each of these regions is separated from neighboring regions by changes in cytoarchitecture, which appear to be consistent across both hemispheres of the brain. The most caudal area of SII is OP1, which is included in the inferior parietal cortex OP1 is in the superficial region of the parietal operculum, immediately ventral to the inferior parietal cortex. In OP1, cortical layer II contains granular cells that spread into upper layer III. The size of pyramidal cells in layer III increases in deeper regions. Layer IV of OP1 is characterized by a layer of granular cells that are distinct from the low density of medium-sized pyramids in layer V. Layer VI contains pyramidal cells that are larger in size and density than the cells in layer V of OP1. OP2 is located deep within the Sylvian fissure, borders on the ventral edge of OP1 and on the dorsal edge of the granular insular cortex. Along with OP1, OP2 extends into the retroinsular area of the Sylvian fissure but does not extend as far as its caudal end. As in OP1, the granular cells in layer II of OP2 spread into cortical layer III (Murray et al., 1984). However, unlike OP1, the pyramidal cells in layer III of OP2 are mostly small, with a uniform density throughout the layer. Layer IV of OP2 contains a very high density of cells, which separates it from layers III and V. Layer V of OP2 is mostly composed of small pyramidal cells. OP3 is located within the Sylvian fissure, although its area spreads over most of the parietal operculum. It is located rostral to OP2, with its ventral end bordering on the insular granular area 1 (Ig1) and its dorsal side bordering on both OP1 and OP4. Like OP1 and OP2, cells of layer II extend into layer III, which is composed of a small number of large pyramidal cells. The granular cells of layer IV are sparse, similar to layer V, in which small pyramidal cells are present at low densities. The most rostral area of SII is OP4. OP4 is present immediately rostral to OP1 on the superficial area of the parietal operculum. There are some individual differences with respect to the borders of OP4. In OP4, the superficial portion of layer III is composed of many small pyramidal cells. Cell density increases from layer IV to layer V. Medium pyramids at a high density make up the superficial portion of layer V, and cells in layer VI extend into the white matter (Murray et al., 1984).

THE SOMATOSENSORY CORTEX

As we have seen in the previous paragraph, the primary somatic sensory cortex comprises four cytoarchitectural areas: Brodmann's areas 3a, 3b, 1, and 2. In a series of pioneering studies of the cerebral cortex, Mountcastle et al., 1975, 1995) discovered that SI cortex is organized into vertical columns or slabs (Westling et al., 1987). Each column is 300–600 μm wide and spans all six cortical layers from the pial surface to the white matter. Neurons within a column receive inputs from the same local area of skin and respond to the same class or classes of touch receptors. A column therefore comprises an elementary functional module of the neocortex; it provides an anatomical structure that organizes sensory inputs to convey related information about location and modality (Oberlaender et al., 2012).

The columnar organization of the cortex is a direct consequence of intrinsic cortical circuitry, the projection patterns of thalamocortical axons, and migration pathways of neuroblasts during cortical development. The pattern of connections within a column is oriented vertically, perpendicular to the cortical surface (Flanagan et al., 2003). Thalamocortical axons terminate primarily on clusters of stellate cells in layer IV, whose axons project vertically toward the surface of the cortex, as well as on star pyramid cells. Thus, thalamocortical inputs are relayed to a narrow column of pyramidal cells that are contacted by the layer IV cell axons. The apical dendrites and axons of cortical pyramidal cells in other cortical layers are also largely oriented vertically, parallel to the thalamocortical axons and stellate cell axons. This allows the same information to be processed by a column of neurons throughout the thickness of the cortex. Pyramidal neurons form the principal excitatory class of somatosensory cortex; they compose approximately 80% of SI neurons (Benedetti et al., 2014). Pyramidal neurons in each of the six cortical layers project to specific targets. Recurrent horizontal connections link pyramidal neurons in the same or neighboring columns, allowing them to share information when activated simultaneously by the same stimulus. Neurons in layers II and III also project to layer V in the same column, to higher cortical areas in the same hemisphere and to mirror-image locations in the opposite hemisphere (Zeilhofer et al., 2012). These feedforward connections to higher cortical areas allow complex signal integration, as described later in this chapter. Pyramidal neurons in layer V provide the principal output from each column. They receive excitatory inputs from neurons in layers II and III in the same and adjacent columns as well as sparse thalamocortical inputs. Neurons in the superficial portion of layer V (layer V-A) send feedforward outputs bilaterally to layer IV of higher-order cortical areas as well as to the striatum. Neurons deeper in layer V (layer V−B) project to subcortical structures, including the basal ganglia, superior colliculus, pontine and other brainstem nuclei, the spinal cord, and dorsal column nuclei. Layer VI neurons project to local cortical neurons and back to the thalamus, particularly to regions of the ventral posterior nuclei providing inputs to that column (Wimmer et al., 2010). In addition to feedforward signals of information from touch receptors, feedback signals from layers II and III of higher somatosensory cortical areas are provided to layer I in lower cortical areas, regulating their excitability (Zeilhofer et al., 2012). Such feedback signals originate not only in somatosensory cortical areas but also in sensorimotor areas of the posterior parietal cortex, frontal motor areas, limbic areas, and regions of the medial temporal lobe involved in memory formation and storage. These feedback signals are thought to play a role in the selection of sensory information for cognitive processing (by the mechanisms of attention) and in short-term memory tasks. Feedback pathways may also gate sensory signals during motor activity. Various local inhibitory interneurons within each column serve to focus columnar output (Caterina et al., 1997).

Somatotopic Organization of the Cortical Columns

The columns within the primary somatic sensory cortex are arranged topographically such that there is a complete somatotopic representation of the body in each of the four areas of SI (Fig. 17.2). The cortical map of the body corresponds roughly to the spinal dermatomes. Sacral segments are represented medially, lumbar and thoracic segments centrally, cervical segments more laterally, and the trigeminal representation of the face at the most lateral portion of SI cortex. Knowledge of the neural map of the body in the brain is important for localizing damage to the cortex from stroke or head trauma (Mountcastle, 1997).

The body surface is represented in at least 10 distinct neural maps in the parietal lobe: four in SI, four in SII, and at least two in the PPC. As a result, these regions mediate different aspects of tactile sensation. Neurons in areas 3b and 1 of SI process details of surface texture, whereas those in area 2 represent the size and shape of objects. These attributes of somatic sensation are further elaborated in SII and the PPC, where neurons are engaged in object discrimination and manipulation, respectively (Oberlaender et al., 2012). Another important feature of somatotopic maps is the amount of cerebral cortex devoted to each body part.

The neural map of the body in the human brain, termed the *homunculus*, does not duplicate exactly the spatial topography of the skin rather, each part of the body is represented in proportion to its importance to the sense of touch. Disproportionately large areas are devoted to certain body regions, particularly the hand, foot, and mouth, and relatively smaller areas to more proximal body parts. In humans and monkeys, more cortical columns are devoted to the fingers than to the entire trunk.

The amount of cortical area devoted to a unit area of skin, called the cortical magnification, varies by more than a hundredfold across different body surfaces. It is closely correlated with the innervation density and thus the spatial acuity of the touch receptors in an area of skin. The areas with the greatest magnification in

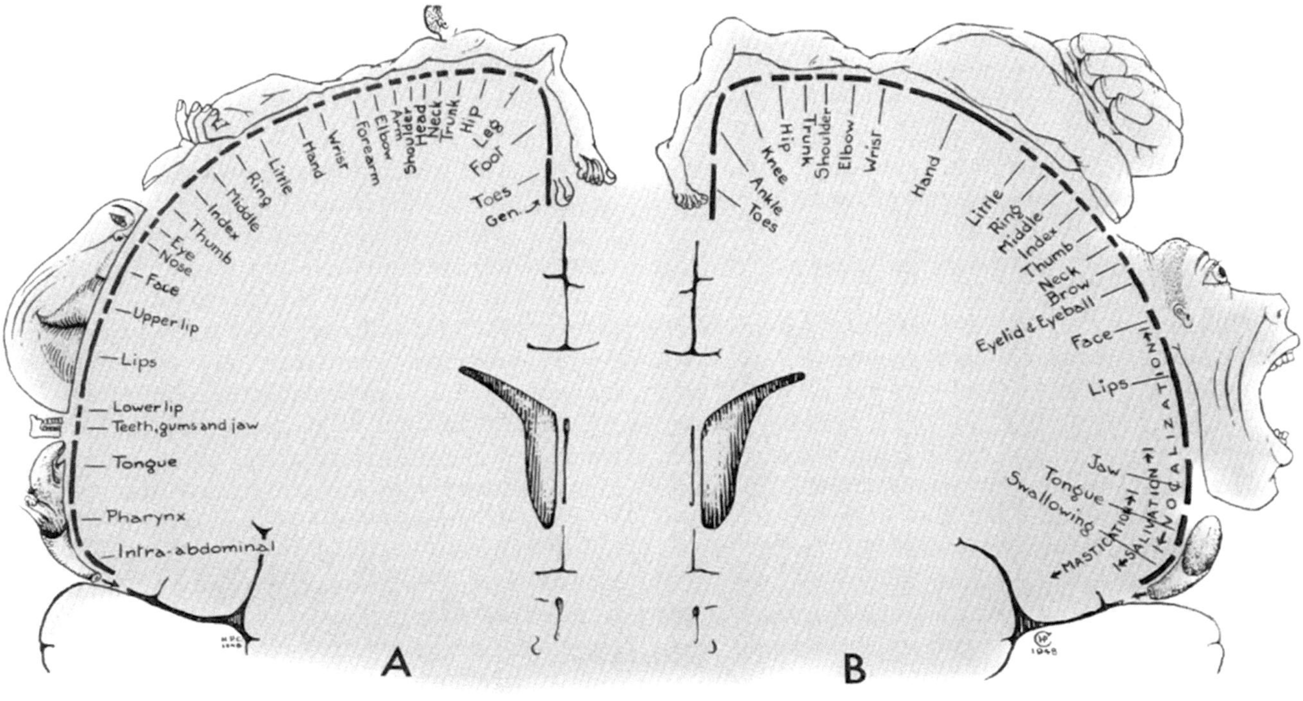

FIGURE 17.2 **Penfield's motor (A) and sensitive (B) homunculi.** This hand-drawn illustration rendered in black ink shows two mirror image outlines of cross sections of the cerebral cortex (primary motor cortex A and primary somatosensory cortex) comprised of segmented lines each of which is labeled with the part of the body to which that part of the brain corresponds. Curving around the surface of each outline are the sensory (on the left) and motor (on the right) homunculi, distorted nude human figures with elongated feet, massive hands, massive faces with huge lips and mouth separated from their bodies, and various parts of the alimentary system represented below the chin. The sensory homunculus also has a scrotum and uncircumcised penis beneath its feet. The heads of each seem to have shorn hair, and musculature and features coded as male I. *Image courtesy of Penfield Archives, Osler Library of the History of Medicine. (Public domain, modified by Brandon M. Ascenzi).*

the human brain—the lips, tongue, fingers, and toes—have tactile acuity thresholds of 0.5, 0.6, 1.0, and 4.5 mm, respectively. Rodents and other mammals that probe the environment with their whiskers have a large number of columns in SI, named barrels, which receive inputs from individual vibrissae on the face. Barrel cortex provides a widely used experimental preparation for studying cortical circuitry (Nelson et al., 1980).

The reason we feel a particular area of skin being touched is because specific populations of neurons in the cortex are activated. This experience can be induced experimentally by electrical or optogenetic stimulation of the same cortical neurons.

The receptive field of a neuron in area 3b represents a composite of inputs from 300 to 400 nerve fibers and typically covers a single phalanx or palm pad. Inputs from SA1 and RA1 touch receptors in the same skin region converge on common neurons in area 3b. Receptive fields in higher cortical areas are even larger, spanning functional regions of skin that are activated simultaneously during motor activity (Peters et al., 2009). These include the tips of several adjacent fingers, or an entire finger, or both the fingers and the palm. Neurons in areas 1 and 2 of SI are concerned with information more abstract than just their innervation sites on the body. Neurons whose receptive fields include more than one finger fire at higher rates when several fingers are touched simultaneously and, in this way, signal the size and shape of objects held in the hand. These large receptive fields allow cortical neurons to integrate the fragmented information from individual touch receptors, enabling us to recognize the overall shape of an object (Quilliam et al., 1978). For example, such neurons may distinguish the handle of a screwdriver from its blade.

Convergent inputs from different sensory receptors in SI may also allow individual neurons to detect the size and shape of objects. Whereas neurons in areas 3b and 1 respond only to touch and neurons in area 3a respond to muscle stretch, many of the neurons in area 2 receive both inputs (Pons et al., 1992). Thus, neurons in area 2 can integrate information about the hand shape used to grasp an object, the grip force applied by the hand and the tactile stimulation produced by the object; this integrated information may be sufficient to recognize the object. The receptive fields of cortical neurons usually have an excitatory zone surrounded by or superimposed upon inhibitory zones. Stimulation of regions of skin outside the excitatory zone may reduce the neuron's responses to tactile stimulation within the receptive field. Similarly, repeated stimulation within the receptive field may also decrease neuronal responsiveness because the excitability of the pathway is diminished by longer-lasting inhibition mediated by local interneurons (Romo et al., 2002).

Inhibitory receptive fields result from feedforward and feedback connections through interneurons in the dorsal column nuclei, the thalamus and the cortex itself that limit the spread of excitation. Inhibition generated by strong activity in one circuit reduces the output of nearby neurons that are only weakly excited. The inhibitory networks ensure that the strongest of several competing responses is transmitted, permitting a winner-take-all strategy (Warren et al., 1986). These circuits prevent blurring of tactile details such as texture when large populations of touch neurons are stimulated. In addition, higher centers in the brain use inhibitory circuits to focus attention on relevant information from the hand when it is used in skilled tasks, by suppressing unwanted, distracting inputs. If a peripheral nerve is injured or transected, its cortical projection targets acquire new receptive fields from less effective sensory inputs that are normally suppressed by inhibitory networks, or from newly developed connections from neighboring skin areas that retain innervation. Likewise, extensive stimulation of afferent pathways through repeated practice may strengthen synaptic inputs, improving perception and thereby performance (Bennet-Clarke et al., 1997).

Cognitive Touch and Secondary Somatosensory Cortex

An SI neuron's response to touch depends primarily on input from within the neuron's receptive field. This feedforward pathway is often described as a bottom-up process because the receptors in the periphery are the principal source of excitation of SI cortical neurons (Chapman et al., 2005).

Higher-order somatosensory areas not only receive information from peripheral receptors but are also strongly influenced by top-down cognitive processes, such as goal-setting and attentional modulation. Data obtained from a variety of studies—single-neuron studies in monkeys, neuroimaging studies in humans and clinical observations of patients with lesions in higher-order somatosensory areas—suggest that the ventral and dorsal regions of the parietal lobe serve complementary functions in the touch system similar to the "what" and "where" pathways of the visual system (see Chapter 20). SII is located on the upper bank and adjacent parietal operculum of the lateral sulcus in both humans and monkeys (Romo et al., 2002). Like SI, the SII cortex contains four distinct anatomical subregions with separate maps of the body. The central zone—consisting of SII proper and the adjacent parietal ventral area—receives its major input from areas 3b and 1, largely tactile information from the hand and face. A more rostral region, the parietal rostroventral area, receives information from area 3a about active hand movements as well as tactile information from areas 3b and. The most caudal somatosensory region of the lateral sulcus extends onto the parietal operculum. This region abuts the posterior parietal cortex and plays a role in integrating somatosensory and visual properties of objects. Physiological studies indicate that SII plays key roles in tactile recognition of objects placed in the hand (stereognosis), distinguishing spatial features, such as shape and texture and temporal properties, such as vibratory frequency (Chapman et al., 2005). The receptive fields of neurons in SII are larger than those in SI, covering the entire surface of the hand, and are often bilateral, representing symmetric, mirror-image locations on the contralateral and ipsilateral hands. Such large receptive fields enable us to sense the shape of an entire large object grasped in one hand, allowing us to integrate the overall contours of a tool as it contacts the palm and different fingers. Bilateral receptive fields enable us to perceive still larger objects with two hands, such as a watermelon or basketball, sharing the load between them (Di Carlo et al., 1998). The large receptive fields of SII neurons also influence their physiological responses to motion and vibration. SII neurons do not represent vibration as periodic spike trains linked to the oscillatory frequency, as do the sensory fibers from the skin or SI neurons. Instead, SII neurons abstract temporal or intensive properties of the vibratory stimulus, firing at different mean rates for different frequencies (Di Carlo et al., 1998). A similar frequency-dependent transition from temporal-to rate-coding neurons underlies sound processing in primary auditory cortex, a brain region juxtaposed to SII cortex in the parietal operculum. Importantly, the firing rates of SII neurons depend on the behavioral context or motivational state of the subject. In elegant recent studies, Ranulfo Romo (Romo et al., 2002) and his colleagues compared responses to vibratory stimuli of neurons in SI, SII, and various regions of the frontal lobe of monkeys while the animals performed a two-alternative forced-choice task. The animals were rewarded if they correctly recognized which of two vibratory stimuli was higher in frequency. Neurons in SI faithfully represent the vibratory cycles of each stimulus using a temporal code: they fire brief spike bursts in phase with each cycle. In contrast, SII neurons respond to the first stimulus with nonperiodic spike trains in which their mean firing rates are directly or inversely correlated with the vibratory frequency. Their responses to the second stimulus are even more abstract. SII spike trains combine the frequencies of both stimuli. In other words, SII responses to vibration depend on the stimulus context: the same vibratory stimulus can evoke different firing rates depending on whether the preceding stimulus is higher or lower in frequency. Even more interesting, Romo's group (Romo et al., 2002) found that neurons in SII send copies of the spike trains evoked by the first stimulus to the prefrontal and premotor cortex to preserve a memory of that response. Neurons in these frontal cortical areas continue to fire during the delay period after the first stimulus ends. Romo and colleagues proposed that these regions in the frontal lobe send the memory signal back to SII when the second stimulus occurs, thereby modifying the response of SII neurons to the direct tactile signals from the hand. In this manner, sensorimotor memories of previous stimuli influence sensory processing in the brain, allowing subjects to make cognitive judgments about newly arriving tactile stimuli. SII is the gateway to the temporal lobe via the insular cortex. Regions of the medial temporal lobe, particularly the hippocampus, are vital to the storage of explicit memory. We do not store in memory every scintilla of tactile information that enters the nervous system, only that which has some behavioral significance (Sripati et al., 2006). In light of the demonstration that the firing patterns of SII neurons are modified by selective attention, SII could make the decision whether a particular bit of tactile information is remembered.

Sensorimotor Circuits in the Posterior Parietal Cortex

Studies in the mid-1970s by Vernon Mountcastle, Juhani Hyvärinen and others demonstrated (Mountcastle et al., 1975) that regions of the PPC, surrounding the intraparietal sulcus, play an important role in the sensory guidance of movement rather than in discriminative touch. These regions include 5 and 7 BA in monkeys and the superior parietal lobule and inferior parietal cortex (39 and 40 BA) in humans (Gardner et al., 2007). These and subsequent studies demonstrated that neural activity in the PPC during reaching and grasping coincides with activation of neurons in motor and premotor areas of the frontal cortex and precedes activity in SI. 5 and 7 BA are postulated to be engaged in the planning of hand actions, because the PPC receives convergent central and peripheral signals that allow it to compare central motor commands with somatosensory feedback during reaching and grasping behaviors (see Chapter 21). Sensory feedback from SI to the PPC is used to confirm the goal of the planned action, thereby reinforcing previously learned skills or correcting those plans when errors occur. Predicting the sensory consequences of hand actions is an important component of active touch. For example, when we view an object and reach for it, we predict how heavy it should be and how it should feel in the hand; we use such predictions to initiate grasping (Pandya et al., 1982).

Daniel Wolpert and Randy Flanagan (Flanagan et al., 2003) have proposed that during active touch the motor system controls the afferent flow of somatosensory information to the brain so that subjects can predict when tactile information should arrive in SI and reach consciousness. Convergence of central and peripheral signals allows neurons to compare planned and actual movements. Corollary discharge from motor areas to somatosensory regions of the cortex may play a key role in active touch. It provides posterior PPC with information on intended actions, allowing them to learn new skills and perform them smoothly.

TACTILE DEFICITS

Patients with lesions in SI cortex have difficulty responding to simple tactile tests: touch thresholds, vibration and joint position sense, and two-point discrimination. Loss of tactile sensation in the hand produces significant motor as well as sensory deficits. With the loss of tactile sensibility, one is completely reliant on vision for directing the hand. Loss of touch does not cause paralysis or weakness because much of skilled movement is predictive, relying on sensory feedback for adjustment if necessary. The motor system in these subjects compensates for the absence of tactile information by generating more force than necessary. Patients with lesions in the PPC usually have only mild difficulty with simple tactile tests. However, they have profound difficulty with complex tactile recognition tasks and use few exploratory and skilled movements (Freund, 2003).

They display kinematic deficits when interacting with objects, failing to shape and orient the hand properly to grasp them and misdirecting the arm during reaching. They typically use too much grip force when an object is placed in their hand and are unable to direct the fingers properly when asked to evaluate its size and shape. These deficits are described clinically as the "useless hand" syndrome (tactile apraxia). Experimental ablation of somatosensory areas of the cortex in monkeys has provided valuable information about the function of these areas. Small lesions limited to area 3b produce major deficits in touch sensation from a particular part of the body. Lesions in 1 BA produce a defect in the assessment of the texture of objects, whereas lesions in 2 BA alter the ability to differentiate the size and shape of objects (Carlson et al., 1981). The damage to tactile function is less severe when such lesions are made in infant animals, apparently because in the developing brain SII cortex may take over functions normally assumed by SI. Removal of SII cortex in monkeys causes severe impairment in the discrimination of both shape and texture and prevents the animals from learning new tactile discriminations. Ablation or inhibition of 2 or 5 BA produces deficits in roughness discrimination but few other alterations in passive touch. However, motor performance is impaired as these animals misdirect reaching toward objects, fail to preshape the hand to grasp objects skillfully, and have difficulty coordinating finger movements because tactile feedback is absent (Freund, 2003).

PAIN

Pain is an unpleasant sensation and emotional experience associated with actual or potential tissue damage or described in terms of such damage. In children born with insensitivity to pain, severe injuries often go unnoticed and can lead to permanent tissue damage. The perception of pain is subjective and is influenced by many factors (De Felice et al., 2013). An identical sensory stimulus can elicit quite distinct responses in the same individual under different conditions. Pain is not the direct expression of a sensory event but rather the product of elaborate processing in the brain of a variety of neural signals. When pain is experienced, it can be acute, persistent or, in extreme cases, chronic. Persistent pain characterizes many clinical conditions and is usually the reason that patients seek medical attention. In contrast, chronic pain appears to have no useful purpose; it only makes patients miserable. Pain's highly individual and subjective nature is one of the factors that make it so difficult to define objectively and to treat clinically (Fields et al., 1987).

THERMAL, MECHANICAL AND POLYMODAL NOCICEPTION

Many organs in the periphery, including skin and subcutaneous structures such as joints and muscles, possess specialized sensory receptors that are activated by noxious insults. Unlike the specialized somatosensory receptors for light touch and pressure, most of these nociceptors are simply the free nerve endings of primary sensory neurons (Han et al., 2013). There are three main classes nociceptors—thermal, mechanical, and polymodal— as well as a more enigmatic fourth class, termed silent nociceptors.

Thermal nociceptors are activated by extremes in temperature, typically greater than 45°C (115°F) or less than 5°C (41°F) (Tominaga et al., 2004). They include the peripheral endings of small diameter, thinly myelinated Aδ axons that conduct action potentials at speeds of 5–30 m/s and unmyelinated C-fiber axons that conduct at speeds less than 1.0 m/s. Mechanical nociceptors are activated optimally by intense pressure applied to the skin; they too are the endings of thinly myelinated Aδ axons. Polymodal nociceptors can be activated by high-intensity mechanical, chemical, or thermal (both hot and cold) stimuli (Jordt et al., 2003). This class of nociceptors consists predominantly of unmyelinated C fibers. These three classes of nociceptors are widely distributed in skin and deep tissues and are often coactivated. When a hammer hits your thumb, you initially feel a sharp pain ("first pain") followed by a more prolonged aching and sometimes burning pain ("second pain"). The fast sharp pain is transmitted by Aδ fibers that carry information from damaged thermal and mechanical nociceptors. The slow dull pain is transmitted by C fibers that convey signals from polymodal nociceptors. Silent nociceptors are found in the viscera. This class of receptors is not normally activated by noxious stimulation; instead, inflammation and various chemical agents dramatically reduce their firing threshold. Their activation is thought to contribute to the emergence of secondary hyperalgesia and central sensitization, two prominent features of chronic pain (Craig et al., 1994). Noxious stimuli depolarize the bare nerve endings of afferent axons and generate action potentials that are propagated centrally. How is this achieved? The membrane of the nociceptor contains receptors that convert the thermal, mechanical, or chemical energy of noxious stimuli into a depolarizing electrical potential.

One such protein is a member of a large family of so-called transient receptor potential (TRP) ion channels (Dhaka et al., 2006). This receptor channel, TRPV1, is expressed selectively by nociceptive neurons and mediates the pain-producing actions of capsaicin, the active ingredient of hot peppers and many other pungent chemicals. The TRPV1 channel is also activated by noxious thermal stimuli, with a threshold for activation around 45°C (115°F), the temperature that provokes heat pain. Importantly, TRPV1-mediated membrane currents are enhanced by a reduction in pH, a characteristic of the chemical milieu of inflammation. Other receptor channels of the TRP channel family are expressed by nociceptors and underlie the perception of a wide range of temperatures, from cold to intense heat (Bautista et al., 2006).

Very recently, a family of mechanical transducers (Piezo1 and Piezo2) was described (Woo et al., 2014).

These channels may be important contributors to the mechanical hypersensitivity that is a prominent feature of many chronic pain conditions. Nociceptors selectively express many different voltage-gated Na^{2+} channels, which are the target of local anesthetics that so effectively block pain. Uncontrolled activation of nociceptors is associated with several pathological conditions. Two common pain states that result from alterations in nociceptor activity are allodynia and hyperalgesia (Woo et al., 2014). Patients with allodynia feel pain in response to stimuli that are normally innocuous: a light stroking of sunburned skin, the movement of joints in patients with rheumatoid arthritis, and even the act of getting out of bed in the morning after a vigorous workout. Nevertheless, patients with allodynia do not feel pain constantly; in the absence of a peripheral stimulus, there is no pain. In contrast, patients with hyperalgesia—an exaggerated response to noxious stimuli—typically report persistent pain in the absence of sensory stimulation. Persistent pain can be subdivided into two broad classes: nociceptive and neuropathic. Nociceptive pain results from the activation of nociceptors in the skin or soft tissue in response to tissue injury, and it usually occurs with inflammation. Sprains and strains produce mild forms of nociceptive pain, whereas arthritis or a tumor that invades soft tissue produce a much more severe nociceptive pain (Bautista et al., 2006, 2007). Typically, nociceptive pain is treated with nonsteroidal antiinflammatory drugs (NSAIDs) or, when severe, with opiates such as morphine.

Neuropathic pain results from direct injury to nerves in the peripheral or central nervous system and is often accompanied by a burning or electric sensation (Colloca et al., 2017). Neuropathic pains include complex regional pain syndrome, which can follow even very minor damage to a limb peripheral nerve; postherpetic neuralgia, the severe pain experienced by many patients after a bout of shingles; or trigeminal neuralgia, an intense, shooting pain in the face that results from a yet unknown pathology of the trigeminal nerve. Other neuropathic pains include phantom limb pain, which can occur after limb amputation (Woolf, 1983). Neuropathic pains do not respond to NSAIDs and are generally poorly responsive to opiates. Since loss of inhibitory controls (as occurs in epilepsy) is an important contributor to neuropathic pain, the first-line therapy for neuropathic pain, not surprisingly, involves anticonvulsants, notably the gabapentinoids (Kuner et al., 2016). (The reference to γ-aminobutyric acid [GABA] was based on a structural similarity of gabapentin to GABA. However, gabapentin in fact exerts its action by binding to the $\alpha2\delta$-subunit of voltage-gated Ca^{2+} channels, ultimately decreasing neurotransmitter release).

Recently a new specialized type of glial cell, called nociceptive Schwann cells, which form a sensory organ in the skin and contribute to the perception of pain has been describe (Abdo et al., 2019). The nociceptive Schwann cells were found to be closely associated with nociceptive nerves and have a mesh-like network of cytoplasmic sheaths around the nerves in the subepidermal border. The study used genetic labeling techniques and optogenetic stimulation to investigate the role of nociceptive Schwann cells in pain perception. It was found that activation of these cells led to pain-like behavioral responses and increased electrical activity in the nerves. The researchers also observed that nociceptive Schwann cells modulated responses to cold, heat, and mechanical stimuli. The study provides new insights into the cellular mechanisms involved in pain sensation and highlights the importance of glial cells in sensory processing (Abdo et al., 2019).

Nociceptive Information: From the Spinal Cord to the Brain

Four major ascending pathways—the spinothalamic, spinoreticular, spinoparabrachial, and spinohypothalamic tracts—contribute sensory information to the central processes that generate pain. The spinothalamic tract is the most prominent ascending nociceptive pathway in the spinal cord. It includes the axons of nociceptive-specific, thermosensitive, and wide-dynamic-range neurons in laminae I and V through VII of the dorsal horn. These axons cross the midline of the spinal cord near their segment of origin and ascend in the anterolateral white matter before terminating in thalamic nuclei (Talbot et al., 1991). The spinothalamic tract has a crucial role in the transmission of nociceptive information. Cells at the origin of this tract typically have discrete, unilateral receptive fields that underlie our ability to localize painful stimuli. Not surprisingly, electrical stimulation of the tract is sufficient to elicit the sensation of pain; conversely, lesioning this tract (anterolateral cordotomy), a procedure that is generally only used for intractable pain in terminal cancer patients, can result in a marked reduction in pain sensation on the side of the body contralateral to that of the lesion (Todd et al., 2010).

The spinoreticular tract contains the axons of projection neurons in laminae VII and VIII. This tract ascends in the anterolateral quadrant of the spinal cord with spinothalamic tract axons and terminates in both the reticular formation and the thalamus. As neurons at the origin of the spinoreticular tract generally have large, often bilateral receptive fields, this pathway has been implicated more in the processing of diffuse, poorly localized pains. The spinoparabrachial tract contains the

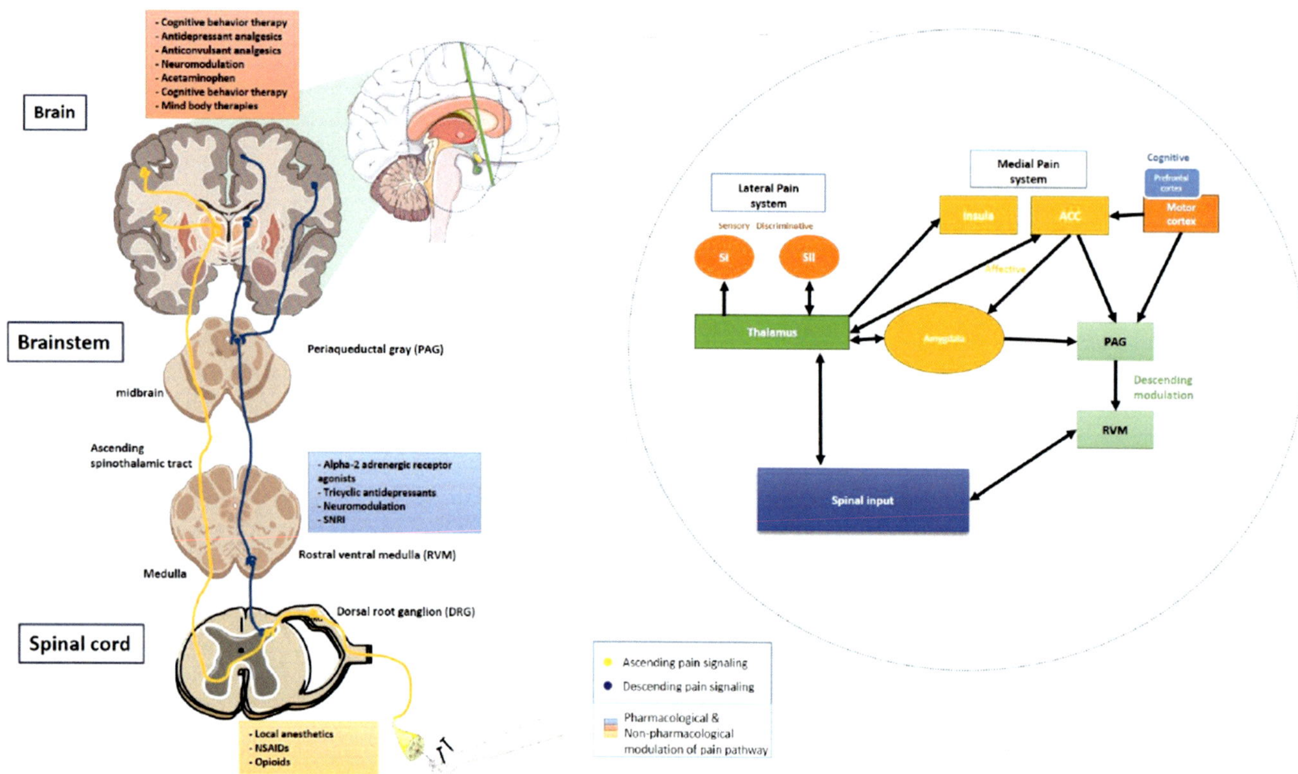

FIGURE 17.3 **Pain modulation scheme.** To the left, ascending and descending pain signaling and modulation of pain at each level. Inset shows a schematic of cortical modulation of pain. To the right, neural networks related to pain. The primary cortical pain matrix (thalamus, S1, S2, posterior insula, and parietal operculum) contributes to pain perception and location. The secondary (ACC, INS, AMY, and hippocampus) represents common structures identified in the affective motivational pain pathway, such as empathy for pain. The third (PFC, MCC, and PCC) represents one component of the cognitive evaluative pain system. The arrows represent multiple cortical connections between regions and systems indicating the complex interconnectedness of brain regions involved with pain. *ACC*, anterior cingulate cortex; *AMY*, amygdala; *DH*, dorsal horn; *DRG*, dorsal root ganglion; *INS*, insula; *MCC*, medial cingulate cortex; *NSAIDs*, nonsteroidal antiinflammatory drugs; *PAG*, periaqueductal gray; *PB*, parabrachial nucleus; *PCC*, posterior cingulate cortex; *PFC*, prefrontal cortex; *RVM*, rostroventral medulla; *SNRI*, serotonin and norepinephrine reuptake inhibitor; *SI/S1*, primary somatosensory cortex; *SII/S2*, secondary somatosensory cortex. *Image courtesy of Khera, T., Rangasamy,V., 2021. Cognition and pain: a review. Front. Psychol. 12. https://doi.org/10.3389/fpsyg.2021.673962.*

axons of projection neurons in laminae I and V (Mantyh et al., 1997). Information transmitted along this tract is thought to contribute to the affective component of pain (Tracey et al., 2007). This tract projects in the anterolateral quadrant of the spinal cord to the parabrachial nucleus at the level of the pons. This pathway has extensive collaterals to the mesencephalic reticular formation and periaqueductal gray matter (PAG) (Tracey et al., 2007). Parabrachial neurons project to the amygdala, a critical nucleus of the limbic system, which regulates emotional states. The PAG has been described as an important node for top-down pain modulation and threat response (Fig. 17.3). Data gathered from a recent human study have shown that PAG has a role in regulating physical responses and sensory decisions in response to aversive or noxious events' stimuli, by establishing connections with the dorsolateral prefrontal cortex and the anterior cingulate cortex (Wang et al., 2022).

Another recent study collected data through functional connectivity patterns within the descending pain modulatory network involving the PAG, amygdala, parabrachial nucleus, and anterior cingulate cortex. PAG and amygdala within the descending pain modulatory network at rest were associated with subsequently experienced pain and modulated by tonic pain (Meeker et al., 2022). The spinohypothalamic tract contains the axons of neurons found in spinal cord laminae I, V, VII, and VIII. These axons project to hypothalamic nuclei that serve as autonomic control centers involved in the regulation of the neuroendocrine and cardiovascular responses that accompany pain syndromes.

Thalamus and Nociceptive Information

The thalamus contains several relay nuclei that participate in the central processing of nociceptive information. Two of the most important regions of the thalamus are the lateral and medial nuclear groups. The lateral nuclear group comprises the ventroposterolateral (VPL), ventroposteromedial (VPM), and posterior/pulvinar nuclei (Lenz et al., 1995). The VPL and VPM, respectively, receive inputs via the spinothalamic tract from nociception-specific and wide-dynamic-range neurons in laminae I and V of the dorsal horn and via the trigeminothalamic tract from the trigeminal nucleus caudalis, the trigeminal homolog of the dorsal horn that processes nociceptive information from orofacial regions. The lateral thalamus processes information about the precise location of an injury, information usually conveyed to consciousness as acute pain. Consistent with this view, neurons in the lateral thalamic nuclei have small receptive fields, matching those of the presynaptic spinal neurons (Melzack et al., 1965). The medial nuclear group of the thalamus comprises the medial dorsal and central lateral nucleus of the thalamus and the intralaminar complex. Its major input is from neurons in laminae VII and VIII of the dorsal horn. The pathway to the medial thalamus was the first spinothalamic projection evident in the evolution of mammals and is therefore known as the paleospinothalamic tract. It is also sometimes referred to as the spinoreticulothalamic tract because it includes indirect connections through the reticular formation of the brainstem. The projection from the lateral thalamus to the ventroposterior lateral and medial nuclei is most developed in primates, and thus is termed the neospinothalamic tract. Many neurons in the medial thalamus respond optimally to noxious stimuli and project to many regions of the limbic system, including the anterior cingulate cortex (Perl et al., 2007).

Nociception and Limbic System

Imaging studies now show that no single area of the cortex is responsible for pain perception (Bliss et al., 2016). Rather, many regions are activated when an individual experiences pain. The anterior cingulate gyrus (ACC) and insular cortex also contain neurons that are activated strongly and selectively by noxious somatosensory stimuli. The anterior cingulate gyrus is part of the limbic system and is involved in processing emotional states associated with pain (Bliss et al., 2016). The insular cortex receives direct projections from the thalamus as well as from the amygdala. Neurons in the insular cortex process information about the internal state of the body and contribute to the autonomic component of pain responses. Importantly, neurosurgical procedures that ablate the cingulate cortex or the pathway from the frontal cortex to the cingulate cortex reduce the affective features of pain without eliminating the ability to recognize the intensity and location of the injury. Patients with lesions of the insular cortex present the striking syndrome of asymbolia for pain. They perceive noxious stimuli as painful and can distinguish sharp from dull pain but fail to display appropriate emotional responses. These observations implicate the insular cortex as an area in which the sensory, affective, and cognitive components of pain are integrated.

Regulation of Pain Perception

Many projection neurons in the dorsal horn of the spinal cord respond selectively to noxious inputs, but others receive convergent inputs from both nociceptive and nonnociceptive afferents. The concept that the convergence of sensory inputs onto spinal projection neurons regulates pain processing first emerged in the 1960s. Ronald Melzack and Patrick Wall (Melzack et al., 1965) proposed that the relative balance of activity in nociceptive and nonnociceptive afferents might influence the transmission and perception of pain. In particular, they proposed that activation of nonnociceptive sensory neurons, by engaging inhibitory interneurons in the dorsal horn, closes a "gate" for afferent transmission of nociceptive signals that can be opened by the activation of nociceptive sensory neurons.

The concept of convergence of different sensory modalities has provided an important basis for the design of new pain therapies (Yaksh et al., 2017). Viewed in its broadest sense, the convergence of high- and low-threshold inputs at spinal or supraspinal sites provided a plausible explanation for several empirical observations about the perception of pain. The shaking of the hand that follows a hammer blow or burn is a reflexive behavior and may alleviate pain by activating large diameter afferent fibers that suppress the transmission of information about noxious stimuli (Tso et al., 2017).

The Peptidergic (Opioid) Modulation of Pain

The descending pain modulatory circuit, which consists of the ventrolateral periaqueductal gray (PAG) projections to the rostral ventromedial medulla (RVM), modulates incoming nociceptive afferents at the level of the spinal cord. Electrical stimulation within the PAG or RVM typically results in analgesia (Barbaro, 1988; Fardin et al., 1984; Heinricher and Ingram, 2008; Hosobuchi, 1986; Mayer, 1984). Neurons in the caudal ventrolateral PAG contribute to stimulation-induced analgesia and opioids microinjected into the PAG produce potent antinociception (Heinricher and Morgan, 1999; Adams, 1976; Akil et al., 1976; Bach and Yaksh, 1995; Hosobuchi et al., 1977). The behavioral antinociception produced by these agents is mediated in turn by projections to the RVM (Sandkuhler and Gebhart, 1984; Tortorici and Morgan, 2002). These early experiments suggested that the PAG to RVM circuit was an "analgesia" circuit. In support of this, human imaging studies find activation of this circuit is associated with opioid analgesia, reward, and placebo responses (Eippert et al., 2009; Pecina et al., 2013; Wanigasekera et al., 2012).

However, substantial evidence has been mounting that this circuit provides bidirectional modulation of pain and that facilitation of pain is also an important function of the circuit (Heinricher, 2003; Porreca et al., 2002). Opto- and chemogenetic circuit mapping strategies have identified a circuit from the amygdala through the PAG-RVM that contributes to hyperalgesia in neuropathic pain states (Huang et al., 2019). The PAG also mediates hyperalgesia associated with alcohol withdrawal (Avegno et al., 2018). Functional imaging studies in humans have implicated dysfunction in descending control through the PAG-RVM circuit in the etiologies of chronic pain syndromes (Bushnell et al., 2013; Yarnitsky, 2015). The anatomical distribution of endogenous opioids was the focus of early studies following the discovery of the endogenous opioid peptide gene family in the 1970s. Subsequent characterization of the receptors for these peptides, as well as for morphine, led to an intense focus on exogenous opioid regulation of the opioid receptor family of G protein–coupled receptors. Although these studies have characterized myriad physiological functions of opioids, there are many basic questions about endogenous opioids in the descending pain modulatory circuit that have yet to be answered. These questions include what are the relevant stimuli that induce their release, what are the temporal and spatial constraints on their release, and how chronic pain states alter the release of endogenous opioids. New methods are being developed and adopted to measure endogenous opioids and define specific neuronal circuits that are modulated by opioids (Miller, 2006; Al-Hasani et al., 2018).

Early studies defining the behavioral consequences of removing opioid peptides and receptors have been carefully reviewed by Corder et al. (2018), Gaveriaux-Ruff and Kieffer (2002), including discussion of developmental adaptations in peptides and receptors in each knockout line. Recent studies in the amygdala have begun to precisely define the release and actions of endogenous opioids in neural circuits activated by nociceptive stimuli (Winters et al., 2017), so we have included the amygdala, a brain structure that coordinates emotional aspects of pain with sensory information, in the discussion of the descending pain modulatory circuit.

Periaqueductal Gray Matter

The PAG integrates information from cortical and subcortical areas to modulate many different behaviors, including defensive responses to pain, threat, and stress (Bandler and Keay, 1996; Keay and Bandler, 2001), as well as cardiovascular control (Carrive et al., 1987), and control of respiration (Sessle et al., 1981), lactation (Lonstein and Stern, 1997) and feeding (Sukikara et al., 2006). Heterogenous cell populations within the PAG surround the cerebral aqueduct and are organized in rostral–caudal columns that mediate distinct functions (Bandler and Shipley, 1994; Silva and McNaughton, 2019). Stimulation of the ventrolateral column of the PAG elicits analgesia in humans and antinociception in rats (Adams, 1976; Akil et al., 1976; Bach and Yaksh, 1995; Barbaro, 1988; Hosobuchi et al., 1977) that is sensitive to naloxone (Adams, 1976; Akil et al., 1976; Hosobuchi et al., 1977) indicating that stimulation of the ventrolateral PAG induces release of endogenous opioids. The term antinociception is used to describe opioid effects in rats because we can only measure their response to noxious stimuli, not their perception of pain. The behavioral antinociception produced by opioids is mediated by activation of PAG output neurons projecting to the RVM (Gebhart et al., 1984; Tortorici and Morgan, 2002). It should be noted that the descending pain modulatory circuit has been shown to be sexually dimorphic (Loyd and Murphy, 2009, 2014). Males rats have significantly higher levels of the MOR (or μ opioid receptor) in the ventrolateral PAG than cycling females, and selective lesions of MOR disrupt morphine analgesia in males, but not females (Loyd et al., 2008). Female rats have a higher number of PAG to RVM projections but a lower percentage of these are activated by morphine (Loyd et al., 2007) suggesting that there are fundamental sex-dependent differences in circuitry. Indeed, differential responses to GABA were noted in the PAG in female compared with male rats (Tonsfeldt et al., 2016). The majority of early studies on endogenous opioid peptide localization and function used male rats so there is still a large gap in knowledge about endogenous opioid peptides in this circuit in females.

Endogenous Opioids in PAG

Endogenous opioids, including met- and leu-enkephalin and β-endorphin, are widely expressed in the brain (Bodnar, 2013; Gu et al., 2017), including the descending pain modulatory circuit (Fig. 17.4). Enkephalin-containing terminals are observed throughout the PAG but are notably most dense in the ventrolateral PAG. These terminals are opposed to

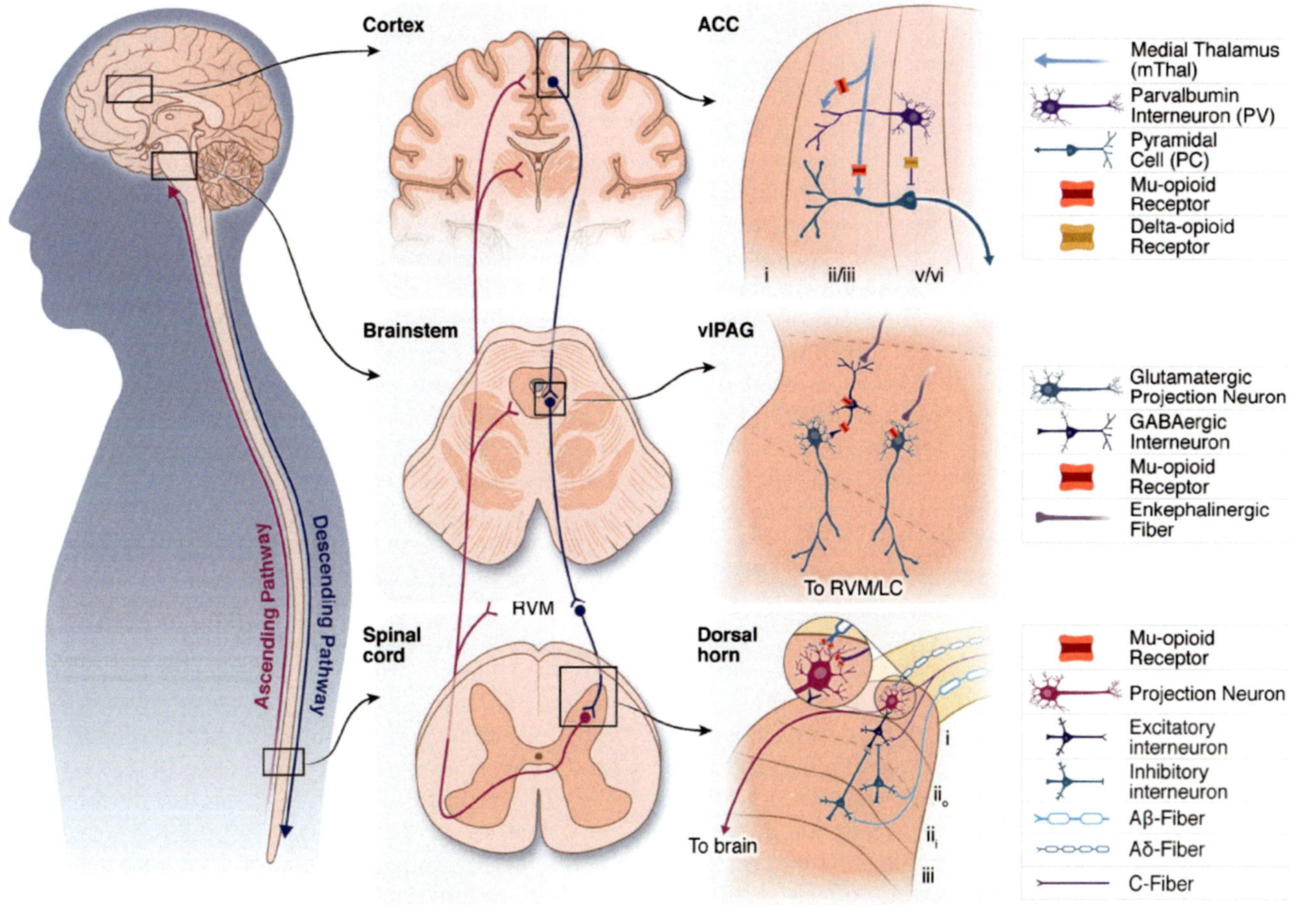

FIGURE 17.4 **Overview of three neural structures that have been targeted by neurostimulation therapies. Left:** Schematic of ascending (purple) and descending (blue) pain modulatory pathways. **Middle:** Macrolevel anatomy of the cortex, brainstem, and spinal cord, showing key nodes in the ascending and descending pain modulatory pathways. Connections between the brainstem and spinal cord via the RVM are indicated. **Right:** Select synaptic connections and microcircuitry of the ACC, vlPAG, and DH are shown. Mu-and delta-opioid receptors are expressed on cell bodies and presynaptic terminals of neurons throughout the pain neuraxis to modulate ascending and descending pain pathways. *ACC*, anterior cingulate cortex; *DH*, dorsal horn; *LC*, locus coeruleus; *RVM*, rostroventromedial medulla; *vlPAG*, ventrolateral periaqueductal gray. *Image courtesy of Lubejko, S.T., Graham, R.D., Livrizzi, G., Schaefer, R., Banghart, M. R., Creed, M.C., 2022. The role of endogenous opioid neuropeptides in neurostimulation-driven analgesia. Front. Sys. Neurosci. 16. https://doi.org/10.3389/fnsys.2022.1044686.*

both GABA and non-GABA-containing dendrites, as well as PAG output neurons that project to the RVM (Williams and Beitz, 1990). A portion of the PAG-RVM projection neurons express MOR and DOR (δ opioid receptors) (Osborne et al., 1996; Wang and Wessendorf, 2002) indicating that endogenous opioids directly inhibit some PAG-RVM output neurons. The source of enkephalin terminals is not completely characterized but the enkephalin-expressing neurons are distributed in discrete populations throughout the PAG (Moss et al., 1983; Williams and Dockray, 1983). Enkephalin staining in the monkey PAG resembles that of rat (Haber and Elde, 1982). Interestingly, some of the enkephalin-containing neurons in the PAG send projections to the amygdala (Li et al., 1990) and the nucleus accumbens (Li et al., 1990) indicating that opioid release in these areas may help to coordinate the response to pain in higher structures.

β-Endorphin-containing fibers from the arcuate nucleus of the hypothalamus project heavily to the PAG (Finley et al., 1981; Ibata et al., 1985; Sim and Joseph, 1991). Stimulation of the arcuate nucleus increases the release of β-endorphin in the PAG (Sun et al., 2007), but stimulation of the PAG predominately increases release of met-enkephalin (Bach and Yaksh, 1995). Both met-enkephalin and β-endorphin are full agonists at MORs (Alt et al., 1998). Endomorphin-2 (Tyr-Pro-Phe-Phe-NH2) is potentially another endogenous opioid peptide (Zadina et al., 1997) that is found at high levels within the PAG (Martin-Schild et al., 1999). Endomorphin 2-containing neurons from the hypothalamus project to PAG (Chen et al., 2009) and RVM (Gu and Wessendorf, 2007). This peptide is a partial agonist at MORs in the PAG (Narita et al., 2000). However, the status of endomorphin as an endogenous opioid has not been fully established as the gene responsible for the production of endomorphin or precursor peptides has not been found. Nociceptive stimulation increases the release of opioid peptides in the ventrolateral PAG (Del Río et al., 1983; Williams et al., 1995).

β-Endorphin release in the PAG is associated with stress-induced analgesia (Kulling et al., 1989), as well as peripheral injury (Nakamura et al., 2013). Similar increases in endomorphin 2 levels were reported following neuropathic pain. Stimulation of the amygdala induces release of the KOR (κ opioid receptor) agonist dynorphin in the PAG (Sun et al., 2001; Nakamura et al., 2013), but dynorphin does not elicit analgesia when microinjected into the PAG (Fang et al., 1989).

Thus, the endogenous opioid system responds to painful situations by activating opioid receptors in the PAG, but the physiological roles of the different peptides are not clearly understood. It should be noted that the majority of studies to date have used adult rats. In neonatal rats, activation of opioid receptors in the PAG actually elicit excitatory responses in the spinal cord that shifts to an inhibitory response in adulthood (Kwok et al., 2014) suggesting that endogenous opioids role in organizing the descending pain modulatory pathway is age-dependent. Neonatal injury enhances opioid tone, which presents as hypoalgesia to painful stimuli in later life that is exacerbated in females (LaPrairie and Murphy, 2007, 2009) indicating that the descending pain modulatory pathway can undergo experience dependent plasticity that can affect overall pain thresholds. These findings may extend to humans where preterm infants with prior NICU (newborn intensive care unit) experience have decreased pain responses as teenagers (Hermann et al., 2006). This plasticity may play an important role in individual differences in the development of chronic pain.

Opioid Receptors in Ventrolateral PAG

The MOR (or μ opioid receptor) is the primary opioid receptor that mediates analgesia (Matthes et al., 1996). In addition to high levels of MOR, neurons in the PAG also express DOR (or δ opioid receptor) and KOR (or κ opioid receptor), as well as orphanin FQ (OFQ) receptors (Bobeck et al., 2016; Chiou et al., 2002; Gutstein et al., 1998; Kalyuzhny and Wessendorf, 1998; Mansour et al., 1995). DORs are localized to enkephalin-containing terminals (Commons et al., 2001) and may act as autoreceptors. DOR-labeled non-GABA terminals synapse onto GABAergic neurons in the vlPAG suggesting that they inhibit glutamate inputs onto GABAergic neurons (Commons et al., 2001). Thus, they could contribute to analgesia by decreasing excitation of intrinsic GABAergic neurons within the PAG. MORs are expressed both pre- and postsynaptically in the ventrolateral PAG. Opioid binding to postsynaptic MORs activates GIRK channels that hyperpolarize a subpopulation of PAG neurons including some PAG to RVM output neurons. Postsynaptic MORs also couple to calcium channels and inhibit their activation; in contrast, MOR inhibition of GABA and glutamate release is observed in all PAG neurons (Chieng et al., 1994; Ingram et al., 1998, 2007; Osborne et al., 1996; Connor and Christie, 1998; Vaughan et al., 2003). Opioid receptor expression is actually markedly different in rats and mice. In mice, agonists of MOR, DOR, and KOR receptors activate GIRK channels, albeit in different subpopulations of neurons (Vaughan et al., 2003). KOR agonists inhibit presynaptic GABA release in both mouse and rat PAG (Lau et al., 2020; Vaughan et al., 2003). Opioid agonist inhibition of glutamate release has largely been ignored because it is difficult to reconcile with the disinhibition hypothesis (see the following), but a recent study finds that MOR inhibition of GABA is more efficient than inhibition of glutamate onto PAG-RVM projections neurons (Lau et al., 2020). These results provide evidence that individual circuits and excitatory/inhibitory input balance are differentially affected by opioids. Further studies examining the spatial and temporal release of endogenous opioids in MORs are expressed both pre- and postsynaptically in the ventrolateral PAG. Opioid binding to postsynaptic MORs activates GIRK channels that hyperpolarize a subpopulation of PAG neurons (Chieng and Christie, 1994; Ingram et al., 2007; Vaughan et al., 2003), including some PAG to RVM output neurons (Osborne et al., 1996). Postsynaptic MORs also couple to calcium channels and inhibit their activation (Connor and Christie, 1998). In contrast, MOR inhibition of GABA and glutamate release is observed in all PAG neurons (Chieng and Christie, 1994; Ingram et al., 1998). Opioid receptor expression is actually markedly different in rats and mice. In mice, agonists of MOR, DOR, and KOR receptors activate GIRK channels, albeit in different subpopulations of neurons (Vaughan et al., 2003). KOR agonists inhibit presynaptic GABA release in both mouse and rat PAG (Lau et al., 2020; Vaughan et al., 2003). Opioid agonist inhibition of glutamate release has largely been ignored because it is difficult to reconcile with the disinhibition hypothesis (see the following), but a recent study finds that MOR inhibition of GABA is more efficient than inhibition of glutamate onto PAG-RVM projection neurons (Lau et al., 2020). These results provide evidence that individual circuits and excitatory/inhibitory input balance are differentially affected by opioids. Further studies examining the spatial and temporal release of endogenous opioids in response to different stimuli will undoubtedly significantly enhance our understanding of these circuits.

Disinhibition Hypothesis of Descending Pain Modulation

PAG output neurons to the RVM are inhibited by GABA under normal conditions. Removal of this inhibition, termed disinhibition, results in activation of the descending pain modulatory circuit and analgesia reviewed in Heinricher and Ingram (2008), Lau and Vaughan (2014). This hypothesis is supported by studies showing that GABAA receptor antagonists increase firing of ~75% of PAG neurons in vivo (Behbehani et al., 1990) and direct injections of GABAA receptor antagonists or glutamate agonists into the PAG elicit antinociception (Bobeck et al., 2014; Moreau and Fields, 1986; Morgan et al., 2003). Inhibition of GABA release was shown to be the primary driver of PAG neuron excitability when compared with opioid-induced hyperpolarization (Chiou and Huang, 1999). In addition, met-enkephalin release is increased after an injection of morphine and the GABAA receptor antagonist bicuculline (Williams et al., 1995) providing evidence that intrinsic enkephalin containing neurons in the PAG are disinhibited by morphine and increased endogenous opioid release in turn disinhibits PAG output neurons. These results were further solidified with chemogenetic studies in transgenic mice where selective activation of PAG glutamatergic neurons decreased nociceptive responses and activation of GABAergic neurons facilitated nociceptive responses (Samineni et al., 2017). Thus, disinhibition plays a key role in the net effect of opioids in the PAG. However, outstanding questions remain in regard to the role of endogenous opioids in the PAG, including the role for endogenous opioids in regulating glutamatergic inputs and the temporal and spatial distribution of the release. In the rat, microinjections of opioids directly into the PAG elicit antinociception (Siuciak and Advokat, 1987; Tortorici and Morgan, 2002; Tortorici et al., 2001; Yaksh et al., 1976) through activation of MORs, not KORs (Fang et al., 1989). It is clear that activation of PAG output neurons leads to antinociception, but the exact circuitry between these PAG output neurons and well-described populations of RVM neurons, namely RVM ON- and OFF-cells (Heinricher and Ingram, 2008), is not yet understood. A simple model has been proposed in which opioids inhibit PAG GABAergic interneurons to disinhibit glutamatergic PAG projections to the RVM (Basbaum and Fields, 1984; Behbehani and Fields, 1979; Lau and Vaughan, 2014; Samineni et al., 2017). However, there is evidence that the circuit is much more complex due to the bidirectional nature of the descending pain modulatory circuit (Barbaro et al., 1986, 1989; Burgess et al., 2002; Porreca et al., 2002; Urban and Gebhart, 1999; Wei and Pertovaara, 1999). Anatomical studies in the rat describe both glutamatergic and GABAergic projections from the PAG to the RVM (Morgan et al., 2008). The GABAergic inputs predominately impinge on GABAergic RVM neurons that project to the spinal cord. Morphine microinjections into the PAG block glutamate activation of RVM neurons supporting the anatomical data that there is an inhibitory connection between PAG and RVM (Morgan et al., 1992). Thus, PAG to RVM circuitry is more complicated than simply disinhibition of excitatory descending projections and probably reflects the existence of parallel circuits contributing to the bidirectional control of pain mediated by the RVM (Lau and Vaughan, 2014; Williams and Beitz, 1990). In addition, PAG neurons project to the locus coeruleus, and this circuit can also modulate nociceptive responses at the level of the spinal cord (Kim et al., 2020) suggesting that there are multiple reciprocal circuits engaged in response to pain.

Results suggest that there are no GABAergic projections from PAG to RVM in mice (Park et al., 2010), but this is certainly not the case in rats (Morgan et al., 2008). It is possible that specific GABAergic neuron markers label only a specific subpopulation of GABAergic neurons, and this explains the difference in anatomical data between rats and mice. In vivo characterization of RVM neurons in mice with respect to responses to nociceptive stimuli is limited (Hellman et al., 2007), while in vivo and in vitro studies are ongoing in rat models to further define this circuitry (Chen et al., 2017), although the genetic tools to study circuits in rats lag behind the plethora of transgenic mouse models.

Rostroventral Medulla

The PAG sends a dense projection to the RVM, which is a brain structure that also integrates information from both cortical and subcortical areas of the brain. The RVM provides the predominant output from the descending pain modulatory circuit to the spinal cord (Fields et al., 2006; Heinricher and Ingram, 2008). RVM neurons respond to nociceptive input through an ascending relay from the parabrachial area (Chen et al., 2017). There are twocell classes identified via their responses to noxious stimuli: OFF-cells that pause firing and ON-cells that fire a burst of action potentials in response to noxious input. Opioids in the RVM elicit antinociception (Azami et al., 1982; Dickenson et al., 1979; Heinricher et al., 2001) via reducing the pause in RVM OFF-cells (Heinricher et al., 1992, 1994). ON-cell firing is correlated with hyperalgesia and systemic morphine reduces ON-cell firing (Barbaro et al., 1986; Heinricher et al., 1992) consistent with the inhibition of ON-cell firing by iontophoresis of MOR agonists directly onto ON cells in vivo (Heinricher and Neubert, 2004; Heinricher et al., 2004). Dense enkephalin containing fibers impinge on RVM ON-cells (Mason et al., 1992) providing an anatomical substrate for the modulation of ON-cells by endogenous opioid peptides.

Endogenous Opioids in RVM

Similar to observations in the PAG, enkephalins are also prevalent in RVM (Guthrie and Basbaum, 1984; Khachaturian et al., 1983; Williams and Dockray, 1983). Additionally, endomorphin fibers in the PAG impinge on serotonergic neurons expressing MOR (Gu and Wessendorf, 2007). Neurons in the RVM express MOR, DOR, and KOR at lower levels than in the PAG (Drake et al., 2007; Gutstein et al., 1998; Kalyuzhny et al., 1996; Wang and Wessendorf, 1999). Activation of these receptors results in both pre- and postsynaptic actions in the RVM in vitro (Pan and Fields, 1996; Pan et al., 1990) supporting the in vivo electrophysiological studies showing direct inhibition of RVM ON-cells (Heinricher and Neubert, 2004; Neubert et al., 2004) and indirect activation of OFF-cells (Cheng et al., 1986; Heinricher et al., 1987, 1989; Morgan et al., 1992). There is evidence for interactions between MOR and DOR in the RVM (Marinelli et al., 2005), especially in chronic pain states (Hurley and Hammond, 2000, 2001). Recent work has used elegant retrograde tracing combined with Cre-drivers in transgenic mouse lines to identify descending inputs from RVM to spinal cord (Cai et al., 2014; Francois et al., 2017; Zhang et al., 2015). Selective expression of markers and opsins in enkephalin-expressing RVM neurons find heterogenous descending RVM circuits that are involved in modulating responses to nociceptive stimuli. Some of the RVM neurons coexpressed GABA and enkephalin and removing these neurons increased hypersensitivity to nociceptive stimuli in the mice (Zhang et al., 2015). The actions of KOR in the RVM are controversial. KOR agonists inhibit neurotransmitter release and activate GIRK currents in a subpopulation of RVM neurons (Ackley et al., 2001; Bie and Pan, 2003; Marinelli et al., 2002; Pan et al., 1997). The endogenous KOR agonist dynorphin is found in some terminals in the RVM (Fallon and Leslie, 1986; Menetrey and Basbaum, 1987). Although direct injections of KOR agonists into the RVM do not elicit antinociception (Meng et al., 2005), antagonists or loss of the either the dynorphin or KOR genes all result in hyperalgesia indicating there is a role for endogenous kappa agonists in the RVM in the modulation of pain (Schepers et al., 2008a,b; Bagley and Ingram, 2020) (Table 17.2).

Amygdala

The central nucleus of the amygdala (CeA), through its dense synaptic outputs to the PAG, plays a role in the "top-down" modulation of pain. The PAG sends projections to the lateral and medial subregions of the central nucleus of the amygdala (Li and Electrical stimulation of the CeA produces analgesia Oliveira and Prado, 2001). In addition, opioid analgesia requires an intact CeA in rodents and primates (Manning, 1998; Johansen et al., 2001) and injection of MOR and DOR agonists into the CeA, such as morphine (Helmstetter et al., 1993; Pavlovic et al., 1996) or β-endorphin (Pavlovic et al., 1996) produce analgesia. Endogenous opioids acting in the CeA produce moderate analgesia (Valverde et al., 1996). Both the analgesia resulting from electrical stimulation of the CeA and opioid actions in the amygdala result from the CeA output neurons stimulating

TABLE 17.2 Most used opioid drugs in neurosurgery and pain relief therapy (Balyan et al., 2020).

Drug	Target receptor	Mechanism of action
Morphine	MOR	Agonist
Fentanyl	MOR	Agonist
Oxycodone	Primarily MOR, but it has affinity also for KOR	Agonist
Hydromorphone	MOR	Agonist
Tramadol	MOR, but it also has additional mechanisms of action involving norepinephrine and serotonin pathways	Agonist

opioid release in the PAG (Oliveira and Prado, 2001; Pavlovic et al., 1996).

Nociceptive Activation of the Amygdala

The amygdala is activated by acute (Bornhovd et al., 2002) and chronic pain (Baliki et al., 2006). It receives nociceptive information through multiple pathways. First, the amygdala receives nociceptive or threat information through a synaptic input to the capsular division of the CeA (Padilla et al., 2018; Palmiter, 2018). This information is delivered by synaptic inputs from the external lateral region of the parabrachial nucleus (PBel) that relays nociceptive information received from the spinal cord (Bernard et al., 1993). This capsular subdivision of the CeA has been termed the "nociceptive amygdala" as it is preferentially activated by noxious stimuli (Neugebauer and Li, 2002). Further, nociceptive inputs are potentiated after acute (Kissiwaa and Bagley, 2018) and chronic (Fu and Neugebauer, 2008) noxious stimuli. Second, the amygdala receives polymodal sensory information, including nociceptive information, from multiple brain regions such as the thalamus (Moga et al., 1995) and cortex (McDonald and Mascagni, 1997). This polymodal nociceptive information is delivered to the basolateral amygdala (BLA) and likely results in the activation of a subpopulation of BLA pyramidal neurons (Corder et al., 2019). The BLA may also receive purely nociceptive information as there is a population of neurons that are selectively activated by nociceptive stimuli (Corder et al., 2019). Third, the intercalated cells receive sensory, including nociceptive, information from the thalamus and sensory cortices (Asede et al., 2015; Bienvenu et al., 2015). Thus, the BLA neurons that code for negative valence, the capsular CeA neurons that are activated by nociceptive stimuli, and the intercalated cells receiving noxious sensory information, all make synaptic connections with the medial and lateral CeA (Beyeler et al., 2016; Cassell et al., 1999; Gregoriou et al., 2019).

Endogenous Opioids in the Amygdala

The level of activity of the CeA neurons that project to the PAG is determined by both their activation by glutamatergic synaptic inputs and their inhibition by GABAergic synaptic inputs. Activation of MOR inhibits GABA release onto ~60% of CeA neurons projecting to the PAG while leaving glutamate release mostly unchecked, with inhibitory effects observed in only 23% of cells (Finnegan et al., 2005). This suggests that, similar to the PAG (Vaughan and Christie, 1997), MOR agonists activate the descending pain modulatory pathway through disinhibition of CeA neurons projecting to the PAG.

The question then arises, how do endogenous opioids activate this descending pathway? Both enkephalin and β-endorphin are expressed in the amygdala. Very modest levels of β-endorphin positive fibers, likely from hypothalamic inputs, are found throughout the amygdala (Gray et al., 2014). Given that β-endorphin is released from neurons in the arcuate nucleus of the hypothalamus, this peptide may participate in stress-induced analgesia. Enkephalin is expressed at high levels in the CeA, with higher expression in the CeL (Gray et al., 2014; Poulin et al., 2006), and one of its major input zones, the intercalated cells of the amygdala (Gray et al., 2014; Jacobsen et al., 2006; Poulin et al., 2006). It is possible that either endogenous opioid could participate in activation of the descending pathway. However, two lines of evidence suggest that enkephalin is more important for activating CeA-mediated projections to the PAG. First, there is significantly greater expression of enkephalin in the CeA and the intercalated cells. Second, endogenous analgesia in the CeA is potentiated by preventing peptidase activity, and enkephalin is much more sensitive to peptidase degradation than β-endorphin (McKnight et al., 1982). Endogenous opioids, presumably dynorphin, also inhibit GABAergic inhibition onto CeM neurons through their actions at the KOR (Gilpin et al., 2014) although it is unknown whether these CeM neurons project to the PAG as only the MOR sensitivity of this projection was determined (Finnegan et al., 2005). The endogenous opioids that produce analgesia in the CeA could be released from either CeA cells themselves or intercalated neurons. When the opioid sensitivity of GABAergic inputs onto CeA-PAG neurons was assessed, GABA release was stimulated indirectly by BLA stimulation (Finnegan et al., 2005). Finnegan and colleagues suggested that the source of this GABA could be from the intercalated neurons, which are activated by the BLA neurons (Winters et al., 2017) that send a strong GABAergic projection to the CeA (Gregoriou et al., 2019). Likewise, endogenous opioid control of feedforward inhibition plasticity in the CeM was suggested to rely on intercalated neurons (Blaesse et al., 2015). Given this suggestion and the fact that endogenous opioids have been shown to be readily released from intercalated cells (Winters et al., 2017), one possible scenario is that intercalated cells release enkephalin, which disinhibits CeA neurons that project to the PAG and activates the descending pain modulatory pathway.

Endogenous Opioids in the Intercalated Cells

The intercalated neurons express high levels of enkephalin (Jacobsen et al., 2006; Poulin et al., 2006) which is packed into dense core vesicles ready for release (Winters et al., 2017). The intercalated cells receive a dense glutamatergic input from the BLA and pairs of electrical stimuli in the BLA excite the intercalated neurons sufficiently to release low levels of enkephalin (Winters et al., 2017). Moderate trains of five stimuli in the BLA stimulate enough enkephalin release to overcome the activity of peptidases and produce multiple cellular effects. The released enkephalin acts as a retrograde neuromodulator and inhibits release of glutamate from BLA terminals through activation of DOR. Of particular note is that while exogenously applied enkephalin inhibits glutamate release at this synapse through activation of DOR or MOR, endogenously released enkephalin only acts via DOR. This is an important distinction that highlights the complex actions of opioids in circuits and suggests care should be taken when inferring the actions of endogenous opioids from studies using application of exogenous agonists. Additionally, a DOR-positive allosteric modulator enhanced opioid inhibition at this synapse, but only at lower levels of opioid release (Winters et al., 2017). This is consistent with the positive allosteric modulators enhancing the affinity for agonists at the receptor and suggests that opioid positive allosteric modulators may be useful in pathologies when endogenous opioid signaling is reduced. The released enkephalin also acts via MOR on intercalated cells themselves and inhibits their release of GABA and directly inhibits their excitability through activation of a G protein–coupled inward rectifier potassium (GIRK) channel (Winters et al., 2017). This combination of reduced glutamatergic drive and direct inhibition would be expected to produce an overall inhibition of intercalated cell activity and their release of GABA in their target zones, including the CeA. Consistent with this, intercalated cell-mediated GABAergic inhibition of CeA neurons is reduced by exogenous enkephalin (Gregoriou et al., 2019). It would be very interesting to know whether intercalated neurons directly inhibit the CeA neurons projecting to the PAG. Thus, it is possible that nociceptive activation of the BLA produces feedforward activation of the intercalated cells to stimulate endogenous opioid release, which then reduces intercalated cell-mediated GABAergic inhibition of the CeA neurons projecting to the PAG. Although endogenous opioids could act at other sites in the amygdala, there are a couple of factors that suggest that endogenous opioids are strong regulators of the intercalated cells. Enkephalin is released in response to no or low levels of electrical stimulation. This is distinct from other sites of peptide release where endogenous opioid actions at other synapses require intense stimulation paradigms (Iremonger and Bains, 2009; Wagner et al., 1993; Weisskopf et al., 1993). It is possible that release of dense core vesicles is differently regulated in the intercalated cells, perhaps through differences in release machinery or intracellular calcium handling. Alternatively, we may be able to detect low concentrations of peptide that is released in response to low stimulation due to the high expression of opioid receptors in the intercalated cells (Poulin et al., 2006).

Role of Endogenous Opioids in Cortical Processing of Pain

Top-down afferents arise from a variety of cortical and subcortical brain regions, including the anterior cingulate (ACC) and amygdala (Fields et al., 2006; Keay and Bandler, 2001; Silva and McNaughton, 2019). Changes in connectivity between the ACC and the PAG are prominent in fMRI studies in chronic pain patients (Kong et al., 2010; Mainero et al., 2011; Mills et al., 2018; Truini et al., 2016). In addition, lesions of the ACC are generally agreed to reduce nociception in human patients (Davis et al., 1994; Foltz and White, 1962; Talbot et al., 1995). The prefrontal cortex (PFC)/ACC–amygdala–PAG circuit has been implicated in processing of aversive prediction-error signals (Roy et al., 2014). Inactivation of the PAG decreases the acquisition of fear conditioning and expectation blocks evoked responses due to aversive unconditioned stimuli in both the amygdala and PAG (Johansen et al., 2010). Interestingly, endogenous opioids have been shown to play important roles in conditioned fear for decades in studies utilizing naloxone to inhibit endogenous opioid signaling (Fanselow and Bolles, 1979; Helmstetter and Fanselow, 1987; Cole et al., 2011). Studies in humans find that naloxone induces sustained responses to aversive unconditioned stimuli indicating that endogenous opioids dampen acquisition of fear (Eippert et al., 2008). The use of systemic naloxone precludes determination of the sites of action of endogenous opioids in the human studies, but MORs in the PAG are critical for fear learning (Cole and McNally, 2007; McNally and Cole, 2006). Collectively, these results suggest that pain may be a teaching signal and endogenous opioids allow for reduced responses to expected pain (Eippert and Tracey, 2014).

More recently, additional projections from the PAG to the ventral tegmental area (VTA) (Omelchenko and Sesack, 2010; Suckow et al., 2013) have been implicated in avoidance behaviors in rodents associated with headache (Waung et al., 2019). Both glutamatergic and GABAergic projections from PAG (Waung et al., 2019) impinge on both DA and GABAergic neurons in VTA (Breton et al., 2019). Interestingly, although dopaminergic neurons comprise the majority of the neurons within the VTA and dopamine in the PAG modulates pain thresholds (Flores et al., 2004; Meyer et al., 2009), the VTA sends primarily GABAergic inputs to PAG (Ingram lab, unpublished observations). The role of endogenous opioids in modulating this circuit is yet to be established.

TASTE

The sensing of food and fluids, which get into our system to inform the brain about the content of nutrients and poisons through flavorings, is the main biological function of the taste. The combination of gustatory, somatosensorial, and olfactory stimuli creates representations of taste, which influence over visceral states, memory, affective, and cognitive states. In this section, we will mainly focus on the gustatory pathway.

There are five taste qualities, which from the flavor arises: sweet, salty, acid, bitter, and umami. The sweet taste identifies highly energetic nutrients, the salty taste helps with the electrolytic balance, acid and bitter tastes prevents our organism from nocuous substances, and umami identifies nutrients rich in amino acids.

In general, for explaining the neural representation of the gustatory information, there have raised two main theories: the "labeled-line" theory and the "across-neuron pattern" theory. The former proposes a binary fashion for taste codifying in neurons, where the neurons when activated by a particular stimulus signalize a single taste quality, and when quiescent is because of the absence of this particular stimulus, in this conception, each structure of the gustatory pathway is tuned to signalize a specific flavor. The latter theory explains that a population of individual neurons contribute to the representation of several qualities of taste, signalizing the response of a whole neuronal assembly.

Peripheral receptors: Sugars, salts, acids, alkaloids, and amino acids are dissolved in the saliva to bind to specific receptors and activate the gustatory buds. The taste receptor cells (TRC) are modified epithelial cells that conform the units and anatomical substrate of taste detection; these are grouped into 50–150 units to compose the taste buds and create the taste pore by projecting microvilli to the apical surface of this bud, where the tastants mixed with saliva interact with the receptors (Dutta et al., 2023).

The taste buds can be isolated on the mucosa of the oral cavity or typically be in the papillae of the tongue, palate, and larynx. There are four kinds of papillae: *circumvallate*, which have thousands of taste buds and span the most posterior region of the tongue; *foliate* that possess hundreds of taste buds and cover the posterior-lateral edge of the tongue; *fungiform* that have one or few taste buds and are located over the two anterior thirds of the tongue; and *palatine* having one taste bud and are placed on the soft palate. In this way, currently it has been demystified the concept of a "taste map" in the tongue, and it has been demonstrated that the five basic qualities of taste are present all over the tongue (Dutta et al., 2023).

Sweet: The sweet flavor receptor is part of a family of G protein–binding receptors, for this quality of taste, specifically the receptor expresses T1R2+3 dimeric complexes that respond to all of the types of sweet tastants from natural to artificial (Wu et al., 2023).

Umami: The umami name was coined by a Japanese term, which means "savory" or "delicious" and frequently describes the meaty or amino acidic savor. Nevertheless, this flavor is less easy to identify compared with the rest of the flavors and is usually foxed with salty taste. As for the sweet flavor, umami flavor receptor is also a G protein–binding receptor, but this one expresses T1R1+3 dimeric complexes, which respond to monosodium glutamate and aspartate, the only two amino acids that evoke umami taste (McMahon et al., 2023).

Bitter: In contrast to umami and sweet flavors that are evoked by a limited group of substances, bitter taste needs to be triggered by several different compounds, as this gustatory information should prevent our organism to be poisoned. The receptors destined to this task are a family of about 30 different G protein–binding receptors expressing T2R dimers, which can detect lots of bitter substances, but do not discriminate between them (Woo et al., 2023).

Salty: The salty flavor is mediated by the entrance of sodium through sodium sensible amiloride channels located on the apical surface of the TRC (Xie et al., 2023).

Acid: Similarly to salty, acid flavor is evoked through channels located on the apical surface of the TRC; however, for this quality of taste, the channels are much more variated and include a family, which interacts with hydrogen molecules such as cyclic-nucleotide-gated channels activated by hyperpolarization, ionic channels sensible to acids, H-gated potassium, and calcium channels; also it is important for the involvement of Na^+/H^+ interchangers and the acid inactivation of potassium channels (Woo et al., 2023) (to see the cortical representation of the five gustatory qualities, see Fig. 17.5).

Cranial nerves: In response to tastants, the TRC releases neurotransmitters to the synapse created between the basolateral membrane of the cell and the nerve terminals of the branches of three cranial nerves that convey the information produced to the brainstem; these branches are the chorda timpani and major superficial petrosum branches of the facial nerve (VII c.n.), the lingual/tonsillar branch of the glossopharyngeal nerve (VIII c.n.), and the superior internal laryngeal branch of the vagus nerve (X c.n.).

These branches are categorized as special visceral afferent fibers and represent the peripheral processes of the first-order neurons of the gustatory pathway, which cell bodies are in ganglions of each nerve and contain pseudo-unipolar neurons (Rhoton, 2019).

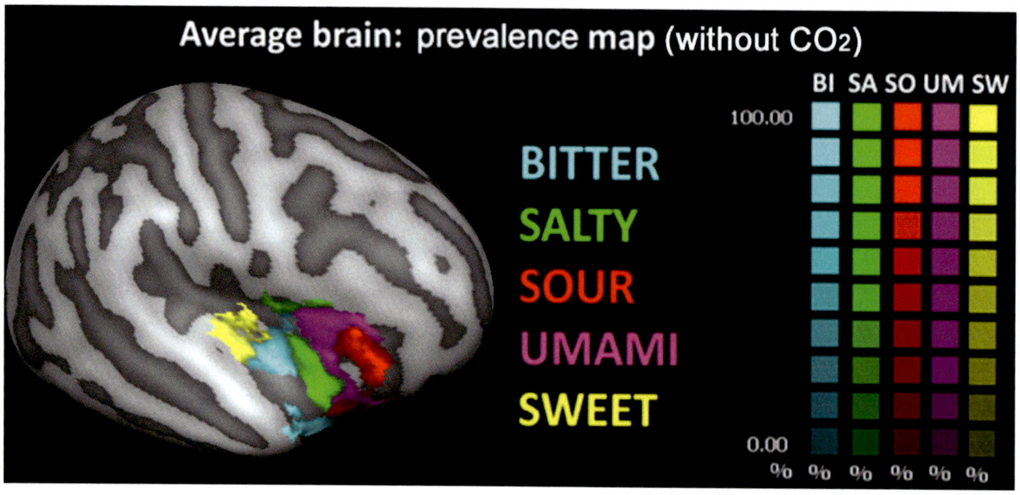

FIGURE 17.5 **Cortical surface representation of right insular activation of the standardized five basic tastes without CO_2.** In these maps, each vertex of the CBA alignment is colored with the taste that has the biggest response averaged across all subjects and its color intensity indicate how strong is the taste prevalence (a 100% value indicate that the taste prevalence is twice the second taste represented on that vertex). *Image courtesy of Prinster, A., Cantone, E., Verlezza, V., Magliulo, M., Sarnelli, G, et al., 2017. Cortical representation of different taste modalities on the gustatory cortex: a pilot study. PLoS One 12 (12), e0190164. https://doi.org/10.1371/journal.pone.0190164.*

Facial Nerve: Chorda Timpani and Major Superficial Petrosum Nerve + Geniculate Ganglion

The chorda timpani (CT) starts with the fibers in the taste buds of the two anterior thirds of the tongue; from here, the nerve fibers leave the tongue traveling within the lingual branch of the trigeminal nerve (this nerve provides the tongue of somatosensorial fibers, and is homonymous but a different structure from the lingual branch of the glossopharyngeal nerve), which loops inferiorly the submandibular duct, crosses the submandibular gland, and ascends anteriorly between the lateral aspect of the medial pterygoid muscle and the mandible, keeps ascending and at the inferior edge of the lateral pterygoid muscle in the infratemporal fossa both nerves split to take different roads. The CT angles and ascends medial to the sphenoid spine and enters to the anterior canaliculus (a separated bony channel within the medial wall of the petrotympanic fissure) to get into the tympanum, traverses it, and perforates the tympanic membrane, running to the stylomastoid foramen to join just superior to it the facial nerve, as a single bundle, the facial nerve penetrates the facial canal (Rhoton, 2019).

Meanwhile, other gustatory fibers raise from the taste buds in the soft palate, which travel within the major and minor palatine nerves, crosses the pterygopalatine ganglion without synapsing there and keep traveling within the nerve of the pterygopalatine canal to finally reach the superficial major petrosum nerve, which also penetrates the facial canal, here both bundles (facial nerve and superficial major petrosum nerve) meet and create the geniculate ganglion, and together as postganglionic fibers leave the acoustic pore, where the gustatory fibers create the intermediate nerve penetrating the lateral surface of the medulla at the pontomedullary joint separated into five to six little fascicles, immediately ventral to the eight cranial nerve, crossing obliquely the trigeminal tract, travel caudally and converge in the rNST, just dorsal to the oralis portion of the trigeminal spinal nucleus.

Glossopharyngeal Nerve: Lingual Branch and Petrosum Ganglion

The lingual nerve innervates the posterior third of the tongue, after carrying the fibers from the taste buds, the nerve joins the branch for the pharynx at the level of the middle constrictor, courses posterior to the stylopharyngeus muscle and travels to the jugular foramen as a single nerve to reach its cellular bodies in the inferior ganglion of the glossopharyngeal nerve, or petrosum ganglion, which is located in the skull base inside the jugular foramen, the postganglionar fibers ascend lying on the jugular tuberculum of the occipital bone, continue lateral to the anterior surface of the floccule, and spread into three to five filaments that penetrate the superior third of the retroolivary sulcus of the medulla, immediately rostral to the entering of the vagus nerve, the fascicles run within the dorsal half of the trigeminal nucleus and tract, and the parvicellular reticular formation, turn caudally over the dorsal edge of the trigeminal spinal tract, reach the NST, and travel inside the nucleus to synapse with its rostral portion (Rhoton, 2019).

Vagus Nerve: Superior Internal Laryngeal Branch and Nodosum Ganglion

The vagus nerve innervates the taste buds of the epiglottis through two to five twigs, which emerge from the larynx perforating the thyrohyoid membrane and travel beneath the thyrohyoid muscle to pass inferior to the major horn of the hyoid bone to meet the external branch of the superior laryngeal nerve to ascend and join to the pharyngeal branches, continuing inside the carotid sheath to reach its cellular bodies located in the inferior ganglion of the vagus nerve, or nodosum ganglion, which lies in the skull base immediately inferior to the jugular foramen, from where the central fibers of these neurons travel, reach the lateral medulla oblongata and perforate it; after this, the nerve splits into two robust bundles that pass immediately dorsal to the trigeminal spinal tract, overpass the parvicellular reticular formation, and create the solitary tract, travel rostrally and synapse with the rNST (Rhoton, 2019).

Nucleus of the Solitary Tract

The gustatory nucleus in the medulla oblongata is the rostral portion of the nucleus of the solitary tract (rNST), which has the second-order neurons of the gustatory pathway.

The NST is in the posterior—lateral portion of the medulla, immediately dorsal to the principal trigeminal nucleus, and has its rostral pole situated in a rostrocaudal axis at the half of this nucleus. The name of the nucleus comes from the fact that the solitary tract runs in its core (Rhoton, 2019).

The three cranial nerves described previously contact the rNST in a rostrocaudal fashion and topographically, where the representation of the of the tongue and palate are located rostrally to the representation of the larynx, meaning the facial nerve synapses more rostrally than the glossopharyngeal nerve, which in turn synapses more rostrally than the vagus nerve (de Araujo and Simon, 2009).

The rNST projects ipsilaterally to the thalamus via the central tegmental tract or medial lemniscus. It is important to highlight that this connection from medulla to the thalamus in humans and primates is direct, bypassing the parabrachial nucleus of the pons, whereas in rodents and inferior mammals, there is an obligatory relay in this nucleus, thus indicating that these species are not optimal for describing the gustatory pathway in humans (Rhoton, 2019).

Thalamus

The parvi or paucicellular portion of the ventral posteromedial nucleus of the thalamus (VPMpc) composes the gustatory thalamus, lodging the third order neurons of the taste pathway. The principal function of this relay is to integrate taste and somatosensorial inputs (Rhoton, 2019). The projection from the rNST is topographically organized here, where the more caudal parts representing the vagus nerve project to the most medial portion of the VPMpc and the facial nerve to the most lateral district of this nucleus, locating the glossopharyngeal projection between them (de Araujo and Simon, 2009).

From this area, efferent fibers run anteriorly and laterally through VPM and are mainly sent to the ipsilateral primary gustatory cortex conveyed by the posterior limb of the internal capsule within the thalamic radiations, dorsal to the claustrum and putamen. Other modest projections are from the central and lateral nuclei of the amygdala, hypothalamus, and innominate substance (Rhoton, 2019).

Cortex

Primary gustatory cortex: The anterodorsal insular cortex and adjacent frontal operculum have been recognized as the primary taste cortex. These regions detect oral gustatory and somatosensorial impulses related with the codification of texture, viscosity, temperature, and capsaicin and thus represents the identity and intensity of flavors, and codifies the physiological significance of flavors right in the anteroventral insula.

The topography carried from the tongue to the brainstem and thalamus is not preserved in the cortex, indicating that probably the localization of the gustatory information is not a function of this region (Dietsch et al., 2023).

Secondary Gustatory Cortex

The caudolateral orbitofrontal cortex (OFC) is known as the secondary taste cortex since it receives afferences from the primary cortex, which carries information that will be modified and refined here. This cortex produces the representation of reward and subjective pleasure related with gustatory stimulus influenced by hunger, satiety, cognition, and selective attention, guiding hedonically our alimentary behavior (de Araujo and Simon, 2009). Furthermore, in this area is where the gustatory, olfactory, and visual inputs meet each other, creating flavors through learned associations between these sensorial qualities. The first region in which this interaction occurs corresponds to an agranular cortex, which cytoarchitectically is the same as the insular cortex but topographically is that of the orbitofrontal cortex; therefore, it belongs to the anterior insula and to the posterior orbitofrontal area.

Tertiary gustatory cortex: The anterior (pregenual) cingular cortex houses the tertiary taste cortex since the orbitofrontal region projects to it.

Other regions: The amygdala capture primary cortex afferences and propagates the information to the hypothalamus, which also receives afferences from the OFC about gustatory inputs to control the food intake by reflexing satiety or hunger. Besides these regions, part of the caudate head and ventral striatum also react to gustatory stimulus; and its hypothesized functions are related to the creation of habits based on stimulus response guided by the taste (de Araujo and Simon, 2009).

References

Abdo, H., Calvo-Enrique, L., Lopez, J.M., Song, J., Zhang, M.D., Usoskin, D., El Manira, A., Adameyko, I., Hjerling-Leffler, J., Ernfors, P., 2019. Specialized cutaneous Schwann cells initiate pain sensation. Science (New York, N.Y.) 365 (6454), 695–699. https://doi.org/10.1126/science.aax6452.

Ackley, M.A., Hurley, R.W., Virnich, D.E., Hammond, D.L., 2001. A cellular mechanism for the antinociceptive effect of a kappa opioid receptor agonist. Pain 91 (3), 377–388. https://doi.org/10.1016/S0304-3959(00)00464-4.

Adams, J.E., 1976. Naloxone reversal of analgesia produced by brain stimulation in the human. Pain 2 (2), 161–166.

Al-Hasani, R., Wong, J.T., Mabrouk, O.S., McCall, J.G., Schmitz, G.P., Porter-Stransky, K.A., Aragona, B.J., Kennedy, R.T., Bruchas, M.R., 2018. In vivo detection of optically-evoked opioid peptide release. Elife 7, e36520. https://doi.org/10.7554/eLife.36520.

Alt, A., Mansour, A., Akil, H., Medzihradsky, F., Traynor, J.R., Woods, J.H., 1998. Stimulation of guanosine-5′-O-(3-[35S]thio) triphosphate binding by endogenous opioids acting at a cloned mu receptor. J. Pharmacol. Exp. Therapeut. 286 (1), 282–288.

Akil, H., Mayer, D.J., Liebeskind, J.C., 1976. Antagonism of stimulation-produced analgesia by naloxone, a narcotic antagonist. Science 191, 961–962.

Asede, D., Bosch, D., Lüthi, A., Ferraguti, F., Ehrlich, I., 2015. Sensory inputs to intercalated cells provide fear-learning modulated inhibition to the basolateral amygdala. Neuron 86 (2), 541–554. https://doi.org/10.1016/j.neuron.2015.03.008.

Avegno, E.M., Lobell, T.D., Itoga, C.A., Baynes, B.B., Whitaker, A.M., Weera, M.M., Edwards, S., Middleton, J.W., Gilpin, N.W., 2018. Central amygdala circuits mediate hyperalgesia in alcohol-dependent rats. J. Neurosci. 38 (36), 7761–7773. https://doi.org/10.1523/JNEUROSCI.0483-18.2018.

Azami, J., Llewelyn, M.B., Roberts, M.H.T., 1982. The contribution of nucleus reticularis paragigantocellularis and nucleus raphe magnus to the analgesia produced by systemically administered morphine, investigated with the microinjection technique. Pain 12 (3), 229–246. https://doi.org/10.1016/0304-3959(82)90155-5.

Bach, F.W., Yaksh, T.L., 1995. Release of beta-endorphin immunoreactivity into ventriculo-cisternal perfusate by lumbar intrathecal capsaicin in the rat. Brain Res. 701 (1–2), 192–200. https://doi.org/10.1016/0006-8993(95)01003-11.

Bagley, E.E., Ingram, S.L., 2020. Endogenous opioid peptides in the descending pain modulatory circuit. Neuropharmacology 173, 108131. https://doi.org/10.1016/j.neuropharm.2020.108131.

Baliki, M.N., Chialvo, D.R., Geha, P.Y., Levy, R.M., Harden, R.N., Parrish, T.B., Apkarian, A.V., 2006. Chronic pain and the emotional brain: specific brain activity associated with spontaneous fluctuations of intensity of chronic back pain. J. Neurosci. 26 (47), 12165–12173. https://doi.org/10.1523/JNEUROSCI.3576-06.2006.

Balyan, R., Hahn, D., Huang, H., Chidambaran, V., 2020. Pharmacokinetic and pharmacodynamic considerations in developing a response to the opioid epidemic. Expert Opin. Drug Metab. Toxicol. 16 (2), 125–141. https://doi.org/10.1080/17425255.2020.1721458. Epub 2020 Feb 24. PMID: 31976778; PMCID: PMC7199505.

Bandler, R., Keay, K.A., 1996. Columnar organization in the midbrain periaqueductal gray and the integration of emotional expression. Prog. Brain Res. 107, 285–300. https://doi.org/10.1016/s0079-6123(08)61871-3.

Bandler, R., Shipley, M.T., 1994. Columnar organization in the midbrain periaqueductal gray: modules for emotional expression? Trends Neurosci. 17 (9), 379–389. https://doi.org/10.1016/0166-2236(94)90047-7.

Barbaro, N.M., 1988. Studies of PAG/PVG stimulation for pain relief in humans. Prog. Brain Res. 77, 165–173. https://doi.org/10.1016/s0079-6123(08)62783-1.

Barbaro, N.M., Heinricher, M.M., Fields, H.L., 1986. Putative pain modulating neurons in the rostral ventral medulla: reflex-related activity predicts effects of morphine. Brain Res. 366 (1–2), 203–210. https://doi.org/10.1016/0006-8993(86)91296-5.

Barbaro, N.M., Heinricher, M.M., Fields, H.L., 1989. Putative nociceptive modulatory neurons in the rostral ventromedial medulla of the rat display highly correlated firing patterns. Somatosens. Motor Res. 6 (4), 413–425. https://doi.org/10.3109/08990228909144684.

Basbaum, A.I., Fields, H.L., 1984. Endogenous pain control systems: brainstem spinal pathways and endorphin circuitry. Annu. Rev. Neurosci. 7, 309–338. https://doi.org/10.1146/annurev.ne.07.030184.001521.

Bautista, D.M., Jordt, S.E., Nikai, T., et al., 2006. TRPA1 mediates the inflammatory actions of environmental irritants and proalgesic agents. Cell 124, 1269–1282.

Bautista, D.M., Siemens, J., Glazer, J.M., et al., 2007. The menthol receptor TRPM8 is the principal detector of environmental cold. Nature 44, 204–208.

Behbehani, M.M., Jiang, M., Chandler, S.D., Ennis, M., 1990. The effect of GABA and its antagonists on midbrain periaqueductal gray neurons in the rat. Pain 40 (2), 195–204. https://doi.org/10.1016/0304-3959(90)90070-T.

Benedetti, F., 2014. Placebo effects: from the neurobiological paradigm to translational implications. Neuron 84, 623–637.

Bennett-Clarke, C.A., Chiaia, N.L., Rhodes, R.W., 1997. Contributions of raphe-cortical and thalamocortical axons to the transient somatotopic pattern of serotonin immunoreactivity in rat cortex. Somatosens. Mot. Res. 14, 27–33.

Bernard, J.F., Alden, M., Besson, J.M., 1993. The organization of the efferent projections from the pontine parabrachial area to the amygdaloid complex: a *Phaseolus vulgaris* leucoagglutinin (PHA-L) study in the rat. J. Comp. Neurol. 329 (2), 201–229. https://doi.org/10.1002/cne.903290205.

Beyeler, A., Namburi, P., Glober, G.F., Simonnet, C., Calhoon, G.G., Conyers, G.F., Luck, R., Wildes, C.P., Tye, K.M., 2016. Divergent routing of positive and negative information from the amygdala during memory retrieval. Neuron 90 (2), 348–361. https://doi.org/10.1016/j.neuron.2016.03.004.

Bie, B., Pan, Z.Z., 2003. Presynaptic mechanism for anti-analgesic and anti-hyperalgesic actions of kappa-opioid receptors. J. Neurosci. 23 (19), 7262–7268. https://doi.org/10.1523/JNEUROSCI.23-19-07262.2003.

Bienvenu, T.C., Busti, D., Micklem, B.R., Mansouri, M., Magill, P.J., Ferraguti, F., Capogna, M., 2015. Large intercalated neurons of amygdala relay noxious sensory information. J. Neurosci. 35 (5), 2044–2057. https://doi.org/10.1523/JNEUROSCI.1323-14.2015.

Birznieks, I., Macefield, V.G., Westling, G., Johansson, R.S., 2009. Slowly adapting mechanoreceptors in the borders of the human fingernail encode fingertip forces. J. Neurosci. 29, 9370–9379.

Blaesse, P., Goedecke, L., Bazelot, M., Capogna, M., Pape, H.C., Jüngling, K., 2015. μ-Opioid receptor-mediated inhibition of intercalated neurons and effect on synaptic transmission to the central amygdala. J. Neurosci. 35 (19), 7317–7325. https://doi.org/10.1523/JNEUROSCI.0204-15.2015.

Bliss, T.V., Collingridge, G.L., Kaang, B.K., Zhuo, M., 2016. Synaptic plasticity in the anterior cingulate cortex in acute and chronic pain. Nat. Rev. Neurosci. 17, 485–496.

Bobeck, E.N., Chen, Q., Morgan, M.M., Ingram, S.L., 2014. Contribution of adenylyl cyclase modulation of pre- and postsynaptic GABA neurotransmission to morphine antinociception and tolerance. Neuropsychopharmacology 39 (9), 2142–2152. https://doi.org/10.1038/npp.2014.62.

Bobeck, E.N., Ingram, S.L., Hermes, S.M., Aicher, S.A., Morgan, M.M., 2016. Ligand-biased activation of extracellular signal-regulated kinase 1/2 leads to differences in opioid induced antinociception and tolerance. Behav. Brain Res. 298 (Pt B), 17–24. https://doi.org/10.1016/j.bbr.2015.10.032.

Bodnar, R.J., 2013. Endogenous opiates and behavior: 2012. Peptides 50, 55–95. https://doi.org/10.1016/j.peptides.2013.10.001.

Bolanowski, S.J., Pawson, L., 2003. Organization of Meissner corpuscles in the glabrous skin of monkey and cat. Somatosens. Mot. Res. 20, 223–231.

Bornhövd, K., Quante, M., Glauche, V., Bromm, B., Weiller, C., Büchel, C., 2002. Painful stimuli evoke different stimulus-response functions in the amygdala, prefrontal, insula and somatosensory cortex: a single-trial fMRI study. Brain 125 (Pt 6), 1326–1336. https://doi.org/10.1093/brain/awf137.

Breton, J.M., Charbit, A.R., Snyder, B.J., Fong, P.T.K., Dias, E.V., Himmels, P., Lock, H., Margolis, E.B., 2019. Relative contributions and mapping of ventral tegmental area dopamine and GABA neurons by projection target in the rat. J. Comp. Neurol. 527 (5), 916–941. https://doi.org/10.1002/cne.24572.

Brisben, A.J., Hsiao, S.S., Johnson, K.O., 1999. Detection of vibration transmitted through an object grasped in the hand. J. Neurophysiol. 81, 1548–1558.

Brochier, T., Boudreau, M.-J., Par, M., Smith, A.M., 1999. The effects of muscimol inactivation of small regions of motor and somatosensory cortex on independent finger movements and force control in the precision grip. Exp. Brain Res. 128, 31–40.

Burgess, S.E., Gardell, L.R., Ossipov, M.H., Malan Jr., T.P., Vanderah, T.W., Lai, J., Porreca, F., 2002. Time-dependent descending facilitation from the rostral ventromedial medulla maintains, but does not initiate, neuropathic pain. J. Neurosci. 22 (12), 5129–5136. https://doi.org/10.1523/JNEUROSCI.22-12-05129.2002.

Bushnell, M.C., Ceko, M., Low, L.A., 2013. Cognitive and emotional control of pain and its disruption in chronic pain. Nat. Rev. Neurosci. 14 (7), 502–511. https://doi.org/10.1038/nrn3516.

Cai, Y.Q., Wang, W., Hou, Y.Y., Pan, Z.Z., 2014. Optogenetic activation of brainstem serotonergic neurons induces persistent pain sensitization. Mol. Pain 10, 70. https://doi.org/10.1186/1744-8069-10-70.

Carlson, M., 1981. Characteristics of sensory deficits following lesions of Brodmann's areas 1 and 2 in the postcentral gyrus of *Macaca mulatta*. Brain Res. 204, 424–430.

Carrive, P., Dampney, R.A., Bandler, R., 1987. Excitation of neurones in a restricted portion of the midbrain periaqueductal grey elicits both behavioural and cardiovascular components of the defence reaction in the unanaesthetised decerebrate cat. Neurosci. Lett. 81 (3), 273–278. https://doi.org/10.1016/0304-3940(87)90395-8.

Cassell, M.D., Freedman, L.J., Shi, C., 1999. The intrinsic organization of the central extended amygdala. Ann. N. Y. Acad. Sci. 877, 217–241. https://doi.org/10.1111/j.1749-6632.1999.tb09270.x.

Caterina, M.J., Schumacher, M.A., Tominaga, M., Rosen, T.A., Levine, J.D., Julius, D., 1997. The capsaicin receptor: a heat-activated ion channel in the pain pathway. Nature 389, 816–824.

Chapman, C.E., Meftah el, -M., 2005. Independent controls of attentional influences in primary and secondary somatosensory cortex. J. Neurophysiol. 94, 4094–4107.

Chen, Y.L., Li, A.H., Yeh, T.H., Chou, A.H., Wang, H.L., 2009. Nocistatin and nociceptin exert opposite effects on the excitability of central amygdala nucleus-periaqueductal gray projection neurons. Mol. Cell. Neurosci. 40 (1), 76–88. https://doi.org/10.1016/j.mcn.2008.09.003.

Chen, Q., Roeder, Z., Li, M.H., Zhang, Y., Ingram, S.L., Heinricher, M.M., 2017. Optogenetic evidence for a direct circuit linking nociceptive transmission through the parabrachial complex with pain-modulating neurons of the rostral ventromedial medulla (RVM). eNeuro 4 (3). https://doi.org/10.1523/ENEURO.0202-17.2017. ENEURO.0202-0217.2017.

Cheng, Z.F., Fields, H.L., Heinricher, M.M., 1986. Morphine microinjected into the periaqueductal gray has differential effects on 3 classes of medullary neurons. Brain Res. 375 (1), 57–65. https://doi.org/10.1016/0006-8993(86)90958-3.

Chieng, B., Christie, M.J., 1994. Hyperpolarization by opioids acting on mu-receptors of a sub-population of rat periaqueductal gray neurones in vitro. Br. J. Pharmacol. 113 (1), 121–128. https://doi.org/10.1111/j.1476-5381.1994.tb16183.x.

Chiou, L.C., Huang, L.Y., 1999. Mechanism underlying increased neuronal activity in the rat ventrolateral periaqueductal grey by a mu-opioid. J. Physiol. 518 (Pt 2), 551–559. https://doi.org/10.1111/j.1469-7793.1999.0551p.x.

Cole, S., McNally, G.P., 2007. Opioid receptors mediate direct predictive fear learning: evidence from one-trial blocking. Learn. Mem. 14 (4), 229–235. https://doi.org/10.1101/lm.489507.

Cole, S., Richardson, R., McNally, G.P., 2011. Kappa opioid receptors mediate where fear is expressed following extinction training. Learn. Mem. 18 (2), 88–95. https://doi.org/10.1101/lm.2049511.

Colloca, L., Ludman, T., Bouhassira, D., et al., 2017. Neuropathic pain. Nat. Rev. Dis. Prim. 3, 17002.

Commons, K.G., Beck, S.G., Rudoy, C., Van Bockstaele, E.J., 2001. Anatomical evidence for presynaptic modulation by the delta opioid receptor in the ventrolateral periaqueductal gray of the rat. J. Comp. Neurol. 430 (2), 200–208.

Connor, C., Hsiao, S.S., Phillips, J., Johnson, K.O., 1990. Tactile roughness: neural codes that account for psychophysical magnitude estimates. J. Neurosci. 10, 3823–3836.

Connor, M., Christie, M.J., 1998. Modulation of Ca^{2+} channel currents of acutely dissociated rat periaqueductal grey neurons. J. Physiol. 509 (Pt 1), 47–58. https://doi.org/10.1111/j.1469-7793.1998.047bo.x.

Corder, G., Castro, D.C., Bruchas, M.R., Scherrer, G., 2018. Endogenous and exogenous opioids in pain. Annu. Rev. Neurosci. 41, 453–473. https://doi.org/10.1146/annurev-neuro-080317-061522.

Corder, G., Ahanonu, B., Grewe, B.F., Wang, D., Schnitzer, M.J., Scherrer, G., 2019. An amygdalar neural ensemble that encodes the unpleasantness of pain. Science (New York, N.Y.) 363 (6424), 276–281. https://doi.org/10.1126/science.aap8586.

Costanzo, R.M., Gardner, E.P., 1980. A quantitative analysis of responses of direction-sensitive neurons in somatosensory cortex of alert monkeys. J. Neurophysiol. 43, 1319–1341.

Craig, A.D., Bushnell, M.C., 1994. The thermal grill illusion: unmasking the burn of cold pain. Science 265, 252–255.

Davis, K.D., Hutchison, W.D., Lozano, A.M., Dostrovsky, J.O., 1994. Altered pain and temperature perception following cingulotomy and capsulotomy in a patient with schizoaffective disorder. Pain 59 (2), 189–199. https://doi.org/10.1016/0304-3959(94)90071-X.

Del Río, J., Naranjo, J.R., Yang, H.Y., Costa, E., 1983. Substance P-induced release of Met5-enkephalin from striatal and periaqueductal gray slices. Brain Res. 279 (1–2), 121–126. https://doi.org/10.1016/0006-8993(83)90169-5.

de Araujo, I.E., Simon, S.A., 2009. The gustatory cortex and multisensory integration. Int. J. Obes. 33 (Suppl. 2), S34–S43. Suppl 2.

De Biasi, S., Rustioni, A., 1990. Ultrastructural immunocytochemical localization of excitatory amino acids in the somatosensory system. J. Histochem. Cytochem. 38, 1745–1754.

De Felice, M., Eyde, N., Dodick, D., et al., 2013. Capturing the aversive state of cephalic pain preclinically. Ann. Neurol. 74, 257–265.

Dhaka, A., Viswanath, V., Patapoutian, A., 2006. Trp ion channels and temperature sensation. Annu. Rev. Neurosci. 29, 135–161.

DiCarlo, J.J., Johnson, K.O., Hsaio, S.S., 1998. Structure of receptive fields in area 3b of primary somatosensory cortex in the alert monkey. J. Neurosci. 18, 2626–2645.

Dickenson, A.H., Fardin, V., Le Bars, D., Besson, J.M., 1979. Antinociceptive action following microinjection of methionine-enkephalin in the nucleus raphe magnus of the rat. Neurosci. Lett. 15 (2–3), 265–270. https://doi.org/10.1016/0304-3940(79)96124-x52.

Dietsch, A.M., Westemeyer, R.M., Schultz, D.H., 2023. Brain activity associated with taste stimulation: a mechanism for neuroplastic change? Brain Behav. 13 (4), e2928. https://doi.org/10.1002/brb3.2928. Epub 2023 Mar 1. PMID: 36860129.

Drake, C.T., De Oliveira, A.X., Harris, J.A., Connor, D.M., Winkler, C.W., Aicher, S.A., 2007. Kappa opioid receptors in the rostral ventromedial medulla of male and female rats. J. Comp. Neurol. 500 (3), 465–476. https://doi.org/10.1002/cne.21184.

Edin, B.B., Abbs, J.H., 1991. Finger movement responses of cutaneous mechanoreceptors in the dorsal skin of the human hand. J. Neurophysiol. 65, 657–670.

Eippert, F., Tracey, I., 2014. Pain and the PAG: learning from painful mistakes. Nat. Neurosci. 17 (11), 1438–1439. https://doi.org/10.1038/nn.3844.

Eippert, F., Bingel, U., Schoell, E., Yacubian, J., Büchel, C., 2008. Blockade of endogenous opioid neurotransmission enhances acquisition of conditioned fear in humans. J. Neurosci. 28 (21), 5465–5472. https://doi.org/10.1523/JNEUROSCI.5336-07.2008.

Eippert, F., Bingel, U., Schoell, E.D., Yacubian, J., Klinger, R., Lorenz, J., Büchel, C., 2009. Activation of the opioidergic descending pain control system underlies placebo analgesia. Neuron 63 (4), 533–543. https://doi.org/10.1016/j.neuron.2009.07.014.

Fallon, J.H., Leslie, F.M., 1986. Distribution of dynorphin and enkephalin peptides in the rat brain. J. Comp. Neurol. 249 (3), 293–336. https://doi.org/10.1002/cne.902490302.

Fang, F.G., Haws, C.M., Drasner, K., Williamson, A., Fields, H.L., 1989. Opioid peptides (DAGO-enkephalin, dynorphin A(1-13), BAM 22P) microinjected into the rat brainstem: comparison of their antinociceptive effect and their effect on neuronal firing in the rostral ventromedial medulla. Brain Res. 501 (1), 116–128. https://doi.org/10.1016/0006-8993(89)91033-0.

Fanselow, M.S., Bolles, R.C., 1979. Naloxone and shock-elicited freezing in the rat. J. Comp. Physiol. Psychol. 93 (4), 736–744. https://doi.org/10.1037/h0077609.

Fardin, V., Oliveras, J.L., Besson, J.M., 1984. A reinvestigation of the analgesic effects induced by stimulation of the periaqueductal gray matter in the rat. II. Differential characteristics of the analgesia induced by ventral and dorsal PAG stimulation. Brain Res. 306 (1–2), 125–139. https://doi.org/10.1016/0006-8993(84)90361-5.

Felleman, D.J., Van Essen, D.C., 1991. Distributed hierarchical processing in the primate cerebral cortex. Cerebr. Cortex 1, 1–47.

Fields, H., 1987. Pain. McGraw-Hill, New York.

Fields, H.L., Basbaum, A.I., Heinricher, M.M., 2006. Wall and Melzack's Textbook of Pain. In: McMahon, S.B., Koltzenburg, M. (Eds.), Wall and Melzack's Textbook of Pain. UK: Elsevier Churchill Livingstone, 125–130.

Finley, J.C., Lindström, P., Petrusz, P., 1981. Immunocytochemical localization of beta-endorphin-containing neurons in the rat brain. Neuroendocrinology 33 (1), 28–42. https://doi.org/10.1159/000123197.

Finnegan, T.F., Chen, S.R., Pan, H.L., 2005. Effect of the {mu} opioid on excitatory and inhibitory synaptic inputs to periaqueductal gray-projecting neurons in the amygdala. J. Pharmacol. Exp. Therapeut. 312 (2), 441–448. https://doi.org/10.1124/jpet.104.074633.

Fitzgerald, P.J., Lane, J.W., Thakur, P.H., Hsiao, S.S., 2006. Receptive field properties of the macaque second somatosensory cortex: representation of orientation on different finger pads. J. Neurosci. 26, 6473–6484.

Flanagan, J.R., Vetter, P., Johansson, R.S., Wolpert, D.M., 2003. Prediction precedes control in motor learning. Curr. Biol. 13, 146–150.

Flores, J.A., El Banoua, F., Galán-Rodríguez, B., Fernandez-Espejo, E., 2004. Opiate anti-nociception is attenuated following lesion of large dopamine neurons of the periaqueductal grey: critical role for D1 (not D2) dopamine receptors. Pain 110 (1–2), 205–214. https://doi.org/10.1016/j.pain.2004.03.036.

Fogassi, L., Luppino, G., 2005. Motor functions of the parietal lobe. Curr. Opin. Neurobiol. 15, 626–631.

Foltz, E.L., White Jr., L.E., 1962. Pain 'relief' by frontal cingulumotomy. J. Neurosurg. 19, 89–100. https://doi.org/10.3171/jns.1962.19.2.0089.

François, A., Low, S.A., Sypek, E.I., Christensen, A.J., Sotoudeh, C., Beier, K.T., Ramakrishnan, C., Ritola, K.D., Sharif-Naeini, R., Deisseroth, K., Delp, S.L., Malenka, R.C., Luo, L., Hantman, A.W., Scherrer, G., 2017. A brainstem-spinal cord inhibitory circuit for mechanical pain modulation by GABA and enkephalins. Neuron 93 (4), 822–839.e6. https://doi.org/10.1016/j.neuron.2017.01.008.

Freund, H.J., 2003. Somatosensory and motor disturbances in patients with parietal lobe lesions. Adv. Neurol. 93, 179–193.

Fu, Y., Neugebauer, V., 2008. Differential mechanisms of CRF1 and CRF2 receptor functions in the amygdala in pain-related synaptic facilitation and behavior. J. Neurosci. 28 (15), 3861–3876. https://doi.org/10.1523/JNEUROSCI.0227-08.2008.

Gardner, E.P., Costanzo, R.M., 1981. Properties of kinesthetic neurons in somatosensory cortex of awake monkeys. Brain Res. 214 (2), 301–319. https://doi.org/10.1016/0006-8993(81)91196-3.

Gardner, E.P., 1988. Somatosensory Cortical Mechanisms of Feature Detection in Tactile and Kinesthetic Discrimination. Can. J. Physiol. Pharmacol., 66 439, 454.

Gardner, E.P., Babu, K.S., Ghosh, S., Sherwood, A., Chen, J., 2007. Neurophysiology of prehension: III. Representation of object features in posterior parietal cortex of the macaque monkey. J. Neurophysiol. 98, 3708–3730.

Gardner, E.P., 2008. Dorsal and ventral streams in the sense of touch. In: Kaas, J.H., Gardner, E.P. (Eds.), The Senses: A Comprehensive Reference, Somatosensation, vol. 6. Elsevier, Oxford, pp. 233–258.

Gavériaux-Ruff, C., Kieffer, B.L., 2002. Opioid receptor genes inactivated in mice: the highlights. Neuropeptides 36 (2–3), 62–71. https://doi.org/10.1054/npep.2002.0900.

Gebhart, G.F., Sandkühler, J., Thalhammer, J.G., Zimmermann, M., 1984. Inhibition in spinal cord of nociceptive information by electrical stimulation and morphine microinjection at identical sites in midbrain of the cat. J. Neurophysiol. 51 (1), 75–89. https://doi.org/10.1152/jn.1984.51.1.75.

Gilpin, N.W., Roberto, M., Koob, G.F., Schweitzer, P., 2014. Kappa opioid receptor activation decreases inhibitory transmission and antagonizes alcohol effects in rat central amygdala. Neuropharmacology 77, 294–302. https://doi.org/10.1016/j.neuropharm.2013.10.005.

Gray, M., Innala, L., Viau, V., 2014. Central vasopressin V1A receptor blockade alters patterns of cellular activation and prevents glucocorticoid habituation to repeated restraint stress exposure. Int. J. Neuropsychopharmacol. 17 (12), 2005–2015. https://doi.org/10.1017/S1461145714000935.

Gregoriou, G.C., Kissiwaa, S.A., Patel, S.D., Bagley, E.E., 2019. Dopamine and opioids inhibit synaptic outputs of the main island of the intercalated neurons of the amygdala. Eur. J. Neurosci. 50 (3), 2065–2074. https://doi.org/10.1111/ejn.14107.

Haber, S., Elde, R., 1982. The distribution of enkephalin immunoreactive fibers and terminals in the monkey central nervous system: an immunohistochemical study. Neuroscience 7 (5), 1049–1095. https://doi.org/10.1016/0306-4522(82)91118-6.

Gu, M., Wessendorf, M., 2007. Endomorphin-2-immunoreactive fibers selectively appose serotonergic neuronal somata in the rostral ventral medial medulla. J. Comp. Neurol. 502 (5), 701–713. https://doi.org/10.1002/cne.21343.

Gu, Z.H., Wang, B., Kou, Z.Z., Bai, Y., Chen, T., Dong, Y.L., Li, H., Li, Y.Q., 2017. Endomorphins: promising endogenous opioid peptides for the development of novel analgesics. Neurosignals 25 (1), 98–116. https://doi.org/10.1159/000484909.

Guthrie, J., Basbaum, A.I., 1984. Colocalization of immunoreactive proenkephalin and prodynorphin products in medullary neurons of the rat. Neuropeptides 4 (6), 437–445. https://doi.org/10.1016/0143-4179(84)90087-8.

Gutstein, H.B., Mansour, A., Watson, S.J., Akil, H., Fields, H.L., 1998. Mu and kappa opioid receptors in periaqueductal gray and rostral ventromedial medulla. Neuroreport 9 (8), 1777–1781. https://doi.org/10.1097/00001756-199806010-00019.

Han, L., Ma, C., Liu, Q., et al., 2013. A subpopulation of nociceptors specifically linked to itch. Nat. Neurosci. 16, 174–182.

Heinricher, M.M., 2003. Orphanin FQ/nociceptin: from neural circuitry to behavior. Life Sci. 73 (6), 813–822. https://doi.org/10.1016/s0024-3205(03)00412-0.

Heinricher, M.M., Ingram, S.L., 2008. The brainstem and nociceptive modulation. Pain 5, 593–626. https://doi.org/10.1016/B978-012370880-9.00183-3.

Heinricher, M.M., Neubert, M.J., 2004. Neural basis for the hyperalgesic action of cholecystokinin in the rostral ventromedial medulla. J. Neurophysiol. 92 (4), 1982–1989. https://doi.org/10.1152/jn.00411.2004.

Heinricher, M.M., Cheng, Z.F., Fields, H.L., 1987. Evidence for two classes of nociceptive modulating neurons in the periaqueductal gray. J. Neurosci. 7 (1), 271–278. https://doi.org/10.1523/JNEUROSCI.07-01-00271.1987.

Heinricher, M.M., Barbaro, N.M., Fields, H.L., 1989. Putative nociceptive modulating neurons in the rostral ventromedial medulla of the rat: firing of on- and off-cells is related to nociceptive responsiveness. Somatosens. Motor Res. 6 (4), 427–439. https://doi.org/10.3109/08990228909144685.

Heinricher, M.M., Morgan, M.M., Fields, H.L., 1992. Direct and indirect actions of morphine on medullary neurons that modulate nociception. Neuroscience 48 (3), 533–543. https://doi.org/10.1016/0306-4522(92)90400-v.

Heinricher, M.M., Morgan, M.M., Tortorici, V., Fields, H.L., 1994. Disinhibition of off-cells and antinociception produced by an opioid action within the rostral ventromedial medulla. Neuroscience 63 (1), 279–288. https://doi.org/10.1016/0306-4522(94)90022-1.

Heinricher, M.M., McGaraughty, S., Tortorici, V., 2001. Circuitry underlying antiopioid actions of cholecystokinin within the rostral ventromedial medulla. J. Neurophysiol. 85 (1), 280–286. https://doi.org/10.1152/jn.2001.85.1.280.

Heinricher, M.M., Martenson, M.E., Neubert, M.J., 2004. Prostaglandin E2 in the midbrain periaqueductal gray produces hyperalgesia and activates pain-modulating circuitry in the rostral ventromedial medulla. Pain 110 (1–2), 419–426. https://doi.org/10.1016/j.pain.2004.04.026.

Hellman, K.M., Brink, T.S., Mason, P., 2007. Activity of murine raphe magnus cells predicts tachypnea and on-going nociceptive responsiveness. J. Neurophysiol. 98 (6), 3121–3133. https://doi.org/10.1152/jn.00904.2007.

Helmstetter, F.J., Fanselow, M.S., 1987. Effects of naltrexone on learning and performance of conditional fear-induced freezing and opioid analgesia. Physiol. Behav. 39 (4), 501–505. https://doi.org/10.1016/0031-9384(87)90380-5.

Helmstetter, F.J., Bellgowan, P.S., Tershner, S.A., 1993. Inhibition of the tail flick reflex following microinjection of morphine into the amygdala. Neuroreport 4 (5), 471–474. https://doi.org/10.1097/00001756-199305000-00002.

Hermann, C., Hohmeister, J., Demirakça, S., Zohsel, K., Flor, H., 2006. Long-term alteration of pain sensitivity in school-aged children with early pain experiences. Pain 125 (3), 278–285. https://doi.org/10.1016/j.pain.2006.08.026.

Hikosaka, O., Tanaka, M., Sakamoto, M., Iwamura, Y., 1985. Deficits in manipulative behaviors induced by local injections of muscimol in the first somatosensory cortex of the conscious monkey. Brain Res. 325, 375–380.

Hinkley, L.B., Krubitzer, L.A., Nagarajan, S.S., Disbrow, E.A., 2007. Sensorimotor integration in S2, PV, and parietal rostroventral areas of the human Sylvian fissure. J. Neurophysiol. 97, 1288–1297.

Hosobuchi, Y., Wemmer, J., 1977. Disulfiram inhibition of development of tolerance to analgesia induced by central gray stimulation in humans. Eur. J. Pharmacol. 43 (4), 385–387. https://doi.org/10.1016/0014-2999(77)90046-2.

Hosobuchi, Y., 1986. Subcortical electrical stimulation for control of intractable pain in humans: report of 122 cases 1970–1984. J. Neurosurg. 64, 543–553.

Huang, J., Gadotti, V.M., Chen, L., Souza, I.A., Huang, S., Wang, D., Ramakrishnan, C., Deisseroth, K., Zhang, Z., Zamponi, G.W., 2019. A neuronal circuit for activating descending modulation of neuropathic pain. Nat. Neurosci. 22 (10), 1659–1668. https://doi.org/10.1038/s41593-019-0481-5.

Hurley, R.W., Hammond, D.L., 2000. The analgesic effects of supraspinal mu and delta opioid receptor agonists are potentiated during persistent inflammation. J. Neurosci. 20 (3), 1249–1259. https://doi.org/10.1523/JNEUROSCI.20-03-01249.2000.

Hurley, R.W., Hammond, D.L., 2001. Contribution of endogenous enkephalins to the enhanced analgesic effects of supraspinal mu opioid receptor agonists after inflammatory injury. J. Neurosci. 21 (7), 2536–2545. https://doi.org/10.1523/JNEUROSCI.21-07-02536.2001.

Ibata, Y., Kawakami, F., Okamura, H., Obata-Tsuto, H.L., Morimoto, N., Zimmerman, E.A., 1985. Light and electron microscopic immunocytochemistry of beta-endorphin/beta-LPH-like immunoreactive neurons in the arcuate nucleus and surrounding areas of the rat hypothalamus. Brain Res. 341 (2), 233–242. https://doi.org/10.1016/0006-8993(85)91062-5.

Ingram, S.L., Vaughan, C.W., Bagley, E.E., Connor, M., Christie, M.J., 1998. Enhanced opioid efficacy in opioid dependence is caused by an altered signal transduction pathway. J. Neurosci. 18 (24), 10269–10276. https://doi.org/10.1523/JNEUROSCI.18-24-10269.1998.

Ingram, S.L., Fossum, E.N., Morgan, M.M., 2007. Behavioral and electrophysiological evidence for opioid tolerance in adolescent rats. Neuropsychopharmacology 32 (3), 600–606. https://doi.org/10.1038/sj.npp.1301139.

Iremonger, K.J., Bains, J.S., 2009. Retrograde opioid signaling regulates glutamatergic transmission in the hypothalamus. J. Neurosci. 29 (22), 7349–7358. https://doi.org/10.1523/JNEUROSCI.0381-09.2009.

Jacobsen, K.X., Höistad, M., Staines, W.A., Fuxe, K., 2006. The distribution of dopamine D1 receptor and mu-opioid receptor 1 receptor immunoreactivities in the amygdala and interstitial nucleus of the posterior limb of the anterior commissure: relationships to tyrosine hydroxylase and opioid peptide terminal systems. Neuroscience 141 (4), 2007–2018. https://doi.org/10.1016/j.neuroscience.2006.05.05410.

Jenmalm, P., Birznieks, I., Goodwin, A.W., Johansson, R.S., 2003. Influence of object shape on responses of human tactile afferents under conditions characteristic of manipulation. Eur. J. Neurosci. 18, 164–176.

Johansson, R.S., Landstr, m U., Lundstr, m R., 1982. Responses of mechanoreceptive afferent units in the glabrous skin of the human hand to sinusoidal skin displacements. Brain Res. 244, 17–25.

Johnson, K.O., Phillips, J.R., 1981. Tactile spatial resolution: I. Two-point discrimination, gap detection, grating resolution and letter recognition. J. Neurophysiol. 46, 1177–1191.

Johansen, J.P., Fields, H.L., Manning, B.H., 2001. The affective component of pain in rodents: direct evidence for a contribution of the anterior cingulate cortex. Proc. Natl. Acad. Sci. USA 98 (14), 8077–8082. https://doi.org/10.1073/pnas.141218998130.

Johansen, J.P., Tarpley, J.W., LeDoux, J.E., Blair, H.T., 2010. Neural substrates for expectation-modulated fear learning in the amygdala and periaqueductal gray. Nat. Neurosci. 13 (8), 979–986. https://doi.org/10.1038/nn.2594103.

Jones, E.G., Powell, T.P.S., 1969. Connexions of the somatic sensory cortex of the rhesus monkey. I. Ipsilateral cortical connexions. Brain 92, 477–502.

Jordt, S.E., McKemy, D.D., Julius, D., 2003. Lessons from peppers and peppermint: the molecular basis of thermosensation. Curr. Opin. Neurobiol. 13, 487–492.

Kaas, J.H., Qi, H.X., Burish, M.J., Gharbawie, O.A., Onifer, S.M., Massey, J.M., 2008. Cortical and subcortical plasticity in the brains of humans, primates, and rats after damage to sensory afferents in the dorsal columns of the spinal cord. Exp. Neurol. 209 (2), 407–416. https://doi.org/10.1016/j.expneurol.2007.06.014.

Kalyuzhny, A.E., Wessendorf, M.W., 1998. Relationship of mu- and delta-opioid receptors to GABAergic neurons in the central nervous system, including antinociceptive brainstem circuits. J. Comp. Neurol. 392 (4), 528–547.

Kalyuzhny, A.E., Arvidsson, U., Wu, W., Wessendorf, M.W., 1996. mu-Opioid and delta-opioid receptors are expressed in brainstem antinociceptive circuits: studies using immunocytochemistry and retrograde tract-tracing. J. Neurosci. 16 (20), 6490–6503. https://doi.org/10.1523/JNEUROSCI.16-20-06490.1996.

Keay, K.A., Bandler, R., 2001. Parallel circuits mediating distinct emotional coping reactions to different types of stress. Neurosci. Biobehav. Rev. 25 (7–8), 669–678. https://doi.org/10.1016/s0149-7634(01)00049-5.

Khachaturian, H., Lewis, M.E., Watson, S.J., 1983. Enkephalin systems in diencephalon and brainstem of the rat. J. Comp. Neurol. 220 (3), 310–320. https://doi.org/10.1002/cne.902200305.

Kim, J.J., Cunnington, R., Kirby, J.N., 2020. The neurophysiological basis of compassion: an fMRI meta-analysis of compassion and its related neural processes. Neurosci. Biobehav. Rev. 108, 112–123. https://doi.org/10.1016/j.neubiorev.2019.10.023.

Kissiwaa, S.A., Bagley, E.E., 2018. Central sensitization of the spino-parabrachial-amygdala pathway that outlasts a brief nociceptive stimulus. J. Physiol. 596 (18), 4457–4473. https://doi.org/10.1113/JP273976.

Koch, K.W., Fuster, J.M., 1989. Unit activity in monkey parietal cortex related to haptic perception and temporary memory. Exp. Brain Res. 76, 292–306.

Kong, J., Tu, P.C., Zyloney, C., Su, T.P., 2010. Intrinsic functional connectivity of the periaqueductal gray, a resting fMRI study. Behav. Brain Res. 211 (2), 215–219. https://doi.org/10.1016/j.bbr.2010.03.042.

Kuner, R., Flor, H., 2016. Structural plasticity and reorganization in chronic pain. Nat. Rev. Neurosci. 18, 20–30.

Külling, P., Siegfried, B., Frischknecht, H.R., Messiha, F.S., Pasi, A., 1989. Beta-endorphin-like immunoreactivity levels in the hypothalamus, the periaqueductal grey and the pituitary of the DBA mouse: determination by ELISA and relationship to nociception. Physiol. Behav. 46 (1), 25–28. https://doi.org/10.1016/0031-9384(89)90315-6.

Kwok, C.H.T., Devonshire, I.M., Bennett, A.J., Hathway, G.J., 2014. Postnatal maturation of endogenous opioid systems within the periaqueductal grey and spinal dorsal horn of the rat. Pain 155 (1), 168–178. https://doi.org/10.1016/j.pain.2013.09.022.

LaMotte, R.H., Mountcastle, V.B., 1979. Disorders in somethesis following lesions of parietal lobe. J. Neurophysiol. 42, 400–419.

Lederman, S.J., Klatzky, R.L., 1987. Hand movements: a window into haptic object recognition. Cognit. Psychol. 19, 342–368.

Lenz, F.A., Gracely, R.H., Romanoski, A.J., Hope, E.J., Rowland, L.H., Dougherty, P.M., 1995. Stimulation in the human somatosensory thalamus can reproduce both the affective and sensory dimensions of previously experienced pain. Nat. Med. 1, 910–913.

LaPrairie, J.L., Murphy, A.Z., 2007. Female rats are more vulnerable to the long-term consequences of neonatal inflammatory injury. Pain 132 (Suppl. 1), S124–S133. https://doi.org/10.1016/j.pain.2007.08.010.

Laprairie, J.L., Murphy, A.Z., 2009. Neonatal injury alters adult pain sensitivity by increasing opioid tone in the periaqueductal gray. Front. Behav. Neurosci. 3, 31. https://doi.org/10.3389/neuro.08.031.2009.

Lieber, J.D., Xia, X., Weber, A.I., Bensmaia, S.J., 2017. The neural code for tactile roughness in the somatosensory nerves. J. Neurophysiol. 118, 3107–3117.

Lau, B.K., Vaughan, C.W., 2014. Descending modulation of pain: the GABA disinhibition hypothesis of analgesia. Curr. Opin. Neurobiol. 29, 159–164. https://doi.org/10.1016/j.conb.2014.07.010.

Lau, B.K., Winters, B.L., Vaughan, C.W., 2020. Opioid presynaptic disinhibition of the midbrain periaqueductal grey descending analgesic pathway. Br. J. Pharmacol. 177 (10), 2320–2332. https://doi.org/10.1111/bph.14982.

Li, Y.Q., Jia, H.G., Rao, Z.R., Shi, J.W., 1990. Serotonin-, substance P- or leucine-enkephalin-containing neurons in the midbrain periaqueductal gray and nucleus raphe dorsalis send projection fibers to the central amygdaloid nucleus in the rat. Neurosci. Lett. 120 (1), 124–127. https://doi.org/10.1016/0304-3940(90)90184-b.

Lonstein, J.S., Stern, J.M., 1997. Role of the midbrain periaqueductal gray in maternal nurturance and aggression: c-fos and electrolytic lesion studies in lactating rats. J. Neurosci. 17 (9), 3364–3378. https://doi.org/10.1523/JNEUROSCI.17-09-03364.1997.

Loyd, D.R., Murphy, A.Z., 2009. The role of the periaqueductal gray in the modulation of pain in males and females: are the anatomy and physiology really that different? Neural Plast. 2009, 462879. https://doi.org/10.1155/2009/462879.

Loyd, D.R., Murphy, A.Z., 2014. The neuroanatomy of sexual dimorphism in opioid analgesia. Exp. Neurol. 259, 57–63. https://doi.org/10.1016/j.expneurol.2014.04.004.

Loyd, D.R., Morgan, M.M., Murphy, A.Z., 2007. Morphine preferentially activates the periaqueductal gray-rostral ventromedial medullary pathway in the male rat: a potential mechanism for sex differences in antinociception. Neuroscience 147 (2), 456–468. https://doi.org/10.1016/j.neuroscience.2007.03.053.

Loyd, D.R., Wang, X., Murphy, A.Z., 2008. Sex differences in micro-opioid receptor expression in the rat midbrain periaqueductal gray are essential for eliciting sex differences in morphine analgesia. J. Neurosci. 28 (52), 14007–14017. https://doi.org/10.1523/JNEUROSCI.4123-08.2008.

Mainero, C., Boshyan, J., Hadjikhani, N., 2011. Altered functional magnetic resonance imaging resting-state connectivity in periaqueductal gray networks in migraine. Ann. Neurol. 70 (5), 838–845. https://doi.org/10.1002/ana.22537.

Manfredi, L.R., Saal, H.P., Brown, K.J., et al., 2014. Natural scenes in tactile texture. J. Neurophysiol. 111, 1792–1802.

Manning, B.H., Franklin, K.B., 1998. Morphine analgesia in the formalin test: reversal by microinjection of quaternary naloxone into the posterior hypothalamic area or periaqueductal gray. Behav. Brain Res. 92 (1), 97–102. https://doi.org/10.1016/s0166-4328(97)00130-7.

Mantyh, P.W., Rogers, S.D., Honore, P., et al., 1997. Inhibition of hyperalgesia by ablation of lamina I spinal neurons expressing the substance P receptor. Science 278, 275–279.

Marinelli, S., Vaughan, C.W., Schnell, S.A., Wessendorf, M.W., Christie, M.J., 2002. Rostral ventromedial medulla neurons that project to the spinal cord express multiple opioid receptor phenotypes. J. Neurosci. 22 (24), 10847–10855. https://doi.org/10.1523/JNEUROSCI.22-24-10847.2002.

Marinelli, S., Connor, M., Schnell, S.A., Christie, M.J., Wessendorf, M.W., Vaughan, C.W., 2005. delta-opioid receptor-mediated actions on rostral ventromedial medulla neurons. Neuroscience 132 (2), 239–244. https://doi.org/10.1016/j.neuroscience.2005.01.015.

Martin-Schild, S., Gerall, A.A., Kastin, A.J., Zadina, J.E., 1999. Differential distribution of endomorphin 1- and endomorphin 2-like immunoreactivities in the CNS of the rodent. J. Comp. Neurol. 405 (4), 450–471.

Mason, P., Back, S.A., Fields, H.L., 1992. A confocal laser microscopic study of enkephalin-immunoreactive appositions onto physiologically identified neurons in the rostral ventromedial medulla. J. Neurosci. 12 (10), 4023–4036. https://doi.org/10.1523/JNEUROSCI.12-10-04023.1992.

Mayer, D.J., 1984. Analgesia produced by electrical stimulation of the brain. Prog. Neuropsychopharmacol. Biol. Psychiatry 8 (4–6), 557–564. https://doi.org/10.1016/0278-5846(84)90015-0.

McDonald, A.J., Mascagni, F., 1997. Projections of the lateral entorhinal cortex to the amygdala: a *Phaseolus vulgaris* leucoagglutinin study in the rat. Neuroscience 77 (2), 445–459. https://doi.org/10.1016/s0306-4522(96)00478-2.

McKnight, A.T., Corbett, A.D., Paterson, S.J., Magnan, J., Kosterlitz, H.W., 1982. Comparison of in vitro potencies in pharmacological and binding assays after inhibition of peptidases reveals that dynorphin (1-9) is a potent kappa-agonist. Life Sci. 31 (16–17), 1725–1728. https://doi.org/10.1016/0024-3205(82)90195-3.

McNally, G.P., Cole, S., 2006. Opioid receptors in the midbrain periaqueductal gray regulate prediction errors during pavlovian fear conditioning. Behav. Neurosci. 120 (2), 313–323. https://doi.org/10.1037/0735-7044.120.2.313.

Matthes, H.W., Maldonado, R., Simonin, F., et al., 1996. Loss of morphine-induced analgesia, reward effect and withdrawal symptoms in mice lacking the μ-opioid-receptor gene. Nature 383, 819–823.

McMahon, D.B., Jolivert, J.F., Kuek, L.E., Adappa, N.D., Palmer, J.N., Lee, R.J., 2023. Savory signaling: T1R umami receptor modulates endoplasmic reticulum calcium store content and release dynamics in airway epithelial cells. Nutrients 15 (3), 493.

Meeker, T.J., Schmid, A.C., Keaser, M.L., Khan, S.A., Gullapalli, R.P., Dorsey, S.G., Greenspan, J.D., Seminowicz, D.A., 2022. Tonic pain alters functional connectivity of the descending pain modulatory network involving amygdala, periaqueductal gray, parabrachial nucleus and anterior cingulate cortex. Neuroimage 256, 119278. https://doi.org/10.1016/j.neuroimage.2022.119278.

Melzack, R., Wall, P.D., 1965. Pain mechanisms: a new theory. Science 150, 971–979.

Menetrey, D., Basbaum, A.I., 1987. The distribution of substance P-, enkephalin- and dynorphin-immunoreactive neurons in the medulla of the rat and their contribution to bulbospinal pathways. Neuroscience 23 (1), 173–187. https://doi.org/10.1016/0306-4522(87)90281-8.

Meng, I.D., Johansen, J.P., Harasawa, I., Fields, H.L., 2005. Kappa opioids inhibit physiologically identified medullary pain modulating neurons and reduce morphine antinociception. J. Neurophysiol. 93 (3), 1138–1144. https://doi.org/10.1152/jn.00320.2004.

Merzenich, M.M., Jenkins, W.M., 1993. Reorganization of cortical representations of the hand following alterations of skin inputs induced by nerve injury, skin island transfers, and experience. J. Hand Ther. 6, 89–104.

Meyer, P.J., Morgan, M.M., Kozell, L.B., Ingram, S.L., 2009. Contribution of dopamine receptors to periaqueductal gray-mediated antinociception. Psychopharmacology 204 (3), 531–540. https://doi.org/10.1007/s00213-009-1482-y.

Miller, M.M., 2006. Heroin detoxification. JAMA 295 (8), 885–886. https://doi.org/10.1001/jama.295.8.885-b.

Mills, E.P., Di Pietro, F., Alshelh, Z., Peck, C.C., Murray, G.M., Vickers, E.R., Henderson, L.A., 2018. Brainstem pain-control circuitry connectivity in chronic neuropathic pain. J. Neurosci. 38 (2), 465–473. https://doi.org/10.1523/JNEUROSCI.1647-17.2017.

Moga, M.M., Weis, R.P., Moore, R.Y., 1995. Efferent projections of the paraventricular thalamic nucleus in the rat. J. Comp. Neurol. 359 (2), 221–238. https://doi.org/10.1002/cne.903590204.

Moreau, J.L., Fields, H.L., 1986. Evidence for GABA involvement in midbrain control of medullary neurons that modulate nociceptive transmission. Brain Res. 397 (1), 37–46. https://doi.org/10.1016/0006-8993(86)91367-33.

Morgan, M.M., Heinricher, M.M., Fields, H.L., 1992. Circuitry linking opioid-sensitive nociceptive modulatory systems in periaqueductal gray and spinal cord with rostral ventromedial medulla. Neuroscience 47 (4), 863–871. https://doi.org/10.1016/0306-4522(92)90036-2.

Morgan, M.M., Clayton, C.C., Lane, D.A., 2003. Behavioral evidence linking opioid-sensitive GABAergic neurons in the ventrolateral periaqueductal gray to morphine tolerance. Neuroscience 118 (1), 227–232. https://doi.org/10.1016/s0306-4522(02)00822-9.

Morgan, M.M., Whittier, K.L., Hegarty, D.M., Aicher, S.A., 2008. Periaqueductal gray neurons project to spinally projecting GABAergic neurons in the rostral ventromedial medulla. Pain 140 (2), 376–386. https://doi.org/10.1016/j.pain.2008.09.009.

Moss, M.S., Glazer, E.J., Basbaum, A.I., 1983. The peptidergic organization of the cat periaqueductal gray. I. The distribution of immunoreactive enkephalin-containing neurons and terminals. J. Neurosci. 3 (3), 603–616. https://doi.org/10.1523/JNEUROSCI.03-03-00603.1983153.

Mountcastle, V.B., 1997. The columnar organization of the neocortex. Brain 120 (Pt 4), 701–722. https://doi.org/10.1093/brain/120.4.701.

Mountcastle, V.B., 2005. The Sensory Hand: Neural Mechanisms of Somatic Sensation. Harvard University Press, p. 156.

Mountcastle, V.B., LaMotte, R.H., Carli, G., 1972. Detection thresholds for stimuli in humans and monkeys: comparison with threshold events in mechanoreceptive afferent fibers innervating the monkey hand. J. Neurophysiol. 35, 122–136.

Mountcastle, V.B., Lynch, J.C., Georgopoulos, A.P., Sakata, H., Acuna, C., 1975. Posterior parietal association cortex of the monkey: command functions for operations within extrapersonal space. J. Neurophysiol. 38, 871–908.

Muniak, M.A., Ray, S., Hsiao, S.S., Dammann, J.F., Bensmaia, S.J., 2007. The neural coding of stimulus intensity: linking the population response of mechanoreceptive afferents with psychophysical behavior. J. Neurosci. 27, 11687–11699.

Murray, E.A., Mishkin, M., 1984. Relative contributions of SII and area 5 to tactile discrimination in monkeys. Behav. Brain Res. 11, 67 83.

Nakamura, T., Tomida, M., Yamamoto, T., Ando, H., Takamata, T., Kondo, E., Kurasawa, I., Asanuma, N., 2013. The endogenous opioids related with antinociceptive effects induced by electrical stimulation into the amygdala. Open Dent. J. 7, 27–35. https://doi.org/10.2174/1874210601307010027.

Narita, M., Mizoguchi, H., Narita, M., Dun, N.J., Hwang, B.H., Endoh, T., Suzuki, T., Nagase, H., Suzuki, T., Tseng, L.F., 2000. G protein activation by endomorphins in the mouse periaqueductal gray matter. J. Biomed. Sci. 7 (3), 221–225. https://doi.org/10.1007/BF02255469.

Neugebauer, V., Li, W., 2002. Processing of nociceptive mechanical and thermal information in central amygdala neurons with knee-joint input. J. Neurophysiol. 87 (1), 103–112. https://doi.org/10.1152/jn.00264.2001.

Nelson, R.J., Sur, M., Felleman, D.J., Kaas, J.H., 1980. Representations of the body surface in postcentral parietal cortex of *Macaca fascicularis*. J. Comp. Neurol. 192, 611–643.

Nolano, M., Provitera, V., Crisci, C., et al., 2003. Quantification of myelinated endings and mechanoreceptors in human digital skin. Ann. Neurol. 54, 197–205.

Oberlaender, M., de Kock, C.P., Bruno, R.M., et al., 2012. Cell typespecific three-dimensional structure of thalamocortical circuits in a column of rat vibrissal cortex. Cerebr. Cortex 22, 2375–2391.

Oliveira, M.A., Prado, W.A., 2001. Role of PAG in the antinociception evoked from the medial or central amygdala in rats. Brain Res. Bull. 54 (1), 55–63. https://doi.org/10.1016/s0361-9230(00)00420-2.

Padilla, L.M., Creem-Regehr, S.H., Hegarty, M., Stefanucci, J.K., 2018. Decision making with visualizations: a cognitive framework across disciplines. Cogn. Res. Princ. Implic. 3 (1), 29.

Palmiter, R.D., 2018. The parabrachial nucleus: CGRP neurons function as a general alarm. Trends Neurosci. 41 (5), 280–293. https://doi.org/10.1016/j.tins.2018.03.007.

Pan, Z.Z., Fields, H.L., 1996. Endogenous opioid-mediated inhibition of putative pain-modulating neurons in rat rostral ventromedial medulla. Neuroscience 74 (3), 855–862. https://doi.org/10.1016/0306-4522(96)00179-0.

Pan, Z.Z., Williams, J.T., Osborne, P.B., 1990. Opioid actions on single nucleus raphe magnus neurons from rat and Guinea-pig in vitro. J. Physiol. 427, 519–532. https://doi.org/10.1113/jphysiol.1990.sp018185.

Pan, Z.Z., Tershner, S.A., Fields, H.L., 1997. Cellular mechanism for anti-analgesic action of agonists of the kappa-opioid receptor. Nature 389 (6649), 382–385. https://doi.org/10.1038/38730.

Pandya, D.N., Seltzer, B., 1982. Intrinsic connections and architectonics of posterior parietal cortex in the rhesus monkey. J. Comp. Neurol. 204, 196–210.

Pause, M., Kunesch, E., Binkofski, F., Freund, H.-J., 1989. Sensorimotor disturbances in patients with lesions of the parietal cortex. Brain 112, 1599–1625.

Park, C., Kim, J.H., Yoon, B.E., Choi, E.J., Lee, C.J., Shin, H.S., 2010. T-type channels control the opioidergic descending analgesia at the low threshold-spiking GABAergic neurons in the periaqueductal gray. Proc. Natl. Acad. Sci. USA 107 (33), 14857–14862. https://doi.org/10.1073/pnas.1009532107.

Pavlovic, Z.W., Cooper, M.L., Bodnar, R.J., 1996. Opioid antagonists in the periaqueductal gray inhibit morphine and beta-endorphin analgesia elicited from the amygdala of rats. Brain Res. 741 (1–2), 13–26. https://doi.org/10.1016/s0006-8993(96)00880-3.

Peciña, M., Azhar, H., Love, T.M., Lu, T., Fredrickson, B.L., Stohler, C.S., Zubieta, J.K., 2013. Personality trait predictors of placebo analgesia and neurobiological correlates. Neuropsychopharmacology 38 (4), 639–646. https://doi.org/10.1038/npp.2012.227.

Perl, E.R., 2007. Ideas about pain, a historical review. Nat. Rev. Neurosci. 8, 71–80.

Peters, R.M., Hackeman, E., Goldreich, D., 2009. Diminutive digits discern delicate details: fingertip size and the sex difference in tactile spatial acuity. J. Neurosci. 29, 15756–15761.

Phillips, J.R., Johansson, R.S., Johnson, K.O., 1990. Representation of braille characters in human nerve fibres. Exp. Brain Res. 81, 589–592.

Pons, T.P., Garraghty, P.E., Ommaya, A.K., Kaas, J.H., Taub, E., Mishkin, M., 1991. Massive cortical reorganization after sensory deafferentation in adult macaques. Science 252, 1857–1860.

Pons, T.P., Garraghty, P.E., Mishkin, M., 1992. Serial and parallel processing of tactual information in somatosensory cortex of rhesus monkeys. J. Neurophysiol. 68, 518–527.

Porreca, F., Ossipov, M.H., Gebhart, G.F., 2002. Chronic pain and medullary descending facilitation. Trends Neurosci. 25 (6), 319–325. https://doi.org/10.1016/s0166-2236(02)02157-4.

Poulin, J.F., Chevalier, B., Laforest, S., Drolet, G., 2006. Enkephalinergic afferents of the centromedial amygdala in the rat. J. Comp. Neurol. 496 (6), 859–876. https://doi.org/10.1002/cne.20956.

Pruszynski, J.A., Johansson, R.S., 2014. Edge-orientation processing in first-order tactile neurons. Nat. Neurosci. 17, 1404–1409.

Quilliam, T.A., 1978. The structure of finger print skin. In: Gordon, G. (Ed.), Active Touch. Pergamon Press, Oxford, pp. 1–18.

Raja, S.N., Campbell, J.N., Meyer, R.A., 1984. Evidence for different mechanisms of primary and secondary hyperalgesia following heat injury to the glabrous skin. Brain 107, 1179–1188.

Rhoton Jr, A.L., 2019. Rhoton's Cranial Anatomy and Surgical Approaches. Oxford University Press, p. 174.

Robinson, C.J., Burton, H., 1980. Somatic submodality distribution within the second somatosensory (SII), 7b, retroinsular, postauditory and granular insular cortical areas of *M. fascicularis*. J. Comp. Neurol. 192, 93–108.

Romo, R., Hernandez, A., Zainos, A., Lemus, L., Brody, C.D., 2002. Neuronal correlates of decision-making in secondary somatosensory cortex. Nat. Neurosci. 5, 1217–1235.

Roy, M., Shohamy, D., Daw, N., Jepma, M., Wimmer, G.E., Wager, T.D., 2014. Representation of aversive prediction errors in the human periaqueductal gray. Nat. Neurosci. 17 (11), 1607–1612. https://doi.org/10.1038/nn.3832.

Saal, H.P., Bensmaia, S.J., 2014. Touch is a team effort: interplay of submodalities in cutaneous sensibility. Trends Neurosci. 37, 689–697.

Salinas, E., Hernandez, A., Zainos, A., Romo, R., 2000. Periodicity and firing rate as candidate neural codes for the frequency of vibrotactile stimuli. J. Neurosci. 20, 5503–5515.

Samineni, V.K., Grajales-Reyes, J.G., Copits, B.A., O'Brien, D.E., Trigg, S.L., Gomez, A.M., Bruchas, M.R., Gereau, R.W., 2017. Divergent modulation of nociception by glutamatergic and GABAergic neuronal subpopulations in the periaqueductal gray. eNeuro 4 (2). https://doi.org/10.1523/ENEURO.0129-16.2017. ENEURO.0129-16.2017.

Sandkühler, J., Gebhart, G.F., 1984. Relative contributions of the nucleus raphe magnus and adjacent medullary reticular formation to the inhibition by stimulation in the periaqueductal gray of a spinal nociceptive reflex in the pentobarbital-anesthetized rat. Brain Res. 305 (1), 77–87. https://doi.org/10.1016/0006-8993(84)91121-1.

Schepers, R.J., Mahoney, J.L., Gehrke, B.J., Shippenberg, T.S., 2008a. Endogenous kappa-opioid receptor systems inhibit hyperalgesia associated with localized peripheral inflammation. Pain 138 (2), 423–439. https://doi.org/10.1016/j.pain.2008.01.023.

Schepers, R.J., Mahoney, J.L., Shippenberg, T.S., 2008b. Inflammation-induced changes in rostral ventromedial medulla mu and kappa opioid receptor mediated antinociception. Pain 136 (3), 320–330. https://doi.org/10.1016/j.pain.2007.07.010.

Sessle, B.J., Ball, G.J., Lucier, G.E., 1981. Suppressive influences from periaqueductal gray and nucleus raphe magnus on respiration and related reflex activities and on solitary tract neurons, and effect of naloxone. Brain Res. 216 (1), 145–161. https://doi.org/10.1016/0006-8993(81)91283-x.

Silva, C., McNaughton, N., 2019. Are periaqueductal gray and dorsal raphe the foundation of appetitive and aversive control? A comprehensive review. Progr. Neurobiol. 177, 33–72. https://doi.org/10.1016/j.pneurobio.2019.02.001.

Sim, L.J., Joseph, S.A., 1991. Arcuate nucleus projections to brainstem regions which modulate nociception. J. Chem. Neuroanat. 4 (2), 97–109. https://doi.org/10.1016/0891-0618(91)90034-a.

Siuciak, J.A., Advokat, C., 1987. Tolerance to morphine microinjections in the periaqueductal gray (PAG) induces tolerance to systemic, but not intrathecal morphine. Brain Res. 424 (2), 311–319. https://doi.org/10.1016/0006-8993(87)91476-4.

Snider, W.D., 1998. How do you feel? Neurotrophins and mechanotransduction. Nat. Neurosci. 1, 5–6.

Srinivasan, M.A., Whitehouse, J.M., LaMotte, R.H., 1990. Tactile detection of slip: surface microgeometry and peripheral neural codes. J. Neurophysiol. 63, 1323–1332.

Sripati, A.P., Yoshioka, T., Denchev, P., Hsiao, S.S., Johnson, K.O., 2006. Spatiotemporal receptive fields of peripheral afferents and cortical area 3b and 1 neurons in the primate somatosensory system. J. Neurosci. 26, 2101–2114.

Suckow, S.K., Deichsel, E.L., Ingram, S.L., Morgan, M.M., Aicher, S.A., 2013. Columnar distribution of catecholaminergic neurons in the ventrolateral periaqueductal gray and their relationship to efferent pathways. Synapse (New York, N.Y.) 67 (2), 94–108. https://doi.org/10.1002/syn.21624.

Sukikara, M.H., Mota-Ortiz, S.R., Baldo, M.V., Felício, L.F., Canteras, N.S., 2006. A role for the periaqueductal gray in switching adaptive behavioral responses. J. Neurosci. 26 (9), 2583–2589. https://doi.org/10.1523/JNEUROSCI.4279-05.2006.

Sun, R.Q., Wang, Y., Zhao, C.S., Chang, J.K., Han, J.S., 2001. Changes in brain content of nociceptin/orphanin FQ and endomorphin 2 in a rat model of neuropathic pain. Neurosci. Lett. 311 (1), 13–16. https://doi.org/10.1016/s0304-3940(01)02095-x.

Sun, Y.G., Gu, X.L., Yu, L.C., 2007. The neural pathway of galanin in the hypothalamic arcuate nucleus of rats: activation of beta-endorphinergic neurons projecting to periaqueductal gray matter. J. Neurosci. Res. 85 (11), 2400–2406. https://doi.org/10.1002/jnr.21396.

Talbot, W.H., Darian-Smith, I., Kornhuber, H.H., Mountcastle, V.B., 1968. The sense of flutter-vibration: comparison of the human capacity with response patterns of mechanoreceptive afferents from the monkey hand. J. Neurophysiol. 31, 301–334.

Talbot, J.D., Marrett, S., Evans, A.C., Meyer, E., Bushnell, M.C., Duncan, G.H., 1991. Multiple representations of pain in human cerebral cortex. Science 251, 1355–1358.

Talbot, J.D., Villemure, J.G., Bushnell, M.C., Duncan, G.H., 1995. Evaluation of pain perception after anterior capsulotomy: a case report. Somatosens. Motor Res. 12 (2), 115–126. https://doi.org/10.3109/08990229509101503.

Todd, A.J., 2010. Neuronal circuitry for pain processing in the dorsal horn. Nat. Rev. Neurosci. 11, 823–836.

Tominaga, M., Caterina, M.J., 2004. Thermosensation and pain. J. Neurobiol. 61, 3–12.

Tonsfeldt, K.J., Suchland, K.L., Beeson, K.A., Lowe, J.D., Li, M.H., Ingram, S.L., 2016. Sex differences in GABAA signaling in the periaqueductal gray induced by persistent inflammation. J. Neurosci. 36 (5), 1669–1681. https://doi.org/10.1523/JNEUROSCI.1928-15.2016.

Tortorici, V., Morgan, M.M., 2002. Comparison of morphine and kainic acid microinjections into identical PAG sites on the activity of RVM neurons. J. Neurophysiol. 88 (4), 1707–1715. https://doi.org/10.1152/jn.2002.88.4.1707.

Tortorici, V., Morgan, M.M., Vanegas, H., 2001. Tolerance to repeated microinjection of morphine into the periaqueductal gray is associated with changes in the behavior of off- and on-cells in the rostral ventromedial medulla of rats. Pain 89 (2–3), 237–244. https://doi.org/10.1016/s0304-3959(00)00367-5.

Tracey, I., Mantyh, P.W., 2007. The cerebral signature for pain perception and its modulation. Neuron 55, 377–391.

Truini, A., Tinelli, E., Gerardi, M.C., Calistri, V., Iannuccelli, C., La Cesa, S., Tarsitani, L., Mainero, C., Sarzi-Puttini, P., Cruccu, G., Caramia, F., Di Franco, M., 2016. Abnormal resting state functional connectivity of the periaqueductal grey in patients with fibromyalgia. Clin. Exp. Rheumatol. 34 (2 Suppl. 96), S129–S133.

Tso, A.R., Goadsby, P.J., 2017. Anti-CGRP monoclonal antibodies: he next era of migraine prevention? Curr. Treat. Options Neurol. 19, 27.

Urban, M.O., Gebhart, G.F., 1999. Central mechanisms in pain. Med. Clin. 83 (3), 585–596. https://doi.org/10.1016/s0025-7125(05)70125-5.

Valverde, O., Fournié-Zaluski, M.C., Roques, B.P., Maldonado, R., 1996. Similar involvement of several brain areas in the antinociception of endogenous and exogenous opioids. Eur. J. Pharmacol. 312 (1), 15–25. https://doi.org/10.1016/0014-2999(96)00437-2.

Vaughan, C.W., Christie, M.J., 1997. Presynaptic inhibitory action of opioids on synaptic transmission in the rat periaqueductal grey in vitro. J. Physiol. 498 (Pt 2), 463–472. https://doi.org/10.1113/jphysiol.1997.sp021872.

Vaughan, C.W., Bagley, E.E., Drew, G.M., Schuller, A., Pintar, J.E., Hack, S.P., Christie, M.J., 2003. Cellular actions of opioids on periaqueductal grey neurons from C57B16/J mice and mutant mice lacking MOR-1. Br. J. Pharmacol. 139 (2), 362–367. https://doi.org/10.1038/sj.bjp.0705261.

Vega-Bermudez, F., Johnson, K.O., 1999. Surround suppression in the responses of primate SA1 and RA mechanoreceptive afferents mapped with a probe array. J. Neurophysiol. 81, 2711–2719.

Wagner, J.J., Terman, G.W., Chavkin, C., 1993. Endogenous dynorphins inhibit excitatory neurotransmission and block LTP induction in the hippocampus. Nature 363 (6428), 451–454. https://doi.org/10.1038/363451a0.

Wang, H., Wessendorf, M.W., 1999. Mu- and delta-opioid receptor mRNAs are expressed in spinally projecting serotonergic and nonserotonergic neurons of the rostral ventromedial medulla. J. Comp. Neurol. 404 (2), 183–196. https://doi.org/10.1002/(sici)1096-9861(19990208)404:2<183::aid-cne4>3.0.co;2-n.

Wang, H., Wessendorf, M.W., 2002. Mu- and delta-opioid receptor mRNAs are expressed in periaqueductal gray neurons projecting to the rostral ventromedial medulla. Neuroscience 109 (3), 619–634. https://doi.org/10.1016/s0306-4522(01)00328-1.

Wanigasekera, V., Lee, M.C., Rogers, R., Kong, Y., Leknes, S., Andersson, J., Tracey, I., 2012. Baseline reward circuitry activity and trait reward responsiveness predict expression of opioid analgesia in healthy subjects. Proc. Natl. Acad. Sci. USA 109 (43), 17705–17710. https://doi.org/10.1073/pnas.1120201109.

Warren, S., H.m.l.inen, H.A., Gardner, E.P., 1986. Objective classification of motion- and direction-sensitive neurons in primary somatosensory cortex of awake monkeys. J. Neurophysiol. 56, 598–622.

Waung, M.W., Margolis, E.B., Charbit, A.R., Fields, H.L., 2019. A midbrain circuit that mediates headache aversiveness in rats. Cell Rep. 28 (11), 2739–2747.e4. https://doi.org/10.1016/j.celrep.2019.08.009.

Weber, A.I., Saal, H.P., Lieber, J.D., et al., 2013. Spatial and temporal codes mediate the tactile perception of natural textures. Proc. Nat. Acad. Sci. U. S. A. 110, 17107–17112.

Weinstein, S., 1968. Intensive and extensive aspects of tactile sensitivity as a function of body part, sex, and laterality. In: Kenshalo, D.R. (Ed.), The Skin Senses. Thomas, Springfield, IL, pp. 195–222.

Wei, H., Pertovaara, A., 1999. MK-801, an NMDA receptor antagonist, in the rostroventromedial medulla attenuates development of neuropathic symptoms in the rat. Neuroreport 10 (14), 2933–2937. https://doi.org/10.1097/00001756-199909290-00011.

Weisskopf, M.G., Zalutsky, R.A., Nicoll, R.A., 1993. The opioid peptide dynorphin mediates heterosynaptic depression of hippocampal mossy fibre synapses and modulates long-term potentiation. Nature 362 (6419), 423–427. https://doi.org/10.1038/362423a0.

Westling, G., Johansson, R.S., 1987. Responses in glabrous skin mechanoreceptors during precision grip in humans. Exp. Brain Res. 66, 128–140.

Williams, F.G., Beitz, A.J., 1990. Ultrastructural morphometric analysis of GABA-immunoreactive terminals in the ventrocaudal periaqueductal grey: analysis of the relationship of GABA terminals and the GABAA receptor to periaqueductal grey-raphe magnus projection neurons. J. Neurocytol. 19 (5), 686–696. https://doi.org/10.1007/BF01188037.

Williams, R.G., Dockray, G.J., 1983. Distribution of enkephalin-related peptides in rat brain: immunohistochemical studies using antisera to met-enkephalin and met-enkephalin Arg6Phe7. Neuroscience 9 (3), 563–586. https://doi.org/10.1016/0306-4522(83)90175-6.

Williams, F.G., Mullet, M.A., Beitz, A.J., 1995. Basal release of Met-enkephalin and neurotensin in the ventrolateral periaqueductal gray matter of the rat: a microdialysis study of antinociceptive circuits. Brain Res. 690 (2), 207–216. https://doi.org/10.1016/0006-8993(95)00554-4.

Winters, B.L., Gregoriou, G.C., Kissiwaa, S.A., Wells, O.A., Medagoda, D.I., Hermes, S.M., Burford, N.T., Alt, A., Aicher, S.A., Bagley, E.E., 2017. Endogenous opioids regulate moment-to-moment neuronal communication and excitability. Nat. Commun. 8, 14611. https://doi.org/10.1038/ncomms14611211.

Wimmer, V.C., Bruno, R.M., de Kock, C.P., Kuner, T., Sakmann, B., 2010. Dimensions of a projection column and architecture of VPM and POm axons in rat vibrissal cortex. Cerebr. Cortex 20, 2265–2276.

Woo, S.H., Ranade, S., Weyer, A.D., et al., 2014. Piezo2 is required for Merkel-cell mechanotransduction. Nature 509, 622–626.

Woo, J.A., Castaño, M., Kee, T.R., Lee, J., Koziol-White, C.J., An, S.S., Kim, D., Kang, D.E., Liggett, S.B., 2023. A Par3/LIM kinase/cofilin pathway mediates human airway smooth muscle relaxation by TAS2R14. Am. J. Respir. Cell Mol. Biol. 68 (4), 417–429.

Woolf, C.J., 1983. Evidence for a central component of postinjury pain hypersensitivity. Nature 306, 686–688.

Wu, X., Cui, L., Wang, H., Xu, J., Zhong, Z., Jia, X., Wang, J., Zhang, H., Shi, Y., Tang, Y., Yang, Q., Liang, Q., Zhang, Y., Li, J., Jiang, X., 2023. Impact of dietary sucralose and sucrose-sweetened water intake on lipid and glucose metabolism in male mice. Eur. J. Nutr. 62 (1), 199–211.

Xie, X., Dang, Y., Pan, D., Sun, Y., Zhou, C., He, J., Gao, X., 2023. The enhancement and mechanism of the perception of saltiness by umami peptide from *Ruditapes philippinarum* and ham. Food Chem. 405 (Pt A), 134886. https://doi.org/10.1016/j.foodchem.2022.134886. Epub 2022 Nov 8.

Yaksh, T.L., Fisher, C., Hockman, T., Wiese, A., 2017. Current and future issues in the development of spinal agents for the management of pain. Curr. Neuropharmacol. 15, 232–259.

Yarnitsky, D., 2015. Role of endogenous pain modulation in chronic pain mechanisms and treatment. Pain 156 (Suppl. 1), S24–S31. https://doi.org/10.1097/01.j.pain.0000460343.46847.58.

Yoshioka, T., Gibb, B., Dorsch, A.K., Hsiao, S.S., Johnson, K.O., 2001. Neural coding mechanisms underlying perceived roughness of finely textured surfaces. J. Neurosci. 21 (17), 6905–6916. https://doi.org/10.1523/JNEUROSCI.21-17-06905.2001104.

Zadina, J.E., Hackler, L., Ge, L.J., Kastin, A.J., 1997. A potent and selective endogenous agonist for the mu-opiate receptor. Nature 386 (6624), 499–502. https://doi.org/10.1038/386499a0.

Zeilhofer, H.U., Wildner, H., venes GE, Y., 2012. Fast synaptic inhibition in spinal sensory processing and pain control. Physiol. Rev. 92, 193–235.

Zhang, Y., Zhao, S., Rodriguez, E., Takatoh, J., Han, B.X., Zhou, X., Wang, F., 2015. Identifying local and descending inputs for primary sensory neurons. J. Clin. Investig. 125 (10), 3782–3794. https://doi.org/10.1172/JCI81156.

Further Reading

Cox, J.J., Reimann, F., Nicholas, A.K., et al., 2006. An SCN9A channelopathy causes congenital inability to experience pain. Nature 444, 894–898.

Darland, T., Heinricher, M.M., Grandy, D.K., 1988. Orphanin FQ/nociceptin: a role in pain and analgesia, but so much more. Trends Neurosci. 21, 215–221.

Dutta Banik, D., Medler, K.F., 2023. Defining the role of TRPM4 in broadly responsive taste receptor cells. Front. Cell. Neurosci. 17, 1148995.

Flor, H., Nikolajsen, L., Jensen, T.S., 2006. Phantom limb pain: a case of maladaptive CNS plasticity? Nat. Rev. Neurosci. 7, 873–881.

Günther, T., Dasgupta, P., Mann, A., et al., 2017. Targeting multiple opioid receptors—improved analgesics with reduced side effects? Br. J. Pharmacol. 2857–2868.

Johansson, R.S., Vallbo, B., 1983. Tactile sensory coding in the glabrous skin of the human hand. Trend. Neurosci. 6, 27–32.

Jordt, S.E., Bautista, D.M., Chuang, H.H., et al., 2004. Mustard oils and cannabinoids excite sensory nerve fibres through the TRP channel ANKTM1. Nature 427, 260–265.

Kelleher, J.H., Tewari, D., McMahon, S.B., 2017. Neurotrophic factors and their inhibitors in chronic pain treatment. Neurobiol. Dis. 97, 127–138.

Lane, N.E., Schnitzer, T.J., Birbara, C.A., et al., 2010. Tanezumabfor the treatment of pain from osteoarthritis of the knee. N. Engl. J. Med. 363, 1521–1531.

McDonnell, A., Schulman, B., Ali, Z., et al., 2016. Inherited erythromelalgia due to mutations in SCN9A: natural history, clinical phenotype and somatosensory profile. Brain 39, 1052–1065.

Ross, S.E., Mardinly, A.R., McCord, A.E., et al., 2010. Loss of inhibitory interneurons in the dorsal spinal cord and elevated itch in Bhlhb5 mutant mice. Neuron 65, 886–898.

Sorge, R.E., Mapplebeck, J.C., Rosen, S., et al., 2015. Different immune cells mediate mechanical pain hypersensitivity in male and female mice. Nat. Neurosci. 18, 1081–1083.

Wercberger, R., Basbaum, A.I., 2019. Spinal cord projection neurons: a superficial, and also deep analysis. Curr. Opin. Physiol. 11, 109–115.

18

The Vestibular System

Jonathan Ortiz Rafael[1,5], *I. Zoe Chavez Valderrama*[2], *Srikant S. Chakravarthi*[3], *Giovanni Paternò*[6], *Brandon M. Ascenzi*[7] *and Alejandro Monroy-Sosa*[3,4]

[1]Department of Neurosurgery, Hospital Juarez de Mexico, Mexico City, Mexico [2]Department of Neurosurgery, XXI Century National Medical Center, Mexico City, Mexico [3]Skull Base and Cerebrovascular Lab, Aurora Research Institute, Milwaukee, WI, United States [4]Department of Neurosurgery, Institute for Social Security and Services for State Workers, (ISSSTE) Tlahuac, Center for Research and Innovation in Neuroscience, Instituto Politécnico Nacional, Superior School of Medicine, Mexico City, Mexico [5]Center for Research and Innovation in Neuroscience, Superior Medicine School, National Polytechnic Institute, Mexico City, Mexico [6]Department of Neurosurgery, Istituto Clinico Città Studi, Milan, Italy [7]Independent Neuroresearcher-Member of Marie Curie Alumni Association (MCAA)

OUTLINE

INTRODUCTION

Anatomy of the Vestibular System—Peripheral Vestibular System

The peripheral vestibular system is composed of the vestibular portion of the eight cranial nerve and five sensory structures. The sensory structures are located in the vestibule of the inner ear in the petrous portion of the temporal bone and consist of the following: two otolith organs (the utricle and saccule) and the three semicircular canals (anterior or superior, lateral or horizontal, and posterior) (Fig. 18.1); each of structures contains sensory hair cells specialized in translation of head motion inputs to neural impulses (Jones et al., 2009; Khan and Chang, 2013). This sensory system is composed of a bony and a

From Anatomy to Function of the Central Nervous System
https://doi.org/10.1016/B978-0-12-822404-5.00005-X

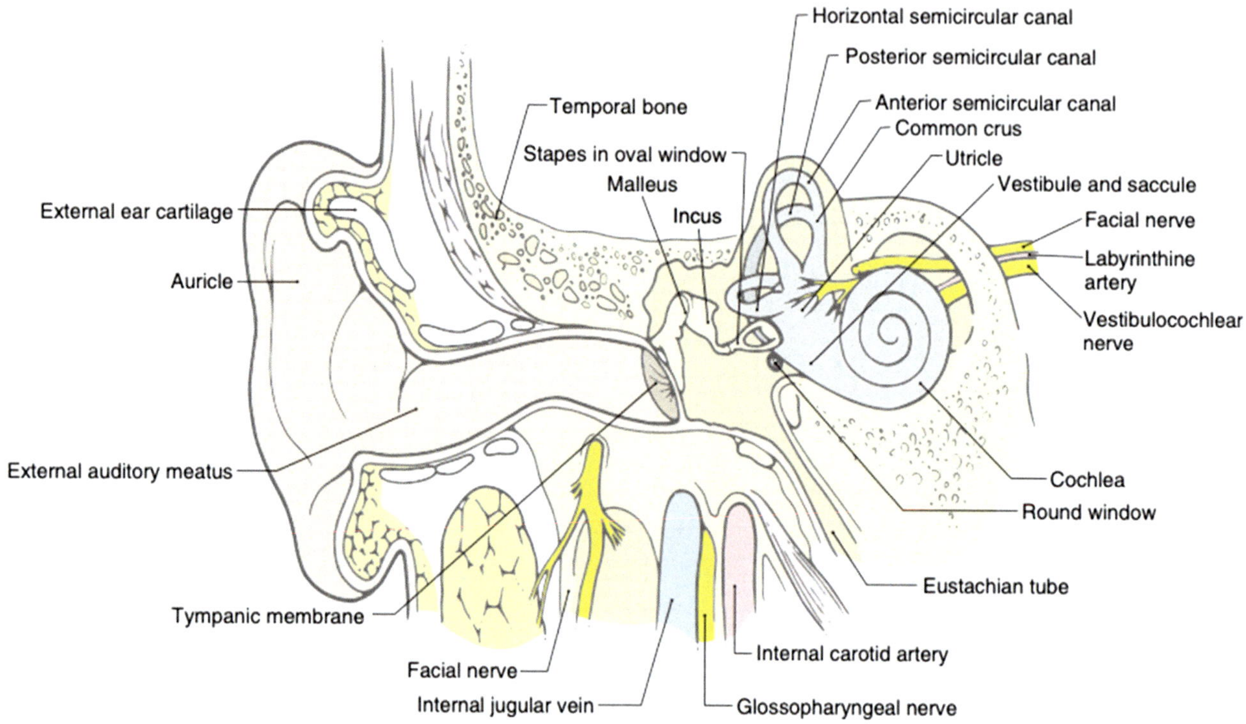

FIGURE 18.1 **A cross section of the outer, middle, and inner ear.** *Image courtesy of Haines, D.E., Mihailoff, G.A. (Eds.), 2018. In Fundamental Neuroscience for Basic and Clinical Applications (Fifth Edition). Elsevier, 321. https://doi.org/10.1016/B978-0-323-39632-5.15001-7.*

membranous labyrinth that do not communicate each other. The labyrinth itself is essentially made up of: vestibule, semicircular canals and cochlea (see later). The vestibule is located posterolateral to the internal acoustic meatus and medial to the middle ear cavity, while the cochlea is situated anteromedial to the vestibule and connects to the vestibule by the ductus reuniens. The floor of the vestibular cavity is separated from the apex of the jugular bulb by a segment of bone approximately 6 mm in thickness (Lyskowski, 2006; Rhoton, 2007).

The cavity of the bony labyrinth, which contains fluid called perilymph, is similar to composition as the cerebral spinal fluid, containing concentration of sodium (140 mEq/L) and a reduced concentration of potassium (10 mEq/L). The stria vascularis has a high concentration of Na+/K+ ATPase along its membrane surface in order to maintain this concentration. The membranous labyrinth is filled with endolymph, which has a higher concentration of potassium (144 mEq/L) and lower concentration of sodium (5 mEq/L), similar to other intracellular fluids (Wangemann, 2006; Baloh and Kerber, 2011). Endolymph is produced by capillaries in the stria vascularis of the cochlear duct, and is reabsorbed to venous blood flow via the endolymphatic sac, which is located on the posterior petrous surface inferolateral to the internal acoustic meatus and near the sigmoid sinus (Rhoton, 2007; Corrales and Mudry, 2017). The endolymphatic sac is connected to the utricle by the endolymphatic duct and contains columnar cells specialized in absorption. Perilymph production is not completely understood, as it in part contains both filtrate of cerebrospinal fluid and vessels of the inner ear, and is drained to the subarachnoid space by the perilymphatic or vestibular duct, which is located inferolateral to the aperture of the internal acoustic meatus (Lyskowski, 2006; Rhoton, 2007; Baloh and Kerber, 2011).

CELLS

Vestibular system has two types of sensory hair cell neuroepithelium, goblet-shaped Type I and cylindrical-shaped Type II cells (Fig. 18.2). Each hair cell is formed by a body, which is a single kinocilium positioned toward one edge of the cell and approximately 70–100 stereocilia at its apical end; the largest stereocilia are located closest to the kinocilium and progressively decrease in length as they get further to the kinocilium (Khan and Chang, 2013). "Tip links" are protein filaments that connect the tips of shorter stereocilia to the body of their adjacent and taller stereocilia; these links are attached to mechanically sensitive potassium channels. Bending of stereocilia toward the kinocilium results in hair cell depolarization, initiating the process of transduction of impulses representing head velocity and acceleration; these signals are then transported by the VIII cranial nerve to central nervous components (Baloh and Kerber, 2011; Highstein and Holstein, 2012; Swenson, 2017). Dendrites of afferent vestibular nerves

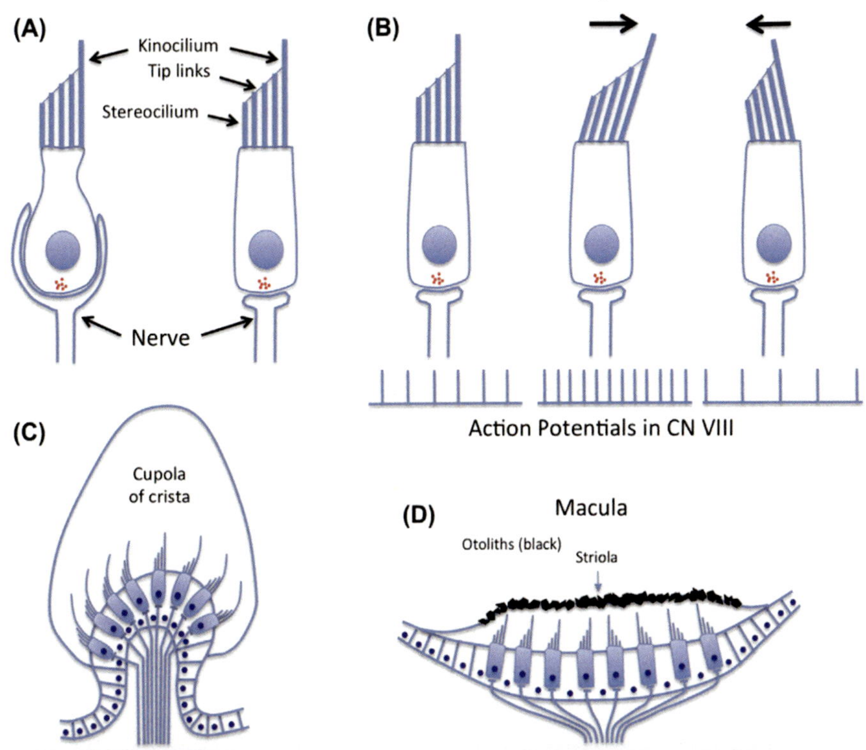

FIGURE 18.2 **Cartoons depicting aspects of hair cell structure, function, and organization.** (A) Type I (left) and type II (right) hair cells. Notice that all hair cells are polarized by the eccentric position of the kinocilium. (B) The effect of hair cell deflection on transmitter release and nerve activity. When hairs are deflected toward the kinocilium, tension is placed on the tip links. This tension will open potassium channels, resulting in an increase in transmitter release and in action potentials in the nerve (below). Also note that there is a baseline resting level of action potential generation. (C) Structure of a cristae, which is the receptive organ found in the ampulla of semicircular ducts. Note that all of the hair cells are polarized in the same direction. (D) Structure of a macula, a receptive organ in the wall of the utricle and saccule. Note that the hair cells have different polarization on either side of the striola (a *curved line* in the middle of the receptive organ). *Image courtesy of Swenson, R.S., 2017. The Vestibular System, Conn's Translational Neuroscience. Elsevier Inc. https://doi.org/10.1016/B978-0-12-802381-5.00014-2. Chapter 8 - The Vestibular System. In P. M. Conn (Ed.), Conn's Translational Neuroscience, Academic Press, 167–183. https://doi.org/10.1016/B978-0-12-802381-5.00014-2.*

make contact with the basal portion of the hair cell, forming ribbon synapses; there are approximately 10–20 synapses per hair cell, and each synapse contains 100 to 200 synaptic vesicles (Baloh and Kerber, 2011).

Hair cells are immersed in endolymph, which is electrically positive (+5 to +10 mV) in relation to the negative charges inside the hair cell. This results in an electrochemical drive that forces entry of potassium into the hair cells when the channels are open by bending the stereocilia toward the kinocilium, with subsequent depolarization of the hair cell. The resting potential of hair cells is between −50 and −70 mV. When the potential rises above −50 mV, the hair cells is depolarized, causing an opening of voltage-gated calcium channels, allowing calcium ions to flow inside the cell (Baloh and Kerber, 2011). As calcium enters the hair cell, it gets bound to proteins in the ribbon synapse, which activate proteins of the SNARE complex, resulting in the release of glutamate into the synaptic cleft. Glutamate activates AMPA receptors (a non-NMDA glutamate receptor) in the dendrites (Baloh and Kerber, 2011; Khan and Chang, 2013). Hair cells generate a tonic and spontaneous neural action potential of 70–90 Hz in the absence of stimulus; in the presence of depolarization this tonic action potential rises to up to 400 Hz. Vestibular afferents can be classified according to the variability in their intervals in the absence of stimulation; type I hair cells are associated with irregular afferent activity and high variability in resting discharges, and type II hair cells have regular afferent activity and a low variably resting discharge rate. Both hair cell types have efferent connections from vestibular nuclei in mediating sensory modulation (Mitchell and Cullen, 2004; Zaleski-king et al., 2019). The outcome of larger defections of stereocilia toward the kinocilium is the opening of a greater number of potassium channels, and consequently, a greater depolarization with a higher frequency of action potentials into the vestibular nerve.

Bending of the stereocilia away from the kinocilium reduces tension on tip links, closing potassium channels and hyperpolarizing hair cells. This change in electrochemical activity can reduce the tonic action potential of 70–90 Hz to virtually zero (Swenson, 2017; Zaleski-king et al., 2019). The opposite end of the tip link in the longer stereocilium is bound to motor proteins that can move toward or away the tip link, based on the relative intracellular calcium concentration. When intracellular calcium is low, motor proteins will pursue movement toward the tip link by inducing potassium channel activation and subsequent calcium inflow. On the contrary, with higher concentrations of intracellular calcium, motor proteins will move away from the tip link, resulting in a lower flow of calcium into the cell. This mechanism results in new baseline hair cell activity based on static positions of the head (Swenson, 2017).

SEMICIRCULAR CANALS

Semicircular canals, which form two-thirds of a circle, are attached to the body of the utricle (Figs. 18.1 and 18.3). These canals include a bony labyrinth portion filled with perilymph, and a membranous labyrinth, referred to as the semicircular ducts, which has the same basic structure as the bony labyrinth and is filled with endolymph (Swartz et al., 1996; Khan and Chang, 2013). As stated before, the three semicircular canals in each side of the head are the anterior (or superior), the lateral (or horizontal) and the posterior canals (Fig. 18.3); the semicircular canals are localized posterosuperior to the utricle (Figs. 18.1 and 18.3). The superior semicircular canal marks an upwardly directed bulge upward into the middle cranial fossa, referred to as the arcuate eminence. This is located in the anterior part of the petrous portion of the temporal bone, lateral to the trigeminal impression (Rhoton, 2007). The semicircular canals sense angular acceleration and rotation of the head in a specific vector, and are oriented in contralateral pairs following the same angle in approximately parallel lines; superior canals are paired with the contralateral posterior canal as they are aligned at a 45 degree angle to the sagittal plane, and lateral canals pair to each other as they are aligned in a 30 degree angle in regards to the axial plane. Any rotation in a plane that will be excitatory to a semicircular canal will be inhibitory to the contralateral one (Lyskowski, 2006; Baloh and Kerber, 2011; Khan and Chang, 2013).

At the end of each semicircular canal there is an opening known as the ampulla which contains the neuroepithelium, and thus, the sensory hair cells. This sensory structure is suspended within the ampulla, forming the crista ampullaris. The ampulla is located in an anterolateral position on the anterior semicircular canals, in a posterolateral position on the posterior semicircular canals, and in anterior position on the lateral canals toward the utricle. The orientation of these end organs is analogous to the orientation of the kinocilium and stereociliary

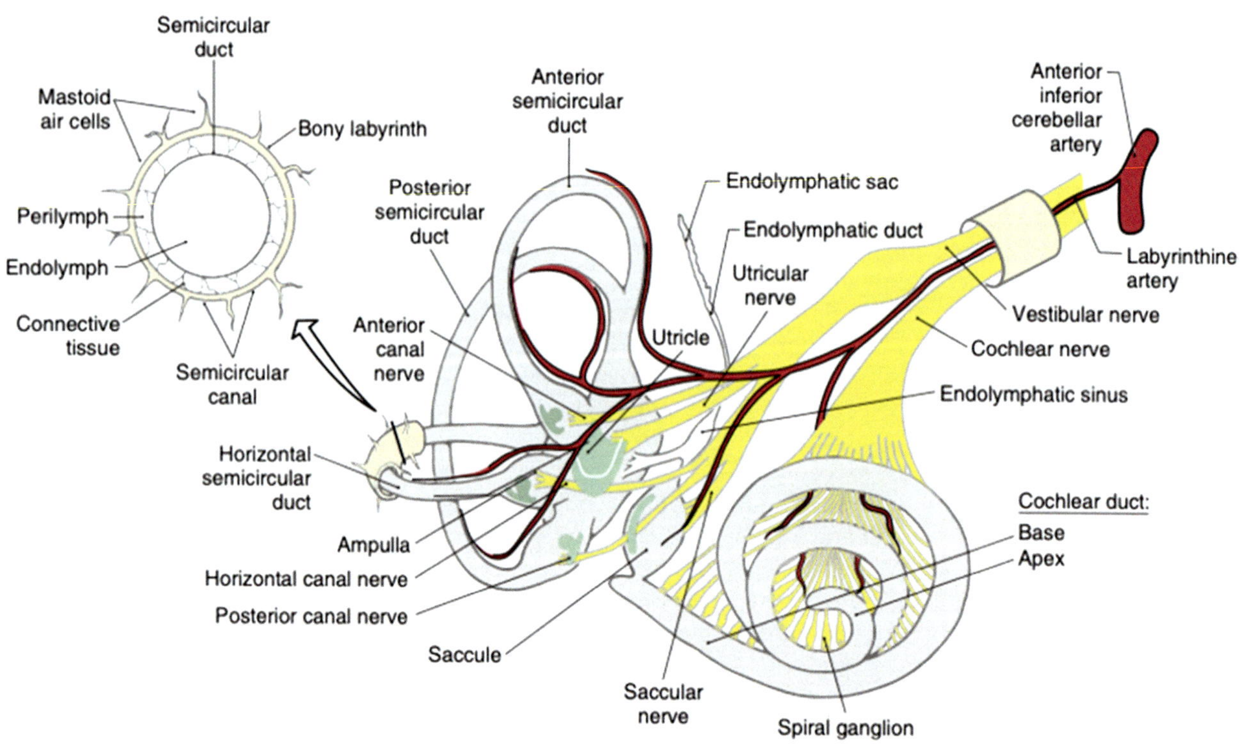

FIGURE 18.3 **The membranous labyrinth and associated vessels and nerves.** The approximate configuration of the receptor sites in the ampulla, utricle, and saccule are shown in green. The detail shows the relationship between bony and membranous labyrinths. *Image courtesy of Haines, D.E., Mihailoff, G.A. (Eds.), 2018. In Fundamental Neuroscience for Basic and Clinical Applications (Fifth Edition). Elsevier, 321. https://doi.org/10.1016/B978-0-323-39632-5.15001-7.*

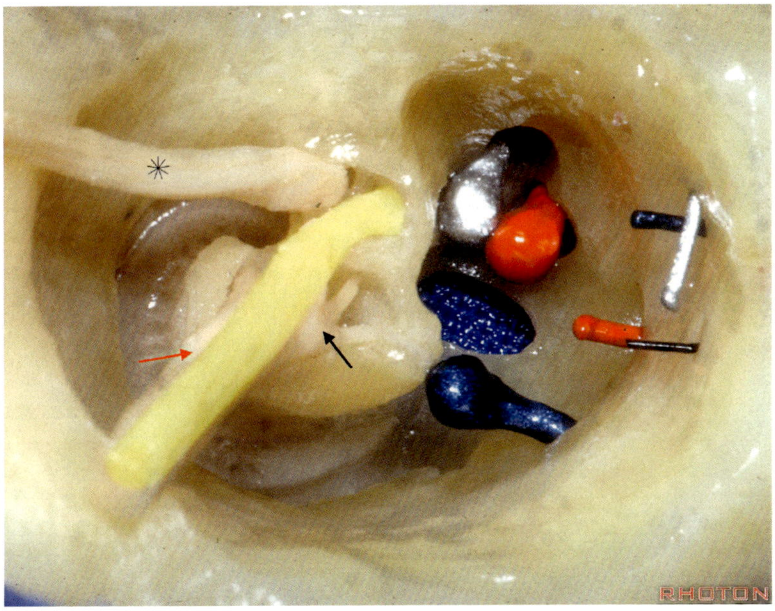

FIGURE 18.4 **Right medial view of the labyrinth.** In the center the labyrinth; behind the latter is the cochlea (the light gray structure). Proximal, from left to right, is: the auditory nerve (red arrow), the superior vestibular nerve (in yellow) and the inferior vestibular nerve (black arrow). Horizontally on the left side is the facial nerve (black star). *Image courtesy of The Rhoton Collection® http://rhoton.ineurodb.org/. Drawing by Jeffrey Sorenson.*

bundles. The nonampullary end of the semicircular canals open up to the utricle (Khan and Chang, 2013; Lyskowski, 2006). Nonampullary ends of the posterior canal, in a superior position, and the superior canal, in a posterior position, share a single canal known as the common crus, which connects them to the utricle (Rhoton, 2007).

OTOLITH ORGANS

The utricle and saccule are the two otolith organs of the inner ear, which are sensitive to linear acceleration of the head activated by gravity. The utricle is oval shaped and adjacent to the semicircular canals and is positioned in a horizontal plane, the saccule is spherical in shape and oriented perpendicularly vertical to the utricle, in a sagittal plane. The otolith organs communicate with the medial wall of the tympanic cavity via the oval window, which is located posterosuperior to the promontory of the cochlea, and is occupied by the footplate of the stapes bone (Rhoton, 2007; Zaleski-king et al., 2019). The sensory receptors of the otolith organs are located in thickened curved patches referred to as maculae, and contain dense layers of supporting cells covering the hair receptor cells, referred to as overhead covers, which serve as hairy cells as well interlace forming loops. The maculae of the otolith organs are covered by gelatinous otolithic membrane that contains calcium carbonate particles called otoliths or otoconia. Otoconia is denser than endolymph, thus can be attracted by gravity; this motion deflects stereocilia toward the kinocilia depending on head position, starting the transduction of linear acceleration or position in reference to gravity. The macula of the utricle senses motion in the horizontal plane while the macula of the saccule senses vertical plane motion (Highstein and Holstein, 2012; Khan and Chang, 2013; Illich Zabolotnyi and Serhiivna Mishchanchuk, 2020).

Sensory areas of the maculae contain a curved central zone referred to as striola. Sterocilliary bundles and kinocilia of the otolith organs are oriented in opposite directions in relation to this zone. Contrarily, the utricle is positioned toward the striola and in the saccule faces away from it. Since the striola and maculae form curved areas, head tilts in any direction cause a fraction of sensor hair cells to activate, and others to be inhibited in one or both maculae. As both otolith maculae hair bundles deflect in vertical and horizontal planes, the ability to distinguish linear motion resides in the central nervous system (Lyskowski, 2006; Khan and Chang, 2013).

VESTIBULAR NERVE

Vestibular (or Scarpa's) ganglion is located deep in the internal auditory canal of the petrous portion of the temporal bone (Fig. 18.5), and its dendrites carry action potentials produced by the excitation of the maculae and the cristae ampullaris.

Vestibular ganglion is composed of approximately 20,000 cell bodies of bipolar cells on each side, and is

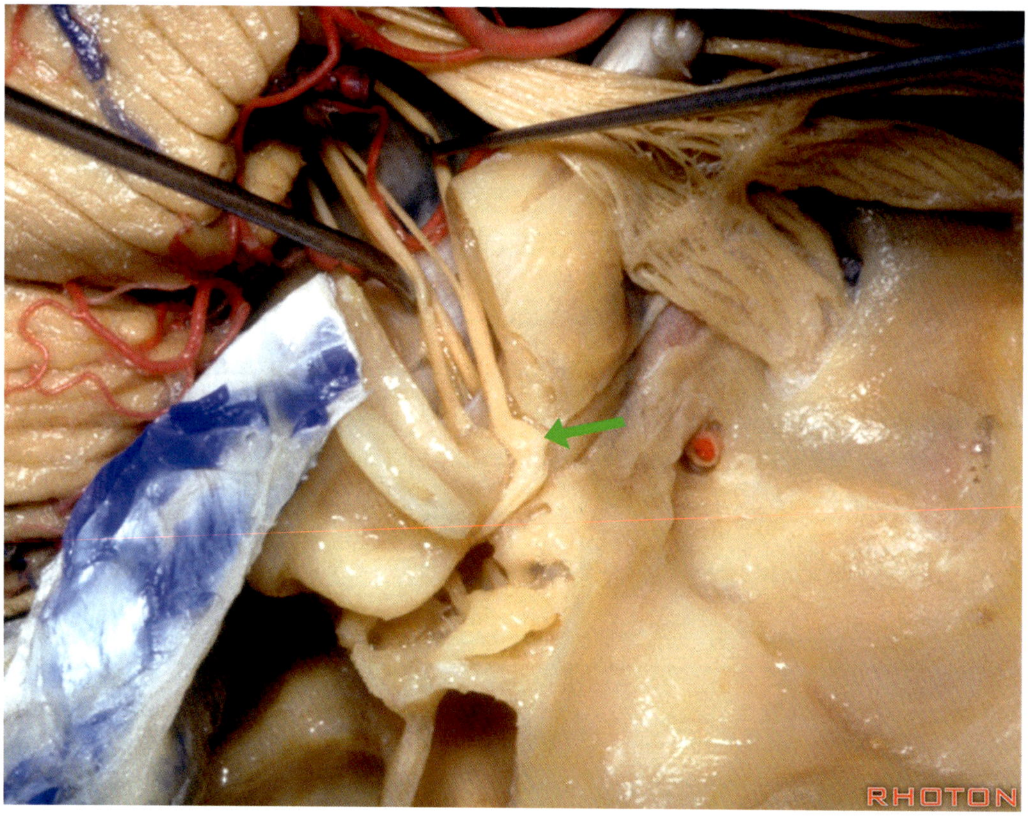

FIGURE 18.5 **Superior-right view of the tympanic cavity.** Eccentrical is the labyrinth (it is easily appreciable the superior semicircular canal). The green arrow indicates the geniculate ganglion. Attached to the latter is the facial nerve and medially is the superior vestibular nerve that joins the labyrinth by a nodular formation: The Scarpa's ganglion. Between the two nerves aforementioned is visible the auditory nerve. In blue is the superior petrosal sinus. *Image courtesy of The Rhoton Collection® http://rhoton.ineurodb.org/. Drawing by Jeffrey Sorenson.*

divided into two portions connected by an isthmus: (1) a superior segment that carries impulses from the superior and lateral crista ampullaris and from the maculae of the utricle, by the superior vestibular nerve; and (2) an inferior segment conveying impulses from the posterior crista ampullaris by the singular nerve, and the saccule's maculae by the saccular nerve. The inferior vestibular nerve commonly has a utricular nerve branch (Rhoton, 2007; Illich Zabolotnyi and Serhiivna Mishchanchuk, 2020). Neuronal diameters are variable inside the vestibular ganglion, with the largest neurons receiving impulses from type I hair cells in the center of the cristae and maculae, and smaller neurons receiving information from type II hair cells in the periphery of the vestibular organs (Baloh and Kerber, 2011).

Axons originating from the vestibular ganglion accompany the cochlear nerve to form the vestibulocochlear nerve, and cross the internal acoustic meatus in close relationship to the facial nerve and the labyrinth artery, to emerge in the posterior cranial fossa. In the skull base, the vestibulocochlear nerve enters the brainstem in the cerebellopontine angle, located superior to the medullary olives in the pontomedullary junction. Posterior to entering the brainstem (Fig. 18.6), fibers from the vestibular nerve pass between the inferior cerebellar peduncle and the spinal tract of the trigeminal nerve, to synapse the four ipsilateral vestibular nuclei; some fibers make direct synapses with the flocculonodular lobe of the cerebellum and paramedian cerebellar cortex, through the inferior cerebellar peduncle. In the cerebellum, primary vestibular afferent terminate in the granulate layer and send collateral projections to the fastigial nerve through mossy fibers (Carpenter, 1989b; Lyskowski, 2006; Khan and Chang, 2013). When axons originating from the primary vestibular neurons enter the brainstem, they are divided into two branches: an ascending branch ends in the rostral vestibular nuclei and the cerebellum, and a descending branch terminates in the caudal vestibular nuclei (Walker et al., 2014).

The internal acoustic meatus is a tubular excavation located in the posterior surface of the petrous bone pyramid, connecting the posterior cranial fossa with the bony labyrinth in the inner ear. The internal acoustic meatus opens to the posterior cranial fossa in the porus acousticus and communicates to the bony labyrinth by a thin perforated bony plate referred to as lamina cribrosa, where nerve fibers from the vestibular and cochlear nerve pass through. The superior vestibular nerve and the facial nerve are separated at the fundus of the internal acoustic meatus by a bony plate called the vertical

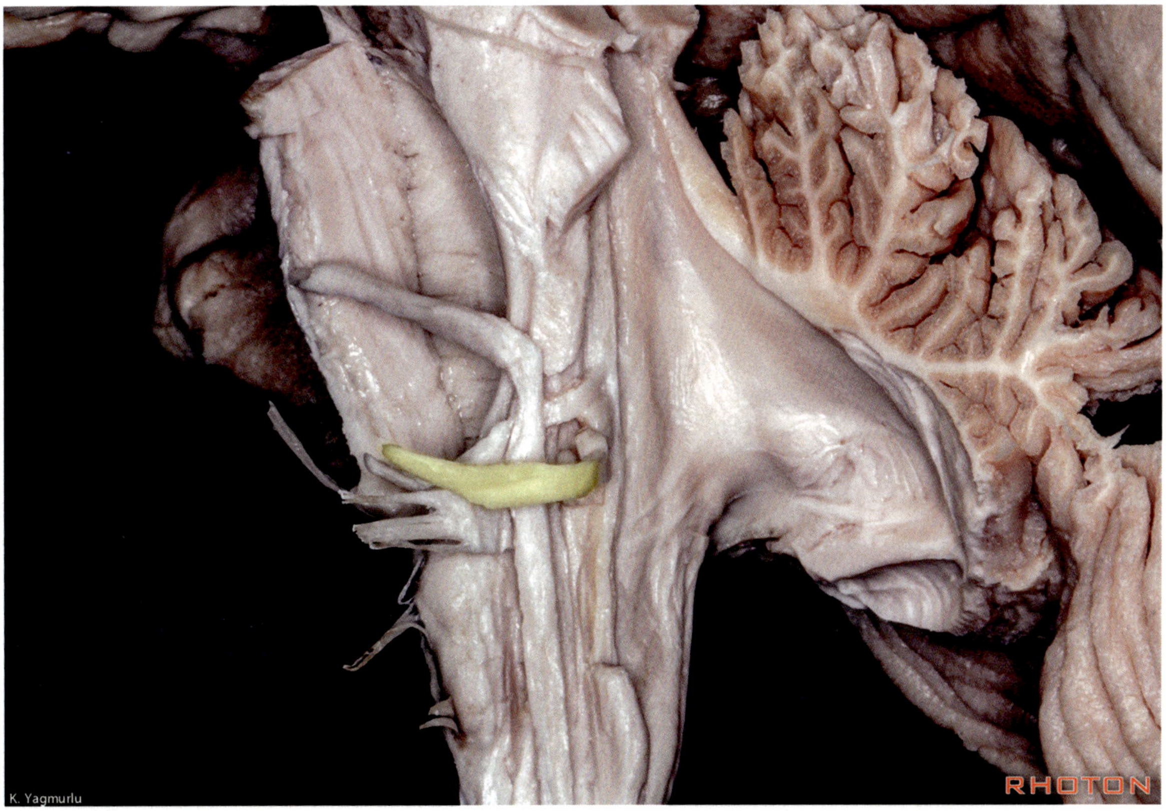

FIGURE 18.6 **Deep dissection of brainstem; lateral view.** In yellow, the vestibular nerve; below the latter vertically (from bottom to top) is the trigeminal lemniscus which ends supero-medially with the trigeminal nerve. *Image courtesy of The Rhoton Collection® http://rhoton.ineurodb.org/. Dissector: K.Yagmurlu. Drawing by Garrett Venable.*

crest, or Bill's bar. The transverse crest separates the superior vestibular nerve from the inferior vestibular nerve and the cochlear nerve. The internal acoustic meatus has a length of approximately 8.39 mm, from the porus acousticus to the lamina cribrosa (Farahani et al., 2007; Rhoton, 2007; Walker et al., 2014).

As the VII and VIII nerve face the same trajectory from the cerebellopontine angle into the internal acoustic meatus, they have been often referred as a VII/VIII neural complex, and are in close relationship with the anteroinferior cerebellar artery, which courses inferior to the facial and vestibulocochlear nerves in their cisternal portion; in some cases, this artery can also pass above or between these nerves (Rhoton, 2000).

CENTRAL VESTIBULAR SYSTEM

Vestibular Nuclei

The vestibular nuclear complex (Fig. 18.7) is the primary processor of vestibular input and consists of second-order neurons located in the lateral aspect of the fourth ventricle. Vestibular nuclei are grouped into two major columns that extend from the rostral medulla to the caudal pons, spreading from the rostral level of the hypoglossal nucleus and slightly exceeding the superior margin of the abducens nucleus. The vestibular nuclear complex is comprised by four major nuclei groups on each side: the superior, medial, inferior (or descending), and lateral vestibular nuclei. The medial vestibular nucleus is the largest and forms the medial cell column, while the lateral, superior, and inferior nuclei form the lateral column (Carpenter, 1989b; Barmack, 2003; Khan and Chang, 2013).

The medial vestibular nucleus, which forms the medial cell column, extends from the superior pole of the hypoglossal nucleus to the level of the abducens nucleus; the medial nucleus is composed of medium and small sized cells. The medial and superior vestibular nuclei are fused in the superior pole of the medial nucleus and the inferior pole of the superior nucleus. In the lateral column, the lateral vestibular nucleus is located at the level of entrance of the vestibular nerve, and is composed of giant cells. The superior vestibular nucleus is the smallest of the main complex, and is located dorsal and superior to the lateral nucleus and medial to the superior cerebellar peduncle. The superior vestibular nucleus is composed of a central part of large neurons surrounded by small neurons. The inferior vestibular nucleus is located in the medulla oblongata, medial to the accessory cuneatus nucleus, and extends

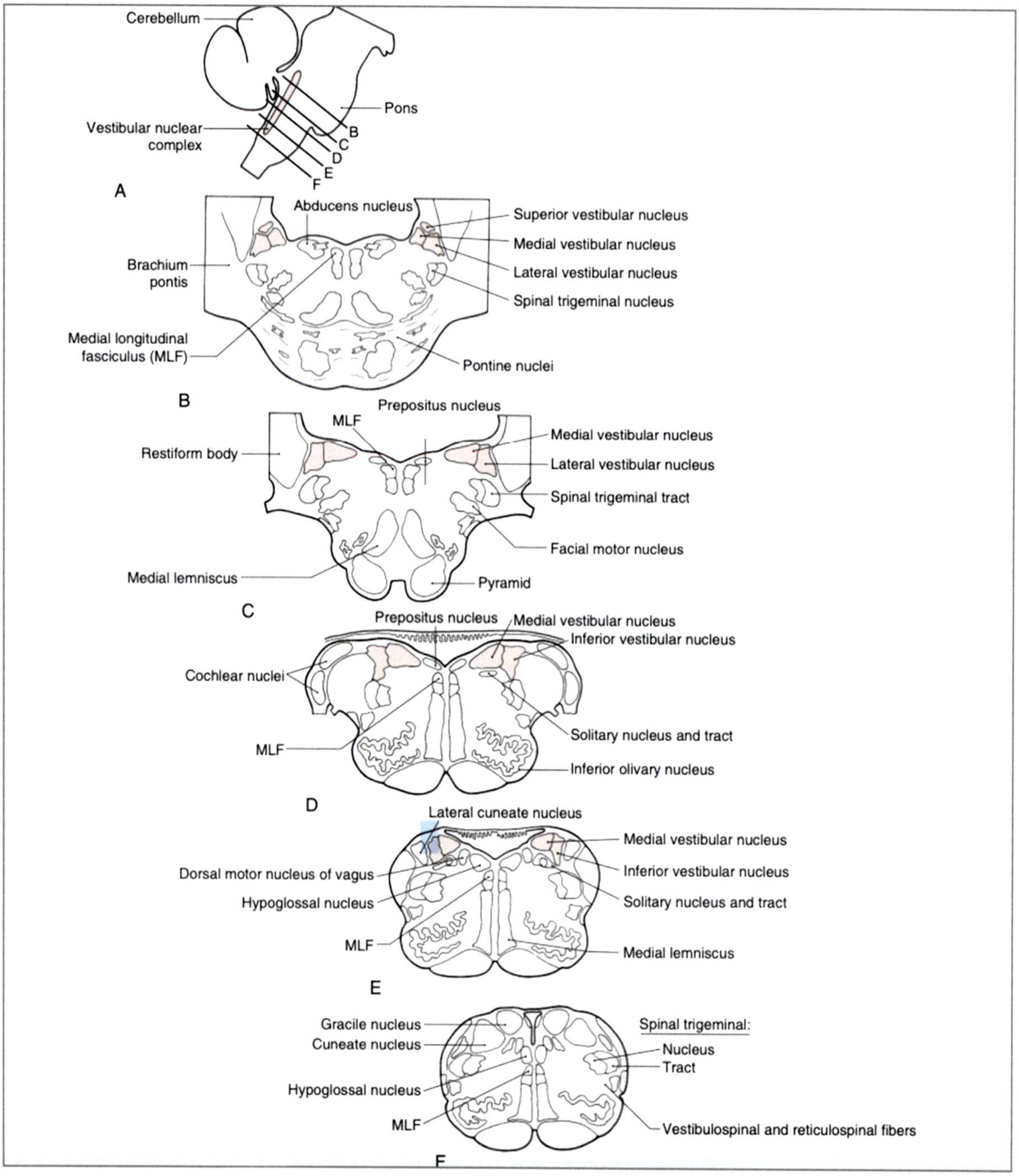

FIGURE 18.7 **Location of the vestibular nuclei in the brainstem in sagittal (A) and representative cross-sectional planes (B–F) from levels indicated in (A).** *Image courtesy of Haines, D.E., Mihailoff, G.A. (Eds.), 2018. In Fundamental Neuroscience for Basic and Clinical Applications (Fifth Edition). Elsevier, 328. https://doi.org/10.1016/B978-0-323-39632-5.15001-7.*

rostrally to the level of the entrance of the vestibular nerve in the pontomedullary junction; the inferior nucleus is composed of medium and small sized neurons (Carpenter, 1989b; Baloh and Kerber, 2011).

The superior vestibular nucleus receives primary afferent projections from the crista ampullaris of the semicircular canals and the maculae of the utricle and saccule. Fibers from the anterior semicircular canal are placed medially, whereas fibers from the lateral and posterior canals are found in the lateral portion of the superior nucleus. Fibers from the maculae are found in the lateral margins of the nucleus. The superior nucleus receives another group of afferent fibers from the flocculonodular lobe and fastigial nucleus from the cerebellum (Baloh and Kerber, 2011). The lateral vestibular nucleus receive primary afferents from the maculae of the saccule and the utricle, in its rostroventral portion, and from the vermis and fastigial nucleus of the cerebellum, in the dorsocaudal portion of the nucleus (Barmack, 2003; Baloh and Kerber, 2011). Neurons in the upper segment of the medial vestibular nucleus acquire primary afferents from the crista ampullaris of the semicircular canals, and from the fastigial nucleus of the cerebellum, while the middle segment receives primary afferents from the macula, and the caudal segment receives fibers from the ipsilateral and contralateral fastigial nucleus and ipsilateral nodule of the vermis (Baloh and Kerber, 2011; Khan and Chang, 2013). The inferior vestibular nucleus gathers primary afferent fibers from the crista ampullaris in the center of the nucleus, and from the maculae in the periphery. Cerebellar afferents from the flocculonodular lobe and uvula are distributed diffusely in the inferior nucleus, together with the fibers from the vestibular afferents (Baloh and Kerber, 2011).

Vestibular nuclei hold connections between the same complex, by internuclear projections, and to the contralateral complex, by commissural projections. Internuclear projections include reciprocal fibers from the superior and inferior vestibular nucleus, and from the superior and medial nucleus. The medial vestibular nucleus receives nonreciprocal fibers from the inferior nucleus (Büttner-Ennever, 1992; Barmack, 2003). Vestibular nuclei on each side, with the exception of the lateral vestibular nucleus, operate in an integrated system using commissural vestibular projections, which connect either nucleus to its contralateral counterpart; the connection between both medial vestibular nuclei is the most prominent. Nevertheless, commissural projections are not limited to the corresponding contralateral nucleus, with one nucleus having connections to the homologous nucleus and the remainder of the nuclei of the vestibular complex, with the exclusion of the lateral vestibular nucleus. Functionally, commissural projections seem to have a critical role in the compensation of the vestibular system function in unilateral lesion to the vestibule (Büttner-Ennever, 1992; Barmack, 2003; Lyskowski, 2006).

It should be noted that several studies have described the patterns of afferent connectivity in the vestibular nuclear complex, and not all of them are in agreement with each other.

Other nuclei with vestibular functions not included in the classical four major groups are smaller nuclei: the parasolitary nucleus, the nucleus intercalatus, and nucleus from the Y-group. The parasolitary nucleus is situated between the medial and the inferior vestibular nucleus in the caudal part of the vestibular complex, and the nucleus intercalatus sits adjacent to the nucleus tractus soitarus. The Y-group is divided into dorsal and ventral, the dorsal group is situated near the nucleus interpositus caudal to the middle cerebellar peduncle, the ventral group is located lateral to the caudal portion of the lateral vestibular nucleus, close to the inferior cerebellar peduncle. Two smaller nuclei sit adjacent to the vestibular complex, it is known as the X-group, which lies lateral to the inferior vestibular nucleus, and the Z-group located rostral to the nucleus gracilis. The X and Z-group nuclei do not appear to receive direct afferents from vestibular primary neurons or end organs, but it is still unclear if they are related to vestibular function (Barmack, 2003; Yates et al., 2018). The parasolitary nucleus is composed of GABAergic neurons that convey impulses from the anterior and posterior semicircular canals and otoliths to the β-nucleus and dorsomedial cell column subnuclei of the inferior olivary complex. The β-nucleus and the dorsomedial cell column project to the contralateral uvula and nodulus (Barmack, 2003; Barmack and Yakhnitsa, 2013). The nucleus intercalatus receives secondary projections from the medial and inferior vestibular nuclei, and the nucleus praepositus hypoglossi. Neurons of the dorsal Y-group receive ipsilateral primary projections and bilateral secondary vestibular projections from the superior vestibular nucleus, and project to the contralateral β-nucleus, ventral Y-group neurons receive information from primary vestibular fibers of the saccule and project to the ipsilateral flocculus and nodulus, and to contralateral oculomotor complex nuclei (Mergner et al., 1977; Büttner-Ennever, 1992; Barmack, 2003).

Vestibulocerebellum

Phylogenetically, the vestibular cerebellum is considered the most ancient division of the cerebellum, being this the reason for its other denomination, the archicerebellum. Vestibulocerebellum is composed of the flocculonodular lobe, the vermian cortex, the fastigial nucleus, and their peduncular connections (Carpenter, 1989a). The main function of the vestibular division of the cerebellum is sensorimotor integration and motor control through the addition of accuracy and modifiability of movement by the employment of sensory feedback for the generation of reflexive activities. Another important function of the cerebellum is working as an adaptative processor of vestibular performance, adjusting sensory integration to facilitate distinction of passive and self-generated head movement (Barmack, 2003; Barmack and Yakhnitsa, 2013; Khan and Chang, 2013; Walker et al., 2014). The primary afferent vestibular fibers in the vestibulocerebellum terminate in the ipsilateral granulate cell layer of the uvula and nodulus as mossy fibers. Primary afferents maintain main regions of linking, but are not exclusive to a particular area, as these connections show a diffuse topographical pattern. The main afferent projections of the saccule and utricule are directed to the ventral uvula, whereas the main projection of the semicircular canal afferents is in the nodulus (Barmack, 2003; Barmack and Yakhnitsa, 2013). Secondary afferent fibers from the medial, inferior, and superior vestibular nuclei, and the Y-group nuclei, also make connections to the granulate layer of the ipsilateral uvula and nodulus, and correspondingly, these secondary afferents likewise project to the flocculus, paraflocculus, and the anterior lobe. Other type of afferent axons in the cerebellum are the climbing fibers, which are considered tertiary projections and originate from the two subnuclei of the inferior olivary complex with vestibular function, from the β-nucleus and the dorsomedial cell column (Fig. 18.8). Climbing fibers cross midline to synapse in dendrites of contralateral Purkinje cells of the uvula and nodulus. The lateral vestibular nucleus and the parasolitary nucleus do not project to the cerebellum (Büttner-Ennever, 1992; Barmack, 2003; Barmack and Yakhnitsa, 2013). An additional group of neurons that project to the flocculus are located in the midline of the medulla and the pons, and are recognized as a group as the nuclei of the paramedian tracts, which in turn, receive afferents from premotor regions of the oculomotor system, providing the cerebellum with eye movement information necessary for eye position regulation and the associated vestibular reflexes (Büttner-Ennever, 1992).

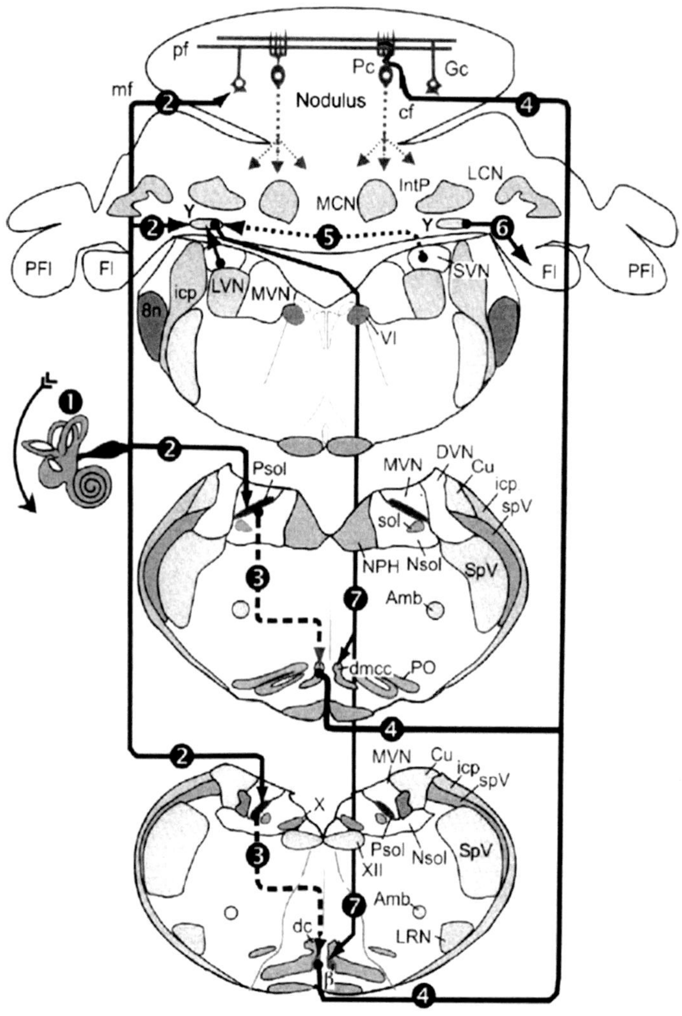

FIGURE 18.8 **Circuitry for activation of vestibular climbing fiber projections to cerebellar nodulus.** Sequences in activation are indicated by *solid lines* for excitatory pathways and *dashed lines* for inhibitory pathways. Pathways are listed numerically. (1) Roll tilt onto the left side increases vestibular primary afferent discharge. (2) Primary afferents project as mossy fibers to ipsilateral granule cell layer of nodulus, parasolitary nucleus (Psol), and dorsal Y-group. (3) Descending GABAergic projection from Psol inhibits activity of olivary neurons in ipsilateral β-nucleus and dorsomedial cell column (dmcc). (4) Climbing fibers from β-nucleus and dmcc project to contralateral nodulus. (5) Vestibular nuclei bilaterally project to dorsal Y-group. (6) Dorsal Y-group projects to ipsilateral flocculus. (7) Dorsal Y-group projects to contralateral dorsal cap, β-nucleus, and dmcc. Abbreviations: *Amb*, nucleus ambiguus; *β*-nucleus; *cf*, climbing fiber; *Cu*, cuneate nucleus; *dc*, dorsal cap of Kooy; *dmcc*, dorsomedial cell column; *DVN*, descending vestibular nucleus; *Fl*, flocculus; *Gc*, granule cell; *icp*, inferior cerebellar peduncle; *IntP*, interpositus nucleus; *LRN*, lateral reticular nucleus; *mf*, mossy fiber; *LCN*, lateral cerebellar nucleus; *LVN*, lateral vestibular nucleus; *MCN*, medial cerebellar nucleus; *MVN*, medial vestibular nucleus; *NPH*, nucleus prepositus hypoglossi; *Nsol*, nucleus solitarius; *Pc*, Purkinje neuron; *pf*, parallel fiber; *PFl*, paraflocculus; *PO*, principal olive; *Psol*, parasolitary nucleus; sol, tractus solitarius; *SVN*, superior vestibular nucleus; *SpV*, spinal trigeminal nucleus; *spV*, spinal trigeminal tract; *X*, dorsal motor nucleus of the vagus; *XII*, hypoglossal nucleus; *VI*, abducens nucleus; *Y*, Y-group: 8n, auditory-vestibular nerve. *Image courtesy of Barmack, N.H., 2003. Central vestibular system: vestibular nuclei and posterior cerebellum. Brain Res. Bull. 60 (5), 511–541. https://doi.org/10.1016/S0361-9230(03)00055-8.*

CORTICAL PROJECTIONS

The ascending vestibulocortical system (Fig. 18.9) includes three identified synaptic stations: the vestibular nuclei, the thalamus, and the parietoinsular vestibular cortex (PIVC), somatosensory, or inferior parietal cortices. These trisynaptical connections receive information from the vestibular endorgans (maculae and crista ampullaris). As with most information that reaches the cerebral cortex, vestibular afferent fibers must first pass through the thalamus, particularly at the ventrobasal, ventrolateral, and intralaminar nuclei, and the geniculate bodies, which consequently projects to the previously described cortical areas (Walker et al., 2014; Wijesinghe et al., 2015). The PIVC is located in and around the lateral sulcus (parietal operculum), and the posterior insula and retroinsular cortex, and receives fibers from the brainstem and transcallosal connections from bilateral motion sensitive visual cortex areas. Hence, cortical vestibular areas receive converging vestibular, somatosensory and visual afferents, creating a multimodal cortical system with the function of postural and eye movement control (De Waele et al., 2001; Dieterich and Brandt, 2018).

Structural and functional hemispheric vestibular dominance has been documented, being the only sensory modality with dominance. Right-handed dominance is concordant to right hemispheric vestibular dominance, and left-handed dominance for the left hemisphere. This dominance is demonstrated by caloric stimulation in positron emitted tomography (PET), galvanic stimulation, and auditory evoked vestibular otolith stimulation, presenting a higher activation of the right insula, retroinsular cortex (RIC), superior temporal gyrus (STG), and inferior parietal lobule (IPL) in right-handed patients, and activation of the same areas of the left hemisphere in left-handed patients. This dominance could explain the presence of hemispatial neglect in dominant parietal lesions, which involves vestibular function (Dieterich and Brandt, 2018).

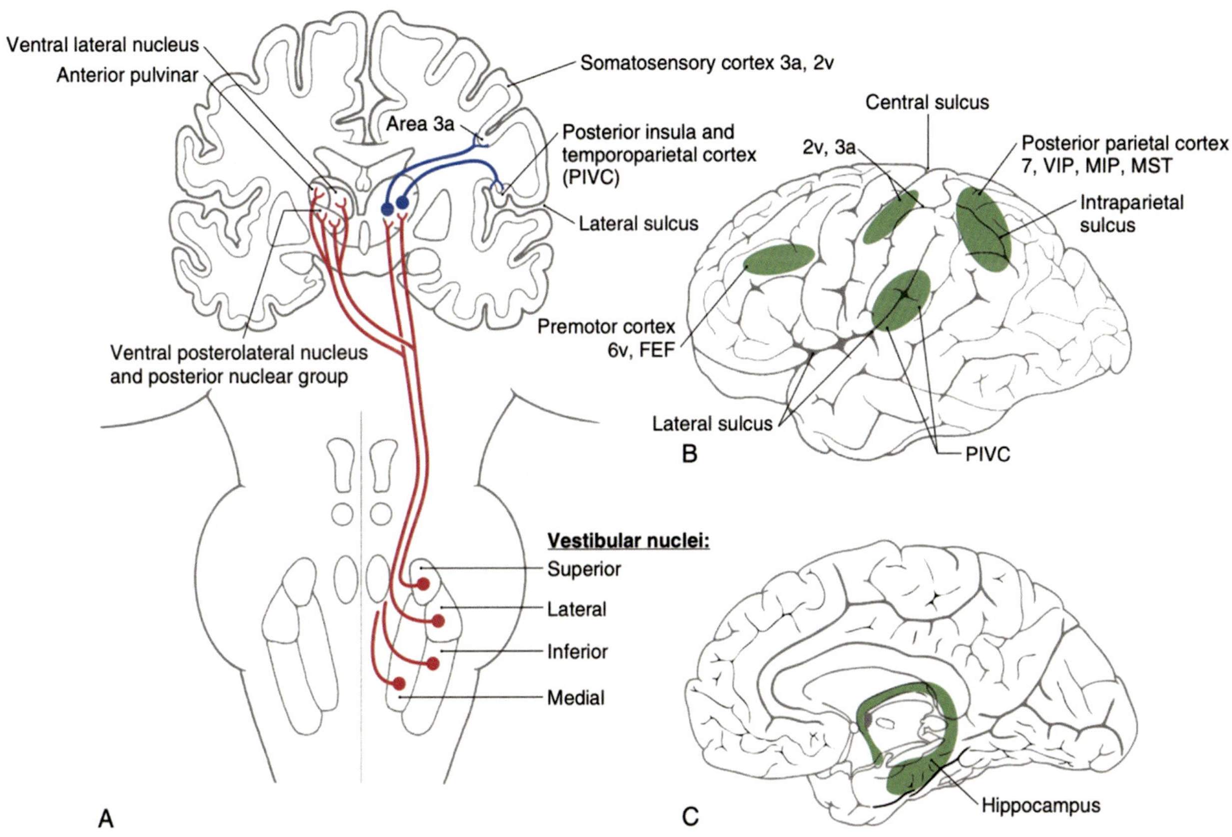

FIGURE 18.9 **The vestibulothalamocortical pathway.** (A), Vestibular nuclei send bilateral projections to the thalamus and then to the cortex. (B) Four main cortical regions receive vestibular signals: Areas 2v and 3a, the posterior insular and temporoparietal cortex (PIVC), the posterior parietal cortex, and the premotor cortex. FEF, frontal eye field; MIP, medial intraparietal area; MST, medial superior temporal area; PIVC, parietoinsular vestibular cortex; VIP, ventral intraparietal area. (C) Vestibular signals are required for spatial navigation and spatial memory in the hippocampus. *Image courtesy of Haines, D.E., Mihailoff, G.A. (Eds.), 2018. In Fundamental Neuroscience for Basic and Clinical Applications (Fifth Edition). Elsevier, 332. https://doi.org/10.1016/B978-0-323-39632-5.15001-7.*

VESTIBULAR EFFERENTS AND VESTIBULAR REFLEXES SYNAPTOLOGY

Although the vestibular reflexes mediated by these projections are highly stereotyped, the vestibular system interacts extensively with the visual and proprioceptive systems, and its activity is influenced by the cerebellum and cortex to provide a high degree of adaptability.

Vestibulo-ocular reflexes (VOR) involve the activation of the motoneurons that innervate the extra-ocular muscles (Fig. 18.10). The function of these reflexes is to maintain the orientation of the eyes in space during head movements so that the visual image remains stable on the retina. It involves a three neuron reflex arc from the semicircular ducts to the vestibular nuclei and then to the extraocular muscles to cause conjugate eye motion in a direction opposite to head turning (Khan and Chang, 2013).

Essentially, two types of reflex responses are produced by the VOR, compensatory and orienting. Compensatory responses can be further divided into two subclasses. Angular rotation of the head is sensed by the semicircular canals, which drive the eyes to maintain gaze along a given direction in space by activating the compensatory angular vestibulo-ocular reflex. Similarly, the linear acceleration associated with side to side, front-to-back, and up–down translation of the head activates the otolith organs, which initiate the compensatory linear vestibulo-ocular reflex (lVOR), to maintain gaze fixed on an invariant point

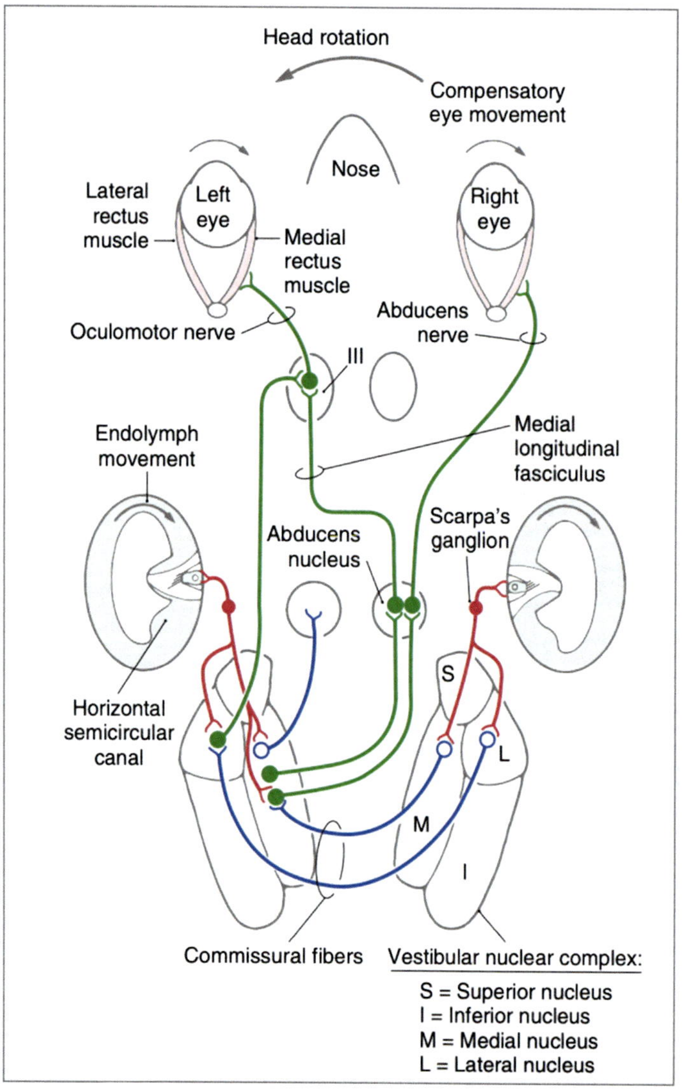

FIGURE 18.10 **Open cell bodies represent inhibitory projections. III, oculomotor nucleus. The connections subserving the horizontal vestibuloocular reflex.** *Image courtesy of Haines, D.E., Mihailoff, G.A. (Eds.), 2018. In Fundamental Neuroscience for Basic and Clinical Applications (Fifth Edition). Elsevier, 330. https://doi.org/10.1016/B978-0-323-39632-5.15001-7.*

in space. The central vestibular system decodes for head and body three-dimensional space information and coordinates in spatial responses to maintain position and posture, which are realized by convergence of otolith and canal-based signals. These mechanisms are paralleled by vestibulocollic reflexes which perform analogous head rotations to stabilize head pointing in space and orient the head to correspond to alterations in the gravitoinertial acceleration in space as well as by vestibulospinal reflexes that readjust the position of the limbs to counter the accelerations to maintain postural stability. This results in ipsilateral medial rectus and contralateral lateral rectus contraction which produces eye movement to the left (opposite to head turning). If eye velocity and head velocity do not match, input from the cerebellar flocconodular lobe is sent to the vestibular nuclei to modify their firing rate to correct this discrepancy (Khan and Chang, 2013).

Because the circuitry underlying the vestibulo-ocular reflexes is interconnected with networks that mediate other ocular reflexes as well as voluntary eye and head movements, these various movements are normally well coordinated. The beauty of this organization is that the central nervous system can code all angular head movements with only three vectors, greatly simplifying central processing. This planar organization is highly conserved and is present over a wide range of phylogeny. Thus, the same canal structure is present in cartilaginous fish such as sharks, rays and mammals (Cohen and Raphan, 2004).

Vestibulospinal reflexes involve the activation of spinal motor neurons that innervate neck, trunk, and limb muscles. The function of these reflexes is to maintain balance and body orientation in the face of ongoing perturbations. The vestibulospinal reflexes are fast, providing compensatory movements that counteract and protect against the potentially injurious results of tripping or falling. The vestibulospinal reflexes are coordinated with other spinal reflexes and the vestibulo-ocular reflexes. This coordination is an essential element of our daily activities. An example is the coordination of the vestibulospinal reflex and the cervicospinal reflex (a proprioceptive reflex that generates limb movements in response to activation of neck proprioceptors). Each of these reflexes initiates unilateral limb extension when the head is tilted appropriately, but on opposite sides of the body.

Another reflex that is related to the vestibulospinal reflex is the vestibulocollic reflex. This activates neck muscles that function to stabilize the head and keep it properly oriented in space. The exact pathways that contribute to this reflex are not completely understood (Hain and Helminski, 2007).

The vestibular reflexes, though highly stereotyped, exhibit plasticity at several levels, from hair cells to central synapses, allowing for adaptive changes in response (Glover, 2004).

Many tests can help determine if the vestibular system is functioning properly. The oculocephalic reflex is a simple test used to determine if the brainstem of a comatose patient is intact using the reflexes of the vestibular system. The test involves rotating a patient's head horizontally, which should activate the vestibular system on the ipsilateral side of rotation. This results in the patient's eyes slowly deviating to the side opposite head movement, if the brainstem is intact. If the brainstem is not intact, the eyes will follow the movement of the head to the ipsilateral side (doll's eyes reflex). Caloric testing is a test that uses differences in temperature to diagnose damage to the acoustic nerve. A small amount of cold water or air is gently delivered into one of your ears. Your eyes should show an involuntary movement called nystagmus. Then they should turn away from that ear and slowly return back. If water is used, it is allowed to drain out of the ear canal. Next, a small amount of warm water or air is gently delivered into the same ear. Again, your eyes should show nystagmus. Then they should turn toward that ear and slowly back. More specific tests for components of vestibular function includes videonystagometry, the most common vestibular function test. The test is divided into three parts including ocular motor, positional testing, and caloric testing. Other diagnostic modalities include rotation tests and video head impulse testing (VHIT). Both of these tests use devices to monitor eye movements when the head is rotated in various directions to test the integrity of the vestibulo-ocular reflex (Zalewski, 2015).

There is first evidence that vestibular subfunctions are mediated in separate and distinct parts of the multisensory cortical network of vestibular and ocular motor areas. Recently, a functional separation was proposed in a review by Ventre-Dominey, who distinguished between two separate cortical vestibular subsystems: a velocity pathway including the middle superior temporal area (MST) and direct descending tracts to the vestibular nuclei, which receives visual and vestibular velocity signals, and is involved in head direction and acceleration perception, with rapid regulation of eye–head coordination; and an inertial processing pathway involving the parietal cortex in connection with the subcortical vestibular nuclei complex accountable for velocity perception integration (Ventre-dominey, 2014). This inertial processing was interpreted to be a perisylvian network active in higher-order cortical processes related to spatial referencing and interpretation (Dieterich and Brandt, 2015).

VERTIGO AND VESTIBULAR SYSTEM SYMPTOMS

Damage to the vestibular nerve or lesions within the brain stem can all cause vertigo and loss of balance. Dizziness designates an imprecise term used by patients to describe sensations of imbalance, lightheadedness, syncope, etc. The character of the sensation described by patients may be used to define dizziness into two categories: vertigo, a sensation of self-rotary movement when no self-motion is present, that indicates disorders of the vestibular system; and nonvertiginous dizziness, that can be described as imbalance, lightheadedness, syncope, and other similar ailments. The "internal" sensation of vertigo should be distinguished from the "external" visual sensation of motion referred as oscillopsia (Bisdorff et al., 2009). Symptoms in peripheral vertigo generally occur on an acute onset, are suppressed by visual fixation, are often accompanied by other otological symptoms such as hearing loss, are fatigable, and go together with mild to moderate instability to stand, and visceral symptoms as nausea, vomiting, and bradycardia. Manifestations nonspecific to peripheral vestibular disease, for instance dizziness, a subacute onset, nonsuppression of vertigo with visual fixation, severe instability (patient is unable to stand), sensation of "drunkenness," stumbling, nonfatigability, and display of other focal neurological symptoms (most commonly ataxia), are expected to be caused by nonvestibular or central vertigo (Bisdorff et al., 2009; Lee, 2012; Ciuman, 2013).

Vertigo caused by vestibular pathology can be classified in peripheral vestibular disorders, including Meniere's disease, viral labyrinthitis, vestibular neuritis, labyrinthine syphilis, trauma, vascular loops or neoplasia at the cerebellopontine angle, perilymphatic fistula, intoxication, alcohol, vascular disorder, benign paroxysmal positional vertigo, vestibular paroxysmia, bilateral vestibulopathy, or residual peripheral vestibular deficit; and central vestibular disorders such as: Brain stem lesions or neoplasia, vertebrobasilar insufficiency, vertebrobasilar anomalies, basilar artery migraine, or vestibular epilepsy (Ciuman, 2013).

NYSTAGMUS

As the vestibulocerebellar system is involved in eye motion regulatory mechanisms, it is expected to find abnormal eye movements in peripheral and central vestibular injury. Nystagmus consists of slow movement of the eyes from the resting position, followed by an immediate rapid return to the initial position (Fluur, 1962) (Fig. 18.11). The fast corrective phase defines the

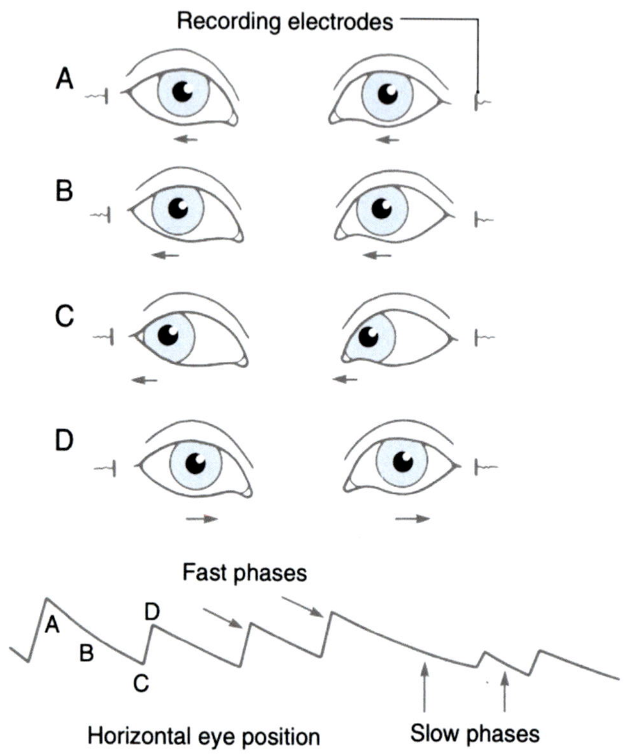

FIGURE 18.11 **Vestibular nystagmus in the horizontal plane showing slow-phase (A–C) and fast-phase (D) eye movements.** *Image courtesy of Haines, D.E., Mihailoff, G.A. (Eds.), 2018. In Fundamental Neuroscience for Basic and Clinical Applications (Fifth Edition). Elsevier, 330. https://doi.org/10.1016/B978-0-323-39632-5.15001-7.*

direction of the nystagmus. This movement can be present in a horizontal, vertical, rotatory, or mixed direction. As the maintenance of the ocular posture and alignment of the eyes depends on retinal and labyrinthine input and on their central connections to the cerebral cortex, brainstem, and cerebellum, nystagmus can be caused by retinal or labyrinthine disease, or disorders affecting the cerebellum or vestibular portions of the brainstem (Lindsay et al., 2011). Therefore, vestibular originated nystagmus can be classified into peripheral or central nystagmus. Peripheral nystagmus appears as horizontal, or a combination of torsional and horizontal movement, with contralateral direction to the affected side. Peripheral nystagmus is fixed in a single direction, visual fixation suppresses it, as it improves with gaze toward the affected side, and worsens directing the gaze opposite to the lesion (Baloh, 1998; Thompson and Amedee, 2009). The eye movement direction in nystagmus of central origin can be horizontal, vertical, rotatory, or dissociated (only present in one eye). Nystagmus of central pathology has a tendency to change direction with gaze, is unaffected by visual fixation, is not fatigable with repetitive testing, and may be associated with hearing loss and other focal neurological symptoms (Thompson and Amedee, 2009; Lindsay et al., 2011).

Since vestibular afferent nuclear neurons have a spontaneous firing rate even in absence of head movement, and most vestibulocerebellar functions are executed bilaterally, the unilateral damage of the vestibular nuclei could result in the central perception of head movement, with consequential disfunction in reflex pathways, as seen in vestibular or cerebellar nystagmus (Lyskowski, 2006).

Peripheral nystagmus can appear as physiological and pathological. Physiological peripheral nystagmus expresses after rotational acceleration, producing movement of the eyes in the plane of rotation, with a slow phase in a direction tending to maintain the primary visual position, and a fast phase in the opposite direction (the direction of the rotation of the head that produced the nystagmus). Another mean to produce physiological vestibular nystagmus is through caloric testing with elevation of the head at 30 degrees from the horizontal plane (positioning the lateral semicircular canal in a vertical angle). The application of cold water to the external auditory meatus produces a reduction of vestibular signal output from the tested side within 20 s after instillation, and consequential imbalance between both labyrinths. Then, cold water causes eye movement toward the irrigated ear, with rapid corrective movement to the opposite ear. Irrigation of hot water increases vestibular output to the tested ear and changes the direction of nystagmus (Lindsay et al., 2011).

INNER EAR PATHOLOGY

Classical inner ear disease is presented by the triad of sensorineural hearing loss, tinnitus, and vestibular vertigo, as it affects the entire membranous labyrinth in its vestibular and cochlear portions, involving hair cells, supporting cells, or aberrant inner ear fluid homeostasis. On acute hearing loss accompanied by vertigo, this tends to be typically presented as rotational vertigo, with accompanied manifestations of peripheral vestibular injury, such as lateropulsion, tilting swaying, and lift sensation (Ciuman, 2013). A common convention is to use the term vestibular neuritis when only vestibular nerve affection is present, and labyrinthitis when the peripheral vestibular symptoms are accompanied by ipsilateral hearing loss. Sudden unilateral loss of vestibular function is most commonly observed in viral and inflammatory disease, as seen in systemic viral illness, such as mumps, measles, and mononucleosis. Reactivation of herpes zoster virus in Ramsay–Hunt's syndrome can cause a combination of facial weakness and acute vestibulopathy, caused by affection to the facial and vestibulocochlear nerves (Walker et al., 2014).

VESTIBULAR NEURONITIS

Vestibular neuronitis is the second most common cause after benign paroxysmal positional vertigo for vestibular disease, manifested by vertigo and horizontally rotatory nystagmus in direction of the healthy inner ear. The most common presentation of vestibular neuronitis is that of viral etiology usually from the herpes virus family. Vestibular neuronitis can be caused by bacterial infections, such a borreliosis (Thompson and Amedee, 2009; Walker et al., 2014). Similar to that of the facial nerve in peripheral facial palsy, vestibular nerve affection is secondary to compressive edema in the epithelium of the narrow internal acoustic meatus. Clinical presentation is characterized by a rapid onset of severe rotatory vertigo with vegetative symptoms as nausea and vomiting, accompanied by severe imbalance. Symptoms last several days and no other otological manifestations are present. Upper respiratory tract infection previous to the development of vestibular symptoms is present with an incidence that varies from 23% to up to 100% (Thompson and Amedee, 2009; Lee, 2012). Vertigo diminishes over days to weeks, with imbalance persisting for months after resolution of acute disease. Recurrence may occur several times per year. Physical examination demonstrates nystagmus and caloric test with diminished responses in the affected side (Thompson and Amedee, 2009). Treatment is primarily supportive and consist in administration of antiemetics, and vestibular sedatives in the acute phase. Use of vestibular sedatives in a prolonged lapse is not recommended as it prevents vertigo compensation by inhibiting central regulation and can slow recovery. Randomized trials have supported brief use of steroids as high-dose methylprednisone for acute vestibular neuritis, as it has been shown to accelerate improvement of symptoms. Antiviral agents have failed to demonstrate added benefit in vestibular neuritis treatment, however, are still commonly used when herpes zoster virus is suspected (Thompson and Amedee, 2009; Lee, 2012; Walker et al., 2014).

BENIGN PAROXYSMAL POSITIONAL VERTIGO

Benign paroxysmal positional vertigo is the most common cause of vertigo, with a lifetime prevalence of 2.4% (Riga et al., 2011). It is caused by canalithiasis, affection that is caused when otoconia becomes displaced from its normal location in the utricle's gelatinous membrane, and migrates into the semicircular ducts, interfering with normal endolymph motion (Mitchell and Cullen, 2004). The process of detachment of otoconia is not fully understood, the presumed primary mechanism is through gravity caused movement of otoconial debris into the vestibular canal semicircular. The most commonly recognized conditions associated with secondary detachment of otoconia are trauma, vestibular neuritis, Menière's disease, idiopathic sensorineural hearing loss, and surgical procedures (Riga et al., 2011; Walker et al., 2014). The underlying pathophysiology of secondary benign paroxysmal positional caused by Menière's disease seems to comprise endolymphatic hydrops, with subsequent destruction of the maculae of the otolith organs by vascular compromise or direct distortion of its surface. In the case of vestibular neuritis, the pathogenic mechanism is seemed to be resultant from superior vestibular nerve lesion deriving in abnormal perception of utricle, and anterior and lateral semicircular canal function. Traumatic injury to the utricle may detach otoconia resulting in canalithiasis (Riga et al., 2011).

The main manifestation of benign paroxysmal positional vertigo is short-lasting vertigo episodes, with predominant appearance in abrupt changes of head position in relation to the gravity vector, for instance, changes from vertical onto horizontal position, or vice versa. These episodes appear predominantly after sleep, and are weakened by repeated head movement. Hearing loss and tinnitus are absent (Lee, 2012; Illich Zabolotnyi and Serhiivna Mishchanchuk, 2020). Vertigo tends to be side specific and last for less than 1 min, as patients have a tendency to lie down and position the head over the affected ear, this change of position results in displacement of the canalith due to gravitational movement to its most dependent position. Nausea can be present, but the transient nature of the vertigo is considerably brief to evoke vomiting (Thompson and Amedee, 2009; Lee, 2012). The posterior canal is involved in 85%–95% of the cases, lateral canal function ins compromised in less than 10% of the cases (Lee, 2012).

Diagnosis of benign paroxysmal positional vertigo is easily made by the Dix-Halpie positional test, which examines the posterior semicircular canal affection. In this test, the patient is transported from sitting to a supine head-hanging position, with the head turned 45 degrees toward the affected ear, after a short latency, the patient will develop a brief vertical-torsional nystagmus with a duration of less than 30 s. In the subsequent return to neutral position, nystagmus will change direction (Walker et al., 2014; Illich Zabolotnyi and Serhiivna Mishchanchuk, 2020). Treatment is primarily supportive, as a large proportion of patients will experience a spontaneous resolution of symptoms. A number of positional maneuvers have been developed in the treatment of persistent symptoms in benign paroxysmal positional vertigo, with the purpose of canalith reposition from the semicircular canals to the utricle. A common procedure for canalith reposition is Epley's maneuver, which begins with placement into Dix-Halpie position to evoke vertigo, after initial nystagmus decreases, the head is rolled in 180 degrees opposite from the affected ear, this position is maintained for 30 s and afterward the patient gets lifted to sitting position with the head turned to the healthy ear. This maneuver can be repeated until no nystagmus is provoked (Walker et al., 2014; Gaur et al., 2015). These repositioning maneuvers are reported to have a 91% success rate. Canalith repositioning maneuvers are less effective in secondary benign paroxysmal positional vertigo, compared to primary vertigo. In some cases, a single canalith repositioning maneuver is not effective and multiple procedures must be performed. Patients with symptoms completely refractory to repositioning maneuvers may be candidates for a singular neurectomy or posterior semicircular canal (Thompson and Amedee, 2009; Riga et al., 2011).

LABYRINTHITIS

Labyrinthitis is an uncommon disorder of the membranous labyrinth that affects both cochlear and vestibular organs. Similar to vestibular neuronitis, it is often preceded by an upper respiratory tract infection. Invasion of microorganisms or inflammatory mediator through otogenic or meningitis infection causes the damage to the labyrinth (Thompson and Amedee, 2009; Lee, 2012). This infection can be caused by bacterial pathogens, viral infection, bacterial toxins, and systemic disease (Thompson and Amedee, 2009). Conditions that affect the cerebrospinal fluid, such as meningitis, can represent an insult to the inner ear, because of the continuity of the cerebrospinal fluid and the perilymph through the perilymphatic duct (Swenson, 2017). Bacterial labyrinthitis may result from both meningitic or otogenic infection. Labyrinthitis from otogenic origin can be caused by cholesteatoma or otitis media, while infection of meningitis origin are more common in children less than 2 years of age, because of susceptibility to central nervous infections (Thompson and Amedee, 2009).

Patients with labyrinthitis present with both vestibular vertigo and hearing loss. To control acute symptoms of the disease treatment consists in vestibular sedatives and antibiotics. Surgical treatment of cholesteatoma or chronic suppurative otitis media may be incited in select cases (Lee, 2012).

MENIÈRE'S DISEASE

Menière's disease courses with endolymphatic hydrops, an increase in endolymphatic pressure by overaccumulation of this fluid in the inner ear, that causes inappropriate nerve excitation and consequent rotational vertigo, fluctuating sensorineural hearing loss, and tinnitus, the classic triad of Menière's disease. The exact etiology for endolymphatic hydrops is currently unknown (Thompson and Amedee, 2009; Lee, 2012). Vertiginous episodes may last from a few minutes to up to 2 h. Hearing loss is typically of low frequency spectrum. A fourth symptom can often be added, as patients can experience aural fullness and pressure, which can precede disease's episodes up to 20 min (Vassiliou et al., 2011). As most peripheral vestibular afflictions, Mêniere disease can present with nausea and vomiting when vertigo is long-lasting. Natural disease history involves spontaneous remission of symptoms in 60%–80% of patients, whereas 10%–20% of patients present with an intractable ailment (Lee, 2012).

Menière's disease tends to present episodical relapses. Hearing capacity is expected to be fully restored after episodes in the initial stage of the ailment. However, with recurring episodes, the patter of hearing loss becomes fluctuating and may become severe and progressive without restoration of hearing capability (Vassiliou et al., 2011).

Medical management of Menière's disease consist of salt restriction, and diuretics, intending to reduce perilymphatic hydrops, in combination with antiemetics, vasodilators, and vestibular sedatives, such as meclizine or promethazine (Walker et al., 2014). Local intratympanic or oral steroids can worsen the perilymphatic hydrops as it increases the expression of aquaporins (Ciuman, 2013). Intratympanic administration of gentamycin, an ototoxic, has shown to be useful in systematic reviews for severe disease (Lee, 2012). Patients with intractable symptoms that conserve serviceable hearing can be candidates to endolymphatic decompression surgery, intratympanic aminoglycoside infusion, or vestibular neurectomy. Patients without functional hearing and intractable symptoms can be candidates to complete labyrinthectomy (Thompson and Amedee, 2009; Lee, 2012).

SCHWANNOMA OF THE VESTIBULOCOCHLEAR NERVE

Schwannoma of the eighth cranial nerve (acoustic neuroma or vestibular schwannoma) is the most common neoplasm of the cerebellopontine angle, as this lesions account for 6%–10% of all intracranial tumors and up to 60%–90% of tumors in the cerebellopontine angle. It is usually a unilateral and sporadic benign nerve sheath tumor, nonetheless it can be found bilaterally and in larger dimensions in cases of Neurofibromatosis 2. Despite the more common name of acoustic neuroma, vestibulocochlear schwannomas grow from vestibular nerve Schwann cells in approximately 85% of the patients, with the remaining arising from cochlear branches (Swartz, 2008; Jung and Ramina, 2019). Vestibulocochlear nerve schwannomas are to some extent more common and tend to be larger in women, and a relationship between pregnancy and accelerated tumor growth has been stablished (Swartz, 2008).

Patients present with progressive unilateral episodic positional vertigo, horizontal nystagmus, disequilibrium, tinnitus, and asymmetrical high-frequency hearing loss, resulting from compression and/or infiltration of the vestibulocochlear nerve (Swartz, 2008; Thompson and Amedee, 2009). Sudden worsening of hearing loss can occur on 26% of patients, and is apparently associated with anteroinferior cerebellar artery, as this vessel is in close relationship to the vestibulocochlear nerve in the cerebellopontine angle (Swartz, 2008). Vertigo has the greatest negative impact on quality of life among patients, compared to tinnitus and hearing loss (Hizli et al., 2016). Slow growth of the tumor often allows central compensation of vertigo. Sustained progress con conditionate compression of the facial nerve in the VII/VIII cranial nerve complex, which occurs in 6% of the patients, and is presented as ipsilateral peripheral facial palsy. In larger and mainly in cystic tumors, compression of the trigeminal nerve and the brainstem can occurs, being manifested as ipsilateral facial numbness and truncal or gait ataxia (Thompson and Amedee, 2009; Jung and Ramina, 2019). The presence of vertical nystagmus implies large tumors with brainstem compression. Obstructive hydrocephalus is relatively common in large tumors that compress the cerebellum and displace the fourth ventricle (Swartz, 2008; Jung and Ramina, 2019).

Audiological examination through audiograms with sound discrimination (including pure-tones, and speech discrimination test) is helpful to determine the grade of remaining hearing function. Brainstem evoked response audiometry provides information on hearing function and is useful to determine the nerve of origin of vestibulocochlear schwannomas. Auditory brainstem response may be normal in 25% of cerebellopontine angle tumors (Kaul and Cosetti, 2018; Jung and Ramina, 2019).

Magnetic resonance imaging with fine cuts of the internal auditory canal, with or without gadolinium enhancement, represents the gold standard for diagnosis of internal auditory canal or cerebellopontine angle pathology (Fig. 18.12). T1, T2, FLAIR, and DWI images are usually sufficient for the diagnosis of vestibulocochlear schwannoma (Kaul and Cosetti, 2018; Jung and Ramina, 2019). These tumors present most commonly as combined internal acoustic meatus/cerebellopontine angle lesions, nevertheless, presentation can be entirely intracanicular (Swartz, 2008). Over 50% of

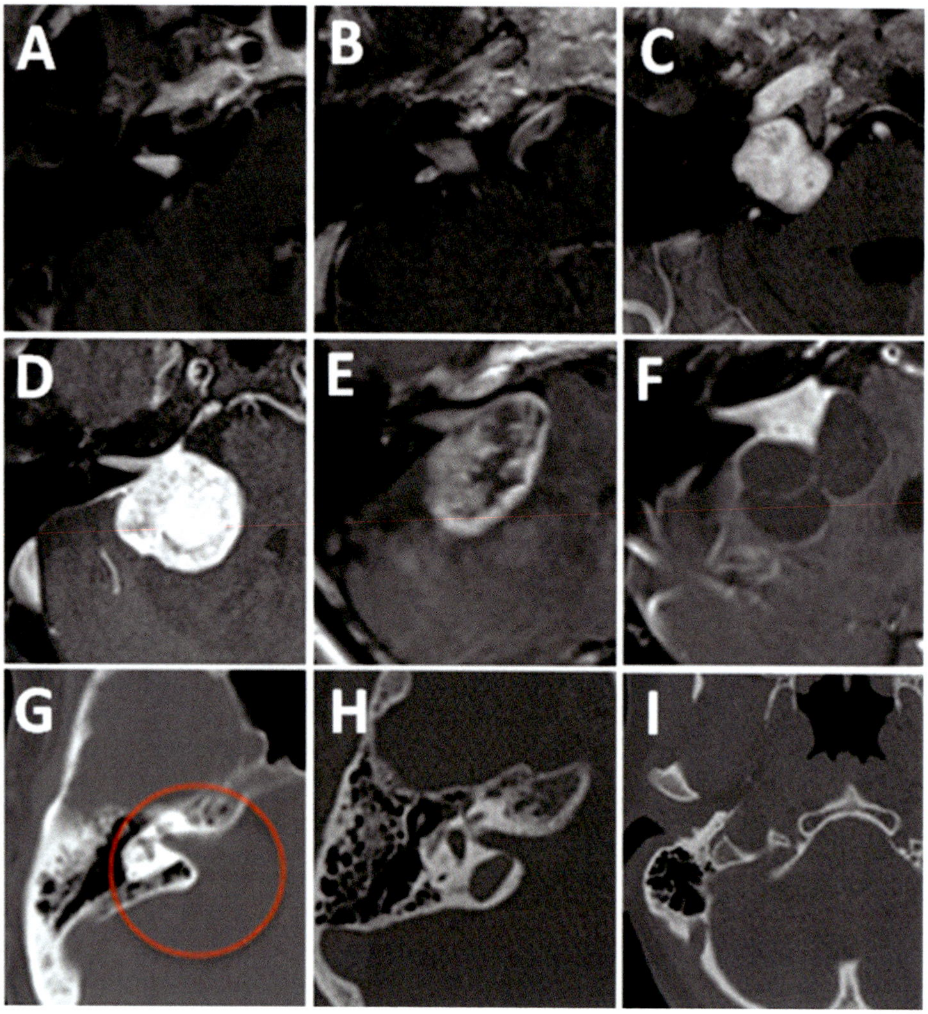

FIGURE 18.12 **Imaging in vestibular schwannoma (VS).** Radiological MR presentation (A–F) of VS may vary considerably: From large to small, from solid to cystic, from compact with high cellular density to soft and well vascularized, from medial located to invading the petrous bone and filling the entire internal auditory canal (IAC). Preoperatively, high-resolution bone window CT is recommended to evaluate the pneumatization of the petrous bone (G) as well as the presence of a high jugular bulb (H) and large emissary veins (I). *Image courtesy of Tatagiba, M., Ebner, F.H., 2021. Nakamura, T. et al. Evolution in Surgical Treatment of Vestibular Schwannomas. Current Otorhinolaryngology Reports 9, 467–476. https://doi.org/10.1007/s40136-021-00366-2.*

vestibulocochlear schwannomas present as isointense in T1-weighted images, as hypointensities may represent a cystic component (Jung and Ramina, 2019). Vestibular schwannomas are typically hyperintense in T2-weighted images. On postcontrast images, these tumors will typically demonstrate intense enhancement. Tumors with a mixed internal acoustic meatus/cerebellopontine angle produce an image of "ice cream on cone" appearance (Farid, 2014).

VESTIBULOCOCHLEAR NERVE SCHWANNOMA TREATMENT

Observation, microsurgical resection, radiotherapy, and radiosurgery are the most common options for management of vestibulocochlear schwannoma, depending of the hearing status, growth velocity, size, and characteristics of the tumor (Jung and Ramina, 2019).

Surgery is generally performed under continuous intraoperative neuromonitoring (IONM). IONM encompass the somatosensory- (SSEP) and motor-evoked potentials (MEPs) to the upper and lower extremities, corticobulbar motor-evoked potentials (CoMEPs), and free-running electromyography (EMG) of the affected cranial nerves (VII, XI, XII c.n.) as well as the brainstem auditory–evoked potentials (BAEPs). In case of semisitting position, we suggest SSEPS during the positioning maneuver in order to avoid any damage to the cervical spinal cord. In large vestibular schwannoma, additional recording of lower cranial nerves (CN XI, XII) is carried out to protect swallowing function.

Observation with serial imaging is reserved for asymptomatic patients with small- to medium-sized tumor, or those with mild hearing loss, advanced age, or comorbidities that could exclude surgical intervention. Imaging it's recommended after first presentation, 6 months later, and then annually for the next 3 year, then every 2 years for a minimal surveillance of 10 years (Kaul and Cosetti, 2018). Reported annual growth rates of vestibular schwannoma vary from 0.3 to 4.8 mm (Rosahl and Eβer, 2017).

SURGICAL RESECTION

VS surgery is performed either in a supine (SP) or a semisitting position (SSP) depending on tumor size and risk profile of the patient (Fig. 18.13).

For large tumors, with brainstem or cerebellum compression, with demonstrable growth on imaging >2 mm per year, or cystic degeneration of tumor in magnetic resonance, craniotomy with microsurgical resection is the modality of choice. Vestibulocochlear schwannomas approach can be through translabyrinthine, retrosigmoid, or middle fossa craniotomy, based on extension to internal acoustic meatus and/or cerebellopontine angle, tumor dimensions, and hearing preservation (Kaul and Cosetti, 2018; Jung and Ramina, 2019; Mastronardi et al., 2019). Medial cranial fossa and retrosigmoid approaches offer the possibility to preserve hearing, whereas translabyrinthine approach intentionally opens through the inner ear leading to profound hearing loss (Kaul and Cosetti, 2018).

The middle fossa approach has long been used to expose the internal auditory canal, mainly for resection of small vestibulocochlear schwannomas located at the porus acousticus, limited to the internal acoustic canal or with minimal (up to 1 cm) extension to the cerebellopontine angle (You et al., 2013; Kaul and Cosetti, 2018). The middle fossa approach to the internal acoustic meatus is selected when there exists probability to preserve hearing. With this approach a temporal craniotomy above the ear and zygoma is made, through the elevation of the inferior dura of the temporal lobe, and via extradural dissection, the internal acoustic meatus is exposed in its superior wall. The preferred method for internal acoustic canal exposition is by drilling at the petrous bone edge above the fundus of the canal in the area medial to the arcuate eminence. This drilling is directed anterolateral from the meatal porus to the meatal fundus, where the vertical crest is located (Rhoton, 2007). Bone removal in the internal acoustic canal should be limited to avoid lesions of important structures: posteriorly is limited by the superior semicircular canal and vestibule, located a few millimeters behind the canal and oriented parallel to the labyrinthine segment of the facial nerve; laterally, the vertical crest marks the position of the facial nerve; anteromedial edge of exposure is the cochlea, which should be avoided if hearing preservation is pursued (Rhoton, 2007). Hearing preservation rates are variable in a range from 64% to 82.7%. As this approach carries extradural dissection, rates of cerebrospinal fluid fistula are low. Because of temporal love retraction, seizures and temporal lobe injury can occur (Kaul and Cosetti, 2018).

Retrosigmoid approach is the most used access in surgical management of vestibulocochlear schwannomas when attempting hearing preservation, as it offers excellent exposition of anatomical structures of the posterior fossa and a panoramic visualization of the cerebellopontine angle. This approach is suitable for tumors of any size regardless of preoperative hearing status. Nevertheless, fundus exposition of the internal acoustic meatus is limited (You et al., 2013; Jung and Ramina, 2019; Mastronardi et al., 2019). Retrosigmoid approach requires a posterior fossa craniotomy that allow access to the cerebellopontine angle, petrous apex, clivus, and the porus acousticus (Kaul and Cosetti, 2018). The technique for this approach involves placement of a burr hole below the asterion and a craniectomy that exposes the lower margin of the transverse sinus in superior position, the posterior margin of the sigmoid sinus in lateral position, and the inferior portion of the squamous portion of the occipital bone in inferior position (Rhoton, 2000). Dural exposure is performed in a semicircular shape, covering part of the cerebellar hemisphere to serve as a protective sheath during retraction, dural adhesion posterior to the porus acousticus is removed or incised. Anteroinferior cerebellar artery can be found firmly adherent to the petrous dura mater, in which case both the dura and the artery should be elevated in conjunction to expose the internal acoustic meatus without arterial injury risk. Drilling of the canal starts from the upper and lower corners of the porus acousticus to remove the posterior meatal wall and expose cochlear and facial nerves (Mastronardi et al., 2019). Posterior and superior semicircular canal with the common crus should be preserved when removing the posterior meatal wall when hearing preservation is pursued. The endolympatich sac may be entered in removing the dura from the posterior meatal lip. Endolympathic duct is the structure at major risk in the meatal exposure. A V-cut is performed in the dorsal surface of the tumor and debulking of the center of the tumor is made to be removed piece by piece from the inside. Removal of the capsule is executed to separate it from the brainstem and cranial nerves (Rhoton, 2000; Mastronardi et al., 2019). If the tumor extends to the vestibule, it can be exposed by drilling the posterior wall lateral to the meatal fundus. Reconstruction of the meatus is important to avoid cerebrospinal fluid

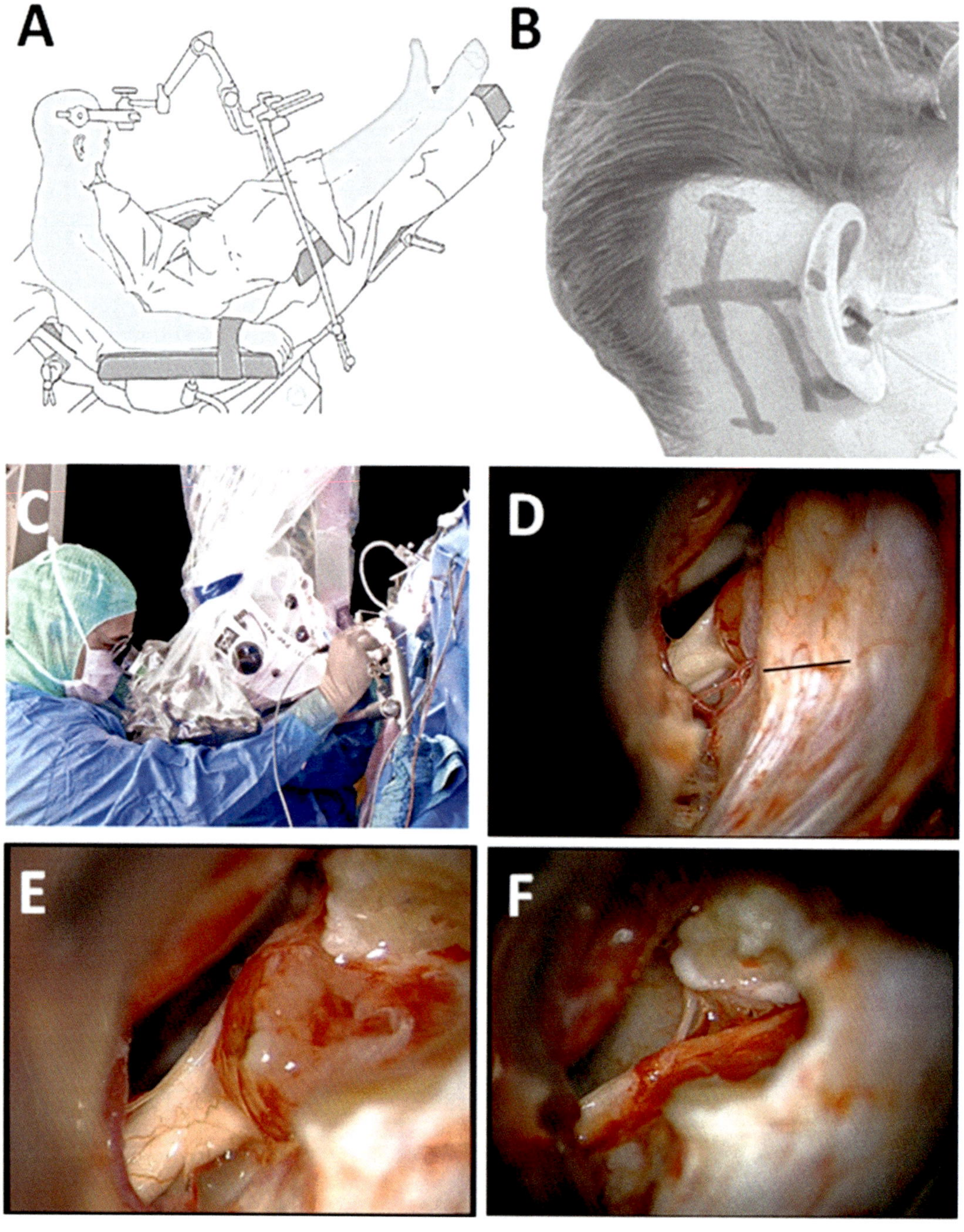

FIGURE 18.13 **A new surgical approach for vestibular schwannoma (VS).** (A) Schematic presentation of the semisitting position in VS surgery. Please note the position of the legs in relation to the head. Illustrations of the skin incision (B) and the position of the surgeon during semisitting surgery (C). After dura opening the Tübingen line (black) is identified (D) representing the position of the internal auditory canal (IAC). Afterward, the IAC is opened with a burr (E) enabling early identification and preservation of the auditory and facial nerves (F). *Image courtesy of Tatagiba, M., Ebner, F.H., Nakamura, T. et al., 2021. Evolution in Surgical Treatment of Vestibular Schwannomas. Curr Otorhinolaryngol Rep 9, 467–476. https://doi.org/10.1007/s40136-021-00366-2.*

fistula and infection, and should be carried by the placement of a fat or muscle graft to seal the defect (Rhoton, 2000; Jung and Ramina, 2019). Cerebrospinal fistula is the most common complication, reason to use continuous lumbar drainage for the majority of cases, except in intrameatal tumors (Nonaka et al., 2013).

The main advantage of translabyrinthine approach is the minimization of brain retraction, and offers an excellent visualization of the lateral portion of the internal acoustic meatus and its fundus, allowing facial nerve identification and a more complete resection of tumors in this region. This transtemporal approach accesses the meatus through the labyrinth, leading to an intentional and profound sensorineural hearing loss. Translabyrinthine approach is preferred in large tumors that grow toward the internal acoustic meatus without preoperative serviceable hearing (You et al., 2013; Kaul and Cosetti, 2018). A mastoidectomy and labyrinthectomy is performed, bone removal should be made searching exposure of the Trautman's triangle (delimited by the sigmoid and superior petrosal sinuses, the otic capsule, and the jugular bulb), and to expose the nerves lateral to the tumor on the course of the internal acoustic meatus (Rhoton, 2007). The cortex over the mastoid bone is removed with a drill in two lines from the top of the external auditory meatus to the mastoid apex, and from the root of the zygoma to the asterion (Mastronardi et al., 2019). Posterior to cortex removal, air cells are removed exposing the mastoid antrum, cortical bone covering the labyrinth, and the digastric groove that marks the mastoid segment of the facial nerve as it exits the stylomastoid foramen (Rhoton, 2007). Bone removal continues to expose the semicircular canals and to skeletonize the sigmoid sinus, middle fossa dura (temporal tegment), mastoid segment of facial nerve, and upper surface of the jugular bulb (Rhoton, 2007). Labyrinthectomy involves removal of the semicircular canals and the vestibule to expose the dura in the internal auditory canal. Bone is removed medial and posterior of the vestibule to expose the facial nerve. As Trautman's triangle is exposed by bone drilling, a V-shaped incision is made to expose the structures of the cerebellopontine angle and the internal acoustic meatus (Rhoton, 2007). Contents of the internal acoustic meatus are separated from the facial nerve and removal of tumor is performed. To reduce possibility of cerebrospinal fistula, incus is removed and an autologous fat or muscle graft is placed occluding the origin of the auditive tube, dural incision is closed and autologous graft is placed (Mastronardi et al., 2019).

Independent of surgical approach, reported resection rates are higher than 95% even with preservation of function as the primary goal. Facial nerve preservation can be achieved in more than 90% of the cases. With functional hearing and small tumors, hearing preservation success exceeds 50% (Rosahl and Eβer, 2017). Balance disturbance is a common symptom after surgery in the three approaches, which in the majority of cases will improve during subsequent follow-ups (Nonaka et al., 2013).

RADIOSURGERY FOR VESTIBULAR SCHWANNOMA

Microsurgery (MS) is the method of choice for large tumors with radiological or neurological signs of brainstem compression. Nowadays, MS can be performed with low mortality, but it is still associated with a significant risk of neurological sequelae such as hearing loss and facial nerve palsy.

Single-dose stereotactic radiosurgery (SRS) and fractionated stereotactic radiotherapy (FSRT) aiming to arrest tumor growth are the most commonly used noninvasive alternatives. In case series of patients suitable for radiosurgery/radiotherapy, the progression-free survival rates are comparable to those of case series of microsurgical resection of VS (Persson et al., 2017). RS and FSRT entail a lower incidence of most common side effects.

The first minimally invasive treatment of vestibular schwannomas was pioneered by Leksell at the Karolinska Hospital in 1969. A single fraction of radiotherapy was delivered using the Gamma Knife and this was named "stereotactic radiosurgery" (SRS). Gamma Knife uses some 200 radiation beams from cobalt-60 sources that converge with high accuracy on the target. By moving the patient's head in relation to the beams' isocenter, the radiation dose can be optimized in relation to the shape and size of the target.

The CyberKnife system uses X-ray image-based feedback and consists of a small linear accelerator mounted on a robotic arm. This permits automated patient movement compensation during radiation delivery. CyberKnife does not require use of a rigid frame and therefore is sometimes referred to as "frameless SRS."

Fractionation of the radiation dose produces, in most situations, less normal tissue toxicity at the same level of tumor control than a single large dose. The basis of fractionation is that dividing a dose into a number of fractions spares normal tissues because of repair of sublethal DNA damage during the interval between fractions and repopulation of cells. At the same time fractionation increases damage to the tumor because of reoxygenation and reassortment of cells into radiosensitive phases of the cell cycle (Mulder et al., 2012).

There are some different advantages/disadvantages related to the radiation techniques in SRS and FSRT. Linear accelerator (LINAC)-based systems for FSRT are more available, and due to the low fractionation dose, larger tumors can be treated. The theoretical rationale for dividing the prescribed total dose into 25–30 fractions is mainly to increase the chance of targeting the tumor cells in the most radiation-sensitive phase of the cell cycle. On the contrary, SRS relies on a single high radiation dose, potentially affecting tumor cells also in the nondividing phase. Physical properties of Gamma Knife SRS allow a steep radiation gradient on the tumor margin, which is important for keeping the dose to adjacent structures low. While it is known that some tumors resistant to fractionated radiotherapy may responds to radiosurgery, the precise differences of the radiobiological properties of the two modalities are to some extent still an issue of discussion.

In a systematic review of both techniques, the hearing deterioration showed no substantial difference between the FSRT group (45%) compared to SRS (49%). Contrary to what has previously been reported in some single-center comparative studies, an increased risk of facial and trigeminal nerve deterioration was found in the included FRST studies compared to studies on SRS (Persson et al., 2017).

While most small tumors are managed conservatively and larger tumors with surgery and/or radiotherapy, there is a gray area surrounding best management options for tumors falling between categories (Hentschel et al., 2018; Kumar Gupta et al., 2020).

Long-term evidence (>10 years) regarding hearing preservation following radiotherapy is limited. Yang et al. reported an average hearing preservation rate of 57% from data derived from 74 articles; however, the average follow-up was only 41.2 months (Coughlin et al., 2018). Cyst formation and malignant transformation should be taken into consideration as late adverse radiation effects (>10 years after radiosurgery), although they appear to be rare (Hasegawa et al., 2013).

A more recent case-controlled study found that hearing preservation among patients decreases from 53% at 5 years to 34% at 10 and 15 years among all tumor grades (Frischer et al., 2018). This emerging evidence at increased time points suggests hearing may not be as well preserved as once thought, and loss occurs due to longer-term nerve damage as a result of radiation exposure. These studies also found the tumor Koos grade (Grade I: Intracanalicular; Grade II Extension into cerebellopontine angle <2 cm; Grade III: Occupies cerebellopontine angle, no brainstem displacement, <3 cm; Grade IV Brain displacement >3 cm) to be an independent predictor of hearing loss, representing a possible confounder (Kumar Gupta et al., 2020).

Some progress has been made in medical therapies for treatment-resistant VS and applies most usefully to patients with NF2 where preservation of function is key. Bevacizumab has made the most progress and appears to be a viable treatment option for patients with NF2 and growing tumors or loss of hearing. In these patients, bevacizumab recovers some useful hearing function and results in tumor reduction; however, the effect is ultimately lost with time succumbing to the natural tendency of the tumor to grow, opening doors for combination or other therapies (Rosahl et al., 2017; Gompel et al., 2018).

When dealing with large VSs, a treatment dilemma arises: a wait-and-scan policy is not recommended given the possibility of life-threatening complications due to further VS progression. The complications of radio surgery are related to dose and volume, and the procedure can cause (transient) VS swelling, which is one reason that this therapy is generally avoided in large VSs. Controversy surrounds complete microsurgical removal for tumor control versus subtotal removal for the preservation of facial nerve function in large VSs. If removed subtotally, these lesions do not usually stop growing. Given the good growth control and facial nerve function preservation, the possibility of preserving serviceable hearing, and the low number of complications, a subtotal resection followed by GKS can be the treatment option of choice in large VS (Van de Langenberg et al., 2011; Rosahl et al., 2017).

Some patients may present with a large VS on MR imaging but without disabling symptoms and mass effect. In addition, serviceable hearing may be present. Other patients may be physically unfit to undergo VS microsurgery as a result of severe comorbidity. In this selected group of patients presenting with large VSs without symptoms of mass effect, microsurgical removal can be avoided, and primary GKS can control lesion growth and better preserve cranial nerve function. Primary GKS for large VS leads to acceptable radiological growth control rates and clinical control rates, with the chance of hearing preservation.

Although VSs are the most common lesion of the cerebellopontine angle CPA (70%–90%), other pathology may present in this location. Meningiomas are the second most common lesion, occurring between 5% and 15% of cases followed by epidermoids and lipomas. Arachnoid cysts and other tumors can extend to CPA compromising vestibular system and have to be in mind with atypical symptom or radiologic presentation (Friedmann et al., 2015).

References

Baloh, R.W., 1998. Differentiating between peripheral and central causes of vertigo. Otolaryngol. Head Neck Surg. 119 (1), 55–59. https://doi.org/10.1016/S0194-5998(98)70173-1.

Baloh, R.W., Kerber, K.A., 2011. Clincal Neurophisiology of the Vestibular System, fourth ed.

Barmack, N.H., 2003. Central vestibular system: vestibular nuclei and posterior cerebellum. Brain Res. Bull. 60 (5), 511–541. https://doi.org/10.1016/S0361-9230(03)00055-8.

Barmack, N.H., Yakhnitsa, V., 2013. Vestibulocerebellar connections. In: Handbook of the Cerebellum and Cerebellar Disorders. Springer Science, pp. 357–376. https://doi.org/10.1007/978-94-007-1333-8.

Bisdorff, A., et al., 2009. Classification of vestibular symptoms: towards an international classification of vestibular disorders. J. Vestib. Res.: Equil. Orientat. 19 (1–2), 1–13. https://doi.org/10.3233/VES-2009-0343.

Büttner-Ennever, J.B., 1992. Patterns of connectivity in the vestibular nuclei. Ann. N. Y. Acad. Sci. 656 (1), 363–378.

Carpenter, M.B., 1989a. The cerebellum. In: Core Text of Neuroanatomy, eighth ed. Williams & Wilkins, pp. 220–243.

Carpenter, M.B., 1989b. The pons. In: Core Text of Neuroanatomy, eighth ed. Williams & Wilkins, pp. 347–358.

Ciuman, R.R., 2013. Inner ear symptoms and disease: pathophysiological understanding and therapeutic options. Med. Sci. Mon. Int. Med. J. Exp. Clin. Res. 19, 1195–1210. https://doi.org/10.12659/MSM.889815.

Cohen, B., Raphan, T., 2004. The Physiology of the Vestibuloocular Reflex (VOR). Springer Handbook of Auditory Research, pp. 235–285.

Corrales, C.E., Mudry, A., 2017. History of the endolymphatic sac: from anatomy to surgery. Otol. Neurotol. 38 (1), 152–156. https://doi.org/10.1097/MAO.0000000000001273.

Coughlin, A.R., Willman, T., Gubbels, S.P., 2018. Systematic review of hearing preservation after radiotherapy for vestibular schwannoma. Otol. Neurotol. 39. https://doi.org/10.1097/MAO.0000000000001672.

Dieterich, M., Brandt, T., 2015. The Bilateral Central Vestibular System: Its Pathways, Functions, and Disorders. Annals of the New York Academy of Sisences, pp. 1–17. https://doi.org/10.1111/nyas.12585.

Dieterich, M., Brandt, T., 2018. The parietal lobe and the vestibular system. In: Handbook of Clinical Neurology, first ed. Elsevier B.V. https://doi.org/10.1016/B978-0-444-63622-5.00006-1.

Farahani, R.M., Nooranipour, M., Nikakhtar, K.V., 2007. Anthropometry of internal acoustic meatus. Int. J. Morphol. 25 (4), 861–865.

Farid, N., 2014. Imaging of vestibular schwannoma and other cerebellopontine angle tumors. Operat. Tech. Otolaryngol. Head Neck Surg. 25 (1), 87–95. https://doi.org/10.1016/j.otot.2013.11.011.

Fluur, E., 1962. The mechanism of nystagmus. Acta Otolaryngol. 54 (1–6), 181–188. https://doi.org/10.3109/00016486209126936.

Friedmann, D.R., et al., 2015. Nonschwannoma tumors of the cerebellopontine angle. Otolaryngol. Clin. 48, 461–475. https://doi.org/10.1016/j.otc.2015.02.006.

Frischer, J.M., et al., 2018. Long-term outcome after Gamma Knife radiosurgery for acoustic neuroma of all Koos grades: a single-center study. J. Neurosurg. https://doi.org/10.3171/2017.8.JNS171281. Published.

Gaur, S., et al., 2015. Efficacy of Epley's maneuver in treating BPPV patients: a prospective observational study. Int. J. Otolaryngol. 1–5. https://doi.org/10.1155/2015/487160, 2015.

Glover, J.C., 2004. Vestibular system. Encyclop. Neurosci. 127–132.

Gompel, J. J. Van, Agazzi, S., Adewumi, D.A., 2018. Congress of neurological surgeons systematic review and evidence-based guidelines on emerging therapies for the treatment of patients with vestibular schwannomas. Neurosurgery 82 (2), 2017–2019. https://doi.org/10.1093/neuros/nyx516.

Hain, T., Helminski, J., 2007. Anatomy and physiology of the normal vestibular system. In: Vestibular Rehabilitation, third ed. Mcgraw Hill, pp. 1–18.

Hasegawa, T., et al., 2013. Long-term safety and efficacy of stereotactic radiosurgery for vestibular schwannomas: evaluation of 440 patients more than 10 years after treatment with gamma knife surgery. J. Neurosurg. 118 (March), 557–565.

Hentschel, M., et al., 2018. An International Comparison of Diagnostic and Management Strategies for Vestibular Schwannoma. European Archives of Oto-Rhino-Laryngology. https://doi.org/10.1007/s00405-018-5199-6. Published.

Highstein, S.M., Holstein, G.R., 2012. The anatomical and physiological framework for vestibular prostheses. Anat. Rec. 295 (11), 2000–2009. https://doi.org/10.1002/ar.22582.

Hizli, Ö., Cureoglu, S., Kaya, S., 2016. Quantitative vestibular labyrinthine otopathology in temporal bones with vestibular schwannoma. Otolaryngol. Head Neck Surg. 154 (1), 150–156. https://doi.org/10.1177/0194599815601070.

Illich Zabolotnyi, D., Serhiivna Mishchanchuk, N., 2020. Vestibular system: anatomy, physiology, and clinical evaluation. In Somatosens. Motor Res. 1–20. https://doi.org/10.5772/intechopen.90538.

Jones, S., Jones, T., Mills, K., 2009. Anatomical and physiological considerations in vestibular dysfunction and compensation. Semin. Hear. 30 (4), 231–241. https://doi.org/10.1038/jid.2014.371.

Jung, G., Ramina, R., 2019. Vestibular schwannomas: diagnosis and surgical treatment. In: Primary Intracranial Tumors. Intech Open, pp. 1–9. https://doi.org/10.1016/j.colsurfa.2011.12.014.

Kaul, V., Cosetti, M.K., 2018. Management of vestibular schwannoma (including NF2): facial nerve considerations. Otolaryngol. Clin. 51 (6), 1193–1212. https://doi.org/10.1016/j.otc.2018.07.015.

Khan, S., Chang, R., 2013. Anatomy of the vestibular system: a review. NeuroRehabilitation 32 (3), 437–443. https://doi.org/10.3233/NRE-130866.

Kumar Gupta, V., Thakker, A., Kumar Gupta, K., 2020. Vestibular schwannoma: what we know and where we are heading. Head Neck Pathol. https://doi.org/10.1007/s12105-020-01155-x.

Van de Langenberg, R., et al., 2011. Management of large vestibular schwannoma. Part I. Planned subtotal resection followed by Gamma Knife surgery: radiological and clinical aspects. J. Neurosurg. 115 (November), 875–884. https://doi.org/10.3171/2011.6.JNS101958.

Lee, A.T.H., 2012. Diagnosing the cause of vertigo: a practical approach. Hong Kong Med. J. 18 (4), 327–332.

Lindsay, K.W., Bone, I., Fuller, G., 2011. Nystagmus. In: Neurology and Neurosurgery Illustrated, fifth ed. Churchill Livingstone, Elsevier, pp. 184–187. https://doi.org/10.1016/B978-0-7020-3369-8.00001-X.

Lyskowski, A., 2006. Anatomy of vestibular end organs and neural pathways. In: Cummings, C.W., et al. (Eds.), Otolaryngology – Head & Neck Surgery, pp. 3089–3114. https://doi.org/10.1017/CBO9781107415324.004.

Mastronardi, L., Fukushima, T., Campione, A., 2019. Advances in Vestibular Schwannoma Microsurgery. Springer Nature Switzerland.

Mergner, T., Pompeiano, O., Corvaja, N., 1977. Vestibular projections to the nucleus intercalatus of staderini mapped by retrograde transport of horseradish peroxidase. Neurosci. Lett. 5 (6), 309–313.

Mitchell, D.E., Cullen, K.E., 2004. Vestibular system. In: Reference Module in Neuroscience and Biobehavioral Psychology, pp. 127–132. https://doi.org/10.1016/B978-0-12-809324-5.02888-1.

Mulder, J.J.S., Kaanders, J.H., Overbeeke, J. J. Van, 2012. Radiation therapy for vestibular schwannomas. Curr. Opin. Otolaryngol. Head Neck Surg. 20, 367–371. https://doi.org/10.1097/MOO.0b013e328357d337.

Nonaka, Y., et al., 2013. Contemporary surgical management of vestibular schwannomas: analysis of complications and lessons learned over the past decade. Neurosurgery 72 (June). https://doi.org/10.1227/NEU.0b013e3182752b05.

Persson, O., et al., 2017. Stereotactic radiosurgery vs. fractionated radiotherapy for tumor control in vestibular schwannoma patients: a systematic review. Acta Neurochirurgica. https://doi.org/10.1007/s00701-017-3164-6.

Rhoton, A.L., 2000. The cerebellopontine angle and posterior fossa cranial nerves by the retrosigmoid approach. In: Rhoton's Cranial Anatomy and Surgical Approaches, first ed., pp. 526–562 Neurosurgery.

Rhoton, A.L., 2007. Overview of the temporal bone. Neurosurgery 61 (4), 7–60. https://doi.org/10.1227/01.NEU.0000280024.07630.65.

Riga, M., et al., 2011. Inner ear disease and benign paroxysmal positional vertigo: a critical review of incidence, clinical characteristics, and management. Int. J. Otolaryngol. 2011 (2014), 1–7. https://doi.org/10.1155/2011/709469.

Rosahl, S., et al., 2017. Diagnostik und Therapie des Vestibularisschwannoms – eine interdisziplinäre Herausforderung Diagnosis and Management of Vestibular Schwannomas – An Inter- disciplinary Challenge.

Rosahl, S., Eßer, D., 2017. Vestibularisschwannom – management und mikrochirurgische Ergebnisse. HNO 65 (5), 381–387. https://doi.org/10.1007/s00106-016-0252-5.
Swartz, J.D., et al., 1996. Anatomic moment balance and equilibrium I: the vestibule and semicircular canals. Am. J. Neuroradiol. 17, 17–21.
Swartz, J.D., 2008. Pathology of the vestibulocochlear nerve. Neuroimag. Clin. 18 (2), 321–346. https://doi.org/10.1016/j.nic.2008.02.001.
Swenson, R.S., 2017. The Vestibular System, Conn's Translational Neuroscience. Elsevier Inc. https://doi.org/10.1016/B978-0-12-802381-5.00014-2.
Thompson, T.L., Amedee, R., 2009. Vertigo: a review of common peripheral and central vestibular disorders. Ochsner J. 9 (1), 20–26.
Vassiliou, A., et al., 2011. Meniere's disease: still a mystery disease with difficult differential diagnosis. Ann. Indian Acad. Neurol. 14 (1), 12–18. https://doi.org/10.4103/0972-2327.78043.
Ventre-dominey, J., 2014. Vestibular function in the temporal and parietal cortex: distinct velocity and inertial processing pathways. Front. Integr. Neurosci. 8 (July), 1–13. https://doi.org/10.3389/fnint.2014.00053.
De Waele, C., et al., 2001. Vestibular projections in the human cortex. Exp. Brain Res. 141 (4), 541–551. https://doi.org/10.1007/s00221-001-0894-7.
Walker, M., et al., 2014. Vestibular System 4. https://doi.org/10.1016/B978-0-12-385157-4.01185-4.
Wangemann, P., 2006. Supporting sensory transduction: cochlear fluid homeostasis and the endocochlear potential. J. Physiol. 576 (1), 11–21. https://doi.org/10.1113/jphysiol.2006.112888.
Wijesinghe, R., Protti, D.A., Camp, A.J., 2015. Vestibular interactions in the thalamus. Front. Neural Circ. 9 (December), 1–8. https://doi.org/10.3389/fncir.2015.00079.
Yates, B.J., et al., 2018. Organization of vestibular inputs to nucleus tractus solitarius and adjacent structures in cat brain stem. Am. J. Physiol. 36, 974–983.
You, Y.P., et al., 2013. Vestibular schwannoma surgical treatment. CNS Neurosci. Ther. 19 (5), 289–293. https://doi.org/10.1111/cns.12080.
Zaleski-king, A.C., Lai, W., Sweeney, A.D., 2019. Anatomy and physiology of the vestibular system. In: Diagnosis and Treatment of Vestibular Disorders. Springer Nature Switzerland, pp. 3–16. https://doi.org/10.1007/978-3-319-97858-1.
Zalewski, C.K., 2015. Aging of the human vestibular system. Semin. Hear. 36 (212), 175–196.

19

The Auditory System

Hideki Derek Kawai

Department of Biosciences, Graduate School of Science and Engineering, Soka University, Tokyo, Japan

OUTLINE

Abbreviations

A1 Primary auditory field
ACx Auditory cortex
AI Primary auditory cortex
AM Amplitude modified
AVCN Anteroventral cochlear nucleus
BF Best frequency
BOLD Blood oxygen level dependent
CD Characteristic delay
CF Characteristic frequency
CN Cochlear nucleus
DCN Dorsal cochlear nucleus
FTS First transverse sulcus
GBC Globular bushy cell
HG Heschl's gyrus
HS Heschl's sulci
IC Inferior colliculus
ICP Inferior cerebellar peduncle
IHC Inner hair cell
ILD Interaural level difference
IPD Interaural phase difference
ITD Interaural time difference
LL Lateral lemniscus
LSO Lateral superior olivary nucleus
MGB Medial geniculate body
MNTB Medial nucleus of the trapezoid body
MRI Magnetic resonance imaging
MSO Medial superior olivary nucleus
OHC Outer hair cell
PAC Primary auditory cortex
PAR Posterior ascending ramus
PET Positron emission tomography
PO Periolivary region
PP Planum polare
PT Planum temporale
PVCN Posteroventral cochlear nucleus
R Rostral field
RF Reticular formation
RT Rostrotemporal field
SBC Spherical bushy cell
SC Superior colliculi
SF Sylvian fissure
SG Spiral ganglion
SI Sulcus intermediate
SOC Superior olivary complex
SPL Sound pressure level
SPON Superior paraolivary nucleus
STG Superior temporal gyrus
STP Superior temporal plane
STS Superior temporal sulcus
VCN Ventral cochlear nucleus

INTRODUCTION

The central auditory system is comprised of the connection of neurons that process auditory information in the brain. This system processes sound information

From Anatomy to Function of the Central Nervous System
https://doi.org/10.1016/B978-0-12-822404-5.00018-8

received by two cochleae located bilaterally in the inner ears. Within a cochlea is the organ of Corti, which converts the frequency and amplitude of sound waves contained within tones, voices, music into neural codes for sound content analyses. The neural codes are then transmitted through the auditory system comprised of the central auditory pathway—connected centers or nuclei located along the pathway of auditory information processing from the spiral ganglia (SG), whose nerves terminate on the hair cells of the organ of Corti, to the cerebral cortex. The accurate understanding of the anatomical centers and cell–cell connections in the central auditory pathway is essential to identify their roles in auditory functions.

The anatomical and functional studies of the human auditory system typically rely on multiple techniques, including histology and structural magnetic resonance imaging (MRI) of postmortem brain tissues, invasive or noninvasive electrophysiological recordings, and noninvasive functional MRI (fMRI) in living subjects. Structural in vivo MRI has been used to investigate the auditory cortex (ACx) (Morosan et al., 2001; Rademacher et al., 2001; Abdul-Kareem and Sluming, 2008; Saenz and Langers, 2014; Wasserthal et al., 2014; Moerel et al., 2019), the thalamus (Devlin et al., 2006; De Martino et al., 2013; Moerel et al., 2015), acoustic radiations containing thalamocortical and corticothalamic axons (Devlin et al., 2006; Behrens et al., 2007; Javad et al., 2014; Maffei et al., 2018), and the inferior colliculus (IC) (Moerel et al., 2015; Sitek et al., 2019). Investigation of deeper structures such as the superior olivary complex (SOC) and the cochlear nucleus (CN) in the brainstem has been challenging with this technique. Noninvasive, high-resolution fMRI has been utilized to identify various aspects of auditory function. A large amount of literature is available for the study of human ACx using this technique, while the studies of brainstem and midbrain early in the auditory pathway are much less due largely to methodological issues and difficulties in monitoring activities in the deep brain areas. Recently, however, an improved signal-to-noise ratio of ultra-high magnetic field allowed higher-resolution functional imaging and better localization of midbrain structures (De Martino et al., 2013; Moerel et al., 2015). Although these imaging techniques provide information about important aspects of the central auditory system in living humans, these techniques alone can not reveal precise and detailed anatomy and functional mechanisms of the auditory system. For better structural understanding, these MRI techniques must be combined with histological analyses for neuronal distribution and morphology. And, to fully understand precise mechanisms of neuronal circuit function, electrophysiological recordings within the brain are imperative. However, both histology and intra-brain electrophysiology have disadvantages in human studies because they are not possible to perform on healthy individuals.

Most of our understanding of the neural circuit connectivity and the cellular properties of auditory neurons, therefore, must rely on animal studies. Animal studies have additional advantages in enabling direct anatomical and functional studies in the same subjects, providing critical evidence for the link between neuronal structure and function. Since the organization of the auditory system is mostly conserved in mammals, we could make good assessments on human auditory system based on animal research, though the use of the animal data requires caution since the closer look reveals differences among mammals (Moore, 1987).

In this chapter, the anatomy and function of the central auditory system will be reviewed aiming to overview the recent studies of the human auditory system. Where information is lacking in the human system, we will refer to animal studies. Readers are requested to refer to many excellent reviews for more details on the topics covered here.

FROM THE PERIPHERY TO THE CENTRAL AUDITORY SYSTEM

Transformation of Sound to Neural Signals

The cochlea is the peripheral organ that transforms sound waves into electrical nerve pulses using its unique structure that is connected to sensory nerve fibers (dendrites) of the SG, initiating the sound information processing in the auditory system. Sound travels through the auditory canal and vibrates the tympanic membrane and the ossicles. The vibration of the ossicles attached to the oval window of the cochlea elicits waves in the perilymph that fills the linked space of the Scala vestibuli and the Scala tympani along the length of the cochlea. The magnitude and extent of the lymph wave, which depend on sound frequency and intensity, induce spatially restricted vibration of the basilar membrane, a soft tissue that separates endolymph-filled Stria vascularis and perilymph-filled Scala tympani from the base to the apex of the cochlea. High frequency sound vibrates the basilar membrane near the base, while low frequency sound vibrates it near the apex. Fig. 19.1 shows the sequential frequency gradient along the basilar membrane within the winding structure of the cochlea (Li et al., 2021), forming a frequency map (or tonotopy). The mean length of the basal membrane averages 34.6 mm, ranging from 31.5 to 36.9 mm, in adult humans. The segmented size of frequency specific area is also variable among individuals (Li et al., 2021). Nevertheless, the cochlear frequency map

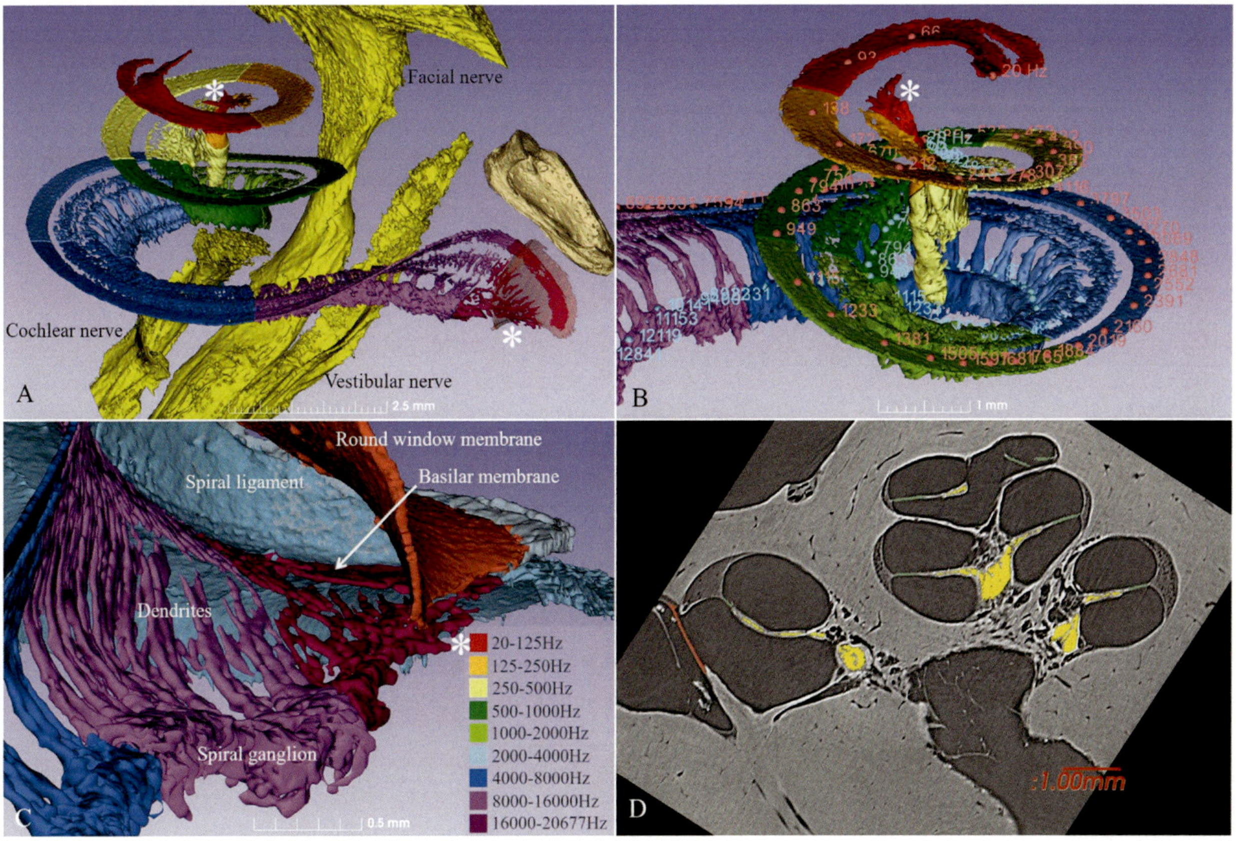

FIGURE 19.1 **The structure of the human cochlea.** (A–C) A left human intra-cochlear image rendered by synchronization radiation phase-contrast imaging (SR-PCI) data, showing simultaneous bone and soft tissue visualization. The basilar membrane and spiral ganglion were segmented with color for corresponding frequency ranges (legend in C). The dendrites of the spiral ganglion were traced from the basilar membrane. Note that the angle of dendritic connections is not radial to the mid-modiolar axis in the apical and basal region denoted by asterisks on the left and right in (A), respectively. (D) Representative tomographic X-ray section showing the segmented round window (*red*), neural elements (*yellow*), and basilar membrane (*green*). *Image courtesy of Li, H., Helpard, L., Ekeroot, J., Rohani, S.A., Zhu, N., Rask-Andersen, H., Ladak, H.M., Agrawal, S., 2021. Three-dimensional tonotopic mapping of the human cochlea based on synchrotron radiation phase-contrast imaging. Sci. Rep. 11, 1–8. https://doi.org/10.1038/s41598-021-83225-w.*

formation requires the organ of Corti situated on the basilar membrane to mediate the mechano-electrical transduction.

The organ of Corti contains about 3500 inner hair cells (IHCs) and 12,000 outer hair cells (OHCs) in a human cochlea (Kim, 1985). Single IHCs are typically connected with 10–20, myelinated auditory nerves of bipolar sensory cells, which occupy 90%–95% of the SG. Low and high frequency responding IHCs are connected to only a few nerves. The basilar membrane vibration elevates the membrane potential (receptor potential) of IHCs and results in the release of neurotransmitter glutamate stored in the synaptic vesicles of the IHCs into synaptic cleft and the activation of ionotropic glutamatergic receptors expressed in the auditory nerve terminals of the SG cells, eliciting the action potentials that propagate the axons toward the CN. Thus, the basilar membrane vibration system in the cochlea works as a frequency bandwidth filter, where a specific component of the bandwidth activates some IHCs to transmit sound frequency information. The auditory nerves respond to specific frequency and intensity of sound, giving rise to a V-shaped tuning curve and a receptive field within the curve, which is the area of auditory stimulus combinations of sound frequency and intensity that elicit action potentials. The frequency with the lowest intensity at the bottom tip of the tuning curve is called characteristic frequency (CF). The shape of tuning curves for IHCs and auditory nerves is identical and corresponds to the vibration pattern at the location of the basilar membrane (i.e., place coding of sound frequency).

Meanwhile, multiple OHCs are contacted by unmyelinated nerves branched from single pseudo-unipolar cells (Carricondo and Romero-Gómez, 2019). At the synaptic terminals of OHCs, glutamate is released and elicits only weak membrane potential changes. To elicit action potentials in the nerves of pseudo-unipolar cells, high intensity sound is required to recruit multiple OHCs (Weisz et al., 2009). This property to use multiple OHCs against very loud sound may be the physiological design to protect ears. A normal function of OHCs is to modify the sensitivity of IHCs. The receptor potential of OHCs generated by the vibratory movement of the basilar membrane quickly modifies the lateral surface of rod-like cell bodies of OHCs, expanding or contracting about 100 nm. This physical movement affects the extent of the basilar membrane vibration, influencing the sensitivity of IHCs and consequently altering the spike rates and spike intervals of the auditory nerves in response to sounds.

Spike rate and spike interval encode sound frequency and intensity information (Rutherford et al., 2021). As sound intensity in sound pressure level (SPL) increases, the spike rate increases and saturates at some point above a specific intensity. The dynamic range is several tens of dB SPL. Because an IHC is connected to auditory nerves with specific SPL sensitivity, the dynamic intensity range is likely conferred by an IHC. Due to the membrane properties of IHCs to transduce sound waves, the intensity of sound signal within a sound frequency bandwidth is expressed by the spike rate (rate code). Since auditory nerve spiking tends to occur at the largest amplitude of the basilar membrane vibration, spiking synchronizes with the cycle of input sound wave on a tonotopic position of the membrane. That is, the spike interval of auditory nerve firing (temporal code) represents the cycle of sound wave or periodicity. But this type of temporal coding is limited up to around 1–3 kHz for most species. It should be noted that spike rates cannot exceed 1 kHz in a single auditory nerve (i.e., an auditory nerve can only synchronize its spikes with a sound wave cycle of several hundred Hz) because the absolute refractory period of an auditory nerve firing—the period during which next spike cannot occur—is a few milliseconds. However, if multiple auditory nerves are combined, auditory nerves can synchronize with the sound wave cycle up to about 3 kHz, even if sound intensity is low. As frequency increases, spike timing becomes random in relation to the periodicity of the sound wave stimulus. However, humans can hear higher frequencies up to ~20 kHz since harmonics combine temporally to produce amplitude-modulation at a fundamental frequency. The amplitude-modified (AM) stimulus likely modulates IHC membrane properties to encode high frequencies, but further studies are necessary for precise mechanisms. In any case, input sound is analyzed in terms of sound frequency and intensity, which are encoded by the spike firing patterns containing a combination of temporal and rate codes of auditory nerves. The central auditory system then works to extract and analyze the spike firing patterns for sound perception.

Tonotopy in the Auditory System

Tonotopy, established first in the cochlea (Nuttall and Dolan, 1996; Ruggero et al., 1997; Narayan, 1998; Cooper, 1999), is preserved through various nuclei in the brainstem, midbrain, thalamus, and cerebral cortex along the central auditory pathway as illustrated in Fig. 19.2 (Saenz and Langers, 2014; Thomas et al., 2015). In human studies, tonotopy has been demonstrated by presenting pure tones carrying a single sound wave frequency with various sound intensity to the ears of subjects while monitoring brain activities using various techniques such as magnetoencephalography (Romani et al., 1982a,b; Pantev et al., 1988, 1989; Yamamoto et al., 1988; Cansino et al., 1994; Huotilainen et al., 1995; Lütkenhöner and Steinsträter, 1998), positron emission tomography (PET) (Lauter et al., 1985; Lockwood et al., 1999), electrophysiology (Howard et al., 2000), and fMRI (Formisano et al., 2003). Tonotopy is at present the only parameter that is clearly identified to form a topographical map in the auditory pathway. Other parameters such as sound intensity (Pantev et al., 1989; Bilecen et al., 2002), tuning bandwidth (Seifritz et al., 2006; Moerel et al., 2012), and amplitude modulation rate (Langner et al., 1997; Barton et al., 2012; Herdener et al., 2013) have been proposed to form complemental topographic maps.

The Ascending and Descending Auditory Pathways

Fig. 19.3 illustrates an overview of ascending and descending pathways of the auditory system. The ascending pathway begins with the cochlea, and SG neurons that receive chemical transmission from IHCs transfer the signals to the CN to initiate the central auditory pathway. Anterior ventral CN (AVCN) transmits monoaural information mostly to contralateral lateral superior olivary nucleus (LSO) via the medial nucleus of the trapezoid body (MNTB) and also to ipsilateral superior olivary complex (SOC), which includes LSO and medial superior olivary nucleus (MSO). Information in the dorsal CN (DCN) and posterior ventral CN (PVCN) is primarily transferred contralaterally to lateral lemniscus (LL) and IC. SOCs are the nuclei that integrate auditory signals from two ears. SOCs feed the integrated signals to the IC via the LL for further integration. The IC sends the signals to the medial geniculate body (MGB) for further analyses. MGB then relays the signals to the ACx. Meanwhile, the descending pathway starts from the cortex that sends projections to various subcortical nuclei, including MGB and IC ipsilaterally, and to SOC bilaterally. The cortex could directly modulate even CN. MGB feeds back to IC, which then sends signals to contralateral SOC as well as ipsilateral LL. In addition

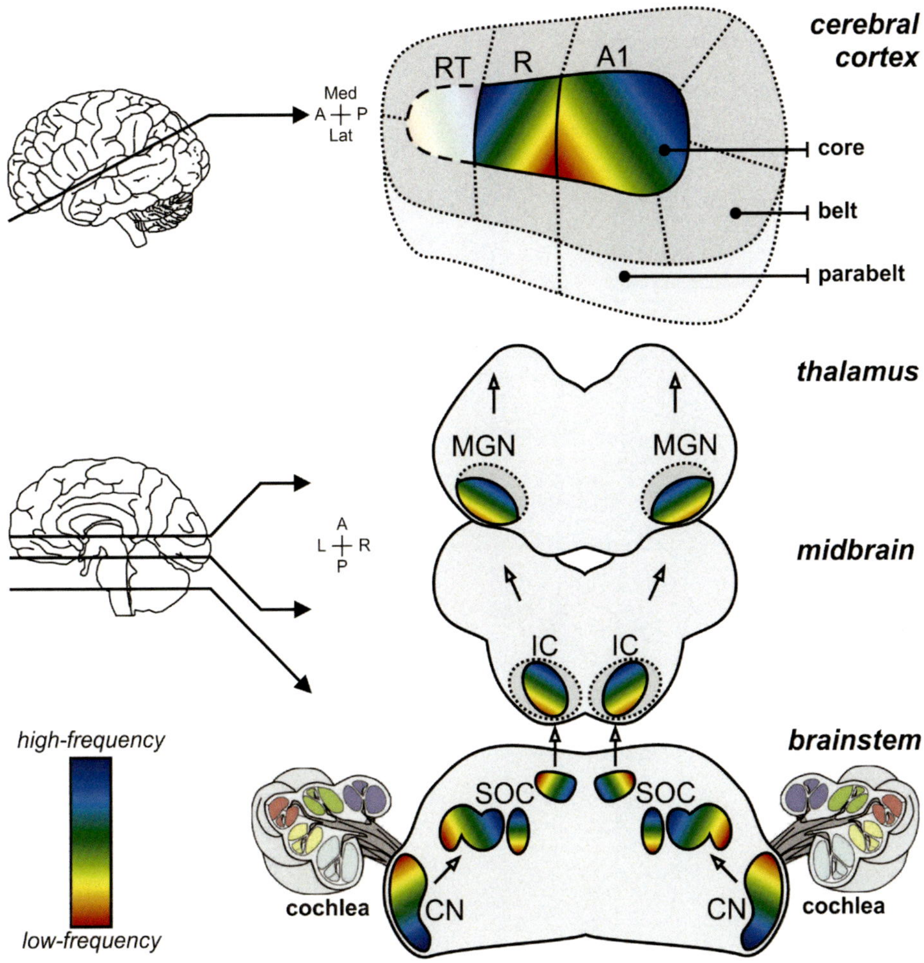

FIGURE 19.2 **Nuclei with tonotopy in the central auditory pathway.** Tonotopically organized nuclei in the classical lemniscal auditory pathway include the cochlear nucleus (CN), superior olivary complex (SOC), inferior colliculus (IC), and medial geniculate body or nucleus (MGN). In the auditory cerebral cortex, core, belt, and parabelt divisions are expected based on the nonhuman primate model. The core has well-recognized two mirror-like high-low-high tonotopy in primary auditory field (A1) and rostral field (R). Rostrotemporal field (RT) may be considered as part of the core or primary auditory cortex. *Image courtesy of Saenz, M., Langers, D.R.M., 2014. Tonotopic mapping of human auditory cortex. Hear. Res. 307, 42–52. https://doi.org/10.1016/j.heares.2013.07.016.*

to the cortical input, the CN also receives inputs from the contralateral IC and bilateral SOCs. Lastly, LSO could modulate the cochlea. Thus, the descending signals could modulate the ascending signals at various nuclei.

Below, we will describe the anatomy and function of these major nuclei along the central auditory system in more detail.

THE COCHLEAR NUCLEUS

Anatomy of the CN

Human CN is localized lateral to the inferior cerebellar peduncle (ICP) (Saini et al., 2019). The CN can be divided into a ventral CN (VCN) and a DCN based on histology in postmortem brains (Moore and Osen, 1979; Adams, 1986). Due perhaps to less clear interareal boundaries, the two divisions have not yet been observed using structural MRI, though the surface CN reconstruction combined with postmortem images support the existence of the two CN divisions (Tarabichi et al., 2020). A recent MRI study at submillimeter resolution indicated that it has a lamellar structure with a fiber running in a ventral–dorsal orientation (Epprecht et al., 2020).

Within the VCN, an anterior area mainly has spherical cells, a central area has multipolar and globular cells, a posterior area has octopus cells, and a laterodorsal cap has small neurons (Moore and Osen, 1979). The bulk of the VCN comprises of the central area and the cap area. The DCN has a structure like cerebellum in terms of anatomical features and gene expression patterns in animals (Mugnaini and Morgan, 1987; Berrebi

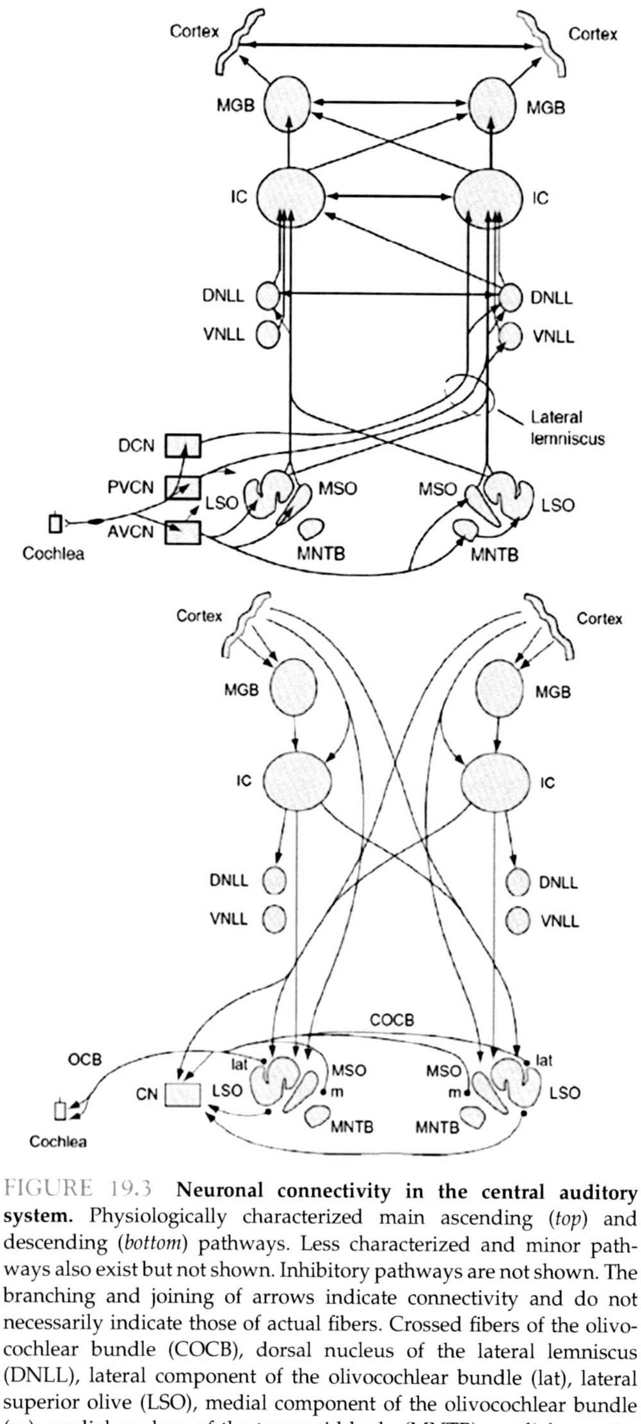

FIGURE 19.3 **Neuronal connectivity in the central auditory system.** Physiologically characterized main ascending (*top*) and descending (*bottom*) pathways. Less characterized and minor pathways also exist but not shown. Inhibitory pathways are not shown. The branching and joining of arrows indicate connectivity and do not necessarily indicate those of actual fibers. Crossed fibers of the olivocochlear bundle (COCB), dorsal nucleus of the lateral lemniscus (DNLL), lateral component of the olivocochlear bundle (lat), lateral superior olive (LSO), medial component of the olivocochlear bundle (m), medial nucleus of the trapezoid body (MNTB), medial superior olive (MSO), olivocochlear bundle (OCB), ventral nucleus of the lateral lemniscus (VNLL). *Image courtesy of Pickles, J.O., 2015. Auditory pathways: anatomy and physiology. Handb. Clin. Neurol. 129, 3–25. http://www.ncbi.nlm.nih.gov/pubmed/25726260.*

et al., 1990; Oertel and Young, 2004). The laminar organization contains three layers in humans: molecular, granular, and deep layers (Moore and Osen, 1979; Wagoner and Kulesza, 2009), which resemble to some extent the organization in animals (Rubio et al., 2008). The external, molecular layer is a fibrous layer with many axons carrying multimodal sensory information with few neurons. The middle, principal layer is the granular layer with mostly small granule cells that send outputs to the IC. In rhesus monkey and other lab animals, a lamina with large fusiform (pyramidal) cells extending dendrites into layers above and below has been observed at the molecular/granular layer interface, but only a few clusters were found in humans. The deep layer contains vertically oriented, elongated glycinergic tuberculovenral cells with occasional excitatory giant stellate neurons.

The Two Auditory Streams in the CN

The functional connectivity of the human CN is largely unclear but may resemble, to some extent, that of mammals. The CN in cats can be divided into three subdivisions: DCN, PVCN, and AVCN cochlear nuclei (Fig. 19.4). The auditory nerve of the SG bifurcates its axon into anterior and posterior branches upon entering the brainstem. The anterior branch innervates the AVCN and forms the ventral stream, carrying information of sound source or location. The posterior branch terminates in the DCN and the PVCN and constitutes the dorsal stream, carrying complex acoustic information. It should be noted that neurons in each CN division are distributed topographically, preserving tonotopy, since they receive the auditory nerves that are tonotopically arranged.

The Ventral Auditory Stream of the Brainstem

The AVCN neurons are considered to play important roles in reducing the variability of spike intervals of the auditory nerves. The AVCN has two types of neurons, spherical bushy cells (SBCs) and globular bushy cells (GBCs). SBCs receive 1 ~ 4 fibers of the auditory nerves, forming large synaptic terminals called the endbulbs of Held. They project to the ipsilateral LSO and both ipsi- and contralateral sides of the MSO. A single endbulb of Held could activate an SBC and therefore could pass on its transmitting information directly to the cell. However, at higher intensities of sound, SBCs are inhibited by neurons located elsewhere in the nucleus, and its action potentials require synchronized inputs from a large number of auditory nerve inputs (Kopp-Scheinpflug et al., 2002). Temporal averaging of the inputs determines the temporal accuracy of spiking in SBCs. Meanwhile, GBCs are innervated by 4 ~ 40 fibers of the auditory nerves, forming smaller synaptic terminals, and project to the contralateral MNTB. Because of a large number of auditory nerve inputs, the impact of temporal averaging on the accuracy of action potential

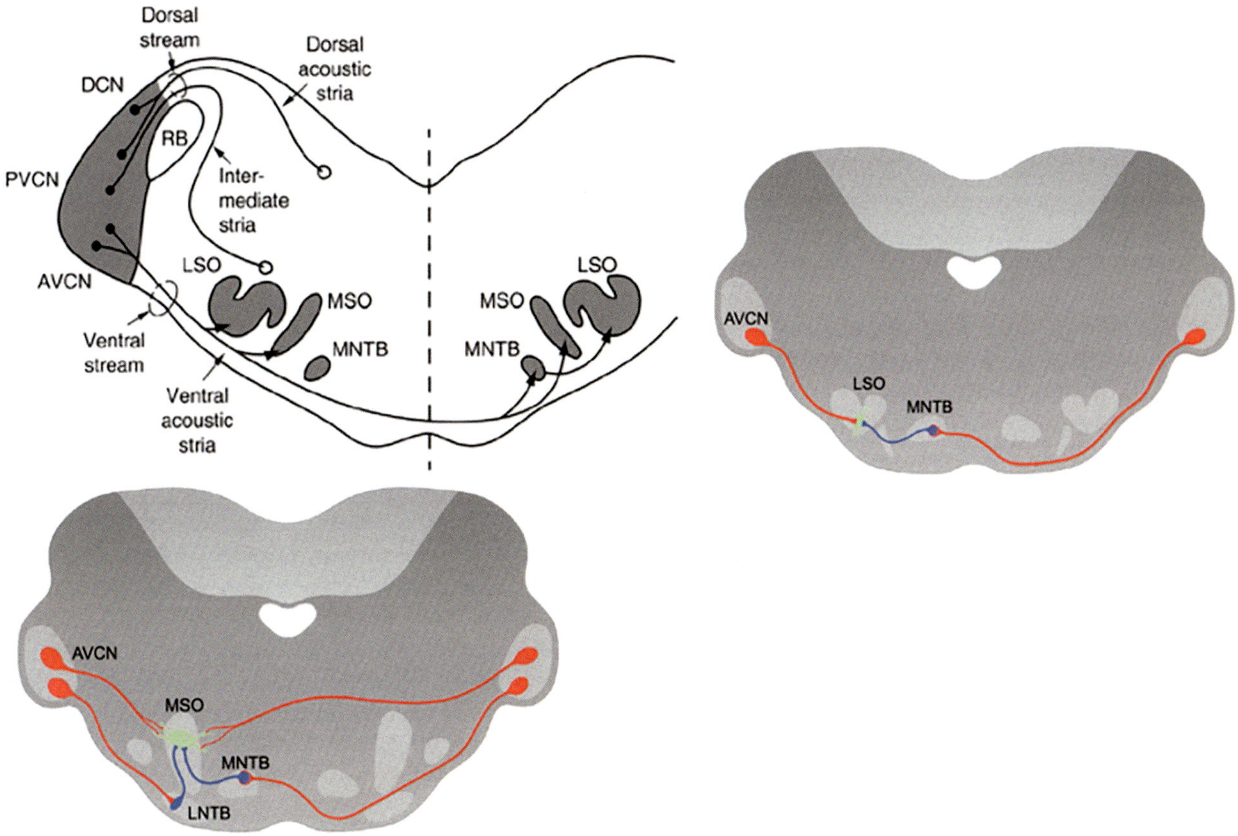

FIGURE 19.4 **The ascending projection in the brainstem.** (*Top left*) The major projections from the cochlear nuclei. The ventral stream arises from AVCN with a contribution from PVCN and runs through the ventral acoustic stria to the SOC of both sides. The dorsal stream arises from DCN, runs through the dorsal and intermediate acoustic striae, and passes dorsally over the restiform body (RB) or inferior cerebellar peduncle. Most fibers of the dorsal stream leave the plane of the section (*small circle*) to run to the cells surrounding the LSO. The branching and joining represent connectivity and do not necessarily indicate those of actual fibers. AVCN and PVCN are superimposed on this plane. (*Top right*) LSO receives direct excitatory input from ipsilateral AVCN and inhibitory input from ipsilateral MNTB driven by contralateral AVCN. (*Bottom*) MSO receives direct excitatory inputs from ipsilateral and contralateral AVCNs, inhibitory input from ipsilateral MNTB driven by contralateral AVCN and an inhibitory input from ipsilateral lateral nucleus of the trapezoid body (LNTB) driven by ipsilateral AVCN. *Image courtesy of Pickles, J.O., 2015. Auditory pathways: anatomy and physiology. Handb. Clin. Neurol. 129, 3–25. http://www.ncbi.nlm.nih.gov/pubmed/25726260; Pecka, M., Encke, J., 2020. Coding of spatial information. In: The Senses: A Comprehensive Reference, vol. 2, pp. 713–731.*

induction is greater in GBCs (Joris et al., 1994a). GBCs extend their heavily myelinated thick axons and terminate on the cell body of inhibitory neurons in the MNTB with each of their presynaptic terminals wrapping single cell bodies in a structure called the calyx of Held, which is the largest presynaptic terminals in the brain (Spirou et al., 1990; Kandler and Friauf, 1993; Sätzler et al., 2002). In addition, there are inhibitory neurons in the AVCN, which serve to preserve tonotopic properties of the bushy cells over a wide range of sound intensities.

Functional Studies in Human CN

In human fMRI studies, obtaining the reliable responses to auditory stimuli against the scanner noise has been challenging. However, researchers have succeeded in identifying the centers of the auditory pathway and provided evidence that human CN responded to broadband noise monotonically with increasing bandwidth (Hawley et al., 2005) and increasing intensity (Sigalovsky and Melcher, 2006). The CN responded continuously throughout a sustained sound stimulus (Dewey et al., 2020). Although Hawley et al. suggested a functional separation of the dorsal and ventral CN divisions based on broadband noise stimulation (Hawley et al., 2005), further studies are needed for its definitive separation and identification of their functions.

The Dorsal Auditory Stream of the Brainstem

The DCN and the PVCN extract and enhance the frequency and timing properties of input sound from the auditory nerves and send them to the contralateral IC. These nuclei constitute the dorsal stream of auditory processing. The dorsal stream has multiple functions in animals, including extraction of signals related to complex sound frequency and sound source of the ventral stream and preservation of the accurate timing of broadband frequency stimuli.

The DCN integrates auditory information as well as nonauditory sensory and motor signals. DCN neurons exhibit a variety of tuning curves. Some of them only respond to a limited frequency and intensity range of external sound inputs due to strong intranucleus inhibition. DCN neurons are responsive to external sound stimuli but insensitive to self-generated sounds due to cancellation by nonauditory inputs of mossy fibers terminating on the granule cells from a number of brain regions (Singla et al., 2017). These features of DCN neurons allow processing of spectrally complex aspects of auditory information. In particular, the DCN is important for sound source location. It has sensitivity to the position of the head with respect to sound source (Rice et al., 1992; Sutherland et al., 1998). Varying spectral notches due to the shadowing effects of the pinna on the broadband sound frequency generated by vertical sound sources affects neuronal firing in the DCN. Cats utilize this feature to correct the ear movement upon the detection of spectral dip (response reduction in a frequency range).

The PVCN is a defined area in the most caudal and dorsal part of VCN. It is considered to analyze the timing and spectral information important for speech perception. The PVCN contains octopus cells, which receive a large number of perpendicularly oriented auditory nerve inputs carrying a wide range of frequencies (Rhode and Smith, 1986; Oertel et al., 1990). Their cell membrane properties make them fire spikes rapidly and accurately in response to the onset of synchronous broadband auditory nerve inputs (Golding et al., 1995; Oertel et al., 2000). This feature is important for understanding speech as it involves broadband transients and gaps of consonants (Shannon et al., 1995). Speech and sound pattern recognition are not impaired upon a loss of hearing in one ear. Therefore, octopus cells are largely involved in monoaural neuronal circuits. In this respect, it would make sense that the excitatory target of the octopus cells is the superior paraolivary nucleus (SPON), which plays an important role in extracting the sound features relevant to vocal communication (Felix et al., 2017). In addition to working as a coincidence detector of synchronous auditory nerve inputs, octopus cells also work as a sequence detector of auditory nerve inputs, exhibiting frequency sweep direction selectivity (Lu et al., 2022). This direction specificity depended on the amplitude and activation sequence of the inputs of auditory nerves, which by themselves do not show any direction selectivity. Octopus cells project mainly to the ventral nucleus of the lateral lemniscus (VNLL) of the opposite side, where they form calyceal-type synapses (Smith et al., 2005) on inhibitory neurons for secure and fast onset responses (Nayagam et al., 2005).

The PVCN also has multipolar cells in the rostroventral region. Primary multipolar cells are the T-stellate cells, which are the excitatory neurons that receive tonotopically organized auditory nerve inputs, interconnected monosynaptic excitatory inputs with other T-stellate cells (Lin et al., 2022), and three glycinergic inhibitory inputs from tuberculoventral neurons in deep layer of the DCN (Wickesberg et al., 1991), D-stellate cells (Doucet and Ryugo, 1997; Oertel et al., 2011; Xie and Manis, 2014) and L-stellate cells (Ngodup et al., 2020) of the VCN. T-stellate cells project their axons to the IC with their collateral branches terminating on multiple brainstem nuclei on the way (Schofield and Cant, 1996). They carry the information of amplitude-modulated sounds (Rhode and Greenberg, 1994; Laudanski et al., 2010) and form a representation of vowel formant structure of various intensities, which is crucial for speech perception (Blackburn and Sachs, 1990; May et al., 1998).

THE SUPERIOR OLIVARY COMPLEX

Anatomy of the SOC

Based on the knowledge gained by tracing and histological analyses in animals and postmortem human brains, several nuclei of the SOC, which includes MNTB, MSO, LSO, and SPON, are considered as the important relay centers of ascending information, sending sensory and other signals to the IC via the LL (Moore, 1987; Kulesza, 2007). The SOC is found in the inferior pons in human brains. Fig. 19.5 illustrates the distance and location of CN, SOC, LL, and IC in humans and cats. Recently, Sitek et al. (2019) provided three-dimensional views of various auditory subcortical nuclei based on previous histological data combined with structural MRI data in postmortem human brains (Fig. 19.6). It should be noted that the size of SOC measured by structural MRI is variable but is typically much larger than that measured by histology, and the size measured by MRI shows much larger volume (García-Gomar et al., 2019; Sitek et al., 2019).

Physical connectivity between these and other subcortical nuclei of the auditory pathway has also been described in vivo using the modern technology of 200 μm diffusion-weighted MRI (Sitek et al., 2019). According to their studies, MSO, the largest SOC nucleus in humans, receives ascending inputs from the right and left AVCN and sends outputs to the ipsilateral LL.

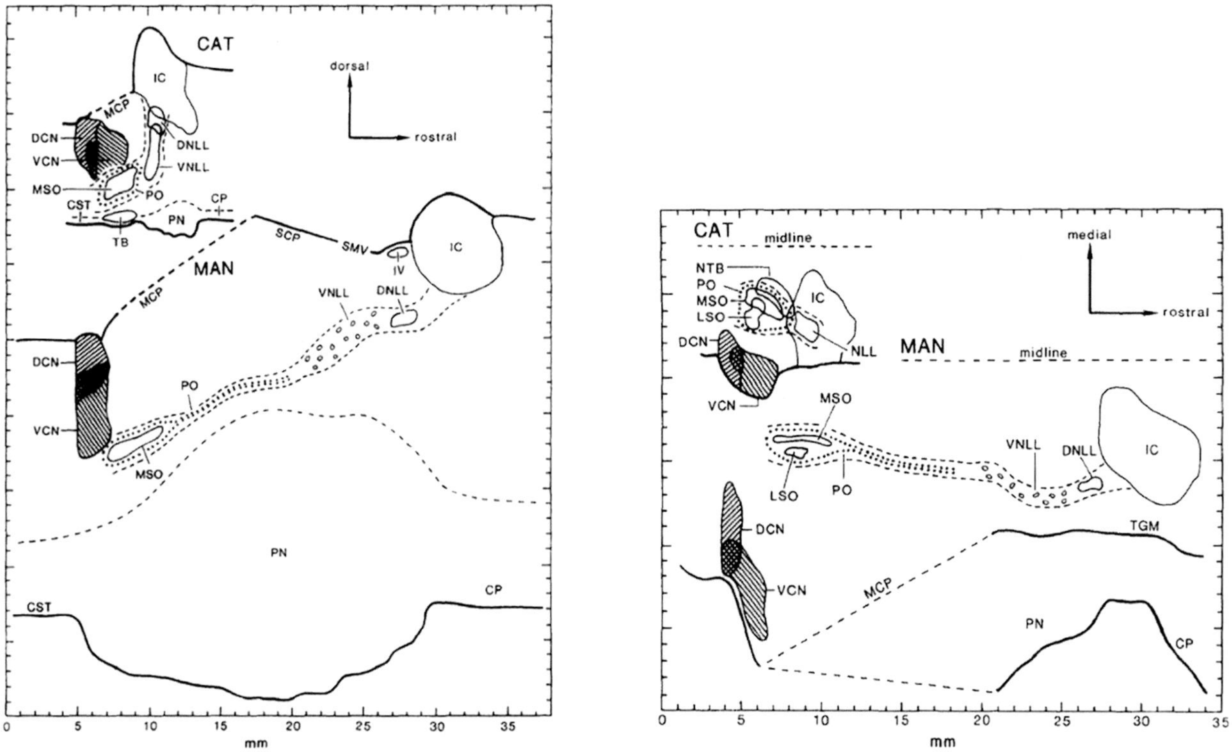

FIGURE 19.5 **The major nuclei and tracts of the auditory brainstem in sagittal projection (*left*) and horizontal plane (*right*) in humans and cats.** (*Left*) *Thick lines* mark the external brain stem surface, with a *dotted line* indicating the point where the middle cerebellar peduncle (MCP) was cut during removal of the cerebellum. Anterior (rostral) to the middle peduncle, the dorsal surface of the human pons is formed by the superior cerebellar peduncle (SCP) and by the superior medullary velum (SMV) containing the trochlear nerve (IV). Rostral to the pontine nuclei (PN), the ventral surface of the midbrain is formed by the cerebral peduncles (CP): caudal to the pons, the ventral surface of the medulla is formed by the corticospinal tract (CST). The cochlear nuclei are shown in diagonal hatching, with the crosshatched area indicating the region where the DCN overlaps the VCN. The SOC contains an MSO surrounded by periolivary cell groups (PO: *dotted line*). Rostrally directed axons (*dashed lines*) form a capsule around the olivary complex and continue forward to form the lateral lemniscus (LL). In man, PO cell groups are present in the axonal pathway rostral to the main body of the SOC. Further rostrally, the axons surround cell groups forming the VNLL and DNLL. In the cat, the large lemniscal nuclei are located in the LL as it runs straight dorsally. In the human pons, the rostrodorsally directed lemniscal pathway contains a distinct DNLL but only scattered cell groups ventrally (VNLL). In both species, the LL enters the IC. TB, trapezoid body. (*Right*) A *dashed line* marks the brain stem midline, while a *heavy line* indicates the lateral surface of the brain stem. In the human brain stem, this figure shows the lateral extent of both the upper tegmental part of the pons (TGM) and the lower half of the brain stem formed by the PN and the CP. In horizontal view, the human cochlear nuclei (DCN, VCN) are seen to be much more elongated along their mediolateral axis than those of the cat. The SOC contains LSO, MSO, and nucleus of the trapezoid body (NTB) in the cat, while it contains MSO and LSO in the human. PO (*dotted line*) surround the MSO and LSO and extend rostrally in the axonal pathway (*dashed lines*) in the human. In the cat, the position of the vertically oriented lemniscal nuclei is indicated (NLL). In the human brain stem, the LL moves laterally to lie along the surface of the pontine tegmentum. It contains a well-developed DNLL and scattered cell clusters constituting the VNLL. In both cat and man. the LL terminates by entering the IC. *Image courtesy of Moore, J.K., 1987. The human auditory brain stem: a comparative view. Hear. Res. 29, 1–32. http://www.ncbi.nlm.nih.gov/pubmed/3654394.*

LSO receives inputs from the ipsilateral AVCN and MNTB and sends outputs to LL bilaterally. Animal studies show that SOC receives most of the ascending information from the contralateral CN. This contralateral bias continues until the primary auditory cortex (PAC).

MSO extends about 6 mm rostro-caudally viewing from above in the left brain (Fig. 19.7) and contains neurons with high density (about 13,000/mm^3), totaling approximately 15,500 cells, which are clearly distinguished into fusiform (long slender soma or slightly round soma) and stellate cells (triangular soma) at nearly equal proportions. Fusiform neurons are bipolar cells with two main thick dendrites extending from the soma rostrally and caudally. Stellate neurons are multipolar cells with a single primary dendrite and various multiple dendrites.

LSO has a skewed Y structure positioned lateral to MSO with a large main body and a small medial limb extending caudally (Fig. 19.7). The rostro-caudal total length of the main body is approximately 2,800 μm. It is estimated to contain approximately 5,600 neurons with a density of about 2,680 neurons/mm^3. There are three morphologically distinct neurons—fusiform, round, and stellate cells—with fusiform and round cells representing approximately one-half of the total

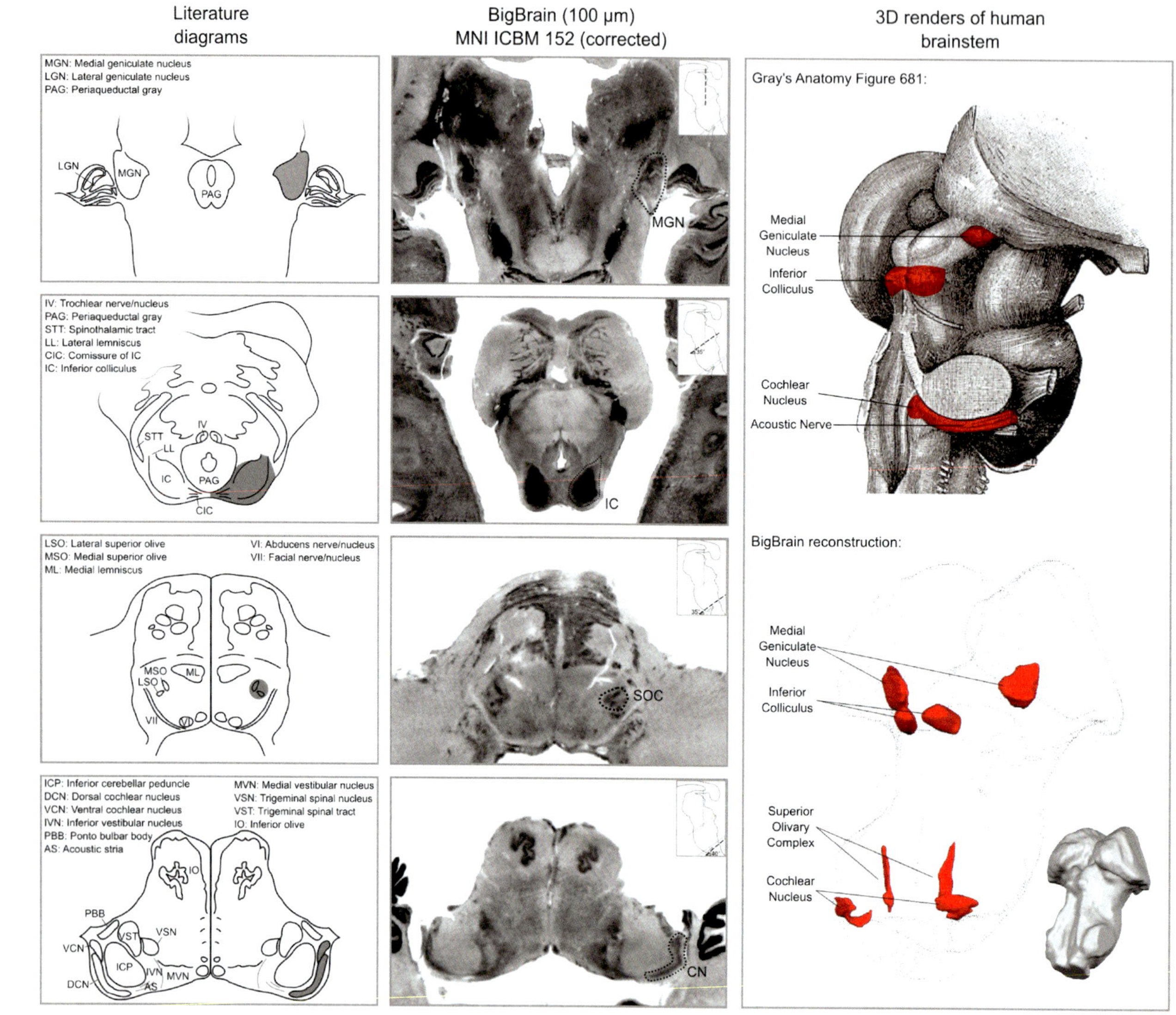

FIGURE 19.6 **Auditory brainstem and midbrain.** Literature diagrams (*left columns*) redrawn from Moore (1987) for the CN, SOC, IC, and from the Allen Human Brain Atlas (Hawrylycz et al., 2012) for the MGB compared to similar cuts from histology (BigBrain) in Montreal Neurological Institute (MNI, central column) and 3D reconstructions of the segmented structures from the histology (*right column, bottom*). The auditory structures are highlighted in *gray* in the *left column*, by a *dotted line* in the *central column*, and in *red* on the modified Gray's anatomy Plate 719 (Gray and Lewis, 1918, anatomy of the human body) and rendered as *solid red* surface meshes within the surface point cloud render of BigBrain MNI brainstem (*right column*). *Image courtesy of Sitek, K.R., Gulban, O.F., Calabrese, E., Johnson, G.A., Lage-Castellanos, A., Moerel, M., Ghosh, S.S., De Martino, F., 2019. Mapping the human subcortical auditory system using histology, postmortem MRI and in vivo MRI at 7T. Elife 8, 1–36.*

population. The fusiform cells extend two primary dendrites in the parasagittal plane (i.e., rostro-caudal direction). The stellate neurons possess at least three primary dendrites, extending mainly in the rostro-caudal direction. The round neurons extend dendrites perpendicular to the rostro-caudal axis of the LSO, extending them in various directions in the coronal plane (latero-medial direction). The somata of these three neuron types appear mostly flattened in sagittal plane, while the stellate neurons appear the largest in the traverse plane. In animals, the principal neurons, which comprise >80% of LSO neurons, are bipolar neurons with elongated soma (Helfert and Schwartz, 1986, 1987; Kandler and Friauf, 1995; Franken et al., 2018). They are excitatory neurons projecting to the contralateral dorsal nuclei of the lateral lemniscus (DNLL) and the IC of both hemispheres. Some LSO neurons send glycinergic inhibitory projections to the ipsilateral DNLL and IC (Brunso-Bechtold et al., 1981; Saint Marie and Baker, 1990; Glendenning et al., 1992). Somatic morphology in humans may resemble that of animals, but dendritic directions appear different (Rietzel and Friauf, 1998).

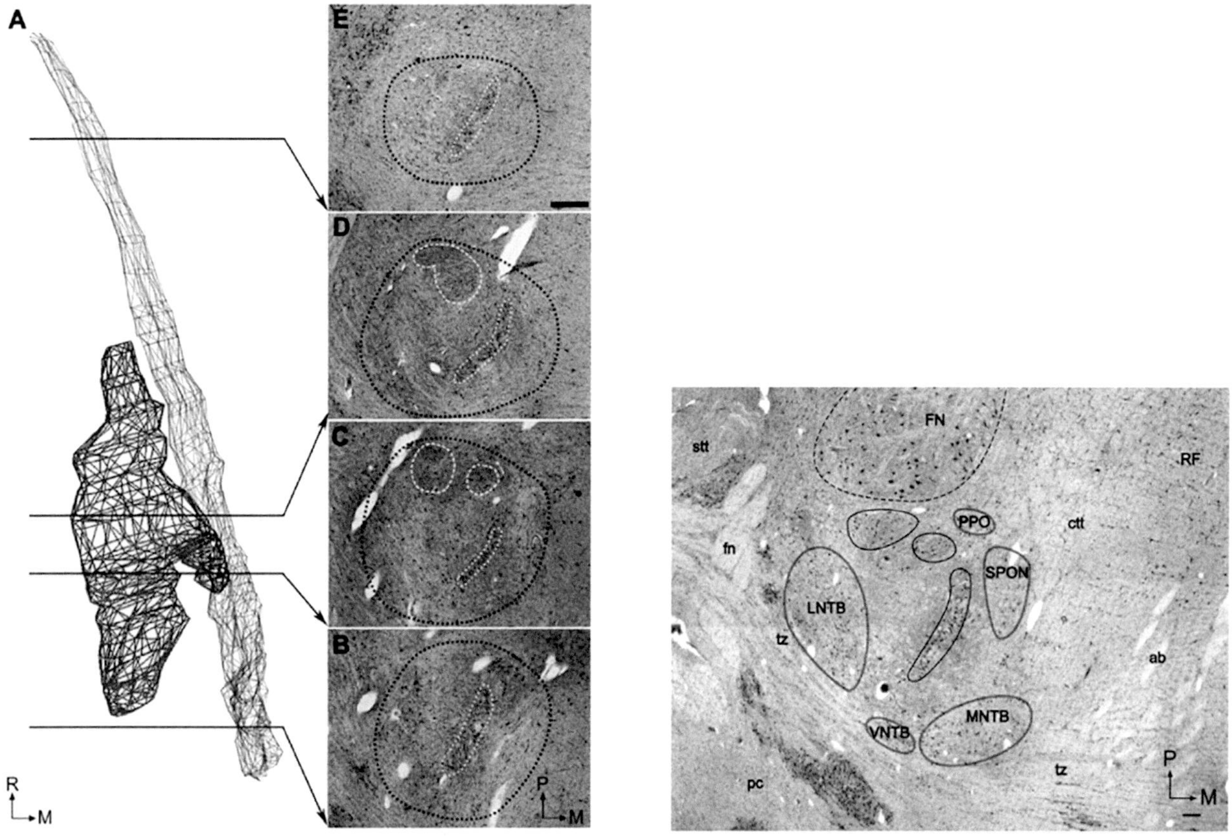

FIGURE 19.7 **Histological analysis of human SOC.** (*Left*) A wire frame, three-dimensional reconstruction of the human MSO (A, *light gray*) and LSO (A, *dark gray*) based on Giemsa-stained brain sections (not to scale). The schematic is the view of the *left side* of the brain from above. Note that the LSO has a large main body and a small medial limb. The MSO spans nearly 6 mm rostro-caudally, while the LSO extends about 3 mm. *White dotted area* is the MSO, and *white dashed area* is the LSO. *Arrows* indicate the approximate locations of the histological sections in the model. The scale bar in (E) represents 500 μm. (*Right*) A Nissl stained section through the caudal aspect of the human SOC. abducens nerve (Ab), central tegmental tract (Ctt), facial nerve (fn), facial nucleus (FN), lateral nucleus of the trapezoid body (LNTB), medial nucleus of the trapezoid body (MNTB), pontocerebellar fibers (pc), posterior periolivary nucleus (PPO), reticular formation (RF), superior paraolivary nucleus (SPON), spinal trigeminal tract (stt), trapezoid body (tz), ventral nucleus of the trapezoid body (VNTB). *Solid lines* indicate the MSO and LSO. At this level, the LSO is composed of two limbs. Note how the fibers of the trapezoid body (tz) at this level form the border between the pontine tegmentum and basilar pons. The scale bar equals 200 μm. *Image courtesy of Kulesza, R.J., 2007. Cytoarchitecture of the human superior olivary complex: medial and lateral superior olive. Hear. Res. 225, 80–90; Kulesza, R.J., 2008. Cytoarchitecture of the human superior olivary complex: nuclei of the trapezoid body and posterior tier. Hear. Res. 241:52–63.*

Functional Studies in Human SOC

In human fMRI studies, the SOC responded to broadband sound stimuli in a bandwidth- (Hawley et al., 2005) and level- (Sigalovsky and Melcher, 2006) dependent manner. However, the SOC responses exhibited much smaller blood oxygen level–dependent (BOLD) responses compared to the CN or the IC. The subdivision of the SOC has not yet been discriminated in human fMRI studies so far.

The Medial Nucleus of the Trapezoid Body

The existence of the MNTB in humans has been questioned for over half a century until recent postmortem histological studies confirming its existence (Kulesza and Grothe, 2015). MNTB is in the medial SOC and contains large glycinergic inhibitory neurons that project to neurons in the SOC (binaural LSO and MSO, and monaural SPON) and VNLL. The glycinergic inhibitory neurons receive excitatory inputs from GBCs of the contralateral AVCN (Cant and Casseday, 1986; Friauf and Ostwald, 1988; Kandler and Friauf, 1993). The GBC-to-inhibitory neuron connection is typically one-to-one at least at the onset of hearing (Forsythe, 1994; Borst et al., 1995), forming the calyx of Held that covers about 25%–60% of the somatic surface of an inhibitory neuron (Smith et al., 1991; Sätzler et al., 2002; Taschenberger et al., 2002). The GBC axons conduct action potentials fast and securely in the most part and transmit tonotopically organized sound information from the contralateral ear faithfully in a temporally precise manner. Therefore, the inhibitory neurons fire action potentials concomitant with synaptic release of neurotransmitter glutamate at the calyx of Held (Englitz et al., 2009; Lorteije et al., 2009; Sonntag et al., 2009). The firing rates of the inhibitory neurons correspond to those of GBCs (Smith et al., 1998). These inherent properties confer the transfer of tonotopic properties. The distribution of the inhibitory neurons within MNTB shows a clear tonotopic organization with low frequency–tuned neurons located laterally and high frequency–tuned neurons located medially (Sommer et al., 1993; Smith et al., 1998).

The frequency tuning curve of MNTB neurons is sharply V-shaped with a low-frequency tail. This tuning curve resembles that of GBCs (Friauf and Ostwald, 1988; Smith et al., 1991; Kopp-Scheinpflug et al., 2008; Tolnai et al., 2008). At sound frequencies below 3 kHz, MNTB neurons respond in phase with tones (i.e., phase locked) (Smith et al., 1998). At higher frequencies, tone responses shift from a transient phasic, high discharge rate at the onset of the stimulus to a tonic, low discharge rate during the stimulus (Kadner and Berrebi, 2008; Kopp-Scheinpflug et al., 2003, 2008; Smith et al., 1991, 1998; Sommer et al., 1993; Tolnai et al., 2008; Tsuchitani, 1997). Phase locking at higher frequencies occur to the envelop of AM stimuli up to 1 kHz. This feature of MNTB neurons makes them the detector of fast transient amplitude changes (Joris and Yin, 1998; Paolini et al., 2001; Kadner and Berrebi, 2008).

Another aspect of MNTB neurons is that most of them have monotonic firing rate-intensity functions with maximal rates up to about 500 Hz (Guinan and Li, 1990; Kopp-Scheinpflug et al., 2003, 2008; Lorteije et al., 2009). Some, however, exhibit nonmonotonicity with reduced spike rates at increasing sound intensities. This nonmonotonicity is likely due to the properties of MNTB, involving inhibitory inputs from the ventral nucleus of the trapezoid body (VNTB) and axon collaterals from other MNTB neurons (Kuwabara and Zook, 1991; Albrecht et al., 2014; Dondzillo et al., 2016) rather than projecting GBC neurons (Adams and Mugnaini, 1990; Banks and Smith, 1992; Sommer et al., 1993; Wu and Kelly, 1995; Awatramani et al., 2004). However, sound-level intolerance—the invariance of spectral tuning at different sound levels—may also contribute to sound-evoked inhibition (Koka and Tollin, 2014).

MNTB neurons reliably convert the excitatory input from the contralateral AVCN into a short latency, precisely timed, and highly reliable inhibitory output to its binaural (LSO, MSO) and monaural (SPON) targets. The different target nuclei of the MNTB make different use of this inhibition. LSO and MSO utilize the inhibition for sound localization by extracting interaural time difference (ITD) and interaural intensity (or level) difference, which are the timing and intensity differences of sound arriving at two ears, respectively (see below for details).

Sound Processing in the Lateral Superior Olivary Nucleus

LSO neurons mainly receive excitatory (glutamatergic) inputs from SBCs in the ipsilateral AVCN and inhibitory (glycinergic) inputs from the ipsilateral MNTB neurons, which are driven by highly myelinated and rapidly conducting axons of the contralateral AVCN at the calyx of Held. The excitatory inputs terminate glutamatergic synapses mainly on the dendrites, while inhibitory inputs terminate glycinergic synapses on the soma via highly myelinated axons. LSO neurons also receive inputs from the contralateral VNTB (Glendenning et al., 1985; Cant and Casseday, 1986), ipsilateral PVCN (Vater and Feng, 1990; Thompson and Schofield, 2000; Gómez-Álvarez and Saldaña, 2016), and ipsilateral DCN (Doucet and Ryugo, 2003).

LSO exhibits peculiar properties for its role in sound localization. The LSO is tonotopically organized as its neurons tuned to lower and higher frequencies are localized to the lateral and medial regions of LSO, respectively. The tonotopic organization is reflected on the close match between the CF of SBCs in the AVCN and the CF of the inhibitory inputs from MNTB. Notably many LSO neurons are preferentially tuned to frequencies above 2 kHz (Tsuchitani, 1977). ILDs are most prominent in these higher frequencies due to sound wave interference by the head (Wightman and Kistler, 1992; Macpherson and Middlebrooks, 2002). The minimal sound intensity required to elicit action potentials increases at higher CFs. There are also lower frequency (<2 kHz)-tuned LSO neurons. Whether these neurons take part in ILDs remain to be clarified (Lingner et al., 2018).

Mechanisms of sound localization involve synaptic activities in LSO neurons. Most LSO neurons are ipsilaterally excited and contralaterally inhibited. This excitation–inhibition balance underlies sound intensity dependence of LSO neurons. As nicely described by Pecka and Encke (2020) using Fig. 19.8, LSO spike rate (response) is the highest for ipsilateral sound as higher sound intensities in the ipsilateral ear activate the excitatory pathway (EPSP), while lower sound intensities in the contralateral ear elicit little activation of the inhibitory pathway (IPSP). Importantly, the spike rate is modulated as a function of ILDs. Most LSO neurons are strongly inhibited since ILDs favor the contralateral ear. When the binaural balance of sound intensity is shifted strongly to the contralateral ear over the ipsilateral ear, their response is very weak. On the other hand, when the balance is shifted to the ipsilateral ear, their response becomes stronger. If the balance is in between, the response follows a sigmoidal curve.

LSO must also integrate the timing of excitatory and inhibitory inputs associated with sound localization (Tollin, 2003). The binaural balance must be understood

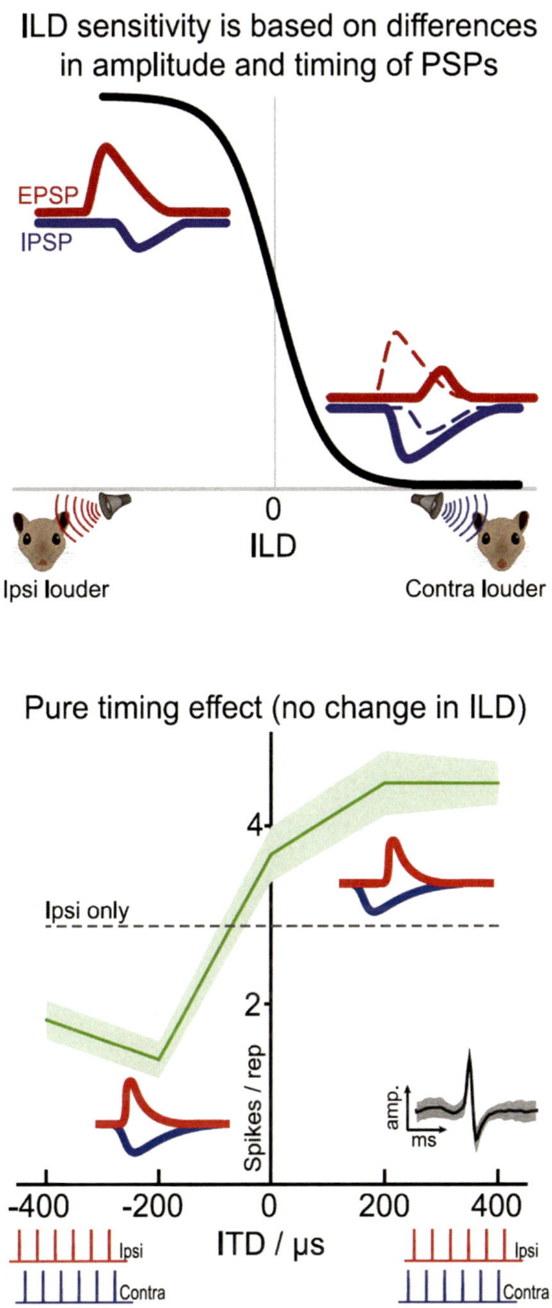

FIGURE 19.8 **Binaural processing in the LSO.** (*Top*) Changes in the sound source location alter the ILD (i.e., the relative sound intensity at the two ears), which cause significant changes in the relative amplitude, but also in timing of the inputs to the LSO: higher absolute sound intensity causes postsynaptic potentials (PSPs) with larger amplitude and shorter latencies. These latency changes largely determine the range of input timing changes associated with changes in the location of a sound source. Thus, ILD computation inherently involves gauging of the relative timing of excitatory PSPs (EPSPs, *red*) and inhibitory PSPs (IPSPs, *blue*). (*Bottom*) Changing the relative timing of ipsi- and contralateral stimulation (i.e., the ITD) in a click train while holding the ILD constant reveals diverse timing effects in the integration of EPSPs and IPSPs. The inhibitory function is observed when IPSPs and EPSPs functionally coincide (here at negative ITDs). When the IPSPs preceded the EPSPs, a postinhibitory facilitation, i.e., an increase of responses above the level for ipsilateral stimulation alone (*dotted line*), can be observed. Inset shows spike waveform of the recorded LSO neuron. *Image courtesy of Pecka, M., Encke, J., 2020. Coding of spatial information. In: The Senses: A Comprehensive Reference, vol. 2, pp. 713–731.*

with natural sounds in mind. Notably, natural sounds are spatiotemporally dynamic and transient in nature (Młynarski and Jost, 2014). Any changes in the sound source will involve changes in the arrival time of neuronal inputs in the LSO, which depends not only on ITDs but also on the change in the input latency due to changes in the ILDs (Grothe and Park, 1995; Irvine et al., 2001). Therefore, LSO neurons must continuously and rapidly process binaural information (Tollin and Yin, 2005; Joris and Trussell, 2018). Indeed, LSO neurons are remarkably sensitive to binaural timing (Beiderbeck et al., 2018; Franken et al., 2018). As Fig. 19.8 (bottom) shows, very brief ITD changes (a few hundred μs) could alter their spiking rates by modulating the contralateral inhibitory inputs, highlighting the speed of conduction of the inhibitory circuit and temporal precision of its input (Ford et al., 2015). Interestingly, the inhibitory input can promote spiking if it precedes the excitatory input within about 200 μs, due in part to the membrane property of LSO neurons (i.e., short membrane time constants) (Franken et al., 2018). Thus, LSO is recognized as an ILD detector but also contributes to ITD detection (Lingner et al., 2018). However, the spike rate of LSO neurons remains low due to the limitation in the modulation of physiologically relevant ITDs, requiring another circuit for ITD processing.

Sound Processing in the Medial Superior Olivary Nucleus

MSO, as LSO does, receives an excitatory input from the ipsilateral AVCN and an inhibitory input from the ipsilateral MNTB excited by the contralateral AVCN (Fig. 19.4). Unlike the LSO, however, the MSO also receives additional contralateral excitatory inputs from the contralateral AVCN and an inhibitory input from the ipsilateral lateral nucleus of trapezoid body (LNTB), which is driven by the ipsilateral AVCN. MSO receives these multiple synaptic inputs for ITD detection.

MSO neurons in cats have laterally- and medially-extending dendrites (Stotler, 1953), which are terminated by excitatory axons of ipsilateral and contralateral bushy neurons of AVCNs, respectively (Cant and Casseday, 1986). These excitatory synaptic currents produce large amplitude with short duration (Couchman et al., 2010), allowing stable, repetitive large inputs during high frequency stimuli (Mathews et al., 2010; Khurana et al., 2011). The bushy cells of AVCNs provide very fast and precise spike conduction and stable synaptic transmission to MNTB and LNTB to permit the microsecond processing for ITDs (Roberts et al., 2014; Ford et al., 2015; Stange-Marten et al., 2017). The inhibitory glycinergic inputs from the ipsilateral MNTB and the contralateral LNTB neurons also provide large amplitude currents of short duration to MSO neurons (Brand et al., 2002). These inhibitory inputs arrive MSO neurons almost simultaneously with similar jitter despite the difference in physical distances of their driving excitatory axons (Roberts et al., 2014). Importantly, these inhibitory inputs arrive MSO neurons earlier than the excitatory inputs do (Roberts et al., 2013). This inhibition reduces excitation and lowers spike probability of MSO neurons, creating a condition to allow coincidence detection of the binaural excitatory inputs from AVCNs.

MSO neurons respond better to binaural inputs. The auditory nerve of the SG transmits the action potentials that are phase-locked to sound waves of the best frequency below about 0.5 kHz. At such low frequencies, the MSO neurons receive phase-locked excitatory and inhibitory inputs from AVCN and evoke spikes with high probability at a specific phase (Joris et al., 1994a,b). These properties along with the characteristics of MSO neurons of low membrane resistance, fast membrane time constant, low rheobases (Mathews et al., 2010) make the circuit an excellent coincidence detector of microsecond ITDs. As illustrated in the ITD function shown in Fig. 19.9, the spike firing of MSO neurons is sensitive to the phase of binaural inputs with the highest frequency of firing observed when the phase of contralateral excitatory inputs and that of ipsilateral inputs coincide. Thus, MSO neurons are sensitive to interaural phase differences (IPDs). To predict the preferred IPD, the phase delay between monoaural contra- and ipsi-lateral inputs can be used (Yin and Chan, 1990). However, the sum of monoaural inputs is generally not enough to elicit action potentials, and therefore an internal mechanism to compensate the binaural inputs by MSO neurons is required. This mechanism involves internal delay (as opposed to the external delay of ITD), which in turn determines the range of ITD sensitivity. The IPD sensitivity is usually the highest at the best frequency (BF) of a neuron, which is the frequency that produces maximum responses and will be the CF if it is below 1 kHz (Pecka et al., 2008; Franken et al., 2015). If tone frequency deviates from the BF, spike firing rates in MSO neurons decline, and the IPD sensitivity is reduced.

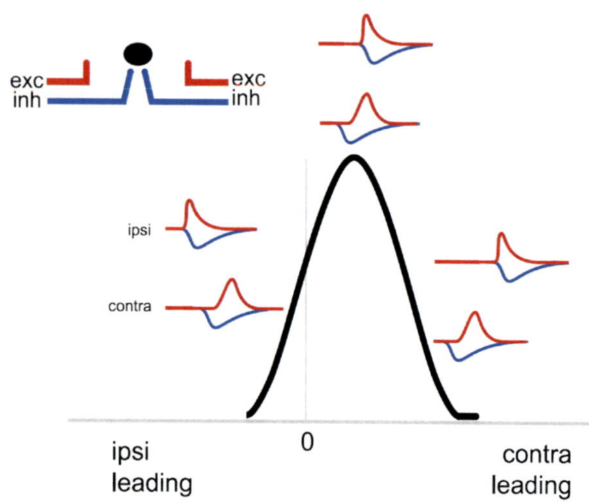

FIGURE 19.9 **A schematic model of ITD sensitivity in the MSO.** ITD sensitivity as measured by spike firing in MSO neurons is based on the complex integration in the timing and relative amplitude of EPSPs (*red waveforms*) and IPSPs (*blue waveforms*) from both the ipsi- and contralateral side. The best ITD is shifted toward the contra-leading ITD. (*Insert*) A simplified schematic for the inputs to an MSO neuron. *Image courtesy of Pecka, M., Encke, J., 2020. Coding of spatial information. In: The Senses: A Comprehensive Reference, vol. 2, pp. 713–731.*

MSO preferentially responds to sound to the opposite ear. In the ITD function of Fig. 19.9, the best ITD is shifted toward the contra-leading ITD, suggesting that the compensatory internal mechanism involves the physical distance of AVCN axons and their input properties to MSO neurons. The ITD for which relative firing rates are identical for different frequencies is called the characteristic delay (CD) (Yin and Kuwada, 1983). In theory, the CD corresponds to the ITD at which spike firing rates peak at any frequencies. However, in most MSO neurons, the CD tends to align along the slope, not the peak, of the ITD function (Spitzer and Semple, 1995; Pecka et al., 2008), suggesting that the ITD sensitivity does not depend simply on the time delay mechanism. In fact, a phase delay mechanism contributes to the coincidence detection since most MSO neurons exhibit a nonzero CD value of IPD at the intercept (i.e., at 0 Hz) in the relationship between IPDs and sound frequencies (Pecka et al., 2008; Stange et al., 2013; Dietz et al., 2014). It should be noted that, in addition to the physical distance, the ITD sensitivity depends on the timing of excitatory and inhibitory synaptic inputs as discussed above. As these inputs are determined by the phase-locked auditory nerves, lower frequencies necessarily produce broader peaks in the ITD function as wavelengths become larger. Further, best ITDs, on average, depended on BFs (McAlpine et al., 2001; Brand et al., 2002; Pecka et al., 2008). These characteristics of ITDs observed in mammalian MSOs are in line with the study of perceptional sound localization in humans (Lingner et al., 2012).

THE NUCLEUS OF LATERAL LEMNISCUS

There are two ascending pathways in the central auditory system. One is the lemniscal pathway, which primarily processes auditory unimodal information, and the other is the nonlemniscal pathway, which handles complex multimodal information. The lemniscal pathway ascends through the LL. The LL is the fiber track from the SOC to the IC. This track contains two nuclei: DNLL and VNLL.

DNLL modulates sound localization signals transmitted from the SOC to the IC to increase the contrast and dynamic range of the signals for accurate sound source detection (Pecka et al., 2010). The DNLL can achieve this by receiving inputs from the ipsilateral MSO and bilateral LSOs as well as the contralateral CN and sending inhibitory inputs to the ICs of both sides and the contralateral DNLL.

VNLL receives its inputs from all cell types of the contralateral VCN, including the octopus cells of PVCN, and from the ipsilateral MNTB (Kelly et al., 2009). Although the VNLL is located between the SOC and the IC, it does not receive any input from the LSO and MSO, which contain binaurally processed information. Therefore, the VNLL essentially processes monoaural sound information as part of the dorsal auditory stream. The tuning curve of many VNLL neurons is complex, covering broad frequency areas with multiple threshold peaks. Some VNLL neurons faithfully respond to sound stimulus timing. Therefore, the VNLL appears to extract temporal information from complex, mosaic sounds with multiple frequencies and transmits it to the ipsilateral IC (Malmierca et al., 1998).

Sound evoked BOLD responses in the nucleus of the LL have been detected using fMRI in humans (Dewey et al., 2020). The magnitude of the fMRI responses and the amplitude of auditory brainstem responses, which measure acoustic responses more directly, were not correlated, however. The VNLL and DNLL were not differentiated in this study, and further studies are needed to better understand the roles of these nuclei in humans.

THE INFERIOR COLLICULUS

Anatomy of the IC

Human IC is found rostral to the trochlear nerve and caudal to superior colliculi (SC) in the midbrain of both hemispheres. The IC volume measured using fMRI ranged from 115 to 200 mm^3 (Kang et al., 2008; Amaral et al., 2016; Cecchetti et al., 2016). These values were much larger than the histological data which ranged from 20.7 to 65 mm^3 (Glendenning and Masterton, 1998; Sitek et al., 2019). These histologically estimated volumes were consistent with the volumes measured using structural MRI (Bestmann et al., 2004; García-Gomar et al., 2019; Sitek et al., 2019). Postmortem studies identified the central nucleus (ICc), dorsal nucleus (ICd), external nucleus (ICx), and other nuclei (Fig. 19.10)

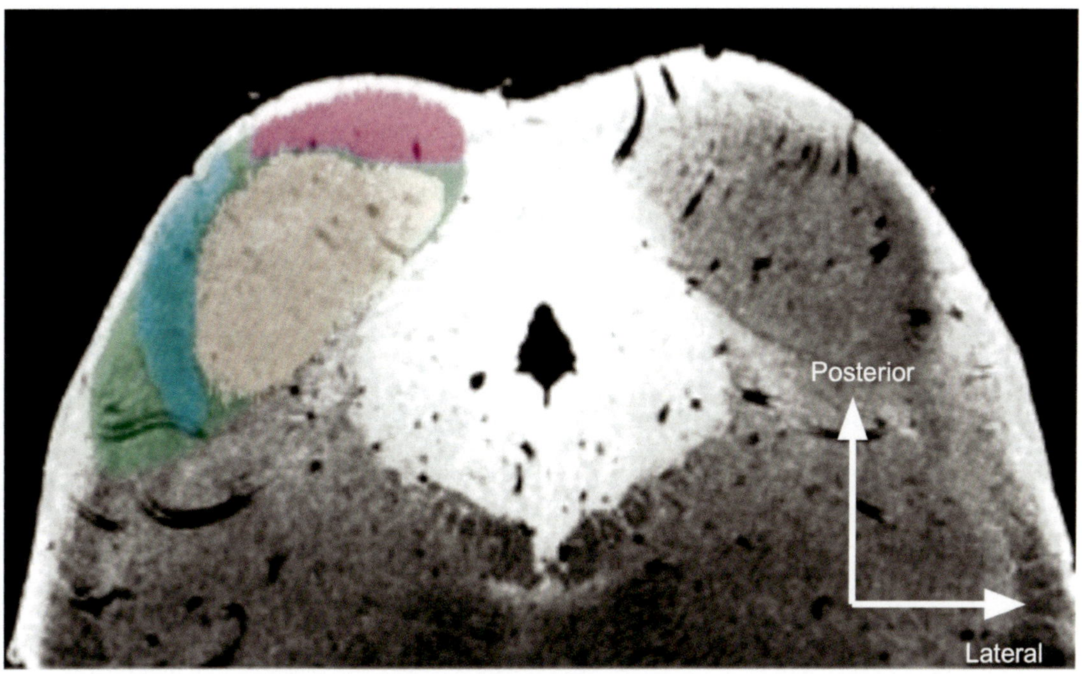

FIGURE 19.10 **Segmentation of postmortem IC and its major subdivisions.** Colored structures in the *left side* are overall IC (*green*), ICc (*gray*), ICd (magenta), and ICx (turquoise). The right IC is unsegmented to show magnetic resonance contrast. Segmentations were performed manually on the 50 μm isotropic T2*-weighted anatomical dataset and transformed to diffusion MRI space (200 μm isotropic). *Image courtesy of Sitek, K.R., Calabrese, E., Johnson, G.A., Ghosh, S.S., Chandrasekaran, B., 2022. Structural connectivity of human inferior colliculus subdivisions using in vivo and post mortem diffusion MRI tractography. Front. Neurosci. 16, 1–11.*

(Sitek et al., 2022). Human fMRI studies, however, could not differentiate the three nuclei (De Martino et al., 2013).

The connectivity of human IC subdivisions is largely unclear. Recent work using diffusion-weighted MRI tractography indicated the connectivity of the LL projections through the ICc and ICx, the brachium of the IC, and interhemispheric commissural connections through the ICd (Sitek et al., 2022). In animal models, the ICc receives the majority of ascending glutamatergic inputs from the auditory brainstem via the LL (Winer et al., 1998; Markovitz et al., 2013). The ICc neurons then project topographically organized efferent fibers to the MGB through the branchium of the IC. ICx and ICd receive nonlemniscal pathways from nonprimary brainstem structures and from top-down inputs from various regions, including the auditory cortex (Straka et al., 2015; Carbajal and Malmierca, 2018). The ICx neurons also project to the SC (Doubell et al., 2000; Bednárová et al., 2018).

The IC contains the excitatory projection neurons and GABAergic inhibitory neurons. GABAergic inhibitory neurons constitute the minority (20%–25%) in animals (Oliver et al., 1994; Merchán et al., 2005; Mellott et al., 2014). Despite similar proportion of inhibitory neurons among different animals, the proportion of projection neurons among inhibitory neurons differs among species (~40% in rats, ~20% in cats and guinea pigs). Each subdivision of IC contains inhibitory neurons, more in the ICc than other IC subdivisions. GABAergic neurons in all IC subdivisions project to both ipsilateral MGB and contralateral IC (Winer et al., 1996; Peruzzi et al., 1997). They were proposed to refine the auditory information transmitted to the higher processing structures. Another major output projects to the contralateral IC and the LL (Ito and Oliver, 2012).

The IC receives descending glutamatergic inputs mainly from layer V but also from layer VI of the auditory cortex. These cortical axonal projections provide top-down signaling to modulate IC responses to frequency, intensity, duration, and location of sound stimuli. The top-down regulation may also involve in the discrimination of sound sources (Bajo and King, 2012).

Tonotopy in the IC

In the ICc, the tonotopic gradient of low-to-high frequency has been detected in humans using fMRI (Fig. 19.11) (De Martino et al., 2013; Ress and Chandrasekaran, 2013; Moerel et al., 2015). Tonotopy in the ICc has been supported by many animal studies using invasive, electrophysiological recordings (Merzenich and Reid, 1974; Schreiner et al., 1997; Malmierca et al., 2008) and noninvasive fMRI (Cheung et al., 2012a,b). The orientation of the low-to-high frequency gradient runs in the dorsolateral to ventromedial direction in the ICc and is orthogonal to its fibrodendritic laminae (Malmierca et al., 1995). In the fMRI study of squirrel monkey, the IC was organized as a spectro-temporal map (Baumann et al., 2011). The systemic representation of frequency and amplitude modulation rates was in perpendicular gradients. This anatomical feature allows the spectro-temporal modulation in the ICc to be matched to the modulation power of natural sounds, suggesting that the ensemble filtering of the ICc provides the basis for biologically relevant perceptual sensitivity (Rodríguez et al., 2010).

The Tuning Curve of IC Neurons

Tuning width, the width of tuning curve, indicates the selectivity of spectral tuning in neurons (Rauschecker et al., 1995). The tuning curve of neurons within an isofrequency band was location-dependent in the IC, where it was narrower in the center and broader toward the edges of the band (De Martino et al., 2013). The tuning width was narrower in the IC than in the ACx. Spectral tuning width (about half an octave) estimated by fMRI was broader than that acquired by electrophysiological recordings in humans (about 1/12 octave) (Bitterman et al., 2008) and monkeys (Bartlett et al., 2011).

Function of the IC

The IC relays ascending information from the brainstem nuclei to auditory thalamus and cortex. The IC is the first site that receives the information on horizontal and vertical sound sources converging from each ear and integrates binaural information for sound localization, as demonstrated in the cat (Aitkin and Martin, 1987, 1990; Ehret and Merzenich, 1988). The ICc contains a map of frequency sensitive location (Merzenich and Reid, 1974) and has frequency-based laminar organization for the psychophysical "critical band" (Schreiner et al., 1997). The IC exhibited close correspondence between ILD sensitivity and perceptual adaptation (Dahmen et al., 2010). Two concurrent frequencies can be discriminated in the IC by a temporal integration mechanism (Chen et al., 2019). These properties of the IC likely contributes to generating the startle response, orienting the body toward relevant stimuli and discriminating pitch (Gruters and Groh, 2012). In human studies, the perception and dislike of harmonic dissonance, which are related to pitch discrimination, are modulated by differences in the activity level of the IC (Fritz et al., 2013; Kim et al., 2017). The IC may also integrate locomotion and auditory signals (Yang et al., 2020).

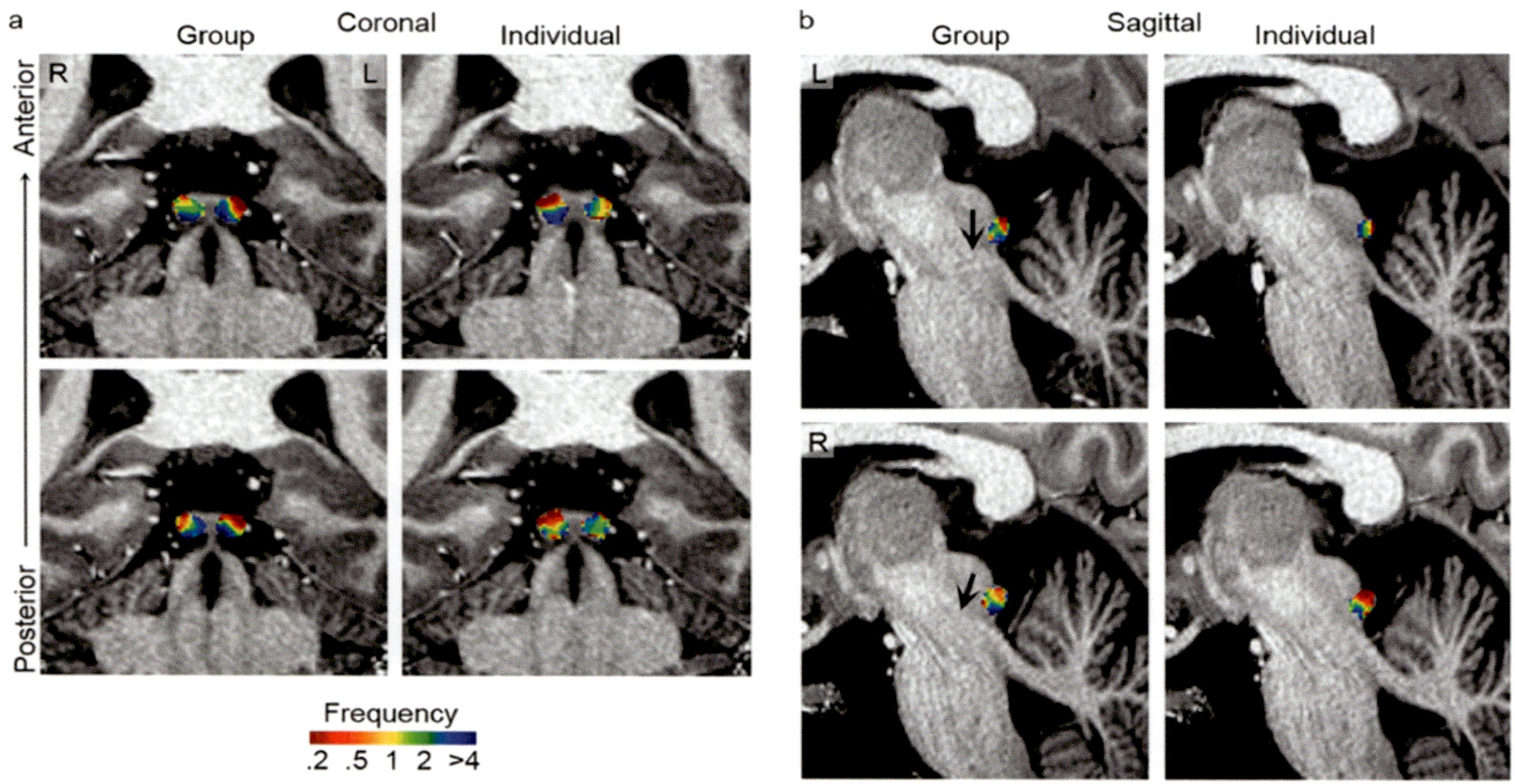

FIGURE 19.11 **Tonotopic maps in the human IC.** Coronal (A) and sagittal (B) views of tonotopic maps of the IC in a group (*left column*) and in an individual (*right column*). Coronal views at an anterior location and a posterior location are shown. Sagittal views of *left* (L, *top*) and *right* (R, *bottom*) IC are displayed. Color frequency maps are superimposed on high-resolution anatomical images of an individual subject. *Image courtesy of Moerel, M., De Martino, F., Ugurbil, K., Yacoub, E., Formisano, E., 2015. Processing of frequency and location in human subcortical auditory structures. Sci. Rep. 5, 1–15.*

SOUND LOCALIZATION

Finding the source of sound in space is a feat we accomplish with relative ease in our daily life. We possess a highly sophisticated neural system to reconstruct sound source by listening to sound using two separate ears. Unlike photoreceptors in our eyes, our ears do not possess sensory cells to encode spatial location of sound wave source. This ability extends our perceptual range of spatial information beyond the visual system to 360 degrees even without vision. Where in the auditory system can we find such a demanding and highly sophisticated system? Encoding the sound location is thought to require specialized neuronal circuits in the brainstem. The encoded neural information is then processed along the central auditory pathway, while being modified at various stages along the way, and decoded or perceived by auditory and other yet unknown brain structures.

For sound localization, our ear (pinna) and cochlea transform the frequency of acoustic waveforms arriving in the vertical plane and from the front and rear sound sources, while sound waves arriving in the horizontal plane are analyzed by the cochlea in the two ears (Oertel and Young, 2004; Macpherson and Sabin, 2013).

Monoaural spectral cues from various vertical sources are mainly detected by the DCN and the IC (Imig et al., 2000; Davis et al., 2003). The DCN specializes in spectra-dependent directionality, which depends on the response reduction produced by co-stimulation of BF and spectral-notch center frequency sounds. Meanwhile, the IC neurons respond sensitively to a single notch frequency flanked by inhibitory regions generated by all other notch frequencies. This feature of IC neurons suggests that they are involved in spectral notch detection and specialized in vertical location-dependent spectral detection.

For localizing sound source in the azimuthal plane (horizontal direction), we utilize the difference in the time and level (intensity) of arriving sounds at the two ears: the ITD and the ILD. The strategy to use these two properties to localize sounds is famously called the "Duplex Theory." These two properties are likely detected by distinct neural circuits and are sensitive to sound frequency. Low frequency sounds are localized primarily by the ITD. Its detection decreases greatly below 1.3 kHz sound (Brughera et al., 2013; Verschooten et al., 2019). The source of high frequency sounds is detected by the ILD as only small differences in ILDs can be found for low frequencies. However, some studies show the sensitivity to ILDs is similar between low and high frequencies (Yost and Dye, 1988), and consequently low frequency ILDs apparently support lateralization of tones (Keating et al., 2014; Hartmann et al., 2016). This is largely due to our head structure with two ears separated by the head. Even if sound is generated from a single source, two ears detect different intensity due to the separation of two ears by the head. However, two ears could not differentiate a sound wave if its wavelength is longer than the distance of the ears. If the distance of two ears were, say, 0.2 m, a sound wave that could overarch the two ears would have the wavelength of 0.2 m or longer. Since the speed of sound is 340 m/s, the frequency limit for two-ear differentiation would be $340\ \mathrm{m/s} / 0.2\ \mathrm{m} = 1{,}700\ \mathrm{Hz}$. Thus, the frequency of sound whose source location could be detected by two ears would be 1.7 kHz or smaller. ILDs increase in magnitude with increasing frequency.

How a representation of horizontal space forms in the auditory system is yet unclear. The earliest analysis of ILDs and ITDs is implemented in the LSO and the MSO, respectively (Middlebrooks, 2015). MSO neurons encode the ITDs over a wide range of sound locations, modulating spike rates widely. LSO neurons similarly encode the ILDs and the ITDs. Since the spike rate modulation is almost linearly correlated with the ILD or the ITD, the contralateral preference of the MSO and the ipsilateral preference of the LSO appear counterintuitive for sound localization. However, this apparent discrepancy could be overcome by the IC, where the excitatory neurons of LSO project to the contralateral IC and those of the MSO target it ipsilaterally, generating coherent representation of binaural sound source location. Based on the bilateral hemispheric balance and the input conversion from the dual nuclei of opposite preference, a new sound source-decoding model ("four-channel code model") has been proposed recently (Lingner et al., 2018). This model enabled to correct localization errors and control spatial resolution in humans. Thus, this model may supersede the long-standing, distributed labeled-line encoder model (Jeffress, 1948) at least in humans.

Interestingly, the IC, as well as the MGB, possesses a relative preference for contralateral and peripheral (vs. central) sound locations, while showing no topographic map of preferred azimuth (Moerel et al., 2015). The ILD cues could provide high frequency information due to the existence of two ears, while ILD tuning may confer the peripheral tuning preference in low frequency bands. Meanwhile, ITDs could be encoded by either place coding or rate coding (McAlpine et al., 2001; Werner-Reiss and Groh, 2008). In place coding, a population of neurons responds strongly to a specific sound location. In the rate coding, a neuronal population responds differentially to various sound locations, and therefore the activity rate estimates the relative location of sound source. The rate coding model of ITDs in a population with low frequency preference has been supported using the high field fMRI (Moerel et al., 2015). This is further supported by human psychophysical experiments, where the spatial location monitored by ITDs is altered by prior stimulation using a lateralized adapter sound (Phillips and Hall, 2005; Vigneault-MacLean et al., 2007). Sound localization experiments using complex sound features also revealed that auditory spatial coding is modulated by preceding stimuli (Magnusson et al., 2008; Park et al., 2008; Stange et al., 2013; Gleiss et al., 2019). In some of these experiments, the LSO and the MSO have been shown to play major roles in the modulation of neuronal sensitivity to ILDs and ITDs, and importantly, both LSO and MSO neurons exhibited hemispheric differences in activity-dependent rate adaptation as a function of sound source locations. Therefore, the hemispheric asymmetry may underlie the shift in the perceived location of subsequently presented sound observed in the human psychophysical experiments. The mechanism for the rate adaptation in the MSO likely involves activity-dependent GABA release from the SPON to affect $GABA_B$ receptors for this shift in perceptual changes (Stange et al., 2013).

Monoaural adaptation of sound level inputs to the LSO neurons alters ILDs, changing their rate functions (Gleiss et al., 2019). This adaptation appears to involve the activity-dependent release of GABA from neuronal dendrites, which activates presynaptic $GABA_B$ receptors to suppress excitatory and inhibitory inputs for short-term synaptic gain modulation. This gain adaptation also involves the auditory nerves of the SG, which is known to exhibit the dynamic range adaptation—the concurrent stimulus probability-dependent spiking to respond maximally during changes in the stimulus range (Wen et al., 2009). This adaptation changes the relative strength of the inputs from the two ears. The monoaural adaptation does not contain sound source information, and as a consequence of this adaptation, individual LSO neurons responded less to spatial stimuli due to high variability in responses. However, as a population average of individual stimuli across multiple LSO neurons, the adaptation enhanced space coding efficiency particularly for the concurrently most probable source locations. This efficient, cooperative population coding improved separation of ILDs not only in animals but also in humans. Thus, the efficient population coding in the LSO, involving dynamic range adaptation may allow better sound source localization via detection and processing of ITDs and ILDs. How the IC and the MGB contribute to this adaptation and sound localization will need to be investigated further for better understanding of the mechanisms underlying sound localization.

THE MEDIAL GENICULATE BODY

Anatomy of the MGB

The MGB is found in the ventromedial thalamus. The histological estimate of human MGB size is in the range of 46.6–90 mm^3 (Winer, 1984; Sitek et al., 2019). Structural MRI estimated it with smaller size 20–65 mm^3 (Amaral et al., 2016; Sitek et al., 2019), while fMRI indicated much larger size: 185 mm^3 (Bestmann et al., 2004) and 285 mm^3 (Sitek et al., 2019). The MGB can be divided into the ventral division (MGv), the dorsal division (MGd), and the medial or magnocellular division (MGm) (Winer, 1984). Human MGv is organized into a laminar structure with laminae mostly representing iso-frequency bands, where dorsal-to-ventral laminae are aligned in low-to-high frequency tonotopic gradient as shown in Fig. 19.12 (Moerel et al., 2015). Tonotopic organization has also been detected in MGm (Moerel et al., 2015) as found in some animals using electrophysiological studies (Rouiller et al., 1989; Hackett et al., 2011) but not in others (Ryan et al., 1982; Calford, 1983). The low frequency area in the dorsomedial region outside MGv may belong to either MGd, a part of the tonotopic map of MGm, or the lateral region of the posterior thalamic nucleus (Pol). Pol was tonotopically organized in humans as it has been found in cats (Imig and Morel, 1984, 1985). Pol is located superior to MGB and is traditionally considered as the secondary (nonlemniscal) auditory thalamus (Jones, 1985). However, it might be considered as a part of lemniscal pathway because of the resemblance of its physiological properties to MGv (Anderson and Linden, 2011).

The MGB is the relay center of auditory information between the IC and the ACx, regulating auditory as well as nonauditory information. As in the IC, the MGB has a contralateral bias for sound responses in humans (Scheffler et al., 1998; Jäncke et al., 2002; Schönwiesner et al., 2007; Moerel et al., 2015), which is also observed in animals based on electrophysiological recordings (Aitkin et al., 1984; Irvine and Gago, 1990; Palmer et al., 1990; Stanford et al., 1992; Groh et al., 2003). The three MGB divisions independently integrate the segregated information in parallel to present it to the ACx as refined auditory information. MGv is considered as a lemniscal pathway, while MGd and MGm are considered as part of the extra-lemniscal auditory pathways, processing complex sensory information. Each division integrates the spatial, temporal, and spectral information of sound that has been partially integrated but separately processed in the lower nuclei of the auditory pathway. They also reciprocally interact with the ACx, forming a tightly coupled tandem integration system.

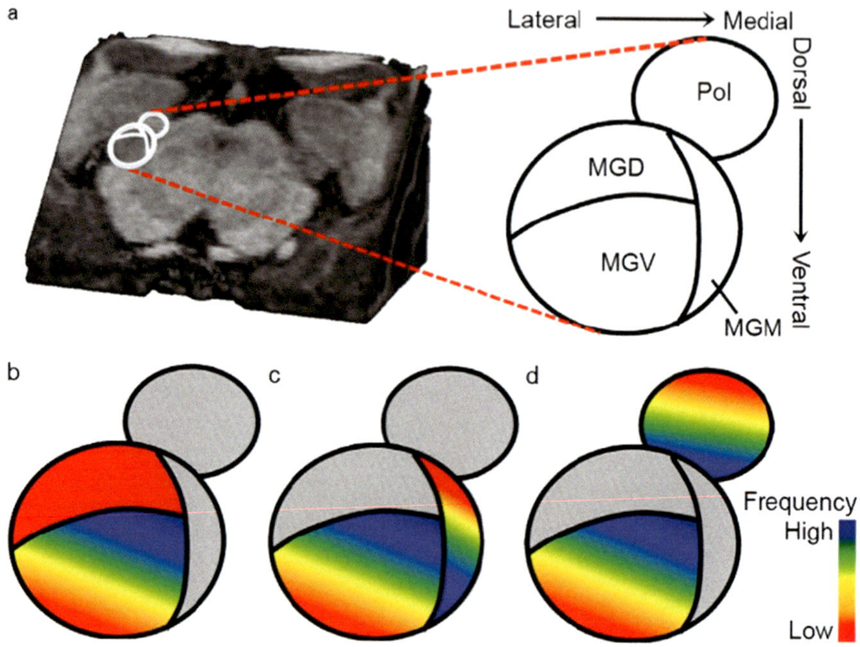

FIGURE 19.12 **fMRI analysis of the human MGB tonotopic gradient.** The cartoon reflects the right auditory thalamic regions as a 45 degrees cut through the brainstem. (A) Dorsal and ventral directions reflect posterior−superior and inferior−anterior locations, respectively. (B−D) Schematic representations of tonotopic maps in MGB and posterior thalamic nucleus (Pol) based on the interpretation of observed tonotopic gradient. (B) The dorsomedial to ventrolateral tonotopic map in MGV. The dorsal low frequency region could reflect low frequency tuning in MGD. (C) A low-to-high dorsomedial to ventrolateral tonotopic map in MGM. (D) A low-to-high dorsomedial to ventrolateral tonotopic map in Pol. *Image courtesy of Moerel, M., De Martino, F., Ugurbil, K., Yacoub, E., Formisano, E., 2015. Processing of frequency and location in human subcortical auditory structures. Sci. Rep. 5, 1–15.*

The Ventral Division

MGv has two types of neurons: large neurons with bushy dendrites and tufted branches and smaller stellate cells with radiating, spherical dendrites (Winer, 1984). The dendrites of the large neurons with terminating neuropils forms medio-laterally oriented fibro-dendritic laminae in parallel rows, which may correspond to the isofrequency bands. MGv receives excitatory and inhibitory inputs mainly from the ipsilateral ICc (Winer et al., 1996; Peruzzi et al., 1997) and smaller input from the IC shell regions (Kudo and Niimi, 1980; LeDoux et al., 1985). MGv neurons, which are tonotopically distributed in the caudal-to-rostral direction, project to primary auditory cortex (AI) in the ventral-to-dorsal direction with little overlap in animals (Storace et al., 2010, 2011; Read et al., 2011). This projection of auditory specific information constitutes the lemniscal auditory pathway in each hemisphere. The axons of MGv neurons terminate mainly to layers 3/4 of AI and less so in other layers in animals (Huang and Winer, 2000; Kimura et al., 2003; Smith et al., 2012; Vasquez-Lopez et al., 2017; Xie et al., 2018; Ishizu et al., 2021). MGv neurons also innervate other core regions and, to a lesser extent, the core-surrounding belt areas of the ACx (Andersen et al., 1980b; Morel and Imig, 1987; Redies et al., 1989; Velenovsky et al., 2003; De La Mothe et al., 2006; Polley et al., 2007; Lee and Winer, 2008; Hackett et al., 2011; Saldeitis et al., 2014). The tuning curve of MGv neurons is relatively sharp (Bartlett and Wang, 2011; Bartlett et al., 2011), perhaps due to external inhibitory input (Suga et al., 1997) or due to corticofugal modulation (Zhang and Suga, 2000).

The Dorsal Division

MGd contains at least three populations of neurons in humans (Winer, 1984). Neurons with small and medium-sized somata are dominant populations, while those with large somata are less populated. Large soma neurons include stellate neurons with radiating dendrites and a dichotomous branching pattern, and elongated neurons with bushy dendrites and tufts. These neurons are mixed with many smaller neurons with tiny round or drumstick-like soma with limited arbor of thin dendrites. MGd receives nontonotopic inputs from the dorsal IC and from the somatosensory system, and projects mainly to layer 1, layers 3/4, and layer 6 of the belt region of the ACx, which surrounds the core region in animals (Andersen et al., 1980a; Kudo and Niimi, 1980; LeDoux et al., 1985).

The Medial Division

The MGm in humans contains large soma neurons more commonly than in other MGB divisions (Winer, 1984). They are present along with smaller neurons. In animals, MGm neurons are diverse in shape and include very large "magnocellular" cells (Winer et al., 1999; Mylius et al., 2013; Saldeitis et al., 2014). Most MGm neurons receive binaural inputs and can therefore exhibit spatial selectivity. MGm neurons typically respond to a wide range of frequencies with short onset (Anderson et al., 2006). Many MGm neurons reliably fire action potentials at the stimulus onset (Anderson and Linden, 2011). Their responses show strong adaptation to ongoing stimuli, amounting to habituation to frequent sounds (Anderson et al., 2009). Consistent with this, in a recent human fMRI study, the MGm exhibited a strong preference toward transient rather than sustained sounds (Meng and Schneider, 2022). Because of these properties, the MGm has been suggested to mediate alerting responses to auditory stimuli by suppressing responsiveness to common sounds.

The MGm has been implicated in many other important behaviors due to its properties to process multiple information. MGm neurons integrate multisensory signals, receiving ascending afferents from the ipsilateral ICc as well as ICx and ICd, the LL, the lateral tegmental system, the SC, the vestibular system, and the spinal cord (Winer and Morest, 1983; Ledoux et al., 1987). The MGm receives descending inputs from the auditory and other cortexes. MGm neurons are inhibited by GABAergic neurons in the IC (Peruzzi et al., 1997) and the DNLL (Whitley and Henkel, 1984; Bajo et al., 1993) and project to all layers of the core and other areas of the ACx as well as the striatum and the lateral nucleus of amygdala (LeDoux et al., 1991). Auditory signals through the MGm could elicit emotional reactions via the amygdala. Indeed, MGm neurons are sensitive to associative learning, where association between sounds and negative stimuli that induce fear such as an air puff to the cornea (McEchron et al., 1996) or an electrical shock to the limbs (Duvel et al., 2001) can modify their responses.

The proportion of GABAergic interneurons in human MGB is unclear. There are >25% of interneurons in cats and monkeys (Winer and Larue, 1996), while only few (<1%) are found in rodents (Winer and Larue, 1988). Such species difference in interneuron proportions is absent in the IC and the ACx. In addition, this species dependence is peculiar to the auditory system, as the lateral geniculate body of the visual system has similar proportions of inhibitory neurons among these species. Based on these observations, one could speculate that auditory processing in higher mammals required intranuclear inhibition in MGB for auditory processing during evolution.

THE AUDITORY CORTEX

The human auditory cortex (ACx) can be considered as the interface between subcortical sound processing of peripherally transformed signals and central processes for perception and cognition. Acoustic sounds such as speech and music contain spectral (frequency) component as well as temporal (timing) component. These spectro-temporal signals are converted to higher-order information by the ACx. Proper understanding of its functional topographic map is necessary to understand the nature of signal transformation in the cortex. For finer maps, direct access to cortical circuits using invasive techniques would be important but is not feasible as the human cortex is convoluted with valleys (sulci) and hills (gyri). However, the ACx is the most accessible structure in the central auditory pathway using noninvasive functional monitoring techniques such as fMRI as it is located on the outer most surface of the brain directly under the skull. Using such noninvasive techniques combined with histological techniques on postmortem brains, scientists have proposed topographical maps critical for our understanding of the structure and function of the ACx.

Anatomy of the ACx

Human ACx is located on the superior temporal plane (STP) (Fig. 19.13). The ACx consists of two-thirds of superior temporal gyrus (STG), which is divided into planum polare (PP), Heschl's gyrus (HG), and planum temporale (PT) (Celesia, 1976; Galaburda and Sanides, 1980; Kim et al., 2000; Moerel et al., 2014). The PP is the anterior region of the ACx, which includes the first transverse sulcus (FTS) and its anterior area. It is delimited medially by the insula, anteriorly by the frontal operculum, and posteriorly by the HG. The HG is in a convoluted region of STP, hidden deep in the Sylvian fissure (SF). It extends postero-medially and is surrounded anteriorly by the PP and postero-laterally by the PT. It includes primary auditory field (A1). The individual variability of the HG is apparent in terms of the number, size, and shape of convolutions created by one or two Heschl's sulci (HS) and, if exists, the sulcus intermediate (SI) formation in each hemisphere (Campain and Minckler, 1976; Penhune et al., 1996; Rademacher et al., 2001). The average size of the HG is approximately 3,200 mm^3. The primary auditory cortex (PAC or core) region is about 1,650 mm^3, about a half the size of the HG (Rademacher et al., 2001). The PT is bordered medially by the HG, laterally by the STP rim, and posteriorly by the horizontal limb of the SF, where it either bifurcates or angulates as the posterior ascending ramus (PAR) (Rubens et al., 1976). The size of the PT is asymmetric with the left hemisphere having a larger size (Geschwind and Levitsky, 1968; Galaburda et al., 1978; Bonte et al., 2013). Although these major anatomical landmark regions are well-recognized, cytoarchitectural studies of the ACx presented different subregions of the cortex using a variety of architectonic schema (Fig. 19.14). Individual variability and interhemispheric anatomical variability likely contributed to the different schema. However, several consensus findings arose from these studies.

Postmortem human anatomy shows a hierarchical organization with the core of ACx receiving ascending

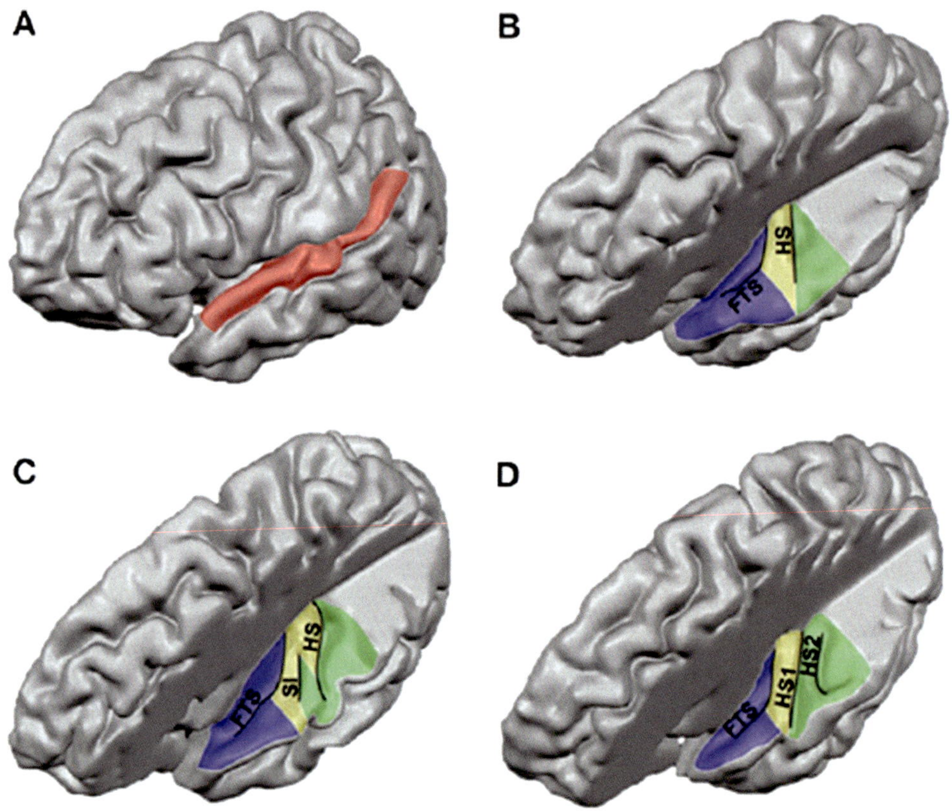

FIGURE 19.13 **Anatomical landmarks on the human supratemporal plane.** (A) Lateral view of the left hemisphere with STG indicated in *red*. (B–D) Top view of left supratemporal plane, after removal of a large part of the parietal cortex. PP, HG, and PT are indicated in *blue*, *yellow*, and *green*, respectively. Major sulci are outlined in *black*. First transverse sulcus (FTS), sulcus intermediate (SI), Heschl's sulcus (HS), first Heschl's sulcus (HS1), second Heschl's sulcus (HS2). Hemispheres with one HG (B), an incomplete separation of HG (C), and two HG in (D) are shown. *Image courtesy of Moerel, M., De Martino, F., Formisano, E., 2014. An anatomical and functional topography of human auditory cortical areas. Front. Neurosci. 8, 1–14.*

projections from the auditory thalamus (Hackett et al., 1998a, 2001). The core is surrounded by nonprimary "belt" and "parabelt" areas (Morosan et al., 2001; Moerel et al., 2014) (see the parcellation made by solid black lines in Fig. 19.15) as found in the monkey, where each of the areas is divided into subregions that can be differentiated based on their anatomical architecture, functional properties, and connectivity with subcortical nuclei and other cortical regions (Kaas and Hackett, 2000). The human PAC is found approximately in the medial two-thirds of HG. The PAC has a highly granular layer 4 formed with thalamocortical terminals of myelinated axons (Saenz and Langers, 2014). Detailed histochemical and architectonic studies showed that this koniocortical structure was heterogenously distributed within the PAC (Morosan et al., 2001; Wallace et al., 2002; Sweet et al., 2005; Fullerton and Pandya, 2007). These and other studies led to the consensus that the PAC is elongated along the crest of the HG and could be divided into two or three fields: A1 and R with or without RT (the "Classical" configuration) as illustrated in Fig. 19.16 (Baumann et al., 2013).

Myeloarchitecture of the ACx

Histological studies showed a heavily myelinated area was the medial HG among all temporal regions (Nieuwenhuys, 2013). Cortical myelin density can be mapped noninvasively in vivo using MRI (Glasser and van Essen, 2011; Dick et al., 2012), where, as in the postmortem analysis, the HG was most myelinated, the PT was modestly myelinated, and the PP was least myelinated (Sigalovsky et al., 2006; Glasser and van Essen, 2011). The highly myelinated HG region overlapped with the symmetric (mirror) tonotopic maps along the HG, which likely represent A1 and R regions (Dick et al., 2012). The most heavily myelinated region was interpreted to be the A1 (De Martino et al., 2015) (see also Fig. 19.18). Consistent with this interpretation, A1 is more myelinated than the rostral field (R) and the rostrotemporal field (RT) within the core in the macaque (Hackett et al., 1998a). As in tonotopic maps, though, in vivo myelo-structure varied between hemispheres and among individuals, pointing out the need for multi-parametric analyses to identify the structure of the ACx.

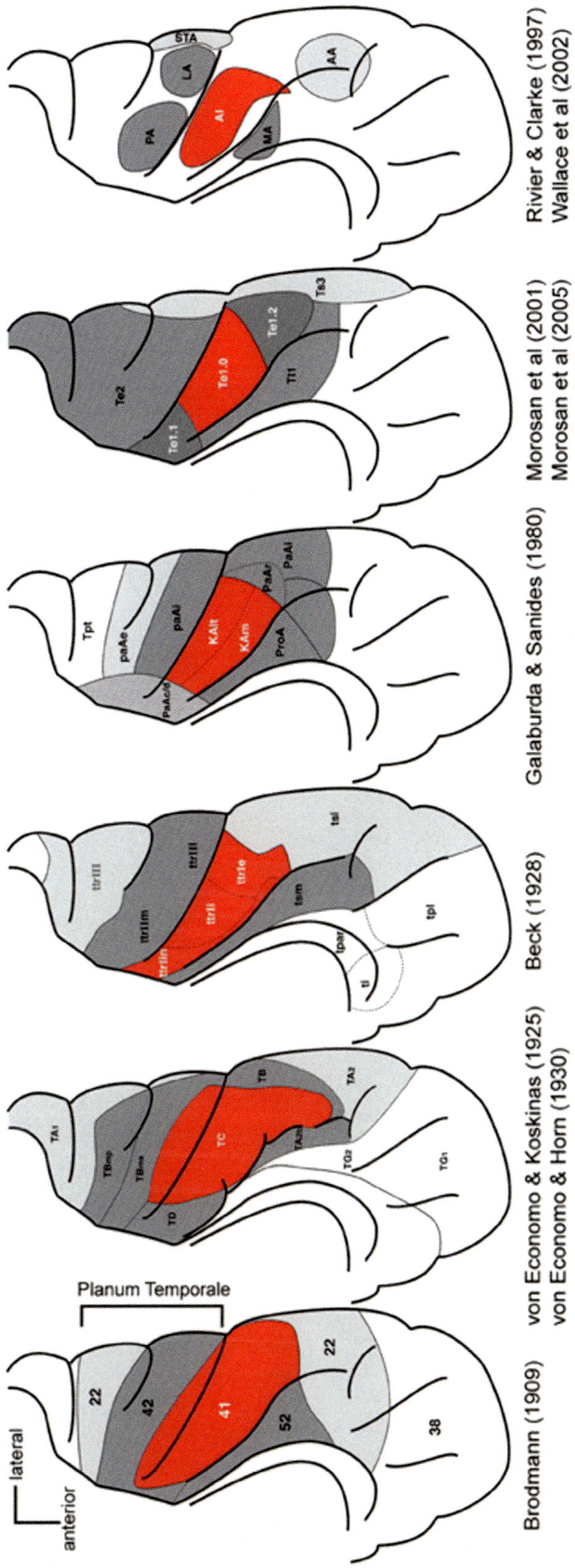

FIGURE 19.14 **Parcellations of the human superior temporal cortex by different investigators.** For each panel, the locations of major auditory cortical regions are drawn on a template schematic of the superior temporal plane. The STG is not visible. *Red*, core region; *dark gray*, belt region, *light gray* parabelt, and possibly other regions. *Image courtesy of Baumann, S., Petkov, C.I., Griffiths, T.D., 2013. A unified framework for the organisation of the primate auditory cortex. Front. Syst. Neurosci. 7, 1–19.*

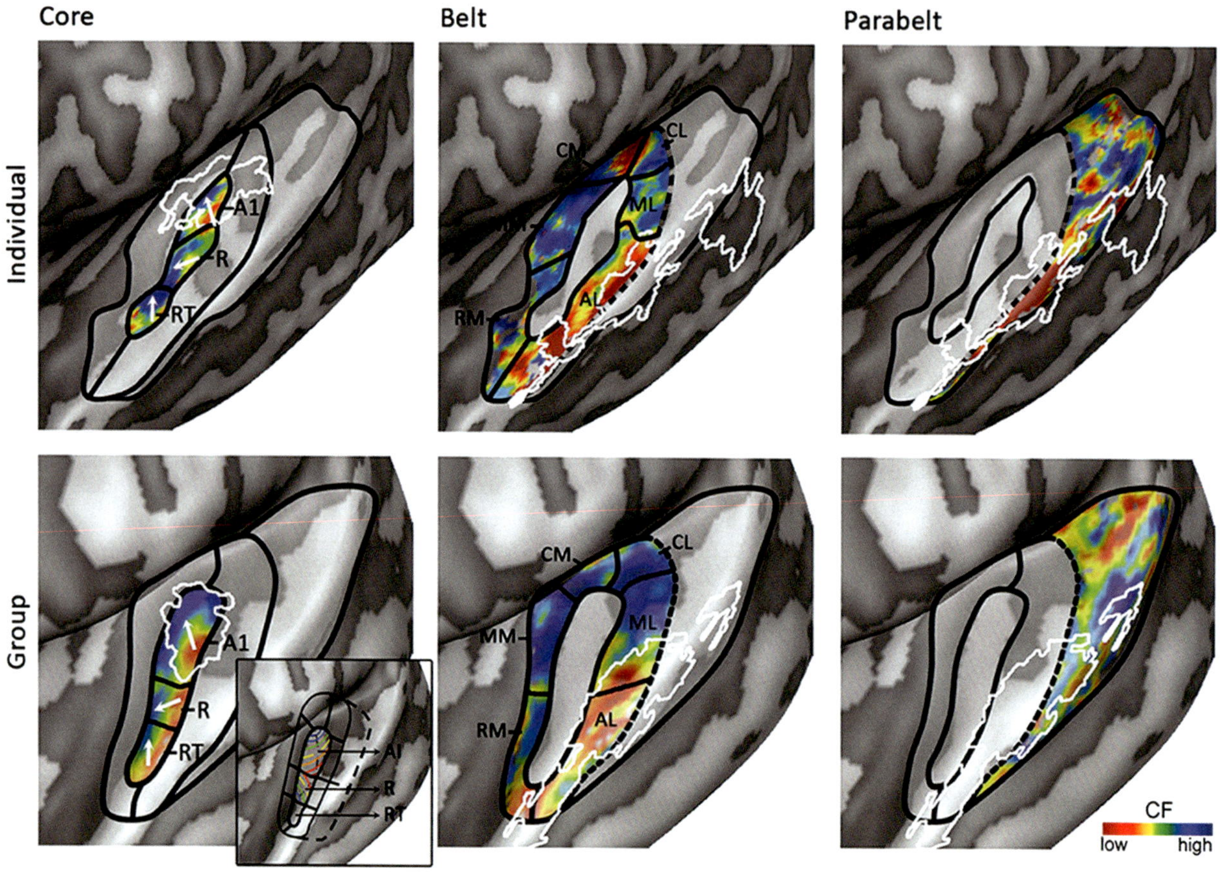

FIGURE 19.15 **Working model of human auditory cortex.** Parcellation of the ACx into core (*left*), medial and lateral belt (*middle*), and parabelt (*right*) based on individual (*top*) and group (*bottom*) data. Solid black lines indicate boundaries between auditory fields based on maps of tonotopy, tuning width, or myelin. *Dotted black lines* indicate boundaries assumed based on literature, but for which no objective measure is available. *White outlines* indicate maps of myelin (*left*) or speech/voice selectivity (*middle* and *right*). The *white arrows* in the first column show the main direction of the frequency gradient within the three core fields. The *inset* in the *bottom left panel* shows the primate model of ACx as interpreted. The model is rotated so that the core aligns with HG. Isofrequency gradients are color-coded to match the color scale in the tonotopy maps. *Image courtesy of Kaas, J.H., Hackett, T.A., 2000. Subdivisions of auditory cortex and processing streams in primates. Proc. Natl. Acad. Sci. U. S. A. 97, 11793–11799; Moerel, M., De Martino, F., Formisano, E., 2014. An anatomical and functional topography of human auditory cortical areas. Front. Neurosci. 8, 1–14.*

Tonotopy of the ACx

Recent studies provided a consistent view of human tonotopic organization in the HG, where the tonotopic gradient is laid in a distinctive, flattened V-shaped manner with a high frequency area just anterior to the HG along the FTS, a wedge-shaped low frequency area located in the lateral HS posterior to the anterior high frequency area, and another high frequency area posteromedial to the low frequency area in the medial HS (Formisano et al., 2003; Humphries et al., 2010; Woods et al., 2010; Da Costa et al., 2011; Striem-Amit et al., 2011; Langers and Van Dijk, 2012; Baumann et al., 2013; Allen et al., 2022). However, there remains controversy over the exact orientation of the core with anterior R and posterior A1 either confined within the rostral and caudal banks of the HG (Classical configuration), elongated perpendicular to the HG with the R mostly in the HG and the A1 mostly beyond the HS (Perpendicular configuration), or extended along the anterior-posterior axis with the A1 crossing the HS (Oblique configuration) (Galaburda and Sanides, 1980; Morosan et al., 2001; Sweet et al., 2005; Hackett et al., 2011; Saenz and Langers, 2014; Allen et al., 2022). The oblique configuration shares the features that are consistent with reports on nonhuman primates (Jones et al., 1995; Hackett, 2011). In addition to the tonotopic map in the anteroposterior gradient, a medial-to-lateral gradient has also been reported (Leaver and Rauschecker, 2016).

Individual tonotopic maps exhibit finer tonotopic gradients with a large number of low-high frequency gradient reversals as shown in Fig. 19.17 (Allen et al., 2022). The finer maps are different among individuals as frequency area size and its precise location are somewhat variable due to the extent of imaged neural

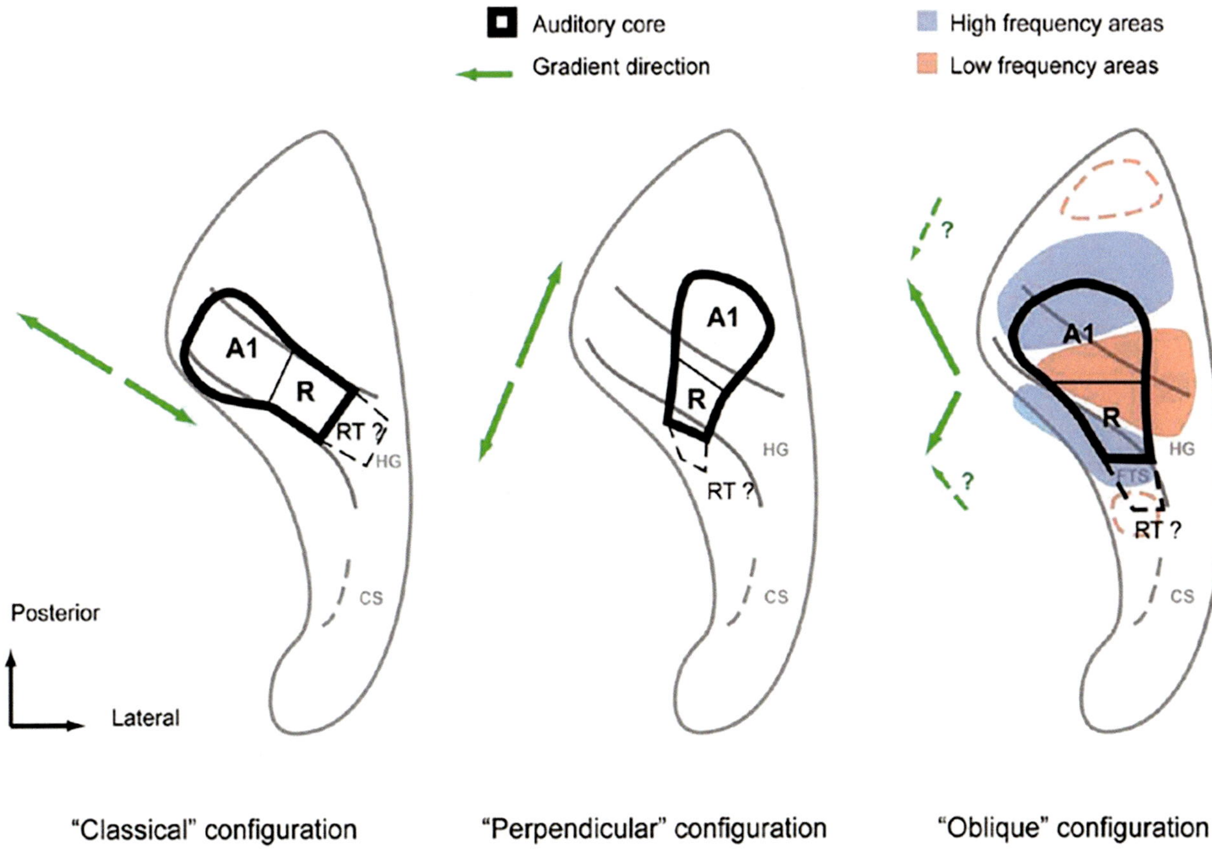

FIGURE 19.16 **Proposed configurations of auditory cortical organization in humans and nonhuman primates.** Three configurations of auditory core fields (*solid line*). The main frequency response areas (mirror-image high-low-high tonotopy) are superimposed over the standard superior temporal plane (*right*). The suggested directions of the main tonotopic gradient axes are indicated with *green arrows* next to each configuration. Additional anterior and posterior low frequency preference areas suggested by some studies are marked by *red dashed lines*. *Image courtesy of Baumann, S., Petkov, C.I., Griffiths, T.D., 2013. A unified framework for the organisation of the primate auditory cortex. Front. Syst. Neurosci. 7, 1–19.*

activities, though general gradient distribution typically agree based on anatomical architecture. Allen et al. (2022) describe five patterns that can be detected in a consistent basis: (1) A long low-frequency area that runs along the HG belongs to the A1 in the core. This area is the part of the main high-low-high flattened V-shaped tonotopic gradient of the auditory core. (2) A large low frequency area, which is located on the lateral HG in the middle STG, belongs to the lateral belt. (3) An anterior high-frequency region that likely belongs to the medial belt, which can be divided into two smaller regions divided by a frequently detected low frequency area, forming two tonotopic maps. (4) A low-frequency area at the postero-medial end of the HG is found in the parakoniocortex. (5) A posterior region that extends from the HS and covers the posterior HG and PT. This region contains a low frequency area and the clusters of high frequency areas of variable sizes. These areas may belong to the lateral belt and parabelt cortex.

Advantages and Disadvantages of Electrophysiological Functional Studies

Our understanding of ACx functions will benefit from more direct methods of neuronal activity detection rather than the detection of blood flow changes as in fMRI, magnetic field changes at distant locations as in magentoencephalography (MEG), or electrocorticography (ECoG). One such method would be intracortical electrophysiological recordings. Though invasive, requiring craniotomy, the direct intracortical placement of sensitive electrodes at an identified brain area allows monitoring of neuron-derived electrical potential changes outside a group of neurons or single neurons. Temporal resolution is more precise, allowing the detection of sound-evoked response latency along the medial-lateral axis of the HG (Liégeois-Chauvel et al., 1994), and frequency selectivity and hierarchical organization of cortical processing in a relatively wide surface area of the cortex (Nourski et al., 2014).

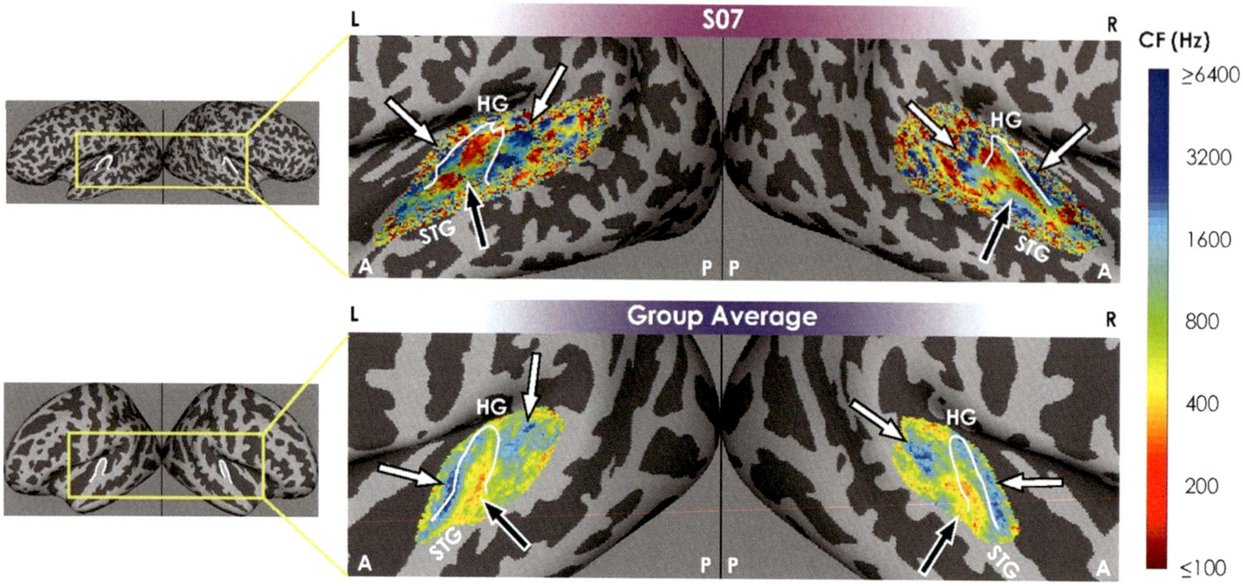

FIGURE 19.17 **Tonotopy elicited by pure tones.** Frequency tuning model characteristic frequency (CF) parameter maps for the pure-tone condition shown on inflated brains for one subject (*top*) and transformed to average space and averaged across all 10 participants (*bottom*) within the intersection of all participants' ROIs. HGs denoted by *white lines. White and black arrows* indicate high (*blue*) and low (*red*) frequency regions, respectively. *Image courtesy of Allen, E.J., Mesik, J., Kay K.N., Oxenham, A.J., 2022. Distinct representations of tonotopy and pitch in human auditory cortex. J. Neurosci. 42, 416–434.*

Recording of action potentials from single cells (single unit recordings) provide ultimate spatial and temporal resolution. Spike rate measurements in response to tone stimuli of various frequencies and intensities could reveal a receptive field of a neuron with a CF, the most sensitive frequency with the lowest intensity of a neuron. A human study showed that the CF of neurons increased toward the postero-medial location in the HG (Howard et al., 1996), consistent with the fMRI studies described above. This technique, however, has its own limitations. To obtain complete tonotopic maps, many neurons from various locations of the ACx must be recorded, and the location of recorded sites must be identified. A multiple array of electrodes with defined distance will help in this regard, but recorded neurons cannot be identified with this method. Other disadvantages in invasive recordings will include limited availability of experimental subjects, restricting to volunteer patients who are undergoing neurological surgeries for epilepsy or brain tumor, and recording locations are limited due to clinical criteria. Therefore, a precise picture of tonotopic maps is not currently possible to obtain in healthy human ACx.

The goal to define precise tonotopic maps would be to understand how sound information is processed for perception and cognition in the ACx, given the fact that the tonotopy is the basis of sound information processing in the central auditory pathway. Defining tonotopic maps has its own value for systemic analysis of the ACx. As the first site of early sound processing in the cortex, the auditory core would naturally attract attention in research community. However, it should be noted that tonotopic maps are not enough to determine the orientation of the auditory core with respect to the HG. It is also the case that tonotopic maps cannot define the size of the core. To achieve systemic understanding of the human ACx, however, accurate estimation of its structure and function will be imperative.

Tuning Width and Spectral Filtering

The tuning width in human ACx has been computationally modeled based on fMRI data as shown in Fig. 19.18 (Moerel et al., 2012; De Martino et al., 2013). When main spectral peak was used to analyze for tuning width, the map of the tuning width showed that the HG was most narrowly tuned in both hemispheres. Consistently, tuning width is narrower in neurons of the core than those of the belt areas in the macaque (Hackett et al., 1998b; Rauschecker and Tian, 2004; Kajikawa et al., 2005). A matching analysis of the tonotopic and tuning width topographic maps showed that narrowly tuned areas in the HG contained the high-low-high tonotopic gradient, corresponding to the PAC (Moerel et al., 2012). This topographic matching extended beyond the PAC though tuning width was broader. These findings suggest that the PAC works as a finer spectral filter for sound signal integration.

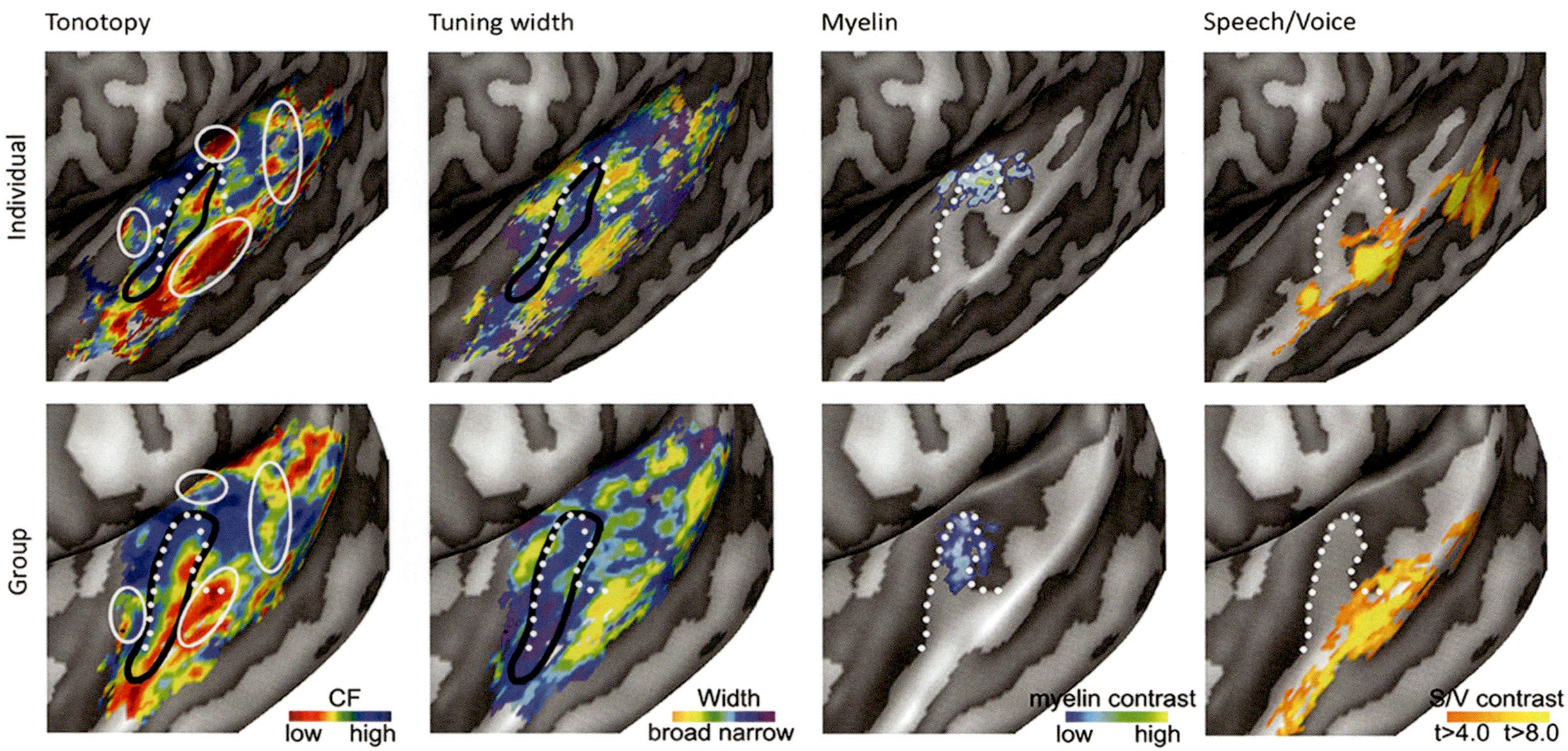

FIGURE 19.18 **Anatomical and functional characterization of human auditory cortex.** The four columns show tonotopy, tuning width, myelin contrast, and speech/voice selectivity in the left hemisphere of a single subject (*top*) and as group results (*bottom*). The *black lines* in the first and second column outline PAC, and the *white circles* in the first column indicate frequency regions beyond the main high-low-high tonotopic gradient. The *white dotted lines* show the location of HG. *Image courtesy of Moerel, M., De Martino, F., Formisano, E., 2014. An anatomical and functional topography of human auditory cortical areas. Front. Neurosci. 8, 1–14.*

Periodotopy (Periodicity Map)

For complex sound processing, a topographical map of the ACx may require not only single dimension but also multiple dimensions combined. Periodotopy or periodicity map may be such a topographic map. Periodicity is the rate of temporal modulations in a sound. The periodotopy characterizes the relationship between the neural activities monitored by fMRI and the rate of amplitude-modulation (AM rate) of carrier tones. There are currently contradicting periodotopy maps. A periodotopy map in humans showed modulation rate gradient that is orthogonal to tonotopic map (Barton et al., 2012). Based on the combinational analysis of these two maps, it was suggested that there were 11 different subfields in the core and belt regions. Another periodotopy map (Fig. 19.19) shows a high-to-low temporal modulation in the HG in the medial-to-lateral direction (Herdener et al., 2013). A large scale map across the ACx like these two maps was not detected in another study (Leaver and Rauschecker, 2016). The use of sounds with simultaneously combined spectral and temporal modulations did not resolve the contrasting results. Whether periodotopy is indeed present and how it contributes to auditory processing will require further investigation.

Speech/Voice Regions

Various species has regions that process conspecific vocalizations. In humans, speech and voice elicit neural activities outside the PAC at the low frequency area in the STG, which is possibly utilized for voice recognition (Fig. 19.18) (Zatorre et al., 2002a; De Lucia et al., 2010; Moerel et al., 2012; Talkington et al., 2012). The speech-activated region overlaps with the low frequency area of the tonotopic map, suggesting that this area works to filter low frequency sounds, which are commonly embedded in human voices. This region

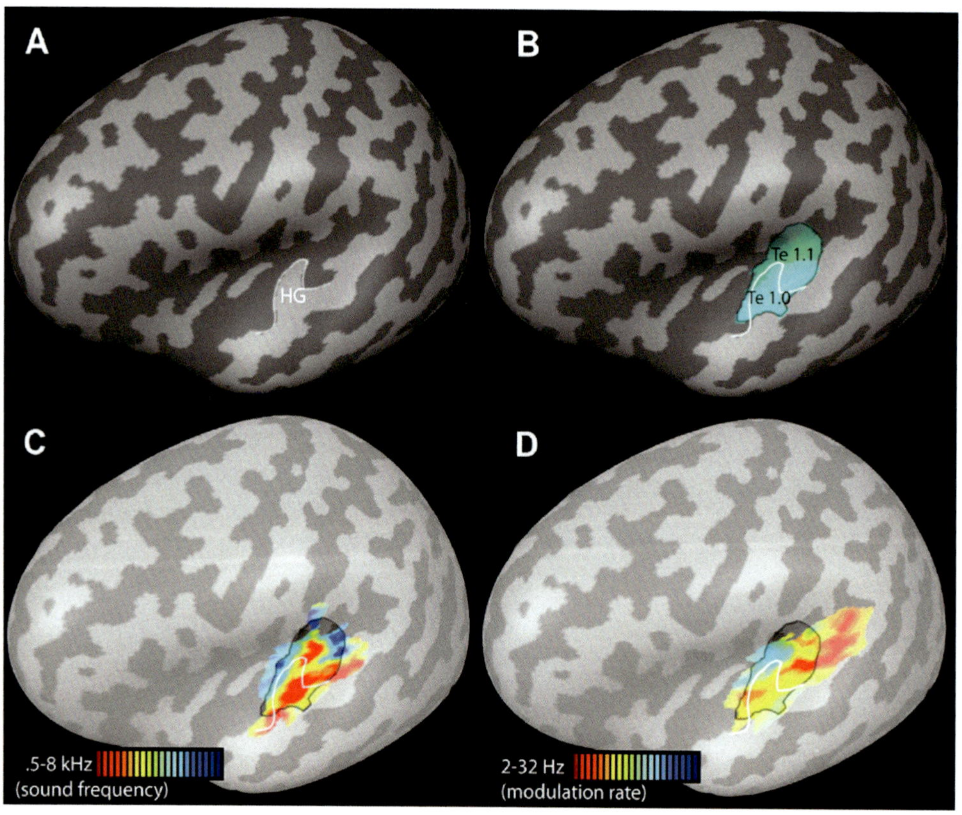

FIGURE 19.19 **Group-average maps of tonotopic and modulation-rate representations for the left hemisphere.** Group averages were obtained after cortex-based alignment of the individual hemispheres to achieve maximal anatomical correspondence of the observed activations across subjects and then rendered onto the reconstructed cortical surface of the Colin 27 standard brain (A, with HG outlined in *white*). In addition, cytoarchitectonic probability maps of auditory core regions (areas Te1.0 in *bluish colors* and Te1.1 in *greenish colors*) are displayed to allow for a comparison of anatomical (B) and functional measures (C, D). (C) The group best-frequency (tonotopic) maps, with *red colors* representing low spectral frequencies, and *blue* for high frequencies (false discovery rate <0.01). (D) The best-modulation rate maps, with *red colors* representing slow temporal modulation rates, and *blue* for high rates of amplitude modulation (false discovery rate <0.01). *Image courtesy of Herdener, M., Esposito, F., Scheffler, K., Schneider, P., Logothetis, N.K., Uludag, K., Kayser, C., 2013. Spatial representations of temporal and spectral sound cues in human auditory cortex. Cortex 49, 2822–2833. https://doi.org/10.1016/j.cortex.2013.04.003.*

may correspond to the pitch region (Griffiths and Hall, 2012) and responds stronger to emotional voices as compared to neutral voices (Frühholz and Grandjean, 2013). Therefore, the sound information contained within the emotional voices will be available to the ACx for neural processing at the auditory-specific perceptual level.

Perception and production of spoken language are reportedly facilitated by predictive mechanisms involving early human ACx (Forseth et al., 2020). The HG is bilaterally involved in speech perception as they are recruited to predict the timing of sound cues. Speech perception was interrupted by chronometric stimulation of the HG. In addition, the language-dominant PT is involved in speech production as a transient response to acoustic stimuli was suppressed during speech production, and its stimulation selectively disrupted speech production. Thus, the neural processing in the core of the ACx is likely critical for language recognition and speech perception.

Hemispheric Differences

Tonotopic maps of the HG are clearly detected in both hemispheres in some fMRI studies (Woods et al., 2009; Da Costa et al., 2011; Striem-Amit et al., 2011; Moerel et al., 2012). The tonotopic gradients in the middle STG, FTS, and the posteromedial end of HG are consistently detected in both hemispheres. Hemispheric differences were also found (Wessinger et al., 1997; Bilecen et al., 1998; Langers et al., 2007). The hemispheric bias may be restricted to the area posterior to the PAC, extending from the HS through the PT and posterior STG. In the right hemisphere, there may be an additional tonotopic reversal with a low frequency area on the posterior STG or lateral PT, while medial PT may not have a high frequency area. Individual variability in the tonotopic map is more apparent in the right hemisphere (Moerel et al., 2013).

As in the anatomical differences, functional hemispheric differences are also well recognized. Left hemispheres are relatively more dedicated to language processing, while right hemispheres are dedicated to tonal, music, and voice processing (Zatorre, 1988; Belin et al., 2000; Hickok and Poeppel, 2000; Scott et al., 2000). These hemispheric biases could reflect neural circuit specialization for auditory processing in each hemisphere. Temporal processing may be optimized in the left (Shannon et al., 1995; Liégeois-Chauvel et al., 1999; Zatorre et al., 2002b), while spectral processing is better performed in the right hemisphere (Liégeois-Chauvel et al., 2001; Zatorre and Belin, 2001). Anatomical substrates for these differences remain to be determined.

CONCLUSION

Recent advances in the fMRI allowed much clearer delineation of human anatomy and function in the central auditory system. More technological improvement will permit us to noninvasively discover more accurate anatomical description, access to unexplored structures, and clearer definition and anatomical assignment of specific auditory functions. Meanwhile, given the limitation of human studies, further investigation of anatomy and function of the auditory system in animals would be imperative to understand detailed mechanisms of auditory processing and anatomical features. Since the central auditory system confers us the critical physiological ability to recognize external auditory events such as alarming sound, baby's cry, verbal communication, attending lectures, watching news and movies, the research to discover fundamental mechanisms of auditory processing would be needless-to-say important to understand human nature, especially for the people who lack the ability due to chronic disorders or acquired disabilities.

References

Abdul-Kareem, I.A., Sluming, V., 2008. Heschl gyrus and its included primary auditory cortex: structural MRI studies in healthy and diseased subjects. J. Magn. Reson. Imag. 28, 287–299.

Adams, J.C., 1986. Neuronal morphology in the human cochlear nucleus. Arch. Otolaryngol. Head Neck Surg. 112, 1253–1261.

Adams, J.C., Mugnaini, E., 1990. Immunocytochemical evidence for inhibitory and disinhibitory circuits in the superior olive. Hear. Res. 49, 281–298.

Aitkin, L., Martin, R., 1990. Neurons in the inferior colliculus of cats sensitive to sound-source elevation. Hear. Res. 50, 97–105.

Aitkin, L.M., Gates, G.R., Phillips, S.C., 1984. Responses of neurons in inferior colliculus to variations in sound-source azimuth. J. Neurophysiol. 52, 1–17.

Aitkin, L.M., Martin, R.L., 1987. The representation of stimulus azimuth by high best-frequency azimuth-selective neurons in the central nucleus of the inferior colliculus of the cat. J. Neurophysiol. 57, 1185–1200.

Albrecht, O., Dondzillo, A., Mayer, F., Thompson, J.A., Klug, A., 2014. Inhibitory projections from the ventral nucleus of the trapezoid body to the medial nucleus of the trapezoid body in the mouse. Front. Neural Circ. 8, 1–15.

Allen, E.J., Mesik, J., Kay, K.N., Oxenham, A.J., 2022. Distinct representations of tonotopy and pitch in human auditory cortex. J. Neurosci. 42, 416–434.

Amaral, L., Ganho-Ávila, A., Osório, A., Soares, M.J., He, D., Chen, Q., Mahon, B.Z., Gonçalves, O.F., Sampaio, A., Fang, F., Bi, Y., Almeida, J., 2016. Hemispheric asymmetries in subcortical visual and auditory relay structures in congenital deafness. Eur. J. Neurosci. 44, 2334–2339.

Andersen, R.A., Knight, P.L., Merzenich, M.M., 1980a. The thalamocortical and corticothalamic connections of AI, AII, and the anterior auditory field (AAF) in the cat: evidence for two largely segregated systems of connections. J. Comp. Neurol. 194, 663–701.

Andersen, R.A., Roth, G.L., Aitkin, L.M., Merzenich, M.M., 1980b. The efferent projections of the central nucleus and the pericentral nucleus of the inferior colliculus in the cat. J. Comp. Neurol. 194, 649–662.

Anderson, L.A., Christianson, G.B., Linden, J.F., 2009. Stimulus-specific adaptation occurs in the auditory thalamus. J. Neurosci. 29, 7359–7363.

Anderson, L.A., Linden, J.F., 2011. Physiological differences between histologically defined subdivisions in the mouse auditory thalamus. Hear. Res. 274, 48–60. https://doi.org/10.1016/j.heares.2010.12.016.

Anderson, L.A., Malmierca, M.S., Wallace, M.N., Palmer, A.R., 2006. Evidence for a direct, short latency projection from the dorsal cochlear nucleus to the auditory thalamus in the guinea pig. Eur. J. Neurosci. 24, 491–498.

Awatramani, G.B., Turecek, R., Trussell, L.O., 2004. Inhibitory control at a synaptic relay. J. Neurosci. 24, 2643–2647.

Bajo, V.M., King, A.J., 2012. Cortical modulation of auditory processing in the midbrain. Front. Neural Circ. 6, 1–12.

Bajo, V.M., Merchán, M.A., López, D.E., Rouiller, E.M., 1993. Neuronal morphology and efferent projections of the dorsal nucleus of the lateral lemniscus in the rat. J. Comp. Neurol. 334, 241–262.

Banks, M.I., Smith, P.H., 1992. Intracellular recordings from neurobiotin-labeled cells in brain slices of the rat medial nucleus of the trapezoid body. J. Neurosci. 12, 2819–2837.

Bartlett, E.L., Sadagopan, S., Wang, X., 2011. Fine frequency tuning in monkey auditory cortex and thalamus. J. Neurophysiol. 106, 849–859.

Bartlett, E.L., Wang, X., 2011. Correlation of neural response properties with auditory thalamus subdivisions in the awake marmoset. J. Neurophysiol. 105, 2647–2667.

Barton, B., Venezia, J.H., Saberi, K., Hickok, G., Brewer, A.A., 2012. Orthogonal acoustic dimensions define auditory field maps in human cortex. Proc. Natl. Acad. Sci. U. S. A. 109, 20738–20743.

Baumann, S., Griffiths, T.D., Sun, L., Petkov, C.I., Thiele, A., Rees, A., 2011. Orthogonal representation of sound dimensions in the primate midbrain. Nat. Neurosci. 14, 423–425.

Baumann, S., Petkov, C.I., Griffiths, T.D., 2013. A unified framework for the organisation of the primate auditory cortex. Front. Syst. Neurosci. 7, 1–19.

Bednárová, V., Grothe, B., Myoga, M.H., 2018. Complex and spatially segregated auditory inputs of the mouse superior colliculus. J. Physiol. 596, 5281–5298.

Behrens, T.E.J., Berg, H.J., Jbabdi, S., Rushworth, M.F.S., Woolrich, M.W., 2007. Probabilistic diffusion tractography with multiple fibre orientations: what can we gain? Neuroimage 34, 144–155.

Beiderbeck, B., Myoga, M.H., Müller, N.I.C., Callan, A.R., Friauf, E., Grothe, B., Pecka, M., 2018. Precisely timed inhibition facilitates action potential firing for spatial coding in the auditory brainstem. Nat. Commun. 9. https://doi.org/10.1038/s41467-018-04210-y.

Belin, P., Zatorre, R.J., Lafaille, P., Ahad, P., Pike, B., 2000. Voice-selective areas in human auditory cortex. Nature 403, 309–312.

Berrebi, A.S., Morgan, J.I., Mugnaini, E., 1990. The Purkinje cell class may extend beyond the cerebellum. J. Neurocytol. 19, 643–654.

Bestmann, S., Baudewig, J., Siebner, H.R., Rothwell, J.C., Frahm, J., 2004. Functional MRI of the immediate impact of transcranial magnetic stimulation on cortical and subcortical motor circuits. Eur. J. Neurosci. 19, 1950–1962.

Bilecen, D., Scheffler, K., Schmid, N., Tschopp, K., Seelig, J., 1998. Tonotopic organization of the human auditory cortex as detected by BOLD-FMRI. Hear. Res. 126, 19–27.

Bilecen, D., Seifritz, E., Scheffler, K., Henning, J., Schulte, A.-C., 2002. Amplitopicity of the human auditory cortex: an fMRI study. Neuroimage 17, 710–718.

Bitterman, Y., Mukamel, R., Malach, R., Fried, I., Nelken, I., 2008. Ultra-fine frequency tuning revealed in single neurons of human auditory cortex. Nature 451, 197–201.

Blackburn, C.C., Sachs, M.B., 1990. The representations of the steady-state vowel sound/e/in the discharge patterns of cat anteroventral cochlear nucleus neurons. J. Neurophysiol. 63, 1191–1212.

Bonte, M., Frost, M.A., Rutten, S., Ley, A., Formisano, E., Goebel, R., 2013. Development from childhood to adulthood increases morphological and functional inter-individual variability in the right superior temporal cortex. Neuroimage 83, 739–750. https://doi.org/10.1016/j.neuroimage.2013.07.017.

Borst, J.G., Helmchen, F., Sakmann, B., 1995. Pre- and postsynaptic whole-cell recordings in the medial nucleus of the trapezoid body of the rat. J. Physiol. 489, 825–840.

Brand, A., Behrend, O., Marquardt, T., McAlpine, D., Grothe, B., 2002. Precise inhibition is essential for microsecond interaural time difference coding. Nature 417, 543–547.

Brughera, A., Dunai, L., Hartmann, W.M., 2013. Human interaural time difference thresholds for sine tones: the high-frequency limit. J. Acoust. Soc. Am. 133, 2839–2855.

Brunso-Bechtold, J.K., Thompson, G.C., Masterton, R.B., 1981. HRP study of the organization of auditory afferents ascending to central nucleus of inferior colliculus in cat. J. Comp. Neurol. 197, 705–722.

Calford, M.B., 1983. The parcellation of the medial geniculate body of the cat defined by the auditory response properties of single units. J. Neurosci. 3, 2350–2364.

Campain, R., Minckler, J., 1976. A note on the gross configurations of the human auditory cortex. Brain Lang. 3, 318–323.

Cansino, S., Williamson, S.J., Karron, D., 1994. Tonotopic organization of human auditory association cortex. Brain Res. 663, 38–50.

Cant, N.B., Casseday, J.H., 1986. Projections from the anteroventral cochlear nucleus to the lateral and medial superior olivary nuclei. J. Comp. Neurol. 247, 457–476.

Carbajal, G.V., Malmierca, M.S., 2018. The neuronal basis of predictive coding along the auditory pathway: from the subcortical roots to cortical deviance detection. Trends Hear. 22, 1–33.

Carricondo, F., Romero-Gómez, B., 2019. The cochlear spiral ganglion neurons: the auditory portion of the VIII nerve. Anat. Rec. 302, 463–471.

Cecchetti, L., Ricciardi, E., Handjaras, G., Kupers, R., Ptito, M., Pietrini, P., 2016. Congenital blindness affects diencephalic but not mesencephalic structures in the human brain. Brain Struct. Funct. 221, 1465–1480. https://doi.org/10.1007/s00429-014-0984-5.

Celesia, G.G., 1976. Organization of auditory cortical areas in man. Brain J. Neurol. 99, 403–414.

Chen, C., Read, H.L., Escabí, M.A., 2019. A temporal integration mechanism enhances frequency selectivity of broadband inputs to inferior colliculus. PLoS Biol. 17, 1–28.

Cheung, M.M., Lau, C., Zhou, I.Y., Chan, K.C., Cheng, J.S., Zhang, J.W., Ho, L.C., Wu, E.X., 2012a. BOLD fMRI investigation of the rat auditory pathway and tonotopic organization. Neuroimage 60, 1205–1211.

Cheung, M.M., Lau, C., Zhou, I.Y., Chan, K.C., Zhang, J.W., Fan, S.J., Wu, E.X., 2012b. High fidelity tonotopic mapping using swept source functional magnetic resonance imaging. Neuroimage 61, 978–986.

Cooper, N.P., 1999. Vibration of beads placed on the basilar membrane in the basal turn of the cochlea. J. Acoust. Soc. Am. 106, L59–L64.

Couchman, K., Grothe, B., Felmy, F., 2010. Medial superior olivary neurons receive surprisingly few excitatory and inhibitory inputs with balanced strength and short-term dynamics. J. Neurosci. 30, 17111–17121.

Da Costa, S., van der Zwaag, W., Marques, J.P., Frackowiak, R.S.J., Clarke, S., Saenz, M., 2011. Human primary auditory cortex follows the shape of Heschl's gyrus. J. Neurosci. 31, 14067–14075.

Dahmen, J.C., Keating, P., Nodal, F.R., Schulz, A.L., King, A.J., 2010. Adaptation to stimulus statistics in the perception and neural representation of auditory space. Neuron 66, 937–948. https://doi.org/10.1016/j.neuron.2010.05.018.

Davis, K.A., Ramachandran, R., May, B.J., 2003. Auditory processing of spectral cues for sound localization in the inferior colliculus. JARO J. Assoc. Res. Otolaryngol. 4, 148–163.

De La Mothe, L.A., Blumell, S., Kajikawa, Y., Hackett, T.A., 2006. Thalamic connections of the auditory cortex in marmoset monkeys: core and medial belt regions. J. Comp. Neurol. 496, 72–96.

De Lucia, M., Clarke, S., Murray, M.M., 2010. A temporal hierarchy for conspecific vocalization discrimination in humans. J. Neurosci. 30, 11210–11221.

De Martino, F., Moerel, M., Van De Moortele, P.F., Ugurbil, K., Goebel, R., Yacoub, E., Formisano, E., 2013. Spatial organization of frequency preference and selectivity in the human inferior colliculus. Nat. Commun. 4, 1–8.

De Martino, F., Moerel, M., Xu, J., Van De Moortele, P.F., Ugurbil, K., Goebel, R., Yacoub, E., Formisano, E., 2015. High-resolution mapping of myeloarchitecture in vivo: localization of auditory areas in the human brain. Cerebr. Cortex 25, 3394–3405.

Devlin, J.T., Sillery, E.L., Hall, D.A., Hobden, P., Behrens, T.E.J., Nunes, R.G., Clare, S., Matthews, P.M., Moore, D.R., Johansen-Berg, H., 2006. Reliable identification of the auditory thalamus using multi-modal structural analyses. Neuroimage 30, 1112–1120.

Dewey, R.S., Francis, S.T., Guest, H., Prendergast, G., Millman, R.E., Plack, C.J., Hall, D.A., 2020. The association between subcortical and cortical fMRI and lifetime noise exposure in listeners with normal hearing thresholds. Neuroimage 204, 116239. https://doi.org/10.1016/j.neuroimage.2019.116239.

Dick, F., Tierney, A.T., Lutti, A., Josephs, O., Sereno, M.I., Weiskopf, N., 2012. In vivo functional and myeloarchitectonic mapping of human primary auditory areas. J. Neurosci. 32, 16095–16105.

Dietz, M., Marquardt, T., Stange, A., Pecka, M., Grothe, B., McAlpine, D., 2014. Emphasis of spatial cues in the temporal fine structure during the rising segments of amplitude-modulated sounds II: single-neuron recordings. J. Neurophysiol. 111, 1973–1985.

Dondzillo, A., Thompson, J.A., Klug, A., 2016. Recurrent inhibition to the medial nucleus of the trapezoid body in the Mongolian gerbil (Meriones Unguiculatus). PLoS One 11, e0160241.

Doubell, T.P., Baron, J., Skaliora, I., King, A.J., 2000. Topographical projection from the superior colliculus to the nucleus of the brachium of the inferior colliculus in the ferret: convergence of visual and auditory information. Eur. J. Neurosci. 12, 4290–4308.

Doucet, J.R., Ryugo, D.K., 1997. Projections from the ventral cochlear nucleus to the dorsal cochlear nucleus in rats. J. Comp. Neurol. 385, 245–264.

Doucet, J.R., Ryugo, D.K., 2003. Axonal pathways to the lateral superior olive labeled with biotinylated dextran amine injections in the dorsal cochlear nucleus of rats. J. Comp. Neurol. 461, 452–465.

Duvel, A.D., Smith, D.M., Talk, A., Gabriel, M., 2001. Medial geniculate, amygdalar and cingulate cortical training-induced neuronal activity during discriminative avoidance learning in rabbits with auditory cortical lesions. J. Neurosci. 21, 3271–3281.

Ehret, G., Merzenich, M.M., 1988. Complex sound analysis (frequency resolution, filtering and spectral integration) by single units of the inferior colliculus of the cat. Brain Res. 472, 139–163.

Englitz, B., Tolnai, S., Typlt, M., Jost, J., Rübsamen, R., 2009. Reliability of synaptic transmission at the synapses of held in vivo under acoustic stimulation. PLoS One 4.

Epprecht, L., Qureshi, A., Kozin, E.D., Vachicouras, N., Huber, A.M., Kikinis, R., Makris, N., Brown, M.C., Reinshagen, K.L., Lee, D.J., 2020. Human cochlear nucleus on 7 tesla diffusion tensor imaging: insights into micro-anatomy and function for auditory brainstem implant surgery. Otol. Neurotol. 41, e484–e493.

Felix Ii, R.A., Gourévitch, B., Gómez-Álvarez, M., Leijon, S.C.M., Saldaña, E., Magnusson, A.K., 2017. Octopus cells in the posteroventral cochlear nucleus provide the main excitatory input to the superior paraolivary nucleus. Front. Neural Circ. 11, 37.

Ford, M.C., Alexandrova, O., Cossell, L., Stange-Marten, A., Sinclair, J., Kopp-Scheinpflug, C., Pecka, M., Attwell, D., Grothe, B., 2015. Tuning of Ranvier node and internode properties in myelinated axons to adjust action potential timing. Nat. Commun. 6, 8073.

Formisano, E., Kim, D.S., Di Salle, F., Van De Moortele, P.F., Ugurbil, K., Goebel, R., 2003. Mirror-symmetric tonotopic maps in human primary auditory cortex. Neuron 40, 859–869.

Forseth, K.J., Hickok, G., Rollo, P.S., Tandon, N., 2020. Language prediction mechanisms in human auditory cortex. Nat. Commun. 11, 1–14. https://doi.org/10.1038/s41467-020-19010-6.

Forsythe, I.D., 1994. Direct patch recording from identified presynaptic terminals mediating glutamatergic EPSCs in the rat CNS, in vitro. J. Physiol. 479, 381–387.

Franken, T.P., Joris, P.X., Smith, P.H., 2018. Principal cells of the brainstem's interaural sound level detector are temporal differentiators rather than integrators. Elife 7, 1–25.

Franken, T.P., Roberts, M.T., Wei, L., Golding, N.L., Joris, P.X., 2015. In vivo coincidence detection in mammalian sound localization generates phase delays. Nat. Neurosci. 18, 444–454.

Friauf, E., Ostwald, J., 1988. Divergent projections of physiologically characterized rat ventral cochlear nucleus neurons as shown by intra-axonal injection of horseradish peroxidase. Exp. Brain Res. 73, 263–284.

Fritz, T.H., Renders, W., Müller, K., Schmude, P., Leman, M., Turner, R., Villringer, A., 2013. Anatomical differences in the human inferior colliculus relate to the perceived valence of musical consonance and dissonance. Eur. J. Neurosci. 38, 3099–3105.

Frühholz, S., Grandjean, D., 2013. Multiple subregions in superior temporal cortex are differentially sensitive to vocal expressions: a quantitative meta-analysis. Neurosci. Biobehav. Rev. 37, 24–35. https://doi.org/10.1016/j.neubiorev.2012.11.002.

Fullerton, B.C., Pandya, D.N., 2007. Architectonic analysis of the auditory-related areas of the superior temporal region in human brain. J. Comp. Neurol. 504, 470–498.

Galaburda, A., Sanides, F., 1980. Cytoarchitectonic organization of the human auditory cortex. J. Comp. Neurol. 190, 597–610.

Galaburda, A.M., Sanides, F., Geschwind, N., 1978. Human brain. Cytoarchitectonic left-right asymmetries in the temporal speech region. Arch. Neurol. 35, 812–817.

García-Gomar, M.G., Strong, C., Toschi, N., Singh, K., Rosen, B.R., Wald, L.L., Bianciardi, M., 2019. In vivo probabilistic structural atlas of the inferior and superior colliculi, medial and lateral geniculate nuclei and superior olivary complex in humans based on 7 tesla MRI. Front. Genet. 10, 1–15.

Geschwind, N., Levitsky, W., 1968. Human brain: left-right asymmetries in temporal speech region. Science (New York, N.Y.) 161, 186–187.

Glasser, M.F., van Essen, D.C., 2011. Mapping human cortical areas in vivo based on myelin content as revealed by T1- and T2-weighted MRI. J. Neurosci. 31, 11597–11616.

Gleiss, H., Encke, J., Lingner, A., Jennings, T.R., Brosel, S., Kunz, L., Grothe, B., Pecka, M., 2019. Cooperative population coding facilitates efficient sound-source separability by adaptation to input statistics. PLoS Biol. 17, 1–24.

Glendenning, K.K., Baker, B.N., Hutson, K.A., Masterton, R.B., 1992. Acoustic chiasm V: inhibition and excitation in the ipsilateral and contralateral projections of LSO. J. Comp. Neurol. 319, 100–122.

Glendenning, K.K., Hutson, K.A., Nudo, R.J., Masterton, R.B., 1985. Acoustic chiasm II: anatomical basis of binaurality in lateral superior olive of cat. J. Comp. Neurol. 232, 261–285.

Glendenning, K.K., Masterton, R.B., 1998. Comparative morphometry of mammalian central auditory systems: variation in nuclei and form of the ascending system. Brain Behav. Evol. 51, 59–89.

Golding, N.L., Robertson, D., Oertel, D., 1995. Recordings from slices indicate that octopus cells of the cochlear nucleus detect coincident firing of auditory nerve fibers with temporal precision. J. Neurosci. 15, 3138–3153.

Gómez-Álvarez, M., Saldaña, E., 2016. Different tonotopic regions of the lateral superior olive receive a similar combination of afferent inputs. J. Comp. Neurol. 524, 2230–2250.

Gray, H., Lewis, W.H., 1918. Anatomy of the Human Body, twentieth ed. Lea & Febiger, Philadelphia.

Griffiths, T.D., Hall, D.A., 2012. Mapping pitch representation in neural ensembles with fMRI. J. Neurosci. 32, 13343–13347.

Groh, J.M., Kelly, K.A., Underhill, A.M., 2003. A monotonic code for sound azimuth in primate inferior colliculus. J. Cognit. Neurosci. 15, 1217–1231.

Grothe, B., Park, T.J., 1995. Time can be traded for intensity in the lower auditory system. Naturwissenschaften 82, 521–523.

Gruters, K.G., Groh, J.M., 2012. Sounds and beyond: multisensory and other non-auditory signals in the inferior colliculus. Front. Neural Circ. 6, 1–39.

Guinan Jr., J.J., Li, R.Y.-S., 1990. Signal processing in brainstem auditory neurons which receive giant endings (calyces of Held) in the medial nucleus of the trapezoid body of the cat. Hear. Res. 49 (1–3), 321–334.

Hackett, T.A., 2011. Information flow in the auditory cortical network. Hear. Res. 271, 133–146.

Hackett, T.A., Preuss, T.M., Kaas, J.H., 2001. Architectonic identification of the core region in auditory cortex of macaques, chimpanzees, and humans. J. Comp. Neurol. 441, 197–222.

Hackett, T.A., Rinaldi Barkat, T., O'Brien, B.M.J., Hensch, T.K., Polley, D.B., 2011. Linking topography to tonotopy in the mouse auditory thalamocortical circuit. J. Neurosci. 31, 2983–2995.

Hackett, T.A., Stepniewska, I., Kaas, J.H., 1998a. Thalamocortical connections of the parabelt auditory cortex in macaque monkeys. J. Comp. Neurol. 400, 271–286.

Hackett, T.A., Stepniewska, I., Kaas, J.H., 1998b. Subdivisions of auditory cortex and ipsilateral cortical connections of the parabelt auditory cortex in macaque monkeys. J. Comp. Neurol. 394, 475–495.

Hartmann, W.M., Rakerd, B., Crawford, Z.D., Zhang, P.X., 2016. Transaural experiments and a revised duplex theory for the localization of low-frequency tones. J. Acoust. Soc. Am. 139, 968–985.

Hawley, M.L., Melcher, J.R., Fullerton, B.C., 2005. Effects of sound bandwidth on fMRI activation in human auditory brainstem nuclei. Hear. Res. 204, 101–110.

Hawrylycz, M.J., et al., 2012. An anatomically comprehensive atlas of the adult human brain transcriptome. Nature 489, 391–399.

Helfert, R.H., Schwartz, I.R., 1986. Morphological evidence for the existence of multiple neuronal classes in the cat lateral superior olivary nucleus. J. Comp. Neurol. 244, 533–549.

Helfert, R.H., Schwartz, I.R., 1987. Morphological features of five neuronal classes in the gerbil lateral superior olive. Am. J. Anat. 179, 55–69.

Herdener, M., Esposito, F., Scheffler, K., Schneider, P., Logothetis, N.K., Uludag, K., Kayser, C., 2013. Spatial representations of temporal and spectral sound cues in human auditory cortex. Cortex 49, 2822–2833. https://doi.org/10.1016/j.cortex.2013.04.003.

Hickok, G., Poeppel, D., 2000. Towards a functional neuroanatomy of speech perception. Trends Cogn. Sci. 4, 131–138.

Howard, M.A., Volkov, I.O., Mirsky, R., Garell, P.C., Noh, M.D., Granner, M., Damasio, H., Steinschneider, M., Reale, R.A., Hind, J.E., Brugge, J.F., 2000. Auditory cortex on the human posterior superior temporal gyrus. J. Comp. Neurol. 416, 79–92.

Howard, R.J., Brammer, M., Wright, I., Woodruff, P.W., Bullmore, E.T., Zeki, S., 1996. A direct demonstration of functional specialization within motion- related visual and auditory cortex of the human brain. Curr. Biol. 6, 1015–1019.

Huang, C.L., Winer, J.A., 2000. Auditory thalamocortical projections in the cat: laminar and areal patterns of input. J. Comp. Neurol. 427, 302–331.

Humphries, C., Liebenthal, E., Binder, J.R., 2010. Tonotopic organization of human auditory cortex. Neuroimage 50, 1202–1211. https://doi.org/10.1016/j.neuroimage.2010.01.046.

Huotilainen, M., Tiitinen, H., Lavikainen, J., Ilmoniemi, R.J., Pekkonen, E., Sinkkonen, J., Laine, P., Näätänen, R., 1995. Sustained fields of tones and glides reflect tonotopy of the auditory cortex. Neuroreport 6, 841–844.

Imig, T.J., Bibikov, N.G., Poirier, P., Samson, F.K., 2000. Directionality derived from pinna-cue spectral notches in cat dorsal cochlear nucleus. J. Neurophysiol. 83, 907–925.

Imig, T.J., Morel, A., 1984. Topographic and cytoarchitectonic organization of thalamic neurons related to their targets in low-, middle-, and high-frequency representations in cat auditory cortex. J. Comp. Neurol. 227, 511–539.

Imig, T.J., Morel, A., 1985. Tonotopic organization in lateral part of posterior group of thalamic nuclei in the cat. J. Neurophysiol. 53, 836–851.

Irvine, D.R., Gago, G., 1990. Binaural interaction in high-frequency neurons in inferior colliculus of the cat: effects of variations in sound pressure level on sensitivity to interaural intensity differences. J. Neurophysiol. 63, 570–591.

Irvine, D.R.F., Park, V.N., McCormick, L., 2001. Mechanisms underlying the sensitivity of neurons in the lateral superior olive to interaural intensity differences. J. Neurophysiol. 86, 2647–2666.

Ishizu, K., Shiramatsu, T.I., Hitsuyu, R., Oizumi, M., Tsuchiya, N., Takahashi, H., 2021. Information flow in the rat thalamo-cortical system: spontaneous vs. stimulus-evoked activities. Sci. Rep. 11, 1–13. https://doi.org/10.1038/s41598-021-98660-y.

Ito, T., Oliver, D.L., 2012. The basic circuit of the IC: tectothalamic neurons with different patterns of synaptic organization send different messages to the thalamus. Front. Neural Circ. 6, 1–9.

Jäncke, L., Wüstenberg, T., Schulze, K., Heinze, H.J., 2002. Asymmetric hemodynamic responses of the human auditory cortex to monaural and binaural stimulation. Hear. Res. 170, 166–178.

Javad, F., Warren, J.D., Micallef, C., Thornton, J.S., Golay, X., Yousry, T., Mancini, L., 2014. Auditory tracts identified with combined fMRI and diffusion tractography. Neuroimage 84, 562–574. https://doi.org/10.1016/j.neuroimage.2013.09.007.

Jeffress, L.A., 1948. A place theory of sound localization. J. Comp. Physiol. Psychol. 41, 35–39.

Jones, E.G., 1985. The Thalamus. Plenum Press, New York.

Jones, E.G., Dell'Anna, M.E., Molinari, M., Rausell, E., Hashikawa, T., 1995. Subdivisions of macaque monkey auditory cortex revealed by calcium-binding protein immunoreactivity. J. Comp. Neurol. 362, 153–170.

Joris, P.X., Carney, L.H., Smith, P.H., Yin, T.C.T., 1994a. Enhancement of neural synchronization in the anteroventral cochlear nucleus. I. Responses to tones at the characteristic frequency. J. Neurophysiol. 71, 1022–1036.

Joris, P.X., Smith, P.H., Yin, T.C.T., 1994b. Enhancement of neural synchronization in the anteroventral cochlear nucleus. II. Responses in the tuning curve tail. J. Neurophysiol. 71, 1037–1051.

Joris, P.X., Trussell, L.O., 2018. The calyx of held: a hypothesis on the need for reliable timing in an intensity-difference encoder. Neuron 100, 534–549. https://doi.org/10.1016/j.neuron.2018.10.026.

Joris, P.X., Yin, T.C.T., 1998. Envelope coding in the lateral superior olive. III. Comparison with afferent pathways. J. Neurophysiol. 79, 253–269.

Kaas, J.H., Hackett, T.A., 2000. Subdivisions of auditory cortex and processing streams in primates. Proc. Natl. Acad. Sci. U. S. A. 97, 11793–11799.

Kadner, A., Berrebi, A.S., 2008. Encoding of temporal features of auditory stimuli in the medial nucleus of the trapezoid body and superior paraolivary nucleus of the rat. Neuroscience 151, 868–887.

Kajikawa, Y., De La Mothe, L., Blumell, S., Hackett, T.A., 2005. A comparison of neuron response properties in areas A1 and CM of the marmoset monkey auditory cortex: tones and broadband noise. J. Neurophysiol. 93, 22–34.

Kandler, K., Friauf, E., 1993. Pre- and postnatal development of efferent connections of the cochlear nucleus in the rat. J. Comp. Neurol. 328, 161–184.

Kandler, K., Friauf, E., 1995. Development of electrical membrane properties and discharge characteristics of superior olivary complex neurons in fetal and postnatal rats. Eur. J. Neurosci. 7, 1773–1790.

Kang, D.H., Kwon, K.W., Gu, B.M., Choi, J.S., Jang, J.H., Kwon, J.S., 2008. Structural abnormalities of the right inferior colliculus in schizophrenia. Psychiatr. Res. Neuroimaging 164, 160–165. https://doi.org/10.1016/j.pscychresns.2007.12.023.

Keating, P., Nodal, F.R., King, A.J., 2014. Behavioural sensitivity to binaural spatial cues in ferrets: evidence for plasticity in the duplex theory of sound localization. Eur. J. Neurosci. 39, 197–206.

Kelly, J.B., Van Adel, B.A., Ito, M., 2009. Anatomical projections of the nuclei of the lateral lemniscus in the Albino rat (Rattus norvegicus). J. Comp. Neurol. 512, 573–593.

Khurana, S., Remme, M.W.H., Rinzel, J., Golding, N.L., 2011. Dynamic interaction of Ih and IK-LVA during trains of synaptic potentials in principal neurons of the medial superior olive. J. Neurosci. 31, 8936–8947.

Kim, D.O., 1985. Functional roles of the inner and outer hair cell subsystems in the cochlea and brainstem. In: Berlin, C. (Ed.), Hearing Science. Taylor and Francis, Philadelphia.

Kim, J.J., Crespo-Facorro, B., Andreasen, N.C., O'Leary, D.S., Zhang, B., Harris, G., Magnotta, V.A., 2000. An MRI-based parcellation method for the temporal lobe. Neuroimage 11, 271–288.

Kim, S.G., Lepsien, J., Fritz, T.H., Mildner, T., Mueller, K., 2017. Dissonance encoding in human inferior colliculus covaries with individual differences in dislike of dissonant music. Sci. Rep. 7, 1–10.

Kimura, A., Donishi, T., Sakoda, T., Hazama, M., Tamai, Y., 2003. Auditory thalamic nuclei projections to the temporal cortex in the rat. Neuroscience 117, 1003–1016.

Koka, K., Tollin, D.J., 2014. Linear coding of complex sound spectra by discharge rate in neurons of the medial nucleus of the trapezoid body (MNTB) and its inputs. Front. Neural Circ. 8, 1–19.

Kopp-Scheinpflug, C., Dehmel, S., Dörrscheidt, G.J., Rübsamen, R., 2002. Interaction of excitation and inhibition in anteroventral cochlear nucleus neurons that receive large endbulb synaptic endings. J. Neurosci. 22, 11004–11018.

Kopp-Scheinpflug, C., Lippe, W.R., Dörrscheidt, G.J., Rübsamen, R., 2003. The medial nucleus of the trapezoid body in the gerbil is more than a relay: comparison of pre- and postsynaptic activity. JARO J. Assoc. Res. Otolaryngol. 4, 1–23.

Kopp-Scheinpflug, C., Tolnai, S., Malmierca, M.S., Rübsamen, R., 2008. The medial nucleus of the trapezoid body: comparative physiology. Neuroscience 154, 160–170.

Kudo, M., Niimi, K., 1980. Ascending projections of the inferior colliculus in the cat: an autoradiographic study. J. Comp. Neurol. 191, 545–556.

Kulesza, R.J., 2007. Cytoarchitecture of the human superior olivary complex: medial and lateral superior olive. Hear. Res. 225, 80–90.

Kulesza, R.J., Grothe, B., 2015. Yes, there is a medial nucleus of the trapezoid body in humans. Front. Neuroanat. 9, 1–9.

Kuwabara, N., Zook, J.M., 1991. Classification of the principal cells of the medial nucleus of the trapezoid body. J. Comp. Neurol. 314, 707–720.

Langers, D.R.M., Backes, W.H., van Dijk, P., 2007. Representation of lateralization and tonotopy in primary versus secondary human auditory cortex. Neuroimage 34, 264–273.

Langers, D.R.M., Van Dijk, P., 2012. Mapping the tonotopic organization in human auditory cortex with minimally salient acoustic stimulation. Cerebr. Cortex 22, 2024–2038.

Langner, G., Sams, M., Heil, P., Schulze, H., 1997. Frequency and periodicity are represented in orthogonal maps in the human auditory cortex. J. Comp. Physiol. 181, 665–676.

Laudanski, J., Coombes, S., Palmer, A.R., Sumner, C.J., 2010. Mode-locked spike trains in responses of ventral cochlear nucleus chopper and onset neurons to periodic stimuli. J. Neurophysiol. 103, 1226–1237.

Lauter, J.L., Herscovitch, P., Formby, C., Raichle, M.E., 1985. Tonotopic organization in human auditory cortex revealed by PET. Hear. Res. 20, 199–205.

Leaver, A.M., Rauschecker, J.P., 2016. Functional topography of human auditory cortex. J. Neurosci. 36, 1416–1428.

LeDoux, J.E., Farb, C.R., Romanski, L.M., 1991. Overlapping projections to the amygdala and striatum from auditory processing areas of the thalamus and cortex. Neurosci. Lett. 134, 139–144.

Ledoux, J.E., Ruggiero, D.A., Forest, R., Stornetta, R., Reis, D.J., 1987. Topographic organization of convergent projections to the thalamus from the inferior colliculus and spinal cord in the rat. J. Comp. Neurol. 264, 123–146.

LeDoux, J.E., Ruggiero, D.A., Reis, D.J., 1985. Projections to the subcortical forebrain from anatomically defined regions of the medial geniculate body in the rat. J. Comp. Neurol. 242, 182–213.

Lee, C.C., Winer, J.A., 2008. Connections of cat auditory cortex: I. Thalamocortical system. J. Comp. Neurol. 507, 1879–1900.

Li, H., Helpard, L., Ekeroot, J., Rohani, S.A., Zhu, N., Rask-Andersen, H., Ladak, H.M., Agrawal, S., 2021. Three-dimensional tonotopic mapping of the human cochlea based on synchrotron radiation phase-contrast imaging. Sci. Rep. 11, 1–8. https://doi.org/10.1038/s41598-021-83225-w.

Liégeois-Chauvel, C., De Graaf, J.B., Laguitton, V., Chauvel, P., 1999. Specialization of left auditory cortex for speech perception in man depends on temporal coding. Cerebr. Cortex 9, 484–496.

Liégeois-Chauvel, C., Giraud, K., Badier, J.M., Marquis, P., Chauvel, P., 2001. Intracerebral evoked potentials in pitch perception reveal a functional asymmetry of the human auditory cortex. Ann. N. Y. Acad. Sci. 930, 117–132.

Liégeois-Chauvel, C., Musolino, A., Badier, J.M., Marquis, P., Chauvel, P., 1994. Evoked potentials recorded from the auditory cortex in man: evaluation and topography of the middle latency components. Electroencephalogr. Clin. Neurophysiol. 92, 204–214.

Lin, L., Campbell, J., Oertel, D., Smith, P.H., 2022. Local targets of T-stellate cells in the ventral cochlear nucleus. J. Comp. Neurol. 530, 2820–2834.

Lingner, A., Pecka, M., Leibold, C., Grothe, B., 2018. A novel concept for dynamic adjustment of auditory space. Sci. Rep. 8, 1–12. https://doi.org/10.1038/s41598-018-26690-0.

Lingner, A., Wiegrebe, L., Grothe, B., 2012. Sound localization in noise by gerbils and humans. JARO J. Assoc. Res. Otolaryngol. 13, 237–248.

Lockwood, A.H., Salvi, R.J., Coad, M.L., Arnold, S.A., Wack, D.S., Murphy, B.W., Burkard, R.F., 1999. The functional anatomy of the normal human auditory system: responses to 0.5 and 4.0 kHz tones at varied intensities. Cerebr. Cortex 9, 65–76.

Lorteije, J.A.M., Rusu, S.I., Kushmerick, C., Borst, J.G.G., 2009. Reliability and precision of the mouse calyx of Held synapse. J. Neurosci. 29, 13770–13784.

Lu, H.-W., Smith, P.H., Joris, P.X., 2022. Mammalian octopus cells are direction selective to frequency sweeps by excitatory synaptic sequence detection. Proc. Natl. Acad. Sci. U. S. A. 119, e2203748119.

Lütkenhöner, B., Steinsträter, O., 1998. High-precision neuromagnetic study of the functional organization of the human auditory cortex. Audiol. Neuro. Otol. 3, 191–213.

Macpherson, E.A., Middlebrooks, J.C., 2002. Listener weighting of cues for lateral angle: the duplex theory of sound localization revisited. J. Acoust. Soc. Am. 111, 2219–2236.
Macpherson, E.A., Sabin, A.T., 2013. Vertical-plane sound localization with distorted spectral cues. Hear. Res. 306, 76–92. https://doi.org/10.1016/j.heares.2013.09.007.
Maffei, C., Jovicich, J., De Benedictis, A., Corsini, F., Barbareschi, M., Chioffi, F., Sarubbo, S., 2018. Topography of the human acoustic radiation as revealed by ex vivo fibers micro-dissection and in vivo diffusion-based tractography. Brain Struct. Funct. 223, 449–459.
Magnusson, A.K., Park, T.J., Pecka, M., Grothe, B., Koch, U., 2008. Retrograde GABA signaling adjusts sound localization by balancing excitation and inhibition in the brainstem. Neuron 59, 125–137.
Malmierca, M.S., Izquierdo, M.A., Cristaudo, S., Hernández, O., Pérez-González, D., Covey, E., Oliver, D.L., 2008. A discontinuous tonotopic organization in the inferior colliculus of the rat. J. Neurosci. 28, 4767–4776.
Malmierca, M.S., Leergaard, T.B., Bajo, V.M., Bjaalie, J.G., Merchán, M.A., 1998. Anatomic evidence of a three-dimensional mosaic pattern of tonotopic organization in the ventral complex of the lateral lemniscus in cat. J. Neurosci. 18, 10603–10618.
Malmierca, M.S., Rees, A., Le Beau, F.E., Bjaalie, J.G., 1995. Laminar organization of frequency-defined local axons within and between the inferior colliculi of the guinea pig. J. Comp. Neurol. 357, 124–144.
Markovitz, C.D., Tang, T.T., Lim, H.H., 2013. Tonotopic and localized pathways from primary auditory cortex to the central nucleus of the inferior colliculus. Front. Neural Circ. 7, 1–11.
Mathews, P.J., Jercog, P.E., Rinzel, J., Scott, L.L., Golding, N.L., 2010. Control of submillisecond synaptic timing in binaural coincidence detectors by K v 1 channels. Nat. Neurosci. 13, 601–609.
May, B.J., Le Prell, G.S., Sachs, M.B., 1998. Vowel representations in the ventral cochlear nucleus of the cat: effects of level, background noise, and behavioral state. J. Neurophysiol. 79, 1755–1767.
McAlpine, D., Jiang, D., Palmer, A.R., 2001. A neural code for low-frequency sound localization in mammals. Nat. Neurosci. 4, 396–401.
McEchron, M.D., Green, E.J., Winters, R.W., Nolen, T.G., Schneiderman, N., McCabe, P.M., 1996. Changes of synaptic efficacy in the medial geniculate nucleus as a result of auditory classical conditioning. J. Neurosci. 16, 1273–1283.
Mellott, J.G., Foster, N.L., Nakamoto, K.T., Motts, S.D., Schofield, B.R., 2014. Distribution of GABAergic cells in the inferior colliculus that project to the thalamus. Front. Neuroanat. 8, 1–9.
Meng, Q., Schneider, K.A., 2022. A specialized channel for encoding auditory transients in the magnocellular division of the human medial geniculate nucleus. Neuroreport 33, 663–668.
Merchán, M., Aguilar, L.A., Lopez-Poveda, E.A., Malmierca, M.S., 2005. The inferior colliculus of the rat: quantitative immunocytochemical study of GABA and glycine. Neuroscience 136, 907–925.
Merzenich, M.M., Reid, M.D., 1974. Representation of the cochlea within the inferior colliculus of the cat. Brain Res. 77, 397–415.
Middlebrooks, J.C., 2015. Sound Localization, first ed. Elsevier B.V. https://doi.org/10.1016/B978-0-444-62630-1.00006-8.
Młynarski, W., Jost, J.J., 2014. Statistics of natural binaural sounds. PLoS One 9, e108968.
Moerel, M., De Martino, F., Formisano, E., 2012. Processing of natural sounds in human auditory cortex: tonotopy, spectral tuning, and relation to voice sensitivity. J. Neurosci. 32, 14205–14216.
Moerel, M., De Martino, F., Formisano, E., 2014. An anatomical and functional topography of human auditory cortical areas. Front. Neurosci. 8, 1–14.
Moerel, M., De Martino, F., Santoro, R., Ugurbil, K., Goebel, R., Yacoub, E., Formisano, E., 2013. Processing of natural sounds: characterization of multipeak spectral tuning in human auditory cortex. J. Neurosci. 33, 11888–11898.
Moerel, M., De Martino, F., Ugurbil, K., Yacoub, E., Formisano, E., 2015. Processing of frequency and location in human subcortical auditory structures. Sci. Rep. 5, 1–15.
Moerel, M., De Martino, F., Uğurbil, K., Yacoub, E., Formisano, E., 2019. Processing complexity increases in superficial layers of human primary auditory cortex. Sci. Rep. 9, 1–9.
Moore, J.K., 1987. The human auditory brain stem: a comparative view. Hear. Res. 29, 1–32.
Moore, J.K., Osen, K.K., 1979. The cochlear nuclei in man. Am. J. Anat. 154, 393–418. https://doi.org/10.1002/aja.1001540306.
Morel, A., Imig, T.J., 1987. Thalamic projections to fields A, AI, P, and VP in the cat auditory cortex. J. Comp. Neurol. 265, 119–144.
Morosan, P., Rademacher, J., Schleicher, A., Amunts, K., Schormann, T., Zilles, K., 2001. Human primary auditory cortex: cytoarchitectonic subdivisions and mapping into a spatial reference system. Neuroimage 13, 684–701.
Mugnaini, E., Morgan, J.I., 1987. The neuropeptide cerebellin is a marker for two similar neuronal circuits in rat brain. Proc. Natl. Acad. Sci. U. S. A. 84, 8692–8696.
Mylius, J., Brosch, M., Scheich, H., Budinger, E., 2013. Subcortical auditory structures in the Mongolian gerbil: I. Golgi architecture. J. Comp. Neurol. 521, 1289–1321.
Narayan, S.S., 1998. Frequency tuning of basilar membrane and auditory nerve fibers in the same cochleae. Science 282, 1882–1884.
Nayagam, D.A.X., Clarey, J.C., Paolini, A.G., 2005. Powerful, onset inhibition in the ventral nucleus of the lateral lemniscus. J. Neurophysiol. 94, 1651–1654.
Ngodup, T., Romero, G.E., Trussell, L.O., 2020. Identification of an inhibitory neuron subtype, the l-stellate cell of the cochlear nucleus. Elife 9, 1–40.
Nieuwenhuys, R., 2013. The myeloarchitectonic studies on the human cerebral cortex of the Vogt-Vogt school, and their significance for the interpretation of functional neuroimaging data. Brain Struct. Funct. 218, 303–352.
Nourski, K.V., Steinschneider, M., Oya, H., Kawasaki, H., Jones, R.D., Howard, M.A., 2014. Spectral organization of the human lateral superior temporal gyrus revealed by intracranial recordings. Cerebr. Cortex 24, 340–352.
Nuttall, A.L., Dolan, D.F., 1996. Steady-state sinusoidal velocity responses of the basilar membrane in guinea pig. J. Acoust. Soc. Am. 99, 1556–1565.
Oertel, D., Bal, R., Gardner, S.M., Smith, P.H., Joris, P.X., 2000. Detection of synchrony in the activity of auditory nerve fibers by octopus cells of the mammalian cochlear nucleus. Proc. Natl. Acad. Sci. U. S. A. 97, 11773–11779.
Oertel, D., Wright, S., Cao, X.J., Ferragamo, M., Bal, R., 2011. The multiple functions of T stellate/multipolar/chopper cells in the ventral cochlear nucleus. Hear. Res. 276, 61–69. https://doi.org/10.1016/j.heares.2010.10.018.
Oertel, D., Wu, S.H., Garb, M.W., Dizack, C., 1990. Morphology and physiology of cells in slice preparations of the posteroventral cochlear nucleus of mice. J. Comp. Neurol. 295, 136–154.
Oertel, D., Young, E.D., 2004. What's a cerebellar circuit doing in the auditory system? Trends Neurosci. 27, 104–110.
Oliver, D.L., Winer, J.A., Beckius, G.E., Saint Marie, R.L., 1994. Morphology of GABAergic neurons in the inferior colliculus of the cat. J. Comp. Neurol. 340, 27–42.
Palmer, A.R., Rees, A., Caird, D., 1990. Interaural delay sensitivity to tones and broad band signals in the guinea-pig inferior colliculus. Hear. Res. 50, 71–86.
Pantev, C., Hoke, M., Lehnertz, K., Lütkenhöner, B., Anogianakis, G., Wittkowski, W., 1988. Tonotopic organization of the human

auditory cortex revealed by transient auditory evoked magnetic fields. Electroencephalogr. Clin. Neurophysiol. 69, 160–170.
Pantev, C., Hoke, M., Lütkenhöner, B., Lehnertz, K., 1989. Tonotopic organization of the auditory cortex: pitch versus frequency representation. Science 246, 486–488.
Paolini, A.G., FitzGerald, J.V., Burkitt, A.N., Clark, G.M., 2001. Temporal processing from the auditory nerve to the medial nucleus of the trapezoid body in the rat. Hear. Res. 159, 101–116.
Park, T.J., Brand, A., Koch, U., Ikebuchi, M., Grothe, B., 2008. Dynamic changes in level influence spatial coding in the lateral superior olive. Hear. Res. 238, 58–67.
Pecka, M., Brand, A., Behrend, O., Grothe, B., 2008. Interaural time difference processing in the mammalian medial superior olive: the role of glycinergic inhibition. J. Neurosci. 28, 6914–6925.
Pecka, M., Encke, J., 2020. Coding of spatial information. In: The Senses: A Comprehensive Reference, vol. 2, pp. 713–731.
Pecka, M., Siveke, I., Grothe, B., Lesica, N.A., 2010. Enhancement of ITD coding within the initial stages of the auditory pathway. J. Neurophysiol. 103, 38–46.
Penhune, V.B., Zatorre, R.J., MacDonald, J.D., Evans, A.C., 1996. Interhemispheric anatomical differences in human primary auditory cortex: probabilistic mapping and volume measurement from magnetic resonance scans. Cerebr. Cortex 6, 661–672.
Peruzzi, D., Bartlett, E., Smith, P.H., Oliver, D.L., 1997. A monosynaptic GABAergic input from the inferior colliculus to the medial geniculate body in rat. J. Neurosci. 17, 3766–3777.
Phillips, D.P., Hall, S.E., 2005. Psychophysical evidence for adaptation of central auditory processors for interaural differences in time and level. Hear. Res. 202, 188–199.
Polley, D.B., Read, H.L., Storace, D.A., Merzenich, M.M., 2007. Multiparametric auditory receptive field organization across five cortical fields in the albino rat. J. Neurophysiol. 97, 3621–3638.
Rademacher, J., Morosan, P., Schormann, T., Schleicher, A., Werner, C., Freund, H.J., Zilles, K., 2001. Probabilistic mapping and volume measurement of human primary auditory cortex. Neuroimage 13, 669–683.
Rauschecker, J.P., Tian, B., 2004. Processing of band-passed noise in the lateral auditory belt cortex of the rhesus monkey. J. Neurophysiol. 91, 2578–2589.
Rauschecker, J.P., Tian, B., Hauser, M., 1995. Processing of complex sounds in the macaque nonprimary auditory cortex. Science 268, 111–114.
Read, H.L., Nauen, D.W., Escabí, M.A., Miller, L.M., Schreiner, C.E., Winer, J.A., 2011. Distinct core thalamocortical pathways to central and dorsal primary auditory cortex. Hear. Res. 274, 95–104. https://doi.org/10.1016/j.heares.2010.11.010.
Redies, H., Brandner, S., Creutzfeldt, O.D., 1989. Anatomy of the auditory thalamocortical system of the guinea pig. J. Comp. Neurol. 282, 489–511.
Ress, D., Chandrasekaran, B., 2013. Tonotopic organization in the depth of human inferior colliculus. Front. Hum. Neurosci. 7, 1–10.
Rhode, W.S., Greenberg, S., 1994. Encoding of amplitude modulation in the cochlear nucleus of the cat. J. Neurophysiol. 71, 1797–1825.
Rhode, W.S., Smith, P.H., 1986. Encoding timing and intensity in the ventral cochlear nucleus of the cat. J. Neurophysiol. 56, 261–286.
Rice, J.J., May, B.J., Spirou, G.A., Young, E.D., 1992. Pinna-based spectral cues for sound localization in cat. Hear. Res. 58, 132–152.
Rietzel, H.J., Friauf, E., 1998. Neuron types in the rat lateral superior olive and developmental changes in the complexity of their dendritic arbors. J. Comp. Neurol. 390, 20–40.
Roberts, M.T., Seeman, S.C., Golding, N.L., 2013. A mechanistic understanding of the role of feedforward inhibition in the mammalian sound localization circuitry. Neuron 78, 923–935. https://doi.org/10.1016/j.neuron.2013.04.022.
Roberts, M.T., Seeman, S.C., Golding, N.L., 2014. The relative contributions of MNTB and LNTB neurons to inhibition in the medial superior olive assessed through single and paired recordings. Front. Neural Circ. 8, 1–14.
Rodríguez, F.A., Chen, C., Read, H.L., Escabí, M.A., 2010. Neural modulation tuning characteristics scale to efficiently encode natural sound statistics. J. Neurosci. 30, 15969–15980.
Romani, G.L., Williamson, S.J., Kaufman, L., 1982a. Tonotopic organization of the human auditory cortex. Science (New York, N.Y.) 216, 1339–1340.
Romani, G.L., Williamson, S.J., Kaufman, L., Brenner, D., 1982b. Characterization of the human auditory cortex by the neuromagnetic method. Exp. Brain Res. 47, 381–393.
Rouiller, E.M., Rodrigues-Dagaeff, C., Simm, G., De Ribaupierre, Y., Villa, A., De Ribaupierre, F., 1989. Functional organization of the medial division of the medial geniculate body of the cat: tonotopic organization, spatial distribution of response properties and cortical connections. Hear. Res. 39, 127–142.
Rubens, A.B., Mahowald, M.W., Hutton, J.T., 1976. Asymmetry of the lateral (sylvian) fissures in man. Neurology 26, 620–624.
Rubio, M.E., Gudsnuk, K.A., Smith, Y., Ryugo, D.K., 2008. Revealing the molecular layer of the primate dorsal cochlear nucleus. Neuroscience 154, 99–113.
Ruggero, M.A., Rich, N.C., Recio, A., Narayan, S.S., Robles, L., 1997. Basilar-membrane responses to tones at the base of the chinchilla cochlea. J. Acoust. Soc. Am. 101, 2151–2163.
Rutherford, M.A., von Gersdorff, H., Goutman, J.D., 2021. Encoding sound in the cochlea: from receptor potential to afferent discharge. J. Physiol. 599, 2527–2557.
Ryan, A.F., Woolf, N.K., Sharp, F.R., 1982. Tonotopic organization in the central auditory pathway of the Mongolian gerbil: a 2-deoxyglucose study. J. Comp. Neurol. 207, 369–380.
Saenz, M., Langers, D.R.M., 2014. Tonotopic mapping of human auditory cortex. Hear. Res. 307, 42–52. https://doi.org/10.1016/j.heares.2013.07.016.
Saini, S., Kaur, C., Pal, I., Kumar, P., Jacob, T.G., Thakar, A., Roy, K.K., Roy, T.S., 2019. Morphological development of the human cochlear nucleus. Hear. Res. 382, 107784. https://doi.org/10.1016/j.heares.2019.107784.
Saint Marie, R.L., Baker, R.A., 1990. Neurotransmitter-specific uptake and retrograde transport of [3H]glycine from the inferior colliculus by ipsilateral projections of the superior olivary complex and nuclei of the lateral lemniscus. Brain Res. 524, 244–253.
Saldeitis, K., Happel, M.F.K., Ohl, F.W., Scheich, H., Budinger, E., 2014. Anatomy of the auditory thalamocortical system in the Mongolian gerbil: nuclear origins and cortical field-, layer-, and frequency-specificities. J. Comp. Neurol. 522, 2397–2430.
Sätzler, K., Söhl, L.F., Bollmann, J.H., Borst, J.G.G., Frotscher, M., Sakmann, B., Lübke, J.H.R., 2002. Three-dimensional reconstruction of a calyx of Held and its postsynaptic principal neuron in the medial nucleus of the trapezoid body. J. Neurosci. 22, 10567–10579.
Scheffler, K., Bilecen, D., Schmid, N., Tschopp, K., Seelig, J., 1998. Auditory cortical responses in hearing subjects and unilateral deaf patients as detected by functional magnetic resonance imaging. Cerebr. Cortex 8, 156–163.
Schofield, B.R., Cant, N.B., 1996. Projections from the ventral cochlear nucleus to the inferior colliculus and the contralateral cochlear nucleus in guinea pigs. Hear. Res. 102, 1–14.
Schönwiesner, M., Krumbholz, K., Rübsamen, R., Fink, G.R., Von Cramon, D.Y., 2007. Hemispheric asymmetry for auditory processing in the human auditory brain stem, thalamus, and cortex. Cerebr. Cortex 17, 492–499.
Schreiner, C.E., Langner, G., Institut, Z., Darmstadt, T.H., Darmstadt, D., 1997. Letters to nature Laminar fine structure of frequency organization in auditory midbrain the main auditory midbrain station, the central nucleus of the. Nature 383–386.

Scott, S.K., Catrin Blank, C., Rosen, S., Wise, R.J.S., 2000. Identification of a pathway for intelligible speech in the left temporal lobe. Brain 123, 2400–2406.

Seifritz, E., Di Salle, F., Esposito, F., Herdener, M., Neuhoff, J.G., Scheffler, K., 2006. Enhancing BOLD response in the auditory system by neurophysiologically tuned fMRI sequence. Neuroimage 29, 1013–1022.

Shannon, R.V., Zeng, F.G., Kamath, V., Wygonski, J., Ekelid, M., 1995. Speech recognition with primarily temporal cues. Science (New York, N.Y.) 270, 303–304.

Sigalovsky, I.S., Fischl, B., Melcher, J.R., 2006. Mapping an intrinsic MR property of gray matter in auditory cortex of living humans: a possible marker for primary cortex and hemispheric differences. Neuroimage 32, 1524–1537.

Sigalovsky, I.S., Melcher, J.R., 2006. Effects of sound level on fMRI activation in human brainstem, thalamic and cortical centers. Hear. Res. 215, 67–76.

Singla, S., Dempsey, C., Warren, R., Enikolopov, A.G., Sawtell, N.B., 2017. A cerebellum-like circuit in the auditory system cancels responses to self-generated sounds. Nat. Neurosci. 20, 943–950.

Sitek, K.R., Calabrese, E., Johnson, G.A., Ghosh, S.S., Chandrasekaran, B., 2022. Structural connectivity of human inferior colliculus subdivisions using in vivo and post mortem diffusion MRI tractography. Front. Neurosci. 16, 1–11.

Sitek, K.R., Gulban, O.F., Calabrese, E., Johnson, G.A., Lage-Castellanos, A., Moerel, M., Ghosh, S.S., De Martino, F., 2019. Mapping the human subcortical auditory system using histology, postmortem MRI and in vivo MRI at 7T. Elife 8, 1–36.

Smith, P.H., Joris, P.X., Carney, L.H., Yin, T.C., 1991. Projections of physiologically characterized globular bushy cell axons from the cochlear nucleus of the cat. J. Comp. Neurol. 304, 387–407.

Smith, P.H., Joris, P.X., Yin, T.C.T., 1998. Anatomy and physiology of principal cells of the medial nucleus of the trapezoid body (MNTB) of the cat. J. Neurophysiol. 79, 3127–3142.

Smith, P.H., Massie, A., Joris, P.X., 2005. Acoustic stria: anatomy of physiologically characterized cells and their axonal projection patterns. J. Comp. Neurol. 482, 349–371.

Smith, P.H., Uhlrich, D.J., Manning, K.A., Banks, M.I., 2012. Thalamocortical projections to rat auditory cortex from the ventral and dorsal divisions of the medial geniculate nucleus. J. Comp. Neurol. 520, 34–51.

Sommer, I., Lingenhöhl, K., Friauf, E., 1993. Principal cells of the rat medial nucleus of the trapezoid body: an intracellular in vivo study of their physiology and morphology. Exp. Brain Res. 95, 223–239.

Sonntag, M., Englitz, B., Kopp-Scheinpflug, C., Rübsamen, R., 2009. Early postnatal development of spontaneous and acoustically evoked discharge activity of principal cells of the medial nucleus of the trapezoid body: an in vivo study in mice. J. Neurosci. 29, 9510–9520.

Spirou, G.A., Brownell, W.E., Zidanic, M., 1990. Recordings from cat trapezoid body and HRP labeling of globular bushy cell axons. J. Neurophysiol. 63, 1169–1190.

Spitzer, M.W., Semple, M.N., 1995. Neurons sensitive to interaural phase disparity in gerbil superior olive: diverse monaural and temporal response properties. J. Neurophysiol. 73, 1668–1690.

Stanford, T.R., Kuwada, S., Batra, R., 1992. A comparison of the interaural time sensitivity of neurons in the inferior colliculus and thalamus of the unanesthetized rabbit. J. Neurosci. 12, 3200–3216.

Stange-Marten, A., Nabel, A.L., Sinclair, J.L., Fischl, M., Alexandrova, O., Wohlfrom, H., Kopp-Scheinpflug, C., Pecka, M., Grothe, B., 2017. Input timing for spatial processing is precisely tuned via constant synaptic delays and myelination patterns in the auditory brainstem. Proc. Natl. Acad. Sci. U. S. A. 114, E4851–E4858.

Stange, A., Myoga, M.H., Lingner, A., Ford, M.C., Alexandrova, O., Felmy, F., Pecka, M., Siveke, I., Grothe, B., 2013. Adaptation in sound localization: from GABA B receptor-mediated synaptic modulation to perception. Nat. Neurosci. 16, 1840–1847.

Storace, D.A., Higgins, N.C., Read, H.L., 2010. Thalamic label patterns suggest primary and ventral auditory fields are distinct core regions. J. Comp. Neurol. 518, 1630–1646.

Storace, D.A., Higgins, N.C., Read, H.L., 2011. Thalamocortical pathway specialization for sound frequency resolution. J. Comp. Neurol. 519, 177–193.

Stotler, W.A., 1953. An experimental study of the cells and connections of the superior olivary complex of the cat. J. Comp. Neurol. 98, 401–431.

Straka, M.M., Hughes, R., Lee, P., Lim, H.H., 2015. Descending and tonotopic projection patterns from the auditory cortex to the inferior colliculus. Neuroscience 300, 325–337. https://doi.org/10.1016/j.neuroscience.2015.05.032.

Striem-Amit, E., Hertz, U., Amedi, A., 2011. Extensive cochleotopic mapping of human auditory cortical fields obtained with phase-encoding fMRI. PLoS One 6.

Suga, N., Zhang, Y., Yan, J., 1997. Sharpening of frequency tuning by inhibition in the thalamic auditory nucleus of the mustached bat. J. Neurophysiol. 77, 2098–2114.

Sutherland, D.P., Masterton, R.B., Glendenning, K.K., 1998. Role of acoustic striae in hearing: reflexive responses to elevated sound-sources. Behav. Brain Res. 97, 1–12.

Sweet, R.A., Dorph-Petersen, K.A., Lewis, D.A., 2005. Mapping auditory core, lateral belt, and parabelt cortices in the human superior temporal gyrus. J. Comp. Neurol. 491, 270–289.

Talkington, W.J., Rapuano, K.M., Hitt, L.A., Frum, C.A., Lewis, J.W., 2012. Humans mimicking animals: a cortical hierarchy for human vocal communication sounds. J. Neurosci. 32, 8084–8093.

Tarabichi, O., Kanumuri, V.V., Klug, J., Vachicouras, N., Duarte, M.J., Epprecht, L., Kozin, E.D., Reinshagen, K., Lacour, S.P., Brown, M.C., Lee, D.J., 2020. Three-dimensional surface reconstruction of the human cochlear nucleus: implications for auditory brain stem implant design. J. Neurol. Surg. Part B Skull Base 81, 114–120.

Taschenberger, H., Leão, R.M., Rowland, K.C., Spirou, G.A., Von Gersdorff, H., 2002. Optimizing synaptic architecture and efficiency for high-frequency transmission. Neuron 36, 1127–1143.

Thomas, J.M., Huber, E., Stecker, G.C., Boynton, G.M., Saenz, M., Fine, I., 2015. Population receptive field estimates of human auditory cortex. Neuroimage 105, 428–439. https://doi.org/10.1016/j.neuroimage.2014.10.060.

Thompson, A.M., Schofield, B.R., 2000. Afferent projections of the superior olivary complex. Microsc. Res. Tech. 51, 330–354.

Tollin, D.J., 2003. The lateral superior olive: a functional role in sound source localization. Neuroscientist 9, 127–143.

Tollin, D.J., Yin, T.C.T., 2005. Interaural phase and level difference sensitivity in low-frequency neurons in the lateral superior olive. J. Neurosci. 25, 10648–10657.

Tolnai, S., Englitz, B., Kopp-Scheinpflug, C., Dehmel, S., Jost, J., Rübsamen, R., 2008. Dynamic coupling of excitatory and inhibitory responses in the medial nucleus of the trapezoid body. Eur. J. Neurosci. 27, 3191–3204.

Tsuchitani, C., 1977. Functional organization of lateral cell groups of cat superior olivary complex. J. Neurophysiol. 40, 296–318.

Tsuchitani, C., 1997. Input from the medial nucleus of trapezoid body to an interaural level detector. Hear. Res. 105, 211–224.

Vasquez-Lopez, S.A., Weissenberger, Y., Lohse, M., Keating, P., King, A.J., Dahmen, J.C., 2017. Thalamic input to auditory cortex is locally heterogeneous but globally tonotopic. Elife 6, 1–18.

Vater, M., Feng, A.S., 1990. Functional organization of ascending and descending connections of the cochlear nucleus of horseshoe bats. J. Comp. Neurol. 292, 373–395.

Velenovsky, D.S., Cetas, J.S., Price, R.O., Sinex, D.G., McMullen, N.T., 2003. Functional subregions in primary auditory cortex defined by thalamocortical terminal arbors: an electrophysiological and anterograde labeling study. J. Neurosci. 23, 308–316.

Verschooten, E., Shamma, S., Oxenham, A.J., Moore, B.C.J., Joris, P.X., Heinz, M.G., Plack, C.J., 2019. The upper frequency limit for the use of phase locking to code temporal fine structure in humans: a compilation of viewpoints. Hear. Res. 377, 109–121. https://doi.org/10.1016/j.heares.2019.03.011.

Vigneault-MacLean, B.K., Hall, S.E., Phillips, D.P., 2007. The effects of lateralized adaptors on lateral position judgements of tones within and across frequency channels. Hear. Res. 224, 93–100.

Wagoner, J.L., Kulesza, R.J., 2009. Topographical and cellular distribution of perineuronal nets in the human cochlear nucleus. Hear. Res. 254, 42–53. https://doi.org/10.1016/j.heares.2009.04.008.

Wallace, M.N., Johnston, P.W., Palmer, A.R., 2002. Histochemical identification of cortical areas in the auditory region of the human brain. Exp. Brain Res. 143, 499–508.

Wasserthal, C., Brechmann, A., Stadler, J., Fischl, B., Engel, K., 2014. Localizing the human primary auditory cortex in vivo using structural MRI. Neuroimage 93, 237–251. https://doi.org/10.1016/j.neuroimage.2013.07.046.

Weisz, C., Glowatzki, E., Fuchs, P., 2009. The postsynaptic function of type II cochlear afferents. Nature 461, 1126–1129.

Wen, B., Wang, G.I., Dean, I., Delgutte, B., 2009. Dynamic range adaptation to sound level statistics in the auditory nerve. J. Neurosci. 29, 13797–13808.

Werner-Reiss, U., Groh, J.M., 2008. A rate code for sound azimuth in monkey auditory cortex: implications for human neuroimaging studies. J. Neurosci. 28, 3747–3758.

Wessinger, C.M., Buonocore, M.H., Kussmaul, C.L., Mangun, G.R., 1997. Tonotopy in human auditory cortex examined with functional magnetic resonance imaging. Hum. Brain Mapp. 5, 18–25.

Whitley, J.M., Henkel, C.K., 1984. Topographical organization of the inferior collicular projection and other connections of the ventral nucleus of the lateral lemniscus in the cat. J. Comp. Neurol. 229, 257–270.

Wickesberg, R.E., Whitlon, D., Oertel, D., 1991. Tuberculoventral neurons project to the multipolar cell area but not to the octopus cell area of the posteroventral cochlear nucleus. J. Comp. Neurol. 313, 457–468.

Wightman, F.L., Kistler, D.J., 1992. The dominant role of low-frequency interaural time differences in sound localization. J. Acoust. Soc. Am. 91, 1648–1661.

Winer, J.A., 1984. The human medial geniculate body. Hear. Res. 15, 225–247.

Winer, J.A., Kelly, J.B., Larue, D.T., 1999. Neural architecture of the rat medial geniculate body. Hear. Res. 130, 19–41.

Winer, J.A., Larue, D.T., 1988. Anatomy of glutamic acid decarboxylase immunoreactive neurons and axons in the rat medial geniculate body. J. Comp. Neurol. 278, 47–68.

Winer, J.A., Larue, D.T., 1996. Evolution of GABAergic circuitry in the mammalian medial geniculate body. Proc. Natl. Acad. Sci. U. S. A. 93, 3083–3087.

Winer, J.A., Larue, D.T., Diehl, J.J., Hefti, B.J., 1998. Auditory cortical projections to the cat inferior colliculus. J. Comp. Neurol. 400, 147–174.

Winer, J.A., Morest, D.K., 1983. The medial division of the medial geniculate body of the cat: implications for thalamic organization. J. Neurosci. 3, 2629–2651.

Winer, J.A., Saint Marie, R.L., Larue, D.T., Oliver, D.L., 1996. GABAergic feedforward projections from the inferior colliculus to the medial geniculate body. Proc. Natl. Acad. Sci. U. S. A. 93, 8005–8010.

Woods, D.L., Herron, T.J., Cate, A.D., Yund, E.W., Stecker, G.C., Rinne, T., Kang, X., 2010. Functional properties of human auditory cortical fields. Front. Syst. Neurosci. 4, 1–13.

Woods, D.L., Stecker, G.C., Rinne, T., Herron, T.J., Cate, A.D., Yund, E.W., Liao, I., Kang, X., 2009. Functional maps of human auditory cortex: effects of acoustic features and attention. PLoS One 4.

Wu, S.H., Kelly, J.B., 1995. Inhibition in the superior olivary complex: pharmacological evidence from mouse brain slice. J. Neurophysiol. 73, 256–269.

Xie, F., You, L., Cai, D., Liu, M., Yue, Y., Wang, Y., Yuan, K., 2018. Fast inhibitory decay facilitates adult-like temporal processing in layer 5 of developing primary auditory cortex. Cerebr. Cortex 28, 4319–4335.

Xie, R., Manis, P.B., 2014. GABAergic and glycinergic inhibitory synaptic transmission in the ventral cochlear nucleus studied in VGAT channelrhodopsin-2 mice. Front. Neural Circ. 8, 1–15.

Yamamoto, T., Williamson, S.J., Kaufman, L., Nicholson, C., Llinas, R., 1988. Magnetic localization of neuronal activity in the human brain. Proc. Natl. Acad. Sci. U. S. A. 85, 8732–8736.

Yang, Y., Lee, J., Kim, G., 2020. Integration of locomotion and auditory signals in the mouse inferior colliculus. Elife 9, 1–17.

Yin, T.C., Chan, J.C., 1990. Interaural time sensitivity in medial superior olive of cat. J. Neurophysiol. 64, 465–488.

Yin, T.C., Kuwada, S., 1983. Binaural interaction in low-frequency neurons in inferior colliculus of the cat. III. Effects of changing frequency. J. Neurophysiol. 50, 1020–1042.

Yost, W.A., Dye, R.H., 1988. Discrimination of interaural differences of level as a function of frequency. J. Acoust. Soc. Am. 83, 1846–1851.

Zatorre, R.J., 1988. Pitch perception of complex tones and human temporal-lobe function. J. Acoust. Soc. Am. 84, 566–572.

Zatorre, R.J., Belin, P., 2001. Spectral and temporal processing in human auditory cortex. Cerebr. Cortex 11, 946–953.

Zatorre, R.J., Belin, P., Penhune, V.B., 2002a. Structure and function of auditory cortex: music and speech. Trends Cognit. Sci. 6, 37–46.

Zatorre, R.J., Bouffard, M., Ahad, P., Belin, P., 2002b. Where is "where" in the human auditory cortex? Nat. Neurosci. 5, 905–909.

Zhang, Y., Suga, N., 2000. Modulation of responses and frequency tuning of thalamic and collicular neurons by cortical activation in mustached bats. J. Neurophysiol. 84, 325–333.

20

The Visual System

Barbara Buccilli[1,2] *and Brandon M. Ascenzi*[3]

[1]University of Rome "La Sapienza", Rome, Italy [2]Department of Neurosurgery, Cerebrovascular and Neurologic Critical Care, Icahn School of Medicine at Mount Sinai, New York, NY, United States [3]Independent Neuroresearcher-Member of Marie Curie Alumni Association (MCAA)

OUTLINE

INTRODUCTION

As human beings we rely on sight more than most other animals, but how does the visual system carry the information on a visual stimulus from the eye to our brain? How does this stimulus become conscious and how do our brains make sense of it in context? In this chapter, we will find the answer to these and other questions revising the classical anatomy of the visual system and looking into some of the most recent discoveries on its functional connectivity. We do not know everything there is to know about the visual system yet, but we hope the following pages will guide you through what we know and what we are learning and prepare you to sail toward new discoveries.

RETINA

The visual system is composed of several parallel pathways, each dedicated to a specific function. The parallel processing of visual information is already present at the level of the retina. The structure and function of the retina have been extensively analyzed (Masland, 2001a,b; Dacey, 1996).

From Anatomy to Function of the Central Nervous System
https://doi.org/10.1016/B978-0-12-822404-5.00003-6

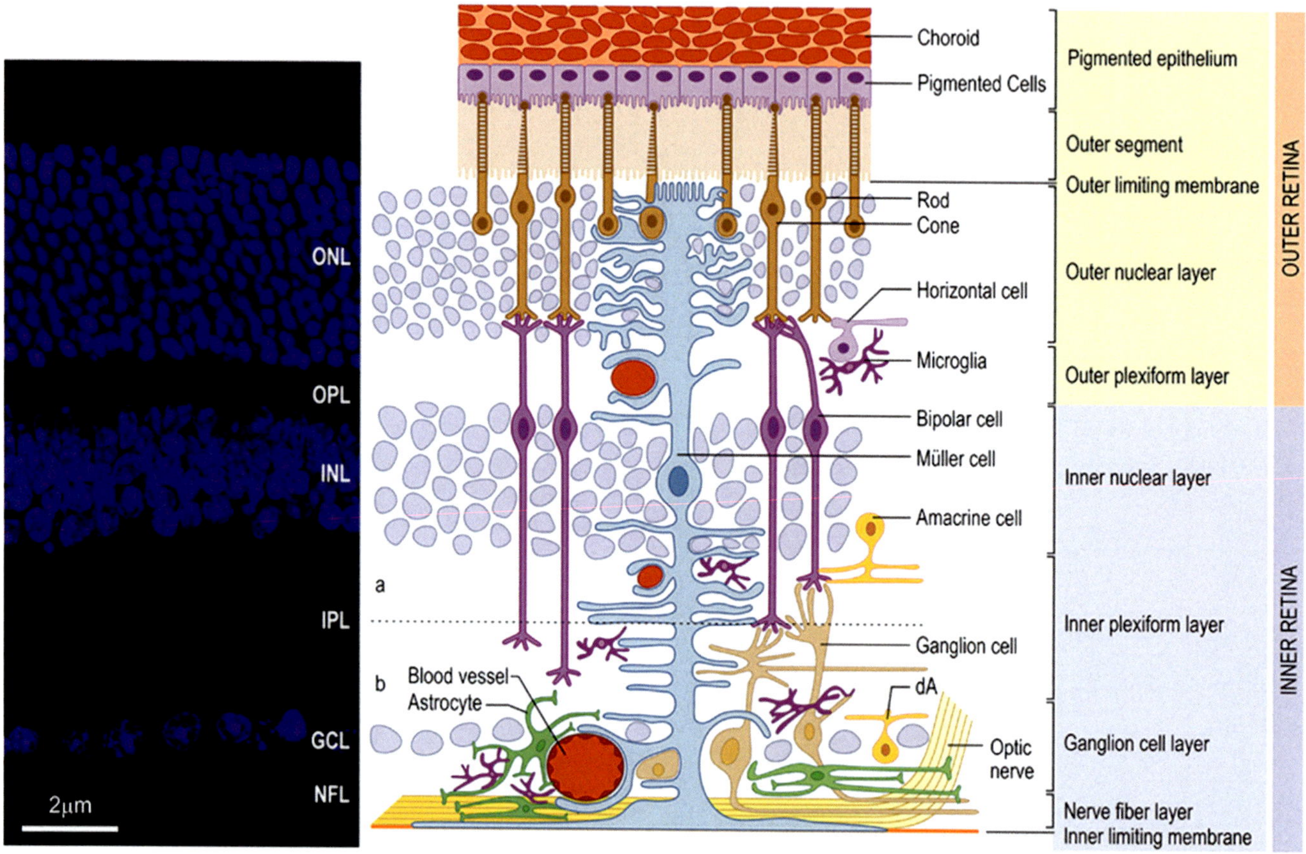

FIGURE 20.1 **Schematic organization of the vertebrate retina.** (Left) Vertical section of mouse retina labeled with nuclear marker DAPI revealing the organization of the retina in layers of cell bodies and process named as follows: ganglion cell layer (GCL), inner plexiform layer (IPL), inner nuclear layer (INL), outer plexiform layer (OPL), outer nuclear layer (ONL), and inner/outer segment of photoreceptors (IS/OS). (Right) The retina consists of different cell types located in specific layers. Cell bodies of rod and cone photoreceptors (brown) are located in the ONL. Both photoreceptors perform synapses in the OPL with bipolar (purple) and horizontal (lilac) cells, which, in turn, show the cell bodies in the outer portion of INL, as well as amacrine (yellow) cells and Müller (blue) cells. Bipolar and amacrine cells arborize in the IPL, contacting ganglion cell dendrites. Rod bipolar cells arborize in the inner portion of IPL. Cone bipolar cells (right pair of purple cells) are subdivided in ON and OFF bipolar cell contacting the same cone in the OPL and synapsing, respectively, in the ON (B) or OFF portion of the IPL (dashed line shows the functional division of the IPL in OFF (A) and ON (B) circuitries). Ganglion (orange) and displaced amacrine (dA) cell bodies are situated in the GCL. Axons of ganglion cells form the nerve fiber layer (NFL), leave the retina through the optic head nerve taking the information to visual brain nuclei by optic nerve (ON). The three types of glial cells (Müller, astrocytes, and microglia) are found in different retinal layers. Astrocytes (green) are restricted to the inner portion of the retina, in the NFL and GCL, and have a close relation to blood vessels. Microglia (dark purple) appear preferentially in the plexiform layers (IPL and OPL). Finally, Müller cells, the predominant glia in the retina, extend their processes radially throughout the retina forming the inner limiting membrane (ILM) and the outer limiting membrane (OLM). Muller glia processes interact with almost all retinal cell types, including blood vessels, displaying a crucial role in the physiology of this tissue. In mice and humans, the inner retina is vascularized by three capillary branches of central retinal artery. The outer retina, with the avascular photoreceptor region, relies on the choriocapillaris (Ch) lying beneath the retinal pigment epithelium (RPE). Scale bar: 20 μm. *Image courtesy of Campos, V. S. de, Calaza, K. C., Adesse, D., 2020. Implications of TORCH diseases in retinal development—special focus on congenital toxoplasmosis. Front. Cell. Infect. Microbiol. 10. https://doi.org/10.3389/fcimb.2020.585727.*

The retina is a laminated structure. Its layers are confined between the outer and inner limiting membranes. The outer limiting membrane is composed of extensions of the radial glia, the Müller cells of the retina, which are connected via adheres junctions to the inner segments of the photoreceptors. The foot processes of Müller cells and the associated basal membrane form the inner limiting membrane.

Generally, the following nine layers are identified (Fig. 20.1):

1. Layer of cones and rods
2. Outer limiting membrane (OLM)
3. Outer nuclear layer (ONL), containing the nuclei of the photoreceptors
4. Outer plexiform layer (OPL), containing the synaptic connections of the photoreceptors and horizontal cells
5. Inner nuclear layer (INL), containing the nuclei of bipolar cells, horizontal cells, and numerous amacrine cells

6. Inner plexiform layer, divided into an off (vitreous) sublayer and an on inner sublayer
7. Ganglion cell layer (GCL)
8. Optic nerve fiber layer (LON)
9. Inner limiting membrane (ILM)

Photoreceptors (Fig. 20.2) are composed of an outer and inner segment, a cell body with a nucleus, and a synaptic terminal. The outer and inner segments protrude through the outer limiting membrane into the space between the retina and the pigment epithelium. The outer segment contains photopigments adhered to the membrane, while the inner segment is full of mitochondria. The outer and inner segments are connected by a modified cilium. The cell body with the nucleus is located in the outer nuclear layer. Synaptic contacts are established in the outer plexiform layer. The inner segment of cones is cone-shaped, while rods are elongated structures. Three types of cones can be distinguished. They differ in their sensitivity to different wavelengths of light. Cones for long wavelengths (l) express maximum sensitivity to red light, cones for medium wavelengths (m) express maximum sensitivity to green light, and cones for short wavelengths (s) express maximum sensitivity to blue light. Cones perceive light of higher intensity (photopic vision), while rods are sensitive to low light intensities (scotopic vision) and can faithfully translate the absorption of single photons.

In the dark, photoreceptors are depolarized and release glutamate from their synaptic terminals. When light hits the photoreceptors, they hyperpolarize, and

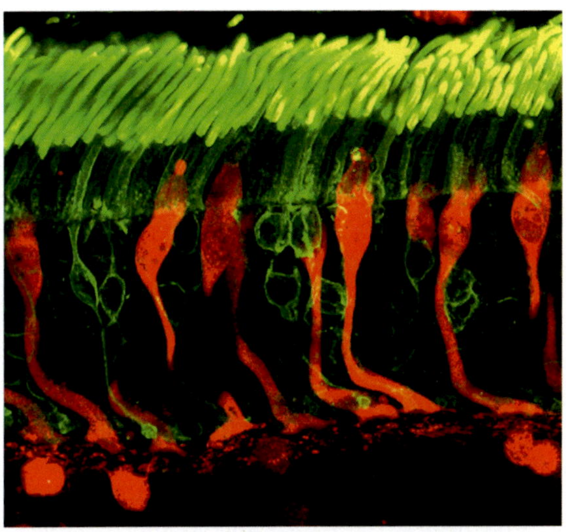

FIGURE 20.2 **Confocal microscope image of rod and cone photoreceptors in a human retina.** Fluorescent probes have been used to identify rod photoreceptors (green) and cone photoreceptors and horizontal cells (red). The cones looks like candle sticks, warped as if viewed through a fun house mirror. The horizontal cells are the round cells in the lower portion of the image. *Image courtesy of Dr. Robert Fariss, National Eye Institute, NIH.*

the release of neurotransmitter to their second-order neurons ceases. The synaptic terminals of cones and rods are arranged in the outer plexiform layer and are called rod spherules and cone pedicles. The main cell types and fundamental circuits of the retina have been described by Cajal et al. (1972) and Polyak (1941). Bipolar neurons, whose cell bodies are located in the inner nuclear layer, connect cones with ganglion cells, which are located in the vitreal inner layer of the retina. From these ganglion cells, the axons of the optic nerve originate (Rockhill et al., 2002). These simple connections are exemplified by the systems of Midget and multipolar ganglion cells.

Horizontal cells (Fig. 20.2), located in the inner nuclear layer, synapse with photoreceptors in the outer plexiform layer; the amacrine cells, arranged in the vitreous part of the inner nuclear layer, establish connections between bipolar cells, ganglion cells, and other amacrine cells in the inner plexiform layer. Horizontal and amacrine cells are inhibitory neurons that do not generate action potentials and conduct impulses in all directions (Lee et al., 1999; Macneil et al., 1999). There are numerous different types of horizontal and amacrine cells. Horizontal cells form an electrotonic connected circuit. Their dendrites contact the pedicles of cones in triad synapses. The L, M, and S cones connect to two different types of horizontal cells. The axons of horizontal cells with axons terminate on the spherules of rod cells. The long axons of these cells are electrotonically isolated from the cone's feedback dendritic system. Horizontal cells are considered important for contrast enhancement and the formation of the antagonistic surround of the ganglion cell's receptive field (Bloomfield et al., 2001; Lee et al., 1999).

The essential features of the retina are given by the existence of on (light-activated) and off (dark-activated) pathways, as well as the center–periphery structure of the receptive fields of numerous ganglion cells, with antagonism between the effects of light on the central and peripheral parts of these receptive fields. On-bipolar cells are depolarized by increased light, while off-bipolar cells are hyperpolarized by light. On- and off-bipolar cells differ in the morphology of their synaptic contacts with photoreceptors, the type of glutamate receptors present at these synapses, and the sublayer of the inner plexiform layer where they establish connections with on and off ganglion cells and amacrine cells. The dendrites of bipolar cells invaginate into the synaptic terminals of photoreceptors and are positioned in front of the synaptic ribbon. Off-bipolar cells make synapses with the flattened base of the synaptic terminals.

The horizontal cells make synaptic contacts on both sides of the invaginated dendrite. The combination of the invaginated dendrite and the terminals of the horizontal cells forms the triad. The metabotropic glutamate receptors expressed by on-bipolar cells reverse the polarity of a light-hyperpolarized photoreceptor into the depolarization of an on-bipolar cell. On the other hand, the ionotropic glutamate receptors maintain the polarity of the light-hyperpolarized photoreceptor in the flat synapses with off-bipolar cells. Off-bipolar cells synapse in the outer off sublayer of the inner plexiform layer, while on-bipolar cells synapse in the inner vitreal on sublayer.

The connections of on- and off-bipolar cells with the dendrites of ganglion cells in the on and off sublayers of the inner plexiform layer explain the presence of ganglion cells with on and off receptive field centers.

The horizontal and amacrine cells form a network of inhibitory connections that modulate the signals transmitted by bipolar cells. They provide lateral inhibition, which means that they suppress the activity of neighboring bipolar cells. This lateral inhibition enhances the contrast between different visual stimuli and contributes to edge detection and spatial filtering. The specific circuitry and synaptic connections involved in these circuits can vary depending on the retinal region and the type of ganglion cell. Different types of horizontal and amacrine cells exist, and they exhibit diverse connectivity patterns within the retina. By providing feedback and feed-forward inhibition, lateral inhibition in the outer retina may support the receptive field surrounds of retinal neurons at photoreceptor terminals and at bipolar cell dendrites. Through gap junction coupling with adjacent photoreceptors and horizontal cell feedback, light falling on nearby photoreceptors can influence the photoreceptor response itself. The hyperpolarizing membrane potential response of horizontal cells exhibits a slow depolarizing sag or roll back in the membrane potential as the size of a spot of light increases and the horizontal cell feedback network is activated, which is partly related to this feedback. The synaptic and cellular mechanisms underlying this process are poorly understood, despite the fact that it is commonly acknowledged that horizontal cells give inhibitory feedback onto photoreceptor terminals. There are a number of GABA, protons, ATP, and an ephaptic mechanisms that have been suggested as synaptic transmission mechanisms. The discrepancy between results from different studies could be due to technological challenges, biological strategies that differ between mammalian and nonmammalian species, or even different synaptic mechanisms that differ between the various horizontal cell subtypes. It is also possible that various cellular processes encode certain components of visual information or may function in different conditions of light. The modification of the voltage-gated L-type Ca^{2+} channels in photoreceptors appears to be the primary consequence of horizontal cell feedback onto cone and rod photoreceptor terminals, though. It has been proposed that the inhibitory neurotransmitter GABA serves as the neurotransmitter of horizontal cells. In nonmammalian retinas, GABA is thought to be released through the voltage-modulated action of a plasma membrane GABA transporter (GAT), whereas in mammalian retinas, it is thought to be released by vesicular release. Byzov et al. (2020) first proposed an ephaptic model of feedback onto cone terminals, which was later developed to include hemichannels as current sinks. Protons have recently emerged as a crucial regulating component, as evidenced, for example, by changes in the pH of the synaptic cleft. Since the ON and OFF bipolar cells depolarize or hyperpolarize in response to light-mediated reductions in glutamate release from photoreceptors, it would seem simplest that the modulatory effect would be integrated at the presynaptic photoreceptor terminal by a feedback mechanism for surround antagonism to affect the two types of bipolar cell responses in a coherent manner. It is necessary for the horizontal cell transmitter to affect the postsynaptic ON and OFF bipolar cells via mechanisms of opposite polarity in order to activate feed-forward surround inhibition. This can happen when the two types of bipolar cells have distinct chloride concentrations. Bipolar cells have a peripheral antagonistic receptive field that is mediated to some extent by horizontal cell input to bipolar cells. At the ultrastructural level, synapses between horizontal cell processes and bipolar cell dendrites have been reported. $GABA_A$ and $GABA_C$ receptor immunoreactivity is localized to bidirectional cell dendrites adjacent to horizontal cell processes. Extensive data supporting GABA-mediated inhibitory feedback from horizontal cells to cones come from studies of turtle, fish, salamander, and some mammalian retinas. The nonzero potential of this feedback synapse on the cones of lower vertebrates suggests that the underlying synaptic mechanism is probably chemical rather than electrical. In addition, bathing the retina in GABA inhibits feedback from cone photoreceptors, as would be expected if a reduction in GABA release signals horizontal cell feedback. In lower vertebrates, GABA has a feedback role, which is compatible with the location of GABA receptors to photoreceptor terminals. The expression of $GABA_A$ receptor subunit cDNA by photoreceptors, as revealed by in situ hybridization immunohistochemistry and single-cell RT-PCR, provides morphological and molecular support for a feedback role. GABA acts on autoreceptors on adjoining horizontal cells as well as on horizontal cells themselves. These investigations collectively point to a variety of GABA targets in the outer plexiform layer, which may mediate GABA's intricate effects in the outer retina. There is also research that contradicts the idea that GABA is necessary to mediate horizontal cell feedback onto cones. In general, surround-evoked depolarization of cones does not always seem to be impacted by GABA agonists and antagonists. However, the interpretation of tests with GABA antagonists is made more difficult by the fact that horizontal cells, bipolar cells, and cone photoreceptors all have GABA receptors. Furthermore, it is evident that modest variations in the Ca^{2+} currents caused by feedback do not result in significant variations in the photoreceptor's membrane potential.

The microcircuitry of rods and cones differs in the central and peripheral retina. The highest concentration of cones is found in the fovea, while rods are absent from this region. Rods are more numerous in the peripheral retina, where cone density rapidly decreases. Photoreceptors are absent in the region of the optic disc, the origin of the optic nerve.

The following parallel pathways with different central connections originate from the retina:

1. L and M cones are densely packed in the fovea, each with their own connections through on and off midget bipolar cells, respectively, to on and off midget ganglion cells (Rodieck et al., 1993; Watanabe et al., 1989). These ganglion cells project to the four upper parvocellular layers of the lateral geniculate body. They are responsible for optimal resolution in the visual system, determined by the density of L and M cones in the fovea. They also participate in the perception of red and green. The receptive fields of on and off midget ganglion cells are color-opponent, with a red or green center and an antagonistic green or red periphery (Dacey, 1996; Dacey & Packer, 2003).
2. Large on and off parasol ganglion cells receive input from numerous diffuse on and off bipolar cells (Rodieck et al., 1993; Watanabe et al., 1989). Diffuse bipolar cells are contacted by a random collection of L, M, and S cones and are therefore not color-sensitive. Multipolar ganglion cells project to two magnocellular layers of the lateral geniculate body and provide them with a luminance contrast signal (Perry et al., 1984; Leventhal et al., 1981).
3. A separate retinal system is responsible for the blue/yellow opponent perception (Calkins et al., 1998). S-on bipolar cells selectively synapse with different S cones and terminate in the on sublayer of the inner plexiform layer on the dendrites of small bistratified S ganglion cells. L and M cones project, through off diffuse bipolar cells, to a group of dendrites of these cells in the off sublayer. The on-type bistratified ganglion cells are excited by blue and inhibited by a combination of red and green, therefore yellow. They project to the interlaminar (koniocellular) layers of the lateral geniculate nucleus. The spatial resolution of this system is relatively low and does not contribute to this property of the visual system, which is determined by L and M midget ganglion cells and their projections to the parvocellular lateral geniculate nucleus. S-off ganglion cells are rare or absent.
4. The central connections of rod photoreceptors include an additional synapse with an amacrine cell (the AII amacrine cell). Rod photoreceptors utilize cone bipolar and ganglion cells for their central projections (Bloomfield et al., 2001). Numerous rods synapse with an on-type rod bipolar cell; off-type rod bipolar cells do not exist. Rod bipolar cells contact the basal dendrites of AII amacrine cells in the on inner plexiform layer. The AII amacrine cells are electrically connected. Dopamine, released by other amacrine cells, regulates the connection of AII amacrine cells and adapts their response in different light adaptation states. Gap junctions connect the dendrites of AII amacrine cells with on-type cone bipolar cells in the on-type inner plexiform layer. Inhibitory (glycinergic) synapses connect the globular appendages of AII amacrine cells with off-type cone bipolar cells in the off sublayer. The rod system, therefore, utilizes the cone system as its efferent system. Consequently, the ganglion cells to which the rod system projects send their projections to the parvocellular and/or magnocellular layers of the lateral geniculate nucleus.
5. The large light-sensitive ganglion cells contain the photopigment melanopsin. Their dendrites form a dual-layered photoreceptive network in the inner off and on sublayers of the plexiform layers (Berson et al., 2002; Hattar et al., 2002; Provencio et al., 2002). Furthermore, these cells are guided by bipolar cells. The axons of melanopsin-containing ganglion cells terminate in the suprachiasmatic nucleus, intergeniculate leaflet, and pretectal olivary nucleus (Lucas et al., 2001; Morin et al., 2003).
6. Several types of ganglion cells are direction-selective. One of them is a bistratified ganglion cell, which receives inputs from starburst amacrine cells (Famiglietti et al., 1992). The cell bodies of starburst amacrine cells are located in the inner nuclear layer and, as displaced amacrine cells, are also present in the ganglion cell layer. Through excitatory and inhibitory signaling, these cells release both acetylcholine and GABA. Starburst amacrine cells are important for direction sensitivity.

Direction-sensitive ganglion cells project to the accessory optic system and the optic tract nucleus. Retinal ganglion cells can be divided both structurally and functionally into different categories. The main populations are represented by midget and parasol cells, as mentioned before. Midget ganglion cells are red-green antagonistic, with small receptive fields and a low luminance contrast gain. Midget ganglion cells respond to luminance gradients with sustained tonic responses. Their axons are small and conduct signals slowly. Parasol ganglion cells are broad-band, meaning they are not selective to a particular wavelength, with large receptive fields and a high luminance contrast gain. Their axons are large, with high conduction velocity. They respond to luminance gradients with transient phasic responses (Kaplan et al., 2003). Midget cells project to the four upper parvocellular layers of the lateral geniculate body, while parasol cells project to the two lower magnocellular layers. It has been proposed that the population of midget cells represents a neoacquisition in primates, serving higher resolution and color vision. Another group consists of ganglion cells that are different in shape and functional properties, with projections to multiple visual centers. These include small bistratified ganglion cells that are blue-yellow antagonistic, with projections to the intercalated layers of the lateral geniculate body, and melanopsin-containing cells. Only a small proportion of ganglion cells project to the lateral geniculate body, with the rest projecting to one or more of these other projection sites.

THE OPTIC NERVE AND ITS PATHWAY

The axons of ganglion cells run along the inner surface of the retina and converge toward the posterior pole of the eye, where they pierce the sclera through the scleral foramina, forming the optic nerve (Fig. 20.3). Only after leaving the retina do the axons of ganglion cells acquire myelin sheaths. The optic nerves extend to the optic chiasm (Fig. 20.4), located at the base of the brain in the most rostral part of the hypothalamus. At the level of the optic chiasm, partial decussation occurs: fibers from the nasal hemiretinas, including the projection of the monocular visual field, cross to the opposite side, while those from the temporal hemiretinas remain uncrossed. After this partial decussation, the axons of retinal ganglion cells continue uninterrupted beyond the chiasm as the two diverging optic tracts. These tracts curve around the lateral surfaces of the diencephalon until they reach the lateral geniculate bodies, where most of the fibers terminate. However, some fibers continue in the mediocaudal direction to the superior colliculus, pretectal region, and terminal nuclei of the accessory optic system.

In the optic tract, retinal fibers with different destinations are mixed. The smaller fibers, likely from midget cells, occupy a deeper position, while the larger fibers, including the axons of multipolar cells and fibers destined for the superior colliculus, are located superficially (Reese et al., 1990). It is important to note that some retinofugal fibers or their collaterals leave the optic chiasm and enter the anterior part of the hypothalamus, where they terminate in the suprachiasmatic nucleus (Dai et al., 1998). This is a small group of periventricular cells located directly above the chiasm. Through this retinohypothalamic projection, the visual system participates in the regulation of various behavioral rhythms. These fibers belong to the system of melanotropic ganglion cells, which in rodents also terminate in the intergeniculate leaflet, pretectal olivary nucleus, various hypothalamic nuclei, amygdala, superior colliculus, and periaqueductal gray (Hattar et al., 2006; Morin et al., 2003). The retinohypothalamic fibers terminate extensively in the hypothalamus.

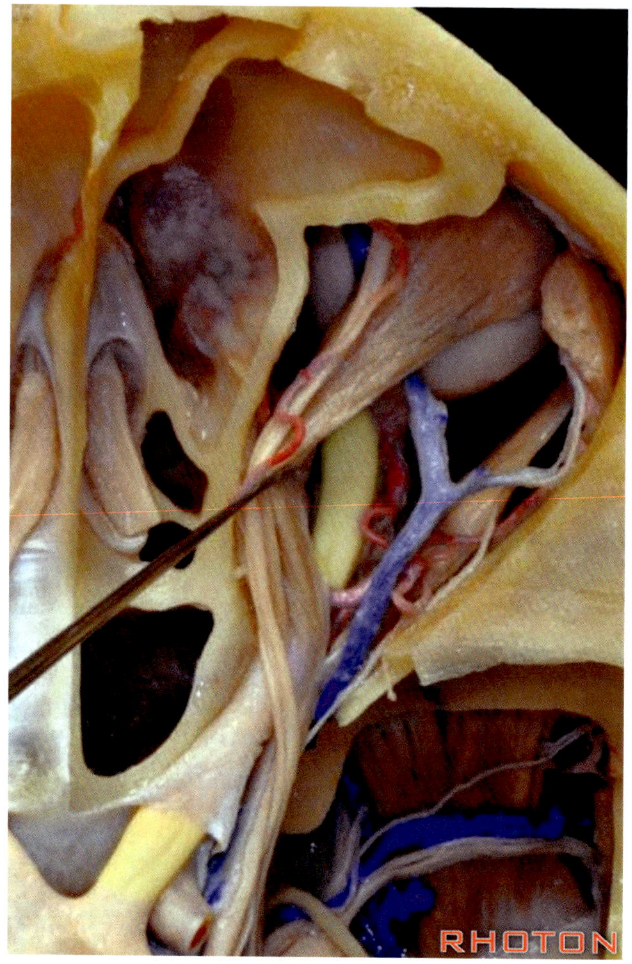

FIGURE 20.3 **Orbital cavity; right-superior view.** In yellow is the optic nerve that reaches, anteriorly, the ocular globe. To the latter are attached: superiorly, the superior rectus muscle joined medially to the oculomotor nerve; laterally, the lateral rectus muscle. In blue, is the ophthalmic vein and in red is the ophthalmic artery. It can be seen that the optic nerve posteriorly continues with the optic chiasm. *Image courtesy of The Rhoton Collection® http://rhoton.ineurodb.org/. Drawing by Jeffrey Sorenson.*

FROM THE LATERAL GENICULATE BODY TO THE VISUAL CORTEX

As previously mentioned, the majority of fibers in the optic tract terminate in the lateral geniculate nucleus. This thalamic nucleus has a clear laminar structure composed of six concentric layers. These layers, originating from the concave ventral surface, are typically

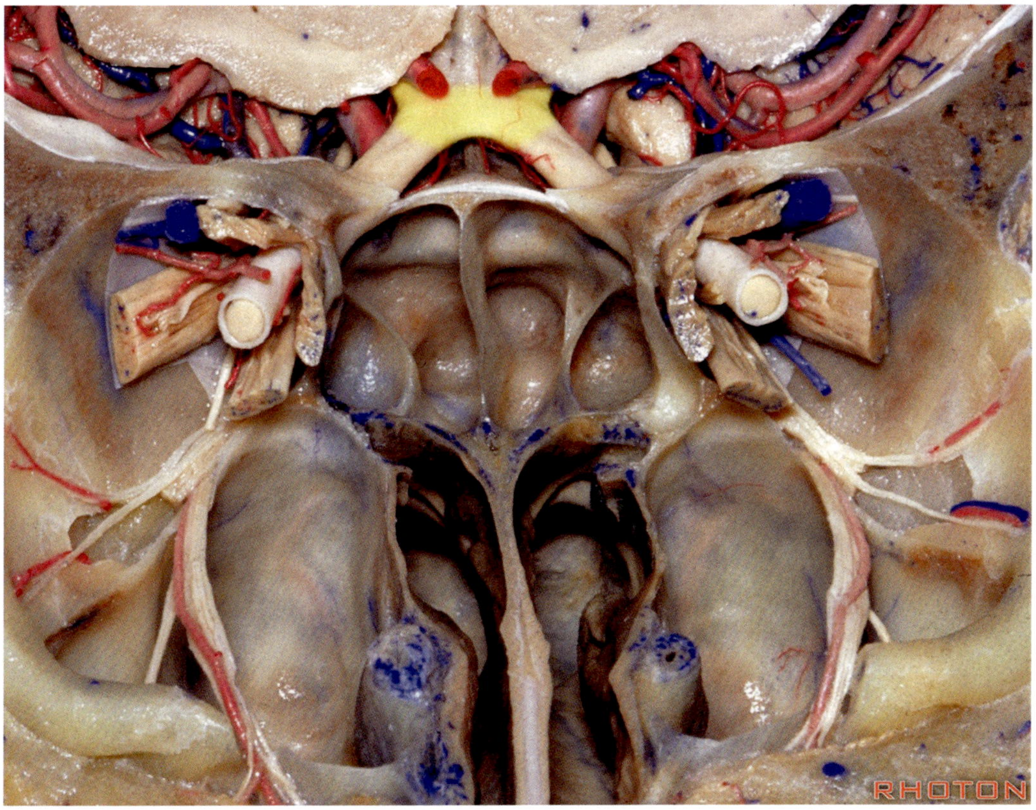

FIGURE 20.4 **Optic chiasm and middle cranial fossa.** In yellow, the optic chiasm located in the anterior cranial fossa whose margins are formed: in the center by the groove of the optic chiasm and laterally by the little wings of the sphenoid. Below the optic chiasm, at the base of the tuberculum sellae (resected), the open sphenoid sinus can be seen. The two posterior ends of the optic nerve, right and left, are visible below the two optic canals. Surrounding the optic nerves are the eye muscles and the ophthalmic artery. *Image courtesy of The Rhoton Collection® http://rhoton.ineurodb.org/. Drawing by Jeffrey Sorenson.*

numbered from 1 to 6. The magnocellular layers (M) 1 and 2 occupy the hilar region. The parvocellular layers 3–6 consist of relatively small cells. Even smaller cells are found in the intercalated or koniocellular layers (K). The K1–K6 layers are situated beneath the six main layers of the nucleus. Retinal fibers terminate in a highly ordered manner in the layers of the lateral geniculate nucleus. Layer M1 and layers P4 and P6 receive fibers from the contralateral eye, while layer M2 and layers P3 and P5 receive fibers from the ipsilateral eye.

Furthermore, multipolar cells project to the M layers, and midget ganglion cells project to the P layers. The K layers are innervated by the same retinal structures that distribute to the overlying main layers. The small, bistratified ganglion cells antagonistic for blue/yellow project to layers K3 and K4, while other types of retinal ganglion cells innervate other K layers.

The ventral K layers receive an extraretinal projection from the superior colliculus and cholinergic input from the parabigeminal nucleus (Hendry & Reid, 2000; Lachica et al., 1992). The efferent fibers of the lateral geniculate nucleus form the optic radiation (Fig. 20.5a and 20.5b), or geniculocalcarine tract, which terminates in the primary visual cortex.

The primary visual area (V1) corresponds to Brodmann's area 17. It is also referred to as the striate cortex (striate area) because it contains the stria of Gennari, a highly characteristic layer of myelinated fibers that is visible microscopically and divides the inner granular layer (sublayer 4B). The striate cortex surrounds the calcarine sulcus on the medial surface of the occipital lobe. The fibers of the optic radiation first pass through the retrolenticular part of the internal capsule, then curve around the lateral ventricle, and finally run posteriorly toward the occipital cortex.

Numerous fibers of the optic radiation do not reach their destination through the shortest pathway. Those originating from the lateral part of the lateral geniculate nucleus and terminating ventrally to the calcarine fissure project forward into the temporal lobe and run laterally on the inferior horn before turning back. There is a precise point-to-point projection from the retina to the lateral geniculate nucleus and from the latter to the visual cortex. In the lateral geniculate nucleus, the contralateral hemifield is represented in each successive layer, with the fovea represented caudally (Van Essen et al., 1984).

Let us analyze some features of the retina's projection to V1 (Van Essen, 2004):

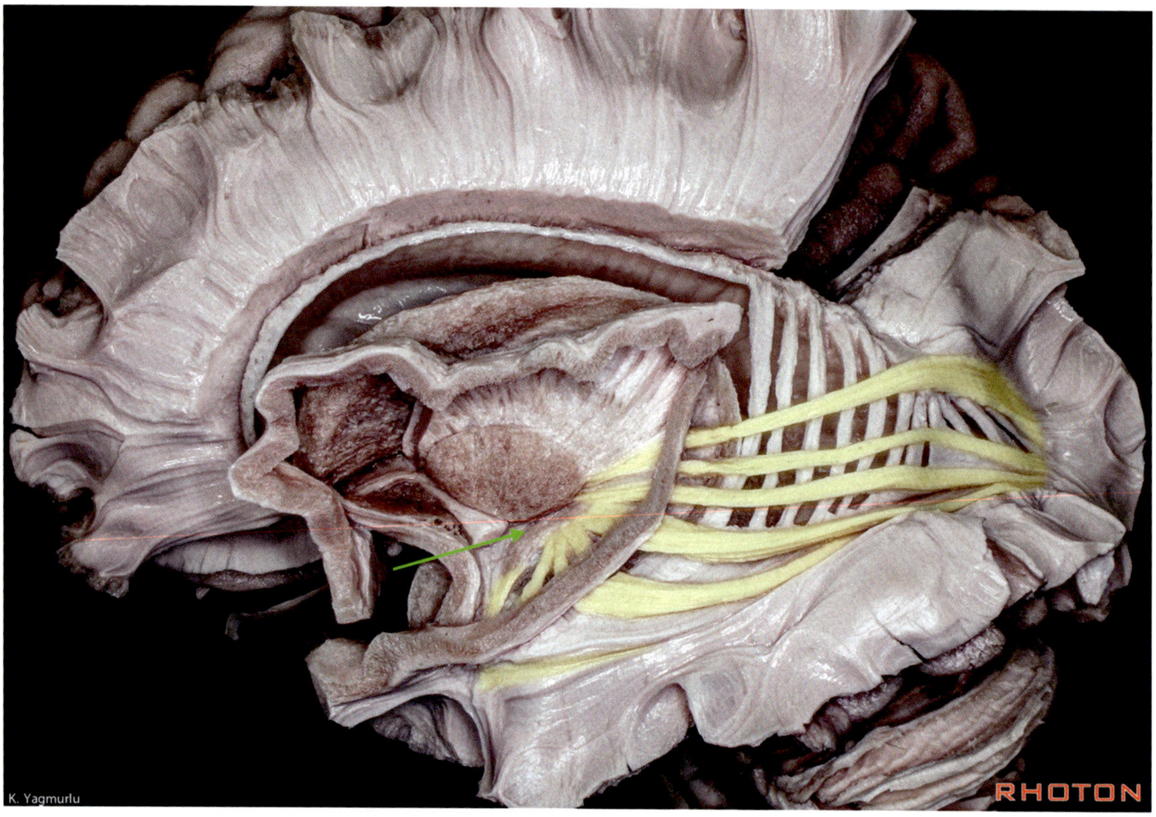

FIGURE 20.5A **Optic radiation.** The optic radiation fibers arise from the lateral geniculate body (LGB, green arrow) and travel in the roof and lateral wall of the temporal horn and then lateral to the atrium to reach the calcarine sulcus. Notable, the upper one-third of the atrium is free from the optic radiation fibers. The optic radiations are separated from the temporal horn and atrium of the lateral ventricle by the tapetum.

1. Due to the partial decussation of retinofugal fibers at the chiasm, the right half of the visual field, which projects onto the right nasal half of the retina and the left temporal half of the retina, is represented in the left occipital lobe, and vice versa.
2. The upper quadrant of the visual field is represented in the lower lip of the calcarine sulcus of the contralateral V1 cortex, while the lower quadrant is represented in the upper lip of this sulcus.
3. The fovea, which corresponds to the areas of the retina responsible for central vision, projects to relatively large areas in the posterior part of the primary visual cortex. The parts of the retina corresponding to the periphery of the binocular field project to a smaller intermediate portion. The nasal periphery of the retina, related to the temporal monocular periphery of the visual field, projects to the anterior end of the visual cortex, near the splenium of the corpus callosum.
4. The segregation of projections from the right and left retinas is preserved in the projection from the lateral geniculate nucleus to the V1 cortex. The projections from the geniculate layers innervated by the contralateral retina, directed to the layer IV of the V1 cortex, are organized in columns, separated from other columns that receive input from the geniculate layers innervated by the ipsilateral retina. The boundaries between these ocular dominance columns are quite distinct at the level of the IV cortical layer, which receives primary afferents, where most cells are monocularly driven, although to a lesser extent than the more superficial cortical layers, where binocularly driven cells predominate. In humans, the ocular dominance columns are about 0.5 mm wide and oriented at right angles to the V1/V2 boundary. They are more regular in the representation of the peripheral visual field and absent from the representation of the monocular crescent. Their width does not vary significantly between the central and peripheral representations of the visual field.

MNEMONIC REPRESENTATIONS IN THE LATERAL GENICULATE NUCLEUS

The importance of the thalamus in higher-order cognition is becoming more widely recognized. Rahmati et al. (2023) looked into the function of the lateral geniculate nucleus (LGN) in working memory (WM) after

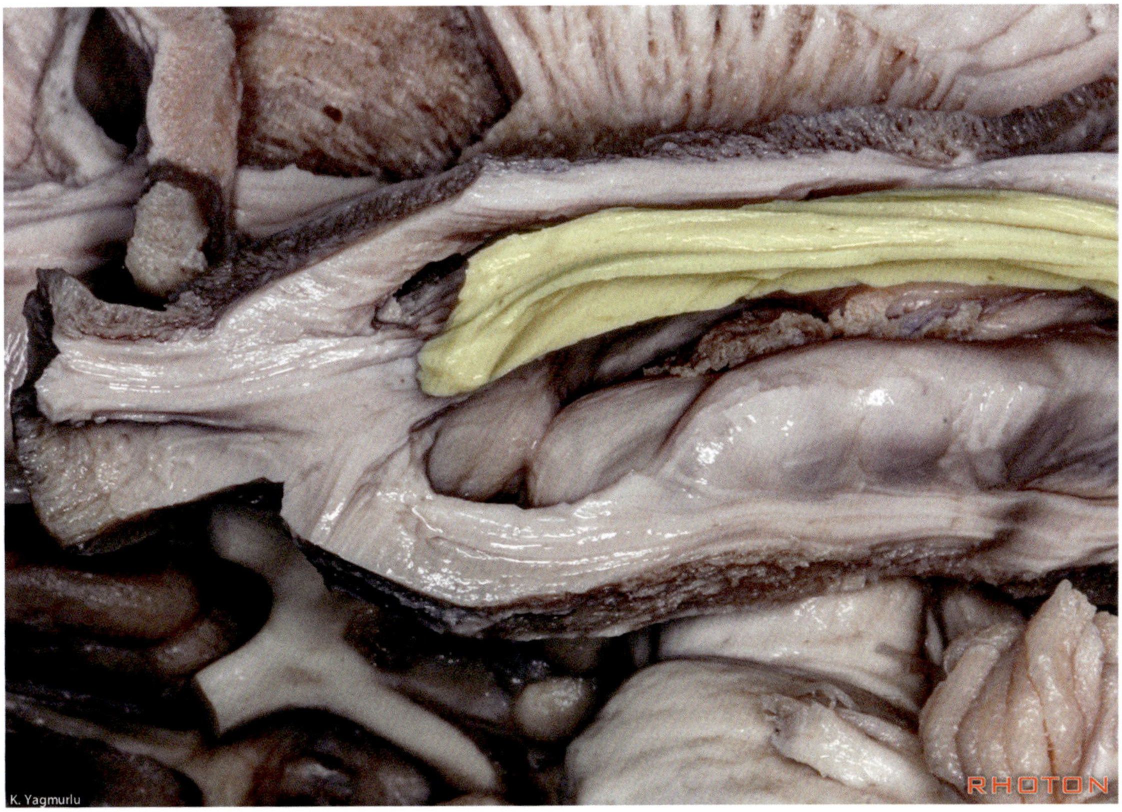

FIGURE 20.5B **Meyer's loop.** The anterior bundle of optic radiations are called Meyer's loop (in yellow), and it extends up to the tip of the temporal horn. The amygdala forms the anterior wall of the temporal horn. The hippocampus is located in the floor of the temporal horn (sausage-shaped structure). *Both images are courtesy of The Rhoton Collection® http://rhoton.ineurodb.org/. Dissector: K.Yagmurlu. Drawings by Jeffrey Sorenson, Kaan Yagmurlu and Brandon M. Ascenzi.*

discovering that activity in feedback layers of primary visual cortex (V1) that targets the LGN is driven by internal cognitive state. They specifically used model-based neuroimaging techniques to investigate the idea that the human LGN contains spatial information that is momentarily stored in WM.

To reconstruct spatial locations stored in WM as subjects underwent modified memory-guided saccade tasks, the researchers used models built on the spatial preferences of LGN populations. They discovered that population LGN activity faithfully encoded the spatial locations held in memory in all subjects. This was significant because it made it possible to separate the sites of ocular stimulation and the motor metrics of memory-guided saccades from the maintained spatial positions, proving that the human LGN accurately represents WM information. These results show a crucial channel through which memories may affect incoming processing at the lowest levels of the visual hierarchy and add LGN to the growing list of subcortical areas involved in WM.

VISUAL AREAS AND THEIR ROLE IN STIMULUS LOCALIZATION

In the cerebral cortex, numerous visual areas have been distinguished based on architecture, connections, visual topography, and/or functional characteristics (Van Essen, 2004). The primary visual area or striate cortex, V1, corresponds to BA17. Layer 4 can be divided into four sublayers (4A, 4B, 4Cα, and 4Cβ). The stria of Gennari, composed of myelinated fibers, is located in sublayer 4B. Spiny stellate neurons, which receive afferents from the lateral geniculate nucleus, are present in sublayers 4A and 4Cα and β. Pyramidal cells are absent from both sublayers 4C but are present in 4A and 4B. The dendrites of pyramidal neurons from the lower layers traverse layer 4 (Lund et al., 1988). Large pyramidal cells (Meynert's cells) are present at the border of layers 5/6. V1 is delimited by the secondary visual area C2. V2 corresponds to BA 18. At the boundary of areas 17/18, the stria of Gennari terminates, and the sublayers of layer 4 in V1 merge into a single inner granular layer.

In the layers 2/3 of V1, populations of neurons expressing cytochrome oxidase (CO) are grouped into clusters or blobs, with the non-CO cells arranged in the interblobs. The blobs continue as distinct columns extending into the deep layers of the striate cortex. Series of blobs occupy the centers of ocular dominance columns. Sublayers 4A and 4C, which receive direct projections from the lateral geniculate nucleus, show strong CO activity. A specific localization of CO-immunoreactive cells in thin and thick alternating stripes, separated by nonimmunoreactive interstripes, is also present in V2. These stripes are oriented perpendicularly to the border of V1/V2. It is difficult to separate the areas of Brodmann 18 and 19. Therefore, most of the higher-order visual areas in area 19 and beyond have been delineated based on noncytoarchitectonic criteria. The presence of a visuotopic organization in many of these areas constitutes a valid tool to define their boundaries.

The contiguous visual areas V1, V2, and V3 contain symmetric-specular visuotopic maps of the contralateral hemifield. Consequently, the vertical meridian forms the boundary between the representations of the visual field in V1 and V2. The horizontal meridian separates areas V2 and V3. V2 and V3, like V1, consist of ventral and dorsal parts that contain paired representations of the upper and lower contralateral quadrants. It is interesting to note that some boundaries of the visual areas could also be established in the human brain following callosal afferents from large lesions in the opposite occipital cortex. This is possible because the termination of callosal fibers in V1—V4 shows a strong preference for representations of the vertical meridian (for V1/V2 and the outer boundaries of V3) (Clarke et al., 1990). Visual areas beyond V3 have been organized into a dorsal stream directed to the parietal lobe and a ventral stream directed temporally. The areas V3a, V6, MT, MST, and the areas of the intraparietal sulcus belong to the dorsal stream. The accessory area V3a is located dorsally, near the parietooccipital sulcus. It shares the representation of the vertical meridian with the dorsal area V3. It contains a complete representation of the contralateral hemifield, divided by the horizontal meridian (Tootell et al., 1989). V6 is located at the dorsal apex of the cuneus and in the posterior occipitoparietal sulcus. The middle temporal area (MT or V5) is located in the ascending branch of the inferior temporal sulcus.

MT is characterized by its strong myelination and can, therefore, be localized in the human brain. The medial superior temporal area (MST) is located anteriorly to MT.

Stereoelectroencephalography recordings from patients with pharmacologically resistant epilepsy performed by Qilin Zhou et al. (2023) give us an inside on the pathways for the identification of psychophysical colors and emotional faces in the human brain.

To study how humans interpret both simple colors and complex emotional faces, delayed match-to-sample paradigms were used. The middle frontal gyrus (MFG) and postcentral gyrus, as well as the middle temporal gyrus (MTG) and posterior superior temporal sulcus (pSTS), were involved in processing incongruent colors and faces. The angular-cuneus gyrus serves as a pioneer in discriminating the two features. A different pathway of incongruency processing is tuned by the beta and gamma band interactions between the cuneus and MTG and between the cuneus and pSTS. Fusiform, MFG, and the posterior insular gyrus were discovered to be involved in the attentional modulation of the two aspects via alpha band activity. These findings provide evidence for the neurological underpinnings of the human brain's discriminative circuits for perception and cognition.

The areas of the ventral stream V4v and V4d, the ventral and dorsal divisions of area V4, are likely independent cortical areas. V4d is included in the lateral occipital area (LO) located between V3 and MT on the convexity of the occipital lobe (Tootell et al., 1996). V4 shares its vertical meridian with ventral V3. Area V8 is located in the collateral sulcus and the lingual gyrus (medial occipitotemporal), rostral to V4d, with which it shares its horizontal meridian. V8 contains a complete representation of the contralateral visual field (Hadjikhani et al., 1998). Visual areas have been organized into different functional streams by tracing the connections of their fibers and characterizing their electrophysiological and psychophysical properties in hierarchies of increasing complexity and specificity. Analyzing the complexity of corticocortical connections between visual areas is less useful for this purpose. Distinguishing anterograde and retrograde connections based on their origin and termination in different layers is not very helpful as there are numerous exceptions to this rule.

THALAMOCORTICAL PROJECTION FROM THE LATERAL GENICULAR BODY

Parallel processing in the visual cortex begins with differential projections from the lateral geniculate body to the primary visual cortex (Fig. 20.6). The projection from the magnocellular layers is concentrated in sublayer 4C, with a minor contribution to layer 6. The parvocellular layers innervate 4C and 4A (Lund et al., 1988). Projections to these cortical layers from the layers that receive their retinal input from the contralateral or ipsilateral eye terminate in alternating stripes and are responsible for the organization of V1 into ocular dominance columns. The K pathway, originating from the koniocellular layers, projects to the cytochrome oxidase (CO) immunoreactive blobs in layers 2/3 and layer 1 (Fitzpatrick et al.,

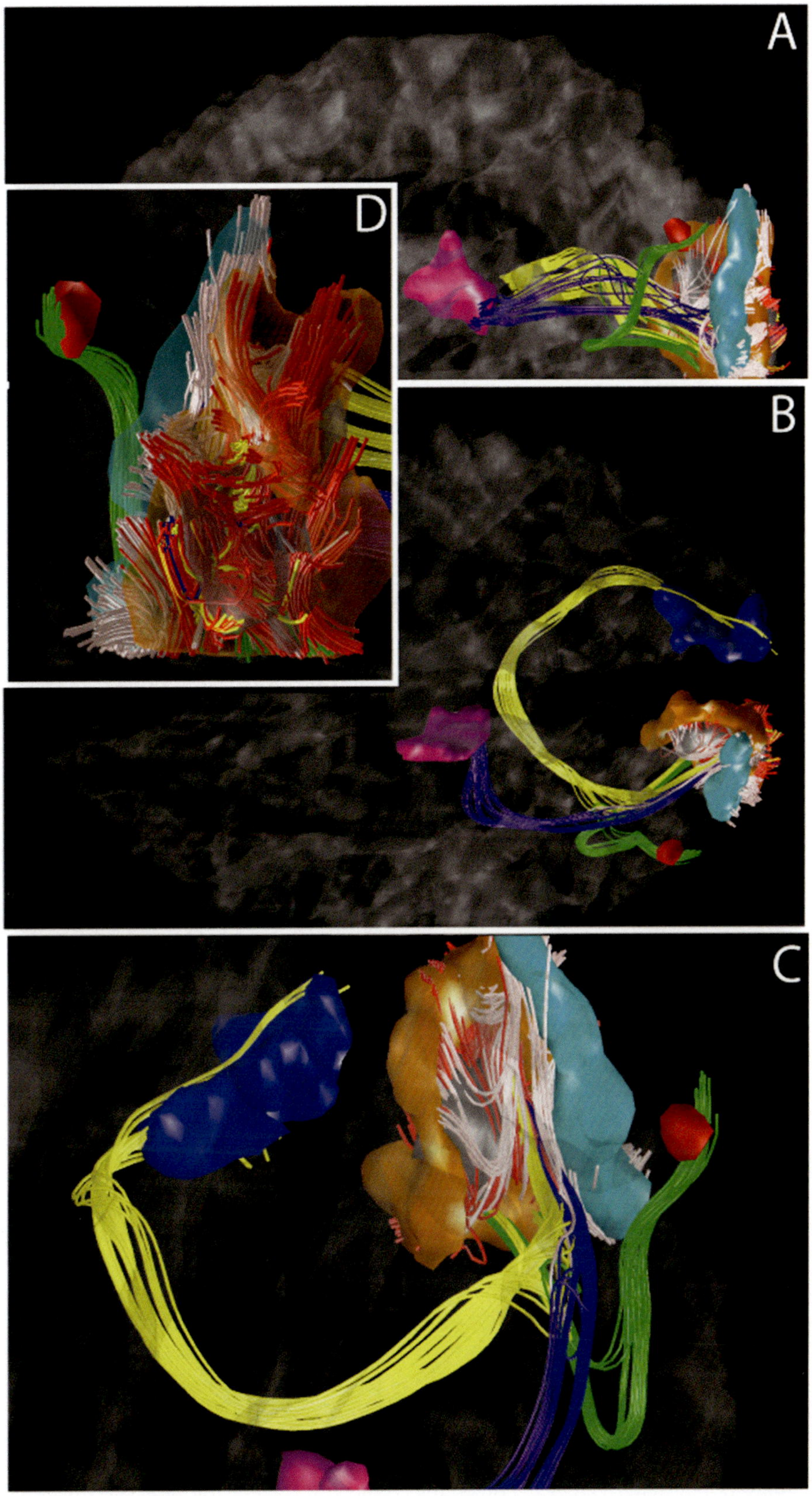

FIGURE 20.6 **Visual system white matter connectivity derived from tractography.** Views from the left (A), from the top (B), zoomed anterior (C), and posterior (D). White matter–gray matter interfaces: Magenta = posterior part of the thalamus, blue and gray = right and left V1; orange = V2; cyan = V3; red = V5. Fibers: Yellow = homotopic V1; red = V1–V2; white = V2–V3; green = V2–V5; blue = lateral geniculate body–V1. *Image courtesy of Hagmann, P., Kurant, M., Gigandet, X., Thiran, P., Wedeen, V. J., Meuli, R., Thiran, J.-P. (2007). Mapping human whole-brain structural networks with diffusion MRI. PLoS One, 2(7), 1–9. https://doi.org/10.1371/journal.pone.0000597.*

1983; Lacica et al., 1992). Layers K1, K4, and K6 innervate blobs arranged in ocular dominance columns, while layers K2 and K3 innervate blobs in ipsilateral columns (Hendry & Reid, 2000). The segregation of projections from different populations of ganglion cells in different layers of the lateral geniculate body and the projection of these layers to the cortex enables the distinction of different functional streams in visual information processing. The parvocellular (P) stream, named after the corresponding layers in the lateral geniculate body from which it originates, is a slow system monopolized by input from midget ganglion cells. This system carries both chromatic and achromatic information. Chromatic information is provided by the red/green color opponent response of midget ganglion cells and enables red–green discrimination in color vision. The L and M channels work synergistically for the detection of achromatic contrast, distinguishing light from dark in object vision. The magnocellular (M) stream is a fast system that receives its retinal input from parasol ganglion cells. These cells are also involved in detecting achromatic contrast and motion perception. The koniocellular stream is directed by small, blue/yellow opponent bistratified ganglion cells. These cells are solely dedicated to color and do not carry achromatic contrast information.

The koniocellular stream is not exclusively related to color vision. Projections from other types of ganglion cells with widely distributed dendrites prevail in some of the koniocellular layers. Cells sharing the immunocytochemical characteristics of the koniocellular population and their projection to CO blobs in layers 2/3 are present in the parvocellular layers of the lateral geniculate body as displaced koniocellular elements (Hendry, 2015). Gouras (2005) suggested that the chromatic and achromatic information carried by midget ganglion cells is divided at the level of the primary visual cortex. Chromatic information regarding red/green discrimination would be carried by the displaced koniocellular elements in the parvocellular layers. The koniocellular projection to blobs in layers 2/3, therefore, would include information regarding the entire chromatic spectrum.

Contrast-related information would be conveyed through other channels, namely the direct projection to layers 4Cβ and 4A. As a useful simplification, the projection of the parvocellular stream from the lateral geniculate body to V1 and beyond can be considered as representing the shape of the object, while the koniocellular stream represents the color of the object. The information carried by the magnocellular stream is used to generate physiological properties related to orientation specificity, binocular fusion, and motion detection.

According to research by Otazu (2022), chromatic assimilation is driven by the interplay between luminance and chromatic channels, primarily the magno- and koniocellular pathways (Fig. 20.7), and does not depend on a chromatic–chromatic interaction. Additionally, these findings demonstrate that chromatic induction is significantly reduced when all three visual pathways are active at the same time and that chromatic pathways can interact with one another via the magnocellular (luminance) pathway. Additionally, we see that chromatic channels can affect the luminance channel, resulting in a slight induction of brightness.

Color induction is a well-known occurrence. This effect occurs whenever the perceived color of a target is affected by the colors around it. Color assimilation and color contrast are the two categories of color induction that can be separated based on the direction of the color change. The former happens when the target's color diverges from the color of the inducer, the nearest nearby object. The latter, on the other hand, happens when the target's color turns more in the direction of the inducer. For instance, color contrast is produced when a gray target is surrounded by a green inducer and is viewed as reddish (green's opposite hue). Assimilation of color is instead caused if the gray target is seen as having a greenish hue. The spatiochromatic characteristics of the inducers and the target primarily determine the kind of induction (contrast or assimilation). According to several research (Monnier and Shevell, 2004; Shevell and Monnier, 2005; Otazu et al., 2010), uniform surrounds typically induce color contrast, while striped surrounds typically induce color assimilation.

Single- (SO), double- (DO), and nonopponent (NO) neurons are three important types of cells. Low spatial frequencies and chromatic stimuli (color-sensitive cells) are the optimal stimulus types for SO neurons, sometimes referred to as color cells. DO neurons, also known as color-brightness cells, have a band pass and are sensitive to the borders of both color and luminance. Band pass neurons with achromatic sensitivity are NO neurons, also known as luminance cells. It is interesting to note that DO neurons may have a significant impact on how colors seem (Nunez et al., 2018). DO and SO activity may be inhibited by NO and DO cells. In a recent work (Cerda-Company et al., 2018), where color assimilation only manifested when there were luminance differences, the observed luminance–chromatic interaction was explained by this process. The M pathway's impact on the P and K chromatic pathways was only examined independently in the earlier study. This raises a fresh query: does this mechanism exist across the P and K pathways as well? A positive answer would suggest that the P pathway suppresses the K pathway and vice versa when both chromatic pathways are active. The observed color induction does not outweigh the JND, contradicting the hypothesis. As a result, the influence of the chromatic pathways on one another is insufficient to cause color assimilation. In actuality, color assimilation was only seen when the brightness M route

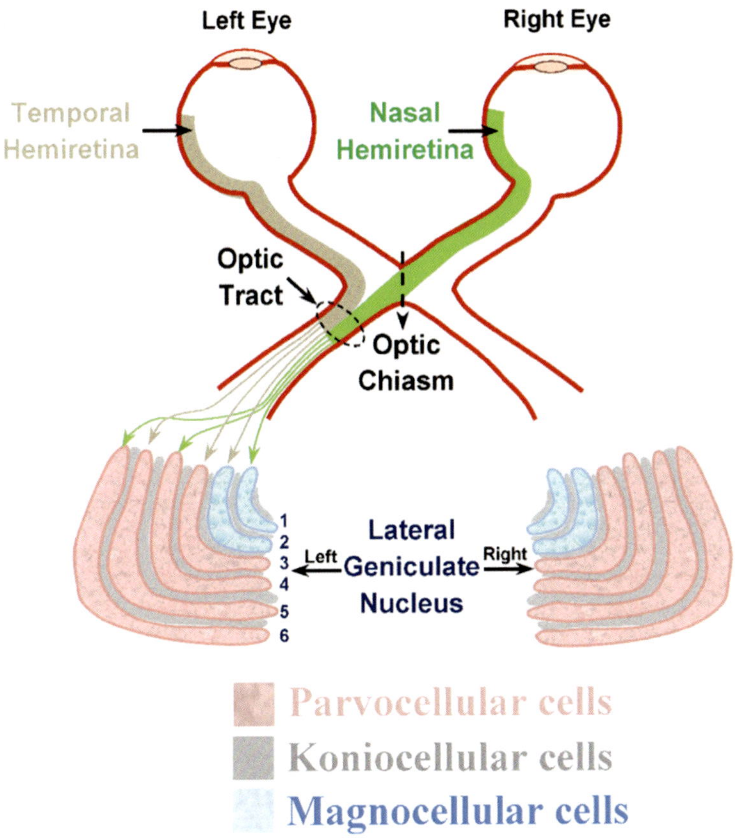

FIGURE 20.7 **Retinal inputs to different layers of LGN in old world monkeys, apes, and humans.** Axons from retinal ganglion cells (RGC) reach the optic chiasm. Axons from the nasal halves of the two retinas cross the optic chiasm and go into the contralateral optic tract and those from the temporal halves go into the ipsilateral optic tract. For example, in the left LGN, retinal afferents from nasal hemiretina project to layers 1, 4, and 6 of the LGN, and those coming from temporal hemiretina go to layers 2, 3, and 5. Under this arrangement, the right visual field is processed by the left LGN, and the left visual field is processed by the right LGN. *Image courtesy of Ghodrati, M., Khaligh-Razavi, S.-M., Lehky, S. R. (2017). Toward building a more complex view of the lateral geniculate nucleus: recent advances in understanding its role. Progr. Neurobiol. 156, 214–255. https://doi.org/10.1016/j.pneurobio.2017.06.002.*

was engaged. The lack of a mutual inhibition-like mechanism between chromatic pathways is one hypothesis that could apply. However, it is conceivable that they have an indirect influence on one another via the luminance pathway, with one chromatic pathway influencing the luminance pathway, which in turn affects the other chromatic pathway. This explanation implies that when the P pathway was activated, it blocked the M pathway. As a result, the M pathway could not as effectively control the K pathway. This suggests that the P pathway has more influence on the M pathway and, in turn, on the K pathway, than vice versa.

We can draw the conclusion that the brightness disparities play a significant role in the color assimilation effect, as do the chromatic and luminance conditions.

The chromatic pathway influences the luminance pathway through the mutual inhibition process when all of the visual pathways are active, and the luminance pathway influences the other chromatic pathway through the same mechanism. Additionally, it appears that the parvocellular pathway has a higher influence on the koniocellular pathway via the magnocellular pathway than the reverse and that color assimilation is stronger along the s than along the l chromatic axis of the MacLeod–Boynton colorspace.

Analysis of steady-state visual evoked potentials (SSVEPs) generated by chromatic stimuli, which typically involve extended stimulation periods, is a promising strategy for examining adaptation effects in the visual cortex.

Zhang et al. (2023) discovered that chromatic SSVEP responses declined with increasing stimulation duration and reached a lower asymptote within a minute of stimulation by evaluating SSVEPs with small time windows. The brightness SSVEPs failed to exhibit any kind of organized adaptation. An exponential decay function with a half-life of roughly 20 s can accurately characterize the time history of chromatic SSVEPs and is extremely similar to earlier psychophysical studies. Despite the fact that the current study's stimuli were different from those used in earlier research, the consistent time course may point to a more widespread adaptation mechanism in the early visual cortex.

FUNCTIONAL CONNECTIONS OF THE VISUAL AREAS

The magnocellular pathway can be traced from the 4Cα sublayer to layer 4B (Fig. 20.8). The blobs in layer 3 receive magnocellular inputs directly from 4Cα and

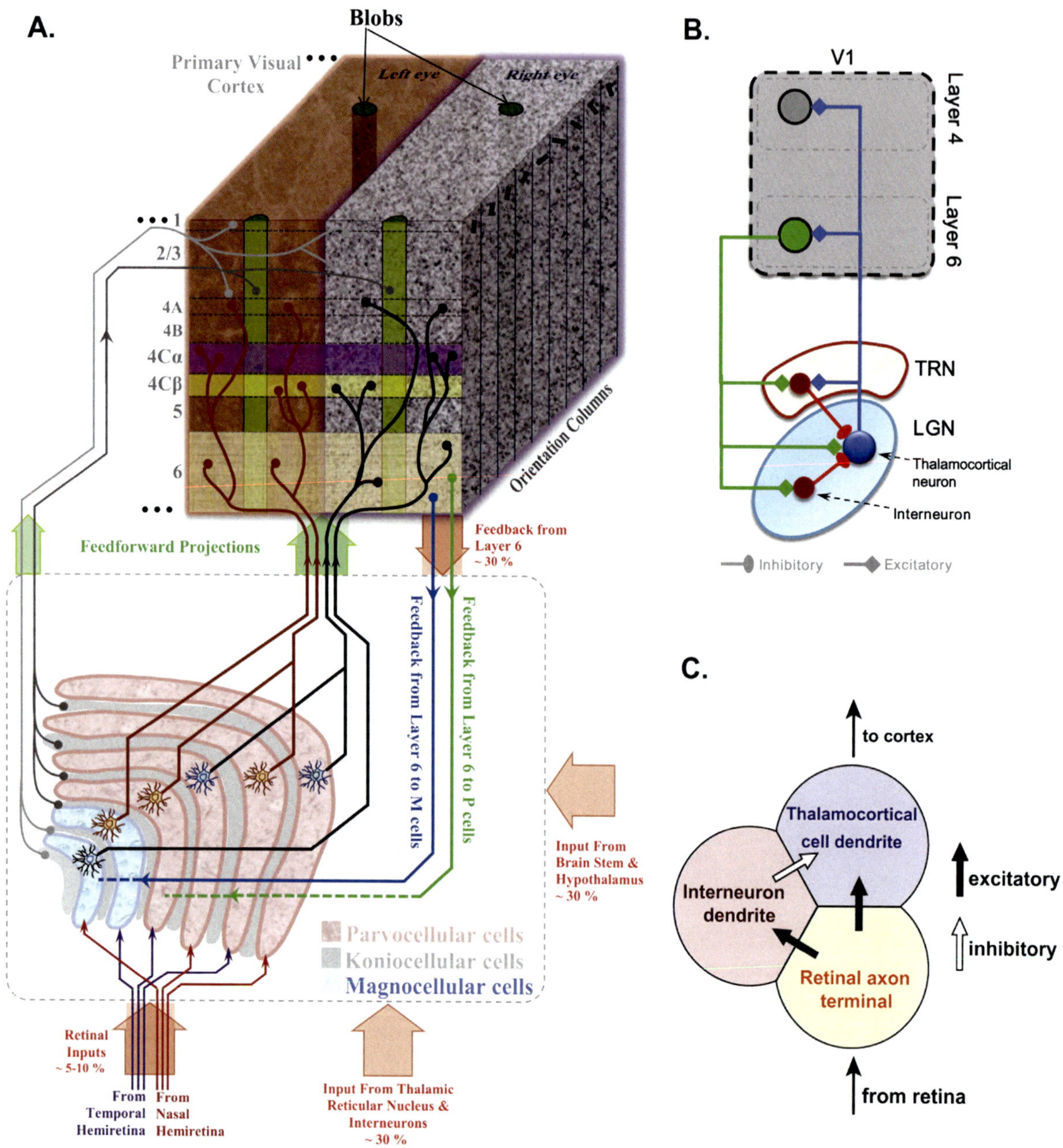

FIGURE 20.8 **LGN feedback and feedforward connections.** (A) Thalamocortical cells in LGN project to layer 4 of primary visual cortex (V1), sending collaterals to layer 6 along the way. Cells are color-coded by eye of origin of visual input. Signals are projected from parvocellular layers of the LGN onto layer 4Cb of V1 and magnocellular layers send their projections to layer 4Ca. Koniocellular layers of the LGN project to layers 1–3. Cortical feedback to LGN is provided by neurons in layer 6 of V1. The layer 6 feedback targets LGN thalamocortical cells and also contact inhibitory interneurons (not shown in the figure). In addition to receiving input from retina and cortical feedback from V1, LGN receives feedback from the thalamic reticular nucleus (TRN) as well as multiple neuromodulatory inputs from the brainstem. There are other afferents to the LGN not shown in the figure, such as neurons from the nucleus of the optic tract in the midbrain and from the tuberomamillary nucleus in the hypothalamus. (B) The LGN-TRN-V1 circuitry. LGN thalamocortical neurons en route to V1 send excitatory collaterals to TRN. Feedback from layer 6 of V1 en route to LGN also sends excitatory collaterals to TRN. The output of TRN is an inhibitory projection back to LGN thalamocortical cells. (C) Synaptic triad organization in the LGN. Within a glomerulus, an incoming axon terminal from the retina makes excitatory synaptic contacts both with a dendrite from an LGN thalamocortical cell (predominantly X cells) and a dendrite from an LGN interneuron. The interneuron dendrite makes an inhibitory synaptic contact with the thalamocortical cell dendrite, forming the third synapse of the triad. The interneuron dendrite is thus presynaptic to the thalamocortical cell dendrite, and under some circumstances, a signal may be transmitted from dendrite to dendrite without passing through the soma of the interneuron. Alternatively, rather than being excited locally by the retinothalamic synapse within the triad, the presynaptic interneuron dendrite may also be activated by backpropagation of spike activity from the axonal output of the interneuron. *Image courtesy of Ghodrati, M., Khaligh-Razavi, S.-M., Lehky, S. R. (2017). Toward building a more complex view of the lateral geniculate nucleus: Recent advances in understanding its role. Progr. Neurobiol. 156, 214–255. https://doi.org/10.1016/j.pneurobio.2017.06.002.*

indirectly through 4B. Magnocellular afferents reach the interblobs of layer 3 from starburst cells in the bottom of 4Cα and from layer 6. Parvocellular afferents are distributed from the 4Cα sublayer to the blobs and interblobs of layer 3 and the interblobs of layer 4A, which also receive a direct projection from the parvocellular layers of the lateral geniculate body. The interarea projection of layer 4B pyramidal cells to V2, V6, and the MT area can be considered as an almost pure extension of the magnocellular pathway (Sincich & Horton, 2003). The magnocellular and parvocellular streams mix at the level of the interblobs in layer 3, and all M, P, and K streams innervate the blobs in this layer. These intrinsic projections to blobs and interblobs terminate deep in layer 3 (sometimes referred to as sublayer 3b). Interarea connections of blobs and interblobs with the V2 area originate from a population of pyramidal cells in the more superficially located layers 2 and 3B. It has been reported that blobs project to thin CO-immunoreactive stripes and interstripes in V2, while interblobs are connected to thick CO-immunoreactive stripes and interstripes in the visual area (Sincich & Horton, 2002, 2005; Xiao et al., 2004). The thick stripes also receive afferents from layer 4B. The thick stripes of V2 project to the MT visual area, which also receives an M afferent from layer 4B of V1. Furthermore, Meynert's pyramidal cells at the border of layers 5/6 project to MT, as well as to the superior colliculus. MT is a key area in the occipitoparietal dorsal processing stream. Thin stripes and interstripes are connected to visual areas belonging to the ventral occipitotemporal processing stream, which will be discussed in more detail in subsequent paragraphs. The thick stripe pathway 4Cα-4B-V2-MT is clearly dominated by magnocellular inputs, with some mixing of the parvocellular stream at the interblob level.

Cells in 4Cα are orientation-specific elements. Cells in layer 4B are specific for both orientation and direction, and direction specificity is preserved in MT and the subsequent stages of dorsal stream processing. The converging pathways onto the ventral occipitotemporal processing stream, which includes V1 blobs and V2 thin stripes, rely on inputs from all layers of the lateral geniculate body and are therefore more difficult to characterize. Studies using the uptake of deoxyglucose by sets of active neurons may have solved this problem. In the original studies by Livingstone and Hubel (1984) and Livingstone and Hubel (1988), it was found that blobs contain higher-order color opponent cells for red/green or blue/yellow. Cells in the interblobs were orientation-selective and organized in long linear sequences (orientation columns), which were separated from the blobs. This preferential location of color and orientation-selective cells has been challenged (Ringach et al., 2002; Xu et al., 2005). However, it has been confirmed by studies using deoxyglucose in the macaque visual cortex for orientation columns and color analysis (Vanduffel, 2002; Tootell et al., 2004). In the primary visual area, a repeated system of orientation columns was first described, composed of neurons that respond to a stimulus presented in the visual field produced by similarly oriented bars. These orientation columns intersect with the ocular dominance columns at right angles. Marked cells in studies based on deoxyglucose uptake, following the presentation of two orthogonal stimuli, are arranged in two groups of orientation columns that are maximally separated. In V1, these columns were well distinguished in the 4Cα sublayer, absent in 4Cβ, and prominent in the more superficial layers where they were absent in the blobs. In V2, they appeared centered on thick CO-immunoreactive stripes, extending to the interstripes but absent in the thin stripes (Vanduffel, 2002). On the other hand, deoxyglucose uptake following the presentation of colored stimuli was stimulated in the blobs and concentrated in CO-immunoreactive thin stripes in V2 (Tootell et al., 2004). The preferential localization of orientation columns in the interblobs and thick stripes of V2 is only true for the peripheral visual field. In the representations of the fovea in V1 and V2, orientation columns extend into the blobs, and the preference for thick stripes in V2 decreases (Vanduffel, 2002). The processing of color information, which relies on afferents from the parvocellular and koniocellular layers of the lateral geniculate body, appears to be segregated in the V1 blobs and thin stripes of V2 and continues in the ventral processing stream. Similarly, orientation selectivity, which depends on magnocellular afference, is segregated in the V1 interblobs and thick stripes of V2 in the peripheral visual field and is transmitted to the dorsal stream, which relies on the peripheral rather than central regions of the visual. The information used in object shape perception by the ventral processing stream may depend on afferents from all layers of the lateral geniculate body. This information can be transmitted by the blobs in the central representation of the visual field, where orientation-selective cells are present.

DORSAL AND VENTRAL PROCESSING STREAMS (FIGS. 20.9 AND 20.10)

The dorsal occipitoparietal stream carries information regarding spatial relationships to the parietal lobe. The ventral occipitotemporal stream conveys information about object properties to the inferotemporal cortex (Fig. 20.9 shows a simplified schematization of the classic visual path and a modified model recently proposed by Choi, S.-H 2020. Fig. 20.10, in turn, shows by tractography (DTI) what is schematically explained in Fig. 20.9. However, in case the reader finds the captions

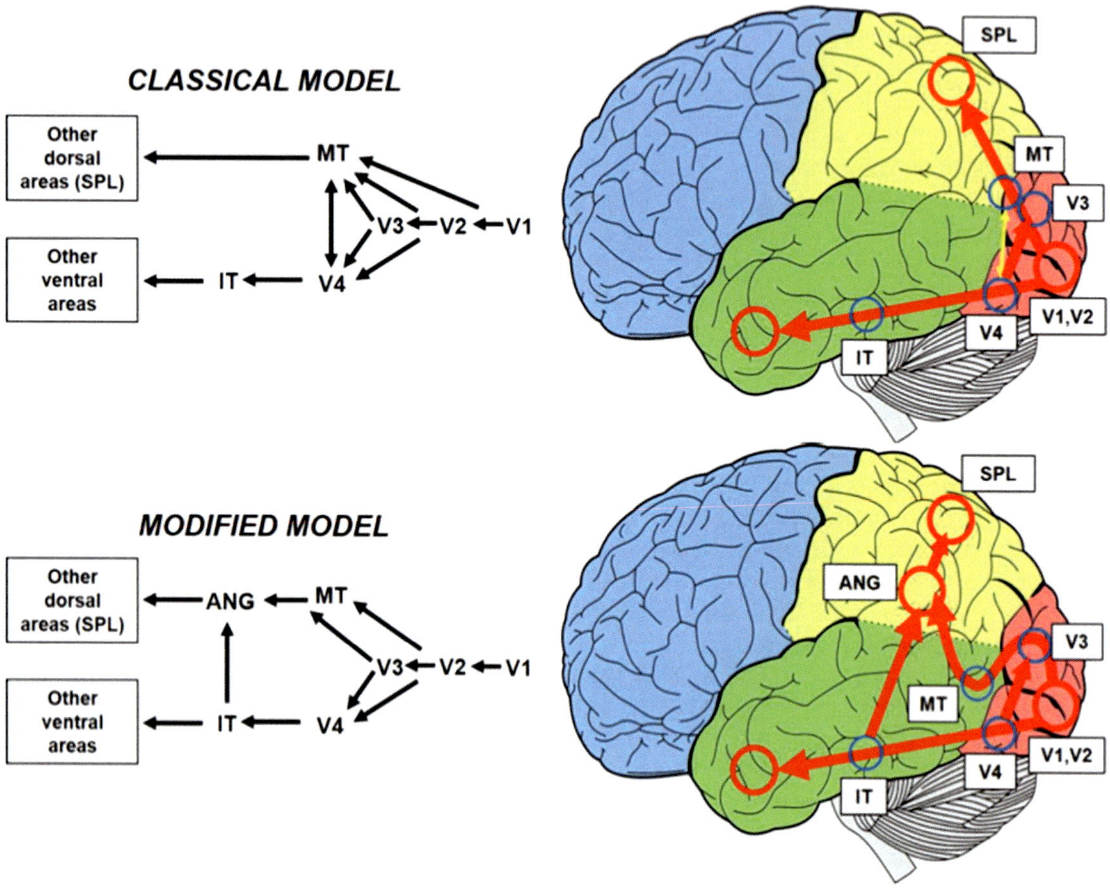

FIGURE 20.9 **The classical visual pathway and Choy, S.-H modified visual pathway.** The upper and lower images represent the classical model and the Choi, S.-H proposed model, respectively. The left and right images are simplified schematic diagram and putative visual pathway on the conceptual brain image, respectively. *ANG*, angular gyrus; *IT*, inferior temporal area; *MT*, middle temporal visual area; *SPL*, superior parietal lobule. The passage from the main visual region to the SPL in the classical model travels via the dorsal stream, but in the modified model, it travels almost in the other direction through the ANG. Additionally, the V3 or MT and the IT are both used to link the ANG fibers. In the modified visual pathway, the IT-ANG provided a further interstream link between the dorsal and ventral streams. The outcome is that the visual system has two interstream connections: ITANG, or the aVOF, in the high-tier visual region, and V4V3MT in the low-tier visual area. The aVOF may be a visual processing fiber bundle like the VOF. *Image courtesy of Choi, S.-H., Jeong, G., Kim, Y.-B., Cho, Z.-H. (2020). Proposal for human visual pathway in the extrastriate cortex by fiber tracking method using diffusion-weighted MRI. NeuroImage, 220, 117145. https://doi.org/10.1016/j.neuroimage.2020.117145.*

too technical and not easy to understand, we recommend supplementing the reading by consulting the article Choi et al. (2020) from which the figures were taken). Generally, the size and latency of receptive fields increase in successive stages of a stream, and eventually, a retinotopic representation is achieved, regardless of the position in the visual field. Areas V4, V8, and the inferotemporal area TE are included in the ventral stream. Areas V3a (accessory V3), V6, V6a, MT, MST (medial superior temporal area), STP (polysensory superior temporal area), and a series of visual areas in the intraparietal sulcus and inferior parietal lobule belong to the dorsal stream. This list is not exhaustive, as other areas not mentioned in this chapter have been identified within each stream. The dorsal and ventral streams also differ based on their preferences for peripheral or central vision and their involvement, respectively, in scotopic or photopic vision. In monkeys, retrograde tracer injections into the parietal projection areas of the dorsal stream result in retrograde labeling of cells in V6, MT, V3A, MST, and STP, primarily in the peripheral field representations of V2, V3, and V4. Injections into the ventral stream in the inferior temporal areas label cells predominantly in the foveal representations of these areas, but primarily in V4 (Baizer et al., 1991). Scotopic vision, which is dependent on rod activity, activates visual areas in the dorsal stream but does not affect the foveal representations of V1, V2, V3, and the color-sensitive area V8. As noted earlier, the afferents from V1 and V2 to MT, V3A, and V6, key areas of the dorsal stream, are dominated by the magnocellular layers of the lateral geniculate body. MT, V3A, and V6 receive their afferents from layer 4B of V1 through heavily myelinated fibers of the stria of Gennari.

The retinotopic localization in MT is not well developed. Its cells have small receptive fields and are selective for direction and orientation. Illusory movements also activate its cells. Cells with a similar preference for motion direction are segregated into columns, which alternate with columns showing a preference for the opposite direction. Estimates of target speed, used in slow pursuit movements of objects, are made by mixed populations of numerous neurons (Priebe et al., 2004). MT is connected to visuomotor areas in the lateral and medial walls of the intraparietal sulcus (Galletti et al., 2001; Rozzi et al., 2006). Area V3A is involved in distinguishing real motion from self-motion. Selective cells for motion and direction are also present in area V3A. In addiction there are connections to the dorsal visual stream and the visuomotor areas, located in the dorsal wall of the intraparietal sulcus.

Although V6 is known to play a role in visual motion, its involvement in navigation and how sensory input affects its functional characteristics are unknown. Aggius-Vella et al. (2023) investigated the role of V6 in egocentric navigation in participants who were either sighted or congenitally blind (CB) and used the EyeCane as a distance-to-sound sensory substitution device (SSD). On two different datasets, they ran two fMRI experiments. In the initial test, both CB and sighted individuals had to find their way through mazes. The mazes were conducted visually by the sighted and auditorily by the CB. Using the EyeCane SSD, the CB ran the mazes both before and after a training session. A group of sighted individuals in the second experiment completed a motor topography test.

Results indicate that, regardless of the chosen sensory modality, the right V6 (rhV6) is selectively involved in egocentric navigation. In fact, just as rhV6 in the sighted, rhV6 in CB is selectively recruited for auditory navigation following training. Additionally, the study discovered body movement activity in region V6, which would suggest that it plays a role in egocentric navigation. All things considered, the research points to region rhV6 as a special hub that converts spatially pertinent sensory data into an egocentric picture for navigation. Although vision is undoubtedly the main modality, rhV6 is a supramodal area that can establish its selectivity for navigation even when it is not exposed to visual information.

MT is connected to the more rostrally located MST and area V6. MST lacks a retinotopic organization, and the receptive fields of its cells are large. Its cells are sensitive to differences, or to the increase or decrease of an image's size. It is believed that this area is involved in depth perception, optic flow, and, like MT, has been implicated in slow pursuit movements (Huk et al., 2002; Luppino et al., 2001). Cells with preferences for image's size reduction or enlargement are grouped into alternating columns. MST is connected to the polysensory superior temporal area (STP) in the dorsal wall of the superior temporal sulcus. This area responds to visual, auditory, and somatosensory stimuli and interacts with the inferotemporal area TE, the final stage in the ventral visual stream, and the inferior parietal lobule, one of the projection sites of the dorsal stream (Rizzolatti et al., 2003). Different visuomotor areas innervated by the dorsal visual stream have been found in the parietooccipital sulcus in monkeys, and tentatively in humans as well (Culham et al., 2001) (Fig. 20.10).

These areas function as a visuomotor interface. They have strong connections with the superior colliculus and pontine nuclei and are connected through various reciprocal parallel pathways with premotor and ocular fields in the frontal lobe (Lynch et al., 2006). Cells in the lateral intraparietal area (LIP or frontal eye field) are modulated by eye position. LIP, located in the lateral wall of the intraparietal sulcus, encodes the positions occupied by the fovea and guides eye movements. The anterior intraparietal area (AIP) plays an essential role in sensorimotor transformation for grasping. The medial intraparietal area (MIP), located in the medial wall, encodes arm-related spatial information and is involved in reaching for a target. Area V6A in the caudal intraparietal sulcus and the ventral intraparietal area (VIP), located in the bottom of the intraparietal sulcus, are head-centered areas that encode head position-related stimuli and are involved in mouth and head movements. In the ventral visual processing stream, the nomenclature and connections are also complex. Area V4A has often been considered the gateway to the ventral stream. Originally, it was divided into a dorsal part (V4d) and a ventral part (V4v), with representations of the ventral and dorsal quadrants of the contralateral hemifield, respectively. A high percentage of its cells are color-selective (Zeki et al., 1991). However, in subsequent studies using fMRI, the dorsal V4 was not recognized (Sereno et al., 1995). Instead, in the same lateral occipital region (LO), between V3 and MT, an area was identified that was activated to a degree comparable with that of object scanning, faces, common objects, and abstract 3-D objects. The retinotopy in LO was coarse or absent, and color selectivity was absent (Tootell et al., 2001). A color-selective area called V8, with a complete representation of the contralateral hemifield, was located rostrally to V4v, in the collateral sulcus (Hadjikhani et al., 1998). Higher-order object-selective areas, selective for specific objects such as faces or buildings, have been described laterally in the collateral sulcus in the fusiform gyrus (lateral occipitotemporal) or in more rostral and medial inferior temporal regions.

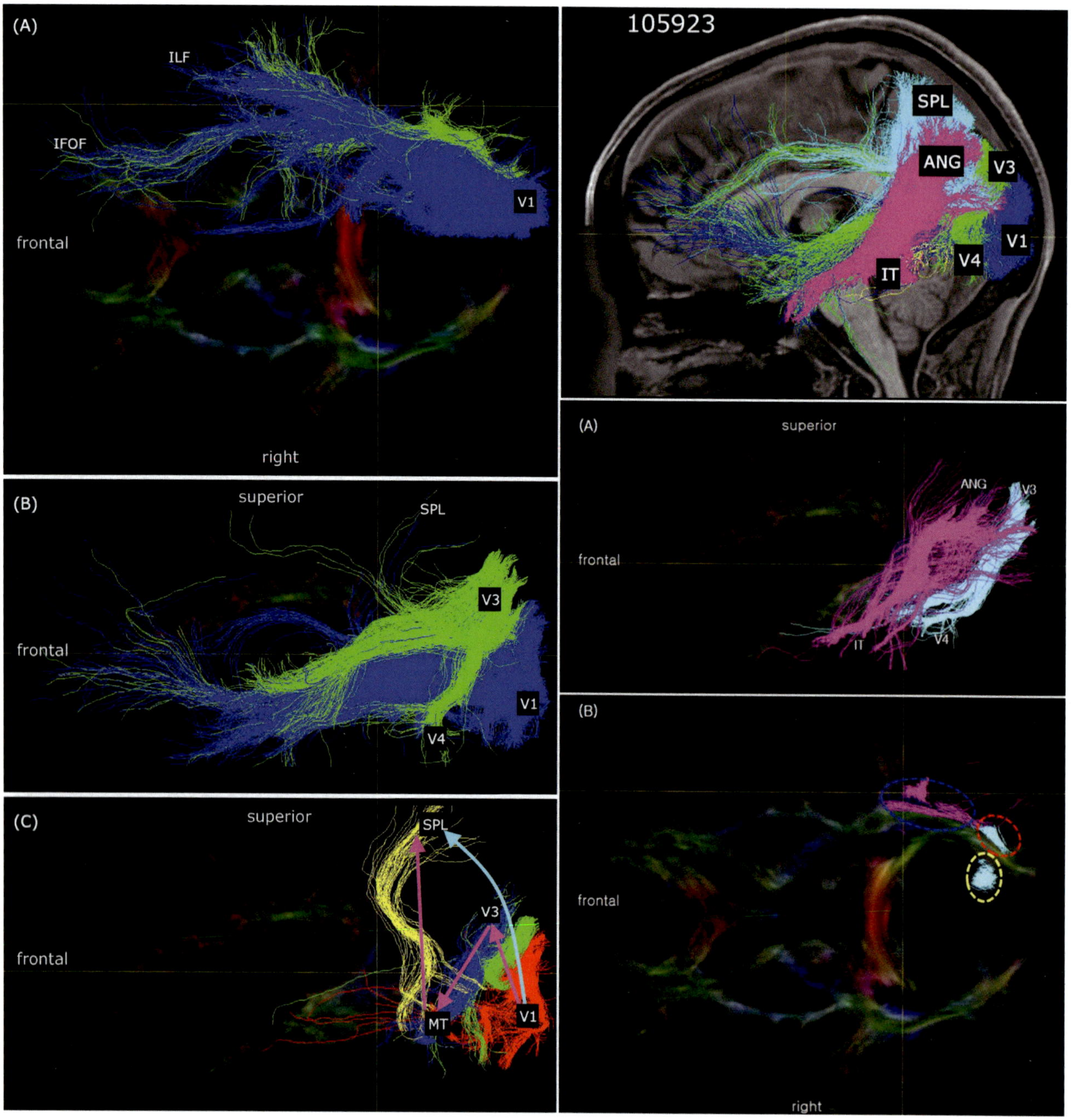

FIGURE 20.10 **(Left) Fiber tracking results of the classical visual pathway of a single subject from left hemisphere on the TDI image.** Axial view (A) and sagittal view (B) image of the fiber tracking result from seed position V1 (blue) and V3 (green) are shown. The generated fiber number from V1 and V3 in Figs. (A) and (B) are 10,000 and 1000, respectively. (C) Sagittal view image of the fiber tracking result according to the classical dorsal pathway from maximum 100,000 generated fibers in each seed position. The colors of each fiber bundle in Figs. (C) are as follows: V1→V2 (red), V2→V3 (green), V3→MT (blue), MT→SPL (yellow), and their conceptual trajectory is represented as a magenta-colored arrow in this figure. The cyan-colored arrow represents the putative pathway of the conventional dorsal stream. ILF, inferior longitudinal fasciculus; IFOF, inferior fronto-occipital fasciculus; MT, middle temporal visual area; SPL, superior parietal lobule. **(Top-right). The fiber tracking results of the main visual pathways in this study from another 9-subject HCP dataset in left hemisphere on the MRI-T1 image.** V1 fiber (blue), V3 fiber (green), VOF (yellow), aVOF (magenta), ANG→SPL (cyan). The center of the crossed yellow lines in each image represents the location of the anterior commissure. ANG, angular gyrus; aVOF, anterior vertical occipital fasciculus; IT, inferior temporal area; SPL, superior parietal lobule; VOF, vertical occipital fasciculus. **(Bottom-right). Fiber tracking results of the aVOF (magenta) and VOF (cyan) of a single subject (on-site scanned data) from maximum 3,000,000 and 300,000 generated fiber numbers, respectively**. They displayed on the sagittal (A) and axial (B) views on TDI image with the labeled colors. The dotted circles in Fig. (B) indicate the aVOF (blue), perpendicular occipital fasciculus (yellow), VOF (red). ANG, angular gyrus; aVOF, anterior vertical occipital fasciculus; IT, inferior temporal area; VOF, vertical occipital fasciculus. *Image courtesy of Choi, S.-H., Jeong, G., Kim, Y.-B., Cho, Z.-H., 2020. Proposal for human visual pathway in the extrastriate cortex by fiber tracking method using diffusion-weighted MRI. NeuroImage, 220, 117145. https://doi.org/10.1016/j.neuroimage.2020.117145 (rearranged).*

The fusiform face area in the posterior fusiform gyrus is better known, activated during face comparison or recognition and the site of lesions causing prosopagnosia, the inability of patients to recognize faces.

It has been emphasized that face recognition, like number and word recognition, is associated with the central visual field, while the recognition of buildings, in a more medial region bordering the parahippocampal gyrus, depends on peripheral vision (Malach et al., 2002). Attributes such as colors and shape are seemingly carried by different divisions of the ventral stream to be used for semantic purposes in higher-order areas, which may rely on information from the central or peripheral visual field. The representation of objects in the ventral temporal cortex is not restricted to small selective areas but should rather be considered as a distributed system that extends from LO to the rostral part of TE (Ishai et al., 2000a,b; Tsao et al., 2003). Functions in this region may be lateralized: responses to object perception are greater on the right side, while responses related to imagining the same objects are greater on the left side. Reciprocal connections between the rostral inferior temporal areas and the inferior orbital and medial prefrontal cortex are a key feature of the ventral visual stream (Saleem et al., 2000). The original claim by Ungerleider and Mishkin (1982) that the dorsal and ventral streams differ in the content of the information they carry (spatial information from the dorsal stream, object properties from the ventral stream) has been criticized by Goodale and Milner (1992), and Rizzolatti and Matelli (2003). According to these authors, the main feature that distinguishes the two streams is the utilization of the information they carry, rather than its content. Parietal areas use visual information for action control, while the inferotemporal cortex uses it for object perception. This difference is well illustrated by one of Goodale's patients, who had a lesion that disrupted the ventral stream in the lateral occipital area (James et al., 2003). This patient was unable to recognize drawings of presented objects but retained her visuomotor abilities in grasping similar objects, with concomitant activation of regions predicted in the dorsal stream in an fMRI study. The dorsal stream was further divided by Rizzolatti and Matelli into a dorsal component, consisting of areas V6, V6A, and the visuomotor area MIP centered on the superior parietal lobule, and a ventral component consisting of areas MT, LIP, AIP, VIP, and other visuomotor areas of the inferior parietal lobule. The dorsal component provides spatial information to the superior parietal lobule used for direct action, information that is utilized in visuomotor transformations by the AIP area for object grasping. The ventral component also receives object-related information from the ventral stream through its connections with the STP, used by the lower parietal area PF to imitate an action from the premotor cortex. The localization of parietal syndromes such as optic ataxia, a disorder of visually guided arm movements toward a point, in the superior parietal lobule, and ideomotor apraxia, the inability to replicate a gesture following a verbal command, in the inferior parietal lobule, is consistent with this dorsoventral subdivision. The projection from the lateral geniculate body to the striate cortex appears essential for conscious vision. However, there are pathways to the higher visual areas that bypass both the striate cortex and the lateral geniculate body and are thought to function in blindsight, which is the unconscious perception of motion, and in conscious perception of motion in patients with damage to the striate area.

These pathways include:

1. Projections from the superficial layers of the superior colliculus, which receive a substantial input from the optic tract, to the pulvinar. The pulvinar innervates the extrastriate visual areas in the occipital, temporal, and parietal lobes.
2. Projections from the superficial layers of the superior colliculus to the MT area, through the koniocellular layers of the lateral geniculate body (Sincich et al., 2004).

VISUOMOTOR PATHWAYS

The visuomotor pathways consist of pathways for reflexes related to accommodation, convergence, and eye stabilization, as well as pathways involved in voluntary saccades and slow pursuit movements of moving objects. There are two main systems that guide conjugate eye movements. One is represented by the three-neuron arc of the vestibuloocular reflex (VOR), which connects each semicircular canal, through the vestibular nuclei, to pairs of muscles that move the eye in the plane of that semicircular canal. This system has developed in lower vertebrates to stabilize the position of the eye in space and has been preserved, with little modification, in the mammalian brain. The pathways used by the VOR are shared by the optokinetic reflex, another reflex that stabilizes the eye, and likely by slow pursuit movements as well, although the latter involve areas of the vestibular nuclei different from those involved in the elementary VOR. The VOR and optokinetic reflex receive their retinal error information from the accessory optic system and the optic tract nucleus. Slow pursuit involves the processing of information regarding the direction and velocity of an object to be tracked by cortical visual areas and the concomitant suppression of the VOR. The second system generates saccades and is controlled by the superior colliculus, various cortical visual areas, and the cerebellum.

SACCADES SYSTEM

Saccades are rapid eye movements that direct the line of sight to an area of interest, allowing the target to fall onto the tiny fovea. The generator that activates the brainstem for saccadic eye movements has been studied in detail by Scudder et al. (2002) and Buttner and Buttner-Ennever (1988). The motor neurons that innervate the extrinsic eye muscles exhibit either burst or tonic activity. Burst activity moves the eye, while tonic activity holds it in position. Burst activity is generated by cells with brief activity that establish excitatory connections with the motor neurons. For horizontal eye movements, these cells are located in the medial part of the caudal reticular formation of the pons, in the so-called paramedian pontine reticular formation (PPRF), rostrally to the abducens nucleus. Excitatory burst cells (EBCs) are glutamatergic. Inhibitory burst cells (IBCs) are located in the medial part of the caudal paragigantocellular reticular nucleus of the medulla, caudally to the abducens nucleus. IBCs utilize glycine as their neurotransmitter. EBCs project to the ipsilateral abducens nucleus, while the projection of IBCs is crossed. The projection of IBCs may also involve nuclei on both sides of the oculomotor and trochlear nerves. The abducens nerve innervates the lateral rectus muscle, and the oculomotor nerve innervates the medial rectus muscle. Their motor neurons receive excitatory input through an internuclear pathway originating from the interneurons of the abducens nucleus. Short-latency neurons for the motor neurons of the oculomotor and trochlear nerves reside in the rostral interstitial nucleus of the medial longitudinal fasciculus (RIMLF). The RIMLF is a wing-shaped nucleus located dorsally to the rostral pole of the red nucleus and rostrally to the interstitial nucleus of Cajal. Neurons in the RIMLF can be identified by their high content of parvalbumin. They share this property with short-latency neurons, both excitatory and inhibitory, found in the PPRF. The projection of the RIMLF to the motor neurons of the oculomotor and trochlear nuclei is primarily ipsilateral. Inhibitory neurons projecting to the motor groups of the inferior rectus and inferior oblique muscles have been located in the interstitial nucleus of Cajal (Horn et al., 2003). The RIMLF projects to the ipsilateral interstitial nucleus. Clinical and pathological evidence indicates that Parinaud's syndrome (vertical gaze palsy) is caused by damage to the RIMLF or its efferent fibers (Pierrot-Deseilligny et al., 1982). The tonic discharge that maintains eye position after the completion of a saccade is generated by two neural integrators (a position signal derived from integrating the velocity signal): the nucleus prepositus of the abducens for horizontal eye movements, and the interstitial nucleus of Cajal for vertical and rotational eye movements. The projections from the nucleus prepositus to the abducens nucleus are bilateral; the ipsilateral projection is excitatory, while the contralateral projection is crossed and also utilizes glycine as a neurotransmitter. Both the nucleus prepositus of the abducens and the interstitial nucleus of Cajal rely on reciprocal connections with the vestibular nuclei for their function, and the nucleus prepositus also connects with the cerebellum. The large fibers originating from the interstitial nucleus descend as the interstitial spinal tract in the medial longitudinal fasciculus. Omnipause neurons, located in the pontine raphe interpositus nucleus, are glycinergic elements that become active when the eye is fixating and pause before and during a saccade, inhibiting EBC neurons in the PPRF and riMLF (Horn et al., 1994). Afferents to the omnipause neurons include the superior colliculus and the frontal and supplementary eye fields.

The system that generates saccades (Fig. 20.11) is used by the superior colliculus and the frontal eye fields to generate goal-directed saccades. The projections from the contralateral fastigial nucleus determine the size of the saccade, defining the acceleration, deceleration, and long-term adaptation of saccades. The superior colliculus plays a major role in orienting behavior, directing the gaze toward objects of interest through eye movements of the head and body. A retinotopic map is present in the superficial layers; auditory, somatosensory, and visual signals that give rise to orienting movements converge onto deeper layers onto premotor neurons that are connected to saccade and head movement circuits. The anatomy of the superior colliculus has recently been reviewed by May (2006). The superior colliculus contains alternating gray and white layers. From the surface to the depths, these layers are:

1. The fibrous zonal layer
2. The superficial gray layer
3. The optic layer
4. The middle gray layer
5. The middle white layer or lemniscal stratum
6. The deep gray layer
7. The deep white layer

Based on the connections and functions, the superior colliculus can be divided into a superficial zone and a deep zone. The superficial zone, consisting of layers 1–3, primarily receives visual afferents and gives rise to ascending fibers. The deep zone, comprising layers 4–7, receives heterogenous multimodal afferents and projects both ascending and descending fibers to brainstem centers, thalamus, and spinal cord. The dendritic trees of neurons in the superficial layers are typically confined to these layers. The large multipolar neurons in the deep layers extend their dendrites into the superficial layers. Connections between the superficial and

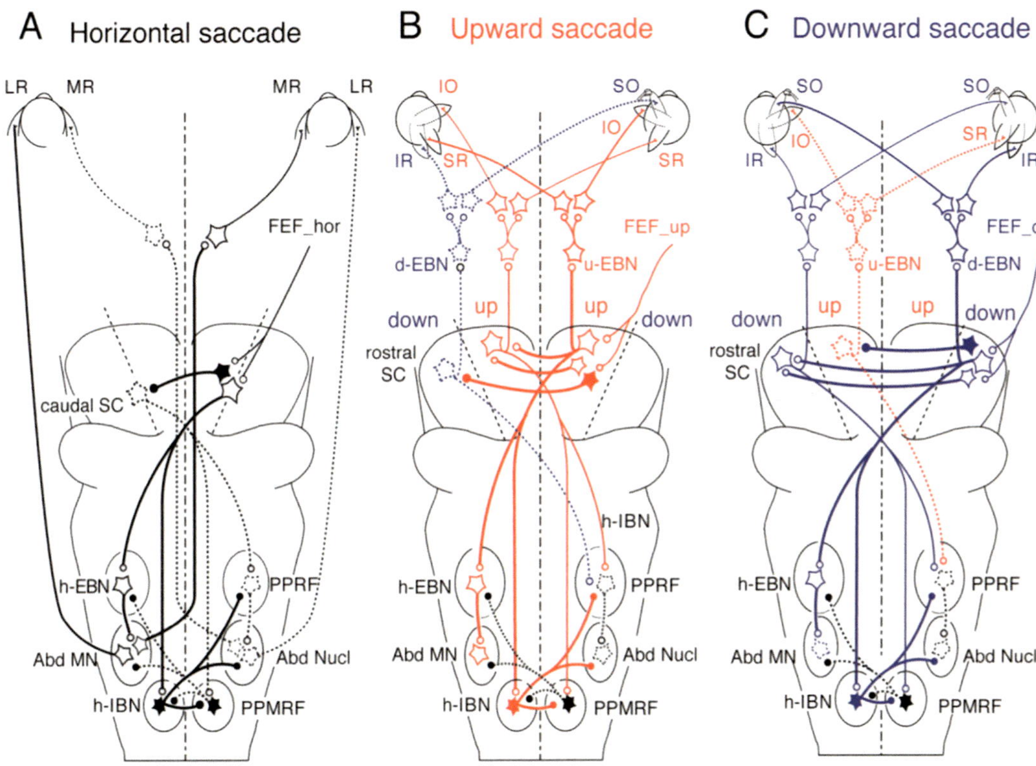

FIGURE 20.11 **Summary diagrams of commissural excitation and inhibition of tectoreticular neurons in the SCs for horizontal (A), upward (B) and downward saccades (C).** (A) horizontal saccades. TRNs in the SC on one side receive inhibition from the SC on the opposite side. Only inhibition from the opposite SC is observed especially in TRNs in the caudal SC. (B) Upward oblique or pure upward saccades. TRNs in the rostromedial SC receive excitation from the rostromedial part of the contralateral SC, and TRNs in the rostrolateral SC receive inhibition from the rostromedial part of the contralateral SC. (C) Downward oblique or pure downward saccades. TRNs in the rostrolateral SC receive excitation from the rostrolateral part of the contralateral SC, and TRNs in the rostromedial SC receive inhibition from the rostrolateral part of the contralateral SC. Note that many TRNs in the rostral part of the SC send their axons to both premotor neurons for vertical and horizontal ocular MNs. Open and filled neurons indicate excitatory and inhibitory neurons, respectively. Solid and broken lines indicate excited and suppressed pathways, respectively. *From Takahashi, M., Sugiuchi, Y., Shinoda, Y., 2007. Commissural mirror-symmetric excitation and reciprocal inhibition between the two superior colliculi and their roles in vertical and horizontal eye movements. J. Neurophysiol. 98 (5), 2664–2682. https://doi.org/10.1152/jn.00696.200711. Image courtesy of Takahashi, M., Shinoda, Y. (2018). Brain stem neural circuits of horizontal and vertical saccade systems and their frame of reference. Neuroscience, 392, 281–328. https://doi.org/10.1016/j.neuroscience.2018.08.027.*

deep layers are formed by collateral axons of small neurons in the superficial layers. The terminations of retinotectal fibers in the superficial layers and the various afferent systems terminating in the intermediate layers of the colliculus show a characteristic mediolateral periodicity. Cholinergic markers (choline acetyltransferase, CHAT, and acetylcholinesterase, AChE) are uniformly distributed in the neuropil of the superficial layers. Some of this activity is of intrinsic origin, but it is mainly due to the bilateral cholinergic projection from the parabigeminal nucleus to these layers. In the intermediate gray layer, CHAT and AChE are distributed in patches in relation to the cholinergic afferents originating from the pedunculopontine nucleus (Graybiel et al., 1994). This patchy distribution is a similar pattern for several substances in this layer, such as enkephalin, and for the discontinuous distribution of most afferent systems and the origin of certain efferent systems in the intermediate layers of the colliculus. This periodicity has been observed in most mammals, including humans (Wallack et al., 1988). Afferent connections to the superficial layers of the superior colliculus originate from both retinas and the ipsilateral cerebral cortex. Retinal fibers pass through the optic tract via the brachium of the superior colliculus and the superficial optic layer to terminate in the superficial gray and zonal layers. The retinocollicular projection is bilateral with contralateral dominance. Retinal fibers distribute according to a retinotopic pattern. The map of the contralateral hemifield is somewhat obliquely oriented. The vertical meridian coincides with the rostral border of the superior colliculus, and the horizontal meridian divides the colliculus into lateral and medial halves.

The fovea is represented rostrolaterally, the peripheral regions caudally, the upper visual field medially, and the lower field laterally. The monocular field is represented only as a contralateral projection to the caudal colliculus. Projections to the foveal representation are relatively few. The fovea expands and contains gaps in the contralateral projection, which are filled by islands of ipsilateral projections. In the more posterior–medial representation of the periphery, the afferents from the contralateral eye become more continuous, but the gaps filled by ipsilateral projections remain present in the deeper portions of the superficial gray (Tiggers et al., 1981). This distribution resembles the columns of ocular dominance in the primary visual cortex, but it is less regular. Several types of ganglion cells, accounting for less than 10% of the total population, contribute to the retinotectal projection. Only a small minority of multipolar cells share projections with the lateral geniculate body; midget cells do not project to the superior colliculus (Leventhal et al., 1981; Perry et al., 1984). Corticotectal fibers originate from pyramidal cells in layer V, except for the primary visual area where they also originate from Meynert's pyramidal cells at the border of layers V/VI. Areas 17 (V1) and 18 (V2) project to the zonal layer and superficial gray, constituting the main corticotectal afferent input. The corticotectal V2 projection originates from band-like groups composed of pyramidal cells in layer V, which in the superficial layers are aligned with thick cytochrome oxidase-positive stripes selective for direction, meaning that this projection originates from a specific direction and orientation region in V2 (Abel et al., 1997). Retinotopy is maintained in the projections from V1 and V2. Area 19 (which includes several visual areas) projects to the deepest portion of the superficial gray layer. Visual area MT, located in the ascending arm of the inferior temporal sulcus, projects to both the superficial and intermediate layers. The neighboring area MST does not project to the superior colliculus. Parietal and frontal visual areas project to the superficial, intermediate, and deep layers of the superior colliculus. The frontal eye field innervates the AChE-positive areas in this region, which contain premotor neurons that generate commands for saccades in the contralateral visual field (Leichnetz et al., 1981). Parietal afferents terminate more deeply and originate from the lateral intraparietal area (LIP) in the lateral wall of the intraparietal sulcus, an area involved in saccade control.

Subcortical afferents directly targeting the superficial layers of the colliculus include the pretectum, parabigeminal nucleus, optic tract nucleus, intercalated nucleus, and substantia nigra (Nieuwenhuys et al., 1998). Afferents to the intermediate and deep layers of the superior colliculus receive somatosensory, auditory, striatal, and cerebellar inputs. Several of these systems terminate in parts of the intermediate gray layer. Somatosensory afferents are derived from all levels of the spinal cord, from the nuclei of the dorsal column, as well as from the interpolaris nucleus of the spinal trigeminal nucleus.

All these projections are crossed and terminate in areas of the intermediate gray layer. The trigeminal projection is limited to the rostral portion of the superior colliculus. The rostral position of trigeminal afferents is consistent with the somatosensory map of the contralateral head in the intermediate layers of the superior colliculus. The head and arm occupy most of the space in the colliculus, respectively, rostral and caudolateral. The smaller representation of the lower limb overlaps with that of the upper limb. The receptive fields of neurons in these layers are wide (Meredith et al., 1991). Auditory projections originate from the dorsal nucleus of the periolivary nuclei, the lateral lemniscus, the external nucleus of the inferior colliculus, and the brachium of the inferior colliculus (Wallace et al., 1989). A map of the contralateral auditory field in the intermediate layers of the superior colliculus has the same orientation as the visual field map. The receptive fields of neurons that respond to auditory stimuli are wide; an auditory stimulus can activate the entire population of neurons that respond to acoustic stimuli, with a maximum response in a more restricted region. Neurons in the rostrolateral pars reticulata of the substantia nigra send an important direct GABAergic projection to the superior colliculus (Huerta et al., 1991). The main projection is directed to the ipsilateral intermediate gray layer where it distributes in longitudinal bands, overlapping with the projection from the pedunculopontine nucleus, frontal eye fields, and AChE-positive areas, and alternating with somatosensory projections. In these areas, nigral afferents establish contacts with premotor cells (Harting et al., 1991). It should be noted that the pars reticulata innervates both the pedunculopontine nucleus, which constitutes the major source of the cholinergic projection directly to the AChE-positive areas in the intermediate gray layer, and these same areas. The basal ganglia control saccadic eye movements by disinhibiting the continuous inhibitory afference exerted by the substantia nigra, thus selecting an appropriate impulse from the frontal eye fields (Hikosaka et al., 2000). Another source of direct inhibitory projection to the intermediate layers of the colliculus is represented by the zona incerta. Efferent connections of the superior colliculus within the superior colliculus have distinguished different types of projection neurons based on their laminar distribution, their involvement in one or more different pathways, and their functional characteristics. Small neurons located in the superficial layers give rise to an ascending dorsal pathway. These neurons have been designated as type L neurons (Moschovakis et al., 1988). The collaterals of type L neurons terminate in the deep layers of the superior colliculus and constitute the main connection between the superficial and deep layers of the colliculus. The ascending dorsal pathway traverses it and terminates in the pretectum. It proceeds to the ventral geniculate nucleus, the inferior and lateral pulvinar, and the lateral geniculate body, where it terminates in the intercalated layers (koniocellular). The axons of neurons belonging to the predorsal fascicle cross in the dorsal tegmental decussation and descend ventrolaterally to the medial longitudinal fasciculus, laterally to the superior central nucleus. Only a few predorsal fibers reach the spinal cord. Predorsal neurons constitute a heterogenous population that mainly resides in the intermediate layers of the superior colliculus. Axons destined for the predorsal fascicle emit an ascending collateral before their decussation. These ascending collaterals form the ventral ascending bundle, which terminates in the Cajal interstitial nucleus, the RIMLF (rostral interstitial nucleus of the medial longitudinal fasciculus), the mesencephalic reticular formation laterally to the interstitial nucleus, the supraoculomotor central gray, the Edinger–Westphal nucleus, the zona incerta, various intralaminar and midline nuclei of the thalamus, and the paralaminar portion of the dorsomedial nucleus of the thalamus. The predorsal bundle terminates in the oral and caudal reticular nuclei of the pons and the paragigantocellular nucleus, regions that contain excitatory and inhibitory cells with short-duration activity in the horizontal saccadic system, arranged rostrally and caudally to the abducens nucleus. Terminations are present in two precerebellar nuclei, the dorsomedial reticular nucleus of the pontine tegmentum and the caudal medial accessory olive. In addition, direct projections to the prepositus nucleus of the hypoglossal nerve and the interpositus nucleus of the raphe have been described (Grantyn et al., 1982). A direct projection to the olive was found to be absent in studies based on individually injected predorsal axons (Scudder et al., 1996). Therefore, tectoolivary fibers may originate from a separate population of tectal neurons.

A noncrossed descending pathway leaves the colliculus laterally, running superficially along the parabigeminal nucleus, and terminates in this nucleus, the lateral pontine gray, and the pontine reticular formation. Known as the tectopontine tract, it was first described by Munzer and Wiener (1902). It originates from a population of small neurons in all layers of the colliculus, including the L neurons (Moschovakis et al., 1988; Redgrave et al., 1996). Three types of projection neurons have been distinguished in the tectum: the L superficial cells and the X and T cells (Moschovakis et al., 1988). Functional profiles have been described for the X and T cell types. The T cells are small- to medium-sized neurons located in the optic layer and the intermediate gray, participating in the formation of ipsilateral and crossed descending bundles and/or dorsal ascending bundles. These correspond to long-activity vector neurons that are active up to 20 ms before a saccade in their movement field. The X cells are large multipolar cells of the predorsal bundle, mainly located in the intermediate gray. Their axons emit commissural collaterals that ascend ventrally. Activity in X cells is not related to spontaneous saccades.

Another group of cells, located in the rostral colliculus at the level of the foveal representation, is composed of tectoreticular cells that are tonically active during the fixation of a remembered target. These cells may correspond to the cells of the predorsal bundle that project to omnipause neurons in the interpositus nucleus of the raphe, which inhibit short-activity excitatory neurons and are mainly located in the rostral superior colliculus. Neurons with long-term activity in the superior colliculus, whose maximum activity is closely related to the onset of a saccade, have been localized in the intermediate gray layer. These cells are topographically organized according to a motor (saccadic) space map that specifies the direction and amplitude of a saccade, preceded by the discharge of a cell at a particular location. In the colliculus, cells that discharge before small contralateral-directed saccades are located rostrally, before large saccades caudally, before upward component saccades medially, and before downward component saccades laterally. The site of increased metabolic activity during saccades, studied using the deoxyglucose method, has shown that this map remains stable. According to the foveation hypothesis, redirecting the line of sight toward objects of interest is achieved by the spreading of excitation from a defined target in the superficial retinotopic map to the underlying presaccadic cells. Since the superficial retinotopic map and the deep saccadic maps are aligned, this would result in the generation of an appropriate saccade that brings the target onto the fovea.

This simple scheme has been developed based on numerous experimental studies of this system:

1. The connections between the superficial and intermediate layers of the colliculus, which contain the retinotopic and saccadic maps, have remained enigmatic for a long time. However, the descending collaterals of the l cells constitute a probable morphological substrate for this connection (Moschovakis et al., 1988).
2. Saccades toward a particular target, specified by premotor neurons in the motor map, can be interrupted by the appearance of another target in a different position. Correction for the direction and amplitude of the intended saccade can be performed by the interposition of another neuron in the direct pathway from the superficial layers to the premotor neurons. These visuoresponsive neurons receive retrograde information related to the expected saccade from cells present in the mesencephalic

reticular formation, innervated by collaterals of the premotor cells. These neurons calculate the direction and amplitude of the corrected saccade based on the information from the new target and measurements of the initial saccade. The visuoresponsive cells have not yet been anatomically confirmed.

3. In models of saccade generation by the superior colliculus, long-term activity neurons are interposed between the premotor cells of the superior colliculus and the short-term activity neurons present in the gaze centers. Few of these cells have been localized in the pontine reticular formation (Scudder et al., 1996, 2002).
4. Saccades can be initiated by visual, auditory, and somatosensory stimuli. Visual targets can be located in the superficial retinotopic map of the contralateral visual field, but a coarse visual map is also present in the intermediate layers of the colliculus (Meredith et al., 1985). Auditory and somatosensory maps in the intermediate layers are based on systematic variations in the receptive fields of neurons, presumably premotor neurons, in this layer. The map of the contralateral auditory field localizes potential targets for a saccade in coordinates centered on the head, meaning their localization is independent of the position of the eyes in the orbits. However, the amplitude and direction of saccades in the saccadic motor map are encoded in eye-centered coordinates, taking into account the initial position of the eye in the orbit. Therefore, calculating the correct direction and amplitude of a saccade for an auditory target should take into account the initial position of the eye. This is achieved by modifying the location in the auditory map relative to the stabilized saccadic map, along with changes in eye position within the orbit. Systematic variations in the organization of auditory receptive fields in premotor neurons, accompanying variations in eye position within the orbit, have been described in awake animals (King et al., 2004; Peck et al., 1993). Similar transformations can occur in the tectal somatosensory map.
5. A peculiar characteristic of the intermediate layers of the superior colliculus, often overlooked in determining the maps in this region, is represented, in many species, by the discontinuous projection of most afferent systems and their relationship with the distribution of AChE and other substances in this layer.
6. It is believed that the superior colliculus also initiates head and body orienting movements. Typically, the tectospinal pathways and/or direct projections to reticular areas with spinal projections are considered responsible (Shinoda et al., 2006). The latter include the paramedian pontine neurons in the oral and caudal pontine nuclei and in the gigantocellular reticular nucleus, with their ponto- and bulbospinal crossed tracts and medial bulbospinal tract, as well as regions in the mesencephalic reticular formation (Cowie et al., 1994; Robinson et al., 1994; Shinoda et al., 2006).

These systems likely exhibit the same level of complexity as the vestibulospinal pathways mediating spinal labyrinthine reflexes.

ROLE OF SUPERIOR COLLICULI ACTIVITY IN THE AFFECTIVE VISUAL PROCESSING

Employing high-field strength (7 T) fMRI techniques, Wang et al. (2020) shone some light on the role of the superior colliculus in affective visual processing.

When viewing emotionally distressing images, the superior colliculus was more activated than when watching neutral images. These findings close the gap between affective studies on humans and nonhuman animals by indicating that the superior colliculus may influence subjective emotional experiences in addition to its visuomotor functions.

Behavioral and pharmaceutical studies have shown that the SC is also involved in approach and defense behaviors, which are modulated by the nigrotectal connections to the substantia nigra. Similar to the periaqueductal gray (PAG), a nearby brainstem nucleus implicated in protective behaviors and fear processing, electrical and chemical stimulation of the rodent SC causes freezing and running behaviors. As a component of the SC—pulvinar—amygdala subcortical magnocellular network, the SC has been postulated to play a function in fast visual danger detection. It has been demonstrated that the SC is involved in processing visually discernible ecological threats. When snake images are shown in the foveal visual area as opposed to the peripheral visual field, a recent study using human volunteers observed increased activity in the SC. This finding is consistent with the central bias in processing ecologically relevant threat stimuli. Additionally, studies using probabilistic tractography and in vivo diffusion tensor imaging have predicted fiber connections between the human SC and amygdala via the pulvinar, offering anatomical proof in favor of the SC's function in threat sensing.

When viewing unpleasant versus neutral images, the majority of fMRI techniques cannot distinguish between the PAG and SC responses because common normalization and smoothing techniques cause serious partial-volume problems.

The differentiation is conceivable with 7 T fMRI, which has an ultrahigh field, and SC activation can be studied as subjects look at a collection of photographs of natural scenes. To test for visual processing specificity, BOLD signals from the inferior colliculus (IC), an auditory midbrain area chosen for similar size and location, can be compared with those from the SC. It has been investigated whether activation in the SC was higher during blocks of viewing aversive images as opposed to neutral images.

The exact anatomical segmentation of the SC and functional data showed that the SC was more active during the processing of visual stimuli than the IC, a control region of comparable size and proximity.

With descending corticotectal pathways providing information on which contexts and features are more or less likely to be relevant, the SC may facilitate responding to threat by quickly identifying context-salient visual features (such as threats arriving from different locations in the environment) and preparing the rapid coordination of behaviors accordingly (fleeing from or orienting toward these locations). These behaviors are known to be supported by relationships between the SC and the PAG.

RESPONSES TO FACES AND THREATENING FACIAL EXPRESSIONS

The state, status, and intentions of others are conveyed through faces, which are among the most significant inputs for humans. As a result, the study of the human brain and behavior is very interested in how people perceive and pay attention to faces.

The human visual system's parallel pathways are tuned differently to brightness, color, and spatial frequency. The differential processing of higher-order stimuli, such as facial threat cues, in the magnocellular (M) and parvocellular (P) pathways has recently been demonstrated to be mediated by these attunements, with greater sensitivity to unambiguous and ambiguous danger, respectively. But there is still some mystery surrounding the third pathway—the koniocellular (K) pathway—which is involved in interpreting facial threats.

In a study by Kveraga et al. (2007), participants were asked to press a key to indicate whether a face was furious or neutral, with their eye movements and manual actions being recorded. They discovered that faces shown in the K channel were targeted by short-latency saccades more frequently than faces in the P or M channels. Expressive and identification cues did not significantly affect saccade latencies. In contrast, route biasing and interactions between facial expression and facial masculinity affected the latencies and accuracy of manual responses, with furious male faces evoking the quickest replies and angry female faces the slowest.

According to research, the M pathway is more activated by obvious threat cues including clear threat in visual settings and clear threat cues such as congruent facial expression and eye gaze (Cushing et al., 2019; Im et al., 2018).

The P pathway, which is consisting of midget retinal ganglion cells that integrate the L and M cone inputs in a color-opponent manner, is sensitive to color in the green to red spectrum and is capable of resolving greater spatial and low temporal frequencies. It was discovered that it was more sensitive to incongruent threat cue combinations involving facial expression and gaze.

Given that mammals with dichromatic vision use the K route, a phylogenetically older pathway, it would be reasonable to assume that it would also be tuned to identify biologically relevant stimuli, particularly those presenting threat signs.

According to Kveraga et al. (2007), the K route is capable of participating in rapid targeting for complex stimuli like human faces since it can elicit faster saccadic responses in response to stimuli presented to it. However, the saccadic latency was comparable for angry and neutral faces, and it did not significantly vary based on the affective nature of the stimulus.

The pathways through which the faces were presented affected the responses. Responses to stimuli presented to the K pathway were predominantly influenced by facial expression, with anger eliciting faster reactions, whereas responses to M-biased faces were influenced by interactions between expression and identification cues. Expressional or identification signals had no discernible impact on how people responded to P-biased faces.

With a relatively short latency, saccades can be elicited by higher-level events projected to the K pathway. Although the K and P saccadic latency distributions are relatively comparable, there was a considerably larger tendency to produce saccades in the 200 ms time range in response to matching K-biased stimuli than for corresponding P-biased (or M-biased) stimuli.

A modest luminance contrast on a rapidly varying EEG backdrop must be resolved without the aid of color opposition signals from either the K or P pathways, which is likely the cause of the extended latencies of sRTs triggered by M-biased stimuli. Although the superior colliculus responds strongly to luminance stimuli, the luminance contrast required to produce strong bursts of activity in the SC neurons responsible for producing short-latency saccades must be quite high. This is because the SC's saccade generation latency scales strongly with target luminance both physiologically (Kveraga and Hughes, 2005) and behaviorally (Marino et al., 2015).

In summary, high-level stimuli, such faces, delivered to the K-pathway are able to elicit quicker saccadic and manual reactions than those provided to the P and M pathways. Manual response latencies are sensitive to facial expression, and accuracy patterns demonstrated modification by both expressive and identity cues and their interaction, despite the saccadic responses not being highly influenced by facial expression and identity cues. We can infer that the koniocellular route is crucial in guiding attention and sight toward complex biological cues such as human faces.

ACCESSORY OPTIC SYSTEM AND OPTIC TRACT NUCLEUS

The accessory optic system (AOS) consists of the lateral and medial terminal nuclei located in a collateral branch of the optic tract at the periphery of the rostral midbrain. These nuclei record the speed and direction of global visual movements. This information is transmitted to the cerebellum via the inferior olive and is used in the long-term adaptation of the VOR (vestibuloocular reflex). The low-speed preference of AOS neurons, along with their wide receptive fields, suggests that they serve as complementary elements to the vestibular system in detecting self-motion. The optic tract nucleus, located in the pretectum, also receives afferents from the optic tract and projects to the inferior olive. Although it is not a terminal nucleus of the accessory optic system, it is adjacent to the dorsal terminal nucleus. The dorsal terminal nucleus and the optic tract nucleus are difficult to separate. The optic tract nucleus is an important station in the subcortical pathway that mediates the horizontal optokinetic reflex. This reflex is particularly effective at low speeds, thus complementing the VOR in stabilizing the image on the retina. The AOS receives fibers from the optic tract; these fibers branch off from their direct route to the superior colliculus in the superior colliculus arm. They descend ventrally along the surface of the midbrain and form the superior fascicle of the AOS, which crosses the cerebral peduncle as Von Gudden's transpeduncular tract. This tract terminates in the medial terminal nucleus of the AOS, located immediately medial to the peduncle.

The dorsal terminal nucleus and the lateral terminal nucleus are included among the fibers of the superior fascicle. The inferior fascicle of the AOS branches off from the optic tract near the chiasm to terminate in the medial terminal nucleus. The terminal nuclei of the AOS and the optic tract nucleus are bilaterally connected by mainly GABAergic pathways. The commissural projections decussate in the posterior commissure (Giolli et al., 1992; Van der Want et al., 1992). The medial and lateral terminal nuclei of the AOS are connected to a group of cells located in the ventral tegmentum, dorsally to the medial terminal nucleus, through a pathway that crosses in the posterior commissure (Blanks et al., 1995; Giolli et al., 1984, 1985). The connections of this area, known as the ventral tegmental reflex zone, are very similar to those of the medial terminal nucleus. The projections from the retina to the optic tract nucleus and the terminal nuclei of the AOS are primarily crossed and originate from direction-selective ganglion cells discussed in the retina section of this chapter. Neurons in the medial and lateral terminal nuclei detect movements close to the vertical plane through the contralateral eye's retina, preferably in the plane of one of the semicircular canals. Similar, but binocular, neurons are found in the ventral tegmental relay zone. These nuclei project to precerebellar, accessory optic, vestibular, and reticular nuclei. Their main precerebellar target is a small subnucleus of the inferior olive: the ipsilateral ventrolateral olive (VLO). From the VLO, a projection of climbing fibers directly targets the contralateral nodulus, flocculus, and ventral paraflocculus. Other projections from the medial terminal nucleus are found in the contralateral dorsolateral pontine nuclei; the lateral terminal nucleus innervates the same structure on the same side. A direct projection to the lateral pole of the pontine tegmental reticular nucleus has been observed only for the lateral terminal nucleus. Most optokinetic centers project to the medial and dorsomedial parts of this nucleus, which also receives numerous other visuomotor afferents.

The accessory optic nuclei project to the interstitial nucleus of Cajal and the Darkschewitsch and Bechterew nuclei. Some of these projections originate from commissural systems in the posterior commissure. The projections to the vestibular nuclei are modest or few. All three terminal nuclei project to the prepositus nucleus (Büttner-Ennever et al., 1996). The dorsal terminal nucleus of the AOS and the optic tract nucleus both encode horizontal movements of the visual periphery. Both nuclei project to the dorsal dome of the inferior olive. However, the two nuclei are distinguishable as the response properties of neurons in the optic tract nucleus differ from those of the dorsal terminal nucleus (Gamlin et al., 2006). Furthermore, the efferent connections of the optic tract nucleus are believed to be more extensive than those of the terminal nuclei of the AOS (Büttner-Ennever et al., 1996). Direct ascending projections to the magnocellular layers of the lateral geniculate body, the intralaminar nuclei, the pulvinar, and the zona incerta, as well as the Darkschewitsch nucleus, have been described. The direct descending projections to the superficial layers of the superior colliculus are GABAergic. Other projections are directed to the ipsilateral hypoglossal prepositus nucleus, the abducens nucleus, caudally to the adjacent reticular formation, and to the same precerebellar nuclei that receive projections from the terminal nuclei of the AOS, namely the dorsolateral and medial pontine nuclei and the dorsal dome of the inferior olive. Terminations are also found in the pontine tegmental reticular nucleus. Through the dorsal dome of the inferior olive, the optic tract nucleus and the dorsal terminal nucleus of the AOS project to the cerebellar nodulus, the flocculus, and the ventral paraflocculus. Cortical regions projecting to the optic tract nucleus include the MT and MST areas in the inferior temporal sulcus, the frontal and supplementary eye fields, and the caudal walls of the superior temporal sulcus (Hoffmann et al., 1992; Shook et al., 1991; Lui et al., 1995). This latter region is also connected with the terminal nuclei of the AOS (Hoffmann et al., 1991). The optokinetic reflex likely utilizes the vestibulooculomotor pathway as a terminal station. Some direct connections have been described, but the hypoglossal prepositus nucleus and the pontine and olivary subnuclei innervated by this nucleus are considered probable stations.

In the frontal lobe, three frontal eye fields have been distinguished, from which eye movements originate when stimulated, and one has been identified in the parietal lobe. In humans, the frontal eye field is located in the precentral sulcus, in the transition zone between areas 6 and 8, and extends into area 8. The supplementary eye field occupies a more dorsomedial position in the interhemispheric fissure. A prefrontal eye field has been distinguished in BA 46 (Luna et al., 1998; Rosano et al., 2002, 2003). In patients, lesions involving the frontal eye field produce conjugate gaze palsy toward the opposite side, with deviation of the eyes and head toward the side of the lesion. Spontaneous recovery typically occurs within a few days. Experimental inactivation of the frontal eye field profoundly alters ocular saccadic movements and slow pursuit movements and causes contralateral visual neglect, which resolves over days or weeks. Lesion studies involving the supplementary frontal eye field have reported only a moderate alteration in oculomotor function. The influence of the prefrontal eye field on voluntary eye movements appears to be inhibitory (Lynch and Tian, 2006). The parietal eye field is located in the lateral wall of the intraparietal sulcus. It is known as the lateral intraparietal area (LIP). Presaccadic activity has been observed in the parietal eye field. It encodes expected positions relative to the fovea and guides eye movements (Culham et al., 2001). The intraparietal sulcus contains several subregions with different connections and functional properties. Damage to this region causes visual—spatial neglect, constructional apraxia, gaze apraxia, and spatial cognition disorders. The three frontal eye fields, the parietal visual fields, and the occipitotemporal MT and MST fields are intensely interconnected. The specificity of these connections is clarified by the observation that the parietal eye field (LIP) is connected to the frontal eye fields, while the nearby MIP and VIP areas are not (Clower et al., 2001). In addition to the reciprocal connections with the frontal eye fields, the parietal eye field is connected to visual areas V2, V3, V4, MT, and MST. Most of these connections are reciprocal. Furthermore, most of the visual areas projecting to the parietal eye field also project directly to the frontal eye field (Lynch and Tian, 2006). Separate divisions of the frontal eye field promote saccades and slow pursuit movements, and activity in other frontal and parietal fields is associated with both types of eye movements (Petit et al., 1999).

In the frontal eye field, neural signals related to saccades are located more dorsally than those related to slow pursuit movements. The frontal eye fields are connected to the striatum, claustrum, thalamus, and centers such as the superior colliculus, vertical and horizontal gaze centers, the prepositus hypoglossal nucleus, and the precerebellar reticular nucleus of the pontine tegmentum and pontine nuclei. Thalamic connections of the parietal eye fields, like those of MT and ST areas, are established with the inferior, lateral, and medial pulvinar (Giolli et al., 2001). The frontal and supplementary eye fields are connected to the ventral anterior mediodorsal and intralaminar nuclei. The supplementary eye field receives greater projections from the medial portion of the ventrolateral nucleus and its nearby medially disposed X subnucleus. The saccade and slow pursuit movement regions of the frontal eye field receive nonoverlapping thalamocortical projections (Huerta et al., 1990). These and other studies allow for predictions of possible cerebellar, striatal, and nigral afferents directed to the frontal and parietal eye fields. It has been concluded that both the saccade and pursuit movement regions of the frontal field receive afferents from the substantia nigra and cerebellar nuclei (Lynch et al., 1994). The pursuit movement region and the supplementary eye field also constitute the site of projection of the internal segment of the globus pallidus. In the supplementary eye field, this combines with a cerebellar afferent. The parietal eye field lacks afferent connections from the cerebellum or striatum. The nearby MIP area (7B) receives thalamic projections mediating cerebellar, pallidal, and nigral nuclear afferents to this lobe (Clower et al., 2001, 2005). Subcortical connections of the eye fields include the superior colliculus and other saccadic premotor centers and various precerebellar nuclei. Parietal afferents from LIP terminate in the intermediate layers of the superior colliculus; the area MT projects to both the superficial and deep layers; the area MST lacks a tectal projection. The frontal and supplementary eye fields project to the zonal layer; their terminations in the deep layers have been described in various ways, occupying areas in the intermediate gray layer or limited to the optic layer. Among the gaze centers, the rostral interstitial nucleus of the medial longitudinal fasciculus (RIMLF) is innervated by both divisions of the frontal eye field and the supplementary eye field. The innervation of the paramedian pontine reticular formation (PPRF) is weak and does not reach the saccade generator. Omnipause neurons of the interstitial nucleus of the raphe receive a projection from the supplementary eye field either in isolation or in combination with the frontal eye field (Huerta et al., 1990; Stanton et al., 1988).

The optic tract nucleus and the prepositus hypoglossal nucleus have occasionally been described as sites of projection for the frontal eye fields; the subthalamic nucleus and the zona incerta are projection sites shared by all frontal eye fields. Projection sites of visual fields directed to precerebellar nuclei include pontine nuclei and the reticular nucleus of the pontine tegmentum, as well as the nucleus of Darkschewitsch and the dorsomedial parvocellular red nucleus, both of which project to the inferior olive. The Darkschewitsch nucleus receives a projection from the frontal eye field and posterior parietal area 7. Both the frontal and supplementary eye fields and parietal area 5 project to the dorsomedial parvocellular red nucleus (Burman et al., 2000). Corticopontine fibers from the frontal eye field have been reported to predominate in the dorsolateral pontine nucleus or the medial pons, where they overlap with a projection from the supplementary eye field (Leichnetz et al., 1996; Shook et al., 1990; Huerta et al., 1990). In the lateral and dorsolateral pontine nuclei, projections from the frontal eye field can overlap with projections from parietal areas 5, 7a, 7B, and LIP, MT and MST areas, and peripheral representations of V1, V2, V3, and V3A, from the pretectum, from the medial and lateral terminal nuclei, and from the ventral tegmental relay zone of the AOS, the optic tract nucleus, and, at least caudally, from the uncrossed tectopontine projection (Baizer et al., 1993 Fries 1990; Schmahmann et al., 1993). Both frontal eye fields project quite diffusely to the entire rostrocaudal extent of the reticular nucleus of the pontine tegmentum. Convergence with the crossed tectopontine projection of the predorsal bundle is possible. The reticular nucleus of the pontine tegmentum does not receive a projection from postrolandic visual areas. Various eye fields are reciprocally connected with the striatum through the globus pallidus or substantia nigra, as well as with the superior colliculus and the cerebellum.

The reciprocal pathways from the superior colliculus to the frontal eye field and the parietal eye field are different. Premotor cells, located in the intermediate layers of the superior colliculus, receive a direct input from the frontal eye field. Their axons split before entering the dorsal tegmental decussation and give rise to an ipsilateral ascending collateral pathway, which terminates in the paralaminar portion of the mediodorsal thalamic nucleus, which in turn projects to the saccadic region of the frontal eye field. This pathway provides the frontal eye field with accessory activity that encodes for proximal eye movements (Sommer et al., 2000). A recurrent system targeting the parietal visual field originates from the superficial layers, while parietal afferents terminate in the intermediate layers of the superior colliculus. These fibers ascend in the ipsilateral dorsal pathway and make a stop in the pulvinar (Clower et al., 2001).

Can the questions regarding the pathways used for slow pursuit movements be explained by this description of the connections of the cortical eye fields? Undoubtedly, these pathways pass through the early visual areas 1 and 2 and the MT and MST areas, with the MT area being responsible for the discrimination of direction in pursuit behavior (Osborne et al., 2004). The parietal visual area LIP may be involved in encoding the localization of an event relative to the fovea and in guiding eye movements. The frontal area of the frontal eye field for pursuit movements processes information in terms of image and eye velocity and plays a role in modulating the amplification of visuomotor responses, using feedback information from the cerebellum (Chou et al., 2004; Tanaka et al., 2002). There are several alternatives for the connections of the frontal eye field with oculomotor neurons. These connections resemble a circuit, in which the optic tract nucleus, the superior colliculus, the prepositus hypoglossal nucleus, and the cerebellum constitute the nodal points. Several of these pathways are shared with the vestibuloocular and optokinetic reflexes.

Access to the system that generates saccades as a final station can be achieved through the superior colliculus or through the cerebellum. Access to the vestibulooculomotor pathways is possible through the prepositus hypoglossal nucleus, a projection site of the frontal eye field, the optic tract nucleus, and the superior colliculus, or through the cerebellum.

The pretectal region is located immediately rostral to the superior colliculus at the level of the posterior commissure. This complex can be divided into five different cell groups: the optic tract nucleus, the pretectal olivary nucleus, and the medial, anterior, and posterior pretectal nuclei. A similar division is also possible in the human brain (Borostyánkői-Baldauf et al., 2002).

Fibers from the retina run to the pretectal region, primarily contralaterally, where they terminate in the optic tract nucleus. The pretectal olivary nucleus receives a bilateral projection. The posterior and medial pretectal nuclei receive diffuse projections from the retina. The posterior commissure nucleus is sometimes contained within the pretectum. The posterior commissure itself contains decussating fibers and commissural fibers from the optic tract nucleus, terminal nuclei of the AOS, and various long ascending systems of the brainstem. The pretectal olivary nucleus is associated with the pupillary light reflex. The anterior and posterior pretectal nuclei receive afferents from contralateral somatosensory nuclei of the brainstem and projections from the primary and secondary somatosensory cortices (Wilberg et al., 1984, 1987). The descending pathways terminate in the superior colliculus, ipsilateral pontine nuclei, and the inferior olive (Bull et al., 1990). The noncrossed projection from the anterior and posterior pretectal nuclei to the inferior olive is distinct from its termination in the dorsal accessory olive, an olivary subnucleus that serves as the main station in the pathways of spinocerebellar climbing fibers. In rodents, the equivalent of the medial pretectal nucleus contains cells that represent the source of a centrifugal projection directed to the retina (primarily contralateral) (Kitao et al., 1989).

The light reflex, which is the constriction of the pupil following eye illumination, is mediated by a pathway involving the following stations:

1. Axons of retinal ganglion cells, which run through the optic nerve and tract to the pretectal region, particularly to the pretectal olivary nucleus (Trejo et al., 1984).
2. Axons of neurons from the pretectal olivary nucleus project to the Edinger–Westphal nuclei of both sides. Their fibers decussate at the posterior commissure (Young et al., 1994; Clarke et al., 2003).
3. Axons of preganglionic parasympathetic neurons from the accessory oculomotor nuclei run with the oculomotor nerves and terminate at the ciliary ganglia, where they synapse with postganglionic neurons.
4. Postganglionic neurons whose axons innervate the sphincter muscle of the iris.

Pretectal efferents directed to the ipsilateral accessory oculomotor nucleus provide for the direct pupillary light reflex, while those from the contralateral nucleus contribute to the consensual pupillary light reflex. In addition to the light response, pupillary constriction also occurs with the onset of eye convergence. The neural pathway for this convergence-induced pupillary constriction is believed to be independent of the pupillary constriction induced by light, as the two reflexes are dissociated in the clinical phenomenon known as Argyll Robertson's sign. This pathway terminates on a group of small neurons outside the classic oculomotor nuclei (Büttner-Ennever et al., 1996).

The center for pupillary dilation is located in the intermediolateral cell column of the upper thoracic spinal cord. The preganglionic sympathetic fibers originating from the so-called ciliospinal center ascend through the sympathetic trunk and synapse in the superior cervical ganglion with postganglionic elements. The axons of the latter run with the branches of the internal carotid artery and pass through the ciliary ganglion before innervating the dilator muscle of the iris. Descending fibers through the lateral part of the medulla and spinal cord synapse in the sympathetic ciliospinal center.

The pathway for accommodation, which involves the increase in lens curvature for near vision, is (regarding its mesencephalic centers and efferent arms) closely comparable with the pathway for the light reflex. In both pathways, the pretectal region, the accessory oculomotor nucleus, and the ciliary ganglion represent the subsequent relay stations. However, the afferent arm of the accommodation reflex includes the visual cortex and is therefore much longer and more complex than that of the light reflex. The final station of the efferent pathway of the accommodation pathway consists of postganglionic fibers directed to the ciliary muscle.

VISUAL SYSTEM AND CEREBELLUM

The oculomotor cerebellum is made of the following regions: the flocculus, the ventral and dorsal paraflocculus, the visual vermis, and the nodulus/uvula. Parts of the ansiform lobe should probably be added to this list. In the human cerebellum, the dorsal paraflocculus is represented by the tonsil, the ventral paraflocculus the accessory paraflocculus, and the ansiform lobe by the superior and inferior semilunar lobules.

To determine which part of the cerebellum is involved in a particular visual subsystem, it is important to remember that the function of the cerebellum depends on the innervation of Purkinje cells by two afferent systems: the system composed of mossy fibers/parallel fibers and the climbing fibers. Climbing fibers always innervate a sagittal strip of Purkinje cells and the cerebellar nuclei cells innervated by these Purkinje cells, and are closely associated with the efferent systems of the cerebellar cortex, both anatomically and functionally. Mossy fibers are more widely distributed. The mossy fiber/parallel fiber systems typically innervate multiple sagittal stripes of Purkinje cells. The main sources of mossy fibers in the visual system are represented by the pontine nuclei, the reticular nucleus of the pontine tegmentum, the prepositus nucleus of the hypoglossal, and possibly the cellular groups of the pontine and bulb formation of the reticular formation, collectively referred to as the paramedian cell groups (Büttner-Ennever et al., 1996). The pontine nuclei provide the majority of mossy fibers that innervate the paraflocculus. These fibers originate from the lateral and medial cell columns, each of which may include several of the ill-defined nuclei that are distinguished in the pons. Neurons that subsequently innervate the more distal folia of the paraflocculus are arranged in more peripheral regions of the pontine gray (Gayer et al., 1988). The lateral column receives a corticopontine projection from the postrolandic visual areas 7A, 7B, and LIP (the parietal visual area), from the occipitoparietal areas MT and MST, and from the V1–V3A regions representing the periphery of the visual field (Fries 1990). These cortical afferents overlap with projections from the ipsilateral superior colliculus, the optic tract nucleus, and the medial and lateral terminal nuclei of the AOS. These projections from the lateral column terminate on neurons from which mossy fibers directed to the dorsal paraflocculus originate (Gayer et al., 1988; Nagao et al., 1997; Xiong et al., 2002). Units in the dorsolateral pons are active during slow pursuit movements or in response to the movement of large pattern drawings. Most of these units are direction-selective. Lesions in the dorsolateral pons cause deficits in slow pursuit movements and optokinetic reflex-induced eye movements, but these deficits disappear within a few days (May et al., 1988). Frontal eye fields project to the medial pontine nuclei. Here, the frontopontine projection overlaps with neurons projecting to the ansiform lobe but not to neurons projecting to the paraflocculus (Xiong et al., 2002). The pontine gray provides only a minor contribution to the mossy innervation of the ventral paraflocculus. The reticular nucleus of the pontine tegmentum is composed of different afferent territories. Projections from premotor cortical areas, including the frontal and supplementary eye fields, primarily terminate in the medial and dorsomedial regions of the nucleus. The same region of the nucleus receives a contralateral projection from the superior colliculus, which may originate as a collateral system from the predorsal bundle, vestibular afferents, a projection from the optic tract nucleus, and a projection from the caudal oculomotor portion of the fastigial nucleus (Sugita et al., 1991). The central portion of the nucleus receives fibers from the primary motor cortex and the dentate and anterior interpositus (emboliform) cerebellar nuclei. Parietal afferents terminate more laterally, but this projection does not include the parietal and parietooccipital visual areas. Physiological studies support a rostrocaudal division of the nucleus, with a rostral nucleus involved in slow pursuit movements. Stimulation of its caudal portion induces eye movements similar to saccades (Suzuki et al., 1999; Yamada et al., 1996). The reticular nucleus of the pontine tegmentum projects bilaterally to large regions of the cerebellar cortex. It projects collaterally to the cerebellar nuclei, including the caudal oculomotor region of the fastigial nucleus, but not to the vestibular nuclei. This projection is complementary to the collateral projection of mossy fibers from the lateral reticular nucleus. Direct projections to the oculomotor cerebellum originate from the medial and dorsal periphery of the nucleus with a visual predominance and from adjacent regions of the dorsomedial pontine gray. These projections include the flocculus, ventral paraflocculus, and the visual vermis (lobule VII) (Nagao et al., 1997; Osanai et al., 1999). The prepositus hypoglossal nucleus receives afferents from the superior colliculus via the predorsal bundle, the terminal nuclei of the accessory optic system, the optic tract nucleus, and the frontal eye fields (but not the postrostral ones). With its mossy fiber projection directed to the flocculus, ventral paraflocculus, lobule VII, adjacent ansiform lobule, and uvula nodulus, it delimits the oculomotor cerebellum, except for the dorsal paraflocculus. The reticular nucleus of the pontine tegmentum corresponds to only a small portion of the afferents to the flocculus and presumably to the ventral paraflocculus. Most of the mossy fibers terminating in these lobules derive from the reticular formation (50%), vestibular nuclei (20%), and the prepositus hypoglossal nucleus (20%).

The role of the vestibular nuclei as a station in the afferent optokinetic pathway is difficult to assess: the interstitial nucleus of Cajal and the prepositus hypoglossal nucleus serve as a visual/vestibular interface, but direct projections from the optic tract nucleus, AOS, superior colliculus, and frontal eye fields to the vestibular nuclei are few or entirely absent. The mossy fiber projections directed to the uvula/ventral nodulus derive almost exclusively from the vestibular nerve and nuclei. The climbing fiber input to the oculomotor cerebellum originates from the subnuclei of the principal and medial accessory olives. The dorsal cap and VLO are often considered as two caudal subnuclei of the principal olive. These two subnuclei are part of a system that provides the cerebellum with information about retinal error in the planes of the horizontal and anterior vertical semicircular canals, which are used in the calibration of vestibuloocular and optokinetic reflexes. Each subnucleus innervates corresponding zones of Purkinje cells extending from the flocculus to the ventral paraflocculus and emits a collateral projection to zones of Purkinje cells in the nodulus (Ruigrok 2003). The lateral and medial terminal nuclei of the AOS provide the VLO and its climbing fibers with information about retinal error in a vertical plane collinear with the ipsilateral semicircular canal plane. The optic tract nucleus and dorsal terminal nucleus of the AOS provide the dorsal cap with information about similar movements of the visual contour in the horizontal plane. Purkinje cells in the zones innervated by the VLO inhibit the vestibuloocular relay cells of the anterior canal in the superior vestibular nucleus. Purkinje cells in the zones innervated by the dorsal cap inhibit the horizontal canal cells in the medial vestibular nuclei. The climbing fiber input to the flocculus and ventral paraflocculus is involved in controlling the input of mossy fibers to these lobules. The molecular basis of this process is represented by long-term depression (LTD) of synapses established between parallel fibers and Purkinje cells, resulting from the coincident activity of climbing fibers and parallel fibers terminating on the same Purkinje cell.

The flocculus and ventral paraflocculus, and their connections with the vestibulooculomotor pathways, can therefore be considered as stations of a cerebellar circuit, parallel and modifiable, superimposed on the three neural reflex arcs of the VOR. However, these lobules also serve as stations for the optokinetic reflex and slow pursuit movements. A specific role has been proposed for the flocculus in long-term adaptation of the VOR and for the ventral paraflocculus in slow pursuit movement. Studies have shown that the flocculus receives visual input from the superior colliculus and vestibular input from the vestibular nuclei, and it integrates these inputs to generate appropriate modifications in the VOR response. Plastic changes synaptic connections within the flocculus are thought to underlie the long-term adaptation of the VOR.

A specific pathway of climbing fibers originating from the superior colliculus synapses in the contralateral caudal medial accessory olive and projects to the ipsilateral lobule VII of the visual vermis, with a collateral projection to its corresponding cerebellar nucleus: the caudal portion of the fastigial nucleus. This pathway may carry a copy of the motor command efferences generated by the intermediate layers of the superior colliculus. In lobule VII, it coincides with the projections of mossy fibers originating from the reticular nucleus of the pontine tegmentum and the adjacent medial pontine gray and prepositus hypoglossal nucleus, which relay information from the superior colliculus, optic tract nucleus, and frontal eye fields. Like the flocculus, lobule VII receives a robust reticular input. Its corticonuclear projection site, the caudal fastigial nucleus, innervates the system that generates saccades in the brainstem and provides reciprocal projections to the superior colliculus, reticular nucleus of the pontine tegmentum, and pontine nuclei (May et al., 1990; Sugita et al., 1991).

The coincident projections of mossy and climbing fibers directed to lobule VII contribute to the long-term adaptation of saccades (Optican et al., 1980). The involvement of the visual vermis in slow pursuit movements has been extensively analyzed (Robinson et al., 2001).

The frontal and parietal eye fields have access to the ventral lamina of the principal olive through the dorsomedial parvocellular red nucleus and the central tegmental tract (Kebschull et al., 2023). The afferents of the Darkschewitsch nucleus include the same visual fields, optic tract nucleus, and medial and lateral terminal nuclei of the AOS. The Darkschewitsch nucleus innervates the rostral medial accessory olive via the medial tegmental tract. Climbing fibers from the ventral lamina of the principal olive innervate a zone of Purkinje cells in the dorsal paraflocculus (zone D1) and its projection nucleus, the ventral and caudal dentate nucleus. The zone of Purkinje cell innervation from the rostral medial accessory olive (zone C2) extends to the flocculus, ventral and dorsal paraflocculus, and beyond. Its projection nucleus, the lateral pole of the posterior interpositus nucleus, receives a collateral projection of climbing fibers (Kebschull et al., 2023). The third and most lateral zone, D2, of the paraflocculus receives its climbing fibers from the dorsal lamina of the principal olive. The afferents of this system originate from nonvisual premotor and motor areas. In the dorsal paraflocculus, climbing fibers coincide with the projections of mossy fibers originating from the dorsolateral pontine nuclei, specifically the postrolandic and tectal afferents. Unlike the climbing fibers innervating zones C2 and D1 of the dorsal paraflocculus, these mossy fibers do not carry a signal from the frontal eye fields. The visual significance of the ventrocaudal dentate and lateral posterior interpositus nuclei, the target nuclei of zones D1 and C2 of the dorsal paraflocculus, has been recognized (Van Kan et al., 1993; Xiong et al., 2002). These cerebellar nuclei provide reciprocal connections to the superior colliculus, parvocellular red nucleus, Darkschewitsch nucleus, and, through the thalamus, to the frontal eye fields and parietal area 7B (Clower et al., 2001, 2005; Kebschull et al., 2023).

The afferents of the climbing fibers to the uvula/nodulus is mainly vestibular. They receive optokinetic afferents from the collaterals of climbing fibers from the dorsal cupola and VLO, which innervate the flocculus, and labyrinthine afferents from the group and dorsomedial cell column, subnuclei of the medial accessory olive that also receive a projection from the optic tract nucleus (Ruigruk 2003). The afferents of vestibular mossy fibers do not seem to encode specific parameters of eye movements. They depend on head position and serve as a mediator for the velocity memory mechanism of the optokinetic reflex in the vestibular nuclei. We can conclude that there are four cerebellar areas involved in visuomotor control. The flocculus and ventral paraflocculus enable adaptation of the vestibuloocular and optokinetic reflexes, and possibly slow pursuit movements. The dorsal paraflocculus provides retrograde information to the frontal eye fields from the postrolandic visual areas. Lobule VII adapts eye movements through its connections with the saccade-generating system in the brainstem. The nodulus influences reflex postural mechanisms through its connections with the vestibular nuclei. A role of the ansiform lobule (corresponding to the semilunar lobes of the human cerebellum) in the regulation of eye movements is supported by the presence of afferent mossy fibers from the frontal eye fields directed to these lobules (Hayakawa et al., 2002; Xiong et al., 2002).

RETINOTOPICAL ORGANIZATION IN THE WHITE MATTER CONNECTING AREA PROSTRIATA AND THALAMUS

In the area prostriata, which is positioned ventrodorsally and anterior to the calcarine sulcus along the parietooccipital sulcus and exhibits a high preference for peripheral and wide-field stimulation, recent fMRI investigations have shown a complete and thorough retinotopic map.

White matter fibers are visible as a continuous, prolonged bundle from which two subcomponents can be separated: one that passes ventrally parallel to the optic radiations (OR) and another that passes dorsally around the lateral ventricle. It is interesting to note that one loop travels more ventrally and connects the LGN with the more periphery visual field representation placed posteriorly, while the second loop travels more dorsally and connects the thalamus with the central visual field representation of prostriata located anteriorly.

Recent research shows for the first time a retinotopic segregation regarding the trajectory of a fiber bundle between the thalamus and an associative visual area, which is congruent with a retinotopic segregation previously revealed in the OR linking the LGN and V1.

Area prostriata possesses a comprehensive representation of the visual field and preferentially responds to very fast motion, making it unique compared with the nearby area V1. It exhibits little to no cortical amplification and instead has a uniform distribution of receptive field centers throughout eccentricity. The far peripheral vision field is represented by the posterior–dorsal border of the human prostriata, which abuts V1. Eccentricities become smaller as you get closer to the anterior boundary in the calcarine sulcus.

The area's functional characteristics imply that it might alert the brain to quick visual events, especially in the peripheral visual field, and it might act as a crucial node for temporary attention.

Comparatively to nearby visual areas, the area prostriata has lesser myelination and higher cortical thickness. The region exhibits traits resembling those of the limbic cortex because it lacks the cerebral cortex's well-defined laminar profile, which has a layer 4 that is thinner and layer 2 that is thicker.

Using functional neuroimaging, region prostriata's estimated retinotopic organization exhibits a typical pattern. According to reports, the visual field periphery is situated next to the V1 periphery in the most posterior part of the prostriata. The anterior part of the prostriata houses the center of the visual field. This suggests that the region prostriata has a gradient of posterior to anterior eccentricity. This reversal is a canonical characteristic of the way visual fields are organized in the human visual cortex.

In contrast to the posterior section of the prostriata, which processes the peripheral visual field, the most anterior region of the prostriata processes the center visual field and has lower myelin content and higher cortical thickness. Other visual areas have also shown a similar systematic link between eccentricity, cortical thickness, and myelin content maps.

The complicated white matter pathway bridging the region prostriata and the visual thalamus was described in terms of its architecture and microstructure. Based on the cortical locus (anterior/posterior) within the prostriata, we separated the tract into two subcomponents. The investigation uncovered two subcomponents with distinct anatomical trajectories and microstructural characteristics.

The dorsal white matter subcomponent of the tract (LGN-Pro-1) connects the front region of the prostriata, which receives information from the central visual field and has less myelin and a higher cortical thickness, with the thalamus. On the other hand, the ventral subcomponent of the tract (LGN-Pro-2) connects the posterior part of the prostriata, which receives information from the periphery of the visual field and has enhanced myelin and low cortical thickness. More intriguingly, the thalamus exhibits a comparable retinotopic structure of fibers.

It has been documented that the LGN has a retinotopic arrangement, with a foveal–peripheral gradient extending from the medial to the lateral direction. When we look at the LGN-Pro-1 and LGN-Pro-2 endpoints, we notice a similar arrangement.

In synthesis, there is a white matter bundle that connects the area prostriata to the LGN, and from this bundle, two subcomponents can be isolated: one that travels along a path similar to that of the optic radiations and another that curves around the lateral ventricle and approaches the LGN from the top. The fibers that make up LGN-Pro-2 are located more posteriorly in prostriata than those that make up LGN-Pro-1, which are located more anteriorly. When comparing our anatomical findings to eccentricity maps from earlier reports, we find that the central visual field representation is connected through LGN-Pro-1, which goes around the lateral ventricle, while the peripheral part of the prostriata is connected with the LGN through LGN-Pro-2, which travels along a path similar to that of the OR.

The examination of endpoints inside the LGN also supports the segmentation of the bundle into fibers that transmit information from various visual fields. While LGN-Pro-2 innervates more lateral regions of the LGN, often processing the visual periphery, LGN-Pro-1 reaches the most medial part of the LGN, which processes the foveal part of the visual field. The subdivision between LGN-Pro-1 and LGN-Pro-2 is somehow anatomically arbitrary.

Recent research reveals a white matter bundle that encompasses the ventricle, crosses the splenium, and extends to the lateral occipital lobe. This bundle resembles LGN-Pro-1 in morphology. This white matter bundle may be a component of the tapetum's lateral region, descending to the temporal lobe close to the OR and IFOF. Such research suggests that the tapetum might be able to transmit information between the occipital lobe and other structures even if prostriata are situated medially in the occipital lobe.

The tapetum, which transmits information between the hemispheres and is a component of the splenium, has been identified by tractography research. It is made up of callosal fibers. Regardless of its independence from the splenium, the tapetum may be sending visual information to the visual cortex. It is interesting to note that a patient with a temporary splenium lesion near the tapetum has been documented to experience a temporary loss of vision without experiencing any other neurological symptoms.

The LGN and area prostriata in the human brain are structurally related. Tractography cannot determine whether this connection is transmitting a thalamic input and communicating information to the prostriata. The multiple feedback projections that emerge from V1 and other visual associative areas and project monosynaptically back to the thalamus in a retinotopic-specific way can also be linked to the LGN–prostriata connection. Nevertheless, electrophysiological research using animal models does point to the possibility of a direct thalamic input. Thaw-mount autoradiography has anatomically shown a direct thalamic projection in the tree shrew to the retrosplenial brain, more precisely to area prostriata. If a direct anatomical connection exists between the LGN and prostriata in humans as well, it has yet to be established, but if it does, it may be of particular significance in the presence of V1 lesions as it may be supporting surviving visual talents.

CORTICOOCULAR COUPLING AND EPISODIC MEMORY FORMATION

For the majority of episodic memory types, visual information must be encoded. Amplitude modulation of brain activity has consistently been found to correlate with and to be functionally involved in successful memory encoding in the searching for a neural signature of memory formation.

Memory and brain activity are related, suggesting that corticoocular interactions have a functional role in the creation of episodic memory. Popov and Staudigl (2023) showed that gaze variability and amplitude modulations of alpha/beta oscillations in the visual cortex can predict subsequent memory performance by simultaneously recording magnetoencephalography and eye tracking in human volunteers.

Echoing the covariation seen during scene encoding, amplitude change during prestimulus baseline was linked to gaze direction variation. Encoding of visual information utilizes oculomotor and visual areas in harmony connection for memory formation.

Humans frequently shift their eyes to examine their surroundings because they rely so largely on vision. These frequently involve, but are not restricted to, micro- and macrosaccadic movements as well as fixations that work in concert to form the overall gaze pattern. According to research (Damiano and Walther, 2019; Fehlmann et al., 2020; Kragel and Voss, 2022; Broers et al., 2022), there is a significant correlation between gaze patterns and the development of episodic memory. According to Voss et al. (2017), visual scenes that were studied with more eye movements are more likely to be remembered than images that were explored with less eye movements. Additional research demonstrating that gaze pattern reinstatement during retrieval improves good remembrance adds to the evidence of links between memory encoding and gaze. In the same way, gaze patterns are restored during visual vision, even when there is no retinal guidance, like in complete darkness (Johansson et al., 2006).

The "subsequent memory effect" (SME) (Paller and Wagner, 2002) is the term for the observation that the alpha/beta power during later remembered items is reduced as compared with later forgotten items. Electro- and magnetoencephalographic studies have confirmed a robust power modulation of alpha/beta oscillatory activity during the encoding of items that is predictive of later memory performance.

To what extent the inverse direction of high versus low gaze variability will predict alpha/beta SMEs?

Popov and Staudigl (2023) found that, corresponding to the alpha/beta SME topography, trials with significant gaze variability were similarly linked with higher modulation of alpha/beta power over occipital sensors. Furthermore, relative to low gaze variability during encoding, greater gaze variability was linked to a rise in the number of later remembered items. Forgotten objects showed the reverse association. The extent to which the gaze was explored during encoding determined the effect size of the memory contrast (number of items remembered vs. forgotten); compared with low gaze variability, increasing gaze variability is associated with improved memory performance. Despite the fact that SME effects are normally predicated on task-specific contrasts, like a difference between the task and prestimulus baseline, the findings show a connection between spontaneous alpha power fluctuations and spontaneous shifts in gaze. The results show that, despite the temptation to assume that the gaze-alpha power correlations are just task-specific, they are actually spontaneous and independent of any particular activity and are nonetheless used to help people build episodic memories.

During the prestimulus baseline, gaze covaries with spontaneous alpha/beta fluctuations, and higher variation in gaze was correlated with larger declines in alpha/beta power. Even further removed from specific, task-related activity, it has been demonstrated that during the entire experiment, alpha/beta activity and gaze covariate. The authors demonstrate significant task-independent covariation of alpha power and gaze using an hour of concurrent eye tracking and MEG recordings for each participant. Together, these findings lend credence to the hypothesis that visual alpha/beta activity and oculomotor activity are fundamentally connected in a domain-general manner.

VISUAL CORTEX AND WORKING MEMORY

Yan et al. (2023) used fMRI and multivariate encoding modeling to show that, even when subjects are not overtly or implicitly urged to categorize the stimuli, categorical codes are already present at the level of extrastriate visual cortex (V4 and VO1) for color stimuli. The categorical coding was evident during working memory but not perception, which is significant. As a result, category representations are probably used in visual working memory, at least in part.

According to the findings, category representations that are already present in the visual cortex are able to retain representations of remembered colors. This implies that distinct parts of the visual cortex already preserve representations in a different neural coding.

These results seem to suggest that representation of an individual item in numerous concurrent coding schemes is necessary for memory storage of that item.

Only when working memory is being used is the preference for categorical tuning evident; it is not present when doing an immediate, perceptual task. This shows that throughout the working memory processing, initially more uniform perception representations change to become more categorical.

There are two scenarios that could account for this variation. One explanation is that the ability to remember items for a long time causes the stimulus to be encoded more categorically. Alternatively, throughout the delay, feedback from more anterior, nonsensory regions might gradually modify representations. In earlier research (Bae et al., 2015; Panichello et al., 2019), the difference in the amount of biasing of behavioral reactions in immediate and delayed memory is directly explained by both options.

Working memory representations could be found in the left ventrolateral prefrontal cortex and left premotor cortex of the frontal lobes.

Earlier studies looked into categorical representations when respondents were specifically told to classify objects. Areas V1 and V4, for instance, show categorical clustering when naming colors, whereas no such clustering was shown when attention was distracted from the displayed colors (Brouwer and Heeger, 2013). Prior research (Ester et al., 2020; Freedman et al., 2003; Freedman and Assad, 2006) has demonstrated that neural representations across visual, parietal, and prefrontal cortices can exhibit categorical properties when subjects are trained to categorize stimuli into two arbitrary classes.

According to one fMRI study in particular (Ester et al., 2020), orientation selective responses in early visual cortex have a bias toward the category's center. This finding raises the question of whether categorical responses are the result of entrained categorical tuning or a rotation-like mental operation that maximizes the discriminability of the stimulus with respect to category (see discussion in Ester et al., 2020).

However, Yan et al. showed that category mnemonic representations exist in the visual cortex even when people are not explicitly or implicitly instructed to categorize the colors and are instead given a task that actively fosters accuracy.

They observed that neither model performed better than the other in V1. According to earlier research (Brouwer and Heeger, 2009), V1 appears to have a fundamentally distinct discontinuous coding, with respect to hue, based on opponent colors that neither model seems to reflect.

PROSOPAGNOSIA

Prosopagnosia is defined as the inability to recognize known and new faces. There often are not any broader issues with vision, object recognition, or memory that could account for this. These individuals are aware that a face is a face and not a dog or a tree, but they are unable to identify the face or recall having seen it before. These subjects make mistakes if other cues, such as haircut, voice, and gait, change, because they rely on them to identify people.

The right or bilateral fusiform—lingual gyrus, which contains neuron pathways that regulate facial perception and memory, may have been damaged or developed abnormally; the amygdala and temporal cortex may also be involved. Fusiform lesions are more likely to cause an apperceptive variant, whereas anterior temporal lobe lesions are more likely to cause an amnesia variety. A gap between the anterior and posterior face networks or a decreased functional activation of the areas in charge of facial recognition and identification appears to be the cause of development prosopagnosia. Modern imaging findings back up these ideas.

A deficit of perceptual encoding or perceptual maintenance? According to several findings, some individuals with developmental prosopagnosia (DP) exhibit broadly typical discrimination of unfamiliar faces presented simultaneously (Ulrich et al., 2017), and apparently normal recognition of facial emotion (Lee et al., 2010). These behavioral findings suggest that in many DP cases, early structural encoding may still be intact. Instead of categorizing DP as an apperceptive disorder, some researchers have suggested that in many cases, DP may be brought on by defective short-term face memory (STFM); many DPs may be able to produce appropriate perceptions, but be unable to maintain such memories over time. Although this mnemonic theory is a fascinating concept, it is not currently supported by strong scientific evidence. In instance, numerous group investigations have so far been unable to identify any impact of short-term memory strain on DPs' capacity for face matching. On the other hand, an investigation by Biotti et al. (2019) shows that perceptual encoding deficiencies are pervasive in this population. This does not mean that all DP cases are perceptive since neurodevelopmental populations are rarely this uniform.

Jackson et al. (2017) investigated the nature of face-specific WM deficits in DPs, with a particular emphasis on the role of perceptual encoding. The results cannot be fully explained by a face encoding deficit, at least not one that is sensitive to the cumulative effect of growing WM load and serial updating, even though they do not dispute that perceptual processing of faces is impaired among DPs and that encoding of faces into WM is insufficient to some extent. The observation that DPs are notably impaired on nonmatch trials brings attention to an essential aspect of how faces are less distinguishable in WM among this group of people when familiarity signals are lacking. It is crucial to demonstrate that DPs are less adept at nonmatch face recognition in WM because false positives can be just as harmful to social interaction as a lack of positive identification. There might also be a deficit in the face comparison process during retrieval given the evidence that the WM deficit among DPs may be caused by factors more than only issues with perceptual face encoding. The resemblance between the physically present test item and the stored item representation(s) has been shown to effect retrieval accuracy in previous studies utilizing healthy young adults and nonface stimuli. When the test object was similar to but different from one of the faces during encoding (i.e., a similar nonmatch), Jackson et al. (2017) found reduced WM than when the test item was a dissimilar nonmatch for complicated (but not basic) abstract shapes (Jackson et al., 2015). Towler and colleagues review the cognitive and neural foundations of DP and report that neuroimaging and electroencephalography studies of DP show intact early, structural perceptual processing of faces among DPs, suggesting that face recognition impairments may be more influenced by postperceptual and higher cognitive processes. According to Towler et al. (2016), there is evidence that DPs have defective configural processing of faces and are unable to properly interpret the visuospatial link between facial characteristics.

The inability to compare integrated, holistic representations of the test face with face representations stored in WM may be what causes the WM deficit for faces.

The more widespread face memory deficiencies among DPs imply that configural information may be particularly important in helping us compare the faces we see to the many face representations we have stored in our minds. According to studies, DP people perceive nonface stimuli holistically less well than non-DP people (Tanzer et al., 2013). As a result, reduced configural/holistic processing and an increased dependence on feature processing generally may cause faces to look more same.

The flexible resource theory states that increasing the resource allotted to an item can reduce representation noise and boost fidelity of encoding and storage. As a result, it is possible that DPs' signal-to-noise ratio for WM face representations is lower. Stronger coupling between features and their configurations may be reflected by a greater dependence on featural than configural processes, which could subsequently impair the processing of individual faces and the recognition of faces through memory comparison processes. It is also feasible, if we take into account the resource-based model of WM, that featural face processing will use up more WM encoding resources than holistic face processing. Binding numerous bits of data together into a single object can help to minimize or compress the amount of data that needs to be processed or encoded, allowing for the retention of objects with multiple features in the same way as objects with a single feature. As a result, ineffective face feature binding across DPs may result in higher encoding requirements per face and a corresponding decrease in signal-to-noise ratio. Alternately, the interference theory of WM (Oberauer and Lin, 2017) would contend that DPs experience more face interference than controls, which worsens the comparison process during retrieval by increasing confusion between the contents of memory.

Are individuals with DP unable to utilize holistic representations of individual faces? Towler et al. (2018) tried to find an answer to this question. To determine where the perceptual deficits in developing prosopagnosia are located, researchers used event-related brain potentials and performance measurements in a face identification matching test. They showed how identification matching processes involved comprehensive face representations. Their findings offer direct evidence from electrophysiology and behavior that the occipital–temporal face processing system in developmental prosopagnosia represents face identity in qualitatively different ways. As anticipated, DPs typically underperformed compared with controls in the face identity matching test. These limitations did not, however, apply to all trial types equally. On partial repetition trials, in particular on trials when the external facial features were repeated while the interior features varied, disproportionate performances were seen. Participants with DP misreported the repetition of the complete face on about one-third of these trials. RTs were slower when an internal feature change was accurately reported. These results indicate that during face matching tasks, people with DP have an unusual bias toward encoding facial features that are included in the external elements of the face, like haircut, and less emphasis on the representation of internal aspects.

What are the implications of the current findings for the neurological and cognitive mechanisms that lead to reduced face recognition in DP, and in particular for the function of holistic processing? These findings imply that, despite the presence of specialized face processing in people with DP, the face representations produced during face perception are abnormal in at least two ways. On the one hand, these representations seem to be more segmented than comprehensive. On the other hand, they could be comparatively insensitive to signals from the eye area that pertain to contrast and offer identity-related information about interior facial characteristics.

Deficits in holistic face processing in DP have a potential explanation, according to a research by Witthoft et al. (2014). From functional MRI data, receptive field sizes were calculated for distinct visual areas along the ventral route. Smaller receptive fields were reported in face- and extra object-selective cortical areas in DP patients. The combined inputs from lower-level visual feature analyzers may not reach face-selective brain regions due to smaller receptive fields. The collection of visual data from a large visual domain that spans the entire face is an essential requirement for creating holistic face representations. This need might not be reached in people with DP if they have face-selective brain areas with narrow receptive fields, leading to face representations that are heavily part-based.

Towler et al. (2018) study has, in conclusion, identified two connected areas where face identification matching processes between DPs and control participants differ. First, and most importantly, those with DP appear to engage in this process considerably more partially, and holistic face representations may not even be engaged. This is despite the fact that when control participants match two consecutively presented face pictures, holistic face representations are definitely implicated. Additionally, DPs show a significant propensity to encode exterior rather than interior face traits first.

References

Abel, P.L., O'Brien, B.J., Lia, B., Olavarria, J.F., 1997. Distribution of neurons projecting to the superior colliculus correlates with thick cytochrome oxidase stripes in macaque visual area V2. J. Comp. Neurol. 377 (3), 313–323. https://doi.org/10.1002/(SICI)1096-9861(19970120)377:3<313::AID-CNE1>3.0.CO;2-5.

Aggius-Vella, E., Chebat, D.R., Maidenbaum, S., Amedi, A., 2023. Activation of human visual area V6 during egocentric navigation with and without visual experience. Curr. Biol. 33 (7), 1211–1219.e5. https://doi.org/10.1016/j.cub.2023.02.025.

Bae, G.Y., Olkkonen, M., Allred, S.R., Flombaum, J.I., 2015. Why some colors appear more memorable than others: a model combining categories and particulars in color working memory. J. Exp. Psychol. Gen. 144 (4), 744–763. https://doi.org/10.1037/xge0000076.

Baizer, J., Ungerleider, L., Desimone, R., 1991. Organization of visual inputs to the inferior temporal and posterior parietal cortex in macaques. J. Neurosci. 11 (1), 168–190. https://doi.org/10.1523/JNEUROSCI.11-01-00168.1991.

Baizer, J.S., Desimone, R., Ungerleider, L.G., 1993. Comparison of subcortical connections of inferior temporal and posterior parietal cortex in monkeys. Vis. Neurosci. 10 (1), 59–72. https://doi.org/10.1017/S0952523800003229.

Berson, D.M., et al., 2002. Phototransduction by retinal ganglion cells that set the circadian clock. Science 295(5557), 1070–1073. https://doi.org/10.1126/science.1067262.

Biotti, F., Gray, K.L.H., Cook, R., 2019. Is developmental prosopagnosia best characterised as an apperceptive or mnemonic condition? Neuropsychologia 124, 285–298. https://doi.org/10.1016/j.neuropsychologia.2018.11.014.

Blanks, R.H.I., Clarke, R.J., Lui, F., Giolli, R.A., Van Pham, S., Torigoe, Y., 1995. Projections of the lateral terminal accessory optic nucleus of the common marmoset (*Callithrix jacchus*). J. Comp. Neurol. 354 (4), 511–532. https://doi.org/10.1002/cne.903540404.

Bloomfield, S.A., Dacheux, R.F., 2001. Rod vision: pathways and processing in the mammalian retina. Prog. Retin. Eye Res. 20 (3), 351–384. https://doi.org/10.1016/S1350-9462(00)00031-8.

Borostyánkői-Baldauf, Z., Herczeg, L., 2002. Parcellation of the human pretectal complex: a chemoarchitectonic reappraisal. Neuroscience 110 (3), 527–540. https://doi.org/10.1016/S0306-4522(01)00462-6.

Broers, N., Bainbridge, W.A., Michel, R., Balestrieri, E., Busch, N.A., 2022. The extent and specificity of visual exploration determines the formation of recollected memories in complex scenes. J. Vis. 22 (11), 9. https://doi.org/10.1167/jov.22.11.9.

Brouwer, G.J., Heeger, D.J., 2009. Decoding and reconstructing color from responses in human visual cortex. J. Neurosci. 29 (44), 13992–14003. https://doi.org/10.1523/JNEUROSCI.3577-09.2009.

Brouwer, G.J., Heeger, D.J., 2013. Categorical clustering of the neural representation of color. J. Neurosci. 33 (39), 15454–15465. https://doi.org/10.1523/JNEUROSCI.2472-13.2013.

Bull, M.S., Mitchell, S.K., Berkley, K.J., 1990. Convergent inputs to the inferior olive from the dorsal column nuclei and pretectum in the cat. Brain Res. 525 (1), 1–10. https://doi.org/10.1016/0006-8993(90)91314-7.

Burman, K., Darian-Smith, C., Darian-Smith, I., 2000. Macaque red nucleus: origins of spinal and olivary projections and terminations of cortical inputs. J. Comp. Neurol. 423 (2), 179–196. https://doi.org/10.1002/1096-9861(20000724)423:2<179::aid-cne1>3.0.co;2-#.

Büttner, U., Büttner-Ennever, J.A., 1988. Present concepts of oculomotor organization. Rev. Oculomot. Res. 2, 3–32.

Büttner-Ennever, J.A., Cohen, B., Horn, A.K.E., Reisine, H., 1996a. Efferent pathways of the nucleus of the optic tract in monkey and their role in eye movements. J. Comp. Neurol. 373 (1), 90–107. https://doi.org/10.1002/(SICI)1096-9861(19960909)373:1<90::AID-CNE8>3.0.CO;2-8.

Büttner-Ennever, J.A., Cohen, B., Horn, A.K.E., Reisine, H., 1996b. Pretectal projections to the oculomotor complex of the monkey and their role in eye movements. J. Comp. Neurol. 366 (2), 348–359. https://doi.org/10.1002/(SICI)1096-9861(19960304)366:2<348::AID-CNE12>3.0.CO;2-L.

Byzov, E.V., Kravchenko, S.V., Moiseev, M.A., Bezus, E.A., Doskolovich, L.L., 2020. Optimization method for designing double-surface refractive optical elements for an extended light source. Opt. Express 28 (17), 24431. https://doi.org/10.1364/OE.400609.

Cajal, S.R., Thorpe, S., Glickstein, M., 1972. The Structure of the Retina.

Calkins, D.J., Tsukamoto, Y., Sterling, P., 1998. Microcircuitry and mosaic of a blue–yellow ganglion cell in the primate retina. J. Neurosci. 18 (9), 3373–3385. https://doi.org/10.1523/JNEUROSCI.18-09-03373.1998.

Cerda-Company, X., Otazu, X., Sallent, N., Parraga, C.A., 2018. The effect of luminance differences on color assimilation. J. Vis. 18 (11), 10. https://doi.org/10.1167/18.11.10.

Chou, I han, Lisberger, S.G., 2004. The role of the frontal pursuit area in learning in smooth pursuit eye movements. J. Neurosci. 24 (17), 4124–4133. https://doi.org/10.1523/JNEUROSCI.0172-04.2004.

Clarke, R.J., Blanks, R.H.I., Giolli, R.A., 2003. Midbrain connections of the olivary pretectal nucleus in the marmoset (*Callithrix jacchus*): implications for the pupil light reflex pathway. Anat. Embryol. 207 (2), 149–155. https://doi.org/10.1007/s00429-003-0339-0.

Clarke, S., Miklossy, J., 1990. Occipital cortex in man: organization of callosal connections, related myelo- and cytoarchitecture, and putative boundaries of functional visual areas. J. Comp. Neurol. 298 (2), 188–214. https://doi.org/10.1002/cne.902980205.

Clower, D.M., Dum, R.P., Strick, P.L., 2005. Basal ganglia and cerebellar inputs to 'AIP.'. Cerebr. Cortex 15 (7), 913–920. https://doi.org/10.1093/cercor/bhh190.

Clower, D.M., West, R.A., Lynch, J.C., Strick, P.L., 2001. The inferior parietal lobule is the target of output from the superior colliculus, Hippocampus, and cerebellum. J. Neurosci. 21 (16), 6283–6291. https://doi.org/10.1523/JNEUROSCI.21-16-06283.2001.

Cowie, R.J., Robinson, D.L., 1994. Subcortical contributions to head movements in macaques. I. Contrasting effects of electrical stimulation of a medial pontomedullary region and the superior colliculus. J. Neurophysiol. 72 (6), 2648–2664. https://doi.org/10.1152/jn.1994.72.6.2648.

Culham, J.C., Kanwisher, N.G., 2001. Neuroimaging of cognitive functions in human parietal cortex. Curr. Opin. Neurobiol. 11 (2), 157–163. https://doi.org/10.1016/S0959-4388(00)00191-4.

Cushing, C.A., Im, H.Y., Adams Jr., R.B., Ward, N., Kveraga, K., 2019. Magnocellular and parvocellular pathway contributions to facial threat cue processing. Soc. Cognit. Affect Neurosci. 14 (2), 151–162. https://doi.org/10.1093/scan/nsz003.

Dacey, D.M., 1996. Circuitry for color coding in the primate retina. Proc. Natl. Acad. Sci. USA 93 (2), 582–588. https://doi.org/10.1073/pnas.93.2.582.

Dacey, D.M., Packer, O.S., 2003. Colour coding in the primate retina: diverse cell types and cone-specific circuitry. Curr. Opin. Neurobiol. 13 (4), 421–427. https://doi.org/10.1016/S0959-4388(03)00103-X.

Dai, J., Van Der, V.J., Swaab, D.F., Buijs, R.M., 1998. Human retinohypothalamic tract as revealed by in vitro postmortem tracing. J. Comp. Neurol. 397 (3), 357–370. https://doi.org/10.1002/(SICI)1096-9861(19980803)397:3<357::AID-CNE4>3.0.CO;2-1.

Damiano, C., Walther, D.B., 2019. Distinct roles of eye movements during memory encoding and retrieval. Cognition 184, 119–129. https://doi.org/10.1016/j.cognition.2018.12.014.

Ester, E.F., Sprague, T.C., Serences, J.T., 2020. Categorical biases in human occipitoparietal cortex. J. Neurosci. 40 (4), 917–931. https://doi.org/10.1523/JNEUROSCI.2700-19.2019.

Famiglietti, E.V., 1992. Dendritic Co-stratification of ON and ON-OFF directionally selective ganglion cells with starburst amacrine cells in rabbit retina. J. Comp. Neurol. 324 (3), 322–335. https://doi.org/10.1002/cne.903240303.

Fehlmann, B., Coynel, D., Schicktanz, N., et al., 2020. Visual exploration at higher fixation frequency increases subsequent memory recall. Cereb. Cortex Commun. 1 (1). https://doi.org/10.1093/texcom/tgaa032.

Fitzpatrick, D., Itoh, K., Diamond, I., 1983. The laminar organization of the lateral geniculate body and the striate cortex in the squirrel monkey (*Saimiri sciureus*). J. Neurosci. 3 (4), 673–702. https://doi.org/10.1523/JNEUROSCI.03-04-00673.1983.

Freedman, D.J., Assad, J.A., 2006. Experience-dependent representation of visual categories in parietal cortex. Nature 443 (7107), 85–88. https://doi.org/10.1038/nature05078.

Freedman, D.J., Riesenhuber, M., Poggio, T., Miller, E.K., 2003. A comparison of primate prefrontal and inferior temporal cortices during visual categorization. J. Neurosci. 23 (12), 5235–5246. https://doi.org/10.1523/JNEUROSCI.23-12-05235.2003.

Fries, W., 1990. Pontine projection from striate and prestriate visual cortex in the macaque monkey: an anterograde study. Vis. Neurosci. 4 (3), 205–216. https://doi.org/10.1017/S0952523800003357.

Galletti, C., Gamberini, M., Kutz, D.F., Fattori, P., Luppino, G., Matelli, M., 2001. The cortical connections of area V6: an occipito-parietal network processing visual information. Eur. J. Neurosci. 13 (8), 1572–1588. https://doi.org/10.1046/j.0953-816x.2001.01538.x.

Gamlin, P.D.R., 2006. The Pretectum: Connections and Oculomotor-Related Roles, pp. 379–405. https://doi.org/10.1016/S0079-6123(05)51012-4.

Gayer, N.S., Faull, R.L.M., 1988. Connections of the paraflocculus of the cerebellum with the superior colliculus in the rat brain. Brain Res. 449 (1–2), 253–270. https://doi.org/10.1016/0006-8993(88)91042-6.

Giolli, R.A., Blanks, R.H.I., Torigoe, Y., 1984. Pretectal and brain stem projections of the medial terminal nucleus of the accessory optic system of the rabbit and rat as studied by anterograde and retrograde neuronal tracing methods. J. Comp. Neurol. 227 (2), 228–251. https://doi.org/10.1002/cne.902270208.

Giolli, R.A., Gregory, K.M., Suzuki, D.A., Blanks, R.H.I., Lui, F., Betelak, K.F., 2001. Cortical and subcortical afferents to the nucleus reticularis tegmenti pontis and basal pontine nuclei in the macaque monkey. Vis. Neurosci. 18 (5), 725–740. https://doi.org/10.1017/S0952523801185068.

Giolli, R.A., Peterson, G.M., Ribak, C.E., McDonald, H.M., Blanks, R.H.I., Fallon, J.H., 1985. GABAergic neurons comprise a major cell type in rodent visual relay nuclei: an immunocytochemical study of pretectal and accessory optic nuclei. Exp. Brain Res. 61 (1). https://doi.org/10.1007/BF00235635.

Giolli, R.A., Torigoe, Y., Clarke, R.J., Blanks, R.H.I., Fallon, J.H., 1992. GABAergic and non-GABAergic projections of accessory optic nuclei, including the visual tegmental relay zone, to the nucleus of the optic tract and dorsal terminal accessory optic nucleus in rat. J. Comp. Neurol. 319 (3), 349–358. https://doi.org/10.1002/cne.903190304.

Goodale, M.A., Milner, A.D., 1992. Separate visual pathways for perception and action. Trends Neurosci. 15 (1), 20–25. https://doi.org/10.1016/0166-2236(92)90344-8.

Gouras, P., 2005. In: Kolb, H., Nelson, R., Fernandez, E., Jones, B. (Eds.), The Organization of the Retina and Visual System. Webvision.

Grantyn, R., Shapovalov, A.I., Shiriaev, B.I., 1982. Combined morphological and electrophysiological description of connections between single primary afferent fibres and individual motoneurons in the frog spinal cord. Exp. Brain Res. 48 (3). https://doi.org/10.1007/BF00238624.

Graybiel, A.M., Illing, R.B., 1994. Enkephalin-positive and acetylcholinesterase-positive patch systems in the superior colliculus have matching distributions but distinct developmental histories. J. Comp. Neurol. 340 (3), 297–310. https://doi.org/10.1002/cne.903400302.

Hadjikhani, N., Liu, A.K., Dale, A.M., Cavanagh, P., Tootell, R.B.H., 1998. Retinotopy and color sensitivity in human visual cortical area V8. Nat. Neurosci. 1 (3), 235–241. https://doi.org/10.1038/681.

Harting, J.K., Van Lieshout, D.P., 1991. Spatial relationships of axons arising from the substantia nigra, spinal trigeminal nucleus, and pedunculopontine tegmental nucleus within the intermediate gray of the cat superior colliculus. J. Comp. Neurol. 305 (4), 543–558. https://doi.org/10.1002/cne.903050403.

Hattar, S., 2002. Melanopsin-containing retinal ganglion cells: architecture, projections, and intrinsic photosensitivity. Science 295(5557), 1065–1070. https://doi.org/10.1126/science.1069609.

Hattar, S., Kumar, M., Park, A., et al., 2006. Central projections of melanopsin-expressing retinal ganglion cells in the mouse. J. Comp. Neurol. 497 (3), 326–349. https://doi.org/10.1002/cne.20970.

Hayakawa, Y., Nakajima, T., Takagi, M., Fukuhara, N., Abe, H., 2002. Human cerebellar activation in relation to saccadic eye movements: a functional magnetic resonance imaging study. Ophthalmologica 216 (6), 399–405. https://doi.org/10.1159/000067551.

Hendry, S.H., 2015. A neurochemically distinct third channel in the macaque dorsal lateral geniculate nucleus. Science 264, 575–577. https://doi.org/10.1126/science.8160015.

Hendry, S.H.C., Reid, R.C., 2000. The koniocellular pathway in primate vision. Annu. Rev. Neurosci. 23 (1), 127–153. https://doi.org/10.1146/annurev.neuro.23.1.127.

Hikosaka, O., Takikawa, Y., Kawagoe, R., 2000. Role of the basal ganglia in the control of purposive saccadic eye movements. Physiol. Rev. 80 (3), 953–978. https://doi.org/10.1152/physrev.2000.80.3.953.

Hoffmann, K.P., Distler, C., Erickson, R., 1991. Functional projections from striate cortex and superior temporal sulcus to the nucleus of the optic tract (NOT) and dorsal terminal nucleus of the accessory optic tract (DTN) of macaque monkeys. J. Comp. Neurol. 313 (4), 707–724. https://doi.org/10.1002/cne.903130413.

Hoffmann, K.P., Distler, C., Ilg, U., 1992. Callosal and superior temporal sulcus contributions to receptive field properties in the macaque monkey's nucleus of the optic tract and dorsal terminal nucleus of the accessory optic tract. J. Comp. Neurol. 321 (1), 150–162. https://doi.org/10.1002/cne.903210113.

Horn, A., Buttner-Ennever, J., Wahle, P., Reichenberger, I., 1994. Neurotransmitter profile of saccadic omnipause neurons in nucleus raphe interpositus. J. Neurosci. 14 (4), 2032–2046. https://doi.org/10.1523/JNEUROSCI.14-04-02032.1994.

Horn, A.K.E., Helmchen, C., Wahle, P., 2003. GABAergic neurons in the rostral mesencephalon of the macaque monkey that control vertical eye movements. Ann. N. Y. Acad. Sci. 1004 (1), 19–28. https://doi.org/10.1196/annals.1303.003.

Huerta, M.F., Kaas, J.H., 1990. Supplementary eye field as defined by intracortical microstimulation: connections in macaques. J. Comp. Neurol. 293 (2), 299–330. https://doi.org/10.1002/cne.902930211.

Huerta, M.F., Van Lieshout, D.P., Harting, J.K., 1991. Nigrotectal projections in the primate Galago crassicaudatus. Exp. Brain Res. 87 (2), 389–401. https://doi.org/10.1007/BF00231856.

Huk, A.C., Dougherty, R.F., Heeger, D.J., 2002. Retinotopy and functional subdivision of human areas MT and MST. J. Neurosci. 22 (16), 7195–7205. https://doi.org/10.1523/JNEUROSCI.22-16-07195.2002.

Im, H.Y., Adams, R.B., Cushing, C.A., Boshyan, J., Ward, N., Kveraga, K., 2018. Sex-related differences in behavioral and amygdalar responses to compound facial threat cues. Hum. Brain Mapp. 39 (7), 2725–2741. https://doi.org/10.1002/hbm.24035.

Ishai, A., Ungerleider, L.G., Haxby, J.V., 2000a. Distributed neural systems for the generation of visual images. Neuron 28 (3), 979–990. https://doi.org/10.1016/S0896-6273(00)00168-9.

Ishai, A., Ungerleider, L.G., Martin, A., Haxby, J.V., 2000b. The representation of objects in the human occipital and temporal cortex. J. Cognit. Neurosci. 12 (Suppl. 2), 35–51. https://doi.org/10.1162/089892900564055.

Jackson, M.C., Counter, P., Tree, J.J., 2017. Face working memory deficits in developmental prosopagnosia: tests of encoding limits and updating processes. Neuropsychologia 106, 60–70. https://doi.org/10.1016/j.neuropsychologia.2017.09.003.

Jackson, M.C., Linden, D.E.J., Roberts, M.V., Kriegeskorte, N., Haenschel, C., 2015. Similarity, not complexity, determines visual working memory performance. J. Experiment. Psychol. 41 (6), 1884–1892. https://doi.org/10.1037/xlm0000125.

James, T.W., Culham, J., Humphrey, G.K., Milner, A.D., Goodale, M.A., 2003. Ventral occipital lesions impair object recognition but not object-directed grasping: an fMRI study. Brain 126 (11), 2463–2475. https://doi.org/10.1093/brain/awg248.

Johansson, R., Holsanova, J., Holmqvist, K., 2006. Pictures and spoken descriptions elicit similar eye movements during mental imagery, both in light and in complete darkness. Cognit. Sci. 30 (6), 1053–1079. https://doi.org/10.1207/s15516709cog0000_86.

Kaplan, E., 2003. The M, P, and K pathways of the primate visual system. In: The Visual Neurosciences, 2-Vol. Set. The MIT Press, pp. 481–493. https://doi.org/10.7551/mitpress/7131.003.0036.

Kebschull, J.M., Casoni, F., Consalez, G.G., et al., February 13, 2023. Cerebellum lecture: the cerebellar nuclei—core of the cerebellum. Cerebellum. https://doi.org/10.1007/s12311-022-01506-0.

King, A.J., 2004. The superior colliculus. Curr. Biol. 14 (9), R335–R338. https://doi.org/10.1016/j.cub.2004.04.018.

Kitao, Y., Nakamura, Y., Kudo, M., Moriizumi, T., Tokuno, H., 1989. The cerebral and cerebellar connections of pretecto-thalamic and pretecto-olivary neurons in the anterior pretectal nucleus of the cat. Brain Res. 484 (1–2), 304–313. https://doi.org/10.1016/0006-8993(89)90374-0.

Kragel, J.E., Voss, J.L., 2022. Looking for the neural basis of memory. Trends Cognit. Sci. 26 (1), 53–65. https://doi.org/10.1016/j.tics.2021.10.010.

Kveraga, K., Boshyan, J., Bar, M., 2007. Magnocellular projections as the trigger of top-down facilitation in recognition. J. Neurosci. 27 (48), 13232–13240. https://doi.org/10.1523/JNEUROSCI.3481-07.2007.

Kveraga, K., Hughes, H.C., 2005. Effects of stimulus-response uncertainty on saccades to near-threshold targets. Exp. Brain Res. 162 (3), 401–405. https://doi.org/10.1007/s00221-004-2155-z.

Lachica, E.A., Casagrande, V.A., 1992. Direct W-like geniculate projections to the cytochrome oxidase (CO) blobs in primate visual cortex: axon morphology. J. Comp. Neurol. 319 (1), 141–158. https://doi.org/10.1002/cne.903190112.

Lee, B.B., Dacey, D.M., Smith, V.C., Pokorny, J., 1999. Horizontal cells reveal cone type-specific adaptation in primate retina. Proc. Natl. Acad. Sci. USA 96 (25), 14611–14616. https://doi.org/10.1073/pnas.96.25.14611.

Lee, Y., Duchaine, B., Wilson, H.R., Nakayama, K., 2010. Three cases of developmental prosopagnosia from one family: detailed neuropsychological and psychophysical investigation of face processing. Cortex 46 (8), 949–964. https://doi.org/10.1016/j.cortex.2009.07.01221.

Leichnetz, G.R., Gonzalo-Ruiz, A., 1996. Prearcuate cortex in the cebus monkey has cortical and subcortical connections like the macaque frontal eye field and projects to fastigial-recipient oculomotor-related brainstem nuclei. Brain Res. Bull. 41 (1), 1–29. https://doi.org/10.1016/0361-9230(96)00154-2.

Leichnetz, G.R., Spencer, R.F., Hardy, S.G.P., Astruc, J., 1981. The prefrontal corticotectal projection in the monkey; an anterograde and retrograde horseradish peroxidase study. Neuroscience 6 (6), 1023–1041. https://doi.org/10.1016/0306-4522(81)90068-3.

Leventhal, A.G., et al., 1981. Retinal ganglion cell classes in the Old World monkey: morphology and central projections. Science 213 (4512), 1139–1142. https://doi.org/10.1126/science.7268423.

Livingstone, M., Hubel, D., 1984. Anatomy and physiology of a color system in the primate visual cortex. J. Neurosci. 4 (1), 309–356. https://doi.org/10.1523/JNEUROSCI.04-01-00309.1984.

Livingstone, M., Hubel, D., 1988. Segregation of form, color, movement, and depth: anatomy, physiology, and perception. Science 240 (4853), 740–749. https://doi.org/10.1126/science.3283936 (1979).

Lucas, R.J., Douglas, R.H., Foster, R.G., 2001. Characterization of an ocular photopigment capable of driving pupillary constriction in mice. Nat. Neurosci. 4 (6), 621–626. https://doi.org/10.1038/88443.

Lui, F., Gregory, K.M., Blanks, R.H.I., Giolli, R.A., 1995. Projections from visual areas of the cerebral cortex to pretectal nuclear complex, terminal accessory optic nuclei, and superior colliculus in macaque monkey. J. Comp. Neurol. 363 (3), 439–460. https://doi.org/10.1002/cne.903630308.

Luna, B., 1998. Dorsal cortical regions subserving visually guided saccades in humans: an fMRI study. Cerebr. Cortex 8 (1), 40–47. https://doi.org/10.1093/cercor/8.1.40.

Lund, J.S., 1988. Anatomical organization of macaque monkey striate visual cortex. Annu. Rev. Neurosci. 11 (1), 253–288. https://doi.org/10.1146/annurev.ne.11.030188.001345.

Luppino, G., Calzavara, R., Rozzi, S., Matelli, M., 2001. Projections from the superior temporal sulcus to the agranular frontal cortex in the macaque. Eur. J. Neurosci. 14 (6), 1035–1040. https://doi.org/10.1046/j.0953-816x.2001.01734.x.

Lynch, J.C., Hoover, J.E., Strick, P.L., 1994. Input to the primate frontal eye field from the substantia nigra, superior colliculus, and dentate nucleus demonstrated by transneuronal transport. Exp. Brain Res. 100 (1). https://doi.org/10.1007/BF00227293.

Lynch, J.C., Tian, J.R., 2006. Cortico-cortical networks and cortico-subcortical loops for the higher control of eye movements. Progr. Brain Res 151, 461–501. https://doi.org/10.1016/S0079-6123(05)51015-X15.

Macneil, M.A., Heussy, J.K., Dacheux, R.F., Raviola, E., Masland, R.H., 1999. The shapes and numbers of amacrine cells: matching of photofilled with Golgi-stained cells in the rabbit retina and comparison with other mammalian species. J. Comp. Neurol. 413 (2), 305–326. https://doi.org/10.1002/(SICI)1096-9861(19991018)413:2<305::AID-CNE10>3.0.CO;2-E.

Malach, R., Levy, I., Hasson, U., 2002. The topography of high-order human object areas. Trends Cognit. Sci. 6 (4), 176–184. https://doi.org/10.1016/S1364-6613(02)01870-3.

Marino, R.A., Levy, R., Munoz, D.P., 2015. Linking express saccade occurrence to stimulus properties and sensorimotor integration in the superior colliculus. J. Neurophysiol. 114 (2), 879–892. https://doi.org/10.1152/jn.00047.2015.

Masland, R., 2001a. Neuronal diversity in the retina. Curr. Opin. Neurobiol. 11 (4), 431–436. https://doi.org/10.1016/S0959-4388(00)00230-0.

Masland, R.H., 2001b. The fundamental plan of the retina. Nat. Neurosci. 4 (9), 877–886. https://doi.org/10.1038/nn0901-877.

May, J.G., Keller, E.L., Suzuki, D.A., 1988. Smooth-pursuit eye movement deficits with chemical lesions in the dorsolateral pontine nucleus of the monkey. J. Neurophysiol. 59 (3), 952–977. https://doi.org/10.1152/jn.1988.59.3.952.

May, P.J., 2006. The Mammalian Superior Colliculus: Laminar Structure and Connections, pp. 321–378. https://doi.org/10.1016/S0079-6123(05)51011-2.

May, P.J., Hartwich-Young, R., Nelson, J., Sparks, D.L., Porter, J.D., 1990. Cerebellotectal pathways in the macaque: implications for collicular generation of saccades. Neuroscience 36 (2), 305–324. https://doi.org/10.1016/0306-4522(90)90428-7.

Meredith, M.A., Stein, B.E., 1985. Descending efferents from the superior colliculus relay integrated multisensory information. Science 227 (4687), 657–659. https://doi.org/10.1126/science.3969558 (1979).

Meredith, M.A., Clemo, H.R., Stein, B.E., 1991. Somatotopic component of the multisensory map in the deep laminae of the cat superior colliculus. J. Comp. Neurol. 312 (3), 353–370. https://doi.org/10.1002/cne.903120304.

Monnier, P., Shevell, S.K., 2004. Chromatic induction from S-cone patterns. Vis. Res. 44 (9), 849–856. https://doi.org/10.1016/j.visres.2003.11.004.

Morin, L.P., Blanchard, J.H., Provencio, I., 2003. Retinal ganglion cell projections to the hamster suprachiasmatic nucleus, intergeniculate leaflet, and visual midbrain: bifurcation and melanopsin immunoreactivity. J. Comp. Neurol. 465 (3), 401–416. https://doi.org/10.1002/cne.10881.

Moschovakis, A.K., Karabelas, A.B., Highstein, S.M., 1988. Structure-function relationships in the primate superior colliculus. I. Morphological classification of efferent neurons. J. Neurophysiol. 60 (1), 232–262. https://doi.org/10.1152/jn.1988.60.1.232.

Nagao, S., Kitamura, T., Nakamura, N., Hiramatsu, T., Yamada, J., 1997. Differences of the primate flocculus and ventral paraflocculus in the mossy and climbing fiber input organization. J. Comp. Neurol. 382 (4), 480–498. https://doi.org/10.1002/(SICI)1096-9861(19970616)382:4<480::AID-CNE5>3.0.CO;2-Z.

Nieuwenhuys, R., ten Donkelaar, H.J., Nicholson, C., 1998. The Central Nervous System of Vertebrates. Springer Berlin Heidelberg. https://doi.org/10.1007/978-3-642-18262-4.

Nunez, V., Shapley, R.M., Gordon, J., 2018. Cortical double-opponent cells in color perception: perceptual scaling and chromatic visual evoked potentials. Iperception 9 (1). https://doi.org/10.1177/2041669517752715, 204166951775271.

Oberauer, K., Lin, H.Y., 2017. An interference model of visual working memory. Psychol. Rev. 124 (1), 21–59. https://doi.org/10.1037/rev0000044.

Optican, L.M., Robinson, D.A., 1980. Cerebellar-dependent adaptive control of primate saccadic system. J. Neurophysiol. 44 (6), 1058–1076. https://doi.org/10.1152/jn.1980.44.6.1058.

Osanai, R., Nagao, S., Kitamura, T., Kawabata, I., Yamada, J., 1999. Differences in mossy and climbing afferent sources between flocculus and ventral and dorsal paraflocculus in the rat. Exp. Brain Res. 124 (2), 248–264. https://doi.org/10.1007/s002210050620.

Osborne, L.C., Bialek, W., Lisberger, S.G., 2004. Time course of information about motion direction in visual area MT of macaque monkeys. J. Neurosci. 24 (13), 3210–3222. https://doi.org/10.1523/JNEUROSCI.5305-03.2004.

Otazu, X., Cerda, X., 2022. The contribution of luminance and chromatic channels to color assimilation. J. Vis. 22 (6). https://doi.org/10.1167/jov.22.6.10.

Otazu, X., Parraga, C.A., Vanrell, M., 2010. Toward a unified chromatic induction model. J. Vis. 10 (12). https://doi.org/10.1167/10.12.5, 5-5.

Paller, K.A., Wagner, A.D., 2002. Observing the transformation of experience into memory. Trends Cognit. Sci. 6 (2), 93–102. https://doi.org/10.1016/S1364-6613(00)01845-3.

Panichello, M.F., DePasquale, B., Pillow, J.W., Buschman, T.J., 2019. Error-correcting dynamics in visual working memory. Nat. Commun. 10 (1), 3366. https://doi.org/10.1038/s41467-019-11298-3.

Peck, C.K., Baro, J.A., Warder, S.M., 1993. Chapter 9 Sensory Integration in the Deep Layers of Superior Colliculus, pp. 91–102. https://doi.org/10.1016/S0079-6123(08)60360-X.

Perry, V.H., Oehler, R., Cowey, A., 1984. Retinal ganglion cells that project to the dorsal lateral geniculate nucleus in the macaque monkey. Neuroscience 12 (4), 1101–1123. https://doi.org/10.1016/0306-4522(84)90006-X.

Petit, L., Haxby, J.V., 1999. Functional anatomy of pursuit eye movements in humans as revealed by fMRI. J. Neurophysiol. 82 (1), 463–471. https://doi.org/10.1152/jn.1999.82.1.463.

Pierrot-Deseilligny, C.H., Chain, F., Gray, F., Serdaru, M., Escourolle, R., Lhermitte, F., 1982. Parinaud's syndrome. Brain 105 (4), 667–696. https://doi.org/10.1093/brain/105.4.667.

Polyak, S., 1941. The Retina. University of Chicago Press.

Popov, T., Staudigl, T., 2023. Cortico-ocular coupling in the service of episodic memory formation. Prog. Neurobiol. 227, 102476. https://doi.org/10.1016/j.pneurobio.2023.102476.

Priebe, N.J., Lisberger, S.G., 2004. Estimating target speed from the population response in visual area MT. J. Neurosci. 24 (8), 1907–1916. https://doi.org/10.1523/JNEUROSCI.4233-03.2004.

Provencio, I., Rollag, M.D., Castrucci, A.M., 2002. Photoreceptive net in the mammalian retina. Nature 415 (6871). https://doi.org/10.1038/415493a, 493-493.

Rahmati, M., Curtis, C.E., Sreenivasan, K.K., 2023. Mnemonic representations in human lateral geniculate nucleus. Front. Behav. Neurosci. 17. https://doi.org/10.3389/fnbeh.2023.1094226.

Redgrave, P., Telford, S., Wang, S., McHaffie, J.G., Stein, B.E., 1996. Functional anatomy of nociceptive neurones in rat superior colliculus. Prog. Brain Res. 107, 403–415. https://doi.org/10.1016/s0079-6123(08)61878-6.

Reese, B.E., Cowey, A., 1990. Fibre organization of the monkey's optic tract: I. Segregation of functionally distinct optic axons. J. Comp. Neurol. 295 (3), 385–400. https://doi.org/10.1002/cne.902950304.

Ringach, D.L., Shapley, R.M., Hawken, M.J., 2002. Orientation selectivity in macaque V1: diversity and laminar dependence. J. Neurosci. 22 (13), 5639–5651. https://doi.org/10.1523/JNEUROSCI.22-13-05639.2002.

Rizzolatti, G., Matelli, M., 2003. Two different streams form the dorsal visual system: anatomy and functions. Exp. Brain Res. 153 (2), 146–157. https://doi.org/10.1007/s00221-003-1588-0.

Robinson, F.R., Fuchs, A.F., 2001. The role of the cerebellum in voluntary eye movements. Annu. Rev. Neurosci. 24 (1), 981–1004. https://doi.org/10.1146/annurev.neuro.24.1.981.

Robinson, F.R., Phillips, J.O., Fuchs, A.F., 1994. Coordination of gaze shifts in primates: brainstem inputs to neck and extraocular motoneuron pools. J. Comp. Neurol. 346 (1), 43–62. https://doi.org/10.1002/cne.903460104.

Rockhill, R.L., Daly, F.J., MacNeil, M.A., Brown, S.P., Masland, R.H., 2002. The diversity of ganglion cells in a mammalian retina. J. Neurosci. 22 (9), 3831–3843. https://doi.org/10.1523/JNEUROSCI.22-09-03831.2002.

Rodieck, R.W., Watanabe, M., 1993. Survey of the morphology of macaque retinal ganglion cells that project to the pretectum, superior colliculus, and parvicellular laminae of the lateral geniculate nucleus. J. Comp. Neurol. 338 (2), 289–303. https://doi.org/10.1002/cne.903380211.

Rosano, C., Krisky, C.M., Welling, J.S., et al., 2002. Pursuit and saccadic eye movement subregions in human frontal eye field: a high-resolution fMRI investigation. Cerebr. Cortex 12 (2), 107–115. https://doi.org/10.1093/cercor/12.2.107.

Rosano, C., Sweeney, J.A., Melchitzky, D.S., Lewis, D.A., 2003. The human precentral sulcus: chemoarchitecture of a region corresponding to the frontal eye fields. Brain Res. 972 (1–2), 16–30. https://doi.org/10.1016/S0006-8993(03)02431-4.

Rozzi, S., Calzavara, R., Belmalih, A., et al., 2006. Cortical connections of the inferior parietal cortical convexity of the macaque monkey. Cerebr. Cortex 16 (10), 1389–1417. https://doi.org/10.1093/cercor/bhj076.

Ruigrok, T.J.H., 2003. Collateralization of climbing and mossy fibers projecting to the nodulus and flocculus of the rat cerebellum. J. Comp. Neurol. 466 (2), 278–298. https://doi.org/10.1002/cne.10889.

Saleem, K.S., Suzuki, W., Tanaka, K., Hashikawa, T., 2000. Connections between anterior inferotemporal cortex and superior temporal sulcus regions in the macaque monkey. J. Neurosci. 20 (13), 5083–5101. https://doi.org/10.1523/JNEUROSCI.20-13-05083.2000.

Schmahmann, J.D., Pandya, D.N., 1993. Prelunate, occipitotemporal, and parahippocampal projections to the basis pontis in rhesus monkey. J. Comp. Neurol. 337 (1), 94–112. https://doi.org/10.1002/cne.903370107.

Scudder, C., Kaneko, C., Fuchs, A., 2002. The brainstem burst generator for saccadic eye movements. Exp. Brain Res. 142 (4), 439–462. https://doi.org/10.1007/s00221-001-0912-9.

Scudder, C.A., Moschovakis, A.K., Karabelas, A.B., Highstein, S.M., 1996. Anatomy and physiology of saccadic long-lead burst neurons recorded in the alert squirrel monkey. I. Descending projections from the mesencephalon. J. Neurophysiol. 76 (1), 332–352. https://doi.org/10.1152/jn.1996.76.1.332.

Sereno, M.I., Dale, A.M., Reppas, J.B., et al., 1995. Borders of multiple visual areas in humans revealed by functional magnetic resonance imaging. Science 268 (5212), 889–893. https://doi.org/10.1126/science.7754376 (1979).

Shevell, S.K., Monnier, P., 2005. Color shifts from S-cone patterned backgrounds: contrast sensitivity and spatial frequency selectivity. Vis. Res. 45 (9), 1147–1154. https://doi.org/10.1016/j.visres.2004.11.013.

Shinoda, Y., Sugiuchi, Y., Izawa, Y., Hata, Y., 2006. Long Descending Motor Tract Axons and Their Control of Neck and Axial Muscles, pp. 527–563. https://doi.org/10.1016/S0079-6123(05)51017-3.

Shook, B.L., Schlag-Rey, M., Schlag, J., 1990. Primate supplementary eye field: I. Comparative aspects of mesencephalic and pontine connections. J. Comparat. Neurol. 301 (4), 618–642. https://doi.org/10.1002/cne.903010410.

Shook, B.L., Schlag-Rey, M., Schlag, J., 1991. Primate supplementary eye field. II. Comparative aspects of connections with the thalamus, corpus striatum, and related forebrain nuclei. J. Comp. Neurol. 307 (4), 562–583. https://doi.org/10.1002/cne.903070405.

Sincich, L.C., Horton, J.C., 2002. Divided by cytochrome oxidase: a map of the projections from V1 to V2 in macaques. Science 295 (5560), 1734–1737. https://doi.org/10.1126/science.1067902 (1979).

Sincich, L.C., Horton, J.C., 2003. Independent projection streams from macaque striate cortex to the second visual area and middle temporal area. J. Neurosci. 23 (13), 5684–5692. https://doi.org/10.1523/JNEUROSCI.23-13-05684.2003.

Sincich, L.C., Horton, J.C., 2005. Input to V2 thin stripes arises from V1 cytochrome oxidase patches. J. Neurosci. 25 (44), 10087–10093. https://doi.org/10.1523/JNEUROSCI.3313-05.2005.

Sincich, L.C., Park, K.F., Wohlgemuth, M.J., Horton, J.C., 2004. Bypassing V1: a direct geniculate input to area MT. Nat. Neurosci. 7 (10), 1123–1128. https://doi.org/10.1038/nn1318.

Sommer, M.A., Wurtz, R.H., 2000. Composition and topographic organization of signals sent from the frontal eye field to the superior colliculus. J. Neurophysiol. 83 (4), 1979–2001. https://doi.org/10.1152/jn.2000.83.4.1979.

Stanton, G.B., Goldberg, M.E., Bruce, C.J., 1988. Frontal eye field efferents in the macaque monkey: I. Subcortical pathways and topography of striatal and thalamic terminal fields. J. Comp. Neurol. 271 (4), 473–492. https://doi.org/10.1002/cne.902710402.

Sugita, S., Noda, H., 1991. Pathways and terminations of axons arising in the fastigial oculomotor region of macaque monkeys. Neurosci. Res. 10 (2), 118–136. https://doi.org/10.1016/0168-0102(91)90035-W.

Suzuki, D.A., Yamada, T., Hoedema, R., Yee, R.D., 1999. Smooth-pursuit eye-movement deficits with chemical lesions in macaque nucleus reticularis tegmenti pontis. J. Neurophysiol. 82 (3), 1178–1186. https://doi.org/10.1152/jn.1999.82.3.1178.

Takahashi, M., Sugiuchi, Y., Shinoda, Y., 2007. Commissural mirror-symmetric excitation and reciprocal inhibition between the two superior colliculi and their roles in vertical and horizontal eye movements. J. Neurophysiol. 98 (5), 2664–2682. https://doi.org/10.1152/jn.00696.200711.

Tanaka, M., Lisberger, S.G., 2002. Role of arcuate frontal cortex of monkeys in smooth pursuit eye movements. I. Basic response properties to retinal image motion and position. J. Neurophysiol. 87 (6), 2684–2699. https://doi.org/10.1152/jn.2002.87.6.2684.

Tanzer, M., Freud, E., Ganel, T., Avidan, G., 2013. General holistic impairment in congenital prosopagnosia: evidence from Garner's speeded-classification task. Cognit. Neuropsychol. 30 (6), 429–445. https://doi.org/10.1080/02643294.2013.873715.

Tigges, J., Tigges, M., 1981. Distribution of Retinofugal and Corticofugal Axon Terminals in the Superior Colliculus of Squirrel Monkey.

Tootell, R., Hamilton, S., 1989. Functional anatomy of the second visual area (V2) in the macaque. J. Neurosci. 9 (8), 2620–2644. https://doi.org/10.1523/JNEUROSCI.09-08-02620.1989.

Tootell, R.B.H., 2004. Search for color "center(s)" in macaque visual cortex. Cerebr. Cortex 14 (4), 353–363. https://doi.org/10.1093/cercor/bhh001.

Tootell, R.B.H., 2001. Where is "dorsal V4" in human visual cortex? Retinotopic, topographic and functional evidence. Cerebr. Cortex 11 (4), 298–311. https://doi.org/10.1093/cercor/11.4.298.

Tootell, R.B.H., Dale, A.M., Sereno, M.I., Malach, R., 1996. New images from human visual cortex. Trends Neurosci. 19 (11), 481–489. https://doi.org/10.1016/S0166-2236(96)10053-9.

Towler, J., Fisher, K., Eimer, M., 2018. Holistic face perception is impaired in developmental prosopagnosia. Cortex 108, 112–126. https://doi.org/10.1016/j.cortex.2018.07.019.

Towler, J., Parketny, J., Eimer, M., 2016. Perceptual face processing in developmental prosopagnosia is not sensitive to the canonical location of face parts. Cortex 74, 53–66. https://doi.org/10.1016/j.cortex.2015.10.018.

Trejo, L.J., Cicerone, C.M., 1984. Cells in the pretectal olivary nucleus are in the pathway for the direct light reflex of the pupil in the rat. Brain Res. 300 (1), 49–62. https://doi.org/10.1016/0006-8993(84)91340-4.

Tsao, D.Y., Freiwald, W.A., Knutsen, T.A., Mandeville, J.B., Tootell, R.B.H., 2003. Faces and objects in macaque cerebral cortex. Nat. Neurosci. 6 (9), 989–995. https://doi.org/10.1038/nn1111.

Ulrich, P.I., Wilkinson, D.T., Ferguson, H.J., Smith, L.J., Bindemann, M., Johnston, R.A., Schmalzl, L., 2017. Perceptual and memorial contributions to developmental prosopagnosia. Quarter. J. Experiment. Psychol. 70 (2), 298–315. https://doi.org/10.1080/17470218.2016.1177101.

Ungerleider, L., Mishkin, M., 1982. Analysis of Visual Behavior: Two Cortical Visual Systems. MIT Press.

van der Want, J.J.L., Nunes Cardozo, J.J., van der Togt, C., 1992. Chapter 14 GABAergic Neurons and Circuits in the Pretectal Nuclei and the Accessory Optic System of Mammals, pp. 283–305. https://doi.org/10.1016/S0079-6123(08)63619-5.

Vanduffel, W., 2002. The organization of orientation selectivity throughout macaque visual cortex. Cerebr. Cortex 12 (6), 647–662. https://doi.org/10.1093/cercor/12.6.647.

Van Essen, D.C., Newsome, W.T., Maunsell, J.H.R., 1984. The visual field representation in striate cortex of the macaque monkey: asymmetries, anisotropies, and individual variability. Vis. Res. 24 (5), 429–448. https://doi.org/10.1016/0042-6989(84)90041-5.

Van Essen, D.C., 2004. Surface-based approaches to spatial localization and registration in primate cerebral cortex. NeuroImage 23 (Suppl. 1), S97–S107. https://doi.org/10.1016/j.neuroimage.2004.07.024.

van Kan, P.L., Houk, J.C., Gibson, A.R., 1993. Output organization of intermediate cerebellum of the monkey. J. Neurophysiol. 69 (1), 57–73. https://doi.org/10.1152/jn.1993.69.1.57.

Voss, J.L., Bridge, D.J., Cohen, N.J., Walker, J.A., 2017. A closer look at the Hippocampus and memory. Trends Cognit. Sci. 21 (8), 577–588. https://doi.org/10.1016/j.tics.2017.05.008.

Wallace, M.N., Fredens, K., 1989. Relationship of afferent inputs to the lattice of high NADPH-diaphorase activity in the mouse superior colliculus. Exp. Brain Res. 78 (2). https://doi.org/10.1007/BF00228917.

Wallack, M.N., 1988. Lattices of high histochemical activity occur in the human, monkey, and cat superior colliculus. Neuroscience 25 (2), 569–583. https://doi.org/10.1016/0306-4522(88)90259-X.

Wang, Y.C., Bianciardi, M., Chanes, L., Satpute, A.B., 2020. Ultra high field fMRI of human superior colliculi activity during affective visual processing. Sci. Rep. 10 (1). https://doi.org/10.1038/s41598-020-57653-z.

Watanabe, M., Rodieck, R.W., 1989. Parasol and midget ganglion cells of the primate retina. J. Comp. Neurol. 289 (3), 434–454. https://doi.org/10.1002/cne.902890308.

Wiberg, M., Blomqvist, A., 1984. The projection to the mesencephalon from the dorsal column nuclei. An anatomical study in the cat. Brain Res. 311 (2), 225–244. https://doi.org/10.1016/0006-8993(84)90086-6.

Wiberg, M., Westman, J., Blomqvist, A., 1987. Somatosensory projection to the mesencephalon: an anatomical study in the monkey. J. Comp. Neurol. 264 (1), 92–117. https://doi.org/10.1002/cne.902640108.

Witthoft, N., Nguyen, M.L., Golarai, G., et al., 2014. Where is human V4? Predicting the location of hV4 and VO1 from cortical folding. Cerebr. Cortex 24 (9), 2401–2408. https://doi.org/10.1093/cercor/bht092.

Xiao, Y., Felleman, D.J., 2004. Projections from primary visual cortex to cytochrome oxidase thin stripes and interstripes of macaque visual area 2. Proc. Natl. Acad. Sci. USA 101 (18), 7147–7151. https://doi.org/10.1073/pnas.0402052101.

Xiong, G., Hiramatsu, T., Nagao, S., 2002. Corticopontocerebellar pathway from the prearcuate region to hemispheric lobule VII of the cerebellum: an anterograde and retrograde tracing study in the monkey. Neurosci. Lett. 322 (3), 173–176. https://doi.org/10.1016/S0304-3940(02)00108-8.

Xu, X., Bosking, W.H., White, L.E., Fitzpatrick, D., Casagrande, V.A., 2005. Functional organization of visual cortex in the prosimian bush baby revealed by optical imaging of intrinsic signals. J. Neurophysiol. 94 (4), 2748–2762. https://doi.org/10.1152/jn.00354.2005.

Yamada, T., Suzuki, D.A., Yee, R.D., 1996. Smooth pursuitlike eye movements evoked by microstimulation in macaque nucleus reticularis tegmenti pontis. J. Neurophysiol. 76 (5), 3313–3324. https://doi.org/10.1152/jn.1996.76.5.3313.

Yan, C., Christophel, T.B., Allefeld, C., Haynes, J.D., 2023. Categorical working memory codes in human visual cortex. Neuroimage 274, 120149. https://doi.org/10.1016/j.neuroimage.2023.120149.

Young, M.J., Lund, R.D., 1994. The anatomical substrates subserving the pupillary light reflex in rats: origin of the consensual pupillary response. Neuroscience 62 (2), 481–496. https://doi.org/10.1016/0306-4522(94)90381-6.

Zeki, S., Watson, J., Lueck, C., Friston, K., Kennard, C., Frackowiak, R., 1991. A direct demonstration of functional specialization in human visual cortex. J. Neurosci. 11 (3), 641–649. https://doi.org/10.1523/JNEUROSCI.11-03-00641.1991.

Zhang, Y., Valsecchi, M., Gegenfurtner, K.R., Chen, J., 2023. The time course of chromatic adaptation in human early visual cortex revealed by SSVEPs. J. Vis. 23 (5), 17. https://doi.org/10.1167/jov.23.5.17.

Zhou, Q., Du, J., Gao, R., et al., 2023. Discriminative neural pathways for perception-cognition activity of color and face in the human brain. Cerebr. Cortex 33 (5), 1972–1984. https://doi.org/10.1093/cercor/bhac186.

Further Reading

Büttner-Ennever, J.A., Horn, A.K.E., 1996. Pathways from cell groups of the paramedian tracts to the floccular region. Ann. N. Y. Acad. Sci. 781 (1 Lipids and Sy), 532–540. https://doi.org/10.1111/j.1749-6632.1996.tb15726.x.

Choi, S.H., Jeong, G., Kim, Y.B., Cho, Z.H., 2020. Proposal for human visual pathway in the extrastriate cortex by fiber tracking method using diffusion-weighted MRI. Neuroimage 220, 117145. https://doi.org/10.1016/j.neuroimage.2020.117145.

Cowie, R.J., Smith, M.K., Robinson, D.L., 1994. Subcortical contributions to head movements in macaques. II. Connections of a medial pontomedullary head-movement region. J. Neurophysiol. 72 (6), 2665–2682. https://doi.org/10.1152/jn.1994.72.6.2665.

Ester, E.F., Sprague, T.C., Serences, J.T., 2015. Parietal and frontal cortex encode stimulus-specific mnemonic representations during visual working memory. Neuron 87 (4), 893–905. https://doi.org/10.1016/j.neuron.2015.07.013.

Kurzawski, J.W., Mikellidou, K., Morrone, M.C., Pestilli, F., 2020. The visual white matter connecting human area prostriata and the thalamus is retinotopically organized. Brain Struct. Funct. 225 (6), 1839–1853. https://doi.org/10.1007/s00429-020-02096-5.

Kveraga, K., Im, H.Y., Ward, N., Adams, R.B., 2020. Fast saccadic and manual responses to faces presented to the koniocellular visual pathway. J. Vis. 20 (2). https://doi.org/10.1167/jov.20.2.9.

Kveraga, K., De Vito, D., Cushing, C., Im, H.Y., Albohn, D.N., Adams, R.B., 2019. Spatial and feature-based attention to expressive faces. Exp. Brain Res. 237 (4), 967–975. https://doi.org/10.1007/s00221-019-05472-8.

Leventhal, A., Thompson, K., Liu, D., Zhou, Y., Ault, S., 1995. Concomitant sensitivity to orientation, direction, and color of cells in layers 2, 3, and 4 of monkey striate cortex. J. Neurosci. 15 (3), 1808–1818. https://doi.org/10.1523/JNEUROSCI.15-03-01808.1995.

Monnier, P., Shevell, S.K., 2003. Large shifts in color appearance from patterned chromatic backgrounds. Nat. Neurosci. 6 (8), 801–802. https://doi.org/10.1038/nn1099.

Nagao, S., 1992. Different roles of flocculus and ventral paraflocculus for oculomotor control in the primate. Neuroreport 3 (1), 13–16. https://doi.org/10.1097/00001756-199201000-00003.

Nouri, A., Ester, E.F., 2020. Recovery of information from latent memory stores decreases over time. Cognit. Neurosci. 11 (1–2), 101–110. https://doi.org/10.1080/17588928.2019.1617258.

Perry, V.H., Cowey, A., 1984. Retinal ganglion cells that project to the superior colliculus and pretectum in the macaque monkey. Neuroscience 12 (4), 1125–1137. https://doi.org/10.1016/0306-4522(84)90007-1.

Rambold, H., Churchland, A., Selig, Y., Jasmin, L., Lisberger, S.G., 2002. Partial ablations of the flocculus and ventral paraflocculus in monkeys cause linked deficits in smooth pursuit eye movements and adaptive modification of the VOR. J. Neurophysiol. 87 (2), 912–924. https://doi.org/10.1152/jn.00768.2000.

Redgrave, P., Odekunle, A., Dean, P., 1986. Tectal cells of origin of predorsal bundle in rat: location and segregation from ipsilateral descending pathway. Exp. Brain Res. 63 (2). https://doi.org/10.1007/BF00236845.

Van Essen, D.C., 2002. Organization of visual areas in macaque and human. Cereb. Cortex. https://www.cns.nyu.edu/csh/csh04/Articles/Vanessen-03.pdf.

Xiong, G., Nagao, S., 2002. The lobulus petrosus of the paraflocculus relays cortical visual inputs to the posterior interposed and lateral cerebellar nuclei: an anterograde and retrograde tracing study in the monkey. Exp. Brain Res. 147 (2), 252–263. https://doi.org/10.1007/s00221-002-1241-3.

21

Motor Systems

Leonardo Fogassi and Stefano Rozzi
Department of Medicine and Surgery, Unit of Neuroscience, University of Parma, Parma, Italy

OUTLINE

INTRODUCTION

The term "motor cortex" implies the existence of a brain cortical region involved in movement execution. The term derives from the observation of motor deficits following brain lesions and also from the pioneering studies on cortical electrical stimulation in mammals, capable of eliciting observable movements (Ferrier, 1873; Fritsch and Hitzig, 1870). To address the issue of the meaning of motor control, and in particular of the control of intentional movements, it is necessary to define what is the main function of the motor system and to introduce some terminological distinctions.

First of all, it is important to define the term "motor intention." In this respect, Benjamin Libet defined an act as intentional *"when it has an endogenous origin, it does not depend, for its starting, from external restrictions or instructions, and is felt as performed by agent's own initiative* (Libet, 1985)." Furthermore, as suggested by many philosophers and neurophysiologists, the intention to act precedes its execution and is directly related to the awareness of being the agent of the performed act (sense of agency). Many philosophers of the last 3 decades made also distinctions between different types/phases of intention, such as, for example, the distinction between "prior intention," denoting the idea to perform a certain action later in time, and "intention-in-action," that refers to the individual's state driving and controlling action implementation (Searle and Willis, 1983). Most of these authors agree on the concept of motor goal at the basis of intentional behavior (Pacherie, 2008).

Neurophysiologists mostly addressed the concept of intention in terms of motor preparation of the next act to be voluntarily executed (Andersen and Buneo, 2002; Andersen and Cui, 2009; Boussaoud, 2001; Deecke and Kornhuber, 1978; Rizzolatti et al., 1990).

Independently of the details characterizing different definitions, there is agreement that intention coding is at the basis of any voluntary motor behavior. In this chapter, we will adopt definitions that allow to relate the concept of goal with neurophysiological underpinnings. First of all, we will define the meaning of action, motor act, and movement. The action corresponds to a sequence of motor acts that allow the agent to achieve a given final goal. For example, if an individual grasps a bottle of water, opens it, pours water in a glass and then grasps the glass, brings it to the mouth and opens the mouth to drink, the action goal is "drinking," while all the described intermediate steps are motor acts. The second important point is that each motor act has its own goal, that is, in some way, a

From Anatomy to Function of the Central Nervous System
https://doi.org/10.1016/B978-0-12-822404-5.00015-2

subgoal with respect to the final goal of the action. However, to perform the action, all these motor acts must be chained one with the other in the right order. Since each motor act is fluently linked to the next, it is plausible that this fluent organization is planned before action starts. Each motor act is composed of a synergy of movements; each movement consists in the displacement of a body part. Finally, movements are executed through the activation of a certain muscle synergy.

This type of organization, originally proposed simply on the basis of behavioral video recordings, has been demonstrated by more sophisticated kinematic studies (Bernshtein, 1947).

There are different anatomical and functional levels underlying the organization of actions. In fact, action goal and motor acts, as well as the final command for movement execution, are represented at the cortical level, while movement implementation occurs through cortico-descending, direct or indirect, pathways to spinal cord and brainstem, whose motor neurons are responsible of muscle control and, as a consequence, of joint displacement.

At the cortical level, action organization requires several areas involved in action coding, planning, and preparation. These aspects also take into consideration the environment (space, objects, other individuals) in which the action will be performed and the individual's state before action is planned (motivation, attention, emotion). All these factors will be considered in later sections, in which their neuroanatomical substrates will be described.

Furthermore, action organization must also take into account the programming of movement parametric aspects, such as force, speed, direction, amplitude, and so on. Actually, when microelectrodes recording techniques have been introduced, the first studies on the cortical motor control addressed the issue of the coding of movement parameters by motor cortex, studying the neural correlates of movement direction and force (Evarts, 1968; Georgopoulos et al., 1992).

Action execution also requires the control of several motor aspects, both before and during execution, such as sequence, timing, adaptation to perturbation, etc. These aspects involve important subcortical structures, that are linked to motor cortex by re-entrant loops. These structures are basal ganglia and cerebellum that receive, directly or indirectly, projections from motor cortex and send back to this latter their efferents through thalamic nuclei (see Chapter 8).

The final output to muscles for movement execution is provided by motor neurons, both in the brainstem and in the spinal cord. It is interesting to note that these neurons are recruited not only by voluntary commands but also from peripheral sensory afferents, thus controlling reflexes, or from brainstem structures, such as reticular formation, vestibular nuclei and red nucleus, that play a crucial role in postural control. Thus, motor neurons act as final motor output for both voluntary and involuntary motor behavior.

ANATOMY OF THE MOTOR CORTEX

The first architectonic map of the human cerebral cortex was published by Campbell (1905). He described two areas located in front of the central sulcus: the *precentral cortex*, caudally, and the *intermediate precentral cortex*, rostrally. According to him, precentral cortex would be involved in motor control, while intermediate precentral cortex in higher order aspects of the motor function. A few years later, Brodmann (1909) identified the motor cortex as the agranular sector of the frontal lobe, and described two distinct cytoarchitectonic areas: area 4 and area 6. Based on the demonstration that lesions of the area 6 in primates produce deficits of skilled movements, Fulton defined this latter cortical sector as *premotor cortex* (Fulton, 1935), proposing that different architectonic organizations underpin different functions.

The existence of a high-order motor area rostral to area 4 was challenged by electrophysiological studies. Woolsey et al. (1952), by employing surface electrodes capable of electrically stimulating the cortex, described a somatotopic map on the agranular frontal cortex, and proposed that area 4 and the caudal part of area 6 form a unitary functional entity, called primary motor cortex (MI). Furthermore, they identified a second complete motor representation of the body, also somatotopically organized, on the mesial surface of area 6, called secondary motor cortex (MII). In their study, the electrical stimulation of the rostral part of lateral area 6 did not elicit body movements and was excluded by these authors from the motor cortex. Thus, Woolsey's vision was in open contrast with the idea of a correspondence between architectonic and functional areas.

The above mentioned early parcellations of the motor cortex has been criticized by several authors (e.g., Vogt and Vogt, 1919; von Bonin and Bailey, 1947), who described a more complex anatomical organization of the motor cortex (Fig. 21.1). In recent years, the combination of classic techniques with newer architectonic approaches (e.g., immuno-histochemistry, neurochemistry, receptor autoradiography) has proven to be extremely useful for a more objective and reliable assessment of anatomical borders (see Geyer et al., 2000; Belmalih et al., 2007). This multiarchitectonic approach to the monkey motor cortex resulted in the early parcellation proposed by Matelli et al. (1985), and developed in subsequent studies (Belmalih et al., 2009; Matelli et al., 1991). In this map, in analogy with von Economo and Koskinas (1925) nomenclature for

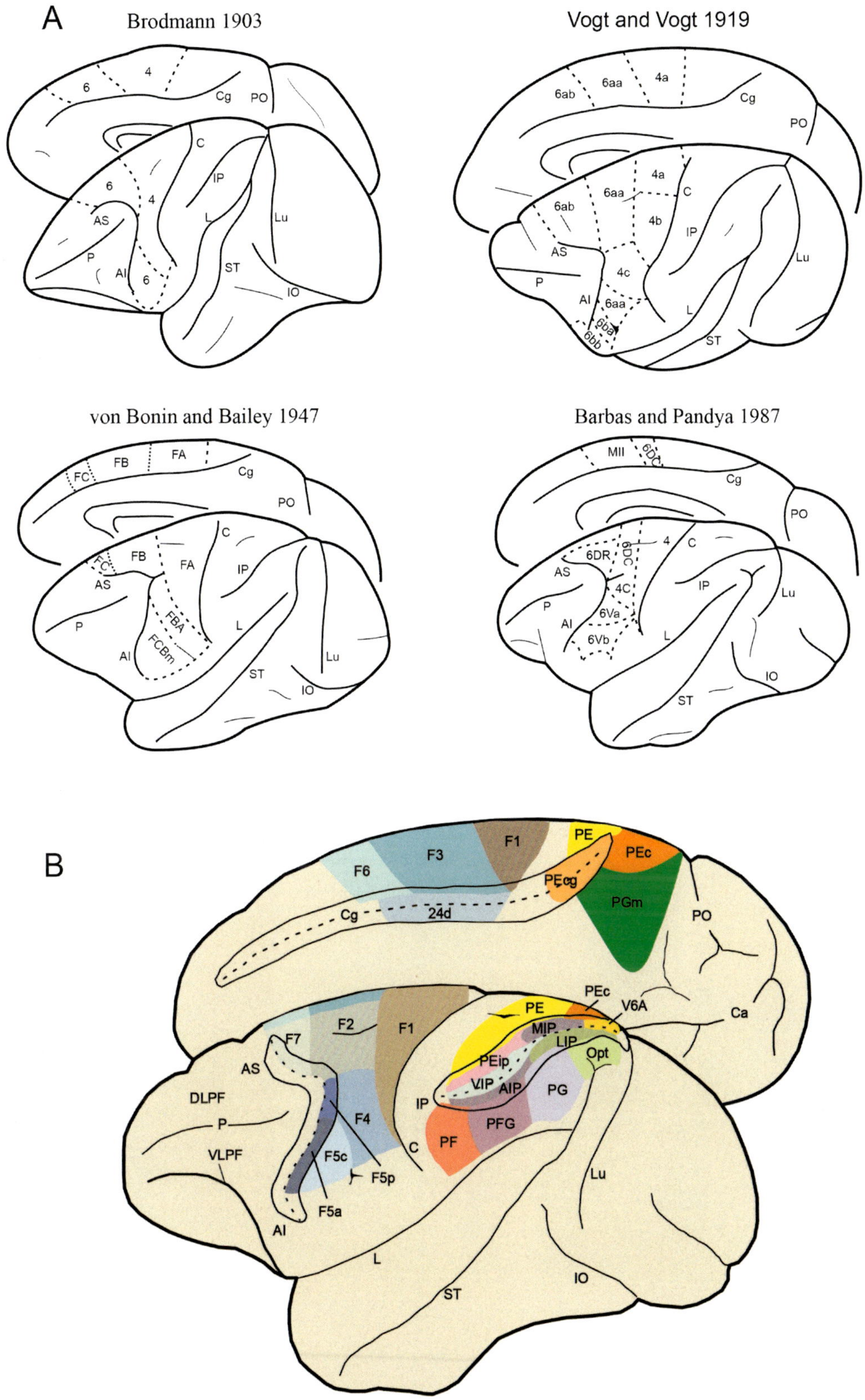

FIGURE 21.1 **Architectonic parcellations of the motor cortex.** (A) Lateral and mesial views of the monkey brain showing classical maps of the agranular frontal cortex. *Dashed lines* indicate borders between different areas. *AI*, inferior arcuate sulcus; *AS*, superior arcuate sulcus; *C*, Central sulcus; *Cg*, cingulate sulcus; *IO*, Inferior occipital sulcus; *IP*, intraparietal sulcus; *L*, lateral fissure; *Lu*, lunate sulcus; *P*, Principal sulcus; *PO*, Parieto-occipital sulcus; *ST*, superior temporal sulcus. (B) Lateral and mesial view of the monkey brain showing a modern anatomical and functional parcellation of the agranular frontal cortex and the posterior parietal cortex. Intraparietal, arcuate, and cingulate sulci are shown unfolded to reveal areas buried inside them. Frontal and parietal areas are classified as in Rizzolatti et al. (2014). *Ca*, calcarine fissure; *DLPF*, Dorsolateral prefrontal cortex; *VLPF*, Ventrolateral prefrontal cortex. Other abbreviations as in (A). *Modified from Rizzolatti, G., Cattaneo, L., Fabbri-Destro, M., Rozzi, S., 2014. Cortical mechanisms underlying the organization of goal-directed actions and mirror neuron-based action understanding. Physiol. Rev. 94, 655–706.*

human cortex, each area is referred to with the letter F (frontal), but differently from these authors, numbers instead of letters indicate the various areas. In this parcellation scheme, area F1 roughly corresponds to area 4 (primary motor cortex), whereas each of the three main sectors of Brodmann area 6 (mesial, dorsal and ventral), is formed by a caudal and a rostral premotor subdivision. Fig. 21.1 shows a modern map of the monkey agranular frontal cortex based on Matelli and coworkers studies (Matelli et al., 1985, 1991; see Belmalih et al., 2007; Rizzolatti et al., 2014). Over the years, this map has been validated by converging evidence showing that each architectonic subdivision has distinct patterns of connections and functions (see below), fulfilling all the criteria required for the definition of a cortical area (Van Essen, 1985). Altogether, these considerations show that the macaque motor cortex has a much more complex organization than previously thought: it is a sort of mosaic of distinct areas and contains a multiplicity of body movement representations, participating to different aspects of motor control (see Rizzolatti et al., 1998, 2014).

The Architectonics of the Macaque and the Human Motor Cortex

Fig. 21.2 shows examples of cytoarchitectonic features of the different monkey motor areas. A common feature of all motor areas is the lack of the IV layer (agranular cortex).

Area **F1** roughly corresponds to Brodmann's area 4, has poor lamination, low cell density, and a very

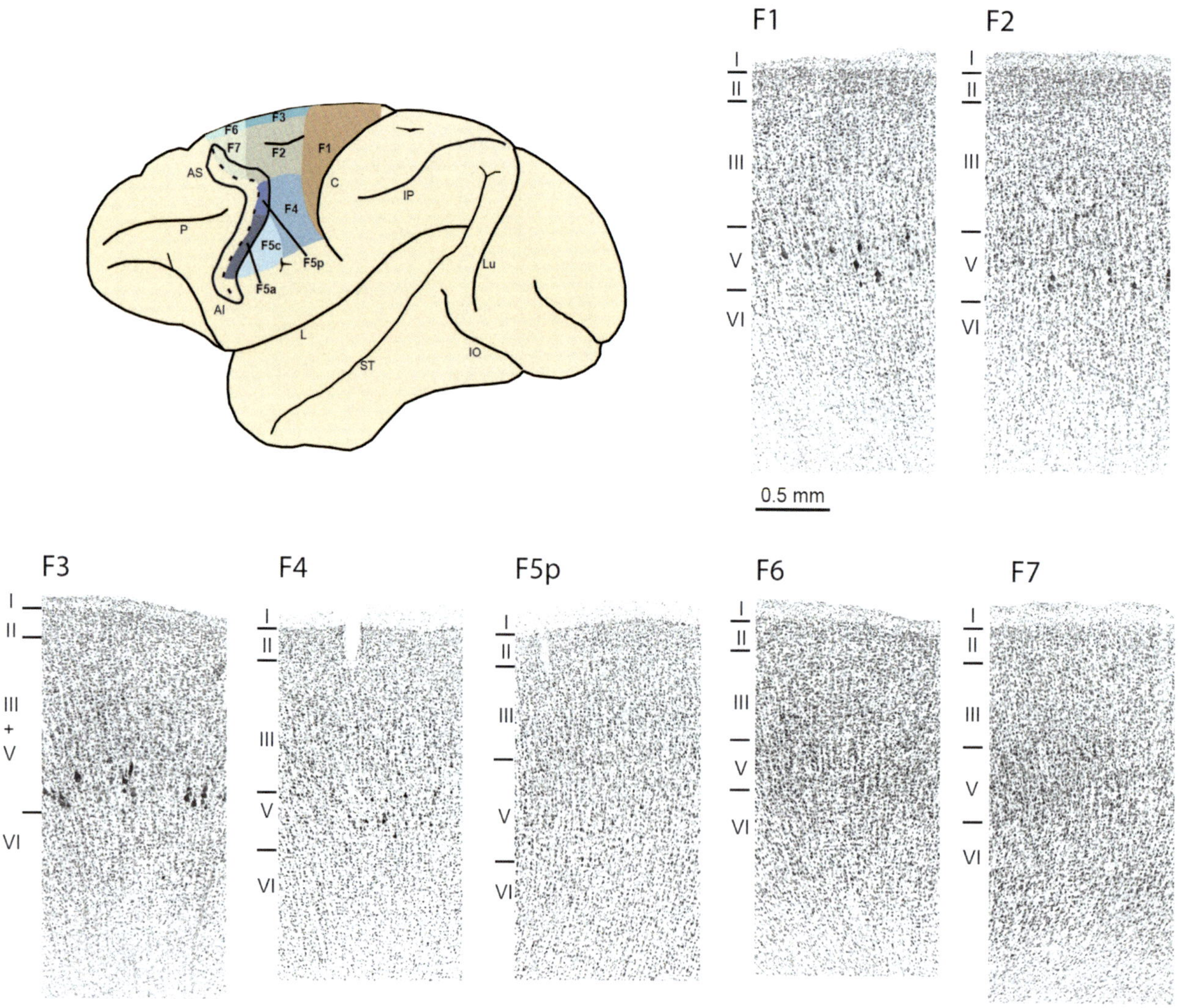

FIGURE 21.2 **Architectonic features of the motor areas.** Photomicrographs of representative fields of cytoarchitectonic areas F1 to F7. Scale bar, shown in F1, applies to all photomicrographs. *Modified from Belmalih, A., Borra, E., Contini, M., Gerbella, M., Rozzi, S., Luppino, G., 2007. A multiarchitectonic approach for the definition of functionally distinct areas and domains in the monkey frontal lobe. J. Anat. 211, 199–211; Belmalih, A., Borra, E., Contini, M., Gerbella, M., Rozzi, S., Luppino, G., 2009. Multimodal architectonic subdivision of the rostral part (area F5) of the macaque ventral premotor cortex. J. Comp. Neurol. 512, 183–217.*

prominent layer V, characterized by giant pyramids arranged in multiple rows. Similar to F1, area **F2** has a poor lamination, but differently from it, giant pyramidal cells are less numerous and more scattered; the lower part of layer III is characterized by medium-sized pyramidal cells distributed in a thin row, and layer V is denser than in F1. Also, area **F3** is poorly laminated, but shows an increase in cellular density in lower layer III and in layer Va; giant pyramidal cells are almost absent. In area **F4,** cell density is low and pyramid size increases from superficial to deep layer III; large pyramids are scattered in layer V, especially near the spur of the arcuate sulcus in the dorsal part of area F4. Area **F5** is clearly laminated and has a cell density higher than F4. This area is formed by three subdivisions: F5p, F5a, and F5c (Belmalih et al., 2007). **F5p** has a relatively homogeneous layer III, a cell dense layer Va, and is characterized by the presence of relatively large pyramids in layer Vb. **F5c** is poorly laminated, due to an overall homogeneity in cell population; **F5a** is characterized by the presence of relatively large pyramids in the lower part of layer III and a prominent and homogeneous layer V. Area **F6** is clearly laminated and is characterized by a dark and prominent layer V, in strong contrast to the less dense layers III and VI. Finally, area **F7** is clearly laminated, has a prominent layer V and layer VI is subdivided into two sublayers.

The cytoarchitectonic properties of the human motor cortex are similar to those of the macaque, though in humans many areas differentiate in further subareas, that could be new emergent areas.

Poor lamination, low cell density, absence of layer IV, and Betz cells in layer V characterize the human Brodmann area 4, while Brodmann area 6 is characterized by scattered or absent giant pyramidal cells and larger, more densely packed pyramids in lower layer III. Human area 4 is constituted by two regions: area 4 posterior (4p) and area 4 anterior (4a), that differ in terms of both cytoarchitectonics and neurotransmitter binding sites (Geyer et al., 1996). In particular, in area 4p lower layer III pyramids are small and loosely aggregated, while in area 4a these cells result larger and more densely packed. Concerning neurotransmitter binding sites, a change in the laminar binding patterns is evident and closely matches the cytoarchitectonic border, especially considering muscarinic cholinergic (labeled with [3H] oxotremorine-M) and serotoninergic (labeled with [3H]ketanserine) binding sites.

In area 6, the most evident cytoarchitectonic similarities between humans and macaques have been described in the mesial cortex, as shown by Zilles and coworkers (Zilles et al., 1995, 1996), who described a clear similarity between monkey areas F3 and F6 and the mesial parts of human areas 6aα and 6aβ as described by Vogt and Vogt (1919). Neurochemical studies revealed further similarities (Geyer et al., 1998; Zilles et al., 1995, 1996), strongly suggesting an homology between the areas forming the humans' and monkeys' mesial motor cortex. The border between primary motor cortex and supplementary motor area (SMA proper; F3) in humans coincides approximately with the vertical plane through the posterior commissure (VCP), while the border between area SMA proper and pre-SMA (F6) with the vertical plane through the anterior commissure (VCA, Zilles et al., 1996). The border between SMA proper and pre-SMA as well as its position at the antero-posterior level of the VCA line has been confirmed by means of diffusion magnetic resonance imaging studies (Crippa et al., 2011; Johansen-Berg et al., 2004; Lehéricy et al., 2004).

In the macaque, the arcuate sulcus represents a clear anatomical landmark separating the motor (agranular) and the prefrontal (granular) cortex. In the human brain, a similar landmark is not evident, but the spatial distribution of the agranular cortex (Brodmann areas 4, 6) is similar. This leads to the proposal of a homology between human and macaque motor cortex (Rizzolatti and Arbib, 1998). In this view, the superior frontal and superior precentral sulci would be, together, the human homolog of the macaque superior arcuate sulcus, and areas 6aα and 6aβ would correspond to macaque areas F2 and F7, respectively. In the ventral part of the premotor cortex, the inferior frontal sulcus and the ascending branch of the inferior precentral sulcus would correspond together to the macaque inferior arcuate sulcus and the descending branch of human inferior precentral sulcus to the inferior precentral dimple of the macaque. Thus, the inferior part of area 6aα and area 44 would be homologous of macaque areas F4 and F5, respectively. An open issue is the macaque homolog of human Brodmann area 45. In 1940, Walker (1940) described an area occupying the rostral bank of the inferior arcuate sulcus of the monkey. Subsequently, Petrides and Pandya (1994) proposed an homology between this latter area and human area 45. Monkey area 45, however, has an oculomotor function (Bruce et al., 1985), while human area 45 is part of Broca's speech region and, to the best of our knowledge, does not participate to oculomotor functions. The architectonic of human areas 44 and 45 is quite complex. The description of the architectonics of this region is out of the general aim of this chapter, for a thorough discussion of this issue see Amunts and Zilles (2012).

Altogether, the previous observations indicate that the human agranular frontal cortex is constituted by a mosaic of architectonic areas. However, the spatial congruence between areal borders and macroanatomic landmarks is much less consistent in humans than in monkeys. Thus, we must consider this variability when, for example, interpreting the activations in brain imaging studies. In addition, classical architectonic maps, based on observations of one or few brains, do not provide information on interindividual variability. This problem could be in part solved by adopting a probabilistic approach to brain mapping, allowing a more reliable comparison between the observed activations and their architectonic localization (Amunts and Zilles, 2012).

CORTICAL AND SUBCORTICAL CONNECTIONS OF THE MOTOR AREAS

The frontal motor cortex is mainly connected with the parietal cortex, the prefrontal lobe and cingulate cortex (see Rizzolatti et al., 1998, 2014). Outside the motor cortex, the strongest connections of the primary motor (F1) and the caudal premotor areas (F3, F4, F5c, and F5p) are with the parietal cortex. This cortical region, similarly to the motor cortex, is formed by a mosaic of areas (Fig. 21.1B), each involved in processing specific aspects of sensory information and in controlling different effectors (e.g., hand, eyes). The areas located in most of inferior parietal lobule (IPL) and in the posterior half of the superior parietal lobule (SPL) process visual and somatosensory information, those located in the most rostral sector of IPL and in the anterior half of SPL are mainly involved in processing somatosensory information (Hyvärinen, 1982; Caminiti et al., 1996; Rizzolatti et al., 1997, 2014; Wise et al., 1997; Rozzi et al., 2008).

In most cases, posterior parietal and agranular frontal areas are connected with each other in a very specific way, and it is therefore possible to identify several dedicated circuits, each constituted by a pair of parietal and frontal areas. Interestingly, parietal and frontal areas forming these circuits share similar functional properties (see below) and are at the basis of sensorimotor transformations for action.

Typically, the parieto-premotor circuits link SPL with dorsal and mesial premotor areas and IPL with ventral premotor areas. This is particularly interesting when considering that IPL and caudal SPL belong to the dorsal visual pathway (Goodale and Milner, 1992; Rizzolatti and Matelli, 2003), and that visual information is particularly relevant for movement guidance in "visual" animals as monkeys and humans. The visual input to IPL and SPL originate mainly from extrastriate areas MT and V6, respectively (Andersen et al., 1990a; Colby et al., 1988; Galletti et al., 2001; Lewis and Van Essen, 2000; Rozzi et al., 2006). Taking into consideration these anatomical data and the evidence deriving from the study of the clinical effects of IPL and SPL lesions in monkeys and humans, Rizzolatti and Matelli proposed that the dorsal stream is subdivided into two separate substreams: the dorso-dorsal visual stream, mainly involved in visuomotor transformation for the online control of movements and the ventro-dorsal visual stream, deeply involved in the organization of goal-directed motor act and in various cognitive functions (Rizzolatti and Matelli 2003).

The connections of prefrontal cortex with the agranular frontal cortex mainly involve the rostral premotor areas F5a, F6, and F7 (Barbas, 1988; Borra et al., 2011; Gerbella et al., 2010, 2013; Lu et al., 1994; Preuss and Goldman-Rakic, 1989; Saleem and Kondo, 2008). Area F5a is connected only with the ventral part of the lateral prefrontal cortex (VLPF), area F7 is mainly connected with the dorsal part of the prefrontal cortex (DLPF), F6 and the Supplementary Eye Field (SEF), that are a subsector of F7, are connected with both DLPF and VLPF. Further characterizing connections of rostral premotor areas (F6 and SEF) are those with the rostral cingulate cortex (area 24c), that is a high order motor sector of area 24.

Concerning subcortical connections, the organization of corticospinal and corticobulbar projections is in strong agreement with the subdivision of the premotor areas into a rostral and a caudal group. In fact, the corticospinal projections originate almost exclusively from F1 and caudal premotor areas (Dum and Strick, 1991; He et al., 1993, 1995). In addition, each caudal premotor area is connected with the corresponding somatotopic field of other premotor areas and F1. This confirms the importance of caudal premotor areas in movement execution. Thus, although F1 is the only cortical sector directly connected to the bulbar and spinal motor neuron pools (Porter and Lemon, 1993), all caudal premotor areas have access to the spinal cord and brainstem circuitry and can be involved in the generation and control of movements, both through F1, or in parallel with it.

On the other side, rostral premotor areas project to several parts of the brainstem (Keizer and Kuypers, 1989), but not to the spinal cord. In addition, they are not directly connected with F1 and are characterized by a widespread pattern of connections with several motor areas. The only exception to this general rule is constituted by SEF which is strongly interconnected with the frontal eye field (FEF) and has weak connections with other motor areas (Schall et al., 1993).

Altogether, these data suggest that parieto-premotor and prefronto-premotor circuits play different functional roles in motor control. The parieto-premotor circuits would be mainly involved in the transformation of sensory information into potential motor acts, while the prefronto-premotor circuits would play a higher hierarchical role in the control of movements. In fact, these latter process information about memorized spatial and nonspatial information, motivation, and others' behavior that could be conveyed to the caudal premotor areas determining if, when and which potential motor acts, coded in the parieto-premotor circuits, might become actual actions. These mechanisms will be outlined in the next sections.

THE CONCEPT OF GOAL

Coding the Goal of Motor Acts

For a long time, in the past century, it has been maintained that the flow of information in the cerebral cortex follows a serial pathway, going from a high-order elaboration of sensory information, named perception, occurring in the posterior part of the cortex (parietal and temporal lobe) to the exploitation of this elaboration for movement execution, occurring in the frontal cortex. According to this view, perception always precedes action. However, this view has been challenged, first of all, by neuroanatomical and behavioral data. As described in the previous sections, neuroanatomical connections between posterior and anterior cortical sectors are reciprocal, making it difficult to identify the direction of information flow. Second, behavioral studies demonstrated that action influences perception, in the same way as perception can drive action. In other words, action and perception are two aspects of the same process and cannot be conceived as independent functions. This concept has important consequences for understanding the role of motor cortex. In the traditional view, the frontal motor cortex would play a major role in movement execution. Accordingly, the first single neuron studies on primary motor cortex tried to verify how this latter codes movement parameters such as force, direction, amplitude, etc. A specific section of this chapter will be dedicated to this issue. However, very soon it was apparent that coding of motor parameters is only a late aspect of movement organization, while the most important, and earlier, role of motor cortex is that of coding motor goals such as reaching, grasping, orienting, gazing, climbing, etc. It has been demonstrated, in particular for forelimb motor acts, that this role is a prerogative of premotor cortex, while in MI it is present at a lower degree. Since this concept will come back several times in the next sections, associated with the distinct cortical circuits, we will describe, as an example, the neuronal coding of hand motor acts.

The rostral part of ventral premotor cortex, named area F5, contains the somatotopic representations of the hand and the mouth, as revealed by electrical microstimulation and single neuron studies (Gentilucci et al., 1988; Maranesi et al., 2012; Rizzolatti et al., 1988). Neurons of this area activate during execution of hand or mouth goal-related motor acts such as grasping, manipulating, holding, tearing, biting, sucking, etc. Grasping is the most represented motor act. Some neurons code the goal of the motor act to an abstract level, for example, there are neurons activated during grasping, when it is executed with the contralateral or ipsilateral hand or the mouth (Rizzolatti et al., 1988) (Fig. 21.3A). This means that the neurons activation is not due to movement execution but to the common goal: "grasping." A further support to this interpretation is the evidence that when the monkey moves the same effectors with a different goal (e.g., scratching) the same neuron does not activate. This "abstract" coding of the goal was also confirmed by studies in which monkeys were instructed to perform grasping acts with "normal" or "inverted" pliers. These latter require opening of the fingers in order to close the plier on the object. After training, F5 neurons respond equally well when the monkey takes possession of a piece of food with the hand or with the two types of pliers (Umiltà et al., 2008), independent of the type of movements (finger flexion or extension) needed to achieve the goal (Fig. 21.3B).

In area F5 there are other purely motor neurons that activate for the specific way in which the motor act is performed. For example, an F5 neuron can activate when the monkey executes a precision grip with either the left or right hand, but not when it executes a whole hand grasp. Note that while in the two types of grips there are many common movements, what is different is the specific type of goal. Thus, neurons of F5 have the role to code several types of motor acts, appropriate to the individual's intention, and also the way in which this act must be executed. The role of translating this high-level representation of the goal into specific movements is assigned to MI, thanks to its direct anatomical connections with premotor cortex. This type of organization is very economic, because it allows to store, in premotor cortex, a limited number of motor representations, representing the individual's motor repertoire and, according to the context and individual's motivation, to choose the representation suitable for a given action. Then this representation is implemented through specific connections with motor neurons commanding the displacement of specific joints. For example, if an individual wants to grasp a small object, the representation of precision grip will be activated, and this grip will be executed through the neurons commanding the extension and flexion of the thumb and index finger, while the other fingers are kept flexed.

THE PARIETO-PREMOTOR CIRCUIT FOR SPACE CODING

Our subjective conception of space is that of a homogeneous entity; however, if we consider it as an interactive entity we can, first of all, define three types of space: (i) a personal space corresponding to the cutaneous

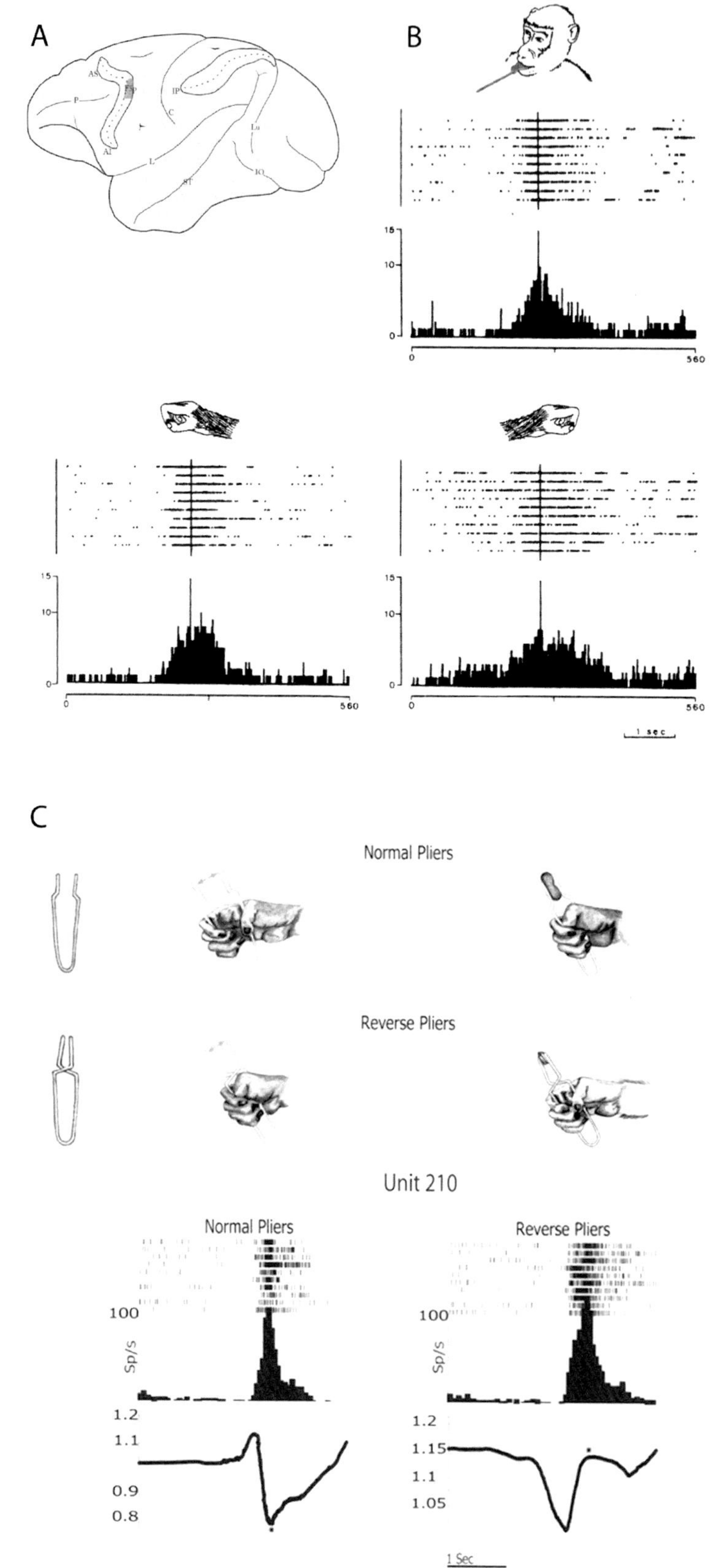

FIGURE 21.3 **Goal coding by F5 grasping neurons.** (A) Lateral view of the monkey brain showing the location of area F5p where grasping neurons were recorded. (B) Discharge of a purely motor neuron active during grasping with the mouth, the right hand and the left hand. Abscissae: time; ordinates: spikes per bin; bin width: 20 ms. (C) Example of an F5 neuron discharging during grasping with normal and reverse pliers. Upper part: Pliers and hand movements necessary for grasping with the two types of pliers. Lower part: Rasters and histograms of the neuron discharge during grasping with pliers. The alignments are with the end of the grasping closure phase (asterisks). The traces below each histogram indicate the hand position, recorded with a potentiometer, expressed as function of the distance between the pliers handles. When the trace goes down, the hand closes, when it goes up, it opens. The values on the vertical axes indicate the voltage change measured with the potentiometer. Abscissae: Time; ordinates: Spikes per sec. *(B) Modified from Rizzolatti, G., Camarda, R., Fogassi, L., Gentilucci, M., Luppino, G., Matelli, M., 1988. Functional organization of inferior area 6 in the macaque monkey. II. Area F5 and the control of distal movements. Exp. Brain Res. 71, 491–507; (C) Modified from Umiltà, M.A., Escola, L., Intskirveli, I., Grammont, F., Rochat, M., Caruana, F., Jezzini, A., Gallese, V., Rizzolatti, G., 2008. When pliers become fingers in the monkey motor system. Proc. Natl. Acad. Sci. U. S. A. 105, 2209–2213.*

surface; (ii) a peripersonal space, that is the space around our body in which we can interact with external stimuli by means of our forelimb; and (iii) an extrapersonal space, that is the space outside our reaching, that can be reached by locomotion or oculomotion. Neurophysiology clearly demonstrates that cerebral cortex processes the different space sectors through different dedicated functional modules, involved in several sensory-motor transformations. Here, we will describe the functional properties of the neurons, belonging to parieto-premotor pathways, coding peripersonal and extrapersonal space.

Coding Peripersonal Space: the VIP-F4 Circuit

The frontal node of this circuit is area F4. The type of movement somatotopy contained in this area has been demonstrated through microstimulation and single neuron studies. Both approaches show that F4 is related to neck, arm, face, and mouth movements (Fogassi et al., 1996; Gentilucci et al., 1988; Godschalk et al., 1981). Microstimulation carried out using standard parameters produces, very often, combined movements at relatively low-threshold, suggesting that this premotor area controls motor synergies. The concept of synergies is confirmed when longer stimulation trains are used, that determine motor responses that resemble voluntary movements, such as protective or reaching-grasping movements (see Graziano and Cooke, 2006). Single neuron studies show that, as in area F5, F4 motor neurons code goal-directed motor acts, such as reaching or bringing to the mouth (Fogassi et al., 1996; Gentilucci et al., 1988; Maranesi et al., 2012). Although until the beginning of the nineties only proprioceptive neurons had been described in motor cortex (Murphy et al., 1978), in area F4 two types of neurons with sensory responses have been discovered: purely somatosensory and bimodal (somatosensory and visual) neurons. The somatosensory responses are similar in both neurons categories: light touch responses with quite large RFs on the same body parts corresponding to the somatotopical motor representation: face, mouth, forelimb (without hand), neck and trunk (Fogassi et al., 1996; Gentilucci et al., 1988). The visual responses of bimodal neurons are elicited by objects introduced and moved toward the tactile RFs (Fig. 21.4). The visual RFs are three-dimensional since, differently from typical visual neurons, the neuron response begins only at a certain distance from the monkey skin, always within the reaching space. These RFs are in register with the tactile fields, independently on eye position or shift of the body part to which the visual RF is related (Fogassi et al., 1996; Gentilucci et al., 1983; Graziano et al., 1994, 1997).

A very interesting feature of the visual RFs of F4 bimodal neurons is that they can expand or shrink when the speed of the approaching object is increased or decreased, respectively (Fogassi et al., 1996). This plastic change can be explained in terms of the behavioral meaning of this visual response: the expansion of the RFs can express the anticipation necessary for the possible reaction to the incoming object, for example, in order to better reach it or avoid it. This explanation leads to the conclusion that the visual responses of F4 bimodal neurons are already potential motor acts in the peripersonal space. This means that the space near the individual is coded, in this area, not in visual but in motor terms. When an object is presented in this space and the individual intends to reach it, this potential motor act is released and become an actual goal-related act.

Area VIP is the parietal node of the circuit, strongly connected with visual area MT. Single neuron studies show that a large percentage of VIP neurons respond to visual stimuli, better if in motion (Colby et al., 1993; Colby and Duhamel, 1996; Duhamel et al., 1998). However, in this area, similarly to area F4, there are also somatosensory neurons, responding to tactile stimuli. Furthermore, there are bimodal (somatosensory and visual) neurons, with properties very similar to those of F4 bimodal neurons. However, many of them respond to tactile stimuli applied to a specific body part (head, mouth, neck, arm) and to visual stimuli presented in a corresponding sector of the visual space, presented on a tangential screen. Only a smaller percent of bimodal VIP neurons have tactile RFs and visual responses to stimuli approaching the tactile RF(Duhamel et al., 1998).

Electrical stimulation of this area evokes head and arm movements, although using long stimulation trains and high-intensity current (Cooke et al., 2003; Thier and Andersen, 1998).

The discovery of the properties of F4 and VIP had a profound impact on the conceptualization of how space is represented in the brain. They indicate that space is not represented as a unitary map in the brain, but derives from independent circuits relating different space sectors to different effectors (Colby, 1998; Graziano et al., 1994; Rizzolatti et al., 1997, 1998). They also indicate that the organization of movements in space is carried out not according to a single system of coordinates (e.g., the body or the head axes), but according to a multiplicity of coordinates located on different body parts (Fogassi et al., 1996; Graziano et al., 1997; Rizzolatti et al., 1997).

Noteworthy, bimodal neurons have also been described in other premotor and parietal areas (Duhamel et al., 1998; Hyvärinen, 1982; Iriki et al., 1996; Ishida et al., 2010; Rozzi et al., 2008), suggesting that also neurons of these areas may participate to peripersonal space coding.

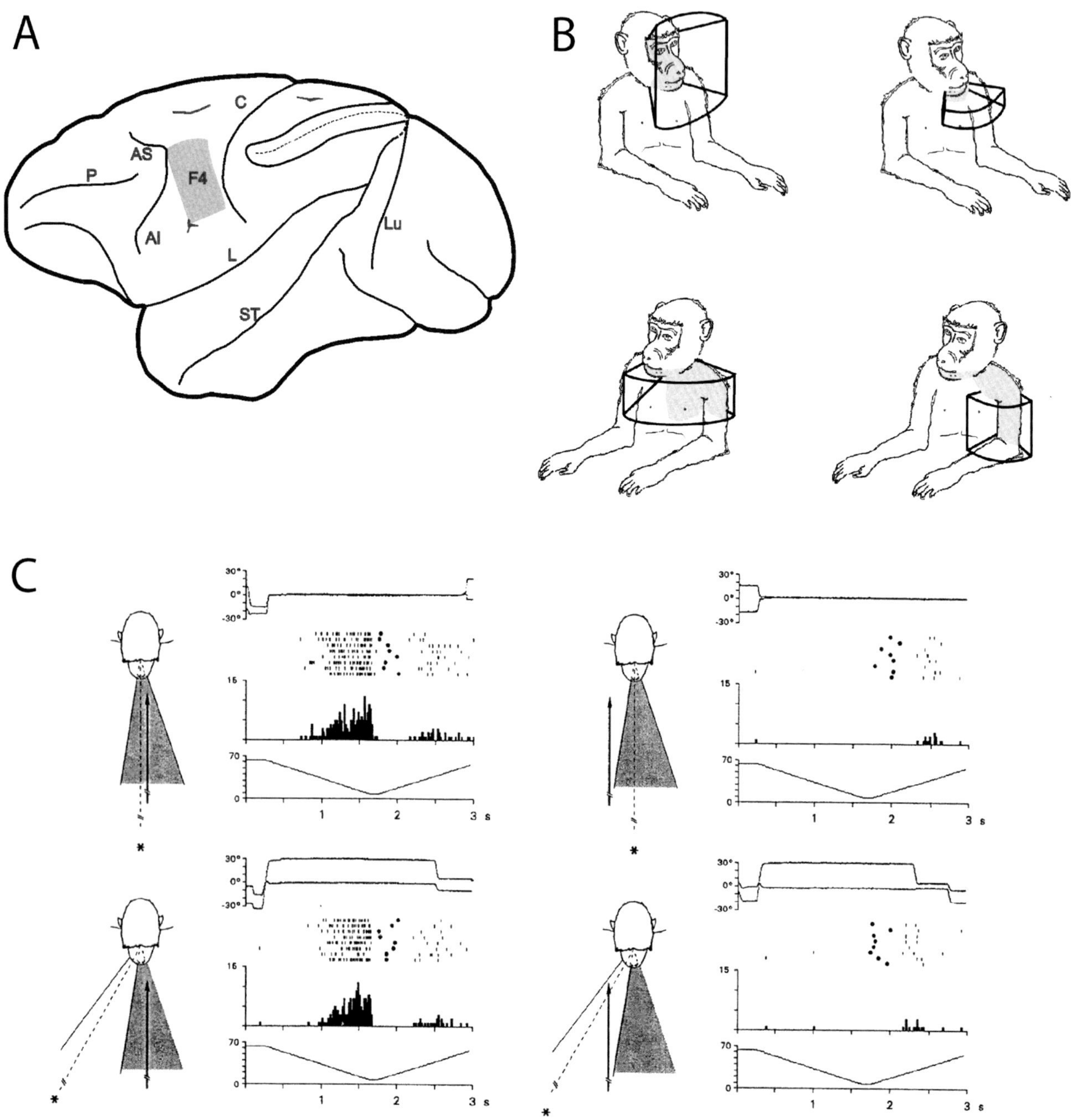

FIGURE 21.4 **Space coding in area F4 bimodal neurons.** (A) Lateral view of the monkey brain showing the location of area F4 where bimodal neurons were recorded. (B) Examples of different types of tactile and visual receptive fields (RF) of F4 bimodal neurons. Solids around different body parts indicate the 3D visual RFs. (C) Example of an F4 bimodal neuron with a visual RF independent of eye position. The monkey had to fixate an LED (fixation position indicated by the asterisks) and release a bar upon its dimming. Top panels indicate central fixation, bottom panels rightward fixation. During fixation, a 3D stimulus was moved toward and away from the monkey face at constant velocity, reversing when the stimulus was a few cm from the tactile RF. In the top panels the stimulus is moved inside (left panel) and outside (right panel) the visual RF in both somatocenteric and retinocentric coordinates. In the bottom panels, the somatocentric and retinocentric coordinates are decoupled. Bottom left: the stimulus is moved inside the somatocentric RF and outside the retinocentric one. Bottom right: The stimulus is moved inside the retinocentric RF and outside the somatocentric one. Each panel shows horizontal and vertical eye position, rasters and histograms representing the neuronal response and, under the histogram, the variation in time and space between the stimulus and the monkey. When the line goes down, the stimulus approaches monkey face, when the line goes up, the stimulus goes away from it. Large dots in the rasters indicate dimming of the fixation point. The neuron tactile RF was located on the lower face. Note that the discharge is present when the visual stimulus is moved toward the tactile RF independent of eye position, and consequently is not related to a retinocentric frame of reference. *Modified from Fogassi, L., Gallese, V., Fadiga, L., Luppino, G., Matelli, M., Rizzolatti, G., 1996. Coding of peripersonal space in inferior premotor cortex (area F4). J. Neurophysiol. 76, 141–157.*

Coding Extrapersonal Space: The LIP-FEF Circuit

The frontal node of the circuit, frontal eye field (FEF) is, anatomically, part of prefrontal cortex but for its properties can be considered a premotor sector. Indeed, its electrical microstimulation (Bruce et al., 1985) elicits mostly saccadic but also slow eye movements, suggesting that this area plays an important role in the generation of voluntary eye movements and, likely, of smooth pursuit of moving visual targets. These motor effects can be explained by FEF connections with superior colliculus and the subcortical, brainstem circuits involved in eye movements control. FEF neurons respond both to the presentation of visual stimuli (also 2D) and during eye movements triggered by the presentation of such stimuli in several space sectors. Interestingly, the same neurons are not active when the monkey makes spontaneous eye movements(Bruce et al., 1985; Bruce and Goldberg, 1985).

The parietal node, i.e., the lateral intraparietal area (LIP), receives its main visual information from the dorsal visual stream, but also from the ventral visual stream, namely from inferotemporal cortex (Andersen et al., 1990a; Blatt et al., 1990; Cavada and Goldman-Rakic, 1989).

As for FEF, also stimulation of LIP produces directional eye movements (Thier and Andersen, 1998). Neural recording shows that its neurons are activated by visual stimuli presented in different spatial positions and by saccades toward the same spatial positions for which the same neurons are selective (Andersen et al., 1985). Even for LIP, the superior colliculus is the major subcortical target of its descending projections, thus voluntary eye movements can, in principle, be driven by both FEF and LIP.

A peculiarity of LIP neurons is that many of them, though keeping directional preference, are modulated by the starting eye position (orbital position), suggesting their capacity to encode stimulus position with respect to the subject's head (Andersen et al., 1985, 1990b). This property can be particularly important in the process leading to the coding of space in a head-centered coordinate system.

Furthermore, it has also been shown that LIP neurons get activated when the monkey has to remember the position of a previously presented visual stimulus that will later become the target of monkey eye movements (Barash et al., 1991).

All features of LIP neurons suggest that they can play the role of transforming visual spatial information in saccadic eye movements and that they can be at the basis of the intention to execute an eye movement toward an interesting object. Another, not necessarily alternative proposal, is that LIP neurons are modulated by the attention the monkey is paying to a spatially organized visual stimulus, in order to enhance the perceptual relevance of that stimulus (Duhamel et al., 1992; Robinson et al., 1978). We will see later in this section how this capacity could be related to spatial deficits in neglect patients.

Interestingly, Duhamel et al. (1992) showed that LIP neurons can shift the location of their receptive field before an eye movement toward that location is performed, suggesting that these neurons can "update the retinal coordinates of remembered stimuli to generate a continuously accurate representation of visual space."

In conclusion, the LIP-FEF circuit is directly involved in the visuomotor transformation of spatial positions of visual targets coded in retinocentric *coordinates* in saccadic eye movements directed to the spatial position in which the visual stimulus is presented.

The Circuits for Space Coding and Neglect Syndrome

Neglect has been classically defined as a deficit in perceiving and conceiving the left side of the body and of space, not due to pure sensory or motor disorders. It can be described in relation to different sensory modalities, such as visual, acoustic, or tactile, not all necessarily present together. Typical symptoms of the syndrome can consist in neglecting to dress or to wash the left part of own body, not responding to presentation of an object in the left visual space, or to an acoustic stimulus, even a word, introduced in the left hemispace. Typical tests for visual neglect are, for example, to ask the patient to tick lines presented on a sheet of paper or to collect coins put on a table. The patient typically marks only the lines presented on the right part of the sheet or collect the coins on the right side of the table. Another classical test for visual neglect is line bisection, that consists in asking the patient to mark the half of a line presented in front of him. The patient always shifts the midline to the right. If asked to reproduce a wall watch shown in front of him, he draws the circle, but then put the numbers indicating the hours only on the right part of the circle and often continue outside it, inserting all numbers, but keeping them always on the right side. When presented with a sound, the patient either neglect it or orient toward the right side.

There is also a motor aspect of neglect. In fact, the patient is reluctant to move both forelimbs toward the left visual space. Furthermore, the eyes are often deviated toward the right side of the space.

Many of the tests here described have typically been tested in the subacute phase at the patient's bed, therefore in her/his peripersonal space. However, there are patients in which neglect is present both near and far from the subject, others in which there is a dissociation (Halligan and Marshall, 1991). This was clearly demonstrated using the line bisection test. In fact, patients who, asked to bisect with a pencil a line drawn on a sheet, shift the half to the right, but they can be perfectly correct when performing a bisection of a line drawn far from them using a light-pen (Halligan and Marshall, 1991). Other patients behave exactly in the opposite way (Cowey et al., 1994; Vuilleumier et al., 1998).

Finally, there are patients showing the so-called allocentric neglect. In these patients, neglect is referred to the object, for example if requested to copy an object based on a visual model, they draw only the left half of it (Marshall and Halligan, 1993). If they are asked to observe a symmetrical object differing only for a detail on the left part, they report no difference between the two sides of the object.

Another space-related disorder is extinction. This syndrome is different from neglect because, when patients are presented with single stimuli in the left hemispace, they react to them. However, when presented with a pair of stimuli, one to the left, the other to the right, they always extinguish the reaction to the leftmost stimulus.

All these descriptions remind some of the characteristics of space coding by the different parieto-premotor circuits described above. In fact, looking at different patients, it is clear that their deficit allows to recognize and distinguish personal, peripersonal, and extrapersonal space, confirming the presence of separate modules related to the processing of different space sectors. Thus, one could imagine that different types of neglect can be due to lesions of different circuits probably homolog to those found in monkeys. There are several types of lesions determining the neglect syndrome in humans, involving most often parietal, frontal, or temporal cortex, but sometimes related to subcortical structures, such as pulvinar or basal ganglia. Classically, there was agreement that the inferior parietal cortex is the most frequently damaged sector in neglect patients. However, there are more recent reports showing that this latter can also be due to more ventral lesion, including temporal cortex (Corbetta and Shulman, 2011; Karnath and Rorden, 2012). One point that receives most consensus is that lesions are usually on the right side of the brain, thus justifying that the impairment is mostly on the left side of the space, that is contralateral to the lesion. Left brain lesion in right-handed patients usually do not cause neglect, but rather other types of impairments. The right-sidedness of the lesion in right-handed patients is one of the examples of function lateralization in humans, different from what described in animals. In monkeys, parietal or frontal lesions do not produce a real neglect, but rather an extinction contralateral to the inactivated or lesioned area (see below).

Different types of signs of neglect in humans have been attributed to different lesion sides. For example, visuo-spatial aspects would be more related to lesions of the posterior parietal lobule, while exploratory and visuomotor signs to frontal lesions. Allocentric signs would instead be related more to the ventral pathway (Lunven and Bartolomeo, 2017). The personal neglect would be associated with the parietal lobe and the posterior part of the superior temporal lobe; extrapersonal neglect would be more related to lesions of the frontal lobe and of the anterior and middle part of superior temporal gyrus. As it can be seen, although many metanalysis studies tried to match lesions and neglect signs, still there are no straightforward conclusions on this correspondence. Some of the reasons can rely on the fact that the lesions causing neglect deficits can derive from both lesions of connected cortical areas and of anatomical connections. In particular, it has been suggested that the three recently described components of superior longitudinal fasciculus (SLFI, SLFII, and SLFIII, Thiebaut de Schotten et al., 2012) can be involved in the functional network subserving the encoding of the different space sectors, in relation to both sensory and motor components.

Concerning the neglect syndrome, it cannot be forgotten that many influencing theories considered it as an attentional deficit. The more classical ones attributed to a conflict between left and right hemispheric control of the space the occurrence of an inattention for the left visual space (Heilman et al., 2000; Kinsbourne, 2013), or to a difficulty to disengage the attention from the right visual hemifield (Posner et al., 1984). A more recent attentional interpretation distinguishes between a dorsal attention network (DAN), related to voluntary direction of attention, and a ventral attentional network (VAN), involved in orienting attention toward new and unexpected spatially located events. The first would be subserved by a parieto-frontal circuit including superior and inferior parietal cortex and FEF, the second by temporo-parietal junction and middle and inferior frontal gyri. Only VAN would be right lateralized and the two networks would interact one with the other (Lunven and Bartolomeo, 2017).

In monkeys, we have described two parieto-premotor circuits for extrapersonal and peripersonal space. In both of them, the parietal node lays inside the intraparietal sulcus. Comparing these circuits with those described with tractography (and compared to the corresponding tracts of humans), it appears that the most likely fascicle that contains both circuits is the SLF II (Thiebaut de Schotten et al., 2012). Thus, if the SLFII is the monkey homolog of that of humans, in our species this bundle would subserve networks involved in both peripersonal and extrapersonal neglect.

As mentioned above, it is difficult to demonstrate a clear neglect syndrome in the monkey. However, it was demonstrated an extinction syndrome, with a dissociation between near and far extinction. More specifically, after lesion of ventral premotor cortex the monkey does not make blinking reactions or grimace to objects moved toward the hemiface contralateral to the lesion side and do not lick drops of juice delivered on the contralateral lips. On the contrary, promptly reacts to approaching stimuli presented in the far space. The opposite effect is observed when the lesion is applied to prearcuate cortex, involving FEF (Rizzolatti and Camarda, 1987).

Interestingly, in patients with tactile extinction, who recognize only the right tactile stimulation when two tactile stimuli are applied contemporarily to right and left hand, the extinction decreases when the patients sees the experimenter hand approaching his left hand (Làdavas et al., 1998). This is a clear demonstration of the role of bimodal neurons that, with their activation, allows the enhancement of tactile stimulation.

THE PARIETO-PREMOTOR CIRCUITS FOR GRASPING

Hand grasping actions represent one of the main behaviors through which primates interact with the external world, thus it is not surprising that a wide cortical network is devoted to the coding of several aspects of grasping actions. In particular, it is very important to consider how object properties (see also the concept of "affordance" of Gibson, 1979) are translated in the appropriate hand configuration and how grasping execution is controlled, based on visual and somatosensory input. We will first focus on the circuit involved in visuomotor transformation for grasping, then on the circuit involved in the on-line control of this action. Finally, we will try to briefly describe the extended grasping network.

The AIP-F5 Circuit: Visuomotor Transformation for Grasping

The frontal node of the circuit, area F5, contains a somatotopic representation of the hand and the mouth. This is clearly shown by electrical stimulation that, using standard parameters, elicits movements of fingers, wrist, jaw, lips, and tongue (Godschalk et al., 1981; Maranesi et al., 2012; Rizzolatti et al., 1988). The hand representation is more dorsal, while the mouth representation is more ventral, though they show a large overlap. As described in a previous section, F5 motor neurons code the goal of hand and mouth motor acts. Beside purely motor neurons, F5 also contains neurons responding not only during movement execution, but also during sensory stimulation, most often responding to somatosensory stimuli, but also to visual stimuli. Among these latter, called visuo-motor neurons, particularly relevant for the issue of the coding of grasping are those responding to the presentation of 3D objects (Rizzolatti et al., 1988; Murata et al., 1997; Raos et al., 2006). These neurons ("canonical" neurons) are mostly located in the dorsal part of the posterior bank of inferior arcuate sulcus (F5p). They respond during grasping execution and to objects presentation. Using a controlled visuomotor paradigm, it has been demonstrated that the visual response is often specific for objects of a certain size, shape, and orientation and that, typically, the visual specificity is congruent with the motor one (Fig. 21.5) (Murata et al., 1997). For example, a neuron specifically activated by presentation of a small object fires during precision grip, while a neuron specifically activated by presentation of a large object fires during whole hand prehension.

Although canonical neurons respond to the vision of an object, even when the monkey is not going to grasp it, on the basis of the congruence between visual and motor response, it has been proposed that object presentation activates a representation of the observed stimulus in a motor format. In other words, when an object is shown, the discharge of specific canonical neurons code a *potential* motor act congruent with the physical features of the object, independently of whether an action will be executed or not. Note that this concept is similar to that proposed for the visual response of F4 bimodal neurons.

F5p is strongly connected with area AIP (Borra et al., 2008; Gerbella et al., 2011; Luppino et al., 1999). Interestingly enough, similar to F5, AIP contains neurons responding to 3D objects and motor neurons (Murata et al., 2000; Sakata et al., 1995; Taira et al., 1990). In particular AIP neurons, using a paradigm similar to that employed to study F5 neurons, have been categorized in three main classes: "motor-dominant," "visual-dominant," and "visual and motor. "Motor-dominant" neurons discharge when the grasping action is performed in light and when it is executed in darkness, but not during object fixation. "Visual-dominant" neurons discharge during grasping in light and object fixation, but not during grasping in dark. "Visual and motor" neurons discharge both during grasping in light and in dark, as well as during object fixation.

The same object can be grasped with different grips according to its possible affordances, but also according to different behavioral contexts. This latter concept has

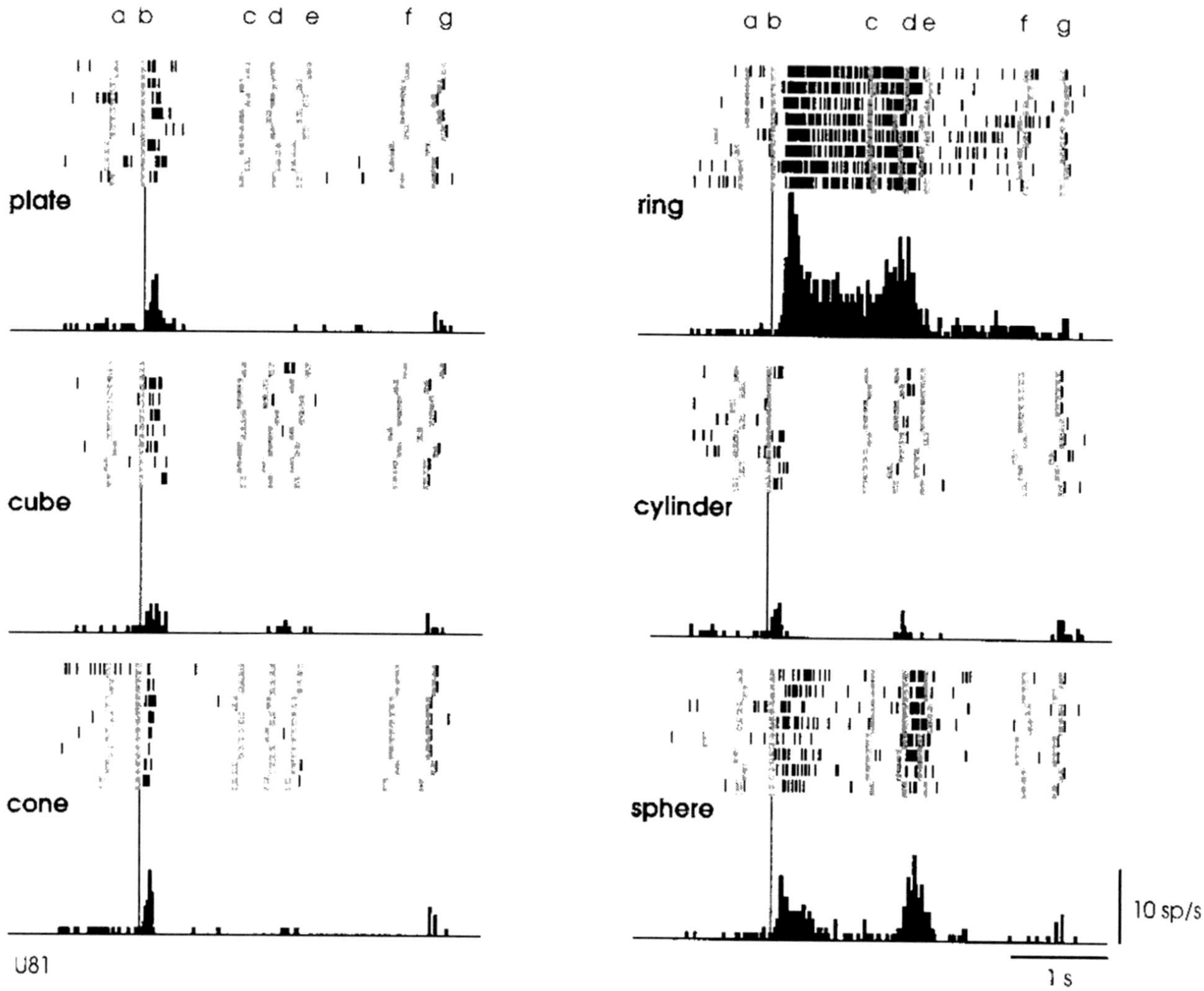

FIGURE 21.5 **Visuo-motor transformation for grasping.** Example of a selective F5 canonical neuron. Panels show neural activity recorded during a grasping in light task with six objects. Rasters and histograms are aligned (vertical line) with key press (onset of object presentation). Small gray bars in each raster indicate the various events of the task: onset of fixation (red LED, (A)), object presentation (key press, (B)), go signal (green LED, (C), start of movement (key release, (D)), onset of object pulling (E). During the period between object presentation (B) and go signal (C) the monkey has to keep fixating the object refraining from grasping it. Note that the neuron has a preferential discharge during observation and grasping of a ring. Note also the sustained discharge during waiting period between object presentation and movement execution. *Modified from Murata, A., Fadiga, L., Fogassi, L., Gallese, V., Raos, V., Rizzolatti, G., Object, G.R., 1997. Object representation in the ventral premotor cortex (area F5) of the monkey. J. Neurophysiol. 78, 2226–2230.*

been addressed by Scherberger et al., who trained monkeys to grasp a handle with two different grips based on different abstract cues (Baumann et al., 2009; Fluet et al., 2010). Their results show that part of AIP and F5 neurons differentially activate after cue presentation, showing context-dependent grasp planning activity. This further confirms the idea that the visual response of visuomotor neurons consists in a potential motor act.

The properties of AIP and F5 neurons suggested their strong role in the transformation of objects visual features in the suitable type of grasping actions. This was further confirmed by transient inactivation studies obtained by injection of the GABA agonist muscimol in macaque area AIP (Gallese et al., 1994) or F5 (Fogassi et al., 2001). Both inactivations produce a dramatic deficit in the ability to shape the hand according to the object's physical features. Interestingly, once touched, the object could be in some trials correctly grasped, likely based on somatomotor transformation, depending on another circuit. This last result shows that these inactivations do not produce purely motor deficits.

All models trying to explain the role of AIP and F5 in visuomotor transformations for grasping (Taira et al., 1990; Jeannerod et al., 1995; Fagg and Arbib, 1998; Rizzolatti and Luppino, 2001; Fluet et al., 2010) and share the common idea that, when we observe an object, AIP neurons extract specific aspects of its physical properties and provide F5 with the description of the possible ways in which it could be to grasped. Based on the subject's intention and the context, prefrontal cortex, through the described connections with premotor and parietal cortex (see above), activates the AIP-F5 circuit in order to implement grasping execution. This latter is allowed by the connection of F5 with F1 and, in part by the descending pathway from F5p to the propriospinal system (Borra et al., 2010).

The MIP/V6A-F2 Circuit for the On-line Control of Reaching–Grasping

Electrical stimulation of the frontal node (area F2) of this circuit, demonstrated a somatotopic organization of movements, with the leg and the arm represented in the dorso-ventral direction around the superior precentral dimple (Dum and Strick, 1991; He et al., 1993; Kurata, 1989; Raos et al., 2003). The electrophysiological studies on sensory properties of F2 neurons showed that many of them respond to somatosensory (mainly proprioceptive) stimulation. Neurons in its ventro-rostral sector (F2vr) also respond to visual stimuli (Fogassi et al., 1999; Raos et al., 2004). The first studies on the role of F2 in motor control, aimed to disentangle between movement planning and execution, adopting paradigms in which a delay was inserted between a stimulus instructing a motor behavior and the associated motor response. In this way, Wise and coworkers (Weinrich and Wise, 1982) identified three types of neurons in area F2: *signal-related*, discharging during the observation of the instructing cue, *set-related*, discharging during the delay period, and *movement-related*, discharging on movement onset (see also Wise et al., 1997). Note that the signal-related activity, although due to a visual stimulus, is not present when it does not instruct a subsequent movement, thus it cannot be interpreted as a true sensory response. Particularly interesting is the neural activity recorded in the delay period of reaching tasks since it conveys information relevant for controlling the direction of the arm movement necessary to correctly reach an object (Crammond and Kalaska, 2000; Cisek and Kalaska, 2010).

The role of F2 in planning reaching actions is further supported by evidence showing that its neurons also integrate information about the location of a target and the arm to use in order to plan actions (Hoshi and Tanji, 2000). More recent studies clearly indicate that neurons in the ventralmost sector of F2 discharge during hand and arm movements. In particular, using the same paradigm employed for studying grasping movements in AIP and F5, it was demonstrated that in this area there are both motor and visuomotor neurons involved in the control of the configuration of the hand grasping actions. The comparison between F2 and F5 shows that during grasping in the dark F5 neurons maintain grip and wrist orientation selectivity, while F2 neurons, although keeping a sustained discharge, tend to become more unselective for both properties (Raos et al., 2004).

The dorsal (F2d) and ventro-rostral (F2vr) sectors of area F2 have partly different patterns of cortical connections: F2d has strong connections with parietal areas PEip and PEc; F2vr is mainly connected with parietal areas MIP and V6A (Bakola et al., 2010; Gamberini et al., 2009; Matelli et al., 1998). In addition, only F2vr is target of a minor but consistent projection from the dorsal part of prefrontal BA 46 (Luppino et al., 2003). The parietal areas connected with area F2vr share with this motor area several functional properties: in particular, they contain neurons that respond to somatosensory and visual stimuli; typically, visual neurons tend to have large and peripheral receptive fields, and these responses can be modulated by gaze. Furthermore, in these areas there are many neurons activated during reaching and grasping movements (Breveglieri et al., 2006; Colby et al., 1988; Galletti et al., 1999; Iwamura and Tanaka, 1996; Kalaska et al., 1990; Marconi et al., 2001).

Altogether, the described evidence indicates that caudal SPL computes object position in static and dynamic conditions (e.g., during self-motion), thus it is crucial for coordinating eye and arm movements for reaching or avoiding objects in space. This information, together with that on intrinsic objects properties and arm-related somatosensory and motor signals, can be exploited in the V6A-F2vr circuit for guidance and on-line monitoring of reaching-grasping motor acts (Breveglieri et al., 2015; see Fattori et al., 2017).

The Dorsal Circuit for Reaching–Grasping and Optic Ataxia

Optic ataxia is a disorder of visually guided movements of the arms toward a goal, independent of visual space misperception (Perenin and Vighetto, 1988). Typically, when patients try to reach for an object, their arm "gropes" for the target and they make errors in the frontal and in the sagittal plane, until they finally touch the object, almost by chance. Both the proximal and the distal components of movements are affected, as shown by specific reaching and hand orientation tasks. The deficit is severe when the target is in the peripheral portion of the visual field, while it is less strong when the target is in parafoveal vision, and it almost disappears when the patient fixates the target. Interestingly, usually patients can correctly point to targets on their own body (De Renzi, 1982). Optic ataxia patients have no perceptual deficits, they are aware of objects presence and location and can correctly describe their position in space. They have problems in producing and on-line correcting reaching movements toward objects under visual control. For example, Pisella et al. (2000) report the case of a patient with optic ataxia who, differently from healthy subjects, does not show an automatic correction of arm movement, in the presence of a spatial target perturbation and suggest that the deficit impairs an "automatic pilot," normally involved in unconscious correction of the direction of visually guided reaching movement.

The brain region whose lesion is most often associated with optic ataxia is the inferior part of the SPL and the superior part of the IPL. Lesions in both hemispheres produce optic ataxia, though with small variations.

Altogether, these observations indicate that the core deficit of optic ataxia is in motor control and that this syndrome derives from lesions of the parietal node of the dorsal circuit, supporting the idea that, as in the monkey, also in humans SPL-dorsal premotor connection is crucial for the visually based programming and control of reaching-grasping movements.

THE PARIETO-PREMOTOR CIRCUIT FOR UNDERSTANDING OTHERS' ACTIONS

Mirror Neurons in Monkeys

In our environment, we have frequent and fundamental interaction with other individuals. The interaction, to be effective requires, first of all, the capacity to rapidly understand others' behavior. This capacity is provided by the mirror neurons mechanism, originally described in monkeys and later in humans (di Pellegrino et al., 1992). "Mirror" neurons are a category of visuomotor neurons recorded in the premotor and the parietal cortex, respectively, areas F5c and PFG. We will begin to describe the properties of F5 mirror neurons, even because the first discovery of this peculiar neuronal class occurred in this area. Mirror neurons are activated during execution of a motor act, e.g., grasping an object, and during observation of the same or a similar motor act performed by another individual (Gallese et al., 1996; Rizzolatti et al., 1996a), as shown in Fig. 21.6A. Simple presentation of an object, or the pantomime of the motor act without the object do not activate these neurons. The former control is important, because it allows to distinguish mirror neurons from the "canonical" neurons present in area F5 (see above), that share the same motor response, but are activated visually by an object presentation.

The most effective observed motor acts activating these neurons are grasping, manipulating, biting, tearing an object; grasping acts are by far the most represented ones. Mirror neurons code the goal of the observed motor act and not the simple pictorial aspect of it; this is demonstrated by a study in which the final part of the observed motor act is covered by an opaque screen (Kohler et al., 2002). The discharge of mirror neurons occurs when the observed agent's hand disappears behind the screen, with the same timing of the response to the fully visible version of the motor act, suggesting that this response depends on a "mental" reconstruction of the other's motor act performed by the monkey brain, built on the memorized motor representation of it. A subcategory of mirror neurons also respond to the sound produced by the motor act, such as paper breaking, noisy manipulation of objects, or dropping of a steak (Umiltà et al., 2001). This means that, independently of the used modality, witness of other's behavior allows to retrieve the goal, that is the meaning of the observed motor act. Note that to recognize others' motor acts does not mean to recognize who is the agent performing the observed act: this latter property probably depends on the processing of further different signals, which allows to attribute agency to the observed or to the performing individual.

Mirror neurons are activated not only by observation and execution of hand-related motor acts but also by those made with the mouth, such as biting or sucking. Among them, a subcategory of these neurons respond specifically during observation and execution of communicative monkey mouth gestures, such as lip-smacking, or lip protrusion (Ferrari et al., 2003).

In mirror neurons, the correspondence between the effective observed and executed motor act can be very strict or broader (Gallese et al., 1996). While in the first case the observed and the executed motor act are the same, in the second case the observed and executed motor act match in terms of goal, but the visual response can be less specific than the motor one. Since in both cases there is a congruence in the goal of the observed and executed motor acts, this can constitute an observation/execution matching mechanism that enables an individual to automatically understand the goal of the other's motor act. This mechanism is allowed by a temporo-parietal-premotor circuit. In fact, an area hidden in the anterior part of the superior temporal sulcus contains neurons responding only to the visual presentation of motor acts (Perrett et al., 1989). This information is likely sent to the inferior parietal lobule, that in turn send it to area F5. The convexity of the inferior parietal lobule is made by several cytoarchitectonic areas

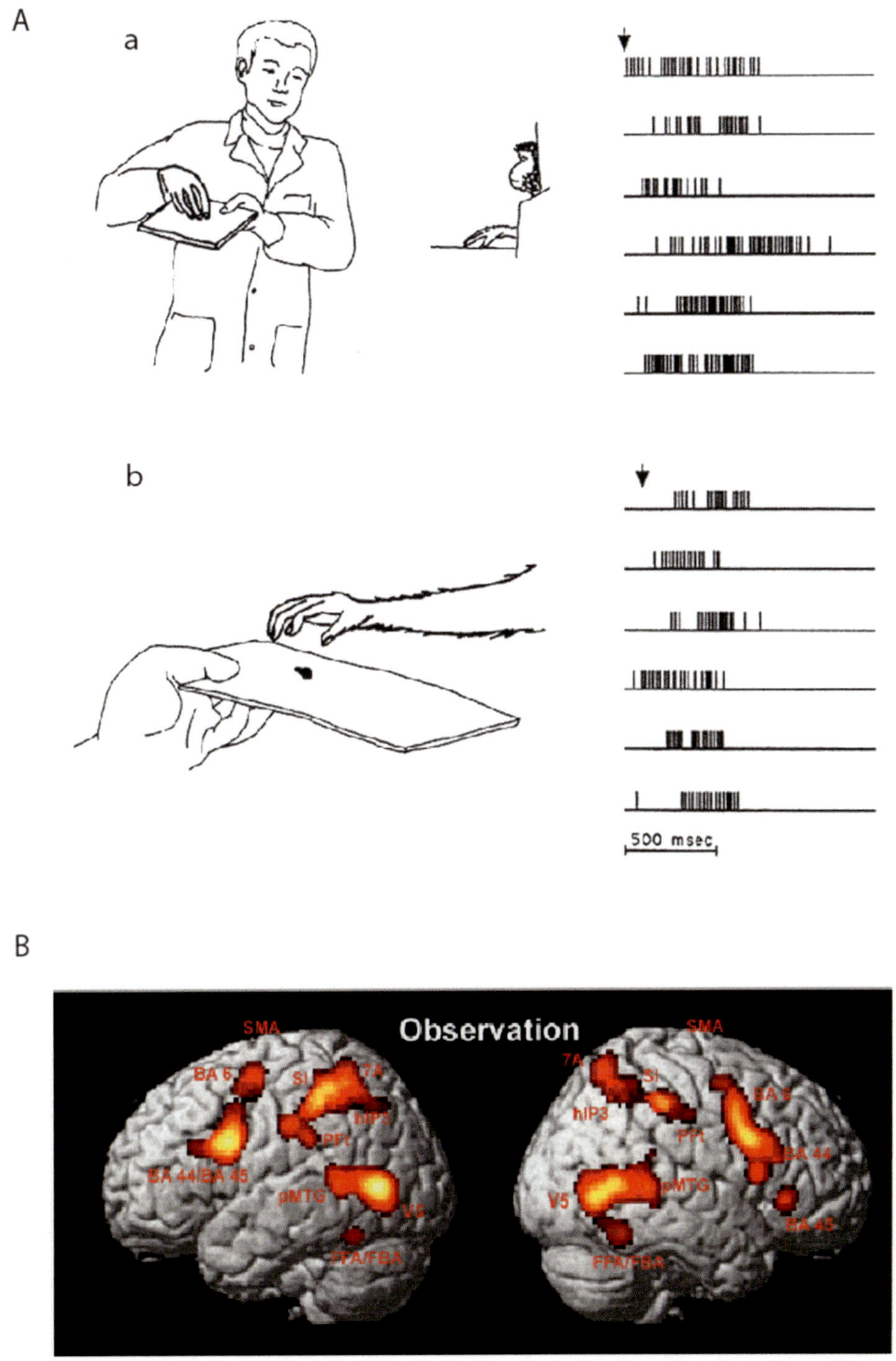

FIGURE 21.6 **Mirror neurons system in monkeys and humans.** (A) Example of an F5 mirror neuron. (A) Grasping observation; (B) grasping execution. On the right of each panel, five trials showing the neuron response are shown. The arrows indicate movement beginning. (B) Meta-analysis showing the brain activations during action observation. Activations are displayed on the left and right lateral surface view of the MNI single subject template. *BA44*, 45, Broca's area; *BA6*, lateral premotor cortex; *hIP3*, intraparietal area; *PFt*, Inferior parietal area; *pMTG*, Posterior middle temporal gyrus; *SI*, Primary somatosensory cortex (BA2); *SMA*, Supplementary motor area (hidden within the interhemispheric fissure); *V5*, Extrastriate visual area; *7A*, Superior parietal area. *(A) From di Pellegrino, G., Fadiga, L., Fogassi, L., Gallese, V., Rizzolatti, G., 1992. Understanding motor events: a neurophysiological study. Exp. Brain Res. 91, 176–180; (B) From Caspers, S., Zilles, K., Laird, A.R., Eickhoff, S.B., 2010. ALE meta-analysis of action observation and imitation in the human brain. Neuroimage 50, 1148–1167.*

(Opt, PG, PFG, PF), corresponding to different somatotopic fields (eyes, arm, hand, and mouth; Pandya and Seltzer, 1982; Gregoriou et al., 2006; Rozzi et al., 2008). Parietal mirror neurons, with properties very similar to those of F5, have been mainly found in area PFG, in agreement with the presence in it of a hand motor representation (Fogassi et al., 2005; Rozzi et al., 2008). More recently, neurons with properties similar to those of mirror neurons have also been described in AIP (Lanzilotto et al., 2019; Maeda et al., 2015).

The first studies on mirror neurons showed that many of them are invariant with respect to many visual details of the observed motor act, although some neurons presented specific visual responses for the direction of the observed act, space position (left, center, right), and the hand used by the observed agent (Gallese et al., 1996). More recent systematic investigation showed that, although the coding of the goal of others' motor acts remain the main function of mirror neurons, they can also be modulated by a certain number of details of the observed act. For example, mirror neurons of monkey area F5 showed differential responses when the monkey observed videoclips of grasping motor acts presented from third person (frontal and side), and first person views: only 25% of neurons were invariant, while the other three-quarter were selectively or preferentially activated, in almost equal percentage, by the three viewpoints (Caggiano et al., 2011). Another example of modulation is that related to the distance at which the observed motor act is performed. Half of studied mirror neurons in F5 were activated differentially when an experimenter grasped a piece of food within the monkey peri-personal space (corresponding to the reaching space) or in the extra-personal space (Caggiano et al., 2009). Half of the recorded neurons discharged strongest when the observed motor act was presented in the monkey peri-personal space, the other half when it was presented in the extra-personal space. This spatially related response could prompt the possibility to socially interact with others in different space sectors, for competing or cooperating (see Jeannerod, 2001 for a speculation on the role of simulation of observed actions for action anticipation).

Understanding Others' Intentions

As largely described, neurons of premotor cortex mainly code the goals of different motor acts and the timing of their activation is generally correlated to the duration of the motor act. It is much less common to record neurons discharging for a whole action. According to the definition of "action" given at the beginning of this chapter, this is composed by a sequence of motor acts aimed to achieve a final goal that coincides with the agent's motor intention. Neurophysiologically, the neural mechanism associated to motor intention is not so clearly elucidated, as well as that involved in decoding others' intentions. Recent studies tried to investigate these mechanisms starting from the consideration that a grasping act can belong to distinct actions relate to different intentions (Bonini et al., 2010; Fogassi and Luppino, 2005). In these studies, in the "motor" condition, monkeys were trained to execute the same grasping act that was included in two different actions related to two different intentions (grasping to eat or to place). In the "visual" condition, the same monkeys had to simply observe an experimenter grasping an object to eat it or to place it. During the performance of the two tasks, hand-related motor and mirror neurons, respectively, were recorded from both F5 and PFG. These experiments showed, first of all, that a high percentage of motor neurons in both areas had a preferential response, during grasping execution, for one of the two actions, while other neurons did not show any difference in discharge. Thus, modulated neurons appear to reflect the agent's intention. Secondly, a high percentage of both parietal and premotor mirror neurons presented a differential discharge also during observation of grasping, depending on the final goal of the observed action. Noteworthy, the visual and motor preference of most mirror neurons was congruent. From these data, a model has been suggested, proposing that each intentional action is coded by a dedicated chain of neurons, each corresponding to a motor act of the action sequence (e.g., reaching-grasping-bringing to the mouth, for the eating chain; Fogassi et al., 2005; Chersi et al., 2011). Therefore, during both execution and observation of a grasping act belonging to a given intentional chain, neurons coding this motor act are recruited. In this way, the mirror neuron mechanism, besides allowing the observing individual to understand the goal of another's motor act, also allows to decode the motor intention of the observed agent.

The Mirror System in Humans

In humans, except in a limited number of cases such as in neurosurgical patients, it is not possible to record from single neurons, for ethical reasons. Thus, also the evidence for the presence of a mirror neuron system in our species must derive from electrophysiological techniques that have a lower spatial or temporal resolution with respect to single neurons recording or are based on indirect measures of neuronal activity, as in the case of neuroimaging methodologies. This does not mean that the data are not reliable, but simply that it is more difficult, for example, to demonstrate the neuronal mechanism of sensorimotor integration, because this latter requires necessarily to assess the single neuron response.

Since the literature on action observation and on mirror neuron system in humans is now quite large, it is necessary to describe findings related to the fundamental concepts that had to be addressed in these investigations. The first concept to be demonstrated was that observation of others' motor acts could activate the motor cortex in healthy subjects. In this respect, the first study addressing this issue is that of Fadiga et al. (1995). They used transcranial magnetic stimulation (TMS) to verify whether during observation of grasping acts performed by an experimenter, there was an activation of motor cortex in observing participants. Since TMS magnetic pulses are given with threshold intensity that does not produce overt movements, it is possible to see whether applying this stimulation together with the visual stimulus can produce a specific enhancement of the electromyographic activation in terms of motor evoked potentials (MEPs). The study demonstrated that grasping observation produced a specific pattern of MEPs enhancement in those hand muscles that are normally used to perform grasping. Furthermore, when the same participants executed a grasping act, there was an activation of the same muscles, thus suggesting the presence of a mirror mechanism. These data demonstrated that during observation there is an activation of excitable cortex, very likely due to activation of premotor mirror neurons. Since, however, TMS studies do not allow to see which areas are directly activated during observation, other types of studies were needed to determine the anatomical location of the areas. Thus, neuroimaging studies were performed to identify the putative mirror neuron system. These studies allowed to show that, when participants observe motor acts done by other agents, three main cortical sectors are activated, one in the STS, a second one in the inferior parietal cortex (supramarginal gyrus), the third one in the ventral premotor cortex plus the caudal sector of the inferior frontal gyrus (IFG) (Grafton et al., 1996; Iacoboni, 1999; Koski et al., 2003; Frey and Gerry, 2006; Caspers et al., 2010; Rizzolatti et al., 2014, Fig. 21.6B). These sectors can be considered the homologs of monkey areas containing mirror neurons (ventral premotor area F5 and inferior parietal area PFG) or simply responding to the visual presentation of motor acts (STS). These first studies were mainly limited to the observation of hand motor acts, while other studies verified the activation related to the observation of acts done with different effectors, namely mouth, hand and foot (Buccino et al., 2001). The results showed that observation of biting with the mouth activated IFG and the ventral premotor cortex (PMV), that of grasping with the hand activated the dorsal part of PMV, that of feet motor acts (e.g., pressing a pedal) activated dorsal premotor cortex (PMD). Similarly, mouth, hand, and foot motor acts activated, in a kind of sequential order, also the posterior parietal cortex, with the mouth motor act activating the rostral part of IPL, followed caudally by the hand and, even more caudally, by the foot. This latter activated also the superior parietal lobule. Overall, this study demonstrated that observation of others' actions imply not only the coding of a goal but also of the effector used by the observed agent. Interestingly, this type of somatotopically organized activation was also found only in the premotor cortex for the pantomimes of the previously described motor acts. This second finding shows that in humans also the pantomime of a goal-directed motor act, without the object, activates the motor cortex, suggesting that participants understand the goal and the effector even in absence of a target. Later studies confirmed the activation of a dorsal circuit, including SPL and PMD, during observation of proximal arm movements (Filimon et al., 2007) or whole-body dancing movements (Calvo-Merino et al., 2005). In the study of Abdollahi et al. (2013), where participants had to observe manipulation, locomotion and climbing, it was shown an anatomical difference in activation, since hand manipulation tended to activate more the IPL, while climbing activated more the SPL. Altogether, the described neuroimaging studies allowed to draw a kind of map of the areas/circuits activated during observation of motor acts.

Another issue to be addressed was whether the human mirror neuron system codes the goal of observed acts. This type of coding was demonstrated, for example, by a series of neuroimaging studies by Gazzola and coworkers (Gazzola et al., 2007a, b). In the first fMRI study (Gazzola et al., 2007a) it was demonstrated that when participants watched videos of grasping performed by a human or a robot hand, there was a comparable activation of the parietal and frontal areas of the mirror system, in spite of many differences between the two effectors. In the second fMRI study, two aplasic patients (experimental group) and a group of healthy control participants had to observe videoclips of hand grasping acts. Then, both patients and control subjects had to perform actions with their feet and mouth and only controls with their hand. First of all, it was demonstrated that also in aplasic individuals, as in control participants, there was a parieto-frontal activation during observation of hand grasping acts. Even more interestingly, when the overlap between activations during observation and execution was analyzed, in aplasic patients there was an overlap with those premotor and parietal sectors active during the execution of foot and mouth movements, meaning that the matching that allows understanding others' motor acts occurs, in these patients, on the basis of the goal (grasping), independent of the effector (Gazzola et al., 2007b).

Electrophysiological studies, carried out by means of electroencephalographic (EEG) and magnetoencephalographic (MEG) techniques, allowed to include the temporal issue to the study of the mirror system. In this respect, an important finding was the demonstration that during both action observation and action execution the μ rhythm, belonging to the EEG alpha band, becomes desynchronized both when a participant performs a motor act and when she/he observes the same act done by another person (see, e.g., Cochin, 1999). Since this rhythm is recorded at the level of primary motor cortex (BA 4), it is likely that the desynchronization during observation is due to the preceding activation of premotor cortex (BA6) that in turn modulates primary motor cortex to which, at least in monkeys, it is directly connected. In terms of timing of activation during action observation, Nishitani and Hari (2000) showed a sequence of activation, starting from the visual cortex, followed by IFG and finally by primary motor cortex (BA 4).

Interestingly, electrophysiological recording is one of the best, noninvasive techniques, to be used in children. Thus, the application of the EEG studies to children of several ages, also as early as 9 months allowed to demonstrate that a desynchronization of the μ/α rhythm occurs during observation of grasping (Lepage and Théoret, 2006; Southgate et al., 2010), drawing (Fecteau et al., 2004), and reaching (Marshall et al., 2011). Also with the Near Infrared Spectroscopy (NIRS) Shimada and Hiraki (2006) were able to demonstrate a frontal activation during grasping observation in children as old as 6 months, suggesting that the mirror mechanism is hardwired, as a product of primates phylogenesis.

Subcortical Structures With Mirror Properties

In monkeys, while cortical mirror neurons have been widely investigated, it is still unknown if neurons with similar properties do also exist in subcortical structures, such as the two major structures with motor properties, cerebellum and basal ganglia. Their possible involvement in a "mirror" circuit is suggested by anatomical data showing that frontal and parietal areas containing mirror neurons project to putaminal regions, different from those connected with MI (Gerbella et al., 2016). The description of subcortical activation during action observation is beyond the aim of this chapter.

Plasticity of the Mirror Neuron System

Neural plasticity is a property of the whole nervous system, very important in the development period, where the role of experience is crucial for the correct maturation of many functions. It is also fundamental for learning, leading to long-lasting changes of the neural structures involved in it.

It is well known that motor training determines plastic changes in motor cortex. Particularly evident are those related to specific skills, such as sports, dance, music performance.

Since the mirror system for actions relies on the motor system also this latter undergoes plastic changes.

Single neuron studies showed that when monkeys learn to use pliers to grasp food, after training F5 grasping neurons become able to activate not only during hand grasping but also during grasping with pliers (Umiltà et al., 2008). Similarly, mirror neurons activated by observation of hand grasping also activate when the observed agent picks up the objects with the same pliers the monkeys learned to use (Rochat et al., 2010; Umiltà et al., 2008).

In humans, plasticity of the mirror neuron system has been shown, for example, in dance. Calvo Merino et al. (2005) performed an fMRI study in which dancers of classic ballet, dancers of capoeira and naïve subjects had to observe videoclips showing steps of classical dance or capoeira. First of all, the activation of the parietal and premotor areas of the mirror system was stronger in dancers, but more interestingly the two groups of dancers showed a higher activation during observation of their type of ballet, while naïve participants did not show differential activations between the two conditions. Thus, motor expertise very likely induced plastic changes in the experts' motor system, so that during observation of their own expertise there is a higher "mirror activation."

This "motor resonance" is very important for imitation learning in fact, during imitation, the trainee observes a model and then tries to reproduce the same movements/motor acts. It is quite reasonable that the matching between the motor acts of the observed model and the observer's repertoire can occur through the mirror neuron system. This has been shown, for example, by an event-related fMRI study in which participants had to observe and imitate novel guitar chords played by an expert guitarist (Buccino et al., 2004). The results showed that during observation and even more during the imitation phase, the parietal, premotor and IFG areas of the mirror system were strongly activated. In this study there was a period of pause between observation of the model and imitation; interestingly, in this period, there was a strong activation of prefrontal cortex (area 46), suggesting that beyond the mirror matching mechanism, a further mechanism is required for imitation, consisting in the sequential organization of the acts to be reproduced. In fact, prefrontal cortex was not active anymore during actual imitation.

Apraxia and the Mirror Neuron System

Apraxia is a high order motor disorder in absence of pure sensory or motor deficits and is caused by acquired brain diseases. Depending on the nature deficit, it can be subdivided in ideomotor and ideational (or conceptual) apraxia (De Renzi and Faglioni, 1999). In the former, the patient is not able to perform a gesture on instruction, while in the latter the patient is not able to associate a tool or an object with the corresponding action. The lesions leading to these two forms of apraxia related to the forelimb usually involve the left inferior parietal cortex, with more anterior lesions mostly producing the ideomotor apraxia and more posterior ones producing the ideational apraxia. Oro-facial apraxia is, instead, usually produced by premotor lesions. The highest frequency of right-handed patients with left parietal lesions brought to the idea of a left specialization for limb praxis, with left cerebral cortex controlling both hands.

Patients with ideomotor apraxia are typically impaired in reproducing meaningful and meaningless hand and arm gestures performed by another individual, acting as a model. However, this form of apraxia is characterized by the surprising phenomenon of automatic-voluntary dissociation, because of which the patient does not perform the gesture when requested, but he/she performs it when the same gestures is spontaneously elicited by the context. This means that the patient still possesses motor representation of meaningful gestures, but cannot retrieve them by imitation. Since it is known that the mirror neuron system is crucial for imitation and that one of its nodes is inferior parietal cortex, it is likely that a damage to this cortical sector produces an impairment in this imitation mechanism. In fact, although in these patients theoretically the model's gesture can be processed by the high order visual system, the visual description of the gesture probably cannot access to the mirror neuron circuitry, thus impairing the matching between the model's gesture and its motor representation. On the contrary, when the patient must execute the same gesture by using external information on objects or context, a different pathway is involved, thus bypassing the parietal lesion.

THE VOLUNTARY CONTROL OF MOVEMENT EXECUTION

Movements Evoked by Electrical Stimulation

The anatomical description of descending projections from agranular frontal cortex is the basis for understanding how motor cortex controls movements execution.

The first type of investigation that paved the way to our understanding of how the motor cortex command the movements of different effectors was the electrical stimulation of agranular frontal cortex by means of surface electrodes put on the dura mater, with current intensity in the range of milliamps. With this technique, Fritsch and Hitzig (1870) could stimulate dog motor cortex, Ferrier (1873) monkey cortex and Sherrington chimpanzee cortex. In all these experiments stimulation evoked overt movements of contralateral body parts. Only in the fifties of the last century could the group of Woolsey (Woolsey and Settlage, 1952) in monkeys (Fig. 21.7A) and that of Penfield (Penfield and Welch, 1951) in neurosurgical patients demonstrate the elicitations of specific movements from surface stimulation of frontal cortex. Both groups showed a similar somatotopical organization of the excitable cortex in front of the central sulcus, with the lower limb represented in the medialmost part, the forelimb in the central part and the mouth in the lateralmost part. The different somatotopic sectors in this map did not show the same relative proportions as in the physical body, those having more descending projections being the most represented. Thus, the hand and the mouth occupy a larger sector of motor cortex with respect to the lower limb. This organization corresponds to a higher sophistication in movement control, as for example the independent movement of single fingers.

Surface stimulation allowed to demonstrate the presence of a single detailed motor homunculus in primary motor cortex and of another one, smaller, in the mesial surface, originally named secondary motor area (MII), then Supplementary Motor Area (SMA).

The introduction of the technique of microelectrodes, capable of recording and stimulating in the cortical depth, opened the possibility to explore more in the details the organization of agranular areas, showing that each of them contains a movements map, with a partial somatotopy. Furthermore, also MI organization was reevaluated, leading to different proposals. For example, Kwan et al. (Murphy et al., 1978) proposed a ring-like organization, in which the representation of more proximal movements surrounds that of more distal movements. Another proposal, that of mosaic organization, maintains that every motor field (e.g., that of the hand) contains several representation of the same movement (e.g., finger flexion, Gould et al., 1986; Fig. 21.7B). Independently of the type of proposal, it has been clarified that primary motor cortex (F1) is completely excitable at low intensity threshold (as low as 3 μAmps) and that its stimulation elicits simple movements. On the contrary, microstimulation of premotor areas reveals that they are in average less excitable and their stimulation very often elicits combination of movements of contiguous or distant articulations. For example, from area F4 it is possible to elicit, at the same intensity threshold, movements of the shoulder, neck, and mouth. Although electrical stimulation is an artificial method, it already allows to conclude that F1 controls discrete movements, while premotor cortex controls movement synergies. Interestingly, Graziano and coworkers (Graziano et al., 2002), by stimulating premotor cortex with longer stimulation trains, could obtain movement synergies that resembled goal-directed movements.

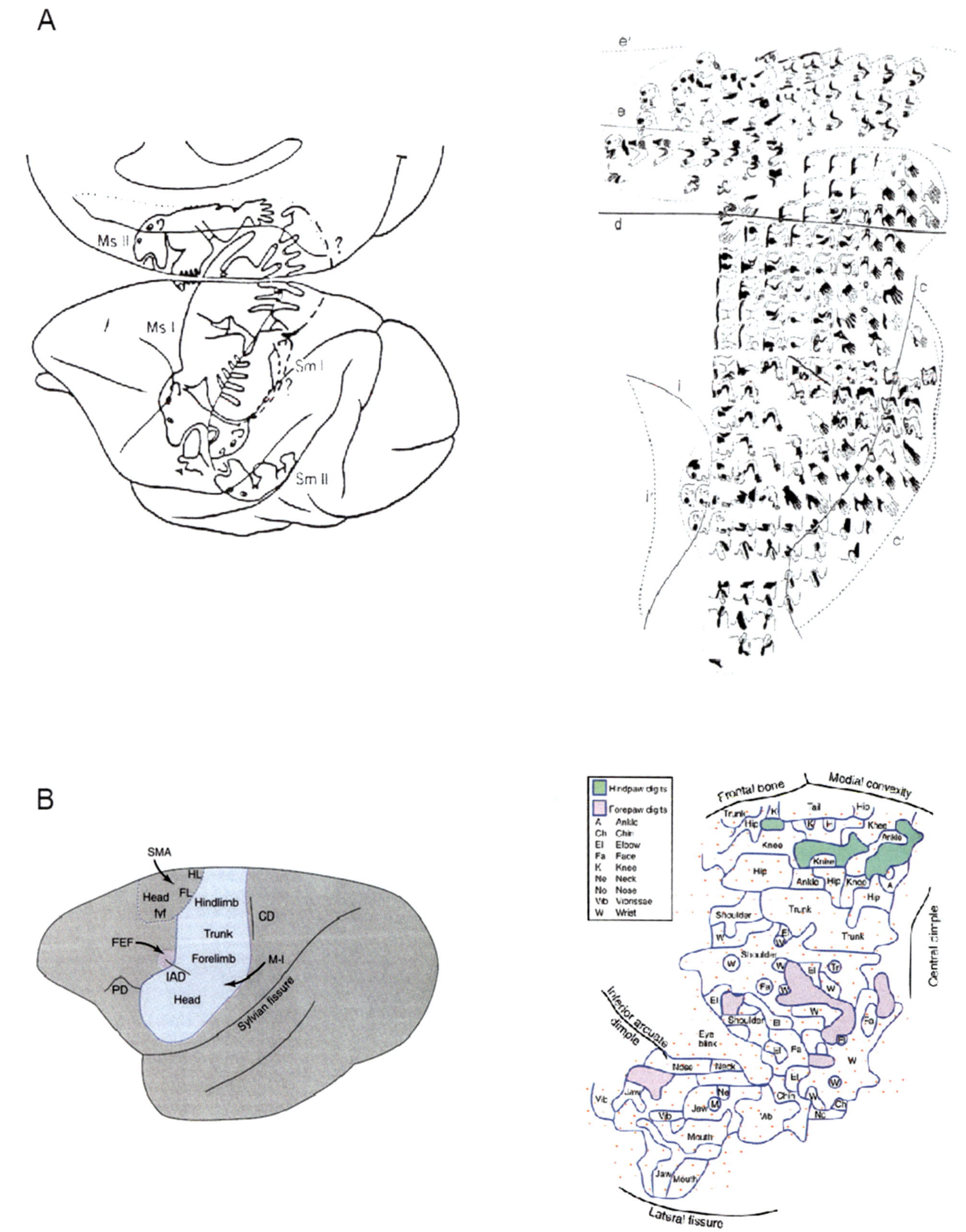

FIGURE 21.7 **Maps of monkey movements elicited by electrical stimulation of the motor cortex.** (A) Map of the movements evoked by superficial electrical stimulation of the macaque monkey motor cortex. Left part: cartoon showing the motor simiusculus of the primary (Ms I) and secondary (Ms II) motor cortex. The primary (Sm I) and secondary (Sm II) somatosensory simiusculi are also shown, in the parietal cortex. Right

Control of Movement Parameters by Primary Motor Cortex

In the introductory chapter, it has been claimed that the primary task of the motor cortex is that of coding motor goals. This is the main task for neurons of premotor areas, while MI must translate motor goals in discrete movements. Single neurons studies, since the first pioneering experiments of Evarts and coworkers in awake monkeys (Evarts, 1968), addressed the issue of if and how MI neurons code movement parameters. They trained monkeys to perform simple voluntary extension or flexion wrist movements to oppose to passively applied flexion and extension movements, respectively. During task execution, they recorded the EMG activity of wrist flexors and extensors muscles and the neural discharge of identified corticospinal neurons of the hand region of MI. First of all, they showed that each neuron discharges before voluntary flexion or extension. Then, they recorded the activity of the same neurons when monkeys had to exert more force to oppose passive movements induced by a load and observed that the discharge increased with the increase in force. In this way, they demonstrated that neurons of MI code movement force.

Subsequently, Georgopoulos et al. (1992) studied the coding of movement direction by neurons of the arm field of MI. Control of arm direction is a complicated task, because to obtain a specific direction, a synergy of different shoulder movements must be coded. The task they employed required the monkey to move a lever in eight different directions and, at the same time, single neurons activity was recorded. They showed that each neuron was not able to code a single direction, but rather a broad direction field. For example, a neuron could discharge almost equally well for three different directions, each differing 45 degrees from the closest one. In order to understand how neurons of this field could code a precise direction, they suggested a mathematic algorithm in which a specific arm direction is achieved through the weighted contribution of all recorded neurons. The application of the model produced, for each direction, an almost accurate prediction of the monkey arm movement, therefore indicating that direction of reaching is the result of a population code.

Further studies (Sergio and Kalaska, 2003) demonstrated that there are primary motor neurons that can integrate force and direction. If the monkey, to keep a given direction, had to oppose to externally applied forces, neuronal activity increased or decreased according to whether external force opposed or facilitated the monkey active movement.

Effects of Lesions of Primary Motor Cortex

The type of coding by MI neurons shown by single neurons recording studies suggests that damage of a somatotopic field should lead to a force or a direction deficit. The former type of impairment has been demonstrated by inactivation experiments in monkeys, showing that when the hand field is temporarily knocked down, there is a flaccid paralysis, with loss of force, making the monkey unable to oppose to a passive joint mobilization and to exert force in order to, for example, grasp an object (Lawrence and Kuypers, 1968).

In humans, damage to motor cortex is hardly confined to a little sector of MI. Thus, it is very frequent to see, after a vascular accident, a spastic paralysis. This latter is not due to a pure damage to MI, but rather also encompasses premotor cortex, thus resulting in muscle hypercontraction. In the cases in which the patient had a focused lesion, the resulting paralysis is flaccid, with loss of force.

Movement Control by Supplementary Motor Area

Similarly to primary motor cortex, SMA (or area F3 in the monkey) contains a complete motor map of the contralateral body and, using intracortical microstimulation, is excitable at low intensity currents. However, there are specific features that allow one to distinguish F3 from F1. First, electrical stimulation of F1 typically evokes single joint movements, while that of F3 induces most often movements of two or more joints (Luppino et al., 1991; Mitz and Wise, 1987). Second, while proximal and distal movements are anatomically segregated in F1, in F3 typically movements of the hand, arm, head, or trunk are coordinated (Luppino et al., 1991). Third,

part: Detailed map of the movements evoked by electrical stimulation. *C*, central sulcus; *c′*, floor of the central sulcus; *d*, midline; *e*, cingulate sulcus; *e′*, floor of the cingulate sulcus; *I*, arcuate sulcus; *I′*, floor of the arcuate sulcus. (B) Map of the movements evoked by intracortical electrical microstimulation of the owl monkey motor cortex. Left part: Lateral view of the left hemisphere of the owl monkey. The light blue color indicates the motor cortex, with the different motor fields revealed by microstimulation. *Cd*, central sulcus; *FEF*, Frontal eye field; *SMA*, Supplementary motor area. Right part: detailed representation of the movements elicited by microstimulation. The sectors in violet indicate subfields in which microstimulation evokes the same movement. *(A) From Woolsey, C., Settlage, P., 1952. Patterns of localization in precentral and "supplementary" motor areas and their relation to the concept of a premotor area. Res. Publ. Assoc. Res. Nerv. Ment. Dis. 30, 238–264; (B) From Gould, H.J., Cusick, C.G., Pons, T.P., Kaas, J.H., 1986. The relationship of corpus callosum connections to electrical stimulation maps of motor, supplementary motor, and the frontal eye fields in owl monkeys. J. Comp. Neurol. 247, 297–325; Credit: Sponsored by Italian PRIN 2015AWSW2Y_005 and by "FIL-Quota Incentivante" of University of Parma and co-sponsored by Fondazione Cariparma.*

proximal movements are more represented than distal ones in F3 (Woolsey and Settlage, 1952; Luppino et al., 1991). Single-neuron studies showed that the large majority of F3 neurons have a discharge time-locked to movement onset. Furthermore, many F3 neurons respond to sensory stimulation, most often to somatosensory stimuli, a minority of neurons also respond to visual stimuli (Tanji and Shima, 1994). To study the motor properties of mesial premotor cortex, Tanji et al. (1996) tested F3 neurons in a task requiring monkeys to perform different sequences of three motor acts (push, pull and turn a manipulandum). They showed that part of F3 neurons discharge in relation to a specific sequence of movements and concluded that this area plays a role in the organization of learned sequential movements (Tanji, 2001; Tanji and Shima, 1996).

F3 is target of strong projections from the caudal and dorsal part of area 24 (area 24d) and from parietal area PECg, located in the caudalmost part of the cingulate sulcus (Luppino et al., 1993). Area 24d roughly corresponds to the "caudal cingulate motor area" (Dum and Strick, 1993). Area PECg was defined by Pandya and Seltzer (1982) as an architectonic area connected with areas PE, PEC, and PGm, containing a complete somatosensory map of the body and thus designated as "supplementary sensory area" (Murray and Coulter, 1981).

Altogether, these data indicate that area F3 controls motor activity in a more global way than area F1, by using information about body-parts position to prepare the posture adjustments necessary for the forthcoming movements and playing a role in predictive postural adjustments (Massion, 1992). Furthermore, F3 plays a relevant role in the late stages of motor sequences learning (Hikosaka et al., 1999, see below).

Association Learning and Control of Action Onset. The Role of Area F6

Area F6 when microstimulated, is weakly excitable and overt movements can be only elicited by applying relatively high and prolonged currents. The motor responses typically consist of slow and complex movements involving the forelimb (Luppino et al., 1991). F6 is strongly connected with the dorsal and ventral parts of prefrontal 46 BA and with the rostral cingulate area 24c BA (Lu et al., 1994; Luppino et al., 1993). It is considered a site of convergence of information related to the working memory of objects or locations in space, to the temporal planning of actions or to motivation, which is used to actually trigger action execution. Accordingly, single neuron studies in monkeys mainly focused on two aspects of F6 functions: the involvement of F6 in executing and learning sequential behaviors and its role in starting an intended action. Shima and Tanji (2000) mainly focused on the first issue. They trained monkeys to perform three movements, push, pull or turn a manipulandum, in different orders and showed that F6 neurons reflected either the occurrence of a particular movement within the sequence or its position in numerical order.

While Tanji et al. focused on the study of specific motor acts, Hikosaka et al. studied the role of F6 neurons in sequential learning and execution of multiple reaching movements directed at different spatial positions (Nakamura et al., 1998). In these tasks, the monkeys faced a panel hosting several LEDs. Two of them could be turned on, and monkeys had to press them in a specific order learned by trial-and-error procedure. When the first sequence ("set") was learned, another pair of LEDs was turned on and the monkey had to press them again in the correct order. Finally, a "hyperset" formed by five different sets was presented. Typically, F6 neurons activated during sequence learning and ceased to respond during the execution of the well-learned hyperset.

Hikosaka et al. also performed an fMRI experiment in humans, aimed to assess whether pre-SMA (the human equivalent of F6) plays a specific role in learning motor sequences or is simply involved in visuomotor association learning (Sakai et al., 1999). The authors examined the activity of pre-SMA in three learning paradigms, including one in which subjects had to press the testing buttons on the basis of learned color associations. Interestingly, they found that pre-SMA is active in all three conditions, including that of color association and concluded that the main role of this area is to exert control over visuomotor transformations, regardless of the type of transformations involved (spatial or nonspatial). This conclusion was confirmed by a subsequent study on monkey (Hoshi and Tanji, 2004).

Concerning the role of F6 in the starting of an intended action, it has been shown that many F6 neurons become active largely in advance of movement onset (Alexander and Crutcher, 1990; Matsuzaka et al., 1992). This behavior is radically different from that of F1 and F3 neurons, whose discharge only slightly precedes movement onset. Rizzolatti and coworkers (Rizzolatti et al., 1990) studied the properties of monkey F6 neurons in a naturalistic context and showed that they do not control specific distal or proximal movements, but rather global actions involving reaching and grasping movements. In addition, they showed that F6 is not directly involved in visuo-motor transformations as posterior premotor areas, e.g., F4 or F5, since F6 neurons activity does not strongly depend on object size, shape, or location. In contrast, they described a clear relation between neuron discharge and action onset. Some neurons discharge on object presentation and the firing rate increases when the graspable stimulus enters into monkey's peripersonal space, others are not active upon object presentation, but discharge when the object become graspable. The authors concluded that F6 plays a role in specifying "whether" and "when" an action can be performed. According to this view, F6 neuron activity is instrumental in releasing the motor program encoded in the parieto-premotor circuits. This hypothesis was recently tested with a controlled paradigm (Lanzilotto et al., 2016). The authors trained monkeys to perform a visuomotor task in which three objects could be grasped using three different grips. They identified two main types of neurons: purely motor and visuomotor. By comparing the properties of these neurons with those previously studied in area F5, they found many common features in neuron activity in the two areas, supporting the idea of a functional interplay between them. However, the neural activity of visuomotor F6 neurons was characterized by an earlier onset and quicker decrease in the discharge with respect to F5, being F6 neurons discharge more closely related to object presentation than to the movement epochs of the task. In addition, F6 motor neurons activate earlier than those of F5 and, in F6, there are less neurons showing object or grip-selectivity.

There is recent evidence that area F6 also plays a role in social behavior. The discovery in this area of neurons that selectively encode the action of the agent ("self" neuron), the action of others ("other" neurons) and the action of both the agent and the observer ("mirror" neurons) is a strong argument in favor of this role (Yoshida et al., 2011, 2012). These neurons can likely interact with those belonging to the mirror neuron circuit (see above), playing a role in the identification of the action agent.

There is rich evidence from human studies that the prefrontal cortex and the pre-SMA are involved in action control. Most of this evidence comes from clinical data and provides a global framework for understanding the relationship between the prefrontal and the motor cortex. Accordingly, in this paragraph we will briefly describe the syndromes producing motor deficits after lesions of the prefrontal and mesial motor cortex. As a premise, we have to note that, typically, the large extent of brain lesions in humans prevents a precise identification of the cortical areas whose damage leads to specific symptoms.

Lesions of the prefrontal lobe can produce several motor syndromes which some are characterized by complex motor behaviors produced by the patients independent of their intention to act (*environmental dependency syndromes*). Among them, particularly interesting are *imitation behavior* and *utilization behavior* (Lhermitte et al., 1986). *Imitation behavior* consists in the reproduction of gestures or complex behaviors performed by another person in the absence of a specific request: often patients continue imitating after being asked to stop (Lhermitte et al., 1986). Similarly, patients showing *utilization behavior*, grasp visible objects with the appropriate grip, but also in this case in the absence of the need to do it (Ishihara et al., 2002). Typically, these two syndromes are observed in patients with bilateral prefrontal lesions, most often located on the mesial or lateral cortical surface, rather than on the orbital one (De Renzi et al., 1996; see Iaccarino et al., 2014).

Another clinical phenomenon related to a release from the executive control of the prefrontal lobe is the syndrome known as *alien hand*, also dubbed *anarchic hand* (Marchetti and Della Sala, 1998) to distinguish it from the alien hand observed in patients with neglect. Patients produce goal-directed and well executed complex movements with the upper limb, but the movements are unintended.

The common feature of all these syndromes is that the patient produces movements in an inappropriate context and timing, this is in agreement with the idea, suggested by monkey studies, that the prefrontal territories connected to the parieto-frontal grasping circuits have a crucial role in controlling the achievement of the goal of an action on the basis of the context (see Rozzi and Fogassi, 2017).

Other more elementary release phenomena appear following lesion of the mesial prefrontal and the anterior cingulate cortex (e.g., *grasp reflex, forced groping*, and *forced grasping*). *Grasp reflex* is a stereotyped prehensile reaction to tactile and proprioceptive stimuli applied to the palm. *Forced groping* and *forced grasping* are more complex and less stereotyped movements of the forelimb in response to a light touch on a patient's hand or to visual presentation of an object (Hashimoto and Tanaka, 1998).

Considering the anatomical connections and the functional properties of monkey area F6, we can hypothesize that these more elementary release phenomena are due to the pre-SMA disinhibition consequent to the damage of the prefrontal cortex. In line with this view, it is known that selective lesions of pre-SMA produce negative symptoms, such as akinesia, arrest of speech and reduction of spontaneous motor activity, in particular when there is no external stimulus to prompt performance (Laplane et al., 1977). Indeed, a recent fMRI study showed that akinetic parkinsonian patients have a hypoactivation of pre-SMA during execution of hand motor tasks (Sabatini et al., 2000).

Finally, the notion that pre-SMA plays an important role in the interplay of action commands and action stops (Hikosaka et al., 1999; Rizzolatti et al., 1996b) has recently been supported by clinical data. In particular, it has been described the case of a patient with a lesion including the prefrontal, the anterior cingulate cortex (ACC) and the pre-SMA but sparing the SMA-proper (Nachev et al., 2008; Sumner et al., 2007) showing a selective, strong impairment in rapidly changing from a motor plan to another. This is a further support to the proposed role of the pre-SMA in controlling "when to go."

References

Abdollahi, R.O., Jastorff, J., Orban, G.A., 2013. Common and segregated processing of observed actions in human SPL. Cerebr. Cortex 23, 2734–2753.

Alexander, G.E., Crutcher, M.D., 1990. Preparation for movement: neural representations of intended direction in three motor areas of the monkey. J. Neurophysiol. 64, 133–150.

Amunts, K., Zilles, K., 2012. Architecture and organizational principles of Broca's region. Trend. Cognit. Sci. 16, 418–426.

Andersen, R.A., Asanuma, C., Essick, G., Siegel, R.M., 1990a. Corticocortical connections of anatomically and physiologically defined subdivisions within the inferior parietal lobule. J. Comp. Neurol. 296, 65–113.

Andersen, R.A., Bracewell, R.M., Barash, S., Gnadt, J.W., Fogassi, L., 1990b. Eye position effects on visual, memory, and saccade-related activity in areas LIP and 7a of macaque. J. Neurosci. 10, 1176–1196.

Andersen, R.A., Buneo, C. a, 2002. Intentional maps in posterior parietal cortex. Annu. Rev. Neurosci. 25, 189–220.

Andersen, R.A., Cui, H., 2009. Intention, action planning, and decision making in parietal-frontal circuits. Neuron 63, 568–583.

Andersen, R.A., Essick, G.K., Siegel, R.M., 1985. Encoding of spatial location by posterior parietal neurons. Science 230, 456–458.

Bakola, S., Gamberini, M., Passarelli, L., Fattori, P., Galletti, C., 2010. Cortical connections of parietal field PEc in the macaque: linking vision and somatic sensation for the control of limb action. Cerebr. Cortex 20, 2592–2604.

Barash, S., Bracewell, R.M., Fogassi, L., Gnadt, J.W., Andersen, R.A., 1991. Saccade-related activity in the lateral intraparietal area. I. Temporal properties; comparison with area 7a. J. Neurophysiol. 66, 1095–1108.

Barbas, H., 1988. Anatomic organization of basoventral and mediodorsal visual recipient prefrontal regions in the rhesus monkey. J. Comp. Neurol. 276, 313–342.

Baumann, M. a, Fluet, M.-C., Scherberger, H., 2009. Context-specific grasp movement representation in the macaque anterior intraparietal area. J. Neurosci. 29, 6436–6448.

Belmalih, A., Borra, E., Contini, M., Gerbella, M., Rozzi, S., Luppino, G., 2007. A multiarchitectonic approach for the definition of functionally distinct areas and domains in the monkey frontal lobe. J. Anat. 211, 199–211.

Belmalih, A., Borra, E., Contini, M., Gerbella, M., Rozzi, S., Luppino, G., 2009. Multimodal architectonic subdivision of the rostral part (area F5) of the macaque ventral premotor cortex. J. Comp. Neurol. 512, 183–217.

Bernshtein, N.A., 1947. Bernstein's Construction of Movements: The Original Text and Commentaries, 2020th ed. Routledge, New York.

Blatt, G.J., Andersen, R.A., Stoner, G.R., 1990. Visual receptive field organization and cortico-cortical connections of the lateral intraparietal area (area LIP) in the macaque. J. Comp. Neurol. 299, 421–445.

Bonini, L., Rozzi, S., Serventi, F.U.F.U., Simone, L., Ferrari, P.F.P.F., Fogassi, L., 2010. Ventral premotor and inferior parietal cortices make distinct contribution to action organization and intention understanding. Cerebr. Cortex 20, 1372–1385.

Borra, E., Belmalih, A., Calzavara, R., Gerbella, M., Murata, A., Rozzi, S., Luppino, G., 2008. Cortical connections of the macaque anterior intraparietal (AIP) area. Cerebr. Cortex 18, 1094–1111.

Borra, E., Belmalih, A., Gerbella, M., Rozzi, S., Luppino, G., 2010. Projections of the hand field of the macaque ventral premotor area F5 to the brainstem and spinal cord. J. Comp. Neurol. 518, 2570–2591.

Borra, E., Gerbella, M., Rozzi, S., Luppino, G., 2011. Anatomical evidence for the involvement of the macaque ventrolateral prefrontal area 12r in controlling goal-directed actions. J. Neurosci. 31, 12351–12363.

Boussaoud, D., 2001. Attention versus intention in the primate premotor cortex. Neuroimage 14, S40–S45.

Breveglieri, R., Galletti, C., Bosco, A., Gamberini, M., Fattori, P., 2015. Object affordance modulates visual responses in the macaque medial posterior parietal cortex. J. Cognit. Neurosci. 27, 1447–1455.

Breveglieri, R., Galletti, C., Gamberini, M., Passarelli, L., Fattori, P., 2006. Somatosensory cells in area PEc of macaque posterior parietal cortex. J. Neurosci. 26, 3679–3684.

Brodmann, K., 1909. Vergleichende Lokalisationslehre der Großhirnrinde. Barth, Leipzig.

Bruce, C.J., Goldberg, M.E., 1985. Primate frontal eye fields. I. single neurons discharging before saccades. J. Neurophysiol. 53, 603–635.

Bruce, C.J., Goldberg, M.E., Bushnell, M.C., Stanton, G.B., 1985. Primate frontal eye fields. II. Physiological and anatomical correlates of electrically evoked eye movements. J. Neurophysiol. 54, 714–734.

Buccino, G., Binkofski, F., Fink, G., 2001. Action observation activates premotor and parietal areas in a somatotopic manner: an fMRI study. Eur. J. Neurosci. 13, 400–404.

Buccino, G., Vogt, S., Ritzl, A., Fink, G., Zilles, K., 2004. Neural circuits underlying imitation learning of hand actions: an event-related fMRI study. Neuron 42, 323–334.

Caggiano, V., Fogassi, L., Rizzolatti, G., Pomper, J.K., Thier, P., Giese, M. a, Casile, A., 2011. View-based encoding of actions in mirror neurons of area F5 in macaque premotor cortex. Curr. Biol. 21, 144–148.

Caggiano, V., Fogassi, L., Rizzolatti, G., Thier, P., Casile, A., 2009. Mirror neurons differentially encode the peripersonal and extrapersonal space of monkeys. Science 324, 403–406.

Calvo-Merino, B., Glaser, D.E., Grèzes, J., Passingham, R.E., Haggard, P., 2005. Action observation and acquired motor skills: an fMRI study with expert dancers. Cerebr. Cortex 15, 1243–1249.

Caminiti, R., Ferraina, S., Johnson, P.B., 1996. The sources of visual information to the primate frontal lobe: a novel role for the superior parietal lobule. Cerebr. Cortex 6, 319–328.

Campbell, A.W., 1905. Histological Studies on the Localisation of Cerebral Function, by Alfred W. Campbell. By Aid of a Subsidy from the Royal Society of London. University Press, Cambridge.

Caspers, S., Zilles, K., Laird, A.R., Eickhoff, S.B., 2010. ALE meta-analysis of action observation and imitation in the human brain. Neuroimage 50, 1148–1167.

Cavada, C., Goldman-Rakic, P.S., 1989. Posterior parietal cortex in rhesus monkey: II. Evidence for segregated corticocortical networks linking sensory and limbic areas with the frontal lobe. J. Comp. Neurol. 287, 422–445.

Chersi, F., Ferrari, P.F., Fogassi, L., 2011. Neuronal chains for actions in the parietal lobe: a computational model. PLoS One 6.

Cisek, P., Kalaska, J.F., 2010. Neural mechanisms for interacting with a world full of action choices. Annu. Rev. Neurosci. 33, 269–298.

Cochin, S., 1999. Observation and execution of movement: similarities demonstrated by quantified electroencephalography. Eur. J. Neurosci. 11, 1839–1842.

Colby, C.L., 1998. Action-oriented spatial reference frames in cortex. Neuron 20, 15–24.

Colby, C.L., Duhamel, J.-R., 1996. Spatial representations for action in parietal cortex. Brain Res. Cogn. Brain Res. 5, 105–115.

Colby, C.L., Duhamel, J.-R., Goldberg, M.E., 1993. Ventral intraparietal area of the macaque: anatomic location and visual response properties. J. Neurophysiol. 69, 902–914.

Colby, C.L., Gattass, R., Olson, C.R., Gross, C.G., 1988. Topographical organization of cortical afferents to extrastriate visual area PO in the macaque: a dual tracer study. J. Comp. Neurol. 269, 392–413.

Cooke, D.F., Taylor, C.S.R., Moore, T., Graziano, M.S.A., 2003. Complex movements evoked by microstimulation of the ventral intraparietal area. Proc. Natl. Acad. Sci. U. S. A. 100, 6163–6168.

Corbetta, M., Shulman, G.L., 2011. Spatial neglect and attention networks. Annu. Rev. Neurosci. 34, 569–599.

Cowey, A., Small, M., Ellis, S., 1994. Left visuo-spatial neglect can be worse in far than in near space. Neuropsychologia 32, 1059–1066.

Crammond, D.J., Kalaska, J.F., 2000. Prior information in motor and premotor cortex: activity during the delay period and effect on pre-movement activity. J. Neurophysiol. 84, 986–1005.

Crippa, A., Cerliani, L., Nanetti, L., Roerdink, J.B.T.M., 2011. Heuristics for connectivity-based brain parcellation of SMA/pre-SMA through force-directed graph layout. Neuroimage 54, 2176–2184.

De Renzi, E., 1982. Disorders of Space Exploration and Cognition. J. Wiley.

De Renzi, E., Cavalleri, F., Facchini, S., 1996. Imitation and utilisation behaviour. J. Neurol. Neurosurg. Psychiatr. 61, 396–400.

De Renzi, E., Faglioni, P., 1999. Apraxia. In: Pizzamiglio, D.G. (Ed.), Clinical and Experimental Neuropsychology. Psychology Press, East Sussex, UK.

Deecke, L., Kornhuber, H.H., 1978. An electrical sign of participation of the mesial "supplementary" motor cortex in human voluntary finger movement. Brain Res. 159, 473–476.

di Pellegrino, G., Fadiga, L., Fogassi, L., Gallese, V., Rizzolatti, G., 1992. Understanding motor events: a neurophysiological study. Exp. Brain Res. 91, 176–180.

Duhamel, J.-R., Colby, C.L., Goldberg, M.E., 1998. Ventral intraparietal area of the macaque: congruent visual and somatic response properties. J. Neurophysiol. 79, 126–136.

Duhamel, J.R., Colby, C.L., Goldberg, M.E., 1992. The updating of the representation of visual space in parietal cortex by intended eye movements. Science 255, 90–92.

Dum, R.P., Strick, P.L., 1991. The origin of corticospinal projections from the premotor areas in the frontal lobe. J. Neurosci. 11, 667–689.

Dum, R.P., Strick, P.L., 1993. Cingulate motor areas. In: Vogt, B. a, Gabriel, M. (Eds.), Neurobiology of Cingulate Cortex and Limbic Thalamus: A Comprehensive Handbook. Birkhäuser, Boston, pp. 415–441.

Evarts, E.V., 1968. Relation of pyramidal tract activity to force exerted during voluntary movement. J. Neurophysiol. 31, 14–27.

Fadiga, L., Fogassi, L., Pavesi, G., Rizzolatti, G., 1995. Motor facilitation during action observation: a magnetic stimulation study. J. Neurophysiol. 73, 2608–2611.

Fagg, A.H., Arbib, M.A., 1998. Modeling parietal-premotor interactions in primate control of grasping. Neural Network 11 (7–8), 1277–1303. https://doi.org/10.1016/s0893-6080(98)00047-1.

Fattori, P., Breveglieri, R., Bosco, A., Gamberini, M., Galletti, C., 2017. Vision for prehension in the medial parietal cortex. Cerebr. Cortex 27, 1149–1163.

Fecteau, S., Carmant, L., Tremblay, C., Robert, M., Bouthillier, A., Théoret, H., 2004. A motor resonance mechanism in children? Evidence from subdural electrodes in a 36-month-old child. Neuroreport 15, 2625–2627.

Ferrari, P.F., Gallese, V., Rizzolatti, G., Fogassi, L., 2003. Mirror neurons responding to the observation of ingestive and communicative mouth actions in the monkey ventral premotor cortex. Eur. J. Neurosci. 17, 1703–1714.

Ferrier, D., 1873. Experimental researches in cerebral physiology and pathology. West Rid. Lunatic Asylum Med. Rep. 30–96.

Filimon, F., Nelson, J.D., Hagler, D.J., Sereno, M.I., 2007. Human cortical representations for reaching: mirror neurons for execution, observation, and imagery. Neuroimage 37, 1315–1328.

Fluet, M.-C., Baumann, M. a, Scherberger, H., 2010. Context-specific grasp movement representation in macaque ventral premotor cortex. J. Neurosci. 30, 15175–15184.

Fogassi, L., Ferrari, P.F., Gesierich, B., Rozzi, S., Chersi, F., Rizzolatti, G., 2005. Parietal lobe: from action organization to intention understanding. Science 308, 662–667.

Fogassi, L., Gallese, V., Buccino, G., Craighero, L., Fadiga, L., Rizzolatti, G., 2001. Cortical mechanism for the visual guidance of hand grasping movements in the monkey: a reversible inactivation study. Brain 124, 571–586.

Fogassi, L., Gallese, V., Fadiga, L., Luppino, G., Matelli, M., Rizzolatti, G., 1996. Coding of peripersonal space in inferior premotor cortex (area F4). J. Neurophysiol. 76, 141–157.

Fogassi, L., Luppino, G., 2005. Motor functions of the parietal lobe. Curr. Opin. Neurobiol. 15, 626–631.

Fogassi, L., Raos, V., Franchi, G., Gallese, V., Luppino, G., Matelli, M., 1999. Visual responses in the dorsal premotor area F2 of the macaque monkey. Exp. Brain Res. 128, 194–199.

Frey, S.H., Gerry, V.E., 2006. Modulation of neural activity during observational learning of actions and their sequential orders. J. Neurosci. 26, 13194–13201.

Fritsch, G., Hitzig, E., 1870. Ueber die elektrishe Erregarkeit des Grosshirns. In: Nowinski, W.W. (Ed.), The Cerebral Cortex. Springfield, IL, pp. 73–96.

Fulton, J., 1935. A note on the definition of the "motor" and "premotor" areas. Brain 58, 311–316.

Gallese, V., Fadiga, L., Fogassi, L., Rizzolatti, G., 1996. Action recognition in the premotor cortex. Brain 119, 593–609.

Gallese, V., Murata, A., Kaseda, M., Niki, N., Sakata, H., 1994. Deficit of hand preshaping after muscimol injection in monkey parietal cortex. Neuroreport 5, 1525–1529.

Galletti, C., Fattori, P., Kutz, D.F., Gamberini, M., 1999. Brain location and visual topography of cortical area V6A in the macaque monkey. Eur. J. Neurosci. 11, 575–582.

Galletti, C., Gamberini, M., Kutz, D.F., Fattori, P., Luppino, G., Matelli, M., 2001. The cortical connections of area V6: an occipito-parietal network processing visual information. Eur. J. Neurosci. 13, 1572–1588.

Gamberini, M., Passarelli, L., Fattori, P., Zucchelli, M., Bakola, S., Luppino, G., Galletti, C., 2009. Cortical connections of the visuomotor parietooccipital area V6Ad of the macaque monkey. J. Comp. Neurol. 513, 622–642.

Gazzola, V., Rizzolatti, G., Wicker, B., Keysers, C., 2007a. The anthropomorphic brain: the mirror neuron system responds to human and robotic actions. Neuroimage 35, 1674–1684.

Gazzola, V., van der Worp, H., Mulder, T., Wicker, B., Rizzolatti, G., Keysers, C., 2007b. Aplasics born without hands mirror the goal of hand actions with their feet. Curr. Biol. 17, 1235–1240.

Gentilucci, M., Fogassi, L., Luppino, G., Matelli, M., Camarda, R., Rizzolatti, G., 1988. Functional organization of inferior area 6 in the macaque monkey. I. Somatotopy and the control of proximal movements. Exp. Brain Res. 71, 475–490.

Gentilucci, M., Scandolara, C., Pigarev, I.N., Rizzolatti, G., 1983. Visual responses in the postarcuate cortex (area 6) of the monkey that are independent of eye position. Exp. Brain Res. 50, 464–468.

Georgopoulos, A.P., Ashe, J., Smyrnis, N., Taira, M., 1992. The motor cortex and the coding of force. Science 256, 1692–1695.

Gerbella, M., Belmalih, A., Borra, E., Rozzi, S., Luppino, G., 2010. Cortical connections of the macaque caudal ventrolateral prefrontal areas 45A and 45B. Cerebr. Cortex 20, 141–168.

Gerbella, M., Belmalih, A., Borra, E., Rozzi, S., Luppino, G., 2011. Cortical connections of the anterior (F5a) subdivision of the macaque ventral premotor area F5. Brain Struct. Funct. 216, 43–65.

Gerbella, M., Borra, E., Mangiaracina, C., Rozzi, S., Luppino, G., 2016. Corticostriate projections from areas of the "lateral grasping network": evidence for multiple hand-related input channels. Cerebr. Cortex 26, 3096–3115.

Gerbella, M., Borra, E., Tonelli, S., Rozzi, S., Luppino, G., 2013. Connectional heterogeneity of the ventral part of the macaque area 46. Cerebr. Cortex 23, 967–987.

Geyer, S., Ledberg, A., Schleicher, A., Kinomura, S., Thorsten, S., Brugel, U., Kilnengberg, T., Larsson, J., Zilles, K., Roland, P.E., 1996. Two different areas within the primary motor cortex of man. Nature 382, 805–807.

Geyer, S., Matelli, M., Luppino, G., Schleicher, A., Jansen, Y., Palomero-Gallagher, N., Zilles, K., 1998. Receptor autoradiographic mapping of the mesial motor and premotor cortex of the macaque monkey. J. Comp. Neurol. 397, 231–250.

Geyer, S., Matelli, M., Luppino, G., Zilles, K., 2000. Functional neuroanatomy of the primate isocortical motor system. Anat. Embryol. 202, 443–474.

Gibson, J.J., 1979. The Ecological Approach to Visual Perception. Houghton Mifflin, Boston.

Godschalk, M., Lemon, R.N., Nijs, H.G., Kuypers, H.G., 1981. Behaviour of neurons in monkey peri-arcuate and precentral cortex before and during visually guided arm and hand movements. Exp. Brain Res. 44, 113–116.

Goodale, M.A., Milner, a D., 1992. Separate visual pathways for perception and action. Trend. Neurosci. 15, 20–25.

Gould, H.J., Cusick, C.G., Pons, T.P., Kaas, J.H., 1986. The relationship of corpus callosum connections to electrical stimulation maps of motor, supplementary motor, and the frontal eye fields in owl monkeys. J. Comp. Neurol. 247, 297–325.

Grafton, S.T., Arbib, M.A., Fadiga, L., Rizzolatti, G., 1996. Localization of grasp representations in humans by positron emission tomography. 2. Observation compared with imagination. Exp. Brain Res. 112, 103–111.

Graziano, M.S.A., Cooke, D.F., 2006. Parieto-frontal interactions, personal space, and defensive behavior. Neuropsychologia 44, 845–859.

Graziano, M.S.A., Hu, X.T., Gross, C.G., 1997. Visuospatial properties of ventral premotor cortex. J. Neurophysiol. 77, 2268–2292.

Graziano, M.S.A., Taylor, C.S.R., Moore, T., 2002. Probing cortical function with electrical stimulation. Nat. Neurosci. 5, 921.

Graziano, M.S.A., Yap, G.S., Gross, C.G., 1994. Coding of visual space by premotor neurons. Science 266, 1054–1057.

Gregoriou, G.G., Borra, E., Matelli, M., Luppino, G., 2006. Architectonic organization of the inferior parietal convexity of the macaque monkey. J. Comp. Neurol. 496, 422–451.

Halligan, P.W., Marshall, J.C., 1991. Spatial compression in visual neglect: a case study. Cortex 27, 623–629.

Hashimoto, R., Tanaka, Y., 1998. Contribution of the supplementary motor area and anterior cingulate gyrus to pathological grasping phenomena. Eur. Neurol. 40, 151–158.

He, S.Q., Dum, R.P., Strick, P.L., 1993. Topographic organization of corticospinal projections from the frontal lobe: motor areas on the lateral surface of the hemisphere. J. Neurosci. 13, 952–980.

He, S.Q., Dum, R.P., Strick, P.L., 1995. Topographic organization of corticospinal projections from the frontal lobe: motor areas on the medial surface of the hemisphere. J. Neurosci. 15, 3284–3306.

Heilman, K.M., Valenstein, E., Watson, R.T., 2000. Neglect and related disorders. Semin. Neurol. 20(4), 463–470.

Hikosaka, O., Sakai, K., Lu, X., Nakahara, H., Rand, M.K., Nakamura, K., Miyachi, S., Doya, K., 1999. Parallel neural networks for learning sequential procedures. Trend. Neurosci. 22(12), 568–573. https://doi.org/10.1016/s0166-2236(99)01439-3.

Hoshi, E., Tanji, J., 2000. Integration of target and body-part information in the premotor cortex when planning action. Nature 408, 466–470.

Hoshi, E., Tanji, J., 2004. Differential roles of neuronal activity in the supplementary and presupplementary motor areas: from information retrieval to motor planning and execution. J. Neurophysiol. 92, 3482–3499.

Hyvärinen, J., 1982. Posterior parietal lobe of the primate brain. Physiol. Rev. 62, 1060–1129.

Iaccarino, L., Chieffi, S., Iavarone, A., 2014. Utilization behavior: what is known and what has to be known? Behav. Neurol. 2014, 297128.

Iacoboni, M., 1999. Cortical mechanisms of human imitation. Science 286, 2526–2528.

Iriki, A., Tanaka, M., Iwamura, Y., 1996. Coding of modified body schema during tool use by macaque postcentral neurones. Neuroreport 7, 2325–2330.

Ishida, H., Nakajima, K., Inase, M., Murata, A., 2010. Shared mapping of own and others' bodies in visuotactile bimodal area of monkey parietal cortex. J. Cognit. Neurosci. 22, 83–96.

Ishihara, K., Nishino, H., Maki, T., Kawamura, M., Murayama, S., 2002. Utilization behavior as a white matter disconnection syndrome. Cortex 38, 379–387.

Iwamura, Y., Tanaka, M., 1996. Representation of reaching and grasping in the monkey postcentral gyrus. Neurosci. Lett. 214, 147–150.

Jeannerod, M., 2001. Neural simulation of action: a unifying mechanism for motor cognition. In: NeuroImage. Academic Press, pp. S103–S109.

Johansen-Berg, H., Behrens, T.E.J., Robson, M.D., Drobnjak, I., Rushworth, M.F.S., Brady, J.M., Smith, S.M., Higham, D.J., Matthews, P.M., 2004. Changes in connectivity profiles define functionally distinct regions in human medial frontal cortex. Proc. Natl. Acad. Sci. U. S. A. 101, 13335–13340.

Kalaska, J.F., Cohen, D.A.D., Prud'homme, M., Hyde, M.L., 1990. Parietal area 5 neuronal activity encodes movement kinematics, not movement dynamics. Exp. Brain Res. 80, 351–364.

Karnath, H.O., Rorden, C., 2012. The anatomy of spatial neglect. Neuropsychologia 50, 1010–1017.

Keizer, K., Kuypers, H.G., 1989. Distribution of corticospinal neurons with collaterals to the lower brain stem reticular formation in monkey (*Macaca fascicularis*). Exp. Brain Res. 74, 311–318.

Kinsbourne, M., 2013. Orientational bias model of unilateral neglect: evidence from attentional gradients within hemispace. In: Unilateral Neglect: Clinical and Experimental Studies. Psychology Press, pp. 63–86.

Kohler, E., Keysers, C., Umiltà, M.A., Fogassi, L., Gallese, V., Rizzolatti, G., 2002. Hearing sounds, understanding actions: action representation in mirror neurons. Science 297, 846–848.

Koski, L., Iacoboni, M., Dubeau, M.-C.C., Woods, R.P., Mazziotta, J.C., Koski, L., Iacoboni, M., Iacoboni, M., Dubeau, M.-C.C., Dubeau, M.-C.C., Woods, R.P., Woods, R.P., Mazziotta, J.C., Mazziotta, J.C., 2003. Modulation of cortical activity during different imitative behaviors. J. Neurophysiol. 89, 460–471.

Kurata, K., 1989. Distribution of neurons with set- and movement-related activity before hand and foot movements in the premotor cortex of rhesus monkeys. Exp. Brain Res. 77, 245–256.

Làdavas, E., Di Pellegrino, G., Farnè, A., Zeloni, G., 1998. Neuropsychological evidence of an integrated visuotactile representation of peripersonal space in humans. J. Cognit. Neurosci. 10, 581–589.

Lanzilotto, M., Ferroni, C.G., Livi, A., Gerbella, M., Maranesi, M., Borra, E., Passarelli, L., Gamberini, M., Fogassi, L., Bonini, L., Orban, G.A., 2019. Anterior intraparietal area: a hub in the observed manipulative action network. Cerebr. Cortex 29, 1816–1833.

Lanzilotto, M., Livi, A., Maranesi, M., Gerbella, M., Barz, F., Ruther, P., Fogassi, L., Rizzolatti, G., Bonini, L., 2016. Extending the cortical grasping network: pre-supplementary motor neuron activity during vision and grasping of objects. Cerebr. Cortex 26, 4435–4449.

Laplane, D., Talairach, J., Meininger, V., Bancaud, J., Orgogozo, J.M., 1977. Clinical consequences of corticectomies involving the supplementary motor area in man. J. Neurol. Sci. 34, 301–314.

Lawrence, D.G., Kuypers, H.G., 1968. The functional organization of the motor system in the monkey. I. The effects of bilateral pyramidal lesions. Brain 91, 1–14.

Lehéricy, S., Ducros, M., Krainik, A., Francois, C., Van de Moortele, P.-F., Ugurbil, K., Kim, D.-S., 2004. 3-D diffusion tensor axonal tracking shows distinct SMA and pre-SMA projections to the human striatum. Cereb. Cortex 14, 1302–1309.
Lepage, J.F., Théoret, H., 2006. EEG evidence for the presence of an action observation-execution matching system in children. Eur. J. Neurosci. 23, 2505–2510.
Lewis, J.W., Van Essen, D.C., 2000. Corticocortical connections of visual, sensorimotor, and multimodal processing areas in the parietal lobe of the macaque monkey. J. Comp. Neurol. 428, 112–137.
Lhermitte, F., Pillon, B., Serdaru, M., 1986. Human autonomy and the frontal lobes. Part I: imitation and utilization behavior: a neuropsychological study of 75 patients. Ann. Neurol. 19, 326–334.
Libet, B., 1985. Unconscious cerebral initiative and the role of conscious will in voluntary action. Behav. Brain Sci. 8, 529–539.
Lu, M.T., Preston, J.B., Strick, P.L., 1994. Interconnections between the prefrontal cortex and the premotor areas in the frontal lobe. J. Comp. Neurol. 341, 375–392.
Lunven, M., Bartolomeo, P., 2017. Attention and spatial cognition: neural and anatomical substrates of visual neglect. Ann. Phys. Rehabil. Med. 60(3), 124–129. https://doi.org/10.1016/j.rehab.2016.01.004.
Luppino, G., Matelli, M., Camarda, R., Rizzolatti, G., 1993. Corticocortical connections of area F3 (SMA-proper) and area F6 (pre-SMA) in the macaque monkey. J. Comp. Neurol. 338, 114–140.
Luppino, G., Matelli, M., Camarda, R.M., Gallese, V., Rizzolatti, G., 1991. Multiple representations of body movements in mesial area 6 and the adjacent cingulate cortex: an intracortical microstimulation study in the macaque monkey. J. Comp. Neurol. 311, 463–482.
Luppino, G., Murata, A., Govoni, P., Matelli, M., 1999. Largely segregated parietofrontal connections linking rostral intraparietal cortex (areas AIP and VIP) and the ventral premotor cortex (areas F5 and F4). Exp. Brain Res. 128, 181–187.
Luppino, G., Rozzi, S., Calzavara, R., Matelli, M., 2003. Prefrontal and agranular cingulate projections to the dorsal premotor areas F2 and F7 in the macaque monkey. Eur. J. Neurosci. 17, 559–578.
M, J., Ma, A., H, G.R.S., Jeannerod, M., Arbib, M.A., Rizzolatti, G., Sakata, H., 1995. Grasping objects: the cortical mechanisms of visuomotor transformation. Trends Neurosci. 18, 314–320.
Maeda, K., Ishida, H., Nakajima, K., Inase, M., Murata, A., 2015. Functional properties of parietal hand manipulation-related neurons and mirror neurons responding to vision of own hand action. J. Cognit. Neurosci. 27, 560–572.
Maranesi, M., Rodà, F., Bonini, L., Rozzi, S., Ferrari, P.F., Fogassi, L., Coudé, G., 2012. Anatomo-functional organization of the ventral primary motor and premotor cortex in the macaque monkey. Eur. J. Neurosci. 36, 3376–3387.
Marchetti, C., Della Sala, S., 1998. Disentangling the alien and anarchic hand. Cognit. Neuropsychiatr. 3, 191–207.
Marconi, B., Genovesio, A., Battaglia-Mayer, A., Ferraina, S., Squatrito, S., Molinari, M., Lacquaniti, F., Caminiti, R., 2001. Eye-hand coordination during reaching. I. Anatomical relationships between parietal and frontal cortex. Cereb. Cortex 11, 513–527.
Marshall, J.C., Halligan, P.W., 1993. Neuropsychology. Imagine only the half of it. Nature 364, 193–194.
Marshall, P.J., Young, T., Meltzoff, A.N., 2011. Neural correlates of action observation and execution in 14-month-old infants: an event-related EEG desynchronization study. Dev. Sci. 14, 474–480.
Massion, J., 1992. Movement, posture and equilibrium: interaction and coordination. Prog. Neurobiol. 38, 35–56.
Matelli, M., Govoni, P., Galletti, C., Kutz, D.F., Luppino, G., 1998. Superior area 6 afferents from the superior parietal lobule in the macaque monkey. J. Comp. Neurol. 402, 327–352.
Matelli, M., Luppino, G., Rizzolatti, G., 1985. Patterns of cytochrome oxidase activity in the frontal agranular cortex of the macaque monkey. Behav. Brain Res. 18, 125–136.
Matelli, M., Luppino, G., Rizzolatti, G., 1991. Architecture of superior and mesial area 6 and the adjacent cingulate cortex in the macaque monkey. J. Comp. Neurol. 311, 445–462.
Matsuzaka, Y., Aizawa, H., Tanji, J., 1992. A motor area rostral to the supplementary motor area (presupplementary motor area) in the monkey: neuronal activity during a learned motor task. J. Neurophysiol. 68, 653–662.
Mitz, A.R., Wise, S.P., 1987. The somatotopic organization of the supplementary motor area: intracortical microstimulation mapping. J. Neurosci. 7, 1010–1021.
Murata, A., Fadiga, L., Fogassi, L., Gallese, V., Raos, V., Rizzolatti, G., Object, G.R., 1997. Object representation in the ventral premotor cortex (area F5) of the monkey. J. Neurophysiol. 78, 2226–2230.
Murata, A., Gallese, V., Luppino, G., Kaseda, M., Sakata, H., 2000. Selectivity for the shape , size , and orientation of objects for grasping in neurons of monkey parietal area AIP. J. Neurophysiol. 83, 2580–2601.
Murphy, J.T., Kwan, H.C., MacKay, W.A., Wong, Y.C., 1978. Spatial organization of precentral cortex in awake primates. III. Input-output coupling. J. Neurophysiol. 41, 1132–1139.
Murray, E. a, Coulter, J., 1981. Supplementary sensory area. In: Woolsey, C. (Ed.), Cortical Sensory Organization I. Multiple Somatic Areas. Humana Press, Clifton, NJ, pp. 167–195.
Nachev, P., Kennard, C., Husain, M., 2008. Functional role of the supplementary and pre-supplementary motor areas. Nat. Rev. Neurosci. 9, 856–869.
Nakamura, K., Sakai, K., Hikosaka, O., 1998. Neuronal activity in medial frontal cortex during learning of sequential procedures. J. Neurophysiol. 80, 2671–2687.
Nishitani, N., Hari, R., 2000. Temporal dynamics of cortical representation for action. Proc. Natl. Acad. Sci. U. S. A. 97, 913–918.
Pacherie, E., 2008. The phenomenology of action: a conceptual framework. Cognition 107, 179–217.
Pandya, D.N., Seltzer, B., 1982. Intrinsic connections and architectonics of posterior parietal cortex in the rhesus monkey. J. Comp. Neurol. 204, 196–210.
Penfield, W., Welch, K., 1951. The supplementary motor area of the cerebral cortex: a clinical and experimental study. Arch. Neurol. Psychiatr. 66, 289–317.
Perenin, M., Vighetto, A., 1988. Optic ataxia: a specific disruption in visuomotor mechanisms. I. Different aspects of the deficit in reaching for objects. Brain 111, 643–674.
Perrett, D.I., Harries, M.H., Bevan, R., Thomas, S., Benson, P.J., Mistlin, A.J., Chitty, A.J., Hietanen, J.K., Ortega, J.E., 1989. Frameworks of analysis for the neural representation of animate objects and actions. J. Exp. Biol. 146, 87–113.
Petrides, M., Pandya, D.N., 1994. Comparative architectonic analysis of the human and the macaque frontal cortex. In: Boller, F., Grafman, J. (Eds.), Handbook of Neuropsychology. Elsevier, Amsterdam, pp. 17–58.
Pisella, L., Gréa, H., Tilikete, C., Vighetto, A., Desmurget, M., Rode, G., Boisson, D., Rossetti, Y., 2000. An "automatic pilot" for the hand in human posterior parietal cortex: toward reinterpreting optic ataxia. Nat. Neurosci. 3, 729–736.
Porter, R., Lemon, R.N., 1993. Corticospinal Function and Voluntary Movement. Oxford University Press, Oxford.
Posner, M.I., Walker, J.A., Friedrich, F.J., Rafal, R.D., 1984. Effects of parietal injury on covert orienting of attention. J. Neurosci. 4, 1863–1874.
Preuss, T.M., Goldman-Rakic, P.S., 1989. Connections of the ventral granular frontal cortex of macaques with perisylvian premotor and somatosensory areas: anatomical evidence for somatic representation in primate frontal association cortex. J. Comp. Neurol. 282, 293–316.
Raos, V., Franchi, G., Gallese, V., Fogassi, L., 2003. Somatotopic organization of the lateral part of area F2 (dorsal premotor cortex) of the macaque monkey. J. Neurophysiol. 89, 1503–1518.

Raos, V., Umiltà, M.A., Gallese, V., Fogassi, L., Umiltá, M.-A., Gallese, V., Fogassi, L., 2004. Functional properties of grasping-related neurons in the dorsal premotor area F2 of the macaque monkey. J. Neurophysiol. 92, 1990–2002.

Raos, V., Umiltà, M.A., Murata, A., Fogassi, L., Gallese, V., 2006. Functional properties of grasping-related neurons in the ventral premotor area F5 of the macaque monkey. J. Neurophysiol. 95, 709–729.

Rizzolatti, G., Arbib, M.A., 1998. Language within our grasp. Trend. Neurosci. 2236, 1667–1669.

Rizzolatti, G., Camarda, R., 1987. Neural circuits for spatial attention and unilateral neglect. Adv. Psychol. 45, 289–313.

Rizzolatti, G., Camarda, R., Fogassi, L., Gentilucci, M., Luppino, G., Matelli, M., 1988. Functional organization of inferior area 6 in the macaque monkey. II. Area F5 and the control of distal movements. Exp. Brain Res. 71, 491–507.

Rizzolatti, G., Cattaneo, L., Fabbri-Destro, M., Rozzi, S., 2014. Cortical mechanisms underlying the organization of goal-directed actions and mirror neuron-based action understanding. Physiol. Rev. 94, 655–706.

Rizzolatti, G., Fadiga, L., Gallese, V., Fogassi, L., 1996a. Premotor cortex and the recognition of motor actions. Brain Res. Cogn. Brain Res. 3, 131–141.

Rizzolatti, G., Fogassi, L., Gallese, V., 1997. Parietal cortex: from sight to action. Curr. Opin. Neurobiol. 7, 562–567.

Rizzolatti, G., Gentilucci, M., Camarda, R.M., Gallese, V., Luppino, G., Matelli, M., Fogassi, L., 1990. Neurons related to reaching-grasping arm movements in the rostral part of area 6 (area 6a beta). Exp. Brain Res. 82, 337–350.

Rizzolatti, G., Luppino, G., 2001. The cortical motor system. Neuron 31, 889–901.

Rizzolatti, G., Luppino, G., Matelli, M., 1996b. The classic supplementary motor area is formed by two independent areas. Adv. Neurol. 70, 45–56.

Rizzolatti, G., Luppino, G., Matelli, M., 1998. The organization of the cortical motor system: new concepts. Electroencephalogr. Clin. Neurophysiol. 106, 283–296.

Rizzolatti, G., Matelli, M., 2003. Two different streams form the dorsal visual system: anatomy and functions. Exp. Brain Res. 153, 146–157.

Robinson, D.L., Goldberg, M.E., Stanton, G.B., 1978. Parietal association cortex in the primate: sensory mechanisms and behavioral modulations. J. Neurophysiol. 41, 910–932.

Rochat, M., Caruana, F., Jezzini, A., Escola, L., Intskirveli, I., Grammont, F., Gallese, V., Rizzolatti, G., Umiltà, M.A., 2010. Responses of mirror neurons in area F5 to hand and tool grasping observation. Exp. Brain Res. 204, 605–616.

Rozzi, S., Calzavara, R., Belmalih, A., Borra, E., Gregoriou, G.G., Matelli, M., Luppino, G., 2006. Cortical connections of the inferior parietal cortical convexity of the macaque monkey. Cerebr. Cortex 16, 1389–1417.

Rozzi, S., Ferrari, P.F.P.F., Bonini, L., Rizzolatti, G., Fogassi, L., 2008. Functional organization of inferior parietal lobule convexity in the macaque monkey: electrophysiological characterization of motor, sensory and mirror responses and their correlation with cytoarchitectonic areas. Eur. J. Neurosci. 28, 1569–1588.

Rozzi, S., Fogassi, L., 2017. Neural coding for action execution and action observation in the prefrontal cortex and its role in the organization of socially driven behavior. Front. Neurosci. 11, 492.

Sabatini, U., Boulanouar, K., Fabre, N., Martin, F., Carel, C., Colonnese, C., Bozzao, L., Berry, I., Montastruc, J.L., Chollet, F., Rascol, O., 2000. Cortical motor reorganization in akinetic patients with Parkinson's disease: a functional MRI study. Brain 123 (Pt 2), 394–403.

Sakai, K., Hikosaka, O., Miyauchi, S., Sasaki, Y., Fujimaki, N., Pütz, B., 1999. Presupplementary motor area activation during sequence learning reflects visuo-motor association. J. Neurosci. 19, RC1.

Sakata, H., Taira, M., Murata, A., Mine, S., 1995. Neural mechanisms of visual guidance of hand action in the parietal cortex of the monkey. Cerebr. Cortex 5, 429–438.

Saleem, K.S., Kondo, H., 2008. Complementary circuits connecting the orbital and medial prefrontal networks with the temporal, insular, and opercular cortex in the macaque monkey. J. Comp. Neurol. 693, 659–693.

Schall, J.D., Morel, A., Kaas, J.H., 1993. Topography of supplementary eye field afferents to frontal eye field in macaque: implications for mapping between saccade coordinate systems. Vis. Neurosci. 10, 385–393.

Searle, J., Willis, S., 1983. Intentionality: An Essay in the Philosophy of Mind.

Sergio, L.E., Kalaska, J.F., 2003. Systematic changes in motor cortex cell activity with arm posture during directional isometric force generation. J. Neurophysiol. 89, 212–228.

Shima, K., Tanji, J., 2000. Neuronal activity in the supplementary and presupplementary motor areas for temporal organization of multiple movements. J. Neurophysiol. 84, 2148–2160.

Shimada, S., Hiraki, K., 2006. Infant's brain responses to live and televised action. Neuroimage 32, 930–939.

Southgate, V., Johnson, M.H., Karoui, I. El, Csibra, G., 2010. Motor system activation reveals infants' on-line prediction of others' goals. Psychol. Sci. 21, 355–359.

Sumner, P., Nachev, P., Morris, P., Peters, A.M., Jackson, S.R., Kennard, C., Husain, M., 2007. Human medial frontal cortex mediates unconscious inhibition of voluntary action. Neuron 54, 697–711.

Taira, M., Mine, S., Georgopoulos, A.P., Murata, A., Sakata, H., 1990. Parietal cortex neurons of the monkey related to the visual guidance of hand movement. Exp. Brain Res. 83, 29–36.

Tanji, J., 2001. Sequential organization of multiple movements: involvement of cortical motor areas. Annu. Rev. Neurosci. 24, 631–651.

Tanji, J., Shima, K., 1994. Role for supplementary motor area cells in planning several movements ahead. Nature 371, 413–416.

Tanji, J., Shima, K., 1996. Supplementary motor cortex in organization of movement. Eur. Neurol. 36 (Suppl. 1), 13–19.

Tanji, J., Shima, K., Mushiake, H., 1996. Multiple cortical motor areas and temporal sequencing of movements. Cognit. Brain Res. 5, 117–122.

Thiebaut de Schotten, M., Dell'Acqua, F., Valabregue, R., Catani, M., Dell'Acqua, F., Valabregue, R., Catani, M., 2012. Monkey to human comparative anatomy of the frontal lobe association tracts. Cortex 48, 82–96.

Thier, P., Andersen, R.A., 1998. Electrical microstimulation distinguishes distinct saccade-related areas in the posterior parietal cortex. J. Neurophysiol. 80, 1713–1735.

Umiltà, M.A., Escola, L., Intskirveli, I., Grammont, F., Rochat, M., Caruana, F., Jezzini, A., Gallese, V., Rizzolatti, G., 2008. When pliers become fingers in the monkey motor system. Proc. Natl. Acad. Sci. U. S. A. 105, 2209–2213.

Umiltà, M.A., Kohler, E., Gallese, V., Fogassi, L., Fadiga, L., Keysers, C., Rizzolatti, G., 2001. I know what you are doing. a neurophysiological study. Neuron 31, 155–165.

Van Essen, D., 1985. Functional organization of primate visual cortex. Cereb. Cortex 259–329.

Vogt, C., Vogt, O., 1919. Allgemeinere ergebnisse unserer hirnforschung. J. Psychol. Neurol. 25, 279–461.

von Bonin, G., Bailey, P., 1947. The Neocortex of Macaca Mulatta. University of Illinois Press, Urbana, Illinois.

von Economo, K., Koskinas, G., 1925. Die cytoarchitektonik der hirnrinde des erwachsenen menschen.

Vuilleumier, P., Valenza, N., Mayer, E., Reverdin, A., Landis, T., 1998. Near and far visual space in unilateral neglect. Ann. Neurol. 43, 406–410.

Walker, A., 1940. A cytoarchitectural study of the prefrontal areas of the macaque monkey. J. Comp. Neurol. 73, 59–86.

Weinrich, M., Wise, S.P., 1982. The premotor cortex of the monkey. J. Neurosci. 2, 1329–1345.

Wise, S.P., Boussaoud, D., Johnson, P.B., Caminiti, R., 1997. Premotor and parietal cortex: corticocortical connectivity and combinatorial computations. Annu. Rev. Neurosci. 20, 25–42.

Woolsey, C., Settlage, P., 1952. Patterns of localization in precentral and" supplementary" motor areas and their relation to the concept of a premotor area. Res. Publ. Assoc. Res. Nerv. Ment. Dis. 30, 238–264.

Yoshida, K., Saito, N., Iriki, A., Isoda, M., 2011. Representation of others' action by neurons in monkey medial frontal cortex. Curr. Biol. 21, 249–253.

Yoshida, K., Saito, N., Iriki, A., Isoda, M., 2012. Social error monitoring in macaque frontal cortex. Nat. Neurosci. 15, 1307–1312.

Zilles, K., Schlaug, G., Geyer, S., Luppino, G., Matelli, M., Qü, M., Schleicher, A., Schormann, T., 1996. Anatomy and transmitter receptors of the supplementary motor areas in the human and nonhuman primate brain. Adv. Neurol. 70, 29–43.

Zilles, K., Schlaug, G., Matelli, M., Luppino, G., Schleicher, A., Qü, M., Dabringhaus, A., Seitz, R., Roland, P.E., 1995. Mapping of human and macaque sensorimotor areas by integrating architectonic, transmitter receptor, MRI and PET data. J. Anat. 187 (Pt 3).

22

The Limbic System

Barbara Buccilli[1,7], *Giovanni Paternò*[2], *Ana Paula Cortes-Contreras*[3], *Maria Fernanda Tejada Pineda*[4], *Alejandro Monroy-Sosa*[5] *and Brandon M. Ascenzi*[6]

[1]University of Rome "La Sapienza", Rome, Italy [2]Department of Neurosurgery, Istituto Clinico Città Studi, Milan, Italy [3]Faculty of Medicine, Universidad Nacional Autónoma de México (UNAM), Mexico City, Mexico [4]Instituto Tecnológico de Estudios Superiores de Monterrey, School of Medicine and Health Sciences, Mexico City, Mexico [5]Department of Neurosurgery, Institute for Social Security and Services for State Workers, (ISSSTE) Tlahuac, Center for Research and Innovation in Neuroscience, Instituto Politécnico Nacional, Superior School of Medicine, Mexico City, Mexico [6]Independent Neuroresearcher-Member of Marie Curie Alumni Association (MCAA) [7]Department of Neurosurgery, Cerebrovascular and Neurologic Critical Care, Icahn School of Medicine at Mount Sinai, New York, NY, United States

O U T L I N E

INTRODUCTION

James Papez in 1937 presented the limbic system as an integrated system of cortical and subcortical structures involved in linking actions and perceptions to emotions. Papez's circuit is believed to be involved in controlling both memory and emotion (Papez, 1937; Roxo et al., 2011). Papez first proposed the idea of a link between the hippocampus, hypothalamus (mammillary bodies), thalamus, cingulate gyrus, and parahippocampal gyrus (Papez, 1937). He further integrated the anatomical definition of limbic models into a functional model and, according to him, two circuits are involved in the emotional processing: one, the medial circuit, via the hippocampus and cingulate cortex, is directly involved with the hypothalamus; the other, via the lateral cortex, is involved in sensory activities mediated by the dorsal thalamus (Papez, 1937). Paul MacLean integrated the Papez and Yakovlev models (for details on the latter see Yakovlev, 1948) as part of the limbic system, and further modified and expanded the Papez circuit to include the prefrontal cortex, the septum, and the amygdalae (MacLean, 1949, 1952). MacLean's model further emphasized that the limbic cortex and the subcortical limbic structures were a functionally integrated system, interconnected by both short-and-long range fiber bundles (MacLean, 1949, 1952). Some of the known limbic system functions in the human brain include learning, memory, olfaction, adjusting visceral functions, emotional processing, attention, and cognition (Nolte, 1993).

Studies have demonstrated the direct involvement of the prefrontal cortex with the limbic system in primates, including humans (Robinson and Kolb, 1997; Whalen

From Anatomy to Function of the Central Nervous System
https://doi.org/10.1016/B978-0-12-822404-5.00019-X

et al., 1998; Barbas et al., 2003; Bryant et al., 2020; Rapan et al., 2021). The cerebral cortex as a whole is also indirectly connected to the anterior limbic nuclei, including the amygdalae, hypothalamic nuclei, and septal nuclei, through extensive cortico-thalamic connectivity mediated by the thalamus (Haines, 2007).

In this chapter, we will describe the anatomo-functional characteristics of the limbic system illustrating what has emerged from various studies in recent years. We will also analyze the relationship between psychiatric disorders and neuroanatomical functional correlates in the light of new discoveries in the field of reticulum (subcortical)-cortical connections.

THE PAPEZ CIRCUIT

The Papez circuit is composed of two parallel loops, generated mainly by the cingulum bundle (CB) and fornix (FX) as follows:

Prefrontal cingulate gyrus

⇐ (CB) ⇒ Parahippocampal gyrus (temporal lobe)
⇐ (FX) ⇒ Mammillary bodies
⇐ (MTT) ⇒ Ventral thalamus
⇐ (TC) ⇒ Prefrontal cingulate gyrus

The CB (Fig. 22.1A and B) is the longest C-shaped limbic pathway and serves as the core of the Papez circuit, connecting prefrontal and parietal projections of the cingulate gyrus to the hippocampus, in the parahippocampal gyrus of the temporal lobe (Figures *green fibers* Fig. 22.1A and B).

The cingulum bundle establishes connections between various regions of the brain. It links the anterior cingulate, dorsolateral and medial prefrontal, and orbitofrontal cortices to the posterior cingulate, parietal, insular, and parahippocampal cortices. Ventrally, it also connects to the amygdala (Crosby, 1963; Nieuwenhuys et al., 1988; Wakana et al., 2004; Schmahmann et al., 2007; Lawes et al., 2008). The CB is composed of a dorsal segment and a parahippocampal segment, which are contiguous. The dorsal segment, from the subgenual area, goes around the corpus callosum and then above the splenium of the corpus callosum, alongside the superior longitudinal fasciculus (Jang and Hong, 2012; Kamali et al., 2014; Wang et al., 2016). The parahippocampal segment, after a turn just posterior to the splenium of the corpus callosum, continues ventrally, goes toward the anterior temporal lobe, along the ventral aspect of the hippocampus, and terminates in the temporal lobe (Catani et al., 2013; Pascalau et al., 2018).

The hippocampus in turn connects to the mammillary bodies via the fornix (FX). The latter is a long C-shaped tract connecting the hippocampus and the amygdala, in the mesial temporal lobe, to the hypothalamic and septal nuclei, centrally at the base of the brain (Figs. 22.1A and B, 22.2A and B, and 22.3).

More in detail, after exiting the hippocampus, the forniceal crura loops anteriorly alongside the lateral ventricles and forms the forniceal body (Fig. 22.2A). The body of the fornix is situated on the midline, coursing anteriorly, and ultimately divides into the two fornix columns at its anterior end. The fornix columns course inferiorly and subsequently split into the precommissural and postcommissural fibers (Figs. 22.2A and 22.3). The precommissural fibers travel anteriorly to the anterior commissure, while the postcommissural fibers run behind the anterior commissure. The precommissural fibers descend anteriorly to the lamina terminalis and end in the anterior septal nuclei. The postcommissural fibers have a C-shape course traveling through the posterior septal nuclei, into the bed nucleus of the stria terminalis, alongside the stria terminalis, and continue inferiorly to reach the hypothalamic nuclei (Fig. 22.3) (Lemaire et al., 2013; Kamali et al., 2015). Two ventral forniceal arms emerge from the junction of the forniceal crura and fimbria, extending away from the hippocampus and coursing medially along the lateral margin of the cerebral peduncles toward the mammillary bodies and other hypothalamic nuclei (Fig. 22.2A), where they terminate (Lemaire et al., 2013; Kamali et al., 2015).

The fornix is the only limbic pathway which generates a complete loop on its own, a hypothalamo-septo-hippocampo-hypothalamic loop, via the forniceal columns, the body, the crura, and the ventral arms (Figs. 22.2A and 22.3).

The mammillary bodies (part of the hypothalamic nuclei), involved in many limbic circuitry, are directly or indirectly connected to the amygdala, the hippocampus, and the thalamus (Yamada et al., 1998; Kamali et al., 2018a,b). They project to the ventral thalamus via the mammillothalamic tract (MTT).

In the end, the ventral thalamus links back into the prefrontal cortex and the anterior cingulate gyrus via the thalamo-cingulate projections (TC) to complete the Papez circuit (Fig. 22.1A and B).

In recent years, diffusion-weighted imaging (DWI) techniques have unveiled additional limbic fiber connectivity. Dissection studies in primates have provided evidence of the equivalent fiber connectivity of these pathways. Some of these tracts include the *amygdalofugal tract* (AFT), the *amygdalothalamic tract* (ATT), the *dorsal thalamo-hypothalamic tract* (DTH), the *prefronto-caudo-thalamic tract* (PFCT), and the *parieto-occipito-hypothalamic tract* (POHT) (Kamali et al., 2015, 2016, 2018a,b, 2020a,b). These newly described tracts add multiple additional limbic circuits to the already-known complex limbic network. It is evident that, similar to other neuroanatomical circuits, the Papez circuit consists of

(A)

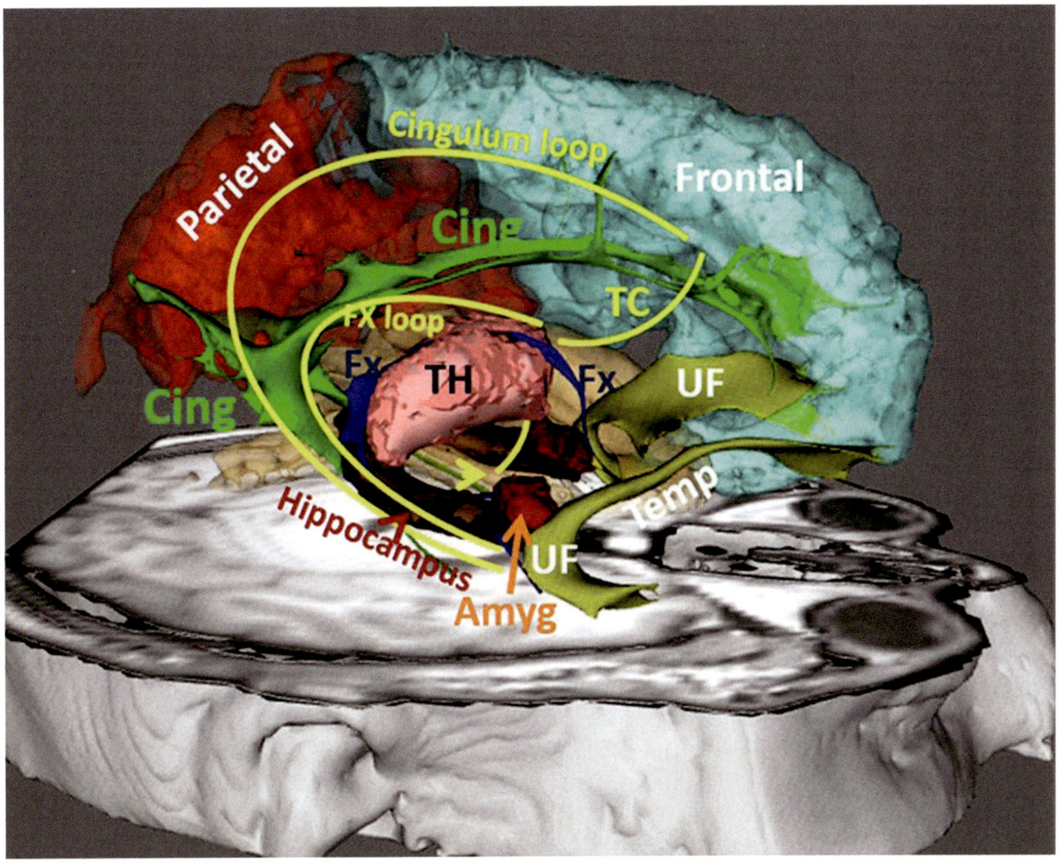

(B)

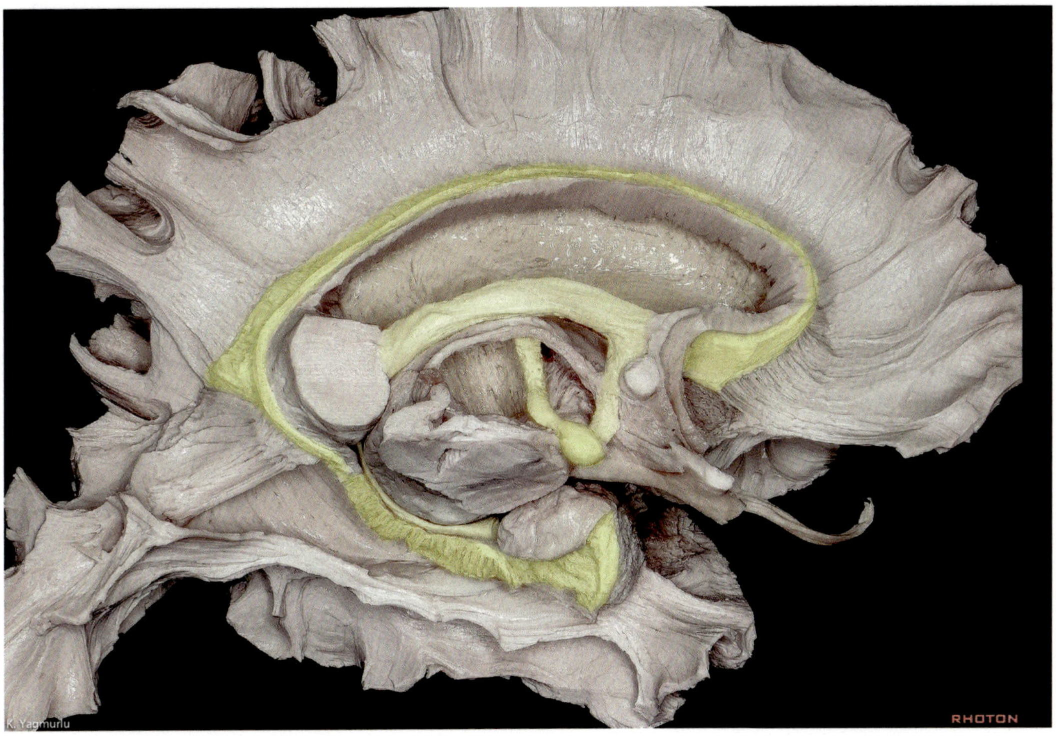

FIGURE 22.1 **(A) 3-D reconstruction of the major gray and white matter structures of the limbic system including the Papez circuit.** Frontal (cyan), parietal (red), and temporal (creamy color) cortices are shown in the background. Major gray matter nuclei of the limbic system are also shown, including the thalamus (TH in pink), hippocampus (brown), and amygdala (amyg in red). Please note that the hypothalamus including the mammillary bodies is not shown, however, the mammillo-hypothalamic tract is shown by an arrowhead connecting the hypothalamus to the

interconnected nuclei and fibers that form a series of reverberant loops. These loops represent the primary anatomical structure underlying emotional and behavioral functions.

In the following pages, we will provide a description of the new thalamic tracts, which can be considered as an anatomical update of Papez's circuit. Additionally, we will focus on the subcortico-reticulo ascending pathways that play a role in regulating the neural activity of the limbic structures.

THE DORSAL THALAMO-HYPOTHALAMIC TRACT

In addition to the MTT, which serves as the major ventral thalamo-hypothalamic pathway, the dorsal thalamohypothalamic tract (DTH) was recently described in the human brain as a dorsal thalamo-hypothalamic connection (Fig. 22.2D) (Kamali et al., 2020a). The DTH connects the dorsal thalamus to the hippocampus and to the anterior hypothalamic nuclei. The DTH is a much greater pathway in volume compared to the MTT. Unlike the MTT, which connects the dorsal hypothalamus (mammillary bodies) to the anterior thalamus, the DTH connects the dorsal thalamus to the anterior hypothalamus (Fig. 22.2D) (Aggleton et al., 2010; Peterson et al., 2023; Kamali et al., 2020a,b). Several animal studies have revealed an equivalent tract connecting the thalamus to the hippocampus and the hypothalamus (Conrad and Pfaff, 2004; Pasquier and Reinoso-Suarez, 1976; Swanson, 1976; Herkenham, 1978; Cornwall and Phillipson, 1988; Groenewegen, 1988; Chen and Su, 1990; Krout and Loewy, 2000).

Aggleton and Mishkin (1982) proposed a dual circuit for memory in monkeys. The first circuit involves the amygdala and the dorsomedial nucleus of the thalamus. The second circuit connects the hippocampus to the anterior limbic nuclei, including the amygdala and the hypothalamus. The DTH, along with the fornix/stria terminalis, may play a role in the second circuit and the thalamo-hippocampo-hypothalamic pathway proposed by Aggleton and Mishkin (1982). The DTH (Dorsal Thalamus-Hippocampus) pathway adds a parallel route to the thalamo-hippocampo-hypothalamic connectivity, running alongside the Papez circuit. It directly connects the thalamus to the hippocampus and the hypothalamic nuclei. Previously, this connectivity was known to be possible via the fornix (hippocampo-hypothalamic) and MTT (hypothalamo-thalamic) pathways in the Papez circuit. Therefore, the DTH may contribute to multiple limbic circuits alongside the Papez circuit. The DTH and the MTT together may generate a thalamo-hippocampo-hypothalamo-thalamic circuit (Fig. 22.8).

THE AMYGDALOTHALAMIC TRACT

The ATT is the largest direct amygdalo-diencephalic connection in primates, providing a direct connection between the amygdala and the dorsomedial nucleus of the thalamus (Figs. 22.2A and C and 22.8) (Kamali et al., 2018a,b). The ATT serves as the amygdalothalamic branch of the ventral amygdalofugal complex, which includes the amygdalo-hypothalamic, amygdaloseptal, amygdalo-mesencephalic, and amygdalo-thalamic tracts (Noback et al., 2005; Haines, 2007). The amygdalofugal complex establish a direct connection between the amygdala and the brainstem (Noback et al., 2005; Nolte, 1993). The amygdala projects directly to the orbitofrontal cortex and to the anterior temporal lobes via the uncinate fasciculus (UF) (Fig. 22.3) (Di et al., 2017); it also indirectly connects with the prefrontal cortex via the Anterior Thalamic Radiation (ATR) (Fig. 22.8). Some of the fibers of the ATT continue beyond the amygdala into the anterior temporal lobe (Klingler and Gloor, 1960; Kiernan, 2012; Kamali et al., 2018a,b). The ATT may participate in multiple circuits of the limbic system, some of which are listed below (Fig. 22.8).

THE PREFRONTO-CAUDO-THALAMIC TRACT (PFCT)

This pathway was first described in 2010 (Kamali et al., 2010). The PFCT is a large projection bundle connecting the thalamus to the basal ganglia, including the caudate nucleus and the lentiform nuclei, and

ventral thalamus. Major white matter pathways of the limbic system are shown, including the cingulum bundle (cing in green), fornix (FX in blue), and uncinate fasciculus (UF in yellow). The Papez circuit is shown in yellow curved lines including the two parallel loops of cingulum loop and forniceal loop (FX loop). The two loops communicate with one another mediated by the thalamus via the mammillothalamic tract (shown by yellow arrowhead) and the thalamo-cortical fibers (TC). The Papez circuit consists of the frontoparietal connectivity with the parahippocampal gyrus/hippocampal formations via the cingulum bundle, hippocampal formations to mammillary bodies via the fornix, mammillary bodies to the thalamus via the mammillothalamic tract, and the thalamus to the frontoparietal lobes and cingulum bundle via the thalamocortical fibers (TC). (B) Klingler's dissection of Papez's circuit (in yellow). For a detailed description of the structures see the caption of (A) and the text. *(A) Image courtesy of Kamali, A., Milosavljevic, S., Gandhi, A., et al., 2023. The cortico-limbo-thalamo-cortical circuits: an update to the original Papez circuit of the human limbic system. Brain Topogr. 36, 371–389. https://doi.org/10.1007/s10548-023-00955-y. (B) Image courtesy of The Rhoton Collection® http://rhoton.ineurodb.org/. Dissector: K. Yagmurlu. Drawings by Jeffrey Sorenson, Nickalus Khan, Kaan Yagmurlu and Alessandro Basso.*

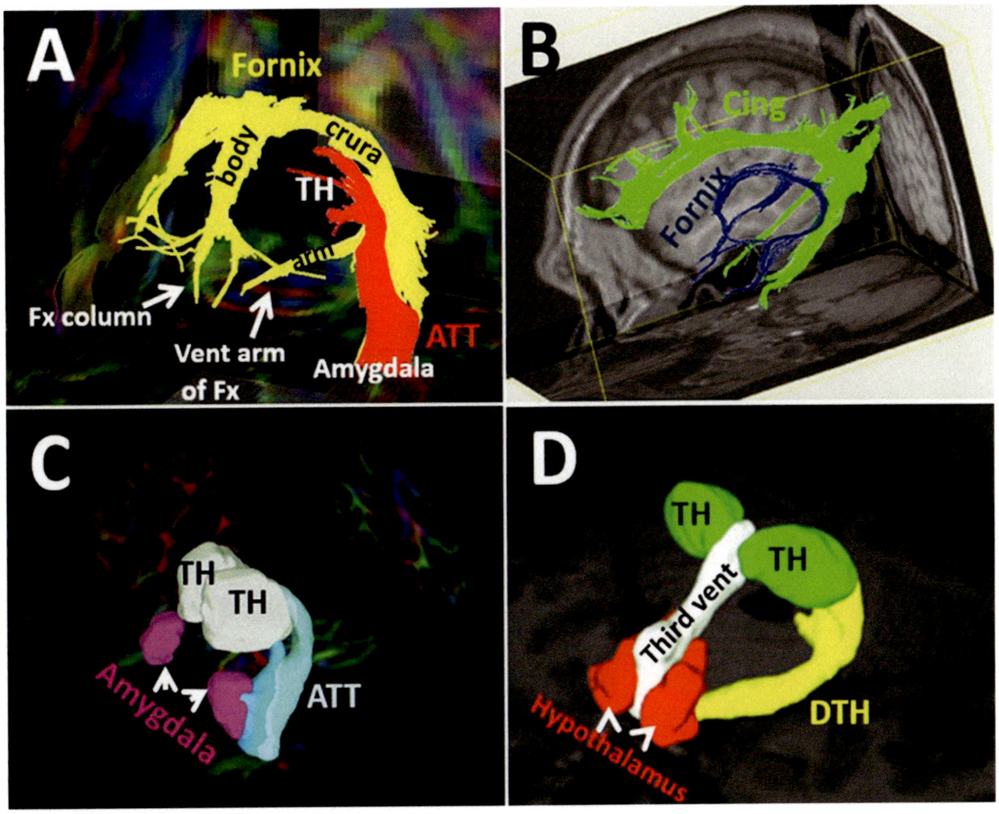

FIGURE 22.2 **(A–D) represents the 3-D fiber tract reconstructions of the fornix, cingulum bundle, amygdalothalamic, and dorsal thalamo-hypothalamic tract.** (A) shows the relationship between the fornix (yellow) and the amygdalothalamic tract (ATT in red). Forniceal arms, crura, body, and columns are shown. (B) shows the relationship of the fornix (blue) and cingulum bundle (green). (C) illustrates the ATT (in cyan), arising from the dorsomedial thalamus and inserting into the amygdala. (D) illustrates the dorsal thalamo-hypothalamic tract (DTH in yellow) arising from the dorsomedial thalamus and inserting into the anterior hypothalamic nuclei. ATT, amygdalothalamic tract; Cing, cingulum bundle; DTH, dorsal thalamo-hypothalamic tract; TH, thalamus; Third vent, third ventricle. *Image courtesy of Kamali, A., Milosavljevic, S., Gandhi, A., et al., 2023. The cortico-limbo-thalamo-cortical circuits: an update to the original Papez circuit of the human limbic system. Brain Topogr. 36, 371–389. https://doi.org/10.1007/s10548-023-00955-y*

eventually projecting into the prefrontal cortex (Figs. 22.4 and 22.6). The PFCT contains some of the cortico-striatal pathways between the putamina and caudate nuclei to the prefrontal cortex (Fig. 22.4C). The PFCT directly connects the thalamus with the Bed Nucleus of the Stria Terminalis (BNST) and the septal nuclei (Figs. 22.4C and 22.5B). The PFCT communicates with multiple limbic nuclei, including the hypothalamic nucleus, the septal nucleus, the BNST, and the nucleus accumbens. The ATR and the PFCT both project from the thalamus to the prefrontal cortex. The ATR travels through the anterior limb of the internal capsule, while the PFCT travels more medially, through the caudate nucleus, to reach the prefrontal cortex (Kamali et al., 2010). At its thalamic origin, the PFCT has a segment with convexity toward the midline and concavity toward the lateral thalamus, which arises from the dorsomedial (DM) nucleus of the thalamus. After arising from the DM nucleus of the thalamus, the PFCT projects to the BNST and receives projections from the septal nuclei and the lentiform nuclei (Figs. 22.4 and 22.5). Aside from the septal nuclei sending projections to the PFCT medially, there are other projections joining the PFCT along the way. These projections are joining the PFCT from the caudate body superiorly (Fig. 22.5B), from the lenticular nuclei (putamen and globus pallidus) laterally (Figs. 22.4C and 22.5B), and from the hypothalamic nuclei inferiorly (Fig. 22.5A).

The PFCT connects all these structures to the caudate head, the nucleus accumbens, and the prefrontal cortex. The PFCT then runs through the caudate head and the nucleus accumbens (Fig. 22.5C), and eventually projects to the medial prefrontal cortex (Haber et al., 1995; Kamali et al., 2010). The trajectories of the PFCT and medial forebrain bundle (MFB) are shown in Fig. 22.6 for a better understanding of their routes. The ATR, similarly to the MFB, travels through the anterior limb of the internal capsule (ALIC). The PFCT establishes connections between the thalamus, basal ganglia, and other gray matter nuclei, enabling communication between them and ultimately connecting them to the prefrontal cortex. Therefore, the PFCT may serve as a

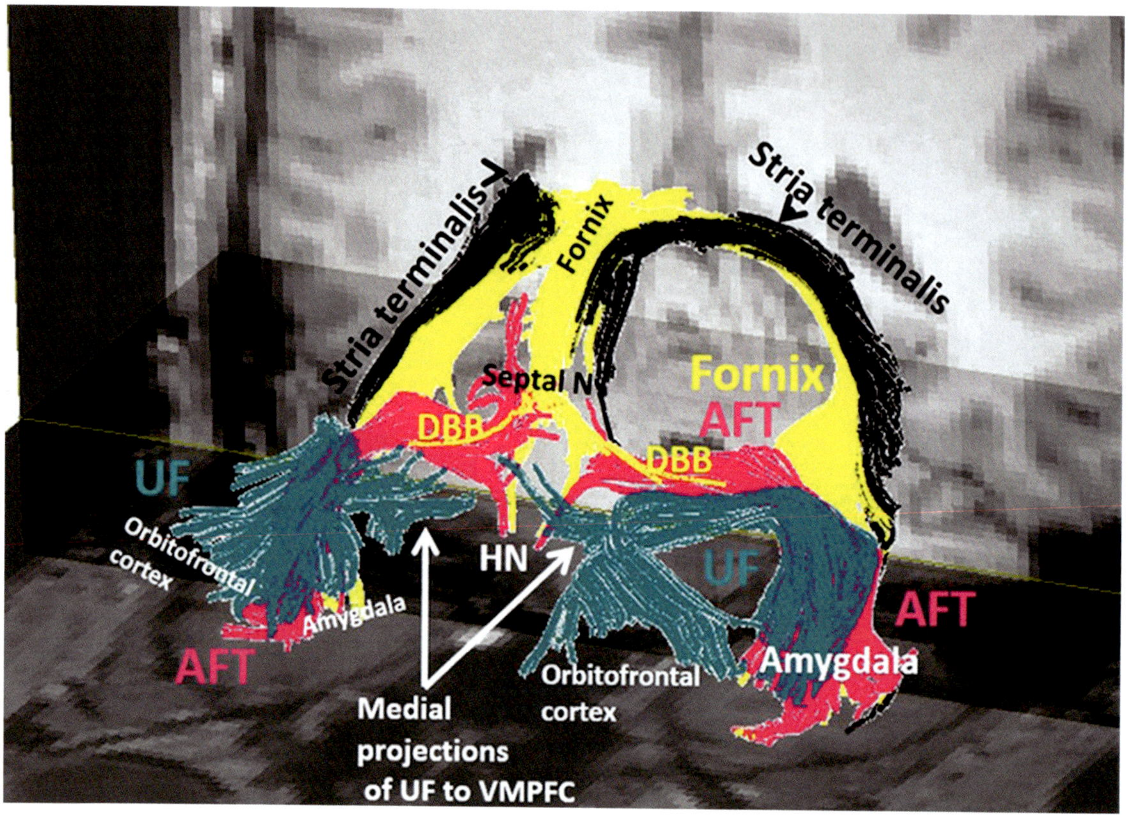

FIGURE 22.3 **3-D fiber reconstruction of the fornix (yellow), stria terminalis (black), amygdalofugal tract (AFT in pink), and uncinate fasciculus (UF in green).** The stria terminalis is shown connecting the amygdala to the region of septal nuclei (Septal N). The amygdalofugal tract is shown connecting the amygdala to the hypothalamic nuclei (HN) and septal nuclei (Septal N). The uncinate fasciculus is shown connecting the amygdala to the orbitofrontal cortex and ventromedial prefrontal cortex (vMPFC). The white arrows demonstrate the medial projections of the UF to the vMPFC. Diagonal band of broca (DBB) are shown as the mustache like yellow fibers arising from the septal nuclei and course toward the para olfactory regions bilaterally. *Image courtesy of Kamali, A., Milosavljevic, S., Gandhi, A., et al., 2023. The cortico-limbo-thalamo-cortical circuits: an update to the original Papez circuit of the human limbic system. Brain Topogr. 36, 371–389. https://doi.org/10.1007/s10548-023-00955-y*

thalamo-prefrontal, an hypothalamo prefrontal, a septo-prefrontal, and a BNST-prefrontal conduit within the cortico-thalamo-limbo-cortical circuit. Through its extensive connectivity, the PFCT is involved in numerous limbic circuits, such as the prefronto-thalamic, prefronto-hypothalamic, prefronto-septal, and prefronto-striatal circuits (Fig. 22.8). Some of the circuits involving the PFCT are shown below (Fig. 22.8).

THE VENTRAL AMYGDALOFUGAL TRACT (AFT)

The ventral amygdalofugal tract consists of afferent and efferent connections to the basal forebrain, the hypothalamus, and the thalamus (Mori et al., 2017). The ventral amygdalofugal complex encompasses the essential amygdalo-diencephalic connections in the human brain. It includes the amygdalo-hypothalamic, amygdalo-septal, amygdalo-thalamic, and amygdalo-mesencephalic tracts. These tracts play a significant role in the communication between the amygdala and various regions of the diencephalon (Figs. 22.3 and 22.7) (Noback et al., 2005; Kamali et al., 2016). The AFT plays a pivotal role in central limbic connections, and allows the amygdala to control the hypothalamic/septal nuclei and BNST via the amygdalo-hypothalamic, amygdalo-septal, and amygdalo-BNST connections (Noback et al., 2005; Nolte, 1993; Usunoff et al., 2009; Kamali et al., 2015). The amygdalo-hypothalamic and amygdalo-septal components of the AFT are hook-like structures in the basal forebrain, running parallel to the posterior-inferior aspect of the anterior commissure (Fig. 22.7) (Kamali et al., 2016). On the midline, some of the superior projections of the AFT project toward and terminate into the BNST, providing a direct amygdalo-BNST connectivity (Weller and Smith, 1982).

Unlike the anterior commissure, which is continuous across the midline, connecting the amygdalae to one another, the AFT runs just posteriorly and parallel to the anterior commissure. The AFT originates from the basolateral nucleus and the central nucleus of the

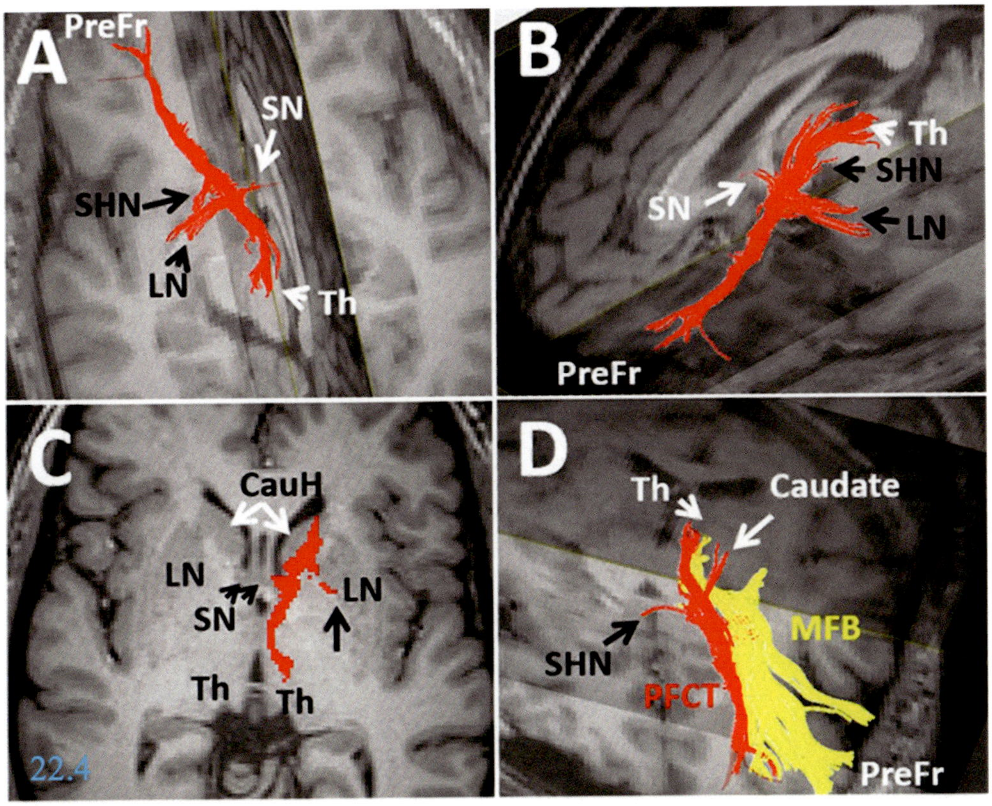

FIGURE 22.4 **Different views of 3-D reconstructions of the prefronto-caudothalamic tract (PFCT in red) on T1 weighted backgrounds.** (A–D) shows the connectivity of the PFCT with multiple gray matter nuclei, including the CaudH, caudate head; LN, lentiform nuclei; SHN, superior hypothalamic nuclei; SN, septal nuclei; TH, the thalamus. The PFCT projects to the medial prefrontal cortex (PreFr). (D) shows the medial forebrain bundle (MFB in yellow), running side by side along the lateral and superior aspect of the PFCT inserting into the lateral prefrontal cortex (PreFr). *Image courtesy of Kamali, A., Milosavljevic, S., Gandhi, A., et al., 2023. The cortico-limbo-thalamo-cortical circuits: an update to the original Papez circuit of the human limbic system. Brain Topogr. 36, 371–389. https://doi.org/10.1007/s10548-023-00955-y.*

amygdala, and projects posteriorly and superiorly toward the striatum (Noback et al., 2005, Nolte, 1993). The AFT then turns medially and enters the gray matter structure of the basal nucleus of Meynert, before reaching the midline (Fig. 22.7A and C). The AFT continues through the nucleus basalis, toward the hypothalamic nuclei. Near the midline, the AFT splits into three groups of fibers. Some of the fibers course cranially and anteriorly and terminate in the septal nuclei (Fig. 22.7C). The second group of cranial fibers runs cranially and slightly posteriorly to the BNST (Fig. 22.7B–D). A third group of fibers splits near the midline and runs caudally to the hypothalamic nuclei and the ventral tegmentum (Fig. 22.7A). Some of the circuits involving the AFT are listed below.

$$\text{amygdala} \Rightarrow (\text{AFT}) \Rightarrow \text{hypothalamus/septalnuc}$$

$$\Leftarrow (\text{PFCT/MFB}) \Rightarrow \text{prefrontalcortex}$$

$$\Leftarrow (\text{UF,CB}) \Rightarrow \text{amygdala}$$

$$\text{orbitofrontal/prefrontalcortex} \Leftarrow (\text{UF,CB}) \Rightarrow \text{amygdala}$$

$$\Rightarrow (\text{AFT}) \Rightarrow \text{hypothalamic/septal} \Leftarrow (\text{PFCT})$$

$$\Rightarrow \text{caudate/nucaccubmbens} \Leftarrow (\text{PFCT})$$

$$\Rightarrow \text{orbitofrontal/prefrontalcortex}$$

THE PARIETO-OCCIPITO-HYPOTHALAMIC TRACT (POHT)

The POHT is a direct corticolimbic connection of the human hypothalamus, connecting the parieto-occipital cortices directly to the hypothalamic nuclei (Kamali et al., 2020b). This pathway is one of the connections between the somatosensory and visual cortices and the limbic system, and it is thought to play a role in the cortico-limbic-cortical loop. Along its path toward the anterior hypothalamic nuclei, the POHT runs through the nucleus basalis of Meynert, alongside the ventral amygdalofugal tract. They both continue medially and project to the anterior hypothalamic nuclei. The POHT may provide a conduit for visual and sensory

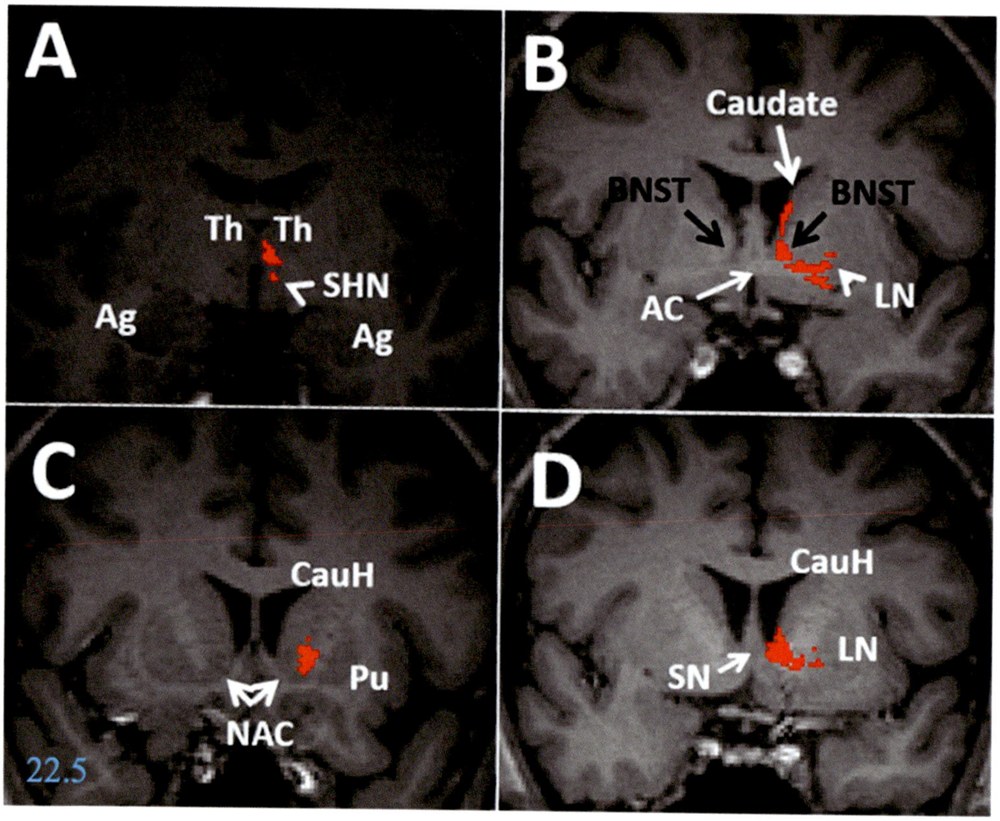

FIGURE 22.5 **Four consecutive coronal T1 weighted anatomical views demonstrating the trajectory of the PFCT (red) from posterior (A) to anterior (D).** The trajectory of the PFCT is shown arising from the dorsomedial thalamus (TH), projecting laterally and anteriorly along the lateral aspect of the third ventricle toward the caudate head (CauH, *white arrow* in C). The PFCT passes through the bed nucleus of stria terminalis (BNST), septal nuclei (SN), nucleus accumbens (NAC), and caudate head before projecting into the medial prefrontal cortex. Along the way, the PFCT receives projections from the superior hypothalamic nuclei (SHN, arrowhead in A) and lentiform nuclei (LN, arrowhead in B). Ag, amygdala; Pu, putamen; Th, thalamus. *Image courtesy of Kamali, A., Milosavljevic, S., Gandhi, A., et al., 2023. The cortico-limbo-thalamo-cortical circuits: an update to the original Papez circuit of the human limbic system. Brain Topogr. 36, 371–389. https://doi.org/10.1007/s10548-023-00955-y.*

information to both the nucleus basalis and the anterior hypothalamic nuclei. The POHT may contribute to multiple cortico-limbo-cortical circuits described in the following image (Fig. 22.9).

FUNCTIONAL-CONNECTOMICS OF THE LIMBIC STRUCTURES

The Influence of Ascending Reticular and Nonreticular Pathways on the Limbic System

It is well documented that at least two of the raphe nuclei are implicated in the functional influence of the limbic system, these are: the paramedian raphe nucleus (PMnR) and the median raphe nucleus (MnR) (Fig. 22.10). The latter, according to Singh et al. (2022), has sparser fMRI connectivity to the hippocampus, thalamus, hypothalamus, cuneiform nucleus (CnF), Pontine Reticular Formation (PnO-PnC), paramedian raphe (PMnR), nucleus raphe pallidus (RPa), nucleus raphe obscurus (ROb), nucleus of the solitary tract, prepositus, interpeduncular, and frontal and temporal cortices. In accordance with the literature, PMnR showed a denser fMRI connectivity pattern (Fig. 22.10) (Singh et al., 2022).

Singh et al. (2022)'s study also highlighted that dorsal raphe nucleus (DRN) is functionally related to other raphe nuclei, limbic forebrain regions, VTA, PAG, LC, and nucleus prepositus (a component of the VSM nuclei complex), but it lacked connections with the Raphe Magnus (RMg), previously reported in rats by Kalén et al. (1985) (Fig. 22.11). fMRI demonstrated that PAG is extensively connected to limbic and reticular structures, including the lateral parabrachial nucleus (LPBN), the medial parabrachial nucleus (MPBN), the viscero-sensory motor nuclei complex (VSM), and the Raphe Pallidus (RPa), but not the amygdala. These results highlight that PAG is also part of the subcortical structure involved in the modulation of the "superior limbic areas" (several cortices, see later) (Fig. 22.11).

In turn, the amygdala has been shown to be connected to basal ganglia, hypothalamus, substantia nigra

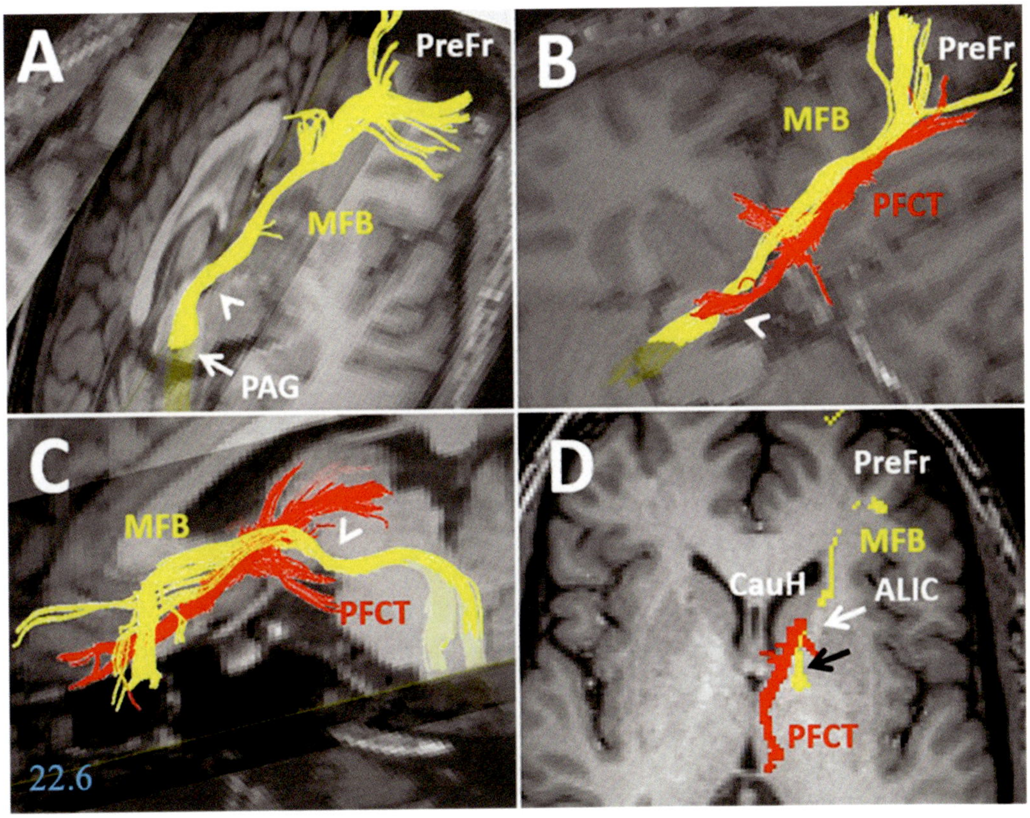

FIGURE 22.6 **Different views of 3-D reconstructions of the prefronto-caudothalamic tract (PFCT in red) and the medial forebrain bundle (MFB in yellow) on T1 weighted imaging backgrounds.** The MFB connects the midbrain's periaqueductal gray matter (PAG) to the ventral tegmental areas and the prefrontal cortex (PreFr). The MFB projects from the PAG anteriorly near the midline into the ventral tegmental area toward the interpeduncular cistern (marked by arrowhead in A–C). The MFB then courses superiorly toward the genu of the internal capsule (*black arrow* in D), where it enters the anterior limb of the internal capsule (ALIC). Unlike the MFB which courses through the ALIC into the prefrontal cortex, the PFCT courses outside the ALIC, more medially and through the caudate head before projecting to the medial prefrontal cortex (B–D). *Image courtesy of Kamali, A., Milosavljevic, S., Gandhi, A., et al., 2023. The cortico-limbo-thalamo-cortical circuits: an update to the original Papez circuit of the human limbic system. Brain Topogr. 36, 371–389. https://doi.org/10.1007/s10548-023-00955-y.*

pars compacta (SNpc), VTA, DRN and MnN, and caudal lineal raphe nucleus (CLi), a nucleus equally implicated in the control of arousal (Fig. 22.12).

Limbic System and Thalamus

The thalamus and isocortex are connected via the septum, the hippocampus, and the amygdala. Isocortical regions, such as the orbital frontal cortex, the cingulate gyrus, and the hippocampal gyrus, connect the limbic system to various corticothalamic systems. Strong connections are present between the cingulate gyrus and the thalamic nuclei, in addition to several cortical areas. These characteristics, together with a robust thalamic connectivity from the hippocampus, the septum, and the amygdala, give rise to the belief that the limbic system serves as a link between the thalamic structures and the overlaying isocortex (Erro et al., 1999; Cornwall et al., 1988; Bourgeais et al., 2001).

Cortical Elaboration of the Limbic Processes

Both the medial orbitofrontal cortex and the pregenual cingulate cortex (32BA, which is a part of the ACC), which participate in rewards, have robust functional connections. The lateral orbitofrontal cortex and the inferior frontal gyrus, which are both stimulated by adverse stimuli, have strong functional connections with the dorsal ACC area.

The parts of the inferior frontal gyrus and the lateral orbitofrontal cortex that are likewise triggered by unpleasant stimuli have substantial functional connectivity with the supracallosal division, which is triggered by unpleasant stimuli and nonreward. These connectivities lend credence to the theory that the medial orbitofrontal cortex, which is associated with rewards, supplies inputs to the pregenual cingulate cortex, which also is stimulated by rewards, and that the lateral orbitofrontal cortex, which is associated with nonreward and punishments, contributes inputs to the supracallosal part of the

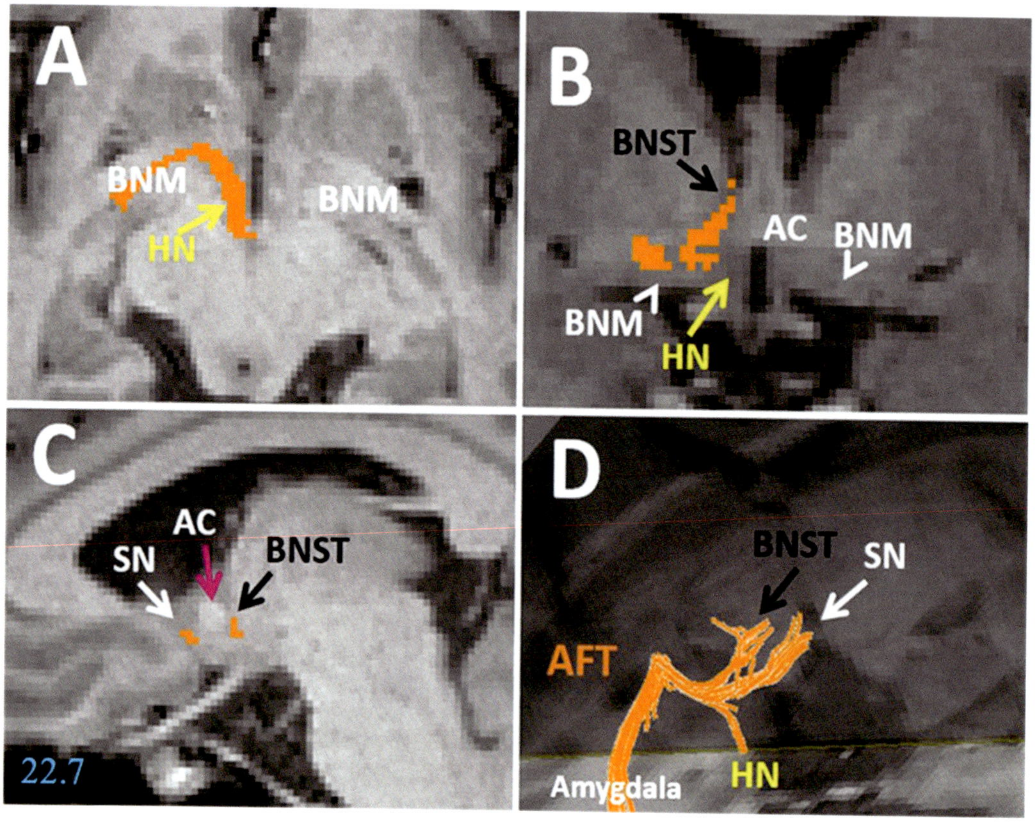

FIGURE 22.7 **(A–C), trajectory of the amygdalofugal tract (orange) on T1 weighted backgrounds. (D), 3-D reconstruction of the AFT is demonstrated.** The AFT originates from the amygdala and projects superiorly and medially toward the midline. While projecting toward the midline (B), the AFT courses through the basal nucleus of meynert (BNM, arrowheads in B) along the inferior aspect of the anterior commissure (AC). The AFT splits into three groups of fibers near the midline (D). Two groups of fibers project cranially toward the septal nuclei (SN, superior anterior marked by *white arrows* in C and D) and the bed nucleus of stria terminals (BNST, superior posterior marked by black arrows in B–D). The third group of fibers projects caudally into the hypothalamic nuclei (HN, *yellow arrow* in A) and ventral tegmental area. *Image courtesy of Kamali, A., Milosavljevic, S., Gandhi, A., et al., 2023. The cortico-limbo-thalamo-cortical circuits: an update to the original Papez circuit of the human limbic system. Brain Topogr. 36, 371–389. https://doi.org/10.1007/s10548-023-00955-y*

ACC, which is also triggered by unpleasant stimuli (Rolls, 2019a,b; Rolls and Wirth, 2018). When choosing between several incentives, the ventromedial prefrontal cortex may be particularly important (Gläscher et al., 2012; Rolls, 2019a). With connections from the ventral-stream processing regions, which focus on "what" the stimulus is, the amygdala and orbitofrontal cortex are engaged in emotion and reward value. Inputs concerning space, action, and "where" events occur come from the dorsal stream cortical regions.

The orbitofrontal cortex, the amygdala, and the ventral stream regions all provide inputs to the anterior cingulate cortex. In addition to receiving input from the parietal cortex and other dorsal stream regions, the posterior cingulate cortex is also connected to the hippocampus memory network. The cingulate cortex contributes to action-outcome learning, which is the process by which behaviors to achieve objectives are learned based on the results, or rewards and punishments, obtained after different behaviors (Kolling et al., 2016; Rolls, 2019a,b).

The perirhinal and entorhinal cortices serve as a major source of "what" and "reward" inputs to the hippocampal memory system.

Contrary to common belief, there is a connection between the orbitofrontal cortex and the posterior cingulate cortex (Vogt and Laureys, 2009), which creates a pathway for information about rewards and punishments to reach the hippocampus, both dorsally and ventrally (Rolls and Wirth, 2018).

The neurons in the orbitofrontal cortex are mostly only aware of the reward value representations for stimuli. The orbitofrontal cortex transmits information to the ACC regarding objectives, including the value of the outcomes and the anticipated value, as well as the value of the stimuli. In addition to connecting behaviors to outcomes using action-outcome learning, the ACC works in tandem with the midcingulate motor area, containing models of actions, and considers the cost of actions to reach the desired outcome when selecting behaviors (Kolling et al., 2016; Rolls, 2019).

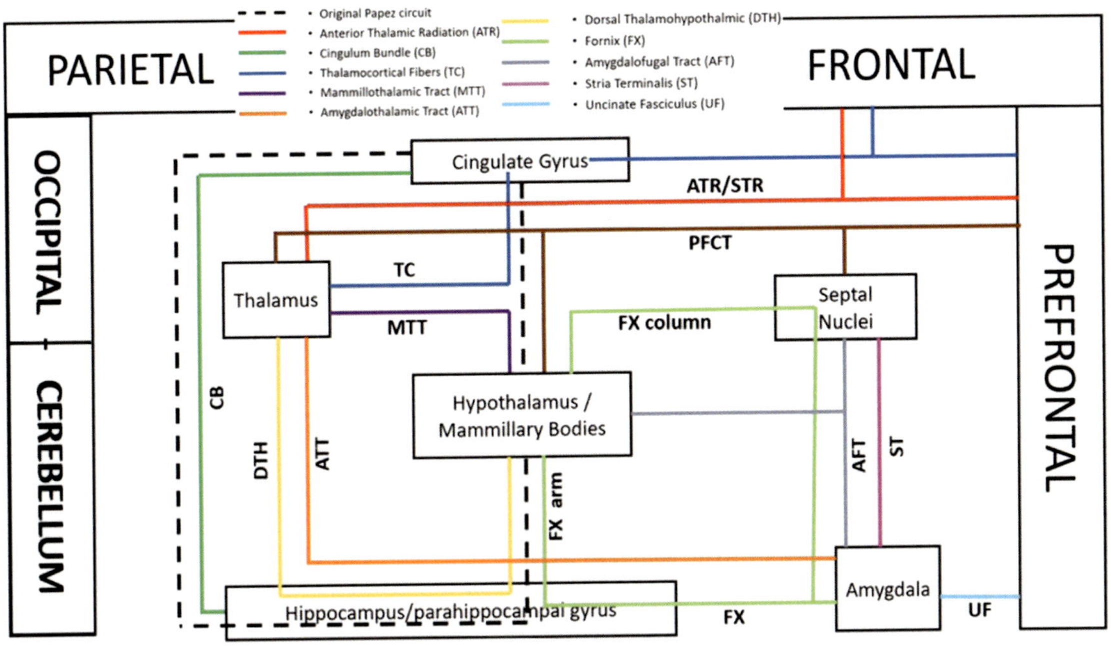

FIGURE 22.8 **Schematic view of the major anterior neuronal circuits (frontotemporal connectivity) of the limbic system mostly connected to the frontal and temporal cortices.** The Papez circuit is shown by dotted box. The Papez circuit consists of the frontoparietal connectivity with the parahippocampal gyrus via the cingulum bundle, hippocampal formations to mammillary bodies via the fornix, mammillary bodies to the thalamus via the mammillothalamic tract (MTT), thalamus to the frontal lobe and cingulum bundle via the thalamocortical fibers (TC). By considering other limbic pathways such as the PFCT, ST, AFT, UF, ATT, and DTH, several neuronal loops are added to the limbic system which were detailed in the text. *Image courtesy of Kamali, A., Milosavljevic, S., Gandhi, A., et al., 2023. The cortico-limbo-thalamo-cortical circuits: an update to the original Papez circuit of the human limbic system. Brain Topogr. 36, 371–389. https://doi.org/10.1007/s10548-023-00955-y*

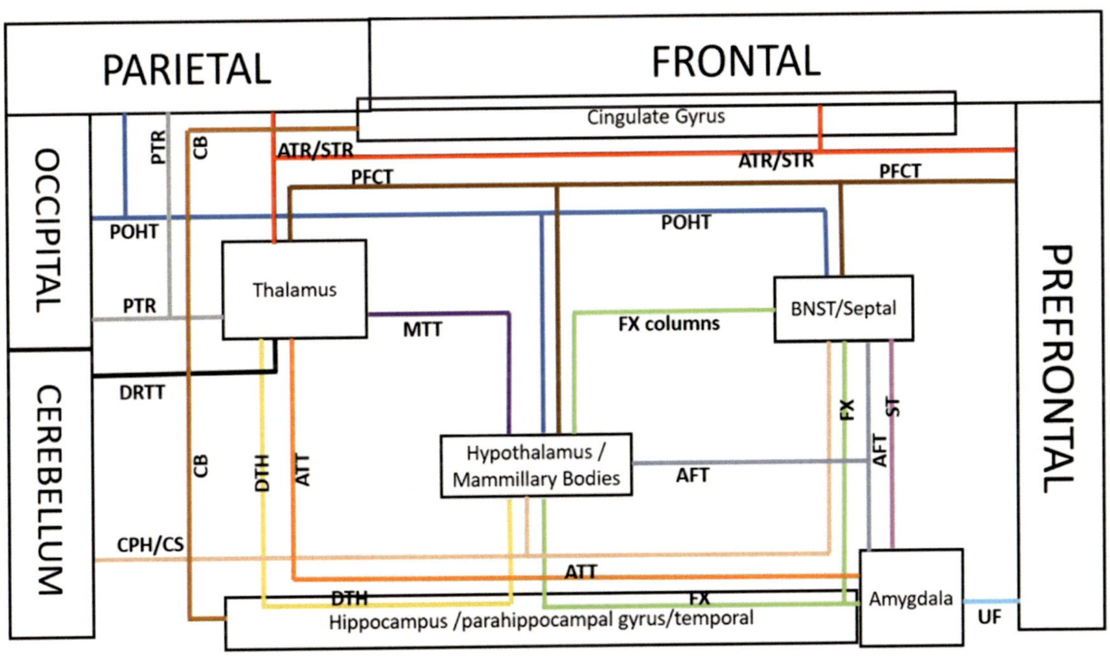

FIGURE 22.9 **Schematic view of the major anterior (fronto-temporal connectivity) and posterior (parieto-occipito-cerebellar connectivity) neuronal circuits of the limbic system connected to the frontal, parietal, occipital and temporal cortices as well as cerebellar hemispheres.** By introducing posterior limbic pathways such as POHT, cerebello-hypothalamic and cerebello-septal/BNST, DRTT, and PTR, several additional neuronal circuits are added to the limbic network. This schematic view shows the tight connectivity of the limbic gray matter nuclei to the posterior cerebral and cerebellar cortices including the parietal, occipital lobes and cerebellar hemispheres. The circuits are detailed in the text. *Image courtesy of Kamali, A., Milosavljevic, S., Gandhi, A., et al., 2023. The cortico-limbo-thalamo-cortical circuits: an update to the original Papez circuit of the human limbic system. Brain Topogr. 36, 371–389. https://doi.org/10.1007/s10548-023-00955-y*

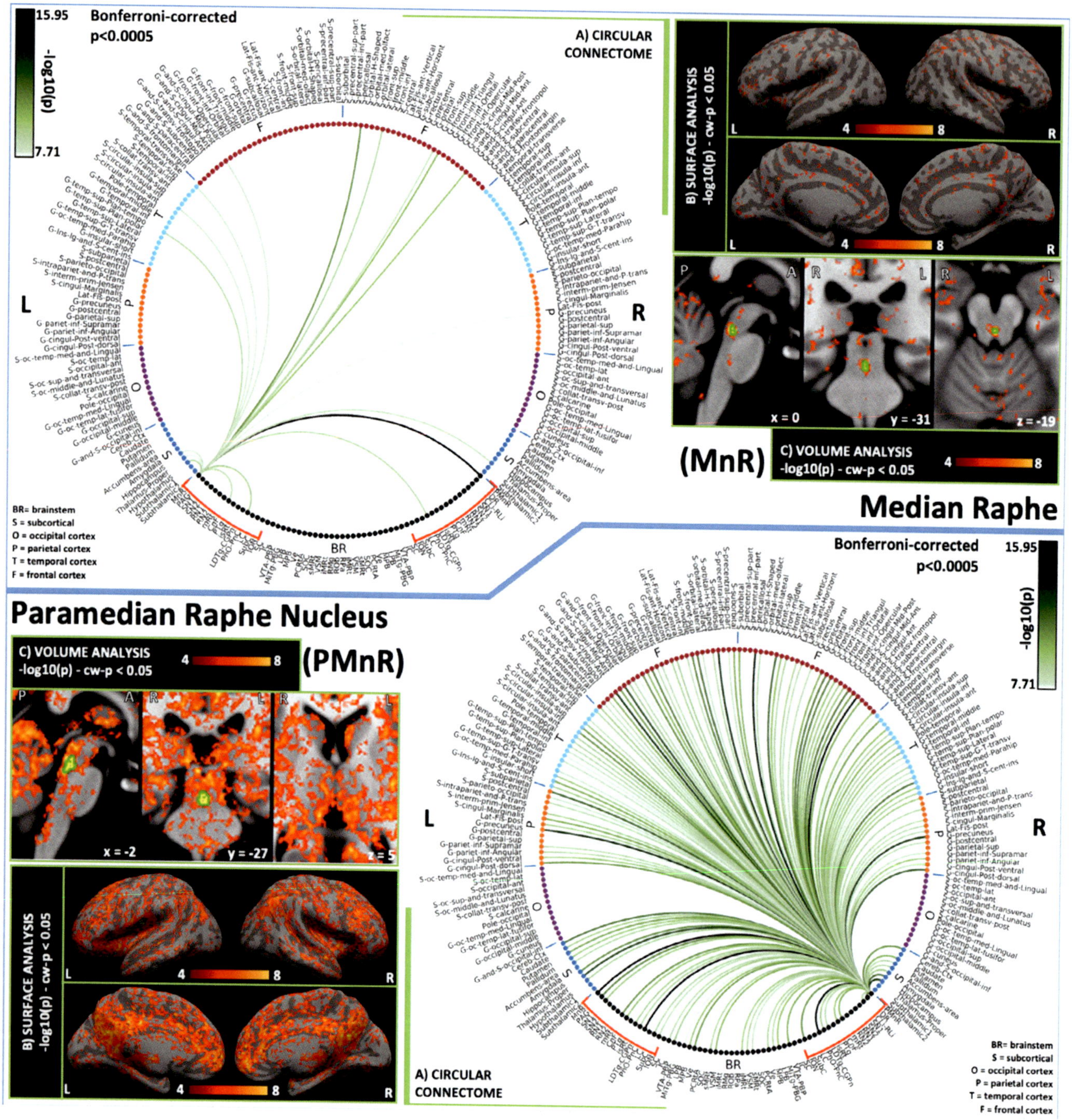

FIGURE 22.10 **2D circular connectome (A), voxel-based connectivity maps in the cortex (B) and subcortex (C) of (*top*) MnR and (*bottom*) PMnR nucleus.** MnR showed significant connections with arousal network regions such as hypothalamus, thalamus, along with arousal brainstem nuclei (PMnR, CnF, and PnO-PnC). Similarly, PMnR showed significant connectivity with several arousal regions including the cortex, forebrain, hypothalamus, and thalamus, along with all major brainstem arousal nuclei (MnR, PAG, SN1, DR, CLi-RLi, mRt, CnF, isRt, VTA, LPB-MPB, LDTG-CGPn, PnO-PnC), except for LC, SN2, PTg. List of abbreviations (brainstem nuclei used as seeds are marked with red brackets): CLi-RLi, caudal-rostral linear raphe; CnF, cuneiform; DR, dorsal raphe; ION, inferior olivary nucleus; isRt, isthmic reticular formation; LC, locus coeruleus; LDTg-CGPn, laterodorsal tegmental nucleus-central gray of the rhombencephalon; MnR, median raphe nucleus; mRt, mesencephalic reticular formation; PAG, periaqueductal gray; PMnR, paramedian raphe nucleus; PnO-PnC, pon-tine reticular nucleus, oral partpontine reticular nucleus caudal part; PTg, pedunculotegmental nuclei; RN1, red nucleus-subregion1; RN2, red nucleus-subregion2; SN1, substantia nigra-subregion1; SN2, substantia nigra-subregion2; SubC, subcoeruleus nucleus. Additional brainstem nuclei used as targets: IC, inferior colliculus; iMRt, inferior medullary reticular formation; LPB, lateral parabrachial nucleus; MiTg-PBG, microcellular tegmental nucleus-parabigeminal nucleus; MPB, medial parabrachial nucleus; PCRtA, parvicellular reticular nucleus alpha-part; RMg, raphe magnus; ROb, raphe obscurus; RPa, Raphe Pallidus; SC, superior colliculus; sMRt, superior medullary reticular formation; SOC, superior olivary complex; Ve, vestibular nuclei complex; VSM, viscero-sensory motor nuclei complex; VTA-PBP, ventral tegmental area-parabrachial pigmented nucleus. *Image courtesy of Singh, K., Cauzzo, S., García-Gomar, M.G., et al., 2022. Functional connectome of arousal and motor brainstem nuclei in living humans by 7 Tesla resting-state fMRI. Neuroimage 249, 118865. doi:10.1016/j.neuroimage.2021.118865.*

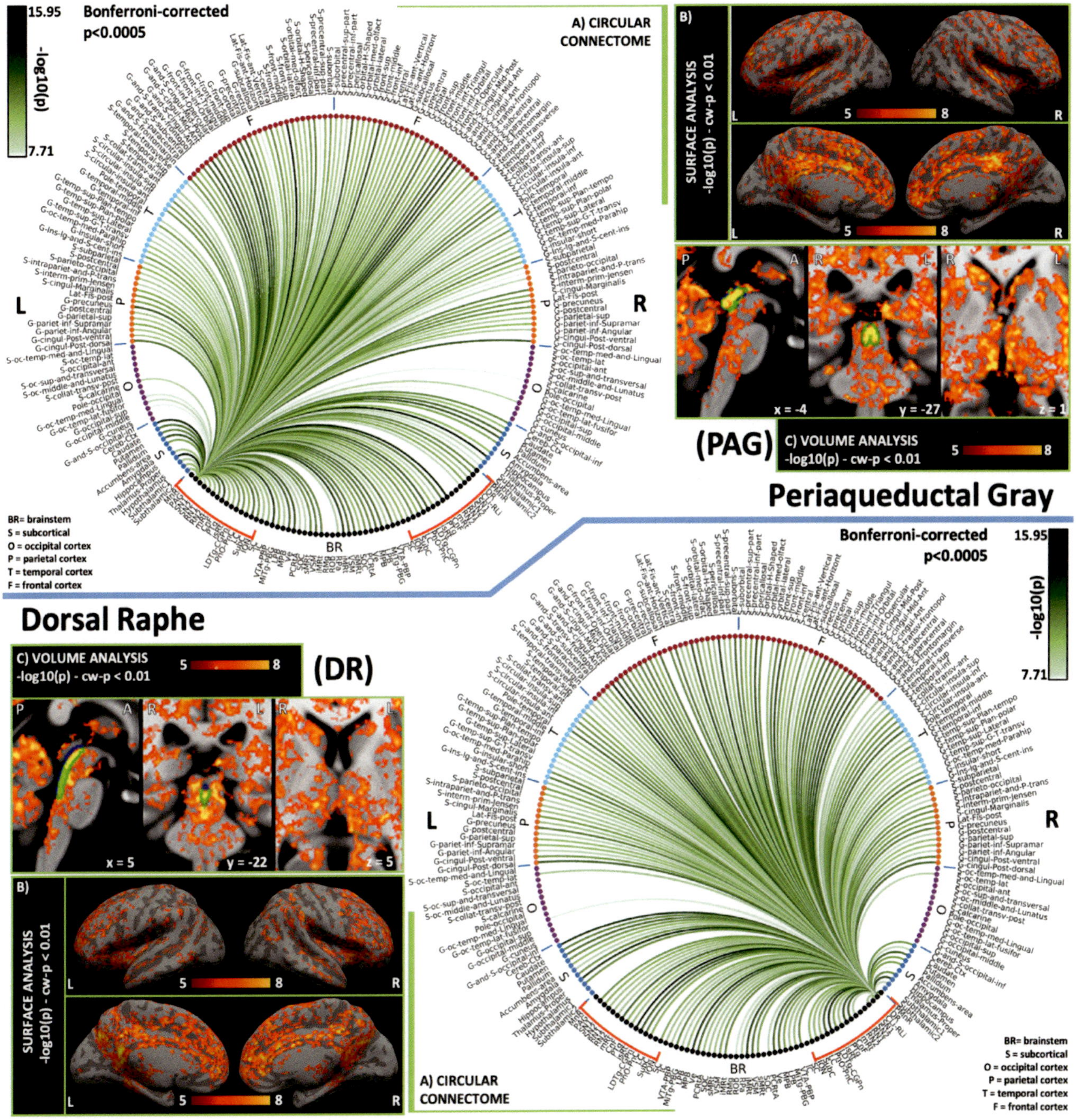

FIGURE 22.11 **2D circular connectome (A), voxel-based connectivity maps in the cortex (B) and subcortex (C) of (*top*) PAG and (*bottom*) DRN.** PAG showed strong connectivity with arousal regions (basal forebrain, hypothalamus, thalamus, VTA-PBP, LC, CnF, LDTg-CGPn, and mRt), REM-sleep brainstem nuclei (SubC, sMRt, iMRt), autonomic and limbic regions (hypothalamus, prefrontal cortex, insular cortex, cingulate cortex, LPB, MPB, VSM, and RPa), yet it was not connected with the amygdala. DR showed high connectivity with all cortical regions except occipital regions. It showed expected connectivity with other arousal network nuclei except MnR although it showed connectivity with PMnR. List of abbreviations (brainstem nuclei used as seeds are marked with red brackets): CLi-RLi, caudal-rostral linear raphe; CnF, cuneiform; DR, dorsal raphe; ION, inferior olivary nucleus; isRt, isthmic reticular formation; LC, locus coeruleus; LDTg-CGPn, laterodorsal tegmental nucleus-central gray of the rhombencephalon; MnR, median raphe nucleus; mRt, mesencephalic reticular formation; PAG, periaqueductal gray; PMnR, paramedian raphe nucleus; PnO-PnC, pontine reticular nucleus, oral part-pontine reticular nucleus caudal part; PTg, pedunculotegmental nuclei; RN1, red nucleus-subregion1; RN2, red nucleus-subregion2; SN1, substantia nigra-subregion1; SN2, substantia nigra-subregion2; SubC, subcoeruleus nucleus. Additional brainstem nuclei used as targets: IC, inferior colliculus; iMRt, inferior medullary reticular formation; LPB, lateral parabrachial nucleus; MiTg-PBG, microcellular tegmental nucleus-prabigeminal nucleus; MPB, medial parabrachial nucleus; PCRtA, parvicellular reticular nucleus alpha-part; RMg, raphe magnus; ROb, raphe obscurus; RPa, raphe pallidus; SC, superior colliculus; sMRt, superior medullary reticular formation; SOC, superior olivary complex; Ve, vestibular nuclei complex; VSM, viscero-sensory motor nuclei complex; VTA-PBP, ventral tegmental area-parabrachial pigmented nucleus. *Image courtesy of Singh, K., Cauzzo, S., García-Gomar, M.G., et al., 2022. Functional connectome of arousal and motor brainstem nuclei in living humans by 7 Tesla resting-state fMRI. Neuroimage 249, 118865. doi:10.1016/j.neuroimage.2021.118865.*

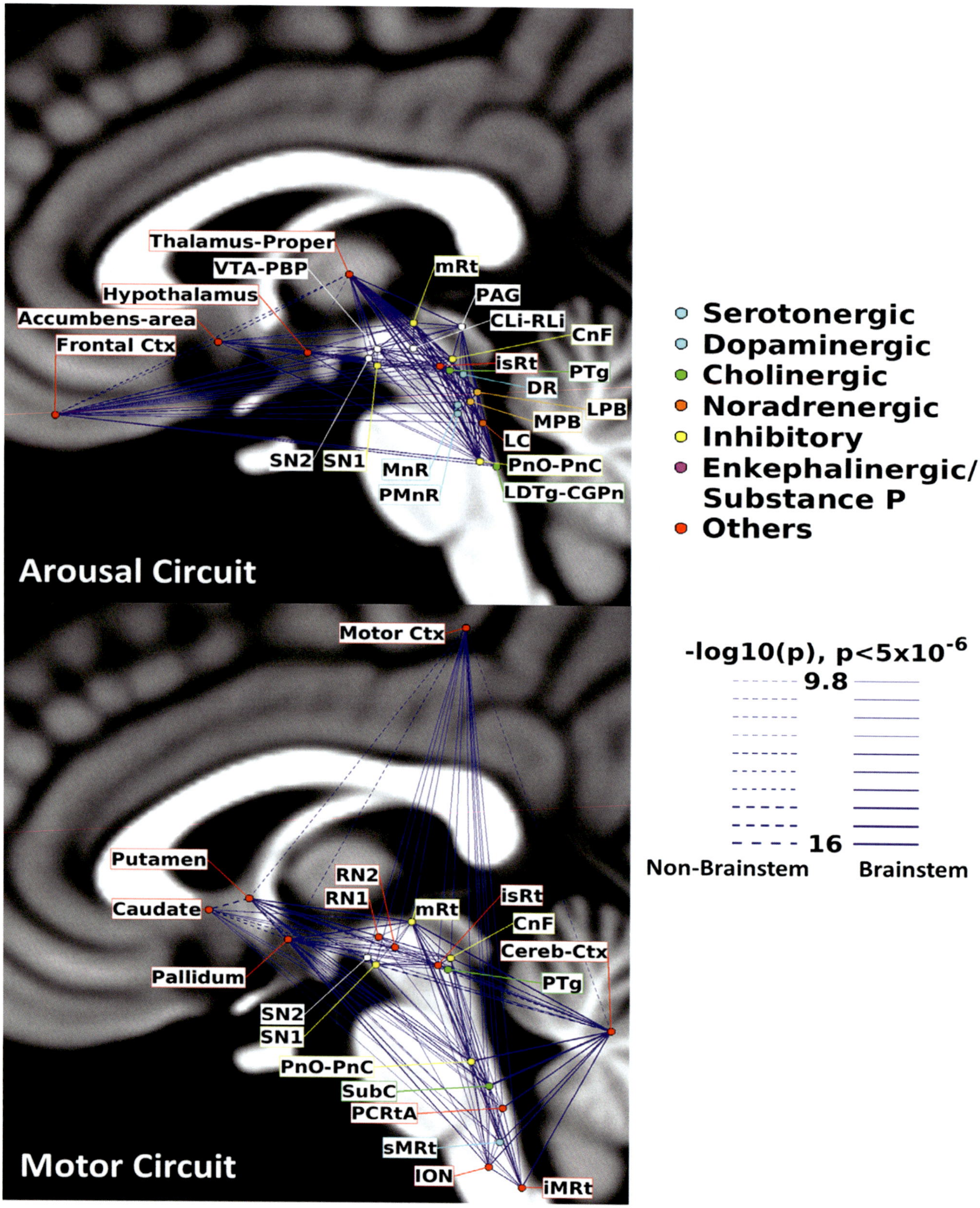

FIGURE 22.12 **Circuit diagram of the arousal (*top*) and motor (*bottom*) network derived from resting-state 7 T fMRI functional connectivity.** The diagram of the arousal network displayed high node interconnectivity. Interestingly, several brainstem arousal nuclei displayed connectivity to the thalamus, hypothalamus, basal forebrain, and frontal cortex, as expected for arousal regions. The diagram of the motor network matched expected connectivity of motor regions, including the motor cortex, basal ganglia, and cerebellum. Note that the link thickness was varied based on the statistical significance of the connectivity strength. For the arousal circuit, the brainstem nodes were color-coded based on the main neurotransmitter employed. *Image courtesy of Singh, K., Cauzzo, S., García-Gomar, M.G., et al., 2022. Functional connectome of arousal and motor brainstem nuclei in living humans by 7 Tesla resting-state fMRI. Neuroimage 249, 118865. doi:10.1016/j.neuroimage.2021.118865.*

Cingulate and Prefrontal Cortex Connectivity in Decision-Making

Both exploring or looking for alternatives in decision-making and dedication to a behavior have been linked to activation of the anterior cingulate cortex (ACC). The ACC integrates a variety of inputs, which represent the modification of assumptions and internal models of the environment, and stores information about the value of alternatives, including options provided by the environment. The human ventromedial prefrontal cortex (vmPFC) is active while making decisions. The subjective values of choices and items have been connected to vmPFC activation. A function in decision-making or attentional selection is suggested by a positive association between activity and the significance of selected alternatives and a negative correlation with the values of rejected alternatives. The vmPFC keeps records of the significance of things even if no decision is made or while observing someone else making a choice.

According to Neubert et al. (2015), the vmPFC displayed significant functional connectivity with the amygdala, the hypothalamus, and the ventral striatum, and also showed positive coupling with the OFC, the retrosplenial, lateral occipital, inferior temporal, posterior temporo-parietal junction (TPJp), and the perirhinal cortex. The ACC, midinferior parietal lobule (IPL), and lateral FP (FPl) cortex, on the other hand, are negatively coupled with it.

Values of abstract items and options are thought to be represented by a more anterior vmPFC area. It showed some connectivity to the amygdala, the hypothalamus, the ventral striatum, the medial temporal cortex, and the temporal pole, similarly to nearby areas, yet it demonstrated stronger connections with the posterior cingulate cortex (PCC), the precuneus, and the head of the caudate. This region did not show negative coupling with the intraparietal sulcus (IPS), the supplementary motor area (SMA), the pre-SMA, and the supplementary motor area (SMA). There is an area where activity matches the perceived values of delayed monetary rewards. In relation to the value-comparison areas covered thus far, it is more dorsal and situated closer to the genu of the corpus callosum. This area is connected to the TPJp, the temporal pole, the PCC, the precuneus, the dorsolateral PFC-(dlPFC), the FPl, and the head of the caudate. The SMA, the dorsal premotor area (PMd), the M1, and the IPS showed a negative coupling with this region. Similar coupling patterns were found in a more dorsal perigenual region, which is related to individual differences in cost–benefit assessments. However, this area demonstrated greater coupling with premotor regions, like the SMA, as well as decreased coupling with medial temporal regions, like the amygdala and the dorsal and ventrolateral PFC.

The altered pattern of self-valuation linked to depression is attributed to a different, more posterior and subgenual vmPFC area. In individuals with depression resistant to therapy, deep brain stimulation is used to target this region (Drobisz et al., 2019; Koenigs et al., 2009).

The area in question displayed a coupling pattern similar to other vmPFC areas, but with greater coupling to ventral striatum, amygdala, and hypothalamus.

When unfavorable consequences, such as punishment, cause behavioral changes, the role of the human lOFC, particularly the posterior lOFC, is crucial. An alternative theory suggests that the lOFC may be involved in processing various consequences, both positive and negative, to guide changes in behavior, rather than solely unfavorable outcomes. The posterior IOFC shows positive coupling with vmPFC, ventrolateral PF (vlFC), ventral striatum and temporal cortex; it shows negative coupling with ACC, sensorimotor, and parietal areas.

A frontoparietal control system was described by Seeley et al. (2007). The system was divided into two components, one focused on "executive control," involving the dorsolateral frontal and parietal cortices, and the other linked with "salience," involving the orbital fronto-insular cortex. The frontoparietal control system is frequently activated by tasks involving controlled processing associated with the concurrent assessment of several interdependent conditions (Kroger et al., 2002). Additionally, numerous frontoparietal control system's regions exhibit sustained activity during a task block (Dosenbach et al., 2006; Yarkoni et al., 2005).

Affective Salience Network Specialization

The Affective Salience Network (ASN), which includes limbic and prefrontal brain regions, encodes emotions. There are still many questions concerning the ASN's processing of valence and emotional intensity. In a study by Metzger et al. human cerebral electrical data was used to extract dominating spectral characteristics, indicating emotional specialization within certain nodes. The dorsal anterior cingulate (dACC), the anterior insula (aINS), and the ventro-medial prefrontal cortex (vmPFC) are responsive to valence and intensity, but the amygdala is specifically sensitive to intensity. The anatomical distribution of Von Economo neurons (VENs), a population of neurons with a distinctive bipolar morphology and a spindle-shaped body, expressed almost exclusively in the human anterior cingulate cortex (concentrated in subgenual portions of the ACC), and the anterior insula may play a role in integrating divergent information streams to influence socio-affective processessing, in addition to their role in facilitating quick communication across distant brain areas. Deficits in social communication, such as those seen in autism and frontotemporal dementia, are associated with abnormalities in Von Economo neurons, which frequently translate to regional dysfunction in the aINS and the ACC (Allman et al., 2005).

Thalamus Influence in Monitoring Viscerosensory Guidance in Anxiety

In classical explanations of the limbic system, including the Papez circuit, it is understood that the mediodorsal nucleus of the thalamus (MD) plays a significant role in its regulation and control.

The MD, which is related to both anterior temporal regions and orbital frontal areas in a special triangle circuit (Jones, 2009), clearly serves as a source for thalamic regulation of the ventral paleocortical limbic division (Aggleton et al., 1986). The triangular circuit could promote reverberating processing between the adaptive motor control in the orbital frontal cortex and the sensory-visceral integration in the amygdala, with the mediodorsal nucleus of the thalamus serving as the network's coordination center.

Inhibitory neurons in a variety of cortical layers are the target of direct connections from the amygdala to the caudal orbital frontal cortex.

In a study looking at potential deep brain stimulation locations for treating anxiety and OCD, Rotge et al. (2010) found evidence of the involvement of MD's in the control of visceral and emotional components of anxiety. The MD thalamus may be a vital component of the dopaminergic, collothalamic midbrain regulation of anxiety and its visceral manifestations, because it receives collothalamic projections from the midbrain, which controls the dopamine modulation of both mesolimbic and frontal networks (Tucker and Luu, 2023).

The terminals of the amygdalar projections to the mediodorsal nucleus (MD) of the thalamus were investigated by Timbie et al. (2020), revealing their similarity to the complex terminals observed in motor regions that indicate self-initiated movements. As a consequence the orbital frontal lobe's involvement in action might be controlled by the amygdala through a reentrant monitoring circuit involving both visceral and affectively relevant sensory data via the triangle circuit through the MD in the thalamus (Timbie et al., 2020).

The ventral limbic-MD-frontal network may have a role in the beta rhythms that control striatal-thalamo-cortical loops to maintain the stability in present cognitive and motor processing. The "redundancy bias" which permits the stabilization of object memory through thalamic mediation of the ventral corticolimbic and striatal networks may be mediated by the limbic, MD thalamic, and frontal components of the triangular circuit (Tucker and Luu, 2012).

With regard to the role of anxiety in behavior regulation, including the exaggerated regulation of redundant actions in chronic and severe anxiety, such as in obsessions and compulsions, the combined ventral limbic and striatal circuits may offer a visceral constraint on working memory (Shapiro, 1981; Tucker and Williamson, 1984). However, as Sanides (1964) notes, there are disorders of inadequate anxiety as well, such as those brought on by orbital frontal lesions, and resulting in personality disinhibition and psychopathy. In order to regulate the adaptive focus of cognition and maintain an important emotional quality of critical conscious assessment, allocentric awareness, and appropriate action in the social environment, anxiety serves as a necessary personality regulator. This regulatory process is mediated by the ventral limbic division.

Nicotinic Acetylcholine Receptors and Their Role in Affective Disorders

ACh signaling via nAChRs clearly plays a role in the control of behavioral responses to stress. It is believed that the psychopathology observed in anxiety and depressive disorders is associated with the disruption of nAChR homeostasis.

As a crucial neurotransmitter and neuromodulator, acetylcholine (ACh) is engaged in a variety of neuronal processes which underlie cognitive and emotional regulation. For effective learning and memory processes to occur, the signaling of acetylcholine (ACh) is essential. Recent research has highlighted the crucial role of ACh in emotional and mood processes. The relationship between nicotine, nAChRs, and depression is complicated, and a number of studies suggest that this correlation is bidirectional, meaning that nicotine use may be started as a way to treat one's own affective disorder symptoms, but that long-term nicotine use also changes the neuronal circuits which control mood, making symptoms of depression worse.

The hypercholinergic hypothesis of depression, which claims that affective regulation depends on a balance of ACh and noradrenergic signaling and that depression is caused by a shift in the balance toward more sustained ACh signaling, may help explain the startling relationship between nicotine use and depression (Dulawa and Janowsky, 2019; Mineur et al., 2018).

Rather than potentiating ACh signaling, nicotine may exert its antidepressant effects by desensitizing nAChRs. The hypercholinergic signaling observed in a subgroup of individuals who are depressed may thus be an effort to be countered by nAChR desensitization through smoking.

In a study by Mineur et al. (2018) was observed that while there are no discernible anxiolytic or antidepressant-like effects from suppression of 2-containing (alpha 2) nAChRs in the hippocampus, fear memory recall in mice can be impaired. The effects of cholinesterase blockage on stress-related behaviors in male mice, on the other hand, are reversed by hippocampal reduction of seven nAChRs.

One of the several human brain regions displaying depression-related variations in nAChR activity is the hippocampus. Mecamylamine, a wide nicotinic antagonist, can be administered systemically or directly infused into the hippocampus to reduce increases in depressive and anxious behaviors, which are caused by a raise in cholinergic activity in the hippocampus in mice. These findings provide compelling evidence that hippocampus' nAChRs are involved in stress-related behaviors.

Research on transgenic mice deficient in the nAChR2 subunit revealed that nAChR2 is essential for adult hippocampus neurogenesis, a crucial process that is impaired in stress-induced depression models and is needed for the response to both fluoxetine and tricyclic antidepressants. According to some data, the hippocampus's seven nAChR antagonism may also have antidepressant effects through alteration of the glutamatergic activity, a key factor in the control of affective behaviors.

Pain-Related Cortico-Limbic Plasticity

Clinical and preclinical research have linked neuroplasticity in cortico-limbic circuits to the persistence and regulation of pain (Fig. 22.13). In the emotional-affective dimension of pain and pain regulation, the amygdala has become a significant actor. In the presence of pain, the reciprocal relationships of the limbic system with medial prefrontal cortical areas change.

As mentioned above in this chapter, the Anterior Cingulate Cortex (ACC) gets thalamo-cortical inputs from all sensory modalities, but the infra- and prelimbic cortices receive inputs largely from limbic rather than somatosensory areas, including pain-related information from the thalamus and the amygdala (Hoover and Vertes, 2007). There have also been reports of direct spinal cord inputs (Cliffer et al., 1991). In addition to intra- and inter-cortical projections mainly from layer II/III, the mPFC subregions project to subcortical targets, including mediodorsal thalamus, lateral hypothalamus, amygdala's nuclei, nucleus accumbens, periaqueductal gray, ventral tegmental area, parabrachial nucleus, nucleus of the solitary tract, rostral ventromedial medulla, and spinal cord (Gabbott et al., 2003, 2005). These targets include areas involved in controlling pain.

According to Ji and Neugebauer (2011), Kiritoshi et al. (2013), and Ji et al. (2010), the infralimbic and prelimbic pyramidal cells have broad receptive fields and react to external painful stimuli.

The neurons in the ACC exhibit several characteristics related to pain processing. These include large receptive fields, encoding of unpleasant somatic and visceral responses, and activation during pain anticipation or pain avoidance behaviors. When measured with electrophysiology, ACC pyramidal layer V cells and some layer II/III neurons (intermediate type), but not interneurons, show increased activity.

The parabrachial nucleus (PB), the thalamic nuclei (midline, posterior intralaminar/paraventricular, and posterior areas), the cortical regions (mPFC, IC, and sensory association cortices), and the thalamic nuclei all contribute nociceptive information to the amygdala (Neugebauer, 2020).

The lateral amygdala-basolateral amygdala (LA-BLA) network is the primary focus of thalamic and cortical inputs, which may also contain nociceptive information (Fu et al., 2020). This information converges with the PB input into the CeA neurons to generate a complex network of GABAergic and peptidergic neurons, and connections that are primarily inhibitory.

CeA suppression reduces behaviors in models of inflammatory and neuropathic pain, while CeA activation produces sensory and emotional pain-like behaviors under normal circumstances (see Neugebauer, 2020).

In rats studies, stimulation of CeA-CRF or BLA-CeA neurons in normal circumstances resulted in mechanical hypersensitivity, emotional reactions, anxiety-like behaviors, and a rise in the activity of nociceptive spinal dorsal horn neurons (Mazzitelli et al., 2021, 2022).

According to Reiner et al. (2019), processing of opioids has been linked to actions in the prelimbic and infralimbic mPFC areas. There is proof that the mPFC's infra- and prelimbic areas, as well as the connections between them, contain opioid receptors involved in modulating pain. The plasticity of the interneurons in the mPFC can be related to both endogenous and external MOR signaling. Mice mPFC dendritic branching of SOM interneurons and dendritic elongation of parvalbumin (PV) interneurons were both induced by chronic systemic morphine treatment (MOR agonist) (Wang et al., 2019). MOR signaling in the mPFC circuitry may affect pain-related processing in the same way as PL and IL inhibitory signaling, which includes SOM and PV interneurons (Jones and Sheets, 2020; Kiritoshi et al., 2016), and is involved in cognitive deficits related to chronic pain. Presynaptic inhibition of glutamatergic transmission, specifically targeting NMDA receptor-expressing neurons, is observed in the mPFC as a result of KOR signaling. The pain in Fabry disease may be caused by an early disruption of the KOR dynorphinergic system. According to several studies (Cahill et al., 2022; Gavériaux-Ruff and Kieffer, 2011; Nadal et al., 2006), the DOR system is involved in the control of both emotional and sensory pain. Similarly to MOR signaling, interneuron plasticity in the mPFC may be impacted by DOR-related processes. However, DOR but not MOR knockdown showed impaired dendritic morphology following a single morphine injection in male mice (Wang et al., 2019). SOM interneurons showed reduced dendritic length and complexity following DOR downregulation (Wang et al., 2019).

The ACC endogenous opioid neurotransmission is crucial for the regulation of pain. MOR, DOR, and to a

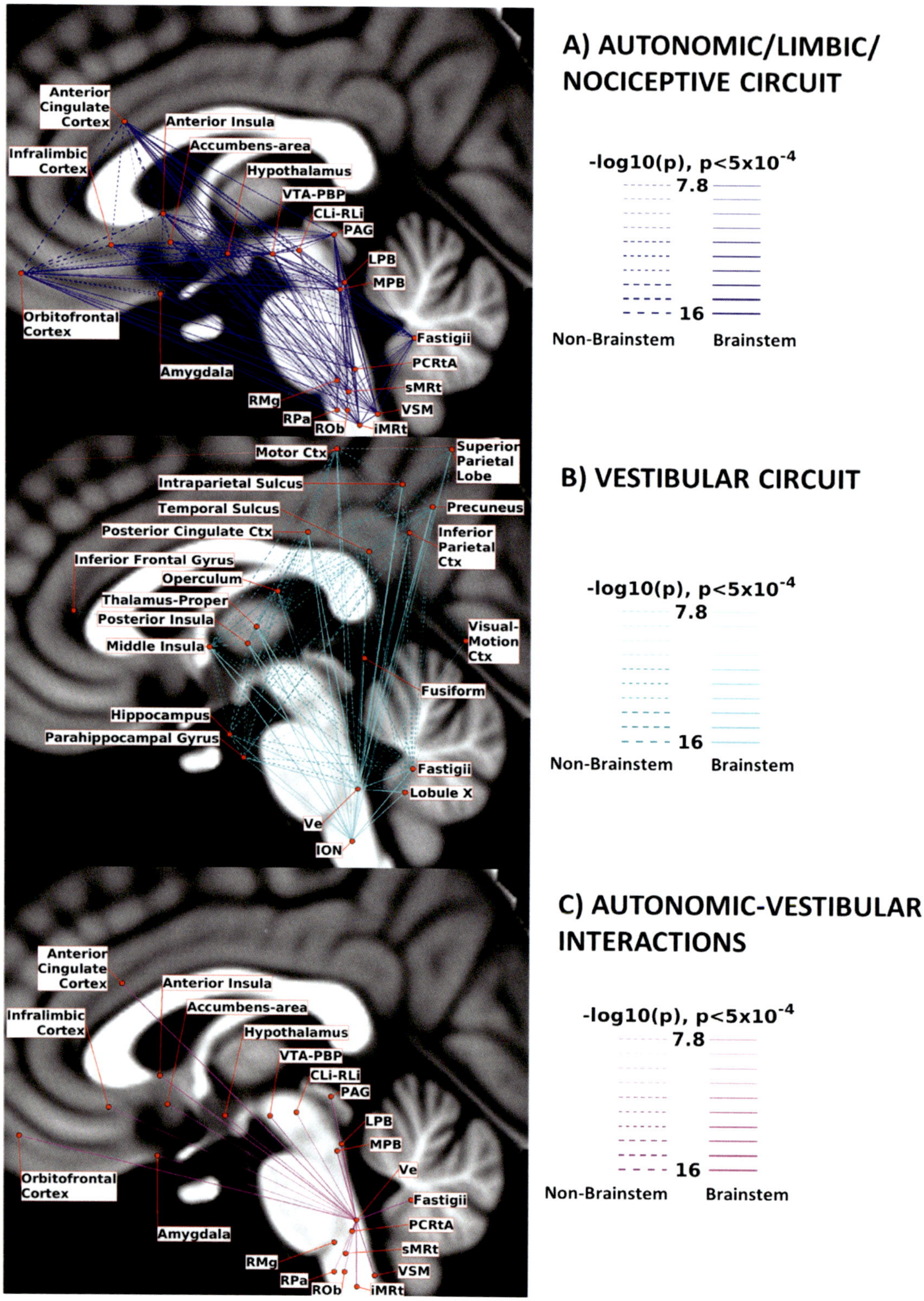

FIGURE 22.13 **Circuit diagrams: we summarize here the links among nodes of interest for (A) the autonomic/limbic/nociceptive circuit, (B) the vestibular circuit, and (C) the interactions between Ve and autonomic nodes.** Brainstem-to-brainstem connections are displayed with solid lines, while other connections are displayed with dashed lines. Line thickness is proportional to its −log10(p) value, thresholded at $P < .0005$, Bonferroni corrected. The nodes positions (region centroids) were projected on the sagittal slice at coordinate x = −2 of the T 1-weighted MNI template, with small adjustments in their positions to avoid node/link crowding. *Image courtesy of Singh, K., Cauzzo, S., García-Gomar, M.G., et al., 2022. Functional connectome of arousal and motor brainstem nuclei in living humans by 7 Tesla resting-state fMRI. Neuroimage 249, 118865. https://doi.org/10.1016/j.neuroimage.2022.118925*

lesser extent KOR are also located in the ACC, which has one of the highest levels of opiate ligand binding in the CNS (Vogt et al., 2001, 2019; Zubieta et al., 2001).

PET studies in humans have shown increased endogenous opioid release in the ACC during acute experimental pain in healthy subjects (Scott et al., 2007), as well as in patients with fibromyalgia (Harris et al., 2007), migraine (DaSilva et al., 2014), and peripheral or central neuropathic pain (Maarrawi et al., 2007). According to preclinical research, endogenous opioids need to be released in the rodent ACC in order to reduce the unpleasantness of continuing pain (Navratilova et al., 2015a).

Although MOR, DOR, and KOR are all expressed at different levels in the amygdala, they could each have a different purpose. Analgesia, reward, and the control of fear, anxiety, and associated behaviors are all correlated with MOR signaling (Bagley and Ingram, 2020; Bodnar, 2022; Tershner and Helmstetter, 2000; Wilson and Junor, 2008; Zhang et al., 2013). Anxiolysis is linked to DOR (Klenowski et al., 2015a,b). Amygdala's KOR signaling is responsi for inducing aversive and anxiogenic behavioral effects (Vialou et al., 2014; Limoges et al., 2022). Negative affective behaviors, like anxiety and depression, are also produced when KOR is activated in NAc. The mesolimbic KOR system has been linked to nociception, stress, and reward. The endogenous ligand dynorphin or pharmaceutical drugs that activate KOR in NAc decrease the motivating impact of rewards and cause aversion and depression-like behaviors, which are caused by suppression of dopamine release.

Limbic System and Olfactory Cortex

Odorant receptors (ORs) located on olfactory sensory neurons in the nose are responsible for detecting odorants (Zou et al., 2001). These signals are transmitted through the olfactory bulb which lies in the olfactory groove within the anterior cranial fossa.

The olfactory bulb is an ovoid structure which contains specialized neurons called mitral cells. The olfactory nerve fibers synapse with the mitral cells, forming collections known as synaptic glomeruli. From the glomeruli, second-order nerves then pass posteriorly into the olfactory tract.

The latter travels posteriorly on the inferior surface of the frontal lobe. As the tract reaches the anterior perforated substance (an area at the level of the optic chiasm) it divides into medial and lateral stria:

- **Lateral stria**—carries the axons to the primary olfactory cortex (located within the uncus of temporal lobe);
- **Medial stria**—carries the axons across the medial plane of the anterior commissure, where they meet the olfactory bulb of the opposite side.

The primary olfactory cortex sends nerve fibers to many other areas of the brain, constituting the rhinencephalon that is rudimental in the human and it includes some of the structure above mentioned more the anterior olfactory nucleus, anterior perforated substance some part of the amygdala, olfactory tubercle, and the secondary olfactory cortex. These areas are involved in the memory and appreciation of olfactory sensations.

The olfactory cortex reaches higher cortical regions as well as limbic regions that are believed to be involved in the emotional and physiological effects of odors (Zou et al., 2001). The olfactory cortex is located in the inferior surface of the temporal lobe and is part of the rhinencephalon (Fig. 22.14).

The olfactory cortex can be divided into multiple subregions based on anatomical characteristics, the largest of which is the piriform cortex (PCx, also known as the pyriform or prepyriform cortex).

The pyriform cortex is characterized by a three-layered structure:

- the outer molecular layer or layer 1;
- the superficial pyramidal layer or layer 2 displaying a dense population of cell bodies and, deep to this layer;
- the deep pyramidal cell layer, or layer 3, displaying a relatively low density of large pyramids (Fig. 22.15).

It receives direct input from the olfactory bulb and is involved in the initial processing of olfactory information.

Adjacent to the primary olfactory cortex, we find the secondary olfactory areas:

- periamygdaloid cortex;
- entorhinal cortex;
- anterior olfactory nucleus.

These areas are involved in further processing and integration of olfactory data. They receive input from the primary olfactory cortex and send projections to other brain regions, such as the amygdala and hippocampus.

Anatomical investigations have revealed that the olfactory cortex does not replicate the map found in the olfactory bulb (Scott, 1986; Shipley et al., 1996). While there may be little topographical organization of sensory data in the olfactory brain, other studies suggest that nearby neurons in the bulb may form synapses at the same locations, indicating that cortical inputs may have an ordered arrangement that has yet to be fully understood (Ojima et al., 1984; Buonviso et al., 1991).

There are multiple regions within the olfactory cortex that can receive input from a single OR, enabling the transmission of this input to different regions responsible for various activities. While the frontal cortex and lateral hypothalamus receive signals from most olfactory cortical areas (Scott, 1986; Shipley et al., 1996), the anterior and posterior piriform cortex send signals to

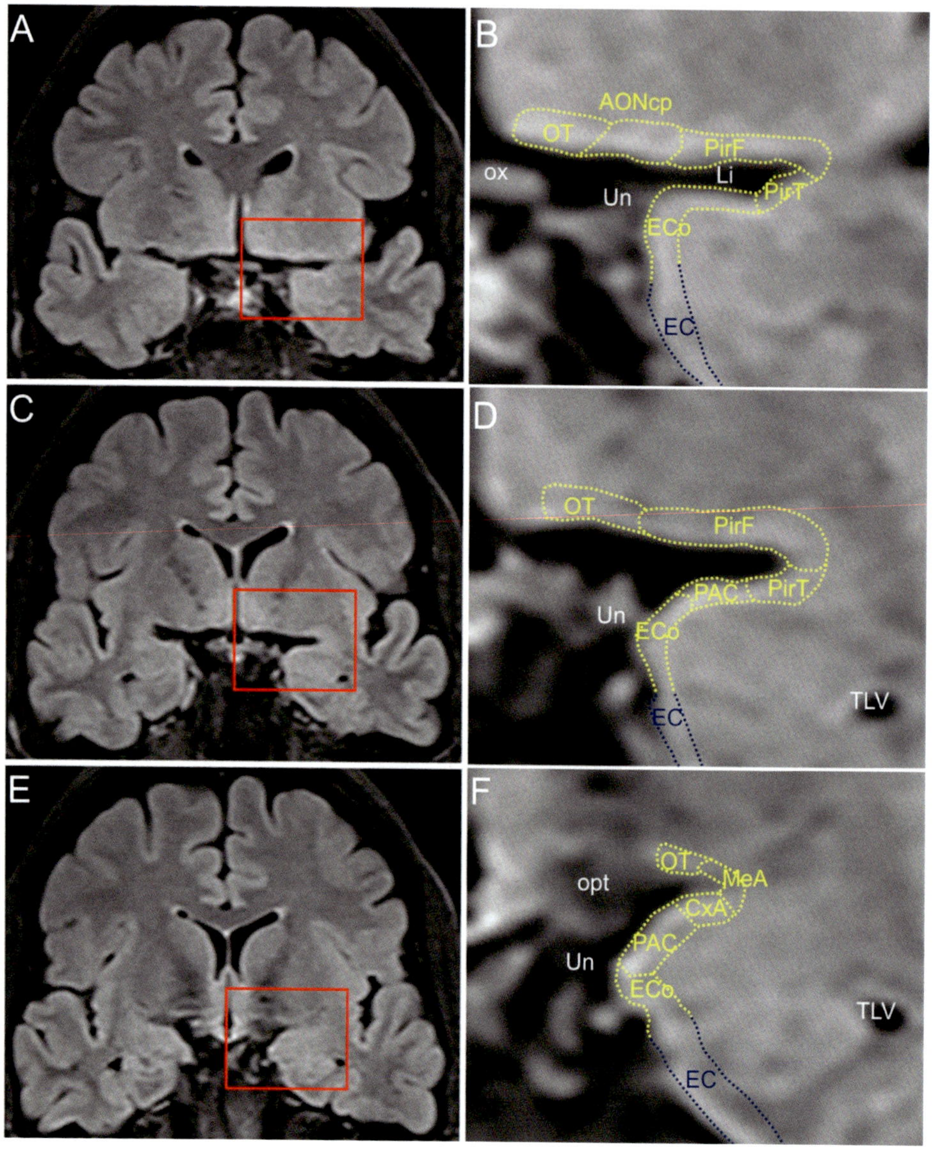

FIGURE 22.14 **Olfactory cortex.** Magnetic resonance images of the human brain illustrating the localization of caudal olfactory structures. Images correspond to a coronal FLAIR sequence in the coronal plane. Calibration bar: 10,000 μm for (A, C, E) and 2600 μm for (B, D, F).*Abbreviations*: AONcp, Anterio olfactory nucleus, cortical posterior; CxA, Cortex–amygdala transition zone; EC, Entorhinal cortex; ECo, Entorhinal cortex, olfactory; Li, Limen insulae; MeA, Medial amygdala; opt, Optic tract; ot, Olfactory tract; OT, Olfactory tubercle; ox, Optic chiasm; PAC, Periamygdaloid cortex; Pir, Piriform cortex; PirF, Piriform cortex, frontal; PirT, Piriform cortex, temporal; TLV, Temporal horn of the lateral ventricle; Un, Uncus. *Image courtesy of Ubeda-Bañon, I., Saiz-Sanchez, D., Flores-Cuadrado, A., et al., 2020. The human olfactory system in two proteinopathies: Alzheimer's and Parkinson's diseases. Transl. Neurodegener. 9, 22. https://doi.org/10.1186/s40035-020-00200-7*

different frontal cortex regions (Datiche et al., 1996), and the olfactory tubercle sends signals to the nuclei gemini of the hypothalamus (Price et al., 1991). This diverse input to various parts of the olfactory cortex allows for simultaneous and differentiated processing of OR inputs, which can then be combined and manipulated in different ways for transmission to the limbic and neocortical systems.

Considering the role of the amygdala in emotional states and its projection to hypothalamic regions not receiving input from other olfactory cortical areas, it is possible that this region specifically receives signals from ORs relevant to specific emotional states, instinctive behaviors, or physiological states controlled by the hypothalamus (McDonald et al., 1998; Kandel et al., 2000; Zigmond et al., 1999; Simerly, 1995).

Olfaction has a role in the control of ingestive behavior, environmental orientation, and interpersonal communication (Stevenson, 2010). Body odors influence the dynamics of close interpersonal relationships by providing information on an individual's emotional state, disease (de Groot et al., 2015; de Groot and Smeets,

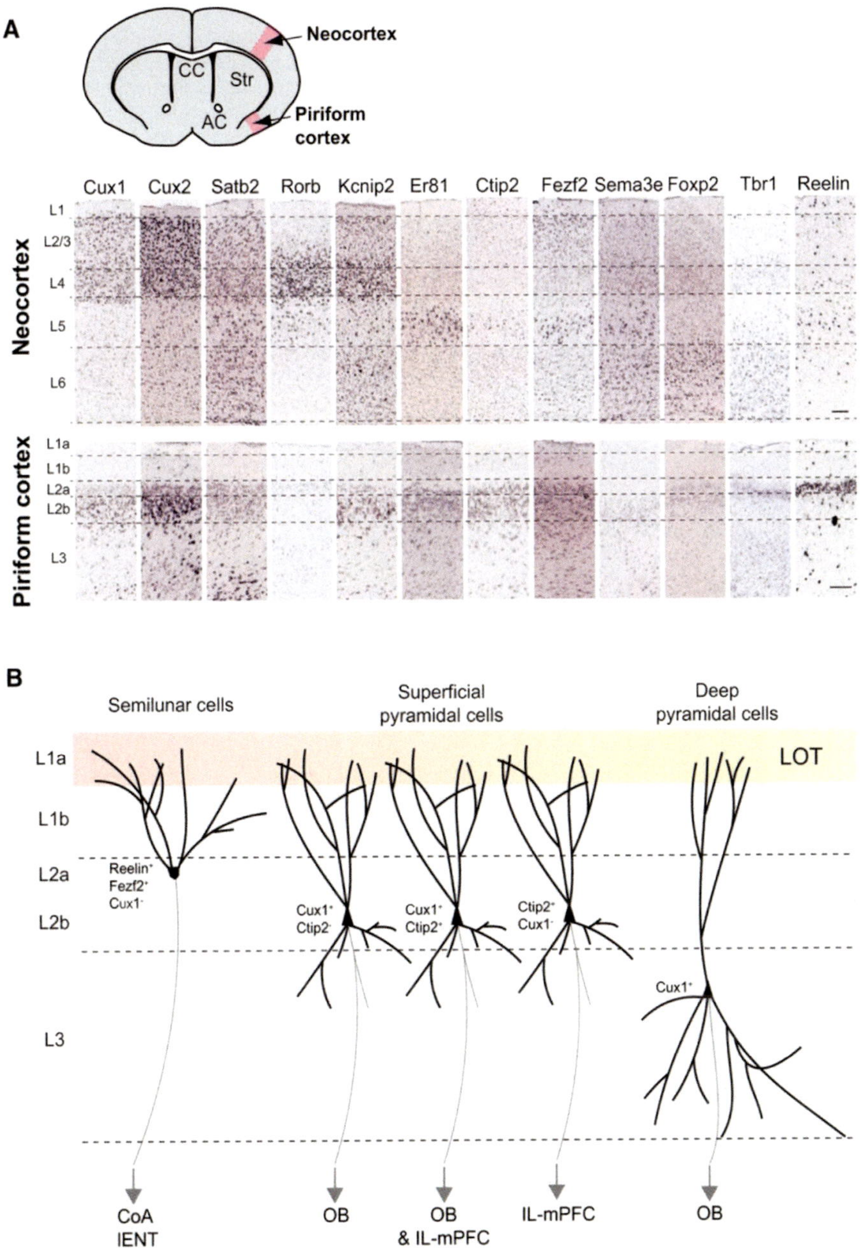

FIGURE 22.15 **Piriform cortex neuron identities.** (A), expression of principal neocortical layer markers in the piriform cortex. In situ hybridizations from Allen Brain Atlas database (postnatal day 56). AC, anterior commissure; CC, corpus callosum; L, layer; Str, striatum. Scale bar, 200 mm. (B) organization, molecular identities and known targets of projection neurons in the piriform cortex. CoA, posteromedial cortical amygdala nucleus; IL-mP-FC, infralimbic subdivision of the medial prefrontal cortex; lENT, lateral entorhinal cortex; OB, olfactory bulb; LOT, lateral olfactory tract. *Image courtesy of Klingler, E., 27 January 2017. Development and organization of the evolutionarily conserved three-layered olfactory cortex. eNeuro 4 (1), ENEURO.0193-16.2016. doi:10.1523/ENEURO.0193-16.2016.*

2017), psychological traits (Sorokowska et al., 2012; Tognetti et al., 2022), and physical characteristics (Zucco et al., 2014).

The limbic system is intricately connected to the olfactory cortex (Rolls et al., 1994; Wilson and Stevenson, 2006). The basolateral amygdala (BLA), in particular, gets strong inputs from the olfactory brain. Through this link, olfactory information can be integrated with emotional and fear-related reactions. The entorhinal cortex, in particular, transmits projections to the hippocampus. The development of olfactory memory and spatial navigation depend on this link. The OFC, which is involved in decision-making, emotion control, and reward processing, has reciprocal connections with the olfactory cortex. Through this link, olfactory information may be integrated with cognitive and emotional processes. Additionally, the olfactory cortex interacts with the hypothalamus, which controls a variety of physiological and behavioral responses. The olfactory cortex has direct connections with the limbic system, particularly the hippocampus and amygdala. These connections allow for the integration of olfactory information with emotional and memory-related processes. As a result, smells can evoke strong emotional responses and trigger vivid memories. The amygdala, in particular, plays a crucial role in associating odors with emotional experiences, influencing emotional responses and behaviors. The limbic system can modulate the processing of olfactory information, influencing how odors are perceived and interpreted. Conversely, olfactory inputs can activate and influence the functioning of the limbic system, affecting emotional and behavioral responses. These connections contribute to the powerful impact of smells on our emotions, memories, and overall subjective experiences.

References

Aggleton, J.P., Mishkin, M., 1982. A comparison of amygdaloid and hippocampal projections to the thalamus in monkeys. SOC. Neurosci. Abst 8 (240.6).

Aggleton, J.P., Desimone, R., Mishkin, M., 1986. The origin, course, and termination of the hippocampothalamic projections in the macaque. J. Comparat. Neurol. 243 (3), 409–421. https://doi.org/10.1002/cne.902430310.

Aggleton, J.P., O'Mara, S.M., Vann, S.D., Wright, N.F., Tsanov, M., Erichsen, J.T., 2010. Hippocampal-anterior thalamic pathways for memory: uncovering a network of direct and indirect actions. Eur. J. Neurosci. 31 (12), 2292–2307. https://doi.org/10.1111/j.1460-9568.2010.07251.x.

Allman, J.M., Watson, K.K., Tetreault, N.A., Hakeem, A.Y., 2005. Intuition and autism: a possible role for Von Economo neurons. Trends Cognit. Sci. 9 (8), 367–373. https://doi.org/10.1016/j.tics.2005.06.008.

Bagley, E.E., Ingram, S.L., 2020. Endogenous opioid peptides in the descending pain modulatory circuit. Neuropharmacology 173, 108131. https://doi.org/10.1016/j.neuropharm.2020.108131.

Barbas, H., Saha, S., Rempel-Clower, N., Ghashghaei, T., 2003. Serial pathways from primate prefrontal cortex to autonomic areas may influence emotional expression. BMC Neurosci. 4 (1), 25. https://doi.org/10.1186/1471-2202-4-25.

Bodnar, R.J., 2022. Endogenous opiates and behavior: 2020. Peptides 151, 170752. https://doi.org/10.1016/j.peptides.2022.170752.

Bourgeais, L., Gauriau, C., Bernard, J.F., 2001a. Projections from the nociceptive area of the central nucleus of the amygdala to the forebrain: a PHA-L study in the rat. Eur. J. Neurosci. 14 (2), 229–255. https://doi.org/10.1046/j.0953-816x.2001.01640.x.

Bourgeais, L., Monconduit, L., Villanueva, L., Bernard, J.F., 2001b. Parabrachial internal lateral neurons convey nociceptive messages from the deep laminas of the dorsal horn to the intralaminar thalamus. J. Neurosci. 21 (6), 2159–2165. https://doi.org/10.1523/JNEUROSCI.21-06-02159.2001.

Bryant, K.L., Li, L., Eichert, N., Mars, R.B., 2020. A comprehensive atlas of white matter tracts in the chimpanzee. PLoS Biol. 18 (12), e3000971. https://doi.org/10.1371/journal.pbio.3000971.

Buonviso, N., Revial, M.F., Jourdan, F., 1991. The projections of mitral cells from small local regions of the olfactory bulb: an anterograde tracing study using PHA-L (*Phaseolus vulgaris Leucoagglutinin*). Eur. J. Neurosci. 3 (6), 493–500. https://doi.org/10.1111/j.1460-9568.1991.tb00836.x.

Cahill, C.M., Holdridge, S.V., Liu, S., et al., 2022. Delta opioid receptor activation modulates affective pain and modality-specific pain hypersensitivity associated with chronic neuropathic pain. J. Neurosci. Res. 100 (1), 129–148. https://doi.org/10.1002/jnr.24680.

Catani, M., Dell'Acqua, F., Thiebaut de Schotten, M., 2013. A revised limbic system model for memory, emotion and behaviour. Neurosci. Biobehav. Rev. 37 (8), 1724–1737. https://doi.org/10.1016/j.neubiorev.2013.07.001.

Chen, S., Su, H.S., 1990. Afferent connections of the thalamic paraventricular and parataenial nuclei in the rat — a retrograde tracing study with iontophoretic application of Fluoro-Gold. Brain Res. 522 (1), 1–6. https://doi.org/10.1016/0006-8993(90)91570-7.

Cliffer, K., Burstein, R., Giesler, G., 1991. Distributions of spinothalamic, spinohypothalamic, and spinotelencephalic fibers revealed by anterograde transport of PHA-L in rats. J. Neurosci. 11 (3), 852–868. https://doi.org/10.1523/JNEUROSCI.11-03-00852.1991.

Conrad, L.C.A., Pfaff, D.W., 2004. Efferents from medial basal forebrain and hypothalamus in the rat. I. An autoradiographic study of the medial preoptic area. J. Comp. Neurol. 169 (2), 185–219. https://doi.org/10.1002/cne.901690205.

Cornwall, J., Phillipson, O.T., 1988. Afferent projections to the dorsal thalamus of the rat as shown by retrograde lectin transport. II. The midline nuclei. Brain Res. Bull. 21 (2), 147–161. https://doi.org/10.1016/0361-9230(88)90227-4.

Crosby, E.C., 1963. Correlative anatomy of the nervous system. J. Med. Educ. 38 (6), 526.

DaSilva, A.F., Nascimento, T.D., DosSantos, M.F., Zubieta, J.K., 2014. Migraine and the mu-opioidergic system—can we directly modulate it? Evidence from neuroimaging studies. Curr. Pain Headache Rep. 18 (7), 429. https://doi.org/10.1007/s11916-014-0429-0.

Datiche, F., Cattarelli, M., 1996. Reciprocal and topographic connections between the piriform and prefrontal cortices in the rat: a tracing study using the B subunit of the cholera toxin. Brain Res. Bull. 41 (6), 391–398. https://doi.org/10.1016/S0361-9230(96)00082-2.

Di, X., Huang, J., Biswal, B.B., 2017. Task modulated brain connectivity of the amygdala: a meta-analysis of psychophysiological interactions. Brain Struct. Funct. 222 (1), 619–634. https://doi.org/10.1007/s00429-016-1239-4.

Dosenbach, N.U.F., Visscher, K.M., Palmer, E.D., et al., 2006. A core system for the implementation of task sets. Neuron 50 (5), 799–812. https://doi.org/10.1016/j.neuron.2006.04.031.

Drobisz, D., Damborská, A., 2019. Deep brain stimulation targets for treating depression. Behav. Brain Res. 359, 266–273. https://doi.org/10.1016/j.bbr.2018.11.004.

Dulawa, S.C., Janowsky, D.S., 2019. Cholinergic regulation of mood: from basic and clinical studies to emerging therapeutics. Mol. Psychiatr. 24 (5), 694–709. https://doi.org/10.1038/s41380-018-0219-x.

Erro, E., Lanciego, J.L., Giménez-Amaya, J.M., 1999. Relationships between thalamostriatal neurons and pedunculopontine projections to the thalamus: a neuroanatomical tract-tracing study in the rat. Exp. Brain Res. 127 (2), 162–170. https://doi.org/10.1007/s002210050786.

Fu, J.Y., Yu, X.D., Zhu, Y., et al., 2020. Whole-brain map of long-range monosynaptic inputs to different cell types in the amygdala of the mouse. Neurosci. Bull. 36 (11), 1381–1394. https://doi.org/10.1007/s12264-020-00545-z.

Gabbott, P.L.A., Warner, T.A., Jays, P.R.L., Bacon, S.J., 2003. Areal and synaptic interconnectivity of prelimbic (area 32), infralimbic (area 25) and insular cortices in the rat. Brain Res. 993 (1–2), 59–71. https://doi.org/10.1016/j.brainres.2003.08.056.

Gabbott, P.L.A., Warner, T.A., Jays, P.R.L., Salway, P., Busby, S.J., 2005. Prefrontal cortex in the rat: projections to subcortical autonomic, motor, and limbic centers. J. Comp. Neurol. 492 (2), 145–177. https://doi.org/10.1002/cne.20738.

Gavériaux-Ruff, C., Kieffer, B.L., 2011. Delta opioid receptor analgesia. Behav. Pharmacol. 22 (5 and 6), 405–414. https://doi.org/10.1097/FBP.0b013e32834a1f2c.

Gläscher, J., Adolphs, R., Damasio, H., et al., 2012. Lesion mapping of cognitive control and value-based decision making in the prefrontal cortex. Proc. Natl. Acad. Sci. USA 109 (36), 14681–14686. https://doi.org/10.1073/pnas.1206608109.

Groenewegen, H.J., 1988. Organization of the afferent connections of the mediodorsal thalamic nucleus in the rat, related to the mediodorsal-prefrontal topography. Neuroscience 24 (2), 379–431. https://doi.org/10.1016/0306-4522(88)90339-9.

de Groot, J.H., Smeets, M.A., Rowson, M.J., Bulsing, P.J., Blonk, C.G., Wilkinson, J.E., Semin, G.R., 2015. A sniff of happiness. Psychol. Sci. 26 (6), 684–700. https://doi.org/10.1177/0956797614566318.

de Groot, J.H.B., Smeets, M.A.M., 2017. Human fear chemosignaling: evidence from a meta-analysis. Chem. Sens. 42 (8), 663–673. https://doi.org/10.1093/chemse/bjx049.

Haber, S., Kunishio, K., Mizobuchi, M., Lynd-Balta, E., 1995. The orbital and medial prefrontal circuit through the primate basal ganglia. J. Neurosci. 15 (7), 4851–4867. https://doi.org/10.1523/JNEUROSCI.15-07-04851.1995.

Haines, D.E., 2007. Neuroanatomy: An Atlas of Structures, Sections, and Systems, sixth ed.vol. 153. Lippincott Williams & Wilkins.

Harris, R.E., Clauw, D.J., Scott, D.J., McLean, S.A., Gracely, R.H., Zubieta, J.K., 2007. Decreased central μ-opioid receptor availability in fibromyalgia. J. Neurosci. 27 (37), 10000–10006. https://doi.org/10.1523/JNEUROSCI.2849-07.2007.

Herkenham, M., 1978. The connections of the nucleus reuniens thalami: evidence for a direct thalamo-hippocampal pathway in the rat. J. Comp. Neurol. 177 (4), 589–609. https://doi.org/10.1002/cne.901770405.

Hoover, W.B., Vertes, R.P., 2007. Anatomical analysis of afferent projections to the medial prefrontal cortex in the rat. Brain Struct. Funct. 212 (2), 149–179. https://doi.org/10.1007/s00429-007-0150-4.

Jang, S.H., Hong, J.H., 2012. The anatomical characteristics of superior longitudinal fasciculus I in human brain: diffusion tensor tractography study. Neurosci. Lett. 506 (1), 146–148. https://doi.org/10.1016/j.neulet.2011.10.069.

Ji, G., Neugebauer, V., 2011. Pain-related deactivation of medial prefrontal cortical neurons involves mGluR1 and GABA A receptors. J. Neurophysiol. 106 (5), 2642–2652. https://doi.org/10.1152/jn.00461.2011.

Jones, A.F., Sheets, P.L., 2020. Sex-specific disruption of distinct mPFC inhibitory neurons in spared-nerve injury model of neuropathic pain. Cell Rep. 31 (10), 107729. https://doi.org/10.1016/j.celrep.2020.107729.

Ji, G., Sun, H., Fu, Y., Li, Z., Pais-Vieira, M., Galhardo, V., Neugebauer, V., 2010. Cognitive impairment in pain through amygdala-driven prefrontal cortical deactivation. J. Neurosci. 30 (15), 5451–5464. https://doi.org/10.1523/JNEUROSCI.0225-10.2010.

Jones, E.G., 2009. Synchrony in the interconnected circuitry of the thalamus and cerebral cortex. Ann. N. Y. Acad. Sci. 1157 (1), 10–23. https://doi.org/10.1111/j.1749-6632.2009.04534.x.

Kalén, P., Karlson, M., Wiklund, L., 1985. Possible excitatory amino acid afferents to nucleus raphe dorsalis of the rat investigated with retrograde wheat germ agglutinin and D-[3H]aspartate tracing. Brain Res. 360 (1–2), 285–297. https://doi.org/10.1016/0006-8993(85)91244-2.

Kamali, A., Ghazi Sherbaf, F., Rahmani, F., et al., 2020a. A direct visuosensory cortical connectivity of the human limbic system. Dissecting the trajectory of the parieto-occipito-hypothalamic tract in the human brain using diffusion weighted tractography. Neurosci. Lett. 728, 134955. https://doi.org/10.1016/j.neulet.2020.134955.

Kamali, A., Hasan, K.M., Adapa, P., et al., 2014. Distinguishing and quantification of the human visual pathways using high-spatial-resolution diffusion tensor tractography. Magn. Reson. Imaging 32 (7), 796–803. https://doi.org/10.1016/j.mri.2014.04.002.

Kamali, A., Karbasian, N., Ghazi Sherbaf, F., et al., 2020b. Uncovering the dorsal thalamo-hypothalamic tract of the human limbic system. Neuroscience 432, 55–62. https://doi.org/10.1016/j.neuroscience.2020.02.021.

Kamali, A., Kramer, L.A., Hasan, K.M., 2010. Feasibility of prefronto-caudate pathway tractography using high resolution diffusion tensor tractography data at 3 T. J. Neurosci. Methods 191 (2), 249–254. https://doi.org/10.1016/j.jneumeth.2010.06.026.

Kamali, A., Riascos, R.F., Pillai, J.J., et al., 2018a. Mapping the trajectory of the amygdalothalamic tract in the human brain. J. Neurosci. Res. 96 (7), 1176–1185. https://doi.org/10.1002/jnr.24235.

Kamali, A., Sair, H.I., Blitz, A.M., et al., 2016. Revealing the ventral amygdalofugal pathway of the human limbic system using high spatial resolution diffusion tensor tractography. Brain Struct. Funct. 221 (7), 3561–3569. https://doi.org/10.1007/s00429-015-1119-3.

Kamali, A., Yousem, D.M., Lin, D.D., et al., 2015. Mapping the trajectory of the stria terminalis of the human limbic system using high spatial resolution diffusion tensor tractography. Neurosci. Lett. 608, 45–50. https://doi.org/10.1016/j.neulet.2015.09.035.

Kamali, A., Zhang, C.C., Riascos, R.F., et al., 2018b. Diffusion tensor tractography of the mammillothalamic tract in the human brain using a high spatial resolution DTI technique. Sci. Rep. 8 (1), 5229. https://doi.org/10.1038/s41598-018-23452-w.

Kamali, A., Milosavljevic, S., Gandhi, A., et al., 2023. The cortico-limbo-thalamo-cortical circuits: an update to the original Papez circuit of the human limbic system. Brain Topogr. 36 (3), 371–389. https://doi.org/10.1007/s10548-023-00955-y.

Kandel, E., Schwartz, J., Jessell, T., 2000. Principles of Neural Sciences. McGraw-Hill.

Kiernan, J.A., 2012. Anatomy of the temporal lobe. Epilepsy Res. Treat. 2012, 1–12. https://doi.org/10.1155/2012/176157.

Kiritoshi, T., Ji, G., Neugebauer, V., 2016. Rescue of impaired mGluR5-driven endocannabinoid signaling restores prefrontal cortical output to inhibit pain in arthritic rats. J. Neurosci. 36 (3), 837–850. https://doi.org/10.1523/JNEUROSCI.4047-15.2016.

Klenowski, P., Morgan, M., Bartlett, S.E., 2015a. The role of δ-opioid receptors in learning and memory underlying the development of addiction. Br. J. Pharmacol. 172 (2), 297–310. https://doi.org/10.1111/bph.12618.

Kiritoshi, T., Sun, H., Ren, W., Stauffer, S.R., Lindsley, C.W., Conn, P.J., Neugebauer, V., 2013. Modulation of pyramidal cell output in the medial prefrontal cortex by mGluR5 interacting with CB1. Neuropharmacology 66, 170–178. https://doi.org/10.1016/j.neuropharm.2012.03.024.

Klenowski, P.M., Fogarty, M.J., Belmer, A., Noakes, P.G., Bellingham, M.C., Bartlett, S.E., 2015b. Structural and functional characterization of dendritic arbors and GABAergic synaptic inputs on interneurons and principal cells in the rat basolateral amygdala. J. Neurophysiol. 114 (2), 942–957. https://doi.org/10.1152/jn.00824.2014.

Klingler, J., Gloor, P., 1960. The connections of the amygdala and of the anterior temporal cortex in the human brain. J. Comp. Neurol. 115 (3), 333–369. https://doi.org/10.1002/cne.901150305.

Koenigs, M., Grafman, J., 2009. The functional neuroanatomy of depression: distinct roles for ventromedial and dorsolateral prefrontal cortex. Behav. Brain Res. 201 (2), 239–243. https://doi.org/10.1016/j.bbr.2009.03.004.

Kolling, N., Wittmann, M.K., Behrens, T.E.J., Boorman, E.D., Mars, R.B., Rushworth, M.F.S., 2016. Value, search, persistence and model updating in anterior cingulate cortex. Nat. Neurosci. 19 (10), 1280–1285. https://doi.org/10.1038/nn.4382.

Kroger, J.K., 2002. Recruitment of anterior dorsolateral prefrontal cortex in human reasoning: a parametric study of relational complexity. Cerebr. Cortex 12 (5), 477–485. https://doi.org/10.1093/cercor/12.5.477.

Krout, K.E., Loewy, A.D., 2000. Parabrachial nucleus projections to midline and intralaminar thalamic nuclei of the rat. J. Comp. Neurol. 428 (3), 475–494. https://doi.org/10.1002/1096-9861(20001218)428:3<475::AID-CNE6>3.0.CO;2-9.

Lawes, I.N.C., Barrick, T.R., Murugam, V., et al., 2008. Atlas-based segmentation of white matter tracts of the human brain using diffusion tensor tractography and comparison with classical dissection. Neuroimage 39 (1), 62–79. https://doi.org/10.1016/j.neuroimage.2007.06.041.

Lemaire, J.J., Nezzar, H., Sakka, L., et al., 2013. Maps of the adult human hypothalamus. Surg. Neurol. Int. 4 (4), 156. https://doi.org/10.4103/2152-7806.110667.

Limoges, A., Yarur, H.E., Tejeda, H.A., 2022. Dynorphin/kappa opioid receptor system regulation on amygdaloid circuitry: implications for neuropsychiatric disorders. Front. Syst. Neurosci. 16. https://doi.org/10.3389/fnsys.2022.963691.

Maarrawi, J., Peyron, R., Mertens, P., et al., 2007a. Differential brain opioid receptor availability in central and peripheral neuropathic pain. Pain 127 (1), 183–194. https://doi.org/10.1016/j.pain.2006.10.013.

Maarrawi, J., Peyron, R., Mertens, P., et al., 2007b. Motor cortex stimulation for pain control induces changes in the endogenous opioid system. Neurology 69 (9), 827–834. https://doi.org/10.1212/01.wnl.0000269783.86997.37.

Maclean, P.D., 1949. Psychosomatic disease and the "visceral brain". Psychosom. Med. 11 (6), 338–353. https://doi.org/10.1097/00006842-194911000-00003.

MacLean, P.D., 1952. Some psychiatric implications of physiological studies on frontotemporal portion of limbic system (Visceral brain). Electroencephalogr. Clin. Neurophysiol. 4 (4), 407–418. https://doi.org/10.1016/0013-4694(52)90073-4.

Mazzitelli, M., Marshall, K., Pham, A., Ji, G., Neugebauer, V., 2021. Optogenetic manipulations of amygdala neurons modulate spinal nociceptive processing and behavior under normal conditions and in an arthritis pain model. Front. Pharmacol. 12. https://doi.org/10.3389/fphar.2021.668337.

Mazzitelli, M., Yakhnitsa, V., Neugebauer, B., Neugebauer, V., 2022. Optogenetic manipulations of CeA-CRF neurons modulate pain- and anxiety-like behaviors in neuropathic pain and control rats. Neuropharmacology 210, 109031. https://doi.org/10.1016/j.neuropharm.2022.1090318.

McDonald, A.J., 1998. Cortical pathways to the mammalian amygdala. Prog. Neurobiol. 55 (3), 257–332. https://doi.org/10.1016/S0301-0082(98)00003-3.

Mineur, Y.S., Cahuzac, E.L., Mose, T.N., et al., 2018. Interaction between noradrenergic and cholinergic signaling in amygdala regulates anxiety- and depression-related behaviors in mice. Neuropsychopharmacology 43 (10), 2118–2125. https://doi.org/10.1038/s41386-018-0024-x.

Mori, S., Kageyama, Y., Hou, Z., et al., 2017. Elucidation of white matter tracts of the human amygdala by detailed comparison between high-resolution postmortem magnetic resonance imaging and histology. Front. Neuroanat. 11. https://doi.org/10.3389/fnana.2017.00016.

Nadal, X., Baños, J.E., Kieffer, B.L., Maldonado, R., 2006. Neuropathic pain is enhanced in δ-opioid receptor knockout mice. Eur. J. Neurosci. 23 (3), 830–834. https://doi.org/10.1111/j.1460-9568.2006.04569.x.

Navratilova, E., Atcherley, C.W., Porreca, F., 2015a. Brain circuits encoding reward from pain relief. Trends Neurosci. 38 (11), 741–750. https://doi.org/10.1016/j.tins.2015.09.003.

Neubert, F.X., Mars, R.B., Sallet, J., Rushworth, M.F.S., 2015. Connectivity reveals relationship of brain areas for reward-guided learning and decision making in human and monkey frontal cortex. Proc. Natl. Acad. Sci. USA 112 (20). https://doi.org/10.1073/pnas.1410767112.

Neugebauer, V., Mazzitelli, M., Cragg, B., Ji, G., Navratilova, E., Porreca, F., 2020. Amygdala, neuropeptides, and chronic pain-related affective behaviors. Neuropharmacology 170, 108052. https://doi.org/10.1016/j.neuropharm.2020.108052.

Nieuwenhuys, R., Voogd, J., van Huijzen, C., 1988. Vessels and meninges. In: The Human Central Nervous System. Springer Berlin Heidelberg, pp. 33–64. https://doi.org/10.1007/978-3-662-10343-2_3.

Noback, C.R., Strominger, N.L., Demarest, R.J., Ruggiero, D.A., 2005. He Human Nervous System: Structure and Function, sixth ed. Springer Science & Business Media.

Nolte, J., 1993. The Human Brain: An Introduction to its Functional Anatomy. Mosby.

Ojima, H., Mori, K., Kishi, K., 1984. The trajectory of mitral cell axons in the rabbit olfactory cortex revealed by intracellular HRP injection. J. Comp. Neurol. 230 (1), 77–87. https://doi.org/10.1002/cne.902300107.

Papez, J.W., 1937. A proposed mechanism of emotion. Arch. Neurol. Psychiatr. 38 (4), 725. https://doi.org/10.1001/archneurpsyc.1937.02260220069003.

Pascalau, R., Popa Stănilă, R., Sfrângeu, S., Szabo, B., 2018. Anatomy of the limbic white matter tracts as revealed by fiber dissection and tractography. World Neurosurg. 113, e672–e689. https://doi.org/10.1016/j.wneu.2018.02.121.

Pasquier, D.A., Reinoso-Suarez, F., 1976. Direct projections from hypothalamus to hippocampus in the rat demonstrated by retrograde transport of horseradish peroxidase. Brain Res. 108 (1), 165–169. https://doi.org/10.1016/0006-8993(76)90172-4.

Peterson, D.C., Reddy, V., Mayes, D.A., 2023. Neuroanatomy. Mammillary Bodies.

Price, J.L., Slotnick, B.M., Revial, M.F., 1991. Olfactory projections to the hypothalamus. J. Comp. Neurol. 306 (3), 447–461. https://doi.org/10.1002/cne.903060309.

Rapan, L., Froudist-Walsh, S., Niu, M., et al., 2021. Multimodal 3D atlas of the macaque monkey motor and premotor cortex. Neuroimage 226, 117574. https://doi.org/10.1016/j.neuroimage.2020.117574.

Robinson, T.E., Kolb, B., 1997. Persistent structural modifications in nucleus accumbens and prefrontal cortex neurons produced by previous experience with amphetamine. J. Neurosci. 17 (21), 8491–8497. https://doi.org/10.1523/JNEUROSCI.17-21-08491.1997.

Rolls, E., Baylis, L., 1994. Gustatory, olfactory, and visual convergence within the primate orbitofrontal cortex. J. Neurosci. 14 (9), 5437–5452. https://doi.org/10.1523/JNEUROSCI.14-09-05437.1994.

Rolls, E.T., 2019a. The cingulate cortex and limbic systems for emotion, action, and memory. Brain Struct. Funct. 224 (9), 3001–3018. https://doi.org/10.1007/s00429-019-01945-2.

Rolls, E.T., 2019b. The orbitofrontal cortex and emotion in health and disease, including depression. Neuropsychologia 128, 14–43. https://doi.org/10.1016/j.neuropsychologia.2017.09.021.

Rolls, E.T., Wirth, S., 2018. Spatial representations in the primate hippocampus, and their functions in memory and navigation. Prog. Neurobiol. 171, 90–113. https://doi.org/10.1016/j.pneurobio.2018.09.004.

Rotge, J.Y., 2010. Individual determination of the surgical target for deep brain stimulation in obsessive–compulsive disorder. Med. Hypotheses 75 (6), 684. https://doi.org/10.1016/j.mehy.2010.08.025.

Roxo, M.R., Franceschini, P.R., Zubaran, C., Kleber, F.D., Sander, J.W., 2011. The limbic system conception and its historical evolution. Sci. World J. 11, 2427–2440. https://doi.org/10.1100/2011/157150.

Sanides, F., 1964. The cyto-myeloarchitecture of the human frontal lobe and its relation to phylogenetic differentiation of the cerebral cortex. J. Hirnforsch. 7, 269–282.

Schmahmann, J.D., Pandya, D.N., Wang, R., et al., 2007. Association fibre pathways of the brain: parallel observations from diffusion spectrum imaging and autoradiography. Brain 130 (3), 630–653. https://doi.org/10.1093/brain/awl359.

Scott, D.J., Domino, E.F., Heitzeg, M.M., et al., 2007. Smoking modulation of μ-opioid and dopamine D2 receptor-mediated neurotransmission in humans. Neuropsychopharmacology 32 (2), 450–457. https://doi.org/10.1038/sj.npp.1301238.

Scott, J.W., 1986. The olfactory bulb and central pathways. Experientia 42 (3), 223–232. https://doi.org/10.1007/BF01942502.

Seeley, W.W., Menon, V., Schatzberg, A.F., et al., 2007. Dissociable intrinsic connectivity networks for salience processing and executive control. J. Neurosci. 27 (9), 2349–2356. https://doi.org/10.1523/JNEUROSCI.5587-06.2007.

Shapiro, B.E., Alexander, M.P., Gardner, H., Mercer, B., 1981. Mechanisms of confabulation. Neurology 31 (9), 1070. https://doi.org/10.1212/WNL.31.9.1070.

Shipley, M.T., Ennis, M., 1996. Functional organization of olfactory system. J. Neurobiol. 30 (1), 123–176. https://doi.org/10.1002/(SICI)1097-4695(199605)30:1<123::AID-NEU11>3.0.CO;2-N.

Simerly, R., 1995. The Rat Nervous System. Paxinos, G. Academic.

Singh, K., Cauzzo, S., García-Gomar, M.G., et al., 2022. Functional connectome of arousal and motor brainstem nuclei in living humans by 7 Tesla resting-state fMRI. Neuroimage 249, 118865. https://doi.org/10.1016/j.neuroimage.2021.118865.

Sorokowska, A., Sorokowski, P., Szmajke, A., 2012. Does personality smell? Accuracy of personality assessments based on body odour. Eur. J. Pers. 26 (5), 496–503. https://doi.org/10.1002/per.848.

Stevenson, R.J., 2010. An initial evaluation of the functions of human olfaction. Chem. Sens. 35 (1), 3–20. https://doi.org/10.1093/chemse/bjp083.

Swanson, L.W., 1976. An autoradiographic study of the efferent connections of the preoptic region in the rat. J. Comp. Neurol. 167 (2), 227–256. https://doi.org/10.1002/cne.901670207.

Tershner, S.A., Helmstetter, F.J., 2000. Antinociception produced by mu opioid receptor activation in the amygdala is partly dependent on activation of mu opioid and neurotensin receptors in the ventral periaqueductal gray. Brain Res. 865 (1), 17–26. https://doi.org/10.1016/S0006-8993(00)02179-X.

Timbie, C., García-Cabezas, M.Á., Zikopoulos, B., Barbas, H., 2020. Organization of primate amygdalar–thalamic pathways for emotions. PLoS Biol. 18 (2), e3000639. https://doi.org/10.1371/journal.pbio.3000639.

Tognetti, A., Durand, V., Dubois, D., Barkat-Defradas, M., Hopfensitz, A., Ferdenzi, C., 2022. The smell of cooperativeness: Do human body odours advertise cooperative behaviours? Brit. J. Psychol. (London, England: 1953) 113 (2), 531–546. https://doi.org/10.1111/bjop.12544.

Tucker, D., Luu, P., 2012. Cognition and Neural Development. Oxford University Press, USA.

Tucker, D.M., Luu, P., 2023. Adaptive control of functional connectivity: dorsal and ventral limbic divisions regulate the dorsal and ventral neocortical networks. Cerebr. Cortex. https://doi.org/10.1093/cercor/bhad085.

Tucker, D.M., Williamson, P.A., 1984. Asymmetric neural control systems in human self-regulation. Psychol. Rev. 91 (2), 185–215.

Usunoff, K.G., Schmitt, O., Itzev, D.E., et al., 2009. Efferent projections of the anterior and posterodorsal regions of the medial nucleus of the amygdala in the mouse. Cells Tissues Organs 190 (5), 256–285. https://doi.org/10.1159/000209233.

Vialou, V., Bagot, R.C., Cahill, M.E., Ferguson, D., Robison, A.J., Dietz, D.M., Fallon, B., Mazei-Robison, M., Ku, S.M., Harrigan, E., Winstanley, C.A., Joshi, T., Feng, J., Berton, O., Nestler, E.J., 2014. Prefrontal cortical circuit for depression- and anxiety-related behaviors mediated by cholecystokinin: role of ΔFosB. J. Neurosci. 34 (11), 3878–3887. https://doi.org/10.1523/JNEUROSCI.1787-13.2014.

Vogt, B.A., 2019. Cingulate Cortex in the Three Limbic Subsystems, pp. 39–51. https://doi.org/10.1016/B978-0-444-64196-0.00003-0.

Vogt, B.A., Laureys, S., 2009. The Primate Posterior Cingulate Gyrus: Connections, Sensorimotor Orientation, Gateway to Limbic Processing. Oxford University Press.

Vogt, L.J., Sim-Selley, L.J., Childers, S.R., Wiley, R.G., Vogt, B.A., 2001. Colocalization of mu-opioid receptors and activated G-proteins in rat cingulate cortex. J. Pharmacol. Exp. Therapeut. 299 (3), 840–848.

Wakana, S., Jiang, H., Nagae-Poetscher, L.M., van Zijl, P.C.M., Mori, S., 2004. Fiber tract–based atlas of human white matter anatomy. Radiology 230 (1), 77–87. https://doi.org/10.1148/radiol.2301021640.

Wang, X., Liu, P., Ma, L., Wang, F., 2019. Opioid signal transduction regulates the dendritic morphology of somatostatin and parvalbumin interneurons in the medial prefrontal cortex. Neuroreport 30 (8), 592–599. https://doi.org/10.1097/WNR.0000000000001254.

Wang, X., Pathak, S., Stefaneanu, L., Yeh, F.C., Li, S., Fernandez-Miranda, J.C., 2016. Subcomponents and connectivity of the superior longitudinal fasciculus in the human brain. Brain Struct. Funct. 221 (4), 2075–2092. https://doi.org/10.1007/s00429-015-1028-5.

Weller, K.L., Smith, D.A., 1982. Afferent connections to the bed nucleus of the stria terminalis. Brain Res. 232 (2), 255–270. https://doi.org/10.1016/0006-8993(82)90272-4.

Whalen, P.J., Rauch, S.L., Etcoff, N.L., McInerney, S.C., Lee, M.B., Jenike, M.A., 1998. Masked presentations of emotional facial expressions modulate amygdala activity without explicit knowledge. J. Neurosci. 18 (1), 411–418. https://doi.org/10.1523/JNEUROSCI.18-01-00411.1998.

Wilson, D.A., Stevenson, R.J., 2006. Learning to Smell: Olfactory Perception from Neurobiology to Behavior. Johns Hopkins University Press.

Wilson, M.A., Junor, L., 2008. The role of amygdalar mu-opioid receptors in anxiety-related responses in two rat models. Neuropsychopharmacology 33 (12), 2957–2968. https://doi.org/10.1038/sj.npp.1301675.

Yakovlev, P.I., 1948. Motility, behavior and the brain. J. Nerv. Ment. Dis. 107 (4), 313–335. https://doi.org/10.1097/00005053-194810740-00001.

Yamada, K., Shrier, D.A., Rubio, A., et al., 1998. MR imaging of the mamillothalamic tract. Radiology 207 (3), 593–598. https://doi.org/10.1148/radiology.207.3.9609878.

Yarkoni, T., Gray, J.R., Chrastil, E.R., Barch, D.M., Green, L., Braver, T.S., 2005. Sustained neural activity associated with cognitive control during temporally extended decision making. Cognit. Brain Res. 23 (1), 71–84. https://doi.org/10.1016/j.cogbrainres.2005.01.013.

Zhang, W., Schneider, D.M., Belova, M.A., Morrison, S.E., Paton, J.J., Salzman, C.D., 2013. Functional circuits and anatomical distribution of response properties in the primate amygdala. J. Neurosci. 33 (2), 722–733. https://doi.org/10.1523/JNEUROSCI.2970-12.2013.

Zigmond, M., Bloom, F., Landis, S., Roberts, J., Squire, L., 1999. Fundamental Neural Science. Academic.

Zou, Z., Horowitz, L.F., Montmayeur, J.P., Snapper, S., Buck, L.B., 2001. Genetic tracing reveals a stereotyped sensory map in the olfactory cortex. Nature 414 (6860), 173–179. https://doi.org/10.1038/35102506.

Zubieta, J.K., Smith, Y.R., Bueller, J.A., Xu, Y., Kilbourn, M. R., Jewett, D.M., Meyer, C.R., Koeppe, R.A., Stohler, C.S., 2001. Regional mu opioid receptor regulation of sensory and affective dimensions of pain. Science (1979) 293 (5528), 311–315. https://doi.org/10.1126/science.1060952.

Zucco, G.M., Schaal, B., Olsson, M.J., Croy, I., 2014. Applied olfactory cognition. Front. Psychol. 5, 873. https://doi.org/10.3389/fpsyg.2014.00873.

Further Reading

Allison, A.C., 1954. The secondary olfactory areas in the human brain. J. Anat. 88 (4), 481–488.

Cahill, C., Tejeda, H.A., Spetea, M., Chen, C., Liu-Chen, L.Y., 2021. Fundamentals of the dynorphins/kappa opioid receptor system: from distribution to signaling and function. Handb. Exp. Pharmacol. 3–21. https://doi.org/10.1007/164_2021_433.

Derbyshiré Vogt, S.B., Jones, A., Derbyshire, S., Vogt, B., 1998. Pain and Stroop Interference Tasks Activate Separate Processing Modules in Anterior Cingulate Cortex, vol. 118. Springer-Verlag.

Gaveriaux-Ruff, C., Nozaki, C., Nadal, X., et al., 2011. Genetic ablation of delta opioid receptors in nociceptive sensory neurons increases chronic pain and abolishes opioid analgesia. Pain 152 (6), 1238–1248. https://doi.org/10.1016/j.pain.2010.12.031.

Haberly, L.B., 2001. Parallel-distributed processing in olfactory cortex: new insights from morphological and physiological analysis of neuronal circuitry. Chem. Senses 26 (5), 551–576. https://doi.org/10.1093/chemse/26.5.551.

Haberly, L.B., Price, J.L., 1978a. Association and commissural fiber systems of the olfactory cortex of the rat. I. Systems originating in the piriform cortex and adjacent areas. J. Comp. Neurol. 178 (4), 711–740. https://doi.org/10.1002/cne.901780408.

Haberly, L.B., Price, J.L., 1978b. Association and commissural fiber systems of the olfactory cortex of the rat II. Systems originating in the olfactory peduncle. J. Comp. Neurol. 181 (4), 781–807. https://doi.org/10.1002/cne.901810407.

Luskin, M.B., Price, J.L., 1983. The topographic organization of associational fibers of the olfactory system in the rat, including centrifugal fibers to the olfactory bulb. J. Comp. Neurol. 216 (3), 264–291. https://doi.org/10.1002/cne.902160305.

Malnic, B., Hirono, J., Sato, T., Buck, L.B., 1999. Combinatorial receptor codes for odors. Cell 96 (5), 713–723. https://doi.org/10.1016/S0092-8674(00)80581-4.

Mineur, Y.S., Soares, A.R., Etherington, I.M., Abdulla, Z.I., Picciotto, M.R., 2023. Pathophysiology of nAChRs: limbic circuits and related disorders. Pharmacol. Res. 191, 106745. https://doi.org/10.1016/j.phrs.2023.106745.

Navratilova, E., Xie, J.Y., Meske, D., et al., 2015b. Endogenous opioid activity in the anterior cingulate cortex is required for relief of pain. J. Neurosci. 35 (18), 7264–7271. https://doi.org/10.1523/JNEUROSCI.3862-14.2015.

Neugebauer, V., Presto, P., Yakhnitsa, V., Antenucci, N., Mendoza, B., Ji, G., 2023. Pain-related cortico-limbic plasticity and opioid signaling. Neuropharmacology 231, 109510. https://doi.org/10.1016/j.neuropharm.2023.109510.

Pieniak, M., Oleszkiewicz, A., Avaro, V., Calegari, F., Hummel, T., 2022. Olfactory training – thirteen years of research reviewed. Neurosci. Biobehav. Rev. 141, 104853. https://doi.org/10.1016/j.neubiorev.2022.104853.

Price, J.L., 1973. An autoradiographic study of complementary laminar patterns of termination of afferent fibers to the olfactory cortex. J. Comp. Neurol. 150 (1), 87–108. https://doi.org/10.1002/cne.901500105.

Rolls, E.T., 2015. Limbic systems for emotion and for memory, but no single limbic system. Cortex 62, 119–157. https://doi.org/10.1016/j.cortex.2013.12.005.

Vogt, B.A., Vogt, L., Farber, N.B., Bush, G., 2005. Architecture and neurocytology of monkey cingulate gyrus. J. Comp. Neurol. 485 (3), 218–239. https://doi.org/10.1002/cne.20512.

Wilson, D.A., 2000a. Comparison of odor receptive field plasticity in the rat olfactory bulb and anterior piriform cortex. J. Neurophysiol. 84 (6), 3036–3042. https://doi.org/10.1152/jn.2000.84.6.3036.

Wilson, D.A., 2000b. Odor specificity of habituation in the rat anterior piriform cortex. J. Neurophysiol. 83 (1), 139–145. https://doi.org/10.1152/jn.2000.83.1.139.

Yokoi, M., Mori, K., Nakanishi, S., 1995. Refinement of odor molecule tuning by dendrodendritic synaptic inhibition in the olfactory bulb. Proc. Natl. Acad. Sci. USA 92 (8), 3371–3375. https://doi.org/10.1073/pnas.92.8.3371.

Appendix

A

The Noradrenergic Modulation of Attention and Memory: The Attentional Boost Effect (ABE)

Giulia Bechi Gabrielli[1,2,3], *Pietro Spataro*[4] *and Clelia Rossi-Arnaud*[2]

[1]Neuroimaging Laboratory, IRCCS Santa Lucia Foundation, Rome, Italy [2]Department of Psychology, University of Rome "La Sapienza", Rome, Italy [3]Department of Human Neurosciences, University of Rome "La Sapienza", Rome, Italy [4]Faculty of Society and Communication Sciences, Universitas Mercatorum, Rome, Italy

THE LOCUS COERULEUS–NOREPINEPHRINE SYSTEM

The Locus Coeruleus (LC) is a small nucleus in the brainstem, which represents the sole source of norepinephrine (NE) in the brain and has widespread connections with most brain regions, including the cerebellum, the forebrain (except the basal ganglia), the hippocampus, and the occipital cortices (Aston-Jones and Cohen, 2005; Nieuwenhuis et al., 2005). Furthermore, LC exhibits substantial regional and laminar specificity in its efferent projections (Morrison et al., 1982), and a single neuron can innervate different regions (Foote and Morrison, 1987). In particular, brain areas that are associated with attentional processing as well as motor responding receive a particularly dense LC-NE innervation (Foote and Morrison, 1987). This anatomical organization has aroused much interest, due to the possible involvement of this system in high-order cognitive processes.

Previous studies based on lesion and pharmacologic manipulations have allowed a great development of our knowledge of how NE acts in the brain (See Berridge and Waterhouse, 2003; Sara and Bouret, 2012, for comprehensive reviews). Early studies indicated a primary role of the LC-NE system in arousal and behavioral states, due to its connections with the thalamus and the neocortex (Berridge et al., 1993; Berridge and Waterhouse, 2003). In particular, the NE action was strongly involved in sleep–waking cycle (Jouvet, 1969), certain autonomic functions (Miyawaki et al.,1991; Miyawaki et al., 1993; Ward and Gunn, 1976), affective states (Siever and Davis, 1985; Valentino and Curtis, 1991), and vigilance (Aston-Jones, 1985; Aston-Jones and Bloom, 1981a).

Moreover, by acting through thalamic nuclei and sensory cortices, the LC activity provides gating and tuning influences on sensory processing in all modalities (e.g., Bouret and Sara, 2002; Devilbiss and Waterhouse, 2011; Lecas, 2004; McLean and Waterhouse, 1994; Sara, 2009; Waterhouse et al., 1998). Several studies showed that an increase in extracellular norepinephrine improved the signal/background noise ratio, reducing spontaneous activity while at the same time preserving or enhancing stimulus-evoked responses in auditory, visual and somatosensory networks; in some cases, nonresponsive neurons shifted to a responsive mode, a phenomenon defined as sensory gating. Moreover, increased levels of norepinephrine sharpen the firing selectivity of neurons. In the auditory cortex, neurons respond preferentially to a "best frequency" and, in a lesser degree, to the neighboring frequencies: NE inhibited response to neighbor frequencies while selectively sparing responses evoked by the best frequencies (i.e., NE "tuned" neurons to their best frequency); similar results were found in the visual cortex: increased NE levels altered specific receptive field properties (e.g., direction selectively, velocity tuning, response threshold) (McLean and Waterhouse, 1994; Waterhouse et al., 1990). In somatosensory and olfactory modalities, the stimulation of LC was related to improvements in spike timing and rhythmicity in neuronal responses to sensory stimuli. In humans, increasing NE levels (obtained through the injection of a reuptake blocker) reduced the threshold for the detection of both sour and bitter tastes (Heath et al., 2006). Finally, as in the sensory neuronal pathways discussed above, also the motor responses were demonstrated to be facilitated by the activation of LC-NE system at multiple level (Berridge and Waterhouse, 2003).

Taken together, these studies indicated that LC phasic activity facilitates the processing and responses to behaviorally relevant stimuli: increased NE levels in target regions enhance perceptual acuity—suppressing the response to irrelevant sensory information and facilitating the processing of relevant one (modulating the amplitude of neural responses and altering the selectively of that response)—and regulate the speed and the efficiency of motor responses to relevant stimuli.

In addition, through its projections to the hippocampus, the NE is implicated in cellular excitability and intracellular cascades and regulates synaptic plasticity and long-term potentiation (LTP) in this region (for reviews see Harley, 1987, 2007). Indeed, due to this hippocampal action, the NE is involved in learning and memory, particularly by interacting with other neuromodulators and hormones (Gallagher et al., 1985; McIntyre et al., 2012). Other pharmacological approaches have revealed a noradrenergic influence in frontal cortical regions that are engaged in attention and working memory functions (reviewed in Arnsten and Li, 2005; Arnsten et al., 2012; Robbins and Roberts, 2007). In particular, a growing body of evidence from animal and human studies showed that the LC-NE system plays an important role in attentional shifting and behavioral flexibility (Angela and Dayan, 2005; Aston-Jones and Cohen, 2005; Bouret and Sara, 2005; Devauges and Sara, 1990; McGaughy et al., 2008).

LC-noradrenergic neurons exhibit two distinct activity modes: tonic and phasic. Tonic activity is characterized by relatively low-frequency, sustained, and highly regular discharge patterns. This modality is state-dependent, displaying a discharge rate that decreases from waking, when is highest (quiet waking <2 Hz; active waking >2 Hz), to slow–wave sleep, when is lower (<1 Hz), and becoming virtually absent during REM or paradoxical sleep (Foote et al., 1980; Hobson et al., 1975). For our purpose, particularly interesting is the fact that changes in LC discharge rates anticipate changes in behavioral state: during waking, environmental stimuli that elicit sustained increases in arousal or attentiveness bring to sustained increases in tonic discharge (Aston-Jones and Bloom, 1981a; Foote et al., 1980).

On the other side, during waking, LC neurons exhibit phasic modulations in discharge rates also in response to salient sensory stimuli (Aston-Jones and Bloom, 1981b; Foote et al., 1980); these modulations are characterized by a relatively short latency (15–70 ms in rats) and by a brief burst of two to three action potentials followed by a more prolonged period of suppression of discharge activity (approximately 300–700 ms). These phasic responses are observed during an overt orienting response to a novel stimulus within a specific environmental location.

In fact, LC phasic discharges seem to be related to tonic activity levels. In particular, phasic responses to correct target detection are observed only when tonic discharge rates fall within a relatively narrow intermediate range. On the contrary, phasic activity is less robust under conditions of high arousal (for example, during a stressful condition) or low vigilance (such as sleep, grooming, and eating), that are associated with higher or lower levels of tonic discharge, respectively (Aston-Jones and Bloom, 1981b; Valentino and Foote, 1987, 1988; Valentino and Wehby, 1988).

Aston-Jones and Choen (2005) argued that these different modes of LC activity have adaptive advantages under different environmental circumstances. The phasic mode (related to an intermediate tonic activity) may support cortically driven behaviors optimized to specific stable environments (e.g., tasks requiring focused attention). Instead, the higher tonic LC activity (related to low phasic discharge) may be adaptive in rapidly changing, unpredictable environments. That is, the behavioral variability associated with high tonic LC activity may be critical for behavioral flexibility and responsiveness to unexpected events.

In line with this interpretation, there is an inverted U-shaped relationship between LC discharge rates (both tonic and phasic discharge) and levels of task performance (Aston-Jones and Cohen, 2005; Berridge and Waterhouse, 2003). In the phasic mode (intermediate tonic activity plus stimulus-evoked phasic activity), the LC system becomes responsive to specific task-relevant stimuli and is associated with highly accurate behavioral performance. In the tonic LC mode (elevated LC baseline activity and reduction in phasic activity with little or no LC response to target stimuli or other tasks), such system increases responsivity to a broader class of stimuli, until the behavioral and attentional responses become relatively indiscriminate (they are more distractible) and labile (they are associated with poor task performance because of frequent false alarm errors).

As the phasic LC response occurs only for the identification of task-relevant stimuli, and not for task-irrelevant stimuli, Aston-Jones and Cohen (2005) proposed the *Adaptive Gain Theory of locus coeruleus–norepinephrine (LC–NE)* function, in which they characterize the phasic response of LC neurons as a temporal attentional filter. This theory suggests that once a task-relevant event has been detected, LC neurons fire to increase the sensitivity (or gain) of target neurons, leading to a transient increase in perceptual processing. In their hypothesis, this attentional filter is temporally specific but, given the broad projections of LC neurons, spatially global. In addition, the link to decision outcomes, rather than to stimulus presentation, indicates that this LC response primarily modulates specific behaviors rather than sensory processing. Therefore, Aston-Jones and Cohen (2005) proposed that the LC phasic response provides a temporal attentional filter that selectively facilitates task-relevant behaviors. In addition, a theory of event perception suggests that changes in observed activities trigger additional perceptual processing, updating internal representations of the current event (Zacks et al., 2007). These theories suggest that increasing attention in response to task-relevant changes in target events may facilitate cognitive processing at the moment of the change (*facilitation hypothesis*). They further suggest that this facilitation may result from orienting attention to the moment in time in which the change occurred, perhaps through the opening of an attentional gate (Olivers and Meeter, 2008).

Supporting this interpretation of LC activity, the LC-NE system has been linked to cortical control of attention by Corbetta et al. (2008). They described two different functional anatomical networks underlying attention: a dorsal and a ventral fronto-parietal network. The dorsal fronto-parietal network is responsible for directing attention to expected stimuli and linking them to appropriate responses. The ventral fronto-parietal network interrupts and resets ongoing activity by detecting behaviorally relevant or unexpected stimuli (Corbetta and Shulman, 2002; Shulman et al., 2002). Usually, the ventral network is actively suppressed. When a salient or behaviorally relevant stimuli occurs, the ventral network becomes activated, sending a "circuit-breaking" signal to the dorsal region that results in a shift of attention to the novel stimulus. The reorienting response from one task state to another could occur in the environment, as between internally and externally directed activities, and is driven by LC-NE system (Corbetta et al., 2008). Indeed, this cortical control of attention reflects the mechanisms underlying the activation of LC by biologically imperative stimuli. Bouret and Sarah (2005) proposed that the activation of the LC-NE system induces a "reset" in its target structures, by interrupting the existing functional networks and facilitating the emergence of new ones (the *network-reset hypothesis*). According to these interpretations, the NE, acting as a signal from LC, would drive the ventral network to "reset" the dorsal region to promote the shift in attention. In conclusion, activity in the dorsal network reflects the configuration of task processes (stimulus and motor representations) in response to the new contingency, while activity in the ventral network facilitates the reset (or reconfiguration) process.

THE ATTENTIONAL BOOST EFFECT

In the last decade, research on the links between attention and memory showed the existence of a paradoxical phenomenon, the *Attentional Boost Effect*, that was hypothesized to be based on a temporary increase in the release of norepinephrine (NE) from the locus coeruleus (LC). This effect was first presented by Swallow and Jiang and by Lin, Pype, Murray, and Boynton in 2010, demonstrating that dividing attention during the encoding phase does not impair but, on the contrary, improves performance in a later memory test. Specifically, they demonstrated that images presented together with a target stimulus that required a motor response were later recognized better than images presented together with a distractor stimulus that required no response.

In the standard version of the ABE paradigm (Swallow and Jiang, 2010), participants encoded a long series of scenes paired with either white or black squares (placed in the center of each scene). The instructions were to memorize the scenes and press the spacebar whenever the central square was white (no response was requested when the square was black). Immediately after the end of the encoding phase, they were administered a four-choice recognition test examining scenes' memory (Fig. A.1). The results showed that the images associated with target white squares were recognized more accurately than the images associated with distractor black squares. The authors referred to this memory advantage of target-paired images as the "*Attentional Boost Effect*" (ABE).

Swallow and Jiang (2010, 2013) hypothesized that detecting an occasional target induced a transient attentional response that temporarily increased the amount of cognitive resources available to process the unrelated scenes that co-occurred with the target stimuli (the so-called *dual-task interaction model*). According to the authors, the neural mechanisms involved in increasing attention to the stimuli associated with the targets would be based on a temporary increase in the release of norepinephrine (NE) from the locus coeruleus (LC). When the target was detected, a transient phasic discharge from the LC enhanced the perceptual

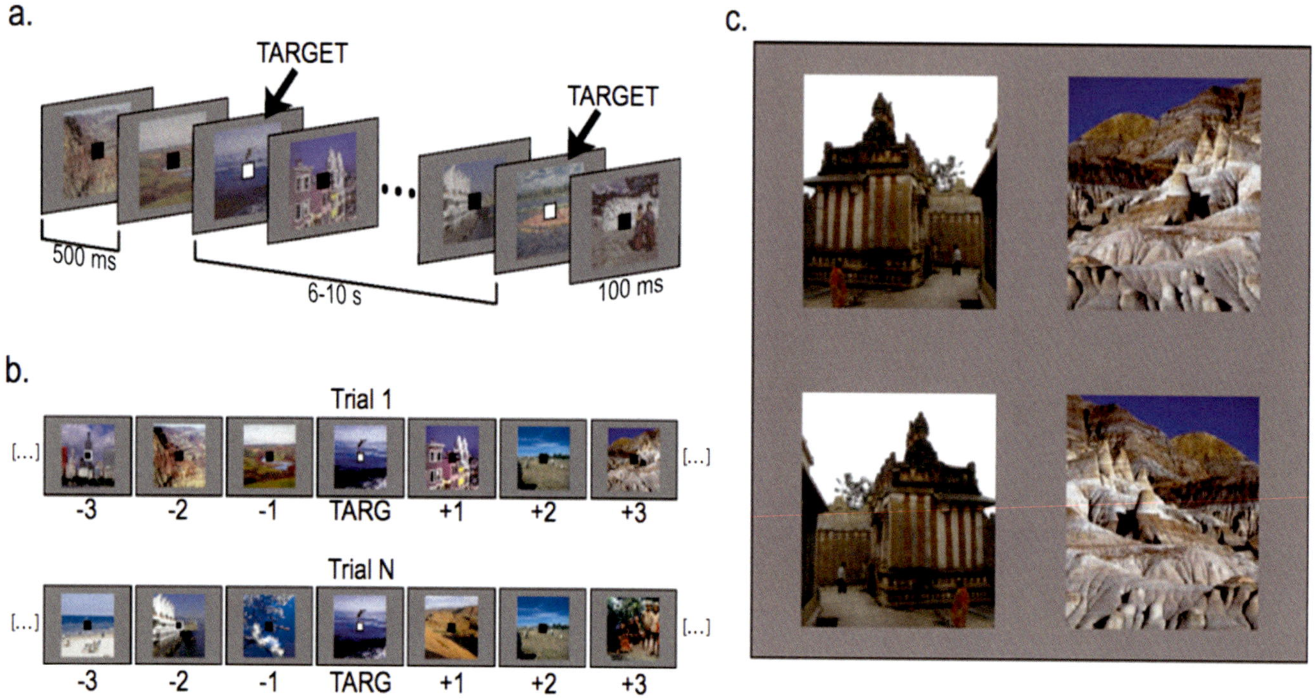

FIGURE A.1 **Representation of dual task encoding used in experiment 1 by Swallow and Jiang (2010).** (A) participants saw a long sequence of scenes with a white or black square in the center. They had to memorize the scenes for a later test and press a key when they saw the white square. (B) target was always in the same position within the trial. (C) example of recognition test: participant had to choose the picture corresponding to the one seen in the encoding phase. *Image courtesy of Swallow, KM., Jiang, YV., 2010. The attentional boost effect: transient increases in attention to one task enhance performance in a second task. Cognition 115 (1), 118–132. https://doi.org/10.1016/j.cognition.2009.12.003. Epub 2010 Jan 18. PMID: 20080232; PMCID: PMC2830300.*

processing of all the scenes presented at the same time of the target stimuli.

They defined the ABE as a compromise between attentional competition (resulting from the interference between the detection and encoding tasks) and attentional facilitation (due to the transient increase in attention following target detection). The authors compared a divided attention (DA) condition (in which participants had to encode the background stimuli and simultaneously detect the targets) with a full attention (FA) condition (in which the sole task was to encode the background stimuli). Results showed that the recognition of distractor-paired items was worse in the DA condition than in the FA condition, confirming the classical negative effects of DA on memory encoding; in contrast, the recognition of target-paired items in the DA condition was boosted to the same level of the FA (Swallow and Jiang, 2010; see also Mulligan et al., 2014; Prull, 2019). Thus, when the attentional requests needed to detect the targets are relatively low, as they are in the standard version of the ABE paradigm illustrated above, the combination between attentional competition and attentional facilitation produces a clear boost in memory performance. However, when the requests of the detection task are increased (for example, when the target stimuli were defined by the conjunction of two features or when participants were asked to press one key for the target stimulus and another key for the distractor stimulus), then attentional competition offset attentional facilitation, reducing or eliminating the ABE (Swallow and Jiang, 2010, exp. 4 and 5). On the basis of this evidence, Swallow and Jiang (2010) defined the ABE as a relative facilitation in memory performance. Later studies demonstrated that an absolute facilitation, in which the DA performance exceeds the FA levels, can be detected under specific conditions, for example when using implicit memory tasks (Mulligan and Spataro, 2015; Prull, 2019; Spataro et al., 2013).

In 2014, Swallow and Jiang proposed a slightly different paradigm to study this phenomenon. The authors presented a series of faces flanked by two target squares (e.g., orange), two distractor squares (e.g., blue) or no squares—the latter condition representing the baseline. The instructions were to study the faces and simultaneously press the spacebar when the target squares appeared on the screen (no action was required in response to the distractor squares). In the test phase, participants performed an old-new recognition task. Results showed that the performance was significantly better for the faces encoded in the target condition than for the faces encoded in the distractor or no-square conditions, which did not differ between them.

According to Swallow and Jiang (2014), these results demonstrate that the ABE represents a memory enhancement triggered by facilitatory mechanisms due to target detection, rather than by inhibitory mechanisms due to distractor rejection (the latter hypothesis predicted that performance in the distractor condition should have been significantly lower than that in the baseline condition).

Later studies have consistently expanded our knowledge about this phenomenon and several regularities have been demonstrated. Specifically, the memory advantage for items presented at the same time of target stimuli has been obtained with different types of materials and different types of memory tasks. Indeed, the ABE is not limited to visual stimuli or explicit long-term memory tasks. Rather, it extends to verbal stimuli (words: Mulligan et al., 2014; Spataro et al., 2013) and has been reported in a wide variety of tasks, including free or cued recall (Mulligan et al., 2014; Spataro et al., 2017), implicit memory tests such as lexical decision and word-fragment completion (Spataro et al., 2013), and several short-term memory paradigms (Lin et al., 2010; Makovski et al., 2011; Spataro et al., 2020).

Moreover, the ABE has been studied in different types of populations. Most of the above evidence has been obtained in studies of healthy young participants, typically university students. Bechi Gabrielli et al. (2018) and Prull (2019) investigated the presence of the ABE in healthy older adults. Bechi Gabrielli et al. (2018) found that the ABE was abolished in older adults between 60 and 75 years using a 20-min interval to separate the encoding and test phases. On the other hand, Prull (2019), with a much shorter study-test interval (2 min), reported that the ABE exhibited by young-old adults (between 60 and 75 years) did not differ from that of young adults, although a significant decrease was observable in older-old adults (from 75 to 94 years). Prull (2019) speculated that the cognitive decline associated with healthy aging resulted in the production of vulnerable boosted memories that would be more easily disrupted over time, compared to the boosted memories of young adults; that is, long study-test delays would create a selective interference that impairs the boosted memories of older adults.

The ABE was also studied in clinical population (Bechi Gabrielli et al., 2021; Kéri et al., 2013; Levy-Gigi and Kéri, 2012; Rossi-Arnaud et al., 2014; Szamosi et al., 2013). Studies on patients with both amnestic mild cognitive impairment and posttraumatic stress disorder showed that these patients were severely impaired in the recognition of scenes encoded with target stimuli. In contrast, these same patients performed like controls, or even displayed an enhanced performance, in the recognition of scenes encoded with distractor stimuli (Levy-Gigi and Kéri, 2012; Szamosi et al., 2013). A different pattern was obtained in patients with Parkinson's disease before and after the administration of dopaminergic medications. At the baseline, there was no difference in performance between drug-naive patients and controls. However, after receiving dopamine agonists for 12 weeks, patients outperformed controls in the recognition of both target-paired and distractor-paired stimuli (Kéri et al., 2013). It must be underlined that these studies used the Rapid Serial Visual Presentation (RSVP) paradigm (Leclercq et al., 2014; Lin et al., 2010; Makovski et al., 2011), in which participants encoded brief sequences of stimuli and the recognition task was administered immediately, at the end of each sequence. More recently, Rossi-Arnaud et al. (2014) and Bechi Gabrielli et al. (2021) investigated the presence of the ABE in schizophrenic patients and bipolar patients in the euthymic phase, using a paradigm similar to that illustrated by Swallow and Jiang (Swallow and Jiang, 2010, 2014). Collectively, the results showed that schizophrenic or bipolar patients exhibited no memory advantage for target-paired stimuli, unlike controls who displayed the typical ABE.

Interestingly for the present purposes, both healthy aging and psychiatric disorders have been associated with LC alterations. In particular, healthy aging is accompanied by increased tau pathology, progressive reduction of hippocampal brain volume, and neuronal loss in the LC, which are already evident from 20 years onwards (Mather and Harley, 2016; Raz et al., 2005; Shibata et al., 2006). Moreover, LC atrophy is an early event in Alzheimer's Disease (AD) and correlates with cognitive performance (Grudzien et al., 2007; Weinshenker, 2008). For example, Jacobs et al. (2015), using resting-state functional connectivity analysis, demonstrated that LC and MTL regions are integrated functionally and that this connectivity is relevant for memory (Jacobs et al., 2015), consistent with animal, pharmacologic, and postmortem studies (see Berridge and Waterhouse, 2003; Sara, 2009 for a review). The evidence of tau pathology in the LC, even before any clinical symptoms or concomitant cerebral amyloid pathology was evident (Braak et al., 2011), led Braak and Del Tredici (2012) to propose a novel model of AD: they hypothesized that the tau pathology in the LC would initiate the pathologic process of AD and would trigger the subsequent neuropathologic changes associated with amyloid deposition, through its anatomical connections to the MTL (Braak and Del Tredici, 2012).

Abnormalities in the noradrenergic system, often in association with alterations of the dopaminergic system, have been implicated in the pathophysiology of mood disorders, major depressive disorder (MDD), bipolar disorder (BD), and schizophrenic pathology (Anand

and Charney, 2000; Lohr and Jeste, 1988; Mäki-Marttunen et al., 2020; Ressler and Nemeroff, 1999; van Enkhuizen et al., 2015).

Finally, it is well known that LC is robustly activated by both visceral and environmental stressors (Korf et al., 1973; Valentino and Van Bockstaele, 2008). LC is under the direct control of primary structures responding to stress from the periphery via the vagus nerve and the paraventricular nucleus of the hypothalamus. The NE released by LC neurons would then mediate, or at least modulate, the stress or arousal responses in forebrain regions. In agreement, several studies indicated that NE activity may contribute to hyperarousal and reexperiencing symptoms associated with PTSD (George et al., 2013; O'donnel, Hegadoren and Coupland, 2004; Southwick et al., 1997). That is, the excessive noradrenergic activation following a traumatic event may enhance memory consolidation and increase the frequency of reexperiencing episodes (Gören and Cabadak, 2014).

THE LC-NE SYSTEM AT THE BASE OF THE ATTENTIONAL BOOST EFFECT

To account for the rapidly increasing amount of findings related to the ABE, Swallow and Jiang (2013) proposed the so-called *dual-task interaction model* (Fig. A.2). This framework assumes that, on each trial, information from the squares and the background stimuli compete for representation in the primary visual cortex, producing the first source of interference. In particular, processes specific to the square detection (i.e., processes triggered by the recognition that an item is a target, and therefore require a response, or a distractor, which therefore require no response) interfere with the ability to encode the concurrent image (*item-specific interference*). In addition, the second source of interference results from the need to maintain two simultaneous goals in working memory: encoding the background stimuli into memory and generating an appropriate response to the target squares (*task-specific interference*). To resolve

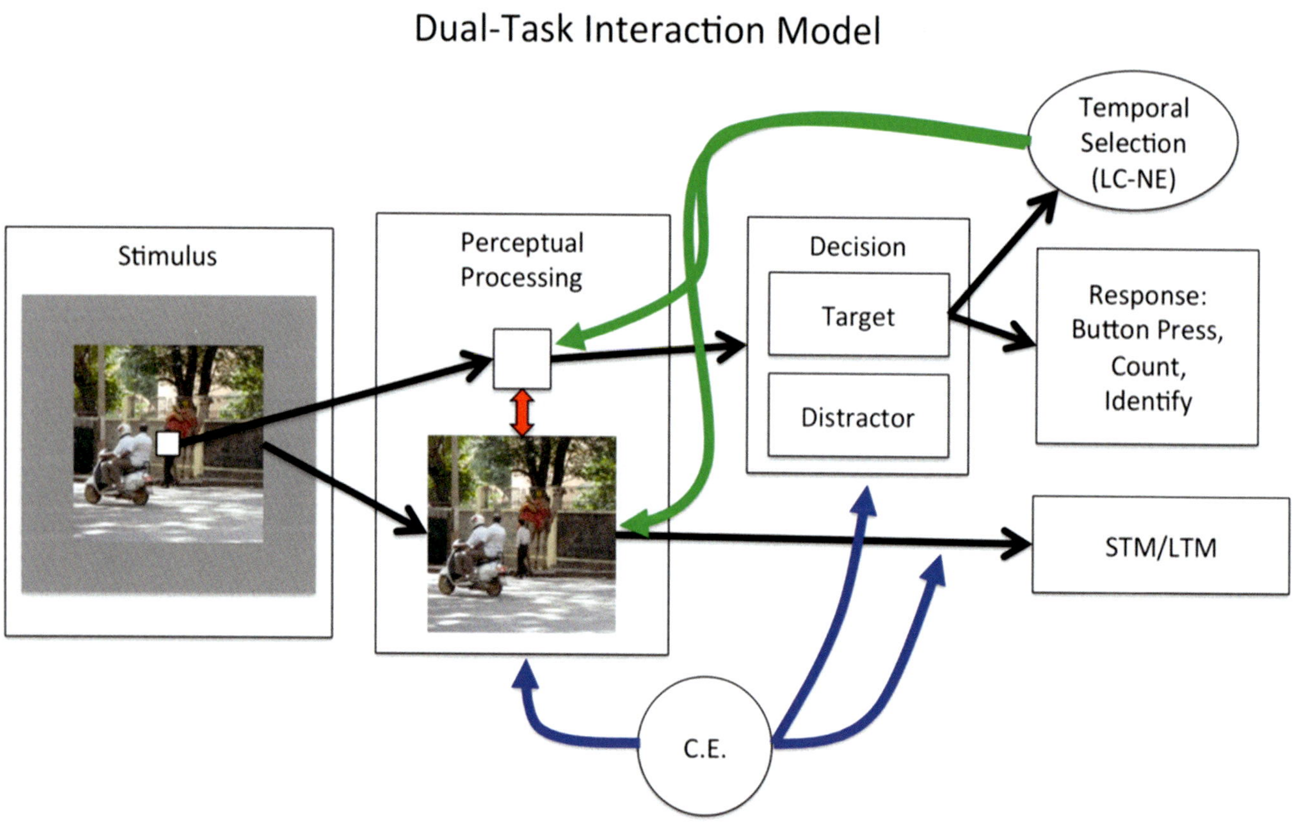

FIGURE A.2 **The dual-task interaction model.** On each trial, sensory information from the two task-relevant stimuli is selected to undergo perceptual processing (early selection), as directed by the central executive (C.E.). Within perceptual processing areas, the two stimuli compete for representation, producing dual-task interference. Dual-task interference also arises from the need to maintain multiple goals simultaneously: the encoding task and the detection task. Perceptual evidence that the square is a target or distractor is accumulated, and the square is categorized once a threshold has been reached. The item may then be selected to guide behavior and be maintained in memory if necessary (late selection). *Image courtesy of Swallow, K.M., Jiang ,Y.V., 2013. Attentional load and attentional boost: a review of data and theory. Front. Psychol. 20(4) 274. https://doi.org/10.3389/fpsyg.2013.00274. PMID: 23730294; PMCID: PMC3657623.*

this competition, a cognitive control system like the central executive prioritizes the elaboration of the goal-relevant items (i.e., the to-be-responded squares). The rapid categorization of targets as stimuli that require an overt response triggers *temporal selective attention* (Olivers and Meeter, 2008), a mechanism that is temporally, but not modality or spatially selective. Temporal selective attention enhances the processing of all perceptual information that is present when a target occurs. The authors hypothesized that this mechanism is based on a transient increase in the release of norepinephrine from the LC.

According to the *dual-task interaction model*, the neural mechanisms underlying the ABE rely on a temporary increase in the release of norepinephrine (NE) from the locus coeruleus (LC). As soon as the target is detected, a transient phasic discharge from the LC triggers temporal selective attention, which is temporally (but not modality or spatially) selective: it, therefore, enhances the perceptual processing of any stimulus occurring at the same time as the target.

First of all, LC phasic activation (and more generally the noradrenergic system) has been often associated with target detection processes in both monkeys and humans (Aston-Jones and Cohen, 2005; Robertson et al., 2014). More specifically, phasic LC activity is elicited by sensory stimuli that are relevant to goal achievement and is associated with focused attention to these stimuli (Aston-Jones and Cohen, 2005).

In one series of experiments (Aston-Jones et al., 1994; Aston-Jones et al., 1999; Rajkowski et al., 2004), LC activity was recorded while monkeys performed a simple detection task in which they were required to release a lever immediately following a specific infrequent visual target (e.g., a vertical bar) to receive a juice reward. Results of these studies showed that LC neurons were physically activated by target stimuli, and only weakly or not at all activated by distractor stimuli, in line with the interpretation of the ABE as a consequence of target detection (rather than distractor rejection: Swallow and Jiang, 2014; Yebra et al., 2019).

But which characteristic of the "action" of detecting a target is related to LC activation? Both studies about LC functioning in animals and ABE experiments in humans have ruled out a number of potential explanations, including accounts based on target distinctiveness and reinforcement learning. Target stimuli are usually novel (infrequent), salient and distinctive stimuli. However, several neurophysiological and behavioral results indicated that LC activation was not mediated by the distinctiveness of target stimuli, but rather by its relevance for the ongoing task (i.e., by the necessity to perform an overt response: Makovski et al., 2013). Recordings of LC activation during experiments on reversal training, where the distractor became the target and vice versa, indicated that the LC responses were not specific to the particular sensory attributes of the target stimuli (Aston- Jones et al., 1994; Aston-Jones et al., 1997). Indeed, LC phasic responses became quickly and selectively activated by the new targets and extinguished for the new distractors. These results indicate that LC activation was not limited by the perceptual salience of the to-be-detected stimuli. Coherently, it has been demonstrated that the perceptual salience of targets does not drive the ABE. Swallow and Jiang (2010, Exp. 3) showed that the ABE was eliminated if participant focused only on the background items and did not pay attention to the stimuli in the detection task, even if they were perceptually salient: in this experiment, participants were instructed to memorize the pictures and to ignore both target (white) and distractor (black) squares—recreating a full attention (FA) condition. In this paradigm, no memory advantage was observed for scenes that were presented at the same time as the infrequent white square. The absence of the ABE in this condition indicated that the effect was not due to the perceptual distinctiveness of the infrequent target squares, but rather that these squares must be processed as to-be-responded targets to produce the ABE (see also Makovski et al., 2013).

Interestingly, while the distinctiveness of the target stimulus plays no role in the ABE, a couple of studies demonstrated that the distinctiveness of the background stimuli modulated the effect. Studies with verbal materials (Mulligan et al., 2014; Spataro et al., 2015) showed a strong ABE with high-frequency and orthographically common words; conversely, the ABE was absent or strongly reduced in size for low-frequency and orthographically distinctive words (but see Prull, 2019, for different results). The latter types of words attract participants' attention independently from their association with target stimuli (Criss and Malmberg, 2008; Malmberg and Nelson, 2003) and are usually remember better than high-frequency and orthographically common words (de Zubicaray et al., 2005; Glanzer and Adams, 1990; Guttentag and Carroll, 1997; Marsh et al., 2006). Both Mulligan et al. (2014) and Spataro et al. (2015) hypothesized that the encoding benefit produced by target condition may be largely redundant with the processing advantage that typically happens with low-frequency and orthographically distinctive words: that is, due to their distinctiveness, these words attract heightened attention during the encoding phase, offsetting any additional processing advantage induced by the ABE.

Turning to the properties of target stimuli, another factor that could make them distinctive and salient is relative frequency. It has long been known that rare, new, and distinctive events, even those that are not perceptually distinctive, affect long-term encoding and memory (Fabiani and Donchin, 1995; Geraci and

Rajaram, 2004). Swallow and Jiang (2012) performed a series of experiments to investigate if the ABE was modulated by the frequency of targets. In one experiment, they manipulated the frequency of target stimuli in two different conditions: a rare condition in which the target-to-distractor ratio was 1:6 and a common condition in which the ratio was 1:1 (i.e., targets were as common as distractors). The results showed that pictures presented at the same time as targets were better recognized than those presented with distractors in both conditions (with no difference in the size of the ABE). This is a clear demonstration that the frequency of target squares did not affect the boost effect. Moreover, the authors demonstrated that also the frequency of distractor did not influence the ABE. Indeed, they performed another experiment in which two different types of distractors were used: one that was common (appearing on 40% of the trials) and another that was rare (appearing on 10% of trials). The idea was that if rare or distinctive stimuli enhance the processing of concurrently presented information, then images presented at the same time as rare distractors should be recognized better than images presented at the same time as common distractors. The results showed that the ABE was present independently of the distractor condition: images presented at the same time as targets were recognized better than images in the two distractor conditions; most important, the images in the rare-distractor condition were recognized more poorly than images in the common-distractor condition—a finding opposite to the a-priori predictions. As stated above, studies on animals reached similar conclusions: LC did not respond physically to distractors even if they are infrequent, and LC responses occur even when targets are presented on every trial. This confirms that the frequency of occurrence of target stimuli is not the determining factor for LC activation (Aston- Jones et al., 1994).

Taken together, these results indicate that the ABE is not triggered by the perceptual distinctiveness or the frequency of target stimuli, but rather by the response associated with their detection. To further support this conclusion, animal studies demonstrated that LC activation did not occur on trials in which the animal made no response despite viewing the cue, and there was no LC response associated with spurious lever responses that occasionally occurred between trials when no stimulus was present. These findings suggest that LC activation is not related to simple motor actions or the isolated perception of target stimuli: the animal has to pay attention to the target and process it as a stimulus that requires a response.

But is a motor response strictly necessary for the ABE to occur? From the behavioral point of view, several studies indicated that the facilitation in the encoding of target-paired stimuli is not limited to the occurrence of an explicit motor response. In fact, the effect is robust even when participants are simply asked to covertly count the number of targets and report it at the end of the encoding phase (Swallow and Jiang, 2010; see also Mulligan et al., 2016). Makovski et al. (2013) replicated and extended these results. They showed that, when participants pressed a button for most images, images that required withholding a button press were remembered better than the others. Their results confirmed that the action itself was not critical to the memory advantage: rather, they interpreted the ABE-related enhancement as being the consequence of the need to modify an ongoing activity.

Other behavioral results showed that paying attention to the target stimuli is a necessary factor for the ABE to occur (Swallow and Jiang, 2010). Swallow and Jiang (2010, Exp. 3) showed that the ABE was eliminated if participant focused only on the background items, not paying attention to the stimuli in the detection task: in this experiment, participants were instructed to memorize the pictures and ignore the squares—recreating a full attention (FA) condition. In this case, no memory advantage was observed for scenes that were presented at the same time as target squares (see also Makovski et al., 2013). Likewise, the ABE is eliminated also when participants focus only on the detection task, ignoring the background stimuli. Swallow and Jiang (2011, Exp. 4) used the standard paradigm but divided participant in two groups: one group was told that they would make judgments about the background images after they completed the detection task (*image-relevant group*); the other group was told to ignore the background images and that attending to them could hurt their performance on the detection task (*image-irrelevant group*). The ABE was present only in the *image-relevant group*, suggesting that the memory facilitation produced by target detection was modulated by attention to the background images (see also Bechi Gabrielli et al., 2019, for a similar interpretation).

A final factor that could mediate LC activation is the fact that detecting a target is in itself rewarding (Raymond et al., 2005; Seitz and Watanabe, 2009). However, the registration of LC response in monkeys was not linked to a specific reward because similar responses were observed for different juice rewards or for water rewards in fluid-restricted animals (Aston- Jones et al., 1994). At the behavioral level, Swallow and Atir (2019) examined the question of whether the ABE was due to an increase in the perceived value of background items that were paired with targets. They presented participants with a series of briefly presented images that depicted valuable (e.g., food) or neutral (e.g., children's toys) items. Participants pressed a button if the square associated with the images was a target color but not if it was a distractor color. Consistent with previous literature, the results replicated the standard ABE, with target-paired images being recognized better than distractor-paired images. On the other hand, valuable images were recognized no better than neutral images, and the ABE did not increase when valuable images were paired to target stimuli (i.e., target detection and image value did not interact). Swallow and Atir (2019) concluded that the ABE was not the result of changes in the perceived value of target-paired items.

In conclusion, neurophysiological and behavioral studies converge in showing that both LC activation and the ABE-related facilitation are not triggered by specific properties of target stimuli (such as perceptual distinctiveness or their infrequent occurrence): rather, they are triggered by the necessity to modify the ongoing activity in order to respond to the target appearance (Makovski et al., 2013).

In human studies, the lack of noninvasive instruments capable of measuring changes in the activity of the LC-NE system did not allow the same theoretical understanding of the noradrenergic functions as that obtained in animal studies.

Because of its small size and difficult localization, LC activity was often assessed using indirect measures, such as electroencephalography (EEG) and pupil recording. More recently, functional magnetic resonance imaging (fMRI) studies allowed to directly assess LC functioning in humans. Murphy et al. (2014), in particular, combined pupillometry and blood-oxygen-level-dependent (BOLD) functional magnetic resonance imaging (fMRI) to investigate the relationship between fluctuations in pupil diameter and human LC activity, both at rest and during a simple detection task. Their results showed a positive correlation between pupil diameter and BOLD activity in the rostral LC. In this way, they found for the first time a functional connection between LC neurons' activation and pupil diameter modifications. More important for our purposes, the region that correlated with pupil diameter during task performance exhibited a greater evoked response to target stimuli than to standards stimuli (distractors). In addition, between-subjects' correlations revealed that larger target-evoked BOLD responses within this region were robustly associated with faster response times (RTs). This observation constitutes the first demonstration in humans of a cardinal characteristic of LC activity: phasic modulation by stimulus-relevance in the oddball task (Aston-Jones et al., 1994; Rajkowski et al., 1994; Rajkowski et al., 2004). The link between LC activation and target detection was replicated by Krebs et al. (2018), who found that LC responses were modulated by stimulus novelty rather than by other stimulus characteristics (i.e., emotional saliency).

In the last few years, several studies used neuroimaging to understand the neural bases of the ABE (Bechi Gabrielli et al., 2019; Moyal et al., 2020; Swallow et al., 2012; Yebra et al., 2019). More specifically, two fMRI studies focused on LC activation during the encoding of stimuli in the ABE paradigm, demonstrating a link between the noradrenergic system and boosted memory, albeit still indirectly (Swallow et al., 2012; Yebra et al., 2019).

Swallow et al. (2012) investigated for the first time the brain activations occurring in the ABE paradigm using functional magnetic resonance imaging (fMRI). They presented a stream of auditory stimuli together with images of faces, scenes, and scrambled images (Exp. 1). Participants were instructed to memorize faces and scenes for a later memory test. They also had to press a button as quickly as possible whenever a prespecified target occurred (no action was request for distractors). The imaging results showed the activation of the primary auditory (A1) and visual (V1) cortices in the target condition, together with several cortical and subcortical areas, and importantly for our discussion, with the posterior brain stem in the vicinity of the LC. However, an important limitation of their work was that data from different experiments were pooled together to obtain the ABE and the effect was small (2.8%), relative to previous reports. Swallow and Jiang (2012) focused their attention on the finding that the detection of target stimuli influenced activity in regions not directly involved in their elaboration. Therefore, they run a second experiment using visual targets and found that the detection of visual targets enhanced activation not only in the primary visual cortex (V1) but also in the primary auditory cortex (A1). The authors concluded that the detection of targets produced a widespread perceptual enhancement that boosted activity in sensory regions of the brain that were not directly involved in their processing. Interestingly, the authors demonstrated that this effect was independent of the memory task, since the same pattern of brain activations was obtained when participants performed only the auditory detection task, without the presentation of background images.

A more recent study by Yebra et al. (2019, exp. 5), using a modified remember/know paradigm, examined the hypothesis that target detection boosted episodic memory by engaging the noradrenergic system. These authors presented a series of greyscale photographs of objects surrounded by colored frames. Participants were instructed to press a button when the pictures were presented with a particular color frame (Go trials). After 1 h, they performed a surprise recognition task, where they were required to indicate whether they remember, were familiar or did not remember the pictures presented. Consistent with the dual-task interaction model of the ABE, the fMRI scanning found a significant interaction between target detection and subsequent memory in a region of the dorsal pons consistent with the LC. Moreover, a psychophysiological interaction analysis revealed that the successful encoding of target-paired images was associated with enhanced connectivity between LC and parahippocampal gyrus. Taken together, these results suggest that target detection triggers the release of norepinephrine from the LC, which in turn promotes memory formation by influencing activity in the medial temporal lobe. However, even in this work, the authors did not find a significant remember advantage for Go versus NoGo stimuli in the context of fMRI scanning (Exp.5). In a subsequent experiment, Yebra et al. (2019, Exp. 6) registered pupil diameter during the same Go/NoGo paradigm. In this case, the results showed that stimuli encoded during Go trials were remembered better than stimuli encoded during NoGo trials. Analysing pupil diameter responses, they found a greater pupil dilatation in Go trials, compared to NoGo trials. More important, they found a significant interaction, indicating a greater pupil dilation to stimuli associated with Go responses that were subsequently remembered in a correct way (see Swallow et al., 2019, for similar results).

Starting from these previous results, a recent study by Bechi Gabrielli et al. (2019) investigated the role of the attention during the encoding and test phases in determining the ABE, using functional magnetic resonance imaging (fMRI). Neuroimaging studies indicated that the detection of salient, behaviorally relevant targets in odd-ball tasks engaged a fronto-parietal ventral network, including the temporoparietal junction (TPJ), the anterior insula, the frontal operculum, and the anterior cingulate cortex (Corbetta et al., 2008; Corbetta and Shulman, 2002). This attentional ventral network is not only involved in the detection of salient stimuli (e.g., rare targets among more frequent distractors) but its activity reflects a dynamic interplay between stimulus saliency and internal goals (Corbetta et al., 2008). Evidence indicated that this neural system was more strongly activated by task-relevant stimuli (targets) than by task-irrelevant stimuli (distractors). This is important because, in the ABE paradigm, targets represent salient changes in the environment, as they require a behavioral response. Coherently, Bechi Gabrielli et al. (2019) found that, compared to distractor condition, the target condition in the ABE paradigm involved a stronger activation of a cerebral system that largely overlapped with the fronto-parietal ventral network typically recruited by detection tasks (Bledowski et al., 2004; Corbetta et al., 2008; Corbetta and Shulman, 2002), including bilateral insula, cingulate cortex, supplementary motor area (SMA), frontal operculum, and supramarginal gyrus (Fig. A.3). Other activation clusters were located in the cerebellum and in the precentral and postcentral gyri and were probably related to the motor response performed by participants. Specifically, the activation of the precentral gyrus (M1 Brodmann area) might reflect the motor component of the response, while the involvement of the postcentral gyrus (S1 Brodmann area) might reflect the engagement of sensorimotor processes (Kim, 2014; Linden et al., 1999).

Contrary to previous fMRI studies (Swallow et al., 2012; Yebra et al., 2019), the authors found a strong ABE using a paradigm in which targets and distractors appeared with the same frequency (see Swallow and Jiang, 2012, for similar results): participants recognized the pictures encoded with target squares significantly better than the pictures encoded with distractor squares. However, when they tested for the interaction between trial type and recognition accuracy to investigate the influence of this attentional network on the recognition of background images, the results did not show any significant activation cluster. This result replicates Swallow et al. (2012), suggesting that the activity of the attentional network engaged by target detection was not modulated by the memory encoding of co-occurring pictures. In other words, at the neural level, target-paired pictures correctly recognized at test did not receive a differential attentional encoding, as compared to distractor-paired pictures correctly recognized, as well as to target- and distractor-paired pictures not recognized. Bechi Gabrielli et al. (2019) proposed that the sustained activity of the fronto-parietal network produced by target detection increased the possibility to recognize the images that were associated with that stimulus in the following memory task. In fact, the activation of different portions of the frontoparietal attention network is related to the way in which visual salience biases the probability for an item to be encoded and remembered (see Santangelo, 2015, for a review). In particular, it has been proposed that the ventral regions of the frontoparietal network support the assignment of selection/encoding priorities (Bouret and Sara, 2005; Corbetta et al., 2008). The to-be-tested hypothesis is that the contemporary activation of the LC during oddball tasks (Krebs et al., 2018; Murphy et al., 2014), due to its connectivity with cortical and subcortical regions, allowed the memorization of stimuli that have been assigned high priority by the ventral attentional network.

Previous behavioral studies demonstrated that target detection is not enough to get the ABE: participant had to pay attention to the background pictures to correctly recognize them in a later test (Swallow and Jiang, 2011, Exp. 4). In agreement, Bechi Gabrielli et al. (2019) found

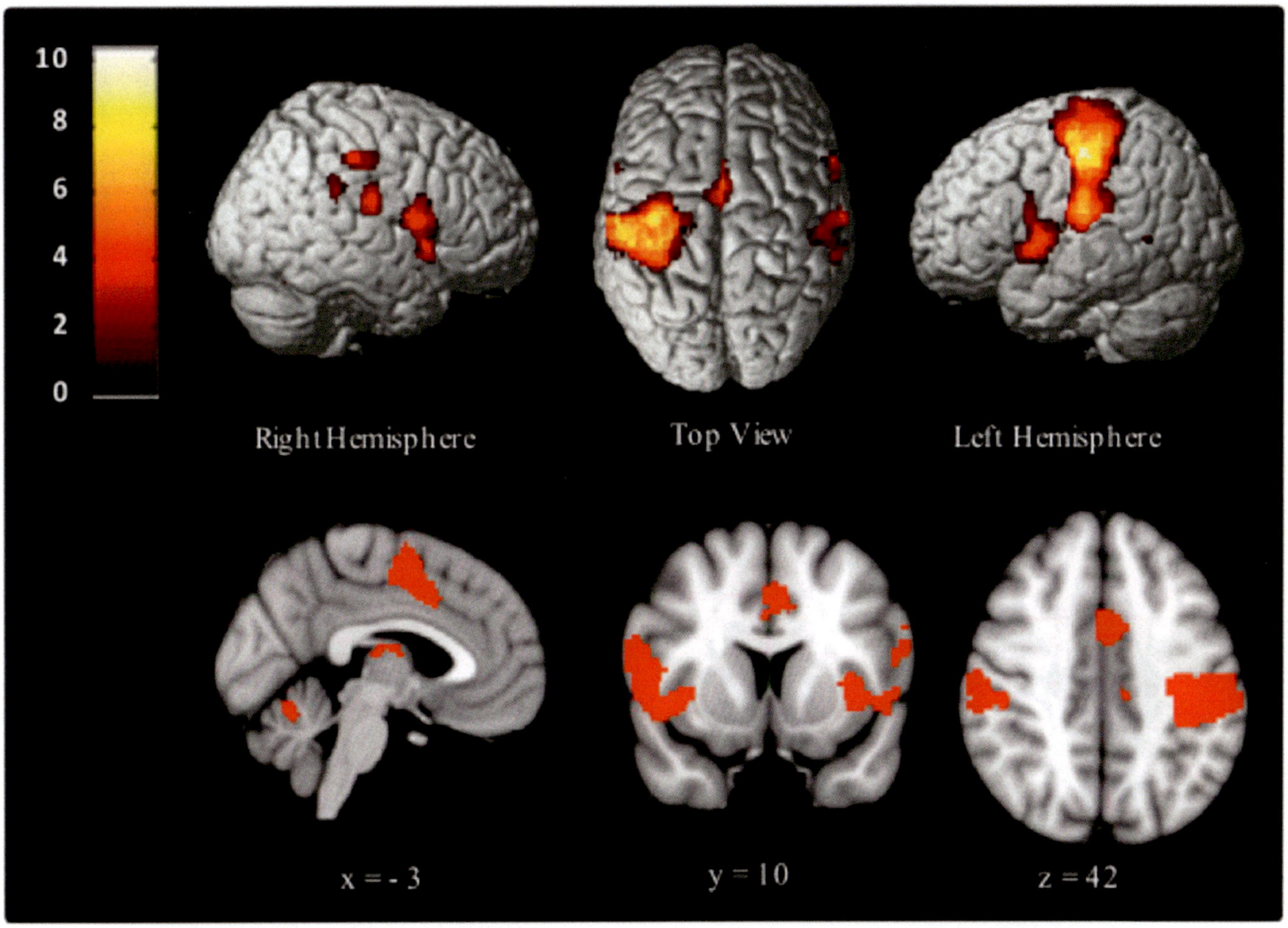

FIGURE A.3 **Differences between target and distractor trials.** The top panel illustrates the three-dimensional rendered projections associated with the detection task (T > D). The bottom panel shows the regions that showed increased activation during target detection in the encoding phase. Activations are rendered at FWE voxel level = 0.05 (cluster size threshold for the display = 100 voxels). *Image courtesy of Bechi Gabrielli, G., Fagioli, S., Spataro, P., Macaluso, M., Serra, L., Bozzali, M., Rossi-Arnaud C., 2019. The Attentional Ventral Network at the Base of the Attentional Boost Effect. Poster session. OHBM Annual Meeting - Organization of Human Brain Mapping. Rome, Italy.*

that the encoding of the images not recognized in the recognition task (both target- and distractor-paired) was associated with an activation pattern compatible with the Default Mode Network (DMN) (Fig. A.4).

The DMN is commonly active when persons are not involved in perceptually demanding tasks, such as autobiographical planning and internally guided thoughts (Andrews-Hanna et al., 2010; Buckner et al., 2008; Smallwood et al., 2013). Several works suggested an interplay between cognitive states that may involve fluctuations in the activity of the DMN and attention networks (Smallwood and Schooler, 2006). For example, Hasenkamp et al. (2012) demonstrated that fluctuations between mind-wandering and an attentional state were associated, respectively, with neural activations in regions corresponding to the DMN network and the attentional networks, including the ventral fronto-parietal system. Mittner et al. (2016) proposed that the DMN and the LC systems interact to give rise to task-unrelated thought (TUT, or mind-wandering). In this model, the norepinephrine fluctuations determine a global processing state that influences the efficiency in solving tasks or engaging in mind wandering. The phasic activation of the LC, according to Swallow and Jiang (2010, 2013); see also Aston-Jones and Choen (2005), induces a state of attentional activation which enhances the processing of all the information present in the moment in which a target stimulus appears on the screen. On the other hand, the tonic activation of the LC (when the task is too simple or too difficult, or when the participant loses interest in the task) determines a more exploratory state in which the activation of task-unrelated network becomes dominant, shifting the focus of attention away from the task. Our data suggest that if participants are engaged in mind wandering during the encoding of pictures, they will not recognize pictures in the later memory task, despite the correct target detection.

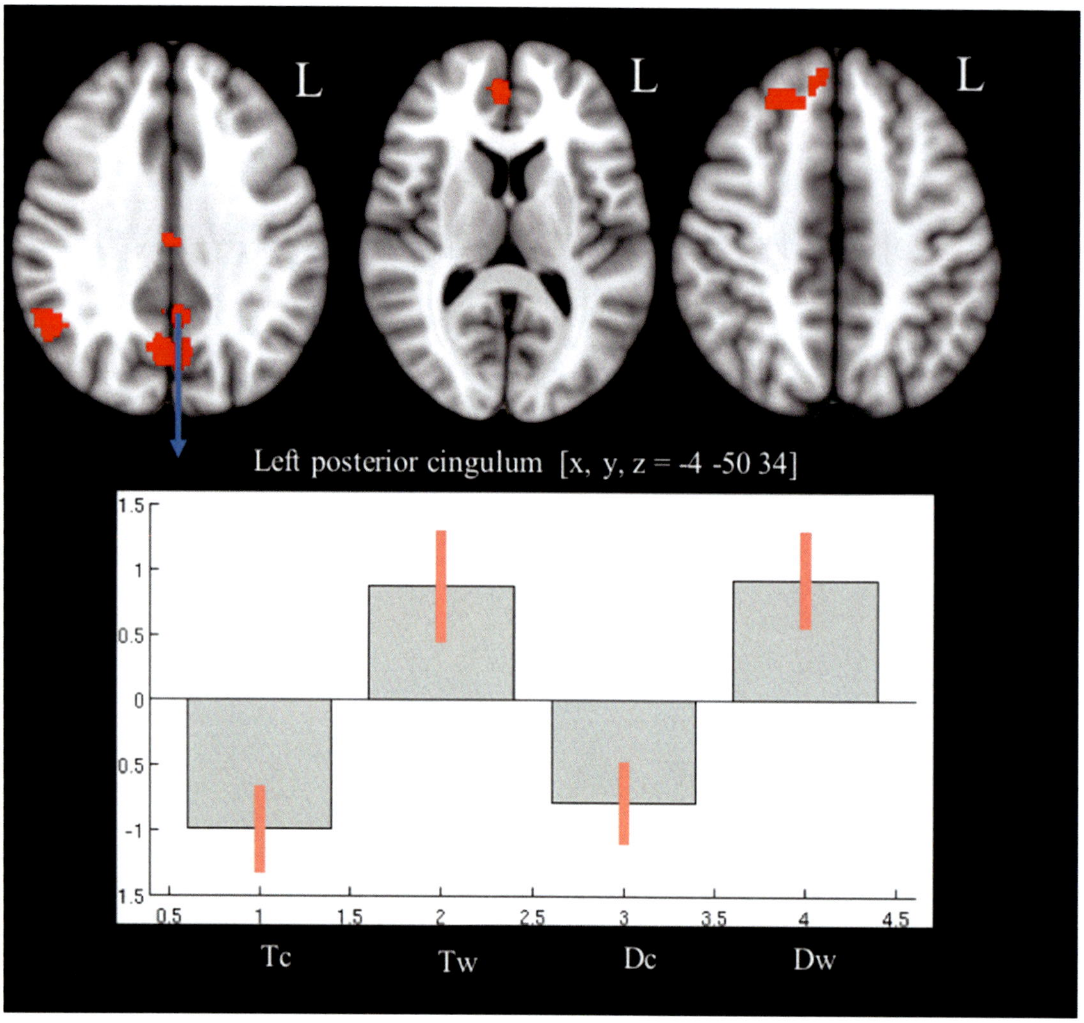

FIGURE A.4 **Differences between recognized and nonrecognized trials.** The top panel illustrates horizontal sections for the regions that showed increased activation during the encoding of nonrecognized pictures (nR > R). The bottom shows the signal plot for the left posterior cingulum. The pattern of activation shows increased activity during the encoding of nonrecognized pictures, independently of trial type. Activations are rendered at FWE voxel level = 0.05 (cluster size threshold for the display = 100 voxels). *Image courtesy of Bechi Gabrielli, G., Fagioli, S., Spataro, P., Macaluso, M., Serra, L., Bozzali, M., Rossi-Arnaud C., 2019. The Attentional Ventral Network at the Base of the Attentional Boost Effect. Poster session. OHBM Annual Meeting - Organization of Human Brain Mapping. Rome, Italy.*

CONCLUSIONS

In conclusion, the findings briefly reviewed in this chapter indicate that the LC activity is highly plastic and that it is not rigidly linked to specific sensory attributes of the target stimulus: rather, it responds to events in a task-sensitive manner. For these reasons, several investigators suggested that the LC-NE system facilitates attentional and cognitive shifts and behavioral adaptation to changes in environmental imperatives (Angela and Dayan, 2005; Bouret & Sara, 2004, 2005; Devauges and Sara, 1990; Sara and Segal, 1991).

This is coherent with the hypothesis that the LC-NE system would be at the base of the ABE, a paradoxical phenomenon in which images or words presented with targets (visual or auditory) are recognized better than stimuli presented with distractors. According to Swallow and Jiang (2010), detecting an occasional target in a secondary task would induce a transient attentional response when the target appears. This attentive orientation response could lead to an increase in the amount of attentional resources available, facilitating the encoding of the background stimuli in memory. They also hypothesized that the ABE represents a compromise between attentional competition, due to the interference effect of the detection task, and attentional facilitation, due to the transient increase in attention following target detection (Swallow and Jiang, 2010, 2013). When the attentional requests needed to detect the target are relatively low, the combination between the two effects produces a clear facilitation. However, as the attentive requests for performing the secondary task increase, interference overcomes facilitation, eliminating the effect (Swallow and Jiang, 2010, Experiment 5).

The dual-task interaction model accounts for the ABE by proposing that the categorization of an item as a target triggers a temporal selection mechanism that enhances the processing of all the information present during behaviorally relevant moments (Swallow and Jiang, 2013). Consistent with this account, detecting a target elicits a temporary increase in the release of norepinephrine (NE) from the locus coeruleus (LC) and the activation of the ventral attention network (Bechi Gabrielli et al., 2019; Corbetta et al., 2008), interrupting the ongoing processes and reorienting attention to salient events (Corbetta et al., 2008; Sara and Bouret, 2012). Furthermore, when a target is detected, a transient phasic discharge from the LC enhances the perceptual processing of the stimuli occurring at the same time as the target. In fact, target detection produces a transient increase in the activation of perceptual brain regions, even when no background stimulus was presented (Swallow et al., 2012). Taken together, the activation of the ventral attentional network (Bechi Gabrielli et al., 2019) and the enhanced perceptual processing mediated by phasic LC discharge (Swallow et al., 2012) might increase the encoding and later memorization of target-paired stimuli.

References

Anand, A., Charney, D.S., 2000. Norepinephrine dysfunction in depression. J. Clin. Psychiatr. 61, 16–24.

Andrews-Hanna, J.R., Reidler, J.S., Huang, C., Buckner, R.L., 2010. Evidence for the default network's role in spontaneous cognition. J. Neurophysiol. 104 (1), 322–335.

Angela, J.Y., Dayan, P., 2005. Uncertainty, neuromodulation, and attention. Neuron 46 (4), 681–692.

Arnsten, A.F., Li, B.M., 2005. Neurobiology of executive functions: catecholamine influences on prefrontal cortical functions. Biol. Psychiatr. 57 (11), 1377–1384.

Arnsten, A.F., Wang, M.J., Paspalas, C.D., 2012. Neuromodulation of thought: flexibilities and vulnerabilities in prefrontal cortical network synapses. Neuron 76 (1), 223–239.

Aston-Jones, G., 1985. Behavioral functions of locus coeruleus derived from cellular attributes. Physiol. Psychol. 13 (3), 118–126.

Aston-Jones, G., Bloom, F.E., 1981. Activity of norepinephrine-containing locus coeruleus neurons in behaving rats anticipates fluctuations in the sleep-waking cycle. J. Neurosci. 1 (8), 876–886.

Aston-Jones, G., Bloom, F.E., 1981. Nonrepinephrine-containing locus coeruleus neurons in behaving rats exhibit pronounced responses to non-noxious environmental stimuli. J. Neurosci. 1 (8), 887–900.

Aston-Jones, G., Cohen, J.D., 2005. An integrative theory of locus coeruleus–norepinephrine function: adaptive gain and optimal performance. Annu. Rev. Neurosci. 28, 403–450.

Aston-Jones, G., Rajkowski, J., Cohen, J., 1999. Role of locus coeruleus in attention and behavioral flexibility. Biol. Psychiatr. 46 (9), 1309–1320.

Aston-Jones, G., Rajkowski, J., Kubiak, P., 1997. Conditioned responses of monkey locus coeruleus neurons anticipate acquisition of discriminative behavior in a vigilance task. Neuroscience 80 (3), 697–715.

Aston-Jones, G., Rajkowski, J., Kubiak, P., Alexinsky, T., 1994. Locus coeruleus neurons in monkey are selectively activated by attended cues in a vigilance task. J. Neurosci. 14 (7), 4467–4480.

Bechi Gabrielli, G., Rossi-Arnaud, C., Spataro, P., Doricchi, F., Costanzi, M., Santirocchi, A., Cestari, V., 2021. The attentional boost effect in young and adult euthymic bipolar patients and healthy controls. J. Personalized Med. 11 (3), 185.

Bechi Gabrielli, G., Fagioli, S., Spataro, P., Macaluso, M., Serra, L., Bozzali, M., Rossi-Arnaud, C., 2019. The Attentional Ventral Network at the Base of the Attentional Boost Effect. Poster session. OHBM Annual Meeting - Organization of Human Brain Mapping, Rome, Italy. Rome (italy).

Bechi Gabrielli, G., Spataro, P., Pezzuti, L., Rossi-Arnaud, C., 2018. When divided attention fails to enhance memory encoding: the attentional boost effect is eliminated in young-old adults. Psychol. Aging 33 (2), 259.

Berridge, C.W., Waterhouse, B.D., 2003. The locus coeruleus–noradrenergic system: modulation of behavioral state and state-dependent cognitive processes. Brain Res. Rev. 42 (1), 33–84.

Berridge, C.W., Page, M.E., Valentino, R.J., Foote, S.L., 1993. Effects of locus coeruleus inactivation on electroencephalographic activity in neocortex and hippocampus. Neuroscience 55 (2), 381–393.

Bledowski, C., Prvulovic, D., Goebel, R., Zanella, F.E., Linden, D.E., 2004. Attentional systems in target and distractor processing: a combined ERP and fMRI study. Neuroimage 22 (2), 530–540.

Bouret, S., Sara, S.J., 2002. Locus coeruleus activation modulates firing rate and temporal organization of odour-induced single-cell responses in rat piriform cortex. Eur. J. Neurosci. 16 (12), 2371–2382.

Bouret, S., Sara, S.J., 2004. Reward expectation, orientation of attention and locus coeruleus-medial frontal cortex interplay during learning. Eur. J. Neurosci. 20 (3), 791–802.

Bouret, S., Sara, S.J., 2005. Network reset: a simplified overarching theory of locus coeruleus noradrenaline function. Trends Neurosci. 28 (11), 574–582.

Braak, H., Del Tredici, K., 2012. Alzheimer's disease: pathogenesis and prevention. Alzheimer's Dementia 8 (3), 227–233.

Braak, H., Thal, D.R., Ghebremedhin, E., Del Tredici, K., 2011. Stages of the pathologic process in Alzheimer disease: age categories from 1 to 100 years. J. Neuropathol. Exp. Neurol. 70 (11), 960–969.

Buckner, R.L., Andrews-Hanna, J.R., Schacter, D.L., 2008. The brain's default network. Ann. N. Y. Acad. Sci. 1124 (1), 1–38.

Corbetta, M., Shulman, G.L., 2002. Control of goal-directed and stimulus-driven attention in the brain. Nat. Rev. Neurosci. 3 (3), 201–215.

Corbetta, M., Patel, G., Shulman, G.L., 2008. The reorienting system of the human brain: from environment to theory of mind. Neuron 58 (3), 306–324.

Criss, A.H., Malmberg, K.J., 2008. Evidence in favor of the early- phase elevated-attention hypothesis: the effects of letter frequency and object frequency. J. Mem. Lang. 59, 331–345.

de Zubicaray, G.I., McMahon, K.L., Eastburn, M.M., Finnigan, S., Humphreys, M.S., 2005. fMRI evidence of word frequency and strength effects during episodic memory encoding. Cognit. Brain Res. 22 (3), 439–450.

Devauges, V., Sara, S.J., 1990. Activation of the noradrenergic system facilitates an attentional shift in the rat. Behav. Brain Res. 39 (1), 19–28.

Devilbiss, D.M., Waterhouse, B.D., 2011. Phasic and tonic patterns of locus coeruleus output differentially modulate sensory network function in the awake rat. J. Neurophysiol. 105 (1), 69–87.

Fabiani, M., Donchin, E., 1995. Encoding processes and memory organization: a model of the von Restorff effect. J. Exp. Psychol. Learn. Mem. Cognit. 21 (1), 224–240.

Foote, S.L., Morrison, J.H., 1987. Extrathalamic modulation of cortical function. Annu. Rev. Neurosci. 10 (1), 67–95.

Foote, S.L., Aston-Jones, G., Bloom, F.E., 1980. Impulse activity of locus coeruleus neurons in awake rats and monkeys is a function of sensory stimulation and arousal. Proc. Natl. Acad. Sci. U. S. A. 77 (5), 3033–3037.

Gallagher, M., Rapp, P.R., Fanelli, R.J., 1985. Opiate antagonist facilitation of time-dependent memory processes: dependence upon intact norepinephrine function. Brain Res. 347 (2), 284–290.

George, S.A., Knox, D., Curtis, A.L., Aldridge, J.W., Valentino, R.J., Liberzon, I., 2013. Altered locus coeruleus–norepinephrine function following single prolonged stress. Eur. J. Neurosci. 37 (6), 901–909.

Geraci, L., Rajaram, S., 2004. The distinctiveness effect in the absence of conscious recollection: evidence from conceptual priming. J. Mem. Lang. 51, 217–230.

Glanzer, M., Adams, J.K., 1990. The mirror effect in recognition memory: data and theory. J. Exp. Psychol. Learn. Mem. Cognit. 16 (1), 5.

Gören, M.Z., Cabadak, H., 2014. Noradrenaline and post-traumatic stress disorder. Comprehen. Guide Post-Traumat. Stre. Disor. 1–16.

Grudzien, A., Shaw, P., Weintraub, S., Bigio, E., Mash, D.C., Mesulam, M.M., 2007. Locus coeruleus neurofibrillary degeneration in aging, mild cognitive impairment and early Alzheimer's disease. Neurobiol. Aging 28 (3), 327–335.

Guttentag, R.E., Carroll, D., 1997. Recollection-based recognition: word frequency effects. J. Mem. Lang. 37 (4), 502–516.

Harley, C.W., 1987. A role for norepinephrine in arousal, emotion and learning?: limbic modulation by norepinephrine and the Kety hypothesis. Progr. Neuro-psychopharmacol. Biol. Psychiat. 11 (4), 419–458.

Harley, C.W., 2007. Norepinephrine and the dentate gyrus. Prog. Brain Res. 163, 299–318.

Hasenkamp, W., Wilson-Mendenhall, C.D., Duncan, E., Barsalou, L.W., 2012. Mind wandering and attention during focused meditation: a fine-grained temporal analysis of fluctuating cognitive states. Neuroimage 59 (1), 750–760.

Heath, T.P., Melichar, J.K., Nutt, D.J., Donaldson, L.F., 2006. Human taste thresholds are modulated by serotonin and noradrenaline. J. Neurosci. 26 (49), 12664–12671.

Hobson, J.A., McCarley, R.W., Wyzinski, P.W., 1975. Sleep cycle oscillation: reciprocal discharge by two brainstem neuronal groups. Science 189 (4196), 55–58.

Jacobs, H.I., Wiese, S., van de Ven, V., Gronenschild, E.H., Verhey, F.R., Matthews, P.M., 2015. Relevance of parahippocampal-locus coeruleus connectivity to memory in early dementia. Neurobiol. Aging 36 (2), 618–626.

Jouvet, M., 1969. Biogenic amines and the states of sleep. Science 163 (3862), 32–41.

Kéri, S., Nagy, H., Levy-Gigi, E., Kelemen, O., 2013. How attentional boost interacts with reward: the effect of dopaminergic medications in Parkinson's disease. Eur. J. Neurosci. 38 (11), 3650–3658.

Kim, H., 2014. Involvement of the dorsal and ventral attention networks in oddball stimulus processing: a meta-analysis. Hum. Brain Mapp. 35 (5), 2265–2284.

Korf, J., Aghajanian, G.K., Roth, R.H., 1973. Increased turnover of norepinephrine in the rat cerebral cortex during stress: role of the locus coeruleus. Neuropharmacology 12 (10), 933–938.

Krebs, R.M., Park, H.R., Bombeke, K., Boehler, C.N., 2018. Modulation of locus coeruleus activity by novel oddball stimuli. Brain Imag. Behav. 12 (2), 577–584.

Lecas, J.C., 2004. Locus coeruleus activation shortens synaptic drive while decreasing spike latency and jitter in sensorimotor cortex. Implications for neuronal integration. Eur. J. Neurosci. 19 (9), 2519–2530.

Leclercq, V., Le Dantec, C.C., Seitz, A.R., 2014. Encoding of episodic information through fast task-irrelevant perceptual learning. Vis. Res. 99, 5–11.

Levy-Gigi, E., Kéri, S., 2012. Falling out of time: enhanced memory for scenes presented at behaviorally irrelevant points in time in posttraumatic stress disorder (PTSD). PLoS One 7 (7), e42502.

Lin, J.Y., Pype, A.D., Murray, S.O., Boynton, G.M., Fahle, M., 2010. Enhanced memory for scenes presented at behaviorally relevant points in time. PLoS Biol. 8 (3), 1–6.

Linden, D.E., Prvulovic, D., Formisano, E., Völlinger, M., Zanella, F.E., Goebel, R., Dierks, T., 1999. The functional neuroanatomy of target detection: an fMRI study of visual and auditory oddball tasks. Cerebr. Cortex 9 (8), 815–823.

Lohr, J.B., Jeste, D.V., 1988. Locus ceruleus morphometry in aging and schizophrenia. Acta Psychiatr. Scand. 77 (6), 689–697.

Mäki-Marttunen, V., Andreassen, O.A., Espeseth, T., 2020. The role of norepinephrine in the pathophysiology of schizophrenia. Neurosci. Biobehav. Rev.

Makovski, T., Jiang, Y.V., Swallow, K.M., 2013. How do observer's responses affect visual long-term memory? J. Exp. Psychol. Learn. Mem. Cognit. 39 (4), 1097.

Makovski, T., Swallow, K.M., Jiang, Y.V., 2011. Attending to unrelated targets boosts short-term memory for color arrays. Neuropsychologia 49 (6), 1498–1505.

Malmberg, K.J., Nelson, T.O., 2003. The word frequency effect for recognition memory and the elevated-attention hypothesis. Mem. Cognit. 31 (1), 35–43.

Marsh, R.L., Cook, G.I., Hicks, J.L., 2006. The effect of context variability on source memory. Mem. Cognit. 34, 1578–1586.

Mather, M., Harley, C.W., 2016. The locus coeruleus: essential for maintaining cognitive function and the aging brain. Trend. Cognit. Sci. 20 (3), 214–226.

McGaughy, J., Ross, R.S., Eichenbaum, H., 2008. Noradrenergic, but not cholinergic, deafferentation of prefrontal cortex impairs attentional set-shifting. Neuroscience 153 (1), 63–71.

McIntyre, C.K., McGaugh, J.L., Williams, C.L., 2012. Interacting brain systems modulate memory consolidation. Neurosci. Biobehav. Rev. 36 (7), 1750–1762.

McLean, J., Waterhouse, B.D., 1994. Noradrenergic modulation of cat area 17 neuronal responses to moving visual stimuli. Brain Res. 667 (1), 83–97.

Mittner, M., Hawkins, G.E., Boekel, W., Forstmann, B.U., 2016. A neural model of mind wandering. Trend. Cognit. Sci. 20 (8), 570–578.

Miyawaki, T., Kawamura, H., Hara, K., Suzuki, K., Usui, W., Yasugi, T., 1993. Differential regional hemodynamic changes produced by L-glutamate stimulation of the locus coeruleus. Brain Res. 600 (1), 56–62.

Miyawaki, T., Kawamura, H., Komatsu, K., Yasugi, T., 1991. Chemical stimulation of the locus coeruleus: inhibitory effects on hemodynamics and renal sympathetic nerve activity. Brain Res. 568 (1), 101–108.

Morrison, J.H., Foote, S.L., O'Connor, D., Bloom, F.E., 1982. Laminar, tangential and regional organization of the noradrenergic innervation of monkey cortex: dopamine-β-hydroxylase immunohistochemistry. Brain Res. Bull. 9 (1), 309–319.

Moyal, R., Turker, H.B., Luh, W.M., Swallow, K.M., 2020. Auditory Target Detection Enhances Visual Processing and Hippocampal Functional Connectivity. bioRxiv.

Mulligan, N.W., Spataro, P., 2015. Divided attention can enhance early-phase memory encoding: the attentional boost effect and study trial duration. J. Exp. Psychol. Learn. Mem. Cognit. 41 (4), 1223–1228.

Mulligan, N.W., Smith, S.A., Spataro, P., 2016. The attentional boost effect and context memory. J. Exp. Psychol. Learn. Mem. Cognit. 42 (4), 598.

Mulligan, N.W., Spataro, P., Picklesimer, M., 2014. The attentional boost effect with verbal materials. J. Exp. Psychol. Learn. Mem. Cognit. 40 (4), 1049–1063.

Murphy, P.R., O'connell, R.G., O'sullivan, M., Robertson, I.H., Balsters, J.H., 2014. Pupil diameter covaries with BOLD activity in human locus coeruleus. Hum. Brain Mapp. 35 (8), 4140–4154.
Nieuwenhuis, S., Gilzenrat, M.S., Holmes, B.D., Cohen, J.D., 2005. The role of the locus coeruleus in mediating the attentional blink: a neurocomputational theory. J. Exp. Psychol. Gen. 134 (3), 291–307.
O'donnell, T., Hegadoren, K.M., Coupland, N.C., 2004. Noradrenergic mechanisms in the pathophysiology of post-traumatic stress disorder. Neuropsychobiology 50 (4), 273–283.
Olivers, C.N.L., Meeter, M., 2008. A boost and bounce theory of temporal attention. Psychol. Rev. 115 (4), 836.
Prull, M.W., 2019. The attentional boost effect for words in young and older adults. Psychol. Aging.
Rajkowski, J., Kubiak, P., Aston-Jones, G., 1994. Locus coeruleus activity in monkey: phasic and tonic changes are associated with altered vigilance. Brain Res. Bull. 35 (5), 607–616.
Rajkowski, J., Majczynski, H., Clayton, E., Aston-Jones, G., 2004. Activation of monkey locus coeruleus neurons varies with difficulty and performance in a target detection task. J. Neurophysiol. 92 (1), 361–371.
Raymond, J.E., Fenske, M.J., Westoby, N., 2005. Emotional devaluation of distracting patterns and faces: a consequence of attentional inhibition during visual search? J. Exp. Psychol. Hum. Percept. Perform. 31 (6), 1404.
Raz, N., Lindenberger, U., Rodrigue, K.M., Kennedy, K.M., Head, D., Williamson, A., Acker, J.D., 2005. Regional brain changes in aging healthy adults: general trends, individual differences and modifiers. Cerebr. Cortex 15 (11), 1676–1689.
Ressler, K.J., Nemeroff, C.B., 1999. Role of norepinephrine in the pathophysiology and treatment of mood disorders. Biol. Psychiatr. 46 (9), 1219–1233.
Robbins, T.W., Roberts, A.C., 2007. Differential regulation of frontoexecutive function by the monoamines and acetylcholine. Cerebr. Cortex 17 (Suppl. 1_1), i151–i160.
Rossi-Arnaud, C., Spataro, P., Saraulli, D., Mulligan, N.W., Sciarretta, A., Marques, V.R., Cestari, V., 2014. The attentional boost effect in schizophrenia. J. Abnorm. Psychol. 123 (3), 588.
Santangelo, V., 2015. Forced to remember: when memory is biased by salient information. Behav. Brain Res. 283, 1–10.
Sara, S.J., 2009. The locus coeruleus and noradrenergic modulation of cognition. Nat. Rev. Neurosci. 10 (3), 211.
Sara, S.J., Bouret, S., 2012. Orienting and reorienting: the locus coeruleus mediates cognition through arousal. Neuron 76 (1), 130–141.
Sara, S.J., Segal, M., 1991. Plasticity of sensory responses of locus coeruleus neurons in the behaving rat: implications for cognition. Prog. Brain Res. 88, 571–585.
Seitz, A.R., Watanabe, T., 2009. The phenomenon of task-irrelevant perceptual learning. Vis. Res. 49 (21), 2604–2610.
Shibata, E., Sasaki, M., Tohyama, K., Kanbara, Y., Otsuka, K., Ehara, S., Sakai, A., 2006. Age-related changes in locus ceruleus on neuromelanin magnetic resonance imaging at 3 Tesla. Magn. Reson. Med. Sci. 5 (4), 197–200.
Shulman, G.L., Tansy, A.P., Kincade, M., Petersen, S.E., McAvoy, M.P., Corbetta, M., 2002. Reactivation of networks involved in preparatory states. Cerebr. Cortex 12 (6), 590–600.
Siever, L.J., Davis, K.L., 1985. Overview: toward a dysregulation hypothesis of depression. Am. J. Psychiatr. 142 (9), 1017–1031.
Smallwood, J., Schooler, J.W., 2006. The restless mind. Psychol. Bull. 132 (6), 946.
Smallwood, J., Tipper, C., Brown, K., Baird, B., Engen, H., Michaels, J.R., Schooler, J.W., 2013. Escaping the here and now: evidence for a role of the default mode network in perceptually decoupled thought. Neuroimage 69, 120–125.
Southwick, S.M., Morgan, C.A., Nicolaou, A.L., Charney, D.S., 1997. Consistency of memory for combat-related traumatic events in veterans of Operation Desert Storm. Am. J. Psychiatr. 154 (2), 173–177.
Spataro, P., Mulligan, N.W., Rossi-Arnaud, C., 2013. Divided attention can enhance memory encoding: the attentional boost effect in implicit memory. J. Exp. Psychol. Learn. Mem. Cognit. 39 (4), 1223–1231.
Spataro, P., Mulligan, N.W., Rossi-Arnaud, C., 2015. Limits to the attentional boost effect: the moderating influence of orthographic distinctiveness. Psychon. Bull. Rev. 22 (4), 987–992.
Spataro, P., Mulligan, N.W., Bechi Gabrielli, G., Rossi-Arnaud, C., 2017. Divided attention enhances explicit but not implicit conceptual memory: an item-specific account of the attentional boost effect. Memory 25 (2), 170–175.
Spataro, P., Saraulli, D., Cestari, V., Mulligan, N.W., Santirocchi, A., Borowiecki, O., Rossi-Arnaud, C., 2020. The attentional boost effect enhances the recognition of bound features in short-term memory. Memory 28 (7), 926–937.
Swallow, K.M., Atir, S., 2019. The role of value in the attentional boost effect. Q. J. Exp. Psychol. 72 (3), 523–542.
Swallow, K.M., Jiang, Y.V., 2010. The attentional boost effect: transient increases in attention to one task enhance performance in a second task. Cognition 115 (1), 118–132.
Swallow, K.M., Jiang, Y.V., 2011. The role of timing in the attentional boost effect. Atten. Percept. Psychophys. 73 (2), 389–404.
Swallow, K.M., Jiang, Y.V., 2012. Goal-relevant events not to be rare to boost memory performance for concurrent images. Atten. Percept. Psychophys. 74 (1), 70–82.
Swallow, K.M., Jiang, Y.V., 2013. Attentional load and attentional boost: a review of data and theory. Front. Psychol. 4, 274.
Swallow, K.M., Jiang, Y.V., 2014. The attentional boost effect really is a boost: evidence from a new baseline. Atten. Percept. Psychophys. 76 (5), 1298–1307.
Swallow, K.M., Jiang, Y.V., Riley, E.B., 2019. Target detection increases pupil diameter and enhances memory for background scenes during multi-tasking. Sci. Rep. 9 (1), 1–13.
Swallow, K.M., Makovski, T., Jiang, Y.V., 2012. Selection of events in time enhances activity throughout early visual cortex. J. Neurophysiol. 108 (12), 3239–3252.
Szamosi, A., Levy-Gigi, E., Kelemen, O., Kéri, S., 2013. The hippocampus plays a role in the recognition of visual scenes presented at behaviorally relevant points in time: evidence from amnestic mild cognitive impairment (aMCI) and healthy controls. Cortex 49 (7), 1892–1900.
Valentino, R.J., Curtis, A.L., 1991. Antidepressant interactions with corticotropin-releasing factor in the noradrenergic nucleus locus coeruleus. Psychopharmacol. Bull. 27 (3), 263–269.
Valentino, R.J., Foote, S.L., 1987. Corticotropin-releasing factor disrupts sensory responses of brain noradrenergic neurons. Neuroendocrinology 45 (1), 28–36.
Valentino, R.J., Foote, S.L., 1988. Corticotropin-releasing hormone increases tonic but not sensory-evoked activity of noradrenergic locus coeruleus neurons in unanesthetized rats. J. Neurosci. 8 (3), 1016–1025.
Valentino, R.J., Van Bockstaele, E., 2008. Convergent regulation of locus coeruleus activity as an adaptive response to stress. Eur. J. Pharmacol. 583 (2–3), 194–203.
Valentino, R.J., Wehby, R.G., 1988. Corticotropin-releasing factor: evidence for a neurotransmitter role in the locus ceruleus during hemodynamic stress. Neuroendocrinology 48 (6), 674–677.
van Enkhuizen, J., Janowsky, D.S., Olivier, B., Minassian, A., Perry, W., Young, J.W., Geyer, M.A., 2015. The catecholaminergic–cholinergic balance hypothesis of bipolar disorder revisited. Eur. J. Pharmacol. 753, 114–126.
Ward, D.G., Gunn, C.G., 1976. Locus coeruleus complex: elicitation of a pressor response and a brain stem region necessary for its occurrence. Brain Res. 107 (2), 401–406.
Waterhouse, B.D., Azizi, S.A., Burne, R.A., Woodward, D.J., 1990. Modulation of rat cortical area 17 neuronal responses to moving visual

stimuli during norepinephrine and serotonin microiontophoresis. Brain Res. 514 (2), 276–292.

Waterhouse, B.D., Moises, H.C., Woodward, D.J., 1998. Phasic activation of the locus coeruleus enhances responses of primary sensory cortical neurons to peripheral receptive field stimulation. Brain Res. 790 (1–2), 33–44.

Weinshenker, D., 2008. Functional consequences of locus coeruleus degeneration in Alzheimer's disease. Curr. Alzheimer Res. 5 (3), 342–345.

Yebra, M., Galarza-Vallejo, A., Soto-Leon, V., Gonzalez-Rosa, J.J., de Berker, A.O., Bestmann, S., Strange, B.A., 2019. Action boosts episodic memory encoding in humans via engagement of a noradrenergic system. Nat. Commun. 10 (1), 1–12.

Zacks, J.M., Speer, N.K., Swallow, K.M., Braver, T.S., Reynolds, J.R., 2007. Event perception: a mind-brain perspective. Psychol. Bull. 133 (2), 273.

B

Neuroanatomical and Neurophysiological Bases of the Anesthesia

Nimet Şenoğlu[1] *and Yağmur Çınar*[2]

[1]Çiğli Education and Training Hospital Anesthesia and Reanimation, İzmir Bakırçay University Faculty of Medicine, Izmir, Turkey [2]Anesthesiology and Reanimation Department, İzmir Tepecik Training and Research Hospital, İzmir, Turkey

GENERAL ANESTHESIA

General anesthesia is a drug-induced reversible depression of the central nervous system (CNS) resulting in the loss of response and perception of all external stimuli (Barash, 2013). One can argue that the previous definition is not fully comprehensive, as it does not emphasize the importance of unconsciousness, amnesia, and analgesia, which are widely accepted as key components of general anesthesia, alongside immobility and attenuation of autonomic responses to noxious stimulation. The Lancet reports more than 300 million people to receive general anesthesia under various conditions every year all over the world. These 300 million people share a common experience of undergoing general anesthesia, but the specific drugs used to induce this state can vary. General anesthetics can be broadly categorized into two groups: inhalational anesthetics and intravenous anesthetics. The mechanisms underlying consciousness are still an ongoing subject of research, and this chapter will delve into the effects of anesthetics on consciousness in greater detail.

QUANTITATIVE ASSESSMENT OF ANESTHETIC DEPTH

Minimum alveolar concentration (MAC) is the alveolar partial pressure of a gas at which 50% of patients do not respond to a surgical incision (Quasha et al., 1980). MAC provides valuable information for inhalation anesthetics, but it cannot be applied to parenteral anesthetics. MAC represents the average response of a whole population of subjects rather than the response of a single subject. To date, patient safety standards require more precise measurements than measurements of delivered concentrations, such as a processed EEG value (Palanca et al., 2009), an evoked potential (Bruhn et al., 2006), or a functional neuroimaging signal indicative of integrated cortical activity (Alkire et al., 2008).

Electroencephalography records electrical activity in the cerebral cortex and can detect ischemia and seizure activity, and it can also assess the effect of anesthetic agent on the brain (Koht et al., 2022). All anesthetic drugs affect the electroencephalogram. Inhalation anesthetics, including isoflurane, sevoflurane, desflurane and halothane, cause a dose-dependent decrease in amplitude and an increase in latency of evoked responses, to varying degrees (Fig. B.1). Volatile agents produce an initial increase in frequency of the frontal EEG, with decrease in frequency and amplitude at higher doses, burst suppression at about 1.5 MAC and electrical silence at high doses.

ANESTHETIC INTERACTIONS WITH THE ELECTROPHYSIOLOGIC FUNCTION OF NERVOUS SYSTEM

Anesthetics show their action by altering the intrinsic firing rate of individual neurons, therefore by changing

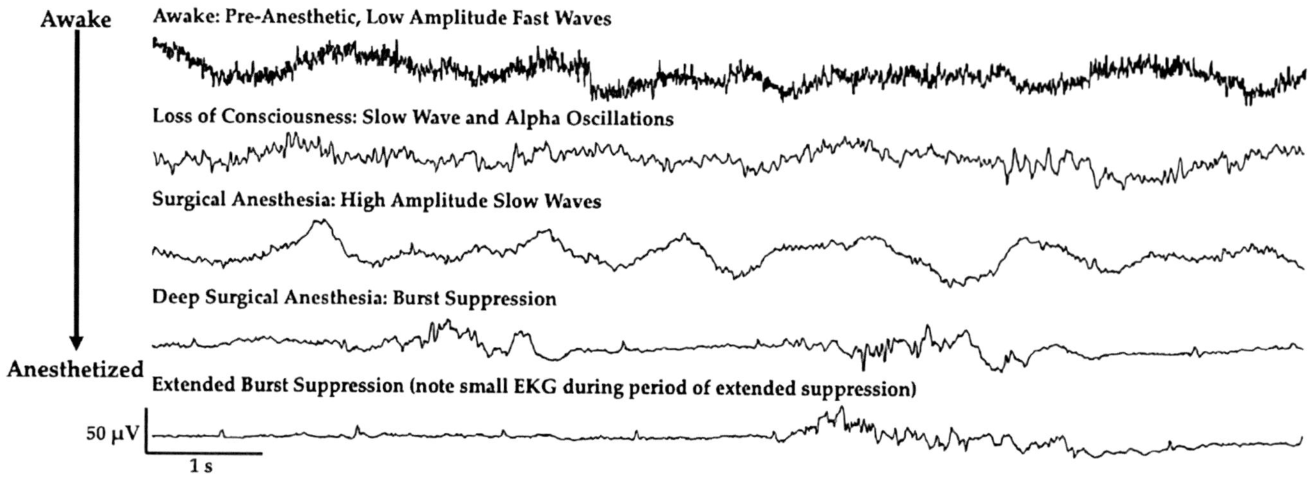

FIGURE B.1 **Patient electroencephalogram (EEG) signals change with increasing anesthetic depth when administered propofol; the image above shows traces from frontal electrode site (F7) from a patient anaesthetized with propofol.** High frequency, low amplitude activity slows to include predominantly alpha and slow oscillations with anesthetic administration. Activity slows more to include a slow-wave dominant pattern, much like during natural sleep. At profound levels of unconsciousness (bottom two traces) a burst suppression pattern emerges, and suppressed periods extend as patients reach more profound levels of unconsciousness. Note the electrical activity from heart beats (electrocardiogram, EKG) which can be seen in the prolonged suppressed period. *Image courtesy of Eagleman, S., MacIver, M.B., 2021. Molecular diversity of anesthetic actions is evident in electroencephalogram effects in humans and animals. Int. J. Mol. Sci. 22 (2), 495. https://doi.org/10.3390/ijms22020495.*

neuronal excitability and/or by altering communication between neurons by disrupting synaptic transmission (Barash, 2013).

The term "neuronal excitability" refers to the propensity of a neuron to generate and propagate action potentials from its cell body to the nerve terminals. Intrinsic neuronal excitability is determined by factors such as resting membrane potential, the threshold potential required for action potential initiation, and the size and propagation of action potentials.

Anesthetics have the ability to hyperpolarize both spinal motor neurons and cortical neurons, and the extent of this effect is correlated with the potency of the anesthetic (MacIver et al., 1991; Madison et al., 1982). While anesthetics generally induce a small hyperpolarization, it is typically insufficient to completely prevent the propagation of an already initiated action potential. However, they can hinder the initiation of action potentials.

For example, isoflurane, an inhalational anesthetic, has been found to hyperpolarize thalamic neurons, leading to the inhibition of tonic firing of action potentials. This interference with thalamic neuronal activity can contribute to the overall effects of anesthesia (Ries and Puil, 1999).

Synaptic transmission is believed to be the primary level of action for general anesthetics. These anesthetics have been shown to inhibit excitatory synaptic transmission in various preparations, including sympathetic ganglia (Larrabee and Posternak, 1952), olfactory cortex (Richards et al., 1975), hippocampus (Richards and White, 1975), and spinal cord (Kullmann et al., 1989). Not all excitatory synapses are equally sensitive to anesthetics; transmission across some hippocampal excitatory synapses is augmented by inhalational anesthetics. MacIver and Roth (1988) and Nicoll et al. (1975) showed that barbiturates enhanced inhibitory synaptic transmission by prolonging the decay of the GABAergic inhibitory postsynaptic current. The same effect is also seen with etomidate (Proctor et al., 1986), propofol (Collins, 1988), and inhalation anesthetics (Nicoll, 1972). Synaptic effects can be divided into two groups: presynaptic and postsynaptic.

Neurotransmitter release at glutamatergic synapses has been found to be inhibited by clinical concentrations of volatile anesthetics. Perouansky et al. (1995) showed in a mouse hippocampus that halothane inhibits excitatory postsynaptic potentials induced by presynaptic electrical stimulation. Reduction of glutamate release by intravenous anesthetics has also been demonstrated. The data for anesthetic effects on inhibitory neurotransmitter release are more complex.

ANESTHETIC ACTIONS ON ION CHANNELS

Anesthetic Effect on Voltage-Dependent Ion Channels

Voltage-dependent sodium, potassium, and calcium channels can sense a change in membrane potential and respond by either opening or closing their pores. These channels share important structural homologies.

Voltage-dependent sodium and potassium channels are remarkably involved in generating and shaping action potentials. Recent studies done by Rehberg et al. (1996) show that clinically relevant concentrations of a variety of inhalational anesthetics suppressed voltage elicited sodium currents. Another study by Ratnakumari and Hemmings (1998) showed that clinical concentrations of halothane inhibit sodium channels in rat brain. Wu et al. (2004) found that isoflurane causes sodium current inhibition in a rat brainstem neuron, and it results in inhibition of synaptic activity. Supraclinical concentrations of intravenous anesthetics have also been shown to inhibit sodium channels (Frenkel et al., 1998; Rehberg and Duch, 1999).

Voltage-gated calcium channels (VGCCs) are located at presynaptic terminals in the nervous system, and they respond to action potentials by opening. Opening of VGCCs results in calcium entry into the cell which causes secretion of neurotransmitters into the synaptic cleft. Many types of calcium channels are identified depending on their electrophysiologic properties or amino acid sequences. L-type calcium channels are best known for their roles in cardiac, skeletal, and smooth muscle excitation-contraction coupling. General anesthetics mostly affect L-, T-, N-, P-type of calcium channels. Volatile anesthetics inhibit VGCCs at concentrations two to five times those required to produce anesthesia in humans. Takenoshita and Steinbech (1991) reported a T-type calcium current in the dorsal root ganglion's neurons that was inhibited by subanesthetic concentrations of halothane. French-Mullen et al. (1993) found that pentobarbital at anesthetic concentrations inhibits an unspecified type of VGCCs in guinea pig hippocampus. One can conclude that even though VGCCs take some part in general anesthetic action mechanisms, they are not the major targets.

There are various types of potassium channels, which can be classified into different categories based on their properties and modes of activation. These include voltage-gated potassium channels, background or leak channels, second messenger-activated channels, ligand-activated channels, and inward rectifying channels. Each type of potassium channel has its own unique characteristics and plays a specific role in regulating the flow of potassium ions across cell membranes. Voltage-gated K^+ channels are affected by high concentrations of both volatile and intravenous anesthetics.

Background or leak K^+ channels are named after their feature that allows opening at even resting membrane potential, and they can be activated by volatile anesthetics (Patel and Honoré, 2001; Steinberg et al., 2015). The first observations about leak channels were made on a ganglion of the pond snail, *Lymnaea stagnalis* (Franks and Lieb, 1988). When this channel is activated by halothane, it results in the silencing of the spontaneous bursting of the neurons. Mammalian potassium channels differ in structure, having two pore-forming domains and four transmembrane segments (Honoré, 2007). Patel et al. (1999) showed that TREK-1 channels are activated by chloroform, diethyl ether, halothane, and isoflurane. TRAAK channels are activated by halothane and isoflurane. TRAAK channels are closely related potassium channels which are insensitive to all the volatile anesthetics. TASK channels are activated by halothane and isoflurane, inhibited by diethyl ether, and unaffected by chloroform. It is known that c-terminal regions of TASK and TREK-1 contain aminoacids essential for the anesthetic action.

Anesthetic Effects on Ligand-Gated Ion Channels

Ligand-gated ion channel proteins, including nicotinic acetylcholine receptors, glycine receptors, and 5-HT_3 receptors, open when neurotransmitters such as glutamate or GABA are released into the synaptic cleft. Thus, excitatory and inhibitory neurotransmission is mediated by the actions of these ion channels.

Glutamate-Activated Ion Channels

There are three types of glutamate-activated ion channels: AMPA receptors, kainite receptors, and NMDA receptors. AMPA and kainite receptors play a role in fast excitatory synaptic transmission and can be described as relatively nonselective monovalent cation channels. NMDA channels, which are involved in long-term potentiation, conduct multiple ions, including Na^+, K^+, and Ca^{2+}. Studies on mouse and rat brain back in the early 1980s revealed that AMPA and kainate-activated currents are insensitive to clinical concentrations of halothane (Wakamori et al., 1991) and enflurane (Lin et al., 1992), but they are sensitive to barbiturates. Pentobarbital at anesthetic doses inhibits kainate and AMPA receptor responses by 50% (Weight et al., 1991). In other studies, submaximal agonist responses of GluR3 (AMPA-type) receptors are inhibited by fluorinated volatile anesthetics, but agonist responses of GluR6 (kainate-type) receptors are enhanced (Dildy-Mayfield et al., 1996). Pentobarbital inhibits both GluR3 and GluR6 receptors.

Electrophysiologic studies show no effects of clinical concentrations of volatile anesthetics (Wakamori et al., 1991; Lin et al., 1992) or barbiturates (Weight et al., 1991) on NMDA receptors, but biochemical flux studies show volatile anesthetics may inhibit NMDA-activated channels. A rat study showed that anesthetic concentrations of halothane and enflurane inhibit NMDA-activated calcium flux by 50% (Aronstam et al., 1994). Ketamine is a well-known NMDA receptor antagonist, and it binds to the phencyclidine site on the NMDA receptor protein (Anis et al., 1983; Zeilhofer et al., 1992). Studies show that nitrous oxide and xenon also target NMDA receptors. They show their action through the inhibition of NMDA-activated currents.

GABA-Activated Ion Channels

GABA is the most important inhibitory neurotransmitter in the central nervous system. When GABA is released into synaptic cleft and binds to $GABA_A$ receptor, chloride ions enter the cell and hyperpolarization occurs. $GABA_A$ receptors pentameric multisubunit proteins and their function is altered by a variety of pharmacologic agents such as anticonvulsants, sedatives, anxiolytics, and anesthetics (Olsen and Li, 2011).

Barbiturates, benzodiazepines, propofol, etomidate, and volatile anesthetics all modulate $GABA_A$ receptor function (Jones et al., 1992). Electrophysiologic effects caused by anesthetics can be examined under three headings: potentiation, direct gating, and inhibition. Potentiation is the ability of an anesthetic agent to strengthen the current elicited by low concentrations of GABA. Anesthetics potentiation of $GABA_A$ currents generally occur at clinical concentration of anesthetics. Direct gating refers to the ability of anesthetics to activate $GABA_A$ channels in the absence of $GABA_A$. Doses higher than the clinical range are required for this effect. Inhibition refers to the ability of anesthetics to prevent GABA from initiating current flow through $GABA_A$ channels and has been observed at high concentrations of both GABA and anesthetics (Hall et al., 1994; Nakahiro et al., 1989).

The studies show that barbiturates (Macdonald et al., 1989), propofol (Hales and Lambert, 1991), and volatile anesthetics (Yeh et al., 1991) do not alter the conductance of the channel; instead, they increase the frequency or duration with which the channel opens or remains open. Anesthetics act on distinct subtypes of $GABA_A$ receptors. Benzodiazepines require the presence of the $\gamma 2$ subunit subtype to show their effect (Pritchett et al., 1989). Etomidate requires the presence of $\beta 2$ or $\beta 3$ subunit (Hill-Venning et al., 1997).

Other Ligand-Activated Ion Channels

Nicotinic acetylcholine receptors (muscle and neuronal types), glycine receptors, and $5HT_3$ receptors are modulated by general anesthetics (Forman and Miller, 2011). Muscle nicotinic receptors are inhibited by anesthetic concentrations in the clinical dosage (Dilger et al., 1994) and is desensitized by higher concentrations of anesthetics (Firestone et al., 1986). Studies done on molluscan neurons (Franks and Lieb, 1991) and bovine chromaffin cells (Charlesworth and Richards, 1995) show that neuronal nicotinic receptors are inhibited by clinical concentrations of volatile anesthetics.

Glycine is an important inhibitory neurotransmitter, especially in the spinal cord and brainstem. Studies show that clinical concentrations of volatile anesthetics potentiate glycine-activated currents in intact neurons (Wakamori et al., 1991). They do so by increasing the affinity of the receptor for glycine, much like in $GABA_A$ receptors (Downie et al., 1996). Propofol, alfaxalone, and pentobarbital also potentiate glycine-activated currents, whereas etomidate and ketamine do not (Mascia et al., 1996).

Another ligand-gated receptor channel is $5\text{-}HT_3$. Clinical concentrations of volatile anesthetics potentiate currents activated by 5-hydroxytryptamine in intact cells (Jenkins et al., 2010). Thiopental inhibits $5\text{-}HT_3$ receptor currents which can be related to anesthetic side effects such as nausea and vomiting, whereas propofol has no effect.

Although the hypothesis that anesthetics interact with the sleep–wake center in the last few decades has been supported, precise interactions and contributions for the general anesthesia state have not been disclosed yet.

The sleep and wake cycle are mediated by selected nuclei in the brain stem and hypothalamus.

NEUROANATOMICAL BASES OF THE GENERAL ANESTHESIA

Brainstem

The Locus Coeruleus (LC) is situated in the pons and has extensive projections throughout the cortex. It is responsible for synthesizing norepinephrine. LC activity is linked to cortical arousal during wakefulness and is hyperpolarized by halothane, an inhalation anesthetic (Takahashi et al., 2010). Additionally, certain intravenous anesthetics have been found to impact the release of norepinephrine in this region, thereby contributing to the depth of anesthesia.

Dexmedetomidine, one of the new generation drugs used for sedation, is an alpha agonist and mimics NREM sleep. The decrease consciousness it causes can be prevented by the alpha 2 antagonist atipemazole (Correa-Sales et al., 1992). This feature of dexmedetomidine reveals the role of LC and norepinephrine.

LATERODORSAL/PEDUNCULOPONTINE TEGMENTUM

The laterodorsal tegmentum (LDT) in the pons and the pedunculopontine tegmentum (PPT) serves as the primary sources of acetylcholine in the brain (Woolf and Butcher, 2011). These regions have direct projections to the thalamus, indicating that the transmission of information to the cortex is likely blocked. Similar to the locus coeruleus (LC), which synthesizes norepinephrine, the LDT/PPT exhibits high activity during wakefulness and reduced activity during NREM sleep (Jones, 2011). However, unlike other monoaminergic neurons, the cholinergic-acting LDT/PPT remains active during REM sleep, when the cortex is stimulated.

PONTINE RETICULAR FORMATION

The PRF (Pons Reticular Formation) is indeed a component of the reticular activating system, which is involved in regulating sleep and wakefulness. While GABA (gamma-aminobutyric acid) serves as the primary inhibitory neurotransmitter in the brain and plays a role in modulating sleep—wake states, its effect on the PRF is associated with cortical arousal. In other words, GABAergic transmission in the PRF leads to an increase in cortical activity, promoting wakefulness (Vanini et al., 2011).

VENTRAL TEGMENTAL AREA

The VTA (Ventral Tegmental Area) is located in the midbrain and consists of dopaminergic neurons that play a role in various behaviors associated with heightened arousal. However, the specific effects of the VTA on the generation and maintenance of wakefulness are still unknown. Recent research, conducted by Eban-Rothschild et al. (2016), has demonstrated a causal role for VTA dopaminergic neurons in the regulation of sleep—wake states and sleep-related behaviors. According to their findings, VTA dopaminergic neurons are capable of sustaining long-term wakefulness during periods of high sleep pressure and are essential for maintaining wakefulness even during periods of low sleep pressure.

HYPOTHALAMUS

Ventrolateral preoptic nucleus. The anterior hypothalamus and the VLPO (ventrolateral preoptic nucleus) in this region are crucial for regulating sleep—wake states. GABA (gamma-aminobutyric acid) and galanin are involved in the transmission of signals in these areas (Sherin et al., 1996). Neurons in the anterior hypothalamus and VLPO exhibit high activity during both NREM (non-rapid eye movement) and REM (rapid eye movement) phases of sleep. During sleep, the VLPO activity inhibits other arousal centers in the brainstem and hypothalamus, contributing to the induction and maintenance of unconsciousness, including during anesthesia. Therefore, the VLPO plays a significant role in promoting unconsciousness during anesthesia.

OREXINERGIC NEURONS

Orexinergic neurons, also known as hypocretin neurons (producing orexin A and B), located in the lateral hypothalamus, project to other arousal-related areas in the brainstem and basal forebrain (Zhang et al., 2016). These neurons exhibit heightened activity during wakefulness but are suppressed during NREM sleep, with intermittent bursts occurring during REM sleep. Dysfunction of the orexinergic system leads to the development of narcolepsy. Orexins play a facilitating role in the emergence from anesthesia by influencing the basal forebrain (Lee et al., 2005).

TUBEROMAMMILLARY NUCLEUS

The TMN (tuberomammillary nucleus) is located in the caudal hypothalamus, and it is the exclusive site of histamine-releasing neurons within the brain. These neurons project to various brain regions, with a particular emphasis on the ventral areas. Histamine's effect is thought to be crucial for maintaining normal wakefulness (Saper et al., 2005). There exists a reciprocal inhibition relationship between histaminergic neurons in the TMN and sleep-promoting GABAergic neurons in the VLPO (ventrolateral preoptic nucleus). TMN activity is highest during wakefulness and lowest during sleep. Although numerous studies have been conducted to investigate the mechanism, the role of the TMN and histaminergic transmission in the mechanism of anesthesia remains unclear (Liu et al., 2010).

ROLE OF THALAMUS

The thalamus plays a critical role in normal consciousness by transmitting arousal signals from the brainstem. It consists of more than 50 nuclei and subnuclei, broadly classified as sensory relays or integrative regions that receive input from the cortex.

Anesthetics have regionally reducing effects on thalamic metabolism and blood flow, making the thalamus a key player in the switch to consciousness. This phenomenon, known as the ON/OFF switch, has been observed in animal experiments through various thalamic manipulations and in clinical studies. However, it's important to note that anesthetics can have diverse effects on thalamic activity. For instance, intravenous anesthetic ketamine increases global metabolism, particularly in the thalamic region (Alkire et al., 2008). On the other hand, some anesthetics, at sedation doses, can reduce thalamic activity. For example, during sevoflurane sedation, thalamic activity can decrease by approximately 23%, even when patients are not fully unconscious. While cortical arousal can occur without thalamic involvement, the thalamus remains crucial for consciousness.

Anesthetic effects on the thalamus are often indirect, and spontaneous thalamic firing during anesthesia is largely driven by feedback from cortical neurons. Therefore, thalamic activity alone may not be sufficient to explain consciousness. Many of these thalamic neurons also extend to arousal centers in the brainstem, meaning that cortical deactivation can result in reduced thalamic activity and arousal (French et al., 1955).

It is shown that, metabolic and electrophysiological effects of anesthetics on the thalamus are abolished after removal of the cortex in animal studies. On the other hand, the cortex can still produce an activated EEG after thalamic ablation.

CORTICAL EFFECTS OF ANESTHETICS

Unconsciousness during anesthesia is associated with specific brain regions such as the mesial parietal cortex, posterior cingulate cortex, and precuneus. These areas are deactivated in vegetative patients but show activation during the recovery phase. Changes in neural activity in these regions are also observed during seizures associated with unconsciousness and during sleep.

The mesial cortical areas, located in the central core of the brain's connections, are strategically positioned and play a role in global monitoring of the internal environment and various self-related functions. However, during REM sleep when individuals experience vivid dreams, these mesial cortical areas are deactivated. In a study by Gyulai et al. (1996), human limbic responses to nitrous oxide inhalation were investigated using Positron Emission Tomography. The study revealed localized and specific changes in neuronal activity in the limbic area during low concentrations of nitrous oxide inhalation. These neuroanatomical sites may be related to behavioral disturbances associated with the effects of nitrous oxide. Overall, the findings suggest that specific brain regions, including the mesial cortical areas, play a crucial role in unconsciousness during anesthesia and are involved in the modulation of conscious states during sleep and other altered states of consciousness. Some anesthetics have the effect of disabling or disconnecting the lateral temporo-parieto-occipital complex, which comprises multimodal connecting areas primarily located in the lower parietal cortex. Consequently, the ultimate common target for anesthesia and the induction of unconsciousness is likely the lateral parieto-temporo-occipital junction, along with a rear mesial cortical area that encompasses a complex core (Alkire et al., 2008).

SUPPLEMENTARY MATERIAL

Supplementary data related to chapter 9 (Subthalamus) can be found online at https://doi.org/10.1016/B978-0-12-822404-5.00012-7

References

Alkire, M.T., Hudetz, A.G., Tononi, G., 2008. Consciousness and anesthesia. Science 322 (5903), 876–880. https://doi.org/10.1126/science.1149213.

Anis, N.A., Berry, S.C., Burton, N.R., Lodge, D., 1983. The dissociative anaesthetics, ketamine and phencyclidine, selectively reduce excitation of central mammalian neurones by N-methyl-aspartate. Br. J. Pharmacol. 79 (2), 565–575. https://doi.org/10.1111/j.1476-5381.1983.tb11031.x.

Aronstam, R.S., Martin, D.C., Dennison, R.L., 1994. Volatile anesthetics inhibit NMDA-stimulated45Ca uptake by rat brain microvesicles. Neurochem. Res. 19 (12), 1515–1520. https://doi.org/10.1007/BF00968999.

Barash, P.G., Cullen, B.F., Stoelting, R.K., et al., 2013. Clinical Anesthesia, eighth ed. Wolters Kluwer Health/Lippincott Williams and Wilkins.

Bruhn, J., Myles, P.S., Sneyd, R., Struys, M.M.R.F., 2006. Depth of anaesthesia monitoring: what's available, what's validated and what's next? Br. J. Anaesth. 97 (1), 85–94. https://doi.org/10.1093/bja/ael120.

Charlesworth, P., Richards, C.D., 1995. Anaesthetic modulation of nicotinic ion channel kinetics in bovine chromaffin cells. Br. J. Pharmacol. 114 (4), 909–917. https://doi.org/10.1111/j.1476-5381.1995.tb13290.x.

Collins, G.G.S., 1988. Effects of the anaesthetic 2,6-diisopropylphenol on synaptic transmission in the rat olfactory cortex slice. Br. J. Pharmacol. 95 (3), 939–949. https://doi.org/10.1111/j.1476-5381.1988.tb11724.x.

Correa-Sales, C., Rabin, B.C., Maze, M., 1992. A hypnotic response to dexmedetomidine, an α2 agonist, is mediated in the locus coerüleus in rats. Anesthesiology 76 (6), 948–952. https://doi.org/10.1097/00000542-199206000-00013.

Dildy-Mayfield, J.E., Eger, E.I., Harris, R.A., 1996. Anesthetics produce subunit-selective actions on glutamate receptors. J. Pharmacol. Exp. Ther. 276 (3), 1058–1065.

Dilger, J.P., Vidal, A.M., Mody, H.I., Liu, Y., 1994. Evidence for direct actions of general anesthetics on an ion channel protein. Anesthesiology 81 (2), 431–442. https://doi.org/10.1097/00000542-199408000-00022.

Downie, D.L., Hall, A.C., Lieb, W.R., Franks, N.P., 1996. Effects of inhalational general anaesthetics on native glycine receptors in rat medullary neurones and recombinant glycine receptors in Xenopus oocytes. Br. J. Pharmacol. 118 (3), 493–502. https://doi.org/10.1111/j.1476-5381.1996.tb15430.x.

Eban-Rothschild, A., Rothschild, G., Giardino, W.J., Jones, J.R., de Lecea, L., 2016. VTA dopaminergic neurons regulate ethologically relevant sleep—wake behaviors. Nat. Neurosci. 19 (10), 1356—1366. https://doi.org/10.1038/nn.4377.

French-Mullen, J., Barker, J., Rogawski, M., 1993. Calcium current block by (−)-pentobarbital, phenobarbital, and CHEB but not (+)-pentobarbital in acutely isolated hippocampal CA1 neurons: comparison with effects on GABA-activated Cl- current. J. Neurosci. 13 (8), 3211—3221. https://doi.org/10.1523/JNEUROSCI.13-08-03211.1993.

Firestone, L.L., Sauter, J.F., Braswell, L.M., Miller, K.W., 1986. Actions of general anesthetics on acetylcholine receptor-rich membranes from Torpedo californica. Anesthesiology 64 (6), 694—702. https://doi.org/10.1097/00000542-198606000-00004.

Forman, S.A., Miller, K.W., 2011. Anesthetic sites and allosteric mechanisms of action on Cys-loop ligand-gated ion channels. Canad. J. Anesthesia 58 (2), 191—205. https://doi.org/10.1007/s12630-010-9419-9.

Franks, N.P., Lieb, W.R., 1988. Volatile general anaesthetics activate a novel neuronal K+ current. Nature 333 (6174), 662—664. https://doi.org/10.1038/333662a0.

Franks, N.P., Lieb, W.R., 1991. Stereospecific effects of inhalational general anesthetic optical isomers on nerve ion channels. Science 254 (5030), 427—430. https://doi.org/10.1126/science.1925602.

French, J.D., Hernandez-Peon, R., Livingston, R.B., 1955. Projections from cortex to cephalic brain stem (reticular formation) in monkey. J. Neurophysiol. 18 (1), 74—95. https://doi.org/10.1152/jn.1955.18.1.74.

Frenkel, C., Weckbecker, K., Wartenberg, H.C., Duch, D.S., Urban, B.W., 1998. Blocking effects of the anaesthetic etomidate on human brain sodium channels. Neurosci. Lett. 249 (2—3), 131—134. https://doi.org/10.1016/S0304-3940(98)00412-1.

Gyulai, F.E., Firestone, L.L., Mintun, M.A., Winter, P.M., 1996. In vivo imaging of human limbic responses to nitrous oxide inhalation. Anesth. Analg. 83 (2), 291—298. https://doi.org/10.1097/00000539-199608000-00016.

Hales, T.G., Lambert, J.J., 1991. The actions of propofol on inhibitory amino acid receptors of bovine adrenomedullary chromaffin cells and rodent central neurones. Brit. J. Pharmacol. 104 (3), 619—628. https://doi.org/10.1111/j.1476-5381.1991.tb12479.x.

Hall, A.C., Lieb, W.R., Franks, N.P., 1994. Stereoselective and non-stereoselective actions of isoflurane on the GABAA receptor. Br. J. Pharmacol. 112 (3), 906—910. https://doi.org/10.1111/j.1476-5381.1994.tb13166.x.

Hill-Venning, C., Belelli, D., Peters, J.A., Lambert, J.J., 1997. Subunit-dependent interaction of the general anaesthetic etomidate with the γ-aminobutyric acid type A receptor. Br. J. Pharmacol. 120 (5), 749—756. https://doi.org/10.1038/sj.bjp.0700927.

Honoré, E., 2007. The neuronal background K2P channels: focus on TREK1. Nat. Rev. Neurosci. 8 (4), 251—261. https://doi.org/10.1038/nrn2117.

Jenkins, T.A., et al., 2010. Tryptophan depletion impairs object-recognition memory in the rat: reversal by risperidone. Behav. Brain Res. 208 (2), 479—483. https://doi.org/10.1016/j.bbr.2009.12.030.

Jones, B.E., 2011. Neurobiology of waking and sleeping. Handb. Clin. Neurol. 131—149. https://doi.org/10.1016/B978-0-444-52006-7.00009-5.

Jones, M.V., Brooks, P.A., Harrison, N.L., 1992. Enhancement of gamma-aminobutyric acid-activated Cl- currents in cultured rat hippocampal neurones by three volatile anaesthetics. J. Physiol. 449 (1), 279—293. https://doi.org/10.1113/jphysiol.1992.sp019086.

Koht, A., Sloan, T.B., Hemmer, L.B., 2022. Neuromonitoring in Surgery and Anesthesia. UpToDate. Published online.

Kullmann, D.M., Martin, R.L., Redman, S.J., 1989. Reduction by general anaesthetics of group Ia excitatory postsynaptic potentials and currents in the cat spinal cord. J. Physiol. 412 (1), 277—296. https://doi.org/10.1113/jphysiol.1989.sp017615.

Larrabee, M.G., Posternak, J.M., 1952. Selective action of anesthetics on synapses and axons in mammalian sympathetic ganglia. J. Neurophysiol. 15 (2), 91—114. https://doi.org/10.1152/jn.1952.15.2.91.

Lee, M.G., Hassani, O.K., Jones, B.E., 2005. Discharge of identified orexin/hypocretin neurons across the sleep-waking cycle. J. Neurosci. 25 (28), 6716—6720. https://doi.org/10.1523/JNEUROSCI.1887-05.2005.

Lin, L.H., Chen, L.L., Zirrolli, J.A., Harris, R.A., 1992. General anesthetics potentiate gamma-aminobutyric acid actions on gamma-aminobutyric acidA receptors expressed by Xenopus oocytes: lack of involvement of intracellular calcium. J. Pharmacol. Exp. Therapeut. 263 (2), 569—578.

Liu, Y.W., Li, J., Ye, J.H., 2010. Histamine regulates activities of neurons in the ventrolateral preoptic nucleus. J. Physiol. 588 (21), 4103—4116. https://doi.org/10.1113/jphysiol.2010.193904.

MacDonald, R.L., Rogers, C.J., Twyman, R.E., 1989. Barbiturate regulation of kinetic properties of the GABAA receptor channel of mouse spinal neurones in culture. J. Physiol. 417 (1), 483—500. https://doi.org/10.1113/jphysiol.1989.sp017814.

Maciver, M.B., Roth, S.H., 1988. Inhalation anaesthetics exhibit pathway-specific and differential actions on hippocampal synaptic responses in vitro. Br. J. Anaesth. 60 (6), 680—691. https://doi.org/10.1093/bja/60.6.680.

Maclver, M.B., Kendig, J.J., 1991. Anesthetic effects on resting membrane potential are voltage-dependent and agent-specific. Anesthesiology 74 (1), 83—88. https://doi.org/10.1097/00000542-199101000-00014.

Mascia, M.P., Machu, T.K., Harris, R.A., 1996. Enhancement of homomeric glycine receptor function by longchain alcohols and anaesthetics. Br. J. Pharmacol. 119 (7), 1331—1336. https://doi.org/10.1111/j.1476-5381.1996.tb16042.x.

Nakahiro, M., Yeh, J.Z., Brunner, E., Narahashi, T., 1989. General anesthetics modulate GABA receptor channel complex in rat dorsal root ganglion neurons. Faseb. J. 3 (7), 1850—1854. https://doi.org/10.1096/fasebj.3.7.2541038.

Nicoll, R.A., 1972. The effects of anaesthetics on synaptic excitation and inhibition in the olfactory bulb. J. Physiol. 223 (3), 803—814. https://doi.org/10.1113/jphysiol.1972.sp009875.

Nicoll, R.A., Madison, D.V., 1982. General anesthetics hyperpolarize neurons in the vertebrate central nervous system. Science 217 (4564), 1055—1057. https://doi.org/10.1126/science.7112112.

Nicoll, R.A., Eccles, J.C., Oshima, T., Rubia, F., 1975. Prolongation of hippocampal inhibitory postsynaptic potentials by barbiturates. Nature 258 (5536), 625—627. https://doi.org/10.1038/258625a0.

Olsen, R.W., Li, G.D., 2011. GABAA receptors as molecular targets of general anesthetics: identification of binding sites provides clues to allosteric modulation. Canad. J. Anesthesia 58 (2), 206—215. https://doi.org/10.1007/s12630-010-9429-7.

Palanca, B.J.A., Mashour, G.A., Avidan, M.S., 2009. Processed electroencephalogram in depth of anesthesia monitoring. Curr. Opin. Anaesthesiol. 22 (5), 553—559. https://doi.org/10.1097/ACO.0b013e3283304032.

Patel, A.J., Honoré, E., 2001. Anesthetic-sensitive 2P domain K+Channels. Anesthesiology 95 (4), 1013—1021. https://doi.org/10.1097/00000542-200110000-00034.

Patel, A.J., Honoré, E., Lesage, F., Fink, M., Romey, G., Lazdunski, M., 1999. Inhalational anesthetics activate two-pore-domain background K+ channels. Nat. Neurosci. 2 (5), 422—426. https://doi.org/10.1038/8084.

Perouansky, M., Baranov, D., Salman, M., Yaari, Y., 1995. Effects of halothane on glutamate receptor-mediated excitatory postsynaptic currents. Anesthesiology 83 (1), 109—119. https://doi.org/10.1097/00000542-199507000-00014.

Pritchett, D.B., Sontheimer, H., Shivers, B.D., et al., 1989. Importance of a novel GABAA receptor subunit for benzodiazepine pharmacology. Nature 338 (6216), 582–585. https://doi.org/10.1038/338582a0.

Proctor, W., Mynlieff, M., Dunwiddie, T., 1986. Facilitatory action of etomidate and pentobarbital on recurrent inhibition in rat hippocampal pyramidal neurons. J. Neurosci. 6 (11), 3161–3168. https://doi.org/10.1523/JNEUROSCI.06-11-03161.1986.

Quasha, A.L., Eger, E.I., Tinker, J.H., 1980. Determination and applications of MAC. Anesthesiology 53 (4), 315–334. https://doi.org/10.1097/00000542-198010000-00008.

Ratnakumari, L., Hemmings, H.C., 1998. Inhibition of presynaptic sodium channels by halothane. Anesthesiology 88 (4), 1043–1054. https://doi.org/10.1097/00000542-199804000-00025.

Rehberg, B., Duch, D.S., 1999. Suppression of central nervous system sodium channels by propofol. Anesthesiology 91 (2), 512–520. https://doi.org/10.1097/00000542-199908000-00026.

Rehberg, B., Xiao, Y.H., Duch, D.S., 1996. Central nervous system sodium channels are significantly suppressed at clinical concentrations of volatile anesthetics. Anesthesiology 84 (5), 1223–1233. https://doi.org/10.1097/00000542-199605000-00025.

Richards, C.D., White, A.E., 1975. The actions of volatile anaesthetics on synaptic transmission in the dentate gyrus. J. Physiol. 252 (1), 241–257. https://doi.org/10.1113/jphysiol.1975.sp011142.

Richards, C.D., Russell, W.J., Smaje, J.C., 1975. The action of ether and methoxyflurane on synaptic transmission in isolated preparations of the mammalian cortex. J. Physiol. 248 (1), 121–142. https://doi.org/10.1113/jphysiol.1975.sp010965.

Ries, C.R., Puil, E., 1999. Mechanism of anesthesia revealed by shunting actions of isoflurane on thalamocortical neurons. J. Neurophysiol. 81 (4), 1795–1801. https://doi.org/10.1152/jn.1999.81.4.1795.

Saper, C.B., Scammell, T.E., Lu, J., 2005. Hypothalamic regulation of sleep and circadian rhythms. Nature 437 (7063), 1257–1263. https://doi.org/10.1038/nature04284.

Sherin, J.E., Shiromani, P.J., McCarley, R.W., Saper, C.B., 1996. Activation of ventrolateral preoptic neurons during sleep. Science (New York, N.Y.) 271 (5246), 216–219. https://doi.org/10.1126/science.271.5246.216.

Steinberg, E.A., Wafford, K.A., Brickley, S.G., Franks, N.P., Wisden, W., 2015. The role of K2P channels in anaesthesia and sleep. Pflügers Archiv. 467 (5), 907–916. https://doi.org/10.1007/s00424-014-1654-4.

Takahashi, K., Kayama, Y., Lin, J.S., Sakai, K., 2010. Locus coeruleus neuronal activity during the sleep-waking cycle in mice. Neuroscience 169 (3), 1115–1126. https://doi.org/10.1016/j.neuroscience.2010.06.009.

Takenoshita, M., Steinbach, J., 1991. Halothane blocks low-voltage-activated calcium current in rat sensory neurons. J. Neurosci. 11 (5), 1404–1412. https://doi.org/10.1523/JNEUROSCI.11-05-01404.1991.

Ulrich Zeilhofer, H., Swandulla, D., Geisslinger, G., Brune, K., 1992. Differential effects of ketamine enantiomers on NMDA receptor currents in cultured neurons. Eur. J. Pharmacol. 213 (1), 155–158. https://doi.org/10.1016/0014-2999(92)90248-3.

Vanini, G., Wathen, B.L., Lydic, R., Baghdoyan, H.A., 2011. Endogenous GABA levels in the pontine reticular formation are greater during wakefulness than during rapid eye movement sleep. J. Neurosci. 31 (7), 2649–2656. https://doi.org/10.1523/JNEUROSCI.5674-10.20115.

Wakamori, M., Ikemoto, Y., Akaike, N., 1991. Effects of two volatile anesthetics and a volatile convulsant on the excitatory and inhibitory amino acid responses in dissociated CNS neurons of the rat. J. Neurophysiol. 66 (6), 2014–2021. https://doi.org/10.1152/jn.1991.66.6.2014.

Weight, F.F., Lovinger, D.M., White, G., Peoples, R.W., 1991. Alcohol and anesthetic actions on excitatory amino Acid? Activated ion channels. Ann. N. Y. Acad. Sci. 625 (1), 97–107. https://doi.org/10.1111/j.1749-6632.1991.tb33833.x.

Woolf, N.J., Butcher, L.L., 2011. Cholinergic systems mediate action from movement to higher consciousness. Behav. Brain Res. 221 (2), 488–498. https://doi.org/10.1016/j.bbr.2009.12.046.

Wu, X.S., Sun, J.Y., Evers, A.S., Crowder, M., Wu, L.G., 2004. Isoflurane inhibits transmitter release and the presynaptic action potential. Anesthesiology 100 (3), 663–670. https://doi.org/10.1097/00000542-200403000-00029.

Yeh, J.Z., Quandt, F.N., Tanguy, J., Nakahiro, M., Narahashi, T., Brunner, E.A., 1991. General anesthetic action on ? Aminobutyric Acid? Activated channels. Ann. N. Y. Acad. Sci. 625 (1), 155–173. https://doi.org/10.1111/j.1749-6632.1991.tb33839.x.

Zhang, L.N., Yang, C., Ouyang, P.R., et al., 2016. Orexin-A facilitates emergence of the rat from isoflurane anesthesia via mediation of the basal forebrain. Neuropeptides 58, 7–14. https://doi.org/10.1016/j.npep.2016.02.003.

Further Reading

Aston-Jones, G., Bloom, F., 1981. Nonrepinephrine-containing locus coeruleus neurons in behaving rats exhibit pronounced responses to non-noxious environmental stimuli. J. Neurosci. 1 (8), 887–900. https://doi.org/10.1523/JNEUROSCI.01-08-00887.1981.

Jenkins, A., Franks, N.P., Lieb, W.R., 1996. Actions of general anaesthetics on 5-HT3 receptors in N1E-115 neuroblastoma cells. Br. J. Pharmacol. 117 (7), 1507–1515. https://doi.org/10.1111/j.1476-5381.1996.tb15314.x.

Jones, M.V., Harrison, N.L., Pritchett, D.B., Hales, T.G., 1995. Modulation of the GABAA receptor by propofol is independent of the gamma subunit. J. Pharmacol. Exp. Ther. 274 (2), 962–968.

Sherin, J.E., Elmquist, J.K., Torrealba, F., Saper, C.B., 1998. Innervation of histaminergic tuberomammillary neurons by GABAergic and galaninergic neurons in the ventrolateral preoptic nucleus of the rat. J. Neurosci. 18 (12), 4705–4721. https://doi.org/10.1523/JNEUROSCI.18-12-04705.1998.

Un, L., Chen, L.L., Harris, R.A., 1993. Enflurane inhibits NMDA, AMPA, and kainate-induced currents in *Xenopus oocytes* expressing mouse and human brain mRNA. Faseb. J. 7 (3), 479–485. https://doi.org/10.1096/fasebj.7.5.7681790.

Vannini, P., Hedden, T., Becker, J.A., et al., 2012. Age and amyloid-related alterations in default network habituation to stimulus repetition. Neurobiol. Aging 33 (7), 1237–1252. https://doi.org/10.1016/j.neurobiolaging.2011.01.003.

Index

Note: Page numbers followed by *f* indicate figures, *t* indicate tables and *b* indicate boxes.

D

N

O

P

R

S